DEVELOPMENTS IN QUATERNARY SCIENCE 15
SERIES EDITOR: JAAP J.M. VAN DER MEER

QUATERNARY GLACIATIONS - EXTENT AND CHRONOLOGY

Developments in Quaternary Science
Series editor: Jaap J.M. van der Meer

Volumes in this series

1. The Quaternary Period in the United States
   Edited by A.R Gillespie, S.C. Porter, B.F. Atwater
   0-444-51470-8 (hardbound); 0-444-51471-6 (paperback) – 2004

2. Quaternary Glaciations – Extent and Chronology
   Edited by J. Ehlers, P.L. Gibbard
   Part I: Europe ISBN 0-444-51462-7 (hardbound) – 2004
   Part II: North America ISBN 0-444-51592-5 (hardbound) – 2004
   Part III: South America, Asia, Australasia, Antarctica ISBN 0-444-51593-3 (hardbound) – 2004

3. Ice Age Southern Andes – A Chronicle of Paleoecological Events
   By C.J. Heusser
   0-444-51478-3 (hardbound) – 2003

4. Spitsbergen Push Moraines – Including a translation of K. Gripp: Glaciologische und geologische Ergebnisse der Hamburgischen Spitzbergen-Expedition 1927
   Edited by J.J.M. van der Meer
   0-444-51544-5 (hardbound) – 2004

5. Iceland – Modern Processes and Past Environments
   Edited by C. Caseldine, A. Russell, J. Hardardóttir, Ó. Knudsen
   0-444-50652-7 (hardbound) – 2005

6. Glaciotectonism
   By J.S. Aber, A. Ber
   978-0-444-52943-5 (hardbound) – 2007

7. The Climate of Past Interglacials
   Edited by F. Sirocko, M. Claussen, M.F. Sánchez Goñi, T. Litt
   978-0-444-52955-8 (hardbound) – 2007

8. Juneau Icefield Research Project (1949–1958) – A Retrospective
   By C.J. Heusser †
   978-0-444-52951-0 (hardbound) – 2007

9. Late Quaternary Climate Change and Human Adaptation in Arid China
   Edited by David B. Madsen, Chen Fa-Hu, Gao Xing
   978-0-444-52962-6 (hardbound) – 2007

10. Tropical and Sub-Tropical West Africa – Marine and Continental Changes During the Late Quaternary
    By P. Giresse
    978-0-444-52984-8 – 2008

11. The Late Cenozoic of Patagonia and Tierra del Fuego
    Edited by J. Rabassa
    978-0-444-52954-1 – 2008

12. Advances in Quaternary Entomology
    By S.A. Elias
    978-0-444-53424-8 – 2010

13. The Mýrdalsjökull Ice Cap, Iceland. Glacial Processes, Sediments and Landforms on an Active Volcano
    Edited by A. Schomacker, J. Krüger, K.H. Kjær
    978-0-444-53045-5 – 2010

14. The Ancient Human Occupation of Britain
    Edited by Nick Ashton, Simon Lewis, Chris Stringer
    978-0-444-53597-9 – 2010

15. Quaternary Glaciations – Extent and Chronology. A Closer Look
    Edited by Jürgen Ehlers, Philip L. Gibbard, Philip D. Hughes
    978-0-444-53447-7 – 2011

For further information as well as other related products, please visit the Elsevier homepage (http://www.elsevier.com)

Developments in Quaternary Science, 15
Series editor: Jaap J.M. van der Meer

QUATERNARY GLACIATIONS - EXTENT AND CHRONOLOGY

A Closer Look

Edited by

Jürgen Ehlers

Geologisches Landesamt, Billstraße 84, D-20539 Hamburg, Germany

Philip L. Gibbard

Cambridge Quaternary, Department of Geography,
University of Cambridge, Downing Place, Cambridge CB2 3EN, England

Philip D. Hughes

Quaternary and Geoarchaeology Research Group, Geography,
School of Environment and Development, The University of Manchester,
Manchester M13 9PL, United Kingdom

ELSEVIER

Amsterdam • Boston • Heidelberg • London • New York • Oxford
Paris • San Diego • San Francisco • Singapore • Sydney • Tokyo

Elsevier
Radarweg 29, PO Box 211, 1000 AE Amsterdam, The Netherlands
Linacre House, Jordan Hill, Oxford OX2 8DP, UK

Copyright © 2011, Elsevier B.V. All Rights Reserved.

No part of this publication may be reproduced, stored in a retrieval system, or transmitted in any form or by any means, electronic, mechanical, photocopying, recording, or otherwise, without the prior written permission of the publisher.

Permissions may be sought directly from Elsevier's Science & Technology Rights Department in Oxford, UK: phone: (+44) 1865 843830, fax: (+44) 1865 853333, E-mail: permissions@elsevier.com. You may also complete your request online via the Elsevier homepage (http://elsevier.com), by selecting "Support & Contact" then "Copyright and Permission" and then "Obtaining Permissions."

Library of Congress Cataloging-in-Publication Data
A catalog record for this book is available from the Library of Congress

British Library Cataloguing-in-Publication Data
A catalogue record for this book is available from the British Library.

ISBN: 978-0-444-53447-7
ISSN: 1571-0866

For information on all Elsevier publications
visit our web site at www.elsevierdirect.com

Printed and bound in China

11 12 13 10 9 8 7 6 5 4 3 2

Working together to grow
libraries in developing countries

www.elsevier.com | www.bookaid.org | www.sabre.org

ELSEVIER   BOOK AID International   Sabre Foundation

Contents

Contributors   xi
Preface   xvii

1. **Introduction**   1
   Jürgen Ehlers, Philip L. Gibbard and Philip D. Hughes

2. **Quaternary Glaciations in Austria**   15
   Dirk van Husen

3. **The Pleistocene Glaciations in Belarus**   29
   A.K. Karabanov and A.V. Matveyev

4. **Pleistocene Glaciations of Czechia**   37
   Daniel Nývlt, Zbyněk Engel and Jaroslav Tyráček

5. **Pleistocene Glaciations in Denmark: A Closer Look at Chronology, Ice Dynamics and Landforms**   47
   Michael Houmark-Nielsen

6. **The Glacial History of the British Isles during the Early and Middle Pleistocene: Implications for the long-term development of the British Ice Sheet**   59
   Jonathan R. Lee, James Rose, Richard J.O. Hamblin, Brian S.P. Moorlock, James B. Riding, Emrys Phillips, René W. Barendregt and Ian Candy

7. **Pleistocene Glaciation Limits in Great Britain**   75
   Philip L. Gibbard and Chris D. Clark

8. **Pleistocene Glaciations in Estonia**   95
   Volli Kalm, Anto Raukas, Maris Rattas and Katrin Lasberg

9. **The Glaciation of Finland**   105
   Peter Johansson, Juha Pekka Lunkka and Pertti Sarala

10. **Quaternary Glaciations in the French Alps and Jura**   117
    Jean-François Buoncristiani and Michel Campy

11. **Recent Advances in Research on Quaternary Glaciations in the Pyrenees**   127
    Marc Calvet, Magali Delmas, Yanni Gunnell, Régis Braucher and Didier Bourlès

12. **Late Pleistocene (Würmian) Glaciations of the Caucasus**   141
    Ramin Gobejishvili, Nino Lomidze and Levan Tielidze

13. **Pleistocene Glaciations of North Germany—New Results**   149
    Jürgen Ehlers, Alf Grube, Hans-Jürgen Stephan and Stefan Wansa

14. **Pleistocene Glaciations of Southern Germany** 163

    Markus Fiebig, Dietrich Ellwanger and Gerhard Doppler

15. **Glaciation in Greece: A New Record of Cold Stage Environments in the Mediterranean** 175

    Jamie C. Woodward and Philip D. Hughes

16. **Pliocene and Pleistocene Glaciations of Iceland: A Brief Overview of the Glacial History** 199

    Áslaug Geirsdóttir

17. **Middle Pleistocene to Holocene Glaciations in the Italian Apennines** 211

    Carlo Giraudi

18. **Pleistocene Glaciations in Latvia** 221

    Vitālijs Zelčs, Aivars Markots, Māris Nartišs and Tomas Saks

19. **Pleistocene Glaciations in Lithuania** 231

    Rimantė Guobytė and Jonas Satkūnas

20. **Pleistocene Glaciation in the Netherlands** 247

    Cees Laban and Jaap J.M. van der Meer

21. **The Pleistocene Glaciations of the North Sea Basin** 261

    Alastair G.C. Graham, Martyn S. Stoker, Lidia Lonergan, Tom Bradwell and Margaret A. Stewart

22. **Glacial History of Norway** 279

    Jan Mangerud, Richard Gyllencreutz, Øystein Lohne and John Inge Svendsen

23. **Quaternary Glaciations in Poland** 299

    Leszek Marks

24. **New Evidence on the Quaternary Glaciation in the Romanian Carpathians** 305

    Petru Urdea, Alexandru Onaca, Florina Ardelean and Mircea Ardelean

25. **Ice Margins of Northern Russia Revisited** 323

    Valery Astakhov

26. **Glaciations of the East European Plain: Distribution and Chronology** 337

    A. A. Velichko, M.A. Faustova, V.V. Pisareva, Yu. N. Gribchenko, N.G. Sudakova and N.V. Lavrentiev

27. **Glacial History of the Barents Sea Region** 361

    Tore O. Vorren, Jon Y. Landvik, Karin Andreassen and Jan Sverre Laberg

28. **Glacial History of the Taymyr Peninsula and the Severnaya Zemlya Archipelago, Arctic Russia** 373

    Per Möller, Christian Hjort, Helena Alexanderson and Florian Sallaba

29. **Glacial History of Slovenia** 385

    Miloš Bavec and Tomaž Verbič

30. **Quaternary Glaciations of Turkey** 393

    Mehmet Akif Sarıkaya, Attila Çiner and Marek Zreda

31. **Limits of the Pleistocene Glaciations in the Ukraine: A Closer Look** 405

    Andrei V. Matoshko

32. **Chronology and Extent of Late Cenozoic Ice Sheets in North America: A Magnetostratigraphical Assessment** 419

    René W. Barendregt and Alejandra Duk-Rodkin

Contents

33. **Alaska Palaeo-Glacier Atlas (Version 2)** 427
    Darrell S. Kaufman, Nicolás E. Young, Jason P. Briner and William F. Manley

34. **Glaciations of the Sierra Nevada, California, USA** 447
    Alan R. Gillespie and Douglas H. Clark

35. **Pleistocene Glaciation of Hawaii** 463
    Stephen C. Porter

36. **Quaternary Glaciations in Illinois** 467
    B. Brandon Curry, David A. Grimley and E. Donald McKay III

37. **Ice-Margin Fluctuations at the End of the Wisconsin Episode, Michigan, USA** 489
    Grahame J. Larson

38. **The Quaternary of Minnesota** 499
    C.E. Jennings and M.D. Johnson

39. **Pleistocene Glaciation of Ohio, USA** 513
    John P. Szabo, Michael P. Angle and Alex M. Eddy

40. **The Glaciation of Pennsylvania, USA** 521
    Duane D. Braun

41. **Glaciation of Western Washington, USA** 531
    Stephen C. Porter

42. **The Quaternary of Wisconsin: An Updated Review of Stratigraphy, Glacial History and Landforms** 537
    Kent M. Syverson and Patrick M. Colgan

43. **Summary of Early and Middle Pleistocene Glaciations in Northern Missouri, USA** 553
    Charles W. Rovey II and Greg Balco

44. **Pleistocene Glaciation of British Columbia** 563
    John J. Clague and Brent Ward

45. **Limits of Successive Middle and Late Pleistocene Continental Ice Sheets, Interior Plains of Southern and Central Alberta and Adjacent Areas** 575
    Lionel E. Jackson, Laurence D. Andriashek and Fred M. Phillips

46. **Magnetostratigraphy of Quaternary Sections in Eastern Alberta, Saskatchewan and Manitoba** 591
    René W. Barendregt

47. **Late Pleistocene–Early Holocene Decay of the Laurentide Ice Sheet in Québec–Labrador** 601
    Serge Occhietti, M. Parent, P. Lajeunesse, F. Robert and É. Govare

48. **The Appalachian Glacier Complex in Maritime Canada** 631
    Rudolph R. Stea, Allen A. Seaman, Toon Pronk, Michael A. Parkhill, Serge Allard and Daniel Utting

49. **Stratigraphical Record of Glacials/Interglacials in Northwest Canada** 661
    Alejandra Duk-Rodkin and René W. Barendregt

50. **The Greenland Ice Sheet During the Past 300,000 Years: A Review** 699
    Svend Funder, Kristian Kjellerup Kjeldsen, Kurt Henrik Kjær and Colm Ó Cofaigh

51. **Pleistocene Glaciations in Southern Patagonia and Tierra del Fuego**   715

Andrea Coronato and Jorge Rabassa

52. **Pleistocene Glaciations in Northern Patagonia, Argentina: An Updated Review**   729

Oscar Martínez, Andrea Coronato and Jorge Rabassa

53. **The High-Glacial (Last Glacial Maximum) Glacier Cover of the Aconcagua Group and Adjacent Massifs in the Mendoza Andes (South America) with a Closer Look at Further Empirical Evidence**   735

Matthias Kuhle

54. **The Pleistocene Glaciations of Chile**   739

Stephan Harrison and Neil F. Glasser

55. **Late Quaternary Glaciations in Bolivia: Comments on Some New Approaches to Dating Morainic Sequences**   757

Klaus Heine

56. **Ecuador, Peru and Bolivia**   773

Jeff La Frenierre, Kyung In Huh and Bryan G. Mark

57. **Late Quaternary Glaciations of Ecuador**   803

Klaus Heine

58. **Quaternary Glaciations of Colombia**   815

Karin F. Helmens

59. **Late Quaternary Glaciations in the Venezuelan (Mérida) Andes**   835

Volli Kalm and William C. Mahaney

60. **Costa Rica and Guatemala**   843

Matthew S. Lachniet and Alex J. Roy

61. **Late Quaternary Glaciation in Mexico**   849

Lorenzo Vázquez-Selem and Klaus Heine

62. **Late Pleistocene Glaciation of the Hindu Kush, Afghanistan**   863

Stephen C. Porter

63. **Late Pleistocene Glaciations in North-East Asia**   865

O.Yu. Glushkova

64. **Extent and Timing of Quaternary Glaciations in the Verkhoyansk Mountains**   877

Georg Stauch and Frank Lehmkuhl

65. **Glaciation in the High Mountains of Siberia**   883

Vladimir S. Sheinkman

66. **Late Quaternary Glaciation of Northern Pakistan**   909

Ulrich Kamp and Lewis A. Owen

67. **Quaternary Glaciation of Northern India**   929

Lewis A. Owen

Contents

68. **The High Glacial (Last Ice Age and Last Glacial Maximum) Ice Cover of High and Central Asia, with a Critical Review of Some Recent OSL and TCN Dates** — 943

    Matthias Kuhle

69. **The Extent and Timing of Late Pleistocene Glaciations in the Altai and Neighbouring Mountain Systems** — 967

    Frank Lehmkuhl, Michael Klinge and Georg Stauch

70. **Quaternary Glaciations: Extent and Chronology in China** — 981

    Zhou Shangzhe, Li Jijun, Zhao Jingdong, Wang Jie and Zheng Jingxiong

71. **Late Pleistocene and Early Holocene Glaciations in the Taiwanese High Mountain Ranges** — 1003

    Margot Böse and Robert Hebenstreit

72. **Late Quaternary Glaciations in Japan** — 1013

    Takanobu Sawagaki and Tatsuto Aoki

73. **The Glaciation of the South-East Asian Equatorial Region** — 1023

    M.L. Prentice, G.S. Hope, J.A. Peterson and Timothy T. Barrows

74. **The Glaciation of Australia** — 1037

    Eric A. Colhoun and Timothy T. Barrows

75. **Quaternary Glaciers of New Zealand** — 1047

    D.J.A. Barrell

76. **Quaternary Glaciations of the Atlas Mountains, North Africa** — 1065

    Philip D. Hughes, C.R. Fenton and Philip L. Gibbard

77. **Quaternary Glacial Chronology of Mount Kenya Massif** — 1075

    William C. Mahaney

78. **Glaciation in Southern Africa and in the Sub-Antarctic** — 1081

    Kevin Hall and Ian Meiklejohn

Index — 1087

High resolution map datasets are available at: http://booksite.elsevier.com/9780444534477/

Contributors

Numbers in paraentheses indicate the pages on which the authors' contrbutions begin.

Helena Alexanderson (373), Department of Earth and Ecosystem Sciences, Division of Geology, Lund University, Sölvegatan 12, SE-223 62 Lund, Sweden

Serge Allard (631), New Brunswick Department of Natural Resources, Fredericton, New Brunswick, Canada E3B 5H1

Karin Andreassen (361), Department of Geology, University of Tromsø, N-9037 Tromsø, Norway

Laurence D. Andriashek (575), E.R.C.B., Alberta Geological Survey, 4th Floor, 4999-98th Avenue, Edmonton, Alberta, Canada T6B 2X3

Michael P. Angle (513), Ohio Department of Natural Resources, Division of Geological Survey, 2045 Morse Road, Building C-1, Columbus, Ohio 43229-6693, USA

Tatsuto Aoki (1013), School of Regional Development Studies, Kanazawa University, Ishikawa, Japan

Florina Ardelean (305), Department of Geography, West University of Timişoara, V. Pârvan Nr.4, 300223 Timişoara, Romania

Mircea Ardelean (305), Department of Geography, West University of Timişoara, V. Pârvan Nr.4, 300223 Timişoara, Romania

Valery Astakhov (323), Geological Faculty, St. Petersburg University, Universitetskaya 7/9, 199034 St. Petersburg, Russia

Margot Böse (1003), Freie Universität Berlin, Institute of Geographical Sciences, Malteserstraße 74-100, 12249 Berlin, Germany

Greg Balco (553), Berkeley Geochronology Center, 2455 Ridge Road, Berkeley, California, USA

René W. Barendregt (59, 419, 591, 661), Department of Geography, The University of Lethbridge, 4401 University Drive, Lethbridge, Alberta, Canada T1K 3M4

D.J.A. Barrell (1047), GNS Science, Private Bag 1930, Dunedin, New Zealand

Timothy T. Barrows (1023, 1037), School of Geography, University of Exeter, Exeter EX4 4RJ, United Kingdom

Miloš Bavec (385), Geological Survey of Slovenia, Dimičeva 14, 1000 Ljubljana, Slovenia

Didier Bourlès (127), UMR 6635 CNRS, Aix Marseille Université, BP 80, 13545 Aix-en-Provence cedex 04, France

Tom Bradwell (261), British Geological Survey, Murchison House, West Mains Road, Edinburgh EH9 3LA, United Kingdom

Régis Braucher (127), UMR 6635 CNRS, Aix Marseille Université, BP 80, 13545 Aix-en-Provence cedex 04, France

Duane D. Braun (521), Geography and Geosciences, Bloomsburg University of Pennsylvania, Mount Desert, Maine, USA

Jason P. Briner (427), Department of Geological Sciences, University at Buffalo, Buffalo, New York 14260, USA

Jean-François Buoncristiani (117), Laboratoire Biogeosciences, Université de Bourgogne, UMR CNRS 5561, 6 Boulevard Gabriel, 21000 Dijon, France

Marc Calvet (127), JE 2522 Médi-Terra, Université de Perpignan-Via Domitia, 52 Avenue Paul Alduy, F 66860 Perpignan cedex, France

Michel Campy (117), Laboratoire Biogeosciences, Université de Bourgogne, UMR CNRS 5561, 6 Boulevard Gabriel, 21000 Dijon, France

Ian Candy (59), Department of Geography, Royal Holloway, University of London, Egham, Surrey TW20 0EX, United Kingdom

John J. Clague (563), Department of Earth Sciences, Simon Fraser University, Burnaby, British Columbia, V5A 1S6 Canada

Chris D. Clark (75), Department of Geography, University of Sheffield, Sheffield S10 2TN, England

Douglas H. Clark (447), Department of Geology, Western Washington University, Bellingham, Washington 98225, USA

Attila Çiner (393), Department of Geological Engineering, Hacettepe University, Ankara, Turkey

Colm Ó. Cofaigh (699), Department of Geography, Durham University, South Road, Durham DH1 3LE, United Kingdom

Patrick M. Colgan (537), Department of Geology, Padnos Hall of Science, Grand Valley State University, Allendale, Michigan 49401, USA

Eric A. Colhoun (1037), Earth Sciences, The University of Newcastle, Callaghan, NSW 2308, Australia

Andrea Coronato (715, 729), CONICET-CADIC, B. Houssay 200, 9410 Ushuaia, Tierra del Fuego, Argentina; Universidad Nacional de la Patagonia San Juan Bosco, Sede Ushuaia, Darwin y Canga, 9410 Ushuaia, Tierra del Fuego, Argentina

B. Brandon Curry (467), Illinois State Geological Survey, Prairie Research Institute, 615 E. Peabody Drive, Champaign, Illinois 61820, USA

Magali Delmas (127), JE 2522 Médi-Terra, Université de Perpignan-Via Domitia, 52 Avenue Paul Alduy, F 66860 Perpignan cedex, France

Gerhard Doppler (163), Bayerisches Landesamt für Umwelt, Lazarettstr. 67, D-80636 München

Alejandra Duk-Rodkin (419, 661), Geological Survey of Canada, 3303-33rd St. NW, Calgary, Alberta, Canada T2L 2A7

Alex M. Eddy (513), Ohio Department of Natural Resources, Division of Geological Survey, 2045 Morse Road, Building C-1, Columbus, Ohio 43229-6693, USA

Jürgen Ehlers (1, 149), Geologisches Landesamt, Billstr. 84, D-20539 Hamburg, Germany; Geologisches Landesamt, Billstraße 84, D-20539 Hamburg, Germany

Dietrich Ellwanger (163), Landesamt für Geologie, Rohstoffe und Bergbau im Regierungspräsidium Freiburg, Freiburg i.Br., Germany

Zbyněk Engel (37), Department of Physical Geography and Geoecology, Faculty of Science, Charles University in Prague, Albertov 6, 128 43 Praha 2, Czech Republic

M.A. Faustova (337), Laboratory of Evolutionary Geography, Institute of Geography, Russian Academy of Sciences, Staromonetny 29, Moscow 109017, Russia

C.R. Fenton (1065), NERC Cosmogenic Isotope Analysis Facility, Scottish Universities Environmental Research Centre, Scottish Enterprise Technology Park, Rankine Avenue, East Kilbride, Glasgow G75 0QF, United Kingdom

Markus Fiebig (163), Institute of Applied Geology, Department of Structural Engineering+Natural Hazards, University of Natural Resources and Applied Life Sciences, Vienna, Peter Jordan-Str. 70, A 1190 Wien, Austria

Svend Funder (699), Centre for GeoGenetics, Natural History Museum, University of Copenhagen, Øster Voldgade 5-7, DK-1350 Copenhagen K, Denmark

Áslaug Geirsdóttir (199), Department of Earth Sciences and Institute of Earth Sciences, University of Iceland, Askja, Sturlugata 7, 101 Reykjavík, Iceland

Philip L. Gibbard (1, 75, 1065), Cambridge Quaternary, Department of Geography, University of Cambridge, Downing Place, Cambridge CB2 3EN, United Kingdom

Alan R. Gillespie (447), Quaternary Research Center, University of Washington, Seattle, Washington 98195-1310, USA

Carlo Giraudi (211), ENEA, C.R. Saluggia, strada per Crescentino, 41, 13040 Saluggia (VC), Italy

Neil F. Glasser (739), Institute of Geography and Earth Sciences, Aberystwyth University, United Kingdom

O.Yu. Glushkova (865), North-eastern Interdisciplinary Research Institute FEB RAS, 16 Portovaya St., 685000 Magadan, Russia

Ramin Gobejishvili (141), Ivane Javkhishvili Tbilisi State University, Vakhushti Bagrationi Institute of Geography, department geomorphology and geoecology, Merab Aleksidze str. Block 1.0193, Tbilisi, Georgia

É. Govare (601), Hydro-Québec Équipement, Expertise Géomatique, 855 rue Sainte-Catherine Est, Montréal, QC, Canada H2L 4P5

Alastair G.C. Graham (261), Ice Sheets Programme, British Antarctic Survey, High Cross, Madingley Road, Cambridge CB3 0ET, United Kingdom

Yu.N. Gribchenko (337), Laboratory of Evolutionary Geography, Institute of Geography, Russian Academy of Sciences, Staromonetny 29, Moscow 109017, Russia

David A. Grimley (467), Illinois State Geological Survey, Prairie Research Institute, 615 E. Peabody Drive, Champaign, Illinois 61820, USA

Alf Grube (149), Landesamt für Landwirtschaft, Umwelt und ländliche Räume, Hamburger Chaussee 25, D-24220 Flintbek, Germany

Yanni Gunnell (127), UMR 5600 CNRS, Université Lumière-Lyon 2, 5 avenue Pierre Mendès-France, 69676 Bron cedex, France

Rimantė Guobytė (231), Lithuanian Geological Survey S. Konarskio 35, LT-03123; Vilnius University, Universiteto 3, LT-01513 Vilnius, Lithuania

Contributors

Richard Gyllencreutz (279), Department of Earth Science and Bjerknes Centre for Climate Research, University of Bergen, Allégt. 41, NO-5007 Bergen, Norway

Kevin Hall (1081), Department of Geography, Geoinformatics and Meteorology, University of Pretoria, Pretoria 0002, South Africa; Department of Geography, University of Northern British Columbia, 3333 University Way, Prince George, Canada BC V2N 4Z9

Richard J.O. Hamblin (59), British Geological Survey, Keyworth, Nottingham NG12 5GG; Department of Geography, Royal Holloway, University of London, Egham, Surrey TW20 0EX, United Kingdom

Stephan Harrison (739), College of Life and Environmental Sciences, University of Exeter, Penryn, Cornwall, United Kingdom

Robert Hebenstreit (1003), Freie Universität Berlin, Institute of Geographical Sciences, Malteserstraße 74-100, 12249 Berlin, Germany

Klaus Heine (757, 803, 849), Geographisches Institut der Universität Regensburg, Universitätsstr. 31, D-93040 Regensburg, Germany

Karin F. Helmens (815), Department of Physical Geography and Quaternary Geology, Stockholm University, 106 91 Stockholm, Sweden

Christian Hjort (373), Department of Earth and Ecosystem Sciences, Division of Geology, Lund University, Sölvegatan 12, SE-223 62 Lund, Sweden

G.S. Hope (1023), Department of Archaeology and Natural History, CAP, Australian National University, Canberra 0200, Australia

Michael Houmark-Nielsen (47), Centre for GeoGenetics, Natural History Museum, University of Copenhagen, Øster Voldgade 5-7, Dk-1350 København K, Denmark

Philip D. Hughes (1, 75, 1065), Quaternary Environments and Geoarchaeology Research Group, Geography, School of Environment and Development, The University of Manchester, Manchester, M13 9PL, United Kingdom

Kyung In Huh (773), Department of Geography, The Ohio State University, Derby Hall, North Oval Mall, Columbus, Ohio 43210, USA

Lionel E. Jackson (575), Geological Survey of Canada, Pacific Division, 625 Robson Street, Vancouver, British Columbia, Canada V6B 5J3

C.E. Jennings (499), Minnesota Geological Survey, University of Minnesota, Minnesota, USA

Wang Jie (981), Department of Geography, Lanzhou University, Lanzhou, China

Li Jijun (981), Department of Geography, Lanzhou University, Lanzhou, China

Zhao Jingdong (981), State Key Laboratory of Cryospheric Sciences, Cold and Arid Regions Environmental and Engineering Research Institute, Chinese Academy of Sciences, Lanzhou, China

Zheng Jingxiong (981), School of Geographical Science, South China Normal University, Guangzhou, China

Peter Johansson (105), Geological Survey of Finland, P.O. Box 77, FI-96101 Rovaniemi, Finland

M.D. Johnson (499), Earth Science Institute, University of Gothenburg, Göteborg, Sweden

Volli Kalm (95, 835), Department of Geology, Institute of Ecology and Earth Sciences, University of Tartu, Ravila 14a, Tartu 50411, Estonia

Ulrich Kamp (909), Department of Geography, The University of Montana, Missoula, Montana 59812-5040, USA

A.K. Karabanov (29), Institute for Nature Management of the National Academy of Sciences of Belarus, F.Skoriny Str., 10 Minsk 220141, Belarus

Darrell S. Kaufman (427), School of Earth Sciences and Environmental Sustainability, Northern Arizona University, Flagstaff, Arizona 86011-4099, USA

Kurt Henrik Kjær (699), Centre for GeoGenetics, Natural History Museum, University of Copenhagen, Øster Voldgade 5-7, DK-1350 Copenhagen K, Denmark

Kristian Kjellerup Kjeldsen (699), Centre for GeoGenetics, Natural History Museum, University of Copenhagen, Øster Voldgade 5-7, DK-1350 Copenhagen K, Denmark

Michael Klinge (967), Rischenangerweg 10, 37139 Adelebsen / Lödingsen

Matthias Kuhle (735, 943), Department of Geography and High Mountain Geomorphology, Geographisches Institut, University of Göttingen, Goldschmidtstr. 5, D-37077 Göttingen, Germany

Jeff La Frenierre (773), Department of Geography, The Ohio State University, Derby Hall, North Oval Mall, Columbus, Ohio 43210, USA

Cees Laban (247), Deltares, PO Box 85467, 3508 AL Utrecht, The Netherlands

Jan Sverre Laberg (361), Department of Geology, University of Tromsø, N-9037 Tromsø, Norway

Matthew S. Lachniet (843), Department of Geoscience, University of Nevada, Las Vegas, 4505 Maryland Parkway, Las Vegas, Nevada 89154, USA

P. Lajeunesse (601), Centre d'études nordiques & Département de géographie, Université Laval, Québec, QC, Canada G1V 0A6

Jon Y. Landvik (361), Department of Plant and Environmental Sciences, Norwegian University of Life Sciences, N-1432 Aas, Norway

Grahame J. Larson (489), Department of Geological Sciences, Michigan State University, East Lansing, Michigan 48834, USA

Katrin Lasberg (95), Department of Geology, Institute of Ecology and Earth Sciences, University of Tartu, Ravila 14a, 50411 Tartu, Estonia

N.V. Lavrentiev (337), Laboratory of Evolutionary Geography, Institute of Geography, Russian Academy of Sciences, Staromonetny 29, Moscow 109017, Russia

Jonathan R. Lee (59), British Geological Survey, Keyworth, Nottingham NG12 5GG; Department of Geography, Royal Holloway, University of London, Egham, Surrey TW20 0EX, United Kingdom

Frank Lehmkuhl (877, 967), Department of Geography, RWTH Aachen University, Templergraben 55, 52056 Aachen, Germany

Øystein Lohne (279), Department of Earth Science and Bjerknes Centre for Climate Research, University of Bergen, Allégt. 41, NO-5007 Bergen, Norway

Nino Lomidze (141), Ivane Javkhishvili Tbilisi State University, Vakhushti Bagrationi Institute of Geography, department geomorphology and geoecology, Merab Aleksidze str. Block 1.0193, Tbilisi, Georgia

Lidia Lonergan (261), Department of Earth Science and Engineering, Imperial College London, South Kensington Campus, London SW7 2AZ, United Kingdom

Juha Pekka Lunkka (105), Institute of Geosciences, University of Oulu, P.O. Box 3000, FI-90014, Oulu, Finland

Per Möller (373), Department of Earth and Ecosystem Sciences, Division of Geology, Lund University, Sölvegatan 12, SE-223 62 Lund, Sweden

William C. Mahaney (835, 1075), Quaternary Surveys, 26 Thornhill Avenue, Thornhill, Ontario, Canada L4J 1J4

Jan Mangerud (279), Department of Earth Science and Bjerknes Centre for Climate Research, University of Bergen, Allégt. 41, NO-5007 Bergen, Norway

William F. Manley (427), Institute of Arctic and Alpine Research, University of Colorado, Boulder, Colorado 80309-0450, USA

Bryan G. Mark (773), Department of Geography, The Ohio State University, Derby Hall, North Oval Mall, Columbus, Ohio 43210, USA

Aivars Markots (221), Faculty of Geographical and Earth Sciences, University of Latvia, Rainis Boulevard 19, Riga LV-1586, Latvia

Leszek Marks (299), Polish Geological Institute, Rakowiecka 4, 00-975 Warsaw, Poland; Institute of Geology, Warsaw University, Żwirki i Wigury 93, 02-089 Warsaw, Poland

Oscar Martínez (729), Universidad Nacional de la Patagonia San Juan Bosco, Sede Esquel, Ruta 259, km 16,5, (9200) Esquel, Chubut, Argentina

Andrei V. Matoshko (405), Department of Geomorphology the Ukrainian National Academy of Sciences, 44, Volodymyrska Street, 01034 Kyiv, Ukraine

A.V. Matveyev (29), Institute for Nature Management of the National Academy of Sciences of Belarus, F.Skoriny Str., 10 Minsk 220141, Belarus

E. Donald McKay (467), Illinois State Geological Survey, Prairie Research Institute, 615 E. Peabody Drive, Champaign, Illinois 61820, USA

Ian Meiklejohn (1081), Department of Geography, Rhodes University, Grahamstown 6140, South Africa

Brian S.P. Moorlock (59), British Geological Survey, Keyworth, Nottingham NG12 5GG; Department of Geography, Royal Holloway, University of London, Egham, Surrey TW20 0EX, United Kingdom

Daniel Nývlt (37), Czech Geological Survey, Brno branch, Leitnerova 22, 658 69 Brno, Czech Republic

Māris Nartišs (221), Faculty of Geographical and Earth Sciences, University of Latvia, Rainis Boulevard 19, Riga LV-1586, Latvia

Serge Occhietti (601), Département de géographie and Geotop, UQAM CP 8888 Centre-ville, Montréal, QC, Canada H3C 3P8; CERPA, Université de Lorraine, Nancy, France

Alexandru Onaca (305), Department of Geography, West University of Timişoara, V. Pârvan Nr.4, 300223 Timişoara, Romania

Lewis A. Owen (909, 929), Department of Geology, University of Cincinnati, Cincinnati, Ohio 45221-0013, USA

M. Parent (601), Geological Survey of Canada, 490 rue de la Couronne, Québec, QC, Canada G1K 9A9

Michael A. Parkhill (631), New Brunswick Department of Natural Resources, Bathurst, New Brunswick, Canada E2A 3Z1

J.A. Peterson (1023), Centre for GIS School of Geography and Environmental Science, Monash University, Clayton 3800, Australia

Contributors

Emrys Phillips (59), British Geological Survey, Murchison House, West Mains Road, Edinburgh EH9 3LA, United Kingdom

Fred M. Phillips (575), Department of Earth and Environmental Science, New Mexico Tech, Sorcorro, New Mexico 87801, USA

V.V. Pisareva (337), Laboratory of Evolutionary Geography, Institute of Geography, Russian Academy of Sciences, Staromonetny 29, Moscow 109017, Russia

Stephen C. Porter (463, 531, 863), Department of Earth and Space Sciences, University of Washington, Seattle, Washington 98195-1360, USA

M.L. Prentice (1023), Indiana Geological Survey, Indiana University, 611 North Walnut Grove, Bloomington, Indiana 47405-2208, USA

Toon Pronk (631), New Brunswick Department of Natural Resources, Fredericton, New Brunswick, Canada E3B 5H1

Jorge Rabassa (715, 729), CONICET-CADIC, B. Houssay 200, 9410 Ushuaia, Tierra del Fuego, Argentina; Universidad Nacional de la Patagonia San Juan Bosco, Sede Ushuaia, Darwin y Canga, 9410 Ushuaia, Tierra del Fuego, Argentina

Maris Rattas (95), Department of Geology, Institute of Ecology and Earth Sciences, University of Tartu, Ravila 14a, 50411 Tartu, Estonia

Anto Raukas (95), Institute of Geology, Tallinn Technical University, Ehitajate tee 5, 19086 Tallinn, Estonia

James B. Riding (59), British Geological Survey, Keyworth, Nottingham NG12 5GG, United Kingdom

F. Robert (601), Groupe Omégalpha, Crabtree, Québec, QC, Canada J0K 1B0

James Rose (59), British Geological Survey, Keyworth, Nottingham NG12 5GG, United Kingdom; Department of Geography, Royal Holloway, University of London, Egham, Surrey TW20 0EX, United Kingdom

Charles W. Rovey (553), Department of Geography, Geology, and Planning, Missouri State University, 901 S. National, Springfield, Missouri 65897, USA

Alex J. Roy (843), Department of Geoscience, University of Nevada, Las Vegas, 4505 Maryland Parkway, Las Vegas, Nevada 89154, USA

Tomas Saks (221), Faculty of Geographical and Earth Sciences, University of Latvia, Rainis Boulevard 19, Riga LV-1586, Latvia

Florian Sallaba (373), Department of Earth and Ecosystem Sciences, Division of Physical Geography and Ecosystem Analysis, Lund University, Sölvegatan 12, SE-223 62 Lund, Sweden

Mehmet Akif Sarıkaya (393), Geography Department, Fatih University, Istanbul, Turkey

Pertti Sarala (105), Geological Survey of Finland, P.O. Box 77, FI-96101 Rovaniemi, Finland

Jonas Satkūnas (231), Lithuanian Geological Survey S. Konarskio 35, LT-03123; Vilnius University, Universiteto 3, LT-01513 Vilnius, Lithuania

Takanobu Sawagaki (1013), Faculty of Environmental Earth Science, Hokkaido University, Sapporo, Japan

Allen A. Seaman (631), New Brunswick Department of Natural Resources, Fredericton, New Brunswick, Canada E3B 5H1

Zhou Shangzhe (981), School of Geographical Science, South China Normal University, Guangzhou, China

Vladimir S. Sheinkman (883), Institute of the Earth Cryosphere, Malygina Street, 86, Russian Academy of Sciences, Siberian Branch, Tyumen 62500, Russia; V.B. Sochava Institute of Geography, Ulan-Batorskaya Street, 1, Russian Academy of Sciences, Siberian Branch, Irkutsk 664033, Russia

Georg Stauch (877, 967), Department of Geography, RWTH Aachen University, Templergraben 55, 52056 Aachen, Germany

Rudolph R. Stea (631), Stea Surficial Geology Services, 851 Herring Cove Road, Halifax, Nova Scotia, Canada B3R 1Z1

Hans-Jürgen Stephan (149), Köhlstraße 3, D-24159 Kiel, Germany

Margaret A. Stewart (261), Department of Earth Science and Engineering, Imperial College London, South Kensington Campus, London SW7 2AZ, United Kingdom; Present address: Neftex Petroleum Consultants, 97 Milton Park, Abingdon, Oxfordshire OX14 4RY, United Kingdom.

Martyn S. Stoker (261), British Geological Survey, Murchison House, West Mains Road, Edinburgh EH9 3LA, United Kingdom

N.G. Sudakova (337), Department of Geography, Lomonosov Moscow State University, Vorob'evy Gory, Moscow 119899, Russia

John Inge Svendsen (279), Department of Earth Science and Bjerknes Centre for Climate Research, University of Bergen, Allégt. 41, NO-5007 Bergen, Norway

Kent M. Syverson (537), Department of Geology, University of Wisconsin, Eau Claire, Wisconsin 54702, USA

John P. Szabo (513), Department of Geology & Environmental Science, University of Akron, Akron, Ohio 44325-4101, USA

Levan Tielidze (141), Ivane Javkhishvili Tbilisi State University, Vakhushti Bagrationi Institute of Geography, department geomorphology and geoecology, Merab Aleksidze str. Block 1.0193, Tbilisi, Georgia

Jaroslav Tyráček (37), Czech Geological Survey, Klárov 3, 118 21 Praha, Czech Republic; This study is dedicated to the memory of our colleague Jaroslav Tyráček (1931-2010) for his substantial contribution to the study of Pleistocene glaciations in our territory

Petru Urdea (305), Department of Geography, West University of Timişoara, V. Pârvan Nr.4, 300223 Timişoara, Romania

Daniel Utting (631), Nova Scotia Department of Natural Resources, 1701 Hollis Street, Halifax, Nova Scotia, Canada B3J 2T9

Lorenzo Vázquez-Selem (849), Instituto de Geografía, Universidad Nacional Autónoma de México, Ciudad Universitaria, 04510 México, D.F., Mexico

Jaap J.M. van der Meer (247), Department of Geography, Queen Mary University of London, Mile End Road, London El 4NS, United Kingdom

Dirk van Husen (15), Simetstraße 18, A-4813 Altmünster, Austria

A.A. Velichko (337), Laboratory of Evolutionary Geography, Institute of Geography, Russian Academy of Sciences, Staromonetny 29, Moscow 109017, Russia

Tomaž Verbič (385), Arhej d.o.o., Institute of geoarchaeology, Rimska 1, 1000 Ljubljana, Slovenia

Tore O. Vorren (361), Department of Geology, University of Tromsø, N-9037 Tromsø, Norway

Stefan Wansa (149), Landesamt für Geologie und Bergwesen Sachsen-Anhalt, Postfach 156, D-06035 Halle, Germany

Brent Ward (563), Department of Earth Sciences, Simon Fraser University, Burnaby, British Columbia, V5A 1S6 Canada

Jamie C. Woodward (175), Quaternary Environments and Geoarchaeology Research Group, Geography, School of Environment and Development, The University of Manchester, Manchester, United Kingdom

Nicolás E. Young (427), Department of Geological Sciences, University at Buffalo, Buffalo, New York 14260, USA

Vitālijs Zelčs (221), Faculty of Geographical and Earth Sciences, University of Latvia, Rainis Boulevard 19, Riga LV-1586, Latvia

Marek Zreda (393), Department of Hydrology and Water Resources, University of Arizona, Tucson, Arizona, USA

Preface

This new volume presents evidence of glaciations from around the world. This compilation of 78 chapters includes new data, views and debates on the nature of global glaciations and aims to represent the current position of glacial research in Quaternary studies.

At the XIV Congress of the International Quaternary Association (INQUA) in Berlin in 1995, the INQUA Commission on Glaciation decided to form a new working group entitled 'Extent and Chronology of Glaciations'. The aim of this project was to provide a comprehensive synthesis of the extent and chronology of Quaternary glaciations as currently known throughout the world.

Following on from a series of presentations beginning at the XV INQUA Congress in Durban, South Africa, in 1999, the first edition of our compilation was published shortly after the XVI INQUA Congress in Reno, Nevada (2003), in a set of three volumes:

Ehlers, J., Gibbard, P.L. (Eds.), 2004a. Quaternary Glaciations—Extent and Chronology, Part I: Europe. Developments in Quaternary Science, vol. 2a, Elsevier, Amsterdam.

Ehlers, J., Gibbard, P.L. (Eds.), 2004b. Quaternary Glaciations—Extent and Chronology, Part II: North America. Developments in Quaternary Science, vol. 2b, Elsevier, Amsterdam.

Ehlers, J., Gibbard, P.L. (Eds.), 2004c. Quaternary Glaciations—Extent and Chronology, Part III: South America, Asia, Africa, Australasia, Antarctica. Developments in Quaternary Science, vol. 2c, Elsevier, Amsterdam.

These volumes contained 111 chapters from different parts of the world, and each volume also included CDs on which the compiled map information was displayed.

Following the success of these first volumes, this new volume of *Quaternary Glaciations—Extent and Chronology: A Closer Look* takes the development of the project a stage further. We have incorporated many new details. The evidence of glaciations from around the world is presented in 78 chapters, all of which have been revised, whilst many have been completely rewritten. Indeed, some chapters are entirely new, sometimes with new or additional authors, and including titles that did not appear in the first editions. A small number of regions that were represented in the first editions have not made it to the second edition. This occurred where authors did not submit a new chapter and probably reflects limited progress since the first edition in these particular areas.

The book includes all regions of the world in a single volume, providing a focus on the extents, timing and wider significance of Quaternary glaciations. The main advances reported in this book include better understanding of the timings of glaciations as well as the detailed extent and subdivision of different glaciations. These advances are supported by new digital maps plotted at a larger scale. With these maps, the accuracy of the information is increased by a factor of 16 on the previous plots. These are available for downloading from the website http://booksite.elsevier.com/9780444534477/.

Unlike in the previous books, this new volume includes an introductory chapter which we include in response to some reviews of the earlier compilation. This introduction summarises the compilation's key findings and raises some of the main common themes that are apparent from study of the glacial records around the world. Thus, we deliberately wish to keep this opening preface short and further background to the book can be found in Chapter 1.

As before, we hope that this compilation, like its predecessors, will prove to be useful to the broad Quaternary community, will stimulate further research in critical regions, by highlighting those areas where more needs to be known or where controversies need to be settled, and ultimately will allow a better reconstruction of earth's glacial history.

This work would not have been possible without the substantial and enthusiastic support and cooperation of all our colleagues who have contributed to this book. This has enabled a particularly efficient process from chapter submission to final publication. Moreover, we must thank all the staff at Elsevier BV involved in the rapid and professional publication of this volume. Finally, this new revised volume would not have been begun without the enthusiastic support of our friend and colleague Professor Jaap van der Meer. To him, and the many others that are too numerous to be acknowledged here, we express our sincere thanks.

Jürgen Ehlers
Philip Gibbard
Philip Hughes

Chapter 1

Introduction

Jürgen Ehlers[1,*], Philip L. Gibbard[2] and Philip D. Hughes[3]

[1]*Geologisches Landesamt, Billstraße 84, D-20539 Hamburg, Germany*
[2]*Cambridge Quaternary, Department of Geography, University of Cambridge, Downing Place, Cambridge CB2 3EN, United Kingdom, plg1@cam.ac.uk*
[3]*Quaternary Environments and Geoarchaeology Research Group, Department of Geography, School of Environment and Development, Arthur Lewis Building, The University of Manchester, Manchester M13 9PL, United Kingdom, philip.hughes@manchester.ac.uk*
*Correspondence and requests for materials should be addressed to J. Ehlers. E-mail: juergen.ehlers@bsu.hamburg.de

1.1. INTRODUCTION

This new edition of *Quaternary Glaciations—Extent and Chronology* gathers together the evidence of glaciation from around the world in 77 chapters. The book combines all regions of the world into one volume and provides a closer look at the extents, timing, and wider significance of Quaternary glaciations. The main advances reported in this book include better understanding of the timings of glaciations—especially through the application of cosmogenic nuclide analyses—and better understanding of the detailed extent and subdivision of different glaciations. These advances are supported by new digital maps, the detailed production of which is presented below. In addition, this introduction summarises the compilation's key findings and raises some of the main common themes that are apparent from the glacial records around the world. However, given the large number of excellent contributions, it is not possible to mention each chapter and it is recommended that readers explore the volume and discover for themselves the new findings from specific areas.

The editors would have liked to produce the volume in full colour. Unfortunately, that was not possible. However, this volume contains far more colour illustrations than the previous edition.

1.2. THE DIGITAL MAPS

For more than 150 years, geologists have mapped the traces of Quaternary glaciations. One main purpose of creating the maps is to obtain reproducible results. Where scale and coordinates are given, the results are considered to be reproducible. But different scales and different projections hinder comparison. In High Asia, for instance, the printed maps do not show much more than that there are very different views on that topic. However, where exactly the ice margins are supposed to be located is very difficult to reconstruct. True comparison is only possible where digital maps in a digital Geographic Information System (GIS) are employed.

At the INQUA Congress in Berlin 1995, a new investigation of the glacial limits seemed desirable. The last such attempt, coordinated by Denton and Hughes ('The Last Great Ice Sheets', 1981), was by that time almost 25 years old and much was out of date. During the congress, a working group on the *Extent and Chronology of Quaternary Glaciations* of the *Commission on Glaciation* was initiated. The first question that work group had to solve was to find a suitable base map. A scale of 1:1,000,000 seemed desirable. However, the printed World Map of that scale was compiled over a period of about 70 years (since 1913) and it was still incomplete. Fortunately, a digital map had become available just in time that seemed ideal for the purpose: the Digital Chart of the World (DCW).

1.2.1. The Digital Chart of the World

The DCW consists of 2094 tiles, each of which (except near the South Pole) is $5 \times 5°$ in size. To identify the tiles, the globe had been subdivided into 12 latitudinal and 24 longitudinal stripes, named using the letters of the alphabet, starting from the south and the west. The resulting $15 \times 15°$ areas were again subdivided into nine tiles, which were numbered. A tile named NK11 (Paris) would be between $0°$ and $15°$E (letter N), between $45°$ and $60°$N (letter K) and the southeasternmost tile of that section (number 11)—thus located between $0°$ and $5°$E and $45°$ and $50°$N (Fig. 1.1). Six hundred of these original tiles are required to cover the formerly glaciated parts of the globe. Gaps in the hypsometry layer of the DCW could be filled with information from the GTOPO30 terrain model (see below).

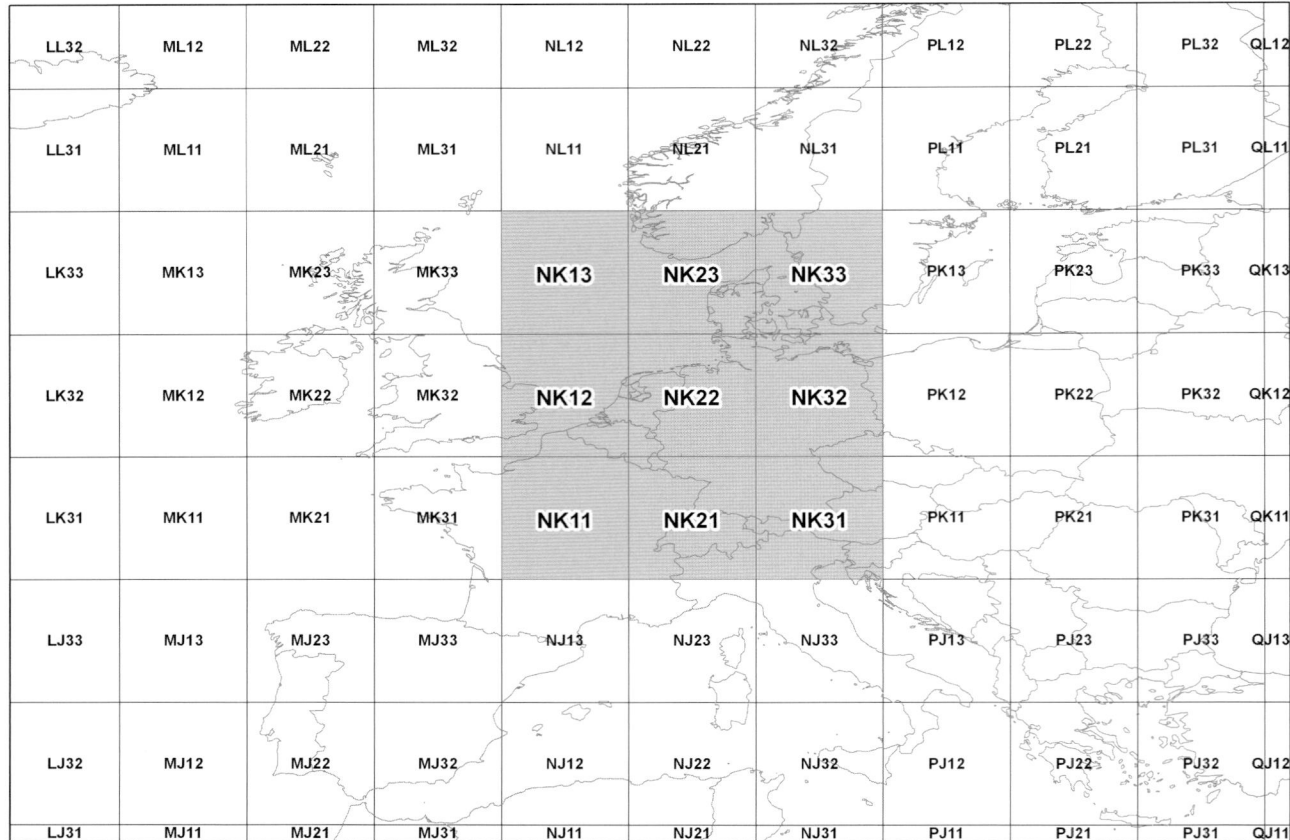

FIGURE 1.1 Naming scheme of the digital maps.

In 1999, when the coordinators of the project started to work with the DCW, the correctness of their base map was not in doubt. However, as soon as better digital maps got available, the shortcomings of the DCW became obvious. The horizontal accuracy was insufficient. The position of each individual line, point or polygon was found to be several hundred metres out of place, in extreme cases even more. At 1:1,000,000, where 1 km is equal to 1 mm, that is still acceptable. However, as soon as comparison with higher resolution topographic data is attempted, the error can no longer be tolerated. The present maps aim to be accurate at a scale of 1:250,000 (1 km equal to 4 mm). Freely available satellite-based topographic data have a much higher accuracy and allow mapping down to 1:75,000.

1.2.2. The GTOPO30 Terrain Model

The GTOPO30 terrain model had been compiled by the United States Geological Survey (USGS) in 1998 and can today be downloaded without charge from various websites. It has a horizontal resolution of ca. 1 km and is very suitable for overview maps, although at a scale of 1:1,000,000, it is at its limit. GTOPO30 was compiled from published sources. For most parts of the world, it is based on the *Digital Terrain Elevation Data* (DTED). Whilst the terrain model seems flawless in global overviews, at a higher resolution artefacts appear. One of them consists of diagonal stripes, especially in Africa and the Middle East. More disturbing is the dissolution of the image into $1 \times 1°$ blocks. This is strongest in areas of low relief, such as the lowlands of West Siberia. When used as an image, this effect can be masked by using the right altitudinal intervals. When trying to reconstruct the extent of the former ice-dammed lakes, however, the effects of this fault are clearly visible.

1.2.3. The Vector Map Level 1 (VMAP1)

Today, the DCW has been re-named Vector Map Level 0 (VMAP0), but the contents have not been updated. The VMAP1 is far more accurate, it is intended for a scale of 1:250,000. Unfortunately, only part of this comprehensive map system, that like the DCW covers the whole globe, has been made available to the public. The published parts include most of the United States, about half of the Russian Arctic, parts of the Central Asian lowlands, Japan, parts of South America and a few other regions such as the Kerguelen Islands. For the present publication, the VMAP1 layers have been adjusted the original DCW tiles format.

The VMAP1, like the DCW before it, is a map-based map. To create it, published maps have been digitised. This results in occasional 'faults' where adjoining map sheets do not match. There are also occasional gaps where either the published map was not available (for instance in minor parts of the former Soviet Union), or it was available but did not contain certain types of information (e.g. contours).

The VMAP1 dataset includes contours, although there is no altitudinal colouring. To make the information easier to read, the contours can be underlain by the 'old' DCW hypsometric tints.

1.2.4. GeoBase (Canada)

Additional topographic data and terrain models are available for several countries. For example, it is possible to download digital files of the Canadian terrain model (scale either 1:250,000 or 1:50,000) from GeoBase, as well as the roads, lakes, rivers and a few other layers without charge. The data have to be transformed to WGS84. The data come in $1 \times 1°$ tiles. This means that for one tile of our digital maps, 25 tiles of the coarser version of the GeoBase terrain model are required. This makes the application of GeoBase possible for the present purpose, but very slow.

1.2.5. The Shuttle Radar Topography Mission (SRTM)

For most parts of the globe, the easier to use and higher resolution SRTM terrain model is available. It covers the area between 60°N and 60°S. It was created by radar measurements from the Space Shuttle in 2000. The data have undergone various corrections. The voids in the original coverage have been filled, the coastlines have been corrected and the surfaces of lakes and the sea have been levelled. The data are easy to download and to import into ArcView. They can be used to calculate contours, so that a highly reliable and very accurate base topography is available. The SRTM topography is good enough to show individual drumlins. However, because it is radar-based and largely uncorrected for contents, it shows forest areas as elevations. This can be seen nicely, when the SRTM model is combined with satellite imagery.

When trying to identify the glacial features shown in the maps, such as key sites, additional information is desirable.

The river courses are visible in the SRTM terrain model, at least in hilly terrain, but there is no possibility to colour them blue, unless information from VMAP1 can be used. Lakes and rivers can be calculated from satellite imagery by isolating the black areas in infrared channel 4, but this method is very time-consuming and also hindered by cloud shadows and—more importantly—by shadows in narrow valleys.

1.2.6. The Open StreetMap

One possible additional source of information is the Open StreetMap (OSM). Particularly in densely populated areas such as in Central Europe, where no official topographic data are available without costs, the roads and coastlines and places given are a valuable source of information. And at a scale of 1:250,000, the OSM data are very reliable. The only problem is that the information is incomplete. In contrast to the several decades old data from DCW and VMAP1, the OSM is continuously updated. In some respects, this goes far beyond what is required in the current project (e.g. by including the street names). In other respects, however, it is still imperfect. Places like Hamburg, that are completely mapped, are the exception.

It should be kept in mind that where other ('official') data sources are available, such as in Canada, they may contain better quality information than the independent free sources.

1.2.7. The ASTER Terrain Model

In 2009, the altimetric data from the Advanced Spaceborne Thermal Emission and Reflection Radiometer (ASTER) instrument were made available to the public. ASTER is an imaging instrument on the Terra satellite launched in December 1999. On its sun synchronous, north to south path around the globe, it covers the whole world and through stereoscopic imagery that allows the construction of a terrain model. The nominal resolution is slightly better than that of the SRTM. Unfortunately, the images are marred by numerous artefacts—hills, holes and 'mole runs'—which cannot be eliminated at present (Fig. 1.2). This misinformation is a major nuisance in regions such as the Arctic, where natural isolated hills (e.g. pingos), holes (e.g. thermokarst lakes) and 'mole runs' (e.g. eskers) are frequent. Therefore, the ASTER data have not been used in this project.

1.2.8. Printed Maps

Scanned topographic maps are freely available at various scales, especially for Canada (Geogratis Canada), the

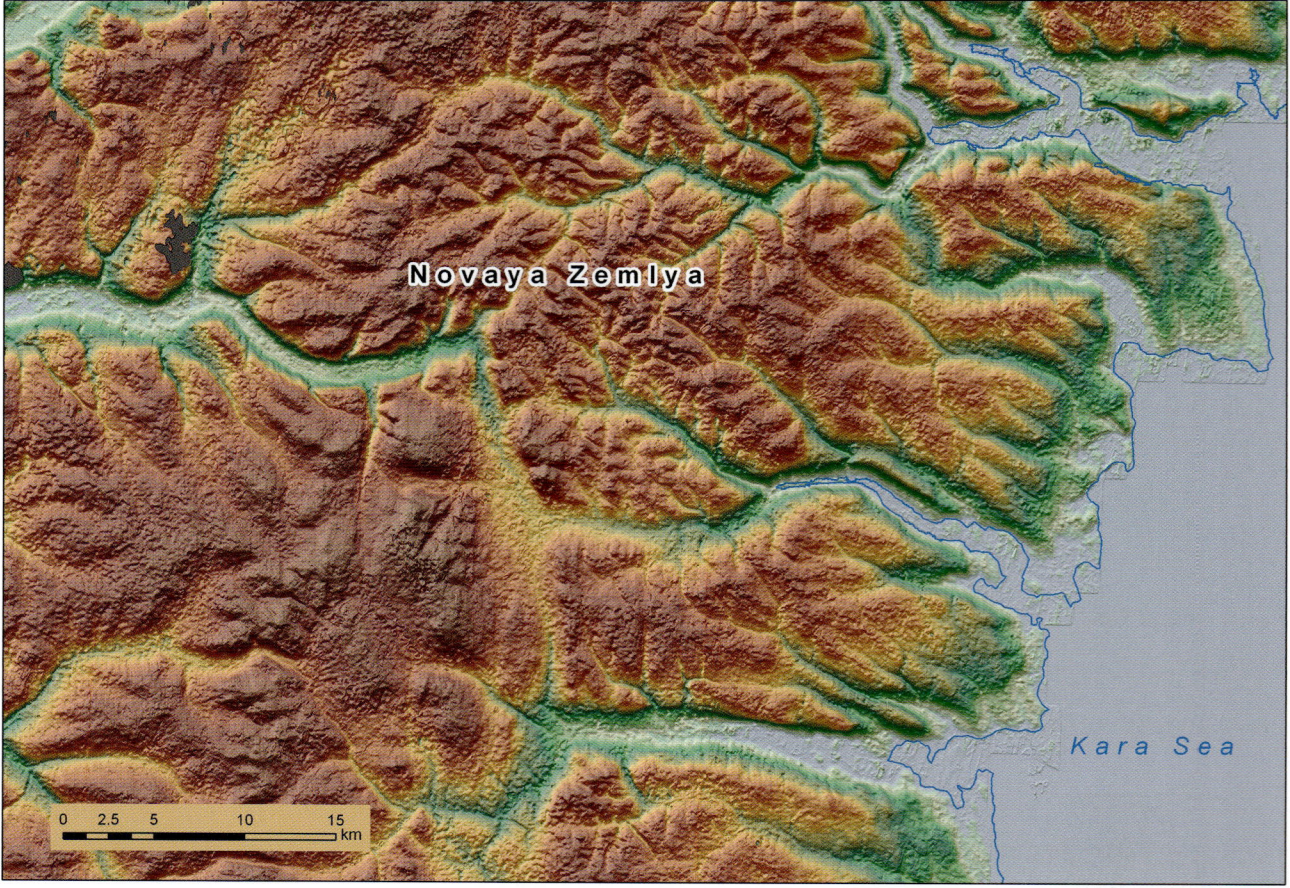

FIGURE 1.2 ASTER terrain model of part of Novaya Zemlya, showing the typical holes, hills and 'mole runs'.

United States (USGS and other sources) and the former Soviet Union (mapstor). However, it must be borne in mind that their projection can differ from the WGS84 system used in the present maps and in all satellite data. For instance, the highly accurate, large-scale Russian topographic maps use a different datum (Pulkovo), which means that the maps would be up to several hundred metres offset, in the absence of mathematical transformation.

1.2.9. Digital Maps: Conclusions

With the larger scale digital base maps used in the present publication, the accuracy of the information is increased by a factor of 16. The difference is striking (Fig. 1.3). Unfortunately, no single, uniform base map can be used at the 1:250,000 scale (Fig. 1.4). Therefore, in order to cope with the different levels of accuracy encountered, some of the layers of information have been grouped, so that only consistent information is shown together.

The SRTM terrain model, VMAP1 data and the OSM information are regarded as 'true' at a scale of 1:250,000. The DCW and GTOPO30 are only 'true' at the 1:1,000,000 scale. Some other layers of information, such as Art Dyke's deglaciation maps of North America from the first edition of this work (Dyke, 2004), are only 'true' at 1:10,000,000 or even smaller scales.

In this publication, all the spatial data are given in shapefiles to allow for easy import into other GIS systems. To take full advantage of the information collected in this project, a GIS is required. The authors are fully aware of the fact that not all readers will have ArcGIS at their disposal. To allow the non-GIS community a glimpse at the maps, some are available as KMZ files that can be read using Google Earth. Nevertheless, the user must be aware of the fact that those files are only images, and their contents are restricted to what has been selected by the author.

The authors hope that the current map set will be widely used, and that it will form the basis for further investigations.

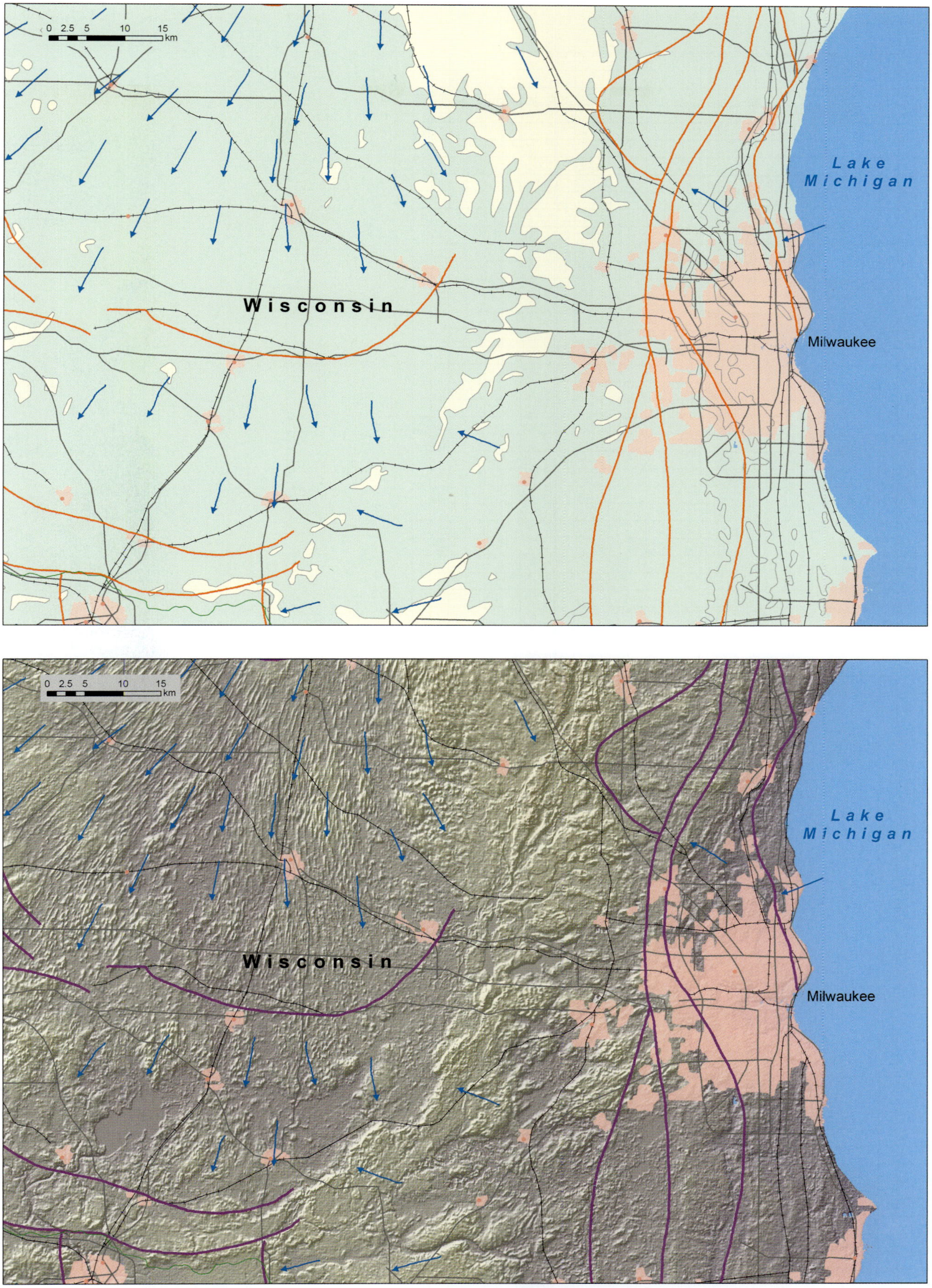

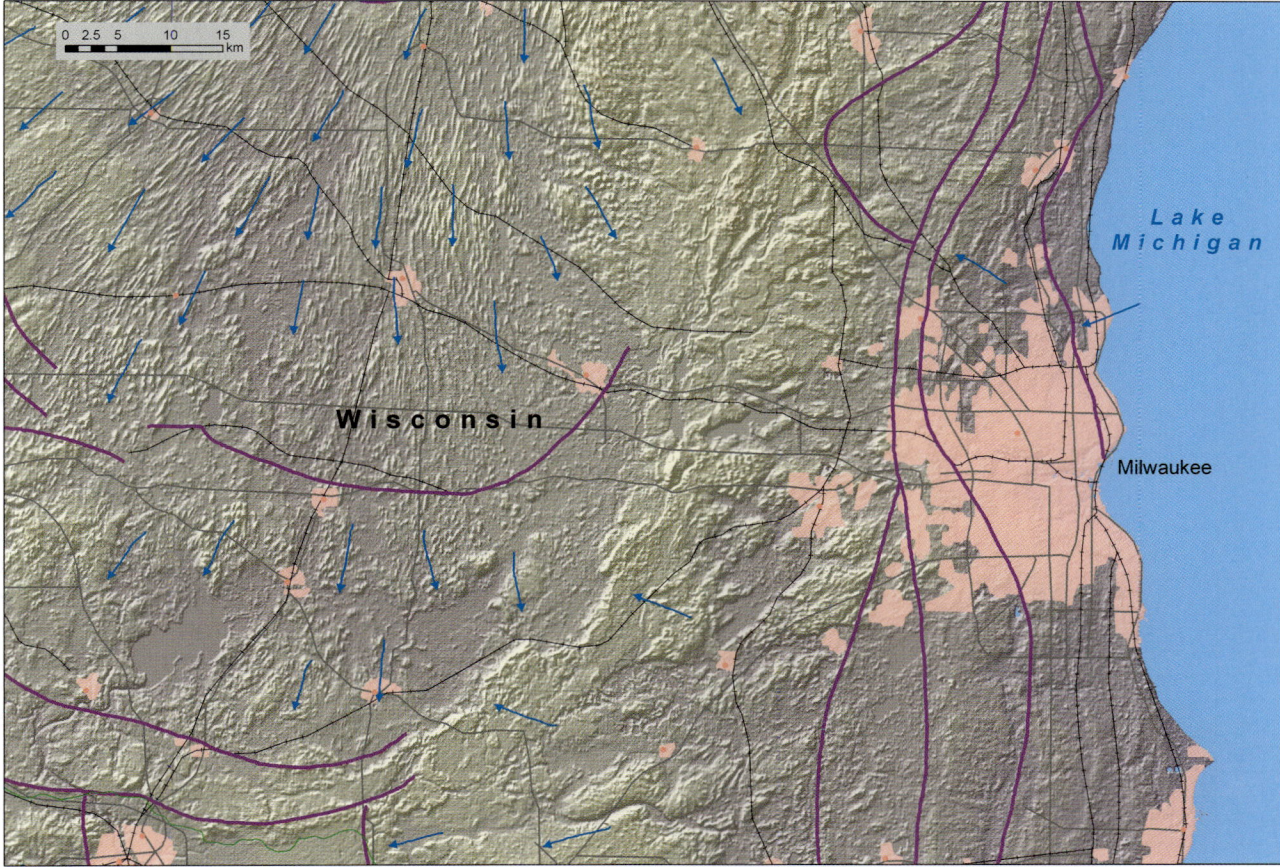

FIGURE 1.3 Comparison of the digital maps from the first edition, based on the Digital Chart of the World, with a digital map from the present edition, using the SRTM terrain model.

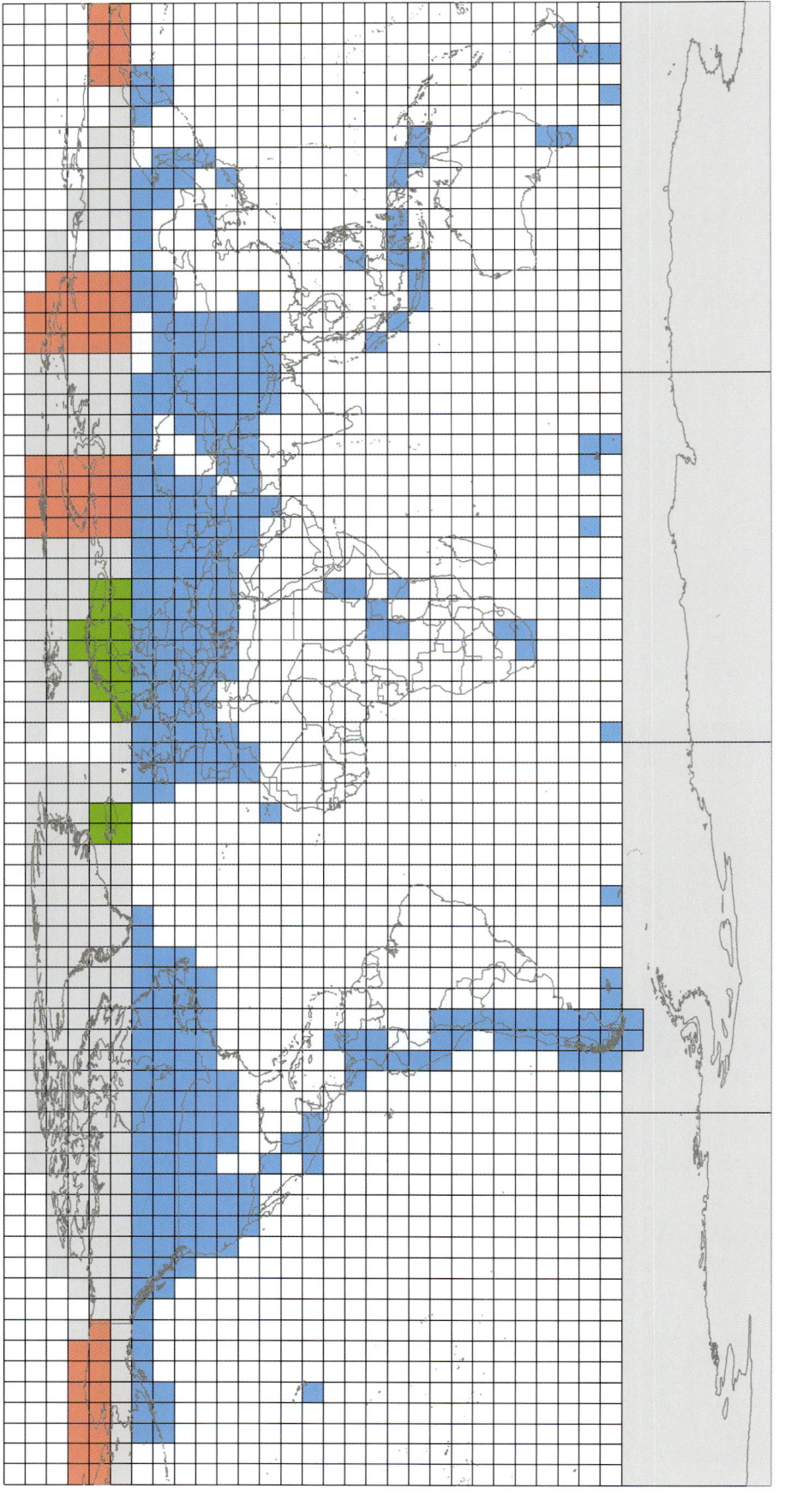

FIGURE 1.4 The different base maps used in the present volume.

1.3. THE GEOLOGY

1.3.1. Prelude to the Quaternary

The initiation of conditions that resulted in Quaternary (Pleistocene and Holocene series) glaciation resulted from the long-term declining cooling trend in world climates that began early in the previous Tertiary Period. Apart from some limited activity in the Eocene both in the southern and northern hemisphere (Stickley et al., 2009; Ingólfsson, 2004; Tripati et al., 2005), significant glaciation began in the late Oligocene (ca. 35 Ma) in eastern Antarctica (Ingólfsson, 2004). It was followed by mountain glaciation through the Miocene (23–5.3 Ma) in Alaska, Greenland, Iceland and Patagonia, and later in the Pliocene (5.3–2.6 Ma) in the Alps, the Bolivian Andes and possibly in Tasmania (see respective chapters in this book). From the Neogene, glacially derived ice-rafted debris is found in ocean-sediment cores from the North Atlantic region, including the Barents Sea (Moran et al., 2006) and areas adjacent to Norway, northern and south-eastern Greenland, Iceland and northern North America, and in the Southern Ocean off-Antarctica.

1.3.2. Glaciation During the Quaternary

Since the recognition in the mid-nineteenth century that glaciers had been considerably more extensive than at present, the Quaternary has been synonymous with glaciation of the mid-latitudes. Today evidence from both the land and ocean-floor sediment sequences demonstrates that the major continental glaciations occurred repeatedly over what are now temperate regions of the Earth's surface. Extensive ice-rafting, an indication that glaciers had reached sea level, is found from the earliest cold stage (2.6–2.4 Ma) in both the North Atlantic and North Pacific oceans (cf. Haug et al., 2005). Extensive glaciations were not restricted to the high latitudes. In Africa, Pleistocene glaciations were common to many mountains over 4000 m a.s.l. in the East African mountains such as Mount Kilimanjaro, Mount Kenya and the High Atlas. Glaciations also occurred on much lower mountains, such as in Algeria where large glaciers formed on mountains as low as 2300 m a.s.l. (Chapter 76). Thus, Pleistocene glaciation was a global affair—with significant presence in the tropics as well as the mid- and high latitudes.

The climatic variations that have characterised the Earth's climate during the late Cenozoic and indeed before are controlled by variations of the Earth's orbit around the Sun which affects the receipt of solar energy at the Earth's surface. These Milankovitch variations, named after their discoverer, are responsible for the cyclic climate changes that characterise the Quaternary and indeed much of Earth's history. One of the most critical ways they are expressed is through the development of 'Ice Ages' or periods when glaciation extended across large areas of the Earth. The Early Pleistocene (2.6–0.8 Ma) was characterised by climatic fluctuations dominated by the 41 ka precession cycle, during which relatively few cold periods were sufficiently cold and long to allow the development of substantial ice sheets. Only 14 of the 41 cold stages of that period display evidence of major glaciation (Ehlers and Gibbard, 2011). They include the Plio-Pleistocene boundary events Marine Isotope Stage (MIS) 104, 100 and 98, together with Early Pleistocene MIS 82, ?78, 68, 60, 58, 54, 52, 36, 34, ?30 and 26 which reach ^{18}Oocean‰ of ca. 4.6–5. It is not until the transition in dominant orbital cyclicity to the 100 ka cycles, that began ca.1.2 Ma and was fully established by about 800 ka ('mid-Pleistocene transition' Tziperman and Gildor, 2003), that the cold periods (glacials) were regularly cold and long enough to allow ice sheet development on a continental scale, outside the polar regions. However, it is during MIS 22 (ca. 870–880 ka) that the first of the 'major' cold events reached critical values of ca. 5.5 or above ^{18}Oocean‰ equivalent to substantial ice volumes that typify the glaciations of the later Pleistocene (i.e. MIS 16, 12, 10, 6, 4–2). Potentially, therefore, it is likely that there were a minimum of 20 periods during which extensive glaciation could have developed during the past 2.6 million years, with the most extensive (ca. 5–6 periods) being limited to the last 900 ka (Gibbard and Cohen, 2008) In fact, recent findings imply that significant glacial expansion even occurred during MIS 14 (Chapter 17 and Giraudi et al., 2010).

Precisely where these glaciations occurred and how far they extended is very difficult to determine, given that the remnants of less extensive early glaciation tend to be obliterated and mostly removed by later, more extensive advances. Although this is so in all terrestrial areas, it is especially difficult in mountain regions where the preservation potential of older sequences rapidly diminishes with time and subsequent glaciation. However, examination of the frequency of glaciation through the Cenozoic indicates that glaciation in the southern hemisphere having been established first, principally in Antarctica and southern South America, occurred continually from the early Neogene to the present day (Ehlers and Gibbard, 2008). By contrast, northern-hemisphere glaciation, although initially somewhat restricted, increased markedly at the beginning of the Quaternary, increasing again in frequency in the latest Early Pleistocene and reaching very high levels in the Middle to Late Pleistocene. Whilst this pattern is not unexpected, the striking increase in ice sheets through the Quaternary clearly emphasises that worldwide glaciation is in effect a northern-hemispheric phenomenon.

1.3.3. Plio-Pleistocene Glaciation

Evidence of glaciation is widespread from throughout the Quaternary and indeed the Neogene in the northern Hemisphere. The longest sequences are restricted to Alaska, and the adjacent North-West Territories of Canada which, together with Greenland and the Rockies, preserve evidence of glaciation from the Neogene to the present. In northern Canada and Alaska, the oldest till and accompanying ice-rafted detritus in marine settings date from the Early Miocene, with regionally widespread glaciation occurring in the Pliocene and regularly throughout the Pleistocene (cf. Haug et al., 2005; Chapter 32). In adjacent British Columbia, a comparable sequence is found, particularly in the north (Chapter 44. Similarly, in Greenland and Iceland, glaciation began in the Miocene, occurring regularly through the Pliocene and onwards to the present day in the mountains (Chapters 50 and 16). Likewise, in Norway, its adjacent offshore and the neighbouring Barents Sea, glaciation is recorded from the Early Miocene, Early Pliocene and Plio-Pleistocene (Knies et al., 2009; Chapters 22 and 27). In the Rockies of the United States, a much shorter glacial sequence occurs, although a Plio-Pleistocene-aged till is known from California (Chapter 34). In Europe, glaciation before the Middle Pleistocene is generally represented only by ice-rafted material, outside the mountain regions (e.g. in the Netherlands, lowland Germany, European Russia and Britain) (Chapters 7, 13, 20 and 26).

In the southern hemisphere, glaciation is much longer established, as noted above. Here the ice already formed in the Late Eocene—Early Oligocene in East Antarctica (Ingólfsson, 2004; Tripati et al., 2005; Miller et al., 1987) and built-up in a step-like pattern through the Neogene. The present polar conditions were already established by the Early Pleistocene after 2.5 Ma (Ingólfsson, 2004). A similar history is known from the Piedmont areas of Argentina and Chile, where substantial ice caps were established by 14 Ma (Heusser, 2003; Rabassa, 2008). Till deposits interbedded with basalt flows indicate the occurrence of glaciation even before the Pliocene–Pleistocene boundary (ca. 2.6 Ma) and widespread lowland glaciation became established between 2.05 and 1.86 Ma (?ca. MIS 68–78), followed by the 'Great Patagonian Glaciation' that took place at 1.15–1.00 Ma (ca. MIS 30–34) (Chapters 51 and 52). Further north, there is little documented evidence of tropical Andean glaciers from the Plio-Pleistocene. It is estimated that the tropical Andes have attained most of their present elevation only during the past 5–6 Ma, However, the earliest glaciation recorded in the Bolivian Andes dates from at least 3.25 Ma (Chapter 56). Further, in Colombia, the first glaciations are dated to near the Gauss/Matuyama magnetic reversal at 2.6 Ma (Chapter 58). The earliest records in Australasia are found in New Zealand from the Plio-Pleistocene (2.6 Ma: MIS 98–104; Chapter 75).

1.3.4. Early and Middle Pleistocene Glaciations

The 'glacial' Pleistocene effectively begins with extensive glaciation of lowland areas, particularly around the North Atlantic region, and the intensification of global cold period (glacial) climates, in general. It coincides with the 'middle Pleistocene transition' (1.2–0.8 Ma) when the 100 ka Milankovitch cycles became dominant and caused the cold periods to become sufficiently cold for long enough to allow the development of continental-scale ice sheets.

The till sheets of the major glaciations of the 'glacial Pleistocene' are found throughout Europe. In northern Europe, till sheets characterise large areas of the lowlands and they are also found at the floors of the adjacent seas (Chapters 3, 6, 7, 8 and 9). New investigations have revealed to what a large degree the North Sea floor was shaped by repeated glaciation (Chapter 21). Further south, in central and southern Europe, till is restricted to the mountains and piedmonts (Chapters 2, 4, 11, 15, 17 and 24). In northern Europe, widespread lowland glaciation began in the early Middle Pleistocene shortly after the Brunhes/Matuyama palaeomagnetic reversal (780 Ka). The phases represented include the Weichselian (Valdaian, MIS 4–2), Saalian (Dniepr and Moscovian, MIS 6, 8 and 10), Elsterian (Okan, MIS 12) and the Donian (Narevian, Sanian, MIS 16). More limited glaciation may also have occurred in the circum-Baltic region during the latest Early Pleistocene (MIS 20 and 22). The evidence for Early Pleistocene glaciations in this region is restricted to Latvia (Chapter 18), Poland (Chapter 23) and possibly Lithuania (Chapter 19), although current research in central Jylland, Denmark, may also reveal evidence for pre-Cromerian glaciation in this area (Chapter 5). Curiously, evidence for early Middle Pleistocene glaciation is absent from the North Atlantic and Norway, whilst it is certainly present in Denmark, the Baltic region and European Russia. In the Italian Dolomites, glaciation becomes established in MIS 22 (Muttoni et al., 2003). Comparable evidence is also found from north of the Alps in Switzerland and southern Germany (Chapter 14). Further to the west, in the Pyrenees, the oldest glaciation identified is of late Cromerian age (MIS 16 or 14) (Calvet, 2004). Widespread lowland glaciation again is first seen in North America in MIS 22 or 20 (Chapters 32 and 49). From this point onwards, major ice sheets covered large regions of the continent during the Middle Pleistocene pre-Illinoian events MIS 16, 12, 8 and 6 (Illinoian s.s.) and the Late Pleistocene MIS 4–2 (Wisconsinan). In Mexico, the oldest moraines on volcanoes have been dated at 205 to 175 ka and probably correspond to an advance early in MIS 6 (Chapter 61). Evidence from East Greenland suggests that the southern dome of the ice sheet may have almost disappeared during the Eemian Stage interglacial (ca. MIS 5e). However, dating of the basal ice in the Dye 3 ice core has shown that ice has been present over this locality since at least MIS 11 (Chapter 50).

According to Chinese investigations, glaciation of Tibet and Tianshu is not recorded before the Middle Pleistocene, of which the MIS 12 glaciation was the most extensive. Four discrete Pleistocene glaciations have been identified on the Qinghai-Tibetan Plateau and the bordering mountains. These four main Pleistocene glaciations are correlated to MIS 18–16, 12, 6 and 4–2. This apparently delayed glaciation of the Himalayan chain might reflect late uplift of high Asia. The Kunlun Glaciation (MIS 18–16) was the most far reaching. Subsequent glaciations have been successively less extensive, probably caused by increasingly arid climates resulting from the progressive Quaternary uplift of the plateau (Chapter 70). On the contrary, Kuhle (Chapter 68) maintains that the last glaciation was most extensive when he envisages an ice sheet covered Tibet. However, this is increasingly a minority view, especially the concept that the Qinghai-Tibetan Plateau was almost entirely covered by an ice sheet. Many workers rigorously maintain that Pleistocene glaciations over the Qinghai-Tibetan Plateau were much more restricted (including Chapter 70; cf. numerous references in Lehmkuhl and Owen, 2005). Indeed, these authors argue that it is now generally well established that a large ice sheet did not cover the Tibetan Plateau. Nevertheless, this book provides both sides of the argument for this particular region and indeed for several other parts of the world where there is debate over the glacial history.

As in Europe and North America, glaciation increased in intensity throughout the South American Andes from 800 ky to the present day. In the northern Andes, there is some evidence of glacial deposition prior to the Late Pleistocene glaciations in Venezuela and Colombia (Chapters 59 and 58). However, in the southern Andes, Late Pleistocene glaciations were less extensive than during the Early Pleistocene events (Chapters 51, 52 and 54). Both easternmost Tierra del Fuego and the Falkland Islands are thought to have remained largely unglaciated during the Pleistocene (Clapperton, 1993), though there are traces of glacial scouring (Ehlers and Gibbard, 2008). In Australasia, following a 1 Ma break, the glacial record continues in MIS 12, followed by MIS 6, 4 and 3. In Tasmania, the earliest Pleistocene ice advances are thought to have been about 1 million years old but may be older. Middle Pleistocene ice advances occurred during MIS 10, 8 and 6 (Chapter 74). The presence of glaciations in this area during MIS 8, revealed by cosmogenic exposure dating, is interesting because a record of glaciation during this interval is lacking in many other parts of the world.

The succession of glaciations in MIS 16, 12, 6 and 4–2 is striking in that it is repeatedly found in numerous areas of the world, and the absence of records of glaciations during MIS 18, 10 or 8 probably reflect the fact that later glaciations were more extensive. However, it is possible that glaciations during MIS 18, 10 and 8 were, in fact, more widespread than currently realised simply because many dating techniques provide minimum ages for glacial deposits and landforms. This is true for U-series dating of cemented moraines (Chapter 15) and also when applying cosmogenic nuclide analyses to date 'ancient' surfaces (e.g. Chapter 74). Lack of precision when dating glaciations of the Middle Pleistocene and earlier in terrestrial settings means that some uncertainty will remain despite the development of better dating techniques. It is likely that the application of multiple methods of developing glacial geochronologies will be required in order to gain greater precision, and, crucially, to test the validity of the various dating techniques (cf. discussion in Chapters 55 and 57).

1.3.5. Late Pleistocene Glaciations

During the last glacial cycle of the Late Pleistocene, the extent of the glaciation of the Southern Hemisphere differed very little from that of the Pleistocene glacial maximum. Glaciers in Antarctica still reached to the shelf edge, and in New Zealand, Tasmania and South America, the glacier tongues were only slightly smaller than during earlier events (Chapters 51, 74 and 75). On mainland Australia, local mountain glaciation occurred (Chapter 74). In many parts of the Northern Hemisphere, glacial ice reached an extent very similar to the Quaternary glacial maximum. In North America, the differences are very small. Again, most parts of Canada were ice covered, including the shelf areas (Chapter 48). It is the same in Greenland (Chapter 50) and Iceland (Chapter 16). In many of these areas, ice reached maximum extents close to the trough of global ice volume recorded in the marine isotope record (ca. 21 ka). This is instructive because it involved the largest ice masses on Earth at the time: the Laurentide and Antarctic ice sheets.

In Europe, and on mid-latitude mountains around the world, however, the situation was different. New evidence suggests that the North Sea was not fully glaciated during the Weichselian glacial maximum but slightly earlier during MIS 4 (Chapter 21). A similar situation has been invoked for the southern Irish Sea area of the British Isles (Bowen et al., 2002), although recently, several studies have presented evidence that contradicts this view (e.g. Ó Cofaigh and Evans, 2007; Ballantyne, 2010). Further north, over the Barents Sea, glaciation during MIS 4 was more extensive than the later Weichselian glaciations (Fig. 1.5). During the Late Weichselian, an ice sheet covered the Barents Sea and extended well into the Kara Sea but hardly touched the Russian mainland and did not reach onto the Severnaya Zemlya islands (Chapter 27). In glaciated mountain areas outside the major ice sheets, such as in Italy and Greece, the maximum Middle Pleistocene glaciations were sometimes markedly greater extent than the local last glacial maxima (Chapters 15 and 17). This is attributed to a change in equilibrium-line altitude (ELA) because it has a much bigger impact on glacier size in areas characterised only by mountain glaciation than in areas where ice covered the lowlands during multiple glaciations. However, this was not the case everywhere. For example, in

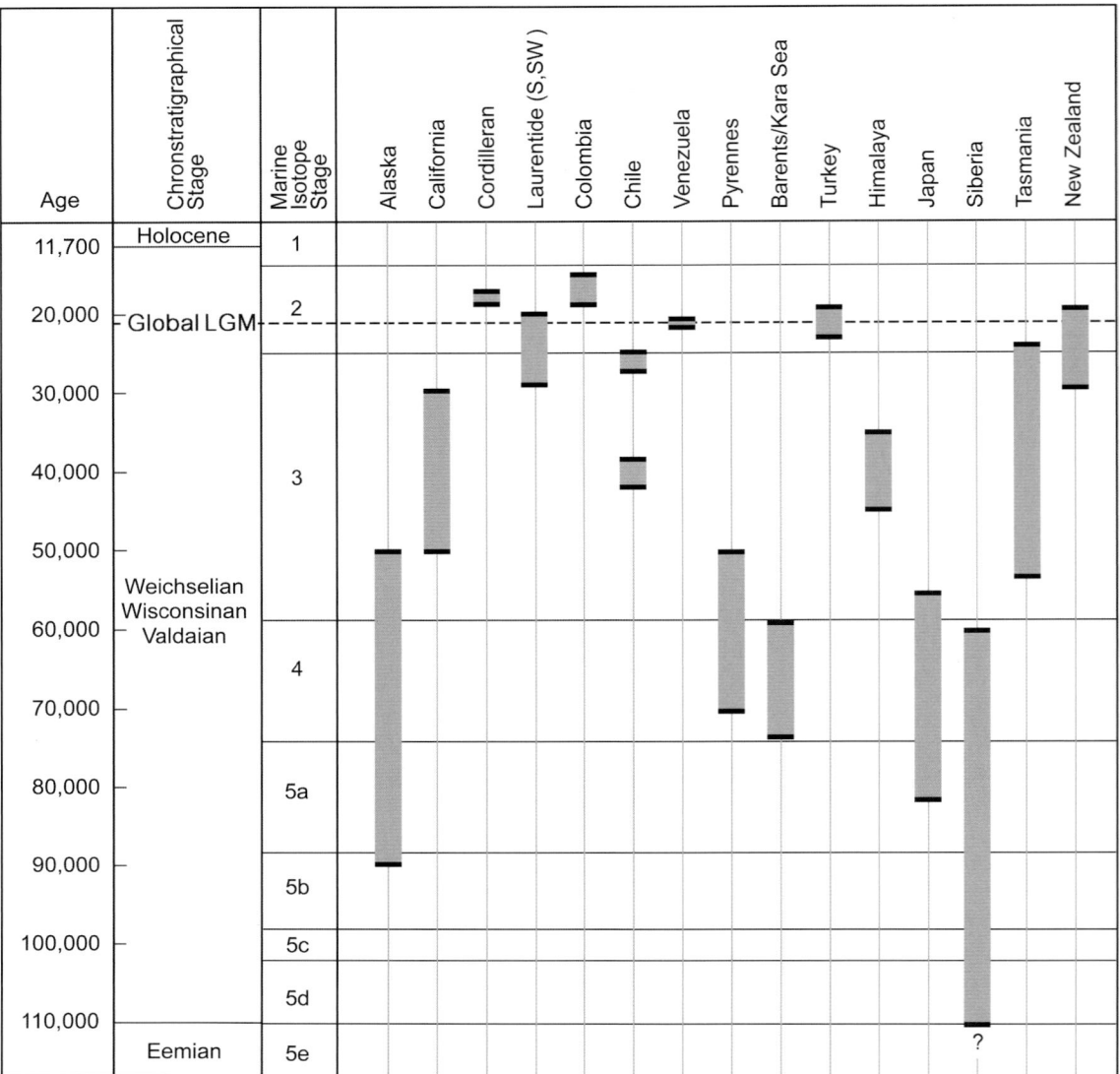

FIGURE 1.5 Timings of maximum extents of glaciers in different parts of the world. The global LGM in this diagram refers to the trough in the marine isotope record at (Martinson et al., 1987) and the associated global eustatic sea level low (Yokoyama et al., 2000) dated to ca. 21 ka cal. BP. This figure just represents a selection of records from around the world and illustrates the differing timing of glacier maxima in different places.

Romania and Turkey, only Late Pleistocene glaciations have been found (Chapters 24 and 30). Mountains often display contrasting geochronologies between areas and sometimes even within the same mountain range (Hughes and Woodward, 2008). This former situation is likely to reflect regional differences in moisture supply, whilst the latter situation, whereby different geochronologies are presented for the same mountain area, are more likely to reflect problems with the reliability or interpretation of dating techniques or results—as has been discussed in this book for the Pyrenees (Chapter 11).

As in parts of Europe, the maximum ice advance of the last glaciation occurred in MIS 4 (and in some places through to MIS 3) in the mid-latitude mountains of Japan (Chapter 72) (Fig. 1.5) and Taiwan (Chapter 71). However, at mid-latitudes in the southern hemisphere, there is still debate as to whether MIS 4 was characterised by larger glaciation than MIS 2 (Chapter 75). Nevertheless, it is clear that global glaciations during the last glacial cycle, whether characterised by lowland ice sheets or mountain glaciers, at high-, mid- or low-latitude, were asynchronous (Fig. 1.5). The challenge now is to understand better why this was the case and to further constrain the timings and extents of glaciers in the different parts of the world if the implications of these apparently different timings of glacial maxima are to be more fully appreciated.

1.3.6. Stratigraphical Terminology

During the editing of this book, a proliferation of stratigraphical terms became apparent (some formal, many not) many of which were used to describe glaciations and their associated deposits. Many workers have devised quasi-stratigraphical terms that are not comparable with formally established chronostratigraphical units. This is perpetuated by the use of previously published terminologies that have no formal stratigraphical meaning, yet which are used as if they were equal in status to formally defined units such as the Würmian, Wisconsinan, etc. This was especially the case for glacial events recognised during the last glacial cycle and, as described above, the use of the term and acronym Last Glacial Maximum (LGM) is often very loosely applied (as noted above). Authors also often interchange between climatostratigraphical and chronostratigraphical terms, especially for Stadials and Chrons, and Glaciations and Stages. Future development of glacial records should endeavour to develop mutually consistent terminologies, preferably with formal stratigraphical meaning, in order to aid correct correlation and efficient communication.

As previously noted (Ehlers and Gibbard, 2007; Thackray et al., 2008), the term LGM is widely accepted as referring to the maximum global ice volume during the last glacial cycle corresponding with the trough in the marine isotope record at ca. ^{14}C ka BP (Martinson et al., 1987) and the associated global eustatic sea level low also dated to 18 ^{14}C ka BP (Yokoyama et al., 2000). It was first applied by the CLIMAP group (CLIMAP, 1981) who adopted the date of 18 ^{14}C ka BP based on the implicit assumption that globally a glacial maximum could be recognised from deep-sea sediment sequences which they used as a time-marker for the reconstruction of global climate at that time—effectively to use the maximum glacial conditions, to contrast with those of today's interglacial situation.

The boundaries of MIS 2 have been calculated by Martinson et al. (1987) to ca. 24 ka for the base and ca. 12 ka cal. BP for the base of MIS 1 (the top of the preceding stage). It is important to realise that these boundaries are not necessarily based on natural events but on graphical plots arising from extrapolation from albeit closely spaced samples. In principle, in ocean-sediment cores, stage boundaries are placed at mid-points between temperature maxima and minima. The boundary points thus defined in ocean sequences are assumed to be globally isochronous. This is reasonable because of the extremely slow sedimentation rate of ocean-floor deposits and the relatively rapid mixing rate of oceanic waters.

However, the whole period of MIS 2 is not necessarily equivalent to the maximum development of glaciation, the LGM being effectively a point in time in the stage rather than a period. This problem has been recognised by others, although many authors who use the term LGM in chronological sense appear unaware that it has not been originally formally defined, apart from in a very general sense. Moreover, many appear to consider that the term should be taken to mean the point at which glaciation all over the world reached its maximum extent. Yet this is not the correct assumption. Since the LGM marks a trough or point in the isotope curves, it is not measuring ice extent but ice volume. Until recently, it was reasonable to assume that the two were one and the same, although evidence was always available to suggest that was an oversimplification. As evidence has accumulated from around the world, particularly through these compilations, it has become apparent that the isotope signal is not reflecting the maximum extent of glaciation worldwide but predominantly the signal from eastern North America's Laurentide Ice Sheet. Even this signal appears to represent only the southern side of the ice sheet, whilst the north achieved its maximum extent later.

In this context, Clarke et al. (2009) have argued that most ice sheets around the world reached their maximum between ca. 26.5 and 19 ka, preceded by a maximum extent of smaller mountain glaciers between ca. 33 and 26.5 ka. This conclusion was reached based on analysis of an extensive data set covering the period 50–10 ka. However, by limiting their study to the period between 10 and 50 ka, Clarke et al. (2009) did not consider the possibility that, in many parts of the world, the maximum extent of glaciers was reached earlier in the glacial cycle, For example, in this volume, Owen notes that glaciation throughout the Himalaya was probably most extensive during the early part of the last glacial cycle and also likely during MIS 3 (Fig. 1.5). Examples of a similar situation are reported elsewhere in this book such as in the Verkhoyansk Mountains, Russia (Chapter 64), Japan (Chapter 72), Australia (Chapter 74), Arctic Russia (Chapter 28) and in some parts of the Pyrenees (Chapter 11) (Fig. 1.5). There is also limited evidence of early glaciation during the last glacial cycle in Morocco, although more dates are needed from this area (Chapter 76).

In Arctic Russia, Möller et al. (Chapter 28) note that the most extensive glaciation dates from the Early Weichselian, cumulate at ca. 100 ka, and a Middle Weichselian event of intermediate extent dates from 70 to 60 ka. The last and least extensive glaciation in Arctic Russia is contemporaneous with the so-called LGM. Even in Alaska, just west of the huge Laurentide ice sheet, the largest glaciers of the last glacial cycle formed during the early Wisconsin and were significantly larger than late Wisconsinan glaciers (Chapter 33). In many other regions, the timing of the largest phases of glaciation during the last glacial cycle occurred later, but still well before the global LGM represented in the marine isotope record. For example, in Australia, the outermost moraines of the Snowy Mountains represent an ice advance at ca. 59 ka during MIS 4 (Chapter 74). There is no doubt that in some areas glaciers did reach their maximum during the last glacial cycle close in time to the record of maximum global ice volume indicated in the marine isotope record. However, as becomes clear from this book, as more and more dated sequences are presented, it appears that in large parts of the world the classical LGM was a period of ice recession or with major ice maxima occurring either before or after the point itself. In this sense, it could be argued that the continued use of the term LGM is becoming a hindrance, rather than a practical 'shorthand term' for referring to the maximum of the last glaciation.

However, there are two points to consider. As already noted, the LGM has been taken to mean the maximum extent of glacial ice globally and has been used not only by glacial geologists across the world but also by many others who have taken it to be a time-marker for subdivision of the late Weichselian and its equivalents. The desire for fine-scale time divisions for the latter half of the last glacial period is strong. Witness the development of the event-timescale from the Greenland ice-core sequence (Björck et al., 1998). However, as also noted above, the LGM is not a time division, although it has been used as such in a quasi-chronostratigraphical sense by many workers.

Some practitioners, especially those who attempt to correlate marine and terrestrial successions, have recognised this problem. Some have even attempted to assign chronozone status. The most useful discussion of this point was made by Mix et al. (2001) who considered the event should be centred on the calibrated date at 21 ka cal. BP and should span the period 23–19 or 24–18 cal. ka BP dependent on the dating applied (e.g. MARGO project members, 2009). This useful proposal seems to have been taken up by some workers whereas others have simply ignored the problem. The problem itself is that the present concept of an event centred on a dated point infringes the rules of stratigraphical nomenclature. The question is whether the LGM represents an event (an event stratigraphical unit), which can have boundaries that vary through time, comparable to the system proposed by Björck et al. for the NGRIP sequence, or whether it is a time interval (a chronostratigraphical unit), in which case the boundaries must be fixed in time and moreover, the unit must be defined in a rock or sediment sequence. As used at present, the term LGM fulfils neither criterion. Its usefulness for communication is therefore questionable.

Until the LGM and similar divisions (e.g. 'the mystery interval': Denton et al., 1999; Broecker and Barker, 2007) are formally defined as chronozones, or as chrons if based on dating (geochronology), the meaning of the terms is in danger of becoming increasingly blurred; a matter that could hinder rather than help the establishment of formal divisions of Weichselian time and simplify communication. Formalisation requires acceptance and ratification by the International Commission on Stratigraphy, and at the time of writing, the Subcommission on Quaternary Stratigraphy is in the process of establishing a working group to consider the whole problem of these fine-scale time divisions.

1.3.7. Developments in Geochronology

Some of the biggest advances made in the work presented in this book have resulted from developments in geochronological understanding. In the first edition, only a few areas of the world had well-constrained glacial geochronologies. However, in this second edition, a majority of the chapter contributions present new dating evidence derived from the application of a range of new or established techniques. For some regions, this is the first time glacial geochronological data have been published (e.g. Chapter 76), whilst for many other areas, authors add to an ever-growing dataset. Future advancement in glacial research is likely to be constrained by availability of dating techniques for researchers in different countries, theoretical advancements in the dating techniques themselves and also the application and testing of multiple dating techniques to the same glacial successions. Most ages in this book are presented following the guidelines outlined in *Quaternary Science Reviews* (Rose, 2007) and *Quaternary Geochronology* (Grün, 2008).

1.4. SUMMARY

The Quaternary is synonymous with extensive glaciation of the Earth's mid- and high latitudes. Although there were local precursors, significant glaciation began in the Oligocene in eastern Antarctica. It was followed by glaciation in mountain areas through the Miocene (in Alaska, Greenland, Iceland and Patagonia), later in the Pliocene (e.g. in the Alps, the Bolivian Andes and possibly in Tasmania) and in the earliest Pleistocene (New Zealand, Iceland and Greenland). Today, evidence from both the land and the ocean floors demonstrates that the major continental glaciations, outside the polar regions, rather than occurring throughout the 2.6 Ma of the Quaternary, were markedly restricted to the last 1 Ma—800 ka or less. MIS 22 (ca. 870–880 ka) included the first of the 'major' worldwide events with substantial ice volumes that typify the later Pleistocene glaciations (i.e. MIS 16, 12, 10, 6, 4–2).

The growing evidence for the asynchroneity of glacial maxima across the world during the last glacial cycle (i.e. the Weichselian, Würmian, Wisconsinan Stage, MIS 5d-early 1) emphasises the need to re-evaluate the concept of an 'LGM' representing a single global event when glaciers reached their maximum extent. Indeed, it appears that glaciations on a global scale were considerably more complex in terms of the timing and extent than previously appreciated. Apart from the obvious implications, this holds for climate and oceanic circulation reconstruction in the past, also implicit in this emerging pattern is that earlier glaciations could well have seen asynchronous maximal extents.

One possible consequence of this could be why some glaciations are represented in one region of the world, whilst in others, there is little or no record of their occurrence. In particular, this could explain why the glaciations in periods like MIS 8, 10, 14 or 18 are apparently absent from regions such as north-western Europe, when others (e.g. MIS 6, 12) are strongly represented. In the absence of a sufficiently fine-resolution geochronology, it might be difficult to examine or differentiate, but it is certain that earlier glaciations were unlikely to have been any less subject to the variable interactions of the controlling variables than the last glacial period. Nevertheless, the mutual timing of advances and retreats could potentially vary to a greater or lesser extent. The consequences could go some way to explaining the variability in extent, ice volume and chronology of glaciation during the Quaternary and earlier.

REFERENCES

Ballantyne, C.K., 2010. Extent and deglacial chronology of the last British–Irish Ice Sheet: implications of exposure dating using cosmogenic isotopes. J. Quatern. Sci. 25, 515–534.

Björck, S., Walker, M.J.C., Cwynar, L.C., Johnsen, S., Knudsen, K.-L., Lowe, J.J., et al., 1998. An event stratigraphy for the last termination in the North Atlantic region based on the Greenland ice-core record: a proposal by the INTIMATE group. J. Quatern. Sci. 13, 283–292.

Bowen, D.Q., Philips, F.M., McCabe, A.M., Knutz, P.C., Sykes, G.A., 2002. New data for the Last Glacial Maximum in Great Britain and Ireland. Quatern. Sci. Rev. 21, 89–101.

Broecker, W., Barker, S., 2007. A 190‰ drop in atmosphere's $\partial14C$ during the "Mystery Interval" (17.5 to 14.5 kyr). Earth Planet. Sci. Lett. 256, 90–99.

Calvet, M., 2004. The Quaternary glaciation of the Pyrenees. In: Ehlers, J., Gibbard, P. (Eds.), Quaternary Glaciations—Extent and Chronology, Part I: Europe. Developments in Quaternary Science, 2a. Elsevier, Amsterdam, pp. 119–128.

CLIMAP, 1981. Seasonal reconstruction of the Earth's surface at the last glacial maximum. Geological Society of America, Map and Chart Series MC-36: 1-18.

Clapperton, C., 1993. Nature and environmental changes in South America at the Late Glacial Maximum. Palaeogeogr. Palaeoclimatol. Palaeoecol. 101, 189–208.

Clarke, P.U., Dyke, A.S., Shakun, J.D., Carlson, A.E., Clark, J., Wohlfarth, B., et al., 2009. The Last Glacial Maximum. Science 325, 710–714.

Denton, G.H., Heusser, C.J., Lowell, T.V., Moreno, P.I., Andersen, B.G., Heusser, L.E., et al., 1999. Interhemispheric linkage of Paleoclimate during the last glaciation. Geogr. Ann. 81A (2), 107–153.

Dyke, A., 2004. An outline of North American deglaciation with emphasis on central and northern Canada. In: Ehlers, J., Gibbard, P.L. (Eds.), Quaternary Glaciations—Extent and Chronology. Part II: North America. Developments in Quaternary Science, vol. 2b. Elsevier, Amsterdam, pp. 1–7.

Ehlers, J., Gibbard, P.L., 2007. The extent and chronology of Cenozoic global glaciation. Quaternary International, 164–165, 6–20.

Ehlers, J., Gibbard, P.L., 2008. Extent and chronology of Quaternary glaciation. Episodes 31, 211–218.

Ehlers, J., Gibbard, P.L., 2011. Quaternary glaciation. In: Singh, V.P., Singh, P., Haritashya, U.K. (Eds.), Encyclopedia of Snow, Ice and Glaciers. Springer.

Gibbard, P.L., Cohen, K.M., 2008. Global chronostratigraphical correlation table for the last 2.7 million years. Episodes 31, 243–247.

Giraudi, C., Bodrato, G., Ricci Lucchi, M., Cipriani, N., Villa, I.M., Giaccio, B., et al., 2010. Middle and late Pleistocene glaciations in the Campo Felice Basin (central Apennines, Italy). Quatern. Res. 75, 219–230.

Grün, R., 2008. Editorial. Quatern. Geochronol. 3 (1–2), 1.

Haug, H.H., Ganopolski, A., Sigman, D.M., Rosell-Mele, A., Swann, G.E.A., Tiedemann, R., et al., 2005. North Pacific seasonality and the glaciation of North America 2.7 million years ago. Nature 433, 821–825.

Heusser, C.J., 2003. Ice Age Southern Andes – A Chronicle of Paleoecological Events. Developments in Quaternary Science, 3. Elsevier, Amsterdam.

Hughes, P.D., Woodward, J.C., 2008. Timing of glaciation in the Mediterranean mountains during the last cold stage. J. Quatern. Sci. 23, 575–588.

Ingólfsson, Ó., 2004. Quaternary glacial and climate history of Antarctica. In: Ehlers, J., Gibbard, P.L. (Eds.), Quaternary Glaciations—Extent and Chronology, Part III. Elsevier, Amsterdam, pp. 3–43.

Knies, J., Matthiessen, J., Vogt, C., Laberg, J.S., Hjelstuen, B.O., Smelror, M., et al., 2009. The Plio-Pleistocene glaciations of the Barents Sea-Svalbard region: A new model based on revised chronostratigraphy. Quaternary Science Reviews 28, 812–829.

Lehmkuhl, F., Owen, L., 2005. Late Quaternary glaciation of Tibet and the bordering mountains: a review. Boreas 34, 87–100.

MARGO Project Members, 2009. Constraints on the magnitude and patterns of ocean cooling at the Last Glacial Maximum. Nat. Geosci. 2, 127–132.

Martinson, D.G., Pisias, N.G., Hays, J.D., Imbrie, J., Moore, T.C., Shackleton, N.J., 1987. Age dating and the orbital theory of the ice ages: development of a high resolution 0–300,000 year chronostratigraphy. Quatern. Res. 27, 1–29.

Miller, K.G., Fairbanks, R.G., Mountain, G.S., 1987. Tertiary oxygen isotope synthesis, sea level history, and continental margin erosion. Paleoceanography 2, 1–19.

Mix, A.C., Bard, E., Schneider, R., 2001. Environmental processes of the ice age: land, oceans, glaciers (EPILOG). Quatern. Sci. Rev. 20, 627–657.

Moran, K., Backman, J., Brinkhuis, H., Clemens, S., Cronin, T., Dickens, G., et al., 2006. The Cenozoic palaeoenvironment of the Arctic Ocean. Nature 441, 601–605.

Muttoni, G., Carcano, C., Garzanti, E., Ghielmi, M., Piccin, A., Pini, R., et al., 2003. Onset of Pleistocene glaciations in the Alps. Geology 31, 989–992.

Ó Cofaigh, C., Evans, D.J.A., 2007. Radiocarbon constraints on the age of the maximum advance of the British–Irish Ice Sheet in the Celtic Sea. Quatern. Sci. Rev. 26, 1197–1203.

Rabassa, J., 2008. Late Cenozoic glaciations in Patagonia and Tierra del Fuego. In: Rabassa, J. (Ed.), Late Cenozoic of Patagonia and Tierra del Fuego. Developments in Quaternary Science, 11. Elsevier, Amsterdam, pp. 151–204.

Rose, J., 2007. The use of time units in Quaternary Science Reviews. Quatern. Sci. Rev. 26, 1193.

Stickley, C.E., St. John, K., Koç, N., Jordan, R.W., Passchier, S., Pearce, R.B., Kearns, K.E., 2009. Evidence for middle Eocene Arctic sea ice from diatoms and ice-rafted debris. Nature 460, 376–379.

Thackray, G.D., Owen, L.A., Yi, C., 2008. Timing and nature of late Quaternary mountain glaciation. J. Quaternary Sci. 23, 503–508.

Tripati, A., Backman, J., Elderfield, H., Ferretti, P., 2005. Eocene bipolar glaciation associated with global carbon cycle changes. Nature 436, 341–346.

Tziperman, E., Gildor, H., 2003. On the mid-Pleistocene transition to 100-kyr glacial cycles and the asymmetry between glaciation and deglaciation times. Paleoceanography 18 (1), 8.

Yokoyama, Y., Lambeck, K., DeDeckker, P., Johnston, P., Fifield, K., 2000. Timing of the last glacial maximum from observed sea-level minima. Nature 406, 713–716.

Chapter 2

Quaternary Glaciations in Austria

Dirk van Husen

Simetstraße 18, A-4813 Altmünster, Austria, aon.913165083@aon.at

2.1. INTRODUCTION

From the onset of rhythmic loess accumulation, at the turn from the Gauss to the Matuyama magnetic Chron, a frequent variation from humid warm to dry cool climate with loess and occasional gravel accumulation occurred until the end of the Matuyama Chron. Remnants of four glaciations (Günz, Mindel, Riss, Würm) within the Eastern Alps and their foreland have long since been known. More recently, evidence for an additional cold stage between the two older ones was found. As a result of a major cooling and build-up of piedmont glaciers in the foreland, the four glaciations show a complete succession of terminal moraines with glaciofluvial terraces connected to them. The last interglacial/glacial cycle can easily be reconstructed climatologically and by sediment development. It may serve as a model for understanding the climatic conditions of the older ones, which had very similar successions.

2.2. THE COURSE OF THE QUATERNARY

Around the Neogene/Quaternary boundary (Gibbard et al., 2009; Mascarelli, 2009), at 2.58 Ma BP, the drainage pattern of the Eastern Alps was generally fully developed. Loess accumulation began along the Danube north of the Eastern Alps at that time (Frank et al., 1997), at the Gauss/Matuyama Chron boundary (Fig. 2.1). According to palaeontological investigations, the Neogene to the Quaternary boundary period was characterised by a moderately warm, humid climate alternating with cool, dry periods. There were frequent repeated changes of climate, but lacking any reference to glacial events. Within the Eastern Alps, no sediments from this early phase of the Quaternary have been preserved, due to the high relief energy and the later glaciations. Sediments from this period, mostly loess, are found only in the Alpine foreland along the Danube and its tributaries.

The loess section at Krems shooting range (Fink et al., 1976) reveals a repeated change of dry-cold (*Pupilla* fauna), warm dry (*Striata* fauna) and warm humid (*Chilostoma* fauna) conditions at the beginning of the Quaternary (Fink et al., 1976; Frank and van Husen, 1995). The sequence contains evidence of 17 interglacials. They post-date the Olduvai Event and thus are younger than Marine Isotope Stage (MIS) 63. The loess was certainly deposited under dry, cold conditions. However, no evidence was found for any climatic deterioration strong enough to generate glaciation.

2.3. THE FOUR ALPINE GLACIATIONS

Following younger interpretations of the marine oxygen isotope record, MIS 24 at about 0.9 Ma represents a clear transition to a different style of climatic regime characterised by more extreme glaciations than the previous period (Shackleton, 1995). This may explain why within the Eastern Alps and their foreland in Austria no glacial deposits older than the Günz Glaciation, *sensu* Penck and Brückner (1909/11), have been found.

In the Salzach, Traun and Krems river areas, terminal moraines of this glaciation are connected to terrace bodies (Weinberger, 1955; Kohl, 1974). These *Ältere Deckenschotter sensu* Penck and Brückner (1909/11) form part of a widespread gravel cover that occurs between the rivers Traun and Enns. In terms of gravel composition and age, this unit is a genetically polymict body (van Husen, 1980, 1981). It probably accumulated over a long time spanning several cold stages. Reworking and lateral erosion by the rivers resulted in the incorporation of older deposits into what appears now as a single terrace accumulation.

In the Alpine foreland, east of the Salzach River (Fig. 2.2 and 2.3), remnants of three glaciations younger than the oldest represented above can be easily identified (Weinberger, 1955; Del Negro, 1969; Kohl, 1976; van Husen, 1977, 1996; Sperl, 1984). Knowledge regarding the extent, development of tills, terminal moraines, glaciofluvial sediments and weathering is good enough to allow the reconstruction of the individual ice streams and tongues of the piedmont glaciers (Penck and Brückner, 1909/11; Sperl, 1984).

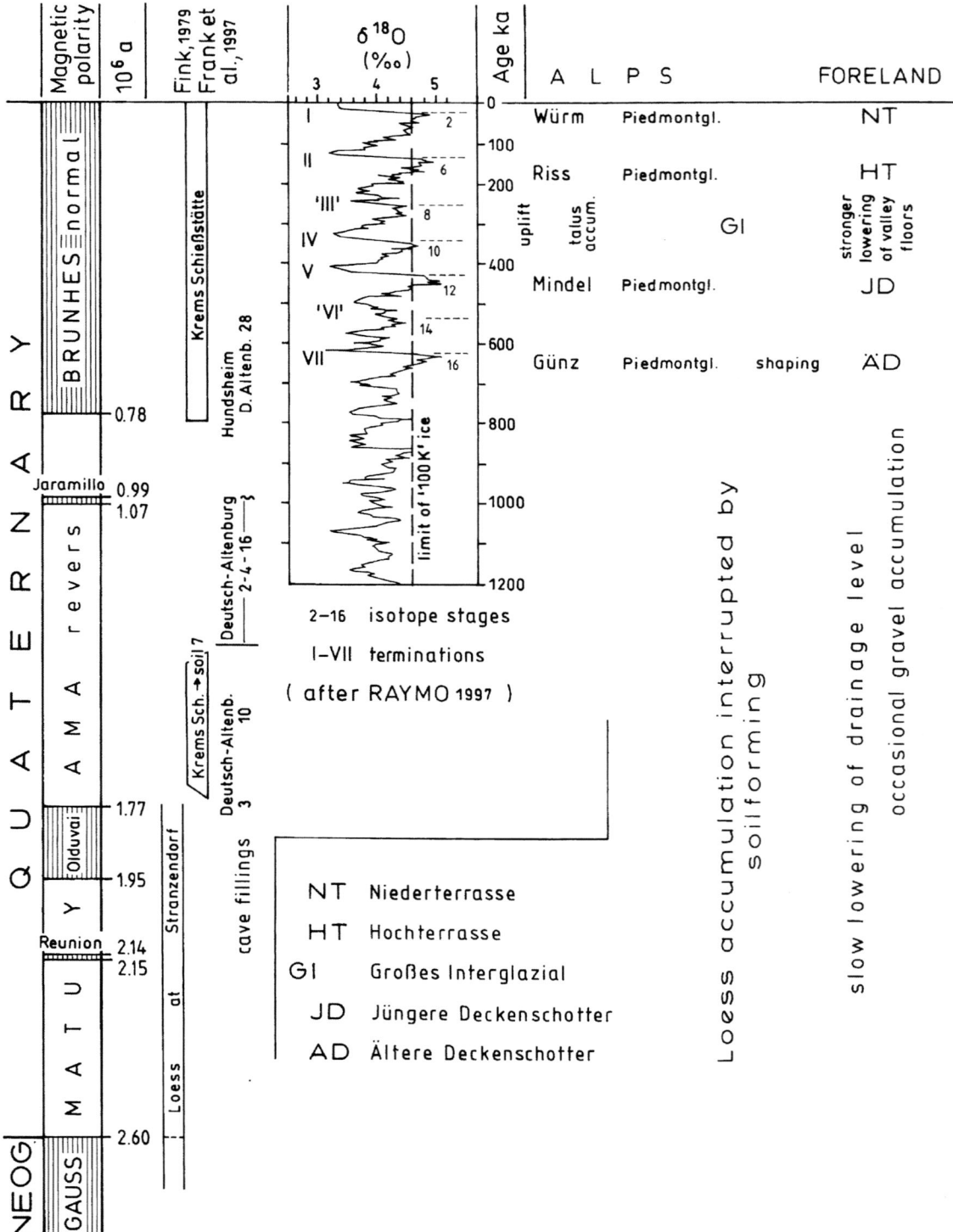

FIGURE 2.1 Geological time scale of the Quaternary events in Austria.

Palaeomagnetic investigations (Fink, 1979) suggest that the tills of these glaciations and their connected outwash terraces all fall into the Brunhes Chron. The marine oxygen isotope record suggests that four major glaciations affected the northern hemisphere during the Brunhes Chron (Shackleton, 1987; Raymo, 1997), generally correlated with MIS

Chapter | 2   Quaternary Glaciations in Austria

FIGURE 2.2 Location map showing the main drainage systems of Austria. Squares mark figures 2.3, 2.4, 2.5.

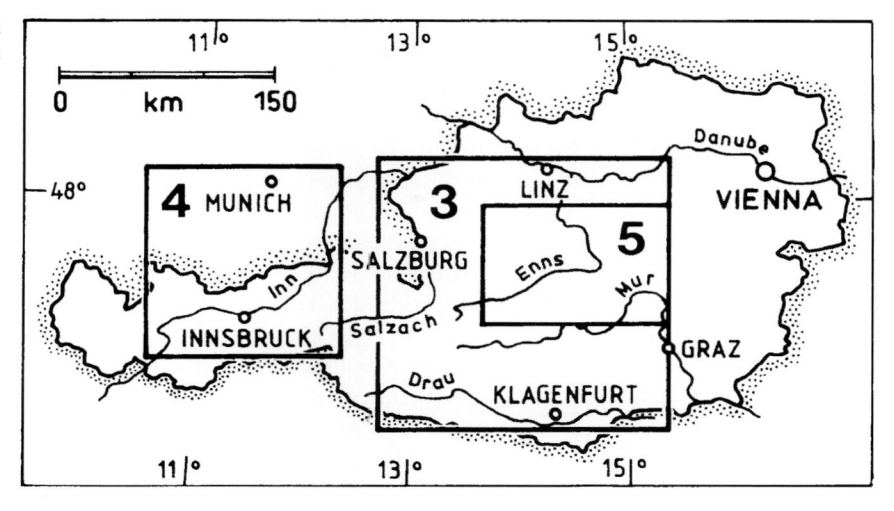

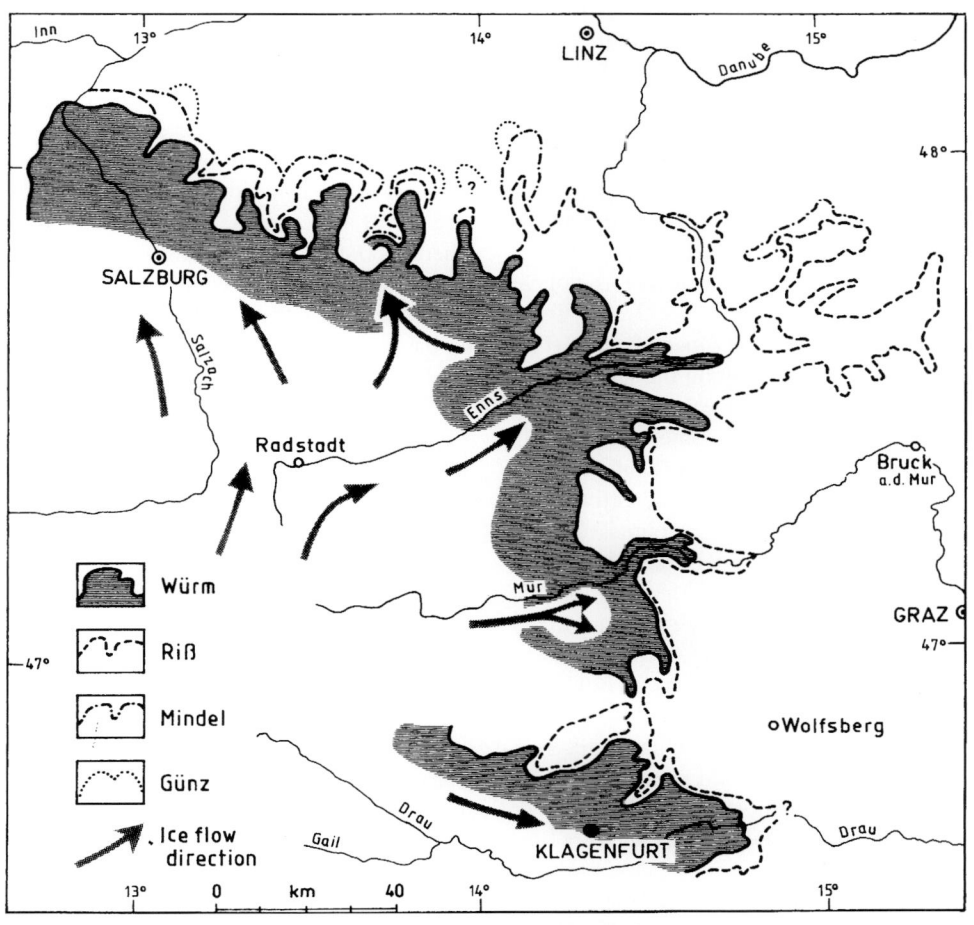

FIGURE 2.3 The extent of the glaciers of the four glaciations known in Austria. Only glaciers of the main valleys are indicated. Ice flow direction is rather similar during each glaciation.

2, 6, 12 and 16 (Raymo, 1997). According to radiometric dating, stratigraphical position as well as weathering, the two youngest Alpine glaciations are correlated with MIS 2 and 6. However, the much stronger weathering, cementation and periglacial modification of all the Mindelian deposits, in comparison to those of the Rissian Stage, indicate that a longer time span must have separated these glaciations. This *Großes Interglazial* (great interglacial) was first postulated by Penck and Brückner (1909/11). Thus, the Mindelian probably correlates with MIS 12 and the Günzian with MIS 16 (Fig. 2.1).

Remnants of another major cold period with gravel deposition in the Alpine foreland that fall between depositions of the Günzian and Mindelian stages have been found more recently (Kohl, 1976). They are separated from both by weathering. This cold stage probably correlates with MIS 14. Extents of the glaciers of these two oldest glaciations were controlled by partly different valley drains leading to locally singular and longer toungues (Fig. 2.3).

2.4. TECTONIC ACTIVITY

The incision of the circum-alpine drainage system as far as the level of the *Ältere Deckenschotter* and beyond seems to have been of an intensity similar to that observed in the Danube and its tributaries (Graul, 1937; Fuchs, 1972; Fischer, 1977). This suggests that no differential uplift has occurred and that there are no traces of other tectonic activity, such as faulting within the northern foreland of the Alps. However, age control of the remnants of terrace systems older than the *Ältere Deckenschotter* is not yet possible.

On the basis of the loess sequence near Krems (Fink et al., 1976), the amount of erosion along the Danube system was about 50 m during the Quaternary period (Fig. 2.1). This supposedly undisturbed development along the northern edge of the Alps ends where the Danube enters the Vienna Basin. Clear evidence of tectonic activity that continues well into the Holocene is recognisable here (Plachy, 1981). The glaciofluvial terraces dating from the last two glaciations (*Hochterrasse, Niederterrasse*, as well as Holocene deposits) show clear evidence of tectonic displacement; indeed parts of the *Hochterrasse* are displaced by as much as ca. 10 m along major faults (Kröll et al., 1993; Decker et al., 2005).

Traces of recent tectonic activities in Quaternary sediments within the Eastern Alps, as would be expected from Miocene and Pliocene development (Peresson and Decker, 1996), have yet not been proved. Only along the rivers Enns and Steyr, where glaciers have not obscured evidence, a displacement of the *Jüngere Deckenschotter* at the northern edge of the Alpine nappe system indicates a tectonic activity before the penultimate glaciation (van Husen, 2000).

This may have been associated with and caused by an uplift of the whole mountain chain (Fig. 2.1).

2.5. DEVELOPMENT OF GLACIERS

The Alpine drainage system was filled by dendritic glaciers during all four major glaciations. The topography of the longitudinal valleys (e.g. the Inn, Salzach, Enns, Mur and Drau; Fig. 2.2) controlled glacier build-up in their tributaries and surrounding areas. On the one hand, this explains the outstanding differences in glacier extent between some of the catchment areas. On the other, it explains the strongly accelerated ice build-up towards the glacial maximum. Two examples are given here.

2.5.1. Inn Valley

The Inn valley is the most extensive drainage system in the Eastern Alps. Its source areas are located south of the valley and in the Engadine, areas almost exclusively underlain by crystalline rocks. North of the Inn valley, the Northern Calcareous Alps have no major valley tributary of the Inn, which follows the border between these two tectonic units east of Landeck (Fig. 2.4).

The valleys in the Northern Calcareous Alps are orientated to the north (Fig. 2.4), forming the heads of the rivers Isar and Loisach and other small rivers. The Inn valley is separated from them by mountain chains rising up to 2500 m a.s.l. or more. Three gaps in the mountain chain are found at Fernpass, Seefeld and Achensee, with watersheds 400–600 m above the Inn valley. During the major glaciations, these cols were crossed by ice (Fig. 2.4). A similar behaviour can also be reconstructed by erratic material in basal tills in the valleys of Salzach, Enns, Mur and Drau (see map).

Till composition reflects the petrographic composition of the catchment areas. In the basal tills of the Inn valley, crystalline components are dominant; only on the northern slope is there a considerable admixture of limestones, derived from the local bedrock. However, up to 35% of crystalline erratics (Fig. 2.4) are found in the glacial and glacigenic material of the Würmian Isar and Loisach glaciers (Dreesbach, 1983). Most of the crystalline material is fresh and unweathered except for a small amount affected by multiple reworking. Such a large amount of unweathered crystalline rocks has to be attributed to extensive ice transport from the Inn valley to the Isar and Loisach systems, across the Fernpass and the Seefeld passes (Penck and Brückner, 1909/11). This transfluence reflects the gradient of the ice surface. The ice filling the Inn valley reached a much higher elevation than in the valleys in the north. The extraordinarily high crystalline boulder content in the sediments in the advance phase of the ice streams suggests a strong and rapid ice build-up in the Inn valley between

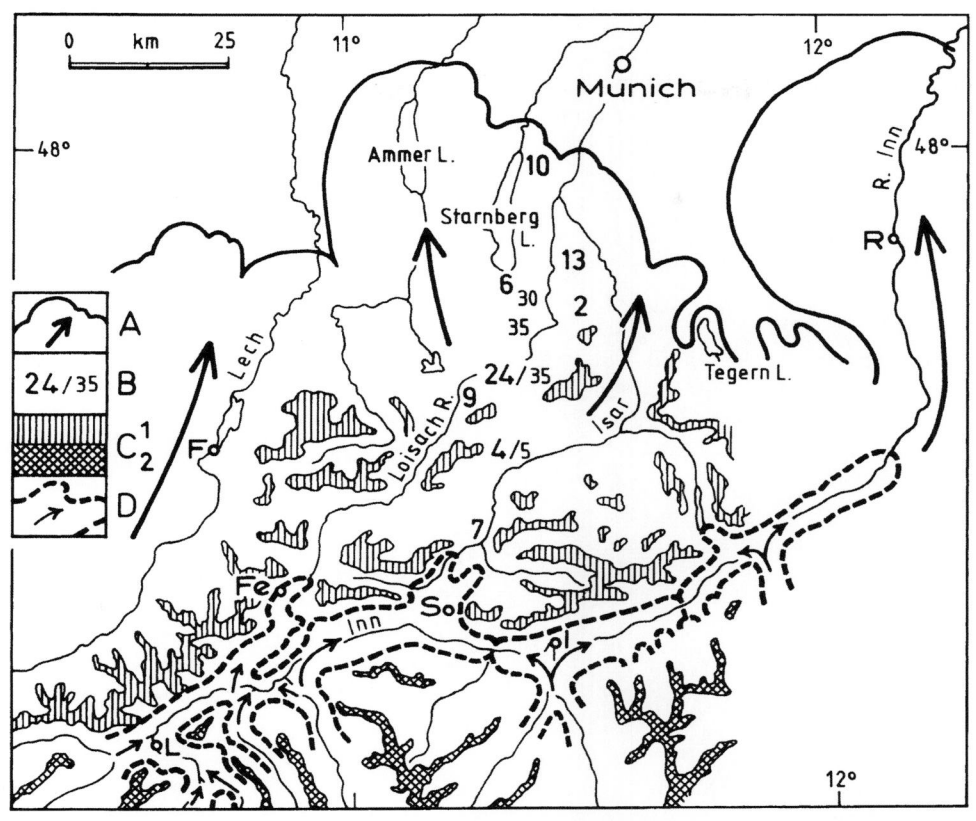

FIGURE 2.4 Ice streams of the Inn, Lech, Isar and Loisach valleys. (A) Glacier extent during the Würm MIS 2. (B) Percentage of crystalline boulders. The larger numbers in basal till. (C) Nunataks formed of sedimentary rocks (C1) and crystalline rocks (C2). (D) Probable ice extent in the Inn valley around the beginning of the final ice build-up during MIS 2. F: Füssen, Fe: Fernpass, I: Innsbruck, L: Landeck, R: Rosenheim, S: Seefeld.

Landeck and Innsbruck. This would allow strong transfluence from the Inn valley even at an early phase of the glaciation, suppressing the influence of the local glaciers and affecting both till and outwash gravel composition. Such an early overflow may result from the internal ice flow conditions within the Inn valley system. The large tributaries from the south, with their extensive source areas at high altitudes, carried huge glaciers during the final approach of the last glaciation, gradually filling the whole Inn valley. Hence, five glaciers finally merged around Landeck. At the same time, the ice flowing out of the Sill and Ziller valleys reached the main valley and thus the glaciers may have blocked each other creating a conspicuous rise in the ice surface. Therefore, crossing the watersheds, the ice of the Inn valley system reached the drainage systems of the Isar and Loisach (with smaller catchment areas at lower altitudes) early enough before these valleys were filled with local ice strong enough to hinder this process.

However, ice congestion in the narrow Inn valley also resulted in a high ice table leading to a rapid expansion of the feeding area, which again favoured rapid ice build-up. Thus, the high gradient to the north persisted throughout the entire glaciation. The volume of ice discharge north across the watersheds (crystalline material included) can be judged from the size of the Isar and Loisach piedmont glaciers as compared with their neighbours, which were fed only by their own source areas within the Northern Calcareous Alps (Fig. 2.4).

2.5.2. Enns Valley

The Enns valley glacier (Fig. 2.2) was connected to the Salzach glacier in the west and the Traun glacier in the north (Fig. 2.5). The Enns valley follows the border between the crystalline zone in the south and the Northern Calcareous Alps in the north. The topography differs slightly from that of the Inn valley. In the western part, the tributaries from the south originate in extensive areas of high elevation, and the wall-like Northern Calcareous Alps drain mainly towards the north. Further to the east, however, the elevation of the crystalline zone declines abruptly over a short distance, and the Northern Calcareous Alps changes from continuous chains and plateaux to more isolated mountains surrounded by much lower terrain.

During the Würmian, the Enns valley was occupied by a glacier extending down to the area of more isolated high

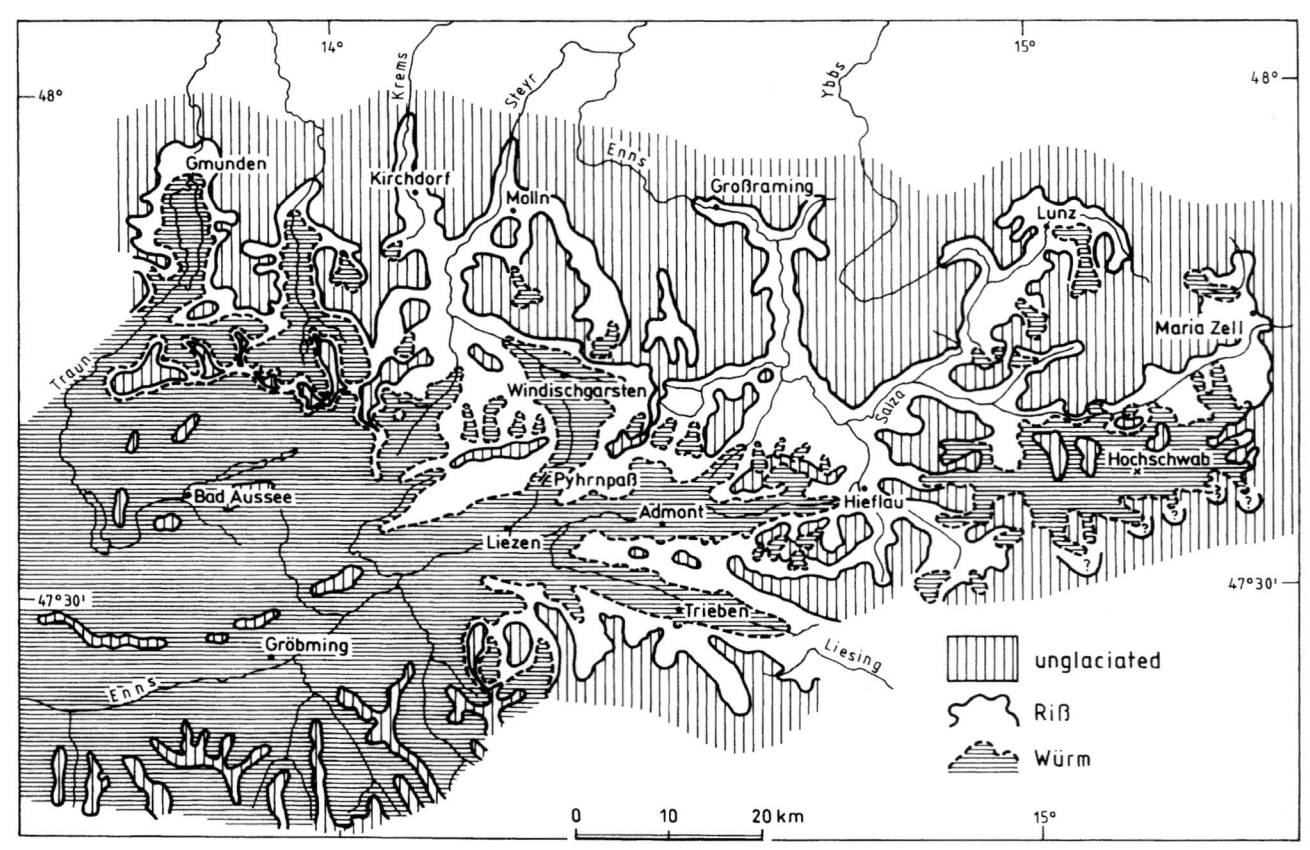

FIGURE 2.5 Glacier extent during the Würmian (MIS 2) and Rissian (MIS 6) in the Enns valley and environs and in the Steyr and Ybbs valleys.

mountains. The latter supported local glaciers (Fig. 2.5). Both the local glaciers and the valley glacier were in contact but had little influence on one another with regard to the ice discharge. To the south, a 20 km long tongue of the valley glacier entered the otherwise unglaciated valley around Trieben. To the north, ice crossed the Pyhrn Pass, entered the drainage system of the Steyr river and filled the small basin at Windischgarsten. The content of crystalline boulders in the tills is around 2–3% in this area. East of the Enns glacier only more or less small local glaciers developed. Their extents depended strongly on snowdrift by prevailing westerly winds.

During the penultimate glaciation (Rissian), the equilibrium line of the glacier system was about 100 m lower than during the Würmian (Penck and Brückner, 1909/11). This led to more extended piedmont glaciers on the northern rim of the Alps and longer valley glaciers within the mountains. The difference in length was in most cases about 5–6 km (Fig. 2.3). The situation was very different in the Enns and Salza valleys and the Steyr-Krems drainage system (Fig. 2.5). In the former, the end moraines at Grossraming (van Husen, 1999) show a glacier extending ca. 40 km further than during the Würmian. The tills throughout the valley and in the immediate vicinity north of Hieflau contain abundant crystalline boulders, indicating that the Enns valley was filled by a much larger Enns glacier with little influence from the local glaciers.

With respect to the lower position of the equilibrium line during the Riss glaciation, both the local glaciers and the valley glacier grew much larger. Thus, the local glaciers were powerful enough to block the valley glacier in the very narrow portion of the valley west of Hieflau. This impediment to ice flow caused a higher ice surface in the Enns valley west of Admont. This area, that formed part of the ablation area during the Würmian, consequently became an accumulation area in the Rissian Stage. This considerable addition to the accumulation area, together with the ice supply from the local glaciers, affected the extent of the glacier, as explained above.

The glaciers on the north slope of Hochschwab grew in the same way (Fritsch, 1993; Kolmer, 1993), filling the Salza valley, which became part of the accumulation area. As a result, an extended ice stream formed here. This process may have been accelerated by the additional cooling effect of the glaciated Enns valley (Fig. 2.5).

The same feedback mechanism occurred in the Windischgarsten basin. The stronger transfluence of ice from the Enns valley in the south caused a larger glacier tongue to develop and to extend further and further into the basin. In the same way, the larger local glaciers contributed to the overall ice volume. Thus, the whole basin became part of the accumulation area. This greatly enlarged feeding area enabled the glacier to advance down valley over 30 km beyond its Würmian limits. It was even powerful enough to cross the watershed into the Krems river system (at Kirchdorf on the Krems), generating a much greater ice stream in that valley (Kohl, 1976).

2.6. OVERDEEPENED VALLEYS

Since the beginning of research on ancient glaciers and their palaeogeographical distribution, overdeepened tongue basins have been known (Fig. 2.6). They seem to form predominantly in the ablation area, where the higher ice velocity increases the basal debris load and where basal meltwater drainage under hydrostatic pressure occurs (van Husen, 1979). They were shaped successively during each glaciation because the tongue areas always developed more or less in the same positions (cf. Fig. 2.3). Hydrogeological investigations in the longitudinal valleys provide new data on the shape and depth of these basins. Geophysical investigations, together with boreholes, also provide good evidence on the sediment filling of the basins and the position of the underlying bedrock.

Thus, on the one hand, the underlying bedrock in the tongue basins of the Salzach glacier was cored repeatedly between 160 and 340 m, on the other, the base in the Gail and Drau valleys was found at 200–240 m (Kahler, 1958). These investigations did show that the overdeepening may be limited to about 400 m in this part of the Austrian Alps. Similar depths are also found in some lakes (e.g. Traunsee), which have a depth of nearly 200 m and a thick sediment fill at the bottom. A depth of ca. 200 m was also determined for the Steyrtal, which was affected by the most extensive glaciations (Fig. 2.6, M) (Enichlmaier, personal communication).

Stronger overdeepening is reported from the Inn valley. East of Innsbruck, for example, the pre-Quaternary basement lies at ca. 180 m a.s.l., about 400 m below the valley floor (Aric and Steinhauser, 1977). Further to the east, seismic investigations revealed that the bedrock lies at about 500 m below sea level (1000 m below the valley floor). The latter was proved by a drill hole 900 m deep that passed through unconsolidated gravel, silt and clay without reaching the bedrock (Weber et al., 1990). This excessive amount of erosion and overdeepening may result from the stronger linear ice and meltwater discharge along valleys with

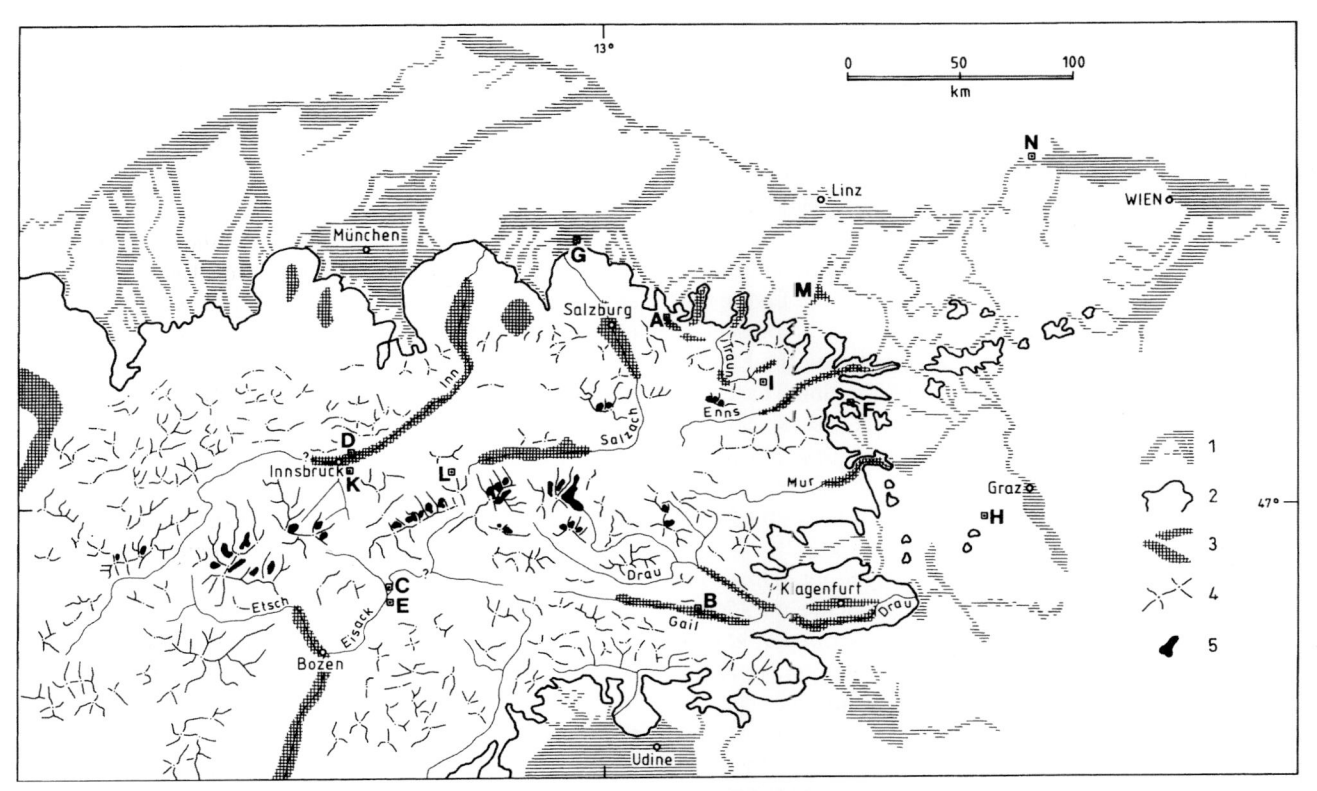

FIGURE 2.6 Sketch map of the Eastern Alps during Würmian (MIS 2); 1 Terrace 'Niederterrasse'; 2 Maximum extent of glaciers; 3 Overdeepened parts of the valleys; 4 Nunataks; 5 Glacier extent of the Holocene. Localities mentioned in the text: A: Mondsee; B: Nieselach; C: Schabs; D: Baumkirchen; E: Albeins; F: Hohentauern; G: Duttendorf; H: Neurath; I: Mitterndorf; K: Lans; L: Gerlos; M: overdeepened area at Molln.

extensive catchment areas. It is in close accordance with the erosion of about 600–1000 m, recently reported from the longitudinal valleys of Rhine and Rhône in Switzerland (Pfiffner et al., 1997).

The filling of the overdeepened valleys depends strongly on the relation of the major rivers to their tributaries, in terms of water and debris discharge, as well as on the size of the basin. Large basins, with a strong main river and small tributaries, were often filled with a thick sequence of fine-grained bottom-set beds that interfinger with coarse delta deposits (e.g. Salzachgletscher: Brandecker, 1974). Strong input of coarse gravel and sand from major tributaries creates a more inhomogeneous valley fill with alternating layers of gravel, sand and silt all over the area.

2.7. LAST INTERGLACIAL-GLACIAL CYCLE

According to oxygen isotope records, all climatic cycles within the Brunhes Chron and in the Late Matuyama Chron show a similar pattern. This is particularly true of the four major cycles before the Terminations I, II, V and VII, following the major glaciations at MIS 2, 6, 12 and 16 (Raymo, 1997) each of which is characterised by a step by step cooling, interrupted by short phases of climatic amelioration, leading eventually to the very cold short true glaciation period (Fig. 2.1). The last, Eemian/Würmian, climatic cycle is relatively well investigated. Therefore, it may serve as a model for the reconstruction of the other cycles in terms of climatically induced sedimentation and facies diversification in the Eastern Alps (van Husen, 1989).

On the northern edge of the Eastern Alps (Fig. 2.6, A), the Mondsee sedimentary sequence has given good, continuous evidence on the climatic development from Termination II, through the Eemian, and well into the first half of the Würmian Stage. The fine-grained sediments, north of the shoreline of the present lake Mondsee, were first investigated by Klaus (1987) and believed to represent a complete sequence covering the time between the Rissian and Würmian glaciations. Recent investigations, based on three long cores, have revealed a delta structure in an ancient lake with a water surface around 50 m above the present lake level. Bottom set, foreset and thin topset beds of a classical Gilbert-delta structure were covered by the till of the last glaciation (Krenmayr, 2000).

The results of palynological investigations (Drescher-Schneider and Papesch, 1998; Drescher-Schneider, 2000) suggest that the Eemian in this area was a warm period with temperatures averaging 2–3 °C above the current Holocene values. The valleys and foreland of the Alps were densely forested at this time with well-developed mixed oak forest with a high content of *Abies* (fir). This phase ended with an abrupt climatic deterioration that affected the forest on the northern edge of the Alps and brought coarser sediments to the delta.

During the Early Würmian, forests recovered twice, showing some elements of mixed oak forest, with a cold stadial intervening between these periods. During the cold stadial, the treeline descended close to the lake level. The cold period at the beginning of the Middle Würmian saw the treeline more or less at the level of the Alpine foreland. This was followed by a slight warming, allowing a forest dominated by *Larix* (larch) and *Picea* (spruce) to grow around the lake. This series of events corresponds closely to the Samerberg sequence from east of the Inn valley (Grüger, 1979), as well as with that at La Grande Pile in the Vosges (Woillard and Mook, 1982).

2.8. CHRONOLOGY

The final climatic deterioration and glacier advance phase is represented at the Würmian Albeins and Baumkirchen (Fig. 2.6, D and E) sites where till overlies gravel and lake deposits, respectively. Radiometric dating indicates that the glaciers of the tributary valleys reached the main longitudinal valleys at about 25–24,000 years BP. The rate at which further ice build-up occurred is unknown because of the total lack of chronological evidence. During this final climatic decay (Fig. 2.8), the valleys were filled with coarse gravel, *Vorstoßschotte* at many places to high elevations, as a result of progressive overloading of the main river with debris (van Husen, 1989). The *Vorstoßschotter* extend laterally into the terraces in the foreland, which accumulated at the same time as the thick gravel bodies along the rivers (Fig. 2.7). These were developed also in non-glaciated areas by periglacial activity (congelifraction). In only two places, Duttendorf and Neurath (Fig. 2.6, G and H), in the Eastern Alps, can radiocarbon dates constrain the climax of the climatic deterioration and the maximum extent of glaciers and periglacial activity, as well as strong congelifraction and periglacial downwash around 20,000 years BP (van Husen, 1989).

Around the Eastern Alps, detailed mapping of terminal moraines and outwash terraces (e.g. the Traun, Enns, Mur, Drau) has shown that most glaciers behaved in a similar way. First of all, the greatest extent of the glacier tongues is marked by small morainic ridges connected to outwash fans. After this, the glacier fronts retreated some hundreds of metres and formed distinctive, high and wide end moraines, also connected to outwash, which grades morphologically into the downstream fluvial terraces. These outwash fans, and their transition into downstream terraces, have a lower gradient than the earlier ones. Within 1–2 km, they are on the same level, merging to form the *Niederterrasse* (Lower Terrace), which continues downstream and can also be traced along the Danube to Vienna (van Husen, 1987). The terraces also correspond to those of the unglaciated tributary valleys.

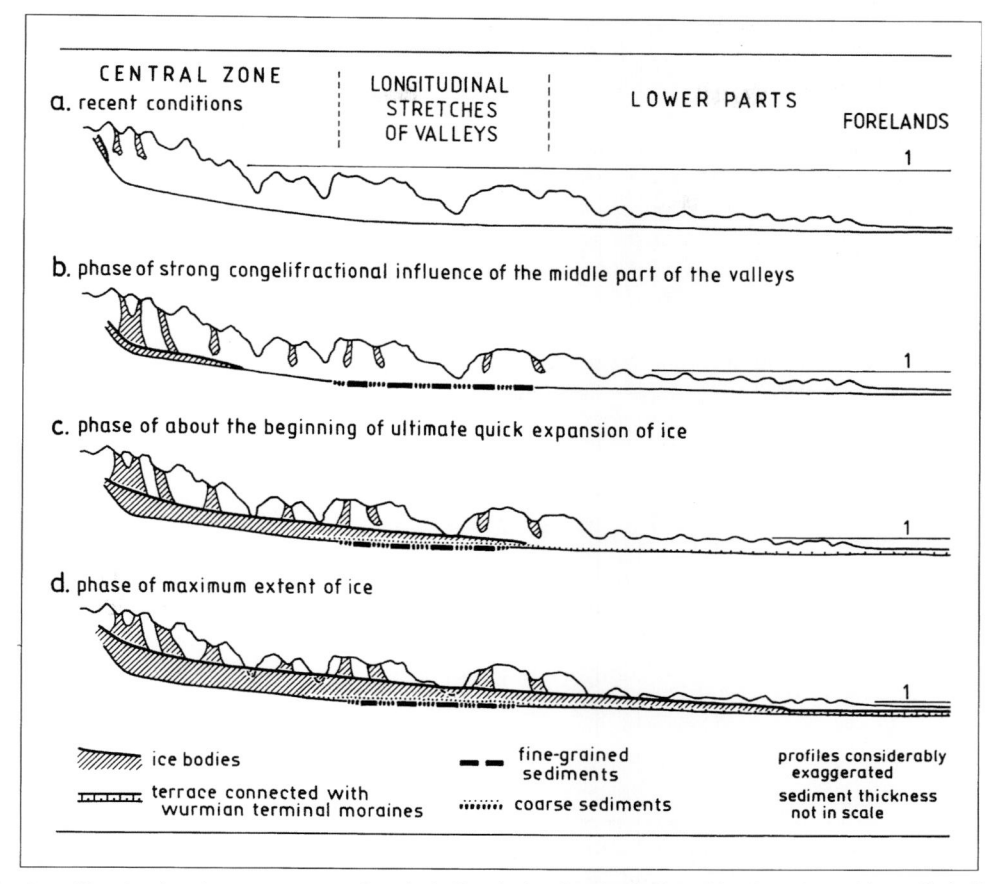

FIGURE 2.7 Sketch profiles showing the consequences of gradual climatic deterioration indicated by the estimated lower limit (l) of strong congelifraction. It shows that only a little lowering of this limit and the equilibrium line (C to D) is necessary to cause a very rapid and substantial expansion of the valley glaciers, such as during the Würmian maximum (see Fig. 2.8).

Evidence of weathering has been found neither between the sediments of these different Würmian terminal moraines nor within the outwash gravel sequence. This suggests that very little time elapsed between these two events, the *Maximalstand* and the *Hochstand* (van Husen, 1977, 2000). It is not known how long the glaciers were in place to form the 20–40 m high moraines of the *Hochstand*, as no datable material has yet been found. An identical differenciation was recently observed at the Tagliamento where organic material allowing to determine time slots of 26.5–23.0 and 24–21 cal. ka BP for these events (Monegato et al., 2007).

The first retreat from these terminal moraines was in the order of some hundreds of metres to several kilometres, depending on the size of the glacier. This generally led to a drainage concentration to only one or two outlets, and an initial minor incision into the outwash fan and terrace. This stage is characterised by small morainic ridges and kettle holes, indicating continuing permafrost conditions. Large blocks of ice were preserved below the sediments during the whole time span from the glacial maximum to its early retreat phase (van Husen, 1977). The kettle holes formed after the downmelting of the glacier tongue, when a single deeply incised outlet drained the overdeepened glacier basin.

The duration of the maximum extent of glaciation and climatic deterioration during the LGM (Last Glacial Maximum) can only be tentatively estimated. After the ice advance around 21 ka BP, the glacier front may have remained in these three maximal positions for about 4000 years, on the basis of data from the subsequent deglaciation phases.

2.9. PHASE OF ICE DECAY

Following the LGM, large-scale retreat and downwasting of the glacier tongues began in the Alpine foreland, as well as in the valleys. In all the great valleys of the Eastern Alps (Inn, Salzach, Drau, Mur, Enns and Traun), no sequences of end moraines or equivalent evidence of former ice margins have been found from this phase. Only kames and ice-marginal terraces that formed in temporary lakes have been identified. These ice-contact sediments,

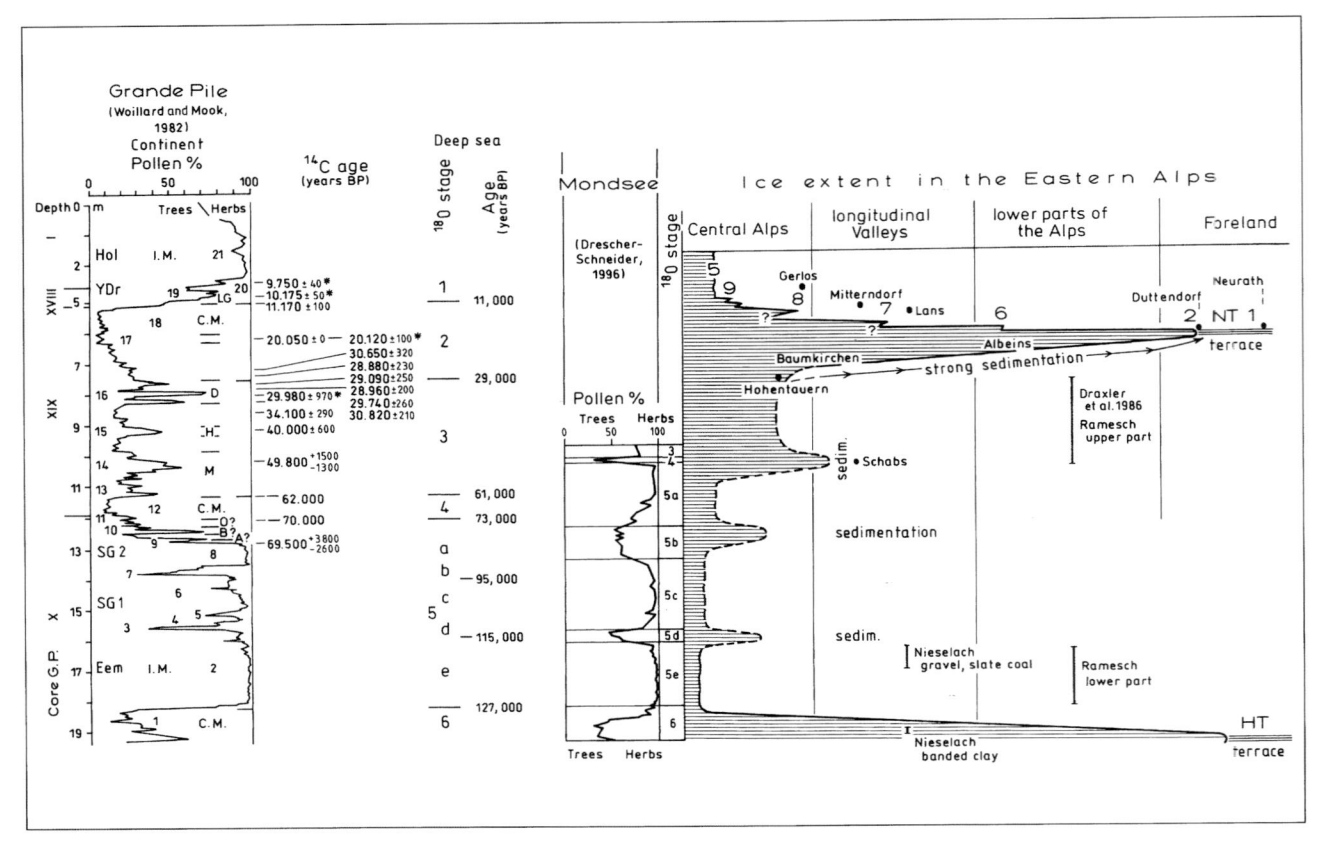

FIGURE 2.8 Temporal development of the ice extent in the Eastern Alps over the past 140,000 years (after van Husen, 2000).

developed especially around the overdeepened parts of the valleys (Fig. 2.8), indicate continuous downmelting, without glacier stillstands or readvances. With respect to the distribution and internal structure of the sediments which formed in temporary lakes, downmelting was rapid. Some hundreds to one thousand years may have been all that was necessary for the loss of about 50% of the glacier lengths in the Eastern Alps. Ice lakes first formed at the glacier margins and then extended over the entire area of the overdeepened basins. This probably resulted in glacier calving, which would have enhanced the rate of ice recession.

Knowledge on the further recessional phases is mainly based on investigations in two valley systems. One is the Traun valley, where the complete sequence of retreat and readvance phases from the LGM to the beginning of the Holocene was mapped in this relatively short valley and investigated by sediment analysis, palynology and radiocarbon dating (Draxler, 1977; van Husen, 1977). A second is the Inn valley where all the type localities of these events are situated (Mayr and Heuberger, 1968), and where intensive investigations have recently been undertaken by Bortenschlager and Patzelt. The close correspondence of the sequences in both areas, in terms of sediment and vegetation development and radicarbon dating, allows the use of classical terms for easier understanding. The following paragraphs describe key sites that provide some evidence about the deglaciation of the Eastern Alps.

2.10. THE BÜHL PHASE

The first sign of a halt in the downmelting of the glaciers is marked around the intramontane basin of Bad Ischl. Here extensive kame deposits, partly covered by a thin layer of till, are connected to small morainic ridges, and this assemblage suggests a stillstand of the glacier margin with minor oscillations (van Husen, 1977). This phase is comparable to the 'Bühl Stage' of Penck and Brückner (1909/11), as shown by the more detailed investigation of the type locality by Mayr and Heuberger (1968). Recent geological re-investigations of the type locality revealed no remarkable readvances there. Predominantly, the forming of extended kame terraces connected with stagnant and downmelting icebodies probably occurred during the period of the Greenland Stadial 2c 21.2–19.5 ka BP (Reitner, 2007). The lithology of pebbles and boulders in the till suggests that a dendritic ice stream in the main valley was still connected to glaciers in all of the tributary valleys at this time. Large kame terraces also exist in other valleys, such as the Drau valley, but they have not been mapped and studied in detail. Nevertheless, their general distribution suggests that they were associated with ice streams comparable to those typical of the Bühl Phase (Fig. 2.8).

2.11. THE STEINACH PHASE

The phase of deglaciation that followed the Bühl was characterised by a minor readvance of the by then much smaller glaciers, again linked with kame terraces and inactive ice masses. Thus, the glacier tongue in the Traun valley near Bad Goisern had advanced over lacustrine and fluvial sediments deposited high above the valley bottom, apparently formed when drainage in the valley to the north was still blocked by stagnant ice masses (van Husen, 1977). A similar situation was described for the Steinach Phase (a term introduced by Senarclens-Grancy, 1958) from the Sill valley, south of Innsbruck (Mayr and Heuberger, 1968). Here, a thick sequence of gravels was also deposited in contact with stagnant ice masses and was covered by the till of an ice readvance. Thus, after the Bühl Phase, the main valleys had become free of active ice and the glaciers had retreated into the tributary valleys. The time interval between the two phases could not have been very long, as large inactive ice masses lingered in the valleys, despite climatic amelioration and large meltwater lakes (Fig. 2.8 and 2.9).

2.12. THE GSCHNITZ PHASE

The next phase in the deglaciation sequence is marked by well-developed blocky end moraines. In the Traun valley, these are present around Bad Goisern and in all of the other source areas of the Würmian ice stream. The morainic ridges are connected to outwash gravels almost everywhere. In the Traun valley north of Bad Goisern and around Bad Aussee, these deposits form terraces extending about 10 km downstream. This indicates that the valleys were free of dead-ice, permitting free drainage along the valley bottoms (van Husen, 1977). This suggests a comparatively long period of time, probably climatic amelioration between the Gschnitz and Steinach Phases and considerable ice recession (Fig. 2.8). The Gschnitz moraines are relatively unmodified by slope processes suggesting that little or no solifluidal shaping has occurred since their deposition. The moraines of the Steinach Phase, in contrast, were clearly smoothed by solifluction during the following (Gschnitz) glacial event. This Gschnitz Phase is also well developed at Trins, south of Innsbruck (Mayr and Heuberger, 1968). Moreover, similar distinctive moraines can be recognised in many of the large tributary valleys draining from higher parts of the central Eastern Alps, as well as in the high cirques to the north and south. This implies that the glacial event was regionally extensive, reflecting a uniform lowering of the equilibrium line (ELA) to a position about 600 m lower than that of the Little Ice Age (Gross et al., 1978).

2.13. CHRONOLOGY

These glacial phases have been indirectly dated by palynological studies of bogs in the Traun valley (Draxler, 1977, 1987). During the early phase, after the main LGM deglaciation, some depressions were slowly filled with varved clay. The pollen record of this time is dominated by *Artemisia*, *Helianthemum*, *Ephedra*, *Hippophaë* and *Juniperus*, in addition to *Pinus*; the *Juniperus* becomes important towards the end of the sequence. This vegetational assemblage, especially the high content of *Artemisia*, is typical of the pioneering phase under dry, cold conditions (Draxler, 1987). The same feature of this early phase has also been described from the western part of the Eastern Alps (Bortenschlager, 1984). The phase terminates with the rapid increase of *Pinus* pollen to values of 70–80%. This interval is well dated in the Traun area to around 12.3 ^{14}C ka BP. The following dates are reported in van Husen (1977):

Moos Alm: 730 m a.s.l., 12.52 ± 0.18 ^{14}C ka BP. VRI-431;

Ödensee: 770 m a.s.l., 12.22 ± 0.18 ^{14}C ka BP. VRI-433;

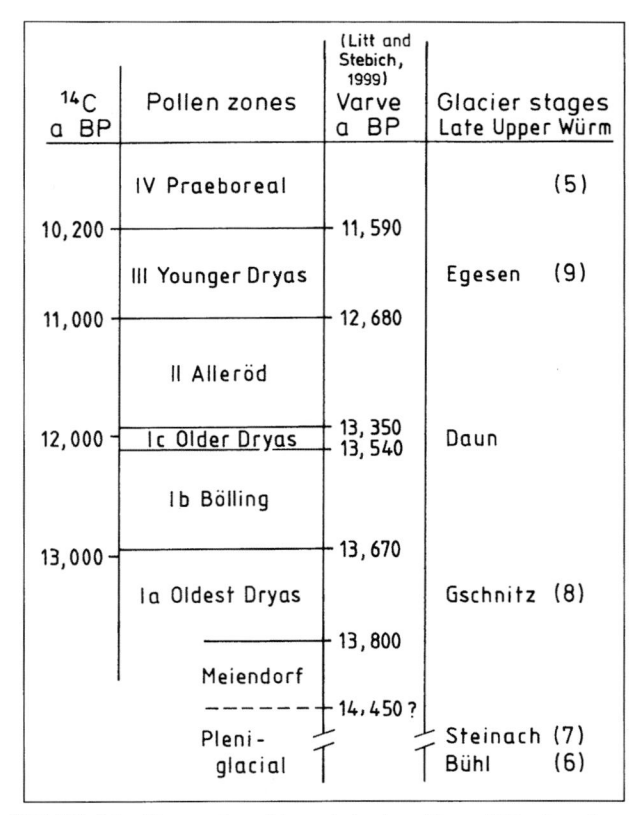

^{14}C a BP	Pollen zones	Varve a BP (Litt and Stebich, 1999)	Glacier stages Late Upper Würm
	IV Praeboreal		(5)
10,200		11,590	
	III Younger Dryas		Egesen (9)
11,000		12,680	
	II Alleröd		
12,000	Ic Older Dryas	13,350 / 13,540	Daun
	Ib Bölling		
13,000		13,670	
	Ia Oldest Dryas		Gschnitz (8)
		13,800	
	Meiendorf		
		14,450 ?	
	Pleni-glacial		Steinach (7) / Bühl (6)

FIGURE 2.9 Temporal position of the late Upper Würmian phases (Termination 1). For explanation see Fig. 2.8.

Plakner: 550 m a.s.l., 12.41 ± 0.19 ^{14}C ka BP. VRI 430;
Ramsau: 515 m a.s.l., 11.97 ± 0.2 ^{14}C ka BP. VRI-432;
Rödschitz: 790 m a.s.l., 12.44 ± 0.42 ^{14}C ka BP. VRI-485.

Dates for the equivalent interval from the Tyrol, reported by Bortenschlager (1984), include:

Lanser See: 840 m a.s.l., 13.23 ± 0.19 ^{14}C ka BP. HV 5269, and
Gerlos: 1590 m a.s.l., 12.155 ± 0.21 ^{14}C ka BP. HV 5284.

The difference of some 100 years between sites may be due to the contrasting rates of soil-forming processes on limestone and crystalline bedrock, as well as to differences in plant immigration. However, the dates suggest that this event occurred during or at the end of the Bølling Ib Chronozone (Fig. 2.9). Bogs documenting this event in the Traun valley lie both outside and inside end moraines of the Gschnitz Phase. Thus, this glacial advance occurred no later than the Oldest Dryas.

This event can be dated more precisely in the pollen profile from Rödschitz in front of the Gschnitz moraines, in which the cooling event is marked by a strong increase in *Artemisia* at 6.40 m depth. Radiocarbon dates from gyttja at 5.40 m depth (12.42 ± 0.44 ^{14}C ka BP. VRI485) and organic detritus (pieces of shrub and herbs) at 7.20–7.00 m depth yield an age of 15.4 ± 0.47 ^{14}C ka BP. VRI-484 suggest that the Gschnitz cold phase had occurred around 14 ^{14}C ka BP, assuming an approximately constant rate of sedimentation in the lake. A similar estimate was made by Patzelt (1975).

Recently, it was attempted to date this event by surface exposure dating (^{10}Be and ^{26}Al) at the type locality (Ivy Ochs et al., 1997, 2000) indicating a final forming of the terminal moraine before 16 ka (Ivy Ochs et al., 2006). This probably would push it into a temporal position before the Meiendorf Stadial at the end of the Pleniglacial (Fig. 2.9).

The periglacial modification of end moraines of the earlier Steinach advance, and the lack of such reshaping on the Gschnitz moraines, agrees well with an Oldest Dryas age for the Gschnitz event, immediately preceding the climatic improvement at Termination 1. After the Bølling Interstadial, there is no evidence for permafrost conditions on the floors of the main valleys in the Eastern Alps.

Based on evidence from the Rödschitz site (basal date of ca. 15.4 ^{14}C ka BP), the Steinach event may have occurred at around 16 ^{14}C ka BP (Fig. 2.9). Thus, the earlier Bühl Phase possibly culminated shortly before this date. During the warmer conditions of the Bølling Chron, the valley bottoms became ice free. Only high parts of the limestone plateaux of the Northern Calcareous Alps and valleys at higher elevations in the Central Alps remained covered with ice.

2.14. THE DAUN PHASE

During the subsequent short, cold Older Dryas Chron, enhanced ice accumulation stimulated small glaciers on the higher limestone plateaux, like the Dachstein Plateau. These small ice masses formed blocky end moraines, including boulders the size of houses. However, in general, this and the following (Egesen) event are not well marked on the comparatively low terrain of the limestone plateaux. By contrast, in the higher parts of the Eastern Alps, moraines of Older and Younger Dryas age are well developed. The older event shows the ELA at a position more than 300 m lower than that of the AD 1850 (Little Ice Age) snowline (Gross et al., 1978).

2.15. THE EGESEN PHASE

This event is marked by well-developed end moraines, which, according to palynological records and radiocarbon (Patzelt and Bortenschlager, 1978) and exposure dating (Ivy Ochs et al., 1996), are believed to have formed during the Younger Dryas Chron of NW Europe. The ELA was lowered by about 300 m (Gross et al., 1978) at this time, but arising from precipitation differences across the mountains, this value varied between 280 m in the drier continental part and 400 m in the more oceanic northern ranges (Kerschner, 1980). Generally, the Younger Dryas was characterised by drier, more continental conditions (around 70% of modern precipitation), with a lowering of the mean annual temperature by some 2.5–4 °C. The climatic deterioration was felt most strongly in the drier central parts of the Alps. Many rock glaciers were reactivated under these cold and dry conditions (Kerschner, 1980). With the onset of the Holocene, the glaciers receded behind their recent limits beginning with a sequence of readvances and retreats (Patzelt, 1995). The last significant ice advance occurred during the Little Ice Age.

REFERENCES

Aric, K., Steinhauser, P., 1977. Geophysikalische Untersuchungen des Inntaluntergrundes bei Thaur östlich von Innsbruck. Zeitschrift für Gletscherkunde und Glazialgeologie 12, 37–54.

Bortenschlager, S., 1984. Beiträge zur Vegetationsge schichte Tirols 1. Inneres Ötztal und unteres Inntal. Berichte des naturwissenschaftlich-medizinischen Vereins Innsbruck 71, 19–56.

Brandecker, H., 1974. Hydrogeologie des Salzburger Beckens. Steirische Beiträge zur Hydrogeologie 26, 26–39.

Decker, K., Peresson, H., Hinsch, R., 2005. Active tectonics and Quaternary basin formation along the Vienna Basin Transform fault. Quatern. Sci. Rev. 24, 307–322.

Del Negro, W., 1969. Bemerkungen zu den Kartierungen L. Weinbergers im Traungletschergebiet (Attersee- und Traunseebereich). Verhandlungen 1969 der Geologischen Bundesanstalt 112–115.

Draxler, I., 1977. Pollenanalytische Untersuchungen von Mooren zur Spät- und postglazialen Vegetationsge-schichte im Einzugsgebiet der Traun. Jahrbuch der Geologischen Bundesanstalt 120, 131–163.

Draxler, I., 1987. Zur Vegetationsgeschichte und Stratigraphie des Würmspätglazials des Traunglet-schergebietes. In: van Husen, D. (Ed.), Das Gebiet des Traungletschers, O.Ö. Eine Typregion des Würm-Glazials, Mitteilungen der Kommission für Quartärforschung der österreichischen Akademie der Wissen-schaften, 7 19–35.

Dreesbach, R., 1983. Sedimentpetrographische Untersuchungen zur Stratigraphie des Würmglazials im Bereich des Isar-Loisachgletschers. Dissertation, München, 176 pp.

Drescher-Schneider, R., Papesch, W., 1998. A contribution towards the reconstruction of Eemian vegetation and climate in central Europe: first results of pollen and oxygen-isotope investigations from Mondsee, Austria. Veg. Hist. Archaeobot. 7, 235–240.

Drescher-Schneider, R., 2000. Die Vegetations- und Klimaentwicklung im Riß/Würm- Interglazial und im Früh- und Mittelwürm in der Umgebung von Mondsee. Mitt. Komm. Quartärforsch. Österr. Akad. Wiss. 12, 37–92.

Fink, J., 1979. Stand und Aufgaben der österreichischen Quartärforschung. Innsbrucker Geogr. Stud. 5, 79–104.

Fink, J., Fischer, H., Klaus, W., Koci, A., Kohl, H., Kukla, J., et al., 1976. Exkursion durch den österreichischen Teil des nördlichen Alpenvorlandes und den Donauraum zwischen Krems und Wiener Pforte. Mitteilungen der Kommission für Quartärforschung der österreichischen Akademie der Wissenschaften. 1, 113 pp.

Fischer, H., 1977. Tal- und Hangentwicklung in pleistozänen Schotterterrassen (Enns Ybbs-Platte). Zeitschrift für Geomorphologie N F Suppl. Band 28, 161–180.

Frank, C., van Husen, D., 1995. Eastern Alps traverse. In: Schirmer, W. (Ed.), Quaternary Field Trips in Central Europe International Union for Quaternary Research, XIV International Congress Berlin, 1. Friedrich Pfeil, München, pp. 417–418.

Frank, Ch., Nagel, D., Rabeder, G., 1997. Chronologie des österreichischen Plio-Pleistozäns. In: Döppes, D., Rabeder, G. (Eds.), Pliozäne und pleistozäne Faunen Österreichs, Wien. Österreichische Akademie der Wissenschaften, pp. 359–374.

Fritsch, A., 1993. Das Quartär Wien, der westlichen Hochschwab-Nordabdachung unter Berücksichtigung des Bergsturzes von Wildalpen. Unpublished Dipl.-Arbeit, Naturwissenschaftliche Fakultät Universität Wien, 122 pp.

Fuchs, W., 1972. Tertiär und Quartär am Südostrand des Dunkelsteiner Waldes. Jahrbuch der Geologischen Bundesanstalt 115, 205–245.

Gibbard, P.L., Head, M.J., Walker, M.J.C., the Subcommission on Quaternary Stratigraphy, 2009. Formal ratification of the Quaternary system/period and the Pleistocene series/epoch with a base at 2.58 Ma. J. Quatern. Sci.

Graul, H., 1937. Untersuchungen über Abtragung und Aufschüttung im Gebiet des unteren Inn und des Hausruck. Mitteilungen der Geographischen Gesellschaft 30, 181–256.

Gross, G., Kerschner, H., Patzelt, G., 1978. Methodische Untersuchungen über die Schneegrenze in alpinen Gletschergebieten. Z. Gletscherk. Glazialgeol. 18, 249–286.

Grüger, E., 1979. Spätriß, Riß/Würm und Frühwürm am Samerberg in Oberbayern—ein vegetationsgeschichtlicher Beitrag zur Gliederung des Jungpleistozäns. Geol. Bavarica 80, 5–64.

Ivy Ochs, S., Schlüchter, Ch., Kubik, P., Synal, H.A., Beer, J., Kerschner, H., 1996. The ^{10}Be, ^{26}Al und ^{36}Cl exposure age of an Egesen moraine at Julier Pass, Swit-zerland, measured with the cosmogenic radionuclides. Eclogae Geol. Helv. 89, 1049–1063.

Ivy Ochs, S., Kerschner, H., Kubik, P.W., Synal, H.-A., Patzelt, G., Schlüchter, Ch., 1997. Moraine formation in the European Alps mirrors North Atlantic Heinrich events. Lunar and Planetary Institute Contribution 921. (Seventh annual V.M. Goldschmidt Conference, Houston Ivy Ochs, S., Kerschner, H., Kubik, P.W., TX), 103.

Ivy Ochs, S., Kerschner, H., Kubik, P.W., Schlüchter, Ch., 2000. Expositionsalter und paläoklimatische Interpreta-tion der Gschnitz-Moräne in Trins, Stubaier Alpen. DEUQUA 2000 (Abstract Band) Bern.

Ivy Ochs, S., Kerschner, H., Kubik, P.W., Schlüchter, Ch., 2006. Glacier response in the European Alps to Heinrich event I cooling: the Gschnitz stadial. J. Quatern. Sci. 21, 115–130.

Kahler, F., 1958. Die Tiefe der Felsoberfläche in den Senken des Klagenfurter Beckens. Carinthia 11 (68), 5–8.

Kerschner, H., 1980. Outlines of the climate during the Egesen advance (Younger Dryas, 11,000-10,000 years BP) in the Central Alps of the Western Tyrol, Austria. Z. Gletscherk. Glazialgeol. 16, 229–240.

Klaus, W., 1987. Das Mondsee-Profil; R/W Interglazial und vier Würm-Interstadiale in einer geschlossenen Schichtfolge. Mitteilungen der Kommission für Quartärforschung der österreichischen Akademie der Wissenschaften 7, 3–17.

Kohl, H., 1974. Die Entwicklung des quartären Flußnetzes im Bereich der Traun Enns-Platte/Oberösterreich. Heidelberger Geographische Arbeiten. H. Graul Festschrift 31–44.

Kohl, H., 1976. In: Fink, J. (Ed.), Exkursion durch den österreichischen Teil des nördlichen Alpenvorlandes und den Donauraum zwischen Krems und Wiener Pforte, Mitteilungen der Kommission für Quartärforschung der österreichischen Akademie der Wissenschaften. 1, 37–48.

Kolmer, Ch., 1993. Die quartäre Landschaftsentwicklung der östlichen Hochschwab Nordabdachung. Diplom-Arbeit, Naturwissenschaftliche Fakultät der Universität Wien, 111 pp.

Krenmayr, H.G., 2000. Sedimentologie der letztinterglazialen limnischen Deltasedimente von Mondsee. Mitteilungen der Kommission für Quartärforschung der österreichischen Akademie der Wissenschaften. 12, 13–37.

Kröll, A., Gnojek, I., Heinz, H., Jirecek, R., Meurers, B., Seiberl, W., et al., 1993. Karten über den Untergrund des Wiener Beckens und der angrenzenden Gebiete (+Erläuterungen). Geologische Bundesanstalt, Wien.

Mascarelli, A.L., 2009. Quaternary geologists win timescale vote. Nature 459, 624.

Mayr, F., Heuberger, H., 1968. Type areas of late glacial and post-glacial deposits in Tyrol, Eastern Alps. Proceedings VII. INQUA Congress 14, University of Colorado Studies, Series in Earth Sciences Nr. 7. Glaciation of the Alps, Boulder/Colorado, pp. 143–165.

Monegato, G., Ravazzi, C., Donagana, M., Pini, R., Calderoni, G., Wick, l., 2007. Evidence of a two-fold glacial advance during the last glacial maximum in the Tagliamento end moraine system (eastern Alps). Quatern. Res. 68, 284–302.

Patzelt, G., 1975. Unterinntal-Zillertal Pinzgau, Kitzbühel. Spät- und postglaziale Landschaftsentwicklung. Innsbrucker geographische Stud. 2, 309–329.

Patzelt, G., 1995. Eastern Alps Traverse. In: Schirmer, W. (Ed.), Quaternary Field Trips in Central Europe. Inter-national Union for Quaternary Research, XIV Interna-tional Congress, 1, 385, Berlin.

Patzelt, G., Bortenschlager, S., 1978. Spät- und nacheiszeitliche Gletscher- und Vegetationsentwicklung im inneren Otztal. Führer zur Tirol-

Exkursion 'Innsbrucker Raum und Ötztal'. DEUQUA, 13-25, Innsbruck.

Penck, A., Brückner, E., 1909. Die Alpen im Eiszeit-alter. Tauchnitz, Leipzig 1199 pp.

Peresson, H., Decker, K., 1996. Early to Middle Miocene tectonics of the eastern part of the North Calcareous Alps. Mitt. Gesell. Geol. Bergbaustud. 41, 53–63.

Pfiffner, O.A., Heitzmann, P., Lehner, P., Frei, W., Pugin, A., Felber, M., 1997. Dynamic Alps. Incision and backfilling of Alpine valleys: Pliocene, Pleistocene and Holocene processes. In: Pfiffner, O.A., Lehner, P., Heitz-mann, P., Mueller, S.T., Steck, A. (Eds.), Deep structure of the Swiss Alps. Birkhäuser, Boston, Berlin, pp. 265–288.

Plachy, H., 1981. Neue Erkenntnisse zur Tektonik im Wiener Raum. Mitteilungen der Österreichischen Geologischen Gesellschaft. 74/75, 231–243.

Raymo, M.E., 1997. The timing of major climatic terminations. Paleoceanography 12, 577–585.

Reitner, J.M., 2007. Glacial dynamics at the beginning of Termination I in the Eastern Alps and their stratigraphic implications. Quatern. Int. 164–165, 64–84.

Senarclens-Grancy, W., 1958. Zur Glazialgeologie des Ötztales und seiner Umgebung. Mitteilungen der Österreichischen Geologischen Gesellschaft. 49, 275–314.

Shackleton, N.J., 1987. Oxygen isotopes, ice volume and sea level. Quatern. Sci. Rev. 6, 183–190.

Shackleton, N.J., 1995. The deep sea sediment record and the Pliocene-Pleistocene boundary. Quatern. Int. 40, 33–35.

Sperl, H., 1984. Geologie und Sedimentologie des Quartärs im Attergau/Oö. Dissertation, naturwissen-schaftliche Fakultät der Universität Wien, 251 pp.

van Husen, D., 1977. Zur Fazies und Stratigraphie der jungpleistozänen Ablagerungen im Trauntal. Jahrbuch der Geologischen Bundesanst. 120, 1–130.

van Husen, D., 1979. Verbreitung, Ursachen und Füllung glazial übertiefter Talabschnitte an Beispielen in den Ostalpen. Eiszeit. Ggw 29, 9–22.

van Husen, D., 1980. Bericht 1979 über Aufnahmen im Quartär auf Blatt 66 Gmunden. Verhandlungen 1980 der Geologischen Bundesanstalt 1980, A42–A43.

van Husen, D., 1981. Geologisch-sedimentologische Aspekte im Quartär von Österreich. Mitteilungen der Österreichischen Geologischen Gesellschaft 74/75, 197–230.

van Husen, D., 1987. Die Entwicklung des Traunglet-schers während des Würm-Glazials. In: van Husen, D. (Ed.), Das Gebiet des Traungletschers, O.Ö. Eine Typ-region des Würm-Glazials, Mitteilungen der Kommission für Quartärforschung der Österreichischen Akademie der Wissenschaften, 7 19–35.

van Husen, D., 1989. The last interglacial glacial cycle in the Eastern Alps. Quatern. Int. 3/4, 115–121.

van Husen, D., 1996. LGM and late glacial fluctuations in the Eastern Alps. Quatern. Int. 38, 109–118.

van Husen, D., 1999. In: Egger, H. (Ed.), Geologische Karte der Republik Österreich 1:50,000, Blatt 69 Großraming. Geologische Bundesanstalt, Wien.

van Husen, D., 2000. Geological Processes during the Quaternary. Mitteilungen der Österreichischen Geologischen Gesellschaft. 92, 135–156.

Weber, F., Schmid, Ch., Figala, G., 1990. Vorläufige Ergebnisse reflexionsseismischer Messungen im Quartär des Inntals/Tirol. Zeitschrift für Gletscherkunde und Glazialgeologie. 26, 121–144.

Weinberger, L., 1955. Exkursion durch das österreichische Salzachgletschergebiet und die Moränengürtel der Irrsee- und Attersee-Zweige des Traungletschers. Verhandlun der Geologischen Bundesanstalt Sonderheft D 1955, 7–34.

Woillard, G.M., Mook, W.G., 1982. Carbon-14 dates at Grande Pile: correlation of land and sea chronologies. Science 215, 159–161.

Chapter 3

The Pleistocene Glaciations in Belarus

A.K. Karabanov* and A.V. Matveyev

Institute for Nature Management of the National Academy of Sciences of Belarus, F.Skoriny Str., 10 Minsk 220141, Belarus
*Correspondence and request for materials should be addressed to A. K. Karabanov. E-mail: nature@ecology.basnet.by

3.1. INTRODUCTION

This interpretation of the glacial limits in Belarus is based on information from the *geomorphological maps of the Belarussian SSR* (Gurski, 1980, 1986) which follows the stratigraphical scheme of the Pleistocene that was accepted by the majority of geologists (Makhnach, 1971; Gursky, 1974; Matveyev et al., 1988; Matveyev, 1995). The aim of this chapter is to present the latest updates to the glacial pattern in Belarus in comparison to the 1980 geomorphological map. These modifications are based on the results of geological and geomorphological investigations undertaken during the past decade. The present authors follow the new stratigraphical scheme of the Pleistocene in Belarus as elaborated by Sanko et al. (2005). The 1:1,000,000 scale digital maps presented is an attempt to integrate the revised ice-marginal positions into the previously mapped glacial limits. On the basis of present knowledge, all known glacial complexes in Belarus were formed during the Middle and Late Pleistocene.

The present-day landscape of Belarus was shaped by repeated Pleistocene glaciations. Six major ice sheets affected the country during the Pleistocene: the Varyazh, Narev, Berezina, Dnieper, Sozh and Poozerian glaciations (Table 3.1). The existence of the most ancient Varyazh Glaciation is still questioned because of the lack of geological information. The sequence of glacial/interglacial deposits reaches a maximum thickness of 325 m, with an average thickness of 75–80 m. All data on the thickness of glacial deposits are cited after Matveyev (1995). To specify the limits of glaciers, their stage and phase complexes, geomorphological and geological information was used for the Pripyat and Poozerian glaciations (including the distribution of end moraines, eskers, kames, fluvioglacial deltas, glaciolacustrine

TABLE 3.1 Glaciations, Interglacials and Warm Intervals (grey) in the Middle to Late Pleistocene of Belarus

Gurski (1986)		This chapter (after Sanko et al., 2005)	
Poozierie	Late Pleistocene	Late Pleistocene	Poozierie
Murava			Murava
Sozh	Middle Pleistocene	Middle Pleistocene	Pripyat Sozh stage Dnieper stage
Shklov			
Dnieper			
Aleksandriya			Aleksandriya
Berezina	Early Pleistocene		Berezina
Beloviezha			Beloviezha
Narev			Narev
Brest			Brest II Ruzhany warm interval
			Brest I Varyazh cold interval (glaciation?)
		Early Pleistocene	Gomel warm interval

sediments, etc.), whilst for the Berezina, Narev and Varyazhsk glaciations only geological data were available (i.e. the distribution of glacial sediments, their stratigraphical position within the Pleistocene sequence, etc.).

3.2. VARYAZH GLACIATION

The oldest known glaciation in Belarus is the Varyazh glaciation. Palaeobotanical evidence indicates a period of cold climate at the beginning of the Brest chronostratigraphical unit (Table 3.1). This evidence is also supported by the distribution of pebbles and gravels of gneiss in the basal parts of the corresponding sediments in some sections in western Belarus (fluvioglacial sediments?). It is also indicated by till-like sediment about 4 m thick found within the Naroch-Vilija lowland and in the western part of the Oshmyany highland. The distribution of these sediments provides the basis for the limit of this most ancient glaciation in the northwest part of Belarus (Fig. 3.1). To the south of this line, a network of boreholes has indicated that fluvioglacial streams distributed erratic materials from Scandinavia.

3.3. NAREV GLACIATION

The greater part of Belarus has been covered by the Narev ice sheet (Fig. 3.1). Deposits of this glaciation have been mostly eroded and disturbed during the subsequent ice advances. Today, the Narev glacial sediments are not exposed anywhere, but they underlie the Berezina deposits. The Narev till varies from 0.2–0.5 m to up to 40–50 m in thickness. The limit of the maximum ice advance is debatable (Matveyev, 1990), however, based on a complex of geological evidence (the distribution of till, pre-glacial and glacial relief), it most likely was reached approximately along the line from Brest to Kobrin, Luninets, Retchitsa, Gomel and Vetka. Three large Narev ice streams have been identified: the Neman, Viliya and Dnieper, the prevailing movement direction of which was towards the south in the west, whereas elsewhere it was towards the south-south-east. Glacial complexes of particular events and phases within the Narev glaciation have been distinguished near the cities of Luninets, Lyakhovichi, Mar'ina Gorka and Kopyl'.

3.4. BEREZINA GLACIATION

The Berezina ice sheet covered almost all of Belarus (except the southernmost part of the country). Generally, the Berezina glacial sediments are 10–15 m thick in the north, 50–70 m in the centre and 15–25 m in the southern part of the country. One of the peculiarities of the Berezina deposits is the abundance of associated glaciolacustrine sediments which are found widespread, especially in western and eastern Belarus. The other is the presence of an intricate

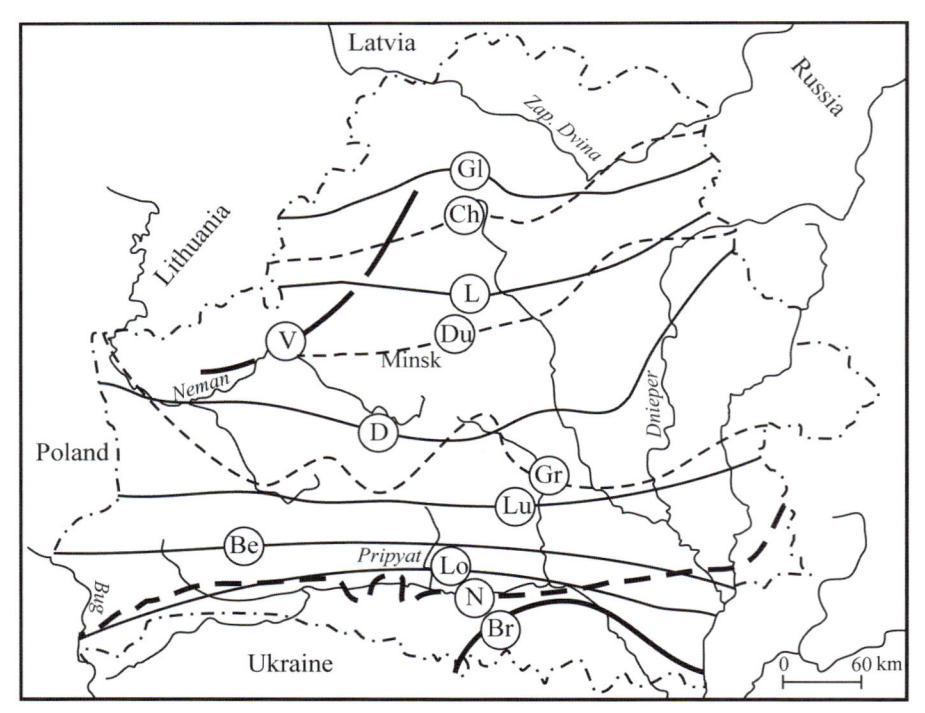

FIGURE 3.1 The major limits of glacial stages represented by buried glacial complexes or deposits not distinctly expressed at the present ground surface. The maximum extent of the glaciations: V, Varyazh; N, Narev; B, Berezina. Stages of the Narev glaciation; Gr, Grodno; Du, Dubrovno; Ch, Chashniki. Stages of Berezina glacier; Lo, Logoshin; Be, Beryoza; Lu, Luban'; D, Dubrovno; L, Logoisk; Gl, Glubokoye.

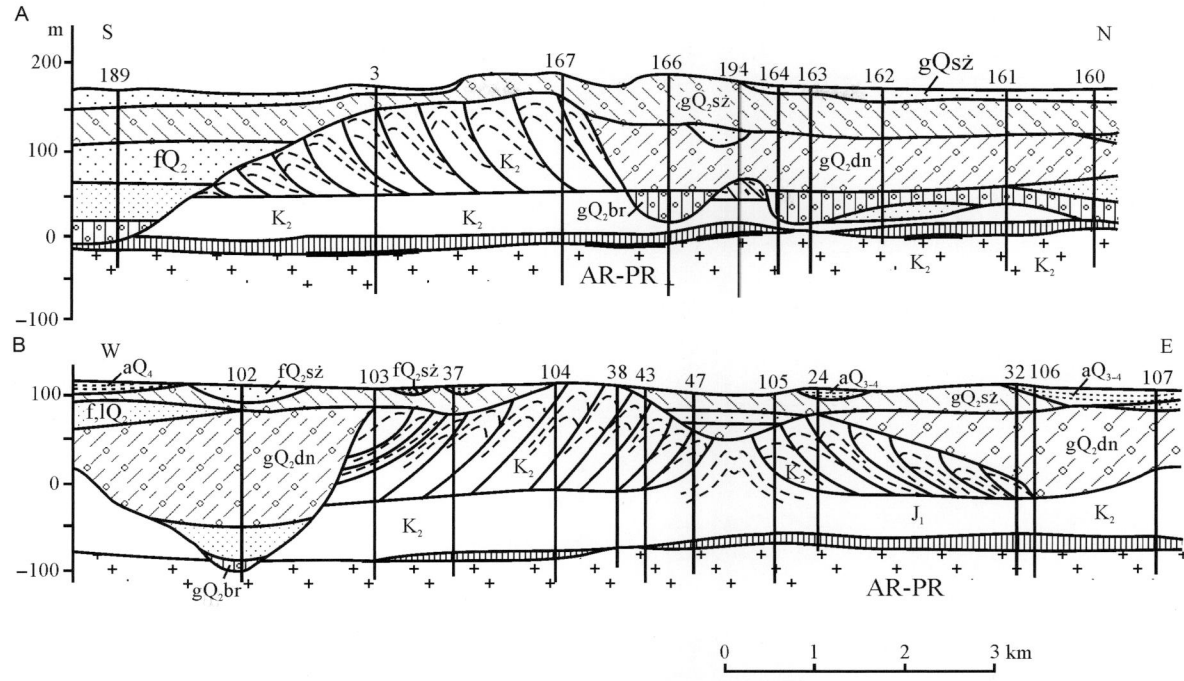

FIGURE 3.2 Glaciotectonic dislocations in western Belarus in (A) the Volkovysk highland and (B) the Novogrudok highland.

network of glacial channels (tunnel valleys) which are deeply incised into the pre-Quaternary bedrock (up to 125 m b.s.l.). After deglaciation, the landscape was dominated by relatively flat surfaces with very few highlands and with large morainic plateaux dissected by the deep glacial valleys.

The most significant buried glacial landforms occur near the cities of Sharkovshchina, Novogrudok, Osipovichi, Gantsevichi and Starobin. The Berezina ice sheet, as was the case during the Narev glaciation, was divided in three glacial lobes: the Neman, Viliya and Dnieper ice streams.

3.5. PRIPYAT' GLACIATION

The main morainic highlands of Belarus were originally formed during the largest, or Pripyat' glaciation. This glaciation can be divided into two main stages: the Dnieper stage and the Sozh stage.

3.5.1. Dnieper Stage

The Dnieper stage of the Pripyat' glaciation saw the most extensive Pleistocene glaciation which advanced well beyond the limits of the country. The thickness of glacial sediments deposited during this phase reaches over 100 m and is generally 40–50 m. The most remarkable result of this glaciation was the formation of the morainic cores of the Grodno, Minsk, Oshmiany, Mozyr and other morainic highlands. Numerous glaciotectonic dislocations are recognised in the structure of these highlands (Fig. 3.2). In many respects, these uplands controlled the dynamics of the subsequent Sozh stage as well as of the last (Poozerian) glaciation.

To map the stages of the Pripyat' glacial cover, the distribution of glacial–morphological complexes has been widely used. A 'glacial–morphological complex' is regarded as an aggregate of genetically linked glacigenic landforms, expressed in the modern landscape and generated between two consecutive stages or phases of glaciation. The oldest surviving 'glacial–morphological complexes' are regional elements of the Dnieper stage of the Pripyat glaciation and its recession phases: the Stolin, Mozyr' and Chechersk complexes (Fig. 3.3). The Dnieper ice advances were characterised by the Neman, Lansk, Dnieper and Verkhnedvinsk glacial streams. During the glacier retreat, the glacial complexes underwent intensive washout by snowmelt, and during the subsequent Poozerian stage, they were partially buried under periglacial alluvial and lacustrine sediment accumulations. Under the influence of erosional and accumulative processes, the original surface was largely altered. Prominent landforms still visible today include the Zagorod'e, Mozyr and Yurovichi ridges. Morphological fragments are also preserved in the areas of Brest, Malorita, Retchitsa and Chechersk. The Dnieper-stage glaciation also formed the morainic cores of the large Grodno, Novogrudok, Oshmyany and Minsk highlands.

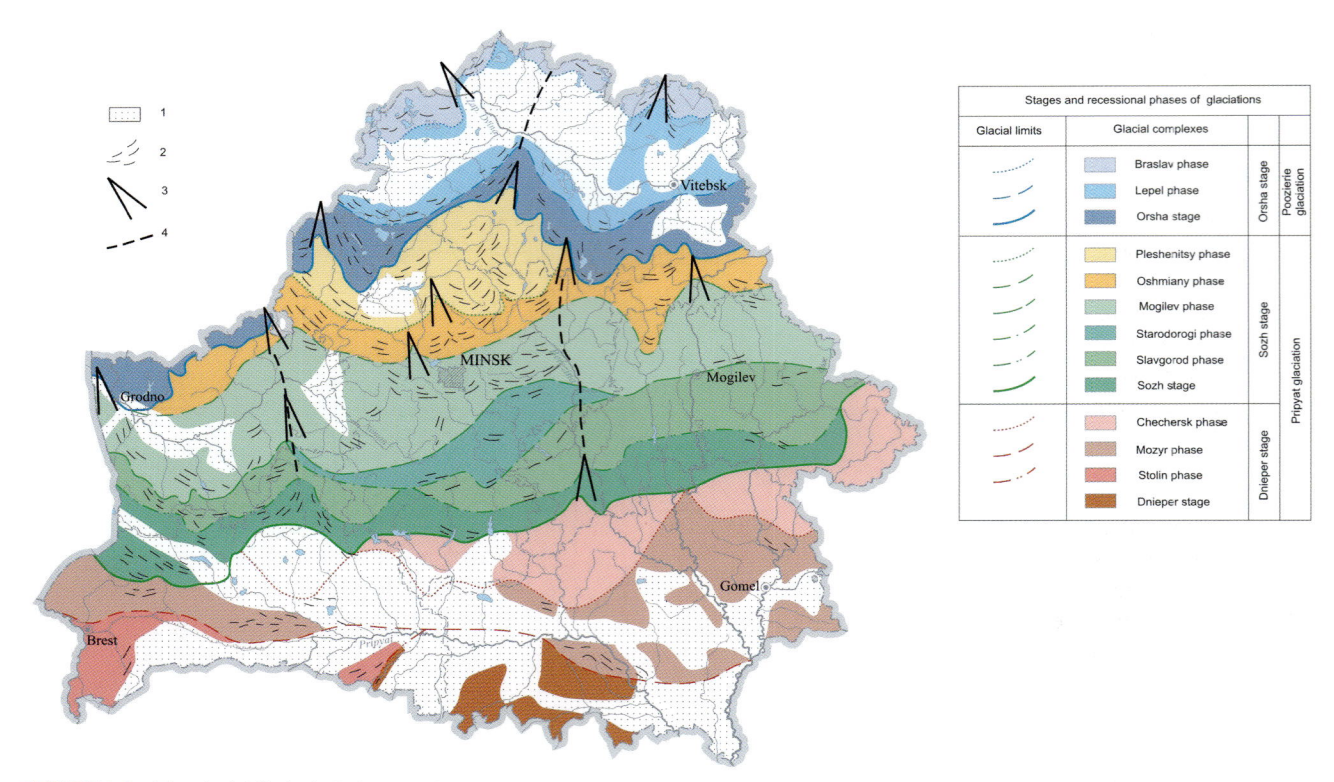

FIGURE 3.3 The glacial limits in Belarus. 1, lacustrine deposits; 2, end moraine ridges; 3, interlobate massifs; 4, main ice divides.

3.5.2. Sozh Stage

The Sozh glacial complex is on average 10–20 m thick but reaches a maximum thickness of about 100 m (in the Minsk highland area). The distribution of the Sozh till is irregular. In the northern part of Belarus, it has been almost completely eroded. Some patches of the original till cover have been preserved within the largest glacial basin in Belarus, the Polotsk glacial depression. Recent investigations have shown that the Sozh deposits are absent from eastern Belarus (Pavlovskaya et al., 1997). This discovery has significantly changed opinions concerning the age of the upper till in that area but confirm the view of Velichkevich, et al. (1996) that only the Dnieper and the Sozh deposits represent glacial stage sediments in the strict stratigraphical sense of this term.

For the purposes of mapping the Sozh glacial complexes (Slavgorod, Starodorogy, Mogilov, Oshmyany, Pleshchenitsy phases), the authors mainly followed previous interpretations regarding the distribution of Sozh deposits and glacial landforms (Karabanov, 1987; Matveyev et al., 1988; Valchik et al., 1990). Synthesis of the geological evidence has allowed the differentiation of several complexes which form morphologically distinct belts within the Grodno, Novogrudok, Oshmyany and Minsk highlands, and also on the Orsha-Mogilov plateau. These glacial complexes were formed by three glacial streams (Neman, Minsk and Dnieper), the ice-divide zones of which, in contrast to those of the Dnieper, are still distinctly visible in the present landscape. Glacial morphology and features of geological structure of angular massifs within the glacial highlands allow the reconstruction of the succession of main ice divides throughout all stages of deglaciation of the Sozh ice sheet. This evidence suggests that no major readvance of the ice front occurred during the recession and that short-term, small oscillations dominated the ice-front behaviour.

A new interpretation of the Sozh-stage limit has been established in the southern part of the Orsha-Mogilov plateau. In recent years, the new sites of Alexandria and Murava interglacial sediments have been studied in eastern Belarus (Savchenko and Pavlovskaya, 1999). The geological situation in this area suggests that there was only one glacial sedimentation cycle between the Alexandria and Murava units. The ratio of glacial and interglacial sediments in these profiles testifies to continuous sediment accumulation during the late Alexandrian to early Pripyat time. At the same time, a break has been found in the sediments between the late Pripyat and early Murava strata. The geological structure of this area probably excludes the erosion of the Sozh deposits by meltwater, since the Dnieper-stage till, throughout practically all the Orsha-Mogilov plateau, forms the present-day land surface, and only in some cases, it is overlain by fluvioglacial deposits.

Therefore, it is safe to assume that the Sozh-stage limit in East Belarus was positioned much further north than was previously thought.

3.6. POOZERIAN GLACIATION

Deposits of the subsequent Poozerian glaciation (Weichselian, Valdaian, Wisconsinan) underlie the northern and north-western part of Belarus. The thickness of the glacial sequence varies from a few metres to 70 m, but on average, the deposits reach about 20–30 m. During the Poozerian glaciation, the Svir, Braslav, Nevel, Gorodok, Vitebsk and Orsha glacial highlands were formed. Similarly, large glacial lakes, such as the Polotsk, Surazh and Luchosa lakes, came into existence as a result of blocked southward meltwater drainage. The maximum extent of the last ice sheet was controlled by the pre-existing topography and depended on the distribution of the morainic highlands and lowlands remaining from the Dnieper and Sozh ice sheets. The Poozerian ice sheet extended the furthest south in the Vilia, Berezina and Luchosa drainage basins.

New evidence obtained in the past few years from north-western and northern Belarus has resulted in a partial revision of the previously mapped glacial limits and maximum extent of the last glaciation. The limits of substages and phases of the Poozerian glaciation are based on the geomorphological analysis of glacial landform complexes, taking into account the results of geological investigations.

Interpretation of aerial photographs has been applied to the study of landforms. The dominant role of the morphological approach was adopted for two main reasons. Firstly, there is a lack of sections exposing interstadial deposits in northern and north-western Belarus. Secondly, the landform assemblages of corresponding phases of the last glaciation exhibit essential morphological differences. The availability of satellite images and aerial photographs has allowed the application of remote-sensing techniques in a new approach to large-scale geomorphological mapping.

New geomorphological evidence have been obtained, and these data combined with a spatial analysis of the distribution and structure of the main ice-marginal complexes in the Belarussian–Lithuanian border region have led to the conclusion that the limits of the main phases of the last glaciation in NW Belarus required revision (Guobyte and Pavlovskaya, 1998). The maximum extent of the last glaciation in Belarus is based not only on the study of glacial landforms but also on till mineralogy and petrography, investigated by Baltrunas et al. (1985) and Gaigalas (1995) in Lithuania, and Karabanov (1987), Gurski et al. (1990) and Sanko (1987) in Belarus. The mineralogical and petrographical data also allow clarification of the course of the deglaciation.

The distribution of ice-marginal formations clearly indicates that the outermost limit of the last ice sheet in NW Belarus occurred south of the Vilia valley where several glaciofluvial deltas and outwash cones have been recognised on aerial photographs. This limit lies 10 km (in some places 20 km) further to the south than previously assumed (Geomorphological Map of the Belarussian SSR, 1980; Geomorphological Map of the Belarussian SSR, 1986). This glacial limit corresponds closely with its equivalent in Lithuania (Guobyte, 1999). Further, field investigations by Pavlovskaya in 1998–1999 in NW Belarus (central part of the Vilia drainage basin) confirmed the assumed position of the maximum extent of the Poozerian glaciation. Two lithologically different tills have been identified here. Clast orientation and till fabrics at the Zamok and Belaya Gora outcrops have revealed differences between the lower and the upper till. The lower till is interpreted as the Dnieper (Saalian) till, on the basis of its clast orientation and the upper is interpreted as a Poozerian (Weichselian) till.

The process of deglaciation in the marginal zone of the Poozerian maximum ice advance created vast fields of dead ice, especially in NW Belarus. This partly explains why there is no massive end-moraine complex, but instead a great number of kames, eskers and pitted outwash plains are found. The occurrence of Poozerian end moraines is limited to the Svir highland in Belarus, as well as to the Mickunai glacial depression in Lithuania. The distribution of ice-marginal complexes and the orientation of glacial dislocations suggest that the Weichselian ice-cover was separated into two lobes: the Disna lobe in NW Belarus and the Zeimena lobe in NE Lithuania. This division probably took place during the Gruda-Ozerskaya phase, which probably equates to the Brandenburg phase. The lobate pattern of ice flow is indicated by the occurrence of almost longitudinal (N–S trending) ice-marginal and glaciofluvial formations. It is very likely that the basal topography affected or even controlled glacier dynamics and played a major role in the formation of the two-lobate pattern.

The two ice lobes were separated by the Shvencionys (Lithuania) and Lyntupy (Belarus) highland areas. Such a supposition is not new. Baltrunas et al. (1985) also concluded that the ice margin was separated into two parts on both sides of these highlands. However, they assumed that the Sventiany highland was formed during their Baltija substage. The present authors consider that this ice-marginal complex is older and that the limit of the Baltija-Braslav stage (which corresponds with the Pomeranian stage) was situated much further north.

Both ice lobes occurred under a relatively passive glacial regime. However, the Disna lobe decayed faster. This might have been caused by climatic factors, but it seems to be more likely that the basal conditions of the ice played a decisive role. For example, an abundant inflow of water at the ice base might be a possible explanation of this mode of deglaciation. Moreover, the faster retreat of the

Disna lobe might have resulted from the wide distribution of ice-dammed lakes, which began to form in this area in the Shvencionys-Sventiany phase (possibly to be correlated with the Frankfurt ice-marginal line). These lakes might have encouraged more rapid decay of the ice margin. The largest water body in the region was the Disna glacial lake, situated between the ice-marginal complexes of the Shvencionys-Sventiany phase and the Baltuja-Braslav substage. The existence of such an extensive water basin provoked a relatively high rate of melting, accelerated by calving. Traces of calving, in the form of ice-rafted detritus, are found in numerous sections of glaciolacustrine deposits.

The passive regime of ice dynamics changed during the Baltuja-Braslav substage, when a prolonged stagnation phase of the Weichselian ice sheet occurred, probably interrupted by short ice-marginal advances which pushed up a series of end-moraine ridges (Pavlovskaya, 1994). A great number of glacial dislocations and the morphology of ridges point to the existence of an active ice margin and a change in the direction of glacial pressure from the NNW to NNE—the latter direction being peculiar of the Baltija substage (Gaigalas, 1995). It is very likely that there were three short pulses of local ice advance which formed the northern, central and southern ridges of the Braslav highland.

In the central and eastern parts of Belarus, the limits of the Poozerian glaciation and the Sozh substage of the Dnieper glaciation have been based on morphological evidence from the geomorphological and Quaternary geological maps of Belarus. In general, they correspond to the position of ice-marginal glacial highlands. Particular investigations of tills have determined that the Poozerian ice sheet in this area extended further south than that mapped previously.

Detailed investigations of the outcrops along the Berezina river valley, between Lake Palik and Borisov, have revealed the following stratigraphical sequence:

1. A brown and greenish-grey till which occurs within the end moraines and is partially exposed within glaciofluvial plains.
2. A thin reddish-brown till. It is distributed mainly in the vast Upper Berezina lowlands and appears to be a flow till.
3. Glaciolacustrine deposits which overlie the upper till. In the vicinity of Borisov, a transition from these glaciolacustrine sediments into the alluvium of the second Berezina river terrace has been seen at an altitude of about 11–13 m above the modern water level.
4. Late Poozerian and Holocene alluvium of the first Berezina river terrace and the floodplain.

Analysis of the pebble fraction (2–5 cm) in both tills indicates their different composition. Clasts of sedimentary rocks (mainly dolomite, limestone, sandstone, marl, flint) dominate the lower till assemblage, whilst crystalline rocks dominate in the upper till (Gurski et al., 1990).

Since the same characteristics have been found in other sites within the Poozerian limits, it has been possible to identify the upper reddish-brown unit as the Poozerian till. Therefore, this evidence suggests that the Upper Berezina lowland, north of Borisov, was overridden by the Poozerian glaciation. The maximum limit of the glaciation can be identified 40–50 km to the south along the Berezina river valley up to its bend at Bolshoye Stakhovo (some 5–7 km north of Borisov). This area was previously attributed to the Sozh glaciation.

Further downstream, in the section of the Berezina river valley between Bobruisk and Parichi and also in the Sozh river valley downstream of Slavgorod, investigations of the till clast petrography (using the 1–2, 2–3, 3–5, 5–7, 7–10 cm fractions) have been carried out (Gurski et al., 1990). The results of these investigations suggest that separate glacial tongues might have extended significantly further to the south (30–40 km in this area) than previously assumed.

3.7. CONCLUSIONS

On the basis of the results of recent investigations, the glacial pattern in Belarus was more complicated than previously thought. The most significant recent modifications of this pattern concern the limits of the Poozerian glaciation in the north-western and northern parts of the country. The maximum limit of this glaciation in NW Belarus was located south of the Vilia valley, and the ice sheet also advanced further south in the Upper Berezina lowland.

The modified limits of the older glaciations (the Dnieper glaciation and the Sozh substage) are mainly based on the results of till-lithological studies and palaeobotanical investigations (Makhnach, 1971; Gursky, 1974; Velichkevich, 1982; Velichkevich, et al., 1993; Pavlovskaya et al., 1997; Savchenko and Pavlovskaya, 1999). The main modifications concern the position of the glacial limit of the Sozh substage in eastern Belarus, and the maximum extent of the Dnieper glaciation in southern Belarus.

On the basis of new geological evidence, the ideas of the processes of ice-sheet degradation within the Belarus territory are revised. Glacier retreat was mainly recessional with short-term readvance movements and widespread areal deglaciation, especially during the Poozerian glaciation. In addition, the limit of the most ancient Varyazh glaciation has been identified for the first time. In the future, the research into the dynamics of the former glaciers should be concentrated on answering the following problems:

- Specify the genesis and the distribution of the Varyazh deposits.

- Improve the understanding of the deglaciation characteristics of the Narev and Berezina glaciers.
- Estimate the impact and significance of tectonic features on the glacial dynamics.

REFERENCES

Baltrunas, V., Krupickas, R., Jurgaitis, A., 1985. Possibility of stadial and phasial till correlation by means of lithological and geochemical methods. Nauchnye trudy VUZ Litovskoi SSR Geologia 6, 102–111 Vilnius [in Russian].

Gaigalas, A., 1995. Glacial history of Lithuania. In: Ehlers, J., Gibbard, P.L., Kozarski, S. (Eds.), Glacial Deposits in North-East Europe. Balkema, Rotterdam, pp. 127–155.

Guobyte, R., 1999. Quaternary Geological Map of Lithuania. Scale 1:200 000. Geological Survey of Lithuania, Vilnius.

Guobyte, R., Pavlovskaya, I., 1998. The stadial limits and maximum extent of the Weichselian glaciation in northeastern Lithuania and north-western Belarus. Field Symposium on Glacial Processes and Quaternary Environment in Latvia, 25–31 May, 1998, Abstracts, Riga, p. 23.

Gurski, B.N. (Ed.), 1980. Geomorphological Map of Belarussian SSR, Scale 1:500 000.

Gurski, B.N. (Ed.), 1986. Geomorphological Map of Belarussian SSR, Scale 1:500 000.

Gurski, B.N., Levkov, E.A., Karabanov, A.K., Bessarab, D.A., 1990. New data about glaciations limits within the Belarus territory. Dokl. Akad. Nauk Belarusi 34 (4), 345–348 [in Russian].

Gursky, B.N., 1974. Lower and Middle Antropogene of Belarus. Nauka i tekhnika, Minsk [in Russian], 144 pp.

Karabanov, A.K., 1987. Grodno Highland. Nauka I tekhnika, Minsk [in Russian], 108 pp.

Makhnach, N.A., 1971. Stages of the Quaternary Evolution of Flora in Belarus. Nauka i tekhnika, Minsk [in Russian], 212 pp.

Matveyev, A.V., 1995. Glacial history of Belarus. In: Ehlers, J., Gibbard, P.L., Kozarski, S. (Eds.), Glacial Deposits in North-East Europe. Balkema, Rotterdam, pp. 267–276.

Matveyev, A.V., 1990. Istoriya formirovaniya reliefa Belorussii (History of forming of the relief of Belarus). Nauka i tekhnika, Minsk.

Matveyev, A.V., Gurski, B.N., Levitskaya, R.I., 1988. Relief of Belarus. Universiteskoye, Minsk 320 pp.

Pavlovskaya, I., 1994. The Polotsk Glaciolacustrine Basin: Structure, Relief and History. Nauka i tekhnika, Minsk [inRussian], 128 pp.

Pavlovskaya, I., Simakova, G., Yakubovskaya, T., 1997. New sections with Alexandrian interglacial deposits in eastern Belarus. Dokl. Akad. Nauk Belarusi 41 (5), 95–99 [in Russian with English summary].

Sanko, A.F., 1987. Neopleistocene of North-Eastern Belarus and Adjacent Areas. Nauka i tekhnika, Minsk [in Russian], 178 pp.

Sanko, A.F., Velichkevich, F.Yu, Rylova, T.B., Khursevich, G.K., Matveyev, A.V., Karabanov, A.K., et al., 2005. Stratigraphic scheme of the quarternary deposits of Belarus. Lithosphere 1 (25), 146–156 [in Russian].

Savchenko, I., Pavlovskaya, I., 1999. Muravian (Eemian) and Early Poozerian (Weichselian) deposits at Azarichi (eastern Belarus). Proceedings of the Fifth European Palaeobotanical and Palynological Conference, 523–527 Cracow.

Valchik, M., Zyc, M., Fedenja, V., Karabanov, A., 1990. Ice-marginal formations of Belarussian Range. Academy of Sciences, Minsk. 161 pp. (in Russian).

Velichkevich, F.Yu., 1982. Pleistocene Flora of Glaciated Areas of the East-European Plain. Nauka I tekhnika, Minsk [in Russian], 239 pp.

Velichkevich, F.Yu, Rylova, T.B., Sanko, A.F., Fedenia, B.M., 1993. The Pleistocene Bereza Stratoregion in Belarus. Nauka I tekhnika, Minsk [in Russian], 147 pp.

Velichkevich, F.Yu., Sanko, A.F., Rylova, T.B., Nazarov, V.I., Khursevich, G.K., Litviniuk, G.I., 1996. Stratigraphic scheme of the Pleistocene deposits of Belarus. Stratigraphiya Geologicheskaya korrelatsiya 4 (6), 75–87. Moscow [in Russian].

Chapter 4

Pleistocene Glaciations of Czechia

Daniel Nývlt[1,*], Zbyněk Engel[2] and Jaroslav Tyráček[3,†]

[1]Czech Geological Survey, Brno branch, Leitnerova 22, 658 69 Brno, Czech Republic
[2]Department of Physical Geography and Geoecology, Faculty of Science, Charles University in Prague, Albertov 6, 128 43 Praha 2, Czech Republic
[3]Czech Geological Survey, Klárov 3, 118 21 Praha, Czech Republic
*Correspondence and requests for materials should be addressed to Daniel Nývlt. E-mail: daniel.nyvlt@geology.cz
[†]This study is dedicated to the memory of our colleague Jaroslav Tyráček (1931-2010) for his substantial contribution to the study of Pleistocene glaciations in our territory

4.1. INTRODUCTION

This is a review of current evidence of continental and mountain glaciation in Czechia and provides a closer look at the Pleistocene glacial limits and chronology for the region. It is based on numerous works published during more than a century of research in the Czech territory. The reference list, therefore, can include only the most important studies or those published in world languages. This chapter is divided into two main parts. The first part focuses on the products of the continental glaciations, whilst the second part deals with the local mountain glaciations in the Sudetes and the Šumava Mountains. Three Northern European ice sheets affected Czech territory during both the Elsterian (marine isotope stage (MIS) 16 and 12) and older Saalian (MIS 6) glaciations. They reached as far as northern Bohemia, northern Moravia and Czech Silesia. In contrast, mountain glaciation was limited to the Krkonoše (Giant), the Šumava (Bohemian Forest) and Hrubý Jeseník (Altvatergebirge in German) Mountains. The timing of mountain glaciation is still poorly known. Sedimentary and numerical evidence is available for both the Weichselian-age glaciations, but the chronostratigraphical framework of the pre-Weichselian glaciation remains questionable. Recent geomorphological, sedimentological and chronological studies have substantially improved the understanding of Pleistocene glaciations in Czechia, providing new information on their extent and chronology.

4.2. CONTINENTAL GLACIATION

The first evidence of continental glaciation in Czechia was already established by the end of the nineteenth century (e.g. Danzig, 1886; Tausch, 1889; Cammerlander, 1891). Czech territory was marginally overridden by northern European ice sheets during three separate events. They left behind considerable masses of glacial sediments, arranged mostly in a relatively narrow belt parallel to the general trend of the northern frontier mountain ridges. The maximum glacial extent is marked either by glacial deposits or by Nordic rocks spread on the land surface. This line has been called the 'flint-line' (*Feuersteinlinie*) because the most commonly represented lithology is flint derived from the Baltic region. More recent studies have focused not only exclusively on glacial sediments but also on glacial erosional morphological features. They have provided better control on the exact glacial limits, particularly in the granite landscapes of northern Bohemia. Numerous sedimentological and petrological studies of glacial deposits from Czechia (e.g. Růžička, 1980, 1995; Nývlt and Hoare, 2000, 2011; Sikorová et al., 2006; Víšek and Nývlt, 2006) are not discussed here, except for those which are directly connected to the topic of this contribution. Dating of individual ice-sheet advances has also been the subject of recent studies, but only a few results have been published to date (e.g. Nývlt, 2008; Nývlt et al., 2008). However, further studies are in progress and will be published in the near future.

Czech territory contains evidence of three Middle Pleistocene glaciations during which glaciers advanced into two regions, northern Bohemia and northern Moravia, and to a larger extent into Czech Silesia. The area was especially invaded by ice of the Polish Odra lobe (e.g. Hesemann, 1932; Piotrowski, 1998). Only the eastern glaciated part of northern Moravia and Czech Silesia was glaciated by ice sheets that combined both the Vistula and Odra lobes (e.g. Marks, 2005). The local morphology at the glacier margin affected the extent of the ice sheets, such that the glaciers penetrated further to the south in the low-lying Zittau and Ostrava Basins or in the Moravian Gate. The fronts of the advancing ice-sheet lobes crossed the local or regional watersheds in some places. The main European watershed was crossed by the ice at the Poruba Gate, and the associated proglacial outwash drained into the Danube River catchment (e.g. Tyráček, 1961, 1963, 2011). The Sudetes mountain belt was crossed by an outlet of the ice sheet in the Jítrava Saddle, thereby penetrating into the inner part of the Bohemian Highlands, that is, to the Labe (Elbe) River catchment (e.g. Danzig, 1886; Šibrava, 1967; Macoun and Králík, 1995). These key areas are important for direct correlation of glacial sequences with fluvial terrace-deposit systems of the rivers in the Labe and Danube catchments (e.g. Šibrava, 1967; Macoun and Králík, 1995; Tyráček and Havlíček, 2009; Tyráček, 2011). Morphostratigraphical correlation was formerly the only available approach for dating of individual glacial deposits and landforms. Individual stratigraphical systems, based on these correlations, were developed for the continental glaciation in northern Bohemia and northern Moravia and Czech Silesia (the most recent are given by Macoun and Králík, 1995 or Růžička, 2004).

4.3. ELSTERIAN GLACIATIONS

Traditionally two separate advances of Northern European ice sheets reaching the northern parts of Czechia were linked to the Elsterian Stage (e.g. Šibrava, 1967; Macoun, 1985). They are correlated with MIS 16 and 12 (e.g. Šibrava et al., 1986; Lindner et al., 2004) and locally to Elsterian 1 and 2 in Germany (Eissmann, 1997, 2002) or the Sanian 1 and 2 in Poland (Lindner et al., 2004; Marks, 2004, 2005). Individual glaciations are described separately for both glaciated areas of Czechia, they could be correlated via the area of Poland.

4.3.1. Northern Bohemia

During the Elsterian 1 (MIS 16 = Donian glaciation of Russia; Chapter 26), the Northern European ice sheet covered Šluknov Hilly land, Frýdlant Hilly land and Zittau Basin, the frontal lobe crossed the Jítrava Saddle in the Ještěd Range and entered the Ralsko Hilly land, that is, the catchments of Ploučnice and Jizera Rivers (Šibrava, 1967; Králík, 1989; Macoun and Králík, 1995; Nývlt, 2008).

The Elsterian 1 glaciation was the most extensive continental glaciation in northern Bohemia. The glacial limit has been mapped in detail in the Šluknov Hilly land (Nývlt, 2001, 2008) district. In contrast to the adjoining parts, this area was only affected by the older Elsterian 1 glacier. This glaciation originated by the confluence of two individual ice-sheet lobes that advanced from the North and East to join along the watershed between the Labe and Oder River catchments (Fig. 4.1). The maximum advance of this ice sheet has been dated on the depth profile of proglacial glaciofluvial deposits using ^{10}Be cosmogenic nuclide to 606 ± 53 ka BP (Nývlt, 2008). This should correspond to MIS 16, which culminated ~ 620–635 ka BP (EPICA Community Members, 2004; Jouzel et al., 2007).

The best-preserved glacial sequence for study of glacial stratigraphy of these Middle Pleitocene deposits is in the Zittau Basin, where both Elsterian tills are preserved in superposition at different sites (Šibrava, 1967). Here the Elsterian 1 glaciation was the most extensive, the ice-sheet front crossed the Jítrava and Horní Saddle and penetrated further to the Ralsko Hilly land (Šibrava, 1967; Králík, 1989). The Elsterian 2 ice front crossed also the Jítrava Saddle and deposited a terminal moraine at Jítrava (Šibrava, 1967; Králík, 1989). Proglacial glaciofluvial accumulations deposited in the Ploučnice river catchment have been correlated with the terrace deposits sequence of the Labe River (Grahmann, 1933; Šibrava, 1967; Tyráček and Havlíček, 2009).

Remnants of both Elsterian glaciations are also found in the Frýdlant Hilly land (Králík, 1989; Macoun and Králík, 1995). Here it seems that the Elsterian 1 glaciation was less extensive than that during the subsequent Elsterian 2 (Králík, 1989; Nývlt, 2003). The ice sheet crossing of the Jizerské hory mountain belt, through the Oldřichov col at 478 m a.s.l. (Králík, 1989), took place during the Elsterian 2 Stage (Nývlt, 2003; Nývlt and Hoare, 2011). The glaciation limit in the northern slope of the Jizerské hory, which is connected with the Elsterian 2 ice sheet, was recently mapped by Černá 2011. Here the glaciation limit was delimited using a combination of different geomorphological methods to the altitude of 470–490 m a.s.l. (Traczyk and Engel, 2006; Černá 2011; Černá and Engel, 2011; Fig. 4.2). The highest site with glacial sediments in northern Bohemia lies in the Anděl Saddle at 522 m (Janásková and Engel, 2009), in the northern part of the Jizerské hory. Here the glaciation is represented by proglacial glaciofluvial sediments deposited in front of the ice-sheet advancing to the saddle from the east. The precise position of the glacier has not been determined, but the glacier's surface must have been higher than that of the saddle, because its melt waters were draining to the west (Janásková and Engel, 2009).

Eissmann (1997, 2002) proposed the concept of a huge ice-dammed lake in the Labe River valley upstream of the present Bad Schandau, where, according to him, the Elsterian ice sheet terminated. Some rhythmically bedded clays occur at the Foksche Höhe in Děčín (Šibrava, 1967) or in Ctiněves near the Říp Hill (Žebera, 1961), but their glacial meltwater origin has never been proved (e.g. Žebera, 1974; Růžička, 2004). Other clay beds have been identified in several river terrace sequences of the Vltava and Labe (Elbe) rivers at various altitudes in the Říp area. However, they do not match any contingent ice-dammed lake. Indeed, no traces of such a huge ice-dammed lake have been found in the Czech part of the Labe River valley, so far. Therefore, Eissmann's (1997, 2002) interpretation of the occurrence of a huge ice-dammed lake in the inner Bohemian Massif during the Elsterian cannot be accepted. If some temporary glacial lake existed around Bad Schandau, then it must have drained beneath the glacier, through the fissures within the ice body or along the glacier front on German territory (Ehlers et al., 2004).

Chapter | 4 Pleistocene Glaciations of Czechia

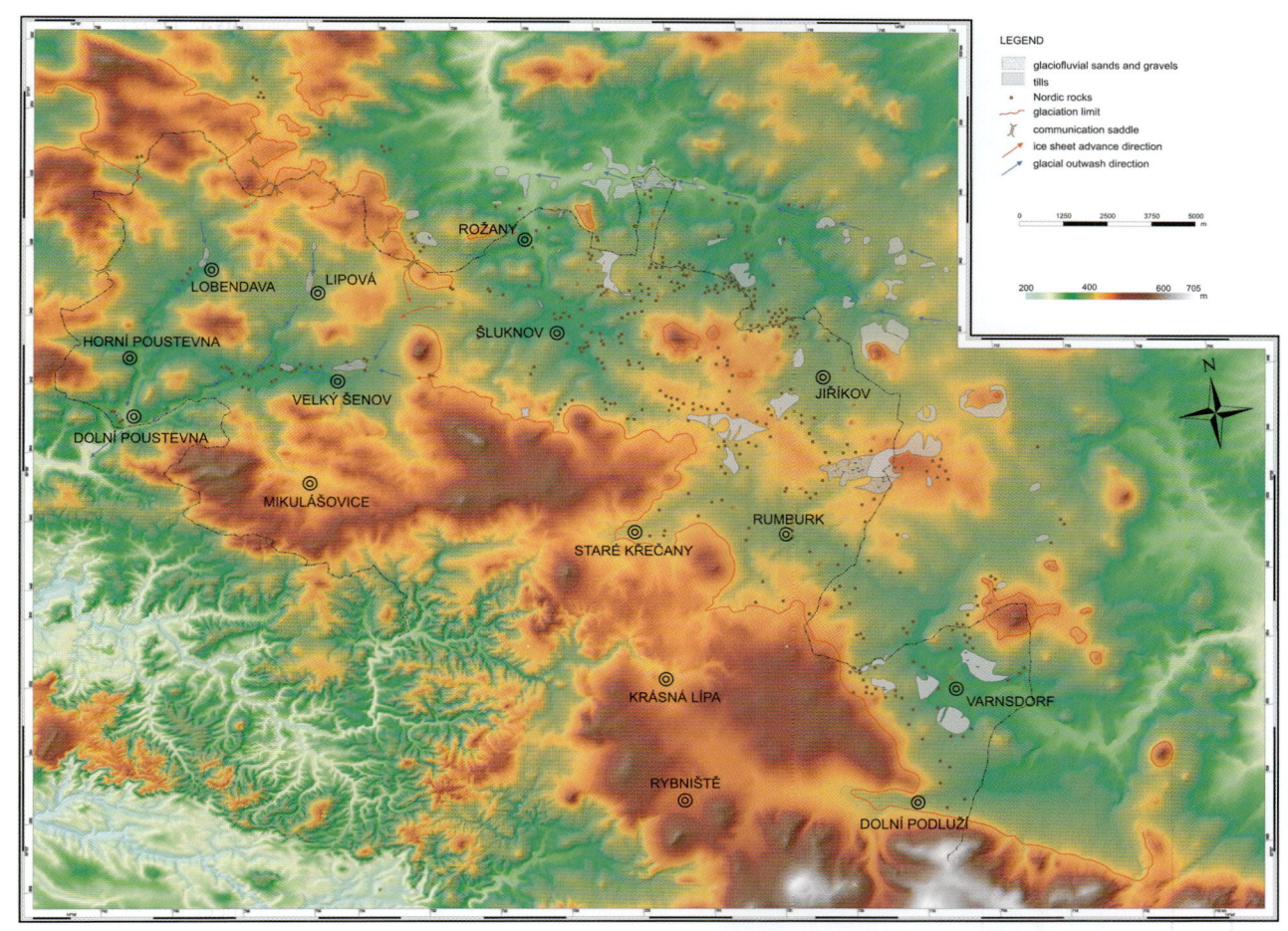

FIGURE 4.1 Palaeogeographical reconstruction of the Elsterian glaciation in the Šluknov Hilly land (modified from Nývlt, 2008).

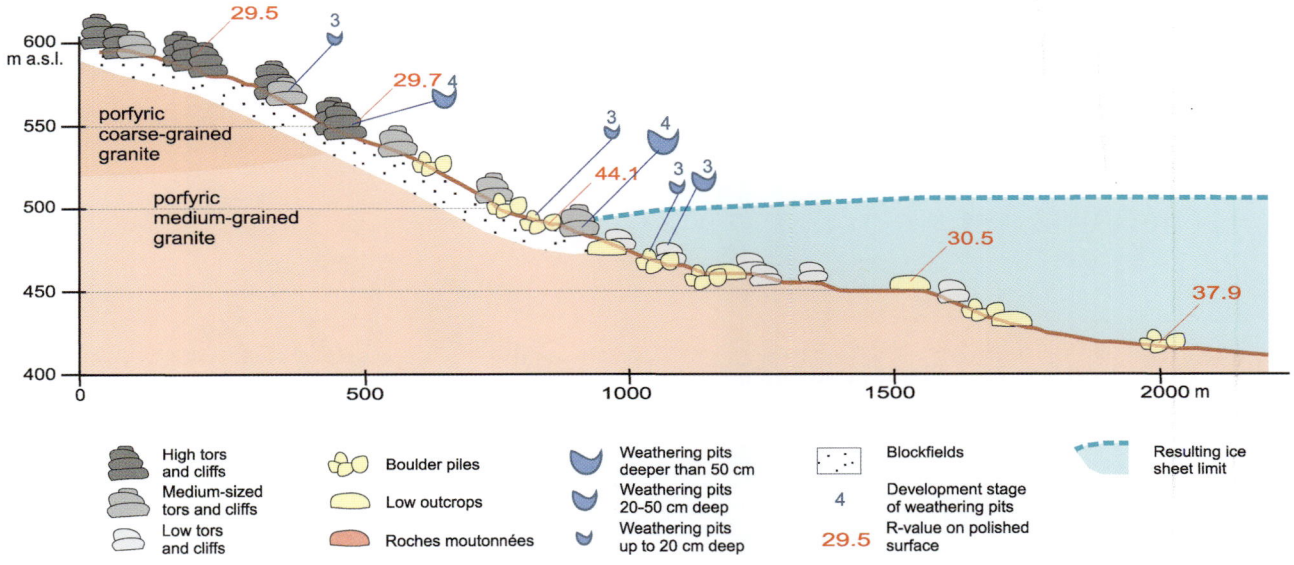

FIGURE 4.2 Profile line through the northern slope of the Jizerské hory reconstructing glaciation limit using Schmidt hammer measurements and geomorphological analyses of glacial and periglacial landforms (from Černá, 2011).

4.3.2. Northern Moravia and Czech Silesia

The Elsterian glaciation also extended into northern Moravia and Czech Silesia, where from the west to east the foothills of the Rychlebské, the Hrubý Jeseník and the Zlatohorská Highlands were glaciated (e.g. Macoun et al., 1965; Prosová, 1981; Macoun and Králík, 1995; Sikorová et al., 2006). The ice sheet advanced into the Ostrava Basin, the Opava Hilly land and the Moravian Gate (Macoun et al., 1965; Macoun, 1980; Macoun and Králík, 1995; Tyráček, 2011) penetrating up to 40 km towards the south of the prevalent maximum glacial limit making an outlet ice-sheet lobe (Fig. 4.3). Because most of the glaciated areas in northern Bohemia and Czech Silesia were overridden by the Saalian ice sheet, most of the Elsterian glacial deposits are buried by younger sediments. In the Ostrava Basin, their greatest thicknesses (over 100 m) are preserved in tunnel valleys or isolated enclosed depressions, which are thought to typify the Elsterian glaciation (Van der Wateren, 2003) and also occur in other European regions.

Morphologically, the most important are two polycyclic push moraines (the northern one at Chuchelná and the southern one at Kravaře) in the Opava Hilly land. They are composed of tills and terminoglacial deposits of both the Elsterian and older Saalian glaciations (Macoun, 1980; Macoun and Králík, 1995). Tills of Elsterian 1-age are known from the Opava Hilly land and the Ostrava Basin. The advance of Elsterian ice sheet into the Moravian Gate was the object of fruitful discussions for decades (e.g. Tyráček, 1963, 2011; Macoun, 1989; Czudek, 2005). On the basis of the recent studies by Tyráček (2011), it seems that the Elsterian ice sheet did not advance closer to the main European watershed and proglacial glaciofluvial outwash was not drained via the Moravian Gate to the Danube River catchment. However, the U-shaped northern termination of the Poruba Gate has not been satisfactorily explained so far (Tyráček, 2011). However, it seems very probable that this morphological feature is of pre-glacial origin. The frontal part of the Elsterian ice sheet was drained towards the North below the glacier. This is supported by numerous north-sloping overdeepened tunnel valleys, sometimes up to

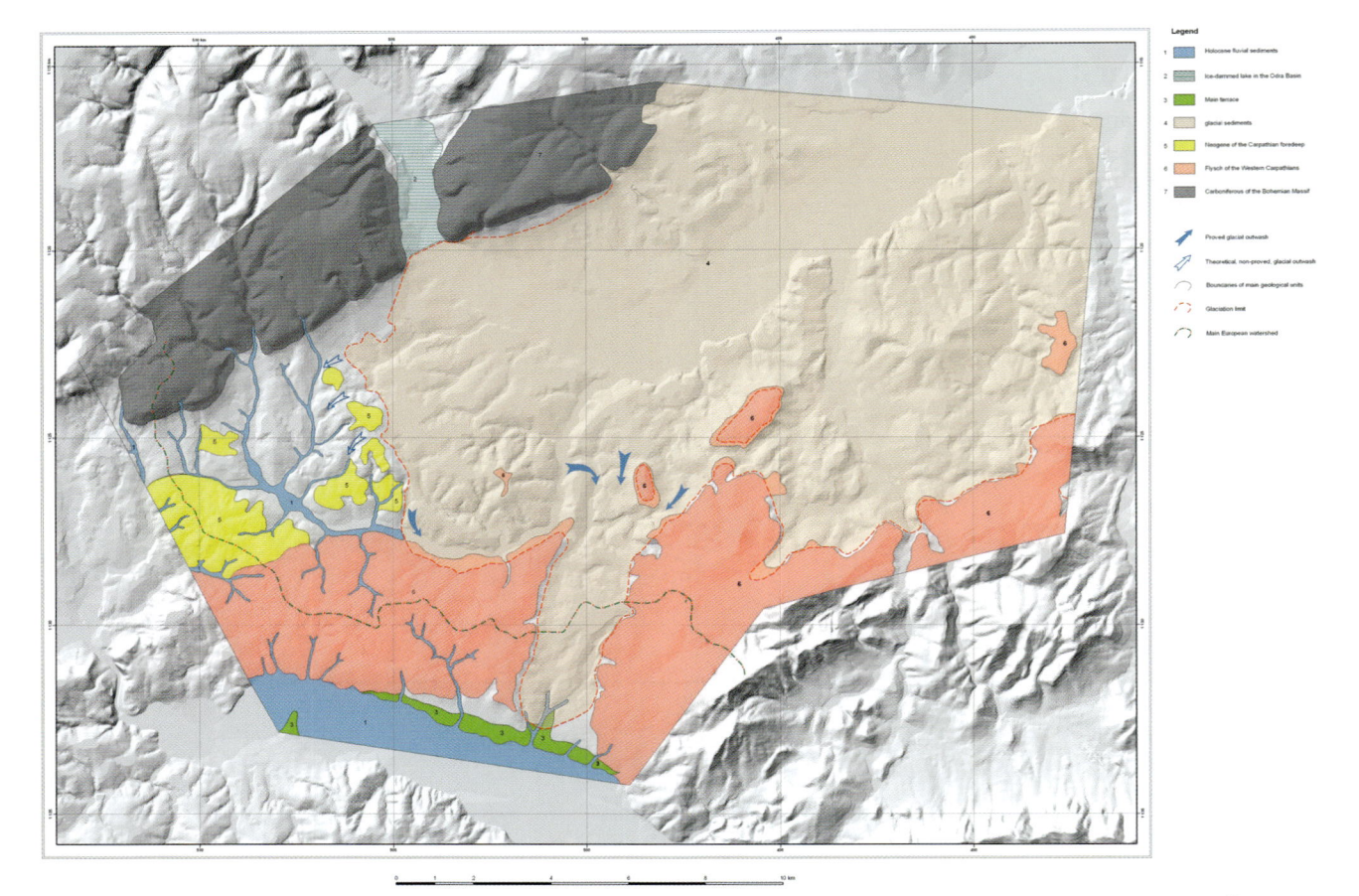

FIGURE 4.3 Palaeogeographical reconstruction of the maximum Saalian glaciation in the Moravian and Poruba Gates (modified from Tyráček, 2011).

100 m deep (Macoun et al., 1965; Czudek, 2005), which developed due to soft bedrock and dynamic advancing ice-sheet front. Local short-lived ice-dammed lakes developed randomly in front of Elsterian ice sheet in the Ostrava Basin and Opava Hilly land (Macoun, 1980).

Shallow lakes formed in the Ostrava Basin during the subsequent Holsteinian Stage interglacial (Stonava Lake, e.g. Menčík et al., 1983). These lacustrine sediments, with abundant organic remains, have been studied in detail (e.g. Kneblová-Vodičková, 1961; Břízová, 1994) and represent the only genuine interglacial separating the Elsterian and Saalian glacial sequences in northern Moravia and Czech Silesia.

4.4. SAALIAN GLACIATION

The northern European ice sheet also advanced into the marginal northern parts of Czechia during the Saalian. This happened during the first Saalian glacial (Drenthe: Woldstedt, 1954, Ehlers, et al., 2004; Odranian: Lindner, et al., 2004, Marks, 2004, 2005), which is dated to the first glacial maximum during MIS 6. The former correlation of the Saalian 1 advance of Northern European ice sheet with MIS 8 (e.g. Šibrava et al., 1986) is no longer valid.

4.4.1. Northern Bohemia

The advance of Saalian ice sheet into the Northern Bohemia has been discussed in the literature for decades (e.g. Šibrava, 1967; Králík, 1989; Macoun and Králík, 1995; Czudek, 2005). Recent studies from neighbouring Poland (e.g. Badura and Przybylski, 1998; Berger et al., 2002; Marks, 2005) show that the Saalian ice-sheet extent was significantly smaller in the Western Sudetes. The Odra ice lobe advanced only into the northernmost part of the Frýdlant Hilly land and deposited a vast accumulation of proglacial glaciofluvial sediments in the surroundings of Horní Řasnice and Háj villages in the Frýdlant Hilly land.

4.4.2. Northern Moravia and Czech Silesia

The main terrace of individual fluvial sequences in the Bohemian and Moravian rivers provides an important stratigraphical marker in the Czechia. This allows reliable correlation of the terrace sequences of individual rivers and glacial sequences throughout northern Moravia and Czech Silesia (Tyráček and Havlíček, 2009). In spite of 'double terrace' origin of most of the main terraces of Czech rivers (e.g. Šibrava, 1964; Tyráček, 2001), the upper accumulation developed during the long pre-glacial period of Saalian glaciation (Tyráček, 2001; Tyráček and Havlíček, 2009), possibly during MIS 8. Because it is developed along most of the larger streams in northern Moravia, this unit may be used (from a stratigraphical point of view) as an important marker for classification of Elsterian and Saalian glacial deposits.

During the Saalian 1 event, the Northern European ice sheet advanced into the Ostrava Basin, the Opava Hilly land and the Moravian Gate (Macoun et al., 1965; Macoun, 1980; Macoun and Králík, 1995). Two oscillations of the Saalian 1 ice sheet are recorded in the Opava Hilly land and the Moravian Gate (Macoun, 1985, 1989; Růžička, 2004). The northern push moraine (at Chuchelná), in the Opava Hilly land, was finally morphologically sculptured (Macoun, 1980) and represents the best-preserved glacial depositional landform connected with continental glaciation in Czechia (Růžička, 2004). The formerly presumed huge ice-dammed lake that was thought to have occupied the whole Ostrava Basin (Žebera, 1964) has not been confirmed. However, local short-lived small lakes developed irregularly during the early Saalian within the basin instead (Růžička, 2004; Czudek, 2005; Tyráček, 2011).

The deposits representing the maximum Pleistocene glaciation in the Moravian Gate are of Saalian 1 age (Tyráček, 2011). Glaciofluvial and glaciolacustrine deposits overlying sediments of the main terrace in the Poruba Gate provide supporting evidence that the Saalian meltwaters crossed the main European watershed. The Moravian Gate was drained during the maximum Saalian 1 ice-sheet extent through the Poruba Gate at the level of the 15 m terrace (Tyráček, 2011); nordic rocks are found in the equivalent terrace deposits of the Bečva River (Tyráček and Havlíček, 2009). Retreating proglacial glaciofluvial sands overlain by glaciofluvial silts and silty muds were dated at Kunín in the Odra part of the Moravian Gate. They directly overlie the fluvial sediments of the main Odra fluvial terrace and have been dated by optically stimulated luminescence (OSL) to 162.0 ± 9.4 ka BP (Nývlt et al., 2008). This corresponds to the end of the first cool period during MIS 6, when the Saalian 1 (Odranian) glaciation occurred (e.g. Lindner et al., 2004; Marks, 2004).

4.5. MOUNTAIN GLACIATIONS

The mountainous regions in the Czech Republic comprise two different geological units. The Hercynian ranges in the western part border an ancient crystalline massif, whereas the mountains on the eastern boundary belong to the West Carpathians, formed by the Alpine orogeny. The geological conditions and tectonic activity controlled the mountain development, resulting in a diverse relief pattern of the two mountain regions. The Hercynian Mountains comprise a series of dissected plateau surfaces forming large areas above 1000 m a.s.l. Ridges and summit plateaux rise up to 1600 m a.s.l. and have a generally low relief with broad valleys. The relief of the Carpathians is more dissected by deep valleys and only individual summits rise above 1200 m a.s.l. During the Quaternary, a periglacial environment prevailed in these mountain areas. However, small mountain glaciers developed in the Krkonoše Mountains, the Šumava Mountains and the Hrubý Jeseník Mountains (Partsch, 1882; Prosová, 1973). The existence of local mountain glaciers in the Krušné hory, the Jizerské hory, the Králický Sněžník and the Beskydy Mountains remains controversial (Pelíšek, 1953; Král, 1968; Demek and Kopecký, 1998; Pilous, 2006; Pánek et al., 2009).

4.6. PRE-WEICHSELIAN GLACIATION

4.6.1. Krkonoše Mountains

At present there is no numerical age evidence for glaciation prior to Weichselian. Understanding the timing and extent of glaciations has historically been limited to the identification of two generations of moraines, originally associated with Riss and Würm glaciations (Partsch, 1894). Results of numerical and relative-age dating reject the idea of pre-Weichselian origin of well-preserved moraines within the Krkonoše Mountains (Braucher et al., 2006; Engel et al., 2011). A pre-Weichselian glacial episode is indicated by the Weischselian moraines embedded in the pre-existing troughs of the upper Labe and the upper Úpa valleys (Engel, 2007). There is some field evidence to suggest that the large high altitude plateau areas may have been glaciated by small plateau ice-fields prior to the last glaciation (Sekyra and Sekyra, 2002). Additionally, recent sedimentological investigations in the Úpa Valley revealed till deposit that probably represents pre-Weichselian glaciation (Carr et al., 2002). If the interpretation is valid, then the Quaternary glaciations were probably more extensive than previously considered.

4.6.2. Šumava Mountains

The chronology and extent of pre-Weichselian glaciations remains unknown. Field evidence appears to conflict with the hypothetical ice-sheet and ice-cap glaciation proposed by Bayberg (1886) and Preihäusser (1934) for the Middle Pleistocene. However, a spatial relationship of glacial landforms in the Bavarian part of the mountains together with glacial transformation of the valleys suggests that the Šumava Mountains had been glaciated prior to Weichselian (Hauner, 1980). Glacial deposits preserved downvalley from the Weichselian moraines in the Kleiner Arbersee region indicate older glaciation preceding the deposition of moraines (Raab, 1999).

4.7. WEICHSELIAN GLACIATION

4.7.1. Krkonoše Mountains

Chronological studies and numerical evidence concerning the Weichselian glaciation are still scarce. Geomorphological and sedimentary evidence, as well as few numerical data, imply glacial advances in the Krkonoše Mountains during the early Weichselian. Thermoluminescence (TL) dating of sediments in a glacial environment below the Sněžné jámy Cirques has provided the first chronological indication of pre-late Weichselian glaciation. The data from two cores suggest that the maximum glacial advance occurred well before 90 ka (Chmal and Traczyk, 1999). The idea of an early Weichselian glaciation was supported by tentative correlation of glacial sediments in the Úpa Valley (Sekyra, 1964; Carr et al., 2002) and by restricting the extent of Late Weichselian moraines in the Łomnica valley (Traczyk, 1989; Engel et al., 2011). The lowermost preserved terminal moraines below Sněžné jámy Cirques, together with indistinct relics of glacial deposits in the Łomnica and Úpa valleys, indicate a more extensive glaciation preceding the Late Weichselian glacier advance. The glaciation on the northern flank of the mountains took the form of broad wall-side glaciers, whereas alpine valley glaciers extended down the valleys on the southern flank.

The Late Weichselian glaciation was reconstructed using numerical and relative-age dating techniques in the Labe, Úpa and Łomnica valleys. The sedimentary record from the Labe valley indicates that the cirque was ice-free around 30 ka BP, suggesting the limited timing of the last glaciation to MIS 2 (Engel et al., 2010). ^{10}Be exposure ages from moraine boulders deposited by the Labe and Łomnica valley glaciers imply a deposition of preserved moraines between 17.0 ± 0.4 ka and 12.1 ± 0.8 ka (Braucher et al., 2006; Engel et al., 2011). The lowermost moraines in the Łomnica valley have been tentatively associated with the Last Glacial Maximum (LGM). The exposure ages show that the glacial recession began around 14 ka and glaciers persisted in suitably orientated parts of cirques until the beginning of the Holocene (Braucher et al., 2006). The retreat of the Labe glacier into the cirque during the Late Glacial period was followed by a re-advance and subsequent recession of the glacier (Engel et al., 2010). A radiocarbon date of around 10 ka BP provides a maximum age for the presence of local cirque glacier in the Łomnica Valley (Chmal and Traczyk, 1999). The extent of the late Weichselian glaciation is indicated by moraines deposited during the LGM by valley glaciers and later by cirque-type glaciation. The extent of glaciers and locations of the dating sites is indicated in Fig. 4.4.

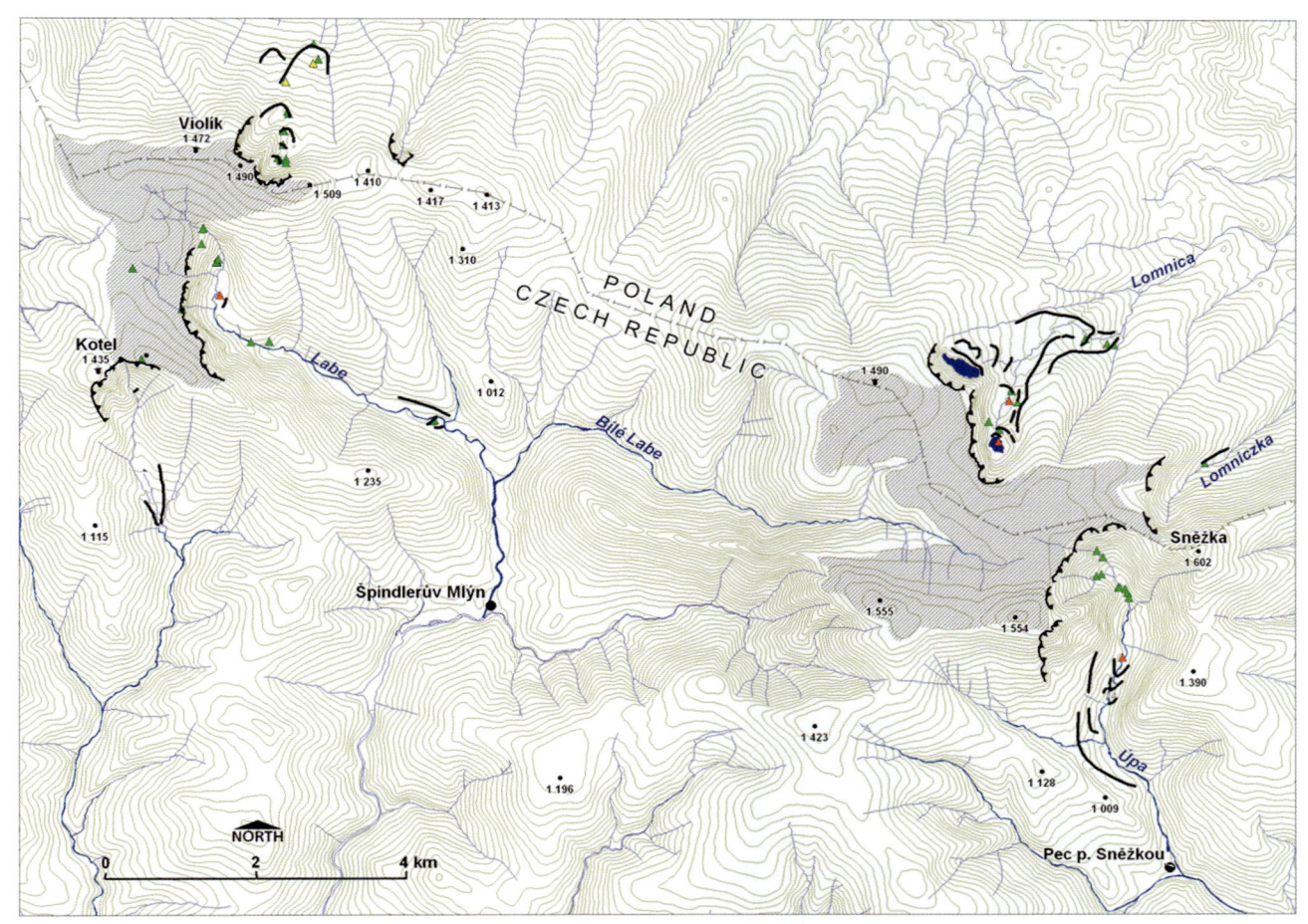

FIGURE 4.4 Late Weichselian moraines in the Krkonoše Mountains. Green, red and yellow triangles represent sites of ^{10}Be, ^{14}C and TL dating.

4.7.2. Šumava Mountains

A reliable glacial stratigraphy and chronology of Weichselian glaciation has yet to be established. Thus, it is difficult to describe the stratigraphy, timing and nature of pre-Late Weichselian glaciation in the region. A pre-Weichselian glacial episode is indicated by the presence of moraine ridges downvalley from the LGM moraines and by exposure data from the Bavarian part of the mountains. An exposure age of 61.5 ± 3.1 ka from a bedrock surface in the Grosser Arber summit region has been interpreted as a result of a more extensive glaciation during the early Würmian (Reuther, 2007). However, these data provide little evidence that may be used to determine the ages of the moraines located in the valleys beyond the Late Weichselian moraines. The location of glacial landforms suggests that the early Weichselian glaciation was the most extensive during the late Quaternary. Valley glaciers up to 7 km long have been reported from the Bavarian side of the mountains (Hauner, 1980; Pfaffl, 1998).

Consistent chronological data for Late Weichselian glaciation have only been derived from the Bavarian side of the mountains. Here the maximum age of the last glaciation was determined to 32.4 ± 9.4 ka using infrared stimulated luminescence dating (Raab and Völkel, 2003). ^{10}Be exposure ages show that the initial advance of the Kleiner Arbersee glacier at 20.7 ± 2.0 ka was followed by several readvances until 15.5 ± 1.7 ka and then by the recession of the glacier into the cirque after 14.5 ± 1.8 ka (Reuther, 2007). The only numerical age, which can constrain the Late Weichselian timing of local glaciation on the Czech side of the mountains, was reported from the closure of the Černý potok River (Vočadlová et al., 2009). The beginning of the deglaciation period is supported by a radiocarbon age of ~ 14 ka cal. BP from cirques in the Plechý and Poledník Mountains. This was interpreted as indicating the termination of the last glaciation on the Czech side of the mountains (Pražáková et al., 2006; Mentlík et al., 2010). On the basis of the position and volume of moraines, the extent of Late Weischselian glaciation was limited only to valleys and cirques.

4.8. CONCLUSIONS

The study of individual glacial limits in the Czech Republic is still under debate. This is because the glacial deposits are mostly correlated using a morphostratigraphical approach with fluvial sequences of inner Czechia. The cross-border correlation with Poland is still not established. However, it seems well established that the maximum continental glaciation in the Sudetes (from northern Bohemia in the west, to the foothills of the Rychlebské hory and Hrubý Jeseník Mountains in the east) was of Elsterian age (e.g. Šibrava, 1967; Králík, 1989; Nývlt, 2003, 2008). On the other hand, the maximum advance in the Ostrava Basin and the Moravian Gate, where the ice sheet and its melt waters, respectively, crossed the main European watershed between the Odra and Danube Rivers, took place during the first Saalian glaciation (e.g. Tyráček, 1963, 2011; Nývlt et al., 2008). The ice-sheet advance into the Moravian Gate represents the southernmost glaciation limit in Central Europe. A rather complex stratigraphy of continental glaciations (Macoun and Králík, 1995) with five 'glaciations' (of the formation rank and 10 oscillations of the member rank), copying partly the German and Polish schemes, is based upon breaks in deposition indicated by palaeosols. None of these divisions have been tested by any modern method (e.g. soil micromorphology, dating techniques, etc.). Therefore, the character and duration of the 'temperate periods' identified are insufficiently defined. So the Stonava interglacial (Holsteinian) is the only genuine interglacial event known to separate the Elsterian and Saalian glaciations in Czechia. Exposure and luminescence dating of individual ice-sheet advances currently in progress (e.g. Nývlt, 2008; Nývlt et al., 2008) will substantially improve our knowledge of the timing of continental glaciations at their southern Central European limits.

Mountain glaciations have been studied in detail in the Krkonoše and Šumava Mountains. Glacial landforms have been identified, the extent of glaciations outlined and preconditions of the glaciation described (Partsch, 1882; Migoń, 1999). However, consistent and precise chronology of the Quaternary glaciations within these ranges and their correlation with chronostratigraphical systems of other glaciated areas in continental Europe are not yet established. Attempts to solve the question of chronology of the local glaciations were usually based on morphological evaluation of the glacial landforms, and this has proven to be insufficient. Some progress has been made with the presentation of the first geochronological data (Chmal and Traczyk, 1999; Mercier et al., 2000). A further step towards the determination of local glaciation chronology has been made recently in the Krkonoše Mountains, although only in these mountains (Carr et al., 2007; Engel et al., 2011). Therefore, it is not possible to include the local mountain glaciations into the chronostratigraphical framework of the Quaternary glaciation of continental Europe. In addition, it is not yet possible to utilise the local knowledge from the mountains for solving problems of environmental changes in the Late Quaternary.

REFERENCES

Badura, J., Przybylski, B., 1998. Extent of the Pleistocene ice sheets and deglaciation between the Sudeten and the Silesian Rampart. Biul. Państwowego Inst. Geol. 385, 9–28.

Bayberg, F., 1886. Geographisch-geologische Studien aus dem Böhmerwalde. Die Spuren alter Gletscher, die Seen und Thäler des Böhmerwaldes. Petermanns Geogr. Mitt. Ergänzungsheft 81, 1–63.

Berger, H.-J., Zitzmann, A., Opletal, M., Nývlt, D., Valečka, J., Prouza, V., et al., 2002. Geologische Übersichtskarte 1:200000, CC 5550 Görlitz. BGR, Hannover.

Braucher, R., Kalvoda, J., Bourlès, D., Brown, E., Engel, Z., Mercier, J.-L., 2006. Late Pleistocene deglaciation in the Vosges and the Krkonoše mountains: correlation of cosmogenic ^{10}Be exposure ages. Geogr. časopis 58 (1), 3–14.

Břízová, E., 1994. Vegetation of the Holsteinian interglacial in Stonava-Horní Suchá (Ostrava region). Sbor. geol. Věd, Antropozoikum 21, 29–56.

Cammerlander, C.v., 1891. Geologische Aufnahmen in den mährisch-schlesischen Sudeten. Jb K. k. Geologischen Reichsanst. 40, 103–316.

Carr, S.J., Engel, Z., Kalvoda, J., Parker, A., 2002. Sedimentary evidence to suggest extensive glaciation of the Úpa valley, Krkonoše Mountains, Czech Republic. Z. Geomorphol. N. F. 46 (4), 523–537.

Carr, S.J., Engel, Z., Kalvoda, J., Parker, A., 2007. Towards a revised model of Late Quaternary mountain glaciation in the Krkonoše Mountains, Czech Republic. In: Goudie, A.S., Kalvoda, J. (Eds.), Geomorphological Variations. P3K, Praha, pp. 253–268.

Černá, B., 2011. Reconstruction of the continental glaciation in the northern slope of the Jizera Mountains. J. Geol. Sci. Anthropozoic 27, 23–38.

Černá, B., Engel, Z., 2011. Surface and sub-surface Schmidt hammer rebound value variation for a granite outcrop. Earth Surf. Process. Landforms 36 (2), 170–179.

Chmal, H., Traczyk, A., 1999. Die Vergletscherung des Riesengebirges. Z. Geomorphol. N. F. Suppl. Band 113, 11–17.

Czudek, T., 2005. Vývoj reliéfu krajiny České republiky vkvartéru. Moravské zemské muzeum, Brno, 238 pp.

Danzig, E., 1886. Bemerkungen über das Diluvium innerhalb des Zittauer Quadergebirges. Gesell. Isis Abh. 4, 30–32 Dresden.

Demek, J., Kopecký, A., 1998. Mt. Kralický Sněžnník (Czech Republic): Landforms and problem of Pleistocene glaciation. Moravian Geogr. Rep. 6 (2), 18–37.

Ehlers, J., Eissmann, L., Lippstreu, L., Stephan, H.-J., Wansa, S., 2004. Pleistocene glaciations of Northern Germany. In: Ehlers, J., Gibbard, P.L. (Eds.), Quaternary Glaciations—Extent and Chronology. Part I: Europe. Developments in Quaternary Science. vol. 2. Elsevier B.V., Amsterdam, pp. 135–146.

Eissmann, L., 1997. Das quartäre Eiszeitalter in Sachsen und Nordostthüringen. Altenburger Naturwiss. Forsch. 8, 98 pp., Altenburg.

Eissmann, L., 2002. Quaternary geology of eastern Germany (Saxony, Saxon–Anhalt, South Brandenburg, Thüringia), type area of the Elsterian and Saalian Stages in Europe. Quatern. Sci. Rev. 21, 1275–1346.

Engel, Z., 2007. Late Pleistocene glaciations in the Krkonoše Mountains. In: Goudie, A.S., Kalvoda, J. (Eds.), Geomorphological Variations. P3K, Praha, pp. 269–286.

Engel, Z., Nývlt, D., Křížek, M., Treml, V., Jankovská, V., Lisá, L., 2010. Sedimentary evidence of landscape and climate history since the end of MIS 3 in the Krkonoše Mountains, Czech Republic. Quatern. Sci. Rev. 29, 913–927. doi:10.1016/j.quascirev.2009.12.008.

Engel, Z., Traczyk, A., Braucher, R., Woronko, B., Křížek, M., 2011. Use of [10]Be exposure ages and Schmidt hammer data for correlation of moraines in the Krkonoše mountains. Z. Geomorph. N.F. 55 (2), 175–196.

EPICA Community Members, 2004. Eight glacial cycles from an Antarctic ice core. Nature 429, 623–628.

Grahmann, R., 1933. Die Geschichte des Elbetales von Leitmeritz bis zu seinem Eintritt in das norddeutsche Flachland. Mitt. Ver. Erdkunde N. F. 132–194.

Hauner, U., 1980. Untersuchungen zur klimagesteuerten tertiären und quartären Morphogenese des Inneren Bayerischen Waldes (Rachel–Lusen) unter besonderer Berücksichtigung pleistozän kaltzeitlicher Formen und Ablagerungen. Bayerisches Staatsministerium für Ernährung, Landwirtschaft und Forsten, München, 198 pp.

Hesemann, J., 1932. Zur Geschiebeführung und Geologie des Odergletschers. 1. Äussere, Rosenthaler und Velgaster Randlage. Jb. Preussischen geologischen Landesanstalt 53, 70–84.

Janásková, B., Engel, Z., 2009. Altitudinal limit of the ice-sheet glaciation in Northern Bohemia—new evidence from the saddle under Andělský vrch hill. In: Mentlík, P., Hartvich, F. (Eds.), Geomorfologický sborník 8. Czech Association of Geomorphologists, Plzeň, pp. 21.

Jouzel, J., Masson-Delmotte, V., Cattani, O., Dreyfus, G., Falourd, S., Hoffmann, G., et al., 2007. Orbital and millennial antarctic climate variability over the past 800,000 years. Science 317, 793–797.

Kneblová-Vodičková, V., 1961. Entwicklung der Vegetation im Elster/Saale-Interglazial im Suchá-Stonava-Gebiet (Ostrava Gebiet). Anthropozoikum 9, 129–174.

Král, V., 1968. Geomorfologie vrcholové oblasti Krušných hor a problém paroviny. Rozpravy Československé Akad. Věd 78 (9), 65.

Králík, F., 1989. Nové poznatky o kontinentálních zaledněních severních Čech. Sbor. geol. Věd, Antropozoikum 19, 9–74.

Lindner, L., Gozhik, P., Marciniak, B., Marks, L., Yelovicheva, Y., 2004. Main climatic changes in the quaternary of Poland, Belarus and Ukraine. Geol. Q. 48, 97–114.

Macoun, J., 1980. Paleogeografický a stratigrafický vývoj Opavské pahorkatiny vpleistocénu. Časopis Slezského muzea řada A 29, 113–132 193–222.

Macoun, J., 1985. Stratigrafie středního pleistocénu Moravy ve vztahu kevropskému kvartéru. Časopis Slezského muzea řada A 34, 125–143 219–237.

Macoun, J., 1989. Kontinentalvereisungen in der Mährischen Pforte. Sbor. geol. Věd, Antropozoikum 19, 75–104.

Macoun, J., Králík, F., 1995. Glacial history of the Czech Republic. In: Ehlers, J., Kozarski, S., Gibbard, P.L. (Eds.), Glacial Deposits in North-East Europe. A.A. Balkema, Rotterdam, pp. 389–405.

Macoun, J., Šibrava, V., Tyráček, J., Kneblová-Vodičková, V., 1965. Kvartér Ostravska a Moravské brány. Nakladatelství ČSAV, Praha, 419 pp.

Marks, L., 2004. Pleistocene glacial limits in Poland. In: Ehlers, J., Gibbard, P.L. (Eds.), Quaternary Glaciations—Extent and Chronology. Part I: Europe. Developments in Quaternary Science. vol. 2. Elsevier B.V., Amsterdam, pp. 295–300.

Marks, L., 2005. Pleistocene glacial limits in the territory of Poland. Przegl. Geol. 53 (10/2), 988–993.

Menčík, E., et al., 1983. Geologie Moravskoslezských Beskyd a Příborské pahorkatiny. ÚÚG, Praha, 304 pp.

Mentlík, P., Minár, J., Břízová, E., Lisá, L., Tábořík, P., Stacke, V., 2010. Glaciation in the surroundings of Prášilské Lake (Bohemian Forest, Czech Republic). Geomorphology 117, 181–194.

Mercier, J.-L., Kalvoda, J., Bourlès, D.L., Braucher, R., Engel, Z., 2000. Preliminary results of [10]Be dating of glacial landscape in the Giant Mountains. Acta Univ. Carol. Geogr. 35 (Suppl.), 157–170.

Migoń, P., 1999. The role of 'preglacial' relief in the development of mountain glaciation in the Sudetes, with the special reference to the Karkonosze Mountains. Z. Geomorphol. N. F. Suppl. Band 113, 33–44.

Nývlt, D., 2001. Main advance directions and maximum extent of Elsterian ice sheet in the eastern part of the Šluknov Hilly Land, Northern Bohemia, Czechia. Slovak Geol. Mag. 7, 231–235.

Nývlt, D., 2003. Geomorphological aspects of glaciation in the Oldřichov Highland, Northern Bohemia, Czechia. Acta Univ. Carol. Geogr. 35 (Suppl.), 171–183.

Nývlt, D., 2008. Paleogeografická rekonstrukce kontinentálního zalednění Šluknovské pahorkatiny. Ph.D. thesis, 103 pp., Faculty of Science, Charles University, Praha.

Nývlt, D., Hoare, P.G., 2000. Valounové analýzy glacifluviálních sedimentů severních Čech. Bull. Czech Geol. Surv. 75 (2), 121–126.

Nývlt, D., Hoare, P.G., 2011. Petrology, provenance and shape of clasts in the glaciofluvial sediments of the Mníšek member, northern Bohemia, Czechia. J. Geol. Sci. Anthropozoic 27, 5–22.

Nývlt, D., Jankovská, V., Víšek, J., Franců, E., Franců, J., 2008. Deglaciační fáze prvního sálského zalednění v Moravské bráně. In: Roszková, A., Vlačiky, M., Ivanov, M. (Eds.), 14. KVARTÉR 2008, 14–15. Brno, 27.11.2008—Sborník abstrakt, Brno.

Pánek, T., Hradecký, J., Šilhán, K., 2009. Geomorphic evidence of ancient catastrophic flow type landslides in the mid-mountain ridges of the Western Flysch Carpathian Mountains (Czech Republic). Int. J. Sediment Res. 24, 88–98.

Partsch, J., 1882. Die Gletscher der Vorzeit in den Karpathen und den Mittlegebirgen Deutschlands, first ed. Wilhelm Koebner, Breslau, 198 pp.

Partsch, J., 1894. Die Vergletscherung des Riesengebirges zur Eiszeit. J. Engelhorn, Stuttgart, 194 pp.

Pelíšek, J., 1953. K otázce zalednění Moravskoslezských Beskyd. Sb. Československé Společnosti Zeměpisné 57, 60–65.

Pfaffl, F., 1998. Glazialmorphologische Untersuchungen am Rachel-Nordkar und am Grossen Arbersee im Bayerischen Wald. Geol. Bl. NO Bayern 38 (1–2), 7–26.

Pilous, V., 2006. Pleistocénní glacigenní a nivační modelace Jizerských hor. Opera Corcontica 43, 21–44.

Piotrowski, J., 1998. Development of the Odra Lobe. INQUA Commission on Glaciation, The Peribaltic Group: Field Symposium on Glacial Geology at the Baltic Sea Coast in Northern Poland, Excursion Guide, 16 pp., Warsaw.

Pražáková, M., Veselý, J., Fott, J., Majer, V., Kropáček, J., 2006. The long-term succession of Cladoceran fauna and palaeoclimate forcing: a 14,600 year record from Plešné Lake, the Bohemian Forest. Biologia 61 (Suppl. 20), 387–400.

Preihäusser, G., 1934. Über Kare und Karseen des Bayrischen Wales. Der Bayerwald 32, 65–71.

Prosová, M., 1973. Zalednění Hrubého Jeseníku. Campanula 4, 115–123.

Prosová, M., 1981. Oscilační zóna kontinentálního ledovce. Jesenická oblast. Acta Univ. Carol. Geol. 25, 265–294.

Raab, T., 1999. Würmzeitliche Vergletscherung des Bayerischen Waldes im Arbergebiet. Regensb. Geogr. Schriften 32, 327 pp.

Raab, T., Völkel, J., 2003. Late Pleistocene glaciation of the Kleiner Arbersee area in the Bavarian Forest, south Germany. Quatern. Sci. Rev. 22, 581–593.

Reuther, A.U., 2007. Surface exposure dating of glacial deposits from the last glacial cycle. Relief Boden Palaeoklima 21, 213 pp.

Růžička, M., 1980. Sedimenty sálského zalednění na Opavsku a Hlučínsku. Sbor. geol. Věd, Antropozoikum 13, 127–148.

Růžička, M., 1995. Genesis and petrography of glacial deposits in the Czech Republic. In: Ehlers, J., Kozarski, S., Gibbard, P.L. (Eds.), Glacial Deposits in North-East Europe. A. A. Balkema, Rotterdam, pp. 407–420.

Růžička, M., 2004. The Pleistocene glaciation of Czechia. In: Ehlers, J., Gibbard, P.L. (Eds.), Quaternary Glaciations—Extent and Chronology. Part I: Europe. Developments in Quaternary Science. vol. 2. Elsevier B.V., Amsterdam, pp. 27–34.

Sekyra, J., 1964. Kvartérně geologické a geomorfologické problémy krkonošského krystalinika. Opera Corcontica 1, 7–24.

Sekyra, J., Sekyra, Z., 2002. Former existence of a plateau icefield in Bílá louka Meadow, eastern Giant Mountains: hypothesis and evidence. Opera Corcontica 39, 35–43.

Šibrava, V., 1964. Double fluvial accumulation in the area of the Bohemian Massif and the Carpathian foredeep. Sbor. geol. Věd, Antropozoikum 2, 57–68.

Šibrava, V., 1967. Study on the Pleistocene of the glaciated and non-glaciated area of the Bohemian Massif. Sb. Geologických Věd Antropozoikum 4, 7–38.

Šibrava, V., Bowen, D.Q., Richmond, G.M. (Eds.), 1986. Quaternary glaciations in the Northern Hemisphere, Quatern. Sci. Rev. 5, 1–514.

Sikorová, J., Víšek, J., Nývlt, D., 2006. Texture and petrography of glacial deposits in the northern foothill of the Hrubý Jeseník and Rychlebské Mts., Czechia. Geol. Q. 50, 345–352.

Tausch, L.v., 1889. Bericht über die geologische Aufnahme der Umgebung von Mähr.-Weisskirchen. Jb. K. k. Geologischen Reichsanst. 39, 405–416.

Traczyk, A., 1989. Zlodowacenie doliny Łomnicy w Karkonoszach oraz poglady na ilość zlodowaceń pleistoceńskich w średnich górach Europy. Czasopis Geograficzny 60 (3), 267–286.

Traczyk, A., Engel, Z., 2006. Maximální dosah kontinentálního zalednění na úpatí Ořešníku a Poledníku vseverním svahu Jizerských hor. Geografie Sb. ČGS 111, 141–151.

Tyráček, J., 1961. Nové názory na rozšíření maximálního zalednění v Moravské bráně. Přírodovědný Časopis slezský 22, 247–254.

Tyráček, J., 1963. On the problem of the parallelization of the continental and the Alpine glaciation on the territory of Czechoslovakia. Report of VIth International Congress on Quaternary Research, Warszaw 1961, 3, 375–384. Lódź.

Tyráček, J., 2001. Upper Cenozoic fluvial history in the Bohemian Massif. Quatern. Int. 79, 37–53.

Tyráček, J., 2011. Continental glaciation of the Moravian Gate (Czech Republic). J. Geol. Sci. Anthropozoic 27, 39–50.

Tyráček, J., Havlíček, P., 2009. The fluvial record in the Czech Republic: a review in the context of IGCP 518. Glob. Planet. Change 68, 311–325.

Van der Wateren, F., 2003. Ice-marginal terrestrial landsystems: Southern Scandinavian ice sheet margin. In: Evans, D.J.A. (Ed.), Glacial Landsystems. Arnold, London, pp. 166–203.

Víšek, J., Nývlt, D., 2006. Leitgeschiebestatistische Untersuchungen im Kontinentalvereisungsgebiet Nordböhmens. Arch. Geschiebekunde 5 (1–5), 229–236.

Vočadlová, K., Křížek, M., Petr, L., Žáčková, P., 2009. Paleogeographical evidence of the sediments from the Černé jezero Lake area, the Bohemian Forest. In: Mentlík, P., Hartvich, F. (Eds.), Geomorfologický sborník, 8. Czech Association of Geomorphologists, Plzeň, pp. 65–66.

Woldstedt, P., 1954. Saaleeiszeit, Warthestadium und Weichseleiszeit in Norddeutschland. Eiszeit. Ggw 4/5, 34–48.

Žebera, K., 1961. Altpleistozäne Bändertone (Warven) in Ctiněves am Říp. Věstník Ústředního Ústavu Geologického 36, 457–460.

Žebera, K., 1964. Kvartér Českého masivu. In: Svoboda, J. (Ed.), Regionální geologie ČSSR, I. Český masiv. Nakladatelství ČSAV, Praha, pp. 443–509.

Žebera, K., 1974. Kvartér Podřipska (III.část). Sbor. geol. Věd, Antropozoikum 10, 23–40.

Chapter 5

Pleistocene Glaciations in Denmark: A Closer Look at Chronology, Ice Dynamics and Landforms

Michael Houmark-Nielsen

Centre for GeoGenetics, Natural History Museum, University of Copenhagen, Øster Voldgade 5-7, Dk-1350 København K, Denmark

5.1. INTRODUCTION

Following the first contribution on the Pleistocene of Denmark in 'Quaternary Glaciations—Extent and Chronology' (Ehlers and Gibbard, 2004), new light has been shed on the chronology, dynamics and landforms of Quaternary glaciations in the south-western part of Scandinavia. Denmark and surrounding areas (Fig. 5.1) were situated in the marginal regions of the former Scandinavian Ice Sheet (SIS) and have frequently been occupied by glaciers but only for short intervals during the Pleistocene. Since new data on pre-Marine Isotope Stage (MIS) 6 glaciations are sparse, glaciations during the period from MIS 6 to 1 (Late Saalian and Weichselian) are given the most attention here. A more comprehensive reference to older glacial-interglacial episodes can be found in the previous edition (Houmark-Nielsen, 2004). Firstly, the present contribution provides a 'closer look' at recent achievements on the extent and chronology

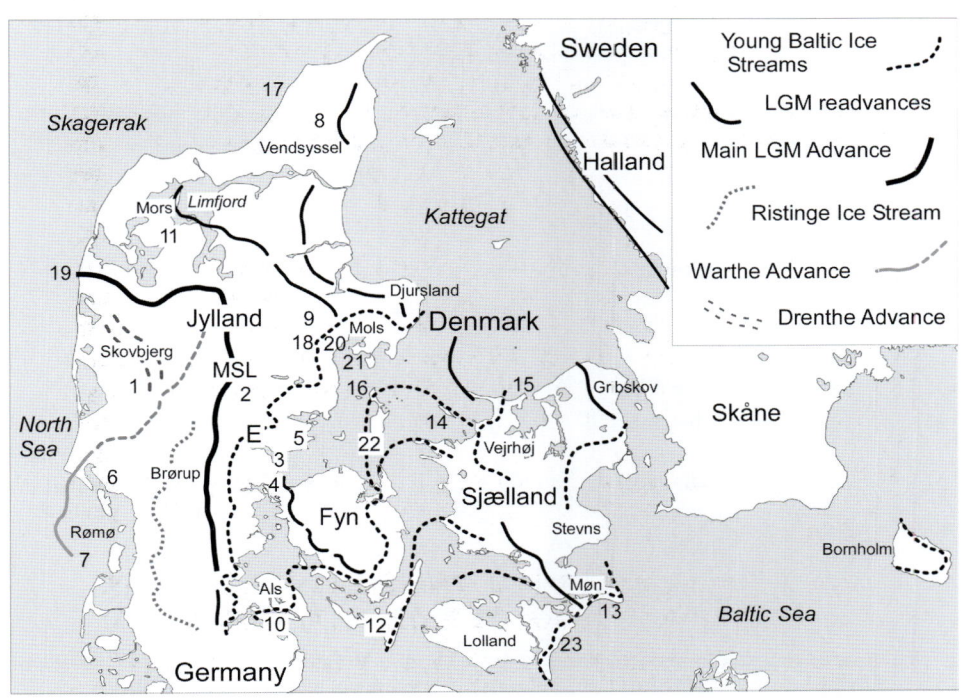

FIGURE 5.1 Key map of Denmark and surrounding countries and the distribution of end moraines of Saalian (Drenthe and Warthe), Middle Weichselian (Ristinge ice stream) and the Late Weichselian (Main advance and Young Baltic ice streams) ages. The numbers refer to key sites; 1: Harreskov; 2: Sønder Vissing; 3: Treldenæs; 4: Røgle; 5: Ashoved; 6: Esbjerg; 7: Lister Deep; 8: Skærumhede; 9: Hinnerup; 10: Stensigmose-Gammelmark; 11: Sundsøre; 12: Ristinge; 13: Klintholm; 14: Sejerø; 15: Hundested; 16: North Samsø; 17: Lønstrup; 18: Fårup; 19: Bovbjerg; 20: Højvang; 21: Mols Hoved; 22: Visborg and 23: Korselitse.

of glaciations of Denmark, in particular, and south-western Scandinavia and the southern circum-Baltic region, in general, over the past ca. 150,000 years (ka).

Northern Hemisphere ice sheets attained their largest volumes during the Saalian and Weichselian glaciations, when global climate achieved periods of extreme cold climate. In south-west Scandinavia, glaciers flowed radially from a main ice dome of the SIS which led to deposition of tills of central and south Swedish provenance during the Drenthe (Late Saalian) and Jylland (Late Weichselian) stadials (Fig. 5.2). Under milder climatic episodes with relatively high sea level, rapid and channelled land-based ice streams were flowing along the Skagerrak–Kattegat Trough and the Baltic depression towards Denmark (Wohlfarth et al., 2008). Glaciation chronologies suggest that episodes of ice streaming seem to have been out of phase with the global ice-volume evolution, including that of the SIS.

Rapidly flowing ice in the land-based sectors of the SIS, which reached far south of the glaciated uplands, could have been initiated by the development of an expanding zone of basal melting beneath a steep gradient ice sheet (Boulton et al., 2001). The growth of the SIS caused the formation of water-filled proglacial depressions which favoured the development of outlet surges and the formation of ice streams that could be channelled through shallow basins in the Baltic depression and the Skagerrak–Kattegat troughs (Fig. 5.1). Ice-sheet instability was enhanced by meltwater and caused ice-bed decoupling over large areas. It might have eventually led to marginal collapse and

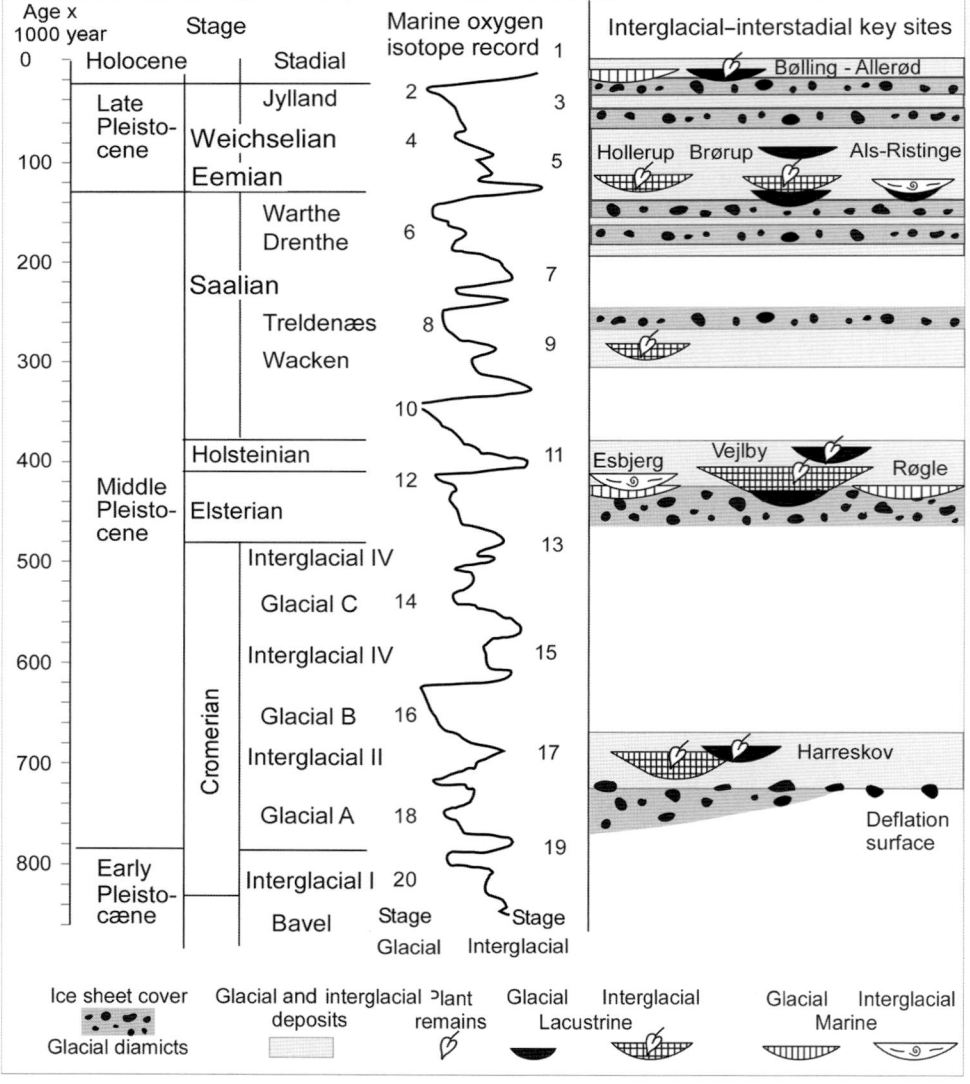

FIGURE 5.2 Event–stratigraphical chart for the Middle and Late Pleistocene, Marine Isotope Stages (MIS) and chronology of Danish interglacial and glacial deposits. Compiled from the International Commission on Stratigraphy (ICS) (2009) and Houmark-Nielsen et al. (2006).

surging through soft, water-saturated sediments bordering the SIS. The ice-marginal collapse and outflow of streaming ice could have been triggered by abrupt warming, which was responsible for an ameliorated climate allowing low Arctic terrestrial biota and boreal-Arctic marine waters to occupy north-western Europe. Stratigraphical evidence suggests at least four episodes of rapid ice-sheet expansion occurred between MIS 4 and 2 (Larsen et al., 2009; Houmark-Nielsen, 2010). Clast provenances in tills indicate glacier flow twice via the Baltic Basin and twice via the Skagerrak–Kattegat trough.

Glacial landforms, including arch-shaped end moraines and mega-scale lineation together with stratigraphically controlled ice-flow indicators of advances from Norwegian and Baltic uplands into Denmark, indicate fan-shaped flow patterns in terminal zones fed by a narrow ice-stream zones further 'up glacier'. Further, discrimination between the flow of rapid streaming ice and ice flow with moderate velocities has been attained through estimations of ice-bed interaction from analyses of dispersal trains of fine gravel-sized erratic clasts (Kjær et al., 2003; Houmark-Nielsen, 2010). Consequently, fan-shaped patterns of movement and indications of limited ice-bed interaction suggest glacier flow in the marginal zones of land-based ice streams (*sensu* Stokes and Clark, 2001). The regional distribution of former land-based ice streams in southern Scandinavia and in the North European lowlands appears to be closely related to the location of an easily deformable substrate.

Because periglacial landscape transformation and post-formational glacier cover increasingly have obscured glacio-morphological features when going back in time only those landforms developed during and after the LGM are well preserved and suitable for detailed analyses of ice-sheet related landforms. However, it has been possible to recognise glacial morphologies much older than the LGM possibly dating back to the Late Saalian and Middle Weichselian. High resolution digital elevation models combined with surface geological mapping have disclosed intimate relationships between landforms, stratigraphy and glacial tectonic architecture, which have revealed a complexity of superimposed landforms not previously acknowledged. The extent of glaciations based on the position of end moraines, vortex of outwash plains, the distribution of stream lined ground moraine and dead-ice features and the presence of till formations holding specific provenance dependent indicator erratics are shown in the digital base maps and palaeogeographical reconstructions.

5.2. THE PLEISTOCENE OF DENMARK

The stratigraphy, lithology and distribution of Pleistocene deposits in Denmark are highly controlled by the geological setting of the pre-Quaternary bedrock. Permian rifting and regional subsidence, connected with the break-up of Pangaea, laid down the foundation to the present-day bedrock geology. The Fennoscandian border zone separates the Baltic shield from the Mesozoic Danish-Norwegian and German basins, the sediments of which form the soft and readily eroded and deformed substratum for Quaternary glaciers. During the Neogene and possibly reaching into the Middle Pleistocene, westward tilting and subsidence, together with uplift and erosion in the east, took place and gave rise to a highly diverse bedrock distribution which is reflected in the glacial-flow dispersal of erratics (Håkansson and Surlyk, 1997; Japsen and Chalmers, 2000; Kjær et al., 2003). Accordingly, the oldest Quaternary deposits are preserved in the North Sea, while Early and Middle Pleistocene are accessible the western part of Denmark, and Late Pleistocene deposits blanket the whole region but dominate the eastern part of the country. Detailed descriptions and references to key sites for till beds of Pleistocene ice advances are given by Houmark-Nielsen (2004, 2007, 2010) and Larsen et al. (2009) and references therein.

A dense network of incised, mainly subglacially formed valleys, now infilled with Quaternary sediments, is reported from the Danish North Sea area, as well as the mainland (Huuse et al., 2003; Kristensen et al., 2008; Krohn et al., 2009). Often the valleys show a composite history of excavation and burial; they most probably date from the Middle and Late Pleistocene and their distribution and orientation have been used as proxies for estimates of the extent and flow pattern of glaciations especially beneath the North Sea floor. In the Late Pleistocene, the regional tectonic activity in the Danish basin shifted eastwards controlled by crustal movements along the Fennoscandian border zone, and the Kattegat region was subject to intensified subsidence.

The age of the oldest glaciations is still being disputed (Houmark-Nielsen et al., 2006). Early Pleistocene glaciations have not been convincingly recognised in Denmark. However, since incised valleys seem to post-date this period, the first Pleistocene glaciations most likely reached Denmark during the 'Cromerian Complex' Stage; in a period when the rhythm of glaciation and interglacials changed from 41 ka astronomical cycles to domination of the 100 ka astronomical cycles. However, new key sites are currently being studied in central Jylland, and the results may push back the age of the first glaciations into the pre-Cromerian.

5.3. MIDDLE PLEISTOCENE GLACIATIONS

The interglacial deposits at the key site Harreskov (Fig. 5.1, site 1) can most likely be correlated with Cromerian Interglacial II (Fig. 5.2) that corresponds to MIS 17. Between the Harreskovian interglacial deposits and Neogene substratum, glaciolacustrine and glaciofluvial deposits overlie a clayey glacial diamicton. Thus, in the older part of the Cromerian Complex, an ice sheet from Scandinavia penetrated southwards leaving the oldest record of glaciation in

Denmark. It is likely that the wind-abraded quartzite and quartzitic sandstone deflation surface between the Neogene and the Pleistocene in Jylland represents the residual of till dating from Cromerian or older glaciations (Fig. 5.2). No limits for these glaciations or morphological features have been recognised, just as the age is only indirectly estimated on the relationship of these sediments to pollen sequences from overlying interglacial deposits.

A regional hiatus separates the Cromerian glaciation from the younger glaciations of the Middle Pleistocene (Fig. 5.2). Elsterian glacigenic deposits comprise tills of Norwegian, Middle Swedish and Baltic provenance, possibly deposited in this order during MIS 12 (i.e. the Elsterian Stage). Key sites include those at Sønder Vissing, Treldenæs and Røgle (Fig. 5.1, sites 2, 3 and 4). Meltwater streams emerging from the first Elsterian ice sheet (Norwegian Elster Advance) probably deposited outwash sand and gravel that overlie solifluction material and interstadial deposits of doubtful age. The ice invaded the country from the north and deposited a sandy till, which is characterised by large amounts of Norwegian indicators (Houmark-Nielsen, 2004). This ice advance terminated south of Denmark.

Whether or not the Norwegian Elster Advance and the advance which deposited the Elsterian sandy to clayey till of Middle Swedish composition were separated by interstadial conditions is uncertain, neither is the distribution of this Middle Elsterian Advance known in any detail. The latter advance was succeeded by a Baltic ice advance, which deposited a clayey till in central Denmark. Key sites for these tills comprise Røgle Klint, Treldenæs, Sønder Vissing (Fig. 5.1, sites 2, 3 and 4). This Baltic Elster Advance probably terminated in northern Germany and the North Sea. Meltwater clay was washed down into local depressions, as well as into the large buried valleys in Jylland. It could well be equivalent to the so-called Lauenburg Clay from northern Germany. These basins and lakes were later to be filled with lacustrine and marine Holsteinian Stage (MIS 11) deposits (Fig. 5.2).

5.4. THE SAALIAN GLACIATION

An upper major hiatus separates pre-Holsteinian glaciations from those of the Saalian (MIS 10–6). It has been proposed that the interglacial lacustrine deposits of Treldenæs (Fig. 5.1) may belong to an inter-Saalian event, that is, the Wacken Interglacial (Houmark-Nielsen et al., 2006) and, not as previously assumed, tied to the Elsterian. Three major ice advances separated in time by deposition of glaciofluvial material dominate the Saalian Stage (Fig. 5.2).

5.4.1. The First Saalian Glaciation

Glaciofluvial sand and gravel locally filled the lake basins and formed floodplains deposited by westward flowing meltwater streams which covered large parts of central and southern Jylland. These waterlain deposits, referred to as the 'Norwegian' gravel, emerged from the first Saalian ice advance probably during MIS 8. This Norwegian Saale Advance deposited a sandy, quartz-rich Treldenæs till which includes a readily apparent Norwegian indicator erratic content and other characteristic rock fragments of northerly provenance (Houmark-Nielsen, 2004). The Norwegian Saale Advance invaded the country from a northerly direction and probably terminated south of the Danish-German borderland. There are apparently no morphological features left from this glaciation. The type section is Treldenæs (Fig. 5.1, site 3).

5.4.2. The Drenthe Phase

The second Saalian glaciation began with the deposition of fine-grained glaciofluvial material and, as the glacier approached, ended with proximal outwash deposits and flow till sediments. The ice sheet invaded the country from the north-east during the early part of MIS 6 and deposited the sandy, quartz-rich Ashoved till, named after the type section on the east coast of Jylland (Fig. 5.1). The till is characterised by Middle Swedish indicators and rock fragments from the Jurassic deposits along the Fennoscandian border zone in Kattegat, as well as by re-deposited Holsteinian-aged Foraminifera (Houmark-Nielsen, 2004). Key sites for this phase include Røgle and Ashoved (Fig. 5.1, sites 2 and 5). The deglaciation of this ice sheet was accompanied by deposition of the glaciofluvial material observed at several places in the central Danish region. End moraines and other elongate features are recognised in the Skovbjerg area in western Jylland (Larsen and Kronborg, 1994; Sahl, 2004) forming linear hill complexes (Fig. 5.1, the digital map).

5.4.3. The Warthe Phase

The youngest of the Saalian glaciers that covered Denmark was a Baltic advance that corresponds to the Warthe Stadial on the North European mainland (Figs. 5.1 and 5.2). This ice sheet invaded the central Danish region from easterly to east-south-easterly directions and deposited a clayey, quartz-poor, chalk-rich till of Baltic clast provenance (Houmark-Nielsen, 2004). This Baltic Advance terminated in western Jylland (Fig. 5.1, the digital map), however, the terminal moraines are sparsely preserved as a consequence of later periglacial landform transformation during the Weichselian Stage. A connection of Warthe end moraines in Schleswig-Holstein via buried submarine push-moraines and thrust interglacial marine deposits south of Esbjerg and offshore by the Lister Deep west of Rømø (Fig. 5.1, sites 6 and 7) to the maximum extent in Jylland (*sensu* Sjørring, 1983) was proposed by Andersen (2004). The proposed ice margin is marked by elongate hills north of

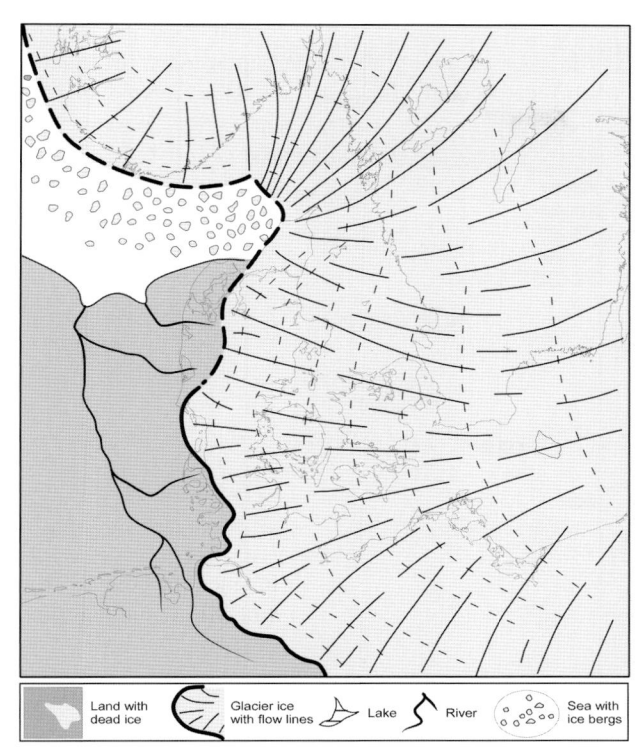

FIGURE 5.3 Palaeogeographical reconstruction of the the Warthe Substage glaciation ca. 150–140 ka ago. From Houmark-Nielsen et al. (2005). Hypothetical ice margins are indicated by the dotted heavy line. Mapped margins are shown by the solid heavy line.

Esbjerg (Figs. 5.1 and 5.3, the digital map) and by the distribution of Baltic indicator erratics on the ground surface. The presence of a Warthe till in the Vendsyssel area (Larsen et al., 2009) and in the area south and east of Esbjerg (Houmark-Nielsen, 2007) suggests that the advance terminated in a large glacier lobe across most of Jylland after flowing through eastern Denmark. It advanced at least as far west as the present North Sea coast of southern Jylland and transgressed the present marsh-land between the Esbjerg area and the northern part of Schleswig-Holstein. It also calved in the newly opened Arctic sea in the Kattegat and Skagerrak (Fig. 5.3). Key sites in Jylland for this unit include Skærumhede, Hinnerup and Stensigmose Klint (Fig. 5.1, sites 8, 9 and 10).

5.4.4. Problems

The ages of the glacial phases which deposited the two older Saalian tills are rather uncertain. They are younger than the Holsteinian, and possibly also the Wacken interglacial. Luminescence dates suggest invasion of the Treldenæs advance between 310 and 245 ka, whilst dating of interstadial deposits in Jylland indicates an age older than 155 ka and younger than 230 ka for the Drenthe advance (Pedersen and Jakobsen, 2005; Houmark-Nielsen et al., 2006; Houmark-Nielsen, 2007). Thus, at present MIS 8 and 6 are the best candidates for these ice advances (Fig. 5.2).

Other problems concern the morphological fingerprint of the younger of the two glaciations. In the Skovbjerg area (Fig. 5.1, the digital map), elongate hills, mostly trending in NW–SE directions, may expose strong tectonic deformation. Some of these are definitely of glaciotectonic origin indicating ice deformation from north-easterly directions (Skibsted, 1990; Larsen and Kronborg, 1994; Houmark-Nielsen, unpublished). However, some of these hill and valley trends, that are aligned parallel to major sub-Quaternary fault systems and some tectonic structures in Saalian sediments, may support the interpretation that some landform features could have been caused by crustal movements (Sahl, 2004; Pedersen and Jakobsen, 2005).

The distribution in north-westernmost Denmark of the Late Saalian, Warthe, glaciation is disputed. Terminal moraines and cliff sites exposing glaciotectonic deformations are present in south-western Jylland (Andersen, 2004). In central Jylland, the morphology is dominated by landforms subjected to re-modelling during the Weichselian. However, in a distinct zone trending SW–NE through central Jylland, the surface distribution of Baltic indicators abruptly drops from being dominant to almost zero. Thus, the terminal zone of the Warthe advance is placed along this zone, following Sjørring (1983). However, Pedersen and Jakobsen (2005) advocated a more westerly course, reaching well inside the Skovbjerg area. Whether the Lillebælt advance crossed the western Limfjord area is still an open question, and Larsen and Kronborg (1994) advocate a far more westerly configuration despite the lack of characteristic provenance dependent clast and boulder contents indicating Baltic origin.

Although a few dates suggest advance in the latter part of MIS 6, ages between 17 and 150 ka have been obtained from a significant number of deposits overlying the Lillebælt till in southern Jylland. These ages could suggest advance before 170–150 ka. However, deposition of sediment grains that were not subjected to sunlight exposure could be responsible for ages older than the internationally accepted age of ca. 140 ka (Houmark-Nielsen, 2007, 2008). Another problem is to distinguish the Warthe Till from Middle Weichselian Baltic tills of similar lithology. This also complicates comparison with the North German and Polish stratigraphy. Only when the till can be stratigraphically related to deposits of the Eemian Stage interglacial will proper correlation be achieved.

5.5. LATE PLEISTOCENE GLACIATIONS

Glaciations during the Late Pleistocene reached Denmark during the Middle and Late Weichselian substages, that is, MIS 4, 3 and 2 (Fig. 5.4). Luminescence

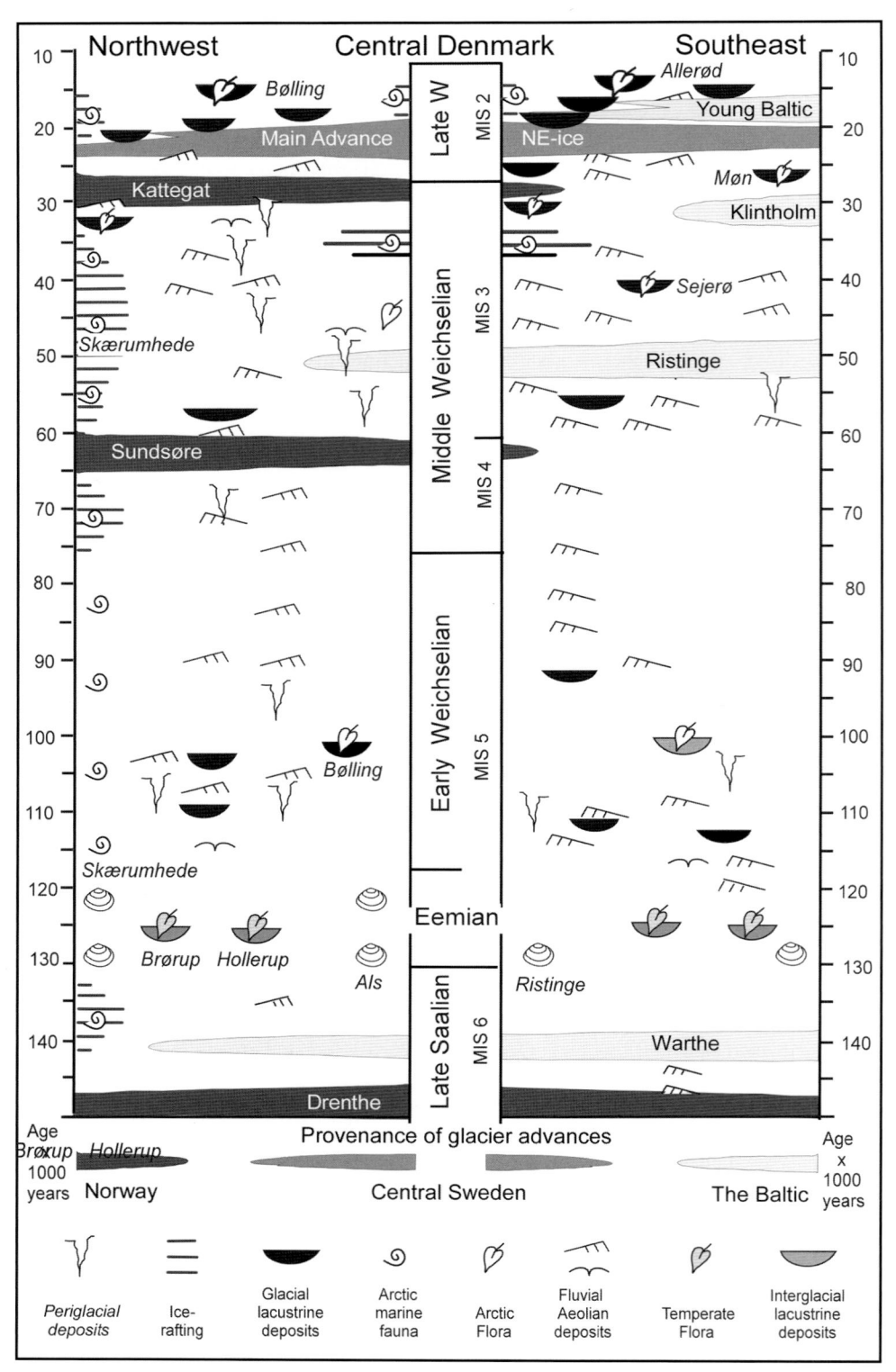

FIGURE 5.4 Chronology of glaciations in Denmark with age indications of dated sediments interstratified between till units.

and radiocarbon from different sources (Eiríksson et al., 2006; Larsen et al., 2009; Houmark-Nielsen, 2010) suggest that six major ice advances transgressed the country comprising the Sundsøre Advance at ca. 65–60 ka; the Ristinge, the Klintholm and the Kattegat Ice Streams at 54–50, 34–29 and 29–27 ka; the Main

Advance at 23–19 and the Young Baltic Ice Streams at 18–16 ka.

5.5.1. The Early Weichselian Substage

Knudsen et al. (2009) report a rapid change from boreolusitanian deep water to shallow water, boreal conditions in Vendsyssel which marks the boundary between the Eemian and the Early Weichselian. Up-sequence, the microfauna changes rapidly from boreal to boreo-Arctic conditions up-sequence. On land, the Early Weichselian is characterised by solifluction events and periglacial conditions dominated by niveo-fluvial and aeolian processes (Houmark-Nielsen, 2007). One major climatic amelioration, namely the Brørup Interstadial, is found in central Denmark (Fig. 5.1). At Brørup, tundra was replaced by low Arctic open forests with birch and juniper succeeded by spruce and pine. In the Skærumhede Group, the Brørup Interstadial has been identified as an alternation towards more boreal conditions (Kristensen et al., 1998). At the end of MIS 5, strong cooling and shallow water Arctic conditions mark the transition to the Middle Weichselian Substage time in the marine sequence.

5.5.2. Middle Weichselian Glaciations

The Middle Weichselian in the Skærumhede Group shows alternating glacigenic and Arctic marine environments with ice rafting (Knudsen et al., 2009). MIS 4 in the Skærumhede Group began with a fall in sea level and deposition of ice-rafted debris in the marine silty clays. Calving glaciers presumably extended beyond the Swedish west Coast and into the Norwegian channel. Eventually, the SIS transgressed the Skagerrak–Kattegat depression from South Norwegian uplands (Fig. 5.4); the Sundsøre Advance covered parts of northern Denmark around 65–60 ka ago during the closure of MIS 4 (Larsen et al., 2009). This glaciation left behind dead ice and proximal outwash sediments in northern Jylland. Key sites for this event include Skærumhede and Sundsøre (Fig. 5.1, sites 8 and 11). Involuted and wind-abraded soil horizons and fluvial and aeolian sand with plant detritus indicate sparsely vegetated Arctic landscapes with floodplains and drifting dunes prior to the first Weichselian glaciation in southern Denmark (Houmark-Nielsen, 2007).

5.5.3. The Ristinge and Klintholm Ice Streams

Major ice-sheet expansion in the southern Baltic region induced an ice stream to spread fan-like across the country (Houmark-Nielsen, 2010), and eventually glaciers covered most of the country early in Middle Weichselian MIS 3 between 55 and 48 ka ago (Figs. 5.4 and 5.5). Key sites include Skærumhede, Ristinge Klint and Klintholm (Fig. 5.1, sites 8, 12 and 13) In south and central Jylland west of the Late Weichselian marginal moraines (Fig. 5.1, MSL), periglacially smoothened end moraines (the digital map) and kettle holes most likely constitute morphological features of the Ristinge Ice Stream and mirror the distribution of this ice advance supported by stratigraphical evidence.

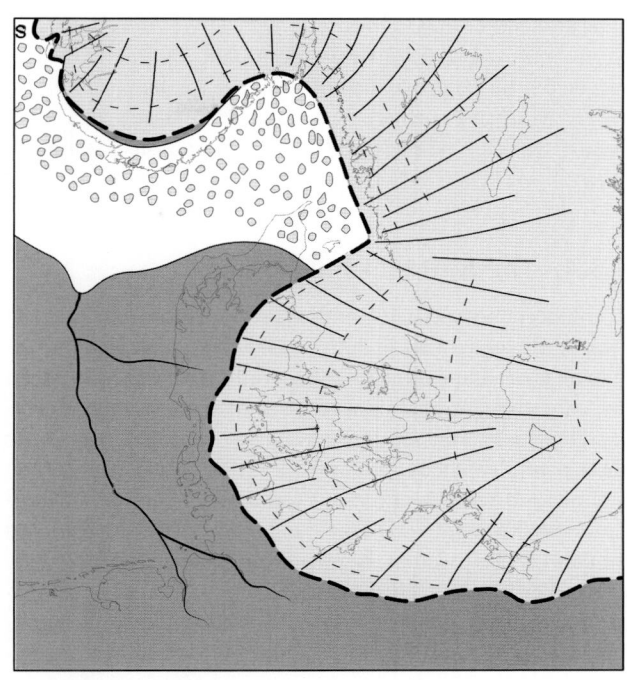

FIGURE 5.5 Palaeogeographical reconstruction of the Ristinge Ice Stream ca. 54–46 ka ago. From Houmark-Nielsen et al. (2005). For explanation of symbols: Fig. 5.3.

Subsequently, active glaciers retreated to the Scandinavian mountains and the central Swedish lakeland, while the former glaciated areas were partly left covered by dead ice. Interstadial environments replaced proglacial regimes. Arctic marine conditions were restored in the Kattegat–Skagerrak sea and this was accompanied by ice rafting due to rapid calving of glaciers as recorded in the Skærumhede Group (Knudsen et al., 2009). The Baltic lake was re-established, and on land, a mammoth steppe with shrub tundra existed from ca. 45 to 30 ka during the middle part of MIS 3 (Bennike et al., 2007; Houmark-Nielsen, 2007).

Renewed ice growth caused the ice sheet to advance onto the coast in southern Norway (Houmark-Nielsen and Kjær, 2003), and glaciers possibly streamed southwards through the Baltic depression feeding the Klintholm Ice Stream (Fig. 5.4). The latter seems only to have covered the eastern most part of the country around 34–29 ka ago (Houmark-Nielsen, 2010). Key sites are located at Klintholm and Sejerø (Fig. 5.1, sites 13 and 14). Glaciomarine clay was deposited proglacially in northern Sjælland caused by glacioisostatic subsidence. Glaciers extended to the coasts of Skagerrak and Kattegat, and in the Upper

Skærumhede Group, ice-rafted material was deposited and Arctic marine conditions were dominant. Deglaciation was succeeded by deposition of interstadial freshwater beds on Skåne, Møn and northern Jylland containing remains of an almost treeless, heather and shrub vegetation (Houmark-Nielsen and Kjær, 2003).

5.5.4. The Kattegat Ice Stream

In southern Norway and south-west Sweden, the SIS transgressed the coast late in MIS 3. Glaciers filled the Norwegian channel and calved into the North Sea and Kattegat. At the same time, an incipient regression is recorded in northern Denmark. The marine conditions in the Skærumhede Group were replaced by deposition of lacustrine fines and fluvial sand with Arctic plant remains in a partly glacier-dammed Kattegat ice lake. In northern Denmark and south-western Sweden, deposits of the Kattegat ice lake were overridden by the Kattegat Ice Stream around 30–27 ka ago (Figs. 5.4 and 5.6). The latter reached its maximum across central Denmark and in northwest Skåne and deposited till of Norwegian provenance (Houmark-Nielsen, 2003; Larsen et al., 2009), while periglacial conditions with Arctic treeless vegetation were maintained in southern Denmark. Deglaciation caused the Kattegat ice lake to re-open, however, still dammed off from the Atlantic Ocean by glacier ice flowing out of the Norwegian Channel. The Kattegat Ice Stream apparently has left no recognisable morphological features, thus, its distribution is founded on stratigraphical evidence alone. Type sections for the Kattegat till comprise Hundested, North Samsø and Lønstrup (Fig. 5.1, sites 15, 16 and 17)

5.5.5. Late Weichselian Substage Glaciations

After a period of glacier retreat, which left most of northern Denmark and parts of Skåne free of active ice, the main ice advance of the Late Weichselian invaded this region from the north-east and deposited till of Middle Swedish provenance. Type sections of this phase include Ristinge, North Samsø, Fårup and Bovbjerg (Fig. 5.1, sites 12, 16, 18 and 19). Strong ice-bed interaction, as deduced from fine-gravel dispersal in the till, suggests steady flow under non-streaming conditions, according to Kjær et al. (2003). The ice reached the Main Stationary Line (Fig. 5.1, MSL) around 22–20 ka BP (Figs. 5.4 and 5.7). The MSL is marked by the vortex of numerous proximal outwash plains that very steadily descend towards the west and south away from the former ice margin. Often, their proximal parts are elevated compared to the ground moraine behind the former termini. The marginal zone of the ice sheet is characterised by few and poorly developed push-moraines suggesting the absence of deformable sediments beneath large sections of the ice sheet that could serve as *decollément*.

The main stationary state was followed by general retreat towards the north-east leaving behind isolated areas

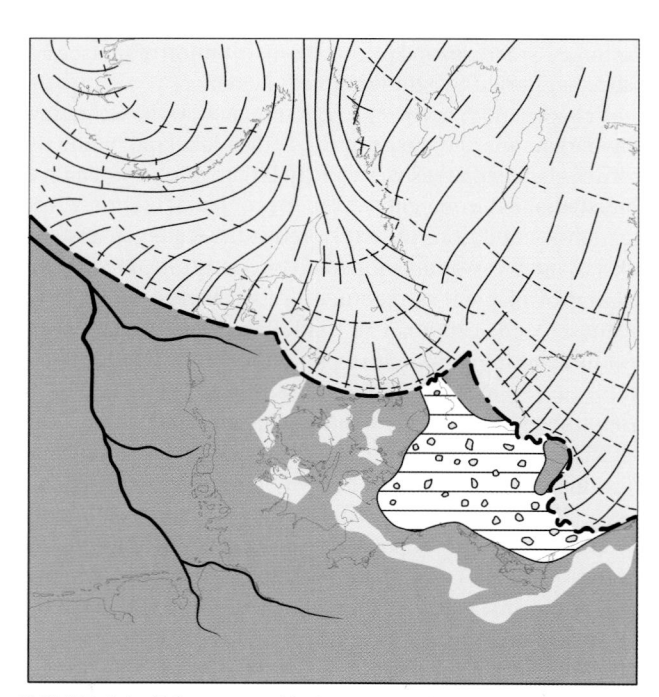

FIGURE 5.6 Palaeogeographical reconstruction of the Kattegat Ice Stream ca. 29–27 ka ago. From Houmark-Nielsen et al. (2005). For explanation of symbols, see Fig. 5.3.

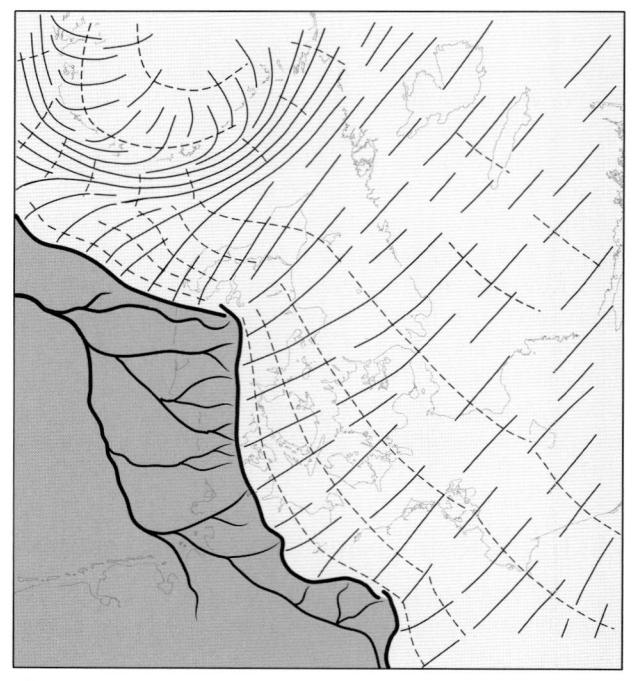

FIGURE 5.7 Palaeogeographical reconstruction of the Main ice advance ca. 22–20 ka ago. From Houmark-Nielsen et al. (2005). For explanation of symbols, see Fig. 5.3.

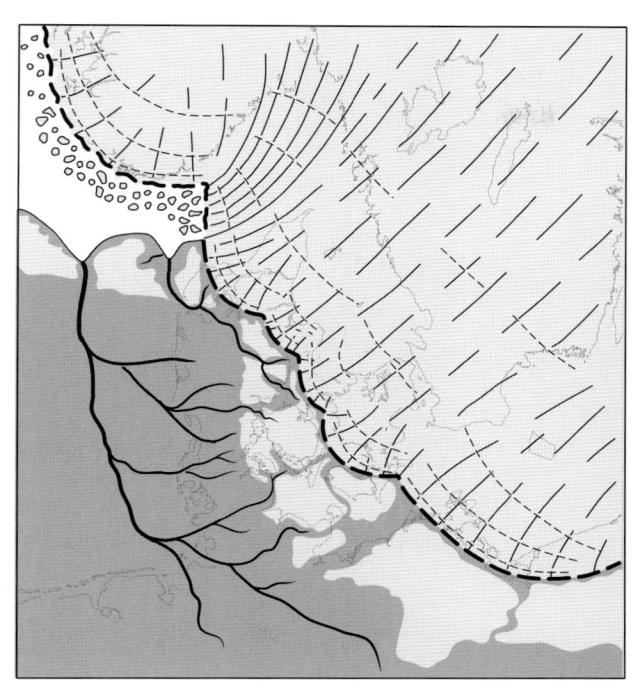

FIGURE 5.8 Palaeogeographical reconstruction of the deglaciation re-advance from main glacial stage ca. 20–19 ka ago. From Houmark-Nielsen et al. (2005). For explanation of symbols, see Fig. 5.3.

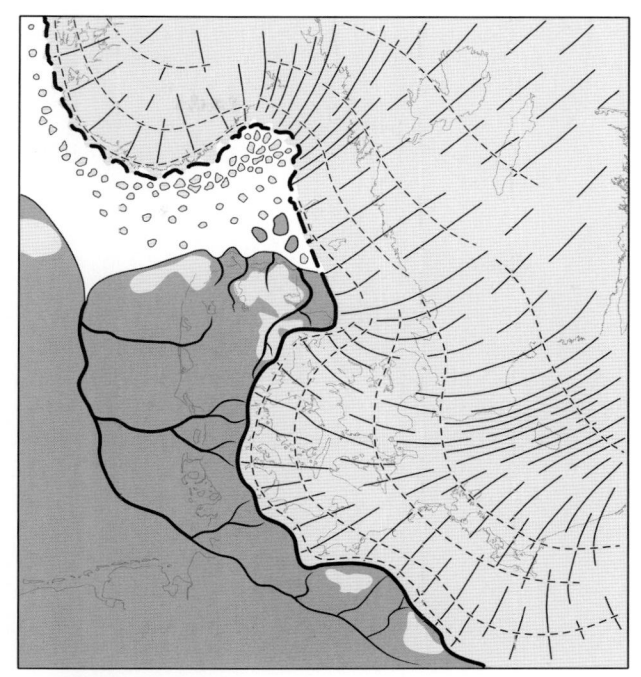

FIGURE 5.9 Palaeogeographical reconstruction of the Young Baltic Ice Streams, East Jylland phase 19–18 ka ago. From Houmark-Nielsen et al. (2005). For explanation of symbols, see Fig. 5.3.

of dead-ice south-west of the retreating ice sheet. However, phases of still-stands and readvances are recorded throughout most of Denmark (Figs. 5.1 and 5.8, the digital map). Reduced ice-bed interaction under the general retreat, as indicated by Kjær et al. (2003), suggests a change towards streaming flow conditions which is supported by morphology. Readvance related landforms may include streamlined ground moraine located behind large arch-shaped push moraines. Directional properties from till beds and glaciotectonic deformation indicate a fan-like flow pattern, which suggest a change in the basal glacier dynamics during the deglaciation of the main advance. Terminal moraines are especially prominent on Mors, in central North Jylland, Djursland, Vendsyssel, southern Kattegat, Gribskov, south-western Fyn, South Sjælland and Møn (Fig. 5.1), as shown in the digital map.

After an interlude of periglacial conditions and dead-ice downwasting, the final expansion of the SIS from the Baltic uplands, around 20–18 ka ago, brought Weichselian ice advances to an end. The Young Baltic ice streams pushed up prominent marginal moraines during the East Jylland phase, the Bælthav phase and succeeding recessional phases (Fig. 5.1, the digital map). Clayey tills of Baltic provenance were deposited. Key sites for this phase include Ristinge, Klintholm, Sejerø, Højvang, Mols Hoved, Visborg and Korselitse (Fig. 5.1, sites 12, 13, 14, 20, 21, 22 and 23). The terminal zones from the maximum extent (Fig. 5.9) are especially well developed on Mols (Fig. 5.1). Here arch-shaped end moraines cross-cut the NW–SE trending terminal moraines of the NE-ice readvances and overprint the prominent streamlined terrain to the north (Houmark-Nielsen, 2008; Fig. 5.7, the digital map). In many areas, especially Mols, Gribskov, southern Sjælland and Møn (Fig. 5.1), the older push moraines are still visible beneath a cover of Young Baltic ground moraine and dead-ice landforms, just as cliff exposures reveal large scale ice-push structures related to the older readvances. After reaching its maximum extent (Fig. 5.1: E), the ice sheet split into several individual ice streams during the Bælthav phase, ca. 18–17 ka ago, forming arcuate end moraines with complex cross-cutting relationships (Fig. 5.10). However, in many cases, these landforms are drowned and therefore not visible above present sea level. The Vejrhøj end moraine system (Fig. 5.1 and Houmark-Nielsen, 2008: fig. 5.8) is an outstanding on-shore example of small-scale ice-stream perturbation forming several generations of arcuate end moraines visible in the digital map. The high end moraines up to 100 m were stacked by an ice advance that penetrated 15 km onto the western Sjælland foreland that later to split into smaller lobes. Although they are covered by Post-Glacial marine deposits in the deepest parts, the hinterland exhibits streamlined ground moraine spreading out from a narrow ice-stream zone to the east. The most clearly expressed stream lines are those formed by kames smeared out in a fan-like pattern towards the west and south-west. In front of the end moraines, proximal outwash fans

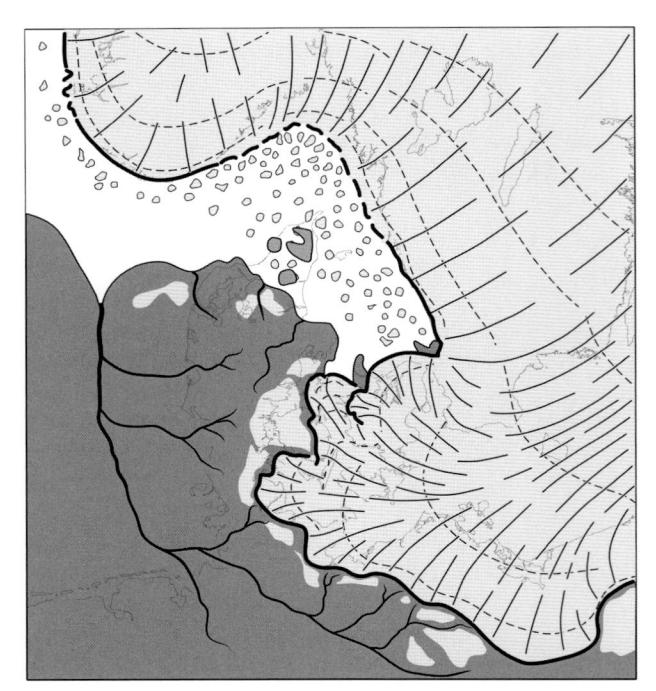

FIGURE 5.10 Palaeogeographical reconstruction of the Young Baltic Ice Streams, Bælthav phase 18–17 ka ago. From Houmark-Nielsen et al. (2005). For explanation of symbols, see Fig. 5.3.

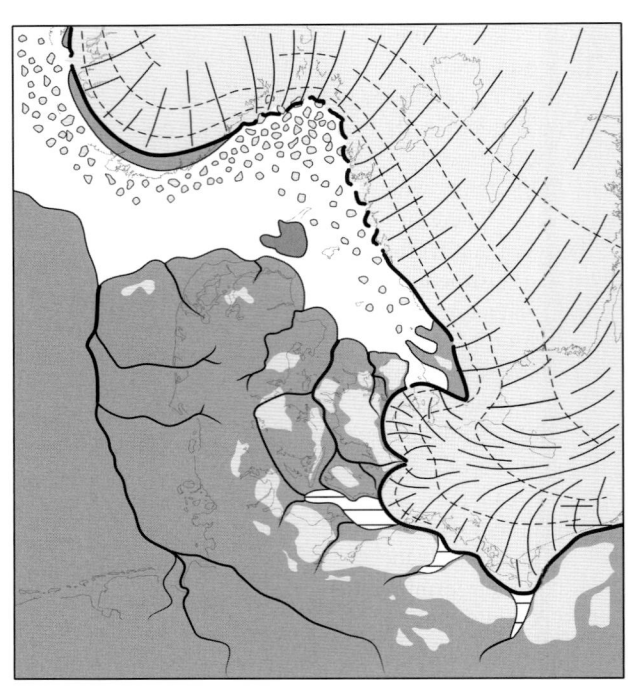

FIGURE 5.11 Palaeogeographical reconstruction of the Young Baltic Ice Streams, Møn phase 17–16 ka ago. From Houmark-Nielsen et al. (2005). For explanation of symbols, see Fig. 5.3.

interfinger with a generation of older arch-shaped push moraines to the west, now mostly situated below sea level and clearly visible on bathymetric maps. Dead-ice fields were successively isolated from the retreating ice front, and prominent arch-shaped end moraines were built in the Storebælt region between Fyn and Sjælland, finally to sculpt the out standing push moraines on Møn along the Baltic Coast. The decay of the Ice Stream and the massive downwasting, connected to deglaciation, took place while the moraine systems in Halland on the Swedish west coast were built along the active ice margin at around 17–16 ka (Figs. 5.1 and 5.11). At the same time, an ice lobe calved in the northern Øresund between Sjælland and Skåne and embryos of what later merged into the Baltic Ice Lake began to develop. As deglaciation proceeded, traces of the youngest terminal positions of the SIS in Denmark are marked by proximal sandar, dump moraines and the distribution of proglacial lakes dammed up along the coast of the island of Bornholm (Fig. 5.1, the digital map). The age of this stage is between 21 and 15 ka (Kjær et al., 2006; Houmark-Nielsen et al., 2010).

From the peak of the Young Baltic Ice Streams and onwards, an Arctic sea transgressed Northern Denmark and penetrated deeply into the Kattegat depression. At about 17 ka ago, Arctic glaciomarine conditions had reached the Øresund region (Lagerlund and Houmark-Nielsen, 1993; Richardt, 1996). Because glacioisostatic rebound caused regression, marine environments were maintained only in the eastern and deeper part of Kattegat, where a transition to marine Holocene deposits is recorded (Knudsen et al., 2009). On land, the Late Weichselian glacial Substage included interstadial conditions with low Arctic pioneer vegetation and progressively reduced ice cover in SW Scandinavia during the Bølling–Allerød event, although a climatic reversal to ice-age conditions occurred during the Younger Dryas Stadial.

5.5.6. Problems

The dynamics of the Sundsøre advance remain clear, even though the younger Kattegat advance also of Norwegian origin has been categorised as ice streams on well-argued grounds (Wohlfarth et al., 2008). Another problem is the estimation of the extent of the Sundsøre and Kattegat advances since their morphological features apparently are not preserved in the present landscape. Alternatively, however, there may be features not acknowledged at present that may relate to these glaciations. Because successive Late Weichselian ice advances produced superimposed landscapes, specific morphologies may have been subjected to misinterpretation both in time and genesis. The extent of glaciers where none or very sparse morphological features are preserved must rest solely on stratigraphical interpretation. However, stratigraphical prerequisites such as distinctive lithological features, robust dating procedures and reliable correlation are not straightforward when dealing with the Middle Weichselian or earlier events. Therefore, the distribution of pre-LGM ice advances may not always be known in detail.

The timing of glacier expansion and retreat is substantially embedded in the part of the numerical age model where luminescence is supported by radiocarbon, especially in the range where ^{14}C ages can be calibrated to calendar ages. Therefore, the ages of Middle Weichselian glaciations are still crude, and the chronologies may be subjected to adjustments or corrections when a denser network of dated glacial deposits becomes available. Environmental interpretation of sedimentary facies and knowledge of the circumstances under which transport and deposition took place is crucial when evaluating luminescence ages (Kjær et al., 2006; Houmark-Nielsen, 2008), just as application of different luminescence methods from region to region makes correlation difficult (Houmark-Nielsen, 2010). Therefore, assessing the synchronicity of not only Middle Weichselian ice advances but also Late Weichselian glaciations and their terminal positions marked by end moraines may not be straightforward.

Compilations of the rapidly changing Scandinavian palaeoenvironments indicate that the repeated outflow of glaciers from the Norwegian, Swedish and Baltic uplands across Denmark during most of the Weichselian was interrupted by episodes of strongly reduced ice-sheet distribution accompanied by the colonisation of low Arctic flora and fauna in central Scandinavia (Ukkonen et al., 2007; Wohlfarth, 2009; Houmark-Nielsen, 2010). Whether ice-sheet fluctuations reflect an immediate adjustment to rapid climate change or whether expansion and withdrawal represent a more complex response remains uncertain. Ice streams flowing from the south-western periphery of the SIS possibly reflect the combined effects of change in regional glacier dynamics caused by ice-bed instabilities, location of shallow proglacial basins containing sediments of low permeability and recurrent external climatic forcing acting upon the ice sheet.

5.6. CONCLUSIONS

- The age of the first Pleistocene ice cover in Denmark is uncertain, however, the Cromerian Glacial A seems to be the most likely candidate.
- The succession of Middle Pleistocene glaciations is not known in detail because of the reduced preservation potential and the presence of regional haiti. However, four major glaciations from the Elsterian Stage, and the Middle and Late Saalian (Drenthe-Warthe advances) Substages can be recognised.
- Late Pleistocene glaciations invaded Denmark from different Scandinavian uplands during the Middle and Late Weichselian Substages. One occurred during late MIS 4 (Sundsøre Advance) and three MIS 3 glaciations (Ristinge, Klintholm and Kattegat Ice Streams) can be recognised, plus two glaciations during MIS 2 (Main Advance and Young Baltic Ice Streams).
- Discrimination between inter-stream and ice-stream flow has added substantially to the understanding of the dynamic behaviour of the Scandinavian Ice Sheet.
- Glacier expansion resulting from ice streaming seems to be out of phase with global ice-sheet volume and appears to correspond to episodes of regional North Atlantic climate amelioration.

REFERENCES

Andersen, L.T., 2004. A study of large scale glaciotectonics using high resolution seismic data and numerical modelling. Danmarks og Grønlands Geologiske Undersøgelse Rapport 2004/30. Ph.D. thesis University of Aarhus, 141 pp.

Bennike, O., Houmark-Nielsen, M., Wiberg-Larsen, P., 2007. A Middle Weichselian interstadial lake deposit on Sejerø, Denmark: macrofossil studies and dating. J. Quatern. Sci. 22, 647–651.

Boulton, G.S., Dongelmans, P., Punkari, M., Broadgate, M., 2001. Palaeoglaciology of an ice sheet through a glacial cycle: the European ice sheet through the Weichselian. Quatern. Sci. Rev. 20, 591–625.

Ehlers, J., Gibbard, P.L. (Eds.), 2004. Quaternary Glaciations: Extent and Chronology, Part 1 Europe. Elsevier, Amsterdam, pp. 34–46.

Eiríksson, J., Kristensen, P.H., Lykke-Andersen, H., Brooks, K., Murray, A., Knudsen, K.L., 2006. A sedimentary record from a deep Quaternary valley in the southern Lillebælt area, Denmark: Eemian and Early Weichselian lithology and chronology at Mommark. Boreas 35, 320–331.

Håkansson, E., Surlyk, F., 1997. Denmark. In: Moores, E.M., Fairbridge, R.W. (Eds.), Encyclopedia of European and Asian Geology. Chapman & Hall, London, pp. 183–192.

Houmark-Nielsen, M., 2003. Signature and timing of the Kattegat Ice Stream: onset of the LGM-sequence in the southwestern part of the Scandinavian Ice Sheet. Boreas 32, 227–241.

Houmark-Nielsen, M., 2004. The Pleistocene of Denmark: a review of stratigraphy and glaciation history. In: Ehlers, J., Gibbard, P.L. (Eds.), Quaternary Glaciations—Extent and Chronology. Part I, Europe. Elsevier, Amsterdam, pp. 35–46.

Houmark-Nielsen, M., 2007. Extent and age of Middle and Late Pleistocene glaciations and periglacial episodes in southern Jylland, Denmark. Bull. Geol. Soc. Denmark 55, 9–35.

Houmark-Nielsen, M., 2008. Testing OSL failures against a regional Weichselian glaciation chronology from south Scandinavia. Boreas 37, 660–677.

Houmark-Nielsen, M., 2010. Extent, age and dynamics of Marine Isotope Stage-3 glaciations in the south western Baltic Basin. Boreas 39, 343–359.

Houmark-Nielsen, M., Kjær, K.H., 2003. Southwest Scandinavia 40-15 ka BP: paleogeography and environmental change. J. Quatern. Sci. 18, 769–786.

Houmark-Nielsen, M., Kjær, K.H., Krüger, J., 2005. De seneste 150.000 år i Danmark, istidslandskabets og naturens udvikling. Geoviden 2, 20.

Houmark-Nielsen, M., Björck, S., Wohlfarth, B., 2006. Cosmogenic ^{10}Be ages on the Pommeranian Moraine, Poland: comments. Boreas 35, 600–604.

Houmark-Nielsen, M., Linge, H., Fabel, D., Xu, S., 2010. Cosmogenic exposure dating of boulders and bedrock in Denmark: wide range in ages reflect strong dependence of post-depositional stability related to specific glacial landforms. Geophys. Res. Abstr. 12, EGU2010-8968.

Huuse, M., Lykke-Andersen, H., Piotrowski, J., 2003. Geophysical investigations of buried Quaternary valleys in the formerly glaciated NW European lowland: significance for groundwater exploration (editorial). J. Appl. Geophys. 53, 153–157.

International Commission on Stratigraphy (ICS), 2009. Global chronostratigraphical correlation table for the last 2.7 million years. http://www.stratigraphy.org/.

Japsen, P., Chalmers, J.A., 2000. Neogene uplift and tectonics around the North Atlantic: overview. Glob. Planet. Change 24, 165–173.

Kjær, K.H., Houmark-Nielsen, M., Richardt, N., 2003. Ice-flow patterns and dispersal of erratics at the southwestern margin of the last Scandinavian ice sheet: signature of palaeo-ice streams. Boreas 32, 130–148.

Kjær, K., Lagerlund, E., Adrielsson, L., Thomas, P.J., Murray, A., Sandgren, P., 2006. The first independent chronology for Middle and Late Weichselian sediments from southern Sweden and the Island of Bornholm. GFF 128, 209–220.

Knudsen, K.L., Kristensen, P., Larsen, N.K., 2009. Marine glacial and interglacial stratigraphy in Vendsyssel, northern Denmark: foraminifera and stable isotopes. Boreas 38, 787–810.

Kristensen, P., Knudsen, K.L., Lykke-Andersen, H., Nørmark, E., Peacock, J.D., Sinnott, A., 1998. Interglacial and glacial climate oscillations in a marine shelf sequence from northern Denmark—a multidiciplinary study. Quatern. Sci. Rev. 17, 813–837.

Kristensen, T.B., Piotrowski, J., Huuse, M., Clausen, O.R., Hamberg, L., 2008. Time-transgressive tunnel valley formation indicated by infill sediment structure, North Sea—the role of glaciohydraulic supercooling. Earth Surf. Process. Land. 32, 546–559.

Krohn, C.F., Larsen, N.K., Kronborg, C., Nielsen, O.B., Knudsen, K.L., 2009. Litho- and chronostratigraphy of the Late Weichselian in Vendsyssel, northern Denmark, with special emphasis on tunnel valley infill in relation to a receding ice margin. Boreas 38, 811–833.

Lagerlund, E., Houmark-Nielsen, M., 1993. Timing and pattern of the last deglaciation in the Kattegat region, southwest Scandinavia. Boreas 22, 337–347.

Larsen, G., Kronborg, C., 1994. Det mellemste Jylland, En beskrivelse af områder af national geologisk interesse. Geografforlaget, Brenderup 272 pp.

Larsen, N.K., Knudsen, K.L., Krohn, C.F., Kronborg, C., Murray, A.S., Nielsen, O.B., 2009. Late Quaternary ice sheet, lake and sea history of southwest Scandinavia—a synthesis. Boreas 38, 732–761.

Pedersen, S.A.S., Jakobsen, P.R., 2005. Geological Map of Denmark, 1:50.000, Ringkøbing. Geological Survey of Denmark and Greenland, Copenhagen.

Richardt, N., 1996. Sedimentological examination of the Late Weichselian sea-level history following deglaciation of northern Denmark. Geological Society of London, Special Publication, 111.

Sahl, B., 2004. Den kvartærgeologiske udvikling af Skovbjerg Bakkeø—belyst ved området omkring Fjaldene. Unpublished M.Sc. thesis, Geological Institute, University of Copenhagen, 105 pp.

Sjørring, S., 1983. The glacial history of Denmark. In: Ehlers, J. (Ed.), Glacial Deposits in North-West Europe. Balkema, Rotterdam, pp. 163–179.

Skibsted, S., 1990. Mellem-Pleistocæn tillstratigrafi på Skovbjerg-, Ølgod- og Esbjerg Bakkeø. Unpublished M.Sc. thesis, Geological Institute, University of Copenhagen, 289 pp.

Stokes, C.R., Clark, C.D., 2001. Palaeo-ice streams. Quatern. Sci. Rev. 20, 1437–1457.

Ukkonen, P., Arppe, L., Houmark-Nielsen, M., Kjær, K.H., Karhu, J.A., 2007. MIS 3 mammoth remains from Sweden—implications for faunal history, palaeoclimate and glaciations chronology. Quatern. Sci. Rev. 26, 3081–3098.

Wohlfarth, B., 2009. Ice-Free Conditions in Fennoscandia During Marine Oxygene Isotope Stage 3? Swedish Nuclear Fuel and Waste Management Co., Stockholm Technical Report TR-09-12, 48 pp.

Wohlfarth, B., Funder, S., Björck, S., Houmark-Nielsen, M., Ingólfsson, O., Lunkka, J.P., 2008. Quaternary of Norden. Episodes 31, 73–81.

The Glacial History of the British Isles during the Early and Middle Pleistocene: Implications for the long-term development of the British Ice Sheet

Jonathan R. Lee[1,2,*], James Rose[1,2], Richard J.O. Hamblin[1,2], Brian S.P. Moorlock[1,2], James B. Riding[1], Emrys Phillips[3], René W. Barendregt[4] and Ian Candy[2]

[1] British Geological Survey, Keyworth, Nottingham NG12 5GG, United Kingdom
[2] Department of Geography, Royal Holloway, University of London, Egham, Surrey TW20 0EX, United Kingdom
[3] British Geological Survey, Murchison House, West Mains Road, Edinburgh EH9 3LA, United Kingdom
[4] University of Lethbridge, Lethbridge, Alberta, Canada TK1 3M4

*Correspondence and requests for materials should be addressed to Jonathan R. Lee. E-mail: jrlee@bgs.ac.uk

6.1. INTRODUCTION

The Early and Middle Pleistocene (ca. 2.6–0.125 Ma) was a period of major climate and earth system change driven by a progressive trend of global cooling. This includes the so-called Middle Pleistocene Transition (MPT) (Head and Gibbard, 2005a) which represents a shift (1.2–0.7 Ma) from global climate forcing driven by 41 ka (obliquity) cycles to a pattern of forcing driven by 100 ka (eccentricity) cyclicity (Clark et al., 2006). The global effects of the MPT on various physical, chemical and biological systems have been documented by a number of papers within Head and Gibbard (2005b). However, one of its most obvious global consequences was to produce longer and more intense 'cold' periods, which changed both the long- and short-term development of ice sheets and the timing and scale of glaciation (Ehlers and Gibbard, 2007).

A large number of Pleistocene ice sheets, including the Scandinavian (SIS), Laurentide and Greenland ice sheets, plus several smaller maritime ice sheets, lie adjacent to the climatically sensitive North Atlantic region. Recent research has demonstrated that each of these ice sheets underwent a step-wise amplification in both the frequency and scale of glaciation across the MPT (Jansen et al., 2000; Sejrup et al., 2005; Knies et al., 2009). However, it remains unclear how the British Ice Sheet (BIS) fits into this regional trend as its long-term development is poorly understood. This is due largely to the fragmentary and sometimes ambiguous nature of the evidence for glaciation in Britain, difficulties with correlating evidence between terrestrial and offshore sequences and major problems regarding chronology and correlation with geological sequences in other countries in northwestern Europe (Clark et al., 2004; Ehlers and Gibbard, 2004, 2007; Rose, 2009; Lee et al., 2010).

The scope of this chapter is to examine spatial and temporal patterns in the scale of glaciation within the Early and Middle Pleistocene British Isles. This will provide a greater understanding behind the controls of the long-term development of glaciation in Britain, and whether it was in-phase with global climate change and other Northern Hemisphere Ice Sheets. This has important implications for understanding the sensitivity of the British land-mass and the level of coupling to climate change. Particular focus is placed upon: (1) reviewing and evaluating the evidence for Early and Middle Pleistocene glaciations; (2) identifying long-term trends in the development of glaciation in Britain; (3) drivers of glaciations and their wider context.

6.2. EARLY AND MIDDLE PLEISTOCENE GLACIAL HISTORY OF THE BRITISH ISLES

6.2.1. Long-Term Sedimentary Archives

Long-term sedimentary archives provide the best means by which the Quaternary evolution of glaciation in Britain can be evaluated. However, records that span the Early and Middle Pleistocene are relatively few in number and discontinuous in nature (Fig. 6.1).

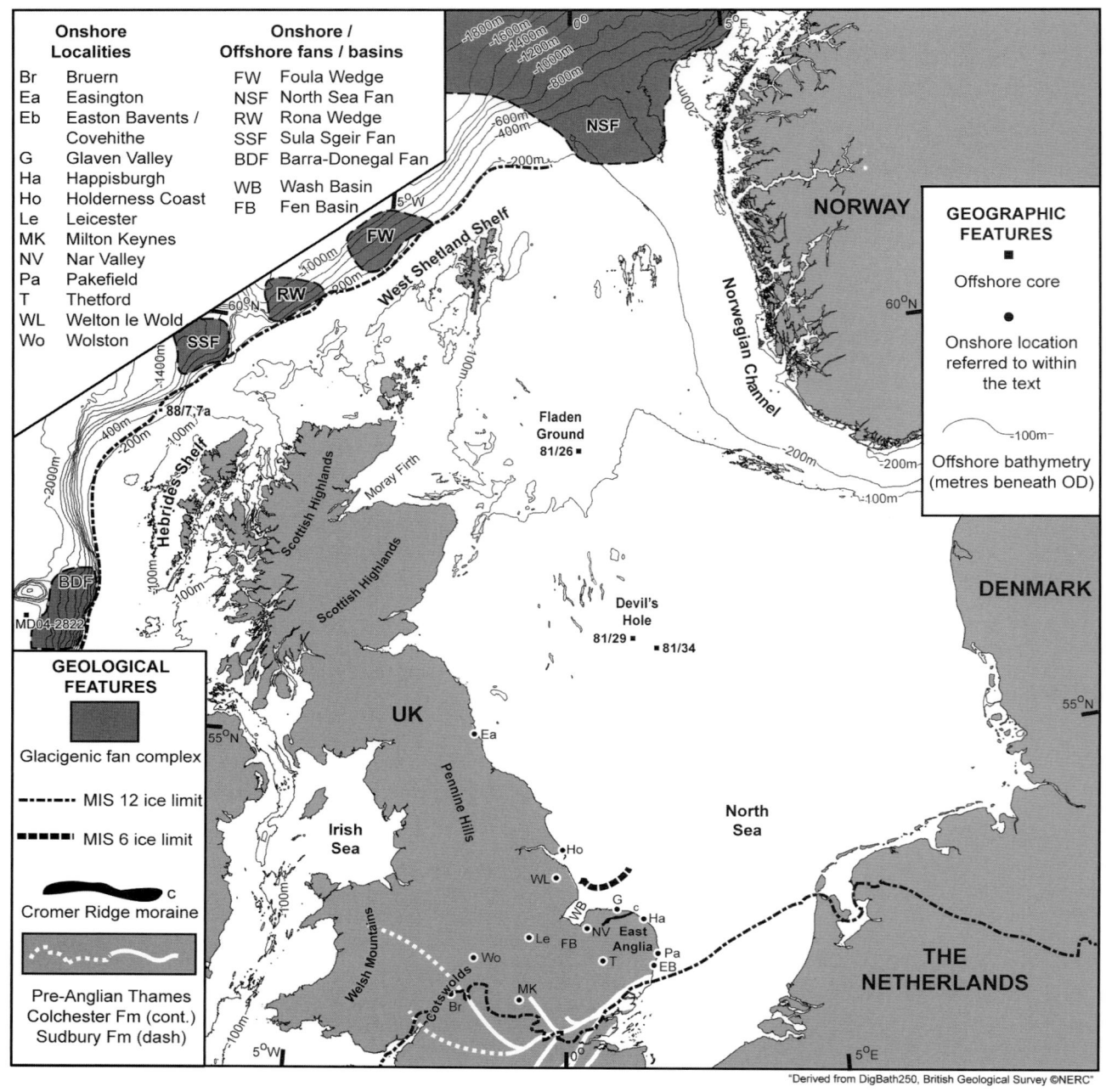

FIGURE. 6.1 Map of the British Isles, North Sea Basin and northwest European margin showing the location of sites and features referred to within the text. The MIS 12 and 6 ice limits are shown after Bowen et al. (1986) and Ehlers and Gibbard (2004).

6.2.2. UK Continental Shelf and Margin

Offshore sediment sequences developed along the Northwestern European Continental Margin offer a detailed insight into glacial processes operating both on and adjacent to the continental shelf, and at the shelf edge with the development of glaciogenic fan complexes (Sejrup et al., 2005). Extending westwards from the Norwegian Channel to Ireland, the continental margin to the north of Scotland reveals evidence for several Pleistocene expansions of the Scottish sector of the BIS (Sejrup et al., 2005; Bradwell et al., 2007). BGS borehole 88/7,7a from the Hebrides Slope records layers of Scottish-derived dropstones between the Gauss–Matuyama palaeomagnetic boundary (ca. 2.6 Ma) and the base of the Anglian (Marine Isotope Stage (MIS) 12) (Stoker et al., 1994, 1993). Whilst the specific number of ice-rafting events is unclear, the fact that they span several different normalised and reversed palaeomagnetic horizons suggests multiple ice-rafting events associated with the extension of Scottish ice into coastal areas. The first shelf-edge glaciation, associated with MIS 12, is represented by a major seismic unconformity that can be traced across the Hebrides Shelf, and several of the glaciogenic fan complexes including the Sula Sgeir Trough Mouth Fan and the Rona Wedge (Stoker et al., 1994, 1993; Stoker, 1995; Holmes et al., 2003; Sejrup et al., 2005). A further expansion of Scottish ice onto the continental shelf during MIS 6 has been recognised by Hibbert et al. (2009) form core MD04-2822 near the Barra-Donegal Fan. Sejrup et al. (2005) has suggested additional shelf-edge glaciations of Scottish ice during MIS 10 and 8 although these require further verification.

6.2.3. The North Sea and 'Crag' Basins

The North Sea is a marine basin fed during the Quaternary by rivers and ice sheets that emanated from Britain and continental Europe (Zagwijn, 1989, 1996; Gibbard, 1995; Sejrup et al., 2000; Rose et al., 2001; Carr, 2004; Busschers et al., 2008). The 'Crag' Basin represents the pre-MIS 12 extension of the North Sea Basin into parts of eastern and southern East Anglia (Funnell, 1995; Rose et al., 2001; Rose, 2009).

The first possible evidence for glaciation within the North Sea region corresponds to 'Baventian' (ca. MIS 68/1.8 Ma) marine clays from Easton Bavents and Covehithe in the 'Crag' basin of East Anglia. These clays exhibit a 'semi-glacial' fauna (West et al., 1980) and contain possible ice-rafted hornblende heavy minerals from Scandinavia (Soloman, 1935). However, the provenance and significance of the hornblende identified by Soloman is ambiguous due to the absence of a full suite of provenance-diagnostic heavy minerals. Alternatively, the hornblende could be ice rafted from northern Britain; or introduced by rivers eroding hornblende-bearing igneous rocks or Tertiary strata in Europe (Berger in Gibbard et al., 1991a) or southern Britain.

The earliest *direct* evidence for the expansion of British ice into the margins of the northern North Sea coincides with the first shelf-edge expansion of the SIS during the Fedje Glaciation (ca. MIS 34/1.1 Ma) (Sejrup et al., 1987) (Fig. 6.2). Evidence consists of large influxes of outwash-derived pre-Quaternary palynomorphs within BGS boreholes 81/29 and 81/34 (Devil's Hole). These palynomorphs are believed to be eroded from lowland bedrock strata along the western fringes of the North Sea offshore from northern Britain (Ekman, 1999). Several additional influxes of glacial outwash into the northern North Sea from northern Britain occurred during the Jaramillo Subchron dated to ca. 1.07–0.99 Ma (Ekman, 1999).

Further small expansions of the BIS in northern Britain during the 'Cromerian Complex' (ca. MIS 22–13/0.9–0.48 Ma) are indicated by several strands of evidence. First,

far-travelled erratics (Green and McGregor, 1990; Larkin, Lee and Connell, unpublished data) and heavy minerals (Lee et al., 2006; Lee, 2009) from northern Britain have been found at several different localities and stratigraphical levels within coastal deposits (Wroxham Crag) in northern East Anglia. Many erratics exhibit a fresh morphology and weigh several kilos and are believed to be the product of ice rafting from a calving BIS in the northern North Sea with localised reworking by coastal processes (Larkin et al., 2011). The precise timing of these ice-rafting events is tenuous, however, their stratigraphical position relative to temperate organic facies of the Cromer Forest-bed Formation suggests deposition during several different cold stages. Further north within the Moray Firth and Firth Approaches, glaciomarine deposits occur in the Aberdeen Ground Formation that contain dropstones (Stoker and Bent, 1985). They directly overlie the Brunhes–Matuyama (B/M) palaeomagnetic boundary that is dated to ca. 0.78 Ma (MIS 19), suggesting that they may be as old as MIS 18 (Merritt et al., 2003). Additional evidence for a pre-MIS 12 expansion of ice into the northern North Sea is suggested by a series of subglacial 'tunnel valleys' and channels cut into the top of the Aberdeen Ground Formation that are partially infilled by 'Cromerian Complex' sediments (Holmes, 1997).

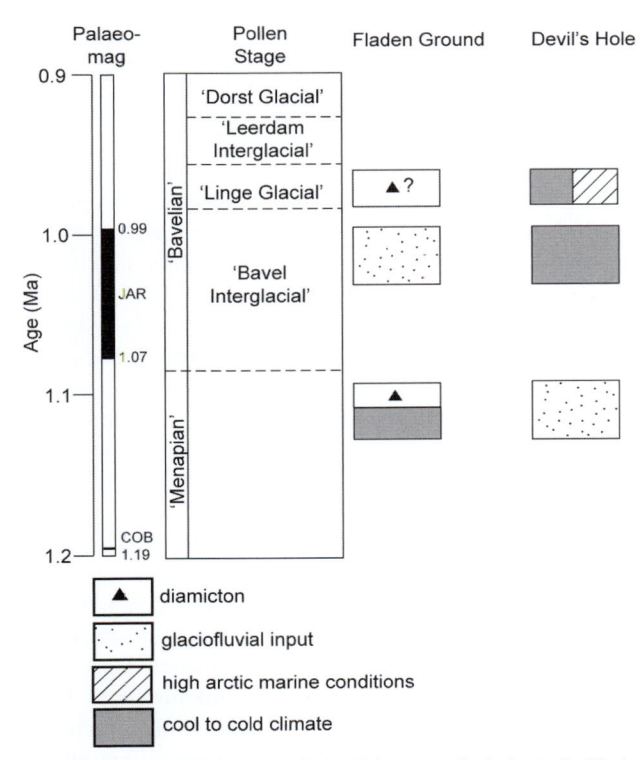

FIGURE. 6.2 Evidence for late Early Pleistocene glaciation in the Fladen Ground and Devil's Hole areas of the North Sea. Adapted from Sejrup et al. (1987) and Ekman (1999). Palaeomagnetism is from Funnell (1995).

Large-scale expansion of the BIS into the North Sea region during the Middle Pleistocene occurred during MIS 12 (Balson and Jeffrey, 1991; Carr, 2004). Evidence for this glaciation is generally in the form of an extensive seismic unconformity that has been interpreted as a glacial erosion surface (Cameron et al., 1992), and a series of over-deepened subglacial tunnel valleys produced by coeval glaciofluvial erosion and deposition during ice-marginal retreat (Praeg, 2003). Sedimentary evidence for the Anglian expansion of the BIS consists of tills, glaciolacustrine and glaciomarine deposits of the Swarte Bank and Aberdeen Ground formations (Cameron et al., 1992; Gatliff et al., 1994). The configuration of the BIS and SIS within the North Sea during this glaciation is unclear. It is possible that many of the glacial features and deposits assigned to the Elsterian in the Dutch Sector (Laban, 1995) may not be equivalent to the MIS 12 Anglian glaciation of the UK, but instead, a younger MIS 10 glaciation (see Lee et al., 2010 for an overview of this subject).

The extent of the BIS within the North Sea during post-Anglian/pre-Ipswichian cold stages is poorly understood. A till of Scottish origin was recognised within core 81/26 from the Fladen Ground area and assigned to the Late Saalian (ca. MIS 6) (Sejrup et al., 1987). However, no intervening British glacial deposits have been recognised in the North Sea suggesting that the BIS did not extend far eastwards during these cold stages (Long et al., 1988).

6.2.4. Kesgrave Proto-Thames Terrace Sequence

The pre-Anglian terraces of the proto-Thames sequence contain evidence for several possible phases of mountain-scale glaciation in western highland and lowland Britain during the Early and early Middle Pleistocene (Whiteman and Rose, 1992; Rose et al., 1999a).

Deposits of the proto-Thames are collectively called the Kesgrave Sands and Gravels (Kesgrave Group). They have been subdivided into the Sudbury (older) and Colchester (younger) Formations based upon lithology and individual terrace aggradations determined by geomorphology and elevation (Whiteman and Rose, 1992). Within the Early Pleistocene, during the deposition of the Sudbury Formation, the headwaters of the Thames drained parts of southern and central Wales. However, during the late Early Pleistocene, the headwaters of the proto-Thames catchment were beheaded and the subsequent terrace aggradations represent deposition by a much smaller river with headwaters to the east of the Cotswold escarpment (Whiteman and Rose, 1992). Far-travelled erratics from north Wales, including frequently outsized and subangular clasts of acid porphyry, tuff and banded rhyolite (Hey and Brenchley, 1977; McGregor and Green, 1978), and glacially-abraded sand grains (Hey, 1980), have been found throughout both the Sudbury and Colchester formations (Whiteman and Rose, 1992), and recently, a glacially striated oversized block of rhyolitic tuff has been discovered *in situ* in the Colchester Formation (Rose et al., 2010). The lithology of the Sudbury Formation terraces, with high quantities (up to 35%) of far-travelled quartz and quartzite (Whiteman, 1992), suggests a direct link to north and south Wales, the Welsh borders and the West Midlands. However, with the exception of the Ardleigh Terrace which shows an increased level of far-travelled material (Whiteman and Rose, 1992), the lithology of the Colchester Formation terraces, with much lower quantities (15–20%) of far-travelled quartz and quartzite (Whiteman, 1992), supports the idea of a more restricted catchment with the far-travelled component of these gravels being recycled from older terrace aggradations (Rose et al., 1999a).

From a temporal perspective, the presence of lithologies from north Wales in Sudbury Formation gravels has been interpreted as reflecting a glaciofluvial input into the proto-Thames catchment from a Welsh ice cap (Whiteman and Rose, 1992; Rose et al., 1999a; Clark et al., 2004). This would indicate the existence of restricted mountain-scale glaciation during the Early Pleistocene. In terms of major landscape evolution, one of the most significant events during the history of the proto-Thames was the beheading and truncation of its headwaters to the west of the Cotswold Hills prior to the deposition of the Colchester Formation terraces. Multiple explanations could be proposed for this event including, for instance, enhanced soft-rock erosion within the headwaters of the Bytham River catchment, however, several authors have indicated that it could reflect a major glaciation (see Whiteman and Rose, 1992). In such a model, the advance of a major ice sheet into Midland England would override the upper reaches of the proto-Thames, disrupting the drainage system and truncating the catchment in the region of the Cotswold escarpment. This is a plausible explanation but is difficult to prove, largely due to the absence of unambiguous glacigenic deposits within the pre-Anglian Thames catchment (see Clark et al., 2004 for a review of the evidence). The increased influx of north Welsh erratic lithologies into the Ardleigh Terrace aggradation has been attributed to a glaciation reaching the Cotswold Hills after the truncation of the Thames catchment, and this is supported by the recent discovery of a striated, oversized clast of rhyolitic tuff in these terrace gravels (Rose et al., 2010).

Developing a chronology for these major events and spatial changes within the pre-Anglian Thames catchment is thus crucial to developing an understanding of the long-term evolution of the Welsh sector of the BIS. This must be viewed as tentative due to the absence of *in situ* biostratigraphical evidence and the dependence upon geomorphological geochronometry although this has similar validity, resolution and possibly more robustness than other chronometers for the period of time under consideration (Lee et al., 2004a). The crude chronological framework developed by Whiteman and Rose (1992), Rose et al. (1999a) and Westaway et al. (2002) is based upon climate forcing of river aggradation and incision, linked to shallow crustal adjustment and, where possible, correlation with First Appearance Datum (FAD) and Last Appearance Datum (LAD) mammal and mollusc assemblage biostratigraphy. Within this chronological model, the earlier Sudbury Formation was deposited between ca. 1.87 and 0.87 Ma, with individual terrace aggradations tentatively assigned to MIS 68 (Stoke Row terrace), MIS 62-54 (Westland Green terrace), MIS 54-36 (Satwell terrace), MIS 36-32 (Beaconsfield terrace) and MIS 22 (Gerrards Cross terrace); providing the first evidence for a Welsh ice cap at around MIS 68. Terraces within the Colchester Formation were, by contrast, formed during single cold stages between ca. 0.87 and 0.48 Ma, with individual aggradations and corresponding Welsh glaciations occurring during MIS 20 (Waldringfield terrace), MIS 18 (Ardleigh terrace, with ice reaching the Bruern region in the Cotswold Hills), MIS 16 (Wivenhoe terrace), MIS 12 (St Osyth terrace), although see Rose et al. (2010) for a more recent assessment of these age attributions. Critically, this chronological interpretation places the first lowland expansion of Welsh ice and the beheading of the Welsh headwaters of the Thames, into MIS 20, although there is no specific influx of erratic material into the Thames at this time to suggest that ice reached the present Thames catchment.

6.2.5. Middle Pleistocene Glacial Sequence of Eastern England and the Midlands

During the Middle Pleistocene, it was widely accepted that East Anglia and the Midlands were glaciated extensively on two occasions—during the Anglian Glaciation, and a later pre-Devensian glaciation known as the 'Wolstonian' (West and Donner, 1956; Mitchell et al., 1973; Shotton, 1983; Straw, 1983). However, detailed examinations of the preglacial and till successions subsequently led to the proposal that there was only one extensive Middle Pleistocene glaciation—the Anglian (Perrin et al., 1979; Bowen et al., 1986; Rose, 1987, 1989a, 1992). Evidence from the Holderness and County Durham areas also suggests a later minor extension of ice along the North Sea coast of England believed to be equivalent to the late Saalian (MIS 6) glaciation (Catt and Penny, 1966; Bateman and Catt, 1996) which is equated to the 'Wolstonian'. Ice is believed to have extended to the margins of the Wash Basin depositing a complex sequence of outwash (Gibbard et al., 1991b, 1992, 2009; Lewis and Rose, 1991). Recently, the glacial model for East Anglia and the Midlands has been challenged by new work that challenges: (1) the regional glacial succession; (2) the provenance of some of the deposits; (3) the number of Middle Pleistocene glaciations and their ages. Each of these issues has courted controversy, and the evidence and debates are presented within the following sections.

6.2.6. Regional Glacial Lithostratigraphy

As stated above, 'conventional' interpretations of the Middle Pleistocene glacial succession of East Anglia and the Midlands provide evidence for two separate glaciations. It was proposed that during the Anglian Glaciation, the BIS extended across much of southern Britain north of the present river Thames. It deposited Mesozoic-rich tills in the Midlands that include the Thrussington and Oadby tills of the Wolston Formation, and the Lowestoft Till of the Lowestoft Formation in central and southern East Anglia (Rice, 1968; Perrin et al., 1979; Bowen et al., 1986). In northern East Anglia, the three 'Cromer Tills' ('North Sea Drift Formation') were laid down contemporaneously by the SIS (Reid, 1882; Banham, 1968; Lunkka, 1994), with a intermediate-stage retreat enabling the BIS to extend eastwards across the region depositing a chalk-rich facies of the Lowestoft Formation called the 'Marly Drift' (Figs. 6.3A and 6.4) (Perrin et al., 1979; Ehlers et al., 1991). This was subsequently covered in northeast Norfolk by a re-advance of the SIS.

Over the past 10 years, the robustness of the regional lithostratigraphy—especially the complex relationship between the 'North Sea Drift' and Lowestoft formations in northern East Anglia—has been challenged by new geological mapping, sedimentological and stratigraphical studies and subsurface modelling of borehole data. Within northern East Anglia, the local stratigraphical schemes of Reid (1882), Banham (1968) and Lunkka (1994) developed largely on coastal and occasional inland sections are robust. However, when attempting to map this stratigraphical scheme inland, it became clear that the geometric relationship between the North Sea Drift and Lowestoft formations as defined by the conventional stratigraphical scheme (i.e. Fig. 6.3A) was erroneous. Common artefacts of mapping to this scheme included various stratigraphical spirals of tills and unit geometries that were geologically implausible.

Based upon new field data collected across several thousand square kilometres of northern East Anglia, a new lithostratigraphy was presented for the region (Hamblin et al., 2000; Lee et al., 2004b) and later extended more widely (Hamblin et al., 2005) across into the Midlands (Clark et al., 2004; Lee et al., 2008a; Rose, 2009) (Fig. 6.2B). The revised glacial succession consists of four broad mappable stratigraphical subdivisions. The lowest subdivision is the Happisburgh Formation of northern East Anglia which includes two tills—the Happisburgh and Corton tills, formerly attributed to the First Cromer Till. The second subdivision is the redefined Lowestoft Formation with some elements of the Wolston Formation.

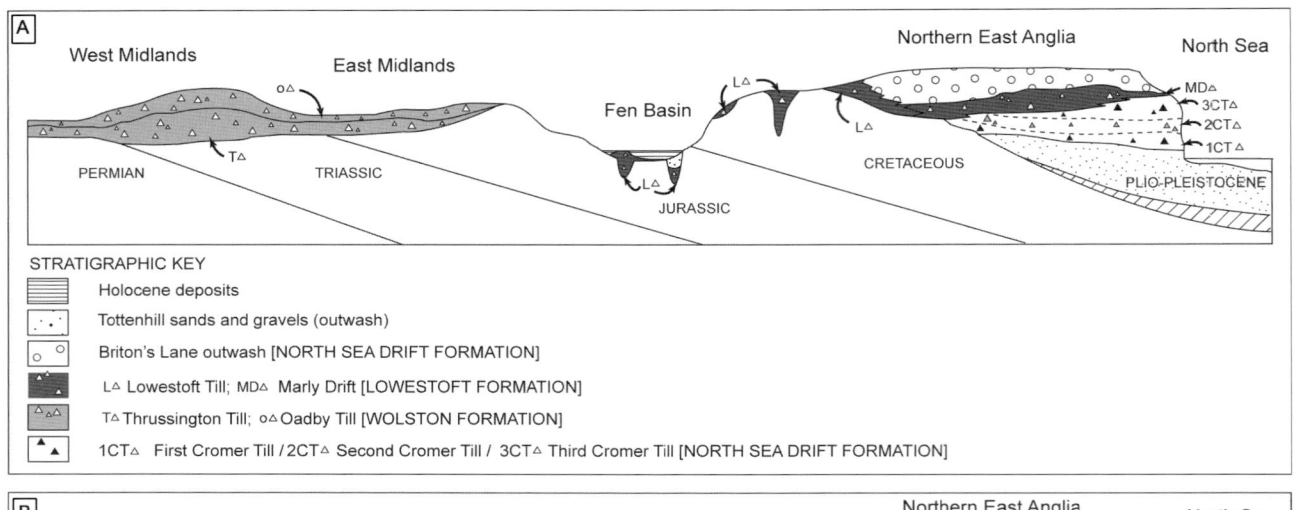

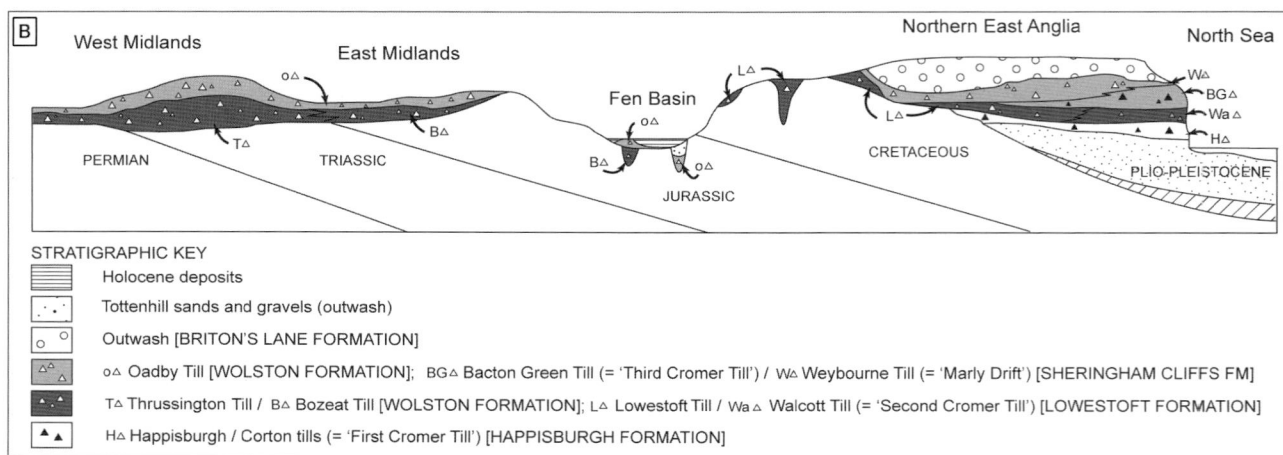

FIGURE. 6.3 Schematic cross-section across the Midlands and East Anglia showing the distribution, geometry and stratigraphy of major till units. (A) The 'conventional' interpretation of the glacial succession showing the interdigitation of the Lowestoft and 'North Sea Drift' in northern East Anglia after Bowen et al. (1986), Ehlers and Gibbard (1991) and Bowen (1999). (B) The 'new' interpretation of the glacial succession based on extensive geological mapping, borehole evidence and lithological analyses—after Lee et al. (2004a), Hamblin et al. (2005), Rose (2009).

It consists of a laterally persistent till sheet that can be traced from the Midlands into central and northern East Anglia to the Midlands and includes: (a) the Thrussington Till rich in Carboniferous, Permian and Triassic materials (Shotton, 1953; Rose, 1989a); (b) a Lias and Early-Middle Jurassic-rich/chalk-free till (Rose, 1992) now called the Bozeat Till (Barron et al., 2006; Carney and Ambrose, 2007); (c) a Late Jurassic (Kimmeridgian)-rich till with chalk clasts—the Lowestoft Till (*s.s.*) (Pointon, 1978; Perrin et al., 1979) and local equivalents in southern East Anglia (Allen et al., 1991); (d) the Walcott Till (Second Cromer Till) of northern East Anglia (Banham, 1968; Lunkka, 1994). The third stratigraphical subdivision in northern East Anglia is the Sheringham Cliffs Formation which includes two till facies, the Bacton Green Till (Third Cromer Till) and the Weybourne Town Till (Marly Drift) (Banham, 1968; Perrin et al., 1979; Ehlers et al., 1987; Lunkka, 1994). It is believed, on the basis of lateral continuity, that this formation is equivalent to the Oadby Till of the Wolston Formation of the Midlands. The fourth, and stratigraphically highest subdivision is the Briton's Lane Formation of northern East Anglia. This formation consists of a series of outwash sands and gravels deposited during a series of ice-marginal oscillations that formed the Cromer Ridge push moraine complex (Hamblin et al., 2005; Rose, 2009). Whilst, the dominant clast lithologies within the outwash sands and gravels are derived from northern Britain, they also contain a very small (<2%) but persistent clast assemblage derived from Scandinavia (Moorlock et al., 2000; Pawley et al., 2004). On the northern ice proximal side of the Cromer Ridge are a further series of outwash deposits and landforms formed during several phases of ice-marginal standstill and retreat including the Kelling and Salthouse outwash plains (Pawley, 2006), the Blakeney Esker (Gray, 1997; Gale and Hoare, 2007) and some possible kame terraces within the Glaven Valley.

6.2.7. Till Provenance

Whilst many of the tills within the stratigraphical succession of the Midlands and East Anglia are of British provenance, it is widely considered that other tills within the region were deposited by the SIS, due to the reported presence of Norwegian erratics from the Oslofjord area of Norway. This includes tills previously attributed to the 'North Sea Drift' (i.e. 'Cromer Tills') in northern East Anglia (Perrin et al., 1979; Bowen et al., 1986; Ehlers and Gibbard, 1991), the Basement Till of East Yorkshire (Catt and Penny, 1966) and the Warren House Till further north in County Durham (Trechmann, 1916; Catt, 2007).

In northern East Anglia, despite claims that '... [rhomb porphyry and larvikite] are found at all exposures of North Sea Drift Cromer Tills...' (Ehlers and Gibbard, 1991: 18), very few *in situ* Scandinavian erratics have actually been reported from the tills (Moorlock et al., 2001; Hoare and Connell, 2005; Lee et al., 2005). Recent analyses have demonstrated that the tills are actually derived from bedrock sources in northern and eastern Britain eroded by the BIS (Lee et al., 2002, 2004b) with minor reworking of erratics derived from earlier incursions of the SIS into the North Sea (Hoare and Connell, 2005; Lee et al., 2006). Whilst some argue that these tills are still of Scandinavian origin (i.e. Gibbard et al., 2008), ice flow trajectories for the SIS required to incorporate such a geographical range of lithologies from southern Norway, central Scotland and northern and eastern England are glaciologically implausible (Lee et al., 2008b). It also overlooks extensive evidence for the configuration of the Norwegian sector of the SIS during repeated Pleistocene expansions into the North Sea (Sejrup et al., 2000).

The removal of the SIS from northern East Anglia raises considerable questions regarding the configuration of the SIS and BIS during the Middle Pleistocene as it is unclear how far west the SIS extended. For instance, several conceptual models require the SIS to be in close proximity to the British land mass to deflect the BIS through the Wash Basin (i.e. Perrin et al., 1979; Fish and Whiteman, 2001) (Fig. 6.4). Equally, the provenance of other possible

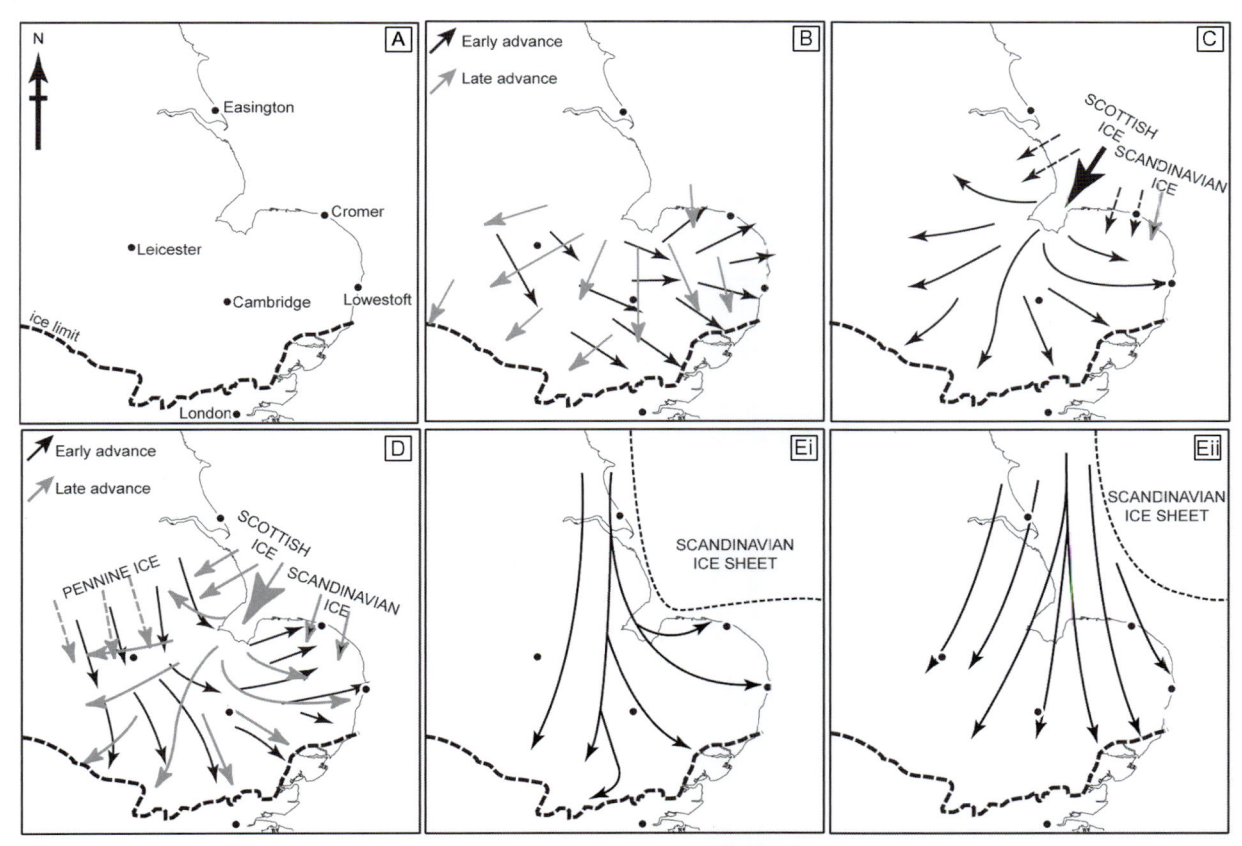

FIGURE. 6.4 Models for the Middle Pleistocene glaciation in the Midlands and Eastern England. (A) Location map. (B) Ice flow model of West and Donner (1956): this model recognises an initial 'Anglian' advance that deposits the Jurassic-rich Lowestoft Till and a later 'Gipping/Wolstonian' advance that deposits the chalky Gipping Till. (C) Ice flow model of Perrin et al. (1979): recognises a single sheet of chalky till including the Calcethorpe Till of Lincolnshire, and a single Lowestoft Till including Jurassic-rich and Chalk-rich facies. (D) Ice flow model of Rose (1992): the 'conventional' model for the Anglian glaciation that recognises an earlier ice advance from the west depositing the Triassic-rich (Thrussington Till) and Jurassic-rich (Lowestoft Till), followed by a later advance and the deposition of the Oadby Till and chalky tills of East Anglia. (E) Ice flow model of Fish and Whiteman (2001): this model recognises two separate ice advances within chalk-bearing tills including an earlier (i) and later (ii) advances.

Scandinavian deposits, including those at Warren House Till in County Durham (Davies et al., 2010), has been discredited, whilst others (e.g. the Basement Till along the Holderness East Yorkshire coast) require verification.

6.2.8. Chronology of the Middle Pleistocene Terrestrial Glacial Record

Whilst the Middle Pleistocene glacial succession presented earlier represents a standalone relative arrangement of the geological units, application of chronology to this succession has courted major controversy with conflicts existing between different types of chronological evidence (see Clark et al., 2004; Hamblin et al., 2005; Hoare et al., 2009; Preece and Parfitt, 2008; Preece et al., 2009 and Rose, 2009 for overviews; Figs. 6.4 and 6.5.

The chronological cornerstone of the Middle Pleistocene terrestrial glacial record is the Anglian Glaciation (MIS 12) (Mitchell et al., 1973; Bowen et al., 1986; Bowen, 1999). Deposits of this age include the Lowestoft Formation of East Anglia and the Bozeat and Thrussington tills of the Wolston Formation in the English Midlands. The age of these tills are constrained by interglacial organic deposits that underlie and overlie them and attributed to the 'Cromerian Complex' and Hoxnian, respectively (West, 1956, 1980; Turner, 1970; Rose, 1989a; Preece et al., 2007; Preece and Parfitt, 2008). Further evidence is the lithostratigraphical relationship of Anglian till to terraces of the River Thames (Gibbard, 1977; Bridgland, 1994); and the post-Anglian Thames terrace chronology based upon temporal patterns of river terrace aggradation and incision, FAD and LAD mammal and molluscan biostratigraphy, and amino-acid racemisation (AAR) (Bowen et al., 1989; Bridgland, 1994; Bridgland and Schreve, 2004). Together,

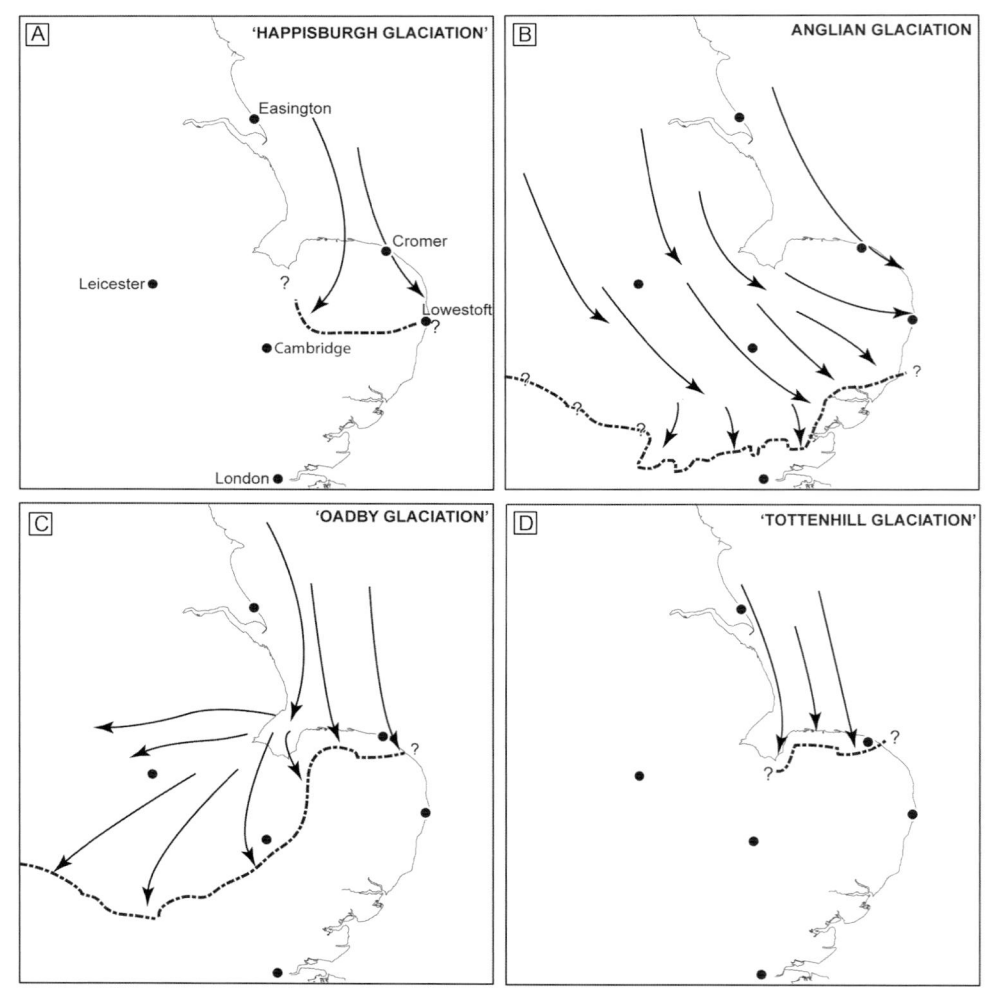

FIGURE. 6.5 Ice flow models based on the four-stage glaciation model after Clark et al. (2004), Hamblin et al. (2005) and Rose (2009). (A) Happisburgh Glaciation. (B) Anglian Glaciation. (C) Oadby Glaciation. (D) Tottenhill Glaciation. Note that the ice flow paths and ice limits are reconstructed on the basis of the distribution of tills and their lithological and provenance properties.

this body of evidence for dating the Anglian Glaciation to MIS 12 must be seen as substantive. However, it needs to be recalled that the original correlation was based upon pollen assemblage biostratigraphy (see Rose, 1989a), and this method has limited chronostratigraphical sensitivity (Tzedakis et al., 2001; Rose, 2009). Further, both MIS 11 and MIS 9 absolute age determinations have been made for organic sediments containing pollen assemblages attributed to the Hoxnian (Bowen et al., 1989; Rowe et al., 1997, 1999; Grün and Schwarz, 2000; Geyh and Müller, 2005; Preece et al., 2007). A systematic dating study of the glaciofluvial deposits in the area using OSL (Pawley et al., 2008) gave a mean age for all the deposits listed above to around MIS 12. It is important to note that this study did not examine material ascribed to MIS 16.

The timing of the deposition of the Happisburgh Formation glacial sediments in northern East Anglia is controversial. Conventional models, based largely upon pollen biostratigraphy, imply that the Happisburgh Formation is of Anglian age (Perrin et al., 1979; West, 1980; Bowen et al., 1986; Bowen, 1999; Gibbard et al., 2008). However, a growing body of evidence demonstrates the existence of a significant time gap and temperate climate(s) between the deposition of the Happisburgh and Lowestoft formations. Evidence includes (a) sedimentological evidence for intervening high sea-level stands at Pakefield (Lee et al., 2006) and Norwich (Read et al., 2007; see Gibbard et al., 2008 and Lee et al., 2008a,b for a debate of this evidence); (b) temperate soil separating till units of the Happisburgh and Lowestoft formations in central East Anglia (Rose, 2009); (c) a further intervening soil that also contains a human artefact (Wymer, 1985). This evidence precludes the Happisburgh Formation from being the product of the same glacial event that deposited the Lowestoft Till, and instead, points to an earlier glacial episode possibly equivalent to MIS 16 (Lee et al., 2004b; Hamblin et al., 2005; Rose, 2009). The rationale for this age is based upon the input of Happisburgh Formation glacial lithologies into the third terrace aggradation of the Bytham River at Leet Hill (Rose et al., 1999b, 2000; Lee, 2001; Lee et al., 2004b). The Bytham River catchment was subsequently overridden and destroyed during the Anglian Glaciation (Rose, 1994). However, between the third terrace and the youngest (first) terrace aggradation that immediately preceded the arrival of Anglian ice within the Bytham catchment (Rose, 1989b), several episodes of terrace aggradation and incision may be delineated. Application of a climate-forced model for river terrace development, where each aggradation/incision cycle represents a ca. 100 ka time-span (c.f. Bridgland, 2000), enables the various stages of aggradation and incision to be counted back from the Anglian placing the 'Happisburgh Glaciation' within MIS 16 (Lee et al., 2004b).

Several recent publications disagree with this evidence and argue for the 'Happisburgh Glaciation' to be still part of the Anglian (MIS 12). Preece and Parfitt (2008), for example, suggest that the glacial input into the Bytham River at Leet Hill generated a 'complex response' within the terrace sequence. Rose (2009) refutes this claim, demonstrating that the 'complex response' occurred later during the deposition of overlying glacial outwash. In a new interpretation of the Bytham terrace sequence, Westaway (2009) presents a terrace model comprising just three terrace aggradations, with the Happisburgh Formation glacial input occurring during the lowest (first/early Anglian) terrace aggradation (cf. Lee et al., 2004b). However, the validity of this terrace model is questionable because individual aggradations are largely unconstrained by palaeosol evidence that is necessary to demonstrate that the projected terrace surfaces correspond to recognisable land surfaces (cf. Lewis, 1993; Lee et al., 2004a). In a different study, Preece et al. (2009) examines the biostratigraphical and aminostratigraphical record from a temperate deposit of the Cromer Forest-bed Formation, the Sidestrand *Unio* Bed, at Sidestrand in northern East Anglia. Critically, the deposit underlies the Happisburgh Formation which would suggest, if the glacial deposits are MIS 16, that its youngest age would be MIS 17. However, the deposit contains the microtine rodent *Arvicola t. Cantiana* (Preece and Parfitt, 2000) which is widely believed to have evolved from *Mimomys savini* and is believe to have first appeared across Europe during MIS 15 (Banham et al., 2001; Preece et al., 2009). Amino acid data from freshwater shells also reveal racemisation rates that are more consistent with younger 'Cromerian Complex' deposits—probably equivalent to either MIS 13 or 15 (Preece et al., 2009). Interestingly, although Preece et al. (2009) argue that this evidence disproves the concept of an MIS 16 glaciation, it does not disprove a pre-Anglian age for the Happisburgh Formation. Further, as the LAD biostratigraphy is empirically derived, the lithostratigraphic interpretation could invalidate the use of this method at this period of time in Britain.

The ages of stratigraphically higher deposits (Oadby Fm, Sheringham Cliffs Fm, Briton's Lane Fm) within the glacial succession are also contentious and highlight further contradictions between different types of chronological evidence (Rose, 2009). Whilst conventional interpretations of their ages imply an Anglian (MIS 12) age (Perrin et al., 1979; Bowen et al., 1986; Banham et al., 2001), other interpretations suggest that they may relate additionally, to further glaciations during MIS 10 and 6 (Rose in Clark et al., 2004; Hamblin et al., 2005; Rose, 2009). In the Nar Valley, organic deposits dated to MIS 9 by U-Series (Rowe et al., 1997) and AAR (Scourse et al., 1999) geochronology form part of a continuum with the Oadby Till suggesting that the till is of MIS 10 age (Ventris, 1996; Scourse et al., 1999; Hamblin et al., 2005). Elsewhere, an MIS 10 age for the Oadby Till is also suggested by the relationship between the till and the terrace aggradations within the Upper Thames (Sumbler, 1995, 2001), and the lower number of terrace aggradations within the river catchments of the Midlands compared to those of the Thames below the Goring Gap (Keen, 1999). Additionally, organic material and a temperate palaeosol have been found in the sequence between a chalk-free till (probably the Anglian-age Bozeat Till) from the chalky Oadby Till in the East Midlands (Rose, 2009).

It has been suggested that the equivalent to the Oadby Till in northern East Anglia is the Sheringham Cliffs Formation, and that this may also equate to MIS 10 (Clark et al., 2004; Hamblin et al., 2005; Rose, 2009). Overlying the Sheringham Cliffs Formation in northern East Anglia is the Briton's Lane Formation. It has been argued by Rose in Clark et al. (2004), Hamblin et al. (2005) and Rose (2009) that the Briton's Lane Formation may correlate with a MIS 6 glaciation for several reasons. First, that the apparent freshness of the Cromer Ridge push moraine and Glaven Valley ice-marginal landforms is akin to the MIS 6 landforms of the Netherlands and north Germany and quite unlike heavily degraded MIS 12 landforms, or any other landforms beyond the MIS 2 ice limit in Britain. Second, an inferred correlation with MIS 6 deposits at Tottenhill (i.e. Gibbard et al., 1991b, 1992, 2009; Lewis and Rose, 1991), and pre-Ipswichian tills in East Yorkshire (Basement Till) and County Durham (Warren House Till) that contain similar reported elevations of Scandinavian lithologies.

The concept of MIS 10 and 6 glaciations in northern East Anglia has been disputed by several different lines of evidence that instead, suggest that the Sheringham Cliffs and Briton's Lane formations belong to the Anglian Glaciation. First, that the apparent freshness of the Cromer Ridge and other ice-marginal landforms could be a function of their sedimentology and later modification by periglacial processes (Banham et al., 2001; Gale and Hoare, 2007). Second, that the push moraine was formed during the Anglian based upon: (a) biostratigraphical and aminostratigraphical age constraint of a Hoxnian-age (MIS 11) kettle-hole infill (Hart and Peglar, 1990; Preece et al., 2009); (b) examination of the polyphase deformation signature within the upper parts of the glaciogenic sequence within the moraine complex that suggest deposition and subsequent deformation within a single glaciation (Lee and Phillips, 2008; Phillips et al., 2008); (c) outwash sands and gravels that lie on top of the Cromer Ridge that are dated by OSL to MIS 12 (Pawley et al., 2008).

6.3. DISCUSSION

6.3.1. Stratigraphical and Correlation Problems

The evidence for the Early and Middle Pleistocene glacial history of the British Isles has been presented within previous sections. It is evident that ice has been active in Britain for 2.6 million years, and that the scale and frequency of glaciation has evolved progressively in response to global-scale climatic trends. It is equally evident that major problems of correlation and chronology still exist as they do elsewhere within Europe (i.e. Ehlers and Gibbard, 2004, 2007). This is largely due to a lack of join-up between the onshore and offshore records, the often limited precision and resolution of dating methods beyond the Late Pleistocene, and the availability of suitable materials for dating (see Rose, 2009).

6.3.2. Glaciations in Britain

Despite the limitations highlighted above, and acknowledging that there are major issues of chronological and stratigraphical debate that require resolving, it is possible to gain a broad overall understanding of how glaciation has evolved in Britain during the Early and Middle Pleistocene and this is outlined in the following sections (Fig. 6.6).

Dropstones within sediments on the Hebrides Shelf provide the first evidence for the inception of the Scottish sector of the BIS at the beginning of the Pleistocene (c. 2.6 Ma) (Stoker et al., 1994). The first evidence for ice within Wales occurs tentatively at c. 1.8 Ma (MIS 68), with erratic input into the oldest Sudbury Formation river terrace of the Kesgrave Thames (Whiteman and Rose, 1992). It may tentatively be correlated with the 'Baventian' cold stage where subarctic conditions prevailed in East Anglia and the North Sea region (West et al., 1980; Gibbard et al., 1991a). Several further mountain-scale glaciations occurred at various times in Wales during the Early Pleistocene (Whiteman and Rose, 1992; Rose et al., 2010)).

Lowland glaciation associated with Scottish ice appears to have occurred at c. 1.1 Ma (MIS 34) with deposition of till (Sejrup et al., 1987) and outwash (Ekman, 1999) within the North Sea (Fig. 6.6). Several further expansions of Scottish into lowland and coastal areas of the North Sea occurred during the Jaramillo Subchron (c. 1.07–0.99 Ma), and during cold stages within the post-B/M 'Cromerian Complex' (c. MIS 18). Icebergs calving from these glacier incursions into the North Sea possibly floated southwards before becoming grounded in the present area of East Anglia shedding the erratics found within the Wroxham Crag (Larkin et al., 2011). The first lowland expansion of Welsh ice may relate to the beheading and truncation of the Thames catchment during the 'Cromerian Complex' at c. 0.8 Ma (MIS 20); with possible lowland glaciation of the Cotswold Hills which deposited a diamicton in that region and an increased influx of far-travelled material into the Ardleigh aggradation of the River Thames during MIS 18 (Whiteman and Rose, 1992).

Terrestrial evidence for the first lowland glaciation to reach eastern England may correspond with the Happisburgh Glaciation of northern East Anglia (Fig. 6.6). It is considered by Lee et al. (2004a) to correlate with MIS 16,

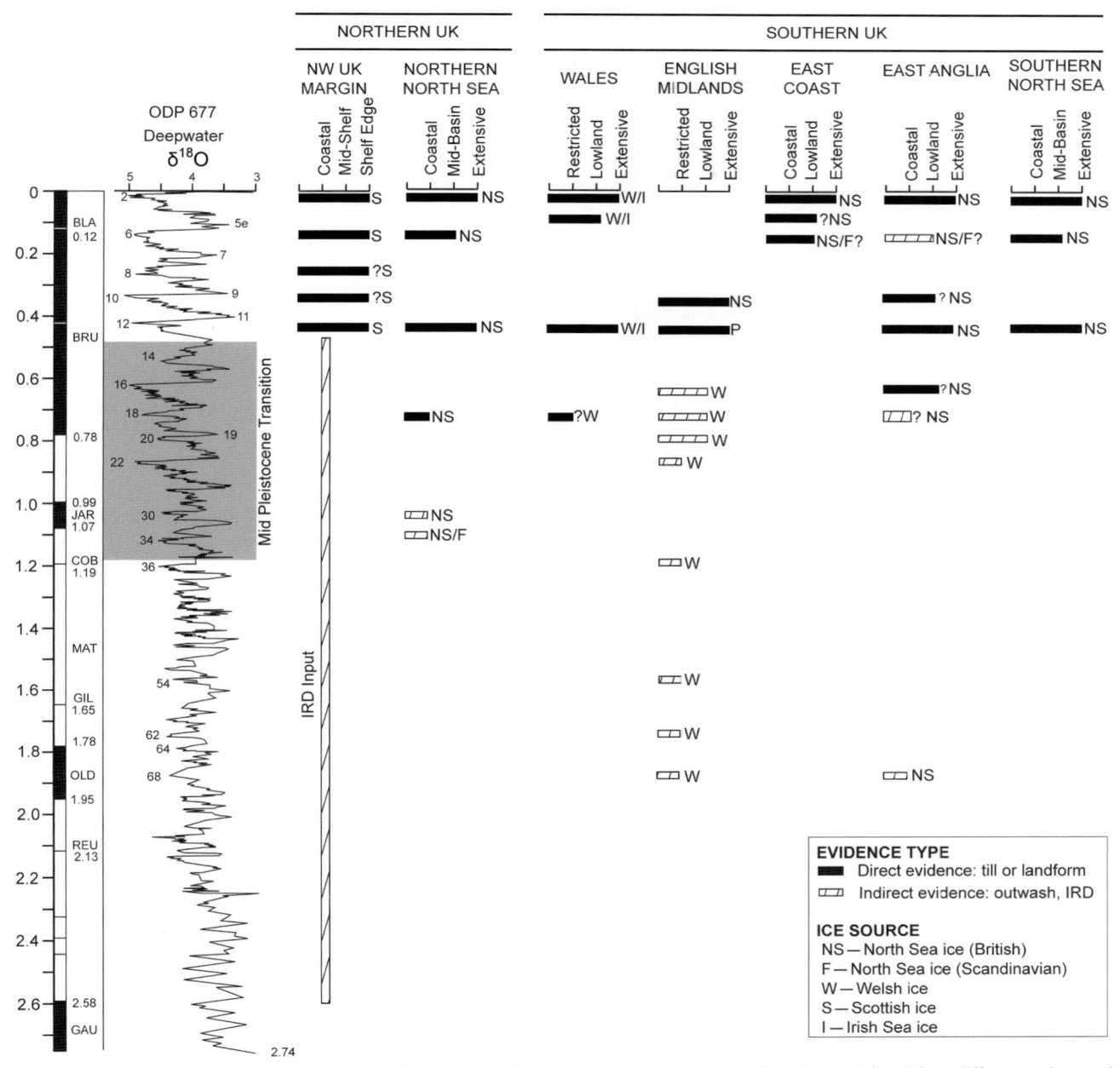

FIGURE. 6.6 Quaternary glacial history of the British Isles with evidence for the scale and sources of glaciations inferred from different onshore and offshore areas. Based upon data and evidence from Bowen et al. (1986), Whiteman and Rose (1992), Stoker et al. (1994), Sejrup et al. (2000), Westaway et al. (2002), Carr (2004), Clark et al. (2004), Hamblin et al. (2005), Catt (2007) and Hibbert et al. (2010).

although others still consider it to be of Anglian (MIS 12) age (Banham et al., 2001; Preece et al., 2009). The most extensive glaciation to affect the British Isles was the Anglian Glaciation of the late Middle Pleistocene (MIS 12). Evidence for this consists of tills, outwash deposits and widespread glacial erosion surfaces throughout both the terrestrial and offshore records (Perrin et al., 1979; Cameron et al., 1992; Fish and Whiteman, 2001; Hamblin et al., 2005). This glaciation also represents the first 'shelf-edge' expansion of the BIS along the northwest European margin (Stoker et al., 1994). The history and extent of the BIS during other late Middle Pleistocene cold stages is vague and open to several different interpretations from the same evidence. Current debate surrounds attempts at trying to establish a reliable chronology for glacial deposits within this part of the stratigraphical record, and whether there are just two Middle Pleistocene glaciations during MIS 12 and 6 four separate glacier expansions during MIS 16, 12, 10 and 6.

6.3.3. Regional and Glaciological Context

Published evidence cited within this review demonstrates a long Quaternary record of glaciation within Britain. The initiation of glaciation within Britain coincides generally with the onset of Northern Hemisphere Glaciation between 3.6 and 2.4 Ma (Mundelsee and Raymo, 2005). Inception of several of the high-latitude polar ice sheets, including the Greenland and Barents Sea ice sheets (Jansen et al., 2000; Knies et al., 2009), occurs slightly earlier during the Late Pliocene. However, major expansions of ice volume within these ice sheets, and the neighbouring Laurentide Ice Sheet, occur at the beginning of the Quaternary (Flesche Kleiven et al., 2002), and these appear to be coeval with the inception of mid-latitude ice sheets in Europe such as the SIS (Jansen et al., 2000; Sejrup et al., 2005) and of course the BIS.

Within Britain, this review also demonstrates that the scale of glaciation progressively stepped-up during the Early and Middle Pleistocene via a series of steps: (Step 1) glacier inception at c. 2.6 Ma and restricted mountain-scale glaciation during the Early Pleistocene; (Step 2) the initiation of lowland glaciation and first incursion of ice into the North Sea during the late Early Pleistocene (c. MIS 34); (Step 3) the first shelf-edge glaciation during the Anglian MIS 12. Interestingly, these 'steps' are broadly synchronous with major increases in the scale of glaciation identified in Scandinavia, and similarities between the two areas are discussed further by Lee et al. (2010). It appears that the principal trigger driving steps '2' and '3' may well have been the MPT and the progressive amplification of the 100 ka 'eccentricity' global climate signal that occurred between 1.2 and 0.7 Ma (Clark et al., 2006). For example, the first expansion of British ice into the North Sea Basin occurs c. 0.1 Ma after the beginning of this transition (1.2 Ma). However, the first extensive expansion of the BIS during MIS 12 demonstrates a lag of c. 0.35 Ma from the end of the MPT. It suggests that while Northern Hemisphere climate may have been modulated by 100 ka forcing, it took a further 0.35 Ma for this forcing to drive the build-up of ice volume within the BIS to sufficient levels to induce its first shelf-edge/continental-scale glaciation.

Within the cited records, it is also possible that there are regional differences between different ice accumulation areas in Britain although it is not clear whether this is real or an artefact of availability of evidence. For example, evidence suggests that the first mountain glaciations in Scotland may have occurred c. 0.8 Ma earlier than in Wales, and that the first significant lowland expansion of ice from Scotland occurred c. 0.3 Ma earlier than in Wales. These differences may involve their respective ability to build-up and retain ice volume over short and long periods of time relative to either the different uplift histories of the two areas, or significant regional differences in elevation/precipitation/relative position to the polar front. Equally, differences in the timing of the first lowland expansion may reflect differences in the nature of the subglacial bed material and topography that partly control the lateral expansion of ice.

6.4. CONCLUSIONS

Britain has a long glacial history with evidence indicating a number of restricted and lowland glaciations during the Early and Middle Pleistocene prior to the Anglian Glaciation. The first evidence for glaciation relates to IRD layers on the Hebrides Shelf tentatively dated to 2.6 Ma.

Whilst the offshore record in Britain provides several lines of evidence for different episodes of glaciation during the Early and Middle Pleistocene, the chronology of these episodes should be viewed with caution as they are largely based on inferred correlations with the terrestrial record, and pollen assemblage biostratigraphy which has limited chronostratigraphical value.

Offshore evidence from the North Sea suggests that the first lowland expansion of Scottish ice occurred during the late Early Pleistocene at c. 1.1 Ma. The onset of lowland glaciations within southern Britain started during the 'Cromerian Complex'. Evidence for these events is linked to the possible beheading of the proto-Thames (MIS 20) and erratic input into the Ardleigh Terrace (possibly MIS 18).

The Anglian Glaciation (c. 0.43 Ma; MIS 12) marks the biggest Quaternary expansion of the BIS and had a marked effect on the landscape of Britain. It radically altered drainage and relief, and initiated the formation of the Straits of Dover which led Britain to be isolated from mainland Europe during various high sea-level stands during the Middle and Late Pleistocene.

The chronology of Middle Pleistocene expansions of the BIS is still contentious, with conflicts existing between different types of evidence, and in some cases, different interpretations of the same evidence.

Despite problems surrounding the precise chronology of expansions of the BIS, it is still possible to demonstrate that the long-term behaviour of glaciation in Britain shows strong links to orbital forcing.

ACKNOWLEDGEMENTS

The authors wish to thank many colleagues for their engaging thoughts and discussions on this subject especially Rodger Connell, Bethan Davies, Peter Hoare, Nigel Larkin, Simon Parfitt, Steven Pawley, Richard Preece, Peter Rowe, Hans Petter Sejrup, Martyn Stoker and Jenni Turner. BGS authors publish with the permission of the Executive Director of the British Geological Survey (NERC).

REFERENCES

Allen, P., Cheshire, D.A., Whiteman, C.A., 1991. The tills of southern East Anglia. In: Ehlers, J., Gibbard, P.L., Rose, J. (Eds.), Glacial Deposits in Great Britain and Ireland. Balkema, Rotterdam, pp. 255–278.

Balson, P.S., Jeffrey, D.H., 1991. The glacial sequence of the southern North Sea. In: Ehlers, J., Gibbard, P.L., Rose, J. (Eds.), Glacial Deposits of Great Britain and Ireland. Balkema, Rotterdam, pp. 245–254.

Banham, P.H., 1968. A preliminary note on the Pleistocene stratigraphy of north-east Norfolk. Proc. Geologists' Assoc. 79, 469–474.

Banham, P.H., Gibbard, P.L., Lunkka, J.P., Parfitt, S.A., Preece, R.C., Turner, C., 2001. A critical assessment of 'A New Glacial Stratigraphy for Eastern England'. Quatern. Newsl. 93, 251–258.

Barron, A.J.M., Morigi, A.N., Reeves, H.J., 2006. Geology of the Wellingborough District—a brief explanation of the geological map. In: Sheet Explanation of the British Geological Survey. British Geological Survey, Nottingham, 34 pp. Wellingborough (England and Wales).

Bateman, M.D., Catt, J.A., 1996. An absolute chronology for the raised beach and associated deposits at Sewerby, East Yorkshire, England. J. Quatern. Sci. 11, 389–395.

Bowen, D.Q. (Ed.), 1999. A Revised Correlation of Quaternary Deposits in the British Isles. Geological Society of London, Bath Special Report No. 23.

Bowen, D.Q., Rose, J., McCabe, A.M., Sutherland, D.G., 1986. Quaternary Glaciations in England, Ireland, Scotland, and Wales. Quatern. Sci. Rev. 5, 299–340.

Bowen, D.Q., Hughes, S., Sykes, G.A., Miller, G.H., 1989. Land-sea correlations in the Pleistocene based on isoleucine epimerization in non-marine molluscs. Nature 340, 49–51.

Bradwell, T., Stoker, M.S., Golledge, N.R., Wilson, C.K., Merritt, J.W., Long, D., et al., 2007. The northern sector of the last British Ice Sheet: maximum extent and demise. Earth Sci. Rev. 88, 207–226.

Bridgland, D.R., 1994. Quaternary of the Thames. Geological Conservation Review Series. Joint Nature Conservation Committee Chapman and Hall, London, pp. 441.

Bridgland, D.R., 2000. River terrace systems in north-west Europe: an archive of environmental change, uplift and early human occupation. Quatern. Sci. Rev. 19, 1293–1303.

Bridgland, D.R., Schreve, D.C., 2004. Quaternary lithostratigraphy and mammalian biostratigraphy of the Lower Thames terrace system, south-east England. Quaternaire 15, 29–40.

Busschers, F.S., Kasse, C., van Balen, R.T., Vandenberghe, J., Cohen, K.M., Weerts, H.J.T., et al., 2008. Late Pleistocene evolution of the Rhine-Meuse system in the southern North Sea Basin: imprints of climate change, sea-level oscillation and glacio-isostacy. Quatern. Sci. Rev. 26, 3216–3248.

Cameron, T.D.J., Crosby, A., Balson, P.S., Jeffrey, D.H., Lott, J.K., Bulat, J., et al., 1992. The Geology of the Southern North Sea. HMSO, London United Kingdom Offshore Regional Report, HMSO for the British Geological Survey.

Carney, J.N., Ambrose, K., 2007. Geology of the Leicester district—a brief explanation of the geological map. In: Sheet Explanation of the British Geological Survey. British Geological Survey, Nottingham Leicester (England and Wales).

Carr, S.J., 2004. The North Sea Basin. In: Ehlers, J., Gibbard, P.L. (Eds.), Quaternary Glaciations—Extent and Chronology, Part I: Europe. Elsevier Publishers, Amsterdam, pp. 261–270.

Catt, J.A., 2007. The Pleistocene glaciations of eastern Yorkshire: a review. Proc. Yorks. Geol. Soc. 56, 177–207.

Catt, J.A., Penny, L.F., 1966. The Pleistocene deposits of Holderness, East Yorkshire. Proc. Yorks. Geol. Soc. 35, 275–420.

Clark, C.D., Gibbard, P.L., Rose, J., 2004. Pleistocene glacial limits in England, Scotland and Wales. In: Ehlers, J., Gibbard, P.L. (Eds.), Quaternary Glaciations—Extent and Chronology, Part I: Europe. Amsterdam, Elsevier, pp. 47–82.

Clark, P.U., Archer, D., Pollard, D., Blum, J.D., Rial, J.A., Brovkin, V., et al., 2006. The Middle Pleistocene transition: characteristics, mechanisms, and implications for long-term changes in atmospheric pCO_2. Quatern. Sci. Rev. 25, 3150–3184.

Davies, B.J., Roberts, D.H., Bridgland, D.R., Ó Cofaigh, C., Riding, J.B., Demarchi, B., Penkman, K.E.H., Pawley, S.M., 2010. Timing and depositional environments of a Middle Pleistocene glaciation of north-east England: New evidence from Warren House Gill, County Durham. Quatern. Sci. Rev. in Press.

Ehlers, J., Gibbard, P.L., 1991. Anglian glacial deposits in Britain and the adjoining offshore regions. In: Ehlers, J., Gibbard, P.L., Rose, J. (Eds.), Glacial Deposits of Great Britain and Ireland. Balkema, Rotterdam, pp. 17–24.

Ehlers, J., Gibbard, P.L. (Eds.), 2004. Quaternary Glaciations—Extent and Chronology, Part I: Europe. Elsevier, Amsterdam, pp. 47–82.

Ehlers, J., Gibbard, P.L., 2007. The extent and chronology of Cenozoic Glaciation. Quatern. Int. 164–165, 6–20.

Ehlers, J., Gibbard, P.L., Whiteman, C.A., 1987. Recent investigations of the Marly Drift of northwest Norfolk, England. In: van der Meer, J.J.M. (Ed.), Tills and Glaciotectonics. Balkema, Rotterdam, pp. 39–54.

Ehlers, J., Gibbard, P.L., Whiteman, C.A., 1991. The glacial deposits of northwest Norfolk. In: Ehlers, J., Gibbard, P.L., Rose, J. (Eds.), Glacial Deposits of Great Britain and Ireland. Balkema, Rotterdam, pp. 223–232.

Ekman, S.R., 1999. Early Pleistocene pollen biostratigraphy in the central North Sea. Rev. Palaeobot. Palynol. 105, 171–182.

Fish, P.R., Whiteman, C.A., 2001. Chalk micropalaeontology and the provenancing of Middle Pleistocene Lowestoft formation till in eastern England. Earth Surf. Process. Land. 26, 953–970.

Flesche Kleiven, H., Jansen, E., Fronval, T., Smith, T.M., 2002. Intensification of Northern Hemisphere glaciations in the circum Atlantic region (3.5–2.4 Ma) - ice-rafted detritus evidence. Palaeogeo. Palaeoclimatol. Palaeoecol. 184, 213–223.

Funnell, B.M., 1995. Global sea-level and the (pen-) insularity of late Cenozoic Britain. In: Preece, R.C. (Ed.), Island Britain: A Quaternary Perspective, Geological Society Special Publication No 96, pp. 3–13.

Gale, S.J., Hoare, P.G., 2007. The age and origin of the Blakeney Esker of north Norfolk: implications for the glaciology of the southern North Sea Basin. In: Hambrey, M.J., Christoffersen, P., Glasser, N.F., Hubbard, B.P. (Eds.), Glacial Sedimentary Processes and Products, International Association of Sedimentologists Special Publication, 39 203–234.

Gatliff, R.W., Richards, P.C., Smith, K., Graham, C.C., McCormac, M., Smith, N.J.P., et al., 1994. The Geology of the Central North Sea. HMSO, London United Kingdom Offshore Regional Report, British Geological Survey.

Geyh, M.A., Müller, H., 2005. Numerical 230Th/U dating and a palynological review of the Holsteinian/Hoxnian Interglacial. Quatern. Sci. Rev. 24, 1861–1872.

Gibbard, P.L., 1977. Pleistocene history of the Vale of St Albans. Philos. Trans. R. Soc. Lond. B280, 445–483.

Gibbard, P.L., 1995. The formation of the Strait of Dover. In: Preece, R.C. (Ed.), Island Britain: A Quaternary Perspective, Geological Society, London, Special Publication, 96 15–26.

Gibbard, P.L., West, R.G., Zagwijn, W.H., Balson, P.S., Burger, A.W., Funnell, B.M., et al., 1991a. Early and early Middle Pleistocene correlations in the Southern North Sea Basin. Quatern. Sci. Rev. 10, 23–52.

Gibbard, P.L., West, R.G., Andrew, R., Pettit, M., 1991b. Tottenhill, Norfolk (TF 636115). In: Lewis, S.G., Whiteman, C.A., Bridgland, D.R. (Eds.), Central East Anglia and the Fen Basin—Field Guide. Quaternary Research Association, London, pp. 131–144.

Gibbard, P.L., West, R.G., Andrew, R., Pettit, M., 1992. The margin of a Middle Pleistocene ice advance at Tottenhill, Norfolk, England. Geol. Mag. 129, 59–76.

Gibbard, P.L., Moscariello, A., Bailey, H.W., Boreham, S., Kock, C., Lord, A.R., et al., 2008. Comment: Middle Pleistocene sedimentation at Pakefield, Suffolk, England. J. Quatern. Sci. 23, 85–92.

Gibbard, P.L., Pasanen, A.H., West, R.G., Lunkka, J.P., Boreham, S., Cohen, K., et al., 2009. Late Middle Pleistocene glaciation in East Anglia, England. Boreas 38, 504–528.

Gray, J.M., 1997. The origin of the Blakeney Esker, Norfolk. Proc. Geologists' Assoc. 108, 177–182.

Green, C.P., McGregor, D.F.M., 1990. Pleistocene gravels of the north Norfolk coast. Proc. Geologists' Assoc. 101, 197–202.

Grün, R., Schwarz, H.P., 2000. Revised open system U-series/ESR age calculations for teeth from Stratum C at the Hoxnian Interglacial type locality, England. Quatern. Sci. Rev. 19, 1151–1154.

Hamblin, R.J.O., Moorlock, B.S.P., Rose, J., 2000. A New Glacial Stratigraphy for Eastern England. Quatern. Newsl. 92, 35–43.

Hamblin, R.J.O., Moorlock, B.S.P., Rose, J., Lee, J.R., Riding, J.B., Booth, S.J., et al., 2005. Revised Pre-Devensian glacial stratigraphy in Norfolk, England, based on mapping and till provenance. Geol. Mijnbouw 84, 77–85.

Hart, J.K., Peglar, S.M., 1990. Further evidence for the timing of the Middle Pleistocene glaciation in Britain. Proc. Geologists' Assoc. 101, 187–196.

Head, M.J., Gibbard, P.L. (Eds.), 2005a. Early-Middle Pleistocene Transitions—The Land-Ocean Evidence, In: Geological Society, London, Special Publication, 247, 326 pp.

Head, M.J., Gibbard, P.L., 2005b. Early-Middle Pleistocene transition: an overview and recommendation for the defining boundary. In: Head, M. J., Gibbard, P.L. (Eds.), Early-Middle Pleistocene Transitions—The Land-Ocean Evidence, Geological Society, London, Special Publication vol. 247. 1–18.

Hey, R.W., 1980. Equivalent of the Westland green gravels in Essex and East Anglia. Proc. Geologists' Assoc. 91, 279–290.

Hey, R.W., Brenchley, P.J., 1977. Volcanic pebbles from the Pleistocene gravels in Norfolk and Essex. Geol. Mag. 114, 219–225.

Hibbert, F.D., Austin, W.E.N., Leng, M.J., Gatliff, R.W., 2009. British Ice Sheet dynamics inferred from North Atlantic ice-rafted debris records spanning the last 175,000 years. J. Quatern. Sci. 25, 461–482.

Hibbert, F.D., Austin, W.E.N., Leng, M.J., Gatliff, R.W., 2010. British Ice Sheet dynamics inferred from North Atlantic ice-rafted debris records spanning the last 175,000 years. J. Quatern. Sci. 25, 461–482.

Hoare, P.G., Connell, E.R., 2005. The first appearance of Scandinavian indicators in East Anglia's glacial record. Bull. Geol. Soc. Norfolk 54, 3–14.

Hoare, P.G., Gale, S.J., Robinson, R.A.J., Connell, E.R., Larkin, N.R., 2009. Marine Isotope Stage 7–6 transition age for the beach sediments at Morston, north Norfolk, UK: implications for Pleistocene chronology, stratigraphy and tectonics. J. Quatern. Sci. 24, 311–316.

Holmes, R., 1997. Quaternary stratigraphy: the offshore record. In: Gordon, J.E. (Ed.), Reflections on the Ice Age in Scotland: An Update on Quaternary Studies. The Scottish Association of Geography Teachers and Scottish Natural Heritage, Glasgow, pp. 72–94.

Holmes, R., Bulat, J., Hamilton, I., Long, D., 2003. Morphology of an ice-sheet limit and constructional glacially-fed slope front, Færoe-Shetland Channel. In: Mienart, J., Weaver, P. (Eds.), European Margin Sediment Dynamics: Side-Scan Sonar and Seismic Images. Springer-Verlag, Berlin, pp. 149–151.

Jansen, E., Fronval, T., Rack, F., Channell, J.E.T., 2000. Pliocene-Pleistocene ice rafting history and cyclicity in the Nordic Seas during the last 3.5 Myr. Paleoceanography 15, 709–721.

Keen, D.H., 1999. The chronology of Middle Pleistocene ('Wolstonian') events in the English Midlands. In: Andrews, P.J., P.H., Banham (Eds.), Late Cenozoic Environments and Hominid Evolution—A Tribute to Bill Bishop. Geological Society of London Special Publication, London, pp. 159–168.

Knies, J., Matthiessen, J., Vogt, C., Laberg, J.S., Hjelstuen, B.O., Smelror, M., et al., 2009. The Plio-Pleistocene glaciation of the Barents Sea-Svalbard region: a new model based on revised chronostratigraphy. Quatern. Sci. Rev. 28, 812–829.

Laban, C., 1995. The Pleistocene glaciations in the Dutch sector of the North Sea. A synthesis of sedimentary and seismic data. Ph.D. thesis, University of Amsterdam.

Larkin, N.R., Lee, J.R., Connell, E.R., 2011. Possible ice-rafted erratics in late Early to early Middle Pleistocene shallow marine and coastal deposits in northeast Norfolk, UK. Proce. Geologists' Assoc. in press.

Lee, J.R., 2001. Genesis and palaeogeographical significance of the Corton Diamicton (basal member of the North Sea Drift Formation), East Anglia, UK. Proc. Geologists' Assoc. 112, 29–43.

Lee, J.R., 2009. Patterns of preglacial sedimentation and glaciotectonic deformation within early Middle Pleistocene sediments at Sidestrand, north Norfolk, UK. Proc. Geologists' Assoc. 120, 34–48.

Lee, J.R., Phillips, E., 2008. Progressive soft sediment deformation within a subglacial shear zone—a hybrid mosaic-pervasive deformation model for Middle Pleistocene glaciotectonised sediments from Eastern England. Quatern. Sci. Rev. 27, 1350–1362.

Lee, J.R., Rose, J., Riding, J.B., Hamblin, R.J.O., Moorlock, B.S.P., 2002. Testing the case for a Middle Pleistocene Scandinavian glaciation in Eastern England: evidence for a Scottish ice source for tills within the Corton Formation of East Anglia, UK. Boreas 31, 345–355.

Lee, J.R., Rose, J., Hamblin, R.J.O., Moorlock, B.S.P., 2004a. Dating the earliest lowland glaciation of eastern England: the pre-Anglian early Middle Pleistocene Happisburgh Glaciation. Quatern. Sci. Rev. 23, 1551–1566.

Lee, J.R., Booth, S.J., Hamblin, R.J.O., Jarrow, A.M., Kessler, H., Moorlock, B.S.P., et al., 2004b. A new stratigraphy for the glacial deposits around Lowestoft, Great Yarmouth, North Walsham and Cromer, East Anglia, UK. Bull. Geol. Soc. Norfolk 53, 3–60.

Lee, J.R., Moorlock, B.S.P., Rose, J., Hamblin, R.J.O., Pawley, S.M., Riding, J.B., et al., 2005. A reply to the paper by Hoare and Connell. "The first appearance of Scandinavian indicators in East Anglia's glacial record" Bull. Geol. Soc. Norfolk 55, 43–62.

Lee, J.R., Rose, J., Candy, I., Barendregt, R.W., 2006. Sea-level changes, river activity, soil development and glaciation around the western margins of the southern North Sea Basin during the Early and early Middle

Pleistocene: evidence from Pakefield, Suffolk, UK. J. Quatern. Sci. 21, 155–179.

Lee, J.R., Pawley, S.M., Rose, J., Moorlock, B.S.P., Riding, J.B., Hamblin, R.J.O., et al., 2008a. Pre-Devensian lithostratigraphy of shallow marine, fluvial and glacial sediments in northern East Anglia. In: Candy, I., Lee, J.R., Harrison, A.M. (Eds.), The Quaternary of Northern East Anglia Field Guide. Quaternary Research Association, London, pp. 42–59.

Lee, J.R., Rose, J., Candy, I., Barendregt, R.W., Moorlock, B.S.P., Riding, J.B., et al., 2008b. Reply: Middle Pleistocene sedimentation at Pakefield, Suffolk, UK. J. Quatern. Sci. 23, 93–98.

Lee, J.R., Busschers, F., Sejrup, H.P., 2010. Pre-Weichselian Quaternary glaciations of the British Isles, The Netherlands, Norway and adjacent marine areas south of 68°N: implications for long-term ice sheet development in northern Europe. Quatern. Sci. Rev. in press.

Lewis, S.G., 1993. The status of the Wolstonian Glaciation in the English Midlands and East Anglia. Unpublished Ph.D. thesis, University of London.

Lewis, S.G., Rose, J., 1991. Tottenhill, Norfolk (TF 639 120). In: Lewis, S. G., Whiteman, C.A., Bridgland, D.R. (Eds.), Central East Anglia and the Fen Basin, Field Guide. Quaternary Research Association, London, pp. 145–148.

Long, D., Laban, C., Streif, H., Cameron, T.D.J., Schüttenhelm, R.T.E., 1988. The sedimentary record of climatic variation in the southern North Sea. Philos. Trans. R. Soc. Lond. B318, 523–537.

Lunkka, J.P., 1994. Sedimentation and lithostratigraphy of the North Sea Drift and Lowestoft Till Formations in the coastal cliffs of northeast Norfolk, England. J. Quatern. Sci. 9, 209–233.

McGregor, D.F.M., Green, C.P., 1978. Gravels of the River Thames as a guide to Pleistocene catchment changes. Boreas 7, 197–203.

Merritt, J.W., Auton, C.A., Connell, E.R., Hall, A.M., Peacock, J.D., Aitken, J.F., et al., 2003. Cainozoic Geology and Landscape Evolution of North-East Scotland: Memoir for the Drift Editions of 1:50,000 Geological Sheets 66E Banchory, 67 Stonehaven, 76E Inverurie, 77 Aberdeen, 86E Turriff, 87W Ellon, 87E Peterhead, 95 Elgin, 96W Portsoy, 96E Banff and 97 Fraserburgh. Memoir of the Geological Survey of Great Britain (Scotland)HMSO, London.

Mitchell, G.F., Penny, L.F., Shotton, F.W., West, R.G., 1973. A Correlation of Quaternary Deposits in the British Isles. Geological Society of London, London Special Report no. 4.

Moorlock, B.S.P., Booth, S.J., Fish, P., Hamblin, R.J.O., Kessler, H., Riding, J., et al., 2000. Briton's Lane Gravel pit, Beeston Regis (TG 168 415). In: Lewis, S.G., Whiteman, C.A., Preece, R.C. (Eds.), The Quaternary of Norfolk and Suffolk: Field Guide. Quaternary Research Association, London, pp. 115–117.

Moorlock, B.S.P., Hamblin, R.J.O., Rose, J., Lee, J.R., 2001. The occurrence of rhomb-porphyry within diamictons in north-east Norfolk. Quatern. Newsl. 95, 49–50.

Mundelsee, M., Raymo, M.E., 2005. Slow dynamics of the Northern Hemisphere Glaciation. Paleoceanography 20, PA4200.

Pawley, S.M., 2006. Quaternary glaciations of north and west Norfolk. Unpublished Ph.D. thesis, Royal Holloway, University of London.

Pawley, S.M., Rose, J., Lee, J.R., Hamblin, R.J.O., Moorlock, B.S.P., 2004. Middle Pleistocene stratigraphy of Weybourne, north-east Norfolk, England. Proc. Geologists' Assoc. 115, 22–42.

Pawley, S.M., Bailey, R.J.J., Rose, J., Moorlock, B.S.P., Hamblin, R.J.O., Booth, S.J., et al., 2008. Age limits on Middle Pleistocene glacial sediments from OSL dating, North Norfolk, UK. Quatern. Sci. Rev. 27, 1363–1377.

Perrin, R.M.S., Rose, J., Davies, H., 1979. The distribution, variation and origins of pre-Devensian tills in eastern England. Philos. Trans. R. Soc. Lond. B287, 535–570.

Phillips, E., Lee, J.R., Burke, H.K., 2008. Progressive proglacial to subglacial deformation and syntectonic sedimentation at the margins of the Mid-Pleistocene British Ice Sheet: evidence from north Norfolk, UK. Quatern. Sci. Rev. 27, 1848–1871.

Pointon, K.W., 1978. The Pleistocene succession at Corton, Suffolk. Bull. Geol. Soc. Norfolk 30, 55–76.

Praeg, D., 2003. Seismic imaging of mid-Pleistocene tunnel valleys in the North Sea Basin—high resolution from low frequencies. J. Appl. Geophys. 53, 273–298.

Preece, R.C., Parfitt, S.A., 2000. The Cromer Forest-bed formation: new thoughts on an old problem. In: Lewis, S.G., Whiteman, C.A., Preece, R.C. (Eds.), The Quaternary of Norfolk and Suffolk, Field Guide. Quaternary Research Association, London, pp. 1–27.

Preece, R.C., Parfitt, S.A., 2008. The Cromer Forest-bed formation: some recent developments relating to early human occupation and lowland glaciation. In: Candy, I., Lee, J.R., Harrison, A.M. (Eds.), The Quaternary of Northern East Anglia, Field Guide. Quaternary Research Association, London, pp. 60–83.

Preece, R.C., Parfitt, S.A., Bridgland, D.R., Lewis, S.G., Rowe, P.J., Atkinson, T.C., et al., 2007. Terrestrial environments during MIS 11: evidence from the Palaeolithic site at West Stow, Suffolk, UK. Quatern. Sci. Rev. 26, 1236–1300.

Preece, R.C., Parfitt, S.A., Coope, G.R., Penkman, K.E.H., Ponel, P., Whittaker, J.E., 2009. Biostratigraphic and aminostratigraphic constraints on the age of the Middle Pleistocene glacial succession in north Norfolk. J. Quatern. Sci. 24, 557–580.

Read, A., Godwin, M., Mills, C.A., Juby, C., Lee, J.R., Palmer, A.P., et al., 2007. Evidence for Middle Pleistocene temperate-climate high sea-level and lowland-scale glaciation, Chapel Hill, Norwich, UK. Proc. Geologists' Assoc. 118, 143–156.

Reid, C., 1882. The Geology of the Country Around Cromer. Memoirs of the Geological Survey of England and Wales.

Rice, R.J., 1968. The Quaternary deposits of central Leicestershire. Philos. Trans. R. Soc. Lond. A262, 459–509.

Rose, J., 1987. Status of the Wolstonian Glaciation in the British Quaternary. Quatern. Newsl. 53, 1–9.

Rose, J., 1989a. Stadial type sections in the British Quaternary. In: Rose, J., Schlüchter, C. (Eds.), Quaternary Type Sections: Imagination or Reality?. Balkema, Rotterdam, pp. 45–67.

Rose, J., 1989b. Castle Bytham. In: Keen, D.H. (Ed.), West Midlands Field Guide. Quaternary Research Association, Cambridge, pp. 117–122.

Rose, J., 1992. High lodge—regional context and geological background. In: Ashton, N.M., Cook, J., Lewis, S.G., Rose, J. (Eds.), High Lodge: Excavations by G. de G. Sieveking, 1962-8 and J. Cook, 1988. British Museum Press, London, pp. 13–24.

Rose, J., 1994. Major river systems of central and southern Britain during the Early and Middle Pleistocene. Terra Nova 6, 435–443.

Rose, J., 2009. Early and Middle Pleistocene landscapes of eastern England. Proc. Geologists' Assoc. 120, 3–33.

Rose, J., Whiteman, C.A., Allen, P., Kemp, R.A., 1999a. The Kesgrave sands and gravels: 'pre-glacial' Quaternary deposits of the River Thames in East Anglia and the Thames Valley. Proc. Geologists' Assoc. 110, 93–116.

Rose, J., Lee, J.A., Candy, I., Lewis, S.G., 1999b. Early and Middle Pleistocene river systems in eastern England: evidence from Leet Hill, southern Norfolk, England. J. Quatern. Sci. 14, 347–360.

Rose, J., Moorlock, B.S.P., Hamblin, R.J.O., 2000. Lithostratigraphy and palaeoenvironments of the pre-Anglian sands and gravels of East Anglia. In: Lewis, S.G., Whiteman, C.A., Preece, R.C. (Eds.), The Quaternary of Norfolk and Suffolk: Field Guide. Quaternary Research Association, London, pp. 35–45.

Rose, J., Moorlock, B.S.P., Hamblin, R.J.O., 2001. Pre-Anglian fluvial and coastal deposits in Eastern England: lithostratigraphy and Palaeoenvironments. Quatern. Int. 79, 5–22.

Rose, J., Carney, J.N., Silva, B.N., Booth, S.J., 2010. A striated, far travelled clast of rhyolitic tuff from Thames river deposits at Ardleigh, Essex, England: evidence for early Middle Pleistocene glaciation in the Thames catchment. Neth. J. Geosci. 89, 137–146.

Rowe, P.J., Richards, D.A., Atkinson, T.C., Bottrell, S.H., Cliff, R.A., 1997. Geochemistry and radiometric dating of a Middle Pleistocene peat. Geochim. Et Cosmochim. Acta 61, 4201–4211.

Rowe, P.J., Atkinson, T.C., Turner, C., 1999. U-series dating of Hoxnian interglacial deposits at Marks Tey, Essex, England. J. Quatern. Sci. 14, 693–702.

Scourse, J.D., Austin, W.E.N., Sejrup, H.P., Ansari, M.H., 1999. Foraminiferal isoleucine epimerization determinations from the Nar Valley Clay, Norfolk, UK: implications for quaternary correlations in the southern North Sea basin. Geol. Mag. 136, 543–560.

Sejrup, H.P., Aarseth, K.L., Ellingsen, K.L., Reither, E., Jansen, E., Løvlie, R., et al., 1987. Quaternary stratigraphy of the Fladen area, central North Sea: a multidisciplinary study. J. Quatern. Sci. 2, 33–58.

Sejrup, H.P., Larsen, E., Landvik, J., King, E.L., Haflidason, H., Nesje, A., 2000. Quaternary glaciations in southern Fennoscandia: evidence from southwestern Norway and the northern North Sea region. Quatern. Sci. Rev. 19, 667–685.

Sejrup, H.P., Hjelstuen, B.O., Dahlgren, K.I.T., Haflidason, H., Kuijpers, A., Nygård, A., et al., 2005. Pleistocene glacial history of the NW European continental margin. Mar. Petrol. Geol. 22, 1111–1129.

Shotton, F.W., 1953. The Pleistocene deposits of the area around Coventry, Rugby and Leamington and their bearing upon topographic development of the Midlands. Philos. Trans. R. Soc. Lond. 237B, 209–260.

Shotton, F.W., 1983. The Wolstonian Stage of the British Pleistocene in and around its type area of the English Midlands. Quatern. Sci. Rev. 2, 261–280.

Soloman, J.D., 1935. The Westleton Series of East Anglia: its ages, distribution and relations. Q. J. Geol. Soc. Lond. 91, 16–238.

Stoker, M.S., 1995. The influence of glacigenic sedimentation on slope-apron development on the continental margin off Northwest Britain. In: Scrutton, R.A., Stoker, M.S., Schimmield, G.B., Tudhope, A.W. (Eds.), The Tectonics, Sedimentation and Palaeooceanography of the North Atlantic Region. Geological Society Special Publications, London, pp. 159–177.

Stoker, M.S., Bent, A., 1985. Middle Pleistocene glacial and glaciomarine sedimentation in the west central North Sea. Boreas 14, 325–332.

Stoker, M.S., Hitchen, K., Graham, C.C., 1993. United Kingdom offshore regional report: the geology of the Hebrides and west Shetland shelves and adjacent deep-water areas. HMSO for the British Geological Survey, London.

Stoker, M.S., Leslie, A.B., Scott, W.D., Briden, J.C., Hine, N.M., Harland, R., et al., 1994. A record of late Cenozoic stratigraphy, sedimentation and climate change from the Hebrides Shelf, NE Atlantic Ocean. J. Geol. Soc. Lond. 151, 235–249.

Straw, A., 1983. Pre-Devensian glaciation of Lincolnshire (Eastern England) and adjacent areas. Quatern. Sci. Rev. 2, 239–260.

Sumbler, M.G., 1995. The terraces of the rivers Thame and Thames and their bearing on the chronology of glaciation in central and eastern England. Proc. Geologists' Assoc. 106, 93–106.

Sumbler, M.G., 2001. The Moreton Drift: a further clue to glacial chronology in central England. Proc. Geologists' Assoc. 112, 13–27.

Trechmann, C.T., 1916. The Scandinavian Drift of the Durham coast and the general glaciology of southeast Durham. Q. J. Geol. Soc. Lond. 71, 53–82.

Turner, C., 1970. Middle Pleistocene deposits at Marks Tey, Essex. Philos. Trans. R. Soc. Lond. B247, 373–437.

Tzedakis, P.C., Andrieu, V., de Beaulieu, J.-L., Birks, H.J.B., Crowhurst, S., Follieri, M., et al., 2001. Establishing a terrestrial chronological framework as a basis for biostratigraphical comparisons. Quatern. Sci. Rev. 20, 1583–1592.

Ventris, P.A., 1996. Hoxnian interglacial freshwater and marine deposits in northwest Norfolk, England and their implications for sea-level reconstruction. Quatern. Sci. Rev. 15, 437–450.

West, R.G., 1956. The Quaternary deposits at Hoxne, Suffolk. Philos. Trans. R. Soc. Lond. B239, 265–356.

West, R.G., 1980. The Pre-Glacial Pleistocene of the Norfolk and Suffolk Coasts. Cambridge University Press, Cambridge.

West, R.G., Donner, J.J., 1956. The glaciations of East Anglia and the East Midlands: a differentiation based on stone orientation measurements of the tills. Q. J. Geol. Soc. Lond. 112, 146–184.

West, R.G., Funnell, B.M., Norton, P.E.P., 1980. An Early Pleistocene cold marine episode in the North Sea: pollen and faunal assemblages at Covehithe, Suffolk, England. Boreas 9, 1–10.

Westaway, R., 2009. Quaternary vertical crustal motion and drainage evolution in East Anglia and adjoining parts of southern England: chronology of the Ingham River terrace deposits. Boreas 10, 261–284.

Westaway, R., Maddy, D., Bridgland, D., 2002. Flow in the lower continental crust as a mechanism for the Quaternary uplift of south-east England: constraints from the Thames record. Quatern. Sci. Rev. 21, 59–603.

Whiteman, C.A., 1992. The palaeogeography and correlation of pre-Anglian Glaciation terraces of the River Thames in Essex and the London Basin. Proc. Geologists' Assoc. 103, 37–56.

Whiteman, C.A., Rose, J., 1992. Thames river sediments of the British Early and Middle Pleistocene. Quatern. Sci. Rev. 11, 363–375.

Wymer, J.J., 1985. Palaeolithic Sites in East Anglia. Geobooks, Norwich.

Zagwijn, W.H., 1989. The Netherlands during the Tertiary and the Quaternary: a case history of Coastal Lowland evolution. Geol. Mijnbouw 68, 107–120.

Zagwijn, W.H., 1996. The Cromerian Complex Stage of the Netherlands and correlation with other areas in Europe. In: Turner, C. (Ed.), The Early Middle Pleistocene in Europe. Balkema, Rotterdam, pp. 145–172.

Chapter 7

Pleistocene Glaciation Limits in Great Britain

Philip L. Gibbard[1],* and Chris D. Clark[2]

[1]Cambridge Quaternary, Department of Geography, University of Cambridge, Downing Place, Cambridge CB2 3EN, United Kingdom
[2]Department of Geography, University of Sheffield, Sheffield S10 2TN, England
*Correspondence and requests for materials should be addressed to Philip L. Gibbard. E-mail: plg1@cam.ac.uk

7.1. INTRODUCTION

As an island off the coast of continental Europe, Great Britain is unique in its geographical setting by comparison to the neighbouring regions. Together with Ireland, Great Britain occupies a position where during interglacial periods it is bathed by the North Atlantic Drift that ensures cool temperate, maritime climates predominate. These conditions provide a stark contrast to those during the cold periods (glacials) when the migration of the polar front, as far south as the Iberian Peninsula, results in establishment of severe, arctic-type conditions in the British Isles. The consequences of this are the appearance of cold winters, the annual development of sea-ice, periglacial conditions on land and when a sufficient supply of precipitation is available, the build-up of glacial ice in mountain regions. The lowering of global sea level, leading to the linking of the British Isles to the Continent, further augments this development of cold-stage continentality. Whilst during warm periods, eustatic sea-level rise ensures the repeated separation of the islands.

The contrasts between the glacial versus interglacial conditions are so strongly marked that the geological record bears a sharply defined imprint of the combined effects of these changes. This provides a fundamental basis for stratigraphical subdivision that is essential for the differentiation of multiple glaciation events. In addition, the proximity of Britain to the North Atlantic Ocean implies that the region is highly sensitive to even minor variations in marine currents and their accompanying wind patterns. Such changes are seen as environmental responses at a range of scales, particularly glacial advance and retreat.

7.2. EVIDENCE FOR GLACIATIONS

Evidence for glaciation is found over most of Britain and is primarily lithological with glacial episodes being identified by till and glaciofluvial sediments and glacial limits being determined by the extent of these deposits. Additionally, geomorphological evidence has played an important role in reconstructing the extent of ice masses in younger glaciations. Moraine ridges and ice-contact landforms, including patterns of glacio-isostatically deformed displaced shorelines, have played an important role in the determination of ice limits of these younger glacial events. Palaeontological evidence, such as pollen, Mollusca and plant macrofossils have also been used to differentiate different stages of the Quaternary, an approach that has been central to the separation of glacial events (West, 1980a,b). Soil evidence, usually in the form of permafrost structures, has been used to indicate cold, periglacial climate conditions, which have been at times linked with the formation of glacier ice elsewhere in England, Wales and Scotland (Ballantyne and Harris, 1994).

However, Britain shows marked physiographical contrasts in geology, Pleistocene history, tectonic setting, relief and glacial regimes. Upland northern and western Britain is dominated by erosional evidence, particularly in the mountainous districts of NW Scotland, the English Lake District and N Wales, which formed the main ice accumulation centres and supported their own ice caps. Where deposition sequences occur, they tend to represent only the most recent glaciation (Devensian = Weichselian, Valdaian, Wisconsinan) and the evidence is often dominated by retreat-phase sequences. In lowland south and eastern Britain (Fig. 7.1), by contrast, depositional sequences predominate with relatively subdued landforms but sediments and associated features indicating glacial expansion and retreat of at least three and potentially more glaciation episodes (Anglian = Elsterian, Wolstonian = Saalian and Devensian; cf. Catt et al., 2006; Fig. 7.2). The extraordinary variety of British substrate geology and the consequent topography developed upon it have given

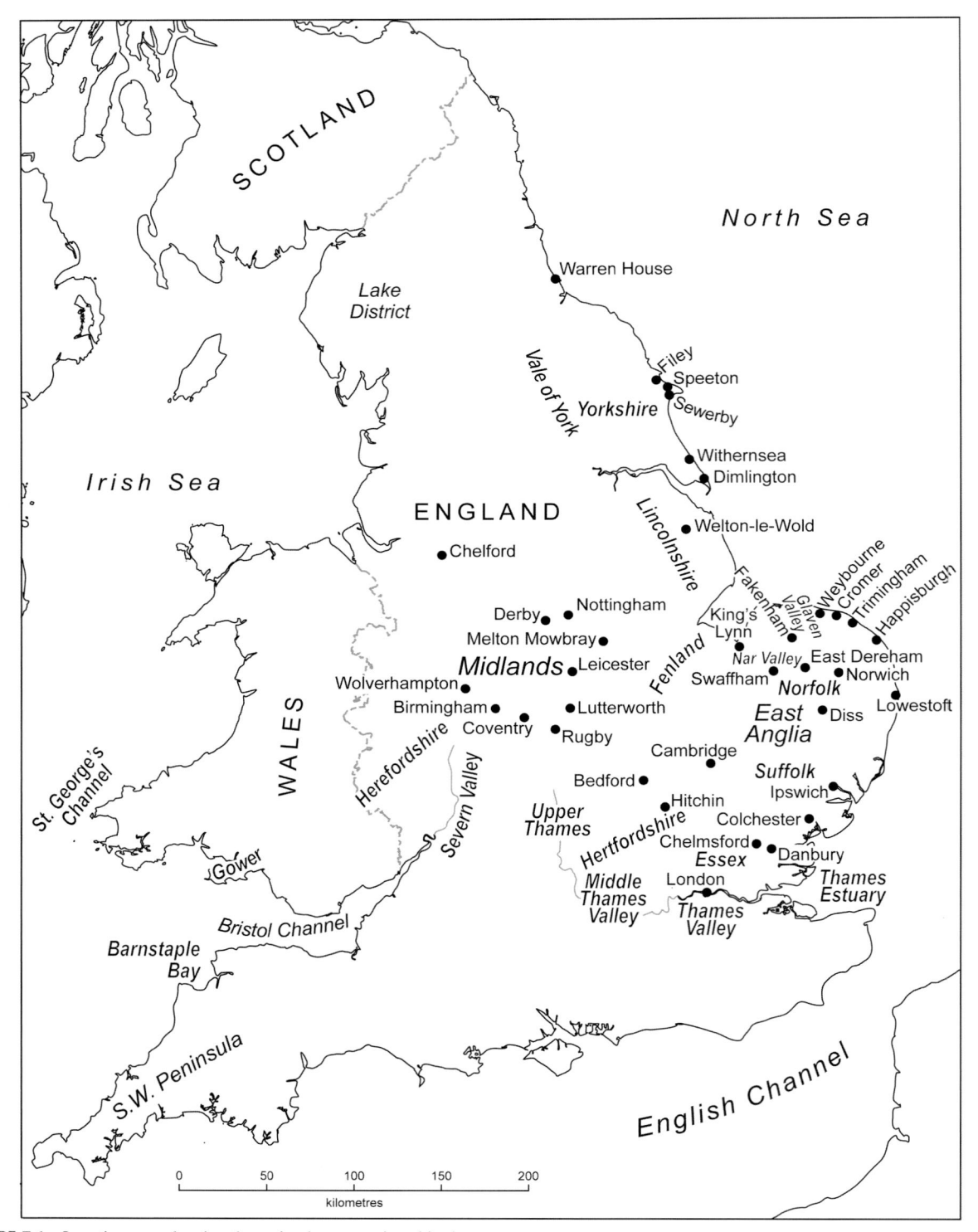

FIGURE 7.1 Location map showing the main places mentioned in the text.

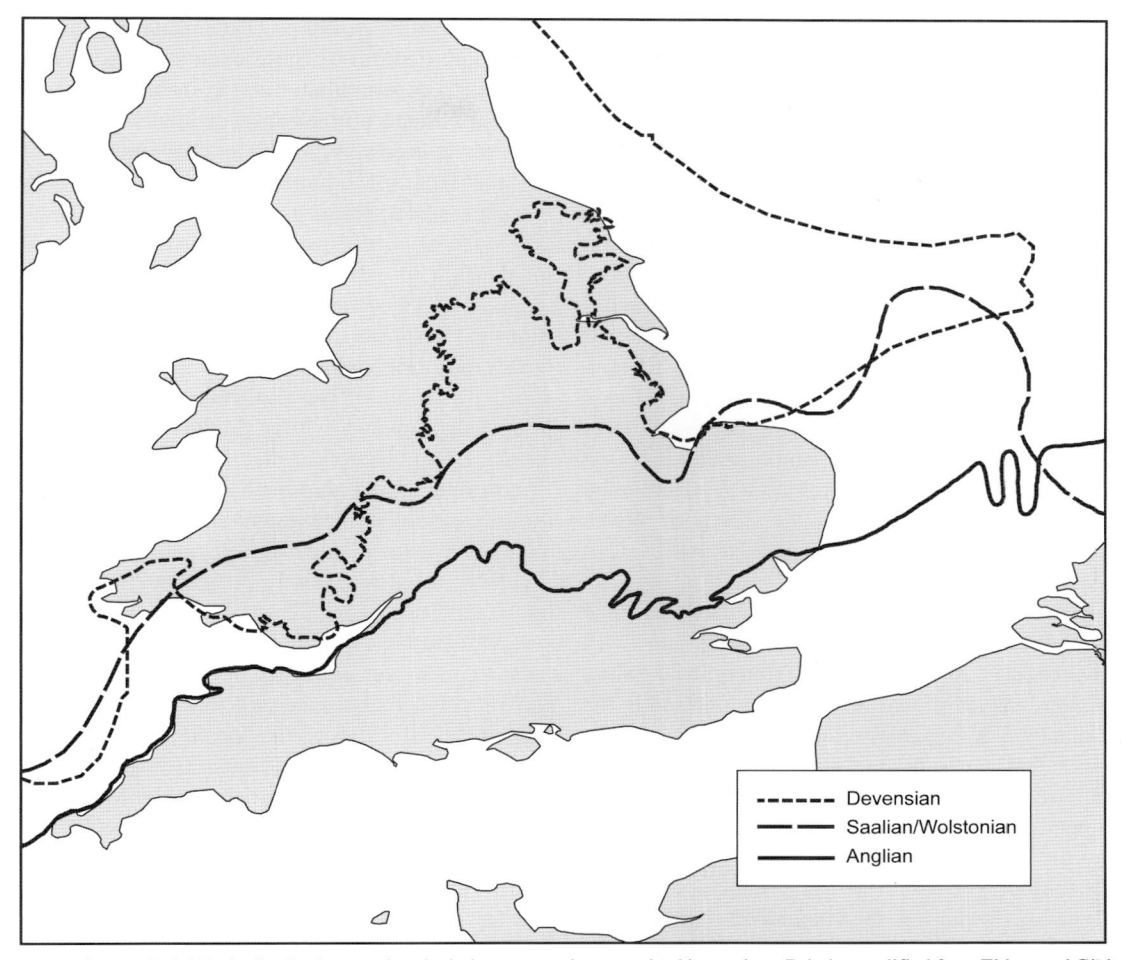

FIGURE 7.2 Maximum glacial limits for the three major glaciations currently recognised in southern Britain, modified from Ehlers and Gibbard (2004).

rise to a vast range of locally variable deposits and landforms across what is after all a relative small area. These factors have together resulted in the variety of glacigenic sedimentation, glaciotectonics, erosion and glacial evolution represented in Britain. Finally, southernmost Britain, broadly south of a line from the Thames to the southern shore of the Bristol Channel, remained unglaciated throughout the Pleistocene but this area has been influenced indirectly by meltwater activity.

7.3. THE BRITISH PLEISTOCENE SUCCESSION

Taking the island of Great Britain as a whole, the Pleistocene sequence can be broadly divided into two major parts: the pre-glacial Pleistocene and the glacial Pleistocene (Fig. 7.1). With some minor exceptions, the bulk of the evidence for the pre-glacial Pleistocene is restricted to southern and eastern England and particularly to East Anglia, remaining areas having been, for the most part, subjected to erosion through the Late Cenozoic. This is because eastern East Anglia forms the western margin of the tectonically downwarping North Sea basin. Sedimentation here, although in a marginal setting, records mostly sea-level highstand sequences of shallow marine, littoral and sublittoral environments. Rapid infilling of this basin led to retreat of the sea through the later Early Pleistocene and its replacement by fluvial and terrestrial sedimentation until the later Middle Pleistocene (ca. 0.45 Ma).

No direct evidence for glaciation is known from these sequences although Solomon (1935) invoked glaciation in the North Sea region to explain the occurrence of anolomous frequencies of hornblende in heavy mineral analyses from cold stage (Baventian = Tiglian C4c Substage) coastal deposits. This is now interpreted as resulting from tidal transport from the northern Netherlands (cf. Burger, in Gibbard et al., 1991). Similar indirect evidence for glaciation is seen as erratic clasts in fluvial sediments, for example, rhyolitic rocks from the North Welsh Berwyn Mountains in Thames' Kesgrave Formation gravels (Rose, 1994; Rose et al., 1999, 2001), are seen as evidence for repeated contemporaneous glaciation that entered the uppermost part of the Thames catchment. Attempts have been made to relate the individual member units within the Kesgrave Formation to the marine isotope-stage chronology dependent on counting backwards from the established age of the Anglian (i.e. Marine Isotope Stage (MIS) 12) and the correlations lack biostratigraphical or similar independent age control.

The earliest undisputed glacial diamicton known on land in Britain is the Happisburgh Diamicton, which is generally regarded as being of Anglian age, although this has been recently questioned (see below). This marks the arrival of the glacial part of the Pleistocene sequence in Britain. Once again evidence for this glaciation is best developed in eastern and southern central England, although sediments are known to extend as far as the Welsh borderlands region near Hereford and south Wales. Anglian-age glaciogenic sediments also occur offshore in the southern North Sea (see below; Carr, 2004) and Irish Sea basins (cf. Catt et al., 2006).

The second glaciation episode is of late Middle Pleistocene age, intermediate between the Hoxnian (= Holsteinian ~ MIS 11c) and Ipswichian (= Eemian ~ MIS 5e) interglacial stages. This Wolstonian glaciation was originally recognised in the English Midlands where glacial sediments underlie a large area in the west and central district. However, they were subsequently referred to the Anglian Stage. Despite this conclusion, a younger Middle Pleistocene glacigenic sequence is known from the Midlands, eastern England (Gibbard et al., 2009). In the light of recent discoveries, it seems conceivable that the Midlands' sequences do not indeed represent the Wolstonian rather than the Anglian events (see below). Because it was apparently of similar extent to that in the subsequent Devensian Stage, this glacial episode is much more poorly represented in the Pleistocene record.

The third or Devensian glaciation is well represented throughout most of glaciated Britain, yet it was possibly the least extensive of the three major events. Although considerable detail is known about its retreat phases, comparatively little evidence is available about the ice sheet's advance. In general, ice extended outwards from the mountain glacial centres, reaching a maximum at about 27 ka and then retreating in an oscillatory manner until 13 ka. Following the Late-glacial Interstadial (dates), the ice readvanced in the Loch Lomond Stadial (= Younger Dryas Stadial).

7.4. THE 'TRADITIONAL MODEL' OF EAST ANGLIAN GLACIAL STRATIGRAPHY

The sequence of glacial episodes described above relates to what has been dubbed the 'traditional model' by Rose and colleagues (e.g. *sensu* Rose in Clark et al., 2004; Lee et al., 2006; Rose, 2009; cf. Chapter 6) during which they reassessed the glacial sequence and differentiated as many as five glaciations in the pre-Devensian Middle Pleistocene. The identification of these events has been based on the separation of individual till or units of assumed glaciogenic or cold-climate origin within sequences. Each of the units they recognised was then assigned to individual MISs, often on the basis of counting backwards or forwards from sequences of apparently known age. The weakness of this approach has been strongly criticised not only on the basis of the lack of acknowledgement of the multi-dimensional depositional systems it displays and the mixing of observation and assumed climatic representation of sediments or erosional horizons but also on correlations in part arising from dismissal of the biostratigraphy of the under- and overlying sequences (e.g. Banham et al., 2001; Gibbard et al., 2006). Therefore, the critical arguments against the 'traditional model' approach can be thought of as disingenuous, especially since they demonstrate a failure to appreciate the nature and significance of biostratigraphical evidence (cf. discussion in Preece and Parfitt, 2000; Preece et al., 2009).

Although it is now generally possible to relate the terrestrial to deep-sea sequences fairly reliably at a coarse scale for the Late Pleistocene and to a lesser extent in the late Middle Pleistocene, earlier the situation was progressively and significantly more difficult (Gibbard and West, 2000). This was especially so in the early Middle Pleistocene (Turner, 1996), the period which Rose (2009) and Lee et al. (2010) not only identify additional pre-Anglian glacial events indirectly from the occurrence of erratic clasts in the fluvial, Kesgrave Formation Thames' deposits, noted above, but also assign the glaciations they identify to specific MISs. In spite of the fact that there is the Matuyama/Brunhes magnetic reversal event marker with which to correlate, the lack of a palaeomagnetic signal preserved in these sediments or any independent means of reliable 'dating' undermines the correlations as currently proposed by these authors. Comparable problems arise when river terraces are assumed to indicate landscape uplift. In this case, when single terrace deposits are assigned to a discrete isotope-stage event, and their age is based on counting backwards or forwards from apparently known-age markers, their elevation has been used to calculate uplift rates. Chronologies relying upon this approach have been used as a basis for questioning existing rigorously constructed stratigraphies, based on assumed uplift (e.g. Westaway et al., 2002), the evidence for which is unsubstantiated and indeed for which there is no independent corroboration.

Thus, whilst there is certainly a longer record of glaciation represented in the British Isles as a whole, the times at which these events occurred remain undetermined at present in any critical sense. Therefore, as elsewhere in the southern North Sea region, no direct evidence is found for pre-Anglian (pre-Elsterian: MIS 12) glaciation (*contra* Rose in Clark et al., 2004; Lee et al., 2004a,b, 2006,

2010; cf. Preece and Parfitt, 2000; Banham et al., 2001; van Gijssel, 2006; Gibbard et al., 2008; Preece et al., 2009).

7.4.1. Anglian Glaciation

The glaciation in the Anglian Stage was the most extensive of the British Pleistocene. However, since for the greater part of its extent its deposits were overridden by younger glaciations, much of the sedimentary record has been lost. There is therefore only fragmentary evidence for reconstructing the glacial history of the Anglian Stage. For this reason, the Anglian glacial sediments are best preserved beyond the margins of later ice advances, and Eastern England and the adjacent offshore region include extensive sequences of deposits and features formed during this event. Therefore, this chapter will not only concentrate on discussion of these areas but will also consider the evidence for the extent of glacial deposits in the English Midlands, the Welsh borderland and south Wales (Fig. 7.1).

7.4.1.1. East Anglia, South-East England and the Thames Valley

The dating of the Anglian glacial deposits in the type area is based on their relation to temperate stage sediments. In north Norfolk, glacial deposits rest on 'Cromerian Complex' Stage and throughout their distribution area are overlain directly by Hoxnian Stage and younger interglacial sediments (Preece and Parfitt, 2000; Banham et al., 2001; Gibbard et al., 2008; Preece et al., 2009). No evidence of interstadial nor higher rank climatic oscillations has been unequivocally identified intervening within the Anglian sequence and therefore it is generally interpreted as representing a single complex glacial event. However, this was recently questioned by Moorlock et al. (2000; see below), in the previous version of this chapter (Rose in Clark et al., 2004) and since (e.g. Rose, 2009; Chapter 6; Lee et al., 2010). On the basis of strong lithological and detailed stratigraphical relationships, the Anglian is equated with the continental Elsterian Stage. Most recently, its equivalence to MIS 12 has been definitively demonstrated both by direct means (Toucanne et al., 2009a) and by numerical dating (Pawley et al., 2008).

The considerable antiquity of the Anglian glaciation means that ice-marginal landforms are difficult to find because of their subsequent modification, and in many cases, total removal during later events. Limits must therefore be determined using a variety of means, including primarily the distribution of diamictons and the occurrence of characteristic erosional and deposition features, such as tunnel valleys, marginal ridges, outwash and glaciolacustrine sequences. The following summary is based on this type of combined evidence, but will emphasise those areas for which particularly characteristic features are known.

The Anglian glaciation began as the ice advanced from the NE and overrode the pre-existing sedimentary sequences throughout the region. This advance deposited a complex suite of interbedded tills and associated meltwater sediments, which are termed the North Sea Drift Formation. These sediments, best exposed on the northern and northeastern coasts of East Anglia, from Weybourne to Lowestoft (Fig. 7.1), can also be traced inland as far as Norwich and south towards Diss (Mathers et al., 1987). Internally, this sequence is highly complex, including evidence for a threefold oscillation of the ice sheet, the intervening phases represented by deltaic outwash and glaciolacustrine sedimentation (Hart, 1987; Hart and Boulton, 1991; Lunkka, 1994; Gibbard, 1995; Gibbard in Clark et al., 2004). The earliest diamicton (the Happisburgh Member) was deposited directly on land, whilst subsequent units were waterlain (e.g. the Cromer Member) being deposited directly from floating ice (Ehlers and Gibbard, 1997). Of particular interest in this context are the Corton Sands, an extensive unit of meltwater deposits derived from the North Sea Drift Formation ice in the Lowestoft area (Bridge and Hopson, 1985; Hopson and Bridge, 1987; Bridge, 1988). Detailed investigation of the sequence indicates that they were deposition by a sandy braided-river delta. The overall vertical and lateral changes in this unit represent the withdrawal of the Scandinavian ice lobe entering an adjacent lacustrine environment (Gibbard et al., 2008). For all but the final stillstand phase during which the delta-morainic Cromer Ridge in northeast Norfolk was laid down, no precise margins are known for these individual advances. Suggestions by Rose in Clark et al. (2004) and Hamblin et al. (2000) that the Briton's Lane Formation from which the Cromer Ridge is formed may relate to a MIS 6 glaciation are untenable. The ridge is clearly intimately related to the glacial sequences on which it rests, the sequence includes the glaciotectonised elements at Trimingham in the east, the sediments are overlain by Hoxnian-age interglacial pond deposits (Preece et al., 2009), and most recently optically stimulated luminescence (OSL) dating of the Morston Raised Beach (Late Saalian; MIS 7-5) which also overlies the deposits (Hoare et al., 2009) again reinforces this interpretation.

The North Sea Drift Formation sediments characteristically contain a suite of igneous and metamorphic erratics of southern Norwegian origin. Some of these rare erratics, rhomb porphyry and larvikite, were derived from Oslofjord area and have therefore been generally accepted as evidence that the ice sheet originated in southern Norway and crossed the North Sea basin (Hoare and Connell, 2004). Although these erratics are found at all exposures of North Sea Drift diamictons, an indication that Scandinavian ice may have reached further south is the present distribution of Norwegian erratics (Boswell, 1914; Slater, 1929). Rhomb porphyries and larvikites can be found as far south as Bedford,

Hitchin, Ipswich and Cambridge (Ehlers and Gibbard, 1991) where they were probably carried by later ice movement. A southern Norwegian/North Sea floor origin for the North Sea Drift Formation materials has recently been questioned by Lee et al. (2004a,b, 2010), these authors preferring to explain the material as being reworked from 'from older pre-existing Scandinavian deposits in the North Sea' (Lee et al., 2010). Apart from evoking the inclusion of deposits which apparently do not exist, this interpretation conflicts with over a century of observations to the contrary (Hoare and Connell, 2004) and is therefore untenable.

During the next phase, ice of British origin advanced through the Vale of York and Lincolnshire into central and western East Anglia. The ice crossed the Fenland basin, possibly exploiting a pre-existing gap in the Lincolnshire–Norfolk Chalk ridge. On the basis of lithology (Perrin et al., 1979), erratic content (Baden-Powell, 1948), detailed fabric measurements (West and Donner, 1956; Ehlers et al., 1987) and fossil assemblages from erratic chalk pebbles (Fish and Whiteman, 2001), it can be shown that the ice radiated outwards from the Fenland basin in a fan-like pattern. It is thought that interaction of this British ice with the Scandinavian ice was responsible for this unusual pattern of ice movement. Moreover, in the eastern part of the area, the British ice seems to have expanded close to its maximal position and replaced the Scandinavian ice. The British ice expanded progressively towards the east and northeast, reaching the present east coast at Lowestoft.

Over a large part of East Anglia, the British ice deposited the so-called Lowestoft Formation till. It has a grey to blue-grey clay matrix (up to 45%), is rich in flint and sub-rounded chalk clasts and includes erratics that originated on the British landmass. Grain-size distribution is remarkably uniform across the region, minor changes being attributed to incorporation of local materials, such as North Sea Drift Formation sediments, etc. (Perrin et al., 1979).

Advance of the British Lowestoft Formation ice into Essex and Hertfordshire brought it into the region influenced by the River Thames and its tributaries. The substantial series of Thames deposits termed the Kesgrave Formation are aligned WSW–ENE to W–E across southern East Anglia that represent the pre-glacial course of the river was overridden by the ice in all but the southernmost part of the region. Oscillations of the ice-sheet margin have been repeatedly identified in the southern marginal zone in the Ipswich, Colchester, Chelmsford and Hertfordshire areas (Allen et al., 1991). Similar frontal movements also occurred in the north and east London areas (Gibbard, 1979, 1994). These oscillations had profound effects on the River Thames since the Lowestoft ice advanced into and overrode the river's valley in Hertfordshire, the till units interdigitating with the fluvial sediments.

Subsequent erosion has largely removed most marginal features left by the ice. The most convincing feature is the marginal moraine-like form in the Finchley–Hendon area of north London. Another landform marking the maximum ice extent is the substantial marginal meltwater–fan complex that occurs south-west of Colchester, Essex, the Stanway and Birch areas (Leszczynska, 2009). Here, over 30 m of meltwater deposits occur on the leeside of the range of Tertiary hills that formed the southern side of the Anglian and immediately pre-Anglian Thames valley (the Danbury–Tiptree Ridge). Although an early oscillation was able to bypass these hills to the south, they apparently later formed an insurmountable barrier to the Anglian ice which rode up against the hills and discharged vast quantities of meltwater eastwards towards the North Sea (P. Gibbard and S. Boreham, unpublished; Leszczynska, 2009). Elsewhere evidence for ice extension several kilometres beyond the maximum mapped extent of glacial diamicton has been identified. For example, small-scale 'tunnel valleys', infilled by glacial sediment, have been identified in Suffolk by Mathers et al. (1993) and Cornwell and Carruthers (1986). The glacial maximum on the map therefore includes these records.

With the continued withdrawal of the Scandinavian ice, the Lincolnshire coast and adjacent offshore area became open for the expanding British ice advancing south-eastwards into Norfolk. Again the Wash and the Breckland Gap directed the ice stream. Ice that flowed over the relatively high ground of the Chalk escarpment and into the Fenland basin led to deposition of the characteristic chalk-rich 'Marly Drift' till facies onto more typical Lowestoft Till near Kings Lynn (Straw, 1991). Ehlers et al. (1987) demonstrated that this facies does not represent a separate glaciation but simply locally derived diamicton deposited during a later phase of the Anglian.

During the 'Marly Drift' phase, chalk-rich till was transported into much of East Anglia, including the Gipping Valley in Suffolk, where it overlies the Jurassic clay-rich tills (Lowestoft facies) of the preceding Lowestoft advance (Ehlers et al., 1987). By this time, however, the ice no longer reached its maximum limits. In East Anglia, the area beyond the Cromer Ridge was no longer covered by ice. As the ice retreated, vast meltwater formations were laid down as sands and gravels, for example, those between East Dereham, Swaffham and Fakenham. Here, the gravels rest on the chalk-rich till of the last Anglian ice advance. Similarly, in the area around the Glaven Valley in North Norfolk kames, a dead-ice topography and an esker are associated with the retreat of the chalk-rich till ice (Sparks and West, 1964; Ehlers et al., 1987).

Throughout East Anglia a series of deep, steep-sided valleys have been found cutting through the Chalk and associated bedrock. These valleys form a radiating pattern, broadly parallel to the dip of the Chalk, that is, they trend east-west in the north and north–south in the south of the region (Fig. 3). Detailed study of these 'tunnel valleys', by Woodland (1970), Cox (1985b), Cox and Nickless

(1972) and van der Vegt et al. (2009) among others, indicates that they are normally filled with glacial sediments, predominantly meltwater sands, gravels or fines. Tills also occur but are less frequent. Closely comparable in form and scale to the *rinnen* of Denmark, Northern Germany, Poland, etc., although of shallower depth, they are undoubtedly of glacial origin and probably result from subglacial drainage discharge under high hydrostatic pressure (Ehlers et al., 1984). Their overall depth appears controlled by substrate and ice-lobe characteristics (GRASP, 2009).

During deglaciation, meltwater rivers and glacial lakes became concentrated into a series of large channel-like valleys the alignment of which, in many places, is determined by the subglacial 'tunnel valley' courses, especially where these valleys were only partially filled subglacially. Valleys such as the Stort, the Gipping, the Waveney and the Wensum are all of this type. However, in northern Norfolk, substantial proglacial outwash sandur plains, apparently independent of tunnel valleys, developed particularly at Kelling and Salthouse Heaths and mark retreat-phase, ice-front stillstand positions (Sparks and West, 1964). The radial pattern of meltwater discharge continued until ice withdrew beyond the Chalk escarpment, north-north-west of which drainage became directed by the regional slope into the Fenland basin.

In the marginal areas, particularly at Corton, and along the Cromer Ridge (already noted), the ice front retreated in standing water (Pointon, 1978; Bridge and Hopson, 1985; Hopson and Bridge, 1987; Bridge, 1988). A similar phenomenon is found in the Nar Valley (Ventris, 1986, 1996) where the 'tunnel valley' deepens towards the west, that is, in the direction of the ice retreat. A proglacial ice-dammed lake seems to have formed in the valley and caused the thinning ice initially to float, depositing subaquatic till, and later to retreat completely leaving laminated clays to accumulate. Subsequent deglaciation gave rise to drainage of the lake and establishment of fluvial sedimentation.

7.4.1.2. English Midlands

The Seisdon Formation consists of 27 m of proximal outwash gravels filling the Seisdon–Stourbridge Channel at Trysull near Wolverhampton, Staffordshire (Worsley, 1991). The gravels are overlain by a thin unconformable cover of Late Devensian Till (the Stockport Formation) and between the two, Morgan (1973) recorded silts (the Trysull Member) containing pollen indicating a Hoxnian interglacial age. At Quinton near Birmingham, the Nurseries Formation, a brown till containing many local erratics mainly from the Coal Measures, is also thought to be Anglian in age because it is overlain by pollen-bearing silts and clays (the Quinton Formation) attributed to the Hoxnian Stage. The occurrence overlying glacial sediments filling channel-like depressions in the underlying bedrock (Horton, 1974) emphasises that Anglian events in the Midlands may have been somewhat similar to those in eastern England.

It is highly likely that it was during the Anglian glaciation of the West Midlands that the pre-existing River Thames' drainage system was decapitated, with the headwaters being realigned into the Severn system to form the River Avon following deglaciation. The latter, probably formed as a meltwater drainageway during this glaciation with the deposition of the Kidderminster gravels (Maddy et al., 1995). The resulting loss of the Thames' headwater system resulted in the loss of access to the local Triassic rocks that had, up to this point, brought quartz pebbles into the upper Thames region and thence into the Middle and Lower Thames system to form the Kesgrave Formation sediments (*contra* Rose, 1987, 1994, etc.; cf. below).

Outwash from an Anglian ice sheet in NE Herefordshire deposited the Risbury Formation, which consists of up to 30 m of glaciofluvial and ice-contact glaciodeltaic gravels (Richards, 1998). The gravels contain clasts of Devonian and Lower Palaeozoic rocks derived from the west and south-west, suggesting that the glacier originated in central Wales. The Anglian age of these glacial deposits is indicated by their stratigraphical relationships to interglacial sequences, paralleling those found in East Anglia. In the Mathon Valley to the west of the Malvern Hills, the formation locally overlies fluvial silts containing a 'Cromerian Complex' age interglacial fauna and flora (Coope et al., 2002), and at the head of the Cradley Brook, it is overlain by silts containing Late Anglian to Early Hoxnian temperate pollen and molluscs (Barclay et al., 1992). Apart from this maximal extent, evidence for four additional retreat phases is also represented in the highly fragmentary sequence. The ice-lobe seems to have adopted virtually the same path as that followed by the Late Devensian ice (Richards, 1998).

7.4.1.3. Wales

In general, the pre-Devensian glaciation of southern Wales is referred to as the 'Irish Sea' Glaciation (Pringle and George, 1961). This glaciation potentially includes evidence for one or more glacial episodes during which complete glaciation of the province occurred by local ice sheets. The only unequivocal Anglian-age unit is the Llandewi Formation of SW Gower which represents the margin of Anglian-age Welsh ice. On south and west of Gower the Paviland Moraine, a marginal complex of meltwater gravels and sands rests on a red clay till (Bowen, 1974, 1999), rich in Millstone Grit clasts derived from southern Welsh Carboniferous rocks. At the same time, this glaciation extended across the Bristol Channel as far as the northern coast of the English South-West Peninsula (Bowen et al., 1986), although there is some dispute over precisely how far it reached. There seems to be doubt whether it reached Cornwall or the Scilly Isles, where deposits previously thought to be Anglian are now assigned to the Devensian (Campbell et al., 1999). If the ice did indeed cross the Bristol Channel during this phase, it would have blocked the River Severn drainage potentially causing a proglacial lake to form in the valley.

7.4.1.4. Offshore Areas

Many of the marginal areas of the Anglian glaciation ice sheet lie not on land but on the sea floor. Therefore, it might be expected that sedimentary basins like the North and Irish Seas would provide excellent sediment traps in which Anglian glaciation sediments would be well preserved. However, because of repeated later erosion, this is not the case, material having been subsequently recycled.

Beneath the St. George's Channel, adjacent to the Welsh landmass, probable Anglian-age diamicton and associated sediments of the Caemarfon Bay Formation have been identified by Tappin et al. (1994), Wingfield (1994, 1995) and Cameron and Holmes (1999) infilling glacial valleys. The southern limit of these valleys occurs at about 51°N and, according to Wingfield (1989), marks the stillstand position of a grounded ice-front in the basin. Floating ice may well have extended further south of this line at times.

The occurrence of Anglian tills is also limited to relatively small areas in the southern part of the North Sea (south of Dogger Bank). As in the St. George's Channel, many buried channels occur beneath the southern and central North Sea floor, the oldest of which have been shown to be of Anglian age (Balson and Cameron, 1985; Cameron et al., 1987; Praeg, 1996, 2003; Moreau et al., 2009; Moreau, 2010). They contain tills and meltwater sediments assigned to the Swarte Bank Formation. These channels are remarkably similar in form and scale to the 'tunnel valleys' of East Anglia or the *rinnen* of the adjacent Continent, already discussed.

This margin is very similar to that suggested by Laban (1995) which is marked by subglacial tongue basins, subglacial 'tunnel' valleys and substantial glaciotectonic deformation structures in the centre of the southern North Sea, east of Texel.

Throughout the North Sea region, the sedimentary reconstructions indicate that the ice dammed the river systems of the Rhine, Meuse, Thames, Scheldt, etc. to form a massive glacial lake that persisted throughout much if not all of the Anglian glaciation (Gibbard, 2007). The rivers and meltwater streams (Gibbard et al., 1996, 2008; Leszczynska, 2009) formed deltaic accumulations built out into the lake through the glacial episode. Overspill from this lake gave rise to the initial breach of the anticlinal Weald-Artois bedrock ridge to initiate the Dover Strait (Gibbard, 1988, 1995; Cohen et al., 2005).

Beyond these areas, the evidence for the extent of the Anglian glaciation remains unestablished.

7.4.2. Wolstonian Stage Glaciation

The second glaciation took place during the late Middle Pleistocene, intermediate between the Hoxnian (Holsteinian; ca. ?MIS 11c; cf. Ashton et al., 2008) and Ipswichian (Eemian; ca. MIS 5e) interglacial Stages. This glacial episode is less well represented in the Pleistocene record and has to date been little studied and weakly defined (Clark et al., 2004).

7.4.2.1. English Midlands

Following the classic work of Shotton (1953, 1968, 1976, 1983a,b), the glacial sequence of the English Midlands in the area around Coventry and Birmingham was considered to represent what was termed the Wolstonian glaciation This name was also selected for the predominantly cold-climate interval or stage in which the glaciation occurred (Mitchell et al., 1973), the Wolstonian (=Saalian) Stage (broadly equivalent to MIS 11b-6). The sequence comprises a series of glacial sediments identified over a large area in the west and central Midlands, (Rice, 1968, 1981; Rice and Douglas, 1991). Of these, diamictons and associated meltwater sediments provide the evidence for an extensive glaciation of the region as far south as the Cotswold Hills at Morton-in-Marsh in Gloucestershire.

The glacial sediments of the Midlands' region overlie deposits of a pre-existing river system, represented by the Baginton–Lillington Gravel. These deposits are characteristically composed of quartz-rich sediment derived from underlying Triassic bedrock. In contrast to normal long-lived fluvial systems in the region, this SW–NE aligned system was apparently relatively short-lived since it lacks a terrace-like system. The river which deposited these sediments appears to have formed after the Hoxnian and was overridden by the Wolstonian ice, and therefore it can only have existed for a relatively limited period (Shotton, 1953, 1983a,b; ~140 ka or thereabouts). The reinterpretation of these sediments as the headwaters of a pre-Anglian 'Bytham river', aligned towards East Anglia across the Fenland, found favour for some years, especially when the Wolston Formation sediments were reassigned to the Anglian Stage (e.g. Rose, 1994). However, recent work (Gibbard et al., 2009) has now shown that the Bytham river did not exist in the form suggested by authors such as Lee et al. (2004a,b) and therefore Shotton's interpretation has regained considerable support in recent years.

As defined by Shotton (1953), the Midlands' Wolston Formation consists of two tills, the Thrussington and Oadby Members, with sands and gravels below, between and above. The Thrussington Till is reddish brown and mainly

Trias-derived, whereas the Oadby Till contains chalk, flint and Jurassic limestone clasts in a grey matrix derived mainly from Liassic clays. These components suggest deposition by ice from western and eastern sources, respectively.

The Thrussington Till was deposited as far east as Melton Mowbray, Rugby and Fenny Compton, and its meltwater formed a large ice-dammed lake between Leicester and Market Bosworth, in which up to 25 m of laminated clays and silts (the Bosworth Member) accumulated. Subsequently, the lake was overridden by meltwater streams derived from the same ice sheet, depositing a similar thickness of sand and gravel (the Wigston Member).

The Oadby Till was then deposited over much of the Eastern Midlands as far north as Nottingham and Derby, and as far west as Stratford-upon-Avon. It also extends southwards to Moreton-in-Marsh, where flint presumed to originate from its associated outwash was incorporated into the Wolvercote Terrace gravels of the Upper Thames valley. As this has been correlated with terraces of the Middle Thames as young as the Lynch Hill/Taplow (?MIS 8-6) Terraces (Gibbard, 1985), whilst Sumbler (1995, 2001) has argued that the Oadby Till at Moreton-in-Marsh might relate to MIS 10. Between Rugby and Lutterworth, the Oadby Till passes laterally into flint-bearing gravels (the Shawell Member), which were deposited as proximal outwash of the eastern ice sheet. Flint outwash was also incorporated into the Dunsmore Member, which is the oldest deposit of the Warwickshire Avon in the area between Hinckley, Leamington Spa and Rugby. The lithological similarity of the Oadby Till to the Lowestoft Till of East Anglia led to suggestions that it was deposited by a westward extension of the same Anglian ice sheet (Perrin et al., 1979), which invaded the E Midlands after deposition of the Thrussington Till by western ice and therefore that the Wolston Formation was itself of Anglian age.

This observation misled geologists for two decades in the 1980–90s, in spite of claims to the contrary, particularly by Straw (1983, 1991), leading to the opinion that glaciation in the Wolstonian made little significant impact on lowland eastern England, south of Yorkshire. Even the stratotype sequences, near Coventry, were thought by some to represent an earlier (i.e. Anglian) age (Sumbler, 1983a,b, 1995; Bowen et al., 1986; Rose, 1988; Rose in Clark et al., 2004). Contrary opinions were held by other authors (e.g. Shotton, 1983a,b; Gibbard 1991; Gibbard and Turner, 1988; Ehlers et al., 1991a,b) who pointed out the apparent disparity with the near Continent, where Scandinavian and Baltic glaciation is very extensive during the Saalian Drenthe Substage (MIS 6). As Gibbard et al. (1992) concluded, it would have been surprising if lowland Britain had not been subjected to significant contemporaneous glaciation.

7.4.2.2. East Anglia

In the past, there has been controversy over the extent of post-Hoxnian/pre-Ipswichian glaciation in eastern England following the work of Bristow and Cox (1973). More recently, following reappraisals of the glacial evidence in northern East Anglia, Rose and colleagues (cf. Clark et al., 2004) proposed that additional glaciations might have occurred early in the Wolstonian (?MIS 8-10). The evidence for the age of this glaciation was based almost exclusively on geochronometry of overlying or underlying non-glacial deposits and long-distance correlation with river sediments in the Upper Thames region of the south Midlands. Evidence considered critical to this interpretation was derived from U-series determinations from the Nar Valley, near Kings Lynn in Norfolk where sediment overlying Lowestoft Formation till at Tottenhill yielded an MIS 9 age Rowe et al. (1997); a date that is close to the maximum that can be resolved using the U-series method. Rose (in Clark et al., 2004), following Scourse et al. (1999), took this to imply that the till was deposited during MIS 10. However, nowhere else in eastern England is a diamicton of this age known, and since this single determination conflicts with both the litho- and biostratigraphy at this site (cf. Ventris, 1985, 1986, 1996), as well as the regional stratigraphy (Gibbard et al., 1992; Gibbard in Clark et al., 2004), its validity is questionable at the least and probably should be rejected.

The presence of glacial-margin delta-like fan sediments of unequivocal post-Anglian/pre-Devensian age at Tottenhill on the NE Fenland basin margin confirmed this view (Gibbard et al., 2009). The dating of this Tottenhill glaciation, in NW Norfolk, is demonstrated by the fact that the Tottenhill Member gravels directly overlie both Anglian-age Lowestoft Formation glacial and the Hoxnian marine Nar Valley interglacial Member deposits (Ventris, 1986, 1996). Later, incision of the Tottenhill sediments left them at an elevation above the adjacent Fenland basin, and the Nar Valley fluvial Pentney Member. The latter includes organic sediments interstratified within the gravels that have been correlated with the first half of the Ipswichian (Eemian) Stage on the basis of their contained pollen assemblages. A palaeosol is also developed on the surface of the Tottenhill sediments which shows only one phase of temperate weathering considered to have formed in the Ipswichian Stage after the deposition of the gravels (Lewis and Rose, 1991). The Tottenhill gravels thus pre-date the stage, demonstrating that they must be of Late Wolstonian age (*sensu* Mitchell et al., 1973; Gibbard and Turner, 1988), as Ventris himself concluded.

The advance of the Tottenhill ice lobe into the East Anglian Fenland basin reached the eastern marginal area and was possibly halted by the rising ground of the Chalk hills to the E and S. Here, a group of landforms and their underlying deposits represent a series of glaciofluvial delta-fan and related sediments deposited as ice-marginal deltas in a lake at the maximum ice-marginal position (the Skertchly Line). This evidence confirms historical descriptions of a glaciation of the Fenland and demonstrates that reinterpretations of the sediments as of fluvial rather than of glacial meltwater origin (e.g. Lee et al., 2004a,b), at sites including Warren Hill, High Lodge, Lakenheath, Feltwell and Shouldham Thorpe, are incorrect. On the basis of regional correlation, supported by OSL dating, the glaciation occurred at ∼160 ka, that is, in the Late Wolstonian (=Saalian) Stage, an age recently independently confirmed by OSL dating (Pawley et al., 2004, 2008; E. Rhodes, unpublished).

This Wolstonian glaciation (Saalian, ca. MIS 6; see below), although originally recognised in the English Midlands, was subsequently identified in Yorkshire, northern East Anglia and the Welsh Borderlands. In many places, however, Wolstonian glacial deposits were subsequently thought by some to pre-date the stage. For example, in the south Midlands, Sumbler (1983a,b) questioned the basis for the relative age of the Wolston Formation sequence, pointing to an apparent continuity with the Anglian deposits further to the east, and therefore concluding that the Midlands' sequence should also be assigned to the Anglian Stage. This conclusion was also supported by Rose (1988). However, now that the latter has been questioned, and glaciation of Wolstonian age has been identified in eastern England, there is a strong likelihood that the Wolston Formation is indeed of Late Wolstonian age (*sensu stricto*, i.e. MIS 6), as Shotton concluded.

The recognition of the eastern Fenland ice margin extends the Tottenhill glacial limit south and south-west (Figs. 7.2 and 7.3) and indicates that, at its maximum extent, the ice lobe must have occupied the entire Fenland basin. Independent confirmation of this interpretation derives from the numerical dates and a possible western limit recognised by H. E. Langford (unpublished, 2007) west of Peterborough, Cambridgeshire and north of Uffington, Lincolnshire. A potential more northerly equivalent in Lincolnshire is the Welton Till, described from the Welton-le-Wold area by Straw (1991, 2005), which in both its stratigraphical position and lithology closely compares with red-brown eastern Fenland diamicton. However, Straw (2000, 2005) favoured an older age for the glaciation which he equated to MIS 8 (i.e. Middle Wolstonian) rather than MIS 6, on the basis of landscape relationships in his area, although he notes that it 'could fall into any of the [MIS] Stages 6, 8 or 10' (Straw, 2005, p. 34). A similar conclusion was reached by White et al. (2010) in the East Midlands, although in both cases the dating of these sequences is equivocal. Therefore, whilst it is potentially possible that these observations might represent a different event to that seen in East Anglia, further work is required to clarify the local correlation question.

Despite these divergent views, on balance, the evidence indicates that during the Late Wolstonian, a substantial ice-lobe advanced down the eastern side of Britain and filled the Fenland basin (Fig. 7.2), where it dammed a series of westward-flowing streams to form shallow glacial lakes that coalesced culminating in an extensive proglacial lake in immediate contact with the ice-front. This lake drained westwards to the North Sea via the river Waveney valley (see below).

7.4.2.3. Miscellaneous Glacial Deposits

In many other parts of the country, poorly dated glacial deposits of restricted lateral extent have also been attributed to a glaciation between the Ipswichian (MIS 5e) and Hoxnian (MIS 11c) interglacial stages, but only because they seem to be either older than Ipswichian or younger than Hoxnian, and are lithologically distinct from extensive local Anglian and Devensian glacial deposits. They are particularly identified where they occur outside the accepted Devensian ice limit. Examples include (a) the Bridlington Member or Basement Till of East Yorkshire which is lithologically similar to the Welton Member in Lincolnshire (cf. above) and Warren House Gill on the coast of County Durham (the Warren House Formation); (b) the Oakwood Formation at Chelford, Cheshire; (c) the Ridgacre Formation overlying the Hoxnian Quinton lacustrine deposits at Quinton, Birmingham; (d) the reddish brown till-like Fremington Member of Barnstaple Bay, N Devon; (e) the Pilkenzane Formation of Lancashire; (f) the Thornsgill Formation of Thornsgill (Lake District); (g) the Balby Formation at Balby and Brayton Barff in the Vale of York and (h) the Bakewell Formation at Shining Bank Quarry, in north Derbyshire.

The Bridlington Member is the most likely candidate for a Wolstonian/Saalian (MIS 6) Till in northern England (Catt, 1991; Catt et al., 2006). This unit underlies the Ipswichian (∼MIS 5e) raised beach at Sewerby near Bridlington on the Yorkshire coast (Catt and Penny, 1966), and at Speeton in Filey Bay overlies the estuarine Raincliff Formation (Speeton Shell Bed), which contains estuarine shells dated to MIS 7 by the amino acid technique (Wilson, 1991). However, it is possible that both the Raincliff Formation and the Bridlington Member at Speeton were glacially rafted to their present position in the Late Devensian (Catt et al., 2006; cf. below).

Offshore a glacial limit based on extent of tunnel valleys and a ridge-like push-morainic structure has been identified north of East Anglia (the Norfolk High) and continuing eastwards towards the centre line and beyond to join the Netherlands' Drenthe glaciation maximum limit by Moreau et al. (2009) and Moreau (2010) (Fig. 7.3). This strongly defined feature can be differentiated from the Anglian/Elsterian limit on the basis of detailed seismic analysis.

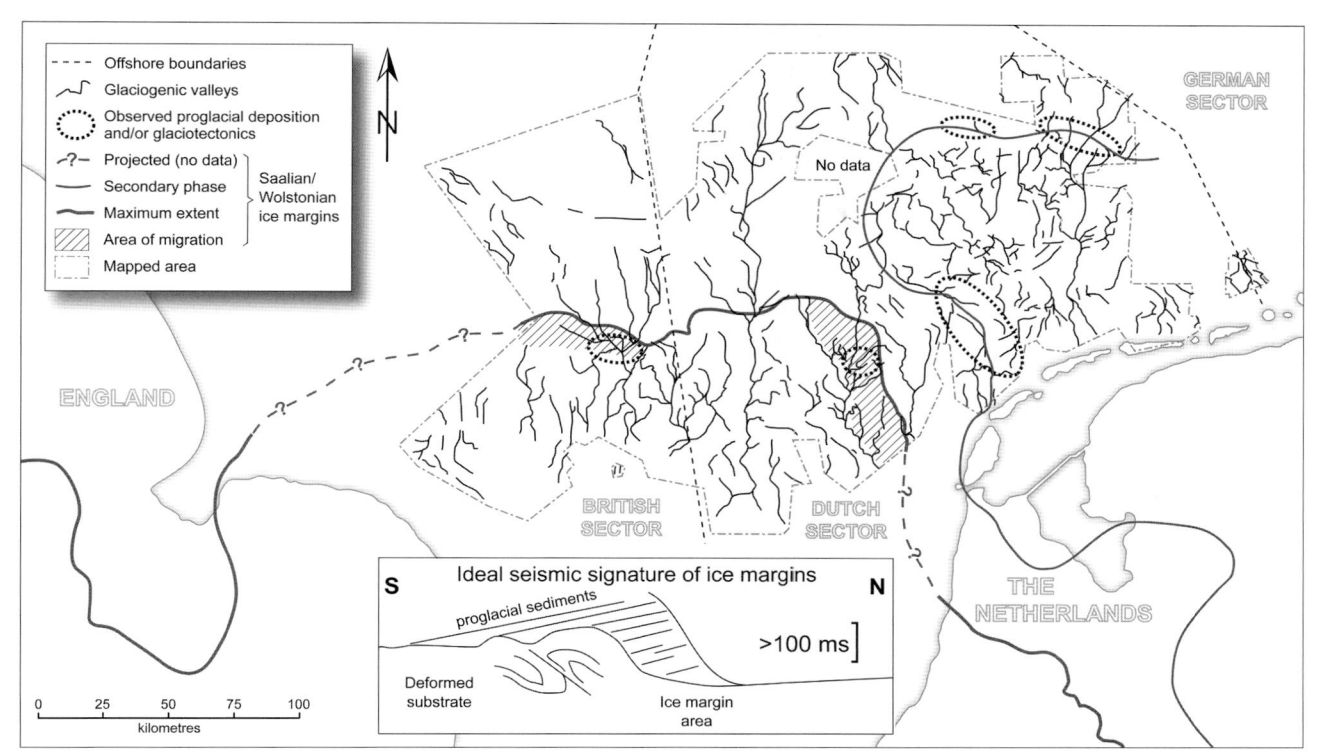

FIGURE 7.3 Ice-front positions recognised for the Wolstonian/Saalian Stage glaciation in the southern North Sea, based on detailed mapping of tunnel-valley networks, marginal formations and associated landforms using seismic data. The lines indicated correspond to regionally correlated fronts between onshore and offshore regions. The correlation derives from this study, Praeg (1996), Clark et al. (2004), Busschers et al. (2008) and Gibbard et al. (2009). The local frontal position does not correlate to a particular glacial phase. The maximum ice front is attributed to the maximum extent of the Wolstonian/Saalian glaciation (180–126 ka, MIS 6—see discussion in the text), whilst the secondary phase position represents either a readvance or a recessional stillstand (after Moreau, 2010).

The identification of this limit confirms that the glacial maximum identified in the Fenland region by Gibbard et al. (1992, 2009) is indeed a continuation of that of the same age as the Drenthe Amersfoort Ice-Push Ridge Complex limit in the Netherlands (Busschers et al., 2008).

Comparison with the Netherlands' sequence shows a remarkable parallelism of events during this glaciation. In particular, the recognition of a major glacial advance during the Drenthe Substage (Stadial) during which the glacier reached a maximum stillstand position in the Central Netherlands (Laban and van der Meer, 2004; Busschers et al., 2007, 2008). It is remarkable that the ice-sheet limit in the Netherlands reaches virtually the same latitude as the Wolstonian ice-lobe, identified in eastern England. Numerical (OSL) dating of the marginal sediments associated with this ice position in the Netherlands indicates a mean date for the glaciation maximum at ca. 160–150 ka (=MIS 6). According to Rappol (1987) and Van den Berg and Beets (1987), the ice-movement direction from the NW requires confluence with British ice. This conclusion has recently been reinforced by the identification of a major proglacial lake that formed in the southern North Sea during this ice advance (Cohen et al., 2005; Busschers et al., 2007, 2008) analogous to that in the Anglain/Elsterian, although of more limited proportions. Such a lake could only have formed if the British and Continental ice sheets were confluent north of the lake. Although an initial lake level close to modern mean sea level is reconstructed for the Saalian southern North Sea lake, this level is lowered as a consequence of spillway erosion in the Dover Strait area during the Saalian glaciation maximum (Busschers et al., 2008). This evidence conclusively demonstrates the time equivalence of the lowland Wolstonian glaciation (Tottenhill glaciation limit) of eastern England and that of the Netherlands during the Late Saalian Drenthe Substage (Amersfoort ice-pushed ridge and sandur complex phase).

The recognition that the most extensive glaciations in lowland Britain occurred in the Anglian, the Late Wolstonian (Saalian-Drenthe) and the Devensian (i.e. MIS 12, 6 and 4-2, respectively) begs the question 'Was there glaciation during intermediate cold periods?' as discussed above. Evidence from the Bay of Biscay sequence offshore of the mouth of the English Channel (Toucanne et al., 2009b) based on 'Fleuve Manche' (Channel River) discharges indicates that glaciation occurred in each event and that these palaeodischarges were directly controlled by the European ice sheets' behaviour during glacial stages MIS 12, 10, 8, 6 and 4-2. However, 'Fleuve Manche' activity during the MIS 10 (Early Wolstonian/Saalian) and MIS 8 (Middle Wolstonian/Saalian) glaciations was significantly less than during MIS 6 and MIS 2 implying that European ice sheets during the penultimate and last glacial periods were more extensive in comparison to the two previous glaciations. Potentially therefore, the ice was markedly more limited in extent during MIS 10 and 8. It also implies that the Scandinavian and British ice masses were almost certainly not confluent across the North Sea basin during these phases (Toucanne et al., 2009a,b).

7.4.3. Extent and Timing of the Devensian (Weichselian) Ice Sheet

During the Devensian Stage, spanning MISs 5d to 2 and generally between ca. 70 and 11.7 ka, ice sheets waxed and waned over the British Isles reaching a culmination in spatial extent during the Late Devensian Substage. Here, this chapter focuses on this latter stage and the maximum extent and timing of this last ice sheet.

Since publication of the last assessment of British glacial limits (Clark et al., 2004) in the European volume of *Quaternary Glaciations—Extent and Chronology* (Ehlers and Gibbard, 2004), much new evidence has been revealed necessitating a revision of the extent and timing of the last ice sheet, and which radically extends both its coverage and timespan. It is mostly marine evidence that has required this change and comes from discoveries of submarine moraines and other landforms on the continental shelf and in the North Sea and from the analysis and dating of ice rafted debris. In Fig. 7.4, the long-held view of the extent of the Late Devensian ice sheet is depicted along with the revised, more extensive version. This changes the reconstruction from a mostly terrestrially constrained and smaller ice sheet (c. 357,000 km^2) to a version twice as large in areal extent and covering extensive areas of current seafloor including the North Sea and continental shelves of Britain and Ireland.

The change in assessment comes from a variety of sources and that seem to be speaking in harmony, always a good sign, and we note that three independent reviews all tackling the issue of maximum ice extent during the Late Devensian reach broadly the same view. Ballantyne (2010) reviewed the burgeoning database of cosmogenic exposure ages and demonstrated that the long-argued Late Devensian 'ice free enclaves' in northeast Scotland and in southernmost Ireland (Fig. 7.4) are in fact erroneous and that the dating evidence is best satisfied by complete terrestrial ice cover here. Chiverrell and Thomas (2010) provide a thorough review of dates and their stratigraphical contexts relevant to the maximum extent, concluding that the ice sheet broadly reached the extents in Fig. 7.4, and they provide a detailed narrative and analysis of dates and sites. Their paper serves as a much more thorough treatment of the topic than exercised here. A third synthesis used all available dates ($n=881$) relevant to the British and Irish ice sheet (build-up, maximum and retreat) to compile a map and GIS database (Hughes et al., 2011), which was then used with landform information to assess the maximum extent of the ice sheet and its pattern and timing of retreat (Clark et al., 2010). The assessment reported below on the timing of maximum extent of ice draws directly from this publication and from Hughes (2008) and Greenwood (2008).

The most prominent discoveries necessitating the reassessment of maximum ice cover are of extensive moraine systems on the continental shelf surrounding northern Scotland (Bradwell et al., 2008) and which continue anticlockwise around the shelf to south-west Ireland (Clark et al., 2010; ÓCofaigh et al., 2010). They clearly demonstrate that, at some time, ice extended to the shelf edge and paused for long enough to build up large moraines. Unfortunately, the moraines have yet to be directly dated. However, moraines of this size and freshness have elsewhere (e.g. Canada, USA, and Fennoscandia), usually been found to be from the last glacial when dated. A more convincing argument for this timing comes from analysis of Ice-Rafted Debris (IRD) concentrations found in cores taken from the trough mouth fans and the seafloor adjacent and beyond the continental shelf edge (Scourse et al., 2009 and references therein). Although there is some uncertainty in how to interpret increases in IRD flux within a core (possibilities include ice advance, retreat, increase in ice streaming, or changes in debris load), leading to varying interpretations (reviewed and explained in Scourse et al., 2009), the first occurrence of significant concentrations of IRD within a core can be taken to indicate that the adjacent ice sheet must, somewhere, have reached a marine-terminating margin from which icebergs were calved. Scourse et al. (2009) report a 'pronounced increase' in IRD at the Rosemary Bank and Barra-Donegal Fan at 29 ka, which requires that the ice sheet must have grown into the sea, but we know not where or how large the marine-terminating margin was. However, it is reasonable to presume that the margin was advancing across the shelf to the NW of Scotland and elsewhere around much of the ice sheet perimeter at this time. By 27 ka, Wilson et al. (2002) have demonstrated that ice must have reached the shelf edge in at least one location because the Barra-Donegal trough mouth fan was being fed with new material. Further south, along the western Irish margin, IRD fluxes also occur at 27 ka (Peck et al., 2006; Scourse et al., 2009). The requisite ice fronts to feed debris-laden icebergs may have existed just at some places along the margin, perhaps as over-extended and isolated lobes (cf. ice streams), or the margin may have reached the shelf edge everywhere from SW Ireland to the Shetlands. Moraines on the shelf and palaeo-margin patterns

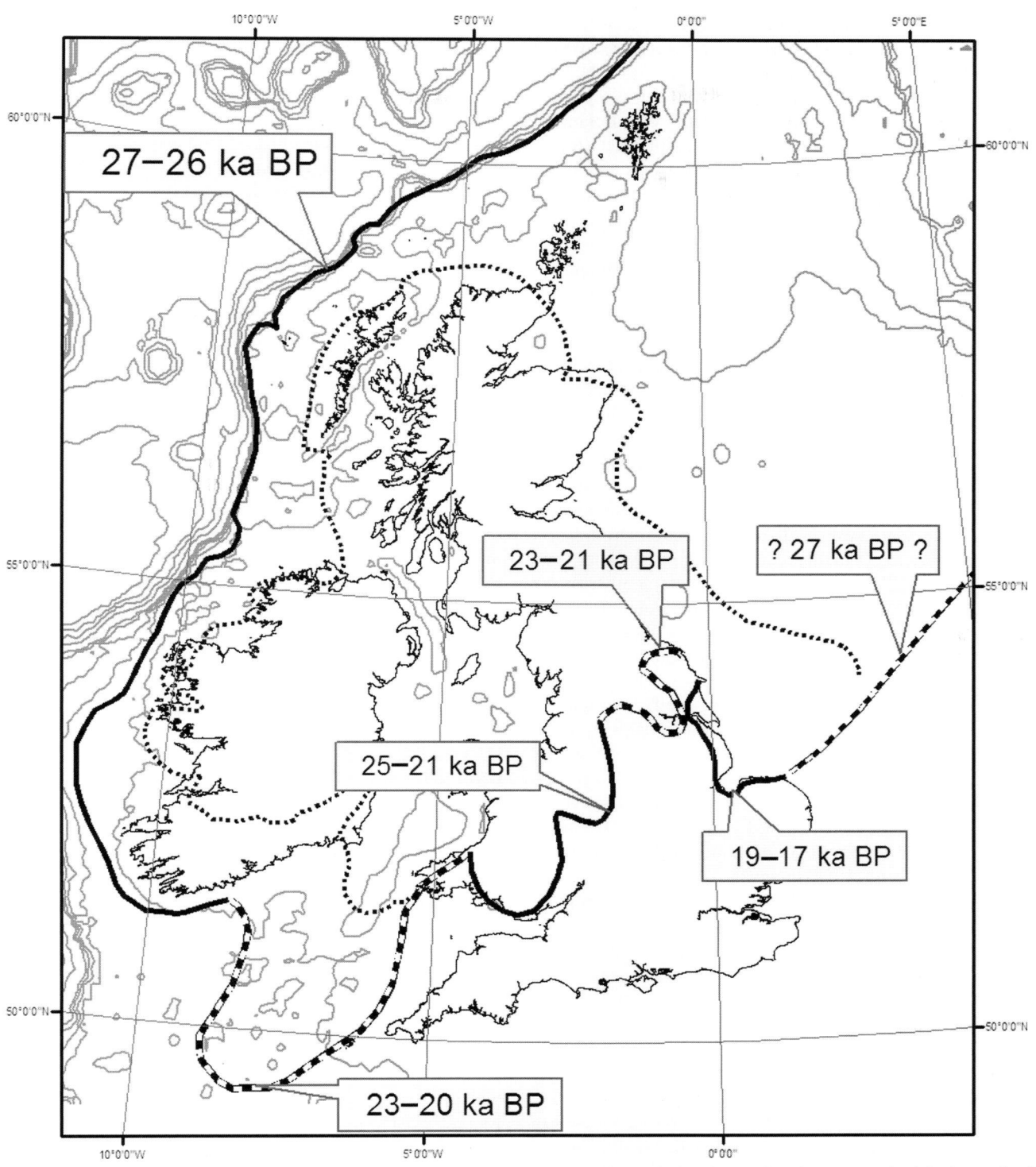

FIGURE 7.4 Maximum limits of the last (i.e. Late Devensian) British-Irish ice-sheet annotated according to when they were attained (in thousands of years before present). The larger extent (thick line) is the assessment reported here and includes complete ice cover of Ireland and Scotland and with ice spreading to the continental shelf edge and covering the North Sea, confluent at some time with the Fennoscandian ice sheet. The smaller extent (dotted line) is the 'traditional' view that held sway for many decades and considered parts of Ireland and Scotland to be ice free at the LGM and with a very limited spread of ice beyond current coastlines, and is depicted here by the assessment of Bowen et al. (1986) and was reported in the first edition of this series (Clark et al., 2004). The new depiction of the extent of the BIIS is twice as large as previously thought, around three quarters of a million km^2 in area and with a probable volume of around a third of that of the current West Antarctic ice sheet. Note that the ice sheet reached its maximum extent at different times in different sectors. The disparity between the maximum extent on the continental shelf, reached as early as 27 ka, against the southern limit of around 25–17 ka might be because the ice sheet could not further expand to the north and west because it ran out of continental shelf and reached water too deep for grounding. Shelf bathymetry is shown with 100 m contours.

reconstructed from them (Clark et al., 2010) appear to indicate the latter scenario because moraines can be traced almost continuously along most of the shelf edge. The simplest inference is that ice reached the shelf edge everywhere by 27 ka along the whole boundary from SW Ireland to the Shetlands. This is consistent with the interpretation of Bradwell et al. (2008), who reconstruct ice at the shelf-break from Scotland and across the North Sea to Norway between 30 and 25 ka and follows Sejrup et al. (1994, 2005, 2009).

The vertical extent of ice, as deduced from trimline evidence, also has a bearing on the lateral extent across the continental shelf because ice thickness at ice divides can be used, along with typical modern ice surface profiles, to estimate lateral spread. Such an (indirect) approach has previously been used to estimate ice extents offshore (e.g. Ballantyne et al., 1998) but is now largely redundant given the discovery of moraines at the shelf edge. Also, the interpretation of many British trimlines representing the upper limit of LGM ice (i.e. demarcating nunataks; Ballantyne et al., 1998) has now been largely rejected (Ballantyne, 2010) and they are best regarded as transitions from warm to cold-based ice, and thus record *minimum* rather than maximum ice thicknesses. This represents another major change from the previous assessment of ice limits (reviewed in Clark et al., 2004) because without trimlines constraining the vertical extent of ice, we can thus envisage a thicker ice sheet.

Having dealt with the continental shelf, we continue anticlockwise around the margin. Advance dates from the Isles of Scilly suggest that these islands were reached after ca. 25 ka (range 26.9–24.6 ka; Scourse, 2006). This is consistent with dates from the Celtic Sea placing ice advance by a short-lived ice stream after 24.2 cal. ka BP (Ó Cofaigh and Evans, 2007). Ice was at the southern Welsh limit by 23 ka (Phillips et al., 1994; range 25.2–21.2 ka). For the rest of the southern margin of the ice sheet, the picture is more complicated. The actual limits recorded by sediment and landform evidence are reviewed in Clark et al. (2004) and Chiverrell and Thomas (2010), here we concentrate on the timing of these limits. The youngest date for advance into the Cheshire Plain suggests that ice advanced inland here after 27 ka (M. Bateman, personal communication, 2009; range 29–25 ka). However, a woolly mammoth bone dated to between 20.0 and 22.9 cal. ka BP (Rowlands, 1971; Bowen, 1974) lying below Irish Sea till could be used to suggest that Irish Sea ice did not advance up the Vale of Clwyd, and potentially the Cheshire Plain, until after 21 cal. ka BP. Although this date is significantly younger than the advance dates in the Cheshire Plain, such a situation is not inconsistent with them. Alternative scenarios include that incursion of ice into the Vale of Clwyd was initially prevented by the presence of Welsh Ice, the location existed as an ice-free enclave until 21 cal. ka BP, the bone date is unreliable or the date reflects an oscillation of the ice margin in this region around 21 cal. ka BP.

Ice advanced down the Vale of York after 23.3 ka (Bateman et al., 2008; range 24.8–21.8 ka) but had retreated to the north by 20.5 ka (range 21.7–19.3). Dates from Dimlington on Holderness suggest that ice did not reach the eastern English coastline until after 22 cal. ka BP (Penny et al., 1969; range 22.5–21.3 ka) and dates from inland Lincolnshire suggest that ice did not progress inland until after ca. 17 ka (Wintle and Catt, 1985; range 19.1–14.9 ka). Ice at this position at this time is consistent with a recently published age for a beach deposit (16.6 ka; Bateman et al., 2008; range 17.8–15.4) related to Glacial Lake Humber, the existence of which requires ice damming the Humber Gap. This new date is significantly younger than the previously quoted maximum age for Lake Humber of 26.2 cal. ka BP (Gaunt, 1974, 1976; range 28.1–24.2 cal. ka BP) and inconsistent with a more recent date for deposition of sands into Lake Humber at ca. 22 ka (Murton et al., 2009) that is difficult to reconcile with the Dimlington date for ice first reaching the eastern coast. It is suggested that the dates at these sites could reflect oscillations of the ice margin, including sporadic damming of the Humber Gap. In the absence of deglacial dates preceding the 'young' advance dates, it is not possible to confirm or disprove this.

If all of the above dates are accepted there are two possible interpretations: (1) Ice did not reach eastern England or the Cheshire Plain until after 17 and 21 ka, respectively; (2) The dates reflect oscillations of the ice margin within the last glaciation. This implies advance into the Cheshire Plain after 27 ka, followed by retreat to an unknown position north of Wales before 21 ka and a subsequent readvance south after 21 ka. This could reflect oscillations of the Irish Sea glacier during uncoupling with Welsh ice. In eastern England, the dates could be interpreted as advance after 25 ka (Ventris, 1985) followed by retreat to an unknown offshore position, followed by a readvance at least as far as Dimlington after 22 ka, with ice reaching the Lincolnshire Wolds after 17 ka. The Dimlington dates have been invoked to support a contemporaneous readvance of the British ice sheet with the Tampen Readvance of the Scandinavian ice sheet (Sejrup et al., 1994; Carr, 2004).

Of the whole perimeter of the ice sheet, the least constrained margin is that in the southern North Sea and for which virtually no information exists. Given good evidence for ice existence in the northern North Sea, then there clearly must have been a southern margin to this, but we know not where. In Fig. 7.4, along with many others, we merely interpolate a line from the eastern England limit to the Main Stationary Line in Denmark (Houmark-Nielsen, 2004).

It is clear that the maximum extent of various ice margins varied with location; the ice sheet did not synchronously reach its most lateral spread everywhere. This is a common finding for other palaeo ice sheets. Fig. 7.4 crudely summarises the maximum limits marked with a date or range of dates suggested by the chronology described above. The disparity between a maximum extent on the continental shelf reached as early as 27 ka against the southern extent of around 23 ka might simply be because the ice sheet may have attempted to continue expanding to the north and west if it had not 'run out' of continental shelf and reached water too deep for grounding (Clark et al., 2010).

The Devensian ice sheet must have been large enough by 70 ka to have at least some marine-terminating margins in order to satisfy offshore evidence from ice rafted debris (Hibbert et al., 2010). It likely experienced considerable waxing and waning during the Devensian culminating in its maximum areal coverage by 27 ka, with asynchronous maximum extents particularly between the NW (marine) margins and southern (mostly terrestrial) margins. It should be noted, however, that much of the margin requires better chronological control, and that large parts are still undated by any direct means.

ACKNOWLEDGEMENT

We thank Philip Stickler (University of Cambridge) for drafting the figures.

REFERENCES

Allen, P., Cheshire, D.A., Whiteman, C.A., 1991. The tills of southern East Anglia. In: Ehlers, J., Gibbard, P.L., Rose, J. (Eds.), Glacial Deposits in Great Britain and Ireland. Balkema, Rotterdam, pp. 255–278.

Ashton, N., Lewis, S.J., Parfitt, S.A., Penkman, K.E.H., Coope, G.R., 2008. New evidence for complex climate change in MIS 11 from Hoxne, Suffolk, UK. Quatern. Sci. Rev. 27, 652–668.

Baden-Powell, D.F.W., 1948. The chalky bounder clays of Norfolk and Suffolk. Geol. Mag. 85, 279–296.

Ballantyne, C.K., 2010. Extent and deglacial chronology of the last British-Irish Ice Sheet: implications of exposure dating using cosmogenic isotopes. J. Quatern. Sci. 25, 515–534.

Ballantyne, C.K., Harris, C., 1994. The Periglaciation of Great Britain. Cambridge University Press, Cambridge, 330pp.

Ballantyne, C.K., McCarroll, D., Nesje, A., Dahl, S.O., Stone, J.O., 1998. The last ice sheet in north-west Scotland: reconstruction and implications. Quatern. Sci. Rev. 17, 1149–1184.

Balson, P.S., Cameron, T.D.J., 1985. Quaternary mapping offshore East Anglia. Modern Geol. 9, 221–239.

Banham, P.H., Gibbard, P.L., Lunkka, J.P., Parfitt, S.A., Preece, R.C., Turner, C., 2001. A critical assessment of 'A New Glacial Stratigraphy for Eastern England'. Quatern. Newslett. 93, 5–14.

Barclay, W.J., Brandon, A., Ellison, R.A., Moorlock, B.S.P., 1992. A Middle Pleistocene palaeovalley-fill west of the Malvern Hills. J. Geol. Soc. 149, 75–92.

Bateman, M.D., Buckland, P.C., Chase, B., Frederick, C.D., Gaunt, G.D., 2008. The Late Devensian proglacial Lake Humber: new evidence from littoral deposits east of Ferrybridge, Yorkshire, England. Boreas 37, 195–210.

Boswell, P.G.H., 1914. On the occurrence of the North Sea Drift (Lower Glacial) and certain other brick-earths in Suffolk. Proc. Geol. Assoc. 25, 121–153.

Bowen, D.Q., 1974. The Quaternary of Wales. In: Owen, T.R. (Ed.), The Upper Palaeozoic and Post-Palaeozoic Rocks of Wales. University of Wales Press, Cardiff, pp. 373–426.

Bowen, D.Q., 1999. Wales. In: Bowen, D.Q. (Ed.), A Revised Correlation of the Quaternary Deposits in the British Isles, 79–90, Geological Society Special Report 23.

Bowen, D.Q., Rose, J., McCabe, A.M., Sutherland, D.G., 1986. Correlation of Quaternary glaciations in England, Ireland, Scotland and Wales. Quatern. Sci. Rev. 5, 299–340.

Bradwell, T., Stoker, M.S., Golledge, N.R., Wilson, C.K., Merritt, J.W., Long, D., et al., 2008. The northern sector of the last British Ice Sheet: maximum extent and demise. Earth Sci. Rev. 88, 207–226.

Bridge, D.M., 1988. Corton cliffs. In: Gibbard, P.L., Zalasiewicz, J.A. (Eds.), Pliocene-Middle Pleistocene of East Anglia, Field Guide. Quaternary Research Association, Cambridge, pp. 119–125.

Bridge, D.M., Hopson, P.M., 1985. Fine gravel, heavy mineral and grain size analysis of mid-Pleistocene deposits in the lower Waveney Valley, East Anglia. Modern Geol. 9, 129–144.

Bristow, C.R., Cox, F.C., 1973. The Gipping Till: a reappraisal of East Anglian glacial stratigraphy. J. Geol. Soc. 129, 1–37.

Busschers, F.S., Kasse, C., Van Balen, R.T., Vandenberghe, J., Cohen, K.M., Weerts, H.J.T., et al., 2007. Late Pleistocene evolution of the Rhine in the southern North-Sea Basin: imprints of climate change, sea-level oscillations and glacio-isostasy. Quatern. Sci. Rev. 26, 3216–3248.

Busschers, F.S., van Balen, R.T., Cohen, K.M., Kasse, C., Weerts, H.J.T., Wallinga, J., et al., 2008. Response of the Rhine–Meuse fluvial system to Saalian ice-sheet dynamics. Boreas 37, 377–398.

Cameron, T.D.J., Holmes, R., 1999. The continental shelf. In: Bowen, D.Q. (Ed.), A Revised Correlation of Quaternary Deposits in the British Isles, 125–139, Geological Society Special Report 23.

Cameron, T.D.J., Stoker, M.S., Long, D., 1987. The history of Quaternary sedimentation in the UK sector of the North Sea basin. J. Geol. Soc. 144, 43–58.

Campbell, S., Hunt, C.O., Scourse, J.D., Keen, D.H., Croot, D.G., 1999. Southwest England. In: Bowen, D.Q. (Ed.), A Revised Correlation of the Quaternary Deposits in the British Isles, 66–78, Geological Society Special Report 23.

Carr, S., 2004. The North Sea basin. In: Ehlers, J., Gibbard, P.L. (Eds.), Quaternary Glaciations—Extent and Chronology, vol. 1. Elsevier, Europe, Amsterdam, pp. 47–82.

Catt, J.A., 1991. The Quaternary history and glacial deposits of East Yorkshire. In: Ehlers, J., Gibbard, P.L., Rose, J. (Eds.), Glacial Deposits in Great Britain and Ireland. Balkema, Rotterdam, pp. 185–191.

Catt, J.A., Penny, L.F., 1966. The Pleistocene deposits of Holderness, East Yorkshire. Proc. Yorkshire Geol. Soc. 35, 375–420.

Catt, J.A., Gibbard, P.L., Lowe, J.J., McCarroll, D., Scourse, J.D., Walker, M.J.C., et al., 2006. Quaternary: ice sheets and their legacy. In: Brenchley, P.J., Rawson, P.F. (Eds.), The Geology of England and Wales. Geological Society, London, 592pp.

Chiverrell, R.C., Thomas, G.S.P., 2010. Extent and timing of the Last Glacial Maximum (LGM) in Britain and Ireland: a review. J. Quatern. Sci. 25, 535–549.

Clark, C.D., Gibbard, P.L., Rose, J., 2004. Glacial limits in the British Isles. In: Ehlers, J., Gibbard, P.L. (Eds.), Extent and Chronology of Glaciation, vol. 1. Elsevier, Europe, Amsterdam, pp. 47–82.

Clark, C.D., Hughes, A.L.C., Greenwood, S.L., Jordan, C., Sejrup, H.P., 2010. Pattern and timing of retreat of the last British-Irish Ice Sheet. Quatern. Sci. Rev.

Cohen, K.M., Busschers, F.S., Gibbard, P.L., Dehnert, A., Preusser, F., 2005. Stratigraphical implications of an elsterian pro-glacial 'North Sea' lake. In: SEQS Subcommission European Quaternary Stratigraphy 2005 Annual Meeting, 22 Bern, Switzerland.

Coope, G.R., Field, M.H., Gibbard, P.L., Greenwood, M., Richards, A.E., 2002. Palaeontology and biostratigraphy of Middle Pleistocene river sediment in the Mathon Member, at Mathon, Herefordshire, England. Proc. Geol. Assoc. 113, 237–258.

Cornwell, J.D., Carruthers, R.M., 1986. Geophysical studies of a buried valley system near Ixworth, Suffolk. Proc. Geol. Assoc. 97, 357–364.

Cox, F.C., 1985. The tunnel valleys of Norfolk, East Anglia. Proc. Geol. Assoc. 96, 357–369.

Cox, F.C., Nickless, E.F.P., 1972. Some aspects of the glacial history of central Norfolk. Bull. Geol. Surv. Great Britain 42, 79–98.

Ehlers, J., Gibbard, P.L., 1991. Anglian glacial deposits in Britain and the adjoining offshore regions. In: Ehlers, J., Gibbard, P.L., Rose, J. (Eds.), Glacial Deposits in Great Britain and Ireland. Balkema, Rotterdam, pp. 17–24.

Ehlers, J., Gibbard, P.L., 1997. The contorted drift of Norfolk, England. In: Festschrift für Professor Dr. Lothar EißmannLeipziger Geowissenschaften 5, 105–113.

Ehlers, J., Gibbard, P.L. (Eds.), 2004. Quaternary Glaciations—Extent and Chronology. Part I: Europe. Elsevier, Amsterdam, 488pp.

Ehlers, J., Meyer, K.-D., Stephan, H.-J., 1984. The pre-Weichselian glaciations of North-West Europe. Quatern. Sci. Rev. 3, 1–40.

Ehlers, J., Gibbard, P.L., Whiteman, C.A., 1987. Recent investigations of the Marly Drift of northwest Norfolk, England. In: Van der Meer, J.J.M. (Ed.), Tills and Glaciotectonics. Balkema, Rotterdam, pp. 39–54.

Ehlers, J., Gibbard, P.L., Rose, J., 1991a. Glacial deposits of Britain and Europe: general overview. In: Ehlers, J., Gibbard, P.L., Rose, J. (Eds.), Glacial Deposits in Great Britain and Ireland. Balkema, Rotterdam, pp. 493–501.

Ehlers, J., Gibbard, P.L., Whiteman, C.A., 1991b. The glacial deposits of northwestern Norfolk. In: Ehlers, J., Gibbard, P.L., Rose, J. (Eds.), Glacial Deposits in Great Britain and Ireland. Balkema, Rotterdam, pp. 223–232.

Fish, P.R., Whiteman, C.A., 2001. Chalk micropalaeontology and the provenancing of Middle Pleistocene Lowestoft formation till in eastern England. Earth Surface Proc. Landforms 26, 953–970.

Gaunt, G.D., 1974. A radiocarbon date relating to Lake Humber. Proc. Yorkshire Geol. Soc. 40, 195–197.

Gaunt, G.D., 1976. The Devensian maximum ice limit in the Vale of York. Proc. Yorkshire Geol. Soc. 40, 631–637.

Gibbard, P.L., 1979. Middle Pleistocene drainage in the Thames valley. Geol. Mag. 116, 35–44.

Gibbard, P.L., 1985. Pleistocene History of the Middle Thames Valley. Cambridge University Press, Cambridge, 155pp.

Gibbard, P.L., 1988. The history of the great north-west European rivers during the past three million years. Philos. Trans. R. Soc. Lond. B318, 559–602.

Gibbard, P.L., 1991. The Wolstonian Stage in East Anglia. In: Lewis, S.G., Whiteman, C.A., Bridgland, D.R. (Eds.), Central East Anglia and the Fen Basin, Field Guide. Quaternary Research Association, London, pp. 7–14.

Gibbard, P.L., 1994. Pleistocene History of the Lower Thames Valley. Cambridge University Press, Cambridge, 229pp.

Gibbard, P.L., 1995. Formation of the Strait of Dover. In: Preece, R.C. (Ed.), Island Britain—A Quaternary Perspective, 15–26, Geological Society Special Publication, 96.

Gibbard, P., 2007. Europe cut adrift. Nature 448, 259–260.

Gibbard, P.L., Turner, C., 1988. In defence of the Wolstonian Stage. Quatern. Newslett. 54, 9–14.

Gibbard, P.L., West, R.G., 2000. Quaternary chronostratigraphy: the nomenclature of terrestrial sequences. Boreas 29, 329–336.

Gibbard, P.L., West, R.G., Andrew, R., Pettit, M., 1991. Tottenhill, Norfolk (TF 636115). In: Lewis, S.G., Whiteman, C.A., Bridgland, D.R. (Eds.), Central East Anglia and the Fen Basin, Field Guide. Quaternary Research Association, London, pp. 131–143.

Gibbard, P.L., West, R.G., Andrew, R., Pettit, M., 1992. The margin of a Middle Pleistocene ice advance at Tottenhill, Norfolk, England. Geol. Mag. 129, 59–76.

Gibbard, P.L., Boreham, S., Burger, A.W., Roe, H.M., 1996. Middle Pleistocene lacustrine deposits in eastern Essex and their palaeogeographical implications. J. Quatern. Sci. 11, 281–298.

Gibbard, P.L., Moscariello, A., Bailey, H.W., Boreham, S., Koch, C., Lord, A.R., et al., 2006. Reply: Middle Pleistocene sedimentation at Pakefield, Suffolk, England: a response to Lee et al. (2006). J. Quatern. Sci. 23, 85–92.

Gibbard, P.L., Pasanen, A., West, R.G., Lunkka, J.P., Boreham, S., Cohen, K.M., et al., 2009. Late Middle Pleistocene glaciation in eastern England. Boreas 38, 504–528.

GRASP 2009. Stratigraphic architecture of the southern North Sea tunnel valleys. Glaciogenic Reservoirs and Hydrocarbon Systems. Abstracts 1-2 December 2009 Petroleum Group Geological Society.

Greenwood, S.L., 2008. A palaeo-glaciological reconstruction of the last Irish Ice Sheet. Unpublished PhD thesis, University of Sheffield, 360pp.

Hamblin, R.J.O., Moorlock, B.S.P., Rose, J., 2000. A new glacial stratigraphy for Eastern England. Quatern. Newslett. 92, 35–43.

Hibbert, F.D., Austin, W.E.N., Leng, M.J., Gatliff, R.W., 2010. British Ice Sheet dynamics from North Atlantic ice-rafted debris records spanning the last 175,000 years. J. Quatern. Sci. 25, 461–482.

Hart, J.K., 1987. The genesis of north-east Norfolk Drift. Unpublished Ph.D. thesis, University of East Anglia.

Hart, J.K., Boulton, G.S., 1991. The glacial drifts of Norfolk. In: Ehlers, J., Gibbard, P.L., Rose, J. (Eds.), Glacial Deposits in Great Britain and Ireland. Balkema, Rotterdam, pp. 233–243.

Hoare, P., Connell, E.R., 2004. The first appearance of Scandinavian indicators in East Anglia's glacial record. Bull. Geol. Soc. Norfolk 54, 3–12.

Hoare, P.G., Gale, S.J., Robinson, R.A.J., Connell, E.R., Larkin, N.R., 2009. Marine Isotope Stage 7-6 transition age for beach sediments at Morston, north Norfolk, UK: implications for Pleistocene chronology, stratigraphy and tectonics. J. Quatern. Sci. 24, 311–316.

Hopson, P.M., Bridge, D.M.C., 1987. Middle Pleistocene stratigraphy in the lower Waveney valley, East Anglia. Proc. Geol. Assoc. 98, 171–185.

Horton, A., 1974. The sequence of Pleistocene deposits proved during the construction of Birmingham motorways. Institute of Geological Sciences Report 74/22, 21pp.

Houmark-Nielsen, M., 2004. The Pleistocene of Denmark: a review of stratigraphy and glaciation history. In: Ehlers, J., Gibbard, P.L. (Eds.),

Quaternary Glaciations Extent and Chronology—Part I: Europe. Elsevier, Amsterdam, pp. 35–46.

Hughes, A.L.C., 2008. The last British Ice Sheet: a reconstruction based on glacial landforms. Unpublished Ph.D. thesis, University of Sheffield, 297pp.

Hughes, A.L.C., Greenwood, S.L., Clark, C.D., 2011. Dating constraints on the last British-Irish Ice Sheet: a map and database. J. Maps 2011, 156–183.

Laban, C., 1995. The Pleistocene glaciations in the Dutch sector of the North Sea. PhD Thesis, University of Amsterdam, 194pp.

Laban, C., van der Meer, J.J.M., 2004. Pleistocene glaciation in The Netherlands. In: Ehlers, J., Gibbard, P.L. (Eds.), Quaternary Glaciations Extent and Chronology—Part I: Europe. Elsevier, Amsterdam, pp. 251–260.

Lee, J.R., Rose, J., Riding, J.B., Moorlock, B.S.P., Hamblin, R.J.O., 2004a. Testing the case for a Middle Pleistocene Scandinavian glaciation in Eastern England: evidence for a Scottish ice source for tills within the Corton Formation of East Anglia. Boreas 31, 345–355.

Lee, J.R., Rose, J., Hamblin, R.J.O., Moorlock, B.S.P., 2004b. Dating the earliest lowland glaciation of eastern England: the pre-Anglian early Middle Pleistocene Happisburgh Glaciation. Quatern. Sci. Rev. 23, 1551–1566.

Lee, J.R., Rose, J., Candy, I., Barendregt, R.W., 2006. Sea-level changes, river activity, soil development and glaciation around the western margins of the Southern North Sea basin during the Early and early Middle Pleistocene: evidence from Pakefield, Suffolk, U.K.. J. Quatern. Sci. 21, 155–179.

Lee, J.R., Busschers, F.S., Sejrup, H.-P., 2010. Pre-Weichselian Quaternary glaciations of the British Isles, The Netherlands, Norway, and adjacent marine areas south of 68N: implications for long-term ice-sheet development in northern Europe. Quatern. Sci. Rev. doi:10.1016/j.quascirev.2010.02.027 (in press).

Leszczynska, K., 2009. Pleistocene glacigenic deposits of the Danbury-Tiptree Ridge, Essex, England: Glaciogenic Reservoirs and Hydrocarbon Systems. The Geological Society Conference Papers, 1–2 December 2009.

Lewis, S.G., Rose, J., 1991. Tottenhill, Norfolk (TF 639120). In: Lewis, S.G., Whiteman, C.A., Bridgland, D.R. (Eds.), Central East Anglia and the Fen Basin, Field Guide. Quaternary Research Association, London, pp. 145–148.

Lunkka, J.P., 1994. Sedimentology and lithostratigraphy of the North Sea Drift and Lowestoft Till Formations in the coastal cliffs of NE Norfolk. J. Quatern. Sci. 9, 209–233.

Maddy, D., Green, C.P., Lewis, S.G., Bowen, D.Q., 1995. Pleistocene Geology of the lower Severn valley, U.K.. Quatern. Sci. Rev. 14, 209–222.

Mathers, S.J., Zalasiewicz, J.A., Bloodworth, A.J., Morton, A.C., 1987. The Banham Beds: a petrologically distinct suite of Anglian glacigenic deposits from central East Anglia. Proc. Geol. Assoc. 98, 229–240.

Mathers, S.J., Zalasiewicz, J.A., Gibbard, P.L., Peglar, S.M., 1993. The Anglian-Hoxnian evolution of an ice-marginal drainage system in Suffolk, England. Proc. Geol. Assoc. 104, 109–122.

Mitchell, G.F., Penny, L.F., Shotton, F.W., West, R.G., 1973. A Correlation of Quaternary eposits in the British Isles. Geological Society of London Special Report 4, 99pp.

Moorlock, B.S.P., Booth, S., Fish, P., Hamblin, R.J.O., Kessler, H., Riding, J., Rose, J. & Whiteman, C.A. 2000. In: Lewis, S.G., Whiteman, C.A. & Preece, R.C. (Eds). The Quaternary of Norfolk and Suffolk, Field Guide, Quaternary Research Association, London, 53–54.

Moreau, J., 2010. The seismic analysis of the southern North Sea glaciogenic record. GRASP report No.1. Delft, The Netherlands.

Moreau, J., Huuse, M., Gibbard, P.L., Moscariello, A., 2009. 3D seismic megasurvey geomorphology of the Southern North Sea, Tunnel valley record and associated ice-sheet dynamic. In: 71st EAGE Conference & Exhibition, 8-11 June 2009, Amsterdam, the Netherlands.

Morgan, A.V., 1973. The Pleistocene geology of the area north and west of Wolverhampton, Staffordshire, England. Philos. Trans. R. Soc. Lond. Ser. B 265, 233–297.

Murton, D.K., Pawley, S.W., Murton, J.B., 2009. Sedimentology and luminescence ages of Glacial Lake Humber deposits in the central Vale of York. Proc. Geol. Assoc. 120, 209–222.

Ó Cofaigh, C., Evans, D.J.A., 2007. Radiocarbon constraints on the age of the maximum advance of the British-Irish Ice Sheet in the Celtic Sea. Quatern. Sci. Rev. 26, 1197–1203.

ÓCofaigh, C., Dunlop, P., Benetti, S., 2010. Marine geophysical evidence for Late Pleistocene ice sheet extent and recession off northwest Ireland. Quatern. Sci. Rev. doi:10.1016/j.quascirev.2010.02.005.

Pawley, S.M., Rose, J., Lee, J.R., Moorlock, B.S.P., Hamblin, R.J.O., 2004. Middle Pleistocene sedimentology and lithostratigraphy of Weybourne, northeast Norfolk, England. Proc. Geol. Assoc. 115, 25–42.

Pawley, S.M., Bailey, R.M., Rose, J., Moorlock, B.S.P., Hamblin, R.J.O., Booth, S.J., 2008. Age limits on Middle Pleistocene glacial sediments from OSL dating, north Norfolk, UK. Quatern. Sci. Rev. 27, 1363–1377.

Peck, V.L., Hall, I.R., Zahn, R., Elderfield, H., Grousset, F., Hemming, S.R., et al., 2006. High resolution evidence for linkages between NW European ice sheet instability and Atlantic meridional overturning circulation. Earth Planet. Sci. Lett. 243, 475–488.

Penny, L.F., Coope, G.R., Catt, J.A., 1969. Age and insect fauna of the Dimlington silts, East Yorkshire. Nature 224, 65–67.

Perrin, R.M.S., Rose, J., Davies, H., 1979. The distribution, variations and origins of pre-Devensian tills in eastern England. Philos. Trans. R. Soc. Lond. B287, 536–570.

Phillips, F.M., Bowen, D.Q., Elmore, D., 1994. Surface exposure dating of glacial features in Great Britain using cosmogenic chlorine-36: preliminary results. Mineral. Mag. 58A, 722–723.

Pointon, K.W., 1978. The Pleistocene succession at Corton, Suffolk. Bull. Geol. Soc. Norfolk 30, 55–76.

Praeg, D., 1996. Morphology, stratigraphy and genesis of buried mid-Pleistocene tunnel-valleys in the southern North Sea Basin. PhD thesis, University of Edinburgh.

Praeg, D., 2003. Seismic imaging of mid-Pleistocene tunnel-valleys in the North Sea Basin—high resolution from low frequencies. J. Appl. Geophys. 53, 273–298.

Preece, R.C., Parfitt, S.A., 2000. The Cromer Forest-bed Formation: new thoughts on an old problem. In: Lewis, S.G., Whiteman, C.A., Preece, R.C. (Eds.), The Quaternary of Norfolk and Suffolk, Field Guide. Quaternary Research Association, London, pp. 1–27.

Preece, R.C., Parfitt, S.A., Coope, G.R., Penkman, K.E.H., Ponel, P., Whittaker, J.E., 2009. Biostratigraphic and aminostratigraphic constraints on the age of the Middle Pleistocene glacial succession in north Norfolk, UK. J. Quatern. Sci. 24, 557–580.

Pringle, J., George, T.N., 1961. British Regional Geology, South Wales. HMSO, London, 152pp.

Rappol, M., 1987. Saalian till in The Netherlands: a review. In: van der Meer, J.J.M. (Ed.), Tills and Glaciotectonics. A.A. Balkema, Rotterdam, pp. 3–21.

Rice, R.J., 1968. The Quaternary deposits of the central Leicestershire. Philos. Trans. R. Soc. Lond. A262, 459–509.

Rice, R.J., 1981. The Pleistocene deposits of the area around Croft in south Leicestershire. Philos. Trans. R. Soc. Lond. B293, 385–418.

Rice, R.J., Douglas, T., 1991. Wolstonian glacial deposits and glaciation in Britain. In: Ehlers, J., Gibbard, P.L., Rose, J. (Eds.), Glacial Deposits in Great Britain and Ireland. Balkema, Rotterdam, pp. 25–36.

Richards, A.E., 1998. Re-evaluation of the Middle Pleistocene stratigraphy of Herefordshire. J. Quatern. Sci. 13, 115–136.

Rose, J., 1987. Status of the Wolstonian glaciation in the British Quaternary. Quatern. Newslett. 53, 1–9.

Rose, J., 1994. Major river systems of central and southern Britain during the Early and Middle Pleistocene. Terra Nova 6, 435–443.

Rose, J., 2009. Early and Middle Pleistocene landscapes of eastern England. Proc. Geol. Assoc. 120, 3–33.

Rose, J., Whiteman, C.A., Allen, P., Kemp, R.A., 1999. The Kesgrave Sands and Gravels: 'pre-glacial' Quaternary deposits of the River Thames in East Anglia and the Thames valley. Proc. Geol. Assoc. 110, 93–116.

Rose, J., Moorlock, B.S.P., Hamblin, R.J.O., 2001. Pre-Anglia fluvial and coastal deposits in Eastern England: lithostratigraphy and palaeoenvironments. Quatern. Int. 79, 5–22.

Rowe, P.J., Richards, D.A., Atkinson, T.C., Bottrell, S.H., Cliff, R.A., 1997. Geochemistry and radiometric dating of a Middle Pleistocene peat. Geochim. Cosmochim. Acta 61, 4201–4211.

Rowlands, B.M., 1971. Radiocarbon evidence of the age of an Irish Sea glaciation in the Vale of Clwyd. Nat. Phys. Sci. 230, 9–11.

Scourse, J.D. (Ed.), 2006. The Isles of Scilly. Field Guide. Quaternary Research Association, London.

Scourse, J.D., Austin, W.E.N., Sejrup, H.-P., Ansari, M.H., 1999. Foraminiferal isoleucine epimerization determinations from the Nar Valley Clay, Norfolk, UK: implications for Quaternary correlations in the southern North Sea basin. Geol. Mag. 136, 543–560.

Scourse, J.D., Haapaniemi, A.I., Colmenero-Hidalgo, E., Peck, V.L., Hall, I.R., Austin, W.E.N., et al., 2009. Growth, dynamics and deglaciation of the last British-Irish ice sheet: the deep-sea ice-rafted detritus record. Quatern. Sci. Rev. 28, 3066–3084.

Sejrup, H.P., Haflidason, H., Aarseth, I., King, E., Forsberg, C.F., Long, D., et al., 1994. Late Weichselian glaciation history of the northern North Sea. Boreas 23, 1–13.

Sejrup, H.P., Hjelstuen, B.O., Dahlgren, K.I.T., Haflidason, H., Kuijpers, A., Nygard, A., et al., 2005. Pleistocene glacial history of the NW European continental margin. Marine Petrol. Geol. 22, 111–1129.

Sejrup, H.P., Nygard, A., Hall, A.M., Haflidason, H., 2009. Middle and Late Weichselian (Devensian) glaciation history of south-western Norway, North Sea and eastern UK. Quatern. Sci. Rev. 28, 370–380.

Shotton, F.W., 1953. The Pleistocene deposits of the area between Coventry, Rugby and Leamington, and their bearing on the topographic development of the Midlands. Philos. Trans. R. Soc. Lond. B237, 209–260.

Shotton, F.W., 1968. The Pleistocene succession around Brandon, Warwickshire. Philos. Trans. R. Soc. Lond. B254, 387–400.

Shotton, F.W., 1976. Amplification of the Wolstonian Stage of the British Pleistocene. Geol. Mag. 113, 241–250.

Shotton, F.W., 1983a. The Wolstonian Stage of the British Pleistocene in and around its type area of the English Midlands. Quatern. Sci. Rev. 2, 261–280.

Shotton, F.W., 1983b. Observations on the type Wolstonian glacial sequence. Quatern. Newslett. 40, 28–36.

Slater, G., 1929. Quaternary period. In: Evans, J.W., Stubblefield, C.J. (Eds.), Handbook of the Geology of Great Britain: A Compilative Work. Thomas Murby & Co., London, pp. 457–498.

Solomon, J.D., 1935. The Westleton Series of East Anglia: its ages, distribution and relations. Q. J. Geol. Soc. Lond. 91, 216–238.

Sparks, B.W., West, R.G., 1964. The drift landforms around Holt, Norfolk. Trans. Inst. British Geograph. 35, 27–35.

Straw, A., 1983. Pre-Devensian glaciation of Lincolnshire (Eastern England) and adjacent areas. Quatern. Sci. Rev. 2, 239–260.

Straw, A., 1991. Glacial deposits of Lincolnshire and adjoining areas. In: Ehlers, J., Gibbard, P.L., Rose, J. (Eds.), Glacial Deposits in Great Britain and Ireland. Balkema, Rotterdam, pp. 213–221.

Straw, A., 2000. Some observations on 'Eastern England' in a revised correlation of Quaternary deposits of the British Isles (ed. Bowen, D.Q.). Quatern. Newslett. 91, 1–10.

Straw, A., 2005. Glacial and Pre-Glacial Deposits at Welton-le-Wold, Lincolnshire. Exeter, Straw, 39pp.

Sumbler, M.G., 1983a. A new look at the type Wolstonian Glacial deposits. Proc. Geol. Assoc. 94, 23–31.

Sumbler, M.G., 1983b. The type Wolstonian sequence—some further observations. Quatern. Newslett. 40, 36–39.

Sumbler, M.G., 1995. The terraces of the rivers Thame and Thames and their bearing on the chronology of glaciation in central and eastern England. Proc. Geol. Assoc. 106, 93–106.

Sumbler, M.G., 2001. The Moreton Drift: a further clue to glacial chronology in central England. Proc. Geol. Assoc. 112, 13–27.

Tappin, D.R., Chadwick, R.A., Jackson, A.A., Wingfield, R.T.R., Smith, N.J.P., 1994. The geology of Cardigan Bay and the Bristol Channel. British Geological Survey, UK Offshore Regional report, London, HMSO, 107pp.

Toucanne, S., Zaragosi, S., Bourillet, J.F., Cremer, M., Eynaud, F., Van Vliet-Lanoe, B., et al., 2009a. Timing of massive 'Fleuve Manche' discharges over the last 350 kyr: insights into the European ice-sheet oscillations and the European drainage network from MIS 10 to 2. Quatern. Sci. Rev. 28, 1238–1256.

Toucanne, S., Zaragosi, S., Gibbard, P.L., Bourillet, J.F., Cremer, M., Eynaud, F., et al., 2009b. A 1.2 my record of glaciation and fluvial discharge from the West European continental margin. Quatern. Sci. Rev. 28, 2974–2981.

Turner, C., 1996. A brief survey of the early Middle Pleistocene in Europe. In: Turner, C. (Ed.), The Early Middle Pleistocene in Europe. Balkema, Rotterdam, pp. 295–317.

van den Berg, M.W., Beets, D.J., 1987. Saalian glacial deposits and morphology in The Netherlands. In: van der Meer, J.J.W. (Ed.), Tills and Glaciotectonics. Balkema, Rotterdam, pp. 235–251.

van der Vegt, P., Janszen, A., Moreau, J., Gibbard, P.L., Huuse, M., Moscariello, A., 2009. Glacial sedimentary systems and tunnel valleys of East Anglia, England. Glaciogenic Reservoirs and Hydrocarbon Systems. Abstracts 1-2 December 2009 Petroleum Group Geological Society.

Van Gijssel, K., 2006. A continent-wide framework for local and regional stratigraphies; application of genetic sequence and event stratigraphy to the Middle Pleistocene terrestrial succession of Northwest and Central Europe. Leiden University, Ph.D. dissertation, 119 pp.

Ventris, P.A., 1985. Pleistocene environmental history of the Nar Valley, Norfolk. Unpublished Ph.D. thesis, University of Cambridge.

Ventris, P.A., 1986. The Nar Valley. In: West, R.G., Whiteman, C.A. (Eds.), The Nar Valley and North Norfolk, Field Guide. Quaternary Research Association, Coventry, pp. 6–55.

Ventris, P.A., 1996. Hoxnian Interglacial freshwater and marine deposits in northwest Norfolk, England and their implications for sea-level reconstruction. Quatern. Sci. Rev. 15, 437–450.

West, R.G., 1980a. Pleistocene forest history in East Anglia. New Phytol. 85, 571–622.

West, R.G., 1980b. The Pre-Glacial Pleistocene of the Norfolk and Suffolk Coasts. Cambridge University Press, Cambridge, 203pp.

West, R.G., Donner, J.J., 1956. The glaciations of East Anglia and the East Midlands: a differentiation based on stone-orientation measurements of the tills. Q. J. Geol. Soc. Lond. 112, 69–91.

Westaway, R., Maddy, D., Bridgland, D.R., 2002. Flow in the lower continental crust as a mechanism for the Quaternary uplift of south-east England: constraints from the Thames terrace record. Quatern. Sci. Rev. 21, 559–603.

White, T.S., Bridgland, D.R., Westaway, R., Howard, A.J., White, M.J., 2010. Evidence from the Trent terrace archive, Lincolnshire, UK, for lowland glaciation of Britain during the Middle and Late Pleistocene. Proc. Geol. Assoc. 121, 141–153.

Wilson, S.J., 1991. The correlation of the Speeton Sheel Bed, Filey Bay, Yorkshire, to an oxygen isotope stage. Proc. Yorkshire Geol. Soc. 48, 223–226.

Wilson, L.J., Austin, W.E.N., Jansen, E., 2002. The last British Ice Sheet: growth, maximum extent and deglaciation. Polar Res. 21, 243–250.

Wingfield, R.T.R., 1989. Glacial incisions indicating Middle and Upper Pleistocene ice limits in Britain. Terra Nova 1, 538–548.

Wingfield, R.T.R., 1994. Pleistocene and Holocene. In: Tappin, D.R., Chadwick, R.A., Jackson, A.A., Wingfield, R.T.R., Smith, N.J.P. (Eds.), The Geology of Cardigan Bay and the Bristol Channel, British Geological Survey, UK Offshore Regional Report. HMSO, London, pp. 76–93.

Wingfield, R.T.R., 1995. Pleistocene and Holocene. In: Jackson, D.I., Jackson, A.A., Evans, D., Wingfield, R.T.R., Barnes, R.P., Arthur, M.J. (Eds.), The Geology of the Irish Sea, British Geological Survey, UK Offshore Regional report. HMSO, London, pp. 85–102.

Wintle, A.G., Catt, J.A., 1985. Thermoluminescence dating of Dimlington Stadial deposits in eastern England. Boreas 14, 231–234.

Woodland, A.W., 1970. The buried tunnel-valleys of East Anglia. Proc. Yorkshire Geol. Soc. 37, 521–578.

Worsley, P., 1991. Glacial deposits of the lowlands between the Mersey and Severn rivers. In: Ehlers, J., Gibbard, P.L., Rose, J. (Eds.), Glacial Deposits in Great Britain and Ireland. Balkema, Rotterdam, pp. 203–211.

Chapter 8

Pleistocene Glaciations in Estonia

Volli Kalm[1],*, Anto Raukas[2], Maris Rattas[1] and Katrin Lasberg[1]

[1]*Department of Geology, Institute of Ecology and Earth Sciences, University of Tartu, Ravila 14a, 50411 Tartu, Estonia*
[2]*Institute of Geology, Tallinn Technical University, Ehitajate tee 5, 19086 Tallinn, Estonia*
*Correspondence and requests for materials should be addressed to Volli Kalm. E-mail: volli.kalm@ut.ee

8.1. INTRODUCTION

The following chapter is a review of evidence of Pleistocene glaciations in Estonia as understood at the beginning of 2010. The knowledge on the spatial pattern and timing of glaciations is drawn from a great number of publications starting from the middle of the nineteenth century. Reviews of the history of glacial geology and Pleistocene stratigraphy in Estonia are available in a number of publications (Raukas and Kajak, 1997a,b; Raukas et al., 2004; Raukas, 2006, 2009).

The most widely accepted model of glaciations distinguishes at least five till beds in superposition and in buried bedrock valleys, occasionally isolated from each other by Holsteinian (Karuküla) and Eemian (Prangli) stage interglacial deposits or interstadial beds (Raukas and Kajak, 1997a; Raukas et al., 2004; Kalm, 2006). From the dense network of boreholes available, it is known that glacial, glaciofluvial and glaciolacustrine deposits comprise about 95% of the Quaternary sediments in Estonia.

In recent years, knowledge of the chronology and palaeogeography of the Pleistocene glaciations in Estonia has developed considerably and has been supported by new, primarily chronological (Kalm, 2005, 2006; Raukas and Stankowski, 2005; Rinterknecht et al., 2006; Molodkov, 2007; Raukas, 2009; Rattas et al., 2010) but also biostratigraphical (Liivrand, 2007; Molodkov et al., 2007; Saarse et al., 2009; Rattas et al., 2010) and palaeogeographical (Kalm, 2010) data. Pleistocene chronostratigraphy in Estonia is based on 69 sites and sections, studied by means of ^{14}C, optically stimulated luminescence/thermoluminescence (OSL/TL), ^{10}Be and varved clay chronology methods. The Late Pleistocene is most extensively studied in Estonia, and a revision of its official stratigraphy (Estonian Land Board: http://geoportaal.maaamet.ee/eng/Maps-and-Data/Geological-data-p317.html) and palaeogeography remains necessary (Kalm, 2006; Molodkov et al., 2007).

8.2. EARLY PLEISTOCENE

Sedimentary records from Early Pleistocene or older glaciations, to which the oldest Pleistocene tills belong in the southern Peribaltic region, Ukraine and western Russia (Glushankova and Sudakova, 1995; Satkunas, 1998; Pavlovskaya, 2000; Matoshko, 2004), have not been preserved in Estonia. Strong erosion by subsequent glaciations has destroyed all the evidence. On the basis of present knowledge, deposits of Lower Pleistocene and older Quaternary ages are absent from Estonia and even the Middle Pleistocene sequence is rather incomplete (Raukas et al., 2004).

8.3. MIDDLE PLEISTOCENE

8.3.1. Elsterian (Sangaste, Oka) Stage Glaciation (Here and Henceforth Correlative Stratigraphical Terms in Estonia and on Russian Plain Are Given in Brackets)

On the basis of its position underlying the Holsteinian (Karuküla) interglacial deposits, the lowermost compact till (Sangaste Till), described in drill-core sections in some buried valleys of central and southern Estonia (Puiestee, Saadjärv, Sudiste, Mägiste; Fig. 8.1), is correlated with the Elsterian glaciation (Raukas, 1978). This brownish or sometimes greenish till, rich in clasts from the Fennoscandian Shield (25–95%), rests directly on bedrock. Within a sequence of Elsterian Till, at a depth of 143.3–169.2 m, clayey sands have been discovered in the Otepää buried valley in southern Estonia (Kajak, 1995). These sands were correlated by Kajak (1995) to the Middle Elsterian Turgeljai and Beloveže beds in Lithuania and Belarus, based on some similarities of the pollen spectra (Kajak and Liivrand, 1967). Glacigenic deposits below these beds were consequently correlated with the Early Elsterian glaciation (Kajak, 1995). However, direct evidence in favour of this conclusion has not been found.

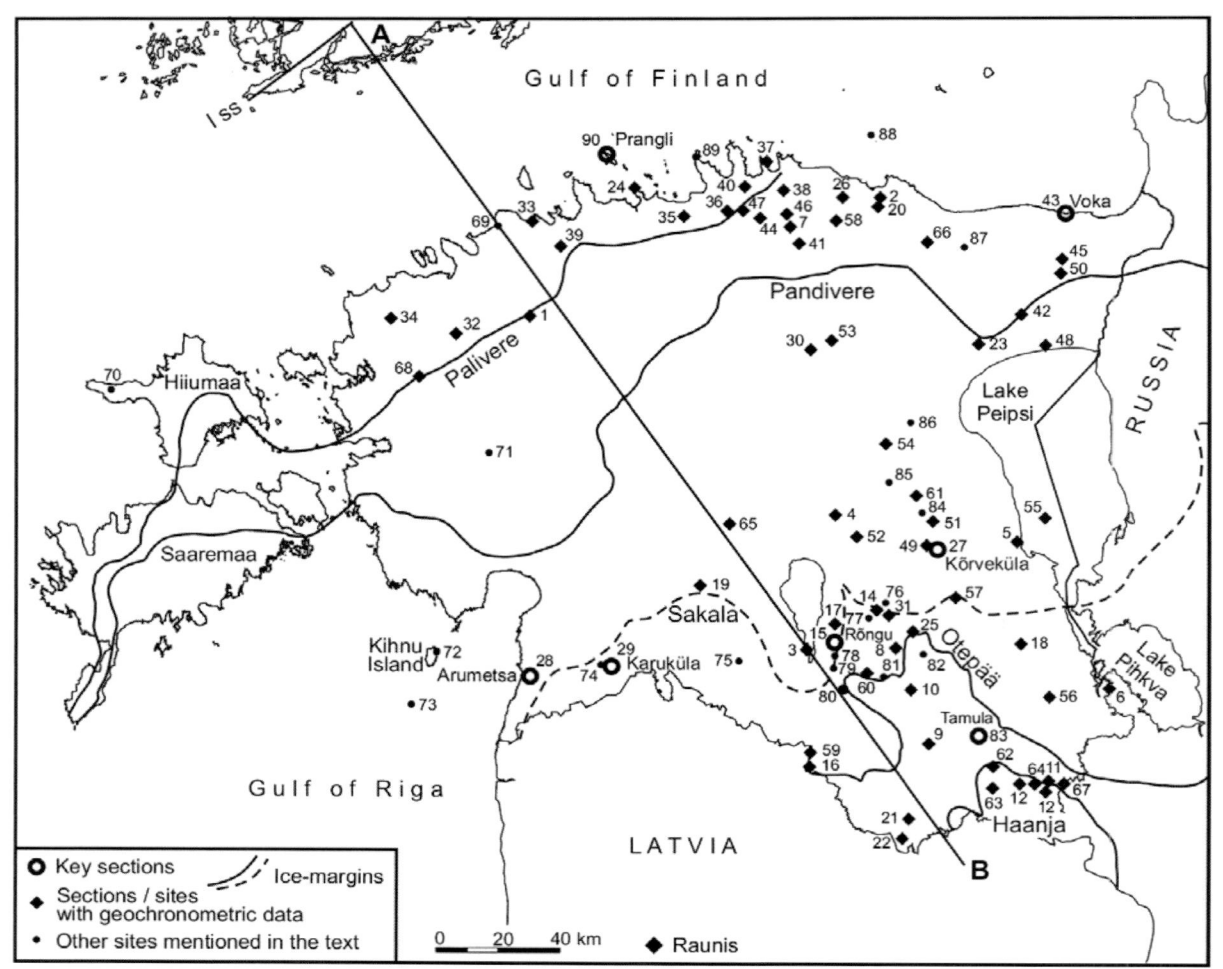

FIGURE 8.1 Location map showing the ice margins (Haanja, Otepää, Sakala, Pandivere, Palivere and Salpausselkä 1 (ISS)) and sites mentioned in the text: 1, Vaharu; 2, Kunda; 3, Naritsa; 4, Puurmani; 5, Saviku; 6, Kuliska; 7, Loobu; 8, Nõuni; 9, Kurenurme; 10, Kaagvere; 11, Remmeski; 12, Viitka; 13, Petruse; 14, Peedu; 15, Rõngu; 16, Valga; 17, Valguta; 18, Mooste; 19, Heimtali; 20, Kunda Lammasmäe; 21, Kallaste; 22, Tahkumägi; 23, Tudulinna; 24, Ihasalu; 25, Kammeri; 26, Pehka; 27, Kõrveküla; 28, Arumetsa; 29, Karuküla; 30, Varangu; 31, Preediku; 32, Valgejärve; 33, Tabasalu; 34, Nõva; 35, Kuusalu; 36, Kemba; 37, Käsmu; 38, Palmse; 39, Männiku; 40, Sillaotsa; 41, Kallukse; 42, Iisaku; 43, Voka; 44, Pikasaare; 45, Pannjärve; 46, Viitna Suurjärv; 47, Valgejõe; 48, Mannjärve; 49, Vasula; 50, Räätsma; 51, Vedu; 52, Laeva; 53, Äntu Sinijärv; 54, Laiuse; 55, Lake Peipsi; 56, Nohipalu; 57, Vana-Kuuste; 58, Haljala; 59, Taimeaia; 60, Äidu; 61, Elistvere; 62, Jaanimäe; 63, Viitina; 64, Pütsepa; 65, Tääksi; 66, Uljaste; 67, Solova; 68, Palivere; 69, Vääna-Jõesuu; 70, Kõpu; 71, Vigala; 72, Kihnu; 73, Gulf of Riga (Core 21); 74, Puiestee; 75, Sudiste; 76, Tõravere; 77, Elva; 78, Kiiti; 79, Aakre; 80, Mägiste; 81, Otepää; 82, Kitse; 83, Tamula; 84, Saadjärve; 85, Visusti; 86, Tõikvere; 87, Savala; 88, Põhja-Uhtju; 89, Juminda; 90, Prangli.

8.3.2. Holsteinian (Karuküla, Likhvin) Stage Interglacial

The two known occurrences of Holsteinian interglacial deposits in Estonia (Karuküla and Kõrveküla; Fig. 8.1) are continental deposits of forest peat, gyttja and lake or river sands (Liivrand, 1984, 1991) described in drill-core sections from southern Estonia. Although both interglacial sections are found in an allochthonous position (Liivrand and Saarse, 1983; Levkov and Liivrand, 1988), the Holsteinian age of peat and gyttja has been established through detailed palynological and carpological (plant macrofossil) investigations (Velichkevich and Liivrand, 1976; Liivrand, 1984, 1991). Because the Holsteinian sediments are of continental origin, they suggest that the Baltic Sea most likely did not reach Estonia (Raukas and Kajak, 1997b) in the stage. The Holsteinian pollen zones in the Kõrveküla and Karuküla sections are identical, and the biostratigraphical correlation to other known Holsteinian sections (Ulmale, Pulvernieki, Butenai, Likhvin) in north-eastern Europe is reliable (Velichkevich and Liivrand, 1976; Liivrand, 1991; Kondratiene, 1993; Donner, 1995). Pollen analysis of the Karuküla interglacial deposits demonstrates widespread conifer forests and a very limited occurrence of *Corylus* (hazel; Liivrand, 1984, 1991) in the region. However, since its discovery in 1939, the Karuküla section had been correlated with the Riss–Würm interglacial (Orviku, 1944), the Brørup interstadial (Orviku and Pirrus, 1965) or even with the Middle Weichselian interstadial (Punning et al., 1969; Serebrjannyi et al., 1969). This last correlation had been previously supported by 20 ^{14}C dates (ages between 33.45 ± 0.8 and ≥ 58.6 ^{14}C ka BP; Kalm, 2005), although half of the samples yielded infinite ages. Recent OSL analysis of fine-grained fluvial and lacustrine sands in the Kõrveküla section (Kalm, 2006) have yielded ages of 68.4 ± 4.2 and 69.5 ± 3.7 ka above and 81.2 ± 7.8 and 93.4 ± 3.7 ka below the interglacial deposits. These ages confirm the allochthonous position of the Holsteinian interglacial deposits and suggest that they have been moved glaciotectonically to their present position. Because the Holsteinian deposits have been displaced from their original location, questions arise concerning the age of the tills below the interglacial deposits, which are considered to be of pre-Holsteinian age.

8.3.3. Saalian (Ugandi, Dniepr) Stage Glaciation

The Saalian glaciation in Estonia is traditionally subdivided into two major substages, the Early Ugandi and the Late Ugandi, correlated, respectively, to Drenthe and Warthe substage glaciations in north-central Europe and to Dniepr and Moscow glaciations on the Russian Plain. However, it is not clear how the two Ugandi till layers in Estonia correspond to the three Saalian tills found in northern Germany (Ehlers et al., 2004) and on the Russian Plain (Early Dniepr, Dniepr Maximum and Moscow; Velichko et al., 2004).

8.3.3.1. The Older Saalian (Drenthe, Early Ugandi, Dniepr) Substage

The older Saalian till occurs in buried valleys in both southern and northern Estonia or in sheltered positions in the cores of the southern Estonian heights. However, at present, there are no known sections where the Saalian complex directly overlies the Holsteinian interglacial deposits. The older Saalian (Early Ugandi) age of the till is deduced from its superposition. The lithologically easily traceable reddish-brown, compact till bed is up to 50 m thick. It is the fourth from the surface and the second complex of glacial deposits below the Eemian Stage interglacial deposits. Its clast composition (the till is rich in Vyborg rapakivi) indicates that the older Saalian till was deposited by a southward moving ice sheet (Raukas, 1978; Raukas and Gaigalas, 1993; Raukas and Kajak, 1997a). Fifteen TL dates (between 43.3 ± 3.0 and 216.0 ± 10.0 ka) were determined for Estonian tills (Kajak et al., 1981) in the late 1970s, a time of over-enthusiasm for the developing TL method. Three of these TL dates (≥ 110.0, ≥ 119.0 and 216.0 ± 10.0 ka) are attributed to the Saalian tills that underlie the Eemian interglacial beds (Kajak et al., 1981). Although the ages obtained may be considered of Saalian age, the TL ages of tills have been omitted from chronological interpretation of Pleistocene stratigraphy in Estonia.

8.3.3.2. Drenthe–Warthe (Middle Ugandi) Interstadial

Possible interstadial deposits are found at a few places (Puiestee, Valguta, Elva, Aakre and Nõuni) in southern Estonia between the two Saalian till strata (Kajak, 1995; Raukas, 1995; Raukas and Kajak, 1997a). Liivrand (1991) has shown that the Middle Ugandi organic sands and sandy loams contain redeposited pollen and their stratigraphical position is not firmly established. However, the 151 ± 8 ka OSL age from the Arumetsa section of Saalian lacustrine sediments (Rattas et al., 2010) correlates well with the mid-marine isotope stage (MIS) 6 discharge event at 155 ka ago interpreted to reflect the retreat of the ice sheet between the Drenthe and Warthe advances (Toucanne et al., 2009).

8.3.3.3. The Younger Saalian (Late Ugandi, Warthe, Moscow) Substage

The younger Saalian till is up to 70 m thick and has been identified in buried valleys in southern and northern Estonia or in the 'cores' of the Otepää and Haanja heights. The till lithology (lacking Vyborg rapakivi and Suursaari quartz-porphyries) and clast fabrics suggest that the ice flow during this glaciation was mostly from the north-west to south-east (Raukas, 1978; Raukas and Gaigalas, 1993; Raukas and Kajak, 1997b). In the Valga borehole, in southern Estonia, the third till unit from the top (the younger Saalian or Moscow Till) was dated using in this case the questionable TL method, which gave an age of 216 ± 10 ka (Kajak et al., 1981).

Recently obtained OSL ages from a 21-m-thick open-pit section of silty clay sediments in the Arumetsa bedrock valley, south-western Estonia, revealed that lacustrine to glaciolacustrine sedimentation at the site began during the Saalian cold Stage prior to 151 ± 8 ka ago. It lasted approximately to the end of MIS 6 at 130 ka (Rattas et al., 2010). Further down from the 151 ka age level to the bottom of the buried valley, there are ca. 60 m of lacustrine fine-grained sediments, the age of which still remain unclear. The late Saalian lacustrine sediments and their respective pollen assemblage zones have also been distinguished in the sections on the Põhja-Uhtju Island (Miettinen et al., 2002) and the Juminda peninsula (Kadastik et al., 2003) in northern Estonia. At both sites, they indicate open late-glacial cold and dry conditions (derived from the presence of herbs *Artemisia* and Chenopodiaceae). The late Saalian silty clay deposits are also found on the Kihnu Island (Liivrand, 1976, 1991, 2007) and in Core 21 in the Gulf of Riga. The latter shows a long continuous vegetation development (Kalnina, 1997, 2001), starting with the Saalian late glacial, continuing through the Eemian Stage interglacial and ending in the Weichselian glaciation. The most complete late Saalian to Eemian pollen succession in Estonia is known from the Prangli core section indicating a gradual change from the late Saalian lacustrine and brackish phase into 'normal marine' conditions in the Early Eemian (Cheremisinova, 1961; Liivrand, 1987, 1991).

8.4. LATE PLEISTOCENE

8.4.1. Eemian (Prangli, Mikulino) Interglacial

Both marine and continental Eemian interglacial deposits have been found in Estonia. In the Gulf of Finland on Prangli Island (Kajak, 1961), and in the Gulf of Riga on Kihnu Island (Liivrand, 1976, 1991, 2007; Raukas, 1978), the Weichselian and Saalian tills are separated by marine clayey deposits of the Eemian Stage interglacial. Diatom analyses of the authochtonous interglacial deposits in the Prangli section (Cheremisinova, 1961; Znamenskaya and Cheremisinova, 1962) have revealed two diatom assemblages that represent two stages in the development of the Eemian Baltic Sea (Raukas, 1991). In some places in south-eastern Estonia (Rõngu, Küti, Kitse, Peedu), peaty gyttja lacustrine deposits occur at the same stratigraphical level (Liivrand, 1977). Although most of the Eemian interglacial deposits in Estonia are found in ice-thrust positions (Liivrand, 1991), both marine and continental sediments may easily be correlated using pollen assemblage zones (Liivrand, 1991; Kondratiene, 1993).

Kajak et al. (1981) reported the first nearly Eemian ages (90.0 and 100.0 TL ka) from these interglacial sediments from alluvial sand and silt in the Peedu section in south-eastern Estonia. The local Estonian stratotype section of the Eemian continental deposits (peat, gyttja, sand) at Rõngu was re-examined recently, and the fine-grained laminated sand above and below the interglacial strata was dated using the OSL method (Kalm, 2006). The dates obtained (32.1 ± 4.4 from above, and 36.7 ± 6.2 and 39.5 ± 2.3 ka from below) appear to verify the earlier ^{14}C ages (≥ 30 ^{14}C ka BP; Liiva et al., 1966) of the organic sediments. However, palynological and diatom investigations have confirmed an Eemian age of these deposits and their correlation to other Eemian sections (Cheremisinova, 1961; Raukas, 1978; Liivrand, 1984, 1991; Kondratiene, 1993; Donner, 1995). Pollen assemblages indicate that the entire Eemian interglacial vegetation cycle is present at Rõngu (Liivrand, 1984, 1991). Assuming that the pollen analysis and bedding conditions in Rõngu site are correct, the OSL ages of deposits containing Eemian erratics reflect glaciotectonic dislocation, in which Eemian deposits have been emplaced into Middle Weichselian sands. The Late Weichselian age of the Rõngu dislocation is supported by the fact that sands with Eemian erratics are covered by only one thin (2 m thick) bed of typical reddish-brown Late Weichselian till.

8.4.2. Weichselian (Järva, Valdai) Glaciation

8.4.2.1. Early Weichselian (Kelnase and Valgjärve, Early Valdai) Substage

The beginning of the Weichselian climatic cooling in Estonia was characterised by deposition of silty clays and sands containing pollen of periglacial vegetation (Liivrand, 1991). Periglacial deposits immediately overlie Eemian interglacial sediments in the Prangli and Kihnu sediment cores, in the Core 21 in the Gulf of Riga, and are probably present between Saalian and Weichselian tills in other core sections (Otepää, Tõravere, Valguta) in southern Estonia (Liivrand, 1995, 2007; Kalnina, 2001). Eight TL dates (Kajak et al., 1981) for a probable Early Weichselian till in south-eastern Estonia (Raukas, 1978; Raukas and Kajak, 1995) range between 46 and 75 ka (and exceptionally 153 ka), which reflect the Middle rather than Early Weichselian time. Recently, Molodkov (2007) and Molodkov et al., 2007 reported from lacustrine sediments in Voka section, north-eastern Estonia, set of OSL ages which span the time period from about 115 to 70 ka and from 44 to 31 ka. If the TL dates from till are excluded, the reliable Early Weichselian ages in Estonia are those from the Voka (115–70 ka; Molodkov, 2007) and the Kõrveküla sections (68–93 ka; Kalm, 2006). These dates suggest an ice-free Estonia during the Early Weichselian. It also corresponds closely to recent evidence from western Finland (Salonen et al., 2008), showing that even Ostrobothnia (Pohjanmaa), located much closer to the glaciation centre, was ice-free throughout the Early Weichselian. These observations support evidence for a lengthy (at least 115–68 ka) ice-free period in Estonia and refute the idea of Early Weichselian glaciation in the country, as suggested earlier (e.g. Raukas et al., 2004).

8.4.2.2. Middle Weichselian (Savala, Middle Valdai) Substage

Palynological and geochronometric investigations have revealed a number of sites (e.g. Savala, Tõravere, Valguta, Peedu, Arumetsa, Vääna-Jõesuu, Pehka, Voka) with Middle Weichselian terrestrial interstadial deposits that yield pollen assemblages representing cold and dry periglacial conditions (Raukas, 1978; Liivrand, 1990, 1991, 1995; Kadastik, 2004; Molodkov, 2007; Molodkov et al., 2007; Rattas et al., 2010). As noted earlier, Molodkov (2007) and Molodkov et al., 2007 reported OSL ages from Voka section which cover the time period from about 44 to 31 ka. Earlier published ^{14}C ages from interstadial organics in the Peedu section (Punning, 1970; Liivrand and Saarse, 1976), ^{14}C accelerator mass spectrometry (AMS) dates from mammoth skeletal finds (Lõugas et al., 2002) and OSL ages from fine-grained sands from the Pehka, Kammeri, Rõngu and Arumetsa sections (Kadastik, 2004; Kalm, 2005, 2006; Rattas et al., 2010) cover time spans of ca. 40–31, >41–31 and 44–27 ka, respectively. Although originating from different materials and different sections, the given age ranges overlap nicely and indicate that Estonia was ice-free at least between 44 and 27 ka. This, together with an ice-free Early Weichselian Substage (between 115 and 68 ka), leaves some 24 ka (68–44 ka) for a possible early Middle Weichselian glaciation, recorded, for example, between 62 and 55 ka in Ostrobothnia, Finland (Salonen et al., 2008) and reached at least south-eastern Finland (Lunkka et al., 2008). However, Molodkov et al. (2007) concluded that there is no evidence in the Voka outcrop, suggesting the presence of glacigenic sediments deposited during the period between about 115 and 31 ka. If that holds true, the question arises how the second till from the surface, widely distributed in central and southern Estonia and in places overlying Eemian interglacial deposits (Raukas, 1978; Kajak, 1995), should be interpreted. Previously, this till complex was presumed to represent the older (Early) Weichselian glaciation (e.g. Raukas, 1978; Raukas et al., 2004).

8.4.2.3. Late Weichselian (Võrtsjärv, Late Valdai) Substage

The beginning of the Late Weichselian glaciation in Estonia has not been directly dated. However, the available data from eastern Finland, located more centrally in relation to the Scandinavian Ice Sheet (SIS), suggest that the onset may have occurred around 22 ka. Lunkka et al. (2001) concluded that ice overrode eastern Finland less than 25 ka ago, while Ukkonen et al. (1999) provided evidence that glaciers did not reach eastern Finland before 22.5 ka. Therefore, the directly dated ice-free period in Estonia (44–27 ka), which began in the Middle Weichselian, can be extended to the beginning of Late Weichselian glaciation some 22 ka ago. Configuration of glacier bedforms and end moraines in south-eastern Lithuania, northern Belarus and on the Valdai Heights in north-western Russia indicate that the earliest advance to the most accepted Last Glacial Maximum (LGM) position was predominantly from the north-west (Kalm, 2010). The Late Weichselian ice sheet deposited tills of different colour and meltwater sediments both above and beneath the till. Moreover, at several sites, intermorainic interstadial or other sediments occur within this till unit. In a number of places, the till also rests directly on Middle Weichselian interstadial deposits.

8.4.2.4. Deglaciation History

During the course of thinning of the ice sheet, the movement of its individual lobes was controlled by the underlying bedrock topography (Tavast and Raukas, 1982). Moreover, the shift in the ice-flow directions might reflect a change in ice pressure/flow gradient between different phases of glaciation (Kalm, 2010). Former ice-marginal positions are represented in the modern topography by discontinuous chains of end moraines and glaciofluvial formations (Raukas and Kajak, 1997b) or marked by overprint of glacier bedforms from consecutive ice flows (Kalm, 2010). Figure 8.2 shows a modelled decay of the Late Weichselian ice in Estonia, with all available dated sites/sections from this time period. Five major ice-marginal zones (in order of decreasing age: Haanja, Otepää, Sakala, Pandivere and Palivere; Fig. 8.1) are usually distinguished in Estonia (Raukas et al., 1971; Raukas and Karukäpp, 1979; Raukas, 1986; Kalm, 2010). They were formed either as the result of stillstands of the ice margin or in some cases, as a result of readvances. At a large scale, the ice-marginal zones in Estonia are sinuous compared to the next, more northerly, relatively linear Salpausselkä end moraines in southern Finland. This suggests a gentler slope of the ice sheet during the deglaciation of Estonia when large ice-stream complexes drained the ice sheet. Ice streams and lobes developed a length of 100–300 km in south-east of the Baltic Sea and are clearly expressed through highly attenuated subglacial bedforms (Kalm, 2010) that correspond to rapid ice flow (Stokes and Clark, 2001).

8.4.2.5. Haanja Ice-Marginal Zone

The Haanja (North Lithuanian, Luga) end moraine zone is the oldest in Estonia. The age of the Haanja zone is traditionally (Pirrus and Raukas, 1996; Raukas, 2009) determined from dating the Raunis interstadial sediments below the Haanja Till (between 15.7 and 15.9 cal. ka BP; Dreimanis and Zelčs, 1995; Zelčs and Markots, 2004) in northern Latvia (Fig. 8.1). After the ice front retreated from the northern slopes of the Haanja Heights, varved clay deposition began in Lake Tamula ca. 14,675 varve years ago (Sandgren et al., 1997; Kalm, 2006; Raukas, 2009). Consequently, dating from below and above the Haanja Stadial Till places the formation of the ice-marginal zone at between ca. 15.7 and 14.7 cal. ka BP. Recently, Raukas (2009) reported two new (10.5 and 10.4 cal. ka BP) dates from the Raunis section and questioned the interstadial age of the sediments there. However, even if the organic sediments in the Raunis section are not interstadial, the Haanja ice-marginal zone did form well before the varved clay

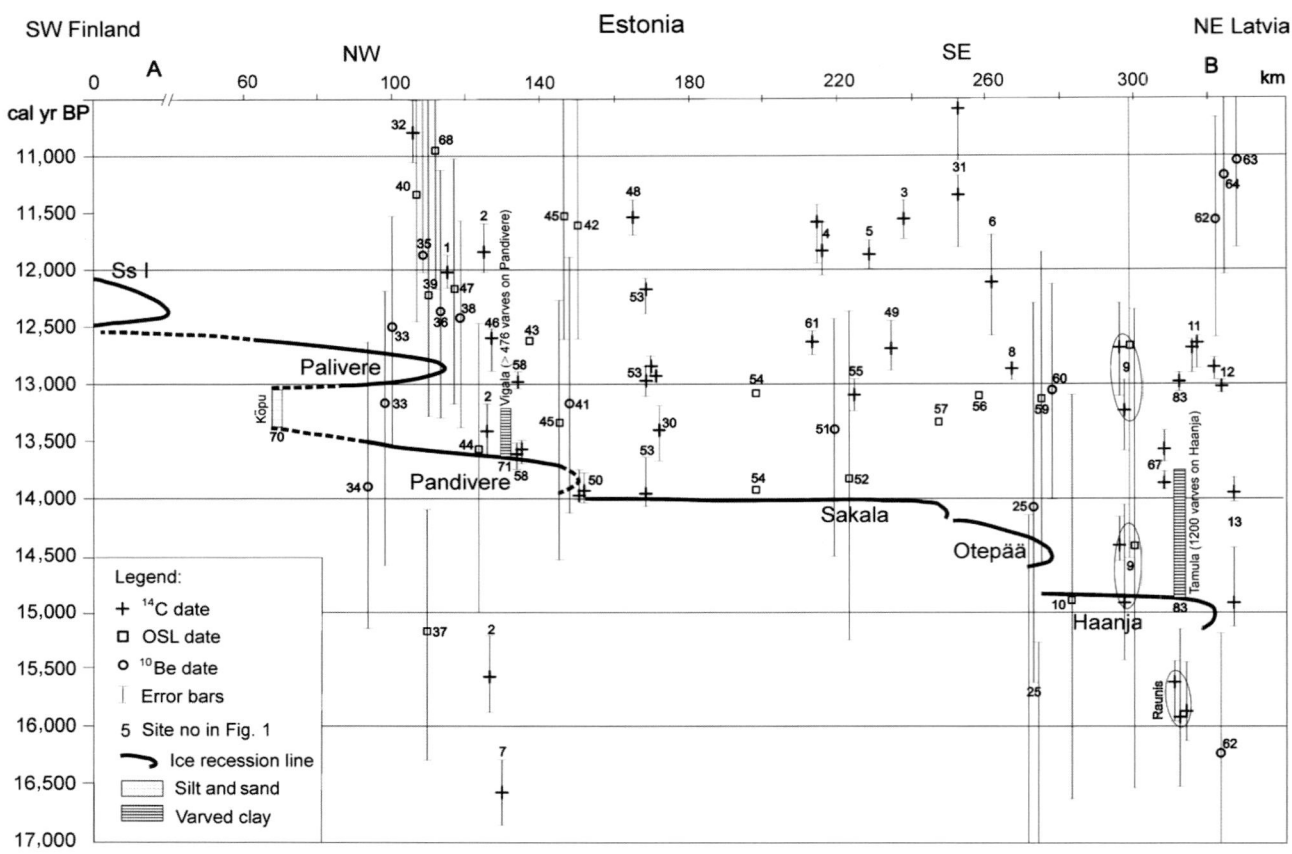

FIGURE 8.2 Modelled decay of the Late Weichselian ice from Estonia along the transect A–B shown in Fig. 8.1 (modified from Kalm, 2006). The numbers denote dated sections indicated in Fig. 8.1.

deposition began in Lake Tamula, that is, before 14.7 ka. If the Estonian mean annual rate of ice recession is estimated (110–120 m/year; Kalm, 2006; Raukas, 2009), it took ca. 800 years for the ice front to recede from the Haanja zone and free the Lake Tamula area for glaciolacustrine deposition. During the Haanja glaciation phase, a small northeast–south-west oriented ice tongue operated between Haanja and the Otepää Heights (Karukäpp, 2004; Kalm, 2010). This was probably responsible for the burial of pieces of wood, gyttja and sand layers under the superficial till in a few localities (Kurenurme, Kaagvere). The oldest ^{14}C ages from buried organic materials under the Haanja Till in Kurenurme (14.5 and 14.9 cal. ka), and OSL ages of buried sand in Kurenurme and Kaagvere (12.7, 14.5 and 14.9 ka), provide evidence that the Haanja zone is older than 14.9 ka.

All five measured ^{14}C ages from Haanja Heights, beyond the Haanja ice-marginal belt (between 12.8 and 14.9 cal. ka BP; Kalm, 2006), are restricted to superficial organic deposits (Remmeski) or plant detritus and peat buried under the slope wash and debris of hummocky morainic relief (Viitka, Petruse). Four ^{10}Be-dated erratics from the Haanja Heights, also occur outside of the Haanja ice margin, have yielded an average age of 12.5 ka, although altogether nine ^{10}Be-dated samples from the Haanja (North Lithuanian) zone in the Baltic region have weighted mean ages of $13.1 + 3$ ka (Rinterknecht et al., 2006). The reported ages are younger or, as a maximum, equal to the minimum age (>14.7 ka) of the Haanja marginal zone, derived from the Lake Tamula varve chronology. According to this, the first organic sediments (gyttja, peat, plant detritus) on the Haanja Heights began accumulating at the time (≤ 14.9 cal. ka BP) when glaciers were surrounding it from the west, north and east (Fig. 8.1). The Haanja Heights, and later also the Otepää Heights, formed ice divides between the individual peripheral ice streams.

8.4.2.6. Otepää and Sakala Ice-Marginal Zones

The age of the Otepää ice-marginal zone is older than its direct dating from the region of Otepää Heights. The 12.9 cal. ka BP age of the first (i.e. basal) late-glacial organic sediments in Nõuni section of lacustrine deposits (Saarse, 1979), which is situated just on the proximal side of the Otepää ice-marginal zone, sets a minimum age for the latter. In addition, two 13.2 ka OSL ages from sand from a glaciofluvial delta (Taimeaia) and a kame (Nohipalu; Raukas and Stankowski, 2005), also located in the proximal position relative to the Otepää ice-marginal zone, indicate an earlier retreat of the ice from the latter. A number of old and recent ^{14}C dates (between 12.9 and 13.9 cal. ka; Saarse and Liiva, 1995; Sohar and Kalm, 2008; Kalm and Sohar, 2010), together with OSL ages (13.1–14.0 ka) from glaciofluvial deposits (Raukas and Stankowski, 2005) from the zone between the Otepää and the Pandivere ice margins, suggest an ice recession from the Otepää marginal well before 14 ka.

Although a Sakala ice-marginal zone has been proposed between the Otepää and Pandivere zones (Pirrus and Raukas, 1996; Raukas et al., 2004; Kalm, 2010), it is morphologically difficult to identify and accurate dating is problematic. End moraines, eskers and kame fields referred to the Sakala zone have been described from the Saadjärve Drumlin Field area between the depressions of Lake Peipsi and Lake Võrtsjärv and on the Sakala Heights in southwestern Estonia (Lõokene, 1961; Raukas et al., 1971).

8.4.2.7. Pandivere Ice-Marginal Zone

Based on varve studies in eastern Estonia, Hang (2003) concluded that deglaciation of the northern part of the glacial Lake Peipsi began around 13,500 varve years ago and ended ca. 370 years later. Hang (2003) also suggested that the first 184 varves in glacial Lake Peipsi were deposited before the ice margin withdrew from the Pandivere marginal zone, and that the start of deposition of distal varves, some 13,300–13,320 varve years ago, corresponds to the withdrawal of the ice from the Pandivere zone. Recently, Saarse et al. (2009) concluded that ice cover of the Pandivere Upland began to perish already about 13.8 cal. ka BP, that is, some 500 years earlier than in the Lake Peipsi depression. One ^{10}Be date from a single boulder (Kallukse) on the Pandivere end moraine yielded an age of 13.1 ^{10}Be ka (Rinterknecht et al., 2006). Geological evidence indicates that the distance of ice recession from the Pandivere ice-marginal zone was considerable. In the Kõpu Peninsula on Hiiumaa Island (Fig. 8.1), lacustrine interstadial sediments are found between the Palivere and Pandivere Stadial Tills, indicating ice-free conditions in north-westernmost Estonia prior to the last Late-Glacial ice readvance (Kadastik and Kalm, 1998).

8.4.2.8. The Palivere Ice-Marginal Zone

Apart from some end moraines on the Island of Saaremaa, the Palivere ice-marginal zone is represented by a curved belt of so-called marginal eskers about 60 km in length. In contrast to the older oscillatory phases in Estonia (Haanja, Otepää, Sakala), the Palivere is clearly of readvance character, as reflected by the distribution of push moraines and by the fact that the older glaciolacustrine sediments are overlain by glaciofluvial material (Raukas and Rähni, 1966). The Palivere ice advance to the limit of its maximum extent must have been at least 35 km (i.e. the distance from the Kõpu Peninsula where interstadial deposits have been found). Hang and Sandgren (1996) counted 476 annual varves at Vigala in western Estonia between the Pandivere and Palivere marginal zones and dated the Palivere marginal formations to about 11.8 ka. These dates include no correction to the Swedish varve chronology, but the number of missing varves is 875 (Andren et al., 1999) and thus the deglaciation of the Palivere marginal zone must be ca. 875 years older, that is, 12.675 ka. From this, an age for the Palivere readvance around 12.7 ka is derived. However, eight ^{10}Be-dated samples from the Palivere ice-marginal zone have weighted mean age of 13.6 ka (Rinterknecht et al., 2006) which predates even the ^{10}Be age (13.1 ka) of the Haanja marginal zone.

8.5. CONCLUSIONS

Taking into account all the lithostratigraphical, biostratigraphical and geochronological information available, the existence of two Pleistocene interglacials—Holsteinian (Karuküla) and Eemian (Prangli)—and three glacials—Elsterian (Sangaste), Saalian (Ugandi) and Weichselian (Järva)—has been determined in Estonia.

The two known occurrences of Holsteinian Stage interglacial sediments (Karuküla and Kõreküla) are continental deposits of forest peat, gyttja and lake or river sands. Although both interglacial sections are found in allochthonous position, the Holsteinian age of the peat and gyttja has been established through detailed palynological and carpological investigations.

There are two Saalian till strata recognised in buried valleys or in sheltered positions in the cores of the southern Estonian heights. Late Saalian lacustrine to glaciolacustrine sedimentation started prior to 151 ka ago and is known from number of sections in northern and western Estonia. A complete Late Saalian to Eemian pollen succession is known from the Prangli core section indicating a gradual change from the Late Saalian lacustrine and brackish phase into the 'normal marine' conditions in the Early Eemian. Both marine and continental Eemian Stage interglacial deposits are present in Estonia, and pollen assemblages indicate that the entire Eemian vegetation cycle is present in the sediments.

Numerous recent OSL ages from Weichselian Stage lacustrine and stream sediments indicate that Estonia was ice-free at least between 115–68 and 44–27 ka ago which leaves in theory 24 ka (68–44 ka) for a possible early Middle Weichselian glaciation.

The beginning of the Late Weichselian glaciation in Estonia has not been directly dated, but the available data from areas located more centrally in relation to the SIS (e.g. Finland) suggest that the onset may have occurred around 22 ka ago. The deglaciation of Estonia from the Haanja ice-marginal zone to the recession of ice from the Palivere zone took place between ca 14.7 and 12.7 ka or between 14,675 and 12,675 varve years ago. The mean annual rate of ice recession in Estonia from the Haanja to the Palivere ice-marginal zone is calculated to be 110 m/year. Hang (2001) calculated the mean rate of ice recession from the Lake Peipsi depression to be 150 m/year. As the onset of the Late Weichselian glaciation in Estonia occurred no earlier than 22 ka, and the ice retreated from the Palivere zone at 12.7 ka, the possible duration of the last glaciation in Estonia ranges between ca. 7.3 and 9.3 ka.

REFERENCES

Andren, T., Björk, J., Johnsen, S., 1999. Correlation of the Swedish glacial varves with the Greenland (GRIP) oxygen isotope stratigraphy. J. Quatern. Sci. 14, 361–371.

Cheremisinova, E.A., 1961. Diatomovye morskikh mezhlednikovykh otlozhenii Estonskoi SSR [Diatoms in the marine interglacial deposits of the Estonian SSR]. Dokl. Akad. Nauk SSSR 141 (3), 698–700 (in Russian).

Donner, J., 1995. The Quaternary History of Scandinavia. Cambridge University Press, Cambridge, 200 pp.

Dreimanis, A., Zelčs, V., 1995. Pleistocene stratigraphy of Latvia. In: Ehlers, J., Kozarski, S., Gibbard, P. (Eds.), Glacial Deposits in North-East Europe. Balkema, Rotterdam, pp. 105–113.

Ehlers, J., Eissmann, L., Lippstreu, L., Stephan, H.-J., Wansa, S., 2004. Pleistocene glaciations of North Germany. In: Ehlers, J., Gibbard, P.L. (Eds.), Quaternary Glaciations—Extent and Chronology. Part I: Europe. Elsevier, Amsterdam, pp. 135–146.

Glushankova, N.I., Sudakova, N.G., 1995. Glacial stratigraphy of the lower Pleistocene in the Oka-Don region. In: Ehlers, J., Kozarski, S., Gibbard, P. (Eds.), Glacial Deposits in North-East Europe. Balkema, Rotterdam, pp. 157–160.

Hang, T., 2001. Proglacial sedimentary environment, varve chronology and Late Weichselian development of the Lake Peipsi, eastern Estonia. Stockholm University, Department of Physical Geography and Quaternary Geology, Quaternaria, Ser. A, No. 11, Doctoral thesis, 44 pp.

Hang, T., 2003. A local clay-varve chronology and proglacial sedimentary environment in glacial Lake Peipsi, eastern Estonia. Boreas 32, 416–426.

Hang, T., Sandgren, P., 1996. Magnetostratigraphy of varved clays. In: Meidla, T., Puura, I., Nemliher, J., Raukas, A., Saarse, L. (Eds.), The Third Baltic Stratigraphical Conference: Abstracts, Field Guide. Tartu University Press, Tartu, pp. 152–154.

Kadastik, E., 2004. Upper-Pleistocene Stratigraphy and Deglaciation History in Northwestern Estonia. Dissertations Geologicae Universitatis Tartuensis XV, PhD Dissertation, 128 pp. Tartu University Press, Tartu.

Kadastik, E., Kalm, V., 1998. Lithostratigraphy of Late-Weichselian tills on West-Estonian Islands. Bull. Geol. Soc. Finland 70 (1–2), 5–17.

Kadastik, E., Kalm, V., Liivrand, E., Mäemets, H., Sakson, M., 2003. Stratigraphy of a site with Eemian interglacial deposits in north Estonia (Juminda Peninsula). GFF 125, 229–236.

Kajak, K., 1961. Kvaternaarsete setete Prangli saare tugiprofiil. [Reference profile of the Quaternary deposits from the Island of Prangli]. In: VI EestiLooduseuurijate päeva ettekannete teesid, pp. 20–21 Tartu (in Estonian).

Kajak, K., 1995. Eesti kvaternaarisetete kaart. Mõõtkava 1:2 500 000. [Map of the Quaternary Deposits of Estonia. In Scale 1:2 500 000]. Eesti Geoloogiakeskus, Tallinn, 26 pp. (in Estonian).

Kajak, K., Liivrand, E., 1967. O nižne-i verhnepleistotsenovykh otlozheniyakh Estonii. [About Lower and Upper Pleistocene deposits of Estonia]. In: Goretskij, G.I., Kriger, N.I. (Eds.), Nizhnii Pleistotsen lednikovyh raionov Russkoi ravniny, 149–157 Moscow (in Russian).

Kajak, K., Raukas, A., Hütt, G., 1981. Opyt issledovaniya raznovozrastnykh estonskikh moren pri pomoshchi termoluminestsentnogo metoda. [Experience of investigation of Estonia's different-aged tills with thermoluminescence method]. In: Jevzerov, V.J. (Ed.), Geologiya Pleistotsena Severo-Zapada SSSR, 3–11 Apatity (in Russian).

Kalm, V., 2005. Chronological data from Estonian Pleistocene. Proc. Estonian Acad. Sci. Geol. 54, 5–25.

Kalm, V., 2006. Pleistocene chronostratigraphy in Estonia, southeastern sector of the Scandinavian glaciation. Quatern. Sci. Rev. 25, 960–975.

Kalm, V., 2010. Ice flow pattern and extent of the last Scandinavian Ice Sheet southeast of the Baltic Sea. Quatern. Sci. Rev. doi:10.1016/j.quascirev.2010.01.019.

Kalm, V., Sohar, K., 2010. Oxygen isotope fractionation in three freshwater ostracod species from early Holocene lacustrine tufa in northern Estonia. J. Paleolimnol. 43, 815–828.

Kalnina, L., 1997. Studies in Eemian deposits and flora history of Latvia. In: Quaternary Deposits and Neotectonics in the Area of Pleistocene Glaciations. Abstract Volume of Field Symposium. Institute of Geological Sciences, Academy of Sciences of Belarus, Minsk, pp. 22–23.

Kalnina, L., 2001. Middle and Late Pleistocene environmental changes recorded in the Latvian part of the Baltic Sea basin. In: Quaternaria Ser. A 9. Department of Physical Geography and Quaternary Geology, Stockholm University, Stockholm 173 pp.

Karuküpp, R., 2004. Late-Glacial ice streams of the southeastern sector of Scandinavian ice sheet and the asymmetry of its landforms. Baltica 17, 41–48.

Kondratiene, O. (Ed.), 1993. Catalogue of Quaternary Stratotypes of the Baltic Region. Baltic Stratigraphical Association, Vilnius 56 pp. (in Russian with English summary).

Levkov, E., Liivrand, E., 1988. On glaciotectonic dislocations of interglacial deposits in the Karuküla and Kõrveküla sections. Proc. Acad. Sci. Estonian SSR Geol. 37 (4), 161–167 (in Russian with English summary).

Liiva, A., Ilves, E., Punning, J.-M., 1966. Verzeichnis der im Institut für Zoologie und Botanik der Akademie der Wissenschaften der Estnischen SSR mittels der Radiokohlenstoff-Methode Datieren Proben. Proc. Acad. Sci. Estonian SSR Biol. 15, 112–121 (in Russian with German summary).

Liivrand, E., 1976. Rebedded pollen and spore in Pleistocene deposits and their role in stratigraphy. In: Bartosh, T., Kabaliene, M., Raukas, A. (Eds.), Palynology in Continental and Marine Geologic Investigations. Zinatne, Riga, pp. 166–178 (in Russian with English summary).

Liivrand, E., 1977. The bedding of Mikulian Interglacial deposits in South-East Estonia. Proc. Acad. Sci. Estonian SSR Geol. Chem. 26 (4), 289–303 (in Russian with English summary).

Liivrand, E., 1984. Biostratificheskaya osnova raschleneniya pleistotsenovykh otlozhenii Estonii [Biostratigraphical base of the Pleistocene deposits subdivision in Estonia]. In: Kondratiene, O.P., Mikalauskas, A.P. (Eds.), Paleogeografiya i stratigrafiya chetvertichnogo perioda Pribaltiki i sopredelnykh raionov [Palaeogeography and Stratigraphy of the Quaternary in the Peribaltic and Adjacent Regions]. Academy of Sciences of Lithuania, Vilnius, pp. 136–139.

Liivrand, E., 1987. Regional type section of the Eemian marine deposits on Suur-Prangli. Proc. Acad. Sci. Estonian SSR Geol. 36 (1), 20–26.

Liivrand, E., 1990. Type section of Lower- and Middle-Weichselian interstadial deposits at Tõravere (South- East Estonia). Proc. Estonian Acad. Sci. Geol. 39 (1), 12–17 (in Russian with English summary).

Liivrand, E., 1991. Biostratigraphy of the Pleistocene deposits in Estonia and correlations in the Baltic Region. Stockholm University. Department of Quaternary Research. Report 19, Doctoral thesis, 114 pp.

Liivrand, E., 1995. South East Estonia as a key area in solution of disagreements in the Pleistocene stratigraphical scheme. In: Meidla, T., Jõeleht, A., Kalm, V., Kirs, J. (Eds.), Liivimaa Geoloogia, 107–112 Tartu (in Estonian with English summary).

Liivrand, E., 2007. Late Saalian—Early Weichselian deposits in the Gulf of Riga (Kihnu Island) and Middle Weichselian deposits in NW Estonia (Vääna-Jõesuu). GFF 129, 325–328.

Liivrand, E., Saarse, L., 1976. Geological and palynological characterization of Upper-Pleistocene deposits in Peedu section. Proc. Acad. Sci. Estonian SSR Geol. Chem. 25 (4), 334–342 (in Russian with English summary).

Liivrand, E., Saarse, L., 1983. Mezhlednikovye otlozheniya razreza Kyrvekyla (Yuzhnaya Estoniya) i yikh stratigraficheskoje znachenie [Interglacial deposits in Kõrveküla section (South Estonia) and their stratigraphical importance]. In: Bartosh, T. (Ed.), Palynological Research in Geological Studies of the Baltic Region and the Baltic Sea, 41–50 Riga (in Russian).

Lõokene, E., 1961. Mandrijää servamoodustistest, fluvioglatsiaalsetest setetest ja mandrijää taandumisest Sakala kõrgustiku põhja- ja keskosas. [About glacial marginal formations, glaciofluvial deposits and retreat of the continental glacier from the northern and central part of Sakala Upland]. In: Orviku, K. (Ed.), Geoloogiline Kogumik. Eesti NSV Teaduste Akadeemia Loodusuurijate Selts, Tartu, pp. 84–105.

Lõugas, L., Ukkonen, P., Jungner, H., 2002. Dating the extinction of European mammoths: new evidence from Estonia. Quatern. Sci. Rev. 21, 1347–1354.

Lunkka, J.P., Saarnisto, M., Gey, V., Demidov, I., Kiselova, V., 2001. Extent and age of the Last Glacial Maximum in the southeastern sector of the Scandinavian Ice Sheet. Glob. Planet. Change 31, 407–425.

Lunkka, J.P., Murray, A., Korpela, K., 2008. Weichselian sediment succession at Ruunmaa, Finland, indicating a Mid-Weichselian ice-free interval in eastern Fennoscandia. Boreas 37, 234–244.

Matoshko, A., 2004. Pleistocene glaciations in the Ukraine. In: Ehlers, J., Gibbard, P.L. (Eds.), Quaternary Glaciations—Extent and Chronology. Part I: Europe. Elsevier, Amsterdam, pp. 431–439.

Miettinen, A., Rinne, K., Haila, H., Hyvarinen, H., Eronen, M., Delusina, I., et al., 2002. The marine Eemian of the Baltic: new pollen and diatom data from Peski, Russia, and Põhja-Uhtju, Estonia. J. Quatern. Sci. 17, 445–458.

Molodkov, A., 2007. IR-OSL dating of uranium-rich deposits from the new late Pleistocene section at the Voka site, North-Eastern Estonia. Quatern. Geochronol. 2, 208–215.

Molodkov, A., Bolikhovskaya, N., Avo Miidel, A., Ploom, K., 2007. The sedimentary sequence recovered from the Voka outcrops, northeastern Estonia: implications for late Pleistocene stratigraphy. Estonian J. Earth Sci. 56 (1), 47–62.

Orviku, K., 1944. Jäävaheaegade geoloogiast Eestis. [About geology of interglacials in Estonia]. Eesti Sõna 138, 4 (in Estonian).

Orviku, K., Pirrus, R., 1965. Mezhmorennye organogennye otlozhenia v Karukyla (Estonskaya SSR) [Intermorainic organogenic deposits of Karuküla (Estonian S.S.R.)]. In: Orviku, K.K. (Ed.), Litologiya i stratigrafiya chetvertichnykh otlozhenii Estonii, 3–21 Tallinn (in Russian with English summary).

Pavlovskaya, I., 2000. The middle and late Pleistocene glacial-interglacial succession of eastern Belarus. Kwart. Geol. 44 (2), 199–203.

Pirrus, R., Raukas, A., 1996. Late-glacial stratigraphy in Estonia. Proc. Estonian Acad. Sci. Geol. 45 (1), 34–45.

Punning, J.-M., Raukas, A., Serebrjannyi, L., 1969. Karukylaskie mezlednikovye otlozheniya Russkoi ravniny (Stratigraphiya i geokronologiya) [Karuküla interglacial deposits in the Russian Plain (Stratigraphy and geochronology)]. Proc. Acad. Sci. USSR: Geology 10, 148–151 (in Russian).

Punning, J.-M., 1970. On possible errors in the determination of sample age by means of radioactive carbon (^{14}C) and on the checking of results. Proc. Acad. Sci. Estonian SSR Geol. Chem. 19 (3), 238–243 (in Russian with English summary).

Rattas, M., Kalm, V., Kihno, K., Liivrand, E., Tinn, O., Tänavsuu-Milkeviciene, K., et al., 2010. Chronology of Late Saalian and Middle Weichselian episodes of ice free lacustrine sedimentation recorded in the Arumetsa section, southwestern Estonia. Estonian J. Earth Sci. 59 (2), 125–140.

Raukas, A., 1978. Pleistocene Deposits of the Estonian SSR. Valgus, Tallinn 310 pp. (in Russian with English summary).

Raukas, A., 1986. Deglaciation of the Gulf of Finland and adjoining areas. Bull. Geol. Soc. Finland 58 (2), 21–33.

Raukas, A., 1991. Eemian interglacial record in the northwestern European part of the Soviet Union. Quatern. Int. 10–12, 183–189.

Raukas, A., 1995. Properties, origin and stratigraphy of Estonian tills. In: Ehlers, J., Kozarski, S., Gibbard, P. (Eds.), Glacial Deposits in North-East Europe. Balkema, Rotterdam, pp. 93–101.

Raukas, A., 2006. Evolution of the theory of continental glaciation in northern Europe. In: Grigelis, A., Oldroyd, D. (Eds.), History of Quaternary Geology and Geomorphology, Abstracts of Papers, INHIGEO Conference, 28–29 July 2006, Vilnius, Lithuania, 47–50.

Raukas, A., 2009. When and how did the continental ice retreat from Estonia? Quatern. Int. 207, 50–57.

Raukas, A., Gaigalas, A., 1993. Pleistocene glacial deposits along the eastern periphery of the Scandinavian ice sheets—an overview. Boreas 22, 214–222.

Raukas, A., Kajak, K., 1995. Quaternary stratigraphy in Estonia. Proc. Estonian Acad. Sci. Geol. 44 (3), 149–162.

Raukas, A., Kajak, K., 1997a. Quaternary cover. In: Raukas, A., Teedumäe, A. (Eds.), Geology and Mineral Resources of Estonia. Estonian Academy Publishers, Estonia, pp. 125–136.

Raukas, A., Kajak, K., 1997b. Ice ages. In: Raukas, A., Teedumäe, A. (Eds.), Geology and Mineral Resources of Estonia. Estonian Academy Publishers, Estonia, pp. 256–262.

Raukas, A., Karukäpp, R., 1979. Eesti liustikutekkeliste akumulatiivsete saarkõrgustike ehitus ja kujunemine. [Geology and formation of the glacial accumulative insular heights of Estonia]. In: Raukas, A. (Ed.), Eesti NSV saarkõrgustike ja järvenõgude kujunemine. Valgus, Tallinn, pp. 9–28 (in Estonian).

Raukas, A., Rähni, E., 1966. K voprosu ob otstupaniy materikovogo l'da poslednego oledeneniya s territorii Estonskoi SSR [On the question about deglaciation history of the last continental glaciation on the territory of Estonian SSR]. In: Kivila, H. (Ed.), Eesti Geograafia Seltsi Aastaraamat 1964/1965. Valgus, Tallinn, pp. 5–17 (in Russian).

Raukas, A., Stankowski, W., 2005. Influence of sedimentological composition on OSL dating of glaciofluvial deposits: examples from Estonia. Geol. Q. 49, 463–470.

Raukas, A., Rähni, E., Miidel, A., 1971. Marginal Glacial Formations in North Estonia. Valgus, Tallinn, 226 pp. (in Russian with English summary).

Raukas, A., Kalm, V., Kaurkäpp, R., Rattas, M., 2004. Pleistocene glaciations in Estonia. In: Ehlers, J., Gibbard, P.L. (Eds.), Quaternary Glaciations—Extent and Chronology. Part I: Europe. Elsevier, Amsterdam, pp. 83–91.

Rinterknecht, V.R., Clark, P.U., Raisbeck, G.M., Yiou, F., Bitinas, A., Brook, E.J., et al., 2006. The last deglaciation of the southeastern sector of the Scandinavian Ice sheet. Science 311, 1449–1452.

Saarse, L., 1979. Rasprostranenye i uslovya zaleganya ozorno-lednikovykh glinistykh otlozhenij yuzhnoi Estonij [On the distribution and bedding conditions of glaciolacustrine clayey sediments in southern Estonia]. Proc. Acad. Sci. Estonian SSR Geol. 28 (4), 145–151 (in Russian with English summary).

Saarse, L., Liiva, A., 1995. Geology of Äntu group of lakes. Proc. Estonian Acad. Sci. Geol. 44, 119–132.

Saarse, L., Niinemets, E., Amon, L., Heinsalu, A., Veski, S., Sohar, K., 2009. Development of the late glacial Baltic basin and the succession of vegetation cover as revealed at Palaeolake Haljala, northern Estonia. Estonian J. Earth Sci. 58 (4), 317–333.

Salonen, V.-P., Kaakinen, A., Kultti, S., Miettinen, A., Eskola, K.O., Lunkka, J.P., 2008. Middle Weichselian glacial event in the central part of the Scandinavian Ice Sheet recorded in the Hitura pit, Ostrobothnia, Finland. Boreas 37, 38–54.

Sandgren, P., Hang, T., Snowball, I., 1997. A Late Weichsealian geomagnetic record from Lake Tamula, SE Estonia. GFF 119, 279–284.

Satkunas, J., 1998. The oldest Quaternary in Lithuania. Mededel. Nederl. Inst. v. Toegepaste Geovetensch. TNO 60, 293–303.

Serebrjannyi, L., Raukas, A., Punning, J.-M., 1969. On the history of the developmnet of glaciation on nortwesterm part of the Russian Plain in the Late Pleistocene. Contrib. Glaciological Investig. 15, 167–181 (in Russian).

Sohar, K., Kalm, V., 2008. A 12.8-ka-long palaeoenvironmental record revealed by subfossil ostracod data from lacustrine freshwater tufa in Lake Sinijärv, northern Estonia. J. Paleolimnol. 40 (3), 809–821.

Stokes, C.R., Clark, C.R., 2001. Palaeo-ice streams. Quatern. Sci. Rev. 20, 1437–1457.

Tavast, E., Raukas, A., 1982. The Bedrock Relief of Estonia. Valgus, Tallinn 192 pp. (in Russian with English summary).

Toucanne, S., Zaragosi, S., Bourillet, J.F., Cremer, M., Eynaud, F., Van Vliet-Lanoë, B., et al., 2009. Timing of massive "Fleuve Manche" discharges over the last 350 kyr: insights into the European ice-sheet oscillations and European drainage network from MIS 10 to 2. Quatern. Sci. Rev. 28, 1238–1256.

Ukkonen, P., Lunkka, J.P., Jungner, H., Donner, J., 1999. New radiocarbon dates from Finnish mammoths indicating large ice-free areas in Fennoscandia during the Middle Weichselian. J. Quatern. Sci. 17, 711–714.

Velichkevich, F., Liivrand, E., 1976. New investigations of the interglacial flora from the Karuküla section in Estonia. Proc. Acad. Sci. Estonian SSR Geol. Chem. 25, 215–221 (in Russian with English summary).

Velichko, A.A., Faustova, M.A., Gribchenko, Y.N., Pisareva, V.V., Sudakova, N.G., 2004. Glaciations of the East European Plain—distribution and chronology. In: Ehlers, J., Gibbard, P.L. (Eds.), Quaternary Glaciations—Extent and Chronology, Part I: Europe. Elsevier, Amsterdam, pp. 337–354.

Zelčs, V., Markots, A., 2004. Deglaciation history of Latvia. In: Ehlers, J., Gibbard, P.L. (Eds.), Quaternary Glaciations—Extent and Chronology. Part I: Europe. Elsevier, Amsterdam, pp. 225–243.

Znamenskaya, O.M., Cheremisinova, E.A., 1962. Rasprostraneniye mginskogo mezhlednikovogo morya i osnovnye cherty paleogeografii [Distribution of the Mga interglacial sea and its main palaeogeographical characteristics]. In: Tokareva, T.N. (Ed.), Voprosy stratigrafii chetvertichnykh otlozhenii severo-zapada evropeiskoi chasti SSSR, 140–160. Gostoptekhizdat. Leningrad, (in Russian).

Chapter 9

The Glaciation of Finland

Peter Johansson[1,*], Juha Pekka Lunkka[2] and Pertti Sarala[1]

[1]Geological Survey of Finland, P.O. Box 77, FI-96101 Rovaniemi, Finland
[2]Institute of Geosciences, University of Oulu, P.O. Box 3000, FI-90014, Oulu, Finland
*Correspondence and requests for materials should be addressed to Peter Johansson. E-mail: peter.johansson@gtk.fi

9.1. INTRODUCTION

The Scandinavian ice sheet, the centre of which is situated in the Scandinavian mountain range, covered Finland and the northwestern Russian Plain several times during the Quaternary cold stages. It is not known precisely how many times Finland and adjacent areas were covered by ice during the Quaternary. This is because the area is situated close to the glaciation centre, and the ice-advances eroded and deformed most of previously deposited interglacial and glacial sediments during the cold stages. Therefore, it is common that only the pre-Quaternary weathered bedrock surface and the sediments deposited during the last cold stage (Weichselian) rest on the pre-Cambrian bedrock. Except for some scattered remnants of the Saalian esker ridges (Kujansuu and Eriksson, 1995) in the ice-divide zone of northern Finland (Finnish Lapland) and in the major river valleys in the Pohjanmaa area, in western Finland (for location: Fig. 9.1), there are no distinct geomorphological landforms related to pre-Weichselian glaciations. However, there are a number of sites where Middle and Late Pleistocene organic and glacial sediments have been preserved, particularly in northern Finland and western Finland (cf. Hirvas, 1991; Nenonen, 1995). These sites provide the basis for the general Quaternary stratigraphy of Finland.

According to the Finnish till stratigraphy, there are six stratigraphically significant till beds in Finnish Lapland, and three till beds in southern Finland. In northern Finland, the key site for the till stratigraphy is the Rautuvaara area in western Finnish Lapland (Hirvas et al., 1977; Hirvas, 1991). The three uppermost till beds are thought to represent Weichselian-age tills (Till Beds I–III), two of these (Till Beds I and II) are thought to have been deposited during the Late Weichselian. The so-called Till Bed IV was laid down during the Saalian glaciation. The two lowermost till beds (Till Beds V and VI) that occur beneath the Holsteinian peat horizon may represent Elsterian or pre-Elsterian tills (cf. Hirvas and Nenonen, 1987; Hirvas, 1991). In the southern part of Finland, south of latitude 65°N, there are three stratigraphically important till beds interbedded with minerogenic sediments (mainly sand and silt) that form the basis of glaciation history of the area (see e.g. Nenonen, 1995; Saarnisto and Lunkka, 2004; Svendsen et al., 2004). At some localities, well-developed palaeosols and peat and gyttja layers are associated within or with these intertill sand and silt units indicating ice-free periods during the Late Pleistocene.

9.2. MIDDLE PLEISTOCENE GLACIATIONS

9.2.1. Pre-Holsteinian Cold Stages

There is hitherto only one site in Finland, Naakenavaara in western central Lapland, where a peat unit, biostratigraphically correlative to the Holsteinian Stage interglacial (Hirvas, 1991; Aalto et al., 1992), is underlain by the sand and gravel unit and the till unit (cf. Hirvas, 1991). According to Hirvas (1991), this till unit was most probably deposited during the Elsterian glaciation. Although Elsterian or pre-Elsterian tills are most likely preserved in scattered localities across northern Finland, there is no conclusive evidence of pre-Saalian tills in southern Finland. However, for example, Nenonen (1995) has tentatively suggested that pre-Holsteinian till occurs beneath Holsteinian lacustrine gyttja unit in Virtasalmi, southeastern Finland although the stratigraphical position of the Virtasalmi lacustrine gyttja is not precisely known.

9.2.2. Saalian Cold Stage

All of Finland was covered by ice during the Saalian Stage glaciation (e.g. Svendsen et al., 2004) although little is known about the Saalian glacial history of Finland. However, there are a number of sites, particularly in Lapland and central western Finland, where till or glaciofluvial

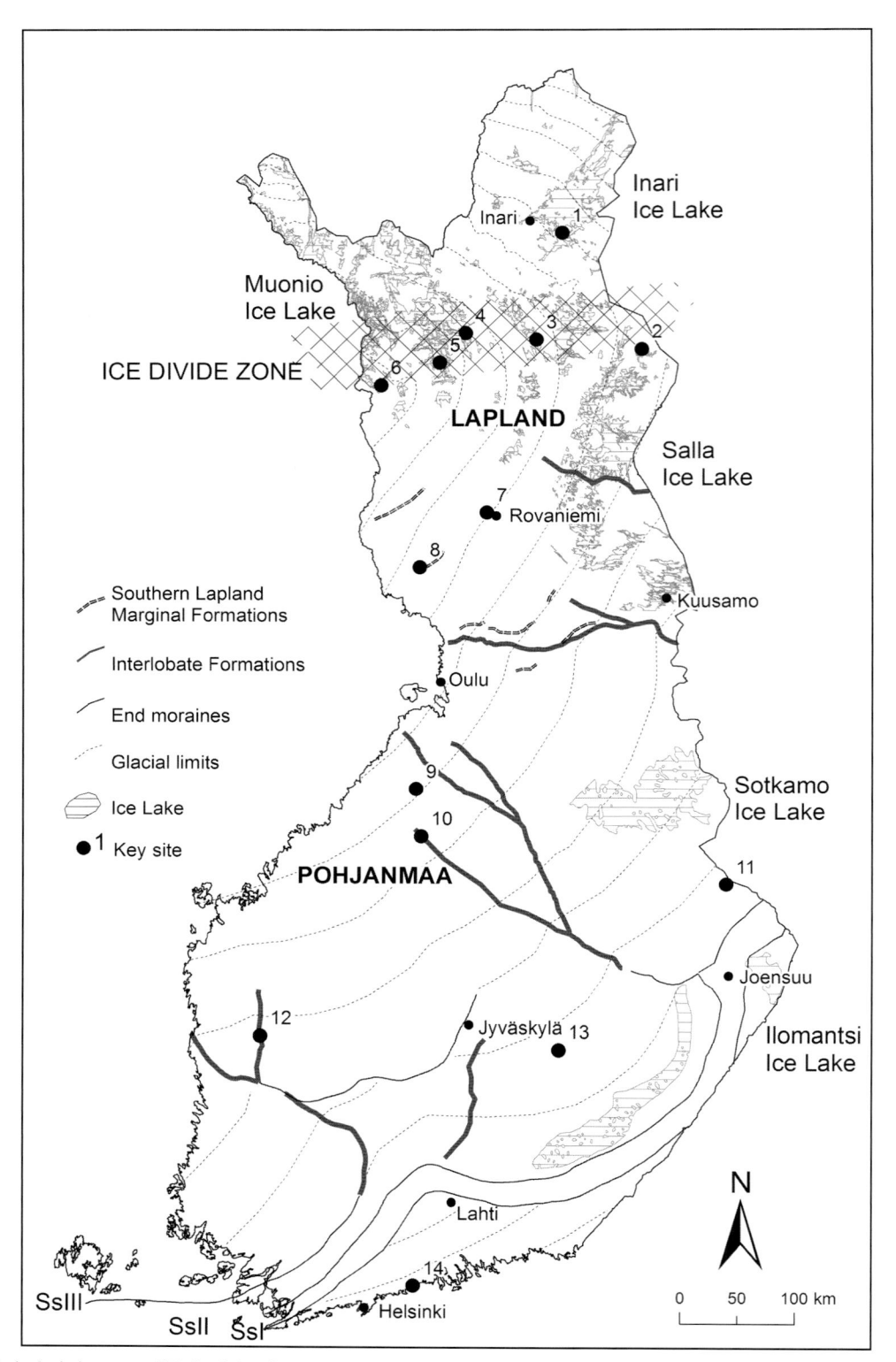

FIGURE 9.1 A deglaciation map of Finland showing the southern Lapland marginal formations and the main Late Weichselian end and interlobate moraines, ice lakes and ice limits. Key sites marked on the map are: 1, Veskoniemi; 2, Sokli; 3, Maaselkä; 4, Tepsankumpu; 5, Naakenavaara; 6, Rautuvaara; 7, Permantokoski; 8, Kauvonkangas; 9, Oulainen; 10, Hitura; 11, Ruunaa; 12, Pohjankangas; 13, Virtasalmi; 14, Vuosaari.

deposits underlie Eemian organic sediments (cf. Grönlund, 1991; Eriksson, 1993; Nenonen, 1995; Donner, 1995).

In northern Finland, observations of the Saalian till bed have been undertaken mainly from the central and southern Lapland (Hirvas, 1991; Sutinen, 1992; Mäkinen, 2005). Saalian till has been preserved in depressions and it is usually highly compacted. Based on till fabrics and striae observations, the ice-flow direction was from the southwest. This indicates that the glaciation centre is situated in the central part of Sweden, west to the Gulf of Bothnia.

Saalian glacigenic sediments in central and southern Finland are mainly preserved in the Pohjanmaa area in central western Finland. In that area, the Eemian interglacial deposits are underlain by the Saalian till and/or glaciofluvial sediments at several sites (cf. Eriksson et al., 1980, 1999; Hirvas and Niemelä, 1986; Forsström et al., 1988; Aalto et al., 1989; Gibbard et al., 1989; Nenonen, 1995; Saarnisto and Salonen, 1995; Salonen et al., 2008). In addition to the Saalian glacial units that underlie Eemian interglacial beds, there are a number of sites where till units are thought to have been deposited during the Saalian glaciation (cf. Nenonen, 1995; Hirvas et al., 1995; Pitkäranta, 2009). The southernmost of these is located south of the Salpausselkä end-moraine zone in Helsinki, where three till beds have been encountered (Hirvas et al., 1995). It was assumed by Hirvas et al. (1995) that the lowermost till of these might have been laid down prior to the Weichselian Stage, most probably during the Saalian glaciation. Ice movement direction, deduced from clast fabric measurements and bedrock cross-striae, indicates that the Saalian ice moved from the northwestern direction across southern Finland towards the southeast. This general flow pattern is also present in the direction of esker ridges of Saalian age although in southern Pohjanmaa, the direction of some esker formations is from north to south.

9.3. LATE PLEISTOCENE GLACIATIONS (CA. 110–11.7 KA)

9.3.1. Weichselian Cold Stage

The stratigraphy of the Early and Middle Weichselian Substages in Finland is based on the correlation of the interstadial organic deposits, till stratigraphy and till-covered glaciofluvial landforms. These data suggest that southern and northern Finland experienced a rather different glaciation history during the Weichselian. Basically, the main difference between these two areas is that northern Finland was already covered by the Scandinavian ice sheet during the Early Weichselian, while there is no firm evidence of Early Weichselian ice-cover in the southern part of Finland at this time (Fig. 9.2). This is probably a result of two separate ice-dome areas in the Scandinavian Mountains that behaved semi-independently in space and time during the Weichselian.

9.3.1.1. Early Weichselian Substage (ca. 110–74 ka)

It is generally assumed that after the Eemian Stage interglacial, ice began building up in the Scandinavian Mountains and spread into the adjacent areas twice during the Early Weichselian stadials (i.e. during the Herning Stadial, marine isotope stage (MIS) 5d and the Rederstall Stadial, MIS 5b). However, during the Early Weichselian interstadials (Brørup, MIS 5c and Odderade, MIS 5a), the ice melted at least once almost completely even in the Scandinavian Mountains (cf. Anderson and Mangerud, 1990; Donner, 1995; Saarnisto and Salonen, 1995).

In northern Finland, many stratigraphically important areas and key sites are located in the ice-divide zone in central Lapland where Pleistocene sediments were preserved as a result of the low ice-velocities and to frozen bed conditions (Hirvas, 1991; Kleman et al., 1999; Sarala, 2005). As mentioned above, six different till units have been discovered in northern Finland and some of these till beds are interbedded with organic layers (cf. Korpela, 1969; Hirvas et al., 1977; Hirvas, 1991; Helmens et al., 2000, 2007). At Tepsankumpu, central Lapland, a type locality for the Tepsankumpu interglacial, a freshwater gyttja unit that occurs stratigraphically between Till Beds IV and III, is characterised by a mixed taiga pollen assemblage (cf. Saarnisto et al., 1999). The Tepsankumpu interglacial deposits and other gyttja and peat deposits between Till Beds IV and III elsewhere in central Lapland (cf. Hirvas et al., 1977; Hirvas, 1991) are correlated with the Eemian Stage (Hirvas, 1991; Saarnisto et al., 1999). Similarly, a continuous sediment succession from Sokli, in northeastern Lapland, that consists of three till beds interbedded with lacustrine, fluvial and glaciolacustrine sediments above the Eemian marker horizon clearly indicates that there were at least three separate ice-advances across Lapland during the Weichselian Stage (cf. Helmens et al., 2000, 2007).

Within these Weichselian sediment successions, there are several sites where interstadial units have been discovered between the Weichselian till units. The interstadial unit at Maaselkä, in central Lapland occurs between Till Beds III and II. In this unit, birch dominates the pollen sequence. The Maaselkä interstadial is considered as the stratotype locality for the Peräpohjola interstadial, an event that was previously correlated with the Brørup interstadial (cf. Donner et al., 1986; Hirvas and Nenonen, 1987). However, the results of recent studies in southern and central Lapland suggest that the southern part of Lapland at least was ice free during the Early Weichselian Herning Stadial (MIS 5d). Likewise, the Peräpohjola interstadial sediments are most probably correlative of either to the Odderade interstadial (MIS 5a) or to one of the Middle Weichselian interstadials (MIS 3; cf. Mäkinen, 2005; Sarala, 2005, 2007, 2008; Sarala et al., 2005).

Results from Sokli, in northeastern Lapland indicate, however, that eastern Lapland experienced a slightly different glaciation history compared to that of western and central Lapland. Results from the Sokli boreholes suggest that the Early Weichselian interstadials, that is, the Brørup (MIS 5c) and Odderade interstadials (MIS 5a), were ice-free intervals. At Sokli, a silt unit, that contains a tundra-type pollen assemblage, separates the Eemian interglacial and the Brørup interstadial units and both the Odderade and the Middle Weichselian interstadial sediments are underlain by till (Helmens et al., 2000). Therefore, it seems that northeastern Finland was unglaciated until Rederstall interstadial (MIS 5b) around 90 ka (cf. Alexanderson et al., 2008). During the Herning Stadial (MIS 5d), the Scandinavian ice sheet margin most probably was located in northwestern Finnish Lapland or in northern Sweden (cf. Lundqvist, 1992) where the Veiki Moraines are thought to represent landforms laid down during the Early Weichselian (cf. Lagerbäck, 1988; Hättestrand, 1997).

Lithostratigraphical and biostratigraphical evidence, combined with a number of optically stimulated luminescence (OSL) dates, suggest that the Scandinavian ice sheet did not cover the western and southern parts of Finland in

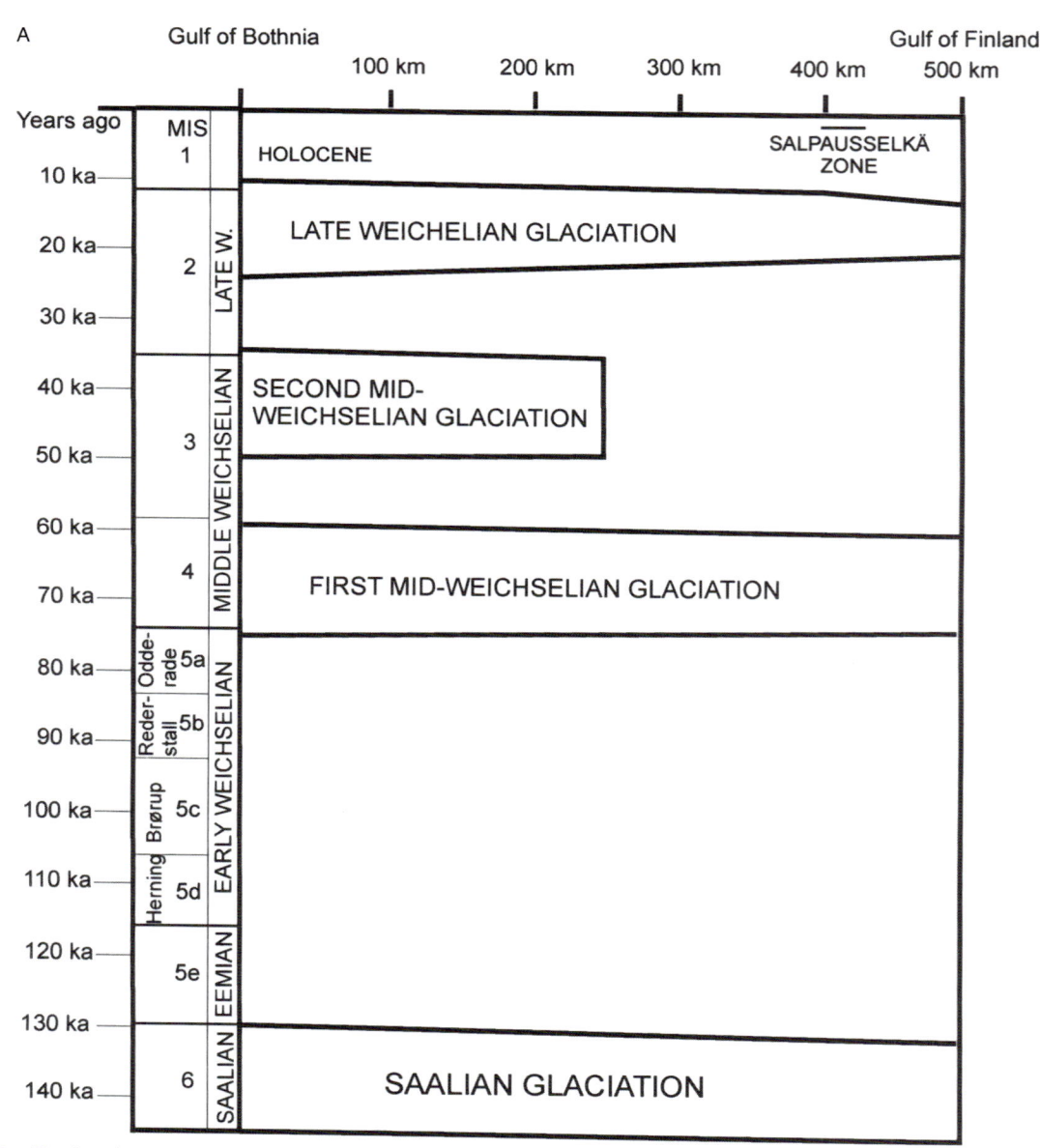

FIGURE 9.2 (Continued)

Chapter | 9 The Glaciation of Finland

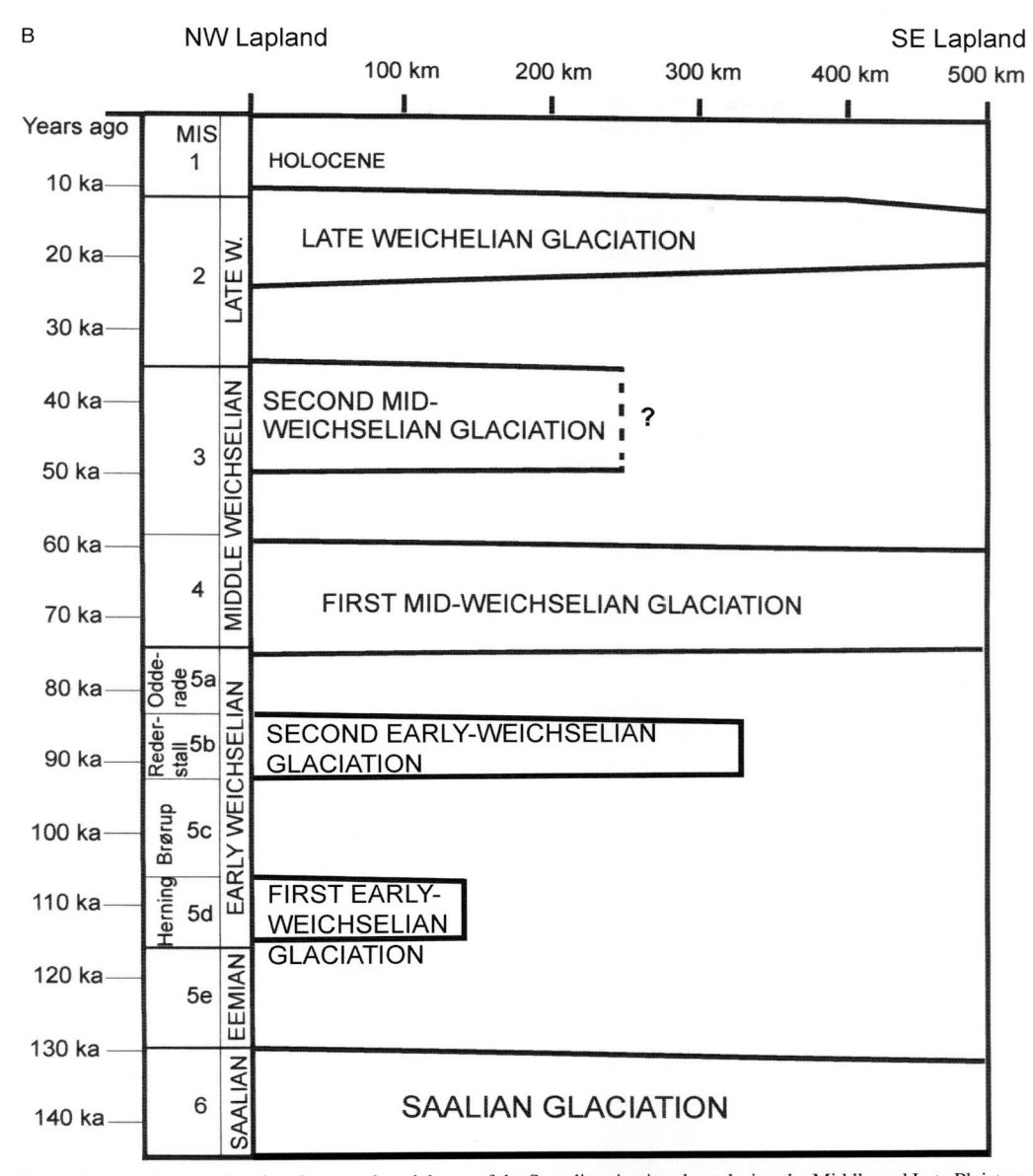

FIGURE 9.2 Time–distance diagram showing the growth and decay of the Scandinavian ice sheet during the Middle and Late Pleistocene. A glaciation curve for the southern and northern Finland from the Gulf of Bothnia to the Gulf of Finland (A) and from northwest to southeast across northern Finland (B).

the Early Weichselian (Salonen *et al.*, 1992; Hütt *et al.*, 1993; Nenonen, 1995; Saarnisto and Salonen;, 1995; Saarnisto and Lunkka, 2004; Svendsen *et al.*, 2004; Lunkka *et al.*, 2008; Salonen *et al.*, 2008). It has been previously suggested that the maximum ice extent in the Early Weichselian was reached during the Herning Stadial (MIS 5d; cf. Lundqvist, 1992; Donner, 1995) when the Scandinavian ice sheet overrode areas north of Oulu (for location see Fig. 9.1), including the present coast of the northern Gulf of Bothnia (Saarnisto and Salonen, 1995). It has further been suggested that the ice extent was more restricted during the Rederstall Stadial (MIS 5b) compared to the Herning Stadial (MIS 5d; cf. Donner, 1995) and did not reach beyond the eastern coast of the Gulf of Bothnia. However, recent results from Sokli (Helmens *et al.*, 2000, 2007) indicate the opposite.

There are a number of landforms that are related to the ice-flow activity in northern Finland during the Early Weichselian glaciations. Ice-marginal landforms from the Pudasjärvi area, northern Pohjanmaa (Ostrobothnia), have been described by Sutinen (1992). These elongate landforms are mostly composed of coarse, glaciofluvial sediment that strikes perpendicular to the Early Weichselian ice-flow direction from northwest to north-north-west. They are buried by Late Weichselian till. On the distal side of the Pudasjärvi end-moraines, the till unit is absent and Eemian deposits lie between the Saalian and Late

Weichselian tills (Sutinen, 1992). In addition to these Early Weichselian landforms, till-covered, ice-marginal landforms in the Tervola–Ylitornio area (Mäkinen, 1985) and till-covered esker sequences (Johansson, 1995) in central Lapland might represent the deglaciation phase of the second Weichselian, Rederstall stadial (MIS 5b).

Although the Early Weichslelian glacial history in Finland is rather well constrained, it is still poorly known in detail. Particular problems are related to the correlation of different till and interstadial units in northern Finland and to the ice extent during the Early Weichselian stadials.

9.3.1.2. Middle Weichselian Substage (ca. 74–25 ka)

At the beginning of the Middle Weichselian (ca. 74–60 ka, MIS 4), the whole of Fennoscandia became covered by the Scandinavian ice sheet (cf. Saarnisto and Lunkka, 2004; Svendsen et al., 2004) (Fig. 9.2). Litho- and biostratigraphical evidence, supported by ^{14}C and OSL dates, suggest that the major part of Finland was ice free at least once or possibly several times during the latter part of the Middle Weichselian Substage (MIS 3) between ca. 54–25 ka ago (cf. Nenonen, 1995; Ukkonen et al., 1999; Lunkka et al., 2001, 2008; Helmens et al., 2007, 2009; Salonen et al., 2008).

The extent of the Middle Weichselian ice sheet in northern Finland is still unknown. According to Hirvas (1991), northern Finland remained ice-covered after the Peräpohjola interstadial, previously correlated with Brørup interstadial, until deglaciation in the Late Weichselian and Early Holocene. However, at several sites in southern Finnish Lapland, silt and sand units between till beds have yielded OSL-ages ranging from 39 to 69 ka (Mäkinen, 2005; Sarala et al., 2005; Sarala and Rossi, 2006). At one site in Kauvonkangas, southern Finnish Lapland peat and gyttja horizons associated with periglacial palaeosols occurring between two till beds have yielded OSL and ^{14}C ages between 39 and 27 ka (Mäkinen, 2005), clearly indicating a Middle Weichselian (MIS 3) ice-free period. Similarly at Sokli, eastern Lapland, a 2-m-thick sequence of laminated sediment is thought to have been deposited in a glaciolacustrine environment during MIS 3 at around 40 ka ago (Helmens et al., 2000, 2009). The laminated unit contains pollen indicating shrub tundra vegetation. Based on relatively high tree pollen percentages in these sediments, it has also been argued that pine and tree birch were probably growing only a few 100 km south or southeast of Sokli (Bos et al., 2009). Recent results have therefore shown that after the Odderade interstadial, the Scandinavian ice sheet advanced across northern Finland during MIS 4. Subsequently, eastern and southern Lapland was deglaciated during the MIS 3 before the last glacial ice-advance in the Late Weichselian (Fig. 9.2).

The glaciation history of southern Finland during the Middle Weichselian is also poorly known in detail. However, at several sites in Pohjanmaa, central western Finland, two till beds are found overlying organic sediments that were deposited in the Eemian interglacial Stage or Early Weichselian stadials (cf. Nenonen, 1995). Moreover, in southernmost part of Finland, the so-called dark till, that underlies the Late Weichselian till, is thought to have been deposited during the Middle Weichselian (cf. Bouchard et al., 1990; Hirvas et al., 1995). Based on this evidence, it is generally accepted that the southern part of Finland was covered by ice during the early Middle Weichselian (MIS 4; Fig. 9.2). Till-clast fabric data indicate that the ice moved from the north and north-north-west across the southern part of Finland (cf. Bouchard et al., 1990; Donner, 1995; Hirvas et al., 1995; Nenonen, 1995; Salonen et al., 2008). The time and the duration of the ice cover over southern Finland is not yet precisely known. It is generally believed that the Scandinavian ice sheet advanced into southern Finland and beyond during MIS 4, some 70–60 ka ago (cf. Saarnisto and Salonen, 1995). However, OSL-ages from southwest Finland (Nenonen, 1995), central western Finland (Salonen et al., 2008) and southeastern Finland (Lunkka et al., 2008) suggest that southern Finland was ice free for several periods during the MIS 3, that is, in the latter part of the Middle Weichselian Substage. In addition, Ukkonen et al. (1999) and Lunkka et al. (2001) have demonstrated that the southern part of Finland was ice free for a minimum of 10 ka between ca. 25 and 35 ka ago.

There is evidence that the behaviour of the ice-streams over northern and southern Finland differed during the late Middle Weichselian (cf. Lunkka et al., 2001; Svendsen et al., 2004). In southern Finland, there are no end moraines related to the Early or Middle Weichselian ice limits. However, there are a number of landforms that are thought to have been deposited by the Middle Weichselian ice sheet in northern Finland. Sarala (2005) has shown the north to northwest–south to southeast trending drumlin field, and associated stratified sediments above in southwestern and western Finnish Lapland were deposited by ice in the early stage of the Middle Weichselian and during its deglaciation. In addition, ice-marginal deposits from Pudasjärvi to Kolari area (cf. Mäkinen, 1985; Sutinen, 1992) were most probably formed during different stages of the Middle Weichselain (MIS 4) ice recession.

At least three cross-cutting esker systems have been mapped in northeastern Finland. Till-covered eskers, with a north–south orientation, were found just south of the Sokli area, in the same region where the 'old northern' till bed possibly deposited during the MIS 4. This till was deposited by ice that moved southwards (Johansson, 1995; Johansson and Kujansuu, 1995). Therefore, the till and esker stratigraphy, together with the interstadial sediments of Middle Weichselian age (^{14}C-AMS age 42 ka) found at Sokli, indicate that the area was deglaciated at least once during the Middle Weichselian, prior to the final build-up of ice at the Late Weichselian maximum (Helmens et al., 2000).

9.3.1.3. Late Weichselian Substage (ca. 25–11.7 ka)

The rapid ice-advance of the Scandinavian ice sheet across southern and central Finland into the Northwestern Russian Plain took place after 25 ka cal BP ago (cf. Lunkka et al., 2001; Svendsen et al., 2004) (Fig 9.2). Based on clast fabric analysis and bedrock striae measurements, the ice-movement direction during this advance phase was from a westerly direction (cf. Salonen et al., 2008). Recent results from Veskoniemi, northern Lapland also indicate that the ice did not cover the area until ca. 22–25 ka ago (Sarala et al., 2010; see Fig. 9.2). The Scandinavian ice sheet reached its maximum extent in the northwestern Russian Plain and the Kanin Peninsula between ca.18.5 and 17 ka ago (Lunkka et al., 2001; Larsen et al., 2006)and began retreating, and Finland was completely deglaciated by ca. 10 ka ago (cf. Saarnisto and Lunkka, 2004).

9.3.1.4. The Late Weichselian ice-Marginal Limits and the Deglaciation of Finland

Two distinct zones of ice-marginal formations occur in southern Finland, the well-known Salpausselkä end-moraines and the central Finland end-moraine (Figs. 9.1 and 9.3) that were deposited at the ice front. The Salpausselkä zone, that runs across southern Finland, belongs to the end-moraine chain that can be traced all around Fennoscandia as well as through Russian Karelia and the Kola Peninsula (cf. Lundqvist and Saarnisto, 1995). This zone consists of three individual end-moraine ridges. The First and the Second Salpausselkäs form two lobe-shaped arcs across southern Finland. The Third Salpausselkä can only be found in southwestern Finland. This complex end-moraine chain is ca. 200 km in length. The central Finland end-moraine, which is also called Näsijärvi-Jyväskylä end-moraine, forms a second lobe-shaped arch complex some 100 km north of the Salpausselkä zone. It is highly continuous, extending for ca. 250 km.

Both the Salpausselkäs and the central Finland end-moraines are mainly composed of glaciofluvial material that was deposited beyond the ice margin as subaquatic fans, deltas and sandur deltas. However, moraine ridges, mainly composed of till, also occur in the end-moraine zones, particularly in the areas where they were deposited above the maximum water level of the Baltic basin.

There are no large end moraines in northern Finland. Instead, small end-moraine ridges are found in places. These end moraines indicate marginal positions of ice-lobes during deglaciation. For example, several well-developed, south–west–north–east trending end moraines in the Pyhä-Luosto fell area (Sarala et al., 2009) indicate different ice-marginal positions approximate about 10,300–10,400 years ago. In addition, Aario (1990) has interpreted small formations in Vikajärvi, close to Rovaniemi, as end moraines that were formed at the final phase of the last deglaciation.

The arcuate forms of the end moraines, with associated feeding esker systems, as well as the so-called interlobate formations, indicate that the Scandinavian ice sheet was subdivided into distinct ice-lobes during deglaciation. A total of four ice-lobes can be identified in southern Finland (Fig. 9.3). Three of these ice-lobes can be related to the Salpausselkä end-moraines (North Karelian ice-lobe, Lake District ice-lobe and Baltic ice-lobe) and one to the central Finland end-moraine (Näsijärvi-Jyväskylä ice-lobe; cf. Salonen, 1998 and references therein). Northeast of the Näsijärvi-Jyväskylä end-moraine, the ice sheet was further divided into two ice-lobes as deglaciation progressed. The prominent interlobate system of Pudasjärvi–Taivalkoski–Hossa was deposited time-transgressively between these two ice-lobes (cf. Aario and Forsström, 1978).

The deglaciation chronology of southern Finland (i.e. from the Gulf of Finland to the Gulf of Bothnia), as well as the dating of the end moraines, is largely based on a varve chronology. Radiocarbon dating, palaeomagnetic methods and the sediment exposure dating (SED) have also been applied to this study. The varve chronology of southern Finland originally spanned 2210 years but was later revised to cover 2800 years (cf. Salonen, 1998 and references therein). In the varved-clay sequences, a distinct marker horizon (the so-called zero year) can be identified. This marker horizon is related to the history of the Baltic basin. It marks the event when the retreating ice front stood at the Second Salpausselkä and the level of the Baltic Ice Lake at the ice front fell to the subsequent Yoldia Sea of the Baltic basin (cf. Donner, 1995).

Repeated attempts have been made to tie the varve chronology to absolute dates (Cato, 1987; Strömberg, 1990) by radiocarbon dating and correlation with the Swedish varve chronology. The recent correlation of Saarnisto and Saarinen (2001) differs from the previous correlation schemes of Cato (1987) and Strömberg (1990) of the ice-marginal formations by ca.1 ka. Saarnisto and Saarinen (2001) based their age determination of the Second Salpausselkä on the results of the varved-clay studies, ^{14}C-AMS dates and palaeomagentic measurements. According to these authors, the First Salpausselkä was formed ca. 12.1–12.3 ka ago, the Second Salpausselkä ca. 11.6–11.8 ka ago and the central Finland end-moraine ca. 11.2–11.1 ka ago. In addition, the third Salpausselkä is thought to have been deposited between ca. 11.4 and 11.5 ka ago. These dates form the foundation of the deglaciation history for southern Finland.

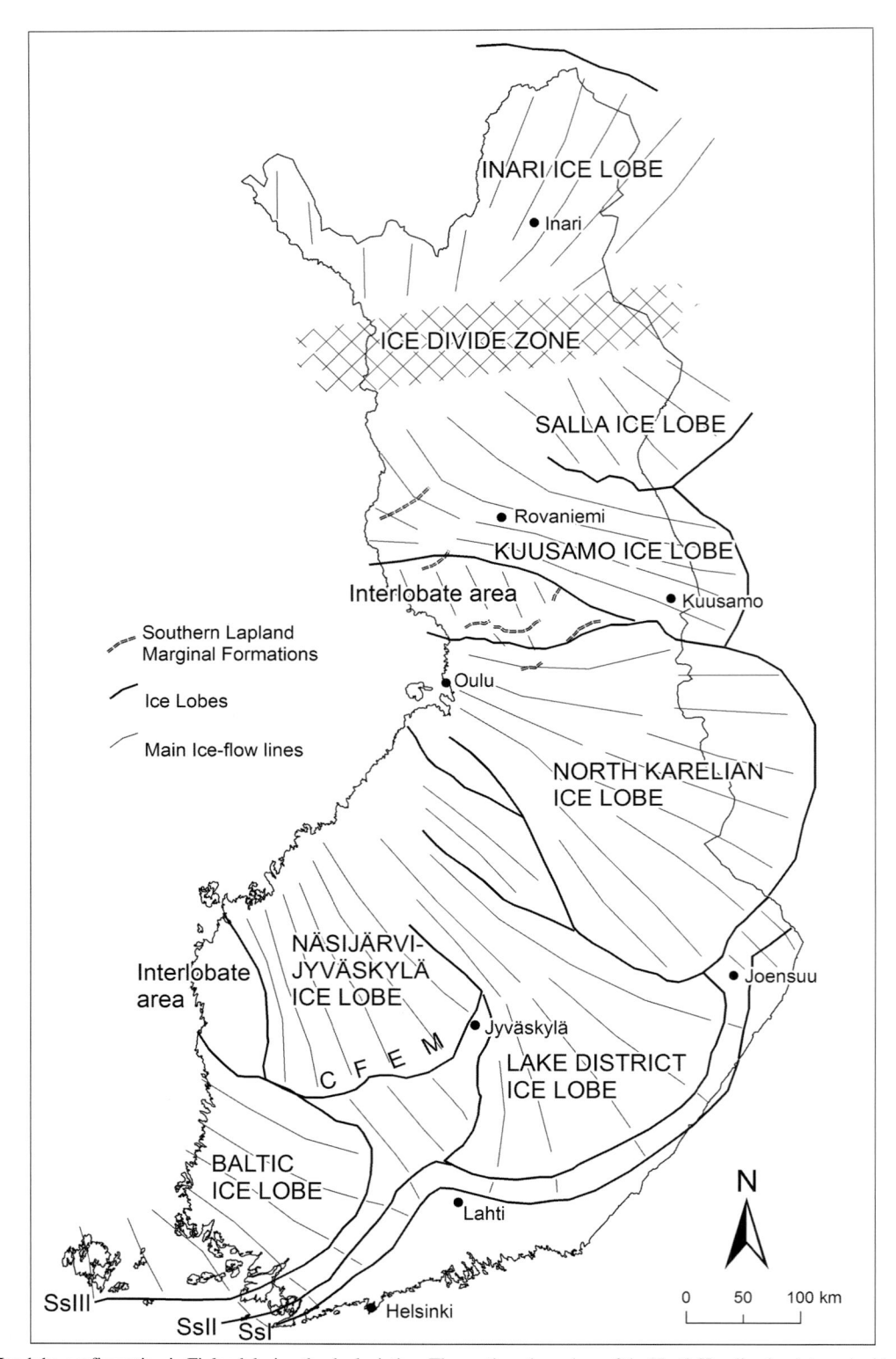

FIGURE 9.3 Ice-lobe configuration in Finland during the deglaciation. The northern boundary of the North Karelian ice-lobe coincides with the Pudasjärvi-Hossa interlobate formation. The Salpausselkä zone (Ss I–Ss III) and the central Finland end-moraine (CFEM) and the Southern Lapland Marginal Formations, as well as the main ice-flow lines are indicated.

The SED results on the age of the First Salpauseselkä of Rinterknecht et al. (2004) are also in good agreement with these results.

The results of Saarnisto and Saarinen (2001), combined with the correlation of the Finnish varve chronology to that in Sweden (Strömberg, 1990), suggest that the retreating ice margin reached the coastal areas of southern Finland ca. 13.1 ka ago. The average retreat rate was 60 m/year south of the Salpausselkä end-moraines (Sauramo, 1924, 1940), while north of the Salpausselkä zone the retreat rate was 260 m/year on average (cf. Donner, 1995). According to the new chronology, the area between the northern shore of the Gulf of Finland and the Gulf of Bothnia became ice free between ca. 13.0 and 10.3 ka ago.

Okko (1962) concluded that the ice sheet first retreated about 30 km north beyond the position of the First Salpausselkä in the Lahti area. Subsequently, the ice readvanced to deposit the First Salpausselkä. This advance took place during the Younger Dryas Stadial (cf. Rainio et al., 1995). Mainly based on till stratigraphy, Rainio (1985) suggested that the ice sheet initially may have retreated as much as ca. 80 km north of the present Salpausselkä zone all over southern Finland prior to its readvance. Although the geological observations support the readvance hypothesis, there is no conclusive evidence regarding the magnitude or the precise age of this oscillation.

As the ice retreated across southern Finland to the Second Salpausselkä, the ice margin terminated into the Baltic Ice Lake in most areas. In addition to the Baltic Ice Lake, separate ice-dammed lakes formed between the ice front and the higher ground during the Early Flandrian (Holocene) deglaciation. There were also ice lake areas in southern Finland which were not directly connected to the Baltic basin after the drop of the Baltic Ice Lake to the Yoldia Sea level. For example, a short-lived ice-lake complex north of the Second Salpausselkä in southeastern Finland was formed at this time (cf. Saarnisto, 1970).

The deglaciation of northern Finland is mainly based on the study of glacigenic deposits and glaciofluvial landforms (Tanner, 1915; Mikkola, 1932; Penttilä, 1963; Kujansuu, 1967; Aario and Forsström, 1978; Johansson, 1988, 1995). The retreating ice sheet melted in a supra-aquatic (terrestrial) environment, the results of which created a range of erosional and depositional landforms. Subaquatic conditions existed only in the southwestern part of Lapland which was covered by the waters of the Ancylus Lake phase of the Baltic basin.

The youngest flow direction can be used to delineate the retreat of the ice sheet because this retreat was usually in the opposite direction to that of the last ice flow. The network of subglacial glaciofluvial systems shows the direction of the retreating ice sheet even more accurately. These systems contain depositional landforms, that is, steep-sided and sharp-crested esker ridges with zones of glaciofluvial erosion between them. The radial pattern of the Late Weichselian subglacial drainage systems reflects the direction of ice-marginal retreat towards the ice-divide zone in central Lapland. In the northern part of Lapland, the ice margin retreated towards the south–southwest and in the southern part of Lapland to the northwest.

Meltwater activity at the boundary of the ice sheet and the exposed terrain produced series of shallow meltwater channels, that is, lateral drainage channels. These channels indicate the surface gradient of the ice sheet and the rate of melting at the end of the deglaciation phase, when the highest mountains emerged from beneath the ice sheet as nunataks. Penttilä (1963), Kujansuu (1967) and Johansson (1995) have used the lateral drainage channels to reconstruct the deglaciation in the mountain areas of Lapland. They found that the surface gradient of ice sheet varied here from 1 to 5 m per 100 m, and the average ice retreat rate was generally between 130 and 170 m/year.

In the supra-aquatic area of northern Finland, the ice margin retreated downslope along the main river valleys. As a result, the meltwaters were unable to drain but formed ice-dammed lakes in the river valleys at lower altitudes. The lake phases are indicated by the presence of ancient shorelines and outlet channels, coarse outwash and fine-grained glaciolacustrine sediments. The largest of these glacial lakes were located in Salla, eastern Lapland, in Muonio, northwestern Lapland and in the Inari Basin, northern Lapland (Kujansuu, 1967; Johansson, 1995; Kujansuu et al., 1998). In eastern Finland, separate ice-dammed lakes were also formed in Ilomantsi and in Sotkamo, eastern Finland (cf. Kilpi, 1937; Hyvärinen, 1971; Eronen and Haila, 1981). Big ice lakes covered thousands of square kilometres. However, most of the ice-dammed lakes were small and short-lived and only occupied the deepest parts of the present river valleys. These ice lakes drained from one river valley to another across the water divides creating erosional landforms, such as deep and narrow gorges. Successive extramarginal meltwater systems formed along the retreating ice-front. Some of these features can be followed hundreds of kilometres from the higher terrain in northwestern Lapland to the lower levels in eastern Lapland (Kujansuu, 1967; Johansson, 2007). Initially, meltwater flow was directed northwards over the main water divide. However, as the ice sheet became smaller and retreated southwestwards, the meltwater flow was redirected eastwards and finally southeastwards along the retreating ice margin towards the Baltic basin. Collecting all the palaeohydrographic information, for example, by mapping the ice-dammed lakes and the extramarginal channels between the ice-dammed lakes, it is possible to reconstruct a reliable picture of successive stages of the ice retreat in the supra-aquatic areas (Johansson, 2007) (Fig. 9.1).

The Younger Dryas ice-marginal landforms are situated in North Norway only 20 km from the northernmost part of Finnish Lapland (Sollid et al., 1973). At that time, the ice-divide zone, that is, the centre of the ice dome located in central Lapland, and the active ice flow was towards the northeast and towards the southeast and east. Extensive drumlin fields in Inari, northernmost Lapland and in Kuusamo, southeasternmost Lapland were formed during this flow phase. As the ice sheet began retreating from the Younger Dryas end-moraine zones ca. 11.6 ka ago (Saarnisto, 2000), the highest mountain tops in northwestern and northern Lapland were the first to emerge as nunataks from beneath the ice (Kujansuu, 1992). The main parts of northern and southeastern Lapland became ice free ca. 10,500 years ago. In southern Lapland, the division of the ice sheet into two ice-lobes, Kuusamo and Oulu ice-lobes, influenced the deglaciation between which the Pudasjärvi–Taivalkoski–Hossa glaciofluvial interlobate system was deposited (Aario and Forsström, 1978) (Fig 9.1), and the passive, interlobate area in Kemi-Ranua-Pudasjärvi-Kuivaniemi was formed (Sarala, 2005). After ice had reached the ice-divide area in central Lapland ca. 10.3 ka ago, the ice margin stagnated in several places and the ice melted *in situ*, partially as separate patches of dead ice. The last remnants of the continental ice sheet melted in western Lapland ca. 10.0 ka ago (Kujansuu, 1967; Saarnisto, 2000; Johansson, 2007; Mäkilä and Muurinen, 2008).

REFERENCES

Aalto, M., Donner, J., Hirvas, H., Niemelä, J., 1989. An interglacial beaver dam deposit at Vimpeli, Ostrobothnia, Finland. Geol. Surv. Fin. Bull. 348, 34pp.

Aalto, M., Eriksson, B., Hirvas, H., 1992. Naakenavaara interglacial—a till-covered peat deposit in western Finnish Lapland. Bull. Geol. Soc. Fin. 64, 169–181.

Aario, R., 1990. Morainic landforms in Northern Finland. In: Aario, R. (Ed.), Glacial Heritage of Northern Finland, Excursion Guide. Nordia tiedonantoja, Sarja A, Nro, 1, pp. 13–27.

Aario, R., Forsström, L., 1978. Koillismaan ja Pohjois-Kainuun deglasiaatiostratigrafia. Summary: deglaciation stratigraphy of Koillismaa and North Kainuu. Geologi 30, 45–53.

Alexanderson, H., Eskola, K.O., Helmens, K.F., 2008. Optical dating of late Quaternary sediment sequence from Sokli, northern Finland. Geochronometria 32, 51–59.

Anderson, B.G., Mangerud, J., 1990. The last interglacial—glacial cycle in Fennoscandia. Quatern. Int. 3/4, 21–29.

Bos, J.A.A., Helmens, K.F., Bohncke, S.J.P., Seppä, H., Birks, H.J.B., 2009. Flora, vegetation and climate at Sokli, north-eastern Fennoscandia, during the Weichselian Middle Pleniglacial. Boreas 38, 335–348.

Bouchard, M.A., Gibbard, P., Salonen, V.-P., 1990. Lithostratotypes for Weichselian and pre-Weichselian sediments in southern and western Finland. Bull. Geol. Soc. Fin. 62, 79–95.

Cato, I., 1987. On the definitive connection of the Swedish time scale with the present. Sver. Geol. Unders. Ca 68, 55pp.

Donner, J., 1995. The Quaternary History of Scandinavia (World and Regional Geology 7). Cambridge University Press, Cambridge, 200pp.

Donner, J., Korpela, K., Tynni, R., 1986. Veiksel-jääkauden ajoitus Suomessa. Terra 98, 240–247.

Eriksson, B., 1993. The Eemian pollen stratigraphy and vegetational history of Ostrobothnia, Finland. Geol. Surv. Fin. Bull. 372, 36pp.

Eriksson, B., Grönlund, T., Kujansuu, R., 1980. Interglasiaalikerrostuma Evijärvellä, Pohjanmaalla. Geologi 6, 65–71.

Eriksson, B., Grönlund, T., Uutela, A., 1999. Biostratigraphy of Eemian sediments at Mertuanoja, Pohjanmaa (Ostrobothnia), western Finland. Boreas 28, 274–291.

Eronen, M., Haila, H., 1981. The highest shore-line in the Baltic in Finland. Striae 14, 157–158.

Forsström, L., Aalto, M., Eronen, M., Grönlund, T., 1988. Stratigraphic evidence for Eemian crustal movements and relative sea-level changes in eastern Fennoscandia. Palaeogeogr. Palaeoclimatol. Palaeoecol. 68, 317–335.

Gibbard, P., Forman, S., Salomaa, R., Alhonen, P., Jungner, H., Peglar, S., et al., 1989. Late Pleistocene stratigraphy at Harrinkangas, Kauhajoki, Western Finland. Ann. Acad. Sci. Fenn. A III 150, 36pp.

Grönlund, T., 1991. The diatom stratigraphy of the Eemian Baltic Sea on the basis of sediment discoveries in Ostrobothnia, Finland. Ph.D. thesis, University of Helsinki. Yliopistopaino, Helsinki, 26pp + appendices.

Hättestrand, C., 1997. Ribbed moraines in Sweden—distribution pattern and paleoglaciological implications. Sed. Geol. 111, 41–56.

Helmens, K.F., Räsänen, M.E., Johansson, P.W., Jungner, H., Korjonen, K., 2000. The last interglacial-glacial cycle in NE Fennoscandia: a nearly continuous record from Sokli (Finnish Lapland). Quatern. Sci. Rev. 19, 1605–1623.

Helmens, K.F., Johansson, P.W., Räsänen, M.E., Alxanderson, H., Eskola, K.O., 2007. Ice-free intervals at Sokli continuing into Marine Isotope Stage 3 in the central area of the Scandinavian glaciations. Bull. Geol. Soc. Fin. 79, 17–39.

Helmens, K.F., Risberg, J., Jansson, K.N., Weckström, J., Berntsson, A., Tillman, P.K., et al., 2009. Early MIS 3 glacial lake evolution, ice-marginal retreat pattern and climate at Sokli (northeastern Fennoscandia). Quatern. Sci. Rev. 28, 1880–1894.

Hirvas, H., 1991. Pleistocene stratigraphy of Finnish Lapland. Geol. Surv. Fin. Bull. 354, 123pp.

Hirvas, H., Nenonen, K., 1987. The till stratigraphy of Finland. In: Kujansuu, R., Saarnisto, M. (Eds.), INQUA Till Symposium, Finland 1985. Geological Survey of Finland Special Paper 3, 49–63.

Hirvas, H., Niemelä, R., 1986. Ryytimaa, Vimpeli. In: Haavisto-Hyvärinen, M. (Ed.), 17th Nordiska Geologmötet 1986. Excursion Guide, Excursion C2, Geological Survey of Finland Guide, 15, pp. 47–50.

Hirvas, H., Alfthan, A., Pulkkinen, E., Puranen, R., Tynni, R., 1977. Raportti malminetsintää palvelevasta maaperätutkimuksesta Pohjois-Suomessa vuosina 1972–1976. Summary: a report on glacial drift investigations for ore prospecting purposes in northern Finland 1972–1976. Geol. Surv. Fin. Rep. Invest. 19, 54pp.

Hirvas, H., Lintinen, P., Lunkka, J.P., Eriksson, B., Grönlund, T., 1995. Sedimentation and lithostratigraphy of the Vuosaari multiple till sequence in Helsinki, southern Finland. Bull. Geol. Soc. Fin. 67, 47–60.

Hütt, G., Junger, H., Kujansuu, R., Saarnisto, M., 1993. OSL and TL dating of buried podsols and overlying sands in Ostrobothnia, western Finland. J. Quatern. Sci. 8, 125–132.

Hyvärinen, H., 1971. Ilomantsi Ice Lake: a contribution to the Late Weichselian history of eastern Finland. Soc. Sci. Fenn. Comment. Physico-Mathematicae 41, 171–178.

Johansson, P., 1988. Deglaciation pattern and ice-dammed lakes along the Saariselkä mountain range in northeastern Finland. Boreas 17, 541–552.

Johansson, P., 1995. The deglaciation in the eastern part of the Weichselian ice divide in Finnish Lapland. Geol. Surv. Fin. Bull. 383, 72pp.

Johansson, P., 2007. Late Weichselian deglaciation in Finnish Lapland. In: Geological Survey of Finland, Special Paper, 46, pp. 47–54.

Johansson, P., Kujansuu, R., 1995. Observations on three subglacial drainage systems (eskers) of different ages in Savukoski, eastern Finnish Lapland. Geological Survey of Finland, Special Paper 20, pp. 83–93.

Kilpi, S., 1937. Das Sotkamo-Gebiet in Spätglazialer Zeit. Bull. Comm. Géol. Fin. 117, 113pp.

Kleman, J., Hättestrand, C., Clarhäll, A., 1999. Zooming in on frozen-bed patches: scale-dependent controls on Fennoscandian ice sheet basal thermal zonation. Ann. Glaciol. 28, 189–194.

Korpela, K., 1969. Die Weichsel-Eiszeit und ihr Interstadial in Peräpohjola, nördliches Nordfinnland im Licht von submoränen sedimenten. Ann. Acad. Sci. Fenn. A III 99, 108pp.

Kujansuu, R., 1967. On the deglaciation of western Finnish Lapland. Bull. Comm. Géol. Fin. 232, 96pp.

Kujansuu, R., 1992. The deglaciation of Finnish Lapland. In: Kauranne, K. (Ed.), Glacial Stratigraphy, Engineering Geology and Earth Construction, Geological Survey of Finland, Special Paper, 15, pp. 21–31.

Kujansuu, R., Eriksson, B., 1995. Pre-Late Weichselian subglacial glaciofluvial erosion and accumulation at Vuotso, Finnish Lapland. Geological Survey of Finland, Special Paper, 20, pp. 75–82.

Kujansuu, R., Eriksson, B., Grönlund, T., 1998. Lake Inarijärvi, northern Finland: sedimentation and late Quaternary evolution. Geol. Surv. Fin. Rep. Invest. 143, 25pp.

Lagerbäck, R., 1988. The Veiki moraines in northern Sweden—widespread evidence of an Early Weichselian deglaciation. Boreas 17, 469–486.

Larsen, E., Kjær, K.H., Demidov, I.N., Funder, S., Grøsfjeld, K., Houmark-Nielsen, M., et al., 2006. Late Pleistocene glacial and lake history of northwestern Russia. Boreas 35, 394–424.

Lundqvist, J., 1992. Glacial stratigraphy in Sweden. In: Kauranne, K. (Ed.), Glacial Stratigraphy, Engineering Geology and Earth Construction, Geological Survey of Finland, Special Paper, 15, pp. 43–59.

Lundqvist, J., Saarnisto, M., 1995. Summary of Project IGPC-253. Quatern. Int. 28, 9–18.

Lunkka, J.P., Saarnisto, M., Gey, V., Demidov, I., Kiselova, V., 2001. The extent and age of the Last Glacial Maximum in the south-eastern sector of the Scandinavian Ice Sheet. Glob. Planet. Change 31, 407–425.

Lunkka, J.P., Murray, A., Korpela, K., 2008. Weichselian sediment succession at Ruunaa, Finland, indicating a Mid-Weichselian ice-free interval in eastern Fennoscandia. Boreas 37, 234–244.

Mäkilä, M., Muurinen, T., 2008. Kuinka vanhoja ovat Pohjois-Suomen suot? Geologi 60, 179–184.

Mäkinen, K., 1985. On the till-covered glaciofluvial formations in Finnish Lapland. Striae 22, 33–40.

Mäkinen, K., 2005. Dating the Weichselian deposits of southwestern Finnish Lapland. Geological Survey of Finland, Special Paper, 40, pp. 67–78.

Mikkola, E., 1932. On the physiography and the late-glacial deposits in northern Lapland. Bull. Comm. Géol. Fin. 96, 88pp.

Nenonen, K., 1995. Pleistocene Stratigraphy and Reference Sections in Southern and Western Finland. Geological Survey of Finland, Regional Office for Mid-Finland, Kuopio, 94pp.

Okko, M., 1962. On the development of the first Salpausselkä, west of Lahti. Bull. Comm. Géol. Fin. 202, 162pp.

Penttilä, S., 1963. The deglaciation of the Laanila area, Finnish Lapland. Bull. Comm. Géol. Fin. 203, 71pp.

Pitkäranta, R., 2009. Lithostratigraphy and age estimations of the Pleistocene erosional remnants near the centre of the Scandinavian glaciations in western Finland. Quatern. Sci. Rev. 28, 166–180.

Rainio, H., 1985. Första Salpausselkä utgör randzonen för en landis som avancerat på nytt. Geologi 4–5, 70–77.

Rainio, H., Saarnisto, M., Ekman, I., 1995. Younger Dryas End moraines in Finland and NW Russia. Quatern. Int. 28, 179–192.

Rinterknecht, V.R., Clark, P.U., Raisbeck, G.M., Yiou, F., Brook, E.J., Tschudi, S., et al., 2004. Cosmogenic 10Be dating of the Salpausselkä I Moraine in southwestern Finland. Quatern. Sci. Rev. 23, 2283–2289.

Saarnisto, M., 1970. The Late Weichselian and Flandrian history of the Saimaa lake complex. Soc. Sci. Fenn. Comment. Physico-Mathematicae 37, 107pp.

Saarnisto, M., 2000. The last glacial maximum and the deglaciation of the Scandinavian Ice Sheet. In: Sandgren, P. (Ed.), Environmental Changes in Fennoscandia During the Late Quaternary, LUNDQUA Report, 37, pp. 26–31.

Saarnisto, M., Lunkka, J.P., 2004. Climate variability during the last interglacial-glacial cycle in NW Eurasia. In: Battarbee, W., Gasse, F., Stickley, C.E. (Eds.), Past Climate Variability Through Europe and Africa. Springer, Dordrecht, pp. 443–464.

Saarnisto, M., Saarinen, T., 2001. Deglaciation chronology of the Scandinavian Ice Sheet from the Lake Onega basin to the Salpausselkä end moraines. Glob. Planet. Change 31, 387–405.

Saarnisto, M., Salonen, V.-P., 1995. Glacial history of Finland. In: Ehlers, J., Kozarski, S., Gibbard, P. (Eds.), Glacial Deposits in North-East Europe. A.A. Balkema, Rotterdam, pp. 3–10.

Saarnisto, M., Eriksson, B., Hirvas, H., 1999. Tepsankumpu revisited—pollen evidence of stable Eemian climates in Finnish Lapland. Boreas 28, 12–22.

Salonen, V.-P., 1998. Isavsmältning och strandlinjer i Finland. In: Andersen, S., Pedersen, S.S. (Eds.), Israndslinier I Norden. Nordisk Ministerråd, København, pp. 83–96. TemaNord 1998:584.

Salonen, V.-P., Eriksson, B., Grönlund, T., 1992. Pleistocene stratigraphy in the Lappajärvi meteorite crater in Ostrobothnia, Finland. Boreas 21, 253–269.

Salonen, V.-P., Kaakinen, A., Kultti, S., Miettinen, A., Eskola, K.O., Lunkka, J.P., 2008. Middle Weichselian glacial event in the central part of the Scandinavian Ice Sheet recorded in the Hitura pit, Ostrobothnia, Finland. Boreas 37, 38–54.

Sarala, P., 2005. Weichselian stratigraphy, geomorphology and glacial dynamics in Southern Finnish Lapland. Bull. Geol. Soc. Fin. 77, 71–104.

Sarala, P., 2007. Glacial morphology and ice lobation in southern Finnish Lapland. Geological Survey of Finland, Special Paper, 46, pp. 9–18.

Sarala, P., 2008. New OSL dating results in the central part of Scandinavian ice sheet. In: Lisicki, S. (Ed.), Quaternary of the Gulf of Gdańsk and Lower Vistula Regions in Northern Poland: Sedimentary Environments, Stratigraphy and Palaeogeography. International Field Symposium of the INQUA Peribaltic Group, Frombork, September 14–19, 2008. Polish Geological Institute, pp. 49–50. Abstracts.

Sarala, P., Rossi, S., 2006. Rovaniemen—Tervolan alueen glasiaalimorfologiset ja—stratigrafiset tutkimukset ja niiden soveltaminen geokemialliseen malminetsintään. Summary: Glacial geological and stratigraphical studies with applied geochemical exploration in the area of Rovaniemi and Tervola, southern Finnish Lapland. Geol. Surv. Fin. Rep. Invest. 161, 115.

Sarala, P.O., Johansson, P.W., Jungner, H., Eskola, K.O., 2005. The Middle Weichselian interstadial: new OSL dates from southwestern Finnish Lapland. In: Kolka, V., Korsakova, O. (Eds.), Quaternary Geology and Landforming Processes. Proceedings of the International Field Symposium, Kola Peninsula, NW Russia, September 4–9, 2005. Kola Science Centre RAS, Apatity, pp. 56–58.

Sarala, P., Johansson, P., Valkama, J., 2009. End moraine stratigraphy and formation in the southwestern Pyhä-Luosto fell area, northern Finland. Quatern. Int. 207, 42–49.

Sarala, P., Pihlaja, J., Putkinen, N., Murray, A., 2010. Composition and origin of the Middle Weichselian interstadial deposit in Veskoniemi, Finnish Lapland. Est. J. Earth Sci. 59, 1–8. doi:10.3176/earth.2010.2.

Sauramo, M., 1924. Lehti B2—Tampere, Maalajikartan selitys. Suomen Geologinen yleiskartta. Suomen Geologinen Komissioni—Geologiska Kommissionen i Finland.

Sauramo, M., 1940. Suomen luonnon kehitys jääkaudesta nykyaikaan. Werner Södeström Oy, Porvoo, 211pp.

Sollid, J.L., Andersen, S., Hamre, M., Kjeldsen, O., Salvigsen, O., Sturød, S., et al., 1973. Deglaciation of Finnmark, North Norway. Nor. Geogr. Tidskr. 27, 233–325.

Strömberg, B., 1990. A connection between the clay varve chronologies in Sweden and Finland. Ann. Acad. Sci. Fenn. A III 154, 32pp.

Sutinen, R., 1992. Glacial deposits, their electrical properties and surveying by image interpretation and ground penetrating radar. Geol. Surv. Fin. Bull. 359, 123pp.

Svendsen, J., Alexanderson, H., Astakhov, V., Demidov, I., Dowdeswell, J., Funder, S., et al., 2004. Late Quaternary ice sheet history of northern Eurasia. Quatern. Sci. Rev. 23, 1229–1271.

Tanner, V., 1915. Studier öfver kvartärsystemet i Fennoskandias nordliga delar III. (Résumé en français: Études sur le système quaternaire dans les parties septentrionales de la Fennoscandie). Bull. Comm. Géol. Fin. 38, 815pp.

Ukkonen, P., Lunkka, J.-P., Jungner, H., Donner, J., 1999. New radiocarbon dates from Finnish mammoths indicating large ice-free areas in Fennoscandia during the Middle Weichselian. J. Quatern. Sci. 14, 711–714.

Quaternary Glaciations in the French Alps and Jura

Jean-François Buoncristiani* and Michel Campy

Laboratoire Biogeosciences, Université de Bourgogne, UMR CNRS 5561, 6 Boulevard Gabriel, 21000 Dijon, France
*Correspondence and requests for materials should be addressed to Jean-François Buoncristiani. E-mail: jfbuon@u-bourgogne.fr

10.1. INTRODUCTION

Today in France the large glaciated areas are only present in the Alps, and they represent approximately 600 km^2 of glaciers (Vivian, 1975). However, during the past two cold periods of the Quaternary, both the Jura and the Alps were covered by major ice sheets. Since Penck and Brückner (1909–1911), two morainic complexes have been recognised in the marginal zone around the Alpine chain and the Jura Mountains (Bourdier, 1961; Monjuvent, 1978, 1984; Campy, 1982; Mandier, 1984). The sedimentary record of the glacial advances is discontinuous. The complexity of the readvance phases caused only the most extensive to be preserved, and there are problems to date these phase. Thus, it was decided to present here the palaeogeography that corresponds to the maximum of the glaciation during each stage. The 'external moraine complex' (EMC; Fig. 10.1) indicates the maximum glacial extension during the Middle Pleistocene from the west and northwest where it reaches the western margin of the Jura, the Lyon region and where it covers the region from the Dombes between Bourg-en-Bresse and Lyon (i.e. the riss *s.l.* after Penck and Brückner, 1909–1911). The 'internal moraine complex' (IMC; Fig. 10.3) can be traced 10–40 km inwards of the preceding limits and has been correlated with the Late Pleistocene glaciations (i.e. Würmian after Penck and Brückner, 1909–1911). The original interpretation was that the glacial advances that laid down the deposits of these two complexes were both derived from inside the Alps. This interpretation is evident for the southern half of the region (as far as approximately the latitude of Geneva) because in this zone there is no mountainous massif separating the Alps from the piedmont. However, further north, the situation is more complex because the Jura formed an obstacle to the advance of the Alpine glaciers.

Nowhere in the Alps have Quaternary deposits been completely preserved. Therefore, the basic stratigraphical subdivision is combined from various regional stratigraphies, though, in many cases, the determination is difficult because age control is lacking. Indications of Early Quaternary cold stages have been reported from the Italian and French Alps. In both areas, an abrupt change from fine-grained sedimentation to the deposition of coarse gravel and conglomerates is manifested in what was regarded as the 'Upper Pliocene'. However, the evidence is ambiguous, because change in sedimentary environment might either be explained by tectonic causes or by climatic changes (Billard and Orombelli, 1986).

On the Chambaran Plateau, west of Grenoble, in a sequence of strata thought to be older than 1.6 million years, a clayey sediment with striated boulders has been found. Bourdier (1961) could not determine with certainty whether this diamicton represented a slope deposit or a till. On the western Chambaran Plateau, the coarse gravels are overlain by strongly cemented loess layers that contain, apart from a fossil soil, a rich mammal fauna (Viret, 1954). The faunal composition suggests that the age of the loess is ca. 2.2 million years (Guerin, 1980).

10.2. MIDDLE PLEISTOCENE GLACIATIONS, THE EXTERNAL MORAINE COMPLEX

10.2.1. Glacial Limits

The Middle Pleistocene glaciation is recognised all around the peripheral Alpine piedmont by a line of end moraines (EMC from Penck and Brückner, 1909–1911) and can be traced over a distance of 200 km from the northern Jura to the Lyon region (Fig. 10.1). Between Ornans in the north and Bourg-en-Bresse in the south, it follows the western slopes of the Jura at altitudes of around 500 m a.s.l. In the extreme north of the Jura, outcrops are much less common, so the deposits' limit is much less precise. In keeping with the work by Hantke (1978), it can be shown that the ice limit occurred at heights of about

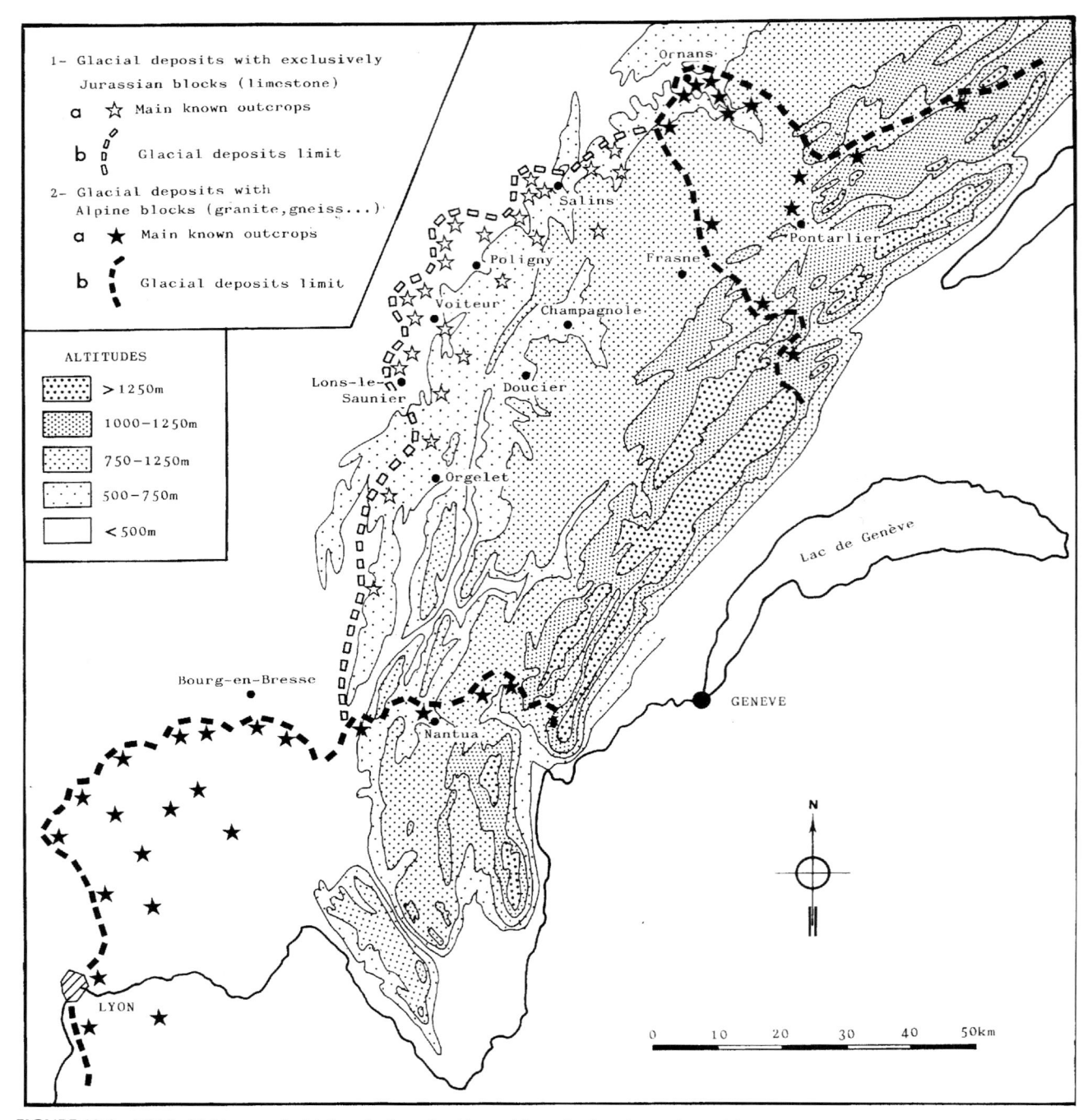

FIGURE 10.1 Middle Pleistocene glacial deposits from the Alps and from the Jura: internal morainic complexes (ICM).

800 m a.s.l., bending towards the east. In the central area, between Bourg-en-Bresse and Lyon, the EMC glacial deposits are very common. There, they have been mapped and recognised by many previous authors, for example, Falsan and Chantre (1879), Delafond and Deperet (1893), Penck and Brückner (1909–1911) and classically termed *Glaciaire de la Dombes*. These deposits consist of sediment several metres thick overlying Pliocene alluvium of the River Bresse (Fleury and Monjuvent, 1984). Studies of these deposits have shown that they were laid down by a vast ice lobe from the Alps (Fleury and Monjuvent, 1984; Mandier, 1984; Monjuvent, 1984). In the south, the Durance glacier advanced as far as Sisteron (Tiercelin, 1974). In the *Alpes provençales*, the glaciers failed to leave very well-developed frontal morainic systems. The main valleys were glaciated, and traces of this glaciation are found in the valleys of the Var, the Bléone and the Verdon.

10.2.2. Morphological Features

In the northwest of the Alps (Fig. 10.2), the Middle Pleistocene glaciation is represented by lines of strongly eroded end moraines sufficiently well preserved to indicate a maximum glacial limit. The main deposits of these moraines are basal tills, ablation tills and glacio-lacustrine deltas identified from outcrops and borehole information (Campy, 1982). The results of field mapping and the petrography of the EMC from the northern Jura to the Lyon region are shown schematically in Fig. 10.2. In the northern and southern zones, where Alpine ice deposited the EMC, it must have crossed the Jura. The central Jura, between Salins in the north and Bourg-en-Bresse in the south, was not overridden by Alpine ice. However, the occurrence of clear frontal moraines in this zone demonstrates that local glaciers from the Jura glaciers were present here.

The limit of these glacial deposits occurs as lobes developed towards the west opposite the main outer Jura valleys, in the Ornans, Salins, Poligny, Voiteur and Lons-le-Saunier regions. Reconstruction of the EMC is sometimes difficult because the moraines are eroded and degraded. However, the available information allows a coherent reconstruction of the Middle Pleistocene glaciation that formed the EMC deposits (Fig. 10.2). In the Geneva basin, the upper limit of the Middle Pleistocene glaciation ice indicates that the glacier was able to partially override the Jura in the relatively lower areas, that is, to the north towards the Ornans and in the south as far as Bourg-en-Bresse and as far as Lyon (Monjuvent, 1984). However, in the central part of the Jura, the petrography of the EMC deposits indicates that this sector had only been occupied by Jura ice (Fig. 10.2).

A large development of the glaciers during the Middle Pleistocene are related to the influence of the mountain barrier witch decrease and allowed convergence between Isère glacier, Arc glacier and Rhône glacier. This increase in the watershed of the Rhône glacier would thus imply an additional alimentation, which could explain the expansion of the lobe of Piedmont resulting from these glaciers confluence during the Middle Pleistocene.

10.2.3. Key Sites

The petrography of the EMC deposits varies from north to south. In the north, in the Ornans and Pontarlier regions, a dozen outcrops have been studied. They expose that basal till, ablation till and glacio-lacustrine delta deposits contain all limestone material from the Jura (40–90%), associated with boulders of Alpine origin (10–60%). In the central zone between Salins and Bourg-en-Bresse, over a distance of about 100 km, the EMC contains only material derived from the Jura (Campy, 1982). South of Bourg-en-Bresse, the Dombes glacial deposits contain Jura limestone material with or without an admixture of Alpine material (20–60%; Fleury and Monjuvent, 1984; Mandier, 1984). The topography of the Jura does explain this separation of glacier flows (Fig. 10.2): In the central zone of the Jura, the eastern slope rises to a height of 1250 m which must have formed a barrier to the Alpine glaciers. However, in the southern and northern zones, the altitudes are lower. Some peaks are over 1250 m high, but there are many cols between them at altitudes of less than 1000 m. In these zones, the Jura would not represent a significant barrier to the advance of the Alpine ice.

10.2.4. Problems

Until the 1980s, most authors, for example, Tricart (1961, 1965) and Jäckli (1962, 1970), considered that also the two morainic complexes present west of the Jura were emplaced under the dominant influence of Alpine glaciers. This hypothesis was based upon the occurrence of erratic boulders of Alpine origin (granite, gneiss, etc.) in certain moraine deposits in the Jura. It was also thought that the Jura Mountains were too small to have supported their own ice-cap. Thus, it was thought that the Alpine glaciers overrode the entire Swiss plain, occupied the Lake Geneva basin and penetrated the Jura through depressions on their eastern side. This question has been partly re-examined in the course of precise mapping and detailed investigation of the glacial deposits on the margin of the Jura massif: The morainic complexes of the western slopes (Campy, 1982, 1992), the glacial deposits of the Geneva Basin and the eastern slope (Arn, 1984; Campy and Arn, 1991) and the moraines south of the Jura (Sbaï, 1986; Monjuvent, 1988).

The palaeogeographic reconstruction shown in the maps is largely based on the EMC (from Penck and Brückner, 1909–1911). However, each morainic complex corresponds to several glacial advances out of the mountain (Billard and Derbyshire, 1985) and has been rearranged each time in the course of numerous climatic fluctuations. Consequently, the maps represent a synthesis of various climatic oscillations during the cold periods. Also, it must be taken into account that the maxima of the glacial advances may not necessarily have been synchronous in all the valleys. According to palynological studies, most of the deposits date from the period immediately before Eemian interglacial (Beaulieu de, 1984; Beaulieu de and Reille, 1989). It therefore seems that the classic '*Complexe des Moraines Externes*' is of Middle Pleistocene (=riss *s.l.*) age.

10.3. LATE PLEISTOCENE GLACIATIONS, THE INTERNAL MORAINE COMPLEX

10.3.1. Glacial Limits

From south of the Alps to the extreme northeast of the Jura Mountains, the IMC has been traced with great precision for over 400 km. It comprises a series of frontal moraines,

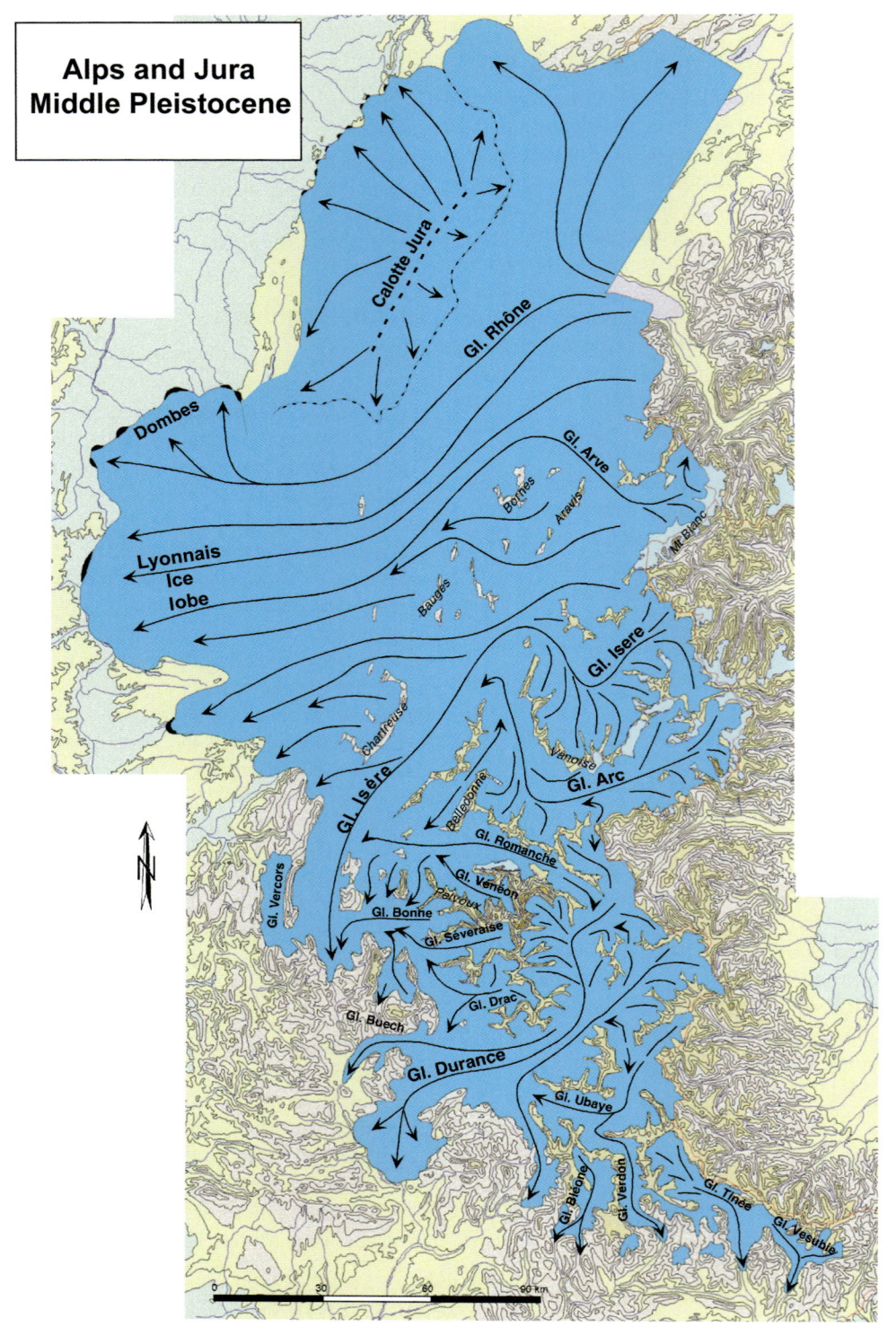

FIGURE 10.2 Middle Pleistocene Alpine and Jura ice sheet extension and ice flow directions.

particularly north of Grenoble. The limit of glacier deposits marking the maximum position of the associated glacial advance is clearly defined (Fig. 10.3). The configuration of the Alpine glacier in the Geneva Basin at the maximum of the Pleistocene was investigated by Swiss geologists since the end of the nineteenth century. The size and surface level of the ice were clearly determined by Jäckli (1962), and his interpretation is still valid today and confirmed

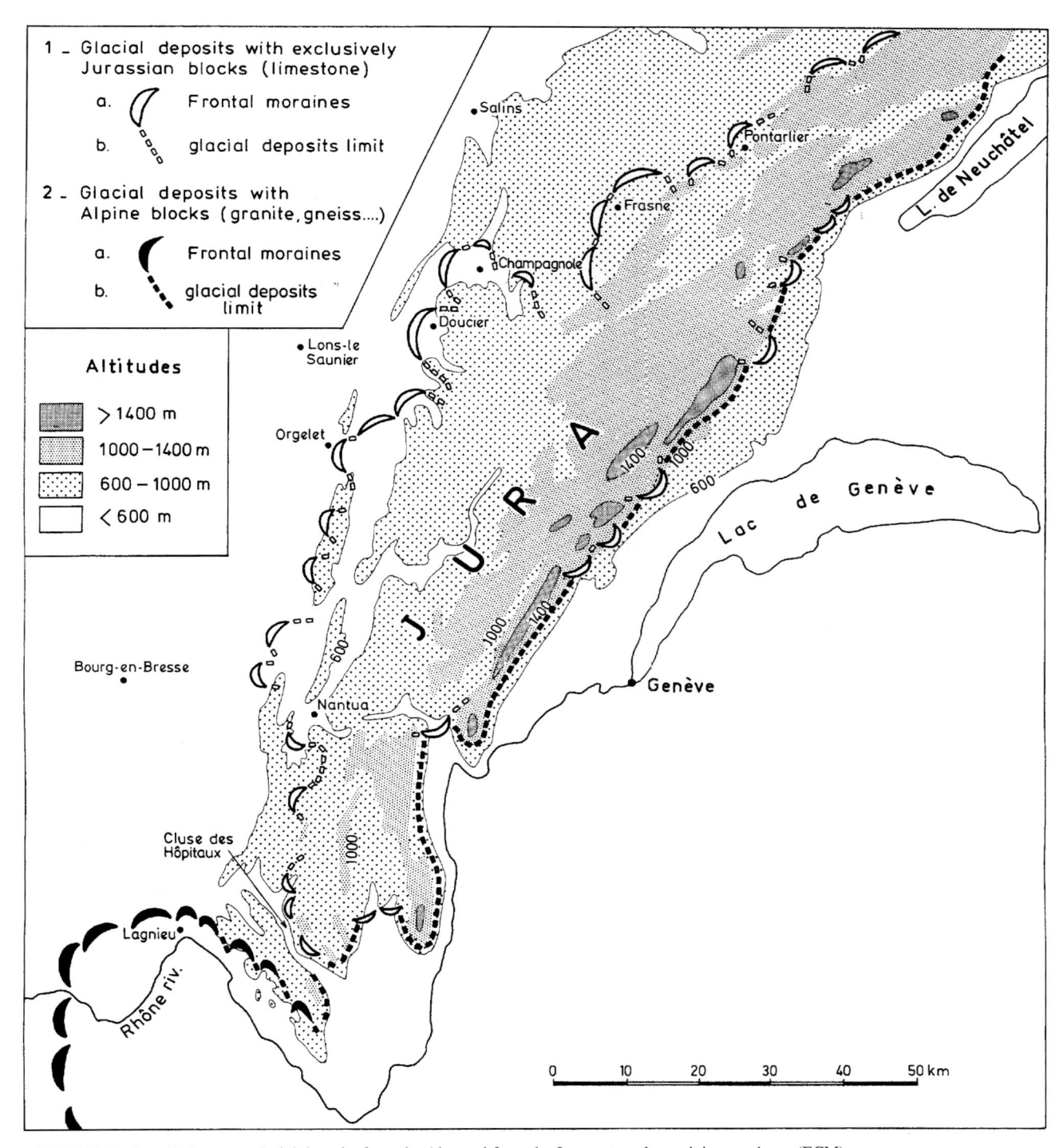

Chapter | 10 Quaternary Glaciations in the French Alps and Jura

FIGURE 10.3 Late Pleistocene glacial deposits from the Alps and from the Jura: external morainic complexes (ECM).

by recent work in the Valais area valleys (Kelly et al., 2004). The maximum height reached by the Alpine ice at its contact with the Jura during the Pleistocene was about 1200 m a.s.l. It descended gradually from this maximum to ca. 400 m a.s.l. at the terminal moraines in the Rhône valley in the southern Jura. In the north and northeast, the ice thinned towards the terminal moraines in the Soleure region to an altitude of about 600 m a.s.l. The level of the Alpine ice-surface only very rarely exceeded that of the Jura massif. Only the valleys were lower than the ice-surface. However, their entrance was blocked by morainic deposits which prevented the Alpine ice from penetrating the Jura.

It was never able to gain sufficient strength to supply sediment to the CMI on the western side of the Jura.

On the inner part of the Alps in the Mont-Blanc area, erosion features allow to reconstruct the palaeogeography of the Late Pleistocene glaciation. The method used here consists to define and map the limit between glacial erosion and atmospheric processes erosion forms, which correspond to the 'trimline'. The cartography of these trimlines and then the interpolation of these data allow the reconstitution of the maximum ice-surface about 2000–2300 m a.s.l. in this area (Coutterand and Buoncristiani, 2006).

10.3.2. Morphological Features

The Pleistocene glacial maximum is well defined by terminal moraines. In North Alps, all the valleys and transfluences were filled by glaciers which receive flow from local glaciers, forming a continuous network (Fig. 10.4). At the glacial maximum, the ice flows from the central Alpine zone and forms a vast piedmont glacier when it entered the Swiss plain. This lobe collided with the Jura at the latitude of Lake Geneva and was forced to flow towards both to the north and south. The southern glacial front stabilised as a piedmont lobe (Lyonnais ice lobe) about 20 km from Lyon.

At the maximum advance of the Late Pleistocene, the Alpine ice did not enter the Jura Mountains, which were covered by a local ice-cap (Fig. 10.4). The assumption of a discrete Jura ice-cap is based on the following evidence. The presence of exclusively local material in the morainic complex on the western slope of the Jura demonstrates that the associated ice flow originated only in the Jura. This morainic complex is particularly developed in the central zone of the western slopes right behind the highest parts of the Jura massif. This is the region where the Alpine ice would have met the greatest obstacle to crossing the mountains. The stratigraphical relationships of the Alpine and Jura tills at the eastern border of the mountains (Fig. 10.3) clearly show that two opposing ice flows occurred in this zone. And finally, the surface level of the ice that overrode the Swiss Basin during the glacial maximum did not allow it to advance far into the Jura massif, even though the Jura was not buried by ice at the same time.

It must be considered, however, that the general form of the Jura ice-cap shown in Fig. 10.4 is more certain in the central zone than further to the north and south. In the latter regions, the moraines are not sufficiently clear to allow a reliable reconstruction but there is no evidence conflicting with the general outline shown in the maps (Campy, 1982; Sbaï, 1986). In the Isère valley, the IMC corresponds to the Bank moraines which result from the confluent glacier of Isère glacier and Arc glacier, in the external area, these glaciers merge together and form a large piedmont glacier. While progressing towards the south, only the valleys become gradually ice field and we find little glaciers: the Bléone glacier, the Verdon glacier and the Var glacier (Fig. 10.4).

10.3.3. Key Sites

Examination of the Combe d'Ain Lake (Fig.10.4) infilling provides insight into the history of the Jura Mountains during the last glacial maximum (Campy, 1982; Buoncristiani and Campy, 2004). By mapping glacial and proglacial deposits on the western slopes and, more particularly, in this area, it has been possible to reconstruct the palaeogeography of the glacial front. Sediments were transported by glacial meltwater and were deposited as different sedimentary complexes in the lake filling. A number of coarse delta-type deposits are in contact with the moraines on the eastern edge of Combe d'A in area, and these form a topographic marker level at 525 m a.s.l. and indicate the former lake level. Laminated fine sediments ranging between 466 and 509 m a.s.l occupy the remainder of Combe d'Ain.

The synthetic section from the Nozon valley (Arn and Aubert, 1984) shows the relations of the two complexes (Fig. 10.5). Resting on tectonised Upper Jurassic and Cretaceous bedrock, the glacial formations reach from an altitude of 650–1000 m a.s.l. In stratigraphical order from the base upwards, four main formations can be described. At the base, horizontally bedded clayey silts with rare stones are present, which are interpreted as glacio-lacustrine sediments. This is overlain by a basal till of essentially Alpine material, very thick in the central part of the section (over 50 m) and thinning up-valley. In the upper part of the section from 950 to 1000 m a.s.l., this Alpine till is overlain by a basal till of exclusively Jura material, 20 m thick, ending in a small morainic ridge at Plan de la Sagne. A number of terraces composed of mixed Jura and Alpine material spread between 950 and 700 m a.s.l. They are interpreted as ice-contact landforms (kame terraces) emplaced during the progressive retreat of the Alpine glacier (Arn, 1984).

This sequence is repeatedly found on the eastern side of the Jura and demonstrates that during the Late Pleistocene glaciation the two ice flows were in contact. Alpine ice occupied the Swiss plain, and a second ice flow from the Jura brought materials that were deposited onto the Alpine sediments. However, the Jura ice sheet did not pass beyond the foot of the Jura chain. The stratigraphical relationship of the two till types shows that the Jura Mountain did support their own ice-cap independent from Alpine glaciers during the Late Pleistocene glaciation, confirming the views of Agassiz (1843), Nussbaum and Gygax (1935) and Aubert (1965).

10.3.4. Problems

The palaeogeographical reconstruction shown in the map (Fig. 10.4) is largely based on IMCs (cf. Penck and Brückner, 1909–1911). However, each morainic complex

FIGURE 10.4 Late Pleistocene Alpine and Jura ice sheet extension and ice flow directions.

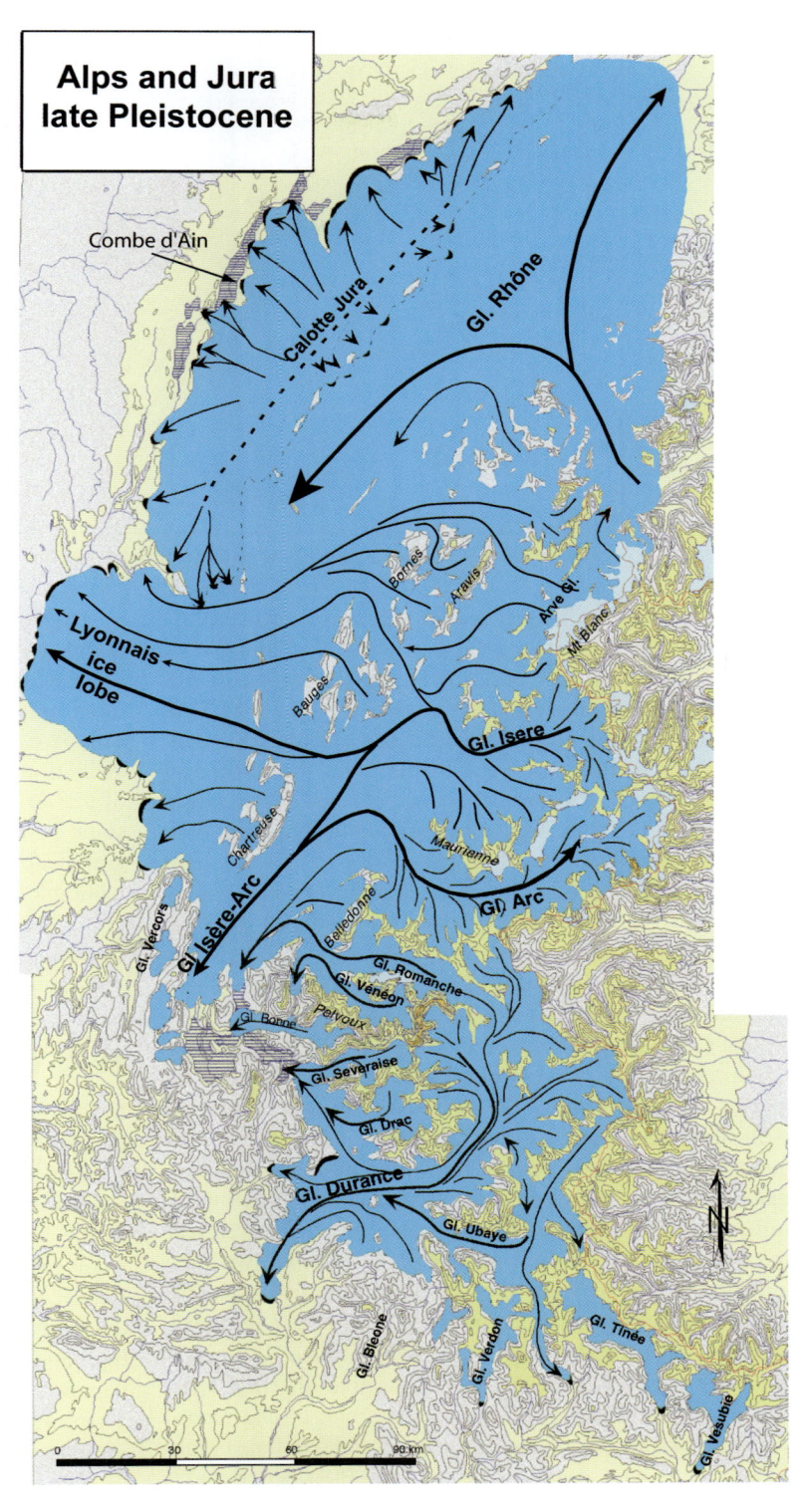

corresponds to several glacial advances out of the mountain massifs and has been rearranged each time in the course of numerous climatic fluctuations. Consequently, the maps represent a synthesis of various climatic oscillations during the Late Pleistocene cold periods. Also, it must be taken into account that the maxima of the glacial advances may not necessarily have been synchronous in all the valleys. Because of the 'freshness' of the deposits and the large number of exposures available, identification of the Late Pleistocene ice advance is easily achieved. On the basis

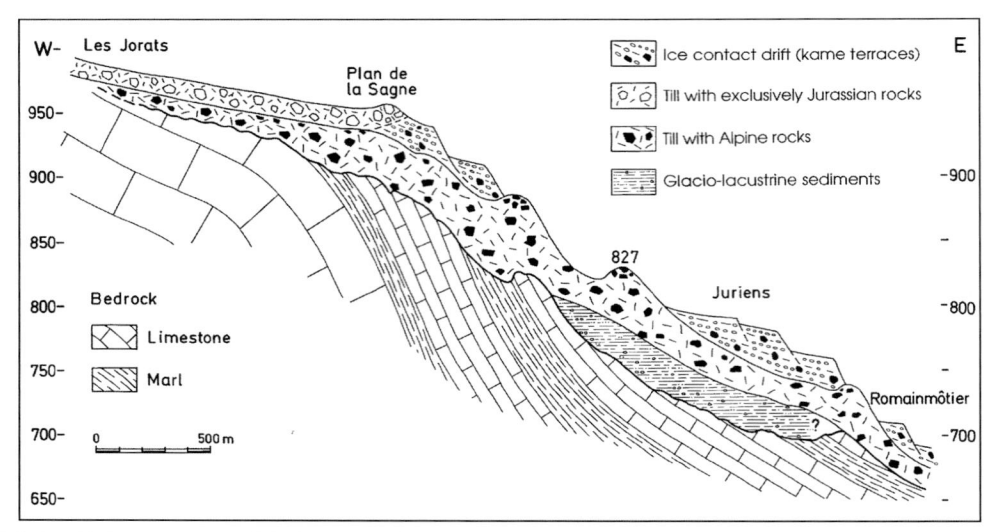

FIGURE 10.5 Relationship between Jura and Alpine glacial deposits along the Nozon valley.

of the ocean core sediments and the ice-core sequences, it is now known that the last cold stage includes two major glacial advances: the first during the Early Pleniglacial (60 ka = MIS 4) and the second in the Late Pleniglacial (20 ka = MIS 2). The problem of the apparent diachroneity of the maximum ice advance is still unsolved. In the Jura, it occurred at 20 ka (Campy and Richard, 1988), in the Isere Glacier region of Lyon prior to 26 ka, with a readvance in the Northern Alps at 20 ka (Monjuvent and Nicoud, 1988). In the Southern Alps, the morainic deposits in the middle Durance valley belonging to the IMC have yielded two samples of fossil wood that have been dated using ^{14}C: 18.6 ± 0.2 ^{14}C ka BP (LY 6338) and 17.68 ± 0.18 ^{14}C ka BP (LY 6387). These indicate that the IMC dates to the Weichselian Upper Pleniglacial (MIS 2) in this region (Jorda et al., 2000).

Initially, some authors correlated the Late Pleistocene glacier maximum with MIS 4, based only on comparison with the ocean isotope curve (Blavoux, 1988; Monjuvent and Nicoud, 1988). However, resolution of the oceanic signal is not sufficiently fine enough to distinguish the weak influence of the Alpine Glaciation in the geochemistry of the oceans. It is therefore preferable to correlate with the temperature curve from the Greenland ice-cap (Grootes et al., 1993). Regarding the ages for the last glacial maximum of the principle Alpine glaciers (Rhône, Linth and Rhine), two series of results available give different ages. Therefore, there are two solutions possible: a long record with the Late Pleistocene maximum glaciation before 27 ka and a short solution with the Late Pleistocene maximum glaciation near 22 ka (Schoeineich, 1998). The shorter variant is in accord with the chronology proposed for the Jura by Campy and Richard (1988) and with that proposed for the Southern Alps by Jorda et al. (2000). And whichever solution may be correct, there can be no doubt that the Late Pleistocene glacier maximum correlates with MIS 2.

However, in all correlation attempts, the erosional effects of the glaciations must be taken into account. Complete sequences of all ice advances are seldom preserved, often only traces of the most extensive phases are found. This is especially true for the older, pre-Late Pleistocene glaciations. Once the glaciers completely retreat, the rivers of the large Alpine valleys continue to erode and remove some or all of the evidence. This implies that the stratigraphical sequences in the large Alpine valleys are very incomplete. It is therefore reasonable to conclude that the preserved traces of glacial advances will not always be synchronous for each reconstructed advance phase in both the Alps and the Jura.

10.4. CONCLUSIONS

In the northwest of the Alps (Figs. 10.1 and 10.3), the two morainic complexes (ICM and ECM) were not entirely formed by ice of Alpine origin. The Jura Mountains had a determining influence on glacial flows and the contents of the Morainic Complexes. It may seem surprising that small mountains like the Jura would be capped during the Late Pleistocene glaciation by a thick ice sheet. In reality, only a few peaks of the mountain chain are over 1500 m high and the elevations over 1000 m are concentrated in a narrow zone 15–30 km wide and 120 km long (Fig. 10.1). There are two reasons for this important glacial build-up in this area: climate and morphology.

The Jura is a particularly cold region; even nowadays, the lowest temperatures in France are always found there. This is because the orientation of the major landforms leaves the Jura open to winds from the northeast derived from the Central European anticyclone. It is also a region of very heavy precipitation. Today, there is about 1800 mm of precipitation annually in the zone above

1000 m a.s.l. which falls as snow between September and June. It seems reasonable to assume that these characteristics also applied during the recent last glacial periods.

The central part of the Jura Mountains is not incised by deep valleys, and the highest zone is characterised by long, synclinal valleys at heights of 900–1000 m a.s.l., flanked by smooth higher ground at altitudes of 1200–1600 m a.s.l. As a result of this morphology, the Jura Mountains can retain snow because of inhibited melting (Aubert, 1965). The poorly drained valley forms efficient traps in which great thicknesses of snow can accumulate and then spring melting is considerably slowed in these closed areas. During full glacial times, the balance between accumulation and ablation would have been positive here and an ice sheet could have formed. From this ice sheet, outlet glaciers formed and flowed down the western and eastern flanks of the Jura chain.

In the Alps, a topographic gradient exists between the north and the south whereby the highest mountains in North Alps sustain glaciers today (Mont Blanc, Vanoise) whilst the lower mountains in the South Alps are cut by steep-sided valleys (the Verdon, Var, Bléone). This north–south topographic gradient can partly explain the limit of the glaciers in the North Alps during Pleistocene ice ages. However, the different positions of the glacier in the North Alps between the Late and the Middle Pleistocene are also functions of the variations of the palaeo-glacial watershed between these glacial periods. During the last cold period of the Late Pleistocene, the Alpine glaciers were less developed than for the cold periods of Middle Pleistocene as illustrated by the theoretical calculation of the Isere glacial topography in Grenoble which was 1100 m a.s.l. during Late Pleistocene and 1500 m a.s.l. during Middle Pleistocene (Monjuvent, 1978). Therefore, the topographic barrier formed by the Western Alps—the Chartreuse, the Bauges, the Bornes and the Aravis—probably controls the configuration of ice masses during Pleistocene glaciations.

REFERENCES

Agassiz, L., 1843. Le Jura a eu ses glaciers propres. Act. Soc. Helv. Sci. Nat. 284–285, 28th session, Lausanne.

Arn, R., 1984. Contribution à l'étude stratigraphique du Pléistocène de la région lémanique. Université de Lausanne, thesis, 307 pp.

Arn, R., Aubert, D., 1984. Les formations quaternaires de l'Orbe et du Nozon, au pied du Jura. Bull. Soc. Vaudoise Sci. Nat. 76 (2), 203–214.

Aubert, D., 1965. Calotte glaciaire et morphologie jurassienne. Eclogae Geol. Helv. 58 (1), 555–578.

Beaulieu de, J.L., 1984. A long upper Pleistocene pollen record from Les Echets, near Lyon, France. Boreas 13, 111–132.

Beaulieu de, J.L., Reille, M., 1989. The transition from temperate phases to stadials in the long upper Pleistocene sequence from Les Echets (France). Palaeogeogr. Palaeoclimatol. Palaeoecol. 72, 147–159.

Billard, A., Derbyshire, E., 1985. Pleistocene stratigraphy and morphogenesis of la Dombes: an alternative hypothesis. Bull. Assoc. Fr. Etude Quatern. 1985/2–3, 85–96.

Billard, A., Orombelli, G., 1986. Quaternary glaciations in the French and Italian piedmonts of the Alps. Quatern. Sci. Rev. 5, 407–411.

Blavoux, B., 1988. L'occupation de la cuvette Lémanique par le glacier du Rhône au cours du Würm. Bull. Assoc. Fr. Etude Quatern. 2/3, 69–81.

Bourdier, F., 1961. Le Bassin du Rhône au Quaternaire. Editions du Centre National de la Recherche Scientifique, Paris, 364 pp.

Buoncristiani, J.F., Campy, M., 2004. Expansion and retreat of the Jura Ice sheet (France) during the last glacial maximum. Sed. Geol. 165, 253–264.

Campy, M., 1982. Le Quaternaire Franc-Comtois. Essai chronologique et paléoclimatique. Université de Besançon, thesis 159, 575 pp.

Campy, M., 1992. Palaeogeographical relationships between Alpine and Jura glaciers during the two last Pleistocene glaciations. Palaeogeogr. Palaeoclimatol. Palaeoecol. 93, 1–12.

Campy, M., Arn, R., 1991. A case study of glacial paleogeography at the würmien circum-alpine zone: the Jura glacier. Boreas 20, 17–27.

Campy, M., Richard, H., 1988. Modalité et chronologie de la déglaciaition würmienne dans la chaine Jurassienne. Bull. Assoc. Fr. Etude Quatern. 2/3, 81–91.

Coutterand, S., Buoncristiani, J.F., 2006. Paléogéographie du dernier maximum glaciaire du pléistocène récentde la région du massif du mont blanc, france. Quaternaire 17 (1), 35–43.

Delafond, F., Deperet, C., 1893. Les terrains tertiaires de la Bresse et leurs gîtes de lignite et de minerais de fer, 1 vol. Mines et travaux publics, Paris, 332 pp.

Falsan, A., Chantre, E., 1879. Monographie géologique des anciens glaciers et du terrain erratique de la partie moyenne du Bassin du Rhône. Annales Société Agrégé Histoire Naturelle Univervité Lyon, 2 vol. 1394 pp.

Fleury, R., Monjuvent, G., 1984. Le glacier alpin et ses implications en Bresse. Géol. Fr. 3, 231–240.

Grootes, P.M., Stuiver, M., White, J.W.C., Johnsen, S.J., Jouzel, J., 1993. Comparison of oxygen isotope records from the GISP2 and GRIP Greenland ice cores. Nature 366, 552–554.

Guerin, C., 1980. Les Rhinoceros (Mammalia perissodactyla) du Miocène terminal au Pleistocène supérieur en Europe Occidentale. Comparaison avec les espèces Actuelles. Doc. Lab. Géol. Lyon 79, 1185 pp.

Hantke, R., 1978. Quartärgeologische Karten in Eiszeitalter, 1 Ott, Thun, 468 pp.

Jäckli, A., 1962. Die Vergletscherung der Schweiz im Würm maximum. Eclogae Geol. Helv. 55 (2), 285–294.

Jäckli, A., 1970. La Suisse durant la derniére période glaciaire. Carte in Atlas de la Suisse, Service topographique fédéral, Wabern-Berne.

Jorda, M., Rosique, T., Evin, J., 2000. Premières datations ^{14}C de dépots morainiques du Pléniglaciaire supérieur de la moyenne Durance (Alpes méridionales, France). Implications géomorphologiques, paléoclimatiques et chronostratigraphiques. Compte rendue Acad. Des Sci. 331/3, 187–188.

Kelly, M.A., Buoncristiani, J.F., Schlüchter, C., 2004. A reconstruction of the last glacial maximum (LGM) ice-surface geometry in the western Swiss Alps and contiguous Alpine regions in Italy and France. Eclogae Geol. Helv. 97, 57–75.

Mandier, P., 1984. Le relief de la moyenne vallée du Rhône au Tertiare et au Quaternaire: essai de synthèse paléogéographique, 3 vols. Thesis, Lyon II, 860 pp.

Monjuvent, G., 1978. Le Drac. Morphologie, stratigraphie et chronologie. Edition du CNRS, Grenoble, 431 pp.

Monjuvent, G., 1984. Quaternaire in Debrand-Passard, Synthèse géologique du Sud-Est de la France. Mémoire B.R.G.M. Fr., n° 125, 230 pp.

Monjuvent, G., 1988. La déglaciaition rhodanienne entre les moraines internes et le Val du Bourget. Géol. Alpine 64, 61–104.

Monjuvent, G., Nicoud, G., 1988. Interprétation de la déglaciation rhôdanienne au Würm, des moraines internes à la cuvette lémanique. Bull. Assoc. Fr. Etude Quatern. 2/3, 129–140.

Nussbaum, F., Gygax, F., 1935. Zur Ausdehnung des risseiszeitlichen Rhonegletschers im französischen Jura. Eclogae Geol. Helv. 28 (2), 659–665.

Penck, A., Brückner, E., 1909–1911. Die Alpen im Eiszeitalter, 3 vols. Tauschnitz, Leipzig, 1199 pp.

Sbaï, A., 1986. Contribution à l'étude géomorphologique de la région d'Oyonnax (Ain-Jura méridional). Lille, Thèse 56, 552 pp.

Schoeineich, P., 1998. Corrélation du dernier maximum glaciaire et de la déglaciation alpine avec l'enregistrement isotopique du Groenland. Quaternaire 9, 203–217.

Tiercelin, J.J., 1974. Le bassin de Larange-Sisteron. Stratigraphie et sédimentologie des dépôts Pleistocènes. Thesis, l'université de Provence, 150 pp.

Tricart, J., 1961. Aperçu sur les formations quaternaires des feuilles de Saint-Claude et Moirans en Montagne. Bull. Serv. Carte Géol. Fr. 58–264, 73–97.

Tricart, J., 1965. Quelques aspects particuliers des glaciations quaternaires du Jura. Rev. géographique Est 4, 499–527.

Viret, J., 1954. Les Loess à banc durcis de Saint-Vallier (Drôme) et sa faune mammifères villafranchiens avec une analyse granulométrique Par E. Schmid et une analyse pollinique par C. Krachenbüchl. Nouvelle Archives du Museum d'histoire naturelle de Lyon 4, 200 pp.

Vivian, R., 1975. Les Glaciers des Alpes Occiedentales. Université de Grenoble, thesis, 575 pp.

Chapter 11

Recent Advances in Research on Quaternary Glaciations in the Pyrenees

Marc Calvet[1,*], Magali Delmas[1], Yanni Gunnell[2], Régis Braucher[3] and Didier Bourlès[3]

[1] JE 2522 Médi-Terra, Université de Perpignan-Via Domitia, 52 Avenue Paul Alduy, F 66860 Perpignan cedex, France
[2] UMR 5600 CNRS, Université Lumière-Lyon 2, 5 avenue Pierre Mendès-France, 69676 Bron cedex, France
[3] UMR 6635 CNRS, Aix Marseille Université, BP 80, 13545 Aix-en-Provence cedex 04, France
*Correspondence and requests for materials should be addressed to Marc Calvet. E-mail: calvet@univ-perp.fr

11.1. INTRODUCTION

In the past decade, particularly since the last summary of the subject (Calvet, 2004), the Quaternary glaciation of the Pyrenees has been the focus of new research. Unequal progress has been achieved on three aspects: mapping the extent of the Pyrenean ice field, quantifying the geomorphological impact of glaciation on the preglacial landscape and refining the chronology of the glacial fluctuations.

Mapping of ice extent has benefited from only minor updates. A relatively accurate 1:100,000-scale illustration of ice extent is only available for the eastern part of the range, the Ariège catchment included. The synthesis by Calvet (2004), who following the minimalist views of Taillefer (1985), at the time underestimated the extent of the Würmian ice field, has been significantly revised (Fig. 11.1). Further west, uncertainty remains concerning the termini of Würmian glaciers in the Salat and Gave d'Oloron catchments; for the last valley, the outermost glacial deposit has been described at Escot, but there is also a presumption that ice advance might have extended some 10 km further out onto the piedmont (Gangloff et al.,

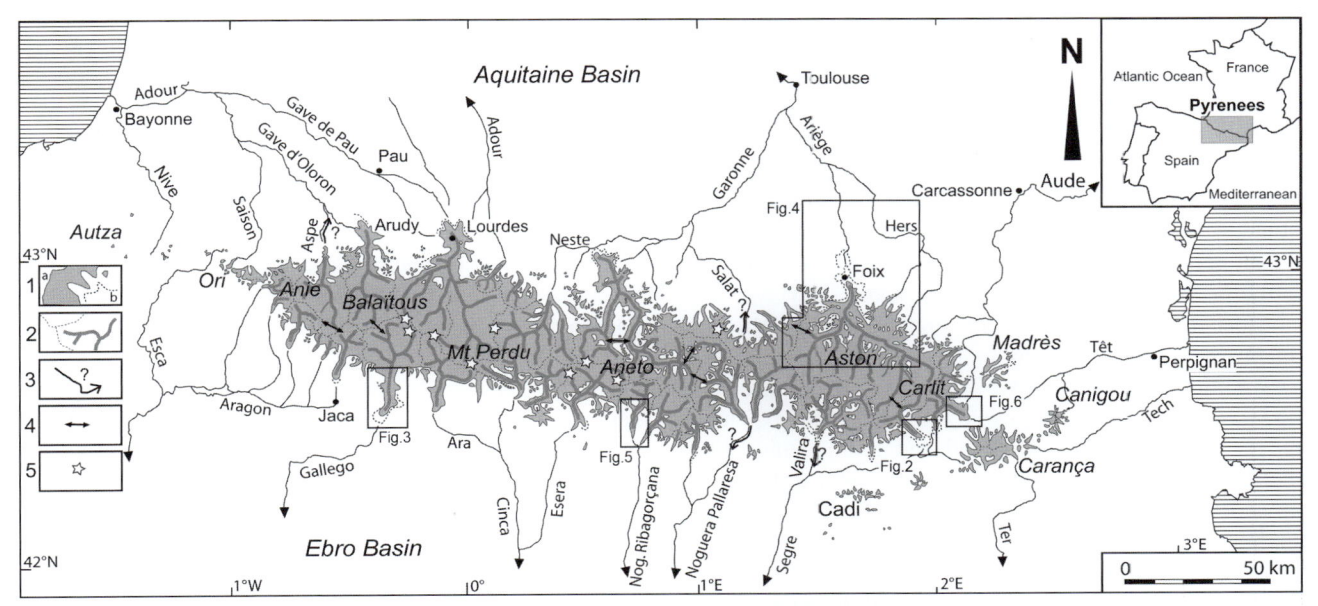

FIGURE 11.1 The Pleistocene glaciation of the Pyrenees: a cartographic synthesis. 1: Glacierised areas, a: during the Würmian; b: during the Middle Pleistocene. 2: Main supraglacial mountain ridges and ice-catchment limits; main ice routes. 3: Possible extent of Pleistocene valley glaciers. 4: Main transfluence cols. 5: Currently glacierised massifs containing small, residual cirque glaciers. The map is modified and updated after Calvet (2004). Note the location of other figures mentioned in the text.

1991). Both for small, transverse valleys and for the main valleys, such as the Noguera Pallaresa or the Valira, the terminal position of the glacier systems is also uncertain in Spain. Work by Serrat et al. (1994) and Turu i Michels and Peña Monné (2006) has claimed that these glacial termini extended ca. 10 km further than thus far accepted. This is in spite of the fact that the valley cross-sections in these areas are V-shaped and sometimes exhibit entrenched meanders. Further, glacial landforms are either scarce or absent, and the interpretation of existing Pleistocene deposits is inconclusive.

The impact of glacial erosion on the landscape has been measured through a combination of approaches: (i) qualitative approaches have endeavoured to list and map all pre- and interglacial landforms and deposits. These features are so well preserved in some places that they necessarily imply extremely limited depths of Pleistocene erosion by glaciers. Relevant markers of limited glacial erosion are thick weathered mantles preserved beneath glacial till or ice-marginal deposits, and relict erosion surfaces the Neogene age of which have been confirmed by apatite fission-track and (U–Th)/He dating (Gunnell et al., 2009). (ii) Quantitative approaches have produced mass-balance studies that exploit the well-preserved glacial sediment repository situated on the southern side of the Carlit massif. Based on research by Delmas et al. (2009), catchment-scale glacial erosion depths have been estimated to have not exceeded 5 m during the last glacial cycle. Cirque morphometry in Aragon ($n=206$; García-Ruiz et al., 2000) and in the Carlit and Ariège catchments ($n=1066$; Delmas, 2009) has also provided some insight into spatial variations in erosional intensity. However, the overwhelming conclusion is that bedrock lithology and structure are the strongest controls on cirque morphology: neither climatic criteria nor the duration of glacier presence (estimated on the basis of cirque floor elevation) have provided clear-cut regional patterns. Overall, even where glacial landforms pervade the crest zone and convey a distinctly alpine character to the mountain scenery (Calvet et al., 2008), evidence points to a relatively limited morphological imprint of glaciation on the pre-Quaternary Pyrenean landscape.

Most progress devoted to understanding Pyrenean Quaternary landscape evolution has been achieved through constraining the chronology of glaciation, and particularly the last Pleistocene glacial cycle (Würmian Stage). This is due to the surge of *in situ*-produced cosmogenic radionuclide (CRN) dating of glacial landforms, together with optically stimulated luminescence (OSL) dating of glaciofluvial outwash deposits, and of the cross-correlation of these new data with existing radiocarbon ages of ice-marginal sediments. It is supported by palynological investigations and by morphostratigraphical studies of glacial deposit sequences. These aspects are the main focus of this review.

11.2. MIDDLE PLEISTOCENE GLACIATIONS

Glaciations pre-dating the Würmian Stage remain poorly documented. Neither their number nor their spatial extent and precise time of formation are well established. This is partially ascribable to the fact that these older glaciers did not advance much further out of the crest zone than those during the subsequent Würmian. It is possible that they exceeded the Würmian maximum by 1.5 km in the Carol (or Querol) valley and by 5–10 km in the case of the Ariège glacier. Further, the last glaciation has often buried, destroyed or reworked pre-existing older deposits.

11.2.1. On the South Side of the Range

The Carol catchment exhibits among the best-preserved glacial deposit sequences of the Pyrenees (Fig. 11.2). It contains three generations of frontal and lateral moraines, each grading to the tread of a glaciofluvial terrace. Clast assemblages in each terrace show distinctive weathering characteristics. Although relative dating of these moraines would have suggested an age sequence bridging the Middle and Late Pleistocene, ^{10}Be dating has provided ages of 65 ka or less, suggesting that all three moraine units formed during the Würmian Stage (Delmas, 2009). Such a young age for deeply weathered glacial and glaciofluvial materials is deceptive. It suggests that the erratic boulders, sampled for CRN dating, have lost significant mass through post-depositional rind weathering and grus formation, hence providing younger than expected exposure ages. Alternatively, it implies that those same erratic boulders were embedded in the moraine and have undergone post-depositional exhumation. In either case, this could explain the relatively uniform age cluster obtained for the Carol moraines.

The Gállego catchment (Fig. 11.3) has provided a puzzling series of OSL ages (Peña et al., 2004; Lewis et al., 2009). The large frontal moraine at Senegüé was until now held to represent the Würmian glacial maximum, but moraines at Aurin, situated 2 km further down-valley, have yielded contrasting ages of 38 and 85 ka. They have therefore also been attributed to the Würmian. Meanwhile, the uppermost glaciofluvial terrace deposits, Qt 5, at Sabiñanigo, which contains glacially striated pebbles and correlates stratigraphically with terrace T2 deposits in other valleys in the eastern Pyrenees (Calvet, 1996), have provided not only two ages, 155 and 156 ka, which are attributable to marine isotope stage 6 (MIS 6; i.e. late Rissian), but also two much younger ages (99 and 84 ka).

The Valira catchment (Turu i Michels and Peña Monné, 2006; Turu i Michels et al., 2007; Turu i Michels, 2009) is reported to contain an Eemian-age sediment fill, discovered during drilling in the central valley of Andorra. Here, deeply weathered moraine deposits, attributable to MIS 6, are also exposed at Sant Julia de Loria and connect down-valley with

FIGURE 11.2 Terminal sequence of glacier deposits in the Carol catchment: age of frontal and lateral Würmian moraines. 1: Pliocene or early Pleistocene alluvium capping erosional surfaces. 2: Proglacial alluvial sheet (T4), deeply weathered. 3: Alluvial sheet (T3). 4: Proglacial alluvial sheet (T2), weathered. 5: Proglacial alluvial sheet (T1), poorly weathered or unweathered. 6: Ice-marginal sediments. 7: Würmian till with crested moraines; materials poorly weathered. 8: Middle Pleistocene (MIS 6) till with physically degraded moraines, connected to T2; materials weathered. 9: Till with highly degraded moraines, connected to T4 (perhaps MIS 12 or MIS 16); materials deeply weathered. 10: ^{10}Be-dated sites; note age dispersion on the more weathered moraines. The map is after Calvet (1996, 2004), with the geochronological data taken from Delmas (2009).

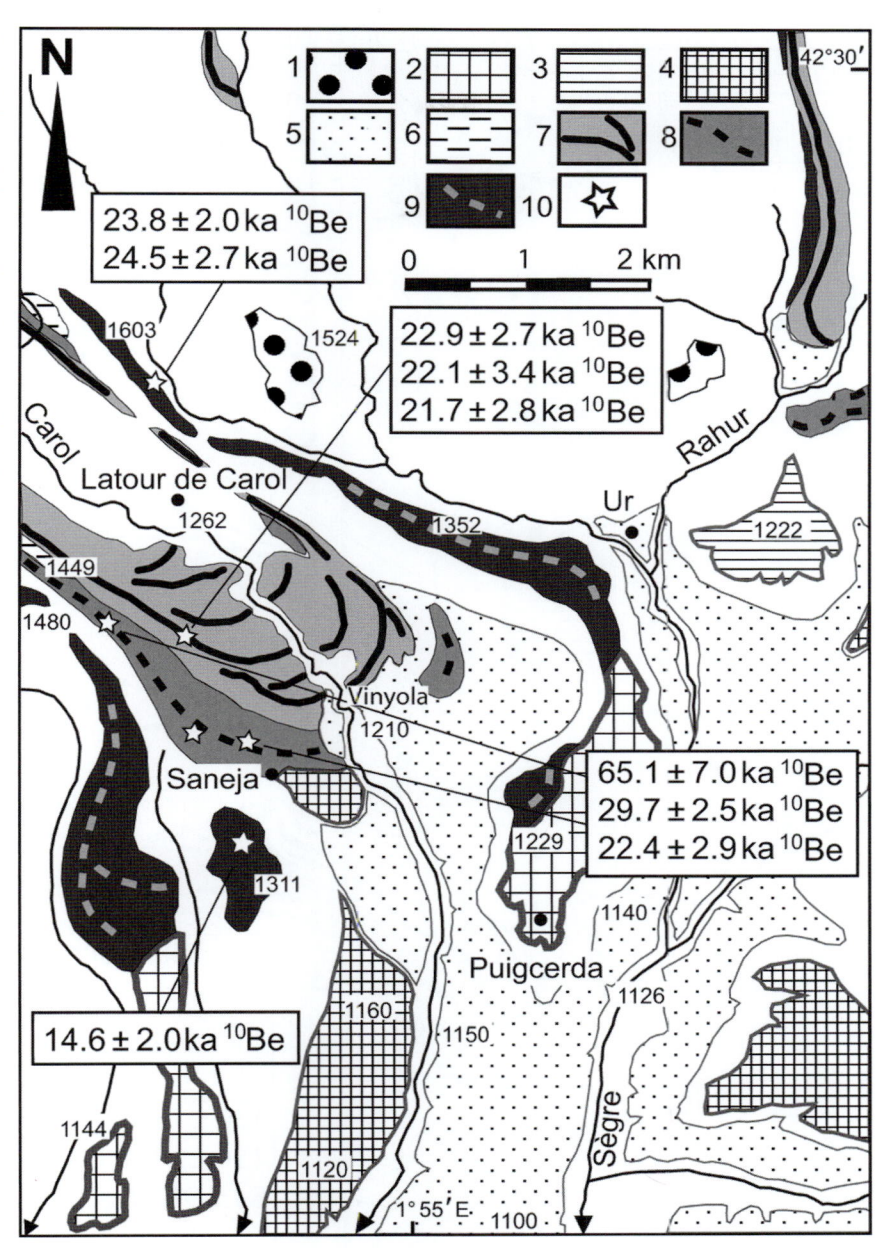

a proglacial alluvial sheet which has yielded an OSL date of 125 ka. A much older ice front position is also reported ca. 10 km further down-valley at Calvinya. This gives an age that is older than 350 ka (the OSL detection limit). However, inspection of exposed sections reveals only very rare striated cobbles, and sedimentological features were indicative of fluvial rather than of glacial deposition.

11.2.2. On the North Side of the Range: Focus on the Ariège Catchment

The Ariège moraine sequence (Fig. 11.4) has raised the same issues as those in the Carol catchment, namely anomalously young CRN ages obtained from erratic boulders preserved on the valley slopes high above the poorly weathered and well-preserved lateral moraines of Würmian age (Delmas, 2009; Delmas et al., 2011). As in the Carol valley, the boulders also show signs of post-depositional weathering, of displacement along fractures and even of *taffoni* development. Following the methodological recommendations of Putkonen and Swanson (2003), out of the sampled population, the oldest ^{10}Be age was considered to be the most accurate, providing a resulting age of 122 ka for the deposits near Caraybat (Fig. 11.4).

Further insight into the Middle Pleistocene glaciations has been gained by examining the relative altitudes of glacial

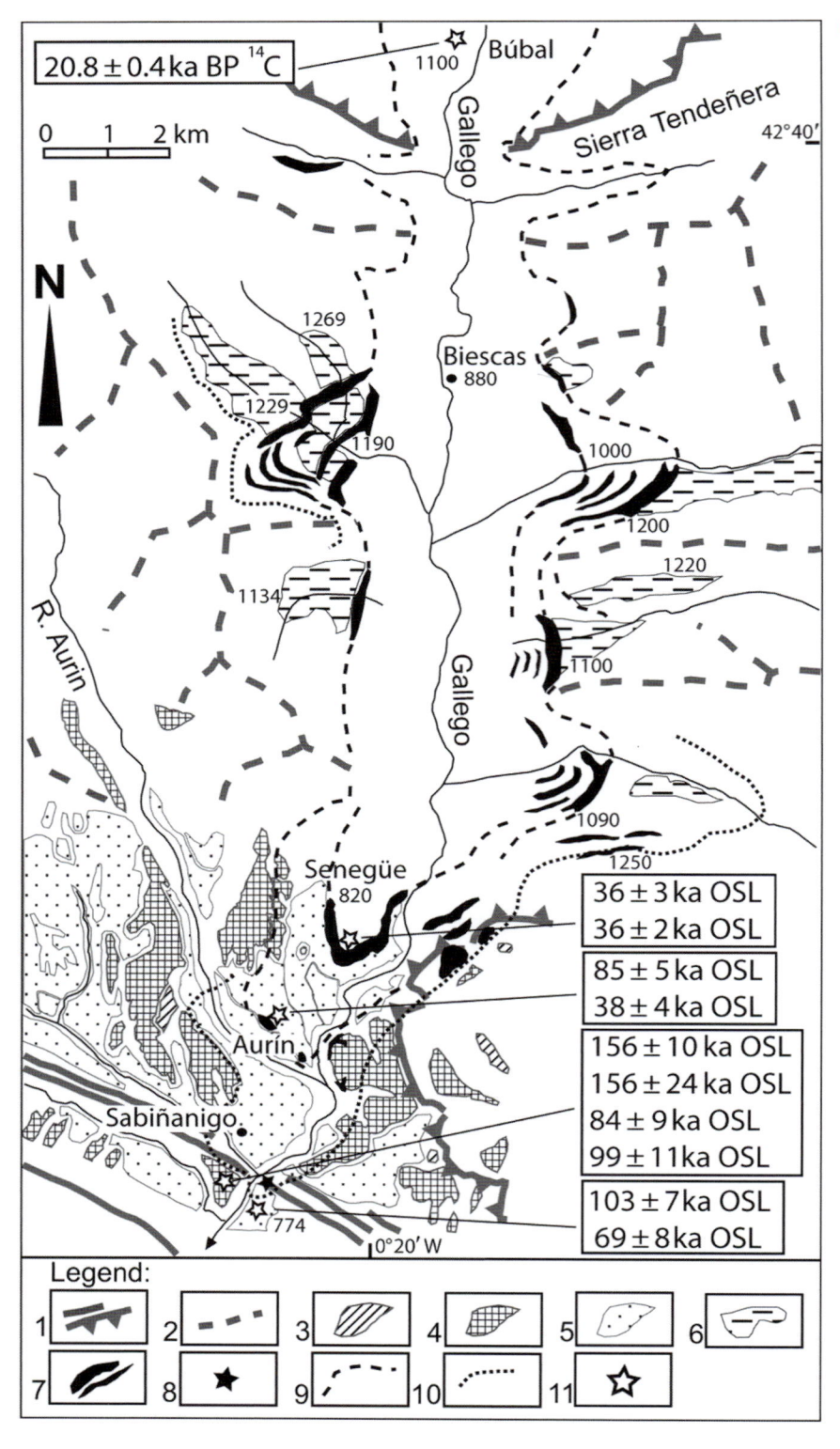

FIGURE 11.3 Age-bracketed terminal sequence of glacier deposits in the Gállego catchment. 1: High-relief landforms, razorback and hogback ridges. 2: arêtes. 3: Tread of highest alluvial deposits (coronas). 4: Proglacial alluvial sheets and mantled wash pediments (weathered and rubified: Qt 5, alias T2). 5: Proglacial alluvial sheets (poorly weathered: Qt 6-7-8, alias T1). 6: Würmian ice-contact sedimentary units (lacustrine or fluvial). 7: Moraine ridges and till deposits. 8: Scoured rock surfaces on bedrock steps. 9: Successive Würmian ice front positions. 10: Middle Pleistocene (MIS 6) ice fronts. 11: OSL and ^{14}C-dated sites. After the geomorphological map by Barrère (1966) and data by Peña et al. (2004), Lewis et al. (2009) and Jalut et al. (1992).

deposits, the advanced weathering state of which indicates that they are older than Würmian. This approach has allowed differentiation between generations of deposits that are legacies of different stadial events within MIS 6, or perhaps even of different glacial cycles: (i) in the Tarascon basin and in the higher Ariège catchment, glacial evidence from MIS 6 occurs as trails of weathered till deposits and erratic boulders situated ca. 50–100 m higher than Würmian

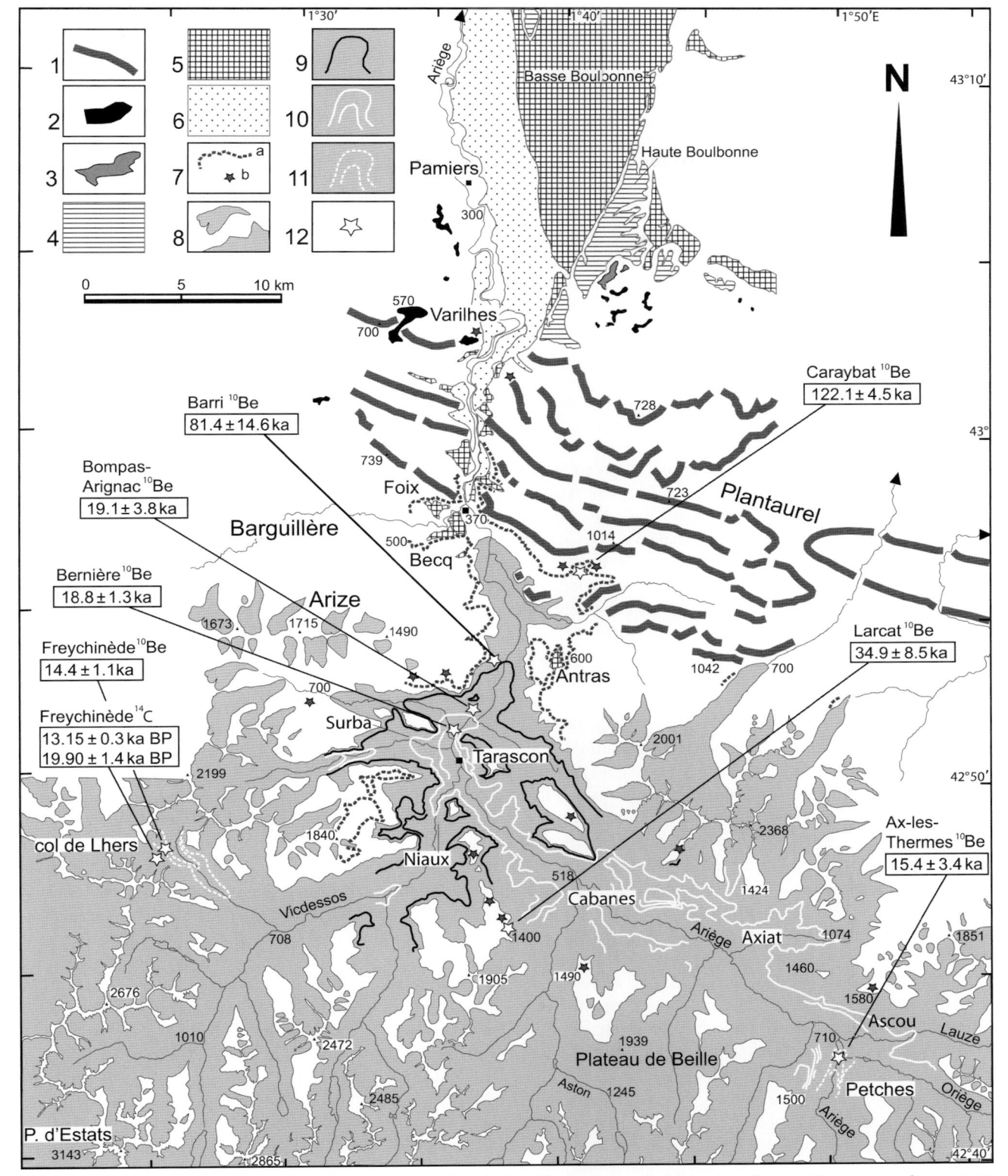

FIGURE 11.4 Pleistocene glaciation in the Ariège catchment. 1: Razorback and hogback scarps in Mesozoic or Cenozoic limestone, sandstone or conglomerates of the outer Pyrenean fold belt. 2: Relict Pliocene to early Pleistocene alluvial fan deposits (formation du Lannemezan). 3: Higher alluvial terrace T4, severely eroded. 4: Intermediate terrace, intensely weathered and rubified (Haute Boulbonne). 5: Lower terraces (generation T2), weathered and rubified (Basse Boulbonne, Vernajoul and ice-marginal terraces of Antras and Foix–Cadirac). 6: Lowermost terrace T1, of Würmian age, unweathered and capped by brown soils. 7: Middle Pleistocene ice boundaries; a: probable MIS 6 ice limits, grading to the T2 terraces; b: erratic boulders or residual morainic deposits dating to MIS 6 or older. 8: Würmian maximum ice extent; the corresponding population of unweathered moraines grades to the T1 alluvial sheets. 9: MIS 2 ice boundaries, that is, Garrabet stadial. 10: Recessional moraines at Bompas–Arignac and Bernière (19–20 ka). 11: Recessional moraines (Petches and Freychinède) dating to the Oldest Dryas. 12: Position of dated sites. Distribution of glaciofluvial deposits in the piedmont zone, based on Hubschman (1975, 1984) and on the geological map of the Pyrenees Quaternary by Barrère et al. (2009). The ice limits and ^{10}Be data are after Delmas (2009) and the ^{14}C ages after Jalut et al. (1982, 1992).

maximum glacial moraines. In the Foix–Montgaillard basin, the Antras ice-contact depositional sequence and the ice-marginal sediment body at Becq-en-Barguillère (Taillefer, 1973) represent two additional sites to which the MIS 6 maximum ice extent can be tethered. Given these new interpretations, the ice-marginal meltwater terrace sequence at Foix–Cadirac was ascribed by Hubschman (1975, 1984) to the Rissian. This was because of the advanced state of weathering of its debris. However, it occurs at a lower elevation than the Becq and other MIS 6 benchmarks and could instead represent the legacy of a deglacial event that occurred toward the end of the MIS 6 cycle. Further down the Ariège valley, the Foix–Cadirac terrace grades to the Basse Boulbonne (T2) alluvial sheet. T2 (mapped by Barrère et al., 2009) is capped by rubified soils displaying a depleted upper horizon. Characteristically evolved soils, such as this, suggest an episode of weathering and eluviation during the Eemian Stage (MIS 5e) climatic optimum. (ii) Erratic boulders sourced by the granitic and gneissic Aston massif have been found resting on the Mesozoic limestone and shale outcrops at Caraybat, down-valley from the Antras delta and 50 m above it. This boulder trail was produced by one or several ancient glaciers of corresponding thickness and has been tentatively correlated with elevated moraine deposits on the outer flanks of the Arize massif and in the Tarascon basin at Quéménailles and Mentiès. In their lower reaches, these ice masses may have extended as far as Pech de Varilles, at the proximal edge of the Pamiers plain. At Pech de Varilles, Faucher (1937) reported erratic boulders of gneiss exceeding 6 m in diameter, sourced from the Aston massif. He hypothesised that they were of glacial origin. However, other geologists later either argued that the boulders had been supplied from the underlying Eocene conglomerate bedrock or else ascribed them to unconformable Pliocene fan deposits that excavated the Eocene conglomerate and locally contain greater than 1 m-sized quartzite boulders. Nevertheless, the boulders described by Faucher are conspicuous by their petrography and provenance, by their large size, and by their extremely limited weathering state. They occur ca. 460 m a.s.l., that is, 90 m above the tread of the deeply weathered Haute Boulbonne (T3) terrace, which occupies the opposite bank of the Ariège river in the same area (Calvet, unpublished). These indications imply that this elevated erratic boulder trail is of early Middle Pleistocene age and correlates, as in the Carol valley, with the older alluvial terrace T4 deposits. This topographical position suggests that the erratics at Pech de Varilles are comparable to the ancient moraine deposits that occur much further west at Arudy. The latter also occur at elevations similar to those of the relict Pliocene to early Pleistocene piedmont deposits of the Gave d'Oloron catchment (Hétu and Gangloff, 1989). CRN dating was not attempted on the Pech de Varilles erratics because field evidence indicates that they have been exhumed from a finer matrix and have been displaced by slope movements.

11.3. THE LAST PLEISTOCENE GLACIAL CYCLE (WÜRMIAN STAGE)

The chronology of the Pyrenean last glacial cycle, correlated with the Würmian Stage of the Alps by most Pyrenean authors, has been mostly based on a large set of ^{14}C ages produced over the past 30 years, supplemented in the last five mostly by new ^{10}Be and OSL ages. The image of the Würmian chronology provided by this body of evidence is still incomplete, with contradictions between the methods employed giving sometimes puzzling contrasts between adjacent valleys. Radiocarbon dating is usually indirect in that it provides an age for proglacial or ice-marginal organic sediments whose relation to adjacent moraines is not always straightforward. Even though a few terminal moraines have been dated directly, most OSL dating raises similar problems. This is because sampling strategies target alluvial sheets that are sometimes situated several tens of kilometres from the nearest ice front deposits. CRN dating, in contrast, systematically allows direct dating of glacial landforms, whether moraines, isolated erratic boulders or ice-scoured bedrock steps. However, CRN exposure ages are, by definition, minimum ages, with the precautions this limitation inevitably brings upon interpretation of the data.

11.3.1. The Age of Würmian Maximum Ice Extent: In or Out of Phase with the Global Last Glacial Maximum?

The mid-latitude Pyrenean mountain range is under direct influence of the North Atlantic weather systems. This makes the Pyrenees an ideal setting for testing whether or not mid-latitude mountain glaciation and global climatic fluctuations were synchronous (Gillespie and Molnar, 1995; Hughes and Woodward, 2008; Thackray et al., 2008). The now classic view that the Würmian Pyrenean maximum ice extent and the global last glacial maximum (LGM-defined as the time period between 23 and 19 ka cal. BP, Mix et al., 2001) did not occur simultaneously was presented in the 1980s on the basis of radiocarbon ages obtained on ice-contact lake sediments on the north side of the range (Andrieu et al., 1988; Jalut et al., 1988, 1992; Andrieu, 1991; detailed synthesis in Delmas, 2009, pp. 96–109). This radiocarbon-based perspective on Pyrenean glaciation advocated a Würmian maximum ice extent between 70 and 50 ka. After this time, the glaciers stagnated slightly up-valley from their maximal advance positions before retreating to the uppermost reaches ca. 29–25 ka, depending on sites. To explain the recessional state of the Pyrenean ice cap by the time of the global LGM (which is broadly correlable to MIS 2), it has been argued that climatic conditions were too dry to sustain extensive ice fields on the Pyrenees. Although data are comparatively scarcer, a similar model has been proposed for the southern side of the Pyrenees. In the Noguera Ribagorçana catchment (Fig. 11.5), the ice-marginal site of Llestui, interpreted as marking a Würmian post-maximum ice position, was initially radiocarbon dated at 31–34 ^{14}C ka BP (Vilaplana, 1983). However, the samples were found to have been contaminated by inorganic carbon and later redated at 18–21 ^{14}C ka BP (Bordonau et al., 1993). For reasons of suspected sample contamination, the 30 ^{14}C ka BP radiocarbon age obtained from the Bassots palaeolake, which is ponded behind the terminal lobe of the Ribagorçana moraine, has also been discounted as unreliable (Bordonau i Ibern, 1992).

Amidst mounting uncertainty over radiocarbon ages of ice-marginal materials, critics have begun to question the asynchroneity of the Pyrenean glacial model. Basing their reassessment on biostratigraphical and geochemical evidence, palynologists in the late 1980s were first to challenge the view that Pyrenean glacier fluctuations were out of phase with the global LGM (Turner and Hannon, 1988; Reille and Andrieu, 1995). The first CRN ages produced for glacial landforms in the Noguera Ribagorçana (Pallàs et al., 2006) prompted a more systematic critique of radiocarbon ages from the Pyrenees, most of them being deemed unreliable and were rejected. The competing hypothesis of synchroneity between the Würmian maximum ice extent and the global LGM was asserted. This was based on ^{10}Be ages of 18.1 ± 1.9 ka obtained from a lateral moraine of the Noguera Ribagorçana trunk glacier and 21.3 ± 4.4 ka obtained from an erratic boulder close to the terminal lobe (Fig. 11.5). These ^{10}Be ages were revised to 19.2 ± 2 of the Noguera Ribagorçana trunk glacier and 22.6 ± 4.7 ka, respectively, by Rodés (2008). Pallàs et al. (2006) have nevertheless not rejected the hypothesis that a relatively protracted 'Pleniglacial' might have prevailed between before 30 ka and ca. 20 ka. This means that glacier snouts may have been stationed far down the valleys not only during the global LGM but also for quite some time before it.

The body of geochronological information available today, however, provides evidence of a somewhat less static history of glacier fluctuation, and consequently, a less serious antagonism between the in-phase and out-phase schools of Pyrenean glaciation. The asynchroneity of maximum advances of mountain and continental glaciers needs no longer to be challenged. The Würmian maximum ice extent now appears to have occurred everywhere quite early during the Würmian Stage, irrespective of the dating methods used to test this hypothesis. More controversial is the relative position of ice fronts during the MIS 2 or more accurately during the global LGM, either because of insufficient data or because of genuine variability between valleys along the mountain range.

11.3.2. Dating of Pyrenean Würmian Terminal Moraine Occurrences

On the Spanish side, the most abundant evidence has been obtained from the Gállego catchment (Fig. 11.3). The terminal moraine at Senegüé has yielded two OSL ages of 36 ± 3 and 36 ± 2 ka, respectively (Peña et al., 2004; Lewis et al., 2009). In the western part of the catchment, elevations do not exceed 2.7 km. As the radiocarbon ages of 30–25 ^{14}C ka BP obtained from proglacial lake sediments around Tramacastilla and Pourtalet demonstrate (García-Ruiz et al., 2003; González-Sampériz et al., 2006) this area was already partially deglaciated in MIS 2 and cut off from transfluent

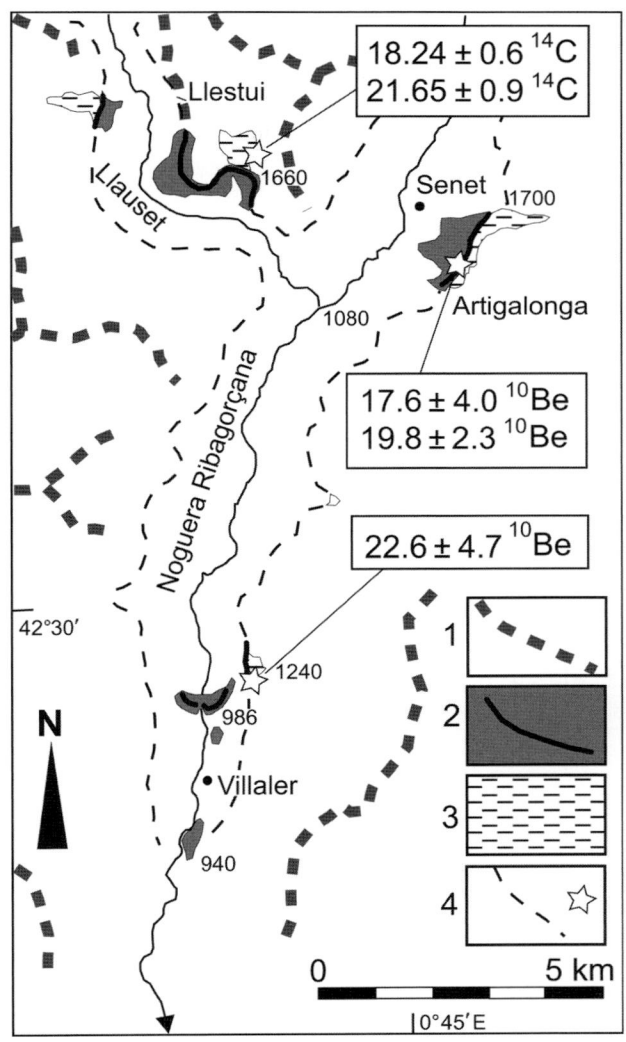

FIGURE 11.5 Terminal sequence of glacier deposits in the Noguera Ribagorçana catchment, age of the Würmian moraines. 1: Mountain ridge. 2: Unweathered Würmian till or moraine. 3: Würmian fluvio-lacustrine ice-contact deposit. 4: Würmian ice front limits and typical U-shaped valley cross-profiles; ^{14}C- and ^{10}Be-dated sites. The map and chronological data are after Bordonau et al. (1993), Pallàs et al. (2006) and Rodés (2008).

sources of northern ice spilling over the Col du Pourtalet. A core obtained from a cirque located 2 km to the east of the Col du Pourtalet indicates a possible readvance between 24.17 and 19.25 ^{14}C ka BP. This advance would have occurred during MIS 2 even though ice had vacated the topography at elevations situated above that site since before 19.25 ^{14}C ka BP. The position of the Gállego ice front in MIS 2 is difficult to establish, but the 20.8 ^{14}C ka BP (i.e. 24.217 ka cal. BP) radiocarbon age obtained from the base of the Bubal lake deposits (Jalut et al., 1992) suggests that the LGM glacier front was stationed as much as 15 km up-valley from Seneguë.

In the Cinca catchment, the maximum extent of the Ara glacier is constrained by ice-contact lake deposits at Linás de Broto. A 50-m drillhole here produced a mid-core AMS ^{14}C age of 30.38 ^{14}C ka BP (Martí-Bono et al., 2002), indicating an early age of Würmian maximum ice extent and possible maintenance of this advanced position until MIS 2 and global LGM time. In the Cinca valley, till at Salinas de Sin is associated with glaciofluvial deposits and has yielded OSL ages of $46 \pm 4, 63 \pm 6$ and 71 ± 15 ka. The Salinas exposure has been subjected to glaciotectonic deformation, proof that it has also been overridden by the glacier since 46 ka (Lewis et al., 2009).

On the Noguera Ribagorçana (Fig. 11.5), MIS 2 ages have been obtained by ^{10}Be dating of deposits situated in the vicinity of the presumed glacier terminus (Pallàs et al., 2006). The terminal lobes, however, have been significantly eroded, making it difficult to draw any firm conclusions from this valley about the age of the Würmian maximum ice extent. In the Valira catchment, most of the data are still unpublished but they clearly confirm that the Würmian maximum ice extent occurred much earlier than the global LGM. The ice-marginal lake sediment sequence at La Massana, which formed where the Valira del Nord and Valira d'Orient glaciers separated, has been subjected to detailed sedimentological and stratigraphical analysis of its progradational deltas. AMS radiocarbon dates have returned ages of 25.63, 21.51 and 17.43 ^{14}C ka BP (29031 to 27820, 24150 to 23310, 21097 to 20271 ka cal. BP) and indirectly confirm the presence of this lake since ~ 41 ka on the basis of sedimentation rate calculations. The lake sediments subsequently recorded a succession of three glacier advances of diminishing intensity, all falling chronostratigraphically within MIS 2 (Turu i Michels, 2002). The maximum ice extent may have been captured by a 59 ± 1.18 ^{21}Ne ka age obtained from an ice-scoured bedrock surface that was sampled on the upper flank of the glacial trough above the village of Canillo (Turu i Michels et al., 2004).

In the eastern Pyrenees, ^{10}Be exposure ages show that the MIS 2 maximum ice advance reached positions within a few hundred metres of the Würmian maximum ice extent. The frontal moraine of the Têt glacier at Mont Louis (Fig. 11.6) has provided an age of 21.4 ± 3.7 ka (Delmas et al., 2008), since then recalibrated as 22 ± 3.6 ka (Delmas, 2009). However, the dated boulders belong to the innermost lobe crest of the Têt Würmian terminal moraine sequence. The two outermost lobes have so far not been dated but could correspond to the sites of maximum ice extent reached earlier during the Würmian stage. In the Carol catchment (Fig. 11.2), the right flank lateral moraine has

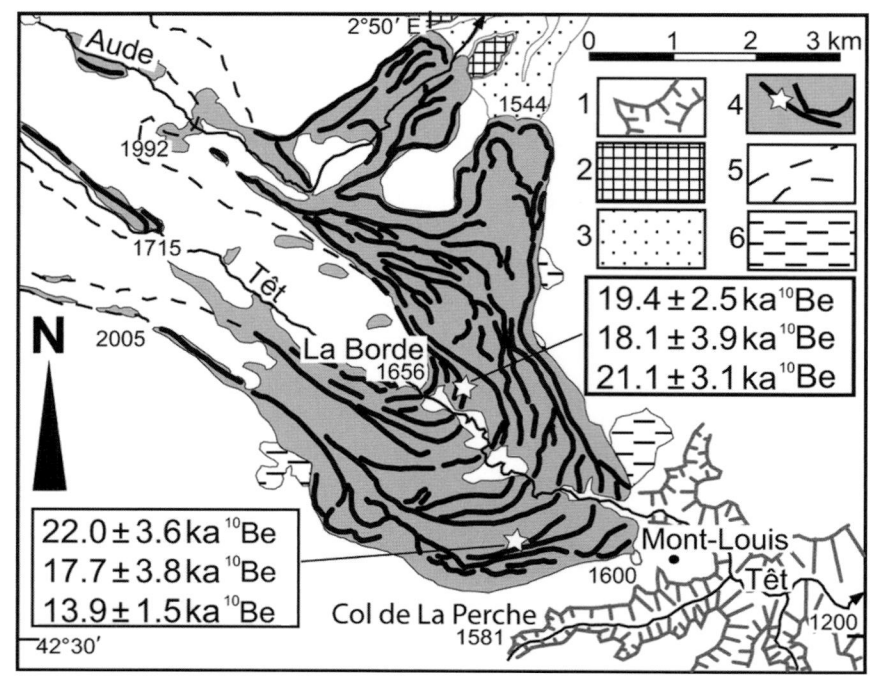

FIGURE 11.6 Terminal sequence of Würmian glacier deposits in the Têt catchment: age of frontal and lateral moraines. 1: Quaternary incision of valleys into the Cenozoic erosion surface of the Col de La Perche. 2: Alluvial sheet T2, weathered (MIS 6). 3: Proglacial alluvial sheet T1, unweathered and connected to the Würmian moraines. 4: Würmian unweathered till and moraines, with the position ^{10}Be exposure-dated sites. 5: Successive Würmian ice front positions. 6: Würmian ice-contact sedimentary units (lacustrine or fluvial). The mapping and geochronological data are after Delmas et al. (2008) and Delmas (2009).

been dated with great precision by three samples: 22.9 ± 2.7, 22.1 ± 3.4 and 21.7 ± 2.8 ka (Delmas, 2009). However, as in the case of the Têt moraine sequence, the dated lateral moraine connects with a frontal lobe that is situated ca. 0.6 km up-valley from the Vinyola terminal lobe. The dated lateral moraine is therefore not the oldest in the Würmian sequence. Finally, the small Malniu glacier, which remained disconnected from the Carol trunk valley throughout the Würmian stage, has yielded ^{10}Be ages of 76.5, 42.3 and 40.6 ka. The post-maximum Malniu frontal lobe, which is situated 0.5 km further up-valley, provided an age of 21.3 ka, which accords with the 23.9 ka age of a coeval lateral moraine preserved on the valley side (Pallàs et al., 2010).

Among the north-facing glacial catchments, only the Ariège basin has recently delivered new data (Delmas, 2009; Delmas et al., 2011). Results, which largely rely on ^{10}Be exposure ages, confirm the classic view of asynchroneity between mountain glaciation and the continental ice-sheet record and are broadly compatible with existing results from the Spanish catchments. The Würmian maximum ice extent (Fig. 11.4) is constrained by two CRN ages that suggest two glacier advances of comparable extent: a 81.4 ± 14.6 ka age obtained from an erratic boulder in the vicinity of the terminal moraine and a younger age of 34.9 ± 8.5 ka obtained on the mid-valley lateral moraine at Larcat. The Ariège glacier advanced to within 2 km of the town of Foix. The older age is in good agreement with the 91.4 ± 2.4 ka U–Th age obtained for the stalagmite floor in the prehistoric cave at Niaux, which occurs 8 km up-valley and is just slightly older than the Niaux glacial deposits (Sorriaux, 1981, 1982; Bakalowicz et al., 1984). The MIS 2 maximum extent ice front has not been directly dated, but its position almost certainly coincides with the Garrabet frontal moraine, which is located 8 km up-valley from the Würmian terminal moraine. It has two recessional fronts at Bompas–Arignac and Bernière, respectively. This chronology was established indirectly by a surface exposure age with a strong component of nuclide inheritance (31.2 ka) on the bedrock step of the Tarascon basin and by the age of an adjacent erratic boulder resting directly on the dated bedrock-step surface (16.7 ka). Such a conjunction suggests so far unreported ice recessions and readvances, that is stadials, within the Pyrenean last glacial cycle. Other bedrock steps have yielded exposure ages of 19 ka, for example, on a scoured rock exposure situated 2.5 km up-valley from Garrabet. These ages equate to a ~ 19 ka age for the Bernière frontal lobe and provide the timing of the deglaciation increment that followed the Garrabet stage.

So far, no new data have been produced that might help to update the radiocarbon-dated terminal lobes of the Garonne, Gave de Pau and Gave d'Ossau catchments (Fig. 11.1). In these western tracts of the Pyrenean range, the glaciers are thought to have receded to the upper valleys as early as 29–25 ^{14}C ka BP, and no post-maximum ice front position corresponding to MIS 2, for example, similar to the Garrabet frontal system, has been clearly identified. The interpretations of Jalut et al. (1992) and Reille and Andrieu (1995), respectively, have diverged on these issues. Accordingly, a discussion of their respective merits in the context of this review seems timely. Assuming first that the 38.4 ± 2 ^{14}C ka BP age obtained at the Biscaye site by Mardonnes and Jalut (1983) is invalid and should be replaced by the AMS ^{14}C ages obtained by Reille and Andrieu (1995) from glaciolacustrine clays at the Lourdes 1 terminal moraine site, a Würmian maximum ice extent older than 20.025 ^{14}C ka BP (22486 to 21562 ka cal. BP) appears correct, with vacation of the piedmont zone between 16.67 and 14.46 ^{14}C ka BP. New ^{10}Be exposure data from the Gave de Pau catchment have been obtained from morainic boulders situated at the Ossen diffluence site, immediately above the town of Lourdes. However, with five Lateglacial and Holocene ages and another of 53.5 ± 38 ka, results remain inconclusive (Rodès, 2008). Likewise, the moraine sequence at Aucun has yielded four ^{10}Be ages ranging between 13 and 10 ka. All these ages, however, are suspiciously young for this small glacier, which at that time was disconnected from the Gave de Pau trunk valley glacier. At the time, as shown by the position of the lateral moraines at Argelès-Gazost (Barrère et al., 1980), this larger glacier still descended to elevations below 400 m a.s.l. and to positions less than 10 km from the outermost Würmian ice front at Lourdes. Assuming now, instead, that the chronology advocated for the Würmian maximum ice extent by Mardonnes and Jalut (1983) is valid, it is feasible that the Argelès-Gazost and Aucun moraine ridges could define the post-maximum glacier position during MIS 2 peak.

If these local insights are assembled into a more coherent Pyrenean-scale picture, it appears that the Würmian maximum ice extent occurred somewhat earlier than the global LGM, perhaps with glaciers advancing out of the crest zone to similar positions repeatedly during MIS 5d, 5b and 4. By definition, all ^{10}Be ages are minimal exposure ages and thus cannot precisely date the moraine ridges that have been sampled. Caution, therefore, must be exercised when determining the exact number and position of glacier advances and retreats. For example, the MIS 2 ice fronts have only been positively identified in the eastern, more Mediterranean part of the mountain range: for example, in the Têt, Carol, Malniu, Ariège, Valira and Ribagorçana valleys. It further appears that, in this region, MIS 2 ice fronts also nearly reached the Würmian maximum glacier positions. The MIS 2 ice fronts reached those positions at the time of, or just before, the global LGM (i.e. between 23–19 ka cal. BP, Mix et al., 2001). In the western catchments (Garonne, gave de Pau, Gave d'Ossau, Gallego), which are under greater direct influence from Atlantic weather conditions, MIS 2 appears to have coincided instead with a recessional period. It is too soon to decide whether such contrasts should be ascribed to methodological artefacts inherent in the dating techniques used, or whether the observed differences in glacier behaviour along the strike of the mountain range truly reflect palaeoclimatic differences. Far from being implausible, such differences could indeed have been controlled over the western Mediterranean by the Balearic low atmospheric pressure centre, which may have been more active than Atlantic weather systems during MIS 2 because of the southerly position of the Polar Front at the time (Florineth and Schlüchter, 2000; Hughes and Woodward, 2008).

11.3.3. Evidence of Rapid Deglaciation from the end of the global LGM

The stages of ice recession are only known for certain valleys, and it is not yet possible to produce a coherent synthesis for the entire Pyrenean range. Evidence of a first recessional stage is provided by frontal moraines situated only a short distance behind the MIS 2 maximum ice advance fronts, with additional evidence of significantly thinner glaciers merely lining the valley floors (Delmas, 2009). In the Ariège catchment (Fig. 11.4), the ice field was dismembered into disjunct trunk and tributary valley glaciers, with, for example, the Courbière, Vicdessos and Lauze glaciers forming disconnected, self-contained systems. In the Cabanes basin, the glacier rapidly melted to half of its thickness. During this time, a number of new ice-contact lakes and deltas also formed, such as at Ascou–Goulours and Axiat–Senconac on the trunk valley right flank, and at Niaux and Surba on the trunk valley left flank. The frontal moraines of this period are situated at Bompas–Arignac (19.1 ± 3.8 ka, a ^{10}Be exposure age obtained indirectly from a ice-scoured bedrock step just down-valley from the Bompas-Arignac ice front) and at Bernière (18.8 ± 1.3 ka, ^{10}Be age obtained on an erratic boulder). The corresponding exposure ages are in remarkable agreement with the 19.1 ± 1.6 ka U–Th age from the younger stalagmite floor that seals the glacial deposits trapped in the Niaux cave (Sorriaux, 1981, 1982; Bakalowicz et al., 1984). This evidence also heralds the end of glacial sedimentation at this site, with the cave becoming abandoned above the level of glacier activity.

A similar story emerges from the Têt ice field, with the Têt glacier snout receding ca. 2.5 km from the Mont Louis terminal lobe to the site of the Borde frontal moraine (Fig. 11.6). Three ^{10}Be exposure ages from the Borde moraine have indicated that recession occurred ca. 18.1 ± 3.9, 19.4 ± 2.5 or 21.1 ± 3.1 ka (Delmas et al., 2008; Delmas, 2009). This period also coincided with the disappearance of major transfluence cols: at the Col du Pourtalet, this occurred before 19.25 ^{14}C ka BP (i.e. 22–23 ka cal. BP; González-Sampériz et al., 2006); at the Col de Lhers (Fig. 11.4), it occurred ca. 22–20 ^{14}C ka BP (Jalut et al., 1982, 1992) and probably likewise at the Col de Puymorens.

After those events, glacier ice rapidly vacated most Pyrenean valleys, although it did not necessarily undergo a steady or uniform decline. In the Têt catchment, for example, four moraine systems, both lateral and frontal, form a regular sequence receding back into the crest zone. Deglaciation also affected cirques, at least in the south-eastwards facing Têt catchment. For example, fossil peat exposed at the Grave-Amont site (elevation: 2150 m) has yielded three ^{14}C ages that cluster around 20 ka cal. BP (from top to bottom: 20.25–19.87, 19.204–18.666 and 20.26–19.37 ka cal. BP; Delmas, 2005; Delmas et al., 2008). The reliability of these ages is reinforced by the fact that they were obtained from well-preserved *Sphagnum*. This material was free from graphite contamination from the Palaeozoic schist outcrops present in the catchment (this was verified by X-ray diffraction analysis of the matrix) and with no possible hard-water effect. In the Têt catchment, the Llat recessional moraine (2200 m a.s.l.) has yielded a ^{10}Be age of 18.4 ± 2.3 ka (Delmas et al., 2008; Delmas, 2009). Likewise, the small Malniu catchment exhibits a recessional moraine, with a ^{10}Be age of 18.2 ± 0.5 ka, situated on the 2200 m a.s.l. contour (Pallàs et al., 2010). Further west, in the Valira catchment, valley glaciers and the Massana moraine-dammed lake seem to have persisted until 17.5 ^{14}C ka BP (Turu i Michels, 2002), and the lower lateral moraine at Engolaster has returned a ^{10}Be age of 18.077 ± 1.31 ka (Turu i Michels et al., 2004; Turu i Michels, 2009). Along the Noguera Ribagorçana, the snout of an 8-km-long valley glacier reaching an elevation of 1560 m a.s.l. formed the Santet moraines, with a mean ^{10}Be age of 13.7 ± 0.9 ka (the oldest boulder being 16.1 ± 2.8 ka: Pallàs et al., 2006). Rodés (2008) has recalibrated the Santet ages to 14.6 ± 1.1 and 17 ± 0.7 ka, respectively. Meanwhile, in the upper subcatchment of the Noguera de Tor, which is the main tributary to the Noguera Ribagorçana river, the Redo d'Aygüestortes lake, dammed by a frontal moraine at an elevation of 2110 m a.s.l. was ice free by 13.47 ± 0.06 ^{14}C ka BP (Copons and Bordonau, 1996).

In the north-facing catchments of the Pyrenees, valley glaciers several kilometres long appear to have persisted until, or even readvanced during, the Oldest Dryas Stadial. This is well observed in the Ariège catchment at Freychinèdes, where a 4- to 6-km-long glacier tongue persisted until 13 ^{14}C ka BP (~16 ka cal. BP) alongside an ice-marginal lake (Jalut et al., 1982, 1992). This has since been confirmed by CRN dating, with a bedrock-step ^{10}Be exposure age of 14.4 ± 1.1 ka (Delmas, 2009; Delmas et al., 2011). In the upper Ariège catchment, the Petches recessional moraine sequence (Fig. 11.4) has been age-bracketed by a ^{10}Be age from the Ax bedrock step of 15.4 ± 3.4 ka (Delmas, 2009; Delmas et al., 2011). This implies the presence of a 26-km-long valley glacier during the Oldest Dryas Stadial. It was well supplied by tributary valleys descending from the elevated crest zone and feeding into the trunk valley just 5 km from its terminus. At the Puymorens col, the lateral moraine of a comparatively shorter glacier that descended to the carol valley yielded three ^{10}Be ages around 15 ka (Pallàs et al., 2010).

11.3.4. The Last Glacial Bastions: Moraines in the Crest Zone Cirques

The chronology of the cirque glaciers is quite variable and depends on their altitude and their position within the mountain mass. As a result, it still remains difficult to produce a definitive synthesis on the latest Pleistocene cirque glaciation of the Pyrenees. In the central Pyrenees, residual glaciers still exist on the most elevated massifs, and evidence that some of these grew during the Little Ice Age has ushered in the idea, still poorly investigated, of a 'neoglacial' resurgence during the Holocene (Gellatly et al., 1992). Other cirques, however, were ice free by the time of the Oldest Dryas as demonstrated, for example, by palynostratigraphy in the cirques of the Madres massif (Reille and Lowe, 1993). In good agreement with ^{10}Be ages obtained from bedrock steps, most cirques in the south-eastern Pyrenees were almost entirely deglaciated by the Allerød Interstadial (Delmas, 2005, 2009), at a time when the tree-line rose rapidly to altitudes of 1800 m a.s.l. on north-facing slopes (Reille and Andrieu, 1993). Nevertheless, in conditions locally favoured either by snow drifts or by limited insolation, such as in the Têt (i.e. upper Grave) valley, some small glaciers persisted or reformed during the Younger Dryas Stadial, as attested by a 11.6 ± 1.8 ka, ^{10}Be-dated moraine (Delmas et al., 2008; Delmas, 2009). This appears to have also occurred in the Lorri valley above Puymorens, supported by a ^{10}Be age of 11.8 ± 0.6 ka (Rodés, 2008; Pallàs et al., 2010). More surprising are the ^{10}Be ages of 10.4 ± 1.2 and 10.3 ± 1.8 ka (Pallàs et al., 2006), recalibrated to 11 ± 1.3 and 11 ± 1.9 ka (Rodés, 2008), that have been attributed to the Mulleres frontal moraine and its neighbouring bedrock step in the Noguera Ribagorçana valley. This would imply the presence, in the Noguera Ribagorçana catchment, of a 3.5-km-long glacier with a front stationed as low as 1700 m a.s.l. at the beginning of the Holocene Period.

11.4. CONCLUSIONS

The chronology and geometry of the Pyrenean ice field during the Quaternary still hold many mysteries. Progress in knowledge of the chronology of glacial cycles older than the Würmian could soon arise from systematic CRN dating of vertical profiles in exposures of glaciofluvial outwash deposits and of terminal moraines. This would advantageously complement, for example, in the Carol and Ariège valleys, the rather haphazard dating of scattered erratic boulders (out of a population of heterogeneous and thus inconclusive CRN ages, a single boulder has provided a suitable age of 122 ka, Delmas et al., 2011). The last glacial cycle, in contrast, has been increasingly well constrained through the use of multiple chronometers. Glaciers during the Würmian appear to have behaved in a more erratic, non-uniform manner than previously believed, particularly by 'mono-glacialists' according to whom the 'Würm' cycle consisted of a single episode of ice advance followed by a single episode of ice retreat (Taillefer, 1985). Nevertheless, the asynchroneity between the Pyrenean mountain glacial chronology and the global LGM seems confirmed. An important interstadial within the Würmian has been revealed in the Ariège catchment, with glaciers having extended repeatedly to approximately the same positions during stadial advances and interstadial recessions. MIS 2, which coincided with significant post-maximum readvances to the outermost terminal moraines in the eastern part of the Pyrenees, remains poorly represented in the western half of the range (Gállego, Garonne, Gave de Pau). In the Ariège basin, perhaps because it is located in the transition zone between east and west, MIS 2 moraines stand distinctly within the earlier Würmian maximum termini. It remains too early to confirm whether this east-to-west gradient is a genuine signal reflecting differences in glacier response to palaeoclimatic differences, or whether it is the consequence of insufficient or insufficiently precise radiometric age brackets. Regional climatic differences certainly became quite sharp after the onset of the latest Pleistocene deglacial stage, and have remained so as a function of topographical setting and local conditions. Finally, although ice receded rapidly everywhere after the global LGM, the detailed chronology of cirque glaciation is still difficult to unravel because of its geographical variability. More work on these aspects is expected in the coming years.

REFERENCES

Andrieu, V., 1991. Dynamique du paléoenvironnement de la vallée montagnarde de la Garonne (Pyrénées centrales, France) de la fin des temps glaciaires à l'actuel. Thèse de Doctorat de 3e cycle, Université de Toulouse-le-Mirail, 311 pp.

Andrieu, V., Hubschman, J., Jalut, G., Hérail, G., 1988. Chronologie de la déglaciation des Pyrénées françaises. Dynamique de sédimentation et contenu pollinique des paléolacs: application à l'interprétation du retrait glaciaire. Bull. AFEQ 34 (35), 55–67.

Bakalowicz, M., Sorriaux, P., Ford, D.C., 1984. Quaternary glacial events in the Pyrenees from U-series dating of speleothems in the Niaux–Lombrives–Sabart caves, Ariège, France. Nor. Geogr. Tidsskr. 38, 193–197.

Barrère, P., 1966. La morphologie quaternaire dans la région de Biescas et de Sabiñanigo (Haut Aragon), carte hors texte en couleur. Bull. AFEQ 2, 83–93.

Barrère, P., Bois, J.P., Soulé, J.C., Ternet, Y., 1980. Notice de la feuille géologique Argelès-Gazost au 1:50,000. BRGM édit, Orléans n° 1070, 46 pp.

Barrère, P., Calvet, M., Courbouleix, S., Gil Peña, I., Martin Alfageme, S., 2009. Courbouleix, S., Barnolas, A. (Eds.), Carte géologique du Quaternaire des Pyrénées, Scale 1:400,000, BRGM and ITGM édit., Orléans-Madrid, International Year of Planet Earth.

Bordonau i Ibern, J., 1992. Els complexos glacio-lacustres relacionats amb el darrer cicle glacial als pirineus. Geoforma Ediciones, Logroño, 251 pp.

Bordonau, J., Vilaplana, J.M., Fontugne, M., 1993. The glaciolacustrine complex of Llestui (Central Southern Pyrenees): a key-locality for the chronology of the last glacial cycle in the Pyrenees. C. R. Acad. Sci. Paris Série II 316, 807–813.

Calvet, M., 1996. Morphogenèse d'une montagne méditerranéenne: Les Pyrénées Orientales. Thèse de Doctorat d'Etat, Document du BRGM, n° 255, 1177 pp.

Calvet, M., 2004. The Quaternary glaciation of the Pyrenees. In: Ehlers, J., Gibbard, P. (Eds.), Quaternary Glaciations—Extent and Chronology, Part I: Europe. Elsevier, Amsterdam, 119–128.

Calvet, M., Gunnell, Y., Delmas, M., 2008. Géomorphogenèse des Pyrénées. In: Canerot, J., Colin, J.P., Platel, J.P., Bilotte, M. (Eds.), Pyrénées d'hier et d'aujourd'hui. Actes du Colloque de Pau, International Year of Planet Earth, 20–21 September 2008. Atlantica, Biarritz, 129–143.

Copons, R., Bordonau, J., 1996. El registro sedimentario del cuaternario reciente en el lago Redó d'Aigües Tortes (Pirineos centrales). In: Grandal d'Anglade, A., Pagés Valcarlos, J. (Eds.), IV Reunión de Geomorfologia. Sociedad Española de Geomorfologia O Castro, La Coruña, Spain, 249–260.

Delmas, M., 2005. La déglaciation dans le massif du Carlit (Pyrénées orientales): approches géomorphologique et géochronologique nouvelles. Quaternaire 16, 45–55.

Delmas, M., 2009. Chronologie et impact géomorphologique des glaciations quaternaires dans l'est des Pyrénées. Thèse doctorat Université de Paris 1 (unpublished), 523 pp.

Delmas, M., Gunnell, Y., Braucher, R., Calvet, M., Bourlès, D., 2008. Exposure age chronology of the last glacial cycle in the eastern Pyrenees. Quatern. Res. 69, 231–241.

Delmas, M., Calvet, M., Gunnell, Y., 2009. Variability of erosion rates in the Eastern Pyrenees during the last glacial cycle—a global perspective on the impact of glacial erosion on mountain landscapes. Quatern. Sci. Rev. 28, 484–498.

Delmas, M., Calvet, M., Gunnell, M., Braucher, R., Bourlès, D., 2011. Palaeogeography and ^{10}Be exposure-age chronology of Middle and Late Pleistocene glacier systems in the northern Pyrenees: implications for reconstructing regional palaeoclimates. Palaeogeography, Palaeoclimatology, Palaeoecology 305, 109–122.

Faucher, D., 1937. Le glacier de l'Ariège dans la basse vallée montagnarde. Rev. Géogr. Pyr. S.O. 8, 335–349.

Florineth, D., Schlüchter, C., 2000. Alpine evidence for atmospheric circulation patterns in Europe during the last glacial maximum. Quatern. Res. 54, 295–308.

Gangloff, P., Courchesne, F., Hétu, B., Jalut, G., Richard, P., 1991. Découverte d'un paléolac sur le piémont des Pyrénées Atlantiques (France). Z. Geomorphol. 35, 463–478.

García-Ruiz, J.M., Gómez-Villar, A., Ortigosa, L., Martí-Bono, C., 2000. Morphometry of glacial cirques in the central Spanish pyrenees. Geografiska Ann. 82A, 433–442.

García-Ruiz, J.M., Valero-Garcés, B.L., Martí-Bono, C., González-Sampériz, P., 2003. Asynchroneity of maximum glacier advances in the central Spanish Pyrenees. J. Quatern. Sci. 18, 61–72.

Gellatly, A.F., Grove, J.M., Switsur, V.R., 1992. Mid-Holocene glacial activity in the Pyrenees. Holocene 2, 266–270.

Gillespie, A., Molnar, P., 1995. Asynchronous maximum advances of mountain and continental glaciers. Rev. Geophys. 33, 311–364.

González-Sampériz, P., Valero-Garcès, B.L., Moreno, A., Jalut, G., García-Ruiz, J.M., Martí-Bono, C., et al., 2006. Climate variability in the Spanish Pyrenees during the last 30,000 yr revealed by the El Portalet sequence. Quatern. Res. 66, 38–52.

Gunnell, Y., Calvet, M., Brichau, S., Carter, A., Aguilar, J.-P., Zeyen, H., 2009. Low long-term erosion rates in high-energy mountain belts: insights from thermo- and biochronology in the Eastern Pyrenees. Earth Planet. Sci. Lett. 276, 302–313.

Hétu, B., Gangloff, P., 1989. Dépôts glaciaires du Pléistocène inférieur sur le piémont des Pyrénées Atlantiques. Z. Geomorphol. 33, 384–403.

Hubschman, J., 1975. Morphogenèse et pédogenèse quaternaire dans le piémont des Pyrénées garonnaises et ariégeoises. Thèse de Doctorat d'Etat Lettres, Université de Toulouse le-Mirail (1974). Atelier de reproduction des Thèses de Lille III, 745 pp.

Hubschman, J., 1984. Glaciaire ancien et glaciaire récent: analyse comparée de l'altération de moraines terminales nord-pyrénéennes. In: Hommage à François Taillefer. Montagnes et Piémonts. Actes du Colloque de Géomorphologie sur les relations entre les montagnes récentes et leurs piémonts, Rev. Géogr. Pyr. S. O., Travaux I, Toulouse, 313–332.

Hughes, P.D., Woodward, J.C., 2008. Timing of glaciation in the Mediterranean mountains during the last cold stage. J. Quatern. Sci. 23, 575–588.

Jalut, G., Delibrias, G., Dagnac, J., Mardones, M., Bouhours, M., 1982. A palaeoecological approach to the last 21,000 years in the Pyrenees: the peat bog of Freychinède (alt. 1350 m, Ariège, South France). Palaeogeogr. Palaeoclimatol. Palaeoecol. 40, 321–359.

Jalut, G., Andrieu, V., Delibrias, G., Fontugne, M., Pagès, P., 1988. Palaeoenvironment of the valley of Ossau (Western French Pyrenees) during the last 27,000 years. Pollenet et Spores 30, 357–394.

Jalut, G., Montserrat, J., Fontugne, M., Delibrias, G., Vilaplana, J.M., Julia, R., 1992. Glacial to interglacial vegetation changes in the northern and southern Pyrenees: deglaciation, vegetation cover and chronology. Quatern. Sci. Rev. 11, 449–480.

Lewis, C.J., McDonald, E.V., Sancho, C., Peña, J.L., Rhodes, E.J., 2009. Climatic implications of correlated Upper Pleistocene and fluvial deposits on the Cinca and Gállego Rivers (NE Spain) based on OSL dating an soil stratigraphy. Glob. Planet. Change 67, 141–152.

Mardonnes, M., Jalut, G., 1983. La tourbière de Biscaye (alt. 409 m, hautes Pyrénées): approche paléoécologique des 45 000 dernières années. Pollenet Spores 25, 163–211.

Martí-Bono, C., González-Sampériz, P., Valero-Garcés, B., García-Ruiz, J.M., 2002. El depósito glaciolacustre de Linás de Broto (Pirineo aragonés) y su implicación paleoambiental. In: Pérez-González, A., Vegas, J., Machado, M.J. (Eds.), Aportaciones a la Geomorfología de España en el Inicio del Tercer Milenio. Actas de la VI Reunión Nacional de Geomorfologia: Madrid, 17–20, September 2000. Publicaciones del Instituto geológico y minero de España, serie Geología, Madrid, 77–83.

Mix, A.C., Bard, E., Schneider, R., 2001. Environmental processes of the ice age: land, oceans, glaciers (EPILOG). Quatern. Sci. Rev. 20, 627–657.

Pallàs, R., Rodés, A., Braucher, R., Carcaillet, J., Ortuno, M., Bordonau, J., et al., 2006. Late Pleistocene and Holocene glaciation in the Pyrenees: a critical review and new evidence from ^{10}Be exposure ages, south-central Pyrenees. Quatern. Sci. Rev. 25, 2937–2963.

Pallàs, R., Rodes, A., Braucher, R., Bourles, D., Delmas, M., Calvet, M., Gunnell, Y., 2010. Small, isolated glacial catchments as priority target for cosmogenic surface dating of Pleistocene climate fluctuations, southeastern Pyrenees. Geology. 38, 891–894.

Peña, J.L., Sancho, C., Lewis, C., McDonald, E., Rhodes, E., 2004. Datos cronológicos de las morrenas terminals del glaciar del Gállego y su relación con las terrazas fluvioglaciares (Pirineo de Huesca). In: Peña, J.L., Longares, L.A., Sánchez, M. (Eds.), Geografía Física de Aragón, Aspectos generales y temáticos. Universidad de Zaragoza e Institución Fernando el Católico, Zaragoza, pp. 71–84.

Putkonen, J., Swanson, T., 2003. Accuracy of cosmogenic ages for moraines. Quatern. Res. 59, 255–261.

Reille, M., Andrieu, V., 1993. Variations de la limite supérieure des forêts dans les Pyrénées (France) pendant le Tardiglaciaire. C. R. Acad. Sci. Paris Série II 316, 547–551.

Reille, M., Andrieu, V., 1995. The late Pleistocene and Holocene in the Lourdes Basin, Western Pyrenees, France: new pollen analytical and chronology data. Veg. Hist. Archaeobot. 4, 1–21.

Reille, M., Lowe, J.J., 1993. A re-evalutation of the vegetation history of the eastern pyrenees (France) from the end of the last glacial to the present. Quatern. Sci. Rev. 12, 47–77.

Rodés, A., 2008. La Última deglaciación en los Pirineos: datación de superficies de exposición mediante ^{10}Be, y modelado numérico de paleoglaciares. Tesis Doctoral, Universitat de Barcelona (unpublished), 238 pp.

Serrat, D., Bordonau, J., Bru, J., Furdada, G., Gomez, A., Marti, J., et al., 1994. Síntesis cartográfica del glaciarismo surpirenaico oriental. In: Marti Bono, C., Garcia Ruiz, J. (Eds.), El glaciarismo surpirenaico: nuevas aportaciones. Geoforma édit. Logroño, 9–15, with 5 maps.

Sorriaux, P., 1981. Etude et datation de remplissages karstiques: nouvelles données sur la paléogéographie quaternaire de la région de Tarascon (Pyrénées ariégeoises). C. R. Acad. Sci. Paris Série II 293, 703–706.

Sorriaux, P., 1982. Contribution à l'étude de la sédimentation en milieu karstique. Le système de Niaux-Lombrives-Sabart (Pyrénées Ariégeoises). Thèse de doctorat de 3e cycle, Université Paul Sabatier de Toulouse (unpublished), 255 pp.

Taillefer, F., 1973. Le glaciaire de Foix. Ann. Féd. Pyrénéenne 'Economie Montagnarde 29, 13–23.

Taillefer, F., 1985. Idées actuelles sur les glaciations dans les Pyrénées de l'Ariège. Rev. Géogr. Pyr. S.O. 56, 323–338.

Thackray, G.D., Owen, L.A., Yi, C., 2008. Timing and nature of late Quaternary mountain glaciation. J. Quatern. Sci. 23, 503–508.

Turner, C., Hannon, G.E., 1988. Vegetational evidence for the late Quaternary climate changes in southwest Europe in relation to the influence of the North Atlantic Ocean. Philos. Trans. R. Soc. Lond. B318, 451–485.

Turu i Michels, V., 2002. Análisis secuencial del delta de Erts. estratigrafía de un valle glaciar obturado intermitentemente. Relación con en último ciclo glaciar. Valle de Arinsal, Pirineos Orientales. In: Estudios recientes (2000–2002) en geomorfología, patrimonio, montaña y dinámica territorial. SEG–Departamento de geografía UVA edit, Valladolid, 555–574.

Turu i Michels, V., 2009. Relief, déglaciation würmienne et effets du paraglaciaire. In: Delmas, M., Turu, B. (Eds.), Le glaciaire des Pyrénées orientales: Ariège et Andorre, Livret-guide de l'excursion annuelle de l'Association Française pour l'Étude du Quaternaire, 5–7 June 2009, 33–85.

Turu i Michels, V., Peña Monné, J.L., 2006. Las terrazas fluviales del sistema Segre-Valira (Andorra-La Seu d'Urgell-Organyà, Pirineos Orientales): relación con el glaciarismo y la tectónica activa. In: Pérez-Alberti, A., López-Bedoya, J. (Eds.), Geomorfología y territorio, IX Reunión Nacional de Geomorfología. Universidad de Santiago de Compostela, Spain, 113–128.

Turu i Michels, V., Vidal-Romani, J.R., Fernandez-Mosquera, D., 2004. Dataciones efectuadas en superficies de erosión y bloques morrénicos mediante Néon cosmogénico, valles del Valira del Nord y Gran Valira, Pirineos orientales. In: Inform Intern de la Fundació Marcel Chevalier Andorra, 64 pp. (unpublished).

Turu i Michels, V., Boulton, G.S., Ros i Visus, X., Peña Monné, J.L., Martí i Bono, C., Bordonau i Ibern, J., et al., 2007. Structure des grands bassins glaciaires dans le nord de la péninsule ibérique: comparaison entre les vallées d'Andorre (Pyrénées orientales), du Gallego (Pyrénées centrales) et du Trueba (Chaîne cantabrique). Quaternaire 18, 309–325.

Vilaplana, J.M., 1983. Quaternary glacial geology of Alta Ribagorça Basin (central southern Pyrenees). Acta Geol. Hisp. 18, 217–233.

Chapter 12

Late Pleistocene (Würmian) Glaciations of the Caucasus

Ramin Gobejishvili*, Nino Lomidze and Levan Tielidze

Ivane Javkhishvili Tbilisi State University, Vakhushti Bagrationi Institute of Geography, department geomorphology and geoecology, Merab Aleksidze str. Block 1.0193, Tbilisi, Georgia

*Correspondence and requests for materials should be addressed to Ramin Gobejishvili. E-mail: goberamin@hotmail.com

12.1. INTRODUCTION

The shaping of the high mountain relief of the Caucasus is closely connected to the geological and geomorphological activity of ancient glaciations. As on other high mountains, the glaciers left deep traces on the relief of the Caucasus in the form of diverse and variously sized landforms. The traces of past glaciations are found in various states of preservation in the river basins. They allow detailed reconstruction of the process of Upper Pleistocene and Holocene glaciations. The investigations of the older glaciations in the Caucasus began in the nineteenth century (Dinnik, 1890). After publication of the works of Penck and Brückner (1901/1909), many scientists attempted to correlate the glaciations of the Caucasus with those in the Alps (e.g. Reingard, 1937; Vardanyants, 1937; Tsereteli, 1966; 'Papers of Caucasus expedition' Kovalev, 1961, 1960–1962; Shcherbakova, 1973). According to these authors, the entire Caucasus was either covered or partially covered by ice, with glaciers descending from the mountains and moving down to very low altitudes, implying a depression of the firn line by some 1100–1300 m. Maruashvili (1956) opposed a mechanical application of the alpine scheme to the Caucasus. On the contrary, he considered that the Late Pleistocene glaciation was of a much more limited character, the depression of the firn line being only about 600–800 m. Since then, a number of new investigations, using different methods for the reconstruction of the old glaciations, have been published (e.g. Kovalev, 1961; Khazaradze, 1968; Gobejishvili, 1995; Bondarev et al., 1997; Serebriani, Orlov, 1989).

12.2. METHODS OF INVESTIGATION

The Caucasus is of interest from a theoretical point of view because it allows the comparison of the great difference in regime and dynamics of glaciers formed under both a maritime and a continental climate. Because the Caucasus occurs midway between the Alps and the mountains of Central Asia, they can provide indispensable information for the modelling of natural processes across vast regions. However, the reconstruction of the extent and morphology of the ancient glaciations in the Caucasus is difficult because the glaciations have often left only weak traces and there is a widespread occurrence of landforms and sediments resembling glacial deposits. In many regions, the original trough forms of formerly glaciated valleys have been greatly changed by weathering, erosion and slope denudation. In these situations, for example, lateral moraines are usually either mostly or completely destroyed. End moraines on the valley floors, which were greatly subjected to glaciofluvial redeposition and mud flows, may be completely removed by streams that redeposited glacial outwash from their original positions to lower altitudinal levels.

In addition, very active exogenic processes may lead to the formation of assemblages of moraine-like landforms, debris cones and mud-flow deposits. Traditional glacial geomorphological methods do not always allow these features to be differentiated and to assign them to a particular age. Consequently, significant discrepancies exist in the reconstructions of the extent of the various glaciation events in practically all the formerly glaciated Eurasian mountain regions. Until recently, the evidence of modern glaciations has been practically ignored in all investigations into these palaeoglaciological problems. However, in more recent studies, the relationship between the feeding area and the length of modern glaciers has been compared to similar parameters reconstructed for the older glaciations. For example, the length of the Late Pleistocene glaciers and their cirque areas have been evaluated by comparison with the data from well-investigated modern glaciated regions. This relationship can be illustrated by the formula:

$$\frac{L_c}{S_c} = \frac{L_d}{S_d}$$

where L_c is the length of modern glaciers, S_c is the snowline altitude of modern glaciers, L_d is the length of Pleistocene glaciers and S_d is the snowline altitude of Pleistocene glaciers.

Using this formula, the definitions of S_c and L_c have been determined by evaluation of topographic maps for all the valley glaciers of Georgia and for some of the larger glaciers in the TienShan, Spitsbergen, Tibet and the Himalayas. According to the relationship

$$\frac{L_c}{S_c} = K$$

four groups of the glaciers could be identified:

- Hanging valley glaciers and simple valley glaciers $K = 0.81$ ($s = 0.96$ correlation coefficient),
- Valley glaciers with multiple source areas $K = 0.50$ ($s = 0.95$),
- Complex valley glaciers $K = 0.33$ ($s = 0.93$),
- Dendritic glaciers $K = 0.13$ ($s = 0.86$).

If the coefficient (K) is known, the following formula can be applied:

$$\frac{L_d}{S_d} = K$$

$$L_d = S_d \times K$$

S_d is the feeding area of the ancient glaciers, which is limited by the lowest boundary of nival zone (Fig. 12.1).

12.3. RESULTS

Based on these criteria, morphological and morphometric analyses, as well as remote sensing techniques, have revealed that the Late Pleistocene glacial cirques formed in crystalline rocks are quite well preserved in the Caucasus. The authors have mapped all glacial cirques in this region and calculated their area. At the same time, the altitude of lower thresholds of the cirques, which is considered to indicate the height of the former firn-line position, was determined. Using this material, it has been possible to calculate the length of the Late Pleistocene glaciers and to determine the altitudes to which the glacier tongues extended (Table 12.1).

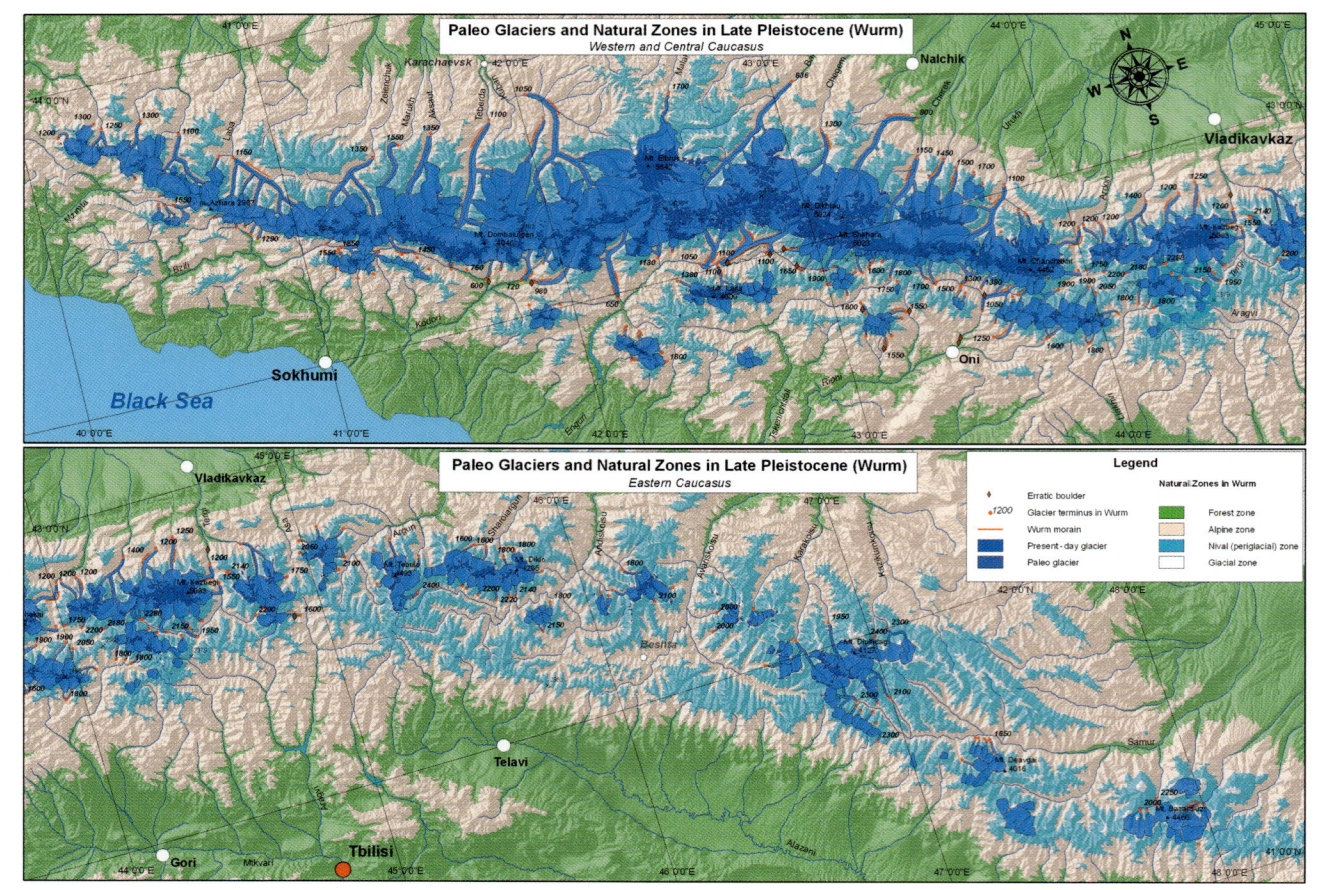

FIGURE 12.1 Extent of the Würmian glaciation and periglacial zone of the Caucasus.

TABLE 12.1 Morphometric Indices of the Late Pleistocene Glaciers of the Central Caucasus

Glaciers	Area of Old Cirque (km^2)	Length (km)	Elevation of Glacier Terminus (m a.s.l.)
Enguri river basin			
Nenskra	275	63.0	680
Dolra	105	34.8	1050
Mulkhura (Lekhziri)	270	35.0	1000
Adishi	24	18.0	1700
Khalde	31	16.0	1650
Shkhara	35	17.0	1900
Lailchala	23	12.0	1100
Laila	17	13.0	1300
Nakra	63	20.0	1180
Tkheishi	7.0	5.6	1450
Kveishkhi	4.5	3.6	1600
Didgali	6.0	4.8	1350
Magana	7.5	6.0	1450
Khobistskali	8.0	6.4	1800
Rioni river basin			
Koruldashi	15.0	11.7	1600
Zeskho	9.0	7.5	1600
Shari	11.5	9.0	1800
Edena	16.5	14.1	1650
Zopkhito	21.5	17.4	1500
Kirtisho	42.0	20.5	1300
Notsara	14.8	12.6	1280
Boko	20.9	18.0	1100
Buba	40.5	23.0	1050
Chanchakhi	13.1	11.0	1750
Garula	17.5	14.0	1250
Jejora	19.8	17.5	1600
Latashuri	11.0	8.5	1500
Sokhurtuli	7.5	6.0	1600
Ghobushuri	8.5	7.0	1800
Shodura	5.4	4.0	1600
Liakhvi river basin			
Zekara	8.5	7.0	1900
Kveshelta	10.0	8.0	1800
Jomagi	11.2	9.1	1550

Continued

TABLE 12.1 Morphometric Indices of the Late Pleistocene Glaciers of the Central Caucasus—cont'd

Glaciers	Area of Old Cirque (km²)	Length (km)	Elevation of Glacier Terminus (m a.s.l.)
Sba	9.2	7.5	1800
Cheliata	8.0	6.5	1850
Kalasani	11.0	9.0	1800
Tergi river basin			
Devdoraki	38.5	14.2	1220
Gergeti	21.0	17.0	1550
Mna	23.0	18.0	1950
Suatisi	32.0	15.0	2150
Tergi	20.0	10.0	2270

12.4. THE WESTERN CAUCASUS

In spite of their much lower elevations in comparison to those in the Central Caucasus, extensive mountain valley glaciation developed here in the Late Pleistocene as a result of the high humidity. The principal glaciation centre was the Main Range. The southern slope of the western Caucasus was a significant glaciation centre during the Würmian (Weichselian, Valdaian, Wisconsinan) Stage. Here, the lower thresholds of former glaciers are found at 1900–2000 m, defining the lower boundary of the nival zone at this time. The largest glaciers occurred in the Kodori river basin, where the glacier tongues from the Chkhalta, Klich and Sakeni valleys converged. The river Chkhalta is the largest right-bank tributary of the Kodori river. Here, field investigations have revealed that the former glaciers, descending through this valley, terminated at different altitudes. Two independent glaciers existed in the upper part of the Chkhalta river valley, the Adange and Marukhi glaciers, which descended from a common source area. Valley-type glaciers formed in the Chkhalta river valley on the Caucasian range, only three of which descended to the valley floor. Two tongues of the Sofruju glacier terminated at 1050 m a.s.l., whereas the tongues of the Aciashi and Ptishi glaciers reached down to 760–600 m.

By contrast, large complex valley-type glaciers in the Kodori river basin descended from the Klich valley. Here, there are well-exposed ancient cirques. The lower thresholds of these cirques occur at a height of 2000 m. Glaciers descending from cirques converged into one tongue and advanced down to the mouth of the river Sakeni (700 m a.s.l.). Remnants of moraines from this advance are found in the villages of Cencvishi and Gvandra (Tsereteli, 1966; Khazaradze, 1968; Gobejishvili, 2000). The Klichi glacier reached a length of 19.5 km during the Würmian.

Another thick glacier formed in the upper reaches of the Sakeni river. Both in the Würmian and at present, the principal glaciation centre was found on the north-west slope of the Kodori range. The glaciers that formed here merged with others and advanced as a single ice stream into the Sakeni River valley, terminating near the village of Sakeni at an elevation of 1000 m a.s.l. This glacier reached a length of 25 km. The fact that, in the Late Pleistocene, glaciers descended down to Sakeni has been confirmed by glaciofluvial terraces, lateral moraines and erratic boulders found in the village (Tsereteli, 1966). In the Late Pleistocene, glaciers not only developed on the southern slope of the Caucasus but also on the Gagra, Bzipi, Chkhalta and Kodori ridges. Here, their crests reached up into the nival zone, and some massifs reached into the glacial zone. During the Late Pleistocene, numerous small corrie-type glaciers developed here, as well as corrie-valley and valley glaciers. Glaciers of 1–3 km length occurred on the northern slope, and their tongues reached down as far as 1600–1700 m a.s.l.

The major ice streams on the north slope of the western Caucasus originated from the Main Range. Small glaciers joined them and descended as a single tongue within the Kubani and Tviberi valleys down to 1100–1200 m a.s.l. These glaciers reached a length of 50–55 km. These valleys are of the trough type, and they contain glaciofluvial terraces that are developed from the glacier-tongue terminus.

In the basins of the rivers Marukhi, Aksaut and Zelenchuk, the glacier tongues reached a maximum length of 30–35 km and terminated at 1500–1700 m a.s.l. The lower threshold of ancient cirques was at 1800–1900 m a.s.l.

12.5. THE CENTRAL CAUCASUS

In the high-altitude longitudinal and traverse valleys and depressions of the Central Cucasasi, firn ice was able to accumulate. This, together with the sufficient supply of precipitation, favoured the development of strong glaciation in this region both during the Late Pleistocene and Holocene. Glacial landforms here are well preserved in the landscape shaped by the Late Pleistocene glaciers. Studies have revealed that the lower thresholds of old cirques occur at different altitudes and rise significantly from west to east from 2100 to 2500 m a.s.l. The largest Late Pleistocene glaciers in the Central Caucasus descended into the Enguri, Rioni, Terek, Baksan, Cherek, Urukh and river valleys, among others.

Older glacial landforms (troughs, moraines, cirques) have been mapped in the Enguri river basin at the scales of 1:50,000 and 1:200,000 on the basis of satellite images of the same scale. This mapping revealed that ancient cirques are fully developed not only on the Main Range but also on the minor ridges. The evaluation of the topographic maps showed that the height of the old cirque thresholds in the river Nenskra basin and on the north slope of the Svaneti range lie at 2100 m a.s.l., while they occur at 2200 m a.s.l. in the basins of the rivers Nakra, Dolra and Mulkhura, as well as in the upper course of the Enguri river. The ancient cirque thresholds in the Samegrelo range are ca. 200 lower. This is thought to reflect the height of the Late Pleistocene firn line. In addition, comparatively well-formed lateral moraines are found in some valleys of the Enguri, Nenskra, Dolra, Mulkhura, Adishura, Khaldechala, Enguri and Lailchala river basins. Where the lateral moraines are absent, they indicate the sites of former end moraines, which were almost completely subsequently removed in some valleys. The altitude of the end moraines has been determined and the length of glaciers measured using published sources and new observations (Table 12.1). Large and complex valley-type glaciers descended from the Main Range to the Rioni river basin. The southern slope of the Main Range from Namkvan to Pasismta borders on the Tskhenistskali river basin (the largest right-bank tributary of the Rioni river). During the largest glaciation, three glaciation centres formed in this area: Koruldashi, Zeskho and the upper reaches of the Tskhenistskali river. All the glaciers were of the simple valley type.

Investigations have shown that the Koruldashi glacier was the largest in the Tskheniskali valley. The length of this glacier was 12 km, and its tongue terminated at 1600 m a.s.l. This is confirmed by reliable valley morphological evidence such as the trough form of the valley itself, as well as fragments of lateral and end moraines. Other glaciers were shorter than the Koruldashi, although their tongues also ended at 1600–1800 m a.s.l.

The morphological valley form of the southern slope, as well as those of the northern slope of the Caucasus, indicates that the glaciers expanded to that height in the Würmian Stage. Here, the alleys have been changed and sometimes preserve well-expressed trough forms. Erosion or erosion-accumulation valleys with well-developed river terraces continue to retain their trough shape.

The local geological and morphological conditions determine the quantity of river terraces. The Würmian-age terraces are mainly well expressed, their relative height varying between 80 and 100 m (Rioni river near Oni and Gebi; Enguri river near to Khaishi, Ipari; Kodori river near to Ajara-Gentsvishi; Bzipi river near Pskhu). In the vicinity of changing valley, morphology of the main rivers of northern slope of Central and Western Caucasus lateral moraine hills are located in the middle or terminal part of the modified trough valley. River ravines are often separated from the main trough valleys. The length of some lateral moraines is 2–8 km, for example, in the valleys of the Enguri river tributaries, Nenskra, Dolra and Mulkhura; upper course of the Enguri river, Lailchala; in the Rioni, Chveshura, Chanchakhi, Gharula, Jejora, Lukhuni and Ritseula river; in the Tskhenistskali, Ghobishura, Akhalchala and Dabieri river; in the rivers Tergi, Juta, Roshka, Arghuni, Bezing, Cherek, Psigansu, Khaznidoni, Laba, Samuri, Akhti and Amsari (Reingard, 1937; Vardanyants, 1937; Maruashvili, 1956; Kovalev, 1961; Tsereteli, 1966; Khazaradze, 1968; Gobejishvili, 1995).

The upper course of the Rioni river drains the part of the Main Range between the town of Pasismta and the Mamisoni Pass. Here, the Würmian-age glaciers descended from the Main Range and terminated at different heights. This suggests that they did not form a united glacier lobe. The Biba-Boko and Kirtisho supported complex valley glaciers. The Kirtisho glacier reached a length of 21 km in the Würmian and terminated at 1300 m a.s.l. at Gebi village. Four cirques and lateral moraines are rather well developed in the river Chveshura basin at Jojokheta. Numerous erratic blocks have been found around this village. Glacial material in the Rioni valley cannot be traced up to Saglolo. Instead, the morainic material in Saglolo derives from glaciers from the Chanchakhi river basin. The length of the Buba-Boko glacier was 23 km; it terminated downvalley of Saglolo at 1100 m a.s.l. In the tributary valleys of the Rioni-Edenura, Zopkhitura and Nitsarula rivers, glaciers of 14–17 km length occurred during the Würmian Stage. The lower threshold of cirques in the Rioni river basin was at 2200 m a.s.l.

In some major river valleys, erratic boulders are preserved at the Würmian glacier termini (Enguri, Rioni, Tergi, Kodori, Cherek, Khaznidoni, Psigansu, Amsari, Bazardiuzu, Roshka and the Juta river valleys). In the Rioni river valley, erratic boulders are encountered in the town of Oni (800 m a.s.l.). The largest of these boulders (14×12 m), with a volume of ca. 670 m^3, is monoclinic granite, derived from the axis zone of the Central Caucasus. The interpretations of these boulders differ greatly. Some consider them to be of glacial origin, while others interpret them as the result of a mud flows. In the authors' opinion, these boulders were brought by an oscillation of the Buba-Boko glacier during the Würmian, which was connected to an eruption of the Tsitela volcano.

The Rioni river basin, on the south slope of the Central Caucasus, was intensely glaciated. Glaciers up to 14–16 km long formed in the Garula and Jejora valleys (both of which are tributaries of the Rioni river) during the Würmian, and their tongues terminated at 1400–1600 m a.s.l. Lateral moraines are well expressed on both slopes of the Garula valley. Some 8–10 km long valley glaciers also formed in the Liakhvi river basin. Their tongues descended down to 1700–1900 m a.s.l., and the lower thresholds of these cirques occurred at 2300–2400 m a.s.l.

The upper valley of the Terek river basin occurs on the north slope of the Main Range of the Caucasus. The lateral Khorkhi ridge was the principal glaciation centre in the Late Pleistocene, as it is today. Here, an ice cap was developed in the eastern part of this range on the Kazbegi massif in the Late Pleistocene (as well as today). From there, hanging-valley glaciers and valley glaciers moved in all directions. The largest of these outlet glaciers, the Devdoraki, Gergeti, Mna and Suatisi glaciers, were up to 14–17 km long (Table 12.1) and descended into the Terek Valley. The Devdoraki glacier tongue terminated at 1200 m a.s.l., while the Gergeti glacier terminated at 1550 m, the Mna glacier at 1950 m and the Suatisi glacier at 2150 m a.s.l. There is no doubt that the erratic block near the village of Larsi, the 'Ermolov Stone', was transported by the Devdoraki glacier during the Late Pleistocene, perhaps during a major oscillation of the ice margin. These ice tongues moved down to the valley floor and crossed the valley, creating favourable conditions for glacial mud flows, for which the headwater region of the river Terek area is famous.

The termini of Late Pleistocene glaciers reached to 1000–1200 m a.s.l. on the northern slope of the Central Caucasus and, as a rule, ended on the south slopes of Skalisti (Rocky) Range, the length of these glaciers being some 34–35 km. Only the Bezing glacier crossed the Skalisti Range and terminated at 836 m a.s.l. This comparatively large ice tongue moved along the Baksan valley for some 70 km. It is obvious that the glaciers on the northern slope were almost twice the size of those on the southern slope. This arose from the trend direction and from the favourable morphological and morphometric conditions found on the north slope.

12.6. THE EASTERN CAUCASUS

The relief topography of the Eastern Caucasus differs greatly from that of the Central Caucasus, both in morphology and altitude. The low altitudinal position and significant aridity restricted the development of glaciers in this region. During the Würmian, like today, the glaciers were largely limited to the highest massifs that rise above 3500 m a.s.l. The authors' calculations suggest that the lower limit of the glacial zone lay at this altitude. The nival zone during the Würmian was well represented in the Eastern Caucasus, especially on the Main Range. Here, numerous corrie glaciers and several valley glaciers formed in the Eastern Caucasus in the Late Pleistocene. The maximum glacier length was ca. 6–8 km, with glacier tongues reaching down to 1700–1800 m a.s.l., and the lower thresholds of the cirques occur at 2400–2500 m a.s.l. On the whole, the extent of glaciation on the Eastern Caucasus during the Late Pleistocene is very similar in comparison to that of the present-day glaciation on the southern slope of the Central Caucasus.

12.7. THE MINOR CAUCASUS

Investigations in the Minor Caucasus have demonstrated that the character of its glaciation was similar to that of the Eastern Caucasus. The Würmian-age firn line was lowest on the ranges located closest to the Black Sea (2200–2300 m) and highest on the ranges situated to the east (2500–2600 m a.s.l.) (Akhalkatsishvili et al., 2003). The nival zone encompassed the crests exceeding 2200–2400 m a.s.l. The strongly degraded cirques in this area are distinguished by their small size. The largest valley glaciers here formed on the Samsari Range, where they were 4–6 km long (Maruashvili, 1956; Akhalkatsishvili et al., 2003). The cirque lower threshold was located at 2500–2700 m a.s.l.

12.8. CONCLUSIONS

The Late Pleistocene (Würmian Stage) glaciation of the Caucasus consisted of valley glaciers, with ice caps only on some peaks. The largest glaciers on the northern slope had a maximum length of 50–70 km, but outlet glaciers of a considerable size (17–35 km length) occurred on the both slopes of the Central Caucasus. The glacier tongues which terminated at the lowest altitudes were those of the Nenskra Glacier on the south slope at 600–680 m a.s.l. and of the Bezingi Glacier on the north slope at 700–750 m a.s.l. Tongues of other large glaciers descended down to 1600–1200 m a.s.l. The firn line in the Central Caucasus was found at 2000–2500 m in the Late Pleistocene, increasing in altitude from west to east. This implies that the firn line in the Caucasus was depressed by some 1200–1300 m during the Late Pleistocene (allowing for neotectonic activity), and the depression increased from west to east. This is equivalent to a decline in mean annual temperature of ca. 7–8 °C (with a gradient of 0.6°/100 m altitude).

REFERENCES

Akhalkatsishvili, M.D., Gobejishvili, R.G., Tutberidze, B.G., 2003. Role of glaciation in transformation of the volcanic relief of Samsari Range, Geography-Geology. In: Proceedings of I. Javakhishvili Tbilisi State University, Tbilisi, pp. 118–125.

Bondarev, L.G., Gobejishvili, R.G., Solomina, O.N., 1997. Fluctuations of local glaciers in the former USSR. Quatern. Int. 38/39, 103–108.

Dinnik, N.Y., 1890. Present and old glaciers of Georgia. KOIRGO 14, 88–102, Tiflis (in Russian).

Gobejishvili, R.G., 1995. Present day glaciers of Georgia and evolution of glaciation in the mountains of Eurasia in late Pleistocene and Holocene. Thesis for a Doctor's degree, Tbilisi, 66 pp. (in Russian).

Gobejishvili, R.G., 2000. Evolution of glaciation of Central Caucasus in the Late Pleistocene and Holocene. MGI, section 89, Moscow, pp. 36–42. (in Russian).

Khazaradze, R.D., 1968. Old Glaciation of the South Slope of the Great Caucasus. Metsniereba, Tbilisi, 135 pp. (in Russian).

Kovalev, P.V., 1961. Present and Old Glaciation of the River Enguri Basin. Materiali Kavkazkoi ekspeditsii (Papers of the Caucasian Expedition) vol. 2. Izv. Kharkoski Gosudarstveni Universitet, Kharkov, pp. 5–32. (in Russian).

Maruashvili, L.I., 1956. The Expediciency of the Revision of Actual Conceptions of the Paleogeographical Conditions of the Glacial Time of the Caucaus. Academy of Science Press, Tbilisi, 126 pp. (in Russian).

Penck, A., Brückner, E., 1901/1909. Die Alpen im Eiszeitalter, 3 vols. Tauchnitz, Leipzig, 1199 pp.

Reingard, A.L., 1937. About the problem of the stratigraphy of the Glacial Time of the Caucasus. In: Transactions of International INQUA Congress Transactions. section M 1, pp. 9–31. (in Russian).

Serebriani, L.P., Orlov, A.V., 1989. Moraines as the source of glacial information. Unpublished manuscript. 235 pp.

Shcherbakova, E.M., 1973. The Ancient Glaciation of the Great Caucasus. Moscow University Press, 272 pp. (in Russian).

Tsereteli, D.V., 1966. Pleistocene Deposits of Georgia. Metsniereba, Tbilisi, 582 pp. (in Russian).

Vardanyants, L.A., 1937. The ancient glaciation of the rivers Iraf (Urukh) and Tsey (Central Caucasus). Proc. State Geogr. Soc. 69 (4), 537–562. (in Russian).

Pleistocene Glaciations of North Germany—New Results

Jürgen Ehlers[1,*], Alf Grube[2], Hans-Jürgen Stephan[3] and Stefan Wansa[4]

[1]Geologisches Landesamt, Billstr. 84, D-20539 Hamburg, Germany
[2]Landesamt für Landwirtschaft, Umwelt und ländliche Räume, Hamburger Chaussee 25, D-24220 Flintbek, Germany
[3]Köhlstraße 3, D-24159 Kiel, Germany
[4]Landesamt für Geologie und Bergwesen Sachsen-Anhalt, Postfach 156, D-06035 Halle, Germany
*Correspondence and requests for materials should be addressed to Jürgen Ehlers. E-mail: juergen.ehlers@bsu.hamburg.de

13.1. INTRODUCTION

The present review is an update of Ehlers et al. (2004). In the past 6 years, the German Pleistocene stratigraphy has been thoroughly revised. Recent overviews had been given by Habbe (2007), Litt et al. (2007) and Ehlers (2011), and an online dictionary of lithostratigraphical terms (Litho-Lex) is currently under construction (http://www.stratigraphie.de/LithoLex/index.html). This chapter deals with the Fenno-Scandinavian ice sheets of north Germany.

In the Late Tertiary, the north-west European Basin was an area of fluvial deposition, dominated by a 'Baltic River System'. This system remained active throughout the Early Pleistocene. Its deposits are found, in particular, in the southern marginal areas of the North Sea Basin, where a large delta was deposited in the North Sea (Westerhoff, 2009). The sediments contain lydites from the southern Central German uplands, as well as large blocks, which must have been transported by drift ice from the eastern Baltic region (Gripp, 1964). The 'Loosen Gravel' of Mecklenburg is regarded as part of this river system (von Bülow, 2000). It remained active until the advance of the first Pleistocene ice sheets altered the drainage system both fundamentally and irreversibly.

During the Pleistocene, Germany was affected by at least three different types of Pleistocene glaciation: the Fenno-Scandinavian ice sheets in the north, the Alpine glaciation in the south and a number of local glaciations in the upland areas. The basic subdivision of the North German Quaternary stratigraphy was established by the end of the nineteenth century. It was originally assumed that north Germany, like the Alps, had been affected by three glaciations (Penck, 1879). The names Elsterian, Saalian and Weichselian first appeared in 1910 on the 1:25,000 map sheets of the 'Königlich Preußische Geologische Landesanstalt' reflecting this view. The glacial limits in north Germany have been largely defined in the course of large-scale geological mapping of the individual till sheets. Only in exceptional cases were maximum glacial positions found to be marked by significant end-moraine landforms.

Since it was established that the Alps had been glaciated (at least) four times, instead of three (Penck and Brückner, 1901/1909), there has been no lack of attempts to identify corresponding till units in north Germany. However, neither the proposed subdivision of the Saalian into two independent glaciations nor the assumed detection of an additional older (Elbe) glaciation could stand up to close scrutiny.

The principal formal lithostratigraphical unit is the formation. It can be divided into subunits (members), and individual strata (beds) can be distinguished. Grube (1981) was one of the first to apply the international principles to the German Quaternary stratigraphy. However, it took a long time before the new approach became fully accepted (Menning and Hendrich, 2005; Litt et al., 2007).

In the course of time, many stratigraphical terms had been coined, the precise meaning of which is only known to a few specialists. To clarify this situation, the Deutsche Stratigraphische Kommission has introduced a new online database 'LithoLex', in which all lithostratigraphical terms that are used in Germany will be registered and defined. The project began in 2006 with about 80 data sets; today the database contains some 400 stratigraphical units, ca. 40 of which are of Quaternary age. However, it will take many years before the collection will be complete.

Because of the limitation of absolute dating for the period beyond 100 ka, and because of the ongoing discussion concerning the validity of 'absolute' numerical ages, very few dates can as yet be given.

13.2. TRACES OF EARLY GLACIATION

The Early Pleistocene deposits of north Germany include at least seven cold and an equivalent number of warm-climatic phases (Menke, 1975). The cold phases show arctic conditions but none show signs of glaciation. The oldest strata that carry major quantities of Scandinavian erratics from East Fennoscandian and central Swedish (Dalarna) source areas (Zandstra, 1983, 1993) in the Netherlands and neighbouring north Germany are the Hattem Beds of the Menapian age. They are exposed in the Ytterbeck-Uelsen push moraine in Niedersachsen (Lower Saxony). They are interpreted as evidence of a first major ice advance beyond the limits of the present Baltic Sea (Ehlers, 1996). Vinx et al. (1997) claim to have found an early, possibly pre-Elsterian till at Lieth, Schleswig–Holstein, Fig. 13.1. However, as there is no possibility to date this till until now, it might as well be a different facies of the Elsterian I till.

13.3. THE ELSTERIAN GLACIATION

The Elsterian deposits in north Germany provide the first unequivocal evidence of a major glaciation during which the ice sheets advanced as far as the margins of the Central German Uplands (Fig. 13.2). There had been earlier glaciations, especially during the latest Early Pleistocene (marine isotope stage (MIS) 24–22) and the early Middle Pleistocene (MIS 16, the Don glaciation), but now the ice sheets extended so far that Scandinavian and British ice met in the North Sea, blocking all drainage to the North. A large ice-dammed lake formed, the water of which eventually crossed the Weald–Artois barrier and created a drainage line towards the south to join the river in the English Channel (Chapter 7). The Strait of Dover, linking the English Channel and the North Sea, was formed.

Recently, Toucanne et al. (2009a,b) were able to date this event by carrying out investigations of deep-sea core

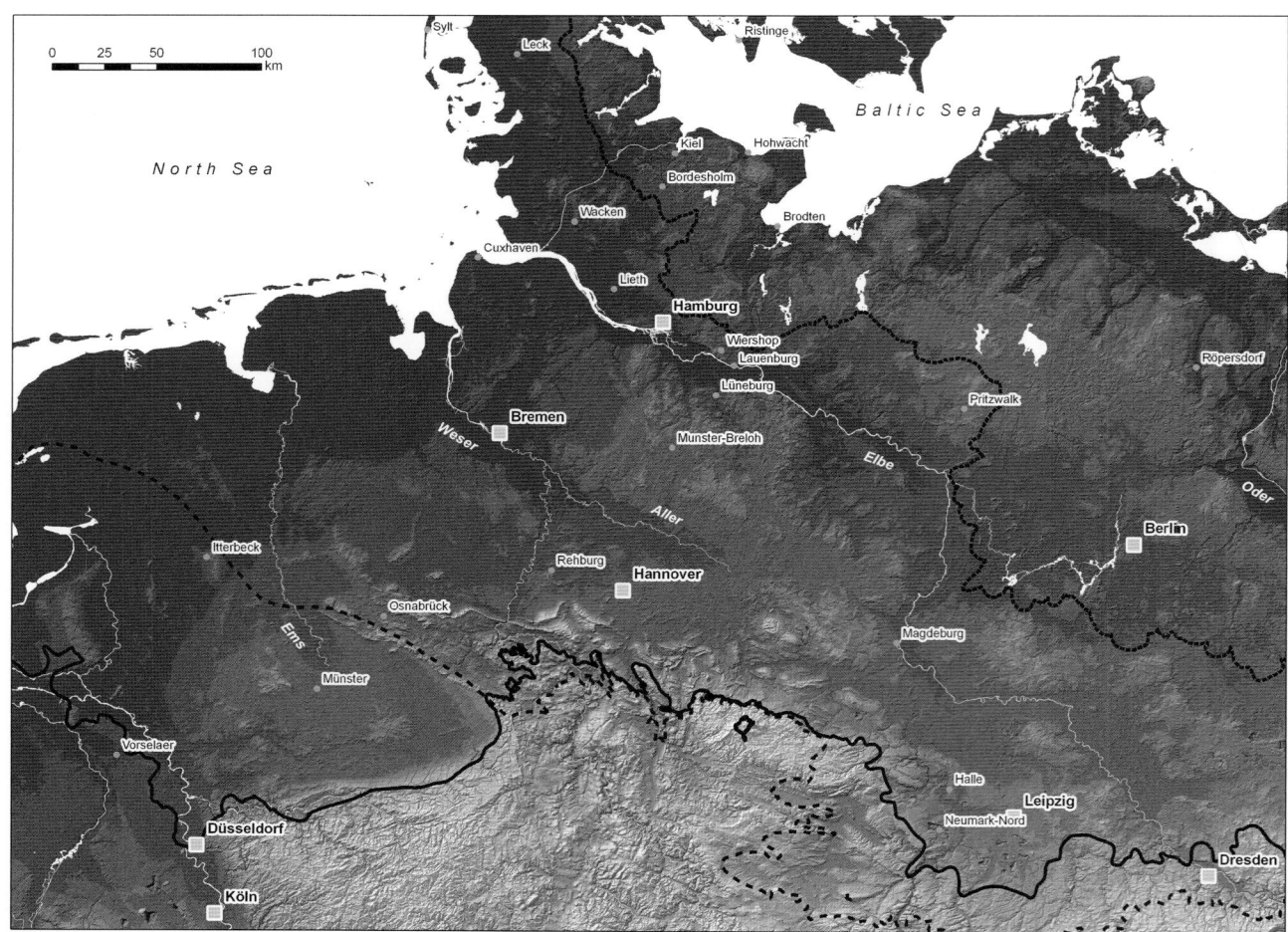

FIGURE 13.1 Location map.

Chapter | 13 Pleistocene Glaciations of North Germany—New Results

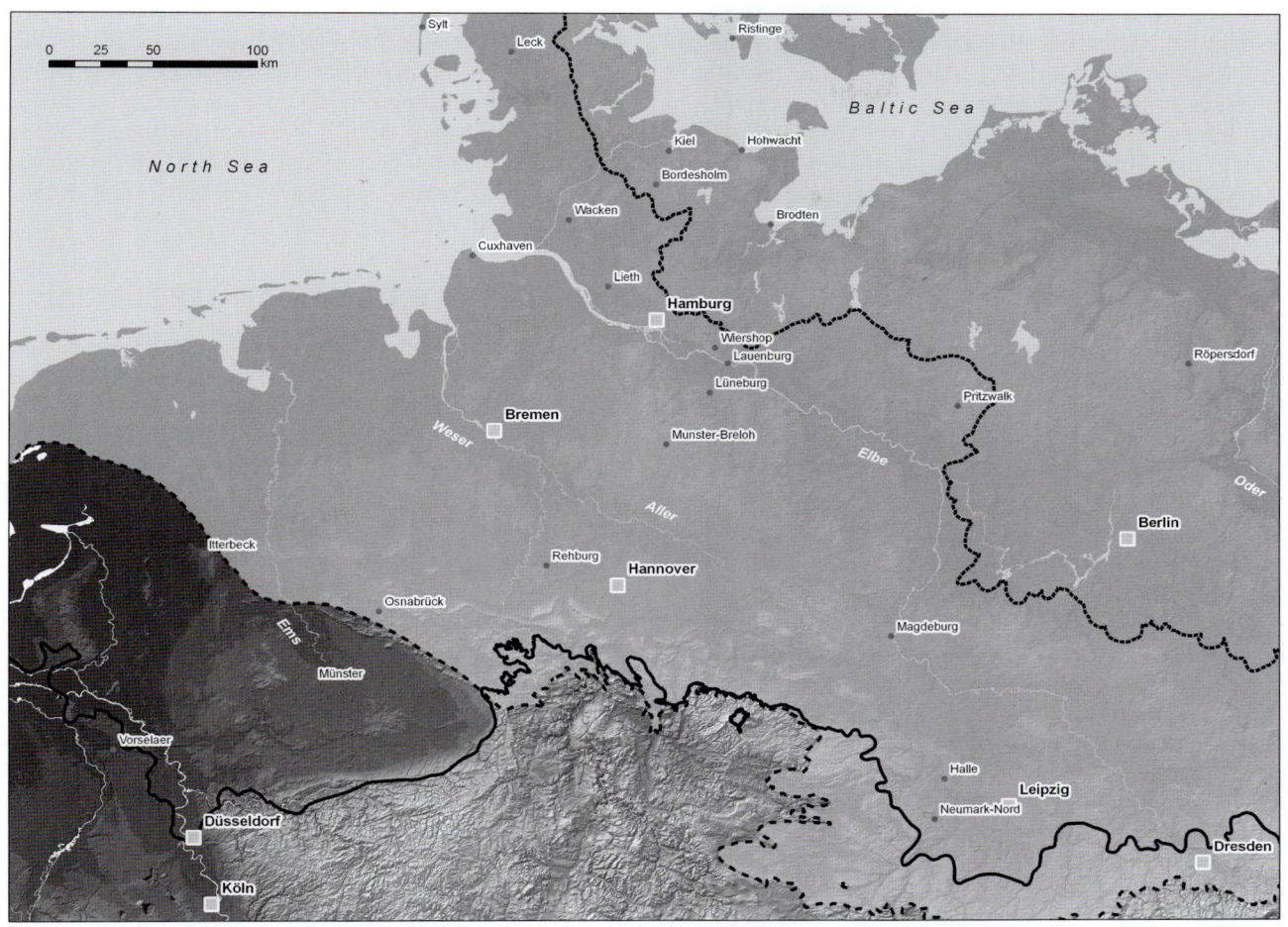

FIGURE 13.2 Maximum extent of the Elsterian glaciation in North Germany.

MD01-2448. This 28.8-m-long core, taken at 44°46,790 N, 11°16,470 S at a water depth of 3460 m on the Charcot Seamount, was ideally positioned to record the sedimentary influx from the Channel as well as the general climatic oscillations of the Ice Age. The sedimentary sequence extends back to MIS 35, that is, to about 1.2 Ma BP. Comparison with the oxygen isotope curve from the LR04-Stack (Lisiecki and Raymo, 2005) shows that the sedimentary record of the core is complete. It demonstrates that glacial drainage through the Channel began at about 455,000 years BP. The Elsterian Stage thus is MIS 12.

During the advance of the Elsterian ice, the drainage system was completely remodelled: rivers that hitherto mostly drained towards the Baltic Sea were partly dammed by the advancing ice sheet and forced to alter their courses. Consequently, the Elsterian tills in major parts of Thüringen, Sachsen and Sachsen-Anhalt are both under- and overlain by varved clays (Eissmann, 1997; Junge, 1998; Junge and Eissmann, 2001). The Elbe River was dammed and formed a large lake south of Dresden, but drainage was orientated westwards rather than to the south. Further to the east, rivers flowed via the 'Moravian Gate' towards the Danube and Black Sea.

In Saxony and Thuringia, where the Elsterian sediments were not overridden by later ice sheets, a fairly accurate reconstruction of the maximum extent of the Elsterian glaciation is possible. Here, the maximum distribution of Nordic erratics, the so-called flint line, indicates the maximum spread of the Elsterian ice sheet. Here, the ice reached up to 450–500 m a.s.l. in the Oberlausitz area (Präger, 1976; Wagenbreth, 1978), 400 m a.s.l. at the foot of the Erzgebirge (Ore Mountains; Eissmann, 1975) and about 300 m a.s.l. in the Thuringian Basin (Unger, 1974). Towards the west, it dips markedly. For example, in the Wesergebirge mountains, the flint line is found at an altitude of 200 m a.s.l. (Kaltwang, 1992). There, however, it marks the upper limit of the Saalian ice sheet; the Elsterian limit was even lower.

In the Saale-Elbe area, two glacial cycles (Zwickau and Markranstädt Phase) can be characteristically identified. The sedimentary sequence of each advance begins with varved clays, overlain by till and ending with various types of meltwater deposits (Eissmann et al., 1995). The advance velocity of the first Elsterian ice sheet ranged between 600 and 900 m/a (Junge, 1998). Many of the distinct thrust moraines in Sachsen and Sachsen-Anhalt (southern Dübener Heide, Dahlener Heide areas) were probably formed during the Elsterian. Even the *Muskauer Faltenbogen* (Muskau arch of folds) might be of Elsterian age (Eissmann, 1997; Kupetz, 1997).

During both Elsterian ice advances, major linear erosional zones formed subglacially, partly by meltwater erosion, partly by glacial erosion. The resulting channels connect with the deeply incised buried 'valleys' beneath the North German Lowland (Eissmann, 1967, 1987; Eissmann and Müller, 1979). In the west, they can be traced through the Netherlands, into the North Sea (Lutz et al., 2009) and into lowland eastern England. Whilst the shallower features that are filled with till (Grube, 1997) normally lack a cover of late Elsterian lacustrine sediments or marine Holsteinian deposits, both are normally present in the deeper channels. That means that the channels remained active until the end of the glaciation. An overview map of the Elsterian channels in north Germany has been given by Stackebrandt (2009). Seismic investigations in the North Sea have clearly demonstrated that the features are not of fluvial origin (Huuse and Lykke-Andersen, 2000; Praeg, 2003; Kristensen et al., 2007, 2008). They are often aligned along older structures so that an influence of neotectonics must be assumed (Schwab and Ludwig, 1996).

When the Elsterian ice melted, large ice-dammed lakes came into existence in which silt and clay were deposited. In north-west Germany, these deposits are referred to as 'Lauenburg Clay'. They are found from the Netherlands (*Potklei* of the Provinces of Friesland, Groningen and Drente) to well into Mecklenburg-Vorpommern. They can reach a thickness of over 150 m.

In the southern parts of Niedersachsen, as well as in Nordrhein-Westfalen, Elsterian till is rarely found so that the extent of the glaciation in most cases can only be assumed from re-deposited Scandinavian erratics in the Middle Terrace. The thrust sequences of the end moraines west of the Weser River contain no Elsterian till (Skupin et al., 1993). In contrast to earlier conclusions, today most workers maintain that the Elsterian maximum lies north of Osnabrück.

Most probably, more than one ice advance occurred during the Elsterian Stage. In Sachsen and Thüringen, where the Elsterian deposits were well exposed in the large opencast brown coal mines, two Elsterian-aged tills are found. They show differences in grain-size composition and quartz content, but over all their differences are rather small. Elsner (2003) has also found two Elsterian tills in the area north of the Harz mountains. The same is true of the Elbe-Weser triangle region (Höfle, 1983b; Wansa, 1994). A specialty there is an upper Elsterian till, which is extremely sandy (only ca. 6% clay and 8% silt). Only its texture reveals that this deposit is a till; in samples from a borehole, it would be impossible to distinguish it from meltwater sands. In Schleswig–Holstein, there is evidence for three different Elsterian glaciation phases, the latest of which brought more easterly (Baltic) ice to north Germany (Stephan, submitted for publication).

13.4. THE HOLSTEINIAN INTERGLACIAL

The Holsteinian Stage Interglacial began with the transition from subarctic to boreal climate at the end of the Elsterian and terminated with the deterioration of climate at the onset of the Saalian Complex Stage (Litt et al., 2007). During this episode, the sea had already invaded parts of Jutland and north Germany at the end of the preceding Elsterian cold stage (Hinsch, 1993; Knudsen, 1993). Marine Late Elsterian deposits can be traced as far inland as Mecklenburg-Vorpommern (Müller and Obst, 2008). This early transgression is interpreted as a result of isostatic depression of the land surface.

During the Holsteinian warm Stage, north-west Europe experienced a second, major transgression. For north Germany, it was the first marine transgression since the Miocene. The Holsteinian lake deposits show a complete interglacial vegetational succession from subarctic-boreal through temperate to boreal conditions. The interglacial climate was not interrupted by any major climatic fluctuations.

Geyh and Müller (2005) have used ^{230}Th/U dating to determine the age of the Holsteinian. They have found that the Holsteinian dates to 320–310 ka and thus should be correlated with MIS 9. This, however, is difficult to reconcile with the Elsterian being correlated with MIS 12 (see above). According to varves counted at the Munster-Breloh site, Müller (1974) determined that the Holsteinian lasted 15,000 years.

13.5. THE SAALIAN COMPLEX

The Saalian was characterised by a number of climatic oscillations. It includes at least one warm event of an interglacial character. Consequently, it should be referred to as the Saalian Complex Stage (Litt and Turner, 1993; Litt et al., 2007).

The early Saalian, that is, the period between the end of the Holsteinian Stage and the first Saalian ice advance, is characterised by extensive valley widening and intensive accumulation of fluvial gravels. These gravel complexes reflect the changing climatic conditions in the Lower Saalian.

According to the results of infrared-radiofluorescence (IR-RF) dates from central Germany, it must be assumed that this period covered about 160 ka and ended at ca. 150 ka (Krbetschek et al., 2008).

13.6. THE DÖMNITZ INTERGLACIAL

The term Dömnitz Interglacial (after a small river in the Prignitz region of Brandenburg) was introduced by Cepek (1965) and Erd (1965) based on the results of boreholes near Pritzwalk. This interglacial was separated from the underlying Holsteinian by sands of the Fuhne cold stage, and it was overlain by till of the first ice advance of the Saalian glacial sequence. The palynological investigations were published by Erd (1970, 1973).

As in Pritzwalk, interglacial deposits have also been found in a stratigraphical position between the Holsteinian and the Saalian s.s. in Wacken (Schleswig–Holstein) (Menke, 1968). Both sites are located in areas that are intensely disturbed by glaciotectonics. Besides the Wacken/Dömnitz Interglacial, two additional Early Saalian warm-climate phases have been claimed by Urban et al. (1991), Urban (1995, 2006) and Stephan et al. (in print), the Reinsdorf and the Leck Interglacials, which separate the Saalian s.s. in a probably non-glacial lower and a glacial upper part (Stephan et al., accepted for publication). According to Litt et al. (2007), however, the Reinsdorf might well be an equivalent of the Holsteinian.

13.7. THE SAALIAN GLACIATION

The Saalian glaciation in north Germany (Fig. 13.3) is traditionally subdivided into two major ice advances, the Drenthe and the Warthe advance. The term Warthe dates from early works by Woldstedt (1927a,b). The Drenthe was introduced much later by Woldstedt (1954), following a suggestion by Van der Vlerk and Florschütz (1950). It is today generally accepted that the upper, glacial part of the Saalian Complex lasted hardly more than 20,000 years and is equivalent to the upper part of MIS 6 (Busschers et al., 2008; Krbetschek et al., 2008).

13.7.1. The Older Saalian Glaciation

Within the Saalian Complex in north Germany, three till sheets separated by meltwater deposits can be distinguished. Because the local stratigraphies still do not match, they are referred to here as 'older', 'middle' and 'younger' Saalian till (Kabel, 1982; Stephan and Kabel, 1982). For local terminology, see discussion in Ehlers (1994).

The older Saalian ice advance is the equivalent of the Dutch Drenthe and the Polish Odra glaciation. Its ice sheet covered almost all of Schleswig–Holstein and Niedersachsen. It intruded into the Münsterland Bight and advanced up to its southern margin (Klostermann, 1992; Skupin et al., 1993). In the west, it reached the Lower Rhine and left behind enormous push moraines. The most prominent thrust moraine of the older Saalian glaciation, however, is located at a distance of some 100 km from the outermost ice margin. The hills of this 'Rehburg Phase' had been originally interpreted as recessional end moraines. Their formation during the advance phase and subsequent overriding by the ice is documented by the occurrence of the same type of sandy basal till in the foreland and on top of the thrust ridges (Meyer, 1987).

Patches of a red-brown till in north-west Germany and the Netherlands are found in the uppermost part of the older Saalian till unit. This till is characterised by East Baltic erratic indicators (much Palaeozoic limestone, little or no flint, usually some dolostones), suggesting a change in ice-movement direction from NNE–SSW to ENE–WSW towards the end of the glaciation.

It is thought that the oldest Saalian till of north-west Germany corresponds to the three tills of the Saale-Elbe area. During the first or Zeitz phase, the Saalian ice sheet reached its maximum extent. This ice advanced to a line from the margin of the Hartz mountains, Eisleben, Freyburg on the Unstrut, Zeitz, Altenburg, Grimma, Döbeln, Kamenz and Görlitz. During the maximum, valley sandurs over 30 m thick accumulated on top of 5–15-m-thick glaciolacustrine deposits (Zeuchfeld Sandur, Großbothen Sandur, Heller Terrace; Eissmann, 1975, 1997). The first halt in the meltdown phase, after the maximum extent, was between the Hornburger Sattel and the Geiseltal marked by the Langeneichstädt end moraine (Meng and Wansa, 2008), that probably continues further to the east in the Dehlitz-Rückmarsdorf end moraine (Eissmann, 1975). The subsequent downwasting of the ice only reached to around Bitterfeld. The ice then readvanced into the Halle-Leipzig area (Leipzig Phase) with two oscillations. The Leipzig Phase ice advances are correlated with the Petersberg end moraine and the end moraines at Breitenfeld and Taucha (Ruske, 1964; Eissmann, 1975, 1987). Also the Jahrstedt-Steimke thrust moraine in the Altmark region was formed during the older Saalian (Stottmeister, 2000). In contrast, the stratigraphical position of other thrust zones in the Altmark (Wiebke-Zichtau and Osterburg) and in the southern Fläming area is still uncertain (von Poblozki, 1995; Litt and Wansa, 2008).

In the forefield of the Petersberg end moraine, that can be traced from Haldensleben via Magdeburg–Bernburg–Halle to Schkeuditz, the oldest known ice-marginal drainage system formed. The meltwaters left the ice margin at Staßfurt and flowed through the present Bode valley and Großer Bruch area into Niedersachsen, where they drained via the Aller-Weser ice-marginal valley (Urstromtal) towards the North Sea.

The tills of both phases are separated by gravel, sand and varved clay. On the basis of clast assemblages, till fabric and directions of striae, the Saalian ice in the Zeitz Phase

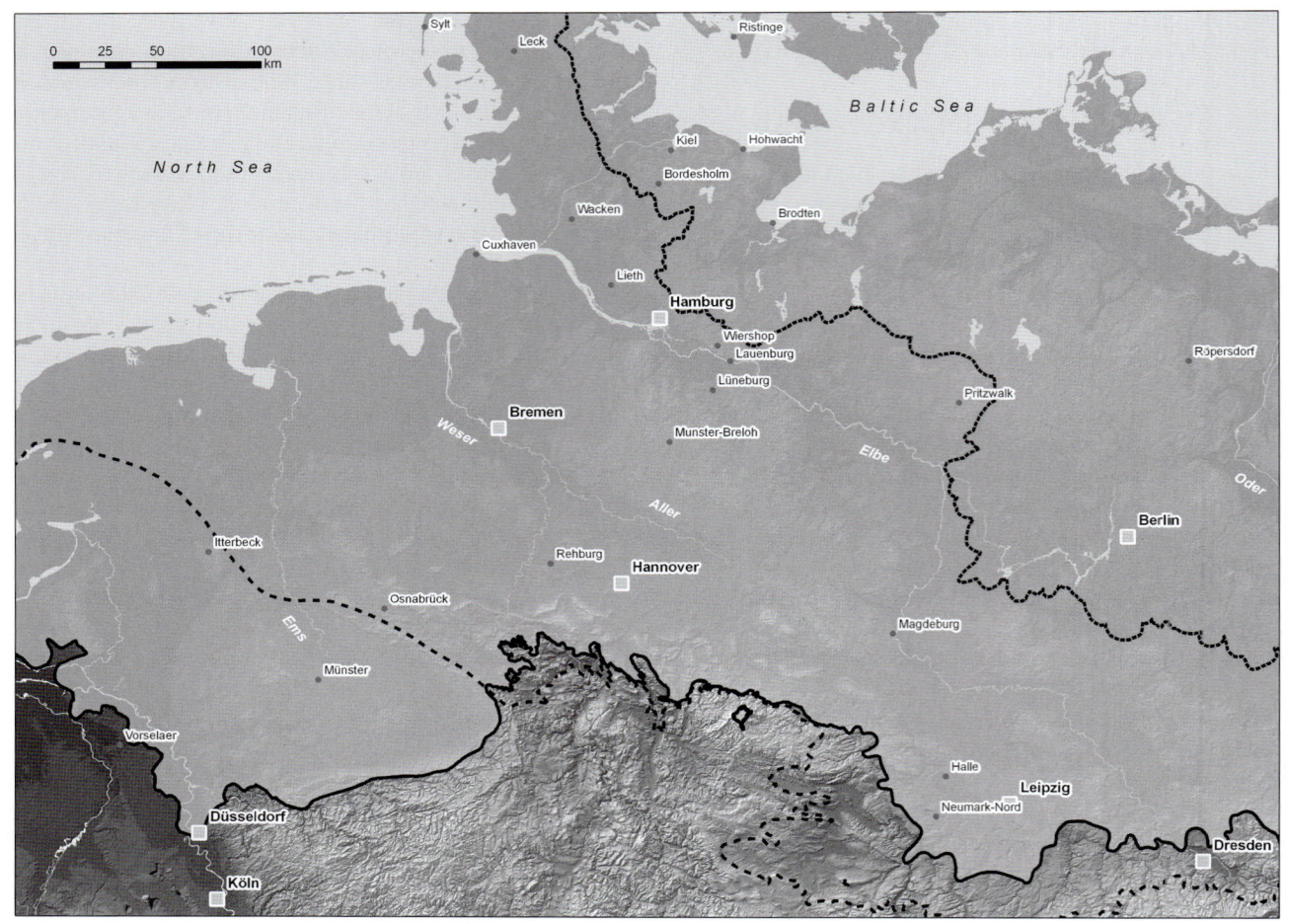

FIGURE 13.3 Maximum extent of the Saalian glaciation in North Germany.

entered the Halle-Leipzig area from the N or NNW. The tills of the Leipzig Phase represent an ice-movement direction from NE to SW (Schulz, 1962; Ruske, 1964; Eissmann, 1986). Thus a similar arrangement of ice-movement directions is met here as in the upper part of the first Saalian till in Niedersachsen and Brandenburg.

At the end of the older Saalian glacial phase, the ice melted back, perhaps beyond the Baltic Sea. Denmark and northern Germany became ice free. Exposures in Dithmarschen and northern Niedersachsen indicate that the older Saalian till at this time was exposed at the land surface. Under the influence of periglacial climate, a gravel lag formed and the till was dissected by a polygonal ice-wedge network (Höfle, 1983a; Stephan et al., 1983).

The warmer intervals between the individual ice advances of the Saalian in north Germany and adjoining areas are characterised by a lack of organic deposits. Apart from the fragile, bryophyte remains described by Grube (1967) (*Calliergon giganteum*), no intra-Saalian autochthonous organic deposits are known, although all three ice advances have left behind kettle holes that would have formed ideal sediment traps for the preservation of such deposits. Only in the subsequent Eemian Stage Interglacial were those depressions foci of peat growth and gyttja, diatomite or lake marl deposition. From the North Sea area, no intra-Saalian high stand of the sea has been reported. Nevertheless, repeatedly, an interglacial status has been postulated for the interval between the Older and middle Saalian ice advances. The supposed warm stages include the 'Uecker Warm Stage' of Röpersdorf (Erd, 1987) and the 'Vorselaer Schichten' (Klostermann et al., 1988). However, the stratigraphic position within the glacial part of the Saalian (i.e. after the Dömnitz/Wacken Interglacial) has not been determined.

According to Hermsdorf and Strahl (2006), the Röpersdorf section consists of glaciotectonically disturbed Eemian sediments. Thus the term 'Uecker Warm Stage' is obsolete. Moreover, the lake sediments of Neumark-Nord (Geiseltal) which were regarded as intra-Saalian deposits by Mania (1990 and later) have now been shown to be of Eemian age (Litt, 1994).

The attempt to define a 'Treene Interglacial' between Drenthe and Warthe Substages in north-west Germany based on palaeosols (mainly by Picard, 1959; Stremme, 1960, 1981) was not successful. Some sites were stratigraphically misinterpreted, others—especially the 'bleached loam' of the Isle of Sylt (Felix-Henningsen, 1983)—were not necessarily formed under interglacial conditions.

13.7.2. The Middle Saalian Glaciation

During the maximum phase of this ice advance, the ice lay south of the present Elbe valley and the meltwater drained southwards and then via the Aller-Weser ice-marginal valley ('urstromtal') towards the North Sea (Meyer, 1983). It is thought that the middle Saalian ice advanced beyond Schleswig–Holstein. It reached at least as far west as the Altenwalder Geest hills, south of Cuxhaven (Höfle and Lade, 1983; Van Gijssel, 1987). In southern Dithmarschen, the youngest fabrics suggest an ice movement from north to south (Stephan, 1980), probably towards the newly developing predecessor of the Elbe urstromtal.

In Brandenburg, equivalent deposits are found in the lower part of the Upper Saalian till. In Sachsen-Anhalt, the second North German Saalian ice advance so far is only known through the occurrence of flint-rich gravels at some places in the Altmark region (Hoffmann and Meyer, 1997). Possibly, some of the Altmark end moraines, especially in the Colbitz-Letzlinger Heide area, must be attributed to this ice advance. In the Altmark and Fläming regions, end moraines and sandurs of different age overlap so that in many cases, the stratigraphy is still unclear.

The middle Saalian glaciation is regarded here as part of the Warthe substage *sensu* Woldstedt (1954). This advance has been referred to as Warthe I by Stephan (1980). In Lower Saxony, the middle Saalian glaciation is regarded as belonging to the Drenthe substage (Younger Drenthe according to Meyer, 1965). Its ice sheet at its maximum terminated over 100 km inside the Drenthe maximum in western Europe. Therefore, meltwater passage towards the west was open. The oldest ice-marginal valley identified in north Germany, the Aller-Weser-Urstromtal, served again as a drainageway for the meltwaters of the middle Saalian glaciation. Towards the east, its catchment area can be traced back to the Letzlinger Heide region (north of Haldensleben) (Glapa, 1971).

13.7.3. The Younger Saalian Glaciation

After the middle Saalian glaciation, the active ice margin retreated far to the north-east and most likely was situated in the present Baltic Sea area. In the proglacial area, widespread dead-ice masses were covered by outwash during the subsequent readvance.

In north Germany, drainage was directed largely towards the west. A line of marked end moraines in the northern Lüneburger Heide area has traditionally been associated with the younger Saalian (Warthe or Warthe II) ice advance. However, more recent investigations show that many of them really have a much older core. The internal structure shows that the thrust occurred from the east. Recent investigations by F. Grube suggest that the youngest Saalian glaciation in northern Niedersachsen indeed reached further than was thought previously (Grube, in preparation).

Based on clast-lithological studies, the youngest Saalian ice advance in Brandenburg comprises both the middle and younger North West German Saalian ice advances. The maximum position in central Germany is marked by the Letzlingen ice-marginal position, the Fläming ice-marginal positions as well as the Niederlausitzer Grenzwall ridge and the Muskauer Faltenbogen (Glapa, 1971; Lippstreu et al., 1995; von Poblozki, 1999). An ice advance beyond the present Elbe valley up to the Schmiedeberg end moraine seems likely (Knoth, 1995; Eissmann, 1997; Büchner, 1999). In the marginal zones of the younger Saalian ice advance, spectacular glacigenic thrust features have been observed, part of which originated already during the earlier glaciation phases.

In the Saale-Elbe region, the first ice advances of both the Elsterian and the Saalian progessed with short oscillations rapidly towards the south, without causing major disturbances of the subsurface. In contrast, the later ice advances in some areas caused deep-penetrating glaciotectonic deformations. The underlying cause for this rule may have been a strongly reduced or total absence of permafrost in the later glaciation phases (Eissmann, 1987; Büchner, 1999).

The youngest Saalian till, which is widespread in eastern Niedersachsen and in the Altmark ('Red Altmark Till'), is characterised by an East Baltic indicator pebble assemblage (Lüttig, 2004). The several tens of metres thick upper Saalian tills of Brandenburg also partially comprise the indicator facies of the middle Saalian till of Niedersachsen (Lippstreu et al., 1995). The youngest Saalian ice advance to north Germany moved in a NE–SW to ENE–WSW direction, in Schleswig–Holstein and around Hamburg eventually from E to W.

A continuous ice-marginal valley of the younger Saalian glaciation can be traced back upstream beyond (Wrocław). It is even possible that it could have extended much further eastwards towards the Warta River.

13.8. THE EEMIAN STAGE INTERGLACIAL

The climatic development of the Eemian very much resembled that of the Holocene (Litt, 1994). Overall, however, the last warm stage seems to have been slightly warmer, and global sea level rose higher than in the Holocene; however, this does not apply to north Germany because of continuous subsidence (Stephan, submitted for publication). The marine transgression occurred later than during the

Holsteinian but considerably earlier than in the Holocene. In contrast to the Holsteinian, a direct connection existed from the North Sea through the English Channel to the south-west so that thermophilous Lusitanian faunal elements could easily immigrate (Hinsch, 1993).

Terrestrial Eemian deposits are found in many places. Strahl and Hermsdorf (2008) have published a map of the Eemian sites in Brandenburg and Berlin, including 566 locations, 202 of which have been palynologically confirmed. In Hamburg, 8235 boreholes have encountered Eemian deposits, although very few of those have been investigated palynologically.

The vegetational and climatic development of the Eemian Interglacial has been reconstructed mainly by palynological investigations of lacustrine deposits from glacigenic kettle holes above Saalian till. Many profiles from northern central Europe represent a complete interglacial cycle without any hints of major climatic deterioration (Litt et al., 1996; Litt and Gibbard, 2008).

On the cliff of the Elbe River in Lauenburg (Schleswig–Holstein) peat and gyttja of the Eemian is exposed. This peat has been investigated by Menke (1992). The vegetational history equals that of other Eemian sites, but important is that in these permanently accessible sites, the peat is not overlain by any younger till. That demonstrates that the Elbe was not crossed by the Weichselian ice sheets.

13.9. THE WEICHSELIAN GLACIATION

The maximum ice advances of the last cold stage both in Britain and continental Europe did not occur before the last part of the Weichselian (Worsley, 1991; cf. Chapter 7). Nevertheless, there can be no doubt that also in the Early Weichselian extensive glaciers existed in northern Europe (Andersen and Mangerud, 1989; Mangerud, 1991; Lundqvist, 1992). The precise extent of the Early Weichselian glaciation in Scandinavia, the deposits of which have been identified in Norway, Finland, northern and central Sweden and in eastern Denmark (Houmark-Nielsen, 1994), is so far unknown. For north Germany, Stephan (2003) claims a first advance around 60 ka BP (Ellund Advance). Müller (2004) even postulates such early ice advance to have reached the Weichselian maximum limit (Warnow Phase).

In Denmark, it had been possible by thermoluminescence (TL) and optically stimulated luminescence (OSL) dating to prove that in the Early Weichselian ice advanced through the depression of the Baltic Sea into the region of the Danish Islands. This 'old Baltic' ice advance deposited the Ristinge Till on Langeland at ca. 50 cal. ka (Houmark-Nielsen, 2010). An ice advance that reached Langeland from the ESE should also have left traces in Schleswig–Holstein. TL dates have shown that an ice advance had occurred that clearly predated the Late Weichselian maximum (Marks et al., 1995). After OSL dating meltwater deposits, Preusser (1999) came to the same result. It turned out that a considerable portion of all Weichselian meltwater deposits in Schleswig–Holstein were deposited during the Early Weichselian (Frechen et al., 2007).

On the basis of present knowledge, the Late Weichselian glaciation of north Germany began around 25 cal. ka in Schleswig–Holstein; for a long time, the Weichselian ice margin was drawn along a line first defined by Gripp (1924). However, later investigations have shown that in a number of places, the ice extended well beyond this morphologically well-defined limit (Stephan, 1997). Nevertheless, since the Eemian deposits at Lauenburg (cf. above; Meyer, 1965) and Wiershop (Stephan, 1997; Strehl, 1997) are not till-covered, the Weichselian ice did not reach the Elbe River. In the area around Kiel, Stephan and Menke (1977) distinguished five Late Weichselian tills. Stephan (1998) associated these tills with only two major ice advances. In contrast to many of the older glaciations, the Weichselian ice margin was strongly fragmented into individual lobes. The overall morphology makes clear that, at least towards the end of the glaciation, the Baltic Sea depression played a decisive role in controlling ice-movement directions. Moreover, till investigations in Schleswig–Holstein have shown that the Weichselian glaciation like its predecessors started with an ice advance from the north-east which later turned to the east (Stephan, 1998).

Whereas in western Germany, the Late Weichselian glaciers only covered a small margin of the coastal zone south of the Baltic Sea, further to the east the ice advanced over 200 km inland (Fig. 13.4). Here, the classic morphostratigraphic subdivision of the Weichselian glaciation was based on the different ice-marginal positions in Mecklenburg-Vorpommern and Brandenburg. Woldstedt (1925, 1928) distinguished a Brandenburg, Frankfurt and Pomeranian Phase, each of which could be subdivided into several arcs of end moraines. The morphological subdivision of the glaciation, however, is not reflected in the stratigraphy. According to clast-lithological investigations by Cepek (1962), only two Weichselian till units could be distinguished that could be correlated with the Brandenburg and Pomeranian Phases. The end moraines of the so-called Frankfurt Stage thus only represent landforms created by an oscillation during the ice retreat from the maximum Brandenburg ice-marginal position (Cepek, 1965). However, minor oscillations of the ice margin have locally deposited more than two till units (e.g. in Berlin; Böse, 1979).

In Mecklenburg, as elsewhere in eastern Germany, originally only two tills of the Weichselian glaciation could be distinguished (Cepek, 1972). However, Heerdt (1965) identified a thin, third till unit which he correlated with the 'Rosenthal end moraine'. This till, according to more recent investigations, can be distinguished from the two older Weichselian tills not only stratigraphically but also based on its clast composition (Rühberg, 1987). The distribution

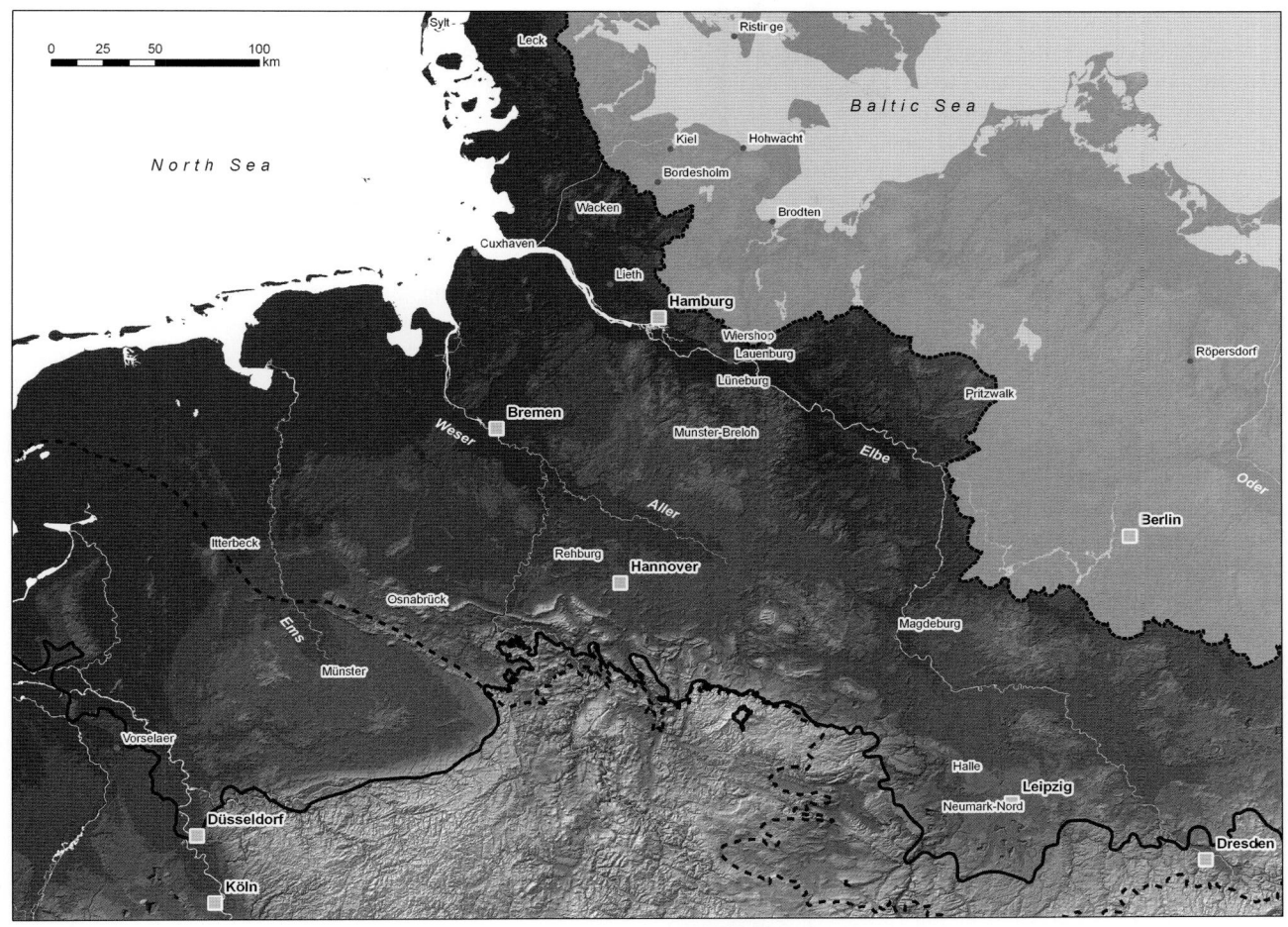

Chapter | 13 Pleistocene Glaciations of North Germany—New Results

FIGURE 13.4 Maximum extent of the Weichselian glaciation in North Germany.

of this youngest Weichselian till is limited to the area north of the Rosenthal end moraine. It is correlated with a Late Weichselian 'Mecklenburg Advance' (Eiermann, 1984). In cases where the ice sheet encountered steep slopes, major push moraines were formed. By geological mapping, it could be demonstrated that this ice advanced south to Malchin and Laage. Most eskers in Mecklenburg-Vorpommern were formed during the 'Mecklenburg Advance', whereas major sandurs of this phase are absent (Rühberg, 1987).

Müller et al. (1995) and Rühberg (1999) described another hitherto oldest known Weichselian till in north Germany. Its composition partly resembles that of the Saalian tills. The correlative ice advance is possibly the early Middle Weichselian (Müller, 2004) and perhaps the Leck Phase of Schleswig–Holstein.

New attempts have been made to date the three Late Weichselian ice advances into north Germany; however, the results are still equivocal. Exposure dating (^{10}Be) of erratics in northern Brandenburg by Heine et al. (2010) suggests that the ice melted back from its maximum position between 21 and 20 ka, and that the area was ice free by

ca. 19 ka. This does not agree with the age estimates of Kozarski (1995), based on radiocarbon dates from underlying sediments of Brandenburg Phase deposits in Pomerania that gave an age of 20 ^{14}C ka BP (calibrated age between 24.07 and 23.875 cal. ka BP; Reimer et al., 2004). Thus, the ages given differ by several thousand years, depending on the method of dating.

The Frankfurt (Oder) Formation must be younger than the Brandenburg Formation. The ^{10}Be results of Heine et al. (2010) suggest that the area was ice free by ca. 18 ka. However, the upper lacustrine sands of the Bordesholm Formation in Schleswig–Holstein, which should be equivalent, have yielded an OSL age of 26.0 ± 4.0 ka (Preusser, 1999). Part of the differences might be explained by the fact that some deposits that were originally assumed to belong to the Frankfurt Formation may really be of a different age (Lüthgens et al., 2009). Rinterknecht et al. (2005) and Heine et al. (2010) have ^{10}Be-dated erratics from the Pomeranian ice-marginal position in Poland and Brandenburg. The Pomeranian dates from Poland gave ages of around 14.8 ± 0.4 ka, whereas the dates from Brandenburg gave 15.8 ± 1.3 ka. Heine et al. (2010) point out that the ages might be slightly too young because they date the time when the boulders came finally to rest, which under periglacial conditions may well postdate the respective glacial maxima. The Mecklenburg Advance, the youngest ice advance into north Germany, is to be correlated with the Young Baltic Advance of Schleswig–Holstein and Denmark (Stephan, 2001). Quartz from the youngest Weichselian ice-contact deposits (kame) of the downwasting glacier in the Lübecker Bucht at Brodten yielded OSL ages between 12.7 ± 1.7 and 18.1 ± 3.8 ka and TL ages between $10.2 \pm .2$ and 21.5 ± 5.8 ka (Preusser, 1999). Preusser concludes that the sands were deposited around ca. 15 ka, which fits well with the end of the last major cold phase of the Weichselian as determined in the Greenland ice cores (Stuiver and Grootes, 2000).

It should be kept in mind that dating of the individual events is just one aspect of Quaternary stratigraphy. The basis must always be a comprehensive survey of the geological and geomorphological evidence, as in the recent works Krienke (2003) or Juschus (2001).

Deposits of the Mecklenburg Advance are overlain by sediments of the Meiendorf Interstadial (Bock et al., 1985). Varve counts from the Meerfelder Maar have shown that this interstadial began at 14,450 varve years ago. The Meiendorf Interstadial is the oldest climatostratigraphical unit in north Germany that has been dated by varve counts (Litt et al., 2007).

The extent of the individual Weichselian ice advances is still a matter of discussion. Based on indicator pebble investigations, Lüttig (2005) has come to the conclusion that the Pomeranian Phase did not reach any further west than the Hohwachter Bucht.

During the Weichselian glaciation, the lower Elbe Valley between the Havel River mouth and the North Sea served continuously as the main drainage path parallel to the ice margin so that no changes of river course had to be made. Further to the east, however, four to five major valley tracts can be distinguished, which as urstromtäler (pradolinas) one after the other drained the southern sector of the Weichselian ice sheet. During the Weichselian, the main watershed occurred considerably further east than during the Saalian glaciation. The easternmost parts of the oldest Weichselian urstromtal, which drained the meltwaters of the Brandenburg Advance, are found at an altitude of about 190 m (Liedtke, 1981). The upper reaches of the most extensive of the four major Weichselian ice-marginal rivers originated in the area around Minsk (Belarus). Further east, the meltwaters drained through the tributaries of the Dniepr towards the east and south.

The Frankfurt ice-marginal position is connected to the Warsaw-Berlin Urstromtal. Liedtke (1957) assumes that during the Pomeranian Advance at first, a Torun-Berlin Urstromtal was formed. At that time, the eastward-aligned drainage via Notec and Warta Rivers and the Oderbruch was already ice free, whereas in the west, the more southerly ice margin forced the meltwaters to drain southwards via the Buckow Gap into the Berlin Urstromtal. This drainage system, however, is not quite safely established. During the Weichselian ice-decay phase, the drainage was first maintained via the Torun-Eberswalde Urstromtal, which continued to function until the formation of the Rosenthal end moraine. During the formation of the Velgast end moraine, according to Liedtke (1981), the ice had melted back far enough that drainage in the west could possibly be redirected via the Mecklenburger Grenztal into the depression of the western Baltic Sea and further via the Belt to the North Sea. This youngest ice-marginal stream, the Notec-Randow Urstrom of Liedtke lost its function when finally the Vistula mouth into the Baltic became ice free.

REFERENCES

Andersen, B.G., Mangerud, J., 1989. The last interglacial-glacial cycle in Fennoscandia. Quatern. Int. 3 (4), 21–29.

Bock, W., Menke, B., Strehl, E., Ziemus, H., 1985. Neuere Funde des Weichselspätglazials in Schleswig-Holstein. Eiszeitalter und Gegenwart 35, 161–180.

Böse, M., 1979. Die geomorphologische Entwicklung im westlichen Berlin nach neueren stratigraphischen Untersuchungen. Berl. Geogr. Abh. 28, 46 pp.

Büchner, L., 1999. Zum geologischen Bau der Schmiedeberger Endmoräne. Mitt. Geol. Sachsen Anhalt 5, 117–134.

Busschers, F.S., Van Balen, R.T., Cohen, K.M., Kasse, C., Weerts, H.J.T., Wallinga, J., et al., 2008. Response of the Rhine-Meuse fluvial system to Saalian ice-sheet dynamics. Boreas 37, 377–398.

Cepek, A.G., 1962. Zur Grundmoränenstratigraphie in Brandenburg. Ber. Geologischen Ges. DDR 7, 275–278.

Cepek, A.G., 1965. Die Stratigraphie der quartären Ablagerungen des Norddeutschen Tieflandes. In: Gellert Hrsg., J.F. (Ed.), Die Weichsel-Eiszeit im Gebiet der Deutschen Demokratischen Republik. Akademie-Verlag, Berlin, pp. 45–65.

Cepek, A.G., 1972. Zum Stand der Stratigraphie der Weichsel-Kaltzeit in der DDR. Wissenschaftliche Zeitschrift der Ernst-Moritz-Arndt-Universität Greifswald, Mathematisch-Naturwissenschaftliche Reihe 21, pp. 11–21.

Ehlers, J., 1994. Allgemeine und Historische Quartärgeologie. Enke, Stuttgart, 358pp.

Ehlers, J., 1996. Quaternary and Glacial Geology. Wiley, Chichester, 578pp.

Ehlers, J., 2011. Das Eiszeitalter. Spektrum, Heidelberg. 372 pp.

Ehlers, J., Eißmann, L., Lippstreu, L., Stephan, H.-J., Wansa, S., 2004. Pleistocene glaciations of North Germany. In: Ehlers, J., Gibbard, P.L. (Eds.), Quaternary Glaciations—Extent and Chronology, Part I: Europe. Elsevier, Amsterdam, pp. 135–146.

Eiermann, J., 1984. Ein zeitliches, räumliches und genetisches Modell zur Erklärung der Sedimente und Reliefformen im Pleistozän gletscherbedeckter Tieflandsedimente—Ein Beitrag zur Methodik der mittelmaßstäbigen naturräumlichen Gliederung. In: Richer, H., Aurada, K. (Eds.), Umweltforschung. Zur Analyse und Diagnose der Landschaft. Haack, Gotha, pp. 169–183.

Eissmann, L., 1967. Glaziäre Destruktionszonen (Rinnen, Becken) im Altmoränengebiet des Norddeutschen Tieflandes. Geologie 16, 804–833.

Eissmann, L., 1975. Das Quartär der Leipziger Tieflandsbucht und angrenzender Gebiete um Saale und Elbe. Modell einer Landschaftsentwicklung am Rand der europäischen Kontinentalvereisung. Schriftenreihe für Geologische Wissenschaften 2, 263pp + Plates.

Eissmann, L., 1986. Quartärgeologie und Geschiebeforschung im Leipziger Land mit einigen Schlußfolgerungen zu Stratigraphie und Vereisungsablauf im Norddeutschen Tiefland. Altenburger Naturwissenschaftliche Forschungen 3, 105–133.

Eissmann, L., 1987. Lagerungsstörungen im Lockergebirge. Exogene und endogene Tektonik im Lockergebirge des nördlichen Mitteleuropa. Geophysik und Geologie. Geophysikalische Veröffentlichungen der Karl-Marx-Universität Leipzig III (4), 7–77.

Eissmann, L., 1997. Das quartäre Eiszeitalter in Sachsen und Nordostthüringen. Altenburger Naturwissenschaftliche Forschungen 8, 98pp.

Eissmann, L., Müller, A., 1979. Leitlinien der Quartärentwicklung im Norddeutschen Tiefland. Z. Geol. Wiss. 7, 451–462.

Eissmann, L., Litt, T., Wansa, S., 1995. Elsterian and Saalian deposits in their type area in central Germany. In: Ehlers, J., Kozarski, S., Gibbard, P.L. (Eds.), Glacial Deposits in North-East Europe. Balkema, Rotterdam, Brookfield, pp. 439–464.

Elsner, H., 2003. Verbreitung und Ausbildung Elster-zeitlicher Ablagerungen zwischen Elm und Flechtinger Höhenzug. Eiszeitalter und Gegenwart 52, 91–116.

Erd, K., 1965. Pollenanalytische Gliederung des mittelpleistozänen Richtprofils Pritzwalk-Prignitz. Eiszeitalter und Gegenwart 16, 252–253.

Erd, K., 1970. Pollenanalytical classification of the Middle Pleistocene in the German Democratic Republic. Palaeogeogr. Palaeoclimatol. Palaeoecol. 8, 129–145.

Erd, K., 1973. Vegetationsentwicklung und Biostratigraphie der Dömnitz-Warmzeit (Fuhne/Saale 1) im Profil von Pritzwalk/Prignitz. Abhandlungen des Zentralen Geologischen Instituts 18, 9–48.

Erd, K., 1987. Die Uecker-Warmzeit von Röpersdorf bei Prenzlau als neuer Interglazialtyp im Saale-Komplex der DDR. Zeitschrift für Geologische Wissenschaften 15, 297–313.

Felix-Henningsen, P., 1983. Palaeosols and their stratigraphhical interpretation. In: Ehlers, J. (Ed.), Glacial Deposits in North-West Europe. Balkema, Rotterdam, pp. 289–295.

Frechen, M., Sierralta, M., Stephan, H.-J., Techmer, A., 2007. Die zeitliche Stellung saale- und weichselzeitlicher Eisrandlagen in Schleswig-Holstein. 74. Tagung der Arbeitsgemeinschaft Norddeutscher Geologen. Tagungsband und Exkursionsführer, S.16.

Geyh, M.A., Müller, H., 2005. Numerical ^{230Th}/U dating and a palynological review of the Holsteinian/Hoxnian Interglacial. Quatern. Sci. Rev. 24, 1861–1872.

Glapa, H., 1971. Warthezeitliche Eisrandlagen im Gebiet der Letzlinger Heide. Geologie 20, 1087–1110.

Gripp, K., 1924. Über die äußerste Grenze der letzten Vereisung in Nordwest-Deutschland. Mitteilungen der Geographischen Gesellschaft in Hamburg 36, 159–245.

Gripp, K., 1964. Erdgeschichte von Schleswig-Holstein. Wachholtz, Neumünster, 411 S.

Grube, F., 1967. Die Gliederung der Saale-(Riß-)Kaltzeit im Hamburger Raum. Fundamenta B2, 168–195.

Grube, F., 1981. The subdivision of the Saalian in the Hamburg Region. Mededelingen Rijks Geologische Dienst 34 (4), 15–25.

Grube, A., 1997. Geologie des Deckgebirges der Struktur Elmshorn (Schleswig-Holstein). (Dissertation). Berichte-Reports des Geologisch-Paläontologischen Institutes, Universität Kiel, vol. 87, pp. 1–169.

Grube, F., in preparation. Saale-Kaltzeit in der Hamburger Region. Eiszeitalter und Gegenwart, Geologian tutkimuskeskus, Espoo.

Habbe, K.A., 2007. Stratigraphische Begriffe für das Quartär des süddeutschen Alpenvorlandes. Eiszeitalter und Gegenwart 56 (1/2), 66–83.

Heerdt, S., 1965. Zur Stratigraphie des Jungpleistozäns im mittleren N-Mecklenburg. Geologie 14, 589–609.

Heine, K., Reuther, A.U., Thieke, H.U., Schulz, R., Schlaak, N., Kubik, P. W., 2010. Timing of Weichselian ice marginal positions in Brandenburg (northeastern Germany) using cosmogenic in situ ^{10}Be. Z. Geomorphol. N.F. 53 (4), 433–454.

Hermsdorf, N., Strahl, J., 2006. Zum Problem der so genannten Uecker-Warmzeit (Intra-Saale)—Untersuchungen an neuen Bohrkernen aus dem Raum Prenzlau. Brandenburgische Geowissenschaftliche Beiträge 13, 49–61.

Hinsch, W., 1993. Marine Molluskenfaunen in Typusprofilen des Elster-Saale-Interglazials und des Elster-Spätglazials. Geologisches Jahrbuch, A 138, 9–34.

Hoffmann, K., Meyer, K.-D., 1997. Leitgeschiebezählungen von elster- und saalezeitlichen Ablagerungen aus Sachsen, Sachsen-Anhalt und dem östlichen Niedersachsen. Leipziger Geowissenschaften 5, 115–128.

Höfle, H.-C., 1983a. Periglacial phenomena. In: Ehlers, J. (Ed.), Glacial Deposits in North-West Europe. Balkema, Rotterdam, pp. 297–298.

Höfle, H.-C., 1983b. Strukturmessungen und Geschiebeanalysen an eiszeitlichen Ablagerungen auf der Osterholz-Scharmbecker Geest. Abhandlungen Naturwiss. Verein Bremen 40, 39–53.

Höfle, H.-C., Lade, U., 1983. The stratigraphic position of the Lamstedter Moraine within the Younger Drenthe substage (middle Saalian). In: Ehlers, J. (Ed.), Glacial Deposits in North-West Europe. Balkema, Rotterdam, pp. 343–346.

Houmark-Nielsen, M., 1994. Late Pleistocene stratigraphy, glaciation chronology and Middle Weichselian environmental history from Klintholm, Møn, Denmark. Bull. Geol. Soc. Denmark 41, 181–201.

Houmark-Nielsen, M., 2010. Extent, age and dynamics of Marine Isotope Stage 3 glaciations in the southwestern Baltic Basin. Boreas 39, 343–359.

Huuse, M., Lykke-Andersen, H., 2000. Overdeepened Quaternary valleys in the eastern Danish North Sea: morphology and origin. Quatern. Sci. Rev. 19, 1233–1253.

Junge, F.W., 1998. Die Bändertone Mitteldeutschlands und angrenzender Gebiete. Altenburger Naturwissenschaftliche Forschungen 9, 210pp.

Junge, F.W., Eissmann, L., 2001. Postsedimentäre Deformationsbilder in mitteldeutschen Vorstoßbändertonen—Hinweise auf den Bewegungsmechanismus des quartären Inlandeises. Brandenburgische Geowissenschaftliche Beiträge 7 (1/2), 21–28.

Juschus, O., 2001. Das Jungmoränenland südlich von Berlin—Untersuchungen zur jungquartären Landschaftsentwicklung zwischen Unterspreewald und Nuthe. Dissertation, Humboldt-Universität Berlin, 245 pp.

Kabel, C., 1982. Geschiebestratigraphische Untersuchungen im Pleistozän Schleswig-Holsteins und angrenzender Gebiete. Dissertation Universität Kiel, 132 pp.

Kaltwang, J., 1992. Die pleistozäne Vereisungsgrenze im südlichen Niedersachsen und im östlichen Westfalen. Mitteilungen aus dem Geologischen Institut der Universität Hannover 33, 161 pp.

Klostermann, J., 1992. Das Quartär der Niederrheinischen Bucht. Ablagerungen der letzten Eiszeit am Niederrhein. Geologisches Landesamt Nordrhein-Westfalen, Krefeld, 200pp.

Klostermann, J., Rehagen, H.-W., Wefels, U., 1988. Hinweise auf eine saalezeitliche Warmzeit am Niederrhein. Eiszeitalter und Gegenwart 38, 115–127.

Knoth, W., 1995. Sachsen-Anhalt. In: Benda, L. (Ed.), Das Quartär Deutschlands. Gebr. Bornträger, Berlin, Stuttgart, pp. 148–170.

Knudsen, K.L., 1993. Late Elsterian-Holsteinian Foraminiferal Stratigraphy in Borings of the Lower Elbe Area, NW Germany. Geologisches Jahrbuch, A 138, 97–119.

Kozarski, S., 1995. Deglacjacja Pólnocno-Zachodniej Polski: Warunki środowiska i transformacja geosystemu (c. 20 ka → 10 ka BP). Dokumentacja Geograficzna 1, 7–82.

Krbetschek, M.R., Degering, D., Alexowsky, W., 2008. Infrarot-Radiofluoreszenz-Alter (IR-RF) unter-saalezeitlicher Sedimente Mittel- und Ostdeutschlands. Zeitschrift der Deutschen Gesellschaft für Geowissenschaften 159 (1), 133–140 Stuttgart.

Krienke, K., 2003. Südostrügen im Weichsel-Hochglazial—lithostratigraphische, lithofazielle, strukturgeologische und landschaftsgenetische Studien zur jüngsten Vergletscherung im Küstenraum Vorpommerns (NE-Deutschland). Greifswalder Geowissenschaftliche Beiträge 12, 148pp.

Kristensen, T.B., Huuse, M., Piotrowski, J.A., Clausen, O.R., 2007. A morphometric analysis of tunnel valleys in the eastern North Sea based on 3D seismic data. J. Quatern. Sci. 22, 801–815.

Kristensen, T.B., Piotrowski, J.A., Huuse, M., Clausen, O.R., Hamberg, I., 2008. Time-transgressive tunnel valley formation indicated by infill sediment structure, North Sea; the role of glaciohydraulic supercooling. Earth Surf. Process. Landforms 33, 546–559.

Kupetz, M., 1997. Geologischer Bau und Genese der Stauchendmoräne Muskauer Faltenbogen. Brandenburger Geowissenschaftliche Beiträge 4 (2), 1–20.

Liedtke, H., 1957. Beiträge zur geomorphologischen Entwicklung des Thorn-Eberswalder Urstromtales zwischen Oder und Havel. Wissenschaftliche Zeitschrift der Humboldt-Universität Berlin, Mathematisch-Naturwissenschaftliche Reihe, vol. 6, pp. 3–49.

Liedtke, H., 1981. Die nordischen Vereisungen in Mitteleuropa, 2. Auflage. Forschungen zur deutschen Landeskunde 204, 307pp.

Lippstreu, L., Brose, F., Marcinek, J., 1995. Brandenburg. In: Benda, L. (Ed.), Das Quartär Deutschlands. Gebr. Bornträger, Berlin, Stuttgart, pp. 116–147.

Lisiecki, L.E., Raymo, M.E., 2005. A Pliocene–Pleistocene stack of 57 globally distributed benthic $\delta^{18}O$ records. Paleoceanography 20, PA1003.

Litt, T., 1994. Paläoökologie, Paläobotanik und Stratigraphie des Jungquartärs im nordmitteleuropäischen Tiefland unter besonderer Berücksichtigung des Elbe-Saale-Gebietes. Dissertationes Botanicae 227, 185pp. (Berlin, Stuttgart, J. Cramer).

Litt, T., Gibbard, P., 2008. A proposed Global Stratotype Section and Point (GSSP) for the base of the Upper (Late) Pleistocene Subseries (Quaternary System/Period). Episodes J. Int. Geosci. 31 (2), 260–263.

Litt, T., Turner, C., 1993. Arbeitsergebnisse der Subkommission für Europäische Quartärstratigraphie, Die Saalesequenz in der Typusregion. Eiszeitalter und Gegenwart 43, 125–128.

Litt, T., Wansa, S., 2008. Quartär. In: Bachmann, G., Ehling, B.-C., Eichner, R., Schwab, M. (Eds.), Geologie von Sachsen-Anhalt. Schweizerbart, Stuttgart, pp. 293–325.

Litt, T., Junge, F.W., Böttger, T., 1996. Climate during the Eemian in north-central Europe—a critical review of the palaeobotanical and stable isotope data from central Germany. Veg. Hist. Archaeobotany 5, 247–256.

Litt, T., Behre, K.-E., Meyer, K.-D., Stephan, H.-J., Wansa, S., 2007. Stratigraphische Begriffe für das Quartär des norddeutschen Vereisungsgebietes. Eiszeitalter und Gegenwart 56 (1/2), 7–65.

Lundqvist, J., 1992. Glacial stratigraphy in Sweden. In: Kauranne, K. (Ed.), Glacial Stratigraphy, Engineering Geology and Earth Construction, Geological Survey of Finland, Special Paper. Geologian tutkimuskeskus, Espoo, 15, 43–59.

Lüthgens, C., Böse, M., Krbetschek, M., 2009. Towards a new understanding of the Last Glacial Maximum (LGM) in NE-Germany—results from optically stimulated luminescence (OSL) dating and their implications. In: Exploratory Workshop on the Frequency and Timing of Glaciations in Northern Europe (including Britain) During the Middle and Late Pleistocene. Abstracts17–18 Berlin (FU).

Lüttig, G., 2004. Ergebnisse geschiebestatistischer Untersuchungen im Umland von Hamburg. Arch. Geschiebekunde 3 (8–12), 729–746.

Lüttig, G., 2005. Geschiebezählungen im westlichen Mecklenburg. Arch. Geschiebekunde 4 (9), 569–600.

Lutz, R., Kalka, S., Gaedicke, C., Reinhardt, L., Winsemann, J., 2009. Pleistocene tunnel valleys in the German North Sea: spatial distribution and morphology. Z. Dtsch Ges. Geowiss. 160, 225–235.

Mangerud, J., 1991. The Scandinavian Ice Sheet through the last interglacial/glacial cycle. Paläoklimaforschung 1, 307–330.

Mania, D., 1990. Stratigraphie, Ökologie und mittelpaläolithische Jagdbefunde des Interglazials von Neumark-Nord (Geiseltal). Veröffentlichungen des Landesmuseums für Vorgeschichte in Halle 43, 9–130.

Marks, C., Piotrowski, J.A., Stephan, H.-J., Fedorowicz, S., Butrym, J., 1995. Thermoluminescence indications of the Middle Weichselian (Vistulian) Glaciation in Northwest Germany. Meyniana 47, 69–82.

Meng, S., Wansa, S., 2008. Sedimente und Prozesse am Außenrand der Saale-Vereisung südwestlich von Halle (Saale). Z. Dtsch. Ges. Geowiss. 159, 205–220.

Menke, B., 1968. Beiträge zur Biostratigraphie des Mittelpleistozäns in Norddeutschland. Meyniana 18, 35–42.

Menke, B., 1975. Vegetationsgeschichte und Florenstratigraphie Nordwestdeutschlands im Pliozän und Frühquartär Mit einem Beitrag zur Biostratigraphie des Weichselfrühglazials. Geologisches Jahrbuch A26, 3–151.

Menke, B., 1992. Eeminterglaziale und nacheiszeitliche Wälder in Schleswig-Holstein. Berichte des Geologischen Landesamtes Schleswig-Holstein 1, 28–101.

Menning, M., Hendrich, A., 2005. Erläuterungen zur Stratigraphischen Tabelle von Deutschland 2005 (ESTD 2005). Newslett. Stratigr. 41, 1–405.

Meyer, K.-D., 1965. Das Quartärprofil am Steilufer der Elbe bei Lauenburg. Eiszeitalter und Gegenwart 16, 47–60.

Meyer, K.-D., 1983. Zur Anlage der Urstromtäler in Niedersachsen. Z. Geomorphol. N. F. 27, 147–160.

Meyer, K.-D., 1987. Ground and end moraines in Lower Saxony. In: van der Meer, J.J.M. (Ed.), Tills and Glaciotectonics. Balkema, Rotterdam, pp. 197–204.

Müller, H., 1974. Pollenanalytische Untersuchungen und Jahresschichtenzählungen an der holsteinzeitlichen Kieselgur von Munster-Breloh. Geologisches Jahrbuch A21, 107–140.

Müller, U., 2004. Weichsel-Frühglazial in Nordwest-Mecklenburg. Meyniana 56, 81–115.

Müller, U., Obst, K., 2008. Junge halokinetische Bewegungen im Bereich der Salzkissen Schlieven und Marnitz in Südwest-Mecklenburg. Brandenburgische Geowissenschaftliche Beiträge 15 (1/2), 147–154.

Müller, U., Rühberg, N., Krienke, H.-D., 1995. The Pleistocene sequence in Mecklenburg-Vorpommern. In: Ehlers, J., Kozarski, S., Gibbard, P.L. (Eds.), Glacial Deposits in North-East Europe. Balkema, Rotterdam, Brookfield, pp. 501–514.

Penck, A., 1879. Die Geschiebeformation Norddeutschlands. Z. Dtsch. Geol. Ges. 31, 117–203.

Penck, A., Brückner, E., 1901. Die Alpen im Eiszeitalter, 3 vols, 1199 pp. Leipzig, Tauchnitz.

Picard, K., 1959. Gliederung pleistozäner Ablagerungen mit fossilen Böden bei Husum/Nordsee. Neues Jb Geol. Paläontol. Monatsh. 6, 259–272.

Praeg, D., 2003. Seismic imaging of mid-Pleistocene tunnel valleys in the North Sea Basin—high resolution from low frequencies. J. Appl. Geophys. 53, 273–298.

Präger, F., 1976. Quartäre Bildungen in Ostsachsen. Abhandlungen des Staatlichen Museums für Mineralogie und Geologie Dresden 25, 125–217.

Preusser, F., 1999. Lumineszenzdatierung fluviatiler Sedimente; Fallbeispiele aus der Schweiz und Norddeutschland. Kölner Forum Geologie Paläontologie 3, 1–62.

Reimer, J.P., Baillie, M.G.L., Bard, E., Bayliss, A., Beck, J.W., Bertrand, C.J.H., et al., 2004. INTCAL04 Terrestrial Radiocarbon Age Calibration, 0–26 Cal KYR BP. Radiocarbon 46 (3), 1029–1058.

Rinterknecht, V.R., Marks, L., Piotrowski, J.A., Raisbeck, G.M., Yiou, F., Brook, E.J., et al., 2005. Cosmogenic ^{10}Be ages on the Pommeranian Moraine, Poland. Boreas 34, 186–191.

Rühberg, N., 1987. Die Grundmoräne des jüngsten Weichselvorstoßes im Gebiet der DDR. Z. Geol. Wiss. 15, 759–767.

Rühberg, N., 1999. Über den Wert der Kleingeschiebezählungen (KGZ). Geschiebekunde aktuell 15, 87–100.

Ruske, R., 1964. Das Pleistozän zwischen Halle (Saale), Bernburg und Dessau. Geologie 13, 570–597.

Schulz, W., 1962. Gliederung des Pleistozäns in der Umgebung von Halle (Saale). Geologie, Beiheft 36, 1–69.

Schwab, G., Ludwig, A.o., 1996. Zum Relief der Quartärbasis in Norddeutschland. Bemerkungen zu einer neuen Karte. Z. Geol. Wiss. 24 (3/4), 343–349.

Skupin, K., Speetzen, E., Zandstra, J.G. (Eds.), 1993. Die Eiszeit in Nordwestdeutschland. Geologisches Landesamt Nordrhein-Westfalen, Krefeld, 143 pp.

Stackebrandt, W., 2009. Subglacial channels of Northern Germany—a brief review. Z. Dtsch Ges. Geowiss. 160, 203–210.

Stephan, H.-J., 1980. Glazialgeologische Untersuchungen im südlichen Geestgebiet Dithmarschens. Schriften des naturwissenschaftlichen Vereins für Schleswig-Holstein 50, 1–36, Kiel.

Stephan, H.-J., 1997. Wie weit reichte die Vergletscherung der letzten Eiszeit in Schleswig-Holstein? Die Heimat 104 (3/4), 52–57.

Stephan, H.-J., 1998. Geschiebemergel als stratigraphische Leithorizonte in Schleswig-Holstein, Ein Überblick. Meyniana 50, 113–135.

Stephan, H.-J., 2001. The Young Baltic advance in the western Baltic depression. Geol. Q. 45 (4), 359–363.

Stephan, H.-J., 2003. Zur Entstehung der eiszeitlichen Landschaft Schleswig-Holsteins. Schriften des naturwissenschaftlichen Vereins für Schleswig-Holstein 67, 101–118.

Stephan, H.-J., submitted for publication. Climato-stratigraphical subdivision of the Pleistocene in Schleswig-Holstein and adjoining areas: status and problems. Quaternary Science Reviews.

Stephan, H.-J., Urban, B., Lüttig, G., Menke, B., Sierralta. M., accepted for publication. Palynologische, petrographische und geochronologische Untersuchungen an Ablagerungen der Leck-Warmzeit (spätes Mittelpleistozän) und begleitender Sedimente. Geologisches Jahrbuch.

Stephan, H.-J., Kabel, Ch., 1982. The subdivision of the glacial Pleistocene in Schleswig-Holstein and problems of correlations with adjacent areas. XI INQUA Congress 1982, Moskva, Abstracts, II, 313.

Stephan, H.-J., Menke, B., 1977. Untersuchungen über den Verlauf der Weichsel-Kaltzeit in Schleswig-Holstein. Z. Geomorphol. N. F. Suppl. 27, 12–28.

Stephan, H.-J., Kabel, Ch., Schlüter, G., 1983. Stratigraphical problems in the glacial deposits of Schleswig-Holstein. In: Ehlers, J. (Ed.), Glacial deposits in North-West Europe. Balkema, Rotterdam, pp. 305–320.

Stottmeister, L., 2000. Zur Geologie der Jahrstedt-Steimker Endmoräne (SachsenAnhalt). Brandenburgische Geowissenschaftliche Beiträge 7, 127–136.

Strahl, J., Hermsdorf, N., 2008. Karte der Eem-Vorkommen des Landes Brandenburg. Brandenburger Geowiss. Beitr. 20 (1–2), 23–55.

Strehl, E., 1997. Zum Verlauf der äußeren Grenze der Weichselvereisung zwischen Schuby und Ellund. Schriften des Naturwissenschaftlichen Vereins für Schleswig-Holstein 67, 27–35.

Stremme, H.E., 1960. Bodenbildungen auf Geschiebelehmen verschiedenen Alters in Schleswig-Holstein. Z. Dtsch. Geol. Ges. 112, 299–308.

Stremme, H.E., 1981. Unterscheidung von Moränen durch Bodenbildungen. Meded. Rijks Geol. Dienst 34, 51–56.

Stuiver, M., Grootes, P., 2000. GSIP2 oxygen isotope ratios. Quaternary Research 53, 277–284.

Toucanne, S., Zaragosi, S., Bourillet, J.F., Cremer, M., Eynaud, F., Turon, J.L., et al., 2009a. Timing of massive 'Fleuve Manche' discharges over the last 350 kyr: insights into the European Ice Sheet oscillations and the European drainage network from MIS 10 to 2. Quatern. Sci. Rev. 28, 1238–1256.

Toucanne, S., Zaragosi, S., Bourillet, J.F., Gibbard, P., Eynaud, F., Turon, J.L., et al., 2009b. A 1.2 Ma record of glaciation and fluvial discharge from the West European Atlantic margin. Quatern. Sci. Rev. 28, 2974–2981.

Unger, K.-P., 1974. Die Elstervereisung des zentralen Thüringer Keuperbeckens. Z. Geol. Wiss. 2, 791–800.

Urban, B., 1995. Palynological evidence of younger Middle Pleistocene Interglacials (Holsteinian, Reinsdorf and Schöningen) in the Schöningen open cast lignite mine (eastern Lower Saxony, Germany). Meded. Rijks Geol. Dienst 52, 175–186.

Urban, B., 2006. Interglacial pollen records from Schöningen, north Germany. In: Sirocko, F., Claussen, M., Sánchez- Goñi, M.F., Litt, T. (Eds.), The Climate of Past Interglacials. Elsevier, Amsterdam, pp. 417–444.

Urban, B., Lenhard, R., Mania, D., Albrecht, B., 1991. Mittelpleistozän im Tagebau Schöningen, Ldkrs. Helmstedt. Z. Dtsch. Geol. Ges. 142, 351–372.

van der Vlerk, I.M., Florschütz, F., 1950. Nederland in het Ijstijdvak. De geschiedenis van flora, fauna en klimaat, toen aap en mammoet ons land bewoonden. Utrecht, de Haan, 287pp.

van Gijssel, K., 1987. A lithostratigraphic and glacio-tectonic reconstruction of the Lamstedt Moraine, Lower Saxony (FRG). In: van der Meer, J.J.M. (Ed.), Tills and Glaciotectonics. Balkema, Rotterdam, Boston, pp. 145–155.

Vinx, R., Grube, A., Grube, F., 1997. Vergleichende Lithologie, Geschiebeführung und Geochemie eines Prä-Elster-I-Tills von Lieth bei Elmshorn. Leipziger Geowissenschaften 5, 83–103.

von Bülow, W. (Ed.), 2000. Geologische Entwicklung Südwest-Mecklenburgs seit dem Ober-Oligozän, In: Schriftenreihe für Geowissenschaften 11, 413 pp.

von Poblozki, B., 1995. Quaternary geology of the Altmark region. In: Ehlers, J., Kozarski, S., Gibbard, P.L. (Eds.), Glacial Deposits in North-East Europe. Balkema, Rotterdam, Brookfield, pp. 473–484.

von Poblozki, B., 1999. Geologische Übersicht-Quartär, - 63. Tagung der Arbeitsgemeinschaft Nordwestdeutscher Geologen, 28. bis 31. Mai 1996 in Helmstedt, Kurzfassung der Vorträge und Exkursionsführer, 56–74.

Wagenbreth, O., 1978. Die Feuersteinlinie in der DDR, ihre Geschichte und Popularisierung. Schriftenreihe für Geologische Wissenschaften 9, 339–368.

Wansa, S., 1994. Zur Lithologie und Genese der Elster-Grundmoränen und der Haupt-Drenthe-Grundmoräne im westlichen Elbe-Weser-Dreieck. Mitteilungen aus dem Geologischen Institut der Universität Hannover 34, 77pp.

Westerhoff, W., 2009. Stratigraphy and sedimentary evolution. The lower Rhine-Meuse system during the Late Pliocene and Early Pleistocene (southern North Sea Basin). Dissertation, Vrije Universiteit Amsterdam.168 S.

Woldstedt, P., 1925. Die großen Endmoränenzüge Norddeutschlands. Z. Dtsch. Geol. Ges. 77, 172–184.

Woldstedt, P., 1927a. Über die Ausdehnung der letzten Vereisung in Norddeutschland. Sitzungsberichte der Preußischen Geologischen Landesanstalt 2, 115–119.

Woldstedt, P., 1927b. Die Gliederung des Jüngeren Diluviums in Norddeutschland und seine Parallelisierung mit anderen Glazialgebieten. Zeitschrift der Deutschen Geologischen Gesellschaft, Monatsberichte 1927 (3/4), 51–52.

Woldstedt, P., 1928. Die Parallelisierung des nordeuropäischen Diluviums mit dem anderer Vereisungen. Zeitschrift für Gletscherkunde 16, 230–241.

Woldstedt, P., 1954. Saaleeiszeit, Warthestadium und Weichseleiszeit in Norddeutschland. Eiszeitalter und Gegenwart 4 (5), 34–48.

Worsley, P., 1991. Possible early Devensian glacial deposits in the British Isles. In: Ehlers, J., Gibbard, P., Rose, J. (Eds.), Glacial Deposits in Great Britain and Ireland. Balkema, Rotterdam, Brookfield, pp. 47–51.

Zandstra, J.G., 1983. Fine gravel, heavy mineral and grain-size analyses of Pleistocene, mainly glacigenic deposits in the Netherlands. In: Ehlers, J. (Ed.), Glacial Deposits in North-West Europe. Balkema, Rotterdam, pp. 361–377.

Zandstra, J.G., 1993. Nördliche kristalline Leitgeschiebe und Kiese in der Westfälischen Bucht und angrenzenden Gebieten. In: Skupin, K., Speetzen, E., Zandstra, J.G. (Eds.), Die Eiszeit in Nordwestdeutschland. Geologisches Landesamt Nordrhein-Westfalen, Krefeld, pp. 43–106.

Chapter 14

Pleistocene Glaciations of Southern Germany

Markus Fiebig[1,*], Dietrich Ellwanger[2] and Gerhard Doppler[3]

[1]Institute of Applied Geology, Department of Structural Engineering+Natural Hazards, University of Natural Resources and Applied Life Sciences, Vienna, Peter Jordan-Str. 70, A 1190 Wien, Austria
[2]Landesamt für Geologie, Rohstoffe und Bergbau im Regierungspräsidium Freiburg, Freiburg i.Br., Germany
[3]Bayerisches Landesamt für Umwelt, Lazarettstr. 67, D-80636 München
*Correspondence and requests for materials should be addressed to Markus Fiebig. E-mail: markus.fiebig@boku.ac.at

14.1. INTRODUCTION

In 1899, Albrecht Penck presented an outline of the Alpine Quaternary stratigraphy which was based on a simple and convincing train of thought: he had found four periods of gravel accumulation separated from each other by valley incision. Because the gravel spreads were in contact with end moraines, Penck inferred that the gravels were meltwater deposits, laid down during periods of cold climate. However, the intervening phases of incision were supposed to relate to periods of warmer climate when glaciers were absent.

Between 1901 and 1909, Penck and Brückner presented an overwhelming quantity of field observations concerning their four gravel spreads. They gathered and discussed extensive evidence from the whole circum-Alpine area. Based on this, Penck and Brückner defined a relative chrono- and climate-stratigraphical system of four glacial units termed *Würm*, *Riss*, *Mindel* and *Günz*. This model was readily accepted by scientists and amateurs alike, for two reasons:

1. Penck and Brückner's definitions could be applied to interdisciplinary evidence since they were valid for many different types of investigations (pedo-, morpho-, litho-, bio-, chrono-, terrace-stratigraphical investigations).
2. Consequently, the model was well-suited for interdisciplinary correlations.

Subsequently, well-known and widely accepted expansions of the terminology of Penck and Brückner were published, for example, by Eberl (1930), Schaefer (1956) and Schreiner and Ebel (1981). They added the terms *Donau*, *Biber* and *Haslach* for three additional gravel accumulations, again separated from the others by periods of valley incision. The relative chronostratigraphy was, as suggested by Penck and Brückner, reflected in the alphabetical sequence of the initial letters of the terms, which are derived from the names of rivers in the northern Alpine Foreland (Fig. 14.1).

More recently, some authors began to focus primarily on sediment successions in the formerly glaciated area instead of the gravel accumulations. In these studies, the classical terminology became partially replaced by lithostratigraphical units. They comprise three generations of glacial basins that were incised by subglacial meltwaters and subsequently refilled by sub- to proglacial sediments, representing three lithostratigraphical units (e.g. Hasenweiler Formation, Illmensee Formation, etc.; Fig. 14.2). With regard to the classical chrono- or climate-stratigraphy, they cover the three glacial events following the MindelDeckenschotter unit (Ellwanger et al., 1995, 2003; Fiebig, 1995). Besides the two classical glaciations of Riss and Würm, a third glacial event had to be inserted that was called 'Hosskirch glaciation' (this term is not derived from the name of a river but from an upper Swabian village).

Obviously, some challenging revision is ongoing regarding the Pleistocene stratigraphy of the circum-Alpine region, as has also been required by Doppler and Jerz (1995). However, it still remains open to what result the revision will finally lead.

14.2. STRATIGRAPHICAL RESULTS AND CONCEPTS

Any parallel use of climate-stratigraphic and lithostratigraphic units may be difficult to reconcile since the two concepts are based upon different ratios of description and interpretation. The standard descriptive stratigraphical category is lithostratigraphy. Lithostratigraphical units are

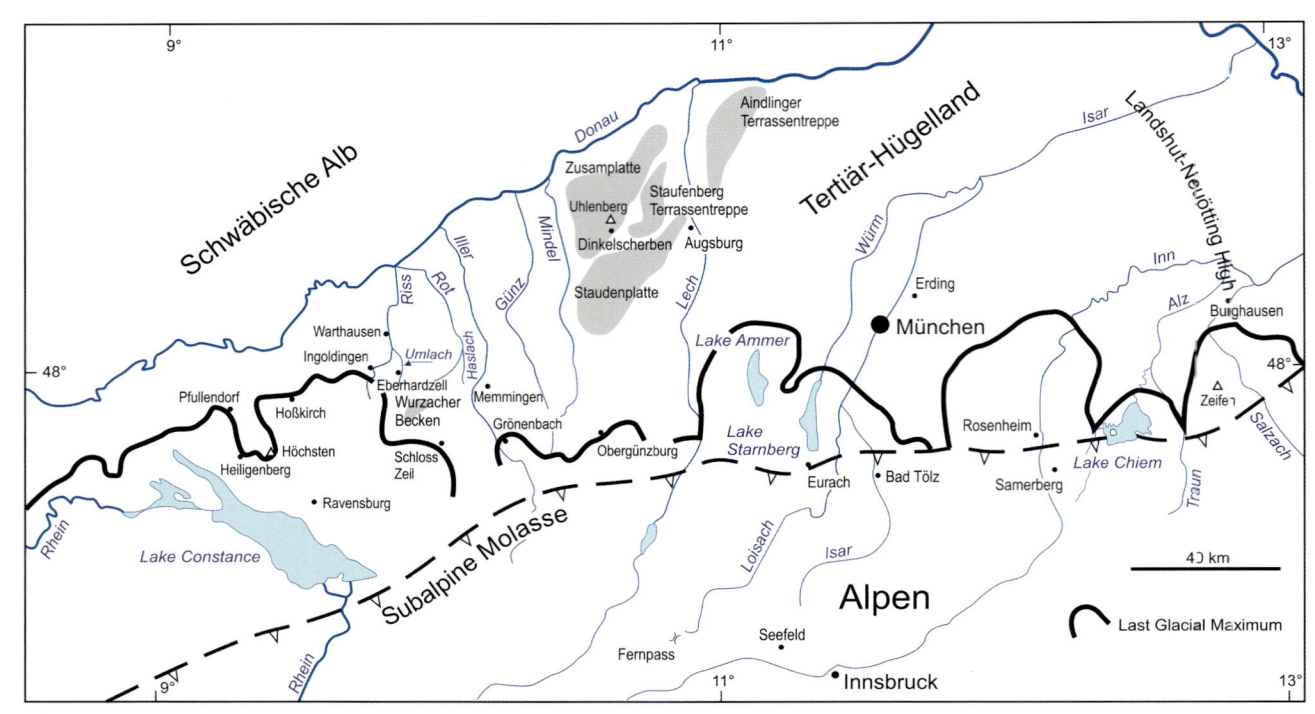

FIGURE 14.1 Overview of the German Alpine Foreland showing the dimensions of the Last Glacial Maximum and localities mentioned in the text.

'characterised on the base of observable lithological properties' of rocks (Salvador, 1994). In contrast, climate is not a directly observable property of a rock. Salvador's International Stratigraphic Guide contains no separate chapter on climatostratigraphy. Instead, he points out that 'climatic changes leave a conspicuous imprint on the geologic record in the form of glacial deposits, evaporites, red beds, coal deposits, palaeontological changes, and such. Since many climatic changes appear to have been regional or worldwide, their effects on the rocks provide valuable information for chronocorrelation' (p. 96). However, climate changes are not equivalent to rock properties, such as colour or lithofacies, but can only be deduced from the interpretation of the properties. Any resulting chronocorrelations of glacial deposits depend on the prior interpretation of the sediments and sedimentary sequences with which they are associated.

The basic units in lithostratigraphy are termed 'formation' (Fm). A formation should be 'mappable' on geological maps at a scale of 1:25 000. A formation or a group of formations may be bounded by unconformities (unconformity-bounded unit), in which some special attention is paid to boundaries and gaps.

Using these considerations, the different stratigraphical concepts of the states of Baden-Württemberg and Bavaria may be introduced. In Bavaria, the classical concept of Penck and Brückner (1901/1909), including numerous updates (e.g. Eberl, 1930; Schaefer, 1953; Schreiner and Ebel, 1981; Doppler and Jerz, 1995; Fiebig and Preusser, 2008), is still successfully applied. As a morphostratigraphical concept, it basically relies on terrace-stratigraphical evidence. This fits well with pedostratigraphical results, and its patterns are often easily transformable into a chronostratigraphical system.

The stratigraphical concept, as applied in Baden-Württemberg, is primarily based on the geometry and sediment infill of the three generations of glacial basins currently known in the Rhine glacier area, as opposed to the *Deckenschotter* units that are interpreted as tectonostratigraphical units (Ellwanger et al., 1995, 2003; Fiebig, 1995). From this evidence, two stratigraphical systems are derived:

- a chronostratigraphical system (Würm, Riss, Hosskirch, Deckenschotter, e.g. STD 2002) and
- a lithostratigraphical system using newly defined unconformity-bounded units (*Symbolschlüssel Geologie BW*).

Both systems can be correlated with one another (Table 14.1). The main advantage of the lithostratigraphical concept is its ability to describe sedimentary bodies as units (not as assemblages of parts of time units). It also enables the correlation of the erosional unconformities of the glacial basins with correlative high-energy deposits in the proximal Upper Rhine Graben (URG) that serve as the main sediment trap for alpine debris along the Rhine.

In addition, there is a chronostratigraphical discrepancy regarding the *Deckenschotter* in Baden-Württemberg and Bavaria. According to palaeomagnetic evidence, the Mindel unit is thought to belong to the Matuyama Chron in Baden-Württemberg but to the Brunhes Chron in Bavaria

Chapter | 14 Pleistocene Glaciations of Southern Germany

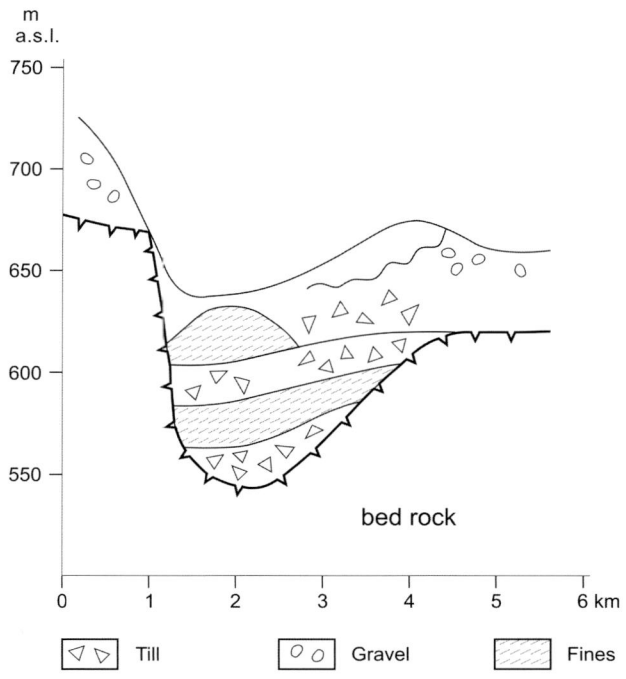

FIGURE 14.2 Section through a glacial basin in the Rhine glacier area (after Ellwanger et al., 1995). The strata between the tills and on top of the uppermost till contains palynological evidence for warm (temperate) climate (Holsteinian, Odderade, cf. Hahne, 2010). Thus the basin fill represents at least three glacial and three interglacial periods. This climatostratigraphy comes close to descriptive lithostratigraphy because the evidence for the glacial and interglacial units is found in a sediment trap in direct superposition. Walther's law of facies applies to stacks like these. In many other cases, Alpine climatostratigraphical interpretation is based on flights of terraces, where the proximity of glaciers is inferred for periods of gravel accumulation, and where evidence of interglacials is widely missing, caused by later erosion. The terrace-based climatostratigraphy is far more interpretative.

(Fromm, 1989; Rolf, 1992; Ellwanger et al., 1995; Strattner and Rolf, 1995; Bibus et al., 1996). According to the Bavarian concept, only the lower part of the Günz unit falls into the Matuyama Chron. This confusion might arise from a correlation error by Penck and Brückner (1901/1909), but it is not evident, where this discrepancy occurs.

The following description is based on the chronostratigraphical system because its terms are more commonly understood by an international audience. The term 'glaciation' will be used in a palaeogeographical sense, referring to glaciers that extended far into the northern Foreland of the Alps. The term 'interglacial' refers to climatic conditions comparable to those of the Holocene.

14.3. REGIONAL DESCRIPTION

The south German Alpine Foreland is subdivided into three large areas (Fig. 14.3):

1. The western area is dominated by the River Rhine and was shaped by former Rhine glacier advances (Ellwanger et al., 1995, 2003). During the Quaternary, the Alpine River Rhine evolved from a tributary of the Danube into a major European stream. This river transported central Alpine material into the URG and further to the North Sea. In the Bodensee (Lake Constance) area, the River Rhine and the Rhine glaciers have lowered the land surface partially by over 700 m during the Quaternary.

2. The central area between the rivers Riss and Lech is characterised by a post-orogenic rebound of the thickest and deeply buried pile of Tertiary sediments of the Northern Alpine Molasse Basin. Continuous uplift has resulted in a landscape consisting of numerous gravel accumulations which are preserved at different altitudes and form a kind of tectonostratigraphical sequence. Here, the classic model of four gravel accumulations (*Ältere Deckenschotter*, *Jüngere Deckenschotter*, *Hochterrasse* and *Niederterrasse*), related to four Alpine glaciations (*Günz*, *Mindel*, *Riss* and *Würm*), was established by Penck (1899) and Penck and Brückner (1901/1909). Later, additional gravel accumulations were identified on top of the highest Molasse ridges (*Älteste Deckenschotter*). *Ältere Deckenschotter* were differentiated into diverse levels and the newly separated accumulations interpreted as evidence of the earliest glaciations (*Donau*, *Biber*).

3. In the east, the Danubian character (superposition of deposits without progressive incision) of the landscape has been partially preserved throughout the Quaternary. Probably, according to tectonic conditions in front of the *Landshut-Neuötting Hoch* (tectonic uplift structure), the Munich gravel plain (*Münchner Schotter-Ebene*) and its southern roots were formed by superposition of several gravel accumulations without any further incisions (Jerz, 1993). Another type of superposition occurs in the Traun-Alz region, with a succession of glacial basin or channel infill followed by *Mindel* gravels and a *Riss*-aged till cover. In other places, this type of accumulation occurs only inside or near the former ice margins.

A first attempt to calculate the maximum deflection of the crust in front of the Inn glacier (Fiebig et al., 2004) during the Last Glacial Maximum (LGM), suggests movements of the whole proglacial area up to the Franconian Jura (Fiebig and Preusser, 2007). Thus glacio-isostasy might have strongly influenced the development of the circum-Alpine rivers, including the Danube, which is separated in the eastern Alpine Foreland from the formerly glaciated terrain by a wide belt of periglacial shaped uplands consisting of tertiary molassic sediments (*Tertiärhügelland*).

TABLE 14.1 Stratigraphical Systems of the German Alpine Foreland and Correlations with the International and Northern German Classifications

Age [ky]	International				Northern German climatic stages	Bavaria			Baden-Württemberg		
	Marine Isotope Stage	Magneto-strat.	Period	(Sub)Epoch		Climatostratigraphy		Terrace stratigraphy	Terrace stratigraphy	Lithostratigraphy	Stratigr. (climatic / morpho-tectonic*)
	1	BRUNHES	QUATERNARY	Holocene	Holozän	Holozän		Postglazialterrasse	Niederterrassenschotter	Hasenweiler-Formation	"Post-würm"
11,5				Upper Pleistocene	Weichsel	Jungpleistozän	Würm — Spät-/Hoch-/Ober. Würm / Mittler. Würm / Unter. Würm	SpätglazT. / Niederterrassenschotter / Übergangsterrassenschotter		Illmensee-Formation	Innen-wall- / Auß.-wall- / Saulgau-Würm — Würm
25	2										
	3–4						Frühwürm				
69											
	5a–5d										
117											
128	5e				Eem		Riß/Würm	--	---		Eem
	6–10			Middle Pleistocene	Saale	Mittelpleistozän	Riß	Hochterrassenschotter	Hochterrassenschotter	Dietmanns-Formation	Innen- / Auß.-wall- / Älteres Riß — Riß
	11				Holstein		Mindel/Riß	---			Holstein
	12–19				Elster		Mindel	Jüngere Deckenschotter		Steinental-Formation	Hoßkirch
780					Cromer	Altpleistozän	Günz	Tiefere Ältere Deckenschotter	Jüngere Deckenschotter		actually without term
		MATUYAMA			Bavel			↑Uhlenberg-Schieferkohle↓	Mittlere Deckenschotter	Jüngere – Deckenschotter - Formation	Mindel* / Haslach* / Haslach–Mindel*
		Jaramillo				Ältest- or Altpleistozän?					
20–103		MATUYAMA		Lower Pleistocene	Menap		Donau or Günz ?		Ältere Deckenschotter	Ältere-Deckenschotter-Formation	Günz*
					Waal						
		Olduvai			Eburon			Höhere Ältere Deckenschotter			
		MATUYAMA			Tegelen	Ältestpleistozän	Biber Donau	Älteste Deckenschotter / Älteste Periglazialschotter	Älteste Deckenschotter	Älteste-Deckenschotter-Formation	Donau* / Biber* / Biber-Donau*
2600					Prätegelen						
	104–	GAUSS	TERTIARY	Pliocene	Reuver	Pliozän					

Chapter | 14 Pleistocene Glaciations of Southern Germany

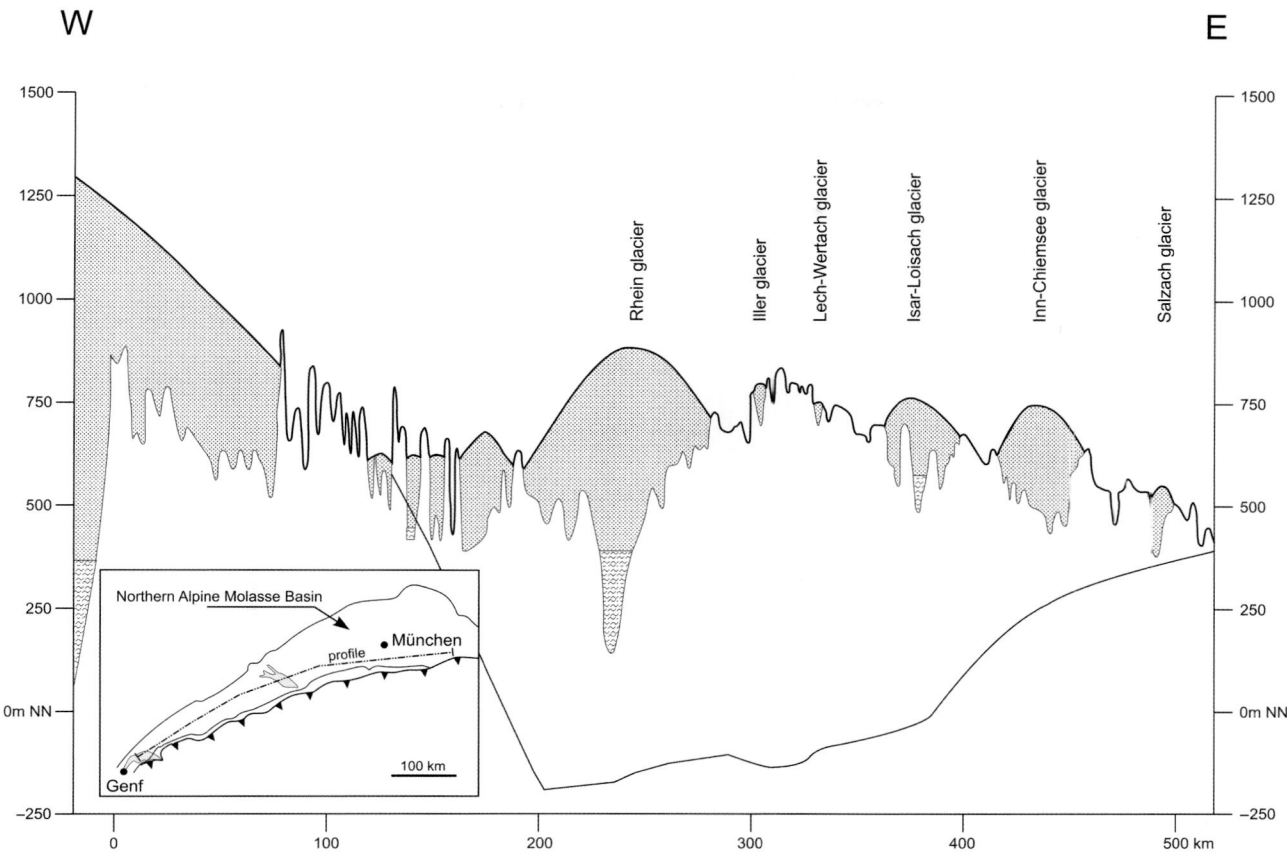

FIGURE 14.3 Cross-section of the northern Alpine Foreland with glaciers of the Last Glacial Maximum (LGM).

14.4. ARGUABLE EARLY PLEISTOCENE GLACIATIONS DERIVED FROM OLDEST GRAVEL SPREADS

In the central south German Alpine Foreland, Eberl (1930) identified ancient, highly elevated gravels in the Iller-Lech region (Fig. 14.3). He interpreted these gravels, widely spread in the *Zusamplatte* region, as deposits of a 'Donau Ice Age' dating back to the time before the *Ältere Deckenschotter* of Penck and Brückner (1901/1909). Schaefer (1953, 1956) finally added still older gravel spreads which were referred to an earlier Biber Ice Age. Its main gravel remnants are located at the *Staudenplatte*, including the *Staufenberg* flight of terraces. These additional gravel spreads have been accepted by most authors as products of cold events (e.g. Brunnacker, 1986). However, the glaciofluvial origin of these gravels is only implied; their correlative tills are still a matter of controversy. Despite of these problems, the terms *Biber* and *Donau* in Southern Germany are used in a chronostratigraphical way to describe the oldest part of Pleistocene.

The *Biber* comprises the time of deposition of *Staufenberg* gravel down to *Staudenplatte* gravels (*Älteste Deckenschotter*) including a terminating interglacial, which may be represented by the 'Bucher Schneckenmergel', an over-bank sediment with an interglacial mollusc fauna (Münzing and Ohmert, 1974). It is lying between gravels some kilometres east of river Iller, which are correlated with *Zusamplatte* gravels and assumed periglacial gravel below it, which is correlated with *Staudenplatte* gravels.

The *Donau* Age is thought to encompass the time since the end of *Biber* to the end of an interglacial period following the deposition of the *Zusamplatte* gravels, representing the upper, dolomite-bearing part of a two- or threefold accumulation south of Danube, between rivers Mindel and Lech. Aeolian and periglacial sediments covering the *Zusamplatte* gravels provide some evidence for stratigraphical correlations, but not without discrepancies.

Häuselmann et al. (2007) tried an absolute dating attempt using cosmogenic isotopes in the gravels of the so-called *Böhener Feld* which is a *Günz* type locality after Penck (1899), but nowadays is classified as *Donau*-aged. This gravel accumulation seems to be quite old (2.35 Ma) but this dating result displays a huge error ($\pm 1.08/0.88$ million years). Further dating attempts are inevitable.

14.5. STRATIGRAPHICAL IMPLICATIONS FROM SEDIMENTS OVERLYING THE *ZUSAMPLATTE* GRAVEL (UHLENBERG SITE AND OTHERS)

Very few sites to yield useful bio- and magnetostratigraphical information have been found in the Alpine foothills. Therefore, an isolated occurrence of dateable sediments on top of the Donau-aged *Zusamplatte* gravel (Deckenschotter) at the Uhlenberg site near Dinkelscherben (west of Augsburg) assumes a key position. The cover sediments start with a 2-m thick fining-upwards sequence containing Arvicola molars and molluscs biostratigraphically correlated with the Neogene Mammal Zone MN17, respectively, Tiglian (Ellwanger et al., 1994; Rähle, 1995). They are overlain by a peat horizon with an Alnus-Tsuga-Pinus pollen spectrum, correlating with Bavelian according to Bludau (1995). On top lies a 3-m thick sequence of solifluidal to alluvial loamy sediments of which some samples show reverse palaeomagnetic polarity (Strattner and Rolf, 1995). But only the nearby localities, Lauterbrunn some 13 km north of Uhlenberg and Roßhaupten east of river Mindel show an unambiguous reverse magnetisation of the lower parts of the loamy cover of the *Zusamplatte* gravels. This implies that the *Zusamplatte* gravel and correlating Deckenschotter deposits date back to the magnetic Matuyama epoch, probably even to the northern European Tiglian (i.e. Gelasian Stage). However, there is presently no solution for the discrepancy in the biostratigraphical classification caused by the palynological results. According to it, the Uhlenberg Schieferkohle (peat) should be at least 700 ka younger than the underlying gravel, but an adequate gap cannot be recognised in the sediments. Moreover it is not clear, whether the Uhlenberg Interglacial should be seen as the final warming period of the Donau time span.

14.6. *ÄLTERE DECKENSCHOTTER* AND POSSIBLY EQUIVALENT GLACIATION

The term *Ältere Deckenschotter* in Bavaria also includes the Donau-aged gravels (*Höhere Ältere Deckenschotter*) such as the *Zusamplatte* gravel, which is dealt with above. In Baden-Württemberg, the term is limited to gravel assigned to the Günz unit.

Until the late 1980s, it had been widely assumed that the Günz unit and connected *Ältere Deckenschotter* post-dated the Matuyama/Brunhes magnetic Chron boundary. Palaeomagnetic investigation of sites on the Höchsten and near Heiligenberg (north of Lake Constance; Rhine glacier) has revealed (Ellwanger et al., 1995; Bibus et al., 1996) that some of the samples from fine lenses within a succession of diamictons, sands and fines that lie beneath a lithostratigraphically identified *Jüngerer Deckenschotter* (of Mindel age) are reversely magnetised. In contrast to this, at some localities in the Traun-Alz region (Salzach glacier) glaciolacustrine fine sediments underlying a Mindel-aged Deckenschotter are normal magnetic orientated (Doppler and Jerz, 1995; Strattner and Rolf, 1995). Hence, the chronostratigraphical position of Günz-aged deposits remains unclear, most likely they do represent several accumulation events of considerably different age.

Eichler and Sinn (1974), Grimm et al. (1979) and Doppler (1980) also reported diamictons, respectively, moraine sediments overlain by Mindel-aged Deckenschotter from the northern and western margin of the Salzach glacier area, König (1979) from the northern margin of the Inn glacier region. These sediments are regarded to be correlatives of Günz-aged *Ältere Deckenschotter* so that one or several glaciation events during its accumulation have to be assumed.

14.7. *MITTLERE DECKENSCHOTTER* AND POSSIBLY EQUIVALENT GLACIATION

The unit of *Mittlere Deckenschotter* has only recently been introduced into the morphostratigraphical scheme of Baden-Württemberg. In the Württemberg Rottal (south of Ulm), Penck and Brückner's (1901/1909) *Jüngerer Deckenschotter* can be subdivided into a higher (i.e. older) Haslach gravel unit and a lower Tannheim gravel unit. They are distinguished using their clast composition. In the proximal area, the terrace-stratigraphical vertical distance between the two units amounts to 10 m. Schreiner and Ebel (1981) assigned the more crystalline-rich Haslach gravel unit to a distinctive *Mittlerer Deckenschotter* and concluded that contemporaneous glaciation had occurred during its accumulation based on their correlation with nearby till sheets. This correlation is not yet generally accepted (e.g. Ellwanger et al., 2003).

14.8. *JÜNGERE DECKENSCHOTTER* AND EQUIVALENT GLACIATION

Jüngerer Deckenschotter firstly was defined by Penck and Brückner (1901/1909) at the Grönenbacher Feld south of Memmingen, hypsographically lying between *Älterer Deckenschotter* and *Hochterrasse*. Correlative deposits are widespread in the northern Alpine Foreland of Germany. The gravels often are conglomerated and sometimes, if not covered by a till, show a very special weathering: circular pipes filled with soil material or washed out. They are reaching down for several metres into the conglomerates and are called *Geologische Orgeln* (Geological Organ Pipes). Gravels of the Grönenbacher Feld have been dated using cosmogenic isotopes (Häuselmann et al., 2007). Penck's type locality for the Mindel glaciation seems to date from ca. 680.0 ± 230.0/240.0 ka. Further age control of these deposits is again essential.

The contact of the *Jüngere Deckenschotter* with moraines north of Obergünzburg caused Penck to propose that a glaciation had occurred during the gravel accumulation (his Mindel glaciation). The terminal moraines south of Memmingen are perfectly formed. However, they do not continue to the west (Sinn, 1972), but can be followed eastwards, as far as Obergünzburg (Penck and Brückner, 1901/1909; Eberl, 1930; Stepp, 1981; Habbe, 1986).

Even further to the east, morphologically correlated landforms have also been identified in the northern parts of Inn glacier and the western and north-western parts of the Salzach glacier region (Eichler and Sinn, 1974; Grimm et al., 1979; König, 1979). In the eastern Rhine glacier area, equivalent glacial landforms are pedologically correlated; they show a weathering depth of 5–8 m. Here, the correlation of the glacial landforms and the Deckenschotter remains doubtful. The age control on the glacial sediments relies primarily on the sediment succession at Unterpfauzenwald (west of Memmingen) that has been attributed on the basis of palynological investigations by different authors at least to the Cromerian (Bibus et al., 1996) or even to the Bavelian stages (Hahne, 2010).

14.9. INCISION BETWEEN JÜNGERE DECKENSCHOTTER AND HOCHTERRASSE, 'THIRD FROM LAST' GLACIATION AND THE SAMERBERG SITE

The incision period between the deposition of the Jüngere Deckenschotter and the Hochterrasse has traditionally been referred to as a 'Great Interglacial'. Today, it is thought that this period actually represents a series of two (or even three) interglacials. In the research borehole Samerberg 2 in southern Bavaria, near the river Inn south-east of Rosenheim, a 'double interglacial' has been discovered beneath a thin 'penultimate glacial' diamicton. On the basis of palynological investigations, Grüger (1983) correlated the interglacials with the Northern German Holsteinian and Wacken/Dömitzian events. Ellwanger et al. (1995) present evidence in the infill of the Hoßkirch basin (Baden-Württemberg) that at least one additional glaciation (a 'third from last' or pre-penultimate glaciation) occurred which post-dates the deposition of the Jüngere Deckenschotter and pre-dates the penultimate glaciation (Riss of the classical Alpine stratigraphy). No correlative basin sediments until now have been identified elsewhere in the Bavarian Alpine Foreland.

The outermost terminal moraine zone in the Southern German Alpine Foreland is interpreted in many regions as a product of a 'third from last' glaciation. In the Rhine glacier area, the period of this pre-penultimate glaciation is called Hoßkirch (Ellwanger in Habbe, 2007) and thought to post-date the Mindel-aged Jüngere Deckenschotter. In contrast in Bavaria, comparable moraines are correlated with Jüngere Deckenschotter and therefore classified to Mindel.

14.10. THE BELT OF MORAINIC AMPHITHEATRES CORRELATED WITH THE PENULTIMATE GLACIATION

Within the Rhine glacier region, the deposits correlated to the penultimate glaciation—that is, Riss *sensu* Penck and Brückner (1901/1909)—are subdivided into three units. The oldest unit is marked pedologically by a greater degree of weathering (Schreiner and Haag, 1982; Schreiner, 1989, 1992). This unit is a diamicton, confined palaeogeographically to the tongue basins in the Riss and neighbouring valleys. The diamicton underlies outwash gravels (Hochterrasse) which are connected to a very distinct double rampart terminal moraine. The extent of the assumed glacial retreat between deposition of the diamicton bed and accumulation of the outwash gravels is unknown.

In the eastern region of the Rhine glacier, the double rampart terminal moraines (*Doppelwall-Riss*) of Schreiner (1989) and Ellwanger (1990) are generally gravelly push moraines 10–30 m in height, formed at a distance of up to 3 km from each other. They are interpreted as the morainic amphitheatres of the penultimate glaciation (see digital map).

The occurrence of additional gravel fields in the Riss and Umlach valleys, as far north as Warthausen, at a lower altitude than the Hochterrasse described above but higher than the Lower Terrace, suggested that a third stratigraphical unit might belong to the penultimate glaciation (Penck and Brückner, 1901/1909; Schreiner, 1989). Possible correlative terminal moraines north of Ingoldingen and south of Eberhardzell are hardly recognisable in the field; therefore, the existence of this third penultimate glacial event remains doubtful.

Older moraines of assumed Riss-age far in front of the terminal moraines of the last glaciation and Hochterrasse gravels following present-day valleys are widespread throughout the Bavarian Alpine Foreland. In different regions, that is around Memmingen, the valley of the Danube or the western Salzach glacier region, Hochterrasse accumulations at two different elevations are found as in the Rhine glacier area. Variations of the weathering profiles of the gravels and overlying solifluidal and aeolian fines give evidence for a warmer, maybe interglacial period between the depositions of these two accumulations (Bibus and Kösel, 1987; Miara, 1995).

14.11. EEMIAN STAGE INTERGLACIAL

In Southern Germany, Beug (1972) investigated the Riss/Würm or Eemian-age interglacial sequences from sites at Zeifen close to the Salzach River, and Eurach, near Lake Starnberg (Beug, 1979). The best-investigated pollen section of the Alpine region comprising the Last Interglacial and the three subsequent interstadials, however, is that at Samerberg in Bavaria, south-east of Rosenheim (Grüger, 1979). Here, the vegetational succession corresponds largely to that of the Northern European Eemian Stage. The Alpine character of the site is expressed by high values of pine pollen. Another locality showing a sequence comparable to the Samerberg has been identified in the Wurzacher Becken, a tongue basin of the penultimate glaciation in the Rhine glacier area (Grüger and Schreiner, 1993). A further though incomplete Eemian sequence at Krumbach near Saulgau contains high-resolution sections that include seasonal patterns (Frenzel and Bludau, 1987) and Füramoos (Müller, 2001).

14.12. LAST GLACIATION (WÜRMIAN)

Cores from the Samerberg and Mondsee (Austria) sites show an undisrupted sequence of lake sediments between undisputed deposits dating from the Last Interglacial and the LGM till (~23.5 ka BP). No evidence is found here of a glaciation intervening between the Eemian and the LGM. The main Würmian ice advance accordingly occurred in the late Upper Pleistocene, similar to that in Northern Europe. Traub and Jerz (1976) obtained an age of 21.65 ± 0.25 ^{14}C ka BP (ca. 25.0 ka) from pleniglacial loess molluscs from the margin of the Salzach glacier. Starnberger et al. (2008) yielded ages between 19.9 ± 2.3 and 21.5 ka by OSL dating; the loess strata, mentioned above, had been deposited immediately before the deposition of the gravels of the Niederterrasse.

The outermost end moraines (*Äußere Jungendmoräne*) do not represent the maximum glacial limit. At various sites (e.g. Ingoldingen, Pfullendorf) diamicton strata are found intercalated into the outwash gravels of the Lower Terrace in front of the terminal moraine zone (see digital map). How far, if at all, the glaciers retreated between deposition of the diamicton and deposition of the Lower Terrace gravels is also unknown. This is similar to the situation found for the penultimate glaciation (see above).

After the advance of the ice to the terminal moraines of the LGM, at least one pronounced readvance is assumed, the *Innere Jungendmoräne* (see digital map). The significance of numerous minor readvance phases (e.g. Rothpletz, 1917; Troll, 1924; Knauer, 1929, 1931; Schreiner, 1974; Grimm et al., 1979; Ellwanger, 1980; Habbe, 1986), assigned to the Würm glaciation, is still a matter of discussion.

During the LGM, as at least during the penultimate glaciation before, transfluences developed at various points of the ice-stream network. In the Inn valley, around Landeck, for instance, five major tributary glaciers grew together to form one major valley glacier that advanced eastwards down the Inn valley. At the same time, around Innsbruck a major tributary glacier from the south had already blocked the Inn valley. Congestion resulted in a rise of the ice surface until finally the threshold to the north was crossed, and ice from the Inn valley flowed over the Fernpass and the Seefelder Senke directly northwards into the Isar and Loisach drainage systems (van Husen, 2000). Such events are reflected in the glacial deposits by far-travelled gravel and major erratics. In the foreland of the Isar-Loisach glacier, Alpine crystalline rocks constitute up to 35% of the total 20–31.5 mm fraction. Further to the east, in the Tölz glacier area, the contents range from 0% to 1% because no ice was supplied from the Inn valley (Dreesbach, 1985).

The Alpine 'Late Glacial' (of the Würm glaciation) comprises the period of recession from the '*Innere Jungendmoräne*' until the end of the Younger Dryas Stadial. The basic subdivision of the Late-glacial ice retreat originates from Penck and Brückner (1901/1909) who distinguished three retreat phases: Bühl, Gschnitz and Daun: Of these, the Bühl is thought to represent the last readvance of the still intact ice-stream network before 17.0 ka (Patzelt, 2002). However, it has been completely abandoned by Reitner (2007) who defined the 'phase of early Late-glacial ice decay'. Gschnitz is an inner-Alpine local readvance between 16.7 and 16.0 ka. The Late-Glacial Daun readvance took place before the Egesen readvance which represents the Younger Dryas Stadial between 12.6 and 11.5 ka (Ivy-Ochs et al., 2008).

Connected to the Würm-aged terminal moraines, the Niederterrasse gravels and their Late-glacial analogues are the widespread in the valleys of the Alpine foothills. Here, they show the flat surface of braided river deposits and sometimes still periglacial features like ice wedges. Conglomerates are rare and a loessic cover is generally missing.

14.13. GLACIATIONS OF LOW MOUNTAIN RANGES

Like the Vosges in neighbouring France (Mercier and Jeser, 2004), on the other side of the URG, the Schwarzwald (Black Forest) was glaciated during the Pleistocene. Here, the first evidence of former glaciation was found by Schimper (1837) in the Titisee region. Later, Steinmann (1892) described moraines from the lower Wehra valley on the southern margin of the Schwarzwald. During the last glaciation, the Schwarzwald was covered by a small ice cap, comparable to those in present-day Norway (e.g.

Jostedalsbreen, Hardangerjökull). Traces of an older glaciation have also been identified in a number of places. Those older moraines are summarily attributed to the penultimate glaciation. It had been assumed that, during the penultimate glaciation, both Alpine and Schwarzwald glaciers might have met (Hantke, 1978). However, according to more recent investigations (Leser, 1979; Wendebourg and Ramshorn, 1987), it is quite clear that this did not happen.

In the Bayerischer Wald (Bavarian Forest), traces of former glaciation have been identified in most areas above 1300 m a.s.l., whilst end moraines of the last glaciation reach down to an altitude of 900 m a.s.l. Traces of older glaciations are also present (Bauberger, 1977; Jerz, 1993). This view has lately been updated by new discoveries by Raab (1999).

The Riesengebirge (Giant Mountains) further to the east was much less glaciated, as a consequence of the more continental climate. Some 15 former glaciers have been identified, the longest of which extended 5.3 km (Schwarzbach, 1942).

The remaining mountain ranges in Germany, like the Thüringer Wald and the Rhön, were probably not glaciated. It seems doubtful whether the Fichtelgebirge (Liedtke, 1981; Jerz, 1993) or even the Schwäbische Alb (Hantke, 1974) had been glaciated.

ACKNOWLEDGEMENTS

This revised, updated and improved chapter would again not have been written without the permanent and tireless encouragement, assistance and advice by Dr. Jürgen Ehlers and Prof. Dr. Philip L. Gibbard, who also corrected and thoroughly improved the chapter (as with the 2004 text; Fiebig et al., 2004). We are very grateful to Helene Pfalz-Schwingenschlögl (Institute of Applied Geology, University of Natural Resources and Applied Life Sciences, Vienna) for help with the drawings and chapter handling.

REFERENCES

Bauberger, W., 1977. Geologische Karte von Bayern 1:25.000, Erläuterungen zum Blatt Nr. 7046 Spiegelau und zum Blatt Nr. 7047 Finsterau (Nationalpark Bayerischer Wald). München, Bayerisches Geologisches Landesamt, 183pp.

Beug, H.J., 1972. Das Riß/Würm-Interglazial von Zeifen, Landkreis Laufen a.d. Salzach. Bayerische Akademie der Wissenschaften, Mathematisch-Naturwissenschaftliche Klasse N.F. 151, 46–75.

Beug, H.J., 1979. Vegetationsgeschichtlich-pollenanalytische Untersuchungen am Riß/Würm-Interglazial von Eurach am Starnberger See/Obb. Geol. Bavarica 80, 91–106.

Bibus, E., Kösel, M., 1987. Paläoböden und periglaziale Deckschichten im Rheingletschergebiet von Oberschwaben und ihre Bedeutung für Stratigraphie, Reliefentwicklung und Standort. Tübinger Geowissenschaftliche Arbeiten D3, 89pp.

Bibus, E., Bludau, W., Ellwanger, D., Fromm, K., Kösel, M., Schreiner, A., 1996. On Pre-Würm glacial and interglacial deposits of The Rhine Glacier (South German Alpine Foreland, Upper Swabia, Baden-Württemberg). In: Turner, C. (Ed.), The Early Middle Pleistocene in Europe. Balkema, Rotterdam, pp. 195–204.

Bludau, W., 1995. Altpleistozäne Warmzeiten im Alpenvorland und im Oberrheingraben.-Ein Beitrag der Palynologie zum Uhlenberg-Problem. Geol. Bavarica 99, 119–134.

Brunnacker, K., 1986. Quaternary stratigraphy in the Lower Rhine Area and Northern Alpine Foothills. Quatern. Sci. Rev. 5, 371–379.

Doppler, G., 1980. Das Quartär im Raum Trostberg an der Alz im Vergleich mit dem nordwestlichen Altmoränengebiet des Salzachvorlandgletschers (Südostbayern). Dissertation, München, 198pp.

Doppler, G., Jerz, H., 1995. Untersuchungen im Alt- und Ältestpleistozän des bayerischen Alpenvorlands—geologische Grundlagen und stratigraphische Ergebnisse. Geol. Bavarica 99, 7–53.

Dreesbach, R., 1985. Sedimentpetrographische Untersuchungen zur Stratigraphie des Würmglazials im Bereich des Isar-Loisachgletschers. Dissertation, München, 176pp.

Eberl, B., 1930. Die Eiszeitenfolge im nördlichen Alpenvorlande—Ihr Ablauf, ihre Chronologie auf Grund der Aufnahmen des Lech- und Illergletschers. Benno Filser, Augsburg, 427pp.

Eichler, H., Sinn, P., 1974. Zur Gliederung der Altmoränen im westlichen Salzachgletschergebiet. Zeitsch. Geomorphol. N.F. 18, 132–158.

Ellwanger, D., 1980. Rückzugsphasen des würmzeitlichen Illergletschers. Arbeiten aus dem Institut für Geologie und Paläontologie der Universität Stuttgart N.F. 76, 93–126.

Ellwanger, D., 1990. Zur Riß-Stratigraphie im Andelsbach-Gebiet (Baden-Württemberg). Jahreshefte des Geologischen Landesamtes Baden-Württemberg 32, 235–245.

Ellwanger, D., Fejfar, O., von Koenigswald, W., 1994. Die biostratigraphische Aussage der Arvicolidenfauna vom Uhlenberg bei Dinkelscherben und ihre morpho- und lithostratigraphischen Konsequenzen. Münchner Geowissenschaftliche Abhandlungen A26, 173–191.

Ellwanger, D., Bibus, E., Bludau, W., Kösel, M., Merkt, J., 1995. Baden-Württemberg. In: Benda, L. (Ed.), Das Quartär Deutschlands, Borntraeger, Berlin, Stuttgart, 255–295.

Ellwanger, D., Lämmermann-Barthel, J., Neeb, I., 2003. Eine landschaftsübergreifende Lockergesteinsgliederung vom Alpenrhein zum Oberrhein. In: Schirmer, W. (Ed.), Landschaftsgeschichte im Europäischen Rheinland. GeoArchaeoRhein 4, 81–124.

Fiebig, M., 1995. Pleistozäne Ablagerungen im süddeutschen und im neuseeländischen Alpenvorland—ein Vergleich. Geologisches Institut, Albert Ludwigs Universität, 122 Seiten, Freiburg.

Fiebig, M., Preusser, F., 2007. Natural changes of river landscapes in the Bavarian Alpine Foreland (Germany). In: Accademia Nazionale dei Lincei, Roma, 3rd International TOPO-EUROPE Workshop, Roma, 3rd International TOPO-EUROPE Workshop, 2.-5.5.2007, Rom.

Fiebig, M., Preusser, F., 2008. Pleistozäne Vergletscherungen des nördlichen Alpenvorlandes. Geogr. Helvetica 3, 145–150.

Fiebig, M., Buiter, S., Ellwanger, D., 2004. Pleistocene glaciations of South Germany. In: Ehlers, J., Gibbard, P.L. (Eds.), Quaternary Glaciations—Extent and Chronology. Part I. Elsevier, Amsterdam, pp. 147–154.

Frenzel, B., Bludau, W., 1987. On the Duration of the Interglacial to Glacial Transition at the End of the Eemian Interglacial (Deep Sea Stage 5): Botanical and sedimentological evidence. In: Berger, W.H., Labeyrie, L.D. (Eds.), Abrupt Climate Change. D. Reidel Publishing Company, Dordrecht, pp. 151–162.

Fromm, K., 1989. Paläomagnetische Altersbestimmungen an eiszeitlichen Quartärablagerungen bei Heiligenberg (Oberschwaben).

Niedersächsisches Landesamt für Bodenforschung—Geowissenschaftliche Gemeinschaftsaufgaben, Bericht (Archiv-Nr. 105 863).

Grimm, W.-D., Bläsig, H., Doppler, G., Fakhrai, M., Goroncek, K., Hintermaier, G., et al., 1979. Quartärgeologische Untersuchungen im Nordwestteil des Salzach-Vorlandgletschers (Oberbayern). In: Schlüchter, C. (Ed.), Moraines and Varves—Origin, Genesis, Classification. Balkema, Rotterdam, pp. 101–119.

Grüger, E., 1979. Spätriß, Riß/Würm und Frühwürm am Samerberg in Oberbayern—ein vegetationsgeschichtlicher Beitrag zur Gliederung des Jungpleistozäns. Geol. Bavarica 80, 5–64.

Grüger, E., 1983. Untersuchungen zur Gliederung und Vegetationsgeschichte des Mittelpleistozäns am Samerberg in Oberbayern. Geol. Bavarica 84, 21–40.

Grüger, E., Schreiner, A., 1993. Riß/Würm- und würmzeitliche Ablagerungen im Wurzacher Becken (Rheingletschergebiet). Neues Jahrbuch für Geologie und Paläontologie Abhandlungen 189, 81–117.

Habbe, K.A., 1986. Bemerkungen zum Altpleistozän des Illergletscher-Gebietes. Eiszeitalter und Gegenwart 36, 121–134.

Habbe, K.A., 2007. Stratigraphische Begriffe für das Quartär des süddeutschen Alpenvorlandes. Eiszeitalter und Gegenwart 56, 66–83.

Hahne, J., 2010. Kommentar zu Pollendiagrammen von W. Bludau. Landesamt für Geologie, Rohstoffe und Bergbau im Regierungspräsidium Freiburg, Bericht (intern) Az. 4762//10_2414.

Hantke, R., 1974. Zur Vergletscherung der Schwäbischen Alb. Eiszeitalter und Gegenwart 25, 214.

Hantke, R., 1978. Eiszeitalter. Die jüngste Erdgeschichte der Schweiz und ihrer Nachbargebiete, Band 1. Ott, Thun, 468pp.

Häuselmann, P., Fiebig, M., Kubik, P.W., Adrian, H., 2007. A first attempt to date the original "Deckenschotter" of Penck and Brückner with cosmogenic nuclides. Quatern. Int. 164–165, 33–42.

Ivy-Ochs, S., Kerschner, H., Reuther, A., Preusser, F., Heine, K., Maisch, M., et al., 2008. Chronology of the last glacial cycle in the European Alps. J. Quatern. Sci. 23 (6–7), 559–573.

Jerz, H., 1993. Das Eiszeitalter in Bayern—Erdgeschichte, Gesteine, Wasser, Boden (Geologie von Bayern, Bd. 2). Schweizerbart, Stuttgart, 243pp.

Knauer, J., 1929. Erläuterungen zur Geognostischen Karte von Bayern 1:100 000. Blatt München West (Nr. XXVII), Teilblatt Landsberg, 47pp.

Knauer, J., 1931. Erläuterungen zur Geognostischen Karte von Bayern 1:100 000. Blatt München West (Nr. XXVII), Teilblatt München-Starnberg, 48pp.

König, W., 1979. Neuere Forschungsansätze aus dem Gebiet der präwürmzeitlichen Moränen und Schotter des pleistozänen Innvorlandgletschers. Heidelberger Geographische Arbeiten 49, 35–36.

Leser, H., 1979. Erläuterungen zur Geomorphologischen Karte der Bundesrepublik Deutschland 1:25.000. GMK 25 Blatt 4, 8313 Wehr, 60pp.

Liedtke, H., 1981. Die nordischen Vereisungen in Mitteleuropa. Erläuterungen zu einer farbigen Übersichtskarte im Maßstab 1: 1 000 000. Forschungen zur deutschen Landeskunde 204, second ed., Trier, 307pp.

Mercier, J.-L., Jeser, N., 2004. The glacial history of the Vosges Mountains. In: Ehlers, J., Gibbard, P.L. (Eds.), Quaternary Glaciations—Extent and Chronology. Part I. Elsevier, Amsterdam, pp. 113–118.

Miara, S., 1995. Gliederung der rißzeitlichen Schotter und ihrer Deckschichten beiderseits der unteren Iller nördlich der Würmendmoränen. Münchner Geographische Abhandlungen 22, 185 S, 33 Abb., 18 Tab., München.

Müller, U., 2001. Die Vegetations- und Klimaentwicklung im jüngeren Quartär anhand ausgewählter Profile aus dem südwestdeutschen Alpenvorland. Tübinger Geowissenschaftliche Arbeiten D7, Tübingen.

Münzing, K., Ohmert, W., 1974. Mollusken aus dem älteren Pleistozän Schwabens. Jahrbuch des Geologischen Landesamts Baden-Württemberg 16, 61–78.

Patzelt, G., 2002. Unterlagen für die Exkursionen der 44. Tagung der Hugo Obermaier-Gesellschaft in Innsbruck. Universität Innsbruck, 21 pp.

Penck, A., 1899. Die vierte Eiszeit im Bereich der Alpen. Schriften der Vereinigung zur Verbreitung naturwissenschaftlicher Kenntnisse 39, 1–20.

Penck, A., Brückner, E., 1901. Die Alpen im Eiszeitalter, 3 Bände. Leipzig, 1199pp.

Raab, T., 1999. Würmzeitliche Vergletscherung des Bayerischen Waldes im Arbergebiet. Regensburger Geographische Schriften 32, 327pp.

Rähle, W., 1995. Altpleistozäne Molluskenfaunen aus den Zusamplattenschottern und ihrer Flussmergeldecke vom Uhlenberg und Lauterbrunn (Iller-Lech-Platte, Bayerisch Schwaben). Geol. Bavarica 99, 103–118.

Reitner, J.M., 2007. Glacial dynamics at the beginning of Termination I in the Eastern Alps and their stratigraphic implications. Quatern. Int. 164–165, 64–84.

Rolf, C., 1992. Statistische Untersuchungen zur Zuverlässigkeit magnetischer Untersuchungen mit dem Spinmagnetometer an Sedimentproben aus Oberschwaben. Niedersächsisches Landesamt für Bodenforschung. Geowissenschaftliche Gemeinschaftsaufgaben, Bericht (Archiv-Nr. 110 230).

Rothpletz, A., 1917. Die Osterseen und der Isar-Vorlandgletscher. Mitt. Geogr. G. München 12, 99–314.

Salvador, A., 1994. International stratigraphic guide. The Geological Society of America, Boulder, 214pp.

Schaefer, I., 1953. Die donaueiszeitlichen Ablagerungen an Lech und Wertach. Geol. Bavarica 19, 13–64.

Schaefer, I., 1956. Sur la division du Quaternaire dans l'avant pays des Alpes en Allemagne. Actes IV. Congress International du Quaternaire (INQUA), Rome/Pise, 1953, vol. 2, pp. 910–914.

Schimper, K., 1837. Über die Eiszeit. Brief an Prof. Agassiz. Actes Societ. Helvetique des Sciences Naturelles 22e Session, Neuchatel, pp. 38–51.

Schreiner, A., 1974. Erläuterungen zur Geologischen Karte des Landkreises Konstanz und Umgebung 1:50.000. 2., berichtigte Auflage. Geologisches Landesamt Baden-Württemberg, Stuttgart, 286pp.

Schreiner, A., 1989. Zur Stratigraphie der Rißeiszeit im östlichen Rheingletschergebiet (Baden-Württemberg). Jahreshefte des Geologischen Landesamtes Baden-Württemberg 31, 183–196.

Schreiner, A., 1992. Einführung in die Quartärgeologie. Schweizerbart, Stuttgart, 257pp.

Schreiner, A., Ebel, R., 1981. Quartärgeologische Untersuchungen in der Umgebung von Interglazial-Vorkommen im östlichen Rheingletschergebiet (Baden-Württemberg). Geol. Jahrbuch A59, 3–64.

Schreiner, A., Haag, Th., 1982. Zur Gliederung der Rißeiszeit im östlichen Rheingletschergebiet (Baden-Württemberg). Eiszeitalter und Gegenwart 32, 137–161.

Schwarzbach, M., 1942. Bionomie, Klima und Sedimentationsgeschwindigkeit im oberschlesische Karbon. Zeitschrift der deutschen geologischen Gesellschaft, 94.

Sinn, P., 1972. Zur Stratigraphie und Paläogeographie des Präwürm im mittleren und südlichen Illergletscher-Vorland. Heidelberger Geographische Arbeiten 37, 159pp.

Starnberger, R., Terhorst, B., Rähle, W., Peticka, R., Haas, J.N., 2008. Palaeoecology of Quaternary periglacial environments during OIS-2 in the forefields of the Salzach Glacier (Upper Austria). Quatern. Int. 198, 51–61.

Steinmann, G., 1892. Mitteilungen, Über Moränen am Ausgang des Wehratales. Bericht der 25. Versammlung des Oberrheinischen Geologischen Vereins zu Basel, pp. 35–39.

Stepp, R., 1981. Das Böhener Feld—Ein Beitrag zum Altquartär im Südwesten der Iller-Lech-Platte. Mitt. Geogr. G. München 66, 43–68.

Strattner, M., Rolf, C., 1995. Magnetostratigraphische Untersuchungen an pleistozänen Deckschicht-Profilen im bayerischen Alpenvorland. Geol. Bavarica 99, 55–102.

Traub, F., Jerz, H., 1976. Ein Lößprofil von Duttendorf (Oberösterreich) gegenüber Burghausen an der Salzach. Zeitschrift für Gletscherkunde und Glazialgeologie 11, 175–193.

Troll, C., 1924. Der diluviale Inn-Chiemseegletscher—Das geographische Bild eines typischen Alpenvorlandgletschers. Forschungen zur deutschen Landes- und Volkskunde 23, 1–121.

van Husen, D., 2000. Geological processes during the Quaternary. Mitteilungen der Österreichischen Geologischen Gesellschaft 92, 135–156.

Wendebourg, J., Ramshorn, C., 1987. Der Verzahnungsbereich alpiner und Südostschwarzwälder Rißvereisung (Baden-Württemberg). Jahreshefte des Geologischen Landesamtes Baden-Württemberg 29, 255–268.

Chapter 15

Glaciation in Greece: A New Record of Cold Stage Environments in the Mediterranean

Jamie C. Woodward* and Philip D. Hughes

Quaternary Environments and Geoarchaeology Research Group, Geography, School of Environment and Development, The University of Manchester, Manchester, United Kingdom

*Correspondence and requests for materials should be addressed to Jamie C. Woodward. E-mail: jamie.woodward@manchester.ac.uk

15.1. INTRODUCTION

In his seminal review of glaciation in the Mediterranean region, Messerli (1967) reported evidence for Pleistocene glacial activity in several areas of upland Greece and assigned all the glacial phenomena to the last (Würmian) cold stage. Messerli's overview brought together some of the earliest work on glaciation in Greece including Niculescu's pioneering (1915) study of Mount Smolikas in northwest Greece and Sestini's (1933) observations on the mountains of southern Epirus. Figure 15.1 shows how the Pindus Mountains, aligned roughly NNW–SSE, extend along the length of mainland Greece and dominate its topography. Much of this mountain chain lies over 1000 m a.s.l., and many peaks exceed 2000 m. Evidence for glacial activity has been identified throughout the Pindus Mountains from Mount Grammos (2523 m) on the Albanian border to Mount Taygetos (2407 m) in the southern Peloponnese. The principal sites in Greece included on Messerli's (1967) map for which there is additional published information on Pleistocene glaciation are shown in Fig. 15.1. Table 15.1 provides summary data for each of the sites shown in Fig. 15.1 and lists the key references in each case.

In general terms, during Pleistocene cold stages, the glaciers of Greece decreased in size with decreasing latitude, although climatic and topographic controls could be locally important. Well-preserved glaciated terrains of Pleistocene age have been observed in the highest parts of the northern Peloponnese (Mastronuzzi et al., 1994), and Fig. 15.2 shows a range of glacial landforms on the high limestone peaks of Mount Chelmos (2341 m) which lies just south of 38°N. The evidence for glacial activity on the mountains of Crete is less clear cut, and further research is needed. Nemec and Postma (1993) have argued that the White Mountains were glaciated during the Pleistocene, and on Mount Idi (also known as Ida or Psiloritis at 2456 m), Fabre and Maire (1983) have reported a cirque with moraines at an altitude of just ca. 1945 m. This claim is out of step with the evidence from mainland Greece. Boenzi and Palmentola (1997), for example, have argued that evidence for glaciation is only present in Greece on mountains that exceed 2200 m. Several publications on the White Mountains of Crete (e.g. Poser, 1957; Bonnefont, 1972; Boenzi et al., 1982) do not report any evidence for the action of glaciers. At some of the Greek sites discussed in Messerli's (1967) review, the evidence for glacial activity is rather limited in areal extent—being confined to the highest peaks—and modern investigations have not yet taken place (Table 15.1). In some of these cases, cirques, ice-steepened cliffs and ice-scoured bedrock surfaces account for most of the evidence for the action of former glaciers, with only very limited evidence for the transport and deposition of glacial sediments to lower elevations. Thus, in some of the localities shown on Messerli's (1967) map and Fig. 15.1, glacial deposits have not been marked on the later 1:50,000 geological sheets produced by the Greek Institute for Geological and Mineral Exploration (IGME).

Recent reviews of glaciation in the mountains of the Mediterranean (Hughes et al., 2006a, Hughes and Woodward, 2009) have shown how the history of research can be broadly classified into three stages of development (either Pioneer, Mapping or Advanced stages) as defined in Table 15.1. The published literature on Pleistocene glaciation in Greece shows that all three of these stages are represented (Table 15.1). By far, the most detailed work

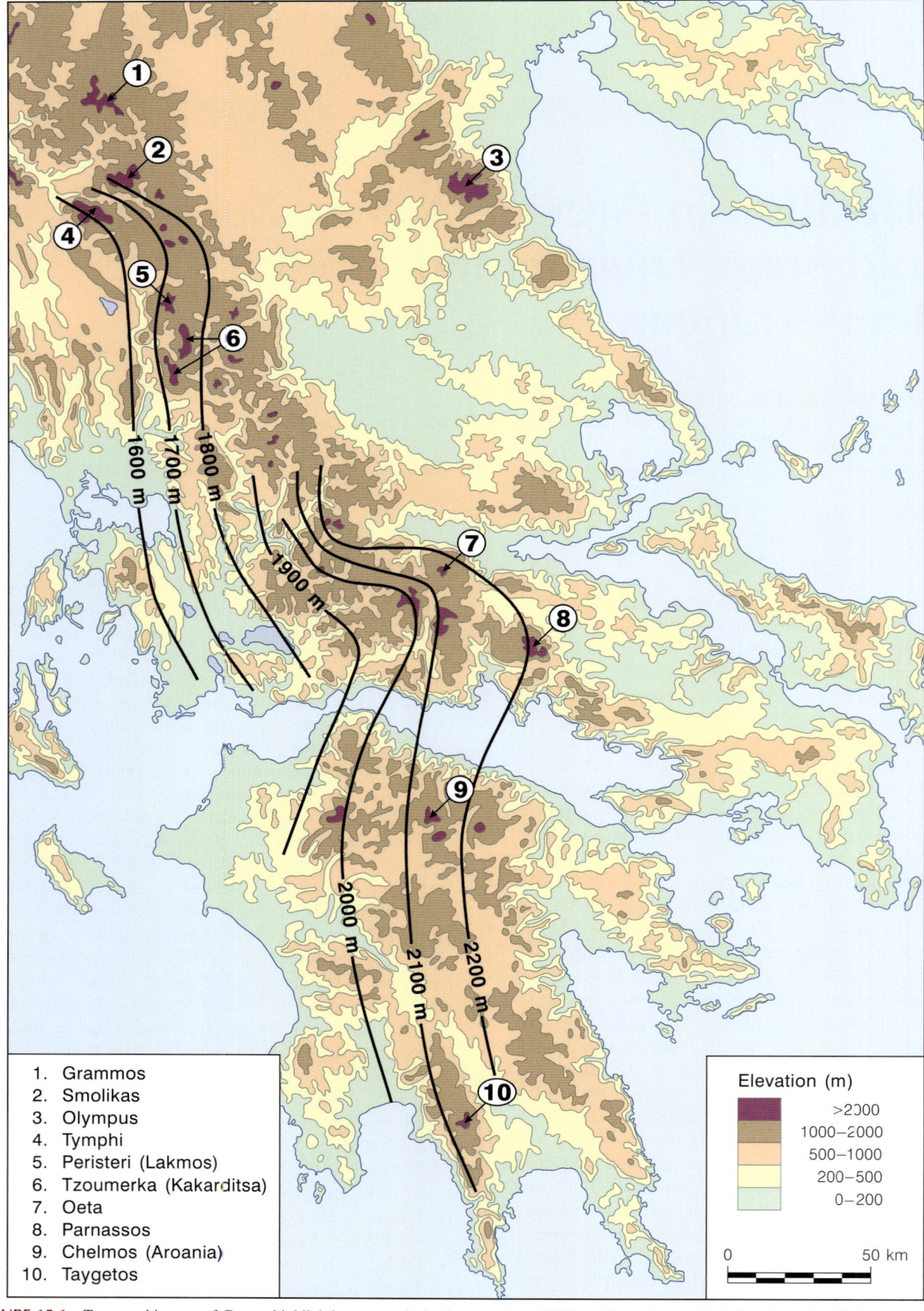

FIGURE 15.1 Topographic map of Greece highlighting mountain landscapes above 2000 m. Most of the published evidence for Pleistocene glacial activity in the mountains of Greece comes from the 10 locations indicated (Table 15.1). Pleistocene snowlines for the Pindus Mountains are also shown (based on Hagedorn, 1969). Glaciers do not exist today in the mountains of Greece, but snowfall is heavy and extensive snow patches are common in the early summer on the highest peaks and ridges. Winters are severe in the mountains and evidence of periglacial processes is widespread. There are marked variations in precipitation across Greece with values in excess of 2000 mm in the northwest and central uplands, falling to <500 mm in the lowland coastal zone of the southeast.

1. Grammos
2. Smolikas
3. Olympus
4. Tymphi
5. Peristeri (Lakmos)
6. Tzoumerka (Kakarditsa)
7. Oeta
8. Parnassos
9. Chelmos (Aroania)
10. Taygetos

TABLE 15.1 Geographical Data for Each of the Glaciated Mountains Shown in Fig. 15.1

Massif and Region	Elevation (m)	Latitude (approx.)	ELA (m)	References	Phase[a]
1. Grammos, Epirus	2523	40°34′N	1800	Louis (1926)	(P)
2. Smolikas, Epirus	2637	40°05′N	1800	Niculescu (1915)	(P)
				Boenzi et al. (1992)	(M)
3. Olympus, Pieria	2917	40°05′N	1030	Faugères (1969)	(M)
				Smith et al. (1997)	(M)
				Manz (1998)	(M/A)
4. Tymphi, Epirus	2497	39°97′N	1600	Palmentola et al. (1990)	(M)
				Smith et al. (1998)	(M)
				Woodward et al. (2004)	(A)
				Hughes et al. (2006b)	(A)
5. Peristeri, Epirus	2295	39°68′N	1750	Sestini (1933)	(P)
6. Kakarditsa, Epirus	2429	39°46′N	1750	Sestini (1933)	(P)
7. Oeta, Sterea	2160	38°75′N	2150	Mistardis (1952)	(P/M)
8. Parnassos, Sterea	2457	38°52′N	2200	Pechoux (1970)	(P/M)
9. Chelmos, Achaia	2341	37°96′N	2150	Mastronuzzi et al. (1994)	(M)
10. Taygetos, Messinia	2404	36°95′N	2150	Mastronuzzi et al. (1994)	(M)

The ELA values for Mount Olympus are from Smith et al. (1997) and the rest are from Hagedorn (1969) as shown in Fig. 15.1.
[a] The letter in brackets after the reference refers to the scheme of Hughes et al. (2006a) for Pioneer, Mapping and Advanced phases of research. Pioneer phase refers to locations where evidence of glaciation has been observed but not mapped. The Mapping phase refers to locations where detailed geomorphological maps of glacial phenomena, and related landforms have been produced but without radiometric dating control. The Advanced phase refers to sites where systematic mapping and detailed stratigraphical assessments have taken place in conjunction with a programme of radiometric dating that has led to the development of a robust chronological framework for the glacial record.

on Pleistocene glaciation has been carried out in northern Greece on Olympus and Tymphi (Fig. 15.1), which both lie close to 40°N. This chapter will therefore review the glacial records from these two areas in detail. A key feature of recent work in Greece has been the development of a dating framework for the glacial record in tandem with systematic geomorphological mapping and formal stratigraphical appraisal of the glacial geology. In the 1960s, when Messerli (1967) produced his synthesis, radiometric dates were not available for any of the glacial records in the Mediterranean and this situation remained largely unchanged for over 30 years. The past decade, however, has seen major advances in geochronology with dating frameworks based on a range of methods (including uranium-series, cosmogenic radionuclides, OSL and tephrostratigraphy) emerging for glacial records across the Mediterranean (e.g. Kotarba et al., 2001; Woodward et al., 2004; Vidal-Romaní and Fernández-Mosquera, 2006; Pallàs et al., 2007; Akçar et al., 2008; Kuhlemann et al., 2008; Sarıkaya et al., 2008, 2009; Lewis et al., 2009; Chapter 30). Research in Greece, using uranium-series dating, has been at the forefront of these developments (Hughes et al., 2006b; Woodward et al., 2008; Hughes and Woodward, 2009). This new chapter builds on an earlier contribution to the first edition of this volume (Woodward et al., 2004) and reports some key advances in our understanding of Pleistocene glaciation in Greece. It can be argued that these findings have important implications for the study of Pleistocene glacial records in upland environments not only across the Mediterranean region but also in glaciated mountain terrains around the world.

15.2. THE GLACIAL RECORD ON MOUNT OLYMPUS, NORTHEAST GREECE

Olympus is the highest mountain in Greece (2917 m) and is located to the east of the main Pindus Mountain chain in the northeast corner of peninsular Greece, close to the Aegean coast (Fig. 15.1). The most recent research on the glacial record on Olympus was conducted by an American team led by the late Geoffrey Smith from the Department of Geological Sciences at Ohio University. Smith and his co-workers (1994, 1997, 2006) argued that the earlier studies in the Mount Olympus area had significantly underestimated the extent of glacial activity, as deposits on the

FIGURE 15.2 Glaciated topography on the upper slopes of Mount Chelmos in the northern Peloponnese (Number 9 in Fig. 15.1). The upper photograph shows moraines and glacially steepened limestone back-wall cliffs behind the dried out lake basin of Mavrolimni. The lower photograph shows a large moraine and active scree slopes in a large cirque on the northwest slopes of Mount Chelmos at an elevation of ca. 1900 m. Note the figure on the moraine crest for scale.

eastern piedmont of the massif—that were previously classified as fluvial sediments—are, in fact, glacial diamictons. As Smith et al. (1997) have pointed out, the early investigations on Mount Olympus did little to establish the sequence and timing of glacial events in the area although the implication was that glaciation was restricted to the latest Pleistocene or Würmian Stage (Messerli, 1967).

15.2.1. The Glacial Sedimentary Sequences on Mount Olympus

The stratigraphic record of Pleistocene and Holocene events on Mount Olympus is most clearly observed on the eastern piedmont within the valleys of the Mavrolongus and Mavroneri Rivers which drain to the Aegean coast (Fig. 15.3). Extensive moraine complexes have been mapped in the Mavroneri valley to the west and southwest of the town of Katerini, and thick proglacial outwash sediments are well preserved on the valley floor (Fig. 15.3A). The deposits in this valley represent the convergence of valley ice from Mount Olympus (to the south) and the High Pieria Mountains (to the north). Three discrete sedimentary packages (units 1–3) have been identified in the area—each capped by a distinctive soil. The deposition of these sediments and the subsequent phases of pedogenic weathering reflect periods of glacial and nonglacial activity, respectively (Smith et al., 1997, p. 809). The deposits of the eastern piedmont comprise a range of glacial, glaciofluvial, fluvial and alluvial fan sediments. Smith et al. (1997, 2006) have argued that each of the three main sedimentary units can be related to a period of glacial activity in the uplands.

During the glacial phases that led to the deposition of the unit 1 and unit 2 sediments, cirque glaciers developed to a size that allowed them to spread out from their basins to form a continuous cover of upland ice (Smith et al., 1997, 2006). The Bara Plateau is part of the Mount Olympus upland that lies a few kilometres south of the Mytikas summit (2917 m). The glacial geology of the plateau is shown in Fig. 15.3B. Here, the unit 2 sediments are the most extensive in this area and can be traced for almost 2 km from the headwaters of the Mavratza River across the Bara Plateau to the head of the Mavrolongus valley (Smith et al., 1997; Fig. 15.3B). Thirteen cirques have been identified on the Bara Plateau with cirque floor elevations ranging from 2180 to 2510 m above present sea level. On the Plateau of the Muses, which includes the Mytikas summit, 11 cirques range in elevation from 2200 to 2660 m a.s.l. (Smith et al., 1997, 2006).

The glacial and proglacial sedimentary records in the Mount Olympus area constitute a potentially important archive of landscape change to the east of the Pindus Mountains and one that could be usefully compared to the long pollen record at Tenaghi Philippon ca. 160 km to the northeast (Wijmstra, 1969; Tzedakis et al., 2006). However, a robust geochronological framework for the glacial sequences on the Olympus massif is still lacking. Smith et al. (1997, p. 820) have conceded that the lack of radiometric dates on deposits of Mount Olympus and the adjacent piedmont precludes the establishment of a numerical chronology for Pleistocene events in this area. Attention was therefore directed towards the study of the weathering profiles associated with soils on the Mount Olympus deposits to develop a relative-age dating framework and to allow comparisons with other Quaternary records in northeast Greece.

Pedogenic maturity indices have been determined for the soil profiles on units 1–3 in an attempt to correlate the deposits of the Olympus piedmont with the sequence of soils developed on dated alluvial sediments in the Larissa Basin immediately to the south (Demitrack, 1986; van Andel et al., 1990). This approach led to the tentative correlation put forward by Smith et al. (1997), shown in Table 15.2. This model places the unit 1 sediments before 200 ka within marine isotope stage (MIS) 8 and the unit 2 deposits within MIS 6. On the basis of this age model, the nature of the unit 1 soil suggests that the last major glaciation of the Olympus massif took place at some time during the last cold stage (MIS 2 or 4) but was restricted to valley heads and glaciers extended to mid-valley positions (Smith et al., 1997).

Smith et al. (1997) presented this correlation as tentative because of the absence of radiometric dates. It is also important to point out that the chronological control for the alluvial sequence and soils of the Peneios River in the Larissa Plain, studied by Demitrack (1986), is relatively poor by the standards of more recent investigations (see Macklin et al., 2002; Macklin and Woodward, 2009) in the Mediterranean, and it can be argued that it does not form a reliable yardstick for regional correlation (Table 15.2). It is often difficult to make valid intersite comparisons of soil profile development when sites are located at different altitudes in contrasting geomorphological settings. In such cases, establishing the long-term constancy of important soil forming factors such as parent material and local climate can be especially problematic (Birkeland, 1984).

To improve the age control for the glacial deposits of Mount Olympus, cosmogenic dating of unit 1 boulder surfaces using chlorine-36 (^{36}Cl) was attempted. This was one of the earliest efforts to use cosmogenic exposure dating in the Mediterranean. Manz (1998) reports cosmogenic dates from two sites within large (unit 1) recessional moraine complexes on the eastern piedmont. Limestone boulders at Site 1 (immediately north of the town of Litochoro) yielded ^{36}Cl ages within the range 32–49 ka. Boulders at Site 2 in the Mavroneri catchment to the west of Katerina (Fig. 15.3B) yielded ages ranging from 43 to 56 ka. One boulder at Site 2 gave an age of 146 ka (MIS 6), and this outlier has been attributed to either previous exposure or

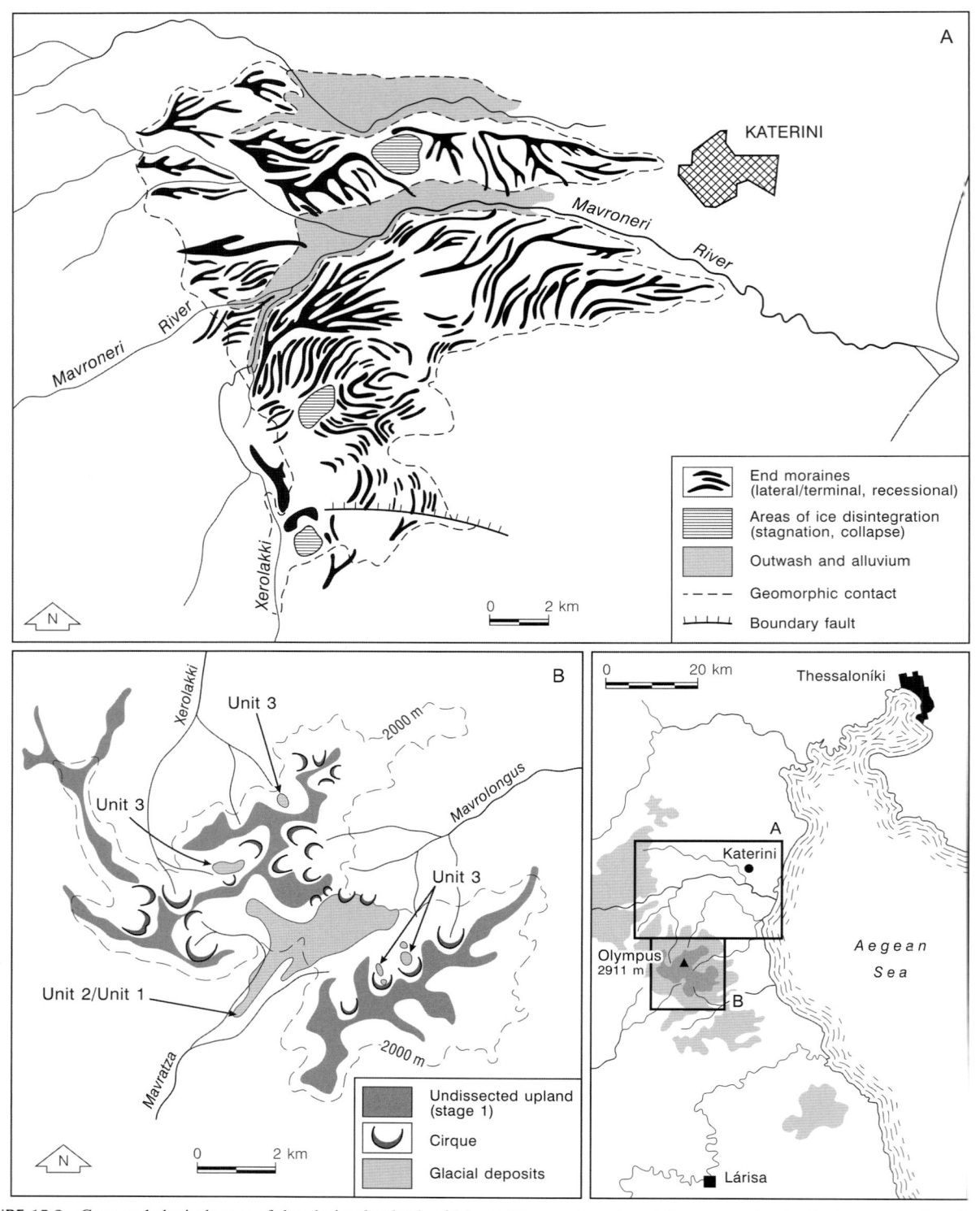

FIGURE 15.3 Geomorphological maps of the glaciated uplands of Mount Olympus in northeast Greece showing the landscape near the village of Katerini (A) and the cirques and glacial deposits close to the summit (B) (modified from Smith et al., 1997).

evidence of an earlier glacial phase. Manz (1998) argued that the Site 2 ages were more reliable because silicate rocks are better suited to this method. The tentative age model for units 1–3 proposed by Smith et al. (1994, 1997, 2006) and shown in Table 15.2 would need to be radically revised if the ^{36}Cl ages for the unit 1 sediments are accepted. Indeed,

TABLE 15.2 A Tentative Correlation of the Olympus Piedmont Soils with the Pleistocene Soils Developed in Terraced Alluvial Sediments on the Larissa Plain Studied by Demitrack (1986) (after Smith et al., 1997)

Larissa Plain Soil name	Depositional Phase	Olympus Piedmont Soil name	Marine Isotope Stage
Pinios Group	<200 years	Unit 3 surface soils	1
Deleria soil	Historical		
Girtoni soil	6–7 ka		
Non-calcareous brown soil	10–14 ka		
Gonnoi Group soils	14–30 ka		
Agia Sophia soil	27–42 ka	Unit 3 truncated (buried) soil	3
Rodia Group soils	≥54–125 ka	Unit 2 soil	5
Deep red soil	≤210 ka	Unit 1 soils	7

Note that the marine isotope stages shown represent warm periods conducive to weathering and soil development. Soil formation began in the marine isotope stage indicated.

it is difficult to reconcile these ages with the extent and thickness of the Pleistocene sediments comprising units 1–3 because the ^{36}Cl ages demand much faster rates of soil development and landscape change in this area than those put forward by Smith et al. (1997) in Table 15.2. The deep red soil developed on the unit 1 sediments is thought to record an extended interval, probably of interglacial duration, during which substantial pedogenesis took place (Smith et al., 1997, p. 811). Manz (1998, p. 56) concluded that additional ^{36}Cl measurements on deposits on the piedmont and upland areas of the mountain might reveal evidence of earlier glacial episodes even though the ages of the deposits (units 1–3) on the eastern Olympus piedmont appear to postdate the last interglacial period.

The work of Manz (1998) represents a pioneering application of the cosmogenic dating method in the Mediterranean. In common with more recent examples of this approach in the region, it highlighted the potential for considerable noise in such datasets and the need for much more extensive sampling and dating programmes. In conclusion, the glacial sediments and landforms of Mount Olympus and its adjacent uplands lack a firm, internally consistent chronological framework and, as Smith et al. (1997, p. 822) pointed out, the sequence of events proposed in Table 15.2 "can easily be shifted backward or forward in time."

15.3. THE GLACIAL RECORD ON MOUNT TYMPHI, NORTHWEST GREECE

Mount Tymphi is located in Epirus in northwest Greece approximately 15 km southwest of Mount Smolikas and 40 km south of Mount Grammos on the Albanian border (Fig. 15.1). The several peaks and ridges of Mount Tymphi form part of the high watershed between the catchments of the Voidomatis (384 km^2) and Aoos (665 km^2) rivers (Fig. 15.4). With its series of jagged limestone peaks, the highest of which exceed 2400 m, Tymphi has been described as both the most extensive and the most majestic of the Greek mountains (Sfikas, 1979). Figure 15.5 is an oblique air photograph that portrays very clearly this distinctive upland landscape and its impressive array of glacial landforms. It also shows some of the major topographic elements of the Voidomatis River basin (see Bailey et al., 1997) that are labelled in Fig. 15.4. The northern side of the Tymphi massif includes a magnificent set of cirques and limestone cliffs dissected by steep ravines. To the south, an extensive tableland is cut by the 900-m-deep Vikos Canyon. Mount Tymphi boasts the most extensive and best preserved glacial terrain on any of the Greek mountains. In common with many of the glaciated mountains on the Balkan Peninsula and wider Mediterranean region, hard limestones are the dominant lithology and the effects of karstic processes are widespread (Waltham, 1978; Lewin and Woodward, 2009).

South of the eastern ridges of Mount Tymphi, around the villages of Tsepelovo (ca. 1100 m) and Skamnelli (ca. 1200 m), extensive spreads of glacial deposits are marked on the 1:50,000 geological sheet (IGME, 1970). Here, numerous moraine ridges are especially well defined as shown in the lower left portion of Fig. 15.5. The glacial deposits south of Tsepelovo reach down to the valley floor of the Voidomatis River to an elevation of around 850 m a.s.l. (Fig. 15.4) (Woodward et al., 1995, 2004; Smith et al., 1998; Hughes et al., 2006b). It is only very recently, however, that the glacial record in this area has been mapped in detail and radiometrically dated. Reconnaissance field

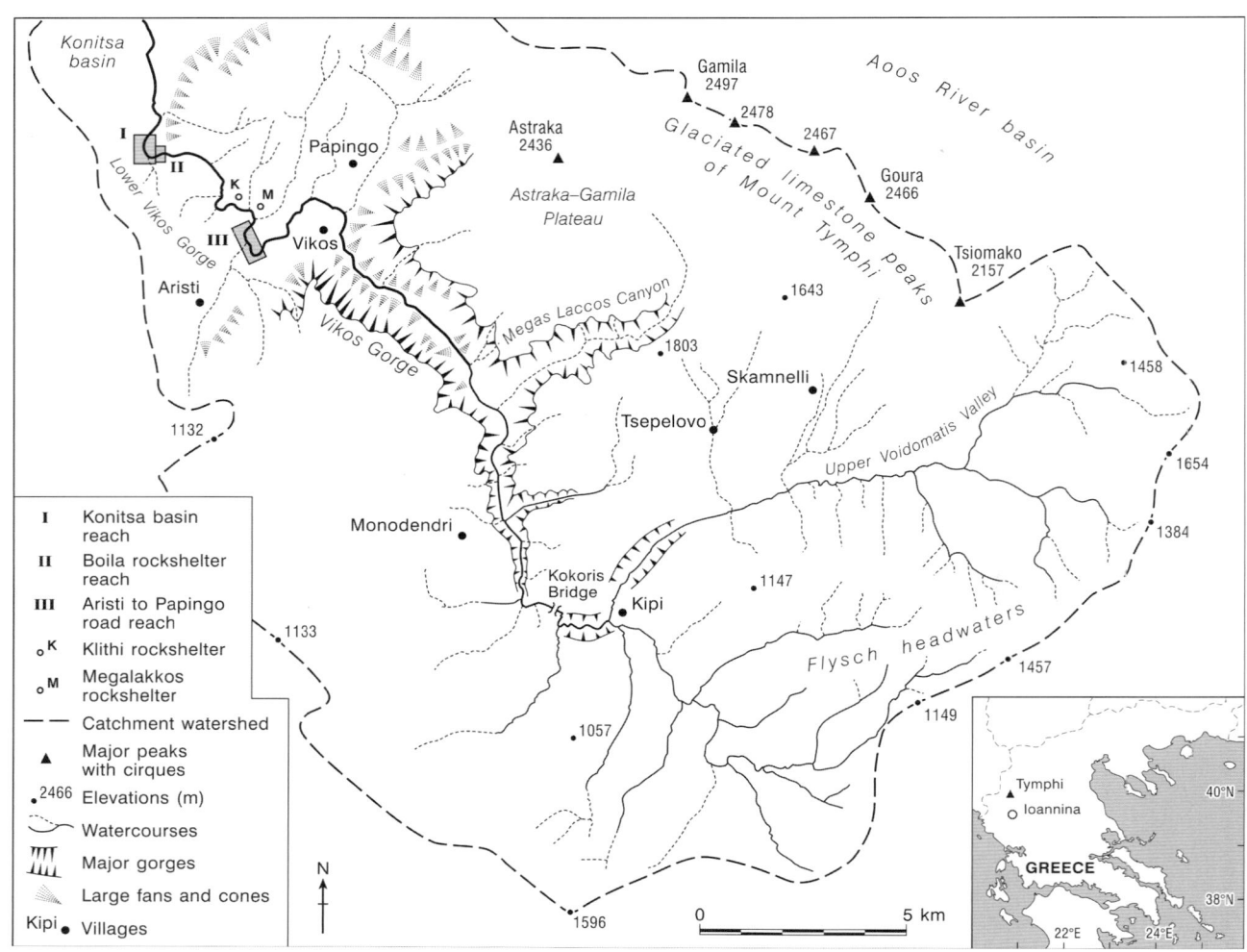

FIGURE 15.4 Map of the Voidomatis River basin of northwest Greece. This basin drains the glaciated limestone uplands of Mount Tymphi (Number 4 in Fig. 15.1). The Pleistocene fluvial record in this basin is particularly well preserved in the lower part of the basin (Reaches I and III) (see Woodward et al., 2008). Flysch rocks are widespread at lower elevations in the Voidomatis River basin and mainly beyond the limits of glacial action (Woodward et al., 1992; Bailey et al., 1997; Hughes et al., 2006b).

studies of this area were conducted in the mid-1980s as part of the first phase of fieldwork investigating the Pleistocene history of the Voidomatis River (Fig. 15.4) and the principal alluvial sediment sources in the basin (Bailey et al., 1990; Lewin et al., 1991; Woodward et al., 1992). This work confirmed that the southern flanks of Mount Tymphi were dominated by glacial sediments and landforms (Woodward, 1990; Woodward et al., 1995; Bailey et al., 1997; Macklin et al., 1997; Woodward et al., 2008).

The assemblage of glacial landscape features on Mount Tymphi includes cirques of various sizes and geometries (Fig. 15.5), deep ice-scoured troughs with stepped long profiles (Fig. 15.6A), limestone bedrock pavements (Fig. 15.6B) and lateral and terminal moraine complexes (Figs. 15.6B, D, and F). Steep limestone cliffs and thick talus deposits are also important landscape features (Fig. 15.6E). As mentioned above, glacial depositional landforms are especially well preserved on the slopes around Tsepelovo and Skamnelli (Fig. 15.5). Good-quality exposures are commonly present in the glacial sediments where the depositional zone on the lower slopes is traversed by mountain tracks or by the Tsepelovo to Skamnelli road. A good example is shown in Fig. 15.6D where a road-side quarry revealed a deep section in till sediments in the large moraine immediately to the east of Tsepelovo (Fig. 15.4). Such exposures in the Mount Tymphi tills show that the glacial sediments contain subrounded limestone boulders, cobbles and gravels commonly dispersed in a fine-grained matrix rich in sands, silts and clays derived from the comminution of the hard limestone bedrock (Fig. 15.7A) (Woodward et al., 1992, 1995, 2004; Hughes et al., 2006b).

During the mid- to late 1980s when Lewin, Macklin and Woodward worked on the alluvial record in the Voidomatis basin, Italian researchers mapped the glacial landforms on

Chapter | 15 Glaciation in Greece: A New Record of Cold Stage Environments in the Mediterranean

FIGURE 15.5 An oblique air photograph of Mount Tymphi looking to the south and southwest. This image shows the steep limestone cliffs and cirques on the northern face of the mountain and the extensive glaciated landscapes that drain southwards to the Vikos Canyon and the upper reaches of the Voidomatis River (Fig. 15.4). Photograph taken by Giannis Tsouratzis in June 2009. T = Tsepelovo and S = Skamnelli.

Tymphi and Smolikas (Palmentola et al., 1990; Boenzi et al., 1992). Geomorphological maps of the major moraines were drawn and, despite the absence of any dating control, it was argued that all the phases of glacial activity took place during the Late Würmian. The next phase of work in the area saw Smith et al. (1998) employ Landsat Thematic Mapper satellite imagery to map the extent of the glacial sediments and landforms on Mount Tymphi. A supervised image classification of the TM scene (using bands 1 to 5 and 7) was carried out using the maximum likelihood classification (MLC) decision rule. The MLC results were evaluated using air photographs and field checking and were shown to provide a good approximation of the geomorphology of the target area. The moraine complex to the north of Skamnelli was correctly classified as being dominated by sharp crested moraines, and the landforms near the village of Tsepelovo and southwards to the Voidomatis River were correctly classified as weathered and degraded moraines (Smith et al., 1998). This contrast in moraine form can be seen in Fig. 15.5.

From this analysis, it became clear that the glacial record on Mount Tymphi consisted of at least two major phases of glaciation, and this pointed towards a more complex Pleistocene history than that proposed just a few years earlier by Palmentola et al. (1990). Indeed, the clear contrasts in weathering depth and landform style suggested that the most extensive moraines were much older than those at higher elevations. Without radiometric dates, however, it was not possible to quantify the difference in age or to establish whether or not the two groups of moraines formed during the same cold stage.

15.4. THE FIRST URANIUM-SERIES AGES FOR THE GLACIAL RECORD IN GREECE

The first radiometric dating programme on Mount Tymphi began in May 1998 when secondary carbonate samples were first recognised in the glacial deposits. At some locations, the road and track cuttings expose cavities within the till matrix where carbonate precipitation has taken place—often in association with subrounded boulders, cobbles and gravels. In places, the fine sediment matrix has been completely removed to leave interclast spaces that have been partially or completely filled with calcite. These secondary carbonates often form a skin on the surface of cobble- and boulder-sized limestone clasts accreting roughly parallel to the host surface (Fig. 15.7b). These secondary carbonates can be dated using uranium-series methods.

The first secondary carbonate samples were collected from five exposures in the glacial sediments on the southern slopes of the Tymphi massif and these yielded nine uranium-series ages (see Woodward et al., 2004 for a detailed discussion of these sites and ages). These ranged in age from 71.1 to >350 ka. In each case, because these ages were obtained from secondary carbonates, they provided *minimum* ages for the host glacial sediments. This set of uranium-series ages lead to a major reinterpretation of the Pleistocene glacial record, not only in Greece but also across the Balkan Peninsula and the wider Mediterranean region. They showed, for the first time, that the most extensive phases of glaciation in Greece took place during the Middle Pleistocene during a cold stage before 350 ka. They also showed that a major phase of glaciation took place before the last interglacial during MIS 6. The nine uranium-series ages published in Woodward et al. (2004) showed that most of the glacial sediments and landforms in the headwaters of the Voidomatis River basin were Middle Pleistocene in age and therefore much older than the last cold stage. Further, two sites revealed multiphase secondary carbonate development demonstrating that the host

FIGURE 15.6 Glaciated landscapes on the slopes of Mount Tymphi in northwest Greece. (A) Limestone pavements and glacially steepened cliffs looking up-valley towards the high cirques between the peaks of Gamila (2497 m) and Goura (2466 m). (B) Well-developed clints and grikes on an elevated limestone pavement with Vlasian Stage moraines in the middle ground.

sediments and their associated landscape features had been stable for an extended period of time. The bulk of the glacial sediments and major glacial depositional landforms in this area are much older than the global Last Glacial Maximum (LGM) of MIS 2, and this was a very significant outcome of the first uranium-series dating programme.

These dates showed that it is unwise to assume that apparently fresh moraines in Mediterranean mountain environments must date to the most recent cold stage (see Palmentola et al., 1990; Boenzi and Palmentola, 1997). Glacial landform stability appears to be a feature of high limestone landscapes especially where secondary carbonate formation (cementation) is widespread. Despite these advances, a major programme of field mapping was still needed to establish the extent of glacial action on Mount Tymphi and the dimensions of the former ice masses during the cold stages of the Middle and Late Pleistocene.

FIGURE 15.6 (C) Large glacially transported boulders on the western side of Tymphi. (D) A deep section in Skamnellian Stage glacial sediments exposed in a quarry on the Tsepelovo to Skamnelli road showing the rounded form of the oldest moraines on Mount Tymphi (MIS 12) and the limestone-rich till sediments.

FIGURE 15.6—cont'd

FIGURE 15.6—cont'd (E) Glaciated terrain and well-developed screes immediately up-valley of Skamnellian Stage end moraines. (F) A Skamnellian Stage moraine crests on the southern slopes of Mount Tymphi.

15.5. TOWARDS A FORMAL STRATIGRAPHICAL FRAMEWORK FOR THE GLACIAL RECORD IN GREECE

It quickly became apparent that the full potential of this glacial record could not be realised without detailed mapping of the entire massif (including the high-level cirques and associated moraines and rock glaciers) allied to a more extensive uranium-series dating programme. This work was completed during the course of a PhD project (Hughes, 2004) and led to a series of papers on the stratigraphy and geochronology of the glacial deposits on Mount Tymphi

(Hughes et al., 2006b) and the palaeoclimatic significance of the glacial record (Hughes et al., 2006c, 2007a,b). This latest phase of research on Mount Tymphi is summarised below. The new record is based on systematic field mapping of the glacial landforms, section logging and sedimentological investigations, detailed examination of soil profiles on moraine surfaces, allied to the most comprehensive programme of uranium-series dating yet conducted on glacial deposits anywhere in the world.

Figure 15.8 provides a useful illustration of the kinds of data that have been compiled to develop a new model of Pleistocene glacier behaviour in Greece. It shows a geomorphological map of the southern slopes of Mount Tymphi detailing a wide range of erosional and depositional features. It also shows 21 of the uranium-series ages and the respective sample locations. This work has led to the formal identification of three glacial phases on Mount Tymphi. Figure 15.9 shows the geography of the ice masses during

FIGURE 15.7 (A) A road-side exposure in till sediments on the north side of Mount Tymphi. Note the abundant limestone-rich fine sediment matrix and the wide range of boulder sizes.

FIGURE 15.7–cont'd

FIGURE 15.7–cont'd (B) An exposure in till on Mount Tymphi showing examples of secondary carbonates that have been sampled for uranium-series dating.

each glacial phase. These stages (Skamnellian, Vlasian and Tymphian) have been formally defined by Hughes et al. (2006b) and correlated with the long lacustrine record from Lake Ioannina that lies about 40 km south of Mount Tymphi (Fig. 15.4). The Mount Tymphi record provides a glacial stratigraphical framework for the Pindus Mountains and is now underpinned by 28 uranium-series ages on secondary carbonates. Based on Hughes et al. (2006b) and Woodward et al. (2008), each phase of glaciation is described in turn below.

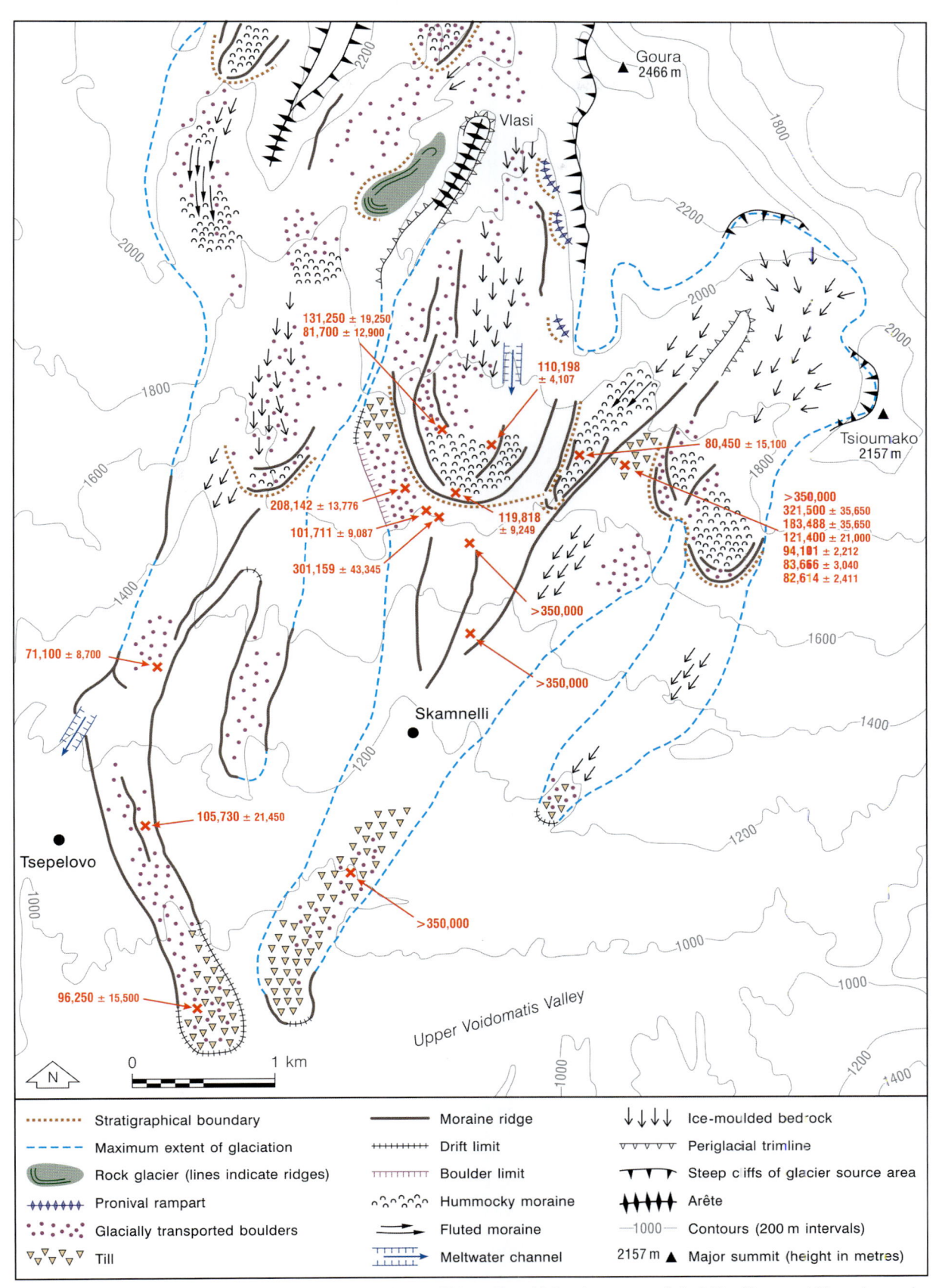

Chapter | 15 Glaciation in Greece: A New Record of Cold Stage Environments in the Mediterranean

FIGURE 15.8 Geomorphological map of part of the southern slopes of Mount Tymphi showing a range of glacial landforms and 21 uranium-series ages obtained from secondary carbonates formed in limestone-rich till (modified from Hughes et al., 2006b). Figure 15.4 shows the location of Tsepelovo and Skamnelli in the Voidomatis River basin.

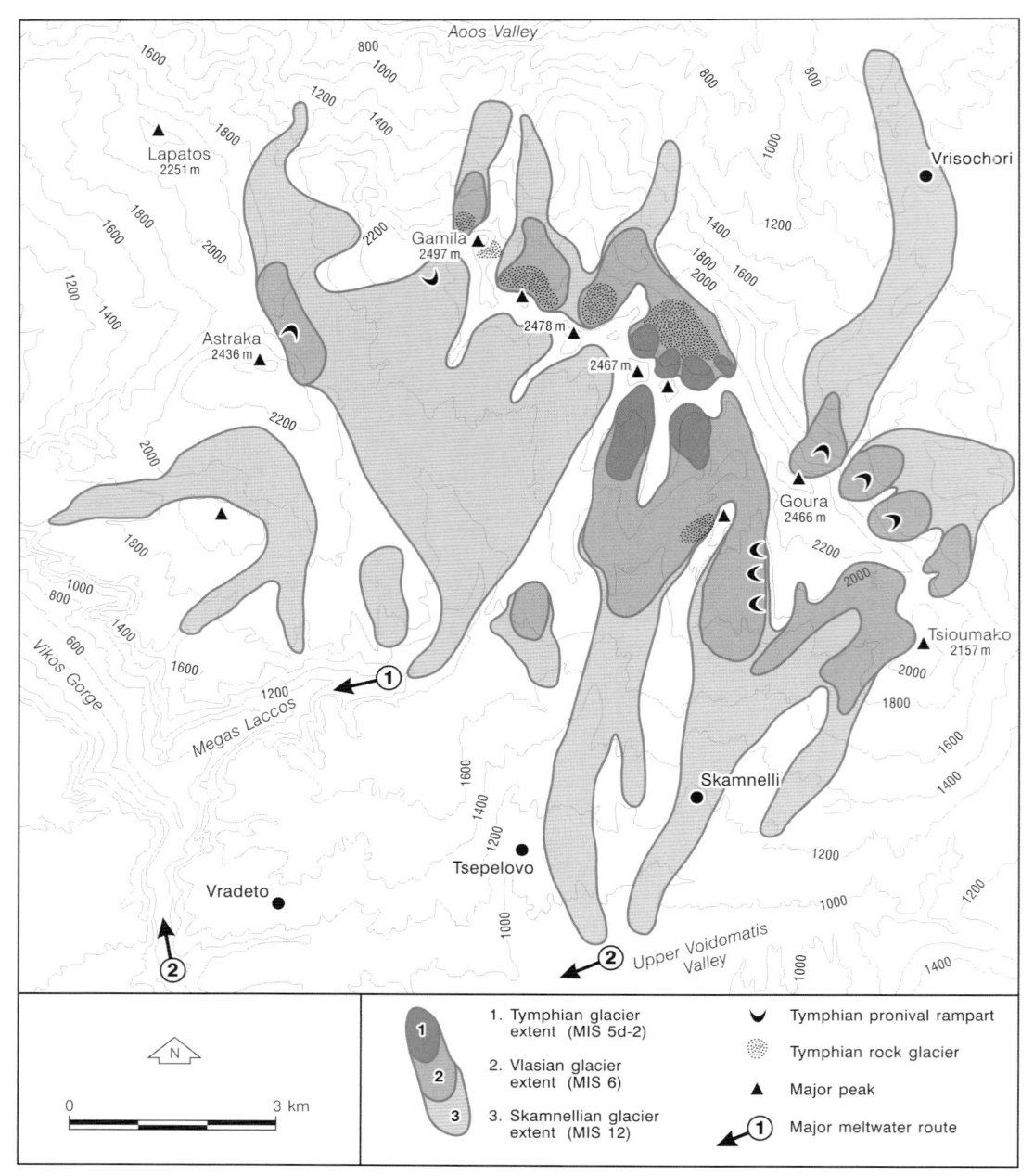

FIGURE 15.9 The extent of glaciation during the Middle and Late Pleistocene on Mount Tymphi (based on Hughes and Woodward, 2009). The rock glaciers on Mount Tymphi and the major meltwater routes to the Voidomatis River are also shown.

15.5.1. The Skamnellian Stage

As Fig. 15.9 shows, the earliest and most extensive phase of glacial activity took place before 350 ka and saw the formation of large ice fields and valley glaciers on the southern slopes of Mount Tymphi and across the Astraka-Gamila plateau. The moraines from this glaciation have deeply weathered soil profiles with profile development indices (PDIs) ranging from 51.8 to 61 (Hughes et al., 2006b).

The glaciers extended beyond the villages of Tsepelovo (1100 m) and Skamnelli (1200 m) down to about 850 m a.s.l. (Fig. 15.4). In this area of Mount Tymphi, these glacial deposits are well exposed on the right bank of the modern Voidomatis River (Woodward et al., 1995; Hughes et al., 2006b). Excellent exposures also exist down to altitudes as low as 1000 m in the Vadulakkos valley on the northern slopes of Mount Smolikas (Hughes et al., 2006d). Ice cover on Mount Tymphi during the Skamnellian glaciation approached 60 km² at its maximum extent, and the mean equilibrium line altitude (ELA) was ca. 1741 m a.s.l. (Hughes et al., 2007b). The oldest secondary carbonates from sediments associated with this phase of glaciation are beyond the range of uranium-series dating and six samples yielded ages >350 ka (Fig. 15.8 and 15.10) (Woodward et al., 2004; Hughes et al., 2006b). The oldest secondary carbonates have been correlated with MIS 11, and the Skamnellian Stage glacial deposits are correlated with the Elsterian Stage of northern Europe and MIS 12 (Hughes et al., 2006b).

15.5.2. The Vlasian Stage

During the Vlasian glaciation, the maximum extent of ice cover (ca. 21 km²) extended over about one-third of the area that was glaciated during the earlier Skamnellian glaciation. At this time, most of the Astraka-Gamila plateau was ice free while glaciers advanced to mid-valley positions above Tsepelovo and Skamnelli (Figs. 15.8 and 15.9). During the maximum extent of this glaciation, the mean ELA was ca. 1862 m a.s.l (Hughes et al., 2007b). Eight uranium-series ages have been obtained from sediments belonging to this

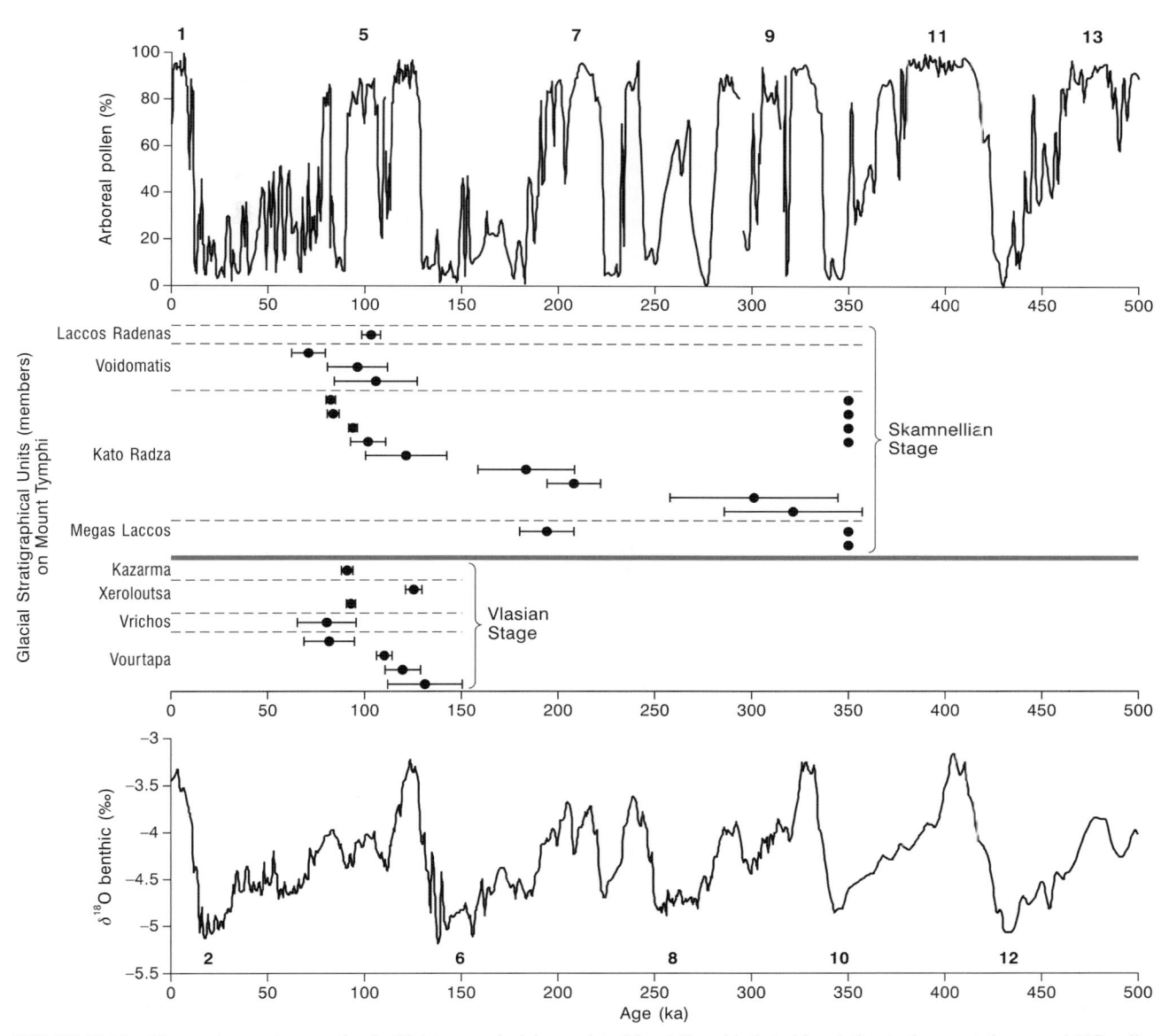

FIGURE 15.10 The uranium-series ages for the Pleistocene glacial record on Mount Tymphi plotted in relation to the vegetation record (AP pollen percentages) from Tenaghi Philippon (top) and the S06 δ^{18}O benthic composite marine isotope record from sites in the equatorial East Pacific (see Tzedakis et al., 2006 for sources). The uranium-series ages are from Woodward et al. (2004) and Hughes et al. (2006b) and have been plotted in relation to their respective glacial stratigraphical units (see Hughes et al., 2006b). The pollen and marine isotope time series come from Tzedakis et al. (2006). See text for discussion.

phase of glaciation. These range in age from 80.45 ± 15.1 to 131.25 ± 19.25 ka and indicate that this phase of glacial activity took place *before* the last interglacial during MIS 6. The Vlasian Stage is correlated with the late Saalian Stage of northern Europe and MIS 6 (Hughes et al., 2006b). It is quite likely that glaciers also developed during the cold stages between the Vlasian and Skamnellian Stages (i.e. during MIS 8 and MIS 10), but deposits from these periods have not been preserved. It appears likely that they were less extensive than the Vlasian Stage glaciers and were overrun and reworked by glaciers during this cold stage.

15.5.3. The Tymphian Stage

The most recent phase of glacial activity took place during the last cold stage, and this is known as the Tymphian Stage. The glacial deposits and landforms of this period are younger than those of the Vlasian (MIS 6) glaciation because they are found at much higher elevations and well within the limits of the Vlasian Stage deposits (Fig. 15.9). Soils on moraine surfaces also show only very limited evidence of pedogenic weathering (PDIs of 7.8–9.0) (Hughes et al., 2006b). However, due to the absence of secondary carbonates in the Tymphian Stage glacial sediments, uranium-series ages have not been obtained. Nonetheless, following the arguments presented in Hughes et al. (2006b), the Tymphian Stage is correlated with the Weichselian/Würmian Stage of northern Europe and the Alps, respectively, and MIS 5d to MIS 2. The climate at the local glacier maximum during this stage would have been wet and cold.

Rock glaciers also formed during this cold stage and provide evidence of cold and dry conditions *after* the local glacial maximum (Hughes et al., 2003, 2006c). The period of rock glacier formation probably correlates with the global LGM around 20–22 ka. About 15 km to the northeast of Mount Tymphi on Mount Smolikas (2637 m), well-developed moraines have been mapped at elevations above the floors of the highest cirques on Mount Tymphi (Hughes et al., 2006e). These glacial landforms postdate the youngest moraines on Mount Tymphi, and it has been argued that they may be Younger Dryas in age, although this has not been confirmed by radiometric dating (Hughes et al., 2006e). Together these records show that, in the Pindus Mountains, climatic conditions were favourable for glacier development at various times during the last cold stage.

15.6. SECONDARY CARBONATE FORMATION AND ENVIRONMENTAL CHANGE

The 28 uranium-series ages from the glacial deposits on Mount Tymphi have been plotted in Fig. 15.10 alongside the tree pollen curve from Tenaghi Philippon and the marine oxygen isotope record. The pollen record shows very rapid expansions and contractions in tree cover during the course of the past 500,000 years (Tzedakis et al., 2003, 2006; Tzedakis, 2005) (Fig. 15.10). The uranium-series ages range from 71 to > 350 ka. As we have argued above, these represent minimum ages for the host glacial sediments because phases of secondary carbonate formation can take place long after the deposition of the primary glacial deposits. The oldest ages obtained for each formation are therefore closest in age to the cold stage in which the glacial deposits were laid down. Figure 15.10 shows that the secondary carbonates in the till sediments form in clusters, with evidence of secondary carbonate formation on Mount Tymphi during four interglacials, namely, MIS 5, 7, 9 and 11. We have obtained eight dates on secondary carbonates from the Vlasian Stage moraines—all of them are associated with calcite formation during the last interglacial (5a–e). The uranium-series database also includes eight ages from within the Skamnellian Stage moraines associated with calcite formation during the last interglacial. This was clearly a very significant period of secondary carbonate formation with prolonged periods of high rainfall, dense forest cover and active soil processes producing high concentrations of CO_2 in percolating vadose waters. The Skamnellian Stage moraines also contain 11 uranium-series ages that predate MIS 6 with six of these being > 350 ka. This earliest phase of secondary carbonate formation probably took place during MIS 11. The host glacial sediments are clearly much older than those of the Vlasian Stage moraines, and we have argued that they were deposited during MIS 12 (Hughes et al., 2006b).

The formation of speleothems and other secondary carbonates is commonly reduced or absent during cold stages but most active during warm intervals when rainfall is higher and biological productivity and soil development increase (Gascoyne, 1992). The pollen records from Ioannina and Tenaghi Philippon show that forest cover in Greece expanded and contracted very rapidly between cold and warm stages, primarily in response to changes in precipitation. These shifts would have dramatically transformed hillslope and soil profile hydrology and the flux of CO_2 in the vadose zone. The dated record of secondary carbonate formation we have assembled from the Mount Tymphi tills shows very clearly how the near-surface karst system switched from cold to warm stage modes. Within the moraines, secondary carbonate formation is influenced by several factors. The water that enters voids in the glacial sediment matrix must have passed through biologically active soil profiles where it acquires elevated concentrations of CO_2, and the ground surface must allow the free passage of soil waters. The combination of low organic productivity, lower rainfall and perhaps also frozen soils effectively shut down secondary carbonate formation during cold stages on Mount Tymphi. Vegetation responded very rapidly to climatic amelioration with a rapid expansion of tree cover at the onset of interglacials and interstadials (Fig. 15.10). This is important for the build-up of organic materials in soils and the enhanced biological productivity that allows elevated concentrations of CO_2 in soil water. The rapid decline in tree cover at this altitude during the onset of cold periods may partly explain the almost complete absence (only 1 out of 28 dates) of secondary carbonate formation during cold stages.

All the uranium-series ages predate the global LGM of MIS 2; they provide compelling evidence for the presence of large ice masses in the Mediterranean during the Middle Pleistocene (Woodward et al., 2004; Hughes et al., 2006a,b,c,d,e). There is no evidence of secondary carbonate formation in the youngest (Tymphian) glacial sediments from the last cold stage. Nor is their evidence for Holocene-age secondary carbonate formation in earlier till sediments. This may be due to the progressive filling of intraclast voids in the oldest glacial sediments and the particularly active phases of secondary carbonate formation and void closure during the last interglacial (Fig. 15.10). The youngest glacial sediments are associated with the smallest moraines on Mount Tymphi, and these sediments do not contain large amounts of fine-grained limestone matrix. There is also some evidence to suggest that secondary carbonate formation is more important towards the end of interglacial periods (Fig. 15.10). It follows therefore that the Holocene may not have been long enough for the development of sufficiently mature soils, and the mid-Holocene removal of much of the tree cover may have reduced biological productivity and the CO_2 concentration of vadose waters. The Mount Tymphi record shows that there was only very limited glaciation during the cold stages that followed MIS 6. The smallest glaciers were restricted to higher elevations, and their deposits contain a much smaller proportion of finely comminuted limestone-derived fine sediments. The dissolution of this carbonate-rich material in the soil profile is the raw material for secondary carbonate accumulation at deeper levels. It may also be significant that all the secondary carbonates we have sampled formed at elevations lower than 1800 m. The Tymphian Stage moraines are all above 1800 m and commonly above 2000 m (Fig. 15.9). In this environment, secondary carbonate formation is more likely at lower elevations where a range of parameters including vegetation cover and deep soil development combined to provide more favourable conditions. These contrasts are also reflected by the PDI values for the three glacial formations (see Hughes et al., 2006b).

summer temperatures at least 11 °C lower than modern values. In the long pollen record at Tenaghi Philippon, the cold and dry interval correlated with MIS 12 was one of the most severe of the Pleistocene (Tzedakis et al., 2003). A dry climate during MIS 12 caused thick loess accumulations to form in the Pannonian basin in the northern Balkans (Marković et al., 2009). During the later Vlasian and Tymphian glaciations, climate would have been significantly warmer and wetter. However, wetter conditions were offset by the warmer summer temperatures resulting in smaller glaciers than during the Skamnellian.

A similar glacier-climate history to that in northern Greece has recently been revealed on Orjen in Montenegro. Here, the most extensive glaciation correlates with the Skamnellian glaciation in Greece during MIS 12 (ca. 480–430 ka). Later and less extensive glaciations are also recorded in the cirques and valleys of Orjen and correlate with glaciations in Greece during the Vlasian (MIS 6; 190–130 ka) and Tymphian Stages (MIS 5d-2; 110–11.7 ka) (Hughes et al., 2010). The Vlasian glaciers in Greece appear to have formed under the wettest conditions of all the three recorded major glaciations. This is in accordance with evidence of higher precipitation during MIS 6, compared with other cold stages, recorded in $\delta^{18}O$ stalagmite records in Italy and sapropel records in the Mediterranean Sea (Bard et al., 2002). For the Tymphian Stage glaciation in Greece, periglacial evidence has been used to reconstruct mean annual temperatures that were 8–9 °C cooler than present (Hughes et al., 2003)—an estimate that is supported by independent evidence from other records (e.g. Peyron et al., 1998). However, this temperature corresponds with the coldest and driest part of the last glacial cycle—close to the global LGM. Glaciers on Mount Tymphi were larger at earlier points of the Tymphian Stage, reaching their maximum between 25 and 30 ka, when climate was wetter—with precipitation > 2000 mm. In fact, glaciers are likely to have oscillated in response to millennial-scale climate fluctuations throughout this cold stage (Hughes et al., 2006c). All the palaeoclimate reconstructions described above were calculated using the Ohmura et al. (1992) regression which relates summer temperatures to values of winter balance + summer precipitation. These palaeoclimate reconstructions were rerun for Mount Tymphi and Smolikas using a degree-day model approach by Hughes and Braithwaite (2008) and produced similar findings.

15.7. PALAEOCLIMATE RECONSTRUCTIONS FROM GLACIAL GEOMORPHOLOGICAL DATA

Hughes et al. (2007b) argued that the large glaciation of the Skamnellian Stage was strongly influenced by very low

15.8. RIVER RESPONSE TO GLACIATION AND THE LAST GLACIAL-TO-INTERGLACIAL TRANSITION

The new age framework for glaciation on Mount Tymphi has allowed fresh insights into the Pleistocene history of the Voidomatis River. The long-term interactions between the glacial and fluvial systems have recently been examined in detail by Woodward et al. (2008). An interesting feature of the alluvial record preserved in the middle and lower reaches (I, II and III in Fig. 15.4) of the Voidomatis River is the evidence for multiple phases of aggradation during the last cold stage. In this context, for the last cold stage at least, the alluvial record may offer a more detailed record of upstream glacial activity than the glacial record itself. The alluvial record can also offer insights into aspects of long-term glacial system behaviour that cannot be obtained from the glacial sediments and landforms themselves. Figure 15.11 is a good example. It presents provenance data from fine-grained slackwater sediments preserved at the downstream end of the Lower Vikos Gorge (Reach II in Fig. 15.4). The lower part of the diagram shows the composition of cold stage meltwater floods towards the end of the last cold stage (Late Tymphian). The suspended load of the full-glacial river was dominated by glacially derived, finely comminuted limestone-rich sediment. The upper part of the diagram shows the composition of flood sediments preserved in Boila Rockshelter (Fig. 15.4). Boila preserves a

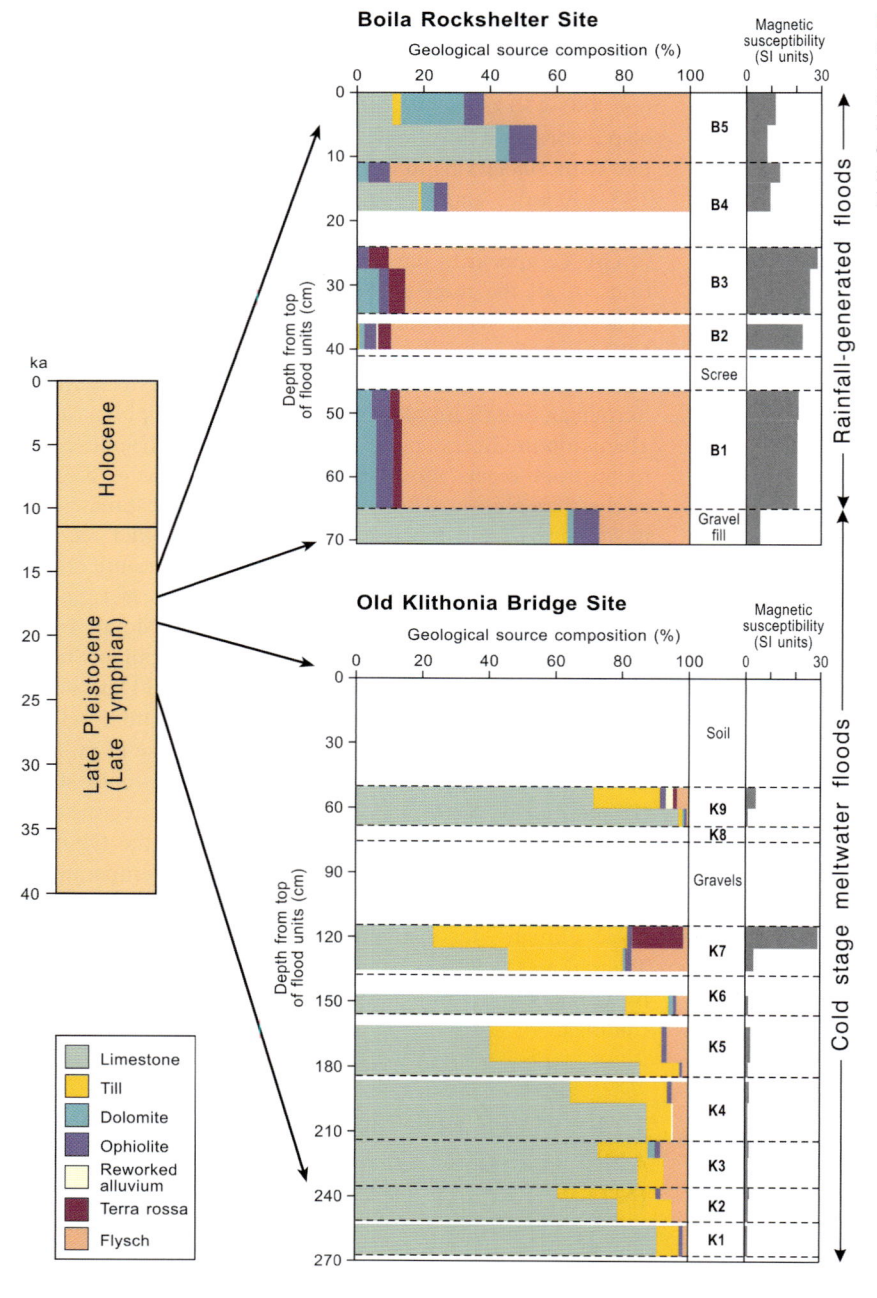

FIGURE 15.11 Slackwater sediment records and suspended sediment source data from two sites in the lower Voidomatis River basin. Together they span the last glacial to interglacial transition in the Pindus Mountains (modified from Woodward et al., 2008). Boila Rockshelter and The Old Klithonia Bridge site are both located in Reach II shown in Fig. 15.4.

Late-glacial record, and these sediments show very clearly how the catchment switched rapidly during the last glacial to interglacial transition to a system dominated by rainfall-generated floods with suspended sediments originating from lower elevation (nonglaciated) flysch terrains (Fig. 15.4). The composition of the flood sediments from Old Klithonia Bridge demonstrates that glaciers were active in the headwaters of the Voidomatis River basin towards the end of the last cold stage. This flood record provides valuable age control for these glacial processes because we do not have any uranium-series ages for the Late Pleistocene Tymphian moraines (Fig. 15.10).

15.9. THE MOUNT TYMPHI AND MOUNT OLYMPUS RECORDS

Smith et al. (1994) have argued that the first phase of glaciation on Olympus was the most extensive, with ice lobes covering part of the eastern, northern and western piedmont of the mountain. This period of glaciation was associated with a snowline as low as 1900 m above present sea level. Thus, in common with the record from Mount Tymphi, glacial activity was not as extensive during the last cold stage. All the uranium-series ages shown in Fig. 15.10 are significantly older than almost all the ^{36}Cl ages reported by Manz (1998) for the unit 1 moraine complexes on Mount Olympus. As Tymphi and Olympus are both located close to 40°N, one may expect a broad correlation between major phases of ice build-up and glacial sediment deposition. Recent work on Mount Orjen in Montenegro, for example, has produced a very similar record of Middle and Late Pleistocene glaciation to that recognised on Tymphi (Hughes et al., 2010). It is not possible to reconcile the extensive Middle Pleistocene glacial sequence we have identified on Mount Tymphi with the compressed sequence of post-MIS 5 glacial cycles and pedogenic weathering phases required on Mount Olympus if the MIS 3 cosmogenic ages preferred by Manz (1998) are accepted. In fact, notwithstanding the weaknesses discussed earlier and in view of the now substantial body of uranium-series ages from Mount Tymphi, the timescales associated with the tentative age model presented by Smith et al. (1997) would appear to be more realistic (Table 15.2). It is important to appreciate that, in common with the uranium-series data from the secondary carbonates on Mount Tymphi, the cosmogenic ages from Mount Olympus also constitute *minimum* ages. Thus, the cosmogenic ages from Site 2 indicate glacial activity at some time before 43–56 ka, while the outlier of 146 ka may represent glacial activity during the Vlasian Stage of MIS 6. This scenario would be much closer to the age model we have proposed for the glacial sediments and landforms on Mount Tymphi. However, before detailed correlations can be attempted, it is clear that more radiometric dates are needed from Mount Olympus.

15.10. PLEISTOCENE ELAs ACROSS GREECE

Hagedorn et al. (1969) was the first to estimate Pleistocene snowline altitudes across Greece, and these are shown in Fig. 15.1. Boenzi and Palmentola (1997) later estimated ELAs for mountains in Greece, Albania and southern Italy. In the northern Pindus Mountains, the estimates of Hagedorn and Boenzi and Palmentola closely correspond with the ELAs calculated for the most extensive Skamnellian Stage glaciers by Hughes et al. (2006c, 2007b). However, in common with earlier workers, they did not provide any radiometric dates for these glacial records. Boenzi and Palmentola (1997, p. 21) stated, "On the highest mountains of the Southern Apennines (Italy) and of the region stretching from Albania as far as Crete, many traces of glacial modelling are preserved, all attributable to the last glacial age (Würmian)." This assumption is now known to be incorrect as the Würmian Stage glaciers (Tymphian) were restricted to the highest cirques, and the ELAs estimated by Hagedorn (1969) and Boenzi and Palmentola (1997) relate to the largest Middle Pleistocene glaciations of the Skamnellian Stage (MIS 12). Nevertheless, as noted for northern Greece, these estimates do represent an accurate reflection of the ELAs during the most extensive glaciation in Greece. Boenzi and Palmentola (1997) have computed an increase in ELA of about 100 m per degree of latitude moving southwards from Albania to Crete. From a palaeoclimatic perspective, it is interesting to note that, in the Pindus Mountains of Greece, glacier formation only took place on peaks higher than 2200 m, whereas in the Southern Appenines of Italy, traces of glacial activity are found at lower elevations on mountains > 1900 m. This observation may reflect regional contrasts in air mass trajectories and other components of cold stage climates.

15.11. CONCLUSIONS AND FUTURE RESEARCH NEEDS

Recent work has shown that the mountains of Greece contain evidence for multiple phases of ice build-up and decay during the Middle and Late Pleistocene. Detailed field-based mapping has only been carried out in two areas: on Mount Olympus and the surrounding piedmont zone (Smith et al., 1997) and on Mount Tymphi in northwest Greece. The morphological and sedimentological evidence for glaciation is less extensive and less well preserved in central and southern Greece. The absence of a reliable and internally consistent chronology for the Mount Olympus record limits its value as a source of palaeoclimatic information and prevents detailed comparison with other sites. Recent work on Mount Tymphi has demonstrated the value of combining detailed field mapping with uranium-series dating in glaciated limestone landscapes (Woodward et al., 2004; Hughes et al., 2006b). The record on Mount Tymphi is proposed as a formal stratigraphical framework for the glacial record in Greece and the wider region. It is quite possible that evidence of glaciation during MIS 8 and 10 has been preserved on some parts of the Pindus Mountains, but there is still a need for detailed field mapping to establish the precise spatial extent and style of glacial activity. The uranium-series ages from Mount Tymphi show that the most extensive glaciations took place during the Middle Pleistocene and not during the most recent cold stage and the global LGM (MIS 2). The alluvial record in the Voidomatis basin shows how the shift from full glacial to interglacial conditions can take place very rapidly.

Large areas of the Mediterranean region are dominated by uplifted carbonate terrains, and the glacial sediments are often cemented by carbonate materials than can be dated by uranium-series methods. The uranium-series ages reported here indicate that glaciation was extensive on Mount Tymphi during MIS 6 and earlier cold stages. We also present the first dated evidence for glacial activity in the Mediterranean region before 350 ka. Together, these ages show that many of the glacial landforms have been very well preserved and that ice build-up on Mount Tymphi (and perhaps in the rest of the Pindus Mountains) was much less extensive during the Late Würmian. Regional comparisons of ELAs in the Mediterranean should *not* assume that fresh glacial landforms relate to the latter part of the most recent cold stage—several reconstructions have made the mistake of using Middle Pleistocene ice limits to reconstruct ELAs for the last cold stage (e.g. Kuhlemann et al., 2008). Data from Mount Tymphi show very clearly that apparently fresh glacial landforms may be much older than previously thought.

Robust geochronologies are required to compare the timings of glaciations in the different mountain areas of the Mediterranean. At present, the geochronological framework for glacial sequences is patchy, although there has been a dramatic increase in the number of studies in the past decade—mainly utilising cosmogenic nuclide analyses to date glacial surfaces (e.g. Vidal-Romaní and Fernández-Mosquera, 2006; Pallàs et al., 2007; Akçar et al., 2008; Kuhlemann et al., 2008; Sarıkaya et al., 2008, 2009; Chapters 11 and 30). Cosmogenic isotope analyses have largely been restricted to noncarbonate lithologies (utilising ^{10}Be, ^{21}Ne and ^{36}Cl isotopes), and different studies have yielded contrasting geochronologies. In Iberia, Vidal-Romaní and Fernández-Mosquera (2006) applied ^{21}Ne and ^{10}Be analyses to date glacial landforms in granitic terrain in Spain and found evidence for Middle and Late Pleistocene glaciations, similar to that found in Greece (although at different times within the Middle Pleistocene). In other areas, such as Turkey, only Late Pleistocene glaciations have been found using cosmogenic dating (Akçar et al., 2008; Sarıkaya et al., 2008, 2009). In theory, carbonate surfaces can also be dated using ^{36}Cl, and it may be possible to develop multiple approaches for dating glacial sequences (by combining U-series and ^{36}Cl). For example, ^{36}Cl analyses have been successfully applied to date landslides on limestone rocks in the Alps (Ivy-Ochs et al., 2009). However, ^{36}Cl analyses have rarely been applied to date glaciated limestone terrains because of problems in estimating surface loss by solution—especially in wet (>2000 mm), highly karstic, mountain regions, such as northwest Greece. Other possibilities include the dating of glaciated ophiolite surfaces, such as those on Mount Smolikas, using ^{3}He. However, there are theoretical and practical obstacles in the effective application of these techniques to the lithologies of the mountains of northwest Greece. Uranium-series dating remains the most effective dating technique for establishing the geochronology of glacial deposits in this area.

There is little doubt that the glacial sediments and landforms in the mountains of southern Europe form an important record of Quaternary environmental change. As we have shown for northwest Greece, where appropriate samples are available, uranium-series methods may provide the best means of developing good age models for sedimentary sequences in glaciated limestone terrains. They can also provide the necessary age control to develop models of ice build-up and decay over several glacial–interglacial cycles (see Hughes et al., 2010). More robust chronological frameworks are now emerging for several parts of the Mediterranean, and this is allowing regional contrasts in glacier dynamics to be explored (e.g. Hughes and Woodward, 2008, 2009; Giraudi et al., 2011).

ACKNOWLEDGEMENTS

We thank Chronis Tzedakis for kindly supplying raw pollen and isotopic data for Fig. 15.10 and Giannis Tsouratzis for allowing us to reproduce Fig. 15.4. We would like to thank the following for their contributions to the work on Mount Tymphi: Mark Macklin, John Lewin, Graham Smith and Philip Gibbard. The first uranium-series age determinations were carried out by Stuart Black (then at the University of Lancaster and now at Reading) and then by collaborators at the NERC Open University Uranium-Series Facility including Mabs Gilmour and Peter van Calsteren (UK Natural Environment Research Council, Award: IP/754/0302). We would also like to thank Nick Scarle in the Cartographic Unit in the School of Environment and Development for producing the diagrams.

REFERENCES

Akçar, N., Yavuz, V., Ivy-Ochs, S., Kubik, P.W., Vardar, M., Schlüchter, C., 2008. A case for a down wasting mountain glacier during Termination I, Verçenik Valley, NE Turkey. J. Quatern. Sci. 23, 273–285.

Bailey, G.N., Lewin, J., Macklin, M.G., Woodward, J.C., 1990. The "Older Fill" of the Voidomatis Valley Northwest Greece and its relationship to the Palaeolithic Archaeology and Glacial History of the Region. J. Archaeol. Sci. 17, 145–150.

Bailey, G.N., Turner, C., Woodward, J.C., Macklin, M.G., Lewin, J., 1997. The Voidomatis Basin: an introduction. In: Bailey, G.N. (Ed.), Klithi:

Palaeolithic Settlement and Quaternary Landscapes in Northwest Greece. Volume 2: Klithi in its Local and Regional Setting. McDonald Institute for Archaeological Research, Cambridge, pp. 321–345.

Bard, E., Delaygue, G., Rostek, F., Antonioli, F., Silenzi, S., Schrag, D.P., 2002. Hydrological conditions over the western Mediterranean during the deposition of the cold sapropel 6 (ca. 175 kyr BP). Earth Planet. Sci. Lett. 202, 481–494.

Birkeland, P.W., 1984. Soils and Geomorphology. Oxford University Press, Oxford.

Boenzi, F., Palmentola, G., Sanso, P., Tromba, F., 1992. Aspetti geomorfologici del massiccio dei Leuka Ori nell'isola di Creta (Grecia), conparticolare reguardo alle forme carsiche. Geologia Applicata e Idrogeologia 17, 75–83.

Boenzi, F., Palmentola, G., 1997. Glacial features and snow-line trend during the last glacial age in the Southern Apennines (Italy) and on Albanian and Greek mountains. Z. Geomorphol. N. F. Suppl 41, 21–29.

Bonnefont, J.C. 1972. La Crète, etude morphologique. Ph.D. thesis. Université de Lille III, France. 845 pp.

Demitrack, A., 1986. The late Quaternary geologic history of the Larissa plain, Thessaly, Greece: Tectonic, climatic and human impact on the landscape. Unpublished PhD dissertation, Stanford University.

Fabre, G., Maire, R., 1983. Néotectonique et morphologénèse insulaire en Grèce: le massif du Mont Ida (Crète). Méditerranée 2, 39–40.

Faugères, L., 1969. Problems created by the geomorphology of Olympus, Greece: Relief, formation, and traces of Quaternary cold periods with discussion. Association Française pour l'Etude du Quaternaire Bulletin 6, 105–127.

Gascoyne, M., 1992. Palaeoclimate determination from cave calcite deposits. Quatern. Sci. Rev. 11, 609–632.

Giraudi, C., Bodrato, G., Lucchi, M.R., Cipriani, N., Villa, I.M., Giaccio, B., et al., 2011. Middle and late Pleistocene glaciations in the Campo Felice Basin (central Apennines), Italy. Quatern. Res. 75, 219–230.

Hagedorn, J., 1969. Beiträge zur Quartärmorphologie griechischer Hochgebirge. Göttinger Geographische Abhandlungen 50, 135 pp.

Hughes, P.D., 2004. Quaternary glaciation in the Pindus Mountains, Northwest Greece. In: PhD thesis. University of Cambridge. 341 pp.

Hughes, P.D., Braithwaite, R.J., 2008. Application of a degree-day model to reconstruct Pleistocene glacial climates. Quatern. Res. 69, 110–116.

Hughes, P.D., Woodward, J.C., 2008. Timing of glaciation in the Mediterranean mountains during the last cold stage. J. Quatern. Sci. 23, 575–588.

Hughes, P.D., Woodward, J.C., 2009. Glacial and periglacial environments. In: Woodward, J.C. (Ed.), The Physical Geography of the Mediterranean. Oxford University Press, Oxford, pp. 353–383.

Hughes, P.D., Gibbard, P.L., Woodward, J.C., 2003. Relict rock glaciers as indicators of Mediterranean palaeoclimate during the Last Glacial Maximum (Late Würmian) of northwest Greece. J. Quatern. Sci. 18, 431–440.

Hughes, P.D., Woodward, J.C., Gibbard, P.L., 2006a. Glacial history of the Mediterranean mountains. Prog. Phys. Geogr. 30, 334–364.

Hughes, P.D., Woodward, J.C., Gibbard, P.L., Macklin, M.G., Gilmour, M.A., Smith, G.R., 2006b. The glacial history of the Pindus Mountains, Greece. J. Geol. 114, 413–434.

Hughes, P.D., Woodward, J.C., Gibbard, P.L., 2006c. Late Pleistocene glaciers and climate in the Mediterranean region. Glob. Planet. Change 46, 83–98.

Hughes, P.D., Gibbard, P.L., Woodward, J.C., 2006d. Middle Pleistocene glacier behaviour in the Mediterranean: sedimentological evidence from the Pindus Mountains, Greece. J. Geol. Soc. Lond. 163, 857–867.

Hughes, P.D., Woodward, J.C., Gibbard, P.L., 2006e. The last glaciers of Greece. Z. Geomorphol. 50, 37–61.

Hughes, P.D., Gibbard, P.L., Woodward, J.C., 2007a. Geological controls on Pleistocene glaciation and cirque form in Greece. Geomorphology 88, 242–253.

Hughes, P.D., Woodward, J.C., Gibbard, P.L., 2007b. Middle Pleistocene cold stage climates in the Mediterranean: new evidence from the glacial record. Earth Planet. Sci. Lett. 253, 50–56.

Hughes, P.D., Woodward, J.C., van Calsteren, P.C., Thomas, L.E., Adamson, K., 2010. Pleistocene ice caps on the coastal mountains of the Adriatic Sea: palaeoclimatic and wider palaeoenvironmental implications. Quatern. Sci. Rev. 29, 3690–3708.

IGME, 1970. 1:50,000 Geological Map of Greece (Tsepelovon Sheet). Institute of Geological and Mineral Exploration, Athens.

Ivy-Ochs, S., Poschinger, A.V., Synal, H.-A., Maisch, M., 2009. Surface exposure dating of the Flims landslide, Graubünden, Switzerland. Geomorphology 103, 104–112.

Kotarba, A., Hercman, H., Dramis, F., 2001. On the age of Campo Imperatore glaciations, Gran Sasso Massif, Central Italy. Geografia Fisica e Dinamica Quaternaria 24, 65–69.

Kuhlemann, J., Rohling, E.J., Krumrei, I., Kubik, P., Ivy-Ochs, S., Kucera, M., 2008. Regional synthesis of Mediterranean atmospheric circulation during the last glacial maximum. Science 321, 1338–1340.

Lewin, J., Woodward, J.C., 2009. Karst geomorphology and environmental change. In: Woodward, J.C. (Ed.), The Physical Geography of the Mediterranean. Oxford University Press, Oxford, pp. 287–318.

Lewin, J., Macklin, M.G., Woodward, J.C., 1991. Late Quaternary fluvial sedimentation in the Voidomatis Basin, Epirus, northwest Greece. Quatern. Res. 35, 103–115.

Lewis, C.J., McDonald, E.V., Sancho, C., Luis Peña, J., Rhodes, E.J., 2009. Climatic implications of correlated Upper Pleistocene glacial and fluvial deposits on the Cinca and Gállego Rivers (NE Spain) based on OSL dating and soil stratigraphy. Glob. Planet. Change 67, 141–152.

Louis, H., 1926. Glazialmorphologishche Beobachtungen im Albanischen Epirus. Z. Ges. Erdkunde 1926, 398–409.

Macklin, M.G., Woodward, J.C., 2009. River Systems and Environmental Change. In: Woodward, J.C. (Ed.), The Physical Geography of the Mediterranean. Oxford University Press, pp. 319–352.

Macklin, M.G., Lewin, J., Woodward, J.C., 1997. Quaternary river sedimentary sequences of the Voidomatis basin. In: Bailey, G.N. (Ed.), Klithi: Palaeolithic Settlement and Quaternary Landscapes in Northwest Greece. Volume 2: Klithi in its Local and Regional Setting. McDonald Institute for Archaeological Research, Cambridge, pp. 347–359.

Macklin, M.G., Fuller, I.C., Lewin, J., Maas, G.S., Passmore, D.G., Rose, J., et al., 2002. Correlation of fluvial sequences in the Mediterranean basin over the Last Interglacial-Glacial Cycle and their relationship to climate change. Quatern. Sci. Rev. 21, 1633–1641.

Manz, L.A., 1998. Cosmogenic ^{36}Cl chronology for deposits of presumed Pleistocene age on the Eastern Piedmont of Mount Olympus, Pieria, Greece. Unpublished MSc thesis, Ohio University, USA.

Marković, S.B., Hambach, U., Catto, N., Jovanović, M., Buggle, B., Machlett, B., et al., 2009. Middle and Late Pleistocene loess sequences at Batajnica, Vojvodina, Serbia. Quatern. Int. 198, 255–266.

Mastronuzzi, G., Sanso, P., Stamatopoulos, L., 1994. Glacial landforms of the Peloponnisos (Greece). Riv. Geogr. Ital. 101, 77–86.

Messerli, B., 1967. Die Eiszeitliche und die gegenwartige Vergletscherung im Mittelmeeraum. Geogr. Helv. 22, 105–228.

Mistardis, G., 1952. Recherches glaciologiques dans les parties supérieures des Monts Oeta et Oxya Grèce Centrale. Zeitschrift für Gletscherkunde und Glazialgeologie 2, 72–79.

Nemec, W., Postma, G., 1993. Quaternary alluvial fans in southwestern Crete: sedimentation processes and geomorphic evolution. Special Publication of the International Association of Sedimentology 17, 256–276.

Niculescu, C., 1915. Sur les traces de glaciation dans le massif Smolica chaîne du Pinde méridional. Bulletin de la Section Scientifique de l'Academie Roumaine 3, 146–151.

Ohmura, A., Kasser, P., Funk, M., 1992. Climate at the equilibrium line of glaciers. J. Glaciol. 38, 397–411.

Pallàs, R., Rodés, Á., Braucher, R., Carcaillet, J., Ortuño, M., Bordonau, J., et al., 2007. Late Pleistocene and Holocene glaciation in the Pyrenees: a critical review and new evidence from ^{10}Be exposure ages, south-central Pyrenees. Quatern. Sci. Rev. 25, 2937–2963.

Palmentola, G., Boenzi, F., Mastronuzzi, G., Tromba, F., 1990. Osservazioni sulle trace glaciali del M. Timfi, Catena del Pindo (Grecia). Geografia Fisica e Dinamica Quaternaria 13, 165–170.

Pechoux, P.Y., 1970. Traces d'activité glaciaire dans les Montagnes de la Grece centrale. Rev. Géogr. Alpine 58, 211–224.

Peyron, O., Guiot, J., Cheddadi, R., Tarasov, P., Reille, R., de Beaulieu, J.-L., Bottema, S., Andrieu, V., 1998. Climatic reconstruction in Europe for 18,000 yr BP from pollen data. Quaternary Research 49, 183–196.

Poser, J., 1957. Klimamorphologische Probleme auf Kreta. Zeitschrift für Geomorphologie 2, 113–142.

Sarıkaya, M.A., Zreda, M., Çiner, A., Zweck, C., 2008. Cold and wet Last Glacial Maximum on Mount Sandıras, SW Turkey, inferred from cosmogenic dating and glacier modelling. Quatern. Sci. Rev. 27, 769–780.

Sarıkaya, M.A., Zreda, M., Çiner, A., 2009. Glaciations and paleoclimate of Mount Erciyes, central Turkey, since the Last Glacial Maximum, inferred from ^{36}Cl dating and glacier modeling. Quatern. Sci. Rev. 23–24, 2326–2341.

Sestini, A., 1933. Tracce glaciali sul Pindo epirota. Bollettino della Reale Società Geografica Italiano 10, 136–156.

Sfikas, G., 1979. The Mountains of Greece. Efstathiadis Group, Athens. 204 pp.

Smith, G.W., Nance, R.D., Genes, A.N., 1994. A re-evaluation of the extent of Pleistocene glaciation in the Mount Olympus region, Greece. Bull. Geol. Soc. Greece 1, 293–305, (Proceedings of the 7th Congress, Thessaloniki, May 1994).

Smith, G.W., Nance, R.D., Genes, A.N., 1997. Quaternary glacial history of Mount Olympus, Greece. Geol. Soc. Am. Bull. 109, 809–824.

Smith, G.R., Woodward, J.C., Heywood, D.I., Gibbard, P.L., 1998. Mapping glaciated karst terrain in a Mediterranean mountain environment using SPOT and TM data. In: Burt, P.J.A., Power, C.H., Zukowski, P.M. (Eds.), RSS98: Developing International Connections, 457–463, (Proceedings of the Remote Sensing Society's Annual Meeting), Greenwich, September 1998.

Smith, G.R., Woodward, J.C., Heywood, D.I., Gibbard, P.L., 2000. Interpreting Pleistocene glacial features from SPOT HRV data using fuzzy techniques. Comput. Geosci. 26, 479–490.

Smith, G.W., Nance, R.D., Genes, A.N., 2006. Pleistocene glacial history of Mount Olympus, Greece: Neotectonic uplift, equilibrium line elevations, and implications for climatic change. In: Dilek, Y., Pavlides, S. (Eds.), Postcollisional Tectonics and Magmatism in the Mediterranean Region and ASIA, vol. 409, Geological Society of America. Special Paper 157–174.

Tzedakis, P.C., 2005. Towards an understanding of the response of southern European vegetation to orbital and suborbital climate variability. Quatern. Sci. Rev. 24, 1585–1599.

Tzedakis, P.C., McManus, J.F., Hooghiemstra, H., Oppo, D.W., Wijmstra, T.A., 2003. Comparsion of changes in vegetation in northeast Greece with records of climate variability on orbital and suborbital frequencies over the last 450,000 years. Earth Planet. Sci. Lett. 212, 197–212.

Tzedakis, P.C., Hooghiemstra, H., Pälike, H., 2006. The last 1.35 million years at Tenaghi Philippon: revised chronostratigraphy and long-term vegetation trends. Quatern. Sci. Rev. 25, 3416–3430.

van Andel, T.H., Zangger, E., Demitrack, A., 1990. Land use and soil erosion in prehistoric Greece. J. Field Archaeol. 17, 379–396.

Vidal-Romaní, J.R., Fernández-Mosquera, D., 2006. Glaciarismo Pleistoceno en el NW de la peninsula Ibérica (Galicia, España-Norte de Portugal). Enseñanza de las Ciencias de la Tierra 13, 270–277. http://www.raco.cat/index.php/ect/article/view/89058/133836.

Waltham, A.C., 1978. The Caves and Karst of Astraka, Greece. Trans. Br. Cave Res. Assoc. 5, 1–12.

Wijmstra, T.A., 1969. Palynology of the first 30 m of a 120 m deep section in northern Greece. Acta Bot. Neerl. 18, 511–527.

Woodward, J.C., 1990. Late Quaternary Sedimentary Environments in the Voidomatis Basin, Northwest Greece. Unpublished PhD thesis, University of Cambridge, Cambridge, UK.

Woodward, J.C., Lewin, J., Macklin, M.G., 1992. Alluvial sediment sources in a glaciated catchment: the Voidomatis basin, northwest Greece. Earth Surf. Process. Land. 16, 205–216.

Woodward, J.C., Lewin, J., Macklin, M.G., 1995. Glaciation, river behaviour and the Palaeolithic settlement of upland northwest Greece. In: Lewin, J., Macklin, M.G., Woodward, J.C. (Eds.), Mediterranean Quaternary River Environments. Balkema, Rotterdam, pp. 115–129.

Woodward, J.C., Macklin, M.G., Smith, G.R., 2004. Pleistocene Glaciation in the Mountains of Greece. In: Ehlers, J., Gibbard, P.L. (Eds.), Quaternary Glaciations—Extent and Chronology. Part I: Europe. Amsterdam, Elsevier, pp. 155–173.

Woodward, J.C., Hamlin, R.H.B., Macklin, M.G., Hughes, P.D., Lewin, J., 2008. Glacial activity and catchment dynamics in northwest Greece: long-term river behaviour and the slackwater sediment record for the last glacial to interglacial transition. Geomorphology 101, 44–67.

Chapter 16

Pliocene and Pleistocene Glaciations of Iceland: A Brief Overview of the Glacial History

Áslaug Geirsdóttir

Department of Earth Sciences and Institute of Earth Sciences, University of Iceland, Askja, Sturlugata 7, 101 Reykjavík, Iceland

16.1. INTRODUCTION

The most direct evidence for the existence of glaciers in the past is the identification of ice-contact sediment. However, reconstruction of the pattern of growth and decay of ice sheets on late Tertiary–Quaternary timescales has proved to be difficult because of the rare preservation of landforms in such erosional systems. Deductions on the extent of glaciers (ice volume) and periodicity of glaciations are, in most cases, derived from indirect proxies such as $\delta^{18}O$ records from microfossils and the distribution of ice-rafted detritus in deep-sea sediment. Available evidence largely extracted from marine sediment cores in the Greenland–Iceland–Norwegian Seas suggests that Northern Hemisphere glaciations began between 10 and 5 Ma, with the onset of large-scale glaciations around 2.75 Ma (cf. Raymo et al., 1986; Jansen et al., 1990; Shackleton et al., 1990; Jansen and Sjøholm, 1991; Wolf and Thiede, 1991; ODP leg 151, 1994; Maslin et al., 1998; Jansen et al., 2000). Based on these proxy data, more than 20 recurrent full glaciations are indicated to have occurred over the approximately 3 Ma.

Iceland occupies an unusual position in glacial and palaeoclimatic research because of its position near the polar front and at a rifting plate boundary. Small changes in global heat and ocean circulation (associated with North Atlantic Deep Water formation) can have a large impact on the terrestrial environment. At present, tropical waters are brought from lower latitudes with the North Atlantic Current, which is diverted up and along the southern and western shelf of Iceland to the Denmark Strait as the Irminger Current. Cold water from the Arctic Ocean influences the northern and north-eastern coasts as the East Icelandic Current, a branch of the East Greenland Current (Fig. 16.1A). This ocean–current system is manifested in a strong north–south temperature gradient across Iceland that has varied in magnitude at certain times in the past, as a function of differential strengths of the cold versus the warm surface currents.

Rifting plate boundaries that divide the island between the Eurasian and North American plate control the geological structure of Iceland (Fig. 16.1B). The configuration of plate margins on the island itself is complex and related to a mantle plume forming a hot spot under Iceland, approximately beneath the Vatnajökull ice sheet in the south-east. As a result of this configuration, the oldest rocks dating from ca. 15 Ma occur in the easternmost and westernmost parts of the island and the youngest occur along the rifting zone (Fig. 16.1B; Saemundsson, 1979; Johannesson and Sæmundsson, 1999). Unusually, high volcanic productivity, caused by the superposition of the mantle plume and mid-ocean ridge magmatism, has made Iceland an environment where constructive geological processes outweigh destructive processes (Geirsdóttir et al., 2007). The results are an excellent preservation of the late Tertiary and Quaternary depositional record, where sedimentary formations are covered and protected from erosion by subsequent basaltic lava flow sequences. Lava flows are deposited in ice-free regions, and the general consensus is that they represent interglacials, interstadials or ice-free periods, whereas hyaloclastites (subglacially formed volcanics) and till represent glaciations (Fig. 16.2). Mapping and dating of remnants of glacial deposits found imbedded within the basaltic lava flows have revealed an intermittent record of glaciation since the late Pliocene, with the earliest glacial deposits found in the south-eastern Iceland and eastern rock sequences and a little later in the western rock sequences (Fig. 16.1; Geirsdóttir and Eiríksson, 1994; Geirsdóttir et al., 2007).

The number of identified ancient glacial deposits in Iceland and comparison with the deep-sea proxy record and studies of glacier cyclicity imply that the late Tertiary

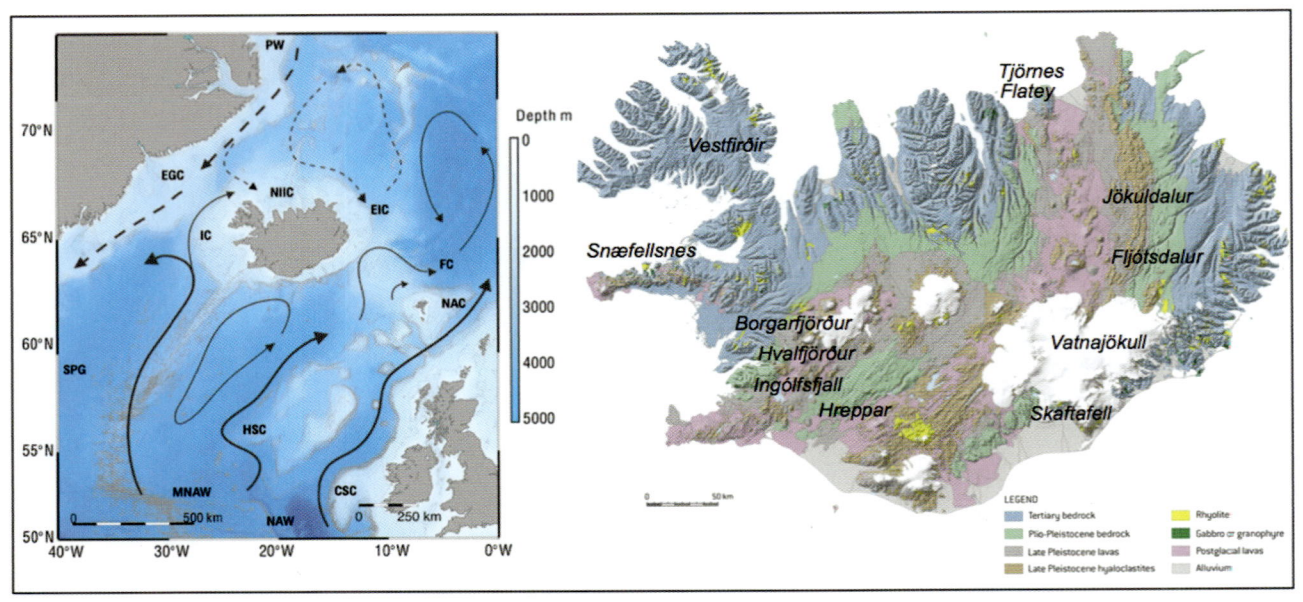

FIGURE 16.1 (A) Surface ocean currents in the North Atlantic Ocean, PW, polar water; EGC, East Greenland Current; EIC, East Iceland Current; IC, Irminger Current; SPG, Subpolar Gyre; HSC, Hatton Slope Current; CSC, Continental Slope Current; NAC, North Atlantic Current; FC, Faroe Current (modified from Hansen and Østerhus, 2000). (B) Geological map of Iceland (Johannesson and Sæmundsson, 1999) with the names of localities mentioned in the text.

and Quaternary terrestrial record captures the large-scale palaeoclimatic processes in the North Atlantic region throughout the late Cenozoic (Figs. 16.3 and 16.4).

16.2. TERTIARY GLACIATIONS—EARLIEST SIGNS OF GLACIATION IN SOUTH-EAST AND EAST ICELAND (>3–2.59 MA)

The Tjörnes beds, exposed in a raised marine section in North Iceland, contain clear evidence of the Pliocene warm period (5–3 Ma) and suggest gradual temperature change (based on oxygen isotope evidence), from warm-water conditions with summer temperatures between 10 and 15 °C at the base to cold-water conditions with temperatures between 5 and 10 °C at the top of the beds (Buchardt and Símonarson, 2003; Símonarson and Eiríksson, 2008). During the same time interval, or between 5 and 4 Ma, emerging but small ice coverage left glacial vestiges in the area of the extant Vatnajökull glacier in south-eastern Iceland (Geirsdóttir and Eiríksson, 1994; Helgason and Duncan, 2001; Geirsdóttir et al., 2007).

The oldest indication of glacier nucleation in Iceland is found in the Miocene (7 Ma) in a deeply eroded rock sequence in the south-east, at the margin of the current Vatnajökull ice cap (Figs. 16.1 and 16.3). Deposits of assumed glacial origin associated with hyaloclastites and pillow lavas, the characteristic rock types of sub-aquatic and/or subglacial eruptions, have been reported from this region (Fridleifsson, 1995). Just to the west in the Skaftafell/Hafrafell sections, Helgason and Duncan (2001) suggest a possible glacial event as old as 4.6–4.7 Ma based on the identification of sub-aquatically formed volcanic rocks (Figs. 16.1 and 16.3). However, the oldest confirmed glacial deposit associated with erosional features in that same section dates from ca. 4 Ma (Fig. 16.3). The age control of these sections is based on palaeomagnetism and K/Ar ages.

In Fljótsdalur, eastern Iceland, the oldest glacial deposit, based on detailed sedimentological analyses, is found interbedded with 3.8–4.0 Ma lava flows (Figs. 16.1 and 16.3; Geirsdóttir and Eiríksson, 1994; Geirsdóttir et al., 2007). Higher in the same stratigraphical sequence, another interpreted glacial bed gives an estimated age of 3.4 Ma also based on synchronisation with the palaeomagnetic timescale. Subglacial activity is further identified in the Skaftafell/Hafrafell sections around that time and correlated with the lowest normal part of the Gauss Chron of the palaeomagnetic timescale between 3.3 and 3.6 Ma (Figs. 16.3 and 16.4; Helgason and Duncan, 2001).

As it has not been possible to trace these oldest glacial beds over a larger area, they are thought to represent only local glacier activity (Fig. 16.5; Geirsdóttir and Eiríksson, 1994; Geirsdóttir et al., 2007). These glacial deposits are all found close to the Vatnajökull ice cap, which covers the highest and most humid region of Iceland, with an annual precipitation of 3600 mm. During Miocene and Pliocene time, this region was within the most active volcanic zone of the island where volcanism was anomalously high and where constructive processes dominated (cf. Harðarson et al., 2008). This gave rise to high-altitude areas which were particularly favourable for the build-up of glaciers.

FIGURE 16.2 Lava flows are deposited in ice-free regions. The general consensus is that they represent interglacials or interstadials, whereas hyaloclastites (subglacially formed volcanics) and tills represent glaciations.

Synchronisation of glacial deposits imbedded within lava flows, found in the three sections Fljótsdalur, Jökuldalur and Skaftafell/Hafrafell between 3 and 2.5 Ma, indicates the initiation of a gradual but progressive expansion of glaciers to ice caps in the south-east, interrupted by interglacial events. Intensification of ice-cap growth during the later part of this time interval is further supported by the discovery of glacial deposits intercalated with lava flows in western Iceland and dated between 2.8 and 2.5 Ma (Fig. 16.3–16.5; Geirsdóttir and Eiríksson, 1994). Intensification of glacial activity at around 2.8 Ma is concomitant with the first indications of continental-scale glaciations in the Northern Hemisphere (cf. Raymo et al., 1986; Jansen and Sjøholm, 1991; Fronval and Jansen, 1996; Jansen et al., 2000; Flesche Kleiven et al., 2002; Ravelo et al., 2004; Mudelsee and Raymo, 2005). If correlations between the sections are correct, at least five glacial advances are recorded in the period between 4 and 2.6 Ma (Fig. 16.4).

16.3. EARLY- TO MID-PLEISTOCENE GLACIATION (2.59–0.78 MA)—AN ICELANDIC ICE SHEET IS ESTABLISHED

The identification of glacial deposits in all parts of the country, dating from 2.5 to 2.4 Ma, suggests that a progressive cooling resulted in incremental north and westward growth of the Quaternary ice sheet from its nucleus in south-east Iceland. By this time, the ice sheet covered more than half of the island (Fig. 16.5). Between the Gauss–Matuyama

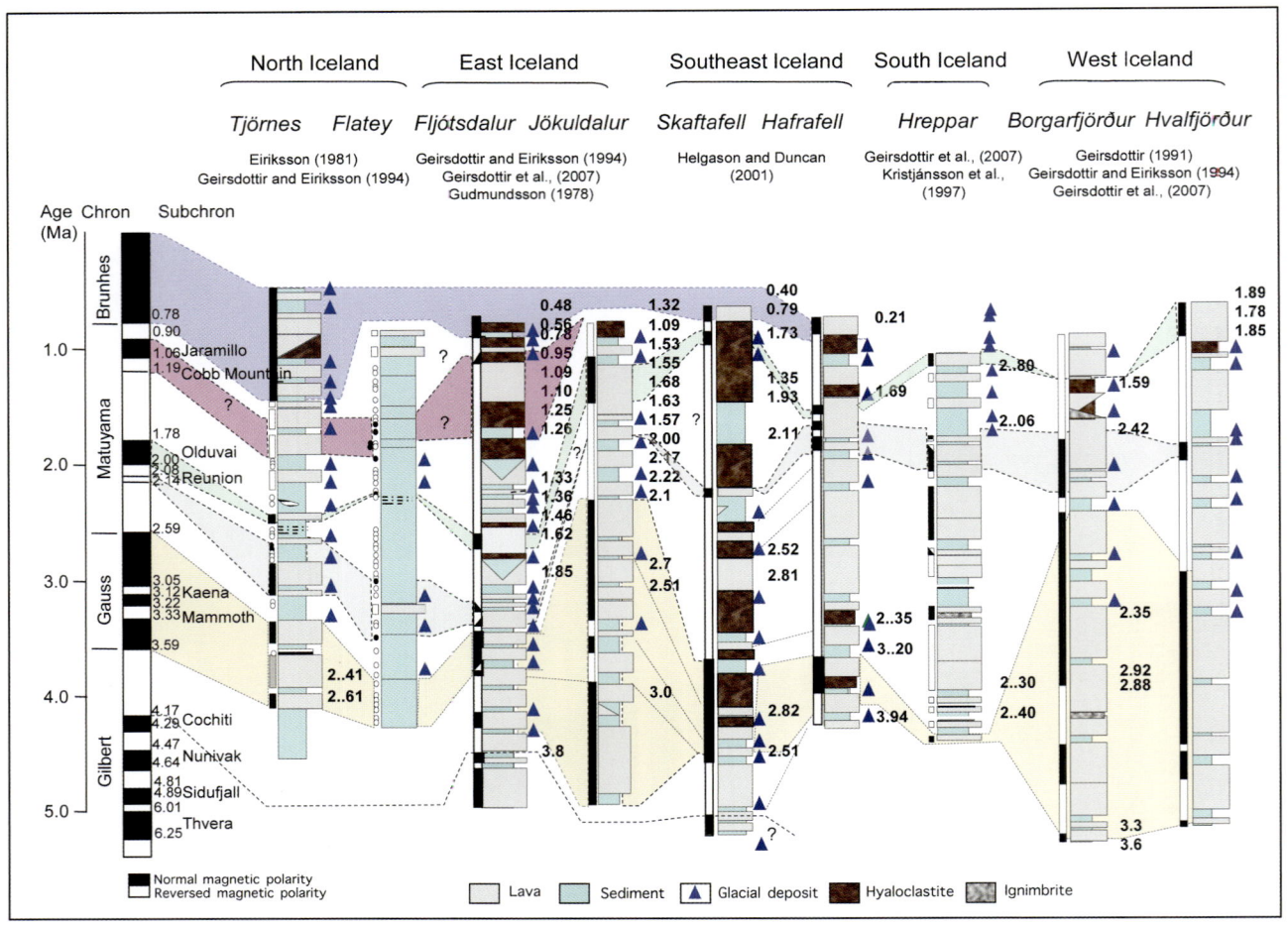

FIGURE 16.3 A proposed correlation between rock sequences from northern, eastern, south-eastern, southern and western Iceland, based on K–Ar dating and palaeomagnetic stratigraphy (Gudmundsson, 1978).

boundary (2.59 Ma) and the Brunhes Chron (0.78 Ma), at least 5 and as many as 13 glacial–interglacial cycles have been recognised (Fig. 16.4). In the Fljótsdalur section in eastern Iceland, 13 glacial deposits have been identified over this time period (Geirsdóttir and Eiríksson, 1994; Geirsdóttir et al., 2007), and in the Ingólfsfjall section, south-central Iceland, 12 strata of glacial deposits or subglacial volcanics have been described (Eiríksson, 2008; Fig. 16.1B). However, Geirsdóttir et al. (1993, 1994) and Kristjánsson et al. (1998) altogether identified only eight tillites during this time span (from ca. 2.5 to 0.78 Ma) in the Hreppar Formation also located in south-central Iceland. The Hreppar Formation is commonly referred to as an anticline between the two neovolcanic zones in southern Iceland showing the oldest date of approximately 2.5 Ma (Aronson and Sæmundsson, 1975). The formation shows a gradual change in the depositional environment through time, which is attributed partly to the climatic cooling and partly to its relative position between the two neovolcanic zones. The Tjörnes section in northern Iceland preserves nine glacial–interglacial cycles between 2.5 and 0.78 Ma, whereas the two western sections (Borgarfjördur and Hvalfjördur) and the south-eastern sections (Skaftafell/Hafrafell) contain seven and eight glacial cycles, respectively (Fig. 16.5). In Vestfirdir, north-western Iceland, a recent study by Roberts et al. (2007) reconstructs glacial activity between 2.4 and 2.2 Ma. There, the chronology is based on fission-track ages of two tephra layers found in raised glacier-dammed lake sediment exposed in the Skagafjall area (Figs. 16.1B and 16.5). This finding further supports interpreted intensification of glacier growth in Iceland around 2.5 Ma. However, it might also suggest a separate nucleation of glacier formation in the Vestfirðir area. Tillites from the Middle Pleistocene have additionally been reported from northern Iceland (Jancin et al., 1985), and on the Snæfellsnes Peninsula in western Iceland (Fig. 16.3; Einarsson, 1994; Leifsdóttir, 1999; Simonarson and Leifsdottir, 2007), where palaeomagnetic correlation suggests that the glacial deposits were probably formed between 2 and 0.7 Ma.

Significant expansion of glaciation by the Scandinavian Ice Sheet has been suggested around 2.57 Ma (Jansen and Sjøholm, 1991) in concert with the expansion of the

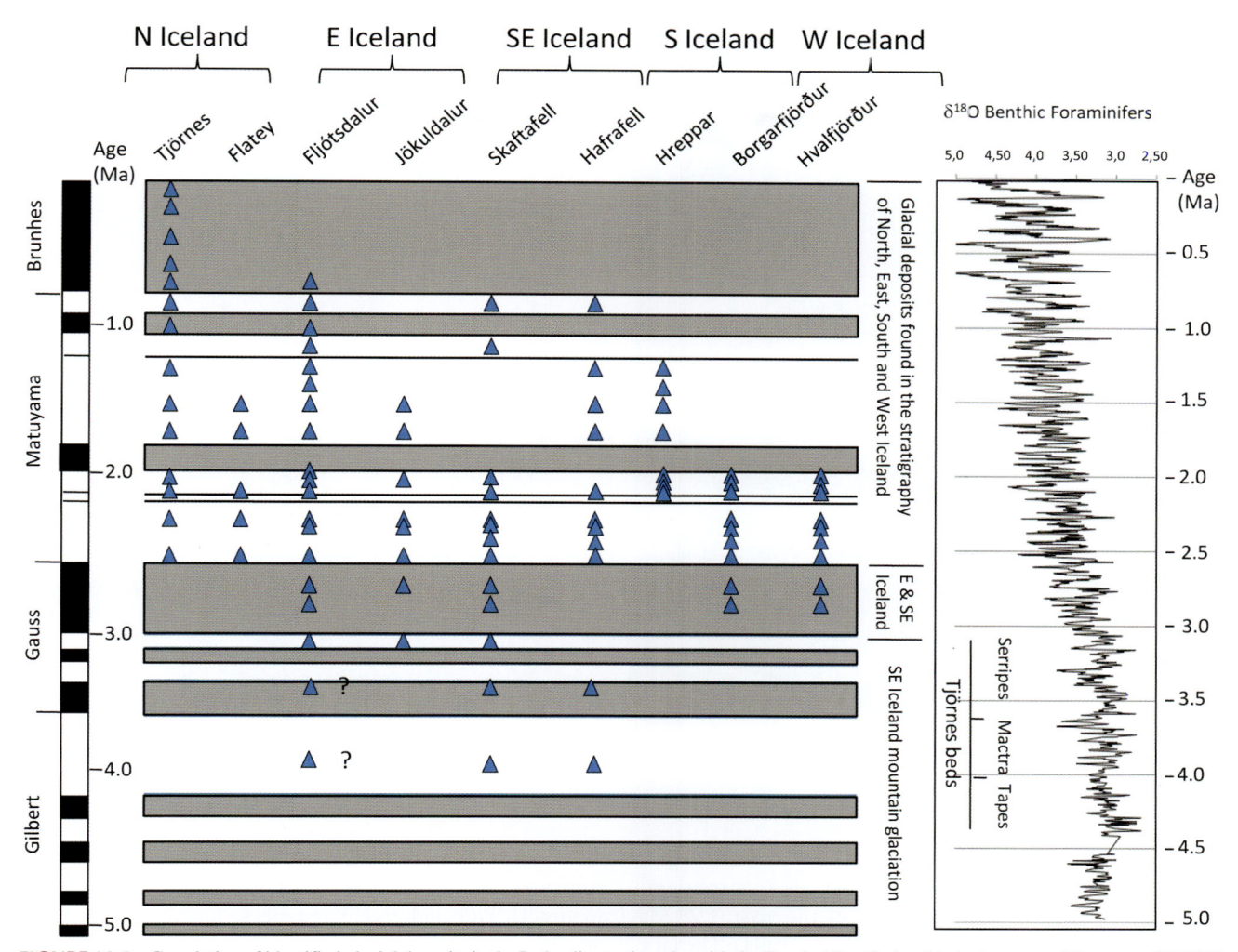

Chapter | 16 Pliocene and Pleistocene Glaciations of Iceland: A Brief Overview of the Glacial History

FIGURE 16.4 Correlation of identified glacial deposits in the Icelandic stratigraphy with the Tropical Pacific benthic isotope record from core ODP849 for 0–5 Ma (Mix et al., 1995a,b). Also shown is the magnetic polarity timescale of Cande and Kent (1995). The triangles show the approximate position of glacial deposits found in the Icelandic record. Molluscan biozones of the Tjörnes beds are also shown on the benthic isotope record based on Símonarson and Eiríksson, 2008.

Icelandic Ice Sheet. Shortly afterwards, or between 2.5 and 2.2 Ma, a single ice sheet covered the whole of Iceland for the first time and full-scale glacial/interglacial cyclicity was established. It is interesting to note that, during the normal palaeomagnetic Olduvai Subchron (2.00–1.78 Ma), glacial deposits are only preserved in the central South Iceland section (Eiríksson, 2008; Fig. 16.3). After that, or around 1.7 Ma, full-scale glacial cyclicity was re-established in Iceland, coinciding with an abrupt change in mineralogical composition and enhanced ice-rafted debris supply interpreted as the onset of large-scale marine-based glaciation in Svalbard (Knies et al., 2009).

Only two of the Icelandic sections contain evidence for full-scale glacial cyclicity between the Olduvai Subchron and the Brunhes Chron. They are the Tjörnes section in North Iceland and Fljótsdalur section in eastern Iceland (Figs. 16.1 and 16.4). Both sections are located at the flank of the rifting zone, where constructive and destructive forces compete continuously, offering ideal preservation conditions for the glacial record for that time. A typical stratigraphical section in Fljótsdalur contains a glacial deposit resting on a striated lava flow but underlying fluvial sediment and lacustrine sediment that is preserved by another lava flow that caps the section. In Tjörnes, this is reflected in glacial deposits underlying kame conglomerates and/or glaciolacustrine sediment and then followed by marine sediment as an indication of glacial retreat.

16.4. MIDDLE TO LATE PLEISTOCENE GLACIATIONS IN ICELAND (0.78–0.126 MA)

The glacial stratigraphy of Iceland from ca. 0.78 Ma to the last deglaciation is fragmentary, apart from in the Tjörnes

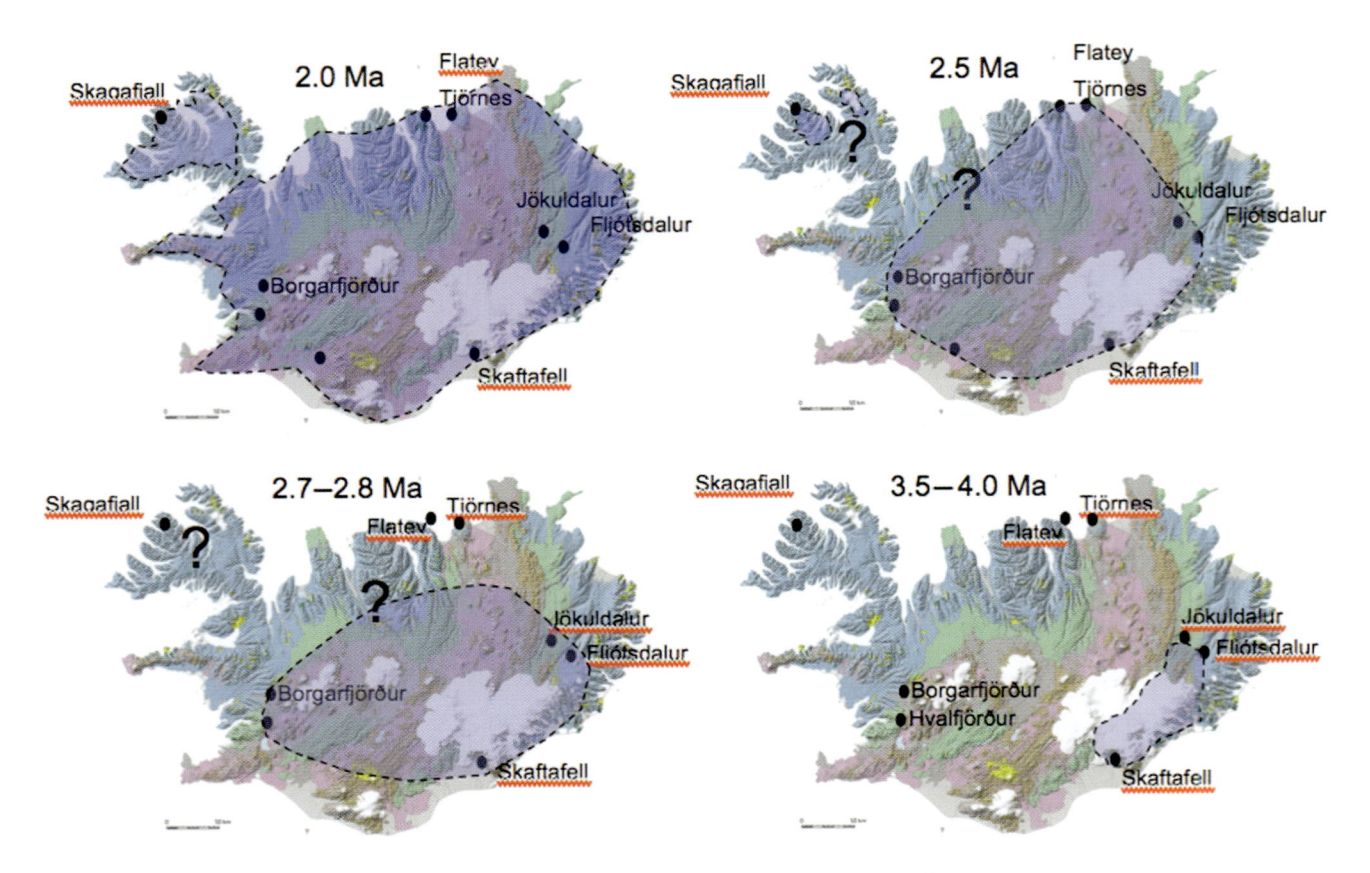

FIGURE 16.5 Hypothetical reconstructions of the Tertiary glaciations on Iceland for different time slices.

section in northern Iceland, which contains five stratigraphically separated glacial deposits in the past 1.78 Ma (Eiríksson, 1985; Geirsdóttir and Eiríksson, 1994). The lack of continuous records elsewhere in Iceland through the Quaternary may be due to a lack of systematic research and radiometric dates; ancient glacial and interglacial deposits of unknown age have been reported from various sites all around the island. However, discontinuities in the glacial record may also reflect differential preservation potential of glacial deposits where constructive forces (production of volcanic material) outweigh destructive forces (erosional) around the plate boundaries (in many cases, below the presumed ice divide)—or they may still be buried beneath younger lava flows, which have still not been eroded away (cf. Geirsdóttir et al., 2007). Between 4 and 0.1 Ma, the focus of volcanic activity was along the northern volcanic zone (Fig. 16.1) but around the Hreppar area (the eastern volcanic zone) in southern Iceland after only 3 Ma.

The Tjörnes section is the only site known in Iceland with distinct glacial–interglacial cyclicity for the past 0.7 Ma when the 100,000 years ice volume cycle developed (Eiríksson, 1985). The lower part of the Tjörnes section (2.5–0.7 Ma) reflects glacial–interglacial units in which glaciation is typically represented by lodgement tillite, followed by kame conglomerate and glaciolacustrine mudrocks and underlying marine sediments that include an interglacial mollusc fauna indicating glacial retreat. The upper part of the Tjörnes section was formed during a phase of local tectonic uplift, which in turn may reflect reduced preservation potential rather than change in the frequency of glacial cyclicity (Eiríksson, 2008). This is in accordance with intensified erosion arising from channelised ice flow and diminishing volcanic build-up in both the Skaftafell area in the south-east of the country (Helgason and Duncan, 2001) and the Fljótsdalur area in eastern Iceland (Geirsdóttir et al., 2007). This could also explain the lack of evidence of repeated early glaciation in Vestfirdir where destructive processes dominated over the constructive forces resulting in erosion of earlier glacial deposits. No volcanic activity has taken place here during the past 6 Ma.

16.5. FROM THE LAST GLACIAL MAXIMUM THROUGH TO THE LAST DEGLACIATION (<0.126 MA)

The full extent of the Iceland ice sheet during the Last Glacial Maximum (LGM) is poorly constrained. However, available evidence indicates that during the LGM, ice streams and outlet glaciers from ice divides in central Iceland reached to, or close to, the shelf edge (Fig. 16.5; Ólafsdóttir, 1975; Kaldal and Víkingsson, 1991; Andrews et al., 2000; Bourgeois et al., 2000; Stokes and Clark, 2001; Geirsdottir et al., 2002; Andrews and Helgadottir, 2003; Helgadottir et al., 2005; Hubbard et al., 2006; Geirsdóttir et al., 2009; Spagnolo and Clark, 2009). Small ice-free areas may have existed along the coastal mountains, particularly in the north-west, north and east (e.g. Einarsson and Albertsson, 1988; Ingólfsson, 1991; Norddahl, 1991; Ingólfsson and Norddahl, 1994; Rundgren and Ingólfsson, 1999; Andrews et al., 2000). The Vestfirdir Peninsula in north-west Iceland might have supported a dynamically independent ice cap and/or valley glaciers with outlet glaciers or ice streams originating within an ice divide near the centre of the peninsula (Hoppe, 1982; Sigurvinsson, 1983; Hjort et al., 1985; Andrews et al., 2002; Geirsdottir et al., 2002; Principato, 2003, 2008; Principato et al., 2006, Roberts et al., 2007; Principato and Johnson, 2009).

Studies based on sites around Iceland indicate that the last deglaciation began between 18 and 15 ka and that it was rapid (e.g. Ingólfsson, 1991; Norddahl, 1991; Ingólfsson and Norddahl, 1994; Ingólfsson et al., 1997; Syvitski et al., 1999; Andrews et al., 2000; Geirsdottir et al., 2002; Andrews and Helgadottir, 2003). Evidence for still-stands in the retreat of the main ice sheet during the last deglaciation is found in the Búdi moraines in south-central Iceland (Hjartarson and Ingólfsson, 1988; Geirsdóttir et al., 1997, 2000), and in moraine complexes in North Iceland (Fig. 16.6; Norddahl, 1991; Norddahl and Pétursson, 2005; Norddahl et al., 2008). The Búdi morainic complex is composed of multiple, discontinuous ridges that parallel the boundary between the interior highlands and the southern lowlands. Most of this morainic complex shows deltaic characteristics with distinct foreset bedding and sandur accumulation, indicating a transition between terrestrial and coastal environments (Geirsdóttir et al., 1997, 2000). The occurrence of the Vedde Ash (11.8 ka, Grönvold

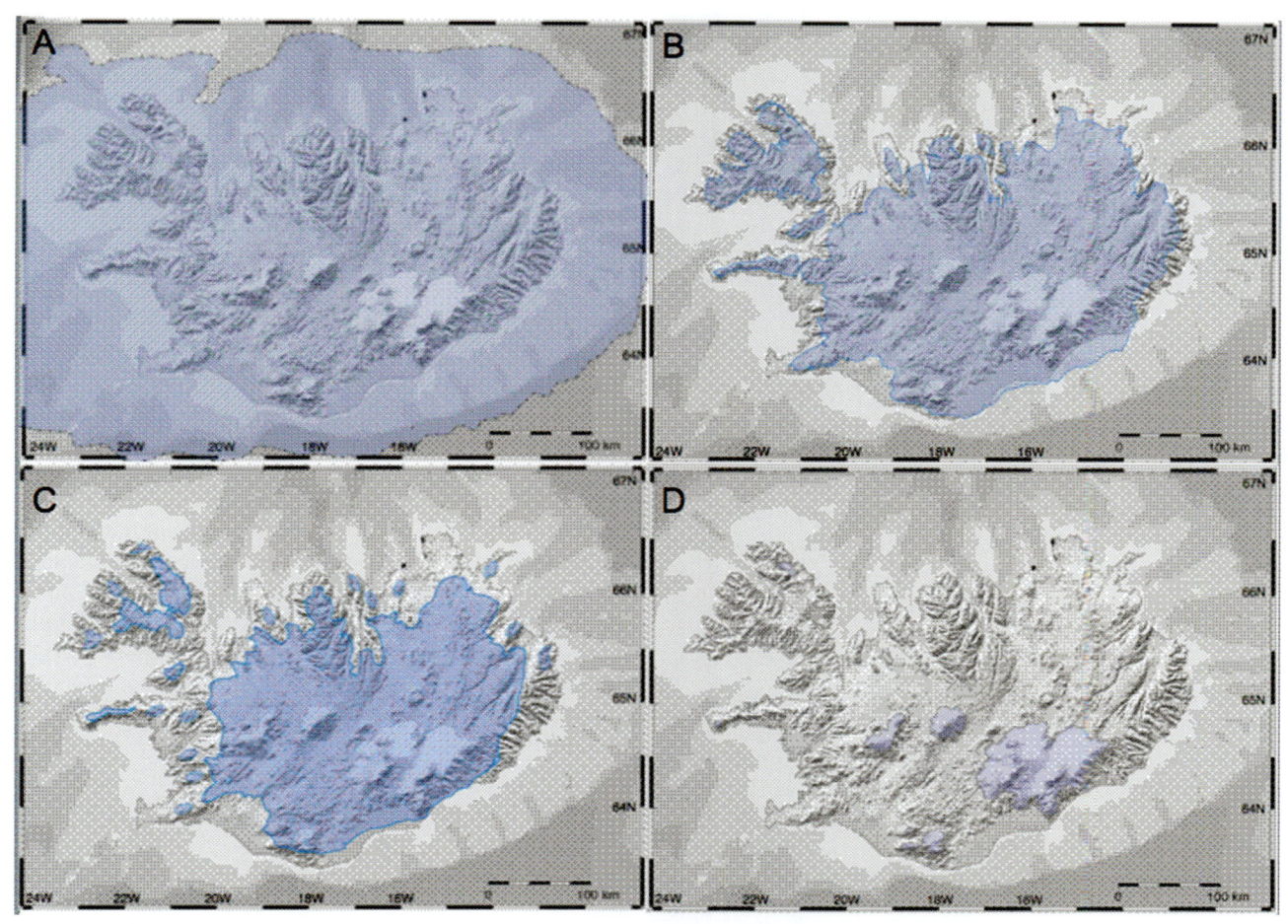

FIGURE 16.6 Interpreted Last Glacial Maximum (LGM: A), Younger Dryas (YD: B), PreBoreal (PB: C) and current glacier extent in Iceland (D). LGM, YD and PB moraines and other physical evidence revealing the maximum size of the Icelandic ice sheet at each time are based on Andrews et al. (2000), Geirsdottir et al. (2002), Andrews and Helgadottir (2003), Norddahl and Pétursson (2005), Norddahl et al. (2008), Hubbard et al. (2006) and Spagnolo and Clark (2009). YD and PB glacier extents are based on Geirsdóttir et al. (1997, 2000), Hannesdottir (2006) and Norddahl and Pétursson (2005).

et al., 1995) in a marine section of Lake Hestvatn sediment, a lake located in front of the Budi moraines, shows that during the Younger Dryas Chronozone (12.9–11.7 ka), outlet glaciers in southern Iceland terminated in a palaeobay inland of the current coastline, at or just beyond the Budi moraines in southern Iceland (Geirsdóttir et al., 2000; Hardardóttir, et al., 2001). The occurrence of the Skógar Tephra, correlated with the Vedde Ash in lake sediments in northern Iceland, constrains the age of assumed ice-dammed lakes and shows that during the Younger Dryas Chronozone, outlet glaciers in northern Iceland extended at least half-way out of Eyjafjörður (Fig. 16.4; Norddahl, 1991; Norddahl and Haflidason, 1992; Norddahl and Pétursson, 2005).

A second period of prominent glacier halt or re-advance in Iceland is dated to the early Preboreal time from 11.5 to 10.1 ka, when most of the inner parts of the Búdi moraines were formed (Fig. 16.6; Hjartarson and Ingólfsson, 1988; Geirsdóttir et al., 1997, 2000; Norddahl and Pétursson, 2005; Norddahl et al., 2008). Marine molluscs found both in front of and behind the Búði moraines range in age from 11.5 to 10.1 ka. The ice margin then retreated rapidly towards the highlands followed by rapid isostatic rebound; Hestvatn (50 m a.s.l) was isolated from the sea around 10.6 ka (Hannesdottir, 2006), indicating a drop in relative sea level of almost 60 m in ca. 1 ka. Although not dated directly, Ingólfsson and Norddahl (1994) assume that the transgression maximum (the highest marine limit is 110 m a.s.l.) occurred during the transition from the Younger Dryas to the Preboreal between 12.2 and 11 ka.

16.6. EARLY HOLOCENE THERMAL MAXIMUM AND THE ONSET OF NEOGLACIATION

Glacial landforms on the central highlands and evidence of ice-marginal lakes offer a general outline of glacial extent in the area during the final deglaciation (Askelsson, 1942; Kjartansson, 1955; Kjartansson, 1964; Kaldal and Vikingsson, 1991; Tómasson, 1993; Möne, 1998; Black et al., 2006; Black, 2008). Radiocarbon dating (all ^{14}C dates are given in calibrated years BP) of lacustrine sediments found near the volcano Hekla, at the boundary between the central highlands and the southern lowlands, suggests the formation of several open lakes as early as ca. 11.0 ka (Fig. 16.1). Further, the presence of the Saksunarvatn Tephra (10.2 ka) in a number of high-elevation sites in north-western, north and central Iceland (Stötter et al., 1999; Caseldine et al., 2003; Johannsdottir, 2007) suggests that the highlands were more or less ice-free at the time of the eruption. The vast subaerially erupted Thjorsa Lava, which originated in the highlands west of Vatnajökull around 8.6 ka (Kjartansson, 1964; Hjartarson, 1988; Hjartarson and Ingólfsson, 1988), confirms an ice-free highland at that time.

The expansion of birch (*Betula*) shortly after 10 ka suggests an early Holocene climate was established characterised by warm, dry summers; a condition that persisted at least until 7 ka (Björck et al., 1992; Hallsdóttir, 1995; Gudmundsson, 1997). The pollen record, summarised by Hallsdóttir (1995), suggests a maximum expansion of birch forests across southern Iceland between about 8 and 5 ka (Geirsdóttir et al., 2009; Axford et al., 2007), suggesting a possible thermal maximum through this interval. This is supported by recent Holocene lacustrine records, which provide evidence for North Atlantic warmth attributed to a stronger Irminger Current coupled with higher summer insolation between ca. 8.5 and 5.0 ka. The lacustrine record from Hvítárvatn by Langjökull, the second largest glacier on Iceland, suggests the disappearance of glacier ice during this period of Holocene warmth (Fig. 16.1; Black, 2008; Geirsdóttir et al., 2009). However, this warmth appears to have been interrupted around 8.5–8 ka by colder summers when the lacustrine records show a decline in climate proxy records (biogenic silica, total organic carbon and diatom assemblages). There is no direct glaciological evidence that glaciers advanced during this cooling event on the island, perhaps because later glacier advances overrode older moraine systems. The onset of Neoglaciation and the decline of tree vegetation remain debated: clear evidence of glacier expansion after 3 ka is available, but some authors have suggested glacier growth began as early as 5 ka (cf. Gudmundsson, 1997; Geirsdóttir et al., 2009). In almost all cases, the Little Ice Age (LIA) moraines (1300–1900 AD) represent the most extensive ice margins since early Holocene deglaciation. During the LIA, Stötter et al. (1999) note numerous glacier advances in the Tröllaskagi area in North Iceland, the oldest of which occurred at ca. AD 1800. Kirkbride and Dugmore (2006) identify two glacial advances in central Iceland, between AD 1690–1740 and AD 1880–1920, although they point out that dating of these advances was hampered by the lack of tephras in the thin and patchy soil cover and few thalli of the lichen genus *Rhizocarpon* on till boulders. Langjökull, Central Iceland, feeds two outlet glaciers that advanced to their largest Holocene extent in Hvítárvatn between AD 1750 and AD 1850 (Black, 2008; Geirsdóttir et al., 2009).

16.7. GLACIAL LIMITS AND QUALITY OF DATA

The changing climate of the past 3–4 Ma has had a marked effect on the landscape of Iceland. Although the mechanism of rifting and volcanism proceeded along the same general lines in the Tertiary and the Quaternary, the face of volcanic activity changed markedly. As the glaciers grew in size, subglacial eruptions became a more frequent occurrence. Móberg ridges (hyaloclastite ridges) and table mountains were formed in ever-increasing numbers. Likewise, while rivers and the glaciers grew larger, their erosional power increased, forming deeper and broader valleys. Consequently, the Plio-Pleistocene landscape was typified by irregular topography where jagged ridges and mountains towered over steep-sided valleys.

The reconstructed Tertiary glacial history of Iceland is based on the use of multiple sedimentological criteria from numerous sites from all parts of the country (Geirsdóttir, 1988, 1991; Geirsdóttir et al., 1993, 1994, 2007; Geirsdóttir and Eiríksson, 1994, 1996; Helgason and Duncan, 2001; Eiríksson, 2008; Roberts et al., 2007). This multiproxy approach includes a comparison with modern depositional environments, lithofacies analysis, clast fabric measurements, textural studies and rock magnetic measurements, including both remanent magnetisation and analysis of anisotropy of magnetic susceptibility. Major emphasis has been placed on detailed mapping and lithofacies analyses where both vertical and lateral changes in facies arrangements could be detected over several kilometres for all sections. The results imply that Iceland has experienced over 20 glaciations since ca. 4 Ma. This is in reasonable agreement with the number of glaciations retrieved from the $\delta^{18}O$ record from the deep sea and suggests that the late Tertiary and Early–Middle Pleistocene terrestrial record in Iceland may be regarded as an example of global palaeoclimatic processes in the North Atlantic region in general (Fig. 16.4).

16.8. DATING OF GLACIAL LIMITS—RELIABILITY OF DATES

Correlations between sections/glacial deposits of Tertiary age are still tentative and mainly based on previous palaeomagnetic work that has been undertaken on lava flows to reconstruct a palaeomagnetic timescale for the Icelandic lava pile (Wensink, 1964; McDougall and Wensink, 1966; McDougall et al., 1976, 1977; Kristjánsson et al., 1980; Eiríksson et al., 1990). Because lava flows chosen for K/Ar dating were not necessarily associated with glacial deposits, some constraints are placed on the reliability of the correlations. Stratigraphical and palaeomagnetical investigations have further revealed some glacial deposits in south-east Iceland (Helgason and Duncan, 2001) and in the Vestfirdir area (Roberts et al., 2007). Their K–Ar dates from identified lava flows and fission-track ages on tephra deposits within the glacial strata confirm previous findings of Geirsdóttir and Eiríksson (1994) on the dating of Tertiary glaciations in the country. Although there is insufficient control on individual glacial units, there does appear to be a variation in the frequency of glaciations with time through the past 2.5 Ma. Changes in the ice cover reasonably correlate with variations in the deep-sea oxygen isotope records from the North Atlantic (Fig. 16.4).

Contentious interpretation of the middle to late Quaternary glacial history of Iceland reflects the lack of continuous chrono- and biostratigraphical records and relatively low age-resolution of the evidence.

16.9. CONCLUSIONS

At least 18 glacial events are found within the stratigraphy of Fljótsdalur in eastern Iceland from ca. 3.4 to 0.7 Ma after which no outcrops are available. In the Tjörnes section in North Iceland, 14 glacial horizons are found throughout the past 2.5 Ma, but differential preservation of sections may explain why all deposits are not found in all sections. Fljótsdalur is close to the probable ice divide, whereas Tjörnes is located at the northern coast of Iceland in a tectonically active setting. A composite stratigraphy combining these two sections reveals at least 22 glacial–interglacial cycles during the past 3 Ma.

Three phases of ice growth in Iceland are indicated: (1) An initial build-up phase, covering mountainous regions in the south-east of Iceland between >4 and 3 Ma. (2) A transitional growth phase (3.0–2.5 Ma), during which the ice sheet expanded towards north and west reaching the sea in the north at Tjörnes by 2.5 Ma. (3) Development of an Icelandic Ice Sheet that left glacial deposits in all the sections studied across the country by 2.5 Ma.

The response of the Icelandic Ice Sheet to Late Pliocene–Pleistocene global change corresponds to studies by Ravelo et al. (2004) and Mudelsee and Raymo (2005), showing that the Northern Hemisphere Glaciation was a gradual climate transition from 3.6 to 2.4 Ma. It is seen as a long-term increase of ice volume superimposed by short-term glacial excursions at 3.34, 3.295, 2.7 Ma, and between 2.523 and 2.433 Ma (Knies et al., 2009).

Pulses of ice-rafted detritus in the North Atlantic Ocean (Shackleton et al., 1984) and on the Vøring Plateau around 2.74 Ma (Jansen and Sjøholm, 1991; Jansen et al., 2000; Flesche Kleiven et al., 2002) indicate synchronous ice sheet development on Iceland, Greenland, Scandinavia and North America from 2.7 to 2.5 Ma (Flesche Kleiven et al., 2002).

Between the Gauss–Matuyama magnetic Chron boundary (2.59 Ma) and the Brunhes Chron (0.78 Ma), at least 5, and as many as 13 glacial–interglacial cycles have been recognised.

The large-scale intensification of glacial expansion in the circum-Atlantic region at ca. 1.0 Ma, and subsequent glacial–interglacial cyclicity of 100,000 year's duration is only preserved in the Tjörnes section, in north Iceland (Eiríksson, 1985). During the Brunhes chron, a total of five glacial–interglacial events are identified. The lack of glacial deposits from this time in other Icelandic sections may reflect reduced preservation potential or even lack of outcrops where constructive processes have out-weighed destructive processes.

REFERENCES

Andrews, J.T., Helgadottir, G., 2003. Late Quaternary ice extent and deglaciation of Hunafloaal, north Iceland: evidence from marine cores. Arct. Antarct. Alpine Res. 35, 218–232.

Andrews, J.T., Hardardóttir, J., Helgadóttir, G., Jennings, A.E., Geirsdóttir, Á., Sveinbjornsdóttir, A.E., et al., 2000. The N and W Iceland shelf: insights into LGM Ice extent and deglaciation based on acoustic stratigraphy and basal radiocarbon AMS dates. Quatern. Sci. Rev. 19, 619–631.

Andrews, J.T., Hardardottir, J., Geirsdottir, A., Helgadottir, G., 2002. Late Quaternary ice extent and glacial history from the Djupall trough, off Vestfirdir peninsula, northwest Iceland: a stacked 36 cal kyr environmental record. Polar Res. 21, 211–226.

Aronson, J.L., Sæmundsson, K., 1975. Relatively old basalts from structurally high area in central Iceland. Earth Planet. Sci. Lett. 28, 83–97.

Askelsson, J., 1942. Jarðfræði og jarðmyndun. In: Eythorsson, J. (Ed.), Arbok Ferðafelags Islands 1942, pp. 22–29 Kerlingarfjoll. Ferðafelag Islands, Reykjavık.

Axford, Y., Miller, G.H., Geirsdottir, A., Langdon, P.G., 2007. Holocene temperature history of northern Iceland inferred from subfossil midges. Quatern. Sci. Rev. 26, 3344–3358.

Björck, S., Ingólfsson, Ó., Haflidason, H., Hallsdóttir, M., Anderson, N.J., 1992. Lake Torfadalsvatn: a high resolution record of the North Atlantic ash zone I and the last glacial-interglacial environmental changes in Iceland. Boreas 21, 15–22.

Black, J., 2008. Holocene climate change in south central Iceland: a multi-proxylacustrine record from glacial lake Hvitarvatn. Ph.D. thesis, University of Colorado, Boulder.

Black, J., Miller, G.H., Geirsdottir, A., 2006. Diatoms as proxies for a fluctuatingHolocene ice cap margin in Hvıtarvatn, central Iceland. EOS 87, Abstract PP41C-08. San Francisco.

Bourgeois, O., Dauteil, O., van Viet-Lanoe, B., 2000. Geothermal control on ice stream formation: flow patterns of the Icelandic Ice Sheet at the Last Glacial Maximum. Earth Surf. Process. Land. 25, 59–76.

Buchardt, B., Símonarson, L.A., 2003. Isotope palaeotemperatures from the Tjörnes beds in Iceland: evidence of Pliocene cooling. Palaeogeogr. Palaeoclimatol. Palaeoecol. 189, 71–95.

Cande, S.C., Kent, D.V., 1995. Revised calibration of the geomagnetic polarity timescale for the late Cretaceous and Cenozoic. J. Geophys. Res. 100, 6093–6095.

Caseldine, C., Geirsdottir, A., Langdon, P., 2003. Efstadalsvatn—a multi-proxy study of a Holocene lacustrine sequence from NW Iceland. J. Paleolimnol. 30, 55–73.

Einarsson, Th., 1994. Geology of Iceland. Rocks and Landscape. Mál & Menning, Reykjavík, 309 pp.

Einarsson, Th., Albertsson, K.J., 1988. The glacial history of Iceland during the past three million years. Philos. Trans R. Soc. Lond. B 318, 637–644.

Eiríksson, J., 1985. Facies analysis of the Breidavík Group sediments on Tjörnes. Acta naturalia Islandica 31, 56.

Eiríksson, J., 2008. Glaciation events in the Pliocene–Pleistocene volcanic succession of Iceland. Jökull 58, 315–329.

Eiríksson, J., Gudmundsson, A.I., Kristjánsson, L., Gunnarsson, K., 1990. Palaeomagnetism of Pliocene–Pleistocene sediments and lava flows on Tjörnes and Flatey, North Iceland. Boreas 19, 39–55.

Flesche Kleiven, H., Jansen, E., Fronval, T., Smith, T.H., 2002. Intensification of Northern Hemisphere Glaciation in the circum Atlantic region (3.5–2.4 Ma)—ice-rafted detritus evidence. Palaeogeogr. Palaeoclimatol. Palaeoecol. 184, 213–223.

Fridleifsson, G.Ó., 1995. Miocene glaciation in SE Iceland. In: Hróarson, B., Jónsson, D., Jonsson, S.S. (Eds.), Eyjar í Eldhafi Gott mál hf., pp. 77–85.

Fronval, T., Jansen, E., 1996. Late Neogene Paleoclimate and Palæoceanography in the iceland-Norwegian Sea. In: Thiede, J., Myhre, A.M., Futh, J.V., Johnson, G.L., Ruddiman, W. (Eds.), Proceedings of the ODP, Scientific Report, vol. 151, pp. 455–468, College Station, TX (Ocean Drilling Project).

Geirsdóttir, Á., 1988. Sedimentologic analysis of diamictites and implications for late Cenozoic glaciation in western Iceland. Ph.D. thesis, University of Colorado, Department of Geological Sciences.

Geirsdóttir, Á., 1991. Diamictites of late Pliocene age in western Iceland. Jökull 40, 3–25.

Geirsdóttir, Á., Eiríksson, J., 1994. Growth of an intermittent ice sheet in Iceland during the late Pliocene and early Pleistocene. Quatern. Res. 42, 115–130.

Geirsdóttir, Á., Eiríksson, J., 1996. A review of studies of the earliest glaciations in Iceland. Terra Nova 8, 400–414.

Geirsdottir, A., Andrews, J.T., Olafsdottir, S., Helgadottir, G., Hardardottir, J., 2002. A 36 Ky record of iceberg rafting and sedimentation from north-west Iceland. Polar Res. 21, 291–298.

Geirsdóttir, Á., Thordarson, Th., Kristjánsson, L., 1993. The interactive impact of climatic deteriorations and volcanic activity on facies assemblages in late Pliocene to early Pleistocene strata of the Hreppar Formation, South Central Iceland. EOS 74 (43), 131

Geirsdóttir, Á., Thórdarson, Th., Kristjánsson, L., 1994. Áhrif eldvirkni og loftslagsbreytinga á upphledslu jardlaga i Gnúpverjahreppi. Ágrip a Jardfraedaradstefnu íslenska Jardfraedafélagsins, 21. Apríl 1994. (Abstract vol. Geological Society of Iceland).

Geirsdóttir, A., Hardardóttir, J., Eiriksson, J., 1997. Depositional history of theYounger Dryas(?)—Preboreal sediments, South Central Iceland. Arct. Alpine Res. 29, 13–23.

Geirsdóttir, Á., Hardardóttir, J., Sveinbjörnsdóttir, A.E., 2000. Glacial extent and catastrophic meltwater events during the deglaciation of Southern Iceland. Quatern. Sci. Rev. 19, 1749–1761.

Geirsdóttir, A., Miller, G.H., Andrews, J.T., 2007. Glaciation, erosion, and landscape evolution of Iceland. J. Geodyn. 43, 170–186.

Geirsdóttir, Á., Miller, G.H., Axford, Y., Ólafsdóttir, S., 2009. Holocene and latest Pleistocene climate and glacier fluctuations in Iceland. Quatern. Sci. Rev. 28, 2107–2118.

Grönvold, K., Óskarsson, N., Johnsen, S., Clausen, H.B., Hammer, C.U., Bard, E., 1995. Ash layers from Iceland in the Greenland GRIP ice core correlated with oceanic and land sediments. Earth Planet. Sci. Lett. 135, 149–155.

Gudmundsson, A., 1978. Austurlandsvirkjun, Mulavirkjun. Frumkönnun á jardfraedi Mula og umhvefis. Orkusofnun, OS-ROD-7818. Reykjavik.

Gudmundsson, H., 1997. A review of the Holocene environmental history of Iceland. Quatern. Rev. 16, 81–92.

Hallsdóttir, M., 1995. On the pre-settlement history of Icelandic vegetation. Icelandic Agricultural Science 9, 17–29.

Hannesdottir, H., 2006. Reconstructing environmental change in South Iceland during the last 12,000 years based on sedimentological and seismostratigraphical studies in Lake Hestvatn. M.S. thesis, University of Iceland, Reykjavik.

Hansen, B., Østerhus, S., 2000. North Atlantic-Nordic Seas exchanges. Prog. Oceanogr. 45, 109–208.

Hardardóttir, J., Geirsdóttir, A., Sveinbjörnsdóttir, A.E., 2001. Seismostratigraphy and sediment studies of Lake Hestvatn, south Iceland: implication for the deglacial history of the region. J. Quatern. Sci. 16, 167–179.

Harðarson, B., Fitton, J.G., Hjartarson, Á., 2008. Tertiary volcanism in Iceland. Jökull 58, 161–178.

Helgadottir, G., Brandsdottir, B., Geirsdóttir, A., Björnsson, H., 2005. Offshore Extent of the Iceland Ice Sheet During the Last Glacial Maximum Inferred from Multibeam Bathymetric Data and Sediment Cores. In: American Geophysical Union (AGU) Fall Meeting, San Francisco, Dec 2005, pp. EOS, 86 (52), Abstract H51G-0449.

Helgason, J., Duncan, R.A., 2001. Glacial-interglacial history of the Skaftafell region, southeast Iceland, 0-5 Ma. Geology 29, 179–182.

Hjartarson, A., 1988. Thjorsar hraunid mikla—stærsta nutimahraun jardar. Natturufraedigurinn 58, 1–16.

Hjartarson, Á., Ingólfsson, Ó., 1988. Preboreal glaciation of Southern Iceland. Jökull 38, 1–16.

Hjort, C., Hreggvidur, N., Ingolfsson, O., 1985. Late Quaternary geology and glacial history of Hornstrandir, northwest Iceland: a reconnaissance study. Jökull 35, 9–29.

Hoppe, G., 1982. The extent of the last inland ice sheet of Iceland. Jökull 32, 3–11.

Hubbard, A., Sugden, D., Dugmore, A., Norddahl, H., Petursson, H., 2006. A modelling insight into the Icelandic Last Glacial Maximum ice sheet. Quatern. Sci. Rev. 25, 2283–2296.

Ingólfsson, Ó., 1991. A review of the late Weichselian and early Holocene glacial and environmental history of Iceland. In: Maizels, J., Caseldine, C. (Eds.), Environmental Change in Iceland: Past and Present. Kluwer Academic Publishers, Dordrecht, Holland, pp. 13–29.

Ingólfsson, Ó., Norddahl, H., 1994. A review of the environmental history of Iceland, 13000-9000 yr BP. J. Quatern. Sci. 9, 147–150.

Ingólfsson, O., Björck, S., Haflidason, H., Rundgren, M., 1997. Glacial and climatic events in Iceland reflecting regional North Atlantic climatic shifts during the Pleistocene-Holocene transition. Quatern. Sci. Rev. 16, 1135–1144.

Jancin, M., Young, K.D., Voight, B., 1985. Stratigraphy and K&Ar ages across the West flank of the Northeast Iceland axial rift zone, in relation to the 7 Ma volcano-Tectonic reorganization of Iceland. J. Geophys. Res. 90, 9961–9985.

Jansen, E., Sjøholm, J., 1991. Reconstruction of glaciation over the past 6 million years from ice-borne deposits in the Norwegian Sea. Nature 349, 600–604.

Jansen, E., Sjøholm, J., Bleil, U., Erichsen, J.A., 1990. Neogene and Pleistocene glaciations in the Northern Hemisphere and late Miocene–Pliocene global ice volume fluctuations: evidence from the Norwegian Sea. In: Bleil, U., Thiede, J. (Eds.), Geological History of the Polar Oceans: Arctic Versus Antarctic. Kluwer Academic, Dordrecht, pp. 677–705.

Jansen, E., Fronval, T., Rack, F., Channell, E.T., 2000. Pliocene-Pleistocene ice rafting history and cyclicity in the Nordic Seas during the last 3.5 Myr. Paleoceanography 15, 709–721.

Johannesson, H., Sæmundsson, K., 1999. Geological Map of Iceland. 1:1 000 000. Icelandic Institute of Natural History, Reykjavík.

Johannsdottir, G.E., 2007. Mid-Holocene to late glacial tephrochronology in West Iceland as revealed in three lacustrine environments. M.S. thesis, University of Iceland, Reykjavik.

Kaldal, I., Vikingsson, S., 1991. Early Holocene deglaciation in Central Iceland. Jökull 40, 51–66.

Kirkbride, M.P., Dugmore, A.J., 2006. Responses of mountain Ice caps in central Iceland to Holocene climate change. Quatern. Sci. Rev. 25, 1692–1707.

Kjartansson, G., 1955. Frodlegar jökulrakir. Natturufraedigurinn 25, 154–171.

Kjartansson, G., 1964. Isaldarlok og eldfjoll á Kili Natturufraedigurinn 34, 9–38.

Knies, J., Matthiessen, J., Vogt, C., Laberg, J.S., Hjelstuen, B.O., Smelror, M., et al., 2009. The Plio-Pleistocene glaciation of the Barents Sea–Svalbard region: a new model based on revised chronostratigraphy. Quatern. Sci. Rev. 28, 812–829.

Kristjánsson, L., Fridleifsson, I.B., Watkins, N.D., 1980. Stratigraphy and paleomagnetism of the Esja, Eyrarfjall and Akrafjall Mountains, SW-Iceland. J. Geophys. 47, 31–42.

Kristjánsson, L., Duncan, R.A., Gudmundsson, Á., 1998. Stratigraphy, palaeomagnetism and age of volcanics in the upper regions of Thjorsardalur valley, central southern Iceland. Boreas 27, 1–13.

Leifsdóttir, Ó.E., 1999. Ísaldarlög á nordanverðu Snæfellsnesi. Setlög og skeldýrafanur. Unpublished M.S. thesis, University of Iceland. 101 pp.

Maslin, M.A., Li, X.S., Loutre, M.-F., Berger, A., 1998. The contribution of orbital forcing to the progressive intensification of northern hemisphere glaciation. Quatern. Sci. Rev. 17, 411–426.

McDougall, I., Wensink, H., 1966. Paleomagnetism and geochronology of the Pliocene-Pleistocene lavas in Iceland. Earth Planet. Sci. Lett. 1, 232–236.

McDougall, I., Watkins, N.D., Kristjánsson, L., 1976. Geochronology and paleomagnetism of Miocene-Pliocene lava sequences at Bessastadaá, eastern Iceland. Am. J. Sci. 276, 1078–1095.

McDougall, I., Watkins, N.D., Walker, G.P.L., Kristjánsson, L., 1977. Potassium-argon and paleomagnetic analysis of Icelandic lava flows: limits on the age of anomaly 5. J. Geophys. Res. 81, 1505–1512.

Mix, A.C., Pisias, N.G., Rugh, W., Wilson, J., Morey, A., Hagelberg, T., 1995a. Benthic foraminifera stable isotope record form Site 849: 0–5 Ma: local and global climate changes. In: Pisias, N.G., Mayer, L., Janecek, T., Palmer-Julson, A., van Andel, T.H. (Eds.), Proceedings of the Ocean Drilling Program, Scientific Results, vol. 138. College Station, TX, USA, pp. 371–412.

Mix, A.C., Le, J., Shackleton, N.J., 1995b. Benthic foraminiferal stable isotope stratigraphy of Site 846: 0–1.8 Ma. In: Fisias, N.G., Mayer, L., Janecek, T., Palmer-Julson, A., van Andel, T.H. (Eds.), Proceedings of the Ocean Drilling Program, Scientific Results, vol. 138. College Station, TX, USA, pp. 839–854.

Möne, C., 1998. Glacial Lake deposits in Kerlingarfjoll, Central Iceland. M.S. thesis, University of Uppsala.

Mudelsee, M., Raymo, M.E., 2005. Slow dynamics of the Northern Hemisphere Glaciation. Paleoceanography 20, PA4022.

Norddahl, H., 1991. Late Weichselian and early Holocene deglaciation history of Iceland. Jökull 40, 27–50.

Norddahl, H., Haflidason, H., 1992. The Skógar Tephra, a Younger Dryas marker in North Iceland. Boreas 21, 23–41.

Norddahl, H., Pétursson, H.G., 2005. Relative sea-level changes in Iceland: new aspects of the Weichselian deglaciation of Iceland. In:

Caseldine, C., Russell, A., Hardardóttir, J., Knudsen, O. (Eds.), Iceland: Modern Processes and Past Environments. Elsevier, Amsterdam, pp. 25–78.

Norddahl, H., Ingólfsson, Ó., Pétursson, H.G., Hallsdóttir, M., 2008. Late-Weichselian and Holocene environmental history of Iceland. Jökull 58, 343–364.

ODP leg 151 Scientific Party, 1994. Exploring Arctic history through scientific drilling. EOS 75 (25), 281–286.

Ólafsdóttir, Th., 1975. Jokulgardar a sjavarbotni ut af Breidafirdi. Náttúrufrædingurinn 45, 31–36.

Principato, S.M., 2003. The late Quaternary history of eastern Vestfirdir, NW Iceland. Unpublished Ph.D. Dissertation. University of Colorado, 258 pp.

Principato, S.M., 2008. Geomorphic evidence for Holocene glacial advances and sealevel fluctuations on eastern Vestfirdir, northwest Iceland. Boreas 37, 132–145.

Principato, S., Johnson, J.S., 2009. Using a GIS to quantify patterns of glacial erosion on northwest iceland: implications for independent ice sheets. Arct. Antarct. Alpine Res. 41, 128–137.

Principato, S.M., Geirsdottir, A., Johannsdottir, G.E., Andrews, J.T., 2006. Late Quaternary glacial and deglacial history of eastern Vestfirdir. J. Quatern. Sci. 21, 271–285.

Ravelo, A.C., Andreasen, D.H., Lyle, M., Olivarez Lyle, A., Wara, M.W., 2004. Regional climate shifts caused by gradual global cooling in the Pliocene epoch. Nature 429, 263–267.

Raymo, M.F., Ruddiman, W.F., Clement, B.M., 1986. Pliocene–Pleistocene paleoceanography of the North Atlantic DSDP Site 609. In: Ruddiman, W.F., Thomas, E. et al., (Eds.), Initial Reports of the Deep Sea Drilling Project 94. U.S. Government Printing Office, Washington, DC, pp. 895–902.

Roberts, S.T., Sigurvinsson, J.R., Westgate, J.A., Sandhu, A., 2007. Late Pliocene glaciation and landscape evolution of Vestfirdir, Northwest Iceland. Quatern. Sci. Rev. 26, 243–263.

Rundgren, M., Ingólfsson, O., 1999. Plant survival in Iceland during periods of glaciation? J. Biogeogr. 26, 387–396.

Saemundsson, K., 1979. Outline of the geology of Iceland. Jökull 29, 7–28.

Shackleton, N.J., Backman, J., Zimmerman, H., Kent, D.V., Hall, M.A., Roberts, D.G., et al., 1984. Oxygen isotope calibration of the onset of ice-rafting and history of glaciation in the North Atlantic region. Nature 307, 620–623.

Shackleton, N.J., Berger, A., Peltier, W.R., 1990. An alternative astronomical calibration of the lower Pleistocene time based on ODP Site 677. Transactions of the Royal Society of Edinburgh Earth Sciences — Trans R Soc Edinburgh Earth Sci 81, 251–261.

Sigurvinsson, J.R., 1983. Weichselian glacial lake deposits in the highlands of north-western Iceland. Jökull 33, 99–109.

Simonarson, L.A., Leifsdóttir, Ó.E., 2007. Early Pleistocene molluscan migration to Iceland - Palaeocenographic implication. Jökull, 57, 1–20.

Símonarson, L.A., Eiríksson, J., 2008. Tjörnes—Pliocene and Pleistocene sediments and faunas. Jökull 58, 331–342.

Spagnolo, M., Clark, C.D., 2009. A geomorphological overview of glacial landforms on the Icelandic continental shelf. J. Maps 2009, 37–52.

Stokes, C.R., Clark, C.D., 2001. Palaeo-ice streams. Quatern. Sci. Rev. 20, 1437–1457.

Stötter, J., Wastl, M., Caseldine, C., Haberle, T., 1999. Holocene palaeoclimatic reconstructions in Northern Iceland: approaches and results. Quatern. Sci. Rev. 18, 457–474.

Syvitski, J.P.M., Jennings, A.E., Andrews, J.T., 1999. High-resolution seismic evidence for multiple glaciation across the southwest Iceland shelf. Arct. Antarct. Alpine Res. 31, 50–57.

Tómasson, H., 1993. Jökulstífluð vötn á Kili og hamfarahlaup í Hvítá, Árnessýslu. Natturufraedingurinn 62, 77–98.

Wensink, H., 1964. Paleomagnetic stratigraphy of younger basalts and intercalated Plio-Pleistocene tillites in Iceland. Geol. Rundsch. 54, 364–384.

Wolf, T.C.W., Thiede, J., 1991. History of terrigenous sedimentation during the past 10 my in the North Atlantic (ODP-leg's 104, 105 and DSDP-Leg 81). Mar. Geol. 101, 83–102.

Chapter 17

Middle Pleistocene to Holocene Glaciations in the Italian Apennines

Carlo Giraudi

ENEA, C.R. Saluggia, strada per Crescentino, 41, 13040 Saluggia (VC), Italy

17.1. INTRODUCTION

The Apennines, a mountain chain formed mainly of Mesozoic and Cenozoic carbonate and flysch sediments extending between 38°N and about 44°30′N, forms the backbone of the Italian Peninsula (Fig. 17.1). The highest peaks in the chain occur in the central sector, between about 42°50′N and 41°30′N where the carbonate rocks predominate: they include the Gran Sasso (2912 m a.s.l.), the Maiella (2793 m a.s.l.), the Velino (2486 m a.s.l.), the Sibillini (2476 m a.s.l.) and the Laga (2458 m a.s.l.) massifs, but numerous other peaks also exceed 2000 m a.s.l. In contrast, the mountains of the Northern Apennines, where the flysch is the dominant lithology, exceed 2000 m a.s.l. only at Mounts Cimone (2165 m a.s.l.), Cusna (2121 m a.s.l.) and Prato (2054 m a.s.l.). The Southern Apennine mountains exceed 2000 m only at Mounts Sirino (2005 m a.s.l.) and Pollino (2266 m a.s.l.), formed of carbonate rocks. At present only a single small, rapidly melting, debris-covered glacier is found in the Apennines: the Calderone Glacier in the Gran Sasso massif.

The traces of the Late Pleistocene glaciations in the Apennines are very abundant and well preserved, whereas evidence of older events is rather scanty. The great majority of the glaciers occupied valleys and glacial cirques orientated northwards, while several developed in valleys orientated towards the west and east. Only a few occurred in valleys that are aligned towards the south (Fig. 17.1).

As summarised in a previous paper (Giraudi, 2004) on the Apennines, the remnants of three glaciations have been recognised, dating from the Middle to Late Pleistocene. In general, the chronology of the Apennine glaciations was assumed on the basis of the degree of weathering of the moraines, or the freshness of the glacial features. Only in a few places has the chronological framework been established using radiometric dates, confirming earlier hypotheses based on field observations. All the radiocarbon dates presented here are calibrated using the CalPal calibration programme (http://www.calpal-online.de).

All the Apennine glaciers disappeared in the Early Holocene, but the Calderone Glacier formed again around ca. 4 ka BP (Giraudi, 2004, 2005) which indicates also a Holocene glacial event (Neoglaciation).

17.2. MIDDLE PLEISTOCENE GLACIATIONS

On the northern and central Apennines, the remnants of the moraines formed during at least two Middle Pleistocene glaciations are known: they are older than the Last Interglacial (i.e. Eemian Stage, ~Marine Isotope Stage, MIS, 5e; Giraudi, 2001, 2004; Kotarba et al., 2001), but their precise age has yet to be established. The authors suggested that this glaciation probably took place during MIS 6.

New researches have been carried out in the Campo Felice plain (Velino Massif, Central Apennine), a tectonic–karstic depression, surrounded by mountains consisting completely of carbonate rocks (Giraudi et al., 2010). Campo Felice is one of the few places in the Apennines where remnants of a Middle Pleistocene glaciation are known (Giraudi, 2004). Detailed geological survey has allowed the recognition of older till deposits, more weathered and pedogenically altered than the morainic sediments presumed to have formed during the MIS 6 event. Other evidence concerning the age of the Campo Felice glaciations has been obtained from investigation of the sediments forming Campo Felice, using nine continuously cored boreholes, 30–120 m in depth (Giraudi et al., 2011a).

The stratigraphy indicated by the cores extended the stratigraphical sequence exposed in the field, or observed in temporary excavations 5–6 m deep. The chronological framework of the Middle Pleistocene sediments is based on ^{39}Ar/^{40}Ar dates on three tephra layers. The recognition of tephras of known age with geochemical methods has

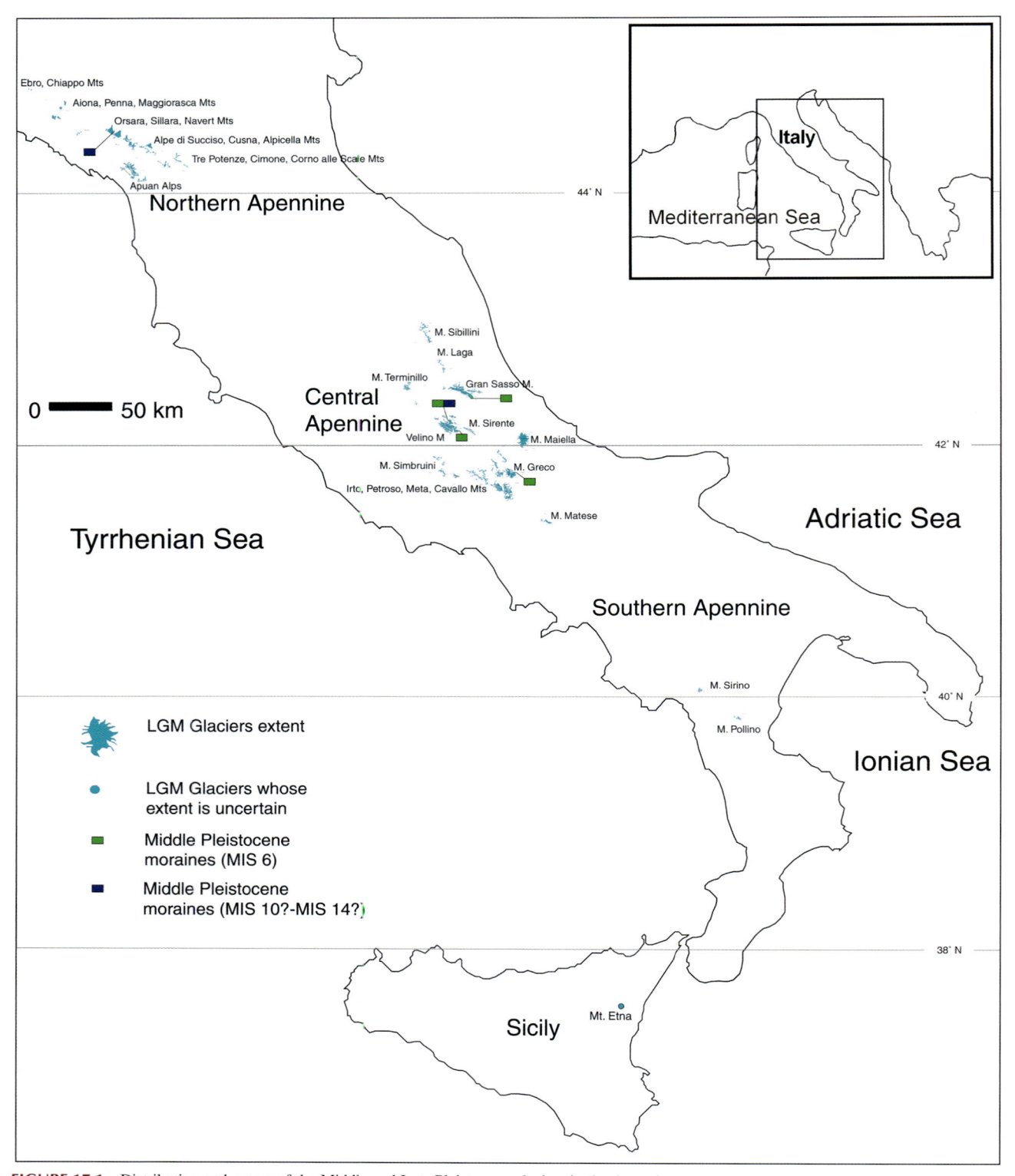

FIGURE 17.1 Distribution and extent of the Middle and Late Pleistocene glaciers in the Apennines.

provided an additional element for dating of the sedimentary sequence.

The borehole sequences indicate five sedimentation–erosion cycles, which allow common features to be identified. Figure 17.2 shows a stratigraphical section through the Campo Felice Plain and the sedimentary cycles.

Each cycle involves an initial phase of erosion indicated by erosion surfaces, which indicate breaks in sedimentation. The erosional phases are followed by sedimentation of low-energy fluvial, colluvial and marsh deposits, separated by minor erosion surfaces. The sediments are pedogenised and are sometimes very rich in organic matter, containing many reworked volcanic minerals. Overall, the composition of the sediments is predominantly non-calcareous. For the uppermost two cycles, the sedimentation of these deposits took place during periods that were characterised by phases of arboreal expansion based on palaeobotantical investigations by Di Rita and Magri (2004) and Chiarini et al. (2007). The geomorphological and sedimentary processes that occurred during the Holocene indicate an environmental development of this type.

In each of the cycles identified, the non-calcareous sediments are mainly overlain by silty clayey lacustrine carbonate deposits, which are extremely poor in organic matter and pollen. In the marginal area of the plain, the lacustrine deposits of the last four cycles interfinger with those of fluvioglacial fans. The lacustrine sediments that accumulated during the last cycle contain pollen attributed to the Last Glacial Maximum (LGM) by Di Rita and Magri (2004).

Each cycle has thus been interpreted as the succession of a warm period, characterised by the development of erosion phases in the initial period and the absence of a lake (or limited episodes of small lakes or marsh formation), followed by a cold period in the course of which the lake was formed, and the development and expansion of the glaciers took place. The uppermost two cycles have been dated to the Late Pleistocene.

The more recent tephra layer was found interbedded with the lacustrine sediments of the third cycle. The chemical composition of the tepha shows that it derives from the Vico volcanic eruption known as 'Ignimbrite B', dated to 157 ± 3 ka ago by Laurenzi and Villa (1987). It is therefore assumed that the sediments of this third cycle formed during MIS 6.

The lacustrine carbonate sediments of the second cycle contain a leucite-bearing tephra stratum, which has been dated to 376 ± 56 ka using ^{39}Ar–^{40}Ar. This age and mineralogical composition are indistinguishable from that of the leucite-bearing Villa Senni Tuff from the Alban Hills Volcanic Complex (AHVC), dated at 350 ± 2 ka by Karner et al. (2001). The Villa Senni Tuff eruption occurred during

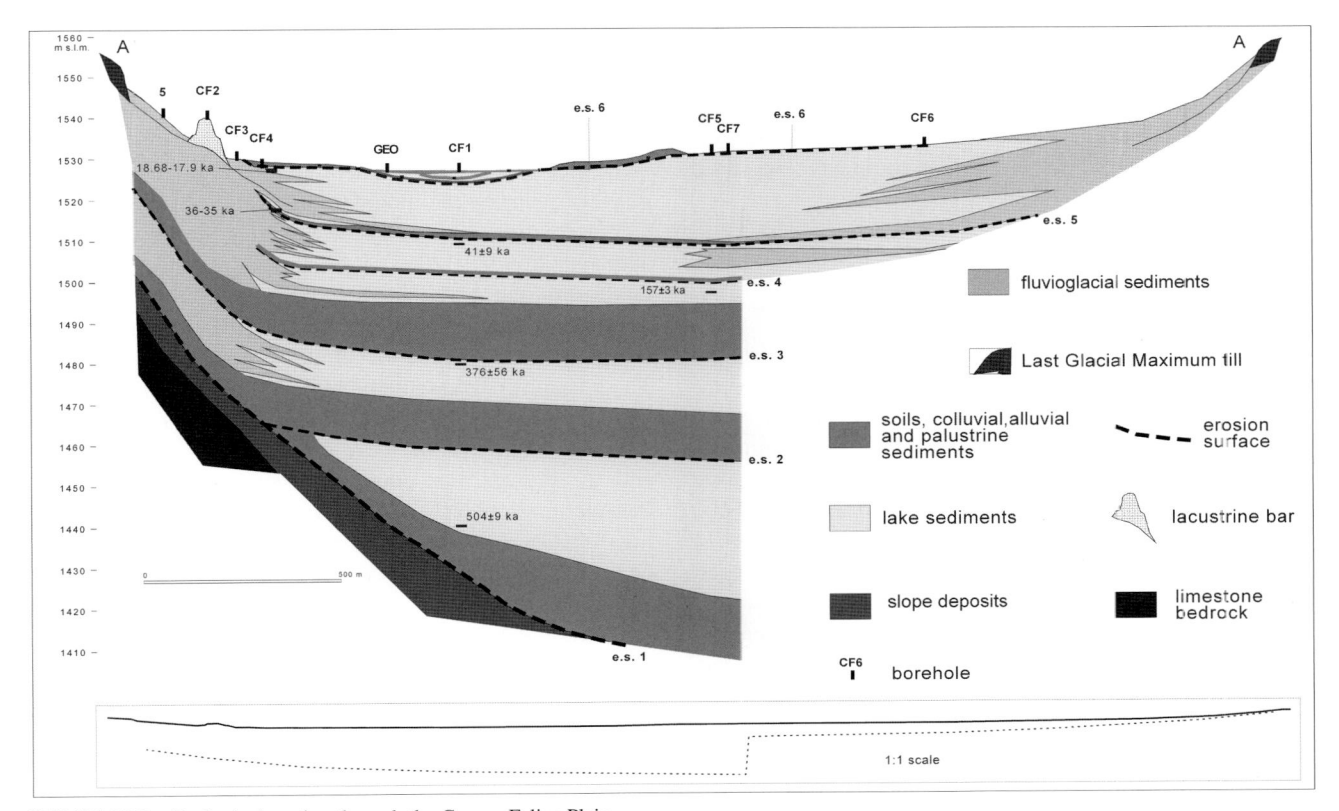

FIGURE 17.2 Geological section through the Campo Felice Plain.

MIS 10 (cf. Giordano et al., 2003). The second cycle sediments are therefore assigned to MIS 10. In the lower part of the first cycle carbonate lacustrine sediment, a leucite-bearing tephra layer occurs, which is dated to 504 ± 9 ka using ^{39}Ar–^{40}Ar. This implies that the first cycle lacustrine sediment accumulated during MIS 14. According to these latest results on Campo Felice, therefore at least three Middle Pleistocene glaciations took place in the Apennines during MIS 6, 10 and 14, respectively.

Regarding the calculations of equilibrium line altitude (ELA) for the glaciations prior to the LGM, it is likely that the Apennines and many other mountains in the Mediterranean area cannot supply reliable values because of the intense tectonic activity. In order to calculate and compare the ELA of the glacial events during MIS 6, 10 and 14, it is necessary to assume that the altitudes of the glacial basins have remained stable for hundreds of thousands of years. This assumption appears implausible for the Apennines. In this zone, rates of uplift can reach and even exceed 1–2 mm/year. It is clear therefore that different parts of the glacial basins are affected by differential tectonic movements (Giraudi, 1995; Giraudi and Frezzotti, 1995), which can modify both the altitude and the dimensions of the glacial valleys. Moreover, the erosion that has taken place between the earliest glacial phases and the most recent ones have both been influenced by climatic variations and tectonics, and these can considerably change the morphology of certain mountain areas.

Judging from the distribution of till in the Central Italian massifs, the extent of glaciers during the MIS 6 glaciation must have exceeded that of the LGM by some 5–10%. The extension of the glaciers that formed the earliest (MIS 10–14) Campo Felice moraines could have exceeded that of the MIS 6 glaciers.

17.3. LATE PLEISTOCENE GLACIATION

Traces of Late Pleistocene glaciation have been reported in a number of papers (e.g. Giraudi, 2004 for an almost complete list of references). All the evidence refers to the Apennine LGM event. In the Central-Southern Apennines, a number of stratigraphical markers have been recognised (tephra and aeolian sediments discussed in Giraudi, 2004) which are fundamental for the dating and correlation of the glacial events on the various massifs. In the same paper, a chronology of the Apennine LGM and stadial and interstadial phases was established using radiocarbon dating of soil and sediments and the chronological data stated above.

Recent studies carried out in the Central Apennines indicate that a Late Pleistocene glaciation, older than the Apennine LGM, took place. The borehole investigation of the Campo Felice plain (noted above) demonstrates that the sediments of the fourth and fifth cycles are of Late Pleistocene age (Giraudi et al., 2011a). The fifth cycle sediments are constrained by two radiocarbon dates. The upper part (cold phase) of this last cycle includes a very thin peat stratum dated to 18.68–17.9 cal. ka BP. The base of the same cycle (warm phase) includes an organic soil dated to 36–35 cal. ka BP. Thus the local LGM is limited between the two dates. Therefore, the age of LGM in the Apennines (between about 30 and 16 cal. ka BP) reported by Giraudi and Frezzotti (1997) and Giraudi (2004) is confirmed. The glacial features observed at the head of the glacial valleys show several moraine ridges formed during the retreat phases following the LGM (Late-glacial substages). The chronological framework of the local LGM and Late-glacial interstadial and stadial events is shown in Fig. 17.3. The fourth cycle sediments studied in the Campo Felice cores include a tephra dated to 41 ± 9 ka by ^{39}Ar–^{40}Ar (Giraudi et al., 2011a). Being younger than the third cycle (MIS 6), it is thought that the fourth cycle dates from MIS 4 to MIS 3. These new data confirm the hypothesis (Frezzotti and Giraudi, 1992) that two glacial expansions occurred during the Late Pleistocene, the first older than a soil dated to 38.0 ± 2.2 ka BP using radiocarbon dating. The ELA variations during the local LGM and Late-glacial stadials, together with the age of the glacial retreat in the southern, central and northern Apennines, modified from data presented by Giraudi (2004), are shown in Fig. 17.4.

In order to complete the review of the glaciations in Mediterranean Italy, the events preserved on the Mount Etna volcano should also be considered. The Etna volcano (Sicily, latitude 37°42′N, maximum elevation 3340 m a.s.l.) is the highest mountain in mediterranean Italy; at present no glaciers occur on the mountain but during LGM some glaciers were almost certainly present on its summit area (Giraudi and Groppelli, 2004). Remnants of glacial deposits and landforms (e.g. terminal moraines) have not been recognised with certainty, but Neri et al. (1995) observed some morphological features that could be produced by glaciers. The present ELA, calculated by Messerli (1967) for eastern Sicily, lies at an altitude of about 3500 m a.s.l.: assuming that the difference of ELA on Etna between the LGM and the present was comparable with the difference on the Apennines (1700–1500 m), the ELA on the volcano during the LGM is estimated at between 1800 and 2000 m a.s.l. on the north-facing slopes. On south-facing slopes, the ELA must undoubtedly have been higher. It is possible, therefore, to suggest that one or more glaciers developed on the volcano during the LGM, although their occurrence would have been reduced by the presence of active craters and the geothermal heat flux. During the LGM, the Etna summit could have been capped by an ice field that was more extensive on the north-facing slope (as observed on some volcanoes in Central and South America, e.g.

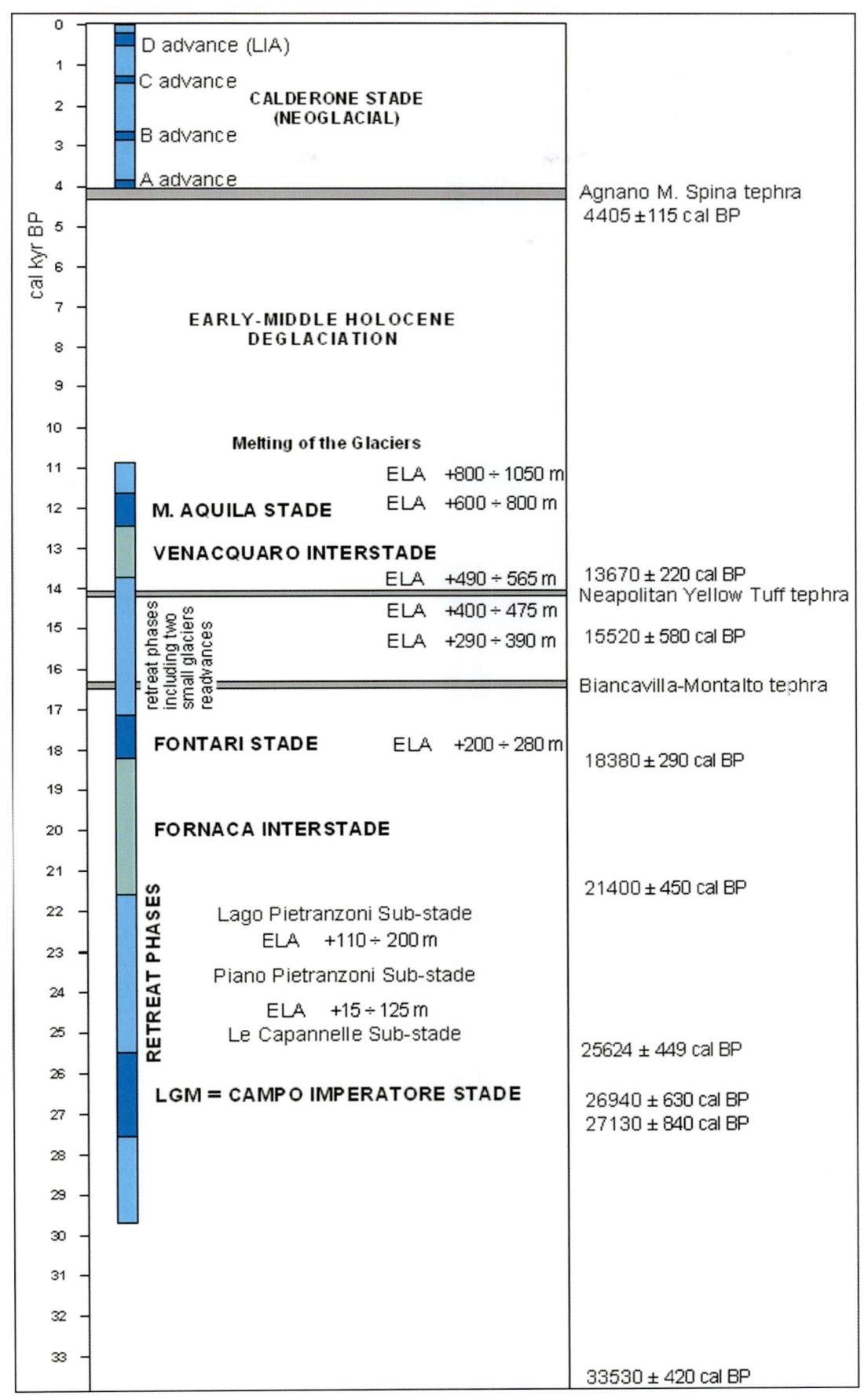

FIGURE 17.3 Synthesis of the chronological evidence delimiting the LGM, Late-glacial and Holocene advance and retreat phases of the Apennine glaciers.

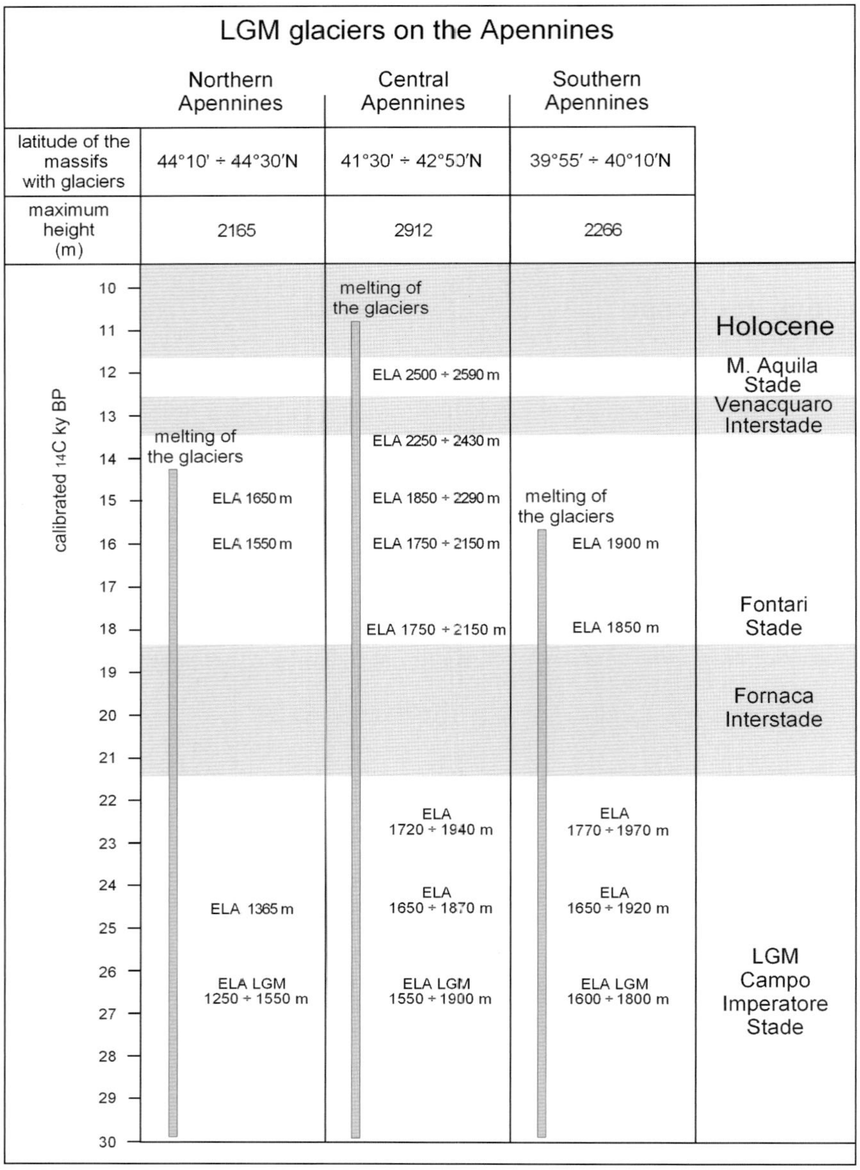

FIGURE 17.4 ELA variations during the LGM, stadial and interstadial phases in the northern, central and southern Apennines.

LGM glaciers on the Apennines			
	Northern Apennines	Central Apennines	Southern Apennines
latitude of the massifs with glaciers	44°10′ ÷ 44°30′N	41°30′ ÷ 42°50′N	39°55′ ÷ 40°10′N
maximum height (m)	2165	2912	2266

Chronology (calibrated ^{14}C ky BP):

- 10–12 ky: melting of the glaciers (Central Apennines); Holocene
- 12 ky: ELA 2500 ÷ 2590 m (Central); M. Aquila Stade
- 13 ky: Venacquaro Interstade
- 14 ky: melting of the glaciers (Northern); ELA 2250 ÷ 2430 m (Central)
- 15 ky: ELA 1650 m (Northern); ELA 1850 ÷ 2290 m (Central); melting of the glaciers (Southern)
- 16 ky: ELA 1550 m (Northern); ELA 1750 ÷ 2150 m (Central); ELA 1900 m (Southern)
- 18 ky: ELA 1750 ÷ 2150 m (Central); ELA 1850 m (Southern); Fontari Stade
- 19–21 ky: Fornaca Interstade
- 22–23 ky: ELA 1720 ÷ 1940 m (Central); ELA 1770 ÷ 1970 m (Southern)
- 24–25 ky: ELA 1365 m (Northern); ELA 1650 ÷ 1870 m (Central); ELA 1650 ÷ 1920 m (Southern)
- 26–27 ky: ELA LGM 1250 ÷ 1550 m (Northern); ELA LGM 1550 ÷ 1900 m (Central); ELA LGM 1600 ÷ 1800 m (Southern); LGM Campo Imperatore Stade
- to 30 ky

Cotopaxi, Equador, and Pico de Orizaba, Mexico). Such an ice field would have covered at least those areas above 2000 m a.s.l. on the north-facing slopes but have been less extensive, and smaller, on the south-facing slopes above 2500 m a.s.l.

In order to discuss the ELA during the local LGM, the geology of the Apennine Chain should be considered. The chain is characterised mainly by NW–SE tectonic structures which give rise to north-eastwards facing fault scarps. This causes the great majority of the glaciers that develop here to be facing northern quadrants. It follows then that the ELA is related mainly to glaciers with northern exposures. In Fig. 17.5, it can be seen that, during LGM, the ELA of the glaciers with northern exposures was increased from north-west to south-east and from south-west to north-east. Only in the southern Apennines did the ELA increase in a west to east direction from the Sirino to the Pollino mountains. The north-west to south-east increase is assumed to be linked to temperature differences resulting from the latitude, whilst the south-west to north-east increase is thought to be a consequence of the southern atmospheric circulation which was predominantly from the south-south-west during the LGM. This circulation carried humidity from the central Mediterranean Sea region onto the Italian landmass. A southerly atmospheric circulation has also been indicated by the sediments dated to the LGM and Late-glacial

FIGURE. 17.5 Map of the ELA in the Apennines. The iso-lines of northern, central and southern Apennine ELA are not extended to the whole peninsula because the glaciated mountains in the three sectors of the chain are separated by large non-glaciated regions. As a consequence, there are no data for the extension of the iso-lines.

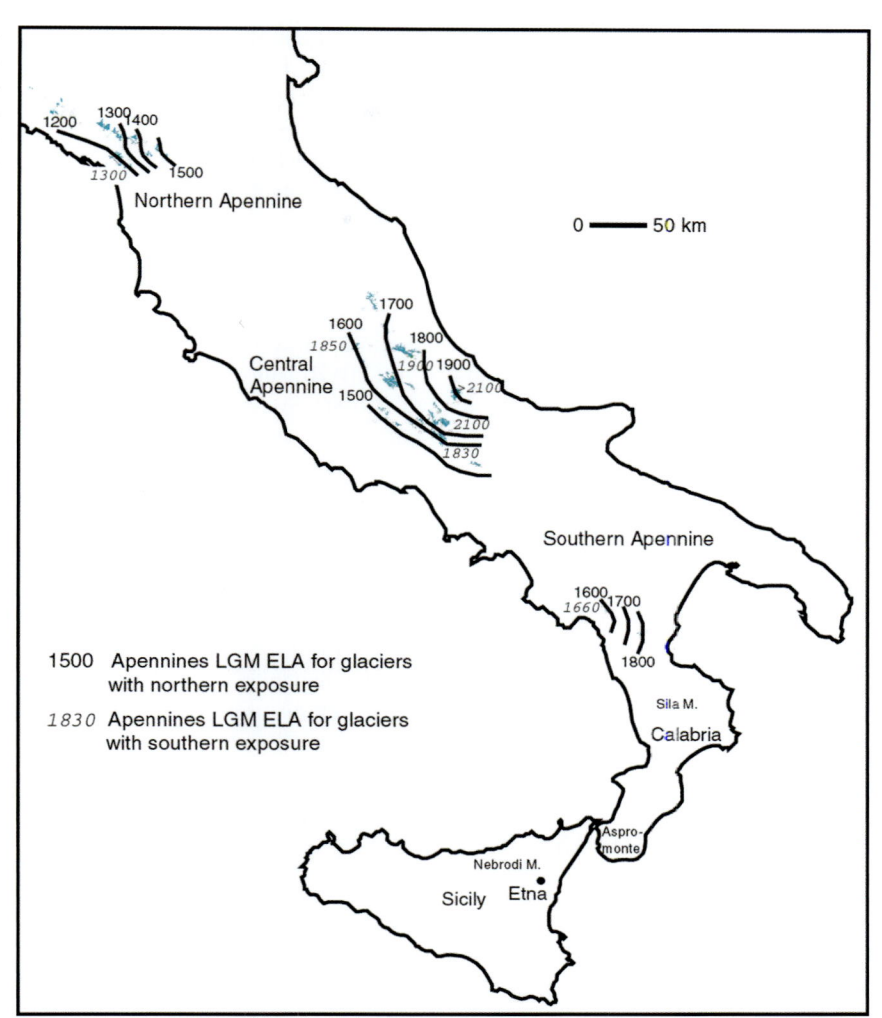

periods, in the central Italian maar lakes (Narcisi, 2000, 2001) and in the glacial environment on the Apennines (Giraudi, 2004), which contain high frequencies of Saharan quartz.

The west–east ELA increase in the southern Apennines could result from the presence of the Etna volcano and mountain ranges reaching altitudes near 2000 m a.s.l. in north-eastern Sicily and in Calabria (Nebrodi, Aspromonte, Sila mountains). These mountains probably caused the air masses from the south and south-east to lose moisture before reaching the Pollino Mountains.

The south-facing valleys preserving glacial remnants that allow an ELA evaluation are five in the central, one in the northern and one in the southern Apennines. In the central Apennines, the ELA was increasing from the south-west to the north-east, and in general, the ELA of glaciers with southern exposures was around 200–250 m higher than that of glaciers with northern exposures.

In the southern and northern Apennines (the Apuan Alps and Sirino mountains), the ELA of glaciers with southern exposures was 60–100 m higher than that of glaciers with northern exposures. Since the mountains occur close to the sea, the southern slopes most probably received more precipitation than their partially wind-shadowed northern slopes and the amount of precipitation partially balanced the temperature differences between their northern and southern exposures.

17.4. HOLOCENE NEOGLACIATION

The presence of glaciers in the Apennines during the Holocene is shown by the debris-covered Calderone Glacier on the Gran Sasso Massif. This glacier occurs at an elevation above ca. 2670 m a.s.l. in a cirque situated on the northern slope of the Corno Grande (2912 m a.s.l.).

On the basis of previous studies (Giraudi, 2004, 2005), some small moraines occur on the Gran Sasso massif, which were formed by glaciers with an ELA 800–1050 m higher than that of the LGM. These glaciers may possibly represent the recessional phases of the Mount Aquila Stadial dated to the Younger Dryas Chronozone (12.9–11.7 ka). However, it cannot be excluded that these moraines formed during the Early Holocene. The Calderone glacier melted during the first half of the Holocene but re-formed (Calderone Stadial A, Fig. 17.3) displacing a soil dated 4.52–4.29 ka BP. The soil developed on parent material containing volcanic ash. Recent chemical analyses (Giraudi et al., 2011b) suggest that this ash correlates to the Agnano–Monte Spina Tephra (AMST) at ca. 4.6 ka BP. Specifically, the uppermost date from this soil just post-dates that of the AMST and therefore the age of the soil corresponds closely to the age of renewed glacier growth.

Subsequently, the glacier experienced several phases of expansion (Calderone Stadials B, C, D, Fig. 17.3) dated ca. 2.88–2.72 ka BP, ca 1.41−1.29 ka BP, ca 0.64−0.58 ka BP, with the largest expansion achieved during the Little Ice Age. Since the end of the nineteenth century, the glacier has progressively retreated and today is close to disappearance.

17.5. CONCLUSIONS

The studies carried out on glacial sediments and features in the Apennine mountains in the past few years, based on dates using the ^{14}C and ^{39}Ar/^{40}Ar methods and tephrachronology, have confirmed previous observations and have allowed the recognition of new Late and Middle Pleistocene glaciations. The dating methods have allowed a direct age assignment that extends further back in time than that using the U–Th disequilibrium method. In summary, there is evidence for at least five pre-Holocene phases of glacial expansion that occurred during MIS 2–4, 6, 10 and 14. If the Alpine glaciation nomenclature and their chronology is considered (Hughes et al., 2006, 2007), it can be stated that sediments were deposited during a pre-Mindel glacial phase, during two Rissian glacial phases and in two Würmian phases. The map of the ELAs suggest that during the LGM the supply of moisture to the Apennine glaciers was carried by air masses linked to south-westerly and southern atmospheric circulation patterns.

ACKNOWLEDGEMENTS

The research was financially supported by ENEA. The studies on Campo Felice were conducted as part of the CNR Special Project 'Lacustrine Sedimentation-Palaeoenvironment-Palaeoclimate'.

REFERENCES

Chiarini, E., Giardini, M., La Posta, E., Papasodaro, F., Sadori, L., 2007. Sedimentology, palinology and new geochronological constrains on Quaternary deposits of the Corvaro intermontane basin (central Italy). Rev. Micropaléontol. 50, 309–314.

Di Rita, F., Magri, D., 2004. Pollen analysis of Upper Pleistocene sediments at Campo Felice, Central Italy. Il Quatern. Ital. J. Quatern. Sci. 17, 117–127.

Frezzotti, M., Giraudi, C., 1992. Evoluzione geologica tardo-pleistocenica ed olocenica del conoide complesso di Valle Majelama (Massiccio del Velino— Abruzzo). Il Quatern 5, 33–50.

Giordano, G., Esposito, A., De Rita, D., Fabbri, M., Mazzini, I., Trigari, A., et al., 2003. The sedimentation along the Roman coast between Middle and Upper Pleistocene: the interplay of eustatism, tectonics and volcanism. Il Quatern. 16, 121–129.

Giraudi, C., 1995. Considerations on the significance of some post-glacial fault scarps in the Abruzzo Apennines (Central Italy). Quatern. Int. 25, 33–45.

Giraudi, C., 2001. Nuovi dati sull'evoluzione tardo-pleistocenica ed olocenica di Campo Felice (L'Aquila, Abruzzo). Il Quatern. 14, 47–54.

Giraudi, C., 2004. The Apennine glaciations in Italy. In: Ehlers, J., Gibbard, P.L. (Eds.), Quaternary Glaciations-Extent and Chronology. Part I: Europe. Elsevier, Amsterdam, pp. 215–224.

Giraudi, C., 2005. Middle to Late Holocene glacial variations, periglacial processes and alluvial sedimentation on the higher Apennine massifs (Italy). Quatern. Res. 64, 176–184.

Giraudi, C., Frezzotti, M., 1995. Palaeoseismicity in the Gran Sasso Massif (Abruzzo, Central Italy). Quatern. Int. 25, 81–93.

Giraudi, C., Frezzotti, M., 1997. Late Pleistocene glacial events in the Central Apennine, Italy. Quatern. Res. 48, 280–290.

Giraudi, C., Groppelli, G., 2004. The problem of the last glaciation on the Mt. Etna volcano. In: Litho-Palaeoenvironmental Maps of Italy During the Last Two Climatic Extremes—Explanatory Notes. Tipografia Artistica Cartografica, Firenze, pp. 29–30.

Giraudi, C., Bodrato, G., Ricci Lucchi, M., Cipriani, N., Villa, I.M., Giaccio, B., et al., 2011a. Middle and late Pleistocene glaciations in the Campo Felice basin (Central Apennines—Italy). Quat. Res. 75, 219–230.

Giraudi, C., Magny, M., Zanchetta, G., Drysdale, R.N., 2011b. The Holocene climatic evolution of the Mediterranean Italy: a review of the continental geological data. The Holocene, 21(1),105–115.

Hughes, P.D., Woodward, J.C., Gibbard, P.L., 2006. Quaternary glacial history of the Mediterranean mountains. Progress Phys. Geogr. 30, 334–364.

Hughes, P.D., Woodward, J.C., Gibbard, P.L., 2007. Middle Pleistocene cold stage climate in the Mediterranean: new evidence from the glacial record. Earth Planet. Sci. Lett. 253, 50–56.

Karner, D., Marra, F., Renne, R., 2001. The history of the Monti Sabatini and Alban Hills volcanoes: groundwork for assessing volcanic-tectonic hazard for Rome. J. Volcanol. Geotherm. Res. 107, 185–215.

Kotarba, A., Hercman, H., Dramis, F., 2001. On the age of Campo Imperatore glaciations, Gran Sasso Massif, Central Italy. Geogr.Fis. Din.Quat. 24, 65–69.

Laurenzi, M.A., Villa, I.M., 1987. ^{40}Ar/^{39}Ar chronostratigraphy of Vico ignimbrites. Per.Miner 56, 285–293.

Messerli, B., 1967. Die eiszeitliche und die gegenwärtige Vergletscherung im Mittelmeerraum. Geogr. Helvet. 22, 105–228.

Narcisi, B., 2000. Late Quaternary eolian deposition in Central Italy. Quatern. Res. 54, 246–252.

Narcisi, B., 2001. Palaeoenvironmental and palaeoclimatic implications of the Late-Quaternary sediment record of Vico volcanic lake (central Italy). J. Quatern. Sci. 16, 245–255.

Neri, M., Coltelli, M., Orombelli, G., Pasquarè, G., 1995. Ghiacciai pleistocenici dell'Etna: un problema aperto. Ist.Lomb. (Rend. Sc.) B128, 103–125.

Chapter 18

Pleistocene Glaciations in Latvia

Vitālijs Zelčs*, Aivars Markots, Māris Nartišs and Tomas Saks

Faculty of Geographical and Earth Sciences, University of Latvia, Rainis Boulevard 19, Riga LV-1586, Latvia
*Correspondence and requests for materials should be addressed to Vitālijs Zelčs.

18.1. INTRODUCTION

Previous stratigraphical investigations of glacial and non-glacial deposits in Latvia suggest that four successive Pleistocene glaciations and three interglacials can be identified (Danilāns, 1973; Meirons and Straume, 1979; Meirons and Juškevičs, 1984; Meirons, 1986, 1992). The internal structure of the Pleistocene sequence has in many places been complicated by glaciotectonic deformation (Āboltinš, 1989; Āboltinš and Dreimanis, 1995; Zelčs and Dreimanis, 1997; Molodkov et al., 1998; Kalnina et al., 2007; Saks et al., 2007). Older sediments have been emplaced above younger ones as megablocks, or as overthrusts and folds (Dreimanis and Zelčs, 1995). As a consequence, solitary cores or narrow exposures may not reveal the true nature of such stratigraphical abnormalities. Large continuous exposures, numerous test drillings in stratotype areas and the implementation of Quaternary dating methods should be applied to solve these stratigraphical puzzles, but, in most places, they have yet not to be undertaken. The Pleistocene sequence has undergone large-scale sediment redistribution and remarkable complication as a result of glaciotectonism and glacial erosion (Āboltinš, 1989; Āboltinš and Dreimanis, 1995; Dreimanis and Zelčs, 1995). In south-eastern Latvia (Latgale and Augšzeme), glacial uplands and buried valleys display stratigraphically complex sections and have a greater than average sediment thickness (Meirons and Straume, 1979; Meirons, 1986, 1992). The thickness of the Pleistocene deposits reaches up to 200 m in the Vidzeme Upland and 310 m in the Aknīste buried valley, in south-eastern Latvia (Fig. 18.1). The glacial lowlands occupy over 60% of the area and are frequently underlain by a thin (less than 20 m) Pleistocene sediment sequence, except for the Baltic bedrock depression in western Latvia, where the Pleistocene sequence is relatively thick (up to 60–70 m) and has a complex stratigraphy.

The extent and timing of Pleistocene glaciations in Latvia has traditionally been interpreted on the basis of palynological studies of separate sections and cross-correlation of lithological data from boreholes. Middle to Late Weichselian-age plant remains, from organic sediments, "sub-fossil" reindeer finds and mammoth teeth have all been radiocarbon dated (Zelčs and Markots, 2004: Table 1; Ukkonen et al., 2006; Arppe and Karhu, 2010). Thermoluminiscence (TL) dating has been applied, with varying success, to some older sediments (Meirons et al., 1981; Meirons, 1986). During recent years, the electron spin resonance (ESR) and optically stimulated luminescence (OSL) dating techniques have also been applied to date basin, glaciofluvial and aeolian sediments underlying and overlying Weichselian till beds (Molodkov et al., 1998; Saks et al., 2007; Nartišs et al., 2009). The chronology of the last deglaciation of the south-eastern sector of the Scandinavian Ice Sheet, including the territory of Latvia, has also been established following the results of 115 cosmogenic ^{10}Be age determinations (Rinterknecht et al., 2006).

18.2. EARLY PLEISTOCENE GLACIATION

The Early Pleistocene sedimentary record has been recovered from the borehole No. II core, in the bottom near the bank of the buried depression, 15 km west of Krāslava town, in south-eastern Latvia (Židiņi in Fig. 18.1). According to Danilāns et al. (1964a), here the pre-Elsterian (Latgale) glacigenic unit consists of grey clayey till at a depth of 106.6–107.1 m (here and henceforth correlative chronostratigraphical terms in Latvia are given in brackets). This diamicton overlies a 12.7-m thick silty sand layer resting on the Middle Devonian sandstone (Danilāns et al., 1964a, p. 69). In later investigations, cores extracted from two boreholes (Nos. 43 and 570) drilled into the deepest part of this buried valley (in 1979 and 1982) revealed that the till bed was absent (Kondratiené et al., 1985; Kalnina et al., 1996; Kondratiené, 1996). In both boreholes, the lowest silty sand is overlain by Cromerian age (Židiņi) interglacial sediments.

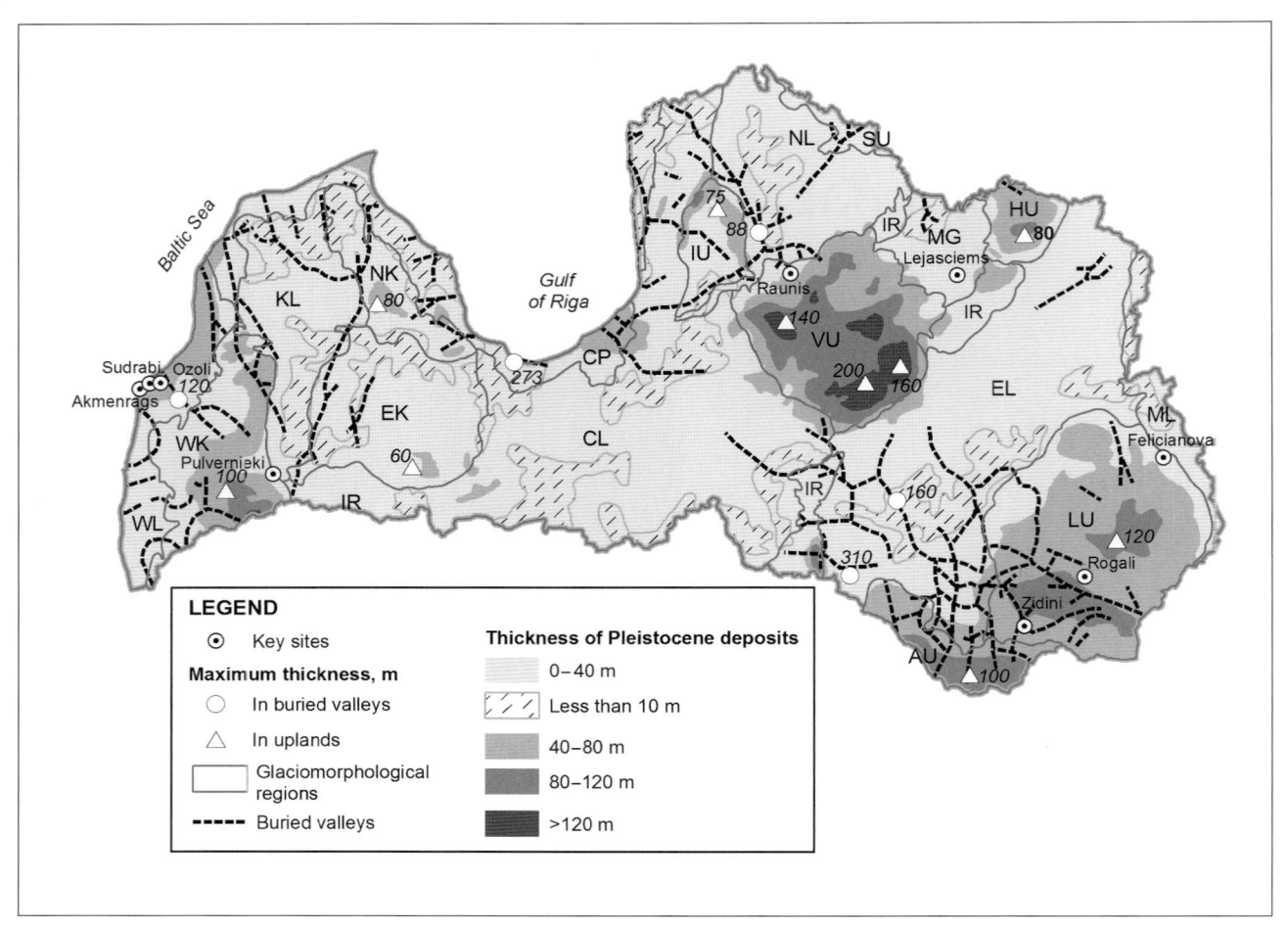

FIGURE 18.1 Thickness of Pleistocene deposits in the glaciomorphological regions of Latvia. Uplands: AU, Augšzeme; LU, Latgale; VU, Vidzeme; HU, Haanja (Alūksne); SU, Sakala; IU, Idumeja; EK, Eastern Kursa; WK, Western Kursa; NK, Northern Kursa. Lowlands: ML, Mudava (Velikoretskaya); EL, Eastern Lanvian; MG, Midle Gauja; NL, North Vidzeme; CL, Central Latvian; KL, Kursa; WL, Western Kursa; CP, Coastal plains. IR, interlobate ridges.

18.3. MIDDLE PLEISTOCENE GLACIATIONS

Elsterian (Lētiža) and Saalian (Kurzeme) till beds comprise the lower part of the Pleistocene sequence in the glacial uplands and buried valleys. Commonly, these till beds are absent from the glacial lowlands where subsequent glacial or post-glacial erosion has occurred. Interglacial sediments at Židiņi in the south-east of the country mark the base of the early Middle Pleistocene.

18.3.1. Cromerian (Židiņi) Stage Interglacial

The Židiņi key site of the early Middle Pleistocene interglacial sediments has been investigated by a number of scientists (Danilāns et al., 1964a; Khursevich, 1984; Meirons and Juškevičs, 1984; Kondratienė et al., 1985; Kalniņa et al., 1996; Kondratienė, 1996). According to Danilāns et al. (1964a) and Danilāns (1981), the Cromerian interglacial sediments here separate a pre-Cromerian diamicton and a silty sand bed and an Elsterian (Lētiža) till. The core extracted from borehole No. II comprises an almost 20-m thick lacustrine sediment sequence of lithified gyttia, silt and clay (Danilāns et al., 1964a). These sediments occur in a buried depression. Palynological investigation has revealed a succession from a cold-climate floral assemblage, with up to 40% content of non-arboreal pollen, to two warm-climatic optima (Danilāns et al., 1964a; Meirons and Juškevičs, 1984; Kondratienė et al., 1985; Kalniņa et al., 1996; Kondratienė, 1996). The first optimum is characterised by *Ulmus*, *Tilia* and *Corylus* maxima, while the second includes *Carpinus*, *Ulmus* and *Tilia* maxima. After the second optimum, the climate deteriorated, as suggested by the presence of a *Pinus*, *Betula* and *Alnus* pollen assemblage. Slight warming is indicated towards the top of this assemblage by the dominance of *Pinus* and *Picea*, followed by further cooling. No similar pollen succession has been encountered in other Latvian or Estonian Pleistocene records. This interglacial sedimentary sequence is considered to be older than the Holsteinian Stage because of its diatom flora (Khursevich, 1984), which includes species that became extinct later in the Middle Pleistocene (Kondratienė et al., 1985).

The Židiņi interglacial sediments have been correlated with the Augustovian interglacial in north-eastern Poland (Ber, 1996). On the basis of its very similar depositional conditions, Chapter 19 suggests that the Vindžiūnai interglacial represents its Lithuanian equivalent, instead of its previous correlation with the Turgeliai interglacial (Danilāns, 1981; Kondratienė et al., 1985; Kondratienė, 1996).

18.3.2. Elsterian (Lētiža) Stage Glaciation

The Elsterian (Lētiža) Stage till mainly rests on the pre-Quaternary bedrock and forms the base of the Pleistocene sequence in the areas of the glacial uplands. Its correlatives in the surrounding territories include the Sangaste-Stage glaciation tills in Estonia, the Dainava-Stage tills in Lithuania and Oka-Stage tills on the Russian Plain. The stratotype area is situated in south-western Latvia where predominantly heavily compacted sandy brownish grey to grey till is widely distributed, particularly in the Lētiža River basin (Danilāns et al., 1964b). These glacial deposits reach up to 60 m in thickness, including associated meltwater sediments. Weathered reddish Elsterian till outcrops on higher areas and Holsteinian (Pulvernieki) Stage interglacial sediments overlie it in several buried valleys. In the coastal lowland area of western Latvia, OSL dating of sediments below the upper Late Weichselian till has revealed that the Saalian till is absent from the Pleistocene sequence (Saks et al., 2007, in press). Here, Elsterian late-glacial marine clay and silt (Sudrabi Member), 1.5–6 m thick, containing lumps of till, the mollusc *Portlandia arctica*, diatoms and subarctic-climate pollen overlie the Elsterian till at a depth of 50–70 m below sea level (Segliņš, 1987; Kalniņa et al., 2000).

Brown sandy till and glaciolacustrine sediments of the Elsterian glaciation are also encountered in the eastern Latvia glacial uplands (Juškevičs, 2000; Juškevičs and Skrebels, 2003; Meirons, 2004). In the Vidzeme Upland, Elsterian till is found immediately resting on the bedrock (Āboltiņš and Dreimanis, 1995). It also fills the deepest parts of pre-Quaternary depressions and buried valleys in other uplands (Meirons, 2004).

18.3.3. Holsteinian (Pulvernieki) Stage Interglacial

The Pulvernieki Stage site consists of up to 8 m of lacustrine and mire sediments (Danilāns, 1973; Meirons, 1986) of Holsteinian interglacial age (Fig. 18.1). Sediments of this age have also been reported from the glacial uplands of south-eastern and east-central Latvia (Krūkle and Stelle, 1964; Savvaitov and Stelle, 1971; Meirons, 1992). However, these sediments are mostly rafted and displaced. At Pulvernieki, the organic sediments contain 52 different seed and fruit species, several of which are typical of the Middle Pleistocene, such as *Caulinia goretskyi* Dorof., *Aracites interglacial is* Wieliczk and *Brasenia borysthenica* var. *nemensis* Wieliczk (Meirons and Ceriņa, 1986). Palynological results demonstrate a sequence of forest development from *Betula–Pinus* to *Picea–Pinus*, followed by a climatic optimum dominated by *Picea–Abies–Carpinus*–Quercetum mixtum. The late interglacial forests consisted of *Pinus–Betula–Picea* and terminated with *Betula–Pinus*. *Pinus* are very abundant throughout the sequence, particularly at the very beginning and the end of the interglacial.

A similar sequence of forest development was also found in cores from Krāslava, Vorzova (Uzvara) and Sarkaņi in south-eastern Latvia (Meirons, 1992, 2004). The uppermost part of the interglacial deposits at Krāslava is absent (Krūkle and Stelle, 1964). The TL age of the sands overlying the Holsteinian sediments at Adamova is greater than 161 ka (Meirons et al., 1981), but the OSL age is 142 ± 26 ka (Hel-TL04150).

The marine equivalent of the Holsteinian interglacial is represented by the Akmeņrags Formation (Segliņš, 1987). Here, along the Baltic Sea coast in western Latvia, green silts and fine sands, up to 16 m thick and rich in organic remains and microfossils, have been encountered in several boreholes (Akmeņrags, Sudrabi, Ozoli a.o.) (Fig. 18.1). *Pinus sect. Strobus* and *Picea sect. Omorica* were present among the evergreens during the climatic optimum (Kalniņa et al., 2000; Kalniņa, 2001). Investigation of the diatom flora suggests that the highest salinity of the sea water was at the beginning of the deposition of the Akmeņrags marine sediments (Konshin et al., 1970).

The Pulvernieki interglacial is correlated with the Karukūla interglacial in Estonia, the Likhvin interglacial in Russia and the Butėnai interglacial in Lithuania (Meirons, 1986, 1992).

18.3.4. Saalian (Kurzeme) Stage

A grey calcareous till and associated meltwater sediments of Kurzeme Stage age are widely distributed throughout Latvia, particularly in the upland areas (Āboltiņš and Dreimanis, 1995; Juškevičs, 2000; Juškevičs and Skrebels, 2003; Meirons, 2004). The Kurzeme Stage is correlated with the Saalian Stage in northern Europe and is equivalent to the Ugandi Stage glaciation in Estonia, the Žeimena Stage glaciation in Lithuania and glaciations in the Russian Plain.

Deposits of the Saalian Stage glaciation were thought to be particularly common in western Latvia. However, previous interpretations of the widespread distribution of the Saalian till in western Latvia (Danilāns, 1973) have been rejected in the light of the recent determination of 13 OSL dates for shallow basin sands that underlie this till unit (Saks et al., 2007, in preparation). These dates cover time interval from 52 to 25 ka and attest that the upper till unit represents the Late Weichselian cold stage succession, not the Saalian (Danilāns, 1973) or combined sediments of the Late Weichselian and Saalian glaciations as suggested earlier (e.g. Juškevičs and Mūrniece, 1998; Kalniņa et al., 2000; Kalniņa, 2001).

Late-glacial freshwater deposits and topographically higher marine sediments, overlying the Saalian Till, have been encountered in boreholes at Zūras, 15 km south of Ventspils in western Latvia (Meirons, 1992).

18.4. LATE PLEISTOCENE GLACIATIONS

The model of a single glaciation in the Late Pleistocene that corresponds to the Last Glacial Maximum has been traditionally accepted in Latvia (Danilāns, 1973). According this model, Latvia was ice free during the Early and Middle Weichselian Substages.

18.4.1. Eemian (Felicianova) Stage Interglacial

Both marine and continental Eemian Stage interglacial sediments have been described from Latvia (Krūkle et al., 1963; Kalniņa et al., 2000; Kalniņa, 2001) and the Gulf of Rīga (Kalniņa, 1997). The key site of the continental Eemian interglacial sediments is at Felicianova in south-eastern Latvia (Fig. 18.1). They comprise up to 7-m thick alluvial and mire sediments (Krūkle et al., 1963; Danilāns, 1973). The climatic optimum of the Felicianova Stage interglacial is characterised by Quercetum mixtum accompanied by *Corylus*, which is followed by the dominance of *Carpinus* (Krūkle et al., 1963). The Felicianova Stage interglacial is correlated with the Merkinė interglacial in Lithuania, the Prangli interglacial in Estonia and the Mikulino interglacial on the Russian Plain (Meirons, 1986). These sediments also occur at Rogaļi, Subate, Kaitra and Skrudaliena in the south-eastern of the country and at Satiķi in the Eastern Kursa Upland (Figs. 18.1 and 18.2). Recently, Kalniņa et al. (2007) reported that pollen data from gyttja and sandy gyttja beds at Satiķi reflect the characteristic forest succession of the Eemian interglacial: *Betula*, *Pinus*, *Ulmus*, *Corylus*, *Quercus*, *Fraxinus*, *Tilia*, *Carpinus*, *Picea* and *Pinus*. Palaeobotanical studies indicate that this sequence is continuous and complete; however, the sediment infill of the Satiki palaeobasin may have been deformed and altered to some extent during the Last glaciation (Kalniņa et al. (2007)).

Based on the detailed palynological studies of the Pleistocene sediments from the coastal lowlands of western Latvia, Kalniņa et al. (2000) and Kalniņa (2001) reported marine Eemian-age sediment beds, distributed in a narrow area. These beds were previously ascribed to the Holsteinian Stage by Danilāns (1973).

18.4.2. Weichselian (Baltija) Stage Glaciation

Deposits of the Weichselian Glaciation are present throughout almost all of Latvia. These deposits play a dominant role in the Pleistocene sequence of the glacial uplands and lowlands and form the glacial landscape of the country (Straume, 1979; Āboltinš, 1989). Glaciotectonic deformational structures are extremely common in the Weichselian glacial deposits and thus complicate their stratigraphical interpretation. Most of the glacigenic landforms were formed during oscillatory retreat of the Late Weichselian Scandinavian ice sheet, when it was divided into several more or less independently flowing ice lobes and glacier tongues (Zelčs and Markots, 2004; Fig. 18.2).

The beginning of the last glaciation is marked by the preglacial deposition of the glaciolacustrine silts at Rogaļi in south-eastern Latvia (Meirons, 1972, 1986). These sediments contain cold-climate vegetational remains (*Betula nana* L., *Dryas* sp. a.o.) and re-deposited Eemian (Felicianova)-age pollen. At Židiņi (Fig. 18.2 for location), lacustrine silt between the Saalian and Weichselian tills has been TL dated to 79 ka (Meirons et al., 1981). TL dates of the lacustrine and fluvial sediments inter-bedded within the till from south-eastern Latvia (Subate, Zvidziena) range between 92 and 97 ka (Meirons, 1986). The ESR dates from *P. arctica* shells from the Pleistocene marine sediments at Līčupe in Central Latvia suggest age of 88.5 ± 7.3 and 97.8 ± 8.2 ka (Molodkov et al., 1998). However, these Early Weichselian interstadial marine sediments were glacially rafted and translocated for some 120 km from its origin in the Gulf of Riga by the Riga Ice Stream during the last glaciation (Dreimanis and Zelčs, 1995). *P. arctica* shells incorporated in the Weichselian till at the Daugmale, Tomēni site (Fig. 18.2) indicate an ESR age of 86.0 ± 7.3 ka, and the same shells collected from the glaciofluvial gravel at the same site have yield an age of 105.0 ± 7.3 ka (Molodkov et al., 1998).

Four OSL dates from the glaciofluvial sands at Lejaslabiņi and Cēre (Fig. 18.2 for location) in north-western Latvia have yielded ages of 85 ± 5, 42 ± 3, 115 ± 10 and 45 ± 3 ka (Risø Nos. 044654, 044654 and 044657, respectively). Taking into account the TL age of glaciofluvial material from Lejaslabiņi (56.056 ka) (Meirons and Juškevičs, 1984), these OSL dates reflect the Middle rather than the Early Weichselian time.

New OSL dating results from the Baltic Sea coastal cliff sections in western Latvia demonstrate the existence of a large inland basin in the Baltic Sea depression during the

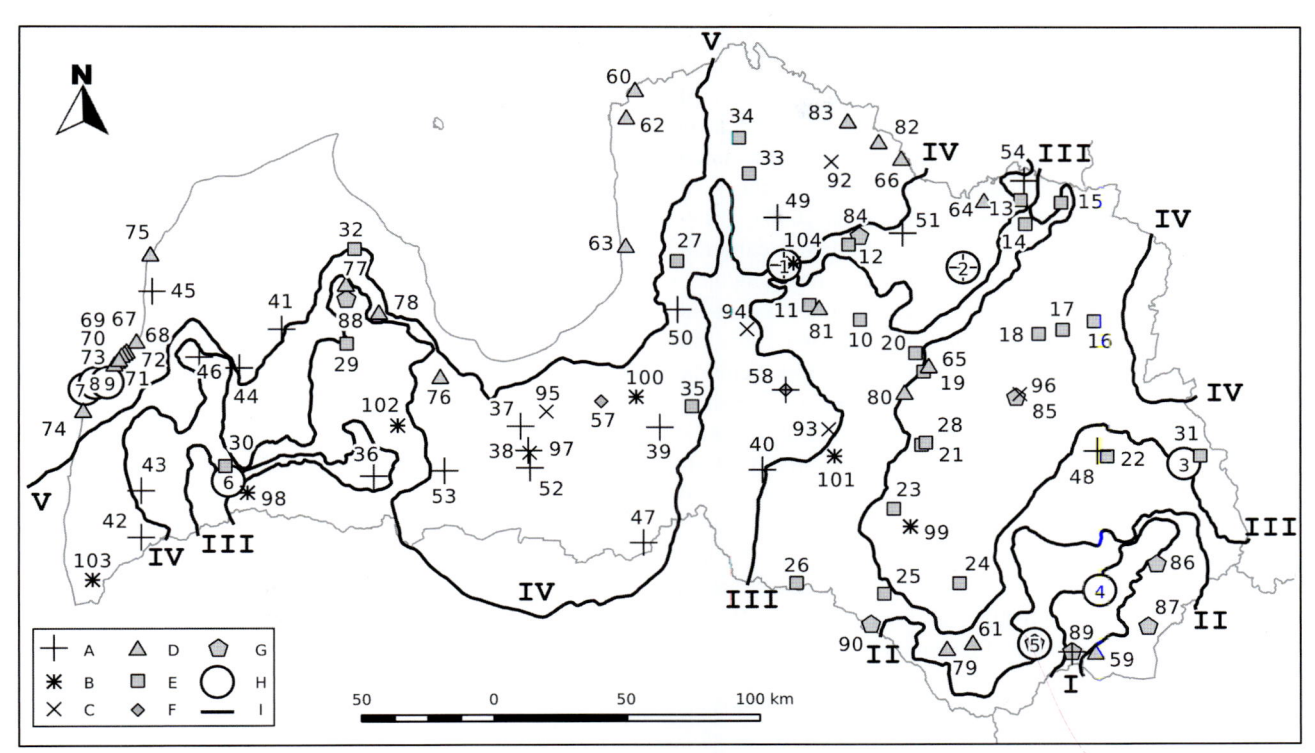

FIGURE 18.2 Location of ice-marginal positions, key and dating sites. A, ^{14}C plant remains; B, ^{14}C mammoth remains; C, ^{14}C reindeer remains; D, OSL; E, ^{10}Be; F, ESR; G, TL; H, Key sites; I–V, ice-marginal positions; I, Dagda; II, Kaldabruņa; III, Gulbene; IV, Linkuva; V, Valdemārpils. Sites: 1, Raunis; 2, Lejasciems; 3, Felicianova; 4, Rogaļi; 5, Židini; 6, Pulvernieki; 7, Akmeņrags; 8, Sudrabi; 9, Ozoli; 10, BALTI-12; 11, LAT-5; 12, MLIT-11; 13, MLIT-13; 14, MLIT-14; 15, MLIT-15; 16, MLIT-16; 17, MLIT-ss BALTI-16; 25, BALTI-17; 26, BALTI-18; 27, LAT-2wm; 28, LAT-7; 29, LAT-12; 30, LAT-13; 31, NLIT-4; 32, LAT-1; 33, LAT-3wm; 34, LAT-4; 35, LAT-8; 36, Lielauce; 37, Ozolnieki; 38, Ane; 39, Kaulezers; 40, Sece; 41, Abavas Rumba; 42, Krikmaņi; 43, Vartaja; 44, Kaltiki; 45, Sārnate; 46, Tireļpurvs; 47, Rakaņi; 48, Burzava; 49, Līči; 50, Viesulēni; 51, Līdumnieki; 52, Progress; 53, Savaiņi; 54, Veclaicene Rugāji; 57, Daugmales Tomēni; 58, Līčupe; 59, Baltica; 60, Ēkaji; 61, Gijantari; 62, Karateri; 63, Lauči; 64, Salkalni; 65, Biksēre; 66, Cepurīte; 67, Ēcenieki; 68, Zitari; 69, Gudenieki; 70, Valcenieki; 71, Baltmuiža; 72, Strante; 73, Ulmale; 74, Ziemupe; 75, Sensala; 76, Kažoki; 77, Lejaslabiņi; 78, Cere; 79, Svente; 80, Smeceres sils; 81, Lodesmuiža; 82, Dores; 83, Pentsils; 84, Mēri; 85, Zvidziena; 86, Ezernieki; 87, Robeznieki; 88, Talsi; 89, Adamova; 90, Subate; 92, Tirelis bog; 93, Odziena; 94, Nītaure; 95, Olaine; 96, Lubāns; 97, Tētele; 98, Nīgrande; 99, Livāni; 100, Ikšķile; 101, Pļaviņas; 102, Jaunpils; 103, Rucava; 104, Veselava.

Middle Weichselian (MIS 3) at least from 52 to 25 ka (Saks et al., in press). Timing of the basin infill is in a good agreement with the oxygen isotope record derived from the mammoth teeth dated to the 28 ka time interval between 52 and 24 ka preceding the Last Glacial Maximum (Arppe and Karhu, 2010). This time span covers earlier published ^{14}C ages from interstadial organics from the Lejasciems section (Arslanov et al., 1975; Meirons, 1986) and TL dates from silty sand sediments inter-bedded within till from Mēri (Meirons, 1986) in north-central Vidzeme (Fig. 18.2). OSL dates from fine-grained sandy sediments at the Silenieki (41.4 ± 8.6 ka, Hel-TL04164), Lejas Zemturi (46.0 ± 8.6 ka, Hel-TL04160) and Kažoki (26.9 ± 4.4 ka, Hel-TL04113; 29.4 ± 4.7 ka, Hel-TL04114) sand pits located in Eastern Latvian Lowland also fall into this Middle Weichselian time interval (Fig. 18.2 for location). OSL dating results from Smeceres sils at the southern slope of the Vidzeme Upland indicate sedimentation in a basin between 19.6 and 26.8 ka that slightly overlaps the subsequent Last Glacial Maximum interval. Summarising the dating results, and taking into account the evidence from Estonia (Kalm, 2006; Chapter 8), a time interval for possible Middle Weichselian glaciation (the Talsi stadial according to Zelčs and Markots, 2004) could tentatively be assigned between 68 and 54 ka in Latvia. The evidence also suggests that Latvian territory was ice free during the Middle Weichselian (Lejasciems Interstadial) during the time interval between 54 and 24 ka.

Meirons (1986, 1992) distinguished three Late Pleistocene glaciation till beds. The lowest, Augšzeme till unit, considered by Meirons (1992) as a formation of the Early Weichselian Substage (the Early Baltija of Meirons), is represented by grey to brownish grey calcareous sandy till. It occurs in the eastern Latvian uplands and the southern part of the western Latvian uplands. Since the Augšzeme till closely resembles the underlying Saalian till, it is possible that this till is composed of re-worked and re-deposited Saalian till material (Dreimanis and Zelčs, 1995). Based on the available regional temporal data (e.g. Kalm, 2006) and the nature of the distribution of the Augšzeme till, Zelčs and Markots (2004) argued that deposition of this till occurred during the major Late Weichselian ice advance. It can be suggested that both overlying upper till beds were deposited during the course of oscillatory deglaciation of the Last Scandinavian Ice Sheet (Zelčs and Markots, 2004). However, stony Weichselian till, that contains weathered erratics and well-rounded cobbles, has also been encountered within 30- to 40-m thick glaciofluvial deposits in the highest part of the Northern Kursa Upland. This till bed thins out in the southern direction and includes clearly expressed features of rewashing accompanied by increasing boulder concentration. It is therefore thought that the Middle Weichselian glaciation probably only reached the Latvian coastal plains, and, possibly, adjoining this upland margins.

The Late Weichselian till and meltwater sediments comprise a sequence up to 90 m thick. The meltwater sediments are widespread and often glaciotectonised. Glaciofluvial sand and gravel often form the core of glacial hills and streamlined landforms, ribbed moraines and ice-marginal formations (Āboltiņš, 1989; Zelčs and Dreimanis, 1997).

18.5. GLACIAL LIMITS

The limits of all the glacial events are located far outside Latvia. Therefore, only zones of ice-marginal formations after the Late Weichselian glacial maximum can be traced. The distribution of the ice-marginal formations has been derived from geomorphological maps at a scale of 1:500,000; in some areas, it has also been taken from 1:50,000 scale geomorphological maps and from Quaternary geology maps at a scale of 1:200,000 and 1:50,000. These data have been compared and corrected with respect to the regular network digital terrain models covering some areas in western and north-eastern Latvia, at increments of 20 m and a digital elevation model. A Shuttle Radar Topographic Mission (SRTM) digital elevation model for Latvia was modified with respect to the Late-glacial (starting with the Baltic Ice Lake) and post-glacial glacioisostatic rebound. As a result, the location of ice-marginal positions has been updated. The deglaciation history is supported by [10]Be dates (Rinterknecht et al., 2006) and recent OSL dating results (Nartišs et al., 2009; Saks et al., in press). Current study and previous investigations (Āboltiņš et al., 1972; Meirons et al., 1976; Straume, 1979; Zelčs and Markots, 2004) enable the differentiation of five distinct zones of ice-marginal formations (Fig. 18.2). Starting from the oldest, they are named the Dagda, Kaldabruņa, Gulbene, Linkuva and Valdemārpils ice-marginal zones. The spatial distribution of the ice-marginal formations and glacial features (drumlins, megaflutes, eskers, plateau-like hills, glacial lakes) and the location of key sections, dating sites and dates are included on the digital map (tiles for Latvia: PK23 and PK33).

18.5.1. Ice-Marginal Zones

The early phases of the deglaciation of the Late Weichselian ice sheet, described by Āboltiņš et al. (1972) as the "insular deglaciation stage", are characterised by prevailing frontal retreat of the ice masses. During this stage, ice-marginal formations of the Dagda, Kaldabruņa, Gulbene phases were formed. As a result, morphologically well-expressed zones of ice-marginal formations (composite marginal ridges, marginal ridges, marginal hilly massifs, shear-margin moraines) were developed at the peripheral part of the isometric insular uplands and Augšzeme marginal upland. The areal deglaciation was common only for the largest lobate and wide linear depressions where, after formation of the streamlined glacial bedforms typically formed below the fast flowing ice, relatively small glacier tongues stagnated in areas now occupied by lakes, for example, lakes Svente, Rāzna and Alauksts, and rivers. In the highest parts of western Latvian (Kursa) uplands and in the lowland areas (Zelčs and Markots, 2004: Fig. 1) where ice masses converged, assemblages of the interlobate glacial landforms, including composite hills, hummocky moraines, cone-shaped hills and even plateau-like hills (limnokames), were formed.

The Dagda phase marginal positions can only be traced in the south-eastern corner of Latvia, in the Latgale Upland (Fig. 18.2). Here, the marginal zone is represented by the composite marginal ridges, which forms the highest part of the upland, and is regarded as interlobate (Fig. 18.2). This phase replaces the initial Indra phase, defined by Āboltiņš et al. (1972), after improved topographical data became available. The Dagda phase is currently correlated to the Baltija (Pomeranian) phase in Lithuania, although there are no reliable numerical age dates on either side of the border to support this assumption.

The Kaldabruņa phase can be also detected in south-eastern part of the country (Fig. 18.2). It can be identified along the slope of the Latgale Upland, and further west in the Augšzeme Upland. This marginal zone was defined by Āboltiņš et al. (1972). Based on improved topographical, geological and numerical age evidence, an updated account can be presented here. Along the Latgale Upland, this marginal zone is represented by fragments of marginal moraines, composite glaciotectonic ridges and meltwater spillways. Further to the west, in the Augšzeme Upland, this zone is composed of rather well-established marginal moraine ridges. On the eastern slope of the Augšzeme Upland, this phase can be drawn along two marginal ridges, interrupted by the meltwater spillway. Further west across the Latvian–Lithuanian border, this marginal zone clearly coincides with the South Lithuanian marginal zone. Both these marginal zones are thought to represent the termination of the Lubāns and Polatsk (both draining from the Peipsi Ice Stream) ice lobes (Zelčs and Markots, 2004). There is only one ^{10}Be date, which corresponds to this glacial phase, and was taken from the boulder near the Latgale Upland northern margin well within the Kaldabruņa glacial limit (Fig. 18.2; Rinterknecht et al., 2006). This date of 15.5 ka probably provides the minimum age of this phase.

During the Gulbene glacial phase, most of the eastern Latvia upland area was ice free, while in western Latvia, only the southern part of the Western Kursa Upland became deglaciated (Fig. 18.2). This phase has been defined by Āboltinš et al. (1972), and in current study, its position has undergone minor changes in eastern Latvia. During this phase in south-eastern Latvia, the Lubāns ice lobe flowed south-westwards from the Russian Plain. Its termination is marked by the spectacular marginal moraine ridge in the Eastern Latvian Lowland (Zelčs and Markots, 2004). During this phase, the lowest marginal landform assemblage of the Latgale Upland was formed, as a termination of Lubāns and Polatsk ice lobes (Zelčs and Markots, 2004). In north-eastern Latvia, this phase can be drawn with some difficulties. During this stage, the Vidzeme Upland became ice free, and this phase is traced by the marginal moraine ridges and heavily glaciotectonised composite marginal ridges on its western, northern and north-eastern margins. Along the slope of Augšzeme Upland, this phase is difficult to trace because no clear ice-marginal landforms are visible. Digital elevation models clearly show the Lubāns ice lobe streamlined bedforms, trending from east-north-east to west-south-west which cross-cut the north–south-trending glacial lineations. Whether these two ice lobes operated asynchronously is still under debate. In central Latvia, during this time, the Zemgale ice lobe was advancing in a highly divergent manner, far into the Middle Lithuanian lowlands. Therefore, the Gulbene phase marginal formations in central Latvia can be traced only along the western margin of the Vidzeme upland. Further south, in the Central Latvian Lowland, the Zemgale ice lobe extent limit can be followed by the assemblage of end-moraine ridges and spillway valleys. Western Latvia during this stage was still covered by active ice, and only in the very south of western Latvia uplands were interlobate ridges formed (Āboltinš et al., 1972; Zelčs and Markots, 2004).

The ^{10}Be dates from within the Gulbene glaciation stage range from (excluding extremes) 12.6 to 14.0 ka (Rinterknecht et al., 2006), giving the average minimum age of the phase of ca. 13.5 ka. This corresponds rather well to the ^{10}Be dates from this phase in neighbouring Lithuania (Rinterknecht et al., 2006; Chapter 19). ^{10}Be dates from outside the Gulbene phase margins range from 10.78 to 15.20 (Rinterknecht et al., 2006), which are inconclusive. The OSL age from the glaciofluvial sands, collected in the kame terraces on the south-eastern slope of the Vidzeme upland, yielded 16.9–26.8, 16.9 ka being the probable maximum age of this phase (Fig. 18.2). OSL dates from the basin sands on the top of the River Daugava ancient delta in the town of Daugavpils, situated approximately 6 km south from the Gulbene phase margin yields an age of 14.5–15.5 ka, further limiting the minimal age of the Gulbene phase. Commonly, this glacial phase is correlated with the Middle Lithuanian phase in Lithuania (Zelčs and Markots, 2004; Chapter 19). The authors suggest to correlate the Gulbene phase ice-marginal zone with the ice-marginal Haanja glacial phase zone in Estonia, but confirmation of this requires more detailed studies.

During the Linkuva phase, glacial ice retreated further north and most of eastern Latvia became ice free (Āboltinš et al., 1972; Zelčs and Markots, 2004). The Lubāns ice lobe disappeared, and only the Mudava (Velikoretski) ice lobe was active in the extreme east of the Eastern Latvia Lowland. The ice-marginal position during this time is marked by the end-moraine chain (Fig. 18.2). In north-eastern Latvia, during this phase, the Burtnieks drumlin field was formed, terminating in the Veselava end-moraine chain (Fig. 18.2). Central Latvia was occupied by the Zemgale ice lobe, producing the well-developed Linkuva end-moraine arch in northern Lithuania and partly in Latvia (Āboltinš et al., 1972; Zelčs and Markots, 2004; Chapter 19). During this phase, the western Latvia uplands became ice free. The Usma ice lobe terminated as a series of glacier tongues, one of which ended in the Venta-Usma glacial lake, therefore, position of the Usma ice lobe margin is very approximate. During this phase, the Baltic Ice Stream in the western Latvia coastal lowlands ceased to exist, and, instead, several ice tongues formed that protruded from west to east, leaving fragmented chains of end moraines on the slopes of the Western Kursa Upland. The ^{10}Be dates from boulders within the margins of the Linkuva phase give a wide distribution of ages (12.0–15.4 ka). Radiocarbon analyses of the plant remains from the lake sediments, approximately 6 km north of the Veselava end-moraine inner margin (Fig. 18.2: Raunis key site), yield ages of 13.2–13.4 ^{14}C ka BP (Vinogradov et al., 1963, Punning et al., 1968; Stelle et al., 1975). They provide the minimum age of this glacial phase (Zelčs and Markots, 2004).

The Linkuva glacial phase can be correlated to the North Lithuanian glacial phase in Lithuania and to the Ottepaa/Sakala ice-marginal formation in Estonia. Cross-border correlation in central Latvia is straightforward, while in western Latvia, the Linkuva ice-marginal formations are thought to correlate with the Pajūris ice-marginal formation zone (Chapter 19) but are hard to trace. In Estonia, the Linkuva glacial phase marginal formations can rather be correlated with the Otepää ice-marginal zone, but here, the cross-border correlation is again difficult.

The Valdemārpils glacial phase is the latest deglaciation stage of the Scandinavian Ice Sheet in Latvia, and its limit stretches along the modern sea coast (Fig. 18.2). Inland, east and west of the Gulf of Riga, the Valdemārpils phase is marked by a chain of relatively low-marginal ridges and end moraines. In central and north-western Latvia, tracing this glacial limit is quite problematic because most of the glacial landscape is smoothed or eroded by Late-glacial meltwater basins and Litorina Sea. There are no reliable dates corresponding to this ice-marginal zone. The Valdemārpils ice-marginal zone can be correlated to the Pandivere ice-marginal zone in Estonia, from which ice started to decay on the Pandivere upland around 13.8 cal. ka BP (Chapter 8).

18.6. CONCLUSIONS

Lithostratigraphical, biostratigraphical and geochronological information available suggest the existence of three Pleistocene interglacials in Latvia—Cromerian (Židiņi), Holsteinian (Pulvernieki) and Eemian (Felicianova), and three glacials—Elsterian (Lētiža), Saalian (Kurzeme) and Weichselian (Baltija). On the basis of present knowledge, there is no conclusive evidence to support the occurrrence of an Early Pleistocene glaciation in the country.

The Late Pleistocene sequence can be tentatively subdivided into the following substages: an Early Weichselian substage (Rogaļi preglacial), which begins from the termination of the Eemian Stage interglacial to 68 ka, an Early/Middle Weichselian substage (Talsi stadial) between 68 and 54 ka, a Middle Weichselian substage (Lejasciems interstadial) between 54 and 24 ka and a Late Weichselian substage. The onset of the Late Weichselian glaciation in Latvia has not been reliably dated, but the available OSL dates from western Latvia suggest that ice masses invaded the country no earlier than 24–25 ka. The possible duration of the Late Weichselian glaciation of Latvia ranges between ca. 9 and 10 ka.

The more easterly lobes retreated earlier than those in the west. Glacier recession from the Kaldabruņa to the Gulbenes zone took place between 15.5 and 13.5 ka.

ACKNOWLEDGEMENTS

We wish to thank Valdis Segliņš and Laimdota Kalniņa for useful additional information on key sites. We express our particular gratitude to our students Ivars Celiņš and Artūrs Putniņš for contributing to the development of the database on linear meltwater features.

REFERENCES

Āboltiņš, O., 1989. Glaciotectonic Structure and Glacial Morphogenesis. Zinatne, Riga, 284 pp. (in Russian).

Āboltiņš, O., Dreimanis, A., 1995. Glacigenic deposits in Latvia. In: Ehlers, J., Kozarski, S., Gibbard, P. (Eds.), Glacial Deposits in North-East Europe. Balkema, Rotterdam/Brookfield, pp. 105–113.

Āboltiņš, O., Veinbergs, I., Danilans, I., Stelle, V., Straume, J., Eberhards, G., et al., 1972. Main features of glacial morphogenesis and peculiarities of deglaciation of the last glaciation in the territory of Latvia. In: Danilans, I., Aboltinš, O. (Eds.), Putyevoditel polevogo simpoziuma III Vsesoyuznogo mezhvedomstvennogo soveschaniya po izucheniyu krayevykh obrazovaniy materikovogo oledeneniya. Zinatne, Riga, pp. 3–16 (in Russian).

Arppe, L., Karhu, J.A., 2010. Oxygen isotope values of precipitation and the thermal climate in Europe during the middle to late Weichselian ice age. Quatern. Sci. Rev. 29 (9–10), 1263–1275.

Arslanov, Kh.A., Filonov, B.A., Chernov, S.B., Makarova, E.V., 1975. Radiocarbon dating of the Geochronology Laboratory of NIGEI at the University of Leningrad. Report No.2. Byul. Komis. po izuch. chetvertich. perioda 43, 212–226 (in Russian).

Ber, A., 1996. Geological situation of Augustovian (Paston ian) Interglacial lake sediments at Szczebra near Augustow and Mazovian Interglacial organogenic sediments at Krzyzewo. Proc. Estonian. Acad. Sci. Geol. 373, 35–48 (in Polish with English summary).

Danilāns, I., 1973. Quaternary Deposits of Latvia. Zinatne, Riga, 312 pp. (in Russian).

Danilāns, I., 1981. Interpretation of data on stratigraphy of Lower Pleistocene of the eastern Baltic area. In: Pleistocene Glaciations of the Eastern European Plain. Nauka, Moscow (in Russian), pp. 114–120.

Danilāns, I., Dzilna, V., Stelle, V., 1964a. Profile of Židiņi. In: Darilāns, I. (Ed.), Questions of Quaternary Geology, vol. III. Publishing House of the Academy of Sciences of the Latvian SSR, Riga, pp. 63–140 (in Russian with English summary).

Danilāns, I., Dzilna, V., Stelle, V., 1964b. Interglacial deposits at Pulvernieki. In: Danilāns, I. (Ed.), Questions of Quaternary Geology, vol. III. Publishing House of the Academy of Sciences of the Latvian SSR, Riga, pp. 141–163 (in Russian with English summary).

Dreimanis, A., Zelčs, V., 1995. Pleistocene stratigraphy of Latvia. In: Ehlers, J., Kozarski, S., Gibbard, P. (Eds.), Glacial Deposits in North-East Europe. Balkema, Rotterdam/Brookfield, pp. 105–113.

Juškevičs, V., 2000. Quaternary deposits. In: Āboltiņš, O., Kuršs, V (Eds.), Geological Map of Latvia. Scale 1:200,000. Sheet 43—Rīga. Sheet 53—Ainaži. Explanatory Text and Maps. State Geological Survey, pp. 10–31 Rīga (in Latvian).

Juškevičs, V., Mūrniece, S., 1998. Quaternary deposits. In: Āboltiņš, O., Kuršs, V. (Eds.), Geological Map of Latvia. Scale 1:200,000. Sheet 41—Ventspils. Explanatory Text and Maps. State Geological Survey, pp. 8–22 Rīga (in Latvian).

Juškevičs, V., Skrebels, J., 2003. Quaternary deposits. In: Āboltiņš, O., Brangulis, A.J. (Eds.), Geological Map of Latvia. Scale 1:200,000. Sheet 34—Jēkabpils & Sheet 24—Daugavpils. Explanatory Text and Maps. State Geological Survey, pp. 10–29 Rīga (in Latvian).

Kalm, V., 2006. Pleistocene chronostratigraphy in Estonia, southeastern sector of the Scandinavian glaciations. Quatern. Sci. Rev. 25, 960–975.

Kalniņa, L., 1997. An outline of the Pleistocene pollen stratigraphy in the Gulf of Riga. In: Cato, I., Klingberg, F. (Eds.), Proceedings of the Forth Marine Geological Conference—The Baltic, Upssala1995, Sveriges Geol. Unders. Ca 86, 67–74.

Kalniņa, L., 2001. Middle and Late Pleistocene environmental changes recorded in the Latvian part of the Baltic Sea basin. Quaternaria A9, 173 pp. Stockholm University.

Kalniņa, L., Juškevics, V., Segliņš, V., 1996. Pleistocene stratigraphy at the Židiņi site, Southeastern Latvia. In: Robertsson, A.-M. (Ed.), Landscapes and life. Studies in honour of Urve Miller. PACT, Rixensart, pp. 31–40.

Kalniņa, L., Dreimanis, A., Mūrniece, S., 2000. Palynology and lithostratigraphy of Late Elsterian to Early Saalian aquatic sediments in the Ziemupe-Jūrkalne area, western Latvia. Quatern. Int. 68–71, 87–109.

Kalniņa, L., Strautnieks, I., Ceriņa, A., 2007. Upper Pleistocene biostratigraphy and traces of glaciotectonics at the Satiki site, western Latvia. Quatern. Int. 164–165, 197–206.

Khursevich, G.K., 1984. Stratigrafija pleistocenovikh otlozheniy Pribaltiki po dannym diatomogo analiza. In: Kondratiené, O., Mikalauskas, A. (Eds.), Paleogeografiya chetvertichnogo perioda Pribaltiki i sopredel'-nikh rayonov. AN Litovskoi SSR, pp. 122–129 (in Russian).

Kondratiené, O., 1996. Stratigrafia i paleogeografia kvartera Litvy po paleo-botanycheskym dannym. Academia, Vilnius, 212 pp. (in Russian).

Kondratiené, O.P., Khursevich, G.K., Loginova, L.P., 1985. Biostratigraficheskoye obosnovaniye vozrasta ozernoj tolschi razreza Zhidini. In: Belzatskaya, L.Y. (Ed.), Problemy Pleistotsena. Navuka i tehnika, Minsk, pp. 86–1001 (in Russian).

Konshin, G., Savvaitov, A., Slobodin, V., 1970. Marine inter-till deposits of Western Latvia and some peculiarities of their development. In: Danilāns, I. (Ed.), Problems of Quaternary Geology, vol. V. Zinātne, Rīga (in Russian with English summary), pp. 37–48.

Krūkle, M., Stelle, V., 1964. Mindel-Rissian deposits in the town of Krāslava. In: Danilāns, I. (Ed.), Questions of Quaternary Geology, vol. III. Publishing House of the Academy of Sciences of the Latvian SSR, Riga, pp. 165–182 (in Russian with English summary).

Krūkle, M., Lūsiņa, L., Stelle, V., 1963. Profile of Pleistocene deposits at Felicianova. In: Danilāns, I. (Ed.), Questions of Quaternary Geology, vol. II. Publishing House of the Academy of Sciences of the Latvian SSR, Riga, pp. 7–34 (in Russian with English summary).

Meirons, Z., 1972. Section of interglacial deposits at Rogali. In: Danilāns, I. (Ed.), Problems of Quaternary Geology, vol. VI. Zinatne, Riga, pp. 7–13 (in Russian with English summary).

Meirons, Z., 1986. Stratigraphy of Pleistocene deposits in Latvia. In: Kondratiene, O., Mikalauskas, A. (Eds.), Issledovaniya lednikovykh obrazovaniy Pribaltiki. AN Litovskoi SSR, pp. 69–81 Vilnius (in Russian).

Meirons, Z., 1992. Stratigraphic scheme of Pleistocene deposits in Latvia. In: Veinbergs, I., Danilans, I., Sorokin, V., Ulst, R. (Eds.), Paleogeografiya i stratigrafiya fanerozoya Latvii i Balstiyskogo morya. Zinatne, Riga, pp. 84–98 (in Russian).

Meirons, Z., 2004. Quaternary deposits. In: Āboltiņš, O., Brangulis, A.J. (Eds.), Geological Map of Latvia. Scale 1:200,000. Sheet 35—Rēzekne & Sheet 25—Indra. Explanatory Text and Maps. State Geological Survey, pp. 9–26 Rīga (in Latvian).

Meirons, Z., Ceriņa, A., 1986. Novye issledovaniya razreza Pulvernieki. In: Āboltiņš, O. (Ed.), Morfogenez relyefa i paleogeografiya Latvii. Latviyskiy gosydarstvenniy universitet, Riga, pp. 4–18 (in Russian).

Meirons, Z., Juškevičs, V., 1984. Quaternary deposits. In: Misans, J., Brangulis, A., Straume, J. (Eds.), Geologija Latviyskoi SSR (Explanatory text with geological maps on scale of 1:500,000). Zinatne, Riga, pp. 89–122 (in Russian).

Meirons, Z., Straume, J., 1979. Cenozoic group. In: Misans, J., Brangulis, A., Danilans, I., Kuršs, V. (Eds.), Geologicheskoe stroyenie i poleznye iskopayemye Latvii. Zinatne, Riga, pp. 176–268 (in Russian).

Meirons, Z., Straume, J., Juškevics, V., 1976. Main varieties of the marginal formations and retreat of the last glaciation in the territory of Latvian SSR. In: Danilans, I. (Ed.), Problems of Quaternary Geology, vol. 9. Zinātne, Rīga, pp. 50–73 (in Russian with English summary).

Meirons, Z., Punning, M.J., Hütt, G., 1981. Results obtained through the TL dating of Southeast Latvian Pleistocene deposits. Proc. Estonian Acad. Sci. Geol. 1, 28–33 (in Russian with English & Estonian summaries).

Molodkov, A., Dreimanis, A., Aboltinš, O., Raukas, A., 1998. The age of *Portlandia arctica* shells from glacial deposits of Central Latvia: an answer to a contraversy onthe age and genesis of their enclosing sediments. Quat. Geochronol. 17, 1077–1094.

Nartišs, M., Celiņš, I., Zelčs, V., Dauškans, M., 2009. Stop 8: history of the development and palaeogeography of ice-dammed lakes and inland dunes at Seda sandy plain, north western Vidzeme, Latvia. In: Kalm, V., Laumets, L., Hang, T. (Eds.), Extent and Timing of Weichselian Glaciation Southeast of the Baltic Sea: Abstracts and Guidebook. The INQUA Peribaltic Working Group Field Symposium in Southern Estonia and Northern Latvia, September 13–17, 2009. Tartu Ülikooli Kirjastus, Tartu, pp. 79–81.

Punning, J.-M., Raukas, A., Serebryanny, L.R., Stelle, V., 1968. Paleogeographical peculiarities and absolute age of the Luga stage of the Valdaian glaciation on the Russian Plain. Dokl. Akad. Nauk SSSR Geol. 178 (4), 916–918 (in Russian).

Rinterknecht, V.R., Clark, P.U., Raisbeck, G.M., Yiou, F., Bitinas, A., Brook, E.J., et al., 2006. The last deglaciation of the southeastern sector of the Scandinavian Ice Sheet. Science 311, 1449–1452.

Saks, T., Kalvāns, A., Zelčs, V., 2007. Structure and micromorphology of glacial and non-glacial deposits in coastal bluffs at Sensala, Western Latvia. Baltica 20 (1–2), 19–27.

Saks, T., Kalvāns, A., Zelčs, V., in press. OSL dating of Middle Weichselian age shallow basin sediments in Western Latvia, Eastern Baltic. Quatern. Sci. Rev, doi:10.1016/j.quascirev.2010.11.004.

Savvaitov, A., Stelle, V., 1971. Lihvinskiye mezhlednikovye otlozheniya na Centralno-Vidzemskoy vozvyshennosti (razrez Klekeri). In: Bartosh, T. (Ed.), Palinologicheskiye issledovaniya v Pribaltike. Zinatne, Riga, pp. 51–55 (in Russian).

Segliņš, V., 1987. Stratigrafiya pleistotsena zapodnoj Latvii: Avtoreferat disertatsii kandidata geologicheskikh i mineralogicheskikh nauk. Academy of Sciences Estonian SSR, Tallinn, 14 pp. (in Russian).

Stelle, V., Savvaitov, A.S., Veksler, V.S., 1975. Dating of Pleistocene deposits in the territory of Latvia. In: Savvaitov, A.S., Veksler, V.S. (Eds.), Opyt i metodika izotopno-geokhimicheskikh issledovaniy v Pribaltike i Belorussii. VNIIMORGEO, Riga, pp. 80–81 (in Russian).

Straume, J., 1979. Geomorphology. In: Misans, J., Brangulis, A., Danilans, I., Kuršs, V. (Eds.), Geologicheskoe stroyenie i poleznye iskopayemye Latvii. Zinatne, Rīga, pp. 297–439 (in Russian).

Ukkonen, P., Lõugas, L., Zagorska, I., Lukševiča, L., Lukševičs, E., Daugnora, L., et al., 2006. History of the reindeer (*Rangifer tarandus*) in the eastern Baltic region and its implications for the origin and immigration routes of the recent northern European wild reindeer populations. Boreas 35, 222–230.

Vinogradov, A.P., Devirts, A.L., Dobkina, E.I., Markova, N.G., 1963. Determination of the absolute age according to ^{14}C. Geochemistry 9, 795–812 (in Russian).

Zelčs, V., Dreimanis, A., 1997. Morphology, internal structure and genesis of the Burtnieks drumlin field. Sed. Geol. 111 (1–4), 73–90.

Zelčs, V., Markots, A., 2004. Deglaciation history of Latvia. In: Ehlers, J., Gibbard, P.L. (Eds.), Quaternary Glaciations—Extent and Chronology. Part I: Europe. Elsevier, Amsterdam, pp. 225–243.

Chapter 19

Pleistocene Glaciations in Lithuania

Rimantė Guobytė[1,2,]* and Jonas Satkūnas[1,2,]*

[1] Lithuanian Geological Survey S. Konarskio 35, LT-03123, Vilnius, Lithuania
[2] Vilnius University, Universiteto 3, LT-01513 Vilnius, Lithuania
*Correspondence and requests for materials should be addressed to Rimantė Guobytė and Jonas Satkūnas. E-mails: rimante@lgt.lt; jonas.satkunas@lgt.lt

19.1. INTRODUCTION

The average thickness of Lithuania's Quaternary deposits is 130 m and varies from 10–30 m in the northern part of country, that is, the area of prevailing glacial erosion, up to 200–300 m in the marginal and insular glacial Uplands and buried valleys or palaeoincisions (Fig. 19.1). A great help for the investigation of the oldest Pleistocene glaciations is the national borehole database (maintained by the Lithuanian Geological Survey) containing over 50,000 boreholes of various depth, among them. However, the most informative are the 4124 boreholes drilled and sampled in the course of geological mapping. Part of these boreholes was recently reinterpreted using the cross section network (Jusienė, 2007), within which all the cross sections were located at a distance of 10–12 km from each other. Therefore, the spatial distribution of stratigraphical units (Table 19.1) in whole area was mapped in a very systematic way.

The ancient Pleistocene Tills, as well as interglacial and interstadial sediments, will be discussed in this chapter. Their stratigraphical position has been constructed on the basis of all available, up to date, relevant material of investigations: borehole and mapping information (that includes cores of deep and shallow drillings, geophysics, examination of outcrops), biostratigraphical, lithostratigraphical and age data.

The modern surface of Lithuania is made up of Upper Weichselian (Nemunas)-age sediments (75.86% of area), late-glacial and Holocene sediments (20.35%) and Saalian-age (Medininkai) deposits (2.25%). The plains are underlain by basal till and glaciolacustrine sediments (Guobytė et al., 2001). The topography of Lithuania is not very imposing because the lowlands, which lie below 100 m a.s.l., occupy a considerable part of the land area, so that only 20% consist of hilly Uplands. Almost all these Uplands are concentrated in the east of country, excluding the Žemaitija (Samogitia) and Western Kuršas (the southern part only) Uplands. The Žemaitija Upland is one of insular-type glacial Uplands in the East Baltic area (Zelčs and Markots, 2004; Raukas, 2009), but the only one situated in Lithuania. The Upland is 80–130 m higher than the surrounding lowlands (Fig. 19.2). A prominent, south-west–north-east orientated belt of hills and ridges, in Eastern Lithuania up to 20–45 km wide, is known as the Baltija marginal Uplands. These hills reach heights of 150–250 m a.s.l. The south-eastern Lithuanian Lowland (140–160 m a.s.l.) separates the Baltija Uplands from North and South Nalšia (previously named Švenčionys and Ašmena) Uplands. These Uplands are branches of a huge ridge that stretches far into neighbouring Belarus. These Uplands are significantly higher, reaching 250–280 m a.s.l. The highest point in Lithuania, the so-called Aukštojas hill (293.8 m a.s.l.), is situated here.

The ubiquitous presence of glacial landforms has long triggered a widespread geological interest in the glacial history that can be traced back to over a century ago. Mapping seems to be the most important issue of the research on geology and geomorphology. The oldest known geomorphological map is that published by A. Misūna in 1903, showing end-moraines of eastern Lithuania and Belarus. Geomorphological sketches, presented by B. Doss (1910) and H. Hausen (1913), as well as Misūna's map were the basis for the first reconnaissance geomorphological map of Lithuania published by H. Mortensen in 1924 (Guobytė, 2004). H. Mortensen's map, reflecting major geomorphological features, served as a basis for later national research by Č. Pakuckas, J. Dalinkevičius and V. Čepulytė (Guobytė, 2002), among others.

Detailed geomorphological investigations of the land surface were carried out by Basalykas (1959) which allowed the identification of glacial limits. It also allowed correlation of these limits with the stadials of the Last Glaciation already recognised in Poland and Germany (Guobytė, 2004). The famous Lithuanian geomorphologist Č. Kudaba considerably improved the research on the age and genesis of the Baltija marginal Uplands. He showed belts of Grūda, Žiogeliai, Baltija stadials and South, Middle

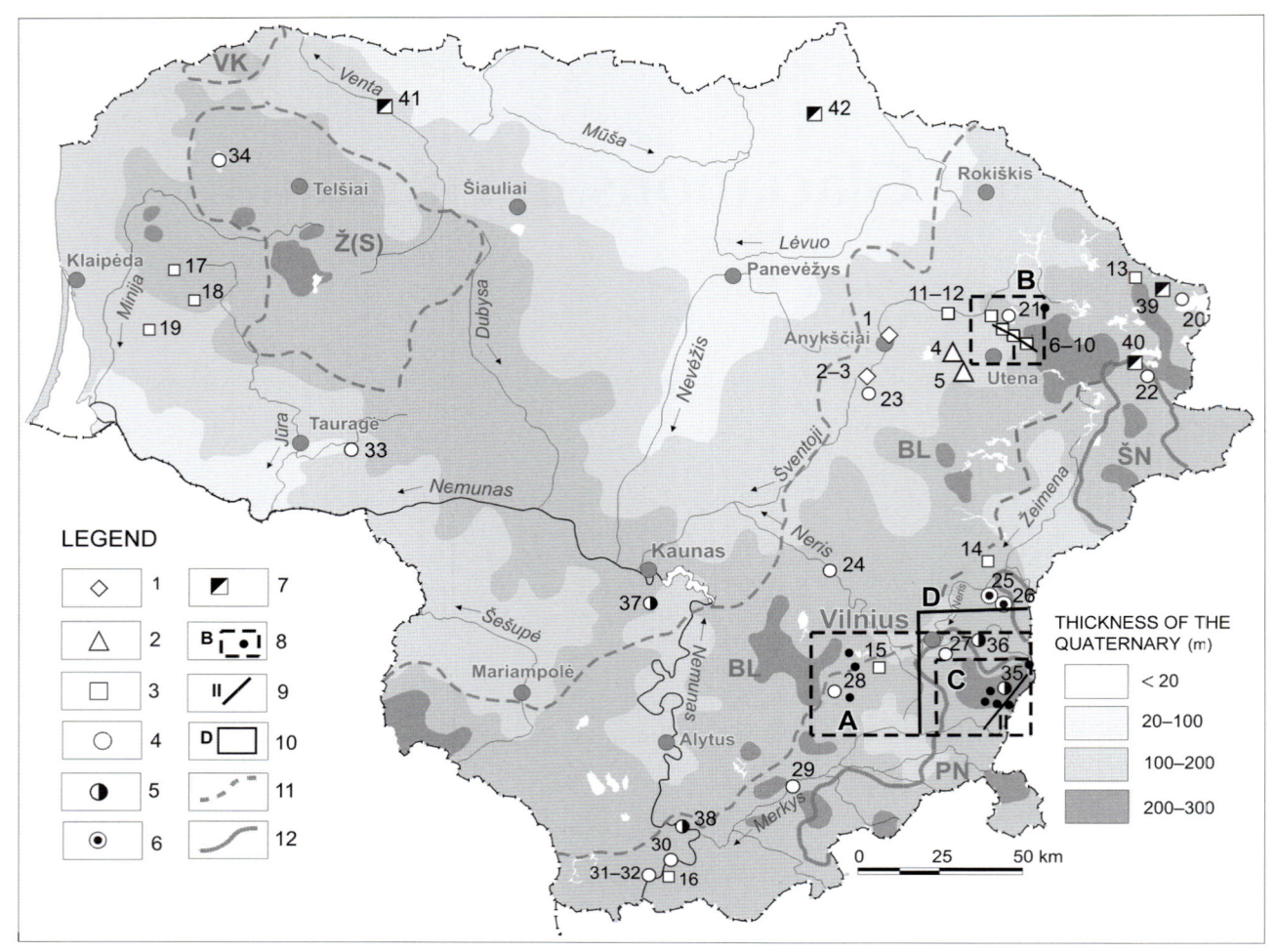

FIGURE 19.1 Thickness of Lithuania's Quaternary and location of the glacial and interglacial stratotypes (areas and sites; after Kondratienė, 1993): 1—Daumantai Formation (sites 1–3), 2—Turgeliai Interglacial (sites 4–5), 3—Būtėnai (Holsteinian) Interglacial (sites 6–19), 4—Merkinė (Eemian) Interglacial (sites 20–34), 5—Merkinė Interglacial and Early–Middle Nemunas Interstadial (sites 35–38), 6—Merkinė and Late Weichselian Grūda Stadial (sites 25–26), 7—Early–Middle Nemunas interstadials (sites 39–42), 8—areal stratotypes with representative boreholes of A—Dzūkija and Dainava (Elsterian) Tills (sites 28, 52*–56*), B—Žemaitija (Lower Saalian) Till (sites 8, 12, 57*), C—Medininkai (Upper Saalian) Till (sites 58*–61*), 9—location of the geological profiles shown in Figs. 19.3 and 19.6, 10—stratotype area of the oldest Quaternary deposits (shown in Fig. 19.3). Contours of glacial Uplands: 10—of the Late Weichselian age, 11—of the Saalian age (for the Uplands name see Fig. 19.2) (*sites shown on the digital map only). Site (stratotype) numbers and names (× dated):

1—Giliai outcrop	13—Minauka-8	23—Kurkliai outcrop	33—Tauragė
2—Vetygala outcrop	14—Pailgė-30	24—Kmitos outcrop	34—Plateliai
3—Daumantai outcrop	15—Račkūnai-685	25—Aukštieji Bezdonys-80	35—Medininkai×
4—Vaitkūnai-914	16—Neravai outcrop	26—Arvydai-79	36—Mickūnai×
5—Kudrė-915	17—Žadeikiai-27	27—Kirtimai	37—Rokai×
6–7—Gaigaliai-921, 876	18—Stančaičiai-25	28—Puponys-674	38—Jionionys×
8—Sudeikiai-936	19—Skomantai-22	29—Smalnykai outcrop	39—Smalvas
9—Stumbrė-878	20—Schedai-3	30—Liškiava outcrop	40—Dysnai
10—Daržiniai-941	21—Bikuškis-25ž	31—Druskininkai outcrop	41—Purviai×
11—Būtėnai-937,938	22—Gervelės-330	32—Ratnyčia outcrop	42—Iržai×
12—Šiekštis-922			

Chapter | 19 Pleistocene Glaciations in Lithuania

TABLE 19.1 Stratigraphical Scheme for the Quaternary of Lithuania, Officially Assigned for State Geological Investigations

System	Division	Subdivision	Stratigraphic units			
				Step Formation	Stadial Subformation	Phasial Member
Quaternary			Holocene			
	Pleistocene	Upper	Nemunas (Weichselian)	Viršutinis Nemunas (Upper Weichselian)	Baltija	Pajūris (Maritime) Šiaurės Lietuvos (North Lithuanian) Vidurio Lietuvos (Middle Lithuanian) Pietų Lietuvos (South Lithuanian)
					Grūda	
				Vidurinis Nemunas (Middle Weichselian)		Mickūnai 4 (thermomer) Nemunas 2e (cryomer) Mickūnai 3 (thermomer) Nemunas 2d (cryomer) Mickūnai 2 (thermomer) Nemunas 2c (cryomer) Mickūnai 1 (thermomer) Nemunas 2b (cryomer) Jonionys 3 (thermomer) Nemunas 2a (cryomer)
				Apatinis Nemunas (Lower Weichselian)		Jonionys 2 (thermomer) Nemunas 1b (cryomer) Jonionys 1 (thermomer) Nemunas 1a (cryomer)
			Merkinė (Eemian) Interglacial			
		Middle	Žeimena Glacial		Pamarys (Medininkai Late Glacial) (thermomer) Medininkai (Warthian) (cryomer) Vilkiškės (thermomer) Žemaitija (Saalian) (cryomer)	
			Būtėnai (Holsteinian) Interglacial			
			Dainava (Elsterian 2) Glacial			
			Turgeliai Interglacial			
			Dzūkija (Elsterian 1) Glacial			
		Lower	Vindžiūnai Interglacial			
			Kalviai Glacial			
	Prepleistocene		Daumantai Preglacial			

and North Lithuanian phases on the geomorphological map at a scale of 1:100,000 (Guobytė, 2002).

After the revision of the Quaternary geological and geomorphological maps at a scale of 1:200,000 in 1998 and 2002, a number of ice-marginal moraine ridges from the Last Glaciation has been reduced because their presence was confirmed neither by the interpretation of large-scale aerial photographs nor by field investigations. Only geomorphologically and geologically well-established marginal ridges and other glacial features marking ice advance and its retreat margins are interpreted as Last Glaciation stadial or phase limits.

The following information, derived from the geomorphological map at a scale of 1:200,000, has been included in the digital map: ice margins, end-moraines, ice-dammed lakes, glacial features (eskers, drumlinoids, subglacial 'tunnel' valleys) and glaciofluvial drainage valleys (lateral ice-marginal rivers and proglacial valleys) of the Last Glaciation.

Ninety key sites were selected and shown on the map including stratotype sections of interglacial sediments and till units which comprise the Quaternary cover. Over 30 sites with the late-glacial sediments dated using optically stimulated luminescence (OSL), thermoluminescence (TL), cosmogenic beryllium (^{10}Be) and radiocarbon (^{14}C) dating methods provide the chronology for the Weichselian deglaciation phases in Lithuania (Figs. 19.1 and 19.2; Table 19.2).

19.2. EARLY PLEISTOCENE GLACIATION

19.2.1. Daumantai Stage

The oldest (pre-Elsterian) Quaternary in Lithuania comprises the Daumantai Formation, Vindžiūnai Interglacial and Kalviai glacial Formation (Table 19.1). The deposits of the Daumantai Formation (have been called by different authors as Pre-Pleistocene, pre-glacial) are widely distributed in eastern Lithuania (Figs. 19.1 and 19.3) where they overlie pre-Quaternary deposits. The Daumantai Formation

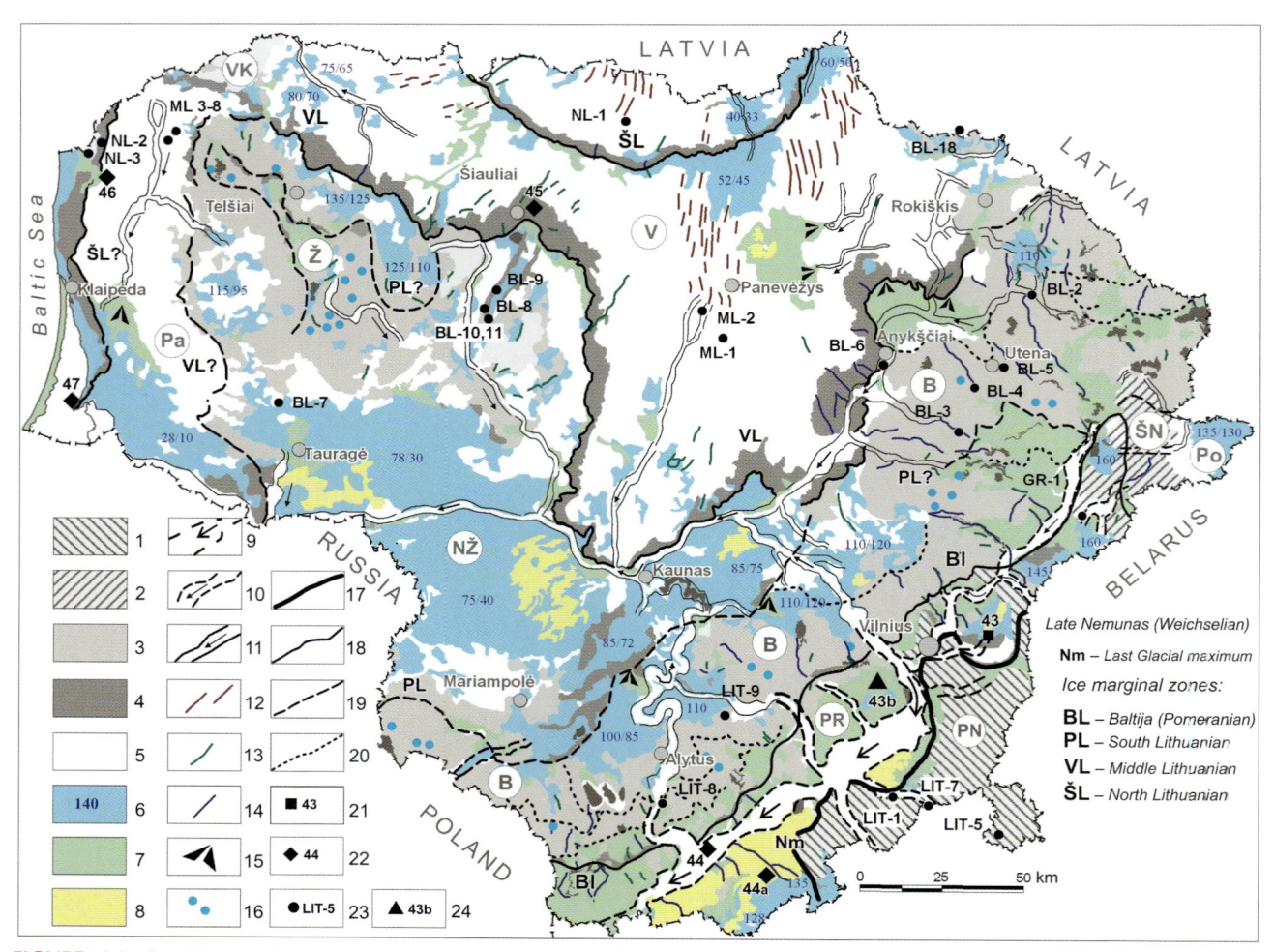

FIGURE 19.2 Late Nemunas (Weichselian) glacial limits and location of dated sites. Glacial Uplands of Saalian age: 1—Southern Nalšia (Ašmena), 2—Northern Nalšia (Švenčionys), 3—Glacial Uplands of Weichselian age: B—Baltija marginal Uplands, Ž—Žemaitija (Samogitia) insular Upland, 4—distinct ice-marginal ridges, 5—glacial lowlands: Pa—Meritime, NŽ—Lower Nemunas, V—Mid-Lithuanian, PR—SE Lithuanian, Po—Polockas, 6—glaciolacustrine plains with maximum/minimum altitude, 7—outwash plains, 8—Aeolian landscape, 9—The Voke–Merkys–Nemunas ice-marginal river valley (urstromtal), 10—The Žeimena–Neris ice-marginal river valley, 11—proglacial valleys, 12—drumlinoid ridges, 13—eskers, 14—subglacial 'tunnel' valleys, 15—deltas, 16—plateau-like hills, 17—LGM limit, limits of ice-marginal zones, 18—definite limit, 19—supposed limit, 20—positions of retreating ice margin, 21—TL dating site, 22—^{14}C dating sites, 23—^{10}Be dating sites, 24—OSL dating sites (Ages are given in Table 19.2).

TABLE 19.2 An Age Data Used for the Determination of the Late Nemunas (Weichselian) Glacial Limits

Site	Site Location in Fig. 19.5	OSL and TL Age (ka)	Dated Material	Depth (m)	References
Mickūnai	43	200 (TL—TLU 652)	Sand	12	Satkūnas (1993)
		25.0 (TL—TLU 650)	Sand	16.5	
		26.0 (TL—TLU 653)	Sand	23.5	
Serafiniškės	43a	15.4 (OSL—TLU R-52)	Sand	2.0	Raukas and Stankowski (2005)
		16.2 (OSL—TLU R-53)	Sand	3.2	
		^{14}C cal. BP			
Pamerkiai outcrop	44	13,710–13,390 (ST-13807)	Wood	5.18	Stančikaitė et al. (2008)
Zervynos	44a	16,470 ± 230 (AA-53595)	Peat	3.5	Rinterknecht et al. (2006, 2008)
Talša	45	14,600–13,950 (Ki-15547)	Bulk carbonates	6.14–6.18	Stančikaitė et al. (pers. com.)
Kašučiai	46	14,140 ± 220 (Ki-12170)	Gyttja	2.75–2.80	Stančikaitė et al. (2008)
Ventės Ragas	47	13,560 ± 180 (Vs-1161)	Gyttja with peat	3.09–3.12	Bitinas et al. (2002)

Boulder or Boulder Field	Site Location in Fig. 19.5	^{10}Be Exposure age (ka)	Boulder or Boulder Field	^{10}Be Exposure Age (ka)	Site Location in Fig. 19.5	References
Navakonys	Lit-1	17.1 ± 1.5	Ružos akmuo-1	14.7 ± 1.0	Bl-10	From Rinterknecht et al. (2006, 2008)
Žirniai	Lit-5	19.1 ± 1.2	Ružos akmuo-2	10.0 ± 0.9	Bl-11	
Kaniūkai	Lit-7	18.4 ± 1.3	Didysis akmuo	15.3 ± 1.1	ML-1	
Didysis Dzūkijos	Lit-8	13.4 ± 1.1	Naujamiestis	11.0 ± 0.9	ML-2	
Dudiškiai	Lit-9	13.5 ± 1.0	Šauklliai-1	15.4 ± 1.1	ML-3	
Didysis akmuo	Gr-1	15.2 ± 1.1	Šauklliai-2	13.4 ± 1.2	ML-4	
Akmuo milžinas	Bl-7	12.6 ± 0.9	Kulaliai-1	13.7 ± 1.0	ML-5	
Puntuko brolis	Bl-6	14.2 ± 1.0	Kulaliai-2	15.8 ± 1.1	ML-6	
Mockėnai	Bl-5	12.0 ± 0.9	Kulaliai-3	9.4 ± 0.9	ML-7	
Utenis	Bl-4	15.0 ± 1.2	Kulaliai-4	14.0 ± 1.1	ML-8	
Mindučiai	Bl-3	14.6 ± 1.1	Bulotiškis	13.6 ± 1.2	NL-1	
Vasaknai	Bl-2	13.8 ± 1.0	'Kulių bobelė'	11.9 ± 1.0	NL-2	
Didysis akmuo	Bl-8	11.3 ± 0.9	Pelėkių akmuo	14.3 ± 1.0	NL-3	
Ertelis	Bl-9	10.5 ± 0.8	In Latvia	13.8 ± 1.0	Balti-18	

is represented by lacustrine, 10–20 m thick, sandy-silt sediments formed in a non-glacial environment (Satkūnas, 1998). On the basis of the palynology, the climate of the Daumantai period was mild; however, some cooler climate intervals were also present (Kondratienė, 1996; Kondratienė and Šeirienė, 2001b; Kondratienė et al., 2001a; Kondratienė et al., 2001b). A more detailed subdivision of the Daumantai period and its correlation with western European stratigraphical divisions remains very problematic. The upper stratigraphical boundary of the Daumantai Stage corresponds to the lower limit of glacial deposits, attributed to the oldest continental glaciation during which ice extended as far south as the latitude of eastern Lithuania.

19.2.2. The Kalviai Glaciation

A possibility that some remnants of an old, pre-Elsterian glaciation existed in Lithuania has been discussed since the 1970s. In Kondratienė's (1971) stratigraphical scheme, the 'First glaciation' unit (Glacial I) has been indicated above

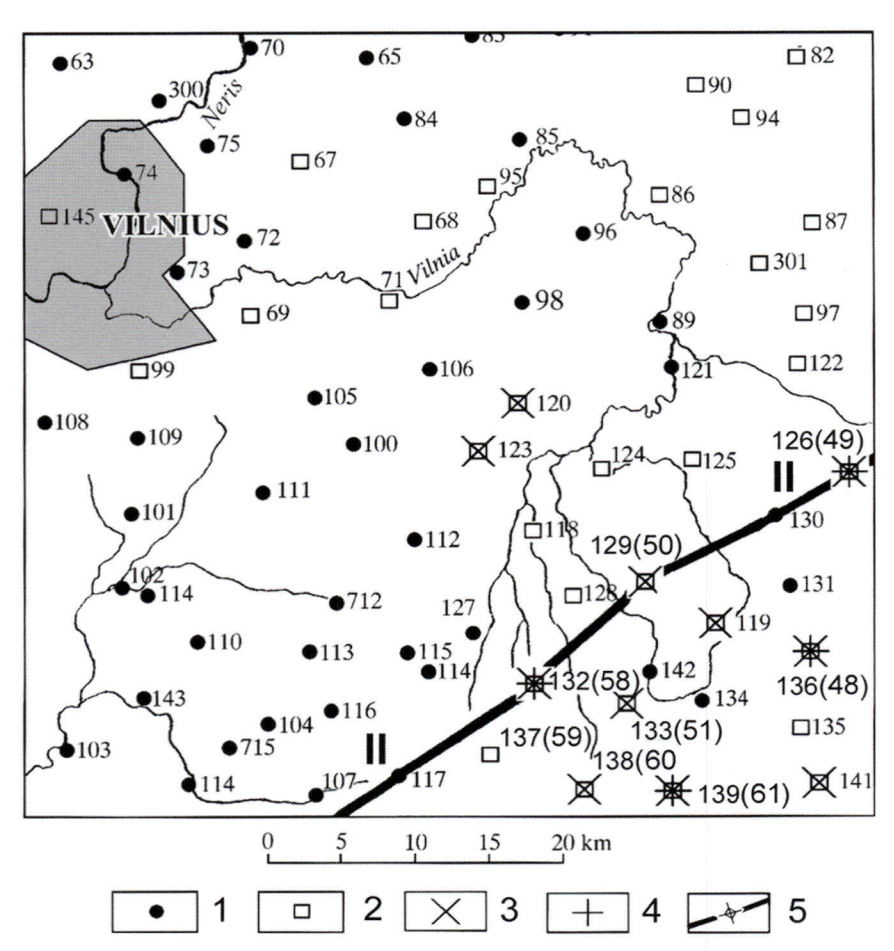

FIGURE 19.3 The stratotype area of the oldest Quaternary deposits: 1—borehole, 2—borehole with Daumantai deposits, 3—section with Kalviai glacial deposits, 4—section with Vindžiūnai Interglacial deposits, 5—location of the cross section II–II (for the location of the area (D), see Fig. 19.1).

the Daumantai Formation. However, this unit was established on the basis of pollen evidence; no deposits of glacial origin had been observed to support this interpretation. As noted by Vaitiekūnas (1977), the Glacial I event did not reflect a climatic deterioration, which could be followed by a glacial advance. Therefore, this Glacial I event was interpreted as a climatic deterioration within the Daumantai Stage.

Traces of a glacial advance, represented by a till older than the Dzūkija unit (generally correlated with the lower Elsterian), have been revealed near Anykščiai in the Šventoji River outcrops, known for its Neogene and Daumantai deposits (Gaigalas, 1987; Fig. 19.1). The till which overlies the Daumantai Formation was named the Katlieriai Till. A similar till was discovered near Kalviai near Vilnius by Gaigalas (1987) who considered it to be a stratigraphical equivalent of the Katlieriai Till. Because of the number of boreholes which revealed a clear section of the till (Fig. 19.3) and demonstrated its position within the Quaternary sequence (Fig. 19.4), the oldest glaciation was assigned the name Kalviai (Satkūnas, 1993, 1998; Gaigalas and Satkunas, 1994). This Kalviai Till was sampled and described from 11 boreholes from an area of ca. 200 km^2

(Satkūnas, 1998) and its position was found to be similar in all the sections examined (Fig. 19.5). The till is a 2–3 m thick loamy sand, of greenish grey colour with a comparatively few pebbles (3–8%), mainly 2–8 mm in diameter. Gravel from the till only rarely includes dolomite and marl, since it is generally carbonate free. These peculiarities are only characteristic of the Kalviai Till and provide a good possibility for distinguishing it from the younger Elsterian Tills (Gaigalas and Satkunas, 1994). The Kalviai Till is underlain or covered by glaciolacustrine deposits in some sections, for example, by massive- or fine-bedded silt with pollen of *Rubus chamaemorus*, *Betula nana*, *Alnaster* and *Selaginella selaginoides* spores. These glaciolacustrine sediments are attributed to the Kalviai glacial cycle.

The independent stratigraphical position of the Kalviai Till as a pre-Elsterian 1 glacial event is supported by presence of the Vindžiūnai Interglacial sediments of which have been encountered in five sections south-east of Vilnius (Figs. 19.1, 19.3 and 19.4). This interglacial is represented by lacustrine silt, silty sand and gyttja. The pollen analysis from these deposits indicated that the vegetation does not represent a complete interglacial climatic cycle (Kondratienė and Satkūnas, 1993). The fragmented sequence can

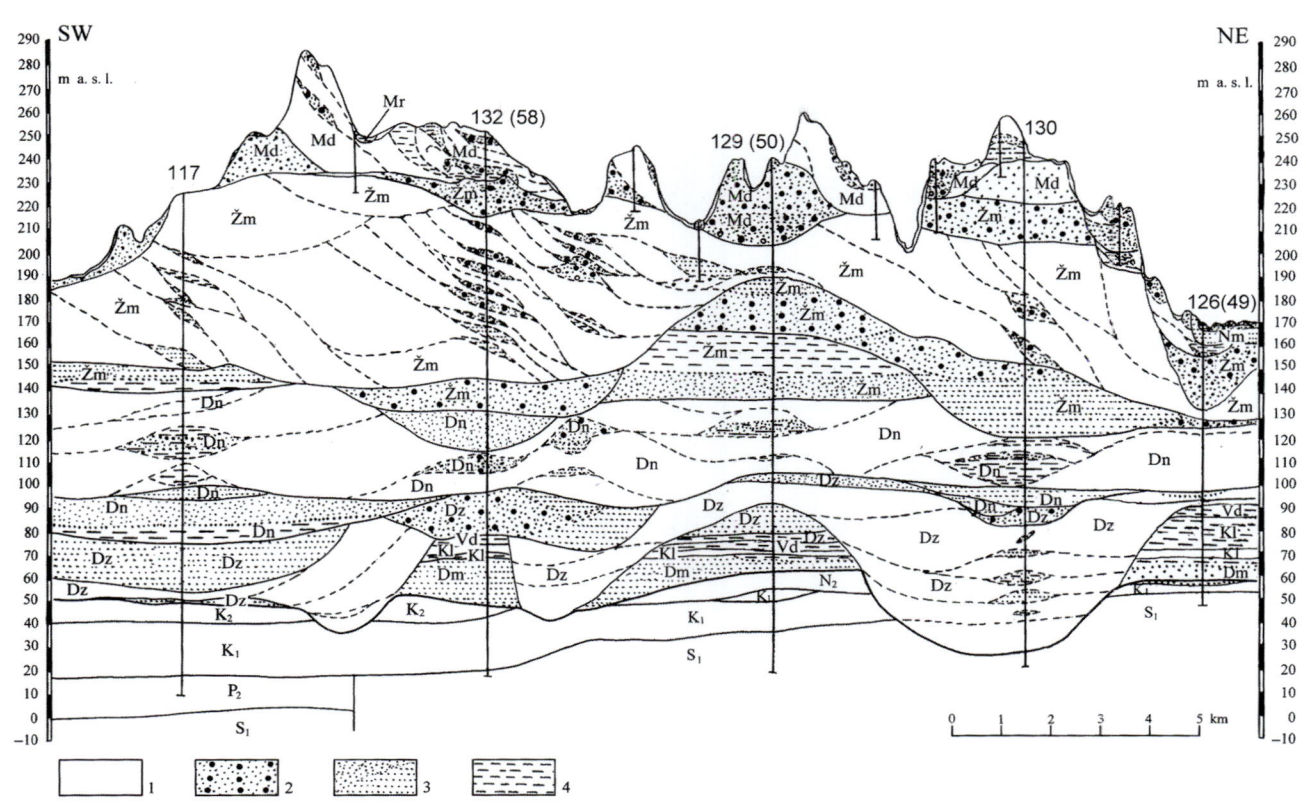

FIGURE 19.4 Cross section of the Quaternary sequence in the Medininkai heights (Southern Nalšia Upland; for location, see Fig. 19.1). Stratigraphical indexes: S_1—Lower Silurian, D_3—Upper Devonian, P_2—Upper Permian, K_1—Lower Creataceous, K_2—Upper Cretaceous, N_2— Upper Neogene, Dm—Daumanta, Kl—Kalviai, Vd—Vindžiūnai, Dz—Dzūkija, Dn—Dainava, Žm—Žemaitija, Md—Medininkai, Mr—Merkinė, Nm—Nemunas. Lithology: 1—till, 2—gravel, 3—sand, 4—silt and clay.

be mainly attributed to the first part of an interglacial sequence. The pollen data indicate forest vegetation composed of *Betula*, *Pinus* and *Picea*, in which *Picea* reaches up to 34% of the total spectrum. Some admixture of pollen of broad-leaved trees, including *Quercus*, *Tilia*, *Alnus*, is also found, and some samples include single grains of exotic taxa such as *Tsuga*, *Pterocarya* and *Pinus Haploxylon*. According to Kondratienė (1996), these finds are most probably redeposited and might reflect the presence of two separate interglacial events. However, the rather similar location (Fig. 19.4) of these deposits and the absence of full vegetation successions cannot definitely support this assumption.

19.2.3. Problems of Stratigraphical Comparison

The absence of complete sections reflecting continuous climatic and palaeoenvironmental records significantly complicates the stratigraphical interpretation and correlation of the oldest Quaternary deposits. Nevertheless, the evidence definitely demonstrates the presence of sediments formed during a period of transition from the end of the Pliocene up to first Pleistocene glaciations. The deposition conditions and characteristic composition of the Kalviai Till suggest that it represents the oldest glaciation to have reached eastern Lithuania. The Narevian Till, known in north-eastern Poland (Ber, 1996), is very probably the lateral equivalent of the Kalviai Till, taking into consideration its depositional conditions and lithostratigraphical indicators.

A correlation of the Vindžiūnai Interglacial with the Cromerian Interglacial might be possible but requires considerably more evidence. However, its correlation with the late Early Pleistocene Bavel Interglacial (Zagwijn, 1992) is also possible. On the basis of its very similar deposition conditions, this interglacial may be correlated with the Zidini Interglacial (Danilans, 1981), found in south-eastern Latvia, and with the Augustovian Interglacial in north-eastern Poland (Ber, 1996; Janczyk-Kopikowa, 1996).

19.3. MIDDLE PLEISTOCENE

The Middle Pleistocene comprises the Dzūkija and Dainava cold-stage glacial units (separated by the Turgeliai Interglacial) and the Žeimena (Žemaitija–Medininkai) cold-stage glacial units. The Būtėnai Interglacial separates the Dainava glacial from the Žeimena. The Middle Pleistocene glacial

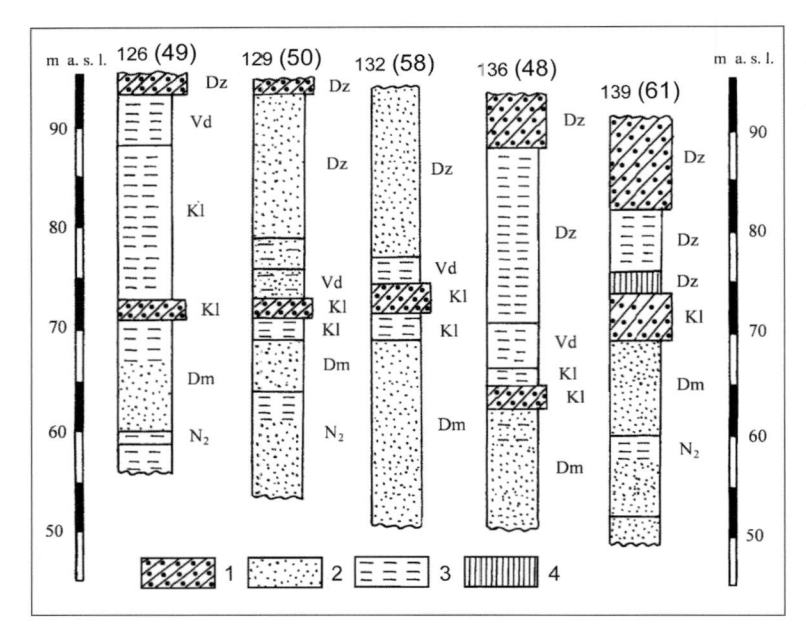

FIGURE 19.5 The Kalviai Till and the Vindžiūnai Interglacial sediments identified in boreholes in eastern Lithuania (for location of boreholes, see Fig. 19.2, site number is written in brackets.). Stratigraphical indexes: N_2—Upper Neogene (Anykščiai Formation of Upper Pliocene); Dm—Daumantai Stage; Kl—Kalviai, glacial stage; Vd—Vindžiūnai, Interglacial stage; Dz—Dzūkija, glacial stage. Lithology: 1—till, 2—sand, 3—silt, 4—gyttja.

deposits are distributed throughout the country and are absent only in areas of subsequent glacial and post-glacial erosion. The limits of glacial extents of the Middle Pleistocene evidently have to be located far outside the country.

19.3.1. Glacial Stages of the Early Middle Pleistocene

The Dzūkija glacial deposits are distributed in the form of discontinuous patches. The stratotype area of the Dzūkija Till is located in south-eastern Lithuania and therefore the name of this glacial stage is derived from the ethnographical region of Dzūkija (Fig. 19.1). The Dzūkija glacial sediments are sandy loam and loam, glaciofluvial sand, sandy gravel, glaciolacustrine sand, silt and clay. It is mainly a grey, greenish grey, greyish brown, predominantly massive till which has an average thickness of 7–10 m. The gravel and pebble of crystalline rocks (presumably from southern Sweden) are dominant in the till. The SiO_2, TiO_2, Ti, Zr and Yb contents are higher than in other tills of the Middle and Late Pleistocene. Respectively, the low contents of CO_2, MgO and FeO are characteristic for the Dzūkija Till.

The Dainava glacial is widely distributed. The stratotype area is in south Lithuania, in the Dainava ethnographical region (Fig. 19.1). The sediments are presented by sandy loam and loam, glaciolacustrine sand, silt, clay, glaciofluvial sand and sandy gravel. Till with an average thickness of 8–12 m is grey, greenish grey and greyish brown. Gravel consists of crystalline rocks (likely from central Sweden), dolomite and Mesozoic sediments. The high Al_2O_3, SiO_2, Y, Yb, Pb, Zr, Cu and low CO_2, CaO and MgO contents characterise the Dainava Till.

The Turgeliai Interglacial (Kondratienė, 1996) separates the Dzūkija and Dainava glacial units. Sediments of this interglacial are represented by lacustrine, bog and fluvial deposits, for example, sand, silt, gyttja and peat, encountered in two boreholes (Fig. 19.1). The thickness of sediments is 5–6 m. The T1–T6 forest development phases are distinguished during the Turgeliai Interglacial. The characteristic flora includes *Pinus* sec. *Cembra*, *Pinus* sec. *Strobus*, *Picea* sec. *Omorica*, *Buxus sempervirens*, *Ilex* sp., *Caulinia* cf. *macrosperma*, *Ranunculus sceleratoides*, *Dulichium* sp. and *Osmunda* sp.

19.3.2. Glacial Events of the Late Middle Pleistocene

Glacial events of this period have been recognised as the Žemaitija and Medininkai stadials, both attributed to the Žeimena Glacial (Table 19.1). This is correlated with the classical Saalian glaciation in Europe. The Žemaitija and Medininkai glacial events have been established between the Būtėnai (Holsteinian) and Merkinė (Eemian) Interglacials. The sediments of these interglacials are quite widely distributed in Lithuania (Fig. 19.1) and are also relatively well studied (Kondratienė and Bitinas, 1989; Baltrūnas, 1995; Kondratienė, 1996). The Būtėnai Interglacial is characterised by the B1–B7 forest development phases and characteristic flora: *Abies alba*, *Taxus baccata* L., *Picea omorica* L., *Caulinia goretskyi*, *Caulinia interglacialis*, *Brasenia borysthenica* var. heterosperma, *Nymphaea cinerea*, *Dulichium arundinaceum*, *Carex paucifloroides*, *Azolla pseudopinnata* and *Selaginella tetraedra* (Kondratienė, 1996). The palaeogeographical environments of these main interglacials has been easily reconstructed by characteristic palynological features in most cases; thus, they provided a reliable biostratigraphical control for the identification of the Žeimena glacial formations. The Žeimena glacial deposits cover a considerable part of Lithuania. Their thickness can reach 80–150 m, or even more in Uplands and palaeoincisions. The significant erosional impact of the Saalian continental ice sheet on the pre-Quaternary bedrock resulted in deep and narrow subglacial valleys as well as broad depressions of the present topography of the pre-Quaternary surface.

It has been established that the Žemaitija Till is generally the thickest, on average it is 23 m thick, but it reaches over 100 m in some sections. The Žemaitija Till is made up of loam and sandy loam with high values of Fe_2O_3, MgO, Ca, Cr, Cr, Co and Ni. It has the maximum value of ratio SiO_2/Al_2O_3 comparing with the other tills (Satkūnas and Bitinas, 1995). The gravel consists mostly of crystalline rocks (derived from southern Finland), dolomite and limestone. The abundant glacial dislocations, predominant dark brown colour and high hardness are characteristic properties of the Žemaitija Till. The stratotype of this unit was recognised in north-eastern Lithuania (Kondratienė and Bitinas, 1989; Fig. 19.1).

The Medininkai (Upper Saalian) Till (loam and sandy loam) texturally and structurally is quite similar to that of the Žemaitija, with the exception of a dominant yellowish brown colour and less mean thickness (~ 15 m). The highest average contents of Cu, Yb, Y, Sc, Zr, Mn and Ti were measured in the Medininkai Till. It is noteworthy that the weathered soil has characteristic association of trace elements. The areal stratotype of the Medininkai glacial unit is the Medininkai Heights in the South Nalšia (Ašmena) Upland, where the Saalian deposits immediately underlie the present-day surface. The Upland is located outside the limit to the Last Glaciation (Fig. 19.2). Sediments of the Merkinė (Eemian Stage) Interglacial are widely distributed in marshy kettle-holes of the Medininkai Heights. Five (M_1–M_5) phases of forest succession from coniferous to broad-leaved and back to coniferous forests have been identified during the Merkinė Interglacial.

The glaciolacustrine and glaciofluvial deposits located between the Medininkai and Žemaitija Tills are comparatively thinner than the inter-till beds separating the Žemaitija (Lower Saalian) and Dainava (Upper Elsterian) Tills or the Medininkai (Upper Saalian) and Nemunas (Weichselian) Tills. Commonly, the Medininkai Till occurs immediately overlying the Žemaitija Till (Fig. 19.6). It seems that lithologically and geochemically the Medininkai and Žemaitija Tills differ from each other less than the Saalian Tills generally differ from those of the Elsterian and Weichselian. Therefore, the lithostratigraphical criterion does not support a higher than stadial rank for the Žemaitija and Medininkai Tills (Satkūnas and Bitinas, 1994). The Vilkiškės interstadial is introduced as an ice-free period separating the Žemaitija and Medininkai glacial events and is comparable to the Late Saalian Drenthe–Warthe Interstadial (Satkūnas and Molodkov, 2005).

19.4. LATE PLEISTOCENE. LOWER AND MIDDLE WEICHSELIAN SUBSTAGE GLACIATIONS (NEMUNAS)

Climatostratigraphical events of the Upper Pleistocene (Fig. 19.7) have been reconstructed on the basis of the all available data from key sections (Fig. 19.1, sites: Jonionys, Medininkai, Mickūnai, Rokai and Purviai; Satkūnas and Grigienė, 2000; Satkūnas et al., 2003, 2009). The previous study results showed the presence of non-glacial (periglacial and interstadial) palaeoenvironments in Lithuania since the end of the Eemian Stage Interglacial, during the Early and Middle Weichselian. It is most likely that the major part of the Eastern Baltic area was not covered with ice until the Late Weichselian Substage.

The environmental history of the Merkinė (Eemian Stage) Interglacial (130–117 ka) is rather well studied (Kondratienė, 1965, 1976, 1996; Kondratienė and Kliveckienė, 1983; Kondratienė and Šeirienė, 2001a; Vaitiekūnas, 1968; Gaigalas et al., 1994, 1994a and 1994b; Baltrūnas, 1995; Gaigalas and Hütt, 1995). The results from the Jonionys site clearly demonstrate the presence of three interstadials following the Eemian (Merkinė) event. The first two, Jonionys 1 and Jonionys 2, correlated, respectively, with the Brørup and Odderade, are rather widely represented in Lithuania and have been identified at several sites including Medininkai, Mickūnai, Dysnai and Smalvos (Fig. 19.1; Satkūnas et al., 1997; Satkūnas et al., 1998). The Jonionys 1 and Jonionys 2 interstadials are characterised by forest flora, for example, birch–pine–spruce–larch, landscapes and comparably mild and humid climate. The Jonionys 3 interstadial was firstly determined from the Jonionys palaeolacustrine site where it was characterised by birch (dwarf birch) forest, that is, tundra palaeoenvironment. This interval is correlated with the Oerel Interstadial in north-west Germany based upon the similarity of the palaeoenvironments.

The Middle Nemunas (Weichselian) is probably one of the most problematic and mostly discussed time intervals of the Late Pleistocene from the point of view of stratigraphy and palaeogeography. Stratigraphical events during this comparatively long period of nearly 50 ka were not expressed as sharp climatic fluctuations during the generally progressing global climate cooling. If the Jonionys 3 warm phase corresponds to the Oerel Interstadial, this means that the cold phase of Nemunas 2a is probably analogous to the Schalkholz Stadial of Behre (1989) which represents remarkable climatic cooling. In turn, this cold phase marks the beginning of the marine isotope stage (MIS) 4 (Mangerud, 1991, 2004).

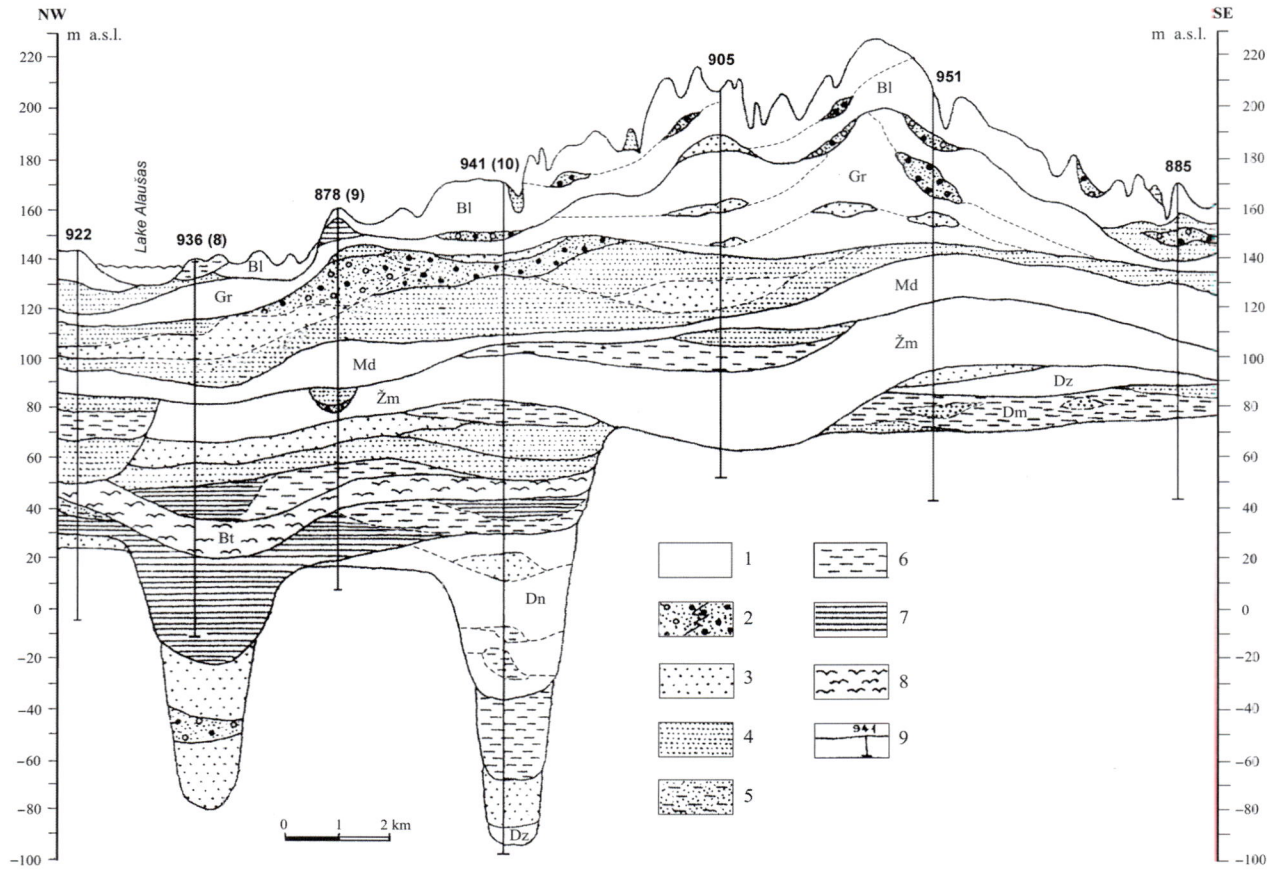

FIGURE 19.6 Geological cross section of Quaternary deposits in the Utena environs (profile I–I). 1—till, 2—gravel, 3—coarse- and medium-grained sand, 4—fine-grained sand, 5—silty sand, 6—silt, 7—clay, 8—sapropelite, 9—borehole with its original and site number shown on the map (for location of cross section, see Fig. 19.1).

The rest of the Middle Weichselian (Pleniglacial) is much more complicated from the point of view of extent of glacial advances (Mangerud, 2004; Kalm, 2006). The glacial deposits attributed to the Middle Weichselian Substage have been described and reported from Estonia (Liivrand, 1991), southern Finland (Nenonen, 1995) and Poland (Marks, 1997, 2004). The Middle Weichselian glacial event is recorded in the central part of the Scandinavian Ice Sheet (Hitura pit, Ostrobothnia), which began after an entirely ice-free Early Weichselian Substage 79 ka ago and was followed by a deglaciation between 62 and 55 ka ago (Salonen et al., 2008). Based on a glaciological modelling, Zelčs and Markots (2004) proposed a possible early Middle Weichselian glaciation in western Latvia (the so-called Talsi Stadial) between 74 and 59 ka; however, no direct evidence was reported. Kalm (2006), summing up recently published ^{14}C dates and OSL ages, indicated that Estonia was ice free at least between 43.2 and 26.8 ka and assumed that a time period including a possible early Middle Weichselian glaciation in Estonia was between 68 and 43 ka. New infrared optically stimulated luminescence (IR-OSL) investigations of tills and inter-till sediments from the Lithuanian maritime region confirm that the south-western margin of the Eurasian ice sheet could have been situated in the Lithuanian coastal area or could have covered all of western Lithuania during the Weichselian Early Pleniglacial maximum (MIS 4; Molodkov et al., 2010).

The recently obtained data from the Venta palaeolacustrine basin (Satkūnas et al., 2009) implies that northern Lithuania was ice free at ca. 33 ka, and that the Late Weichselian glacial advance probably did not reach the north of the country earlier than ca. 25 ka. This conclusion corresponds very well with the general Late Weichselian glaciation chronology as determined in Estonia and Finland (Lunkka et al., 2004; Kalm, 2006).

19.4.1. Late Weichselian (Nemunas) Substage

An important problem in the Quaternary sequence in Lithuania is the climatostratigraphical subdivision of the Upper Weichselian (Upper Nemunas) Substage. On the basis of all

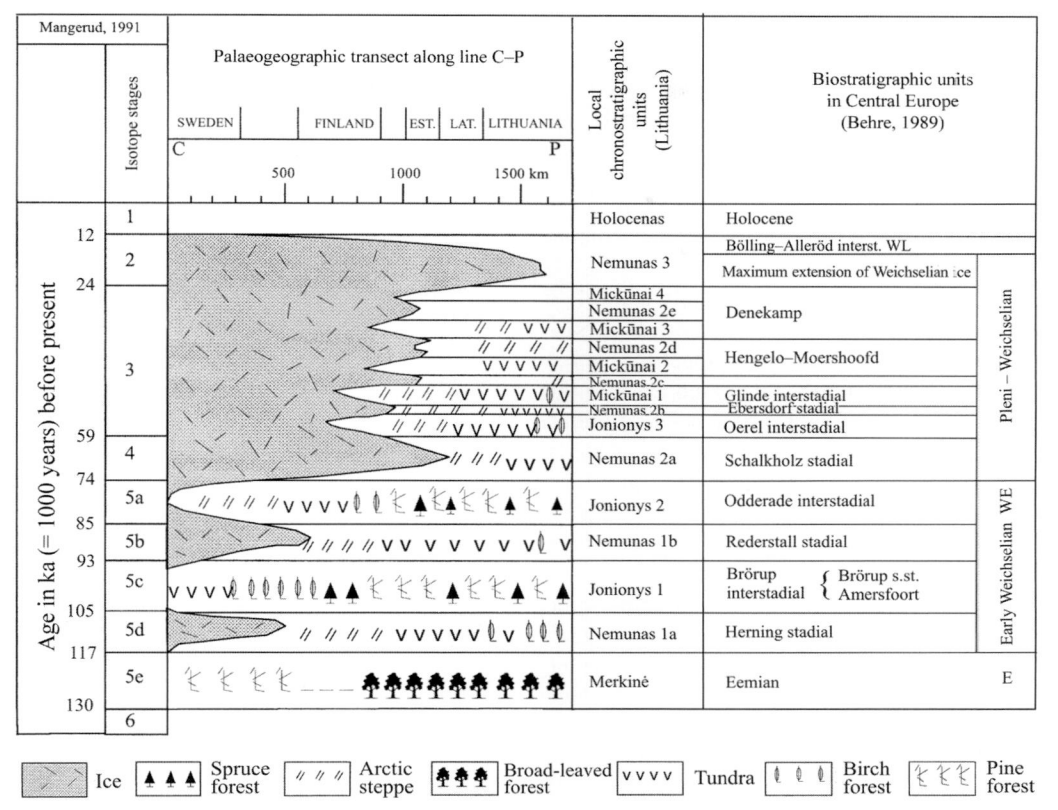

FIGURE 19.7 Climatostratigraphical events of the Late Pleistocene.

the available stratigraphical information (Satkūnas and Hütt, 1999), it can be stated that no site with reliably determined Upper Weichselian interstadial deposits has been found so far in Lithuania. Therefore, 'stadials', 'phases' and 'interstadials' have no proper climatostratigraphical foundation, and the Upper Weichselian at present can be subdivided only into lithostratigraphical units. The Upper Nemunas glacial is subdivided into two stratigraphical units: the Grūda and Baltija Subformations (stadials). The older of the two units, the Grūda Till is composed of brown, dark brown, greyish brown loam and sandy loam and has an average thickness of 5–9 m. The gravel within this unit consists of crystalline rocks, dolomite, Silurian limestone and Mesozoic rocks. The FeO and CO_2 content is comparatively high. The stratotype boreholes of the Grūda Till are situated in south-eastern Lithuania (Fig. 19.1; sites 25 and 26). The till of the Baltija Stadial is represented only by a sandy loam of reddish brown, yellowish brown colour with higher amount of gravel and pebbles of carbonate rocks. The thickness is on average 6–18 m, however, in places it can reach even 80–90 m. The stratotype area is the Baltija marginal Uplands (Fig. 19.2). These till units are either separated by glaciofluvial and glaciolacustrine deposits or immediately overlie one another, which are therefore easily distinguished visually. However, in many places, a 50–100 m thick till bed occurs that can be differentiated neither visually nor by chemical or mineral composition of gravel. The subdivision of Baltija Subformation into South Lithuanian, Middle Lithuanian and North Lithuanian 'phasials' (members; Table 19.2) is based on geomorphological evidence. However, this division is not reflected in the stratigraphy. Although the climatostratigraphical criteria are absent, there are some authors who consider a possibility of recognising even 'phasial' tills in vertical profiles (Baltrūnas et al., 2005).

The limits of the Grūda and Baltija stadials and three ice-front phases, that is, the South Lithuanian, Middle Lithuanian and North Lithuanian, within the Baltija Stadial, have been distinguished in Lithuania. The reconstruction of their glacial limits is based on geomorphological evidence combined with geological data.

19.4.2. Last Glacial Maximum

The limit of the Late Nemunas glacial maximum, which corresponds to the Grūda (Brandenburg) stadial in Germany, has been traced at the extreme south-eastern corner of Lithuania, in the border area with neighbouring Belarus (Fig. 19.2). Here, the absence of continuous ice-marginal ridges of the last ice sheet, as well as the presence of the high Saalian-age Uplands near its margin, makes it difficult to define a limit for the classical Late Weichselian (last glacial maximum, LGM) in the country. Some have even considered possibility that the ice during the LGM reached the country (Guobytė, 2004 and references herein). The Saalian age of the Medininkai Highlands and entire Southern Nalšia Upland was confirmed by discovery of undisrupted lacustrine sequences spanning the period from the Merkinė (Eemian) Interglacial to the Lower and Middle Nemunas (Satkūnas et al., 2003).

The margin of the last ice sheet was strongly controlled by the pre-Weichselian topography. The high ridges of Saalian age caused the margin of the last ice sheet to diverge into separate ice lobes and tongues that extended into the pre-existing depressions. A low but prominent end-moraine ridge which outlines the ice-tongue depression clearly defines the LGM limit in the surroundings of Vilnius. To the south of Vilnius, the LGM limit is marked by the distal margin of a sandur plain near the margin of the Southern Nalšia Upland (Fig. 19.2). The slightly undulating southwestern part of Southern Nalšia Upland is thought by some investigators to have been covered by the last ice sheet, so that the outermost limit of the Last Glaciation must be traced further to the south-east than is shown in Fig.19.2 (Guobytė, 2004 and references herein).

The time of a maximum extent of the Last Glaciation was most reliably defined by the detail investigation of the Mickūnai palaeolake sediments discovered to the east of Vilnius. Here TL dates, obtained from the lacustrine sands underlying the Grūda Till, revealed that the last ice sheet reached its maximum extent no earlier than 22–21 ka (Fig. 19.2; Table 19.2; Satkūnas, 1993).

The newly obtained ^{10}Be dates from erratic boulders show that the ice margin began to retreat from its maximum extent at ca. 18 ± 4 ka (Rinterknecht et al., 2008; Fig. 19.2; Table 19.2). During this retreat, an extensive sandur was accumulated between the Southern Nalšia and Baltija Uplands (Fig. 19.2). Two types of outwash-plain surfaces occur at different altitudes. The higher sandur plain, with a distinct network of dry gullies and ravines, is older (than the Grūda Stage) than the lower plain. According to the OSL dates obtained from the sand of the higher sandur plain, this could have accumulated between 16.2 and 15.4 ka (Raukas and Stankowski, 2005; Fig. 19.2; Table 19.2). The lower outwash plain was formed by active ice melting of the Baltija Stage ice retreat close to the Baltija Uplands. An intense outflow of meltwaters on the existing sandur plain resulted in ice-marginal drainage generally orientated towards the south-west (Blažauskas et al., 2007). The ca. 20 km wide, terraced palaeovalley was formed parallel to the retreating ice margin in the central part of the sandur plain. This, Vokė–Merkys–Nemunas, palaeovalley served as a main drainage artery, which drained the ice meltwater of the retreating Baltija (Pomerania) ice margin (Fig. 19.2).

The Žiogeliai (Frankfurt) phase ice limit was drawn at the foot of the extensive dune fields stretching to the south-east of the palaeovalley (Guobytė, 2004 and references herein).

19.4.3. Baltija (Pomerania) Ice-Marginal Zone

The major part of the present topography of Lithuania was formed during the Baltija Stadial. The belt of the ice-marginal formation, the Baltija marginal Uplands, forms a distinct hilly arc stretch across the country from south-west to north-east. It was formed during the re-advance and long standstill of the ice margin in this position. Subdivision of the ice-sheet front during this advance was strongly controlled by the pre-Weichselian topography. Geomorphological features show that ice margin was divided into several ice lobes, tongues and micro-tongues. Different activities of the ice in the southern, south-eastern and north-eastern parts of the Uplands are reflected in the glacial morphology. The southern and eastern lobes formed a 20 km-wide hummocky moraine belt with chains of push-moraine ridges outlining its southern margin. Deep distinct ice-tongue depressions in the proximal part of the belt are encircled by hummocky ribbed moraines. Plateau-like hills occur on the triangular-shaped inter-tongue massifs at highest altitudes (180, 200, 210, 220, 240 and 260 m a.s.l.). Kame terraces, kames and ice-crevasse infilling forms situated between the hummocks show that the wide marginal zone of the ice sheet downwasted as dead ice.

The twice as wide north-eastern part of the Baltija Uplands is densely resolved into blocks by deep subglacial tunnel valleys and proglacial spillways. Geomorphologically well-expressed ice-marginal formations, as well as eskers and subglacial tunnel valleys, within these blocks indicate the northern ice advance and retreat direction in the area (Fig. 19.2).

Traditionally, the boundary of the Baltija Stage is drawn along the distal (south-eastern) margin of the Baltija Uplands at the foot of the Northern Nalšia (Švenčionys) Upland, which divides the last ice-sheet margin in to two lobes (Fig. 19.2). The Sventiany (Švenčionys) phase limit of the Disna lobe (in Belarus) is most likely correlative of the Baltija ice-marginal limit than the Frankfurt ice-marginal zone (Karabanov et al., 2004).

The time of final ice-sheet retreat from the Baltija Uplands was 14.0 ± 0.4 ka (Table 19.2; Rinterknecht et al., 2006, 2008).

19.4.4. South Lithuanian (Pietų Lietuvos) Ice-Marginal Zone

The south Lithuanian (PL) phase limit has been a prominent feature of the geomorphology of southern Lithuania since

1913, when a fragment of ridge, later assigned to the PL phase, was first mentioned by Hausen. The South Lithuanian ice limit on the proximal slope of the Baltija Uplands might be marked by poorly developed fragments of marginal ridges and kame terraces. In the Marijampolė district, the limit is outlined by recessional end-moraine ridges. Several ice-marginal fans and glaciolacustrine plains, lying 15–20 m higher than those within the ice limit, show an ice-dam position existed to the East of Kaunas (Fig. 19.2). It is almost impossible to trace the South Lithuanian phase limit in the eastern part of the Baltija Uplands because of lack of continuous ice-marginal ridges. The limit can only be identified by ice-contact slopes, kame terraces and by discontinuous hummocky ribbed-moraine fragments. The ^{10}Be boulder exposure age determinations are similar as in the Lithuania-Latvia and Lithuania–Belarus border area. Thus, simultaneous ice melting in a broad ice-marginal zone can be considered. After the careful reinterpretation of geological and geomorphological features, it appears that the Poozerje Stadial limit is much closer to the South Lithuanian phase than to the Baltija (Pomerania) Stadial limit, than previously thought (Guobytė, 2004).

Chains of recessional ridges and swarms of small kames, observed on the north-eastern and south-western slopes of the Žemaitija insular Upland, indicate that retreating ice stayed a long time at the foot of the Upland. Therefore, this ice-marginal zone has been correlated with the South Lithuanian phase limit on the Baltija Uplands (Guobytė, 2004 and references herein). Meanwhile, the PL phase limit is not reliably established in many places, especially in north-eastern Lithuania and Samogitia.

The age of the South Lithuanian phase is uncertain because the ^{10}Be exposure age of erratic boulders from this zone varies from 14.6 to 12.0 ka (Table 19.2; Rinterknecht et al., 2006, 2008). A more reliable age for South Lithuanian ice retreat phase could be the ^{10}Be date obtained from the Braslav Moraine 13.1 ± 0.5 ka in Belarus (Rinterknecht et al., 2007). It seems likely that during the classical the South Lithuanian (phase), the ice-sheet remnant downwasted as dead ice throughout the whole of Lithuania, producing a vast ice-dammed lake between Baltija and Žemaitija Uplands (Fig. 19.2).

19.4.5. Middle Lithuanian (Vidurio Lietuvos) Ice-Marginal Zone

A marginal ridge is prominent on the surface of central Lithuania clearly outlining an ice-lobe advance into the pre-existing Middle Lithuanian topographical lowland. It is known as Middle Lithuanian Ridge. This ice-pushed ridge is up to 8 km wide and stretches ca. 400 km from the northern limits of the Žemaitija Upland in the west to Rokiškis in the north-eastern part of the Baltija Uplands.

Small sandurs, as well as ice-dammed lakes, are frequent on the outer side of the ridge (Fig. 19.2). Remarkable fluted moraine fields and several radial eskers are related to the till plain outlined by the ridge. The ridge is commonly intersected by subglacial channel valleys and proglacial spillways. The Middle Lithuanian (VL) phase limit is drawn along the distal slope of the ridge. To the north-east of Rokiškis, this limit might be formally traced by sporadic hummocky ribbed-moraine areas in the Baltija Uplands and can be correlated to the Gulbene phase in neighbouring Latvia (Zelčs and Markots, 2004). The boundary, however, is less-clearly expressed along the western slope of the Žemaitija Upland. Here, it is marked by a north–south orientated zone of subdued small ridges and sporadic loaf-like hills. The VL ice limit is well defined by the high, so-called Vilkyškiai Ridge in southern Lithuania, near the Russian border (Kaliningrad district; Fig. 19.2).

A detailed study of fragments of glaciolacustrine kame terraces on the distal slope of the Middle Lithuanian Ridge revealed them to have originated between the active ice lobe and blocks of dead ice. It is likely that there was no common boundary of the frontal Middle Lithuanian ice-sheet limit. Moreover, the areal rather than frontal deglaciation dominated the area (Bitinas et al., 2004; Bitinas, 2011, in press). The latter fact could explain the wide spread of boulder ^{10}Be exposure ages between 11.0 and 15.4 ka (Table 19.2; Rinterknecht et al., 2008) obtained from seven dated boulders in the two genuine boulder fields in north-western Lithuania.

A ^{10}Be age of 13.5 ± 0.6 ka, obtained from the 10 dated boulders, is suggested to represent the final deglaciation from the Middle Lithuanian ice margin (Rinterknecht et al., 2008; Table 19.2). A radiocarbon date from the Talša Lake in the Middle Lithuanian Ridge (Fig. 19.2; Table 19.2) shows that terrigenous bottom sediments with negligible organic matter were deposited in the lake before 13.7 cal. ka BP.

19.4.6. North Lithuanian (Šiaurės Lietuvos) Ice-Marginal Zone

The latest ice re-advance in Lithuania was named the North Lithuanian (ŠL) phase (Guobytė, 2004 and references herein). The North Lithuanian phase limit might be traced by two separate distinct push-moraine ridges in the north and west of the country. The Linkuva Ridge, situated in northern Lithuania near the Latvian border (Fig. 19.2), was first mentioned by Doss (1910). It is a ca. 130 km long, 8–10 to 20 m high and 2–4 km wide in its central part, thinning to 0.5 km at the edges where the ridge bends towards the south. Several subglacial tunnel valleys cross-cut the ridge in its central part. Radial eskers and flute-like subglacial streamlined forms in the till plain to north from the ridge suggest the ridge-producing ice lobe to advance from the north. Small sandurs, peat bogs and the clayey ice-dammed lake plain, at an altitude of 52–45 m, occur to the south from the ridge.

Pajūris Ridge is represented by a subdued, low, slightly bend to East Ridge stretching along the Baltic coast. The ca. 100 km long ridge is approximately 3 km wide, reaching 7 km in its central part, and thinning to 30–60 m near the margins. Strong erosion has destroyed the ca. 5–8 m high ridge in some places and smoothed it to the same level as the surrounding till plain. Therefore, the ribbed hummocky surface of the ridge near Nemirseta was erroneously described and later morphometrically analysed as a drumlin landscape (Guobytė, 2004 and references herein).

An ^{10}Be age of 13.3 ± 0.7 ka has been suggested for ice melting from the Linkuva and Pajūris Ridges (Table 19.2; Rinterknecht et al., 2006, 2008). The radiocarbon ages from the two sites Ventės Ragas and Kašučiai, associated with the Pajūris Ridge, indicate the deglaciation age of the area. In the Ventės Ragas, outcrop occurs on the proximal slope of the Pajūris Ridge. Peat from here was dated at 13.6 ± 0.2 cal. ka BP (Rinterknecht et al., 2008). An age of 14.1 ± 0.2 cal. ka BP (Table 19.2; Stančikaitė et al., 2008) was obtained from gyttja from the Kašučiai Lake which occurs to the East of the Pajūris Ridge. The up-to-date results of cosmogenic and radiocarbon dating show that both the ridges may indicate simultaneous ice-lobe retreat, and therefore the ridge belongs to the North Lithuanian phase.

As well as in the Middle Lithuania Ridge, fragments of glaciolacustrine kame terraces have been recognised and described on the distal slopes of the Linkuva and Pajūris ridges (Bitinas et al., 2004). On the basis of the geomorphological data, the ridges were probably produced by fast-moving local ice-lobe re-advances into areas of dead, almost melted ice.

19.5. CONCLUSIONS

An existence of the oldest pre-glacial (Pleistocene) sediments and till of the Early Pleistocene glaciation has been recognised using the biostratigraphical and lithostratigraphical data available. While dating of these sediments remains problematic, the data nevertheless definitely demonstrate the presence of sediments formed during a period of transition from the end of the Pliocene up to first Pleistocene glaciation. Three Middle Pleistocene glaciations (Dzūkija, Dainava and Žeimena), separated by the Turgeliai and Būtėnai (Holsteinian) Interglacial sediments, are distributed throughout the country. The Žeimena glacial event which is correlated with Saalian glaciation in Europe is represented by two substantial stadials—Žemaitija and Medininkai. The Vilkiškės warm phase (interstadial) is introduced as an ice-free period separating the Žemaitija and Medininkai glacial events and is compared with the Drenthe–Warthe Interstadial (Satkūnas and Molodkov, 2005; Table 19.1). Limits of glacial extents of the Middle Pleistocene evidently must have been located far beyond Lithuania.

The presence of non-glacial palaeoenvironments during the Early and Middle Weichselian was determined for almost all of Lithuania, except its coastal zone. New IR-OSL studies of tills and inter-till sediments from the Lithuanian Maritime Region confirm that the south-western margin of the Eurasian ice sheet could have been situated in the coastal region or throughout western Lithuania during the maximum of the Weichselian Early Pleniglacial (MIS 4; Molodkov et al., 2010). Recent information from the Venta palaeolake implies that North Lithuania was ice free at about 33 ka.

The Late Nemunas (Weichselian) ice advance had reached its maximum limits in south-eastern Lithuania at ca. 21–22 ka. The Late Nemunas ice advance is considered to comprise two parts, the Grūda Stadial (the older) and the Baltija Stadial. However, they can only be distinguished using lithostratigraphical criteria. According to ^{10}Be age evidence, the retreat of the Grūda ice margin began no later than 18.3 ± 0.8 ka. The major part of the present topography of Lithuania was formed during the Baltija Stadial which is correlated with Pomeranian Stage. The retreat age of the Baltija moraine is 14.0 ± 0.4 ka (Rinterknecht et al., 2006, 2008). Three ice-marginal zones distinguished to the north-west of the Baltija Uplands indicate South Lithuanian, Mid-Lithuanian and North Lithuanian ice retreat and re-advance phases of the Pomeranian ice sheet. The ^{10}Be ages of these phases are South Lithuanian 13.1 ± 0.5 ka, Middle Lithuanian 13.5 ± 0.6 ka and North Lithuanian 13.3 ± 0.7 ka. A large volume of data, including ^{10}Be ages, show that during the traditionally distinguished South Lithuanian (phase), the ice-sheet remnant downwasted as dead ice throughout the country. More and more data have been acquired showing that the Middle Lithuanian, Linkuva and Pajūris ice-push moraine ridges were probably produced by fast-moving, local ice-lobe re-advances into areas of dead, almost melted ice.

ACKNOWLEDGEMENTS

We would like to thank our colleagues Alma Grigienė and Albertas Bitinas for valuable suggestions and comments on the chapter. For technical assistance, we are grateful to Regina Norvaišienė. We appreciate very much Gražina Skridlaitė's help with English language. Special thanks to J. Ehlers and prof. P. L. Gibbard for their tremendous work while compiling and editing this volume.

REFERENCES

Baltrūnas, V., 1995. Pleistoceno stratigrafija ir koreliacija. Academia, Vilnius, 178, pp. (in Lithuanian).

Baltrūnas, V., Karmaza, B., Dundulis, K., Gadeikis, S., Šinkūnas, P., 2005. Characteristic of till formation during the Baltija (Pomeranian) stage of the Nemunas (Weichselian) glaciation in Lithuania. Geol. Q. 49 (4), 417–428.

Basalykas, A., 1959. Zur Frage der Entwicklung der Oberflächengestaltung Litauens. Moksliniai pranešimai, Geologija, Geografija, IX, 57–86. (Lithuanian with German summary).

Behre, K.-E., 1989. Biostratigraphy of the last glacial period in Europe. Quatern. Sci. Rev. 8, 25–44.

Ber, A., 1996. Geological situation of Augustovian (Pastonian) Interglacial lake sediments at Szczebra near Augustow and Mazovian Interglacial organogenic sediments at Krzyzewo. Biul. Państw. Inst. Geol. 373, 35–48 (in Polish with English summary).

Bitinas, A., 2011. New insights in to the last deglaciation of the south-eastern flank of the Scandinavian ice sheet. Quatern. Sci. Rev. (2011), doi:10.1016/j.quascirev.2011.01.019. in press.

Bitinas, A., Damušytė, A., Stančikaitė, M., Aleksa, P., 2002. Geological development of the Nemunas River Delta and adjacent areas, West Lithuania. Geol. Q. 46 (4), 375–389.

Bitinas, A., Karmaziene, D., Jusiene, A., 2004. Glaciolacustrine kame terraces as an indicator of conditions of deglaciation in Lithuania during the Last Glaciation. Sed. Geol. 165, 285–294.

Blažauskas, N., Jurgaitis, A., Šinkūnas, P., 2007. Patterns of Late Pleistocene proglacial fluvial sedimentation in the SE Lithuanian Plain. Sed. Geol. 193, 193–201.

Danilans, I., 1981. Interpretation of data on stratigraphy of Lower Pleistocene of the eastern Baltic area. In: Faustova, M.A. and Velichko, A. A. (Eds.). Pleistocene Glaciations of the Eastern European Plain. Nauka, Moscow, pp. 114–120 (in Russian).

Doss, B. 1910. Über das Vorkomen einer Endmoräne, sowie Drumlins, Åsar and Bänderton im nördlichen Litauen. Centralblatt für Mineralogie, Geologie and Paläontologie. 723–731.

Gaigalas, A., 1987. Border layers and boundary between Quaternary and Neogene systems in the Baltic region. In: Alekseev, M.N. and Nikiforova, K.B. (Eds.). Boundary Between Neogene and Quaternary in USSR. Nauka, Moscow, pp. 13–26 (in Russian).

Gaigalas, A., Hütt, G., 1995. OSL dating of the Merkinė (Eem) Interglacial (in Jonionys) and the Nemunas Glaciation (Rokai Section) in Lithuania. Landscapes and Life. Studies in honour of Urve Miller. PACT 50.

Gaigalas, A., Satkunas, J., 1994. Evolution of the quaternary stratigraphic scheme in Lithuania. Geologija 17, 152–158.

Gaigalas, A., Hütt, G., Melešytė, M., 1994a. The OSL age of the Merkinė (Mikulino) Interglacial and the Nemunas (Valdaj) glacial period in Lithuania. In: Conference on Geochronology and Dendrochronology of Old Towns and Radiocarbon Dating of Archeological Findings Lithuania, Vilnius, October 31–November 4, 1994. Abstracts and Papers, 15.

Gaigalas, A., Arslanov, Kh., Chernov, T., Tertychnaja, T., Kazarceva, T., 1994b. New radiocarbon dating of the Middle Nemunas (Valdaj) in the Rokai exposure near Kaunas. In: Conference on Geochronology and Dendrochronology of Old Towns and Radiocarbon Dating of Archeological Findings, Lithuania Vilnius, October 31–November 4, 1994. Abstracts and Papers, 16.

Guobytė, R., 2002. Lithuanian surface: geology, geomorphology and deglaciation. Abstract of doctoral dissertation. Vilnius, 31 pp.

Guobytė, R., 2004. A brief outline of the Quaternary of Lithuania and the history of its investigation. In: Ehlers, J., Gibbard, P.L. (Eds.), Quaternary Glaciations—Extent and Chronology, Part I: Europe. Elsevier, Amsterdam, pp. 245–250.

Guobytė, R., Aleksa, P., Satkūnas, J., 2001. Lietuvos paviršiaus genetinių, litologinių ir stratigrafinių tipų gruntų paplitimo analizė. (Distribution of Quaternary deposits in Lithuania according their age, genetic types and lithological varieties). Geografijos metraštis 34 (2), 57–67 (in Lithuanian).

Hausen, H., 1913. Über die Entwiclung der Oberflächenformen in der Russichen Ostseeländern und angegrenzenden Gouvernments in der Quartärzeit. Fennia, 34,3. Helsingfors.

Janczyk-Kopikowa, I., 1996. Temperate stages of the Mesopleistocene in Northeastern Poland. Biul. Państw. Inst. Geol. 373, 49–66 (in Polish with English summary).

Jusienė, A., 2007. Kvartero pjuvių stratigrafinė ir genetinė revizija. Report. Lithuanian Geological Survey (in Lithuanian).

Kalm, V., 2006. Pleistocene chronostratigraphy in Estonia, southeastern sector of the Scandinavian glaciation. Quatern. Sci. Rev. 25 (9–10), 960–975.

Karabanov, A.K., Matveyev, A.V., Pavlovskaya, I.E., 2004. The main glacial limits in Belarus. In: Ehlers, J., Gibbard, P.L. (Eds.), Quaternary Glaciations—Extent and Chronology, Part I: Europe. Elsevier, Amsterdam, pp. 15–18.

Kondratiene, O., 1965. Stratigraficheskoye raschleneniye pleistocenovykh otlozheniy yugo-vostochnoy chasti Litvy na osnove palinologicheskikh dannych. Stratigrafiia chetvertichnykh otlozheniy i paleogeografiia antropogena Yugo-Vostochnoy Litvy. Vilnius, pp. 189–261 (in Russian).

Kondratiene, O., 1971. Paleobotanichekaia kharakteristika opornykh razrezov Litvy. In: Stroenie, litologiia i stratigrafia otlozheniy nyzhnego pleistocena Litvy, 57–116 Vilnius (in Russian).

Kondratiene, O., 1976. Nekotorye metodichieskye osobienosty i pogrieshnosty sporovo-pylcevogo analiza na primere razrieza Kurkliai (Litovskaja SSR). Palinologija v kontinentalnykh i morskykh geologichieskykh issledovanyjakh. Riga, Zinatne, pp. 179–187 (in Russian).

Kondratiene, O., 1993. Catalogue of Quaternary stratotypes of the Baltic region. Baltic Stratigraphic Association & Lithuanian Geological Institute. Vilnius, 55 pp. (in Russian).

Kondratiene, O., 1996. Stratigrafia i paleogeografia kvartera Litvy po paleobotanycheskym dannym. Academia, Vilnius, 212 pp. (in Russian).

Kondratiene, O., Bitinas, A., 1989. Stratigraphy and Palaeogeography of the Middle and Lower Pleistocene of the southern Baltics in view of palaeobotanical data. In: Quaternary Age: Palaeontology and Archaeology, pp. 103–110, Kishinev (in Russian).

Kondratiene, O., Klivečkiene, A., 1983. Novye razriezy miarkinskogo miezhliednikovia v vostochnoy chasti Litvy. Palinologija v geologicheskykh issliedovanyakh Pribaltyki i Baltyiskogo morja. Riga, Zinatne, pp. 51–53 (in Russian).

Kondratiene, O., Satkūnas, J., 1993. Stratigraphy of Early Pleistocene of Lithuania. In: 2nd Baltic Stratigraphic Conference May 11–14, Vilnius. Abstracts, 47.

Kondratiene, O., Šeirienė, V., 2001a. Merkinė (Eemian) deposits at the Kurkliai outcrop. In: Field Symposium on Quaternary Geology in Lithuania, pp. 37–39, Exursion guide, 19–25 May, 2001. Vilnius.

Kondratiene, O., Šeirienė, V., 2001b. Neogene/Pre-Pleistocene sediments at the Giliai outcrop. In: Field Symposium on Quaternary Geology in Lithuania, pp. 37–39, Exursion guide, 19–25 May, 2001. Vilnius.

Kondratiene, O., Šinkūnas, P., Barzdžiuvienė, V., 2001a. Daumantai-1 outcrop—the stratotype of Neogene/Quaternary boundary. In: Field

Symposium on Quaternary Geology in Lithuania, pp. 20–27, Exursion guide, 19–25 May, 2001. Vilnius.

Kondratienė, O., Barzdžiuvienė, V., Šinkūnas, P., 2001b. Neogene/Quaternary boundary at the Vetygala outcrop. In: Field Symposium on Quaternary Geology in Lithuania, pp. 27–30, Exursion guide, 19–25 May, 2001. Vilnius.

Liivrand, E., 1991. Biostratigraphy of the Pleistocene deposits in Estonia and correlations in the Baltic region. Stockholm University. Report 19. Doctoral thesis, 114 pp.

Lunkka, J.P., Johansson, P., Saarnisto, M., Sallasmaa, O., 2004. Glaciation of Finland. In: Ehlers, J., Gibbard, P.L. (Eds.), Quaternary Glaciations—Extent and Chronology. Part I: Europe. Elsevier, Amsterdam, pp. 93–100.

Mangerud, J., 1991. The last interglacial/glacial cycle in Northern Europe. In: Shane, L.C.K. & Cushing, E.J. (eds),Quatern. Landscapes, 38–75, Minneapolis.

Mangerud, J., 2004. Ice sheet limits in Norway and on the Norwegian continental shelf. In: Ehlers, J., Gibbard, P.L. (Eds.), Quaternary Glaciations—Extent and Chronology. Part I: Europe. Elsevier, Amsterdam, pp. 271–294.

Marks, L., 1997. Middle and Late Vistulian glaciation in Poland. In: The Late Pleistocene in Eastern Europe: Stratigraphy, Palaeoenvironment and Climate. Abstract Volume and Excursion Guide of the INQUA-SEQS Symposium Vilnius, Lithuania, p. 36.

Marks, L., 2004. Pleistocene glacial limit in Poland. In: Ehlers, J., Gibbard, P.L. (Eds.), Quaternary Glaciations—Extent and Chronology. Part I: Europe. Elsevier, Amsterdam, pp. 296–300.

Misūna, A. B., 1903. Materjaly k izucheniju ledrikowych otlozhenij Belorusii i Litovskogo kraja. Materialy k poznaniju geol. stroijenia Rosijskoj imperii, vypusk II, 1–72 (in Russian).

Molodkov, A., Bitinas, A., Damušytė, A., 2010. IR-OSL studies of till and inter-till deposits from the Lithuanian Maritime Region. Quat. Geochronol. 5 (2–3), 263–268.

Mortensen, H., 1924. Beiträge zur Entwicklung der glacialen Morphologie Litauens. Geologisches Archiv. 3, 1–2, 1–93.

Nenonen, K., 1995. Pleistocene stratigraphy of southern Finland. In: Glacial Deposits in North-East Europe. Rotterdam, Balkema, pp. 11–28.

Raukas, A., 2009. When and how did the continental ice retreat from Estonia? Quatern. Int. 207, 50–57.

Raukas, A. and Stankowski, W., 2005. Influence of sedimentological composition on OSL dating of glaciofluvial deposits: examples from Estonia. Geol. Q 49 (4), 463–470.

Rinterknecht, V.R., Clark, P.U.M., Raisbeck, G.M., Yiou, F., Bitinas, A., Brook, E.J., et al., 2006. The last deglaciation of the southeastern sector of the Scandinavian Ice Sheet. Science 311, 1449–1452.

Rinterknecht, V.R., Pavlovskaya, I.E., Clark, P.U., Raisbeck, G.M., Yiou, F., Brook, E.J., 2007. Timing of the last deglaciation in Belarus. Boreas 36, 307–313.

Rinterknecht, V.R., Bitinas, A., Clark, P.U., Raisbeck, G.M., Yiou, F., Brook, E.J., 2008. Timing of the last deglaciation in Lithuania. Boreas 37, 426–433.

Salonen, V.-P., Kaakinen, A., Kultti, S., Miettinen, A., Eskola, O.K., Lunkka, J.P., 2008. Middle Weichselian glacial event in the central part of the Scandinavian Ice Sheet recorded in the Hitura pit, Ostrobothnia, Finland. Boreas 37 (1), 38–54.

Satkūnas, J., 1993. Conditions of occurrence, structure and forming peculiarities of the interglacial sediments in eastern Lithuania. Abstract of doctoral thesis. Vilnius, 26 pp. (in Lithuanian with English summary).

Satkūnas, J., 1998. The oldest Quaternary in Lithuania. Mededelingen Nederlands Instituut voor Toegepaste Geowetenschappen TNO 60, 293–304.

Satkūnas, J., Bitinas, A., 1995. The lithological composition of Saalian tills in Lithuania as a criterion for lithostratigraphical subdivision. Acta Geogr. Lodziensia 68, 173–180.

Satkūnas, J., Grigienė, A., 2000. New Eemian–Weichselian sequences from Mickūnai depression (eastern Lithuania). In: Quaternary Geology of Denmark, International Field Symposium, pp. 43–44, August 29–September 3, University of Aarhus: Abstracts of Papers and Posters, The Peribaltic Group, INQUA Commission on Glaciation, University of Aarhus.

Satkūnas, J., Hütt, G., 1999. Stratigraphy of the section Antaviliai, eastern Lithuania, and its implication for the Upper Weichselian climatostratigraphic subdivision. Geol. Q. 43 (2), 213–218.

Satkūnas, J., Molodkov, A., 2005. Middle Pleistocene stratigraphy in the light of data from the Vilkiskes site, eastern Lithuania. Polish Geological Institute Special Papers 16, pp. 94–102.

Satkūnas, J., Grigienė, A., Guobytė, R., Marcinkevičius, V., Račkauskas, V., 1997. Upper Pleistocene stratigraphy based on data from Smalvos and Dysnai localities, North-eastern Lithuania. Geologija 22, 26–35.

Satkūnas, J., Grigienė, A., Robertsson, A.-M., 1998. An Eemian—Middle Weichselian sequence from the Jonionys site, Southern Lithuania. Geologija 25, 82–91.

Satkūnas, J., Grigienė, A., Velichkevich, F., Robertsson, A.-M., Sandgren, P., 2003. Upper Pleistocene stratigraphy at the Medininkai site, eastern Lithuania: a continuous record of the Eemian. Boreas 32, 627–641.

Satkūnas, J., Grigiene, A., Jusiene, A., Damusyte, A., Mazeika, J., 2009. Middle-Weichselian palaeolacustrine basin in the Venta river valley and vicinity (northwest Lithuania), exemplified by the Purviai outcrop. Quatern. Int. 207, 14–25.

Stančikaitė, M., Šinkūnas, P., Šeirienė, V., Kisielienė, D., 2008. Patterns and chronology of the Lateglacial environmental development at Pamerkiai and Kašučiai, Lithuania. Quatern. Sci. Rev. 27 (1–2), 127–147.

Vaitiekūnas, P., 1968. Problems of stratigraphy of the Neopleistocene in Lithuania. Kwart. Geol. 12 (3), 646–664 (in Polish).

Vaitiekūnas, P., 1977. Pre-pleistocene deposits in basin of the river Šventoji. Geogr. Geol. 13, 73–90, Vilnius (in Russian).

Zagwijn, W.H., 1992. The beginning of the Ice Age in Europe and its major subdivisions. Quatern. Sci. Rev. 11, 583–591.

Zelčs, V., Markots, A., 2004. Deglaciation history of Latvia. In: Ehlers, J., Gibbard, P.L. (Eds.), Quaternary Glaciations—Extent and Chronology. Part I: Europe. Elsevier, Amsterdam, pp. 225–243.

Chapter 20

Pleistocene Glaciation in The Netherlands

Cees Laban[1,*] and Jaap J.M. van der Meer[2]

[1]Deltares, PO Box 85467, 3508 AL Utrecht, The Netherlands
[2]Department of Geography, Queen Mary University of London, Mile End Road, London E1 4NS, United Kingdom
*Correspondence and requests for materials should be addressed to Cees Laban. E-mail: cees.laban@deltares.nl

20.1. PRE-ELSTERIAN GLACIATIONS

20.1.1. Introduction

The Netherlands is part of the southern North Sea basin and as such has seen more or less continuous sedimentation since the Tertiary. However, 'cold' marine and fluvial sediments related to Early Pleistocene glaciations are only present at a few locations in boreholes in the Dutch sector of the North Sea or on the Dutch mainland. In the Dutch sector of the North Sea, the oldest indications of glacial influence probably date from the Tiglian C4c cold Stage. Stiff marine clays recorded in a borehole are thought to date from this early glaciation. Indications of sediments deposited during the Menapian Stage in a prodelta and delta-front environment are present in a deep borehole and are referred to the Aurora and Outer Silver Pit Formations. Fluvial sediments deposited by the rivers Elbe and Weser during the Menapian Stage are present in only two boreholes. The oldest glacigenic sediments recorded have been found in a borehole near the Dogger Bank and were probably deposited during the Cromerian Glacial A or B (Laban, 1995). 'Cold' marine sediments of Cromerian age are known to be present in a borehole farther east. Correlation of the 'cold' sediments between boreholes is not possible because of the great distance between localities and also the fact that in most boreholes, sediments of only one cold stage are recorded.

20.1.2. Tiglian C4c Glacial Phase and Eburonian Stage

In the Dutch sector of the North Sea, the oldest deposits indicating a cold Pleistocene phase are marine deposits probably dating from the Late Tiglian (C4c; marine isotope stage (MIS) 68). They have been found in the well-cored borehole G16-22 (Fig. 20.1) in which a deposit of stiff clay is present between 260 and 253 m below MSL (Laban, 1995). Based on the evidence of dinoflagellate cyst analysis, the clay probably settled in a shallow marine (prodelta, delta front to mouth bar environment) low salinity environment during the Late Tiglian and Early Eburonian (MIS 62; Maher and Hallam, 2005). The pollen content of the clay is poor, indicating cold climatic conditions (de Jong, 1991). The foraminiferal fauna is poor to very poor, again pointing to arctic conditions with species such as *Elphidium albiumbilicatum*, *Elphidium excavatum f. clavata* and *Nonion orbiculare* (Neele, 1991b). On top of this clay (between 242.90 and 198.10 m), sand is present, thought to have been deposited in a fluvial environment during the Eburonian Stage (Burger, 1992). During deposition of the fluvial sand, the pollen indicate cold climatic conditions (de Jong, 1991). On land, periglacial structures, like frost wedges and involutions (van Straaten, 1956; Maarleveld, 1960), point at widespread cold conditions during different stages of the Early Pleistocene.

20.1.3. Menapian Stage

Indications of cold, post-Eburonian and pre-Cromerian conditions have been encountered in several boreholes in the Dutch sector of the North Sea. In borehole F8-6, a stiff marine clay occurs between 244.60 and 242.50 m below MSL; its foraminiferal content (dominated by *E. excavatum f. clavata*) represents a rich, high arctic, fauna poor in species. Between 242.56 and 242.50 m, a relatively high percentage of *N. orbiculare* is present (Neele, 1991a), molluscs are absent (Pouwer, 1991), while pollen indicate cold climatic conditions (P. Cleveringa, personal communication). In borehole L10-6 (Fig. 20.1), fluvial sediments are present below Cromerian III/IV deposits (Zagwijn, 1977) between 129.35 and 80.35 m below MSL. They were deposited under cold climatic conditions during the Menapian Stage (MIS 36 or 34). The pollen assemblage contains a mixture of Pleistocene and Tertiary pollen which have been derived from glaciofluvial deposits. Near the base

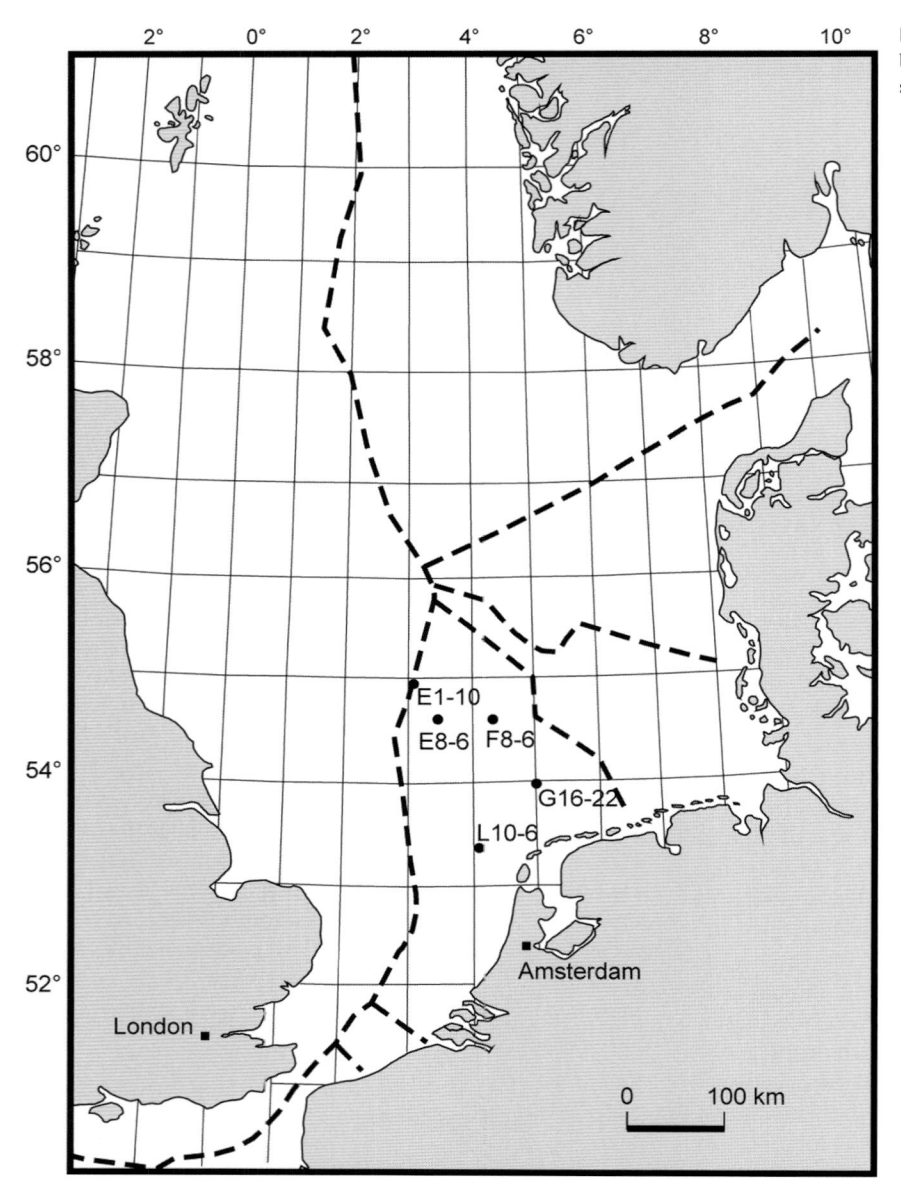

FIGURE 20.1 Map showing the position of boreholes recording Early Pleistocene 'cold' sediments (after Laban, 1995).

of the borehole, at 129.35 m below MSL, the pollen assemblage contains a high percentage of herbaceous pollen also suggesting cold conditions. In borehole G16-22 (Fig. 20.1), fluvial sand is present between 179 and 149 m below MSL. Again the pollen record indicates the sand accumulated under a cold climate (de Jong, 1991) and is again correlated with the Menapian Stage. In the Netherlands, deposition of the Appelscha Formation, of central European origin, continues into the Cromerian. At the top of the Peize Formation, the Hattem Layers (Hattem Bed Complex) are found in the Netherlands and in adjoining parts of Germany. This lag deposit is taken as evidence of glaciation in northern Europe (see Lee et al., 2010) because it contains (weathered) Scandinavian erratics that could not have reached the Netherlands by fluvial transport alone (van der Meer, 1987). Lee et al. (2010) suggest that the coarseness of the material implies that the ice must have been quite proximal and quote a MIS 34 age.

20.1.4. Cromerian Stage

Deposits dating from the 'Cromerian Complex' Stage are found in the Dutch sector in borehole E1-10 (Fig. 20.1) drilled on the south-east flank of the Dogger Bank. In this borehole, very fine, grey, slightly silty sand with sporadic clay laminae and organic matter has been sampled between

153.20 and 130.30 m. Pollen analysis points to a Scandinavian, non-British glacigenic provenance (de Jong, personal communication), while foraminifera or molluscs are absent (Neele, 1986). Zagwijn (1986) suggests that the age of this bed is Cromerian and probably belongs to Glacial B or A (MIS 20–18). The sand layer is overlain by a marine bed which Zagwijn (1986) suggested to be probably of Cromerian III age at the base and Glacial C towards the top. Near borehole E1-10, borehole E8-6 in block E8 was drilled to a depth of 183.50 m below MSL. In this borehole, a clay bed is present between 152.50 and 147.50 m with a high arctic foraminiferal fauna, dominated by *E. excavatum f clavata*, *N. orbiculare* and *Elphidium ustulatum* (Neele, 1991b). The pollen record also shows evidence of cool climatic conditions (J.D. de Jong, personal communication). The glacial, laterally equivalent sand deposit in borehole E1-10 is, however, non-marine, and it is suggested that a coastline lay between these two boreholes at this particular time. Evidence for 'cold' deposits on land during the Cromerian Stage was found in the eastern Netherlands near Emmerschans. Here, the 'Weerdinge Beds' of the Appelscha Formation are regarded as glaciofluvial and deposited close to an ice margin (within the Netherlands) during the Cromerian Glacial C (conventional, MIS 16; Ruegg and Zandstra, 1977; de Jong and Maarleveld, 1983) because like the Hattem Layers they contain Scandinavian erratics. Lee et al. (2010) point to a possible correlation to MIS 12 instead of MIS 16 on the basis that overlying augite minerals may stem from more easterly sources instead of the conventional sourcing in the West Eifel.

20.2. THE ELSTERIAN GLACIATION

20.2.1. The Elsterian Glaciation in The Netherlands

The effect of the Elsterian glaciation (MIS 12) on the northern Netherlands has always been enigmatic, as a morphologically distinct ice margin is absent, and tills have only been recognised relatively recently. This enigma resulted in two types of reconstruction of the maximum extent of the Elsterian Ice Sheet. One type showed the ice margin coming from the east to take a wide swing N of the Dutch mainland. The other followed the southern boundary of the Elsterian Peelo Formation (Fig. 20.2).

At present, the following Elsterian deposits and landforms are known on land: in the northern Netherlands, Elsterian meltwater deposits occur in depressions varying in depth between ca. 10 and over 100 m. They consist of hard, dark brown clay; fine, micaceous, locally laminated, sand; and fine- to medium-grained aeolian sands.

In a pit near Peelo, the type locality of the Elsterian Peelo Formation (Zagwijn, 1973; Ruegg, 1975), typical Scandinavian gravel, has been found at the top of the Elsterian deposits. In two boreholes, notably Den Burg (913-36) and Tzum (6D-49) in Friesland, rhythmites of Elsterian age have been found. The Tzum borehole was drilled in a depression over 100 m deep. ter Wee (1983a) and Bosch (1990) describe 20- to 30-km-long channels, 3–5 km wide and locally up to 350 m deep from the northern Netherlands. Measurements on sedimentary structures point to an infill from north to south (J.H.A. Bosch, personal communication, 1994). In the northern Netherlands, Elsterian till has been found in only three boreholes, while in a few others, rhythmites of Elsterian age have been encountered, for example, 9B-36 (Fig. 20.2). In borehole 5B-7 in the Waddenzee, Oostmeep, 0.30 m of till has been sampled at a depth of between 51.70 and 51.40 m at the base of a depression and in a relatively small glacial valley. The gravel in this till is only of eastern provenance; Scandinavian rock fragments have not been observed. In the Witmarsum borehole (I OB/191) in Friesland, greyish-black, calcareous, very sandy till at a depth of between 226 and 220 m was sampled at about 20 m above the floor of a glacial valley. This till is poor in coarser particles and consists of both fluvial material of eastern provenance and a few fragments of flint and Fennoscandian granite (Zandstra, 1983). In another borehole on the former island of Wieringen (I 4E/I 10, Stroe 11), two till layers occur. The deepest till is present between 60 and 59 m at the base of a glaciolacustrine clay, equated with the Peelo Formation. The upper till between 13.40 and 1.20 m belongs to the Drente Formation (Saale glaciation). The gravel content of the lower Elsterian 'till' is of eastern origin and derived from the fluvial deposits of mid-German rivers. The flint, however, is of typical Scandinavian provenance (Zandstra, 1986). One reason for the lack of observations on Elsterian tills can be their sandy nature which does not help their detection in flush borings. Further, Passchier et al. (2010) point at the permeable nature of the bed of the Elsterian ice and how this leads to thin tills as observed in Iceland (Kjaer et al., 2003).

20.2.2. The Elsterian Glaciation in the Dutch Sector of the North Sea

The first glaciation in the North Sea from which abundant glacial sediments and features are preserved pertains to the Elsterian Stage (MIS 12). The most striking feature of this glaciation is the occurrence of a complex system of deeply eroded anastomosing valleys and isolated oval depressions in the southern North Sea in a broad zone between 53°N and 56°N (Passchier et al., 2010). The valleys have been eroded into pre-Elsterian deposits. The zone containing this type of depressions continues towards the west into the British sector (Cameron et al., 1986, 1992; Jeffery et al., 1989) and towards the east to beyond Germany. The valleys in the

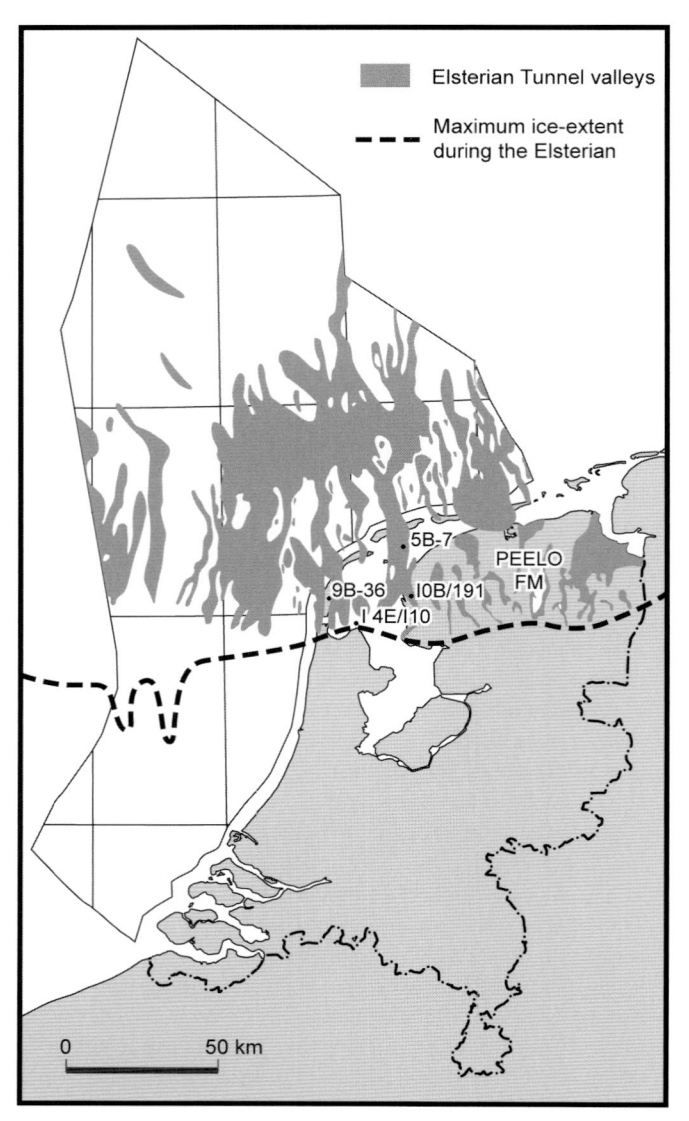

FIGURE 20.2 Map showing the extent of the Elsterian Ice Sheet in the onland and offshore parts of the Netherlands. Numbered boreholes record till and/or rhythmites of Elsterian age (after Laban, 1995; de Mulder et al., 2003).

Dutch sector of the North Sea have been mainly mapped from a regular and dense grid of north-east/south-west and north-west/south-east seismic lines. The valleys appear to occur within the maximum ice limit of the Elsterian glaciation and mainly trend north-north-west/south-south-east. A difficulty in interpreting the valleys from seismic lines is that the shoulders are not always clearly defined or visible. As they occur on the upper part of the profiles, they are often obscured by multiple reflections from the sea bed, even after processing. Most of the valleys have steep slopes in cross section with angles ranging between 5° and 25° in the marine sector (Joon, 1987) and up to 55° on land in northern Germany (Kuster and Meyer, 1979). From the longitudinal profile, it is clear that the base of the valleys is not flat but shows isolated sub-basins and thresholds. The valleys vary in width between < 1 and 23 km and are generally 100–250 m deep, while exceptionally depths of over 500 m below MSL are reached. The valleys are most numerous between 53°N and 54°N.

South of 53°N, comparable valleys are absent. Except that on two seismic lines between 52°50′N and 53°N and ca. 02°27′E, some valleys up to 50 m deep and several kilometres wide are present incised into the Yarmouth Roads Formation of Cromerian age. They form probably the southern extensions of deep valleys north of 53°N (Cameron et al., 1984).

Between 54°N and 56°N, only isolated valleys are present. Based on the presence of these valleys, as well as the presence of glaciotectonic structures, the maximum extent of the Elsterian ice has been reconstructed (Fig. 20.2). It ran from the Dutch coast at 52°50′00″N in a south-west direction towards Ipswich in East Anglia, near the British east coast. It appears to link up well with what is known from the British side.

20.2.3. Drainage History During the Elsterian Glacial Maximum

Gibbard (1988) reconstructed the palaeogeography of the Elsterian (Anglian) Stage at its glacial maximum with an ice-dammed lake in front of the ice sheet in the southern North Sea. Drainage partly took place over the Weald–Artois anticline and over the north-western French coastal area (Fig. 20.3). This is supported by the occurrence of fluvial sediments at Wissant in France (Roep et al., 1975) dating from the Saalian or an older glaciation. Sea level according to Valentin (1954) was 90 m below present level and Fairbridge (1961) calculated a sea level of about 48 m below present level. Praeg (1994) in his reconstruction projected a lake at least 100 m deep in the southern North Sea dammed by the Weald–Artois anticline in the south and the ice margin in the north. Recently, Gupta et al. (2007) described an incised valley system in the Channel which formed by catastrophic flooding due to the sudden drainage of a large meltwater lake in the southern North Sea. Cohen et al. (2008) reconstruct a huge (ca. 250 × 600 km) Elsterian ice-dammed lake which covered, apart from the southern North Sea, northern Belgium, almost all of the Netherlands and parts of northern Germany and which was fed by meltwater as well as by the Rhine and the Thames (Toucanne et al., 2009a,b). However, the drainage associated with the Elsterian glaciation is still problematic. Until now, no depositional evidence has been produced from the southern North Sea basin to support the presence of a continuous lake between the Weald–Artois anticline in the south and an ice sheet in the north. Although the clayey filling of the deep tunnel valleys is seen as lacustrine, they are not evidence of a single, continuous lake. Nevertheless, Toucanne et al. (2009a,b) produce convincing evidence for the overflow from such a lake. In the Dutch sector of the southern North Sea, the pre-Elsterian formations consisted of fluvial deltaic deposits, while Elsterian sediments are fluviatile in origin (Yarmouth Roads Formation). If such a huge lake existed, one would expect the development of deltas, and although Cohen et al. (2008) hint at their existence, we are awaiting the sedimentological evidence.

20.3. THE SAALIAN GLACIATION IN THE NETHERLANDS

20.3.1. Introduction

Many of the sediments and features associated with the cold Saalian Stage are much better known than the preceding stage and are invariably thought to relate to MIS 6. However, Beets et al. (2005) discussed the possibility that there has been an earlier Saale (MIS 8) ice advance in the North Sea basin. This is based on the presence of a till below the common level of the MIS 6 and above the level of the MIS 12, Elsterian glacial sediments. The evidence consists of seismic and micropalaeontological data, and amino acid age determination. Although there is no doubt about the nature of the till (van der Meer, 1992), so far it is the only evidence for such an earlier advance throughout the North Sea basin (Toucanne et al., 2009a,b). It should be noted that Meijer and Cleveringa (2009) support the MIS 8 age of the till on biostratigraphical grounds. But this is by no means a secured age as claimed by Westaway (2010).

As a result of the widespread presence of sediments and landforms, the geometry of the Saalian deposits has been mapped in more detail. From the period immediately preceding the Late Saalian glaciation, the sediments present in the northern part of the Dutch sector of the North Sea are very fine- to fine-grained periglacial sediments. In a valley in the eastern part of the sector, however, marine sediments are overlain by Late Saalian till, which may indicate arctic to boreal marine conditions during an Early Saalian Interstadial. Stiff to very stiff glaciolacustrine clay was deposited in the north and north-west parts of the Dutch sector. Tills are present near the northern coast of the Netherlands, and based on their lithology, two different facies can be distinguished. In the east, the till contains up to 47% of gravel, while in the west only up to 3% of gravel has been recorded. At sea, fine- to medium-grained glaciofluvial sediments are only found very locally. The extent of the formation was probably greater, but much of it was reworked during the Eemian transgression. On land, Saalian deposits mainly consist of coarser fluvial sediments of the Rhine–Meuse system, while glaciofluvial sediments are mainly known from the fringe of the ice-pushed ridges. It is now known that glaciofluvial deposits are also included in the ice-pushed ridges (Ruegg, 1977) and much more widely than previous thought (Bakker, 2004). Outside the ice-pushed ridges, their extent may never have been much greater, as this fringe was in touch with the westward diverted Rhine. In the eastern part of the country, a continuous esker/entrenched valley has been mapped, feeding into the proglacial meltwater-Rhine system (van den Berg and Beets, 1987).

In the North Sea, the occurrence and nature of the Saalian subglacial valleys are problematical. The dimensions of the valleys are much smaller than those of the Elsterian glaciations (Passchier et al., 2010), and they are partly filled with Eemian marine sediments. Since, on 3.5 kHz seismic records, the base of some of the valleys occurs below the penetration depth, it is difficult to distinguish Saalian valleys from those formed during the Eemian transgression. Tongue-shaped basins (with deformation structures along

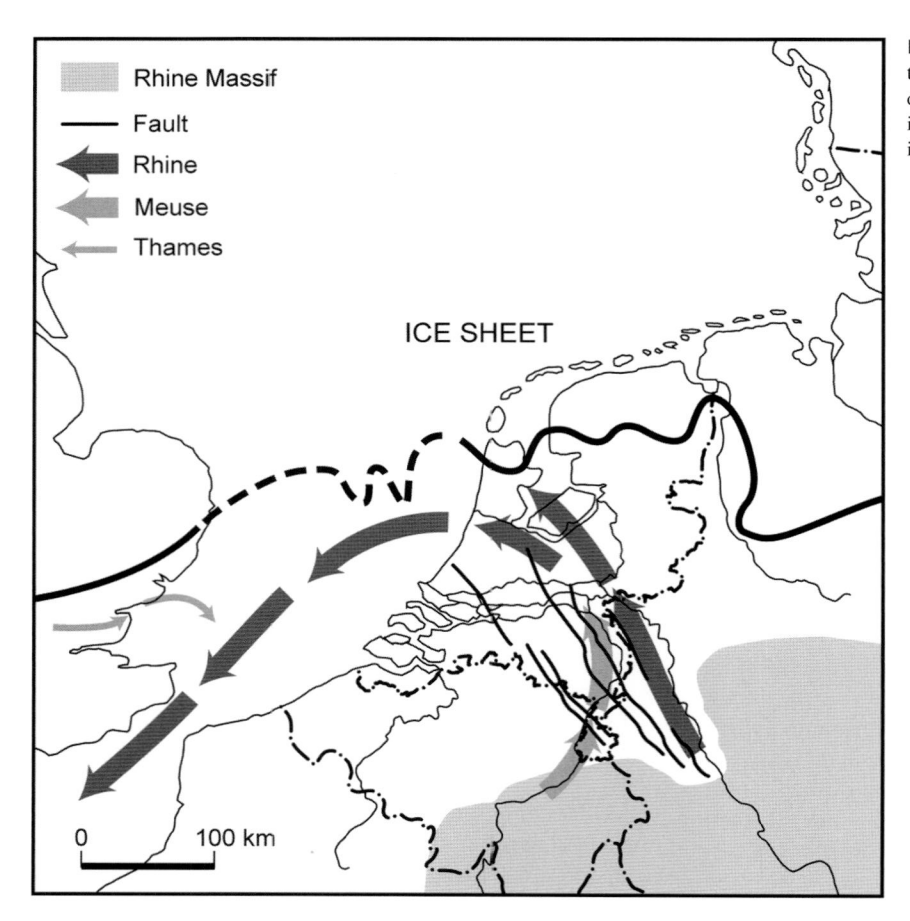

FIGURE 20.3 NW European drainage during the Elsterian. Most likely there has been an ice-dammed lake in the North Sea basin, but geological evidence of its position and size is as yet lacking (after de Mulder et al., 2003).

their flanks) although forming a continuous belt on land (de Gans et al., 1987; van der Wateren, 1992) have only been found at two locations at sea (Laban, 1995).

In the Netherlands, all Saalian glacial sediments are placed in one lithostratigraphic unit, the Drente Formation (Zagwijn, 1961). This formation includes one till sheet, indicating a single ice advance. This contrasts with northern Germany where three Saalian tills occur in certain areas. Correlation with the glacial deposits on land in the northern Netherlands (ter Wee, 1983b; van den Berg and Beets, 1987) and northern Germany (Ehlers, 1990) indicates that the advance of Scandinavian ice in the North Sea took place during the Drenthe Substage (see also Busschers et al., 2008; Lee et al., 2010).

20.3.2. Saalian Glaciation in The Netherlands

The glaciation of the northern half of the country has had a pronounced influence on the geomorphology of the Netherlands. In this relatively short time span, all the relief in this part of the country has been created. The present-day relief has been strongly influenced by younger processes; originally, the largest height differences created by the ice must have been in the order of 250 m. The latest map of the glacial geomorphology of the Netherlands was published in 1987 (van den Berg and Beets, 1987); it shows the extent of ice-pushed ridges surrounding five (known) deep glacial basins as well as the extent of glaciofluvial deposits. The ice-pushed ridges are asymmetric in cross section with steep proximal and more gentle distal slopes, the latter grading into glaciofluvial deposits. The ice-pushed ridges consist mainly of older Pleistocene and Tertiary deposits, with a trend towards increasingly younger materials towards the west. In the eastern part of the country, Tertiary deposits are involved, while in the central part, the pushed sequence consists of Early and Middle Pleistocene deposits. The westernmost basins are deeply buried under younger deposits and are the least known (de Gans et al., 1987); an overview of the shallow glaciotectonics is given by Van der Wateren (1981, 1985, 1992) on the basis of exposures, while the overall, deep tectonic structure is presented by Bakker (2004), who based himself on GPR and other seismics as well as deep boreholes.

The northern part of the country is characterised by the Saalian till plateau, similar to the 'Geest' areas in northern Germany. The most prominent feature on this slightly undulating plateau is a series of streamlined ridges, collectively known as the Hondsrug, and nowadays interpreted as a set of megaflutes (Rappol, 1983, 1984). The strong NW–SE orientation of these ridges, in combination with the till stratigraphy, plays an important role in the reconstruction of events during the Saalian glaciation. Moreover, the stratigraphy of the associated till is quite well known nowadays. The most complicated element is the occurrence of 'red' till floes of completely different composition and derivation (Rappol, 1987; van der Meer, 1987 and references therein). In the Netherlands, the interpretation of these red till floes is difficult because the top of the till is always eroded and the original configuration of the floes is unknown (van der Meer and Lagerlund, 1991). Even when the till surface has been preserved by postglacial burial, as exposed in a deep pit near Grouw in the northern Netherlands, the till had been weathered as shown by decalcification and soil formation (J.J.M. van der Meer, unpublished). In the 1980s, till stratigraphy has been studied in a number of deep exposures related to motorway construction. Provenance and directional studies have demonstrated a more complex stratigraphy than suggested previously, that is, it was found that 'red' till occurs at two stratigraphic levels (Rappol et al., 1989).

The studies also demonstrate that on land, the last recorded ice movement was from the NW (Rappol et al., 1989; Kluiving et al., 1991). As the Scandinavian ice was moving upslope and also created its own relief, relatively small, ice-dammed lakes developed. The most prominent of these must have occurred in the central Veluwe area, around the present-day Leuvenumse Beek valley. From this area, mass-flow deposits have been described (Postma et al., 1983; Postma, 1997). During deglaciation, lakes also developed in the glacial basins, as demonstrated by the deposition of rhythmites (de Gans et al., 1987; Beets and Beets, 2003). The largest of these lakes, in the present-day IJssel valley, was part of the course of the river Rhine and acted as a sediment trap. Because of the sediment depletion, the 'Rhine', on exiting the lake, cut a deep channel in the subsoil of the northern Netherlands (van den Berg and Beets, 1987) while sea level was still low. This channel had previously been interpreted as ice marginal in origin and formed during the glacial advance.

20.3.3. Saalian Glaciation in the Dutch Sector of the North Sea

In the south-western and north-eastern parts of the Dutch sector of the North Sea, a range of Saalian glacial and periglacial sediments and phenomena are found. Based on these, the outer limit of the Saalian glaciation in the Dutch sector of the North Sea can be reconstructed (Fig. 20.4). The most obvious feature of this limit is that it curves towards the NE into the German sector and does not cross the North Sea anywhere in the Dutch sector. This implies that at least in the southern North Sea, there was no connection between Scandinavian and British ice.

20.3.3.1. Reconstruction of the Maximum Extent of the Saalian Ice Sheet in the Dutch Sector of the North Sea

The reconstruction of the maximum extent of the ice sheet by the present authors is based on a study of its legacy (sediments and landforms). Within the maximum extent of the Saalian ice in the Dutch sector of the North Sea, till plateaux and tongue-shaped basins were formed together with deformation structures along their margins. In addition, subglacial channels, esker-like features and ice-contact gravels were formed. The ice margin in the North Sea forms a continuation of the north-west/south-east trending margin on land. Only in block PI 5 has the ice moved further south than on land. During the initial ice advance (from the north-east), and in the north-eastern and western parts of the Dutch sector, tongue-shaped basins were formed pointing to the formation of a lobate ice margin. The subglacial valleys which are present between 53°N and 54°N latitude indicate a straight ice front in that area, while south of the 53°N a lobate margin again appears to be developed as in the eastern part of the Netherlands. In this area, no tills have so far been found, similar to them being rare in the ice-pushed ridges on land. The maximum extent of the ice is based in this area on the presence of the tongue-shaped basins, end-morenes and glaciotectonic structures. In the area west and north of the reconstructed ice margin, none of the above features have been recorded in boreholes or are evident from widespread seismic data. Glaciolacustrine and glaciofluvial deposits, overlying Saalian periglacial sediments, dominate the area west and north of the ice margin (Fig. 20.4). The fact that lacustrine sediments were deposited implies that there must have been a barrier in the north (outside the Dutch sector), blocking direct drainage into the northern North Sea/Atlantic. Busschers et al. (2008) reconstruct a lake level similar to present-day sea level, which provides an indication of the lake configuration. Given the extent of the ice, the lake must have been substantial smaller than the one postulated for the Elsterian (see also Toucanne et al., 2009a,b). Passchier et al. (2010) surmise that the lake sediments may have accumulated subglacially, looking to Antarctic subglacial lakes for comparison. However, Antarctic subglacial lakes occur under very thick ice, unlike the ice margin in the southern North Sea. Given the overflow into the Bay of Biscay (Toucanne et al., 2009a,b), a lake must have been present between the Weald–Artois ridge and the ice and this complicates matters. A lake in front of the ice with a water depth of ca. 100 m; a calving ice front and behind that a subglacial lake, fed through incised valleys perpendicular to the pressure gradient,

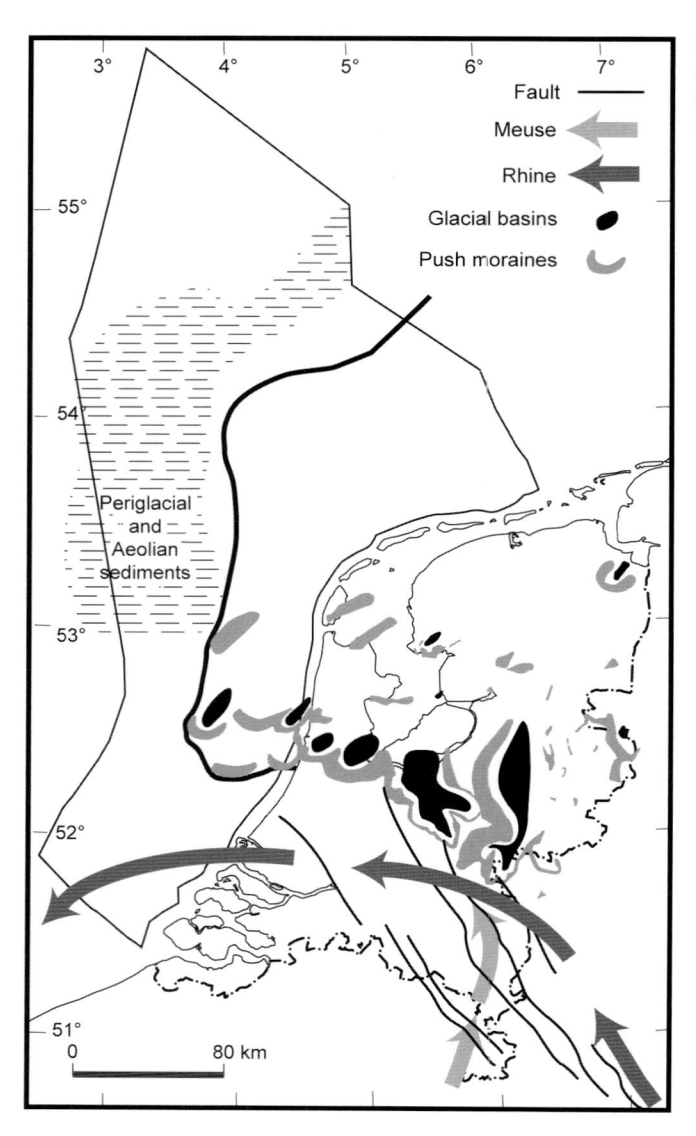

FIGURE 20.4 Ice cover of the Netherlands during the Saalian, demonstrating that in the Dutch sector of the North Sea, there is no evidence for a connection between a British and the Scandinavian Ice Sheet (after Joon et al., 1990; de Mulder et al., 2003).

presents an unlikely reconstruction. Subglacial drainage is directed by pressure gradients (Hooke, 1998), which in this setting would be from underneath the ice towards the front, not towards a pressurised subglacial lake.

The reconstruction of the maximum extent of the ice during the Saalian is thus based on the presence of a variety of sub- and proglacial sediments and landforms on one hand and extra-glacial sediments on the other and these are not overlapping. The reconstruction is further supported by the distribution of Eemian sediments. Postglacial isostatic rebound is reflected in the thickness of Eemian marine sediments in the coastal area of the Netherlands and their progressive thinning in westerly and north-westerly directions outside the Saalian glacial limit.

This reconstruction in the North Sea is contradicting the well-established idea that the Scandinavian Ice Sheet crossed the North Sea and coalesced with the British Ice Sheet, as it had done during the Elsterian. This 'established' reconstruction and especially its position is persistent (Fig. 1 in Lee et al., 2010) even though it is not supported by geological or geomorphological evidence in the Dutch sector. Joint mapping of the UK and the Dutch sectors led to a compatible reconstruction, while in the German and Danish sectors, no further studies have been carried out in this field. Lee et al. (2010) draw the ice margin as turning W instead of NE at a latitude where this is not supported by the geological evidence and their reconstruction is similar to the one adopted by Gibbard et al. (2009), both without supplying the evidence on which this is based. See also Westaway (2010) for the continuing discussion about the Saalian glaciation on the UK side of the North Sea.

However, the complication (see also Busschers et al., 2008) is that according to M. Rappol (personal communication), the maximum extent of the ice sheet in the Dutch sector of the North Sea is too close to the coast of the northern Netherlands to allow the ice to change direction from NE–SW to N–S or even NW–SE. Rappol (1987) and van den Berg and Beets (1987) assumed that the ice sheet extended much further north in the North Sea area in order to explain the origin of the north-west/south-east trending Hondsrug in the eastern Netherlands. Based on data collected by the Geological Survey of the Netherlands, however, the extent of the Saalian Ice Sheet in the North Sea is far enough to the west to allow ice to enter the northern Netherlands from a north-easterly direction. The Hondsrug complex is interpreted as megalineations and as such to have formed under an ice stream (for the latest, see Passchier et al., 2010). At the end of the glaciation, a drawdown system developed in the Norwegian Trench, effectively cutting off the ice flow towards the south-west (Rappol et al., 1989; Lee et al., 2010). In order for the ice to develop a SE flow direction, there must have been a substantial supply in the North Sea basin. Speculatively, a local ice dome with a radial flow pattern could be such a source. It must be clear that for the system to sustain an ice stream, as envisaged by the different authors, this ice dome must have acted as an accumulation and dispersal centre of its own. In that, it could be similar in shape and size to marginal ice domes envisaged for the SW Baltic margin of the Weichselian Ice Sheet (Lagerlund, 1987; Houmark-Nielsen, 2010).

20.3.4. Drainage During the Saale Glaciation

During the Saalian glaciation, the NW European rivers drained through the Straits of Dover which (re-?)opened during this period (Gibbard, 1995; Toucanne et al., 2009a,b). Previous to the presence of ice during MIS 6, the Rhine and Meuse extended in a NW to N direction across the Netherlands. With the ice progressing towards the SW, the rivers were forced to deviate in a W direction. Initially, the rivers—supplemented by meltwater from the ice itself—ran along an ice front in the northern Netherlands. Later on when the ice reached its southernmost position across the central Netherlands, the rivers ran more or less at their current position (Busschers et al., 2008). When the ice retreated, it vacated deep glaciotectonic depressions which were subsequently filled with interconnected lakes (Holland Lake of Beets and Beets, 2003). While the Meuse stayed in its full-glacial position, the Rhine intruded in the ice marginal belt along the line of the current IJssel arm, more or less resuming its preglacial course (van den Berg and Beets, 1987).

20.3.5. Models for the Saale Glaciation of The Netherlands

Since the Second World War, a number of glaciation models (based on glaciotectonic stratigraphy, geomorphology and, more recently, also on till stratigraphy) have been presented. For many years, investigators have worked with the Maarleveld/ter Wee model (ter Wee, 1962; Maarleveld, 1981) in which five retreat phases were distinguished. Although this model was sometimes criticised (Zonneveld, 1975), it has stood for many years. It was replaced by the Jelgersma and Breeuwer model (1975), which to some extent reshuffled some of the previously known five phases, but which did not take into account all that was known about the glaciation. Since then, new models have quickly emerged: in 1987, the van den Berg and Beets model was presented, which only recognised two phases. The youngest phase is characterised by the existence of an ice stream coming from the North Sea basin, across the north-eastern part of the country. A more recent model by Rappol et al. (1989) is similar to the van den Berg and Beets model, but it recognises three phases and the existence of at least two ice streams in the Netherlands and one in the adjoining part of Germany. It also tries to explain ice movement by presenting a model for the behaviour of the south-western part of the Scandinavian Ice Sheet. These last two models rely heavily on the existence of coalescing Scandinavian/British ice in the Dutch sector of the North Sea basin. However, as outlined above, the available evidence does not support this interpretation (Laban, 1995; Busschers et al., 2008). Unfortunately, the German and Danish sectors of the North Sea have not been mapped in the same detail as the British and the Dutch sectors, which means that the NE continuation of the ice margin is virtually unknown. This still leaves much room for speculation about sequences and events of the Saalian glaciation (Fig. 20.4).

20.4. THE WEICHSELIAN GLACIATION IN THE NORTH SEA

20.4.1. Introduction

During the Weichselian, glaciers did not reach the current terrestrial territory of the Netherlands. However, British ice did invade the Dutch sector of the North Sea, and consequently, this section of the report will only deal with the marine area. The authors would like to stress that common opinion is strongly influenced by the modern distribution of land and sea. As sea level is highly variable in time, this notion of 'wet-and-dry' is often irrelevant to the reconstruction of Quaternary glaciation.

A wide range of fluvial, lacustrine, glacial, glaciofluvial, glaciolacustrine and periglacial sediments were formed during the Weichselian glaciation in the Dutch sector of the North Sea. The oldest Early Weichselian sediments are stiff clays, which were deposited in the Southern Bight of the North Sea in a lacustrine environment, implying a low sea-level stand. In the central and southern parts of the Southern Bight, medium- to coarse-grained fluvial sediments were deposited by the rivers Rhine and Meuse in the form of a south-west prograding delta. During the Late Weichselian, an ice sheet from Britain entered the southern North Sea and extended into the Dutch sector. The Scandinavian Ice Sheet, however, did not extend as far south and west as the Dutch sector and did not even deposit fluvioglacial sediments in it. However, fine- to medium-grained glaciofluvial sand was deposited both prior to the advance and during the retreat of the British Ice Sheet and is recorded locally, while stiff glaciolacustrine clays are widespread. In the north—along the median line between the German and Dutch sectors—a glaciomarine facies is present. Tills are present in the southern part of the glaciated area and also as patches locally in the northern part. Microstructures indicate deposition of both flow and subglacial tills (van der Meer and Laban, 1990). Studies of the glacial gravels have identified them as originating from Britain (Veenstra, 1969; Dijkmans, 1981). Two different types of subglacial valleys are present. The first type consists of a braided system of partly open valleys with their floors up to 80 m below MSL and occurring near the southern limit of the ice. They are infilled with soft clays and fine-grained sediments. The second type consists of V-shaped valleys. They are found in the north and are infilled with stiff glaciolacustrine clay. Based on the occurrence of subglacial valleys, two main ice-flow directions of the British ice have been recognised, one from the west and south-west near the southern limit of the British ice and the other from the north-east in the northern part of the Dutch sector. In the Dutch sector, only one tongue-shaped basin has been recorded and is present in the Dogger Bank Formation. This indicates ice cover without accompanying deposition of tills or alternatively, their later erosion. Deformation structures formed by ice push are rare. Very fine- to fine-grained periglacial sediments are widespread, occur mainly in the eastern half of the Dutch sector and were deposited during the Early, Middle and Late Weichselian. Locally, interstadial peat has been sampled and dated by radiocarbon. Various ages, including $45.09 + 3.75/-2.55$ ^{14}C ka BP, 11.28 ± 0.04 ^{14}C ka BP and 10.945 ± 0.05 ^{14}C ka BP have been recorded.

20.4.2. Weichselian Glaciation in the Dutch Sector of the North Sea

Twice during the Weichselian, glaciers extended over large areas of Great Britain. The early 'Ferder' glaciation took place during the Early Weichselian probably around 70.0 ^{14}C ka BP (Carr, 1998; Carr et al., 2006). No evidence for this Early Weichselian glaciation has, however, been found so far in the southern North Sea (Balson and Jeffery, 1991; Cameron et al., 1992). Evidence for an Early Weichselian Scandinavian Ice Sheet in north-western Europe is still a subject of debate (Houmark-Nielsen, 1987), while there is good evidence of mid-Weichselian advances (Houmark-Nielsen, 2010). Recent investigations in the northern North Sea by Sejrup et al. (2000) indicate that the maximum Weichselian glaciation in the northern North Sea took place between 29.4 and ca. 22 ^{14}C ka BP with a coalescing British and Scandinavian Ice Sheet. According to Sejrup et al. (2000), during a second advance of the Weichselian Ice Sheet between 18.5 and 15.1 ^{14}C ka BP, the British and Scandinavian sheets did not coalesce.

During the Weichselian glacial maximum, as well as at ~ 18 ^{14}C ka BP, sea level dropped substantially, probably to more than 130 m below present level. According to Oele and Schüttenhelm (1979), in the Dutch sector of the North Sea, sea level was more than 50 m below the present level during the Early Weichselian. This is based on freshwater deposits sampled at depths up to 48 m below MSL. The freshwater deposits overlie open marine to brackish marine Eemian sediments.

20.4.3. Maximum Extent of the Weichselian Ice Sheet in the Southern North Sea

It is widely accepted that the maximum glaciation occurred around 18 ^{14}C ka BP (Jardine, 1979). In the past, many authors attempted reconstruction of the maximum extent of the Weichselian Ice Sheets in the North Sea, predominantly by using evidence of the gravels exposed on the sea bed. All authors suggest a broadly similar location of the limits (15 reconstructions are presented in Laban, 1995). Some assume a connection between the Scandinavian and British Ice Sheets, whereas others consider such a connection never existed. According to Sejrup et al. (2000), during a second advance of the Weichselian Ice Sheet between 18.5 and 15.1 ^{14}C ka BP, the British and Scandinavian Ice Sheets did not coalesce.

The revised map (Fig. 20.5) shows the extent of the British Ice Sheet in the Dutch sector, based on the occurrence of tills, deformation structures and subglacial valleys as mentioned above. The arrows indicate the suggested directions of ice flows.

20.5. CONCLUSIONS

- Circumstantial evidence indicates that during the Early and part of the Middle Pleistocene, glaciations occurred in northern Europe, but ice did not reach the Netherlands.
- The Weerdinge Beds of Cromerian age suggest that ice may have reached the NE part of the Netherlands.
- During the Elsterian coalescing, British and Scandinavian ice covered the northern part of the Netherlands and the southern North Sea.

FIGURE 20.5 Maximum extent of the Weichselian British Ice Sheet in the Dutch sector of the North Sea (from Laban, 1995).

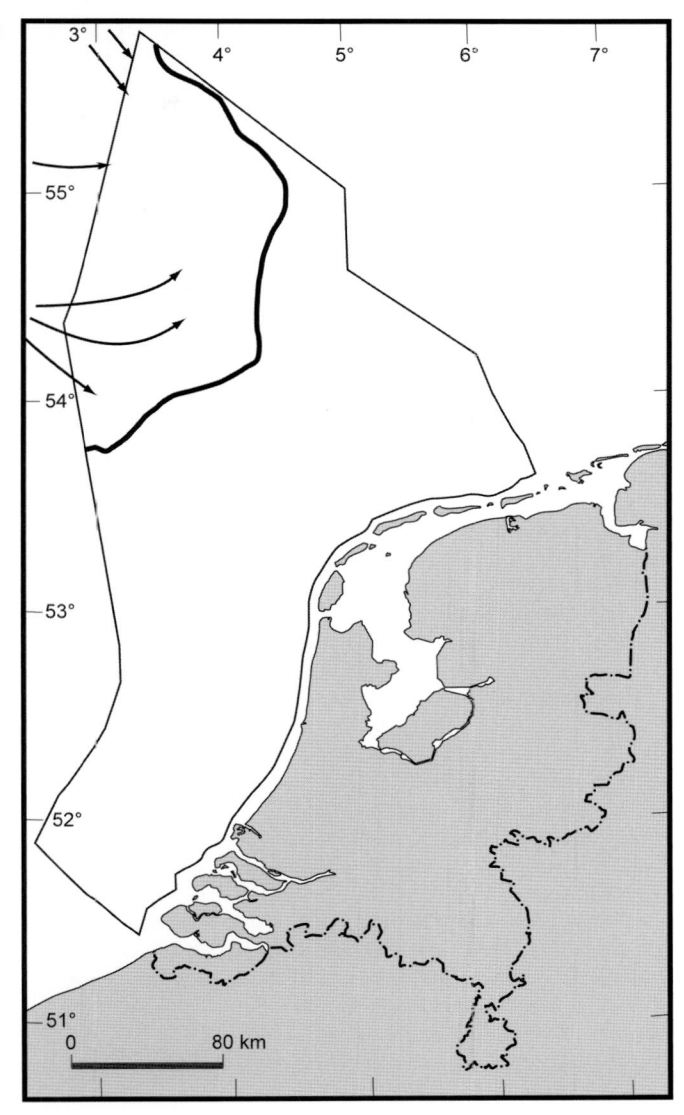

- Reconstructions of the catastrophic formation of the Channel indicate the presence of a large ice-dammed lake in the southern North Sea. However, the geological evidence of such a lake is still lacking.
- During the Saalian, the most extensive glaciation of the Netherlands occurred, giving the country its current structure.
- Events S of the Channel suggest that a smaller ice-dammed lake existed in the southern North Sea, implying coalescing British–Scandinavian ice. However, the geological evidence indicates that this coalescence must have occurred north of the Dutch sector of the North Sea.
- During the Weichselian, British ice reached the Netherlands but only the western part of the Dutch sector of the North Sea.

ACKNOWLEDGEMENTS

The authors would like to thank colleagues for providing relevant, unpublished information as well as Ed Oliver at Queen Mary for swift production of figures.

REFERENCES

Bakker, M.A.J., 2004. The internal structure of Pleistocene push moraines. A multidisciplinary approach with emphasis on ground-penetrating radar. PhD Thesis Queen Mary, University of London, TNO-Geological Survey of the Netherlands, 180pp.

Balson, P.S., Jeffery, D.H., 1991. The glacial sequence of the southern North Sea. In: Ehlers, J., Gibbard, P.L., Rose, J. (Eds.), Glacial Deposits in Great Britain and Ireland. Balkema, Rotterdam/Brookfield, pp. 245–253.

Beets, D.J., Meijer, T., Beets, C.J., Cleveringa, P., Laban, C., van der Spek, A.J.F., 2005. Evidence for a Middle Pleistocene glaciation of MIS 8 age in the southern North Sea. Quatern. Int. 133 (4), 7–19.

Beets, C.J., Beets, D.J., 2003. A high resolution stable isotope record of the penultimate deglaciation in lake sediments below the city of Amsterdam, The Netherlands. Quatern. Sci. Rev. 22, 195–207.

van den Berg, M.W., Beets, D.J., 1987. Saalian glacial deposits and morphology in the Netherlands. In: van der Meer, J.J.M. (Ed.), Tills and Glaciotectonics. Balkema, Rotterdam, pp. 235–251.

Bosch, J.H.A., 1990. Landijs, zee en rivieren als geologische "opbouwwerkers" van het Noorden. Grondboor Hamer 44, 90–94.

Burger, A.W., 1992. Zware mineralenonderzoek aan boring F8-6 in de Noordzee. Internal report 958. Geological Survey of the Netherlands.

Busschers, F.S., van Balen, R.T., Cohen, K.M., Kasse, C., Weerts, H.J.T., Wallinga, J., et al., 2008. Response of the Rhine-Meuse system to Saalian ice-sheet dynamics. Boreas 37, 377–398.

Cameron, T.D.J., Laban, C., Schüttenhelm, R.T.E., 1984. Flemish Bight: sheet 52°N/02°E. Quaternary Geology, 1:250 000 series British Geological Survey and Geological Survey of the Netherlands.

Cameron, T.D.J., Laban, C., Schüttenhelm, R.T.E., 1986. Indefatigable: Sheet 53°N/02°E Quaternary Geology, 1:250 000 series British Geological Survey and Geological Survey of the Netherlands.

Cameron, T.D.J., Crosby, A., Balson, P.S., Jeffery, D.H., Lott, G.K., Bulat, J., et al., 1992. United Kingdom Offshore Regional Report: The Geology of the Southern North Sea. HMSO for the British Geological Survey, London, 150pp.

Carr, S., 1998. The last glaciation in the North Sea Basin. Unpublished Ph.D. Thesis, Royal Holloway, University of London.

Carr, S.J., Holmes, R., van der Meer, J.J.M., Rose, J., 2006. The last glacial maximum in the North Sea Basin. J. Quatern. Sci. 21, 131–153.

Cohen, K.M., Gibbard, P.L., Busschers, F.S., 2008. Middle Pleistocene ice lake high stands in the North Sea: how do they change regional stratigraphical frameworks? In: INQUA-SEQS 2008 Conference.

Dijkmans, J., 1981. Grind- en zware mineralenanalyse aan grindvoorkomens uit het Botney Cutgebied. Unpublished MSc Thesis, Fysisch Geografisch en Bodemkundig Laboratorium, University of Amsterdam.

Ehlers, J., 1990. Untersuchungen zur Morphodynamik der Vereisungen Norddeutschlands unter Berücksichtigung benachbarter Gebiete. Bremer Beitr. Geogr. Raumpl. 19, 166pp.

de Gans, W., de Groot, Th., Zwaan, H., 1987. The Amsterdam basin, a case study of a glacial basin in the Netherlands. In: van der Meer, J.J.M. (Ed.), Tills and Glaciotectonics. Balkema, Rotterdam, pp. 205–216.

Fairbridge, R.W., 1961. Eustatic changes in sea level. Physics and Chemistry of the Earth 4, 99–185.

Gibbard, P.L., 1988. The history of the great northwest European rivers during the past three million years. Philos. Trans. R. Soc. Lond. 318B, 559–602.

Gibbard, P.L., 1995. Formation of the Strait of Dover. In: Preece, R.C. (Ed.), Island Britain a Quaternary perspective, Geological Society Special Publication, 96 15–26.

Gibbard, P.L., Pasanen, A.H., West, R.G., Lunkka, J.P., Boreham, S., Cohen, K.M., et al., 2009. Late Middle Pleistocene glaciation in East Anglia, England. Boreas 38, 504–528.

Gupta, S., Collier, J.S., Palmer-Felgate, A., Potter, G., 2007. Catastrophic flooding origin of shelf valley systems in the English Channel. Nature 448, 342–345.

Hooke, R. LeB, 1998. Principles of Glacier Mechanics. Prentice Hall, Upper Saddle River.

Houmark-Nielsen, M., 1987. Pleistocene stratigraphy and glacial history of the central part of Denmark. Bull. Geol. Soc. Denmark 36, 1–189.

Houmark-Nielsen, M., 2010. Extent, ice and dynamics of Marine Isotope Stage 3 glaciations in the southwestern Baltic Basin. Boreas 39, 343–359.

Jeffery, D.H., Frantsen, P., Laban, C., Schüttenhelm, R.T.E., 1989. Silver Well: sheet 54°N/02°E. Quaternary Geology, 1:250,000 series British Geological Survey and Geological Survey of the Netherlands.

de Jong, J.D., Maarleveld, G.C., 1983. The glacial history of the Netherlands. In: Ehlers, J. (Ed.), Glacial Deposits in North-West Europe. Balkema, Rotterdam, pp. 353–356.

de Jong, J., 1991. Palynological investigation of the borehole 89/2 (EEC). Internal report 1112. Geological Survey of the Netherlands.

Jardine, W.G., 1979. The western (United Kingdom) shore of the North Sea in Pleistocene and Holocene times. In: Oele, E., Schüttenhelm, R.T.E., Wiggers, A.J. (Eds.), The Quaternary history of the North Sea. Acta Universitatis Upsalensis, Uppsala, pp. 159–174.

Jelgersma, S., Breeuwer, J.B. 1975. Toelichting bij de geologische overzichtsprofielen van Nederland. In: Zagwijn, W.H., van Staalduinen, C.J. (Eds.), Geologische overzichtskaarten van Nederland. Rijks Geologische Dienst, Haarlem, 91–93.

Joon, B., 1987. The Tertiary history of the Terschelling Basin. Internal report KSEPL, (Unpublished).

Joon, B., Laban, C., van der Meer, J.J.M., 1990. The Saalian glaciation in the Dutch part of the North Sea. Geol. Mijnbouw 69, 151–158.

Kjaer, K., Krüger, J., van der Meer, J.J.M., 2003. What causes till thickness to change over distance? Answers from Mýrdallsjökull, Iceland. Quatern. Sci. Rev. 22, 1687–1700.

Kluiving, S., Rappol, M., van der Wateren, D., 1991. Till stratigraphy and ice movements in eastern Overijssel, The Netherlands. Boreas 20, 193–205.

Kuster, H., Meyer, K.D., 1979. Glaziäre Rinnen im mittleren und nordöstlichen Niedersachsen. Eisz. Gegenw. 29, 135–156.

Laban, C., 1995. The Pleistocene glaciations in the Dutch sector of the North Sea. A synthesis of sedimentary and seismic data. Ph.D. Thesis, University of Amsterdam, 194pp.

Lagerlund, E., 1987. An alternative Weichselian glaciation model, with special reference to the glacial history of Skåne, south Sweden. Boreas 16, 433–450.

Lee, J.R., Busschers, F.S., Sejrup, H.P., 2011. Pre-Weichselian Quaternary glaciations of the British Isles, The Netherlands, Norway and adjacent marine areas south of 68°N: implications for long-term ice sheet development in northern Europe. Quatern. Sci. Rev. doi:10 1016/j.quascirev.2010.02.027.

Maarleveld, G.C., 1960. Les phénomènes périglaciaires au Pleistocène Ancien et Moyen aux Pays-Bas. Biul. Perygl. 9, 135–141.

Maarleveld, G.C., 1981. The sequence of ice-pushing in the central Netherlands. In: Ehlers, J., Zandstra, J.G. (Eds.), Glacigenic Deposits in the Southwest Part of the Scandinavian Icesheet, Mededelingen Rijks Geologische Dienst vol. 34-1/11. 2–6.

Maher, B.A., Hallam, D.F., 2005. Palaeomagnetic correlation and dating of Plio/Pleistocene sediments at the southern margins of the North Sea Basin. J. Quatern. Sci. 20, 67–77.

van der Meer, J.J.M., 1987. Field trip 'Tills and end moraines in the Netherlands and NW Germany'. In: van der Meer, J.J.M. (Ed.), Tills and Glaciotectonics. Balkema, Rotterdam, pp. 261–268.

van der Meer, J.J.M., Laban, C., 1990. Micromorphology of some North Sea till samples, a pilot study. J. Quatern. Sci. 5, 95–101.

van der Meer, J.J.M., Lagerlund, E., 1991. First report of a preserved Weichselian periglacial surface in NW-Europe—the "van der Lijn," geological reserve in the Netherlands. Eisz. Gegenw. 41, 100–106.

Meijer, T., Cleveringa, P., 2009. Aminostratigraphy of Middle and Late Pleistocene deposits in the Netherlands and the southern part of the North Sea basin. Global Planet. Change 68, 326–345.

de Mulder, E.F.J., Geluk, M.C., Ritsema, I., Westerhoff, W.E., Wong, Th. E., 2003. De ondergrond van Nederland. Nederlands Instituut voor Toegepaste Geowetenschappen TNO, Utrecht.

Neele, N.G., 1986. Micropaleontologisch onderzoek aan het traject 0.00–151.13 m van de boring El-10 en correlatie met de boringen E8-4 (Rapport 1516) en E2-3 (rapport 1495). Internal report 1517. Geological Survey of the Netherlands.

Neele, N.G., 1991a. Verslag van micropaleontologisch onderzoek aan de boring Noordzee F8-89/006, traject 0.00–195.10 m onder zeebodem. Internal report 1620. Geological Survey of the Netherlands.

Neele, N.G., 1991b. Micropaleontological investigation of borehole North Sea 89/2 (A). Internal report 1614. Geological Survey of the Netherlands.

Oele, E., Schüttenhelm, R.T.E., 1979. Development of the North Sea after the Saalian glaciation. In: Oele, E., Schüttenhelm, R.T.E., Wiggers, A.J. (Eds.), The Quaternary History of the North Sea. Acta Universitatis Upsaliensis, Uppsala, pp. 191–216.

Passchier, S., Laban, C., Mesdag, C.S., Rijsdijk, K.F., 2010. Subglacial bed conditions during Late Pleistocene glaciations and their impact on ice dynamics in the southern North Sea. Boreas 39, 663–697.

Postma, G., Roep, Th.B., Ruegg, G.H.J., 1983. Sandy-gravelly mass-flow deposits in an ice-marginal lake (Saalian, Leuvenumse Beek Valley, Veluwe, The Netherlands) with emphasis on plug-flow deposits. Sediment. Geol. 34, 59–82.

Postma, G., 1997. De Leuvenumsche Beek-vallei. Een voormalig door landijs gevormd meer met zandige massatransport-afzettingen. Grondboor Hamer 51, 117–124.

Pouwer, R., 1991. Molluskenonderzoek van Noordzeeboring F8/006. Internal report 1582, Geological Survey of the Netherlands.

Praeg, D., 1994. Quaternary seismic morphology and stratigraphy—glacial and fluvial channels in the southern North Sea. In: 2nd International Conference on the Geology of Siliciclastic Shelf Seas 24–28 May, Gent, 96.

Rappol, M., 1983. Glacigenic properties of till. Studies in glacial sedimentology from the Allgäu Alps and the Netherlands. PhD Thesis, University of Amsterdam, 225pp.

Rappol, M., 1984. Till in southeast Drente and the origin of the Hondsrug Complex, The Netherlands. Eisz. Gegenw. 34, 7–27.

Rappol, M., 1987. Saalian till in the Netherlands: a review. In: van der Meer, J.J.M. (Ed.), Tills and Glaciotectonics. Balkema, Rotterdam/Boston, pp. 3–22.

Rappol, M., Haldorsen, S., Jørgensen, P., van der Meer, J.J.M., Stoltenberg, H.M.P., 1989. Composition and origin of petrographically-stratified thick till in the Northern Netherlands and a Saalian glaciation model for the North Sea basin. Meded. WTKG. 26, 31–64.

Roep, Th.B., Holst, H., Vissers, R.L.M., Pagnier, H., Postma, D., 1975. Deposits of southwardflowing, Pleistocene rivers in the Channel Region, near Wissant, NW France. Palaeogeogr. Palaeoclimatol. Palaeoecol. 17, 289–308.

Ruegg, G.H.J., 1975. Sedimentary structures and depositional environments of Middle and Upper Pleistocene glacial time deposits from an excavation at Peelo, The Netherlands. In: Mededelingen Rijks Geologische Dienst, N.S., vol. 26, pp. 17–37.

Ruegg, G.H.J., 1977. Features of middle Pleistocene sandur deposits in the Netherlands. Geol. Mijnbouw 56, 5–24.

Ruegg, G.H.J., Zandstra, J.G., 1977. Pliozäne und pleistozäne gestauchte Ablagerungen bei Emmerschans (Drenthe, Niederlande). In: Mededelingen Rijks Geologische Dienst, N.S., vol. 28, pp. 65–99.

Sejrup, H.P., Larsen, E., Landvik, J.H., King, E.L., Haflidason, H., Nesje, A., 2000. Quaternary glaciations in southern Fennoscandia: evidence from southwestern Norway and the northern North Sea region. Quatern. Sci. Rev. 19, 667–685.

van Straaten, L.M.J.U., 1956. Structural features of the "Papzand" formation at Tegelen (Netherlands). Geol. Mijnbouw 18, 416–420.

Toucanne, S., Zaragosi, S., Bourillet, J.F., Cremer, M., Eynaud, F., van Vliet-Lanoe, B., et al., 2009a. Timing of massive 'Fleuve Manche' discharges over the last 350 kyr: insights into the European ice-sheet oscillations and the European drainage network from MIS 10 to 2. Quatern. Sci. Rev. 28, 1238–1256.

Toucanne, S., Zaragosi, S., Bourillet, J.F., Gibbard, P.L., Eynaud, F., Giraudeau, J., et al., 2009b. A 1.2 Ma record of glaciation and fluvial discharge from the West European Atlantic margin. Quatern. Sci. Rev. 28, 2974–2981.

Veenstra, H.J., 1969. Gravels of the southern North Sea. Marine Geol. 7, 449–464.

Valentin, H., 1954. Die Grenze der letzten Vereisung im Nordseeraum. Verhandl. Deut. Geogr. Tag Hamburg 30, 359–366.

van der Meer, J.J.M. 1992 Micromorphology of Pleistocene sediments from the southern North Sea: boreholes BH 89.2 and BH 89.3. Report Fysische Geografisch en Bodemkundig Laboratorium, UvA.

van der Wateren, F.M., 1981. Glacial tectonics at the Kwintelooijen sandpit, Rhenen, The Netherlands. In: Ruegg, G.H.J., Zandstra, J.G. (Eds.), Geology and Archeology of Pleistocene Deposits in the Icepushed Ridge near Rhenen and Veenendaal, Mededelingen Rijks Geologische Dienst, vol. 35, 252–268.

van der Wateren, F.M., 1985. A model of glacial tectonics, applied to the ice-pushed ridges in the Central Netherlands. Bull. Geol. Soc. Denmark 34, 55–74.

van der Wateren, F.M., 1992. Structural geology and sedimentology of push moraines. PhD Thesis, University of Amsterdam, 230pp.

ter Wee, M.W., 1962. The Saalian glaciation in the Netherlands. In: Mededelingen Geologische Stichting, N.S., vol. 15, pp. 57–76.

ter Wee, M.W., 1983a. The Elsterian Glaciation in the Netherlands. In: Ehlers, J. (Ed.), Glacial Deposits in North-West Europe. Balkema, Rotterdam, pp. 413–415.

ter Wee, M.W., 1983b. The Saalian glaciation in the northern Netherlands. In: Ehlers, J. (Ed.), Glacial Deposits in North-West Europe. Balkema, Rotterdam, pp. 405–412.

Westaway, R., 2010. Implications of recent research for the timing and extent of Saalian glaciation of eastern and central England. Quatern. Newslett. 121, 3–23.

Zagwijn, W.H., 1986. Preliminary interpretation and summary of the micro- and macropalaeontological and pollen analysis of the boreholes El-10 and E8-4. Internal report. Geological Survey of the Netherlands.

Zagwijn, W.H., 1961. Vegetation, climate and radiocarbon datings in the Pleistocene of the Netherlands. Part 1: Eemian and Early Weichselian. In: Mededelingen Geologische Stichting, N.S. vol. 14, pp. 15–45.

Zagwijn, W.H., 1973. Pollenanalytical studies of Holsteinian and Saalian beds in the northern Netherlands. In: Mededelingen Rijks Geologische Dienst N.S., vol. 24, pp. 139–156.

Zagwijn, W.H., 1977. Stratigrafische interpretatie van boringen tot circa 100 m onder zeeniveau in het Noordzeegebied (blokken P13, P5, J14, K17, K14, K15, L13, LIO, L7). Internal report 769. Geological Survey of the Netherlands.

Zandstra, J.G., 1983. Fine gravel, heavy mineral and grain size analyses of Pleistocene, mainly glacigenic, deposits in the Netherlands. In: Ehlers, J. (Ed.), Glacial Deposits in North-West Europe. Balkema, Rotterdam, pp. 361–377.

Zandstra, J.G., 1986. Korrelgrootte-onderzoek van glacigene zandige klei op Wieringen. Internal report 865. Geological Survey of the Netherlands.

Zonneveld, J.I.S., 1975. Zijn de noordnederlandse stuwwallen overreden of niet? In: Berichten Fysisch Geografische Afdeling, Rijks Universiteit Utrecht, vol. 9, pp. 3–14.

Chapter 21

The Pleistocene Glaciations of the North Sea Basin

Alastair G.C. Graham[1,*], Martyn S. Stoker[2], Lidia Lonergan[3], Tom Bradwell[2] and Margaret A. Stewart[3,†]

[1] Ice Sheets Programme, British Antarctic Survey, High Cross, Madingley Road, Cambridge CB3 0ET, United Kingdom
[2] British Geological Survey, Murchison House, West Mains Road, Edinburgh EH9 3LA, United Kingdom
[3] Department of Earth Science and Engineering, Imperial College London, South Kensington Campus, London SW7 2AZ, United Kingdom
*Correspondence and requests for materials should be addressed to Alastair G.C. Graham. E-mail: alah@bas.ac.uk
[†] Present address: Neftex Petroleum Consultants, 97 Milton Park, Abingdon, Oxfordshire OX14 4RY, United Kingdom.

21.1. INTRODUCTION

The North Sea has had a long and complex geological history, with its present-day structural configuration being largely the result of Late Jurassic–Early Cretaceous rifting, followed by thermal cooling and subsidence (Glennie and Underhill, 1998; Zanella and Coward, 2003). During the Cenozoic, the basin was gently deformed by tectonic inversion and basin-margin uplift driven by intraplate compression, resulting from the interplay between the opening of the northeast Atlantic Ocean and Alpine orogeny. Since the Middle Cenozoic, up to 3000 m of Oligocene to Holocene sediment has accumulated in the central graben region, including, locally, in excess of 800 m of Quaternary sediment (Caston, 1977; Gatliff et al., 1994; Fig. 21.1A). At the present day, the North Sea forms a shallow, epicontinental shelf that is mostly <100 m deep but increases to 200–400 m water depth towards the shelf edge, along its northern margin and in the Norwegian Channel (Fig. 21.1B).

Ice sheets are known to have transgressed into the North Sea at several key stages of the Quaternary, contributing to the episodic erosion and infill of the basin. Traditional models of North Sea Pleistocene glaciations suggest three major glacial episodes during the past 500 ka, known locally, and recorded sequentially, as the Elsterian Stage (Marine Isotope Stage [MIS] 12), Saalian Stage (MIS 10–6) and Weichselian Stage (MIS 5d–2) glaciations. Discrete sets of tunnel valleys have been used as the main criterion for this threefold subdivision, delimiting the broad (submarginal) extents of ice sheets during each of the dominant glaciations in the North Sea (Cameron et al., 1987; Wingfield, 1989, 1990; Ehlers and Wingfield, 1991; Praeg, 2003; Fig. 21.1B). However, this simple three-stage model has come under considerable scrutiny in recent years, and there is now growing evidence, which we will review in this chapter, that many more glacial episodes are preserved in the North Sea sedimentary sequence (Lonergan et al., 2006; Graham, 2007; Stewart, 2009; Stewart and Lonergan, 2011). Nevertheless, in a recent study of the link between glaciation and fluvial discharge from the West European Atlantic margin, Toucanne et al. (2009) have demonstrated that maximum fluvial discharge rates of the Fleuve Manche palaeo-river (through the English Channel) are associated with the Elsterian, Saalian and Weichselian glacial episodes. This implies that North Sea ice sheets were indeed at their maximum extent during these stages.

The recognition of the North Sea as a key 'archive' of information on Quaternary glacial activity has become increasingly apparent in recent times, in conjunction with a mounting interest in the role of ice sheets on the northwest European continental shelves. In particular, a number of the Middle-to-Late Pleistocene ice sheets are now known to have had terminal extents on or at the margins of these shelf regions (Stoker et al., 1993; Sejrup et al., 2005) and were commonly marine based. Thus, the North Sea is likely to preserve significant evidence for former continental glaciation in the bordering regions.

In addition, the basin was an important pathway for large-scale glacial transport to the deeper ocean, as shown by the presence of large glacigenic accumulations (glacial debris fans) on the northwest European continental margin (Stoker et al., 1993, 1994; Stoker, 1995; Sejrup et al., 2005; Bradwell et al., 2008). Ice streams, comparable with those that drain the majority of ice from modern-day Greenland and Antarctica, are known to have fed these fans and were probably a key feature of the North Sea ice sheets

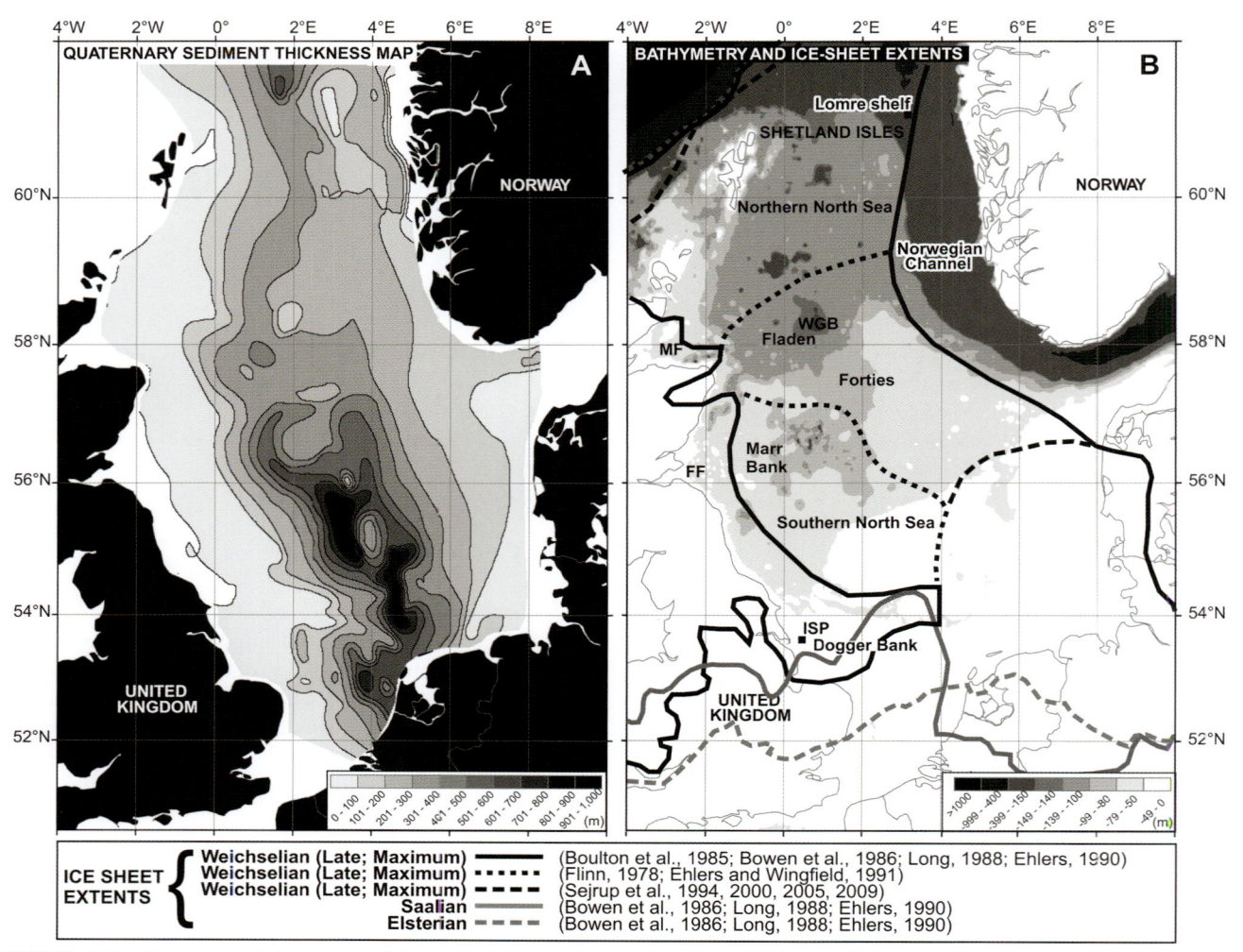

FIGURE 21.1 (A) Isopach (thickness) map of Quaternary sediments in the North Sea basin, derived from interpreted 2D seismic datasets (after Caston, 1977). (B) Bathymetry of the North Sea basin, and Quaternary ice-sheet extents for each of the three major Middle-to-Late Pleistocene northwest European glaciations. Key sites are also shown. WGB, Witch Ground Basin; FF, Firth of Forth; MF, Moray Firth; ISP, Inner Silver Pit.

(Stoker et al., 1993; Graham, 2007). As a result, the North Sea basin is also likely to be an important site for understanding the discharge and stability of the major northern European palaeo-ice masses, including the British and Fennoscandian Ice Sheets (BIS and FIS).

This chapter reviews the evidence for extents and timings of glaciations in the North Sea basin through the Pleistocene, from ~2.58 Ma to the present. We draw upon an extensive body of existing literature, from the work of Valentin (1957), through to state-of-the-art marine geological survey and analytical techniques. The main objectives of the chapter are to (1) review information related to North Sea Quaternary glaciations, in terms of ice-sheet limits, configuration and chronology, and (2) link this information, for the first time, to a geomorphological framework. For the latter, new observations of glacial landforms on commercial three-dimensional (3D) seismic reflection data and single-beam sea-floor mapping comprise novel lines of evidence. Our review highlights that the North Sea has a complex glaciated history, which is likely to have been affected by a range of glacial environments throughout the Pleistocene. As this is a regional review of the North Sea basin, the existing British Geological Survey (BGS) lithostratigraphical nomenclature for the Quaternary units is used (cf. Stoker et al., 2010). As direct correlation between the Quaternary continental sedimentary record and the deep-ocean oxygen isotope sequence (MISs) remains to be fully substantiated, we primarily correlate our glacial events to the northwest European stage nomenclature (Fig. 21.2). It should be noted, however, that the past 2.6 Ma of the continental Quaternary record has recently been correlated to the marine isotope stratigraphy (Gibbard and Cohen, 2008; Toucanne et al., 2009), and we note such correlations where appropriate.

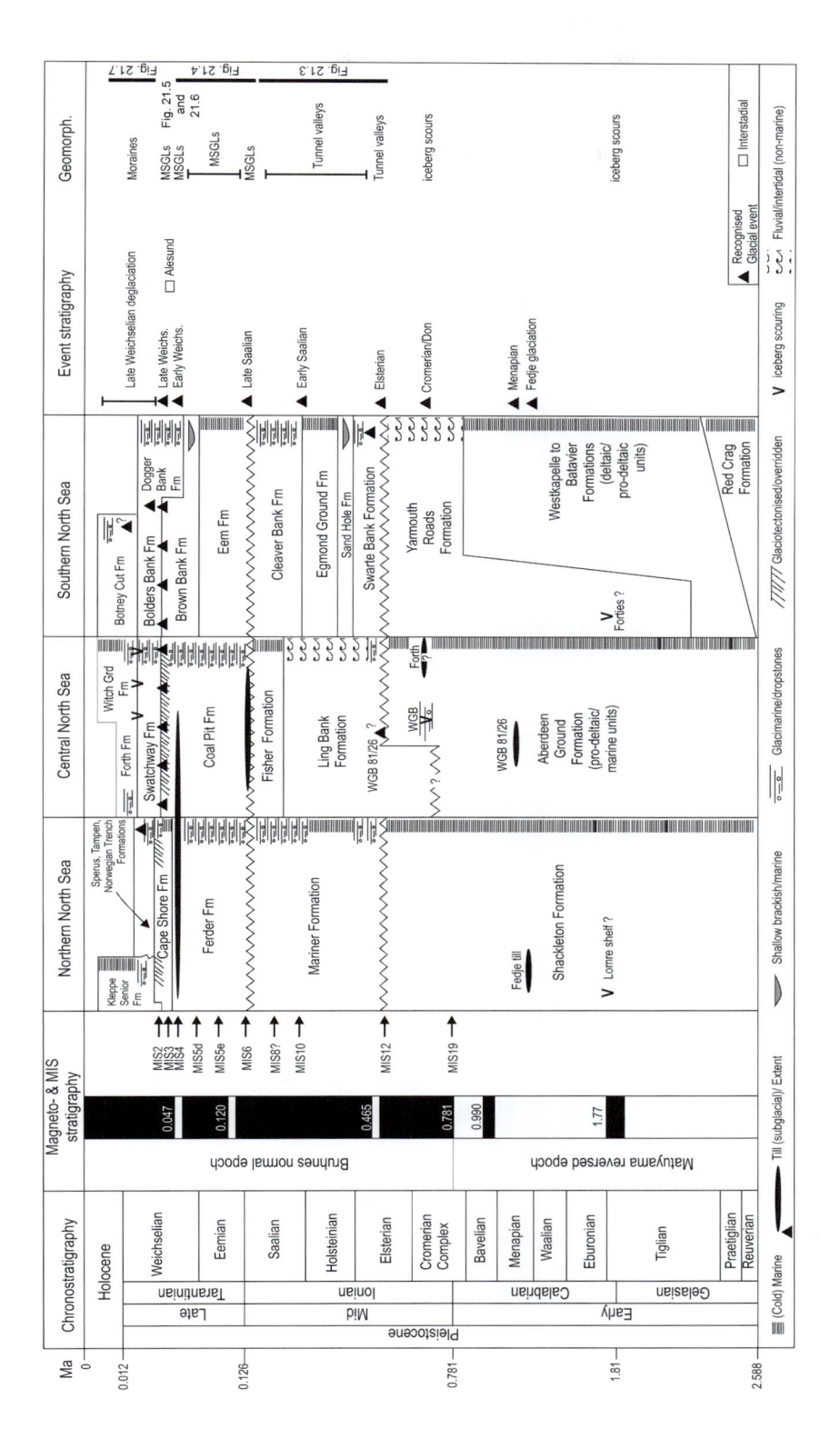

FIGURE 21.2 Summary panel outlining the Quaternary framework for the North Sea basin (after Stoker et al., 1985; Cameron et al., 1987, 1992; Johnson et al., 1993; Gatliff et al., 1994; Scourse et al., 1998; Stoker et al., 2010), showing geomorphological observations, key glacial event stratigraphy, inferred ages and correlation to the regional stratigraphical nomenclature. Palaeomagnetic events (Ma): 0.047, Laschamp; 0.120, Blake; 0.465, Emperor; 0.990, Jaramillo; 1.77, Olduvai.

21.1.1. Evidence for Glaciation: Techniques and Methods

Evidence for glaciations in the North Sea basin, including their extents and timings, were inferred in the past from two main data sources: (1) rotary-drilled sedimentary boreholes and (2) comprehensive networks of two-dimensional (2D) marine reflection seismic surveys. For the most part, these datasets were largely acquired by the BGS, the Dutch Geological Survey and the University of Bergen, Norway, during the 1970s and 1980s (e.g. Sejrup et al., 1987). Traditionally, two types of information were used for interpreting glacial depositional environments from these sources: (1) sedimentary data, including the identification and characterisation of glacigenic sediments (e.g. tills) in association with seismostratigraphical analysis and (2) landform data, specifically the mapping of glacial bedforms (e.g. meltwater valleys and ice-keel ploughmarks).

Many early studies focused primarily on the sedimentary and seismic stratigraphical information. Palaeoenvironmental interpretations were constrained using conventional biostratigraphical, chronostratigraphical and lithostratigraphical tools (Stoker et al., 1983, 1985; Long et al., 1986; Sejrup et al., 1987), and seismic datasets were interpreted in terms of their broad acoustic facies and interrelationships. Together with the core data, these interpretations led to the establishment of a formal seismic stratigraphical nomenclature for the U.K. continental shelf, which remains a key foundation for studies of Quaternary depositional history today (e.g. Stoker et al., 1985).

However, there were several problems with these early investigations: (1) in many cases, sediment sequences were often poorly recovered, giving fragmentary geological insights; (2) the genesis of the sediments was not always clear from the cores alone, and detailed sedimentology was sometimes lacking (e.g. provenance analyses); (3) dating was not well constrained for much of the sequence; (4) interpretation of the high-frequency Boomer and Sparker seismic data was limited by its shallow depth penetration and its 2D nature; and (5) the geomorphological context for the features observed (e.g. landforms) was not easily identifiable except where side-scan data were also available, as in Stoker and Long (1984).

3D seismic reflection data have been acquired for North Sea hydrocarbon exploration for over 30 years, and the application of these types of dataset to understanding former glacial activity from shallow Pleistocene successions has developed significantly over the past decade (e.g. Praeg, 2003; Andreassen et al., 2004; Rise et al., 2004; Lonergan et al., 2006; Kristensen et al., 2007; Lutz et al., 2009). Several recent, in-depth 3D seismic studies have been used to revise understanding of parts of the Quaternary framework of the North Sea basin (Graham, 2007; Stewart, 2009) and to address some of the problems outlined above. Merged commercial datasets (e.g. Petroleum Geo-Services mega-survey) supplemented by higher-resolution commercial 3D seismic volumes provide a basis for updating some of the information reviewed herein. High-resolution 2D seismic datasets and single-beam (fisheries-sourced) seafloor bathymetric compilations have also been utilised to improve understanding of more recent glacial activity (e.g. Bradwell et al., 2008; Sejrup et al., 2009) and to capture geological features beneath the resolution of 3D datasets.

Recent sedimentological studies have also begun to take advantage of higher-precision AMS radiocarbon dating of carbonate material, leading to improved chronological control over the Late Quaternary sequence (Sejrup et al., 2009; Graham et al., 2010). Where problems with the genetic interpretation of sediments once stood, the use of micromorphological techniques to complement macro-scale studies of cores has become an important tool (Carr, 2004; Carr et al., 2006). In addition, new core material has been collected from the basin in recent years (Sejrup et al., 2009; Graham et al., 2010), and 3D seismic data have afforded consideration of these deposits in a glacial geomorphological context. In view of these recent methodological developments, a review of the Quaternary history of the North Sea basin is timely.

21.2. EARLY PLEISTOCENE GLACIATION(S)

Lower to Middle Pleistocene sediments comprise a large proportion of the Quaternary succession in the North Sea region (Fig. 21.2). Existing studies of this succession are generally limited to discrete evidence for interglacials (e.g. Gibbard, 1991; Zagwijn, 1992; Ekman and Scourse, 1993; Sejrup and Knudsen, 1993) or have targeted the non-glacial sequence of Lower to lower Middle Pleistocene sediments, consisting of deltaic sediments and pro-deltaic bottomsets deposited by rivers emanating from continental Europe (Zagwijn, 1974; Cameron et al., 1987; Stoker and Bent, 1987). The latter was associated with the development of the southern North Sea delta, which has been compared in size to the largest modern delta complexes in the world (Ekman and Scourse, 1993), and was accountable for the majority of non-glacial deposition during the Early to Middle-Pleistocene in the southern and central North Sea (Cameron et al., 1992). This delta probably started developing in the Latest Pliocene–Early Pleistocene and comprises a number of well-defined formations that record its progradation northwards towards the central North Sea where it passes into the pro-deltaic–marine Aberdeen Ground Formation (Gatliff et al., 1994; Fig. 21.2).

To date, the earliest known glaciation of the North Sea basin, based on sedimentary records, is found within the Norwegian Channel (Fig. 21.1B). Here, subglacial diamict lies unconformably upon Oligocene rocks at the base of the channel which has been tentatively assigned a 1.1-Ma age based on Sr-isotope, palaeomagnetic and micropalaeontological data (Sejrup et al., 1995, 2000). Deposition of this till (the 'Fedje' till; Fig. 21.2) was followed by a period of extensive marine deposition, interbedded with glacimarine sediments. A thin interglacial layer found within this sequence, and below the 0.78-Ma Bruhnes-Matuyama palaeomagnetic reversal, provides further evidence to justify the till's ca. 1 million year age (related to the Radøy interglacial in the corresponding literature; Sejrup et al., 1995; see chapter 'Pleistocene Glaciations in Norway' in this volume). North of the Norwegian Channel (Lomre shelf; Fig. 21.1B) time-structure maps of the base-Pleistocene unconformity, mapped from 3D seismic data, also imaged localised buried iceberg scours, with a possible age of 1.7–2.6 Ma (Jackson, 2007). If correct, these features indicate relatively proximal marine ice-sheet margins during a period of glaciation pre-dating the 'Fedje' glaciation, although the sources of the icebergs could be distal to the North Sea itself (e.g. northern Norwegian margin).

Similar coeval records are scarce outside of the Norwegian Channel. However, in BGS borehole 81/27 on the western margin of the central North Sea (Marr Bank; Fig. 21.1), Graham (2007) noted glacigenic sediments consisting of dropstone-rich muddy glacimarine sands overlying the Tertiary rockhead. These muddy sands may give fragmentary evidence for Early Pleistocene glacial activity in the North Sea basin because the deposits occur well below the Bruhnes–Matuyama reversal in the core (Stoker et al., 1983).

All other lines of evidence indicate significantly younger Pleistocene glacial activity in the North Sea region. For example, indirect evidence for glaciation of the basin during the Menapian stage of the Early Pleistocene has been described by Bijlsma (1981) and Gibbard (1988). The former suggested that a proto-Baltic basin was scoured by regional glaciation, which resulted in a paucity of pre-Menapian deposits in the North Sea that bear an eastern European provenance (Carr, 2004).

Sejrup et al. (1987) presented the earliest known sedimentary records of glaciation in the central North Sea itself with evidence for Early Pleistocene subglacial tills in borehole records from the Witch Ground area (Figs. 21.1B and 21.2). In BGS 81/26, a Menapian age of between 800 and 900 ka was suggested for a buried subglacial till (diamicton F) using palaeomagnetic, biostratigraphical and amino acid stratigraphical evidence (Fig. 21.2; Sejrup et al., 1987, 2000). Sejrup et al. (1987) originally used these findings to suggest that British–Fennoscandian ice sheets were extensive in the North Sea during this time. Later investigation of the borehole by Ekman and Scourse (1993) identified a cold-stage pollen assemblage for the sediments that correlate with the Menapian till, and they identified extinct pollen taxa (species of Carya and Ostrya) along with abundant reworked Neogene taxa which prove a pre-Cromerian age. Using 3D seismic data Stewart (2009) found a N-S oriented branching tunnel valley system that extends for some 50 km in the northwestern part of the Witch Ground basin. The valley occurs within the Aberdeen Ground Formation and beneath the 780 ka Bruhnes-Matuyama boundary, well below the Elsterian and younger tunnel valleys that dominate the later Quaternary stratigraphy in this area. This discovery provides further evidence for grounded ice in the North Sea Basin in the Early Pleistocene.

21.3. MIDDLE PLEISTOCENE/PRE-ELSTERIAN GLACIATIONS

Evidence for Middle Pleistocene, pre-Elsterian glaciation was suggested by Stoker and Bent (1985) by the presence of subglacial and glacimarine sediments in cores recovered from Firth of Forth (Figs. 21.1 and 21.2). These deposits were assigned a tentative early Cromerian age based on their stratigraphical position and palaeomagnetic evidence (Fig. 21.2). Terrestrial studies of the neighbouring glacial stratigraphy in Norfolk have argued, more recently, for a probable Cromerian (MIS 16) glaciation (the 'Happisburgh' glaciation), on the basis of subglacial diamictons correlated against well-dated fluvial terrace sequences (Lee et al., 2004). However, recent work in the area, including optically stimulated luminescence dating, and detailed biostratigraphical and aminostratigraphical analyses, suggests that these deposits may be younger than first thought, and most likely relate to the later Elsterian glaciation (of MIS 12, or 'Anglian' in the United Kingdom; Preece et al., 2009).

Nevertheless, supporting evidence for Cromerian glaciation comes from the equivalent central European Donian glaciations (Fig. 21.2). A subglacial diamicton termed the Don till is constrained to a probable MIS 16 age by the presence of Pleistocene mammalian faunal remains and by pollen stratigraphy in discrete beds surrounding the deposit (Velichko et al., 2004). Ice sheets are interpreted to have been extensive across mainland Europe during the Donian and to have reached coastal positions in western Norway (Gibbard, 1988). However, no record of tills is present in the Norwegian Channel during equivalent times (between ~1.1 Ma and 500 ka) suggesting that, if present, the Don glaciation was of relatively limited extent (Sejrup et al., 1995, 2000) and ice did not enter the central North Sea during this period.

Graham (2007) suggested evidence for pre-Elsterian glacial influence on the North Sea succession, based on the study of 3D seismic datasets from the Witch Ground basin, central North Sea. Geomorphological evidence for a proximal ice-sheet limit is present in the form of iceberg ploughmarks which are mapped at 130–170 m depth, within layers of pre-glacial strata corresponding to the Aberdeen Ground Formation (Fig. 21.2). Age constraints on the iceberg scours in this locality, as constrained from palaeomagnetic data from BGS borehole 77/02, indicate that the scours probably formed during the Cromerian. The fact that the scours have also been cross-cut by tunnel valleys of a minimum Elsterian age and younger provides a strong support to their pre-Elsterian age (Graham, 2007). In addition, deposition of ice-rafted erratics, within the Wroxham Crag Formation of northeast Norfolk, provides additional persuasive evidence for early-Middle Pleistocene iceberg activity in the southern North Sea during one or more of the pre-Elsterian cold stages (Larkin et al., 2011).

21.4. THE ELSTERIAN (MIS 12)

21.4.1. Glacial Limits

The Elsterian glaciation—unequivocally correlated to MIS 12 by Gibbard and Cohen (2008) and Toucanne et al. (2009)—was probably the most extensive in the North Sea Pleistocene glacial history (Fig. 21.1B), marking the onset of repeated shelf-edge glaciations on the northwest European margin (Stoker et al., 1993, 1994, 2010; Sejrup et al., 2005), and also a major switch in North Sea sedimentation from non-glacial to predominantly glacial deposition (Cameron et al., 1987; Fig. 21.2). Southern ice limits for the Elsterian glaciation have been mapped onshore based on the presence of extant end moraines and incised tunnel valleys (Anglian 'rinnen'; Fig. 21.3A) which are observed throughout continental Europe and into the United Kingdom. Offshore, the southerly Elsterian limit is associated with morphologically similar buried, subglacial tunnel valleys, glaciotectonic deformation structures and subglacial 'till tongues' (Fig. 21.3A and B; Laban, 1995; Praeg, 2003). For example, Praeg (1996) showed unequivocal evidence that south–north-oriented tunnel valleys are associated with an Elsterian margin in the southwestern North Sea, at ~53°N (Fig. 21.3B). Taken together, the various geomorphological elements serve as good indicators for large, coalescent ice sheets in the North Sea basin at this time (Fig. 21.3). A major consequence of this was the southerly redirection of the European drainage network south of the ice margin, with the Fleuve Manche palaeo-river draining into the Bay of Biscay (Toucanne et al., 2009). For the northern ice sheet margin, sedimentary fans on the Atlantic continental margin record elevated rates of glacial sedimentation during the Elsterian (Stoker et al., 1994; Sejrup et al., 2005), consistent with an ice sheet which reached the shelf break during this stage. Ice in the northeastern North Sea (Norwegian Channel), and further northeast along the Norwegian margin, also reached the shelf break at least once during the Elsterian glaciation (Rise et al., 2004).

21.4.2. Morphological Features

The main morphological evidence for Elsterian glaciation in the North Sea is restricted to subglacial tunnel valleys (Fig. 21.3). In the central and southern North Sea, south of ca. 58°N, separate generations of tunnel valleys are relatively easy to distinguish from each other (see Ehlers and Wingfield, 1991); the oldest generation of valleys having often been related to a southern margin of the Elsterian ice sheet in the North Sea (Huuse and Lykke-Andersen, 2000). The timing of incision of this valley network has been inferred from cross-correlation to onshore stratigraphy and features, in the United Kingdom, the Netherlands and Germany (e.g. Kluving et al., 2003; Lutz et al., 2009), although from a careful review of the literature, it appears that no valleys of presumed Elsterian age have been directly dated. On seismic records, these older valleys are associated with a strong glacial unconformity, which can be traced throughout the North Sea basin (Cameron et al., 1987; Huuse and Lykke-Andersen, 2000; Stoker et al., 2010). This unconformity surface is incised into the underlying southern North Sea deltaic units as well as into the laterally equivalent Aberdeen Ground and Shackleton Formations in the central and northern North Seas, respectively (Fig. 21.2). The unconformity overlying each formation is believed to correlate approximately with the Elsterian glacial stage.

From 3D seismic datasets, Lonergan et al. (2006) have mapped, in detail, the geometry of tunnel valleys in the Witch Ground area of the central North Sea and have proposed a complex polygenetic origin for the larger Elsterian valleys, which they attribute to the action of episodic meltwater erosion. The number and complexity of cross-cutting patterns probably suggest that the ice sheet was actively eroding and re-eroding its bed throughout this stage. Stewart and Lonergan (2011) have mapped over 180 tunnel valleys in the central North Sea from 3D seismic data, identifying seven separate phases of valley incision (Fig. 21.3B). These authors related several generations of deeply buried cross-cutting valleys to the Elsterian (at least two phases between MIS 12 and 10) and proposed that it is unlikely that the complex valley sequences observed formed during just two glacial stages (Elsterian and Saalian). Lutz et al. (2009) report at least three generations of cross-cutting tunnel valleys mapped on 3D seismic data from the German North Sea, which they too infer are of Elsterian age, supporting greater complexity to the Elsterian stage than previously thought. Lonergan et al. (2006) and Stewart (2009) also suggest that, based on the orientation and fill of valleys, it is unlikely that all the valleys are ice marginal. However, the overall

Chapter | 21 The Pleistocene Glaciations of the North Sea Basin

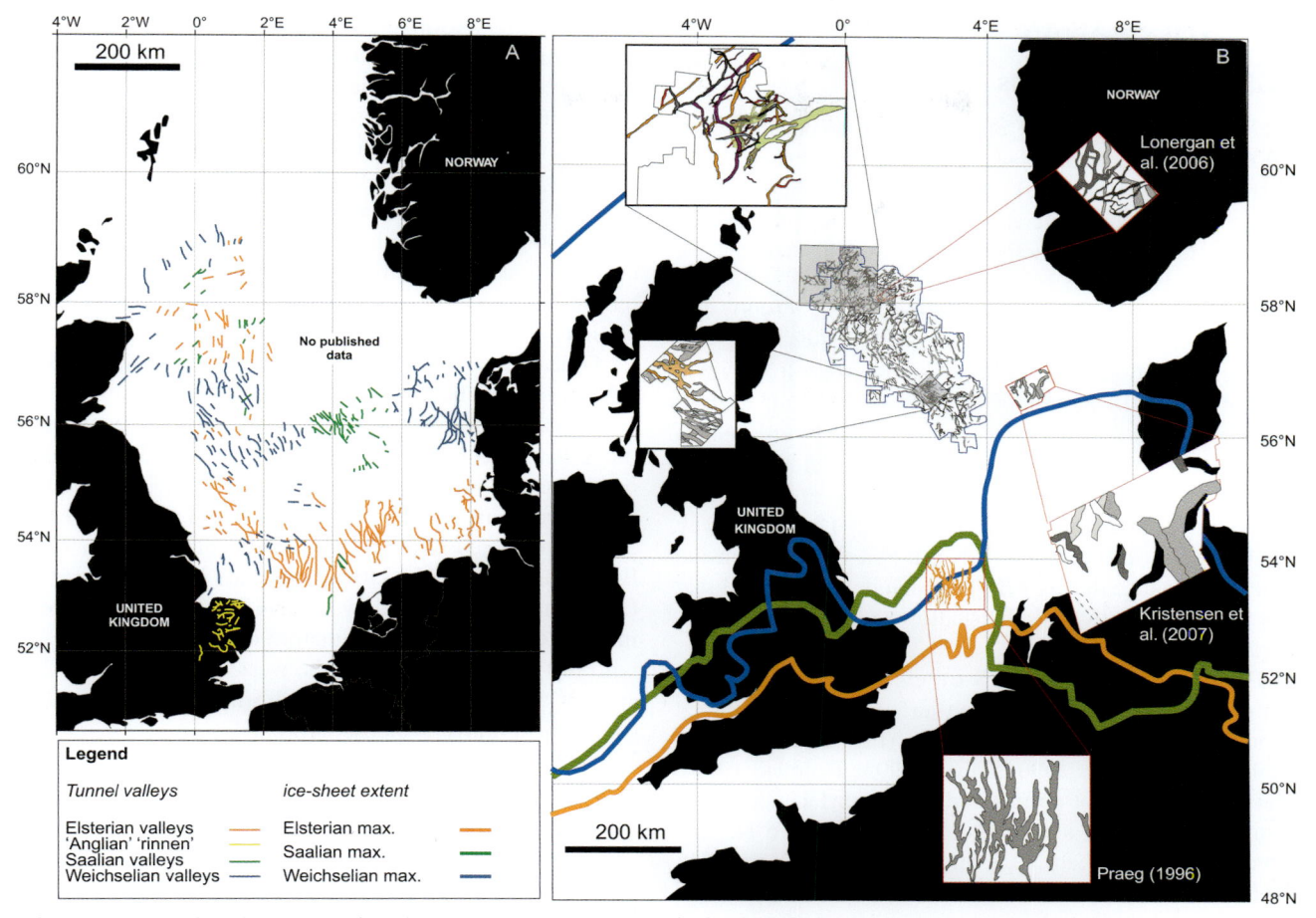

FIGURE 21.3 (A) Compilation map of previously published tunnel valleys in the North Sea basin and their assignment to the major Pleistocene glaciations in the region based on 2D seismic reflection data (modified from Huuse and Lykke-Andersen, 2000). (B) Recently reported distribution of buried tunnel valleys mapped from 3D seismic datasets in the North Sea region (from Stewart, 2009), which updates significantly the older compilation in (A). The new mapping of tunnel valleys in central North Sea (top inset map) illustrates seven generations of tunnel valleys formed from Elsterian to MIS 5e. The most recent tunnel valleys formed during the Weichselian glaciation are not shown on this map.

distribution of valleys (Fig. 21.3) implies that the ice sheet, at its maximum extent, covered the North Sea basin. This is consistent with the southerly deflection of the North Sea fluvial system at this time due to the expansive ice sheet (Toucanne et al., 2009).

21.4.3. Key Sites

Little sedimentary evidence for Elsterian glaciation is forthcoming from the central North Sea (Long, 1988; Carr, 2004), although some upper units of the Aberdeen Ground Formation have been interpreted as an Elsterian till (Fig. 21.2; Sejrup et al., 1987, 1991). Given the apparent size of the ice sheet(s) and the pervasive presence of subglacial meltwater features (which implies a significant bedload), it is likely that the absence of tills in cores either relates to a lack of penetration by existing boreholes or reflects their reworking by ice rather than non-deposition (e.g. Carr, 2004).

In the southern North Sea, south of the Dogger Bank, some authors have suggested that buried channels contain tills and sediments derived from subglacial meltwater that can be assigned to the Swarte Bank Formation of probable Elsterian age (Balson and Cameron, 1985). In the Inner Silver Pit area of the southwestern North Sea, temperate marine sediments belonging to the locally restricted Sand Hole Formation are sandwiched between the Egmond Ground and Swarte Bank Formations (Fig. 21.2). In BGS borehole 81/52a, and from neighbouring vibrocores, Scourse et al. (1998, 1999) reliably correlated the Sand Hole Formation to the Holsteinian interglacial, of MIS 9 (Fig. 21.2), thus proving an Elsterian age for the underlying Swarte Bank diamictons. Corroborating this evidence, recent detailed micromorphological, provenance and sedimentary analyses have interpreted the Swarte Bank Formation as a subglacial till and provide additional sedimentary data in support of the landform record for subglacial environments and extensive Elsterian glaciation (Davies, 2009; Davies et al., 2011).

21.5. THE SAALIAN (MIS 6–10)

21.5.1. Glacial Limits

According to Gibbard and Cohen (2008) and Toucanne et al. (2009), the Saalian cold-stage spans MIS 6–10 (Fig. 21.2). Onshore (e.g. in Denmark, Poland and the Netherlands) evidence for Saalian glacial activity is widespread, and we refer the reader to respective chapters in this volume for further details.

In the central North Sea region, Ehlers (1990) has suggested that it is possible to reconstruct two phases of Saalian glaciation. For the earliest phase, till of early Saalian age (MIS 8), found offshore of the Netherlands, requires British ice-sheet occupation of the North Sea, in order to explain a south-easterly ice-sheet flow onshore (Rappol et al., 1989). In support, Beets et al. (2005) presented convincing evidence from Dutch Survey borehole 89/2 in the southern North Sea for an extensive ice-sheet advance during MIS 8, which deposited a till that was subsequently overlain by shallow marine sands, correlated with MIS 7.

For the later Saalian (MIS 6), evidence for glaciation comprises a single glacial erosion surface that can be traced through large parts of the North Sea (Fig. 21.2; Cameron et al., 1987; Ehlers, 1990; Laban, 1995; Holmes, 1997). Glacial incisions that correspond to this surface suggest a minimum southern ice sheet terminus at ~56°N, and extending to the shelf edge in the northern North Sea (Holmes, 1997; Carr, 2004). This erosion surface is overlain by glacigenic sediments, including till and glacimarine deposits within the Fisher, Coal Pit and Ferder Formations in the central and northern North Sea (Fig. 21.2; Stoker et al., 1985; Cameron et al., 1987; Sejrup et al., 1987; Holmes, 1997). The presence of tills in the Southern North Sea, offshore of the Netherlands and offshore of Denmark has been used in the past to infer more extensive glacial occupation of the North Sea basin during the later Saalian (Carr, 2004). This has been further implied by more recent studies of tunnel valleys and sediments in coastal areas of the southern North Sea (e.g. Kluving et al., 2003; Kristensen et al., 2007; Fig. 21.3B), which provide minimum constraints on southerly Saalian ice-sheet extents at ~54°N (Figs. 21.1B and 21.3B).

21.5.2. Morphological Features

Tunnel valleys of supposed Saalian age are relatively common across the North Sea (Wingfield, 1989; Cameron et al., 1987; Huuse and Lykke-Andersen, 2000), although none of these have been directly dated (Fig. 21.3A). Whereas the central North Sea valleys are deeply buried, some Saalian tunnel valleys lie as relict, filled features at the sea floor, in the southern North Sea (Fig. 21.3A).

In the central North Sea, Stewart and Lonergan (2011) recently mapped up to seven regionally correlatable tunnel valley generations, incising into the Aberdeen Ground Formation (Fig. 21.3B). Some of these are most likely Elsterian as previously discussed, but the authors' correlation of tunnel valley generations to the marine isotope record is consistent with phases of repeated valley incision during each glacial stage of the Saalian, during MIS 10, 8 and 6. The cross-cutting tunnel valleys document a complicated pattern of reoccupation and overprinting during extensive glaciations of the Middle to Late Pleistocene.

Graham et al. (2007) also described localised patches of sub-ice-stream bedforms (mega-scale glacial lineations, MSGLs), which they mapped on 3D seismic datasets in the Witch Ground basin. A small suite of MSGLs occurs on an erosion surface at the base of the Coal Pit Formation in this area (Fig. 21.2), which is thought to range from Late Saalian to Weichselian in age (Graham et al., 2007). These lineations, termed the 'lower surface' by Graham et al. (2007) and shown as flowset 1 in Fig. 21.4, have been interpreted as the buried signature of a palaeo-ice stream, with fast-flow sourced from the west, within the BIS. The authors tentatively relate the bedforms to a Late Saalian (or possibly Early Weichselian) expansion of ice into the Witch Ground basin.

21.5.3. Key Sites

To date, the only record of a central North Sea Saalian-aged till comes from BGS borehole 81/26 where a diamict is found in the Fisher Formation, containing clasts of a probable Scottish source and interpreted as subglacial in origin (Fig. 21.2; Sejrup et al., 1987; Carr, 2004; Davies, 2009; Davies et al., 2011). However, recent reassessment of the borehole site based on 3D seismic observations has indicated that this deposit is probably found only locally, infilling one of the many buried tunnel valleys that characterise the subsurface (Graham, 2007). No other known or published reports of Saalian till have been found from the central and northern North Sea regions (Johnson et al., 1993; Carr, 2004), though tills of both MIS 8 and 6 age appear to be relatively common farther south, recovered in a number of boreholes from several sites in the southern North Sea and northern European coastal regions (e.g. Laban and van der Meer, 2004; Beets et al., 2005).

21.6. THE WEICHSELIAN (MIS 5D–2)

In the North Sea, there is good evidence for at least two phases of extensive Weichselian ice-sheet growth: in the Early Weichselian (MIS 4) and during the Late Weichselian (MIS 3/2; Fig. 21.2; Carr et al., 2006; Graham, 2007). This is consistent with evidence for a two-stage Weichselian ice sheet on the Atlantic margins of northwest Scotland (Stoker

Chapter | 21   The Pleistocene Glaciations of the North Sea Basin

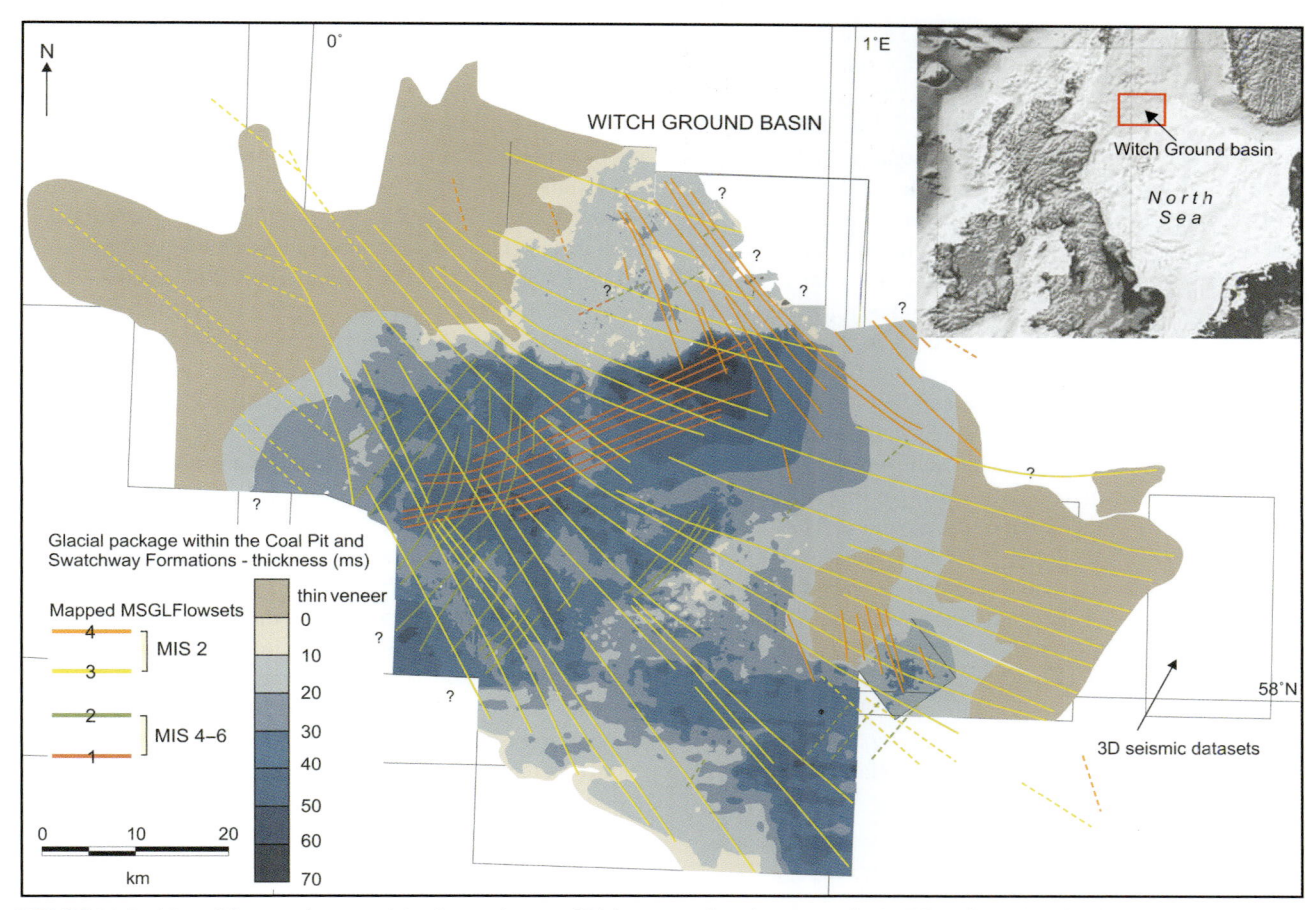

FIGURE 21.4 Map of flowsets that record the flow pathways of Pleistocene palaeo-ice streams in the Witch Ground basin. Flowsets were interpreted from suites of buried, mega-scale glacial lineations corresponding to relict palaeo-ice-stream beds in 3D seismic datasets (Graham et al., 2007, 2010). Three-dimensional datasets are shown as grey boxes. A lower-resolution, regional 3D seismic mega-survey, also used for lineation mapping, covers the majority of the area shown in the figure. The underlying basemap shows the thickness of the glacial package in which bedforms are observed, correlative with the Coal Pit and Swatchway Formations in the central North Sea.

and Holmes, 1991; Stoker et al., 1993) and northern Norway (Mangerud, 2004).

21.6.1. Early Weichselian Glaciation: Glacial Limits, Morphological Features and Key Sites

In the northern North Sea, a till forming the upper part of the Ferder Formation overlies Eemian interglacial deposits and glacimarine sediments, in which the Blake magnetic event has been proven (Fig. 21.2; Stoker et al., 1985; Johnson et al., 1993; Carr, 2004; Carr et al., 2006). Infilled tunnel valleys provide primary evidence for glaciation, which correlate with this early stage (Fig. 21.3A). Evidence for the offshore limits of this stage remain unclear, although it is thought that the northern ice edge reached the shelf break, based on sedimentary evidence from the Norwegian Channel and Atlantic margin (Sejrup et al., 2003; Mangerud, 2004) and from the analysis of microstructures in sediments from the northern North Sea area (Carr et al., 2006). All three sites indicate extensive grounded MIS 4 ice sheets. Southerly ice extents are uncertain, but onshore, Scandinavian and Baltic ice sheets reached at least as far central Denmark, implying significant ice cover in the North Sea also (see relevant chapters in this volume).

In 3D seismic datasets from the central North Sea, Graham (2007) described well-preserved morphological evidence for palaeo-ice-stream activity and inferred extensive glaciation, which may correspond to the Early Weichselian. Graham (2007) mapped at least four separate suites of MSGLs which correspond to palaeo-ice-stream bed signatures (flowsets) within the Coal Pit Formation (Fig. 21.2), infilling the Witch Ground basin (Fig. 21.4). Existing chronostratigraphical constraints on this part of the sequence are poor but suggest at least two of these flowsets correspond to pre-MIS 2 shelf glaciations, between MIS 10–6 and 2. On this basis, at least one of the palaeo-ice streams is thought to have operated during MIS 4 (flowset 2); the other was assigned a tentative

Late Saalian MIS 6 age (flowset 1). The acoustic stratigraphy and bedform record also indicate that ice streams are associated with discrete till horizons in a stacked sedimentary sequence and may be interlayered with glacimarine or proglacial deposits (Graham, 2007). Notably these sediments had previously been ascribed a simple glacimarine–marine genesis, comprising a single formation (Fig. 21.2; the Coal Pit Formation; Stoker et al., 1985; Cameron et al., 1987).

Recent syntheses of marine and terrestrial geological evidence by Svendsen et al. (2004) and Sejrup et al. (2005) as well as offshore evidence of Carr et al. (2006) provide good support to ice sheet occupation of the North Sea basin during MIS 4. Depositional fans located variously along the North Atlantic continental margin provide additional independent evidence for shelf-edge glacial limits revealing dramatic increases in sediment flux to the margin during the MIS 4 glacial period (Elverhøi et al., 1998; Sejrup et al., 2003, 2005; Mangerud, 2004). Diamictons interpreted as subglacial till are also recorded in the neighbouring Norwegian Channel and are assigned an MIS 4 age (Sejrup et al., 1995), while onshore to the west, there is general agreement for two extensive Middle-to-Late Weichselian glaciations corresponding to MIS 4 and 3–2, shown by a two-tiered till stratigraphy separated by organic horizons at Balglass Burn in central Scotland (Brown et al., 2006).

21.6.2. Late Weichselian Maximum Glaciation Limits

The limits of Late Weichselian glaciation in the North Sea basin (MIS 3–2) have been heavily debated over the past two decades due to a lack of information regarding palaeo-ice-flow extents and palaeo-ice-sheet configuration. Numerous ice-sheet reconstructions have been proposed, often based on relatively select pieces of data (e.g. single cores), and for the purposes of this chapter, a range of these are shown (Fig. 21.1B). In some areas, there was a general agreement between ice-sheet limits; however, poor agreement surrounded others, in particular, in the central North Sea where various forms of ice-free, proximal glacial and subglacial environments were interpreted and where ice-sheet reconstructions were clearly at odds (Fig. 21.1B).

Superseding the borehole studies of Sejrup et al. (1987, 1991), which reconstructed an ice-free North Sea at the Late Weichselian maximum, seismic-based studies by Graham et al. (2007) documented that an ice stream occupied the central North Sea at the last maximum ice extent. The main phase of ice cover is associated with subglacial tills recovered in two marine cores that have been related to a period of extensive North Sea glaciation, dated to between 29 and 22 ^{14}C ka BP when ice is thought to have covered the entire North Sea shelf and reached the shelf break (Figs. 21.5 and 21.6; Rise and Rokoengen, 1984; Sejrup et al., 1994, 2000, 2005, 2009; Carr et al., 2006; Bradwell et al., 2008). In this model, the period of maximum areal extent was followed by widespread retreat and a series of subsequent stillstands and possible readvances to inner-shelf limits, which we will discuss below.

Extensive glacial cover followed by at least one localised glacial stillstand or readvance is supported by geomorphological and chronostratigraphical evidence from the onshore record (Merritt et al., 2003; Mangerud, 2004) and the Barents Sea, Norwegian and Atlantic margins (Davison, 2004; Sejrup et al., 2005) as well as recent micromorphological studies on the North Sea deposits themselves (Carr et al., 2006). Based on all these data, the northern extent of the extensive ice sheet is now accepted to have reached the shelf break. Moraines and tills recovered on the shelf to the northwest of Shetland (Stoker and Holmes, 1991; Davison, 2004, Bradwell et al., 2008), and ice-flow patterns mapped across the Shetland Isles themselves, both support this interpretation (Golledge et al., 2008; Fig. 21.7).

In contrast to the northwestern margin, the southern extent of the Late Weichselian maximum ice sheet is less well defined. In the eastern North Sea, ice is known to have filled the Skagerrak and the Norwegian Channel at the last glacial maximum (LGM), based on information from cores and landform data (Sejrup et al., 2003). Farther south Baltic ice extended onshore into Denmark, while to the west, the Dogger Bank remains a likely southernmost limit of the 'North Sea Lobe' part of the last BIS (Figs. 21.1B and 21.5). Evidence for deformation structures on seismic reflection data in this area indicate ice movement from the north, and geomorphological mapping and sediment provenance analyses from cores recovered from the Bolders Bank Formation (Fig. 21.2) show that ice streams emanating from the east of Scotland and northern England were clearly deflected south along the coast by Scandinavian ice occupying the central North Sea basin (Everest et al., 2005; Davies et al., 2011).

Between the Dogger Bank and western Denmark, it is now widely accepted that British and Fennoscandian ice probably coalesced (Sejrup et al., 1994, 2000, 2009; Graham et al., 2007; Bradwell et al., 2008), and an arbitrary southern ice boundary is mapped, broadly coincident with the limit of exposed sea-floor tunnel valleys at $\sim 56°$N (Figs. 21.1B and 21.5). In terms of timing, the period of maximum ice extent appears to have been attained earlier (at ~ 25 cal. ka BP) than the global LGM as conventionally defined by sea-level records (Mix et al., 2001), based on evidence from the Barra–Donegal Fan as well as the North Sea basin itself (Figs. 21.5 and 21.6; Peck et al., 2006; Sejrup et al., 2009; Scourse et al., 2009). A corollary is that the prominent Wee Bankie and Bosies Bank moraines, which were used in the past to demarcate the limits of the LGM east of Britain (Figs. 21.6 and 21.7; Hall and Bent, 1990; Stewart, 1991), probably correlate with the 'Dimlington Stadial' (ca.

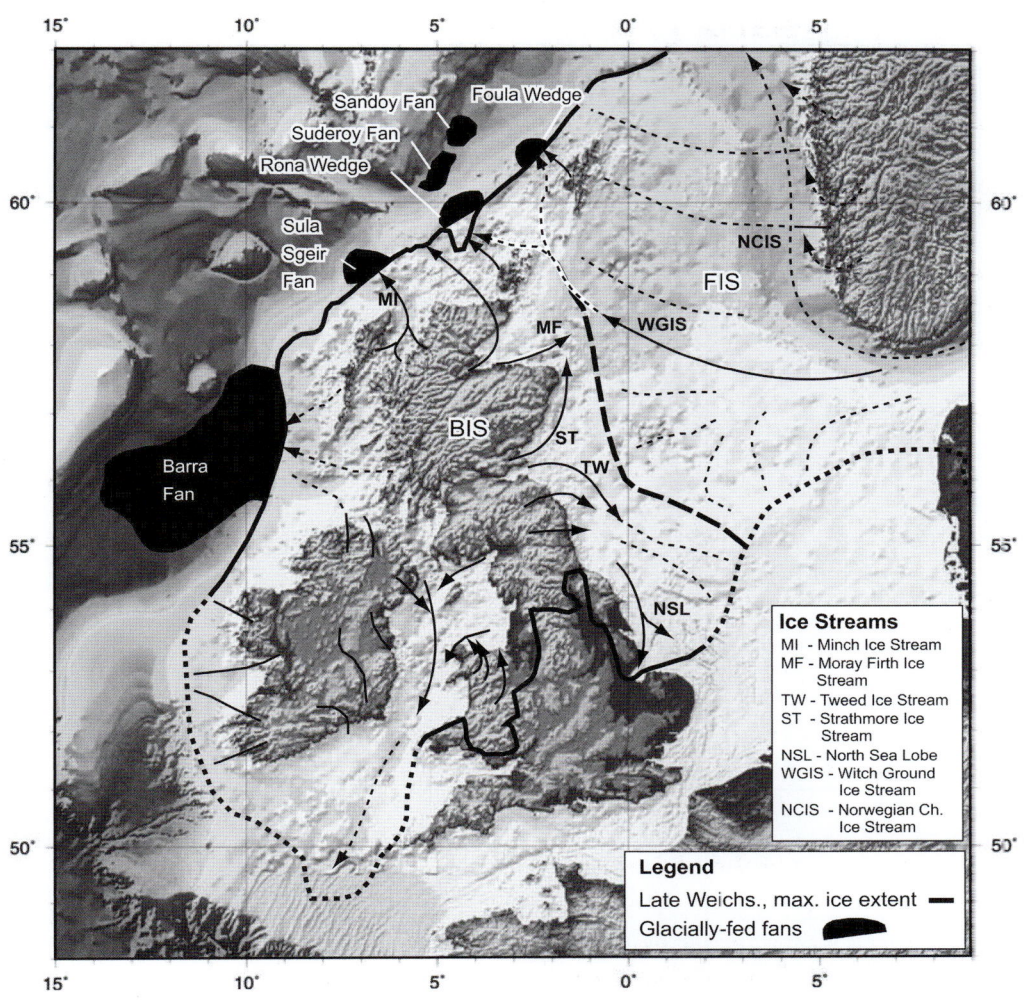

FIGURE 21.5 Reconstruction of ice-sheet extent and configuration for the Late Weichselian glacial maximum (MIS 3–2) in the North Sea basin. The reconstruction is based on the existing literature and is intended to highlight the broad flow patterns recorded within the northwest European ice sheets. It cannot replicate the full dynamics and various advance/retreat configurations that this ice mass certainly possessed. Arrowed flow lines represent fast-flow elements of the ice sheet (N.B. not necessarily ice streams), at certain times during its lifespan, whilst headless black lines show generalised ice-flow characteristics. Ice streams were likely active at different times (see text for details). Confluence is inferred for the British and Scandinavian ice sheets, as shown by the grey stippled area. In this configuration, the majority of ice-flow drainage is directed towards the North Atlantic shelf edge, feeding sedimentary fans at the continental margin. Sizes and locations of glacigenic fans from Stoker et al. (1993), Stoker (1995), Davison (2004) and Sejrup et al. (2005).

21 cal. ka BP) or younger deglacial events but were almost certainly preceded by more extensive North Sea glacial cover and, thus, do not represent Late Weichselian maxima (Carr et al., 2000, 2006; Sejrup et al., 2000, 2009; Bradwell et al., 2008; Graham et al., 2009, 2010).

21.6.3. Late Weichselian Maximum Glaciation: Morphological Features

Morphological features relating to the last main phase of ice-sheet activity are well preserved in the North Sea geological record. Geomorphological evidence for ice flow during the extensive Late Weichselian maximum comes primarily from the central Witch Ground basin. Buried submarine landforms mapped on 3D reflection seismic datasets provided the first glacial geomorphological evidence for glacial occupation of the central North Sea by at least one Late Quaternary palaeo-ice stream (Fig. 21.4; Graham et al., 2007). Streamlined subglacial bedforms (MSGLs) and iceberg ploughmarks, mapped from 40 m below sea bed to near sea floor, record the presence and subsequent break-up of grounded ice in the region. The most extensive and best-preserved lineation flowset is attributed to the action of the Witch Ground Ice Stream, which was probably sourced from the southeast within the FIS (Fig. 21.4, flowset 3 and Fig. 21.5; Graham et al.,

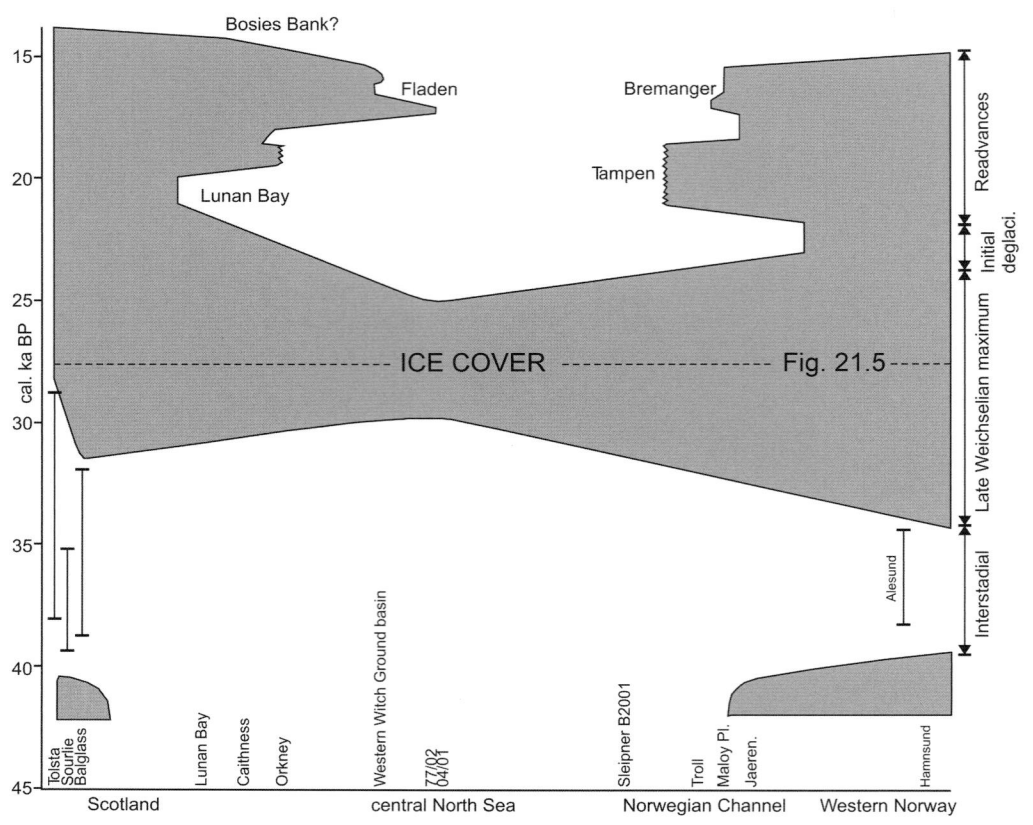

FIGURE 21.6 Simplified glaciation curve for the Late Weichselian in the North Sea basin, showing changing ice-sheet extents through time, constrained by a published radiocarbon chronology. Modified from Sejrup et al. (2009).

2007, 2010). The palaeo-ice stream is imaged over an area at least 30–50 km wide and along-flow for a minimum of 100 km, trending northwest–southeast. Cored sedimentary records tied to the 3D seismic observations support the age, and subglacial interpretation, of the bedforms. Importantly, the lineations provide independent geomorphological evidence in support of previous ice-sheet reconstructions that favoured complete ice coverage of the North Sea between Scotland and Norway during the Late Weichselian (e.g. Figs. 21.5 and 21.6; Sejrup et al., 1994, 2000; Carr et al., 2000).

Shelf-edge moraines probably mark the limit of this extensive ice sheet, which concentrated the delivery of sediment through ice streams (the Witch Ground Ice Stream included) to glacial debris fans on the continental margin (Fig. 21.5; Stoker, 1990; Stoker and Holmes, 1991; Sejrup et al., 2005; Stoker and Bradwell, 2005; Graham et al., 2007). Ice-flow trajectories on Shetland and in northern Scotland support the offshore morphological observations of a dominant north-westerly ice drainage (Bradwell et al., 2008; Golledge et al., 2008), although there remains some contention over the extent to which Scandinavian ice overran these fringing islands (Flinn, 2009).

21.6.4. Late Weichselian Maximum Glaciation: Key Sites

The shallow Quaternary successions in the central and northern North Seas preserve good evidence for extensive glaciation and palaeo-ice-stream activity and include sediments that relate to the Coal Pit and Cape Shore Formations (Fig. 21.2; Carr et al., 2006). These sequences have been cored and were analysed for their sedimentology and chronology. BGS boreholes 77/02 and 04/01 both show evidence for glacial overriding of the Coal Pit Formation and secondary deformation of pre-existing Late Weichselian sediments by the Witch Ground Ice Stream (Sejrup et al., 1994; Graham et al., 2010). Thin-section analysis of the broadly correlative Cape Shore Formation in other BGS boreholes confirms glacial overriding and deformation by grounded ice in the northern North Sea (Fig. 21.2; Carr et al., 2000, 2006). The glacimarine sediments that were deformed by the passage of ice were previously emplaced during the Alesund/Tolsta interstadial, when the North Sea was believed to be largely ice free (Figs. 21.2 and 21.6; Mangerud, 2004). The corresponding sequence of sediments relating to ice-sheet extents along the southern margin of the

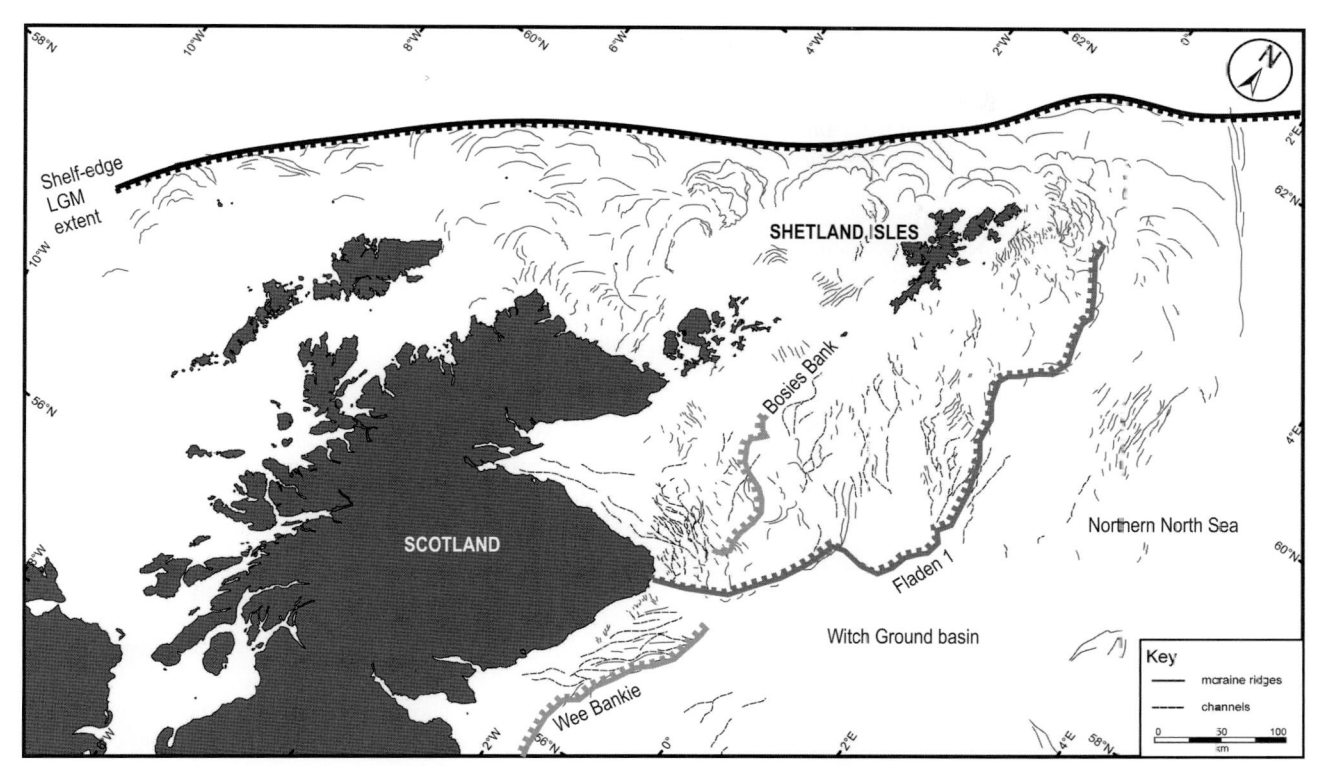

FIGURE 21.7 Map of sea-floor moraine ridges and meltwater channels in the northern North Sea and north of Scotland. The features were mapped from high-resolution sea-floor bathymetry and record the dynamics and decay of British ice during the last deglaciation (Bradwell et al., 2008). Thick grey and stippled lines depict moraines formed during readvances or stillstands of the British Ice Sheet during the last deglaciation. Moraine positions drawn from Stewart (1991), Graham et al. (2009) and Sejrup et al. (2009).

North Sea ice sheet are also heavily deformed but have not been examined in detail. Limited existing micromorphological analyses from this region including samples from the Dogger Bank suggest that the feature may be a terminal moraine formed during the Late Weichselian maximum, corresponding to the Bolders Bank Formation (Fig. 21.2; Carr, 2004). The Dogger Bank was likely shaped further by a more localised and predominantly land-based lobe of the BIS during the later 'Dimlington Stadial', when North Sea ice sheets had receded to coastal fringes (Fig. 21.5; e.g. Davies, 2009; Sejrup et al., 2009; Davies et al., 2011).

21.6.5. Last Deglaciation: Limits, Morphological Features and Key Sites

While the maximum extent of Late Weichselian ice seems clear to the Northwest and largely inferred to the South, simple 'two-stage' models for the deglaciation of the North Sea basin (e.g. Sejrup et al., 1994) have now given way to a model of more complex dynamic and oscillatory ice-margin retreat (Boulton and Hagdorn, 2006; Bradwell et al., 2008; Graham et al., 2009; Hubbard et al., 2009; Sejrup et al., 2009). Details on ice-sheet limits during the last deglaciation have been described by Bradwell et al. (2008), based on mapping from a new fisheries-sourced bathymetric compilation derived from single-beam echo-sounder data (Olex data; Fig. 21.7). The authors showed convincing evidence for coalescent BIS and FIS in the central and northern North Sea, and a subsequent pull-apart or 'unzipping' of the ice sheet, followed by a stepped, landward retreat to coastal positions. The retreat formed abundant hummocky topography, meltwater channels, and terminal moraines that are traceable on the sea bed today (Fig. 21.7). In many cases, the morainic features appear to comprise the sediments that correlate with the Sperus Formation in the northern North Sea (Johnson et al., 1993) and Swatchway Formation in the central North Sea (Stoker et al., 1985); both formations record subglacial-to-glacimarine conditions, from ~ 14 ^{14}C ka onwards (Fig. 21.2; Sejrup et al., 1994, 2000; Carr et al., 2006; Graham et al., 2007). The arrangement and existing age constraints on the sequence led Bradwell et al. (2008) to suggest that initial deglaciation in the northern North Sea may have been forced, at least in part, by rising sea-level, and was focused in the Witch Ground region at the confluence between British and Scandinavian ice (Figs. 21.5 and 21.6). This forcing appears to mirror the pattern of retreat in other major marine ice-sheet systems in northern Europe (e.g. the Barents Sea;

Winsborrow et al., 2010) and has been supported by modelling studies (Hubbard et al., 2009).

Several of the Late Weichselian stillstands or readvances have been studied discretely, including the Tampen (Sejrup et al., 2000), Fladen (Sejrup et al., 2009) and Bosies Bank episodes (Figs. 21.6 and 21.7; Hall and Bent, 1990; Graham et al., 2009). Moraine units relating to these events, including the Tampen Till (northeastern North Sea) as well as tills of the Norwegian Trench Formation (eastern North Sea), are correlated with the Swatchway and Sperus Formations farther seaward (Carr et al., 2006), which indicate ice-free conditions in large parts of the North Sea during their deposition (Fig. 21.2).

The Tampen episode probably marks one of the earliest North Sea deglacial events and is recorded by the presence of a sandy, shelly subglacial diamicton (interpreted as till) in cores from the eastern Witch Ground area (Figs. 21.1B and 21.6). Dates from shell fragments within the till were used to date a major pause or incursion of the FIS on the North Sea plateau, at about 18.6–15 ^{14}C ka BP (Rise and Rokoengen, 1984; Sejrup et al., 1994). The ice-margin terminus is believed to lie east of BGS borehole 77/02, which records marine deposition continuously during the equivalent time period in the Swatchway and Witch Ground Formations (Figs. 21.2 and 21.6).

Until recently, the Bosies Bank readvance (correlated with the Bolders Bank readvance by Carr et al., 2006) of the BIS was believed to correlate broadly with the Tampen readvance, in the east (Sejrup et al., 1994, 2000). Graham et al. (2009) originally supported this argument, showing that the Bosies Bank formed as a redvance or stillstand subsequent to a more extensive phase of ice streaming (Fig. 21.7). At the mouth of the Moray Firth, a morainal suite, consisting of a large terminal bank and superimposed by smaller crescentic ridges formed by ice-push, clearly overrides an older bedform signature of a palaeo-ice stream (Graham et al., 2009; see also Hall and Bent, 1990). Although no chronological data were presented, the authors assigned the main phase of ice streaming to the North Sea Late Weichselian maximum, because the bedforms and moraine unit appeared to override sediments belonging to the Coal Pit Formation (Fig. 21.2), and suggested that the younger moraine-forming event may have been correlative with the Tampen/Dimlington readvance, as described above.

Since then, the work of Bradwell et al. (2008), Sejrup et al. (2009) and Graham et al. (2010) has confirmed that the Bosies Bank is actually significantly younger and was likely formed as part of a relatively late-stage stillstand of the BIS (Fig. 21.6). Graham et al. (2010) suggested that its age may be younger than $\sim$14–13.5 ^{14}C ka BP, based on dates on the Fladen readvances—the evidence for which lies seaward of the Bosies Bank moraine (Fig. 21.7; Sejrup et al., 2009)—and on ^{14}C ages from bivalves (Graham et al., 2010) from an ice-proximal deposit recovered in BGS 04/01, which indicates extensive British ice in the Witch Ground region prior to $\sim$13.9 ^{14}C ka BP. These results would also imply that ice streaming in the Moray Firth, as recorded in the bedform patterns (Graham et al., 2009), relates to a phase of deglacial activity, rather than to the Late Weichselian maximum as originally proposed. The precise stratigraphical context of the Moray Firth ice stream, however, is unclear at present.

To the south of the Bosies Bank, the Wee Bankie moraine may form a lateral equivalent to the Bosies Bank feature and presumably marks an ice-recessional morphological feature too (Fig. 21.7). South of the Wee Bankie, the history of ice recession is poorly understood, but the broadly corresponding sediments, including those of the Botney Cut Formation, record the gradual, ice-recessional infill of subglacial channels cut by the LGM ice sheet in the southern North Sea (Fig. 21.2).

Most recently, cores from the western Witch Ground basin have been studied which support evidence for further oscillations of the BIS during its late-stage retreat. Buried grounding zone wedges (or till tongues) have been mapped from subsurface acoustic profiles to the west of BGS borehole 77/02 (Sejrup et al., 2009). The moraines correlate to glacigenic diamictons recovered in cores, which were deposited during at least two supposed ice-sheet readvances dated to between 17.5 and 15.5 cal. ka BP. These readvances have been termed the Fladen readvances and suggest rapid localised ice advances akin to those modelled by Boulton and Hagdorn (2006) and Hubbard et al. (2009) late on in the deglaciation (Figs. 21.6 and 21.7). They may correspond, chronologically, to a similar advance of Norwegian ice onto the Maløy plateau, west of Norway, which formed streamlined bedforms and large arcuate moraines at the sea bed (Nygård et al., 2004). Stratigraphically, the Fladen moraines form part of the polygenetic Swatchway Formation, which encompasses many of the features formed during the last deglaciation of the central North Sea area (Fig. 21.2).

A compilation of all published offshore chronological data, together with inferred ice-sheet extents for the North Sea, portrays multiple ice-margin oscillations and a stepped pattern of retreat during Late Weichselian deglaciation (Fig. 21.6; Sejrup et al., 2009). The glaciation curve (Fig. 21.6) still lacks ties to many of the features mapped by Bradwell et al. (2008) in Fig. 21.7, and therefore we predict even more complexity to the ice-margin retreat pattern than shown here. One possible clue to the dynamic retreat, however, lies in numerous discrete ice-stream systems that drained into the basin during the last deglaciation (Fig. 21.5; e.g. Moray Firth, Tweed, Strathmore, Witch Ground, North Sea Lobe and Norwegian Channel ice streams). Indeed, recent modelling experiments by Hubbard et al. (2009) seem to confirm that these arteries of flow were influential in controlling the overall dynamics of the decaying BIS.

During the latter stages of deglaciation, ice sheets remained in contact with the open-marine North Sea basin as late on as ~12 ^{14}C ka BP (Graham et al., 2010). Purges of icebergs discharged from the fronts of landward-retreating tidewater glaciers or ice streams, depositing the most distal part of the Swatchway Formation and the lower parts of the Witch Ground Formation in the central North Sea (Fig. 21.2). Icebergs scoured the sea floor, and keel marks are now found as the buried and exposed signatures of deglaciation, between these two formations in the stratigraphy (Stoker and Long, 1984; Graham et al., 2010). Age measurements on the most pervasive and prominent scoured horizon constrain iceberg activity to ~13.9–12 ^{14}C ka BP (Stoker and Long, 1984; Graham et al., 2010). The overlying sediments show that distal-glacimarine conditions persisted in the central North Sea until, and for some time after 12 ^{14}C ka BP, as ice sheets shrank to smaller ice caps and became restricted to the adjacent land masses.

In the northern and central North Sea sediments relating to the Witch Ground, Forth, upper parts of the Botney Cut and the Kleppe Senior Formations record the transition from glacimarine to temperate (shallow) marine conditions through the Late-Glacial and Early Holocene (Fig. 21.2). The connection between the North Sea and the English Channel was only established between 9 and 7 ka BP, and the North Sea only existed as a full marine basin as recently as ~6 ka BP. The southern North Sea was, therefore, likely exposed as a periglacial plain until the Early Holocene.

21.7. SUMMARY

We have presented an updated review of the Quaternary stratigraphy of the North Sea basin, which demonstrates a complicated history influenced by glacial environments throughout the past 2.6 Ma.

The North Sea may have had glacimarine influences from fringing marine ice sheets during times of traditionally non-glacial activity: in the Early Pleistocene, during the Menapian (MIS 36; ~1.1–1 Ma), and in the Middle Pleistocene, during the Cromerian between MIS 19 and 12 (900–450 ka). A switch from a deltaic–marine setting to a glacial setting during the Middle Pleistocene saw the first major expansions of continental ice sheets into the North Sea.

Complete ice cover of much of the North Sea basin occurred during the Elsterian (MIS 12) glaciation, and significant phases of glacial activity are inferred during each stage of the Saalian (MIS 10, 8 and 6) as well as Early Weichselian glaciations, based on information from bedform geomorphology and sediments. Combined 2D and 3D seismic observations and associated geomorphological and sediment core studies suggest that meltwater drainage systems dominated the subglacial environment during the period MIS 12–6. The MIS 12 and 10–6 ice sheets appear to have been particularly erosive, and ice-sheet extents have been determined by tunnel valley networks mapped across the basin, although these do not always demonstrate an ice-marginal association and are more complex than previously indicated. Indeed, based on the generations of buried tunnel valleys mapped in the central North Sea, it is clear that the bedforms record a much more complex glacial history for the North Sea than the conventional three-stage model first proposed (Stewart, 2009). Stewart and Lonergan (2011) note that the number of mapped tunnel valley generations in the central North Sea match in number the cold events in the global marine isotope record within the last 500 ka, providing direct geomorphic evidence for frequent, extensive glaciation of the North Sea during glacial stages of the Pleistocene.

Ice sheets from Norway, Denmark and Scotland coalesced in the North Sea at least once during the Elsterian and Saalian glaciations and during the Weichselian, possibly during MIS 4 and certainly during MIS 2. Palaeo-ice streams drained into, and crossed, the central North Sea, leaving footprints of their flow at least three times, correlated to MIS 10–6, 4 and 2. The best-preserved palaeo-ice-stream bed relates to the Late Weichselian-aged Witch Ground Ice Stream, which was sourced from the southeast Fennoscandian ice sheet, and probably drained to the shelf edge near Shetland.

The break-up of the last ice sheets was probably initiated in the northern North Sea and Witch Ground areas, and the ensuing deglaciation of the North Sea basin was characterised by a dynamic ice-sheet system; the retreat was punctuated by regular readvances and stillstands, which formed buried and sea-floor moraines that document final ice-marginal recession onto land.

21.8. NOTE ON THE MAPS

Although we have provided 'a closer look' at Quaternary glaciations of the North Sea basin in this chapter, the regional picture concerning extents of the main Pleistocene glacial stages has not changed since the first edition of *Quaternary Glaciations—Extent and Chronology*. While new work has focused on the buried geomorphology, such studies show local detail beyond the remit of this project. Also, where the glacial features have been mapped regionally, chronological constraints are often poor, and the features do little to change the broad ice-marginal extents. Hence, for this review, we make no update to the digital maps of ice extents in the North Sea, presented in the first volume.

ACKNOWLEDGEMENTS

A. G. C. G. was funded by a BGS–Imperial College consortium grant whilst a Ph.D. student at Imperial College London (2003–2006). M. A. S. was funded by an Imperial College Janet Watson scholarship. A. G. C. G. and L. L. acknowledge Petroleum Geo-Services (PGS) for providing 3D seismic reflection data and Landmark Graphics for a University Software Grant. A. G. C. G. is grateful to M. S. S. and L. L. for helping formulate many of the ideas presented within. M. S. S. and T. B. publish with the permission of the executive director of the BGS (NERC). Thanks to H. P. Sejrup, S. Carr, A. Nygard, H. Haflidason, J. Rose, C. Ó. Cofaigh, B. Davies and J. Lee for useful advice and discussions.

REFERENCES

Andreassen, K., Nilssen, L.C., Rafaelsen, B., Kuilman, L., 2004. Three-dimensional seismic data from the Barents Sea margin reveal evidence of past ice streams and their dynamics. Geology 32, 729–732.

Balson, P., Cameron, T.D.J., 1985. Quaternary mapping offshore East Anglia. Mod. Geol. 9, 231–239.

Beets, D.J., Meijer, T., Beets, C.J., Cleveringa, P., Laban, C., van der Spek, A.J.F., 2005. Evidence for a Middle Pleistocene glaciation of MIS 8 age in the southern North Sea. Quatern. Int. 133–134, 7–19.

Bijlsma, S., 1981. Fluvial sedimentation from the Fennoscandian area into the North-West European Basin during the Cenozoic. Geol. Mijnbouw 60, 337–345.

Boulton, G.S., Hagdorn, M., 2006. Glaciology of the British Isles Ice Sheet during the last glacial cycle: form, flow, streams and lobes. Quatern. Sci. Rev. 25, 3359–3390.

Boulton, G.S., Smith, G.D., Jones, A.S., Newsome, J., 1985. Glacial geology and glaciology of the last mid-latitude ice sheets. J. Geol. Soc. Lond. 142, 447–474.

Bowen, D.Q., Rose, J., McCabe, A.M., Sutherland, D.G., 1986. Correlation of Quaternary glaciations in England, Ireland, Scotland and Wales. In: Sibrava, V., Bowen, D.Q., Richmond, G.M. (Eds.), Quaternary Glaciations in the Northern hemisphere. Pergamon Press, New York, p. 514.

Bradwell, T., Stoker, M.S., Golledge, N.R., Wilson, C.K., Merritt, J.W., Long, D., et al., 2008. The northern sector of the last British Ice Sheet: maximum extent and demise. Earth Sci. Rev. 88, 207–226.

Brown, E.J., Rose, J., Coope, R.G., Lowe, J.J., 2006. An MIS 3 age organic deposit from Balglass Burn, central Scotland: palaeoenvironmental significance and implications for the timing of the onset of the LGM ice sheet in the vicinity of the British Isles. J. Quatern. Sci. 22, 295–308.

Cameron, T.D.J., Stoker, M.S., Long, D., 1987. The history of Quaternary sedimentation in the UK sector of the North Sea Basin. J. Geol. Soc. Lond. 144, 43–58.

Cameron, T.D.J., Crosby, A., Balson, P.S., Jeffrey, D.H., Lott, G.K., Bulat, J., et al., 1992. United Kingdom Offshore Regional Report: The Geology of the Southern North Sea. HMSO for the British Geological Survey, London.

Carr, S., 2004. The North Sea basin. In: Ehlers, J., Gibbard, P.L. (Eds.), Quaternary Glaciations—Extent and Chronology, Part I. Elsevier, Amsterdam, pp. 261–270.

Carr, S., Haflidason, H., Sejrup, H.P., 2000. Micromorphological evidence supporting Late Weichselian glaciation of the Northern North Sea. Boreas 29, 315–328.

Carr, S.J., Holmes, R., van der Meer, J.J.M., Rose, J., 2006. The last glacial maximum in the north sea basin: micromorphological evidence of extensive glaciation. J. Quatern. Sci. 21, 131–153.

Caston, V.N.D., 1977. A new isopachyte map of the Quaternary of the North Sea. Inst. Geol. Sci. Rep. 10 (11), 3–10.

Davies, B.J., 2009. British and Fennoscandian ice sheet interactions during the Quaternary. Ph.D. thesis, University of Durham, 300pp.

Davies, B.J., Roberts, D.H., Bridgland, D.R., O´Cofaigh, C., Riding, J.B., 2011. Provenance and depositional environments of Quaternary sediments from the western North Sea Basin. J. Quatern. Sci. 26, 59–75.

Davison, S., 2004. Reconstructing the Last Pleistocene (Late Devensian) Glaciation on the Continental Margin of Northwest Britain. (Ph. D. thesis, University of Edinburgh).

Ehlers, J., 1990. Reconstructing the dynamics of the North-West European Pleistocene ice sheets. Quatern. Sci. Rev. 9, 71–83.

Ehlers, J., Wingfield, R., 1991. The extension of the Late Weichselian/Late Devensian ice sheets in the North Sea Basin. J. Quatern. Sci. 6, 313–326.

Ekman, S.R., Scourse, J.D., 1993. Early and Middle Pleistocene pollen stratigraphy from British Geological Survey borehole 81/26, Fladen Ground, central North Sea. Rev. Palaeobot. Palynol. 79, 285–295.

Elverhøi, A., Hooke, R.L., Solheim, A., 1998. Late Cenozoic erosion and sediment yield from the Svalbard-Barents Sea region: implications for understanding erosion of glacierized basins. Quatern. Sci. Rev. 17, 209–241.

Everest, J., Bradwell, T., Golledge, N., 2005. Subglacial landforms of the tweed palaeo-ice stream. Scot. J. Geol. 121, 163–173.

Flinn, D., 1978. On the glaciation of Orkney-Shetland and North Sea east of Orkney. Scot. J. Geol. 14, 109–123.

Flinn, D., 2009. The omission of conflicting evidence from the paper by Golledge et al. (2008). Geogr. Ann. A 91, 253–256.

Gatliff, R.W., Richards, P.C., Smith, K., Graham, C.C., McCormac, M., Smith, N.J.P., et al., 1994. United Kingdom Offshore Regional Report: The Geology of the Central North Sea. HMSO for the British Geological Survey, London.

Gibbard, P.L., 1988. The history of the great northwest European rivers during the past three million years. Philos. Trans R. Soc. Lond. B 318, 559–602.

Gibbard, P.L., 1991. The Wolstonian Stage in East Anglia. In: Lewis, S.G., Whiteman, C.A., Bridgland, D.R. (Eds.), Central East Anglia & the Fen Basin Field Guide. Quaternary Research Association, Cambridge, pp. 7–13.

Gibbard, P.L., Cohen, K.M., 2008. Global chronostratigraphical correlation table for the last 2.7 million years. Epsiodes 31, 243–247.

Glennie, K.W., Underhill, J.R., 1998. Origin, development and evolution of structural styles. In: Glennie, K.W. (Ed.), Petroleum Geology of the North Sea: Basic Concepts and Recent Advances. fourth ed. Blackwell Science Ltd., Oxford, pp. 42–84.

Golledge, N.R., Finlayson, A., Bradwell, T., Everest, J.D., 2008. The last glaciation of Shetland, North Atlantic. Geogr. Ann. 90A, 37–53.

Graham, A.G.C., 2007. Reconstructing Pleistocene Glacial Environments in the Central North Sea Using 3D Seismic and Borehole Data. Ph.D. thesis, University of London, 410pp.

Graham, A.G.C., Lonergan, L., Stoker, M.S., 2007. Evidence for Late Pleistocene ice stream activity in the Witch Ground basin, central North Sea, from 3D seismic reflection data. Quatern. Sci. Rev. 26, 627–643.

Graham, A.G.C., Lonergan, L., Stoker, M.S., 2009. Seafloor glacial features reveal the extent and decay of the last British Ice Sheet, east of Scotland. J. Quatern. Sci. 24, 117–138.

Graham, A.G.C., Lonergan, L., Stoker, M.S., 2010. Depositional environments and chronology of Late Weichselian glaciation and deglaciation in the central North Sea. Boreas 39, 471–491.

Hall, A.D., Bent, A.J., 1990. The limits of the last British Ice Sheet in Northern Scotland and the Adjacent Shelf. Quatern. Newslett. 61, 2–12.

Holmes, R., 1997. Quaternary stratigraphy: the offshore record. In: Gordon, J.E. (Ed.), Reflections on the Ice Age in Scotland. Scottish Natural Heritage, Glasgow, pp. 72–94.

Hubbard, A.L., Bradwell, T., Golledge, N.R., Hall, A., Patton, H., Sugden, D., et al., 2009. Dynamic cycles, ice streams and their impact on the extent, chronology and deglaciation of the British–Irish ice sheet. Quatern. Sci. Rev. 28, 758–776.

Huuse, M., Lykke-Andersen, H., 2000. Overdeepened quaternary valleys in the eastern Danish north sea: morphology and origin. Quatern. Sci. Rev. 19, 1233–1253.

Jackson, C.A.-L., 2007. Application of three-dimensional seismic data to documenting the scale, geometry and distribution of soft-sediment features in sedimentary basins: an example from the Lomre Terrace, offshore Norway. In: Geological Society, London, Special Publications, vol. 277, pp. 253–267.

Johnson, H., Richards, P.C., Long, D., Graham, C.C., 1993. United Kingdom Offshore Regional Report: The Geology of the Northern North Sea. HMSO, London, 111pp.

Kluving, S.J., Bosch, J.H.A., Ebbing, J.H.J., Mesdag, C.S., Westerhoff, R.S., 2003. Onshore and offshore seismic and lithostratigraphic analysis of a deeply incised Quaternary buried valley system in the Northern Netherlands. J. Appl. Geophys. 53, 249–271.

Kristensen, T.B., Huuse, M., Piotrowski, J.A., Clausen, O.R., 2007. A morphometric analysis of tunnel valleys in the eastern North Sea based on 3D seismic data. J. Quatern. Sci. 22, 801–815.

Laban, C., 1995. The Pleistocene Glaciations in the Dutch Sector of the North Sea. Ph.D. thesis, Universiteit van Amsterdam, 200pp.

Laban, C., van der Meer, J.J.M., 2004. Pleistocene glaciation in The Netherlands. In: Ehlers, J., Gibbard, P.L. (Eds.), Quaternary Glaciations—Extent and Chronology, Part I. Elsevier, Amsterdam, pp. 251–263.

Larkin, N.R., Lee, J.R., Connell, E.R., 2011. Possible ice-rafted erratics in late Early to early Middle Pleistocene shallow marine and coastal deposits in northeast Norfolk, UK. P. Geologist Assoc., in press.

Lee, J.R., Rose, J., Hamblin, R.J.O., Moorlock, B.S.P., 2004. Dating the earliest lowland glaciation of eastern England: a pre-MIS 12 early Middle Pleistocene Happisburgh glaciation. Quatern. Sci. Rev. 23, 1551–1556.

Lonergan, L., Maidment, S., Collier, J., 2006. Pleistocene sub-glacial tunnel valleys in the central North Sea: 3D morphology and evolution. J. Quatern. Sci. 21, 891–903.

Long, D., 1988. The sedimentary record of climatic variations in the southern North Sea. Philos. Trans. R. Soc. Lond. B318, 523–537.

Long, D., Bent, A., Harland, R., Gregory, D.M., Graham, D.K., Morton, A.C., 1986. Late Quaternary palaeontology, sedimentology and geochemistry of a vibrocore from the Witch Ground Basin, central North Sea. Mar. Geol. 73, 109–123.

Lutz, R., Kalka, S., Gaedicke, C., Reinhardt, L., Winsemann, J., 2009. Pleistocene tunnel valleys in the German North Sea: spatial distribution and morphology. Z. dt. Ges. Geowiss. 160, 225–235.

Mangerud, J., 2004. Ice sheet limits in Norway and on the Norwegian continental shelf. In: Ehlers, J., Gibbard, P.L. (Eds.), Quaternary Glaciations—extent and chronology, Part I. Elsevier, Amsterdam, pp. 271–294.

Merritt, J.W., Auton, C.A., Connell, E.R., Hall, A.M., Peacock, J.D., 2003. Cainozoic Geology and Landscape Evolution of North-East Scotland: Memoir for the Drift Editions of 1:50 000 Geological Sheets 66E, 67, 76E, 77, 86E, 87W, 87E, 95, 96W, 96E, and 97 (Scotland). British Geological Survey, Edinburgh, 178pp.

Mix, A.C., Bard, E., Schneider, R., 2001. Environmental processes of the ice age: land, oceans, glaciers (EPILOG). Quatern. Sci. Rev. 24, 627–657.

Nygård, A., Sejrup, H.P., Haflidason, H., Cecchi, M., Ottesen, D., 2004. The deglaciation history of the southwestern Fennoscandian Ice Sheet between 15 and 13 ^{14}C ka. Boreas 33, 1–17.

Peck, V.L., Hall, I.R., Zahn, R., Elderfield, H., Grousset, F., Hemming, S.R., Scourse, J.D., 2006. High resolution evidence for linkages between NW European ice sheet instability and Atlantic Meridional Overturning Circulation. Earth Planet. Sc. Lett. 243, 476–488.

Praeg, D., 1996. Morphology, stratigraphy and genesis of buried Mid-Pleistocene tunnel valleys in the southern North Sea. Ph.D. thesis, University of Edinburgh.

Praeg, D., 2003. Seismic imaging of mid-Pleistocene tunnel-valleys in the North Sea basin. J. Appl. Geophys. 53, 273–298.

Preece, R.C., Parfitt, S.A., Coope, R.C., Penkman, K.E.H., Ponel, P., Whittaker, J.E., 2009. Biostratigraphic and aminostratigraphic constraints on the age of the Middle Pleistocene glacial succession in north Norfolk, UK. J. Quatern. Sci. 24, 557–580.

Rappol, M., Haldorsen, S., Jorgensen, P., van der Meer, J.J.M., Stoltenberg, H.M.P., 1989. Composition and origin of petrographically stratified thick till in the northern Netherlands and a Saalian glaciation model for the North Sea Basin. Contrib. Tert. Quatern. Geol. 26, 31–64.

Rise, L., Rokoengen, K., 1984. Surficial sediments in the Norwegian Sector of the North Sea between 60°30′ and 62° N. Mar. Geol. 58, 287–317.

Rise, L., Olsen, O., Rokoengen, D., Ottesen, D., Riis, F., 2004. Mid-Pleistocene ice drainage pattern in the Norwegian Channel imaged by 3D seismic. Quatern. Sci. Rev. 23, 2323–2335.

Scourse, J.D., Ansari, M.H., Wingfield, R.T.R., Harland, R., Balson P.S., 1998. A Middle Pleistocene shallow marine interglacial sequence, Inner Silver Pit, southern North Sea: pollen and dinoflagellate cyst stratigraphy and sea-level history. Quatern. Sci. Rev. 17, 871–900.

Scourse, J.D., Austin, W.E.N., Sejrup, H.P., Ansari, M.H., 1999. Foraminiferal isoleucine epimerization determinations from the Nar Valley Clay, Norfolk, UK: implications for Quaternary correlations in the southern North Sea basin. Geol. Mag. 136, 543–560.

Scourse, J.D., Haapaniemi, A.I., Colmenero-Hidalgo, E., Peck, V.L., Hall, I.R., Austin, W.E.N., et al., 2009. Growth, dynamics and deglaciation of the last British–Irish ice sheet: the deep-sea ice-rafted detritus record. Quatern. Sci. Rev. 28, 3066–3084.

Sejrup, H.P., Knudsen, K.L., 1993. Paleoenvironments and correlations of interglacial sediments in the North Sea. Boreas 22, 223–235.

Sejrup, H.P., Aarseth, I., Ellingsen, E., Reither, E., Jansen, E., Løvlie, R., et al., 1987. Quaternary stratigraphy of the Fladen area, central North Sea: a multidisciplinary study. J. Quatern. Sci. 2, 35–53.

Sejrup, H.P., Aarseth, I., Haflidason, H., 1991. The Quaternary succession in the northern North Sea. Mar. Geol. 101, 103–111.

Sejrup, H.P., Haflidason, H., Aarseth, I., King, E., Forsberg, C.F., Long, D., et al., 1994. Late Weichselian glaciation history of the northern North Sea. Boreas 23, 1–13.

Sejrup, H.P., Aarseth, I., Haflidason, H., Løvlie, R., Bratten, Å., Tjøstheim, G., et al., 1995. Quaternary of the Norwegian channel: glaciation history and palaeoceanography. Nor. Geol. Tidsskr. 75, 65–87.

Sejrup, H.P., Larsen, E., Landvik, J., King, E., Haflidason, H., Nesje, A., 2000. Quaternary glaciations in southern Fennoscandia: evidence from southwestern Norway and the northern North Sea region. Quatern. Sci. Rev. 19, 667–685.

Sejrup, H.P., Larsen, E., Haflidason, H., Berstad, I., Hjelstuen, H.E., 2003. Configuration, history and impact of the Norwegian Channel Ice Stream. Boreas 32, 18–36.

Sejrup, H.P., Hjelstuen, B.O., Torbjørn Dahlgren, K.I., Haflidason, H., Kuijpers, A., Nygård, A., et al., 2005. Pleistocene glacial history of the NW European continental margin. Mar. Petrol. Geol. 22, 1111–1129.

Sejrup, H.P., Nygård, A., Hall, A.M., Haflidason, H., 2009. Middle and late Weichselian (Devensian) glaciation history of south-western Norway, North Sea and eastern UK. Quatern. Sci. Rev. 28, 370–380.

Stewart, F.S., 1991. A reconstruction of the eastern margin of the Late Weichselian ice sheet in northern Britain. Unpublished Ph.D. thesis, University of Edinburgh.

Stewart, M.A., 2009. 3D Seismic Analysis of Pleistocene Tunnel Valleys in the Central North Sea. Ph.D. thesis, University of London, 319pp.

Stewart, M.A., Lonergan, L., 2011. Direct evidence for seven glacial cycles in the Middle-Late Pleistocene of NW Europe. Geology 39, 283–286.

Stoker, M.S., 1990. Glacially-influenced sedimentation on the Hebridean slope, northwestern United Kingdom continental margin. In: Dowdeswell, J.A., Scourse, J.D. (Eds.), Glacimarine Environments: Processes and Sedimentation, Geological Society, London, Special Publications, 53, pp. 349–362.

Stoker, M.S., 1995. The influence of glacigenic sedimentation on slope-apron development on the continental margin off Northwest Britain. In: Scrutton, R.A., Stoker, M.S., Shimmield, G.B., Tudhope, A.W. (Eds.), The Tectonics, Sedimentation and Palaeoceanography of the North Atlantic Region, Geological Society Special publication, 90, pp. 159–177.

Stoker, M.S., Bent, A.J., 1985. Middle Pleistocene glacial and glaciomarine sedimentation in the west central North Sea. Boreas 14, 325–332.

Stoker, M.S., Bent, A.J., 1987. Lower Pleistocene deltaic and marine sediments in boreholes from the central North Sea. J. Quatern. Sci. 2, 87–96.

Stoker, M.S., Bradwell, T., 2005. The Minch palaeo-ice stream: NW sector of the British–Irish ice sheet. J. Geol. Soc. Lond. 162, 425–428.

Stoker, M.S., Holmes, R., 1991. Submarine end-moraines as indicators of Pleistocene ice limits off NW Britain. J. Geol. Soc. Lond. 148, 413–434.

Stoker, M.S., Long, D., 1984. A relict ice-scoured erosion surface in the central North Sea. Mar. Geol. 61, 85–93.

Stoker, M.S., Skinner, A.C., Fyfe, J.A., Long, D., 1983. Palaeomagnetic evidence for early Pleistocene in the central and northern North Sea. Nature 304, 332–334.

Stoker, M.S., Long, D., Fyfe, J.A., 1985. A revised quaternary stratigraphy for the central North Sea. British Geological Survey Research Report, 17, London, HMSO, 35pp.

Stoker, M.S., Hitchen, K., Graham, C.C., 1993. United Kingdom Offshore Regional Report: The Geology of the Hebrides and West Shetland Shelves and Adjacent Deep-Water Areas. HMSO for the British Geological Survey, London.

Stoker, M.S., Leslie, A.B., Scott, W.D., Briden, J.C., Hine, N.M., Harland, R., et al., 1994. A record of Late Cenozoic stratigraphy, sedimentation and climate change from the Hebrides slope, NE Atlantic Ocean. J. Geol. Soc. Lond. 151, 235–249.

Stoker, M.S., Balson, P.S., Long, D., Tappin, D.R., 2010. An overview of the lithostratigraphical framework for the Quaternary deposits on the United Kingdom continental shelf. British Geological Survey Research Report, RR/00/00, 48pp, http://nora.nerc.ac.uk/3241/1/RR04004.pdf.

Svendsen, J.I., Alexanderson, H., Astakhov, V.I., Demidov, I., Dowdeswell, J.A., Funder, S., et al., 2004. Late Quaternary ice sheet history of northern Eurasia. Quatern. Sci. Rev. 23, 1229–1271.

Toucanne, S., Zaragosi, S., Bourillet, J.F., Gibbard, P.L., Eynaud, F., Giraudeau, J., et al., 2009. A 1.2 Ma record of glaciation and fluvial discharge for the west European Atlantic margin. Quatern. Sci. Rev. 28, 2974–2981.

Valentin, H., 1957. Die grenze der letzen Vereisung im Nordseeraum. Ver. Deutschen Geog. 30, 259–366.

Velichko, A.A., Faustova, M.A., Gribchenko, Y.N., Pisareva, V.V., Sudakova, N.G., 2004. Glaciation of the East European Plain—distribution and chronology. In: Ehlers, J., Gibbard, P.L. (Eds.), Quaternary Glaciations—Extent and Chronology—Part 1—Europe. Elsevier, Amsterdam, pp. 337–354.

Wingfield, R., 1989. Glacial incisions indicating Middle and Upper Pleistocene Ice limits off Britain. Terra Nova 1, 528–548.

Wingfield, R., 1990. The origin of major incisions within the Pleistocene deposits of the North Sea. Mar. Geol. 91, 31–52.

Winsborrow, M.C.M., Andreassen, K., Corner, G.D., Laberg, J.S., 2010. Deglaciation of a marine-based ice sheet: late Weichselian palaeo-ice dynamics and retreat in the southern Barents Sea reconstructed from onshore and offshore glacial geomorphology. Quatern. Sci. Rev. 29, 424–442.

Zagwijn, W.H., 1974. The palaeogeographic evolution of the Netherlands during the Quaternary. Geol. Mijnbouw 53, 369–385.

Zagwijn, W.H., 1992. The beginning of the Ice Age in Europe and its major subdivisions. Quatern. Sci. Rev. 11, 583–591.

Zanella, E., Coward, M.P., 2003. Structural framework. In: Evans, D., Graham, C., Atmour, A., Bathurst, P. (Eds.), The Millennium Atlas: Petroleum Geology of the Central and Northern North Sea. The Geological Society of London, London, pp. 45–59.

Chapter 22

Glacial History of Norway

Jan Mangerud*, Richard Gyllencreutz, Øystein Lohne and John Inge Svendsen

Department of Earth Science and Bjerknes Centre for Climate Research, University of Bergen, Allégt. 41, NO-5007 Bergen, Norway
*Correspondence and requests for materials should be addressed to Jan Mangerud. E-mail: Jan.Mangerud@geo.uib.no

22.1. INTRODUCTION

This review is in principle an updated version of a similar chapter in the previous book edition (Mangerud, 2004). This time, we have to a large extent discussed other aspects of the glaciation history and most of the discussions and references to older papers that were included in Mangerud (2004) have been omitted.

Glaciers in the Norwegian and Swedish mountains certainly have acted as nuclei for Scandinavian Ice Sheet growth several tens of times during the Quaternary, and probably even during the Pliocene. The ice sheets, and during milder periods also smaller glaciers, have had a profound impact on the Norwegian landscape. The many deep fjords, long U-shaped valleys, cirques and thousands of lakes in overdeepened bedrock basins are the results of glacial activity. The vertical linear glacial erosion along several fjords amounts to 1500–2000 m, whereas it probably is considerably less than 100 m on some plateaus and summits; a description and discussion of the pattern of glacial erosion is given by Kleman et al. (2008). A mean bedrock lowering of ~520 m through the Quaternary glaciations has been calculated for a large area of Mid-Norway (Dowdeswell et al., 2010). However, the bedrock landforms cannot by themselves be used to unravel the full glaciation history. On the contrary, the glaciation history inferred from other data is used as one of the main factors explaining the observed pattern and inferred rates of glacial erosion.

22.2. EARLY AND MIDDLE QUATERNARY GLACIATIONS

The last ice sheets that covered the Norwegian land masses during the last (Weichselian) ice age have removed most of the older Quaternary deposits on land. Accordingly, pre-Weichselian glacial deposits are found only in a few sites (Mangerud, 2004). The older glacial history has to be deciphered from sea floor records in the deep Norwegian Sea and from deposits along the continental margin and from the North Sea region with the adjacent land masses to the south.

Ice-rafted debris (IRD) dropped from "Norwegian" icebergs into the deep sea beyond the coast represents the most continuous records of former glaciations. However, the ice sheets may have reached a considerable size over Norway–Sweden before they reached the sea and released icebergs. One should also keep in mind that the distribution and amount of IRD are also influenced by other factors, such as ocean currents and sea surface temperatures. IRD in the records from the Nordic Seas may also stem from Greenland or from sea ice. Kleiven et al. (2002) presented a synthesis of IRD results from several cores covering the period 3.5–2.4 Ma ago, and they concluded that there had been marked expansions of the Greenland Ice Sheet at 3.3 Ma and of the Scandinavian Ice Sheet from 2.75 Ma. Cores from the Vøring Plateau (Fig. 22.1) outside the Norwegian coast also contain some minor IRD pulses starting as early about 12.6 Ma, and another at about 7.2 Ma (Fronval and Jansen, 1996).

There are several recent reviews and syntheses of the glaciation history of the Norwegian continental shelf (Dahlgren et al., 2005; Hjelstuen et al., 2005; Nygård et al., 2005; Rise et al., 2005; Sejrup et al., 2005; Ottesen et al., 2009). The architecture of the deposits is somewhat different on the North Sea Fan at the mouth of the Norwegian Channel (Fig. 22.6) as compared with the architecture in areas to the north and south of the fan. The Naust Formation is the youngest formation on the Mid-Norwegian continental shelf in the area between 64 and 68°N, that is, north of the North Sea Fan. It consists of a series of prograding wedges (Fig. 22.2) where the base is considered to have an age of ca. 2.8 Ma (Ottesen et al., 2009), that is, close to the recorded major increase in IRD in the Norwegian Sea and the onset of the Quaternary according to the new definition of the boundary (Gibbard et al., 2009). However, there are still significant uncertainties concerning the exact age of the lower boundary of the Naust Formation as well as the internal subunits. The Naust Formation is more than a

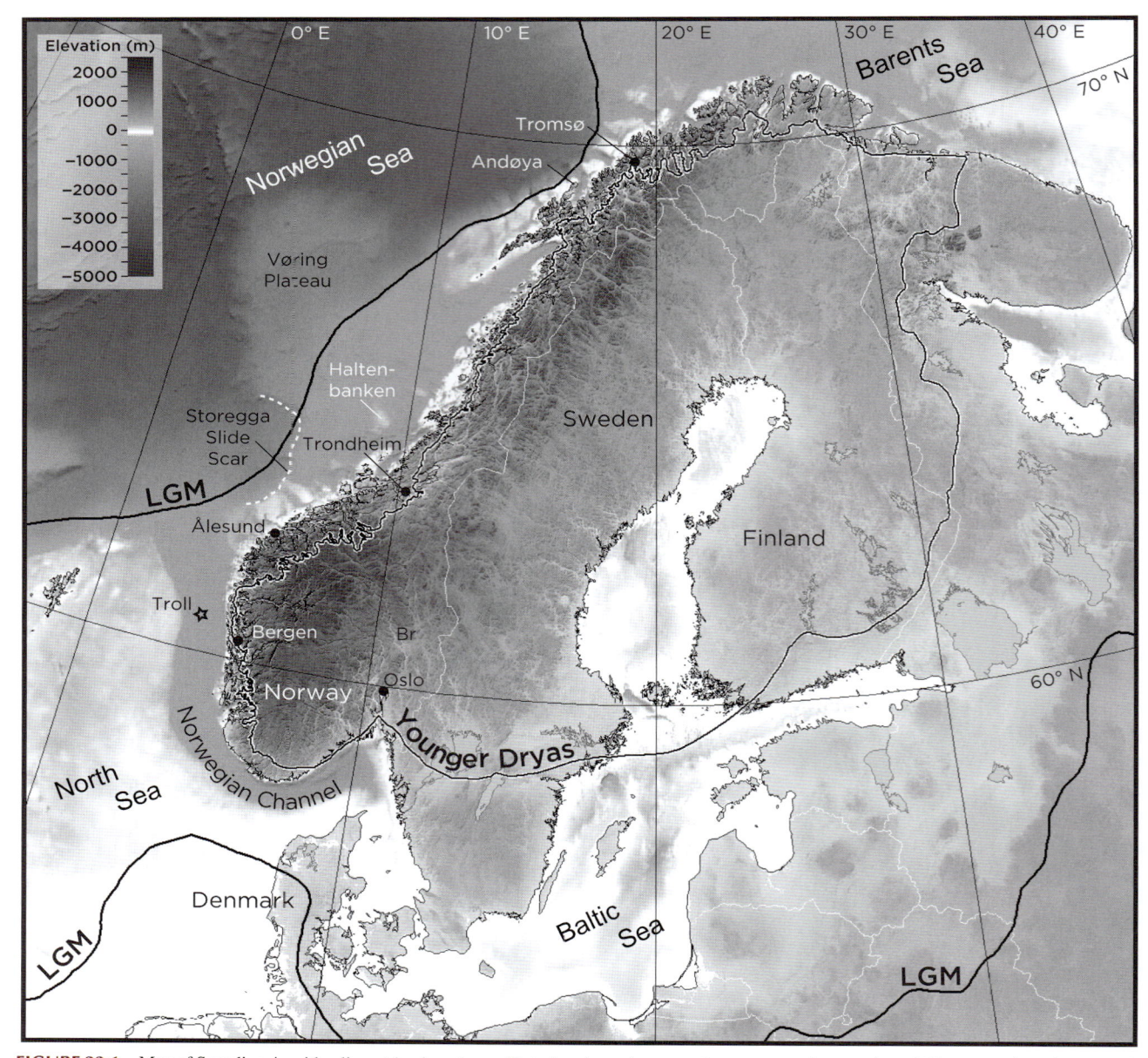

FIGURE 22.1 Map of Scandinavia with adjacent lands and seas. Note that the main mountain range starts in central south Norway and continues northeastwards along the Norwegian–Swedish border. The (asynchronous) ice limit at the Last Glacial Maximum (LGM) (Svendsen et al., 2004) and also the Younger Dryas moraines (Gyllencreutz et al., 2007) are marked. Br is the Brumunddal site. Digital elevation model is based on the SRTM30PLUS-dataset (Becker et al., 2009).

kilometre thick over a large area, and the shelf edge prograded up to 150 km during accumulation of this formation (Dahlgren et al., 2005; Rise et al., 2005; Ottesen et al., 2009). Independent of the remaining chronological uncertainties it is clear that the Naust Formation represents a period with more rapid deposition as compared to the underlying units. Rise et al. (2005) identified till in the oldest part of the Naust in boreholes on Haltenbanken (Fig. 22.1), off Mid-Norway, and Ottesen et al. (2009) described about 1.5 Ma old glacial flutes crossing the shelf. They conclude that large parts of the thick sediment wedge that has been dated to 2.8–1.7 Ma were formed during several ice sheet advances to the palaeo-shelf edge. Contrary to this view, several other investigators (Jansen et al., 2000; Dahlgren et al., 2005; Hjelstuen et al., 2005; Sejrup et al., 2005) consider that this older part of the Naust reflects some smaller mountain-centred ice sheets and that the ice margin reached the shelf edge for the first time about 1.1 Ma. However, there is a full agreement that the Naust Formation reflects a number of glaciations that at least covered a large part

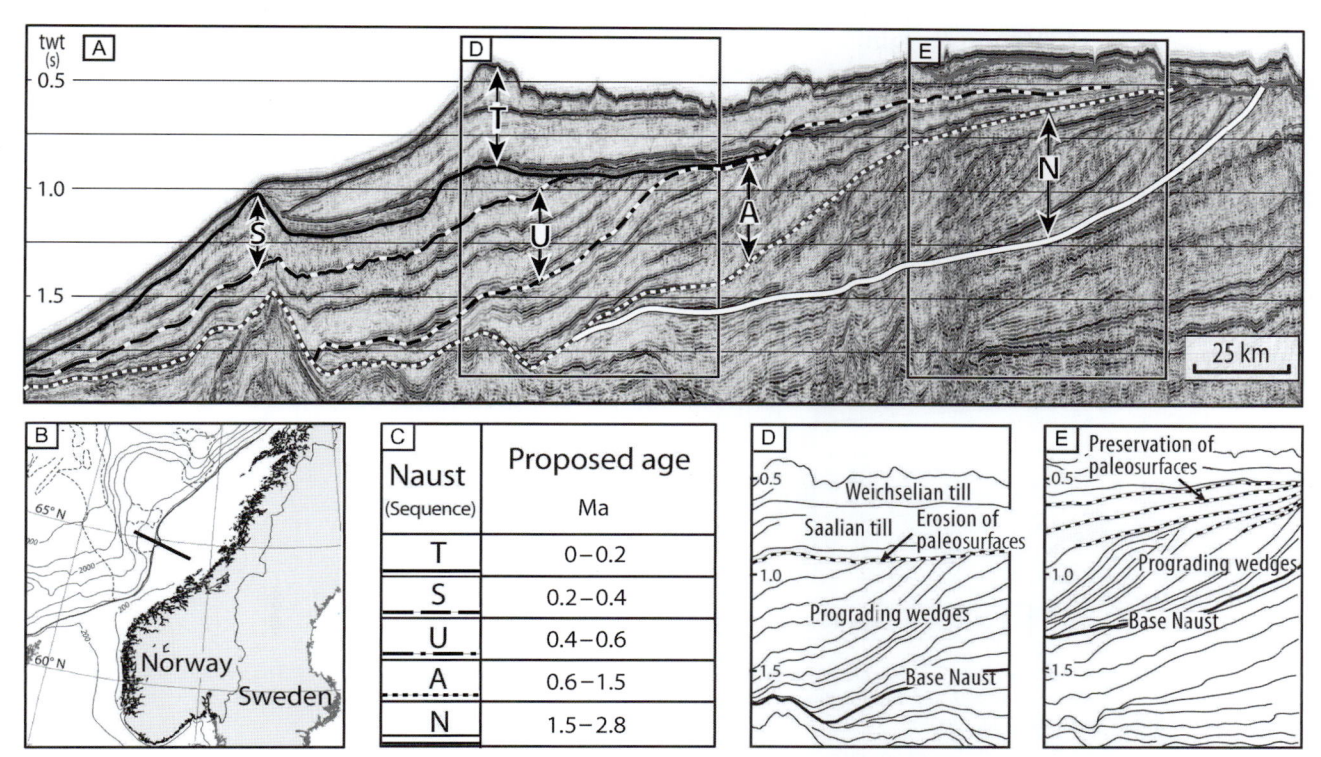

FIGURE 22.2 (A) Seismic-stratigraphical section across the mid-Norwegian continental shelf and slope. The Naust Formation (marked with letters N, A, U, S and T for subunits) is the glacigenic influenced wedge. Twt, two-way travel time; 1 s equals ~1000 m of sediment thickness. (B) Location of the profile. (C) Subdivisions and proposed ages (Ottesen et al., 2009) for the Naust Formation. The boundaries are marked with different line types used in (A). (D) Interpreted part of the section showing truncation of prograding wedges and palaeo-shelf surfaces. (E) Interpreted part showing preservation of wedges and palaeo-shelf surfaces. Stippled lines, see (D) above. Modified from Dowdeswell et al. (2007).

Naust (Sequence)	Proposed age Ma
T	0–0.2
S	0.2–0.4
U	0.4–0.6
A	0.6–1.5
N	1.5–2.8

of mainland Norway. On the European continent further to the south, only one or two glaciations older than 0.5 Ma have been proven (Ehlers et al., 2004; Houmark-Nielsen, 2004; Laban and van der Meer, 2004). During the period 2.7–1.1 Ma ago that was dominated by the 41 ka glaciation cycle, the Scandinavian mountains were repeatedly hosting sizeable ice sheets that eroded Norwegian fjords (Mangerud et al., 1996). Only during the past million years or so, when the 100 ka cycles dominated did the Scandinavian Ice Sheet for short periods expand far south of Scandinavia.

The oldest identified till in the Norwegian Channel is the Fedje Till that is dated to about 1.1 Ma (Sejrup et al., 1995, 2005). In the period from 1.1 Ma to marine isotope stage (MIS) 12 (about 500 ka), there were more limited glaciations. Starting with MIS 12, the North Sea Fan became the main depo-centre of glacial sediments from southern Scandinavia (Nygård et al., 2005; Sejrup et al., 2005). Four pre-Weichselian glacigenic debris-flow packages that were correlated with MIS 12, 10, 8 and 6, respectively, are identified in the fan (Hjelstuen et al., 2005; Nygård et al., 2005; Sejrup et al., 2005), indicating that the ice stream each time reached the mouth of the channel. There is a general agreement that the ice sheet reached the shelf edge further north on the continental shelf several times during the past 0.5 Ma (Fig. 22.3) also. This includes the Elsterian, Saalian and other glaciations that expanded far south on the continent (Dahlgren et al., 2005; Hjelstuen et al., 2005; Rise et al., 2005; Ottesen et al., 2009). It is interesting to note that Dowdeswell et al. (2006) identified a different and longer ice stream path on the shelf off Mid-Norway during the Saalian than during the Late Weichselian, which also implies thicker ice up-flow.

The extension of the Scandinavian Ice Sheet towards west was probably several times limited by the water depth and thus calving and not climate because grounded ice sheets cannot extend into deep open water. Therefore, the 150-km progradation of the Mid-Norwegian shelf during deposition of the Naust Formation made possible a further ice sheet extension during the younger than during the earlier glaciations, also favouring thicker ice sheets over the main land.

22.3. THE LATE QUATERNARY GLACIATIONS—THE WEICHSELIAN

By definition, the Weichselian starts at the end of the Eemian, corresponding approximately with the end of the MIS 5e (or early MIS 5d). The glacial history during this

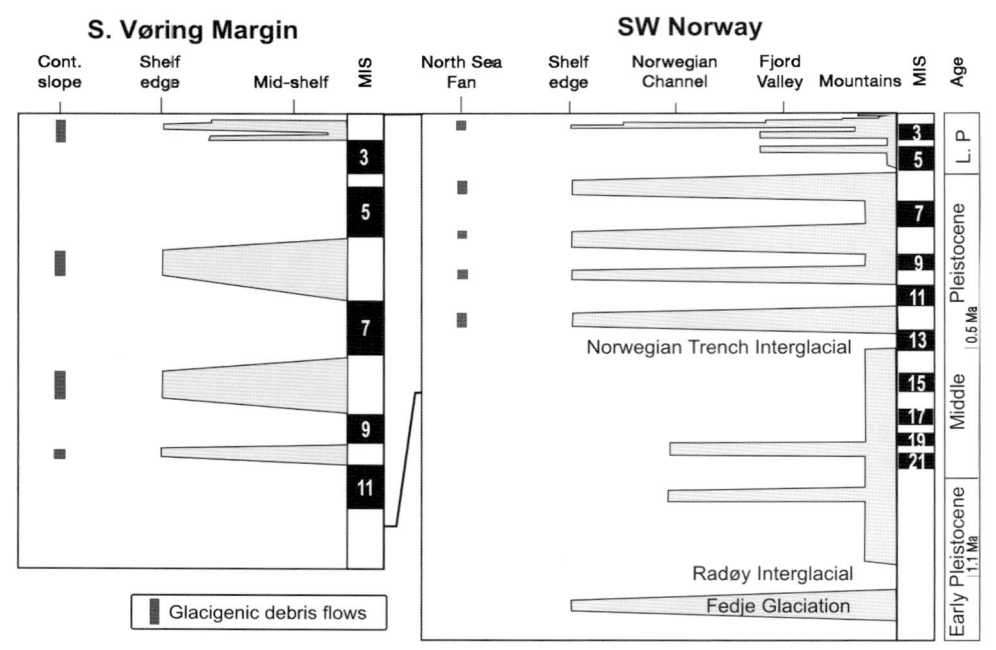

FIGURE 22.3 Schematic time–distance diagrams for the margin of the Scandinavian Ice Sheet through the Quaternary. The left diagram (South Vøring, Fig. 22.1) shows the Mid-Norwegian Shelf, and the right diagram shows SW Norway from the mountains through the Norwegian Channel to the shelf edge. Slightly modified from Hjelstuen et al. (2005).

period is for obvious reasons better known than for older periods. However, considering that the Late Weichselian (Last Glacial Maximum (LGM)) ice sheet advance removed most pre-existing sediments, our knowledge of the foregoing Weichselian remains fragmentary. The interpretation of the timing and extent of these older glaciations are based on observations from some few localities. The available data are insufficient to allow accurate mapping of the ice sheet limits, but they nevertheless give a rough idea of the ice sheet dimensions (Fig. 22.5). We have chosen to subdivide this chapter into two main sections: in the first one, we describe the Early and Middle Weichselian, and in the second, we describe the ice sheet extend during the LGM and the last deglaciation.

22.4. THE EARLY AND MIDDLE WEICHSELIAN

22.4.1. The Period of MIS 5e to 4

Dating events beyond than the range of the radiocarbon method are problematic. Different scientists have therefore disagreed on the age of some of these deposits and thus also the correlation between different sites. We refer to the discussion in Mangerud (2004); here, we will mainly describe some key sites and review some new results. Figure 22.4 shows our favoured glaciation curve for south-western Norway. The oldest part of the curve is based mainly on the stratigraphy at the key site Fjøsanger (Mangerud et al., 1981).

The correlation of the Fjøsangerian Interglacial with the Eemian is well established. Conformably, overlying the interglacial strata is a thick glaciomarine silt unit (Gulstein in Fig. 22.4), indicating that a large glacier ended somewhere in the fjord close to the Fjøsanger site. The glaciomarine silt is covered by a non-glacial beach deposit (Fana interstadial) followed by a thick till bed, the Bønes Till (Fig. 22.4). The till contains a large number of sediment clasts, molluscs and wood remains that must have been picked up from the 80-m-deep fjord basin adjacent to the site. The least certain part of this chronology is the correlation of the Fana interstadial with the Brørup interstadial (MIS 5c). If our correlation should be wrong, we find it most probable that Fana is older than MIS 5c and that the Bønes Till was deposited during MIS 5d, as originally proposed by Mangerud et al. (1981).

Another key site is located in Brumunddal in central parts of southern Norway (Fig. 22.1). At this site, a well-defined layer of peat is interbedded between two till beds (Helle et al., 1981). The pollen stratigraphy of the peat shows a succession starting from open pioneer vegetation to shrubs with some trees including *Larix*, and a reversion to arctic tundra near the top of the peat, that is, a cold–mild–cold climate cycle. The correlation of the peat with the Brørup seems plausible (Helle et al., 1981), although not unambiguously proven. If correct, the site provides evidence for the existence of a pre-Brørup till that most likely postdates the last interglacial and thus suggests a MIS 5d ice sheet advance across the site (Fig. 22.4).

Chapter | 22 Glacial History of Norway

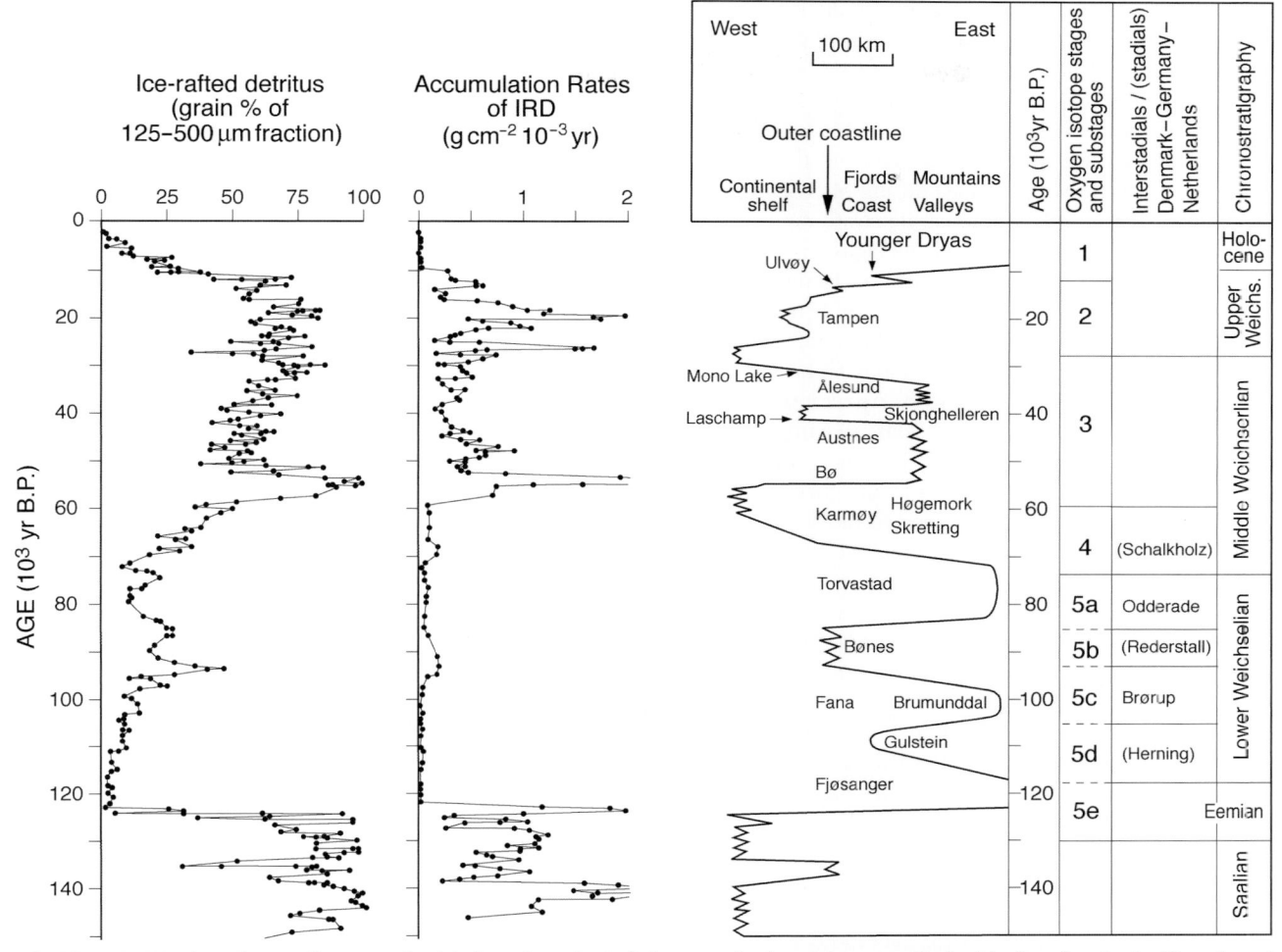

FIGURE 22.4 The time–distance diagram to the right is a schematic glaciation curve for the south-western flank of the Scandinavian Ice Sheet through the Weichselian, modified from Mangerud (2004). Names on the curve represent geological sites. The Laschamp and Mono Lake palaeomagnetic excursions are marked with arrows. The curve to the left is a stacked curved for grain % of IRD in five cores, and the middle curve a stacked record of accumulation rates in three cores; all cores collected from the Vøring Plateau (Fig. 22.1; Baumann et al., 1995). The timescale is (in principle) in calendar years, but the curves are partly dated and correlated with different methods.

The MIS 5e–MIS 4 part of the glaciation curve (Fig. 22.4) is recently supported by peaks in the ice-rafted detritus (IRD) record in a core collected from the Vøring Plateau (Fig. 22.1; Brendryen et al., 2010), in addition to earlier IRD curves shown in Fig. 22.4 (Baumann et al., 1995). Even though IRD is not a simple monitor of ice sheet size, iceberg rafting demonstrates that glaciers at this time extended down to sea level and probably even reached the open sea (i.e. beyond the mouth of the fjords). Brendryen et al. (2010) even correlate some of their IRD peaks with individual Dansgaard–Oeschger events; such a detailed correlation is not yet possible with the more poorly dated land record. We note that the $\delta^{18}O$ in precipitation on Greenland respond on a seasonal scale to weather and climate changes, and also some marine organisms and chemical parameters respond almost that fast. Ice sheets, and especially the build-up of ice sheets, will have a much slower response, and it will take hundreds and probably even thousands of years to build up a large ice sheet covering Norway.

Some of the most important new evidence for the ice sheet development in Norway during MIS 4 and early MIS 3 come from Denmark (Larsen et al., 2009a,b; Houmark-Nielsen, 2010). Based on several occurrences of a till with Fennoscandian erratics, Larsen et al. (2009a, b) describe the Sundsøre glacial advance from Norway which reached almost half-way down Denmark and was dated to 65–60 ka. They correlate this advance with the Karmøy advance in Norway (Mangerud, 2004; Fig. 22.4). If correct, the implication is that all of southern Norway with the adjacent shelves was ice covered during MIS 4, as indeed reconstructed by Carr et al. (2006). We also note that there is evidence to suggest that also the Barents–Kara Ice Sheet expanded at this time reaching the edge of the continental shelf west of Svalbard (Mangerud et al., 1998). Although there at present do not exist observations to test our hypothesis, we find it reasonable to assume that the Scandinavian Ice Sheet reached close to the shelf edge along the entire Norwegian coast during MIS 4 (Fig. 22.5).

Larsen et al. (2009a) demonstrate stratigraphically that the Sundsøre ice margin retreated and lacustrine and marine sediments were deposited in northern Denmark before there was another (Ristinge) ice advance across much of the country. However, the Ristinge advance came from the Baltic Sea, and it is dated to about 50 ka (Houmark-Nielsen, 2010), that is, early MIS 3. Apparently, the main dome of the ice sheet had moved east towards the Baltic Sea from its more westerly position when the Sundsøre advance occurred. There is no site in Norway that can resolve these two ice advances, and in the schematic map for MIS 4 in Fig. 22.5, we have combined the maximum limits for the two, although we acknowledge that they are asynchronous. Svendsen et al. (2004) correlated this Ristinge advance with the MIS 4 glacial maximum in the Barents Sea–Kara Sea Northern Russia area. In the entire Russian sector, the MIS 4 ice sheet was considerably larger than the LGM ice extent. The conclusion now is that the Scandinavian Ice Sheet probably was as (or almost as) extensive west and south of Norway during MIS 4 (and/or early MIS 3) as during MIS 2 (LGM).

22.4.2. Marine Isotope Stage 3

The start of MIS 3 is beyond the reach of radiocarbon dating, but ^{14}C ages are getting more and more trustworthy towards the end of MIS 3. In this section, we will not follow a chronological description, but start with the best dated part in the middle and late MIS 3, and then discuss the older part.

The coastal caves Skjonghelleren (Larsen et al., 1987), Hamnsundhelleren (Valen et al., 1996) and Olahola (Valen et al., 1995) located near Ålesund (Fig. 22.1) are key localities for reconstructing the ice sheet history during the middle part of MIS 3 because the cave sediments have enabled exceptional accurate dating of the ice margin fluctuations. The caves are found in vertical coastal cliffs in crystalline rocks; they are almost horizontal, up to 100 m long, and formed by wave action during an earlier ice-free period. Sediment records from these caves can be utilised to make inferences of the ice sheet fluctuations because when the ice margin expanded westwards and blocked their entrances, then ice-dammed lakes formed in the caves and laminated clay and silt accumulated on the floor. In contrast, when the caves were open, frost-wedged blocks fell from the roof and, at least during the Ålesund interstadial, ten-thousands of bones were brought into the caves by polar foxes. An accurate chronology is established in the following ways.

The Laschamp palaeomagnetic excursion was found in the glaciolacustrine sediments below the Ålesund interstadial strata. This excursion could be correlated with peaks of ^{10}Be and ^{36}Cl in the Greenland ice cores (Mangerud et al., 2003), demonstrating that an advancing ice sheet margin passed by the caves during the Greenland interstadial (GI) 10, or about 41 ka in the GICC05 timescale (Svensson et al., 2008). As many as 32 AMS ^{14}C dates were obtained from very well preserved bones from the Ålesund interstadial beds, yielding ages in the range 34–28 ^{14}C ka (Mangerud et al., 2010). These dating results could also be correlated with Greenland ice core δ^{18}O curves via marine cores collected close to Greenland (Voelker et al., 1998, 2000), indicating that Ålesund interstadial corresponds with the period of GI 8–GI 7, that is, 38.2–34.5 ka in the GICC05 chronology when using 1950 as the zero year. However, there were fewer dates from the colder Greenland stadial (GS) 8 between GI 8 and GI 7, suggesting that it was colder also in western Norway during this cold spell, but the ice margin did apparently not expand across the caves during GS 8. The inferred age of the end of the Ålesund interstadial was confirmed by the occurrence of the Mono Lake excursion in the glaciolacustrine sediments above the Ålesund bone layer. In the Greenland ice cores, the Mono Lake excursion is identified within GS 7, shortly after the end of GI 7.

Reindeer antlers were among the animal remains that were found in the Ålesund interstadial layers, and the presence of this animal indicates that at a broad zone along the coast was ice-free and had enough vegetation for grazing. Finds of assumed Ålesund interstadial age exist from several sites in Norway, but in no other place is the chronology as well established as in the mentioned caves. At Jæren further south-west along the coast (Fig. 22.6), strata that was termed Sandnes interstadial was in fact identified before the Ålesund interstadial was defined (Feyling-Hanssen, 1974; Raunholm et al., 2004). Olsen et al. (2001a,b) obtained a large number of ^{14}C dates of Mid-Weichselian age from Norway. However, many of these are performed on bulk-sediment samples with total organic content <1%, some even <0.1%, and the reliability of samples with such low carbon content is questionable. Olsen et al. (2001a,b) also report a number of similar ^{14}C ages (conventional and AMS) from marine molluscs, although very rarely they have obtained several consistent dates from one section. There are all reasons to assume that some of the ages reported by Olsen et al. (2001a,b) are correct, but presently, we find it very difficult to distinguish reliable dates from those that are more dubious. In Fig. 22.5, we have schematically drawn a very restricted ice sheet extent during the Ålesund interstadial; it may well have been larger. The reconstruction is partly substantiated by published observations from other sites in Scandinavia (Olsen et al., 2002; Mangerud et al., 2010; Wohlfarth, 2010) and also by modelling results (Arnold et al., 2002; Lambeck et al., 2010).

Chapter | 22 Glacial History of Norway

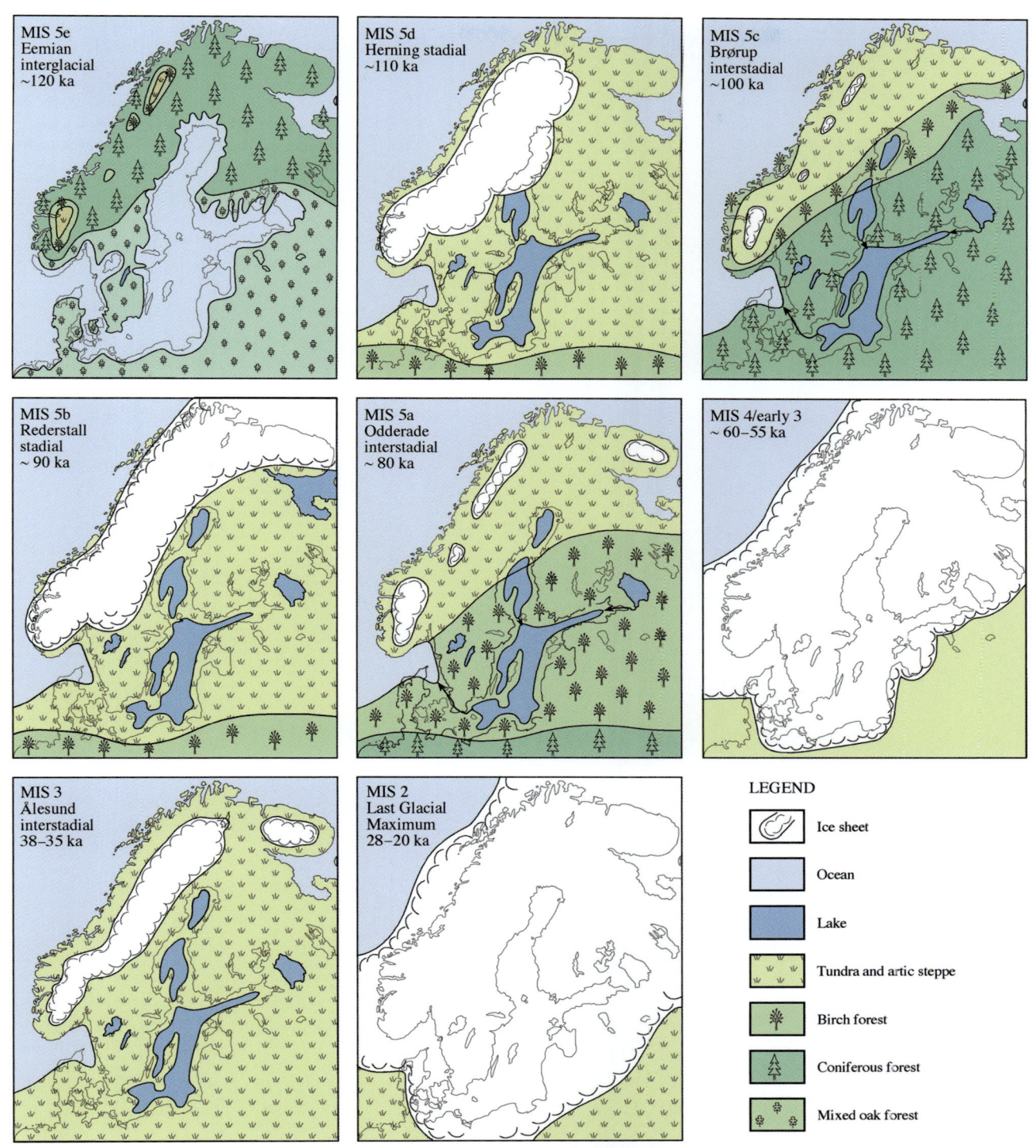

FIGURE 22.5 Conceptual maps illustrating the development of the Scandinavian Ice Sheet and surrounding areas. The original idea and version was by Lundqvist (1992); here, it is developed from Mangerud (2004) and Vorren and Mangerud (2008).

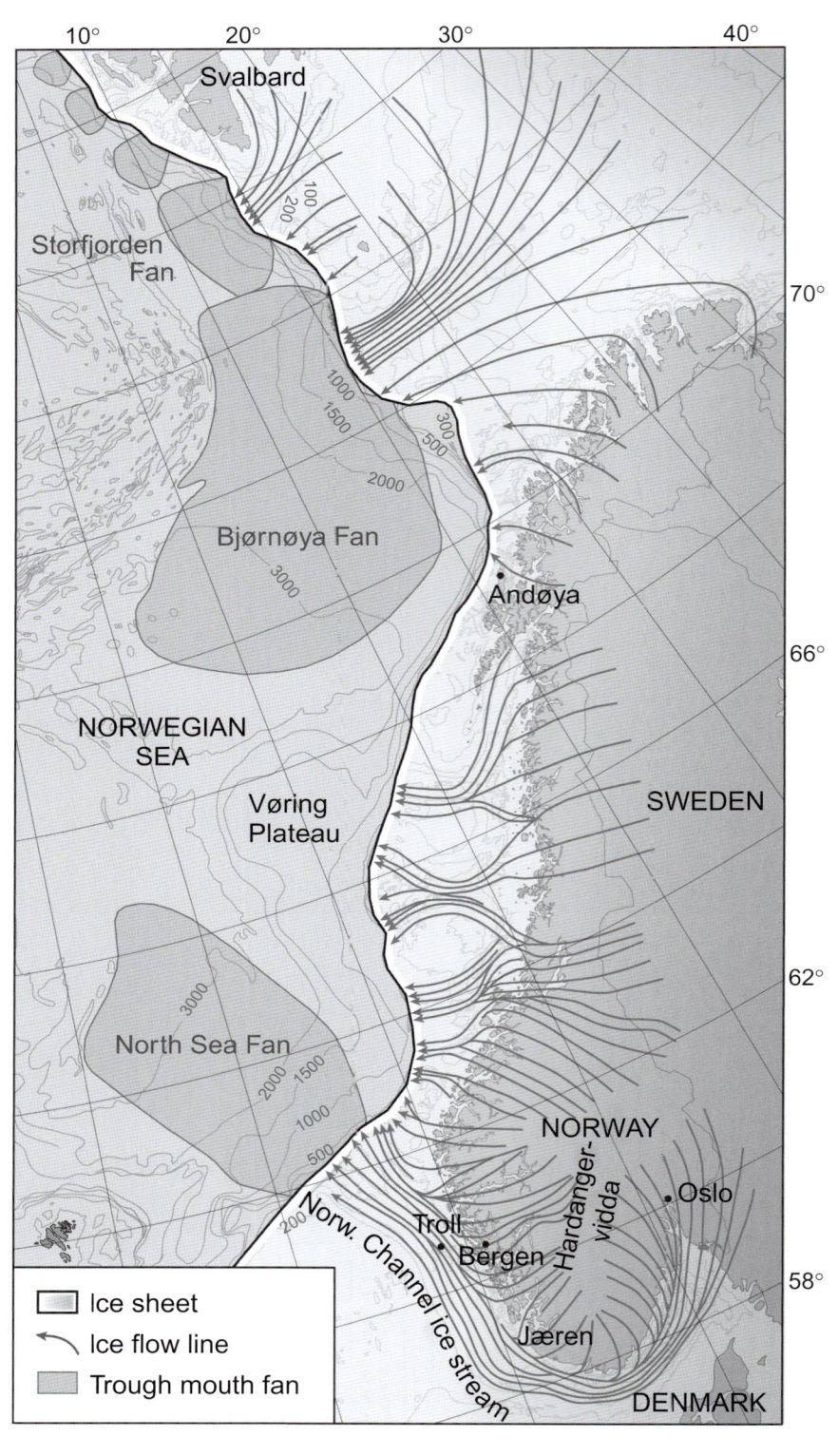

FIGURE 22.6 Reconstruction of ice flow regime of the western margin of the Scandinavian and Barents ice sheets during later phases of the LGM, partly based on Ottesen et al. (2005); here, modified from Vorren and Mangerud (2008). Between the marked ice streams, there was slower-flowing ice. The major trough mouth fans are also marked.

Further back in time, ^{14}C dates become even more uncertain, and their reliability are getting even more questionable. The fact that the caves near Ålesund were blocked in succession from east to west during the Laschamp excursion is therefore important for assessing the finer details of the chronology (Valen et al., 1995; Mangerud et al., 2003) because the implication is that this part of the coast was ice-free shortly before the Laschamp. This older ice-free period is named the Austnes interstadial, and it is supported by some ^{14}C dates from marine shells (Mangerud et al., 2010), but we consider the correlation with Laschamp to be a more reliable dating than the ^{14}C ages. The Austnes interstadial might represent later stages of a long ice-free period starting with the ice margin retreat during the early MIS 3 (Bø interstadial) (Fig. 22.4), or alternatively a shorter ice-free period separated from Bø with one or more ice advance(s).

22.5. THE LATE WEICHSELIAN

22.5.1. The Age of the Last (Local) Glacial Maximum

Here, we use the term LGM in a broad sense, namely as the maximum ice sheet extent in a particular area after the Ålesund interstadial and before about 18 cal. ka. We assume that the maximum extent was not reached simultaneously along the entire coast of Norway. Earlier results were described in some details by Mangerud (2004); here, we will mainly call attention on some more recent findings. Most studies dealing with the Late Weichselian glacial history conclude that there have been (several) fluctuations of the ice margin during the period 30–18 cal. ka BP (Olsen et al., 2001b, 2002; Vorren and Plassen, 2002; Sejrup et al., 2009; Mangerud et al., 2010). Further, marine cores from west of Norway show several peaks of ice-rafted detritus (IRD) that may reflect fluctuations of the ice margin (Dokken and Jansen, 1999; Elliot et al., 2001; Lekens et al., 2009). However, some of the inferred ages for glacial expansions are conflicting; others are difficult to correlate because of uncertain chronology. There seems to be an agreement that the first and largest ice advance from Norway towards south (Denmark) and south-west occurred ~29–27 ka (Houmark-Nielsen and Kjær, 2003; Larsen et al., 2009a; Sejrup et al., 2009), in which case the assumed age (~29 ka) of the Hamnsund interstadial must be wrong (Mangerud et al., 2010). However, Houmark-Nielsen and Kjær (2003) and Larsen et al. (2009a) also find evidence of another re-advance ~23–21 ka from Norway. We emphasise that the maximum extent is not directly dated anywhere north of the mouth of the Norwegian Channel, that is, along most of the ice margin west of Norway, except the Egga II moraine outside Andøya that is dated to about 19 cal. ka BP (Vorren and Plassen, 2002).

22.5.2. The Maximum Extent of the Last (Local) Glacial Maximum

There are no new observations to suggest that the maximum ice sheet extent given in Mangerud (2004) and Svendsen et al. (2004) needs a major revision. However, the glacial morphology of the shelf is now much better mapped, including end moraines and glacial lineations (Ottesen et al., 2001, 2005, 2008; Dowdeswell et al., 2007; Andreassen et al., 2008; Sejrup et al., 2009; Winsborrow et al., 2010); a simplified synthesis map is given in Fig. 22.6. A main result from recent mapping efforts is that a number of fast-flowing ice streams have crossed the continental shelf in between areas of much slower-flowing ice. This would of course also influence the processes and the form of the ice margin. Based on the mentioned mapping, there is now an almost unanimous agreement that the LGM ice sheet reached the shelf edge. This break has for long time been postulated to represent the maximum possible limit because of calving in the deeper water outside.

The only area where there recently has been some debate if the LGM reached the shelf edge or not is around the island Andøya in northern Norway (Figs. 22.1 and 22.11). Vorren and Plassen (2002) mapped two moraines, Egga I and II, along the shelf edge outside Andfjorden just north of Andøya. Their continuation on land was mapped on Andøya and dated to >25 and 19–18 cal. ka BP, respectively (Fig. 22.11). The ages were based on ^{14}C dates from the long and continuous lacustrine records from Andøya (Vorren, 1978; Vorren et al., 1988; Alm, 1993). The reconstructions are reasonable, and they are supported by the exposure dates (although mainly on bedrock) by Nesje et al. (2007). This site demonstrates some of the extreme differences in interpretations, as Lambeck et al. (2002) predicted 1000–1500 m thick ice from isostatic modelling. Considering that the distance to the shelf edge is only 10–20 km outside Andøya, the latter inferred ice thickness sounds unlikely and was also later reduced by Lambeck et al. (2010). We assume that the Egga I (>25 cal. ka BP) advance overran Andøya and reached the shelf edge, partly based on the marine mapping (Ottesen et al., 2005) and partly on the high marine limit of 36 m a. s.l. after Egga I (Vorren et al., 1988); the latter indicating a major glacio-isostatic depression as the eustatic sea level at that time certainly was low.

22.5.3. The Elevation of the LGM Ice Sheet Surface

The existence of possible nunataks during glacial maxima is a more than 100 year's old discussion theme in Norway; a short review is given in Mangerud (2004). Several recent reconstructions of ice thickness are based on glacio-isostatic or glaciological modelling, but here, we will mainly discuss observational data. Geomorphologic features (tors, block fields, trim lines, etc.) have been used to map the upper surface of the LGM ice sheet. In some exceptional cases, that might be right, but it has repeatedly been shown that such forms can survive under (often cold-based parts of) ice sheets (Kleman, 1994; Phillips et al., 2006). We also clarify that any lower limit of such features, including "warm climate clay minerals" (Roaldset et al., 1982) would rather represent the highest level of any Middle- to Late Quaternary glaciation and not necessarily the LGM.

The introduction of exposure dating methods using cosmogenic isotopes (e.g. ^{10}Be and ^{26}Al) opened new possibilities to solve some of the problems. However, for several reasons, it has been a difficult task to obtain reliable dates from isolated mountain summits that potentially were nunataks—or alternatively covered by ice—during LGM: (1) there have often been minimal glacial erosion, as shown by, for example, the comparison of ^{10}Be exposure ages from exposed bedrock surfaces and erratics (Goehring et al., 2008). (2) Erratics may, as the block fields themselves, have survived glaciations after deposition. (3) Erratics from local bedrock or upstream peaks will, because of the limited glacial erosion, more often than erratics from lower terrain have an "inherited age" from earlier exposures. Such peak-source erratics will dominate if the summit is covered by only thin ice. Erratics picked from lower terrain and transported up-hill to a summit are less vulnerable for inheritance, but they would require a thick ice over the summit. Some of the mentioned problems can be solved by using more than one isotope (Fabel et al., 2002), and by obtaining internally and geographically consistent results both concerning exposure dates and glacial reconstructions.

There are few published studies using exposure dating to estimate maximum ice sheet surface elevation near the ice divide. Goehring et al. (2008) present results from two summits, Blåhø and Elgåhogna, located in southern central Norway, slightly north of the assumed ice divide during the LGM (Fig. 22.7). The summit of Blåhø is 1617 m a.s.l. and based on their results, especially a ^{10}Be age of 25.1 ± 1.8 ka from a boulder near the summit; they conclude that the mountain was entirely covered during LGM. Elgåhogna is 1460 m a.s.l., and the results are not unambiguous, but again, based on young ages near the top, they conclude the mountain was completely covered by ice.

Mangerud (2004) considered that the ice sheet probably covered (almost?) all mountains in southern central Norway, and that the ice divide was >2000–2500 m a.s.l., before the Norwegian Channel Ice Stream commenced to operate and lowered the ice surface. We maintain that a low ice surface in central southern Norway as postulated by Nesje and Dahl (1992) is difficult to combine with other

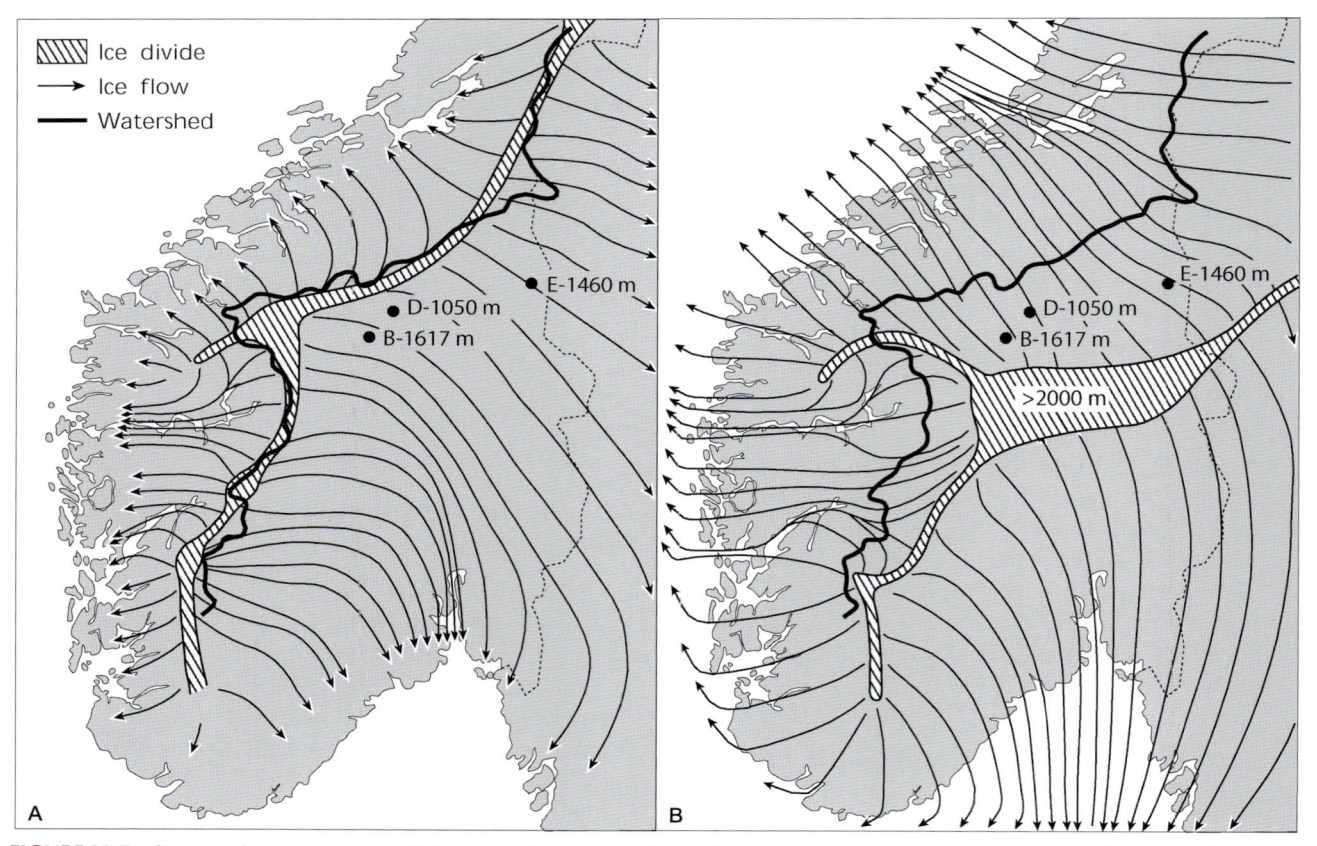

FIGURE 22.7 Conceptual reconstructions of (A) an early phase of a glaciation when the ice divide was located close to the highest mountains and the water divide, and (B) a time around the Last Glacial Maximum when the ice divide was located well south and east of the watershed. Locations and elevations are shown for the summits Blåhø (B in map) and Elgåhogna (E), where exposure dates have been performed and marked, and also for the Dørålen (D) site. Modified from Vorren and Mangerud (2008).

observations such as the geographical extent of the ice sheet to Denmark–Germany and that ice flowed across the water shed from south to north; and even more difficult when Dahl et al. (2010) postulate a maximum altitude for the LGM ice sheet as low as 1050 m a.s.l. in Dørålen. These low ice surfaces are also contradicted by the ^{10}Be dates mentioned earlier (Fig. 22.7). Glaciological and isostatic ice sheet models produce very simplified ice sheets, but they all produce an ice surface up to 2000–3000 m a.s.l. (Näslund et al., 2003; Forsström and Greve, 2004; Peltier, 2004; Siegert and Dowdeswell, 2004; Lambeck et al., 2010), although most allow the highest peaks to project through the surface as nunataks.

Another type of ice-free summits that have been proposed is mountains along the fjords on the west coast. As stated in Mangerud (2004), nunataks in these areas are more feasible from a glaciological point of view than peaks located far inland because the deep fjords would efficiently drain the ice and thus keep the ice surface low. Dahl (1955) provided a strong theoretical argument in favour of such nunataks. He pointed out that the ice sheet margin would calf at the shelf edge because the water depth there increased fast, and that the distance from land to the shelf edge is very short outside Møre (around Ålesund, Fig. 22.1) and Andøya (Fig. 22.1). He further argued that with reasonable ice surface profiles, some of the coast-near mountain peaks would penetrate above the ice surface. This argument is even stronger today when we assume there were fast-flowing ice streams with lower gradients across the shelf. The two mentioned areas are in fact the most studied and discussed areas during the past decades. All scientists who have worked in Nordfjord, just south of Møre, have concluded that, for example, the mountain of Skåla (1848 m a.s.l.) probably remained ice-free, partly based on trimlines and weathering limits and partly on results of exposure dating (Nesje and Dahl, 1992; Brook et al., 1996; Goehring et al., 2008). Similar methods were used along a profile from Andøya and inland (Nesje et al., 2007), although as already mentioned, most exposure dates there were from bedrock exposures. In both areas, the conclusions are geologically reasonable, but not unambiguously demonstrated.

22.5.4. The Deglaciation

A new reconstruction of the deglaciation of the Eurasian ice sheets, based on a large database of radiometric dates, geomorphologic elements and stratigraphy, is presently under construction (Gyllencreutz et al., 2007). Here, we will describe some few examples of deglaciation moraines, chronology and processes.

The deglaciation of Norway can be subdivided into three periods. The first period starts when the ice margin withdraws from its maximum position and lasts until the onset of the Younger Dryas cold reversal (Fig. 22.1). During this first period, the ice sheet melted and calved back so that the continental shelf became completely ice-free and also parts of the coast became deglaciated. Obviously, this ice retreat led to a major thinning of the remaining ice sheet further inland. No end moraines or other ice-marginal features from this period have been mapped over long distances. The next period is the Younger Dryas when the ice margins re-advanced in some areas, notably in the Bergen area (see below), halted in others and even withdrew in some areas as, for example, in the area around Trondheim (Fig. 22.1; Reite, 1994). Moraines from the Younger Dryas are the only deglacial moraines that can be mapped continuously around the Scandinavian Ice Sheet (Fig. 22.1), although the moraines were not formed strictly synchronously (Mangerud, 1980). The third period is the Early Holocene, when the ice sheet disappeared during some 1000–1500 years. There were some distinct halts in most areas, often depending on topographical situations as sills in fjords or palaeo-fjord heads.

22.5.5. The Norwegian Channel and Oslofjord

The Norwegian Channel Ice Stream had a different pattern than the other ice streams on the continental shelf (Fig. 22.6). The ice stream flowed southward in Oslofjorden, continued around the southern coast of Norway and collected ice flowing radially from the SW mountain plateau (Hardangervidda) and ended with a northward flow at the mouth of the channel. The ice stream was not active during the entire Late Weichselian. As discussed in Mangerud (2004), the first MIS 2 ice flow to reach Denmark crossed the Norwegian Channel and transported erratics from Norway to Denmark. This development was recently substantiated by Larsen et al. (2009a) and Houmark-Nielsen (2010). According to Mangerud (2004), the fast ice flow probably started at the mouth of the Norwegian Channel and subsequently propagated up-ice to form a long ice stream. The main point here is that the ice stream was active during the later phases of LGM and that deposition of glacial sediments at the mouth of the channel ceased at 19.0 ± 0.2 cal. ka BP (King et al., 1998; Nygård et al., 2007). Some 220 km upstream the well-dated Troll cores (Fig. 22.6) indicate a deglaciation near 18.5 cal. ka (Sejrup, et al., 2009) implying very fast calving up the deep channel, whether this etreat was triggered by a sea level rise or a climate warming. The oldest obtained ^{14}C ages that show ice-free conditions on land in southern Norway were obtained from Jæren (Fig. 22.6) where several dates indicate a deglaciation between 17 and 16 cal. ka (Knudsen, 2006). At that time, the Norwegian Channel must have been ice-free this far upstream.

Around the Oslofjord, the pattern of ice retreat is known in some more detail. As seen from Fig. 22.8, a major calving

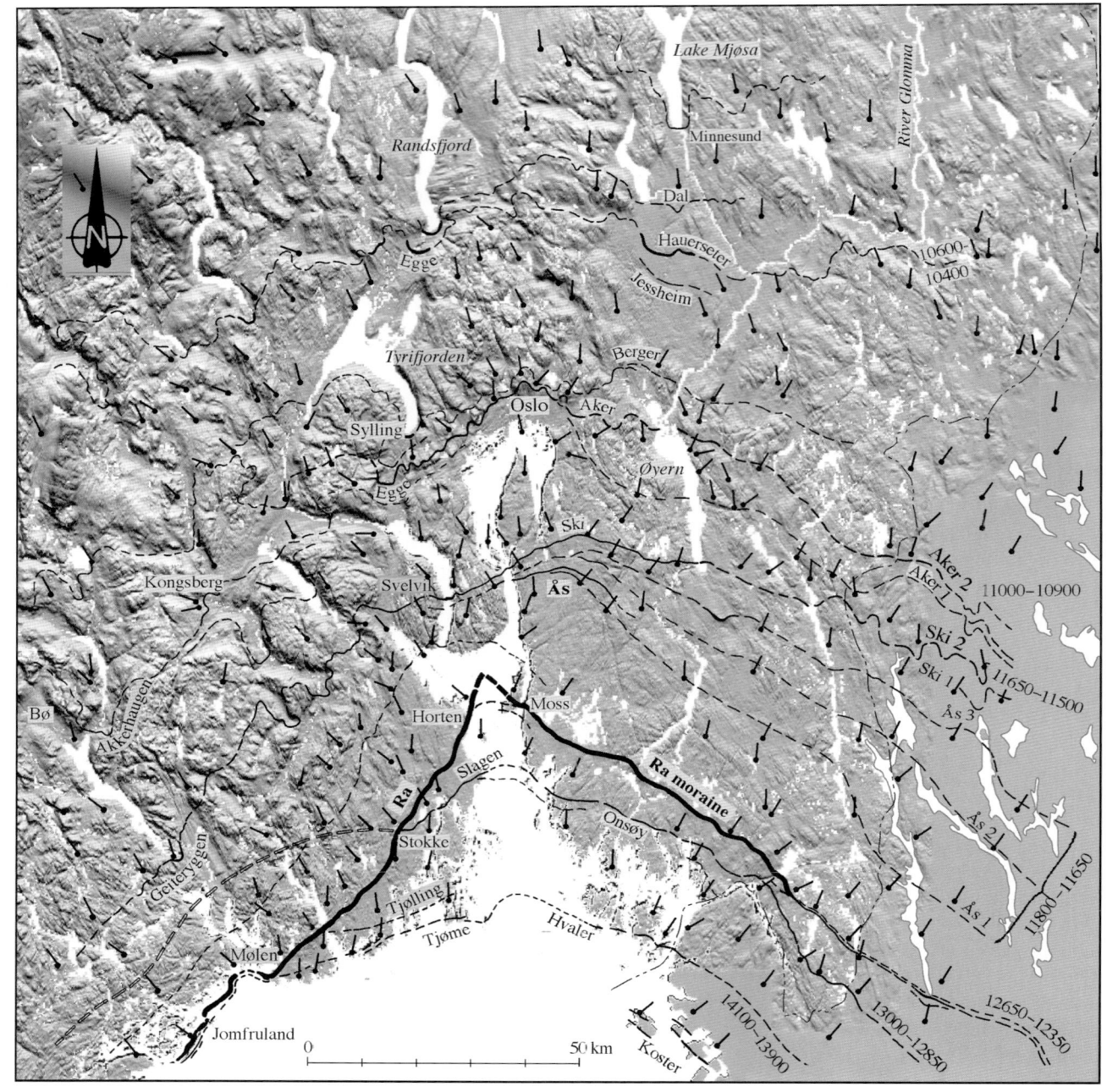

FIGURE 22.8 A map showing ice-marginal deposits and directions of glacial striae around the Oslofjorden. Compiled by Rolf Sørensen (unpublished); here, modified from Vorren and Mangerud (2008). Ages are given in calibrated years BP, mainly based on calibrated ^{14}C dates. Stippled lines indicate less distinct moraines or uncertain correlations.

bay developed along the axis of the deeper part of the fjord, where the main ice stream was flowing some few thousand years earlier. The oldest moraine that can be correlated across the fjord is the Tjøme–Hvaler moraine formed about 14 cal. ka BP (Sørensen, 1979) and correlated with Trollhättan moraine in Sweden (Lundqvist and Wohlfarth, 2001). It should be noted that on the eastern side of the fjord, the younger moraines are oriented in a more or less parallel retreat fashion, whereas on the west side the prominent (Younger Dryas), Ra moraine cut across the older moraines, reflecting a re-advance in this area (Bergstrøm, 1999). Mainly, based on ^{14}C dates of marine molluscs, the prevailing view has been that the Ra moraines around Oslofjorden were deposited during the middle part of the

Younger Dryas, whereas the Ås–Ski moraines were formed during later parts of the Younger Dryas (Mangerud 2004). However, some new dates obtained by Rolf Sørensen (written communication December 2009, Sørensen et al., in press) may indicate that only the Ra moraines are of Younger Dryas age, and that the Ås–Ski moraines are younger. If correct, this would make the age of the Younger Dryas re-advance on the west side of the Oslofjord more in line with the re-advance in the Bergen area (see below) that originated from the same ice culmination over the mountain plateau Hardangervidda (Fig. 22.6).

The Ra moraines consist for a large part of diamicton, mainly till, except for some meltwater deposits that are localised near the valley mouths. The glaciofluvial components increase in the younger moraines, but ridges of till are found also in the Ås, Ski and Aker moraines. The even younger and large ice margin deposits at Berger, Jessheim, Hauerseter, Dal and Minnesund (Fig. 22.8) are totally dominated by glaciofluvial sediments. The marine limit around Oslo is 220 m a.s.l., and in valleys distal to the mentioned ice-marginal deposits, there are up to 100 m thick deposits of glaciomarine clay.

22.5.6. Western Norway

The break-up of the Norwegian Channel Ice Stream must have had a major impact on the ice sheet configuration and ice flow over western Norway. If we consider the Bergen area, the westerly ice flow over land turned northwards into the Norwegian Channel (Fig. 22.6) before the ice masses ended up as icebergs at the shelf edge some 220 km further to the northwest. The mentioned distance to the ice margin suddenly became much shorter when the ice disappeared from the northern segment of the Norwegian Channel. The inferred fast retreat from the mouth of the Norwegian Channel to the location of the Troll cores probably resulted in a much steeper gradient of the ice surface towards the mountain areas further inland, that is, to the east in Fig. 22.9. This may have led to a situation where the ice margin during deglaciation was "hanging" on the outer islands west of Bergen (Fig. 22.9), calving into the Norwegian Channel. As mentioned earlier, the Troll site became ice-free as early as 18.5 cal. ka BP, whereas the oldest dates on land (Blomvåg) have yielded significantly younger ages of about 14.6 cal. ka BP (Fig. 22.10). This implies that the ice margin has been calving just outside the coast near Bergen for an extended period of 4000 years (Fig. 22.10), which is the most long-lasting identified halt during the deglaciation of Norway. The standstill was partly a result of ice sheet dynamics, that is, caused by the fast break-up of the Norwegian Channel. However, important for maintaining this position over so long time was also the cool climate combined with the mountain plateaux extending from the central mountains (Hardangervidda) and almost to the coast (Fig. 22.9), together providing large accumulation areas for snow because the equilibrium line was low (Mangerud, 1980). This can be seen when we compare with Jæren (Fig. 22.6), where the high plateaux are located further away and the coast remained ice-free from about 16 cal. ka BP.

During the Bølling, the outermost coast was ice-free for a short period (Fig. 22.10) before the ice margin re-advanced across the outermost islands and presumably again calved into the Norwegian Channel. These outer islands became finally ice-free during early Allerød when the ice margin retreated well inland. However, this retreat apparently stopped before the end of Allerød and was followed by a lasting period with ice growth (Lohne et al., 2007), resulting in an ice advance that reached its maximal position at the very end of the Younger Dryas (Bondevik and Mangerud, 2002). The corresponding glacio-isostatic rebound commenced shortly after (Lohne et al., 2004). During this advance, the ice front reached almost the extreme west coast once again (Fig. 22.9), and the outlet glaciers in the fjords grew to a thickness of 2000 m in areas that were completely ice-free a few hundred years before (e.g. Hardangerfjorden) (Andersen et al., 1995).

At the Younger Dryas/Preboreal boundary, a very fast melting and calving commenced in the fjords. The pattern can be seen from glacial striae which show that the ice flow changed from being regionally directed towards west to flow directions depending on local topography and directed from land areas towards the fjords. The pattern of the moraines (Fig. 22.9) indicates that calving was the dominating process as the distances between the Younger Dryas and Preboreal moraines (Osa–Eidfjord and correlated moraines, Fig. 22.9) are largest along the deeper fjords (Hardangerfjorden and Sognefjorden). Evidently, the retreat rate slowed down considerably at the fjord heads. The Osa–Eidfjord (Anundsen and Simonsen, 1967) and correlated moraines are dated to about 11 cal. ka BP (Bergstrøm, 1975; Mangerud et al., 2009), indicating that the deglaciation of the up to 1300 m deep fjords only lasted some few hundred years during the earliest Holocene. The break-up of the fjords must have led to a major draw down of the entire ice sheet draining towards the fjords.

22.5.7. Andøya–Andfjord

The oldest organic sediments discovered on land in Norway and that are postdating the LGM are found in lake basins on the island Andøya in northern Norway (Figs. 22.1 and 22.11; Vorren et al., 1988). As discussed earlier, they have also been used to argue that Andøya remained ice-free during the LGM, which possibly is correct. In this area, the early deglaciation is well dated and mapped (Vorren and Plassen, 2002). The Bjerka Moraine in Andfjorden has been overrun by ice, but Vorren and Plassen (2002) argue that it

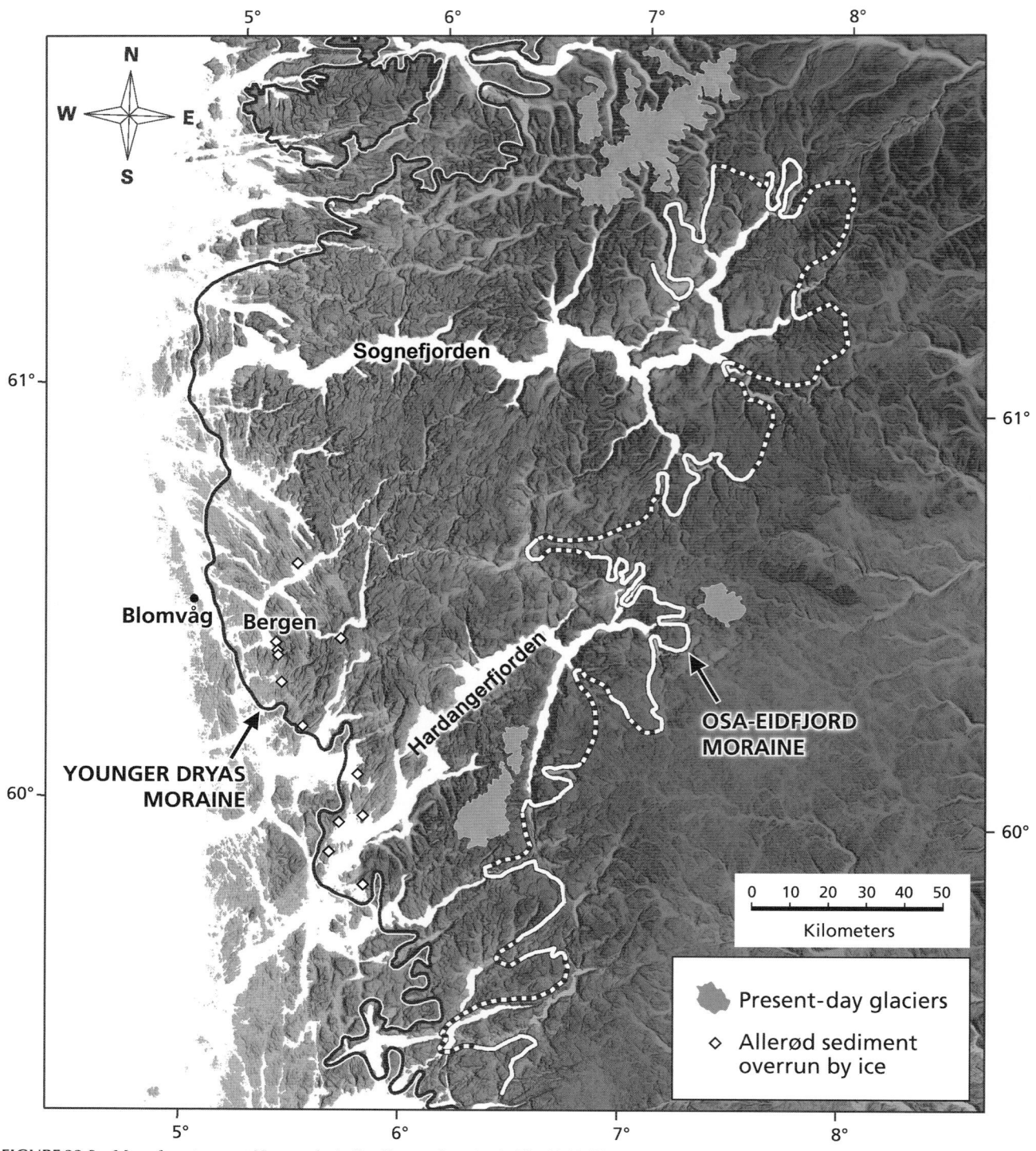

FIGURE 22.9 Map of westernmost Norway, including Bergen (location in Fig. 22.1). The Younger Dryas (Herdla–Halsnøy) moraines are marked. Note an isolated ice cap outside the main ice sheet limits in the upper left corner (Sønstegaard et al., 1999). Some localities around Bergen–Hardangerfjorden with ^{14}C-dated Allerød sediments overrun by the Younger Dryas re-advance are marked with diamonds. The Osa–Eidfjord moraines near the head of Hardangerfjorden and correlated moraines of Preboreal age are also marked. Possible isolated ice caps of Preboreal age outside the Osa–Eidfjord moraines are not indicated.

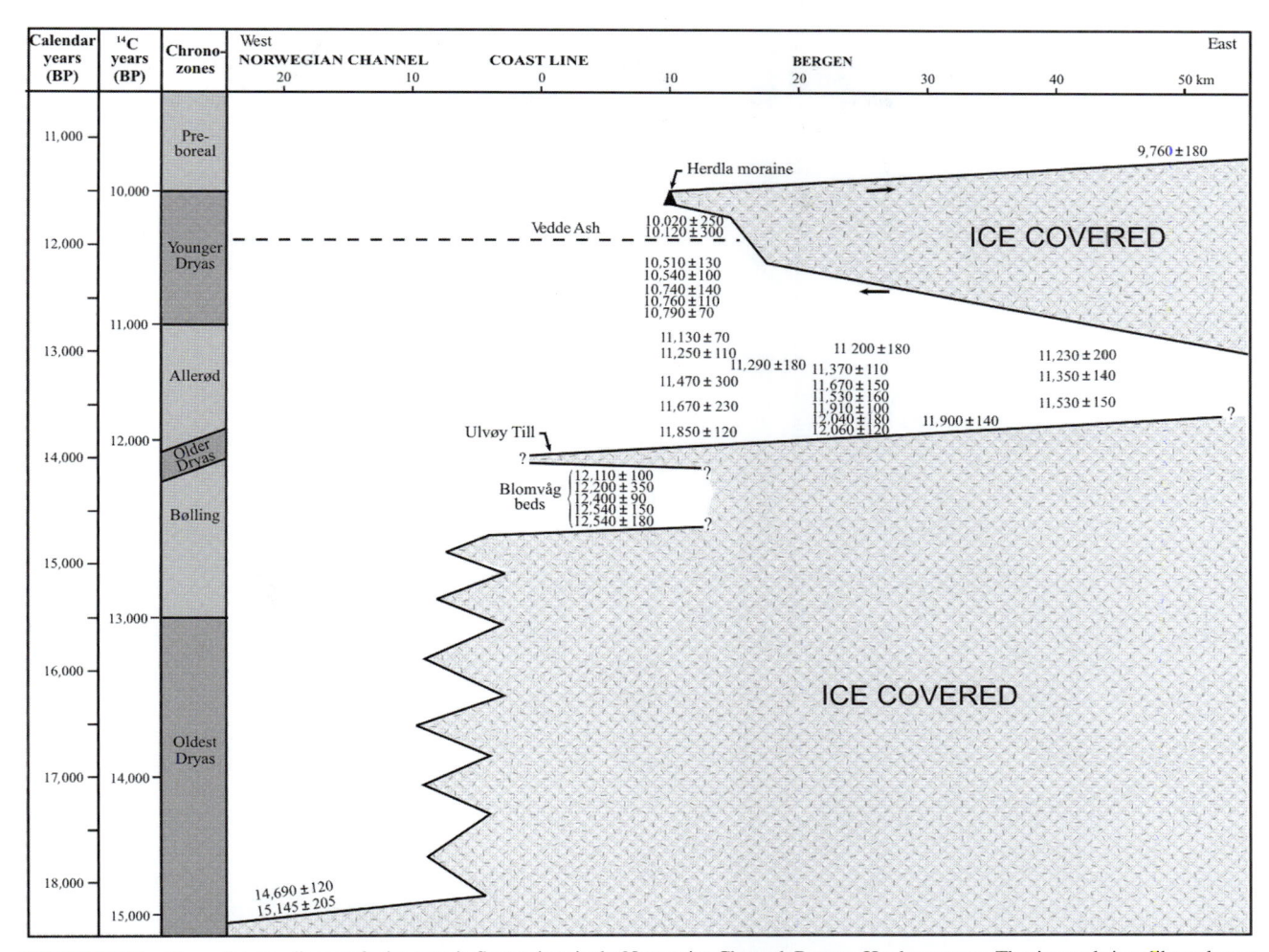

FIGURE 22.10 Time–distance diagram for ice margin fluctuations in the Norwegian Channel–Bergen–Hardanger area. The timescale in calibrated years (Reimer et al., 2009) is linear, but a (non-linear) timescale in ^{14}C years is also shown. Selected representative ^{14}C ages are given in ^{14}C years; for marine samples corrected for a reservoir age of 440 years. The ages plotted in the Norwegian Channel are from the Troll cores (Sejrup et al., 2009), located in the middle part of the channel some 50–60 km from the coast (Fig. 22.1). But in this diagram, these dates are for simplicity plotted in the eastern part of the channel because the deglaciation was from the north with a retreating ice front across the channel. The ages are therefore representative also for the eastern part. Developed from a diagram in Mangerud (2000).

probably has been overrun only once and that its age is between the ages of Egga I and II (Fig. 22.12). If the proposed chronology is correct, then there was a period lasting about 7000 years between two ice margin advances to the shelf edge (Egga I, 26 cal. ka BP and Egga II, 19–18 cal. ka BP), indicating that the ice margin oscillated near the LGM position for an extended period. After 18 cal. ka, there was apparently a fast deglaciation of the shelf and deeper part of Andfjord, until the retreat slowed down when the ice margin reached the coast of about 16 cal. ka. Vorren and Plassen (2002) could not find any end moraine marking the D-event (Figs. 22.11 and 22.12) which is reconstructed along the outer coast; it is marked by a maximum of IRD in cores and a glacial fauna. Moraines of the Skarpnes event, dated to about 14 cal. ka, are found shortly outside or close to the Younger Dryas moraines over large areas in this part of the country (Andersen, 1968), but they are not as continuous as the Younger Dryas moraines. The Tromsø–Lyngen moraines of Younger Dryas age are distinct and well mapped not only in this restricted area but also along most of the coast of Northern Norway (Andersen et al., 1995).

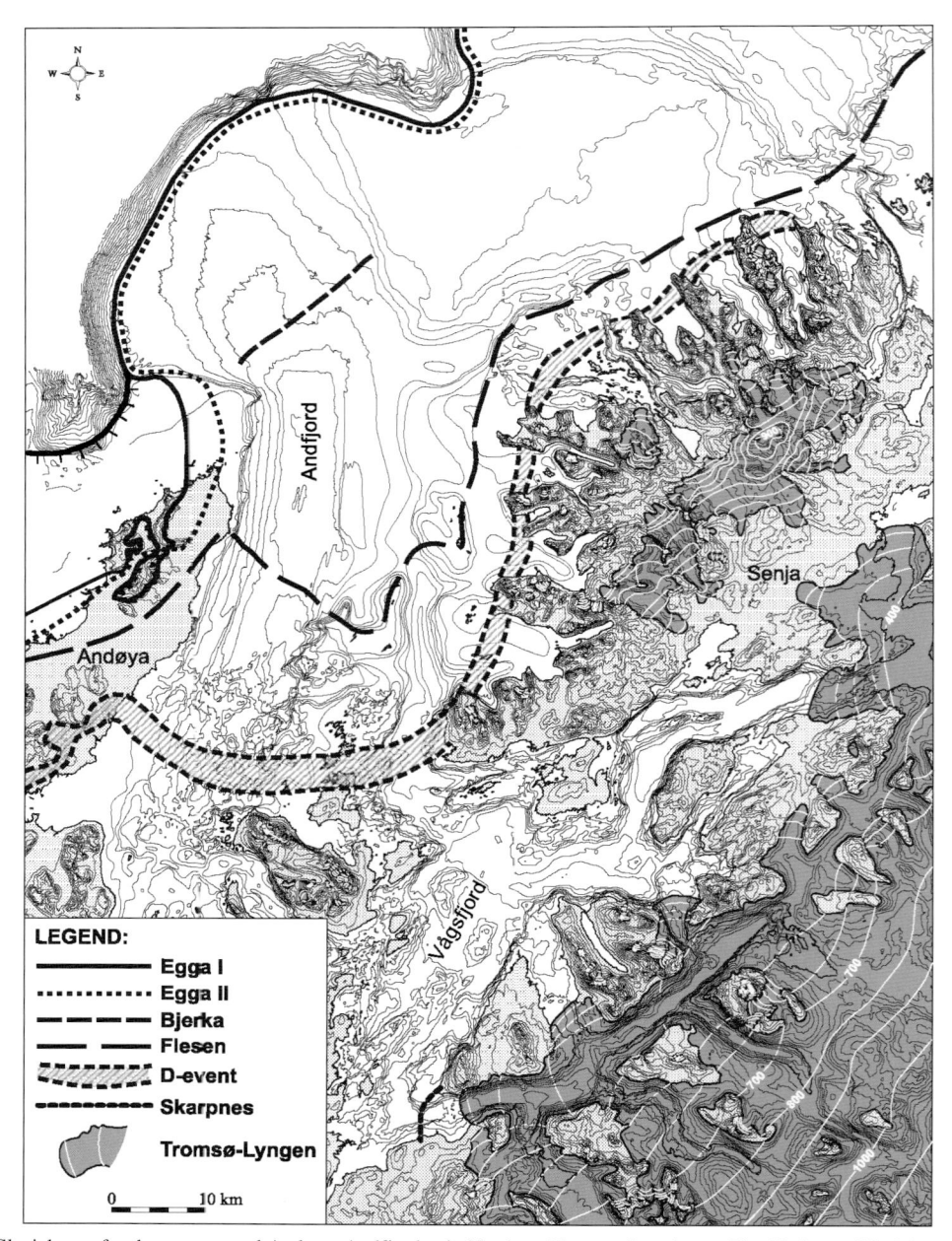

FIGURE 22.11 Glacial map for the area around Andøya–Andfjorden in Northern Norway (location on Fig. 22.1), modified from Vorren and Plassen (2002). Full reconstructions of the ice sheet and glaciers (with white ice surface elevation contours at 100 m interval) are shown for the Younger Dryas. We have added an alternative continuation (marked with tagged line) for the Egga I along the shelf edge towards south (corresponding with the limit shown in Fig. 22.6).

Chapter | 22 Glacial History of Norway

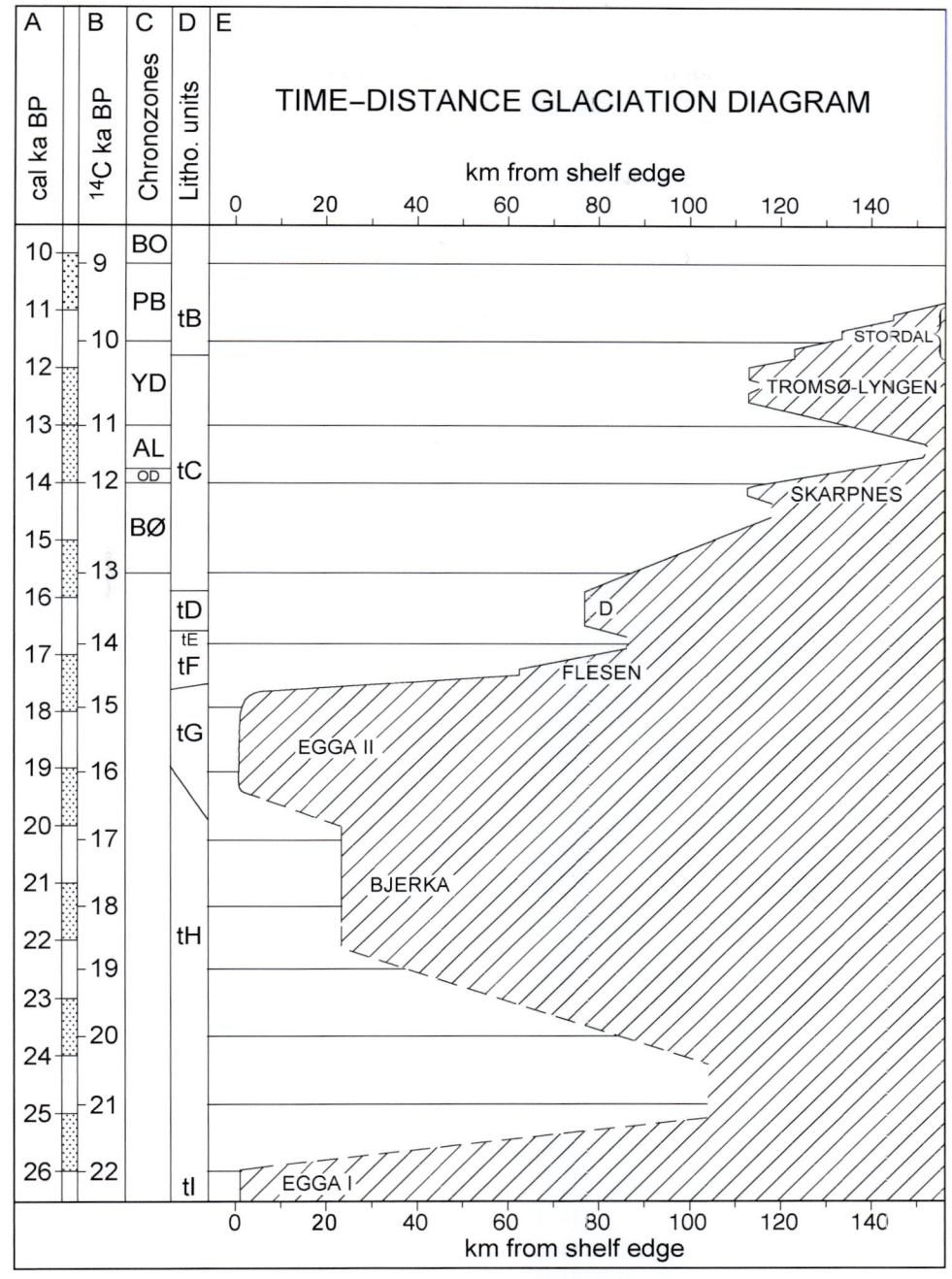

FIGURE 22.12 Time–distance glaciation diagram for the Andfjord area (Fig. 22.11). Modified from Vorren and Plassen (2002). To the left are age scales in both ^{14}C (non-linear) and calibrated (linear) kilo years.

REFERENCES

Alm, T., 1993. Øvre Æråsvatn—palynostratigraphy of a 22,000 to 10,000 BP lacustrine record on Andøya, northern Norway. Boreas 22, 171–188.

Andersen, B.G., 1968. Glacial geology of Western Troms, North Norway. Nor. geol under. 256, 160pp.

Andersen, B.G., Mangerud, J., Sørensen, R., Reite, A., Sveian, H., Thoresen, M., et al., 1995. Younger Dryas ice-marginal deposits in Norway. Quatern. Int. 28, 147–169.

Andreassen, K., Laberg, J.S., Vorren, T.O., 2008. Seafloor geomorphology of the SW Barents Sea and its glaci-dynamic implications. Geomorphology 97 (1–2), 157–177.

Anundsen, K., Simonsen, A., 1967. Et pre-borealt breframstøt på Hardangervidda og i området mellom Bergensbanen og Jotunheimen. Årbok Uni. Bergen. Mat. nat. 7, 1–42.

Arnold, N., van Andel, T., Valen, V., 2002. Extent and dynamics of the Scandinavian Ice Sheet during oxygen isotope stage 3 (65,000-25,000 yr B.P.). Quatern. Res. 57, 38–48.

Baumann, K.-H., Lackschewitz, K.S., Mangerud, J., Spielhagen, R.F., Wolf-Welling, T.C.W., Henrich, R., et al., 1995. Reflection of Scandinavian Ice Sheet fluctuations in Norwegian Sea sediments during the last 150,000 years. Quatern. Res. 43, 185–197.

Becker, J., Sandwell, D., Smith, W., Braud, J., Binder, B., Depner, J., et al., 2009. Global bathymetry and elevation data at 30 arc seconds resolution: SRTM30_PLUS. Mar. Geodesy 32, 355–371.

Bergstrøm, B., 1975. Deglasiasjonsforløpet i Aurlandsdalen og områdene omkring, Vest-Norge. Nor. geol. under. 317, 33–69.

Bergstrøm, B., 1999. Glacial geology, deglaciation chronology and sea-level changes in the southern Telemark and Vestfold counties, south-eastern Norway. Nor. geol. under. 435, 23–42.

Bondevik, S., Mangerud, J., 2002. A calendar age estimate of a very late Younger Dryas ice sheet maximum in western Norway. Quatern. Sci. Rev. 21, 1661–1676.

Brendryen, J., Haflidason, H., Sejrup, H.P., 2010. Norwegain Sea tephrostratigraphy of marine isotope stages 4 and 5: prospects and problems for tephrochronology in the North Atlantic region. Quatern. Sci. Rev. 29, 847–864.

Brook, E.J., Nesje, A., Lehman, S.J., Raisbeck, G.M., Yiou, F., 1996. Cosmogenic nuclide exposure ages along a vertical transect in western Norway: implications for the height of the Fennoscandian ice sheet. Geology 24, 207–210.

Carr, S.J., Holmes, R., Meer, J.J.M.v.d., Rose, J., 2006. The Last Glacial Maximum in the North Sea Basin: micromorphological evidence of extensive glaciation. J. Quatern. Sci. 21 (2), 131–153.

Dahl, E., 1955. Biogeographic and geologic indications of unglaciated areas in Scandinavia during the glacial stages. Bull. Geol. Soc. Am. 66, 1499–1619.

Dahl, S.O., Linge, H., Fabel, D., Murray, A.S., 2010. Extent and timing of the Scandinavian Ice Sheet during Late Weichselian (MIS3/2) glacier maximum in central southern Norway—link to the Norwegian Channel Ice Stream? Abstr. Proc. Geol. Soc. Norway 1–2010, 37–38.

Dahlgren, K.I.T., Vorren, T.O., Stoker, M.S., Nielsen, T., Nygard, A., Sejrup, H.P., 2005. Late Cenozoic prograding wedges on the NW European continental margin: their formation and relationship to tectonics and climate. Mar. Petrol. Geol. 22 (9–10), 1089–1110.

Dokken, T., Jansen, E., 1999. Rapid changes in the mechanism of ocean convection during the last glacial period. Nature 401, 458–461.

Dowdeswell, J.A., Ottesen, D., Rise, L., 2006. Flow switching and large-scale deposition by ice streams draining former ice sheets. Geology 34 (4), 313–316.

Dowdeswell, J.A., Ottesen, D., Rise, L., Craig, J., 2007. Identification and preservation of landforms diagnostic of past ice-sheet activity on continental shelves from three-dimensional seismic evidence. Geology 35 (4), 359–362.

Dowdeswell, J.A., Ottesen, D., Rise, L., 2010. Rates of sediment delivery from the Fennoscandian Ice Sheet through an ice age. Geology 38, 3–6.

Ehlers, J., Eissmann, L., Lippstreu, L., Stephan, H.-J., Wansa, S., 2004. Pleistocene glaciations of North Germany. In: Ehlers, J., Gibbard, P.L. (Eds.), Quaternary Glaciations—Extent and Chronology. Part I: Europe. Elsevier, Amsterdam, pp. 135–146.

Elliot, M., Labeyrie, L., Dokken, T., Manthé, S., 2001. Coherent patterns of ice-rafted debris deposits in the Nordic regions during the last glacial (10–60 ka). Earth Planet. Sci. Lett. 194 (1–2), 151–163.

Fabel, D., Stroeven, A.P., Harbor, J., Kleman, J., Elmore, D., Fink, D., 2002. Landscape preservation under Fennoscandian ice sheets determined from in situ produced ^{10}Be and ^{26}Al. Earth Planet. Sci. Lett. 201 (2), 397–406.

Feyling-Hanssen, R.W., 1974. The Weichselian section of Foss-Eigeland, south-western Norway. Geol. Fören. Stockh. Förhandlinger 96, 341–353.

Forsström, P.-L., Greve, R., 2004. Simulation of the Eurasian ice sheet dynamics during the last glaciation. Glob. Planet. Change 42 (1–4), 59–81.

Fronval, T., Jansen, E., 1996. Late Neogene paleoclimates and paleoceanography in the Iceland-Norwegian Sea: evidence from the Iceland and Vøring plateaus. In: Thiede, J., Myhre, A.M., Firth, J.V., Johnsen, G., Ruddiman, W. (Eds.), Proceedings of the Ocean Drilling Program, Scientific Results, vol. 151. College Station, Texas, USA.

Gibbard, P.L., Head, M.J., Walker, M.J.C., and the Subcommission on Quaternary Stratigraphy, 2009. Formal ratification of the Quaternary System/Period and the Pleistocene Series/Epoch with a base at 2.58 Ma. J. Quatern. Sci. 25, 96–102.

Goehring, B.M., Brook, E.J., Linge, H., Raisbeck, G.M., Yiou, F., 2008. Beryllium-10 exposure ages of erratic boulders in southern Norway and implications for the history of the Fennoscandian Ice Sheet. Quatern. Sci. Rev. 27 (3–4), 320–336.

Gyllencreutz, R., Mangerud, J., Svendsen, J.I., Lohne, Ø., 2007. DATED—a GIS-based reconstruction and dating database of the Eurasian Deglaciation. Geol. Surv. Finland Spec. Pap. 46, 113–120.

Helle, H., Sønstegaard, E., Coope, G., Rye, N., 1981. Early Weichselian peat at Brumunddal, southeastern Norway. Boreas 10, 369–379.

Hjelstuen, B., Sejrup, H., Haflidason, H., Nygård, A., Ceramicola, S., Bryn, P., 2005. Late Cenozoic glacial history and evolution of the Storegga Slide area and adjacent slide flank regions, Norwegian continental margin. Mar. Petrol. Geol. 22 (1–2), 57–69.

Houmark-Nielsen, M., 2004. The Pleistocene of Denmark: a review of stratigraphy and glaciation history. In: Ehlers, J., Gibbard, P.L. (Eds.), Quaternary Glaciations—Extent and Chronology. Part I: Europe. Elsevier, Amsterdam, pp. 35–46.

Houmark-Nielsen, M., 2010. Extent, age and dynamics of Marine Isotope Stage 3 glaciations in the southwestern Baltic Basin. Boreas 39, 343–359.

Houmark-Nielsen, M., Kjær, K.H., 2003. Southwest Scandinavia, 40–15 kyr BP: palaeogeography and environmental change. J. Quatern. Sci. 18 (8), 769–786.

Jansen, E., Fronval, T., Rack, F., Channel, J., 2000. Pliocene-Pleistocene ice rafting history and cyclicity in the Nordic Seas during the last 3.5 Myr. Paleoceanography 15, 709–721.

King, E.L., Haflidason, H., Sejrup, H.-P., Løvlie, R., 1998. Glacigenic debris flows on the North Sea Trough Mouth Fan during ice stream maxima. Mar. Geol. 152, 217–246.

Kleiven, H., Jansen, E., Fronval, T., Smith, T.M., 2002. Intensification of Northern Hemisphere glaciations in the circum Atlantic region (3.5-2.4 Ma)—ice-rafted detritus evidence. Palaeogeogr. Palaeoclimatol. Palaeoecol. 184 (3–4), 213–223.

Kleman, J., 1994. Preservation of landforms under ice sheets and ice caps. Geomorphology 9, 19–32.

Kleman, J., Stroeven, A.P., Lundqvist, J., 2008. Patterns of Quaternary ice sheet erosion and deposition in Fennoscandia and a theoretical framework for explanation. Geomorphology 97 (1–2), 73–90.

Knudsen, C.G., 2006. Glacier dynamics and Lateglacial environmental changes—evidences from SW Norway and Iceland. Department of Earth Science, Ph.D., University of Bergen, Bergen, p. 98.

Laban, C., van der Meer, J.M., 2004. Pleistocene glaciations in The Netherlands. In: Ehlers, J., Gibbard, P.L. (Eds.), Quaternery Glaciations—Extent and Chronology—Part I: Europe. Elsevier, Amsterdam, pp. 251–260.

Lambeck, K., Esat, T.M., Potter, E.-K., 2002. Links between climate and sea levels for the past three million years. Nature 419 (6903), 199–206.

Lambeck, K., Purcell, A., Zhao, J., Svensson, N.-O., 2010. The Scandinavian Ice Sheet: from MIS 4 to the end of the Last Glacial Maximum. Boreas 39, 410–435.

Larsen, E., Gulliksen, S., Lauritzen, S.-E., Lie, R., Løvlie, R., Mangerud, J., 1987. Cave stratigraphy in western Norway; multiple Weichselian glaciations and interstadial vertebrate fauna. Boreas 16, 267–292.

Larsen, N.K., Knudsen, K.L., Krohn, C.F., Kronborg, C., Murray, A.S., Nielsen, O.B., 2009a. Late Quaternary ice sheet, lake and sea history of southwest Scandinavia—a synthesis. Boreas 38, 732–761.

Larsen, N.K., Krohn, C.F., Kronborg, C., Nielsen, O.B., Knudsen, K.L., 2009b. Lithostratigraphy of the Late Saalian to Middle Weichselian Skaerumhede Group in Vendsyssel, northern Denmark. Boreas 38 (4), 762–786.

Lekens, W.A.H., Haflidason, H., Sejrup, H.P., Nygård, A., Richter, T., Vogt, C., et al., 2009. Sedimentation history of the northern North Sea Margin during the last 150 ka. Quatern. Sci. Rev. 28 (5–6), 469–483.

Lohne, Ø., Bondevik, S., Mangerud, J., Schrader, H., 2004. Calendar year age estimates of Allerød-Younger Dryas sea-level oscillations at Os, western Norway. J. Quatern. Sci. 19, 443–464.

Lohne, Ø.S., Bondevik, S., Mangerud, J., Svendsen, J.I., 2007. Sea-level fluctuations imply that the Younger Dryas ice-sheet expansion in western Norway commenced during the Allerød. Quatern. Sci. Rev. 26, 2128–2151.

Lundqvist, J., 1992. Glacial stratigraphy in Sweden. Geol. Surv. Finland Spec. Pap. 15, 43–59.

Lundqvist, J., Wohlfarth, B., 2001. Timing and east-west correlation of south Swedish ice marginal lines during the Late Weichselian. Quatern. Sci. Rev. 20, 1127–1148.

Mangerud, J., 1980. Ice-front variations of different parts of the Scandinavian Ice Sheet, 13,000-10,000 years BP. In: Lowe, J.J., Gray, J.M., Robinson, J.E. (Eds.), Studies in the Lateglacial of North-West Europe. Pergamon Press, Oxford, pp. 23–30.

Mangerud, J., 2000. Was Hardangerfjorden, western Norway, glaciated during the Younger Dryas? Nor. Geol. Tidsskr. 80, 229–234.

Mangerud, J., 2004. Ice sheet limits on Norway and the Norwegian continental shelf. In: Ehlers, J., Gibbard, P.L. (Eds.), Quaternary Glaciations—Extent and Chronology, vol. 1. Elsevier, Europe, Amsterdam, pp. 271–294.

Mangerud, J., Sønstegaard, E., Sejrup, H.-P., Haldorsen, S., 1981. A continuous Eemian-Early Weichselian sequence containing pollen and marine fossils at Fjøsanger, western Norway. Boreas 10, 137–208.

Mangerud, J., Jansen, E., Landvik, J., 1996. Late Cenozoic history of the Scandinavian and Barents Sea ice sheets. Glob. Planet. Change 12, 11–26.

Mangerud, J., Dokken, T., Hebbeln, D., Heggen, B., Ingólfsson, O., Landvik, J.Y., et al., 1998. Fluctuations of the Svalbard-Barents Sea Ice Sheet during the last 150 000 years. Quatern. Sci. Rev. 17, 11–42.

Mangerud, J., Løvlie, R., Gulliksen, S., Hufthammer, A.-K., Larsen, E., Valen, V., 2003. Paleomagnetic correlations between Scandinavian Ice-Sheet fluctuations and Greenland Dansgaard-Oeschger Events, 45,000-25,000 yrs B.P. Quatern. Res. 59, 213–222.

Mangerud, J., Lohne, Ø.S., Goehring, B.M., Svendsen, J.I., Gyllencreutz, R., Schaefer, J.M., 2009. The chronology and rate of ice-margin retreat in the major fjords of Western Norway during the Early Holocene. EOS Trans., AGU, 90 (52), Fall Meet. Suppl., Abstract PP23D-05.

Mangerud, J., Gulliksen, S., Larsen, E., 2010. ^{14}C-dated fluctuations of the western flank of the Scandinavian Ice Sheet 45–25 kyr BP compared with Bølling-Younger Dryas fluctuations and Dansgaard-Oeschger events in Greenland. Boreas 39, 328–342.

Näslund, J.O., Rodhe, L., Fastook, J.L., Holmlund, P., 2003. New ways of studying ice sheet flow directions and glacial erosion by computer modelling—examples from Fennoscandia. Quatern. Sci. Rev. 22 (2–4), 245–258.

Nesje, A., Dahl, S.O., 1992. Geometry, thickness and isostatic loading of the Late Weichselian Scandinavian ice sheet. Nor. Geol. Tidsskr. 72, 271–273.

Nesje, A., Dahl, S.O., Linge, H., Ballantyne, C.K., McCarroll, D., Brook, E.J., et al., 2007. The surface geometry of the Last Glacial Maximum ice sheet in the Andoya-Skanland region, northern Norway, constrained by surface exposure dating and clay mineralogy. Boreas 36 (3), 227–239.

Nygård, A., Sejrup, H.P., Haflidason, H., Bryn, P., 2005. The glacial North Sea Fan, southern Norwegian Margin: architecture and evolution from the upper continental slope to the deep-sea basin. Mar. Petrol. Geol. 22 (1–2), 71–84.

Nygård, A., Sejrup, H.P., Haflidason, H., Lekens, W.A.H., Clark, C.D., Bigg, G.R., 2007. Extreme sediment and ice discharge from marine-based ice streams: new evidence from the North Sea. Geology 35 (5), 395–398.

Olsen, L., Van der Borg, K., Bergstrøm, B., Sveian, H., Lauritzen, S.-E., Hansen, G., 2001a. AMS radiocarbon dating of glacigenic sediments with low organic carbon content—an important tool for reconstructing the history of glacial variations in Norway. Nor. Geol. Tidsskr. 81, 59–92.

Olsen, L., Sveian, H., Bergstrøm, B., 2001b. Rapid adjustments of the western part of the Scandinavian Ice Sheet during the Mid and Late Weichselian—a new model. Nor. Geol. Tidsskr. 81, 93–118.

Olsen, L., Sveian, H., van der Borg, K., Bergstrøm, B., Broekmans, M., 2002. Rapid and rhythmic ice sheet fluctuations in western Scandinavia 15–40 kya—a review. Polar Res. 21, 235–242.

Ottesen, D., Rise, L., Rokoengen, K., Sættem, J., 2001. Glacial processes and large-scale morphology on the mid-Norwegian continental shelf. In: Martinsen, O.J., Dreyer, T. (Eds.), Sedimentary Environments Offshore Norway–Palaeozoic to Recent. Norw. Petroleum Soc. Spec. Publ. 10, 441–449. Elsevier, Amsterdam.

Ottesen, D., Dowdeswell, J.A., Rise, L., 2005. Submarine landforms and reconstruction of fast-flowing ice streams within a large Quaternary ice sheet: the 2500-km-long Norwegian-Svalbard margin (57-80°N). GSA Bull. 117, 1033–1050.

Ottesen, D., Stokes, C.R., Rise, L., Olsen, L., 2008. Ice-sheet dynamics and ice streaming along the coastal parts of northern Norway. Quatern. Sci. Rev. 27 (9–10), 922–940.

Ottesen, D., Rise, L., Andersen, E.S., Bugge, T., Eidvin, T., 2009. Geological evolution of the Norwegian continental shelf between 61°N and 68°N during the last 3 million years. Norw. J. Geol. 89, 251–265.

Peltier, W., 2004. Global glacial isostasy and the surface of the ice age Earth: the ICE-5G(VM2) model and GRACE. Annu. Rev. Earth Planet. Sci. 32, 111–149.

Phillips, W.M., Hall, A.M., Mottram, R., Fifield, L.K., Sugden, D.E., 2006. Cosmogenic 10Be and 26Al exposure ages of tors and erratics, Cairngorm Mountains, Scotland: timescales for the development of a classic landscape of selective linear glacial erosion. Geomorphology 73 (3–4), 222–245.

Raunholm, S., Larsen, E., Sejrup, H.P., 2004. Weichselian interstadial sediments on Jaeren (SW Norway)—paleoenvironments and implications for ice sheet configuration. Norw. J. Geol. 84 (2), 91–106.

Reimer, P., Baillie, M., Bard, E., Bayliss, A., Beck, J., Blackwell, P., et al., 2009. IntCal09 and Marine09 radiocarbon age calibration curves, 0–50,000 years cal BP. Radiocarbon 51, 1111–1150.

Reite, A.J., 1994. Weichselian and Holocene geology of Sør-Trøndelag and adjacent parts of Nord-Trøndelag county, Central Norway. Nor. geol. under. 426, 1–30.

Rise, L., Ottesen, D., Berg, K., Lundin, E., 2005. Large-scale development of the mid-Norwegian margin during the last 3 million years. Mar. Petrol. Geol. 22 (1–2), 33–44.

Roaldset, E., Pettersen, E., Longva, O., Mangerud, J., 1982. Remnants of preglacial weathering in western Norway. Nor. Geol. Tidsskr. 62, 169–178.

Sejrup, H.P., Aarseth, I., Haflidason, H., Løvlie, R., Bratten, Å., Tjøstheim, G., et al., 1995. Quaternary of the Norwegian Channel: glaciation history and palaeoceanography. Nor. Geol. Tidsskr. 75, 65–87.

Sejrup, H.P., Hjelstuen, B.O., Dahlgren, K.I.T., Haflidason, H., Kuijpers, A., Nygård, A., et al., 2005. Pleistocene glacial history of the NW European continental margin. Mar. Petrol. Geol. 22 (9–10), 1111–1129.

Sejrup, H.P., Nygård, A., Hall, A.M., Haflidason, H., 2009. Middle and Late Weichselian (Devensian) glaciation history of south-western Norway, North Sea and eastern UK. Quatern. Sci. Rev. 28 (3–4), 370–380.

Siegert, M.J., Dowdeswell, J.A., 2004. Numerical reconstructions of the Eurasian Ice Sheet and climate during the Late Weichselian. Quatern. Sci. Rev. 23 (11–13), 1273–1283.

Sønstegaard, E., Aa, A.R., Klagegg, O., 1999. Younger Dryas glaciation in the Ålfoten area, western Norway; evidence from lake sediments and marginal moraines. Nor. Geol. Tidsskr. 79, 33–45.

Sørensen, R., 1979. Late Weichselian deglaciation in the Oslofjord area, south Norway. Boreas 8, 241–246.

Sørensen, R., Høeg, H., Henningsmoen, K., Skog, G., Labowsky, S., & Stabell, B., 2011. Utviklingen av det senglasiale og tidlig preboreale landskapet og vegetasjonen omkring steinalderboplassene ved Pauler, Larvik kommune, Vestfold, In Jaksland, L. (Ed.) E18 Brunlanesprosjektet.Varia 79, Kulturhistorisk Mus., University of Oslo.

Svendsen, J.I., Alexanderson, H., Astakhov, V.I., Demidov, I., Dowdeswell, J.A., Funder, S., et al., 2004. Late Quaternary ice sheet history of Northern Eurasia. Quatern. Sci. Rev. 23, 1229–1271.

Svensson, A., Andersen, K., Bigler, M., Clausen, H., Dahl-Jensen, D., Davies, S., et al., 2008. A 60 000 year Greenland stratigraphic ice core chronology. Climate Past 4, 47–57.

Valen, V., Larsen, E., Mangerud, J., 1995. High-resolution paleomagnetic correlation of Middle Weichselian ice-dammed lake sediments in two coastal caves, western Norway. Boreas 24, 141–153.

Valen, V., Larsen, E., Mangerud, J., Hufthammer, A.K., 1996. Sedimentology and stratigraphy in the cave Hamnsundhelleren, western Norway. J. Quatern. Sci. 11, 185–201.

Voelker, A., Sarnthein, M., Grootes, P., Erlenkeuser, H., Laj, C., Mazaud, A., et al., 1998. Correlation of marine ^{14}C ages from the Nordic seas with the GISP2 isotope record: implications for ^{14}C calibration beyond 25 ka BP. Radiocarbon 40, 517–534.

Voelker, A., Grootes, P., Nadeau, M.-J., Sarnthein, M., 2000. Radiocarbon levels in the Iceland Sea from 25–53 kyr and their link to the earth's magnetic field intensity. Radiocarbon 42, 437–452.

Vorren. K.D., 1978. Late and Middle Weichselian stratigraphy of Andøya, North Norway. Boreas 7 (1), 19–38.

Vorren, T., Mangerud, J., 2008. Glaciations come and go. Pleistocene, 2.6 million-11,500 years ago. In: Ramberg, I., Bryhni, I., Nøttvedt, A., Rangnes, K. (Eds.), The Making of a Land—Geology of Norway. Norsk Geologisk Forening, Trondheim, pp. 480–533.

Vorren, T.O., Plassen, L., 2002. Deglaciation and palaeoclimate of the Andfjord-Vågsfjord area, North Norway. Boreas 31, 97–125.

Vorren, T.O., Vorren, K.-D., Alm, T., Gulliksen, S., Løvlie, R., 1988. The last deglaciation (20,000 to 11,000 B.P.) on Andøya, northern Norway. Boreas 17, 41–77.

Winsborrow, M.C.M., Andreassen, K., Corner, G.D., Laberg, J.S., 2010. Deglaciation of a marine-based ice sheet: Late Weichselian palaeo-ice dynamics and retreat in the southern Barents Sea reconstructed from onshore and offshore glacial geomorphology. Quatern. Sci. Rev. 29 (3–4), 424–442.

Wohlfarth, B., 2010. Ice-free conditions in Sweden during Marine Oxygen Isotope Stage 3? Boreas 39, 377–398.

Chapter 23

Quaternary Glaciations in Poland

Leszek Marks

Polish Geological Institute, Rakowiecka 4, 00-975 Warsaw, Poland
Institute of Geology, Warsaw University, Żwirki i Wigury 93, 02-089 Warsaw, Poland

23.1. INTRODUCTION

During the Pleistocene, Poland was occupied by ice sheets of eight Scandinavian glaciations, named after the rivers in the country (Fig. 23.1). Among these glaciations, the Narewian and Nidanian occurred during the Early Pleistocene, the other five glaciations that is Nidanian, Sanian 1, Sanian 2, Liwiecian, Krznanian and Odranian were of the Middle Pleistocene age and the Vistulian was the only Late Pleistocene glaciation. The limits of two glaciations (Odranian and Vistulian) can be detected at the land surface, whereas of the other two (Sanian 1, Sanian 2), although more or less expressed in covering deposits and landforms, are considerably buried by deposits of younger glaciations. The limits of four glaciations (Narewian, Nidanian, Liwiecian and Krznanian) are buried everywhere under younger deposits and, with the exception of the Nidanian and presumably also the Krznanian, they have been noted mostly

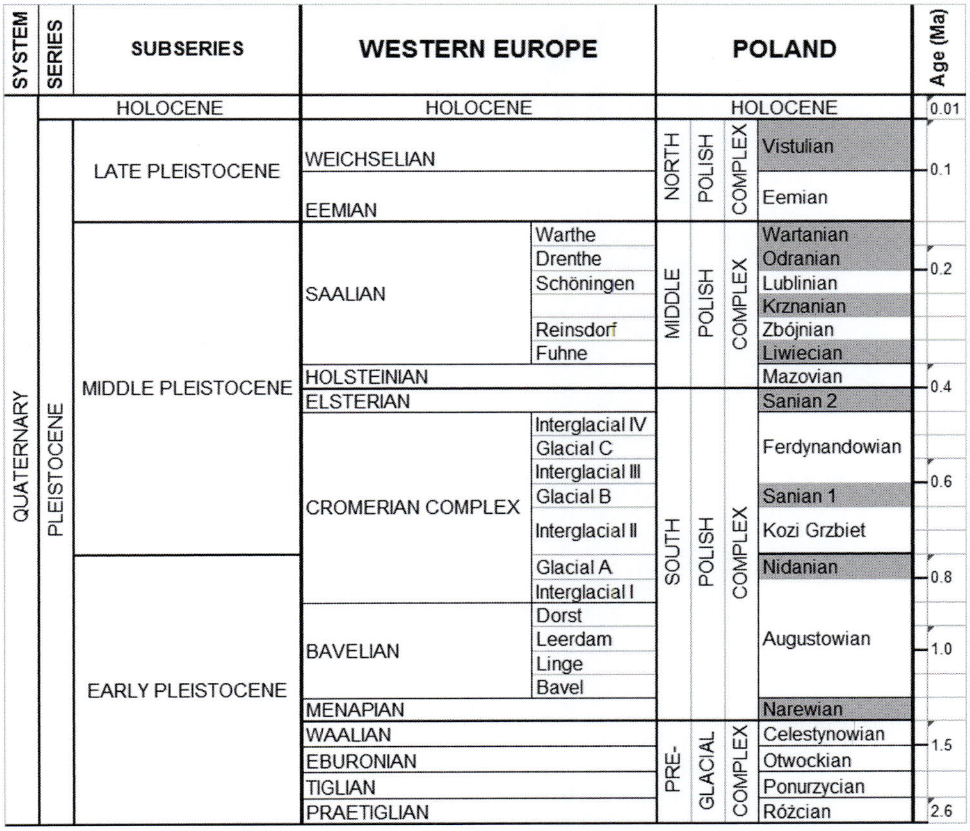

SYSTEM	SERIES	SUBSERIES	WESTERN EUROPE		POLAND			Age (Ma)
		HOLOCENE	HOLOCENE		HOLOCENE			0.01
QUATERNARY	PLEISTOCENE	LATE PLEISTOCENE	WEICHSELIAN		NORTH POLISH COMPLEX		Vistulian	0.1
			EEMIAN				Eemian	
		MIDDLE PLEISTOCENE	SAALIAN	Warthe	MIDDLE POLISH COMPLEX		Wartanian	
				Drenthe			Odranian	0.2
				Schöningen			Lublinian	
							Krznanian	
				Reinsdorf			Zbójnian	
				Fuhne			Liwiecian	
			HOLSTEINIAN				Mazovian	0.4
			ELSTERIAN				Sanian 2	
			CROMERIAN COMPLEX	Interglacial IV	SOUTH POLISH COMPLEX		Ferdynandowian	
				Glacial C				
				Interglacial III				0.6
				Glacial B			Sanian 1	
				Interglacial II			Kozi Grzbiet	
				Glacial A			Nidanian	0.8
				Interglacial I				
		EARLY PLEISTOCENE	BAVELIAN	Dorst			Augustowian	
				Leerdam				1.0
				Linge				
				Bavel				
			MENAPIAN				Narewian	
			WAALIAN		PRE-GLACIAL COMPLEX		Celestynowian	1.5
			EBURONIAN				Otwockian	
			TIGLIAN				Ponurzycian	
			PRAETIGLIAN				Różcian	2.6

FIGURE 23.1 General stratigraphical subdivision of the Quaternary of Poland and its correlation with the Western European scheme (after Wagner, 2008, slightly modified); glacial episodes in Poland are indicated in grey.

in the eastern and north-eastern part of the country. The maximum limit of the Pleistocene glaciations in Poland is indicated by the occurrence of Scandinavian erratics in southern Poland and is represented exclusively by the Sanian 1 and Odranian events.

The present chapter is based on results from the Detailed Geological Map of Poland, scale 1:50,000, the almost complete edition of which has been recently compiled in a more general scale (Marks et al., 2006). Such original field material, supplied with numerous research boreholes, covers the whole territory of the country and enables a profound interpretation of glacial sediments and landforms in Poland. It is much more reliable than it could be presented previously (Marks, 2004, 2005). However, numerous significant references, that were previously cited in the first edition (Marks, 2004), are not repeated here.

The newly presented Pleistocene glacial limits (Fig. 23.2) have also been verified and supplemented by data from numerous recent compilations of the Quaternary stratigraphy and palaeogeography of Poland. Some of them have not been limited to the territory of this country but have also been presented by correlation through the cross-border regions with the neighbouring countries (Marks, 2002, 2010; Wójcik et al., 2004; Ber, 2005; Rinterknecht et al., 2005, 2006a,b; Marks and Pavlovskaya, 2006; Gozhik et al., 2010; among others).

23.2. EARLY PLEISTOCENE GLACIATIONS

The first ice sheet of this Early Pleistocene glaciation, namely the Narewian glaciation (Fig. 23.1), occupied north-eastern Poland and advanced as far south as the South Polish Uplands (Lindner et al., 2004). Its maximum limit has been roughly determined (Fig. 23.2), based upon research borehole sections, some of which include deposits of the Augustowian Interglacial at several sites. The latter overlie the Narewian Till, whereas the Brunhes/Matuyama palaeomagnetic reversal boundary has been noted at the very top of the interglacial sequence (Ber, 2005) or in the lower part of the younger, the Kozi Grzbiet Interglacial (Gozhik et al., 2010). The Narewian glaciation is presumed to be older in Poland than the Narev glaciation in Belarus

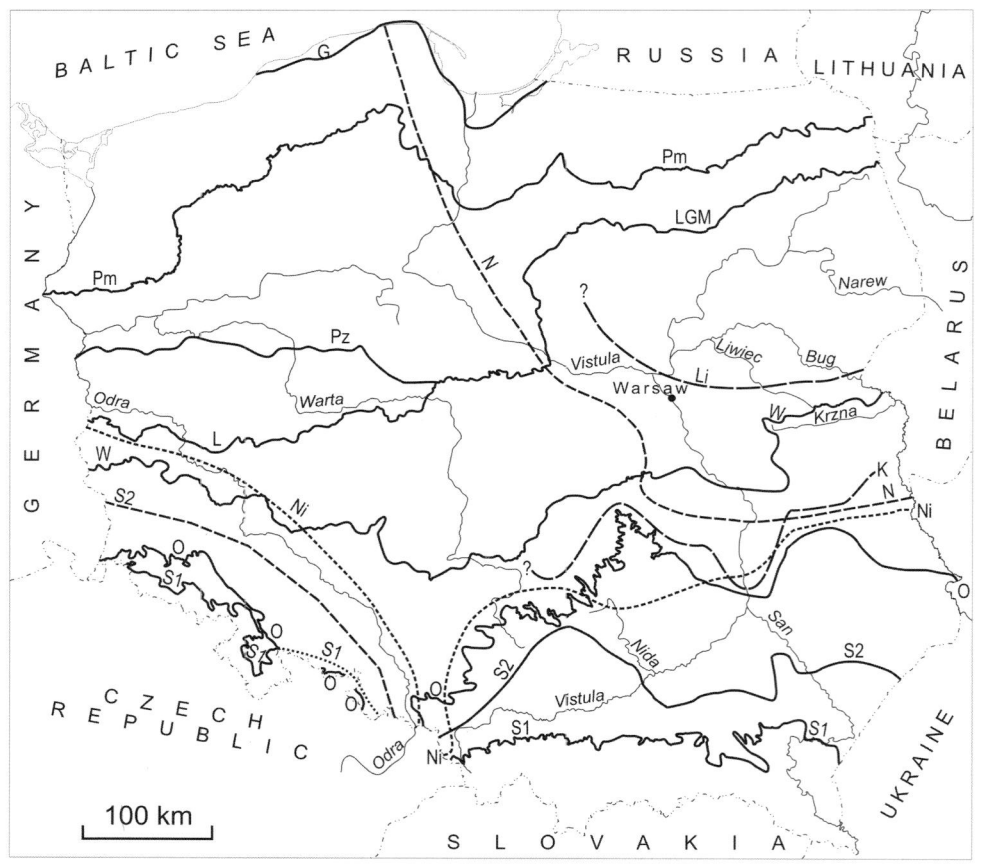

FIGURE 23.2 Limits of Pleistocene glaciations in Poland (after Lindner, 1988, 2001; Lindner and Marks, 1999; Marks, 2004, 2005, 2010; Marks et al., 2006; Gozhik et al., 2010; and others). *Early Pleistocene*: N, Narewian; Ni, Nidanian. *Middle Pleistocene*: S1, Sanian 1; S2, Sanian 2; Li, Liwiecian; K, Krznanian; O, Odranian with the secondary stadial of W, Wartanian. *Late Pleistocene*: Vistulian: LGM, Last Glacial Maximum; L, Leszno Phase; Pz, Poznań Phase; Pm, Pomeranian Phase; G, Gardno Phase. Buried or obscured glacial limits are indicated either with dashed or with dotted lines.

(cf. Voznyachuk, 1985), because of its relation to the Brunhes/Matuyama palaeomagnetic reversal boundary (Sanko and Moiseyev, 1996; Gozhik et al., 2010).

The ice sheet of the Nidanian glaciation (Fig. 23.1) occupied the Wielkopolska Lowland and a considerable part of the Lower Silesia in western Poland, reached the South Polish Uplands and its extensive lobe advanced into the Upper Odra drainage basin, entering the Czech Republic (Lindner et al., 2004; Fig. 23.1). A till of this glaciation is particularly well preserved in north-eastern Poland (Lindner and Astapova, 2000; Ber, 2005, 2009). In the Nida drainage basin in the Holy Cross Mountains, cave deposits of the younger Kozi Grzbiet Interglacial, in which the Brunhes/Matuyama palaeomagnetic reversal boundary has been noted (Głazek et al., 1976, 1977), are underlain by sands with Scandinavian erratic material. This implies that a preceding ice sheet advanced far to the south of the site.

23.3. MIDDLE PLEISTOCENE GLACIATIONS

The ice sheet of the Sanian 1 glaciation (Fig. 23.1) passed over the South Poland Uplands, and its lobe occupied the Upper Odra drainage basin, again entering the Czech Republic (Fig. 23.2). The ice sheet reached the Sudetes and the Carpathians in the south (Lindner, 2001; Lindner et al., 2004), damming proglacial lakes during its advance and producing outwash terraces during its retreat. It entered the Jelenia Góra and Kłodzko Basins in the Sudetes and the Krosno-Jasło Basin in the Carpathians. The Sanian 1 glaciation was the most extensive glaciation in Poland. Tills of this glaciation are underlain by organic deposits of the Kozi Grzbiet Interglacial in the Sandomierz Basin in southern Poland (Dąbrowski, 1967; Laskowska-Wysoczańska, 1967; Wojtanowicz, 1985). Apart from the Scandinavian ice sheet, mountain glaciers developed in the Tatra Mountains and, presumably, also in the Sudetes.

The ice sheet of the Sanian 2 glaciation (Fig. 23.1) occupied a large part of Poland, reaching the Sudetes and the Carpathian Foredeep, and its lobe occupied the Upper Odra drainage basin, entering the Czech Republic (Lindner et al., 2004; Fig. 23.2). However, the ice sheet was less extensive in the Sandomierz Basin than during the previous glaciation (Lindner, 2001). No marginal features are preserved from this event (cf. Lindner, 1988; Dolecki, 2002); therefore, the ice sheet limit can only be detected mostly on the basis of till occurrences.

The subsequent Liwiecian glaciation was the first glacial episode after the Mazovian (Holsteinian Stage) Interglacial and preceded the Zbójnian Interglacial (Fig. 23.1). During this event, the ice sheet occupied the eastern Mazovian and northern Podlasie Lowlands, but it could also have reached the northern foreland of the South Polish Uplands (Lindner et al., 2004; Fig. 23.2).

Recent reinterpreted data (cf. Lindner and Marks, 1999) suggest a possible setting of the 'Krzna Stadial' of Rühle (1970) as a separate glaciation (Fig. 23.1). During the Krznanian glaciation, ice sheet occupied eastern Poland, reaching as far south as the northern foreland of the South Polish Uplands, and presumably, it also approached the Silesian Upland (Fig. 23.2). A till, at present ascribed to the Krznanian glaciation, has been considered previously as evidence for a pre-maximum stadial of the Odranian glaciation (cf. Lindner, 1992).

During the maximum stadial of the Odranian glaciation (Fig. 23.1), the ice sheet reached the South Polish Uplands, a lobe advanced into a gap formed by the Vistula River through to the South Polish Uplands, occupied not only the Silesian Upland and the foreland of the Sudetes but also most of the Upper Odra drainage basin (Lindner et al., 2004). Finally it entered the Czech Republic through the Moravian Gate (Fig. 23.2). Therefore, the proglacial meltwaters during this glaciation also discharged into the Danube drainage basin. In mid-eastern Poland, the advancing ice sheet blocked the valleys of the Vistula, Pilica and Wieprz rivers, thus favouring development of several large ice-dammed lakes. They formed near Koniecpol on the Pilica River and Sandomierz on the upper Middle Vistula River, among others. The ice sheet cannot have been thick in its marginal part, especially in central Poland, because its maximum limit was considerably influenced by the landscape of the forefield and a lobate ice margin typically developed. In the Sudetes and the Holy Cross Mountains, the maximum limit of the ice sheet is indicated by marginal (kame) terraces that developed on mountain slopes and by proglacial meltwater trains in the river valleys. There are only scarce end moraines and eskers that developed during deglaciation, mostly in the central part of the country. The meltwater in the south-eastern part of Poland drained to the east, to the Dniester valley, by a sub-Carpathian ice-marginal streamway.

During the Wartanian, previously treated as a separate glaciation (Lindner, 1992) but at present considered as a post-maximum stadial of the Odranian glaciation (Fig. 23.1), the ice sheet occupied the southern Wielkopolska Lowland, the Łódź Upland and the Mazovian and Podlasie Lowlands (Marks et al., 1995; Lindner and Marks, 1999; Fig. 23.2). The Wartanian ice margin is commonly indicated by end moraines, which are especially impressive in the western part of the country. This is because of their giant glacially dislocated cores that had developed primarily during the earlier events, especially the Sanian 1 glaciation. The Wartanian ice sheet limit indicates the presence of numerous lobes, a distribution of which is underlined by rows of deglacial end moraines and by streamlined features, represented mostly by eskers. The ice sheet margin could be correlated through the cross-border region to Belarus where a lobate pattern of the ice sheet was also a common feature

(Marks and Pavlovskaya, 2006). The western part of the country was drained by the sub-Sudeten ice-marginal streamway towards the Weser drainage basin. The waters in the east were discharged through the Pilica–Wieprz–Krzna ice-marginal streamway. The latter drained mid-eastern Poland to the Prypiat drainage basin in Belarus and further to the Dnieper valley. The mountain glaciers again developed on the Tatra Mountains and the Sudetes.

23.4. LATE PLEISTOCENE GLACIATION

There were at least two major ice sheet advances during the Vistulian glaciation (Weichselian) in Poland. The Middle Vistulian ice sheet advance, namely the Świecie Stadial, occurred at about 60–50 ka ago and mostly occupied mid-northern Poland. All the glacial landforms and sediments of this glacial episode are presumably mantled by younger glacial deposits. However, there could have been also the ice sheet advance sometime at 33–37 ka ago in northern Poland (Marks, 2010). During the Late Vistulian glaciation, the ice sheet occupied the southern Wielkopolska Lowland, Pomerania, Kujawy and the Mazury Lakelands. Based on radiocarbon and cosmic isotope ages (Marks, 2002, 2010; Rinterknecht et al., 2005, 2006a,b), the Last Glacial Maximum (LGM) limit in Poland was determined to have been time-transgressive and occurred at 24–19 ka (the younger in the east). This limit marks the maximum extent during the Leszno (Brandenburg) Phase in the west and the Poznań (Frankfurt) Phase in the Middle Vistula valley region and the east. Both these glacial phases were followed by the less extensive Pomeranian and Gardno Phases, the latter restricted to the northernmost part of the country (Fig. 23.2).

The Late Vistulian ice sheet margins were presumably modelled by palaeo-ice streams, especially in the Vistula and Odra valleys but also in the north-eastern part of the country (Marks, 2002). There are numerous deglacial end moraines that are frequently arranged in rows. They are believed to represent minor and local standstills or even short advances during a general ice sheet retreat. The eskers are mostly but irregularly concentrated within the area that has been occupied by the ice sheet during the Pomeranian Phase. In contrast, the drumlin fields are rather connected with the palaeo-ice stream areas, predominantly with the Vistula Lobe during the Poznań Phase and the Odra Lobe during the Pomeranian Phase. However, a structure of the ice sheet body is best expressed by numerous tunnel valleys, at present partly occupied by lakes.

The advancing ice sheet of the Late Vistulian glaciation blocked numerous river valleys in Poland causing a vast proglacial lake to develop in the Warsaw Basin in central Poland and also in the Middle Neman valley in western Belarus, close to the Polish border. Runoff from these lakes and huge discharge of meltwaters and extraglacial waters merged together and passed through the Narew River in north-eastern Poland initiating the earliest of the eastern complex ice-marginal streamways, referred to as the Warsaw-Berlin and the Toruń-Eberswalde, etc. During deglaciation, a large ice-dammed lake was formed in the Lower Vistula valley and in the Gulf of Gdańsk. Discharge of the lake waters played a significant role in the development of the youngest, Pomeranian ice-marginal streamway.

Valley and cirque glaciers developed in the Tatra Mountains and in the Sudetes during this phase. Indeed, there were at least three separate glacial episodes in the mountains during the Vistulian Stage. The glaciers also occurred in strictly limited areas of the lower ranges of the Outer Carpathians, especially on mounts Pilsko, Babia and Barania in the western Beskids (cf. Lindner and Marks, 1995).

23.5. CONCLUSIONS

Poland was occupied by ice sheets during eight individual Scandinavian glaciations in the Early Pleistocene (Narewian, Nidanian), the Middle Pleistocene (Sanian 1, Sanian 2, Liwiecian, Krznanian and Odranian) and the Late Pleistocene (Vistulian). The limits of the Odranian and Vistulian glaciations can be fully identified at the modern land surface, whereas those of the Sanian 1 and Sanian 2 are more or less buried by deposits of younger glaciations. The limits of the Narewian, Nidanian, Liwiecian and Krznanian are mantled everywhere by younger deposits. The maximum limit of the Pleistocene glaciations in Poland is indicated by the occurrence of Scandinavian erratics in southern Poland, where it is represented exclusively by Sanian 1 and Odranian glacial deposits.

During the Odranian (and its deglacial Wartanian Stadial) and the Vistulian, the ice sheet was dissected into numerous, presumably asynchronous lobes, either adjusted to a configuration of the forefield or being terminations of the palaeo-ice streams in the ice sheet body. The streamlined structure of the ice sheet is expressed principally by occurrences of rows of end moraines and also by tunnel valleys, eskers and drumlins. The LGM limit in Poland has been shown to have been time-transgressive and occurred between 24 and 19 ka. This limit represents the glacial advances during the Leszno Phase in the west and the Poznań Phase in the centre and, presumably, also in the east.

There have been mountain glaciers in the Carpathians and in the Sudetes, presumably since at least the Sanian 1 glaciation.

REFERENCES

Ber, A., 2005. Polish Pleistocene stratigraphy—a review of interglacial stratotypes. Neth. J. Geosci./Geol. Mijnbouw 84–2, 61–76.

Ber, A., 2009. Pleistocene interglacials and glaciations of northeastern Poland compared to neighbouring areas. Quatern. Int. 149, 12–23.

Dąbrowski, M.J., 1967. Pollen analysis of an interstadial profile from Jasionka near Rzeszów. Acta Geol. Pol. 17 (3), 509–520.

Dolecki, L., 2002. Main Profiles of the Neopleistocene Loesses on the Horodło Plateau-Ridge and their Lithological-Stratigraphical Interpretation. Wydawnictwo Uniwersytetu Marii Curie-Skłodowskiej, Lublin.

Głazek, J., Lindner, L., Wysoczański-Minkowicz, T., 1976. Interglacial Mindel I/Mindel II in fossil-bearing karst at Kozi Grzbiet in the Holy Cross Mts. Acta Geol. Pol. 26 (3), 376–393.

Głazek, J., Kowalski, K., Lindner, L., Młynarski, M., Stworzewicz, E., Tuchołka, P., et al., 1977. Cave deposits at Kozi Grzbiet (Holy Cross Mts, Central Poland) with vertebrate and snail faunas of the Mindelian I/Mindelian II interglacial and their stratigraphic correlations. In: Proceedings of the Seventh International Speleological Congress, Sheffield, pp. 211–214.

Gozhik, P., Lindner, L., Marks, L., 2010. Late Early and early Middle Pleistocene limits of Scandinavian glaciations in Poland and Ukraine. Quatern. Int. doi:10.1016/j.quaint.2010.07.027.

Laskowska-Wysoczańska, W., 1967. The interstadial of the Cracovian Glaciation from Jasionka near Rzeszów. Acta Geol. Pol. 17 (3), 495–507.

Lindner, L., 1988. Glacial and interglacial units in the Pleistocene of the Miechów Upland and Nida Basin. Prz. Geol. 419 (3), 140–148.

Lindner, L., 1992. Stratygrafia (klimatostratygrafia) czwartorzędu. In: Lindner, L. (Ed.), Czwartorzęd: osady, metody badań, stratygrafia. Państwowa Agencja Ekologiczna, Warszawa, pp. 441–633.

Lindner, L., 2001. Problems of the age and extent of the Scandinavian glaciations at the margin of the Polish Carpathians (southern Poland). Prz. Geol. 49, 819–821.

Lindner, L., Astapova, S.D., 2000. The age and geological setting of Pleistocene glacigenic beds around the border between Poland and Belarus. Geol. Q. 45 (2), 187–197.

Lindner, L., Marks, L., 1995. Correlation of glacial episodes of the Wisła (Vistulian) Glaciation in the Polish Lowland and mountain regions, and in Scandinavia. Bull. Pol. Acad. Sci. Earth Sci. 43 (1), 5–15.

Lindner, L., Marks, L., 1999. New approach to stratigraphy of palaeolake and glacial sediments of the younger Middle Pleistocene in mid-eastern Poland. Geol. Q. 43 (1), 1–8.

Lindner, L., Gozhik, P., Marciniak, B., Marks, L., Yelovicheva, Y., 2004. Main climatic changes in the Quaternary of Poland, Belarus and Ukraine. Geol. Q. 48 (2), 97–114.

Marks, L., 2002. Last glacial maximum in Poland. Quatern. Sci. Rev. 21, 103–110.

Marks, L., 2004. Pleistocene glacial limits in Poland. In: Ehlers, J., Gibbard, P.L. (Eds.), Quaternary Glaciations—Extents and Chronology, 1. Elsevier, Amsterdam, pp. 295–300.

Marks, L., 2005. Pleistocene glacial limits in the territory of Poland. Przegląd Geol. 53 (10/2), 988–993.

Marks, L., 2010. Timing of the Late Vistulian (Weichselian) glacial phases in Poland. Quatern. Sci. Rev. doi:10.1016/j.quascirev.2010.08.008.

Marks, L., Pavlovskaya, I.E., 2006. Correlation of the Saalian glacial limits in eastern Poland and western Belarus. Quatern. Int. 149, 87–93.

Marks, L., Lindner, L., Nitychoruk, J., 1995. New approach to a stratigraphic position of the Warta stage in Poland. Acta Geogr. Lodz. 68, 135–147.

Marks, L., Ber, A., Gogołek, W., Piotrowska, K. (Eds.), 2006. Geological Map of Poland 1:500 000, with Explanatory Text. Państwowy Instytut Geologiczny, Warszawa.

Rinterknecht, V.R., Marks, L., Piotrowski, J.A., Raisbeck, G.M., Yiou, F., Brook, E.J., et al., 2005. Cosmogenic ^{10}Be ages on the Pomeranian Moraine, Poland. Boreas 34, 186–191.

Rinterknecht, V.R., Clark, P.U., Raisbeck, G.M., Yiou, F., Bitinas, A., Brook, E.J., et al., 2006a. The last deglaciation of the southeastern sector of Scandinavian ice sheet. Science 311, 1449–1452.

Rinterknecht, V.R., Marks, L., Piotrowski, J.A., Raisbeck, G.M., Yiou, F., Brook, E.J., et al., 2006b. 'Cosmogenic dating of the Pomeranian Moraine: adding a regional perspective': reply to comments. Boreas 35 (3), 605–606.

Rühle, E., 1970. Les nouvelles unites stratigraphiques de la glaciation de la Pologne Centrale (Riss) sur le territoire entre la moyene Vistule et le bas Bug. Acta Geogr. Lodz. 24, 389–412.

Sanko, A.F., Moiseyev, E.I., 1996. First definition of the Brunhes-Matuyama boundary of Pleistocene deposits in Belarus. Dokl. Akad. Nauk Bel. 40 (5), 106–109.

Voznyachuk, L.N., 1985. Problemy glyatsyopleystotsena vostochno-evropeiskoi ravniny. Problemy Pleistotsena. Nauka i Tekhnika, Minsk, pp. 8–55.

Wagner, R. (Ed.), 2008. Tabela Stratygraficzna Polski. Państwowy Instytut Geologiczny, Warszawa.

Wójcik, A., Nawrocki, J., Nita, M., 2004. Pleistocene in the Kończyce Profile (Oświęcim Basin)—sediment genesis and age analysis at the background of stratigraphic schemes of the Quaternary. Państw. Inst. Geol. 409, 5–50.

Wojtanowicz, J., 1985. The TL dated profile of the Quaternary deposits at Giedlarowa (Sandomierz Basin) and its palaeogeographic importance. St. Geom. Carp.-Balc. 19, 37–44.

Chapter 24

New Evidence on the Quaternary Glaciation in the Romanian Carpathians

Petru Urdea*, Alexandru Onaca, Florina Ardelean and Mircea Ardelean

Department of Geography, West University of Timişoara, V. Pârvan Nr.4, 300223 Timişoara, Romania
*Correspondence and requests for materials should be addressed to Petru Urdea. E-mail: urdea@cbg.uvt.ro

24.1. INTRODUCTION

The Carpathian Mountains represent a key region for the understanding of the spatial link between the eastern part of the Scandinavian Ice Sheet and Alpine glaciers. Since the second half of the nineteenth century, many authors have discussed the question of the origin and significance of the glacial landforms and the Pleistocene glaciation in the Romanian Carpathians. From a historical point of view, the research on the Romanian glacial relief and the effect of Pleistocene glaciation in this part of Europe has taken place in four distinct periods (see Urdea and Reuther, 2009 for a comprehensive bibliographical list).

Since the previous chapter, presented in Ehlers and Gibbard (2004), new evidence has been collected on the glacial landforms in the Romanian Carpathians, and consequently, new ideas concerning the extension of the Pleistocene glaciations in this area have emerged. However, glacial geomorphological studies have still not been undertaken in many mountain areas, especially for the middle mountains of the Carpathian arch, or the knowledge of the glacial geomorphology and stratigraphy are only recorded in a general way. Although many doctoral theses have focused on the geomorphological aspects of different mountain areas (Ancuţa, 2005; Nedelea, 2006; Murătoreanu, 2009) in the period since our previous paper and have included a chapter on the glacial relief, unfortunately, the authors have not tackled the problem of the Pleistocene glaciation evolution, or this problem is presented in a rather general manner, for example, for Iezer Mountains by Szepesi (2007), or for Piatra Craiului Massif by Constantinescu (2009). In addition, the same situation applies to papers that discuss the glacial relief of some mountain areas (e.g. Simoni and Flueraru, 2006; Simoni, 2008).

Although the existence of Pleistocene glaciers in the Romanian Carpathians was pointed out over 125 years ago—in the Northern Carpathians (Tietze, 1878) and in the Transylvanian Alps (Lehmann, 1881, 1885)—for the remaining mountain areas reaching maximum altitudes below 2000 m a.s.l., the problem is still controversial. The situation is even more surprising because, during the Pleistocene, glaciers descended to 1050–1200 m a.s.l. in the highest Romanian mountains. Examples include the Southern Carpathians (Făgăraş, Retezat, Parâng)—also called the Transsylvanian Alps—and the Rodna Mountains in the Eastern Carpathians.

Moreover, even the existence of the Pleistocene glaciations in the middle mountains of Central Europe and South-Eastern Europe has been accepted unequivocally, for example, the Bavarian Forest (Raab and Völkel, 2003), Bohemian Forest (Vočadlová and Křížek, 2005), Krkonoše (Giant) Mountains (Carr et al., 2007), Risnjak Mountains, Croatia (Bognar and Prugovečki, 1997), etc. For some of the Romanian geomorphologists, this concept is presented in an unsatisfactory way (Naum, 1957a; Mac et al., 1990). In consequence, the authors focused on some problematical Romanian mountain areas for their investigations, including the Eastern Carpathians, Southern Carpathians and the Apuseni Mountains. The aim of this chapter therefore is to present a synthesis including the new observations and images concerning the presence and extent of Pleistocene glaciers in some mountain areas of the Romanian Carpathians (Fig. 24.1; Table 24.1).

24.2. THE SETTING

The Carpathians are a distinct part of the Alpine-Himalayan orogenic chain, and form the backbone of Romania's relief, being situated in the central part of the country (Fig. 24.1). The Carpathians are included in the northern branch of the European system of the Alpides and are the results of Mesozoic and Cenozoic continental collision. They take the form off several Nappe complexes and thrust sheets. In the

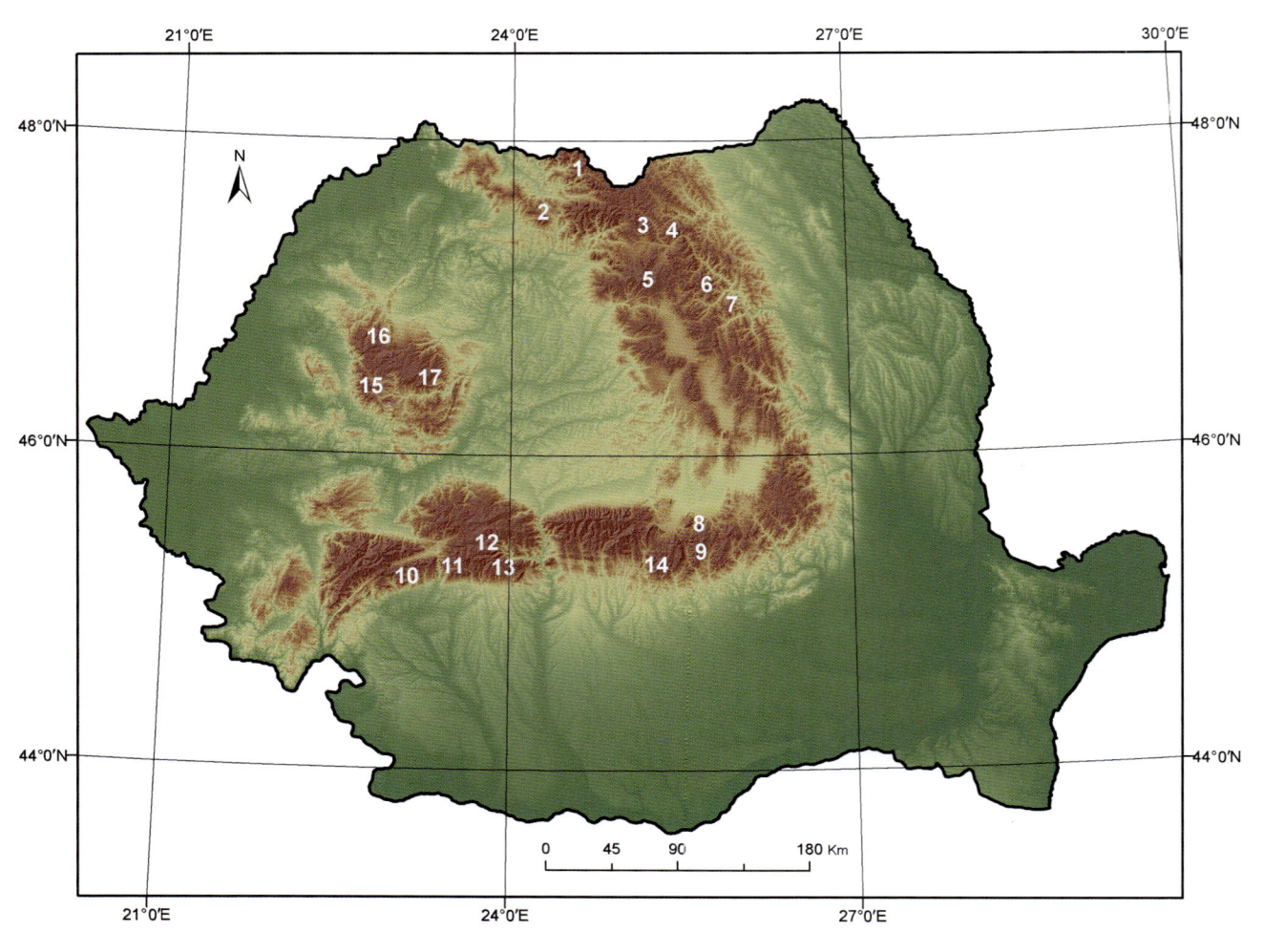

Eastern Carpathians (1-9)

Southern Carpathians (10-14)

Apuseni Mountains (15-17)

FIGURE 24.1 Position of the Carpathian Mountains in Romania and location of the mentioned area: 1, Maramureş Mountains; 2, Ţibleş Mountains; 3, Suhard Mountains; 4, Giumalău Mountains; 5, Călimani Mountains; 6, Bistriţei Mountains; 7, Ceahlău Massif; 8, Piatra Mare Massif; 9, Neamţu Mountains; 10, Vâlcan Mountains; 11, Capra-Buha Massif; 12, Latoriţei Mountains 13, Căpăţânei Mountains; 14, Leaota Mountains; 15, Bihor Mountains; 16, Vlădeasa Mountains; 17, Muntele Mare Mountains.

Apuseni Mountains, Southern Carpathians and in the central axial area of the Eastern Carpathians, in particular, the metamorphic rocks associated with numerous granitoid plutons characterise the chain. By contrast, sedimentary rocks occur in the marginal zones. In the south-eastern part of the Apuseni Mountains and on the eastern side of the Eastern Carpathians, the Flysch zone represents a prominent tectonic unit. In addition, in the southern part of Apuseni Mountains and on the western side of the Eastern Carpathians, to the Transsylvania Basin, Mio-Pliocene volcanic rocks and deposits are present (volcanic chain). A complex inner fault system is connected to the appearance of a series of tectonic basins. Although the mean elevation of these mountains is 840 m a.s.l., only 34% of the chain has altitudes exceeding 1500 m. However, there are 11 peaks higher than 2500 m a.s.l., the highest peak being Moldoveanu, at 2544 m a.s.l. in the Făgăraş Mountains. In the high area of the Romanian Carpathians, the landscape includes evidence of glacial sculpture, with cirques, steep slopes and U-shaped valleys, associated either with sharp peaks or with ridges. This typical alpine landscape of central part of the Făgăraş, Retezat, Parâng and Rodnei Mountains includes either rounded mountain tops or interfluves, integrated on some levelled surfaces with true peneplains, the

TABLE 24.1 Identity of Investigated Mountain Areas

Mountain Groups	Mountains	Latitude/longitude	Maximum altitude (m)
Eastern Carpathians			
1.	Maramureş Mountains	47°53′36″N; 24°27′20″E	1957
2.	Ţibleş Mountains	47°31′22″N; 24°15′11″E	1853
3.	Suhard Mountains	47°30′20″N; 25°05′31″E	1937
4.	Giumalău Massif	47°26′1″N; 25°29′04″E	1857
5.	Căliman Mountains	47°08′12″N; 25°03′46″E	2100
6.	Bistriţei Mountains	47°07′23″N; 25°40′32″E	1853
7.	Ceahlău Massif	46°57′23″N; 25°56′47″E	1907
8.	Piatra Mare Massif	45°32′59″N; 25°38′45″E	1843
9.	Neamţu Mountains	45°28′4″N; 25°41′42″E	1923
Southern Carpathians			
10.	Vâlcan Mountains	45°18′29″ N; 23°15′56″E	1868
11.	Capra-Buha Massif	45°25′41″N; 23°34′30″E	1927
12.	Latoriţei Mountains	45°23′24″N; 23°44′33″E	2055
13.	Căpăţânii Mountains	45°19′39″N; 23°50′48″E	2130
14.	Leaota Mountains	45°19′21″N; 25°18′59″E	2133
Apuseni Mountains			
15.	Bihor Mountains	46°26′26″N; 22°41′16″E	1849
16.	Vlădeasa Mountains	46°46′34″N; 23°44′33″E	1836
17.	Muntele Mare Mountains	46°29′29″N; 23°41′13″E	1825

latter characteristic of the Godeanu, Ţarcu, Şureanu, Cindrel, Latoriţei, Căpăţânii, Leaota or Vâlcan Mountains. At the beginning of twentieth century, Martonne (1907) described three peneplains in the Southern Carpathians, Borăscu at 1800–2200 m a.s.l., Râu Şes at 1400–1600 m a.s.l. and Gornoviţa at 1000–1200 m a.s.l. In the sedimentary area of the mountain massifs, structural and karstic landforms typically occur.

The climatic conditions that occur in the high zone of the Romanian Carpathians are cold, with the mean annual air temperature of 3 °C at Cozia (1577 m a.s.l.), 1.0 °C at Vlădeasa (1836 m a.s.l.), 0.2 °C at Bâlea-Lake (2038 m a.s.l.), −0.5 °C at Ţarcu (2180 m a.s.l.) and −2.5 °C at Omu (2505 m a.s.l.), where the absolute minimum temperature can reach −38 °C. The mean annual precipitation is 8442 mm at Cozia, 1151.3 mm at Vlădeasa, 1246 mm at Bâlea-Lake, 1180 mm at Ţarcu and 1280 mm at Omu, the continentality index Gams (CIG) reaching values over 50°. The thickness of the snow layer can range between 50 and 370 cm and is highly variable depending on the wind action. About 60–75% of the precipitation falls as snow, and the snow cover in the region lasts between 150 and 210 days each year.

24.3. METHODOLOGICAL ASPECTS

The complex Pleistocene cold events have seen the build-up of ice at many scales in the Carpathian Mountains, depending upon the intensity of freezing, the availability of snow fall, etc. This has resulted in a range of glacier types: niche glaciers, cirque glaciers, valley glaciers, snow fields, ice aprons, wall-sided glaciers, mountain ice caps and plateau ice fields. Therefore, although the presence and development of glaciers depend on the snow accumulation-melt balance, which is related to the regional climatic characteristics that control precipitation and temperature (in particular, latitude and air masses), the topographic setting is of paramount importance in determining the location, extent, shape and evolution of each glacial body. Thus, the combined effects of altitude, exposure to incoming solar radiation, slope and mean curvature may explain the main part of the observed variations, related to in the equilibrium-line altitudes (ELAs) of different glaciers in the same region (López-Moreno et al., 2006).

Reconstructing the former glaciers requires field investigations, detailed geomorphological mapping and the analysis of landforms and sediments. The most accurate methods also require that there should be sufficient geomorphological evidence, usually lateral–terminal moraines, trimlines, striated bedrock, roches moutonnées and erratics, to allow the extent and the form of Pleistocene glaciers to be reconstructed. Special attention is to be paid to the reconstruction of atypical or small glacial entities, including ice aprons, wall-sided glaciers, avalanche-cone glaciers, firn and ice fields. Because these smaller bodies occur in extreme settings or particular topographic or climatic situations, they have an equally minor morphological role. They also have a discrete 'morphological fingerprint' in the landscape. Based on the authors' field experience, they consider that the differentiation of discrete, incipient glacial forms; the identification of a small semicircular scarp; an incipient trimline with a knickpoint; a sudden break of slope; and the presence of the knickpoint in the lower part of the longitudinal profile of this hollow are required. They do not ignore the fact that various types of pre-glacial landscape, such as funnel-shaped valley heads, gullies and ravines, have provided the conditions necessary for the early stages of enlargement and modification by snow patches and niche glaciers, which may have eventually led to cirque formation (Grove, 1961). In consequence, the less well developed features are considered to be transitional forms between nivation hollows and incipient cirques.

Although the glacial features are easy to recognise in the main part of the alpine area of the Romania Carpathians, their identification at low altitude and in the more vegetated terrain is more difficult. Frequently, glaciation has often only left weak traces and, in many regions, the original glacial landforms have been greatly changed by postglacial processes. For the reconstruction of Pleistocene glaciers in this geomorphological and biogeographical context, the largest problem remains the identification of terminal moraines, which may or may not be present. Scree, debris flows and landslide deposits, and, sometimes, fluvioglacial and fluvial deposits and landforms, which occur along the valley walls and floor, may be mistakenly interpreted as moraines. In the absence of unequivocal glacial deposits and erosion features, or erratics, the evidence of glacial advance at the lowest elevations must be used to determine the glacial terminal position. This is often where the change from a U- to V-shaped valley cross-section is found. However, while the glacial extents in some valleys have undoubtedly been misidentified, orthorectified aerial photographs and Landsat images can be used to determine the transition from a glaciated to a fluvial landscape throughout the Romanian Carpathians.

Therefore, the database for the glacial geomorphology and topographic information has been compiled using topographic maps (1:25,000), aerial photographs (1:10,000), digitally orthorectified aerial photographs (2005 edition, 0.5 m resolution) and satellite images (2003 edition, 2.5 m resolution). The contours were created digitally from the digital elevation model (DEM) using Adobe Illustrator software and geo-referenced using ESRI ArcView software.

The glaciers have been reconstructed to estimate their areas, lengths, volumes and palaeo-ELAs. Following field investigations and mapping, the reconstructions were based on moraine elevations and the apparent upper elevation limit of glacial features, as determined by glacially sculpted bedrock on the valley walls. The ELA of each palaeo-glacier was estimated using a toe-to-headwall altitude ratio or THAR (cf. Porter, 2001). The toe (minimum) and headwall (maximum) elevation for each glacier was determined, and the ELA was calculated as 0.45 of the vertical distance from the toe to the headwall (ELA = lowest elevation of glacier + vertical range × ratio). The area, length and ELA of each palaeo-glacial unit were determined by overlaying the mapped glacial extents on the best available topographic maps 1:25,000 produced by Direcţia Topografică Militară.

Taking into account that the volume–area scaling relationship gives both a practical and physically based method for estimating glacier volume (Bahr et al., 1997), the volume (V) of each palaeo-glacial entity was calculated, based on Fox (1993) equation $V = 0.048\,S^{1.36}$ (Klein and Isacks, 1998).

24.4. THE ARGUMENTS

Contrary to the ideas promoted during the 1960–1980s and are still advocated by some workers (e.g. Ielenicz and Pătru, 2005), the present authors (Urdea, 2000, 2004; Urdea and Reuther, 2009) consider that the Pleistocene glaciation was more extensive. For example, the north slope of the Rodnei Mountains, in the Pietroasa Valley, is an area in which controversial glacial geomorphological features are found; morainic deposits with a huge gneiss erratic were found at 950–1050 m a.s.l. (Fig. 24.2). Here, the area is dominated by the Palaeogene sedimentary deposits (flysch) of the Maramureş Depression (Sawicki, 1911; Sîrcu, 1978).

Geomorphological investigations in the Bistra Mărului valley (Ţarcu Mountains) have also recorded more extensive Pleistocene glaciers than were previously known. This extent has been demonstrated by huge erratic and lateral and terminal morainic deposits, with striated boulders (Fig. 24.3). The reconstruction of the upper part of these Bistra Mărului glaciers (Fig. 24.4), corresponding to the proposed maximum configuration of the Pleistocene glaciers, occurred at 1150 m a.s.l. This contrasts with the

FIGURE 24.2 Erratic boulder in Pietroasa Valley (Rodnei Mountains).

FIGURE 24.3 Striated boulder in the morainic deposits in the Bistra Marului Valley (Țarcu Mountains).

earlier opinions that the palaeo-glaciers of this area descended to 1310–1350 m a.s.l. In addition, the Varângu-Frâncu glacier covered an area of 6.88 km² and the Dalciu glacier 3.46 km², orientated towards the north-west. The geomorphological features of the Pietrele Albe-Nedeia area suggest that both mentioned glaciers were in connection with the Pietrele Albe-Bistricioara plateau glacier, which covers over 5.5 km².

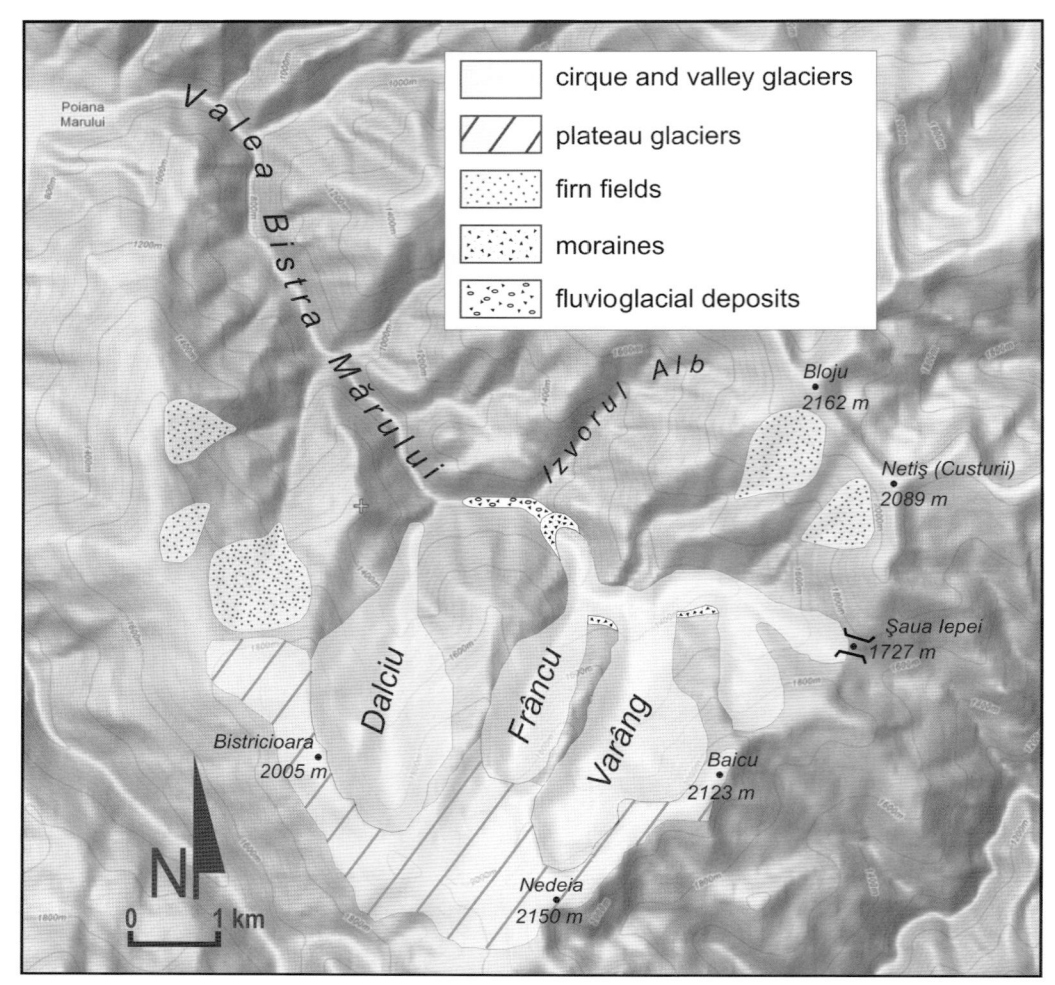

FIGURE 24.4 The reconstructed Pleistocene glaciers on the 3 DEM in the upper part of Bistra Mărului Basin (Țarcu Mountains).

Glaciation that extended further than that previously recorded has also been demonstrated on other mountain massifs, including in the eastern part of the highest mountain of the Romanian Carpathians, Făgăraş Mountains (Urdea and Reuther, 2009).

24.5. EASTERN CARPATHIANS

The Eastern Carpathians are aligned north–south, with their highest area in the north half (the Rodna Mountains—2303 m a.s.l., Călimani Mountains—2100 m a.s.l.). Evidence for Quaternary glaciation on these highlands has been known since the nineteenth century but, apart from some areas of the Rodnei, Maramureş and Călimani Mountains, it has remained understudied and poorly dated. This area has been the focus of some recent investigations.

Geomorphological evidence of past glaciation in the Maramureş Mountains has been known since the work by Zapalowicz (1886). However, the deposits remain poorly dated, or their interpretation has been controversial. For example, witness the conflicting views of Sawicki (1911) versus Sîrcu (1963), or the work on glacial landforms by Mac et al. (1990) or Mîndrescu (1997, 2001–2002).

Based on the identification of fundamental geomorphological features of this mountain area, the authors distinguish between the north-east area, an external part situated along the Ukrainian border, with the highest point at Pop Ivan (1940 m a.s.l.), and inner massifs, Farcău (1957 m a.s.l.), Mihailec (1918 m a.s.l.), Pietrosu Bardăului (1850 m a.s.l.), Toroiaga (1930 m a.s.l.) and Cearcănu (1846 m a.s.l.), well individualised by deep valleys.

In the 'external area' (including the Ukrainian slope), the authors have reconstructed 89 glacial entities, with a total area of 23.05 km^2. Of these, the Jupania glacier was the largest (3.32 km^2), with a maximum length of 4.4 km, reaching down to 1205 m a.s.l., followed by the Kvasny glacier (2.11 km^2), which descended to 1050 m a.s.l.

The 'inner massifs' hosted glaciers especially on the northern slopes. The Farcău–Mihailec Massif supported 14 glacial entities covering 7.9 km^2, the Pietrosu Bardăului with 10 glacial entities covering 2.54 km^2, the Toroiaga with 17 glacial entities covering 73.33 km^2 and a newly identified area, the Cearcănu, with 11 glacial entities covering 3.5 km^2. Based on the authors' work, the total palaeo-glaciated area on Maramureş Mountain reached 40.32 km^2.

Another area which has been the focus of considerable discussion (Athanasiu, 1899; Sawicki, 1911; Someşan, 1933; Naum, 1957b, 1970; Sîrcu, 1964) is the Călimani Mountains, the highest volcanic area of Romania, with the highest peak Pietrosu reaching 2100 m a.s.l. Based on geomorphological analysis, 79 individual palaeo-glacial bodies have been reconstructed: 50 niche glaciers, 14 ice aprons, 14 cirque and valley glaciers and 1 plateau glacier. Of these, 45.4% were orientated towards the north, which covered a total area of 44.35 km^2 (Table 24.2). The largest valley glaciers are present in the inner part of the volcanic caldera, in particular, the Pietricelu-Reţiţiş, which covered an area of 3.03 km^2 and a maximum length of 3.3 km, descending to 1350 m a.s.l. The surface of the upper volcanic shield was occupied by an ice-cap-plateau glacier, with a surface area of 22.7 km^2, with some short outlet ice-tongue glaciers descending from it (Straja, Iezer-Puturosu, Voivodeasa and Secu).

In the Ţibleş Mountains, in the Eastern Carpathians, the presence of glacial landforms is still under discussion. A part of the Northern Group of the Eastern Carpathians (Fig. 24.1) is orientated west–east and reaches 1842 m at Ţibles Peak. Much of the research on this area disregards the existence of glacial landforms, (e.g. Coteţ, 1973; Posea et al., 1974), while others very briefly mention the past occurrence of Pleistocene glaciers in the area. For example, Kubijovici (1934) comments, 'on the northern slope of the Ţibles we may find picturesque glacial landforms,' and Morariu (1942) states, 'Ţibleş had some glaciers, which are noticeable especially on the Maramureş slope, in the wide valleys with glacial aspect situated under the peak,' while others consider them to be nothing more than nival landforms (Posea, 1962). More recent research confirms the existence of two glacial cirques under the ridge between the Ţibleş and the Arcer peaks. These cirques are orientated towards the north-north-east (M. Mândrescu, personal communication 2005), whilst two glacial niches, located north of Arcer (1830 m a.s.l.), are orientated towards the north. Based on the authors' palaeo-glacial reconstruction, the Ţibleş Mountains hosted 13 glaciers, that is, 6 ice aprons, 6 niche glaciers and firn fields and 1 valley glacier. The latter, Izvoru Fundăului, descended to 1250 m a.s.l. and reached a total length of 1.6 km, the total surface of the glaciated area being estimated ca. 2.26 km^2.

Recently, 10 individual palaeo-glacial entities, that is, 7 niche glaciers and 3 ice aprons, have been recognised descending from the highest peak, Omu (1932 m a.s.l.), in the Suhard Mountains (Urdea and Reuther, 2009). The biggest were the niche glaciers Zimu, 0.74 km^2, and Valea Ursului, which climb down northwards, to 1420 and 1430 m, respectively (Table 24.2). The reconstructed surface of all these glaciers amounts only to 2.36 km^2.

Until recently, the Giumalău Mountains (1857 m a.s.l.) were considered to not have been glaciated (Lesenciuc, 2006). However, by careful geomorphological examination, glacio-nival niches have been recognised along both sides of the main north–south-orientated ridge, above 1660 m, the glaciated surface here amounting to only 0.779 km^2.

Also, until recently, the Bistriţei Mountains were also considered not to have been glaciated. It is now considered that the Budacu Mountain (1853 m a.s.l.), the main valley head (Borca, Borcuţa, Neagra, Ortoiţa), situated below the main ridge, between Budacu Peak (1859 m a.s.l.) and Muntele între Borci (1831 m a.s.l.), hosted small niche glaciers. The largest of these glaciers achieved areas of 0.66 km^2, Borcuţa, and 0.53 km^2, Borca. However, other valley heads in the area have the character of nivation hollows. The maximum total surface of the glaciated area is estimated to have been 2.13 km^2 (Table 24.2).

The Ceahlău Massif (1907 m a.s.l.) lies in the middle part of the Eastern Carpathians. According to Macarovici (1963), the Ocolaşu Mare plateau, at 1750–1900 m a.s.l., was covered by ice in the Pleistocene. However, because the evidence for glacial activity was considered equivocal for the main part of the Eastern Carpathians, Sîrcu (1964) excluded the Ceahlău Massif from the glaciated area.

Because the contact between the uppermost structural plateau and structural cliffs and walls is rounded in some areas and is continued with a U-shaped profile, the authors consider that the entire plateau was occupied by ice fields, spreading across 0.84 km^2. These ice fields continued across the abrupt area of the ice aprons, wall-sided glaciers, avalanche-cone glaciers (included in the wider category of cliff glaciers: *sensu* Lewandowski and Zgorzelski, 2004), simple niches and U-shaped niches in the rock wall, filled with snow and ice. A total of 20 glacial entities with a total surface of 3.67 km^2 have been reconstructed in this range (Table 24.2).

With a maximum elevation of 1843 m a.s.l., the Piatra Mare Massif is, from the general geological and geomorphological point of view, similar but smaller than the Ceahlău Massif. The presence of glacial landforms was never considered here, although nivation niches have been identified and mapped (Mihai, 2005). The morphological features suggest the occurrence of nine small niche glaciers and ice aprons/avalanche-cone glaciers and one plateau glacier that occupied a total surface area of 1.17 km^2.

Situated in the neighbourhood and not much higher in altitude, the Gârbova-Baiul Mountains are also included in the unglaciated Romanian Carpathians; only nivation landforms were recognised here (Niculescu, 1981). In the

TABLE 24.2 Summary of the Reconstructed Pleistocene Glaciers of Some Mountain Units of Eastern Carpathians

	Area	No. of Glacial Entities	Surface (km²) Average/Overall		ELA THAR$_{0.45}$	ELA STDEV	ELA THAR$_{0.45}$ N (North)	ELA THAR$_{0.45}$ S (South)	Volume (10⁶m³) Average/Overall		Longest Glacier (km)	Aspect (%)	Type of Glacial Entities[a]
1.	Maramureş Mountains	141	0.31	40.32	1561	80.14	1534	1565	15.34	1644.9	4.4	N-22.2; NE-12.2; E-10.7; SE-7.9; S-15.8; SV-7.2; V-15.8; NV-7.9	80 n.g.; 36 i.a.; 24 g.; 1 p.g.
2.	Ţibleş Mountains	13	0.18	2.26	1573	51.8	1550	1606	5.33	69.4	1.6	N-15.3; NE-38.4; S-30.7; SV-7.7; V-7.7	6 i.a.; 6 n.g.; 1 g.
3.	Suhard Mts.	10	0.23	2.36	1641	47.43	1621	1641	7.23	77.2	—	N-10; NE-10; E-10; SE-10; S-10; SV-20; NV-30	7 n.g.; 3 i.a.
4.	Giumalău Massif	10	0.07	0.79	1673	47	1637	1673	1.57	15.7	—	N-10; NE-30; E-10; SE-10; SV-10; V-30	10 n.g.
5.	Căliman Mountains	79	0.56	44.35	1714	75.45	1674	1714	12.84	2986.8	3.3	N-22.7; NE-16.4; E-10.1; SE-10.1; S-8.8; SV-7.6; V-16.4; NV-6.3	50 n.g.; 14 i.a.; 14 g.; 1 p.g.
6.	Bistriţei Mountains	10	0.21	2.13	1657	46.15	1665	1657	6.89	68.9	—	N-10; NE-10; E-10; SE-10; S-20; SV-20; V-10; NV-10	10 n.g.
7.	Ceahlău Massif	20	0.18	3.67	1655	83.75	1544	1655	5.82	116.4	0.7	N-5; E-35; SE-5; SV-5; V-25; NV-20	6 n.g.; 12 i.a.; 1 g.; 1 p.g.
8.	Piatra Mare Massif	10	0.117	1.17	1712	56.04	1627	1712	1.27	11.5	—	N-44.4; E-11.1; SE-11.1; S-11.1; SV-22.2	6 i.a.; 3 n.g.; 1 p.g.
9.	Neamţu Mountains	23	0.2	4.67	1670	41.02	1651	1670	6.31	145.2	1.9	N-13; NE-8.6; E-13; SE-17.4; S-4.3; SV-4.3; V-26; NV-13	22 n.g.; 1 g.
	Count	315	0.25	101.52	1627	58.75	1587	1627	6.95	5136	4.4	N-16.8; NE-15.6; E-12.2; SE-9; S-11.2; SV-9.1; V-14.5; NV-16.7	194 n.g.; 77 i.a.; 41 g.; 3 p.g.

[a] n.g., niche glacier; i.a., ice apron; g., glacier; p.g., plateau glacier.

northern half of this area, the Neamţu Massif reaches a maximum elevation of 1923 m at the Neamţu Peak. Here, the morphology of the valley heads situated along of the main interfluve, the Paltinu Peak (1900 m a.s.l.) to the Rusului Peak (1903 m a.s.l.), suggests that, in Pleistocene glacial periods, they hosted 22 niche glaciers and/or firn fields, and on the northern slope of the Paltinu Peak a small valley glacier (0.64 km^2, 1.9 km length), which descended to 1400 m a.s.l., with a total surface of 4.67 km^2.

24.6. THE SOUTHERN CARPATHIANS

Although the Southern Carpathians, or the Transylvanian Alps, show the most obvious glacial relief, with typical alpine landforms, in some areas, an indistinct glacial topography leaves the evidence of the Pleistocene glaciations somewhat controversial or it is reduced to the traces of small cirque glaciers. This area includes mountains that reach altitudes of over 2000 m a.s.l., including the Leaota, Căpăţânii and Latoriţei mountains, and therefore, special attention has been given to these areas in recent investigations.

In the Leaota Mountains (2133 m a.s.l.), the previous research by Nedelcu (1964) and Murătoreanu (2009) recognised only small glacial cirques and two glacio-nival cirques. Based on the morphological features identified by the authors, 39 individual palaeo-glacial entities have been reconstructed, that is, 29 niche glaciers, 9 ice aprons and 1 valley glacier, Mitarca (0.67 km^2, 1.7 km), 51.13 % orientated to the north. These ice bodies together covered an area of 7.54 km^2 (Table 24.3).

Although they achieve altitudes higher than 2000 m, the Căpăţânei (Nedeia Peak—2130 m a.s.l.) and Latoriţei Mountains (Bora Peak—2055 m a.s.l.), previous research by Martonne (1907) and Călin (1987) only recognised small glacial cirques on the eastern slope of the Ursu Peak (2125 m a.s.l.), one glacial cirque on eastern slope of the Frătoşteanu Peak (2053 m a.s.l.), and other four glacio-nivation cirques.

In cases where glacial landforms are less clearly expressed, detailed analysis offers a new basis for palaeo-glaciological reconstruction. Thus, in the Căpăţânei Mountains, 104 glacial entities—63 niche glaciers, 32 ice aprons and 9 cirque and valley glaciers—situated generally from west to the east along the main interfluves and having formed a total surface of 27.49 km^2 (Table 24.3) have been reconstructed. The largest glaciers were firn fields situated on the south slope of the Cosa-Ursu Mountain, with a surface of 1.31 km^2. It is important to mention that in the Târnovu Massif and Buila-Vânturariţa Massif, dominated by a calcareous ridge, with expressive cliffs, the ice-aprons, avalanche-con-glaciers and very small composite forms have been typical.

For the Latoriţei Mountains, investigations suggest that in Pleistocene cold periods, they hosted 15 cirque and valley glaciers, 20 niche glaciers and/or firn fields and 12 ice aprons, on a total surface of 4.67 km^2. The largest valley glaciers were present on the northern slope, Miru (2.66 km^2, 4 km), which descends to 1340 m, and Puru (1.78 km^2), which descends to 1470 m a.s.l.

An interesting situation is that of Capra-Buha Massif, situated in the north-eastern part of the Parâng Mountains. Reaching a maximum elevation of 1927 m a.s.l., the morphology is dominated by rounded interfluves. The existence of glacial landforms here had never been considered. Present investigations demonstrate that the morphology of the valley heads, situated along the main interfluve, Tomeşti (1861 m a.s.l.)–Capra (1927 m a.s.l.)–Buha (1905 m a.s.l.)–Cotu Ursului, suggests that, in the Pleistocene, the area hosted eight glacial niche glaciers and/or firn fields; one small valley glacier, Buha Mică (0.31 km^2, 1.2 km) on the northern slope of the Capra Peak, which descended to 1480 m; and one small cirque glacier, Cotu Ursului (0.31 km^2, 0.8 km). All glaciers and firn fields covered an area of 2.09 km^2 in this area.

A new interpretation of the less well developed features, the discrete and incipient glacial forms on the eastern part of the Vâlcan Mountains, the Straja Peak (1868 m a.s.l.)–Drăgoiu Peak (1690 m a.s.l.) area, indicates that 19 glacial entities—18 very small firn fields/niche glaciers, one ice apron-avalanche-cone glacier, Baleia (0.66 km^2), which descends to 1210 m—orientated mainly to the north, with a total surface of 3.86 km^2, occurred here.

24.7. THE APUSENI MOUNTAINS

The Apuseni Mountains are the lowest range in the Romanian Carpathians. Here, only three peaks higher than 1800 m a.s.l., and the issue of Pleistocene glaciation was again not considered, except for on the Bihor Mountains, and the southern part of the Vlădeasa Mountains, the highest mountains, with an altitude of 1849 m a.s.l. at Curcubăta Mare Peak or Bihor Peak and 1836 m a.s.l. at Vlădeasa Peak. The presence of glacial landforms was first debated by Szadeczky (1906), although the frontal moraines, that he described near Stâna de Vale, are in reality a mixture of scree and alluvial fan deposits (Martonne, 1922). Sawicki (1909) described a glacial cirque at the NE of Curcubăta Mare Peak, although Martonne (1922) and later Berindei (1971) considered such landforms to be of cryo-nival origin. However, Mac et al. (1990) identified a glacial cirque under Bihor Peak which generated a 1–1.5 km long glacial valley.

Recent investigations have confirmed the latter interpretation and enabled the precise configuration of the total extent of Pleistocene glaciation in this mountain area (Fig. 24.5). During the global Last Glacial Maximum, the total surface occupied by the ice here was 4.31 km^2. The glacial entities, 1 valley glacier, 3 ice aprons and 14 niche glaciers, were orientated especially towards the east

TABLE 24.3 Summary of the Reconstructed Pleistocene Glaciers of Some Mountain Units of Southern Carpathians

	Area	No. of Glacial Entities	Surface (km²) Average/Overall		ELA THAR$_{0.45}$	ELA STDEV	ELA THAR$_{0.45}$ N (North)	ELA THAR$_{0.45}$ S (South)	Volume (10⁶ m³) Average/Overall		Longest Glacier (km)	Aspect (%)	Type of Glacial Entities[a]
10.	Vâlcan Mts.	19	0.19	3.86	1497	82.65	1516	1522	5.73	108.8	–	N-15.8; NE-5.3; E-15.8; SE-10.5; S-15.8; V-21.1; NV-15.8	18 n.g.; 1 i.a.
11.	Capra-Buha Massif	10	0.2	2.09	1718	42.28	1710	1745	5.96	59.6	1.2	N-30; NE-10; E-10; S-20; SV-10; NV-20	8 n.g.; 2 g.
12.	Latoriței Mts.	47	0.42	20	1701	60.07	1705	1697	18.87	887	4	N-25.5; NE-10.6; E-4.3; SE-19.1; S-29.8; SV-6.4; NV-4.3	20 n.g.; 12 i.a.; 15 g.
13.	Căpățânii Mts.	104	0.26	27.49	1689	89.38	1669	1690	9.84	1024	1.3	N-22.1; NE-5.8; E-8.7; SE-11.5; S-17.3; SV-7.7; V-4.8; NV-22.1	63 n.g.; 32 i.a.; 9 g.
14.	Leaota Mts.	39	0.19	7.54	1719	70.45	1714	1705	5.98	233.4	1.7	N-30.8; NE-7.7; E-5.1; SE-7.7; S-5.1; SV-10.3; V-20.5; NV-12.8	29 n.g.; 9 i.a.; 1 g.
	Count	219	0.25	60.98	1664	68.96	1662	1671	9.27	2312.8	4	N-24.2; NE-7.3; E-7.8; SE-11.9; S-17.8; SV-7.3; V-7.8; NV-16	138 n.g.; 54 i.a.; 27 g.

[a] n.g., niche glacier; i.a., ice apron; g., glacier; p.g., plateau glacier.

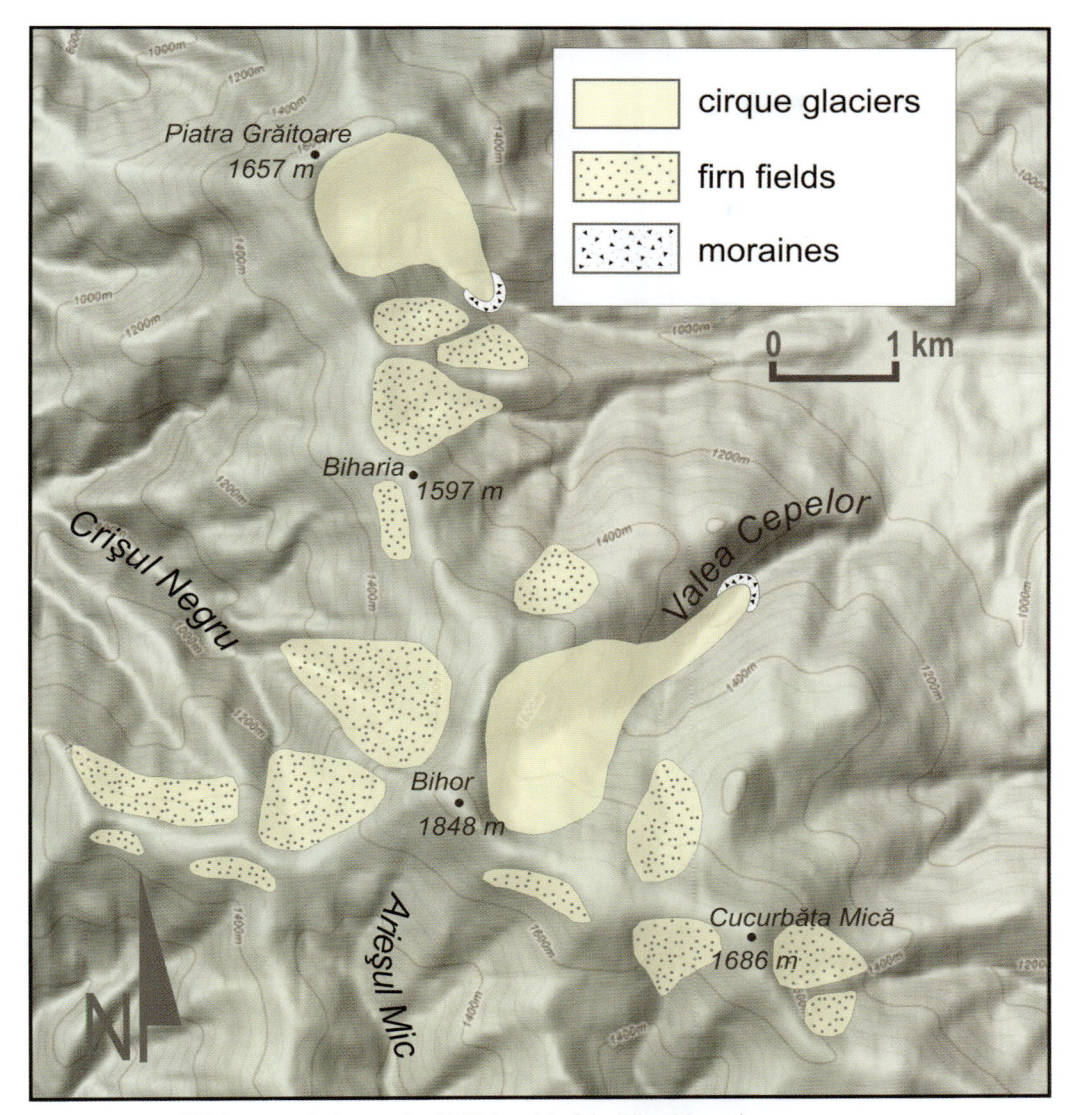

FIGURE 24.5 The reconstructed Pleistocene glaciers on the 3 DEM model of the Bihor Mountains.

(38.9%) (Table 24.4). This depended on the pre-glacial relief with the main ridge orientated north–south, and the accumulation of the drifting snow, blown by westerly winds. The cirque on the Bihor Peak continues into a glacial valley, the Valea Cepelor, which is covered with erratic blocks, down to about 1340 m a.s.l. Remains of morainic arcs are found at 1675, 1630, 1550 and 1340–1350 m and allow the reconstruction of the fluctuation of the glacier which reached a maximum surface area of 0.71 km².

Glacial landforms in Vlădeasa Mountains were first identified here by Szadeczky (1906) and Sawicki (1909) who described a glacial cirque on the north-eastern slope of the Buteasa Peak (1792 m a.s.l.). Modern investigations confirm these glacial features, and the reconstructed glacier achieved a surface area of 0.53 km² and was 1.5 km in length and descended to 1310 m. The writers have also recognised glacial features in the Vlădeasa Peak area and in Bohodei Peak (1654 m a.s.l.)–Cârligatele (1694 m a.s.l.)–Brăiasa (1692 m a.s.l.). Four glacial entities have been differentiated in the Vlădeasa Peak, situated along the main interfluve, especially on the eastern slope. The largest one of these, the Valea Zănoghii, covered an area of 0.72 km² and descended to 1340 m. The glacial and nivation forms on the Bohodei Peak–Cârligatele–Brăiasa are cut into the flat Cârligata peneplain and are the expression of the strong influences of topographic conditions. These landforms occur on the opposite margins of the highest planation surfaces, that is, on sites leeward of the prevailing winds from the west and north-west. They also occur in positions exposed towards the south to south-west. Of these

TABLE 24.4 Summary of the Reconstructed Pleistocene Glaciers of the Highest Mountain Units of Apuseni Mountains

	Area	No. of Glacial Entities	Surface (km^2) Average/Overall		ELA THAR$_{0.45}$	ELA STDEV	ELA THAR$_{0.45}$ N (North)	S (South)	Volume (10^{6}m^3) Average/Overall		Longest Glacier (km)	Aspect (%)	Type of Glacial Entities[a]
15.	Bihor Mountains	18	0.23	4.31	1500	106.5	1523	1609	7.59	136.7	1.5	N-16.7; NE-5.6; E-38.9; S-11.1; SV-5.6; V-16.7; NV-5.6	14 n.g.; 3 i.a.; 1 g.
16.	Vlădeasa Mountains	15	0.44	6.63	1500	64.7	1482	1505	17.99	268.8	1.3	N-13.3; NE-13.3; E-13.3; SE-33.3; S-6.7; SV-6.7; V-13.3	8 n.g.; 3 i.a.; 4 g.
17.	Muntele Mare Mountains	3	3.71	11.15	1622	9.19	1622		228.17	684.5	1.5	N-100	1 n.g.; 1 g.; 1 p.g.
	Count	36	0.46	22.09	1540	60.13	1542	1557	84.58	1090	–	N-20; NE-8.6; E-25.7; SE-14.3; S-8.6; SV-5.7; V-14.3; NV-2.9	23 n.g.; 6 g.; 6 i.a.; 1 p.g.

[a] n.g., niche glacier; i.a., ice apron; g., glacier; p.g., plateau glacier.

reconstructed glaciers, mainly comprising niche glacial/firn fields, two Cârligatele and Valea Voiosu, occupied over 1 km² in area.

An interesting situation is that of the Muntele Mare Mountains, with a maximum elevation of 1825 m a.s.l. The morphology of these uplands is dome shaped as a result of a peneplain cut in the granitic bedrock. Given the medium–high altitude of this mountain area, the existence of some glacial landforms was never considered and, indeed, evidence of glaciation is much less well represented morphologically. But, the presence of some very flared valley heads on the margin of the plateau level continuing as small U-shaped valleys (Crețoaia, Vânătu, Valea Mare, etc.) suggests that a plateau glacier extended here across 9.7 km², with outlet glaciers descending down the valleys over all sides of the mountain. The longest glacier, Vânătu, had a length of 1.5 km, and a total glacier area of 11.15 km² (Table 24.4).

24.8. GLACIER ORIENTATIONS

An important characteristic of the reconstructed glaciers in the areas discussed here is their orientation, or aspect. It is clear that for each massif of the Romanian Carpathians, asymmetry is a typical feature (Fig. 24.6). This is expressed as a pronounced west–east asymmetry in the Eastern Carpathians and the Apuseni Mountains, and a less pronounced north–south asymmetry in the Southern Carpathians. The west–east asymmetry arises from the general north–south orientation of the main interfluves, in conjunction with the position in respect to the main westerly to north-westerly winds and associated precipitation. These directions correspond to that determined from examination of longitudinal and barcane dunes of Würmian age that occur in south-west of the Romanian Plain, more exactly in the Oltenia Plain (Coteț, 1957). However, the major part of the study areas are dominated by rounded and gentle summit interfluves, a factor that on windward slopes would favour drifting snow (cf. Evans, 2005).

For the Southern Carpathians, the north–south asymmetry arises from the north–south asymmetry of the relief, with the main interfluves orientated east–west. This orientation is connected with a tectonic asymmetry, as well as to the north–south climatic asymmetry.

In fact, the asymmetry can be explained through the role of the pre-glacial relief. In particular, the direction of the main interfluves represents the ideal area for the initiation of glacial cirques. A particular importance is not only their configuration, the widespread of the round-shaped interfluves, the latter having been inherited from peneplains, but also the development degree of the reception basins. These basins have a semi-funnel form, which was favourable for snow accumulation and for the processes of glacial ice formation.

24.9. PROBLEMS OF AGE ASSIGNMENT

More recently, an attempt has been made to develop an improved temporal framework for the glaciations of the Romanian Carpathians. This has involved the application of cosmogenic dating of boulders from moraines located in the Pietrele valley (Retezat Mountains) using ^{10}Be (Reuther, 2005; Reuther et al., 2007). Two major glacial advances, M 1 and M 2 (the Lolaia and Judele-Jieț glaciations in local terminology), have been previously recognised (Urdea, 1989, 2000). The most extensive M 1 advance reached an elevation of 1035 m a.s.l. in the Retezat Mountains, some 250 m below the terminal moraines of the younger M 2 advance.

The timing of the M 1 advance is uncertain, but it occurred either during the Early Würmian Substage, that is, approximately marine isotope stage (MIS) 4, or during the Rissian (~MIS 6) glaciation. This correlation is based on the relative chronology derived from pedological investigations (Reuther et al., 2004). The exposure ages show that the younger M 2 glacial advance (the Judele-Jieț) was deposited during the Würmian late glacial at 16.8 ± 1.8 ka in the Retezat Mountains and 17.9 ± 1.6 ka

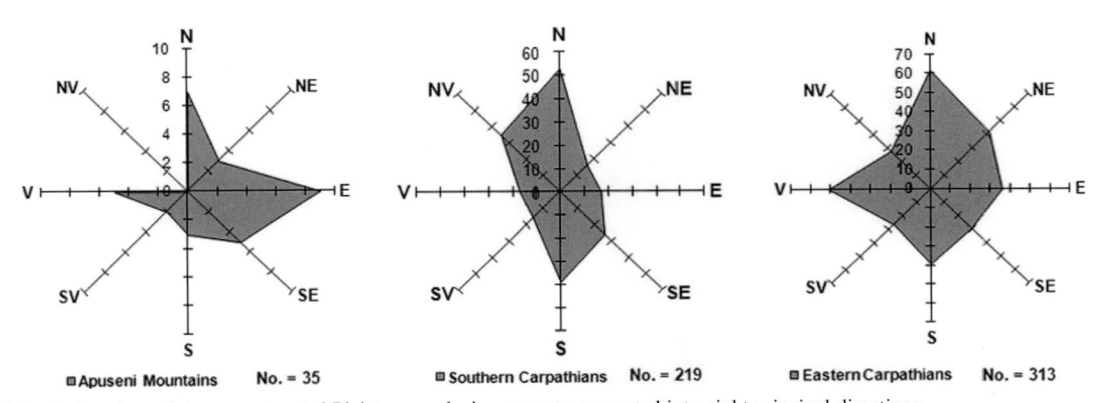

FIGURE 24.6 Radar plots of the reconstructed Pleistocene glaciers aspect aggregated into eight principal directions.

in the Parâng Mountains. Two exposure ages constrain the time of an additional younger glacial advance (M 3) during the Younger Dryas Stadial. This advance, that deposited the moraine between two massive boulders (horizontal distance 350 m), falls in the period between the deposition of the older (13.6 ± 1.5 ka) and the younger (11.4 ± 1.3 ka) of the two boulders situated at 1851 and 1902 m a.s.l, respectively.

For the mountains situated in the northern part of the Eastern Carpathians, the development of the late glaciers is correlated to the period of cooler and drier conditions between 12.9 and 11.5 cal. ka. This development is represented by palaeontological and sedimentological evidence in the Preluca Ţiganului, Gutâi Mountains, situated at 730 m a.s.l. These conditions are interpreted from the recurrence of an open landscape with only scattered trees of *Betula*, *Larix*, *Salix* and *Pinus* (Feurdean, 2005).

It should be noted that in the eastern part of the Făgăraş Mountains, in the upper part of the Dejani valley cirques, four stadial moraines situated between 1860 and 2010 m occur. Two of these moraines are connected to rock glaciers that represent the latest part of the last glaciation (Urdea and Reuther, 2009). These events are tentatively associated with the cooling episodes characterised by an increase of *Pinus* and *Artemisa* pollen during the Younger Dryas-Preboreal interval (LPAZ 6, LPAZ 7 and LPAZ 8) in Avrig marsh (600 m a.s.l.) in the Făgăraş depression (Tantău et al., 2005). In addition, oxygen and carbon stable isotope records from two stalagmites from Bihor Mountains indicate a cold and dry climate between 12.6 and 11.4 (11.7) ka. BP, that is, during the Younger Dryas Chron (GS-1). The ^{18}O profiles also indicate three short, cold intervals during the early Holocene at 11.0–10.6, 10.5–10.2 and 9.4–9.1 ka (Tămaş et al., 2005).

24.10. RECONSTRUCTED PLEISTOCENE ELAs

Palaeo-ELAs, which are sometimes referred to as the snowline, were estimated by using the toe-to-headwall altitude ratio (THAR), with a THAR ratio of 0.45. To explore the nature of intra-regional ELA variability, we have considered different groupings of glaciers and computed both mean ELA and standard deviation (Tables 24.2–24.4).

At 17 sites in the Romanian Carpathians, estimates of average ELA are 1610 m, with an average standard deviation of 62.61. Analysing the characteristic values for each Romanian Carpathian branch, the authors find that there is a low amount of intra-regional variance in the ELA, with values ranging over 100 m, from a minimum value of 1540 m for the Apuseni Mountains, to a maximum of 1664 m, for Southern Carpathians. Most of the variance in ELAs can be explained by the climatic conditions—the differences between the western part, more humid, and the eastern part, and between the south and the north, cooler—and by the local topographic context (cf. Nesje, 1992).

Examination of all the 570 reconstructed palaeo-glaciers with ELAs derived from this study shows an important intra-regional variability in reconstructed ELAs. For example, within the Eastern Carpathians investigated areas, the values range from 1561 m in Maramureş Mountains and 1714 m in Călimani Mountains, or within the Southern Carpathians, between 1497 m in the Straja area (Vâlcan Mountains), situated in the western part, and 1719 m in the Leaota Mountains, situated in the eastern part. Because the ELA value of Straja is the lowest value of all reconstructed ELAs, this can be explained by the topographic-climatic conditions of the composite palaeo-glacier, ice apron and avalanche-cone glacier, orientated towards the north, on a shaded area. The concept of the local topographic temperature-precipitation-wind-ELA (TPW-ELA) can be applied to the particular situations described here, and the cirque glaciers may have existed well below the regional temperature-precipitation-ELA (TPELA) (Dahl and Nesje, 1992).

Taking into account the westwardly air masses circulation during the Pleistocene time (Mîndrescu et al., 2010), in the authors' opinion, in Apuseni Mountains, the contrast of the west to eastern trend in the ELA values—1500 m for the Bihor and Vlădeasa and 1622 m for Muntele Mare—is best explained also by precipitation patterns, with 1631.5 mm at Stâna de Vale (1102 m a.s.l.), situated on the western slope, and only 843 mm at Băişoara (1385 m a.s.l.), situated on the eastern slope and affected by foehnic air masses.

If the average ELAs values specific for the north and south slopes are compared, it can be seen that, in all the Carpathian ranges, the values generally rise to the south (Figs. 24.7–24.9). Those that are the exceptions, that is, the Budacu Massif in the Eastern Carpathians, the Latoriţei and Leaota Mountains in the Southern Carpathians, have a morphological peculiarity, imposed by the pre-glacial landforms.

Sawicki (1911) estimated the full-glacial snowline on the Maramureş Mountains to lie at ca. 1500 m, for the Mihailec-Farcău Massif. This is close to the value estimated by Sîrcu (1963) at 1550 m, at ca. 1480 m for Pop Ivan Massif, and at 1750 m and at ca. 1700 m for the deglaciation period (*Zeit des Rückzuges*), derived using the Kurowski method.

In the Retezat Mountains, the reconstructed ELAs, derived using the accumulation area ratio $AAR_{(0.75)}$ method, for the M 2 and M 3 glacial advances in the Pietrele-Nucşoara area, were modelled to 1770 and 2030 m (Reuther et al., 2004).

In addition, estimated ELAs for the last two glacial stages of Răchitiş and Pietrosu glaciers of Călimani Mountains (Kern et al., 2006) occurred at 1840 and 1915 m and at 1850 and 1925 m, respectively.

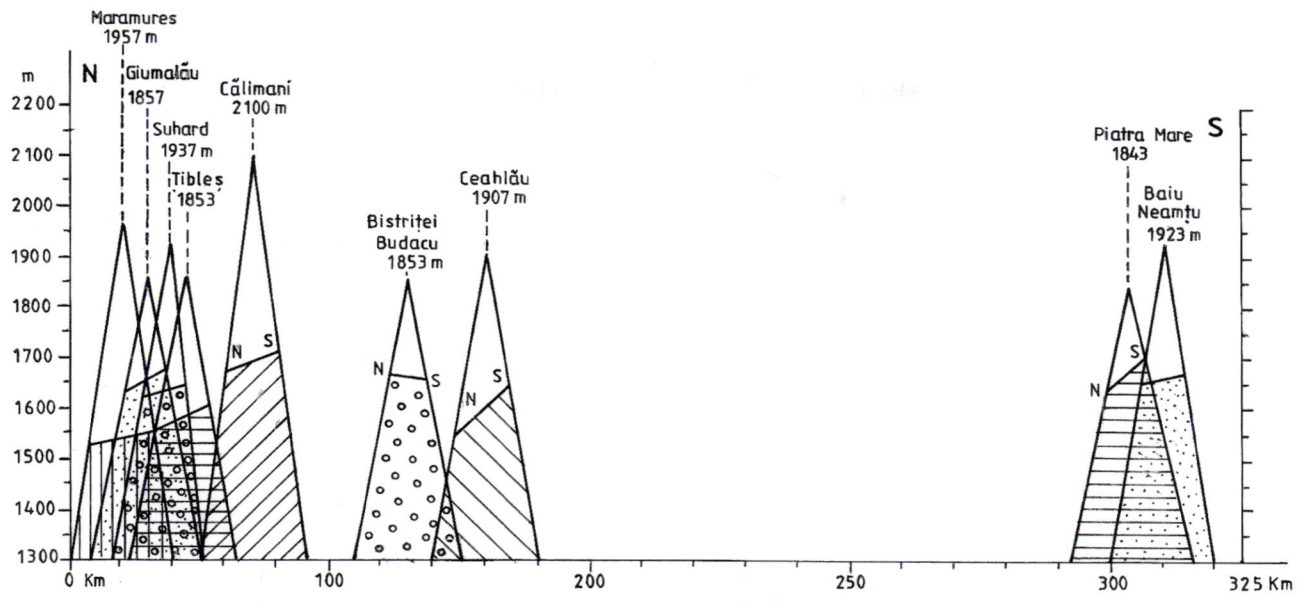

FIGURE 24.7 The full-glacial north and south values of ELAs for the reconstructed Pleistocene glaciers of the Eastern Carpathians.

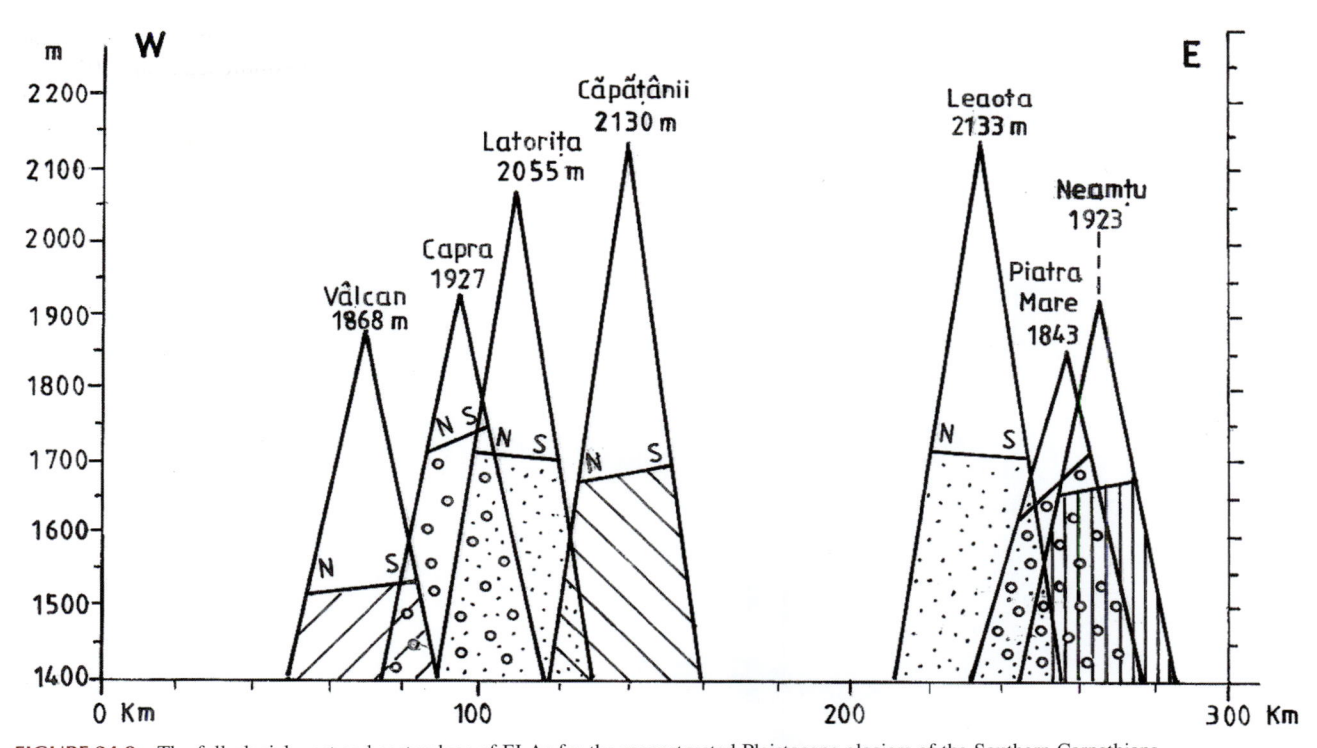

FIGURE 24.8 The full-glacial west and east values of ELAs for the reconstructed Pleistocene glaciers of the Southern Carpathians.

24.11. CONCLUSIONS AND OPEN QUESTIONS

On the basis of the results presented and recent investigations, the Pleistocene glaciation pattern in the Romanian Carpathians was more extensive and complicated than previously thought. In all, 570 glacial entities (glaciers, niche glaciers, ice aprons, plateau ice fields and firn fields) have been reconstructed in the 17 mountain ranges of the Romanian Carpathians (9 in the Eastern Carpathians, 5 in the Southern Carpathians and 3 in the Apuseni Mountains). The most frequent surface orientation of these glacial

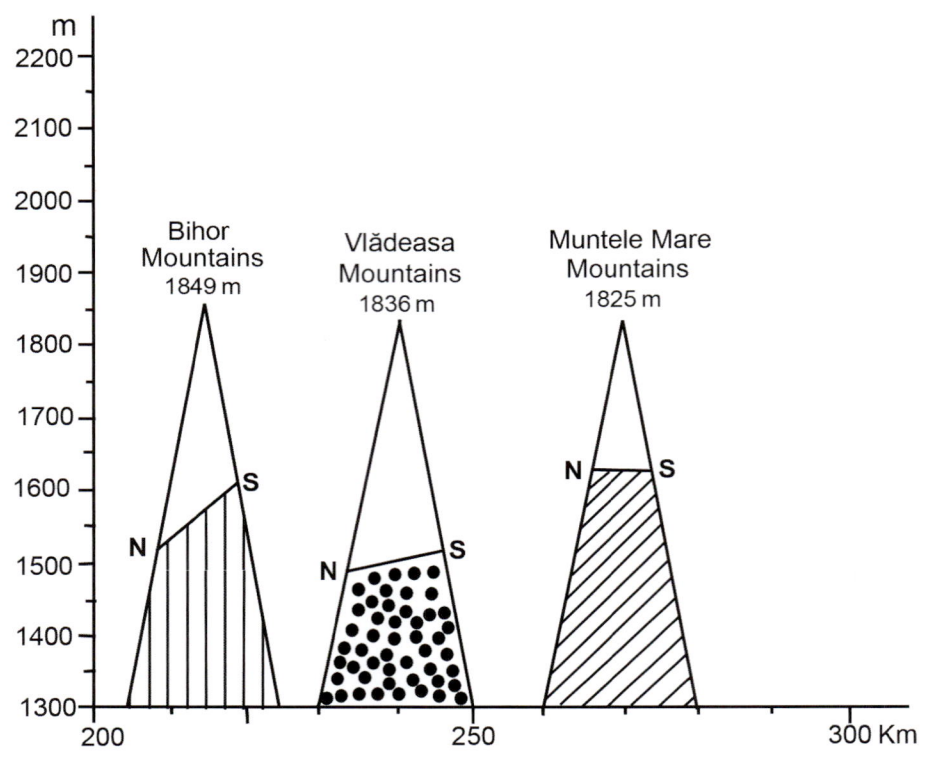

FIGURE 24.9 The full-glacial west and east values of ELAs for the reconstructed Pleistocene glaciers of the Apuseni Mountains.

FIGURE 24.10 Stadial moraines and periglacial landforms in the Mija cirque (Parâng Mountains, Southern Carpathians).

entities was consistently towards the north, north-east and north-west, 42.7%, differentiated by each branches: 49% in the Eastern Carpathians, 47.5% in the Southern Carpathians and 31.5% in the Apuseni Mountains.

Under the general climatic conditions during Pleistocene cold periods, the development of glaciers in the Romanian Carpathians was controlled by the pre-glacial relief, the orientation of the main slopes and interfluves—which had a direct influence on the appearance of the specific topographic–climatic conditions—and the magnitude of climatic continentality.

What is particularly certain is that glaciation in the Romanian Carpathians was more widespread in the Pleistocene than was previously considered and that it had several distinct stages throughout the region. The first radiometric dating from the Pietrele Valley (Retezat Mountains) indicates that, when sufficient numerical dating is available, the timing and extent of the Pleistocene glaciation in Romanian Carpathians will be unravelled. However, it will be possible to draw comparison between the most glaciated areas of Romania and, in consequence, construct an image of the evolution of the Quaternary glaciation in this part of the Alpine-Himalayan orogenic chain.

A comprehensive understanding of the extent and evolution of Pleistocene glaciation requires a detailed geomorphological analysis and mapping, connected to lithostratigraphical and chronological studies, based on surface exposure dating using cosmogenic radionuclides, palynological analysis and radiocarbon dating. This approach will also include the research on the Würmian late glacial and early Holocene (Fig. 24.10). Future international cooperation will help in solving the dating problems and clarify some open questions.

The Carpathians Mountains still hold much promise for research into all forms of glacial and periglacial geomorphology.

REFERENCES

Ancuţa, C., 2005. Munţii Lotrului. Studiu geomorfologic. Teză de doctorat, Universitatea din Oradea, 186pp.

Athanasiu, S., 1899. Morphologische Skizze in den Nordmoldauischen Karpathen. Bul. Soc. de Ştiinţe, Bucureşti 3, 232–277.

Bahr, D.B., Meier, M.F., Peckham, S.D., 1997. The physical basis of glacial volume-area scaling. J. Geophys. Res. 102 (B9), 20355–20362.

Berindei, I., 1971. Microrelieful crio-nival din Masivul Biharea. Lucrările ştiinţifice ale Institutului Pedagogic din Oradea, seria, Geografie, pp. 19–28.

Bognar, A., Prugovečki, I., 1997. Glaciation traces in the area of the Risnjak Mountain Massif. Geol. Croatica 50 (2), 269–278.

Carr, S., Zbyněk, E., Kalvoda, J., Parker, A., 2007. Toward a revised model of Quaternary mountain glaciation in the Krkonoše Mountains, Czech Republic. In: Goudie, A., Kalvoda, J. (Eds.), Geomorphological Variations. P3K, Prague, pp. 253–268.

Călin, D., 1987. Munţii Latoriţei—schiţă geomorfologică. Terra XIX (XXXIX) (2), 39–43.

Constantinescu, T., 2009. Masivul Piatra Craiului. Sudiu geomorfologic. Edit. Universitară, Bucureşti, 163 p.

Coteţ, P., 1957. Câmpia Olteniei. Studiu geomorfologic. Edit. Ştiinţifică, Bucureşti, 271pp.

Coteţ, P., 1973. Geomorfologia României. Edit, Tehnică, Bucureşti, 414 pp.

Dahl, S.O., Nesje, A., 1992. Paleoclimatic implications based on equilibrium-line altitude depressions of reconstructed Younger Dryas and Holocene cirque glaciers in inner Nordfjord, western Norway. Palaeogeogr. Palaeoclimatol. Palaeoecol. 94, 87–97.

Ehlers, J., Gibbard, P.L. (Eds.), 2004. Quaternary Glaciations—Extent and Chronology, Part 1: Europe. Elsevier B.V. 475 p.

Evans, I.S., 2005. Local aspect asymmetry of mountain glaciation: a global survey of consistency of favoured directions for glacier numbers and altitudes. Geomorphology 73 (1–2), 166–184.

Feurdean, A., 2005. Tracking Lateglacial and early Holocene environmental change: a paleo-limnological study od sediments at Preluca Ţiganului, NW Romania, Studia Universitatis, "Babeş-Bolyai" Geologia 50 (1–2), 3–11.

Grove, J.-M., 1961. Some Notes on Slab and Niche Glaciers, and the Characteristic of Proto-Cirque Hollows, vol. 54. International Association of Scientific Hydrology Publication, Helsinki, pp. 281–287.

Ielenicz, M., Pătru, I., 2005. Geografia fizică a României. Edit. Universitară, Bucureşti, 255pp.

Kern, Z., Nagy, B., Kohán, B., Bugya, E., 2006. Glaciological characterization of small palaeo-glaciers from Călimani Mountains. Analele Univ. de Vest din Timişoara. Geografie XVI, 35–44.

Klein, A.G., Isacks, B.L., 1998. Alpine glacial geomorphological studies in the Central Andes using Landsat Thematic Mapper Images. Glacial Geol. Geomorphol. rp01, http://ggg.qub.ac.uk/ggg/papers/full/1998/rp011998/rp01.htm.

Kubijovici, V., 1934. Păstoritul în Maramureş. Bul. Soc. Reg. Rom. Geogr. LII, 215–293.

Lehmann, P.W., 1881. Beobachtungen über Tektonik und Gletscherspuren im Fogaraschen Gebirge. Zeitsch. d. Deutsch. Geol. Gesellsch. XXXIII, 109–117.

Lehmann, P.W., 1885. Die Südkarpaten zwischen Retjezat und Königstein. Zeitsch. d. Geseellsch. F. Erdkunde Berlin, XX 325–336, 346–364.

Lesenciuc, C.D., 2006. Masivul Giumalău. Studiu geomorfologic. Edit. Tehnopress, Iaşi, 268 p.

Lewandowski, W., Zgorzelski, M., 2004. Wall-sided glaciers. Misc. Geogr. 11, 75–80.

López-Moreno, J.I., Nogués-Bravo, D., Chueca-Cía, J., Julián-Andrés, A., 2006. Glacier development and topographic context. Earth Surface Process. Landforms 31, 1585–1594.

Mac, I., Covaci, I., Moldovan, C., 1990. Glaciaţiune şi morfologie glaciară în munţii mijlocii din România. Studia Univ. "Babeş-Bolyai", Geogr. XXXV (2), 3–11.

Macarovici, N., 1963. Unele observaţii în legătură cu problema glaciaţiunii cuaternare din Carpaţii Orientali. Natura, s. Geogr.- Geol. XV, IX, 4, 8–15.

de Martonne, Emm., 1907. Recherches sur l'évolution morphologique des Alpes de Transylvanie (Karpates méridionales). Rev. de géogr. Annuelle Paris I (1906–1907), 286pp.

de Martonne, Emm., 1922. Le Massif du Bihor. In: Lucrările Institutului de Geografie al Universităţii din Cluj, vol. I. Cluj, Inst. de Geografie, Tipografia, Cultura Naţinală, pp. 47–114.

Mihai, B.A., 2005. Munții Timișului (Carpații Curburii). Potențialul geomorfologic și amenajarea spațiului montan. Edit. Universitară, București, 409 p.

Mîndrescu, M., 1997. Perenitatea formelor de relief glaciare din Munții Maramureșului. An. Univ. "Stefan cel Mare" Suceava, Geogr.—Geol. 6, 45–48.

Mîndrescu, M., 2001–2002. Muntele Jupania. Un nou areal glaciat din Carpații Orientali. An. Univ. "Stefan cel Mare" Suceava, Geogr.—Geol. 11–12, 41–47.

Mîndrescu, M., Evans, I.S., Cox, N.J., 2010. Climatic implications of cirque distribution in the Romanian Carpathians: palaeowind directions during glacial periods. J. Quatern. Sci. doi:10.1002/jqs.1363.

Morariu, I., 1942. Vegetația muntelui Țibleș. Bul. Soc. Reg. Rom. Geogr. LXI, 143–180.

Murătoreanu, G., 2009. Munții Leaota. Studiu de geomorfologie. Edit. Transversal, Târgoviște, 182pp.

Naum, T., 1957a. Geomorfologie R.P.R. Probleme speciale, Vol. I. Facultatea de Științe Naturale-Geografie, Tipografia Universității, "C. I. Parhon", București.

Naum, T., 1957b. Observații geomorfologice în Siriu. Anal. Univ., "C.I. Parhon" București, Seria Șt. naturii, geologie-geografie 13, 233–254.

Naum, T., 1970. Complexul de modelare nivo-glaciar din Masivul Căliman. Anal. Univ. București Geogr. XIX, 67–75.

Nedelea, A., 2006. Valea Argeșului în sectorul montan. Studiu geomorfologic. Edit. Universitară, București, 229 p.

Nedelcu, E., 1964. Sur la cryo-nivation actuelle dans les Carpates Méridionales entre les rivieres Ialomița et Olt. Rev. roum. géol. géophys. géogr. Géographie 8, 121–128.

Nesje, A., 1992. Topographical effects on the Equilibrium-Line Altitude on glaciers. Geo J. 27 (4), 383–391.

Niculescu, Gh., 1981. Munții Gârbova—caractere geomorfologice. Stud. cerc. geol. geofiz. geogr. Geografie XXVII, 9–19.

Porter, S.C., 2001. Snowline depression in the tropics during the Last Glaciation. Quatern. Sci. Rev. 20, 1067–1091.

Posea, G., 1962. Țara Lăpușului. Edit, Științifică, București 281 pp.

Posea, G., Popescu, N., Ielenicz, M., 1974. Relieful României. Edit, Științifică, București, 483 pp.

Raab, T., Völkel, J., 2003. Late Pleistocene glaciations of the Kleiner Arbersee area in the Bavarian Forest, south Germany. Quatern. Sci. Rev. 22, 581–593.

Reuther, A., 2005. Surface exposure dating of glacial deposits from the last glacial cycles. Evidence from the Eastern Alps, the Bavarian Forest, the Southern Carpathians and the Altai Mountains. University of Regensburg, PhD thesis, 246pp.

Reuther, A., Geiger, C., Urdea, P., Heine, K., 2004. Determining the glacial equilibrium line altitude (ELA) for the Northern Retezat Mts. Southern Carpathians and resulting paleoclimatic implications for the last glacial cycle. Analele Univ. de Vest din Timișoara, GEOGRAFIE XIV, 11–34.

Reuther, A., Urdea, P., Geiger, C., Ivy-Ochs, S., Niller, H.P., Kubik, P., et al., 2007. Late Pleistocene glacial chronology of the Pietrele Valley, Retezat Mountains, Southern Carpathians constrained by ^{10}Be exposure ages and pedological investigations. Quatern. Int. 164–165, 151–169.

Sawicki, L., 1909. A Biharhegység eljegesedésének kérdéséhez. Földrajzi Közlemenyek, XXXVII 10, 316–325.

Sawicki, L., 1911. Die glazialen Züge der Rodnaer Alpen und Marmaroscher Karpaten. Mitteilungen d.k.k. Geographische Gesellschaft in Wien LIV, IX-X, 510–571.

Simoni, S., 2008. The glacial valleys in the basin of the Doamna river (Făgăraș Massif). Geogr. Phorum 7, 35–51.

Simoni, S., Flueraru, C., 2006. Leaota and Zârna-Ludișor glacial-complexes (Făgăraș Massif). Analele Univ. de Vest din Timișoara, GEOGRAFIE XVI, 61–74.

Sîrcu, I., 1963. Le probleme de la glaciation quaternaire dans les montagnes du Maramureș. Anal. Șt. Univ., Al I. Cuza" Iași, (Serie nouă), secț. II (Șt. nat.) b. Geologie-Geografie IX, 125–134.

Sîrcu, I., 1964. Cîteva precizări în legătură cu glaciația cuaternară din Carpații Orientali Românești. Natura, Seria Geologie-Geografie 3, 24–31.

Sîrcu, I., 1978. Munții Rodnei. Studiu morfogeografic. Edit. Academiei, București, 112pp.

Someșan, L., 1933. Urme glaciare în Munții Călimani. Bul. Soc. Reg. Rom. Geogr. LI, 295–299.

Szadeczky, J., 1906. Glecsernyomok a Biharhegységben. Földrajzi Közlemenyek, XXXIV 8, 299–304.

Szepesi, A., 2007. Masivul Iezer. Elemente de geografie fizică. Edit. Universitară, București, 208 p.

Tantău, I., Reille, M., de Beaulieu, J.-L., Farcas, S., 2005. Late Holocene vegetation history in the southern part of Transylvania (Romania): pollen analysis of two sequence from Avrig. J. Quatern. Sci. 21 (1), 49–61.

Tămaș, T., Onac, B.P., Bojar, A.-V., 2005. Lateglacial-Middle Holocene stable isotope records in two coeval stalagmites from the Bihor Mountains, NW Romania. Geol. Q. 49 (2), 185–194.

Tietze, K., 1878. Über das Vorkommnis der Eiszeitspuren in den Ostkarpathen. Verhandlungen der geologische Reichsanstalt, Wien, pp. 142–146.

Urdea, P., 1989. Munții Retezat. Studiu geomorfologic. Teză de doctorat, Universitatea, "Al.I. Cuza" Iași, 186pp.

Urdea, P., 2000. Munții Retezat. Studiu geomorphologic. Edit. Academiei Române, București, 272pp.

Urdea, P., 2004. The Pleistocene glaciation of the Romanian Carpathians. In: Ehlers, J., Gibbard, P.L. (Eds.), Quaternary Glaciations—Extent and Chronology, Part 1: Europe. Amsterdam, Elsevier, pp. 301–308.

Urdea, P., Reuther, A.U., 2009. Some new data concerning the Quaternary glaciation in the Romanian Carpathians. Geogr. Pannon. 13 (2), 41–52.

Vočadlová, K., Křižek, M., 2005. Glacial landforms in the Černé jezero Lake area. Miscellanea Geographica 11, 45–62, KGE, Z_U v Plzni.

Zapałowicz, H., 1886. Geologische Skizze des östlichen teiles der Pokutisch-Marmaroscher Grenzkarpaten. Jahrb. D. Geol. Reichsanst. Wien 361–394, 580–587.

Chapter 25

Ice Margins of Northern Russia Revisited

Valery Astakhov

Geological Faculty, St. Petersburg University, Universitetskaya 7/9, 199034 St. Petersburg, Russia

25.1. INTRODUCTION

This chapter is a regional explanatory note to the second edition of the electronic database on the world ice limits. The text concerns the huge landmass of northern Eurasia between 48°E and 114°E embracing north-eastern European Russia, Urals, West and Central Siberia. The Pleistocene ice limits of this area have been updated according to recent international research, particularly owing to many new radiocarbon, luminescence, uranium series and cosmogenic exposure datings. The main change since the first edition (Astakhov, 2004a) is the addition of MIS 6 ice limit for Siberia and of area glaciated in the Middle Weichselian time span for the Arctic. The Late Weichselian glaciation has shrunk to the Barents Sea shelf and to hypothetical ice caps over the Putorana Plateau. A corrected and more extensive list of stratigraphical key sites is offered. Figures give a sketch of the renovated ice limits and explain major stratigraphical results in the glaciated area.

Our understanding of the configuration and especially timing of Eurasian Pleistocene ice sheets beyond the realm of Fennoscandian glaciations has considerably changed since the first edition of this compendium (Astakhov, 2004a). The contents of the electronic database are principally the same; it includes (i) the boundaries of ice sheets interpolated between crucial sections with the help of photogeology or, where this tool is not instrumental, by hypothetical evaluation of topographic features; (ii) crests of major ice-pushed ridges and axes of largest glaciotectonic disturbances; and (iii) key stratigraphical sites, which are necessary for determination of relative age of the mapped ice limits. The map pattern presented herein in the form of CD files is modified mostly east of the Urals and partly in the Arctic as a result of the large number of new luminescence, uranium–thorium, radiocarbon and cosmogenic exposure dates on Upper Pleistocene deposits obtained by international teams in the 6 years. These new data have also resulted in a re-evaluation of some former conclusions concerning the late Middle Pleistocene. However, the huge glaciated area between 48°E and 114°E remains poorly investigated and will probably remain so for years to come. To fill the study gap, and in view of possible further developments, the author has added several not strictly proven (Taz and Middle Weichselian glaciations) or provisional (mostly Late Weichselian) ice margins. As a consequence of the new dates, several key sections from the first edition have been eliminated but more are added. Figure 25.1 provides an overview of the latest results, whilst other figures explain the new stratigraphy. An important part of this essay is Table 25.1 where all stratigraphical sites used for determining relative ages of former ice sheets are listed.

25.2. MIDDLE PLEISTOCENE GLACIATIONS

Quaternary sediments with reverse palaeomagnetic polarity have not been reliably identified in the Russian North. Several samples from long drilling cores in the More-Yu river catchment ca. 68°N/61°E showed the reverse polarity from strata very close to the surface (Yakhimovich et al., 1992). However, the measurements were taken from heavily glaciotectonised diamicton formations, and as they contrast strongly with Late Pleistocene luminescence and U/Th ages from the same area (Astakhov and Svendsen, 2002), these magnetic determinations should be treated as spurious. The remaining glacial formations up to 300–400 thick known from the north are thought to belong to the Bruhnes Chron, including the deepest (342 m below sea level) Lebed borehole (Fig. 25.1).

The general succession of diamicton formations in the European North is given by borehole profiles A and B and Lake Chusovskoye (Fig. 25.1) published as Figs. 2 and 8 in Astakhov (2004b). Four thick diamictons recognised in the Arctic (Lavrushin et al., 1989) and three on the Pechora-Volga interfluve (Stepanov, 1974) bear distinct traces of ice flow from the north and northeast along the Urals (Astakhov, 2004b). The upper diamicton of the Arctic region, overlying the marine Boreal Strata containing an Atlantic mollusc fauna such as *Arctica islandica*, should be a counterpart of the Weichselian glacial of the Fennoscandian ice-dispersal centre. The remaining three diamictons underlying the

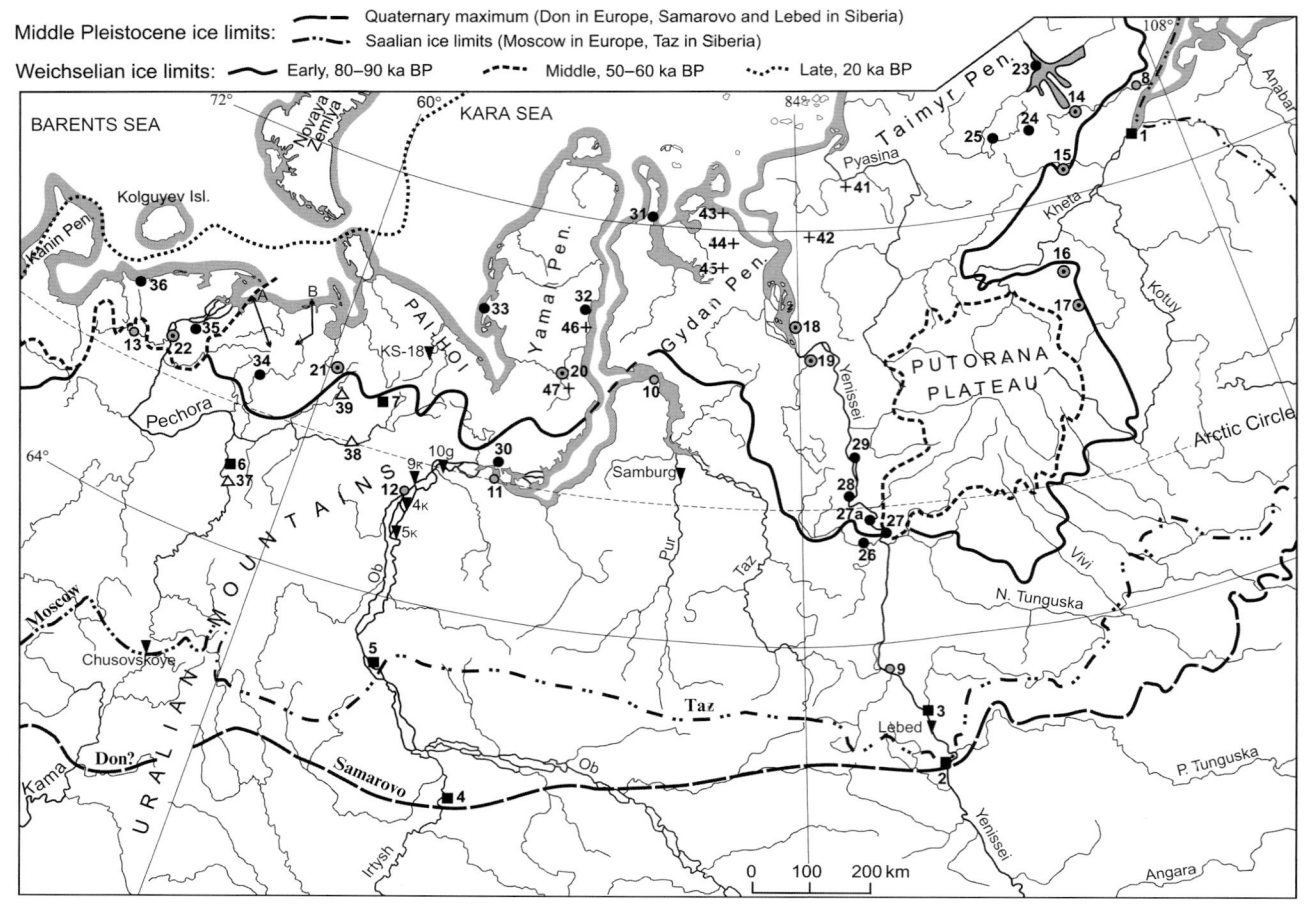

FIGURE 25.1 Selected stratigraphic localities for distinguishing ice limits in northern Russia. Literary sources and more sites and are listed in Table 25.1. Symbols: *Black triangles* are boreholes drilled through several glacial and interglacial formations: Lebed – thick diamicts beneath Turukhan alluvium; Samburg – marine formation overlain by Middle Pleistocene till, 4k, 5k, 9k, 10g – marine formation overlain by thick diamictons. *Arrowhead lines* A and B are borehole profiles which uncovered multiple diamictons and marine formations including Eemian; Chusovskoye lake – borehole profiles through 3 tills and interglacial sequences; KS-18 – till-covered terrestrial Eemian. Black quadrangles are sections of Middle Pleistocene tills overlying interglacial formations: 1– Novorybnoye, 2 – Khakhalevka; 3 – Bakhta, 4 – Semeyka, 5 –Kormuzhikhanka, 6 — Rodionovo, 7 — Seyda. *Grey circles* are Eemian sediments not covered by tills: 8 – N-114, marine, 9 – Alinskoye, alluvium, 10 – Observation Cape, marine, 11 – Pyak-Yakha, alluvium, 12 – Shuryshkary, peat, 13 – Sula 21 and 22, marine. *Dotted grey circles* are Eemian formations overlain by tills: 14 – K-54, marine, 15– IL-254, marine 16 – Maimecha, alluvium, 17 – Amnundakta, alluvium, 18 – Cape Karginsky, 19 – Malaya Kheta, 20 –Nurma-Yakha, Yuribei, 21 – More-Yu, 22 – Vastiansky Kon. *Black circles* are sites of dated Middle and Late Weichselian sediments not covered by till: 23 – Cape Sabler, 24 – R-59, 25 –F-9, 26 – Farkovo, 27 – Bolshoi Shar, 27a – Konoshchelye, 28 – Poloy, 29 – Igarka Shaft, 30 – Sangompan, 31 – Mongotalyang, 32 – Syo-Yakha, 33 – Marresale, 34 – Kolva terrace, 35 – Upper Kuya, 36 – Timan Beach. *Open triangles* are Palaeolithic sites not covered by till: 37 – Byzovaya, 38 – Mamontovaya Kurya, 39 –Pymva-Shor. Crosses indicate locations of radiocarbon dated frozen mammoth remains older than 17 ka BP not overlain by tills: 40 – Pyasina, 41 – Mokhovaya, 42 – Leskino , 43 –Mongoche, 44 –Schmidt, 45 –Masha, 46 – Lyuba.

Boreal strata are obviously Middle Pleistocene, although only one diamicton formation covering vast expanses of subarctic Russia can be readily correlated with the Saalian Stage glaciation (MIS 6). The next oldest glacial formation, termed the Pechora Till in the European North and Samarovo till in Siberia, has been associated with an Early Saalian (MIS 8) for decades. The lowermost double till that underlies the Likhvin (Holsteinian Stage) type interglacial sediments has been attributed to the Elsterian Stage. In Western Siberia, a similar position is occupied by a thick double-diamicton succession termed the Shaitan in the Lower Ob area and Lebed in the Yenissei valley (Fig. 25.2). In the Siberian stratigraphical schemes, the double Shaitan glacial complex is in places interstratified with the Tiltim interstadial beds that include arctic foraminifera, and the similar Lebed Formation also occupy the position of the Elsterian complex (Isayeva et al., 1936; Arkhipov, 1989). However, close to the mouth of the Irtysh river in West Siberia, the oldest till in the Bruhnes Chron—the Mansi Till—was discovered below sea level underlying pre-Holsteinian interglacial alluvium (Arkhipov, 1989).

TABLE 25.1 Key stratigraphical sites for determining ice limits marked in the electronic maps of northern Russia on the CD

Boreholes drilled through several glacial and interglacial formations	
Middle Pleistocene formations overlain by till	
5, 8 (marine, Kazantsevo fauna)	Kind and Leonov (1982)
Samburg (marine)	Zubakov (1972)
4k, 5k, 9k, 10g (Kazantsevo fauna)	Arkhipov et al. (1992, 1994)
2k, 6k, 37 (marine)	Arkhipov et al. (1994)
SDK-80, LK-9, 74, 454 (marine)	Zarkhidze (1981)
DK-13, 501, 502, 709 (marine)	Lavrushin et al. (1989) and Astakhov (2004b)
754, 755 (marine)	Loseva et al. (1992)
Lebed (terrestrial)	Isayeva et al. (1986)
Lake Chusovskoye (terrestrial by borehole profiles)	Stepanov (1974) and Astakhov (2004b)
Eemian formations beneath till	
701, 703, 704, 708, 710 (marine)	Lavrushin et al. (1989) and Astakhov (2004b)
KS-18 = Silova-Yakha (terrestrial)	Loseva and Duryagina (1983)
Sections of Middle Pleistocene interglacials covered by tills	
Marine formations with boreal fauna	
Novorybnoye, A-8, A-66, A-329, B-117 (Kazantsevo fauna)	Kind and Leonov (1982)
Pupkovo	Zubakov (1972)
Limbya-Yakha, Russkaya, Hutty-Yakha (Kazantsevo fauna)	Troitsky (1975)
Bol. Volma, Nyamdayakha, Yangarei	Zarkhidze (1981)
Bagan	Astakhov (in preparation)
Dated terrestrial sequences	
Khakhalevka (Fig. 25.2)	Levina (1964)
Bakhta (Fig. 25.2)	Zubakov (1972)
Semeyka (Fig. 25.2)	Kaplyanskaya and Tarnogradsky (1974)
Kormuzhihanka (Fig. 25.2)	Arkhipov (1989)
Rodionovo	Arslanov et al. (2006)
Seyda	Astakhov (2004b)
Kipiyevo	Guslitser and Isaychev (1983)
Sections of Eemian sediments not covered by till	
Marine formations	
A-434, N-114 (Karginsky fauna)	Kind and Leonov (1982)
Observation Cape, Belaya Yara, Sede-Yakha	Nazarov (2007)
Sula 21, Sula 22	Mangerud et al. (1999)
Terrestrial formations with interglacial flora	
Alinskoye, Mirnoye	Zubakov (1972)
Chembakchino	Laukhin et al. (2008)
Karymkary	Arkhipov (1989)
Pyak-Yakha	Zubakov (1972) and Astakhov et al. (2004)

Continued

TABLE 25.1 Key stratigraphical sites for determining ice limits marked in the electronic maps of northern Russia on the CD—cont'd

Shuryshkary	Astakhov et al. (2005)
Yamozero	Henriksen et al. (2008)
Sections of Eemian sediments overlain by till	
Marine formations	
K-54, IL-254 (Karginsky fauna)	Kind and Leonov (1982)
Romanikha-1, Romanikha-2 (Karginsky fauna)	Isayeva et al. (1976)
Krestyanka, Rogozinka, Olenyi Roga	Sachs (1953)
Cape Karginsky	Kind (1974) and Arkhipov (1989)
Lukova Protoka	Sachs (1953)
Karaul	Astakhov et al. (1986)
Parisento	Astakhov, Nazarov (2010)
Nurma-Yakha, Yuribei (Karginsky fauna)	Dolotov et al. (1981) and ILevchuk (1984)
More-Yu-2, More-Yu-3	Astakhov and Svendsen (2002)
Vashutkiny Lakes	Astakhov (2001)
Golaya Shchel (boreal marine diatoms)	Loseva (1978)
Vastiansky Kon, Sopka	Mangerud et al. (1999)
Hongurei	Henriksen et al. (2001)
Golodnaya Guba	Astakhov (unpublished)
Sula-7, Sula-27	Astakhov and Svendsen (in preparation)
Terrestrial formations	
Kotuy-2, Amnundakta (Karginsky)	Bardeyeva (1986)
Maimecha (Karginsky)	Bardeyeva et al. (1980)
Yermakovo	Zubakov (1972)
Yaran-Musyur	Astakhov and Svendsen (in preparation)
Dated Weichselian glacial deposits	
Bol. Shar (Fig. 25.8)	Astakhov and Mangerud (2007)
Yuribei 9 (sandur)	Nazarov (2008)
Vorga-Yol (sandur)	Astakhov and Svendsen (in preparation)
Bol. Usa (sandur)	Dolvik (2004)
Chernov glacier (boulders)	Mangerud et al. (2008)
Bol. Kara (sandur)	Nazarov et al. (2009)
Ileymusyur (sandur)	Mangerud et al. (2004)
Markhida	Mangerud et al. (1999)
Upper Kuya	Astakhov and Svendsen (2008)
Surficial non-glacial deposits with dates ≥ 17 ka BP	
F-9, F-17, R-10, R-59—45 to 28 ^{14}C ka BP	Fisher et al. (1990)
A-50	Kind and Leonov (1982)
Lake Labaz—>48 to 20 ^{14}C ka BP	Siegert et al. (1999)

TABLE 25.1 Key stratigraphical sites for determining ice limits marked in the electronic maps of northern Russia on the CD—cont'd

Site	Reference
Cape Sabler—39 to 17 ^{14}C ka BP	Kind and Leonov (1982) and Möller et al. (1999)
Kotuy-1	Bardeyeva (1986)
Bakhta—35.2 ± 1.5 and 34.2 ± 1 ^{14}C ka BP	Astakhov et al. (1986)
Farkovo—42 to 34 ^{14}C ka BP	Kind (1974)
Konoshchelye, Poloy—76 to 27 ka BP	Astakhov and Mangerud (2007)
Igarka permafrost shaft—>50 to 35 ^{14}C ka BP	Kind (1974)
Kureika—32 ^{14}C ka BP	Astakhov and Isayeva (1988)
Tab-Yakha—31 ± 2 and 24 ± 2 ka BP	Astakhov, Nazarov (2010)
Lysukansyo—bones 19 to 16 ^{14}C ka BP	Astakhov, Nazarov (2010)
Bogdashka—80 to 64 ka BP	Arkhipov (1989)
Pitlyar, Sangompan—108 to 62 ka BP	Astakhov (2006)
Igorskaya Ob, Yerkata—89 to 59 ka BP	Astakhov et al. (2007)
Marresale—45 to 26 ka BP	Forman et al. (2002)
Syo-Yakha—40 to 17 ^{14}C ka BP	Vasilchuk et al. (2000)
Mongotalyang—31 to 21 ^{14}C ka BP	Vasilchuk (1992)
Syatteityvis—77 and 75 ka BP	Mangerud et al. (2004)
Rogovaya—24 to 22 ka BP	Astakhov and Svendsen (2008)
More-Yu-1, Nguta-Yakha—46 to 22 ka BP	Astakhov and Svendsen (2002)
Korotaikha—mammoth teeth 34 to 36 ^{14}C ka BP	Astakhov and Svendsen (2002)
Urdyuga—102 to 53 ka BP	Mangerud et al. (1999)
Podkova, Yarei-Shor—bones 37 to 26 ^{14}C ka BP	Mangerud et al. (1999)
Timan Beach—52 to 13 ka	Mangerud et al. (1999)
Dates from superficial Palaeolithic sites	
Pymva-Shor—26 to 10 ^{14}C ka BP	Mangerud et al. (1999)
Mamontovaya Kurya—37 to 24, Byzovaya, 33 to 25 ka BP	Mangerud et al. (1999)
Frozen mammoth remains close to surface	
Mammoth by F. Schmidt—33.5 ± 1, mammoths on rivers	
Pyasina—25.1 ± 0.5 and Mokhovaya—35.8 ± 2.7 ^{14}C ka BP	Sulerzhitsky (1995)
Leskino mammoth and plants—30.1 ± 0.3 and 29.7 ± 0.3 ^{14}C ka BP	Astakhov (1998)
Mongoche river—17 ^{14}C ka BP	Gilbert et al. (2007)
Baby mammoth Masha—39.1 ± 1.4 ^{14}C ka BP	Tomirdiaro and Tikhonov (1999)
Baby mammoth Lyuba—41.9 ^{14}C ka BP	Kosintsev (2008)

In the digital maps, the Middle Pleistocene glacial maximum is more or less the same as in the first edition, as the drift limit follows the pattern previously established in the National Geological Map of scale 1:1,000,000 (Rudenko et al., 1984; Bobkova, 1985; Babushkin and Shatsky, 1989; Babushkin, 1997; Potapenko, 1998; Chumakov et al., 1999; Lider et al., 1999). Chronologically, it has been attributed to the Early Saalian or MIS 8 for many decades. However, the most extensive glaciation, indicated by the Don Till of south-eastern Russia, was found sandwiched

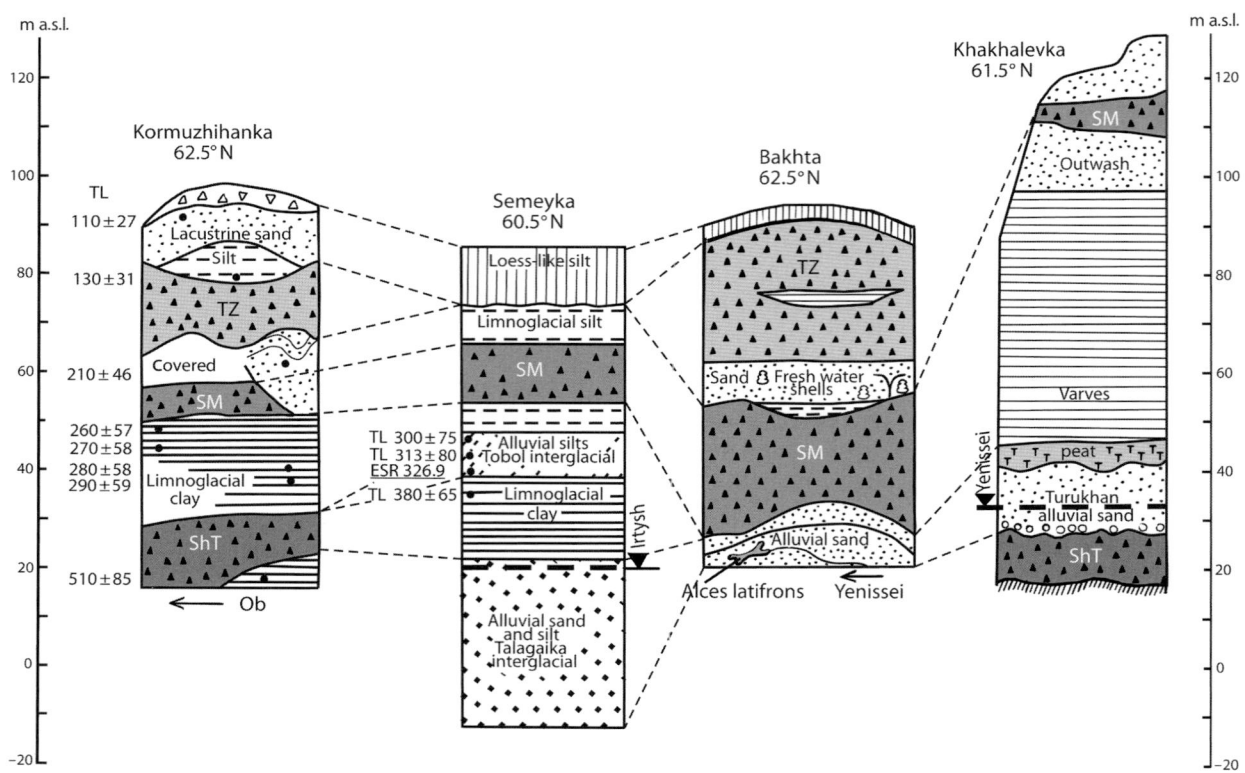

FIGURE 25.2 Key natural sections of Middle Pleistocene glacial formations of Siberia. Indices of glacial diamictons (shaded) according to the formal chronostratigraphic scale of West Siberia: ShT – Shaitan climatolith of early (pre-Holsteinian) Middle Pleistocene, SM – Samarovo climatolith of West Siberian glacial maximum (presumably MIS 8); TZ –Taz climatolith of late Middle Pleistocene (MIS 6). Intervening lacustrine and alluvial sediments contain pollen of rich boreal forest (adapted from Arkhipov, 1989; Kaplyanskaya & Tarnogradsky, 1974; Levina, 1964; Zubakov, 1972).

between strata that yielded Cromerian flora and fauna in the 1970s. Therefore, the glacial drift maximum limit was thus firmly dated as pre-Elsterian, older than 500 ka (Velichko et al., 2004). Consequently, in new Quaternary maps of eastern European Russia, the drift limit is also associated with the Don glaciation (e.g. Chumakov et al., 1999) and is tentatively shown as such in this volume. However, there is still a possibility that the drift limit in the eastern Russian Plain is time transgressive and may be of Elsterian or even of Early Saalian age. In any case, the diamicton formations and the intercalated Likhvin-type interglacial sediments at Lake Chusovskoye (Stepanov, 1974; Astakhov, 2004b) offer no clue to which of the three tills underlying the Eemian interglacial sediments can be traced to the drift limit in the Kama catchment area.

In West Siberia, the drift limit is thought to be the margin of the Samarovo ice sheet which is traditionally correlated with the Early Saalian (MIS 8) (Arkhipov, 1989). The basic fact is that the Samarovo glacial strata imperceptibly grade downwards into the Tobol interglacial sands and silts which contain elements of Pacific flora and shells of freshwater *Corbicula tibetensis*, a Central Asia species. The Tobol interglacial of the Ob and Irtysh area is the closest analogue of the Likhvin interglacial of Central Russia which is reliably correlated with the Holsteinian Stage. This correlation is supported by rare electron spin resonance (ESR) dates on *Corbicula* shells and also by suspiciously old thermoluminescence (TL) dates (from the Semeyka site: Figs. 25.1 and 25.2) (Arkhipov, 1989). No geochronometrical dates are available from the eastern West Siberian Plain, where correlation of the drift limit with the Samarovo glaciation is based solely on the position of the till overlying the so-called Turukhan alluvium. The latter contains pollen spectra of rich boreal forest. This interglacial has repeatedly been equated with the Tobol alluvium in all stratigraphical schemes, although this correlation is not rigorously determined. The Samarovo ice limit is traceable only using rare boreholes and natural sections in the flat and swampy West Siberian Plain. On the Central Siberian uplands, the boundary has been mapped morphostratigraphically from several isolated chains of push moraines overlying the Palaeozoic bedrock and occasionally interglacial alluvium that occurs within buried valleys (Rudenko et al., 1984).

Another morphostratigraphical ice limit appears in the Central Siberian uplands east of 103–104°E beyond the Samarovo moraines. This easternmost glacial maximum is associated with the Lebed diamicton of the Yenissei valley (Bobkova, 1985). Its relation to the Don glaciation of European Russia is unknown because of the lack of correlation tools.

The limit of the Taz glaciation (Fig. 25.1), traditionally associated with the Late Saalian, was omitted from the first edition but is now shown on the Siberian digital maps. The reason that this was previously omitted was the lack of dated interglacial formations directly underlying the glacial complex correlated to the Late Saalian. However, recently, such sequences have now become available. In European Russia, they are Rodionovo and Seyda terrestrial sediments which are interstratified with the till formations. These terrestrial sediments contain pollen indicating a climate warmer than the present (Fig. 25.1). Thick peat between two tills at Seyda has yielded a U/Th date of ca. 200 ± 30 ka and several OSL dates in the range of 180–191 ka (Astakhov, 2004b). At Rodionovo, a peat 3–4 m thick between the superficial Moscow and Pechora Tills has been dated by U/Th method to 250 ka (Arslanov et al., 2006).

A marine formation containing a boreal fauna which underlies the second from the surface till in the Arctic often contains shells of extinct mollusc *Cyrtodaria angusta*, and on this basis, contrary to the opinion of Sachs (1953), they are unlikely to be of Late Pleistocene age. In the Pechora Basin, the *Cyrtodaria* strata, found in borehole 454 and at Bol. Volma, Nyamdayu and Yangarei sections (Table 25.1), have been attributed to the ancient Padimei formation (Zarkhidze, 1981). In West Siberia, a similar position is occupied by the Kazantsevo marine formation with boreal fauna including *Cyrtodaria*. The Kazantsevo strata have, for a long time, been taken as representatives of the first Late Pleistocene warm period analogous to the Eemian (e.g. Arkhipov, 1989). However, on the Lower Ob, there is clear evidence that the formation with Kazantsevo foraminifera assemblage occurs beneath thick Middle Pleistocene glacial sediments and well below superficial Eemian peats and alluvium dated by U/Th and OSL methods (Fig. 25.3; also Astakhov, Nazarov, 2010). Therefore, the sub-till position of the Kazantsevo formations can no longer be taken as an indication of a Late Pleistocene age of the overlying till. This is why many sections that include interglacial marine sediments in the area of Late Pleistocene glaciations of Siberia shown in the maps of the first edition have now been eliminated. Other important localities that include *Cyrtodaria* fauna overlain by tills, such as the deep Samburg borehole (Fig. 25.1), are added. Superficial interglacial marine formations with warm-water Atlantic fauna but without *Cyrtodaria* are left in their places on the map south of the Weichselian limit (e.g. 10 and 13; Fig. 25.1).

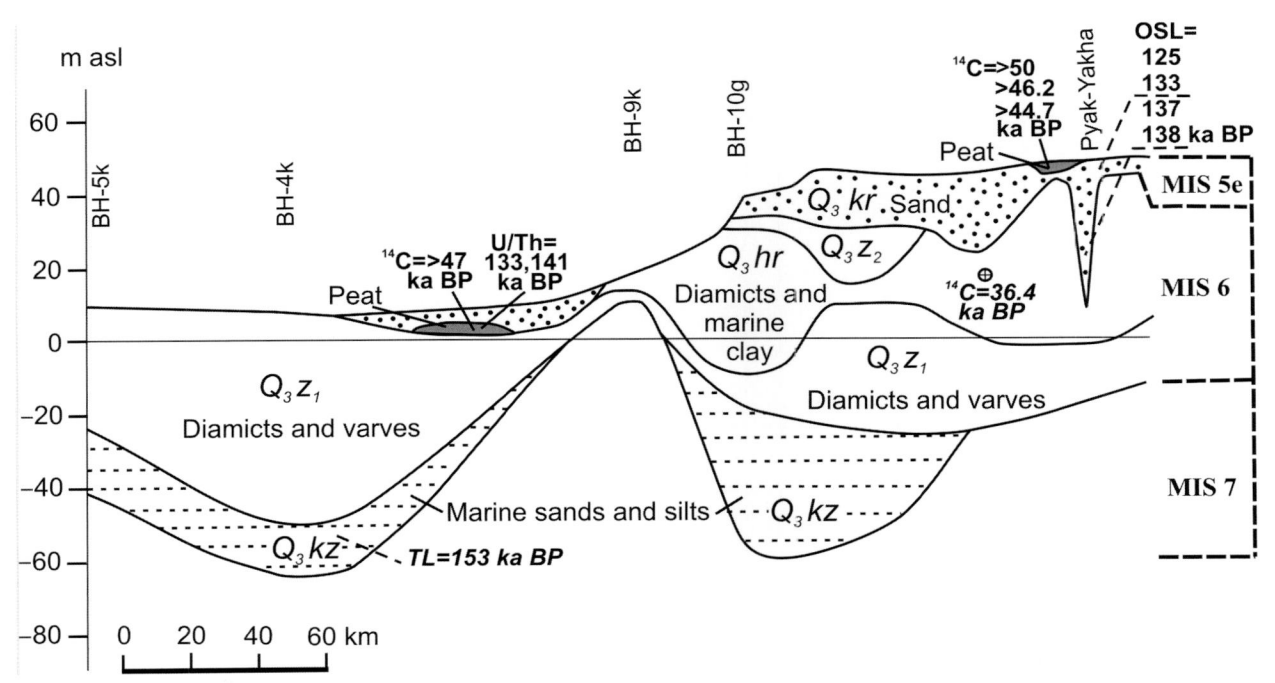

FIGURE 25.3 Modern geochronological results in the Lower Ob valley by ICEHUS (regular print) (Astakhov, Nazarov, 2010) as compared to the results of Arkhipov's et al. (1977, 1992, 1994) (italicized). Indices of Late Pleistocene stratigraphical units according to Arkhipov's interpretation: Q3k, marine sediments with Kazantsevo foraminifera assemblage (Eemian), $Q_3 z_1$ and $Q_3 z_2$, Lower Zyryankian and Middle Zryankian glacial and lacustrine sediments, Q_3 hr, Kharsoim interstadial marine clay, Q_3 kr, Karginskian alluvium (all Weichselian). Note that the Arkhipov's Kazantsevo marine formation with TL date of 153 ka BP occurs beneath a thick glacigenic succession covered by alluvium and peat with dates of the Eemian Stage. See Fig. 25.1 for locations.

25.3. LATE PLEISTOCENE GLACIATIONS

The stepping of the Kazantsevo interglacial down in the chronostratigraphical scale has influenced the mapped limits of the Weichselian glaciations. Thus, the Early Weichselian ice limit on the Taimyr Peninsula has to be moved slightly northwards because the sub-till marine formations encountered in boreholes here are Middle Pleistocene and the overlying Karginsky sediments have proved to be of Eemian age. The most important stratigraphical result has been obtained by the re-dating of Karginsky interglacial stratotype sequences which were used to mark the Middle Weichselian (MIS 3) level in all Siberian schemes. This is because many conventional radiocarbon dates with finite values were obtained from these units (Kind, 1974; Arkhipov, 1989). Later the marine formation at Cape Karginsky (18; Fig. 25.1) has provided an ESR date of 122 ka (Arkhipov, 1989) and six OSL dates with a mean value of 111 ka BP (Astakhov, Nazarov, 2010). Its alluvial counterpart at Malaya Kheta (19; Fig. 25.1) has yielded non-finite radiocarbon dates, three OSL dates in the range of 80–112 ka from alluvium s. stricto and three OSL dates in the range of 78–98 ka from the overlying loess-like silt (Figs. 25.4 and 25.5; Astakhov and Mangerud, 2005).

These ages have been supplemented by old dates from the other classical Karginsky sections; for example, the 1-m-thick peat with interglacial flora at Shuryshkary (12; Fig. 25.1), known as the Lower Karginsky ca. 50 ka old (Arkhipov et al., 1977), yielded U/Th dates 133 and 141 ka (Fig. 25.3; Astakhov et al., 2005). Downstream of this site, the superficial sandy alluvium with peat layers at Pyak-Yakha (11, Fig. 25.1) has been dated by OSL to 125–135 ka (Astakhov et al., 2004, 2007). It is now clear that most of old conventional ^{14}C dates in the Russian North are as spurious as that of 36 ka BP from the 'Early Karginsky' strata by Arkhipov in Fig. 25.3. A warning that too young dates are often obtained on plant detritus from permafrost has long been issued by Sulerzhitsky (Kind and Leonov, 1982) but obviously has not sufficiently influenced the stratigraphical reasoning. Results of the past decade definitely point out to the Karginsky strata correlation with the Eemian. Therefore, their sub-till position can no longer be used as evidence of Late Weichselian glaciation.

Another consequence of the different solution of the Karginsky transgression problem is the re-assessment of age of the so-called Murukta glaciation in the east of Central Siberia (in the Kotuy and Kheta rivers catchment areas, Fig. 25.1). Previously, it was correlated with the Early Weichselian because the Murukta moraines were clearly overlain by Karginsky marine and terrestrial strata containing boreal microfossils (Bardeyeva et al., 1980; Bardeyeva,

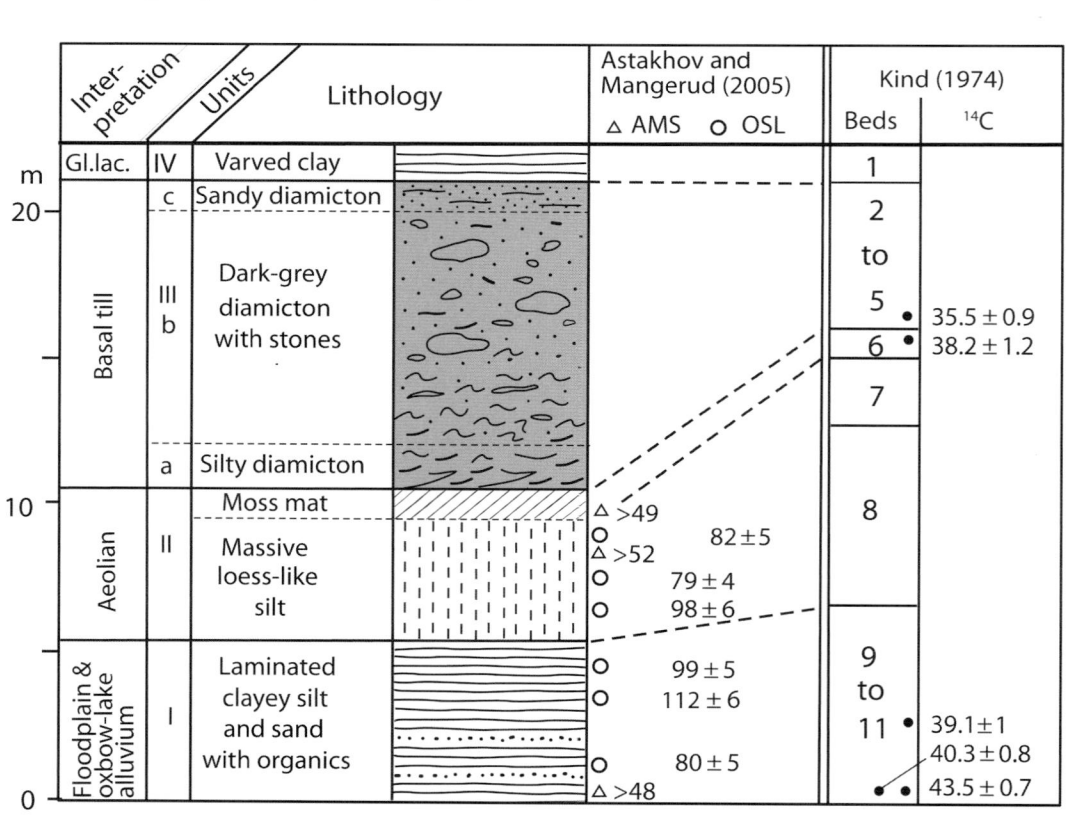

Interpretation	Units	Lithology		Astakhov and Mangerud (2005) △ AMS ○ OSL	Kind (1974) Beds	^{14}C
Gl.lac.	IV	Varved clay			1	
Basal till	c	Sandy diamicton			2 to 5	
	III b	Dark-grey diamicton with stones			6	35.5 ± 0.9
					7	38.2 ± 1.2
	a	Silty diamicton				
Aeolian	II	Moss mat		△ >49	8	
		Massive loess-like silt		○ 82 ± 5 △ >52 ○ 79 ± 4 ○ 98 ± 6		
Floodplain & oxbow-lake alluvium	I	Laminated clayey silt and sand with organics		○ 99 ± 5 ○ 112 ± 6 ○ 80 ± 5 △ >48	9 to 11	39.1 ± 1 40.3 ± 0.8 43.5 ± 0.7

FIGURE 25.4 Malaya Kheta section — terrestrial stratotype of the Karginsky interglacial — MIS 3 according to Kind (1974). New OSL and ^{14}C dates (Astakhov and Mangerud, 2005) suggest correlation of the subtill alluvium with the Eemian.

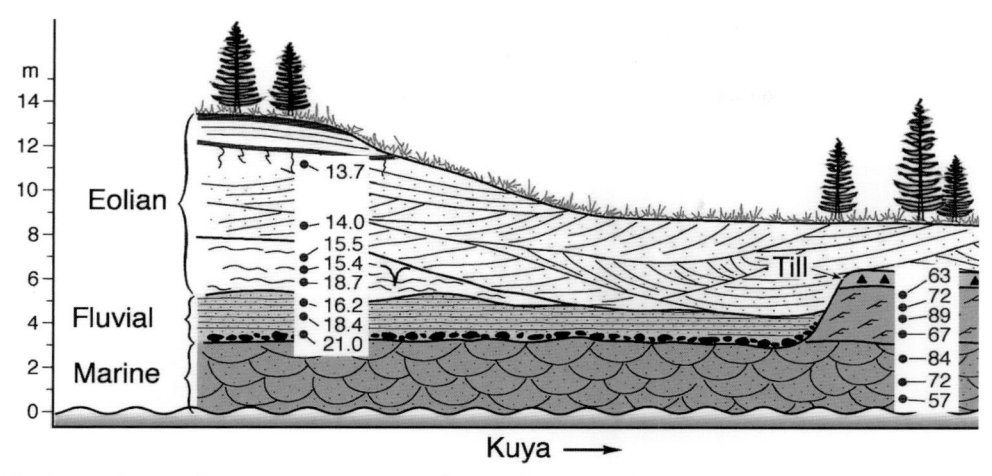

FIGURE 25.5 OSL dated sediments of Upper Kuya section, 67°35′N/53°30′E (35, Fig. 25.1). The black dots are OSL ages, ka BP. Marine sand of average age of 72 ka BP is overlain by Middle Weichselian till. The succession is capped by Late Weichselian fluvial and aeolian sands (Astakhov & Svendsen, 2008).

1986; Isayeva et al., 1986). This means that the Murukta moraines were formed in the Middle Pleistocene and now their area should be included within the limit of the Taz glaciation of the Late Saalian (MIS 6) period (Fig. 25.1; Svendsen et al., 2004).

Moreover, because fairly young OSL ages (younger than the age of the Early Weichselian ice advance) have repeatedly been obtained from beneath the Markhida moraines, the stratigraphical result is the changing of the age of Weichselian ice limits. OSL values ca. 60 ka from the Markhida section were reported already in the 1990s (Mangerud et al., 1999) which was confirmed by numerous sub-till dates of this order from other regions. These dates have led to a tentative reconstruction of the second Weichselian ice sheet that advanced into the Russian mainland ca. 50–60 ka (Svendsen et al., 2004). This reconstruction for the European sector was recently confirmed by OSL dating of sub-till sediments on river Kuya east of Naryan-Mar city (35; Fig. 25.1). There, a Weichselian till is clearly positioned between interstadial with mean OSL age of 72 ka and Late Weichselian dunes (Fig. 25.3). The second Weichselian ice advance in the European Arctic is constrained from above by OSL date of 52 ka at the Timan Beach section (36; Fig. 25.1; Mangerud et al., 1999).

Thus, it appears that the Markhida moraine proper and similar moraines west of the Pechora river (Astakhov et al., 1999) relate to a local Middle Weichselian maximum advance from the Barents Sea shelf (Fig. 25.1). The preceding Weichselian ice sheet, marked by the solid line in Fig. 25.1, culminated earlier, approximately 80–90 ka. The mean weighted value of many OSL dates from beach sediments of the corresponding proglacial Lake Komi gave an age of 82 ka (Mangerud et al., 2004; Svendsen et al., 2004; Astakhov et al., 2007). Beyond the Urals, the base of the proglacial lake sequence in the Sangompan section (30, Fig. 25.1) yielded four OSL dates that gave a mean value of 81 ka. This early ice advance has a minimum age of ca. 70 ka constrained by four OSL dates with a mean value of 65 ka on postglacial sands at Yerkata in southern Yamal (Astakhov, 2006). Outwash sediments inclined to the south and glaciolacustrine sands have lately provided six OSL dates with the mean value of 65 ka in the Gydan Peninsula (Nazarov, 2008).

Similar results have recently been obtained from the eastern border of the West Siberian Plain (Astakhov and Mangerud, 2007). Here, the Bolshoi Shar section (27; Fig. 25.1) exposes periglacial alluvium that grades upwards into varved clay and silts of a proglacial lake dammed by an ice margin along the Arctic Circle (Astakhov and Isayeva, 1988). Eighteen OSL dates from this alluvial/lacustrine sequence (Fig. 25.6) have yielded the mean age of 85 ka which is close to that of the early proglacial lakes on the Ob and in the eastern Pechora Basin (Mangerud et al., 2004; Astakhov, 2006; Astakhov et al., 2007). The Early Weichselian succession is overlain by a till and sandur of eastern provenance. The Bolshoi Shar sandur and that in the nearby Goroshikha section have yielded 10 OSL ages which give a mean value ca. 60 ka (Astakhov and Mangerud, 2007). The age of the main Early Weichselian ice advance is constrained by old OSL values from post-dating fluvial terraces at Poloy and Konoshchelye (Fig. 25.6).

This implies that the major Early Weichselian ice sheet was followed by an independent ice advance from the northeast which did not dam the Yenissei river. Further to NE, the Middle Weichselian ice limit of the Yenissei can only be traced by morainic arcs which block deep valleys dissecting the Putorana Plateau (Fig. 25.1). Traditionally, these moraines, locally called the Norilsk Stade, were attributed to the Late Weichselian, even to the Younger Dryas Stadial (Kind, 1974; Isayeva et al., 1986). In the first

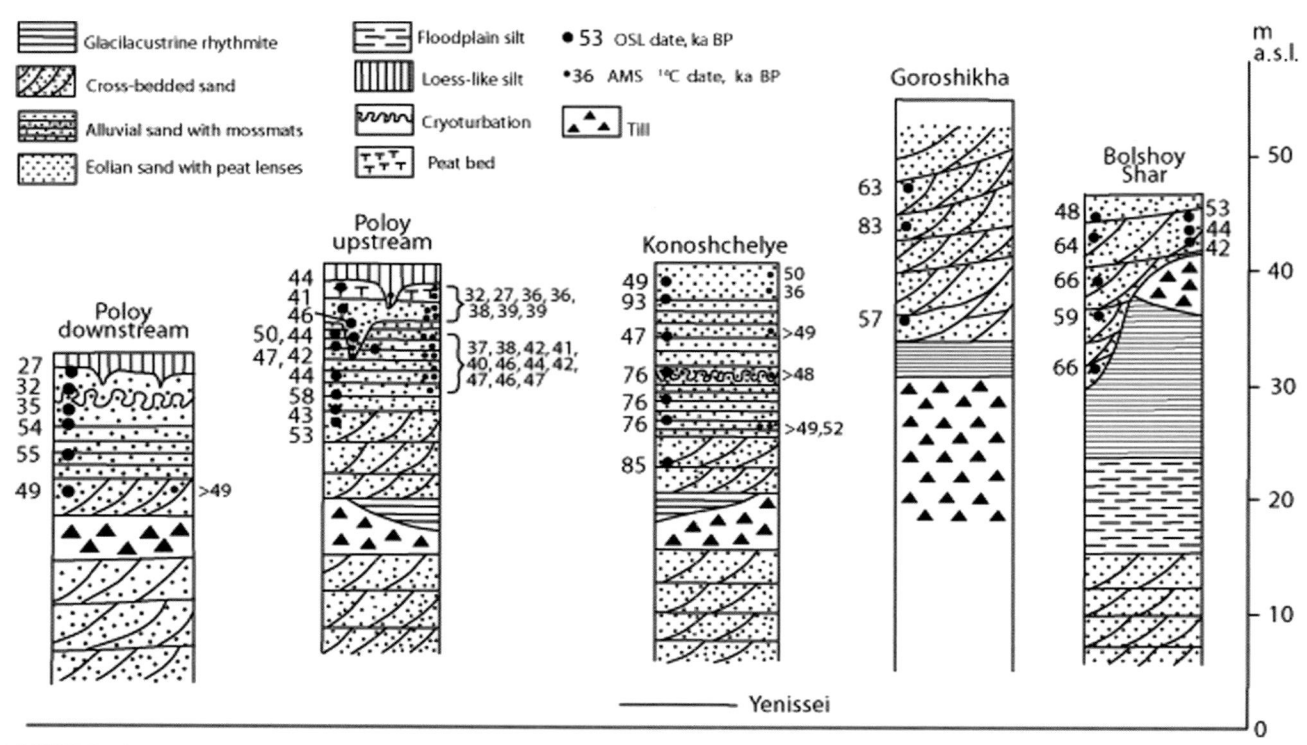

FIGURE 25.6 Dated terraces of the Yenissei near the Arctic Circle between Poloy and Bolshoi Shar sections (27, 27a and 28 in Fig. 25.1). Two Weichselian glacial complexes are identified. The Middle Weichselian ice advance from the northeastern Putorana Plateau (Fig. 25.1) is determined from the upper till of Bolshoi Shar with coarse sandur of a mean OSL age of 60 ka BP on 10 samples (Astakhov & Mangerud, 2007). Underlying periglacial alluvium changing into varves of the first Weichselian ice advance has recently yielded 18 OSL dates with mean value of 85 ka BP (same authors). The incised alluvial terraces constrain the age of the ice dam across the Yenissei.

edition of this work (Astakhov, 2004a), the Norilsk moraines were taken to mark the Last Glacial Maximum (LGM) stillstand position because of their overlying the Karginsky marine formation, the latter giving finite radiocarbon dates (Kind and Leonov, 1982). However, as the Karginsky transgression is now considered as a counterpart of the Eemian (see above), this argument is no longer valid. Moreover, a drilling in thick lake bottom silts, proximal to the Norilsk moraines, makes their Late Weichselian age highly improbable (Svendsen et al., 2004).

The main Late Weichselian glaciation is confined to the Barents Sea shelf by evidence amassed from marine geological investigations (Fig. 25.1; Svendsen et al., 2004). The absence of Late Weichselian glacial features on the northern plains has been proven by well-dated supra-till formations, especially by loess-like sequences (23, 31, 32, 33; Fig. 25.1) and old postglacial fluvial terraces (27, 27a, 28). This event has very good independent chronology based upon radiocarbon dating of (i) megafaunal bones from oldest Palaeolithic sites in European Russia and (ii) frozen mammoth flesh found in fine-grained sediments overlying the uppermost glacial diamictons in Siberia. Such finds have been recovered from Siberian permafrost since the nineteenth century (e.g. the Schmidt mammoth). The dates on the frozen mammoths and bones from Palaeolithic sites span the entire radiocarbon range from 42 to 17 ka BP (Table 25.1). They thus totally exclude any glacial activity within this time range. The sites of these finds obviously lying *in situ* in organic sediments dated by different laboratories are shown in the digital maps and in Fig. 25.1 (open triangles 37–39 and crosses 40–46).

25.4. NEW SOLUTIONS

Some problems of the last ice age mentioned in the first edition now seem to be resolved. Thus, the Early Weichselian age of the Laya-Adzva moraines in the Pechora Basin is now very probable, as six OSL dates in the range of 71–93, mean age 84 ka BP, have been obtained from the coarse sandur at the distal foot of the morainic ridge (Vorga-Yol; Table 25.1). In a proximal section of Yaran-Musyur, OSL ages of greater than 100 ka BP have been obtained on sands of Eemian type (Astakhov and Svendsen, in preparation). The last glaciation of the Polar Urals was long ago firmly attributed to the invasion of thick ice from the Kara Sea shelf (Astakhov et al., 1999). This is indicated by the mapped morainic pattern (Fig. 25.7). A great thickness of Kara Sea ice can be judged from altitudes of end moraines pushed into the Ural valleys, sometimes up to 560 ma.s.l. (Fig. 25.8). Their altitudes decrease rapidly along both flanks of the

FIGURE 25.7 Glacial features on the north-western slope of the Polar Urals (from Astakhov et al., 1999). The arrows are ice flow directions, black broken line is glacial trimline. Gravelly moraines with scarce Uralian pebbles delineate a south-bound stream of the Early Weichselian Kara Sea ice sheet. The stream inserted horseshoe-shaped moraines into the montane valleys as the ridge in Fig. 25.8 (indicated by green arrow). The white arrow indicates Bol. Kara moraine dated to 73 ka BP (Nazarov et al., 2009), blue arrow points to the Bol. Usa sandur dated to 75 ka BP (Dolvik, 2004).

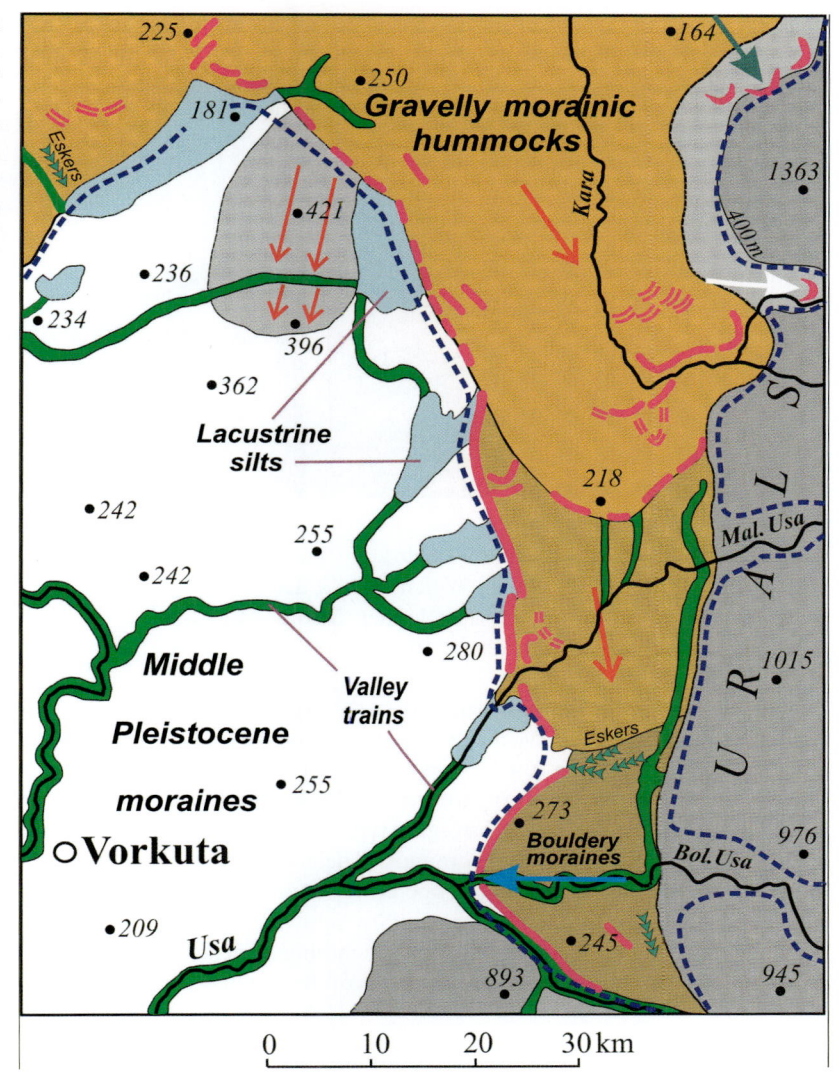

Urals, and close to 67th parallel, all glacial features are now absent. The age of the last glacial invasion into the Urals which left undisturbed morainic ridges can be determined from 13 OSL dates on a sandur distal to the horseshoe moraine formed by an ice advance from the west into the Bol. Kara river valley. The mean value of the OSL ages of 73 ka BP (Nazarov et al., 2009) is by 8–12 ka younger than that of the lowland ice dams. The southernmost ice lobe on the Bol. Usa river has also provided relatively young ages: mean 75 ka BP from eight OSL dates on the coarse frontal sandur (Dolvik, 2004). These younger ages probably reflect final ice surges of the same Early Weichselian Kara ice sheet that dammed the northern rivers already by 81–85 ka BP. Its disintegration is marked by 65 ka BP OSL ages from Yamal and Gydan peninsulas (see above). Notwithstanding these points, it is clear that after 73–75 ka BP no glacial ice filled the major valleys of the Urals.

Very important results have recently also been obtained by cosmogenic exposure dating by ^{10}Be method of large quartzite boulders in the western Polar Urals. Morainic boulders, 1 km east from the present-day Chernov glacier at 67°38′N/65°48′E, yielded 6 ^{10}Be ages that range from 14.7 to 28.9 ka BP, with a mean value of 21.9 ka BP. Down valley, beyond that small moraine, boulders are normally older, four determinations gave ages in the range of 50–60 ka BP. This implies that Late Weichselian ice in the Polar Urals hardly exceeded in size puny present-day glaciers (Mangerud et al., 2008). This probably reflects a moisture deficiency in the region during the last glacial event. Therefore, the features of alpine glaciation seen in many Urals valleys, especially in the highest massif of the Subpolar Urals, probably belong to the Early Weichselian Substage. It is possible that some of the alpine moraines were left by Middle Weichselian glaciers, as apparently

FIGURE 25.8 Morainic ridge at 560 m a.s.l. pushed up into a Uralian valley (for location see Fig. 25.7). The author's hand is pointing to a lowland source of ice flow from NW.

suggested by 50–60 ka ages of the Polar Urals boulders. A very similar situation of no Late Weichselian glaciers has been recently described from the higher Verkhoyansk mountains bounding Central Siberia from the east (Stauch et al., 2007).

These new results implying an extremely dry climate during the continental LGM make the existence of even modest Late Weichselian glaciers on Putorana Plateau at an altitude of 1200–1600 m very problematic. On the digital maps, possible glaciers of the Norwegian type are shown in this area on purely speculative grounds. Their extent is based on calculations of past snow-line elevation assessed from the position of glacial cirques (Shvaryov, 1998). The mountain glacial features have no geochronometrical dates and may easily belong to earlier stages of the last glacial cycle.

25.5. CONCLUSION

The revision of foundations for traditional ice limits in the central sector of northern Eurasia suggests significant changes in the configuration of Late Pleistocene glaciers remarkably but irregularly decreasing in size through the last glacial cycle. These changes are mostly suggested by the improved dating methods. No comparable progress in the understanding of the Middle Pleistocene ice margins has been achieved because of the lack of adequate geochronological techniques.

ACKNOWLEDGEMENTS

This work was supported by the Russian-Norwegian project ICEHUS (Ice Age Development and Human Settlement in Northern Eurasia) sponsored by the Research Council of Norway. This chapter is also a contribution to the European coordination programme APEX (Arctic Palaeoclimate and its Extremes), part of the International Polar Year.

REFERENCES

Arkhipov, S.A., 1989. The chronostratigraphic scale of the glacial Pleistocene of the West Siberian North. In: Skabichevskaya, N.A. (Ed.), Pleistotsen Sibiri. Stratigrafia i mezhregionalnye korrelatsii. Nauka, Novosibirsk, pp. 20–30 (in Russian).

Arkhipov, S.A., Votakh, M.R., Golbert, A.V., Gudina, V.I., Dovgal, L.A., Yudkevich, A.I., 1977. The Last Glaciation in the Lower Ob River Region (Posledneye oledeneniye Nizhniego Priobya). Nauka, Novosibirsk, 215 pp. (in Russian).

Arkhipov, S.A., Levchuk, L.K., Shelkoplyas, V.N., 1992. Marine Quaternary sediments of the Lower Ob. In: Murzayeva, V.E., Punning, J.-M.K., Chichagova, O.A. (Eds.), Geokhronolohgia chetvertichnogo perioda. Nauka, Moscow, pp. 90–101.

Arkhipov, S.A., Levchuk, L.K., Shelkoplyas, V.N., 1994. Stratigraphy and geological structure of the Quaternary in the Lower Ob-Yamal-Taz region of West Siberia. Geologiai geofizika 35 (6), 87–104 (in Russian).

Arslanov, Kh.A., Maksimov, F.E., Kuznetsov, V.Yu., Chernov, S.B., Velichkevich, F.Yu., Razina, V.V., et al., 2006. ^{230}Th/U isochron dating and paleobotanical study of the Middle Pleistocene interglacial section Rodionovo, northeastern European Russia. In: Abstracts of International Workshop 'Correlation of Pleistocene Events in the Russian North'. VSEGEI, St. Petersburg, Russia, p. 12.

Astakhov, V., 1998. The last ice sheet of the Kara Sea: terrestrial constraints on its age. Quatern. Int. 45/46, 19–29.

Astakhov, V., 2001. The stratigraphic framework for the Upper Pleistocene of the glaciated Russian Arctic: changing paradigms. Glob. Planet. Change 31 (1–4), 281–293.

Astakhov, V., 2004a. Pleistocene ice limits in Russian northern lowlands. In: Ehlers, J., Gibbard, P.L. (Eds.), Quaternary Glaciations—Extent and Chronology. Part 1: Europe. Elsevier, Amsterdam, pp. 309–319.

Astakhov, V., 2004b. Middle Pleistocene glaciations of the Russian North. Quatern. Sci. Rev. 23 (11–13), 1285–1311.

Astakhov, V.I., 2006. Evidence of Late Pleistocene ice-dammed lakes in West Siberia. Boreas 35, 607–621.

Astakhov, V.I., Isayeva, L.L., 1988. The 'Ice Hill': an example of retarded deglaciation in Siberia. Quatern. Sci. Rev. 7, 29–40.

Astakhov, V., Mangerud, J., 2005. The age of the Karginsky interglacial strata on the Lower Yenisei. Doklady Earth Sci. 403 (5), 673–676. Translated from Doklady Akademii Nauk, 2005, 401(1), 63–66.

Astakhov, V., Mangerud, J., 2007. The geochronometric age of Late Pleistocene terraces on the Lower Yenisei. Doklady Earth Sci. 1028-334X, 416 (7), 1022–1026. Translated from Doklady Akademii Nauk 2007, 416(4), 509–513.

Astakhov, V., Nazarov, D., 2010. Correlation of Upper Pleistocene sediments in northern West Siberia. Quaternary Science Reviews 29, 3615–3629.

Astakhov, V.I., Svendsen, J.I. 2002. Age of remnants of a Pleistocene glacier in Bol'shezemel'skaya Tundra. Doklady Earth Sciences 384(4), 468–472, ISSN 1028-334X. Translated from Doklady Akademii Nauk, 384(4), 534–538.

Astakhov, V.I., Svendsen, J.I., 2008. Environments during the time of initial human settlement of the northern part of the Urals area. In: Velichko, A.A., Vasil'ev, S.A. (Eds.), Way to North: Paleoenvironment and Earliest Inhabitants of Arctic and Subarctic. Institute of Geography RAS, Moscow, pp. 98–106 (in Russian).

Astakhov, V.I, Isayeva, L.L., Kind, N.V., Komarov, V.V., 1986. On geologic and geomorphic criteria of subdivision of glacial history in the Yenissei North. In: Velichko, A.A., Isayeva, L.L. (Eds.), Chetvertichnye oledeneniya Sredney Sibiri. Nauka, Moscow, pp. 18–28.

Astakhov, V.I., Svendsen, J.I., Matiouchkov, A., Mangerud, J., Maslenikova, O., Tveranger, J., 1999. Marginal formations of the last Kara and Barents ice sheets in northern European Russia. Boreas 28 (1), 23–45.

Astakhov, V.I., Arslanov, K.A., Nazarov, D.V., 2004. The age of mammoth fauna on the Lower Ob. Doklady Earth Sci. 396 (4), 538–542, Translated from Doklady Akademii Nauk 396(4), 253–257.

Astakhov, V.I., Arslanov, Kh.A., Maksimov, F.E., Kuznetsov, V.Yu., Razina, V.V., Nazarov, D.V., 2005. The age of interglacial peat on the Lower Ob. Doklady Earth Sci. 1028-334X, 401 (2), 298–302, Translated from Doklady Akademii Nauk 2005, 400(1), 95–99.

Astakhov, V.I., Mangerud, J., Svendsen, J.I., 2007. Trans-Uralian correlation of the northern Upper Pleistocene. Regionalnaya Geologia i Metallogenia, 30–31, 190–206 (in Russian).

Babushkin, A.Ye., 1997. National Geological Map of Russian Federation, Scale 1:1 000 000 (New Series), Sheet P-44,45 (Verkhneimbatsk). Map of Quaternary Formations. VSEGEI, St. Petersburg.

Babushkin, A.Ye., Shatsky, S.B., 1989. National Geological Map of the USSR, Scale 1:1000 000 (New Series), Sheet P-42,43 (Khanty-Mansiysk). Map of Quaternary Formations. VSEGEI, Leningrad.

Bardeyeva, M.A., 1986. The key sections of Quaternary deposits in the Central Siberian Upland. In: Velichko, A.A., Isayeva, L.L. (Eds.), Chetvertichnye oledeneniya Srednei Sibiri. Nauka, Moscow, pp. 35–52.

Bardeyeva, M.A., Isayeva, L.L., Andreyeva, S.M., Kind, N.V., Nikolskaya, M.V., Pirumova, L.G., et al., 1980. Stratigraphy, geochronology and palaeogeography of the Pleistocene and Holocene of the northern Central Siberian Upland. In: Ivanova, I.K., Kind, N.V. (Eds.), Geokhronologia chetvertichnogo perioda. Nauka, Moscow, pp. 198–207.

Bobkova, Z.S., 1985. National Geological Map of the USSR, Scale 1:1,000 000, New Series, Quadrangle Q-48,49 (Aikhal). Map of Quaternary Deposits. VSEGEI, Leningrad.

Chumakov, O.Ye., Shik, S.M., Kirikov, V.P., 1999. National Geological Map of the Russian Federation, Scale 1:1 000 000, New Series, Quadrangle O-38,39 (Kirov). Map of Quaternary Formations. VSEGEI, St. Petersburg.

Dolotov, M.S., et al., 1981. Photogeological Mapping of the Yamal Peninsula, Scale 1:200000. Aerogeologia, Moscow, manuscript (in Russian).

Dolvik, T., 2004. Rekonstruksjon av weichsel-brelober i Polare Ural. Cand. Scient.-oppgave i geologi. University of Bergen, Geological Institute, 141 pp.

Fisher, E.L., Leonov, B.N., Nikolskaya, M.V., Petrov, O.M., Ratsko, A.P., Sulerzhitsky, L.D., et al., 1990. The Late Pleistocene of the central North Siberian Lowland. Izvestiya Acad. Sci. USSR, seria geograficheskaya 6, 109–118 (in Russian).

Forman, S.L., Ingólfsson, Ó., Gataullin, V., Manley, W.F., Lokrantz, H., 2002. Late Quaternary stratigraphy, glacial limits, and paleoenvironments of the Marresale area, western Yamal Peninsula, Russia. Quatern. Res. 57, 355–370.

Gilbert, M.T.P., Tomsho, L.P., Rendulic, S., Packard, M., Drautz, D.I., Sher, A., et al., 2007. Whole-genome shotgun sequencing of mitochondria from ancient hair shafts. Science 317, 1927–1930.

Guslitser, B.I., Isaychev, K.I., 1983. The age of the Rogovaya formation of the Timan-Pechora region as determined by fossil pied lemmings. Bull. Komissii po izucheniyu chetvertichnogo perioda Acad. Sci. USSR 52, 58–72 (in Russian).

Henriksen, M., Mangerud, J., Maslenikova, O., Matiouchkov, A., Tveranger, J., 2001. Weichselian stratigraphy and glaciotectonic deformation along the lower Pechora river, Arctic Russia. Glob. Planet. Change 31 (1–4), 297–319.

Henriksen, M., Mangerud, J., Matiouchkov, A., Murray, A.S., Paus, A., Svendsen, J.I., 2008. Intriguing climatic shifts in a 90 kyr old lake record from northern Russia. Boreas 37, 20–37.

Isayeva, L.L., Kind, N.V., Kraush, M.A., Sulerzhitsky, L.D., 1976. On the age and structure of marginal formations in the northern foothills of the Putorana Plateau. Bull. Komissii po izucheniyu chetvertichnogo perioda Acad. Sci. USSR 45, 117–123 (in Russian).

Isayeva, L.L., Kind, N.V., Laukhin, S.A., Kolpakov, V.V., Shofman, I.L., Fainer, Yu.B., 1986. The stratigraphic scheme of the Quaternary of Central Siberia. In: Velichko, A.A., Isayeva, L.L. (Eds.), Chetvertichnye oledeneniya Srednei Sibiri. Nauka, Moscow, pp. 4–17.

Kaplyanskaya, F.A., Tarnogradsky, V.D., 1974. The Middle and Lower Pleistocene of the Lower Irtysh Area (Sredny i nizhny pleistotsen nizovyev Irtysha). Nedra, Leningrad, 160 pp. (in Russian).

Kind, N.V., 1974. Late Quaternary Geochronology According to Isotopes Data (Geokhronologia pozdnego antropogena po izotopnym dannym). Nauka, Moscow, 255 pp. (in Russian).

Kind, N.V., Leonov, B.N. (Eds.), 1982. The Anthropogene of Taimyr (Antropogen Taimyra). Nauka, Moscow, 184 pp. (in Russian).

Kosintsev, P.A., 2008. Mammoth fauna of the Yuribei river basin (Yamal Peninsula). In: Kosintsev, P.A. (Ed.), Biota Severnoi Yevrazii v kainozoye, 6. Institute of Plant and Animal Ecology. Uralian Branch Russian Academy of Sciences, Yekaterinburg–Chelyabinsk, pp. 147–157 (in Russian).

Laukhin, S.A., Arslanov, Kh.A., Maksimov, F.Ye, Kuznetsov, V.Yu, Shilova, G.N., Velichkevich, F.Yu., et al., 2008. New outcrop of buried Kazantsevo peat at lower reaches of the Irtysh river. Doklady Earth Sci. 1028-334X, 419 (2), 200–204. Translated from Doklady Akademii Nauk, 2008, 418(5), 650–654.

Lavrushin, Yu.A., Chistyakova, I.A., Gaidamanchuk, A.S., Golubev, Yu. K., Vasilyev, V.P., 1989. The structure and composition of the glacial paleoshelf sediments in Bolshezemelskaya Tundra. In: Lavrushin, Yu.A. (Ed.), Litologia kainozoiskikh shelfovykh otlozheniy. Geological Institute of Academy Sciences USSR, Moscow, pp. 3–51.

Levchuk, L.K., 1984. Biostratigraphy of the Upper Pleistocene of the Siberian North by Foraminifera (Biostratigrafia verkhniego pleistosena severa Sibiri po foraminiferam). Nauka, Novosibirsk, 128 pp. (in Russian).

Levina, T.P. 1964. Pollen spectra of the Quaternary sediments from the proglacial zone of the Samarovo ice sheet (the Yenissei catchment). In: Sachs, V.N. and Khlonova, A.F., eds. Sistematika i metody izucheniya iskopayemykh pyltsy i spor. Moscow, Nauka, 208–217 (in Russian).

Lider, V.A., Stefanovsky, V.V., Vigorova, I.E., 1999. National Geological Map of the Russian Federation, Scale 1:1 000 000 (New Series). Map of the Quaternary Deposits. Yekaterinburg.

Loseva, E.I., 1978. The Middle Valdai sea-lake in western Bolshezemelskaya Tundra. Bull. Komissii po izucheniyu chetvertichnogo perioda 48, 103–112 (in Russian).

Loseva, E.I., Duryagina, D.A., 1983. Palaeobotanic evidence for stratigraphy of Cenozoic sediments of the central Pai-Hoi. Trudy Inst. Geol. Komi Branch Acad. Sci. USSR 43, 56–68. Syktyvkar, (in Russian).

Loseva, E.I., Duryagina, D.A., Andreicheva, L.N., 1992. The Middle Pleistocene of central Bolshezemelskaya Tundra. Trudy Inst. Geol. Russ. Acad. Sci. Komi Branch 75, 113–123, Syktyvkar, 113–123 (in Russian).

Mangerud, J., Svendsen, J.I., Astakhov, V.I., 1999. Age and extent of the Barents and Kara Sea ice sheets in Northern Russia. Boreas 28 (1), 46–80.

Mangerud, J., Jakobsson, M., Alexanderson, H., Astakhov, V., Clarke, G.K.C., Henriksen, M., et al., 2004. Ice-dammed lakes and rerouting of the drainage of northern Eurasia during the Last Glaciation. Quatern. Sci. Rev. 23 (11–13), 1313–1332.

Mangerud, J., Gosse, J., Matiouchkov, A., Dolvik, T., 2008. Glaciers in the Polar Urals, Russia, were not much larger during the Last Global Glacial Maximum than today. Quatern. Sci. Rev. 27, 1047–1057.

Möller, P., Bolshiyanov, D., Bergsten, H., 1999. Weichselian geology and palaeoenvironmental history of the central Taimyr Peninsula, Siberia, indicating no glaciation during the last global glacial maximum. Boreas 28 (1), 92–114.

Nazarov, D.V., 2007. New data on Quaternary sediments in the central part of the West Siberian Arctic. Regionalnaya Geol. Metallogenia 30/31, 213–221 (in Russian).

Nazarov, D., Henriksen, M., Svendsen, J.I., 2009. The age of the last glacier invasion into the Polar Urals. In: Abstracts of Third International Conference on Arctic Palaeoclimate and its Extremes (APEX). University of Copenhagen, Denmark, p. 55.

Potapenko, L.M., 1998. National Geological Map of the Russian Federation, Scale. 1:1 000 000, New Series, Quadrangle P-38,39 (Syktyvkar). Map of Quaternary Formations. VSEGEI, St. Petersburg.

Rudenko, T.A., Fainer, Yu.B., Fainer, T.G., 1984. National Geological Map of the USSR, Scale 1:1 000 000, New Series, Quadrangle P-48,49 (Vanavara). Map of Quaternary Deposits. VSEGEI, Leningrad.

Sachs, V.N., 1953. The Quaternary Period in the Soviet Arctic (Chetvertichny period v Sovietskoi Arktike). Institute of Geology of the Arctic, Leningrad-Moscow, 627 pp. (in Russian).

Shvaryov, S.V., 1998. Reconstruction of Sartan glaciation of Putorana Plateau (by the remote sensing data). Geomorfologia 1, 107–112 (in Russian).

Siegert, C., Derevyagin, A.Y., Shilova, G.N., Hermichen, W.D., Hiller, A., 1999. Paleoclimatic indicators from permafrost sequences in the eastern Taymyr Lowland. In: Kassens, H., Bauch, H., Dmitrenko, I.A., Eicken, H., Hubberten, H.-W., Melles, M., Thiede, J., Timokhov, L.A. (Eds.), Land-Ocean Systems in the Siberian Arctic. Springer, Berlin-Heidelberg, pp. 477–499.

Stauch, G., Lemkuhl, F., Frechen, M., 2007. Luminescence chronology from the Verkhoyansk Mountains (North-Eastern Siberia). Quat. Geochronol. 2, 255–259.

Stepanov, A.N., 1974. Stratigraphy and Sedimentary Environments of the Upper Cenozoic of the Pechora-Kama Interfluve. Resume of PhD thesis. Geological Faculty of Moscow University, 34 pp. (in Russian).

Sulerzhitsky, L.D., 1995. Characteristics of radiocarbon chronology of the woolly mammoth (Mammuthus primigenius) of Siberia and north of Eastern Europe. Trudy Zoologicheskogo instituta 263, 163–183. St. Petersburg (in Russian, English summary).

Svendsen, J.I., Alexanderson, H., Astakhov, V.I., Demidov, I., Dowdeswell, J.A., Funder, S., et al., 2004. Late Quaternary ice sheet history of northern Eurasia. Quatern. Sci. Rev. 23, 1229–1271.

Tomirdiaro, S.V., Tikhonov, A.N., 1999. The Yamal baby mammoth. Paleogeographical situation and burial conditions. Trudy Zoologicheskogo instituta 275, 7–19, St. Petersburg (in Russian).

Troitsky, S.L., 1975. The Modern Antiglacialism: A Critical Essay (Sovremenny Antiglatsialism). Nauka, Moscow, 163 pp. (in Russian).

Vasilchuk, Yu.K., 1992. Oxygen Isotope Composition of Ground Ice: Application to Paleogeocryological Reconstructions, vol. 1. (Izotopno-kislorodny sostav podziemnykh ldov: opyt paleokriologicheskikh rekonstruktsiy). Mosoblpoligrafizdat, Moscow, 420 pp. (in Russian).

Vasilchuk, Yu.K., van der Plicht, J., Jungner, H., Sonninen, E., Vasilchuk, A.K., 2000. First direct dating of Late Pleistocene ice wedges by AMS. Earth Planet. Sci. Lett. 179, 237–242.

Velichko, A.A., Faustova, M.A., Gribchenko, Yu.N., Pisareva, V.V., Sudakova, N.G., 2004. Glaciations of the East European Plain—distribution and chronology. In: Ehlers, J., Gibbard, P.L. (Eds.), Quaternary Glaciations—Extent and Chronology. Part 1: Europe. Elsevier, Amsterdam, pp. 337–354.

Yakhimovich, V.L., Zarkhidze, V.S., Afanasyeva, T.A., 1992. A Key Magnetostratigraphic Section of the Upper Pliocene of the Timan-Uralian Region. Preprint of Bashkir Science Centre RAS, Ufa, 12 pp.

Zarkhidze, V.S., 1981. Stratigraphy and correlation of Pliocene and Pleistocene deposits. In: Kamaletdinov, M.A., Yakhimovich, V.L. (Eds.), Pliotsen i pleistotsen Volgo-Uralskoi oblasti. Nauka, Moscow, pp. 7–28.

Zubakov, V.A., 1972. Recent Sediments of the West Siberian Lowland (Noveishie otlozhenia Zapadno-Sibirskoi nizmennosti). Nedra, Leningrad, 312 pp. (in Russian).

Chapter 26

Glaciations of the East European Plain: Distribution and Chronology

A.A. Velichko[1,*], M.A. Faustova[1], V.V. Pisareva[1], Yu.N. Gribchenko[1], N.G. Sudakova[2] and N.V. Lavrentiev[1]

[1]*Laboratory of Evolutionary Geography, Institute of Geography, Russian Academy of Sciences, Staromonetny 29, Moscow 109017, Russia*
[2]*Department of Geography, Lomonosov Moscow State University, Vorob'evy Gory, Moscow 119899, Russia*
*Correspondence and requests for materials should be addressed to A.A. Velichko. E-mail: paleo_igras@mail.ru

26.1. INTRODUCTION

The East European Plain is one of the regions that have been repeatedly subjected to the expansion of large continental ice sheets. The concept of three ice sheets expanding successively into European Russia developed in the 1930s. They were known as the Oka, Dnieper and Valdai glaciations and were tentatively correlated with the Mindel, Riss and Würm of the Alpine region. Since then, opinions on the number of glaciations and on the periodicity of glacial events have undergone considerable revision. Biostratigraphical and lithostratigraphical evidence suggest that the first strong cooling which might have resulted in glaciation that occurred as early as 2.4–1.8 million years ago (Frenzel, 1967; Grichuk, 1981), although no glacial deposits dating from this interval have been found.

The largest ice sheets, that within the past million years expanded into East European Russia, were developed at the end of the Matuyama and during the Brunhes Chrons. They were associated with the Tiraspolian faunal complex that corresponds to the 'Cromerian Complex' Stage in the West European chronostratigraphy (Figs. 26.1 and 26.2).

26.2. LIKOVO GLACIATION

The oldest glaciation is now correlated with the glacial 'Pre-Cromerian Complex'; it expanded as far south as the latitude of Moscow. The so-called Likovo Till attributed to this ice sheet, was only discovered in the 1980s, after members of the Geological Survey of the Central Regions of the East European Plain had thoroughly examined many boreholes in the Moscow region (Figs. 26.3 and 26.4). It has not yet been found elsewhere (Maudina et al., 1985). Boreholes west of Moscow, near the western margin of the town Odintsovo (Akulovo village), penetrated the till near the base of the Quaternary sequence. It is represented by black or greenish, dark grey loam including abundant gravel and small pebbles of sedimentary rocks (predominantly flint and limestone). Crystalline (augite and rose fine-grained granite with a grain size less than 1–2 mm) gravel is only occasionally found. No rocks of unequivocally Scandinavian origin are recorded in any of the sections. The heavy mineral composition is typically garnet–disthene–tourmaline. In the Don drainage basin, this cryochron (cold stage) is probably represented by the loess unit exposed in the Troitskoye section beneath the oldest soil. The fossil fauna of the latter is dominated by the ground squirrel *Citellus* (Krasnenkov et al., 1997). The Troitskoye fauna's antiquity is suggested by the presence of *Prolagurus pannonicus* Korm and the abundance of European vole *Pitymys hintoni* Kretzoi (Agadjanian, 1992). A similar faunal assemblage has also been recovered from the ancient fluvial sediments near the Karai-Dubina (Dnieper drainage basin) by Markova (1982) who attributed it to the final stage of the Matuyama Chron.

26.3. AKULOVO INTERGLACIAL

The subsequent, Akulovo interglacial was identified in the Akulovo section mentioned above. It is characterised by the oldest plant macrofossil flora (Maudina et al., 1985), including over 30% of local and regional exotic taxa and greater than 11% of extinct species. The antiquity of the Akulovo flora, which bears similarities to undoubtedly Pliocene assemblages, is suggested by the diversity of coniferous pollen present with high percentages of *Pinus sec. Cembra* (up to 30–50%), *P. sec. Strobus* (up to 5–8%), as well as *P. sec. Mirabilis*, *P. sec. Omorica*, *Tsuga* and *Taxus*. The broad-leaved taxa, which comprised 35–50% in the climatic optimum, *Quercus, Ulmus, Carpinus, Zeltis, Zelkova, Fagus, Pterocarya, Juglans, Castanea, Ilex, Morus, Eucommia* and *Vitis* are represented, as well as shrubs *Ligustrum, Corylus, Ostrya* and *Myrica* (Fig. 26.5).

West European glacial region			East European glacial region			East European loess region			Cryogenic horizons
Holocene									
Weichselian glaciation	Late Weichselian		Valdai glaciation	Late Valdai		Altynovo loess			Yaroslavl
						Trubchevsk soil			
						Desna loess			Vladimir
	Middle Weichselian	I/s Denekamp		Middle Valdai	Bryansk megainterval	I/s Dunaevo	I/s Bryansk soil		
		Cool and warm stages				Cool and warm stages			
	Early Weichselian	Cool and warm stages		Early Valdai		Early Valdai stages	Khotylevo loess		Smolensk phase "b"
		I/s Brorup				I/s Upper Volga (Krutitsa)	Mezin soil complex	I/s Krutitsa soil	
		Cool and warm stages				Early Valdai stages		Sevsk loess	Smolensk phase "a"
Eemian Interglacial			Mikulino Interglacial				Salyn interglacial soil		
Saale glaciation	Warthe (Saale III) stage		Dnieper glaciation	Moscow stage		Moscow loess			Moscow
	I/s Treene			I/s Kostroma		I/s Kursk soil			
	Saale II (Drente II) stage			Dnieper stage		Dnieper loess			Dnieper
						I/s soil (Romny?)			
						Loess (Orchik?)			Early Dnieper
	Saale I (Drente I) stage			Interstadial		Kamenka soil complex	I/s Late Kamenka soil		
				Stage			Loess		Igorevka
Demnits, Wacken Interglacial			Kamenka Interglacial				Early Kamenka interglacial soil		
Fuhne glaciation	Stage II		Glatiation (Pechora?)	Stage		Borisoglebsk loess			Stupino, phase "b"
	Interstadial			Interstadials		Inzhavino interglacial soil	I/s Late Inzhavino soil		
	Stage I			Stage			Loess		Stupino, phase "a"
Holstein Interglacial			Likhvin Interglacial				Early Inzhavino intrglacial soil		
Elster glaciation	Elster II stage		Oka glaciation			Korostelevo loess			Oka
	Interstadial								
	Elster I stage								
Cromer complex	Interglacial IV Voigstedt		Ikorets Interglacial ?			Vorona soil complex	Late Vorona interglacial soil		
	Stage C		?				Loess		
	Interglacial III		Muchcap Interglacial				Early Vorona interglacial soil		
	Stage B		Don glaciation				Don loess		Don
	Interglacial II		Okatovo Interglacial				Rzhaksa interglacial soil		
	Stage A		Setun stage				Bobrov loess		B
	Interglacial I		Krasikovo Interglacial			Balashov soil complex	Balashov interglacial soil		M
			?						
			Akulovo Interglacial						
			Likovo glaciation						

FIGURE 26.1 The Pleistocene stratigraphical scheme of European Russia and correlation with the West European units (compiled by A. A. Velichko, V. V. Pisareva, M. A. Faustova, T. D. Morozova, V. P. Nechaev and Yu. N. Gribchenko).

Chapter | 26 Glaciations of the East European Plain: Distribution and Chronology

Glacial Zone		Deposits	Periglacial Zone	
Holocene			Holocene soil	
Late Pleistocene	Late Valdai	Yaroslavl' cryogenic horizon	Altynovo loess	
			Trubchevsk soil	
		Vladimir' cryogenic horizon	Desna loess	
	Middle Valdai		Bryansk soil	
	Early Valdai	Smolensk cryogenic horizon	khctylevo loess	
				Krutitsa soil
			Mezin soil complex	Sevsk loess
	Mikulino Interglacial			Salyn' soil
Middle Pleistocene	Dnieper Glacial — Moscow stage		Dnieper loess complex	Moscow loess
	Interstadial			Kursk soil
	Dnieper stage (maximum)			Dnieper loess
	Interstadial			Romny soil
	Cold Epoch			Orchik loess
	Kamenka Interglacial		Kamenka soil	
	Cold Epoch		Borisoglebsk loess	
Early Pleistocene	Likhvin interglacial		Inzhavino soil	
	Oka Glacial		Korostylevo loess	
	Ikorets interglacial ?		Vorona soil complex	
	Cold Epoch			
	Muchkap (Roslavl) Intergl.			
	Don Glacial		Don loess complex	
	Okatovo Interglacial		Rzhaksa soil	
	Setun' Glacial		B Bobrov loess	
	Krasikovo Interglacial		M Balashov soil complex	
	Glacial			
	Akulovo Interglacial			
	Likovo Glacial			

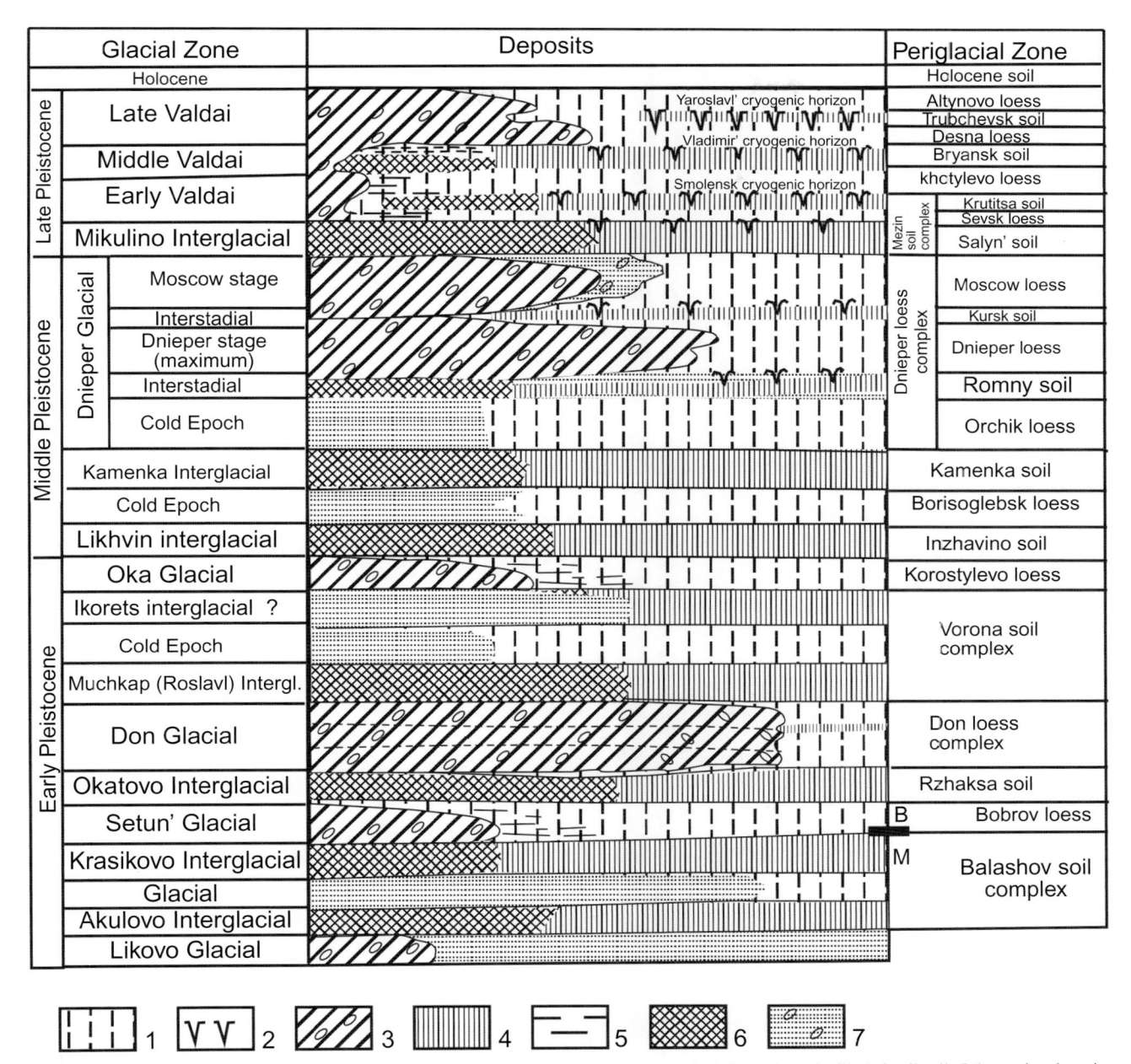

FIGURE 26.2 Correlation between glacial and periglacial stratigraphy. 1, loess; 2, cryogenic deformations; 3, till; 4, fossil soil; 5, lacustrine deposits; 6, peat deposits; 7, glacial deposits (from Velichko et al., 2004, with additions).

The Akulovo interglacial was followed by a strong cooling, which restored tundra-like landscape in the Moscow region. This cooling possibly caused the development of a glaciation, although existing evidence does not allow determination of its boundaries.

26.4. KRASIKOVO INTERGLACIAL

The new records available in recent years suggest the occurrence of an additional Krasikovo Interglacial that was presumably cooler than Akulovo event. The Krasikovo Interglacial has been recognised on the basis of palynological studies of the section near the Krasikovo Village, Konakhovo district in the Tver' region. Here, the 19-m-thick interglacial deposits are exposed beneath the Don Till and are represented by lacustrine clays, loams bearing plant remains and peat beds, and gyttja. They are underlain by glaciofluvial sands with quartz and flint pebble. The whole sequence is normally magnetised with reversed polarity intervals in the lower and upper parts of the lacustrine sediments. The pollen diagram (Fig. 26.6) shows a climatic optimum of the interglacial characterised by

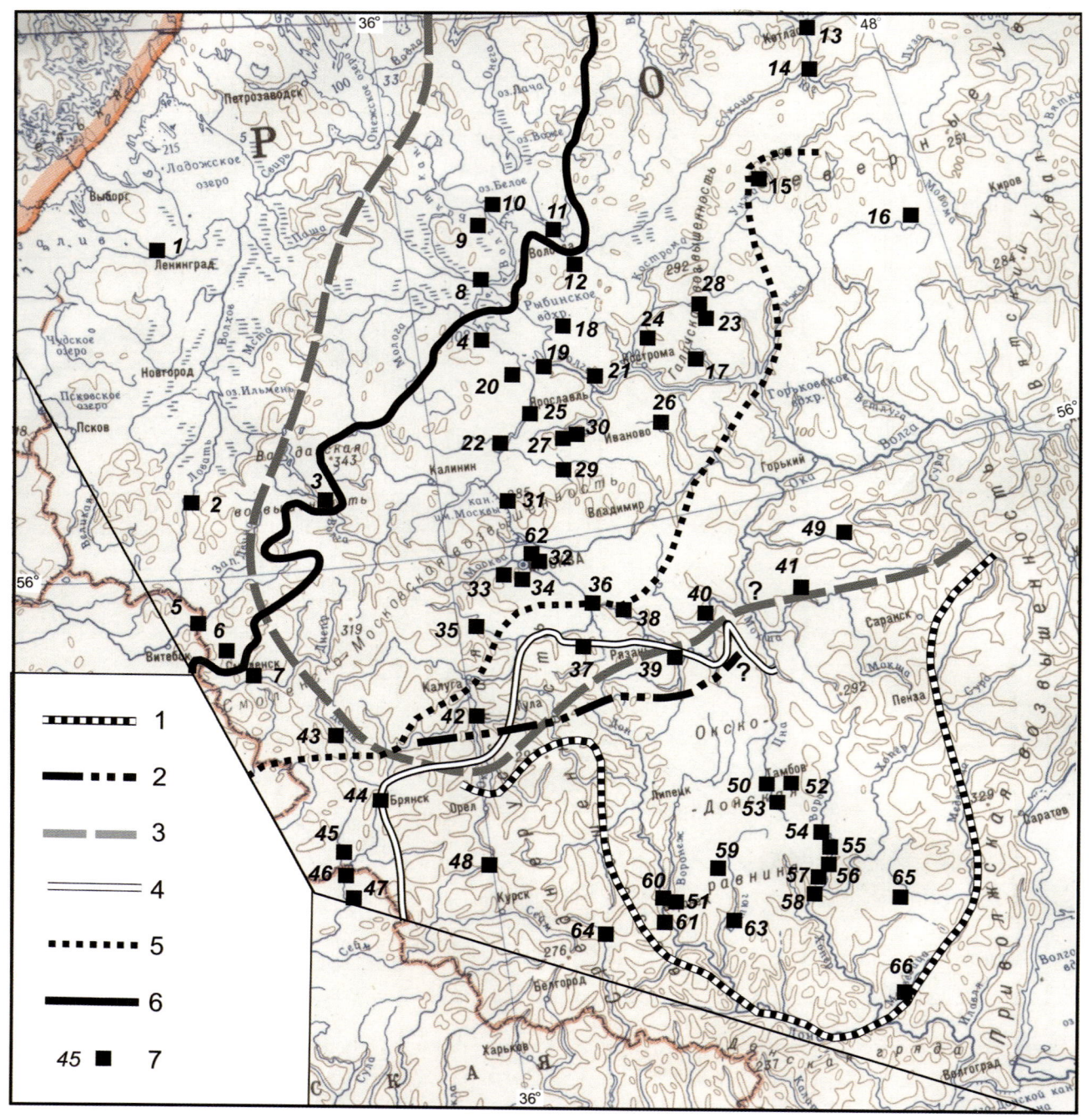

FIGURE 26.3 Glacial limits and key sections in the East European Plain (by A. A. Velichko, V. V. Pisareva and M. A. Faustova). Glacial limits and key sections. The boundaries of ice sheets: 1, Don; 2, Oka; 3, Pechora; 4, Dnieper; 5, Moskow; 6, Valdai; 7, Key sections. Key sections: 1, Grazhdansky Prospekt, Kelkolovo (vd); 2, Dunaevo, Gora (vd); 3, Bulatovo, Tyaglitsy, Bolshaya Kosha (ok, lh, pč, ms); 4, Palnikovo (pč, km, ms); 5, Drichaluki, Shapurovo (ms,vd); 6, Mikulino (mk, vd); 7, Smolensk, Kuchino(dn, ms, mk); 8, Cherepovets (dn, ms, mk); 9, Moeksa (dns, dn, vd); 10, Ferapontovo (dns, dn, vd); 11, Puchka (dn, ms,vd); 12, Molochnoe (dn, ms); 13, Kotlas (dns, dn, ms); 14, Veliky Ustyug, Dymkovo (dns, dn, ms); 15, Anyuk (dn, mc); 16, Vavilyata (pre-dns, dns); 17, Pepelovo (dns, mc); 18, Yakovlevskoe (ok, lh, dn, ms); 19, Rybinsk, Chermenino (pre-dns, dns, lh, dn, ms); 20, Shestikhino (ms); 21, Tutaev, Dolgopolka (ms, mk); 22, Alkhimkovo (pre-dns, dns, ok, lh, dn, ms); 23, Galich (ms, mk); 24, Zakharyino (dns, mc, dn, ms); 25, Verkhniye Ploski, Altynovo (dns, dn, ms); 26, Bibirevo (dns, mc, ms); 27, Cheremoshnik (ms, mk); 28, Gorki, Chelsma (dns, mc, dn, ms); 29, Pereslavl Salessky (ms, mk, vd); 30, Lake Nero (ms, mk, vd); 31, Dmitrov (ms, mk); 32, Balashikha (dns, mc, ms); 33, Akulovo (pre-dns, dns, mc, ms); 34, Okatovo (pre-dns, dns, ok, ms); 35, Borovsk, Satino (pre-dn, dn, ms); 36, Golutvin, Gololobovo (dn, ms); 37, Zaraisk (dns, dn, mk); 38, Alpatyevo (dns, pre-dn, mk); 39, Troitsa, Fatyanovka (dns, ok, pre-dn, dn); 40, Elatma, Kasimov (pre-dn, dn); 41, Narovatovo (dns, ok, lh); 42, Chekalin, Bryankovo (dns, ok, lh, dn); 43, Roslavl, Konakhovka, Malakhovka, Sergeevka (pre-dns, dns, mc, ok, lh, ms); 44, Bryansk (dn, mk); 45, Pogar (dn, mk); 46, Pushkari (dn,mk); 47, Arapovichi, Mezin (dn, mk); 48, Zheleznogorsk (pre-dn, mk); 49, Lukoyanov (dns, pre-dns); 50, Tambov, Preobrazhenie (dns, mc, dn); 51, Demshinsk (dns, mc, lh, dn, km); 52, Rasskazovo (dns, mc, lh, dn); 53, Kotovsk (dns, pre-dn); 54, Inzhavino (dns, mc, lh, dn); 55, Muchkapsky, Korostelevo (dns, mc, lh, pre-dn, mk); 56, Bolshaya Rzhaksa, Perevoz, Posevkino (pre-dns, dns, lh, pre-dn); 57, Borisoglebsk, Volniye Vershiny (pre-dn, mk); 58, Novokhopersk (pre-dns, dns); 59, Polnoye Lapino (dns, mc); 60, Bogdanovka, Verkhnyaya Emancha (dns, lh, pre-dn); 61, Uryv, Mastuyuzhenka, Korotoyak (pre-dns, dns, lh, pre-ok); 62, Krasikovo (kr, dns); 63, Klepki (pre-dns, dns); 64, Stolinsky (pre-dns); 65, Nizhne-Dolgovsky (dns, mc); 66, Mikhailovka (pre-dns, dns, pre-dn, mk).

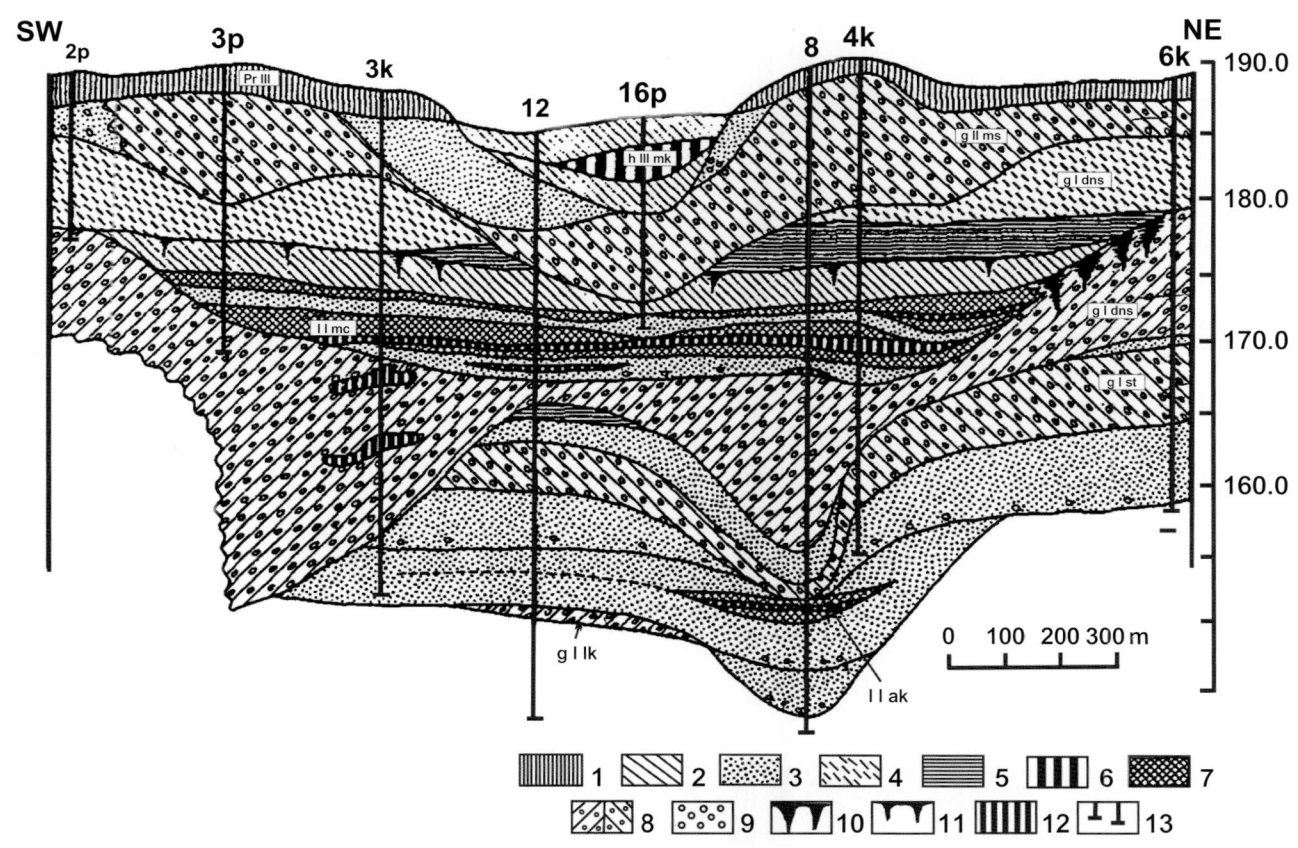

FIGURE 26.4 Longitudinal geological section of the Akulovo site (Maudina et al., 1985). 1, cover loam; 2, lacustrine loam; 3, sand; 4, sandy loam; 5, clay; 6, peat; 7, gyttja; 8, diamicton (till); 9, gravel and pebble; 10, fossil soils; 11, weakly comminuted fossil soils; 12, detached blocks; 13, boreholes.

coniferous–broad-leaved forests. The conifers were dominated by *Pinus* ex gr. *Haploxylon* and include common spruce together with fir at the beginning and at the end of the optimum. The assemblage also contains *Larix*, *Tsuga* and *Taxus*. The broad-leaved species were mainly represented by oak and hornbeam. The complex phytocoenoses included *Pterocarya*, *Juglans* and *Ilex*, and the undergrowth contained diverse bushes. Based on the detailed floral assemblages, the Krasikovo Interglacial can be correlated with the 'Interglacial 1' of the Cromerian Complex Stage.

26.5. SETUN′ GLACIATION

The so-called Setun′ glaciation (named after the Setun′ River in the Moscow region) expanded as far south as the northern margin of the Tula region (Fig. 26.7). The Setun′ Till (described by the Geological Survey team) is represented by massive brownish and greenish dark grey loams and sandy loams. Magmatic rocks of Scandinavian provenance comprise about 40–60% of the clasts. The till is characterised by an amphibole–epidote–garnet–disthene mineral association.

26.6. OKATOVO INTERGLACIAL

The subsequent interglacial is known as the 'Okatovo' Warm Stage'. Its deposits were drilled west of Moscow, 4 km east of Vnukovo railway station near Okatovo village (Fursikova et al., 1992) and near Skhodnya railway station, at Dubrovka village. This younger flora is poorer than that of the Akulovo in every respect. Unlike the previously described interglacial, its climatic optimum featured broad-leaved forests (with several, 5–10 dominant tree species on average), primarily composed of oak, elm and lime trees, and later with hornbeam and other species characteristic of a mild temperate climate (Fig. 26.8).

26.7. DON GLACIATION

The subsequent expansion of the ice sheet known as the Don Glaciation corresponds to the maximum ice extent in the East European Plain. Its age is determined primarily from its relation to palaeontological evidence from over- and underlying sediments. Until the late 1970s, it was generally assumed that the maximum glaciation in the East

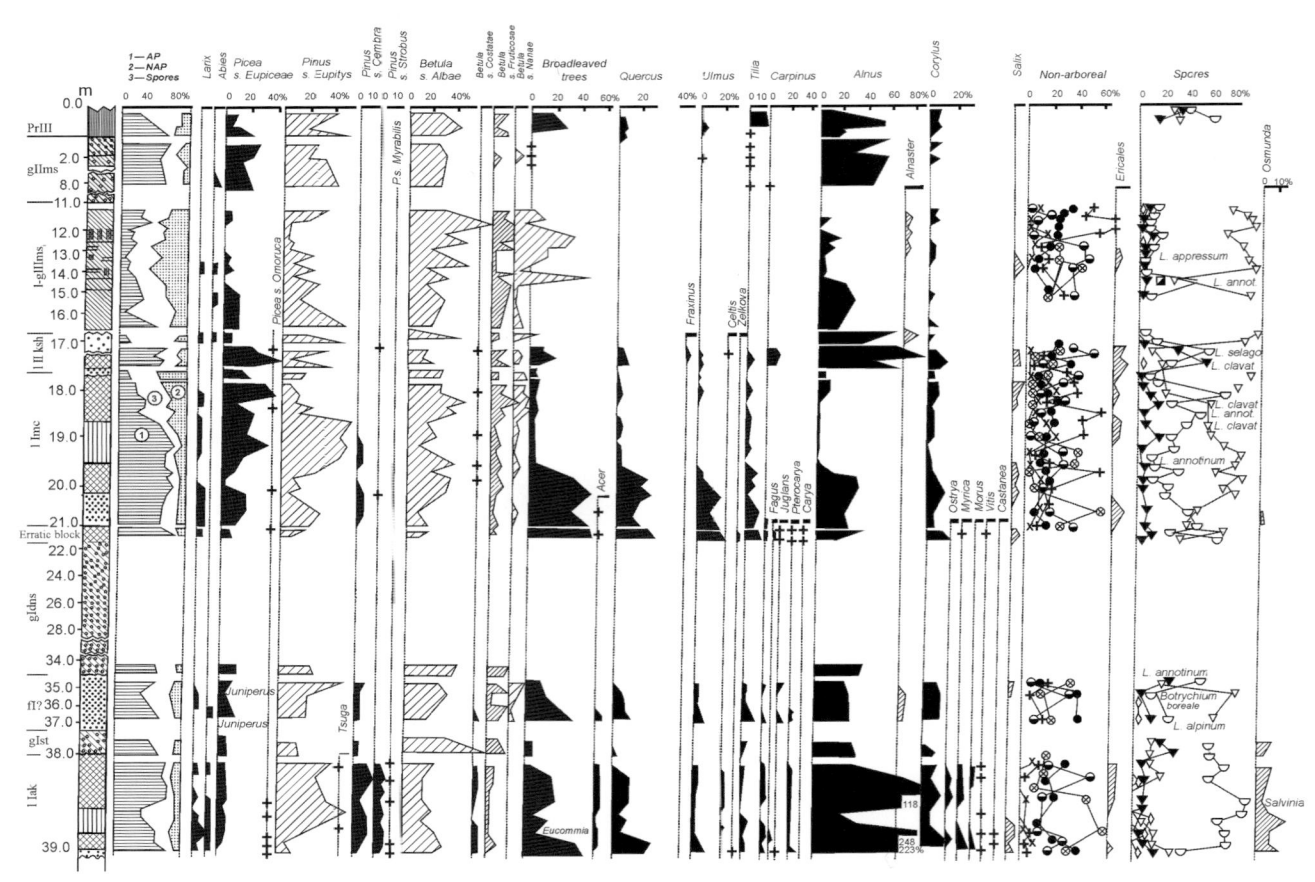

FIGURE 26.5 Pollen diagram of the Akulovo site deposits (Borehole 8k) (Maudina et al., 1985). For symbols of grass pollen and spores, see Fig. 26.14.

European Plain was the late Middle Pleistocene Dnieper (Saalian) Glaciation, and the oldest known Oka ice sheet was correlated to the Elsterian Stage in western and central Europe.

However, multidisciplinary studies and analysis of the relations between tills belonging to the two largest ice lobes in Russia (the Dnieper and the Don lobes) on the one hand, and loess–palaeosol periglacial sequences on the other, revealed that the tills of the two lobes differed substantially in age (Gerasimov and Velichko, 1980). Studies of loess and soil horizons exposed in many sections, and the analysis of palaeontological evidence allowed the identification of several glacial stages during which active loess accumulation occurred (Fig. 26.9). In contrast to the glacial sequence of the Dnieper lobe, which was formed by the most extensive Middle Pleistocene ice sheet, the Don Till was deposited by the largest of all Pleistocene ice sheets dated to the Tiraspolian. The limits of the Don ice sheet are only clearly defined east of the Central Russian Uplands, whereas to the west their position is indistinct. The maximum glaciation of the East European Plain, the Don Glaciation expanded far south, into the drainage basin of the middle and lower Don River.

The correlation of the Don Glaciation with the 'Cromerian Complex' Stage has been substantiated by investigations of glacial and loess–palaeosol sequences in the Don and Dnieper drainage basins. Palaeontological evidence has confirmed beyond doubt the late Tiraspolian age of fossil rodent remains recovered from numerous sections both above and below the Don Till (Gerasimov and Velichko, 1980; Krasnenkov et al., 1997). At most sites (Vol'naya Vershina, Korotoyak-4, Korostelevo-2, Kuznetsovka and others in the Don basin, as well as in the vicinity of Roslavl', at Konakhovka, Podrudnyansky and Sergeevka in the Dnieper basin), the late Tiraspolian small mammal remains recovered from the beds immediately underlying the till are somewhat older than those found in the overlying sediments. Thus a fauna dominated by *Lemmus*, in combination with *Mimomys*, *Pitymys* and *Microtus oeconomus*, has been identified in the overlying sediments. These assemblages compare closely to the Tiraspolian faunas in Moldova, Czechoslovakia, Hungary and France (Aleksandrova, 1982; Agadjanian, 1992). The investigations have provided corroborative evidence for the different age of the ice lobes extending southwards into the Dnieper and Don basins, as well as for the different age of their tills.

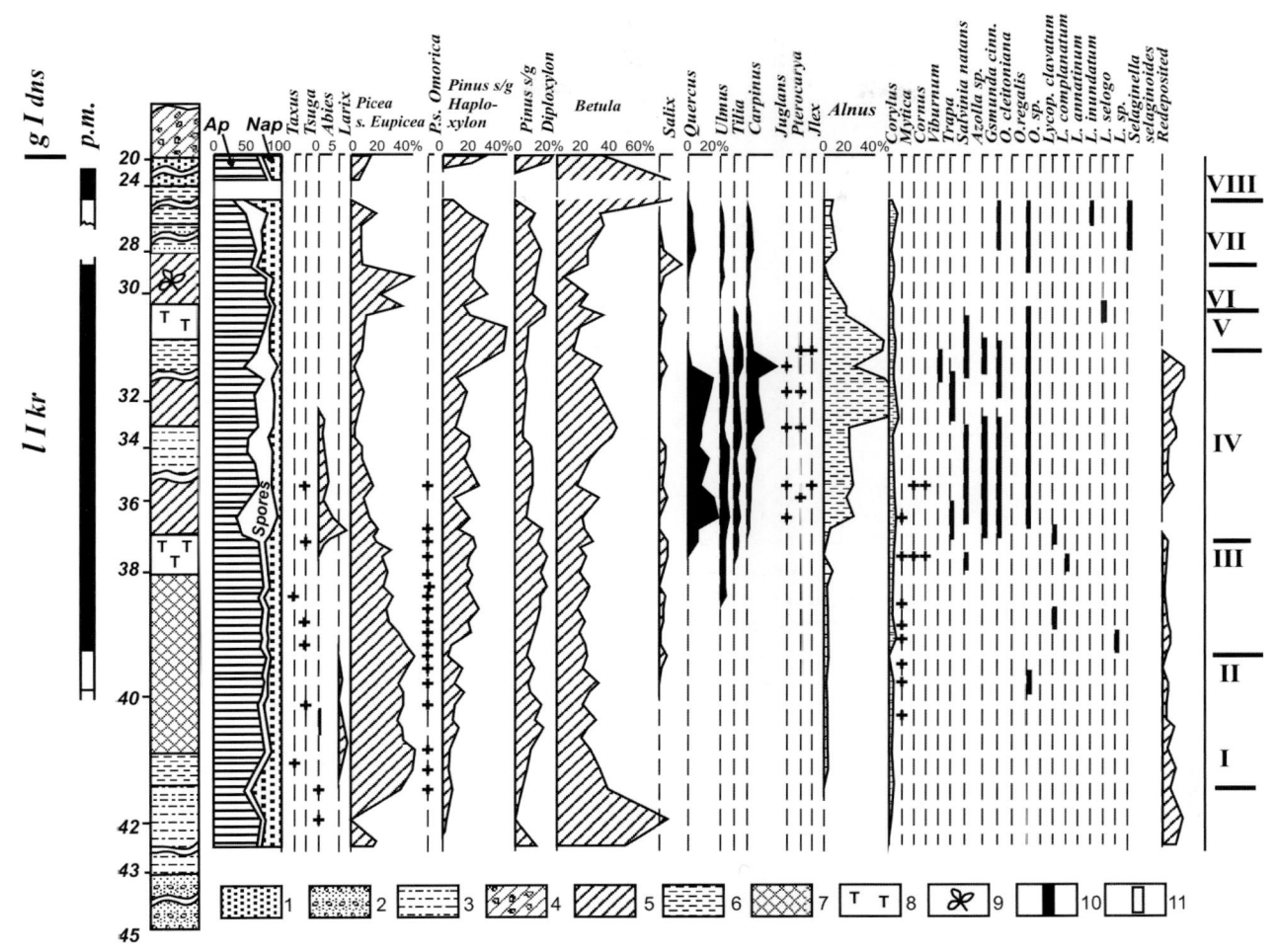

FIGURE 26.6 Pollen diagram of the deposits nearby Krasikovo Village (Shick et al., 2006). 1, sand; 2, gravel and pebbles; 3, loamy sand, loam; 4, till; 5, loam; 6, clay; 7, gyttja; 8, peat; 9, plant remains. Polarity: 10, normal; 11, reversed; 12, I–VIII phases of vegetation evolution.

Consequently, the Cromerian palaeogeography of the East European Plain has had to be revised. The Don Formation glacial sequence is as a rule from 3–5 to 15–20 m thick and consists of several till units which can be traced throughout the Don lobe area. Individual strata differ slightly in both composition and clast orientation (Grishchenko, 1976; Shick, 1984; Sudakova and Faustova, 1995). The most characteristic feature of the Don Till is the abundance of small gravel, whereas large pebbles and boulders are only occasionally found. Most of large particles are derived from local sedimentary rocks varying in composition throughout the area, whilst the far-travelled clasts are chiefly granites, metamorphic rocks and quartzite-like sandstone. The maximum content of the debris 2–5 cm in size (up to 15–20%) has been found in the western portion of the lobe; in the central and eastern parts, the debris content decreases down to 4–7% (Gribchenko, 1980; Maudina et al., 1985).

With regard to crystalline indicator rocks, the Don Glaciation tills differ markedly from all the other tills in the central, northern and north-western Russian Plain (Fig. 26.10). The fact that the Don Till lacks the main index rocks, which Chirvinsky (1914), Yakovlev (1939) and Viiding et al. (1971) have identified in the Middle and Late Pleistocene tills, suggests a more north-easterly Don Glaciation ice-centre. This is supported by a south-western orientation of clasts in many sections (e.g. Gerasimov and Velichko, 1980). The mineralogy of the Don Till matrix varies considerably regionally. In contrast to the younger tills, minerals of local and medium distance provenance dominate, whilst the actual percentages of exotic minerals, such as hornblende, amphibole, pyroxene and biotite, do not exceed 6–10%. Amphiboles are more common in the western sector, whereas epidote, tourmaline, ilmenite and rutile are most typical of the central and eastern sectors. The most conspicuous result of the topography-forming activities of the Don ice sheet is the numerous linear depressions eroded by the moving ice. Later, during the Muchkap Interglacial Complex, they were filled by sediments characterised by a late Tiraspolian rodent fauna (Agadjanian, 1992).

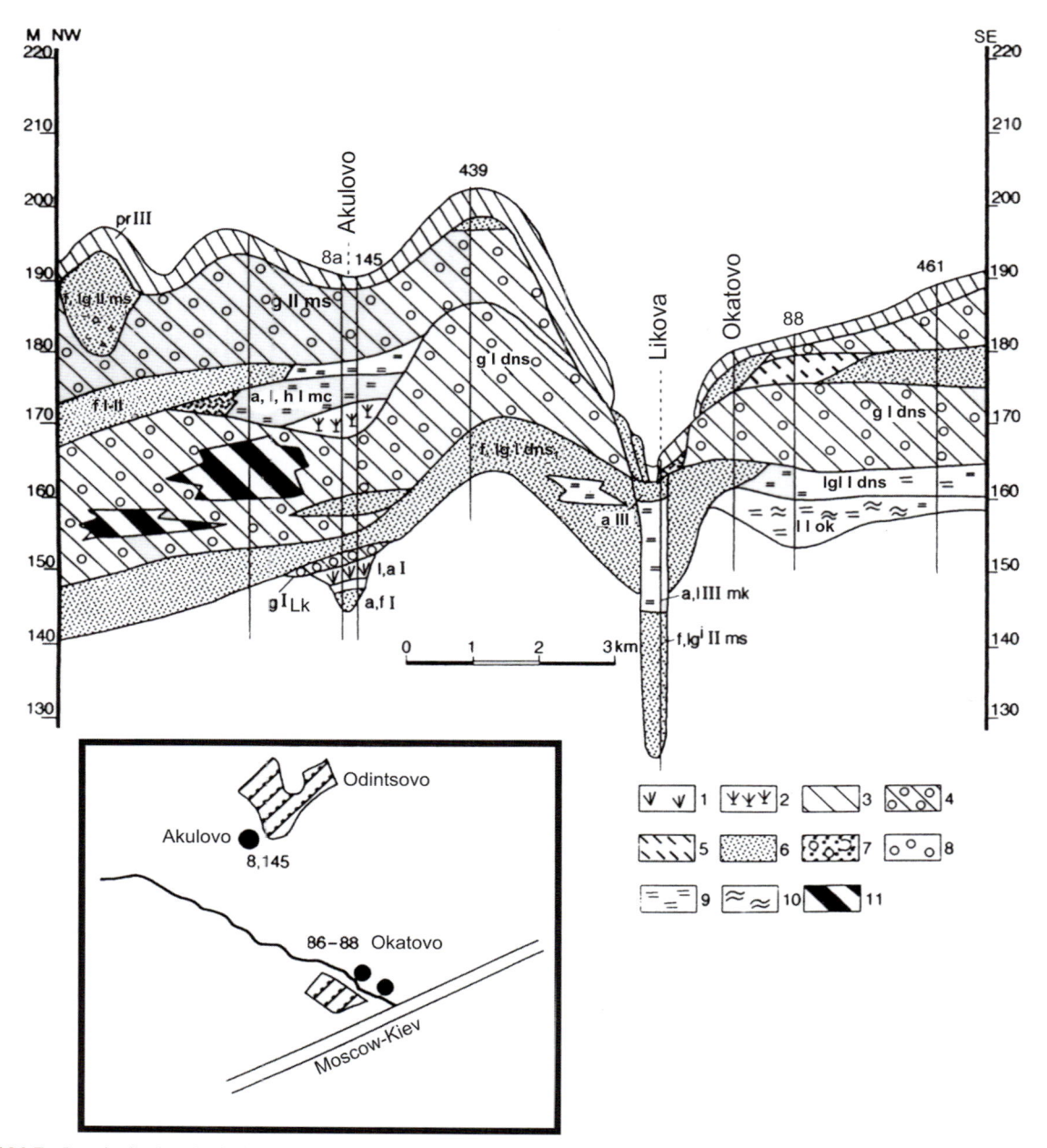

FIGURE 26.7 Longitudinal geological section from Akulovo to Okatovo Village (Fursikova et al., 1992). 1, gyttja, peat; 2, plant remains; 3, loam; 4, till; 5, sandy loam; 6, sand, silt; 7, sand with pebbles; 8, gravel and pebble; 9, silty clay; 10, clay; 11, detached blocks.

26.8. MUCHKAP INTERGLACIAL

The sediments of the Muchkap Interglacial have been identified in the Dnieper basin, in the Smolensk region (Biryukov et al., 1992), Don basin (sites Demshinsk, Pol'noe Lapino, Zapadnaya Starinka, Losino, Preobrazhen'e, etc.) (Iosifova et al., 2006) and at the Akulovo site, in the Moscow region (Maudina et al., 1985).

This Muchkap Interglacial is characterised by two climatic optima that alternated with cooler climate phases. The two optima (Glazov and Konakhovka) were recognised in the sections within the Roslavl' region near Konakhovka Village, in the upper Dnieper drainage basin (Figs. 26.11 and 26.12).

During the first (Glazov) optimum, broad-leaved forests occurred in the area between 59°N and 51°N. Mild temperate species as *Juglans*, *Pterocarya* and *Carya* grew in the upper Dnieper basin. *Pterocarya* even occurred at the latitude of Moscow. Whereas in the second (Konakhovka) optimum, *Quercus* and *Carpinus* forests grew in the upper Dnieper basin and forests of *Pinus* and *Picea* with *Carpinus* were common near Moscow. Pollen spectra and plant

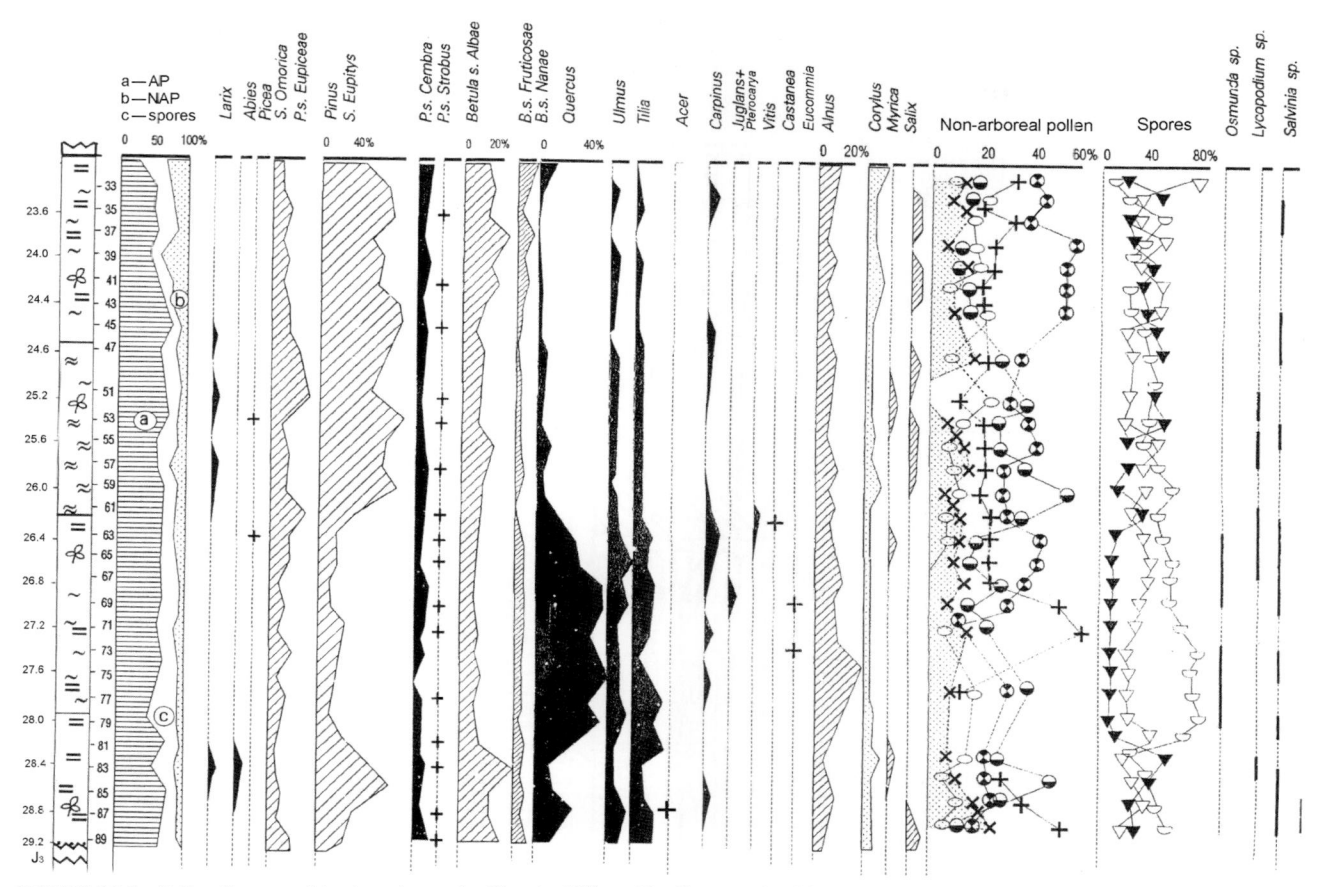

Chapter | 26 Glaciations of the East European Plain: Distribution and Chronology

FIGURE 26.8 Pollen diagram of the deposits nearby Okatovo Village (Fursikova et al., 1992). For symbols of grass pollen and spores, see Fig. 26.14.

macrofossil remains from the Muchkap = Roslavl' sediments at Konakhovka (near Roslavl') and Sergeevka (both in the upper Dnieper basin) have yielded taxa indicative of their antiquity including *Ligustrina amurensis* Rupr., *Pterocarya* sp., *Juglans* sp., *Carya* sp., *Tilia* cf. *amurensis* Rupr. and *Woodsia* cf. *manchuriensis* Hook (Fig. 26.12). Generally, the Muchkap = Roslavl' sediments correlate well with those described from the Ferdynandow section in Poland by Janczyk-Kopikowa (1975).

Both climatic optima can be correlated with 'Interglacial III' of the Netherlands' Cromerian Complex Stage. The cooling between the optima was accompanied by a partial decline of forests; however, no traces of glaciation have been identified during this phase.

Thus the relationship of the Roslavl' to Likhvin (Holsteinian) floras can be clarified: the Roslavl' sediments occur intermediate between the Don and Oka tills, whereas the Likhvin interglacial deposits overlie the Oka Till (Fig. 26.13). In the loess–palaeosol sequences of the periglacial regions, the Muchkap Interglacial is represented by phases of the Vorona soil. It has been extensively investigated in a number of key sections within the Don and Dnieper basins using palaeopedological (Velichko et al., 2007) and palaeofaunal methods (Markova, 1982; Agadjanian, 1992).

26.9. IKORETS INTERGLACIAL

The Ikorets unit, that is nowadays considered to represent an additional interglacial, is recognised in the interval preceding the Oka (Elsterian) glaciation. The so-called Ikorets Interglacial has been distinguished on the basis of faunal studies from the Mastyuzhenka section, in the Ikorets River basin, in the Voronezh region (Iosifova et al., 2009). In this section, the bone-bearing alluvium is overlain by a buried soil interpreted as Inzhavino (Likhvin, Holsteinian) and which includes traces of cryogenic deformations. According to Agadjanian, the faunal assemblage includes *M. oeconomus* Pallas, *Microtus* sp., *Sorex* sp., *Microtus (Stenocranius) gregalis* Pallas and other taxa and is dominated by the archaic *Arvicola mosbachensis* Schmidtgen that first occurred in the Russian Plain (Fig. 26.9). Agadjanian placed this assemblage between the Tiraspolian and Singil faunas and correlated it on the basis of its evolutionary advancement with the type *Arvicola* population from Mosbach, in Germany. The sediments in the Shekhman' section

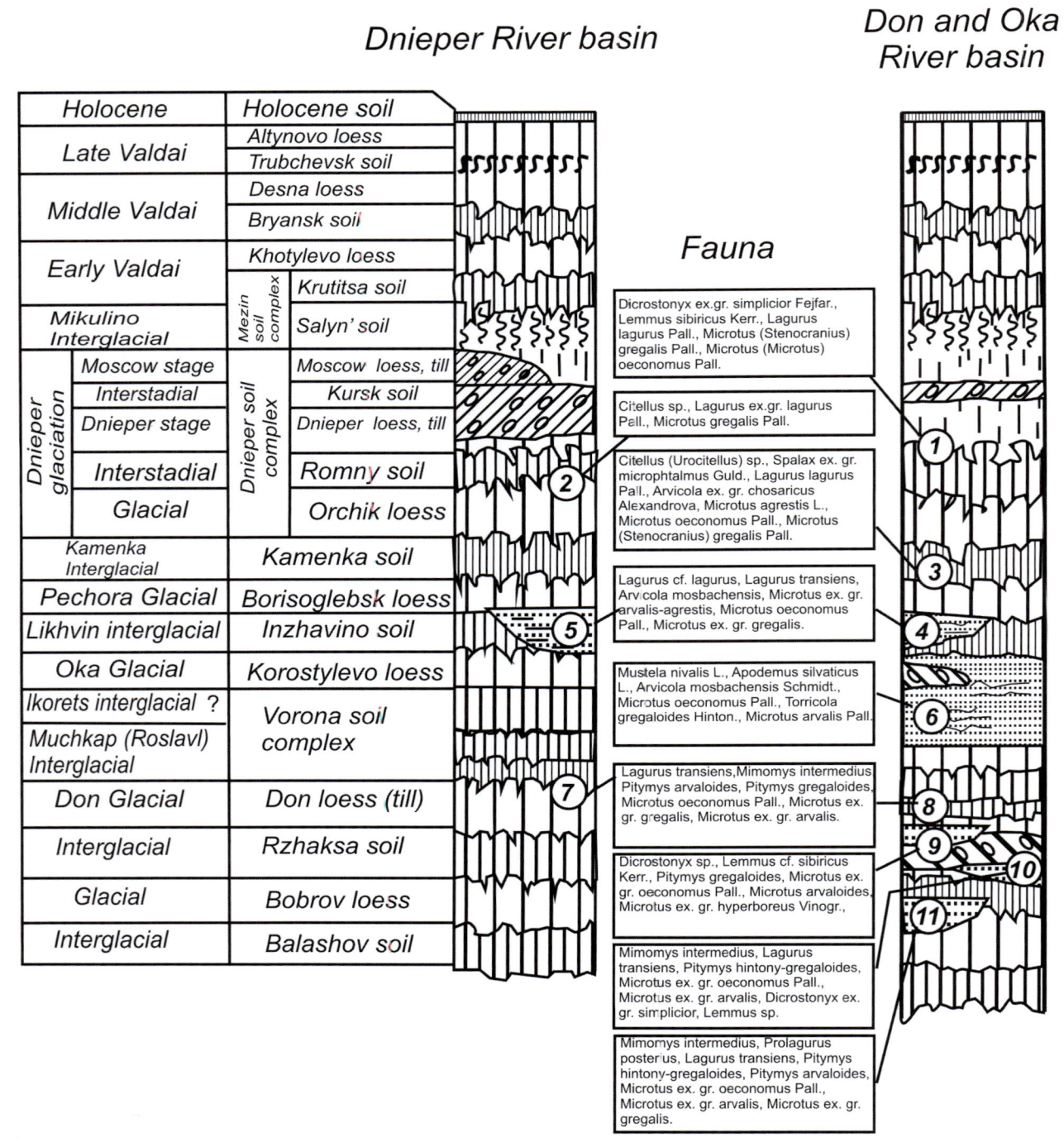

FIGURE 26.9 Correlation between the glacial and interglacial units and stratigraphical position of tills in the Dnieper and Don River basins. 1, Alpat'evo, Troitsa; 2, Priluki; 3, Rasskazovo; 4, Verkhnyaya Emancha; 5, Gunki; 6, Mastyuzhenka; 7, Roslavl; 8, Posevkino, Perevoz, Muchkap; 9, Troitsa; 10, Bogdanovka; 11, Klepki (from Velichko et al., 2004; Iosifova et al., 2009).

south of Michurinsk town, Tambov region, are probably also associated with this interval (Iosifova et al., 2006).

Palaeobotanical evidence from the Ikorets sediments is still fragmentary, but they indicate the expansion of forest vegetation in the upper Don basin during the event.

26.10. OKA GLACIATION

The Oka Glaciation can be correlated with the Elsterian Stage of central Europe. From about 500 to 460 ka BP, the Oka ice sheet extended as far south as the Oka River basin. However, its precise boundaries are not defined yet.

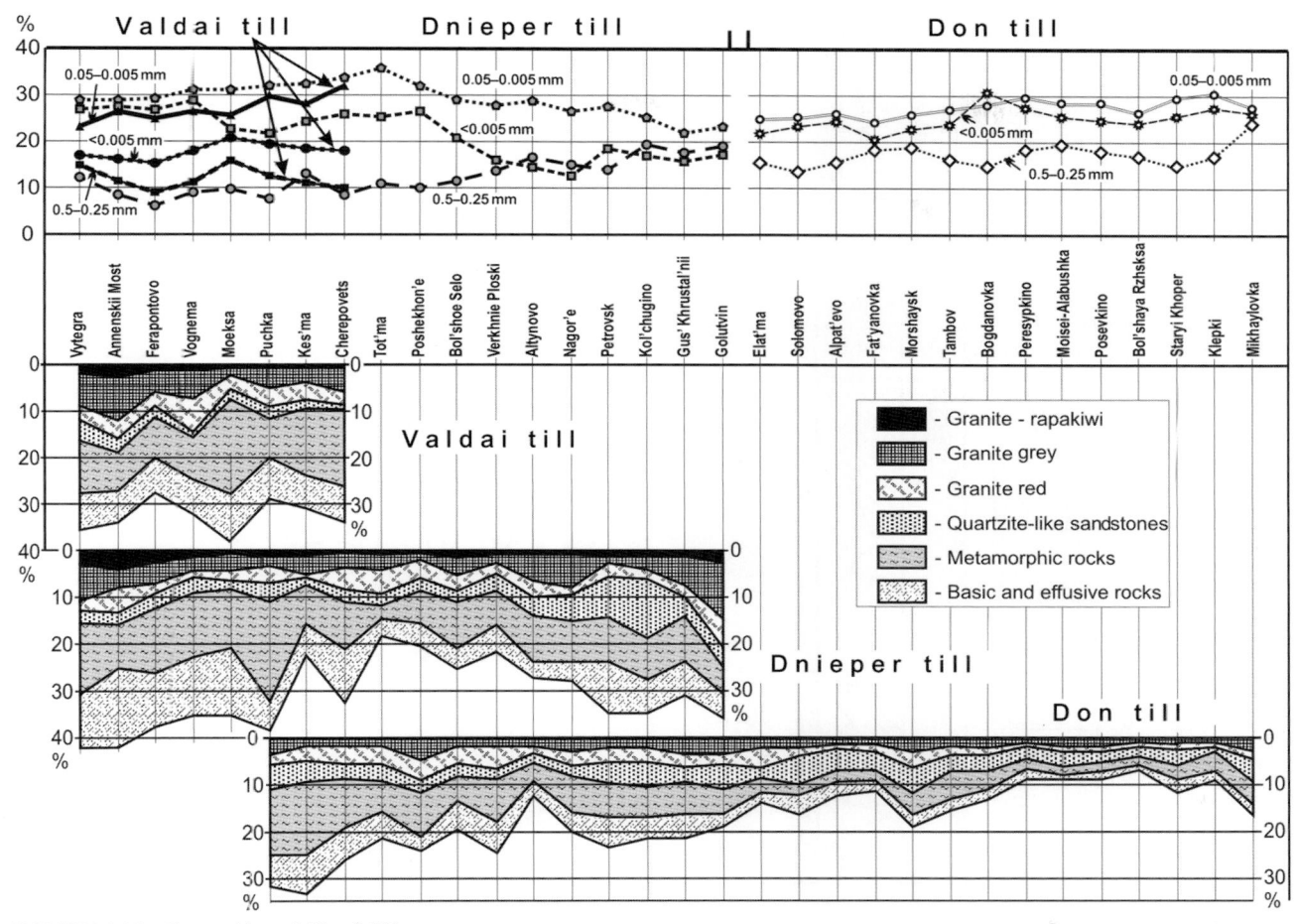

Chapter | 26 Glaciations of the East European Plain: Distribution and Chronology 347

FIGURE 26.10 Composition of tills of different ages.

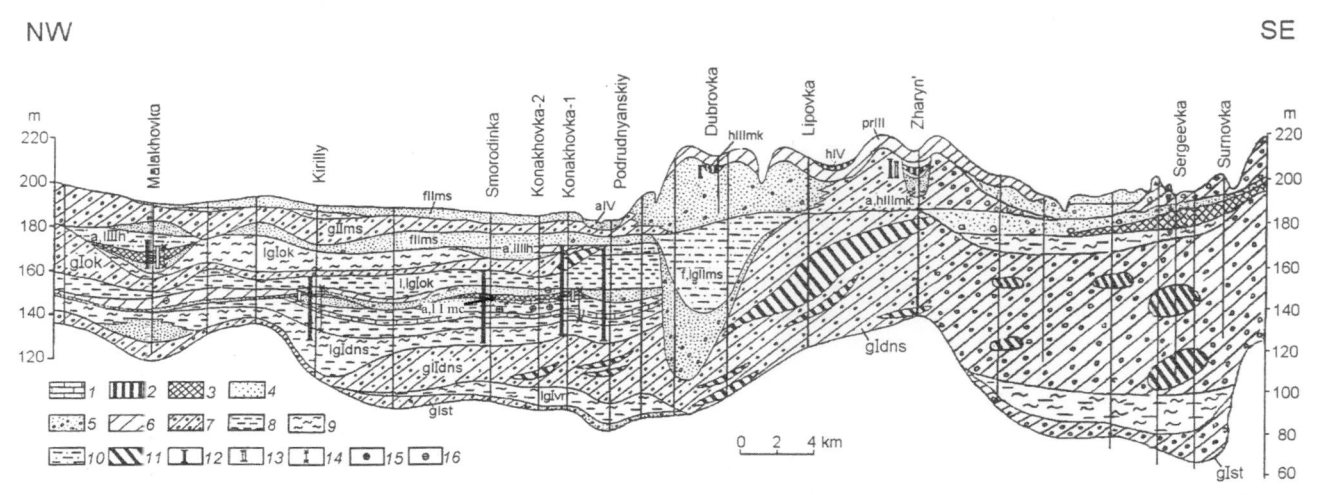

FIGURE 26.11 Longitudinal geological section of the Roslavl area (Konakhovka and Sergeevka sites). Compiled by V.V. Pisareva after Biryukov (1992) and Zarrina (1991). 1, marl; 2, peat; 3, gyttja; 4, sand; 5, sand with pebbles; 6, loam; 7, till; 8, glaciolacustrine clay; 9, glaciolacustrine loam; 10, sandy loam; 11, detached blocks; 12, position of palynological samples; 13, position of carpological samples; 14, position of samples for diatom analysis; 15, thermophilic small mammals; 16, cryophylic small mammals.

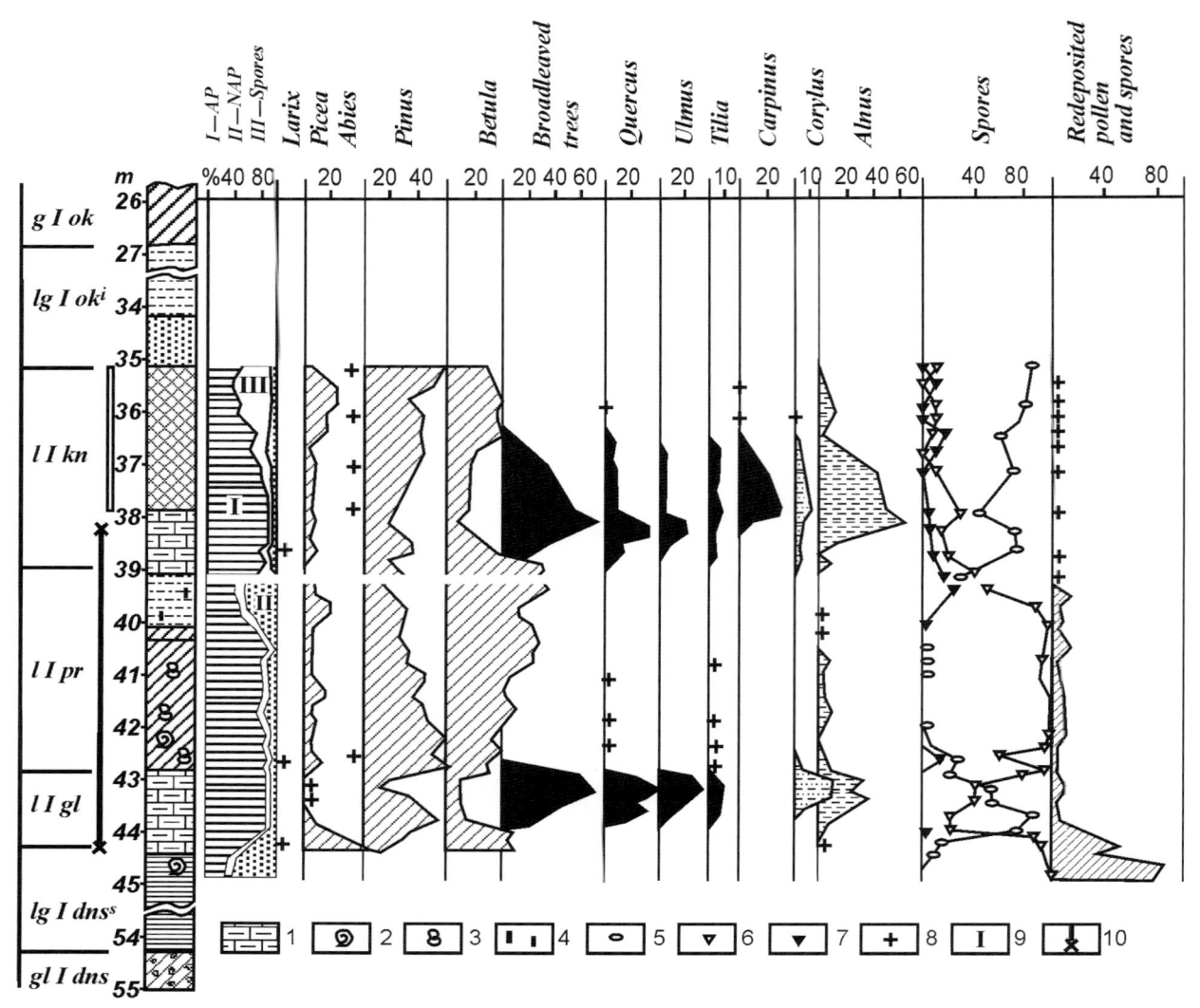

FIGURE 26.12 Pollen diagram of the Muchcap deposits nearby Konakhovka Village in the Roslavl stratotype area (Shick et al., 2006). 1, marl; 2, shells of fresh-water mollusks; 3, vivianite; 4, peat inclusions; 5, Polypodiaceae; 6, Bryales; 7, Sphagnales; 8, rare grains; 9, deposits, where fossil plants were studied; 10, deposits, where the diatoms were studied. Indices: 1, gl dns—Don Till; 2, lgl dns^s—glaciolacustrine clays dated to the Don ice retreat; 3, l I gl- Glazov optimum lacustrine deposits; 4, l I pr- Podrudnyanski cooling lacustrine deposits; 5, l I kn Konakhovka optimum lacustrine deposits; 6, lg I ok— glaciolacustrine and glaciofluvial deposits associated with the advance of the Oka ice sheet; 7, gl ok—detached block of the Oka Till. For symbols of grass pollen and spores, see Fig. 26.14.

Unfortunately, the central regions of the East European Plain generally lack sections where the Oka Till can be unambiguously demonstrated by its relation to overlying and underlying glacial and interglacial strata associated with loess and fossil soils. In a few sections where the till is present, its age is interpreted from its lithology, which is closely similar to that of the other older tills. It is grey, greenish-grey or greyish-brown and contains gravel of local provenance together with rare clasts of exotic rocks. The till can be identified by its relatively low hornblende content (Sudakova, 1990).

Judging from the clast orientation, the ice moved into the East European Plain from north to south. Exotic rock fragments are more abundant in the Oka Till than in the Don Till. Sedimentary rocks are dominant, together with an admixture of erratic boulders of granite, gneiss and igneous rocks. The most typical heavy mineral association is garnet–hornblende. The Oka Till has been most reliably identified in the Upper Dnieper and Upper Volga basins; in the sections of Malakhovka (Smolensk region), Mar'ino and Pan'kovo (north of Moscow), the till grades into glaciolacustrine and glaciofluvial sediments and then into lacustrine and paludal deposits. The severe climate of the Oka Cold Stage is suggested by finds of the bones of cold-tolerant animals (such as *Dicrostonyx simplicior okaensis* Alexandrova) found in the Chekalin section on

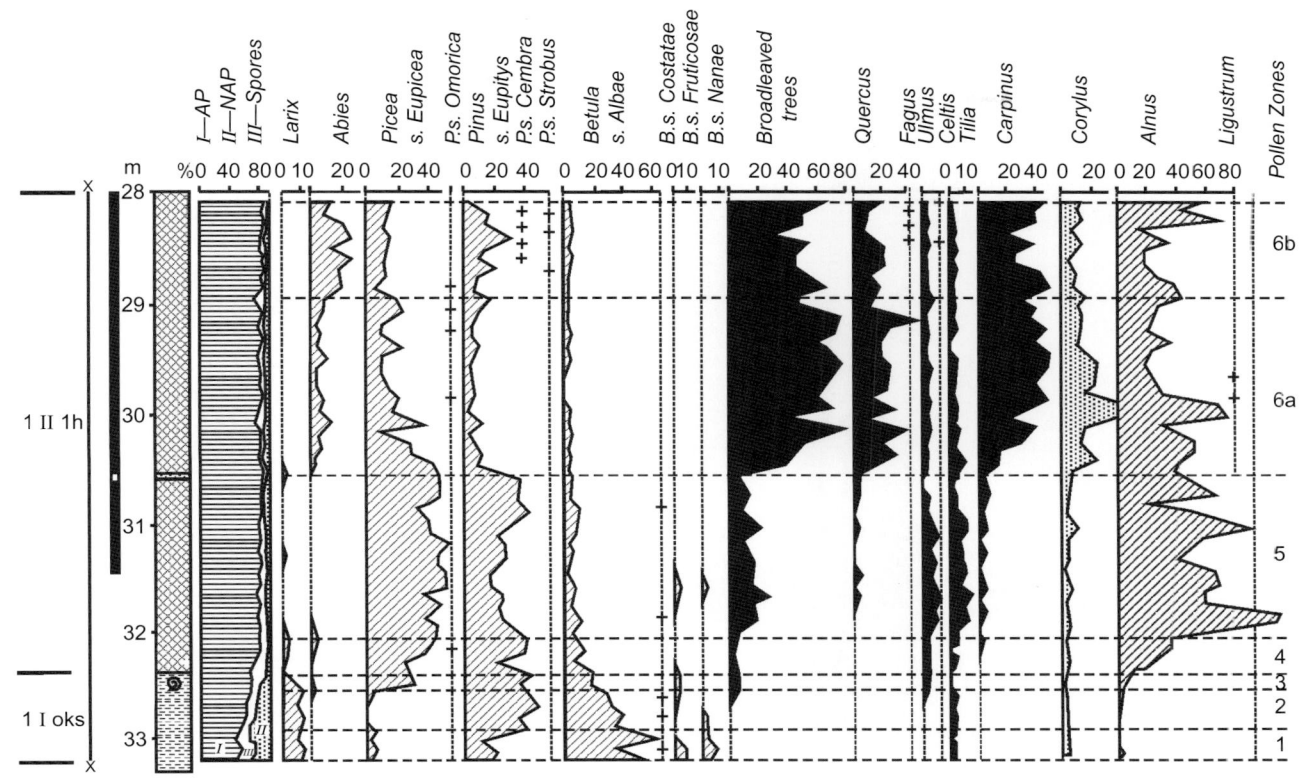

FIGURE 26.13 Pollen diagram of the Likhvin deposits nearby Malakhovka Village in the Roslavl stratotype area (Shick et al., 2006). Indices: 1, lgl ok^s—glaciolacustrine deposits of the time of Oka ice retreat and 2, l II lh—likhvin interglacial lacustrine deposits (by V. V. Pisareva).

the Oka River (Aleksandrova, 1982) and in the Mikhailov quarry in the Kursk region (Agadjanian, 1992). This is reinforced by finds of the teeth of *Microtus* ex gr. *hyperboreus* Vinog at Bogdanovka in the Upper Don basin (Markova, 1982). At present, this vole inhabits tundra and forest-tundra.

26.11. LIKHVIN INTERGLACIAL

The Likhvin Interglacial, that followed the Oka Glaciation, is correlated closely with the Holsteinian Interglacial of central Europe. The interglacial deposits are exposed in the Chekalin stratotype and in other type sections at Yakovlevskoe and Rybinsk in the Yaroslavl' region, at Malakhovka in the Smolensk region (Figs. 26.13 and 26.14), at Mar'ino and Pan'kovo near Moscow, Nyaravai in Lithuania (Voznyachuk et al., 1984), Strelitsa on the Don River (Krasnenkov et al., 1997), Gun'ki in the Dnieper basin (Velichko et al., 1997a), Ozernoe Village in the north-western Black Sea region (Mikhailesku et al., 1991) and others.

Comprehensive faunal and floral evidence from these deposits have been published (Markov, 1977; Grichuk, 1989; Pisareva, 1997). The Likhvin fauna includes archaic *Arvicola mosbachensis* Schmidtgen, which replaced *Mimomys intermedius* Newton. The wealth of evidence permits not only the reconstruction of the interglacial climate and vegetation but also the understanding of the landscape zone dynamics (Grichuk, 1989). The whole area under consideration, including the southernmost regions, was positioned within the forest zone, although the forest differed in composition. South of the latitude of Moscow a mixed coniferous–broad-leaved forest grew, in which first oak and hornbeam and later hornbeam and fir assemblages dominated. Increased rainfall enabled the forests to expand into the Don basin. The temperate, slightly oceanic climate favoured the continued presence of relict plants, such as *Taxus*, *Ilex*, *Castanea*, *Buxus*, *Pterocarya* and *Fagus*; their pollen have been recovered from the Likhvin stratotype section at Chekalin (Grichuk, 1989). Many of the taxa occurred occasionally at the latitude of Moscow but were completely absent further north. The composite spruce forests with *Quercus*, *Carpinus* and *Abies* occurred there.

26.12. PECHORA GLACIATION

The transition from the Likhvin Interglacial to the subsequent glaciation was accompanied by a cooling represented in a number of sections of the Upper Volga, Oka, Dnieper

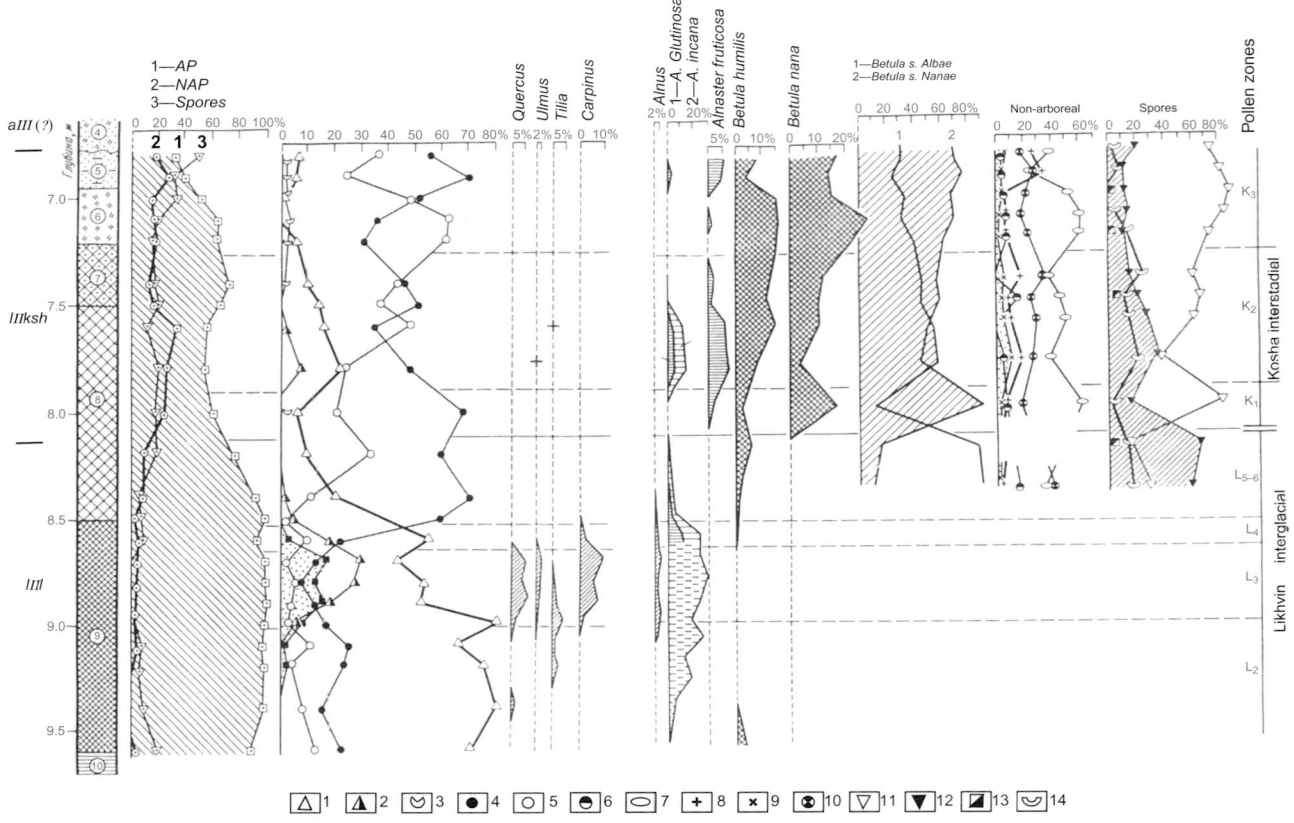

FIGURE 26.14 Pollen diagram of the Likhvin interglacial and Kosha interstadial deposits on the Bolshaya Kosha River (Grichuk, 1989). 1, Picea; 2, Abies; 3, Larix; 4, Pinus; 5, Betula; 6, Gramineae; 7, Cyperaceae; 8, Artemisia; 9, Chenopodiaceae; 10, Varia; 11, Bryales; 12, Sphagnales; 13, Lycopodiaceae; 14, Polypodiaceae.

and Dniester basins. It was repeatedly interrupted by warm-events of interstadial rank. During the Koshin warming, first recognised in the section on the Bol'shaya Kosha River (Volga River tributary) in the Tver' region (Grichuk, 1989), plant associations similar to those of the middle taiga forests of Eastern Europe extended to the Upper Volga basin (Fig. 26.14). The Dnieper and Dniester basins were covered by south taiga forests.

The two following interstadials, that is, the Bulatovo and Mar'ino, recorded in the sections of the Tver' and northern Moscow regions, are correlated with the Hoogeveen and Bantega interstadials of the Netherlands (Zagwijn, 1985). These events were characterised by more continental climate and occurrence of forests similar in composition to the middle and north taiga forests of western Siberia.

These interstadials were followed by the substantial Kaluga cool event, known from the central East European Plain (Markov, 1977). It is not inconceivable that it should be associated with the Pechora advance.

The Middle Pleistocene sequence of the northern East European Plain suggests three large glacial phases could have occurred here; two or even three till units can be distinguished in the composite geological transects across the East European Plain from north-west to south-east. The units differ considerably in their lithological characteristics.

A distinctive feature of the till, associated with the Pechora advance (after the drainage basin of the Pechora River) is a high content of local rocks and minerals in both central and north-eastern regions. In the latter, the till contains sedimentary and metamorphic rocks including agate-bearing basalts, derived from the Urals and Timan. Among the minerals, glauconite, sulphides, siderite and others of the local and transit provenances are recorded. An ilmenite-garnet heavy mineral association predominates and is characterised by a remarkable presence of epidote and a relatively low content of hornblende (less than 20% of the total assemblage). The clast orientation in the till suggests that the ice moved southwards in the western regions and south-eastwards in the central and south-eastern areas. An extensive ice sheet that formed three large ice streams (Belomorsky, Cheshsko-Mezensky and Pomorsky) moved into the East European Plain from the Scandinavian and Ural–Novaya Zemlya glacial centres. Judging from the indicator clast composition, the ice streams were related to different source areas. The Pomorsky stream flowed from the Pai-Khoi and Novaya Zemlya provinces (indicators are

Silurian bituminous limestones and dolomite with coral fauna, as well as Permian and Triassic polymict sandstones). The source area of the Cheshsko-Mezensky stream lay primarily in Novaya Zemlya, as indicated by boulders of rose marble-like crinoid–bryozoan Ordovician limestones. As for the chronological correlation of the Pechora Till, there is still disagreement among specialists. Some researchers (e.g. Andreicheva et al., 1997) correlate it with the Dnieper (Saale) Till, while others object to such long-distance correlations.

The present authors (except N.G. Sudakova) can accept that the Pechora Till could be older than the Dnieper Till of the central regions, considering the possibility of heterochroneity in the development of the Scandinavian and Kara–Novaya Zemlya glacial centres. The Middle Pleistocene age of the Pechora till has been demonstrated by the fauna and flora remains in both underlying and overlying sediments.

26.13. KAMENKA INTERGLACIAL

The Kamenka Interglacial occurred during the interval between the Pechora and Dnieper glaciations. In the country subjected to the Pechora Glaciation, alluvial, lacustrine and paludal sediments were accumulated and a fossil soil was formed in the periglacial zone. The chronostratigraphical position of the Kamenka deposits has been discussed in the monograph *Middle Pleistocene Glaciations* (Velichko and Shick, 2001) and other publications (Pisareva, 1997; Arslanov et al., 2005; Velichko et al., 2005, 2007). In the stratotype section, near Chekalin town on the Oka River, the Chekalin soil (an analogue of the Kamenka soil) is exposed above the Likhvinian sediments (Bolikhovskaya, 1995). The Kamenka soil occurs in the same stratigraphical position in the Narovatovo section on the Moksha River in Mordovia (Runkov et al., 1993). The Uranium–Thorium (U–Th) age of the Kamenka deposits is about 200 ka BP (Arslanov et al., 2005), which corresponds to marine isotope stage (MIS) 7.

In addition to the information from the exposures mentioned above, the vegetational development during this phase in the central East European Plain has been revealed by palaeobotanical studies of the Lipna village section in the Klyaz'ma River basin, Moscow region (Zarrina, 1991), Tyaglitsy and Pal'nikovo in the Tver' region (Shick et al., 2006), Gorka and Rylovo in the Vologda region (Zarrina and Shick, 2000) and others.

According to the palaeobotanical records, all the regions where the sediments of the second Middle Pleistocene Kamenka Interglacial have been recorded, a single zone of mixed coniferous–broad-leaved forests occurred. The fossil dendroflora, together with the widely distributed taxa, included the arboreal species that are absent today from this region. They include *Larix* cf. *decidua*, *Abies* cf. *alba*, *Pinus* s. *Cembra*, *P*. s. *Strobus*, *Picea* s. *Omorica*, *Betula* s. *Costatae*, *Tilia platyphyllos*, *T. tomentosa* and *Carpinus betulus*. Judging from the floral composition, the Kamenka Interglacial was somewhat cooler than the Likhvin time.

26.14. DNIEPER GLACIAL EPOCH

In the East European Plain, there is one glacial complex that belongs to the latter part of the Middle Pleistocene. Its stratigraphical position has been defined by its occurrence between the Kamenka soils and the Mezin soil complex. The earlier part of the latter is attributed to the Mikulino Interglacial.

26.14.1. Dnieper Stage (of the Dnieper Glaciation)

The Dnieper complex can be traced south–westwards into the Upper Dnieper and Upper Oka drainage basins. The age of the till is indicated by the presence of a lemming fauna in the lacustrine deposits underlying the till in the Likhvin stratotype section, nearby Chekalin town on the Oka River (Markov, 1977).

Further north, the Dnieper Till is the second from the ground surface. There the Middle Pleistocene glacial complex often includes thick intermorainic sediments that allow the differentiation of two glacial units. At Rybinsk, in the Yaroslavl' region, the Dnieper Till also overlies the lacustrine silt bearing the analogous lemming fauna of the same evolutionary level as that found in the Likhvin section (Markov, 1977; Borodin et al., 1981).

In some central and northern areas of the plain, the uppermost till unit differs considerably from that beneath in both its lithological and mineralogical characteristics (Velichko and Shick, 2001). The difference is strongly pronounced in the whole granulometric spectrum of the till, from pebbles and boulders to the clay fraction (Fig. 26.10).

The ice sheet that deposited the lower till of the Dnieper Glacial complex extended to the East European Plain after the beginning of MIS 6 (about 200 ka). At that time, the source provenances were engaged, which resulted in the increase of Scandinavian components in the till mineralogical spectra. The ice sheet advanced in five major ice streams, the Baltic, Ladozhsky, Onezhsky, Belomorsky and Kolsky.

26.14.2. Moscow Stage (of the Dnieper Glaciation)

This ice began to draw on other source areas, which were strongly connected with the Scandinavian glacier centre. It advanced on the East European Plain about 190 ka ago, following a relatively prolonged and warm interval.

Whereas during the accumulation of the Dnieper Till the ice moved from the north-west to the south-east, in the course of formation of the Moscow glacial formation, the glacial centre was shifted westwards. The outermost limit of the ice sheet can be traced from the Dnieper basin into the drainage basins of the Upper Volga, Kama and Vyatka rivers. In the lower reaches Pechora and upper Vychegda rivers, the till corresponding to the Moscow unit is referred to as the Vychegda Till (Fig. 26.3).

The marginal landforms of the principal (Dnieper) ice advance were essentially destroyed in the central region of the Russian Plain by the subsequent Moscow advance. The latter left behind a well-defined assemblage of landforms including ice-marginal and radial ridges, such as the Smolensk–Roslavl', Klin–Dmitrov, Tver', Galich–Chukhloma and others. Frontal ice-marginal features, forming at least ten end moraine belts, mark the stages of the ice retreat. They often include push moraines with detached blocks and glaciotectonic dislocations. The radial landforms are highly diverse in structure and morphology and include interstream, interlobate and intertongue hilly massifs and ridges, as well as systems of ridges formed within the ice-divide and ice-contact zones. Many researchers have pointed out that the ice sheet was active even during deglaciation. In fact, landforms of the active ice are more abundant than in the last glaciation area (Goretskii et al., 1982). During the Moscow ice retreat, numerous temporary meltwater drainage valleys and large proglacial lakes were formed, most of which no longer exist.

Since the Dnieper and Moscow ice advances differed in dynamics and direction of ice flow, they might have been separated by a considerable time interval. However, the studies of organic deposits interstratified with the till units in the glacial regions, and the correlative sediments in the periglacial areas, have not revealed interglacial sediments in this chronostratigraphical interval. Some researchers refer the typical interstadial deposits underlying the Moscow Till (in the Kostroma and Yaroslavl' Volga region, Moscow region and Belarus) to the second half of the Middle Pleistocene. The available palaeobotanical records indicate the occurrence of boreal forest vegetation and therefore a considerable ice retreat at that time. These data suggest that during the climatic optimum of the interstadial separating the Dnieper and Moscow Stages, an open spruce forests with the admixture of fir, Siberian pine, Scots pine and birch colonised the region. In the Upper Volga basin, the soil bed froze as indicated by finds of *Alnaster fruticosus* pollen. However, according to Sudakova et al. (2007), the Moscow ice sheet was distinctive and was preceded by a warming similar to an interglacial interval.

Further southwards, in the Dnieper basin, broad-leaved trees and hazel occurred infrequently. In the terminal phase of the interstadial, the area was covered by open birch woodland, but *Artemisia*/Chenopodiaceae associations were common.

26.15. MIKULINO INTERGLACIAL

During the Mikulino Interglacial, the 'Boreal Transgression' flooded the northern part of the East European Plain. Numerous boreholes penetrated its sediments that contain a Lusitanian mollusc fauna in the Kola Peninsula, Finnish Gulf coast, in Karelia, in the lower Severnaya Dvina, Mezen' and Pechora River basins.

The Mikulino Interglacial (=Eemian Stage of central Europe) vegetation has been thoroughly studied. Palaeobotanical remains recovered from about 150 sections allow the identification of seven vegetation phases in the central part of the East European Plain, from pine forests at the beginning to broad-leaved forests in the optimum and later to mixed pine-birch forests at the end of the interglacial. Most of the evidence suggests the occurrence of a single climatic optimum during this stage (Fig. 26.15).

The soils of the Mikulino Interglacial were formed in the forest zone of the Subboreal belt. Their modern analogues are probably the texturally differentiated lessivé soils of the northern Central and Middle Europe (Velichko et al., 2007). Global warming during the Mikulino Interglacial resulted in open birch and pine woodlands expanding into the Pai-Khoi, Bolshezemelskaya and Malozemelskaya tundras, and birch forests with pine and spruce developed in the Kola Peninsula. The Onega, Severnaya Dvina and Mezen' drainage basins were colonised by spruce and birch forests, with oak and elm. Similar forests with hornbeam occupied northern Karelia and partially the Vetluga drainage basin, whereas the northern limit of broad-leaved forests was north of Vologda towards the upper Unzha River (left tributary of the Volga River). The Saratov region east of the Volga River favoured the forest-steppe landscapes, where nowadays only grassland steppe occurs.

26.16. VALDAI GLACIATION

The events during the Valdai (Weichselian) glaciation are the best studied, especially those of its second half. In the early Valdai (MIS 5a–d), several cool and warm intervals have been demonstrated on the basis of chronometric, palaeobotanical, and geological evidence (Figs. 26.16 and 26.17).

The first post-Mikulino cooling (ca. MIS 5d, about 110 ka ago) corresponded to a glaciation limited to the mountain regions of Scandinavia (Baumann et al., 1995; Thiede and Bauch, 1999), whereas south of the Baltic Shield, in north-western Russia, a progressive cooling strongly changed the landscapes of the terminal Mikulino. Open birch forests, alternating with boggy areas of dwarf birch and shrub alder, became widespread (Grichuk, 1989; Borisova and Faustova, 1994; Pisareva, 2001). The subsequent warming (the Upper Volga) was the most pronounced in the early Valdai and is correlated with the

Chapter | 26 Glaciations of the East European Plain: Distribution and Chronology

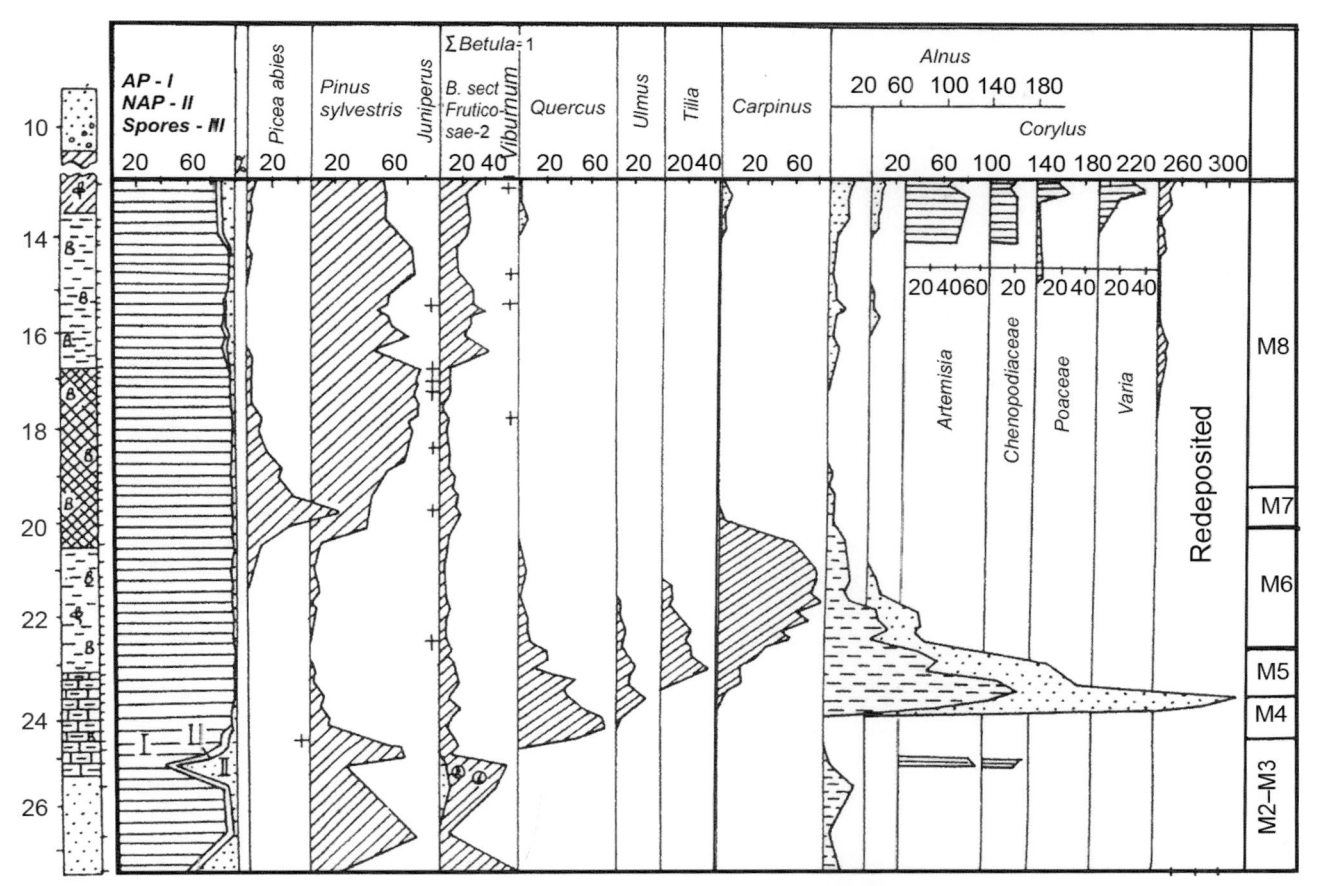

FIGURE 26.15 Pollen diagram of the Mikulino interglacial deposits near Podrudnyanskiy site in the Roslavl stratotype area (Rumyantsev, 1998) M 2–M 8, Polinizones after Grichuk (1989).

Brørup Interstadial and associated warming events corresponding to the terminal MIS 5c. At that time, the extraglacial zone, south of the Baltic Sea depression (Krutitsa interstadial), was covered with coniferous and, in the Dnieper and Upper Volga regions, with pine forests with an admixture of spruce (Velichko, 1999).

The following cooling (MIS 5b), although was considerably strong, did not result in a glaciation beyond the Baltic Shield as might be suggested. According to the latest information, it was associated with local ice cover in mountain regions of the Kola Peninsula. The early Valdai glacial sediments have been described from its south-western part, in the Kovdor vicinities and Lovozerskii massif foothills, where a small ice sheet is thought to have developed (Yevzerov, 2005). Beyond the Baltic Shield, the cooling resulted in wide distributed birch–pine forests alternating with herbaceous areas that graded eastwards into open birch forests with common dwarf-shrub formations (Borisova and Faustova, 1994; Zarrina and Shick, 2000; Velichko, 2009a). At the end of the early Valdai (MIS 5a), the climate become slightly milder. However, beginning with 70 ka ago, that is, in MIS 4, the Scandinavian glaciers again extended south- and south-eastwards beyond the mountain regions (Baumann et al., 1995; Thiede and Bauch, 1999).

For that time, 70–65 ka ago, Larsen et al. (2006) and Demidov et al. (2006) have proposed that traces of the ice advance from the shelf and Timan Peninsula when an ice cover occurred in the White Sea mouth and Mezen' basin.

At the MIS 4/3 boundary, the ice again advanced to the Kola Peninsula. This advance is indicated by the lower till in the sections bearing two tills with the upper one being late Valdai in age (Yevzerov, 2005).

The Late Pleistocene tills often differ in clast composition (Faustova and Gribchenko, 1995). The lower till is distinguished by the predominance of rocks derived from the Belomorsk Formation (garnet-bearing granite–gneiss and amphibolites). Rare pebbles of Kola nepheline syenites are present as well.

The ice lobes did not extend beyond Karelia and the Baltic Sea depression. This is indicated, for instance, by the evidence from a number of sections in the north-western East European Plain where the Mikulino sediments are overlain only by the late Valdai Till, which has been noted

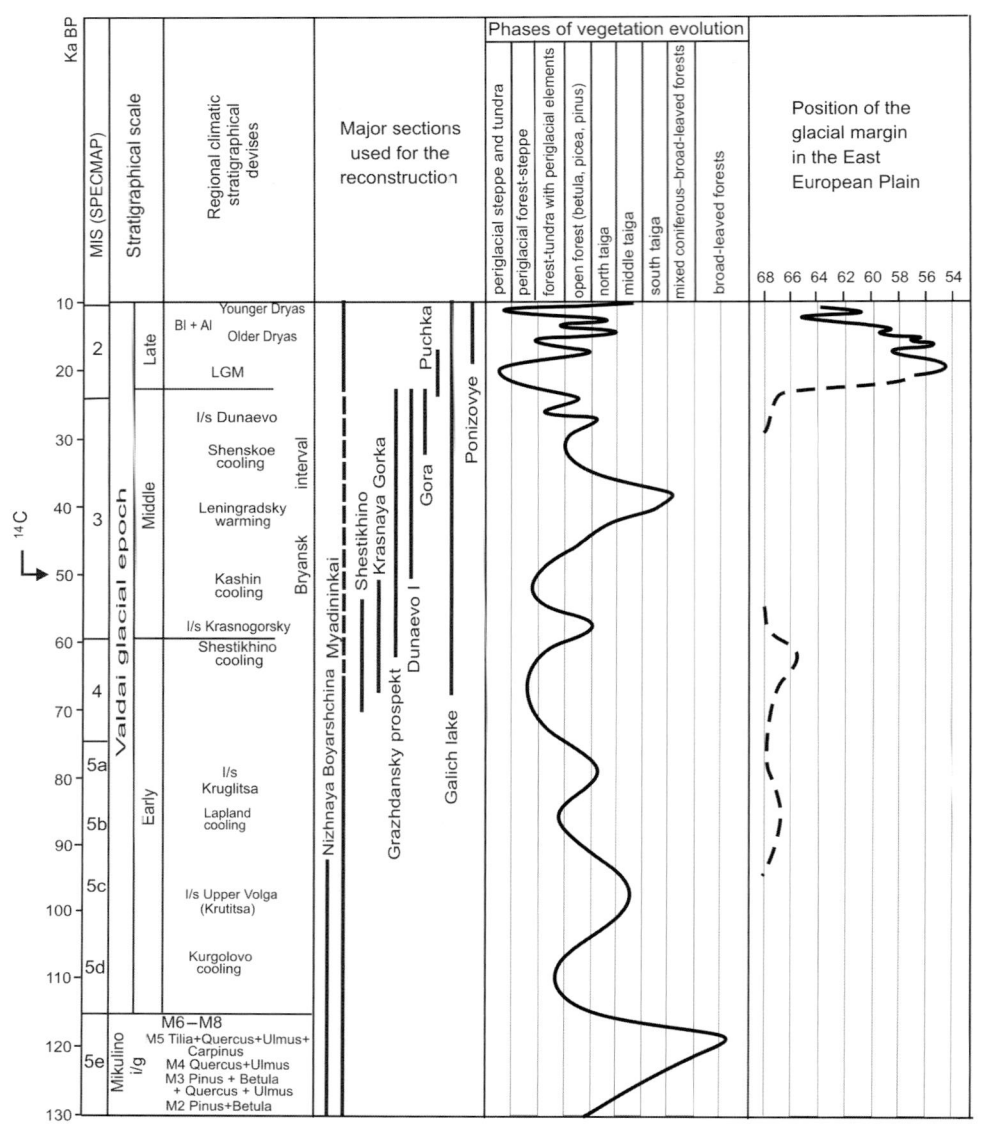

FIGURE 26.16 Changes of the vegetation cover and glacial extension in the East European Plain during the past 130 000 years (Velichko, 2009a,b, with additions).

over the past 20 years (San'ko, 1987; Gei and Malakhovskii, 1998; Gei et al., 2000a,b).

The cooling in the periglacial zone of north-western and central Russia resulted in distribution of tundra and steppe associations with patches of birch forests. Some sections are marked by freezing processes in the form of cryoturbation and small wedges.

The period from 58 to 25 ka ago, corresponding to MIS 3, and designated in Eastern Europe as 'megainterstadial', the 'Leningrad or Bryansk megainterval', is characterised by alternated warm and cool phases (Arslanov et al., 1981; Chebotareva and Makarycheva, 1982; Borisova and Faustova, 1994; Velichko, 1999; Velichko, 2009a,b). During this phase, the general amelioration of the continental climate resulted in glaciation being restricted only to the Scandinavian mountains. Starting from 50 ka BP, most of the sections in the Leningrad, Tver' and Vologda regions display a rhythm of several warming events that have been equated by certain researchers (e.g. Pisareva and Faustova, 2009) with the West European units Moershoofd, Hengelo and Denekamp interstadials, when the open to middle and south taiga forests became established across the region. However, other investigators considered the main warm phase to have occurred within the interval at about 40–34 ka BP and the rhythmic climate changes to be close to that of a interglacial magnitude (Fedorova, 1963, Levkovskaya et al., 2005) or a short interglacial (Grichuk, 1989; Velichko, 2009a,b) when coniferous–broad-leaved forests

Chapter | 26 Glaciations of the East European Plain: Distribution and Chronology

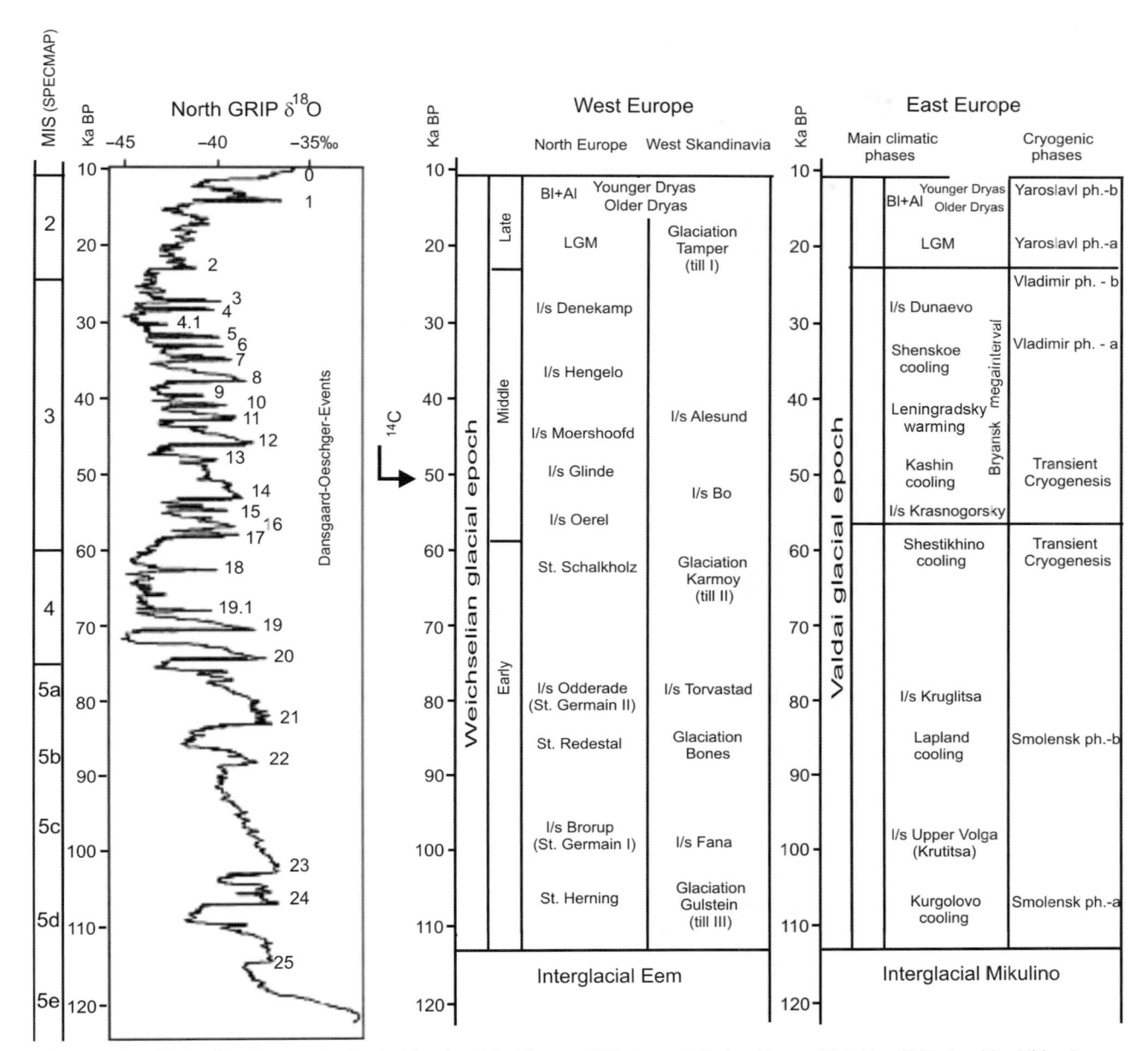

FIGURE 26.17 Major climatic events in the glacial and periglacial areas of Western and Eastern Europe (Velichko, 2009a,b, with additions).

with *Corylus* and *Tilia platyphyllos* are thought to have been established. The terminal Middle Valdai interval corresponds to the Dunaevo warming dated at 31–25 ka. In some sections in the Upper Volga and Sukhona basins, earlier warmings, corresponding to the European Glinde and Oerel intervals, are recorded. Their landscapes were represented by forests similar to the north taiga vegetation (Spiridonova, 1970; Chebotareva and Makarycheva, 1974; Arslanov et al., 1981; Borisova and Faustova, 1994; Pisareva and Faustova, 2008, 2009).

In transition to the last global cooling, the hyperzonal landscapes became widespread and new vegetation types, periglacial–tundra and periglacial–steppe, were formed.

The late Valdai expansion of the ice sheets began 25–23 ka ago. The most favourable conditions for ice advance appeared at the end of the glaciation in the western Eurasia. They resulted in strongest growth of the Scandinavian glacier by the time of maximum cooling (Velichko, 1980, 1999). In recent years, a reasonably reliable reconstruction of the Late Pleistocene glacial system of the Northern Hemisphere has been suggested. The existing disagreement lies in the estimation of the area of shelf glaciation.

The results of international and national research programmes and important data obtained from the drilling on the sea floor and on the land, comprehensive studies of sections, seismo-acoustic profiling, micropaleontological studies and analysis of the glacial isostasy isobases (Velichko and Faustova, 1989; Velichko et al., 1997b, 2000).

According to the proposed reconstruction, the large continental Scandinavian ice cover in the European Arctic sector was characterised by asymmetric slopes and a radial structure that resulted from a series of glacier flows (e.g. Chebotareva, 1977; Gerasimov and Velichko, 1982). Its extension onto the Barents Sea shelf, north of the Kola Peninsula, was limited. In the Russian Arctic sector, separate glacial domes occurred on the Franz Josef Land and on shelf elevations, for instance, on the Central highland (Pavlidis et al., 2005; Velichko, 2009a,b). These ice domes might have been connected to one another for a short time, and in the east, to the local cover of the northern island of Novaya Zemlya. The estimates based on geological, geomorphological and palaeobotanical evidence suggest that the ice sheet occupied an area smaller than the present land area in the archipelago (Krasnozhen et al., 1987; Malyasova and Serebryanny, 1993). No fresh morainic ridges have been found south of the islands, and there is no evidence of the contact between the Scandinavian and Novaya Zemlya ice sheets (neither moraines nor glacial dislocations) (Pavlidis, 1992).

Further eastwards, in the Kara sector of the Arctic, several small glacial domes are interpreted to have formed on the Severnaya Zemlya islands. As suggested previously (Velichko and Faustova, 1989; Velichko et al., 2000), the Kara ice sheet might be developed, not at the glacial maximum but at the beginning of Valdai, which has been subsequently confirmed (Svendsen et al., 2004). The Polar Urals and Taimyr were areas of valley glaciers (Velichko, 2009a,b). From another standpoint (Svendsen et al., 2004), the Barents Sea shelf glaciation during both MIS 4 and the Last Glacial Maximum (LGM) was greater in size.

The Scandinavian ice sheet, controlling the landscape and climatic situation on the European landmass, and the predominance of peculiar periglacial–tundra–steppe landscape beyond the advancing ice (Nazarov, 1984; Velichko, 1993 and others) reached its maximum in the West European sector at about 28–26 ka BP (Boulton et al., 2001) and in East Europe at 24–17 ka BP (Arslanov et al., 1971; Ostanin et al., 1979; Zimenkov, 1985; San'ko, 1987; Zarrina and Shick, 2000; Demidov et al., 2006). At its maximum, this ice sheet approached the upper reaches of the Nieman, Dnieper and Zapadnaya Dvina rivers, invaded the Mologa-Sheksna lowland and extended into the Onega basin.

The East European Plain, within Russia and Belarus, was invaded by the Chudskoe, Ladoga, Onega-Karelia and Belomorsk ice flows that reached the upper reaches of the Dnieper and Volga. The eastern maximum limit of the Scandinavian ice occurred in the middle reaches of the Severnaya Dvina river, the low reaches of the Onega and the mouth of the White Sea. The north-eastern part of the plain was not covered by the Scandinavian ice during the LGM, an observation which was confirmed by comprehensive investigations at the beginning of the twenty-first century during the international QUEEN project.

The late Valdai Till displays regular changes in erratic clast composition over the area, related to the flow pattern of the Scandinavian ice sheet. In comparison to older tills, these deposits include boulders and pebbles derived from more western source areas, namely, the Baltic and Ladoga–Onega provinces. In addition, the late Valdai Till lithology depends to a large extent on the composition of underlying older tills, especially within the ice-marginal formation zones (Faustova and Gribchenko, 1995). The relief-forming activities of the ice streams have been discussed in detail in a number of monographs and papers (e.g. Chebotareva, 1977; Gerasimov and Velichko, 1982). Three major phases of deglaciation have been distinguished differing in the degree of ice-marginal mobility. The initial phase (prior to 16 ka) is noted for the gradual and synchronous retreat of the ice margin (a regressive deglaciation), the integrity of the ice sheet being preserved, together with the outlines of the ice lobes. The well-pronounced marginal deposition left linear or festoon-like (large festoons) landform assemblages. The ice margin retreated spasmodically and phases of rapid ice melt apparently alternated with short cool intervals of stagnation. The receding ice-marginal zones can be easily correlated, thus enabling the reconstruction of the ice-margin configuration. In the outermost, about 30-km-wide zone, the ice thickness presumably did not exceed 100–150 m. The dominant landforms are flat-topped pre-Valdai elevations covered by a thin mantle of clayey Valdai Till in association with the diverse landforms related to dead ice found on the slopes of higher areas and in meltwater valleys. The area north of the peripheral zone was sculpted by thicker ice detached from the thin, stagnant peripheral ice during the course of retreat. The retreating ice tongues, still highly active, formed 2–4 belts of hills and ridges often composed of sand and gravel. At the contacts of adjacent ice lobes, the broad zones of radial deposition developed, typified by isolated uplands and hilly massifs, whereas the zones of ice divergence are marked by angular massifs often associated with deep linear depressions. The proglacial zone included vast ice-dammed basins and smaller glaciofluvial plains.

The type of deglaciation changed after a short but pronounced glacial readvance along the whole ice front; this Vepsovo (=Pomeranian) advance occurred at about 15.5 ka. It was preceded by a cold interval. During the advance, the ice margin became twice as dissected as before and oscillated incessantly. The well-defined end moraines of this phase form a continuous belt from the Baltic ridge in the west to the Onega drainage basin in the east. Numerous minor radial landforms were formed, such as 'intertongue' angular massifs. Twice during the retreat, the ice margin underwent reactivation, as shown by the Luga and Neva ice-marginal formations. As ice thickness decreased, all major oscillations ceased. This accounts for the narrower ice-marginal zones and thinner glacial deposits. The ice itself moulded to the underlying topography, and transgressive–regressive lobate deglaciation gave way to individual decaying glacier tongues. This second stage of deglaciation was characterised by a rapid transition from active to passive ice. The most important of these glacial 'litho–morpho-complexes' are various inversion landforms related to dead ice (e.g. *zvonets*). This is a local name meaning large hills with flattened tops, built of the lake-glacial clay to the depth of several metres starting from the surface; there is usually moraine underlying the clay. These landforms of dead ice are closely genetically associated with active ice landforms, such as terminal moraines and hummocky topography. During the second stage of deglaciation, substantial proglacial lakes appeared, which rapidly changed their margins as the ice retreated; some smaller lakes resulted from thawing of isolated ice blocks together with ice-dammed lakes in the marginal zones. The proglacial zone featured various levels of valley trains and multiple-cone deltas. At the end of that stage, in the Bølling Interstadial, the East European Plain became completely ice free (Faustova, 1994).

The last stage of deglaciation (predominantly areal-type decay) began about 11 ka. Its marginal formations are found in the adjacent Baltic region. In Russia, the last advance of the ice front in the Younger Dryas Stadial only affected Karelia and the Kola Peninsula.

REFERENCES

Agadjanian, A.K., 1992. Stages of small mammals evolution in the central regions of the Russian Plain. In: Velichko, A.A., Shick, S.M. (Eds.), Quaternary Stratigraphy and Paleogeography of Eastern Europe. Institute of Geography, Russian Academy of Sciences, Moscow, pp. 37–49 (in Russian).

Aleksandrova, L.P., 1982. A new species of *Dicrostonyx okaensis* and its significance for Oka till dating in the Likhvin stratotypical section. In: Gerasimov, I.P. (Ed.), Stratigraphy and Palaeogeography of Anthropogene. Nauka, Moscow, pp. 17–21 (in Russian).

Andreicheva, L.N., Nemtsova, G.M., Sudakova, N.G., 1997. Middle Pleistocene Tills of the North and Center of the Russian Plain. Ural Branch of the Komi Research Center, Russian Academy of Sciences, Ekaterinburg 83 pp. (in Russian).

Arslanov, Kh.A., Voznyachuk, L.N., Velichkevich, F.Yu., et al., 1971. Age of the Last Glacial Maximum in the interfluve of the Zapadnaya Dvina and Dnieper rivers. Doklady Akademii Nauk. 196 (1b), 901–909 (in Russian).

Arslanov, Kh.A., Breslav, S.L., Zarrina, E.P., Znamenskaya, O.M., Krasnov, I.I., Spiridonova, E.A., 1981. Climatostratigraphy and chronology of the Middle Valdai in the northwest and center of the Russian Plain.

In: Velichko, A.A., Faustova, M.A. (Eds.), Pleistocene Glaciations of the East European Plain. Nauka, Moscow, pp. 12–27 (in Russian).

Arslanov, Kh.A., Maksimov, F.E., Kuznetsov, V.Yu., Chernov, S.B., Velichkevich, F.Yu., Razina, V.V., et al., 2005. U-Th age and paleobotanical characteristics of the interglacial peat in the Rodionovo reference section. In: Proceedings of the fourth All-Russia Conference on the Quaternary. Quaternary-2005. Geoprint, Syktyvkar, pp. 21–23 (in Russian).

Baumann, K.-H., Lackschtwitz, K.S., Mangerud, J., et al., 1995. Scandinavian ice sheet fluctuations in Norwegian Sea sediments during the past 150,000 years. Quatern. Res. 43, 185–197.

Biryukov, I.P., Agadjanian, A.K., Valueva, M.N., Velichkevich, F.Yu., Shick, S.M., 1992. Quaternary deposits of the Roslavl' stratotype region. In: Velichko, A.A., Shick, S.M. (Eds.), Quaternary Stratigraphy and Palaeogeography of Eastern Europe. Institute of Geography, Russian Academy of Sciences, Moscow, pp. 152–180.

Bolikhovskaya, N.S., 1995. Evolution of the loess-soil formation of Northern Eurasia. In: Kaplin, P.A. (Ed.), Moscow University Press.

Borisova, O.K., Faustova, M.A., 1994. Sequence of natural phases of the Valdai ice age in European Russia. In: Velichko, A.A., Starkel, L. (Eds.), Paleogeographical Basis of the Modern Landscapes. Nauka, Moscow, pp. 17–25 (in Russian).

Borodin, N.G., Danilina, A.A., Kozlov, V.B., Maudina, M.I., 1981. Section of the Likhvin interglacial sediments nearby Yakovlevskoe Village, Poshekhon'e-Volodarsk. Comprehensive studies of the Lower and Middle Pleistocene reference sections in the European USSR. In: To XI INQUA Congress23–28 Moscow (in Russian).

Boulton, G.S., Dongelman, P., Punkari, M., Broadgate, M., 2001. Palaeoglaciology of an ice sheet through glacial cycle: the European ice sheet through the Weichselian. Quatern. Sci. Rev. 20 (4), 521–625.

Chebotareva, N.S. (Ed.), 1977. Structure and Dynamics of the Last Ice Sheet of Europe. Nauka, Moscow 143 pp. (in Russian).

Chebotareva, N.S., Makarycheva, I.A., 1974. Last Glaciation of Europe and Its Geochronology. Nauka, Moscow 254 pp. (in Russian).

Chebotareva, N.S., Makarycheva, I.A., 1982. Geochronology of environmental change in the East European glacial region during the Valdai glaciation. In: Gerasimov, I.P., Velichko, A.A. (Eds.), Paleogeography of Europe During the Last 100,000 Years. Nauka, Moscow, pp. 16–28 (in Russian).

Chirvinsky, V.N., 1914. Materials to understanding of chemical and petrographic composition of glacial deposits in southwestern Russia in connection with the problem of ice sheet motion. Zapiski Kievskogo Obshchestva. Estestvoispytateley 24, 316 pp. (in Russian).

Demidov, I.N., Houmark-Nielsen, M., Kjœr, K.H., Larsen, E., 2006. The last Scandinavian ice sheet in northwestern Russia: ice flow patterns and decay dynamics. Boreas 35, 1–19.

Faustova, M.A., 1994. Deglaciation and types of glacial relief in the European part of Russia. In: Velichko, A.A., Starkel, L. (Eds.), Paleogeographical Basis of the Modern Landscapes. Nauka, Moscow, pp. 30–40 (in Russian).

Faustova, M.A., Gribchenko, Yu.N., 1995. Lithology of glacial deposits of the last Pleistocene glaciation. In: Ehlers, J., Kozarski, S., Gibbard, P.L. (Eds.), Glacial Deposits in North-East Europe. A.A. Balkema, Rotterdam, pp. 183–188.

Fedorova, R.V., 1963. Environmental conditions during the Upper Paleolithic man age in Kostenki, Voronezh region. Mat-ly issled. po arkh. SSSR 121, 220–229 (in Russian).

Frenzel, B., 1967. Die Klimaschwankungen des Eiszeitalters. Vieweg, Braunschweig, 296 pp.

Fursikova, I.V., Pisareva, V.V., Yakubovskaya, T.V., Vlasov, V.K., Kulikov, O.A., Semenenko, L.T., 1992. Pleistocene key section near Okatovo Village west of Moscow. In: Shik, S.M. (Ed.), Phanerozoic Stratigraphy of the Central East European Platform. Tsentrgeologiya, Moscow, pp. 59–82 (in Russian).

Gei, V.P., Malakhovskii, D.B., 1998. On the age and distribution of the maximum Upper Pleistocene glacial advance in the western Vologda region. Izvestiya. Russkogo. Geographicheskogo. Obshchestva, Issue 1, 43–53 (in Russian).

Gei, V.P., Auslender, V.G., Kiseleva, V.B., 2000a. Geomorphology and Quaternary sediments of the central Vologda region. Geomorphological structure of the territory. In: Zarrina, E.P., Shick, S.M. (Eds.), Problems of Quaternary Stratigraphy and Marginal Glacial Forms in the Vologda Region, Northwestern Russia. GEOS, Moscow, pp. 10–17 (in Russian).

Gei, V.P., Saarnisto, M., Lunka, Yu.P., Auslender, V.G., Pleshivtseva, E.S., Kiseleva, V.B., et al., 2000b. Guide-book of excursions. In: Problems of Quaternary Stratigraphy and Marginal Glacial Forms in the Vologda Region, Northwestern Russia. GEOS, Moscow, pp. 71–78 (in Russian).

Gerasimov, I.P., Velichko, A.A. (Eds.), 1980. Age and Distribution of the Maximum Glaciation in Eastern Europe. Nauka, Moscow 212 pp. (in Russian).

Gerasimov, I.P., Velichko, A.A., 1982. Paleogeography of Europe During the Last 100,000 Years (Atlas–Monograph). Nauka, Moscow 156 pp. (in Russian).

Goretskii, G.I., Chebotareva, N.S., Shick, S.M. (Eds.), 1982. Moscow Ice Sheet of Eastern Europe. Nauka, Moscow 240 pp. (in Russian).

Gribchenko, Yu.N., 1980. Petrographic composition of tills of the Dnieper and Don glacial lobes. In: Gerasimov, I.P., Velichko, A.A. (Eds.), Age and Distribution of the Maximum Glaciation in Eastern Europe. Nauka, Moscow, pp. 73–88 (in Russian).

Grichuk, V.P., 1981. The oldest ice sheet in Europe: its indicators and stratigraphic position. In: Velichko, A.A., Grichuk, V.P. (Eds.), Problems of Paleogeography of the Pleistocene in Glacial and Periglacial Regions. Nauka, Moscow, pp. 7–35 (in Russian).

Grichuk, V.P., 1989. History of Flora and Vegetation of the Russian Plain in the Pleistocene. Nauka, Moscow 175 pp. (in Russian).

Grishchenko, M.N., 1976. Pleistocene and Holocene of the Upper Don Drainage Basin. Nauka, Moscow 228 pp. (in Russian).

Iosifova, Yu.I., Agadjanian, A.K., Pisareva, V.V., Semenov, V.V., 2006. The Upper Don basin as stratotype region of the Middle Pleistocene in the Russian Plain. In: Palynological, Climatostratigraphic, and Geoecological Reconstructions. In Memory of E.N. Ananova. Nedra, St.-Petersburg, pp. 41–84 (in Russian).

Iosifova, Yu.I., Agadjanian, A.K., Ratnikov, V.Yu., Sycheva, S.A., 2009. On the Ikorets Formation and Horizon of the uppermost Lower Neopleistocene in the Mastyuzhenka section, Voronezh region. In: Bulletin of the Regional Interdepartmental Stratigraphy Committee on the Center and South of the Russian Plain, 4 Russian Academy of Natural Sciences, Moscow, pp. 89–104 (in Russian).

Janczyk-Kopikowa, Z., 1975. Flora interglacialu Mazowieckiego w Ferdynandowie. Z badan czwartorzedu Polsce 17, 5–70.

Krasnenkov, R.V., Iosifova, Yu.I., Semenov, V.V., 1997. The Upper Don drainage basin—an important stratoregion for climatostratigraphy of the lower Middle Pleistocene (Lower Neopleistocene) of Russia. In: Alekseev, M.N., Khoreva, I.M. (Eds.), Quaternary Geology and Paleogeography. GEOS, Moscow, pp. 82–96 (in Russian).

Krasnozhen, A.S., Baranovskaya, A.F., Zarkhidze, V.S., Malyasova, E.S., Lev, O.M., 1987. Stages of evolution and the Cenozoic mantle structure of the Novaya Zemlya archipelago. In: Cenozoic Deposits of the European Northeast and Placer Geology, Syktyvkar, Proceed. Ural. Branch, Komi Res. Center, USSR Acad. Sci.34–36 (in Russian).

Larsen, E., Kjær, K.N., Demidov, I.N., Funder, S., Grosfjeld, K., Houmark-Nielsen, M., et al., 2006. Late Pleistocene glacial and lake history of northwestern Russia. Boreas 35, 394–424.

Levkovskaya, G.M., Hoffecker, J.F., Anikovich, M.V., et al., 2005. Climatic Stratigraphy of the Earliest Palaeolithic Layers at Kostenki 12 (preliminary results of palynological, palynoteratical, palaeozoological, palaeomagnetic, and SEM-palaeobotanical research on archaeological layer V at Kostenki 12). In: Anikovich, M.V. (Ed.), The Problems of Early Upper Palaeolithic of the Kostenki-Borschevo Region and Adjacent Territories. Institute for the Material Culture History RAS, St.-Petersburg, pp. 119–130.

Malyasova, E.S., Serebryanny, L.R., 1993. Natural history of Novaya Zemlya. In: Bojarsky, V.V. (Ed.), Novaya Zemlya vol. 2. Institute of Culture and Natural Heritage, Moscow, pp. 10–22 (in Russian).

Markov, K.K. (Ed.), 1977. Sections of Glacial Deposits in the Central Russian Plain. Moscow University Press, Moscow, 198 pp. (in Russian).

Markova, A.K., 1982. Pleistocene Rodents of the Russian Plain. Nauka, Moscow 185 pp. (in Russian).

Maudina, M.I., Pisareva, V.V., Velichkevich, F.Yu., 1985. The Odintsovo stratotype in the light of new data. Dokl. Akad. Nauk SSSR 284 (5), 1195–1199 (in Russian).

Mikhailesku, K.D., Markova, A.K., Chepalyga, A.L., et al., 1991. Biostratigraphy of the reference section (lectostratotype) of the Early Euxinian sediments near the Ozernoe Village. Commission for study of the Quaternary. Bulletin. AN SSSR 60, 29–40 (in Russian).

Nazarov, V.I., 1984. Anthropogene Insects of the Northeastern Byelorussia and Adjacent Regions. Nauka I Tekhnika, Minsk, 115 pp. (in Russian).

Ostanin, V.E., Atlasov, R.R., Bukreev, V.A., Levina, N.B., 1979. Marginal forms and the Valdai glaciation boundary in the Vaga River basin. Geomorphology 1, 72–76 (in Russian).

Pavlidis, Yu.A., 1992. The World Ocean Shelf During the Late Quaternary. Nauka, Moscow, 272 pp. (in Russian).

Pavlidis, Yu.A., Bogdanov, Yu.A., Levchenko, O.V., Murdmaa, I.O., 2005. New record on the Barents Sea environment at the end of the Valdai glaciation. Okeanologiya 45 (1), 92–106 (in Russian).

Pisareva, V.V., 1997. Flora and vegetation of the Early and Middle Pleistocene of the Russian Plain. In: Alekseev, M.N., Khoreva, I.M. (Eds.), Quaternary Geology and Paleogeography. GEOS, , pp. 124–133 (in Russian).

Pisareva, V.V., 2001. Reference correlative profile VIII Gryazovets–Bui–Galich–Unzha River. In: Velichko, A.A., Shick, S.M. (Eds.), Middle Pleistocene Glaciations of Eastern Europe. GEOS, Moscow, pp. 87–95 (in Russian).

Pisareva, V.V., Faustova, M.A., 2008. Reconstruction of landscapes of northern Russia during the middle Valdai megainterstadial. In: Velichko, A.A., Vasil'ev, S.A. (Eds.), Way to North: Paleoenvironment and Earliest Inhabitants of Arctic and Subarctic. Russian Academy of Sciences, Inst. of Geography, Moscow, pp. 53–62 (in Russian).

Pisareva, V.V., Faustova, M.A., 2009. Reconstruction of paleoenvironments during the second part of the Late Glacial epoch in Eurasian north. In: Fundamental Problems of the Quaternary: Results and Trends of Further Researches, Proceedings of the fourth All-Russia

Quaternary Conference October 19–23, 2009. Novosibirsk. 470–474 (in Russian).

Rumyantsev, V.A. (Ed.), 1998. History of the Pleistocene Lakes of the East European Plain. Nauka, St.-Petersburg 404 pp. (in Russian).

Runkov, S.I., Bol'shakov, V.A., Nemtsova, G.M., Pisareva, V.V., Sudakova, N.G., 1993. Pleistocene reference section near the Narovatovo Village, Moksha River. In: Bulletin of the Regional Interdepartmental Stratigraphy Committee on the Center and South of the Russian Plain, 2 Rosgeolfond, Moscow, pp. 144–152 (in Russian).

San'ko, A.F., 1987. Neopleistocene of Northeastern Byelorussia and the Adjacent Regions of the RSFSR. Nauka i Tekhnika, Minsk 178 pp. (in Russian).

Shick, S.M. (Ed.), 1984. Lower Pleistocene Key Sections of the Upper Don Drainage Basin. Voronezh University Press, Voronezh, 213 pp. (in Russian).

Shick, S.M., Zarrina, E.P., Pisareva, V.V., 2006. Neopleistocene stratigraphy and paleogeography of central and northwestern European Russia. In: Zubakov, V.A. (Ed.), Palynological, Climatostratigraphic, and Geoecological Reconstructions. In Memory of E.N. Ananova. Nedra, St.-Petersburg, pp. 85–121 (in Russian).

Spiridonova, E.A., 1970. Palynological characteristics of the Valdai interstadial deposits in northwestern Russian Plain and its significance for stratigraphy and paleogeography. In: Candidate Dissertation in Geology and Mineralogy. Leningrad University Press, Leningrad, 22 pp. (in Russian).

Sudakova, N.G., 1990. Paleogeographic Regularities of Glacial Lithogenesis. Moscow University Press, Moscow, 159 pp. (in Russian).

Sudakova, N.G., Faustova, M.A., 1995. Ice sheets of the Middle and Late Pleistocene in the East European Plain. In: Climate and Environmental Changes in Eastern Europe during the Holocene and Late–Middle Pleistocene. Moscow University Press, Moscow, pp. 69–75 (in Russian).

Sudakova, N.G., Rychagov, G.I., Antonov, S.I., 2007. Topical problems of the middle Neopleistocene stratigraphy and paleogeography of the central Russian Plain. In: Gladenkov, Ju, B. (Ed.), Geologic Events of the Neogene and Quaternary of Russia: Modern State of Stratigraphic Schemes and Paleogeographic Reconstructions. GEOS, Moscow, pp. 86–90 (in Russian).

Svendsen, J.I., Alexanderson, H., Astakhov, V.I., et al., 2004. Late Quaternary ice sheet history of northern Eurasia. Quatern. Sci. Rev. 23, 1229–1271.

Thiede, J., Bauch, Y.A., 1999. The Late Quaternary history of northern Eurasia and the adjacent Arctic Ocean: an introduction to QUEEN. Boreas 28 (1), 3–5.

Velichko, A.A., 1980. Latitudinal asymmetry of environmental components during glacial epochs in the Northern Hemisphere. USSR, Academy of Sciences, Izvestiya, Ser. Geogr. 5, 5–23 (in Russian).

Velichko, A.A. (Ed.), 1993. Evolution of landscapes and climates of Northern Eurasia, Issue 1. Regional Paleogeography. Nauka, Moscow, 102 pp.

Velichko, A.A. (Ed.), 1999. Climate and Environmental Changes During the Last 65 Million Years (Cenozoic, from Paleocene to Holocene). GEOS, Moscow 260 pp. (in Russian).

Velichko, A.A. (Ed.), 2009a. Paleoclimates and Paleoenvironments of Extratropical Area of the Northern Hemisphere in the Late Pleistocene–Holocene. Atlas-Monograph. GEOS, Moscow 119 pp.

Velichko, A.A., 2009b. Middle Valdai, Zyryan–Sartan megainterval and climatic rank of its optimum. In: Fundamental Problems of the Quaternary: Results and Trends of Further Researches, Proceedings of the fourth All-Russia Quaternary Conference October 19–23, 2009. Novosibirsk (in Russian).

Velichko, A.A., Faustova, M.A., 1989. Reconstruction of the last Late Pleistocene glaciation in the Northern Hemisphere (18–20 ka BP). Dokl. Akad. Nauk SSSR 309 (6), 1465–1468 (in Russian).

Velichko, A.A., Shick, S.M. (Eds.), 2001. Middle Pleistocene Glaciations in Eastern Europe. GEOS, Moscow 159 pp. (in Russian).

Velichko, A.A., Gribchenko, Yu.N., Gubonina, Z.P., et al., 1997a. Major features and structure of the loess–soil formation. In: Velichko, A.A. (Ed.), Loess–Soil Formation of the East European Plain. Paleogeography and Stratigraphy. Institute of Geography. RAN, Moscow, pp. 5–25 (in Russian).

Velichko, A.A., Kononov, Yu.M., Faustova, M.A., 1997b. The last glaciation of the Earth: size and volume of ice sheets. Quatern. Int. 41/42, 43–51.

Velichko, A.A., Kononov, Yu.M., Faustova, M.A., 2000. Geochronology, distribution, and ice volume on the Earth during the Last Glacial Maximum: inferences from new data. Stratigr. Geol. Correl. 6 (1), 1–12 (in Russian).

Velichko, A.A., Faustova, M.A., Gribchenko, Yu.N., Pisareva, V.V., Sudakova, N.G., 2004. Glaciations of the East European Plain—distribution and chronology. In: Ehlers, J., Gibbard, P.L. (Eds.), Quaternary Glaciations—Extent and Chronology. Elsevier, Amsterdam, pp. 337–354.

Velichko, A.A., Pisareva, V.V., Faustova, M.A., 2005. Early and Middle Pleistocene glaciations and interglacials in the East European Plain. Stratigr. Geol. Correl. 13 (2), 84–102 (in Russian).

Velichko, A.A., Morozova, T.D., Panin, P.G., 2007. Soil polygenetic complexes as a systematic phenomenon of Pleistocene macrocycles. Russian Academy of Sciences, Izvestiya, Ser. Geoger. 2, 44–53 (in Russian).

Viiding, H., Gaigalas, A., Gudelis, V., Raukas, A., Tarvidas, R., 1971. Crystalline Index Boulders of the Baltic Region. Mintis Publishers, Vilnius, 95 pp. (in Russian).

Voznyachuk, L.N., Kondratene, O.P., Motuzko, A.N., 1984. First finding of the Likhvin small mammal fauna in the western glacial area of the East European Plain. In: Quaternary Paleogeography and Stratigraphy of the Baltic Region and Adjacent Areas, 105–121 Vilnius (in Russian).

Yakovlev, S.A., 1939. Index boulders, tills, and limits of the Novaya Zemlya ice sheet in the Russian Plain. Commission for study of the Quaternary. Bulletin. 5, 18–26 (in Russian).

Yevzerov, V.Ya., 2005. Geology and mineralogy of the Quaternary sediments in the northeastern Baltic Shield. Doctoral dissertation in geology and mineralogy (Voronezh University Press) (in Russian).

Zagwijn, W.N., 1985. An outline of the Quaternary stratigraphy of Netherlands. Geologia en. Mijnbouw 64 (1), 17–24.

Zarrina, E.P., 1991. Quaternary Deposits of Northwestern and Central Regions of the European Part of the USSR. Nedra, Leningrad 187 pp. (in Russian).

Zarrina, E.P., Shick, S.M. (Eds.), 2000. Problems of the Quaternary Stratigraphy and Marginal Glacial Formations of the Vologda Region (NW Russia). Proceedings of the International Symposium. GEOS, Moscow, 100 pp. (in Russian).

Zimenkov, O.I., 1985. On the age of maximum stage of the Valdai glaciation in Byelorussia. In: Velichko, A.A. (Ed in Chief) Marginal Forms of Continental Glaciations. Nauka, Moscow, pp. 131–132 (in Russian).

Chapter 27

Glacial History of the Barents Sea Region

Tore O. Vorren[1,*], Jon Y. Landvik[2], Karin Andreassen[1] and Jan Sverre Laberg[1]

[1]*Department of Geology, University of Tromsø, N-9037 Tromsø, Norway*
[2]*Department of Plant and Environmental Sciences, Norwegian University of Life Sciences, N-1432 Aas, Norway*
*Correspondence and requests for materials should be addressed to Tore O. Vorren. E-mail: tore.vorren@uit.no

27.1. INTRODUCTION

During the 1970s and early 1980s, there was much discussion about the possible extent and chronology of glaciers on the Barents Sea shelf and adjoining Arctic areas (Fig. 27.1). Total glaciation of these areas during the Late Weichselian was advocated by Hughes et al. (1977, 1981) and Grosswald (1980). At the other extreme, Boulton (1979) suggested no glaciation in the Barents Sea during the Late Weichselian or in the past two million years. Various intermediate scenarios were also advanced at that time (e.g. Kvasov, 1978; Mathisov, 1980).

During the early 1980s, it was shown that (1) grounded glaciers reached the shelf edge at least four times during the Late Cenozoic (Solheim and Kristoffersen, 1984), (2) the northern Barents Sea was covered by glaciers during the Late Weichselian (Salvigsen, 1981; Salvigsen and Nydal, 1981), (3) Spitsbergenbanken was covered by glaciers, possibly as late as the Late Weichselian (Elverhøi and Solheim, 1983) and (4) the southern Barents Sea as well had been covered by glaciers merging with glaciers further north (Vorren and Kristoffersen, 1986). Thus, ample evidence was established for multiple glaciations during the Late Cenozoic as well as for an ice sheet covering the whole of the Barents Sea during the Late Weichselian Substage. These and later results were summarised by Landvik et al. (1998) and Svendsen et al. (2004b).

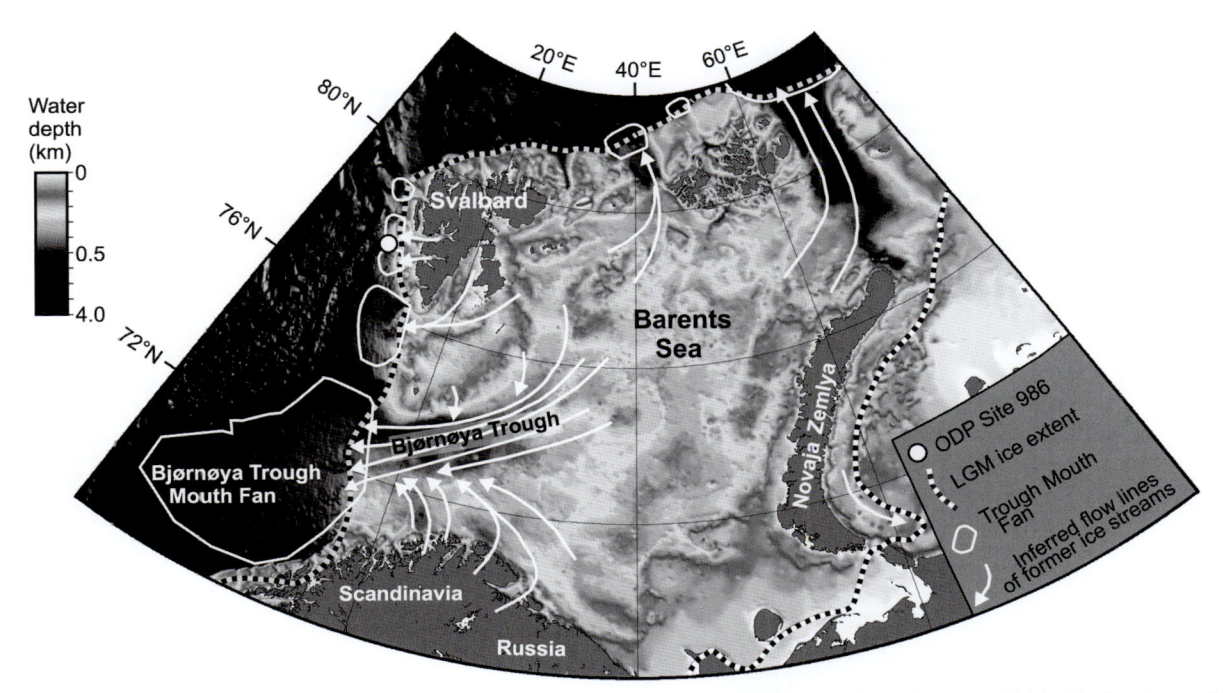

FIGURE 27.1 Overview map indicating the location of trough-mouth fans, flow lines and marginal position of the Late Weichselian ice sheet and ODP site 986. Modified from Andreassen et al. (2007).

During the past decade, much progress has been made through the analysis of high-resolution sea-floor morphology, three-dimensional (3D) seismic investigations and analysis of long cores (e.g. Butt et al., 2000; Andreassen et al., 2004, 2007; Ottesen et al., 2007; Ottesen and Dowdeswell, 2009; Knies et al., 2009; Winsborrow et al., 2010). These and onshore works on the Arctic islands and in north-western Russia (e.g. Svendsen et al., 2004a; Larsen et al., 2006) have contributed to a better understanding of the extent and chronology as well as to the internal dynamics of the Late Cenozoic Ice Sheets covering the Barents Sea region.

There is now a large-scale temporal framework of the glaciations and high-resolution understanding of glacial dynamics at selected time slices, and research approaches have been developed to further develop the reconstructions. However, the present-day records are still to fragmentary for full-scale ice-sheet reconstructions through time. With respect to such reconstructions, the future challenge is to improve correlation of onshore and offshore records, better age control of the individual glacial events, as well as a better understanding of the lateral and vertical extent of the different glaciations. The latter should be integrated with numerical modelling.

In this chapter, the earlier reviews are updated and, in particular, expanded on the development in ice-sheet dynamics and the new knowledge on pre-Late Weichselian glaciations. The authors follow the *International Commission on Stratigraphy* concerning the subdivision of the geological time scale. Thus, the base of the Quaternary and Pleistocene is defined at 2.588 Ma and the base of Holocene to 11.7 Ka. The Pleistocene is divided into Early Pleistocene (2.588–0.781 Ma), Middle Pleistocene (0.781–0.126 Ma) and Late Pleistocene (0.126 Ma–11.7 Ka).

27.2. PHYSIOGRAPHY AND QUATERNARY SEDIMENTS

The epicontinental Barents Sea covers one of the widest continental shelves in the world (Fig. 27.1), where the bathymetry is characterised by banks and troughs/channels opening to the Arctic Ocean in the north and the Norwegian Sea to the west. Typically, the water depths in the troughs range from 300 to 500 m, whereas depths of the banks mostly range from 300 to 50 m.

The major geomorphological feature of the southern Barents Sea is the Bjørnøyrenna Trough that is 750 km long and 150–200 km wide. Two south-east–north-west trending troughs just off the coast of Norway, Ingøydjupet and Djuprenna, reach water depths around 400 m (Fig. 27.1). Cross-shelf troughs separated by shallower banks also characterise the bathymetry of the northern Barents Sea (Franz Victoria Trough and St. Anna Trough) and Svalbard continental shelf (Fig. 27.1).

The most pronounced seismic reflector on the continental shelf is the upper regional unconformity (URU) that separates underlying sedimentary bedrock from overlying glacial deposits (Solheim and Kristoffersen, 1984; Vorren et al., 1986) (Fig. 27.2). A sedimentary sequence of varying

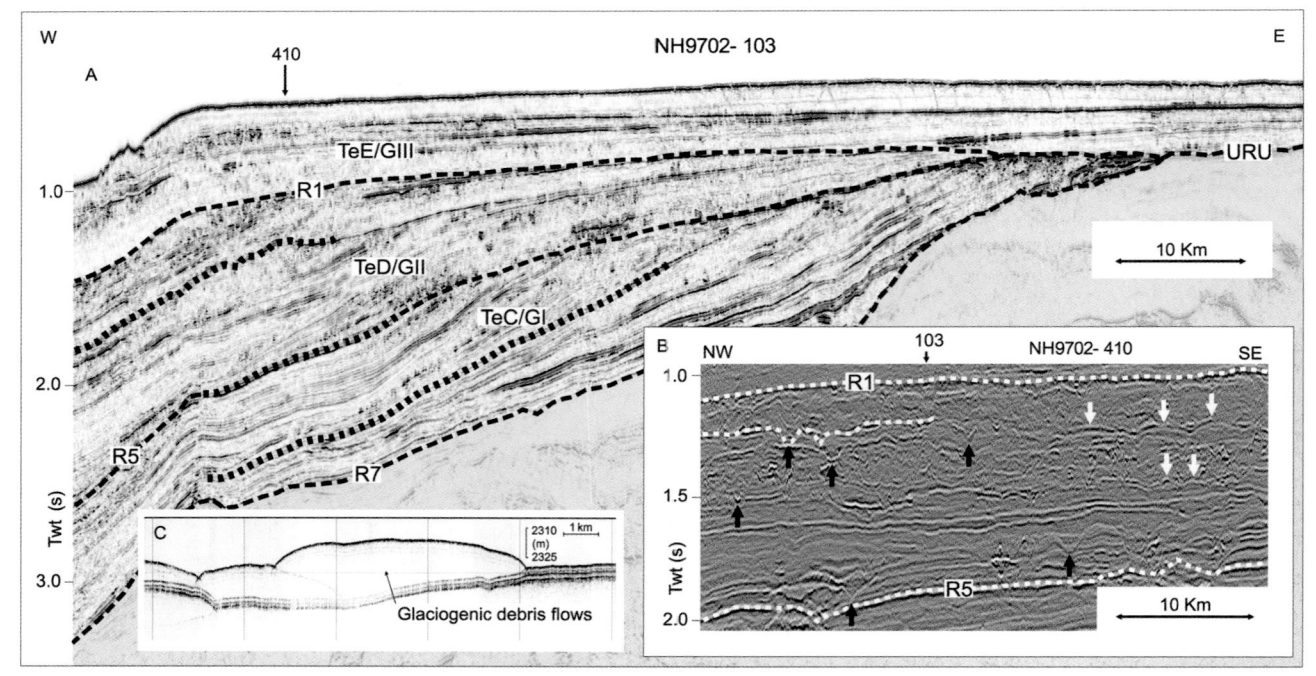

FIGURE 27.2 (A) Seismic line across the Bear Island TMF showing seismostratigraphical units and their bounding reflectors. (B) Part of along-slope orientated line showing channels and debris flows in unit TeD/GII. (C) Late Weichselian debris flow deposits. Modified from Laberg et al. (2010).

thickness (0–300 m) overlies the URU. Its glacigenic origin is recorded by several core and drill samples (e.g. Vorren et al., 1984, 1989, 1990; Sættem and Hamborg, 1987; Elverhøi et al., 1998). The URU probably represents the erosional base for several continental shelf glaciations.

The western and northern continental margins of the Barents Sea region are characterised by fan-shaped protrusions located at the mouth of cross-shelf troughs and therefore named 'trough-mouth fans' (TMFs by Vorren et al., 1988, 1989). Investigations along the Norwegian-Barents Sea-Svalbard continental margin suggest that the TMFs are depocentres of sediments that accumulated in front of ice streams of the former Fennoscandian-Barents Sea-Svalbard Ice Sheets (Vorren and Laberg, 1997; Vorren et al., 1998; Sejrup et al., 2003). The Bjørnøya (Bear Island) TMF (Fig. 27.1), the largest of the Polar North Atlantic TMFs, contains up to 3–4 km of Plio-Pleistocene sediments. As Vorren and Laberg (1997) pointed out, the fans act as the most important archives for the palaeoclimate and ice-sheet evolution in the Barents Sea region.

Several deep-sea boreholes have recently given much information, particularly, concerning the Late Pliocene and Early–Mid-Pleistocene environments. The sites are located (Figs. 27.1 and 27.3) on the Yermak Plateau (Ocean Drilling Program (ODP) sites 910, 911), in the Fram Strait (ODP sites 908, 909), on the western Barents Sea continental slope (ODP site 986) and on the Barents Sea continental shelf (Well 7216/11-15) (e.g. Butt et al., 2000; Knies et al., 2009).

The Barents Sea region includes several archipelagos on the shelf which are partially glaciated at present: Svalbard, Franz Josefs Land and Novaya Zemlya. The stratigraphical records on the islands are fragmentary because of glacial erosion and long periods of non-deposition. Interpretations of glacial history on these islands are mainly based on investigations of exposed tills and uplifted glaciomarine sediments. The best-studied area is Svalbard which has been focus of intensive investigations for several decades (e.g. Landvik et al., 1998; Mangerud et al., 1998; Svendsen et al., 2004b).

Ice sheets from Scandinavia, the Barents Sea and the Kara Sea merged from time to time in north-western Russia. This resulted in large ice-dammed lakes. Thus an intricate stratigraphy comprising shifting beds of fluvial, glaciofluvial, glaciolacustrine, marine sediments and tills exists in this area. The reconstruction of the blocking and re-routing of the large north-flowing Russian rivers, with the formation of large ice-dammed lakes between the glaciers and the water divides to the south, has been studied in great detail along the southern margin of the Barents and Kara Ice Sheets (Chapter 25).

27.3. PALAEOGENE AND NEOGENE

Sediment cores from the Arctic Ocean have provided indirect evidence of Eocene and Miocene glaciations somewhere in the surrounding high Arctic land mass.

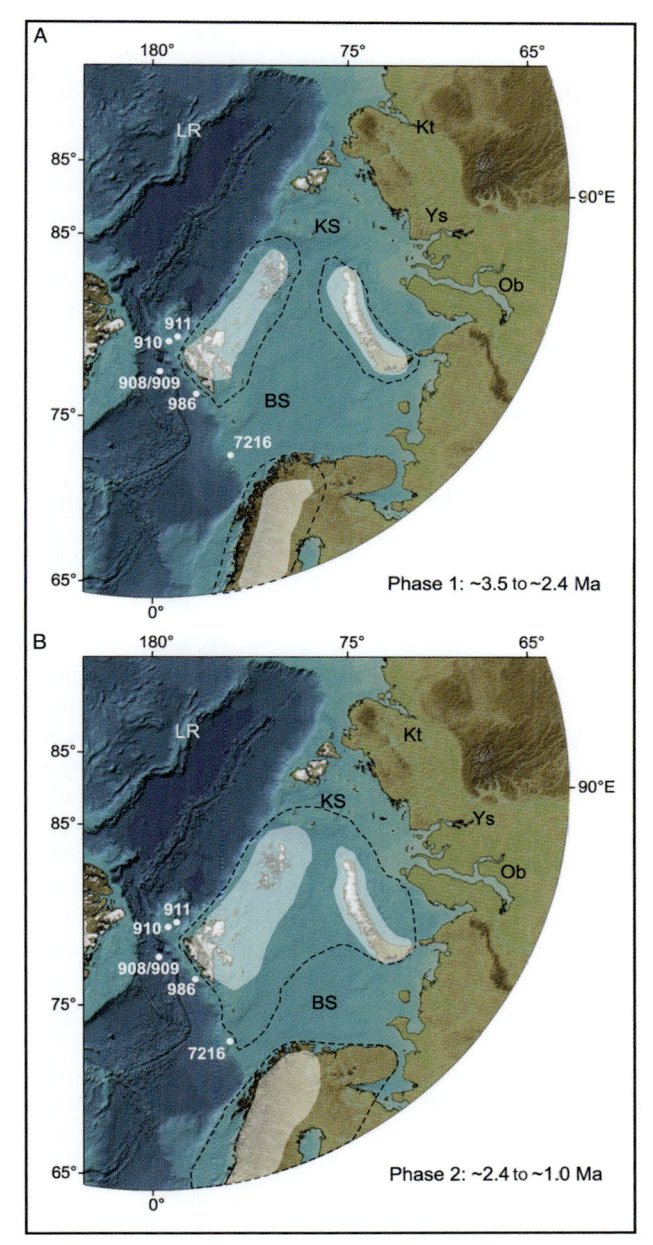

FIGURE 27.3 Schematic minimum/maximum model of glacier extent in the Barents Sea region during (A) 3.5–2.4 Ma and (B) 2.4–1.0 Ma, according to Knies et al. (2009).

According to Stickley et al. (2009), input of iceberg-rafted detritus, occurring at about 46 Ma, suggests the presence of small isolated glaciers in the high Arctic. In the Early Miocene, dropstones and sand imply that sea ice and icebergs calved from glaciers were present in the Arctic Ocean, and at 14 Ma, the abundance of ice-derived sand increases significantly (Moran et al., 2006).

Evidence for Late Pliocene to Early Pleistocene glaciations from marine studies was reviewed by Knies et al. (2009), together with the development of an age model for the glacial events. Based on ice-rafted detritus (IRD) pulses to the ODP drilling sites 910 and 911 on the Yermak Plateau and Fram Strait sites ODP 908 and 909 (Fig. 27.3), they found that there are no indications of any Barents Ice Sheet until the Pliocene glaciations (3.5–2.4 Ma). From the regional distribution of clay minerals, they concluded that the ice sheet was of limited extent, covering mountainous regions over Svalbard and Franz Josef Land and only reaching the coastline of the northern Barents Shelf during short-term glacial intensifications (Fig. 27.3A).

27.4. EARLY PLEISTOCENE (~2.6 TO ~0.7 MA)

27.4.1. Seismic Stratigraphy and Chronology

A seismostratigraphical framework has been established both for the western and the northern Barents Sea margins (e.g. Vorren et al., 1991; Faleide et al., 1996; Fiedler and Faleide, 1996; Hjelstuen et al., 1996; Laberg and Vorren, 1996; Solheim et al., 1998; Ryseth et al., 2003; Geissler and Jokat, 2004). Here, the writers focus on the best-studied area, the Bjørnøya TMF. Three main sequences (TeC/GI, TeD/GII and TeE/GIII) and seven regional seismic reflectors (R7–R1) were identified here (Fig. 27.2) and along the western margin. The deepest reflector, R7, was interpreted to mark the base of the glacial deposits (Faleide et al., 1996).

An age assignment of 2.3 Ma for reflector R7 was derived from the chronostratigraphical framework of IKU's shallow boreholes and exploration wells from the southwestern Barents Sea (e.g. Sættem et al., 1992, 1994; Eidvin et al., 1993; Mørk and Duncan, 1993). R7 was later penetrated at ODP site 986 (Fig. 27.1), and palaeomagnetic and biostratigraphical data, as well as strontium isotope analyses, supported the Early Pleistocene age (2.3–2.5 Ma) of the prominent reflector (e.g. Channell et al., 1999; Eidvin and Nagy, 1999; Smelror, 1999; Butt et al., 2000). Recently, Knies et al. (2009) redated R7 to 2.7 Ma. The R5 reflector is assigned the interpolated age 1.3–1.5 Ma at ODP site 986, supported by biostratigraphical and Sr evidence (Butt et al., 2000). This implies that the seismic unit TeC/GI (bounded by the R7 and R5 reflections) comprises sediments deposited in the period from ~2.7 to ~1.5 Ma.

The age of the base of unit TeE/GIII has variously been suggested to be 0.2 Ma (Andersen et al., 1994; Elverhøi et al., 1995), younger than 0.44 Ma (Sættem et al., 1992, 1994), 0.44 Ma (Faleide et al., 1996; Fiedler and Faleide, 1996) or 0.8 Ma (Vorren et al., 1991, Knutsen et al., 1992; Richardsen et al., 1992; Vorren and Laberg, 1997). The authors favour an age of ~0.7 Ma because unit TeE/GIII seem to comprise at least eight full-scale glacial advances to the shelf break (Vorren and Laberg, 1997).

This implies that the TeD/GII sediments have an age of ~1.5 to ~0.7 Ma, while unit TeE/GIII (R1—sea floor) includes the Middle and Late Pleistocene succession. However, unquestionably, the chronology of the seismostratigraphy will most certainly undergo some revision in the years to come.

27.4.2. The Early Part of Early Pleistocene (~2.6 to ~1.5 Ma)

The precise extent of the glaciations in the early part of Early Pleistocene is largely unknown, except for some phases where it can be shown that it reached the shelf break. However, some generalised patterns of glaciation can be drawn from the available data. According to Knies et al. (2009), during the period 2.4–1.0 Ma, ice sheets over the northern Barents Sea expanded southwards and periodically reached the western Kara Sea (Fig. 27.3B). This growth is suggested by the decrease of Siberian river-supplied smectite, probably caused by ice-sheet growth and blocking of the transport pathway. However, based on IRD data from ODP site 986 and high-resolution seismic data from the Bjørnøya TMF, glaciers did not reach the shelf break until ca. 1.6 Ma, according to Butt et al. (2000), which corresponds to the seismic result from the Bjørnøya TMF (see below).

The unit TeC/GI palaeo-slope sediments in the Bjørnøya TMF are interpreted to be dominantly distal glaciomarine and mostly settled from meltwater overflows. Sediment-laden underflows resulted in turbidity currents forming channels (Laberg et al., 2010). Sættem et al. (1992) identified channels on the outermost part of the palaeo-shelf. The channels are filled by gravelly sand interpreted as of glaciofluvial origin. This indicates that the glaciers did not reach the shelf break.

Thus, during the first part of the Early Pleistocene, the continental shelf was probably subaerial with glaciers terminating on land as also indicated by numerical modelling (Butt et al., 2002) and studies of the ODP sites along the western Barents Sea continental margin (Knies et al., 2009). This probably indicates fluvial and glaciofluvial sediment transport to the western Barents Sea continental palaeo-shelf break. Mountain glaciation is interpreted on Spitsbergen between 2.6 and 1.6 Ma (Forsberg et al., 1999; Butt et al., 2000).

27.4.3. The Later Part of Early Pleistocene (~1.5 to ~0.7 Ma)

The first well-documented shelf-edge glaciation occurred during the start of this period (Andreassen et al., 2007). This resulted in rapid sediment deposition on the continental shelf-break and -slope. Ice streams transported deformation till to the shelf break. These sediments were subsequently remobilised to form glacigenic debris flows. Blocks of glacigenic sediments occurring on the outer shelf probably also formed from subglacial erosion and sediment transport (Andreassen et al., 2004). The increased sediment loads provided conditions for large failure events on the continental slope. Hjelstuen et al. (2007) have mapped several slides occurring during this time, including three buried megaslides, which have left scars containing up to 500 m thick debris units.

At about 1 Ma, that is, during deposition of the upper part of unit TeE/GII, the Barents Sea Ice Sheet is suggested to have transformed from a mainly subaerial to a marine-based ice sheet (e.g. Butt et al., 2002); at that time, glacier ice probably reached as far as the Yermak Plateau, northwest of Svalbard (Knies et al., 2009). This is the largest ice extent shown to have occurred in the Arctic Ocean during the Quaternary. The intensified growth of the prograding sediment wedge along the Barents shelf suggests that ice sheets reached a full-scale Barents/Kara Sea glaciation.

During this time, the glacially sculpted landscape of north-western Svalbard was already formed. Coastal exposures show that till, glaciomarine mud and sublittoral sands were deposited in the Kongsfjorden fjord basin some time prior to 1 Ma as indicated by the mollusc fauna and the reversed palaeomagnetic signal (Houmark-Nielsen and Funder, 1999).

27.5. MIDDLE PLEISTOCENE GLACIATIONS (0.78–0.12 MA)

Unit TeE/GIII mainly comprises large debris flows deposited during full-glacial conditions. Eight stacked units of debris flows indicate eight full-scale glaciations where shelf sediments were eroded and transported subglacially as deformation till by ice streams to the Bjørnøya TMF. Seven similar stacked debris flow units have been identified in the Storfjord TMF (Vorren and Laberg, 1997). Andreassen et al. (2004) have identified five horizons with mega-scale glacial lineations in the proximal part of the TeE/GIII in the Bjørnøya TMF. Sparse occurrence of glaciomarine sediments, as well as the lack of channels, indicates that input from meltwater was of minor importance during the non-debris flow periods (Laberg and Vorren, 1996; Vorren and Laberg, 1997). This indicates more polar conditions, that is, a colder ice sheet with less meltwater production. Erosion and sediment transport mostly occurred below fast-flowing ice streams.

As an attempt to date the various debris flow units, Vorren and Laberg (1997) made the following assumption that (1) all the identified debris flow units on the fan comprise sediments deposited during glacial maxima; (2) the Barents Sea Ice Sheets advanced to the shelf break once during the Weichselian Stage and once during each of the immediately preceding Middle and Late Pleistocene glacial maxima; (3) the Middle and Late Pleistocene fan stratigraphical record is complete; and (4) the glacial maxima correspond to the even marine isotope stages (MISs) 2, 4, 6, etc. Then 'counting-backwards' from the Last Glacial Maximum (LGM, during MIS 2) would give the result that the earliest debris flow unit was between 0.6 and 0.7 Ma. However, two or more glacier advances to the shelf break could have occurred during one and the same MIS, or on the other hand, during one or more of the isotope stages, the ice margin may not have reached the shelf break. Further information is needed before a safer conclusion can be reached.

27.6. LATE PLEISTOCENE GLACIATIONS

The geomorphology and sedimentary records both offshore and onshore in the Barents Sea region provide excellent means for understanding the glacigenic sedimentation, dynamics of glaciations, locations of fast-flowing ice streams and the chronology during the last glacial period (e.g. Vorren and Laberg, 1996; Landvik et al., 1998, 2005; Andreassen et al., 2004, 2007; Ottesen et al., 2005, 2007; Andreassen and Winsborrow, 2009; Winsborrow et al., 2010).

As a consequence of the higher preservation potential of the Late Pleistocene glacial deposits, there is inevitably a better knowledge concerning glaciations during this time interval. However, for the glacial episodes prior to the very last glaciation, glacial limits cannot be morphologically mapped, with the exception of north-western Russia (see Chapter 25). However, glacial deposits from the Early and Middle Weichselian Substages are preserved in stratigraphical successions where the age control is within reach of several dating methods. However, most of the morphological features are related to the Late Weichselian Barents Ice Sheet for which good control of its extent is now available, as well as a steadily increasing understanding of its behaviour and dynamics, which will be discussed below.

The following discussion is intended to update previous reviews. The history of the north-western part of the Weichselian Ice Sheets was reviewed by Mangerud et al. (1998), and later compilations of both the western and the eastern sectors of the ice sheets have been presented by Svendsen et al. (2004a,b) and Larsen et al. (2006), mainly based on new studies in north-western Russia and adjacent offshore areas. The latter study (Larsen et al., 2006) also addressed the interplay between Barents Sea, Kara Sea and Scandinavian-dominated ice sheets in north-western Russia throughout the Weichselian Stage. Models for the extent and timing of the much better known Late Weichselian Ice Sheet were presented by Landvik et al. (1998), Svendsen et al. (2004a,b) and later updated for the Kara region of the ice sheet by Möller et al. (2006) and Ingólfsson et al. (2008).

27.7. THE EARLY AND MIDDLE WEICHSELIAN GLACIAL HISTORY

27.7.1. Introduction

The only area where morphological and stratigraphical evidence for the Early and Middle Weichselian glaciations can be found on land is in northern Russia and the Taimyr Peninsula (see Chapters 25 and 28). The recent development in reconstructions, as outlined below, clearly shows that ice sheets of the region are time-transgressive systems, not regional scale contemporaneous events. There are demonstrably east–west variations and evidence that the Barents and Kara Ice Sheets do not share a synchronous history. Further, the spatial behaviour of the ice sheets was forced by the underlying topography controlling the ice-stream and inter-ice-stream configuration, as is now known from the latest reconstructions of the Late Weichselian Ice Sheet (see below).

In their syntheses, Svendsen et al. (2004a,b) showed that Early and Middle Weichselian Barents and Kara Sea-based ice extended further into northern Russia than during the Late Weichselian. They proposed a maximum Weichselian ice-sheet extent around 90 ka and a later ice advance culminating around 60 ka, an age estimate which agrees with the ice-advance timing proposed by a sea level and isolation-driven numerical model (Siegert et al., 2001). The ice margins for the final phase, the Late Weichselian Ice Sheet, were established with the uncertainty whether the Kara shelf ice reached the northern part of the Taymyr Peninsula (Svendsen et al., 2004a,b).

27.7.2. Kara Sea and North-West Russia

The first glaciation after the Eemian Stage interglacial was more extensive in the Kara Sea region than the later Weichselian glaciations (Svendsen et al., 2004a,b). It was confluent with ice over the Putorana Plateau and covered most of the Taymyr Peninsula. Studies on Severnaya Zemlya (Möller et al., 2006) show that the islands were glaciated by an ice sheet that probably persisted on the northern part of the Kara shelf until the Middle Weichselian MIS 3 (Ingólfsson et al., 2008). Large glacial lakes formed south of the ice margin (Mangerud et al., 2004).

An age for this ice sheet occurring around 90 ka was proposed by Svendsen et al. (2004b). From studies in the White Sea area, Larsen et al. (2006) suggested that the first ice advance occurred slightly earlier (100–90 ka) and was dominated by an ice sheet centred over the Kara Sea (Fig. 27.4). Because no influence by Barents Sea ice was found, they suggested that most of the southern Barents Sea remained ice free during this period, contrary to the extensive southern Barents Sea Ice Sheet suggested by Mangerud et al. (2004).

While an ice sheet persisted on the northern Kara Shelf (see above), most of the Barents Sea was probably ice free until Barents-dominated glacial ice reached the White Sea area at 70–65 ka (Larsen et al., 2006) and the Pechora region ca. 60 ka (Svendsen et al., 2004a). They suggest it was confluent with the Scandinavian Ice Sheet, covering the southern Barents Sea (Fig. 27.4), and a large ice-dammed lake in the White Sea basin has been assumed (Mangerud et al., 2004; Larsen et al., 2006). However, Larsen et al. (2006) proposes that this phase was succeeded by a deglaciation of the Barents Sea before a Kara Sea Ice Sheet advanced into northern Russia by about 55–45 ka.

Dating of 37 mollusc-shell samples from Novaja Zemlya (Mangerud et al., 2008) shows that the island was mostly ice free between 48 and 26 ^{14}C ka BP, suggesting that neither Barents nor Kara Sea Ice Sheets existed at this time. Raised shorelines of up to 140 m a.s.l. from this period indicate a substantial glacial load during the preceding glaciation during the MIS 4/3 interval, compared to the Late Weichselian Ice Sheet which produced marine limits of below 20 m a.s.l.

27.7.3. Svalbard and the Western Barents Sea

Along the western and northern perimeter of the Barents Sea, the shelf break will always constitute a maximum restriction for ice-sheet extent. Recent studies suggest that the ice-sheet marginal advances and retreats to a large degree are controlled by ice-stream fluctuations and thus represent the more dynamic sectors of the past ice sheets (Landvik et al., 2005; Andreassen et al., 2008; Winsborrow et al., 2010). In the review of Svendsen et al. (2004b), the Early and Middle Weichselian glacial history of the northern and north-western part of the Barents Ice Sheet was mainly based on the compilation by Mangerud et al. (1998) determined by sediment successions on the Svalbard archipelago and marine sediment cores.

Two major glacial advances were identified centring around 110 and 60 ka, with a persistent ice cover of at least the northern Barents Sea between. The westward extent of the 110 ka advance can be questioned on the basis of optically stimulated luminescence (OSL) redating (J.Y.L. Landvik, unpublished) of the coastal site of Skilvika (cf. Mangerud et al., 1998). This work discards the published thermoluminescence (TL) ages and shows that the ice advance that was the most extensive westwards is of pre-Eemian age, not Early Weichselian as earlier assumed by Landvik et al. (1992). However, reinvestigation (Alexanderson et al., 2010) of the Leinestranda site (Miller et al., 1989) in the inter-fjord areas of north-west Spitsbergen shows a glacial event succeeded by high (>30 m) relative sea level at 99 ± 8 ka. In this 'inter-ice-stream' setting, the ice cover may have been of longer duration compared to the rather short glacial phase as reconstructed from

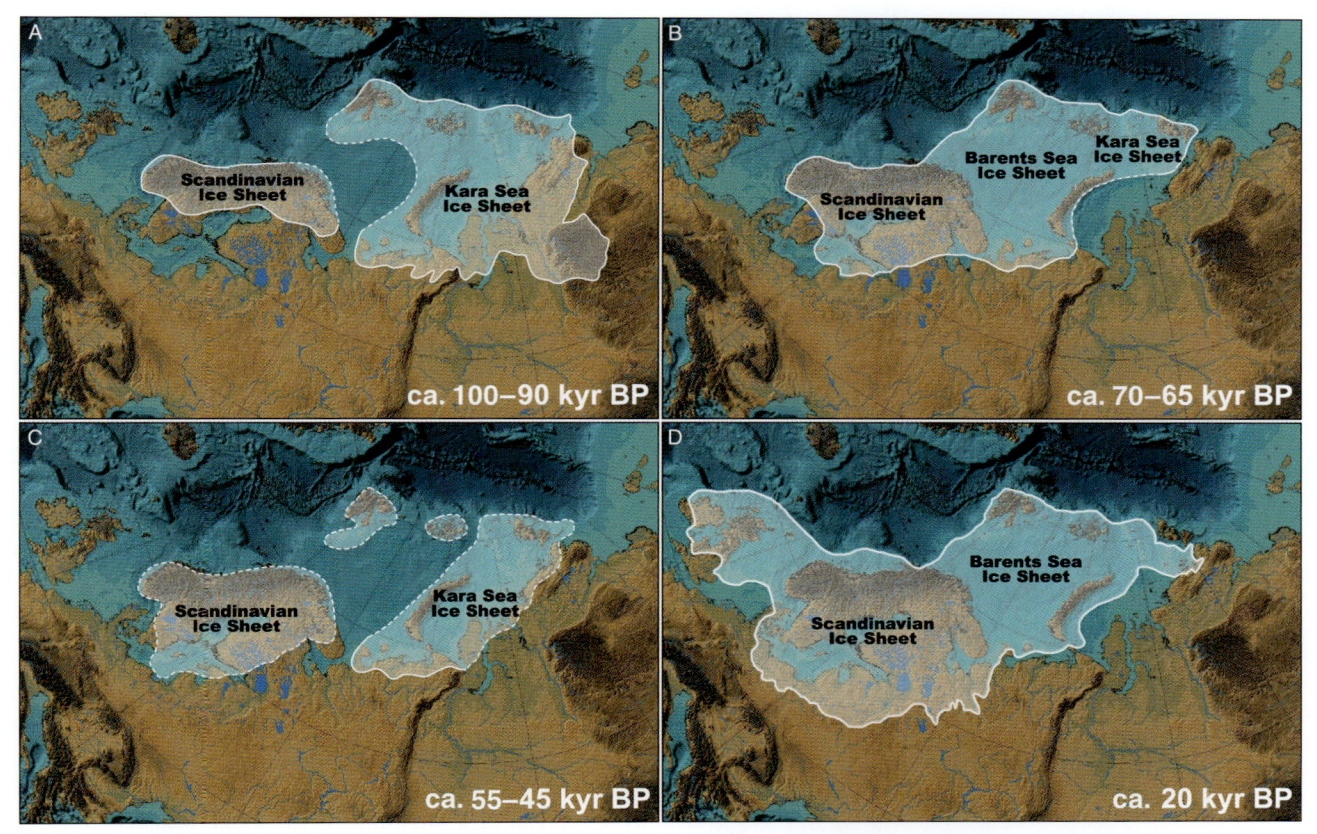

FIGURE 27.4 Reconstruction of Eurasian Weichselian ice sheets during four stages. Ice-dammed lakes in front of the glacier margin are not included. From Larsen et al. (2006).

the palaeo-ice-stream system of Isfjorden (Mangerud and Svendsen, 1992; Mangerud et al., 1998).

The ice advance during MIS 4 represented a significant glacial load, as shown by high deglacial marine levels along the west coast of Svalbard (Mangerud et al., 1998). Exposure-age dating of glacial erratics indicates that the outer islands of north-west Svalbard were covered by a thick ice sheet prior to 70 ka, whereas the Late Weichselian Ice Sheet only inundated the lower ground (Landvik et al., 2003). The Early Weichselian (MIS 4) glaciation of Svalbard was succeeded by ice-free conditions from ca. 50 ka until the onset of the Late Weichselian glaciation, as has also been confirmed for the Barents Sea by the new evidence from Novaja Zemlya (Mangerud et al., 2008).

27.8. THE LATE WEICHSELIAN ICE SHEET

The extent of the Late Weichselian Ice Sheet over the Barents and Kara Sea is now well established, as outlined by Svendsen et al. (2004b). Glaciers reached the shelf break in the west and north, reached the Russian mainland in the White Sea area only, terminated in the Pechora and Kara Sea, east of the Novaja Zemlya coastline, possibly reached the coast of the Taymyr Peninsula (Polyak et al., 2000, 2008) but left the Severnaya Zemlya islands ice free (Möller et al., 2006). The major achievement during the last few years has been the increased understanding of the ice-sheet dynamics from mapping of glacial sea-floor morphology and ice-sheet/ocean interplay in the coastal areas of Svalbard. Most past reconstructions of the ice sheet (e.g. Vorren and Laberg, 1996; Landvik et al., 1998; Stokes and Clark, 2001) have incorporated some generalised ice streams, mostly following bathymetric lows and feeding the TMFs along the Barents Sea margin. With detailed bathymetric data and 3D seismic data, it has been possible to reconstruct ice streams with great precision.

27.8.1. The Barents Sea

Winsborrow et al. (2010) have recently reconstructed the Late Weichselian ice streams and the deglaciation pattern in the southern Barents Sea (Fig. 27.5). At the Late Weichselian maximum, an ice sheet covered the whole of the Barents Sea continental shelf extending to the shelf edge. Ice drainage was dominated by the Bjørnøyrenna Ice Stream fed by large source areas to the north-east and south. Two main southerly tributary ice streams merged with the Bjørnøyrena Ice Stream. In the east, an ice stream flowed

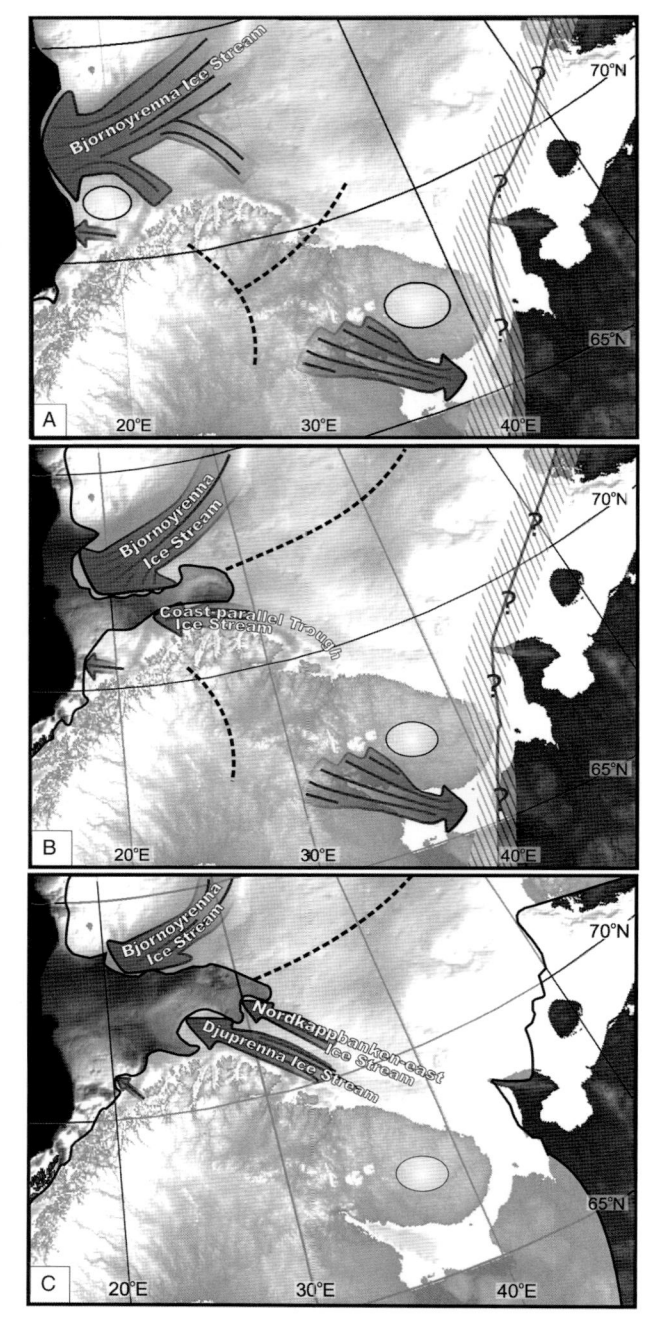

FIGURE 27.5 Reconstruction of the Late Weichselian maximum and subsequent deglaciation of the Barents Sea. Ice streams are shown as large arrows, cold-based ice as white discs and possible ice divides as dashed lines. Modified from Winsborrow et al. (2010).

into the White Sea through Kandalaksha Gulf. The landform record does not indicate whether this ice stream continued south-eastwards terminating in eastern Pechora Sea, as proposed by Demidov et al. (2006), or whether it curved along the coast of Kola Peninsula, perhaps contributing ice to the Bjørnøyrenna Ice Stream.

Deglaciation began with significant retreat of the ice margin in the southern Barents Sea (Fig. 27.5B). A grounding-zone wedge in Bjørnøyrenna indicates retreat and subsequent readvance of this ice stream. The Coast-parallel Trough Ice Stream was fed by tributaries draining through the fjords of mainland Norway. In the east, the ice-sheet extent had not yet reached its maximum (Larsen et al., 1999). The deglaciation occurred early in Storfjordrenna and the western Svalbard margin. Planktonic foraminifera in glaciomarine sediments indicate open-water conditions between 19.4 and 15.3 cal. ka BP (Rasmussen et al., 2007). Based on radiocarbon dates from Andfjorden in northern Norway (Vorren and Plassen, 2002) and Tromsøflaket (Vorren et al., 1978), Winsborrow et al. (2010) assigned an age of ~17 cal. ka BP to this early deglacial stage (Fig. 27.5B).

The next stage reconstructed by Winsborrow et al. (Fig. 27.5C) marks a significant change in the dynamics of the ice sheet. The centre of maximum ice volume shifted eastwards, and much of the south-western Barents Sea was deglaciated. This eastwards shift is indicated by the major readvance of the Djuprenna and Nordkappbanken-east Ice Streams, which were fed by ice flowing north across the Kola Peninsula and north-east across Kandalaksha Gulf and the eastern tip of Kola Peninsula. At this stage, the deepest parts of Bjørnøyrenna were deglaciated, but the Bjørnøyrenna Ice Stream was still active, and Winsborrow et al. identified a grounding-zone wedge 250 km from the shelf break marking its position at this time. Along the western continental shelf, the ice had retreated close to the coast. Winsborrow et al. suggested that this occurred at ~16 cal. ka BP, by which time large parts of the western Barents Sea were ice free indicated by glaciomarine conditions in mid-Ingøydjupet by 15.7 cal. ka BP (Vorren and Kristoffersen, 1986). According to Winsborrow et al. (2010), this stage marks a major shift in the dynamics of the Barents Sea Ice Sheet, with the largest ice volume and major ice streams located in the eastern sector. They linked this stage to the maximum Late Weichselian ice extent in north-west Russia to which Larsen et al. (1999) assigned a maximum age of 17.0 ± 0.5 cal. ka BP and a minimum age of 14.8 ± 0.6 cal. ka BP.

27.8.2. Svalbard

At about 32 cal. ka BP, the Svalbard glaciers began to grow (e.g. Andersen et al., 1996; Landvik et al., 1998). The shelf break west of Svalbard was reached by the glaciers at about 24 cal. ka BP (Jessen et al., 2010). Along the western coast of Svalbard, fast-flowing ice streams drained the major fjords and cross-shelf troughs, whereas the inter-ice-stream areas were dominated by dynamically less active, possibly periodically cold-based portions of the ice sheet (Landvik et al., 2005; Ottesen et al., 2007). This ice sheet left distinct subglacially formed glacial mega-scale lineations in the ice-stream zones, whereas younger transverse ridges related to ice retreat characterise the inter-ice-stream areas (Ottesen and Dowdeswell, 2009). By mapping these features, Ottesen et al. (2007) reconstructed the main ice-stream systems that were active within the Svalbard archipelago during the Late Weichselian (Fig. 27.6). Ice drainage along the major fjords of western Svalbard, as well as the Hinlopen Strait, points to an ice-sheet culmination over the northern Barents Sea. Mega-scale glacial lineations now suggest that this centre of ice flow was situated in the southern part of the Hinlopen Strait (Dowdeswell et al., 2010), not south-east of Kong Karls Land as previously assumed (Lambeck, 1996; Landvik et al., 1998; Siegert et al., 2001).

On the basis of studies of sediment cores recovered west of Svalbard (Jessen et al., 2010), the initial retreat occurred about 20.5 cal. ka BP that is somewhat earlier than previous reconstructions (e.g. Andersen et al., 1996). At that time, the conditions were cold and the influence of Atlantic water reduced; therefore, Jessen et al. speculated that the retreat resulted from reduced accumulation of snow. The mouth of the best-studied fjord, Isfjorden, was deglaciated at ca. 14.1 cal. ka BP, and the final glacier retreat in the inner fjord system terminated around 11.3 cal. ka BP

FIGURE 27.6 Reconstruction of the Late Weichselian ice margin and ice-sheet flow regime in the Svabard area interpreted from sea-floor morphology. After Ottesen et al. (2007).

(e.g. Mangerud et al., 1992, 1998; Elverhøi et al., 1995; Svendsen et al., 1996; Lønne, 2005). The deglaciation pattern in the Spitsbergen fjords and the location of the glacier fronts during the Younger Dryas Stadial cooling are still under discussion. Mangerud and Svendsen (1990), Svendsen and Mangerud (1992) and Mangerud and Landvik (2007) indicated that local glaciers on western Spitsbergen were smaller during the Younger Dryas than during the Little Ice Age, and that no evidence for a Younger Dryas-age readvance exists. It has been suggested that the tributaries of Isfjorden were occupied by outlet glaciers and that their fronts were located either in the inner main fjord (Mangerud et al., 1992) or 'far out in the main fjord' (Svendsen et al., 1996). Retarded glacio-isostatic uplift during the Younger Dryas Chron (Landvik et al., 1987; Lehman and Forman, 1992; Forman et al., 2004) led Svendsen et al. (1996) to conclude that the glaciers in the inner parts of Isfjorden readvanced during this period. Boulton (1979) suggested a marked Younger Dryas glacier readvance in the Billefjorden area.

ACKNOWLEDGEMENTS

This work is a contribution to the 'Democen' project (Depositional models for Cenozoic sandy sediments) funded by the Norwegian Research Council and Statoil, and the International Polar Year project 'Science Pub' (Arctic Natural Climate and Environmental Changes and Human Adaption: From Science to Public awareness) funded by the Research Council of Norway. J. P. Holm helped draft the figures.

REFERENCES

Alexanderson, H., Landvik, J.Y., Ryen, H.T., 2011. Chronology and styles of glaciation in and inter-fjord setting, northwestern Svalbard. Boreas 40, 175–197.
Andersen, E.S., Solheim, A., Elverhøi, A., 1994. Development of a glaciated continental margin: exemplified by the western margin of Svalbard. In: Thunston, D.K., Fujita, K. (Eds.), International Conference on Arctic Margins. Proceedings Anchorage, Alaska 1992, pp. 155–160. U.S. Department of the Interior, Mineral Management Service, Alaska Outer Continental Shelf Region; OCS Study, MMS 94-0040.
Andersen, E.S., Dokken, T.M., Elverhøi, A., Solheim, A., Fossen, I., 1996. Late Quaternary sedimentation and glacial history of the Western Svalbard continental margin. Mar. Geol. 133, 123–156.
Andreassen, K., Winsborrow, M., 2009. Signature of ice streaming in Bjørnøyrenna, Polar North Atlantic, through the Pleistocene and implications for ice-stream dynamics. Ann. Glaciol. 50, 17–26.
Andreassen, K., Nilssen, L.C., Rafaelsen, L., Kuilman, L., 2004. Three-dimensional seismic data from the Barents Sea margin reveal evidence of past ice streams and their dynamics. Geology 32, 729–732.
Andreassen, K., Ødegaard, C.M., Rafaelsen, B., 2007. Imprints of former ice streams, imaged and interpreted using industry three-dimensional seismic data from the south-western Barents Sea. Geological Society London, Special Publication 277, 151–169.
Andreassen, K., Laberg, J.S., Vorren, T.O., 2008. Seafloor geomorphology of the SW Barents Sea and its glaci-dynamic implications. Geomorphology 97, 157–177.
Boulton, G.S., 1979. Glacial history of the Spitsbergen archipelago and the problem of a Barents Shelf ice sheet. Boreas 8, 31–57.
Butt, F.A., Elverhøi, A., Solheim, A., Forsberg, C.F., 2000. Deciphering late Cenozoic development of the western Svalbard margin from ODP site 986 results. Mar. Geol. 169, 373–390.
Butt, F.A., Drange, H., Elverhøi, A., Otterå, O.H., Solheim, A., 2002. Modelling Late Cenozoic isostatic elevation changes in the Barents Sea and their implications for oceanic and climatic regimes: preliminary results. Quatern. Sci. Rev. 21, 1643–1660.
Channell, J.E.T., Smelror, M., Jansen, E., Higgins, S., Lehman, B., Eidvin, T., et al., 1999. Age models for glacial fan deposits off East Greenland and Svalbard (ODP Site 986 and Site 987). In: Raymo, M., Jansen, E., Blum, P., Herbert, T.D. (Eds.), Proceeding Ocean Drilling Program, Scientific Results, 162. Ocean Drilling Program, College Station, Texas, pp. 149–166.
Demidov, I.N., Houmark-Nielsen, M., Kjaer, K.H., Larsen, E., 2006. The last Scandinavian Ice Sheet in northwestern Russia: ice flow patterns and decay dynamics. Boreas 35, 425–443.
Dowdeswell, J.A., Hogan, K.A., Evans, J., Noormets, R., Ó Cofaaigh, C., Ottesen, D., 2010. Past ice sheet flow east of Svalbard inferred from streamlined subglacial landforms. Geology 38, 163–166.
Eidvin, T., Nagy, J., 1999. Foraminiferal biostratigraphy of Pliocene deposits at Site 986, Svalbard margin. In: Raymo, M., Jansen, E., Blum, P., Herbert, T.D. (Eds.), Proceeding Ocean Drilling Program, Scientific Results, 162. Ocean Drilling Program, College Station, Texas, pp. 3–17.
Eidvin, T., Jansen, E., Riis, F., 1993. Chronology of Tertiary fan deposits off the western Barents Sea: implications for the uplift and erosion history of the Barents shelf. Mar. Geol. 112, 109–131.
Elverhøi, A., Solheim, A., 1983. The Barents Ice Sheet, a sedimentological discussion. Polar Res. 1, 23–43.
Elverhøi, A., Svendsen, J.I., Solheim, A., Andersen, E.S., Milliman, J., Mangerud, J., et al., 1995. Late Quaternary sediment yield from the high arctic Svalbard area. J. Geol. 103, 1–17.
Elverhøi, A., Hooke, R.L., Solheim, A., 1998. Late Cenozoic erosion and sedimentation yield from the Svalbard-Barents sea region: implication for understanding erosion of glacierized basins. Quatern. Sci. Rev. 17, 209–241.
Faleide, J.I., Solheim, A., Fiedler, A., Hjelstuen, B.O., Andersen, E.S., Vanneste, K., 1996. Late Cenozoic evolution of the western Barents Sea—Svalbard continental margin. Glob. Planet. Change 12, 53–74.
Fiedler, A., Faleide, J.I., 1996. Cenozoic sedimentation along the southwestern Barents Sea margin in relation to uplift and erosion of the shelf. Glob. Planet. Change 12, 75–93.
Forman, S.L., Lubinski, D.J., Ingólfsson, Ó., Zeeberg, J.J., Snyder, J.A., Siegert, M., et al., 2004. A review of postglacial emergence on Svalbard, Franz Josef Land and Novaya Zemlya, northern Eurasia. Quatern. Sci. Rev. 23, 1391–1434.
Forsberg, C.F., Solheim, A., Elverhøi, A., Jansen, E., Channell, J.E.T., Andersen, E.S., 1999. The depositional environment of the western Svalbard margin during the late Pliocene and Pleistocene: sedimentary facies changes at Site 986. Proceedings Ocean Drilling Program. Sci. Results 162, 233–246.
Geissler, W.H., Jokat, W., 2004. A geophysical study of the northern Svalbard continental margin. Geophys. J. Int. 158, 50–66.

Grosswald, M.G., 1980. Late Weichselian Ice Sheet of northern Eurasia. Quatern. Res. 13, 1–32.

Hjelstuen, B.O., Elverhøi, A., Faleide, J.I., 1996. Cenozoic erosion and sediment yield in the drainage area of the Storfjorden fan. Glob. Planet. Change 12, 95–117.

Hjelstuen, B.O., Eldholm, O., Faleide, J.I., 2007. Recurrent Pleistocene mega-failures on the SW Barents Sea margin. Earth Planet. Sci. Lett. 258, 605–618.

Houmark-Nielsen, M., Funder, S., 1999. Pleistocene stratigraphy of Kongsfjordhallet, Spitsbergen, Svalbard. Polar Res. 18, 39–49.

Hughes, T., Denton, G.H., Grosswald, M.G., 1977. Was there a Late Würm ice sheet. Nature 266, 596–602.

Hughes, T., Denton, G.H., Andersen, B.G., Schilling, D.H., Fastook, F.L., Lingle, C.S., 1981. The last great ice sheet: a global view. In: Denton, G.H., Hughes, T. (Eds.), The Great Ice Sheets. John Wiley & Sons, New York, p. 484.

Ingólfsson, Ó., Möller, P., Lokrantz, H., 2008. Late Quaternary marine-based Kara Sea ice sheets: a review of terrestrial stratigraphic data highlighting their formation. Polar Res. 27, 152–161.

Jessen, S.P., Rasmussen, T.L., Nielsen, T., Solheim, A., 2010. A new Late Weichselian and Holocene marine chronology for the Western Svalbard slope 30,000-0 cal years BP. Quatern. Sci. Rev. 29, 1301–1312.

Knies, J., Matthiessen, J., Vogt, C., Laberg, J.S., Hjelstuen, B.O., Smelror, M., et al., 2009. The Plio-Pleistocene glaciation of the Barents Sea-Svalbard region: a new model based on revised chronostratigraphy. Quatern. Sci. Rev. 28, 812–829.

Knutsen, S.-M., Richardsen, G., Vorren, T.O., 1992. Late Miocene-Pliocene sequence stratigraphy and mass-movements on the western Barents Sea margin. In: Vorren, T.O. et al., (Ed.), Arctic Geology and Petroleum Potential. NPF Special Publication, 2 Elsevier Science, Amsterdam, pp. 573–606.

Kvasov, D.D., 1978. The Barents Ice Sheet as a relay regulator of glacial and interglacial alternation. Quatern. Res. 9, 288–289.

Laberg, J.S., Vorren, T.O., 1996. The Middle and late Pleistocene evolution of the Bear Island Trough Mouth Fan. Glob. Planet. Change 12, 309–330.

Laberg, J.S., Andreassen, K., Knies, J., Vorren, T.O., Winsborrow, M., 2010. Late Pliocene–Pleistocene development of the Barents Sea Ice Sheet. Geology 38, 107–110.

Lambeck, K., 1996. Limits on the areal extent of the Barents Sea Ice Sheet in Late Weichselian time. Glob. Planet. Change 12, 41–51.

Landvik, J.Y. Mangerud, J., Salvigsen, O., 1987. The Late Weichselian and Holocene shoreline displacement on the western central coast of Svalbard. Polar Res. 5 n.s., 29–44.

Landvik, J.Y., Bolstad, M., Lycke, A.K., Mangerud, J., Sejrup, H.P., 1992. Weichselian stratigraphy and palaeoenvironments at Bellsund, western Svalbard. Boreas 21, 335–358.

Landvik, J.Y., Bondevik, S., Elverhøi, A., Fjeldskaar, W., Mangerud, J., Salvigsen, O., et al., 1998. The last glacial maximum of the Barents Sea and Svalbard area: ice sheet extent and configuration. Quatern. Sci. Rev. 17, 43–75.

Landvik, J.Y., Brook, E.J., Gualtieri, L., Raisbeck, G., Salvigsen, O., Yiou, F., 2003. Northwest Svalbard during the last glaciation: ice free areas existed. Geology 31, 905–908.

Landvik, J.Y., Ingólfsson, Ó., Mienert, J., Lehman, S.J., Solheim, A., Elverhøi, A., et al., 2005. Rethinking Late Weichselian ice sheet dynamics in coastal NW Svalbard. Boreas 34, 7–24.

Larsen, E., Lyså, A., Demidov, I., Funder, S., Houmark-Nielsen, M., Kjær, K.H., et al., 1999. Age and extent of the Scandinavian Ice Sheet in northwest Russia. Boreas 28, 115–132.

Larsen, E., Kjær, K.H., Demidov, I.N., Funder, S., Grøsfjeld, K., Houmark-Nielsen, M., et al., 2006. Late Pleistocene glacial and lake history of northwestern Russia. Boreas 35, 394–424.

Lehman, S.J., Forman, S.L., 1992. Late Weichselian glacier retreat in Kongsfjorden, West Spitsbergen, Svalbard. Quatern. Res. 37, 139–154.

Lønne, I., 2005. Faint traces of high Arctic glaciations: an early Holocene ice-front fluctuation in Bolterdalen, Svalbard. Boreas 34, 308–323.

Mangerud, J., Landvik, J., 2007. Younger Dryas cirque glaciers in western Spitsbergen: smaller than during the Little Ice Age. Boreas 36, 278–285.

Mangerud, J., Svendsen, J.I., 1990. Deglaciation chronology inferred from marine sediments in a proglacial lake basin, western Spitsbergen, Svalbard. Boreas 19, 249–272.

Mangerud, J., Svendsen, J.I., 1992. The last interglacial-glacial period on Spitsbergen, Svalbard. Quatern. Sci. Rev. 11, 633–664.

Mangerud, J., Bolstad, M., Elgersma, A., Helliksen, D., Landvik, J.Y., Lønne, I., et al., 1992. The Last Glacial Maximum on Spitsbergen. Quatern. Res. 38, 1–31.

Mangerud, J., Dokken, T.M., Hebbeln, D., Heggen, B., Ingólfsson, Ó., Landvik, J.Y., et al., 1998. Fluctuations of the Svalbard-Barents Sea Ice Sheet during the last 150,000 years. Quatern. Sci. Rev. 17, 11–42.

Mangerud, J., Jakobsson, M., Alexanderson, H., Astakhov, V., Clarke, G.K.C., Henriksen, M., et al., 2004. Ice-dammed lakes and rerouting of the drainage of northern Eurasia during the Last Glaciation. Quatern. Sci. Rev. 23, 1313–1332.

Mangerud, J., Kaufman, D., Hansen, J., Svendsen, J.I., 2008. Ice-free conditions in Novaya Zemlya 35,000 to 30,000 cal years BP, as indicated by radiocarbon ages and amino acid racemization evidence from marine molluscs. Polar Res. 27, 187–208.

Mathisov, G.G., 1980. Geomorphological signs of the action of the Scandinavian, Novaya Zemlya, and Spitsbergen ice sheets on the floor of the Barents Sea. Oceanology 20, 440–447.

Miller, G.H., Sejrup, H.P., Lehman, S.J., Forman, S.L., 1989. Glacial history and marine environmental change during the last interglacial-glacial cycle, western Spitsbergen, Svalbard. Boreas 18, 273–296.

Möller, P., Lubinski, D.J., Ingólfsson, Ó., Forman, S.L., Seidenkrantz, M.S., Bolshiyanov, D.Y., et al., 2006. Severnaya Zemlya, Arctic Russia: a nucleation area for Kara Sea ice sheets during the Middle to Late Quaternary. Quatern. Sci. Rev. 25, 2894–2936.

Moran, K., Backman, J., Brinkhuis, H., Clemens, S.C., Cronin, T., Dickens, G.R., et al., 2006. The Cenozoic palaeoenvironment of the Arctic Ocean. Nature 441, 601–605.

Mørk, M.B.E., Duncan, R.A., 1993. Late Pliocene basaltic volcanism on the western Barents shelf margin—implications from petrology and Ar40-Ar39 dating of volcaniclastic debris from a shallow drill core. Nor. Geol. Tidsskr. 73, 209–225.

Ottesen, D., Dowdeswell, J.A., 2009. An inter-ice-stream glaciated margin: submarine landforms and a geomorphic model based on marine-geophysical data from Svalbard. Geol. Soc. Am. Bull. 121, 1647–1665.

Ottesen, D., Dowdeswell, J.A., Rise, L., 2005. Submarine landforms and the reconstruction of fast-flowing ice streams within a large Quaternary ice sheet: the 2500-km-long Norwegian-Svalbard margin (57 degrees-80 degrees N). Geol. Soc. Am. Bull. 117, 1033–1050.

Ottesen, D., Dowdeswell, J.A., Landvik, J.Y., Mienert, J., 2007. Dynamics of the Late Weichselian ice sheet on Svalbard inferred from high-resolution sea-floor morphology. Boreas 36, 286–306.

Polyak, L., Levitan, M., Gataullin, V., Khusid, T., Mikhailov, V., Mukhina, V., 2000. The impact of glaciation, river discharge and sea-level change on Late Quaternary environments in the southwestern Kara Sea. Int. J. Earth Sci. 89, 550–562.

Polyak, L., Niessen, F., Gataulin, V., Gainanov, V., 2008. The easern extent of the Barents-Kara ice sheet during the Last Glacial Maximum based on seismic reflection data from the eastern Kara Sea. Polar Res. 27, 162–174.

Rasmussen, T.L., Thomsen, E., Slubowska, M.A., Jessen, S., Solheim, A., Koc, N., 2007. Palaeoceanographic evolution of the SW Svalbard margin (76 degrees North) since 20,000 radiocarbon years before present. Quatern. Res. 67, 100–114.

Richardsen, G., Knutsen, S.-M., Vail, P.R., Vorren, T.O., 1992. Mid-late Miocene sedimentation on the southwestern Barents Shelf margin. In: Vorren, T.O. et al., (Ed.), Arctic Geology and Petroleum Potential. NPF Special Publication, 2 Elsevier Science, Amsterdam, pp. 539–571.

Ryseth, A., Augustson, J.H., Charnock, M., Haugerud, O., Knutsen, S.-M., Midbøe, R.S., et al., 2003. Cenozoic stratigraphy and evolution of the Sørvestsnaget Basin, southwestern Barents Sea. Norw. J. Geol. 83, 107–130.

Sættem, J., Poole, D.A.R., Ellingsen, L., Sejrup, H.P., 1992. Glacial geology of outer Bjørnøyrenna, southwestern Barents Sea. Mar. Geol. 103, 15–51.

Sættem, J., Hamborg, M., 1987. The geological implications of the upper seismic unit, southwestern Barents Sea. Polar Research 5, 299–301.

Sættem, J., Bugge, T., Fanavoll, S., Goll, R.M., Mørk, A., Mørk, M.B.E., et al., 1994. Cenozoic margin development and erosion of the Barents Sea: core evidence from southwest of Bjørnøya. Mar. Geol. 118, 257–281.

Salvigsen, O., 1981. Radiocarbon dated raised beaches in Kong Karls Land, Svalbard, and their consequences for the glacial history of the Barents Sea. Geogr. Ann. 63A, 283–291.

Salvigsen, O., Nydal, R., 1981. The Weichselian glaciation in Svalbard before 15,000 B.P.. Boreas 10, 433–446.

Sejrup, H., Larsen, E., Haflidason, H., Berstad, I., Hjelstuen, B.O., Jonsdotter, H.E., et al., 2003. Configuration, history and impact of the Norwegian Channel Ice stream. Boreas 32, 18–36.

Siegert, M.J., Dowdeswell, J.A., Hald, M., Svendsen, J.I., 2001. Modelling the Eurasian Ice Sheet through a full (Weichselian) glacial cycle. Glob. Planet. Change 31, 367–385.

Smelror, M., 1999. Pliocene–Pleistocene and redeposited dinoflagellate cysts from the western Svalbard Margin (Site 986): biostratigraphy, paleoenvironments and sediment provenance. In: Raymo, M., Jansen, E., Blum, P., Herbert, T.D. (Eds.), Proceedings of the Ocean Drilling Program, Scientific Results, 162. Ocean Drilling Program, College Station, TX, USA, pp. 83–97.

Solheim, A., Kristoffersen, Y., 1984. Sediments above the upper regional unconformity: thickness, seismic stratigraphy and outline of the glacial history. Nor. Polarinst. Skr. 179B, 26 pp.

Solheim, A., Faleide, J.I., Andersen, E.S., Elverhøi, A., Forsberg, C.F., Vanneste, K., et al., 1998. Late Cenozoic seismic stratigraphy and glacial geological development of the East Greenland and Svalbard Barents Sea continental margins. Quatern. Sci. Rev. 17, 155–184.

Stickley, C.E., St John, K., Koç, N., Jordan, R.W., Passchier, S., Pearce, R.B., et al., 2009. Evidence for middle Eocene Arctic sea ice from diatoms and ice-rafted debris. Nature 460, 376–379.

Stokes, C.R., Clark, C.D., 2001. Palaeo-ice streams. Quatern. Sci. Rev. 20, 1437–1457.

Svendsen, J.I., Mangerud, J., 1992. Paleoclimatic inferences from glacial fluctuations on Svalbard during the last 20 000 years. Climate Dyn. 6, 213–220.

Svendsen, J.I., Elverhøi, A., Mangerud, J., 1996. The retreat of the Barents Sea Ice Sheet on the western Svalbard margin. Boreas 25, 244–256.

Svendsen, J.I., Alexanderson, H., Astakhov, V.I., Demidov, I., Dowdeswell, J.A., Funder, S., et al., 2004a. Late Quaternary ice sheet history of northern Eurasia. Quatern. Sci. Rev. 23, 1229–1271.

Svendsen, J.I., Gataullin, V., Mangerud, J., Polyak, L., 2004b. The glacial history of the Barents and Kara Sea region. In: Ehlers, J., Gibbard, P.L. (Eds.), Quaternary Glaciations—Extent and Chronology. Part 1: Europe. Developments in Quaternary Science Elsevier, Amsterdam, pp. 359–368.

Vorren, T.O., Kristoffersen, Y., 1986. Late Quaternary glaciations in the south-western Barents Sea. Boreas 15, 51–59.

Vorren, T.O., Laberg, J.S., 1996. Late glacial air temperature, oceanographic & ice sheet interactions in the southern Barents Sea region. In: Andrews, J.T., Austin, W.E.N., Bergsten, H., Jennings, A.E. (Eds.), Late Quaternary Paleoceanography of the North Atlantic Margins, Geological Society, London, Special Publication, 111, London, 303–321.

Vorren, T.O., Laberg, J.S., 1997. Trough mouth fans—palaeoclimate and ice-sheet monitors. Quatern. Sci. Rev. 16, 865–881.

Vorren, T.O., Plassen, L., 2002. Deglaciation and palaeoclimate of the Andfjord-Vågsfjord area, north Norway. Boreas 31, 97–125.

Vorren, T.O., Strass, I.F., Lind-Hansen, O.W., 1978. Late Quaternary sediments and stratigraphy on the continental shelf off Troms & west Finnmark, northern Norway. Quatern. Res. 19, 340–365.

Vorren, T.O., Hald, M., Thomsen, E., 1984. Quaternary sediments and environments on the continental shelf off northern Norway. Mar. Geol. 57, 229–257.

Vorren, T.O., Kristoffersen, Y., Andreassen, K., 1986. Geology of the inner shelf west of North Cape, Norway. Norsk Geologisk Tidsskrift 66, 99–105.

Vorren, T.O., Hald, M., Lebesbye, E., 1988. Late Cenozoic environments in the Barents Sea. Paleoceanography 3, 601–612.

Vorren, T.O., Lebesbye, E., Andreassen, K., Larsen, K.B., 1989. Glacigenic sediments on a passive continental margin as exemplified by the Barents Sea. Mar. Geol. 85, 251–272.

Vorren, T.O., Lebesbye, E., Larsen, K., 1990. Geometry and genesis of the glacigenic sediments in the southern Barents Sea. In: Dowdeswell, J.A., Scource, J.D. (Eds.), Glaciomarine Environments: Processes & Sediments, Geological Society Special Publication, 53, London, 309–328.

Vorren, T.O., Richardsen, G., Knutsen, S.-M., Henriksen, E., 1991. Cenozoic erosion and sedimentation in the western Barents Sea. Mar. Petrol. Geol. 8, 317–340.

Vorren, T.O., Laberg, J.S., Blaume, F., Dowdeswell, J.A., Kenyon, N.H., Mienert, J., et al., 1998. The Norwegian Greenland Sea continental margins: morphology and late Quaternary sedimentary processes and environment. Quatern. Sci. Rev. 17, 273–302.

Winsborrow, C.M., Andreassen, K., Corner, G.D., Laberg, J.S., 2010. Deglaciation of marine based ice sheet: Late Weichelian paleo-ice dynamics & retreat in the southern Barents Sea reconstructed from onshore & offshore glacial geomorphology. Quatern. Sci. Rev.

Chapter 28

Glacial History of the Taymyr Peninsula and the Severnaya Zemlya Archipelago, Arctic Russia

Per Möller[1,]*, Christian Hjort[1], Helena Alexanderson[1] and Florian Sallaba[2]

[1]*Department of Earth and Ecosystem Sciences, Division of Geology, Lund University, Sölvegatan 12, SE-223 62 Lund, Sweden*
[2]*Department of Earth and Ecosystem Sciences, Division of Physical Geography and Ecosystem Analysis, Lund University, Sölvegatan 12, SE-223 62 Lund, Sweden*
*Correspondence and requests for materials should be addressed to Per Möller. E-mail: per.moller@geol.lu.se

28.1. INTRODUCTION

Since the late 1980s, there has been a re-assessment of the glacial history of the Russian Arctic, from the Kola Peninsula in the west to the Lena Delta and beyond in the east (e.g. Svendsen et al., 1999, 2004). In this context, work has been carried out from Lake Taymyr, south of the Byrranga Mountains, northwards to the Arctic coast of the Taymyr Peninsula and on the Severnaya Zemlya archipelago (Fig. 28.1). The present contribution is a QUEEN member's summary, update and extension of the Taymyr Peninsula chapter (Hjort et al., 2004) in the first edition of this book. Here, the authors review the results regarding the glacial and marine history of Taymyr and the Severnaya Zemlya islands in chronological order, with reference to previous as well as to contemporary Russian and other work. It should be noted that the Putorana Plateau at the southern base of the Taymyr Peninsula is not included.

28.2. THE MIDDLE PLEISTOCENE PRE-SAALIAN HISTORY

Summaries on the extensive Russian literature on the Quaternary stratigraphy of the west Siberian Plain and the Taymyr Peninsula, and critical reviews of these, have been provided by Arkhipov et al. (1986) and Astakhov (2001, 2004a,b) among others. In many regions, three to five diamict beds are identified, interbedded with terrestrial interglacial strata in subarctic areas and usually marine deposits within the arctic areas within or just outside the boundaries of the Late Pleistocene glaciations (Astakhov, 2004a). However, such stratigraphical successions are rarely observed in superposition but constructed from borehole correlations. The 'pancake' stratigraphy of the Ozernaya River sections on southern October Revolution Island in the Severnaya Zemlya archipelago, first briefly described by Bolshiyanov and Makeyev (1995) and later more comprehensively by Möller et al. (2007), is thus unique for Arctic Russia in revealing at least four marine units intercalated with tills in superposition, two of which seem to be of Middle Pleistocene age (Fig. 28.2).

The two lowermost marine units, represented by deeper-marine to prograding deltaic sediment successions (marine unit M-I) and a glaciomarine/marine off-shore to shoreface sediment succession (marine unit M-II), are erosionally cut and overlain by tills (till units T-II and T-III; Fig. 28.2). Till fabrics and glaciotectonic structures within underlying marine sediment imply that the ice-movement direction was from a northern sector towards the south (Fig. 28.2), thus suggesting that till deposition during the expansions for T-II–T-IV was from the local ice cap(s) (Möller et al., 2007). From retrieved electron spin resonance (ESR) and green stimulated luminescence (GSL) ages of marine unit M-I (~300–400 ka), it is suggested that it possibly represents a marine isotope stage (MIS) 10–9 transitional phase with a deep-marine setting induced by isostatic loading from the previous glacial phase (T-I). Similar ages for marine deposits have been reported from a few localities on the Taymyr Peninsula (Bolshiyanov et al., 1998; Fig. 28.1), such as along the Shrenk and Mamota Rivers, north of the Byrranga Mountains, at Labaz Lake, south of the Byrrangas and at the Novorybnoe Village, along the Khatanga River. The marine unit M-I could possibly correlate to the *Tobol Interglacial* (Arkhipov, 1989) of the

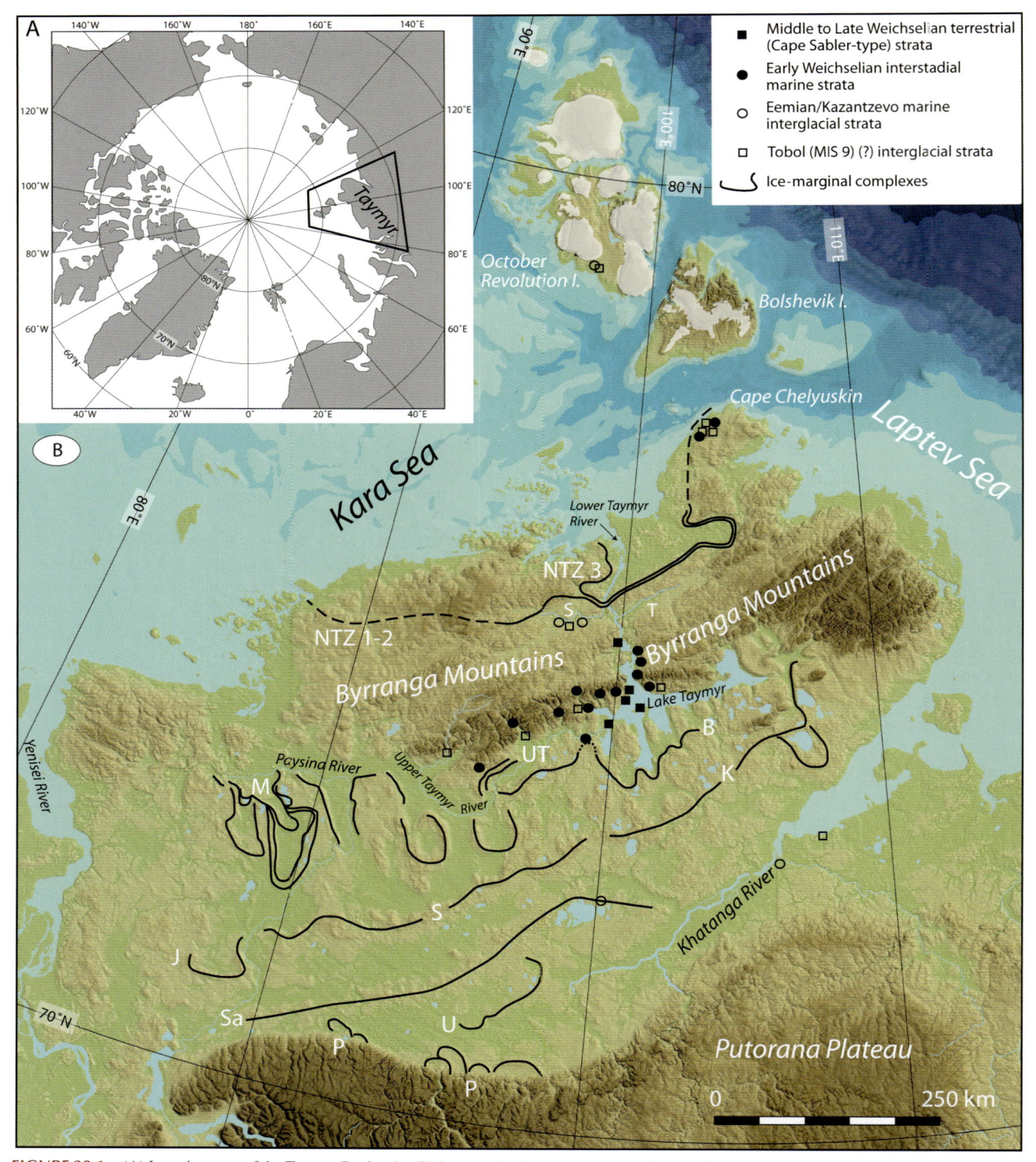

FIGURE 28.1 (A) Location map of the Taymyr Peninsula. (B) Ice-marginal complexes on the Taymyr Peninsula, named according to Kind and Leonov (1982), but partly redrawn after Landsat image interpretation by Möller and Sallaba (2010): U, Urdachsk; Sa, Sampesa; K, Severokokorsk; J, Jangoda; S, Syntabul; B, Baikuronyora; M, Mokoritto; UT, Upper Taymyr; NTZ, North Taymyr ice-marginal zone (Alexanderson et al., 2001). Lines south of the Urdachsk line (marked P) are piedmont glacier moraines, deposited from the Putorana Plateau. The Urdachsk and Sampesa moraines are possibly connected to the Saalian glaciation, while moraines north of these are tentatively from the Weichselian. Note that the Baikuronyora (B) moraine is drawn to connect with the Upper Taymyr (UT) moraine and not towards the SW, connecting to the Syntabul (S) moraine, as proposed by Kind and Leonov (1982). The JSB line, proposed by Svendsen et al. (2004) as marking the Weichselian KSIS maximum on Taymyr seems not so viable today, as these moraine

western Siberian lowlands, capped by till of the *Samarovo glaciation* (MIS 8) which forms the maximum glacial drift boundary in western Siberia. However, such correlations are highly speculative because of the poor age constraints; also older ESR and thermoluminescence (TL) ages have been presented for the Tobol Interglacial (Volkova and Babushkin, 2000; as cited in Astakhov, 2004a), suggesting an MIS 11 age, that is, a correlative of the Holsteinian Stage of western Europe (Astakhov, 2004b).

The ESR and GSL ages recorded in marine unit M-II also suggest that this marine event is of a pre-Saalian age, most probably representing a deglacial sedimentary succession, formed at the transition from MIS 8 to 7 with a relative high sea level stand in excess of 71 m a.s.l. at deglaciation (Fig. 28.2). A direct correlative on Taymyr for this marine event is not recorded. However, Molodkov and Bolikhovskaya (2002) report an ESR age cluster at around 220 ka for a number of raised marine deposits in northern Eurasia. The age envelope for the marine unit M-II possibly fits into the *Shirta Interglacial* (MIS 7), as dated on the north-western Siberian plain (Astakhov, 2004a), sandwiched between the *Taz* and *Samarovo glaciations*.

28.3. THE MIDDLE/LATE PLEISTOCENE TRANSITION: THE SAALIAN AND EEMIAN STAGES

According to Svendsen et al. (2004), the maximum glaciation of the Kara Sea shelf and Taymyr Peninsula, and merging of the ice caps over the Putorana Plateau and the Anabar Uplands, that is, when this whole north-eastern region was covered by an Eurasian ice sheet, took place during the late Saalian (MIS 6), with a maximum spread at ca. 140 ka and with deglaciation of the shelf at around 130 ka (Lambeck et al., 2006). However, the absolute Eurasian ice sheet maximum in this area might have already occurred during MIS 8 (the Samarovo glaciation; cf. Astakhov, 2004a), the Pleistocene maximum glaciation limit thus being spatially diachronous (Astakhov, 2008). The Saalian is considered equal to the *Taz glaciation* of West and North Central Siberia. On Severnaya Zemlya, this glaciation is associated with the highest recorded (up to 140 m a.s.l.; Fig. 28.2, marine unit M-III) beach–ridge complexes formed at deglaciation (Möller et al., 2007), and this only 200 km from the shelf break to the deep Arctic Ocean. This implies an ice-sheet thickness in excess of 3000 m over the Kara Sea (Lambeck et al., 2006) and concurs with possible ice-shelf grounding at 1000 m water depths on the Lomonosov Ridge in the central Arctic Ocean (Jakobsson et al., 2001, 2008b; Polyak et al., 2001). On the southern flank of Taymyr, a number of wide ice-thrust/push-moraine ridges have been documented from earlier mapping, for example, by Kind and Leonov (1982), and from recent LANDSAT satellite imagery interpretation and modelling of digital elevation (ASTER) data (Möller and Sallaba, 2010) (Fig. 28.1). If the suggestion in Svendsen et al. (2004) that the Jangoda–Syntabyl–Baikuronyora zone (the JSB line) marks the maximum Kara Sea Ice Sheet (KSIS) extent during an Early Weichselian glaciation is correct, then the Urdachsk (U), Sampesa (Sa) and Severokokorsk (K) ice-marginal zones (Fig. 28.1) are older and might thus represent recessional phases of the Saalian/Taz glaciation after separation from the ice cap over the Putorana plateau.

The deep isostatic depression of the land, caused by the thick Saalian ice load, led to widespread post-glacial marine inundation at deglaciation and subsequent deposition of marine sediments over large parts of Taymyr and Severnaya Zemlya. Such *Kazantzevo Interglacial* (equivalent to the Eemian Stage in the European terminology; ~ MIS 5e) sediments, typified by an arcto-boreal mollusc assemblage, are described from a large number of sites on the Taymyr Peninsula (Urvantsev, 1931; Saks, 1953; Kind and Leonov, 1982; Möller et al., 2008) and from the Severnaya Zemlya archipelago (Bolshiyanov and Makeyev, 1995; Möller et al., 2007) (Fig. 28.1). Reported Eemian-age sediments range from deep-marine over shoreface and foreshore to beach-face sediment. The latter reach altitudes up to 130 m a.s.l. south of the Byrranga Mountains (Möller et al., 1999a,b) and as high as 140 m a.s.l. on Severnaya Zemlya (Möller et al., 2007) and on Cape Chelyuskin, the northern tip of the Taymyr Peninsula (Miroshnikov, 1959; Shneyder, 1989; Möller et al., 2008).

28.4. THE BUILD-UP OF KSIS: CHANGING PARADIGMS

An early view of ice-sheet inception in north-western Siberia was that ice started to grow over mountains (e.g. the Byrranga Mountains, Ural Mountains, etc.), thereafter

---

complexes do not match geomorphologically (Möller and Sallaba, 2010). The NTZ 1–3 lines are from the Early, Middle and Late Weichselian, respectively. S and T mark the Shrenk and Trautfetter Rivers, respectively. Open circles mark ESR-dated sites with presumably Tobol Interglacial marine sediments (MIS 9?) (Bolshiyanov et al., 1998; Möller et al., 2007). Open squares within and south of the Byrranga Mountains mark ESR-dated sites with Kazantzevo/Eemian marine sediments according to Bolshiyanov et al. (1998), whereas those in the north are ESR- and OSL-dated Eemian marine sediments in the Cape Chelyuskin area (Möller et al., 2008) and on Severnaya Zemlya (Möller et al., 2007). Filled circles mark ESR- and OSL-dated sites with marine deltaic sediments deposited during Early Weichselian ice recession (Möller et al., 1999a, 2008; Hjort et al., 2004). Filled squares mark sites with 'Cape Sabler-type' terrestrial sediments of Middle to Late Weichselian age (Möller et al., 1999a). The basic map is from the International Bathymetric Chart of the Arctic Ocean (IBCAO) (Jakobsson et al., 2008a).

Ice flow Environment	D/L Asp ratio	ESR age ka	GSL age ka	^{14}C age cal.ka	Interpreted age, ka	Interpreted MIS	Highest shoreline m a.s.l.
Marine IV — off-shore ⇨ lagoonal ⇨ foreshore full regressional sequence	0.119 ± 0.014 n=31	20 29 29 32 33	36 49 82 129 144 153	37 39 40 42 43 43 43 43 45 46 46 47	ca. 50 ka	MIS 3	> 40 m
		59	>180	45 47 50		MIS 3 (or 5a?)	>60 m
Till IV						MIS 5d-4 (or MIS 5d-5b; or MIS 4)	
Marine III — off-shore ⇨ foreshore full regressional sequence	0.196 ± 0.015 n=42	77 81 82 83 89 95 97 105 105* 120*	>59 >89 143 156 162 165 167 176 225	47 51 >50 >50	>90 <160	MIS 6/5e	140 m
Till III						MIS 6	
Marine II — off-shore ⇨ shoreface	0.250 ± 0.020 n=14	150 216 220 224 231	>152	>50	>220	MIS 8/7?	>71 m
Till II						MIS >6 MIS 8?	
Marine I — deltaic progradation	0.281 ± 0.011 n=18	269 271 363 400 452 492	>252 >270 >282 >288 >473	>50	~300? ~400?	MIS 10/9? MIS 12/11?	>70 m
Till I						MIS >6 MIS 10? MIS 12?	

FIGURE 28.2 Summary of the glacial stratigraphy, depositional environment of marine strata, ice-flow directions from till units and glaciotectonites and geochronological data for October Revolution Island (modified from Möller et al., 2007). ESR ages marked * are from Bolshiyanov and Makeyev (1995).

flowing into lowland areas (e.g. Urvantsev, 1931; Saks, 1953). From such a scenario, there has been a shift into a paradigm suggesting repeated build-up of thick ice sheets on the shallow shelves of the Barents and Kara Seas (e.g. Astakhov 1976, 1979, 1998; Grosswald, 1980, 1998; Kind and Leonov, 1982). The substantial loading from such continental-scale ice could explain the occurrence of high raised beaches, such as the ~140 m a.s.l. Saalian/Eemian-inundation shorelines on October Revolution Island, which are among the highest described from the Eurasian Arctic. Using such shoreline evidence and dated ice-marginal formations, the maximum KSIS thickness in the Saalian was estimated to 3000–3300 m by earth rheological inverse modelling (Lambeck et al., 2006). More recent reconstructions of Eurasian Ice Sheet accumulations and decays have confirmed this latter paradigm (e.g. Möller et al., 1999a; Alexanderson et al., 2001, 2002; Hjort et al., 2004; Svendsen et al., 2004; Astakhov and Mangerud, 2005; Larsen et al., 2006).

However, growing evidence suggests that the latter paradigm could be combined with the first. Structural and textural evidence from tills and sub-till sediments at a number of sites around the perimeter of the shallow Kara Sea, and within the Kara Sea itself, including Severnaya Zemlya (Möller et al., 2007), at Cape Chelyuskin (Möller et al., 2008), the north-western Taymyr Peninsula (Hjort and Funder, 2008), the Yamal Peninsula (Forman et al., 1999, 2002) and the Yugorski Peninsula (Lokrantz et al., 2003) imply subglacial deformation and deposition during expansions of local ice caps and often suggesting ice-flow directions that are not compatible with ice flows that should have been generated from an ice sheet centred over the Kara Sea shelf (cf. review in Ingólfsson et al., 2008). However, such local ice caps could not have induced sufficient isostatic depression of the crust for explaining the high elevations documented for raised beaches. For resolution of this conflicting evidence, and based on an original concept of Hughes (1987) further modelled by Siegert et al. (2002) and Siegert and Dowdeswell (2004), Möller et al. (2007, Fig. 26) suggested that these peripheral sites were critical as nucleation centres for repeated initiations of large, coherent KSISs. These local ice domes in highland areas and along the perimeter of the Kara Sea were characterised by wet-based thermal regimes in their initiation phases, as indicated by deposition of deformation tills and plastic deformation of pre-existing sediment. They later gradually coalesced on the adjacent shelf with globally falling sea level. The latter facilitated the formation of a stable and thick sea-ice cover and later ice-shelf formation. Eventually, a large ice dome formed as a combination of ice flowing into the Kara Sea basin and snow accumulation on the grounded ice shelf, initiating redistribution of ice drainage and basal thermal regime. A fully developed KSIS preferentially drained towards the Arctic Ocean in the north as ice streams, such as those developed along the St. Anna and Voronin troughs. Towards the south, the ice flow reversed compared to the initiation phase north and west of the Byrranga Mountains, whereas south of these mountains the south-directed flow had already formed during the initiation phase. This continued, eventually leading to the terminal ice-margin positions of the individual glaciation phases in question. Such Kara Sea shelf-based ice sheets would have been sensitive to sea level changes; as global sea level rose when continental ice sheets disintegrated at terminations of glaciations, the KSIS rapidly lost mass, resulting in marine inundation and the deposition of marine sediments (Ingólfsson et al., 2008).

28.5. THE EARLY WEICHSELIAN—THE WEICHSELIAN GLACIATION MAXIMUM ON TAYMYR

From the southern foothills of the Byrranga Mountains and ~250 km southwards, at least eight moraine-ridge complexes have been recorded (e.g. Kind and Leonov, 1982; Fig. 28.1). However, their age of formation remains poorly constrained. Individual ridge complexes are up to 15 km wide and 100–150 m high and can be followed laterally for hundreds of kilometres. Some form distinct, sometimes complex morainic loops, others are more diffuse and ridge trends and connections are lost in large interlobate complexes. According to Kind and Leonov (1982), the ridge complexes consist of active ice-thrust stacks of both Quaternary and pre-Quaternary strata, some ridges being glaciotectonic in origin, whereas others are more of ice disintegration type, lined with kame terraces and hummocky kame topography. Fossil ice is still present in many ridge complexes (e.g. Kind and Leonov, 1982; Siegert et al., 1999), a phenomenon that can also be seen from Landsat images, showing spatially concordant distributions of numerous thermokarst lakes. The glacial geomorphology, the direction of glaciotectonic deformation and the provenance of crystalline erratic boulders clearly indicate that these moraine-ridge complexes were constructed at KSIS marginal positions south of the Byrranga Mountains. Numerous suggestions have been proposed regarding which glacial stages and phases individual ridges represent (Andreyeva, 1978; Andreeva and Isaeva, 1982; Kind and Leonov, 1982; Isayeva, 1984; Fisher et al., 1990; Siegert et al., 1999), varying from Saalian (Taz), Early Weichselian (Lower Zyryanka/Muruktin) to Late Weichselian (Upper Zyryanka/Sartan). The most recent compilation by Svendsen et al. (2004) suggested that the Jangoda–Syntabyl–Baikuronyora zone, the JSB line (Fig. 28.1), should mark the maximum expansion of the KSIS in the Early Weichselian (unfortunately, the JSB line has been incorrectly drawn in Fig. 2 of Svendsen et al. (2004), where it instead follows the Sampesa ridge (line Sa, Fig. 28.1), lying distal to the Syntabul ridge). Major argument for the JSB line being the maximum limit for an Early Weichselian KSIS is that marine deposits assigned to the Eemian/Kazantzevo interglacial are covered by a till north of the JSB (Urvantsev, 1931; Saks, 1953; Kind and Leonov, 1982), but not south of it. However, it must be stressed that no direct age constraints exist for the JSB line as such. Based on the recently performed Landsat mapping (Möller and Sallaba, 2010), the geomorphological connection between the Syntabyl and Baikuronyora ridge complexes seems less probable; the latter is readily traced into the younger Upper Taymyr ridge system (see below), and the former seems to continue into the wide Severokokorsk ridge complex (line K, Fig. 28.1). A connected Jangoda–Syntabyl–Severokokorsk moraine system (a connection already proposed by Isayeva, 1984) might thus be an alternative to the JSB line of Svendsen et al. (2004) for the Early Weichselian KSIS maximum.

28.6. DEGLACIATION FROM THE EARLY WEICHSELIAN MAXIMAL POSITION TO THE BYRRANGA MOUNTAINS

The ice recession from the Early Weichselian maximum towards the Byrranga Mountains seems to have been halted at certain positions and/or interrupted by ice-margin readvances, as indicated from a number of ice-marginal formations, some of them with distinctly lobate planforms and with intricate patterns of secondary ridges and within-lobe younger ridges cutting off, or on-lapping to, older lobes (e.g. Fig. 28.3). Isayeva (1984) named these younger ice-marginal complexes the Mokoritto ridge system in the Pyasina River basin on the western Taymyr Peninsula and the Upper Taymyr Ridge system in the upper Taymyr River basin (ridge system M and UT in Fig. 28.1, respectively). From Landsat imagery interpretation (Möller and Sallaba, 2010), the latter system seems to link to the Baikuronyora ridge system along the southern shore of Lake Taymyr (line B, Fig. 28.1) which, however, was considered to be part of the terminal Early Weichselian ice-marginal zone by Svendsen et al. (2004).

The ice recession from the Early Weichselian maximum took place in a marine basin with water depths up to 100 m, as demonstrated from several localities with marine sediments within the Taymyr Lake basin and along the foothills of the Byrranga Mountains (filled circles, Fig. 28.1). These marine successions usually constitute off-shore glaciomarine to marine clayey–silty deposits. They continue in places into thick deltaic deposits, or at other places into shoreface and foreshore and sometimes also beach-face deposits and thus demonstrate more or less full isostatically driven regressional sediment sequences. The retrieved ages from mollusc shells and sediments from six investigated sites (26 ESR and 5 OSL ages) give a consistent age-frame of 95–70 ka. Radiocarbon dating of mollusc shells from the same localities all give infinite ages. The most prominent of the delta successions occurs in the Ledyanaya River valley, west of Lake Taymyr (Fig. 28.1), the type locality for the 'Ledyanaya Gravel Event' (Möller et al., 1999a,b). Here, delta topset beds reach 100–120 m a.s.l. and the marine basin into which the deltas prograded probably reached 90–100 m a.s.l. This high, post-glacial marine limit, together with the huge accumulations of coarse sediments within the deltas, confirms the substantial glacioisostatic depression and indicate a substantial flow of sediment-laden glacial meltwater southwards through the Byrranga mountain valleys and into the marine basin. This meltwater must have emanated from an ice front which at that time stood along the northern slopes of the mountains. The marine event is probably the equivalent of the Karginsk marine transgression of Andreeva and Kind (1982), in its earliest phase radiocarbon-dated to 50–39 ^{14}C ka BP. However, these mollusc ^{14}C ages stem from conventional, that is, not accelerator mass spectrometry (AMS), dating and must be considered underestimates.

There are no signs that glaciers regrowing during the Middle and Late Weichselian reached south of the Byrranga Mountains since this Early Weichselian deglaciation (e.g. Möller et al., 1999a). The Taymyr Lake basin has thus been continuously ice free ever since. This is shown by a number of sites with 'Cape Sabler-type' sediment successions of laminated fine sand and silt, rich in organic detritus and also thick units of silt-soaked peat, deposited in a terrestrial setting with peat bog and aeolian deposition, interrupted by occasional lacustrine floods (Möller et al., 1999a; Hubberten et al., 2004). AMS ^{14}C dates obtained from Cape Sabler and nearby sites (infilled squares, Fig. 28.1) suggest continuous deposition from before 40 ^{14}C ka BP and into the Holocene. Lake sediment successions from both the Taymyr Lake itself and the adjacent Levinson-Lessing Lake (Ebel et al., 1999; Hahne and Melles, 1999; Niessen et al., 1999) also indicate generally ice-free conditions in the Byrranga Mountains since at least the Middle Weichselian Substage time.

The Early Weichselian KSIS also inundated the Chelyuskin Peninsula on north-easternmost Taymyr. This conclusion is based on till and glaciotectonic deformations overlying and affecting Eemian-age interglacial marine sediments, and from the occurrence of Kara Sea crystalline erratics found on the hilltops (Möller et al., 2008). In addition, the glacioisostatic depression in this area (at least 80 m a.s.l.) resulted in post-glacial marine inundation, as indicated by ESR dating of mollusc shells in the marine sediments to ca. 93–80 ka and thus largely contemporaneous with that in the Taymyr Lake basin.

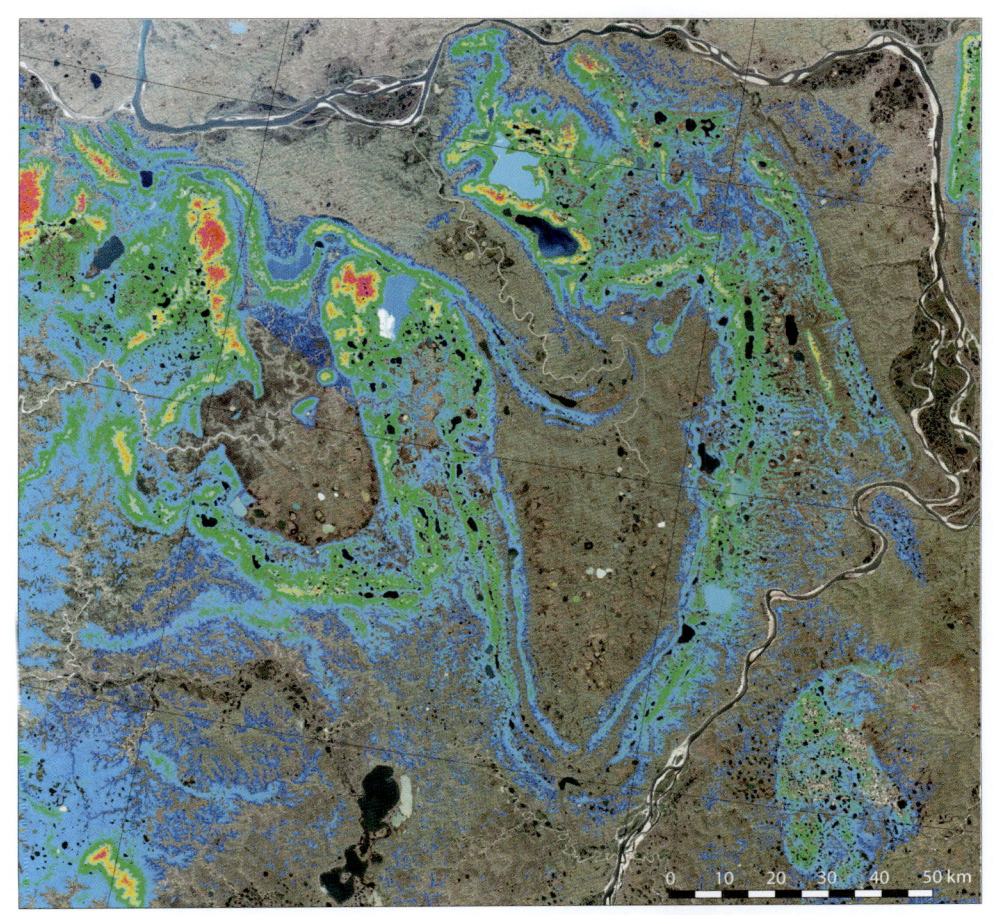

FIGURE 28.3 The Mokoritto ice-marginal complex (complex M, Fig. 28.1), indicating an extremely lobate ice margin with 2–3 older moraines as outer ridges, and then younger moraines from a recessional stage, on-lapping to and/or cutting off older lobes.

28.7. THE NORTH TAYMYR ICE-MARGINAL ZONE

The north-westward retreat of the Early Weichselian ice front from the Byrranga Mountains seems to have proceeded largely by calving into a glacial lake filling the Shrenk, Trautfetter and part of the Lower Taymyr River valleys (Fig. 28.1) and dammed towards the north-west by the ice itself. A new grounding line was reached on the north-western sides of the Shrenk and Trautfetter valleys, causing a temporary still-stand of the ice front and resulting in the formation of the North Taymyr ice-marginal zone, the NTZ (Fig. 28.1, NTZ 1).

The NTZ is a complex of glacial, glaciofluvial and glaciolacustrine deposits, containing large amounts of redeposited Quaternary marine sediments and also glacially displaced, coal-bearing Cretaceous sands. It has now been dated for the first time and described in some detail by Alexanderson et al. (2001, 2002) but had already been broadly mapped and discussed by Kind and Leonov (1982). When the KSIS front stood at this ice-marginal zone, it seems to have crossed the present coastline near the Michailova Peninsula at ca. 75°N. The NTZ can then be roughly (Hjort and Funder, 2008) followed, first eastwards and then northwards for 700–750 km, mostly 80–100 km inland, and seems to recross the present coastline south of the Tessema River, around 77°N (Fig. 28.1). It is best developed in its central parts, ca. 100 km north-east and south-west, respectively, of where it is today cut through by the Lower Taymyr River. The base of the NTZ in these central parts (Alexanderson et al., 2001, 2002) is a series of ridges up to 100 m high and 2 km wide, mainly consisting of, or possibly only covered by, redeposited marine silts. The ridges are still ice cored, but in most parts of the zone, the present active layer only rarely reaches the ice surface. Associated with the NTZ are deltas, abrasion terraces and shorelines corresponding to two generations of ice-dammed lakes, with shore-levels at between 140 and 120 m and at ca. 80 m a.s.l. (Figs. 4 and 5 in Hjort et al., 2004). These lakes drained southwards into the Taymyr Lake basin, as recorded by current directions in fluvial sediment sequences along the Taymyr River valley

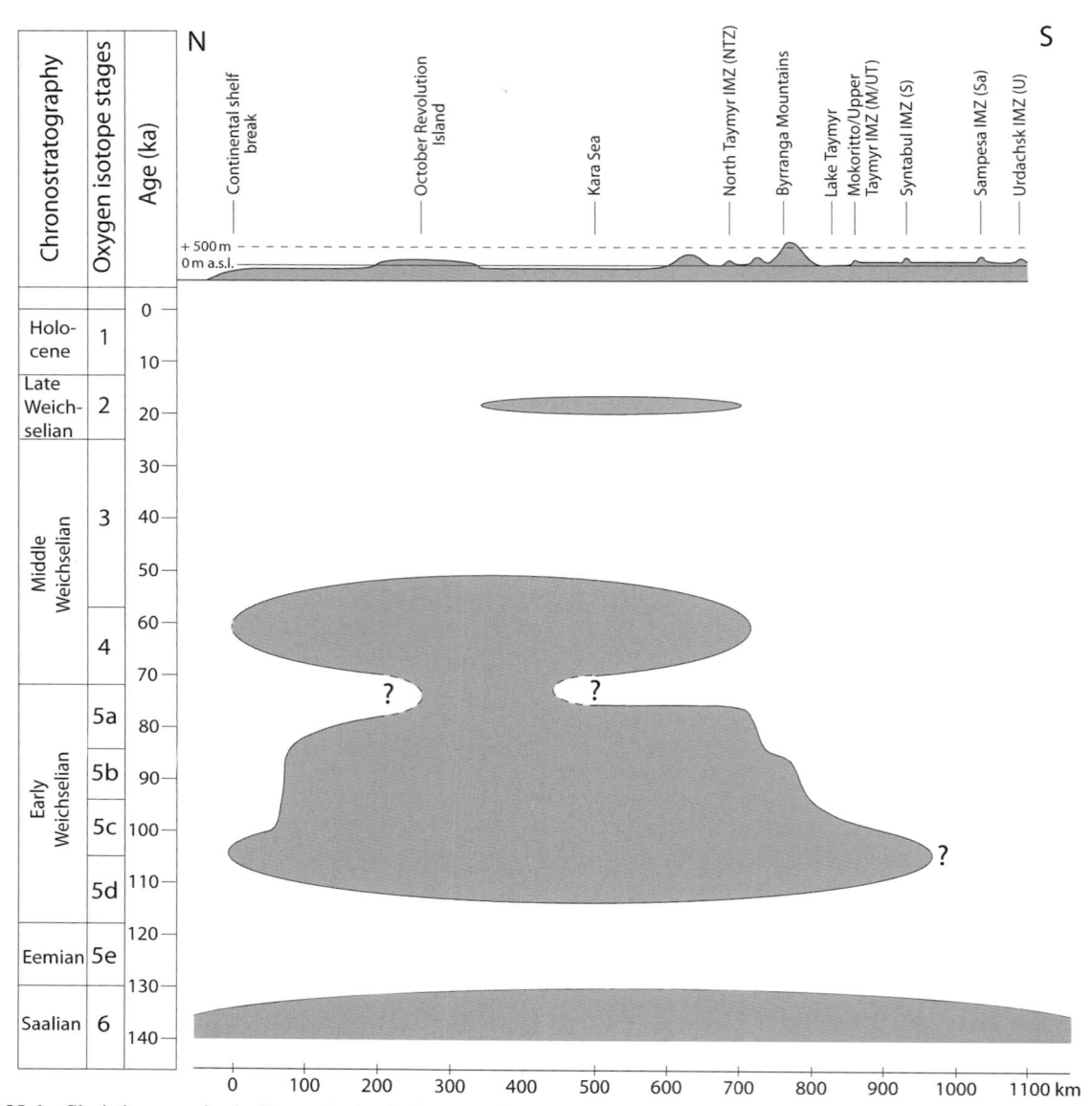

FIGURE 28.4 Glaciation curve for the Taymyr Peninsula, Severnaya Zemlya islands and the Kara Sea shelf. The ice sheets, which originated on the Kara Sea shelf, advanced onto the Taymyr Peninsula from the north to north-west. During the Saalian, the whole of Taymyr seems to have been ice covered. The three Weichselian glaciations were of progressively decreasing amplitude (Fig. 28.1). The maximum limit of the Early Weichselian glaciation is not exactly known, but it did at least reach the Jangoda–Syntabul–Severokokorsk ice-marginal complexes (J, S and K lines, Fig. 28.1). The Middle Weichselian ice front at the NTZ was roughly the same as the Early Weichselian one, whereas the Late Weichselian (LGM) ice cover was considerably smaller and thinner.

where it passes through the Byrranga Mountains (today the river flows northwards). From the Taymyr Lake basin, the water continued either westwards to the south-eastern Kara Sea shelf or eastwards towards Khatanga Bay and the Laptev Sea.

28.8. THE EARLY WEICHSELIAN NTZ STAGE

The NTZ is morphologically very complex and in its central part consists of three generations of ice-marginal deposits (Alexanderson et al., 2001, 2002). The oldest is that formed as the ice front, largely through calving into its frontally dammed lake, had retreated north-westwards from its Byrranga still-stand position to the new grounding line. This stage is associated with the deepest glacial lake, reaching 140–120 m a.s.l. Two OSL dates from an ice-contact glaciofluvial sequence aggradated to the 140 m level gave ages of ca. 80 ka, which combined with the ESR ages obtained for the 'Ledyanaya Gravel Event', indicate its relationship with the Early Weichselian KSIS deglaciation process from its maximum stand south of the Byrrangas and the Taymyr Lake basin.

28.9. THE MIDDLE WEICHSELIAN NTZ STAGE

During the second NTZ generation, the ice front seems to have stood more or less at the same positions as during the older stage (Fig. 28.1: NTZ 2). This caused an overprinting on the previous morphology of a number of over-deepened lake basins and a new system of marginal moraine ridges, associated glacial lake deltas and shorelines, valley fills, etc. The ice-dammed lake was, however, shallower than the previous water body and reached only 80 m a.s.l. It has been OSL-dated at two localities; two delta samples gave an age of ca. 65 ka BP, and two dates from fluvial terrace deposits, connecting the ice front with the lake basin, gave ca. 70–55 ka BP. This stage of glacial lake sedimentation is further supported by three OSL dates of 60–55 ka from glaciolacustrine rhythmites in the Taymyr Lake basin, just south of the Byrranga Mountains, where the glacial damming in the north also led to a rising water level. The available dates thus indicate that an interval of at least 10,000 years occurred between the two oldest NTZ events, and data from the south-westernmost part of this ice-marginal zone indicate that the event around 70–60 ka was the last one to affect the whole NTZ (Hjort and Funder, 2008).

28.10. THE LATE WEICHSELIAN NTZ STAGE

During the third NTZ generation, the KSIS affecting north-western Taymyr was much thinner than previously and inundated a much smaller area (Alexanderson et al., 2001, 2002). Because it did not cross a 300–500 m high range of coastal hills (Fig. 28.1), which were overridden during the two previous stages, its thickness near the present coastline could not have exceeded 500 m. Nonetheless, it penetrated 100 km inland on a 150 km broad front centred along the Lower Taymyr River valley and terminated at altitudes presently below 150 m a.s.l. (Fig. 28.1: NTZ 3). North-east of the valley the front abutted a system of bedrock cuestas, whilst to the south-west, it was in contact with the pre-existing NTZ 1–2 moraines and, in one case, formed an independent lobate moraine (Fig. 4 in Hjort et al., 2004). The area overridden by this ice sheet, the most recent to inundate the Taymyr Peninsula, is to a large extent covered by dislocated marine sediments, identifiable on satellite images by their dendritic erosional pattern. In a 5- to 10-km-wide zone behind the former ice front, where the ice contained most debris, the landscape is patterned by a multitude of shallow slides, exposing remnant glacier ice under a melt-out till cover of only about 0.5 m. (Fig. 6 in Hjort et al., 2004). Further north-west (up-ice), there are fewer indications of the former overriding. However, a boulder-lag on top of glaciolacustrine sediments at the Kara Sea coast (Funder et al., 1999) may date from this glacial event. This youngest ice-sheet advance is pre-dated by two radiocarbon dates of ca. 20 ^{14}C ka BP from mollusc shells (*Hiatella arctica*, *Astarte* sp.), from glacially redeposited marine silt sampled ca. 2 km behind the former ice-front position (Alexanderson et al., 2001, 2002). It is post-dated by a radiocarbon date of ca. 12 ^{14}C ka BP from *in situ* terrestrial material just inside the present coast near the Taymyr River mouth (Bolshiyanov et al., 2000). The glaciation thus dates from the Weichselian global Last Glacial Maximum (LGM). This brief advance (8000 years or less) of a thin ice sheet onto the present land, as marked by the NTZ 3 line, thus dates from the Weichselian Last Glacial Maximum (LGM). However, it is still unclear if the ice advanced from a regional ice cap on the very shallow, and for global eustatic reasons at that time mostly dry shelf in the north-eastern corner of the Kara Sea, or if it was connected westwards to ice centred north of Novaya Zemlya, as suggested by Polyak et al. (2008).

No evidence of any glacial lake dammed by this LGM ice sheet has been found north of the Byrranga Mountains, and it is therefore thought that its meltwater mainly drained southwards via the Lower Taymyr River valley into the Taymyr Lake basin (Alexanderson et al., 2001, 2002). Indications of an increasing rate of sedimentation in the lake around 19 ^{14}C ka BP (Möller et al., 1999a, Hubberten et al., 2004) suggest a causal connection with a meltwater input. Neither have any raised marine shorelines, dated to the LGM or thereafter, been found on Taymyr. This is not surprising considering the thin, short-lived and thus isostatically insignificant ice, as well as the extremely low eustatic sea level at the time that persisted into the Holocene.

28.11. THE SEVERNAYA ZEMLYA ISLANDS DURING THE WEICHSELIAN STAGE

Eemian/Kazantzevo sediments in the Ozernaya River sections on southern October Revolution Island (MIS 5e; marine unit M-III, Fig. 28.2) form a regressional sequence, initiated by off-shore glaciomarine–marine sediments capped by shoreface to foreshore sands and gravels, in turn capped by till (till unit T-IV) of obvious Weichselian age (Möller et al., 2007). This till is in turn at a number of sites overlain by marine unit M-IV sediments, forming a coarsening upwards succession from off-shore marine silty clays to foreshore/shoreface sands and gravels, starting well above 60 m a.s.l. Some M-IV sediments also consist of estuary/lagoonal sediment successions, at one site with a spectacular discovery of at least nine narwhal (*Monodon monoceros*) skeletons on the same depositional surface. These finds have been dated to ~50 ka. A number of ^{14}C ages on marine molluscs, which are finite but close to the analytical limit, suggest an age of 50–37 ^{14}C ka BP. This age span is supported by some of the retrieved GLS ages, whereas others are significantly older. It is assumed that insufficient solar resetting of sediment in these cases is responsible for age overestimates. From all this, Möller et al. (2007) concluded that the M-IV sediments are of Middle Weichselian (MIS 3) age and, because no marine sediments of an undoubtedly interstadial MIS 5d–5a age have been identified, that October Revolution Island (and the whole Severnaya Zemlya archipelago) became glaciated during the initial phase of the Early Weichselian and remained so from MIS 5d to MIS 4. During the same period, the Cape Chelyuskin peninsula south of the islands became glaciated and thereafter deglaciated, experiencing no Middle Weichselian ice advance (Möller et al., 2008). However, as previously described, the other parts of northern Taymyr experienced both an Early Weichselian ice retreat from the NTZ zone and thereafter a readvance to the NTZ in the Middle Weichselian. If ice persisted over Severnaya Zemlya throughout MIS 5a–d, then it could have acted as a nucleation area for the regrowth and expansion of the KSIS on the northern shelf during MIS 4.

The studies on Severnaya Zemlya by Möller et al. (2007) revealed neither any sediments nor raised shorelines that could be tied to a Late Weichselian (MIS 2) glacial/deglacial event, and that Severnaya Zemlya was outside any KSIS during the LGM is further supported by a continuous sediment core from Changeable Lake on southern October Revolution Island. There ^{14}C and luminescence ages from the lake sediments suggest that the latest ice expansion occurred before 53 ka, followed by marine and lacustrine deposition during the LGM and into the Holocene (Raab et al., 2003; Berger et al., 2004). A largely ice-free LGM on Severnaya Zemlya is further supported by mammoth remains that have yielded ages between 12 and 30 ^{14}C ka BP (Velichko et al., 1984; Bolshiyanov and Makeyev, 1995), and seemingly also by the submarine geology of the surrounding parts of the Kara Sea (Polyak et al., 2008).

28.12. SUMMARY OF RESULTS

The main results of this study of the glacial and marine history of the Taymyr Peninsula, the Severnaya Zemlya archipelago and the Kara Sea shelf, as summarised in Fig. 28.4, are as follows:

- The so far documented maximum glaciation of this region seems to have taken place during the Late Saalian (MIS 6), when even the central Arctic Ocean was glaciated, with ice grounding on the Lomonosov Ridge. Thereafter followed the Eemian (Kazantzevo, MIS 5e) interglacial with, extensive marine inundation due to the isostatic effect of the preceding glaciation—the so-called Boreal Transgression.

- During the Weichselian Stage, three phases of successively decreasing ice-sheet expanses and retreats have been mapped and dated. The most extensive glaciation dates from the Early Weichselian, culminating at ca. 100 ka, and a Middle Weichselian event of intermediate extent dates from 70 to 60 ka. The last and least extensive glaciation, contemporaneous with the global LGM, was short, lasting only 8000 years or less. It culminated after 20 ka and had largely disappeared from present onshore areas by 12 ka BP.

- The main (culminating) ice sheets covering the Taymyr Peninsula during the Weichselian on all three occasions did emanate from the Kara Sea continental shelf, from where they advanced south-eastwards across the land. At most, the ice front reached some 400 km from the present coast, leaving a series of more or less distinct zones of ice-marginal features south of the Byrranga Mountains and the Taymyr Lake basin. In the south and east, the ice front reached the Laptev Sea drainage basin.

- The KSISs dammed large glacial lakes, filling the lake- and river basins both north and south of the Byrranga Mountains (Fig. 5 in Hjort et al., 2004) and, during the final stages of the deglaciations, they also developed on the lowland areas along the coast. Water from north of the mountains drained southwards along the Taymyr River valley (where it flows northwards today) into the Taymyr Lake basin and, in most cases, thereafter probably westwards to the Kara Sea shelf.

- An Early Weichselian marine inundation, following the regional Weichselian glaciation maximum, reached ca. 100 m above present sea level. However, the short-lived, thin and comparatively very small Late Weichselian ice cap (with no ice expanding from Severnaya Zemlya), contemporaneous with the global eustatic sea level low around 18 ^{14}C ka BP, did not isostatically influence land sufficiently to create marine shorelines above the present one.

- The concept of a maximum ice cover during the Late Weichselian LGM, in which ice should more or less have totally covered the Eurasian Arctic (and thus the Taymyr Peninsula), for long advocated by some researchers (e.g. Grosswald, 1998) and extensively used by climate modellers (e.g. Budd et al., 1998), is incorrect. As have now been shown (e.g. Svendsen et al., 2004; Jakobsson et al., 2008b), this situation has occurred, but not later than during the Saalian glaciation.

REFERENCES

Alexanderson, H., Hjort, C., Möller, P., Antonov, O., Pavlov, M., 2001. The North Taymyr ice-marginal zone, Arctic Siberia—a preliminary overview and dating. Glob. Planet. Change 31, 427–445.

Alexanderson, H., Adrielsson, L., Hjort, C., Möller, P., Antonov, O., Eriksson, S., et al., 2002. The depositional history of the North Taymyr ice-marginal zone, Siberia—a landsystem approach. J. Quatern. Sci. 17, 361–382.

Andreeva, S.M., Isaeva, L.L., 1982. Muruktin (Nizhne Zyryanka) deposits of the North-Siberian Lowland. (in Russian). In: Kind, N.V., Leonov, B.N. (Eds.), The Antropogen of the Taimyr Peninsula. Nauka, Moscow, pp. 34–46.

Andreeva, S.M., Kind, N.V., 1982. Karginsk deposits. (in Russian). In: Kind, N.V., Leonov, B.N. (Eds.), The Antropogen of the Taimyr Peninsula. Nauka, Moscow, pp. 47–71.

Andreyeva, S.M., 1978. Zyryanka glaciation in north-central Siberia. (in Russian). USSR Academy of Sciences, Izvestiya seriya geograficheskaya 5, 72–78.

Arkhipov, S.A., 1989. A chronostratigraphic scale of the glacial Pleistocene of the West Siberian North. In: Skabichevskaya, N.A. (Ed.), Pleistotsen Sibiri. Stratigrafia i mezhregionalnye korrelatsii. Nauka, Novosibirsk, pp. 19–30 (in Russian).

Arkhipov, S.A., Isayeva, L.L., Bespaly, V.G., Glushkova, O., 1986. Glaciation of Siberia and North-East USSR. Quatern. Sci. Rev. 5, 463–474.

Astakhov, V.I., 1976. Geologičeskie dokazatel´stva centra plejstocenovigo oledenija na Karskom šel´fe. (Geological evidence of a centre of Pleistocene glaciation on the Kara shelf). Dokl. Akad. Nauk SSSR 23, 1178–1181.

Astakhov, V.I., 1979. New data on the largest activity of Kara-shelf glaciers in West Siberia. In: Šibrava, V. (Ed.), IGCP Project 73/1/24 Quaternary Glaciations in the Northern Hemisphere. Czech Geological Survey, Prague, pp. 21–31, Report no. 5.

Astakhov, V.I., 1998. The last ice sheet of the Kara Sea: terrestrial constraints on its age. Quatern. Int. 45/46, 19–28.

Astakhov, A., 2001. The stratigraphic framework for the Upper Pleistocene of the glaciated Russia: changing paradigms. Glob. Planet. Change 31, 283–295.

Astakhov, A., 2004a. Middle Pleistocene glaciations of the Russian North. Quatern. Sci. Rev. 23, 1285–1311.

Astakhov, A., 2004b. Pleistocene ice limits in the Russian northern lowlands. In: Ehlers, J., Gibbard, P.L. (Eds.), Quaternary Glaciations—Extent and Chronology. Part 1: Europe. Developments in Quaternary Science, 2A Elsevier, Amsterdam, pp. 309–319.

Astakhov, V.I., 2008. Geographical extremes in the glacial history of northern Eurasia: post-QUEEN considerations. Polar Res. 27, 280–288.

Astakhov, V.I., Mangerud, J., 2005. The age of the Karginsky interglacial strata on the lower Yenisei. Dokl. Akad. Nauk Earth Sci. 403 (5), 673–676.

Berger, G.N., Melles, M., Banerjee, D., Murray, A.S., Raab, A., 2004. Luminescence chronology of non-glacial sediments in Changeable Lake, Russian High Arctic, and implications for limited Eurasian ice-sheet extent during the LGM. J. Quatern. Sci. 19, 513–523.

Bolshiyanov, D.Yu. Makeyev, V.M., 1995. Arkhipelag Severnaya Zemlya: oledeneniye, istoria razvitia prirodnoi sredy (Severnaya Zemlya Archipelago: Glaciation, Environmental History). Gidrometeoizdat, St. Petersburg, 215 pp. (in Russian).

Bolshiyanov, D.Y., Savatuygin, L.M., Shneider, G.V., Molodkov, A.N., 1998. New data about modern and ancient glaciations of the Taimyro-Severozemlskaya region. Materialny glyatchiologicheskich issledovanii 85, 219–222 (in Russian).

Bolshiyanov, D.Y., Ryazanova, M., Savelieva, L., Pushina, Z., 2000. Peatbog at the shoreline of Cape Oskar (Taymyr Peninsula). In: Abstracts 4th QUEEN Workshop. European Science Foundation, Lund, Sweden, p. 9.

Budd, W.F., Coutts, B., Warner, R.C., 1998. Modelling the Antarctic and Northern Hemisphere ice-sheet changes with global climate through the glacial cycle. Ann. Glaciol. 27, 153–160.

Ebel, T., Melles, M., Niessen, F., 1999. Laminated sediments from Levinson-Lessing Lake, northern Central Siberia—a 30,000 year record of environmental history? In: Kassens, H., Bauch, H.A., Dmitrenko, I.A., Eicken, H., Hubberten, H.-W., Melles, M., Thiede, J., Timokhov, L.A. (Eds.), Land-Ocean Systems in the Siberian Arctic: Dynamics and History. Springer-Verlag, Berlin, pp. 425–435.

Fisher, E.L., Leonov, B.N., Nikolskaya, M.Z., Petrov, O.M., Ratsko, A.P., Sulerzhitsky, L.D., et al., 1990. The Late Pleistocene of the central North-Siberian lowland. Izvestia Acad. Sci. USSR Geogr. 6, 109–118 (in Russian).

Forman, S.L., Ingólfsson, Ó., Gataullin, V., Manley, W.F., Lokrantz, H., 1999. Late Quaternary stratigraphy of western Yamal Peninsula, Russia: new constraints on the configuration of the Eurasian ice sheet. Geology 27, 807–810.

Forman, S.L., Ingólfsson, Ó., Gataullin, V., Manley, W., Lokrantz, H., 2002. Late Quaternary stratigraphy, glacial limits, and paleoenvironments of the Marresale Area, Western Yamal Peninsula, Russia. Quatern. Res. 57, 355–370.

Funder, S., Riazanova, M., Rydlevski, A., Seidenkrantz, M.S., 1999. Late Quaternary events in northern Siberia—preliminary results of field work on coastal Taymyr. In: Abstracts 3rd QUEEN Workshop Öystese, Norway, p. 19.

Grosswald, M.G., 1980. Late Weichselian ice sheets of Northern Eurasia. Quatern. Res. 13, 1–32.

Grosswald, M.G., 1998. Late Weichselian ice sheets in Arctic and Pacific Siberia. Quatern. Int. 45–46, 3–18.

Hahne, J., Melles, M., 1999. Climate and vegetation history of the Taymyr Peninsula since Middle Weichselian time—palynological evidence from lake sediments. In: Kassens, H., Bauch, H.A., Dmitrenko, I.A., Eicken, H., Hubberten, H.-W., Melles, M., Thiede, J., Timokhov, L.A. (Eds.), Land-Ocean Systems in the Siberian Arctic: Dynamics and History. Springer-Verlag, Berlin, pp. 361–376.

Hjort, C., Funder, S., 2008. Mountain derived versus shelf based glaciations on the western Taymyr Peninsula. Polar Res. 27, 152–161.

Hjort, C., Möller, P., Alexanderson, H., 2004. Weichselian glaciation of the Taymyr Peninsula, Siberia. In: Ehlers, J., Gibbard, P.L. (Eds.), Quaternary Glaciations—Extent and Chronology. Part 1: Europe. Developments in Quaternary Science, 2A Elsevier, Amsterdam, pp. 359–367.

Hubberten, H.W., Andreev, A., Astakhov, V.I., Demidov, I., Dowdeswell, J.A., Henriksen, M., et al., 2004. The periglacial climate and environment in northern Eurasia during the last glaciation (LGM). Quatern. Sci. Rev. 23, 1333–1357.

Hughes, T.J., 1987. The marine ice transgression hypothesis. Geogr. Ann. 69, 237–250.

Ingólfsson, Ó., Möller, P., Lokrantz, H., 2008. Late Quaternary marine-based Kara Sea ice sheets: review of terrestrial stratigraphic data highlighting their formation. Polar Res. 27, 152–161.

Isayeva, L.L., 1984. Late Pleistocene glaciation of North-Central Siberia. In: Velichko, A.A. (Ed.), Late Quaternary Environments of the Soviet Union. University of Minnesota Press, Minneapolis, pp. 21–30.

Jakobsson, M., Løvlie, R., Arnold, E.M., Backman, J., Polyak, L., Knutsen, J.-O., et al., 2001. Pleistocene stratigraphy and palaeoenvironmental variation from Lomonosov Ridge sediments, central Arctic Ocean. Glob. Planet. Change 31, 1–22.

Jakobsson, M.R., Macnab, L., Mayer, R., Anderson, M., Edwards, J., Hatzky, H., et al., 2008a. An improved bathymetric portrayal of the Arctic Ocean: implications for ocean modeling and geological, geophysical and oceanographic analyses. Geophys. Res. Lett. 35, L07602.

Jakobsson, M., Polyak, L., Edwards, M., Kleman, J., Coakley, B., 2008b. Glacial geomorphology of the central Arctic Ocean: the Chukchi Borderland and the Lomonosov Ridge. Earth Surf. Process. Land. 33, 526–545.

Kind, N.V., Leonov, B.N., 1982. Antropogen Taimyra (The Antropogen of the Taimyr Peninsula). Nauka, Moscow, 184 pp. (in Russian).

Lambeck, K., Purcell, A., Funder, S., Kjær, K.H., Larsen, E., Möller, P., 2006. Constraints on the Late Saalian to Early Middle Weichselian ice sheet of Eurasia from field data and rebound modelling. Boreas 35, 539–575.

Larsen, E., Kjær, K.H., Demidov, I.N., Grøsfjeld, K., Houmark-Nielsen, M., Jensen, M., et al., 2006. Late Pleistocene glacial and lake history of northwestern Russia. Boreas 35, 394–424.

Lokrantz, H., Ingólfsson, Ó., Forman, S.L., 2003. Glaciotectonised Quaternary sediments at Cape Shpindler, Yugorski Peninsula, Arctic Russia: implications for glacial history, ice-movements and Kara Sea Ice Sheet configuration. J. Quatern. Sci. 18, 527–543.

Miroshnikov, L.D., 1959. Chetvertichnye otlozeniya I nekotorye cherty geomorfologii poluostrova Cheluskin. (Quaternary deposits and some of the geomorphological features of Chelyuskin Peninsula). Bull. Leningrad State Univ. Ser. Geol. Geogr. 2 (12), 11–21.

Möller, P., Sallaba, F., 2010. Ice marginal zones on the Taymyr Peninsula from the last glacial cycles, as interpreted from Landsat and digital elevation (ASTER) data. Abstract. In: Fourth International Conference and Workshop on the Arctic Palaeoclimate and Its Extremes (APEX), Iceland, May 26–30, 2010.

Möller, P., Bolshiyanov, D.Yu., Bergsten, H., 1999a. Weichselian geology and palaeo-environmental history of the central Taymyr Peninsula, Siberia, indicating no glaciation during the last global glacial maximum. Boreas 28, 92–114.

Möller, P., Bolshiyanov, D.Yu., Jansson, U., Schneider, G.V., 1999b. The "Ledyanaya Gravel Event"—a marker of the last glacioisostatic-induced depression along the Byrranga Mountains and south thereof. AbstractIn: Quaternary Environments of the Eurasian North (QUEEN), Third QUEEN Workshop, Øystese, Norway, 17–18, April 1999, p. 41.

Möller, P., Lubinski, D., Ingólfsson, Ó., Forman, S.L., Siedenkrantz, M.-S., Bolshiyanov, D.Yu., et al., 2007. Erratum to: Severnaya Zemlya, Arctic Russia: a nucleation area for Kara Sea ice sheets during the Middle to Late Quaternary. Quatern. Sci. Rev. 26, 1149–1191.

Möller, P., Federov, G., Pavlov, M., Seidenkrantz, M.-S., Sparrenbom, C., 2008. Glacial and palaeo-environmental history of the Cape Chelyuskin area, Arctic Russia. Polar Res. 27, 222–248.

Molodkov, A.N., Bolikhovskaya, N.S., 2002. Eustatic sea-level and climate changes over the last 600 ka as derived from mollusc-based ESR-chronostratigraphy and pollen evidence in Northern Eurasia. Sed. Geol. 150, 185–201.

Niessen, F., Ebel, T., Kopsch, C., Fedorov, G.B., 1999. High-resolution seismic stratigraphy of lake sediments on the Taymyr Peninsula, central Siberia. In: Kassens, H., Bauch, H.A., Dmitrenko, I., Eicken, H., Hubberten, H.-W., Melles, M., Thiede, J., Timokhov, L. (Eds.), Land-Ocean Systems in the Siberian Arctic: Dynamics and History. Springer-Verlag, Berlin, pp. 437–456.

Polyak, L., Edwards, M.H., Coakley, B.J., Jakobsson, M., 2001. Ice shelves in the Pleistocene Arctic Ocean inferred from glaciogenic deep-sea bedforms. Nature 410, 453–457.

Polyak, L., Niessen, F., Gataullin, V., Gainanov, V., 2008. The eastern extent of the Barents-Kara ice-sheet during the Last Glacial Maximum based on seismic-reflection data from the eastern Kara Sea. Polar Res. 27, 162–174.

Raab, A., Melles, M., Berger, G.W., Hagedorn, B., Hubberten, H.-W., 2003. Non-glacial paleoenvironments and the extent of Weichselian ice sheets on Severnaya Zemlya, Russian High Arctic. Quatern. Sci. Rev. 22, 2267–2283.

Saks, V.N., 1953. The Quaternary Period in the Soviet Arctic (Chetvertichny period v Sovietskoi Arktike). Vodtransizdat, Leningrad-Moscow, 627 pp. (in Russian).

Shneyder, G.V., 1989. Stratigrafiya kaynozoyskih otlozeniy i nekotorye cherty rel'efa severo-vostochnoy okonechnosti Taymyrskogo Poluostrova (Stratigraphy of Cenozoic deposits of some topographic features of the Taymyr Peninsula, north-eastern extremity). PGO Sevmorgeologiya 1989, 35–48.

Siegert, M.J., Dowdeswell, J.A., 2004. Numerical reconstructions of the Eurasian Ice Sheet and the climate during the Late Weichselian. Quatern. Sci. Rev. 23, 1273–1283.

Siegert, C., Derevyagin, A.Yu., Shilova, G.N., Hermichen, W.-D., Hiller, A., 1999. Paleoclimatic evidences from permafrost sequences in the Eastern Taymyr Lowlands. In: Kassens, H., Bauch, H.A., Dmitrenko, I.A., Eicken, H., Hubberten, H.-W., Melles, M., Thiede, J., Timokhov, L.A. (Eds.), Land-Ocean Systems in the Siberian Arctic: Dynamics and History. Springer-Verlag, Berlin, pp. 477–499.

Siegert, M.J., Dowdeswell, J.A., Svendsen, J.I., Elverhøi, A., 2002. The Eurasian Arctic during the last ice age. Am. Sci. 90, 32–39.

Svendsen, J.I., Astakhov, V.I., Bolshiyanov, D.Y., Demidov, I., Dowdeswell, J.A., Gataullin, V., et al., 1999. Maximum extent of the Eurasian ice sheets in the Barents and Kara Sea region during the Weichselian. Boreas 28, 234–242.

Svendsen, J.I., Alexanderson, H., Astakhov, V.I., Demidov, I., Dowdeswel, J.A., Funder, S., et al., 2004. Late Quaternary ice sheet history of northern Eurasia. Quatern. Sci. Rev. 23, 1229–1271.

Urvantsev, N.N., 1931. Quaternary glaciation of Taymyr. Bulleten komissii po izucheniyu chetvertichnogo perioda 3, 23–42 (in Russian).

Velichko, A.A., Isayeva, L.L., Makeyev, V.M., Matishov, G.G., Faustova, M.A., 1984. Late Pleistocene glaciation of the Arctic Shelf, and the reconstruction of Eurasian ice sheets. In: Velichko, A.A., Wright, H.E., Barnosky, C.W. (Eds.), Late Quaternary Environments of the Soviet Union. University of Minnesota Press, Minneapolis, pp. 35–41.

Volkova, V.S., Babushkin, A.Ye., 2000. Unifitsirovannaya regionalnaya stratigraficheskaya skhema chetvertichnykh otlozheniy Zapadno-Sibirskoi ravniny (The Unified Regional Stratigraphic Scheme of the Quaternary of the West Siberian Plain). SNIIGGiMS, Novisibirsk, 64 pp. (in Russian).

Chapter 29

Glacial History of Slovenia

Miloš Bavec[1,*] and Tomaž Verbič[2]

[1] *Geological Survey of Slovenia, Dimičeva 14, 1000 Ljubljana, Slovenia*
[2] *Arhej d.o.o., Institute of geoarchaeology, Rimska 1, 1000 Ljubljana, Slovenia*
*Correspondence and requests for materials should be addressed to Miloš Bavec. E-mail: milos.bavec@geo-zs.si

29.1. INTRODUCTION

The Pleistocene glaciers in Slovenia formed an ice cap over the central parts of the Julian Alps, Kamnik-Savinja Alps and Karavanke Mountains and drained into the foreland along the glacial/fluvial valleys (Bavec and Verbič, 2004). The relief in Slovenia is relatively low compared to other parts of the Alps. Only a few mountain peaks in Slovenia exceed 2800 m a.s.l., and intramontane valleys and basins are situated as low as 200 m a.s.l. In trying to reconstruct the extent of glaciers in Slovenia, one should also bear in mind that owing to Quaternary tectonic activity (e.g. Rižnar et al., 2005, 2007; Verbič, 2006; Komac and Bavec, 2007), fluvial erosion and mass movements of sediments have almost entirely removed the glacial sedimentary successions in some places.

Ice from the Karavanke Mountains, and part of the Julian Alps north of the Adriatic/Black Sea divide, fed across cols and smaller side valleys the Bohinj and Sava Dolinka valley glaciers, so that the Sava Dolinka glacier was also a continuation of the ice cover in Italy (Fig. 29.1). Along the present valleys of the Sava Bohinjka and Sava Dolinka rivers, these glaciers advanced to the vicinity of the town of Radovljica (Melik, 1930; Kuščer, 1955; Šifrer, 1969, 1992). The Bohinj glacier was (at least in the Late Pleistocene) the largest glacier on Slovenian territory, and at its maximum, it exceeded 900 m in thickness. Ice from the Julian Alps accumulated south of the continental divide in the Soča glacier following the course of the recent Soča river valley that drains into the Adriatic Sea.

The main ice stream from the Kamnik-Savinja Alps extended down the Savinja river valley. Erosion of the Savinja river carried away practically all the distal glacial sediments; therefore, the maximum extent of the glacial ice can only be reconstructed from sparse geomorphological evidence (Meze, 1966; Mioč, 1983).

In addition to the ice cover of the Slovenian Alps (Southern Alps) and their outlet glaciers reaching down to the foothills, a smaller ice cap also covered the summit and slopes of Mount Snežnik in the Dinarides. The terminus of this ice reached down to approximately 900 m a.s.l. (Šikić et al., 1972).

29.2. EARLY/MIDDLE PLEISTOCENE GLACIATIONS

In addition to some sparse sedimentary evidence for the Early/Middle Pleistocene glaciations, glaciofluvial sediments may be the only evidence available for understanding some aspects of early glaciations in the region. The Sava river drained the bulk of the Alpine sector throughout the Pleistocene, its drainage system having been developed prior to the onset of the pronounced climatic oscillations, that is, most probably during the Pliocene. The pre-Pleistocene origin of the Sava river drainage system has been interpreted from remnants of fluvial sandy gravel preserved at high elevations downstream of Ljubljana (around 700 m a.s.l.; up to 500 m above the modern river bed). Indirect evidence of a substantial age of this sedimentary succession includes the absence of carbonate pebbles. In contrast, the modern sediments of the Sava river contain around 75% of carbonate pebbles. The difference is interpreted as a consequence of multiple consecutive resedimentation cycles during which the carbonate component was leached out (Verbič, 2004). These sediments are most abundant in the Krško Basin and its vicinity and are usually referred to as a Plio-Pleistocene succession (Fig. 29.1). The only attempt to date the sediment so far by thermoluminescence determination has yielded a maximum age of (older than) 306 ka $\pm 2\sigma$ (Bavec, 2000).

29.2.1. Glacial Limits

Sedimentary evidence of Early and Middle Pleistocene glaciation in Slovenia is sparse. Yet by combining certain

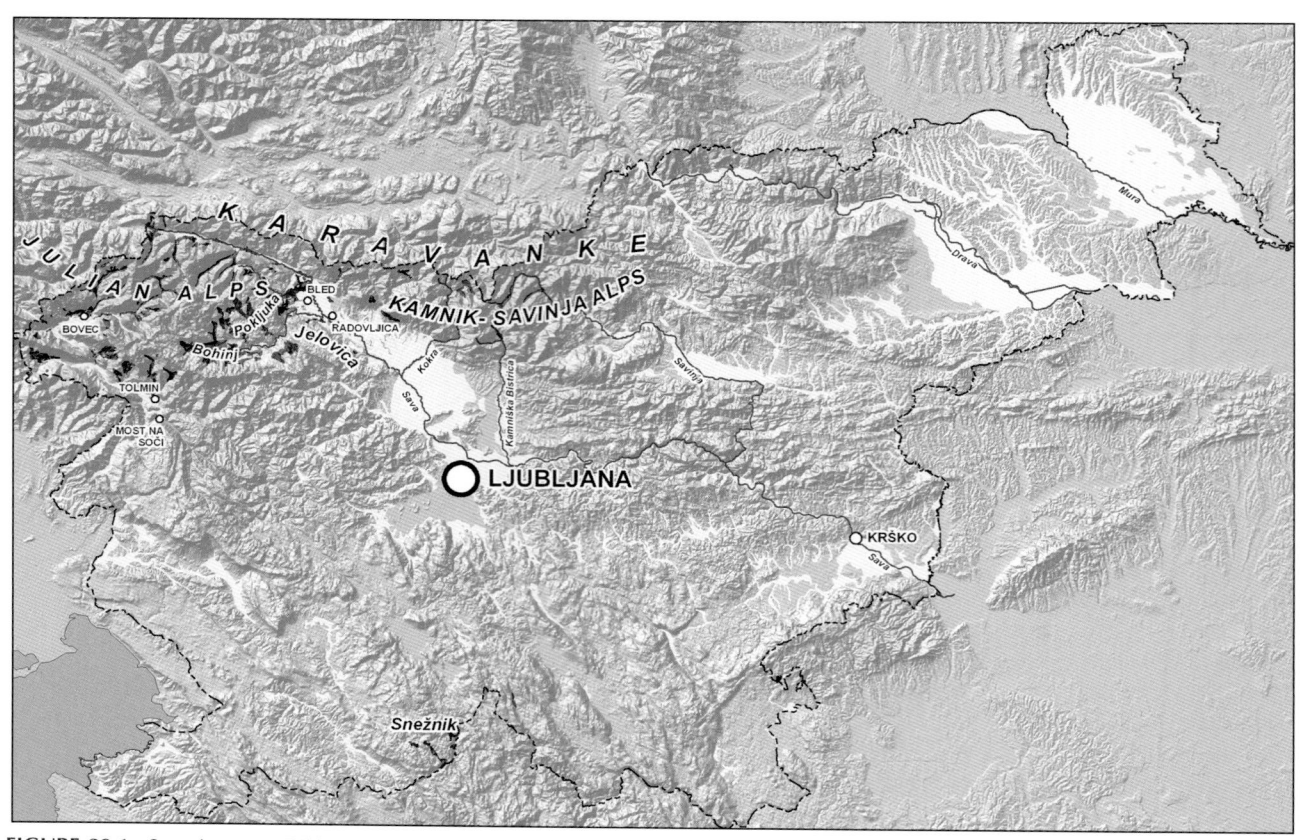

FIGURE 29.1 Location map. White: glaciofluvial and fluvial sediments of Quaternary and 'Plio-Quaternary' age. Black: glacial and paraglacial deposits. Geology taken from Buser (2009).

modelling results, sedimentary records and morphological evidence, the maximum extent of the Soča and Bohinj glaciers can be discussed and evaluated.

Assuming that the topography during the Pleistocene was at least roughly similar to the present day, it is very unlikely that the Soča glacier should have reached beyond the Bovec basin (Bavec et al., 2004). For the Middle, and especially for the Early Pleistocene, such an estimate is based solely on modelling results. Although this extent appears to be overestimated, the writers do not have any firm evidence to reject the interpretation of Penck and Brückner (1909; later also Kuščer et al., 1974; Kunaver, 1975) that the Soča glacier reached as far as the town of Tolmin during earlier Pleistocene glaciations. Early authors interpreted the maximum extent on the basis of relatively ambiguous morphological observations, the occurrence of glacially reworked boulders in the area and a patch of till found about 300 m above the modern Soča river, 3 km north of Tolmin, with the lithology indicating an upstream source.

The maximum extent of the Early/Middle Pleistocene Bohinj glacier was determined by interpreting landforms in the vicinity of the towns of Radovljica and Bled as terminal moraines (Kuščer, 1955, 1990; Žlebnik, 1971). Four generations of topographically relatively well-expressed moraines—all at a distance of less than 4 km apart—have been identified and correlated with four stages of fluvial valley fill, tentatively referred to as 'Würmian, Rissian, Mindelian and Günzian' (Žlebnik, 1971). The chronology of the deposits was based on litho- and morphostratigraphical correlations, partly supported by palynological analyses (Šercelj, 1970). According to this chronology, the most distal moraine is the oldest, while the terminal moraines of the subsequent glacial events follow in a proximal direction. Revisiting the site in 2008 confirmed that morainic remains are, in general, progressively older in the distal direction, yet the details and the absolute ages may not necessarily follow those of the earlier interpretations (see Bohinj glacier below).

29.2.2. Soča Glacier

Regardless of the discussion presented above on the maximum extent of the Soča glacier during the Early and Middle Pleistocene, and regardless of the fact that even in the Bovec basin, some 30 km upstream from Tolmin, no glacial deposits have been identified, the proximity of an Early/Middle Pleistocene glacier is well documented by a sequence of paraglacial deposits in the Bovec basin. The Quaternary sedimentary sequence in this basin consists of

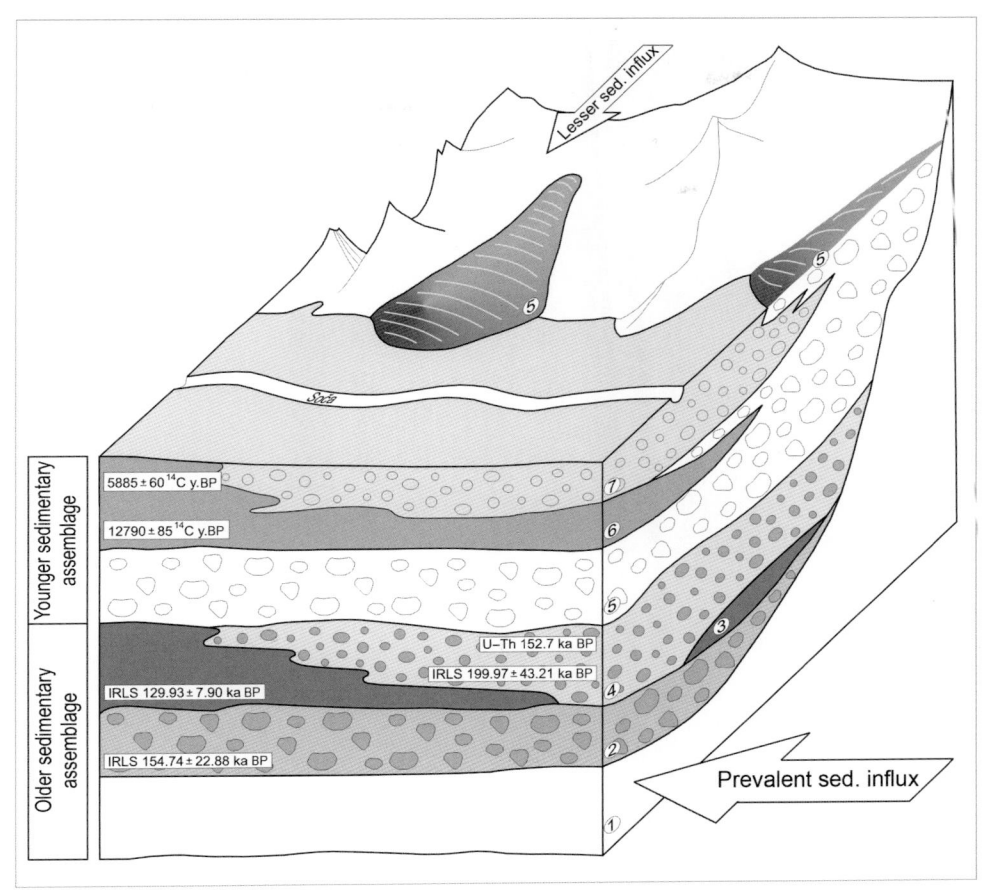

FIGURE 29.2 Model of the Bovec basin paraglacial sedimentation. 1—bedrock, 2—diamictite, 3—lacustrine deposits, 4—glaciofluvial conglomerate, 5—diamicton, 6—lacustrine deposits, 7—glaciofluvial gravel.

two sedimentary assemblages that are genetically similar (Fig. 29.2). The older assemblage is Middle Pleistocene in age and consists of diamictite, overlain by a series of fluvial conglomerate and fine-grained lacustrine sediments.

After the (Early and Middle Pleistocene) glaciers melted, a large amount of poorly consolidated and water-saturated tills they deposited were left unsupported and therefore slid downhill both along the main glacial valley and along the slope flanks. This till, mixed with non-glacial material, thus forms a large volume of the sediments that were often erroneously interpreted as tills or moraines owing to their content of glacially shaped clasts. Typical examples of such remobilised tills, belonging to the older sedimentary assemblage, are found at Ravni Laz near Bovec (Fig. 29.3). Here, the diamictite was dated by optically stimulated luminescence (OSL) to 154.74 ± 22.88 ka. Deposition of the diamicton (now diamictite) was followed by coarse fluvial gravel (now conglomerate) and lake deposits. It has to be noted that the age determination of the whole sedimentary fill (Fig. 29.2) was strongly influenced by the high carbonate content and should therefore be taken merely as a rough estimate. The authors interpret the older assemblage in Bovec as having been deposited at the transition from full glacial to interglacial conditions, that is, between MIS (marine isotope stages) 6 and 5 (Bavec et al., 2004).

29.2.3. Bohinj Glacier

In the vicinity of Radovljica, there is relatively good sedimentary evidence of Early/Middle Pleistocene glaciations. The remains of older morainic arcs are found beyond the most distal limits of the Late Pleistocene glaciations, as was previously clearly noted by early researchers (e.g. Kuščer, 1955, 1990; Žlebnik, 1971). The site was revisited in 2008 during highway construction through the eastern part of the Bled–Radovljica basin (Bavec et al., 2008). The road cutting revealed sedimentary features that enabled some modifications of the glacier dynamics. The four positive topographic features identified east of the Sava river, which correspond to early interpretations as arcs of pre-Late Pleistocene terminal moraines, could thus be related to the deposits revealed in the road section. The sedimentary properties and morphology demonstrate that the plain around Radovljica was formed in a proglacial and ice-marginal

FIGURE 29.3 Paraglacial diamictite (below) and conglomerate in Ravni Laz (Bovec basin).

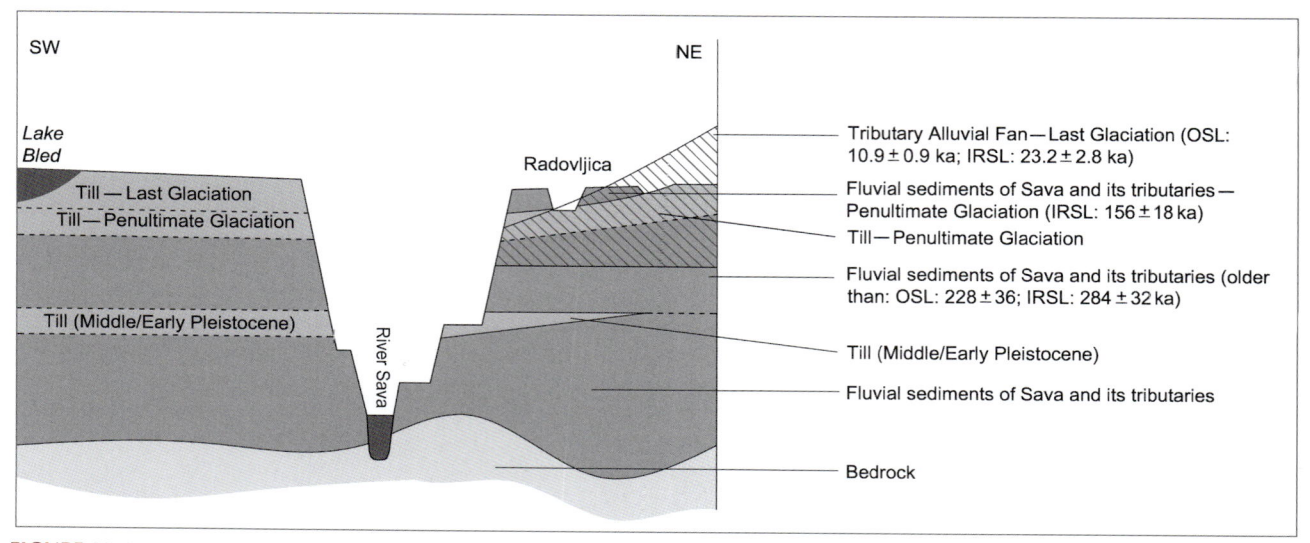

FIGURE 29.4 Model of the sedimentary fill in the Bled–Radovljica basin.

environment where accumulation dominated over erosion. At least two generations of Early/Middle Pleistocene tills are found stacked east of the Sava river. The two generations of till are separated/covered/underlain by at least three generations of glaciofluvial sediments (Fig. 29.4). Based on OSL, and infrared-stimulated luminescence datings (OSL/IRSL), the writers conclude that Early/Middle Pleistocene glaciers crossed the axis of the modern Sava valley at least twice. The same event happened at least once earlier in the Early/Middle Pleistocene, when the glacier reached the plain of Radovljica again during the penultimate glaciation (Bavec et al., 2008). The till of the penultimate glaciation is revealed underlying the latest glaciofluvial sequence and can be linked to topographically expressed end moraines east of Radovljica.

There three positions of glacial termini are reconstructed based on the positions of end moraines, and they are all related to the glacier retreat during the penultimate deglaciation. The most proximal moraine east of the Sava river, however, is now interpreted to be an erosional remnant of an older basal till and is the oldest morainic feature in the area.

29.3. LATE PLEISTOCENE GLACIATIONS

29.3.1. Glacial Limits

Remains of glacial sediments from the Late Pleistocene are mainly limited to higher mountainous regions, although some erosional remnants are only found in alpine valleys where moraines locally also cover the valley floors. In the high mountain regions of the Julian Alps, moraines have been described to some detail, especially from the hinterland of the Soča glacier (Kunaver, 1975, 1980, 1990) and Pokljuka (Melik, 1930; Šifrer, 1952). Here, they are found to be of considerable extent around Jelovica and in places in the hinterland of the Sava Dolinka and Bohinj glaciers.

As earlier authors concluded, the Late Pleistocene Soča glacier should have reached the towns of Tolmin or even Most na Soči (Penck and Brückner, 1909; Kuščer et al., 1974; Kunaver, 1975). Sediment properties and glacier modelling indicate that such an extent is, at least for the time of the last two Alpine glaciations (Rissian, Würmian), most probably overestimated. During these glaciations, the main valley ice stream along the Soča and Koritnica valleys extended no further than the Bovec basin (Bavec et al., 2004).

Ice from the Karavanke Mountains, and part of the Julian Alps north of the Adriatic/Black Sea divide, accumulated across smaller side valleys (such as Radovna) and slope glaciers in the Bohinj and Sava Dolinka valley glaciers, where the Sava Dolinka glacier was also an outlet glacier of the ice cover of the neighbouring Italian Alps (Fig. 29.5). Along the present valleys of the Sava Bohinjka and Sava Dolinka rivers, these glaciers advanced almost to Radovljica (Melik, 1930; Kuščer, 1955; Šifrer, 1969, 1992; Bavec et al., 2008). The Bohinj glacier was (at least in the Late Pleistocene) the largest glacier in Slovenia. At its optimum, it most probably exceeded 900 m in thickness and therefore would have reached and partially covered the Pokljuka plateau. Near the village of Gorje, a well-preserved terminal/medial moraine demonstrates the contact between the Bohinj and Radovna glaciers at the time of their maxima (Fig. 29.6).

Descriptions of glacial sediments from the Karavanke Mountains and the Kamnik-Savinja Alps are rare. Major glaciers from the Kamnik-Savinja Alps included the Jezersko glacier, that extended along the river Kokra to the village of Fužine (Meze, 1974; Buser, 1980), and a glacier in the Kamniška Bistrica river valley, that extended approximately to the confluence of the Kamniška Bistrica and Korošica rivers (Šifrer, 1961).

Meze (1966, 1974) and Buser (1980) refer to some moraines in the high mountain area. Moraines from the Kamniška Bistrica river catchment area have been studied by Šifrer (1961). Meze (1966) and Mioč (1983) describe moraines from the area of Savinja at Robanov Kot and Logar and in the Savinja valleys. They also mention the existence of 'hanging moraines' in smaller side valleys. Meze (1966) attributed all the preserved moraines of the Savinja glacier and its tributaries to the 'Latest' or Late-glacial period.

Evidence of Pleistocene glaciations is also found on the Snežnik Mountain, being the only glaciated area in Slovenia outside the Alps. There sparse till remnants cover a relatively wide area, reaching downslope to the lowest end moraines at about 900 m a.s.l. (Šikić et al., 1972; Šikić and Pleničar, 1975).

FIGURE 29.5 Reconstruction of maximal glacier extent during the Late Pleistocene in the surroundings of Bled and Radovljica. The Julian Alps are to the left (west) and the Karavanke Mountains to the right (east). (Photograph: Bavec and Knific 2004; © National Museum of Slovenia).

FIGURE 29.6 A terminal moraine of both the Bohinj glacier (from the left) and Radovna glacier (from the right) near the village of Gorje (Bled). (Photograph: J. Hanc; © National Museum of Slovenia).

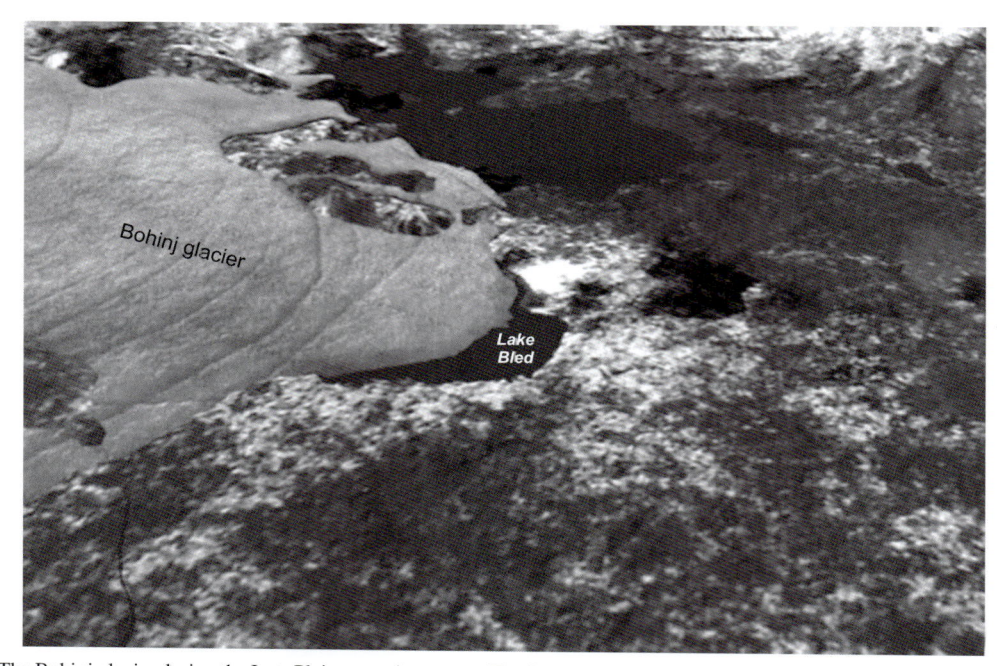

FIGURE 29.7 The Bohinj glacier during the Late Pleistocene ice retreat. The fragmentation into three separate tongues is already visible. In front of the terminus, Lake Bled is being exposed. Meltwater is also forming small water bodies on the Bled plateau (dark patches E and N of the early Lake Bled).

29.3.2. Key Sites

Kuščer (1955) and Šifrer (1969, 1992) consider that almost all terminal moraines of the Bohinj glacier between Bled and Radovljica relate to the retreat of the latest Bohinj glacier. The relative position of newly dated sediments and morphological features (Fig. 29.4) largely support this early interpretation (Bavec et al., 2008). The whole 'Bled Plateau'—the area between today's Lake Bled and the Sava Dolinka river—consists of the Late Pleistocene till and is party covered by fine-grained sediments. The latter were deposited in proglacial water bodies during the ice retreat (Fig. 29.7). Between the villages of Gorje and Zasip, the maximal extent of the last glaciations is marked by one of the best preserved terminal moraines in Slovenia (Fig. 29.6). In fact, it is a terminal moraine of the Bohinj and Radovna glaciers that converged there. Several morainic arcs were deposited during the retreat of the Late Pleistocene glacier west of Lake Bled. During melting, the uniform glacier terminus was gradually fragmented into two and then into three tongues. The central one carved the hollow of the present Lake Bled, the southern the depression at the village of Ribno and the northern one stretched towards the village of Gorje west of Bled Castle. This interpretation by Melik (1930) was recently supported by an animated model of the Bohinj glacier (Bavec and Knific, 2004). Most of the Late Pleistocene sedimentary record is preserved on the Bled Plateau and between the villages of Gorje and Zasip, including the area around the entrance to the Blejski Vintgar gorge.

Sedimentation of Late Pleistocene paraglacial deposits in the Bovec basin followed the same pattern as during deposition of the older sedimentary assemblage (Fig. 29.2). During and after the retreat of the Soča glacier (that never extended beyond the Bovec basin during the Late Pleistocene), the glacial sediments mixed with non-glacial gravel and reworked slope deposits to form various gravity mass flow sequences. This form of resedimentation is, to a certain extent still taking place, causing damage and even claiming lives. Meanwhile, the basin has become filled by fluvial and lacustrine sediments. The majority of the basin infill was transported along the main valley axis, and sedimentation has only partially been balanced during the Holocene (Bavec et al., 2004). An interesting feature in this young sedimentary assemblage is a succession of laminated lacustrine deposits up to 200 m thick (Fig. 29.2). An attempt has been made to obtain palaeoclimatic information from this sequence (Gosar et al., 2008); however, it turned out that the inorganic, detrital component of the sediment was too high to enable this type of analysis. The younger sedimentary assemblage was deposited between MIS 2 (the Late Weichselian) and today.

29.4. CONCLUSIONS

The preservation potential of glacial deposits in the Slovenian glacial valleys is extremely low. This phenomenon can also clearly be attributed to regional active tectonic uplift and displacements that range from several millimetres per year in the Alps, to several tenths of a millimetre in the south of the country. These displacement rates have been quantified either by direct measurements (Rižnar et al., 2005, 2007) or by indirect observations of predominantly glaciofluvial deposits (Verbič, 2005, 2006). Bearing in mind that differential vertical displacement over certain structural boundaries, such as the SW-rim of the Alps, might have reached as much as 4 mm per year, one should be very cautious when comparing geomorphological features only on the basis of their topographical position. This is the reason why the results of many of the early studies have been re-evaluated recently, as discussed in this overview. Another problem for their interpretation is the lack of suitable sediments for dating. Where present, such sediments are rich in carbonate which makes them particularly difficult to date. Recent developments in luminescence dating are opening new possibilities. Ongoing active tectonic studies and absolute numerical dating will provide firmer foundations for future reinterpretations of the glacial history of Slovenia.

REFERENCES

Bavec, M., 2000. Poročilo o določanju starosti kvartarnih sedimentov v Krški kotlini z metodo termoluminescence (TL) in optično stimulirane luminescence (OSL). Geološki zavod Slovenije, Ljubljana, 3 pp.

Bavec, M., Knific, T., 2004. The Bohinj Glacier. DVD-ROM. National museum of Slovenia and Infrastruktura d.o.o., Ljubljana (Animation with spoken text in Slovene, English, German and Italian).

Bavec, M., Verbič, T., 2004. The extent of Quaternary glaciations in Slovenia. In: Ehlers, J., Gibbard, P.L. (Eds.), Quaternary glaciations: extent and chronology, Part I: Europe. Elsevier, Amsterdam, pp. 385–388.

Bavec, M., Tulaczyk, S.M., Mahan, S.A., Stock, G.M., 2004. Late Quaternary glaciation of the Upper Soča River Region (Southern Julian Alps, NW Slovenia). Sed. Geol. 165, 265–283.

Bavec, M., Rižnar, I., Jež, J., Novak, M., 2008. Geološka spremljava v okviru varstva naravne dediščine "Strukturno tektonske značilnosti na območju trase in spremljajočih objektov na AC Vrba—Črnivec (Peračica)" Geološki zavod Slovenije, 64 pp., Ljubljana.

Buser, S., 1980. Tolmač lista Celovec (Klagenfurt) Osnovne geološke karte SFRJ 1: 100.000. Zvezni geološki zavod, Beograd, 62 pp.

Buser, S., 2009. Geological map of Slovenia 1:250.000; adopted after OGK 1: 100.000 and amended. Geological Survey of Slovenia, Ljubljana.

Gosar, M., Kovačič, K., Bavec, M., 2008. Geochemical feature of laminae in Late Quaternary lacustrine sequence in Srpenica, Soča valley. Geologija 51/1, 119–126 (in Slovene with English summary).

Komac, M., Bavec, M., 2007. Application of PSInSAR for observing the vertical component of the recent surface displacement in Julian Alps. Geologija 50/1, 97–110 (in Slovene with English summary).

Kunaver, J., 1975. On the geomorphological development of the Basin of Bovec during the Pleistocene epoch. Geografski Vestnik 47, 11–39 (in Slovene with English summary).

Kunaver, J., 1980. Razvoj in sledovi zadnje stadialne poledenitve v Zgornjem Posočju (I). Geografski Vestnik 52, 17–36.

Kunaver, J., 1990. Poznoglacialne morene v najvišjih delih posoških Julijskih Alp in poskus njihove datacije. In: Natek, K. (Ed.), 5. Znanstveno posvetovanje geomorfologov Jugoslavije, Zbornik referatov, Geomorfologija in geoekologija, ZRC SAZU, 207–215, Ljubljana.

Kuščer, D., 1955. Beitrag zur Pleistozängeologie des Beckens von Radovljica. Geologija 3, 136–150 (in Slovene with German summary).

Kuščer, D., 1990. The Quaternary valley fills of the Sava River and neotectonics. Geologija 33, 299–314 (in Slovene with English summary).

Kuščer, D., Grad, K., Nosan, A., Ogorelec, B., 1974. Geology of the Soča Valley between Bovec and Kobarid. Geologija 17, 425–476 (in Slovene with English summary).

Melik, A., 1930. Bohinjski ledenik. Geografski Vestnik 5–6 1–39.

Meze, D., 1966. La vallée supérieurre de la Savinja nou-velles constatations dans le développement géomorpholo-gique de la région. Dela IV razreda SAZU, 20. 199 pp. (in Slovene with French summary).

Meze, D., 1974. River Basin of Kokra during the Pleistocene Period. Geografski zbornik 14, 5–101 (in Slovene with English summary).

Mioć, P., 1983. Tolmač lista Ravne na Koroškem Osnovne geološke karte SFRJ 1: 100.000. Zvezni geološki zavod, Beograd, 69 pp.

Penck, A., Brückner, E., 1909. Die Eiszeiten in den Südalpen und im Bereich der Ostabdachung der Alpen. Die Alpen im Eiszeitalter 3, 718–1199, Leipzig.

Rižnar, I., Koler, B., Bavec, M., 2005. Identification of potentially active structures along the Sava River using topographic, and leveling line data. Geologija 48 (1), 107–116 (in Slovene with English summary).

Rižnar, I., Koler, B., Bavec, M., 2007. Recent activity of the regional geologic structures in Western Slovenia. Geologija 50 (1), 111–120 (in Slovene with English summary).

Šercelj, A., 1970. Würmeiszeitliche Vegetation und Klima in Slowenien. Razprave IV razreda SAZU, 13, 209–249 (in Slovene with German summary).

Šifrer, M., 1952. Obseg zadnje poledenitve na Pokljuki. Geografski Vestnik 24, 95–114.

Šifrer, M., 1961. Porečje Kamniške Bistrice v pleistocenu. – Dela IV. razreda SAZU, 12, 211 pp.

Šifrer, M., 1969. Kvartarni razvoj Dobrav na Gorenjskem. Geografski zbornik 11, 99–221.

Šifrer, M., 1992. Geomorfološki razvoj Blejsko-radovljiške ravnine in Dobrav v kvartarju. Radovljiški zbornik, 6–14.

Šikić. D., Pleničar, M., 1975. Tumač za list Ilirska Bistrica Osnovne geološke karte SFRJ 1: 100.000. Zvezni geološki zavod, Beograd.

Šikić, D., Pleničar, M., Šparica, M., 1972. Osnovna geološka karta SFRJ 1:100 000, list Ilirska Bistrica. Zvezni geološki Zavod, Beograd, 51 pp.

Verbič, T., 2004. Quaternary stratigraphy and neotectonics of the Eastern Krško Basin, Part 1: Stratigraphy. Razprave IV. razreda SAZU 45, 171–225, (in Slovene with English summary).

Verbič, T., 2005. Quaternary stratigraphy and neotectonics of the Eastern Krško Basin, Part 2: Neotectonics. Razprave IV. razreda SAZU 46, 171–216, (in Slovene with English summary).

Verbič, T., 2006. Quaternary-active reverse faults between Ljubljana and Kranj, Central Slovenia. Razprave IV. razreda SAZU 47, 101–132, (in Slovene with English summary).

Žlebnik, L., 1971. Pleistocene deposits of the Kranj, Sora and Ljubljana fields. Geologija 14, 5–51, (in Slovene with English summary).

Chapter 30

Quaternary Glaciations of Turkey

Mehmet Akif Sarıkaya[1,*], Attila Çiner[2] and Marek Zreda[3]

[1]Geography Department, Fatih University, Istanbul, Turkey
[2]Department of Geological Engineering, Hacettepe University, Ankara, Turkey
[3]Department of Hydrology and Water Resources, University of Arizona, Tucson, Arizona, USA
*Correspondence and requests for materials should be addressed to Mehmet Akif Sarıkaya. E-mail: masarikaya@fatih.edu.tr

30.1. INTRODUCTION

Although Turkey is not the first country that comes to mind when glaciers and glacial deposits are considered, recent research has revealed important new information regarding mountain glaciers developed during the Quaternary (Zreda et al., 2006; Akçar et al., 2007, 2008; Sarıkaya et al., 2008, 2009; Zahno et al., 2009, 2010). Turkey, situated in the Eastern Mediterranean region between 36°N and 42°N latitude, and 26°E and 45°E longitude, is characterised by strong topographic and climatic contrasts. The mean elevation increases eastwards, from less than 500 m a.s.l. in the western lowlands, to ca. 1100 m a.s.l. in the centrally located Anatolian Plateau, to much over 3000 m a.s.l. in the mountains to the south, east and north of the plateau (Fig. 30.1). Many of the mountain tops lie above the modern snowline (Kurter and Sungur, 1980; Kurter, 1991) and have climatic conditions suitable for glaciers and ice caps (Çiner, 2004; Sarıkaya, 2009, 2010a,b).

The existence of glaciers in the Taurus Mountains and the Pontic range (Eastern Black Sea Mountains) was first reported by Ainsworth (1842) and Palgrave (1872), but systematic investigations did not begin until the beginning of the twentieth century. Maunsell (1901), Bobek (1940), Louis (1944), İzbırak (1951), Erinç (1953), Blumenthal (1952, 1958) and Wright (1962) studied glaciers and glacial deposits and collected glacio-geological data from the south-eastern Taurus Mountain range, which is the highest and most mountainous part of the country, where a quarter of the recent glaciers are concentrated (Sarıkaya, 2010a). Sırrı Erinç, the pioneer of the glacial geology in Turkey, published detailed papers on glaciers and variations in snowline altitudes (Erinç, 1944, 1951, 1952, 1953, 1978). Later, Messerli (1964, 1967), Birman (1968) and Atalay (1987) presented summary papers on recent Turkish glaciers. More recently, Landsat (Kurter and Sungur, 1980; Kurter, 1991) and ASTER (Advanced Spaceborne Thermal Emission and Reflection Radiometer; Sarıkaya, 2010a) satellite images were used to report the extent of modern Turkish glaciers. Since the beginning of this century, two research groups, one from Bern in Switzerland and the other from Tucson in the USA, together with their Turkish partners from Istanbul Technical and Hacettepe universities, respectively, conducted research that used cosmogenic dating of moraine boulders to quantify the extent and timing of Quaternary glaciations. While the Swiss group used ^{10}Be and ^{26}Al isotopes, the American group used ^{36}Cl, which allowed an excellent diversification of the target areas. Today, the distribution and extent of Turkish glaciers are well established, and likewise, the knowledge of Pleistocene glaciations is much improved compared to that in the first edition of this book. In this chapter, the authors present the state of research based on the existing literature combined with an evaluation of unpublished evidence and their personal observations.

30.2. QUATERNARY GLACIAL LANDFORMS

Most of Quaternary glacial landforms occur in three regions of Turkey: (1) the Taurus Mountains, along the Mediterranean coast and south-eastern Turkey; (2) the mountain ranges along the Eastern Black Sea Region and (3) the volcanoes and independent mountain chains scattered on the Anatolian Plateau (Kurter and Sungur, 1980; Çiner, 2004; Table 30.1).

30.3. THE TAURUS MOUNTAINS

The Taurus Mountains extend from south-east to south-west Turkey and contain the mountains with the best-studied glaciers. This mountain range can be subdivided into three sections: (1) the south-eastern Taurus, (2) the Central Taurus and (3) the western Taurus.

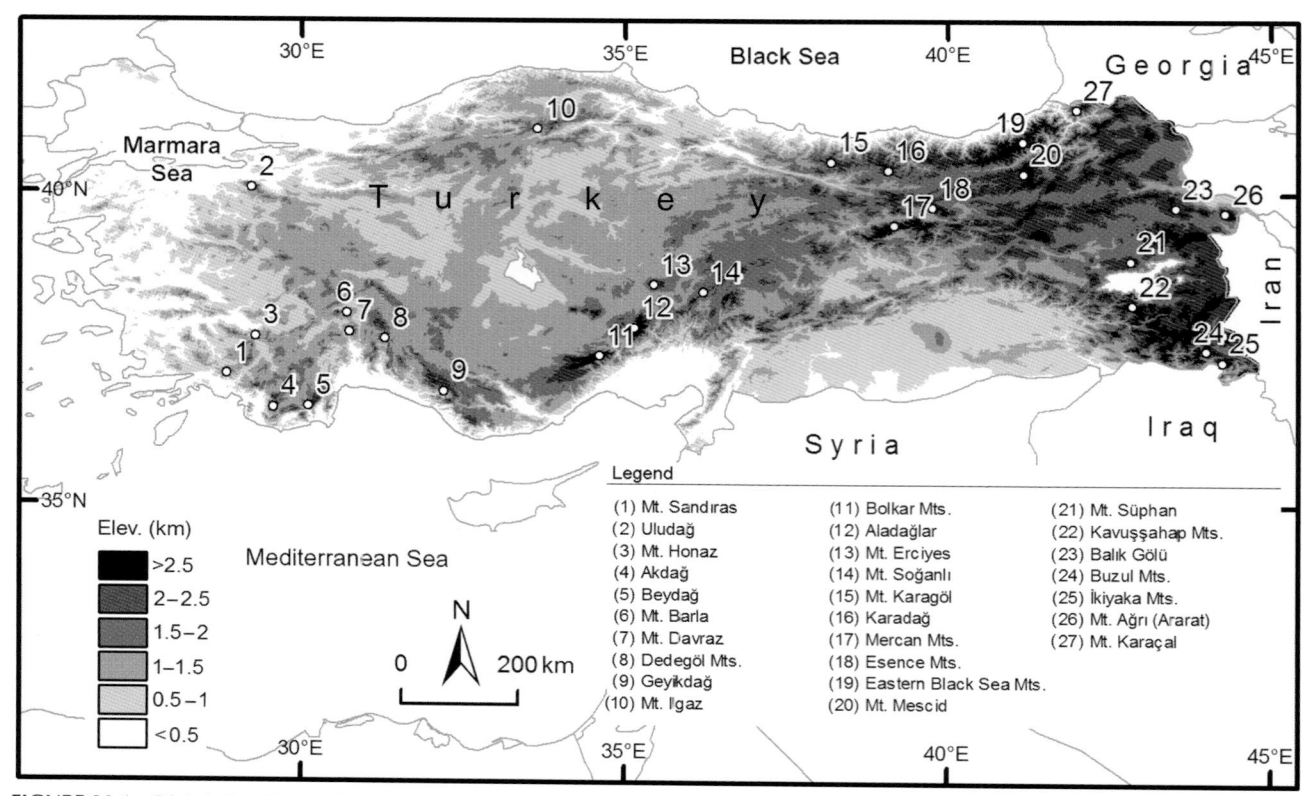

FIGURE 30.1 Digital elevation model indicating the location of glaciated mountains in Turkey.

30.3.1. The South-Eastern Taurus

The south-eastern Taurus Mountains are located on the junction between the Turkey's Taurus Mountains and the Iran's Zagros Mountains. Together, the Taurus–Zagros range separates the Anatolian–Iranian Plateau from the Mesopotamian Lowlands. These mountains contain the highest crests of the region and thus support the largest modern glaciers in the Middle East (Sarıkaya, 2009). The mountains consist largely of Palaeozoic and Mesozoic metamorphic and volcanic rocks (Wright, 1962) together with folded Mesozoic limestones and Tertiary terrestrial sedimentary rocks (Altınlı, 1966). The climate in the region is marked by high winter precipitation arising from the cyclonic disturbances that travel eastwards along the Taurus range (Butzer, 1958) and from the Arabian anticyclones derived from the south (Wright, 1962). Pleistocene glacial features in the south-eastern Taurus are found in three regions: the Buzul and İkiyaka Mountains, on the Iraqi border and the Kavuşşahap Mountains, south of Lake Van.

The Buzul and İkiyaka Mountains, also known as Mount Cilo (37.49°N, 44.00°E, 4135 m a.s.l.) and Mount Sat (37.31°N, 44.25°, 3794 m a.s.l.), respectively, are located at the south-eastern corner of the country and comprise the highest sector of the Taurus–Zagros range (Fig. 30.1). Despite the importance of the region in terms of recent and past glaciations, no research has been conducted there in recent decades because of the political difficulties and physiographic inaccessibility of the region. Only a few studies, which date back to the 1940–1960s, are available. Bobek (1940) mapped the distribution of Pleistocene (Würmian, Last Glacial Maximum, LGM) glaciers, and marked the prominent end moraines in the Buzul and the İkiyaka Mountains. He reported that the former glaciers on the northern side of the Uludoruk Tepe in the Buzul Mountains reached lengths of ca. 9 km from the cirque to the termini at ca. 1800 m a.s.l. In the İkiyaka Mountains, 30 km south-east of the Buzul Mountains, the Pleistocene glaciers reached an elevation of 2100 m a.s.l. and had a length of up to 10 km (Bobek, 1940). Erinç (1953) noted that the lowest terminal moraine in the Buzul Mountains occurs at 1600 m, and concluded that the moraine was deposited during the Last Glaciation. Later, Wright (1962) investigated the area on the north and north-western sides of the Buzul Mountains, and reported that the Pleistocene glacial features in the region were much more extensive than those described by Bobek (1940) and Erinç (1953). He mapped a broader area adjacent to the Greater Zap River Valley, and discovered a few lower elevation cirques. Based on a blanket of till over the 2700–3000 m a.s.l. ridges on the northern side of the Buzul Mountains, he and proposed the lowest terminus elevation of Pleistocene glaciers

Chapter | 30 Quaternary Glaciations of Turkey

TABLE 30.1 Glaciated Mountains in Turkey

Mountain name	Highest peak's Name	Location Latitude (DD°N)	Longitude (DD°E)	Elevation (m)	Snowline (m) LGM/recent
Taurus Mountains					
Southeastern Taurus Mountains					
Buzul Mountains	Uludoruk Tepe	37.4877	44.0012	4135	2100–2800/3600
İkiyaka Mountains	Dolampar Tepe	37.3105	44.2502	3794	2600/3600
Kavuşşahap Mountains	Hasanbeşir Tepe	38.2146	42.8563	3503	unknown/3400
Central Taurus Mountains					
Aladağlar	Demirkazık Tepe	37.8366	35.1453	3756	1900–2200/3450
Bolkar Mountains	Medetsiz Tepe	37.3862	34.6087	3524	1900–2075/3450–3700
Geyikdağ	Geyikdağ	36.8075	32.2021	2850	2000/3200
Mount Soğanlı	Beydağ	38.4084	36.2119	3075	2610/3550
Western Taurus Mountains					
Mount Sardıras	Çiçekbaba Tepe	37.0814	28.8380	2295	2000/3000–3500
Dedegöl Mountains	Dipoyraz Tepe	37.6437	31.2835	2992	2350–2400/3300–3500
Akdağ	Uyluk Tepe	36.5439	29.5674	3016	2200–2400/3500
Beydağ	Beydağ	36.5684	30.1017	3086	2400–2600/3600
Mount Barla	Gelincik Tepe	38.0531	30.7022	2800	2400/3750
Mount Honaz	Honaz Dağı	37.6791	29.2850	2571	2600/3600
Mount Davraz	Davraz Dağı	37.7571	30.7317	2637	2400/3750
Coastal ranges of the Eastern Black Sea					
Eastern Black Sea Mountains	Kaçkar Dağı	40.8354	41.1614	3932	2300–2500/3100–3200
Karadağ	Aptalmusa Tepe	40.3793	39.0710	3331	2600–2850/3500
Mount Karagöl	Karagöl Dağı	40.5101	38.1928	3107	2600–2700/3500
Mount Karaçal	Karaçal Dağı	41.3472	41.9830	3415	unknown/3400
Individual Mountains					
Mount Ağrı (Ararat)	Büyük Ağrı	39.7018	44.2983	5137	3000/4300
Mount Erciyes	Büyük Erciyes D.	38.5318	35.4469	3917	2700/3550
Mount Süphan	Sandık Tepe	38.9309	42.8326	4058	3200/3700–4000
Uludağ	Uludağ Tepe	40.0706	29.2215	2543	2400/3500
Mercan Mountains	Gedik Tepe	39.4934	39.1669	3368	2750/3600–3700
Esence Mountains	Keşiş Dağı	39.7836	39.7548	3477	2750/3600–3700
Mount Mescid	Mescid Dağı	40.3273	41.1673	3239	2750/3600–3700
Mount Ilgaz	Ilgaz Dağı	41.0342	33.6545	2587	unknown/3500
Balık Gölü	Çıplakyurt Tepe	39.7766	43.5274	2804	unknown/4300

at about 1500 m a.s.l. Further, he correlated these glacial features with the Würmian-age deposits in the Alps based upon their degree of weathering and the erosional modifications of glacial deposits. Unfortunately, no numerical ages have yet been obtained from this region, precluding any further interpretation of the glacial stratigraphy.

The Kavuşşahap Mountains, also known as Mount İhtiyarşahap (38.21°N, 42.86°E), located about 20 km south of Lake Van (Fig. 30.1), contain numerous cirques on the northern slopes, many of them (particularly those above 3000 m a.s.l.) littered with rock glaciers (Sarıkaya, 2010a). There have been no reports of late Quaternary glacial activity in these mountains. However, satellite image analysis of Google Earth® and ASTER scenes acquired in 2004 and 2006, respectively, revealed the presence of very well-developed lateral moraines at altitudes as low as 2100 m a.s.l. If confirmed in the field, these lateral moraines will mark past glaciers that were up to 10 km long.

The modern snowline (∼equilibrium line altitude or ELA), calculated from 20 small glaciers in the Buzul Mountains, varies between 3100 and 3600 m a.s.l. (Bobek, 1940; Messerli, 1967), whereas the LGM snowline was estimated to have been either 2100 (Wright, 1962) or 2800 m a.s.l. (Messerli, 1967). The İkiyaka Mountains have similar snowline elevations as the Buzul Mountain. However, the modern snowline on the Kavuşşahap Mountains (∼3400 m a.s.l.; Kurter, 1991) is lower than on their southern counterparts, probably a result of the proximity to the moisture source of Lake Van (Fig. 30.2).

30.3.2. The Central Taurus

The Central Taurus Mountains, located along the Mediterranean coast of central Anatolia, have attracted much scientific interest because of their accessibility. The scientific investigations began in 1927 with German Alpine expeditions (Künne, 1928) to Aladağlar and Bolkar Mountains, where the most prominent glacial deposits are found.

The Aladağlar (37.84°N, 35.15°E, 3756 m a.s.l.) consists mainly of Mesozoic carbonates (Tekeli et al., 1984) with an extensive karstic drainage network (Klimchouk et al., 2006) similar to the other Central Taurus Mountains, and abundant evidence of former glacial activity (Klaer, 1962; Zreda et al., 2006). The Hacer Valley (Fig. 30.3A), a 1.4-km-deep U-shaped glacial valley, located on the eastern side of the mountain, contains several moraine ridges at elevations from 1100 (at the mouth of the valley) to 3100 m a.s.l. (on the Yedigöller Plateau; Fig. 30.3B), just below the summits of Aladağlar, where an ice cap developed. Seven moraines dated using cosmogenic ^{36}Cl on 22 boulders gave ages from 10.2 ± 0.2 ka (1 ka = 1000 calendar years) at the bottom of the valley to 8.6 ± 0.3 ka on the plateau (Zreda et al., 2006), indicating large Holocene glaciers and rapid deglaciation (Fig. 30.4). These ages may become older as

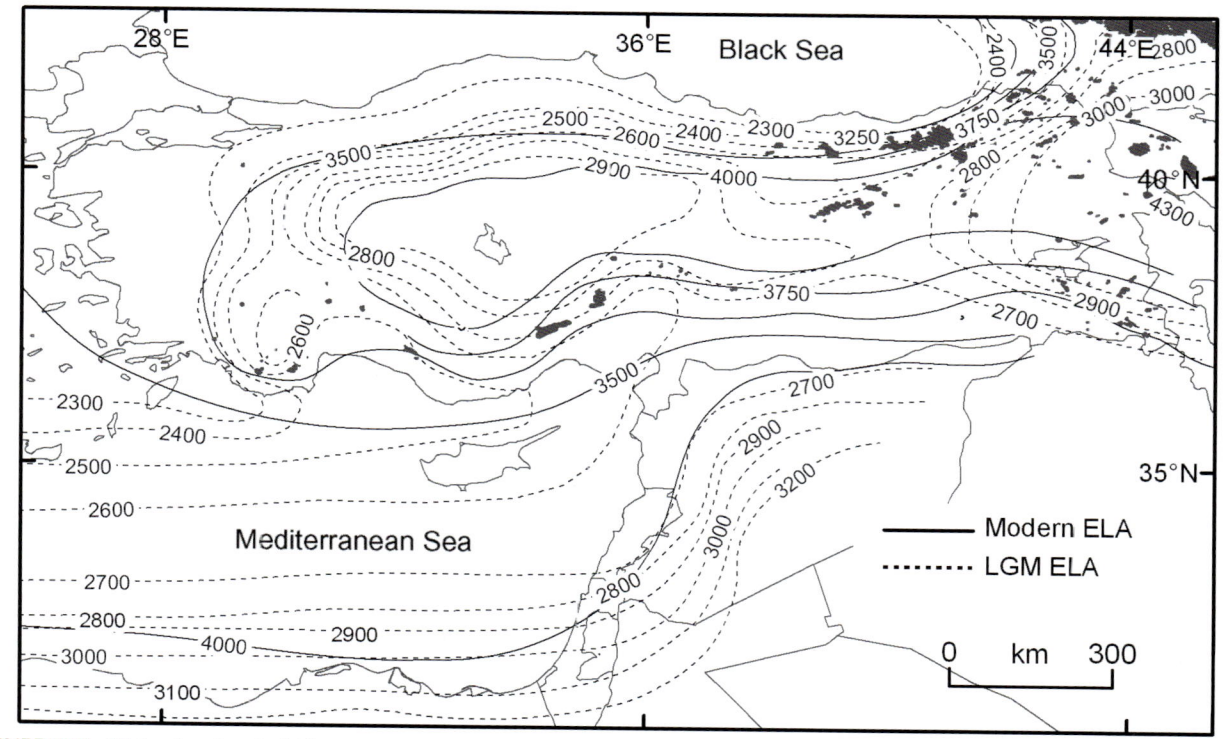

FIGURE 30.2 ELAs of modern (solid lines) and LGM (dotted lines) glaciers, adapted from Messerli (1967). The grey areas show the maximum extent of late Quaternary glacial deposits.

Chapter | 30 Quaternary Glaciations of Turkey

there is evidence that production rates of ^{36}Cl from ^{40}Ca may be lower, by a third or so, than those used by Zreda et al. (2006); if so, the oldest of these moraines will be of latest Pleistocene age or earliest Holocene age (12–13 ka), and the youngest will be of early Holocene age (10–11 ky). Glacial landforms in other valleys of the Aladağlar suggest much older glaciations. To the north and northwest of Yedigöller Plateau, several valleys contain terminal and lateral moraines reaching down to altitudes of about 1850–2100 m a.s.l. (Blumenthal, 1952; Klaer, 1962; Birman, 1968). Preliminary ^{36}Cl dates on moraines in the north-western Maden Valley suggest LGM and older ages of glacial deposits.

Bolkar Mountains (37.39°N, 34.61°E, 3524 m a.s.l.), located 65 km south-west of Aladağlar (Fig. 30.1), also bear evidence of Pleistocene glaciations. Several lateral and terminal moraines are present along the northern Maden (Fig. 30.3C) and the south-eastern Elmalı Valleys

FIGURE 30.3 (Continued)

FIGURE 30.3 Pictures from field studies. (A) Hacer Valley, Aladağlar (August 2008), (B) Yediğoller Plateau, Aladağlar (August 2008), (C) well-preserved lateral moraines in Maden Valley, Bolkar Mountains (July 2001), (D) an erratic boulder with glacial striations in the Maden Valley, Bolkar Mountains (July 2001), (E) cosmogenic sampling from the right lateral moraine in Maden Valley, Bolkar Mountains (July 2001), (F) 'flat iron' and 'bullet' stones from the Namaras Valley, Geyikdağ (August 2001), (G) LGM terminal moraines in Kartal Lake Valley, Mount Sandıras (August 2006), (H) a lateral moraine in the Dedegöl Mountains (August 2007) and (I) The Upper Aksu Valley and Erciyes glacier (July 2005).

(Blumenthal, 1956; Birman, 1968). Messerli (1967) and Klaer (1969) suggest the presence of possible morainic deposits as low as 1650 m a.s.l. in the Maden Valley (Fig. 30.3D). Recent findings from cosmogenic ^{36}Cl analysis of moraine boulders (Fig. 30.3E) from the Maden and Elmalı Valleys indicate several advances of glaciers since the LGM.

Soğanlı Mountains (38.41°N, 36.21°E, 2967 m a.s.l.), located 120 km north-east of Aladağlar (Fig. 30.1), also show evidence of former glaciations. With similar lithologies as in Aladağlar, the Soğanlı Mountains bear extensive glacio-karstic features that developed during the Pleistocene (Ege and Tonbul, 2005). The lowest glacial deposits are found in the Dökülgen Valley at 2250 m a.s.l. on the western and north-western side of the mountain. Geyikdağ, (36.81°N, 32.20°E, 2850 m a.s.l.), situated 100 km north-east of the city of Antalya, is characterised by the presence of glacier-shaped boulders (Fig. 30.3F) and peculiar hummocky topography that covers an area of about 30 km^2 in the Namaras and Susam Valleys (Arpat and Özgul, 1972; Çiner et al., 1999). The hummocky and lateral moraines at an elevation of ca. 2000 m a.s.l. have been interpreted as evidence of Pleistocene glaciations on the basis of relative age determination techniques. Unfortunately, no numerical ages have been reported from these mountains.

The modern snowline in the Central Taurus Mountains varies between 3200 and 3700 m a.s.l., and the LGM snowline was estimated to have been between 1900 and 2600 m a.s.l. (Messerli, 1967; Ege and Tonbul, 2005).

30.3.3. The Western Taurus

Mount Sandıras (37.08°N, 28.84°E, 2295 m a.s.l.), located on the extreme south-western part of the Taurus Mountains (Fig. 30.1), shows evidence of both LGM and Late-Glacial advances in two northern valleys (Fig. 30.3G). Sarıkaya et al. (2008) published cosmogenic ^{36}Cl ages obtained from nine moraine boulders. These ages indicate that LGM glaciers reached a maximum length of. 1.5 km and terminated at an elevation of ca. 1900 m a.s.l. ca. 20.4 ± 1.3 ka ago (Fig. 30.4). The glaciers of Mount Sandıras readvanced and retreated by 19.6 ± 1.6 ka, and then again by 16.2 ± 0.5 ka ago (Sarıkaya et al., 2008).

Similar results were obtained from the Dedegöl Mountains (37.64°N, 31.28°E, 2992 m a.s.l.), ~300 km east of

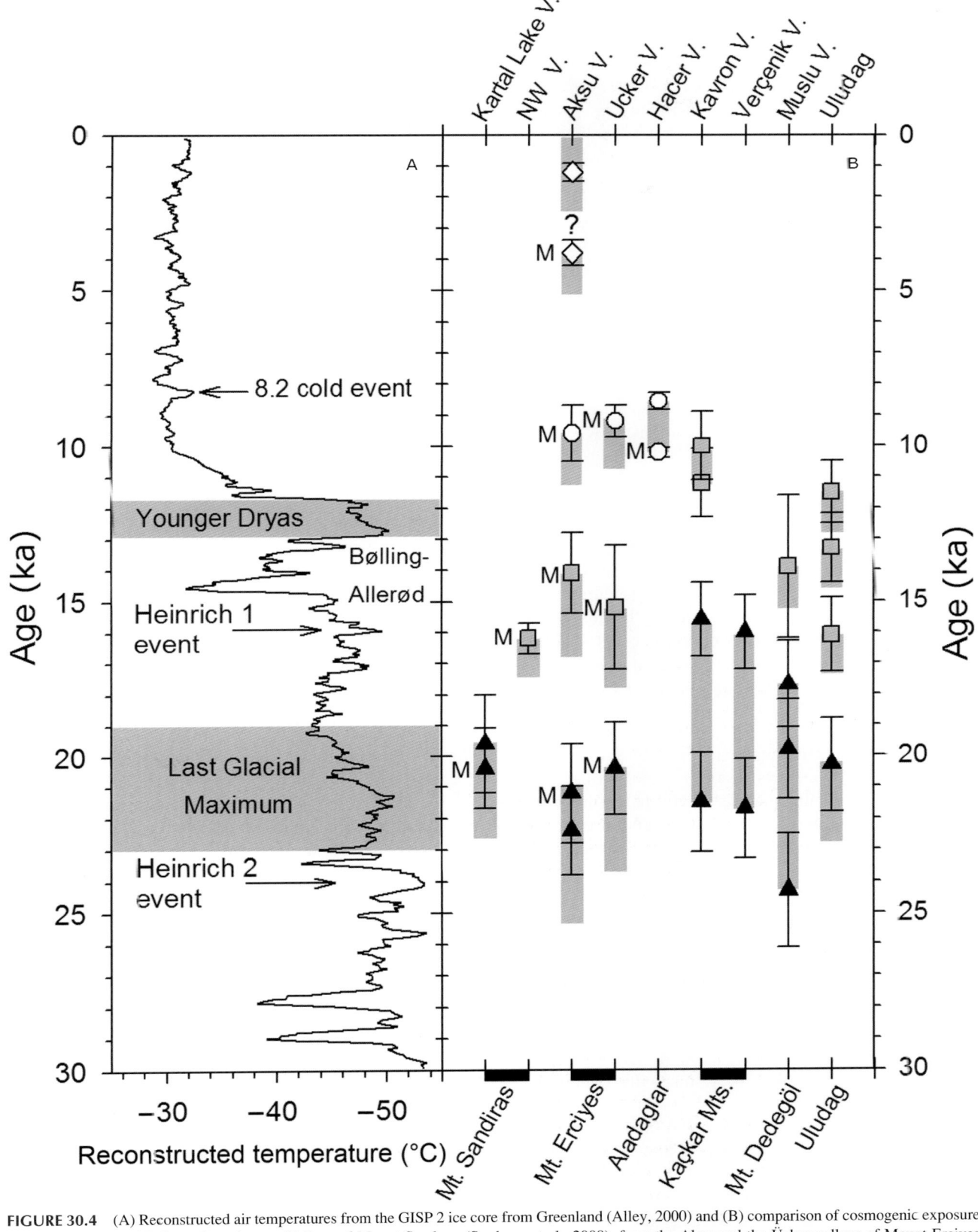

FIGURE 30.4 (A) Reconstructed air temperatures from the GISP 2 ice core from Greenland (Alley, 2000) and (B) comparison of cosmogenic exposure ages from the Kartal Lake and north-west Valleys of Mount Sandıras (Sarıkaya et al., 2008), from the Aksu and the Üçker valleys of Mount Erciyes (Sarıkaya et al., 2009) and from the Hacer Valley of Aladağlar (Zreda et al., 2006), from the Kavron Valley (Akçar et al., 2007) and the Verçenik Valley (Akçar et al., 2008) of the Eastern Black Sea Mountains, from the Muslu Valley of Dedegöl Mountains (Zahno et al., 2009) and from Uludağ (Zahno et al., 2010). The timing of maximum glaciations is indicated by a capital letter M, wherever possible. The vertical grey bars indicate possible timing of glaciations where black triangles represent the LGM, grey squares the Würmian Late Glacial, circles the early Holocene and diamonds the late Holocene.

Mount Sandıras (Fig. 30.1), using cosmogenic ^{10}Be and ^{26}Al on glacial landforms (Fig. 30.3H). In the east-facing Muslu Valley, two 50–100 m high lateral moraines that extend down to ~1400 m a.s.l. and ice-abraded bedrock were sampled by Zahno et al. (2009). The results indicate glacial advances before the LGM (>24.3 ± 1.8 ka ago), 19.8 ± 1.6 ka ago, 17.7 ± 1.4 ka ago and again 13.9 ± 2.3 ka ago (Fig. 30.4). Several other mountains (Fig. 30.1), such as Akdağ (36.54°N, 29.57°E, 3016 m a.s.l.; Onde, 1954; Doğu et al., 1999), Beydağ (36.57°N, 30.10°E, 3086 m a.s.l.; Louis, 1944; Messerli, 1967), Mount Barla (38.05°N, 30.70°E, 2800 m a.s.l.; Ardos, 1977), Mount Honaz (37.68°N, 29.29°E, 2571 m a.s.l.; Yalçınlar, 1954; Erinç, 1955, 1957) and Mount Davraz (37.76°N, 30.73°E, 2637 m a.s.l.; Monod, 1977; Atalay, 1987), in the Western Taurus Mountains have been reported to have evidence of Quaternary glaciations; however, they have not been dated numerically.

The modern snowline elevation in the western Taurus Mountains is estimated to be between 3000 and 3750 m a.s.l., whereas the LGM snowline varied between 2200 and 2600 m a.s.l., except on Mount Sandıras, where the glacial-period snowline stood at 2000 m a.s.l. (Fig. 30.2). This was probably the result of their proximity to the Mediterranean Sea and the favourable humid climatic conditions during the LGM period (Sarıkaya et al., 2008).

30.4. MOUNTAIN RANGES ALONG THE EASTERN BLACK SEA

The series of mountains that extends parallel to the Black Sea coast (Fig. 30.1) shows various signs of former glacial activity. The mountains include several peaks that exceed 3000 m a.s.l. that were well above the glacial-period snowline, their elevations increasing eastwards. These mountains (also known as the Pontic Mountains) act as a physiographic barrier between the Black Sea and the Anatolian interior. They consist largely of Cretaceous volcanic rocks and Tertiary granitic intrusions (Okay and Şahintürk, 1997). The climate of the Black Sea Mountains is affected by northerly incursions of Siberian high-pressure systems and are characterised by year-long orographic precipitation from air masses originating from the Black Sea (Akçar et al., 2007). Coastal areas of Eastern Black Sea region are the wettest parts of Turkey, receiving over 2 m of rainfall annually.

The highest part of the Black Sea Range is the Rize Mountains, which contain the fourth highest peak in Turkey, Mount Kaçkar (40.84°N, 41.16°E, 3932 m a.s.l.; Fig. 30.1). Akçar et al. (2007) dated glacial landforms in the Kavron Valley, a U-shaped valley descending northwards from Mount Kaçkar. The cosmogenic ^{10}Be ages show that the Kavron palaeoglaciers advanced at least 21.5 ± 1.6 ka ago (recalculated by Zahno et al. (2009) to include geomagnetic field variations), with the LGM glaciation continuing until 15.6 ± 1.2 ka (Fig. 30.4). Subsequent to the recession of LGM glaciers in the Kavron Valley, a younger glacial advance took place around 11.2 ± 1.1 to 10.0 ± 1.1 ka (Zahno et al., 2009). Similar results were obtained from the nearby Verçenik Valley (Akçar et al., 2008), 20 km south-east of Kavron Valley, where LGM glaciers advanced before 21.7 ± 1.6 ka ago and continued until 16.0 ± 1.2 ka ago (Akçar et al., 2007; recalculated in Zahno et al., 2009; Fig. 30.4). Doğu et al. (1993) and Gürgen (2003) reported that this mountainous part of the Eastern Black Sea ranges, from Soğanlı Mountains in the west to Bulut-Altıparmak Mountains in the east, contains several large U-shaped valleys and many glacier-related landforms, including terminal, ground, ablation and lateral moraines, roches moutonnées and glacial lakes (Çiner, 2004).

Although lower in altitude, Karadağ (40.38°N, 39.07°E, 3331 m a.s.l.) and Mount Karagöl (40.51°N, 38.19°E, 3107 m a.s.l.), both located on the westernmost part of the Pontic Mountains, on the Gümüşhane and Giresun Mountains (Fig. 30.1), respectively, bear terminal and hummocky moraines assigned to the Pleistocene (Çiner, 2004). On the far east of the Pontic range, close to the Georgian border, Mount Karaçal (or Mount Karçal in some maps; 41.35°N, 41.98°E, 3415 m a.s.l.; Fig. 30.1) carries well-preserved glacial deposits apparent in ASTER satellite images and in Google Earth® images. No published information is available concerning glacial landforms on this mountain. If the remote-sensing result is confirmed in the field, this region will potentially become a newly discovered formerly glaciated region.

The modern snowline in the Eastern Black Sea Mountains changes significantly depending on the aspect, from 3100–3200 m a.s.l. on the north-facing slopes of the Rize Mountains (Erinç, 1952) to 3500–3550 m a.s.l. on the southern side of the mountain (Erinç, 1949a; Çiner, 2004). The difference probably reflects the effects of humid air derived from the Black Sea (Erinç, 1952). The LGM-snowline elevations mimic the modern trends; they were lower on the northern side of the mountains (2300–2500 m a.s.l.) and higher on the southern side (2600–2700 m a.s.l.) (Çiner, 2004; Fig. 30.2).

30.5. VOLCANOES AND INDIVIDUAL MOUNTAINS ON THE ANATOLIAN PLATEAU

High volcanic mountains scattered throughout the Anatolian Plateau show signs of former glacial activity. Among them, Mount Ağrı (also known as Mount Ararat; 39.70°N, 44.30°E, 5137 m a.s.l.), located in the far east of Turkey (Fig. 30.1), the third highest mountain in the

Middle East and the highest in the country, supports an ice cap of 5.5 km² in area (based on 2008 ASTER imagery), the largest single glacier in Turkey, and has a snowline at 4300 m a.s.l. (Sarıkaya and Bishop, 2010). Mount Ağrı is a dormant, composite, calc-alkaline stratovolcano, consisting of two separate cones: Büyük Ağrı (Great Ararat) and Küçük Ağrı (Lesser Ararat, 3896 m a.s.l.). Blumenthal (1958) calculated a Pleistocene snowline elevation of 3000 m on this mountain, which would result in an ice cap of 100 km². So far, no well-preserved moraines have been identified. However, Birman (1968) reported some glacial deposits on the south-facing slope of Büyük Ağrı, about 300 m below the base of the modern ice, and distinguished a possibly glacially sculpted valley as low as 3750 m a.s.l. He noted, however, that his observations were extremely crude, since they were made using clinometers sightings from a distance. Blumenthal (1958) explained the lack of moraines by the absence of confining ridges to support valley glaciers, by insufficient debris load to form moraines and by volcanic eruptions that obliterated older moraines.

Another volcano, Mount Erciyes (38.53°N, 35.45°E, 3917 m a.s.l.), located in central Anatolia (Fig. 30.1), has evidence of four periods of glacial activity during the last 22 ka (Sarıkaya et al., 2009; Fig. 30.4). Cosmogenic ^{36}Cl surface exposure dating results, obtained from 44 samples in two valleys of Mount Erciyes (Sarıkaya et al., 2009), show that LGM glaciers reached 6 km in length and descended to 2150 m a.s.l. (Fig. 30.3I). These glaciers began retreating 21.3 ± 0.9 ka ago, but they readvanced by 14.6 ± 1.2 ka ago (Weichselian Late Glacial) and again by 9.3 ± 0.5 ka ago (early Holocene). The latest advance took place 3.8 ± 0.4 ka ago (Sarıkaya et al., 2009; Fig. 30.4). Three more glaciated valleys, located on the north-east and south of the mountain, have similar glacial records. The LGM-snowline elevation has been calculated at 2700 m a.s.l. on the northern (Sarıkaya et al., 2009) and 3000 m a.s.l. on the southern side (Messerli, 1967). Today, a small glacier ($\sim$260 m long) persists to the north of the peak (Fig. 30.3I), with snowline elevation at 3550 m a.s.l. (Sarıkaya et al., 2009).

Mount Süphan (38.93°N, 42.83°E, 4053 m a.s.l.), situated to the north of Lake Van in Eastern Turkey (Fig. 30.1), is the third highest peak of the country, rising to over 2400 m above the lake level. During the LGM, the summit of Mount Süphan was covered by an ice cap, with outlet glaciers descending to 2650–2700 m a.s.l. on the northern slope, and 2950–3000 m a.s.l. on the south side (Kesici, 2005). Former glaciers that emerged from the ice cap descended about 1.5–2 km from their sources and especially deposited moraines on the northern flank of the mountain. The glacial-period snowline elevation was estimated at 3100 m a.s.l. for the northern side, and 3300 m a.s.l. for the southern side of the mountain (Kesici, 2005). The modern snowline elevation here is $\sim$4000 m a.s.l. (Fig. 30.2).

Apart from the individual volcanoes, a few small mountain ranges in Anatolia also bear evidence of past glacial activity. Mount Uludağ (40.07°N, 29.22°E, 2543 m a.s.l.; Fig. 30.1), situated 30 km south-east of the Marmara Sea, near the city of Bursa, has several cirques and valleys that were previously occupied by glaciers (Erinç, 1949b). Birman (1968) made a rapid survey of the Çayırlı Valley, 4 km east of Uludağ ski resorts, and differentiated four main moraine sets that he attributed to the Early, Middle, Late and Post-'Wisconsinan' (i.e. Weichselian; Würmian Stage) glaciations. These glaciers extended several kilometres from their cirques down to as low as 1850 m a.s.l. (Birman, 1968). Recently, ^{10}Be and ^{26}Al cosmogenic dating of glacial landforms has demonstrated that the local LGM here occurred no later than 20.3 ± 1.5 ka ago (Zahno et al., 2010). Post-LGM fluctuations are also evident on the mountain, and show distinct phases of glacier advances at around 16.1 ± 1.2, 13.3 ± 1.1 and again at 11.5 ± 1.0 ka ago (Zahno et al., 2010; Fig. 30.4). Here, the LGM snowline was estimated at $\sim$2400 m a.s.l. (Messerli, 1967), while the modern snowline elevation is at around 3500 m a.s.l.

In Eastern Anatolia, the Mercan Mountains (39.49°N, 39.17°E, 3368 m a.s.l.; Fig. 30.1), to the south of the city of Erzincan, experienced extensive glaciations during the Pleistocene (Bilgin, 1972). This occurred when the Mercan Valley glacier descended to an altitude of about 1650 m a.s.l. (Atalay, 1987; Türkünal, 1990). Another area glaciated during the Pleistocene is the Esence Mountains (39.78°N, 39.75°E, 3477 m a.s.l.; Fig. 30.1), 50 km north of the Mercan massif. Here, well-developed lateral moraines descend to about 1950 m, $\sim$9 km from their cirques. Within the same region, on Mount Mescid (40.33°N, 41.17°E, 3239 m.s.l.; Fig. 30.1), glaciers were about 3–5 km long (Yalçınlar, 1951; Atalay, 1987). On Mount Mercan, the LGM-snowline elevation was estimated to have occurred at about 2750 m, whereas the modern snowline elevation is between 3600 and 3700 m a.s.l. (Fig. 30.2). Finally, Mount Ilgaz (41.03°N, 33.65°E, 2587 m a.s.l.; Louis, 1944) in north central Anatolia and Balık Gölü (Birman, 1968; Balık Lake, 39.78°N, 43.53°E, 2804 m a.s.l.) near Mount Ağrı (Fig. 30.1) show evidence of previous glacial processes; however, they have not yet been investigated in detail.

30.6. CONCLUSIONS

Recent investigations on the glacial geology and chronology of Turkey, following the first edition of this book, have revealed important information regarding the extent and timing of late Quaternary glaciers. Cosmogenic ages of glacial landforms, the first in the region, are essential to the understanding of past glacial activity and the palaeoenvironmental reconstructions. The oldest and the most extensive glacial advances date to the LGM (Fig. 30.4). ^{10}Be and ^{26}Al exposure ages from the Eastern Black Sea Mountains (Akçar et al., 2007, 2008) provide evidence for glaciations between 21 (recalculated by Zahno et al., 2009) and 16 ka ago. Consistent local chronologies, based upon moraine exposure ages, closely corresponding to the global LGM (between 19 and 23 ka, centred on 21 ka; Martinson et al., 1987; Yokoyama et al., 2000; Mix et al., 2001), were reported from western Taurus Mount Sandıras (Sarıkaya et al., 2008) and Dedegöl Mountains (Zahno et al., 2009) and from Mount Erciyes in central Turkey (Sarıkaya et al., 2009; Fig. 30.4). These results are supported by glacial chronologies from other circum-Mediterranean mountains, such as the central Spanish mountains (Palacios et al., 2007), the Pyrenees (Pallás et al., 2007; Chapter 11) and the Maritime Alps (Granger et al., 2006). Late-glacial advances also occurred between 16 and 11 ka ago in these mountains. Unusual Holocene glaciations, dated to 9–10 ka, were also reported from central Anatolia (Mount Erciyes, Sarıkaya et al. 2009) and the south-central Anatolia (the Aladağlar, Zreda et al., 2006). Late Holocene and Little Ice Age advances were less extensive than older glaciations and were present only at certain sites in the Turkish mountains (Fig. 30.4).

The Quaternary and glacial geology of Turkey will continue developing in a similar fashion to that of the past decade. Advances in geochronology (cosmogenic dating), coupled with glacial modelling and palaeoclimatic reconstructions, have yielded new evidence and information regarding past climates, and further progress will bring a fuller understanding of climatic variations in space and time. This will inevitably lead to a better understanding of the regional palaeoenvironments. This, in turn, will allow us to understand better the physical environment in which humans evolved, from the Palaeolithic to modern times in response to the physical changes.

ACKNOWLEDGEMENTS

Many thanks to Dr. Serdar Bayarı, Dr. Erdal Şen and Ms. Şükran Açıkel (Hacettepe University) and Dr. Bülent Akıl (İller Bankası) for their field assistance. This work was partially supported by the United States National Science Foundation (Grant 0115298) and by the Scientific and Technological Research Council of Turkey (TÜBİTAK; Grants 101Y002 and 107Y069).

REFERENCES

Ainsworth, W.F., 1842. Travels and Researches in Asia Minor, Mesopotamia, Chaldea and Armenia. J.W. Parker, London.

Akçar, N., Yavuz, V., Ivy-Ochs, S., Kubik, P.W., Vardar, M., Schluchter, C., 2007. Paleoglacial records from Kavron Valley, NE Turkey: field and cosmogenic exposure dating evidence. Quatern. Int. 164–165, 170–183.

Akçar, N., Yavuz, V., Ivy-Ochs, S., Kubik, P.W., Vardar, M., Schluchter, C., 2008. A case for a downwasting mountain glacier during Termination I, Vercenik valley, northeastern Turkey. J. Quatern. Sci. 23 (3), 273–285.

Altınlı, İ.E., 1966. Geology of eastern and southeastern Anatolia. Bull. Miner. Res. Explor. Inst. Turk. 66, 35–76.

Alley, R.B., 2000. The Younger Dryas cold interval as viewed from central Greenland. Quatern. Sci. Rev. 19, 213–226.

Ardos, M., 1977. Geomorphology and Pleistocene glaciation of Mount Barla and surrounding (in Turkish). Rev. Geogr. Inst. Univ. Istanbul 20–21, 151–168.

Arpat, E., Özgul, N., 1972. Rock glaciers around Geyikdağ, Central Taurids (in Turkish). Bull. Miner. Res. Explor. Ankara 80, 30–35.

Atalay, I., 1987. Introduction to Geomorphology of Turkey (in Turkish), second ed. Ege University Press, Izmir, Turkey.

Bilgin, T., 1972. Glacial and periglacial morphology of Eastern Munzur mountains (in Turkish). Rev. Geogr. Inst. Univ. Istanbul 1757, 69.

Birman, J.H., 1968. Glacial reconnaissance in Turkey. Geol. Soc. Am. Bull. 79, 10091026.

Blumenthal, M.M., 1952. The high mountains of Taurids Aladağ, recent research on its geography, stratigraphy and tectonics (in German). Bull. Miner. Res. Explor. 6, 136pp.

Blumenthal, M.M., 1956. Geology of northern and western Bolkardağ region (in Turkish). Bull. Miner. Res. Explor. Ankara 7, 153pp.

Blumenthal, M.M., 1958. From Mount Ağrı (Ararat) to Mount Kaçkar (in German). Bergfahrten in nordostanatolischen Grenzlanden. Die Alpen 34, 125–137.

Bobek, H., 1940. Recent and ice time glaciations in central Kurdish high mountains (in German). Z. Gletscherk. 27 (1–2), 50–87.

Butzer, K.W., 1958. Quaternary stratigraphy and climate in the Near East Bonner. Geogr. Abh. 24, 157.

Çiner, A., 2004. Turkish glaciers and glacial deposits. In: Ehlers, J., Gibbard, P.L. (Eds.), Quaternary Glaciations: Extent and Chronology. Part I: Europe. Elsevier, Amsterdam, pp. 419–429.

Çiner, A., Deynoux, M., Çörekçioğlu, E., 1999. Hummocky moraines in the Namaras and Susam Valleys, Central Taurids, SW Turkey. Quatern. Sci. Rev. 18, 659–669.

Doğu, A.F., Somuncu, M., Çiçek, İ., Tuncel, H., Gürgen, G., 1993. Glacier shapes, yaylas and tourism on the Kaçkar Mountains (in Turkish). Turk. Geogr. Bull. Ankara Univ. 157–183.

Doğu, A.F., Çiçek, İ., Gürgen, G., Tuncel, H., 1999. Geomorphology of Akdağ and its effect on human activities (in Turkish). Turk. Geogr. Bull. Ankara Univ. 7, 95–120.

Ege, I., Tonbul, S., 2005. The relationship of karstification and glaciation in Soğanlı Mountain (in Turkish). V. Quaternary Workshop of Turkey, Istanbul Technical University, Istanbul, Turkey.

Erinç, S., 1944. Glazialmorphologische Untersuchungen im Nordostanatolischen Randgebirge. Istanbul University Geography Inst. Pub., Ph.D. dissertation Series, 1, 56pp.

Erinç, S., 1949a. Past and present glacial forms in Northeast Anatolian mountains (in German). Geol. Rundsch. 37, 75–83.

Erinç, S., 1949b. Research on glacial morphology of Mount Uludağ (in Turkish). Rev. Geogr. Inst. Univ. Istanbul 11–12, 79–94.

Erinç, S., 1951. The glacier of Erciyes in Pleistocene and Post-glacial epoch. Rev. Geogr. Inst. Univ. Istanbul 1 (2), 82–90 (in Turkish).

Erinç, S., 1952. Glacial evidences of the climatic variations in Turkey. Geogr. Ann. 34, 89–98.

Erinç, S., 1953. From Van to Mount Cilo (in Turkish). Turk. Geogr. Bull. Ankara Univ. 3–4, 84–106.

Erinç, S., 1955. Periglacial features on the Mount Honaz (SW Anatolia) (in Turkish). Rev. Geogr. Inst. Univ. Istanbul 2, 185–187.

Erinç, S., 1957. About glacial evidences of Honaz and Bozdağ (in Turkish). Turk. Geogr. Bull. 8, 106–107.

Erinç, S., 1978. Changes in the physical environment in Turkey since the end of last glacial. In: Brice, W.C. (Ed.), The Environmental History of the Near and Middle East Since the Last Ice Age. Academic Press, London, pp. 87–110.

Granger, D.E., Spagnolo, M., Federici, P., Pappalardo, M., Ribolini, A., Cyr, A.J. 2006. Last glacial maximum dated by means of ^{10}Be in the Maritme Alps (Italy). Eos Transactions: American Geophysical Union 87, Fall Meet. Suppl. 87, Abstract H53B-0634.

Gürgen, G., 2003. Glacial morphology of North of Çapans Mountains (Rize) (in Turkish). Gazi Univ. Educ. Fac. J. 23, 159–175.

İzbırak, R., 1951. Geographical research in Lake Van and in the Hakkari and Cilo Mountains (in Turkish). Turk. Geogr. Bull. Ankara Univ. 67 (4), 149.

Kesici, O., 2005. Glacio-morphological investigations of Süphan and Cilo Mountains in regard to current global warming trends. TÜBİTAK (The Scientific and Technical Research Council of Turkey) Report No: 101Y131 (in Turkish).

Klaer, W., 1962 Untersuchungen zur klimagenetischen Geomorphologie in den Hochgebirgen Vorderasiens. Heidelb. Geogr. Arb. 11, 1–135.

Klaer, W., 1969. Glacio-morphological problems in the near east high mountains. Erdkunde 23 (3), 192–200.

Klimchouk, A., Bayari, S., Nazik, L., Törk, K., 2006. Glacial destruction of cave systems in high mountains, with a special reference to the Aladağlar massif, Central Taurus, Turkey. Acta Carsol. 35, 111–121.

Künne, G., 1928. Die deutsche alpine Taurus-Expedition 1927 (Ala Dagh in Zilizien). Petermanns Geogr. Mitt. 74, 273–276.

Kurter, A., 1991. Glaciers of Middle East and Africa—Glaciers of Turkey. In: Williams, R.S., Ferrigno, J.G. (Eds.), Satellite Image Atlas of the World, USGS Professional Paper, 1386-G-1, pp. 1–30.

Kurter, A., Sungur, K., 1980. Present glaciation in Turkey. In: Muller, F., Scherle, K. (Eds.), World Glacier Inventory, vol. 126. International Association of Hydrological Sciences, Bartholomew press, Dorking, Surrey, pp. 155–160.

Louis, H.L., 1944. Evidence for Pleistocene glaciation in Anatolia (in German). Geol. Rundsch. 34 (7–8), 447–481.

Martinson, D.G., Pisias, N.G., Hays, J.D., Imbrie, J., Moore, T.C., Shackleton, N.J., 1987. Age, dating and orbital theory of the Ice Ages: development of a high resolution 0–300,000 year chronostratigraphy. Quatern. Res. 27, 1–29.

Maunsell, F.R., 1901. Central Kurdistan. Geogr. J. 18 (2), 121–141.

Messerli, B., 1964. Der gletscher am Erciyes Dagh und das problem der rezenten Schneegrenze im Anatolischen und Mediterranen Raum. Geogr. Helv. 19 (1), 19–34.

Messerli, B., 1967. Die eiszeitliche und die gegenwartige Vergletscherung in Mittelmeerraum. Geogr. Helv. 22, 105–228.

Mix, A., Bard, A., Schneider, R., 2001. Environmental processes of the ice age, land, oceans, glaciers (EPILOG). Quatern. Sci. Rev. 20, 627–657.

Monod, O., 1977. Geological research in the Western Taurides south of Beyşehir, Turkey (in French). Unpublished thesis, University of Paris, 442pp.

Okay, A.I., Şahintürk, Ö., 1997. Geology of the eastern Pontides. In: Robinson, A.G. (Ed.), Regional and Petroleum Geology of the Black Sea and Surrounding Region, AAPG Memoir. 68, 291–311.

Onde, H., 1954. Forms of glaciers in the Lycien Massif of Akdağ (southwest Turkey) (in French). Cong. Géol. Int. 15, 327–335.

Palacios, D., Marcos, J., Andrés, N., Vazquez, L., 2007. Last glacial maximum and deglaciation in central Spanish mountains. Geophys. Res. Abstr. 9, 05634.

Palgrave, W.G., 1872. Vestiges of the glacial period in northeastern Anatolia. Nature 5, 444–445.

Pallás, R., Rodés, Á., Braucher, R., Carcaillet, J., Ortunò, M., Bordonau, J., et al., 2007. Late Pleistocene and Holocene glaciation in the Pyrenees: a critical review and new evidence from ^{10}Be exposure ages, south-central Pyrenees. Quatern. Sci. Rev. 25, 2937–2963.

Sarıkaya, M.A., 2009. Late Quaternary glaciation and paleoclimate of Turkey inferred from cosmogenic ^{36}Cl dating of moraines and glacier modeling. Ph.D. Thesis, University of Arizona, Tucson, AZ, USA.

Sarıkaya, M.A., 2010a. Present glaciers of Turkey. In: Kargel, J.S., Bishop, M.P., Kääb, A., Raup, B.H., Leonard, G. (Eds.), GLIMS: Global Land Ice Measurements from Space: Monitoring the World's Changing Glaciers. Praxis-Springer (in review).

Sarıkaya, M.A., 2010b. Quaternary Glaciers of Turkey: A Glacio-Chronologic and Paleoclimatic View. Lambert Academic Publishing, Germany, 140pp.

Sarıkaya, M.A., Bishop, P.M., 2010. Space-based assessments of the ice cap on Mount Ağrı (Ararat), Eastern Turkey, Associations of American Geographers Annual Meeting, Washington, DC, USA, 14–18. April 2010 Abstract #31922.

Sarıkaya, M.A., Zreda, M., Çiner, A., Zweck, C., 2008. Cold and wet Last Glacial Maximum on Mount Sandıras, SW Turkey, inferred from cosmogenic dating and glacier modeling. Quatern. Sci Rev. 27 (7–8), 769–780.

Sarıkaya, M.A., Zreda, M., Çiner, A., 2009. Glaciations and paleoclimate of Mount Erciyes, central Turkey, since the Last Glacial Maximum, inferred from ^{36}Cl cosmogenic dating and glacier modeling. Quatern. Sci. Rev. 28 (23–24), 2326–2341.

Tekeli, O., Aksay, A., Ürgün, B.M., Işık, A., 1984. Geology of the Aladağ Mountains. In: Tekeli, O., Göncüoğlu, M.C. (Eds.), The Geology of the Taurus Belt. MTA Publications, Ankara, pp. 143–158.

Türkünal, S., 1990, Mountain chains and mountains of Turkey, (in Turkish). Bulletin of the Chamber of Geological Engineers of Turkey, 30, 42.

Wright, H.E., 1962. Pleistocene glaciation in Kurdistan. Eiszeit. Gegenwart 12, 131–164.

Yalçınlar, İ., 1951. Glaciations on the Soğanlı-Kaçkar mountains and Mescid Dağ (in French). Rev. Geogr. Inst. Univ. Istanbul 1–2, 50–55.

Yalçınlar, İ., 1954. On the presence of the Quaternary glacial forms on Honaz Dag-and-Boz Dag (western Turkey) (in French). C. R. Sommaire Soc. Géol. France 13, 296–298.

Yokoyama, Y., Lambeck, K., De Deckker, P.P.J., Fifield, L.K., 2000. Timing of the last glacial maximum from observed sea-level minima. Nature 406, 713–716.

Zahno, C., Akçar, N., Yavuz, V., Kubik, P.W., Schlüchter, C., 2009. Surface exposure dating of Late Pleistocene glaciations at the Dedegöl Mountains (Lake Beyşehir, SW Turkey). J. Quatern. Sci. 24, 1016–1028.

Zahno, C., Akçar, N., Yavuz, V., Kubik, P.W., Schlüchter, C., 2010. Chronology of Late Pleistocene glacier variations at the Uludağ Mountain, NW Turkey. Quatern. Sci. Rev. 29, 1173–1187.

Zreda, M., Zweck, C., Sarıkaya, M.A., 2006. Early Holocene glaciation in Turkey: large magnitude, fast deglaciation and possible NAO connection. EOS Trans. Am. Geophys. Union 87 (52) Fall Meet. Suppl., Abstract PP43A-1232.

Chapter 31

Limits of the Pleistocene Glaciations in the Ukraine: A Closer Look

Andrei V. Matoshko

Department of Geomorphology the Ukrainian National Academy of Sciences, 44, Volodymyrska Street, 01034 Kyiv, Ukraine

31.1. INTRODUCTION

In the first edition of the '*Quaternary Glaciations—Extent and Chronology*', the Pleistocene glaciations in the Ukraine were reviewed (Matoshko, 2004). That overview included a short history of research starting at the beginning of the nineteenth century as well as a discussion of controversial ideas of ice-sheet development within the Ukraine, the number of glaciations and the different versions of the glacial limits. It also contained most comprehensive references to these issues. The evidence cited reliably indicated that the plain region of the Ukraine was glaciated twice in the Middle Pleistocene. This conclusion was first presented by Gozhik (1995). The Oka (Elsterian) and Dnieper (Saalian) glaciations occupied different areas with the exception of West (Volyn') Polissia, where their deposits overlap (Figs. 31.1 and 31.2). Small-scale mountain glaciation also occurred in the Ukrainian Carpathians.

Based on the theses mentioned above and according to the scope of the present volume, this chapter presents a deeper analysis of the primary evidence, the methods of mapping the glacial limits, glacial landforms and geological features. A direct and joint author's use of the medium scale topographic and thematic maps (1:100,000, 1:200,000, 1:500,000) in an ArcMap environment allowed the mapping of glacial limits and other glacial features much more accurately. Additional features (landforms, key sites) associated with the glacial limits have been added to the database. They result from the revision of previous materials and new investigations in the past 5 years. Special emphasis is given to the key sites and chronology of the glacial events. In this work, the author looks beyond the national borders to create a more complete picture of the glacial limits as well as to compare some different points of view regarding the age and position of the glacial limits.

31.2. THE INTERPRETATION OF THE PRIMARY DATA

At this point in the project, the author used a simple but straightforward approach for the interpretation of the primary evidence. The glacial limits were drawn by joining the outermost points, where glacial features were discovered. It might be a landform, exposure or borehole section with glacial deposits or glacial dislocations, or the occurrence of erratic boulders. An exception was made for proglacial valley sandur terraces and deposits, although they also indicate the vicinity of the former glacial margin. According to Matoshko and Chugunny (1993), the proximal edges of valley sandar lie within the glaciated area and not more than 20–30 km from the probable ice-marginal position. In the same work, the altitudinal analysis of the glacial limits showed that its elevation changes only gradually over large distances and therefore may also be used for the control of the position of the glacial limits between established points. In this approach, any speculations concerning the glacial limits without supporting evidence or based on doubtful data are avoided.

31.3. THE OKA GLACIATION IN THE WESTERN AND NORTHERN PARTS OF THE UKRAINE AND IN THE EASTERN AND SOUTHERN PARTS OF BELARUS

31.3.1. Glacial Boundaries

In comparison to the previous version (Matoshko, 2004), the present digital map of the Oka glacial limits has remained largely unchanged (Fig. 31.1). Its outline was only determined according to the applied approach (described above). The southernmost point reached by the Oka glaciation occurs at the main watershed between the Vistula and Dniester Rivers at 49°40′N where the L'viv

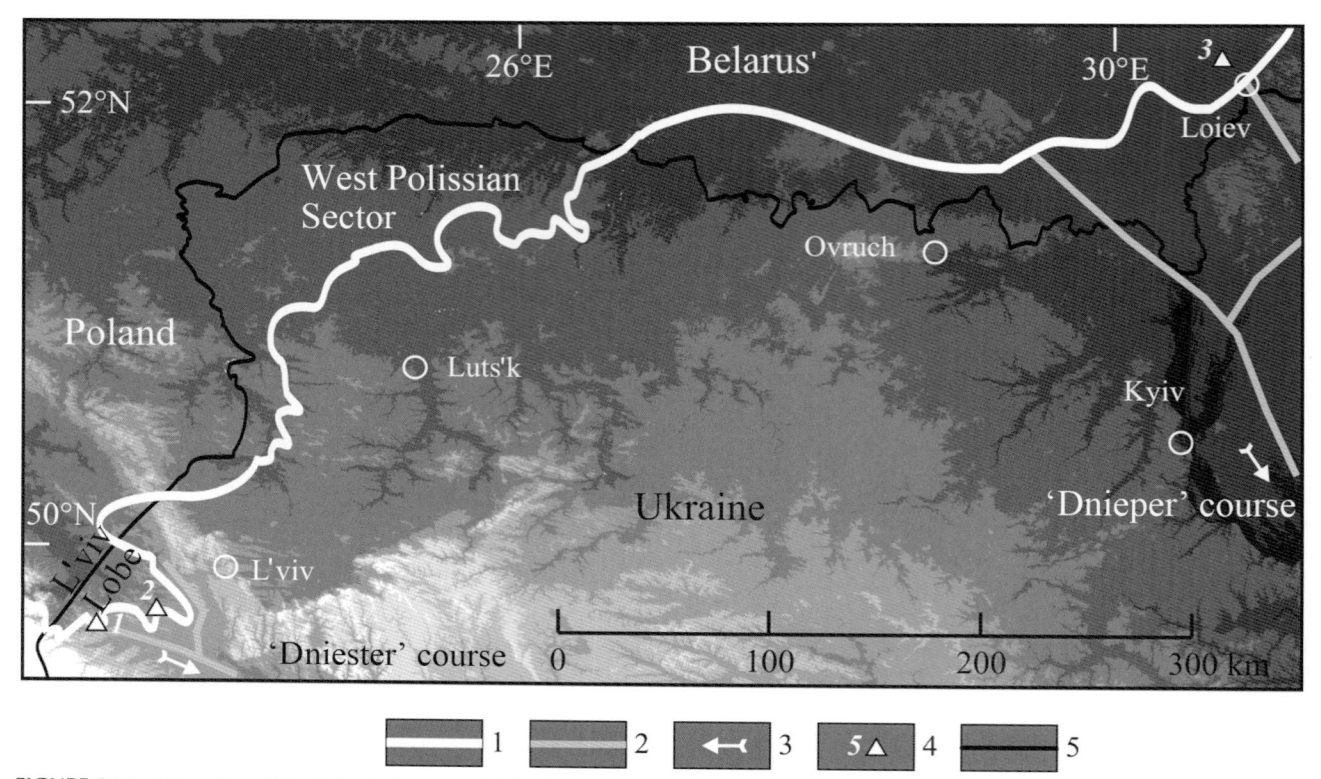

FIGURE 31.1 The extent of the Oka glaciation in the Ukraine. 1—boundary of glaciation; 2—proglacial meltwater runoff; 3—direction of runoff; 4—sites, mentioned in the text and their numbers: 1—Krukenichi, 2—Dubanevichi, 3—Peredelka; 5—state border. Digital hypsometrical map of the modern surface with grey colour scale (step—50 m), in the range from 50–100 (darkest grey) to more 400 m (white), in the background.

Ice Lobe reached 260–280 m a.s.l. The front of the lobe was controlled by the position of local heights (Matoshko, 2004). From this region, the Oka glacial boundary sharply turns to the north, forming the eastern margin of the L'viv lobe and thence to the east-north-east. In this direction, it descends gradually to not more than 140 m a.s.l. at the Ukraine–Belarus border. The further continuation of this glacial limit to the east is at issue. According to the latest publications by Belarus' specialists (Machnach et al., 2001; Karabanov, 2004), it turns to the south into the limits of the Middle Dnieper area. That revives the idea that Oka Till might be present near Kaniv (Ukraine) that was previously disputed (Goretsky, 1970; Matoshko, 2004). According to the state geological survey, the southernmost boreholes encountered Oka Till (Berezina Till in Belarus) at some 30–80 m below the modern surface in south-eastern Belarus. In the Dnieper valley, the Oka Till is exposed in the cliffs on its right side at 105–110 m a.s.l. The author's version of the glacial limits in Belarus (Fig. 31.1) is based on these data. It largely follows the boundary of the Berezina Till as drawn by Gurskii (1974) and Mander (1973). An extreme eastern point of the Oka glacial limit in Belarus lies near its border with Russia at 52°30′N.

31.3.2. Glacial Features

Each region covered by the Oka glaciation possesses a specific set of glacial features marking the glacial limit.

The L'viv lobe is represented by single erratic boulders, rare till remnants and small-scale glacial dislocations (Demedyuk and Demedyuk, 1988a). It was drained by a well-developed proglacial runoff system which included short streams orientated in distal direction from the ice margin, others flowing parallel to the ice margin and the main arteries exporting the meltwater to the Pre-Dniester and further to the Black Sea (Fig. 31.1, map). Today, these courses are found in modern valleys incised into bedrock with several exposures of valley sandur deposits (Fig. 31.3).

To the north from L'viv, the Oka glacial limit can be traced through several outcrops of till and also small-scale dislocations of glacial deposits (Bogutskii, 1967; Gruzman and Chebotareva, 1978).

In the West Polissian sector (Fig. 31.1), the outermost limits of the Oka and Dnieper glaciations are close to each other. Often at the highest points of the pre-Pleistocene land surface, a single layer of silty sands with both local and far-transported debris is found overlying the Cretaceous rocks.

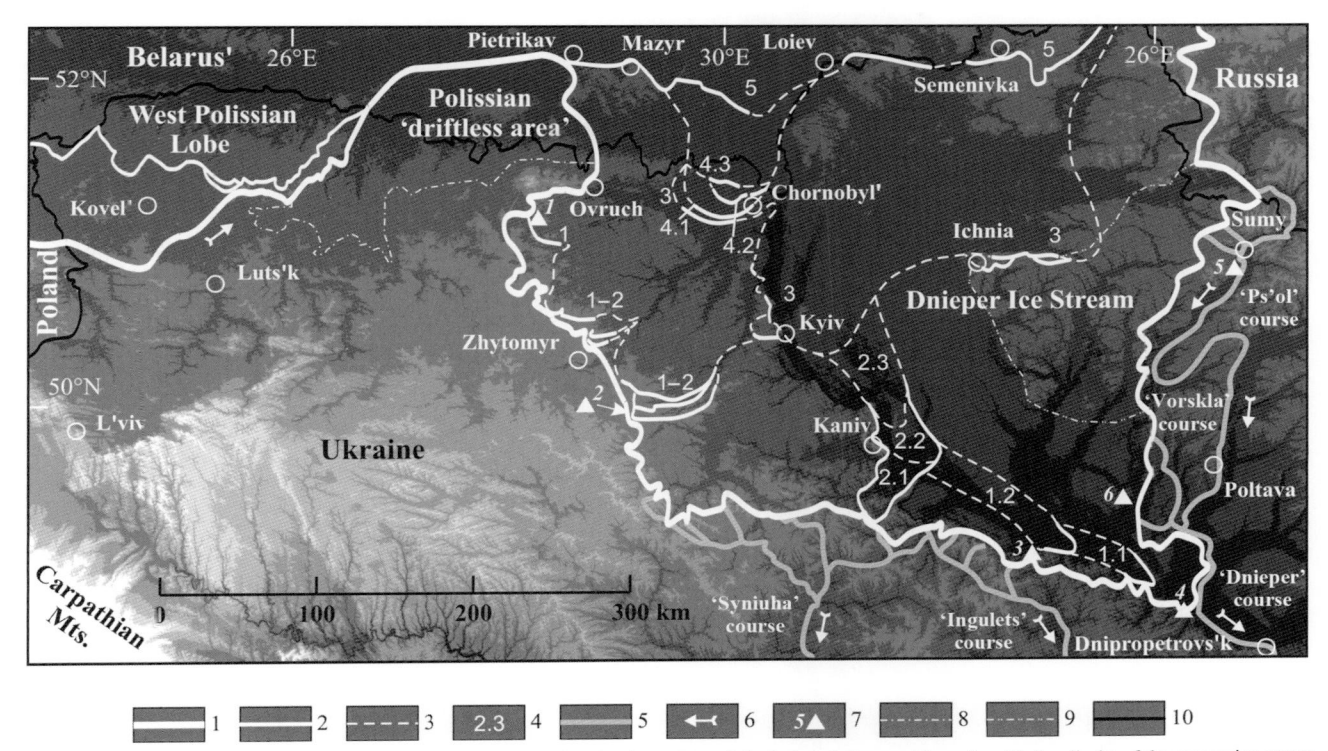

Chapter | 31 Limits of the Pleistocene Glaciations in Ukraine: A Closer Look   407

FIGURE 31.2 The extent of the Dnieper glaciation in the Ukraine. 1—boundary of glaciation; 2–3—position of oscillation limits of the regressive stage: 2—established, 3—inferred; 4—relative numbers of oscillations during regressive stage: first number—phase, second number—recession or re-advance, hyphened numbers—undifferentiated oscillations; 5—proglacial meltwater runoff; 6—direction of runoff; 7—sites, mentioned in the text and their numbers: 1—Tesnovka and Krasnostav, 2—Gorodische, 3—Velyka Andrusovka, 4—Domotkan', 5—Chervone, 6—Hun'ky; 8–9—inferred limits of the ice-dammed lakes: 8—in the Polissian 'driftless area', 9—in the Sula basin; 10—state border. Digital hypsometrical map of the modern surface with grey colour scale (step—50 m), in the range from 50–100 (darkest grey) to more 400 m (white).

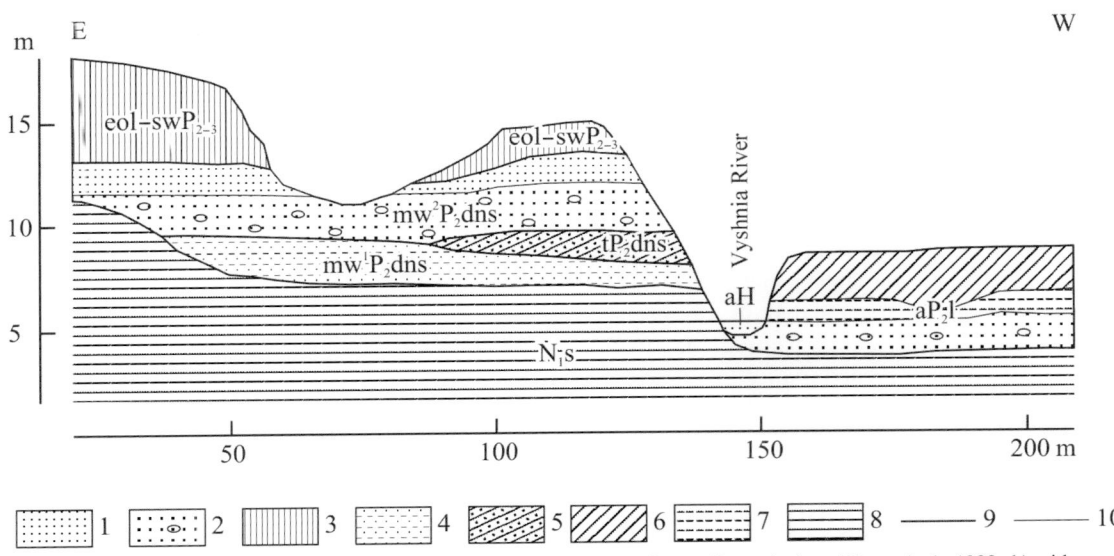

FIGURE 31.3 The section of Pleistocene deposits near Dubanevichi village, according to (Demedyuk and Demedyuk, 1988a,b) with some modifications. 1—silty sands; 2—cross-laminated sands with lenses of gravelly sands, gravel and pebble; 3—loess-like silts; 4—interlayered silty sands and silts; 5—diamicton; 6—loams, peat; 7—sandy clays; 8—hard silty clays deposits; 9 – stratigraphical boundaries; 10 – boundaries between lithological facies: eol–sw—aeolian–slopewash, a—alluvial, a–l—alluvial–lacustrine, mw^2—meltwater deposits overlying the till (valley sandur), mw^1—meltwater deposits underlying the till, t—basal till. International stratigraphy: H—Holocene, P_{2-3}—Middle–Upper Pleistocene undifferentiated, P_2—Middle Pleistocene, N_1—Miocene Local stratigraphy: s - Sarmatian, dns—Dniester glacial complex of the Oka glaciation, l—Likhvian (Holsteinian).

Traditionally, the glacial landforms and glacial deposits (exposed and subsurface) are referred to the Dnieper glaciation (see below), not leaving space for any older glacial formations. In the glacial depressions up to 100 m deep, the stratigraphical position of the Oka Till occurs in their lower part, being separated from the Dnieper Till by interglacial deposits (Zalesskii, 1987; Matoshko, 1990). The distal margins of these depressions are chosen as relative indicators of the Oka glacial limit (Fig. 31.1).

In the south-eastern part of Belarus, only single boreholes have revealed glacial deposits of the Oka formation which occur both in glacial depressions (where their thickness reaches several tens of metres) and within plain areas.

The basal fill of the Dnieper valley (underlying the alluvial deposits of the Middle-Upper Pleistocene and Holocene) contains thick medium to coarse sands with inclusions of gravel and pebbles. They are considered as meltwater deposits of undifferentiated Oka and Dnieper glaciations. It is expected that proglacial drainage existed in the Pre-Dnieper valley—the main drainage pathway to the Black Sea (Fig. 31.1).

31.3.3. Key Sites

The key sites included in the digital map should be spatially related to the glacial limits and contain important information about the occurrence of glacial deposits and (or) their relative (absolute) age. Unfortunately, a very limited number of sites fulfil these conditions. The most important site exposing the Oka glaciation sediments is Krukenichi (Fig. 31.1, map). The characteristics of this site and the main references have been previously presented (Matoshko, 2004). Apart from the palaeontological and petrographical analyses, it is one of only two points in the Ukraine and Belarus where till samples have been dated by the thermoluminescence (TL) method (Butrym et al., 1988). On the basis of these measurements, the till was found to be 510–520 ka. In the present map, this site is supplemented by the neighbouring Dubanevichi site, where a glacial complex (till and meltwater deposits) is exposed in two pits (Fig. 31.3). The TL analysis by the laboratory of Lublin University (Poland) dated the till to 529 ± 72 ka and the overlying non-glacial sandy loams to 241 ± 36 ka (Demedyuk and Demedyuk, 1988b).

The series of good exposures stretches from Loeiv (Fig. 31.1) to the north-west along the Dnieper valley in Belarus. These exposures have been thoroughly investigated, and in part of them, two till horizons, divided by lacustrine, alluvial and meltwater deposits, are exposed (Mander, 1973). The upper till is correlated with the Dnieper glaciation and the lower with the Berezina (Oka) glaciation. The southernmost exposure of this series is found at Peredelka which is included in the present digital map.

In glacial depressions of the West Polissian sector, the interglacial deposits are represented by alluvial–lacustrine deposits rich in plant fossils. In several sections, they overlie the Oka Till. The analogues of these deposits were studied in the same region (but only in sections where the Oka Till was absent) by palaeobotanical analysis (Khursevich and Loginova, 1980; Vevichkevich, 1982; Yelovicheva, 2001). The results indicate a Likhvian (Holsteinian) age. It thus confirms the possible pre-Likhvian age of the Oka glacial complex by implication.

31.3.4. Problems

The name 'Oka glaciation' is taken from the name of the Russian River where its deposits were first identified in the former USSR. The paradox is that a great gap exists in the limits of that glaciation between the basin of the Oka River and the south-eastern regions of Belarus (see above). However, it was shown in the digital maps of Ehlers and Gibbard (2004) that the boundaries of the Oka and South Polish glaciations coincide absolutely at the Polish–Ukrainian border. They can be correlated with the Elsterian glaciation in Germany. However, the precise glacial limits of the Oka glaciation in the Russian Federation are still a matter of speculation (Velichko, et al., 2004).

There are no clearly defined criteria for the identification of Oka glacial deposits within the plain areas of the West Polissian sector. On the whole, the Oka age of the second till from the surface found in the depressions and in the plain areas (the altitudinal difference reaches 70–80 m) has not been proved. Thus, the present glacial limit of the Oka glaciation should be considered as a working hypothesis. Better information will only be obtained from complex studies of the existing and new key sites. The most promising objects for such a project are the glacial depressions of Volyn' Polissia in the Ukraine and the exposures along the Dnieper valley in south-eastern Belarus.

31.4. THE DNIEPER GLACIATION IN THE CENTRAL AND NORTHERN UKRAINE AND ADJACENT REGIONS OF RUSSIA

31.4.1. Glacial Boundaries

Two main glacial structures of the Dnieper ice sheet have been distinguished in the Ukraine: the wedge-shaped Dnieper Ice Stream and West Polissian Lobe, separated by the Polissian 'driftless area' (an analogue of the driftless area in Wisconsin and some adjacent states in the USA). The glacial features occupy the northern and partly central part of Ukraine (Fig. 31.2). As a whole, the position of the glacial limits in the present version of the digital map has changed only in detail. The glacial limit has undergone some shift near Ovruch (Fig. 31.2, map). The latest detailed field investigations within the Ovruch Heights have provided evidence that the glacier went around the elevated terrain. The southernmost part of the Dnieper Ice Stream has been even more reduced owing to the exclusion of several dubious points and areas where concrete evidence of glacial features is lacking. The author suggests his own version of the glacial limit within the Russian Federation near the Ukrainian border up to the latitude of 52°N. It is based on materials from the USSR State Geological Survey (1970–1980 years). The boundary given by Velichko et al., (2004) for this region represents an older, simplified glacial limit from the beginning and middle of the twentieth century. In all other cases, the position of the glacial limit was specified according to the principles cited above.

31.4.2. Glacial Features

The areas of the main glacial structures possess their own characteristics for tracing the extent of the former ice sheets. Instead of the rather provisional previous line (Matoshko, 2004), the West Polissian Lobe glacial limit given in this chapter is based only on fixed glacial features. Among them are boreholes in which Dnieper Till (or, to be exact, the first till from the surface, see also the Section 31.3.2) was encountered, outer esker fans, fluvioglacial fans, ridges of thrust end moraines, erratic boulders and the proximal margin of a glacial depression. The combined information on most of these features has been published elsewhere (Matoshko and Chugunny, 1993; Chugunny and Matoshko, 1995; Matoshko, 1995a,b). It should be considered that in some areas, the distance between the adjacent glacial features reaches 5–15 km and that some of the end moraines might have been created during re-advance phases of the ice sheet at its retreat stage. The same restrictions apply to the area of the Dnieper Ice Stream (see below). In some places, the line of the maximal extent is very close to the retreat phase boundary. The latter has also been changed slightly from that of the previous version.

The extent of the Dnieper Ice Stream as well as its glacial limits is based on much better evidence compared to the area of the West Polissian Lobe. The average distance between points of mapped features varies from hundred metres to a few kilometres. However, to the north of Sumy, the glacial limits are based on rare points. The main evidence of the extent of the Dnieper ice sheet is the Dnieper Till (Fig. 31.4). It is easily identified in boreholes and exposures. Practically, the Dnieper Till can be traced from section to section throughout its distribution area (thousands of points) as a continuous single stratum, apart from a few places where multilayered till ('schollen till') occurred. The latter probably reflects special movement of the basal ice and the entrainment of subglacial deposits (Matoshko, 1995a). Other features are less widely spread. They include ridges of small thrust and push moraines (Fig. 31.5; see also digital map), single esker fans (Fig. 31.6; see also digital map) and lacustrine deposits of ice-dammed lakes.

New exposures of valley sandar deposits on the slopes of the Ps'ol (Fig. 31.7) and Vorskla Rivers, in combination with previously published evidence (Nazarenko, 1968), allowed the redrawing of the system of proglacial runoff of the south-eastern flank of the Dnieper Ice Stream in more detail. The meltwater drainage followed the course of the Ps'ol River, then a section of the Tashan' valley (draining against the modern gradient), dead valley at the Tashan' - Vorskla interfluve and parts of the Vorskla and Dnieper valleys. (Fig. 31.2). This complicated drainage system was probably caused by the ice sheet blocking of the Pra-Ps'ol valley at some part of its lower reaches.

31.4.3. Correlation Between Glacial Limits and Ice-Sheet Evolution

The extent of an ice sheet is an element and also an important indicator of the ice-sheet dynamics with which it should be in accordance at least in the marginal zone of the glaciated area. This connection has been dealt with in several publications (Bondarchuk et al., 1978; Matoshko and Chugunny, 1993, 1995). In the latest publications, the Dnieper Ice Stream and the West Polissian Lobe are regarded as peripheral outlet glaciers of the ice sheet which originated under favourable dynamic conditions and on a favourable preglacial relief (Figs. 31.1 and 31.2). Until recently, they were assumed to have been characterised by free and uniform ice diffluence in distal direction during a single advance stage. New facts compel correction of this thesis.

The presence of relatively small massifs of glaciolacustrine deposits near the south-western limit of the Dnieper Ice Stream is known (Matoshko and Chugunny, 1993). Originally they were thought to be of minor importance (Matoshko and Chugunny, 1995). This view has been revised by the author in the light of new field observations and thorough analysis of the primary borehole descriptions from the Sula River basin (Fig. 31.2) and some adjacent areas. Basically, the thick (30–70 m) strata of uniform silts with clay interlayers, overlain by the Dnieper Till, are interpreted as the glaciolacustrine Sula Suite belonging to the Dnieper glacial complex. Earlier, these strata were either not noticed or considered as a sequence of loess-like deposits and buried soils. The new interpretation suggests the existence of a great ice-dammed lake (more than 20,000 km^2) and some smaller ones on the left bank plateau of the Middle Dnieper area (Fig. 31.2). The dammed pre-rivers Seim and Sula drained into that lake. Such lake could form only at the transgressive stage during a rather long oscillation of the ice stream preceding the final advance.

The altitudinal analysis of the maximum glacial limit (based mostly on the level of the till bed) supplements

FIGURE 31.4 Basal (mature monolithic) till with structures of entrainment of underlying deposits and irregular contact between till and glacially deformed alluvial sands. The cliff of the Kremenchuk Reservoir, 2.5 km to the NW from the Velyka Andrusovka village (Photograph: Matoshko, 2006). Deposits: sw, slopewash; t, till; a, alluvial deposits. P_{2-3}, Middle–Upper Pleistocene; P_2, Middle Pleistocene; P_1, Lower Pleistocene; dn, Dnieper Horizon (Formation).

essential information to the picture of the ice-marginal evolution. In the frontal part of the West Polissian Lobe, the glacial limit lies in the range of 180–200 m a.s.l. and descends along its eastern flank towards Petrikov (Belarus) where it is found at about 130 m a.s.l. Comparison of the hypsometric profiles of the eastern and western flanks of the Dnieper Ice Stream (Fig. 31.8) demonstrates that there are sharp differences in their shapes and in their absolute altitude. It is obvious that the eastern flank, forming an almost straight slope, lies much lower, except its proximal part. To the contrary, the profile of the western flank is convex and strangely descends towards the north (to Pietrikav) in its proximal part. That and some asymmetry of the wedge-shaped area of the Dnieper Ice Stream are explained

FIGURE 31.5 Fragment of glacial diapir in Pliocene—Lower Pleistocene clays and silty meltwater sands. Vertex of glacial marginal ridge near the Gorodische village, western boundary of the Dnieper Ice Stream area (photograph: Matoshko, 2007).

FIGURE 31.6 Sands, gravel and pebbles representing the sedimentary structures of the apex of the esker between the Krasnostav and Tesnovka villages (photograph: Matoshko, 2009).

by the predominant activity of the western sector of the glacier. Such a shift towards the West that existed also during the active retreat stage was and has been attributed to the Coriolis force (Matoshko and Chugunny, 1993, 1995). Later, the same effect with the same explanation was independently revealed for lobes of the Last glaciation in the Baltic countries and in Karelia (Karukäpp, 1999). At the same time, the abnormal northwards descending glacial limits along the flanks of the Polissian 'driftless area' are not causal. They might be associated with a dynamically induced lowering at the ice-sheet margin as result of ice outflow. The latter was caused by peripheral ice structures and first of all the Dnieper Ice Stream. Thus, the 'driftless area' probably reflects a compensatory ice-thickness depression in the marginal zone of the ice sheet.

FIGURE 31.7 Meltwater sands and silts forming a terrace on the right bank of the Ps'ol River near Chervone village. A combination of various fluvial facies of variable thickness and lithological composition is typical to valley sandur deposits near the former ice margin (photograph: Matoshko, 2009).

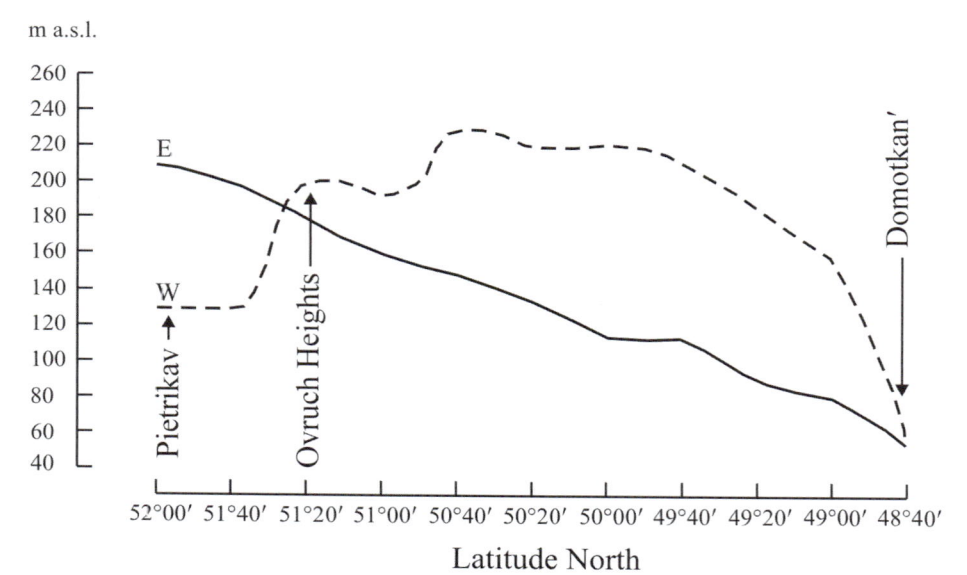

FIGURE 31.8 Altitudinal diagram along the flanks of the Dnieper Ice Stream. E, eastern flank; W, western flank.

An active regressive stage, including recessions and re-advances of the ice margin (combined into phases), is no less important at least to be able to distinguish the limits of maximum glaciation from the limits of retreat phases. Six phases are assumed to have occurred within the central-northern regions of the Ukraine and the southern part of Belarus between the maximum limits of the Dnieper and Sozh (Varta, Moscow) glaciations (Matoshko and Chugunny, 1993, 1995; Karabanov, 2004). Here, they are cited with some modifications, based on new information. Only established limits of oscillations are given in the digital map, while both established and probable outlines are shown in Fig. 31.2. In comparison with the poorly expressed glacial features marking the maximum glacial limit (see above), the retreat complexes are represented by exclusively ice-marginal landforms and deposits of various types (Matoshko and Chugunny, 1993, 1995; Chugunny and Matoshko, 1995; Matoshko, 1995a,b). The greatest glacial 'macro-ridges' in Eastern Europe at Kaniv and Moshnogiria are parts of one of these complexes.

31.4.4. Key Sites and Problems

Within the West Polissian Lobe area, more or less complete sections of the glacial Dnieper Formation, including studied overlying and underlying fossiliferous interglacial units, are unknown. On the contrary, the Middle Dnieper area is characterised by numerous exposures where the stratigraphic relation of the Dnieper Till to the over- and underlying interglacial deposits is clearly defined. Its position has been determined in tens of regional profiles (Goretsky, 1970; Barschevskii, 1977; Yes'kov, 1977; Matoshko et al., 2002, 2004), where the Dnieper Till is considered as second by importance as a regional index bed. Throughout the Middle Dnieper area, the glacial complex occurs on top of the lacustrine–alluvial Krivichi Suite referred to the Likhvian (Holsteinian) stage.

Earlier, several key sites for the determination of the stratigraphical position of the Dnieper Horizon and thus of the Dnieper glacial limits have been indicated (Velichko et al., 2004). However, according to the author's opinion, only one of them, Hun'ky, qualifies as a proper key site. Here (Fig. 31.2), the loess-like deposits with several buried soils underlying the Dnieper Till are characterised by Likhvian (Holsteinian) complex of plant remnants (Gubonina, 1980). The Singil fauna complex of small mammals (Rekovets, 1994) and a corresponding collection of freshwater molluscs (Gozhik, 1992) were both identified in the alluvial deposits occurring at the base of the section.

Most of the other sites contain only indirect stratigraphical evidence such as different lithological units including buried soils, loesses, permafrost features, etc. An example is the Viazivok site that includes one of the most complete sequences of loess-like deposits in the Ukraine (Gozhik et al., 2001). The palaeosols of this section are characterised by poor spore and pollen spectra and by mollusks. One of them, the Kaidaky soil, overlies the Dnieper Till which in turn overlies the Dnieper Loess. The reversed and anomalous polarity revealed at the base of the Viazivok section was referred to the Brunhes–Matuyama boundary (Vigilanskaya, 2001). In this section, the Dnieper Till layer rather serves as the main division of the Pleistocene sequence than the other way round.

The problem of the stratigraphical position of the Dnieper Formation and the age of the glacial limits is redoubled by the lack of reliable absolute dates. As previously mentioned (Matoshko, 2004), the numerous TL dates by the Institute of Geological Sciences of the Ukrainian Academy of Sciences in Kiev (Laboratory of Shelkoplyas), including those applied to the Dnieper Horizon have been shown to be invalid (Zhou et al., 1995). This conclusion is shared by the author. In this situation new, independent investigations are required. Traditionally, the Dnieper unit of the USSR and Ukraine is correlated with the Drenthe advance of the Saalian glaciation (*s.s.*) in Germany and the corresponding Odranian glaciation (stage) in Poland (Gozhik et al., 2001). The glacial limits of the latter coincide with the limit of the West Polissian Lobe at the Polish–Ukrainian border (Fig. 31.2). According to Mojski (1995), the TL dates from the two first advances of the Odranian stage (obtained from samples in south-eastern Poland) range from 250 to 300 ka BP (i.e. marine isotope stage (MIS) 8). However, many specialists put the upper age limit of TL dating closer to 100 ky (Ehlers, 1996) which raised doubts concerning the cited dates. Earlier, Shackleton (1987) had suggested that one of the major Pleistocene glaciations should equate to MIS 6. This opinion was shared by Marks (2004) concerning both the Odranian and the Wartanian stages. It is also advocated by Velichko et al., (2001) for the Dnieper glaciation. Thus, the issue is far from settled.

Recently, an effort has been made to determine the age of the maximum advance of the Dnieper glaciation using ^{10}Be surface exposure dating (SED). The project 'Termination II in central Europe: ice-sheet dynamics and climate variability' is lead by V. Rinterknecht. For this purpose, the unique Domotkan site, at the southernmost point of the Dnieper Ice Stream (Figs. 31.2 and 31.9), was chosen. Here, boulders of local granite lie there on the summit of a small hill, partly washed by the waters of the Dniprodzerzhinsk Reservoir. In the reservoir cliff (up to 18 m high), glacially dislocated alluvial sands are exposed. At the top, they are overlain by a carbonated boulder layer and somewhere at the foothill by basal till. At the time of writing (late 2009), the samples of these boulders were yet to reveal their results.

31.5. MOUNTAIN GLACIATION IN THE UKRAINIAN CARPATHIANS

The Ukrainian Carpathians form part of the great Alpine–Carpathian mountain arc, but information concerning their Pleistocene glaciations is very limited (Ehlers and Gibbard, 2004).

In the Ukraine, real traces of the mountain glaciations are found in the Charnogora, Svidovets and Sivulia massifs (Matoshko, 2004) at an altitudinal range of 1350–1850 m a.s.l. (Fig. 31.10). Here, the mountain peaks and ridges reach an altitude of 1700–2060 m a.s.l. At present, in cold years, long-term snowfields in cirques may remain to the end of the summer.

The lower part of the mountain range is associated with basins, troughs and end moraines of former valley glaciers, whereas glacial cirques (Fig. 31.11) are limited to the upper part of this range. In this edition, the map of several new formerly glaciated areas higher than 1650–1700 m have

FIGURE 31.9 Boulders of local granites glacially transported onto the top of the marginal hill, composed of boulder 'caprock' and intensively glacially deformed underlying alluvial deposits of the Lower–Middle Pleistocene. The Dniprodzerzhinsk Reservoir is seen at the background. Southernmost point of Pleistocene glaciation in the Dnieper basin near Domotkan' village (photograph: Matoshko, 2006).

been added (Fig. 31.10, map). These values are approximate and reflect the assumed snowline of one or possibly several of the Pleistocene glacial events.

It should be noticed that the modern relief of the highest ridges and individual mountains has an alpine appearance and is unfavourable for substantial firn-ice accumulation. At the same time, the existence of former summit and transfluent glaciers cannot be excluded.

For the estimation of the former snowline altitude, neotectonic movements should be taken into account. According to Starostenko (2005), the average velocity of Pliocene–Quaternary uplift in the central zone of the Ukrainian Carpathians is 0.045 mm per year. This is insignificant in the case of the Last glaciation but as for the Middle Pleistocene, the snowline could have reached more than 50 m lower in absolute values than the present evidence would suggest.

The Pozhizhevs'ka site, between the Breskul and Pozhizhevs'ka summits, described in the previous edition (Matoshko, 2004) remains unique in this glaciated region. However, it has gained more importance as a result of the application of the ^{10}Be SED method to large boulders in the end moraines during the 'Termination II' project (see above). The exposure age of the boulders is in the range of 10.8–13.8 ka (Rinterknecht et al., 2010). These dates correspond with the Holocene ^{14}C dates (Tretyak and Kuleshko, 1982) from peat in the tongue basin surrounded by the moraines mentioned above.

They confirm that the glacial features are of late Late Glacial age.

The assumed snowline during this time, according to Urdea (2004; with reference to Halouska, 1977), lay at 1825–2075 m a.s.l. in the Tatra Mountains and at 2035–2203 m a.s.l. in the Southern Carpathians, while altitudes of the terminal moraines are 1825–2075 m a.s.l. and 1840–2150 m a.s.l. correspondingly. Under these conditions, glaciation in the Ukrainian Carpathians would have been unlikely to have occurred. However, the lowest end moraines of the valley glaciers in Romania descend to 1000 m a.s.l. and below. They are considered as of Rissian age (i.e. *c*. Saalian) with a palaeo-snowline depression to 1500–1700 m a.s.l. (Urdea, 2004). The latter would match with the Ukrainian evidence (see above), but it contradicts the dates obtained from the Pozhizhevs'ka site. To solve this paradox, some additional efforts are required. They include radiometric dating of the well-expressed Romanian end moraines at different altitudes as well as—on a larger scale—a modern analysis of all the available data about the Pleistocene mountain glaciations in Europe and adjacent highlands.

31.6. CONCLUSION

The boundaries of the Middle Pleistocene Oka (Elsterian) and Dnieper (Saalian) glaciations, as well as the mountain

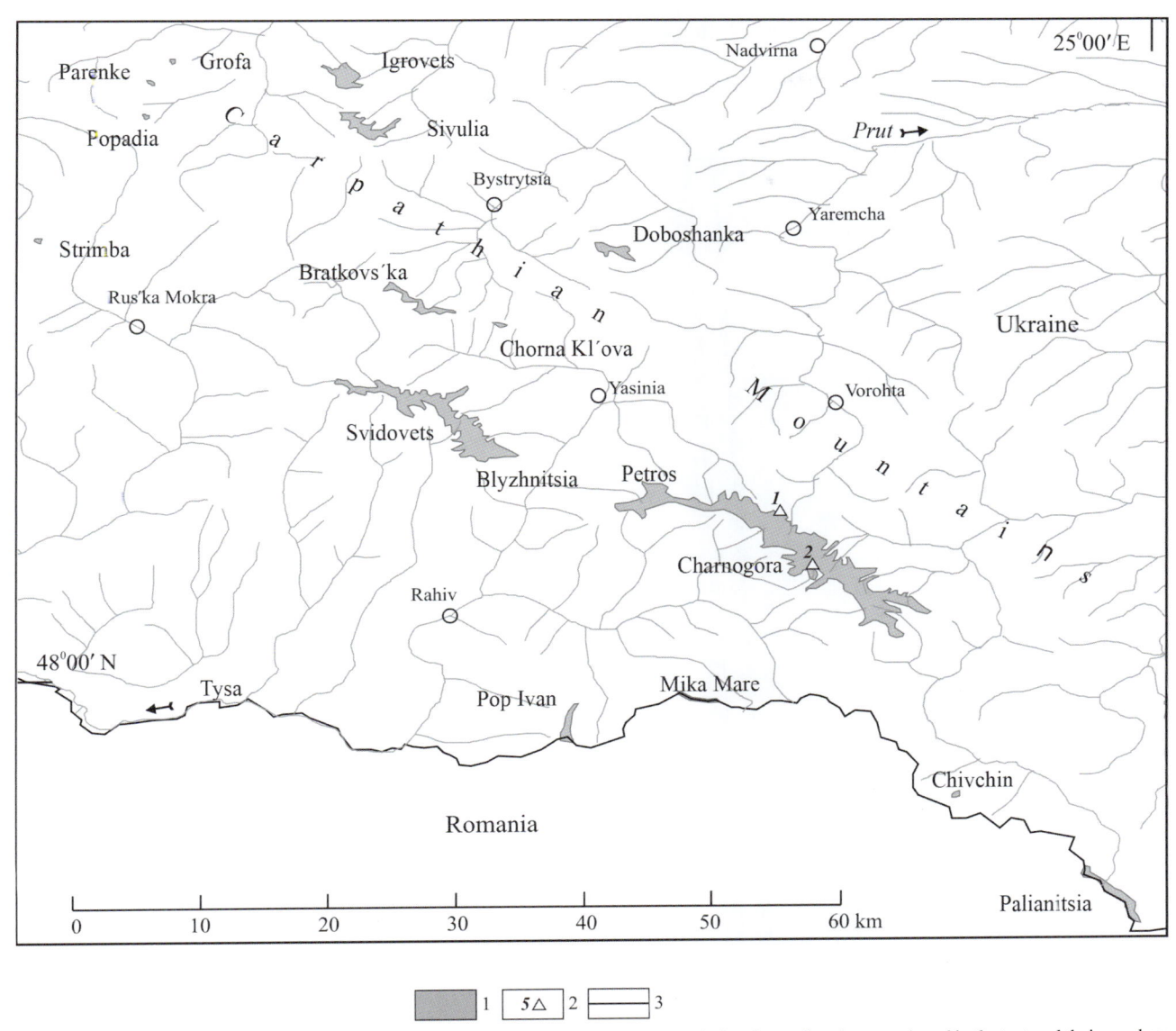

FIGURE 31.10 The extent of the mountain glaciation in the Ukrainian Carpathians. 1—glaciated area; 2—sites mentioned in the text and their numbers: 1—Pozhizhevs'ka, 2—Gutyn Tomnatek; 3—state border.

glaciation in the Carpathians within the Ukraine and in adjacent regions of Poland, Romania, Belarus and Russia, are considered in order to differentiate their positions and age. The focus is on a deeper and more rigorous analysis of the primary evidence, methods of mapping the glacial limits and spatial features of the glacial landscape, deposits and dislocations, using the treatment of the topographical and thematic data maps in an ArcMap environment. This has allowed the drawing of a more accurate version of the digital map of the glacial boundaries and extending its content. New field and laboratory evidence have been included in the study. The first results of a ^{10}Be SED investigation applied to boulders from end moraines in the Charnogora Ridge are discussed. They suggest a Late Glacial age of the glaciation in the Ukrainian Carpathians. A short comparison of the main concepts in dating and correlation of the Middle Pleistocene glacial complexes in the south-western part of the East-European Plain reveals the great gaps and still unsettled issues in the regional stratigraphy.

ACKNOWLEDGEMENTS

The author thanks the editors for considerable editorial work on the chapter and English improvement, and Anton Matoshko for preparing the figures.

FIGURE 31.11 'Armchair-like' glacial cirque on the slopes of the Charnogora Ridge, Ukrainian Carpathians (photograph: Stel'mah, 2008).

REFERENCES

Barschevskii, N.Ye., 1977. Stroenie chetvertichnogo pokrova i istoriia chetvertichnogo osadkonakopleniia na territorii Kievskogo Pridniproviia (Structure of the Quaternary cover and history of the Quaternary sedimentation in Kiev near-Dnieper area). Ph.D. thesis, Institute of Geological Sciences of NAN of Ukraine (Kiev), 161pp.

Bogutskii, A.B., 1967. Morennye otlozheniya na teritorii Volynskoi vozvyshennosti (Till deposits within the Volyn' Upland). Doklady i soobscheniia L'vovskogo otdela Geographicheskogo obschestva UkrSSR za 1965 g. Izdatelstvo L'vovskogo Universiteta, L'vov.

Bondarchuk, V.G., Gozhik, P.F., Chugunny, Yu.G., 1978. Kraevyie obrazovaniia i voprosy dinamiki razvitiia dneprovskogo oledeneniia (Marginal formations and issues of dynamics of Dnieper Glaciation development). In: Bondarchuk, V.G. (Ed.), Kraevyie obrazovaniia materikovyh oledenenii. Naukova Dumka, Kiev, pp. 3–5.

Butrym, J., Maruszczak, H., Wojtanowicz, J., 1988. Chronologia termoluminescencyjna osadow ladolodu Sanian (=Elsterian II) w dorzeczu sanu i gornego Dniestru (Thermoluminescence chronology of the Sanian (=Elsterian II) inland-ice deposits in the San and Upper Dniester River basins). Ann. Soc. Geol. Poloniae 58, 191–205.

Chugunny, Yu.G., Matoshko, A.V., 1995. Dnieper Glaciation—till deposits of ice marginal zones. In: Ehlers, J., Kozarski, S., Gibbard, P. (Eds.), Glacial Deposits in North-East Europe. A.A. Balkema, Rotterdam, Brookfield, pp. 249–256.

Demedyuk, N.S., Demedyuk, Yu.N., 1988a. Rozriz antropogeny poblyzu s. Dubanevichi L'vivs'koi oblasti. Dokl. Akad. Nauk UkrSSR B 5, 6–9.

Demedyuk, N.S., Demedyuk, Y.N., 1988b. Dnestrovskii lednikovyi complex Predkarpatiia (Dniester Glacial Complex of Fore-Part Area of Carpathians). Institut Geologicheskih Nauk AN UkrSSR, Kiev, preprint 88–27, 55pp.

Ehlers, J., 1996. Quaternary and Glacial Geology. John Wiley & Sons, Chichester, 578pp.

Ehlers, J., Gibbard, P.L. (Eds.), 2004. Quaternary Glaciations—Extent and Chronology. Part 1: Europe. Amsterdam, Elsevier, 475pp.

Goretsky, G.I., 1970. Alluvialnaia letopis' Velikogo Pra-Dnepra (Alluvial Record of Great Pra-Dnieper). Nauka, Moscow, 492pp.

Gozhik, P.F., 1992. Presnovodnyie mollusci i korreliatsia verkhnekainozoiskih allyuvialnyh otlozhenii yuga Vostochno-Yevropeiskoi platformy (Freshwater molluscs and correlation of the Upper Cenozoic alluvial deposits of the southern part of the East European Platform). Ph.D. thesis, Institute of Geological Sciences, National Academy of Sciences of Ukraine, Kiev, 287pp.

Gozhik, P., 1995. Glacial history of the Ukraine. In: Ehlers, J., Kozarski, S., Gibbard, P.L. (Eds.), Glacial Deposits in North-East Europe. A.A. Balkema, Rotterdam, Brookfield, pp. 213–215.

Gozhik, P.F., Matviishina, Z., Shelkoplyas, V., Palienko, V., Rekovets, L., Gerasimenko, N., Korniets, N., 2001. The Upper and Middle Pleistocene of Ukraine. In: The Ukraine Quaternary Explored: The Middle and Upper Pleistocene of the Middle Dnieper Area and Its Importance for the East-West European Correlation, Kiev, Ukraine, Conference, 9–14 September 2001. Abstract Volume, 32–33, Institute of Geological Sciences of National Academy of Sciences of Ukraine, Kiev, Ukraine.

Gruzman, G.G., Chebotareva, L.Ye., 1978. Novyie dannyie o lednikovyh otlozheniiah v raione g. Sokal' (New data on glacial deposits in the region of Sokal' Town). Tektonika i Stratigrafiia, 15, 104–105.

Gubonina, Z.P., 1980. Palinologicheskaia harakteristika podmorennyh otlozhenii v basseine Dnepra (po dannym razreza "Hun'ki") (Pollen and spores characteristic of submoraine deposits in Dnieper basin (by data of "Hun'ki" section)). In: Velichko, A.A. (Ed.), Vozrast i rasprostranenie maksimal'nogo oledeneniia Vostochnoi Evropy. Nauka, Moskva, pp. 153–168.

Gurskii, B.N., 1974. Nizhnii I srednii antropogen Belarusi (Lower and Middle Quaternary of Belarus). Nauka i tekhnika, Minsk, 144pp.

Karabanov, A.V., 2004. The main glacial limits in Belarus. In: Ehlers, J., Gibbard, P.L. (Eds.), Quaternary Glaciations—Extent and Chronology. Part 1: Europe. Elsevier, Amsterdam, pp. 15–18.

Karukäpp, R., 1999. Discussion of the observed asymmetrical distribution of landforms of the southeaster sector of the Scandinavian Ice Sheet. In: Mikelson, D., Attig, J.W. (Eds.), Glacial Processes, Past and Present, The Geological Survey of America, Special Paper, 337, 187–192.

Khursevich, G.K., Loginova, L.P., 1980. Vozrast i paleogeographicheskie usloviia nakopleniia drevneozernyh diatomovyh porod Severo-Zapada Volyni (The age and paleogeographic conditions of accumulation of ancient lacustrine diatomaceous rocks in North-West Volyn'). In: Bondarchuk, V.G. (Ed.), Paleolandshafty, fauna i flora lednikovyh i perygliacial'nyh zon pleistocena. Institut Geologicheskih Nauk, Kiev, pp. 32–34, preprint 80-15.

Machnach, A.S, Garetskii, A.V., Matveev, A.V., Anoshko, Yu.I., Il'kevitch, G.I., Konischev, V.S., Kruchek, S.A., Machnach, A.A., Naidenkov, I.V., Pashkevitch, V.I., 2001. Geologiia Belarusi (Geology of Belarus). Institut geologicheskih nauk NAN Belarusi, Minsk, 814pp.

Mander, Ye.P., 1973. Antropogenovyie otlozheniia i razvitiie reliefa Belarusii (Quaternary Deposits and Relief Evolution of Belarus'). Nauka i tekhnika, Minsk 126pp.

Marks, L., 2004. Pleistocene glacial limits in Poland. In: Ehlers, J., Gibbard, P.L. (Eds.), Quaternary Glaciations—Extent and Chronology. Part 1: Europe. Elsevier, Amsterdam, pp. 431–439.

Matoshko, A.V., 1990. Pogrebennyie lednikovyie vrezy na teritorii UkrSSR (Buried glacial incisions in the territory of UkrSSR). Dokl. Akad. Nauk UkrSSR Ser. B 5, 16–18.

Matoshko, A.V., 1995a. Dnieper glaciation—basal till deposits. In: Ehlers, J., Kozarski, S., Gibbard, P. (Eds.), Glacial Deposits in North-East Europe. A.A. Balkema, Rotterdam, Brookfield, pp. 231–239.

Matoshko, A.V., 1995b. Dnieper glaciation—meltwater deposits. In: Ehlers, J., Kozarski, S., Gibbard, P. (Eds.), Glacial Deposits in North-East Europe. A.A. Balkema, Rotterdam, Brookfield, pp. 257–263.

Matoshko, A.V., 2004. Pleistocene glaciations in Ukraine. In: Ehlers, J., Gibbard, P.L. (Eds.), Quaternary Glaciations—Extent and Chronology. Part 1: Europe. Elsevier, Amsterdam, pp. 431–439.

Matoshko, A.V., Chugunny, Yu.G., 1993. Dneprovskoe oledenenie teritorii Ukrainy (geologicheskii aspect) (Dnieper Glaciation of Ukraine (Geological Aspect)). Naukova Dumka, Kiev, 192pp.

Matoshko, A.V., Chugunny, Yu.G., 1995. Geological activity and dynamic evolution of the Dnieper glaciation. In: Ehlers, J., Kozarski, S., Gibbard, P.L. (Eds.), Glacial Deposits in North-East Europe. A.A. Balkema, Rotterdam, Brookfield, pp. 225–230.

Matoshko, A.V., Gozhik, P.F., Ivchenko, A.S., 2002. The fluvial archive of the Middle and Lower Dnieper (a review). Netherlands J. Geosci. 81, 339–355.

Matoshko, A.V., Gozhik, P.F., Danukalova, G., 2004. Key Late Cenozoic fluvial archives of Eastern Europe: the Dniester, Dnieper, Don and Volga. Proc. Geol. Assoc. 115, 141–173.

Mojski, J.E., 1995. Pleistocene glacial events in Poland. In: Ehlers, J., Kozarski, S., Gibbard, P.L. (Eds.), Glacial Deposits in North-East Europe. A.A. Balkema, Rotterdam, Brookfield, pp. 225–230.

Nazarenko, D.P., 1968. Geomorphologicheskaia karta i istoriya formirovaniia neogenovyh i antropogenovyh teras levoberezhiia UkrSSR (Geomorphological map and history of formation of the Neogene and Quaternary terraces on the left bank of the Dnieper within the Ukrainian SSR). Materialy Kharkovskogo otdela geographicheskogo obschestva Ukrainy 6, 44–51.

Rekovets, L.I., 1994. Melkiie mlekopitayushchiie antropogena yuga Vostochnoy Yevropy (Small Mammals of the Quaternary of the Southern Part of Eastern Europe). Naukova Dumka, Kiev, 371pp.

Rinterknecht, V., Matoshko, A., Gorokhovich, Y., Fabel, D., Xu, Sh., 2010. Expression of the Younger Dryas cold event in the Carpathian Mountains, Ukraine. In: EGU General Assembly 2010, Held 2–7 May, 2010 in Vienna, Austria, 9383.

Shackleton, N.J., 1987. Oxygen isotopes, ice volume and sea level. Quatern. Sci. Rev. 6, 183–190.

Starostenko, V.I. (Ed.), 2005. Doslidzhennia suchasnoi dynamiky Ukrains'kyh Karpat (Study of the Modern Dynamics of the Ukraine Carpathians). Naukova Dumka, Kyiv, 256pp.

Tretyak, P.R., Kuleshko, M.P., 1982. Degradatsiya poslednego oledeneniya v Ukrainskih Karpatah (Degradation of the last glaciation in Ukrainian Carpathians). Dokl. Akad. Nauk UkrSSR, Ser. B 8, 26–31.

Urdea, P., 2004. The Pleistocene glaciation of the Romanian Carpathians. In: Ehlers, J., Gibbard, P.L. (Eds.), Quaternary Glaciations—Extent and Chronology. Part 1: Europe. Elsevier, Amsterdam, pp. 301–308.

Velichko, A.A., Morozova, T.D., Gribchenko, Y.N., Nechaev, V.P., Dlussky, K.G., Timireva, S.N., et al., 2001. The Ukraine Quaternary Explored: The Middle and Upper Pleistocene of the Middle Dnieper Area and Its Importance for the East-West European Correlation, Kiev, Ukraine, Conference, 9–14 September 2001. Abstract Volume 95. Institute of Geological Sciences of National Academy of Sciences of Ukraine, Kiev, Ukraine.

Velichko, A.A., Faustova, M.A., Gribchenko, Yu.N., Pisareva, V.V., Sudakova, N.G., 2004. Glaciations of the East European Plain—distribution and chronology. In: Ehlers, J., Gibbard, P. (Eds.), Quaternary Glaciations—Extent and Chronology. Part 1: Europe. Elsevier, Amsterdam, pp. 337–354.

Velichkevich, F.Yu., 1982. Pleistotsenovye flory lednikovyh oblastei Vostochno-Evropeiskoi ravniny (The Floras of the Glaciated Areas of the East-European Plain). Nauka i tekhnika, Minsk, 240pp.

Vigilianskaya, L.I., 2001. Paleomagnetic section and magnetic properties of Quaternary deposits of the Viazivok site. In: The Ukraine Quaternary Explored: The Middle and Upper Pleistocene of the Middle Dnieper Area and Its Importance for the East-West European Correlation, Kiev, Ukraine, Conference, 9–14 September 2001. Abstract Volume 97. Institute of Geological Sciences of National Academy of Sciences of Ukraine, Kiev, Ukraine.

Yelovicheva, Ya., 2001. Middle Pleistocene deposits of Ukraine in the sections Guta and Rudki. In: The Ukraine Quaternary Explored: The Middle and Upper Pleistocene of the Middle Dnieper Area and Its Importance for the East-West European Correlation, Kiev, Ukraine, Conference, 9–14 September 2001. Abstract Volume 106. Institute of Geological Sciences of National Academy of Sciences of Ukraine, Kiev, Ukraine.

Yes'kov, B.G., 1977. Inzhenerno-geologicheskie svoistva alluvia Srednego Dnepra (Engineering-Geological Properties of the Alluvium of the Middle Dnieper Area). Naukova Dumka, Kiev, 196pp.

Zalesskii, I.I., 1987. Okskii etap razvitiia Volynskogo Polesiia (Oka stage of the Volyn' Polessie development). Stratigrafiia I koreliatsiia morskih I kontinentalnyh otlozhenii Ukrainy. Naukova Dumka, Kiev, 43–47.

Zhou, L.P., Dodonov, A.E., Shackleton, N.J., 1995. Thermoluminescence dating of the Orkutsay loess section in Tashkent region, Uzbekistan, Central Asia. Quatern. Sci. Rev. 14, 721–730.

Chronology and Extent of Late Cenozoic Ice Sheets in North America: A Magnetostratigraphical Assessment

René W. Barendregt[1,]* and Alejandra Duk-Rodkin[2]

[1]*Department of Geography, The University of Lethbridge, 4401 University Drive, Lethbridge, Alberta, Canada T1K 3M4*
[2]*Geological Survey of Canada, 3303-33rd St. NW, Calgary, Alberta, Canada T2L 2A7*
*Correspondence and requests for materials should be addressed to René W. Barendregt. E-mail: barendregt@uleth.ca

Studies of $\delta^{18}O$ and $CaCO_3$ content in cores from the deep ocean, and the presence of ice-rafted debris in cores taken from continental margins, have long suggested that earth's past climate has experienced cold periods, some of which produced extensive glaciation on land. While oceanographers have provided evidence for well over 50 such cold periods in the past 3 million years, many of which clearly point to periods of ice sheet development, the terrestrial records initially suggested only relatively few glaciations. The classic fourfold subdivision of the Pleistocene in both North America and Europe dominated terrestrial Quaternary palaeoclimate studies for many decades.

With the exception of the last major continental (Late Wisconsinan) glaciation, our understanding of the extent and timing of ice sheet development in North America has until recently remained uncertain. With the more widespread use of magnetostratigraphy and detailed mapping of surficial deposits, it has become possible to identify the approximate timing (appearance and disappearance) of ice sheets which pre-date the Late Wisconsinan, and to estimate their approximate spatial extent.

With appropriate sampling and analytical techniques, geomagnetic polarity data can be obtained from glacial tills as well as from altered sediments such as palaeosols, to provide a direct assessment of the palaeomagnetism of glacial and interglacial deposits. Such data can be used to assign sedimentary sequences to polarity chrons and subchrons of the global geomagnetic polarity timescale (Fig. 32.1). Where palaeomagnetic data are combined with tephrochronology, palynology and other proxy records of palaeoclimate, magnetostratigraphy affords a robust tool for establishing the timing and extent of glaciations and interglaciations and has substantially enhanced our understanding of Late Cenozoic climate change.

Mapping of surficial geology in northern regions of Canada, as well as remapping of key southern portions of Canada under the NATMAP programme, has provided unique opportunities to study extensive outcrop and borecore records of glacial and interglacial sediments preserved in pre-glacial valleys, in coastal lowlands and interbedded with volcanics in mountainous terrain.

The work described in this volume points to some of the advances which have been made in dating and modelling of past terrestrial climates. It has been the contribution of magnetostratigraphy in particular, which has provided timelines for glacial and interglacial events of the past 3.0 million years, and facilitated the assignment of sediments to the chrons and subchrons (Fig. 32.1) of the geomagnetic polarity timescale (Barendregt and Irving, 1998; Froese et al., 2000; Clague et al., 2003; Huscroft et al., 2004; Roy et al., 2004; Parfitt et al., 2005; Nelson et al., 2009; Barendregt et al., 2010; Mahaney et al., 2011).

In 1998, Barendregt and Irving provided a summary of the magnetostratigraphical data for western Canada and the north-western USA (Fig. 32.2). This summary was updated to include new sampling locations (Table 32.1), and new magnetochron/subchron ages of tills, intertill beds, interglacial sediments, palaeosols and pre-glacial sediments for sites in NW Canada, published in Barendregt et al. (2010), Duk-Rodkin et al. (2010), Roy et al. (2004) and Duk-Rodkin and Barendregt (2011, this volume, see Figs. 49.1–49.3 and 49.23). In these summaries, distribution and extent of ice sheets in western North America are reconstructed, based on some 70 magnetostratigraphical records, and show marked differences between the Matuyama and Brunhes

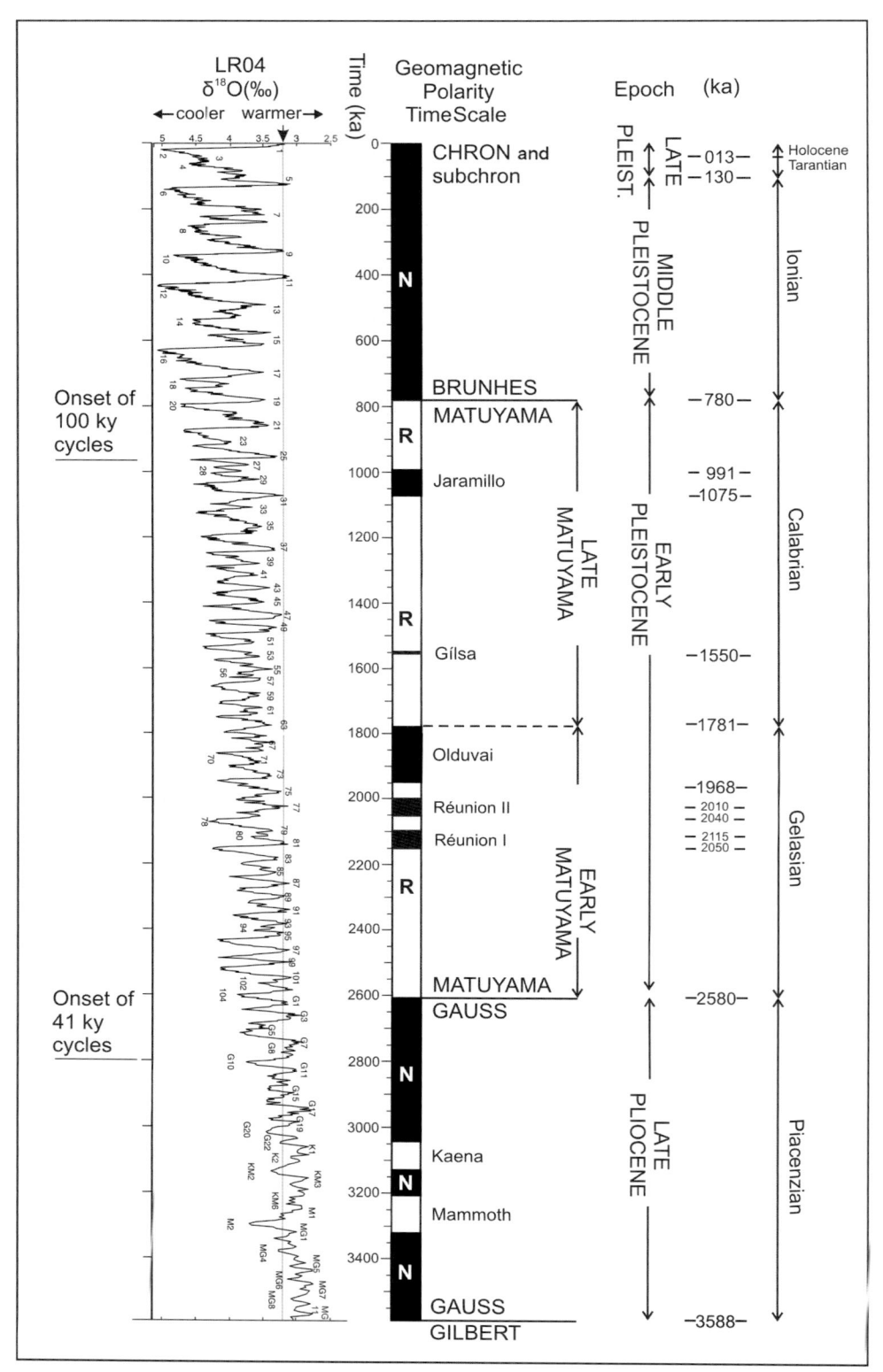

FIGURE 32.1 Geomagnetic polarity timescale (Cande and Kent, 1995) for LR04 benthic $\delta^{18}O$ palaeotemperature profile (Lisiecki and Raymo, 2005). Black/white intervals represent normal/reversed polarity. Marine isotope stages (MIS) are labelled on LR04 (even numbers represent glacials, odd numbers represent interglacials). MIS numbering scheme follows Ruddiman et al. (1986, 1989) and Raymo et al. (1989, 1992) from present to MIS 104, and Shackleton et al. (1995) in the Gauss Chron. Arrow marks Holocene mean $\delta^{18}O$ (Raymo, 1992).

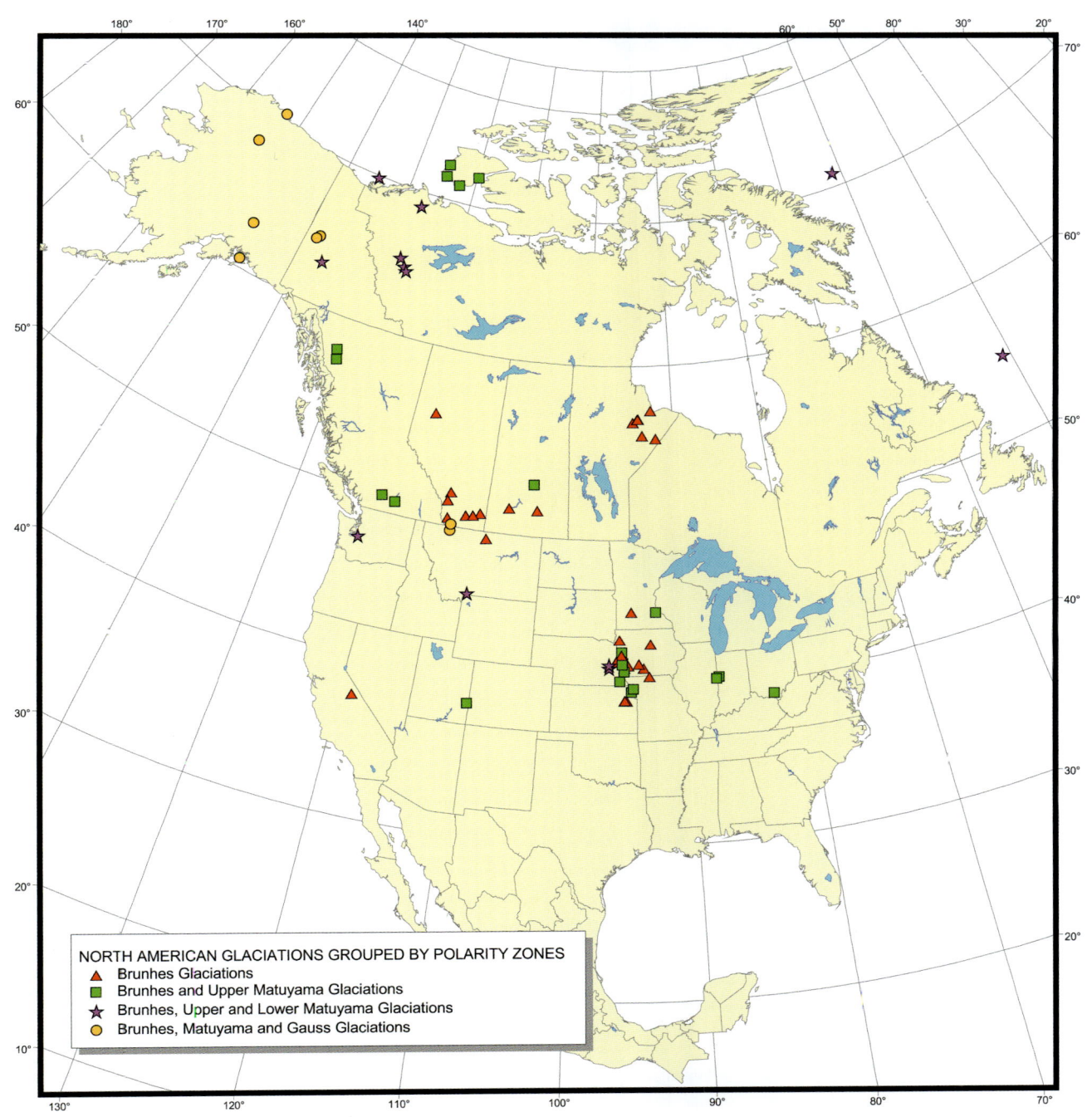

Chapter | 32 Chronology and Extent of Late Cenozoic Ice Sheets in North America

FIGURE 32.2 North American glaciations grouped by polarity chrons.

Chrons (Fig. 32.3). During the Matuyama Chron, ice appears to have been largely absent from large areas of the southern prairie provinces in Canada and the adjacent states of Montana and North Dakota, as well as from much of the Arctic Islands. In the Late Matuyama, a modest Keewatin Ice Centre formed, delivering ice as far distant as Banks Island, North-west Territories (NWT), and south-central Saskatchewan. In contrast, the Labrador/Hudson Bay ice centre delivered ice as far south as Kansas during both the Early and Late Matuyama (Roy et al., 2004). During the Brunhes Chron, ice caps appear for the most part to have been far more extensive than in the Matuyama, and only in the southern Midwestern states did Brunhes-age ice not reach previous limits. During the last major glaciation (Late Wisconsinan), ice cover was continuous from Atlantic to Pacific, with Cordilleran and Keewatin ice sheets in contact in western Alberta, and along the eastern margin of the Mackenzie Mountains in the NWT.

TABLE 32.1 Palaeomagnetic Data Sites

Site name	Symbol	Latitude	Longitude	Site name	Symbol	Latitude	Longitude
Port Nelson, Manitoba	T	57.06	−92.61	Elk Creek, Nebraska	S	40.2	−96.13
Gillam, Manitoba	T	56.34	−94.71	City-Wide Quarry, Nebraska	S	41.26	−95.93
Sundance, Manitoba	T	56.53	−94.07	Atchison, Kansas	S	39.56	−95.13
Henday, Manitoba	T	56.53	−94.18	Wathena, Kansas	S	39.76	−94.95
Limstone, Manitoba	T	56.52	−94.13	Hersey, Minnesota	S	44.5	−92.8
Echoing River, Manitoba	T	55.3	−92.15	West Lebanon, Indiana	S	40.27	−87.38
Stupart Creek, Manitoba	T	55.5	−93.7	Danville, Illinois	S	40.15	−87.63
Prince Albert, Saskatchewan	T	52.3	−105.77	Minford, Ohio	S	38.89	−82.83
Smeaton Saskatchewan	T	53.5	−104.82	La Sal Mountains, Utah	S	38.31	−109.24
Stewart Valley, Saskatchewan	T	50.6	−107.8	Sezill Creek, British Columbia	S	57.9	−31.17
Wascana Creek, Saskatchewan	T	50.6	−104.82	Merrit, British Columbia	S	50.12	−120.78
Medicine Hat, Alberta	T	50.05	−110.67	Kelowna, British Columbia	S	49.88	−119.52
Taber, Alberta,	T	49.78	−112.13	Saskatoon, Saskatchewan	S	52.13	−106.58
Bow Island, Alberta	T	49.87	−111.37	Worth Point, NWT	S	72.25	−125.5
Calgary, Alberta	T	51.05	−114.08	Duck Hawk Bluffs, NWT	S	71.98	−121.5
Pincher Creek, Alberta	T	49.48	−113.95	Nelson River Bluffs, NWT	S	71.22	−122.5
Longview, Alberta	T	50.53	−114.23	Morgan Bluffs, NWT	S	72.2	−119.75
Watino, Alberta	T	55.72	−117.62	West River, NWT	G	69.3	−124.75
Havre, Montana	T	48.54	−109.68	Afton, Iowa	G	41.03	−97.13
Greenfield, Iowa	T	41.31	−94.46	David City, Nebraska	G	41.25	−97.13
Thayer, Iowa	T	41.03	−94.05	Yellowstone Park, Wyoming	G	45.07	−110.76
Macedonia, Iowa	T	41.19	−95.42	Puget Lowland, Washington	G	47.18	−122.28
Glenwood, Iowa	T	41.05	−95.74	Fort Selkirk, Yukon	G	62.78	−137.38
Le Mars, Iowa	T	42.79	−96.17	Katherine Creek, NWT	G	64.95	1127.57
Alden, Iowa	T	42.51	−93.38	Little Bear Creek, NWT	G	64.45	−126.75
County Line, Iowa	T	41.85	−95.99	Inlin Brook, NWT	G	64.33	−126.62
Freemont, Nebraska	T	41.44	−96.49	Mackenzie Delta	G	69.8	−135
Wood Valley Road, Kansas	T	39	−95.75	Labrador Sea, NWT	G	55	−51
Recumseh, Kansas	T	39.02	−95.55	Davis Strait, NWT	G	71	−61.5
Topeka, Kansas	T	39.03	−95.7	Deadman Pass, California	C	37.63	−118.99
Redwood Falls, Minnesota	T	44.54	−95.11	St. Mary, Montana	C	48.74	−113.43
Mercer, Missouri	T	40.51	−93.52	Beazer, Alberta	C	49.12	−113.48
Tahltan, British Columbia	T	57.9	−131.17	Tintina Trench, Yukon	C	64.2	−139.2
Dog Creek, British Columbia	S	51.55	−122.25	Dawson, Yukon	C	64.06	−139.43
Crescent, Iowa	S	41.37	95.86	North Coast, Alaska	C	72	−152
Turin, Iowa	S	42.02	−95.97	Brooks Range	C	68	−154
Thurman, Iowa	S	40.82	−95.75	Southern Alaska Mountains	C	63	−148

T, Brunhes Chron magnetizations only (triangles); S, Brunhes and Upper Matuyama Chrons (squares); G, Brunhes and Upper and Lower Matuyama Chrons (stars); C, Brunhes, Matuyama and Gauss Chrons (circles).

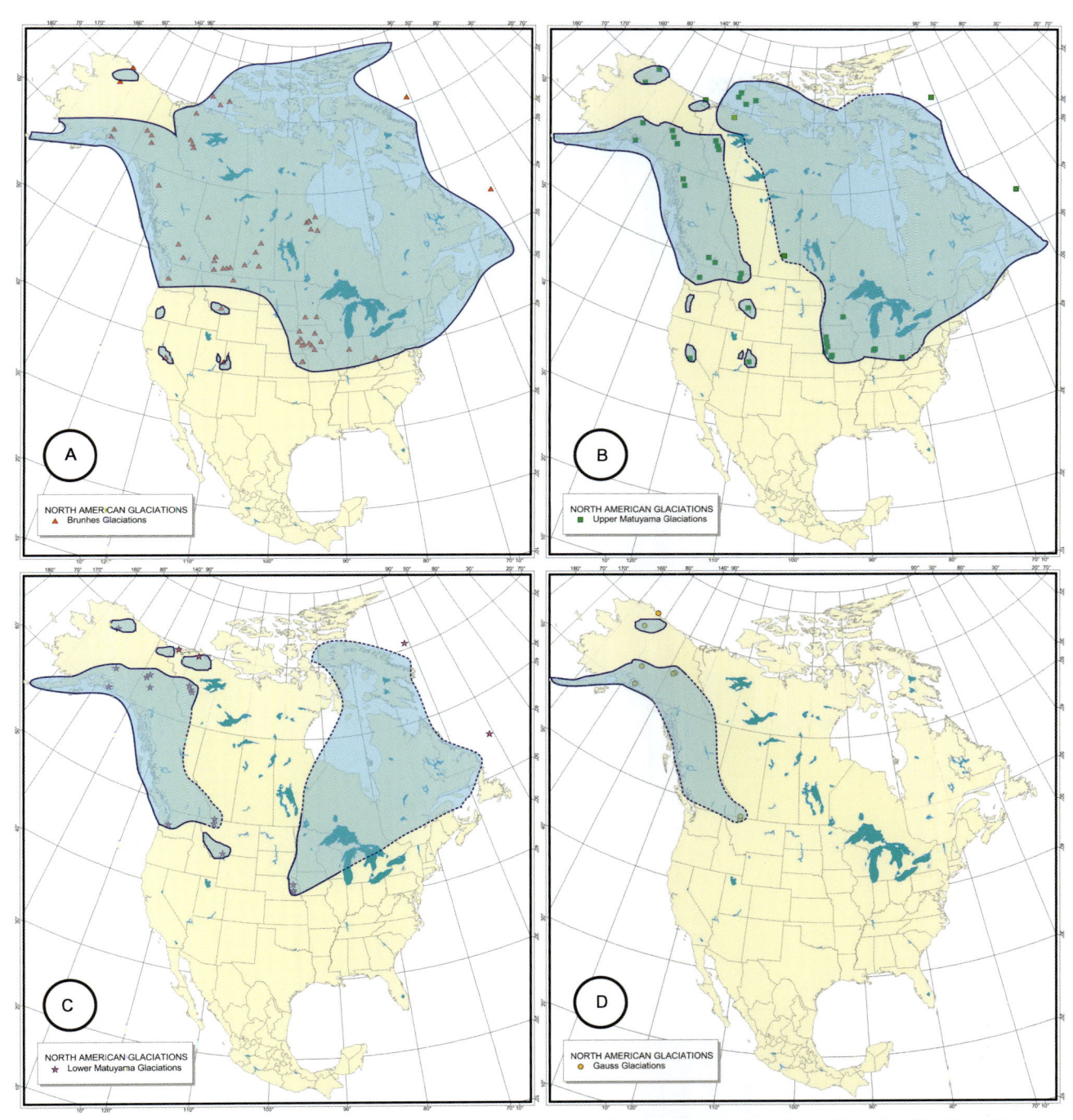

Chapter | 32 Chronology and Extent of Late Cenozoic Ice Sheets in North America

FIGURE 32.3 Proposed maximum ice distribution during late Gauss, early and late Matuyama and Brunhes polarity Chrons (modified from Barendregt and Irving, 1998; Barendregt and Duk-Rodkin, 2004).

Data from the Yukon, NWT and Alaska (Fig. 32.4) have led to considerable refinement of the timing of glaciations in NW North America, and provided a better spatial resolution of the extent of some of the ice sheets (Duk-Rodkin et al., 2004, 2010; Barendregt et al., 2010; Duk-Rodkin and Barendregt, 2011). The most complete terrestrial record of Late Cenozoic glaciations is in the Tintina Trench where multiple tills and palaeosols record six polarity chrons and subchrons, in the Klondike area where loess and glacial gravel sequences record six polarity chrons and subchrons, and along the Yukon River near Fort Selkirk where Quaternary volcanics and interbedded glacial/interglacial sediments record at least five polarity chrons and subchrons (Fig. 32.4). Bore-core data from the Mackenzie Delta provide evidence for

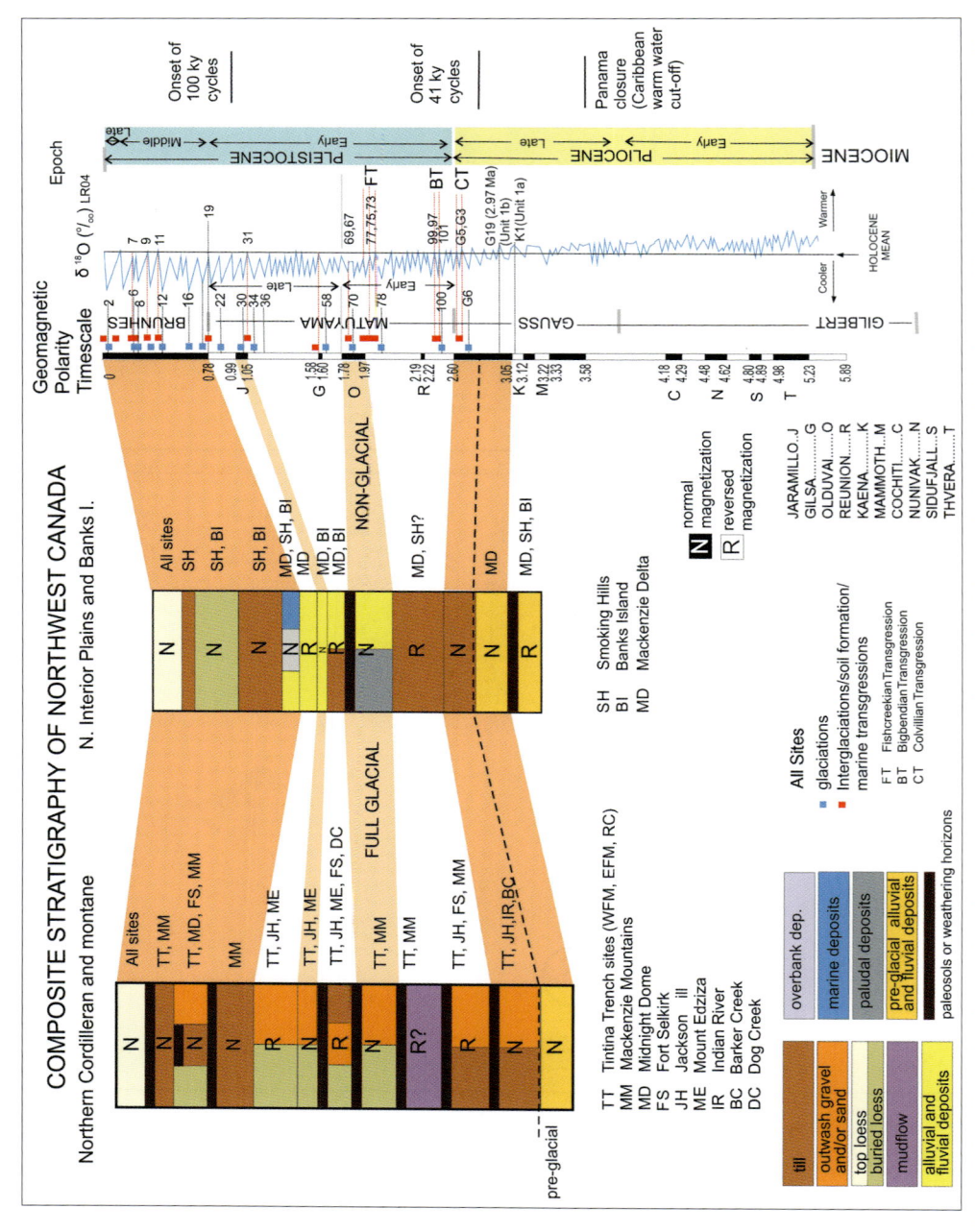

FIGURE 32.4 Correlation of glacial/interglacial events based on a composite stratigraphy developed from multiple sites in the Northern Cordillera, the Northern Interior Plains and Banks Island. The sites included here are discussed in Duk-Rodkin et al. (2010, 2011) and identify glaciations (blue squares) as well as interglacial and pre-glacial sediments containing palaeosols and/or pollen (red squares). See Fig. 32.1 for legend of geomagnetic polarity timescale and composite δ^{18}O LR04 marine isotopic record. The Gauss and Brunhes normal chrons are highlighted in medium brown, and Olduvai and Jaramillo normal subchrons in light brown.

an early Matuyama glaciation only and contain a record of all polarity chrons and subchrons within an extensive nonglacial late Pliocene and Pleistocene sequence. The West River section on the Horton Plateau (Table 32.1) records at least two polarity chrons, in tills and pre-glacial sediments. Shield stones are absent from all the tills, except for the surface till, and therefore a Horton Plateau-centred ice dome is proposed (Fig. 32.3). This ice dome first developed in the early Matuyama, and was restricted to the local highlands. During the late Matuyama, this ice centre encompassed a considerably larger area and may have reached to Banks Island and merged with the more extensive Keewatin Ice Centre to the south and east. From the late Matuyama Chron onwards, the distribution of ice from the Keewatin Ice Centre became progressively more extensive (Fig. 32.3). Previously published data from the Mackenzie Mountains record five mountain glaciations and one continental glaciation, spanning three polarity chrons and subchrons, and include an extensive palaeosol sequence (Barendregt et al., 1996; Duk-Rodkin et al., 1996), while on Banks Island, four widely separated sites record five glaciations spanning four polarity chrons and subchrons, all post-dating the Olduvai subchron (Barendregt et al., 1998).

Recent work in NW Canada has pointed to the importance of tectonics and proximity to open oceans for the development of ice centres (Barendregt et al., 1998; Duk-Rodkin et al., 2004, 2010). The timing of uplift and erosion in the Wrangell/St Elias Mountains, and the Continental Divide (Mackenzie and Selwyn Mountains), appears to have been an important controlling factor in the distribution of moisture supply and the build-up of ice masses in NW Canada. Likewise, periods of low-amplitude variation in benthic $\delta^{18}O$ appear to have seen little or no ice cover in the interior of the North American continent, as suggested by the lack of glacial sediments during these intervals of time (see Fig. 49.24 in Duk-Rodkin and Barendregt (2011, this volume).

In the absence of absolute dating tools, magnetostratigraphy affords a valuable means of assigning terrestrial ice age deposits to the geologic timescale and, most importantly, allows a correlation to be made with the more complete marine record. The distribution of past ice sheets and their timing will undoubtedly be better defined with future magnetostratigraphical work.

REFERENCES

Barendregt, R.W., Irving, Edward, 1998. Changes in the extent of North American ice sheets during the late Cenozoic. Can. J. Earth Sci. 35, 504–509.

Barendregt, R.W., Enkin, R.J., Duk-Rodkin, A., Baker, J., 1996. Paleomagnetic evidence for Late Cenozoic glaciations in the Mackenzie Mountains of the Northwest Territories, Canada. Can. J. Earth Sci. 33, 896–903.

Barendregt, R.W., Vincent, J.-S., Irving, E., Baker, J., 1998. Magnetostratigraphy of quaternary and late tertiary sediments on Banks Island, Canadian Arctic Archipelago. Can. J. Earth Sci. 35, 147–161.

Barendregt, R.W., Duk-Rodkin, A., 2004. Chronology and extent of Late Cenozoic ice sheets in North America: a magnetostratigraphic assessment. In: Ehlers, J., Gibbard, P.L. (Eds.), Quaternary Glaciations - Extent and Chronology: Part II: North America. Elsevier, Amsterdam, pp. 1–7.

Barendregt, R.W., Enkin, R.J., Duk-Rodkin, A., Baker, J., 2010. Paleomagnetic evidence for multiple Late Cenozoic glaciations in the Tintina Trench of west central Yukon, Canada. Can. J. Earth Sci. 47, 987–1002.

Cande, S.C., Kent, D.V., 1995. Revised calibration of the geomagnetic timescale for the Late Cretaceous and Cenozoic. J. Geophys. Res. 100 (B4), 6093–6095.

Clague, J.J., Barendregt, R.W., Enkin, R.J., Foit, N., Jr. 2003. Paleomagnetic and tephra evidence for tens of Missoula floods in southern Washington. Geology 31, 247–250 (plus data repository item 2003023).

Duk-Rodkin, A., Barendregt, R.W., 2010. The glacial history of Northwestern Canada. In: Ehlers, J., Gibbard, P.L., Hughes, P.D. (Eds.), Quaternary Glaciations—Extent and Chronology. Part IV: A Closer Look. Elsevier, Amsterdam.

Duk-Rodkin, A., Barendregt, R.W., Tarnocai, C., Phillips, F.M., 1996. Late Tertiary to late Quaternary record in the Mackenzie Mountains, Northwest Territories, Canada: stratigraphy, paleosols, paleomagnetism, and chlorine-36. Can. J. Earth Sci. 33, 875–895.

Duk-Rodkin, A., Barendregt, R.W., Froese, D.G., Weber, F., Enkin, R.J., Smith, I.R., et al., 2004. Timing and extent of Plio-Pleistocene glaciations in North-Western Canada and East-Central Alaska. In: Ehlers, J., Gibbard, P.L. (Eds.), Quaternary Glaciations—Extent and Chronology. Part II: North America. Elsevier, Amsterdam, pp. 313–345.

Duk-Rodkin, A., Barendregt, R.W., White, J., 2010. An extensive late Cenozoic terrestrial record of multiple glaciations preserved in the Tintina Trench of west central Yukon: stratigraphy, paleomagnetism, paleosols, and pollen. Can. J. Earth Sci. 47, 1003–1028.

Froese, D.G., Barendregt, R.W., Enkin, R.J., Baker, J., 2000. Paleomagnetism of Late Cenozoic terraces of the lower Klondike Valley: evidence for multiple Late Pliocene-Early Pleistocene glaciations. Can. J. Earth Sci. 37, 863–877.

Huscroft, C.A., Ward, B.C., Barendregt, R.W., Jackson, L.E., Jr. 2004. Pleistocene volcanic damming of Yukon River, and age of the Reid glaciation, west central Yukon. Can. J. Earth Sci. 41, 151–164.

Lisiecki, L.E., Raymo, M., 2005. A Pliocene-Pleistocene stack of 57 globally distributed benthic $\delta^{18}O$ records. Paleoceanography 20, PA1003. doi:10.1029/2004PA001071.

Mahaney, W.C., Barendregt, R.W., Villeneuve, M., Dostal, J., Hamilton, T.S., Milner, M.W., 2011. Upper Neogene volcanics and interbedded paleosols near Mt. Kenya. In: Torsvik, T.H., van Hinsbergen, D.J.J., Buiter, S., Webb, S., Gaina, C. (Eds.), Out of Africa: A Synopsis of 3.8 Ga of Earth History. Geological Society of London Special Publication Vol. 357. (in press), Geological Society Publishing House, Bath, UK.

Nelson, Faye E., Barendregt, Rene W., Villeneuve, M., 2009. Stratigraphy of the Fort Selkirk Volcanogenic Complex in central Yukon and its paleoclimatic significance: Ar/Ar and paleomagnetic data. Can. J. Earth Sci. 46, 381–401.

Parfitt, S.A., Barendregt, R.W., Breda, M., Candy, I., Collins, M.J., Coope, G.R., et al., 2005. The earliest record of human activity in northern Europe. Nature 438, 1008–1012.

Raymo, M.E., 1992. Global climate change: a three million year perspective. In: Kukla, G.K., Went, E. (Eds.), Start of a Glacial, Nato ASI Series, Series I. Global Environmental Change 3, 207–223.

Raymo, M.E., Ruddiman, W.F., Backman, J., Clement, B.M., Martinson, D.G., 1989. Late Pliocene variation in Northern Hemisphere ice sheets and North Atlantic deep water circulation. Paleoceanography 4, 413–446.

Roy, M., Clark, P.U., Barendregt, R.W., Glasmann, J.R., Enkin, R.J., Baker, J., 2004. Glacial stratigraphy and paleomagnetism of late Cenozoic deposits of the north-central United States. Geol. Assoc. Am. Bull. 116, 30–41.

Ruddiman, W.F., Raymo, M.E., McIntyre, A., 1986. Matuyama 41,000-year cycles: North Atlantic Ocean and northern hemisphere ice sheets. Earth Planet. Sci. Lett. 80, 117–129.

Ruddiman, W.F., Raymo, M.E., Martinson, D.G., Clement, B.M., Backman, J., 1989. Pleistocene evolution of Northern Hemisphere climate. Paleoceanography 4, 353–412.

Shackleton, N.J., Hall, M.A., Pate, D., 1995. Pliocene stable isotope stratigraphy of site 846. Proc. Ocean Drill. Program Sci. Results 138, 337–353.

Alaska Palaeo-Glacier Atlas (Version 2)

Darrell S. Kaufman[1,*], Nicolás E. Young[2], Jason P. Briner[2] and William F. Manley[3]

[1] School of Earth Sciences and Environmental Sustainability, Northern Arizona University, Flagstaff, Arizona 86011-4099, USA
[2] Department of Geological Sciences, University at Buffalo, Buffalo, New York 14260, USA
[3] Institute of Arctic and Alpine Research, University of Colorado, Boulder, Colorado 80309-0450, USA
*Correspondence and requests for materials should be addressed to Darrell S. Kaufman. E-mail: Darrell.Kaufman@nau.edu
Contributors: T.K. Bundtzen, J. Harvey, R.D. Reger, and A. Werner

33.1. INTRODUCTION

This report is an update of the *Alaska Palaeo-Glacier Atlas* (APG Atlas), a collaborative effort among glacial geologists working in Alaska to compile a coherent map of former glacier limits from across the state. The previous version of the digital atlas (v1) was originally available online in 2002 (Manley and Kaufman, 2002). The evidence used to draw the glacial limits in 14 regions across Alaska was described in a subsequent report (Kaufman and Manley, 2004), which was part of a larger effort to create a global database of Pleistocene glacial extents with a target scale of 1:1,000,000. This update (APG Atlas v2) is based on a more detailed map scale and, therefore, provides more accurately resolved limits of former glacier extents. In addition, the previous version included the only two glacier limits that could be most confidently mapped and correlated around the state: (1) the all-time maximum extent of former glaciers; and (2) the late Wisconsinan extent (= late Weichselian = marine isotope stage (MIS) 2 = spanning the period between about 24,000 and 12,000 years ago). In this update, we added a third mapped limit for areas where it has been sufficiently well studied, namely for the penultimate glaciation, which over most of Alaska occurred during the early Wisconsinan.

This version of the APG Atlas (v2) builds on the pervious version (v1), which synthesized information from 42 published and unpublished source maps (Kaufman and Manley, 2004). Version 2 includes revisions based on information from an additional 12 source maps that were published since v1 was completed in 2002, or that relate to the extent of early Wisconsinan drift, and on additions provided by the four contributing authors whose input is recognized below. This new version of the APG Atlas is available in GIS formats (including kmz for Google Earth at www.ncdc.noaa.gov/paleo/alaska-glacier/). Anyone who has new information on the extent of palaeo-glaciers in Alaska and would like to make that information available, please contact the authors to update the online version of the APG Atlas.

The update also includes a compilation of all cosmogenic exposure ages that have been published from Alaska ($n=249$), with each sample locality linked to the geospatial database. Each sample is identified by a placemark in Google Earth. The placemarks are colour coded according to glaciation; they are labelled according to the cosmogenic exposure age, and include the sample identification and the reference publication. This comprehensive, mapped compilation of ages is useful for visualizing the extent of coverage and the detailed location of cosmogenic exposure ages available from Alaska and in the context of former glacier extents. Combining glacier extents with the locations of cosmogenic exposure ages in GIS format enables easy reference to sample location relative to geomorphic features at a zoomable scale without relying solely on small or schematic figures provided in journal article format. Readers are referred to the original references for calculations used for each of the ages.

For more information about the glacial geology of Alaska and the history of previous investigations, readers are referred to the edited volume of Hamilton et al. (1986), which is the most complete collection of glacial geologic studies in the Alaska, and to more recent summaries by Hamilton (1994), Kaufman et al. (2004) and Kaufman and Manley (2004). A review of the most secure chronologies of Pleistocene mountain glacier advances in Alaska is provided by Briner and Kaufman (2008). Holocene glaciation in Alaska is summarized by Barclay et al. (2009), and historical glacier fluctuations are reviewed by Molnia (2007, 2008)

33.2. GIS PROCEDURES

Standard GIS techniques were used to compile the glacier limits, incorporating information from a variety of sources.

The procedure used to create the digital shapefiles in the original version (v1) of the APG Atlas is briefly described by Kaufman and Manley (2004), with additional details in Manley and Kaufman (2002; overview page and metadata). We converted the v1 polygon shapefiles from datum NAD27 to NAD83 and continue to use the Alaska Albers projection. Working copies were edited in both ArcGIS Desktop with ArcOnline world imagery and Google Earth to overlay the polygons onto a zoomable digital base of orthorectified, high-resolution imagery. A benefit of this high-resolution imagery is that ice-marginal features (e.g. moraines and outwash fans) can often be identified and used to demarcate former ice extent in areas where field-based mapping is unavailable. Within ArcMap (v. 9.3), newly added polygons and refinement of v1 polygons were digitized onscreen at resolutions greater than 1:250,000, which in some cases were at resolutions better than the base map. Spacing between vertices was generally less than 1 km and often < 500 m for regions with detailed mapping. Polygons for the late Wisconsinan reconstruction were converted to lines, and attributed by "certainty" level (see below). The spatial data were then compiled in one dataset, formatted with symbology and converted to a kmz file for distribution and presentation.

33.3. PALAEO-GLACIER LIMITS

Pleistocene glaciers in Alaska included a vast assortment of types and sizes (Fig. 33.1). Their former extents have been mapped at a wide range of scales by many glacial geologists

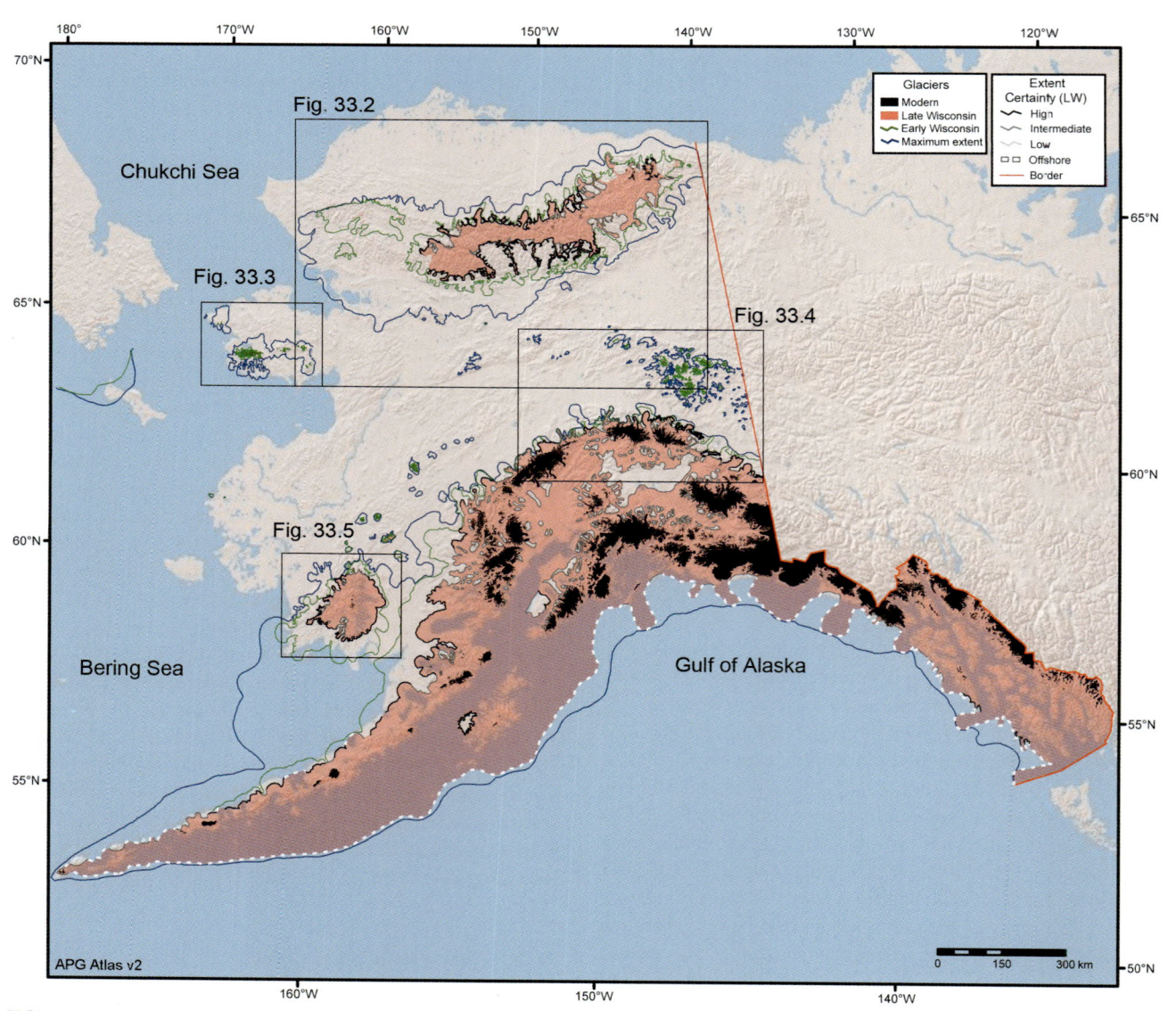

FIGURE 33.1 Alaska Palaeo-Glacier Atlas v2 showing the Pleistocene maximum, early Wisconsinan, late Wisconsinan and modern glacier extents across Alaska.

for different purposes. This results in contrasting styles of mapped limits among regions. For small valley glaciers, the former extent of ice has been delimited precisely at a scale of 1:63,360 or more detailed, and includes the elevational upper limit of the reconstructed glacier. In contrast, the former limits of ice caps and the Cordilleran Ice Sheet are more generalized. They cover large continuous areas with little attention to delimiting the innumerable unglaciated upland areas that extended above the ice sheets, especially around their margins. The largest and most generalized of all the ice limits is along the southern margin of the Cordilleran Ice Sheet where ice terminated on the continental shelf south of Alaska, and little information is available to constrain the former extent.

33.3.1. All-Time Maximum Extent of Glaciers

Pleistocene glaciers in Alaska once covered $>1.2 \times 10^6$ km^2. The reconstructed maximum extent of glaciers in this version (v2) of the APG Atlas is essentially the same as in the original version (v1). The limit is only loosely constrained relative to the high-resolution base. The drift is largely buried by eolian sediment or no longer exhibits diagnostic constructional relief features, and the high-resolution base maps do not afford additional geomorphological evidence for further refinement. The reconstructed maximum extent of glaciers generally coincides with the outer limit of drift mapped by Coulter et al. (1965). It does not represent a single ice advance, but ranges in age from late Tertiary (e.g. the Nenana River Valley) to middle Pleistocene (e.g. the Baldwin Peninsula). The placement of the maximum glacial limit in many places is essentially an educated guess, based on extrapolation of limited data and guided by large-scale geographical patterns and general geomorphology. The limit is depicted as a line rather than a polygon in v1 to emphasize the lack of detailed information about ice-free areas within the area of the maximum extent of glacier ice, and we have omitted certainty estimates (see below) because the limit is generally located only approximately.

33.3.2. The Penultimate Glaciation (Early Wisconsinan)

The first moraines located beyond the geomorphologically well-defined limits of the late Wisconsinan ice have been studied in several areas of Alaska. The glaciation represented by these moraines is known collectively as the "penultimate glaciation", recognizing that, other than their morphostratigraphical position beyond the late Wisconsinan limit, there is often little basis for assigning an age or for correlating deposits regionally. Most previous studies (e.g. Hamilton, 1994) correlated the penultimate glaciation with a global glacial interval younger than the last interglaciation. Several more recent studies (reviewed by Briner and Kaufman, 2008) report geochronological evidence that generally agrees with this age assignment. A growing number of cosmogenic exposure ages (Table 33.1) combined with ^{14}C, luminescence, and tephra-based ages from the few well-dated deposits place the culmination of the penultimate glaciation into MIS 4 or early MIS 3, between around 60 and 50 ka. In the Kvichak Bay area, the penultimate glaciation appears to correlate with late MIS 5 or early MIS 4 (discussed below). In contrast, tephrostratigraphical and other evidence from a few sites suggest that the penultimate drift along the northeast Alaska Range might predate the last interglaciation (discussed below). For simplicity, we refer to the penultimate limit as early Wisconsinan, while recognizing that it probably includes drift older than MIS 4 in places, and might date to MIS 6 in a few places. Version 2 includes mapped limits for the penultimate glaciation where previous studies provide some basis for constraining the former ice extent. We do not attempt to represent the ice-free areas within the limits of early Wisconsinan ice nor to assign certainty levels to the mapped limits, as we do for the late Wisconsinan. The certainties are generally low, but are higher in those areas with geochronological control, which primarily constitutes cosmogenic exposure ages whose locations are viewable in GIS (including kmz format for Google Earth).

33.3.3. Late Wisconsinan Glaciation

Late Wisconsinan glaciers occupied 725,800 km^2 of Alaska. The extent of late Wisconsinan glaciers is delimited more accurately than it is for the other two glacier limits. Generally, late Wisconsinan deposits are easily recognized by their sharply defined moraines on which the details of glacial constructional relief are well preserved. The late Wisconsinan limit on our map roughly coincides with areas covered by ice during the glacial advances of late Pleistocene age, as shown originally on Coulter et al.'s (1965) 1:2,500,000-scale statewide map. In places, however, Coulter et al.'s (1965) late Pleistocene unit included areas that are now assigned to the penultimate glaciation. The APG Atlas v1 refined the limits of late Wisconsinan mountain glaciers based on a compilation of many published maps, and it added palaeo-glacier reconstructions from numerous small mountain ranges that were previously unmapped. Version 2 includes the limits for the small glaciers that emanated from the highest cirques in the Blackburn Hills, Sunshine Mountains, Horn Mountains, Taylor Mountains and Shotgun Hills.

The extent of ice during the late Wisconsinan glaciation in v2 was modified from v1. In areas where the v1 glacier limits were based on detailed (1:63,360) mapping (small mountain glaciers), no adjustments were made. For the Brooks Range, the mapped glacier limits in v1 were largely inferred from a 1:1,000,000 scale, unpublished map

TABLE 33.1 Cosmogenic Exposure Ages from Late and Early Wisconsinan Drift in Alaska (Included as a kmz File for Google Earth)

Sample ID	Latitude (°N)	Longitude (°W)	Age (ka)	Isotope	Region	Reference
Ages from late Wisconsinan (or younger) drift						
FL06-01	63.549	−144.357	22.4±0.6	^{10}Be	Northeast Alaska Range	Young et al. (2009)
FL06-02	63.554	−144.369	9.3±0.4	^{10}Be	Northeast Alaska Range	Young et al. (2009)
FL06-03	63.545	−144.416	16.5±0.5	^{10}Be	Northeast Alaska Range	Young et al. (2009)
FL06-04	63.546	−144.427	16.3±0.4	^{10}Be	Northeast Alaska Range	Young et al. (2009)
FL06-05	63.544	−144.455	16.5±0.5	^{10}Be	Northeast Alaska Range	Young et al. (2009)
FL06-06	63.547	−144.469	16.6±0.4	^{10}Be	Northeast Alaska Range	Young et al. (2009)
FL06-12	63.552	−144.391	9.5±0.3	^{10}Be	Northeast Alaska Range	Young et al. (2009)
US07-09	63.505	−144.526	15.4±0.6	^{10}Be	Northeast Alaska Range	Young et al. (2009)
US07-10	63.505	−144.526	13.9±0.4	^{10}Be	Northeast Alaska Range	Young et al. (2009)
US07-11	63.504	−144.527	13.2±0.3	^{10}Be	Northeast Alaska Range	Young et al. (2009)
FL07-01	63.510	−144.516	6.6±0.2	^{10}Be	Northeast Alaska Range	Young et al. (2009)
FL07-02	63.510	−144.518	11.3±0.3	^{10}Be	Northeast Alaska Range	Young et al. (2009)
FL07-06	63.511	−144.535	5.3±0.2	^{10}Be	Northeast Alaska Range	Young et al. (2009)
FL07-07	63.511	−144.539	8.5±0.4	^{10}Be	Northeast Alaska Range	Young et al. (2009)
FL07-08	63.510	−144.537	11.8±0.7	^{10}Be	Northeast Alaska Range	Young et al. (2009)
US07-04	63.500	−144.523	11.6±0.8	^{10}Be	Northeast Alaska Range	Young et al. (2009)
WM01-1D	61.277	−157.887	15.8±1.8	^{10}Be	Chuilnuk Mountains, Interior Alaska	Briner et al. (2005)
WM01-1C	61.279	−157.889	14.6±1.5	^{10}Be	Chuilnuk Mountains, Interior Alaska	Briner et al. (2005)
WM01-1A	61.296	−157.891	12.1±2.0	^{10}Be	Chuilnuk Mountains, Interior Alaska	Briner et al. (2005)
WM01-1B	61.283	−157.891	9.3±1.8	^{10}Be	Chuilnuk Mountains, Interior Alaska	Briner et al. (2005)
WM01-2E	61.293	−157.862	17.7±1.5	^{10}Be	Chuilnuk Mountains, Interior Alaska	Briner et al. (2005)
WM01-2C	61.294	−157.861	16.4±1.7	^{10}Be	Chuilnuk Mountains, Interior Alaska	Briner et al. (2005)
WM01-2D	61.293	−157.862	15.2±1.5	^{10}Be	Chuilnuk Mountains, Interior Alaska	Briner et al. (2005)
WM01-2B	61.296	−157.859	14.3±1.2	^{10}Be	Chuilnuk Mountains, Interior Alaska	Briner et al. (2005)
NLL-00-2	61.458	−155.359	38.1±2.1	^{10}Be	Lime Hills, western Alaska Range	Briner et al. (2005)
NLL-00-1	61.459	−155.360	25.6±4.5	^{10}Be	Lime Hills, western Alaska Range	Briner et al. (2005)
SR2-00-2	61.481	−154.535	21.3±0.9	^{10}Be	Lime Hills, western Alaska Range	Briner et al. (2005)
SR2-00-5	61.475	−154.504	19.3±1.0	^{10}Be	Lime Hills, western Alaska Range	Briner et al. (2005)
SR2-00-3	61.486	−154.568	19.2±0.8	^{10}Be	Lime Hills, western Alaska Range	Briner et al. (2005)
SR2-00-4	61.459	−154.466	18.2±0.7	^{10}Be	Lime Hills, western Alaska Range	Briner et al. (2005)
KH 2-1	65.142	−154.359	30.0±1.4	^{10}Be	Kokrines Hills, Interior Alaska	Briner et al. (2005)
KH 2-2	65.141	−154.356	23.7±1.0	^{10}Be	Kokrines Hills, Interior Alaska	Briner et al. (2005)
KH 2-3	65.141	−154.355	23.3±1.7	^{10}Be	Kokrines Hills, Interior Alaska	Briner et al. (2005)
KH 2-4	65.142	−154.357	21.1±1.3	^{10}Be	Kokrines Hills, Interior Alaska	Briner et al. (2005)
WM00-09A	64.673	−143.357	28.4±1.1	^{10}Be	Yukon–Tanana Upland, Interior Alaska	Briner et al. (2005)
WM00-09B	64.673	−143.355	25.4±1.8	^{10}Be	Yukon–Tanana Upland, Interior Alaska	Briner et al. (2005)
WM00-09D	64.673	−143.353	18.3±1.2	^{10}Be	Yukon–Tanana Upland, Interior Alaska	Briner et al. (2005)

Continued

TABLE 33.1 Cosmogenic Exposure Ages from Late and Early Wisconsinan Drift in Alaska (Included as a kmz File for Google Earth)—cont'd

Sample ID	Latitude (°N)	Longitude (°W)	Age (ka)	Isotope	Region	Reference
WM00-C9E	64.672	−143.354	~15.7	^{10}Be	Yukon–Tanana Upland, Interior Alaska	Briner et al. (2005)
WM00-08B	64.703	−143.367	28.2±1.2	^{10}Be	Yukon–Tanana Upland, Interior Alaska	Briner et al. (2005)
WM00-03E	64.702	−143.377	20.9±1.2	^{10}Be	Yukon–Tanana Upland, Interior Alaska	Briner et al. (2005)
WM00-08D	64.702	−143.377	17.1±0.9	^{10}Be	Yukon–Tanana Upland, Interior Alaska	Briner et al. (2005)
WM00-08 C	64.703	−143.369	16.4±1.1	^{10}Be	Yukon–Tanana Upland, Interior Alaska	Briner et al. (2005)
BR02-15	69.455	−144.082	76.0±1.9	^{10}Be	Northeastern Brooks Range	Briner et al. (2005)
BR02-16	69.458	−144.081	50.0±1.3	^{10}Be	Northeastern Brooks Range	Briner et al. (2005)
BR02-14	69.454	−144.082	27.3±0.9	^{10}Be	Northeastern Brooks Range	Briner et al. (2005)
BR02-8	69.460	−143.794	55.7±1.4	^{10}Be	Northeastern Brooks Range	Briner et al. (2005)
BR02-7	69.460	−143.800	26.3±0.7	^{10}Be	Northeastern Brooks Range	Briner et al. (2005)
BR02-6	69.458	−143.801	24.8±0.6	^{10}Be	Northeastern Brooks Range	Briner et al. (2005)
BR02-5	69.445	−143.787	21.6±0.6	^{10}Be	Northeastern Brooks Range	Briner et al. (2005)
BR02-9	69.461	−143.796	20.1±0.7	^{10}Be	Northeastern Brooks Range	Briner et al. (2005)
BR02-11	69.461	−143.713	42.3±1.1	^{10}Be	Northeastern Brooks Range	Briner et al. (2005)
BR02-10	69.460	−143.711	26.9±0.7	^{10}Be	Northeastern Brooks Range	Briner et al. (2005)
BR02-12	69.459	−143.715	12.8±0.5	^{10}Be	Northeastern Brooks Range	Briner et al. (2005)
BR02-13	69.449	−143.743	9.2±0.4	^{10}Be	Northeastern Brooks Range	Briner et al. (2005)
BR02-4	69.343	−143.561	21.6±0.4	^{10}Be	Northeastern Brooks Range	Briner et al. (2005)
BR02-1	69.339	−143.578	20.9±0.6	^{10}Be	Northeastern Brooks Range	Briner et al. (2005)
BR02-3	69.349	−143.576	18.1±0.7	^{10}Be	Northeastern Brooks Range	Briner et al. (2005)
BR02-2	69.337	−143.575	15.6±0.5	^{10}Be	Northeastern Brooks Range	Briner et al. (2005)
Denali-15	63.454	−150.741	146.4±10.4	^{10}Be	Denali NP, Central Alaska Range	Dortch et al. (2010a)
Denali-16A	63.449	−150.676	7.4±2.2	^{10}Be	Denali NP, Central Alaska Range	Dortch et al. (2010a)
Denali-16B	63.449	−150.676	5.7±2.5	^{10}Be	Denali NP, Central Alaska Range	Dortch et al. (2010a)
Denali-17	63.444	−150.616	2.5±0.3	^{10}Be	Denali NP, Central Alaska Range	Dortch et al. (2010a)
Denali-18A	63.445	−150.585	11.5±1.0	^{10}Be	Denali NP, Central Alaska Range	Dortch et al. (2010a)
Denali-18B	63.445	−150.585	28.3±2.0	^{10}Be	Denali NP, Central Alaska Range	Dortch et al. (2010a)
Denali-20A	63.454	−150.546	141.7±9.8	^{10}Be	Denali NP, Central Alaska Range	Dortch et al. (2010a)
Denali-20B	63.454	−150.546	27.9±1.9	^{10}Be	Denali NP, Central Alaska Range	Dortch et al. (2010a)
Denali-21	63.457	−150.552	4.3±0.5	^{10}Be	Denali NP, Central Alaska Range	Dortch et al. (2010a)
DFCR-1	63.212	−144.828	11.7±1.2	^{10}Be	Central Alaska Range	Matmon et al. (2006)
DFCR-2	63.211	−144.828	10.5±1.1	^{10}Be	Central Alaska Range	Matmon et al. (2006)
DFCR-3	63.211	−144.828	11.7±1.2	^{10}Be	Central Alaska Range	Matmon et al. (2006)
DFCR-4	63.208	−144.833	10.3±1.0	^{10}Be	Central Alaska Range	Matmon et al. (2006)
DFCR-5	63.208	−144.834	11.3±1.2	^{10}Be	Central Alaska Range	Matmon et al. (2006)
DFCR-6	63.206	−144.835	11.0±1.2	^{10}Be	Central Alaska Range	Matmon et al. (2006)
DFCR-7	63.209	−144.829	10.2±1.1	^{10}Be	Central Alaska Range	Matmon et al. (2006)

Continued

TABLE 33.1 Cosmogenic Exposure Ages from Late and Early Wisconsinan Drift in Alaska (Included as a kmz File for Google Earth)—cont'd

Sample ID	Latitude (°N)	Longitude (°W)	Age (ka)	Isotope	Region	Reference
DFCR-8	63.209	−144.832	11.2±1.2	^{10}Be	Central Alaska Range	Matmon et al. (2006)
DFCR-9	63.209	−144.832	11.2±1.2	^{10}Be	Central Alaska Range	Matmon et al. (2006)
DFCRSD-1	63.212	−144.828	10.9±1.2	^{10}Be	Central Alaska Range	Matmon et al. (2006)
DFCRSD-2	63.209	−144.829	10.9±1.2	^{10}Be	Central Alaska Range	Matmon et al. (2006)
DFMF-1	63.152	−144.591	13.2±1.4	^{10}Be	Central Alaska Range	Matmon et al. (2006)
DFMF-2	63.152	−144.590	12.9±1.4	^{10}Be	Central Alaska Range	Matmon et al. (2006)
DFMF-3	63.150	−144.597	12.4±1.3	^{10}Be	Central Alaska Range	Matmon et al. (2006)
DFMF-4	63.150	−144.596	14.3±1.5	^{10}Be	Central Alaska Range	Matmon et al. (2006)
DFMF-5	63.151	−144.594	11.6±1.2	^{10}Be	Central Alaska Range	Matmon et al. (2006)
DFMFSD-1	63.152	−144.591	10.4±1.1	^{10}Be	Central Alaska Range	Matmon et al. (2006)
DFMFSD-2	63.151	−144.594	10.9±1.1	^{10}Be	Central Alaska Range	Matmon et al. (2006)
DFSC1	63.463	−148.644	14.9±1.6	^{10}Be	Central Alaska Range	Matmon et al. (2006)
DFSC2	63.463	−148.644	16.0±1.7	^{10}Be	Central Alaska Range	Matmon et al. (2006)
DFSC3	63.462	−148.644	15.1±1.6	^{10}Be	Central Alaska Range	Matmon et al. (2006)
DFSC7	63.461	−148.648	15.9±1.7	^{10}Be	Central Alaska Range	Matmon et al. (2006)
DFSC8	63.459	−148.652	14.9±1.6	^{10}Be	Central Alaska Range	Matmon et al. (2006)
DFWC-1	63.488	−148.084	2.1±0.2	^{10}Be	Central Alaska Range	Matmon et al. (2006)
DFWC-2	63.487	−148.085	2.4±0.4	^{10}Be	Central Alaska Range	Matmon et al. (2006)
DFNM-1	62.617	−143.032	10.1±1.1	^{10}Be	Central Alaska Range	Matmon et al. (2006)
DFNM-2	62.617	−143.032	11.5±1.3	^{10}Be	Central Alaska Range	Matmon et al. (2006)
DFNM-1R	62.617	−143.032	12.2±1.3	^{10}Be	Central Alaska Range	Matmon et al. (2006)
DFNM-RG	62.617	−143.032	9.3±1.0	^{10}Be	Central Alaska Range	Matmon et al. (2006)
DFDP-1	62.672	−143.156	10.2±1.1	^{10}Be	Central Alaska Range	Matmon et al. (2006)
DFDP-2	62.673	−143.155	8.0±0.8	^{10}Be	Central Alaska Range	Matmon et al. (2006)
DFDP-3	62.672	−143.153	14.1±1.5	^{10}Be	Central Alaska Range	Matmon et al. (2006)
DFDP-4	62.672	−143.156	12.3±1.3	^{10}Be	Central Alaska Range	Matmon et al. (2006)
DFTR-6	62.674	−142.807	13.0±1.4	^{10}Be	Central Alaska Range	Matmon et al. (2006)
DFTR-7	62.674	−142.807	14.3±1.5	^{10}Be	Central Alaska Range	Matmon et al. (2006)
DFTR-8	62.669	−142.809	16.1±1.7	^{10}Be	Central Alaska Range	Matmon et al. (2006)
DFTR-9	62.670	−142.814	9.2±1.0	^{10}Be	Central Alaska Range	Matmon et al. (2006)
DFTR-10	62.670	−142.814	10.8±1.2	^{10}Be	Central Alaska Range	Matmon et al. (2006)
DFTR-11	62.674	−142.812	13.2±1.4	^{10}Be	Central Alaska Range	Matmon et al. (2006)
DFTR-12	62.674	−142.812	9.9±1.1	^{10}Be	Central Alaska Range	Matmon et al. (2006)
DDDN-1	63.784	−145.742	17.4±1.9	^{10}Be	Delta River Valley, central Alaska Range	Matmon et al. (in press)
DDDN-1-SD	63.784	−145.742	17.2±1.9	^{10}Be	Delta River Valley, central Alaska Range	Matmon et al. (in press)
DDDN-2	63.778	−145.763	23.6±2.6	^{10}Be	Delta River Valley, central Alaska Range	Matmon et al. (in press)
DDDN-2-SD	63.778	−145.763	16.8±1.9	^{10}Be	Delta River Valley, central Alaska Range	Matmon et al. (in press)

Continued

TABLE 33.1 Cosmogenic Exposure Ages from Late and Early Wisconsinan Drift in Alaska (Included as a kmz File for Google Earth)—cont'd

Sample ID	Latitude (°N)	Longitude (°W)	Age (ka)	Isotope	Region	Reference
DDDN-3	63.774	−145.774	18.1±2.0	^{10}Be	Delta River Valley, central Alaska Range	Matmon et al. (in press)
DDDN-3-SD	63.774	−145.774	17.0±1.9	^{10}Be	Delta River Valley, central Alaska Range	Matmon et al. (in press)
DR1-1	63.777	−145.756	13.2±1.5	^{10}Be	Delta River Valley, central Alaska Range	Matmon et al. (in press)
DR1-2	63.777	−145.757	16.8±1.9	^{10}Be	Delta River Valley, central Alaska Range	Matmon et al. (in press)
DR1-3	63.779	−145.753	11.9±1.3	^{10}Be	Delta River Valley, central Alaska Range	Matmon et al. (in press)
DR1-4	63.780	−145.753	13.1±1.5	^{10}Be	Delta River Valley, central Alaska Range	Matmon et al. (in press)
DR1-5	63.779	−145.756	67.6±7.4	^{10}Be	Delta River Valley, central Alaska Range	Matmon et al. (in press)
Ala-126A	63.610	−148.777	15.0±2.3	^{10}Be	Nenana River Valley, central Alaska Range	Dortch et al. (2010b)
Ala-126B	63.610	−148.777	14.4±2.2	^{10}Be	Nenana River Valley, central Alaska Range	Dortch et al. (2010b)
Ala-127	63.606	−148.799	20.4±3.4	^{10}Be	Nenana River Valley, central Alaska Range	Dortch et al. (2010b)
Ala-128	63.605	−148.799	14.9±2.0	^{10}Be	Nenana River Valley, central Alaska Range	Dortch et al. (2010b)
Ala-130	63.603	−148.800	12.5±2.5	^{10}Be	Nenana River Valley, central Alaska Range	Dortch et al. (2010b)
Ala-132	63.599	−148.799	15.9±2.7	^{10}Be	Nenana River Valley, central Alaska Range	Dortch et al. (2010b)
Ala-133	63.598	−148.799	12.9±4.1	^{10}Be	Nenana River Valley, central Alaska Range	Dortch et al. (2010b)
Ala-134	63.597	−148.799	15.6±2.4	^{10}Be	Nenana River Valley, central Alaska Range	Dortch et al. (2010b)
Ala-119	63.675	−148.842	10.9±1.9	^{10}Be	Nenana River Valley, central Alaska Range	Dortch et al. (2010b)
Ala-120	63.675	−148.842	57.3±9.3	^{10}Be	Nenana River Valley, central Alaska Range	Dortch et al. (2010b)
Ala-121	63.676	−148.841	19.1±3.1	^{10}Be	Nenana River Valley, central Alaska Range	Dortch et al. (2010b)
Ala-122	63.701	−148.872	11.3±3.5	^{10}Be	Nenana River Valley, central Alaska Range	Dortch et al. (2010b)
Ala-123	63.699	−148.886	8.4±2.7	^{10}Be	Nenana River Valley, central Alaska Range	Dortch et al. (2010b)
Ala-15	63.736	−148.893	0.8±0.1	^{10}Be	Nenana River Valley, central Alaska Range	Dortch et al. (2010b)
Ala-16	63.736	−148.893	9.3±1.7	^{10}Be	Nenana River Valley, central Alaska Range	Dortch et al. (2010b)
Ala-18	63.735	−148.894	6.9±1.4	^{10}Be	Nenana River Valley, central Alaska Range	Dortch et al. (2010b)
Ala-19	63.735	−148.895	8.1±4.1	^{10}Be	Nenana River Valley, central Alaska Range	Dortch et al. (2010b)
Ala-20	63.735	−148.894	1.0±0.1	^{10}Be	Nenana River Valley, central Alaska Range	Dortch et al. (2010b)
Ala-135	63.897	−149.117	12.7±1.9	^{10}Be	Nenana River Valley, central Alaska Range	Dortch et al. (2010b)
Ala-136	63.897	−149.124	15.2±2.9	^{10}Be	Nenana River Valley, central Alaska Range	Dortch et al. (2010b)
Ala-137A	63.865	−149.132	73.6±7.2	^{10}Be	Nenana River Valley, central Alaska Range	Dortch et al. (2010b)
Ala-137B	63.865	−149.132	97.3±9.5	^{10}Be	Nenana River Valley, central Alaska Range	Dortch et al. (2010b)
Ala-156	63.868	−149.130	43.7±4.5	^{10}Be	Nenana River Valley, central Alaska Range	Dortch et al. (2010b)
Ala-151	63.404	−148.843	12.0±2.5	^{10}Be	Reindeer Hills, central Alaska Range	Dortch et al. (2010b)
Ala-152	63.403	−148.843	12.5±1.8	^{10}Be	Reindeer Hills, central Alaska Range	Dortch et al. (2010b)
Ala-153	63.401	−148.847	14.1±2.0	^{10}Be	Reindeer Hills, central Alaska Range	Dortch et al. (2010b)
Ala-154	63.401	−148.840	12.9±1.9	^{10}Be	Reindeer Hills, central Alaska Range	Dortch et al. (2010b)
Ala-155	63.400	−148.847	13.4±2.0	^{10}Be	Reindeer Hills, central Alaska Range	Dortch et al. (2010b)
Ala-158	63.402	−148.858	11.2±1.4	^{10}Be	Reindeer Hills, central Alaska Range	Dortch et al. (2010b)
Ala-159	63.401	−148.858	15.6±2.5	^{10}Be	Reindeer Hills, central Alaska Range	Dortch et al. (2010b)

Continued

TABLE 33.1 Cosmogenic Exposure Ages from Late and Early Wisconsinan Drift in Alaska (Included as a kmz File for Google Earth)—cont'd

Sample ID	Latitude (°N)	Longitude (°W)	Age (ka)	Isotope	Region	Reference
Ala-160	63.401	−148.858	16.7±4.1	^{10}Be	Reindeer Hills, central Alaska Range	Dortch et al. (2010b)
Ala-161	63.899	−148.866	15.6±3.3	^{10}Be	Reindeer Hills, central Alaska Range	Dortch et al. (2010b)
Ala-162	63.899	−148.866	12.5±2.4	^{10}Be	Reindeer Hills, central Alaska Range	Dortch et al. (2010b)
Ala-164	63.893	−148.850	15.6±3.5	^{10}Be	Reindeer Hills, central Alaska Range	Dortch et al. (2010b)
Ala-165	63.893	−148.860	13.2±2.1	^{10}Be	Reindeer Hills, central Alaska Range	Dortch et al. (2010b)
Ala-166	63.893	−148.860	16.0±3.8	^{10}Be	Reindeer Hills, central Alaska Range	Dortch et al. (2010b)
Ala-41	63.303	−148.211	1.2±0.2	^{10}Be	Monahan Flat, central Alaska Range	Dortch et al. (2010b)
Ala-42	63.303	−148.210	1.6±0.2	^{10}Be	Monahan Flat, central Alaska Range	Dortch et al. (2010b)
Ala-43	63.302	−148.205	12.0±1.5	^{10}Be	Monahan Flat, central Alaska Range	Dortch et al. (2010b)
Ala-145	63.306	−148.212	21.3±2.8	^{10}Be	Monahan Flat, central Alaska Range	Dortch et al. (2010b)
Ala-147	63.305	−148.210	15.0±2.1	^{10}Be	Monahan Flat, central Alaska Range	Dortch et al. (2010b)
Ala-148	63.305	−148.210	31.0±3.5	^{10}Be	Monahan Flat, central Alaska Range	Dortch et al. (2010b)
Ala-140	63.238	−147.778	12.5±1.9	^{10}Be	Monahan Flat, central Alaska Range	Dortch et al. (2010b)
Ala-141	63.238	−147.777	12.2±1.8	^{10}Be	Monahan Flat, central Alaska Range	Dortch et al. (2010b)
Ala-143	63.238	−147.774	11.8±1.9	^{10}Be	Monahan Flat, central Alaska Range	Dortch et al. (2010b)
Wattamuse T-1	59.349	−161.299	26.7±0.9	^{36}Cl	Ahklun Mountains, southwestern Alaska	Briner et al. (2001)
Wattamuse T-2	59.349	−161.298	20.4±0.9	^{36}Cl	Ahklun Mountains, southwestern Alaska	Briner et al. (2001)
Wattamuse T-3	59.350	−161.300	57.7±4.4	^{36}Cl	Ahklun Mountains, southwestern Alaska	Briner et al. (2001)
Wattamuse T-4	59.352	−161.303	30.3±2.5	^{36}Cl	Ahklun Mountains, southwestern Alaska	Briner et al. (2001)
Wattamuse T-5	59.352	−161.302	39.1±2.0	^{36}Cl	Ahklun Mountains, southwestern Alaska	Briner et al. (2001)
Wattamuse T-6	59.351	−161.301	30.3±2.3	^{36}Cl	Ahklun Mountains, southwestern Alaska	Briner et al. (2001)
Wattamuse R-1	59.346	−161.318	25.9±0.8	^{36}Cl	Ahklun Mountains, southwestern Alaska	Briner et al. (2001)
Wattamuse R-2	59.346	−161.318	26.4±1.6	^{36}Cl	Ahklun Mountains, southwestern Alaska	Briner et al. (2001)
Wattamuse R-3	59.363	−161.317	47.0±1.3	^{36}Cl	Ahklun Mountains, southwestern Alaska	Briner et al. (2001)
Kisogle-1	59.445	−161.226	17.9±1.4	^{36}Cl	Ahklun Mountains, southwestern Alaska	Briner et al. (2001)
Kisogle-2	59.444	−161.225	17.1±1.5	^{36}Cl	Ahklun Mountains, southwestern Alaska	Briner et al. (2001)
Kisogle-3	59.444	−161.232	3.1±0.6	^{36}Cl	Ahklun Mountains, southwestern Alaska	Briner et al. (2001)
Kisogle-4	59.449	−161.236	18.6±1.8	^{36}Cl	Ahklun Mountains, southwestern Alaska	Briner et al. (2001)
Kisogle-5	59.446	−161.241	16.6±1.3	^{36}Cl	Ahklun Mountains, southwestern Alaska	Briner et al. (2001)
Cloud Lake-1	59.423	−161.198	17.2±1.1	^{36}Cl	Ahklun Mountains, southwestern Alaska	Briner et al. (2001)
Cloud Lake-2	59.423	−161.198	24.0±1.3	^{36}Cl	Ahklun Mountains, southwestern Alaska	Briner et al. (2001)

Continued

TABLE 33.1 Cosmogenic Exposure Ages from Late and Early Wisconsinan Drift in Alaska (Included as a kmz File for Google Earth)—cont'd

Sample ID	Latitude (°N)	Longitude (°W)	Age (ka)	Isotope	Region	Reference
Cloud Lake-3	59.423	−161.198	18.1±1.4	^{36}Cl	Ahklun Mountains, southwestern Alaska	Briner et al. (2001)
Cloud Lake-4	59.421	−161.198	15.0±1.2	^{36}Cl	Ahklun Mountains, southwestern Alaska	Briner et al. (2001)
Cloud Lake-5	59.419	−161.194	16.0±1.2	^{36}Cl	Ahklun Mountains, southwestern Alaska	Briner et al. (2001)
Chilly Valley-1	59.606	−160.589	26.7±1.0	^{36}Cl	Ahklun Mountains, southwestern Alaska	Briner et al. (2001)
Chilly Valley-2	59.606	−160.586	19.5±1.0	^{36}Cl	Ahklun Mountains, southwestern Alaska	Briner et al. (2001)
Chilly Valley-3	59.608	−160.582	6.2±0.2	^{36}Cl	Ahklun Mountains, southwestern Alaska	Briner et al. (2001)
Chilly Valley-4	59.608	−160.579	16.8±0.6	^{36}Cl	Ahklun Mountains, southwestern Alaska	Briner et al. (2001)
Gusty Lakes-1	59.612	−160.599	19.3±0.4	^{36}Cl	Ahklun Mountains, southwestern Alaska	Briner et al. (2001)
Gusty Lakes-2	59.612	−160.599	20.2±0.6	^{36}Cl	Ahklun Mountains, southwestern Alaska	Briner et al. (2001)
Gusty Lakes-3	59.609	−160.598	17.6±0.5	^{36}Cl	Ahklun Mountains, southwestern Alaska	Briner et al. (2001)
Gusty Lakes-4	59.606	−160.600	20.7±0.4	^{36}Cl	Ahklun Mountains, southwestern Alaska	Briner et al. (2001)
Gusty Lakes-5	59.608	−160.605	9.4±0.6	^{36}Cl	Ahklun Mountains, southwestern Alaska	Briner et al. (2001)
MB1-99-1	59.870	−159.222	11.2±2.0	^{10}Be	Ahklun Mountains, southwestern Alaska	Briner et al. (2002)
MB1-99-2	59.871	−159.224	15.6±1.6	^{10}Be	Ahklun Mountains, southwestern Alaska	Briner et al. (2002)
MB1-99-3	59.873	−159.227	16.1±2.1	^{10}Be	Ahklun Mountains, southwestern Alaska	Briner et al. (2002)
MB1-99-2	59.871	−159.224	14.1±2.2	^{26}Al	Ahklun Mountains, southwestern Alaska	Briner et al. (2002)
MB1-99-3	59.873	−159.227	16.2±2.2	^{26}Al	Ahklun Mountains, southwestern Alaska	Briner et al. (2002)
MB1-00-4	59.868	−159.220	11.0±0.6	^{10}Be	Ahklun Mountains, southwestern Alaska	Briner et al. (2002)
MB4-00-1	59.869	−159.276	11.7±1.2	^{10}Be	Ahklun Mountains, southwestern Alaska	Briner et al. (2002)
MB4-00-2	59.869	−159.276	9.4±1.5	^{10}Be	Ahklun Mountains, southwestern Alaska	Briner et al. (2002)
MB4-00-3	59.870	−159.271	10.1±0.8	^{10}Be	Ahklun Mountains, southwestern Alaska	Briner et al. (2002)
MB6-00-1	59.868	−159.217	9.8±0.7	^{10}Be	Ahklun Mountains, southwestern Alaska	Briner et al. (2002)
MB6-00-2	59.868	−159.218	11.2±0.5	^{10}Be	Ahklun Mountains, southwestern Alaska	Briner et al. (2002)
Ages from early Wisconsinan drift						
WM00-06B	64.661	−143.528	61.4±3.9	^{10}Be	Yukon–Tanana Upland, Interior Alaska	Briner et al. (2005)
WM00-06D	64.662	−143.536	46.4±2.7	^{10}Be	Yukon–Tanana Upland, Interior Alaska	Briner et al. (2005)
WM00-06A	64.661	−143.528	39.2±1.4	^{10}Be	Yukon–Tanana Upland, Interior Alaska	Briner et al. (2005)
WM00-06C	64.662	−143.532	25.3±1.0	^{10}Be	Yukon–Tanana Upland, Interior Alaska	Briner et al. (2005)
WM00-07A	64.637	−143.541	66.9±2.2	^{10}Be	Yukon–Tanana Upland, Interior Alaska	Briner et al. (2005)
WM00-07D	64.633	−143.548	115.9±3.7	^{10}Be	Yukon–Tanana Upland, Interior Alaska	Briner et al. (2005)
WM00-07C	64.633	−143.548	91.4±2.5	^{10}Be	Yukon–Tanana Upland, Interior Alaska	Briner et al. (2005)
WM00-07B	64.634	−143.546	77.6±3.1	^{10}Be	Lime Hills, western Alaska Range	Briner et al. (2005)
SR1-00-2	61.474	−154.540	55.4±1.4	^{10}Be	Lime Hills, western Alaska Range	Briner et al. (2005)
SR1-00-4	61.470	−154.503	51.4±1.6	^{10}Be	Lime Hills, western Alaska Range	Briner et al. (2005)
SR1-00-1	61.474	−154.544	47.9±1.7	^{10}Be	Lime Hills, western Alaska Range	Briner et al. (2005)
SR1-00-3	61.468	−154.489	44.2±1.8	^{10}Be	Lime Hills, western Alaska Range	Briner et al. (2005)
KH 1-3	65.169	−154.422	51.6±1.7	^{10}Be	Kokrines Hills, Interior Alaska	Briner et al. (2005)

Continued

TABLE 33.1 Cosmogenic Exposure Ages from Late and Early Wisconsinan Drift in Alaska (Included as a kmz File for Google Earth)—cont'd

Sample ID	Latitude (°N)	Longitude (°W)	Age (ka)	Isotope	Region	Reference
KH 1-1	65.167	−154.419	23.9±1.1	^{10}Be	Kokrines Hills, Interior Alaska	Briner et al. (2005)
KH 1-2	65.167	−154.419	18.8±1.2	^{10}Be	Kokrines Hills, Interior Alaska	Briner et al. (2005)
DFDD1	63.774	−145.715	43.5±4.8	^{10}Be	Delta River Valley, central Alaska	Matmon et al. (in press)
DFDD2	63.774	−145.715	34.8±3.8	^{10}Be	Delta River Valley, central Alaska	Matmon et al. (in press)
DDOC-1	63.772	−145.715	32.5±3.6	^{10}Be	Delta River Valley, central Alaska	Matmon et al. (in press)
DDOC-2	63.772	−145.715	34.1±3.8	^{10}Be	Delta River Valley, central Alaska	Matmon et al. (in press)
DFDD3	63.802	−145.777	39.7±4.4	^{10}Be	Delta River Valley, central Alaska	Matmon et al. (in press)
DFDD4	63.802	−145.777	28.6±3.2	^{10}Be	Delta River Valley, central Alaska	Matmon et al. (in press)
DDDL-1	63.797	−145.788	18.2±2.0	^{10}Be	Delta River Valley, central Alaska	Matmon et al. (in press)
DDDL-2	63.862	−145.627	57.2±6.3	^{10}Be	Delta River Valley, central Alaska	Matmon et al. (in press)
DDDL-3	63.844	−145.605	55.5±6.1	^{10}Be	Delta River Valley, central Alaska	Matmon et al. (in press)
DDDL-4Q	63.777	−145.778	16.0±1.8	^{10}Be	Delta River Valley, central Alaska	Matmon et al. (in press)
DDDL-4 G	63.777	−145.778	16.0±1.8	^{10}Be	Delta River Valley, central Alaska	Matmon et al. (in press)
DDDL-5	63.804	−145.781	12.4±1.4	^{10}Be	Delta River Valley, central Alaska	Matmon et al. (in press)
DDDL-6	63.763	−145.692	46.3±5.1	^{10}Be	Delta River Valley, central Alaska	Matmon et al. (in press)
DDDL-8	63.766	−145.681	28.2±3.1	^{10}Be	Delta River Valley, central Alaska	Matmon et al. (in press)
DR2-1	63.854	−145.733	51.2±5.6	^{10}Be	Delta River Valley, central Alaska	Matmon et al. (in press)
DR2-2	63.854	−145.726	70.8±7.8	^{10}Be	Delta River Valley, central Alaska	Matmon et al. (in press)
DR2-3	63.856	−145.741	43.8±4.8	^{10}Be	Delta River Valley, central Alaska	Matmon et al. (in press)
DR2-4	63.857	−145.734	57.0±6.3	^{10}Be	Delta River Valley, central Alaska	Matmon et al. (in press)
DR2-5	63.854	−145.724	25.7±2.8	^{10}Be	Delta River Valley, central Alaska	Matmon et al. (in press)
Ala-11	63.845	−149.034	44.8±4.8	^{10}Be	Nenana River Valley, central Alaska Range	Dortch et al. (2010b)
Ala-12	63.848	−149.025	51.8±6.2	^{10}Be	Nenana River Valley, central Alaska Range	Dortch et al. (2010b)
Ala-13	63.848	−149.025	50.7±5.9	^{10}Be	Nenana River Valley, central Alaska Range	Dortch et al. (2010b)
Ala-23	63.834	−149.000	55.5±6.0	^{10}Be	Nenana River Valley, central Alaska Range	Dortch et al. (2010b)
Ala-24	63.833	−149.000	52.8±5.9	^{10}Be	Nenana River Valley, central Alaska Range	Dortch et al. (2010b)
Ala-25	63.834	−149.001	55.8±5.6	^{10}Be	Nenana River Valley, central Alaska Range	Dortch et al. (2010b)
Ala-107	63.837	−148.978	53.2±5.4	^{10}Be	Nenana River Valley, central Alaska Range	Dortch et al. (2010b)
Ala-108	63.834	−148.975	49.6±4.9	^{10}Be	Nenana River Valley, central Alaska Range	Dortch et al. (2010b)
Ala-157	63.855	−149.043	26.6±3.6	^{10}Be	Nenana River Valley, central Alaska Range	Dortch et al. (2010b)
Olympic Creek-1	59.379	−161.202	57.9±1.8	^{36}Cl	Ahklun Mountains, southwestern Alaska	Briner et al. (2001)
Olympic Creek-2	59.378	−161.202	62.3±1.7	^{36}Cl	Ahklun Mountains, southwestern Alaska	Briner et al. (2001)
Olympic Creek-3	59.380	−161.197	57.5±1.6	^{36}Cl	Ahklun Mountains, southwestern Alaska	Briner et al. (2001)
Olympic Creek-4	59.381	−161.202	63.9±2.3	^{36}Cl	Ahklun Mountains, southwestern Alaska	Briner et al. (2001)

Note: Ages are as reported in the original reference. Most 10Be and 26Al ages use the same or very similar production rate. All 36Cl ages use the same production rates. Refer to original reference for all details of age and scaling calculations, in addition to how authors handle erosion rate and snow shielding. The location of all samples can be obtained in the Google Eearth KML file. Latitude/longitude datum WGS84.

compiled by T.D. Hamilton. For APG Atlas v2, we overlaid all of Hamilton's 1:250,000 published surficial geologic maps from across the Brooks Range in Google Earth and used the mapped extent of drift to guide the location of the former ice limits. For the Cordilleran Ice Sheet, which contained the vast majority of Pleistocene glacier ice in Alaska, the original (v1) glacier limits were based mainly on 1:2,500,000-scale statewide map (Coulter et al., 1965), which resulted in significant inaccuracies when superposed onto the high-resolution imagery. The reconstructed glacier extents were adjusted to align with visually obvious geomorphic transitions, including the outer edge of prominent moraines and kettled drift, and they were drawn to fit major topographical constraints while adhering to basic principles of glacier flow. The original mapping was retained where the location of the former ice margin was ambiguous. Additional polygons were added, and some were modified according to the new information since v1, as described below.

Reconstructed margins of late Wisconsinan glaciers were classified according to their relative "certainty": a measure of confidence relating to *both* age determination and geographical position. Well-dated moraines were ascribed a high level of certainty, comparable to a solid line on traditional glacial geologic maps. Because they are based on detailed mapping, the certainty for small mountain glaciers is also generally high. Certainty classifications were omitted from all alpine glaciers, however, because the lines obscure the small polygons when zoomed out. Intermediate certainty was encoded for limits without well-resolved ice-marginal features, or those with little or no direct geochronological control (comparable to dashed lines for "uncertain" contacts on traditional maps). A low level of certainty was attributed to areas lacking detailed air-photo interpretation or substantial, field-based glacial geologic study. This includes the southern margin of the Cordilleran Ice Sheet where the reconstructed ice margin is based on little evidence where ice terminated on the presently submerged continental shelf. The location of the offshore margin is designated as the lowest certainty level.

The ages of the moraines used to delimit the late Wisconsinan glaciers vary by thousands of years across the state. Although still sparse, the chronologies do show some pattern in timing of the maximum extent of mountain glaciers during MIS 2 (Briner and Kaufman, 2008). Many of these chronologies are based on cosmogenic exposure ages of moraine boulders, which likely date the timing of moraine stabilization upon glacier retreat (Briner et al., 2005). In northern Alaska, glaciers retreated from their late Wisconsinan terminal moraines by 25 ka, compared with 22–20 ka in central and southern portions of the state. Following the maximum phase of the late Wisconsinan, glaciers across the state constructed end moraines during subsequent periods of stabilization or re-advance. Although most glaciated valleys across Alaska contain multiple moraines, few have been dated, hampering statewide comparisons; however, glaciers in many valleys built sizeable moraines near terminal moraines shortly following their initial retreat. In the Ahklun Mountains, for example, prominent end moraines were deposited about 20 ka, and in the Alaska Range, end moraines post-dating the terminal moraine formed around 19 ka. In both cases, glaciers stabilized near their former limits for one or two thousand years following the maximum phase.

33.4. REGIONAL UPDATES FOR APG ATLAS V2

The following describes only the specific revisions included in v2 of the APG Atlas. The reader is referred to Kaufman and Manley (2004) for references to source maps used to delimit former glacier extents in v1. The regions are arranged roughly from north to south.

33.4.1. Brooks Range

33.4.1.1. Early Wisconsinan

The extensive suite of moraines in the Itkillik River area, central Brooks Range (Fig. 33.2), serves as the reference locality for late Pleistocene glaciations of the Brooks Range (Hamilton and Porter, 1975; Hamilton, 1982, 1986). The late Pleistocene moraines were subdivided into the Itkillik I (=penultimate glaciation) and Itkillik II (=late Wisconsinan glaciation) advances. Glaciers expanded up to 40 km north of the northern range front during the Itkillik I glaciation, and up to 25 km north of the range front during the Itkillik II glaciation. Recent detailed mapping in the Itkillik River area resulted in further subdivision of the glacial deposits (Hamilton, 2003a). The Itkillik I glaciation was subdivided into two phases based on differences in postglacial modification of moraines; both are older than non-finite ^{14}C ages of 53 ka, and are believed to be younger than the last interglacial maximum (MIS 5e; Hamilton, 1994). In the Noatak River basin of the western Brooks Range, two separate advances are younger than the Old Crow tephra (131 ± 11 ka; Péwé et al., 2009) and older than 36–34 ka (Hamilton, 2001).

In the Brooks Range, the early Wisconsinan glacier limit in v2 of the APG Atlas coincides with the outer extent of drift of the Itkillik I glaciation as mapped by Hamilton (1978, 1979a,b, 1980, 1981, 1984a,b, 2002a,b, 2003b) in his series of 1:250,000-scale surficial geologic maps that extend across much of the Brooks Range, from 162°W longitude in the west to 147°W longitude in the east. In the area east of Hamilton's coverage, we know of only two other studies that have identified the extent of glaciers during the early Wisconsinan glaciation. Namely, moraines have

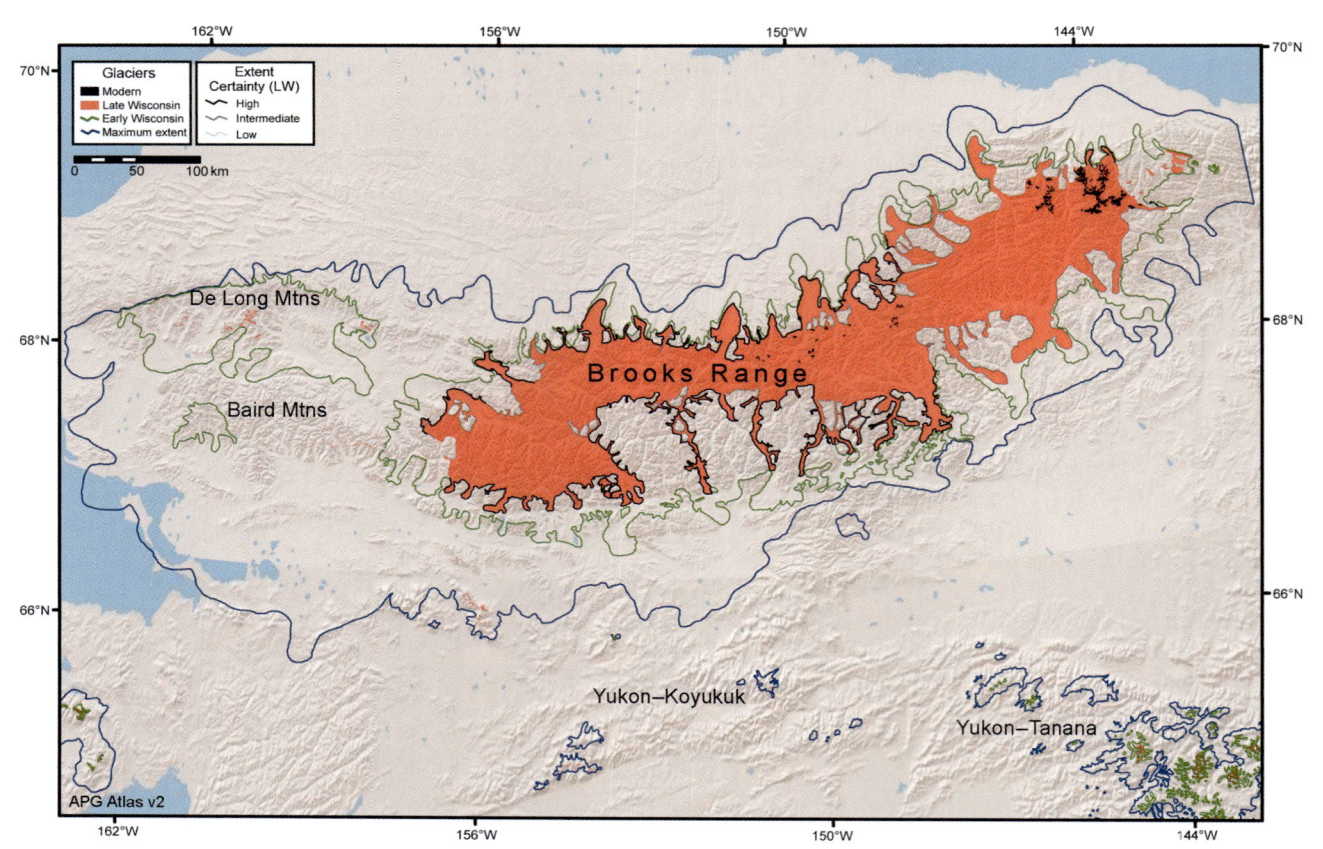

FIGURE 33.2 Pleistocene maximum, early Wisconsinan, late Wisconsinan and modern glacier extents, Brooks Range and Yukon–Koyukuk region.

been correlated with the Itkillik I glaciation both in the Jago River area (Balascio et al., 2005a) and in the Kavik River area (Carson, 2009). The glacier limit outside these areas largely follows the Qg_3 limit of Coulter et al. (1965).

33.4.1.2. Late Wisconsinan

The late Wisconsinan limit in v2 coincides with the outer extent of drift of the Itkillik II (or Walker Lake) glaciation as mapped by T.D. Hamilton in his series of 1:250,000-scale surficial geologic maps referenced above. Westward of the main body of coalescent Brooks Range ice (west of 157°W longitude), summit elevations are lower and late Wisconsinan glaciers were confined to the highest mountain valleys. We retain the reconstruction of late Wisconsinan glaciers in the western Brooks Range (DeLong and Baird Mountains) from the APG Atlas v1, which was based on photo-interpretive mapping at 1:63,360 scale by DSK (Darrell S. Kaufman), and which was used to reconstruct equilibrium-line elevations across the Brooks Range (Balascio et al., 2005b). Our late Wisconsinan limits in the western Brooks Range agree with more recent 1:250,000-scale mapping by Hamilton (2003b). The only significant difference is the moraines near the mouth of Avan River valley, which Hamilton (2003b) ascribed to "an unusually large valley glacier", but which we interpret as a moraine from an older, more extensive glacier advance. Like Hamilton (2003b), we find no evidence for extensive alpine glaciation in the DeLong Mountains during the late Wisconsinan, as was recently suggested by Hill et al. (2007) based on evidence for seismic stratigraphy and sediment coring of meltwater channels that extend offshore of northwest Alaska on the shelf of the Chukchi Sea. In the area east of Hamilton's coverage, we relied on mapping by Balascio et al. (2005a) and Carson (2009).

33.4.2. Seward Peninsula

Moraines of the Salmon Lake (=early Wisconsinan) glaciation have been mapped at 1:63,360 scale across the Kigluaik Mountains (Kaufman et al., 1989), the largest of the four glaciated ranges on Seward Peninsula (Fig. 33.3). In the three other ranges, the limit of Salmon Lake drift mapped by Kaufman (1986) was used to guide the reconstructed ice extent.

33.4.3. Yukon–Koyukuk Region

Little is known about the extent of ice during the early Wisconsinan in the several small mountain ranges located south of the Brooks Range and north of the Yukon River. Relating

Chapter | 33 Alaska Palaeo-Glacier Atlas

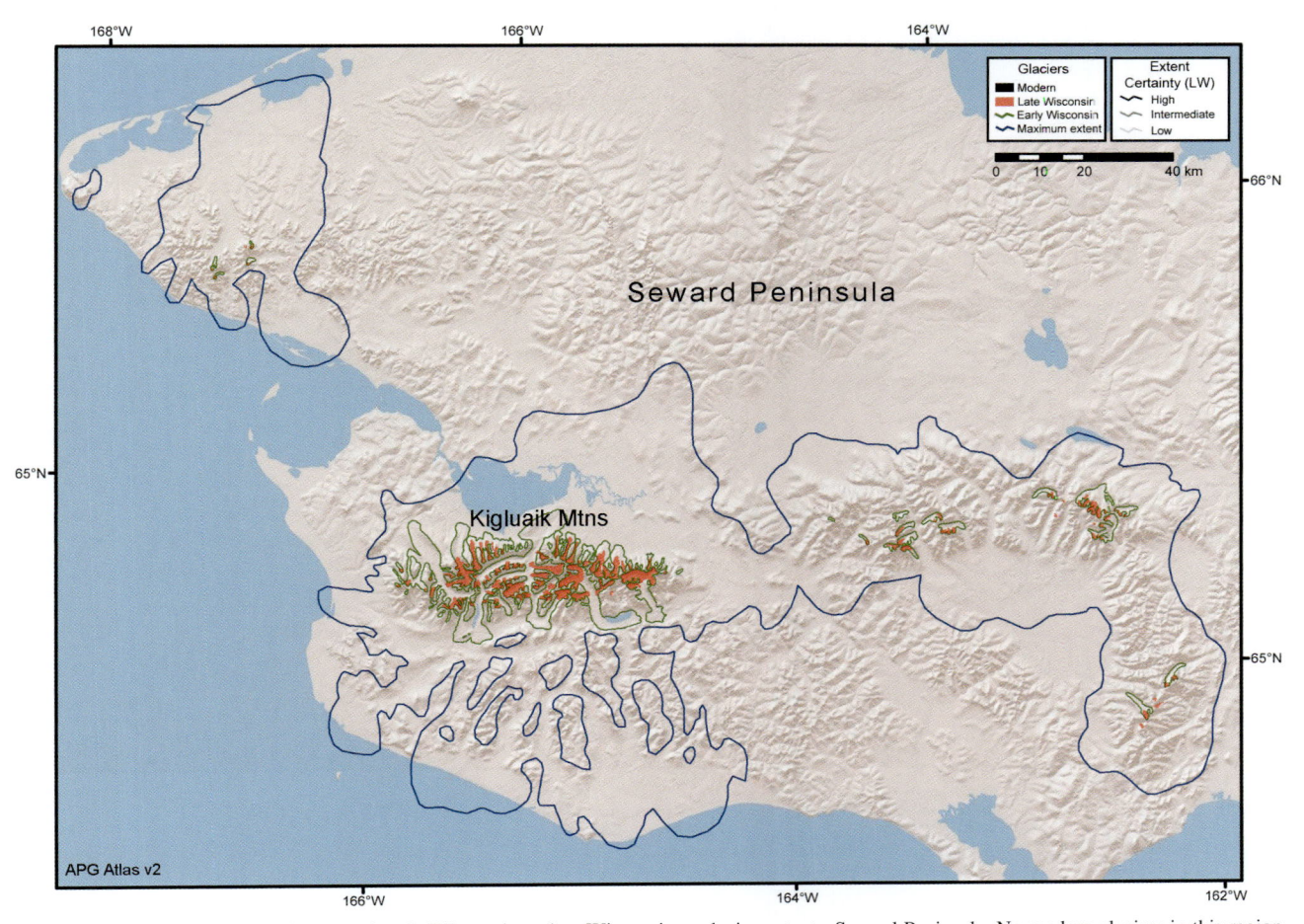

FIGURE 33.3 Pleistocene maximum and early Wisconsinan, late Wisconsinan glacier extents, Seward Peninsula. No modern glaciers in this region.

the mapped extents shown in Coulter et al. (1965) to specific topographical features in high-resolution imagery is difficult and is not included in APG Atlas v2. The early Wisconsinan extent for the Indian Mountain massif is based on Reger (1979). Cosmogenic surface exposure ages on moraines of the penultimate glaciation in the Kokrines Hills indicate an early Wisconsinan age (Briner et al., 2005; Table 33.1 and Fig. 33.2). Mapping of these moraines has not been completed and are not shown in the APG Atlas.

33.4.4. Yukon–Tanana Upland

This region encompasses rolling hills punctuated by several rugged peaks scattered between the Yukon and Tanana rivers in east-central Alaska. The limit of early Wisconsinan glaciers coincides with moraines of the Eagle glaciation as mapped around Mt. Prindle by Weber and Hamilton (1984), modified slightly based on aerial-photo interpretation and field survey by W. F. M. For most of the rest of the Yukon–Tanana upland (Fig. 33.4), the early Wisconsinan reconstruction was taken from Weber (1986) and Hamilton (1994). Cosmogenic surface exposure ages on moraines of the Eagle glaciation indicate an early Wisconsinan age (Briner et al., 2005; Table 33.1), consistent with earlier interpretations (Weber, 1986).

33.4.5. Kuskokwim Mountains

The Kuskowim Mountains region encompasses at least 13 isolated highlands, all located within about 50 km of the Kuskokwim River, that generated alpine glaciers during the Pleistocene. The early Wisconsinan reconstruction in v2 was taken from Waythomas (1990) for the Chuilnuk and Kiokluk Mountains, where it is represented by the Chuilnuk glaciation. We retain this designation despite cosmogenic isotope ages that indicate a younger age of the Chuilnuk glaciations (Briner et al., 2005). In the Russian and Beaver Mountains, it is represented by the Bifurcation Creek glaciation of Kline and Bundtzen (1986), and in the Taylor Mountains, we assumed it coincides with unit Qgu of Coulter et al. (1965).

Two of the small ranges, the Horn and Sunshine Mountains, contained small valley glaciers during the late Pleistocene that were not shown in the APG Atlas v1. For this version, we have added the glacier limits inferred from drift

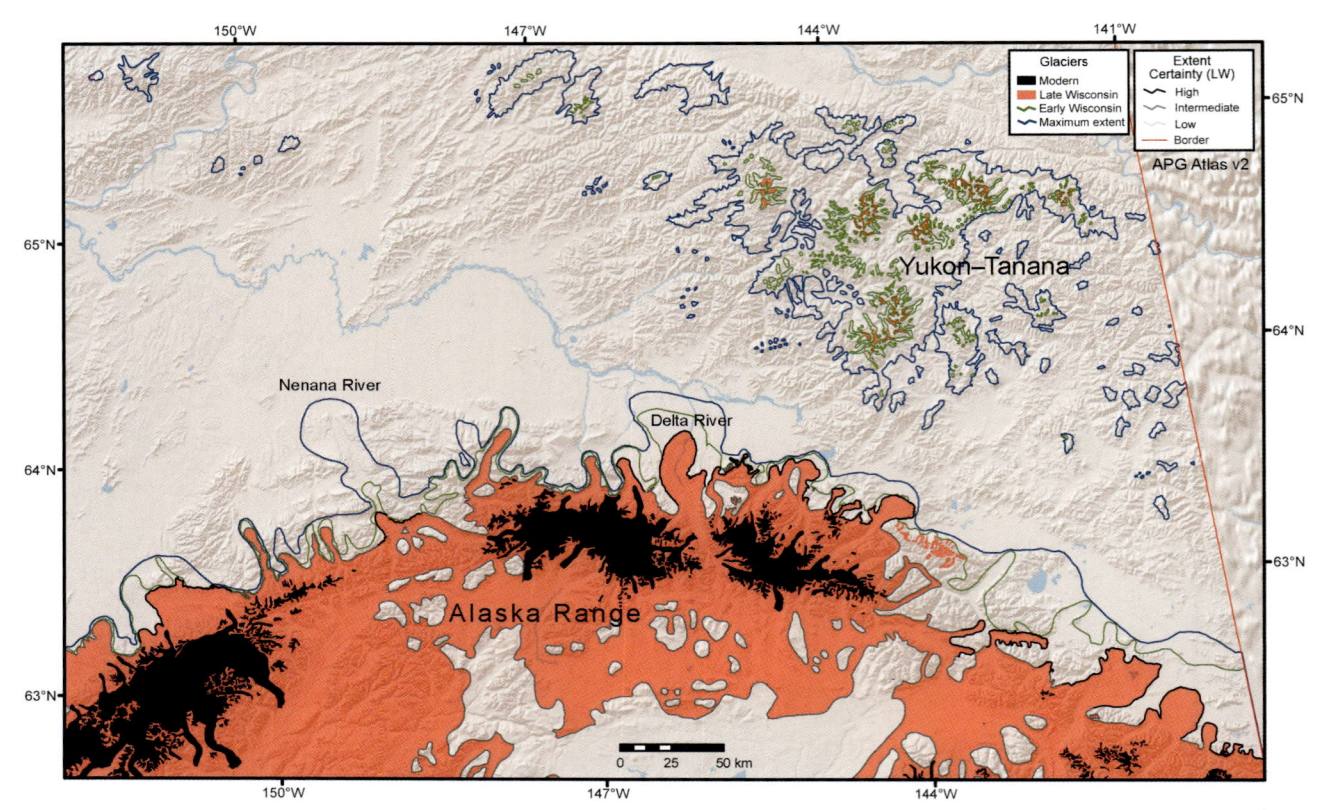

FIGURE 33.4 Pleistocene maximum, early Wisconsinan, late Wisconsinan and modern glacier extents, Cordilleran Ice Sheet.

mapped by Bundtzen et al. (1999) in the Horn Mountains, where till of the Bifurcation Creek glaciation marked the extent of early Wisconsinan ice, and till of the Tolstoi glaciation was used to delimit the extent of late Wisconsinan valley glaciers. Version 2 now includes the small valley glaciers that emanated from several of the highest cirques in the Sunshine Mountains during the late Wisconsinan (D. S. K. unpublished aerial-photo-interpretative mapping). Similarly, two small, isolated plutonic stocks, the Taylor Mountains and the Shotgun Hills, located northeast of the Ahklun Mountains, also supported late Wisconsinan glaciers (D. S. K. unpublished field and aerial-photo-interpretative mapping) and are now shown in v2. Finally, v2 includes new unpublished mapping of late Wisconsinan limits by TKB (T.K. Bundtzen) for the Blackburn Hills adjacent to the lower course of the Yukon River.

33.4.6. Ahklun Mountains

During the late Pleistocene, the Ahklun Mountains hosted an ice cap over its east-central spine that expanded radially, extending farther to the south and west than to the north and east; isolated alpine glaciers occupied the highest valleys beyond the ice cap margin (Fig. 33.5). In most valleys, late Pleistocene drift is composed of several moraine belts formed by outlet glaciers of the central ice cap (Manley et al., 2001). Moraines of the Arolik Lake (=early Wisconsinan) glaciation are dated in several locations across the range. In the southern Ahklun Mountains, Kaufman et al. (2001) reported a thermoluminescence (TL) age of 70 ± 10 ka on lava-baked sediment that underlies penultimate drift and provides a maximum-limiting age on the glaciation. Manley et al. (2001) report a minimum ^{14}C age of 39.9 ^{14}C ka BP on organic material that overlies Arolik Lake drift. In the western Ahklun Mountains, Briner et al. (2001) used four ^{36}Cl exposure ages on erratic boulders deposited in the Goodnews River valley to constrain the age of the Arolik Lake glaciation to between 56 and 53.8 ± 2.6 ka. Thus, the ^{36}Cl ages on boulders deposited during the Arolik Lake glaciation fit well between the TL maximum age of 70 ± 10 ka and the ^{14}C minimum age of 40 ka ^{14}C ka BP. These ages are, in general, agreement with amino acid and luminescence ages from glacial-estuarine sediments of the penultimate glaciation in the Bristol Bay lowland, which ranged between 90 and 55 ka (Kaufman et al., 1996). Collectively, these ages indicate a major glaciation in the Ahklun Mountains roughly coincident with MIS 4; around Nushagak and Kvichak bays, however, it appears that the advance culminated late during MIS 5 (Kaufman and Thompson, 1998).

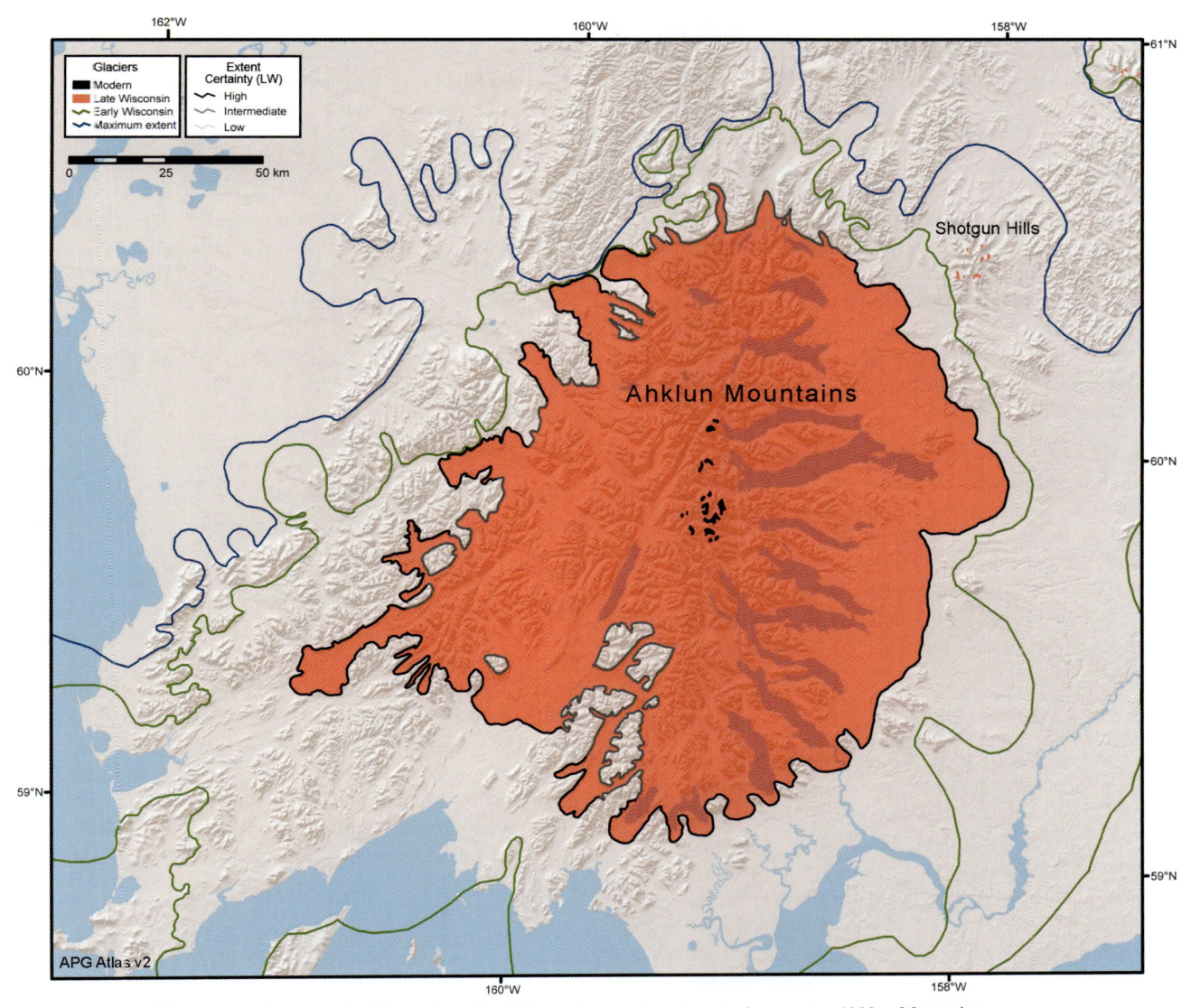

FIGURE 33.5 Pleistocene maximum, early Wisconsinan, late Wisconsinan and modern glacier extents, Ahklun Mountains.

33.4.7. Alaska Range (Fig. 33.4)

33.4.7.1. Early Wisconsinan

The APG Atlas v2 now includes the limits of glaciers during the penultimate glaciation along the north side of the Alaska Range where it largely follows the northern extent of Coulter et al.'s (1965) unit Qg$_3$. The ages of moraines have been determined in three places:

1. A moraine sequence deposited along the Delta River Valley beyond the northern Alaska Range front constitutes the reference locality of the Donnelly (=late Wisconsinan) and Delta (=early Wisconsinan) glaciations (Péwé, 1953). An outwash terrace that grades to the Delta moraine is overlain by the Old Crow tephra (Bégét and Keskinen, 2003), suggesting that it is older than 131 ± 11 ka (Péwé et al., 2009). Matmon et al. (2010) analysed ^{10}Be ages of 15 boulders and sediment samples from the surface of the inner portion of the Delta moraine, near Donnelly Dome. The ^{10}Be ages range from ~70 to ~12 ka, with the majority between ~70 and ~40 ka. The new ^{10}Be ages suggest that this portion of the Delta moraine stabilized during MIS 4/3. Recent mapping indicates that the Delta moraine is a composite feature comprising drift of two ages, with drift just down-valley of the Donnelly terminal moraine significantly younger than the drift farther down valley near Delta Junction, which was probably deposited during MIS 6 (Reger et al., 2008).

2. A more detailed moraine sequence in the Nenana River Valley, north-central Alaska Range (Thorson, 1986), was the focus of a recent exposure-dating study. Dortch

et al. (2010b) obtained nine ^{10}Be ages on boulders from landforms created during the Healy glaciation, which is thought to be the equivalent to the moraine deposited in the Delta River Valley during the Delta glaciation (Hamilton, 1994). The landforms of the Healy glaciation, excluding one young outlier, range between 56 and 51.6 ± 3.8 ka (Dortch et al., 2010b).

3. In the Swift River valley of the western Alaska Range, Briner et al. (2005) mapped a sequence of moraines and correlated them with the Farewell I (=early Wisconsinan) and Farewell II (=late Wisconsinan) moraines in the nearby Farewell region (Kline and Bundtzen, 1986). The Farewell I equivalent moraine, dated by four ^{10}Be ages, stabilized between 58 and 52.5 ± 5.6 ka (Briner et al., 2005).

In sum, ^{10}Be ages from three sites along the northern Alaska Range indicate that moraines of the penultimate glaciation stabilized between 58 and 52 ka, and tephrostratigraphy constrains the outer part of the Delta moraine to >140 ka. A pulse of loess deposition in the Tanana River valley (Begét, 2001) that appears to coincide with MIS 4 supports the notion of a regionally significant early Wisconsinan glacier advance in the north Alaska Range.

33.4.7.2. Late Wisconsinan

New surficial geologic studies associated with sighting the corridor for a gas pipeline has provided new information on the extent of late Wisconsinan glaciation along the northeastern Alaska Range (Reger et al., 2008). In addition, three studies add new chronologic control to late Wisconsinan drift from the north side of the Alaska Range. (1) Dortch et al. (2010b) provide dozens of ^{10}Be ages on moraines in the Nenana River Valley. The ^{10}Be ages are scattered, but indicate that the late Wisconsinan moraines stabilized between ~20 and 10 ka. (2) Matmon et al. (2010) provide 11 new ^{10}Be ages from the Donnelly moraine near Donnelly Dome, which they interpret to indicate that the maximum phase of the Donnelly glaciation culminated by ~17 ka. (3) Young et al. (2009) found a similar age for the culmination of the Donnelly glaciation in the Fish Creek valley, ~70 km east of the Delta River. The mapping of the late Wisconsinan limit in this area is updated according to Young et al. (2009). Finally, v2 of the APG Atlas includes minor modifications of the late Wisconsinan limit based on unpublished mapping by AW (A. Werner) in the Wonder Lake area.

33.4.8. Aleutian Range

During Pleistocene glaciations, the Cordilleran Ice Sheet extended westward from the Alaska Range to the Aleutian Range. The elongate arc of glacier ice encompassed the volcanic peaks of the Alaska Peninsula and the adjacent continental shelf, and extended west some distance along the Aleutian Island archipelago (Mann and Hamilton, 1995). Evidence for the age and extent of Pleistocene ice is sparse along the entire 3000-km-long arc that extends from the eastern Alaska Peninsula west to Kamchatka.

In the east, costal bluffs rimming Kvichak Bay expose superposed drift units of at least two glaciations (Kaufman et al., 1996; Kaufman and Thompson, 1998). The uppermost drift contains clast lithologies from the Alaska Peninsula, and amino acid and luminescence geochronology indicate that the drift is younger than the last interglaciation, possibly 75–90 ka, or perhaps early MIS 4. In this area, the extent of early Wisconsinan (*sensu lato*) shown in v2 follows the mapping of Kaufman et al. (1996); to the west, it merges with northern limit of unit Qg_3 shown by Coulter et al. (1965). The limit of early Wisconsinan ice is unknown westward of Coulter et al.'s mapping, which extends to 163.7°W longitude.

The extent of late Wisconsinan drift has been clarified in a few places along the Alaska Peninsula. In the Pavlof Bay area, Dawson tephra overlies what had previously been mapped as late Wisconsinan drift. The tephra has been dated to 27 ka (Mangan et al., 2003), providing a minimum age for the underlying drift. Therefore, the late Wisconsinan ice limit in v2 was relocated along the outer edge of the next younger moraine in this area, supplemented by mapping by Wilson and Weber (2001). Version 2 also includes minor modifications of the late Wisconsinan limit based on unpublished mapping by JH (J. Harvey) in the Lake Clark area.

33.4.9. Cook Inlet Area

Version 2 includes recent unpublished mapping by RDR (R.D. Reger) in the Cook Inlet area. A volcanic plateau between 600 and 900 m elevation and 22 km southeast of the summit of Mt. Spurr remained unglaciated through the late Pleistocene. The margins of this volcanic plateau are rimmed by the highest granitic erratics of the penultimate glaciation, perched well above the high-relief lateral moraines of the last major (Naptowne) glaciation. In the southeastern Cook Inlet, the southern Kenai lowlands remained unglaciated during the late Wisconsinan (Reger et al., 2007). Fluted features in this area west of the Caribou Hills indicate that ice flow was oriented parallel to the trough axis. High-level massive sand deposits overlain by 1-m-thick diamictons on the northern end of the Caribou Hills are interpreted as early Wisconsinan ice-marginal lake deposits (RDR, unpublished). Apparently, the Cook Inlet trough was filled with ice to the summit of the Caribou Hills during the penultimate glaciation.

33.4.10. Southern Margin of the Cordilleran Ice Sheet

The southern rim of the ice sheet was supplied with ice by the coastal mountains extending from Kodiak Island on the west, to the Kenai, Chugach and St. Elias Mountains, which border the northern Gulf of Alaska, and to the Alexander Archipelago of the southeastern panhandle. Nowhere else in Alaska is the extent of Pleistocene glacier ice more poorly constrained than along the southern margin of the Cordilleran Ice Sheet. We know of no evidence for the extent of ice during the early Wisconsinan anyplace along the southern margin of the Cordilleran Ice Sheet. Even during the late Wisconsinan, the position of the southern margin of the ice sheet is speculative, as discussed by Kaufman and Manley (2004). We retain the restricted-ice model shown by Molnia (1986) around the Gulf of Alaska. The only modification made in v2 was to extend the late Wisconsinan limit to beyond the present-day coast around the Copper River delta because we expect this major trough to have conveyed a major outlet of the Cordilleran Ice Sheet rather than remained free of glacier ice. Previously unpublished mapping in the southern Alexander Archipelago, which was included in v1, is now published by Carrara et al. (2007). Their work highlights evidence of unglaciated refugia along the west coast of several islands. A similar palaeogeography might extend northward along the panhandle, but we are unaware of other studies that map Pleistocene ice extents along the continental shelf in the Gulf of Alaska region.

33.5. DISCUSSION

The updated version of the APG Altas provides the first statewide look at the extent of ice during the penultimate glaciation. This version also includes the first statewide georeferenced compilation of all cosmogenic exposure ages published to date from late and early Wisconsinan drift. Further, several features in Google Earth improve visualization, such as viewing areas of interest obliquely in 3D, and panning around the compass from different vantage points. Version 2 is another step towards a more accurate geospatial database of glacier extents, including the addition of reconstructed mountain glaciers in five small ranges in southwestern Alaska, which are important because they barely intersected the snowline elevation during the late Wisconsinan, as well as some more major refinements of the larger ice mass, which have now been located more accurately because their moraines are visible on high-resolution satellite imagery. APG v2 provides a comprehensive, up-to-date compilation of limits and ages in a format that is publicly available and user-friendly.

The geospatial data in v2 contributes to both education and research. The website for v1 has received more than 20,000 visits to date, with more than 3500 dataset downloads. Beyond visualization, mapping and outreach, the new data will be valuable for modelling and quantitative analysis by glacial geologists, with application more broadly to geological sciences, geomorphology, palaeoclimatology, ecology and land resource management.

FULL CITATION

Kaufman, D.S., Young, N.E., Briner, J.P., Manley, W.F., 2011. Alaska Palaeo-Glacier Atlas (Version 2). In: Ehlers, J. & Gibbard, P.L. (eds), Quaternary Glaciations Extent and Chronology, Part IV: A Closer Look. *Developments in Quaternary Science 15*, Amsterdam, Elsevier, 427–445.

REFERENCES

Balascio, N.L., Kaufman, D.S., Briner, J.P., Manley, W.F., 2005a. Late Pleistocene glacial geology of the Okpilak-Kongkut Rivers region, northeastern Brooks Range, Alaska. Arctic Antarctic Alpine Res. 37, 416–424.

Balascio, N.L., Kaufman, D.S., Manley, W.F., 2005b. Equilibrium-line altitudes during the last glacial maximum across the Brooks Range, Alaska. J. Quatern. Sci. 20, 821–838 (Additional source maps used for v2 and not included in v1).

Barclay, D.J., Wiles, G.C., Calkin, P.E., 2009. Holocene glacier fluctuations in Alaska. Quatern. Sci. Rev. 28, 2034–2048.

Begét, J.E., 2001. Continuous Late Quaternary proxy climate records from loess in Beringia. Quatern. Sci. Rev. 20, 499–507.

Begét, J.E., Keskinen, M.J., 2003. Trace-element geochemistry of individual glass shards of the Old Crow tephra and the age of the Delta glaciation, central Alaska. Quatern. Res. 60, 63–69.

Briner, J.P., Kaufman, D.S., 2008. Late Pleistocene mountain glaciation in Alaska: key chronologies. J. Quatern. Sci. 23, 659–670.

Briner, J.P., Swanson, T.W., Caffee, M., 2001. Cosmogenic ^{36}Cl glacial chronology of the southwestern Ahklun Mountains, Alaska: extensive early Wisconsin ice. Quatern. Res. 56, 148–154.

Briner, J.P., Kaufman, D.S., Werner, A., Caffee, M., Levy, L., Manley, W.F., et al., 2002. Glacier readvance during the late glacial (Younger Dryas?) in the Ahklun Mountains, southwestern Alaska. Geology 30, 679–682.

Briner, J.P., Kaufman, D.S., Manley, W.F., Finkel, R.C., Caffee, M.W., 2005. Cosmogenic exposure dating of late Pleistocene moraine stabilization in Alaska. Geol. Soc. Am. Bull. 117, 1108–1120 (Additional source maps used for v2 and not included in v1).

Bundtzen, T.K., Harris, E.E., Miller, M.L., Layer, P.W., Laird, G.M., 1999. Geology of the Sleetmute C-7, C-8, D-7, and D-8 Quadrangles, Horn Mountains, Southwestern Alaska. Alaska Division of Geological and Geophysical Surveys Report of Investigations, pp. 98–12 (Additional source maps used for v2 and not included in v1).

Carrara, P.E., Ager, T.A., Baichtal, J.F., 2007. Possible refugia in the Alexander Archipelago of southeastern Alaska during the late Wisconsin glaciation. Can. J. Earth Sci. 44, 229–244 (Additional source maps used for v2 and not included in v1).

Carson, E.C., 2009. Surficial-geologic map of the Kavik River area, west-central Mount Michelson Quadrangle, northeastern Brooks Range. Alaska Division of Geological and Geophysical Surveys Report of Investigations, 2009-3 (Additional source maps used for v2 and not included in v1).

Coulter, H.W., Hopkins, D.M., Karlstrom, T.N.V., Péwé, T.L., Wahrhaftig, C., Williams, J.R., 1965. Map showing extent of glaciations in Alaska. U.S. Geological Survey Map, I-415.

Dortch, J.M., Owen, L.A., Caffee, M.W., Brease, P., 2010a. Late Quaternary glaciation and equilibrium line altitude variations of the McKinley River region, central Alaska Range. Boreas 39, 233–246.

Dortch, J., Owen, L., Caffee, M., Li, D., Lowell, T., 2010b. Beryllium-10 surface exposure dating of glacial successions in the central Alaska Range. J. Quatern. Sci. 25, 1259–1269.

Hamilton, T.D., 1978. Surficial geologic map of the Chandalar quadrangle, Alaska. U.S. Geological Survey Miscellaneous Field Investigation Map, MF-878A.

Hamilton, T.D., 1979a. Surficial geologic map of the Chandler Lake quadrangle, Alaska. U.S. Geological Survey Miscellaneous Field Investigation Map, MF-1121.

Hamilton, T.D., 1979b. Surficial geologic map of the Wiseman quadrange, Alaska. U.S. Geological Survey Miscellaneous Field Investigation, Map MF-1122.

Hamilton, T.D., 1980. Surficial geologic map of the Killik River quadrangle, Alaska. U.S. Geological Survey Miscellaneous Field Investigation, Map MF-1234.

Hamilton, T.D., 1981. Surficial geologic map of the Survey Pass quadrangle, Alaska. U.S. Geological Survey Miscellaneous Field Investigation, Map MF-1320.

Hamilton, T.D., 1982. A late Pleistocene glacial chronology for the southern Brooks Range — stratigraphic record and regional significance. Geological Society of America Bulletin 93, 700–716.

Hamilton, T.D., 1984a. Surficial geologic map of the Howard Pass quadrangle, Alaska. U.S. Geological Survey Miscellaneous Field Investigation, Map MF-1677.

Hamilton, T.D., 1984b. Surficial geologic map of the Amber River quadrangle, Alaska. U.S. Geological Survey Miscellaneous Field Investigation Map, MF-1678.

Hamilton, T.D., 1986. Late Cenozoic glaciation of the central Brook Range. In: Hamilton, T.D., Reed, K.M., Thorson, R.M. (Eds.), Glaciation in Alaska—The Geologic Record. Alaska Geological Society, Anchorage, pp. 9–50.

Hamilton, T.D., 1994. Late Cenozoic glaciation of Alaska. In: Plafker, G., Berg, H.C. (Eds.), The Geology of Alaska, The Geology of North America, vol. G-1. Geological Society of America, Boulder, Colorado, pp. 813–844.

Hamilton, T.D., 2001. Quaternary glacial, lacustrine, and fluvial interactions in the western Noatak basin, northwest Alaska. Quatern. Sci. Rev. 20, 371–391.

Hamilton, T.D., 2002a. Surficial geologic map of the Hughes quadrangle, Alaska. U.S. Geological Survey Miscellaneous Field Investigation Map, MF-2408.

Hamilton, T.D., 2002b. Surficial geologic maps of the Bettles quadrangle, Alaska. U.S. Geological Survey Miscellaneous Field Investigation Map, MF-2409.

Hamilton, T.D., 2003a. Surficial geology of the Dalton Highway (Itkillik-Sagavanirktok Rivers) area, southern Arctic Foothills, Alaska. Alaska Department Natural Resources/Division Geological & Geophysical Surveys Professional Report, 121.

Hamilton, T.D., 2003b. Surficial geologic map of parts of the Misheguk Mountain and Baird Mountains Quadrangles, Noatak National Preserve, Alaska. U.S. Geological Survey Open-File Report, 03-367 (Additional source maps used for v2 and not included in v1).

Hamilton, T.D., Porter, S.C., 1975. Itkillik glaciation in the Brooks Range, northern Alaska. Quatern. Res. 5, 471–497.

Hamilton, T.D., Reed, K.M., Thorson, R.M. (Eds.), 1986. Glaciation in Alaska—The Geologic Record. Alaska Geological Society, Anchorage, 265pp.

Hill, J.C., Driscoll, N.W., Brigham-Grette, J., Donnelly, J.P., Gayes, P.T., Keigwin, L., 2007. New evidence for high discharge to the Chukchi shelf since the Last Glacial Maximum. Quatern. Res. 68, 271–279.

Kaufman, D.S., 1986. Surficial geologic map of the Solomon, Bendeleben and southern Kotzebue quadrangles, Alaska. U.S. Geological Survey Miscellaneous Field Studies Map, 1838-A.

Kaufman, D.S., Manley, W.F., 2004. Pleistocene maximum and Late Wisconsin glacier extents across Alaska, USA. In: Ehlers, J., Gibbard, P.L. (Eds.), Quaternary Glaciations—Extent and Chronology, Part II: North America. Developments in Quaternary Science, vol. 2 Elsevier, Amsterdam, pp. 9–27.

Kaufman, D.S., Thompson, C.H., 1998. Re-evaluation of pre-late-Wisconsin glacial deposits, lower Naknek valley, southwestern Alaska. Arctic Alpine Res. 30, 142–153 (Additional source maps used for v2 and not included in v1).

Kaufman, D.S., Calkin, P.E., Whitford, W.B., Przybyl, B.J., Hopkins, D.M., Peck, B.J., et al., 1989. Surficial geologic map of the Kigluaik Mountains area, Seward Peninsula, Alaska. U.S. Geological Survey Miscellaneous Field Studies Map, MF-2074.

Kaufman, D.S., Forman, S.L., Lea, P.D., Wobus, C.W., 1996. Age of pre-late-Wisconsin glacial-estuarine sedimentation, Bristol Bay, Alaska. Quatern. Res. 45, 59–72.

Kaufman, D.S., Manley, W.F., Forman, S.L., Layer, P., 2001. Pre-late-Wisconsin glacial history, coastal Ahklun Mountains, southwestern Alaska—new amino acid, thermoluminescence, and $^{40}Ar/^{39}Ar$ results. Quatern. Sci. Rev. 20, 337–352.

Kaufman, D.S., Porter, S.C., Gillespie, A.R., 2004. Quaternary alpine glaciation in Alaska, the Pacific Northwest, Sierra Nevada, and Hawaii. In: Gillespie, A.R., Porter, S.C., Atwater, B.F. (Eds.), The Quaternary Period in the United States. Developments in Quaternary Science, vol. 1. Elsevier, Amsterdam, pp. 77–103.

Kline, J.T., Bundtzen, T.K., 1986. Two glacial records from west-central Alaska. In: Hamilton, T.D., Reed, K.M., Thorson, R.M. (Eds.), Glaciation in Alaska—The Geologic Record. Alaska Geological Society, Anchorage, pp. 123–150.

Mangan, M.T., Waythomas, C.F., Miller, T.P., Trusdell, F.A., 2003. Emmons Lake volcanic center, Alaska Peninsula: source of the Late Wisconsin Dawson tephra, Yukon Territory, Canada. Can. J. Earth Sci. 40, 925–936 (Additional source maps used for v2 and not included in v1).

Manley, W.F., Kaufman, D.S., 2002. Alaska PaleoGlacier Atlas: Institute of Arctic and Alpine Research (INSTAAR). University of Colorado. http://instaar.colorado.edu/QGISL/ak_paleoglacier_atlasv.1.

Manley, W.F., Kaufman, D.S., Briner, J.P., 2001. Late Quaternary glacial history of the southern Ahklun Mountains, southeast Beringia—soil development, morphometric, and radiocarbon constraints. Quatern. Sci. Rev. 20, 353–370.

Mann, D.H., Hamilton, T.D., 1995. Late Pleistocene and Holocene paleoenvironments of the north Pacific coast. Quatern. Sci. Rev. 14, 441–471.

Matmon, A., Schwartz, D.P., Haeussler, P.J., Finkel, R., Lienkaemper, J.J., Stenner, H.D., et al., 2006. Denali fault slip rates and Holocene-late Pleistocene kinematics of central Alaska. Geology 34, 645–648.

Matmon, A., Briner, J.P., Carver, G., Bierman, P., Finkel, R., 2010. Moraine chronosequence of the Donnelly Dome region, Alaska: implications for the late Pleistocene glacial history of interior Alaska. Quatern. Res. 74, 63–72.

Molnia, B.F., 1986. Glacial history of the northeastern Gulf of Alaska—a synthesis. In: Hamilton, T.D., Reed, K.M., Thorson, R.M. (Eds.), Glaciation in Alaska—The Geologic Record. Alaska Geological Society, Anchorage, pp. 219–236.

Molnia, B.F., 2007. Late nineteenth to early twenty-first century behavior of Alaskan glaciers as indictors of changing regional climate. Glob. Planet. Change 56, 23–56.

Molnia, B.F., 2008. Glaciers of North America—glaciers of Alaska. In: Williams, R.S., Jr., Ferrigno, J.G. (Eds.), Satellite Image Atlas of Glaciers of the World, U.S. Geological Survey Professional Paper, 1386-K 525pp.

Péwé, T.L. (Ed.), 1953. Multiple Glaciations in Alaska—A Progress Report, Geological Survey Circular, 289, 13pp.

Péwé, T.L., Westage, J.A., Preece, S.J., Brown, P.M., Leavitt, S.W., 2009. Late Pliocene Dawson Cut Forest Bed and new tephrochronological findings in the Goold Hill Loess, east-central Alaska. Geol. Soc. Am. Bull. 121, 294–320.

Reger, R.D., 1979. Glaciation of Indian Mountain, west-central Alaska. Alaska Division of Geological and Geophysical Surveys Geologic Report, 61, pp. 15–18.

Reger, R.D., Sturmann, A.G., Berg, E.E., Burns, P.A.C., 2007. A guide to the late Quaternary history of the northern and western Kenai Peninsula, Alaska. Alaska Division of Geological and Geophysical Surveys Guidebook, vol. 8. 112pp (Additional source maps used for v2 and not included in v1).

Reger, R.D., Stevens, D.S.P., Solie, D.N., 2008. Surficial geology of the Alaska highway corridor, Delta Junction to Dot Lake, Alaska. Alaska Division of Geological and Geophysical Surveys Preliminary Interpretive Report, 2008-3a, 48pp (Additional source maps used for v2 and not included in v1).

Thorson, R.M., 1986. Late Cenozoic glaciation of the northern Nenana River valley. In: Hamilton, T.D., Reed, K.M., Thorson, R.M. (Eds.), Glaciation in Alaska—The Geologic Record. Alaska Geological Society, Anchorage, pp. 171–192.

Waythomas, C.F., 1990. Quaternary geology and late Quaternary environments of the Holitna Lowland and Chuilnuk-Kiokluk Mountains region, interior southwest Alaska. Ph.D. dissertation, University of Colorado, Boulder, 268pp.

Weber, F.R., 1986. Glacial geology of the Yukon-Tanana upland. In: Hamilton, T.D., Reed, K.M., Thorson, R.M. (Eds.), Glaciation in Alaska—The Geologic Record. Alaska Geological Society, Anchorage, pp. 79–98.

Weber, F.R., Hamilton, T.D., 1984. Glacial geology of the Mt. Prindel area, Yukon-Tanana upland, Alaska. Alaska Division of Geological and Geophysical Surveys Professional Report, 86, pp. 42–48.

Wilson, F.H., Weber, F.R., 2001. Quaternary geology of the Cold Bay and False Pass Quadrangles. In: Gough, L.P., Wilson, F.H. (Eds.), Geologic Studies in Alaska by U.S. Geological Survey, 1999, 51–71 U.S. Geological Survey Professional Paper, 1633.

Young, N.E., Briner, J.P., Kaufman, D.S., 2009. Late Pleistocene and Holocene glaciation of the Fish Lake valley, northeastern Alaska Range, Alaska. J. Quatern. Sci. 24, 677–689 (Additional source maps used for v2 and not included in v1).

Chapter 34

Glaciations of the Sierra Nevada, California, USA

Alan R. Gillespie[1],* and Douglas H. Clark[2]

[1]*Quaternary Research Center, University of Washington, Seattle, Washington 98195-1310, USA*
[2]*Department of Geology, Western Washington University, Bellingham, Washington 98225, USA*
*Correspondence and requests for materials should be addressed to Alan R. Gillespie. E-mail: arg3@uw.edu

34.1. INTRODUCTION

The Sierra Nevada is a major north–south mountain range in California that separates the internally drained basins and desert ranges to the east from the Central Valley to the west. Although it runs from the Tehachapi Mountains on the northern edge of the Mojave Desert to near Mount Lassen (35–40.5°N), the highest part of the Sierra Nevada (crest elevations of 3400–4300 m) lies between ~36°N and ~38°N, the approximate latitudes of the towns of Olancha (south) and Bridgeport (north). This entire reach was heavily glaciated during the Pleistocene, and some small glaciers still occupy sheltered cirques high in the mountains. Enhanced accumulation and shading from adjacent peaks allow these modern glaciers to exist well below the regional climatic snowline (estimated at ~4500 m elevation at 37°N; Flint, 1957, p. 47). Although the glaciers occur at progressively lower altitudes to the north, the topographic crest north of Bridgeport plunges below the average snowline, or equilibrium-line altitude (ELA), for the modern glaciers. During the major Pleistocene glaciations, however, the ELA was ~820 m lower than for modern glaciers (Warhaftig and Birman, 1965; Gillespie, 1991), and then large glaciers occurred in the northern Sierra Nevada also (Fig. 34.1). Remarkably, the terminal moraines of maximum advances of different ages cluster closely together (Fig. 34.2), such that there appears to be a rough maximum limit to the size of Sierran glaciers over a half million years or more.

In the Sierra Nevada, the advance and retreat of glaciers are especially sensitive to changes in winter precipitation and summer temperature. At present, the Sierra Nevada receives moisture mainly from winter low-pressure systems from the Pacific Ocean guided by the jet stream. Mean annual precipitation on the eastern slope of the Sierra Nevada at ~37°N ranges from ~100 cm a^{-1} at the crest to ~25 cm a^{-1} at the range front (Danskin, 1998); east of the crest it is controlled by a strong rainshadow. During Pleistocene glaciations, the jet stream shifted south so that precipitation increased and temperatures were reduced, although not necessarily in phase. There are two main sources of precipitation: northerly, from the Gulf of Alaska; and southerly (the so-called "Pineapple Express"), from tropical latitudes in the Pacific driven by a southern branch of the jet stream. Therefore, precipitation in the Sierra Nevada and the Cascade Mountains to the north is not necessarily in phase on an annual or even a decadal scale.

Antevs (1938, 1948) first proposed this modern view of the ice-age climate, with Sierra Nevada precipitation controlled by intensified winter Pacific storms, driven farther south than today due to the influence of the Cordilleran and Laurentide ice sheets. However, modelling by Rupper et al. (2009) has shown that except for arid environments with precipitation less than ~150 mm a^{-1}, there is generally enough snowfall to grow glaciers and summertime temperatures have a strong control on their advance and retreat.

The Sierra Nevada consists of a batholith of Mesozoic plutons of intermediate composition, commonly granodiorite, intruded metamorphic marine sedimentary rocks and island arc volcanic rocks, both now preserved as roof pendants. The range is a large crustal block that has been tilted to the west and broken by extensional basin-and-range faulting on the east. The timing of these tectonic events is controversial, but tilted Miocene lava flows west of the crest, and faulted Pliocene basalt flows in the ranges east of the Sierra Nevada indicate that much of the early tectonic activity started before the Quaternary Period, although faulting continues today.

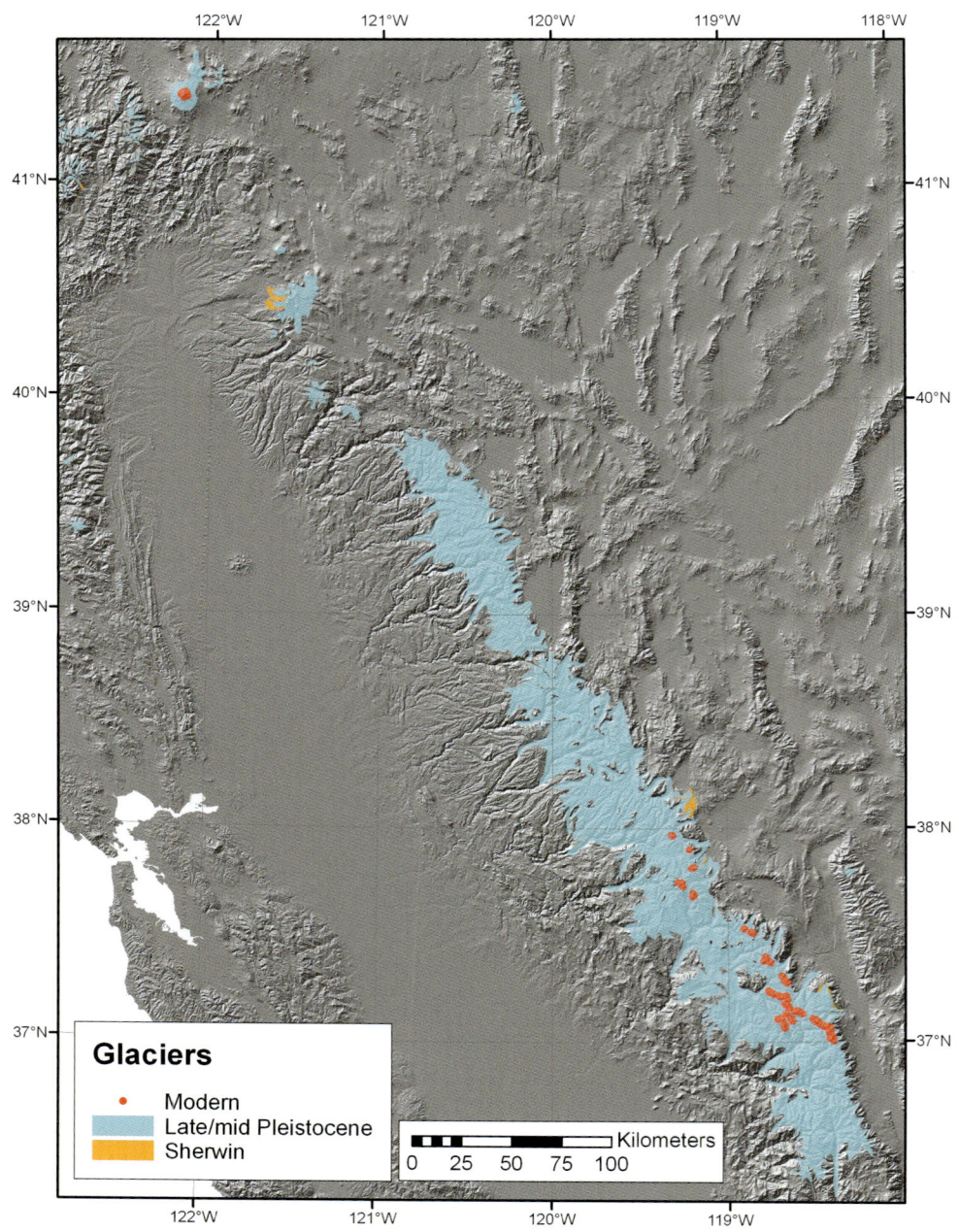

FIGURE 34.1 Extent of Quaternary glaciers in the Sierra Nevada, modified from Warhaftig and Birman (1965) and Clark (1995). Late Pleistocene (Tioga) and older glacial deposits are, in general, so similar in extent that they are difficult to distinguish at this scale; see Figs. 34.2 and 34.3. Base map: USGS NED dataset (3-s resolution, ~90 m).

34.1.1. Background

California in the nineteenth century was a long way from the European centres where geological theory was first developed. Nevertheless, the Sierra Nevada received a surprising amount of attention, and the understanding of mountain glaciation was advanced in no small part there. For example, Josiah Whitney, the California State Geologist, and John Muir, prominent naturalist, debated famously the relative roles of glacial, and tectonic processes in sculpting the spires, ridges, and deep valleys of the Sierra Nevada, with special attention given to Yosemite Valley. LeConte (1873) modified the positions taken earlier and derived an essentially modern view that glaciers had excavated the highlands and widened the upper reaches of river valleys, but that the basic landscape was nevertheless formed largely by fluvial activity. Matthes (1930, 1965) undertook decades of study of the glaciated troughs of the western

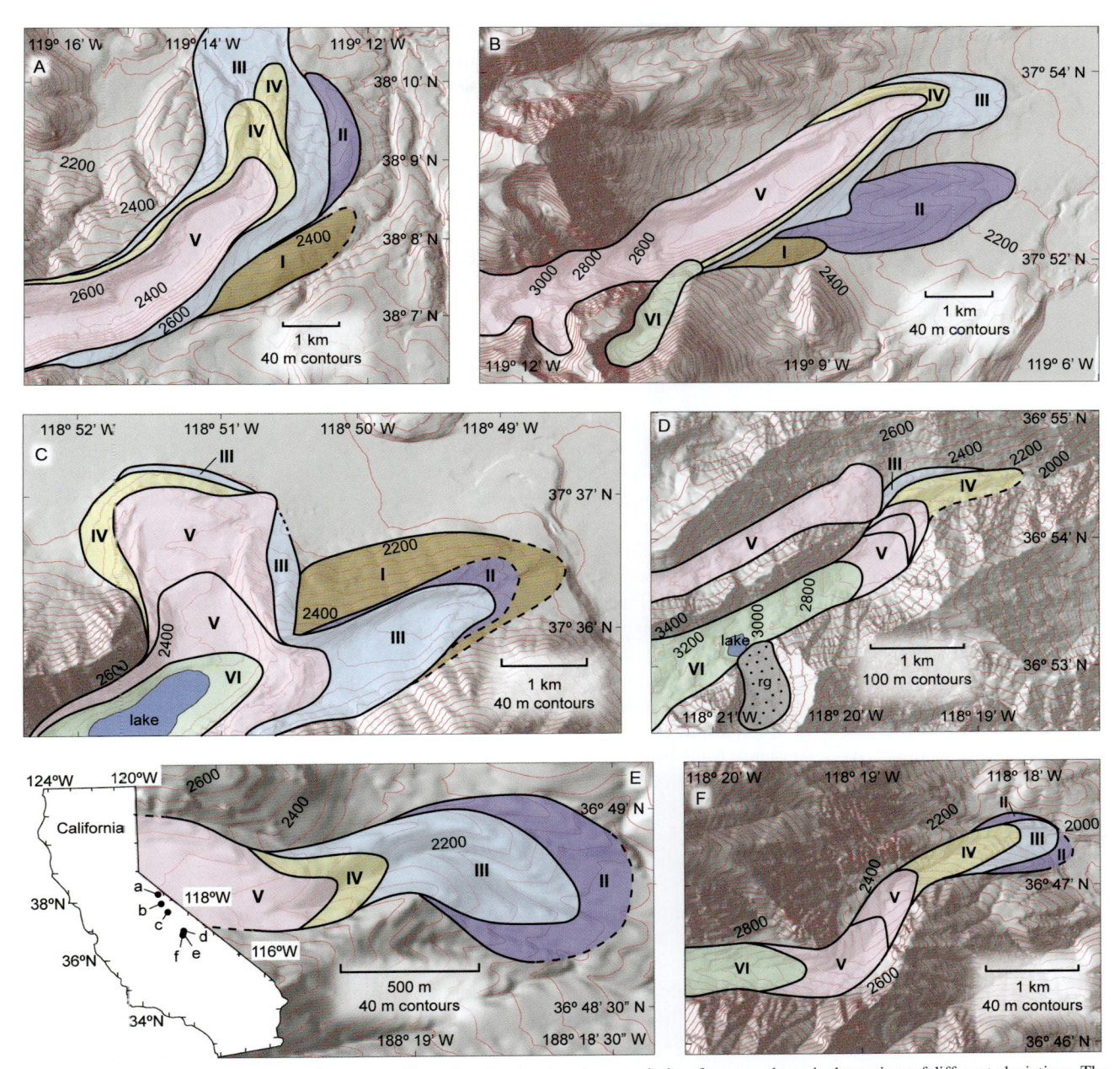

FIGURE 34.2 Six glaciated drainages of the eastern Sierra Nevada, showing close proximity of preserved terminal moraines of different glaciations. The ELAs are also close (Table 34.1). Index map of California shows location of drainages (A–F). (A) Green Creek. (B) Walker Creek/Bloody Canyon/Sawmill Canyon (Mono County). (C) Convict Creek. (D) Sawmill Canyon (Inyo County). (E) South Fork Oak Creek. (F) Independence Creek. The greater dispersion of moraines in (F) arises because the accumulation area there comprises three drainages versus one for the other examples; therefore, minor differences in ELA result in major extension of the ablation area. ΔELA (accumulation-area ratio method) is 60 m or less for the moraines shown. Moraines are identified by Roman numerals and colour coding. Name assignments are: pre-Tahoe moraines (I, brown); Mono Basin moraines (II) and Tahoe moraines (III), blue; Tenaya (IV) moraines and Tioga (V) moraines, yellow and pink and latest Tioga moraines (VI, green). Elevation contours are in m a.s.l. Except for Convict Creek (Gillespie, unpublished map, 2009), mapping follows Gillespie (1982). Base maps are contoured and shaded NASA/ASTER 30-m digital elevation models (DEMs).

slopes of the Sierra Nevada, and Blackwelder (1931) similarly analysed the glacial evidence east of the crest. Blackwelder, in particular, pioneered relative dating in an effort to correlate moraines from valley to valley, and to the glacial stratigraphy developed along the southern margin of the Laurentide Ice Sheet. Matthes and Blackwelder made an early attempt to reconcile their glacial sequences, but in the absence of numerical age control, correlation was difficult, as successive generations of glacial geologists have confirmed while refining the glacial stratigraphy.

By the 1970s, the modern framework of the Sierra Nevada glaciations was well developed (e.g. Warhaftig and Birman, 1965; Fullerton, 1986), but numerical age control was weak or missing entirely. Recent investigations have improved this glacial history. In particular, a more thorough accounting has been developed for glaciations pre-dating marine oxygen isotope stage (MIS) 6 (>186 ka), and the MIS 6–2 (186–12 ka) and Holocene glacial history has been refined by numerical dating. Most of the research has taken place in the eastern Sierra Nevada, although recently new studies have been made west of the crest also (James et al., 2002). The improvements to the glacial history are based on continued exploration and mapping, improved and new methods of numerical and relative dating and extraction and analysis of sediment cores from lakes and bogs.

Many advances in the past few decades have resulted from the development and application of rigorous dating techniques to glacial drift and landforms. Early studies included K/Ar dating of lava flows interbedded with till (Dalrymple, 1963, 1964) and ^{14}C dating of latest Pleistocene sediments in bogs that could be related stratigraphically to glacial deposits (Adam, 1966, 1967). K/Ar dating, however, was problematic and opportunistic, because the glacial deposits and landforms themselves could not be dated. For example, Gillespie et al. (1984) obtained high-precision ^{40}Ar–^{39}Ar dates for basalt flows at Sawmill Canyon (Inyo Country), but in the end could only conclude that the Hogsback moraine was less than 119 ± 3 ka and must postdate MIS 6 (186–128 ka), a conclusion already reached by Burke and Birkeland (1979) on the basis of soil development.

Beginning in the 1980s, advances in accelerator mass spectrometry (AMS) led to the measurement of trace concentrations of exotic isotopes created by cosmic-ray bombardment of rocks exposed at the Earth's surface, and the calculation of exposure ages of these rocks (e.g. Nishiizumi et al., 1989; Gosse and Phillips, 2001). From these dates, landform ages could be inferred if the erosion rate of the dated surfaces could be estimated. The glacial chronology worldwide was a logical and early target of this new technology, and convenient samples for analysis were found in boulders exposed on the crests of moraines.

Evidence from the Sierra Nevada and other glaciated mountain ranges is incomplete, due to the overriding of earlier deposits by later glaciers and the high potential for erosion in the steep canyons. Constrained by the limited availability of local numerical ages, correlation to oxygen isotope data from marine sediment cores and/or to ice cores from Greenland or Antarctica has often been used as a proxy chronology for the local glacial history. In this review, we use MISs as a convenient chronological timescale, recognising that improved numerical dating may someday compel revision, especially for the past 50 ka for which soil erosion and boulder exhumation and erosion do not present such major complications as for dating older rocks.

34.2. GLACIAL ADVANCES

Blackwelder (1931) recognised four main glaciations in the eastern Sierra Nevada: McGee, Sherwin, Tahoe and Tioga. Subsequently, the glacial history has been refined and revised. The recognised glaciations of the Sierra Nevada are summarised by Fullerton (1986, Chart 1, Table 34.1) and those discussed in this review are listed in Table 34.1. Discussion draws on Kaufman et al. (2004) and Clark et al. (2003). The glacial advances are grouped for discussion below into Pliocene and early Pleistocene (MIS 22–6, ~830–186 ka; MIS 6, 186–128 ka), late Pleistocene (MIS 5–latest 2, 128–14 ka) and Post-Tioga glaciations and advances (latest MIS 2 and MIS 1, 14 ka to present).

ELAs summarised in Table 34.1 were calculated using the accumulation/ablation area ratio (AAR, with a ratio of 0.65 giving the ELA) and the highest-moraine techniques for palaeoglaciers in 70 eastern Sierra Nevada valleys between 36.5 and 39.5°N (Gillespie, 1991). The modern ELA was taken to be 3860 at 37°N (Burbank, 1991), but the "true" value may be 100–200 m higher (e.g. Meierding, 1982) or even 640 m higher (Flint, 1957) since anomalously low cirque glaciers can occur in sheltered locations.

34.2.1. Pliocene and Early Pleistocene Glaciations

The earliest glaciations are represented by deeply weathered erratic boulders and diamictons on highland surfaces. These deposits occur in a wide range of topographic settings: mountaintops, arêtes, beheaded valleys hundreds of metres above modern canyons and benches on the eastern escarpment (Gillespie, 1982). Not all of the diamictons are necessarily glacial drift, but some appear to be (e.g. Brocklehurst et al., 2002). Given the range of sites and elevations of these deposits, it is likely that a number of unnamed glaciations are represented. However, our understanding of glacial chronology, extents and ELA depressions for this time is poor. The oldest identified till, the McGee till, is among the deposits in this category.

The McGee till crops out near the summit of McGee Mountain, a large, broad peak south of the town of Mammoth Lakes (37°38'55.93"N, 118°58'23.16"E). Metamorphic rocks and the eroded remnant of a 2.7-Ma basalt flow (Dalrymple, 1963, 1964) here are covered by large, exotic granitic boulders from a source area located on the other side of McGee Canyon. Evidently, the time of transport was long ago (~1.5 Ma; Huber, 1981) because McGee

TABLE 34.1 Recognised Glaciations of the Sierra Nevada, Their Ages and East-Side ELAs Relative to Tioga and Interpolated to 37°N

Glaciation	Mean age (ka)[a]	ΔELA (m)[b]	Age references
Matthes (Little Ice Age)	0.6–0.1[c]	480 ± 25	Wood (1977), Stine (1994)
Recess Peak	14.2–13.1[c]	335 ± 20	Clark (1997)
	13.4–12.0		Phillips et al. (2009)
Tioga (retreat)	15–14	125 ± 30	James et al. (2002), Clark and Gillespie (1997)
Tioga (start)	21–20[c]		Clark et al. (2003)
	<25[c]		Bursik and Gillespie (1993)
Tioga ("Tioga 2–4")	25–16	0	Phillips et al. (1996)[d]
	18.5–14.5		Phillips et al. (2009)
Tenaya ("Tioga 1")	31	−45 ± 15	Phillips et al. (1996)[d], Benson et al. (1996)
	32[c]		Bursik and Gillespie (1993)
	28–24		Phillips et al. (2009)
Tahoe II[††]	50–42	−95 ± 10	Phillips et al. (1996)
Tahoe I[††]	?	∼−100(?)	
	170–130		Phillips et al. (2009)
Casa Diablo	126–62	?	Bailey et al. (1976)
pre-Tahoe (Bloody Canyon)[e]	220–140	?	Phillips et al. (1990), discussed in Clark et al. (2003)
Mono Basin	80–60	−195 ± 50	Phillips et al. (1990)[d]
Walker Creek[e]	∼550	?	Clark (1968), A. M. Sarna-Wojcicki
Sherwin	∼820	∼−200	Sharp (1968), Birkeland et al. (1980), Nishiizumi et al. (1989)
Lower Rock Creek	∼920	?	Sharp (1968), Birkeland et al. (1980)
McGee	2700–1500	?	Huber (1981), Dalrymple (1963, 1964)

[a] Ages are shown as ranges or approximate values. Uncertainties may be found in text and/or references.
[b] Elevation differences are relative to the lowest Tioga ELA (∼3040 ± 150 m) and interpolated to 37°N (Gillespie (1991). Uncertainties are ±1 σ and are random, and do not include the systematic uncertainties of ∼150 m. Accuracy depends critically on assignment of moraines to the correct glaciation. Regression was done on ≤70 glaciated drainages.
[c] cal. ka BP.
[d] Revised (see discussions in James et al., 2002 and Phillips et al., 2001).
[e] Nomenclature from Clark et al. (2003).

Creek has been incised ∼800 m since access to the summit of McGee Mountain for the erratic boulders was last possible (e.g. Putnam, 1962; Gillespie et al., 1999).

The next youngest drift deposits occur in the vicinity of the late Pleistocene moraines, suggesting that deposition occurred when the landscape had much its present appearance. Moraines from this period have been largely eroded, but Sharp (1972) identified a degraded Sherwin moraine in the Bridgeport Basin. Two tills have been described: the "old red till" on Lower Rock Creek and the Sherwin till (Sharp, 1968).

Sharp (1968) observed both tills exposed in a road cutting next to Little Rock Creek, just south of Long Valley. The lower till has deeply weathered, disintegrating granitic boulders and a red palaeosol. Birkeland et al. (1980) estimated from relative dating that the palaeosol on this buried till represents ~100 ka of development. The till is overlain unconformably by a second deeply weathered till that also contains grusy boulders but lacks the distinctive red palaeosol. Sharp (1968) traced this second till ~125 m up the canyon wall to the "type locality" of the Sherwin till, the Big Pumice Cut on U.S. Highway 395. Another till with a distinctive red palaeosol underlies till mapped as Sherwin by Sharp (R. P. Sharp, unpublished data, 1979) near the end of the left-lateral moraine of Big Pine Creek, ~50 km to the south.

Sharp (1968) demonstrated that the Sherwin till underlies and therefore pre-dates the Bishop Tuff at the Big Pumice Cut, near the southern end of Long Valley. Sarna-Wojcicki et al. (2000) dated sanidine crystals in the Bishop Tuff to 759 ± 2 ka ($^{40}Ar/^{39}Ar$). R. P. Sharp (personal communication, 1976) estimated that upon burial, the till was weathered about as much as Tahoe till is today, requiring ~50–100 ka. Other studies support this estimate. From the development of the palaeosol on the buried till, Birkeland et al. (1980) estimated an age at burial of ~50 ka, and Nishiizumi et al. (1989) analysed cosmogenic nuclides to yield an estimate of ~67–53 ka. Thus, the Sherwin glaciation probably occurred at ~820 ka (i.e. late Early Pleistocene). It follows that Sharp's (1968) old red till at Lower Rock Creek is ~920 ka.

Figure 34.3 shows that the Sherwin glaciers of the central Sierra Nevada were more extensive than their successors, extending farther onto the low-angle floors of Bridgeport Basin and down the slopes of the Sherwin Grade at the Big Pumice Cut. Because the topographic gradients are relatively low here ($<5°$), a minor additional depression of ELA could account for the larger length of the Sherwin glaciers. In the Bridgeport Basin, an additional ELA depression of as little as ~100 m from the Tahoe ELAs might explain the difference in extent (Clark et al., 2003).

34.2.1.1. MIS 22–6 Glaciations

The period between ~820 and 186 ka, the beginning of MIS 6, is sparsely populated with glacial evidence. However, two glacial or glaciofluvial deposits, one near the Sonora Junction and the other in Mohawk Valley, are dated to this interval by tephras.

The deposit near Sonora Junction is from the West Walker River (Blackwelder, 1931; Clark, 1967, 1968). A roadcut near U.S. highway 395 exposes fluvial gravel of Wheeler Flats that contains a tephra identified as Rockland Ash-Tuff by Sarna-Wojcicki et al. (1985), now dated at ~550 ka (A. M. Sarna-Wojcicki, quoted in Clark et al., 2003). If the gravel is glaciofluvial, as seems likely, the tephra may date an unnamed, pre-Tahoe glaciation. Mathieson and Sarna-Wojciki (1982) found the Rockland ash in a similar stratigraphic relation in the Mohawk Valley in the northern Sierra Nevada.

Several other tills appear to date from the MIS 8–6 (303–186 ka) interval, but the age control is not compelling. Fullerton (1986) discussed a till from Reds Meadow, Devils Postpile National Monument. This till overlies the Bishop Tuff and underlies an andesite, the age of which has been very loosely constrained by K/Ar to 650 ± 350 ka. Curry (1971) and Sharp (1972) presented evidence for other tills of this same general period from Rock Creek and the Bridgeport Basin. Gillespie (1982) described two moraines outside the Mono Basin moraines at Bloody Canyon ($QpMB_I$ and $QpMB_{II}$ in Table 34.1) that appear to postdate the Sherwin till. Phillips et al. (1996) presented ^{36}Cl exposure ages from an "older Tahoe" moraine at Walker Creek (Bloody Canyon) in the ~200 ka range, but other dates from the same moraine were 130 and 150 ka.

The MIS 22–6 interval appears to have been marked by a number of glacier advances, despite the scarcity of widespread evidence. In the absence of accurate and precise numerical dates, correlation of tills and elucidation of the glacial history from this interval remains problematical.

34.2.2. MIS 6 Glaciations

The record of Sierra Nevada glaciations becomes more detailed with MIS 6, but remains incomplete compared to the generalised history deduced from lake-sediment cores (e.g. Smith et al., 1991; Bischoff et al., 1997). Tills that probably date from MIS 6 include the Mono Basin, pre-Tahoe and Tahoe I tills of Bloody Canyon (Fig. 34.4), the Casa Diablo till of Mammoth Lakes and the pre-Hogsback till of Sawmill Canyon in Inyo County. Moraines from this period are eroded and commonly broad-crested, with few exposed boulders. Boulders that are exposed may be disintegrated, heavily pitted and split.

The type locality for the Mono Basin moraines is at Bloody Canyon, near Mono Lake (Sharp and Birman, 1963; Figs. 34.4 and 34.5). The degree of weathering and erosion is similar to many Tahoe moraines (Burke and Birkeland, 1979), and it is uncertain how many moraines from the two glaciations may have been misclassified. At the same distance from the range front, the Mono Basin moraines at Bloody Canyon are lower in elevation than the Tahoe moraines, possibly as a result of range-front faulting between glaciations (Clark, 1972). They were preserved because the subsequent Tahoe glaciers extended along a more northerly course. Phillips et al. (1990) measured cosmic-ray exposure ages averaging ~103 ka for eight boulders from these moraines (Fig. 34.6). The dates were revised downwards to ~80–60 ka as new estimates for production rates of ^{36}Cl were made (Phillips et al., 2001; revised as discussed in James et al., 2002). Even these ages are minima, as soil and boulder erosion on the moraines (e.g. Birkeland and Burke, 1988; Hallet and Putkonen, 1994) reduce the exposure ages estimated by this technique.

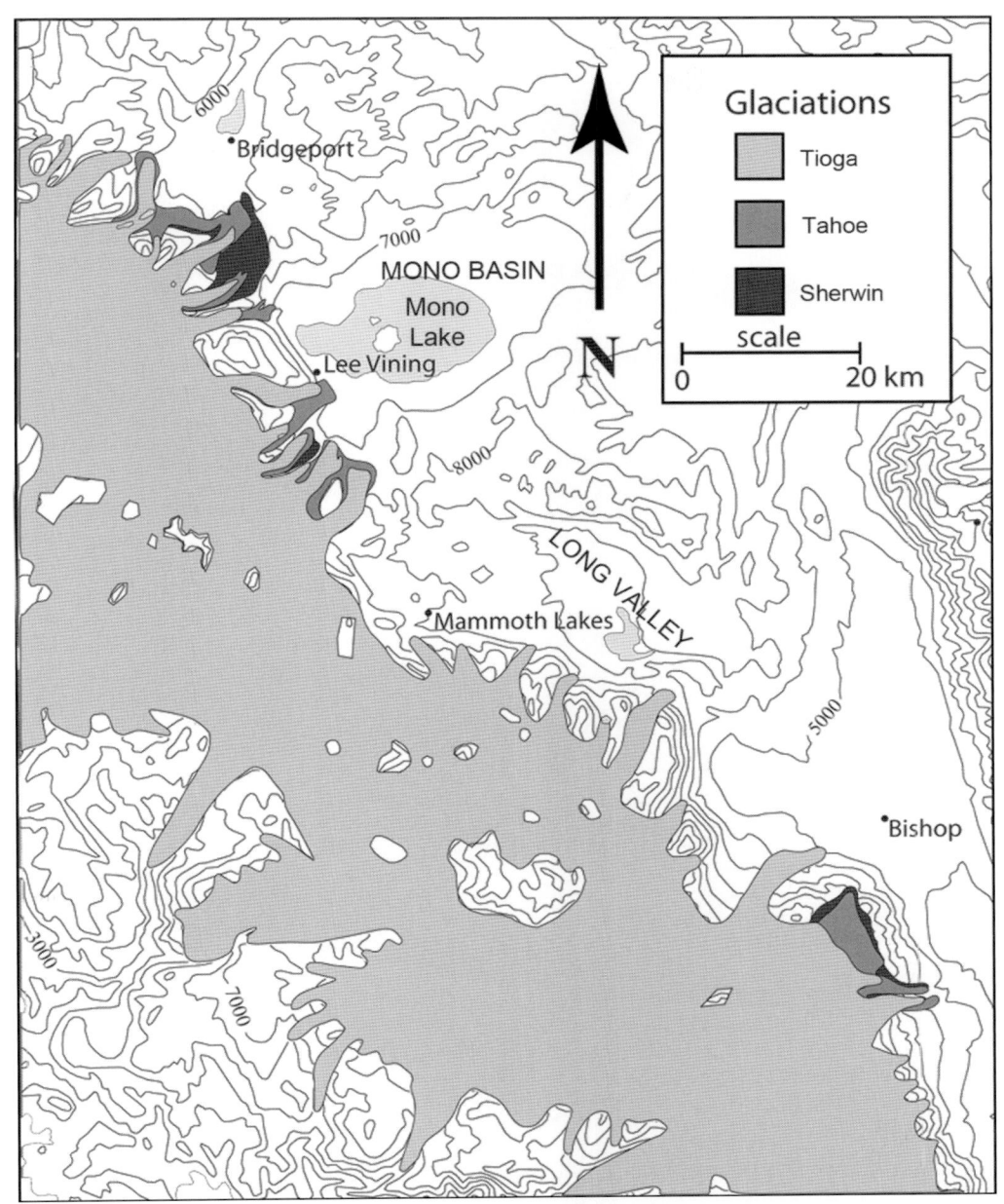

FIGURE 34.3 Extent of the ~820 ka Sherwin drift relative to ~20 ka Tioga drift in the central Sierra Nevada. Contour intervals are 305 m (1000 ft). Modified from Warhaftig and Birman (1965), Kistler (1966), Gillespie (1982) and Huber et al. (1989).

The Tahoe glaciation is one of the four major Sierra Nevada glaciations recognised by Blackwelder (1931). Gillespie (1982) showed that the "Tahoe" moraine of Sharp and Birman (1963) at Bloody Canyon was composite, with a young till comprising the crest and an older till the right-lateral flank (Figs. 34.4 and 34.5). These two tills were called "Tahoe II" and "Tahoe I," respectively. However, they are probably from different glaciations. Phillips et al. (1990) obtained cosmogenic ages for the Tahoe II moraine. These ages averaged ~60 ka, later revised to ~50–42 ka. No age estimate for the Tahoe I moraine was obtained.

Phillips et al. (1990) did measure ^{36}Cl dates for five boulders from the "older Tahoe" moraines, discussed above, that protrude from the right-lateral composite moraine. The stratigraphic relation between this "older Tahoe" moraine and the Mono Basin moraines is unclear, although both are buried by Tahoe I till (see Clark et al., 2003; Kaufman et al., 2004). The ^{36}Cl dates for the "older Tahoe" till, called "pre-Tahoe I, post-Mono Basin" till in Fig. 34.5, are ~220–140 ka (unrevised).

Although these "older Tahoe" values have not been adjusted downward in accordance with new production

FIGURE 34.4 View west from pre-Tahoe end moraines of Bloody Canyon (Fig. 34.2B), showing important among the younger Tahoe and Mono Basin moraines. The left-lateral Tahoe II moraine comprises the high ridge to the right (north). The older Tahoe I moraine crops out part way down the flank of the moraine. Both Tahoe moraines bury the older Mono Basin left-lateral moraine, seen emerging from the composite Tahoe moraine in left centre. Photograph by D. H. Clark.

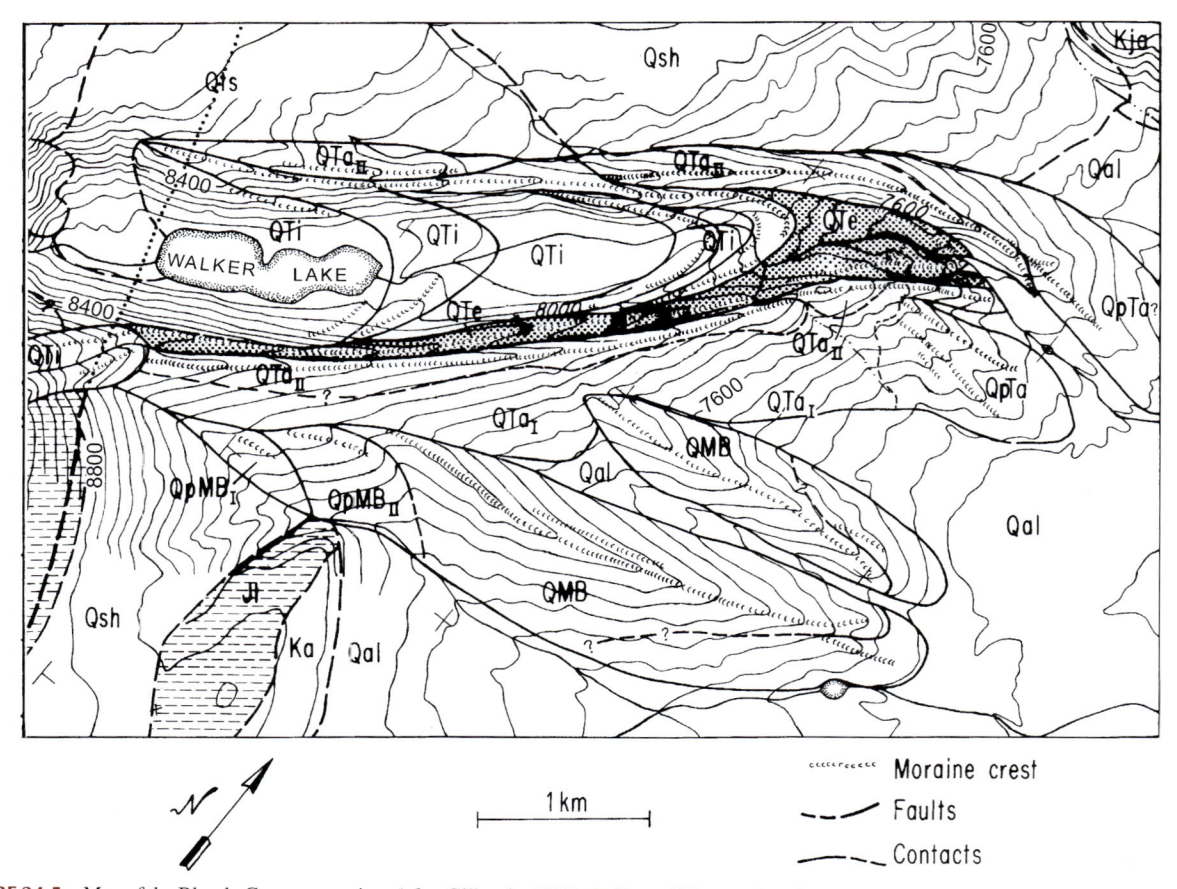

FIGURE 34.5 Map of the Bloody Canyon moraines (after Gillespie, 1982). Jl, Ka and Kja are plutonic rocks; Qal is Quaternary alluvium. Qsh is Sherwin till of Sharp and Birman (1963). $QpMB_I$ and $QpMB_{II}$ are pre-Mono Basin moraines; QMB is Mono Basin moraines. QpTa is the oldest set of moraines (pre-Tahoe) along Walker Creek ("older Tahoe" of Phillips et al., 1990); QTa_I and QTa_{II} are the Tahoe I and II moraines (Gillespie, 1982). QTe (shaded for clarity) is the Tenaya moraine, and QTi are the undifferentiated Tioga moraines. Topographic contour interval is 24 m (80 ft).

FIGURE 34.6 ^{36}Cl dates (Phillips et al., 1990, 1996) for Bloody Canyon moraines with the following independent age control (courtesy of R. M. Burke): (1) 100–600 cal. a BP (Yount et al., 1982; Clark and Gillespie, 1997; Konrad and Clark, 1998). (2) 13.1–14.2 cal. ka BP (11.2–12.2 ^{14}C ka BP; Clark, 1997). (3) 30 cal. ka BP (Bursik and Gillespie, 1993). (4) >25 ka on ^{10}Be and ^{26}Al (Gillespie, unpublished data). See text for discussion of "older Tahoe".

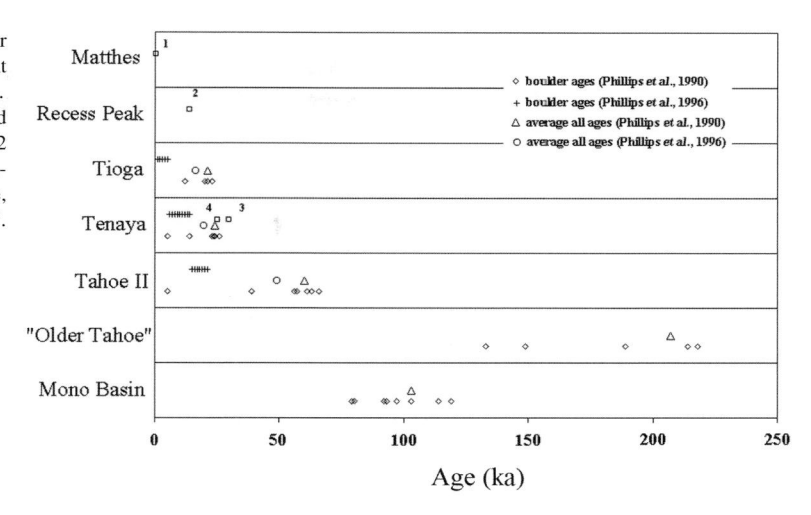

rates, it is clear that they are greater than the ^{36}Cl ages for the Mono Basin moraines ~1 km away. Phillips et al. (1990) regarded the older Tahoe as pre-dating the Mono Basin glaciation. These findings appear to contradict field relations among the moraines. It is noteworthy that only a small number of boulders were dated, and scatter among dates for each till is large. The dating studies have illuminated the conflicts between stratigraphic and chronologic analyses, and the current strengths and deficiencies in each.

The pre-Tahoe Casa Diablo till, near the town of Mammoth Lakes, is weathered to a similar degree as nearby Tahoe till (Burke and Birkeland, 1979; Birkeland et al., 1980). Given the degree of soil development, the Tahoe till was taken to date from MIS 6 by Burke and Birkeland (1979). Although the Casa Diablo till is interbedded with basalt flows that should afford a good dating opportunity, K/Ar analyses by Curry (1971) and Bailey et al. (1976) disagree, constraining the till to either ~453–288 or 126–62 ka, respectively. On the basis of relative dating techniques, Burke and Birkeland (1979) suggested that the Casa Diablo till was correlated with the older Tahoe and/or Mono Basin tills to the north. Fullerton (1986) pointed out that the absence of 185-ka quartz latite boulders in the Casa Diablo till, and their presence in the nearby Tahoe till, suggests that the Casa Diablo till may predate MIS 6. However, recent field work shows that the quartz latite boulders do occur in the Casa Diablo till, and the dates of Bailey et al. (1976) are now generally accepted as the age range for the till (R. M. Burke, personal communication, 2003).

Another till from the same general age range occurs in the oldest moraine in Sawmill Canyon, Inyo County. It is also interbedded with basalt flows, but dates for the flows only loosely constrain the moraine to ~465–130 ka (Gillespie et al., 1984).

On Bishop Creek, an unusually complete sequence of moraines has been preserved. Phillips et al. (2009) measured ^{36}Cl exposure ages for Tahoe moraines of 190–130 ka there (Fig. 34.7).

ELAs for the MIS 6 glaciers were depressed about 100–200 m below the levels for the MIS 2 Tioga glaciers of the last glacial maximum (LGM: Table 34.1).

34.2.3. Late Pleistocene Glaciations

Blackwelder (1931) recognised that there typically were two Tioga moraines in Sierra Nevada valleys. Sharp and Birman (1963) and Birman (1964) recognised a third, "Tenaya" moraine between Tioga and Tahoe in relative age. Lake-core evidence (e.g. Benson et al., 1998b; Fig. 34.7) suggests that a dozen or more short-lived advances occurred between ~50 and 14 ka. Only moraines from the largest advances appear to have been preserved near Mono Lake, and at Bishop Creek, Phillips et al. (2009) found a gap in the record until 26 ka (Fig. 34.7). The Tahoe II and Tenaya advances at Mono Lake seem to have occurred during earlier and later parts of MIS 3, respectively. MIS 2 Tioga advances culminated around 19–23 ka and ended by 15 ka with retreat to the Sierra Nevada crest, or even complete disappearance for a brief period. ELAs for maximum Tioga glaciers were 3040 ± 150 m (Table 34.1), depressed about 820 m relative to modern glaciers and perhaps 900–1500 m below the climatic ELA, depending on how it is estimated. The climatic ELA is above the crest of the Sierra Nevada, and cirque glaciers exist only in sheltered localities.

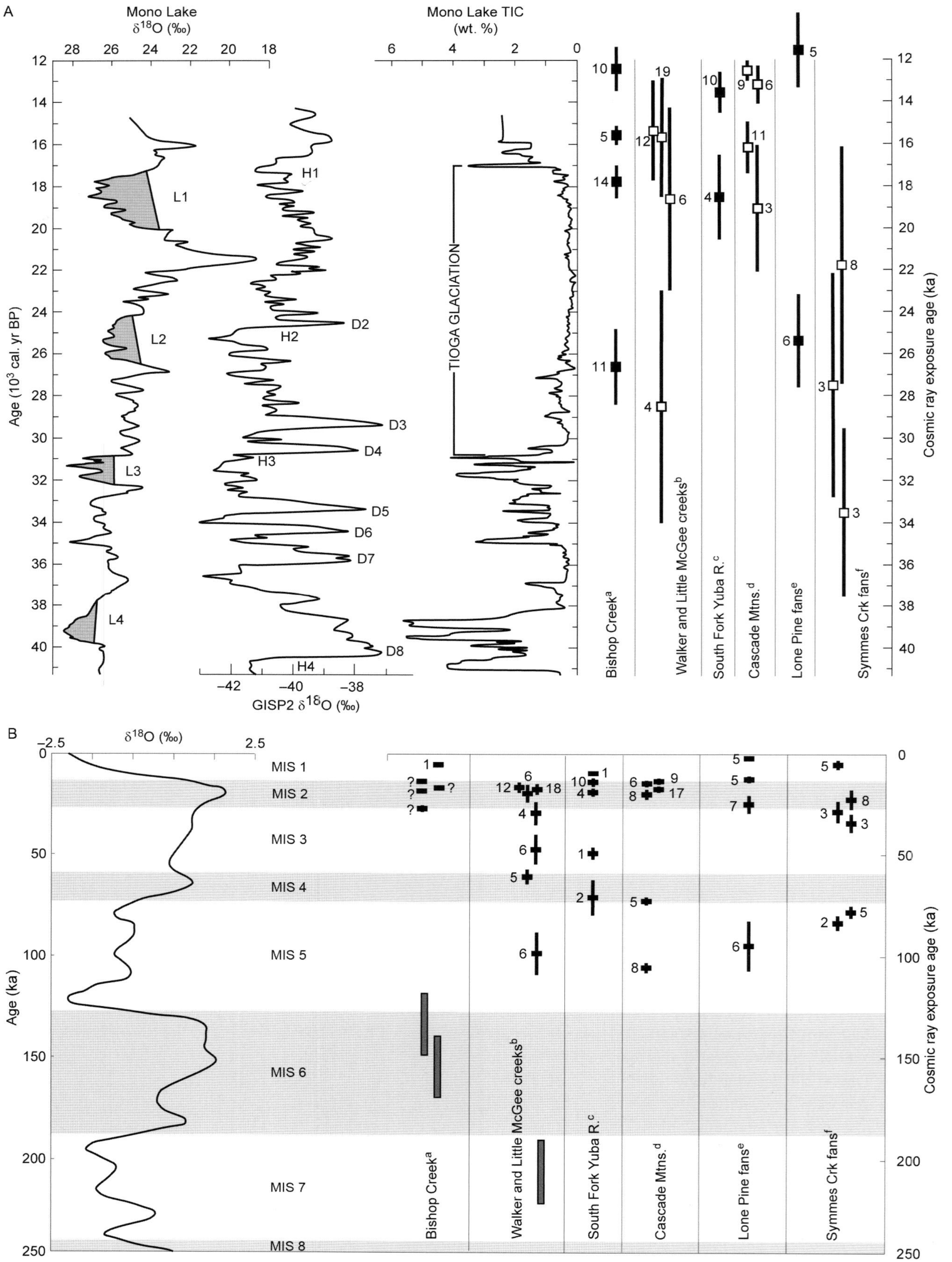

A
Mono Lake δ18O (‰)
Mono Lake TIC (wt. %)
Age (10³ cal. yr BP)
L1
L2
L3
L4
H1
H2
H3
H4
D2
D3
D4
D5
D6
D7
D8
TIOGA GLACIATION
GISP2 δ18O (‰)
Cosmic ray exposure age (ka)
Bishop Creek[a]
Walker and Little McGee creeks[b]
South Fork Yuba R.[c]
Cascade Mtns.[d]
Lone Pine fans[e]
Symmes Crk fans[f]
B
δ18O (‰)
Age (ka)
MIS 1
MIS 2
MIS 3
MIS 4
MIS 5
MIS 6
MIS 7
MIS 8

From ^{36}Cl cosmogenic exposure ages at Bloody Canyon and other canyons, Phillips et al. (1996) inferred four separate Tioga stades ranging in age from 25 to 14 ka, revised for the changes in production rates as discussed above. Their dates failed to resolve the Tenaya as a separate glaciation at Bloody Canyon. James et al. (2002) measured ^{10}Be and ^{26}Al cosmogenic exposure ages in the South Fork of the Yuba River, suggesting that the maximum extent of the Tioga glaciation there occurred $\sim 18.6 \pm 1.2$ ka. Likewise, the record at Bishop Creek gave moraines ranging from 26 to 15.5 ka in age (Phillips et al., 2009), or latest MIS 3–2 (Fig. 34.7). In Humphreys Basin, above the headwaters of Bishop Creek, Phillips et al. (2009) reported ^{36}Cl cosmic-ray exposure ages of 15.2 ± 0.7 ka that suggested to them that even the crest of the Sierra Nevada was nearly free of ice by then. It follows that retreat of glaciers from their maximum Tioga extents was rapid, taking only 1000 or 2000 years.

Based on ^{14}C ages of ostracodes in Mono Lake sediments interbedded with basaltic ash, Bursik and Gillespie (1993) inferred at nearby June Lake an age of $<25.2 \pm 2.5$ cal.ka BP for the maximum Tioga advance, which overrode a cinder cone. They inferred an age of 31.7 cal.ka BP or more for the Tenaya moraine, through which an eruption may have occurred while ice was present. The two cinder cones are the only sources that have been discovered for the ash in the lake sediments. Bursik and Gillespie (1993) regarded the Tenaya as a separate advance.

Other age control for the Tioga glaciation is available from sediment cores collected from bogs and lakes. D. H. Clark cored Grass Lake Bog, south of Lake Tahoe, and dated a sharp transition from Tioga glacial to overlying non-glacial sediments at 21.13–19.85 cal.ka BP. This date may mark the onset of retreat of the Tioga maximum advance, and is consistent with the ^{36}Cl dates from Bishop Creek (Phillips et al., 2009; Fig. 34.7). Basal lake sediment from the west slope of the Sierra Nevada yielded a minimum date for the beginning of Tioga retreat of 15.57 ± 0.82 ^{14}C ka BP (18.84 ± 0.91 cal.ka BP; Wagner et al., 1982, cited in Fullerton, 1986). The basal age from the Greenstone Lake cores instead demonstrates that the area near Tioga Pass was deglaciated by ~ 13.1 ^{14}C ka BP (~ 15.5 cal.ka BP; Clark, 1997). James et al.'s (2002) cosmogenic data show that the Tioga glaciers retreated rapidly from the middle elevations of the Yuba River 15,000–14,000 years ago. Clark and Gillespie (1997) agreed that the Tioga glaciers vanished entirely or were restricted to cirques during this interval.

34.2.3.1. Post-Tioga Advances

The Hilgard advance, proposed by Birman (1964) as a separate post-Tioga glaciation, was considered to be a very late Tioga stade by Birkeland et al. (1976), and a recessional standstill by M. Clark (personal communication, 1988). Thus, the Hilgard glaciation is no longer regarded as a separate stade.

The Recess Peak glaciation is the first post-Tioga advance in the Sierra Nevada for which evidence has been discovered. As first described by Birman (1964), Recess Peak moraines are restricted to the vicinity of Pleistocene cirques. Because of their fresh character, most early workers concluded that the Recess Peak moraines were Neoglacial, constructed within the past 2000–3000 years (Birman, 1964; Curry, 1969; Scuderi, 1987). However, soil work by Yount et al. (1982) suggests that Recess Peak deposits are early Holocene or older. Firm numerical constraints on the moraines from sediment coring of nearby lakes demonstrates that the Recess Peak advance began by ~ 14.2 cal.ka BP and ended before ~ 13.1 cal.ka BP (Clark, 1997; Phillips et al., 2009). Plummer's (2002) ^{36}Cl cosmogenic ages of 12.6 ± 1.3 ka (production rate uncertainties included) overlap Clark's (1997) age range.

ELA estimates for Recess Peak glaciers were only ~ 340 m higher than that for the maximum Tioga glaciers (Table 34.1). This value may probably overestimate the climatic severity during the Recess Peak advance because cirque glaciers can occur at anomalously low elevations.

Curry (1971) recognised two Holocene Neoglacial advances having lichenometric ages equivalent to ~ 1100 and 970 ^{14}C a BP (~ 1015 and 920 cal.a BP). Coring of the Conness Lakes indicates that Neoglaciation began there by ~ 3.2 ^{14}C ka BP (3.4 cal.ka BP, Konrad and Clark, 1998; 3.2 cal.ka BP, Bowerman and Clark, 2011). Curry (1971) dated the Matthes (Little Ice Age) advances at 620 ± 55 ^{14}C a BP ($\sim 610 \pm 40$ cal.a BP). The absence of a ~ 700 ^{14}C a BP (~ 630 cal.a BP) tephra blanketing Matthes moraines coupled with evidence from dendrochronology (Wood, 1977) indicates that Matthes glaciers reached their maximum positions after that eruption (180 cal.a BP: Bowerman and Clark, 2011). Stine (1994) identified two droughts at about AD 1112–900 and 1350–1250,

---

FIGURE 34.7 (A) δ^{18}O record (left) and total inorganic carbon (TIC) from Mono Lake (right), for 41–12 ka. Low stands L1–L4 have been labelled and shaded. Middle curve shows the δ^{18}O record from the GISP2 ice core, Greenland, with Heinrich events H1–H4 and Dansgaard–Oeschger events D2–D8 labelled for reference. From Benson et al. (1998b). ^{10}Be cosmic-ray exposure ages for Bishop Creek compared to three nearby drainages, Walker Creek (Fig. 34.2B), Little McGee Creek and the South Fork of the Yuba River, on the west side of the Sierra Nevada are shown on the far right. Error bars are 1σ and the integer by each indicates the number of samples dated. (B) Dates for the same drainages over an expanded time range (250–0 ka) compared to the SPEC-MAP marine δ^{18}O record (Imbrie et al., 1984; data from Imbrie and McIntyre, 2006). Marine oxygen isotope stages are labelled and shaded for reference. Dating references: [a]Phillips et al. (2009); [b]Phillips et al. (1996); [c]James et al. (2002); [d]Porter and Swanson (2008); [e]Bierman et al. (1995); [f]Dünhnforth et al. (2007).

just before the onset of the Matthes advances in the Sierra Nevada. ELAs for Matthes glaciers were ~480 m higher than ELAs for the maximum Tioga glaciers.

The absence of moraines between the Recess Peak and Matthes moraines, and the absence of outwash deposits between 13.1 and 3.4 cal. ka BP, suggests that no significant glacier advances in the Sierra Nevada occurred during that time, including during the Younger Dryas interval (Clark and Gillespie, 1997). If Younger Dryas glaciers were present in the Sierra Nevada, they must have been smaller than both the Recess Peak and Matthes glaciers and restricted to cirques.

34.3. DISCUSSION

34.3.1. Younger Dryas

The absence of any moraines between the Recess Peak moraines and the Matthes moraines, as well as the absence of any periods of outwash between 13,100 and 3400 cal. a BP, indicates that no significant glacial advances in the Sierra Nevada occurred during that time, which includes during the Younger Dryas interval (Clark and Gillespie, 1997), although alluvial fans have experienced limited Holocene aggradation (e.g. Bierman et al., 1995; Zehfuss et al., 2001). This finding, combined with evidence favouring Younger Dryas advances in the Rocky Mountains and potentially the North Cascades (e.g. Reasoner et al., 1994; Kovanen and Easterbrook, 2001) suggest a complex regional climate during the late-glacial period in western North America. It also indicates that the climate during the Little Ice Age was the coldest and/or wettest (i.e. most glacial) in the Sierra Nevada of the past 13,000 years.

34.3.2. Comparison with the Cascade Range

Porter and Swanson (2008) reported 76 ^{36}Cl ages for a moraine sequence from Icicle Creek, in the Cascade Range. The dates cluster at 12.5 ± 0.5, 13.3 ± 0.8, 16.1 ± 1.1, 19.1 ± 3, 70.9 ± 1.5, 93.1 ± 2.6 and 105.4 ± 2.2 ka (Fig. 34.7). The younger, Tioga-age dates are in good agreement with those from the Sierra Nevada. However, the agreement is less clear for MIS 4 and 5 moraines. It is possible that this is due to incomplete preservation, or due to the noted problems with dating older, eroded deposits, but it is also possible that the glaciations themselves were more synchronous up and down the Pacific coast of North America at the MIS 2 LGM than before.

34.3.3. Lake Records

Direct glacial deposits and landforms present an incomplete record of mountain glaciations. This record has been fleshed out by analysis of lake-sediment cores (e.g. Bischoff et al., 1997; Fig. 34.7). Core data can establish the duration of glaciations, information that is difficult to acquire from moraines and drift alone. The findings of Benson et al. (1996) at Mono Lake suggest that the average duration of glacial advance and retreat during MIS 3–2, period including the Tioga glaciation, was on the order of 3 ka. In fact, the fluctuation of Sierra Nevada glaciers inferred from Owens Lake (Bischoff et al., 1997; Bischoff and Cummins, 2001) and Mono Lake (Benson et al., 1996, 1998a,b) cores seems to show the same three scales of climatic oscillation as the marine/global system: Milankovitch, Heinrich and Dansgaard–Oesger (Benson et al., 1996, 1998a,b). Benson et al. (1998a) interpreted the data to show that glacier activity in the Sierra Nevada was synchronous with cold periods in the North Atlantic.

34.3.4. Reliability of "Older" Cosmic-Ray Exposure Dates

Figure 34.7 suggests that, especially for glacial deposits older than ~50 ka, there may be "geologic" scatter in the dates in excess of analytic and systematic measurement errors and in excess of what might be expected from geologic mapping and relative dating, possibly due to unaccounted-for effects of soil and boulder burial and erosion. It is also possible that our ability to "read" the landscape and correlate landforms on the basis of relative weathering is less than we have thought and hoped; thus, preservation of moraines may have been more erratic than we have assumed. Nevertheless, the spatial richness of the record from glacial deposits and landforms supplies information difficult to glean from a limited number of cores alone, especially in the latest Pleistocene.

34.3.5. Dating Paraglacial Deposits

For a time, it seemed that the record of glaciation might be better read from alluvial or outwash fans downvalley from the glaciers than from the moraines themselves. Gillespie (1982) suspected that paraglacial deposition accounted for much of the Sierran Bajada, and Whipple and Dunne (1992) elaborated a process-based explanation for this synchrony. Bierman et al. (1995) measured pairs of ^{10}Be and ^{26}Al of cosmic-ray exposure dates that seemed to confirm the hypothesis. Dünhnforth et al. (2007) added further dates for the glaciated Shepherd Creek fan, and the adjacent unglaciated Symmes Creek fan, finding more Holocene aggradation in the unglaciated drainage. Nevertheless, comparison of fan and moraine dates (Fig. 34.7) suggests that the glacial record is not much better preserved in the fans than in the moraines. It may be that fan-resurfacing floods that do occur occasionally beneath glaciated canyons (e.g. July 2007, Oak Creek) are sufficient to partially obliterate or at least add scatter to the recovered record, or again it may be that erosion on the fan surface is by itself sufficient to do so.

34.3.6. ELA Depression

ELA depressions for the different Pleistocene Sierra Nevada glacier advances represented in the land record increase with age but show a remarkable consistency. Figure 34.8A shows the trend in Tioga ELAs from North to South for the Sierra Nevada. In Fig. 34.8B, the nearly parallel trend for the youngest Tahoe glaciers is shown. From the Tioga to the Sherwin glaciation, the ELA depression of the maximum advances was within a range of ~200 m.

That ELA depressions should be increasingly greater for older glaciations is expected because moraines for smaller, older glaciers were obliterated by younger glaciers. What is surprising is that the ELAs dropped time and again to within 10% or 20% of their MIS 2 values, even though the pattern of sea-level depression inferred from marine cores suggests that the high-latitude ice sheets were much larger during MIS 2 (Tioga) than, for example, during MIS 4–3 (Tenaya, Tahoe II; e.g. Martinson et al., 1987).

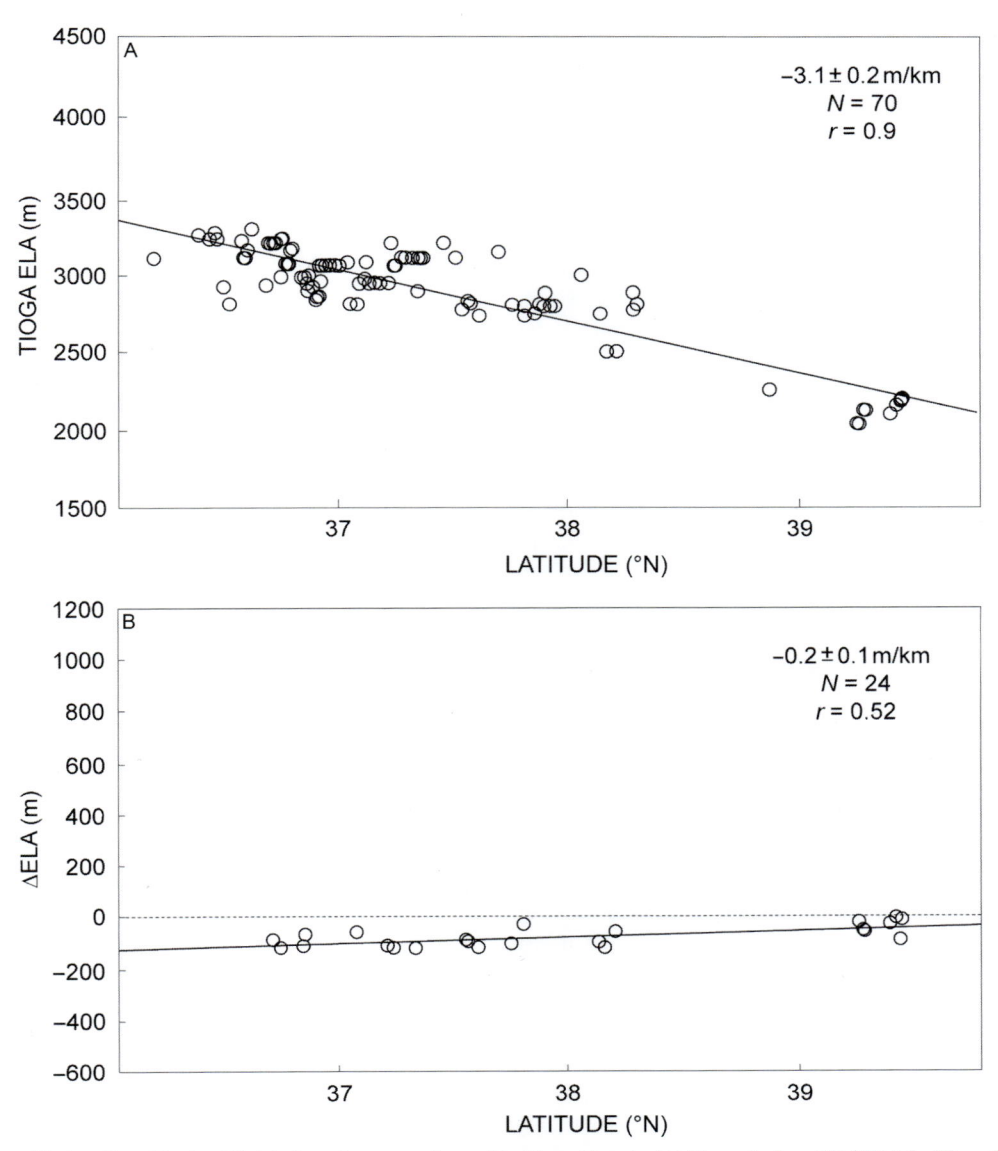

FIGURE 34.8 Equilibrium-line altitudes (ELAs) along the eastern front of the Sierra Nevada. (A) Tioga glaciers. (B) ΔELA for Tioga (Ti) and late Tahoe (Ta) moraines (ELA$_{Ti}$− ELA$_{Ta}$). ELA$_{Ti}$ descends almost linearly from south to north. The cause of the deviation of up to 300 m above the regression line near 37.5°N is unknown. ELA$_{Ta}$ rose slightly higher to the north than ELA$_{Ti}$; the cause for this trend, if real, is also unknown. After Gillespie (1991).

Gillespie and Molnar (1995) emphasised this discrepancy, but Shackleton (2000) pointed out that the sawtooth history of ice volume inferred from the marine cores was due more than previously suspected to effects of cold water. Consequently, James et al. (2002) suggested that the record of mountain and high-latitude glaciations was more similar than Gillespie and Molnar (1995) suspected. Nevertheless, there does appear to some fundamental limit to maximum ELA depressions, at least over the past ~800 ka. This in turn implies a fundamental limit to climatic extremes on the western coast of North America.

34.4. SUMMARY

The Sierra Nevada was repeatedly glaciated during the Quaternary Period. The glacial record on the eastern side of the range includes at least eight Pleistocene glaciations, and multiple stades are known for some of them. Lake-sediment core data suggest that this record is incomplete and that there have been many more advances than have been recognised, no doubt in part due to obliteration of evidence from smaller, older glaciers by larger, younger ones.

The last Pleistocene glacier advance, the Recess Peak advance, pre-dated the Younger Dryas event in the Sierra Nevada, and any Younger Dryas glaciers there must have been restricted to the cirques. Because there is evidence elsewhere in western North America suggesting the presence of Younger Dryas glaciers, it appears that there may have been considerable local variability in the regional response to a "global" climatic event. Nevertheless, the available lake-core data have been interpreted to show a general synchronism of glacier advances in the Sierra Nevada to cold periods in the North Atlantic.

ELA depressions of the largest glaciers were the same within 20%, independent of age. This suggests a remarkable consistency in the extreme climate in California over a time span of 800 ka or more.

Recent efforts to date glacial deposits numerically have added detail to our understanding of Sierra Nevada glacial chronology, and have facilitated correlation with the lake-core records. However, except for results for the late Pleistocene advances, results may serve only as limits to the age of the glaciations, because of erosion and other geological complexities. Further numerical dating will probably be required in order to refine the glacial history, particularly for older glaciations for which great uncertainties remain.

ACKNOWLEDGEMENTS

We thank Laura Gilson, Harvey Greenberg and Paul Zehfuss for assistance in preparing this chapter.

REFERENCES

Adam, D.P., 1966. Osgood Swamp (C-14) date A-545. Radiocarbon 8, 10.

Adam, D.P., 1967. Late Pleistocene and recent palynology in the central Sierra Nevada. In: Cushing, E.J., Wright, H.E., Jr. (Eds.), Quaternary Paleoecology. INQUA Congress VII, Proc. 7. Yale University Press, New Haven, Connecticut, pp. 275–300.

Antevs, E., 1938. Postpluvial climatic variations in the Southwest. Am. Meteorol. Soc. Bull. 19, 190–193.

Antevs, E., 1948. The Great Basin, with emphasis on glacial and post-glacial times—climatic changes and pre-white man. Bull. Univ. Utah Biol. 38, 168–191.

Bailey, R.A., Dalrymple, G.B., Lanphere, M.A., 1976. Volcanism, structure, and geochronology of Long Valley caldera, Mono County, California. J. Geophys. Res. 81, 725–744.

Benson, L., 1999. Records of millennial-scale climate change from the Great Basin of the Western United States. In: Clark, P.U., Webb, R.S., Keigwin, L.D. (Eds.), Mechanisms of Global Climate Change at Millennial Time Scales. Geophysical Monograph 112 American Geophysical Union, Washington, DC, pp. 203–226.

Benson, L.V., Burdett, J.W., Kashgarian, M., Lund, S.P., Phillips, F.M., Rye, R.O., 1996. Climatic and hydrologic oscillations in Owens Lake basin and adjacent Sierra Nevada, California. Science 274, 746–751.

Benson, L.V., May, H.M., Antweiler, R.C., Brinton, T.I., 1998a. Continuous lake-sediment records of glaciation in the Sierra Nevada between 52,600 and 12,500 ^{14}C yr B.P. Quatern. Res. 50, 113–127.

Benson, L.V., Lund, S.P., Burdett, J.W., Kashgarian, M., Rose, T.P., Smoot, J.P., et al., 1998b. Correlation of Late-Pleistocene lake-level oscillations in Mono Lake, California, with North Atlantic climate events. Quatern. Res. 49, 1–10.

Bierman, P.R., Gillespie, A.R., Caffee, M.W., 1995. Cosmogenic ages for earthquake recurrence intervals and debris flow fan deposition, Owens Valley, California. Science 270, 447–450.

Birkeland, P.W., Burke, R.M., 1988. Soil catena chronosequences on eastern Sierra Nevada moraines, California, U.S.A. Arctic Alpine Res. 20, 473–484.

Birkeland, P.W., Burke, R.M., Yount, J.C., 1976. Preliminary comments on late Cenozoic correlations on the Sierra Nevada. In: Mahoney, W.C. (Ed.), Quaternary Stratigraphy of North America. Dowden, Hutchinson and Ross, Stroudsburg, PA, pp. 283–295.

Birkeland, P.W., Burke, R.M., Walker, A.L., 1980. Soils and sub-surface weathering features of Sherwin and pre-Sherwin glacial deposits, eastern Sierra Nevada, California. Bull. Geol. Soc. Am. 91, 238–244.

Birman, J.H., 1964. Glacial geology across the crest of the Sierra Nevada, California. Geological Society of America Special Paper, 75, 80pp.

Bischoff, J.L., Cummins, K., 2001. Wisconsin glaciation of the Sierra Nevada (79,000-15,000 yr B.P.) as recorded by rock flour in sediments of Owens Lake, California. Quatern. Res. 55, 14–24.

Bischoff, J.L., Meisling, K.M., Fitts, J.P., Fitzpatrick, J.A., 1997. Climatic oscillations 10,000-155,000 yr B.P. at Owens Lake, California reflected in glacial rock flour abundance and lake salinity in Core OL-92. Quatern. Res. 48, 313–325.

Blackwelder, E., 1931. Pleistocene glaciation in the Sierra Nevada and Basin Ranges. Geol. Soc. Am. Bull. 42, 865–922.

Bowerman, N. D., Clark, D. H., 2011. Holocene glaciation of the central Sierra Nevada, California. Quatern. Sci. Rev. in press.

Brocklehurst, S.H., Granger, D.E., Whipple, K.X., 2002. Implications of old, glaciated surfaces at high elevations in the Sierra Nevada, California. EOS Trans. Am. Geophys. Union 83 (47) Fall Meeting Supplement, Abstract H22B-0884.

Burbank, D., 1991. Late Quaternary snowline reconstructions for the southern and central Sierra Nevada, California: reassessment of the "Recess Peak glaciation" Quatern. Res. 36, 294–306.

Burke, R.M., Birkeland, P.W., 1979. Re-evaluation of multiparameter relative dating techniques and their application to the glacial sequence along the eastern escarpment of the Sierra Nevada, California. Quatern. Res. 11, 21–51.

Bursik, M.I., Gillespie, A.R., 1993. Late Pleistocene glaciation of Mono Basin, California. Quatern. Res. 39, 24–35.

Clark, M.M., 1967. Pleistocene glaciation of the drainage of the West Walker River, Sierra Nevada, California. Ph.D. dissertation, Stanford, Stanford University, 170pp.

Clark, M.M., 1968. Pleistocene glaciation of the upper West Walker drainage, Sierra Nevada, California. Geological Society of America Special Paper, 115, 317.

Clark, M.M., 1972. Range-front faulting: cause of anomalous relationships among moraines of the eastern slope of the Sierra Nevada, California. Geol. Soc. Am. Abstr. Programs 4, 137.

Clark, D.H., 1995. Extent, timing, and climatic significance of latest Pleistocene and Holocene glaciation in the Sierra Nevada. Ph.D. dissertation, University of Washington, 193pp.

Clark, D.H., 1997. A new alpine lacustrine sedimentary record from the Sierra Nevada: implications for Late-Pleistocene paleoclimate reconstructions and cosmogenic isotope production rates. EOS Trans. Am. Geophys. Union 78, F249.

Clark, D.H., Gillespie, A.R., 1997. Timing and significance of late-glacial and Holocene glaciation in the Sierra Nevada, California. Quatern. Int. 38 (39), 21–38.

Clark, D., Gillespie, A.R., Clark, M.M., Burke, R.M., 2003. Mountain glaciations of the Sierra Nevada. In: Easterbrook, D.J. (Ed.), Quaternary Geology of the United States. International Quaternary Association (INQUA) 2003 Field Guide Volume XVI INQUA Congress. Desert Research Institute, Reno, NV, pp. 287–312.

Curry, R.R., 1969. Holocene climatic and glacial history of the central Sierra Nevada, California. In: Shumm, S.A., Bradley, W.C. (Eds.), United States Contributions to Quaternary Research, Geological Society of America Special Paper, 123, 1–47.

Curry, R.R., 1971. Glacial and Pleistocene History of the Mammoth Lakes Sierra—A Geologic Guidebook. Geological Series Publication No. 11 University of Montana, Department of Geology, Missoula, 49pp.

Dalrymple, G.B., 1963. Potassium-Argon ages of some Cenozoic volcanic rocks of the Sierra Nevada, California. Geol. Soc. Am. Bull. 74, 379–390.

Dalrymple, G.B., 1964. Cenozoic chronology of the Sierra Nevada, California. University of California Publications in Geological Science, 47, 41pp.

Danskin, W.R., 1998. Evaluation of the hydrologic system and selected water-management alternatives in the Owens Valley, California. U.S. Geological Survey Water Supply Paper, 2370, 175pp.

Dünnforth, M., Densmore, A.L., Ivyochs, S., Allen, P.A., Kubik, P.W., 2007. Timing and patterns of debris flow deposition on Shepherd and Symmes creek fans, Owens Valley, California, deduced from cosmogenic 10Be. J. Geophys. Res. 112, F03S15. doi:10.1029/2006JF000562.

Flint, R.F., 1957. Glacial and Pleistocene Geology. Wiley, New York 553pp.

Fullerton, D.S., 1986. Chronology and correlation of glacial deposits in the Sierra Nevada, California. In: Sibrava, V., Bowen, D.Q., Richmond, G.M. (Eds.), Quaternary Glaciation of the Northern Hemisphere, Quaternary Science Reviews, vol. 5, pp. 161–169.

Gillespie, A.R., 1982. Quaternary glaciation and tectonism in the Southeastern Sierra Nevada, Inyo County, California. Ph.D. dissertation, Caltech, 695pp.

Gillespie, A.R., 1991. Testing a new climatic interpretation for the Tahoe glaciation. In: Hall, C.A., Jr., Doyle-Jones, V., Widawski, B. (Eds.), Natural History of Eastern California and High-Altitude Research, Proceedings of the White Mountain Research Station Symposium, vol. 3, pp. 383–398.

Gillespie, A.R., Molnar, P., 1995. Asynchronism of maximum advances of mountain and continental glaciations. Rev. Geophys. 33, 311–364.

Gillespie, A.R., Huneke, J.C., Wasserburg, G.J., 1984. Eruption age of a 100,000-year-old basalt from ^{40}Ar-^{39}Ar analysis of partially degassed xenoliths. J. Geophys. Res. 89, 1033–1048.

Gillespie, A.R., Burke, R.M., Clark, M.M., 1999. Eliot Blackwelder and the Alpine Glaciations of the Sierra Nevada. In: Moores, E.M., Sloan, D., Stout, D.L. (Eds.), Classic California Cordilleran Centennial Concepts: A View from California. Geological Society of America, Special Paper 338, Boulder, Colorado, pp. 443–452.

Gosse, J.C., Phillips, F.M., 2001. Terrestrial in situ cosmogenic nuclides: theory and applications. Quatern. Sci. Rev. 40, 1475–1560.

Hallet, B., Putkonen, J., 1994. Surface dating of dynamic landforms: young boulders on aging moraines. Science 265, 937–940.

Huber, N.K., 1981. Amount and timing of Cenozoic uplift and tilt of the central Sierra Nevada, California—evidence from the upper San Joaquin River. U.S. Geological Survey Professional Paper, 1197, 28pp.

Huber, N.K., Bateman, P.C., Wahrhaftig, C., 1989. Geologic map of Yosemite National Park and Vicinity, California. U.S. Geological Survey, one sheet.

Imbrie, J.D. & McIntyre, A., 2006. SPECMAP time scale developed by Imbrie et al., 1984 based on normalized planktonic records (normalized O-18 vs time, specmap. 017). PANGAEA®Publishing Network for Geoscientific & Environmental Data, Doi: 10.1594/PANGAEA.441706, last accessed 15 April 2011.

Imbrie, J.J.D., Hays, D.G., Martinson, A., McIntyre, A.C., Mix, J.J., Morley, N.G., et al., 1984. The orbital theory of Pleistocene climate: Support from a revised chronology of the marine δ^{18}O record. In, Milankovitch and Climate, Part I, edited by A. Berger et al., pp. 269–305, D. Reidel, Norwell, MA.

James, L.A., Harbor, J., Fabel, D., Dahms, D., Elmore, D., 2002. Late Pleistocene glaciations in the Northwestern Sierra Nevada, California. Quatern. Res. 57, 409–419.

Kaufman, D.S., Porter, S.C., Gillespie, A.R., 2004. Quaternary alpine glaciation in Alaska, the Pacific Northwest, Sierra Nevada, and Hawaii. In: Gillespie, A.R., Porter, S.C., Atwater, B.F. (Eds.), The Quaternary Period in the United States. Developments in Quaternary Science, vol. 1. Elsevier, Amsterdam, pp. 77–103.

Kistler, R.W., 1966. Geologic map of the Mono Craters Quadrangle, Mono and Tuolumne Counties, California. U.S. Geological Survey Geologic quadrangle Map GQ-462, scale 1:62,500, one sheet.

Konrad, S.K., Clark, D.H., 1998. Evidence for an early Neoglacial glacier advance from rock glaciers and lake sediments in the Sierra Nevada, California, USA. Arctic Alpine Res. 30, 272–284.

Kovanen, D.J., Easterbrook, D.J., 2001. Late Pleistocene, post-Vashon, alpine glaciation of the Nooksack drainage, North Cascades, Washington. Geol. Soc. Am. Bull. 113, 274–288.

LeConte, J., 1873. On some of the ancient glaciers of the Sierra. Am. J. Sci. 5, 325–339.

Martinson, D.G., Pisia, N.G., Hayes, J.D., Imbrie, J., Moore, T.C., Jr., Shackleton, N.J., 1987. Age dating and the orbital theory of the ice ages; development of a high-resolution 0 to 300,000-year chronostratigraphy. Quatern. Res. 27, 1–29.

Mathieson, S.A., Sarna-Wojciki, A.M., 1982. Ash layer in Mohawk Valley, Plumas County, California, correlated with the 0.45 M.Y.-old Rockland ash—implications for the glacial and lacustrine history of the region. Geol. Soc. Am. Abstr. Program 14, 184.

Matthes, F., 1930. Geologic history of the Yosemite Valley. U.S. Geological Survey Professional Paper, 160, 137pp.

Matthes, F., 1965. Glacial reconnaissance of Sequoia National Park, California. U.S. Geological Survey Professional Paper, 504-A, A1–A58.

Meierding, T.C., 1982. Late Pleistocene glacial equilibrium-line altitudes in the Colorado Front range: a comparison of methods. Quatern. Res. 18, 289–310.

Nishiizumi, K., Winterer, E.L., Kohl, C.P., Klein, J., Middleton, R., Lal, D., et al., 1989. Cosmic ray production rates of ^{10}Be and ^{26}Al in quartz from glacially polished rocks. J. Geophys. Res. 94, 17907–17915.

Phillips, F.M., Zreda, M.G., Smith, S.S., Elmore, D., Kubik, P.W., Sharma, P., 1990. Cosmogenic chlorine-36 chronology for glacial deposits at Bloody Canyon, eastern Sierra Nevada. Science 248, 1529–1532.

Phillips, F.M., Zreda, M.G., Benson, L.V., Plummer, M.A., Elmore, D., Sharma, P., 1996. Chronology for fluctuations in late Pleistocene Sierra Nevada glaciers. Science 274, 749–751.

Phillips, F.M., Stone, W.D., Fabryka-Martin, J.T., 2001. An improved approach to calculating low-energy cosmic-ray neutron fluxes at the land/atmosphere interface. Chem. Geol. 175, 689–701.

Phillips, F.M., Zreda, M., Plummer, M.A., Elmore, D., Clark, D.H., 2009. Glacial geology and chronology of Bishop Creek and vicinity, eastern Sierra Nevada, California. Geol. Soc. Am. Bull. 121, 1013–1033.

Plummer, M., 2002. Paleoclimatic conditions during the last deglaciation inferred from combined analysis of pluvial and glacial records. Ph.D. dissertation, New Mexico Institute of Mining and Technology.

Porter, S.C., Swanson, T.W., 2008. ^{36}Cl dating of the classic Pleistocene glacial record in the Northern Cascade Range, Washington. Am. J. Sci. 308, 130–166.

Putnam, W.C., 1962. Late Cenozoic geology of McGee Mountain, Mono County, California. Univ. California Publ. Geol. Sci. 40, 181–218.

Reasoner, M.A., Osborn, G., Rutter, N.W., 1994. Age of the Crowfoot advance in the Canadian Rocky Mountains: a glacial event coeval with the Younger Dryas oscillation. Geology 22, 439–442.

Rupper, S., Roe, G., Gillespie, A.R., 2009. Spatial patterns of Holocene glacier advance and retreat in Central Asia. Quatern. Res. 72, 337–346.

Sarna-Wojcicki, A.M., Meyer, C.E., Bowman, H.R., Hall, N.T., Russell, P.C., Woodward, M.J., et al., 1985. Correlation of the Rockland ash bed, a 400,000-year-old stratigraphic marker in northern California and western Nevada, and implications for middle Pleistocene paleogeography of central California. Quatern. Res. 23, 236–257.

Sarna-Wojcicki, A.M., Pringle, M.S., Wijbrans, J., 2000. New ^{40}Ar/^{39}Ar age of the Bishop Tuff from multiple sites and sediment rate calibration for the Matayama-Bruhnes boundary. J. Geophys. Res. 105, 21,431–21,433.

Scuderi, L.A., 1987. Glacier variations in the Sierra Nevada, California, as related to a 1200-year tree-ring chronology. Quatern. Res. 27, 220–231.

Shackleton, N.J., 2000. The 100,000 year ice-age cycle identified and found to lag temperature, carbon dioxide, and orbital eccentricity. Science 289, 1897–1902.

Sharp, R.P., 1968. Sherwin TillBishop Tuff relationship, Sierra Nevada, California. Bull. Geol. Soc. Am. 79, 351–364.

Sharp, R.P., 1972. Pleistocene glaciation, Bridgeport Basin. Bull. Geol. Soc. Am. 83, 2233–2260.

Sharp, R.P., Birman, J.H., 1963. Additions to the classical sequence of Pleistocene glaciations, Sierra Nevada, California. Bull. Geol. Soc. Am. 74, 1079–1086.

Smith, G.I., Bischoff, J.L., Bradbury, J.P., 1991. Synthesis of the paleoclimatic record from Owens Lake core OL-92. Geol. Soc. Am. Spec. Pap. 317, 143–160.

Stine, S., 1994. Extreme and persistent drought in California and Patagonia during mediaeval time. Nature 369, 546–549.

Wagner, D.L., Jennings, C.W., Bedrossian, T.L., Bortugno, E.J., Saucedo, G.J., 1982. Sacramento Quadrangle, California. California Divisions of Mines and Geology, Region Geologic Map Series, Scale 1:250,000.

Warhaftig, C., Birman, J.H., 1965. The Quaternary of the Pacific mountain system in California. In: Wright, H.E., Frey, D.G. (Eds.), The Quaternary of the United States. Princeton University Press, Princeton, pp. 299–340.

Whipple, K.X., Dunne, T., 1992. The influence of debris-flow rheology on fan morphology, Owens Valley, California. Geol. Soc. Am. Bull. 104, 887–900.

Wood, S.H., 1977. Distribution, correlation, and radiometric dating of late Holocene tephra, Mono and Inyo Craters, eastern California. Geol. Soc. Am. Bull. 88, 89–95.

Yount, J.C., Birkeland, P.W., Burke, R.M., 1982. Holocene glaciation, Mono Creek, central Sierra Nevada, California. Geol. Soc. Am. Abstr. Programs 14, 246.

Zehfuss, P.H., Bierman, P.R., Gillespie, A.R., Burke, R.M., Caffee, M.W., 2001. Slip rates on the Fish Springs Fault, Owens Valley, California, deduced from cosmogenic ^{10}Be and ^{26}Al and soil development on fan surfaces. Geol. Soc. Am. Bull. 113, 241–255.

Chapter 35

Pleistocene Glaciation of Hawaii

Stephen C. Porter

Department of Earth and Space Sciences, University of Washington, Seattle, Washington 98195-1360, USA

The islands of Hawaii, the tallest of which rise 8–9 km above the floor of the deep ocean, are formed of overlapping volcanoes constructed during the past several million years. Of the many volcanoes comprising the islands, only Mauna Kea, on the island of Hawaii, has an unequivocal record of multiple glaciation (Porter, 1979a, 2005; Wolfe and Moore, 1996). Haleakala volcano on Maui likely was glaciated repeatedly during the Middle Pleistocene (Moore et al., 1993; Porter, 2005), but details of its glacial history have yet to be worked out. Mauna Loa, Hawaii's second-highest summit, probably had a small ice cap during the last glacial maximum (LGM), but if so, the record of it is buried beneath Holocene lava that mantles the upper slopes and is nowhere exposed (Figs. 35.1 and 35.2).

The extent of the last (Makanaka) ice cap on Mauna Kea (4206 m a.s.l.) is marked by a discontinuous terminal moraine or moraine complex and upper limits of erratic boulders on cinder cones that lie within the glacial limit

FIGURE 35.1 Topographic map showing the five Pleistocene volcanoes that comprise the island of Hawaii, and the location of the ice cap on Mauna Kea.

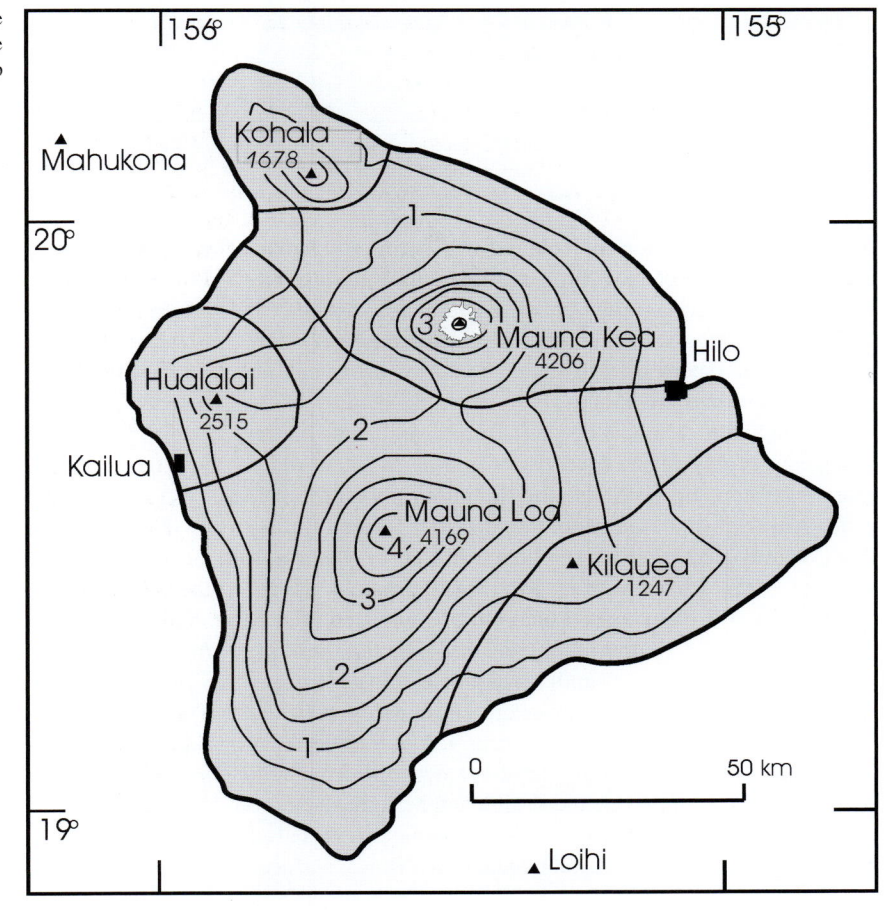

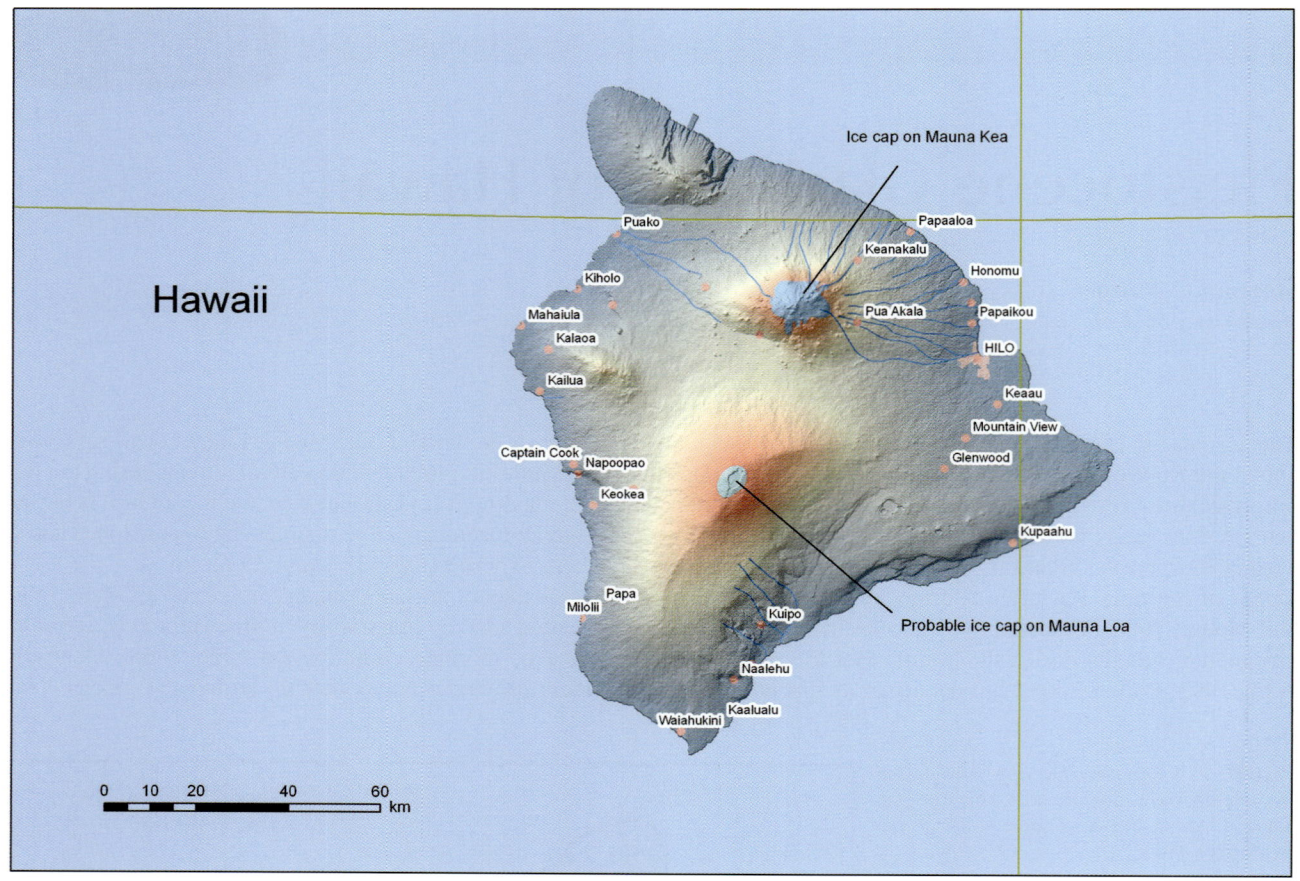

FIGURE 35.2 Relief map showing Late Pleistocene ice caps on Mauna Loa (inferred) and Mauna Kea on island of Hawaii.

(Porter, 1979b). The ice cap covered ca. 70 km² and had an average thickness of ca. 75 m. Its lobate margin was controlled by local topography and reached as low as 3200 m a.s.l. Upslope from the outer moraines, striated bedrock surfaces are interspersed with loose bouldery drift, but no younger moraines are found, implying rapid deglaciation.

Beyond the Makanaka drift limit, and below the youngest pre-Makanaka lava flows, remnants of an older moraine system are visible on the southern flank of the volcano. Although the area of the ice cap that formed during this (Waihu) glaciation was greater than that of Makanaka age, its extent is estimated only approximately as ca. 150 km². A still older drift (Pohakuloa) is exposed in several gullies on the southern slope of the mountain beneath pre-Waihu lavas and in a small kipuka on the eastern slope. This glacier may have had an extent similar to that of the Waihu ice cap. It formed prior to eruption of alkalic lavas and cinder cones in the summit region that supplied the bulk of the sediment comprising the Waihu and Makanaka drifts.

The chronology of Mauna Kea glaciation is based primarily on K/Ar dates of intercalated lava flows, surface-exposure ages (^{36}Cl, ^{3}He and ^{10}Be) of moraine boulders and several limiting ^{14}C ages. Collectively, the dates indicate that Pokhaukloa drift probably dates to marine isotope stage (MIS) 6, Waihu drift either to late MIS 6 or to MIS 5d and Makanaka to MIS 2 (Wolfe et al., 1997). After reaching its maximum limit of the last glaciation ca. 20,600 years ago, the glacier receded and then likely re-advanced about 16,000 years ago (Dorn et al., 1991; Wolfe et al., 1997). ^{3}He and ^{36}Cl ages from lava near the summit and ^{14}C dates from sediments in nearby Lake Waiau suggest that the ice cap had disappeared by ca. 15,000–14,000 years ago.

The reconstructed equilibrium-line altitude (ELA) of the Makanaka ice cap slopes gently towards the south-south-east and lies, on average, ca. 425 m lower than the modern summit; corrected for isostatic subsidence of the volcano, the figure is increased to 470 m (Porter, 1979b). The Waihu ELA can be reconstructed only on the southern flank of the volcano, where it lay ca. 675 m below the summit during that glaciation. The summit is now too low to sustain a glacier under the present climate. If the ELA of the LGM approximated the level of the modern July freezing isotherm (ca. 4715 m a.s.l.; Porter, 2005), then ELA depression during the LGM was about 930 m.

REFERENCES

Dorn, R.I., Phillips, F.M., Zreda, M.G., Wolfe, E.W., Jull, A.J.T., Donahue, D.J., et al., 1991. Glacial chronology of Mauna Kea, Hawaii, as constrained by surface-exposure dating. Natl. Geogr. Res. Explor. 7, 456–471.

Moore, J.G., Porter, S.C., Mark, R., 1993. Glaciation of Haleakala volcano, Hawaii [abst]. Geol. Soc. Am. Bull. 25, 123–124.

Porter, S.C., 1979a. Quaternary stratigraphy and chronology of Mauna Kea, Hawaii: a 380,000-yr record of mid-Pacific volcanism and ice-cap glaciation. Geol. Soc. Am. Bull. 90, Part I: Summary, 609–611; Part II: Complete article, 980–1093.

Porter, S.C., 1979b. Hawaiian glacial ages. Quatern. Res. 12, 161–187.

Porter, S.C., 2005. Pleistocene snowlines and glaciation of the Hawaiian Islands. Quatern. Int. 138/139, 118–128.

Wolfe, E.W., Morris, J. (Eds.), 1996. Geologic map of the island of Hawaii, U.S. Geological Survey Miscellaneous Investigations Map, I-2524-A (1:100,000 scale).

Wolfe, E.W., Wise, W.S., Dalrymple, G.B., 1997. The geology and petrology of Mauna Kean volcano, Hawaii—a study of postshield volcanism. U.S. Geological Survey Professional Paper, 1557, 129pp.

Quaternary Glaciations in Illinois

B. Brandon Curry*, David A. Grimley and E. Donald McKay III

Illinois State Geological Survey, Prairie Research Institute, 615 E. Peabody Drive, Champaign, Illinois 61820, USA
*Correspondence and requests for materials should be addressed to B. Brandon Curry. E-mail: curry@isgs.illinois.edu

36.1. INTRODUCTION

During the Quaternary, Illinois was glaciated by the south-central margin of the Laurentide Ice Sheet at least four times. The latest and penultimate glaciations are the Wisconsin and Illinois Episodes (Hansel and Johnson, 1996). The glaciations and interglacials prior to this and extending back to the onset of the Pleistocene in Illinois are collectively known as the pre-Illinois Episode (PIE). The upper part of the PIE is a long-lived interglacial known as the Yarmouth Episode (Hansel and McKay, 2010). The material referents for glacial episodes include glacigenic diamicton (mostly till but including debris-flow deposits), sand and gravel outwash, laminated or stratified fine sands, silt and clay lake deposits and silty loess. The glacial events are separated by weathering horizons formed in the glacial sediment, which may include a relatively thin mantle (< 4 m) of colluvium and accreted material. Stratigraphically important weathering horizons are classified as geosols (Follmer et al., 1979; Follmer, 1983; Curry and Follmer, 1992; Hansel and Johnson, 1996; Stiff and Hansel, 2004). North America's southernmost Quaternary glacial diamicton of the Laurentide Ice Sheet is mapped just south of Carbondale, Illinois (Fig. 36.1; 37.58°N; Weibel and Nelson, 1993). The sediment was deposited during the Illinois Episode, a time when glaciers advanced about 200 km south of the southernmost Wisconsin Episode till margin in Illinois (39.33°N).

Since publication of the last volume (Stiff and Hansel, 2004), the Illinois State Geological Survey (ISGS) has acquired ^{14}C and optically stimulated luminescence (OSL) ages that have improved the chronological resolution of glacial events. We will focus on recent advancements in the chronology of Illinois Episode materials dated with OSL, and a modified Wisconsin Episode deglacial chronology of the Lake Michigan lobe based chiefly on radiocarbon ages of tundra plant macrofossils preserved in sediments of periglacial ice-walled lakes. Firm chronological ages for PIE glacial deposits remain elusive. Other advancements include findings based on recently published 1:24,000 superficial geology maps. From 2000 to 2009, about 130 new maps have been released by the ISGS, and most are available on-line (http://www.isgs.uiuc.edu/maps-data-pub/).

In Illinois, the lithological and pedological characteristics that define the rock and soil stratigraphical units result from a long tradition of applying stratigraphical principles to the glacial/interglacial successions (Willman and Frye, 1970; Hansel and Johnson, 1996; Stiff and Hansel, 2004; Curry et al., 2010; Hansel and McKay, 2010). In this chapter, we use the diachronic nomenclature formally introduced by Hansel and Johnson (1996 and references therein) and followed in the first edition of this volume (Fig. 36.2; Stiff and Hansel, 2004). We point out, however, that material reference sections for diachronic units have been formally published for only the Wisconsin Episode (Hansel and Johnson, 1996).

36.2. PRE-ILLINOIS EPISODE

36.2.1. Introduction

The PIE in Illinois consists of all Quaternary glaciations, as well as intervening interglacial events, prior to the Illinois Episode (Fig. 36.2). The Yarmouth Episode embraces the time after the last occurrence of glacial PIE conditions up to the onset of the Illinois Episode (Hansel and McKay, 2010). During the PIE, there were probably two or more major glaciations in Illinois, yet confirmation is difficult without additional chronological control. PIE deposits are delineated from Illinois Episode deposits by the Yarmouth Geosol (a strong interglacial palaeosol) developed into the upper portion of PIE deposits (Fig. 36.2; Leverett, 1899; Willman and Frye, 1970; Grimley et al., 2003). Because of erosion and burial by younger glacial deposits (except for a small area in western Illinois; Fig. 36.1), PIE deposits are discontinuously preserved, thus obscuring the details of past glacial lobe configurations. Even so, compositional

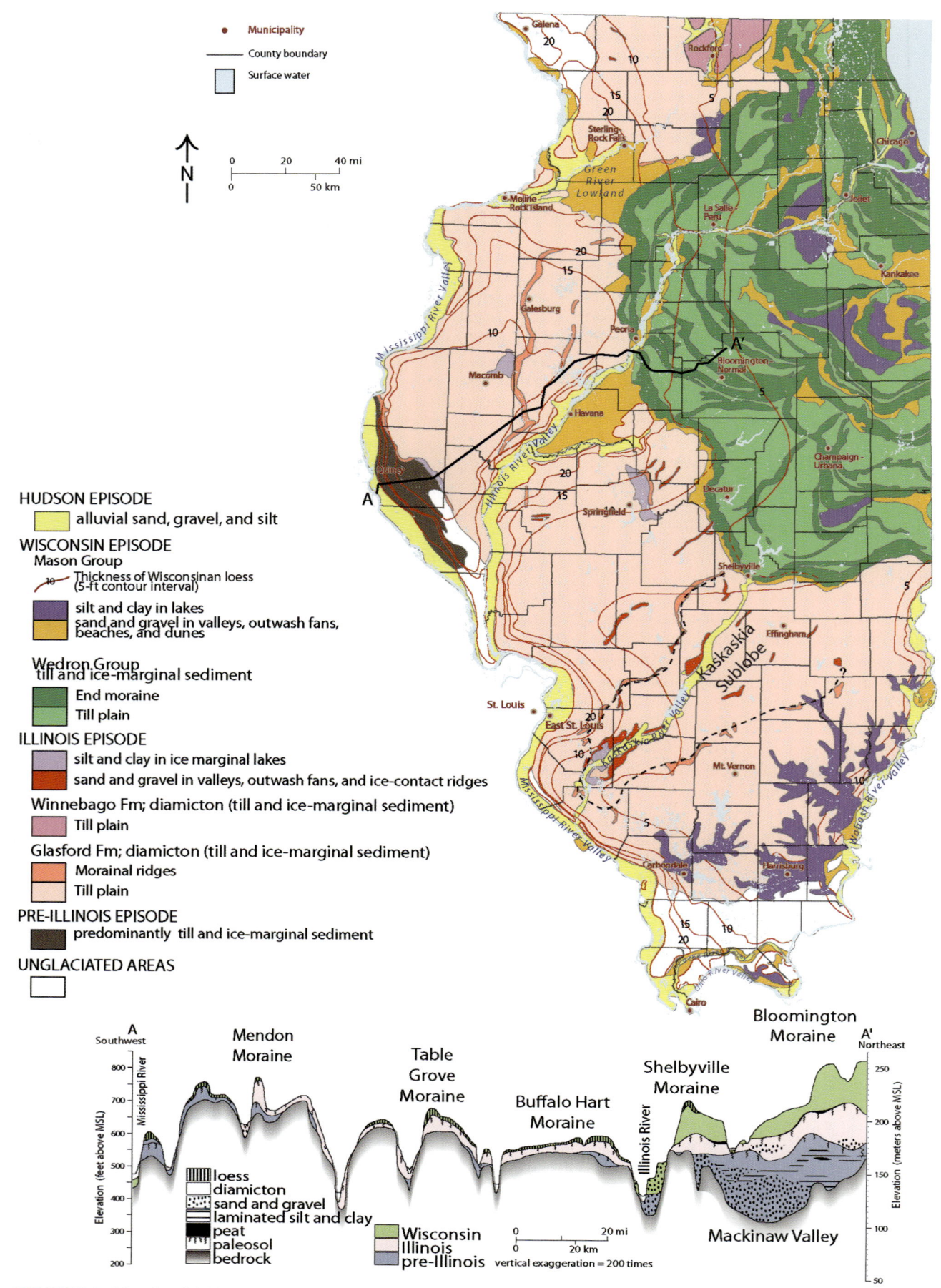

FIGURE 36.1 Map of surficial deposits in Illinois. Cross-section A–A' shows the increased thickness of the glacial sediments approaching Lake Michigan. *Source: Illinois State Geological Survey.*

Chapter | 36   Quaternary Glaciations in Illinois

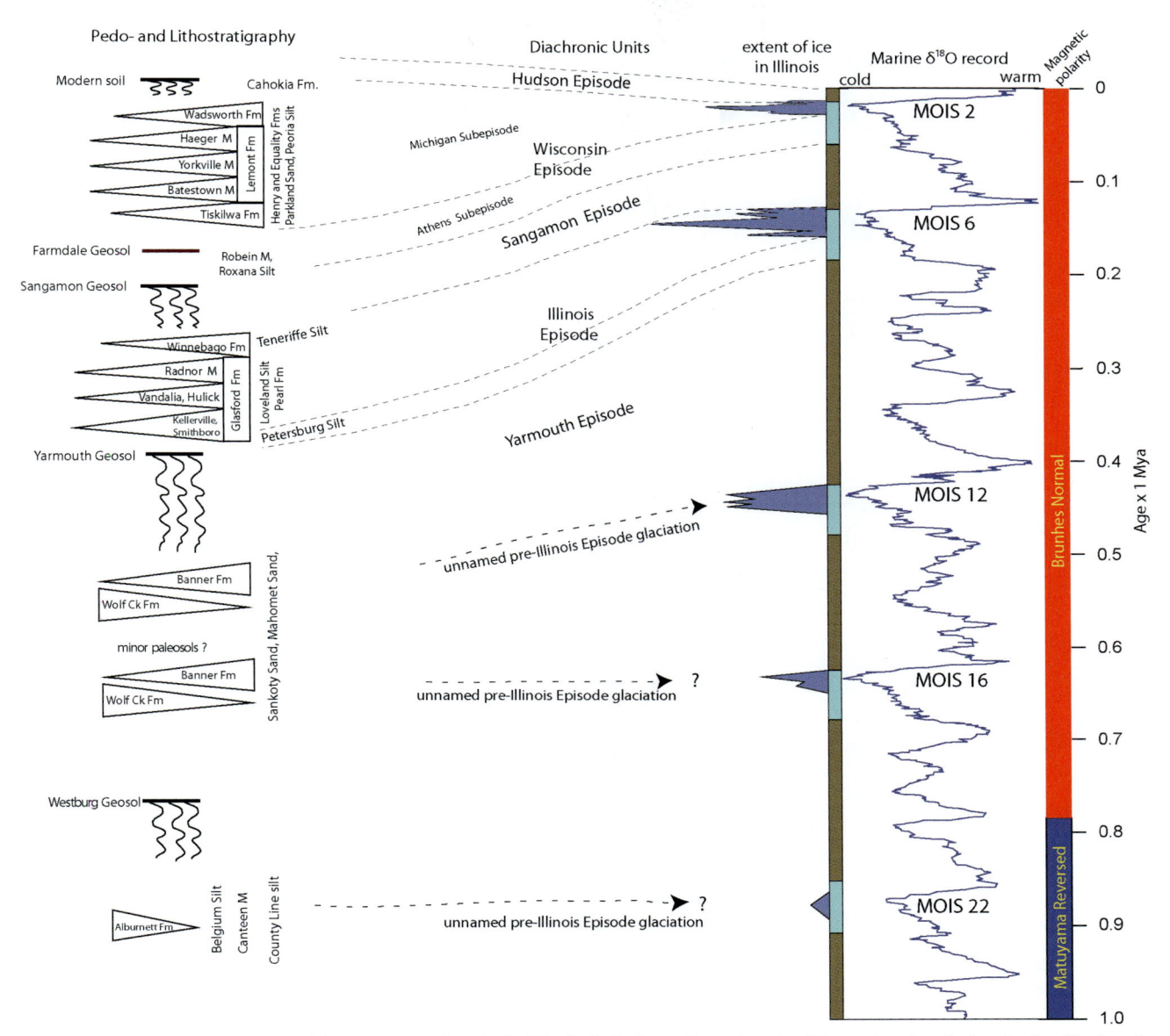

FIGURE 36.2 Schematic diagram of Quaternary stratigraphy in Illinois, including pedostratigraphy, lithostratigraphy, diachronic classification, ice extent in Illinois, the marine $\delta^{18}O$ record (Zachos et al., 2001) and magnetostratigraphy (Rovey and Kean, 2001). Note that pedostratigraphical and lithostratigraphical units are not scaled with respect to time.

data have proven useful in locally distinguishing PIE diamictons from deposits of younger glaciations as well as differentiating among deposits from eastern and western source lobes (Willman and Frye, 1970; Johnson, 1976; Wickham, 1979). At one point, PIE glacial ice covered at least two-thirds of the state (Fig. 36.3) and meltwater inundated many major sluiceways including the Mahomet Valley (now buried, Figs. 36.3 and 36.4) and the ancient valleys of the Mississippi, Kaskaskia and Ohio river systems. Outwash aggraded in these sluiceways, choking them with sediment. Consequently, slackwater lakes were formed in many tributaries to these river systems. Beyond the PIE ice margin in southern Illinois, loess deposits accumulated, sourced from wind-swept valley-train outwash.

36.2.2. Pre-Glacial Alluvium

Pre-glacial alluvium and colluvium occur in buried bedrock valleys of south-western Illinois (Phillips, 2004; Grimley, 2010). Known as the Canteen member, it is greenish-grey, typically non-calcareous and contains lower illite content and lower magnetic susceptibility than overlying glacial

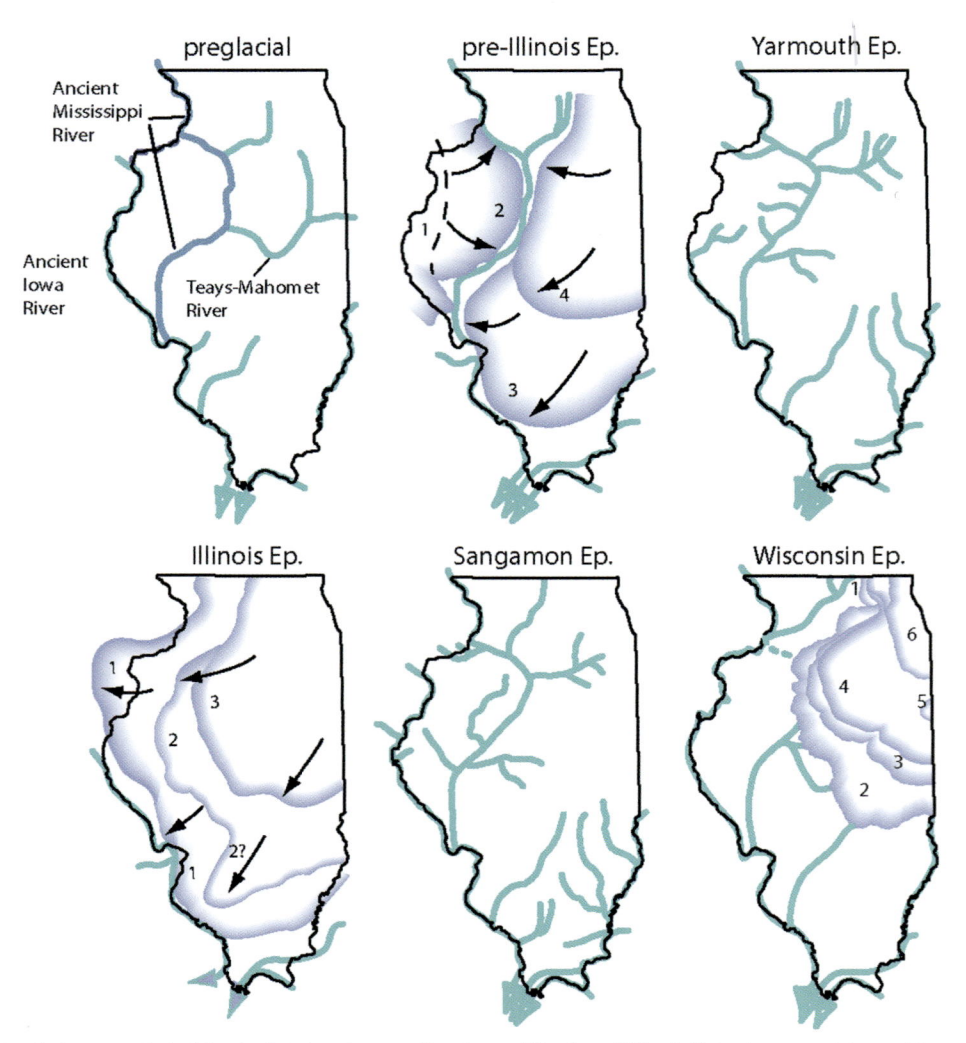

FIGURE 36.3 Major drainages and glacial episode lobe advances. For the pre-Illinois and Illinois Episodes, many lobe positions are inferred and are based on the location of type sections and other key sections. For the pre-Illinois Episode, lobes labelled 1 and 2 are associated with the Keewatin ice centre of the Laurentide Ice Sheet; lobes 3 and 4 are associated with the Labradorean ice centre. All subsequent lobes extended from the Labradorean ice centre. The chronological context of these sites is evolving through age determinations using OSL and other methods. For the Wisconsin Episode, all numbered features are ice margins associated with the following phases of the Michigan Subepisode: (1) Marengo, (2) Shelbyville, (3) Putnam, (4) Livingston, (5a) Nissouri (Elgin Subepisode; Huron-Erie lobe; Karrow et al., 2000), (5b) Woodstock and (6) Crown Point.

sediments. In places, the lower Canteen member is composed of angular fragments of the local bedrock. The landscape and deposits at this time and earlier (>0.7 Mya) were probably analogous to that in the Palaeozoic terrain of central Kentucky and Tennessee today. Similar pre-glacial alluvium, colluvium and lake deposits have also been noted in the basal succession that partially fills the Mahomet Valley in east-central Illinois (Fig. 36.4; A. Stumpf, personal communication, 2010).

Beyond the southern limit of glaciations, 5–24 m of silty and sandy fluvial deposits of the Metropolis Formation occur along the margins of the Cache River Valley where they were originally mapped as high-level outwash terraces (Fig. 36.5; Nelson et al., 1999). The unit is composed largely of quartz sand and fragments of chert derived from reworked pre-glacial Mounds Gravel. The age of the Metropolis Formation and Mounds Gravel is unknown. The upper part contains the Sangamon Geosol, so it is probable that part of the unit is contemporaneous with Illinois and PIE glaciations. The thickest deposits of the Metropolis Formation (15–24 m) occur in grabens associated with Quaternary mid-continental rifting (Nelson et al., 1999).

In many places in unglaciated Illinois, red clayey bedrock residuum with chert fragments and other accreted materials occur below loess. The residuum is named the Oak Formation (Nelson et al., 1991); it is deeply weathered and in most cases has proportionally much more randomly interstratified kaolinite–smectite than does the overlying Pleistocene geosols, including the modern soil. In the driftless area of north-eastern Illinois underlain by dolomite, such residuum is also known as 'geest' (Willman et al., 1989).

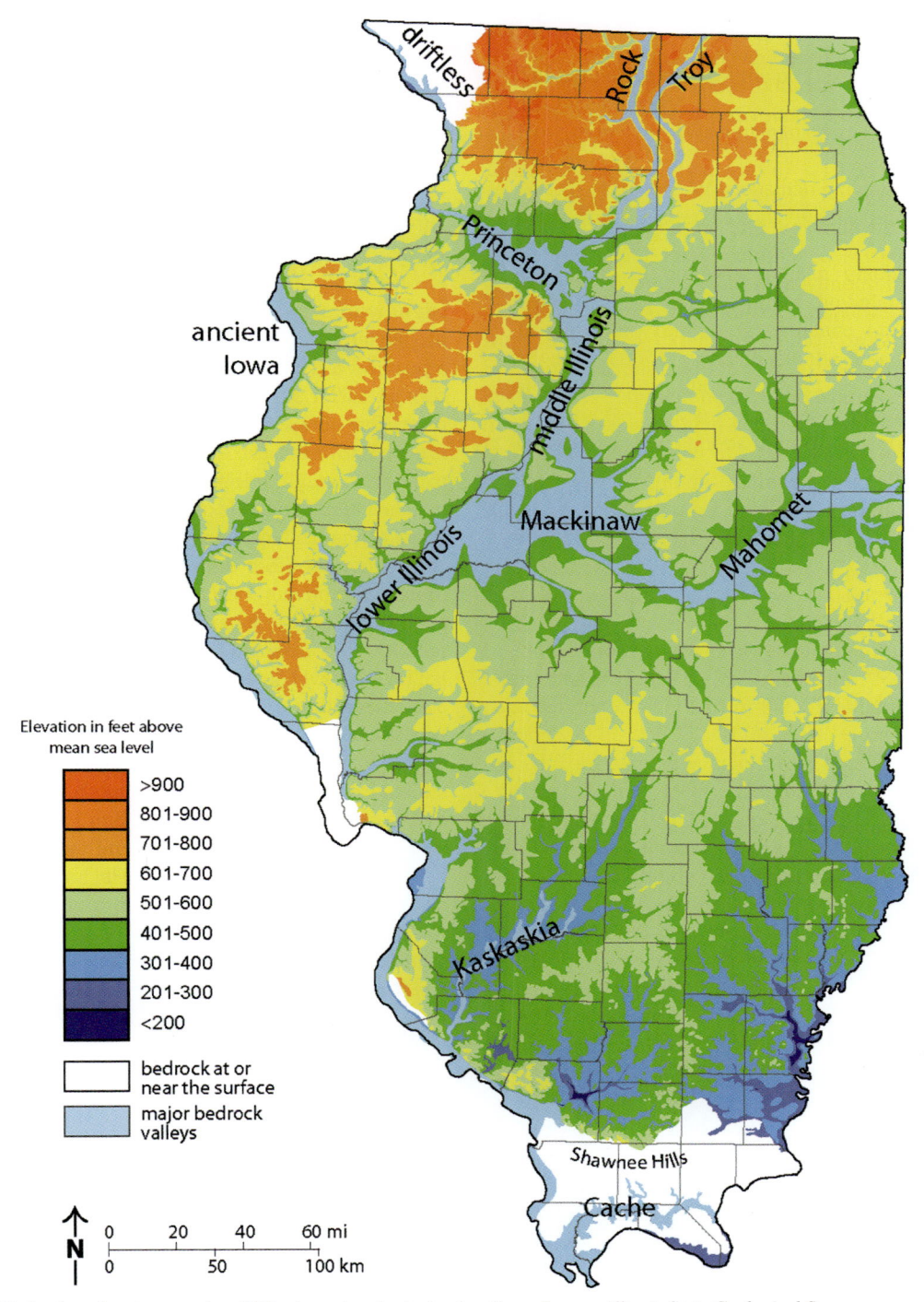

FIGURE 36.4 Bedrock surface topography of Illinois, and major bedrock valleys. *Source: Illinois State Geological Survey.*

36.2.3. Glacial History and Glacial Diamicton Provenance

Mineralogical and lithological evidence have confirmed that PIE glacial ice in Illinois advanced from both the eastern (Labradorean) and western (Keewatin) ice centres (Willman and Frye, 1970; Johnson, 1976; Wickham, 1979) (Fig. 36.3). PIE ice from the Keewatin source advanced south-eastwards through Minnesota and Iowa, whereas ice from the Labradorean source advanced south-westwards from the Lake Michigan, Saginaw or

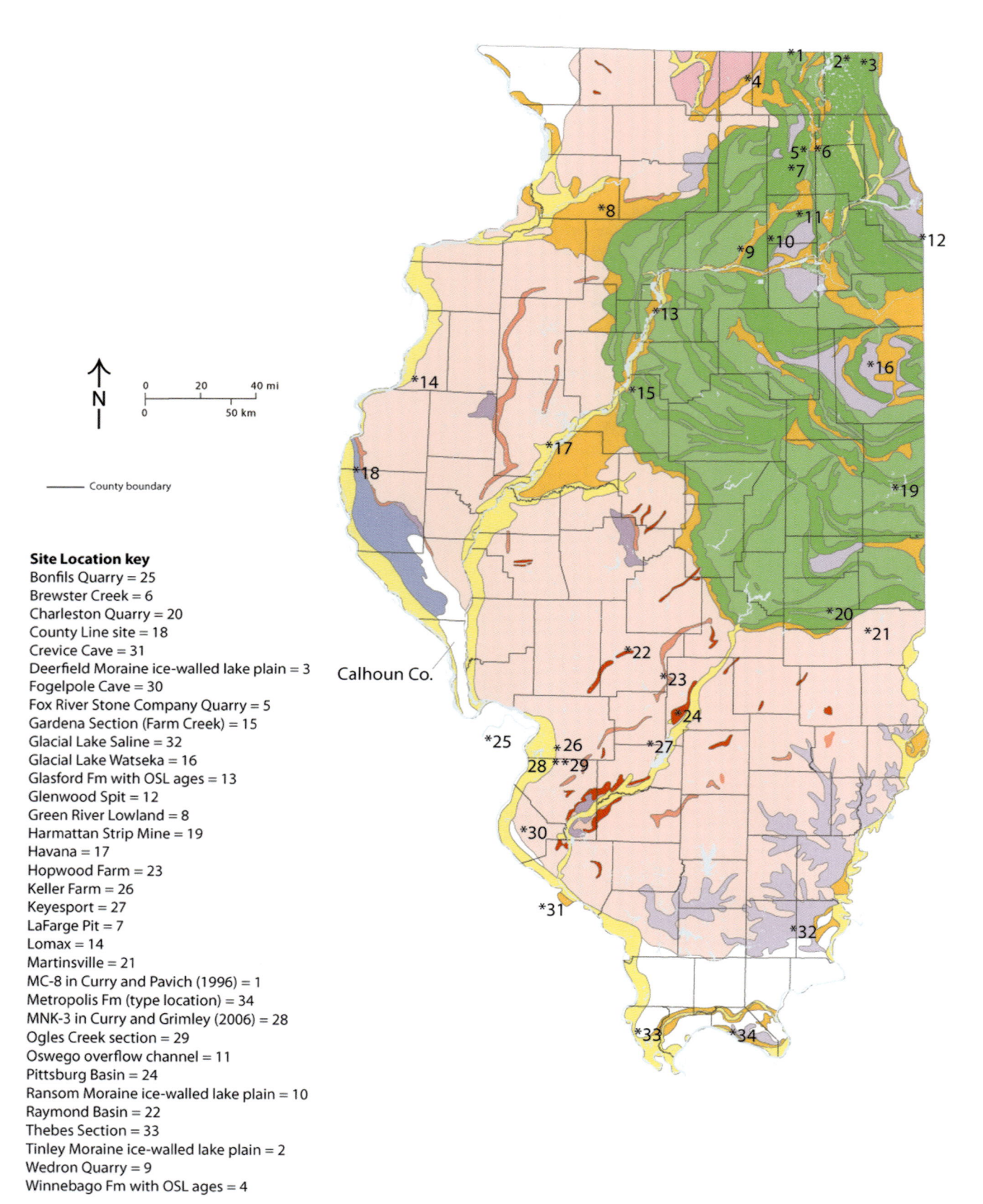

Site Location key
Bonfils Quarry = 25
Brewster Creek = 6
Charleston Quarry = 20
County Line site = 18
Crevice Cave = 31
Deerfield Moraine ice-walled lake plain = 3
Fogelpole Cave = 30
Fox River Stone Company Quarry = 5
Gardena Section (Farm Creek) = 15
Glacial Lake Saline = 32
Glacial Lake Watseka = 16
Glasford Fm with OSL ages = 13
Glenwood Spit = 12
Green River Lowland = 8
Harmattan Strip Mine = 19
Havana = 17
Hopwood Farm = 23
Keller Farm = 26
Keyesport = 27
LaFarge Pit = 7
Lomax = 14
Martinsville = 21
MC-8 in Curry and Pavich (1996) = 1
Metropolis Fm (type location) = 34
MNK-3 in Curry and Grimley (2006) = 28
Ogles Creek section = 29
Oswego overflow channel = 11
Pittsburg Basin = 24
Ransom Moraine ice-walled lake plain = 10
Raymond Basin = 22
Thebes Section = 33
Tinley Moraine ice-walled lake plain = 2
Wedron Quarry = 9
Winnebago Fm with OSL ages = 4

FIGURE 36.5 Location of key sites in Illinois. The sites include: 1, boring MC-8 (Curry and Pavich, 1996); 2, Tinley Moraine ice-walled lake plain (Curry, 2008b); 3, Deerfield Moraine ice-walled lake plain (Curry, 2008b); 4, Winnebago Formation with OSL ages (E.D. McKay and R.C. Berg, personal communication); 5, Fox River Stone Company Quarry (Curry et al., 1999); 6, Brewster Creek (Curry et al., 2007); 7, LaFarge Pit (Jacobs et al., 2009);

Huron-Erie basins. The middle and lower Illinois River Valley systems (part of the ancient Mississippi River Valley system) were the approximate boundary between Labradorean and Keewatin glacial ice in Illinois (Fig. 36.3). It is not known whether glacial ice from either source crossed the ancient Mississippi Valley system to abut the other ice sheet. The comparative age of eastern and western source glacial advances is also not well known, yet it is likely that they were generally in phase.

PIE glaciations sourced from eastern (Labradorean) included three to five advances based on the diamicton record formerly exposed in the Harmattan strip mine in the Danville area (Fig. 36.5; Johnson et al., 1971; Johnson, 1986). The units, locally associated with organic-rich interstadial soils, are classified as members of the Banner Formation (Hansel and McKay, 2010), and are distinguished by contrasting lithological attributes (Johnson et al., 1971; Vonder Haar and Johnson, 1973). High dolomite/calcite values, abundant illite and chlorite relative to expandable clay minerals and low magnetic susceptibility mark materials with provenance similar to the Wisconsin Episode Lake Michigan lobe; low calcite/dolomite values and high magnetic susceptibility suggest provenance similar to the Huron-Erie lobe or Saginaw lobes. In southern and south-western Illinois, a single PIE diamicton unit of the Banner Formation occurs below the Yarmouth Geosol (Maclintock, 1929; McKay, 1979; Grimley et al., 2001) and probably correlates to one of the diamicton units in east-central Illinois (Johnson et al., 1971).

Pre-Matuyama Chron (magnetically reversed) diamictons are known from adjacent west-central Indiana (Bleuer, 1976). In east-central Illinois, the occurrence of magnetically reversed deposits is less certain. A fossiliferous silt unit below three of the Banner diamicton units at the Harmattan Strip Mine in east-central Illinois (Fig. 36.5) had reversed remnant magnetic polarity in one exposure, but normal polarity in another exposure immediately below PIE diamicton (Johnson, 1986; Kempton et al., 1991). A diamicton below the silt, once interpreted as the earliest PIE glacial advance in eastern Illinois (Johnson, 1971), was later reinterpreted as colluvium, perhaps reworked from early, magnetically reversed PIE diamicton deposits (Johnson, 1986).

PIE glaciers originating from the western (Keewatin) source area advanced into western Illinois from Iowa during at least two major glaciations represented by the Alburnett and Wolf Creek Formations (Fig. 36.2). The Westburg interglacial geosol is developed into the upper Alburnett Formation (Wickham, 1979). The type sections of these units are located in eastern Iowa (Hallberg, 1986) where the Alburnett Formation is magnetically reversed (Matuyama Chron) and the Wolf Creek Formation is normally magnetised (Bruhnes Chron). Although interglacial-grade palaeosols separate multiple Wolf Creek diamicton units in eastern Iowa (Hallberg, 1986), only one Wolf Creek diamicton unit is known in Illinois (Wickham, 1979). Diamicton units correlative to the Wolf Creek and Alburnett Formations have been documented in northern Missouri (Rovey and Kean, 2001), and in eastern Kansas and Nebraska, and western Iowa, in addition to older PIE diamicton units (Roy et al., 2004).

36.2.4. Fluvial, Lacustrine and Aeolian Sediments

Most PIE fluvial deposits in Illinois' major valleys are probably outwash, but in some places may be interglacial alluvium. As much as 60 m thick, PIE waterlain sediment partially fills the Mahomet Bedrock Valley (Fig. 36.4; Kempton et al., 1991). The sand and gravel deposits are locally known as the Mahomet Sand and are buried by slowly permeable fine-grained diamicton units (from PIE and younger glaciations). The Sankoty Sand is a fluvial sand unit mapped north of Peoria in the Ancient Mississippi River Valley (Fig. 36.2; Horberg et al., 1950; Horberg, 1953). It is composed of primarily pink-stained quartz grains. Once thought to be a 'key bed' at or near the base of the PIE succession, the status of this unit is currently under investigation (McKay and Berg, 2008; McKay et al., 2008). In south-western Illinois, PIE fluvial sand may be preserved in the deepest troughs of the lower Kaskaskia Bedrock Valley (Fig. 36.4; Grimley and Webb, 2009; Grimley, 2010).

Subsurface fossiliferous PIE slackwater lake deposits are sporadically preserved in Illinois. The thickest known deposit is 18 m thick, and located in an abandoned tributary to the lower Kaskaskia Bedrock Valley (Grimley and Webb, 2009). Fossil gastropods and ostracods suggest there were initially permanent lakes that shoaled to temporary pools as lake sedimentation accelerated with the

---

8, Green River Lowland (Miao et al., 2010); 9, Wedron Quarry (Johnson and Hansel, 1990); 10, Ransom Moraine ice-walled lake plain (Curry, 2008b); 11, Oswego overflow channel (Curry, 2008b); 12, Glenwood Spit (Hansel and Johnson, 1996); 13, Glasford Formation with OSL ages (ISGS, personal communication); 14, Lomax (Curry, 1998); 15, Gardena Section (Farm Creek; Follmer et al., 1979); 16, Glacial Lake Watseka (Willman and Frye, 1970); 17, Havana (Hajic and Curry, 2008); 18, County Line site (Miller et al., 1994); 19, Harmattan Strip Mine (Johnson, 1971); 20, Charleston Quarry (Hansel and Johnson, 1996); 21, Martinsville (Curry et al., 1994); 22, Raymond basin (Zhu and Baker, 1995; Curry and Baker, 2000); 23, Hopwood Farm (King and Saunders, 1986); 24, Pittsburg basin (Grüger, 1972; Teed, 2000); 25, Bonfils Quarry (Forman and Pierson, 2002); 26, Keller Farm (Wang et al., 2000); 27, Keyesport (David Grimley, personal communication); 28, boring MNK-3 (Curry and Grimley, 2006); 29, Ogles Creek section (Kaplan and Grimley, 2006); 30, Fogelpole Cave (Panno et al., 2004); 31, Crevice Cave (Dorale et al., 1998); 32, Glacial Lake Saline (Heinrich, 1982); 33, Thebes Section (Grimley et al., 2003); 34, Metropolis Formation (type locality; Nelson et al., 1999). *Source: Illinois State Geological Survey.*

approaching glacial ice and increased loess accumulation. Thick PIE lacustrine deposits of possible slackwater origin have also recently been documented in successions that partly fill tributaries to the Mahomet Bedrock Valley (Fig. 36.4; A. Stumpf, personal communication, 2010).

In the unglaciated parts of southern and western Illinois, conformable loess–palaeosol sequences inclusive of the PIE are preserved in isolated areas. The Thebes site (Fig. 36.5), for example, revealed 2.4 m of PIE loess encompassing the Yarmouth Geosol solum (Grimley et al., 2003). In unglaciated Calhoun County of western Illinois (Fig. 36.5), probable PIE fluvial, loessal and lacustrine sediments were exposed at the Pancake Hollow Section and sampled by core at the Green Bay Hollow Section (Hajic et al., 1991). In these sections, PIE aeolian sediment interfingers with fluvial and lacustrine sediments in bedrock valleys tributary to the ancient Iowa River system and lower and middle Illinois River Valley (part of the Ancient Mississippi River system).

36.2.5. Age and Palaeoclimate

Based on current knowledge, most glacial PIE sediment in Illinois was probably deposited during marine isotope stages (MIS) 16 ($\sim$620 ka) and 12 ($\sim$430 ka; Johnson, 1986; Hansel and McKay, 2010). These events mark the greatest global ice volumes during the Pleistocene (Fig. 36.2; Shackleton, 1987; Zachos et al., 2001; Pillans and Naish, 2004). The oldest glacigenic diamicton in Illinois, the Alburnett Formation, was deposited prior to the Matuyama Chron (>780 ka; Hallberg, 1986). The Banner and Wolf Creek Formation diamictons, as well as intervening lacustrine and fluvial deposits, are post-Matuyama Chron (>780 ka). At the County Line site in western Illinois (Fig. 36.5), the age of fossiliferous lacustrine silts, below Wolf Creek diamicton, is ca. 800 ka based on amino acid geochronology, palaeomagnetic data and mammalian fossils (Miller et al., 1994). Magnetic remanence and cosmogenic dating indicate that correlative and older diamictons occur in Missouri (Roy et al., 2004; Rovey and Kean, 2001), Nebraska, South Dakota and western Iowa (Roy et al., 2004; Balco et al., 2005).

The Yarmouth Episode covers the period from the end of PIE glaciations (about 430 ka) to the onset of Illinois Episode glaciation, a span of about 230 ka (Hansel and McKay, 2010). The upper age limit of 430 ka for glacial PIE deposits is consistent with the estimated duration of surface exposure, based on multi-proxy weathering indices for the modern soil, and the Sangamon and Yarmouth geosols (Grimley et al., 2003).

Short-lived, poorly dated fossiliferous sites provide glimpses of environmental conditions during the ca. 600 ka-long PIE in Illinois. Pre-glacial PIE conditions are based on interpretations of plant macrofossils, trace fossils, molluscs and ostracods from slackwater lake deposits. Most PIE fossil wood identified from sites in western and southern Illinois is *Picea*. Slackwater molluscan fossils (Leonard et al., 1971, and unpublished findings) suggest a transition from a closed boreal forest to mixed boreal-open prairie as PIE ice approached sites in south-western Illinois. In western Illinois, the flora and fauna preserved in the County Line silt ($\sim$800 ka) has analogues in northern Iowa (Fig. 36.5; Miller et al., 1994).

36.3. ILLINOIS EPISODE

36.3.1. Introduction

The Illinois Episode glaciation (or Illinoian Stage) was named by Leverett (1899) after the state of Illinois, where Illinois Episode deposits are best preserved and exposed. The thickest deposits, about 50 m, occur in bedrock valleys (Curry et al., 1994). About 90% of Illinois was glaciated during this time (Fig. 36.3). Occurring in the upper portion of Illinois Episode deposits is the Sangamon Geosol, an interglacial soil (Fig. 36.2; Willman and Frye, 1970; Follmer et al., 1979). In the 1970s, three major glacial events were recognised in Illinois; they were associated with multiple stages in the marine oxygen isotope record (Willman and Frye, 1970; Johnson, 1976; Lineback, 1979b). The palaeosol record, the primary evidence for multiple, major glacial events, does not support this paradigm. First, the palaeosols are uncommonly observed in key stratigraphical positions. Second, the palaeosols are of interstadial character and possess only thin, organic-rich A-horizons or thin cambic (colour) B horizons with partial leaching of carbonate minerals. These observations and the results of recently developed age dating methods (OSL, ^{10}Be and amino acid geochronology) indicate that the Illinois Episode is restricted to MIS 6, $\sim$190–130 ka (Martinson et al., 1987; Curry et al., 1994; Curry and Pavich, 1996; Oches, 2001; McKay and Berg, 2008; see, however, Forman and Pierson, 2002). Moreover, recently obtained OSL ages suggest that much, if not all of the Winnebago Formation in north-central Illinois, is associated with the Illinois Episode (Figs. 36.2 and 36.5; E.D. McKay and R.C. Berg, personal communication). The age of Winnebago diamicton units had previously been variably assigned to the 'early' Wisconsin, Illinois and Pre-Illinois episodes (Berg et al., 1985; Curry, 1989; Curry and Pavich, 1996).

36.3.2. Glacial History and Diamictons

The direction of flow of Illinois Episode glacial ice in Illinois radiated outwards from the Lake Michigan basin. Orientation of moraines, bedrock striae, diamicton macrofabrics, eskers, megaflutes and rectilinear to parallel drainage patterns indicates that ice-flow directions were generally south-eastwards in south-eastern Illinois (Curry et al., 1994), south-westwards towards St. Louis in south-west Illinois (Lineback, 1971; Heigold et al., 1985), west to south-west in western Illinois (Leverett, 1899; Leighton and Brophy, 1961) and north-westward in north-west Illinois (Grimley, 1997). On a local scale, especially near former ice margins, the ice-flow directions were more variable (Webb, 2009). Notable obstructions to glacier flow included regional cuestas or escarpments of resistant Pennsylvanian sandstone in southernmost Illinois and Mississippian limestone in far south-western and western Illinois.

Several diamicton units of the Illinois Episode are classified with the Glasford Formation (Fig. 36.2; Willman and Frye, 1970). Key lithostratigraphical units were defined by Willman and Frye (1970) including the Kellerville (=Smithboro?), Vandalia (=Hulick) and Radnor (=Sterling) till members. In some cases, units are probably lateral facies and synchronous, such as the informal Fort Russell till of the East St. Louis area (McKay, 1979) and Vandalia till of central Illinois (Lineback, 1979a). In other cases, the relationships are less clear, such as the correlation of the buried Smithboro till member in central Illinois with the Kellerville till member in western Illinois (Fig. 36.2; Curry et al., 1994).

Consistent with the state's geomorphology and other glacial-flow indicators, the mineralogical composition of Glasford diamictons indicates they were deposited by the Lake Michigan lobe. In the <2 mm fraction, key characteristics include abundant dolomite and a high proportion of illite, chlorite and kaolinite with respect to expandable clay minerals (Willman and Frye, 1970; Follmer et al., 1979; McKay, 1979; Berg et al., 1985; Curry et al., 1994; McKay et al., 2008). Diamictons are generally more silty and expandable clay-rich to the south and west as silt-rich lake sediments, loess and palaeosols were overridden and entrained in the glacial ice (McKay, 1979; Glass and Killey, 1988; Curry et al., 1994). Uncommon, but notable boulder and pebble erratics of rhyolite porphyry, jasper conglomerate and Gowganda tillite in Illinois Episode diamictons indicate a distal Canadian bedrock contribution north of Lake Huron (Leverett, 1899; Horberg, 1956). These erratic lithologies are also known from Wisconsin Episode deposits, and their occurrence is attributed to the Lake Michigan lobe entraining materials originally deposited by the Saginaw lobe (Hansel and Johnson, 1996; Fig. 36.6).

The origin of ice-contact ridges in the Kaskaskia River drainage basin has been of interest for over a century (Leverett, 1899). Known as the 'ridged-drift', the ridges rise more than 40 m relative to the flat-lying, but dissected, Illinois Episode till-plain (e.g. Heigold et al., 1985; Grimley and Webb, 2009). Individual ridges are composed of primarily sand and gravel, diamicton or complexes of diamicton and sorted sediment. The heterogeneous textures have led geologists to various hypotheses on the genesis of the ridges ranging from ice marginal processes forming moraines (Leverett, 1899) or interlobate ridges (Willman et al., 1963) to ice-stagnation processes that form crevasses

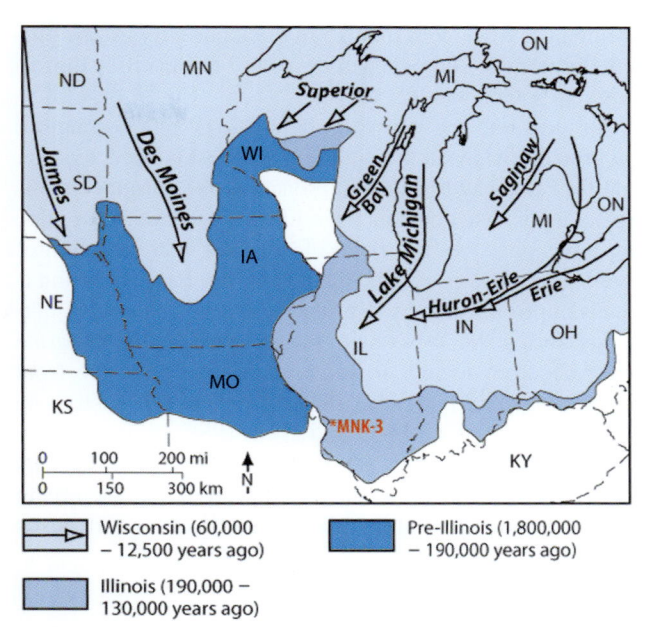

FIGURE 36.6 Approximate flow lines of major lobes of the south-central margin of the Laurentide Ice Sheet during the last glaciation (late Wisconsin Episode), and the location of boring MNK-3. The mapped limits of the Wisconsin, Illinois and pre-Illinois Episode glaciations are also indicated. State and province acronyms include: IA, Iowa; IL, Illinois; IN, Indiana; KS, Kansas; KY, Kentucky; ND, North Dakota; NE, Nebraska; MI, Michigan; MN, Minnesota; OH, Ohio; ON, Ontario; SD, South Dakota; WI, Wisconsin.

and/or eskers (Leighton and Brophy, 1961; Jacobs and Lineback, 1969; Lineback, 1979b). Recent mapping has differentiated the Kaskaskia sublobe in south-central Illinois (Fig. 36.1; Phillips, 2004; Webb, 2009; Grimley, 2010); some ridges in this area formed of chiefly diamicton may be ice-stream shear margin moraines (e.g. Stokes and Clark, 2002), whereas others are ice-walled channels or eskers (Heigold et al., 1985). In this scenario, the sublobe is thought to have been thin and fast-moving owing in part to the low topography of the Kaskaskia basin, and to the soft deformable substrate of shale and proglacial lake sediment. Post-glacial erosion, as well as 'cut-and-fill' events along the Kaskaskia Valley after the Illinois Episode, has left a regionally discontinuous, but locally continuous, system of ridges that are relics of the advance, stagnation and drainage of a downwasting sublobe.

36.3.3. Fluvial, Lacustrine and Aeolian Sediments

Deposits of sand and gravel outwash (Pearl Formation), loess and lacustrine deposits (Loveland Silt, Teneriffe Silt and Petersburg Silt) are intercalated with Glasford Formation diamicton units (Fig. 36.2; Willman and Frye, 1970). The outwash and lacustrine deposits are thickest in valleys. For example, proglacial slackwater lake deposits

(Petersburg Silt) more than 15 m thick partially fill the buried Embarrass Bedrock Valley near Martinsville, Illinois (Fig. 36.5; Curry et al., 1994).

In general, the outwash deposits (Pearl Formation) occur in drainageways, such as the Kaskaskia River system, that flowed away from the approaching ice margin. Illinois Episode outwash deposits were likely abundant at one time but have since been removed from valleys by incision of rivers and erosion of deposits during the Sangamon interglacial episode and by late-stage glacial meltwater outbursts during the Wisconsin glacial episode. Ice-marginal lakes occurred locally in some areas where the ice front blocked streams that were flowing towards it, such as in southern Illinois (Willman and Frye, 1980). Many proglacial or slackwater lakes were subsequently overridden by the advancing ice, preserving fossiliferous lake sediment in the subsurface below glacigenic diamicton deposits (Curry et al., 1994; Grimley et al., 2001).

36.3.4. Age and Palaeoclimate

The age of Illinois Episode deposits is based on ^{10}Be accumulation in palaeosols, OSL ages, amino acid geochronology and correlations to the marine isotope record. A ^{10}Be inventory of the Sangamon and Farmdale geosols in a northern Illinois core (MC-8) indicates an age of $\sim 155 \pm 30$ ka for Illinois Episode colluvium, outwash and diamicton (Fig. 36.5; Curry and Pavich, 1996). In north-central Illinois (north of Peoria), McKay et al. (2008) and McKay and Berg (2008) reported OSL ages between 190 and 130 ka (MIS 6) for Illinois Episode outwash (Fig. 36.5). Three OSL age determinations on fluvial sand immediately underlying the lowermost Illinois Episode diamicton (Kellerville Member in Marshall County) yielded a weighted mean of 155 ± 8 ka (UNL-1668, 1669, 1670) (E.D. McKay, unpublished data). Outwash deposits (Pearl Formation) at the Keyesport sand and gravel pit in south-western Illinois were recently dated by OSL at 154 ± 19 ka (UNL-1872) and 147 ± 19 ka (UNL-1873) (Fig. 36.5; D.A. Grimley, unpublished data). In north-central Illinois, the weighted mean of eight samples of sand interbedded with diamicton of the Winnebago Formation (latest Illinois Episode) is 130 ± 9 ka (UNL-2237 to 2244) (Figs. 36.1 and 36.5; E.D. McKay and R.C. Berg, unpublished data). The collective OSL ages listed above suggest Illinois Episode glaciation and deglaciation occurred primarily during the latter half of MIS 6 ($\sim 160-130$ ka), coincident with a time of high global ice volumes (Fig. 36.2; Zachos et al., 2001). Amino acid ratios for Illinois Episode gastropod shells are greater than those for Wisconsin Episode shells (Curry et al., 1994; Mirecki and Miller, 1994; Oches, 2001) and are consistent with a correlation to MIS 6, approximately 190–130 ka (Fig. 36.2; Martinson et al., 1987).

Illinois Episode fossils (primarily pollen, ostracods, gastropods and plant macrofossils) occur in lacustrine sediment that occurs above, within and beneath Glasford diamicton (till) units. Most sites with fossils are located between about 37°N and 41°N (Leonard et al., 1971). Early Illinois Episode fossils occur in the slackwater and proglacial lacustrine Petersburg Silt (Leonard et al., 1971; Curry et al., 1994; Geiger, 2008). Fossil gastropods, ostracods, wood and spruce needles are especially abundant. The Ogles Creek section (St. Clair County) has yielded more than 15 *Picea* logs, including one in growth position that were planed off by the overriding glacier (Fig. 36.5; Kaplan and Grimley, 2006). A pollen diagram from an 80-cm-thick section (sample spacing = 8 cm) reveals that 60–80% pollen is of boreal trees (mostly *Picea* with some *Pinus*), <5% of deciduous tree elements and 20–40% of herbaceous pollen (Poaceae, *Chenopodium*, Cyperaceae, Caryophyllaceae). Common spores of *Selaginella* and the Cyperaceae pollen indicate an analogue environment at or near the northern extent of boreal forests. These findings are consistent with the fossil gastropod assemblage (Geiger, 2008). The gastropod assemblages of Illinois Episode environments indicate and a dominance of boreal/woodland species in southern Illinois (Geiger, 2008) with some species tolerant of colder and drier conditions in central and northern Illinois (Leonard et al., 1971).

Fossiliferous late Illinois Episode pollen is known from the base of kettle-lake deposits on the Illinois till-plain. These sites are best known for their Sangamon Episode fossils, such as *Geochelone crassiscutata* (giant tortoise) from Hopwood Farm (Fig. 36.5; King and Saunders, 1986) and long pollen and ostracod records that extend to the early Wisconsin Episode, such as in the Pittsburg (Grüger, 1972; Teed, 2000), and Raymond Basins (Fig. 36.5; Zhu and Baker, 1995; Curry and Baker, 2000). These records have yet to yield consistent OSL ages. The most complete record occurs at Raymond basin which includes a lower biofacies that lacks pollen, but contains an ostracod assemblage also known from late Wisconsin Episode ice-walled lake deposits. Overlying biofacies, common to all the other sites listed above, includes abundant *Picea* pollen and an ostracod assemblage consistent with the presence of a closed boreal forest (interpreted as latest Illinois Episode in age). The regional pollen profiles thus record the switchover from tundra-like conditions to a boreal forest. The ensuing switchover to *Quercus*-dominated pollen spectra indicative of the early Sangamon Episode is abrupt in each case (Curry and Baker, 2000).

36.3.5. Wisconsin Episode

In Illinois, the onset of the last glaciation is marked by the first occurrence of loess deposited on the Sangamon Geosol. This loess, named the Roxana Silt (Willman and Frye,

1970), was related to aeolian deflation of the first post-Sangamon Episode outwash associated with the Laurentide Ice Sheet in the alluvial valley of the ancient Mississippi River. The collective evidence indicates that this occurred about $60,000 \pm 5000$ years ago. The evidence derives from OSL ages of distal outwash deposits in the lower Mississippi River Valley and pedogenically complicated upland loess deposits, as well as from proxy records associated with well-dated U-series dated speleothems. In areas where the new loess was deposited above glacigenic diamicton and outwash, bioturbation and colluviation resulted in a transitional zone with upward-decreasing sand, gravel and pedogenic clay contents, and increase silt content, with attendant changes in mineralogy and soil morphology (Frye et al., 1974; Follmer et al., 1979; McKay, 1979; Follmer, 1983; Jacobs et al., 2009). Calibrated radiocarbon ages of this material generally range from about 47.5 to 31.0 cal. ka BP (45.0–29.0 ^{14}C ka BP; Curry, 1989), but the presence of the mixing zone indicates a somewhat earlier age for the first loess. Successions of radiocarbon ages from the Roxana Silt have been used to extrapolate the onset of the last glaciation. Ignoring several outliers, calibrated radiocarbon ages from Leigh and Knox, 1993 indicate that the onset of the Wisconsin Episode is about $58.5 \pm \sim 3.0$ cal. ka BP ($56.37 \pm \sim 2.0$ ^{14}C ka BP).

Waterlain deposits ostensibly related to the loessic Roxana Silt include distal outwash terraces (braid belts) of the lower Mississippi River Valley as well as slackwater lake deposits and alluvium preserved in buried valleys tributary to the Mississippi River Valley. Located in south-eastern Missouri and north-eastern Arkansas, the oldest braid belts yield OSL ages consistent to that of the Roxana Silt, including the Dudley Terrace deposits (63.5 ± 4.8–50.1 ± 4.0 ka), Melville Ridge (41.6 ± 3.0–34.5 ± 2.5 ka) and Lower Macon Ridge (33.3 ± 2.4–30.0 ± 2.1 ka; Rittenour et al., 2007). North of Thebes Gap, Illinois (Fig. 36.5), the Mississippi River Valley narrows considerably, and correlatives of these terraces are buried by younger outwash. Lateral outwash facies extend up tributary valleys of the Mississippi River (and other major sluiceways) where they intercalate with fossiliferous slackwater lake sediment. In core MNK-3, sampled in a tributary to the Mississippi River near St. Louis (Fig. 36.5), the provenance of the lake sediment was determined by clay mineralogy, colour and magnetic susceptibility (Curry and Grimley, 2006). Wood fragments in the lake sediment associated with high proportion of sediment from the Superior and western glacial lobes (Des Moines and James lobes) date as 47.65 cal. ka BP (44.15 ± 1.3 ^{14}C ka BP); one interval dating at about 31.5 cal. ka BP (29.03 ± 0.42 ^{14}C ka BP) has a significant contribution of Superior lobe sediment (Fig. 36.6). The oldest slackwater lake sediment that includes a high proportion of Lake Michigan lobe sediment dates at 28.02 cal. ka BP (25.07 ± 0.34 ^{14}C ka BP) consistent with other lines of evidence regarding initial glacial activity (Fig. 36.7; discussed below). Other significant early Wisconsin Episode slackwater lake deposition occurred on low-gradient tributaries beyond the glacial margin such as in Glacial Lake Saline (Fig. 36.5; Frye et al., 1972; Heinrich, 1982).

The age of the basal 'mixed zone' of the lower Roxana Silt has been investigated in upland environments by OSL techniques in addition to the back-extrapolation of radiocarbon ages discussed earlier. Infrared-stimulated luminescence ages of the lower Roxana Silt at the Bonfils site near St. Louis are as old as 82.5 ± 6.8 ka (Fig. 36.5; Forman and Pierson, 2002); this age is complicated by pedogenically admixed material from the Sangamon Geosol. Other ages from the same section indicate the basal Roxana dates ~ 60.0 ka, an age derived from the mean of the two chronologically 'out-of-order' ages in the lower 0.6 m of the Roxana (54.9 ± 4.6 and 64.0 ± 5.8 ka, Forman and Pierson, 2002).

Poorly dated biozones are based on pollen and ostracod records from lacustrine deposits on the Illinois Episode till plain (Zhu and Baker, 1995; Curry and Baker, 2000). These records have been correlated to the U-series dated δ^{13}C record from Crevice Cave, Missouri (Fig. 36.5; Dorale et al., 1998; Curry et al., 2002). Early and middle Sangamon biozone conditions were consistent with Illinois' modern continental climate with occasional warmer-than-present winters (King and Saunders, 1986; Zhu and Baker, 1995; Curry and Baker, 2000). During the late Sangamon biozone, the collective evidence of non-analogue elements (such admixed *Picea* and *Liquidambar* pollen) and spring-dwelling ostracods that live in the Great basin today suggest cold winters and dry summers (Curry and Baker, 2000). This interval is overlain by Wisconsin Episode sediment that contains cryophyllic tree pollen (*Picea*) and ostracods (*Limnocythere friabilis*). The late Sangamon biozone may date from about 70,000 to 55,000 years ago (Curry et al., 2002).

Important long palaeoenvironmental records are emerging from δ^{13}C records from speleothems from mid-western North America (Dorale et al., 1998). Shifts in the isotope record are likely related to changes in the proportion of 'C$_3$'- and 'C$_4$'-type vegetation (interpreted to be forest and prairie vegetation, respectively; Dorale et al., 1998). U-series ages provide accurate and precise chronology. For example, a well-dated stalagmite from Fogelpole Cave, Illinois, shows a marked change from older, clean travertine to material that is interbedded with silt laminae dating from about 60 to 55 ka (Fig. 36.5; Zhou et al., 2005; Curry et al., 2010). The laminae were deposited in the cave by floods that redeposited silty Illinois Episode lake sediment. The source of the lake sediment was possibly from material introduced in the cave from sinkhole collapse resulting from lowering water tables and landscape instability (Panno et al., 2004). Notably, speleothems from Crevice Cave, Missouri (Fig. 36.5) show a dramatic shift from prairie to

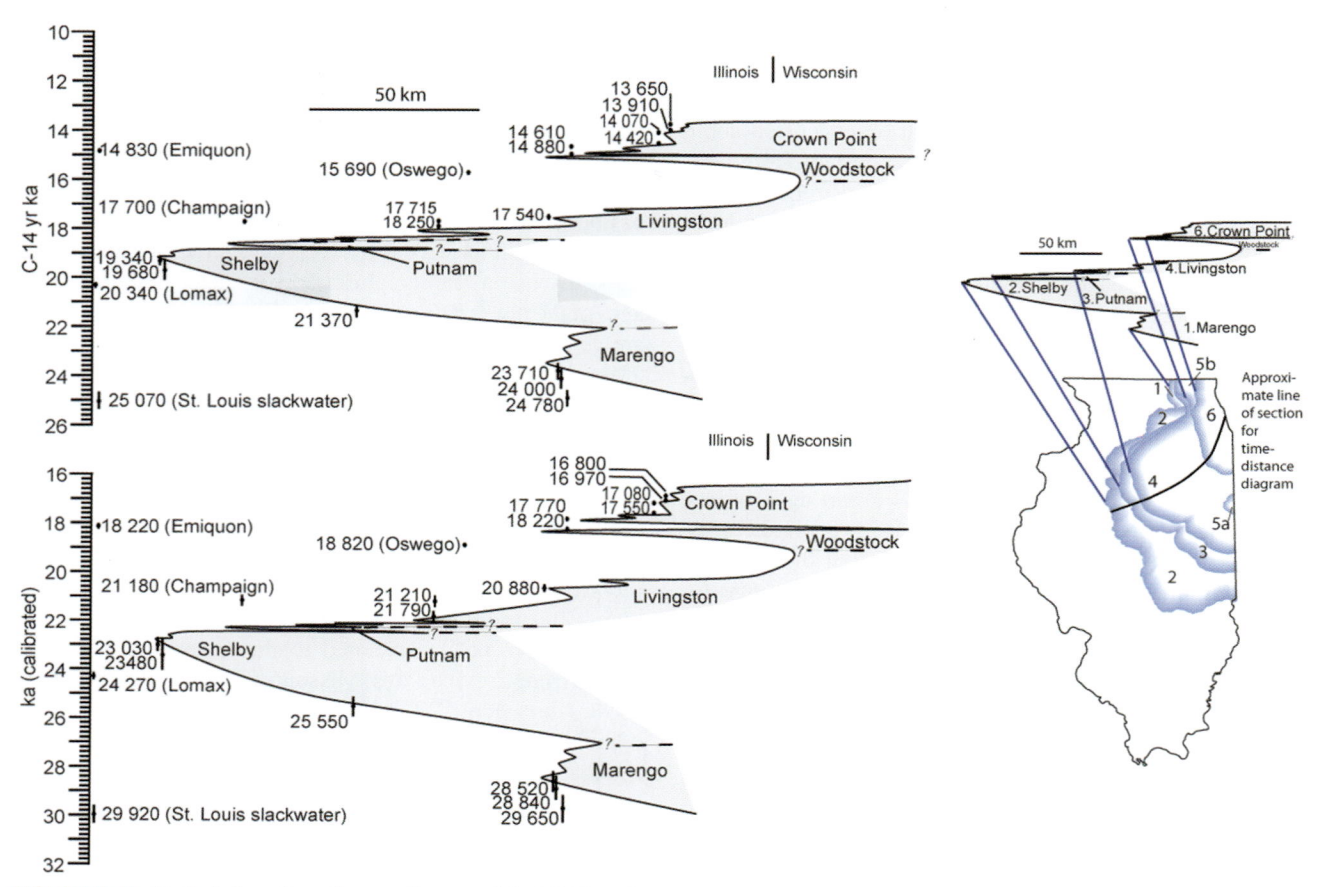

FIGURE 36.7 On the left are time–distance diagrams of the Michigan Subepisode, Lake Michigan lobe, based on radiocarbon ages (top) and calibrated ages (bottom). The calibrations were done using Calib 5.02 and IntCal 8 (Reimer et al., 2004). The inset figure on the right shows the line of section of the time–distance diagrams, and the corresponding ice-margin positions. The phases associated with the numbered ice margins include (1) Marengo, (2) Shelby, (3) Putnam, (4) Livingston, (5a) Nissouri (Huron-Erie lobe; Karrow et al., 2000), (5b) Woodstock and (6) Crown Point. Details on the radiocarbon ages (errors, contexts, references) are provided in Table 36.1.

forest vegetation at 55,000 years ago, which Dorale et al. (1998) interpret as the onset of the Wisconsin Episode.

Material referents of Wisconsin Episode loess deposits are the Roxana Silt and Peoria Silt. The Athens Subepisode of the early Wisconsin Episode is represented by loess (Roxana Silt; Fig. 36.2) which locally has evidence of relatively short periods of minimal soil development and cryoturbation (Curry and Follmer, 1992). A state-wide marker bed, the Robein Member of the Roxana Silt is composed of organic-rich silt that has yielded many radiocarbon ages ranging from about 20.0 to >45.0 ^{14}C ka BP (Curry, 1989; Hansel and Johnson, 1996); several large wood samples have yielded ages of about 24–25 ^{14}C ka BP. Carbonate leaching and weak colour-B horizonation in the Robein Member are attributed to the Farmdale Geosol (Follmer et al., 1979).

The Robein Member is overlain by diamicton, outwash, lake sediment and loess deposited during the Michigan Subepisode. Fossil boreal trees are commonly buried by Michigan Subepisode deposits, and many ^{14}C ages provide minimum estimates for the diachronic contact with the Michigan and Athens Subepisodes (Table 36.1 and Fig. 36.7).

Ice entered northern Illinois during the late Wisconsin Episode at about 29,000 cal. ka BP, about 100,000 years after the end of the Illinois Episode. The Lake Michigan lobe formed many distinctive curvilinear moraines with gently hummocky topography and little relief along their crests. Five sublobes were inferred by Willman and Frye (1970) from moraine patterns, including the Harvard, Princeton, Decatur, Peoria and Joliet sublobes (Fig. 36.8). After reaching its maximum extent, the Lake Michigan lobe retreated in an irregular fashion, forming many recessional moraines composed of lithologically distinctive and mappable diamicton units. Diamicton units are the material referents for several phases (Fig. 36.8). The north–south orientation of the 60 km-long segment of the Marengo Moraine in southernmost Wisconsin and northern Illinois indicates that the early Harvard sublobe of the Lake Michigan lobe flowed unimpeded from Lake Michigan basin. This initial incursion of the Laurentide Ice Sheet into

TABLE 36.1 Important Radiocarbon Ages, Illinois.

C-14 age	±	Laboratory number	cal. year (BP) Sigma 1	cal. year (BP) Mean	cal. year (BP) Sigma 1	References	Site name	Material dated/other notes	Diachronic unit(s)
12,055	45	CAMS-111410	13,830	13,900	13,970	Curry et al. (2007)	Brewster Creek	*Larix* needles in marl	Hudson Episode (Ep)
12,495	45	CAMS-111411	14,260	14,720	14,320	Curry et al. (2007)	Brewster Creek	*Larix* needles and wood frags	Michigan Subep, Wisconsin Ep (Wisc Ep)
13,470	130	ISGS-1378	16,420	16,590	16,840	Monaghan and Hansel (1990)	Riverside, Michigan	Organics	Mackinaw Low Phase, Wisc Ep
13,650	40	UICAMS-46829	16,720	16,800	16,880	Curry (2008a,b)	Deerfield Moraine IWLP	*Dryas integrifolia* leaves, stems	Crown Pt Ph/Glenwd Lk Ph/Mich Subep, Wisc Ep
13,870	170	ISGS-1549	16,800	16,970	17,130	Hansel and Johnson (1992)	Glenwood Spit	Driftwood	Crown Pt Ph/Glenwd Lk Ph/Mich Subep, Wisc Ep
13,870	60	CAMS-105798	16,840	16,950	17,030	Curry et al. (2007)	Brewster Creek	Wood in silt	Unnamed phase
13,890	120	ISGS-1649	16,830	16,980	17,090	Hansel and Johnson (1992)	Glenwood Spit	Driftwood	Crown Pt Ph/Glenwd Lk Ph/Mich Subep, Wisc Ep
13,910	35	UCIAMS-63076	16,870	16,970	17,060	Unpublished	Deerfield Moraine IWLP	*D. integrifolia* leaves, stems	Crown Pt Ph, Mich Subep, Wisc Ep
14,070	40	UCIAMS-26262	16,940	17,080	17,180	Curry (2008a,b)	Tinley Moraine IWLP	*D. integrifolia* leaves, stems	Crown Pt Ph, Mich Subep, Wisc Ep
14,420	40	UCIAMS-26264	17,400	17,550	17,720	Curry (2008a,b)	Tinley Moraine IWLP	*D. integrifolia* leaves, stems	Crown Pt Ph, Mich Subep, Wisc Ep
14,610	110	ISGS-A-0143	17,620	17,770	17,960	Curry (2005, 2008a,b)	Nancy Dr kettle, Crystal L	Plant fragments	Woodstock Ph, Mich Subep, Wisc Ep
14,780	150	AA-4680	17,740	18,000	18,470	Curry et al. (1999)	Nelson Lake	Needles in silt	Unnamed phase
14,830	50	B-207031	17,900	18,050	18,450	Hajic and Curry, 2008	Emiquon Ntl Wldlfe Ref	Plant fragments	Post late-stage Kankakee Torrent
14,860	40	UCIAMS-26265	17,940	18,210	18,460	Curry (2008a,b)	Cranberry Lake IWLP	*Salix herbacea* leaves, stems	Woodstock Ph, Mich Subep, Wisc Ep
14,860	110	ISGS-A-0165	17,910	18,150	18,480	Curry (2005, 2008a,b)	Nancy Dr kettle, Crystal L	Plant fragments	Woodstock Ph, Mich Subep, Wisc Ep

Continued

TABLE 36.1 Important Radiocarbon Ages, Illinois.—cont'd

C-14 age	±	Laboratory number	cal. year (BP) Sigma 1	cal. year (BP) Mean	cal. year (BP) Sigma 1	References	Site name	Material dated/other notes	Diachronic unit(s)
14,880	80	UICAMS-46834	17,940	18,220	18,480	Curry (2008a,b)	Higgins/Sutton IWLP	Tundra plant fossils	Woodstock Ph, Mich Subep, Wisc Ep
15,690	35	_b	18,740	18,820	18,890	Curry (2008a,b)	Oswego overflow	*S. herbacea, Vaccinium* leaves and stems	Post early-stage Kankakee Torrent?
17,540	130	OxA-W814-13	20,540	20,880	21,180	Curry et al. (1999); Grimley and Curry, 2001)	Fox River Stone Co	Tundra plant fossils	Mid-Livingston Phase
17,700	45	UCIAMS-46838	21,030	21,180	21,390	Unpublished	West Champaign	*D. integrifolia* leaves, stems	Putnam Ph, Mich Subep, Wisc Ep
17,700	40	_d	21,040	21,180	21,380	Curry (2008a,b)	Ransom Moraine IWLP	*S. herbacea* leaves, stems	Livingston Phase
18,270	50	_c	21,620	21,810	22,010	Curry (2008a,b)	Ransom Moraine IWLP	*S. herbacea* leaves, stems	Livingston Phase
19,340	180	ISGS-2918	22,660	23,030	23,360	Hansel and Johnson (1996; Liu et al., 1986)	Charleston Quarry South	Wood in silt over Farmdale Geos	Shelby Ph, Mich Subep, Wisc Ep
19,680	460	ISGS-532	23,660	23,870	24,140	Follmer et al. (1979)	Gardena	Small wood fragments and moss	Shelby Ph, Mich Subep, Wisc Ep
19,980	150	ISGS-2842	22,680	23,480	24,040	Hansel and Johnson (1992), Hansel et al. (1999)	Charleston Quarry South	*In situ* stump covered by lake sediment	Shelby Ph, Mich Subep, Wisc Ep
20,345	85	_a	24,130	24,270	24,460	Curry (1998)	Lomax	Wood fragments	Shelby Ph, Mich Subep, Wisc Ep
21,370	240	ISGS-2484	25,180	25,550	25,880	Hansel and Johnson (1996; Liu et al., 1986)	Wedron Quarry	Org debris	Shelby Ph, Mich Subep, Wisc Ep
23,710	320	ISGS-2108	28,080	28,520	28,880	Hansel and Johnson (1996)	Feltes Pit	*In situ* stump in palaeosol@	Marengo Ph, Mich Subep, Wisc Ep
24,000	390	ISGS-2108	28,470	28,840	29,300	Curry et al. (1999)	Feltes Pit	*In situ* stump in palaeosol@	Marengo Ph, Mich Subep, Wisc Ep

24,780	360	ISGS-2601		Hansel et al. (1999)	Hebron Core MC-8	Wood fragments in Morton Silt	Marengo Ph, Mich Subep, Wisc Ep
			29,360	29,650	30,180			
25,070	340	OxA-V-2016-35	29,580	29,920	30,230	Curry and Grimley (2006)	MNK-3	Wood frags in lake sed (Equality Formation), Lake Michigan lobe provenance	Alton Ph, Athens Subep, Wisc Ep
29,030	420	OxA-V-2016-36	33,190	33,650	34,440	Curry and Grimley (2006)	MNK-3	Wood frags in lake sed (Equality Formation), Superior lobe provenance	Alton Ph, Athens Subep, Wisc Ep
44,150	1300	OxA-V-2016-37	46,050	47,420	48,680	Curry and Grimley (2006)	MNK-3	Wood frags in lake sed (Equality Formation), Superior and western lobe provenance	Alton Ph, Athens Subep, Wisc Ep
							IWLP = ice-walled lake plain	

[a] Weighted mean of ISGS-136, ISGS-3463, ISGS-3462 and ISGS-2110 (Curry, 1998).
[b] Weighted mean of UCIAMS-26256, UCIAMS-26257, UCIAMS-26258 and UCIAMS-26259 (Curry, 2008a,b).
[c] Weighted mean of UCIAMS-71284 and UCIAMS-71279 (unpublished).
[d] Weighted mean of UCIAMS-71283, UICAMS-71288 and IICIAMS-71289.

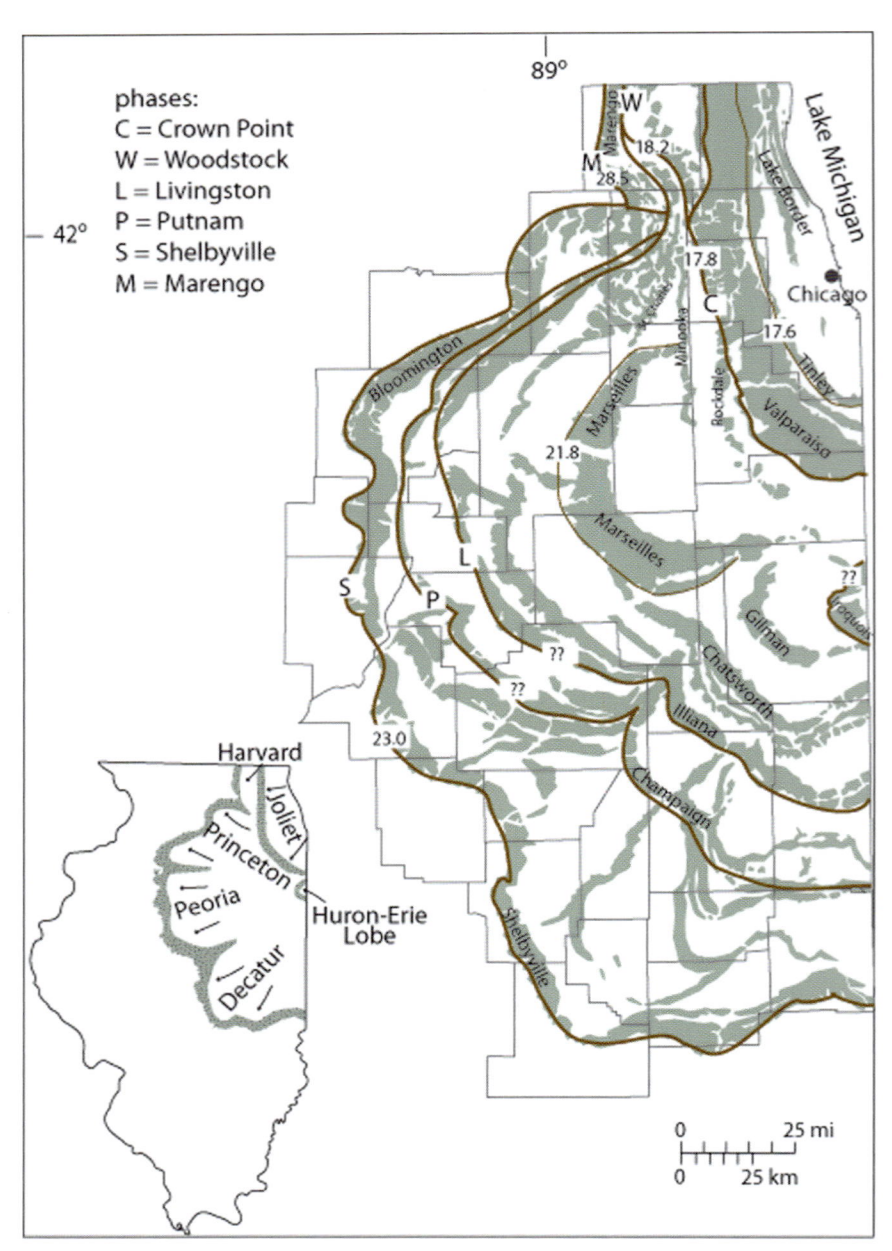

FIGURE 36.8 Location of Wisconsin Episode moraines, and the approximate ages (cal. ka) of major moraines associated with the phases shown in Fig. 36.7. The inset map shows the sublobes of the Lake Michigan lobe. Diagram from Hansel and McKay (2010).

Illinois occurred at ca. 28.67 cal. ka BP (23.71 ^{14}C ka BP) during the onset of the Marengo Phase (Michigan Subepisode; Wisconsin Episode; Figs. 36.3 and 36.7). The radiocarbon ages from the LaFarge Pit provide the youngest ages associated with this event, including tree stumps and organic-rich soil that were enfolded into the basal Wisconsin Episode diamicton (Curry et al., 1999; Figs. 36.5 and 36.7). The Marengo Moraine is formed of chiefly diamicton of the Tiskilwa Formation (Wedron Group), the oldest of several lithostratigraphical units of the Wedron Group (Fig. 36.2). The Marengo Moraine is formed of the thickest known succession (about 100 m) of a single lithostratigraphical unit of the last glaciation (Wickham et al., 1988). Tundra plant fossils dating at 25.54 cal. ka BP (21.37 ± 0.240 ^{14}C ka BP) at Wedron Quarry (Figs. 36.5 and 36.7; see also Garry et al., 1990) provide a minimum date for a major change in flow behaviour of the Lake Michigan lobe as the Princeton, Peoria and Decatur sublobes formed the lobe's outermost moraines during the Shelby Phase (Figs. 36.6–36.8). The moraines are formed primarily of diamicton of the Tiskilwa Formation. The change in flow behaviour has been attributed to interaction of the south-eastern part of the Lake Michigan lobe with the Huron-Erie lobe (Wickham et al., 1988; Fig. 36.6). Data from the Lomax site (Curry, 1998; Fig. 36.5) indicate that the Lake Michigan lobe blocked drainage of the ancient Mississippi River at 24.34 cal. ka BP (20.34 ± 0.085 ^{14}C ka BP) eventually causing the river to flow along its modern course between Moline and Alton, Illinois. The approximate rate of advance of the Lake Michigan lobe from Wedron Quarry (Fig. 36.5) to the glacial maximum, a distance of about 40 km, ranged from about 19 m/year (23 m/^{14}C year) for the Princeton sublobe to 34 m/year (42 m/^{14}C year) for the Peoria sublobe.

The Lake Michigan lobe reached its maximum extent about 1300 years after the diversion of the Mississippi River as ice-flow velocity decreased to about 15 m/year (20 m/^{14}C year) or less (depending on where the width of the Shelbyville Morainic System is measured). Minimum radiocarbon ages of wood fragments in overridden proglacial lake sediment or calcareous loess include 23.03 cal. ka BP (19.34 ± 0.18 ^{14}C ka BP) at Charleston Quarry (Fig. 36.5; Hansel and Johnson, 1996) for the Decatur sublobe; and for the Peoria sublobe, an age of 23.87 cal. ka BP (19.68 ± 0.46 ^{14}C ka BP) obtained from the Gardena Section (near the classic Farm Creek section; Fig. 36.5; Follmer et al., 1979). The two ages are statistically similar at the one-sigma level. At both sites, several other radiocarbon ages are available (Follmer et al., 1979; Hansel and Johnson, 1996), but most are at least 500 ^{14}C years older that the ones reported above. We find no reason to reject the younger ages, and look forward to verifying them in the future.

Details regarding the deglacial history of the Lake Michigan lobe have been revealed from the chronologies of tundra plant fossils encased in sediment deposited in periglacial ice-walled lakes (Curry, 2008a,b; Curry et al., 2010). The basal ages of these features mark when the ice became stagnant after depositing diamicton that formed the underlying moraine. Basal ages of ice-walled lake deposits also provide a minimum date for when the ice readvanced to form the next youngest moraine (Fig. 36.7).

At about 20.7 cal. ka BP (17.54 ± 0.13 ^{14}C ka BP), the Huron-Erie lobe ceased its interaction with the Lake Michigan lobe (Fig. 36.6). The age estimate is derived from radiocarbon ages on tundra plant fossils in lake sediment atop the St. Charles Moraine and buried by ice-marginal deltaic and alluvial deposits adjacent to the Minooka Moraine at the Fox River Stone Company pit (Figs. 36.5 and 36.8; Curry et al., 1999). This time marks when the Princeton, Peoria and Decatur sublobes became stagnant, the advance of the Joliet sublobe, and renewed activity of the Harvard sublobe (Curry and Yansa, 2004; Fig. 36.8). The age is corroborated by basal ice-walled lake ages of 21.81 cal. ka BP (18.27 ± 0.05 ^{14}C ka BP) associated with the Ransom Moraine (Marseilles Morainic System; Figs. 36.5–36.8), assumed to be time equivalent with the St. Charles Moraine (Curry, 2008b). Erosion of the southern portion of the St. Charles Moraine precludes certain correlation with the Marseilles Morainic System (Fig. 36.8).

There is no evidence of an active Lake Michigan lobe in Illinois between about 20.5 and 18.0 cal. ka BP (Fig. 36.7). Instead, there is evidence of large-scale routing of meltwater from north-eastern Indiana along the Kankakee River that probably involved breaching of moraines by proglacial lakes. The most prominent of these, the Marseilles Morainic System, is bisected by three channels eroded by overflow of proglacial Lake Wauponsee; one of the channels developed into the modern valley of the Illinois River. Another (earlier?) breach formed the Oswego channel just prior to 18.82 cal. ka BP (15.69 ± 0.035 ^{14}C ka BP; Figs. 36.5–36.7; Curry, 2008b). Other evidence points to younger large-scale floods along the Kankakee River system. At the Emiquon site (about 40 km south-west of the last glacial maximum along the Illinois River; Fig. 36.5), a large slack-water lake formed laterally against bouldery outwash; organics deposited in the base of the lake sediment date 18.03 cal. ka BP (14.83 ± 0.050 ^{14}C ka BP) (Hajic and Curry, 2008; Fig. 36.7). This age is remarkably similar to radiocarbon ages associated with ice-walled lakes on the Woodstock and West Chicago moraines that indicate ice began stagnating at 18.1 cal. ka BP (Fig. 36.7). Other prominent landscape features may be related to meltwater routing at about this time, but material suitable for radiocarbon dating is scarce (Johnson and Hansel, 1989; Hajic and Curry, 2008).

Between the Woodstock and Crown Point Phases (Fig. 36.7), the Lake Michigan lobe retreated north of Milwaukee (Schneider and Need, 1985; Curry, 2008b). Several radiocarbon ages of tundra plant fossils (primarily those of *Dryas integrifolia*) in ice-walled lake deposits bracket the Crown Point Phase between about 18.15 and 16.80 cal. ka BP (14.86 ± 0.110 and 13.65 ± 0.04 ^{14}C ka BP; Table 36.1; Curry, 2008b); other dates indicate ice stagnation of recessional moraines beginning at 17.55 cal. ka BP (14.42 ± 0.04 ^{14}C ka BP) for the Tinley Moraine and 16.97 cal. ka BP (13.91 ± 0.035 ^{14}C ka BP) for the Deerfield Moraine (Lake Border Morainic system; Figs. 36.5, 36.7 and 36.8; Curry, 2008b). The younger ages are consistent with an early period of beach development by Glacial Lake Chicago in the southern Lake Michigan basin at 16.97 cal. ka BP (13.87 ± 0.17 ^{14}C ka BP; Fig. 36.5; Hansel and Johnson, 1996). At this time, drainage of the lake was to the south through the Chicago outlet. The intra-Glenwood (Mackinaw) low-water phase of Glacial Lake Chicago, dating at about 16.59 cal. ka BP (13.47 ± 0.130 ^{14}C ka BP; Monaghan and Hansel, 1990), marks when the Lake Michigan lobe retreated north to the point that the Chicago outlet was abandoned in favour of drainage to the north across the Straits of Mackinaw or Indian River lowland in northern Michigan. Hence, at some point between about 17.0 and

16.5 cal. ka BP, the Lake Michigan lobe had retreated from Illinois for the last time.

36.3.6. Aeolian and Lacustrine Deposits

The loessial Peoria Silt is the youngest glacial deposit of the Wisconsin Episode (Hansel and Johnson, 1996). Beyond the limit of the last glaciation, the Peoria Silt and Roxana Silt together range in thickness from <2 m to as much as 25 m in areas proximal to the Mississippi and Illinois River Valleys (McKay, 1979; Grimley, 2000; Wang et al., 2000). Within the last glacial limit, Peoria Silt covers the Wedron Group. Along outwash terraces within and beyond the limit, Peoria Silt covers the Henry Formation (Hansel and Johnson, 1996). Like the Wedron Formation, Peoria Silt was deposited during the Michigan Subepisode of the Wisconsin Episode. In kettles and other wetland areas, Peoria Silt was redeposited as lake sediment of the Equality Formation. At the Brewster Creek site (Fig. 36.5; Curry et al., 2007), an abrupt change from deposition of quartz-rich silty sediment to marl occurred at 14.6 cal. ka BP (12.495 ^{14}C ka BP). This age equates with Termination I (Rasmussen et al., 2006). Thus, although ice had largely melted from Illinois' landscape by about 16 cal. ka BP, we recognise the end of the last glaciation (Wisconsin Episode) by the cessation of loess deposition some 1400 years later in north-eastern Illinois. In the Mississippi River drainage, the collective evidence indicates a younger boundary between the interglacial Hudson and glacial Wisconsin Episodes of 10.5 ^{14}C ka BP (∼12.5 cal. ka BP). This age is consistent with that extrapolated for the youngest Peoria Silt at the Keller Farm section (Wang et al., 2000), and with the approximate age of the switch from deposition of sand and gravel outwash and reddish, laminated silty clay (overbank flood deposits) to primarily sand with less gravel (Hajic, 1993). In addition to geomorphological considerations, the change in alluvium composition is attributed to the abrupt change in the Mississippi River from a seasonally variable, meltwater-influenced braided river to a meandering system. The 12.5 cal. ka BP boundary age is consistent with OSL ages on the Morehouse braid belt, the youngest well-dated braid belt of the upper-lower Mississippi River basin (Rittenour et al., 2007).

Tracts of sand dunes cover several places in Illinois, notably in the Green River Lowland (Fig. 36.5; Miao et al., 2010), west of the confluence of the Sangamon and Illinois Rivers in Mason County, and in the broad valley of the Kankakee River. The earliest late-glacial dune activity is indicated by buried dune sand at Lomax (Curry, 1998) that is bracketed by an age on wood fragments of about 20.5 cal. ka BP below, and by terrestrial snail shells above dating at 19.5 cal. ka BP (Oches et al., 1996). The sand was probably sourced from exposed alluvial bars in the upper Mississippi River Valley. The dunes in the Green River Lowland formed during a short period centring on about 17.5 ka cal. BP, and the sand was derived from outwash along the Rock River Valley (Miao et al., 2010). Superficial dune sand is mapped as the Parkland facies of the Henry Formation (Hansel and Johnson, 1996).

Fossiliferous deposits of Wisconsin Episode proglacial lakes are extensive in Illinois (Fig. 36.1). Slackwater lake sediment accumulated in valleys tributary to major glacial sluiceways as outwash aggraded. The earliest evidence of the Laurentide Ice Sheet contributing outwash to the Mississippi River drainage includes buried Alton Subepisode slackwater lake sediment at sites near St. Louis (Grimley et al., 2001; Curry and Grimley, 2006). Younger slackwater lake deposits are mapped throughout the state and form prominent terraces at the mouths of valleys tributary to the Mississippi, Illinois and Wabash Rivers (Hajic et al., 1991). Some slackwater lake terraces, mapped as the Equality Formation, are many kilometres wide, and extend tens of kilometres upstream of the main trunk stream such as the terrace attributed to Glacial Lake Saline (Fig. 36.5; Frye et al., 1972; Heinrich, 1982; Trent and Esling, 1995). Expansive terraces formed of fine-grained, laminated sediment were deposited in moraine-dammed proglacial lakes such as Glacial Lake Watseka (Fig. 36.5; Willman and Frye, 1970; Moore, 1981).

REFERENCES

Balco, G., Rovey, C.W., Stone, J.O.H., 2005. The first glacial maximum in North America. Science 307, 222.

Berg, R.C., Kempton, J.P., Follmer, L.R., McKenna, D.P., 1985. Illinoian and Wisconsinan stratigraphy and environments in northern Illinois: the Altonian revised. Illinois State Geological Survey Guidebook 19, 177 pp.

Bleuer, N.K., 1976. Remnant magnetism of Pleistocene sediments of Indiana. Indiana Acad. Sci. 85, 277–294.

Curry, B.B., 1989. Absence of Altonian glaciation in Illinois. Quatern. Res. 31, 1–13.

Curry, B.B., 1998. Evidence at Lomax, Illinois, for mid-Wisconsin (ca. 40,000 yr B.P.) position of the Des Moines Lobe, and for diversion of the Mississippi River by the Lake Michigan Lobe (20,350 yr B.P.). Quatern. Res. 50, 128–138.

Curry, B.B., 2005. Surficial geology of Crystal Lake Quadrangle, McHenry and Kane Counties, Illinois. Illinois State Geological Survey, Illinois Geological Quadrangle Map, IGQ—Crystal Lake-SG, 1:24,000.

Curry, B.B., 2008a. Surficial geology of Hampshire Quadrangle, Kane and DeKalb Counties, Illinois. Illinois State Geological Survey, Illinois Geological Quadrangle Map, IGQ—Hampshire SG, 1:24,000.

Curry, B.B. (Ed.), 2008b. Deglacial History and Paleoenvironments of Northeastern Illinois, In: Illinois State Geological Survey Open File 2008-1, p. 175.

Curry, B.B., Baker, R.G., 2000. Paleohydrology, vegetation, and climate during the last interglaciation (Sangamon episode) in south-central Illinois. Palaeogeogr. Palaeoclimatol. Palaeoecol. 155 (1–2), 59–81.

Curry, B.B., Follmer, L.R., 1992. The last interglacial–glacial transition in Illinois: 123–25 ka. In: Clark, P.U., Lea, P.D. (Eds.), The Last

Interglacial–Glacial Transition in North America, Geologic Society of America Special Paper 270, 71–88.

Curry, B.B., Grimley, D.A., 2006. Provenance, age, and environment of mid-Wisconsin Episode slackwater lake sediment in the St. Louis Metro East area. Quatern. Res. 65, 108–122.

Curry, B.B., Pavich, M.J., 1996. Absence of glaciation in Illinois during marine isotope stages 3 through 5. Quatern. Res. 31, 19–26.

Curry, B.B., Yansa, C.H., 2004. Evidence of stagnation of the Harvard sub-lobe (Lake Michigan lobe) in northeastern Illinois, USA, from 24,000 to 17,600 BP and subsequent tundra-like ice-marginal palaeoenvironments from 17,600 to 15,700 BP. Géogr. Phys. Quatern. 58, 305–321.

Curry, B.B., Troost, K.G., Berg, R.C., 1994. Quaternary geology of the Martinsville alternative site, Clark County, Illinois, a proposed low level radioactive waste disposal site. Illinois State Geological Survey Circular 556, 85 pp.

Curry, B.B., Grimley, D.A., Stravers, J.A., 1999. Quaternary geology, geomorphology, and climatic history of Kane County, Illinois. Illinois State Geological Survey Guidebook 28, 40 pp.

Curry, B.B., Dorale, J.A., Henson, R.A., 2002. Function-fitting vegetation proxy record profiles of the Sangamon and Wisconsin episodes from Missouri and Illinois. Geol. Soc. Am. Abstr. Programs 35 (6), 199.

Curry, B.B., Grimm, E.C., Slate, J.E., Hansen, B.C., Konen, M.E., 2007. The late glacial and early Holocene geology, paleoecology, and paleohydrology of the Brewster Creek site, a proposed wetland restoration site, Pratt's Wayne Woods Forest Preserve and James "Pate" Philip State Park, Bartlett, Illinois. Illinois State Geological Survey Circular 571, 50 pp.

Curry, B.B., Wang, H., Panno, S.V., Hackley, K.C., 2010. Quaternary paleoclimates. In: Kolata, D.R., Nimz, C.K. (Eds.), Geology of Illinois, Illinois State Geological Survey, Champaign, IL, USA, 248–260.

Curry, B.B., Larson, T.H., Konen, M.E., Alexanderson, H., Yansa, C.H., Lowell, T., submitted for publication. The DeKalb mounds of northeastern Illinois as archives of deglacial history and postglacial environments. Quatern. Res. 74, 82–90.

Dorale, J.A., Edwards, R.E., Ito, E., Gonzalez, L.A., 1998. Climate and vegetation history of the Midcontinent from 75 to 25 ka: a speleothems record from Crevice Cave, Missouri, USA. Science 282, 1871–1874.

Follmer, L.R., 1983. Sangamonian and Wisconsinan pedogenesis in the midwestern United States. In: Porter, S.C. (Ed.), Late Quaternary Environments of the Unites States, vol. 1, the Late Pleistocene University of Minnesota Press, Minneapolis, pp. 138–144.

Follmer, L.R., McKay III, E.D., Lineback, J.A., Gross, D.L., 1979. Wisconsinan, Sangamonian, and Illinoian Stratigraphy in Central Illinois. Illinois State Geological Survey Guidebook 13, 139 pp.

Forman, S.L., Pierson, J., 2002. Late Pleistocene luminescence chronology of loess deposition in the Missouri and Mississippi river valleys, United States. Palaeogeogr. Palaeoclimatol. Palaeoecol. 186 (1–2), 25–46.

Frye, J.C., Leonard, A.B., Willman, H.B., Glass, H.D., 1972. Geology and paleontology of Late Pleistocene Lake Saline, Southeastern Illinois. Illinois State Geological Survey Circular 471, 44 pp.

Frye, J.C., Follmer, L.R., Glass, H.D., Masters, J.M., Willman, H.B., 1974. Earliest Wisconsinan sediments and soils. Illinois State Geological Survey Circular 485, 12 pp.

Garry, C.E., Schwert, D.P., Baker, R.G., Kemmis, T.J., Horton, D.G., Sullivan, A.E., 1990. Plant and insect remains from the Wisconsin interstadial/stadial transition at Wedron, north-central Illinois. Quatern. Res. 33, 387–399.

Geiger, E.C., 2008. Paleoecology of Pleistocene gastropods in glacial lake deposits in southern Illinois/Missouri. Southern Illinois University Master's thesis, Carbondale, Illinois, 138 pp.

Glass, H.D., Killey, M.M., 1988. Principles and applications of clay mineral composition in Quaternary stratigraphy: examples from Illinois, USA. In: van der Meer, J.J.M. (Ed.), Tills and Glaciotectonics. A.A. Balkema, Rotterdam, pp. 117–125.

Grimley, D.A., 1997. Quaternary deposits in Carroll County, Illinois. Illinois State Geological Survey Open File 1997-13b.

Grimley, D.A., 2000. Glacial and nonglacial sediment contributions to Wisconsin Episode loess in the central United States. Geol. Soc. Am. Bull. 112, 1475–1495.

Grimley, D.A., 2010. Surficial geology of Mascoutah quadrangle, St. Clair County, Illinois: Illinois State Geological Survey. Illinois Geologic Quadrangle Map, IGQ—Mascoutah-SG, 1:24,000.

Grimley, D.A., Curry, B.B., 2001. Surficial geology map, Geneva Quadrangle, Kane County, Illinois. Illinois State Geological Survey, Illinois Geological Quadrangle Map, IGQ—Geneva-SG, 1:24,000.

Grimley, D.A., Webb, N.D. (2009). Surficial geology of Red Bud Quadrangle, Randolph, Monroe and St. Clair Counties, Illinois: Illinois State Geological Survey, Illinois *Geologic Quadrangle Map, IGQ—Red Bud-SG*, 1:24,000.

Grimley, D.A., Phillips, A.C., Follmer, L.R., Wang, H., Nelson, R.S., 2001. Quaternary and environmental geology of the St. Louis Metro East area. In: Malone, D. (Ed.), Illinois State Geological Survey Guidebook 33, pp. 21–73.

Grimley, D.A., Follmer, L.R., Hughes, R.E., Solheid, P.A., 2003. Modern, Sangamon, and Yarmouth soil development in loess of unglaciated southwestern Illinois. Quatern. Sci. Rev. 22, 225–244.

Grüger, E., 1972. Late Quaternary vegetation development in south-central Illinois. Quatern. Res. 2, 217–231.

Hajic, E.R., 1993. Geomorphology of the northern American Bottom as context for archeology. Ill. Archeol. 5, 54–65.

Hajic, E.R., Curry, B.B., 2008. Megafloods and evolution of the upper Illinois River valley. In: Curry, B.B. (Ed.), Deglacial History and Paleoenvironments of Northeastern Illinois, Illinois State Geological Survey Open File 2008–1, pp. 62–67.

Hajic, E.R., Johnson, W.H., Follmer, L.R., 1991. Quaternary deposits and landforms, confluence region of the Mississippi, Missouri, and Illinois Rivers, Missouri and Illinois: terraces and terrace problems. In: Midwest Friends of the Pleistocene Guidebook, 38th Field Conference, May 10–12, 1991. Department of Geology, University of Illinois at Urbana-Champaign, 106 pp.

Hallberg, G.R., 1986. Pre-Wisconsin glacial stratigraphy of the central plains region in Iowa, Nebraska, Kansas, and Missouri. Quatern. Sci. Rev. 5, 11–15.

Hansel, A.K., Johnson, W.H., 1992. Fluctuations of the Lake Michigan lobe during the late Wisconsin subepisode. Sver. Geol. Unders. 81, 133–144.

Hansel, A.K., Johnson, W.H., 1996. Wedron and Mason groups: lithostratigraphic reclassification of deposits of the Wisconsin episode, Lake Michigan lobe area. Ill. State Geol. Surv. Bull. 104, 116.

Hansel, A.K., Berg, R.C., Phillips, A.C., Gutowski, V., 1999. Glacial sediments, landforms, paleosols, and a 20,000-year-old forest bed in east-central Illinois. Illinois State Geologic Survey Guidebook, 26, 31 pp.

Hansel, A.K., McKay, E.D., III, 2010. Quaternary period. In: Kolata, D.R., Nimz, C.K. (Eds.), Geology of Illinois. Illinois State Geological Survey, Champaign, IL, USA, pp. 216–247.

Heigold, P.C., Poole, V.L., Cartwright, K., Gilkeson, R.H., 1985. An electrical earth resistivity survey of the Macon-Taylorville ridged-drift aquifer. Illinois State Geological Survey Circular 533, 23 pp.

Heinrich, P.V., 1982. Geomorphology and sedimentology of Pleistocene lake saline, southern Illinois. Unpublished MS thesis, University of Illinois at Urbana-Champaign, 145 pp.

Horberg, C.L., 1953. Pleistocene deposits below the Wisconsin drift in northeastern Illinois. Illinois State Geological Survey Report of Investigations 165, 61 pp.

Horberg, C.L., 1956. Pleistocene deposits along the Mississippi Valley in central-western Illinois. Illinois State Geological Survey Report of Investigations 192, 39 pp.

Horberg, C.L., Larson, T.E., Suter, M., 1950. Groundwater in the Peoria region: part I, geology. Ill. State Geol. Surv. Bull. 75, 13–49.

Jacobs, A.M., Lineback, J.A., 1969. Glacial geology of the Vandalia, Illinois region. Illinois State Geological Survey Circular 442, 23 pp.

Jacobs, P.M., Konen, M.E., Curry, B.B., 2009. Pedogenesis of a catena of the Farmdale–Sangamon Geosol complex in the north central United States. Palaeogeogr. Palaeoclimatol. Palaeoecol. 282, 119–132.

Johnson, W.H., 1971. Old glacial drift near Danville, Illinois. Illinois State Geological Survey Circular 457, 16 pp.

Johnson, W.H., 1976. Quaternary stratigraphy in Illinois—status and current problems. In: Mahaney, W.C. (Ed.), Quaternary Stratigraphy in North America. Dowden, Hutchinson, & Ross, Inc., Stroudsburg, PA, USA, pp. 169–196.

Johnson, W.H., 1986. Stratigraphy and correlation of the glacial deposits of the Lake Michigan lobe prior to 14 ka B.P. Quatern. Sci. Rev. 5, 17–22.

Johnson, W.H., Hansel, A.K., 1989. Age, stratigraphic position, and significance of the Lemont drift, northeastern Illinois. J. Geol. 97, 301–318.

Johnson, W.H., Hansel, A.K., 1990. Multiple Wisconsinan glacigenic sequences at Wedron, Illinois. J. Sed. Petrol. 60, 26–41.

Johnson, W.H., Gross, D.L., Moran, S.R., 1971. Till stratigraphy of the Danville region, east-central IllinoisGoldtwait, R.P. (Ed.), Till, a Symposium. Ohio State University Press, Columbus, OH, USA, pp. 184–216.

Kaplan, S.W., Grimley, D.A., 2006. A multi-proxy reconstruction of Illinois episode glacial maximum environments in the central U.S.. In: American Association of Geographers Annual Meeting, Chicago, IL.

Karrow, P.F., Dreimanis, A., Barnett, P.J., 2000. A proposed diachronic revision of late Quaternary time-stratigraphic classification in the eastern and northern Great Lakes area. Quatern. Res. 54, 1–12.

Kempton, J.P., Johnson, W.H., Heigold, P.C., Cartwright, K., 1991. Mahomet bedrock valley in east-central Illinois; topography, glacial drift stratigraphy and hydrogeology. In: Melhorn, W.N., Kempton, J.P. (Eds.), Geology and Hydrogeology of the Teays-Mahomet Bedrock Valley System, Geological Society of America Special Paper 258, 91–124.

King, J.E., Saunders, J.J., 1986. *Geochelone* in Illinois and the Illinoian–Sangamonian vegetation of the type region. Quatern. Res. 25, 89–99.

Leigh, D.S., Knox, J.C., 1993. AMS radiocarbon age of the upper Mississippi river valley Roxana silt. Quatern. Res. 39, 282–289.

Leighton, M.M., Brophy, J.A., 1961. Illinoian glaciation in Illinois. J. Geol. 69, 11–31.

Leonard, A.B., Frye, J.C., Johnson, W.H., 1971. Illinoian and Kansan molluscan faunas of Illinois. Illinois State Geological Survey Circular 461, 24 pp.

Leverett, F., 1899. The Illinois glacial lobe. U.S. Geological Survey Monograph 38, 817 pp.

Lineback, J.A., 1971. Pebble orientation and ice movement in south-central Illinois. In: Goldtwait, R.P. (Ed.), Till, a Symposium. Ohio State University Press, Columbus, OH, USA, pp. 328–334.

Lineback, J.A., 1979a. Quaternary deposits of Illinois. Illinois State Geological Survey Map, scale 1:500,000.

Lineback, J.A., 1979b. The status of the Illinoian glacial stage. Illinois State Geological Survey Guidebook 13, pp. 69–78.

Maclintock, P., 1929. Recent discoveries of pre-Illinoian drift in southern Illinois. Illinois State Geological Survey Report of Investigation 19. Part II, pp. 27–57.

Martinson, D.G., Pisias, N.G., Hays, J.D., Imbrie, J., Moore, T.C., Jr., Shackleton, N.J., 1987. Age dating and the orbital theory of the Ice ages: development of a high-resolution 0 to 300,000-year chronostratigraphy. Quatern. Res. 27, 1–29.

McKay, E.D., III, 1979. Stratigraphy of Wisconsinan and older loesses in southwestern Illinois. Illinois State Geological Survey Guidebook 14, pp. 37–67.

McKay, E.D., III, Berg, R.C., 2008. Optical ages spanning two glacial–interglacial cycles from deposits of the ancient Mississippi river, north-central Illinois. Geol. Soc. Am. Abstr. Programs 40 (5), 78.

McKay, E.D., III, Berg, R.C., Hansel, A.K., Kemmis, T.J., Stumpf, A.J., 2008. Quaternary deposits and history of the ancient Mississippi river valley, north-central Illinois. Illinois State Geological Survey Guidebook 35, 106 pp.

Miao, X., Hanson, P.R., Wang, H., Young, A.R., 2010. Timing and origin for sand dunes in the Green river lowland of Illinois, upper Mississippi river valley, USA. Quatern. Sci. Rev. 29, 763–773.

Miller, B.B., Graham, R.W., Morgan, A.V., Norton, G.M., McCoy, W.D., Palmer, D.F., et al., 1994. A biota associated with Matuyama-age sediments in west-central Illinois. Quatern. Res. 41, 350–365.

Mirecki, J.E., Miller, B.B., 1994. Aminostratigraphic correlation and geochronology of two Quaternary loess localities, central Mississippi Valley. Quatern. Res. 41, 289–297.

Monaghan, G.W., Hansel, A.K., 1990. Evidence for the intra-Glenwood (Mackinaw) low-water phase of glacial Lake Chicago. Can. J. Earth Sci. 27, 1236–1241.

Moore, D.W., 1981. Stratigraphy of till and lake beds of late Wisconsinan age in Iroquois and neighboring counties, Illinois. Unpublished Ph.D. thesis, University of Illinois at Urbana-Champaign, 211 pp.

Nelson, W.J., Devera, J.A., Jacobson, R.J., Weibel, C.P., Follmer, L.R., Riggs, M.H., et al., 1991. Geology of the Eddyville, Stonefort, and Creal Springs quadrangles, southern Illinois. Ill. State Geol. Surv. Bull. 96, 85 pp.

Nelson, W.J., Denny, F.B., Follmer, L.R., Masters, J.M., 1999. Quaternary grabens in southernmost Illinois: deformation near an active intraplate seismic zone. Tectonophysics 105, 381–397.

Oches, E.A., 2001. Aminostratigraphy of three Pleistocene silts in the Collinsville-Belleville area, Illinois. Illinois State Geological Survey Guidebook 33, pp. 70–72.

Oches, E.A., McCoy, W.D., Clark, P.U., 1996. Amino acid estimates of latitudinal temperature gradients and geochronology of loess deposition during the last glaciation, Mississippi valley, United States. Geol. Soc. Am. Bull. 108, 892–903.

Panno, S.V., Curry, B.B., Wang, H., Hackley, K.C., Liu, C.-L., Lundstrom, C., et al., 2004. Climate change in southern Illinois, USA, based on the age and $\delta^{13}C$ of organic matter in cave sediments. Quatern. Res. 64, 301–313.

Phillips, A.C., 2004. Surficial geology of Collinsville quadrangle, Madison and St. Clair Counties, Illinois. Illinois State Geological Survey. Illinois Preliminary Geologic Map, IPGM Collinsville-SG, 1:24,000.

Pillans, B., Naish, T., 2004. Defining the Quaternary. Quatern. Sci. Rev. 23, 2271–2282.

Rasmussen, S.O., Andersen, K.K., Svensson, A.M., Steffensen, J.P., Vinther, B.M., Clausen, H.B., et al., 2006. A new Greenland ice core chronology for the last glacial termination. J. Geophys. Res. 111, D06102. doi:10.1029/2005JD006079.

Reimer, P.J., Baillie, M.G., Bard, E., Bayliss, A., Beck, J.W., 2004. IntCal04 terrestrial radiocarbon age calibration, 0–26 cal kyr BP. Radiocarbon 46, 1029–1058.

Rittenour, T.M., Blum, M.D., Goble, R.J., 2007. Fluvial evolution of the lower Mississippi river valley during the last 100 k.y. glacial cycle: response to glaciations and sea-level change. Geol. Soc. Am. Bull. 119, 586–608.

Rovey, C.W., II, Kean, W.F., 2001. Palaeomagnetism of the Moberly formation, northern Missouri, confirms a regional magnetic datum within the pre-Illinoian glacial sequence of the midcontinental USA. Boreas 30, 53–60.

Roy, M., Clark, P.U., Barendregt, R.W., Glasmann, J.R., Enkin, R.J., 2004. Glacial stratigraphy and paleomagnetism of late Cenozoic deposits of the north-central United States. Geol. Soc. Am. Bull. 116, 30–41.

Schneider, A.F., Need, E.A., 1985. Lake Milwaukee: an "early" proglacial lake in the Lake Michigan basin. In: Karrow, P.F., Calkin, P.E. (Eds.), Quaternary Evolution of the Great Lakes, Geological Association of Canada Special Paper 30, pp. 55–62.

Shackleton, N.J., 1987. Oxygen isotopes, ice volume and sea level. Quatern. Sci. Rev. 6, 183–190.

Stiff, B.J., Hansel, A.K., 2004. Quaternary glaciations in Illinois. In: Ehlers, J., Gibbard, P.L. (Eds.), Quaternary Glaciations—Extent and Chronology, vol. II, North America. Elsevier, Amsterdam, pp. 71–82.

Stokes, C.R., Clark, C.D., 2002. Ice stream shear margin moraines. Earth Surf. Process. Land. 27, 547–558.

Teed, R., 2000. A > 130,000-year-long pollen record from Pittsburg basin, Illinois. Quatern. Res. 54, 264–274.

Trent, G.C., Esling, S.P., 1995. The Big muddy valley. (Chapter 1)In: Quaternary Sections in Southern Illinois and Southeast Missouri, Midwest Friends of the Pleistocene 42nd Annual Meeting, Southern Illinois University, Carbondale.

Vonder Haar, S.P., Johnson, W.H., 1973. Mean magnetic susceptibility: a useful parameter for stratigraphic studies of glacial till. J. Sed. Petrol. 43, 1148–1151.

Wang, H., Follmer, L.R., Liu, J.C., 2000. Isotope evidence of paleo-El Niño-southern oscillation cycles in the loess-paleosol record in the central United States. Geology 28, 771–774.

Webb, N.D., 2009. An investigation of the origin of the ridged drift of the lower Kaskaskia basin, southwestern Illinois. Masters thesis, University of Illinois at Urbana-Champaign, 139 pp.

Weibel, C.P., Nelson, W.J., 1993. Geology of the lick creek quadrangle, Johnson, union, and Williamson Counties, Illinois. Illinois Geologic Quadrangle Map, IGQ—Lick Creek-G, 1:24,000.

Wickham, J.T., 1979. Pre-Illinoian till stratigraphy in the Quincy, Illinois, area. In: Geology of Western Illinois, Illinois State Geological Survey Guidebook 14, 69–90.

Wickham, S.S., Johnson, W.H., Glass, H.D., 1988. Regional geology of the Tiskilwa Till Member, Wedron Formation, northeastern Illinois. Illinois State Geological Survey Circular 543, 35 pp.

Willman, H.B., Frye, J.C., 1970. Pleistocene stratigraphy of Illinois. Ill. State Geol. Surv. Bull. 94, 204 pp.

Willman, H.B., Frye, J.C., 1980. The glacial boundary in southern Illinois. Illinois State Geological Survey Circular 511, 23 pp.

Willman, H.B., Glass, H.D., Frye, J.C., 1963. Mineralogy of glacial till and their weathering profiles. Part I. Glacial tills. Illinois State Geological Survey Circular 347, 55 pp.

Willman, H.B., Glass, H.D., Frye, J.C., 1989. Glaciation and origin of the geest in the driftless area of northwestern Illinois. Illinois State Geological Survey Circular 535, 44 pp.

Zachos, J., Pagani, M., Sloan, L., Thomas, E., Billups, K., 2001. Trends, rhythms, and aberrations in global climate 65 Ma to present. Science 292, 686–693.

Zhou, J., Lundstrom, C., Fouke, B., Panno, S., Hackley, K., Curry, B., 2005. Geochemistry of speleothems records from southern Illinois: development of ($^{234}U/^{238}U$) as a proxy for paleoprecipitation. Chem. Geol. 221, 1–20.

Zhu, H., Baker, R.G., 1995. Vegetation and climate of the last glacial–interglacial cycle in southern Illinois, USA. J. Paleolimnol. 14, 337–354.

Ice-Margin Fluctuations at the End of the Wisconsin Episode, Michigan, USA

Grahame J. Larson

Department of Geological Sciences, Michigan State University, East Lansing, Michigan 48834, USA

37.1. INTRODUCTION

During the Quaternary, ice sheets extended over Michigan and the surrounding Great Lakes multiple times (Fig. 37.1). Within the state, however, the record of glaciation is mainly limited to the last ice sheet (Larson and Kincare, 2009), whereas in states south and southwest of Michigan, it is much more complete and indicates that the state was glaciated at least six times (Fullerton, 1986) with the earliest glaciation occurring sometime prior to 780,000 years ago (Fullerton, 1986; Johnson, 1986).

In the 1970s, temporal classification of the last glaciation of Michigan (Saarnisto, 1974; Farrand and Eschman, 1974) followed a widely accepted system developed in Illinois (Willman and Frye, 1970). It defined the last glacial interval as the *Wisconsinan Stage* and divided the stage into five *substages*. Several years later, an alternate system was introduced to Michigan (Eschman, 1978, 1980) from Ontario (Dreimanis and Karrow, 1972). It defined the last glacial interval as the *Wisconsin(an) Stage* and divided the stage into three *substages* and 15 *stadial* and *interstadials*.

An inherent difficulty with both the above classification systems is that the Wisconsinan or Wisconsin(an) Stage and its subdivisions have time-parallel boundaries that do not correspond to timing of events across Michigan, let alone the Great Lakes region. For example, the interval of ice-free conditions known as the Twocreekan Substage (Willman and Frye, 1970; Evenson et al., 1976) or Two Creeks Interstadial (Dreimanis and Karrow, 1972) clearly lasted longer in the central part of the Lake Michigan basin than in the northern part, but in both classification systems it represents a fixed interval of time. To address this issue, a new classification system based on diachronic (time transgressive) divisions was introduced by Hansel and Johnson (1992) and Johnson et al. (1997) for the western part of the Great Lakes region and by Karrow et al. (2000) for the eastern part of the region. More recently, the classification system was applied by Larson and Kincare (2009) to the central part of the Great Lakes region (Fig. 37.2).

In this chapter, major ice readvances during the deglaciation of Michigan are reviewed using the diachronic classification system of Hansel and Johnson (1992), Johnson et al. (1997) and Karrow et al. (2000). Also, timing of the readvances is compared to palaeoclimatic records from central Greenland and Crawford Lake in southeast Ontario.

37.2. STRUCTURE OF THE DIACHRONIC CLASSIFICATION

In the diachronic classification system of Hansel and Johnson (1992), Johnson et al. (1997) and Karrow et al. (2000), the highest order time division is an *episode* which is applied regionally (e.g. Great Lakes basin) and defined by one or more referent units. It is subdivided into *subepisodes* which are applied more locally (e.g. Lake Michigan basin) and also defined by one or more referent units. The smallest subdivision is a *phase* which is applied locally (e.g. northwestern lower Michigan) and defined by a single referent such as a unit of till, outwash, lacustrine sediment, loess, peat, or a buried soil. It is important to note that episodes, subespisodes and phases can be glacial or non-glacial intervals.

The Wisconsin Episode (Fig. 37.2) represents a time when climate in the Great Lakes region deteriorated and glacier ice extended into the region. It is preceded by the Sangamon Episode, a time of relative warmth and ice-free conditions, and followed by the Hudson Episode, a time of improved climatic conditions and ice-margin retreat (Johnson et al., 1997; Karrow et al., 2000). In the western area of the Great Lakes region, the Wisconsin Episode is subdivided into two subepisodes, the Athens and Michigan. The Athens is a time characterised by cold dry climate, ice-free conditions and loess deposition, while the Michigan is a time characterised by cold climate and substantial cover of glacial ice (Johnson et al., 1997). In the eastern area

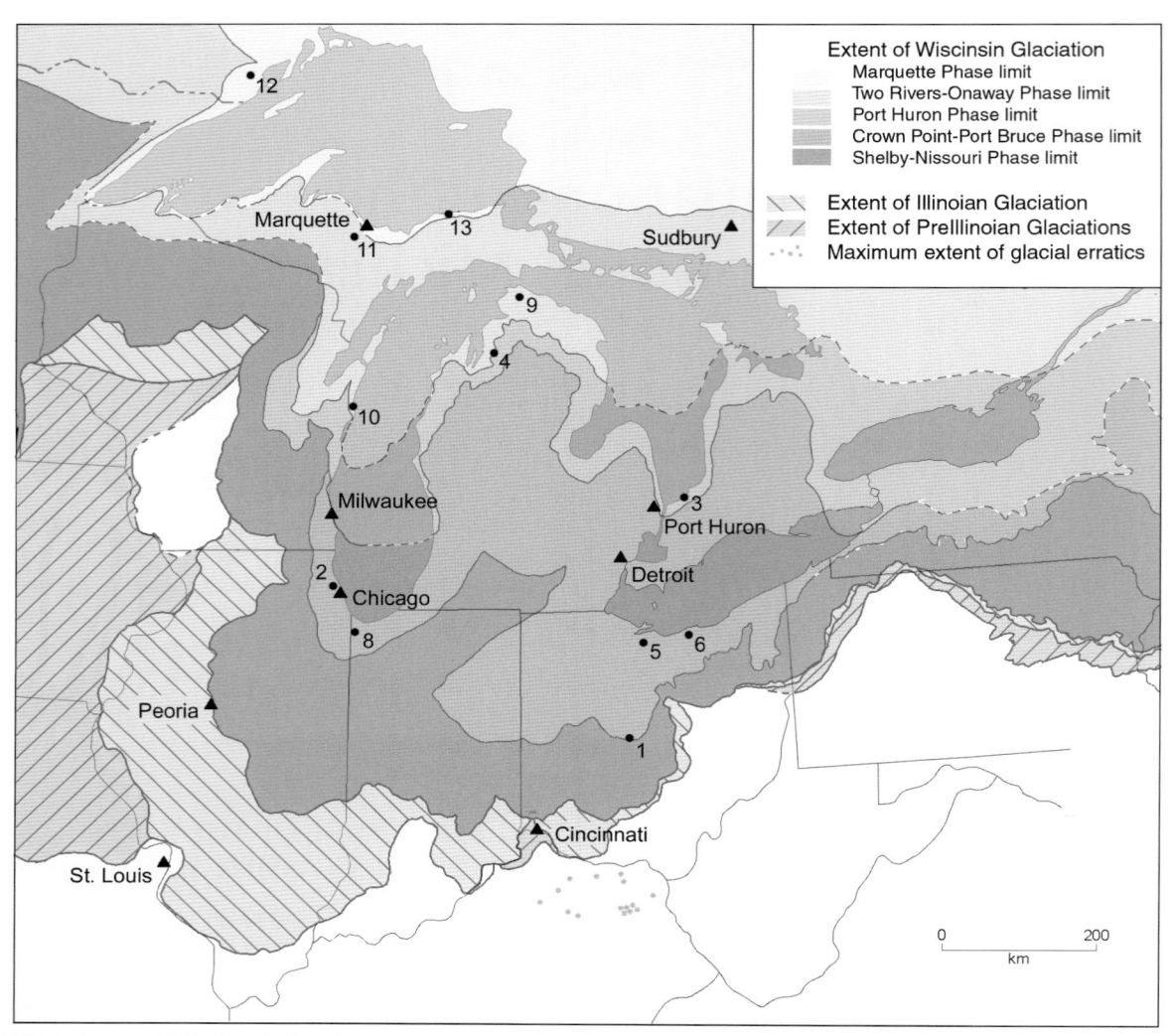

FIGURE 37.1 Extent of drift sheets in the mid-continent of North America (modified from Larson and Kincare, 2009). Numbered points correspond to location of ^{14}C dates listed in Table 37.1.

of the Great Lakes region, the Wisconsin Episode is subdivided into three subepisodes, the Ontario, Elgin and Michigan. The Ontario is a time of cold climate and limited ice cover, whereas the Elgin is a time of moderated climate and reduced ice covert (Karrow et al., 2000). The Michigan is a time of cold climate and extensive ice cover (Karrow et al., 2000). The reason three subepisodes occur in the eastern part of the Great Lakes region while only two occur in the western part is that the glacial record begins earlier in the eastern part.

37.3. MAJOR GLACIAL PHASES IN MICHIGAN

To date, there are no reports of sediments in Michigan deposited prior to or during the Ontario Subepisode, but there are reports of sediments deposited during the Athens-Elgin and Michigan Subepisodes. Sediments associated with the Athens-Elgin Subepisode consist of wood bearing non-glacial sediments (clay, muck and marl) found along cut banks of streams (Eschman, 1980) or encountered while drilling water wells (Winters et al., 1986). They have yielded radiocarbon dates ranging from ~51.9 to ~42.3 ^{14}C ka BP and indicate that the lower peninsula of Michigan was generally ice free during the Athens-Elgin Subepisode. The position of the ice margin north of Michigan during the subepisode is unknown but it probably extended into the Lake Superior basin and perhaps even as far south as the northern end of the Lake Michigan basin (Grimely, 2000). East of Michigan the ice margin extended across the Lake Ontario basin (Karrow et al., 2000). Sediments associated with the Michigan Subepisode (so named because much of Michigan's landscape is developed on glaciogenic sediments deposited near the end of the last glaciations; Johnson et al., 1997) consist of

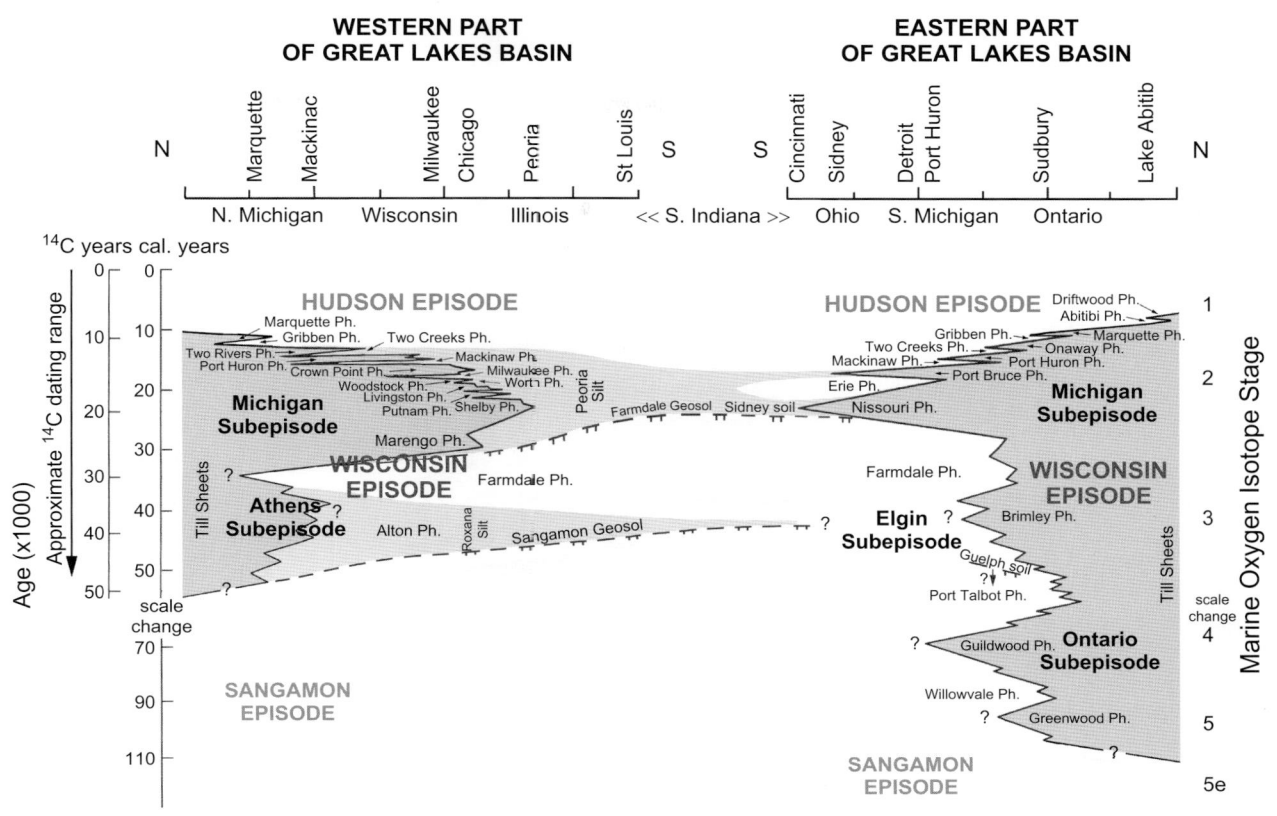

Chapter | 37 Ice-Margin Fluctuations at the End of the Wisconsin

FIGURE 37.2 Comparison of time–distance diagrams for the western part of the Great Lakes basin (modified from Johnson et al., 1997) and the eastern part of the basin (modified from Karrow et al., 2000).

glacial and non-glacial deposits. In Michigan, not all of the 'glacial' phases of the Michigan Subespisode are represented by glacial deposits, but four are. They include the Crown Point-Port Bruce, Port Huron, Two Rivers-Onaway and Marquette phases (Figs. 37.1–37.3).

37.3.1. Crown Point-Port Bruce Phase

The Crown Point-Port Bruce Phase is the time of ice advance to and retreat from the Kalamazoo moraine system (Leverett and Taylor, 1915) in southwestern lower Michigan and the Mississinewa moraine system (Leverett and Taylor, 1915) in southeastern lower Michigan. The phase is named for Crown Point, a town in the Valparaiso moraine area of Indiana (Hansel and Johnson, 1992). In southwestern lower Michigan, the phase is represented by the Saugatuck till (Monaghan et al., 1986) and in south-central lower Michigan by the Bedford till (Monaghan and Larson, 1986). In Illinois and Lake Michigan, it is represented by the Wadsworth Till Member (Hansel and Johnson, 1992) and in Wisconsin by the Oak Creek Till Member (Mickelson et al., 1984). In Indiana, it is represented by part of the New Holland Till Member (Fullerton, 1986). During the Crown Point-Port Bruce Phase, ice advanced out of the southern end of Lake Michigan to the position of the Valparaiso moraine in northeastern Illinois and northwestern Indiana (Hansel and Johnson, 1992) and to the position of the Kalamazoo moraine system in lower Michigan (Monaghan and Larson, 1986). In southeastern lower Michigan, it advanced to the position of the Mississinewa moraine system, and in Indiana and Ohio to the position of the Union City-Powell moraine system (Monaghan and Larson, 1986). In Michigan, there is no information of when the ice margin advanced to the Kalamazoo and Mississinewa moraine systems, but in central Ohio a log incorporated in the Powell moraine (Figs. 37.1 and 37.3, Table 37.1, Location No. 1) has yielded a radiocarbon date of 14.78 ± 0.192 ka BP (Ogden and Hay, 1965). In Illinois, wood samples from lake sediments above the Wadsworth Till Member (Location No. 2) have yielded radiocarbon dates of 14.1 ± 0.64, 13.89 ± 0.12 and 13.87 ± 0.17 ^{14}C ka BP and indicate that the Crown Point Phase in Illinois was waning by ~ 14.0 ^{14}C ka BP (Hansel and Johnson, 1992).

37.3.2. Port Huron Phase

The Port Huron Phase is the time of ice advance to and retreat from the Port Huron moraine system in Michigan (Leverett and Taylor, 1915; Blewett, 1991) and correlative

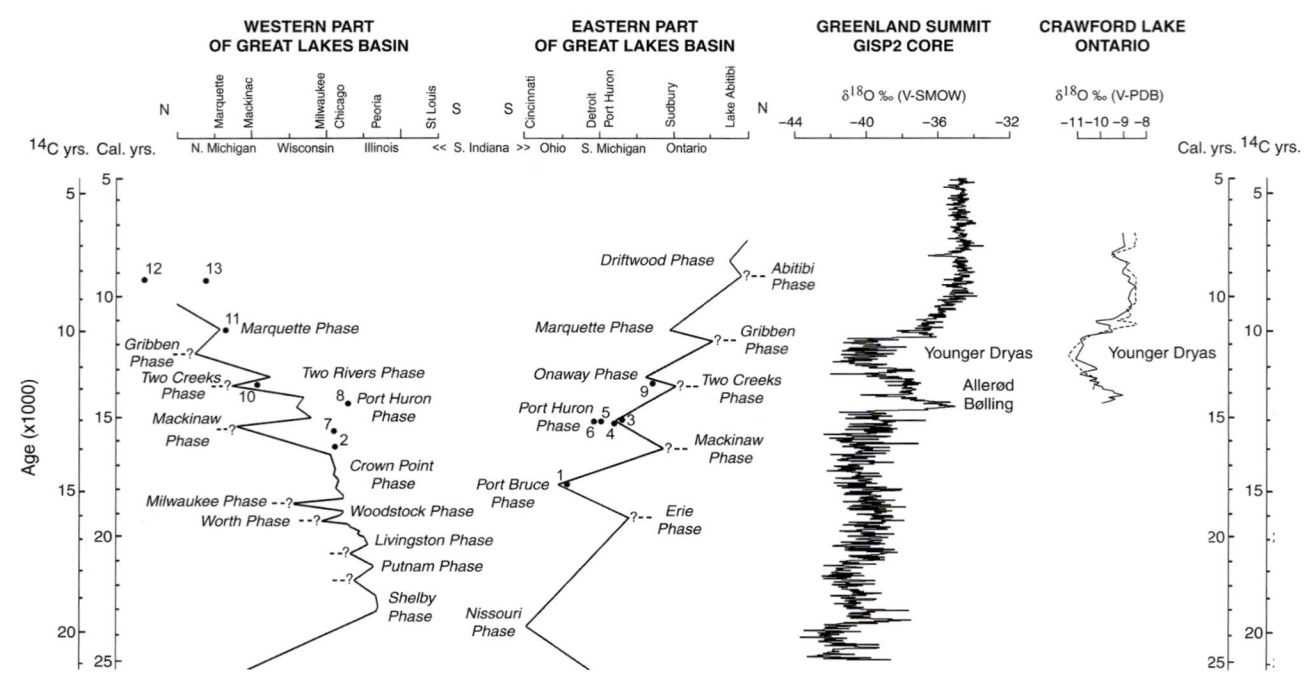

FIGURE 37.3 Correlation of ice-margin oscillations in the Great Lakes region and palaeoclimatic records from Greenland GISP2 ice core (Grootes et al., 1993; GISP2 20-year data set, University of Washington) and Crawford Lake, Ontario (Yu and Eicher, 1998). Numbered points correspond to ^{14}C dates listed in Table 37.1. Calendar ages from GISP2 are from annual ice layer counting.

moraines in Wisconsin (Evenson et al., 1973, 1976) and Ontario (Cowan et al., 1975; Gravenor and Stupavsky, 1976). The phase is named for the Port Huron moraine system in Michigan which is well developed and was interpreted almost a century ago as representing the culmination of a major readvance of the ice margin across much of northern lower Michigan (Leverett and Taylor, 1915). In western lower Michigan, the Port Huron Phase is represented by the Montigue and Riverton tills (Taylor, 1990) and in eastern lower Michigan by the Jeddo till (Eschman and Michelson, 1984). In the Lake Michigan basin, it is represented by the Shorewood and Manitowoc Till Members (Lineback et al., 1974) and in eastern Wisconsin by the Ozaukee and Haven Till Members. Along the eastern shore of Lake Huron, it is represented by the St. Joseph till (Cooper and Clue, 1974).

During the Port Huron Phase, ice advanced across the Straits of Mackinac connecting the Lake Michigan and Lake Huron basins and extended into the southern part of the Lake Michigan basin and to the southern rim of the Lake Huron basin. Exactly when the Port Huron moraine system was built is unknown because of the paucity of datable *in situ* organic material, but it was probably ~13.0 ^{14}C ka BP. This estimate is based on a spruce log found in St. Joseph till along the eastern shore of Lake Huron (Location No. 3) that has yielded a radiocarbon age of 13.1 ± 0.11 ^{14}C ka BP (Gravenor and Stupavsky, 1976), as well as disseminated organic material found in outwash associated with the Port Huron moraine system in northern lower Michigan (Location No. 4) that has yielded a radiocarbon date of 12.96 ± 0.35 ^{14}C ka BP (Blewett et al., 1993). In northern Ohio (Location Nos. 5 and 6), wood fragments have been found in sediments of glacial Lake Whittlesey that developed in the Lake Erie and Lake Huron basins as the ice margin advanced to form the Port Huron moraine system. These fragments have yielded radiocarbon dates of ~12.92 ± 0.40, 12.80 ± 0.25 and 12.90 ± 0.20 ^{14}C ka BP (Barendsen et al., 1957; Goldthwait, 1958; Calkin, 1970). In the Lake Michigan basin, wood fragments have also been found in sediments of Glenwood II phase of glacial Lake Chicago (Location Nos. 7 and 8) that developed as the ice margin advanced across the Straits and to the position of the Port Huron moraine system. These fragments have yielded radiocarbon dates of ~13.47 ± 0.13, 12.65 ± 0.35 and 12.22 ± 0.35 ^{14}C ka BP (Rubin and Suess, 1955; Monaghan and Hansel, 1990). Little is known as to how far north the ice margin retreated at the end of the Port Huron Phase but it was at least to the Straits of Mackinac. Evidence for this is development of an Intra-Glenwood low phase of glacial Lake Chicago that could only have drained eastward through the Straits (Hansel et al., 1985).

37.3.3. Two Rivers-Onaway Phase

The Two Rivers-Onaway Phase is the time of ice advance to and retreat from a position behind the Port Huron moraine

TABLE 37.1 Selected Radiocarbon Dates, Great Lakes Region

Site name	Location No.	Age (^{14}C ka BP)	Lab. No.	Material	Comments	References
Powell moraine, Central Ohio	1	14.78 ± 0.19	OWU-83	Wood	From till north of Powell moraine	Ogden and Hay (1965)
Glenwood spit, Northern Illinois	2	14.10 ± 0.64	ISGS-1570	Wood	From beach sand associated with Glenwood spit	Hansel and Mickelson (1988)
		13.87 ± 0.17	ISGS-1549	Spruce cone	From beach sand associated with Glenwood spit	Hansel and Mickelson (1988)
		13.89 ± 0.12	ISGS-1649	Wood	From lacustrine sediment above Wadsworth Till Member	Hansel and Johnson (1992)
Wyoming moraine Southern Ontario	3	13.1 ± 0.11	GSC-2213	Wood	From St. Joseph till	Gravenor and Stupavsky (1976)
Port Huron moraine, Northern Lower Michigan	4	12.96 ± 0.35	TX-6151	Carbonaceous material	From folded lacustrine silt on proximal slope of Inner Port Huron moraine	Blewett et al. (1993)
Whittlesey beach, Northern Ohio	5	12.92 ± 0.40	W-430	Wood	From beach sand associated with glacial Lake Whittlesey	Goldthwait (1958)
		12.80 ± 0.25	Y-240	Wood	From beach sand associated with glacial Lake Whittlesey	Barendsen et al. (1957)
Whittlesey beach, Northern Ohio	6	12.90 ± 0.20	I-3175	Wood	From beach sand associated with glacial Lake Whittlesey	Calkin (1970)
Riverside, Western Lower Michigan	7	13.47 ± 0.13	ISGS-1378	Wood	From beach gravel on till	Monaghan and Hansel (1990)
Dyer Spit, Northern Indiana	8	12.65 ± 0.35	W-140	Wood	From peat deposit beneath Dyer Spit	Rubin and Suess (1955)
		12.22 ± 0.35	W-161	Wood	From Dyer Spit	Rubin and Suess (1955)
Cheboygan, Northern Lower Michigan	9	12.10 ± 0.10	BETA-50967	Bryophyte	From bryophyte bed beneath Munro Lake till	Larson et al. (1994)
		12.05 ± 0.08	ETH-9241	Wood	From bryophyte bed beneath Munro Lake till	Larson et al. (1994)
		11.75 ± 0.08	BETA-50966	Vole dropping	From bryophyte bed beneath Munro Lake till	Larson et al. (1994)
		11.40 ± 0.29	ISGS-2333	Bryophyte	From bryophyte bed beneath Munro Lake till	Larson et al. (1994)
Two Creeks, Eastern Wisconsin	10	12.03 ± 0.06	ETH-8273 (AMS)	Wood (pith)	From forest bed beneath Two Rivers till	Kaiser (1994)
		12.01 ± 0.09	ETH-8612 (AMS)	Wood (pith)	From forest bed beneath Two Rivers till	Kaiser (1994)
		11.98 ± 0.09	ETH-8611 (AMS)	Wood (tree rings)	From forest bed beneath Two Rivers till	Kaiser (1994)
		11.96 ± 0.09	ETH-8601 (AMS)	Wood (tree rings)	From forest bed beneath Two Rivers till	Kaiser (1994)
		11.91 ± 0.10	ETH-8272 (AMS)	Wood (tree rings)	From forest bed beneath Two Rivers till	Kaiser (1994)
		11.89 ± 0.09		Wood (tree rings)	From forest bed beneath Two Rivers till	Kaiser (1994)
		11.80 ± 0.09		Wood (tree rings)	From forest bed beneath Two Rivers till	Kaiser (1994)
		11.85 ± 0.11		Wood (tree rings)	From forest bed beneath Two Rivers till	Leavitt and Kalin (1992)
		11.80 ± 0.16		Wood (tree rings)	From forest bed beneath Two Rivers till	Leavitt and Kalin (1992)
		11.75 ± 0.09		Wood (tree rings)	From forest bed beneath Two Rivers till	Leavitt and Kalin (1992)

Continued

TABLE 37.1 Selected Radiocarbon Dates, Great Lakes Region—cont'd

Site name	Location No.	Age (^{14}C ka BP)	Lab. No.	Material	Comments	References
		11.64 ± 0.60	ETH-8609 (AMS)	Wood (tree rings)	From forest bed beneath Two Rivers till	Leavitt and Kalin (1992)
		11.89 ± 0.10	ETH-8610 (AMS)	Wood	From forest bed beneath Two Rivers till	Broeker and Farrand (1963)
		11.85 ± 0.10	A-Calib 99	Wood	From lacustrine sediment below forest bed	Broeker and Farrand (1963)
		11.79 ± 0.10	A-5551	Wood	From lacustrine sediment below forest bed	Broeker and Farrand (1963)
			A-5550	Wood (cellulose)		
			A-5552	Wood		
			L-698C	Wood (cellulose)		
			L-607A	Wood (lignin)		
			L-698C			
Lake Gribben, Northern Michigan	11	10.20 ± 0.055	A-7875	Wood	From outwash graded to Grand Marais I moraine	Lowell et al. (1999)
		10.15 ± 0.065	A-7882	Wood	moraine	Lowell et al. (1999)
		10.07 ± 0.095	A-7880	Wood	From outwash graded to Grand Marais I moraine	Lowell et al. (1999)
		10.05 ± 0.055	A-7883	Wood	moraine	Lowell et al. (1999)
		10.40 ± 0.065	A-7881	Wood	From outwash graded to Grand Marais I moraine	Lowell et al. (1999)
		10.40 ± 0.055	A-7879	Wood	moraine	Lowell et al. (1999)
		9.96 ± 0.055	A-7877	Wood	From outwash graded to Grand Marais I moraine	Lowell et al. (1999)
		9.91 ± 0.055	A-7876	Wood	From outwash graded to Grand Marais I moraine	Lowell et al. (1999)
		9.89 ± 0.30	A-7878	Wood	From outwash graded to Grand Marais I moraine	Lowell et al. (1999)
		10.33 ± 0.30	W-3896	Needles	From outwash graded to Grand Marais I moraine	Hughes and Merry (1978)
		10.22 ± 0.21	DAL-338	Wood	From outwash graded to Grand Marais I moraine	Hughes and Merry (1978)
		9.85 ± 0.30	W-3866	Wood	From outwash graded to Grand Marais I moraine	Hughes and Merry (1978)
		9.78 ± 0.25	W-3904	Wood	From outwash graded to Grand Marais I moraine	Hughes and Merry (1978)
		9.54 ± 0.22	DAL-340	Wood	From outwash graded to Grand Marais I moraine	Hughes and Merry (1978)
Rosslyn, Ontario	12	9.38 ± 0.15	GSC-287	Wood	From possible glacial Lake Minong sediments	Drexler et al. (1983)
Grand Marias Moraine, Upper Michigan	13	9.34 ± 0.24	GX-4883	Wood	From glacial Lake Minong sediments	Zoltai (1965)

system. Nowhere in Michigan did the ice leave a recognisable moraine, but in eastern Wisconsin, it left behind hummocky terrain known as the Two Rivers moraine (Evenson et al., 1973). The Two Rivers-Onaway Phase is named after the Two Rivers till in eastern Wisconsin (Evenson et al., 1973, 1976) and the Onaway till in northeastern lower Michigan (Burgis, 1977). In northwestern lower Michigan, the Two Rivers-Onaway Phase is represented by the Orchard Beach till (Taylor, 1990), in northern lower Michigan by the Monro Lake till (Larson et al., 1994) and in northeastern lower Michigan by the Onaway till (Burgis, 1977). In the Lake Michigan basin and in eastern Wisconsin, it is represented by the Two Rivers Till Members (Evenson et al., 1973, 1976; Lineback et al., 1974).

During the Two River-Onaway Phase ice advanced across the Straits of Mackinac and extended to the central part of the Lake Michigan basin and into the northwestern end of the Lake Huron basin. It also advanced over and buried a bryophyte bed near Cheboygan, Michigan (Location No. 9), that has yielded an average radiocarbon date of ~ 11.82 ^{14}C ka BP (Larson et al., 1994), while in eastern Wisconsin, it advanced over and buried a spruce and pine forest near Two Creeks, Wisconsin (Location No. 10), that has yielded an average radiocarbon age of 11.85 ^{14}C ka BP (Broeker and Farrand, 1963). These dates indicate the ice advance culminated ~ 11.7 ^{14}C ka BP. Following retreat of the ice margin from the Straits of Mackinac, glacial Lake Algonquin extended from the Lake Huron basin into the Lake Michigan and Lake Superior basins (Larsen, 1987). Little is known how far north the ice margin eventually retreated at the end of the phase but shorelines associated with glacial Algonquin have been traced to as far north as Alona Bay 30 km north of the Straits (Heath and Karrow, 2007). It may also have retreated to a position somewhere in the northern part of the Lake Superior basin (Farrand and Drexler, 1985; Lowell et al., 1999).

37.3.4. Marquette Phase

The Marquette Phase is the time of ice advance to and retreat from the Porcupine Mountain and Grand Marais moraines in northern Michigan (Drexler et al., 1983; Farrand and Drexler, 1985), the Saxon moraine in northern Wisconsin (Clayton, 1983), and possibly the Cartier and McConnell moraines in southern Ontario (Lowell et al., 1999). It is named after the city of Marquette at the western end of the Grand Marais moraine system in central northern Michigan. The Marquette Phase is not represented by any named till although till has been reported within the Saxon, Porcupine Mountain and Grand Marais (Munising) moraines (Hack, 1965; Levin et al., 1965; Black, 1976; Peterson, 1982).

During the Marquette Phase outwash associated with the Grand Marais moraine system buried a spruce and larch forest at Lake Gribben southwest of Marquette (Location No. 11). Samples of wood and needles from the forest have yielded an average radiocarbon date of ~ 10.02 ^{14}C ka BP (Lowell et al., 1999). Logs found in till associated with the Saxon and Porcupine Mountain moraines have also yielded an average radiocarbon date of ~ 10.05 ^{14}C ka BP (Hack, 1965; Levin et al., 1965; Black, 1976; Peterson, 1982). The dates from the forest and logs indicate ice advance during the Marquette Phase culminated ~ 10.0 ^{14}C ka BP. Following the advance glacial lake Minong in the eastern end of the Superior basin rapidly expanded westward and northward, eventually extending to the northern shore of Lake Superior (Farrand and Drexler, 1985). Driftwood found in late Minong sediment near Grand Marias, Michigan (Location No. 12) has yielded a radiocarbon date of 9.38 ± 0.15 ^{14}C ka BP (Drexler et al., 1983) while wood found from probable late Minong sediment near Rosslyn, Ontario (Location No. 13) has yielded a radiocarbon date of 9.34 ± 0.24 ^{14}C ka BP (Drexler, 1981).

37.4. DISCUSSION

In addition to the record of readvances during the deglaciation of Michigan, Fig. 37.3 shows the ^{18}O/^{16}O ratio (δ^{18}O) palaeoclimatic record from the Greenland GISP2 ice core (Grootes et al., 1993) and the ^{18}O/^{16}O ratio palaeoclimatic record from two cores taken from Crawford Lake, Ontario (Yu and Eicher, 1998). Comparison of the deglaciation record and the GISP2 ice-core record shows little relationship between the two other than a general warming trend through time. The deglaciation record prior to ~ 13.0 ^{14}C ka BP includes a number of major ice-margin oscillations whereas the ice-core record shows only minor climatic oscillations that are not in phase with the ice-margin oscillations. The only exception is the oscillation associated with the Erie Phase (Erie Interstadial) which generally coincides with a slight warming interval recorded in the GISP2 ice core between ~ 16.0 and ~ 15.0 ^{14}C ka BP. This was also noted by Clark (1999) in the Greenland GRIP ice-core record (Table 37.1).

Following ~ 13.0 ^{14}C ka BP, the GISP2 ice-core record shows a significant warm interval (Allrød-Bølling) between ~ 12.6 and ~ 10.7 ^{14}C ka BP followed by a significant cold interval (Younger Dryas) between ~ 10.8 and ~ 10.1 ^{14}C ka BP. The same cold interval and part of the preceding warm interval are also recorded in authigenic lake carbonates from Crawford Lake. The deglaciation record, however, shows an out of phase behaviour of the ice-margin with the Port Huron Phase generally coinciding with the warm interval and the Gribben Phase generally coinciding with the cold interval.

The most likely explanation for the out of phase behavior of the ice margin is the delayed response of the southern part of the Lauentide ice sheet to climatic forcing after

~13.0 ^{14}C ka BP. This was suggested by Lowell et al. (1999) who argued that the ice advance associated with the Marquette Phase was in response to climatic forcing of the Younger Dryas cold event. In support of their argument, they point to the coincidence in timing of the maximum limit of the advance (~10.0 ^{14}C ka BP) and the end of the Younger Dryas (~10.1 ^{14}C ka BP). Presumably, the Gribben, Two Rivers-Onaway and Two Rivers phases are associated with the Allerød-Bølling warm interval.

What caused the ice margin oscillations prior to ~13.0 ^{14}C ka BP is unknown despite the fact that some oscillations where on the order of 250–350 km (Hansel and Johnson, 1992). One explanation proposed is that substantial changes in water level of ice-marginal lakes greatly amplified small fluctuations of the ice margin (Mickelson et al., 1981). Another is that sudden changes in physical properties of subglacial deforming sediments accelerated ice flow and caused the ice margin to advance (Clark, 1994).

37.5. SUMMARY

Diachronic and event classification is a useful way to outline the deglaciation history of a region especially when the timing of events differs locally and regionally. In Michigan, four glacial phases mark the retreat of the last ice sheet: the Crown Point-Port Bruce, Port Huron, Two Rivers-Onaway and Marquette. During each phase, the ice margin advanced and retreated with each advance terminating at a position behind the limit of the previous advance. Radiocarbon dates from the state and surrounding area indicate the four advances culminated at ~15.0, 13.0, 11.7 and 10.0 ^{14}C ka BP. Only the Two Rivers-Onaway Phase is well constrained by ^{14}C age dates from *in situ* organic material.

REFERENCES

Barendsen, G., Deevey, E., Jr., Gralenski, L., Jr. 1957. Yale natural radiocarbon measurements III. Science 126, 908–919.

Black, R.F., 1976. Quaternary geology of Wisconsin and contiguous Upper Michigan. In: Mahaney, W.C. (Ed.), Quaternary Stratigraphy of North America. Dowden, Hutchenson and Ross, Inc., Stroudsburg, Pennsylvania, pp. 93–117.

Blewett, W.L., 1991. Characteristics, correlations, and refinements of Leverett and Taylor's Port Huron Moraine in Michigan. East Lakes Geogr. 26, 52–60.

Blewett, W.L., Winters, H.A., Rieck, R.L., 1993. New age control on the Port Huron moraine in northern Michigan. Phys. Geogr. 14, 131–138.

Broeker, W.S., Farrand, W.R., 1963. Radiocarbon age of the Two Creeks forest bed, Wisconsin. Geol. Soc. Am. Bull. 74, 795–802.

Burgis, W.A., 1977. Late-Wisconsinan History of Northeastern Lower Michigan. Ph.D. Thesis, Ann Arbor, University of Michigan, 396pp.

Calkin, P.E., 1970. Strand lines and chronology of the glacial Great Lakes in northwestern New York. Ohio J. Sci. 70, 78–96.

Clark, P.U., 1994. Unstable behavior of the Laurentide ice sheet over deforming sediment and its implications for climate change. Quatern. Res. 41, 19–25.

Clayton, L., 1983. Chronology of Lake Agassiz drainage to Lake Superior. In: Teller, J.T., Clayton, L. (Eds.), Glacial Lake Agassiz. Geological Association of Canada Special Paper, 26, Ottawa, Ontario, pp. 291–307.

Cooper, A.J., Clue, J., 1974. Quaternary Geology of the Grand Bend area, southern Ontario. Ontario Division of Mine, Preliminary Map, 974.

Cowan, W.R., Karrow, P.F., Cooper, A.J., Morgan, A.V., 1975. Late Quaternary stratigraphy of the Waterloo-Lake Huron area, southwestern Ontario. In: Field Trips Guidebook. Geological Association of Canada, Ottawa, Ontario, pp. 180–222.

Dreimanis, A., Karrow, P.F., 1972. Glacial history of the Great Lakes—St. Lawrence region, the classification of the Wisconsin(an) stage, and its correlatives. In: 24th International Geological Congress, pp. 5–15.

Drexler, C.W., 1981. Outlet channels for the post-Duluth lakes in the Upper Peninsula of Michigan. Ph.D. Dissertation, Ann Arbor, University of Michigan, 295pp.

Drexler, C.W., Farrand, W.R., Hughes, J.D., 1983. Correlation of glacial lakes in the Superior basin with eastward discharge events from Lake Agassiz. In: Teller, J.T., Clayton, L. (Eds.), Glacial Lake Agassiz. Geological Association of Canada Special Paper, 26, pp. Ottawa, Ontario, pp. 309–329.

Eschman, D.E., 1978. Pleistocene geology of the Thumb Area of Michigan. In: Guidebook, North Central Section. Geological Society of America Meeting, Ann Arbor, Michigan, pp. 35–62.

Eschman, D.F., 1980. Some evidence of mid-Wisconsinan events in Michigan. Michigan Acad. 12, 423–436.

Eschman, D.F., Michelson, D.M., 1984. Correlation of glacial deposits of the Huron, Lake Michigan and Green Bay lobes in Michigan and Wisconsin. Quatern. Sci. Rev. 5, 53–57.

Evenson, E.B., Eschman, D.F., Farrand, W.R., 1973. The "Valderan" problem, Lake Michigan basin. In: 22nd Annual Field Conference, Midwest Friends of the Pleistocene, June 1–3, 1–59.

Evenson, E.B., Farrand, W.R., Eschman, D.F., Mickelson, D.M., Maher, L.J., 1976. Greatlakean Substage: a replacement for the Valderan Substage in the Lake Michigan basin. Quatern. Res. 6, 411–424.

Farrand, W.R., Drexler, C.W., 1985. Late Wisconsin and Holocene history of the Lake Superior basin. In: Karrow, P.F., Calkin, P.E. (Eds.), Quaternary Evolution of the Great Lakes, Geological Association of Canada Special Paper, 30, Ottawa, Ontario, pp. 17–32.

Farrand, W.R., Eschman, D.F., 1974. Glaciation of the Southern Peninsula of Michigan: a review. Michigan Acad. 7, 31–56.

Fullerton, D.S., 1986. Stratigraphy and correlation of glacial deposits from Indiana to New York and New Jersey. Quatern. Sci. Rev. 5, 23–37.

Goldthwait, R.P., 1958. Wisconsin-age forests in western Ohio. Part 1—age and glacial events. Ohio J. Sci. 58, 209–219.

Gravenor, C.P., Stupavsky, M., 1976. Magnetic, physical, and lithologic properties of till exposed along the east coast of Lake Huron, Ontario. Can. J. Earth Sci. 13, 1655–1666.

Grimely, D.A., 2000. Glacial and nonglacial sediment contribution to Wisconsin Episode loess in the central United States. Geol. Soc. Am. Bull. 112, 1475–1495.

Grootes, P.M., Stuiver, M., White, J.W.C., Johnsen, S., Jouzel, J., 1993. Comparison of oxygen isotope records from the GISP2 and GRIP Greenland ice cores. Nature 366, 552–554.

Hack, J., 1965. Postglacial drainage evolution and stream geometry of the Ontonagon area, Michigan. U.S. Geological Survey Professional Paper P 0504-B, B1B40pp.

Hansel, A.K., Johnson, W.H., 1992. Fluctuations of the Lake Michigan lobe during the late Wisconsin subepisode. In: Robertson, A., Ringberg, B., Miller, U., Brunnberg, L. (Eds.), Sveriges Geologiska Undersökning, Ser. Ca. 81, pp. 133–144.

Hansel, A.K., Mickelson, D.M., 1988. A reevaluation of the timing and causes of high lake phases in the Lake Michigan basin. Quat. Res. 29, 113–128.

Hansel, A.K., Mickelson, D.M., Schneider, A.F., Larsen, C.E., 1985. Late Wisconsinan and Holocene history of the Lake Michigan basin. In: Karrow, P.F., Calkin, P.E. (Eds.), Quaternary Evolution of the Great Lakes, Geological Association of Canada Special Paper, 30, 39–53.

Heath, A.J., Karrow, P.F., 2007. Northernmost(?) glacial Lake Algonquin series shorelines, Sudbury basin, Ontario. J. Great Lakes Res. 33, 264–278.

Hughes, J.D., Merry, W.J., 1978. Marquette buried forest 9,850 years old. In: American Association for the Advancement of Science, 144th Annual Meeting, Abstract 12–14 February 1978, p.115.

Johnson, W.H., 1986. Stratigraphy and correlation of the glacial deposits of the Lake Michigan lobe prior to 14 ka BP. In: Sibrava, V., Bowen, D.Q., Richmond, G.M. (Eds.), Quaternary Glaciations in the Northern Hemisphere, Quaternary Science Reviews, vol. 5, pp. 17–22.

Johnson, W.H., Hansel, A.K., Bettis III, E.A., Karrow, P.F., Larson, G.J., Lowell, T.V., et al., 1997. Late Quaternary temporal and event classification, Great Lakes region, North America. Quatern. Res. 47, 1–12.

Kaiser, K.F., 1994. Two Creeks interstadial dated through dendrochronology and AMS. Quatern. Res. 42, 288–298.

Karrow, P.F., Dreimanis, A., Barnett, P.J., 2000. A proposed diachronic revision of Late Quaternary time-stratigraphic classification in the eastern and northern Great Lakes area. Quatern. Res. 54, 1–12.

Larsen, C.E., 1987. Geological history of glacial Lake Algonquin. U.S. Geol. Surv. Bull. 1801, 36pp.

Larson, G.J., Kincare, K., 2009. Late Quaternary history of the eastern midcontinent region, USA. In: Schaetzl, R., Darden, J., Brandt, D. (Eds.), Michigan Geography and Geology. Custom Publishing, New York, pp. 69–90.

Larson, G.J., Lowell, T.V., Ostrom, N.E., 1994. Evidence for the Two Creeks interstade in the Lake Huron basin. Can. J. Earth Sci. 31, 793–797.

Leavitt, S.T., Kalin, R.M., 1992. A new tree-ring width, ^{13}C, ^{14}C investigation of the Two Creeks site. Radiocarbon 34, 792–797.

Leverett, F., Taylor, F.B., 1915. The Pleistocene of Indiana and Michigan and history of the Great Lakes. U.S. Geological Survey Monograph 526pp.

Levin, B., Ives, P.C., Oman, C.L., Rubin, M., 1965. U.S. Geological Survey, Radiocarbon Dates VIII. Radiocarbon 7, 376.

Lineback, J.A., Gross, D.L., Meyer, R.P., 1974. Glacial tills under Lake Michigan. Illinois State Geological Survey. Environ. Geol. Notes 69, 48.

Lowell, T.V., Larson, G.J., Hughes, J.D., Denton, G.H., 1999. Age verification of the Lake Gribben forest bed and the Younger Dryas advance of the Laurentide ice sheet. Can. J. Earth Sci. 36, 383–393.

Mickelson, D.M., Acomb, L.J., Bentley, C.R., 1981. Possible mechanism for the rapid advance and retreat of the Lake Michigan lobe between 13,000 and 11,000 years BP. Ann. Glaciol. 2, 185–186.

Mickelson, D.M., Clayton, L., Baker, R.W., Mode, W.H., Schneider, A.F., 1984. Pleistocene stratigraphic units of Wisconsin. Wisconsin Geological and Natural History Survey Miscellaneous Paper, 84-1, 107pp.

Monaghan, G.W., Hansel, A.K., 1990. Evidence for the intra-Glenwood (Mackinaw) low-water phase of glacial Lake Chicago. Can. J. Earth Sci. 27, 1236–1241.

Monaghan, G.W., Larson, G.J., 1986. Late Wisconsinan drift stratigraphy of the Saginaw ice lobe in south-central Michigan. Geol. Soc. Am. Bull. 97, 324–328.

Monaghan, G.W., Larson, G.J., Gephart, G.D., 1986. Late Wisconsinan drift stratigraphy of the Lake Michigan lobe in southwestern Michigan. Geol. Soc. Am. Bull. 97, 329–334.

Ogden, J.G., Hay, R.J., 1965. Ohio Wesleyan University natural radiocarbon measurements II. Radiocarbon 7, 166–173.

Peterson, W.L., 1982. Preliminary surficial geologic map of the Iron river 1° × 2° quadrangle, Michigan and Wisconsin. U.S. Geological Survey Open-File Report, 82–301.

Rubin, M., Suess, H.E., 1955. U.S. Geological Survey radiocarbon dates; (Part) 2. Science 121, 481–488.

Saarnisto, M., 1974. The deglaciation history of the Lake Superior region and its climatic implications. Quatern. Res. 4, 316–339.

Taylor, L.D., 1990. Evidence for high glacial-lake levels in the northeastern Lake Michigan basin and their relation to the Glenwood and Calumet phases of glacial Lake Chicago. In: Geological Society of America Special Paper, 251, 91–109.

Willman, H.B., Frye, J.C., 1970. Pleistocene stratigraphy of Illinois. Illinois State Geol. Surv. Bull. 94, 204pp.

Winters, H.A., Rieck, R.L., Knapp, R.O., 1986. Significance and ages of mid-Wisconsinan organic deposits in southern Michigan. Phys. Geogr. 7, 292–305.

Yu, Z., Eicher, U., 1998. Abrupt climate oscillations during the last deglaciation in central North America. Science 282, 2235–2238.

Zoltai, S.C., 1965. Glacial features of the Quetico-Nipigon area, Ontario. Can. J. Earth Sci. 2, 247–269.

Chapter 38

The Quaternary of Minnesota

C.E. Jennings[1,*] and M.D. Johnson[2]

[1]Minnesota Geological Survey, University of Minnesota, Minnesota, USA
[2]Earth Science Institute, University of Gothenburg, Göteborg, Sweden
*Correspondence and requests for materials should be addressed to C.E. Jennings. E-mail: carrie@umn.edu

38.1. GENERAL CHARACTER AND AGE OF QUATERNARY DEPOSITS IN MINNESOTA

The rich and complex sedimentary record of glaciation in Minnesota includes deposits of glaciers, associated rivers and lakes, as well as windblown and other periglacial deposits that span the Quaternary Period and are at the surface in nearly every part of the state. The thickness of these deposits is commonly more than 100 m. One of the thickest, and perhaps most stratigraphically complete, records of the Quaternary in Minnesota lies beneath the Coteau des Prairies, an inter-ice stream sediment highland that spans the Minnesota–South Dakota border where over 300-m of fine-grained diamictons are preserved (Fig. 38.1).

Though the glacial deposits of the state are dominated by a complex Late Wisconsinan history (marine isotope stage (MIS) 2), Minnesota has many lithostratigraphical units from the Middle and likely Early Pleistocene. For example, Minnesota has units older than the Sangamon Geosol (>125 ka), units older than volcanic ashes derived from Yellowstone (>610 ka) (Boellstorff, 1978), as well as magnetically reversed units from prior to the Brunhes–Matayuma boundary (>788 ka). Recent cosmogenic burial dating of glacigenic sediment (Balco et al., 2005) indicates that numerous glacial stratigraphical units were deposited prior to MIS 14. A rare bedrock exposure in the southern part of the Coteau des Prairie highland was striated as early as 640–740 ka (MIS 16 or 14) based on cosmogenic exposure dating of the quartzite using a paired-isotope system (Bierman et al., 1999). A stack of 12 tills surrounding this isolated bedrock high is therefore most likely a record of glaciation prior to and including MIS 14–16. Ice streams that were active primarily during MIS 2 focused erosion on either side of the Coteau des Prairie leaving it as a remnant between broad erosional unconformities (Fig. 38.1). The southeastern corner of Minnesota was also glaciated many times early in the Quaternary Period but remained ice-free during MIS 2–4, during which time it was affected by strong, northwesterly, periglacial winds and permafrost (Zanner, 1999). Thus, the earlier record of glaciation of this part of Minnesota is obscured and in places, confined to sinkholes and caves (e.g. Milske et al., 1983).

The southern margin of the Late Wisconsinan (MIS 2) Laurentide ice sheet produced many dynamic ice protuberances or lobes that emanated from discrete ice-source areas (Fig. 38.2). Some of the tributary ice sheds had distinctive bedrock geology allowing the provenance of the ice as well as the evolution of ice sheds to be discerned. This condition has produced distinct lithologic compositions for the tills derived from different ice centres and has provided a basis for differentiating and formalising lithostratigraphical units (Johnson et al., in preparation). Four broad source regions have been identified and their characteristics are shown in Table 38.1 and Fig. 38.3. Minnesota's pre-MIS 2 till units share the same broadly defined provenance regions indicating that older glaciations had similar sources areas. However, the shape of the former ice margins is much more difficult to determine from the scattered subsurface information and therefore ice dynamics more difficult to infer. Where the pre-MIS 2 ice limits are at the surface, their breadth and more southerly extent suggest that at the very least, the ice lobes were broader. It remains possible that the ice dynamics were substantially different at times in the past and did not lead to the creation of ice streams and lobes, for example, during the ice-sheet-build-up phase of each glaciation and prior to the Middle Pleistocene transition (when the frequency of glaciation and volume of changed from 41,000-year periodicity to 100,000-year periodicity, e.g. Rovey and Balco, 2010).

38.2. LITHOSTRATIGRAPHY OF QUATERNARY DEPOSITS IN MINNESOTA

Since the initial investigations of Minnesota's surficial deposits in the late 1800s (see Wright, 1972), numerous units of diamicton and other sediment types have

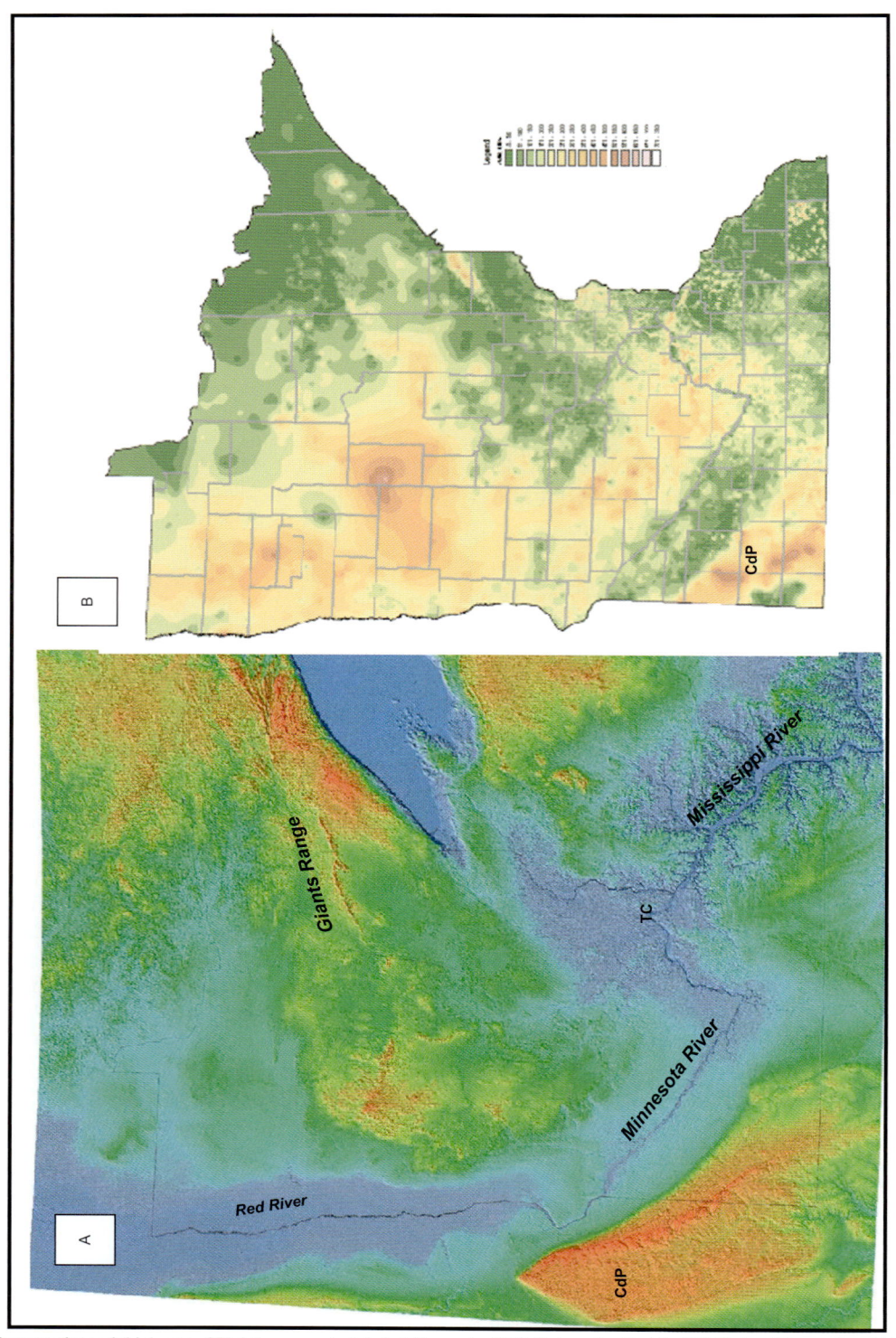

FIGURE 38.1 Topography and thickness of Pleistocene materials in Minnesota. CdP, Coteau des Prairie; TC, Twin Cities region; (A) surface elevation of Minnesota; and (B) depth to bedrock (from Lively et al., 2006).

been described, and some have been named formally (Table 38.2). Formal definition of lithostratigraphical units began with the work of Schneider (1961), Matsch (1972) and Wright et al. (1973). However, most of Minnesota's glacial deposits have been named only informally and characterised by their texture and the provenance of their

Chapter | 38 The Quaternary of Minnesota

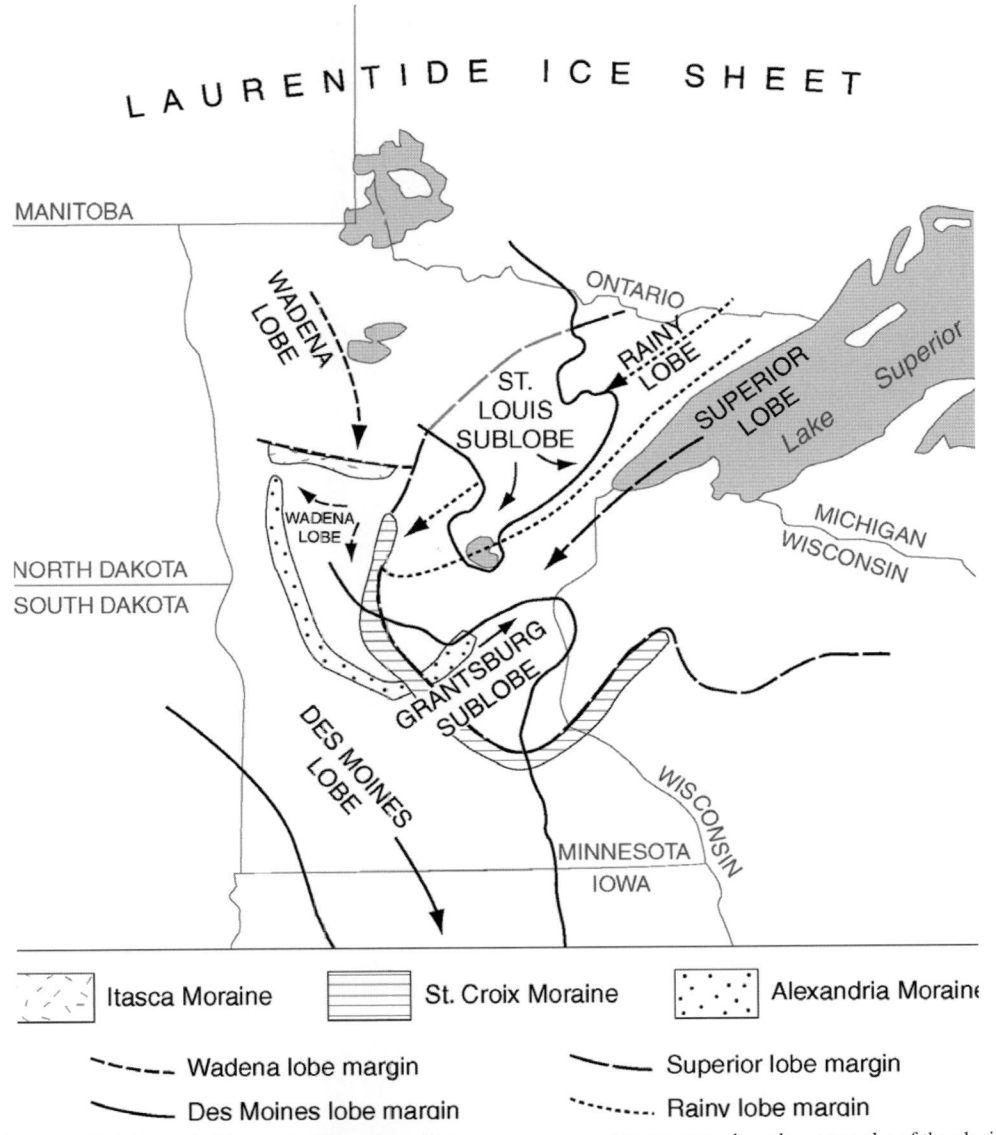

FIGURE 38.2 Names of glacial lobes in Minnesota. These three diagrams are not meant to represent the paleogeography of the glacial lobes at precise times, but rather show the change in names as the geometry of the lobes evolved through time. (A) Lobes during the late glacial maximum, MIS 2, approximately 20–15 ^{14}C ka BP. (B) Lobes following retreat from the late glacial maximum, approximately 15–11 ^{14}C ka BP. (C) Lobes of the latest glacial phases, approximately 11–9 ^{14}C ka BP.

very-coarse-sand lithology. A formal lithostratigraphical framework (Johnson et al., in preparation) is currently being developed to define all of Minnesota's Quaternary stratigraphical units (Fig. 38.4). Here, we describe the local stratigraphical sections for several specific regions of the state. Some subsurface stratigraphical units are found in more than one local sequence but are only extended between the different areas if well supported. It is the goal of future lithostratigraphical work to establish correlations among the local stratigraphical sections.

Regional, lithostratigraphical sections for eight areas of the state (Figs. 38.5–38.9) were developed where control (outcrop or subsurface drilling information) was more abundant, typically as a result of a mapping effort conducted by the Minnesota Geological Survey. Many of the formations shown in Figs. 38.5–38.9 and listed in Table 38.2 have been further subdivided into numerous members, but these are not shown here.

38.2.1. Red River Valley

This low-relief and low-lying area along the northwestern border of Minnesota (Fig. 38.5) is dominated by units of bedded silt and clay associated with glacial Lake Agassiz

TABLE 38.1 Lithologic Characteristics of Glacial Deposits in Minnesota Based on Their Provenance (Modified from Meyer and Knaeble, 1996)

Ice sector	Keewatin (northwest)		Labrador (northeast)	
Provenance	**Riding Mountain**	**Winnipeg**	**Rainy**	**Superior**
Diamicton texture	Loamy to clayey	Loamy to clayey	Sandy	Predominantly sandy, but can be clayey
Colour				
Oxidised	Yellow brown to olive brown	Yellow brown to olive brown	Yellow brown to olive brown	Brown to red brown
Unoxidised	Grey	Grey, dark grey, green grey	Grey to green grey	Grey to red grey
Pebble type				
Carbonate	Common	Uncommon to abundant	Rare to common	Rare to common
Black, grey, green crystalline rocks	Uncommon to common	Uncommon to common	Uncommon to common	Common to abundant
Red felsite and sandstone	Absent to common	Absent to common	Rare to uncommon	Uncommon to common
Grey shale	Common to abundant	Absent to common	Absent to rare	Absent

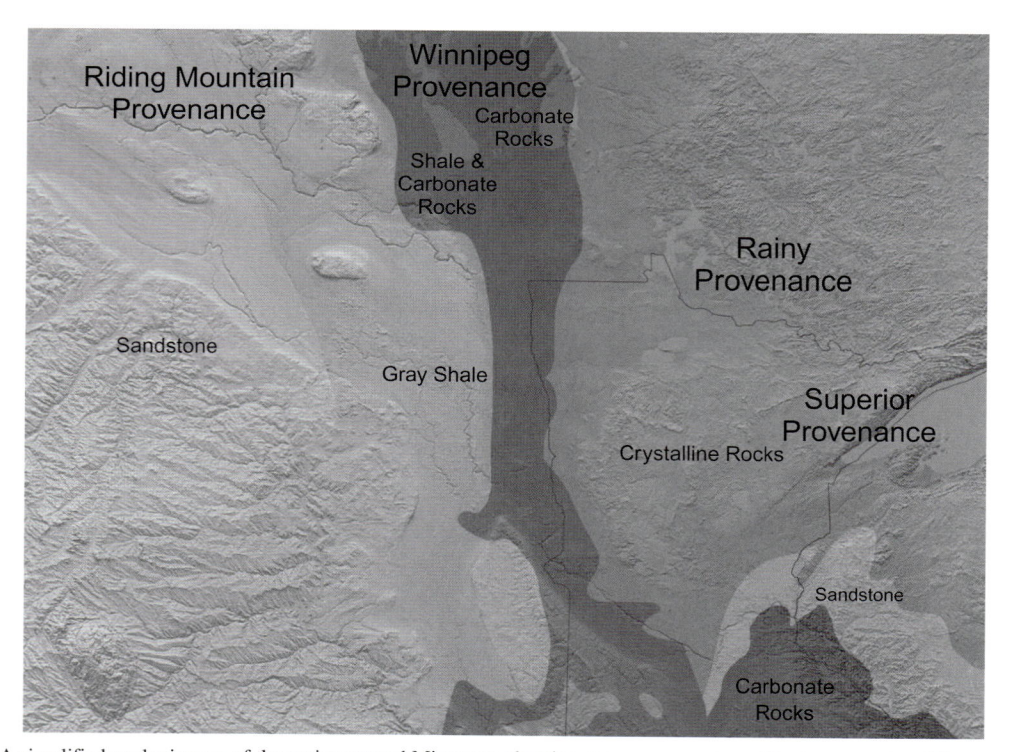

FIGURE 38.3 A simplified geologic map of the region around Minnesota showing provenance regions. Glaciers came into Minnesota from the north, northwest and northeast, and incorporated bedrock from these regions. (From Meyer and Knaeble, 1996.)

TABLE 38.2 Quaternary lithostratigraphic units evaluated by the Minnesota Geological Survey (MGS). (A) Units formally defined in earlier reports and recognized by MGS, (B) Units formally named in earlier reports but revised or redefined by MGS, (C) Units informally named in earlier reports and new units recognized as formal by MGS, (D) Units formally and informally named in earlier reports but no longer considered to be valid formal units by MGS, (E) Units used in earlier reports for which no action in this report. These units was agreed upon on by the MGS in January 2011

A	B	C	D	E
Argusville Fm	Cromwell Fm	Aitkin Fm	Mahtowa Mbr	Arsenal Sand	South Fork Till
Brenna Fm	Dahlen Mbr	Alborn Mbr	Meyer Lake Mbr	Barnesville Fm	SWRA units
Gardar Fm	Falco Mbr	Bennington Mbr	Mille Lacs Mbr	Brainerd Till	UMRB units
Gervais Fm	Falun Mbr	Big Fork Fm	Moland Mbr	Falcon Heights Sand	Windrow Fm
Hillside Sand	Hawk Creek Fm	Browerville Fm	Moose Lake Mbr	Fridley Fm	Yellow Bank Till
Loveland Loess	Huot Mbr	Buffalo River Fm	Mulligan Fm	Granite Falls Till	
Pierce Fm	Independence Fm	Coon Creek Mbr	Nashwauk Mbr	Hugo Sand	
Poplar River Fm	Marcoux Mbr	Crow Wing River Fm	Nelson Lake Mbr	Indus Fm	
Red Lake Falls Fm	New Brighton Fm	Dovray Mbr	New York Mills Mbr	Kandiyohi Till	
River Falls Fm	New Ulm Fm	Eagle Bend Fm	Prairie Lake Mbr	Lake Agassiz Clay	
Roxana Silt	Peoria Fm	Elmdale Fm	Rose Creek Fm	Little Fork Fm	
Sherack Fm	St Hilaire Mbr	Forest River Fm	Sauk Centre Mbr	Mankato Drift	
Wylie Fm	Sebeka Mbr	Funkley Fm	Saum Fm	Pierz Till	
	Sunrise Mbr	Garden City Mbr	Sheyenne Fm	Rainy Drift	
	Twin Cities Mbr	Goose River Fm	Shooks Fm	Red River Valley unit 14	
		Harwood Mbr	South Long Lake Mbr	St. Charles Fm	
		Hawley Mbr	St Francis Fm	Superior Drift	
		Heiberg Mbr	Traverse des Sioux Fm	Superior Till	
		Henderson Fm	Upper Mbr	Toronto Till	
		Hewitt Fm	Verdi Mbr	Turtle Lake Sand	
		Ivanhoe Mbr	Villard Mbr	Wadena Till	
		James River Fm	West Fargo Mbr	Wadena Drift	
		Knife River Mbr	Whetstone Fm	West Campus Sand	
		Lake Henry Fm	Wirt Fm	West Union Gravel	
		Lower Mbr	Wrenshall Mbr		

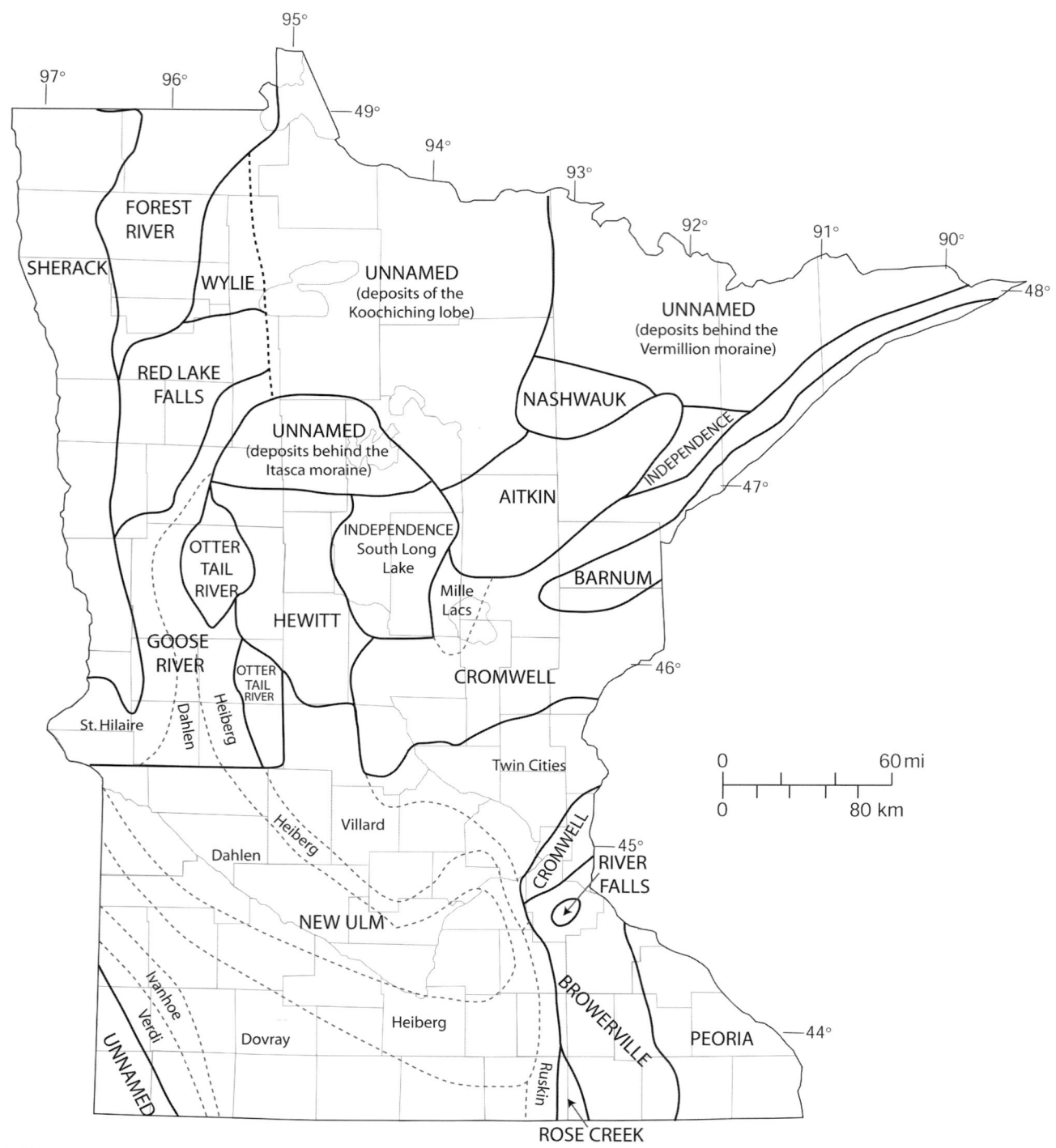

FIGURE 38.4 Formal lithostratigraphical units occurring at the surface in Minnesota. With the exception of the Browerville and River Falls Formations in the southeast, all units are from the Late Wisconsinan glaciations.

which overlie diamicton units of Riding Mountain provenance (Harris et al., 1974; Harris, 1998). These are separated from older diamicton units of Rainy, Winnipeg and mixed provenance by a regional, glacial–erosional unconformity. The Red River, a northward-flowing modern river, has a very low gradient, is still rebounding in the north and therefore losing gradient so there are few, extensive, natural outcrops in the watershed.

Chapter | 38 The Quaternary of Minnesota

Red River Valley

Sherack Fm
Poplar River Fm
Brenna Fm
Forest River Fm
Wylie Fm
Red Lake Falls Fm
Goose River Fm
Otter Tail River Fm
James River Fm
Gardar Fm
Buffalo River Fm
Marcoux Fm
Sheyenne River Fm
Browerville Fm
Gervais Fm

Legend

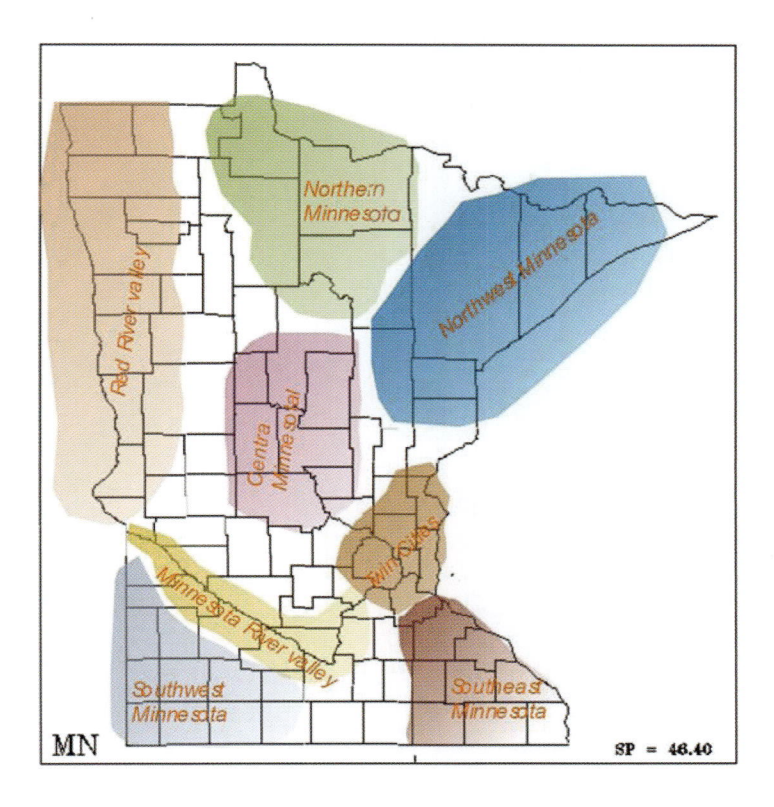

- silt & clay
- sand &/or gravel
- Riding Mtn provenance diamicton
- mixed-provenance diamicton
- Winnipeg provenance diamicton
- Superior provenance diamicton
- Rainy provenance diamicton

FIGURE 38.5 Location map and legend for stratigraphical columns shown in Figs. 38.5–38.9 and stratigraphical column for units in the Red River valley region. Units in bold and underlined are older than MIS 2.

506   Quaternary Glaciations - Extent and Chronology

Upper Minnesota River

| New Ulm Fm |
| 'UMRB Till unit 7' |
| 'UMRB Till unit 8' |
| **Hawk Creek Fm** |
| **Dry Weather Bed** |
| **'UMRB Till unit 10** |
| **'UMRB Till unit 11'** |

Central Minnesota River

| New Ulm Fm |
| unnamed unit(s) |
| **Hawk Creek Fm** |
| **unnamed units** |

River Bend region

| New Ulm Fm |
| Traverse des Sioux Fm |
| **Browerville Fm** |
| **Henderson Fm** |
| **Good Thunder Fm** |

FIGURE 38.6 Stratigraphical column for units in the Minnesota River valley region. See Fig. 38.5 for legend and location. Units in bold and underlined are older than MIS 2.

Southwestern Minnesota

| New Ulm Fm |
| SWRA 1 |
| SWRA 2 |
| SWRA 3 ≈ Crooks till of SD |
| SWRA 4 ≈ Brandon till of SD |
| SWRA-5 = 'Renner till' of S. Dak. |
| SWRA 6 |
| SWRA 7 |
| SWRA 8 |
| SWRA-9 |

Sangamon soil ~125,000 YBP

Volcanic ash 610,000

Brunhes Matuyama boundary 788,000 YBP

Central Minnesota

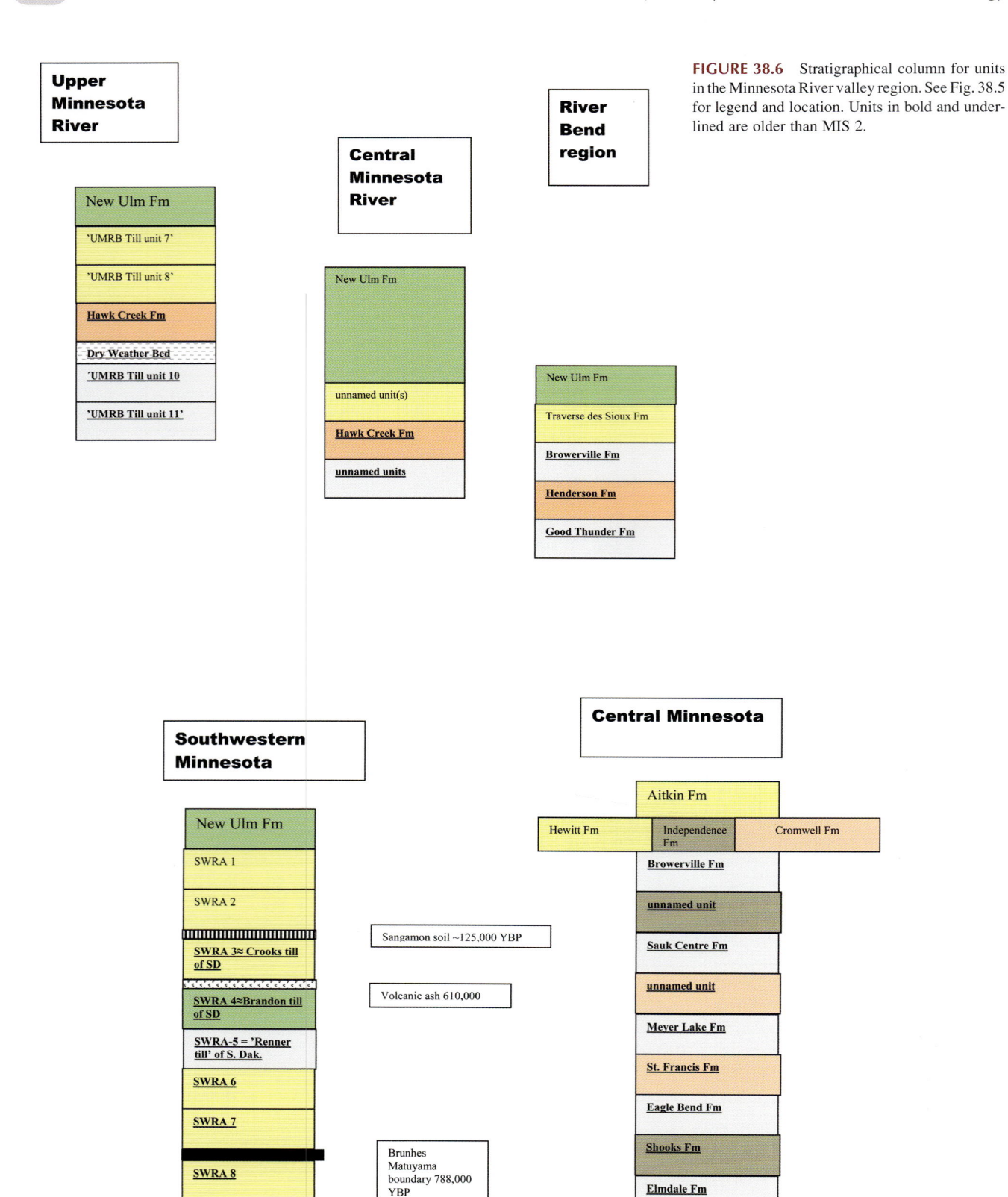

Aitkin Fm		
Hewitt Fm	Independence Fm	Cromwell Fm
Browerville Fm		
unnamed unit		
Sauk Centre Fm		
unnamed unit		
Meyer Lake Fm		
St. Francis Fm		
Eagle Bend Fm		
Shooks Fm		
Elmdale Fm		

FIGURE 38.7 Stratigraphical column for units in the southwestern and central Minnesota region. See Fig. 38.5 for legend and location. Indicated dates are based on correlations to dated units in South Dakota. Units in bold and underlined are older than MIS 2.

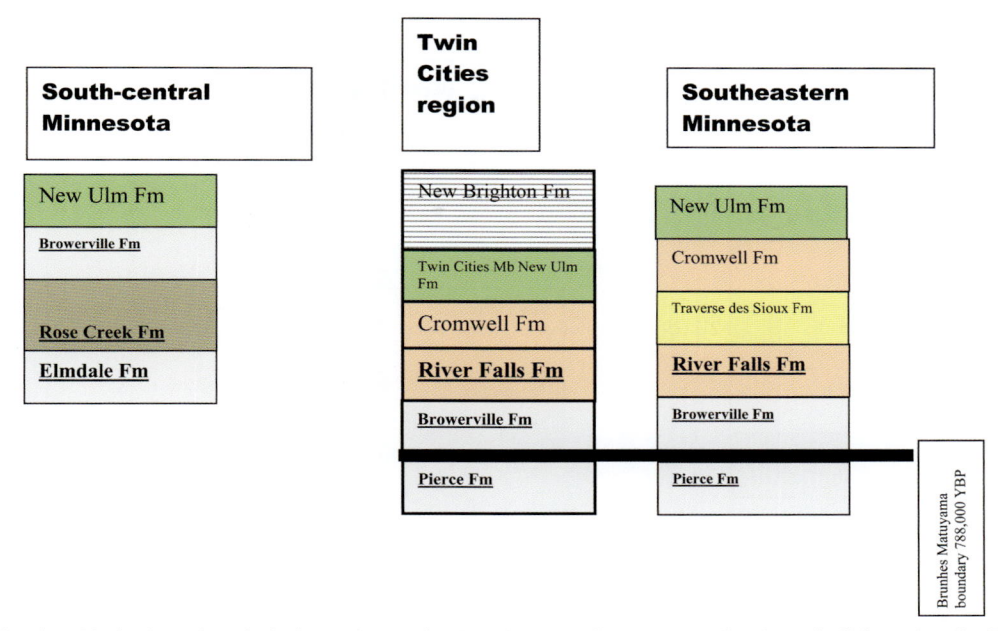

FIGURE 38.8 Stratigraphical column for units in the south-central and southeastern Minnesota as well as the Twin Cities region. See Fig. 38.5 for legend and location. Units in bold and underlined are older than MIS 2.

38.2.2. Minnesota River Valley

Matsch (1972) established southern Minnesota's first formal lithostratigraphy for exposures in the Minnesota River valley. More recent studies have expanded on and modified his work. Multiple diamicton units appear in the Late Wisconsinan, New Ulm Formation of the Des Moines lobe (Fig. 38.6) (Patterson et al., 1999). These units, which are defined as members of the New Ulm Formation, are stacked in places and also occur side by side, suggesting that multiple ice streams from an evolving ice shed contributed to the Des Moines lobe, sometimes simultaneously (Lusardi et al., 2011). The New Ulm Formation also includes sorted sediment deposited in ice-marginal streams and short-lived (decades) proglacial lakes Benson and Minnesota.

At variable depths at the base of the New Ulm Formation lies a regional, erosional unconformity that is in places marked by a boulder pavement. The unconformity complicates the correlation of underlying units because some of them are very low in the stratigraphical sequence. Matsch (1972) described a Superior provenance 'Hawk Creek Till', beneath the New Ulm that is patchy in its distribution, and a mixed provenance 'Granite Falls Till' that has now been recognised as more than one unit. Below the Hawk Creek in western Minnesota and into eastern South Dakota, a rare interglacial bed, the Dry Weather Bed, formerly referred to informally as the 'gastropod silts', contains abundant flora and fauna indicative of a cool, but not glacial climate. It has been dated using amino acid racemisation on shells as 140 ± 70 ka (Gilbertson, 1990; Pirkl et al., 1998). Beneath the Hawk Creek Formation and the Dry Weather Bed, several Winnipeg provenance tills have been described (Gilbertson, 1990; Patterson et al., 1999; Balco et al., 2005). The use of paired, radioactive cosmogenic nuclides with different decay constants to estimate the burial of previously exposed sediment provides limiting ages of 0.5–1.5 Ma for sediment at depths of 20–40 m. Magnetostratigraphical work documents a reversal in this section and therefore ages greater than 788 ka for some of the lower units (Roy et al., 2004a,b). A similar stratigraphical section is present in the Riverbend region, where thick Winnipeg provenance sediment is found below the Henderson Formation, a Superior provenance diamicton with associated lake and stream sediment.

38.2.3. Southwest Minnesota

At least 10 lithostratigraphical units have been described in southwest Minnesota (Patterson, 1997; Fig. 38.7). The three youngest units are members of the New Ulm Formation (D-1, D-2 and D-3 of Patterson, 1997). Below the New Ulm Formation and south and west of the region affected by Des Moines lobe erosion, at least nine other diamicton units occur (SWRA 1 through SWRA 9 of Patterson, 1997), most of which can be correlated to informal till units in southeast South Dakota (the Toronto, Crooks, Brandon and Renner tills of Lineburg, 1993) and to 'drift complexes' in the Prairie Coteau region of eastern South Dakota (Gilbertson and Lehr, 1989; Gilbertson, 1990). It is important to note that in the stratigraphy of southwest Minnesota, if units can be correlated to those in South Dakota, northeastern Nebraska and northeastern Iowa, they can be constrained by other dating methods. A well-developed soil

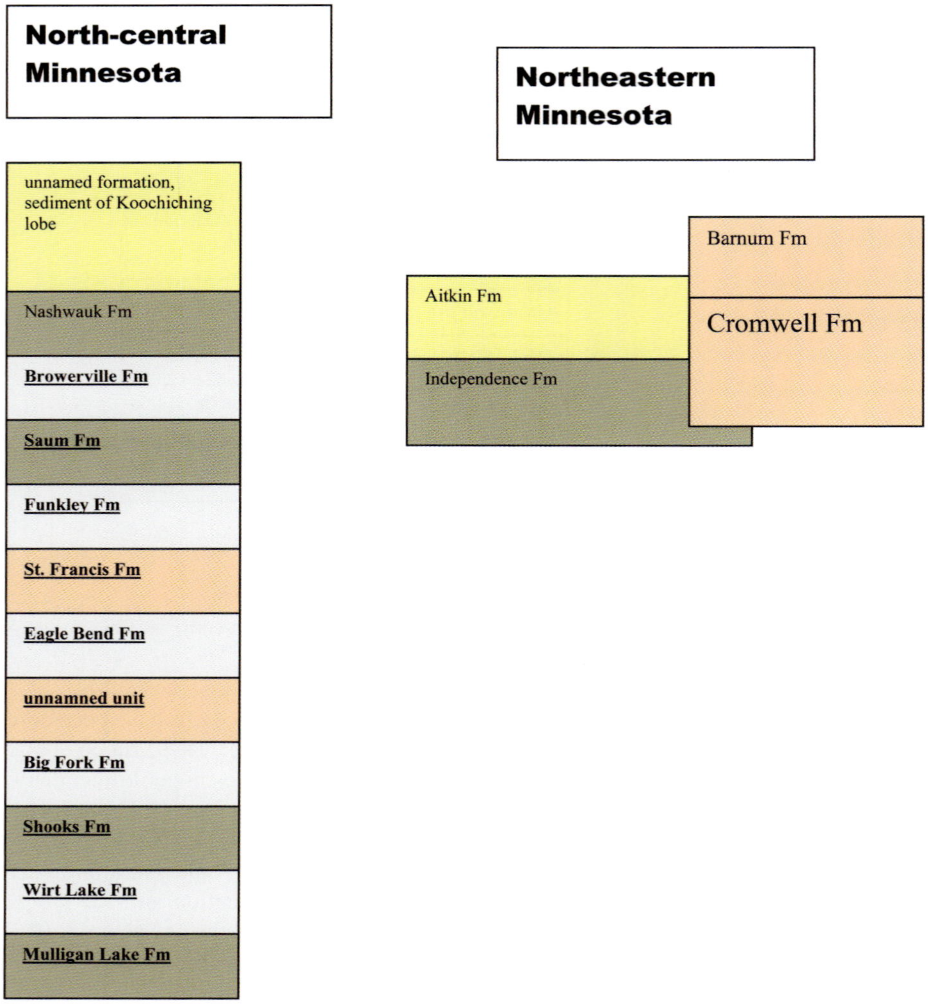

FIGURE 38.9 Stratigraphical column for units in the north-central and northeastern Minnesota region. See Fig. 38.5 for legend and location. Units in bold and underlined are older than MIS 2.

in unit SWRA 3 has been tentatively correlated to the Sangamon Geosol which is most often attributed to a long period of formation spanning oxygen isotope stages 5 and 4, from approximately 135–125 to 65–55 ka (Grimley et al., 2003). This would suggest that SWRA 3 is pre-MIS 5, that is, older than about 125 ka (Fig. 38.7). Units SWRA 4 through 7 are correlated to units that occur below a Yellowstone ash layer dated to 610 ka. Units SWRA 8 and 9 of Patterson (1997) are correlated to units that lie beneath magnetically reversed lake sediment, making them older than 788 ka. Roy et al. (2004a,b) use magnetostratigraphical data along with dated units that are constrained by three volcanic ashes (0.6, 1.2 and 2.0 Ma) derived from eruptions of the Yellowstone Caldera present in the sections (Boellstorff, 1978; Izett, 1981) to create a map of the limit of Early and Middle Pleistocene ice in eastern North America that includes all of Minnesota (Fig. 38.2; Roy et al., 2004a,b). Balco et al. (2005) use cosmogenic burial dating to show that glacial sediment at 60 m depth in a borehole has a minimum limiting age of >1.1 Ma (Balco et al., 2005). Other cosmogenically dated layers include a paleosol and sorted sediment and they also support a regional glacial advance at approximately 1–1.2 Ma. The lowermost tills in the Coteau were deposited between 2 and 1.5 Ma (Balco et al., 2005).

38.2.4. Southeast Minnesota

In the southeast corner of Minnesota, east of the deposits of the New Ulm Formation, older diamicton units of Winnipeg provenance are at the surface. In the counties that have been mapped, several diamicton units are present, sometimes separated by paleosols or cryoturbated horizons, but regional correlations have not been firmly established. The diamicton units are thickest near the contact with the New Ulm Formation and gradually thin and are more deeply buried by silt eastward. Three silt units, the Peoria, Roxana and Loveland Formations, are thickest close to the Mississippi River valley and are interpreted as loess. Much silt has also been redeposited on slopes and in streams (Mason and Knox, 1997).

The overall relief of the landscape increases eastward owing to incision by tributary streams to the Mississippi River. Steep valley slopes plus intensive permafrost conditions have led to widespread erosion of Quaternary units. Gravel derived from Winnipeg provenance diamictons and deposited in Mystery Cave is interpreted as a late-stage fluvial deposit corresponding to MIS 2 glacial retreat. Using the disequilibrium of $^{234}U/^{230}Th$ in encasing speleothems, the gravel was deposited prior to 145 ka (Milske et al., 1983). This confirms that the youngest Winnipeg provenance till in the region is at least MIS 6.

In Mower County in south central Minnesota (Fig. 38.8), the New Ulm Formation overlies the Browerville, Rose Creek and Elmdale Formations. The Browerville and Elmdale Formations are defined in central Minnesota, but stratigraphical position and strong similarity to the diamicton units in Mower County led Meyer (2000) to extend them into Mower County. The Rose Creek Formation occurs stratigraphically between the Browerville and the Elmdale and is of Rainy provenance.

Deposits of Superior provenance are rare in southeast Minnesota, with only a suggestion of a presence in Mower County (Meyer, 2000). However, the so-called 'Hampton moraine' of southern Dakota County in southeastern Minnesota (Ruhe and Gould, 1954) contains diamicton and sand and gravel of Superior provenance. This sediment is included in the River Falls Formation, originally defined in Wisconsin (Baker, 1984b). In Dakota County south of the St. Croix moraine and east of the Bemis moraine, valley bottoms are filled with glacial sand and gravel of the New Ulm and Cromwell Formations (Hobbs, 1998).

The Winnipeg diamicton units of southeastern Minnesota likely correlate to the Wolf Creek and Alburnett Formations of Iowa (Hallberg, 1980) and, in part, to the Pierce Formation of Wisconsin (Baker, 1984a). It is important to note that, if these units can be correlated even farther south with units in Missouri (Rovey and Balco, 2010), then interpretations based on cosmogenic and paleomagnetic signatures (used to distinguish the isolated early ice advances (2.4 and 1.3 Ma) from the more common, post-8 Ma advances in Missouri) could be used to correlate with units in Minnesota.

38.2.5. Twin Cities

The Twin Cities region contains some pre-Wisconsinan Formations that are mainly confined to bedrock valleys (Fig. 38.8). The surface units are dominated by deposits of criss-crossing glacier lobes: the Superior lobe advanced from the northeast and left deposits included in the Cromwell Formation that are, in most of the area, covered by deposits of the Twin Cities member of the New Ulm Formation, which was deposited by the Des Moines lobe and its offshoot, the Grantsburg sublobe.

Deposits of ice-dammed lakes accompanied the retreat of these two lobes and partially overlie one another. Glacial Lake Lind formed during retreat of the Superior lobe, and the sediments deposited in this lake are included in the Sunrise Member (Johnson et al., in 1999). This member was originally defined in Wisconsin as part of the Copper Falls Formation, but in Minnesota, it is included in the Cromwell Formation. A re-advance of the Superior lobe during the existence of glacial Lake Lind is recorded by a more fine-grained Cromwell diamicton that Meyer (1998) referred to as the 'Coon Creek till', but has not been formally defined. As the Grantsburg sublobe advanced, it formed glacial Lake Grantsburg, a short-lived glacial lake, in the same regional lowland. Its sediment is included in the Falun member of the Trade River Formation in Wisconsin (Johnson and Hemstad, 2000). As the Grantsburg sublobe retreated, several stages of glacial Lake Anoka occurred in the Anoka Sand Plain region (Meyer, 1998).

Pre-Wisconsinan, northwest and northeast provenance till units along the St. Croix River have been investigated in Wisconsin and include predominantly the Pierce and River Falls Formations (see Baker et al., 1983; Attig et al., 1988). These names have been used along the Minnesota side as well in Washington (Meyer et al., 1990) and Dakota counties (Hobbs et al., 1990)

38.2.6. Central Minnesota

The many surficial lithostratigraphical units in central Minnesota are associated with several different lobes in the central part of the state (Fig. 38.2) and many occur in what has been referred to as the Alexandria moraine complex (see Wright, 1972). The oldest sediment exposed at the surface in this area (Fig. 38.2A) is the Hewitt Formation, which is the surface diamicton in the Wadena drumlin field, created by the Rainy lobe. The prominent St. Croix Moraine lies east of the drumlin field. The drumlin field shows a gradational, north-to-south change in composition with till of the Independence Formation (associated with the Rainy lobe) in its northern portion and the Cromwell Formation (associated with the Superior lobe) to the south. As the Rainy lobe retreated, glacial Lake Brainerd formed in part of the vacated area. The Independence and Cromwell Formations are roughly the same age as the deposits in the Itasca Moraine, which marks the southern margin of the Itasca lobe (Fig. 38.2).

Diamicton and sand and gravel of the New Ulm Formation occur in the southern part of central Minnesota. For the most part, they form a drape over earlier, more pronounced glacial landforms. To the north, an advance of the Koochiching lobe brought Winnipeg provenance diamictons to the northwestern border of a bedrock high known as the Giants Range. The sediment from this advance is correlated to the Roseau Formation of southern Manitoba. A narrow breach in the linear highland near Grand Rapids allowed ice to squeeze through and spread out on the southeastern side, much like a piedmont lobe (see Fig. 38.2). This resulted in the incorporation, possibly by freezing on, of enough local material to change the colour and texture of the diamicton. On the northwest side, the diamictons resemble the New Ulm Formation but on the southeast side, they contain red, clayey sediment incorporated from underlying, Superior provenance material. These red clayey deposits are included in the Aitkin Formation.

Beneath the surficial units of central Minnesota, a thick sequence of older Quaternary deposits occurs, and these have been studied by Schneider (1961), Meyer (1997, 1986), Meyer and Knaeble (1996) and Goldstein (1998). They include the Browerville, Eagle Bend, Elmdale, Meyer Lake, St. Francis, Sauk Centre and Shooks Formations. Additionally, there are several buried units of Superior provenance diamicton described in earlier reports that are difficult to correlate and name at this time, and these include the 'Red Drift' of Goldstein (1998) and the 'Second Red till' of Meyer (1986). These older formations are of Winnipeg and Superior provenance, and their regular alternation in the local stratigraphy has led Meyer (1997) to postulate a regular pattern of growth and migration of accumulations centres in Canada during the Pleistocene.

38.2.7. Northern Minnesota

Much of north central Minnesota is covered by Holocene peat and lacustrine clay, silt and bedded sand of glacial Lake Agassiz and other glacial lakes. Diamicton of the Koochiching lobe, tentatively thought to be equivalent to the Roseau Formation of southern Manitoba, occurs at the surface in places. Below the surface is a thick sequence of Quaternary lithostratigraphical units of Winnipeg, Rainy and Superior provenance. Similar to the older record in central Minnesota, Winnipeg units alternate regularly with units of Rainy or Superior provenance (Meyer, 1997). These include the Big Fork, Browerville, Eagle Bend, Funkley, Mulligan Lake, Nashwauk, Saint Francis, Saum, Shooks and Wirt Lake Formations.

38.2.8. Northeast Minnesota

Deposits of Rainy and Superior provenance dominate the stratigraphy in northeast Minnesota. Sediment associated with the Rainy lobe in northeast Minnesota includes the Nashwauk Member of the Independence Formation, expressed north of the Giants Range, and the remainder of the Independence Formation, which occurs east of the Giants Range where not covered by younger deposits (Fig. 38.4). Further north, a younger phase of the Rainy lobe formed the Vermillion moraine and its deposits extend north to the Canadian border.

Superior lobe deposits of the Cromwell Formation occur in northeast Minnesota, but these are overlain by finer-grained Superior lobe deposits of the Lakewood Member of the Barnum Formation. The finer sediment was derived from lake sediment deposited in the Superior basin during a retreat of the lobe after deposition of the Cromwell Formation. As the ice re-advanced, it incorporated the fine sediment and deposited the Barnum Formation. Its four recognised members include the Lakewood, Moose Lake and Knife Lake members, which are predominantly diamicton, and the Wrenshall member, which contains clay, silt and bedded sand associated with an early phase of glacial Lake Duluth.

At about the same time that the Superior lobe re-advanced, the St. Louis sublobe, mentioned earlier, advanced. In the northeast part of the 'piedmont' lobe, it deposited red, clay-rich diamicton that was referred to as the 'Alborn Till' by Baker (1964) (see also Wright and Ruhe, 1965; Matsch and Schneider, 1986).

REFERENCES

Attig, J.W., Clayton, Lee, Mickelson, D.M., 1988. Pleistocene stratigraphic units of Wisconsin 1984–87. Wisconsin Geological and Natural History Survey Information Circular, 62, 61 pp.

Baker, R.G., 1964. Late-Wisconsinan glacial geology and vegetation history of the Alborn Area, St. Louis County, Minnesota. Minneapolis, University of Minnesota, MS thesis, 44 pp.

Baker, R.W., 1984a. Pierce formation. In: Mickelson, D.M., Clayton, Lee, Baker, R.W., Mode, W.N., Schneider, A.F. (Eds.), Pleistocene Stratigraphic Units of Wisconsin, Wisconsin Geological and Natural History Survey Miscellaneous Paper 84-1, p. A4-1.

Baker, R.W., 1984b. River falls formation. In: Mickelson, D.M., Clayton, Lee, Baker, R.W., Mode, W.N., Schneider, A.F. (Eds.), Pleistocene Stratigraphic Units of Wisconsin, Wisconsin Geological and Natural History Survey Miscellaneous Paper 84-1, p. A5-1.

Baker, R.W., Diehl, J.F., Simpson, T.W., Zelazny, L.W., Beske-Diehl, S., 1983. Pre-Wisconsin glacial stratigraphy, chronology, paleomagnetics of west-central Wisconsin. Geol. Soc. Am. Bull. 94, 1442–1449.

Balco, G., Stone, J., Jennings, C., 2005. Dating Plio-Pleistocene glacial sediments using the cosmic-174 ray-produced radionuclides ^{26}Al and ^{10}Be. Am. J. Sci. 305, 1–41, 175 pp.

Bierman, P.R., Marsella, K.A., Patterson, C., Thompson Davis, P., Caffee, M., 1999. Mid-Pleistocene cosmogenic minimum-age limits for pre-Wisconsinan glacial surfaces in southwestern Minnesota and southern Baffin Island: a multiple nuclide approach. Geomorphology 27, 25–39.

Boellstorff, J., 1978. North American Pleistocene stages reconsidered in light of probable Pliocene–Pleistocene continental glaciation. Science 202, 305.

Gilbertson, J.P., 1990. Quaternary geology along the Eastern Flank of the Coteau des Prairies, Grant County, South Dakota. Duluth, University of Minnesota, MS thesis, 108 pp.

Gilbertson, J.P., Lehr, J.D., 1989. Quaternary stratigraphy of northeastern South Dakota. In: Gilbertson, J.P. (Ed.), Quaternary Geology of Northeastern South Dakota, South Dakota Geological Survey Guidebook No. 3, 36th Midwest Friends of the Pleistocene, pp. 1–13.

Goldstein, B.S., 1998. Quaternary stratigraphy and history of the Wadena drumlin region, central Minnesota. In: Patterson, C.J., Wright, H.E., Jr. (Eds.), Contributions to Quaternary Studies in Minnesota, Minnesota Geological Survey Report of Investigation, 49, pp. 61–84.

Grimley, D.A., Follmer, L.R., Hughes, R.E., Solheid, P.A., 2003. Modern, Sangamon and Yarmouth soil development in loess of unglaciated southwestern Illinois. Quatern. Sci. Rev. 22, 225–244.

Hallberg, G.R., 1980. Pleistocene stratigraphy in east-central Iowa. Iowa Geological Survey Technical Information Series, 10, 168 pp.

Harris, K.L., 1998. Computer-assisted lithostratigraphy. In: Patterson, C.J., Wright, H.E., Jr. (Eds.), Contributions to Quaternary Studies in Minnesota, Minnesota Geological Survey Report of Investigation, 49, pp. 179–192.

Harris, K.L., Moran, S.R., Clayton, L., 1974. Late Quaternary stratigraphic nomenclature Red River valley, North Dakota and Minnesota. North Dakota Geological Survey, Miscellaneous Series, 52, 47 pp.

Hobbs, H.C., 1998. Quaternary stratigraphy, plate 4. In: Setterholm, D.R. (Ed.), Geologic Atlas Goodhue County, Minnesota, Minnesota Geological Survey County Atlas Series, Atlas C-8, scale 1:100,000.

Hobbs, H.C., Aronow, Saul, Patterson, C.J., 1990. Surficial geology, plate 3. In: Balaban, N.H., Hobbs, H.C. (Eds.), Geologic Atlas Dakota County, Minnesota, Minnesota Geological Survey County Atlas Series, Atlas C-4, scale 1:100,000.

Izett, G.A., 1981. Volcanic ash beds: recorders of Upper Cenozoic silicic pyroclastic volcanism in the western United States. J. Geophys. Res. 86, 10200–10222.

Johnson, M.D., Hemstad, C., 2000. Falun member of the Trade River formation, in Johnson, M.D., Pleistocene geology of Polk County, Wisconsin. Wis. Geol. Nat. Hist. Surv. Bull. 92, 63–67.

Johnson, M.D., Addis, K.L., Ferber, L.R., Hemstad, Chris, Meyer, G.N., Komai, L.T., 1999. Glacial Lake Lind, Wisconsin and Minnesota, USA. Geol. Soc. Am. Bull. 111, 1371–1386.

Johnson, M.D., Harris, K.L., Hobbs, H.C., Jennings, C.E., Knaeble, A.R., Lusardi, B.A., Meyer, G.N., Thorleifson, H., in preparation. Quaternary Lithostratigraphic Units of Minnesota. Minnesota Geological Survey Report of Investigations.

Lineburg, J.M., 1993. Sedimentology and stratigraphy of Pre-Wisconsin drifts, Coteau des Prairie, eastern south Dakota. University of Minnesota-Dukuth, thesis, 122 pp.

Lively, R.S., Bauer, E.J., Chandler, V.M., 2006. Maps of gridded bedrock elevation and depth to bedrock in Minnesota. Minnesota Geological Survey Open File Report, OFR2006_02 .

Lusardi, B.A., Jennings, C.E., Harris, K.L., 2011. Provenance of Des Moines lobe till records ice-stream catchment evolution during Laurentide deglaciation. Boreas 40 (in print).

Mason, J.A., Knox, J.C., 1997. Age of colluvium indicates accelerated late Wisconsinan hillslope erosion in the Upper Mississippi Valley. Geology 25, 267–270.

Matsch, C.L., 1972. Quaternary geology of southwestern Minnesota. In: Sims, P.K., Morey, G.B. (Eds.), Geology of Minnesota: A Centennial Volume. Minnesota Geological Survey, St. Paul, Minnesota, pp. 547–560.

Matsch, C.L., Schneider, A.F., 1986. Stratigraphy and correlation of the glacial deposits of the glacial lobe complex in Minnesota and northwestern Wisconsin. In: Sibrava, V., Bowen, D.Q., Richmond, G.M. (Eds.), Quaternary Glaciations in the Northern Hemisphere. Pergamon Press, Oxford, pp. 59–64.

Meyer, G.N., 1986. Subsurface and till stratigraphy of the Todd County area, central Minnesota: Minnesota Geological Survey Report of Investigations 34, 40 p.

Meyer, G.N., 1997. Pre-late Wisconsinan till stratigraphy of north-central Minnesota. Minnesota Geological Survey, Report of Investigations, 48, 67 pp.

Meyer, G.N., 1998. Glacial lakes of the Stacy basin, east-central Minnesota and northwestern Wisconsin. In: Patterson, C.J., Wright, H.E., Jr. (Eds.), Contributions to Quaternary Studies in Minnesota, Minnesota Geological Survey Report of Investigation 49, pp. 35–48.

Meyer, G.N., 2000. Quaternary geology of Mower County. In: Mossler, J.H. (Ed.), Contributions to the Geology of Mower County, Minnesota, Minnesota Geological Survey Report of Investigations 50, pp. 31–61.

Meyer, G.N., Knaeble, A.R., 1996. Part C, text supplement. In: Meyer, G.N., Swanson, L. (Eds.), Geologic Atlas Stearns County, Minnesota, Minnesota Geological Survey County Atlas Series C-10, p. 63.

Meyer, G.N., Baker, R.W., Patterson, C.J., 1990. Surficial geology, plate 3. In: Swanson, L., Meyer, G.N. (Eds.), Geologic Atlas Washington County, Minnesota, Minnesota Geological Survey County Atlas Series, Atlas C-5, scale 1:100,000.

Milske, J.A., Alexander, E.C. Jr. Lively, R.S., 1983. Clastic sediments in Mystery Cave, Southeastern Minnesota, National Speleological Society Bulletin 45, 55–70.

Patterson, C.J., 1997. Surficial geology of southwestern Minnesota. In: Patterson, C.J. (Ed.), Contributions to Quaternary Geology of Southwestern Minnesota, Minnesota Geological Survey Report of Investigation 47, 1–45.

Patterson, C.J., Knaeble, A.R., Setterholm, D.R., Berg, J.A., 1999. Quaternary stratigraphy, plate 2. In: Patterson, C.J. (Ed.), Regional Hydrogeologic Assessment: Quaternary Geology-Upper Minnesota River Basin, Minnesota, Minnesota Geological Survey Regional Hydrogeological Assessment Series RHA-4, Part A.

Pirkl, M.E., Kuglin, C.L., Cotter, J.F.P., 1998. The 'gastropod silts' of western Minnesota and northeastern south Dakota. In: Patterson, C.J., Wright, H.E., Jr. (Eds.), Contributions to Quaternary Studies in Minnesota, Minnesota Geological Survey Report of Investigation 49, pp. 155–158.

Rovey, C.W., Balco, G., 2010. Periglacial climate at the 2.5 Ma onset of northern hemisphere glaciations inferred from the Whippoorwill formation, northern Missouri, USA. Quatern. Res. 73, 151–161.

Roy, M., Clark, P.U., Barendregt, R.W., Glasmann, J.R., Enkin, R.J., 2004a. Glacial stratigraphy and paleomagnetism of late Cenozoic deposits of the north-central United States. Geol. Soc. Am. Bull. 226 (1–2), 30–41.

Roy, M., Clark, P.U., Raisbeck, G.M., Yiou, F., 2004b. Geochemical constraints on the regolith hypothesis for the middle Pleistocene transition. Earth Planet. Sci. Lett. 227, 281–296.

Ruhe, R.V., Gould, L.M., 1954. Glacial geology of the Dakota County area, Minnesota. Geol. Soc. Am. Bull. 65, 769–792.

Schneider, A.F., 1961. Pleistocene Geology of the Randall Region, Central Minnesota. Minn. Geol. Surv. Bull. 151 pp.

Wright, H.E., Jr., 1972. Quaternary history of Minnesota. In: Sims, P.K., Morey, G.B. (Eds.), Geology of Minnesota: A Centennial Volume. Minnesota Geological Survey, St. Paul, MN, pp. 515–547.

Wright, H.E., Jr., Ruhe, R.V., 1965. Glaciation of Minnesota and Iowa. In: Wright, Jr., H.E., Frey, D.G. (Eds.), The Quaternary of the United States. Princeton University Press, Princeton, pp. 29–41.

Wright, H.E., Jr., Matsch, C.L., Cushing, E.J., 1973. Superior and Des Moines lobes. In: Black, R.F., Goldthwait, R.P., Willman, G.P. (Eds.), The Wisconsinan Stage, Geological Society of America Memoir. 136, 153–185.

Zanner, C.W., 1999. Late-Quaternary landscape evolution in southeastern Minnesota: loess, aeolian sand, and the Periglacial environment. Ph.D. Dissertation, University of Minnesota, St. Paul, Minnesota, 398 pp.

Chapter 39

Pleistocene Glaciation of Ohio, USA

John P. Szabo[1,]*, Michael P. Angle[2] and Alex M. Eddy[2]

[1]Department of Geology & Environmental Science, University of Akron, Akron, Ohio 44325-4101, USA
[2]Ohio Department of Natural Resources, Division of Geological Survey, 2045 Morse Road, Building C-1, Columbus, Ohio 43229-6693, USA
*Correspondence and requests for materials should be addressed to John P. Szabo. E-mail: jpszabo@uakron.edu

39.1. INTRODUCTION

The article of Szabo and Chanda (2004) is herein expanded to provide much more detail into the glaciation of Ohio. Little has changed in our knowledge of Early Pleistocene glaciations, but a summary of extensive research into the stratigraphy of the Illinoian Stage of the Middle Pleistocene (Szabo and Totten, 1995) is included. Additionally, recent work of Thomas Lowell at the University of Cincinnati on dating Late Pleistocene advances in western Ohio is discussed. A brief summary of the history of late-glacial and post-glacial Lake Erie and its effect on drainage systems is also presented. A map of the distribution of beach ridges is included in ArcView in addition to maps of the limits of glaciation and moraines.

Information on the general geology of Ohio is useful in understanding its glaciation. The Portage Escarpment (Fig. 39.1) is the boundary of two major physiographic provinces that occur in Ohio. Eastern Ohio lies within the Allegheny Plateaux (Brockman, 1998), which is capped in most places by resistant Pennsylvanian-aged sandstones. Western Ohio is lower in elevation and is part of the Central Lowlands; bedrock consists predominately of carbonate rocks. As ice flowed into the Erie Basin from the north or at times from the northeast, its spread southward was controlled by these physiographic provinces. Ice split into topographically controlled sublobes; herein referred to as lobes (Fig. 39.1). The Miami and Scioto lobes flowed much farther south through the Central Lowlands, whereas the Killbuck, Cuyahoga and Grand River lobes flowed onto the higher elevations of the plateaux.

39.2. TERTIARY PERIOD

There is no evidence of Tertiary glaciations in Ohio, but there are sands of questionable origin on high divides south of the glacial boundary near Dresden (D, Fig. 39.2). Nine metres of loess and possible aeolian sand overlie at least 6 m of Mn-stained fluvial sands. The intensity and thickness of this staining have not been found in fluvial deposits north of the glacial boundary and may be suggestive of weathering over a long period of time. Other 'black' sands have been reported in water wells south of the glacial boundary. There is insufficient evidence to determine if these sands originated during the Tertiary or are the product of some Early Pleistocene glaciation.

39.3. EARLY PLEISTOCENE GLACIATIONS

Magnetically reversed lacustrine sediments, Minford Silt, in abandoned valleys of the Teays drainage system provide evidence of an Early Pleistocene glaciation in at least the

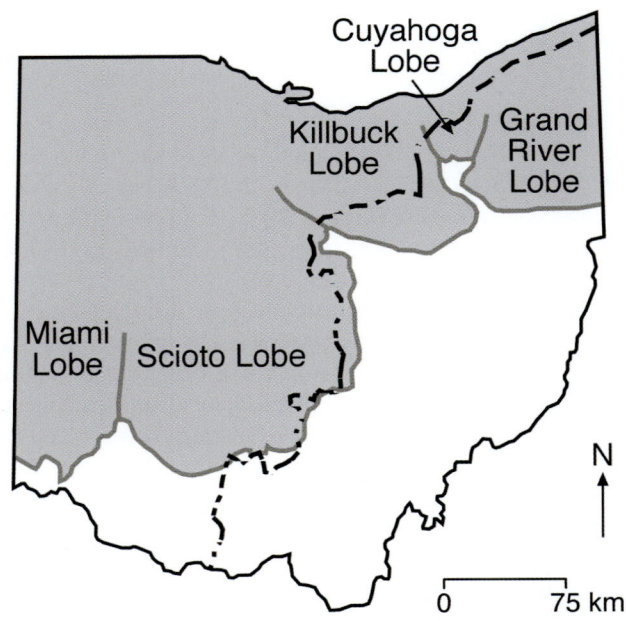

FIGURE 39.1 Topographically controlled sublobes of the Erie Lobe at the southern terminus of the Laurentide ice sheet in Ohio. Dashed line is the boundary between the Central Lowlands to the west and the Allegheny Plateaux to the east.

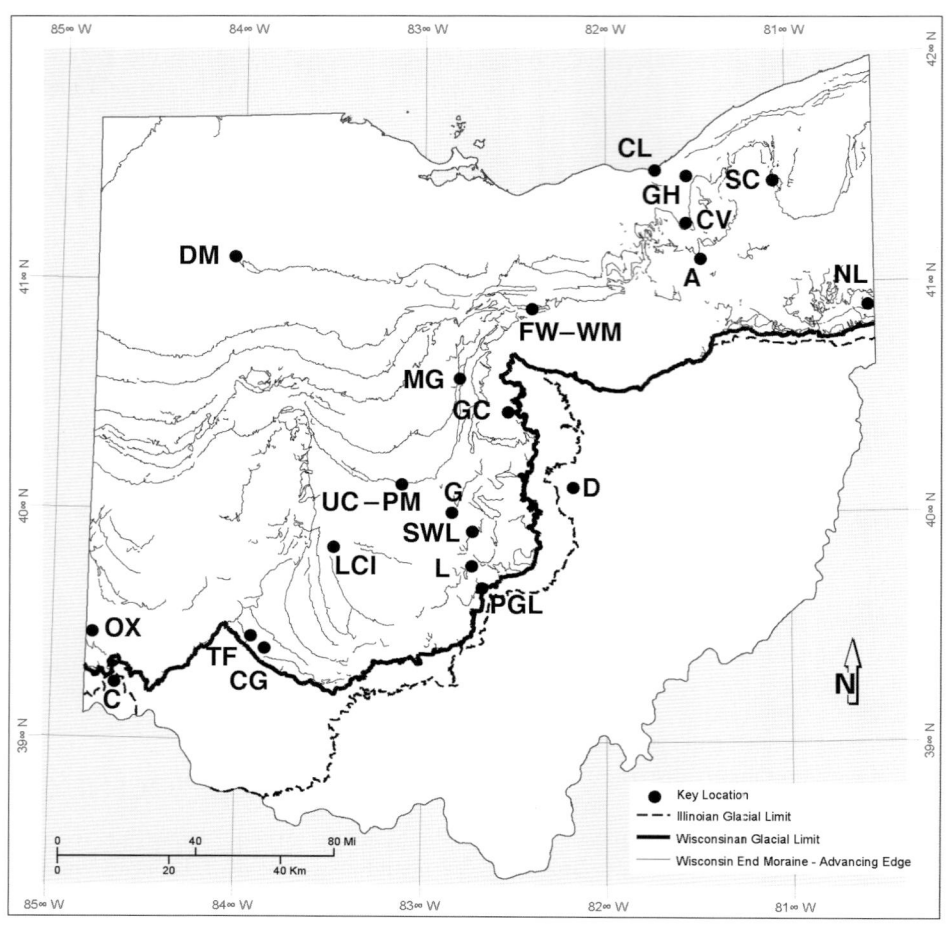

FIGURE 39.2 Glacial limits, end moraines and key locations in Ohio. A, Akron; C, Cincinnati; CG, Cuba Gully; CL, Cleveland; CV, Cuyahoga Valley; D, Dresden; DM, Defiance Moraine; FW–WM, Fort Wayne–Wabash Moraine; G, Gahanna; GC, Granny Creek; GH, Garfield Heights; L, Lancaster; LCI, London Correctional Institute; MG, Mt. Gilead; NL, North Lima; OX, Oxford; PGL, Pingrove Landfill; SC, Swine Creek; SWL, Southwest Licking; TF, Todd Fork; UC-PM, Union City-Powell Moraine.

northern half of Ohio. The Minford Silts are older than 780,000 years (Szabo and Chanda, 2004). A possibly older deposit, Calcutta Silt (Lessig, 1963; Fullerton, 1986), is found at high elevations near the Ohio–Pennsylvania state line.

Attempts have been made to integrate many buried valleys in Ohio into the Teays drainage (Stout et al., 1943). However, a complex network of buried valleys incised into bedrock underlies most of glaciated Ohio; efforts to connect all buried bedrock valleys into one drainage network lead to impossibly complex unnatural patterns (Frolking and Szabo, 1998). Many deep valleys (>70 m) are orientated approximately north–south and are generally filled with lacustrine deposits indicative of ice damming (Szabo, 2006). The lower parts of their cross-sections are generally V-shaped, whereas their upper parts have been glacially modified (Szabo, 1987). These are possible remnants of drainage during Tertiary or Early Pleistocene time. Generally, shallower buried valleys (<70 m) may have formed along ice-front positions or by meltwater draining southward from the ice sheet towards the ancestral Ohio River throughout the latter part of the Pleistocene Epoch (Szabo, 2006).

39.4. MIDDLE PLEISTOCENE GLACIATIONS

Little work has been done in the pre-Illinoian area beyond the Illinoian limit (Fig. 39.2), nor has much been done within the area mapped as Illinoian surficial materials. Within these two areas, possible Middle Pleistocene sediments could be exposed in deeply incised stream valleys or in deeper borings for water wells. The Illinoian end moraine as mapped on the glacial map of Ohio (Pavey et al., 1999) may not represent the actual maximum extent of Illinoian ice; erratic boulders often are found beyond the mapped limit suggesting erosion of either older glacial deposits or thin Illinoian sediments (White, 1982). The Illinoian moraine is generally more subdued than its

Wisconsinan equivalent, consisting of broad, gently rolling hills. Illinoian ground moraine is found between the Illinoian and Wisconsinan terminal moraines and is generally dissected having a well-integrated drainage compared to that of Wisconsinan ground moraine.

Sites exposing sediments from the early part of the Middle Pleistocene are not easily identified (Szabo and Chanda, 2004). In a 100-m deep borehole into a buried bedrock valley in southwestern Licking County (SWL; Fig. 39.2), at least 50 m of Illinoian glacial sediments overlie older tills and laminated fines that appear to have normal magnetic polarity (Frolking and Szabo, 1998). In north-central and northeastern Ohio, multiple Illinoian-age units have been differentiated (Szabo and Totten, 1995), whereas in southwestern Ohio, only one unit, Rainsboro Till, has been identified (Rosengreen, 1974; Quinn and Goldthwait, 1985). At Mount Gilead (MG) in central Ohio (Fig. 39.2), several tills underlie silt (loess?) having a themoluminescence (TL) date from near the end of the Middle Pleistocene (Szabo and Chanda, 2004). Tills beneath this loess were separated based on percentages of calcite and dolomite in their <0.074-mm fractions and were traced northward and eastward to establish the stratigraphy illustrated in Table 39.1. The two lowermost tills at this section are currently inseparable and are referred to as the Chesterville Till best exposed at Granny Creek (GC; Fig. 39.2). This unit contains nearly 2% calcite and 9% dolomite (Szabo, 2006) and is overlain by the Gahanna Till (G; Fig. 39.2) having up to 20% fine carbonate.

The most areally extensive unit found at MG is the Millbrook Till and its equivalents (Table 39.1) containing almost no calcite and 4–6% dolomite depending on location. Its total fine-carbonate content declines eastward in Ohio and allows this unit to be used as a marker bed from central Ohio into adjacent western Pennsylvania (Szabo and Totten, 1995). A representative section of equivalent Titusville Till (Table 39.1) is exposed in the valley of Swine Creek (SC) in northeastern Ohio (Fig. 39.2). Analysis of tills at this section (Szabo, 2006) identified a repetition of textural and compositional properties reflective of thrust stacking as proposed by Moran (1971). Heavy-mineral studies of the tills in this section show that their provenance is in the Madoc–Arnprior area of the Grenville province in Canada (Matz, 1996; Szabo, 2006), whereas analysis of equivalent Millbrook Till at MG shows a source area on the north shore of Lake Huron in the Superior-Southern province (Matz, 1996). Other evidence for thrust stacking is present in strip mines in eastern Ohio; Franko (2008) delineated compositional differences between meltout tills and deformational comminution tills at North Lima (NL; Fig. 39.2).

Other Illinoian units overlie the Millbrook Till and its equivalents (Table 39.1). The moderately calcareous, more clay-rich Northampton Till overlies the Millbrook Till and its equivalent Mogadore Till in north-central and western northeastern Ohio (Szabo and Totten, 1995). This unit is very firm and has intensely oxidised joints exposed where it outcrops in deeply incised tributaries in the valley of the Cuyahoga River (CV; Fig. 39.2). A sandier, less consolidated till, informally referred to as Millbrook Till U is found north of MG (Fig. 39.2) along the eastern edge of the area glaciated by the Scioto lobe and extends into the northern portion of this lobe.

Several problems exist in defining and correlating units of the Middle Pleistocene, more specifically those of the Illinoian Stage. Dating of advances of this stage is not well constrained, relying on established observational criteria summarised in (White, 1982); however, no other dates

TABLE 39.1 Tentative Correlations of Lithologic Units in North-Central and Northeastern Ohio

Time	Scioto lobe		Killbuck lobe	Cuyahoga lobe	Grand River lobe
	Northern	Eastern			
Late Wisconsinan (Weichselian)					Ashtabula Till
	Hiram Till	Hiram Till	Hiram Till	Hiram Till	Hiram Till
	Hayesville Till	Hayesville Till	Hayesville Till	Lavery Till	Lavery Till
	Navarre Till	Navarre Till	Navarre Till	Kent Till	Kent Till
Middle Wisconsinan (Weichselian) through Sangamonian (Eemian)	Millbrook Till U	Millbrook Till U			
	Millbrook Till M	Northampton Till	Northampton Till	Northampton Till	Not found
Illinioan (Saalian)	Millbrook Till L	Millbrook Till	Millbrook Till	Mogadore Till	Titusville Till
		Gahanna Till			Keefus Till
		Chesterville Till			

are available, and a Late Wisconsinan till overlies the TL-dated silt at MG. Another common marker horizon used to differentiate Illinoian from Wisconsinan deposits is the presence of a Sangamonian soil that is missing throughout much of northern Ohio (Szabo, 1997). This soil was formed in Illinoian tills of southwestern Ohio (Hall and Zbieszkowski, 2000) and in Illinoian loess in southeastern Ohio (Frolking and Szabo, 1998), but it is not found in sections in the northern part of the state. Sharp contacts between unweathered Illinoian tills and their overlying Wisconsinan tills suggest erosion before or during the Late Wisconsinan advances. Not much mapping on a countywide basis has occurred since the end of the state-sponsored programme in the late 1980s. Thus, workers have been unable to correlate isolated outcrops in the area glaciated by the Miami and western Scioto lobes in western Ohio with the better-established Illinoian stratigraphy from the eastern Scioto lobe and other lobes in northern Ohio.

39.5. LATE PLEISTOCENE GLACIATIONS

The normal geological processes including pedogenesis that operate on landscapes began during the Sangamonian Stage interglacial and continued through climate oscillations of Early and Middle Wisconsinan time. Calculation of ice volumes based on sea-level curves suggests that there was not enough ice during these parts of the Wisconsinan Stage to glaciate Ohio (Clark and Lea, 1992). Specimens of wood collected from gravels deposited during this time all have 'greater than' dates and contain both deciduous hardwoods from warm periods and spruce from cold periods (Goldthwait, 1958). A few dates in the 40 ^{14}C ka range suggest a warm period during which colluviation of slopes and alluvial fan deposition occurred (Goldthwait, 1992; Hall and Zbieszkowski, 2000).

Additional evidence suggests that baselevels of streams were lower than present, and streams were well incised into Illinoian deposits. Parts of the Teays system may have been exhumed because wood dated at 28.39 ± 0.33 ^{14}C ka BP (ISGS-3224) was found in alluvium (Lloyd, 1998; Frolking and Szabo, 1998) about 75 m below the surface in a buried bedrock valley associated with the Teays system (London Correctional Institute (LCI); Fig. 39.2). A boring near Lancaster (L) in Fairfield County (Fig. 39.2) contained at least 56 m of Late Wisconsinan sediments, and at Pingrove Landfill (PGL) in the same county (Fig. 39.2) wood dated at 26.78 ± 0.43 ^{14}C ka BP (ISGS-3223) was found in the basal Wisconsinan till 35 m below the surface in a buried bedrock valley.

The section at Garfield Heights (GH; Figs. 39.2 and 39.3) is the only location on the Allegheny Plateau in northern Ohio having a nearly complete record of the Late Pleistocene and has been studied periodically for over 50 years (Szabo, 1992). The site is currently covered by vegetated colluvium, but can be excavated to expose most of the younger units described in the literature (Szabo, 1997). Units older than the Illinoian Stage at this site and at an adjacent site are no longer accessible; their age assignments are discussed in Fullerton (1986) and Szabo (1992).

Gravels are the oldest accessible unit at the site (Fig. 39.3) and are interpreted to be Illinoian age (White, 1968; Szabo, 1992). A dark reddish-brown, truncated paleosol is found in the upper part of the gravels, and its formation has been assigned to the Sangamonian interglacial (White, 1968). The matrix of the gravel below the paleosol becomes calcareous 25 cm below the paleosol (Miller and Szabo, 1987); lower in the section carbonate cements layers of gravel.

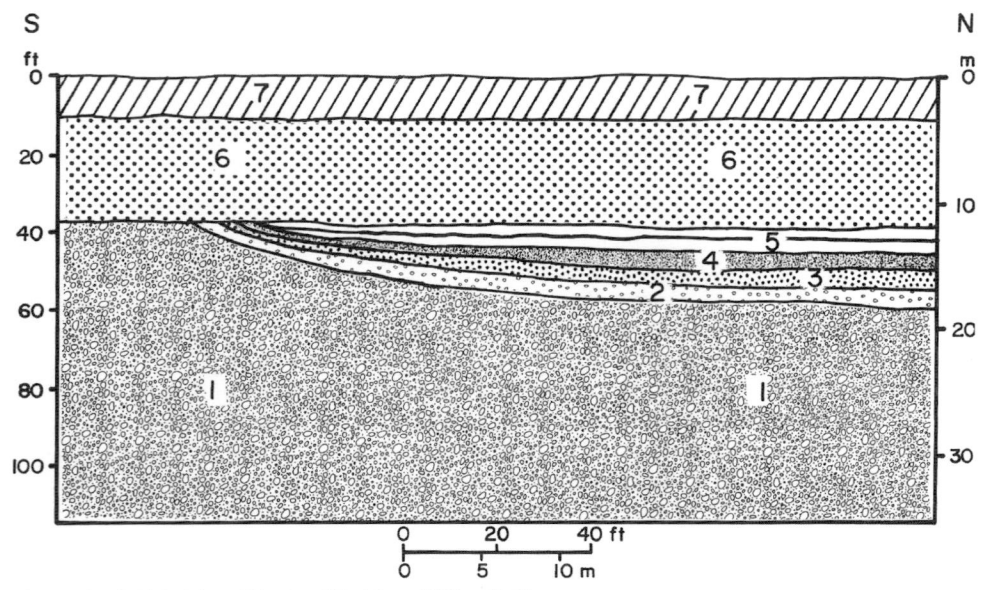

FIGURE 39.3 Section at Garfield Heights, Ohio (modified from White, 1968).

Although loess overlies the paleosol near the south end of the section (Fig. 39.3), a yellowish-brown to greenish-grey, platy to blocky to massive accretion gley of variable texture overlies the gravels through most of the section to the north. Clay mineralogy varies between layers of gley having definable X-ray diffraction peaks associated with vermiculite and montmorillonite and those layers having broad peaks associated with heterogeneous swelling material (Miller and Szabo, 1987), suggestive of episodic erosion upslope.

A bipartite loess sequence (Fig. 39.3) overlies the gley and thickens northward. The older, yellowish-brown, friable loess is non-calcareous, and its clay mineralogy also suggests that this loess has been weathered. A weakly developed soil, 10 cm below the contact with the upper loess unit contains disseminated organic material dated at 27.39 ± 0.35 ^{14}C ka BP (ISGS-1949). A wood fragment from this loess dated at 28.195 ± 0.535 ^{14}C ka BP (K-361-3). At some locations, this loess has involutions accentuated by iron; Fullerton (1986) suggested that this cryoturbated loess implies the presence of ice in the nearby Erie Basin. The upper part of the loess sequence consists of a light olive-brown to light greyish-brown, calcareous, friable silt containing snails and wood fragments (Miller and Szabo, 1987).

Laminated lacustrine silts and clays overlie the loess sequence. The lower part of these brownish-grey, platy, calcareous deposits consists of sandy zone containing snails and wood fragments interpreted as a colluvium (Miller and Szabo, 1987). Several radiocarbon dates on wood in this zone average about 24 ^{14}C ka BP. The fossils in the lacustrine sequence suggest deterioration of local climate as the ice advanced (Szabo, 1997). Dark brown, leached, calcareous till caps the section and may represent deposition by Late Wisconsinan Lavery or Hiram ice (Table 39.1). However, about 50 m to the south, sandy Kent till of the first Late Wisconsinan advance lies between the dark brown till and the lacustrine sediments.

The Wisconsinan limit in southwestern Ohio is defined by well-formed moraines especially in the area covered by the Scioto lobe (Fig. 39.2). The limit in the Miami lobe is less well constrained by moraines but was reconstructed based on boulder density counts (Goldthwait et al., 1961). The terminal moraine near Cincinnati (C; Fig. 39.2) has been breached by meltwater streams and dissected by tributaries of the Miami and Ohio rivers. Figure 39.2 shows that the Allegheny Plateaux deflected the Scioto lobe to the southwest. In areas of both the Scioto and Miami lobes, the topography consists of a succession of recessional moraines (Fig. 39.2) separated by intervening areas of loamy ground moraine. Outwash trains, kames and very short eskers are found in the interlobate area between these lobes. Inter-moraine lake beds are more common and kames less common in the poorly drained, clay-rich tills in the northern areas glaciated by these lobes.

The Late Wisconsinan glacial limit in the Miami and Scioto lobes is well constrained by radiocarbon dates (Lowell et al., 1999). Several dates on organic material and trees 2 km behind the maximum limit of ice near Todd Fork (TF; Fig. 39.2) average about 23.2 ± 0.086 ^{14}C ka BP. Dates of another expansion of ice that overran trees at Cuba Gully (CG; Fig. 39.2) average 20.36 ± 0.084 ^{14}C ka BP (Lowell et al., 1999). Farther west in the Miami lobe at Oxford (OX, Fig. 39.2), ice overrode trees having an average date of 20.77 ^{14}C ka BP and reached its maximum limit north of Cincinnati (C, Fig. 39.2) at an average date of 19.59 ± 0.035 ^{14}C ka BP (Lowell et al., 1999).

The Late Wisconsinan limit in the Killbuck, Cuyahoga and Grand River lobes is less well defined in north-central and northeastern Ohio (Figs. 39.1 and 39.2). Dissected resistant bedrock affected the initial Late Wisconsinan advance onto the plateaux. Loamy tills along the glacial limit occur as discontinuous, hummocky moraine without linear trends in many areas (White, 1982), and the limit is often defined by weathering criteria. Widely spaced recessional moraines of the Scioto lobe are compressed and even merge in the area of the Killbuck lobe; the Defiance Moraine (DM; Fig. 39.2) is the only well-defined recessional moraine. Very extensive outwash trains and kame deposits occupy the re-entrant (Fig. 39.1) among the Killbuck, Cuyahoga and Grand River lobes in northeastern Ohio. Stagnation topography is common along the margins of these lobes, suggesting that ice remained stationary in these areas for long periods of time (Szabo, 2006).

Dating of the Late Wisconsinan advance in northeastern Ohio is not well constrained because very little organic matter has been found in moraines at the limit. Based on dates from wood at the base of the lacustrine part of the section at GH, ice advanced onto the plateaux after 24,000 ^{14}C years ago and had retreated to north-east of Cleveland (CL) by 14,450 ^{14}C years ago (Szabo et al., 2003). By about 16,000 ^{14}C years ago, ice had retreated into the northwards and eastwards during the Erie Interstadial (Dreimanis and Goldthwait, 1973) and then readvanced southwestward from the Niagara Falls area of Ontario (Fullerton, 1986). Ice overrode interstadial lacustrine clays and deposited clay-rich tills over the northern half of Ohio. Probability curves of radiocarbon dates from basal organic matter in 30 basins in Ohio and eastern Indiana show that the majority of basins formed at either 15,500 or 14,800 ^{14}C years ago (Glover et al., 2004). These are dates traditionally assigned to the post-interstadial advances that formed the Union City-Powell Moraine and the Fort Wayne–Wabash Moraine (UC-PM and FW–WM; Fig. 39.2), respectively (Dreimanis and Goldthwait, 1973).

As ice retreated from northeastern Ohio at about 14.5 ^{14}C ka BP (Szabo et al., 2003), a series of lakes formed that were trapped between the ice and higher topography to the south. A series of beach ridges along the south shore of Lake Erie remain as evidence of changing ancestral lake levels as various outlets were exposed by deglaciation

(Szabo et al., 2003). As ice retreated from the Niagara peninsula about 12.5 ^{14}C ka BP (Lewis, 1969), the last ancestral lake in the Erie basin drained. The level of the resultant Early Lake Erie was about 40 m below the present level (Szabo et al., 2003) and causing extensive downcutting in drainage systems that emptied into this lake and shifting the continental divide at Akron (A; Fig. 39.2) southwards. A post-glacial channel is deeply incised into a till in the buried bedrock valley of the lower CV 37 m below the present level of Lake Erie at CL (Fig. 39.2) and possibly was formed by the rapid drop in base level (Szabo et al., 2003). Post-glacial isostatic rebound of the Niagara Falls area raised the outlet of Early Lake Erie, and it filled the basin to modern levels. By about 7.18 ± 0.070 ^{14}C ka BP (ISGS-4439), the level of the floodplain of the CV north of Akron was within 3 m of its present elevation (Szabo et al., 2003).

Several problems are apparent in our understanding of the Late Pleistocene glaciation of Ohio. The timing of the advance of ice to its limit in northeastern Ohio is not well constrained. Was the climate substantially different on the Allegheny Plateau compared to the lowlands of southwestern Ohio? The recessional moraine succession in western Ohio appears to have been driven by relative ablation rates (Lowell et al., 1999), whereas the sequence of glacial retreat on the Allegheny Plateau in northeastern Ohio represents an extended period of stagnation.

39.6. CONCLUSIONS

Consideration of our knowledge of the glaciation of Ohio produces the logical conclusion that the sequence of events creating the Late Pleistocene landscape of Ohio is better understood than the results of earlier glaciations. The identification of Early Pleistocene deposits relies on the acquisition of orientated core samples from deep borings and the use of paleomagnetic analysis. Some outcrop analysis of lacustrine sediments may be of value in the deeply dissected, non-glaciated area of south-eastern Ohio. The stratigraphy of the Middle Pleistocene, especially the last glaciation, the Illinoian, is better understood in north-central and northeastern Ohio then in western Ohio. The timing of Late Wisconsin advances of ice during the Late Pleistocene is better constrained in western Ohio than in eastern Ohio suggesting some fundamental difference not only in topography but also in climate.

REFERENCES

Brockman, C.S., 1998. Physiographic regions of Ohio (Map). Ohio Division of Geological Survey.

Clark, P.U., Lea, P.D. (Eds.), 1992. The Last Interglacial—Glacial Transition in North America, Geological Society of America Special Paper, 270, p. 317.

Dreimanis, A., Goldthwait, R.P., 1973. Wisconsin glaciation in the Huron, Erie, and Ontario lobes. In: Black, R.F., Goldthwait, R.P., Willman, H.B. (Eds.), The Wisconsinan Stage, Geological Society of America Memoir, 136, pp. 71–106.

Franko, B.J., 2008. Mineralogy and provenance of pink inclusions in the Illinoian Titusville Till, Mahoning County, Ohio. Unpublished M.S. thesis, University of Akron, 132 pp.

Frolking, T.A., Szabo, J.P., 1998. Quaternary geology along the Eastern Margin of the Scioto Lobe in Central Ohio. Ohio Division of Geological Survey Guidebook, 16, 40 pp.

Fullerton, D.S., 1986. Stratigraphy and correlations of glacial deposits from Indiana to New York and New Jersey. Quatern. Sci. Rev. 5, 23–37.

Glover, K.C., Lowell, T.V., Pair, D., Wiles, G., 2004. Deglaciation and post-glacial climate change in a regional network of sites, Ohio and eastern Indiana. Geol. Soc. Am. Abstr. Programs 36 (5), 249–250.

Goldthwait, R.P., 1958. Wisconsin age forests in western Ohio: I. Ohio J. Sci., 58, 209–219.

Goldthwait, R.P., 1992. Historical overview of early Wisconsin glaciation. In: Clark, P.U., Lea, P.D. (Eds.), The Last Interglacial—Glacial Transition in North America, Geological Society of America Special Paper, 270, pp. 13–18.

Goldthwait, R.P., White, G.W., Forsyth, J.L., 1961. Glacial map of Ohio. U.S. Geological Survey Miscellaneous Geologic Investigations Map, I-316.

Hall, R.D., Zbieszkowski, D.J., 2000. Glacial and nonglacial Quaternary stratigraphy of eastern Indiana and western Ohio. Field Trip Guidebook: 34th Annual Meeting of the North-Central Section of the Geological Society of America. Department of Geology, Indiana University-Purdue University, Indianapolis, p. 142.

Lessig, H.D., 1963. Calcutta silt, a very early Pleistocene deposit, upper Ohio Valley. Geol. Soc. Am. Bull., 74, 129–139.

Lewis, C.F.M., 1969. Late Quaternary history of lake levels in the Huron and Erie basins. In: Proceedings of 12th Conference on Great Lakes Research. International Association for Great Lakes Research, pp. 250–270.

Lloyd, B.A., 1998. Stratigraphy of late Wisconsinan tills from the London Correctional Institute, Union Township, Madison County, Ohio. Unpublished M.S. thesis, University of Akron, 159 pp.

Lowell, T.V., Hayward, R.K., Denton, G.H., 1999. Role of climate oscillations in determining ice-margin position: hypothesis, examples, and implications. In: Mickelson, D.M., Attig, J.W. (Eds.), Glacial Processes Past and Present, Geological Society of America Special Paper, 337, pp. 193–203.

Matz, J.B., 1996. Determination of heavy-mineral assemblages of pre-Wisconsinan tills in northern Ohio by use of sodium polytungstate and x-ray diffraction. Unpulished M.S. thesis, University of Akron, 138 pp.

Miller, B.B., Szabo, J.P., 1987. Garfield heights: Quaternary stratigraphy of Northeastern Ohio. Geological Society of America Centennial Field Guide, North-Central section, pp. 399–402.

Moran, S.R., 1971. Glacio-tectonic structures in drift. In: Goldthwait, R.P. (Ed.), Till: A Symposium. Ohio State University Press, Columbus, pp. 127–148.

Pavey, R.R., Goldthwait, R.P., Brockman, C.S., Hull, D.N., Swinford, E.M., Van Horn, R.G., 1999. Quaternary geology of Ohio. Ohio Div. Geol. Surv. Map No. 2.

Quinn, M.J., Goldthwait, R.P., 1985. Glacial geology of Ross County, Ohio. Ohio Div. Geol. Surv. Rept. Invest., 127, 42 pp.

Rosengreen, T.E., 1974. Glacial geology of Highland County, Ohio. Ohio Div. Geol. Surv. Rept. Invest., 92, 36 pp.

Stout, W., Ver Steeg, K., Lamb, G.F., 1943. Geology of water in Ohio. Ohio Div. Geol. Surv. Bull., 44, 694 pp.

Szabo, J.P., 1987. Wisconsinan stratigraphy of the Cuyahoga Valley in the Erie Basin, northeastern Ohio. Can. J. Earth Sci., 24, 279–290.

Szabo, J.P., 1992. Reevaluation of early Wisconsinan stratigraphy of Northern Ohio. In: Clark, P.U., Lea, P.D. (Eds.), The Last Interglacial—Glacial Transition in North America, Geological Society of America Special Paper, 270, pp. 88–107.

Szabo, J.P., 1997. Nonglacial surficial processes during the early and middle Wisconsinan substages from the glaciated Allegheny Plateau in Ohio. Ohio J. Sci., 87, 66–71.

Szabo, J.P., 2006. Quaternary geology of the interlobate area between the Cuyahoga and Grand River lobes, northeastern Ohio. Ohio Div. Geol. Surv. Guidebook, 20, 52 pp.

Szabo, J.P., Chanda, A., 2004. Pleistocene glaciation of Ohio, U.S.A.. In: Ehlers, J., Gibbard, P.L. (Eds.), Quaternary Glaciations—Extent and Chronology, Part II, Developments in Quaternary Science, 2, 233–236.

Szabo, J.P., Totten, S.M., 1995. Multiple pre-Wisconsinan glaciations along the northwestern edge of the Allegheny Plateau in Ohio and Pennsylvania. Can. J. Earth Sci., 32, 2081–2089.

Szabo, J.P., Bradley, K.N., Tevesz, M.J.S., 2003. Foundations from the past: clues to understanding late Quaternary stratigraphy beneath Cleveland, Ohio. J. Great Lakes Res., 29, 566–580.

White, G.W., 1968. Age and correlation of Pleistocene deposits at Garfield Heights (Cleveland), Ohio. Geol. Soc. Am. Bull., 79, 749–752.

White, G.W., 1982. Glacial geology of northeastern Ohio. Ohio Div. Geol. Surv. Bull., 68, 75 pp.

Chapter 40

The Glaciation of Pennsylvania, USA

Duane D. Braun

Geography and Geosciences, Bloomsburg University of Pennsylvania, Mount Desert, Maine, USA

40.1. INTRODUCTION

There is evidence of at least four Pleistocene glacial advances into Pennsylvania (Shepps et al., 1959; White et al., 1969; Marchand, 1978; Braun, 1994, 1999, 2008; Braun et al., 2008). From oldest to youngest (using the terminology of Richmond and Fullerton, 1986), these glaciations are Early Pleistocene (pre-Illinioian G—marine isotope stage (MIS) 22 or older), early Middle Pleistocene (probably pre-Illinoian D—MIS 16), mid Middle Pleistocene (pre-Illinioan A or B—MIS 10 or 12) or late Middle Pleistocene (MIS 6—Illinoian or Saalian) and Late Pleistocene (MIS 2—Late Wisconsinan or Weichselian). The Early Wisconsinan (Weichselian) did not reach Pennsylvania (Braun, 1988; Ridge et al., 1990). The oldest glaciation extended the furthest south, and each younger glaciation extended less far to the south (see digital map and Fig. 40.1). The trace of each successive glacial advance's maximum limit is remarkably similar (essentially parallel) to that of the previous advance (see digital map and Fig. 40.1).

Each time the Laurentide ice sheet advances, it enters Pennsylvania from the north-west (the Erie—Ontario lobe) and the north-east (the Lake-Champlain—Hudson River lobe) (Crowl and Sevon, 1999) (Fig. 40.2). The triangular-shaped unglaciated area between the two lobes is called

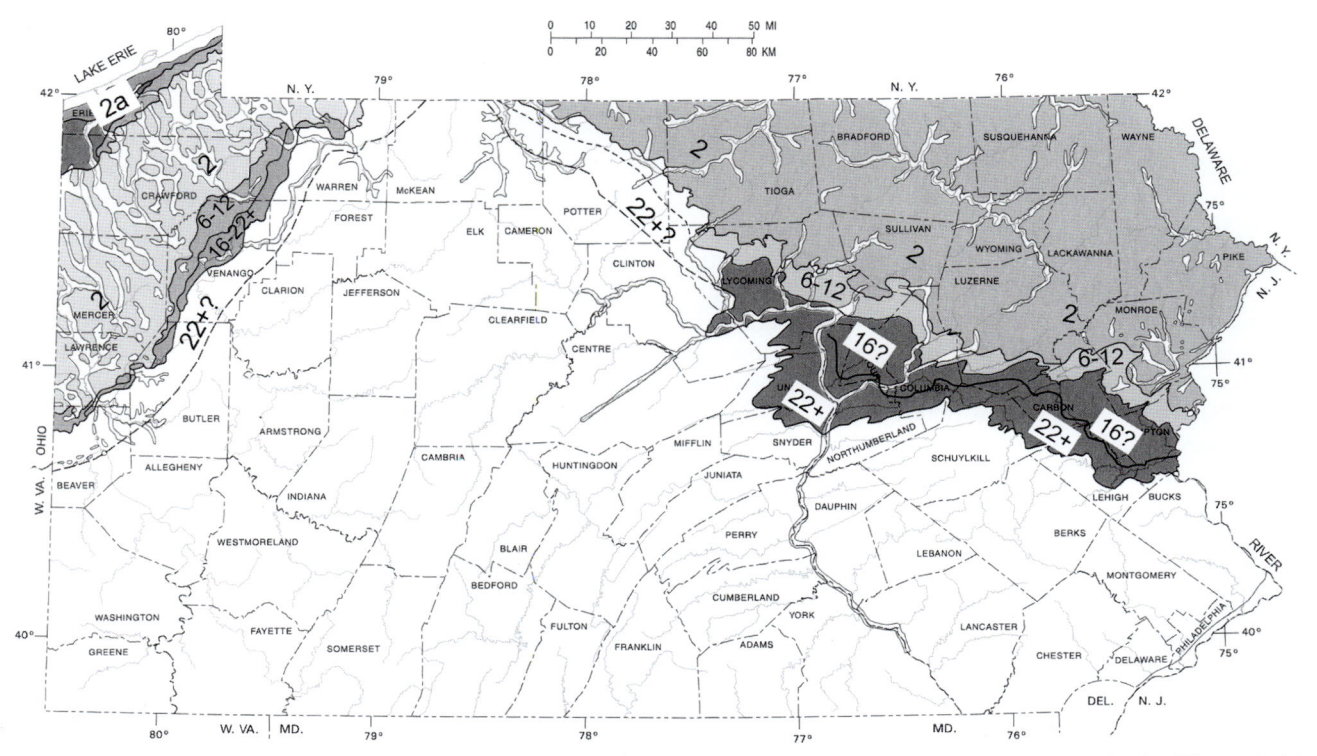

FIGURE 40.1 Map of the extent and age of glacial deposits in Pennsylvania. The numbers are the marine isotope stage ages for the different glacial limits. The older glacial limits have uncertain ages and are either labelled with a question mark with a range of ages like 6–12. Label 2a represents MIS 2 recessional or readvance positons. Narrow strips along stream valleys are glaciofluvial deposits (modified from Sevon and Braun, 1997).

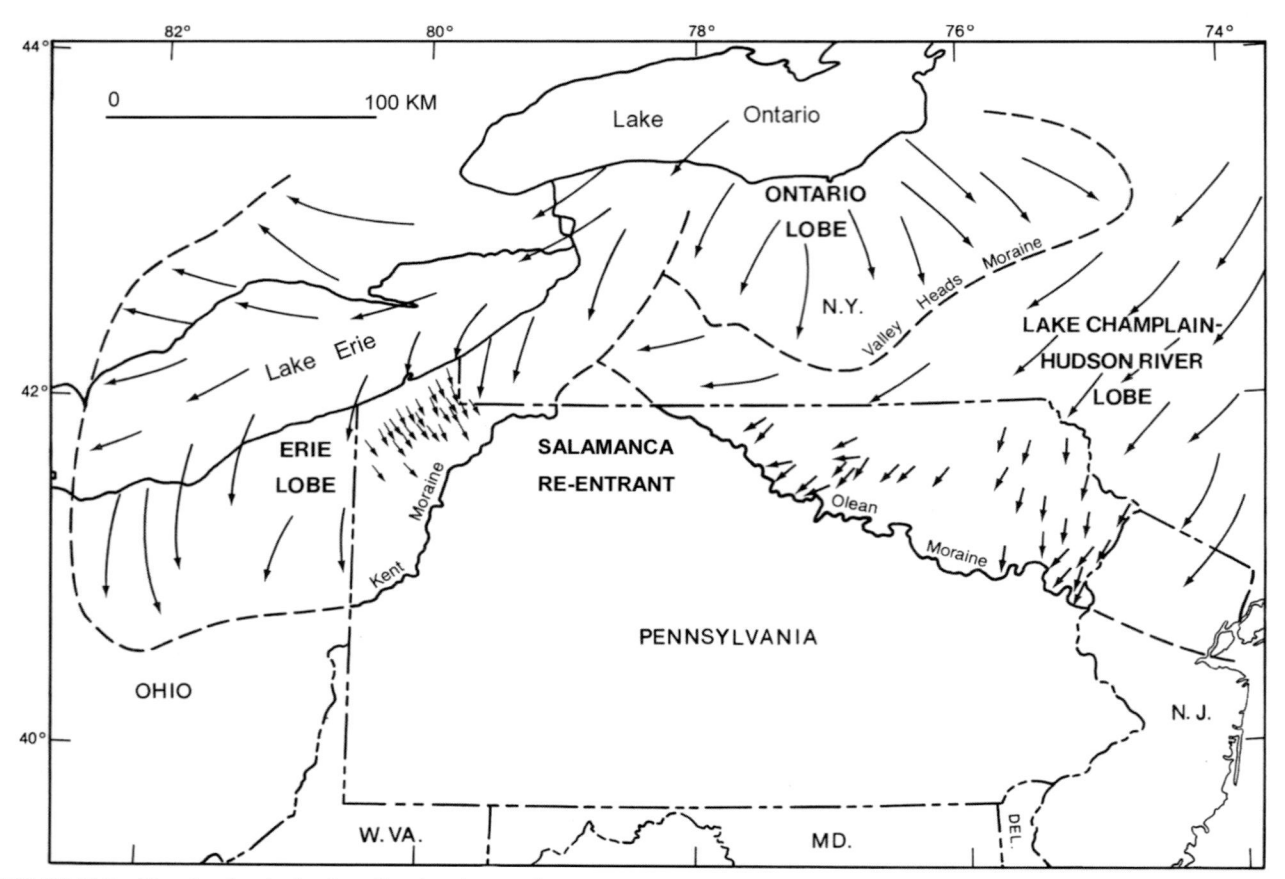

FIGURE 40.2 Map showing the ice flow directions into north-eastern Pennsylvania and north-western Pennsylvania to either side of the Salamanca re-entrant (modified from Crowl and Sevon, 1999, p. 226, Fig. 15-2).

the Salamanca re-entrant (see digital map and Fig. 40.2). In north-western Pennsylvania, the relatively low relief (100–300 m), shaly bedrock, and abundant debris from the Great Lake basins produced a dominance of deposition over erosion in each glacial advance. This produced multiple till sheet sequences whose layers can be separated by partly preserved weathering profiles. These till sequences are well exposed in open pit coal mines (White et al., 1969). The pre-glacial stream drainage in north-western Pennsylvania was to the north-west and was blocked and diverted to the south-west by glaciation to form an ice marginal drainage system, the present Allegheny–Ohio River system (Carll, 1880; Leverett, 1902, 1934; Kaktins and Delano, 1999). The Early Pleistocene and probably early Middle Pleistocene glaciations imponded large proglacial lakes, especially glacial lake Monogahela (see digital map), whose deposits are wide spread in valleys in south-west Pennsylvania and northern West Virginia (White, 1896; Leverett, 1902, 1934; Campbell, 1903; Lessig, 1963; Jacobson et al., 1988). By mid-Pleistocene time at the latest, the Allegheny–Ohio system was fully integrated into its present form (Leverett, 1934).

In north-eastern Pennsylvania, the moderate relief (300–500 m) on sandstone bedrock produced a dominance of erosion over deposition in each glacial advance. Each successive advance almost entirely removed the deposits of the previous advance and some bedrock (Braun, 1989a,b, 1994, 2006a). Described multiple till sites are few and those are all questionable sites (Braun, 1994). Older glacial advances are recognised only because remnants of their deposits are present south of the younger advances. Each advance left thick glacial and proglacial deposits in the valleys, while adjacent ridge crests are essentially bare bedrock. Portions of the pre-glacial stream drainage were to the east or northeast and glacial advance blocked that drainage to form glacial lake Lesley (Williams, 1902) and Packer (Williams, 1894) at the Early Pleistocene glacial limit. Other proglacial lakes were impounded along that limit, along younger limits (Alden and Fuller, 1903; Braun, 1988), and north of the Late Pleistocene limit as ice receded from Pennsylvania (Willard, 1932; Coates, 1966; Braun, 1989a, 1997, 2002; Gardner et al., 1993). Part of the drainage was diverted to form the 'Grand Canyon' of Pennsylvania, a 230-m-deep bedrock gorge (Alden and Fuller, 1903; Crowl, 1981).

The Marine Isotope record and the North American glacial deposit record (Richmond and Fullerton, 1986) suggest that at least 10 Pleistocene glacial advances should have approached close to Pennsylvania and caused periglacial climate conditions there (Braun, 1989b, 1994). The area south of the Late Pleistocene glacial limit is characterised by extensive colluvial deposits and other features of palaeoperiglacial origin (Peltier, 1949; Sevon et al., 1975; Ciolkosz et al., 1986; Clark and Ciolkosz, 1988; Braun, 1989b, 1994, 2006a; Clark et al., 1992; Marsh, 1999). Most of the material mapped as Early and Middle Pleistocene glacial deposits is actually colluvium derived from glacial deposits (Braun, 1994, 1999; Braun et al., 2008). Even north of the Late Pleistocene limit, there are significant periglacial effects and production of colluvium on the slopes (Peltier, 1949; Denny and Lyford, 1963; Coates and King, 1973; Braun, 1997, 2002, 2006a).

40.2. DISCUSSION OF THE GLACIAL LIMITS

The limit of the most recent glaciation or the 'Terminal Moraine' was traced across the entire state of Pennsylvania by Lewis (1884) and assigned a Late Wisconsinan age by Chamberlin (1883). In north-western Pennsylvania, that limit was remapped and refined by Shepps et al. (1959). In north-eastern Pennsylvania, revisions to the limit to better account for the effect of topography on the ice front were made by Leverett (1934) and the entire north-eastern Pennsylvania border was remapped by Crowl and Sevon (1980). Other minor revisions were made to produce the current small scale glacial limit map of Pennsylvania (Sevon and Braun, 1997). Detailed 1:24,000 scale mapping has further refined the position of the Late Wisconsinan terminus in north-eastern Pennsylvania (Braun, 2004a,b, 2006b, 2007a,b, 2008; selected maps with significant change). The Late Wisconsinan limit is a true moraine only in a few places, mainly on the Pocono plateau (Crowl and Sevon, 1980). Elsewhere in north-eastern Pennsylvania, the limit is marked by heads of outwash (frontal kame fans) in the valleys with, at most, an 'indistinct' moraine on adjacent hillsides (Crowl and Sevon, 1980). In north-western Pennsylvania, the limit is marked by an almost continuous belt of morainic topography and heads of outwash in the valleys (Shepps et al., 1959).

The next older glaciation reached only a few kilometres beyond the Late Wisconsinan in both north-western Pennsylvania (see digital map and Fig. 40.1) (Leverett, 1934; Shepps et al., 1959) and north-eastern Pennsylvania (Braun, 1988). This glacial advance has been traditionally thought to be Late Illinoian (Saalian) in age (Leverett, 1934; Sevon et al., 1975; Marchand, 1978; Berg et al., 1980; Braun, 1988). In north-eastern Pennsylvania, the Late Illinoian limit is marked by heads of outwash in the valleys and discontinuous patches of till or colluvium derived from till on the uplands (Braun, 1988). The deposits are deeply weathered (10 m or more) and deeply eroded with essentially no constructional topography remaining (Braun, 1988, 1994). In north-western Pennsylvania, the Late Illinoian deposits are also deeply weathered and eroded (Shepps et al., 1959; White, 1969) but less so than in north-eastern Pennsylvania. Where the Late Illinoian deposits in north-western Pennsylvania are overlain by Late Wisconsinan deposits, a truncated weathering profile separates the two units (Shepps et al., 1959; White, 1969).

Yet older pre-Illinoian glaciations extended a few tens of kilometres beyond the Late Illinoian limit (see digital map and Fig. 40.1). In north-eastern Pennsylvania, two pre-Illinoian advances are indicated by the presence of glacial lake sediments having normal polarity magnetism just south-west of the Illinoian limit and other such sediments having reversed magnetic polarity farther south-west of the Illinoian limit (Sasowsky, 1994). In addition, within the pre-Illinoian glacial limit in the Lehigh Valley region, there are two belts of markedly thicker glacial deposits that suggest two different glacial margins, one at the maximum glacial limit and a second one about 10–20 km north-east of the maximum limit (Braun, 1999). Likewise in the central Susquehanna valley lowland, there are areas of thicker glacial deposits that probably mark pre-Illinoian ice limits (Fig. 40.1, MIS 22 + line and MIS 16? Line) (Braun et al., 2008). The deposits are so weathered and eroded that they generally cover less than 10% of the land surface except where 'trapped' on dissolving carbonate bedrock (Braun, 1996, 1999, 2008; Braun et al., 2008). Most often just a few erratics remain as a scant lag on the ground surface. In the Salamanca re-entrant area, where erratics are very rare on the ground surface on the uplands, valley floor wells show the presence of at least two different clay sediment intervals (Lohman, 1939) that probably represent proglacial lake sediments. In north-western Pennsylvania, two pre-Illincian glacial advances are shown as two belts (Sevon and Braun, 1997), one as a belt of discontinuous deeply weathered glacial deposits (the Mapledale drift of White, 1969) and a second belt of rare erratics on the upland surface, well records of proglacial lake deposits under the valley floors, and drainage anomalies (Leverett, 1934; Leggette, 1936; Sevon, 1992; Braun, unpublished).

40.3. DISCUSSION OF THE DATING OF THE AGE OF THE GLACIAL LIMITS

The Late Wisconsinan limit in both north-eastern and north-western Pennsylvania is considered to be about 20–24,000 radiocarbon years in age from carbon 14 dates from adjacent states (Cotter et al., 1985; Lowell, 1991; Ridge,

2003). It is considered to be 24–28,000 calibrated or calendar years in age (Ridge, 2003). There have been no carbon 14 dates in Pennsylvania that directly provide the age of the Late Wisconsinan maximum. Previously, the difference in expression of the glaciation to either side of the Salamanca re-entrant caused some to think that the Wisconsinan limit in north-eastern Pennsylvania was somewhat older than that in north-western Pennsylvania (Denny and Lyford, 1963; Muller, 1977). That difference resulted from the relatively more weathered and eroded appearance of the 'drab' Olean drift in north-eastern Pennsylvania (MacClintock and Apfel, 1944) in contrast to the 'bright' Kent drift in north-western Pennsylvania (White, 1960). Remapping of the Late Wisconsinan limit by Crowl and Sevon (1980) and the carbon 14 dates mentioned above have convinced all that the Late Wisconsinan limit is the same age in both north-eastern and north-western Pennsylvania and that age is about 24–28 ka.

The next older glacial limit, presently considered to be of questionable Late Illinoian age (Braun, 1999, 2008; Braun et al., 2008), was originally dated by long distance correlation to type localities in the mid-western United States. Leverett (1934) correlated some of the older glacial deposits closest to the Wisconsinan terminus with the Illinoian of Illinois on the basis that they were distinctly more weathered and eroded than the Wisconsinan material but less weathered and eroded than glacial material farther to the south. Others have agreed that is a reasonable correlation (Sevon et al., 1975) and that the Marine Isotope record also supports the correlation (Braun, 1988, 1994), but there are no absolute dates to further support that correlation. Material along this limit in north-eastern Pennsylvania is more weathered and eroded than in north-western Pennsylvania and in the mid-west United States (Braun, 1988, 1994, 1999; Braun et al., 2008). It has been argued that the lithology of the material and the topographic setting in north-eastern Pennsylvania produces more weathering than in north-western Pennsylvania (Braun, 1988, 1994). Still though, the great amount of weathering and erosion in north-eastern Pennsylvania makes it doubtful that just 150,000 years (MIS 6 age) have elapsed since the material was deposited (Braun, 1999; Braun et al., 2008). The next older cold event that appears to be intense enough to bring ice south of the Late Wisconsinan limit is the MIS 10 or 12 (pre-Illinoian A or B) event. So presently, this glacial limit is considered to be of either of MIS 6 or MIS 10 or MIS12 age (see digital map and Fig. 40.1; Braun, 1999; Braun et al., 2008).

The older pre-Illinoian glacial limits are dated by their relative degree of erosion and by whether they have a reversed or normal magnetic polarity. These older materials cover only a few percent of the land surface except where trapped on dissolving carbonates and most of the material is colluvium derived from glacial deposits (Braun, 1996, 1999; Braun et al., 2008). Because of the long continued erosion and redeposition of these materials, weathering profiles are truncated and the overall weathering of these materials is not markedly different from the 'Late Illinoian' (MIS 6 or 10 or 12) materials (Braun, 1994; Braun et al., 2008).

The reversed magnetic polarity of pre-Illinoian proglacial lake deposits near the maximal limit of glaciation (Jacobson et al., 1988; Gardner et al., 1994; Sasowsky, 1994; Marine, 1997; Ramage et al., 1998) is the key evidence that the most extensive glaciation is of Early Pleistocene age. The strong MIS 22 or pre-Illinoian G event at about 850–880 ka was chosen by (Braun 1994, 1999; Braun et al., 2008) for the glacial maximum event. It is considered that with long continued erosion, to preserve any remnants of glaciation on the moderate relief Appalachian landscape, a younger Early Pleistocene event was more likely than an older Early Pleistocene or Pliocene event. For correlative deposits immediately to the east in New Jersey, a Late Pliocene or pre-Illinoian K (2.1 Ma) age has been argued by Stanford (1997) on the basis of Pliocene exotic plant taxa in proglacial lake sediments.

At five sites between the glacial maximum limit and the 'Late Illinoian' (MIS 6 or 10 or 12) limit, proglacial lake sediments have a normal magnetic polarity (Sasowsky, 1994; Braun and Saswosky, unpublished). This indicates that there are remnants of at least one and possibly more other Middle Pleistocene glacial advances in north-eastern Pennsylvania. A probable age for this advance is the especially cold MIS 16 (pre-Illinoian D) 650-ka event. If the 'Late Illinoian' (MIS 6 or 10 or 12) limit is actually Late Illinoian (MIS 6) in age, then there could be as many as three Middle Pleistocene events (MIS 10, 12 and 16) whose materials would retain a normal magnetic polarity.

40.4. EARLY PLEISTOCENE PSEUDO-MORAINE AT SELINSGROVE AND ALLENTOWN PENNSYLVANIA

At the Selinsgrove and Allentown sites, Leverett (1934) observed a subdued knob and kettle-like landscape and assumed that it represented a relict constructional glacial moraine of Late Illinoian (MIS 6) age. The Selinsgrove site caused him to project a 50-km long, few kilometre wide, ice tongue down the Susquehanna valley to Selinsgrove. Long tongues of ice would be a popular way of projecting pre-Wisconsinan glacier borders in the Ridge and Valley Province for later workers up through the 1980s. The problem is that such tongues of ice have nearly zero ice-surface profile gradients and require impossibly fluid glacial ice. Also the Late Wisconsinan border clearly showed that lobation of ice down the strike valleys in the region was only 3–8 km, close to what would be expected from a theoretical one-bar ice-surface profile and from profiles of existing glaciers (Braun, 1988). Surficial geology mapping in the Allentown area in the 1990s indicated that the 'moraine' is actually a pseudo-moraine that has a polygenetic origin only initially related to glaciation (Braun, 1996, 1999).

The glacial deposits under the pseudo-moraine sites have been undergoing weathering and erosion for at least 880,000 years (MIS 22) (Fig. 40.3). The degree of erosion of the deposits is most clearly seen in the slate and shale belt north of Allentown where only the broadest hilltops retain any glacial materials. On a few hilltops where glacial till thicker than 2 m remains, numerous shallow depressions form a 'patterned' ground effect that suggests a periglacial origin for the depressions (Braun, 1994, 1999; Braun et al., 2008). Where less than 1 m of glacial deposits overly the slate or shale bedrock, weathering has penetrated and rubified the bedrock 1–2 m below the glacial material. In most of the slate and shale belt, the glacial materials have been eroded completely from the hills and redeposited as colluvium in the valleys. Any knob and kettle topography has long since been removed from the landscape.

Allentown and Selinsgrove are underlain by carbonate bedrock and glacial diamicton remnants are thick and extensive, having been 'trapped' by 880,000 years of lowering of the landscape by dissolution (Fig. 40.3). Reviews of studies of carbonate denudation rates in Pennsylvania indicate that deposits this old should have been lowered on the order of 30 m or more (Sevon, 1989; Ciolkosz et al., 1995). There should be at least one to several metres of residuum from the dissolution of the carbonates under the 'let down' old glacial material (Sevon, 1989; Ciolkosz et al., 1995; White, 2000), and this is what is observed under the pseudo-moraine (Fig. 40.3) (Braun, 1996, 1999). Also there is a subtle but distinct pattern of thinner diamict remnants on the hilltops and thicker remnants in the valleys and especially in solution depressions (seen in both outcrop and borehole observations), indicating redistribution of the original glacial deposits. The pseudo-moraine does have a relatively thicker and more continuous glacial mantle than elsewhere and probably once had a genuine morainic topography about 30 m above the present landscape (Fig. 40.3). Now though the deeply eroded and 'let down' colluvium (collapse-uvium?) derived from glacial material is draped over underlying bedrock features (Fig. 40.3). In places on the present knob tops in the pseudo-moraine, colluvium derived from carbonate residuum overlies the colluvium derived from the glacial diamict (Braun, 1996, 1999). This suggests repeated episodes of topographic reversal as the landscape was lowered.

The present swell and sag topography is composed of smaller scale periglacial depressions that developed in the reworked glacial material and that are superimposed on larger-scale bedrock solution features (Braun, 1996, 1999). The numerous smaller scale depressions commonly contain wetlands and perennial ponds that make the pseudo-moraine landscape distinctly different from the surrounding more 'karstic' landscape. Thick sand and gravel deposits in ice

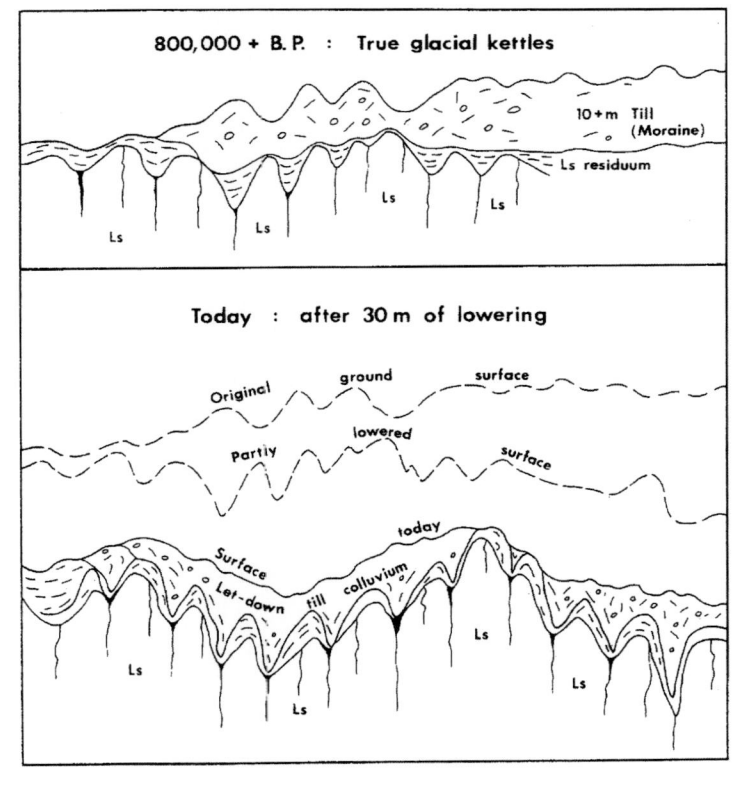

FIGURE 40.3 Schematic cross-section showing the lumpy landscape of the 'pseudo-moraine' developed by 880,000 years of limestone dissolution, as glaciation that has lowered the land surface 30 m or so. The 'kettles' are sinkholes with smaller periglacial landforms on their flanks (Braun, 1999, p. 37, Fig. 20).

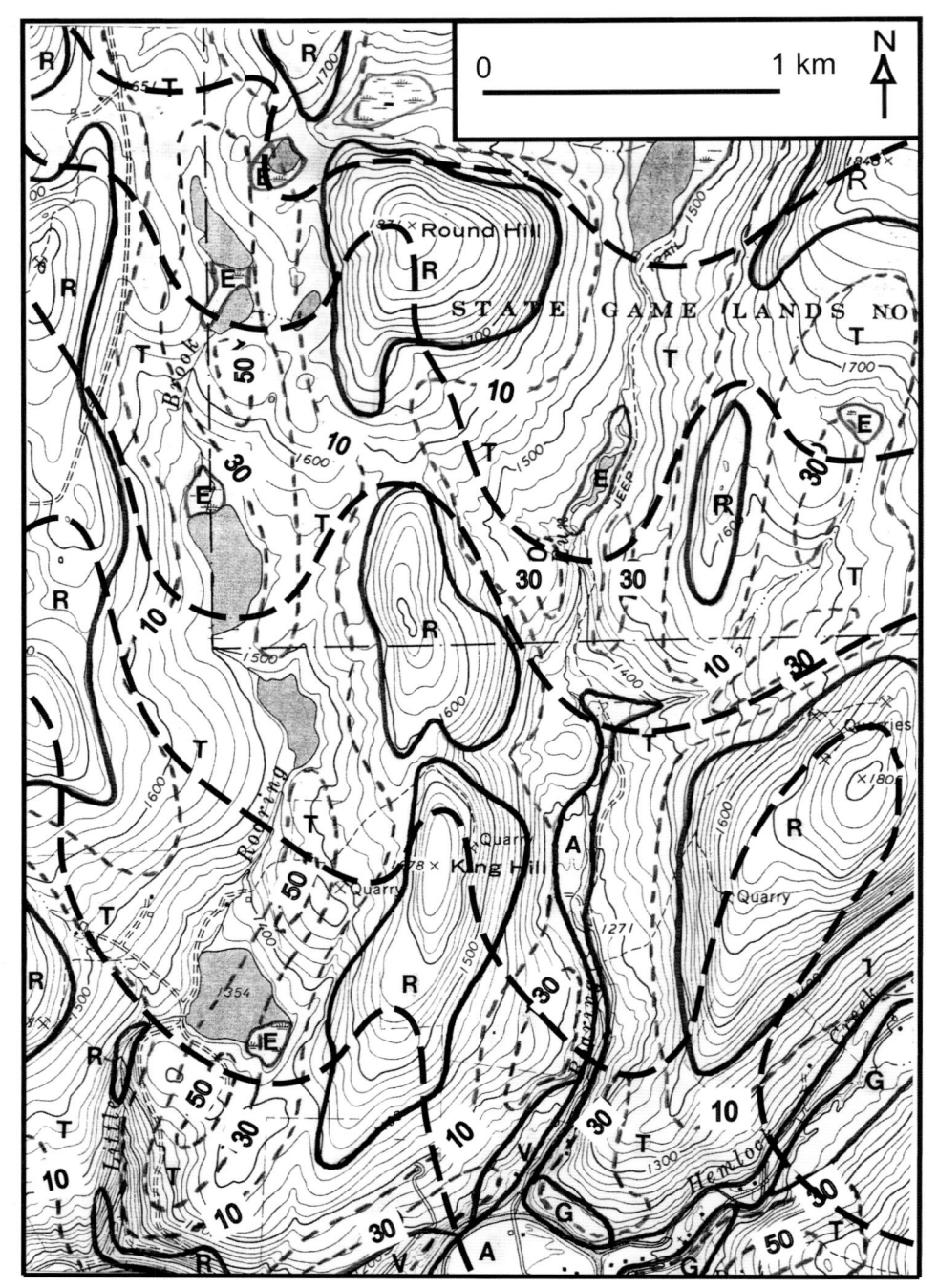

FIGURE 40.4 North-eastern Pennsylvania glacial deposit pattern of bedrock ridges projecting through valleys mantled with thick till. Individual valleys have a 'beaded' appearance from a series of till knobs and intervening lakes or wetlands spaced about 1 km apart. Heavy dashed lines are active ice margin positions thought to be responsible for the deposition of the till knobs. A, Alluvium; E, Wetlands; G, ice-contact stratified drift; R, Bedrock at or within 2 m of the surface; T, Till; V, Till underlain by varves. Glacial deposit thickness contours (isochores) at 10, 30 and 50 m. Topographic contour interval is 20 feet. Area shown is the north-eastern part of the Susquehanna, Pennsylvania 7.5′ quadrangle (Braun, 2006a, p. 252, Fig. 4).

marginal kames at the base of South Mountain at Allentown have been partly buried by extensive periglacially derived boulder colluvium (Braun, 1999) and further attest to the reshaping of the landscape by periglacial activity.

40.5. LATE WISCONSINAN TILL KNOBS FORMING 'BEADED VALLEYS' IN NORTH-EASTERN PENNSYLVANIA

Glacial retreat from valleys near parallel to ice flow in the moderate relief (300–500 m) Appalachian Plateau in north-eastern Pennsylvania was characterised by episodic deposition of till in a series of knobs that form 'beaded valleys' (Braun, 2002, 2006a). Individual valleys have a series of till knobs alternating with wetlands or lakes at a spacing of 1–5 km (Fig. 40.4). The ubiquity of chains of lakes in the region led Coates (1974) to call the area the Small Lakes Section of the Appalachian Plateau. Coates (1974, 1981) also described the common occurrence of similar till knobs in New York just north of Pennsylvania. Braun and 11 field assistants (1996–2001) have mapped, including deposit thicknesses, the entire 10,000 km^2 area characterised by 'beaded valleys'. That mapping has shown that the till knobs form systematic patterns across the region that, when constrained by striation orientations, permit the delineation of ice margin positions like heads-of-outwash do in New England or moraines do in the mid-west. South to north in each 7.5' quadrangle, there are from 5 to 15 till knobs and intervening wetlands or lakes. Outcrop and well data, while small and few in any individual knob, when put together from the entire region show that the till knobs are typically 30–50 m thick (maximum 100 m) (Fig. 40.4 thickness contours). Collectively, the outcrops also show the knobs to be cored by dense subglacial till with a thick wedge of supraglacial till (re-sedimented till) on their south sides, push structures on their north sides and an overall veneer of 'colluviated till' that thickens downslope on all sides. Glaciofluvial deposits are almost non-existent, usually showing as thin lenses or pockets.

The spacing of the till knobs is similar to that of heads of outwash in New England (Koteff, 1974; Koteff and Pessl, 1981) and suggests a similar systematic stagnation-zone retreat origin. The primary difference would be that the ice in north-eastern Pennsylvania would be retreating in a till rich rather than glaciofluvial rich environment. Active ice shearing over stagnate ice would stack till up to form the till knob mass. As the ice continued to thin, a mass of ice 1–5 km long would become stagnate and active ice shearing across the north side of the stagnant ice would start developing a new till knob. Melting stagnate ice on the south side of the original till knob would cause collapse of the south side of the knob and remobilisation of the till into a 're-sedimented' till mass.

40.6. OPEN QUESTIONS

1. The actual age of all of the pre-Late Wisconsinan glacial advances into PA.
2. How many pre-Wisconsinan glacial advances are present in PA?
3. Are all of the pre-Wisconsinan glacial advance found in both north-eastern Pennsylvania and north-western Pennsylvania?
4. The trace of pre-Wisconsinan glacial limits on the Appalachian Plateau south of the Salamanca re-entrant.
5. Is there any remaining evidence of Pliocene glacial advances in PA, particularly, the first and, particularly, strong events at about 2.5 Ma?
6. The precise mechanism of deposition of the Late Wisconsinan age till knobs in north-eastern Pennsylvania.

REFERENCES

Alden, W.C., Fuller, M.L., 1903. Pleistocene geology of the Elkland and Tioga quadrangles. In: Fuller, M.L., Alden, W.C. (Eds.), Elkland-Tioga Folio, Pennsylvania, U.S. Geological Survey Atlas of U.S. Folio, 93, Washington, DC, 6 maps, 9 pp.

Berg, T.M., Edmunds, W.E., Geyer, A.R., et al., (Compilers), 1980. Geologic map of Pennsylvania. Pennsylvania Geological Survey, 4th ser., Map 1, scale 1:250,000, 3 sheets.

Braun, D.D., 1988. Glacial geology of the Anthracite and North Branch Lowland Regions. In: Inners, J.D. (Ed.), Bedrock and Glacial Geology of the North Branch Susquehanna Lowland and the Eastern Middle Anthracite Field, Annual Field Conference of Pennsylvania Geologists, 53rd, Hazleton, PA, Guidebook, pp. 3–25.

Braun, D.D., 1989a. The use of proflacial lake and sluiceway sequences to determine ice margin positions on the Appalachian Plateau of north-central PA. Geol. Soc. Am. Abstr. Programs 21 (2), 6.

Braun, D.D., 1989b. Glacial and periglacial erosion of the Appalachians. In: Gardner, T.W., Sevon, W.D. (Eds.), Appalachian Geomorphology, Geomorphology, 2 (1–3). Elsevier, Amsterdam, pp. 233–258.

Braun, D.D., 1994. Late Wisconsinan to pre-Illinoian (G?) glacial events in eastern Pennsylvania. In: Braun, D.D. (Ed.), Late Wisconsinan to Pre-Illinoian (G?) Glacial and Periglacial Events in Eastern Pennsylvania, Guidebook for the 57th Field Conference, Friends of the Pleistocene, pp. 1–21. U.S. Geologic Survey Open File Rpt. 94-434.

Braun, D.D., 1996. Pseudo-morainic topography of the Allentown area of eastern Pennsylvania. In: Guidebook for the 15th Annual Field Trip of the Harrisburg Area Geological Society (available by writing: Harrisburg Area Geological Society, c/o Pennsylvania Geological Survey, P.O. Box 8453, Harrisburg, PA 17105-8453), 28 pp.

Braun, D.D., 1997. Physiography and Quaternary-history of the Scranton/Wilkes-Barre area. In: Inners, J.D. (Ed.), Geology of the Wyoming-Lackawanna Valley and Its Mountain Rim, Northeastern Pennsylvania, Annual Field Conference of Pennsylvania Geologists, 62nd, Scranton, PA, Guidebook, pp. 1–15.

Braun, D.D., 1999. Pleistocene geology of the Allentown area. In: Sevon, W.D., Fleeger, G.M. (Eds.), Economic and Environmental Geology and Topography in the Allentown Bethlehem Area, Annual Field Conference of Pennsylvania Geologists, 64th, Allentown, PA, Guidebook, pp. 31–40.

Braun, D.D., 2002. Quaternary history of the Tunkhannock—Great Bend region. In: Inners, J.D., Fleeger, G.M. (Eds.), Bedrock, Surficial, Economic, and Engineering Geology of the "Endless Mountains" Region, Susquehanna and Wyoming Counties, Annual Field Conference of Pennsylvania Geologists, 67th, Harrisburg, PA, Guidebook, pp. 32–38.

Braun, D.D., 2004a. Surficial geology of the picture Rocks 7.5-minute quadrangle, Lycoming and Sullivan Counties, Pennsylvania. Pennsylvania Geological Survey, 4th ser., Open-File Report, OFSM 04–01.1, 23 pp., Portable Document Format (PDF).

Braun, D.D., 2004b. Surficial geology of the Sonestown 7.5-minute quadrangle, Lycoming and Sullivan Counties, Pennsylvania. Pennsylvania Geological Survey, 4th ser., Open-File Report OFSM 04–02.1, 21 pp., Portable Document Format (PDF).

Braun, D.D., 2006a. Deglaciation of the Appalachian Plateau, northeastern Pennsylvania—till shadows, till knobs forming "beaded valleys": revisiting systematic stagnation-zone retreat. In: Fleisher, P.J., Knuepfer, P.L.K., Butler, D.R. (Eds.), Ice Sheet Geomorphology—Past and Present Processes and Landforms, Geomorphology, 75 (1–2). Elsevier, Amsterdam, pp. 248–265.

Braun, D.D., 2006b. Surficial geology of the Bodines 7.5-minute quadrangle, Lycoming County, Pennsylvania. Pennsylvania Geological Survey, 4th ser., Open-File Report, OFSM 06–09.1, 23 pp., Portable Document Format (PDF).

Braun, D.D., 2007a. Surficial geology of the Elk Grove 7.5-minute quadrangle, Sullivan, Columbia, and Lycoming Counties, Pennsylvania. Pennsylvania Geological Survey, 4th ser., Open-File Report, OFSM 07–09.0, 21 pp., Portable Document Format (PDF).

Braun, D.D., 2007b. Surficial geology of the Trout Run 7.5-minute quadrangle, Lycoming County, Pennsylvania. Pennsylvania Geological Survey, 4th ser., Open-File Report, OFSM 07-11.0, 26 pp., Portable Document Format (PDF).

Braun, D.D., 2008. Surficial geology of the Sybertsville 7.5-minute quadrangle, Luzerne County, Pennsylvania. Pennsylvania Geological Survey, 4th ser., Open-File Report, OFSM 08-18.0, 22 pp., Portable Document Format (PDF).

Braun, D.D., Ciolkosz, E., Drohan, P., Sevon, W., 2008. The Pleistocene record in the middle and lower Susquehanna River Basin and the longer term evolution of the Susquehanna Basin landscape. In: 20th Biennial Meeting of the American Quaternary Association, University Park, PA, Fieldtrip Guidebook 58 pp.

Campbell, M.R., 1903. Geographic development of northern Pennsylvania and Southern New York. Geol. Soc. Am. Bull. 14, 553–573.

Carll, J.F., 1880. Geology of the oil regions of Warren, Venango, Clarion and Butler Counties. Pennsylvania Geological Survey, 2nd ser., Report, III, xxxiv, 482 pp.

Chamberlin, T.C., 1883. Terminal moraine of the second glacial epoch. U.S. Geological Survey, 3rd Annual Report, 291–402.

Ciolkosz, E.J., Cronce, R.C., Sevon, W.D., 1986. Periglacial features in Pennsylvania. Pennsylvania State University, Agronomy Series, 92, 15 pp.

Ciolkosz, E.J., Cronce, R.C., Sevon, W.D., Waltman, W.J., 1995. Genesis of Pennsylvania's limestone soils. Pennsylvania State University Agronomy Series, 135, 28 pp.

Clark, G.M., Ciolkosz, E.J., 1988. Periglacial geomorphology of the Appalachian Highlands and Interior Highlands south of the glacial border—a review. Geomorphology, 1, 191–220.

Clark, G.M., Behling, R.E., Braun, D.D., Ciolkosz, E.J., Kite, J.S., Marsh, B., 1992. Central Appalachian periglacial geomorphology. In: International Geographical Congress, 27th, Guidebook. Pennsylvania State University Agronomy Series, 120, 248 pp.

Coates, D.R., 1966. Discussion of K.M. Clayton "Glacial Erosion in the Finger Lakes Region (New York, U.S.A.)" Z. Geomorphol., 10, 469–474.

Coates, D.R., 1974. Reappraisal of the glaciated Appalachian Plateau. In: Coates, D.R. (Ed.), Glacial Geomorphology. Annual Geomorphology Symposia, 5th, Publications in Geomorphology, SUNY at Binghamton, New York, pp. 205–244.

Coates, D.R., 1981. Geomorphology of South-Central New York. In: Enos, P.E. (Ed.), Guidebook for Field Trips in South-Central New York, Annual Meeting New York State Geological Association, 53rd, Binghamton, New York, pp. 171–200.

Coates, D.R., King, C.A.M., 1973. Glacial geology of Great Bend and adjacent region. In: Coates, D.R. (Ed.), Glacial Geology of the Binghamton-Western Catskill Region, Contribution 3, Publications in Geomorphology, SUNY at Binghamton, New York, pp. 3–30.

Cotter, J.F.P., Ridge, J.H., Evenson, E.B., Sevon, W.D., 1985. The late Wisconsinan history of the Great Valley, Pennsylvania and New Jersey, and the age of the "terminal moraine" In: Evenson, E.B. (Ed.), Friends of the Pleistocene Field Conference, 48th, Guidebook, pp. 1–59.

Crowl, G.H., 1981. Glaciation in north-central Pennsylvania and the Pine Creek Gorge. In: Berg, T.M. et al., (Ed.), Geology of Tioga and Bradford Counties, Pennsylvania, Annual Field Conference of Pennsylvania Geologists, 46th, Wellsboro, PA, Guidebook, pp. 39–44.

Crowl, G.H., Sevon, W.D., 1980. Glacial border deposits of Late Wisconsinan Age in Northeastern Pennsylvania. Pennsylvania Geological Survey, 4th ser., General Geology Report, 71, 68 pp.

Crowl, G.H., Sevon, W.D., 1999. Quaternary. In: Shultz, C.H. (Ed.), The Geology of Pennsylvania, Pennsylvania Geological Survey and Pittsburgh Geological Society, Special Publication, 1, pp. 225–231.

Denny, C.S., Lyford, W.H., 1963. Surficial geology and soils of the Elmira-Williamsport Region, New York and Pennsylvania. U.S. Geological Survey Professional Paper, 379, 60 pp.

Gardner, T.W., Braun, D.D., Pazzaglia, F.L., Sevon, W.D., 1993. Late Cainozoic landscape evolution of the Susquehanna River basin. In: International Geomorphology Conference, 3rd, Post-Conference Field Trip Guidebook, 288 pp.

Gardner, T.W., Sasowsky, I.D., Schmidt, V.A., 1994. Reversed polarity glacial sediments, West Branch Susquehanna River Valley, Central Pennsylvania. Quatern. Res. 42, 131–135.

Jacobson, R.B., Elston, D.P., Heaton, J.W., 1988. Stratigraphy and magnetic polarity of the high terrace remnants in the upper Ohio and Monongahela Rivers in West Virginia, Pennsylvania, and Ohio. Quatern. Res. 29, 216–232.

Kaktins, U., Delano, H.L., 1999. Drainage basins. In: Shultz, C.H. (Ed.), The Geology of Pennsylvania, Pennsylvania Geological Survey and Pittsburgh Geological Society, Special Publication, 1, pp. 378–390.

Koteff, C., 1974. The morphologic sequence concept and deglaciation of southern New England. In: Coates, D.R. (Ed.), Glacial Geomorphology, Annual Geomorphology Symposia, 5th, Publications in Geomorphology, S.U.N.Y. at Binghamton, New York, pp. 121–144.

Koteff, C., Pessl, F., Jr., 1981. Systematic ice retreat in New England. U.S. Geological Survey Professional Paper, 1179, 20 pp.

Leggette, R.M., 1936. Groundwater in northwestern Pennsylvania. Pennsylvania Geological Survey, 4th ser., W 3, iv, 215 pp.

Lessig, H.D., 1963. Calcutta Silt, a very early Pleistocene deposit, upper Ohio Valley. Geol. Soc. Am. Bull. 74, 129–139.

Leverett, F., 1902. Glacial formations and drainage features of the Erie and Ohio Basins. U.S. Geologic Survey Monograph, 41, 802 pp.

Leverett, F., 1934 Glacial deposits outside the Wisconsin terminal moraine in Pennsylvania. Pennsylvania Geological Survey, 4th ser., General Geology Report, 7, 123 pp.

Lewis, H.C., 1884. Report on the terminal moraine in Pennsylvania and Western New York. Pennsylvania Geological Survey, 2nd ser., Report on Progress, Z. lvi, 299 pp.

Lohman, S.W., 1939. Ground water in North-Central Pennsylvania, with analyses by E.W. Lohr. Pennsylvania Geological Survey, 4th ser., Bulletin, W6, 219 pp.

Lowell, T., 1991. Chronologic constraints on late Quaternary events in Ohio. Geol. Soc. Am. Abstr. Programs 23 (3), 18.

MacClintock, P., Apfel, E.T., 1944. Correlation of the drifts of the Salamanca Re-entrant, New York. Geol. Soc. Am. Bull. 55, 1143–1164.

Marchand, D.E., 1978. Quaternary deposits and Quaternary history. In: Marchand, D.E., Ciolkosz, E.J., Bucek, M.F., Crowl, G.H. (Eds.), Quaternary Deposits and Soils of the Central Susquehanna Valley of Pennsylvania, Friends of the Pleistocene Field Conference, 41st, Pennsylvania State University Agronomy Series, 52, pp. 1–19.

Marine, J.T., 1997 Terrace deposits associated with ancient Lake Monongahela in the Lower Allegheny Drainage, Western Pennsylvania. Unpublished M.S. thesis, University of Pittsburgh, 182 pp.

Marsh, B., 1999. Paleoperiglacial landscapes of Central Pennsylvania. Guidebook for the 62nd Northeast Friends of the Pleistocene Field Trip, 69 pp.

Muller, E.H., 1977. Quaternary geology of New York, Niagara sheet. New York State Museum and Science Service Map and Chart Series, 28, scale 1:250,000.

Peltier, L.C., 1949. Pleistocene terraces of the Susquehanna River. Pennsylvania Geological Survey, 4th ser., General Geology Report, 23, 158 pp.

Ramage, J.M., Gardner, T.W., Sasowasky, I.D., 1998. Early Pleistocene Glacial Lake Lesley, West branch Susquehanna River valley, central Pennsylvania. Geomorphology, 22, 19–37.

Richmond, G.M., Fullerton, D.S., 1986. Summation of Quaternary glaciations in the United States of America. In: Sibrava, V., Bowen, D.Q., Richmond, G.M. (Eds.), Quaternary Glaciations in the Northern Hemisphere, Quater. Sci. Rev. 5. Pergamon Press, Oxford, Great Britian, pp. 183–196

Ridge, J.C., 2003. The last deglaciation of the northeastern United States: a combined varve, paleomagnetic, and calibrated ^{14}C chronology. In: Cremeens, D.I., Hart, J.P. (Eds.), Geoarechaeology of Landscapes in the Glaciated Northeast: New York State Museum Bulletin, 497. Albany, New York, pp. 15–45.

Ridge, J.C., Braun, D.D., Evenson, E.B., 1990. Does Altonian drift exist in Pennsylvania and New Jersey? Quatern. Res. 33 (2), 253–258.

Sasowsky, I.D., 1994. Paleomagnetism of glacial sediments from three locations in eastern Pennsylvania. In: Braun, D.D. (Ed.), Late Wisconsinan to Pre-Illinoian (G?) Glacial and Periglacial Events in Eastern Pennsylvania, Guidebook for the 57th Field Conference, Friends of the Pleistocene, pp. 21–24. U.S. Geologic Survey Open File Report 94-434.

Sevon, W.D., 1989. Erosion in the Juniata River drainage basin, Pennsylvania. In: Gardner. T.W., Sevon, W.D. (Eds.), Appalachian Geomorphology, Geomorphology, 2. Elsevier, Amsterdam, pp. 303–318.

Sevon, W.D., 1992. Surficial Geology and Geomorphology of Warren County, PA. In: Geology of the Upper Allegheny River Region in Warren County, Northwestern Pennsylvania, Annual Field Conference of Pennsylvania Geologists, 57th, Warren, PA, Guidebook, 204 pp.

Sevon, W.D., Braun, D.D., 1997. Glacial deposits of Pennsylvania, second ed. Pennsylvania Geological Survey, 4th ser., Map, 59, scale 1:2,000,000.

Sevon, W.D., Crowl, G.H., Berg, T.M., 1975. The Late Wisconsinan drift border in Northeastern Pennsylvania. In: Annual Field Conference of Pennsylvania Geologists, 40th, Bartonsville, PA, Guidebook, 108 pp.

Shepps, V.C., White, G.W., Droste, J.B., Sitler, R.F., 1959. Glacial geology of Northwestern Pennsylvania. Pennsylvania Geological Survey, 4th ser., General Geology Report, 32, 59 pp.

Stanford, S.D., 1997. Pliocene-Quaternary geology of northern New Jersey: an overview. In: Stanford, S.D., Witte, R.W. (Eds.), Pliocene-Quaternary Geology of Northern New Jersey, Guidebook for the 60th Annual Reunion of the Northeastern Friends of the Pleistocene, 1-1 to 1-26.

White, I.C., 1896. Origin of the high terrace deposits of the Monongahela River. Am. Geol. 18, 368–379.

White, G.W., 1960. Classification of glacial deposits in northeastern Ohio. U.S. Geol. Surv. Bull. 1121-A, A1–A12.

White, G.W., 1969. Pleistocene deposits of the north-western Allegheny Plateau. USA Quaterly Jour. Geo. Soc. London, 124, 131–151.

White, W.B., 2000. Dissolution of limestone from field observation. In: Kimchouk, A.B., Ford, D.C., Palmer, A.N., Dreybrodt, W. (Eds.), Speleogenesis, Evolution of Karst Aquifers. January 2000 ed. National Speleological Society, Huntsville. Alabama, pp. 149–155.

White, G.W., Totten, S.M., Gross, D.L., 1969. Pleistocene stratigraphy of Northwestern Pennsylvania. Pennsylvania Geological Survey, 4th ser., General Geology Report, 55, 88 pp.

Willard, B., 1932. The Devonian section at Selinsgrove Junction, Pennsylvania. Am. Midl. Nat. 13 (4), 222–235.

Williams, E.H., Jr., 1894. The age of the extramoraine fringe in eastern Pennsylvania. Am. J. Sci. 47, 34–37.

Williams, E.H., Jr., 1902. Kansan glaciation and its effects on the river systems of northern Pennsylvania. Proc. Wyo. Hist. Geol. Soc. 7, 21–28.

Chapter 41

Glaciation of Western Washington, USA

Stephen C. Porter
Department of Earth and Space Sciences, University of Washington, Seattle, Washington 98195, USA

41.1. CORDILLERAN ICE SHEET IN WESTERN WASHINGTON

During at least six glaciations, the Puget Lobe of the Cordilleran Ice Sheet advanced southward from British Columbia (BC) and left a record of its passage in the Quaternary stratigraphy and the present landscape. During successive glaciations, the lobe originated in the Coast Mountains and initially consisted of valley and fjord glaciers that coalesced to form a continuous expansive piedmont lobe at the latitude of Vancouver, BC (Clague, 1989). Advancing southward, the ice front encountered the northeastern margin of the Olympic Mountains, where it divided into an eastern lobe that flowed south into the Puget Lowland of Washington, and a western lobe that flowed west along the Strait of Juan de Fuca to the Pacific Ocean (Fig. 41.1). Because the Quaternary record of the southwestern sector of the ice sheet is best documented in the Puget and Fraser lowlands, these areas are emphasised here.

The latest interval of glacier expansion coincided broadly with Marine Isotope Stage (MIS) 2 and is known as the Fraser Glaciation (Armstrong et al., 1965). The glacier left a complex stratigraphical record in the Fraser and Puget lowlands that includes proglacial lake sediments

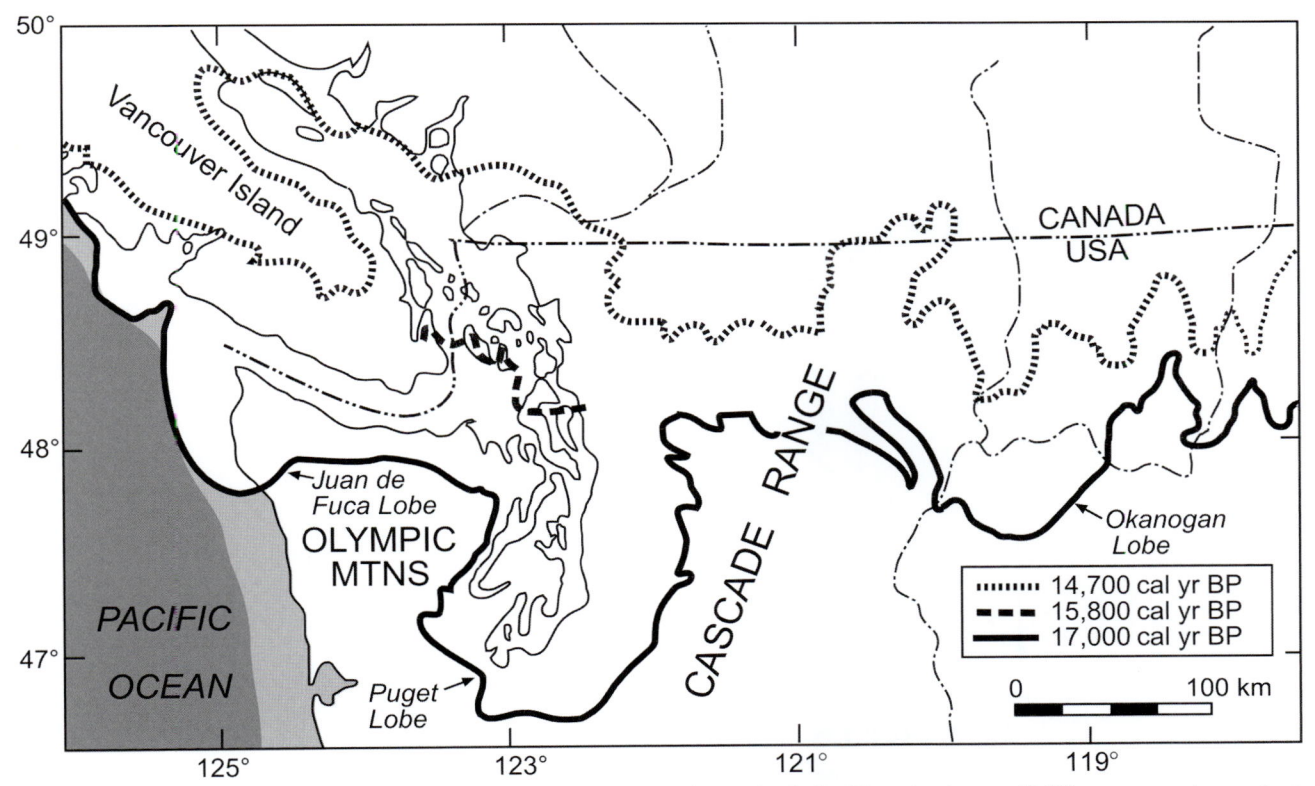

FIGURE 41.1 Map of northern Washington and south-western British Columbia showing limit of Fraser ice sheet ca. 17,000 years ago, and approximate limits during two recessional phases (based on Clague, 1989; Dethier et al., 1995; Thorson, 1980).

and outwash, till, ice-contact stratified drift, postglacial deltas, glacialmarine drift and recessional outwash. The modern landscape is strongly linear. The long axes of ubiquitous drumlinoid hills generally parallel bedrock striations, grooves and till fabrics. No distinct, continuous terminal moraine was built, but a broad belt of dead-ice terrain, major outwash plains and meltwater channels occurs near the southern ice limit. The eastern and western limits of the lobe, where the glacier impinged on the bordering mountain fronts, are identified by crystalline erratics of BC provenance and local delta moraines built into mountain valleys. The ice limits define a lobe that was about 95 km wide and increased in altitude from ca. 200 m at its southern limit to ca. 1000 m near Seattle and 1600 m near Vancouver, BC (Thorson, 1980; Porter and Swanson, 1998).

The advance and early retreat history of the Puget Lobe has been reconstructed using limiting ^{14}C ages (Booth, 1987; Porter and Swanson, 1998). The advancing ice margin passed the latitude of the international boundary about 19,000 years ago and reached its southern limit close to 17,000 years ago (Fig. 41.2). The average rate of advance was about 135 m/a. The retreat of the glacier was more rapid and non-uniform. Sectors of the ice margin fronted on meltwater lakes that would have generated calving embayments, leaving grounded ice remnants on the adjacent uplands. The largest proglacial lake, Lake Russell, reached depths of 300 m or more and was up to 10 km wide. By ca. 16.5 ka ago, the ice front had retreated past the latitude of Seattle, and by 15.8 ka ago, it reached the San Juan Islands, north of Puget Sound, where it briefly stabilised before continuing to retreat into Canada (Dethier et al., 1995; Fig. 41.2).

Evidence of pre-Fraser advances of the Puget Lobe is found on Whidbey Island, near the centre of the Puget Lowland (Easterbrook, 1994). Here, two drifts underlie Fraser drift, from which they are separated by Olympia nonglacial sediments.

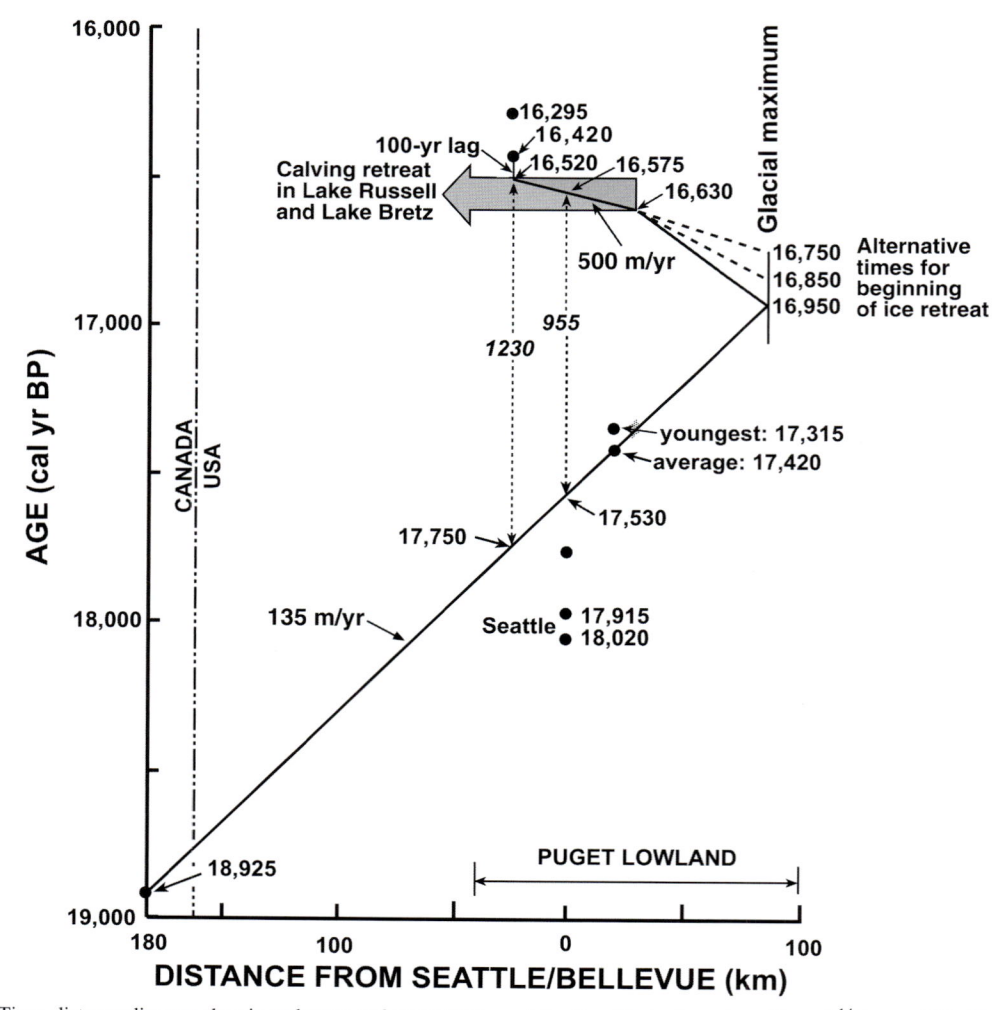

FIGURE 41.2 Time–distance diagram showing advance and retreat history of the Puget Lobe (ages in calibrated ^{14}C years BP). Mean retreat rate was several times that during the advance due to calving of ice margin in a series of proglacial lakes. Residence time of the ice at the latitude of Seattle was ca. 955 years (from Porter and Swanson, 1998; Fig. 41.4).

The younger drift (Possession) overlies peat- and wood-bearing fluvial and lacustrine sediments of the Whidbey Formation, which is presumed to have been deposited during the last-interglacial *sensu lato* (i.e. MIS 5). Available luminescence and amino acid dates imply that Possession Drift was deposited 70,000–80,000 years ago (Easterbrook, 1994; K. Troost, personal communication), which would make it broadly equivalent with early MIS 4. The southern limit of this drift is not known, but it may not extend much beyond the latitude of Seattle (Haase, 1987).

The older drift on Whidbey Island is named for Double Bluff. It antedates the Whidbey Formation and is the oldest glacial deposit exposed in the northern Puget Lowland. Amino acid and luminescence dates for the drift range widely, from ca. 111 to 289 ka ago. Although Double Bluff Drift may correlate broadly with MIS 6, such a correlation has not yet been demonstrated convincingly. Weathered drift beyond the Fraser limit in the southern Puget Lowland has been interpreted to correlate with Double Bluff drift (Lea, 1984), but at present there are no dates to support this correlation.

Still older advances of the ice sheet in western Washington are recorded in glacial drift exposed in the southeastern Puget Lowland (Crandell et al., 1958). There, evidence of three ice-sheet glaciations (Orting (oldest), Stuck and Salmon Springs) is found beneath deposits of the Fraser advance, and all pre-date the Brunhes–Matuyama boundary (ca. 780 ka ago) (Easterbrook, 1994; Fig. 41.3). None apparently extended beyond the limit of the Fraser-age glacier.

Evidence of late-glacial halts or readvances of the retreating ice margin at the end of the last glaciation has been reported on both sides of the international boundary. However, the chronology of the ice limits is disputed. Kovanen and Easterbrook (2002) have described evidence of three late-glacial ice advances in northernmost Washington and argue that one of them correlates with the European Younger Dryas interval (ca. 12.9–11.7 ka ago). Clague et al. (1997) have interpreted evidence in nearby southernmost BC as indicating two advances but conclude that both pre-dated Younger Dryas time. They interpret radiocarbon dates of basal gyttja in two low-altitude post-glacial lakes in the canyon of the Fraser River near Yale, BC, a major source of late-glacial ice in the Fraser Lowland, as implying ice-free conditions well north of the late-glacial moraines by the beginning of Younger Dryas time.

FIGURE 41.3 Pleistocene stratigraphical sequences in the northern and southern Puget Lowland, showing inferred correlation of the uppermost units with marine isotopic stages and approximate ages in millions of years for older units that fall within the Matuyama Polarity Chron.

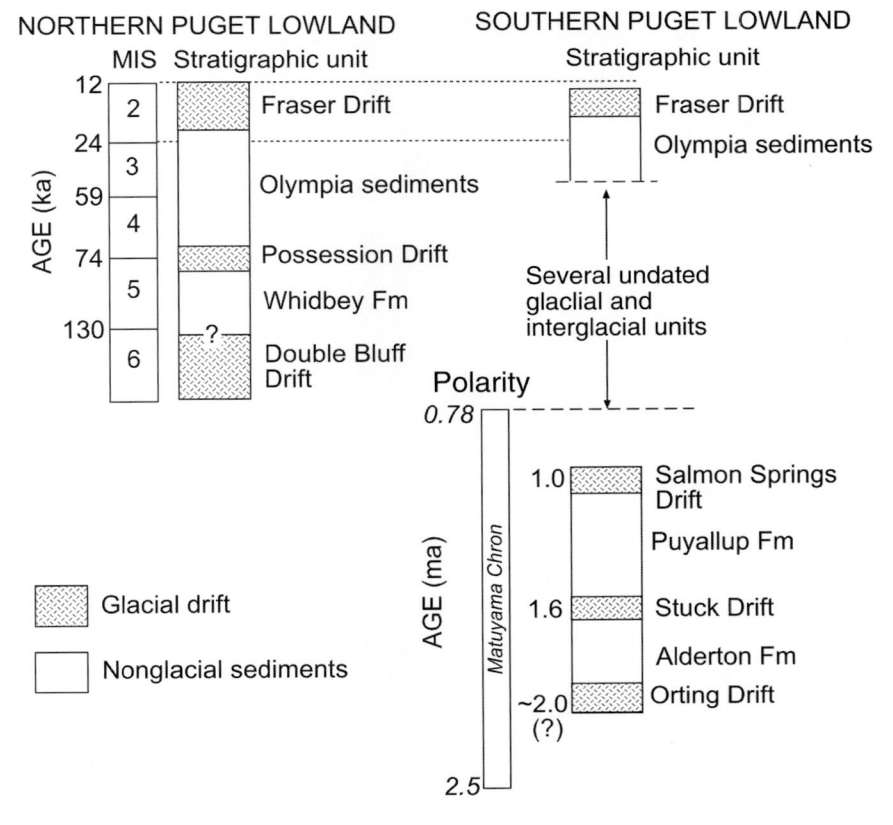

41.2. GLACIATION OF THE CASCADE RANGE

Quaternary glacial limits have been mapped in most Cascade valleys, and the moraine sequences are generally comparable. Among the key areas in the southern Cascades are the major valleys draining Mount Rainier volcano where Crandell and Miller (1974) mapped the distribution of four drifts. The youngest, named Evans Creek, was deposited by valley glaciers 24–64 km long during the Fraser Glaciation. Older deposits lying downvalley [Wingate Hill (oldest) and Hayden Creek] record advances of glaciers that were as much as 105 km long. No closely limiting radiometric ages have yet been obtained for the Evans Creek drift, which is assumed to date to MIS 2, nor for the older moraines, which have been tentatively correlated with MIS 4 (Early Wisconsinan) and pre-Wisconsinan ($\geq$ MIS 6), respectively. A late-glacial ice advance deposited McNeeley drift in the upper valleys. Heine (1998) assigned two McNeeley-age moraines to pre- and post-Younger Dryas advances based on radiocarbon ages and tephrochronology.

The most easily identifiable moraine successions in the Cascades lie east of the range crest in valleys where dense forests of the moist western slope give way to open forests and prairie vegetation of the drier lee-side valleys. Bouldery moraines in the valley of Icicle Creek have been mapped in detail and dated by using ^{36}Cl method (Swanson and Porter, 2000; Fig. 41.4). In this valley, two moraines of the last glaciation (Leavenworth) have mean ages of 19.1 and 16.0 ka, and two late-glacial (pre-Younger Dryas) advances have ages of ca. 13.3 and 12.5 ka. A more extensive moraine

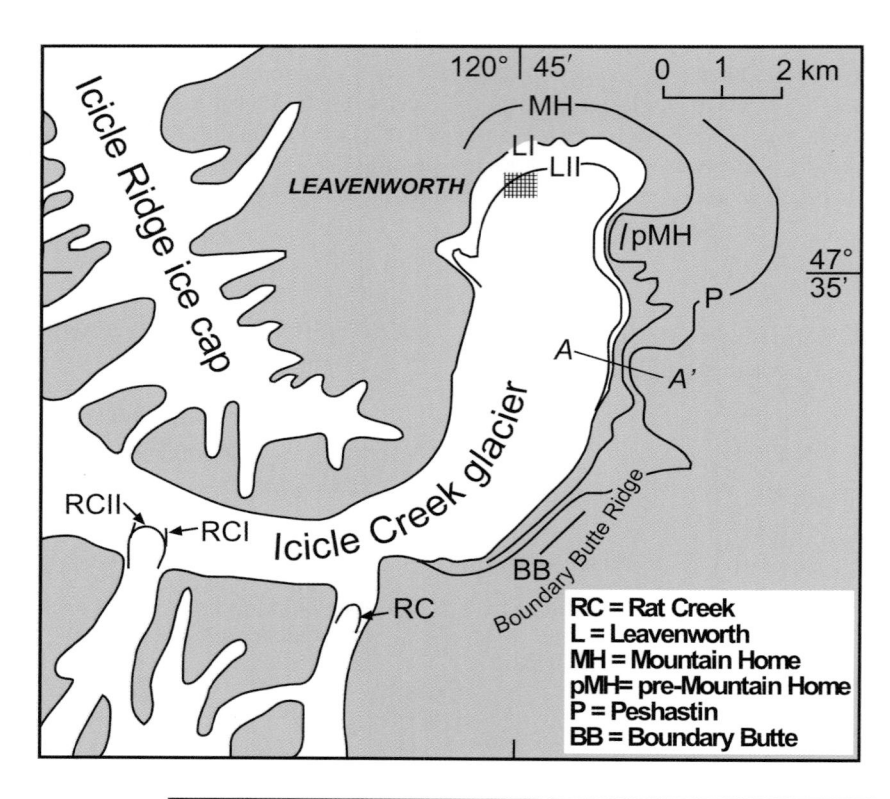

FIGURE 41.4 Map showing the distribution of lateral and terminal moraines of the Icicle Creek glacier in the south-eastern North Cascades. Cosmogenic isotope ages suggest that all but the Boundary Butte moraine post-date the peak of the last interglacial (Marine Isotope Substage 5e).

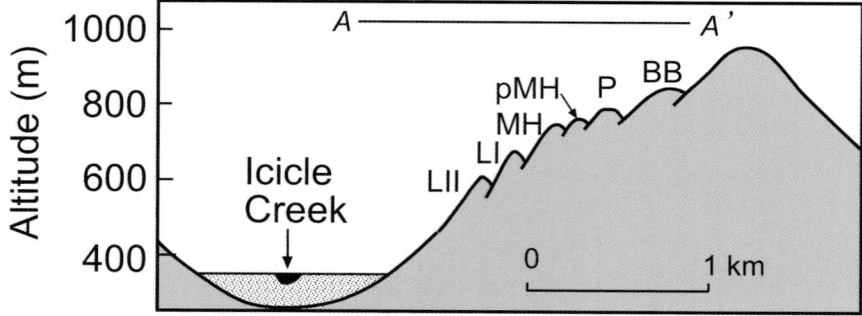

(Mountain Home) has an age of 70.9 ka, and two older moraines date to ca. 93.0 and 108.0 ka ago. The latter represents the most extensive Late Pleistocene ice advance in the valley, and its age implies correlation with MIS 5d. A still-older moraine (Boundary Butte), in which most granite boulders are weathered to grus, has not yet been dated but is older than the last-interglacial and probably is at least 165,000 years old (Middle Pleistocene). A similar moraine succession has been mapped in the upper Yakima Valley to the south (Porter, 1976). The five youngest prominent moraines are comparable in relative position to those in Icicle Creek valley and have similar ^{36}Cl ages (Fig. 41.4).

41.3. GLACIATION OF THE OLYMPIC MOUNTAINS

Long glaciers advanced down valleys of the Olympic Mountains during at least six major glacier advances. The record of alpine advances in the eastern Olympics is poorly known, both because of a dense vegetation cover and because alpine deposits are overlain by Fraser drift along the western margin of the Puget Lobe. Piedmont lobes flowing south from the mountain front have been mapped, but as yet there is no reliable chronology for these glaciations. Thackray (2001) has studied ice limits in two major valleys of the western Olympics, where the longest glaciers nearly reached, or terminated beyond, the present coastline. Abundant organic matters, including wood, peat and lake sediments in the glacial succession, have helped to date the youngest part (<50 ka old) of this record. A key stratigraphical horizon is a marine beach deposit exposed in sea cliffs along the Olympic coast that dates to the last-interglacial, based on an amino acid age of ca. 100 ka obtained from solitary corals in the deposits (Kvenvolden et al., 1979).

Six moraine systems were mapped and dated. The youngest three (Hoh Oxbow 3 and Twin Creeks 1 and 2) fall within MIS 2; the Hoh Oxbow 3 is bracketed by dates of 22.0 and 19.3 ^{14}C ka BP; and the Twin Creeks I is restricted to 19.1–18.3 ^{14}C ka BP. The more extensive Oxbow 1 and 2 date to 42.5–35.0 and 30.8–26.3 ^{14}C ka BP, respectively, implying correlation with MIS 3. An older, more extensive drift (Lyman Rapids) post-dates the MIS 5 shoreline but pre-dates ca. 54.0 ^{14}C ka BP; it may correlate either with MIS 4 or with part of MIS 5. The age of a still older drift, exposed beneath the last-interglacial beach deposits, is unknown.

REFERENCES

Armstrong, J.E., Crandell, D.R., Easterbrook, D.J., Noble, J.B., 1965. Late Pleistocene stratigraphy and chronology in southwestern British Columbia and northwestern Washington. Geol. Soc. Am. Bull. 76, 321–330.

Booth, D.B., 1987. Timing and processes of deglaciation along the southern margin of the Cordilleran ice sheet. In: Ruddiman, W.F., Wright Jr., H.E. (Eds.), North America and Adjacent Oceans During the Last Glaciation, vol. K-3. Geological Society of America, Boulder, pp. 71–90.

Clague, J.J., compiler, 1989. Quaternary geology of the Canadian Cordillera. In: Fulton, R.J. (Ed.), Quaternary Geology of Canada and Greenland, The Geology of North America, vol. K-1. Geological Society of America, Boulder, pp. 16–96.

Clague, J.J., Mathewes, R.W., Guilbault, J.-P., Hutchinson, I., Ricketts, B.D., 1997. Pre-Younger Dryas resurgence of the southwestern margin of the Cordilleran ice sheet, British Columbia, Canada. Boreas 26, 261–278.

Crandell, D.R., Miller, R.D., 1974. Quaternary stratigraphy and extent of glaciation in the Mount Rainier region, Washington. In: U.S. Geological Survey Professional Paper 847, 59pp.

Crandell, D.R., Mullineaux, D.R., Waldron, H.H., 1958. Pleistocene sequence in the southeastern part of the Puget Sound Lowland, Washington. Am. J. Sci. 256, 384–397.

Dethier, D.P., Pessl Jr., F., Kueler, R.F., Balzarini, M.A, Pevear, D.R., 1995. Late Wisconsinan glaciomarine deposition and isostatic rebound, northern Puget Lowland, Washington. Geol. Soc. Am. Bull. 107, 1288–1303.

Easterbrook, D.J., 1994. Chronology of pre-late Wisconsin Pleistocene sediments in the Puget Lowland, Washington. Washington Div. Mines Earth Resour. Bull. 80, 191–206.

Haase, P.C., 1987. Glacial stratigraphy and landscape evolution of the north-central Puget Lowland, Washington. Unpublished M.S. Thesis, University of Washington.

Heine, J.T., 1998. Extent, timing, and climatic implications of glacier advances Mount Rainier, Washington, USA, at the Pleistocene/Holocene transition. Quatern. Sci. Rev. 17, 1139–1148.

Kovanen, D.J., Easterbrook, D.J., 2002. Timing and extent of Allerød and Younger Dryas Age (ca. 12,500-10,000 ^{14}C yr B.P.) oscillations of the Cordilleran Ice Sheet in the Fraser Lowland, western North America. Quatern. Res. 57 (2), 208–224.

Kvenvolden, K.A., Blunt, D.J., Clifton, H.E., 1979. Amino-acid racemization in Quaternary shell deposits at Willapa Bay, Washington. Geochim. Cosmochim. Acta 43 (9), 1505–1520.

Lea, P.D., 1984. Pleistocene glaciation at the southern margin of the Puget Lobe, western Washington. Unpublished M.S. Thesis, University of Washington.

Porter, S.C., 1976. Pleistocene glaciation in the southern part of the North Cascade Range, Washington. Geol. Soc. Am. Bull. 87, 61–75.

Porter, S.C., Swanson, T.W., 1998. Radiocarbon age constraints on rates of advance and retreat of the Puget Lobe of the Cordilleran Ice sheet during the last glaciation. Quatern. Res. 50, 205–213.

Swanson, T.W., Porter, S.C., 2000. ^{36}Cl Evidence for maximum Late Pleistocene glacier extent in the Cascade Range during Marine Oxygen Isotope Substage 5d. Geological Society of America Abstracts with Program, Reno, Nevada, A-472.

Thackray, G.D., 2001. Extensive early and middle Wisconsin glaciation on the western Olympic Peninsula, Washington, and the variability of Pacific moisture delivery to the northwestern United States. Quatern. Res. 55, 257–270.

Thorson, R.M., 1980. Ice-sheet glaciation of the Puget Lowland, Washington, during the Vashon Stade (Late Pleistocene). Quatern. Res. 13, 303–321.

Chapter 42

The Quaternary of Wisconsin: An Updated Review of Stratigraphy, Glacial History and Landforms

Kent M. Syverson[1,]* and Patrick M. Colgan[2]

[1] Department of Geology, University of Wisconsin, Eau Claire, Wisconsin 54702, USA
[2] Department of Geology, Padnos Hall of Science, Grand Valley State University, Allendale, Michigan 49401, USA
*Correspondence and requests for materials should be addressed to Kent M. Syverson. E-mail: syverskm@uwec.edu

42.1. INTRODUCTION

Wisconsin was probably glaciated dozens of times during the Pleistocene Epoch (2.58–0.012 Ma), but stratigraphical units provide direct evidence for at least four glaciations. Even though Wisconsin lies well north of the maximum extent of Quaternary glaciations, the Driftless Area in the south-western part of the state remained unglaciated (Fig. 42.1). Glacial, alluvial and aeolian sediments from several glaciations and interglacials are present, but age control for all except the Late Wisconsinan Glaciation (marine isotope stage 2 or MIS 2) is limited to palaeosols and palaeomagnetic data (Whittecar, 1979; Baker et al., 1983; Jacobs and Knox, 1994; Miller, 2000). Radiocarbon analyses are numerous for deglaciation after 13.0 ^{14}C ka BP, but they are rare for the rest of the Wisconsinan Glaciation (MIS 2–4). Limited optically stimulated luminescence (OSL) and cosmogenic radionuclide (CRN) data are also available for the Late Wisconsinan Glaciation.

In this review, all ages are reported in thousands of calendar years (ka) unless stated otherwise. Radiocarbon ages (^{14}C ka BP) have been converted to calendar years using CALIB v. 5.0 (Stuiver and Reimer, 1993). OSL and CRN age estimates are assumed to be roughly equivalent to calibrated radiocarbon ages. Numerical ages of stratigraphical boundaries are taken from charts produced by the Subcommission on Quaternary Stratigraphy of the International Union of Geological Sciences (Gibbard and Cohen, 2008). We also use the current definition of the base of the Holocene as 11.7 ka (Walker et al., 2009).

Glacial sediment covers approximately three-fourths of the 145,000 km² land surface of Wisconsin (Figs. 42.2–42.4). Ice flowing from three major source regions deposited sediment (Mickelson et al., 1984; Attig et al., 1988; Syverson et al., 2011). Ice from the Keewatin ice dome to the north-west (e.g. the Des Moines Lobe; Fig. 42.2) deposited silt-rich, calcareous tills. Ice from the Labradoran ice dome to the north-east flowed out of the Superior lowland and deposited reddish-brown tills with Precambrian basalt, banded iron formation and reddish sandstone erratics (e.g. Superior Lobe and other smaller lobes; Fig. 42.2). Labradoran ice flowing out of the Green Bay and Lake Michigan lowlands (e.g. Green Bay and Lake Michigan Lobes; Fig. 42.2) deposited calcareous tills whose grain size was strongly influenced by ice-dammed lakes within those lowlands.

Syverson and Colgan (2004) summarised the glacial history of Wisconsin (including an extensive literature review which is not repeated here). Since that publication, the Pleistocene lithostratigraphy of Wisconsin has been reviewed and updated (Syverson et al., 2011) to incorporate more recent research findings. Additionally, our knowledge of ice dynamics and landform genesis has expanded as more has been learned more about modern ice sheet analogs. Here, we summarise our current understanding of the glacial history of Wisconsin and suggest areas of future research.

42.2. EARLY PLEISTOCENE GLACIATIONS

Three formations might represent at least two pre-Illinoian glaciations in Wisconsin. These units have been assigned an Early Pleistocene age based on intense weathering characteristics and reversed remanent palaeomagnetism. The units underlie sediment assigned to the Illinoian Glaciation (MIS 6 or 8) and are sometimes referred to informally as pre-Illinoian (older than MIS 6 or 8). Tills of the Pierce Formation of western Wisconsin (the 'old grey' till of Leverett, 1932)

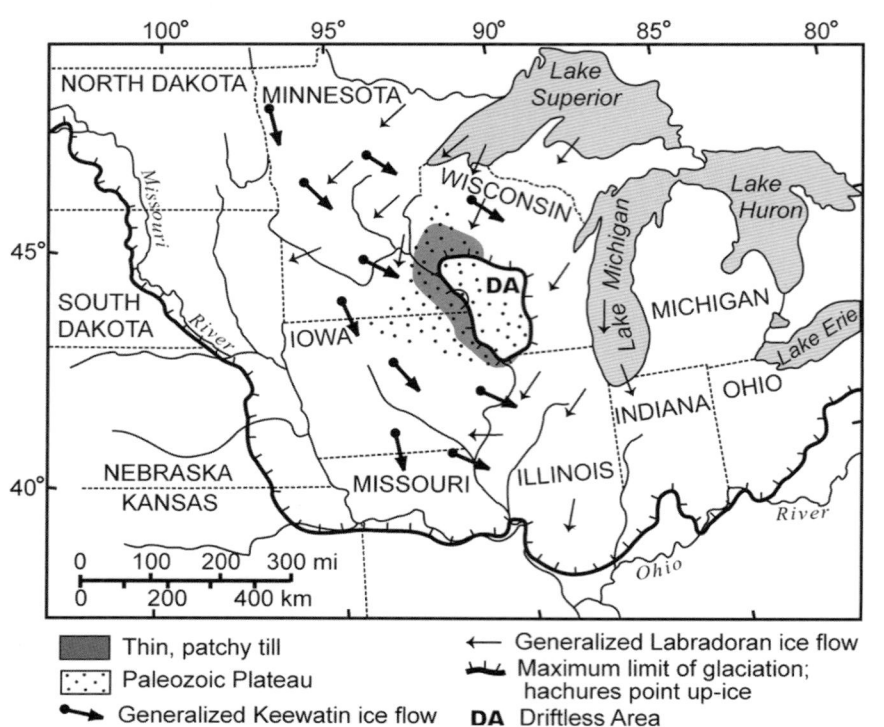

FIGURE 42.1 Map of Wisconsin and surrounding states relative to the maximum extent of glacier ice during the Quaternary (modified from Hobbs, 1999). The Driftless Area in south-western Wisconsin does not show evidence for burial by glacier ice. The shaded area of patchy, eroded till displays the same deeply incised river valleys as the Driftless Area and has been referred to as the 'pseudo-driftless area' by Hobbs (1999).

and Marathon Formation in north-central Wisconsin (Syverson et al., 2011) are the most extensive (Figs. 42.4 and 42.5).

42.2.1. Western Wisconsin

Tills of the Pierce Formation in western Wisconsin are grey to brown, calcareous (where unleached), silt rich, kaolinite rich and they probably represent an ice advance from a Keewatin source during at least two events (Baker et al., 1983; Johnson, 1986; Thornburg et al., 2000; Syverson et al., 2011). Till of the Woodville Member of the Pierce Formation marks the first ice advance (Fig. 42.5). Peat and wood overlie the Woodville Member till at the type section (Attig et al., 1988, p. 8 and 11).

Keewatin-source ice flowed south-east across the Mississippi River during the later Reeve Phase, deposited the Hersey Member of the Pierce Formation in western Wisconsin and dammed the major south-easterly flowing tributaries of the Mississippi River (Baker et al., 1983; Johnson, 1986). The resulting ice-dammed lakes extended at least tens of kilometres east of the modern Mississippi River valley. Silt- and clay-rich lake sediment of the Kinnickinnic Member of the Pierce Formation was deposited in these lakes (Fig. 42.5). Based on lake sediment elevations and varve counts, the lakes might have covered an area of 5800 km^2 for more than 1200 years (Baker, 1984; Syverson et al., 2011).

42.2.2. North-Central Wisconsin

The Wausau Member of the Marathon Formation in north-central Wisconsin is a silt-rich, intensely weathered till (Fig. 42.5). No similar till units have been found in the rest of Wisconsin, and the Wausau Member may be evidence for an extremely old glacial event. The Medford and Edgar Members of the Marathon Formation are calcareous and silt rich (Figs. 42.4 and 42.5). These tills contain less kaolinite than the Hersey Member of the Pierce Formation, but otherwise they are very similar (Muldoon et al., 1988; Attig and Muldoon, 1989; Thornburg et al., 2000; Syverson et al., 2011). The Marshfield moraine contains 30–50 m of Edgar till (Weidman, 1907, p. 452; Clayton, 1991), and this is the only primary glacial landform that remains from a pre-Illinoian ice advance.

42.2.3. Southern Wisconsin

In south-eastern Wisconsin, till is present in erosional remnants outside of end moraines deposited during the Wisconsinan Glaciation (Alden, 1918). Bleuer (1970, 1971) and Whittecar (1979) proposed that some of the silt-rich, calcareous grey till units were deposited during a pre-Illinoian ice advance from the east out of the Lake Michigan lowland. These till units have been observed beneath Illinoian till of the Walworth Formation in southern Wisconsin (Bleuer, 1971, p. 143; Miller, 2000, p. 116).

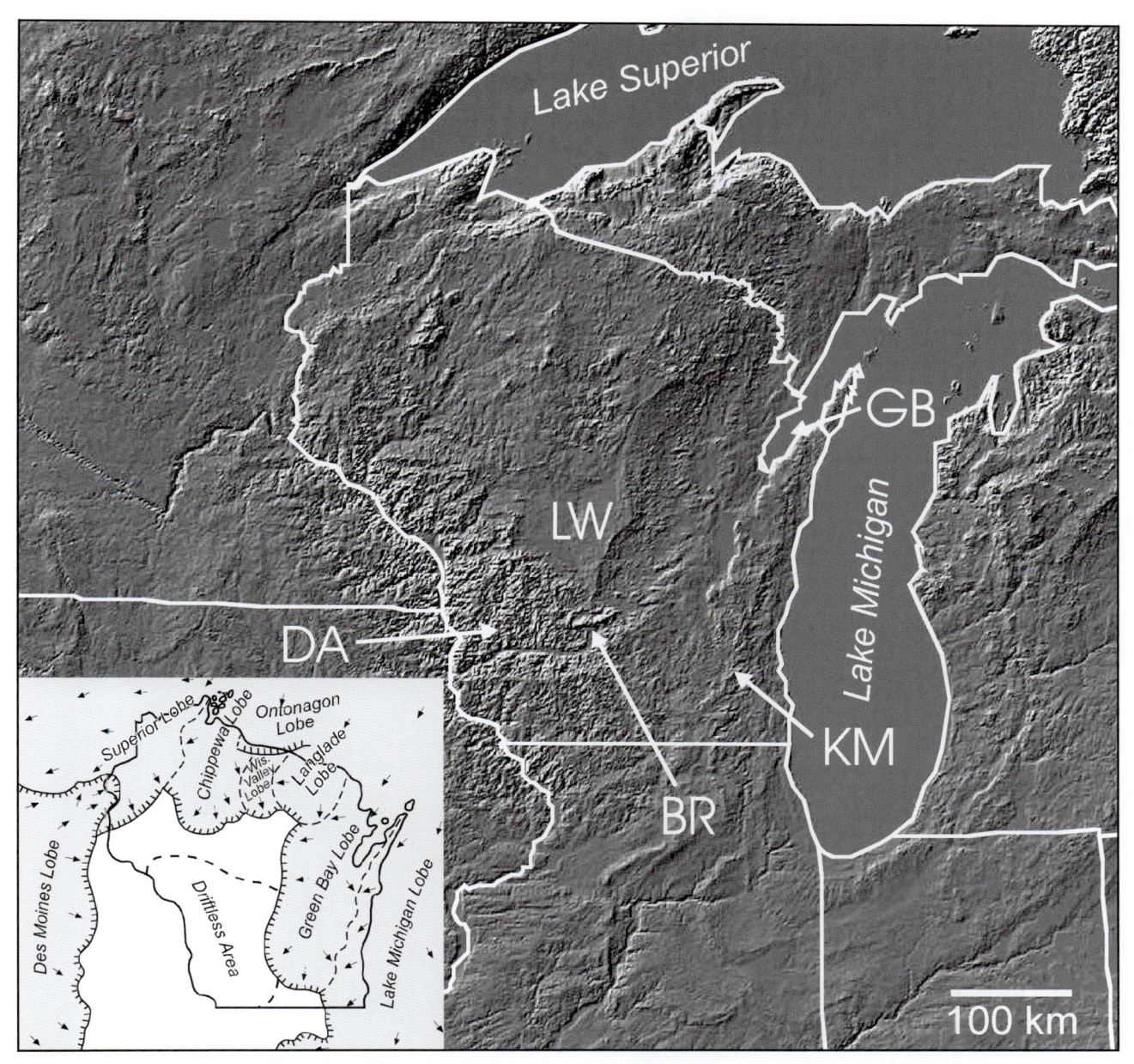

FIGURE 42.2 Major ice lobes during the Late Wisconsinan Glaciation (inset, modified from Clayton et al., 2006). The shaded-relief image of Wisconsin shows the following major features: BR, Baraboo Range; DA, Driftless Area; GB, Green Bay; KM, Kettle Moraine; LW, glacial Lake Wisconsin bed. Image was created from USGS 3 arcsec digital elevation data. Illumination direction is approximately 315°, and sun angle is 25°.

42.2.4. Ice Extent and Chronology

Ice extent and the chronology concerning these pre-Illinoian events are poorly known. Any original glacial landforms other than the Marshfield moraine have been completely removed by extensive erosion, and weathered till remnants are widely scattered and buried by younger sediment. The Powers Bluff chert dispersal fan (Fig. 42.3, location PB) suggests that pre-Illinoian Keewatin ice flowed towards the south-east in central Wisconsin (Weidman, 1907, p. 444; Clayton, 1991). Additionally, Knox and Attig (1988) described till in the Bridgeport terrace of the Wisconsin River (ca. 3 km east of the Mississippi River junction). This till formed during a pre-Illinoian ice advance to the south-east across the Mississippi River (Knox and Attig, 1988) with an ice margin closely following the location of the Mississippi River (Figs. 42.3 and 42.6A; Clayton et al., 2006).

Till units with reversed remanent magnetism provide some age control for pre-Illinoian till units. A Keewatin-source ice lobe may have deposited the Hersey Member of the Pierce Formation and the Medford Member of the

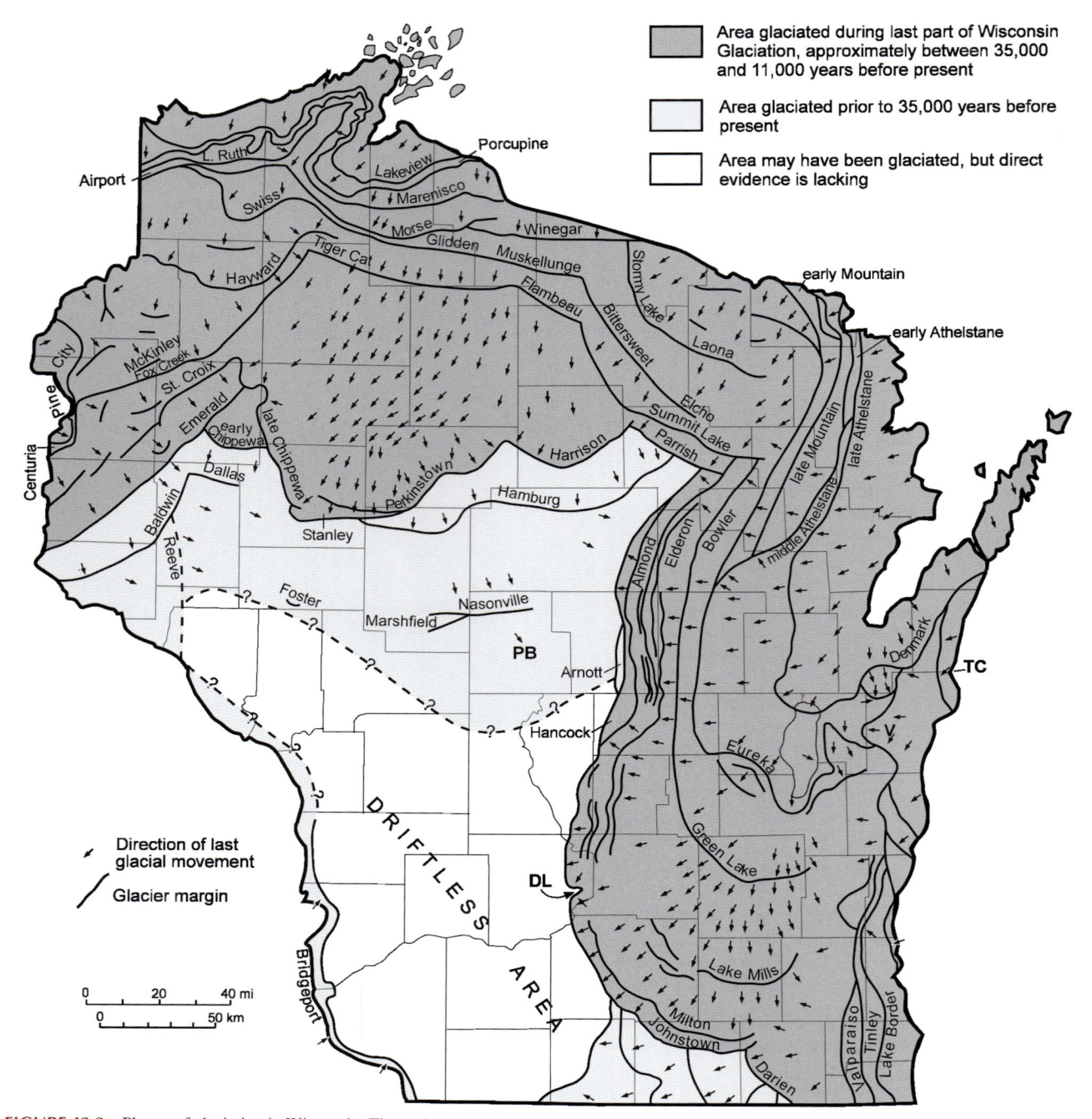

FIGURE 42.3 Phases of glaciation in Wisconsin. These phases are events that probably represent at least a minor advance of the ice sheet. Letters indicate important localities mentioned in the text: DL, Devils Lake; PB, Powers Bluff chert fan; TC, Two Creeks Forest Bed; V, Valders. Age estimates are in calendar years. Modified from Clayton et al. (2006) using information from Hooyer and Mode (2008) and Syverson et al. (2011).

Marathon Formation at the same time (Fig. 42.5; Baker et al., 1987). This interpretation is based on evidence such as the Powers Bluff chert fan, similar grain sizes, similar stratigraphical positions, a common provenance of sedimentary carbonate and black shale sources located to the north-west, and palaeomagnetism. Till of the Hersey Member and the lower part of the Kinnickinnic Member display reversed remanent magnetism, and the uppermost part of the Kinnickinnic Member displays normal remanent magnetism (Baker et al., 1983; Baker, 1984). The Medford Member also has reversed remanent magnetism (Syverson et al., 2005). Thus, the Kinnickinnic, Hersey and Medford Members were probably deposited before the Illinoian Glaciation, perhaps during the Emperor event at 0.42 Ma or more likely before the Matuyama/Brunhes boundary at 0.781 Ma (Baker et al., 1983). Miller (2000) also described till and interbedded lake sediment with reversed remanent magnetism in south-central Wisconsin.

Magnetically reversed till units have been described in Nebraska, Kansas and Iowa (Boellstorff, 1978; Colgan, 1998, 1999a; Roy et al., 2004) and northern Missouri (Rovey and Kean, 1996, 2001; Roy et al., 2004; Rovey and Balco, 2010). Roy et al. (2004) used sedimentology, palaeomagnetism and K–Ar geochronology of volcanic ashes to re-examine the tills first described by Boellstorff (1978) and to propose a revised stratigraphy for pre-Illinoian tills in the midcontinent. They described four 'R' type tills deposited during the Matuyama Reversed Polarity Chron and three 'N' type tills deposited during the Brunhes Normal Polarity Chron. 'R2' tills (rich in sedimentary lithologies, kaolinite and quartz) are older than 2.0 Ma. Two 'R1' tills were deposited between 1.3 and 0.781 Ma.

The Hersey and Medford Member tills might have been deposited at approximately the same time by Keewatin ice (Syverson et al., 2005). If so, the Medford till must have been deposited by a different lobe with a flow line that incorporated less kaolinite from Cretaceous source materials in Minnesota (Morey and Setterholm, 1997). If the Medford and Hersey Member tills are time equivalent, the Reeve Phase and the deposition of the Kinnickinnic Member lake sediment might represent a recessional event. The tills of the Marathon and Pierce Formations also might possibly represent at least two glaciations. Roy et al. (2004) reported seven pre-Illinoian tills in the midcontinent, and following their reasoning, the Hersey Member could represent an older glaciation where more kaolinite-rich saprolith was available for glacial erosion.

Clearly, the maximum ice extent in Wisconsin was reached well before the Illinoian Glaciation (MIS 6 or 8), and most of Wisconsin except the Driftless Area probably was ice covered during these glaciations. Sediment from these older events might be preserved in buried valleys and as remnants below sediment from the Late Wisconsinan Glaciation (MIS 2). Subsurface drilling, cosmogenic burial dating of palaeosols (Balco et al., 2005; Rovey and Balco, 2010) and additional palaeomagnetic data could help decipher this enigmatic part of the glacial record in Wisconsin.

42.3. MIDDLE PLEISTOCENE (ILLINOIAN GLACIATION)

42.3.1. Western Wisconsin

Two pre-Late Wisconsinan tills in western Wisconsin are thought to represent events during the Illinoian Glaciation of the Middle Pleistocene (MIS 6 or 8). The River Falls Formation (the 'old red' till of Leverett, 1932) and the Bakerville Member of the Copper Falls Formation were deposited by ice from the Superior region (Figs. 42.4 and 42.5). The River Falls Formation displays normal remanent magnetism and unconformably overlies deeply weathered Pierce Formation till in western Wisconsin (Baker et al., 1983). Both the River Falls and Bakerville till units contain reddish-brown, sandy till with abundant Precambrian basalt and red sandstone clasts from the Lake Superior region. Both of these tills are extensively eroded and do not display primary glacial topography (Johnson, 1986; Clayton, 1991; Syverson et al., 2011). Proximal stream sediment of the River Falls Formation is common in western Wisconsin and is intensely weathered. Soil-derived clay extends to depths of 5 m and cements the stream sediment in some cases (Syverson, 2007).

42.3.2. Southern Wisconsin

Illinoian tills in south-central Wisconsin (adjacent to northern Illinois where the Illinoian Glaciation was first defined) are members of the Walworth Formation and the Capron Member of the Zenda Formation (Figs. 42.4 and 42.7; Bleuer, 1971; Curry, 1989; Syverson et al., 2011). Walworth Formation tills typically are yellowish-brown, silt rich to sandy in texture, contain abundant dolomite pebbles and are dissected by erosion (Alden, 1918, p. 151; Bleuer, 1970, 1971; Miller, 2000; Syverson et al., 2011). Capron Member till is pink and silt rich. Weathered loess units in the Driftless Area (Wyalusing and Loveland Members of the Kieler Formation) also have been attributed to the Illinoian Glaciation based on stratigraphical position and palaeosols (Leigh and Knox, 1994; Jacobs et al., 1997; Knox et al., 2011).

42.3.3. Ice Extent and Chronology

The River Falls Formation and Bakerville Member of the Copper Falls Formation are associated with the Baldwin/Dallas/Foster and Nasonville Phases, respectively (Figs. 42.3–42.5; Johnson, 1986; Attig and Muldoon, 1989; Syverson, 2007). The lack of primary glacial landforms makes it difficult to determine the maximum extent of glacier ice and the number of depositional events. Baker et al. (1983) and Syverson (2007) attributed the intense weathering of River Falls Formation till and stream sediment to the Sangamonian interglacial (MIS 5e–d) and used this to support an Illinoian age for the sediment, even though any number of older interglacials (MIS 7, 9 or 11) might have weathered these till units. The Bakerville till does not display intense weathering but based on its stratigraphical position and eroded nature, workers such as Johnson (2000) have tentatively correlated the Bakerville and River Falls tills.

In south-central Wisconsin, tills of the Walworth Formation and Capron Member of the Zenda Formation were

FIGURE 42.4 Pleistocene lithostratigraphical units for tills in Wisconsin. Age estimates are in calendar years. Modified from Clayton et al. (2006) using information from Syverson et al. (2011).

probably deposited by ice flowing westward out of the Lake Michigan lowland during several events (Fig. 42.6C). The Capron Member was recently dated using the OSL method at ~130 ka, so the Capron Member (and the older, more eroded Walworth Formation) was deposited before the last part of the Illinoian Glaciation (R. Berg, Illinois State Geological Survey, oral communication, 2009, as reported in Syverson et al., 2011).

42.4. LATE PLEISTOCENE (WISCONSINAN GLACIATION)

Landforms and sediment formed during the Wisconsinan Glaciation (~80–11.7 ka) are abundant within the state. Age control is lacking for events other than the last part of the Wisconsinan Glaciation (MIS 2). Thus, the subdivisions of the Wisconsinan Glaciation proposed by Johnson

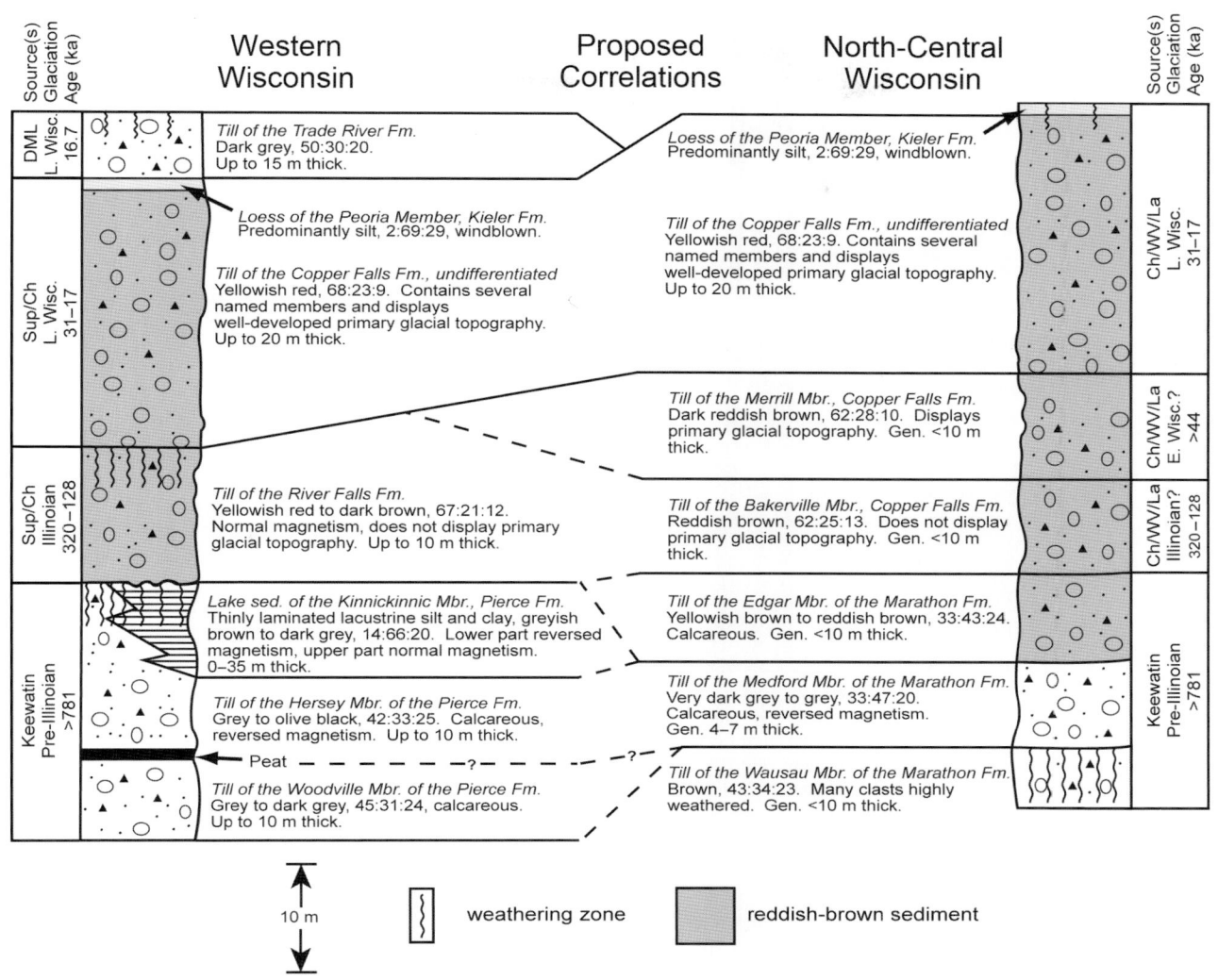

FIGURE 42.5 Glacial lithostratigraphy in western and north-central Wisconsin. Vertical scale is only approximate, and mean grain size is reported as sand:silt:clay percentages. Sources for the sediment are indicated as follows: Ch, Chippewa Lobe; DML, Des Moines Lobe; Keewatin, derived from Keewatin ice dome to the north-west; La, Langlade Lobe; Sup, Superior Lobe; WV, Wisconsin Valley Lobe. Proposed age estimates in calendar years and correlations for glacial units are shown. Modified from Syverson and Colgan (2004) using information from Syverson et al. (2011).

et al. (1997; Athens and Michigan Subepisodes of the Wisconsin Episode) have not been used in Wisconsin. Most workers in Wisconsin use the terms 'pre-Late Wisconsinan Glaciation' (>35 ka) and 'Late Wisconsinan Glaciation' or 'last part of the Wisconsinan Glaciation' (35–11.7 ka) to refer to sediment ages.

42.4.1. Early Wisconsinan Glaciation (MIS 4?)

Glacial tills attributed to the Early Wisconsinan Glaciation are rare in the state. Where present, these till units appear relatively fresh, display little to moderate stream incision and lack well-developed weathering horizons at the surface. The most extensive unit is the reddish-brown, sandy till of the Merrill Member of the Copper Falls Formation in north-central Wisconsin (Figs. 42.4 and 42.5). It displays some streamlined glacial landforms and low-relief hummocky topography. Thus, it is probably younger (MIS 4?) than pre-Illinoian and Illinoian units, which typically display no primary glacial topography. Stewart and Mickelson (1976) presented clay mineral analyses as evidence for greater weathering of the Merrill Member than similar units deposited during the Late Wisconsinan Glaciation. Thornburg et al. (2000) could not reproduce this trend so it is not clear whether these criteria are valid. One radiocarbon analysis of >40.8 ^{14}C ka BP (>44 ka) for organic material overlying till of the Merrill Member supports an Early Wisconsinan age (Stewart and Mickelson, 1976).

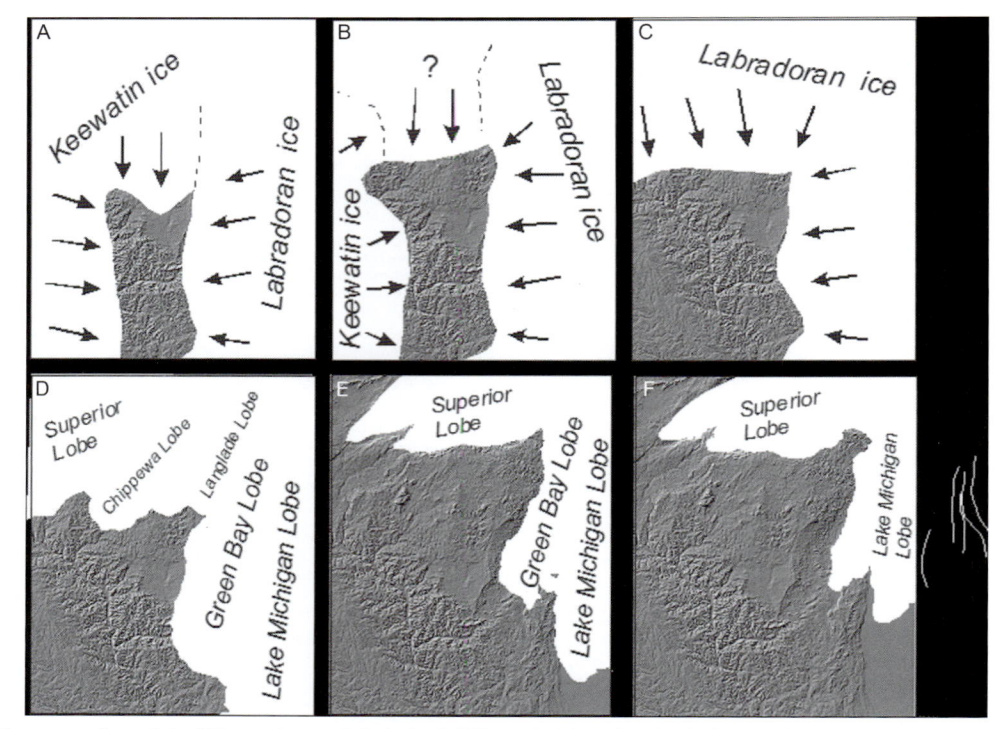

FIGURE 42.6 Reconstructions of six different phases of glaciation in Wisconsin (plotted on the shaded-relief base map from Fig. 42.2). (A) Pre-Illinoian glacial maximum, Stetsonville Phase. Limit based on interpretation of Clayton et al. (2006). The ice margin shown here may have been diachronous. (B) Pre-Illinoian Reeve Phase during the deposition of the Hersey Member (till) and Kinnickinnic Member (lake sediment) of the Pierce Formation. (C) Illinoian Glaciation maximum, Nasonville Phase in north-central Wisconsin. (D) Late Wisconsinan Glaciation maximum ice position at ~21 ka. (E) Late Wisconsinan Glaciation ice-margin reconstruction at ~16 ka during initial deposition of Kewaunee Formation. Ice margin extends south to the vicinity of Milwaukee along the Lake Michigan shoreline. (F) Ice-margin positions at ~13 ka (just after the Two Creeks Forest is covered by ice and the Two Rivers Member of the Kewaunee Formation is deposited).

Evidence for extensive Early Wisconsinan ice is lacking, but loess of this age is common in Wisconsin. Dates from the base of the Roxana Member of the Kieler Formation, a loess unit derived from the Mississippi River valley in the Driftless Area, range from a non-finite analysis of >47.0 to 45.2±2.65 ^{14}C ka BP (Leigh and Knox, 1993, 1994; Knox et al., 2011). Grimley (2000) used the magnetic susceptibility and clay mineralogy of the Roxana Formation to infer that the Superior, Wadena and Des Moines Lobes were contributing glacial meltwater (and fine-grained sediment) to the Mississippi River during the Early and Middle Wisconsinan (MIS 4 and 3).

42.4.2. Late Wisconsinan Glaciation

Ice crossed the drainage divide south of Lake Superior by ~32 ka, shortly before the Late Wisconsinan (MIS 2; ca. 24–12 ka). This inference is based on spruce wood in western Wisconsin buried by 60 m of glacial stream sediment (26.06±0.8 ^{14}C ka BP; Black, 1976b; Attig et al., 1985). The ice may have reached its Late Wisconsinan maximum by 21.3–18.8 ka and remained there until approximately 18.0 ka (Clayton and Moran, 1982; Fig. 42.6D). This is not well constrained in Wisconsin because organic material is rare from 32 to 16 ka. Clayton et al. (2001) attribute this lack of organic material to the presence of permafrost in Wisconsin, yet many areas not subject to permafrost also lack radiocarbon dates from this period. Some workers are starting to use OSL and CRN exposure dating to fill this void in geochronology. The Green Bay Lobe was at its maximum extent in the Devils Lake area (DL; Fig. 42.3) ~20–18 ka based on recent OSL dates on ice-marginal lake sediment (Attig et al., 2009) and CRN exposure dates (Colgan et al., 2002).

Numerous tills were deposited during many phases of the Late Wisconsinan Glaciation (Figs. 42.3 and 42.4). The Copper Falls and Holy Hill Formations are the most extensive units (Figs. 42.4, 42.5 and 42.7). During the earliest phases of the Green Bay Lobe (the Hancock/Johnstown Phases), the ice margin intersected the Baraboo Range in central Wisconsin, blocked the Wisconsin River drainage and formed glacial Lake Wisconsin in the central part of the state (Fig. 42.2; locations BR, LW). This lake was up to 115 km long and 50 m deep, and its outlet was to the west via the Black River. The ice dam broke as the ice thinned, and catastrophic drainage of glacial Lake Wisconsin around

Chapter | 42 The Quaternary of Wisconsin

FIGURE 42.7 Schematic stratigraphical column for tills deposited by the Lake Michigan Lobe in eastern Wisconsin. Shaded units display reddish colours. Estimated ages in calendar years are from Maher and Mickelson (1996) and Mickelson et al. (2007). The leaves at Valders Quarry are located above Holy Hill Formation till and below till of the Valders Member. Modified from Syverson and Colgan (2004) using stratigraphical information compiled from Syverson et al. (2011).

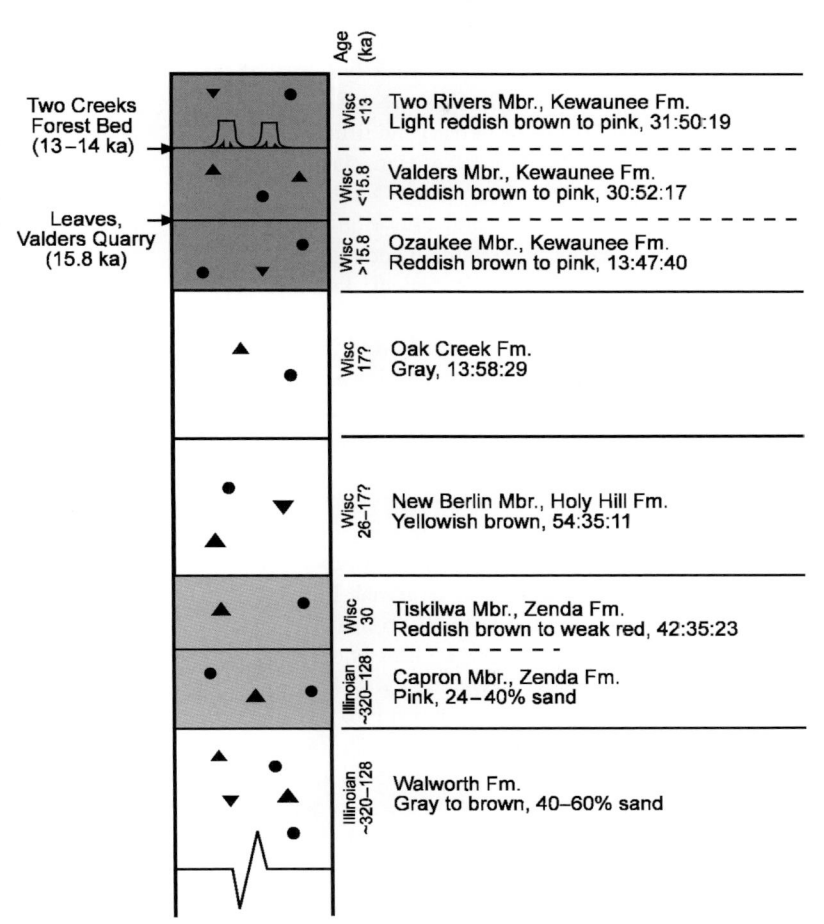

the east side of the Baraboo Range (estimated discharge $\sim 4 \times 10^4$ m^3/s; Clayton and Knox, 2008) incised deep, branching bedrock channels that are now part of the Wisconsin Dells (Clayton and Attig, 1989, p. 44–45; Clayton and Attig, 1990). Clayton and Attig (1989) estimated that glacial Lake Wisconsin drained between 15.5 and 17 ka. Active sand dunes were present on the eastern part of the exposed plain of glacial Lake Wisconsin by 14 ka (Rawling et al., 2008).

The Superior Lobe advanced to its Late Wisconsinan maximum during the Emerald Phase (Figs. 42.2 and 42.3; Johnson, 2000). Emerald Phase ice deposited thin till of the Poskin Member of the Copper Falls Formation, indistinct landforms and small outwash plains. The Superior Lobe then retreated more than 15 km to the St. Croix Phase ice-margin position where a markedly different suite of landforms was deposited (Fig. 42.3). The St. Croix moraine displays numerous large hummocks, ice-walled-lake plains and tunnel channels adjacent to an extensive outwash plain. According to Johnson (2000), the Superior Lobe was a cold, non-surging glacier during the Emerald Phase and a less cold, surging glacier during the St. Croix Phase. This warming effect caused more meltwater erosion and deposition during the St. Croix Phase. The Spooner Hills north-west of the St. Croix moraine may have been incised by subglacial meltwater erosion during the St. Croix Phase (Johnson, 1999, 2000).

A small sublobe of the Des Moines Lobe advanced from the west-southwest into western Wisconsin during the Pine City Phase and deposited till of the Trade River Formation (Figs. 42.3–42.5). This Keewatin-source ice occupied an area previously covered by the Superior Lobe. Trade River Formation till is silt rich, calcareous and similar to till of the pre-Illinoian Pierce and Marathon Formations, but Trade River Formation till exhibits little weathering and erosional modification (Johnson, 2000; Syverson et al., 2011). The Pine City Phase ice margin dammed the St. Croix River drainage along the Minnesota and Wisconsin border and glacial Lake Grantsburg formed. This lake lasted at least 80–100 years based on varve counts (Johnson and Hemstad, 1998; Johnson, 2000). The Pine City Phase might have occurred at ~ 14.0 ^{14}C ka BP (16.7 ka) based on the well-dated advance of the Des Moines Lobe to the Bemis margin in central Iowa (Johnson, 2000).

As ice margins wasted northward across drainage divides for Lake Superior, Green Bay and Lake Michigan, modern outlets towards the east were blocked by ice and lakes formed within these deep bedrock basins. The lakes drained to the south via the St. Croix (e.g. Lakes Nemadji, Duluth), Wisconsin (Lake Oshkosh) and Illinois (Lake Chicago) River systems (Clayton, 1984; Hansel and Mickelson, 1988; Colman et al., 1995; Hooyer and Mode, 2008; Knaeble and Hobbs, 2009). Silt- and clay-rich sediment was deposited in these lakes, and ice readvancing out of the lake basins eroded the fine-grained sediment. Till units deposited by these later advances are clast poor and contain up to 90% silt and clay (Miller Creek Formation in northern Wisconsin, Oak Creek and Kewaunee Formations in eastern Wisconsin; Figs. 42.4 and 42.7; Schneider, 1983; Simpkins, 1989; Simpkins et al., 1990; Ronnert, 1992; Syverson et al., 2011). For these reasons, early workers such as Chamberlin (1883) and Alden (1918) found it difficult to determine if the sediments were lacustrine or till.

The reddish-brown colour of the Kewaunee Formation is quite different from the grey to yellowish-brown till units deposited previously by the Late Wisconsinan Green Bay and Lake Michigan Lobes (Figs. 42.4 and 42.7). Alden (1918, p. 315) proposed that water carried iron-oxide-rich sediment from the Lake Superior basin into the Green Bay and Lake Michigan lowlands during a time of ice-margin retreat, an event that occurred before 15.8 ka. Ice advance subsequently eroded red lake sediment and deposited the reddish-brown tills of the Kewaunee Formation as far south as Milwaukee (Rovey and Borucki, 1995).

The Two Creeks Forest Bed along the western Lake Michigan shoreline (Fig. 42.3, location TC) contains spruce wood and has been recognised as an important interstadial marker bed within the Late Wisconsinan Glaciation sedimentary sequence of the Great Lakes region (Goldthwait, 1907; Black, 1970a, 1974; Kaiser, 1994; Leavitt et al., 2007; Mickelson et al., 2007). The Two Creeks Forest Bed is first overlain by lake sediment and then till of the Kewaunee Formation. Acomb et al. (1982) interpreted this sequence to represent an ice readvance into the Lake Michigan basin that blocked the eastern drainage, raised the lake level, flooded and buried the forest and deposited till of the Two Rivers Member of the Kewaunee Formation as the site was overridden by ice (Figs. 42.6F and 42.7). Kaiser (1994) used dendrochronology and accelerator mass spectrometry dates to delimit the growth period of the Two Creeks Forest to between 12.05 and 11.75 ^{14}C ka BP ($\sim$14.0–13.6 ka). Most ages for Two Creeks Forest wood fall between 11.2 and 12.4 ^{14}C ka BP ($\sim$13–14 ka; Mickelson et al., 2007).

Early workers (e.g. Black, 1976b, 1980) initially thought all reddish-brown Kewaunee Formation till was younger than the Two Creeks Forest (the 'Valders till'; see Acomb et al. (1982) for details). Field mapping and lab analyses later showed some red till units to be older, and others younger, than the Two Creeks Forest (Acomb et al., 1982; McCartney and Mickelson, 1982).

Ice margins in eastern Wisconsin oscillated markedly during the last part of the Wisconsinan Glaciation (Fig. 42.8), although the timing of events is poorly constrained. Some researchers have suggested an early Green Bay Lobe retreat before $\sim$18.8 ka (Colgan, 1996, 1999b), while others have suggested a later retreat beginning $\sim$14.6 ka (Maher, 1982). Hundreds of small moraines, interpreted as annual moraines, are visible on surfaces uncovered by the Green Bay Lobe (Colgan, 1996; Clayton and Attig, 1997). If annual, these moraines suggest that the

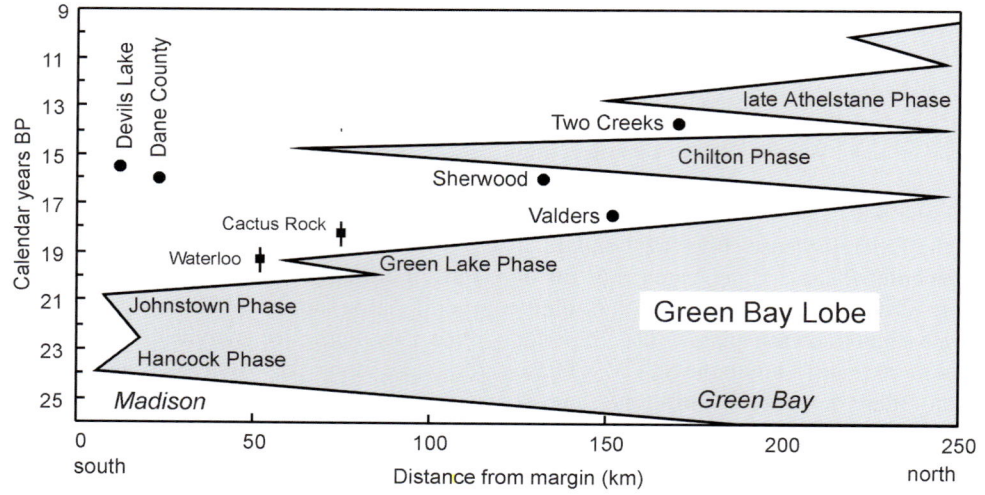

FIGURE 42.8 Time–distance diagram for the Green Bay Lobe (modified from Colgan et al., 2002). Dots are radiocarbon dates converted to calendar years using CALIB 5.0 (Stuiver and Reimer, 1993). Cross bars are cosmogenic isotope dates from Colgan et al. (2002) using the production rates of Stone et al. (1998). Bar lengths represent one sigma analytical uncertainty. Yellowish-brown, sandy till of the Holy Hill Formation was deposited during the Hancock, Johnstown, and Green Lake Phases. Reddish-brown, fine-grained till of the Kewaunee Formation was deposited during the Chilton and late Athelstane Phases.

Green Bay Lobe retreated at a rate of ~50–100 m/year. The Green Bay Lobe was at its maximum position at ~20 ka based on OSL-dated ice-marginal lake sediments high in the Baraboo Hills (Attig et al., 2009). Cosmogenic exposure dates at two sites more than 50 km from the maximum extent of the Green Bay Lobe suggest ice retreat before 17–19 ka (Colgan et al., 2002).

Loess of the Peoria Member of the Kieler Formation was deposited during the Late Wisconsinan Glaciation (Knox et al., 2011). This loess is thickest (up to 10 m) along the Mississippi River valley in the Driftless Area. Outwash streams and sparsely vegetated periglacial landscapes were the most important sources of loess (Leigh and Knox, 1994; Mason et al., 1994). Loess also was derived from sources such as freshly exposed glacial sediment and the exposed beds of recently drained proglacial or ice-walled lakes (Schaetzl et al., 2009).

Glaciers last advanced into Wisconsin south of Lake Superior during the Lakeview Phase (Fig. 42.3; Clayton, 1984). Black (1976b) reported wood dates from the red, clay-rich till of this event in northern Wisconsin (9.73 ± 0.140 and 10.1 ± 0.1 ^{14}C ka BP). Clayton (1984) correlates this 9.9 ^{14}C ka BP advance with the Marquette Phase in the Upper Peninsula of Michigan and estimates that the glacier margin wasted out of Wisconsin for the last time by 9.5 ^{14}C ka BP (10.9 ka).

42.4.3. Late Wisconsinan Landforms and Palaeoglaciology

Based on an extensive geographic information system database of landforms and sediments, Colgan et al. (2003) classified the landforms in the State of Wisconsin into two glacial landsystems. These include (A) a younger landsystem with low-relief till plains and end moraines and (B) an older landsystem with high-relief hummocky moraines, ice-walled-lake plains, drumlins and tunnel channels.

Colgan et al. (2003) proposed that landsystem (B) formed under cold subpolar climate and permafrost conditions during the last glacial maximum and the early part of the deglaciation (~32–16 ka). Evidence for permafrost includes ice-wedge casts and ice-wedge polygons found on landscapes older than 15.4 ka (Black, 1965, 1976a; Mason et al., 1994; Colgan, 1996; Holmes and Syverson, 1997; Clayton et al., 2001).

Moraines marking the ice maximum in northern Wisconsin (St. Croix, late Chippewa, Perkinstown, Harrison Phases; Fig. 42.3) are hummocky complexes 5–20 km wide with numerous, large ice-walled-lake plains (Attig, 1993; Johnson et al., 1995; Ham and Attig, 1996, 1997; Johnson and Clayton, 2003; Syverson, 2007; Clayton et al., 2008). Lake sediment is commonly at least as thick as the height of the ice-walled-lake plain above adjacent depressions; this suggests a supraglacial origin, rather than a subglacial ice-pressing origin (Boone and Eyles, 2001), for the ice-walled-lake plains and the surrounding hummocks in the moraines (Clayton et al., 2008). Prominent tunnel channels cut the outermost hummocky moraines but are generally not associated with recessional moraines (Attig et al., 1989; Clayton et al., 1999; Cutler et al., 2002).

The correlation of cold winter climate, hummocky moraines and tunnel channels has been used to infer a frozen bed zone up to tens of kilometres wide near the ice margins in Wisconsin between 32 and 16 ka (Attig et al., 1989; Winguth et al., 2004). This frozen bed zone would have enhanced compressive ice flow, increased the transport of sediment to the ice surface (as modelled by Moore et al., 2009) and produced broad, high-relief hummocky moraine complexes (Johnson and Clayton, 2003). Behind the frozen zone was an unfrozen wet bed with high subglacial pore-water pressures (Attig et al., 1989; Colgan et al., 2003). Tunnel channels could have been eroded by water escaping through the frozen bed of the glacier margin (Attig et al., 1989; Clayton et al., 1999), perhaps as geothermal heat melted the frozen bed at the glacier margin (Cutler et al., 2000).

Well-developed, radiating drumlin fields are present in areas formerly covered by the Green Bay Lobe (Borowiecka and Erickson, 1985; Colgan and Mickelson, 1997) and Lake Michigan Lobe (Whittecar and Mickelson, 1977, 1979; Stanford and Mickelson, 1985). These drumlins are up-flow from narrow (2–5 km wide), single-crested end moraines that are cut by numerous tunnel channels (Attig et al., 1989; Clayton et al., 1999; Colgan, 1999b; Colgan et al., 2003). According to Mickelson et al. (1983) and Attig et al. (1989), glacial ice with a thawed bed eroded these drumlins up-flow from the ice-marginal frozen bed. Colgan and Mickelson (1997) suggested that the drumlins did not all form during the ice maximum, but rather formed during several deglaciation/readvance events. If so, then ice must have advanced over permafrost redeveloped on recently deglaciated surfaces. Permafrost clearly did redevelop during ice retreat prior to 15.4 ka based on numerous ice-wedge-cast polygons on drumlinized surfaces (Colgan, 1996; Clayton et al., 2001). Colgan and Mickelson (1997) cited the absence of tunnel channels associated with these younger ice-margin positions to argue against a catastrophic subglacial flood origin for the drumlins (Shaw, 2002); if the drumlins formed via subglacial meltwater processes, many flood channels should be associated with the younger drumlin-forming phases.

The Kettle Moraine of eastern Wisconsin trends approximately parallel to the Lake Michigan shoreline for more than 125 km (KM; Fig. 42.2). This time-transgressive ridge complex formed during the time of drumlin formation as meltwater was concentrated in the interlobate region between the Green Bay and Lake Michigan Lobes. Most fluvial sediment was deposited in contact with glacial ice, so the Kettle Moraine contains hummocky stream sediment, pitted outwash plains, outwash plains and eskers (Chamberlin, 1883; Alden, 1918; Black, 1969, 1970a; Syverson, 1988; Mickelson and Syverson, 1997; Clayton, 2001; Carlson et al., 2005).

Following the formation of landsystem (B), ice-surface slopes probably changed from steeply sloping margins during Late Wisconsinan maximum to more gently sloping margins during deglaciation phases (Colgan and Mickelson, 1997). Other palaeo-reconstructions of ice lobes in Wisconsin during retreat from the glacial maximum suggest gentle ice-surface slopes, thin ice and low basal shear stress values typical of fast-moving outlet glaciers or surging glaciers (Clark, 1992; Colgan, 1999b; Socha et al., 1999). This could reflect increasing meltwater generation during deglaciation and/or a rapid ablation-driven lowering of the ice surface during retreat.

The low-relief till plains and end moraines of landsystem (A) of were formed by low-profile ice lobes during retreat, according to Colgan et al. (2003). At this time, the bed of the ice was wet, and ice was sliding from the interior of the ice sheet to the terminus. Colgan et al. (2003) interpreted these low-relief features to indicate englacial and subglacial sediment transport. Some of the sediment may have been transported and deposited by a subglacial deforming bed (Colgan et al., 2003, p. 122). Landsystem (A) is found in areas where tills were deposited after climate became warmer and permafrost conditions diminished at ~16 ka.

Surging glaciers may have modified the landscape during the Late Wisconsinan Glaciation. The Kewaunee Formation in eastern Wisconsin is associated with low-relief moraines and till plains. Low basal shear stress values estimated for the Chilton and Denmark (late Athelstane) Phases of the Green Bay Lobe (2–3 kPa) may indicate a surging ice lobe with elevated subglacial water pressures enhancing basal sliding and/or subglacial sediment deformation (Socha et al., 1999; Figs. 42.3, 42.4 and 42.8). Offset moraine segments associated with the Chilton Phase (Colgan, 1996) and the late Chippewa Phase (Attig et al., 1998) are morphologic evidence for glacial surges during ice-margin retreat.

42.5. DRIFTLESS AREA

The Driftless Area of south-western Wisconsin is characterised by deeply incised, dendritic river valleys developed in relatively flat-lying Palaeozoic bedrock (Figs. 42.1 and 42.2). River valleys were eroded to their present levels as the Mississippi River incised sometime before 0.781 Ma (Baker et al., 1998). Most early workers thought evidence for glaciation was lacking in the Driftless Area (e.g. Chamberlin and Salisbury, 1885), but Black (1970b,c) presented what he thought to be evidence for glaciation of the Driftless Area, perhaps as recently as the Early Wisconsinan Glaciation. Mickelson et al. (1982) evaluated Black's evidence for glaciation in the Driftless Area and concluded that no unequivocal evidence supported glaciation of the entire Driftless Area. However, periglacial processes such as solifluction eroded Driftless Area uplands and slopes during glaciations (Knox, 1989; Clayton et al., 2001). Loess blown out of the Mississippi and Wisconsin River valleys mantles many upland surfaces in the Driftless Area (Knox et al., 2011).

Hobbs (1999) provides a review of different hypotheses used to explain the origin of the Driftless Area. According to Chamberlin and Salisbury (1885), bedrock highlands in northern Wisconsin and the Upper Peninsula of Michigan deflected ice to the east and west and 'protected' the Driftless Area from glaciation. Based on numerical ice sheet modelling, Cutler et al. (2001) argued that the presence of the deep Lake Superior basin retarded ice advance into central Wisconsin and perhaps funnelled ice to the west and east, leaving the Driftless Area unglaciated. If so, the Driftless Area owes its existence to the deep Superior basin to the north. In addition, permeable Palaeozoic bedrock might have dewatered the glacier bed and inhibited ice from flowing into the Driftless Area from the west (Hobbs, 1999). All of these factors may have played a role in preventing the glaciation of the Driftless Area.

42.6. FUTURE WORK

The glacial history of Wisconsin is complex because it reflects ice advances from both the Keewatin and Labradoran ice domes of the Laurentide and earlier ice sheets. Many questions about the glacial history still need to be answered. When did ice first advance into Wisconsin? Glaciers advanced into Missouri as early as ~2.5 Ma (Balco et al., 2005; Rovey and Balco, 2010), so Wisconsin was likely glaciated during the Early Pleistocene as well. The marine oxygen-isotope record suggests dozens of glaciations (Gibbard and Cohen, 2008), far more than can be recognised from the known geological record in Wisconsin. Will it ever be possible to date pre-Late Wisconsinan events with any confidence? OSL is starting to produce new results. Cosmogenic isotope dating of surfaces and buried palaeosols is another technique yet to be applied to pre-Late Wisconsinan events in the state.

Using glacial landforms to determine Quaternary environmental and glaciological conditions is another area of fruitful research. A better understanding of the relationship between proglacial permafrost, ice sheet dynamics and glacial sedimentology in modern environments is likely to inform our interpretations of Quaternary processes and landforms in Wisconsin. The interaction of terrestrial ice lobes with the ice streams that probably fed fast-flowing glaciers is another important area to be explored.

Clearly, new technologies and researchers with new questions are needed. The State of Wisconsin will almost certainly yield new and exciting answers in the future.

ACKNOWLEDGMENTS

This chapter was strengthened by comments from Lee Clayton, Mark Johnson, John Attig and David Mickelson. Gene Leisz (UW-Eau Claire) assisted with figures.

REFERENCES

Acomb, L.J., Mickelson, D.M., Evenson, E.B., 1982. Till stratigraphy and late glacial events in the Lake Michigan Lobe of eastern Wisconsin. Geol. Soc. Am. Bull. 93, 289–296.

Alden, W.C., 1918. The Quaternary geology of southeastern Wisconsin, with a chapter on the older rock formations. United States Geological Survey Professional Paper 106, 356pp.

Attig, J.W., 1993. Pleistocene geology of Taylor County, Wisconsin. Wisconsin Geol. Nat. History Surv. Bull. 90, 25.

Attig, J.W., Muldoon, M.A., 1989. Pleistocene geology of Marathon County, Wisconsin. Wisconsin Geol. Nat. History Surv. Inform. Circular 65, 27pp.

Attig, J.W., Clayton, L., Mickelson, D.M., 1985. Correlation of late Wisconsin glacial phases in the western Great Lakes area. Geol. Soc. Am. Bull. 96, 1585–1593.

Attig, J.W., Clayton, L., Mickelson, D.M. (Eds.), 1988. Pleistocene stratigraphic units of Wisconsin 1984-87, Wisconsin Geol. Nat. History Surv. Inform. Circular 62, 61pp.

Attig, J.W., Mickelson, D.M., Clayton, L., 1989. Late Wisconsin landform distribution and glacier-bed conditions in Wisconsin. Sediment. Geol. 62, 399–405.

Attig, J.W., Ham, N.R., Mickelson, D.M., 1998. Environments and processes along the margin of the Laurentide Ice Sheet in north-central Wisconsin. Wisconsin Geol. Nat. History Surv. Open-File Report, 1998-01, 62pp.

Attig, J.W., Hanson, P.R., Rawling III, J.E., Young, A.R., Carson, E.C., 2009. Optical ages from ice-marginal lake deposits in the Baraboo Hills indicate the Green Bay Lobe was at its maximum extent about 20,000 years ago. Geol. Soc. Am. Abstr. Programs 41 (7), 334.

Baker, R.W., 1984. Pleistocene history of west-central Wisconsin. Wisconsin Geol. Nat. History Surv. Field Trip Guide Book 11, 76pp.

Baker, R.W., Diehl, J.F., Simpson, T.W., Zelazny, L.W., Beske-Diehl, S., 1983. Pre-Wisconsinan glacial stratigraphy, chronology, and paleomagnetics of west-central Wisconsin. Geol. Soc. Am. Bull. 94, 1442–1449.

Baker, R.W., Attig, J.W., Mode, W.N., Johnson, M.D., Clayton, L., 1987. A major advance of the pre-Illinoian Des Moines Lobe. Geol. Soc. Am. Abstr. Programs 19 (4), 187.

Baker, R.W., Knox, J.C., Lively, R.S., Olsen, B.M., 1998. Evidence for early entrenchment of the upper Mississippi River valley. In: Patterson, C.J., Wright Jr., H.E. (Eds.), Contributions to Quaternary Studies in Minnesota. Minnesota Geological Survey Report of Investigations, Minnesota Gelogical Survey Publication, vol. 49, pp. 113–120.

Balco, G., Rovey II, C.W., Stone, J.O.H., 2005. The first glacial maximum in North America. Science 307 (5707), 222.

Black, R.F., 1965. Ice-wedge casts of Wisconsin. Wisconsin Acad. Sci. Arts Lett. Trans. 54, 187–222.

Black, R.F., 1969. Glacial geology of Northern Kettle Moraine State Forest, Wisconsin. Wisconsin Acad. Sci. Arts Lett. Trans. 57, 99–119.

Black, R.F., 1970a. Glacial geology of Two Creeks Forest Bed, Valderan type locality, and Northern Kettle Moraine State Forest. Wisconsin Geol. Nat. History Surv. Inform. Circular 13, 40pp.

Black, R.F., 1970b. Blue Mounds and the erosional history of southwestern Wisconsin. Wisconsin Geol. Nat. History Surv. Inform. Circular 15-H, H1–H11.

Black, R.F., 1970c. Residuum and ancient soils of the Driftless Area of southwestern Wisconsin. Wisconsin Geol. Nat. History Surv. Inform. Circular 15-I, I1–I12.

Black, R.F., 1974. Geology of Ice Age National Scientific Reserve of Wisconsin. National Park Service Scientific Monograph Series 2, 234pp.

Black, R.F., 1976a. Periglacial features indicative of permafrost: ice and soil wedges. Quatern. Res. 6, 3–26.

Black, R.F., 1976b. Quaternary geology of Wisconsin and contiguous upper Michigan. In: Mahaney, W.H. (Ed.), Quaternary Stratigraphy of North America. Dowden, Hutchinson and Ross, Stroudsburg, PA, pp. 93–117.

Black, R.F., 1980. Valders—Two Creeks, Wisconsin, revisited: the Valders Till is most likely post-Twocreekan. Geol. Soc. Am. Bull. 91, 713–723.

Bleuer, N.K., 1970. Glacial stratigraphy of south-central Wisconsin. Wisconsin Geol. Nat. History Surv. Inform. Circular 15-J, J1–J35.

Bleuer, N.K., 1971. Glacial stratigraphy of south-central Wisconsin. Unpublished Ph.D. Thesis, University of Wisconsin, Madison, 173pp.

Boellstorff, J., 1978. Chronology of some late Cenozoic deposits from the central United States and the Ice Ages. Trans. Nebraska Acad. Sci. 6, 35–49.

Boone, S.J., Eyles, N., 2001. Geotechnical model for great plains hummocky moraine formed by till deformation below stagnant ice. Geomorphology 38, 109–124.

Borowiecka, B.Z., Erickson, R.H., 1985. Wisconsin drumlin field and its origin. Z. Geomorphol. 29, 417–438.

Carlson, A.E., Mickelson, D.M., Principato, S.M., Chapel, D.M., 2005. The genesis of the northern Kettle Moraine, Wisconsin. Geomorphology 67, 365–374.

Chamberlin, T.C., 1883. General geology. In: Chamberlin, T.C. (Ed.), Geology of Wisconsin, Wisconsin Geol. Nat. History Surv. 1, pp. 1–300.

Chamberlin, T.C., Salisbury, R.D., 1885. Preliminary paper on the Driftless Area of the Upper Mississippi Valley. United States Geological Survey 6th Annual Report, pp. 199–322.

Clark, P.U., 1992. Surface form of the southern Laurentide Ice Sheet and its implications to ice-sheet dynamics. Geol. Soc. Am. Bull. 104, 595–605.

Clayton, J.A., Knox, J.C., 2008. Catastrophic flooding from Glacial Lake Wisconsin. Geomorphology 93, 384–397.

Clayton, L., 1984. Pleistocene geology of the Superior region, Wisconsin. Wisconsin Geol. Nat. History Surv. Inform. Circular 46, 40pp.

Clayton, L., 1991. Pleistocene geology of Wood County, Wisconsin. Wisconsin Geol. Nat. History Surv. Inform. Circular 68, 18pp.

Clayton, L., 2001. Pleistocene geology of Waukesha County, Wisconsin. Wisconsin Geol. Nat. History Surv. Bull. 99, 33pp.

Clayton, L., Attig, J.W., 1989. Glacial Lake Wisconsin. Geological Society of America Memoir 173, 80pp.

Clayton, L., Attig, J.W., 1990. Geology of Sauk County, Wisconsin. Wisconsin Geol. Nat. History Surv. Inform. Circular 67, 68pp.

Clayton, L., Attig, J.W., 1997. Pleistocene geology of Dane County, Wisconsin. Wisconsin Nat. History Geol. Surv. Bull. 95, 64pp.

Clayton, L., Moran, S.R., 1982. Chronology of late Wisconsinan glaciation in middle North America. Quatern. Sci. Rev. 1, 55–82.

Clayton, L., Attig, J.W., Mickelson, D.M., 1999. Tunnel channels formed in Wisconsin during the last glaciation. In: Mickelson, D.M., Attig, J.W. (Eds.), Glacial Processes: Past and Present. Geological Society of America, Special Paper, 337, 69–82.

Clayton, L., Attig, J.W., Mickelson, D.M., 2001. Effects of late Pleistocene permafrost on the landscape of Wisconsin, USA. Boreas 30, 173–188.

Clayton, L., Attig, J.W., Mickelson, D.M., Johnson, M.D., Syverson, K.M., 2006. Glaciation of Wisconsin. Wisconsin Geol. Nat. History Surv. Educ. Ser. 36, 4pp.

Clayton, L., Attig, J.W., Ham, N.R., Johnson, M.D., Jennings, C.E., Syverson, K.M., 2008. Ice-walled-lake plains: implications for the origin of hummocky glacial topography in middle North America. Geomorphology 97, 237–248.

Colgan, P.M., 1996. The Green Bay and Des Moines Lobes of the Laurentide Ice Sheet: evidence for stable and unstable glacier dynamics 18,000 to 12,000 B.P. Unpublished Ph.D. Thesis, University of Wisconsin, Madison, 293pp.

Colgan, P.M., 1998. Paleomagnetism of pre-Illinoian till near Kansas City, Kansas. Trans. Kansas Acad. Sci. 101 (1–2), 25–34.

Colgan, P.M., 1999a. Early middle Pleistocene Glaciation (0.78 to 0.61 Ma) of the Kansas City area, northwestern Missouri, USA. Boreas 28 (4), 477–489.

Colgan, P.M., 1999b. Reconstruction of the Green Bay Lobe, Wisconsin, United States, from 26,000 to 13,000 radiocarbon years B.P. In: Mickelson, D.M., Attig, J.A. (Eds.), Glacial Processes Past and Present. Geological Society of America, Special Paper 337, 137–150.

Colgan, P.M., Mickelson, D.M., 1997. Genesis of streamlined landforms and flow history of the Green Bay Lobe, Wisconsin, USA. Sediment. Geol. 111, 7–25.

Colgan, P.M., Bierman, P.R., Mickelson, D.M., Caffee, M., 2002. Variation in glacial erosion near the southern margin of the Laurentide Ice Sheet, south-central Wisconsin, USA: implications for cosmogenic dating of glacial terrains. Geol. Soc. Am. Bull. 114, 1581–1591.

Colgan, P.M., Mickelson, D.M., Cutler, P.M., 2003. Landsystems of the southern Laurentide ice sheet. In: Evans, D.J.A. (Ed.), Glacial Landsystems. Arnold Publishing, London, pp. 111–142.

Colman, S.M., Clark, J.A., Clayton, L., Hansel, A.K., Larsen, C.E., 1995. Deglaciation, lake levels, and meltwater discharge in the Lake Michigan basin. Quatern. Sci. Rev. 13, 879–890.

Curry, B.B., 1989. Absence of Altonian glaciation in Illinois. Quatern. Res. 31, 1–13.

Cutler, P.M., MacAyeal, D.R., Mickelson, D.M., Parizek, B.R., Colgan, P.M., 2000. A numerical investigation of ice-lobe–permafrost interaction around the southern Laurentide ice sheet. J. Glaciol. 46 (153), 311–325.

Cutler, P.M., Mickelson, D.M., Colgan, P.M., MacAyeal, D.R., Parizek, B.R., 2001. Influence of the Great Lakes on the dynamics of the southern Laurentide ice sheet: numerical experiments. Geology 29, 1039–1042.

Cutler, P.M., Colgan, P.M., Mickelson, D.M., 2002. Sedimentologic evidence for outburst floods from the Laurentide Ice Sheet margin in Wisconsin, U.S.A: implications for tunnel-channel formation. Quatern. Int. 90, 23–40.

Gibbard, P., Cohen, K.M., 2008. Global chronostratigraphical correlation table for the last 2.7 million years. Episodes 31 (2), 243–247.

Goldthwait, J.W., 1907. The abandoned shore-lines of eastern Wisconsin. Wisconsin Geol. Nat. History Surv. Bull. 17, 134pp.

Grimley, D.A., 2000. Glacial and nonglacial contributions to Wisconsin Episode loess in the central United States. Geol. Soc. Am. Bull. 112, 1475–1495.

Ham, N.R., Attig, J.W., 1996. Ice wastage and landscape evolution along the southern margin of the Laurentide Ice Sheet, north-central Wisconsin. Boreas 25, 171–186.

Ham, N.R., Attig, J.W., 1997. Pleistocene geology of Lincoln County, Wisconsin. Wisconsin Geol. Nat. History Surv. Bull. 93, 31pp.

Hansel, A.K., Mickelson, D.M., 1988. A reevaluation of the timing and causes of high lake phases in the Lake Michigan basin. Quatern. Res. 29, 113–128.

Hobbs, H., 1999. Origin of the Driftless Area by subglacial drainage - a new hypothesis. In: Mickelson, D.M., Attig, J.W. (Eds.), Glacial Processes Past and Present. Geological Society of America, Special Paper 337, 93–102.

Holmes, M.A., Syverson, K.M., 1997. Permafrost history of Eau Claire and Chippewa Counties, Wisconsin, as indicated by ice-wedge casts. Compass 73 (3), 91–96.

Hooyer, T.S., Mode, W.N., 2008. Quaternary geology of Winnebago County, Wisconsin. Wisconsin Geol. Nat. History Surv. Bull. 105, 41pp.

Jacobs, P.M., Knox, J.C., 1994. Provenance and pedology of a long-term Pleistocene depositional sequence in Wisconsin's Driftless Area. Catena 22, 49–68.

Jacobs, P.M., Knox, J.C., Mason, J.A., 1997. Preservation and recognition of middle and early Pleistocene loess in the Driftless Area, Wisconsin. Quatern. Res. 47, 147–154.

Johnson, M.D., 1986. Pleistocene geology of Barron County, Wisconsin. Wisconsin Geol. Nat. History Surv. Inform. Circular 55, 147–154.

Johnson, M.D., 1999. Spooner Hills, northwest Wisconsin: high-relief hills carved by subglacial meltwater of the Superior Lobe. In: Mickelson, D.M., Attig, J.W. (Eds.), Glacial Processes Past and Present. Geological Society of America, Special Paper 337, 83–92.

Johnson, M.D., 2000. Pleistocene geology of Polk County, Wisconsin. Wisconsin Geol. Nat. History Surv. Bull. 92, 70pp.

Johnson, M.D., Clayton, L., 2003. Supraglacial landsystems in lowland terrain. In: Evans, D.J.A. (Ed.), Glacial Landsystems. Arnold Publishing, London pp. 228–258.

Johnson, M.D., Hemstad, C., 1998. Glacial Lake Grantsburg: a short-lived lake recording the advance and retreat of the Grantsburg sublobe. In: Patterson, C.J., Wright Jr., H.E. (Eds.), Contributions to Quaternary studies in Minnesota. Minnesota Geological Survey Report of Investigations 49, 49–60.

Johnson, M.D., Mickelson, D.M., Clayton, L., Attig, J.W., 1995. Composition and genesis of glacial hummocks, western Wisconsin, USA. Boreas 24, 97–116.

Johnson, W.H., Hansel, A.K., Bettis III, E.A., Karrow, P.F., Larson, G.J., Lowell, T.V., et al., 1997. Late Quaternary temporal and event classifications, Great Lakes region, North America. Quatern. Res. 47, 1–12.

Kaiser, K.F., 1994. Two Creeks Interstade dated through dendrochronology and AMS. Quatern. Res. 42, 288–298.

Knaeble, A.R., Hobbs, H.C., 2009. Quaternary Stratigraphy. Geologic Atlas of Carlton County, Minnesota. Minnesota Geological Survey, County Atlas Series, C-19 Part A, Plate 4.

Knox, J.C., 1989. Long- and short-term storage and removal of sediment in watersheds of southwestern Wisconsin and northwestern Illinois. In: Hadley, R.F., Ongley, E.D. (Eds.), Sediment and the Environment, International Association of Scientific Hydrology, Publication 184, 157–164.

Knox, J.C., Attig, J.W., 1988. Geology of the pre-Illinoian sediment in the Bridgeport terrace, lower Wisconsin valley, Wisconsin. J. Geol. 96, 505–513.

Knox, J.C., Leigh, D.S., Jacobs, P.M., Mason, J.A., Attig, J.W., 2011. Introduction: part II: Quaternary loess lithostratigraphy in Wisconsin. In: Syverson, K.M., Clayton, L., Attig, J.W., Mickelson, D.M. (Eds.), Lexicon of Pleistocene Stratigraphic Units of Wisconsin: Wisconsin Geol. Nat. History Surv. Technical Report, 1, 14–17.

Leavitt, S.W., Panyushkina, I.P., Lange, T., Cheng, L., Schneider, A.F., Hughes, J., 2007. Radiocarbon "wiggles" in Great Lakes wood at about 10,000 to 12,000 BP. Radiocarbon 49, 855–864.

Leigh, D.S., Knox, J.C., 1993. AMS radiocarbon age of the Upper Mississippi Valley Roxana Silt. Quatern. Res. 39, 282–289.

Leigh, D.S., Knox, J.C., 1994. Loess of the Upper Mississippi Valley Driftless Area. Quatern. Res. 42, 30–40.

Leverett, F., 1932. Quaternary geology of Minnesota and parts of adjacent states. United States Geological Survey Professional Paper 161, 149pp.

Maher Jr., L.J., 1982. The palynology of Devils Lake, Sauk County, Wisconsin. In: Knox, J.C., Clayton, L., Mickelson, D.M. (Eds.), Quaternary History of the Driftless Area, Wisconsin Geol. Nat. History Surv. Field Trip Guidebook 5, 119–135.

Maher Jr., L.J., Mickelson, D.M., 1996. Palynological and radiocarbon evidence for deglaciation events in the Green Bay Lobe, Wisconsin. Quatern. Res. 46, 251–259.

Mason, J.A., Milfred, C.J., Nater, E.A., 1994. Distinguishing soil age and parent material effects on an Ultisol of north-central Wisconsin, USA. Geoderma 61, 165–189.

McCartney, M.C., Mickelson, D.M., 1982. Late Woodfordian and Greatlakean history of the Green Bay Lobe, Wisconsin. Geol. Soc. Am. Bull. 93, 297–302.

Mickelson, D.M., Syverson, K.M., 1997. Quaternary geology of Ozaukee and Washington Counties, Wisconsin. Wisconsin Geol. Nat. History Surv. Bull. 91, 56pp.

Mickelson, D.M., Knox, J.C., Clayton, L., 1982. Glaciation of the Driftless Area: an evaluation of the evidence. In: Knox, J.C., Clayton, L., Mickelson, D.M. (Eds.), Quaternary History of the Driftless Area, Wisconsin Geol. Nat. History Surv. Field Trip Guidebook, vol. 5, 155–169.

Mickelson, D.M., Clayton, L., Fullerton, D.S., Borns Jr., H.W., 1983. The Late Wisconsin glacial record of the Laurentide Ice Sheet in the United States. In: Porter, S.C. (Ed.), The Late Pleistocene. University of Minnesota Press, Minneapolis, pp. 3–37.

Mickelson, D.M., Clayton, L., Baker, R.W., Mode, W.N., Schneider, A.F., 1984. Pleistocene stratigraphic units of Wisconsin. Wisconsin Geol. Nat. History Surv. Miscellaneous Paper 84-1, 15pp.

Mickelson, D.M., Hooyer, T.S., Socha, B.J., Winguth, C., 2007. Late-glacial ice advances and vegetation changes in east-central Wisconsin. In: Hooyer, T.S. (Ed.), Late-Glacial History of East-Central Wisconsin, 72–87 Wisconsin Geol. Nat. History Surv. Open-File Report, 2007-01.

Miller, J., 2000. Glacial stratigraphy and chronology of central southern Wisconsin, west of the Rock River. Unpublished M.S. Thesis, University of Wisconsin, Madison, 148pp.

Moore, P.L., Iverson, N.R., Cohen, D., 2009. Ice flow across a warm-based/cold-based transition at a glacier margin. Ann. Glaciol. 50 (52), 1–8.

Morey, G.B., Setterholm, D.R., 1997. Rare earth elements in weathering profiles and sediments of Minnesota: implications for provenance studies. J. Sediment. Res. 67, 105–115.

Muldoon, M.A., Bradbury, K.R., Mickelson, D.M., Attig, J.W., 1988. Hydrogeologic and geotechnical properties of Pleistocene materials in north-central Wisconsin. Wisconsin Water Resources Center Technical Report 88-03, 58pp.

Rawling III, J.E., Hanson, P.R., Young, A.R., Attig, J.W., 2008. Late Pleistocene dune construction in the Central Sand Plain of Wisconsin, USA. Geomorphology 100, 494–505.

Ronnert, L., 1992. Genesis of diamicton in the Oak Creek Formation of south-east Wisconsin, USA. Sedimentology 39, 177–192.

Rovey II, C.W., Balco, G., 2010. Periglacial climate at the 2.5 Ma onset of Northern Hemisphere glaciation inferred from the Whippoorwill Formation, northern Missouri, USA. Quatern. Res. 73 (1), 151–161.

Rovey II, C.W., Borucki, M.K., 1995. The southern limit of red till deposition in eastern Wisconsin. Geosci. Wisconsin 15, 15–23.

Rovey II, C.W., Kean, W.F., 1996. Pre-Illinoian glacial stratigraphy in north-central Missouri. Quatern. Res. 45, 17–29.

Rovey II, C.W., Kean, W.F., 2001. Palaeomagnetism of the Moberly formation, northern Missouri, confirms a regional magnetic datum within the pre-Illinoian glacial sequence of the midcontinental USA. Boreas 30, 53–60.

Roy, M., Clark, P.U., Barendregt, R.W., Glasmann, J.R., Enkin, R.J., 2004. Glacial stratigraphy and paleomagnetism of late Cenozoic deposits of the north-central United States. Geol. Soc. Am. Bull. 116 (1/2), 30–41.

Schaetzl, R.J., Stanley, K., Scull, P., Attig, J.W., Bigsby, M., Hobbs, T., 2009. An overview of loess distribution in Wisconsin: possible source areas and paleoenvironments. Geol. Soc. Am. Abstr. Programs 41 (4), 22.

Schneider, A.F., 1983. Wisconsinan stratigraphy and glacial sequence in southeastern Wisconsin. Geosci. Wisconsin 7, 59–85.

Shaw, J., 2002. The meltwater hypothesis for subglacial bedforms. Quatern. Int. 90, 5–22.

Simpkins, W.W., 1989. Genesis and spatial distribution of variability in the lithostratigraphic, geotechnical, hydrogeological, and geochemical properties of the Oak Creek Formation in southeastern Wisconsin. Unpublished Ph.D. Thesis, University of Wisconsin, Madison, 857pp.

Simpkins, W.W., Rodenbeck, S.A., Mickelson, D.M., 1990. Geotechnical and hydrogeological properties of till stratigraphic units in Wisconsin: Proceedings of Symposium on Methods and Problems of Till Stratigraphy. LUNDQUA Rep. 32, 11–15.

Socha, B.J., Colgan, P.M., Mickelson, D.M., 1999. Ice-surface profiles and bed conditions of the Green Bay Lobe from 13,000 to 11,000 ^{14}C-years B.P.. In: Mickelson, D.M., Attig, J.W. (Eds.), Glacial Processes Past and Present. Geological Society of America, Special Paper 337, 151–158.

Stanford, S.D., Mickelson, D.M., 1985. Till fabric and deformational structures in drumlins near Waukesha, Wisconsin, U.S.A. J. Glaciol. 31, 220–228.

Stewart, M.T., Mickelson, D.M., 1976. Clay mineralogy and relative age of tills in north-central Wisconsin. J. Sediment. Petrol. 46, 200–205.

Stone, J.O.H., Evans, J.M., Fifield, L.K., Cresswell, R.G., 1998. Cosmogenic chlorine-36 production in calcite by muons. Geochim. Cosmochim. Acta 62, 433–454.

Stuiver, M., Reimer, P.J., 1993. Extended ^{14}C database and revised CALIB radiocarbon calibration program. Radiocarbon 35, 215–230.

Syverson, K.M., 1988. The glacial geology of the Kettle Interlobate Moraine region, Washington County, Wisconsin. Unpublished M.S. Thesis, University of Wisconsin, Madison, 123pp.

Syverson, K.M., 2007. Pleistocene geology of Chippewa County, Wisconsin. Wisconsin Geol. Nat. History Surv. Bull. 103, 53pp.

Syverson, K.M., Colgan, P.M., 2004. The Quaternary of Wisconsin: a review of stratigraphy and glaciation history. In: Ehlers, J., Gibbard, P.L. (Eds.), Quaternary Glaciations—Extent and Chronology. Part II: North America. Elsevier Publishing, Amsterdam, pp. 295–311.

Syverson, K.M., Baker, R.W., Kostka, S., Johnson, M.D., 2005. Pre-Wisconsinan and Wisconsinan glacial stratigraphy, history, and landscape evolution, western Wisconsin. In: Robinson, L. (Ed.), Field Trip Guidebook for Selected Geology in Minnesota and Wisconsin, Minnesota Geological Survey Guidebook, vol. 21, 238–278.

Syverson, K.M., Clayton, L., Attig, J.W., Mickelson, D.M. (Eds.), 2011. Lexicon of Pleistocene stratigraphic units of Wisconsin, Wisconsin Geol. Nat. History Surv. Technical Report, 1, 180pp.

Thornburg, K.L., Syverson, K.M., Hooper, R.L., 2000. Clay mineralogy of till units in western Wisconsin. Geol. Soc. Am. Abstr. Programs 32 (7), A270.

Walker, M., et al., 2009. Formal definition and dating of the GSSP (Global Stratotype Section and Point) for the base of the Holocene using the Greenland NGRIP ice core, and selected auxiliary records. J. Quatern. Sci. 24 (1), 3–17.

Weidman, S., 1907. The geology of north central Wisconsin. Wisconsin Geol. Nat. History Surv. Bull. 16, 697pp.

Whittecar, G.R., 1979. Geomorphic history and Pleistocene stratigraphy of the Pecatonica River valley, Wisconsin and Illinois. Unpublished Ph.D. Thesis, University of Wisconsin, Madison, 195pp.

Whittecar, G.R., Mickelson, D.M., 1977. Sequence of till deposition and erosion in drumlins. Boreas 6, 213–217.

Whittecar, G.R., Mickelson, D.M., 1979. Composition, internal structures, and a hypothesis for the formation of drumlins, Waukesha County, Wisconsin, U.S.A. J. Glaciol. 22, 357–371.

Winguth, C., Mickelson, D.M., Colgan, P.M., Laabs, B.J.C., 2004. Modeling the deglaciation of the Green Bay Lobe of the southern Laurentide Ice Sheet. Boreas 33, 34–47.

Chapter 43

Summary of Early and Middle Pleistocene Glaciations in Northern Missouri, USA

Charles W. Rovey II[1,]* and Greg Balco[2]

[1] *Department of Geography, Geology, and Planning, Missouri State University, 901 S. National, Springfield, Missouri 65897, USA*
[2] *Berkeley Geochronology Center, 2455 Ridge Road, Berkeley, California, USA*
*Correspondence and requests for materials should be addressed to Charles W. Rovey. E-mail: charlesrovey@missouristate.edu

43.1. INTRODUCTION

This chapter summarises recent work on the stratigraphy of glacial sediments at or near the southern limit of North American Pleistocene glaciation in northern Missouri, USA (Fig. 43.1). This stratigraphical sequence records the most extensive advances of the Laurentide Ice Sheet (LIS) during the Early and Middle Pleistocene.

In this chapter, we use "till" in a dual but related sense. The term always indicates a general texture and mode of deposition—a diamicton generated by and deposited from glacial ice. Additionally, the term may refer to lithostratigraphical units that are comprised predominantly of till. We also use the revised Quaternary time scale (Gibbard et al., 2010) for chronological references. Thus, "Early Pleistocene" denotes ages between 2.6 and 0.78 Ma, "Middle Pleistocene" 0.78–0.13 Ma, and "Late Pleistocene" 0.13–0.012 Ma. Finally, many glacial deposits in central North America are commonly referred to by regional stage/episode names. Although the original conception of four Pleistocene glaciations that led to this naming system is obsolete, the terms "Wisconsinan" (referring to the most recent glaciation), "Illinoian" (referring to the penultimate glaciation), and "pre-Illinoian" (referring to all older glaciations) are still in common use, and we use them in this sense here.

Pre-Illinoian glacial deposits of the Midwestern United States are not, in general, well dated, because they are too old for radiocarbon and luminescence techniques. Moreover, they apparently are too old to preserve ice-constructional topography, so ice margins cannot be reconstructed from moraine systems and other landform assemblages. Therefore, the results and interpretations here are generated by (i) observations at numerous isolated exposures, (ii) correlation between these exposures based on quantitative analyses of lithological properties, (iii) palaeomagnetic measurements, and (iv) direct age measurements of individual tills by the cosmogenic-nuclide burial dating technique. We have applied some or all these methods at 30 sections in this study area which expose two or more superposed tills. Of these, 14 preserve three or more tills separated by mature weathering profiles, four expose at least four such tills, and one preserves five tills (the complete stratigraphical sequence).

The resulting stratigraphical synthesis is summarised in Fig. 43.2. Glaciogenic sediments in Missouri range in approximate age from 2.4 Ma to 20 ka and span pre-Illinoian to Wisconsinan Stages. However, direct glacial deposits (i.e. till), which record glacial advances into the state, are limited to the Early and Middle Pleistocene. With the exception of distal valley-train outwash in the major river valleys (Mississippi and Missouri Rivers), Wisconsinan and most of the Illinoian Stage deposits are limited to loess. The Illinoian glaciation (during Marine Isotope Stage (MIS) 6) extended ~140 km south of the confluence of the Mississippi and Missouri Rivers but generally in Illinois east of the present Mississippi River valley. Locally, as near St. Louis, the present channel may have been established slightly east of this ice-marginal position. Illinoian till is periodically exposed north of St. Louis in a small area just west of the Mississippi River (Whitfield, 1995). The extent of this till is not known accurately, but it probably extends no farther than a few kilometres west of the Mississippi. With this minor exception, all known till in Missouri is pre-Illinoian in age. Thus, five major pre-Illinoian glaciations in Missouri are recorded by six widely distributed tills that are separated vertically by four mature weathering profiles.

43.2. EARLY WORK IN MISSOURI

Early workers in Missouri largely followed nomenclature and concepts established in neighbouring states. Descriptions of tills were qualitative, and correlations were based

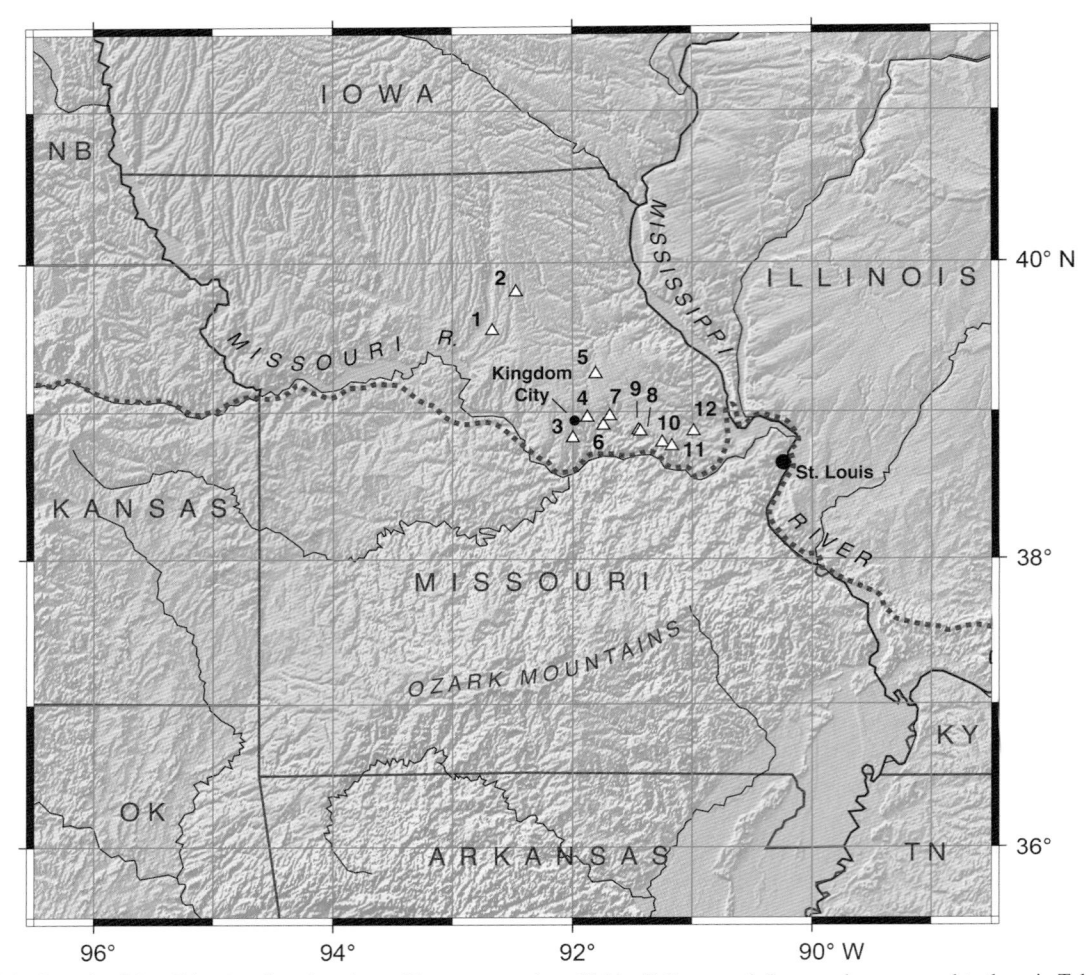

FIGURE 43.1 Location Map. Triangles show locations of important sections (Table 43.1); appended numerals correspond to those in Table 43.1. The dashed line is the (composite) southern margin of glaciation. In Illinois, this line shows the limit of Illinoian (MIS 6) advance; in Missouri and Kansas, this line represents the southern boundary of multiple pre-Illinoian glaciations. "NB" denotes Nebrasksa.

on their position relative to any observed weathering profile, with the assumption that only one such profile occurred within the pre-Illinoian glacial sequence. Apparently, no sections ever exposed more than one weathering profile, and early workers in Missouri followed the now obsolete North American framework of two pre-Illinoian till sheets, consisting of "Kansan" till overlying the older "Nebraskan" (Shipton, 1924; Holmes, 1942; Heim, 1961). This twofold division of "Kansan" and "Nebraskan" influenced perceptions and interpretations of glacial sediments in Missouri into the 1980s. Hallberg (1986, p. 13), in summarising the state of knowledge for Missouri at that time, could only state that "at least two pre-Illinoian tills have been recognised".

Following publication of *The Stratigraphic Succession in Missouri* in 1961 (Koenig, 1961), the Missouri Geological Survey placed little emphasis on glacial and Quaternary sediments. Nevertheless, by the 1970s, various authors had recognised and described (mostly in informal field guides) three or more tills in direct superposition at multiple sites in Missouri (Guccione et al., 1973; Allen and Ward, 1974).

Guccione (1983) introduced an informal lithostratigraphical framework for glaciogenic sediment in Missouri. The present lithostratigraphical nomenclature has grown from Guccione's work with later additions and restrictions by Tandarich (1992) and Rovey and Kean (1996); formation names were formalised by Rovey and Tandarich (2006).

43.3. STRATIGRAPHY AND LITHOLOGIES

Six tills can be recognised and correlated over large portions of northern Missouri (Fig. 43.1, Table 43.1; Rovey and Tandarich, 2006; Balco and Rovey, 2010). These six tills represent five major glaciations, based on the presence of four mature weathering profiles with argillic B horizons developed atop individual tills. Two of the tills are not separated by a significant weathering horizon and appear to

Chapter | 43 Summary of Early and Middle Pleistocene Glaciations in Northern Missouri

Age	Lithostratigraphy		Polarity	Stage/Episode
Late Pleistocene	Peoria Loess and Roxana Silt (0.02 Ma)			Wisconsinan
Middle Pleistocene	Loveland Silt (0.16 Ma)			Illinoian
	McCredie Formation	Macon member (0.2–0.4 Ma)	N	Pre-Illinoian
		Columbia member (0.2–0.4 Ma)	N	
		Fulton member (0.76 Ma)	N	
Early Pleistocene	Moberly Formation (undifferentiated) (1.3 Ma)		R	
			R	
	Atlanta Formation (2.4 Ma)		R	

FIGURE 43.2 Stratigraphy of Glaciogenic Sediment in northeast Missouri. The Roxana and Loveland Silts are mostly loess; the McCredie, Moberly, and Atlanta Formations are mostly till. Members within the McCredie Formation are informal. Approximate ages are given for each stratigraphical unit; see the text for a discussion of these ages and error limits. "N" denotes normal magnetic polarity; "R" reversed.

represent closely spaced ice advances during a single glacial episode. The six tills are grouped into three formations that are recognisable in the field by visual characteristics.

Laboratory characteristics of each till are summarised in Table 43.2. These averages are most representative within the western portion of the study area (Sections 1–9, Fig. 43.1) where the complete stratigraphical sequence was first recognised (Rovey and Kean, 1996). More recently, the Missouri Division of Geology and Land Survey completed a drilling transect between St. Louis and Kingdom City (Fig. 43.1). These results showed that the same stratigraphical sequence is present in the easternmost portion of Missouri as well, and individual tills display the same relative differences in lithology. Nevertheless, analysis of these cores has defined an east–west gradient in some lithological parameters. Tills within the McCredie and Moberly Formations become sandier to the east, and they also contain more igneous-rock fragments within the coarse-sand fraction. The clay mineralogy within these two formations also grades eastwards to compositions with a lower percentage of expandable clay.

The oldest till (Atlanta Formation) is rarely preserved, but it is recognised by a very cobbly texture with a high proportion of sedimentary to igneous clasts. In some cases, a highly weathered residual lag between bedrock and the Moberly Formation is assigned to the Atlanta Formation, based on the presence of erratic lithologies.

The younger Moberly and McCredie Formations have a lower concentration of clasts but a much higher proportion of igneous and metamorphic lithologies. The Moberly Formation usually is the first till above bedrock, and it generally has a higher concentration of organic materials (wood, charcoal, and coal) than the overlying McCredie Formation. The high organic content within the Moberly renders it much more resistant to oxidation than the other tills, and the Moberly is the only till which commonly retains a significant unoxidised zone.

Recently, we recognised two tills with nearly identical lithology within the Moberly Formation (Balco and Rovey, 2010). However, these two tills can be distinguished only where they are in direct superposition and separated by interbedded fluvial/lacustrine sediment and/or a weak weathering profile, marked mainly by a concentration of redox features typical of relatively short (interstadial not interglacial) surface exposure.

The three members of the McCredie Formation represent distinct glaciations, as they are separated vertically by mature weathering profiles. Nevertheless, they are lumped within a single formation, because each one cannot always be identified unambiguously in the field without any stratigraphical context. The (lower) Fulton member is quite distinct, however, based on laboratory parameters. The Fulton has the finest clay-rich texture and also contains a much lower percentage of crystalline rock fragments in the coarse-sand fraction. The youngest (Macon) member generally is preserved in stable landscape positions along interfluves. Even in this position, however, the Macon usually is highly weathered throughout its entire thickness, which precludes an adequate lithological characterisation. Based on three occurrences in the northern portion of the study area, the unaltered Macon till has a sandier texture than the other tills within the McCredie Formation, and it also has a lower percentage of expandable clay minerals (Table 43.2).

Tills within the McCredie Formation have been left as informal members (Rovey and Tandarich, 2006) due to several factors. Among these reasons is some uncertainty in correlating the Macon member among sections. "Macon" is generally applied to a highly weathered till above a discrete weathering profile atop the Columbia member. Thus, superposition, without supporting lithological characterisation, is the only criterion used for its assignment in these cases, and multiple tills of different ages could be lumped within the same stratigraphical unit.

TABLE 43.1 Important Sections in Northeast Missouri

Section/location			
Till unit	Polarity	Age (Ma)	Reference
1. AECI Pit—S.26, T55N, R16W: 39°32.50'N, 92° 40.17'W			Rovey and Kean (1996), Rovey (1997), Rovey and Kean (2001)
Columbia	N		
Fulton			
Moberly	R		
2. SMS92a (core)—S.28, T78N, R14W: 39°48.15'N, 92° 28.45'W			Balco and Rovey (2010), Rovey and Kean (1996)
Macon			
Columbia		0.31±0.50	
Fulton			
Moberly			
3. FU02 (core)—S.24, T47N, R10W: 38°49.30'N, 92°00.00'W			Balco and Rovey (2010), new
Fulton	N	0.84±0.11	
upper Moberly	R		
lower Moberly			
4. Harrison Pit—S.1, T48N, R9W: 38°57.53'N, 91° 52.42'W			Rovey and Kean (1996), Rovey and Kean (2001)
Columbia	N		
Fulton			
Moberly	R		
5. Blum/Sieger Pit—S.33, T52N, R8W: 39°14.88'N, 91° 48.36'W			Balco and Rovey (2010), Rovey (1997), Rovey and Kean (2001), Balco and Rovey (2008)
Macon		0.21±0.18	
Columbia	N		
6. Prairie Fork (core)—S.30, T48N, R7W: 38°54.24'N, 91° 44.58'W			Balco and Rovey (2010), Balco and Rovey (2008), new
Macon		≤0.18±0.15	
Columbia		0.21±0.16	
Fulton			

Section	Unit	Polarity	Age (Ma)	Reference
7. WB 19 (core)—S.3, T48N, R7W: 38°57.84′N, 91°41.22′W				Balco and Rovey (2010)
	Fulton			
	upper Moberly			
	lower Moberly		1.31 ± 0.09	
8. Johnson/Deeker Pit—S.2, T47N, R5W: 38°52.20′N, 91°27.10′W				Rovey and Kean (2001), new
	Columbia			
	Moberly	R		
	Atlanta	R		
9. Musgrove Pit/NF06 (core)—S.1&2, T47N, R5W: 38°51.80′N, 91°26.10′W				Balco and Rovey (2010), Balco et al. (2005), Rovey and Kean (2001), Rovey et al. (2006), Balco and Rovey (2008), Rovey and Balco (2010)
	Columbia		0.74 ± 0.06	
	Moberly	R	1.31 ± 0.15	
	Atlanta	R	2.48 ± 0.15	
10. Pendleton Pit—S.33, T47N, R3W: 38°47.28′N, 91°15.11′W				Balco and Rovey (2010), new
	Fulton	N		
	Moberly	R(?)		
	Atlanta	R	2.28 ± 0.16	
11. Polston Pit—S.7, T49N, R3W: 38°45.79′N, 91°10.45′W				new
	Atlanta	R		
12. WL3 (core)—S.14, T46N, R1W: 38°51.78′N, 90°59.40′W				Balco and Rovey (2010), Balco and Rovey (2008)
	Fulton		0.80 ± 0.06	
	upper Moberly			
	lower Moberly			

This table includes all sections in northeast Missouri with palaeomagnetic and/or age control. Numerals correspond to those in Fig. 43.1; some nearby sections are lumped together, as they cannot be plotted separately at this scale. The lithostratigraphy of till units is shown in Fig. 43.2. Any section with previously unpublished data pertaining to palaeomagnetism or burial dates is denoted by "new."

TABLE 43.2 Chronological, Magnetic, and Lithological Properties of Tills in Northeast Missouri

Age (Ma)	Polarity	Unit	Clay Mineralogy				Texture				Sand-fraction: Lithology			
			E	I	K+C	(n)	Sa	Si	Cl	(n)	Q+F	C+C	I+M	(n)
0.2–0.4	N	Macon	48	23	29	(3)	37	34	29	(3)	61	29	10	(3)
0.2–.04	N	Columbia	54	23	23	(8)	34	34	32	(8)	60	30	10	(8)
0.76	N	Fulton	53	25	22	(6)	26	37	37	(7)	60	36	4	(7)
1.3	R	Moberly	39	34	27	(7)	28	41	31	(7)	53	40	7	(7)
2.4	R	Atlanta	16	49	35	(4)	18	46	36	(4)	16	82	2	(4)

Values within each category are normalised percentages taken from Rovey and Tandarich (2006); these values are means of individual averages at various sites. See Rovey and Tandarich (2006) for procedures, definitions, and site-specific values. Ages for tills are based on the cosmogenic-nuclide burial dating method; see text and Balco and Rovey (2010) for a summary of age measurements and error limits. Rovey and Kean (2001) and Rovey et al. (2006) summarise palaeomagnetic measurements.
Abbreviations: E, expandable clay minerals; I, Illite; K+C, kaolinite+chorite; Sa, sand; Si, silt; Cl, clay; Q+F, quartz+feldspar; C+C, carbonates+chert; I+M, igneous+metamorphic rock fragments; (n) is the number of sites.

43.4. CHRONOLOGY

Prior to the recent advent of the cosmogenic-nuclide burial dating method, only indirect age control could be applied to the pre-Illinoian till sequence in Missouri. The youngest till (Macon member) is present beneath the Loveland Silt (Rovey, 1997), which is widely distributed regionally and has a well-established age of ~ 160 ka at its base (MIS 6). Thus, the entire sequence of Pre-Illinoian till in Missouri must be older than ~ 160 ka.

The Matuyama/Bruhnes boundary is a widespread datum between Early and Middle Pleistocene deposits throughout the midwestern United States (Rovey and Kean, 2001; Roy et al., 2004). Therefore, the palaeomagnetic remanence of these tills and intertill sediments also provides some age control. Specifically, the reversed remanence of the Moberly and Atlanta Formations (Rovey and Kean, 1996, 2001; Rovey et al., 2006) restricts their age to the Early Pleistocene, > 0.78 Ma. The normal depositional remanence within the McCredie Formation, combined with its stratigraphical position beneath Illinoian (MIS 6) sediment, restricts its age to the Middle Pleistocene (0.78–0.13 Ma).

Cosmogenic-nuclide burial dates recently have provided additional age control on the till sequence in Missouri. General aspects of cosmogenic-nuclide dating are summarised in Granger (2006); specific procedures and methods for these till sections are given in Balco and Rovey (2008). Results to date are summarised in Balco et al. (2005), Rovey and Balco (2010), and Balco and Rovey (2010).

The Atlanta till has been dated at two sites (Fig. 43.1, Table 43.1) approximately 20 km apart with excellent consistency. Ages of 2.28 ± 0.16 and 2.48 ± 0.15 Ma agree and yield an error-weighted mean of 2.42 ± 0.14 Ma, which is indistinguishable in age from the first major pulse of isotopically light meltwater in the Gulf of Mexico (Joyce et al., 1993) and the first sustained occurrence of ice-rafted debris in the North Atlantic Ocean (Raymo et al., 1989). The Atlanta Formation appears to record the first major Quaternary expansion of the LIS into central North America.

The Moberly Formation has been dated at two sites approximately 25 km apart, both of which yielded ages of 1.31 Ma with a weighted error limit of 0.089 Ma. Thus, the age of the Moberly is firmly established within the latter portion of the Matuyama Chron.

Three cosmogenic-isotope dates for the Fulton member of the McCredie Formation (Fig. 43.1, Table 43.1) range from 0.74 to 0.84 Ma, with an error-weighted mean of 0.80 ± 0.04 Ma. This mean age is nearly coincident with, but slightly older than, the Matuyama/Bruhnes boundary at 0.78 Ma. Nevertheless, the detrital remanent magnetisation within the Fulton (both till and proglacial-silt facies) is consistently normal, meaning that the true age must be younger than this datum, most likely within MIS 18 (0.71–0.76 Ma, Lisiecke and Raymo, 2005).

Cosmogenic-nuclide burial dating using the ^{10}Be–^{26}Al nuclide pair (as used here) is not well suited for events younger than about 0.5 Ma (Balco and Rovey, 2008), and the relative uncertainty in dates younger than this age increases dramatically. Thus burial dates on the two youngest tills within the McCredie Formation (Macon and Columbia) are relatively imprecise. One analysis for the (younger) Macon member gives an age of 0.21 ± 0.17 Ma, while a second gives a maximum age of $\leq 0.18 \pm 0.15$ Ma. Two dates for the subjacent Columbia member are 0.21 ± 0.16 and 0.31 ± 0.5 Ma. Thus, the burial ages indicate that these two tills are younger than ~ 0.5 Ma but yield little further information. As discussed above, these tills predate the 0.15–0.135 Illinoian glaciation, so they were most likely deposited during MIS 8 (~ 0.25 Ma),

10 (~0.35 Ma), or 12 (~0.45 Ma). This implies that the youngest pre-Illinoian tills in central North America may predate the Illinoian glaciation by only one or two 100,000-year glacial-interglacial cycles, a much shorter time interval than previous assessments (e.g. Richmond and Fullerton, 1986), which are mostly based on the development of weathering profiles. This issue remains to be investigated in more detail.

43.5. GLACIAL BOUNDARIES

Early workers recognised that the pre-Illinoian till sheets do not preserve ice-constructional topography and associated moraine systems. Thus, landforms are of little use in reconstructing ice margins and till-sheet boundaries. Accordingly, various workers produced different interpretations of ice margins within Missouri, but all such reconstructions were based on the old twofold concept of "Nebraskan" followed by "Kansan" glaciation (e.g. Holmes, 1942; Flint, 1957). By the 1960s, a consensus was reached that the composite southern boundary of pre-Illinoian glaciations is nearly coincident with the modern Missouri River valley (e.g. Heim, 1961), and this consensus has endured to the present (Middendorf, 2003). In the eastern portion of the state that boundary is well defined by the common preservation of till along interfluves to within a few kilometres of the valley; we are unaware of any *bona fide* examples of till south of the Missouri River. Reworked erratics within fluvial sediment (Rubey, 1952; Heim, 1961) a few kilometres south of this border may indicate that one or more early glaciations reached a terminus slightly farther south. Nevertheless, the Missouri River is a close approximation to the southern limit of glaciation in eastern Missouri.

In the western portion of the state (west of the "big bend" in the Missouri River, Fig. 43.1), the glacial boundary is more problematic, but most published maps depict it as being a few kilometres to several tens of kilometres south of the Missouri River (Heim, 1961; Middendorf, 2003). Again, till exposures are common just north of the Missouri River valley, but none to our knowledge has been found to the south. Placement of the glacial boundary south of the Missouri River in this area apparently is based on the presence of isolated erratics, but criteria for this placement are vague.

A few reports (e.g. Aber, 1999) have interpreted subtle topographic features and drainage patterns as moraine ice-margin positions in Missouri. Allen (1973) traced curvilinear features on satellite imagery of northeast Missouri and also interpreted these as degraded moraines. However, if these features do represent moraines, they would have to be recessional, because they do not coincide with any till boundaries; all six tills recognised in Missouri (Figs. 43.1 and 43.2) have been found within ~15 km of the Missouri River valley. Therefore, the Missouri River seems to be a good approximation to the ice margin of all the pre-Illinoian glaciations that reached Missouri. Advances farther south of this location may have been prevented by a reversal in the slope of the bedrock topography. The Missouri River channel coincides nearly exactly with the southern margin of the Ozark Dome, and the general bedrock-surface elevation rises rapidly southwards of this location. If a series of thinning ice margins advanced to this boundary, such a reversal could focus the terminus of successive glaciations at nearly the same latitude.

43.6. CORRELATION AND FUTURE WORK

Now that the Pleistocene glacial stratigraphy of northeast Missouri is established, an obvious question for future research is that of the correlation between the Pleistocene section in Missouri and those elsewhere in glaciated central North America. In particular, an informal lithostratigraphy is well established for pre-Illinoian tills northwest of Missouri in western Iowa and eastern Nebraska. The till sequence in these states has some age control based on palaeomagnetic remanence (Easterbrook and Boellstorff, 1984; Roy et al., 2004) and fission-track dates of interbedded tephras (Boellstorff, 1978a,b). Moreover, the stratigraphical, lithological, and palaeomagnetic sequence there is similar to that in northeast Missouri (Rovey and Kean, 1996, 2001; Rovey et al., 2006); three normal-polarity tills overlie three tills with reversed magnetic remanence. The oldest of these reversed-polarity deposits (the "C-Till" of Boellstorff) locally includes a second lower till which likely records a minor ice-margin fluctuation, but this unit generally is represented by a single till.

These observations suggest several possible correlations between the two sequences. First, the oldest glacial deposit in the Nebraska–Iowa sequence is a highly weathered till that at one site underlies tephra dated at ~2 Ma (Boellstorff, 1978a,b; Hallberg, 1986). This till may correlate with the Atlanta Formation in Missouri, but it has not been directly dated. Second, two younger reversed-polarity tills with nearly identical lithology overlie, at one site, a tephra within unweathered silt with a fission-track age of ~1.3 Ma. As these two tills also have similar lithology to the Moberly Formation in Missouri, it is possible that they correlate with the Moberly.

However, available chronological information does not support a one-to-one correlation between the three normal-polarity tills in Nebraska/western Iowa and the three till members of the McCredie Formation in Missouri. Two of the normal-polarity tills in Nebraska/western Iowa are older than a tephra dated at ~0.6 Ma (Boellstorff, 1978a,b). However, the cosmogenic-nuclide burial ages described above indicate that only one of the normal-polarity tills in Missouri is older than this datum. This difference could be explained, if the Middle Pleistocene till sequence is more complex than currently recognised. Specifically, stratigraphical observations in Iowa, Nebraska and Missouri show unambiguously that *at least three* normal-polarity tills are present; however, due to the fragmentary and scattered nature of the exposures and similarity in composition among some normal-polarity tills, it is not possible to prove that *no more than three* such tills are present. Thus, there may be more than three normal-polarity tills within the Iowa–Nebraska sequence, and in Missouri as well.

43.7. SUMMARY AND CONCLUSIONS

Major expansions of the LIS to $\sim 39°$ latitude occurred at about 2.4 and 1.3 Ma during the Early Pleistocene. Another major glaciation reached this same latitude at ~ 0.76 Ma during the Middle Pleistocene shortly after the Matuyama/Bruhnes transition. This episode was followed by at least two more expansions of the LIS to the same position between about 0.2 and 0.4 Ma. Other dating techniques for tills to the north in Nebraska and Iowa have shown that at least two glaciations reached at least as far south as 41° latitude between 0.6 and 0.78 Ma, whereas only one till within this age range is known from Missouri. The relationship between these till sequences and other Early and Middle Pleistocene sections in the region could, in the future, be constrained by cosmogenic-nuclide burial dating of additional glacial sections.

REFERENCES

Aber, J.S., 1999. Pre-Illinoian glacial geomorphology and dynamics in the central United States, west of the Mississippi. In: Mickelson, D.M., Attig, J.W. (Eds.), Glacial Processes Past and Present, Geological Society of America Special Paper, 337, 113–119.

Allen, W.H., Jr., 1973. Pleistocene and engineering geology of north-central Missouri. In: Pleistocene and Engineering Geology of North-Central Missouri, Guidebook, North-Central GSA, April 13, 1973, Miscellaneous Publication No. 26, Missouri Geological Survey, 3–4.

Allen, W.H., Jr., Ward, R.A., 1974. Deeker clay pit. In: Geology of East-Central Missouri with Emphasis on Pennsylvanian Fire Clay and Pleistocene Deposition, Guidebook, Association of Missouri Geologists, 21st Annual Fieldtrip, 19–22.

Balco, G., Rovey, C.W., 2008. An isochron method for cosmogenic-nuclide dating of buried soils and sediments. Am. J. Sci. 308, 1083–1114. doi:10.2475/10.2008.02.

Balco, G., Rovey, C.W., 2010. Absolute chronology for major Pleistocene advances of the Laurentide Ice Sheet. Geology 38 (9), 795–798. doi:10.1130/G30946.1.

Balco, G., Rovey, C.W., Stone, J.O.H., 2005. The first glacial maximum in North America. Science 307, 222. doi:10.1126/science.1103406.

Boellstorff, J., 1978a. Chronology of some Late Cenozoic deposits from the central United States and the ice ages. Trans. Nebr. Acad. Sci. 6, 35–49.

Boellstorff, J., 1978b. North American Pleistocene stages reconsidered in light of probable Pliocene-Pleistocene continental glaciation. Science 202, 305–307. doi:10.1126/science.202.4365.305.

Easterbrook, D.J., Boellstorff, J., 1984. Paleomagnetism and chronology of early Pleistocene tills in the central United States. In: Mahaney, W.C. (Ed.), Correlation of Quaternary Chronologies. GeoBooks, Norwich, England, pp. 73–90.

Flint, R.F., 1957. Glacial and Pleistocene Geology. Wiley and Sons, New York.

Gibbard, P.L., Head, M.J., Walker, M.J.C., 2010. Formal ratification of the Quaternary System/Period and the Pleistocene Series/Epoch with a base at 2.58 Ma. J. Quatern. Sci. 25 (2), 96–102. doi:10.1002/jqs.1338.

Granger, D.E., 2006. A review of burial dating methods using ^{26}Al and ^{10}Be. In: Siame, L.L., Bourlés, D.L., Brown, E.T. (Eds.), In Situ-Produced Cosmogenic Nuclides and Quantification of Geological Processes, Geological Society of America Special Paper, 415, 1–16.

Guccione, M.J., 1983. Quaternary sediments and their weathering history in northcentral Missouri. Boreas 12, 217–226. doi:10.1111/j.1502-3885.1983.tb0034.x.

Guccione, M.J., Davis, S.N., Allen Jr., W.H., Williams, J.H., 1973. Pleistocene and engineering geology of north-central Missouri. Guidebook, North-Central GSA, April 13, 1973, Miscellaneous Publication No. 26, Missouri Geological Survey.

Hallberg, G.R., 1986. Pre- Wisconsin glacial stratigraphy of the central plains region in Iowa, Kansas, and Missouri. In: Richmond, G.M., Fullerton, D.S. (Eds.), Quaternary Glaciations in the Northern Hemisphere, Quatern. Sci. Rev. 5, 11–15.

Heim, G.E., Jr., 1961. Quaternary System. In: Koenig, J.W. (Ed.), The Stratigraphic Succession in Missouri, Division of Geological Survey and Water Resources, 40, pp. 130–136 (2nd Series).

Holmes, C.D., 1942. Nebraskan-Kansan drift boundary in Missouri. Geol. Soc. Am. Bull. 53, 1479–1490.

Joyce, J.E., Tjalsma, L.R.C., Prutzman, J.M., 1993. North American glacial meltwater history for the past 2.3 m.y.: oxygen isotope evidence from the Gulf of Mexico. Geology 21, 483–486. doi:10.1130/0091-7613 (1993)021<0483:NAGMHF>2.3.CO;2.

Koenig, J.W. (Ed.), 1961. The stratigraphic succession in Missouri, In: Missouri Geological Survey and Water Resources, 40 (2nd Series).

Lisiecke, L., Raymo, M., 2005. A Pliocene-Pleistocene stack of 57 globally distributed benthic δ^{18}O records. Paleoceanography 20, doi:10.1029/2004PA001071, PA1003.

Middendorf, M., 2003. Geologic Map of Missouri—Sesquicentennial Edition. Missouri Division of Geology and Land Survey, Rolla, Missouri, USA.

Raymo, M., Ruddiman, W., Backman, J., Clement, B., Martinson, D., 1989. Large Pliocene variation in northern hemisphere ice sheets and north Atlantic deep water circulation. Paleoceanography 4, 413–446. doi:10.1029/89PA00116.

Richmond, G.M., Fullerton, (Eds.), 1986. Quaternary Glaciations in the United States of America, Quaternary Science Reviews, 5.

Rovey, C.W., 1997. The nature and origin of gleyed polygenetic paleosols in the loess covered glacial drift plain of northern Missouri, USA. Catena 31, 153–172. doi:10.1016/S0341-8162(97)00037-4.

Rovey, C.W., Balco, G., 2010. Periglacial climate at the 2.5 Ma onset of Northern Hemisphere glaciation inferred from the Whippoorwill Formation, northern Missouri, USA. Quatern. Res. 73, 151–161. doi:10.1016/j.yqres.2009.09.002.

Rovey, C.W., Kean, W.F., 1996. Pre-Illinoian glacial stratigraphy in north-central Missouri. Quatern. Res. 45, 17–29. doi:10.1006/qres.1996.0002.

Rovey, C.W., Kean, W.F., 2001. Paleomagnetism off the Moberly formation, northern Missouri, confirms a regional magnetic datum within the pre-Illinoian glacial sequence of the midcontinental USA. Boreas 30, 53–60. doi:10.1111/j.1502-3885.2001.tb00988.x.

Rovey, C.W., Tandarich, J.D., 2006. Lithostratigraphy of glacigenic sediments in north-cental Missouri. In: Mandel, R.D. (Ed.), Guidebook of the 18th Biennial Meeting of the American Quaternary Association, Kansas Geological Survey Technical Series 21, pp. 3-A-1–3-A-12.

Rovey, C.W., Kean, W.F., Atkinson, L., 2006. Paleomagnetism of sediments associated with the Atlanta Formation, north-central Missouri, USA. In: Mandel, R.D. (Ed.), Guidebook of the 18th Biennial Meeting of the American Quaternary Association, Kansas Geological Survey Technical Series 21, pp. 3-B-1–3-B-12.

Roy, M., Clark, P.U., Barendregt, R.W., Glasmann, J.R., Enkin, R.J., 2004. Glacial stratigraphy and paleomagnetism of late Cenozoic deposits of the north-central United States. GSA Bull. 116 (1/2), 30–41. doi:10.1130/B25325.1.

Rubey, W.W., 1952. Geology and mineral resources of the Hardin and Brussels Quadrangles (in Illinois). U.S. Geological Survey Professional Paper, 218 .

Shipton, W.D., 1924. The occurrence of Nebraskan drift in northern Missouri. Washington University Studies vol. XII, Scientific Series, No. 1 53–71.

Tandarich, J.D., 1992. A re-evaluation of profile concepts in pedology and geology. Unpublished draft, Ph.D. Dissertation, University of Illinois.

Whitfield, J.W., 1995. Quaternary system (revised). In: Thompson, T.L. (Ed.), The Stratigraphic Succession in Missouri, Missouri Division of Geology and Land Survey, 40, pp. 143–149 (2nd Series)-Revised.

Chapter 44

Pleistocene Glaciation of British Columbia

John J. Clague* and Brent Ward

Department of Earth Sciences, Simon Fraser University, Burnaby, British Columbia, V5A 1S6 Canada
*Correspondence and requests for materials should be addressed to John J. Clague. E-mail: jclague@sfu.ca

44.1. INTRODUCTION

British Columbia lies within the mid-latitudes of the North Hemisphere adjacent to the Pacific Ocean, which is the moisture source for present and past glaciers in the region. The region is dominated by northwest-trending mountain ranges, rolling plateaus, and, on the west, coastal lowlands. The highest peaks in British Columbia are in the St. Elias Mountains (Mt. Fairweather, 4671 m a.s.l.), the southern Coastal Mountains (Mount Waddington, 4016 m a.s.l.), and the southern Rocky Mountains (Mount Robson, 3954 m a.s.l.).

The high mountains of British Columbia today support valley and cirque glaciers and some ice caps. The total current ice cover in British Columbia is about 26,000 km^2, which is probably close to the minimum value for the entire Quaternary.

Most of our knowledge of glaciation in British Columbia comes from the last period of ice-sheet glaciation (the Late Wisconsinan; approximately coincident with Marine Isotope Stage (MIS) 2). Patchy stratigraphical evidence exists for earlier Pleistocene glaciations, but much of the record for these evidence was eroded by Late Wisconsinan glaciers or buried beneath their deposits. Thus our understanding of Pleistocene glaciation in the region may be biased by temporal filtering of events. Nonetheless, the record of the last Cordilleran ice sheet is well preserved, within the range of radiocarbon dating, and readily interpretable. From this record, we now understand that a true ice sheet, formed by coalescence of glaciers over the interior of the province, was a rarely achieved during the Pleistocene. Through nearly all the Pleistocene, the plateaus of interior British Columbia were not covered by an ice sheet; instead, glaciers were confined to high mountains and intermontane valleys. The style of glaciation over these long periods was alpine; the signature of long-lasting alpine glaciation is strong in all high mountain ranges in British Columbia (Fig. 44.1).

The development of the Cordilleran ice sheet is, of course, rooted in alpine glaciation. The current paradigm for establishment of the ice sheet involves its inception in an extended alpine phase that, either progressively or more likely episodically, leads to establishment of a continuous ice cover over the interior of the province during an ice-sheet phase (see below).

44.2. CHARACTER AND EXTENT OF CORDILLERAN ICE SHEET

The Cordillera of western Canada was repeatedly enveloped by a continental ice sheet, known as the Cordilleran ice sheet, during the Pleistocene and latest Pliocene (Flint, 1971; Clague, 1989; Jackson and Clague, 1991). At its maximum extent, the Cordilleran ice sheet and its satellite glaciers covered almost all of British Columbia, as well as southern Yukon Territory and southern Alaska. It extended south into the northwestern conterminous United States (Fig. 44.2). The ice sheet, to a considerable extent, was confined between the high mountain ranges bordering the Canadian Cordillera on the west and east, but large areas on the east flank of the Rocky Mountains and west of the Coast Mountains were also covered by ice. Glaciers in several bordering mountain ranges, such as the Olympics, Cascade, and Mackenzie Mountains and the Queen Charlotte Range, were more or less independent of the ice sheet, even at times of maximum ice cover.

The last Cordilleran ice sheet attained its maximum size in British Columbia where it was up to 900 km wide and reached to 2000–3000 m elevation over the plateaus of the interior (Wilson et al., 1958). When fully formed, the ice sheet in British Columbia probably had the shape of an elongate dish, with gentle slopes in the interior region and steeper slopes at the periphery. It closely resembled the present-day Greenland ice sheet at such times. More commonly, the interior of the ice sheet had an irregular, undulating surface, with several ice divides that shifted through time. These ice divides were subordinate to the main divide along the axis of the Coast Mountains.

FIGURE 44.1 Glaciated landscape, southern Coast Mountains, British Columbia. The erosional landforms visible in this photograph are typical of a landscape sculpted by alpine glaciation and include cirques, arêtes, horns, and U-shaped valleys. Although this area was also covered by the Cordilleran ice sheet for up to a few thousand years during the Late Wisconsinan, alpine glaciers and mountain ice caps were much more important in shaping the landscape during the Quaternary (Province of British Columbia).

In western British Columbia, ice streamed down fjords and valleys in the coastal mountains and covered large areas of the Pacific continental shelf. At these times, parts of the British Columbia continental shelf were exposed due to eustatic lowering of sea level driven by the growth of continental ice sheets on land. Some lobes at the western margin of the ice sheet extended to the shelf edge where they calved into deep water. Glaciers issuing from the southern Coast Mountains and Vancouver Island Ranges coalesced over the Strait of Georgia to produce a great outlet glacier that flowed far into Puget Lowland in Washington State (Waitt and Thorson, 1983). Glaciers streaming down valleys farther to the east likewise terminated as large lobes in eastern Washington, Idaho, and Montana (Waitt and Thorson, 1983).

On the east, ice flowed from the British Columbia interior and the Rocky Mountains and locally coalesced with ice originating in the Keewatin sector of the Laurentide ice sheet over the Rocky Mountain Foothills (Clague, 1989). These ice masses coalesced only rarely, at times of maximum glaciation, most recently about 17,000 years ago (Jackson et al., 1999). At other times, an ice-free zone (the "Ice-free corridor") existed between the Cordilleran and Laurentide ice sheets.

44.3. GROWTH AND DECAY OF CORDILLERAN ICE SHEET

The Cordilleran ice sheet nucleated in the high mountains of British Columbia (Figs. 44.3A and 44.4A). Small mountain ice fields grew, and valley glaciers advanced when climate deteriorated early during each glaciation (Fig. 44.3B; Kerr, 1934; Davis and Mathews, 1944; Fulton, 1991). With continued cooling and an increase in precipitation, glaciers expanded and coalesced to form a more extensive cover of ice in mountains (Fig. 44.4B). The glaciers advanced out of the mountains and across plateaus and lowlands (Fig. 44.4C), eventually coalescing to form an ice sheet that covered most of British Columbia and adjacent areas (Figs. 44.3C and 44.4D). During this period, which spanned thousands of years, the major mountain ranges remained the principal sources of ice, and ice flow was controlled by topography. Ice thickened to such an extent during the final phase of glaciation that one or more domes became established over the interior of British Columbia, with surface flow radially away from their centres. This full continental ice sheet phase of glaciation was rarely achieved. The transition into this final phase was accompanied by a local reversal of ice flow in the Coast Mountains, as the ice divide shifted from the mountain crest eastwards to a position over the British Columbia interior (Stumpf et al., 2000). A comparable westward shift and reversal of flow may also have occurred locally in the Rocky Mountains. The flow reversals resulted from the build-up of ice in the interior to levels higher than the main accumulation areas in the flanking mountains.

The model outlined in the preceding paragraph provides a framework for conceptualising the growth of the Cordilleran ice sheet, but the actual history of the ice sheet is more complicated (Clague, 1989). Ice did not build up in a uniform, monotonic fashion; rather, periods of growth were interrupted by intervals during which glaciers stabilised or receded.

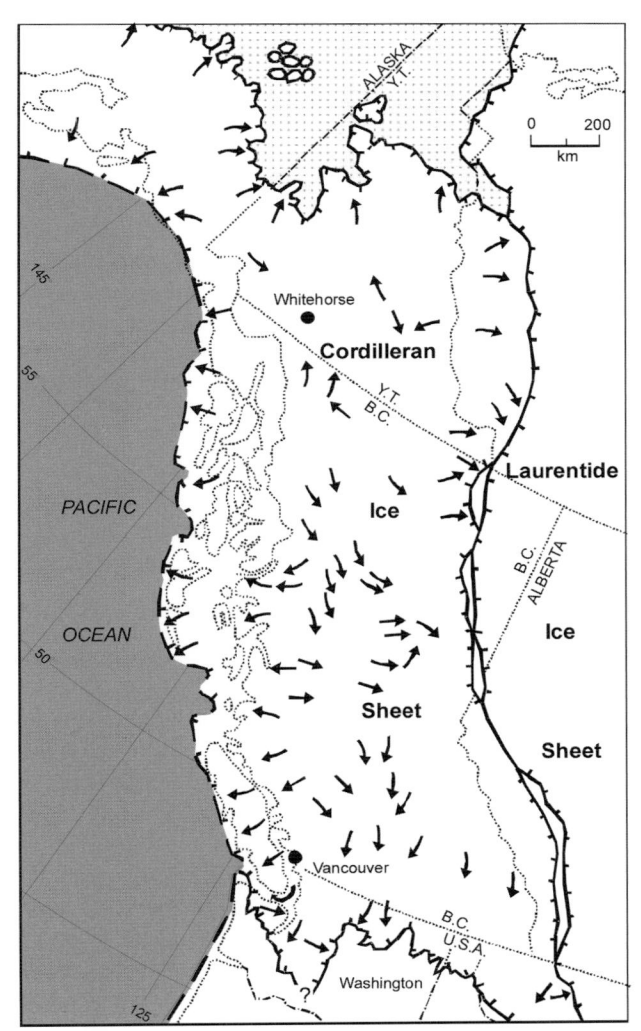

Chapter | 44 Pleistocene Glaciation of British Columbia

FIGURE 44.2 The Cordilleran ice sheet about 17,000 years ago at the maximum of the last glaciation. The upper surface of the ice sheet reached up to 2000–3000 m a.s.l. Arrows indicate directions of ice flow.

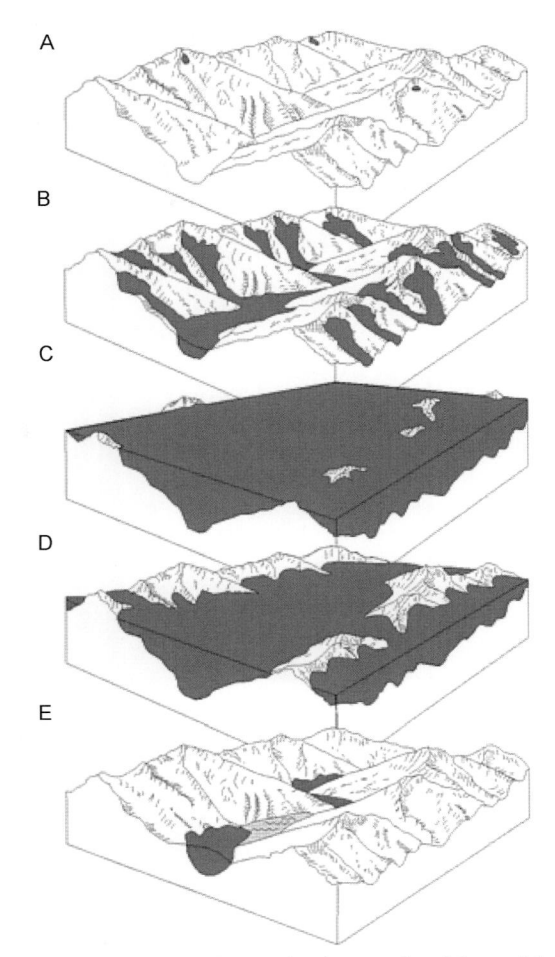

FIGURE 44.3 Schematic diagram showing growth and decay of the Cordilleran ice sheet. (A). Mountain area at the beginning of a glaciation. (B). Development of a network of valley glaciers. (C). Coalescence of valley and piedmont lobes to form an ice sheet. (D) Decay of ice sheet by downwasting; upland areas are deglaciated before adjacent valleys. (E). Residual dead ice masses confined to valleys. (Modified from Clague, 1989, fig. 1.13).

Most glacial cycles terminated with rapid climate warming. Deglaciation was characterised by complex frontal retreat in peripheral glaciated areas and by downwasting accompanied by widespread stagnation throughout the Cordilleran interior (Fulton, 1967). The western periphery of the ice sheet became unstable due, in part, to the global rise in sea level that occurred at such times. The British Columbia continental shelf was rapidly freed of ice, as glaciers calved back to fjord heads and valleys. Frontal retreat also occurred elsewhere along the periphery of the ice sheet, for example, in northern Washington and southern Yukon.

A different style of deglaciation has been documented for areas of low and moderate relief nearer the centre of the ice sheet. Deglaciation in these areas occurred mainly by downwasting and stagnation and proceeded through four stages (Fulton, 1967): (1) active ice phase—regional flow continued but diminished as ice thinned (Fig. 44.3D); (2) transitional upland phase—the highest uplands became ice free but regional flow continued in valleys; (3) stagnant ice phase—ice was confined to valleys but was still thick enough to flow; and (4) dead ice phase—ice tongues in valleys thinned to the point that they no longer flowed (Fig. 44.3E). Geomorphological evidence for this pattern of deglaciation is widespread and convincing: successively lower and younger lake sediments in valleys; flights of ice-marginal channels over vertical ranges of hundreds of metres; and ice-stagnation features at a range of elevations and in a variety of geomorphological positions.

The first areas to become ice-free were those near the periphery of the ice sheet, for example, British

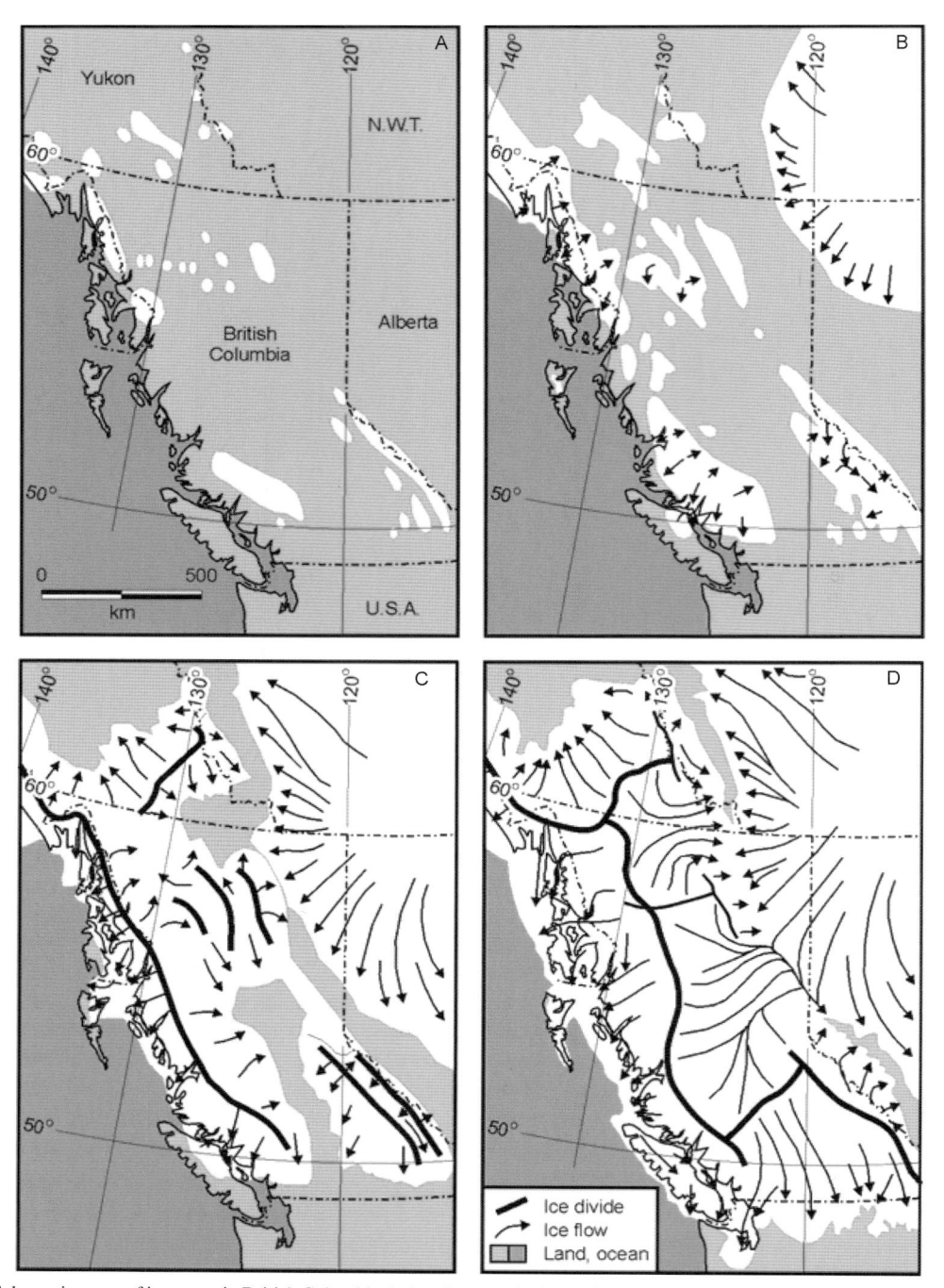

FIGURE 44.4 Schematic maps of ice cover in British Columbia during the growth phase of the last glaciation, about (A) 35,000, (B) 30,000, (C) 25,000, and (D) 18,000 years ago. Ice distributions are inferred from limited data and should only be considered approximations. (Modified from Clague et al., 2004, fig. 4).

Columbia's continental shelf and the plateaus of the northwestern United States (Clague, 1989). Active glaciers probably persisted longest in high mountain valleys, but they may have coexisted with large masses of dead ice on the plateaus of the Cordilleran interior. In general, retreat in the interior proceeded from both southern and northern peripheral areas towards the centre of the ice sheet. In detail, however, the pattern of retreat was complex, with uplands in each region becoming ice-free before adjacent valleys.

Decay of the Cordilleran ice sheet at the close of the last glaciation, and probably during the terminations of earlier glaciations, was interrupted repeatedly by glacier readvances (Alley and Chatwin, 1979; Clague, 1984, 1989; Clague et al., 1997; Friele and Clague, 2002; Kovanen, 2002; Lakeman et al., 2008). Most readvances affected relatively small areas and may not have been synchronous from one region to another.

44.4. GLACIAL EROSION AND DEPOSITION

The ice sheet and the alpine glaciers from which it formed modified the late Tertiary landscape of British Columbia (Mathews, 1989). Mountain areas are dominated by erosional glacial landforms, whereas plateaus, coastal lowlands, and intermontane valleys record both the erosional and depositional effects of glaciers (Clague, 1989). In high mountains, classic alpine forms were created, including cirques and over-deepened valley heads, horns, and comb ridges. Most mountain valleys are typical glacial troughs. Some valleys in the westernmost Cordillera extend into fjords, which attain water depths of up to 750 m. Much of the sediment produced by glacial erosion was transported beyond the periphery of the ice sheet. Large amounts of sediment, however, were deposited in valleys, on coastal lowlands, and on the plateaus of the Cordilleran interior as proglacial and ice-contact sediments, mainly during the advance and recessional phases of the last glaciation. Deposits of older glaciations are less common, because they have been extensively eroded by the last Cordilleran ice sheet. Even in areas where these older deposits are present, they are covered by younger sediments and, consequently, are poorly exposed.

44.5. CRUSTAL DEFORMATION

Growth and decay of the Cordilleran ice sheet triggered crustal movements that were dominantly isostatic in origin (Clague, 1983; Clague and James, 2002; Hetherington et al., 2004). The crust was displaced downward during periods of ice-sheet growth. Initially, the depression was localised beneath the mountain ranges that were loci of glacier growth. The area of crustal subsidence grew larger as glaciers advanced out of mountains and into lowlands. At times of maximum ice cover, the entire area of the ice sheet was displaced downward.

The amount of isostatic depression during times of maximum ice cover depended primarily on the thickness and extent of the ice sheet, the length of time over which it formed, and the structure and composition of the crust and mantle (James et al., 2000). Isostatic depression was greatest beneath the centre of the ice sheet and decreased west of the Coast Mountains and Strait of Georgia towards the continental margin and south into Washington state.

Elevated glaciomarine sediments and shoreline features along the British Columbia and northern Washington coasts provide evidence for isostatic depression at the end of the last glaciation (Clague et al., 1982a; Thorson, 1989). The elevation of the late-glacial marine limit differs in relation to distance from former centres of ice accumulation and time of deglaciation. In general, the marine limit is highest (ca. 200 m a.s.l.) on the British Columbia mainland coast and drops towards the west, southwest, and south. Many mainland fjords, however, were deglaciated after much of the local isostatic rebound had occurred, and consequently, the marine limit in those areas is relatively low. Late-glacial shorelines on Haida Gwaii (Queen Charlotte Islands) were lower than at present, indicating that glacioisostatic depression there was less than the coeval global (eustatic) lowering of sea level (Clague et al., 1982b; Clague, 1983; Josenhans et al., 1997; Barrie and Conway, 2002; Hetherington et al., 2004).

Rapid deglaciation at the end of each glaciation triggered isostatic adjustments that were opposite in direction to those that occurred during ice-sheet growth (Fig. 44.5; Clague, 1983; Clague and James, 2002). Material moved laterally in the mantle from extraglacial regions towards the centre of the decaying ice sheet. Areas at the periphery of the ice sheet, which were deglaciated earliest, rebounded first. The total amount of uplift in these areas, however, was less than at the centre of the ice sheet where ice thicknesses generally were greater. As deglaciation progressed, the zone of rapid isostatic uplift migrated in step with receding glacier margins (Clague, 1983). The rate of uplift in each region decreased exponentially with time, and rebound was largely complete within several thousand years of deglaciation (Clague et al., 1982a; James et al., 2000).

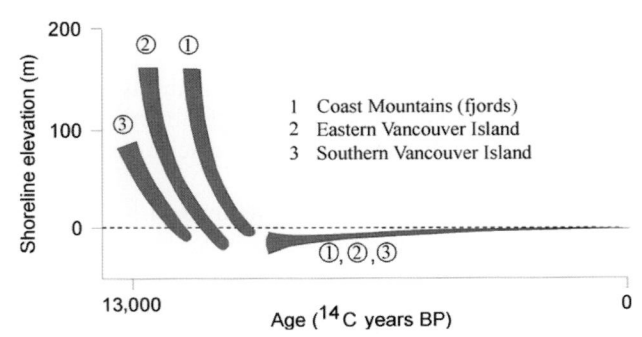

FIGURE 44.5 Generalised patterns of sea-level change on the British Columbia coast since the end of the last glaciation. Deglaciation and isostatic rebound occurred later in the Coast Mountains than on Vancouver Island. (Modified from Muhs et al., 1987, fig. 10).

44.6. STRATIGRAPHICAL RECORD AND CHRONOLOGY

Scattered evidence of Early and Middle Pleistocene glaciations has been reported throughout British Columbia, but reconstructions are sparse and correlations difficult. Near Merritt in southern British Columbia, two sequences of glaciolacustrine sediments separated by a palaeosol are reversely magnetised and thus are older than 780 ka (Fulton et al., 1992). At a site along the Fraser River north of Lillooet, a reversely magnetised till overlies a normally magnetised sequence of glaciofluvial sediments (Lian et al., 1999). This sequence may date to the early part of the Matuyama Chronozone at about 2.6 Ma.

Glacial deposits are interstratified with lavas and pyroclastic sediments on the flanks of Quaternary volcanoes in British Columbia, allowing some older glaciations to be dated. Two basalts overlying tills in the Clearwater River area of east-central British Columbia have been dated by the potassium–argon method to 0.5 and 0.27 Ma (Hickson et al., 1995). At Dog Creek south of Williams Lake, a striated bedrock surface is covered with basalt, which in turn is overlain by a sequence of glacial sediments that is capped by another basalt flow (Mathews and Rouse, 1986). The older basalt has been dated to 2.84 Ma, and the younger one to 1.06 Ma (Graham Andrews, personal communication, 2010). Till in the Stikine River area underlies a basalt flow dated at 330 ka (Spooner et al., 1995).

Volcanic rocks erupted beneath or against glaciers can also provide information on the age of glaciation, but the extent of these glaciers can be difficult to ascertain. Ice-contact volcanic rocks in the Garibaldi Lake area have been dated to 0.75, 0.6, 0.45, 0.13, and 0.090 Ma (Kelman et al., 2002), and others in the Clearwater River area have yielded ages of 1.9, 1.6, and 0.27 Ma (Hickson et al., 1995). In northwest British Columbia, glaciovolcanic eruptions have been documented at >1.1, 0.9, 0.75, 0.43, 0.085, and <0.040 Ma (Spooner et al., 1996; Edwards et al., 2002, 2011; Harder and Russell, 2007). Some of these glacial events have been inferred to be of ice-sheet scale, suggesting that the Cordilleran ice sheet formed and decayed many times during the Pleistocene. The Early and Middle Pleistocene record of glaciation in British Columbia is beginning to approach in completeness that of the marine isotopic record, although the extent of these glaciations remains unclear.

Some subsurface glacial deposits in British Columbia have been assigned a late Middle Pleistocene or early Late Pleistocene age, but because they have not been dated, questions remain as to whether they record MIS 4, MIS 6, or even older glaciation. Some units underlie deposits that are likely Sangamonian (MIS 5) and thus record MIS 6 or older glaciation. One of these units, originally described by Armstrong (1975) in the Fraser Lowland, is Westlynn drift. It comprises a complex sequence of till, glaciofluvial, glaciomarine, and possibly glaciolacustrine sediments. An unnamed till, underlying deposits correlated to MIS 5 on southern Vancouver Island, has been correlated to Westlynn drift by Hicock (1990) and Lian et al. (1995). Bobrowsky and Rutter (1992) describe deposits of an "Early advance" below what they interpreted to be Sangamonian fluvial deposits in the Northern Rocky Mountain Trench in northeast British Columbia. Other till sheets, without stratigraphical relationship to MIS 5 deposits, could also correlate to Westlynn drift, for example two tills beneath MIS 2 till along the Fraser River north of Lillooet (Ryder, 1976).

Glacial deposits assigned to MIS 4 and consisting of till and glaciofluvial and glaciolacustrine sediments have been identified in many parts of British Columbia (Table 44.1). They include the Okanogan Centre Drift in south-central British Columbia (Fulton and Smith, 1978), deposits of the Early Portage Mountain advance in the Northern Rocky Mountain Trench (Bobrowsky and Rutter, 1992), Muchalat River Drift on north-central Vancouver Island (Howes, 1981, 1983), Dashwood Drift (Fyles, 1963) and Mapleguard sediments (Hicock and Armstrong, 1983) on southeast Vancouver Island, and Semiahmoo Drift in the Fraser Lowland (Hicock and Armstrong, 1983). These units are not well dated, although in some cases they appear to be conformably overlain by radiocarbon-dated MIS 3 deposits. There is evidence of extensive MIS 4 glaciation, probably of ice-sheet scale, from an ODP core off the west coast of Vancouver Island. The core records glaciomarine sedimentation similar to that of MIS 2 (Cosma et al., 2008). In addition, an episode of glaciovolcanism in northern British Columbia record probable ice-sheet glaciation during MIS 4 (Edwards et al., 2011).

Deposits and landforms of the MIS 2 Fraser glaciation are ubiquitous and well constrained by radiocarbon ages, allowing robust reconstruction (Table 44.1). Glacier advance in southwestern BC is recorded by a diachronous sheet of outwash termed Quadra Sand. (Clague, 1976; Clague et al., 2005). The unit is up to 50 m thick and is exposed along the margins of the Strait of Georgia and in adjacent mountain valleys. Numerous radiocarbon ages on plant and animal fossils recovered Quadra Sand range from about 29 ^{14}C ka BP at the northern end of the Strait of Georgia to 17 ^{14}C ka BP near Victoria on southeastern Vancouver Island. The outwash was deposited on isostatically depressed braidplains and in deltaic and shallow marine environments. Brackish-water diatoms have been found in laminated silt within Quadra Sand at an elevation of 14 m a.s.l. at Vancouver, indicating isostatic depression of the order of 100 m at 24.6 ^{14}C ka BP (Clague et al., 2005). A similar amount of isostatic depression has been inferred on the west coast of Vancouver Island at ca. 16.5 ^{14}C ka BP (Ward et al., 2003).

TABLE 44.1 Quaternary stratigraphic framework for selected areas of British Columbia

	GEOLOGIC-CLIMATE UNITS	^{14}C years ka BP (not to scale)	FRASER LOWLAND (Armstrong, 1891, 1984; Hicock and Lian, 1995) lowland / mountains	North-Central Vancouver Island (Howes, 1981)	Northern Vancouver Island (Howes, 1983)	South-Central British Columbia (Fulton and Smith, 1978)	Southern Rocky Mountain Trench (Clague, 1975)	Northern Rocky Mountain Trench (Bobrowsky and Rutter, 1992)
HOLOCENE	Postglacial	5	Fraser River sediments, Salish sediments / Sumas Stade	Postglacial sedimets	Postglacial sedimets	Postglacial sedimets	Postglacial sedimets	Postglacial sedimets
PLEISTOCENE — LATE WISCONSINAN (MIS 2)	Fraser Glaciation	10, 11, 14, 15, 17, 19	Capilano sediments / Ft. Langley interval; Vashon Stade; Port Moody Interstade; ? Coquitlam Stade ?	Gold River Drift	Port McNeil Drift	Kamloops Lake Drift	Younger Drift / Interdrift sediments / Older drift	Late Portage Mountain
PLEISTOCENE — MIDDLE WISCONSINAN (MIS 3)	Olympia nonglacial interval	25, 36, 50 — Not calibrated	Quadra Sand; Cowichan Head Formation			Bessette Sediments	'Interglacial' Sediments	Fluvial and lacustrine sediments
PLEISTOCENE — EARLY WISCONSINAN (MIS 4)			Dashwood and Semiahmoo Drift	Muchalat River Drift	Older Drift	Okanagan Centre Drift		Early Portage Mountain
PLEISTOCENE — SANGAMONIUM (MIS 5)			Muir Point Formation and Highbury Sediments			Westwold Sediments		Fluvial sediments
PLEISTOCENE — MIS 6 OR OLDER			Westlynn Drift					Early Advance

Advance outwash deposits are present in interior valleys in British Columbia, but they are generally not well dated. They comprise thick glaciofluvial gravel, conformably overlain by glaciolacustrine sediments deposited in proglacial lakes (Ryder, 1976, 1981; Fulton and Smith, 1978; Howes, 1981; Clague, 1986, 1987). The limited number of radiocarbon ages that bear on the age of these deposits constrain growth of the last Cordilleran ice sheet growth and support the conceptual model outlined above (Fig. 44.4). Glaciers sourced in the mountains of Vancouver Island were advancing 25 ^{14}C ka BP, prior to the island being overtopped by the Cordilleran ice sheet flowing from the east (Howes, 1981). Ice sourced in the high mountains of British Columbia advanced onto the Interior Plateau after 25 ^{14}C ka BP, and parts of the interior remained ice-free until after 19 ^{14}C ka BP (Clague et al., 1980). Glaciomarine sedimentation on the continental slope off central Vancouver Island began about 25 ^{14}C ka BP (Cosma et al., 2008).

The Fraser Lowland and adjacent mountain valleys have been the subject of detailed Quaternary stratigraphical research for a half century. Late Pleistocene events have been detailed at higher resolution here than in other areas of British Columbia (Armstrong et al., 1965; Armstrong, 1975, 1981; Armstrong and Clague, 1977; Clague, 1980, 1981, 1989; Clague et al., 1980, 1988, 1997; Hicock and Armstrong, 1981, 1983, 1985; Hicock et al., 1982, 1999; Miller et al., 1985; Saunders et al., 1987; Hicock and Lian, 1995, 1999; Lian et al., 2001; Kovanen, 2002; Ward and Thomson, 2004). The Fraser glaciation in this region has been divided into the Coquitlam, Vashon, and Sumas stadials (Table 44.1). Nonglacial sediments of the Sisters Creek Formation, radiocarbon dated from 19 to 18 ^{14}C ka BP, separate deposits of the Coquitlam and Vashon stadials (Hicock and Lian, 1995). Similar deposits beneath Vashon till in the eastern Fraser Lowland have been radiocarbon dated from 20 to 19 ^{14}C ka BP (Ward and Thomson, 2004). The Coquitlam and Vashon stadials have been linked to Heinrich events (Hicock et al., 1999), and the former has been correlated to the Evans Creek stadial, an alpine glacier advance in the Cascade Range in Washington State (Armstrong et al., 1965; Hicock et al., 1999; Riedel et al., 2010).

Deposits of the Coquitlam stadial have been identified at seven sites and have maximum and minimum ages of 21.3 and 18.7 ^{14}C ka BP. It was initially proposed to be a brief advance of valley glaciers from the Coast Mountains and a piedmont glacier flowing down the Strait of Georgia (Hicock and Armstrong, 1981). Subsequent work suggested that ice during the Coquitlam stadial was more extensive, with westward flow across the entire Fraser Lowland (Hicock et al., 1999; Hicock and Lian, 1999). Recent work in the Chehalis River watershed indicates that glacier cover during the Coquitlam stadial was not this extensive (Ward and Thomson, 2004). The lower Chehalis River valley was not glaciated until the Vashon stadial, after 19 ^{14}C ka BP.

Ice cover in British Columbia during MIS 2 was greatest during the Vashon stadial (Fig. 44.6). The Cordilleran ice sheet developed to its maximum size at 14.5 ^{14}C ka BP; at that time, the Fraser Lowland was covered by 2 km of ice and ice reached as far south in the Puget Lowland as Olympia, Washington (Armstrong et al., 1965; Clague, 1981; Porter and Swanson, 1998).

The age of maximum ice cover is known in other areas of British Columbia (Fig. 44.6). Glaciers on Haida Gwaii (Queen Charlotte Islands) reached their limit prior to ~16 ^{14}C ka BP and were retreating by ~15 ^{14}C ka BP (Clague et al., 2004). Ice also was thickest on northern Vancouver Island after ~16 ^{14}C ka BP (Ward et al., 2003; Al-Suwaidi et al., 2006). Ice advanced to the edge of the British Columbia continental shelf ca. 16–17 ^{14}C ka BP and retreated shortly thereafter, based on records from two core sites in the eastern North Pacific Ocean (Blaise et al., 1990; Cosma et al., 2008). Based on these data, it appears that the Cordilleran ice sheet achieved its maximum extent when the Laurentide Ice Sheet was retreating.

Deglaciation was rapid, triggered both by climate warming and by calving at the western margin of the ice sheet. By 13 ^{14}C ka BP, what are now Vancouver, Victoria, and Prince Rupert were free of ice (Fulton, 1971; Armstrong, 1981; Huntley et al., 2001). The entire Strait of Georgia was completely deglaciated shortly thereafter.

Early retreat was followed by a period during which glaciers stabilised at pinning positions at the front of the Coast Mountains and fluctuated about those positions for 1500–2000 year. As an example, the glacier in the Fraser Lowland retreated rapidly eastwards across the Fraser Lowland near Vancouver shortly after 13.0 ^{14}C ka BP but stabilised near Abbotsford, where it advanced and retreated several times between 13 and 10.5 ^{14}C ka BP (Armstrong, 1981; Clague et al., 1997; Kovanen, 2002). These late Glacial advances were of the order of several kilometres to 20 km and were caused by climate and non-climatic factors, including perhaps rapid emergence of the Fraser Lowland. At least one of the advances occurred during the Younger Dryas Chronozone.

Deglaciation in the interior also was rapid. The Cordilleran ice sheet persisted until about 11.0–10.5 ^{14}C ka BP; by

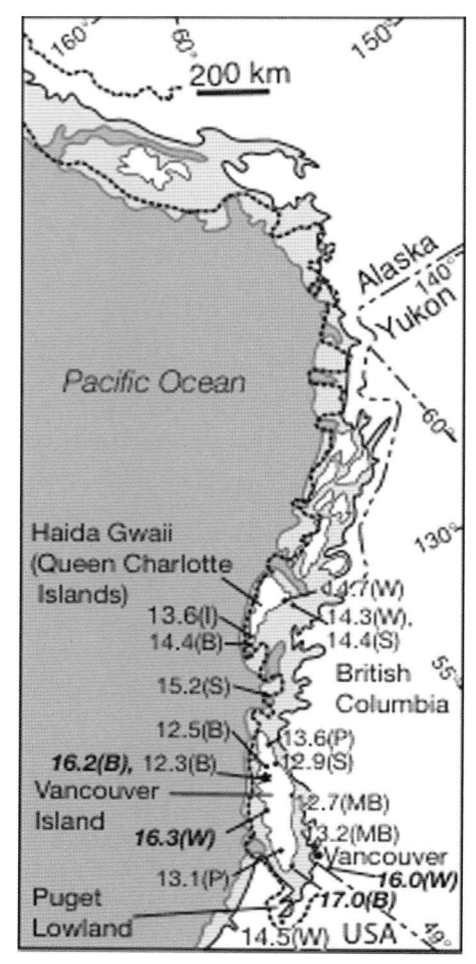

FIGURE 44.6 Time of the Last Glacial Maximum along the British Columbia coast. The continental shelf (<200 m water depth) is shaded light grey. The dotted line delineates the approximate maximum extent of the Cordilleran ice sheet during MIS 2 (Clague, 1989; Blaise et al., 1990; Manley and Kaufman, 2002). Radiocarbon ages (in ^{14}C ka BP) pertaining to maximum extent of the ice sheet are bolded and italicised; limiting minimum radiocarbon ages are shown in plain font (Clague et al., 1982b; 1988; Blaise et al., 1990; Howes, 1997; Nagorsen and Keddie, 2000; Hetherington et al., 2004; Ramsey et al., 2004; Lacourse et al., 2005; Al-Suwaidi et al., 2006). S = shell; W = wood/terrestrial macrofossil; B = bone; P = basal peat; MB = marine mammal bone; I = insect. (Modified from Ward et al., 2003.)

9.5 ^{14}C ka BP, glaciers in the Coast Mountains were probably no more extensive than they are today (Clague, 1981).

REFERENCES

Alley, N.F., Chatwin, S.C., 1979. Late Pleistocene history and geomorphology, southwestern Vancouver Island, British Columbia. Can. J. Earth Sci. 16, 1645–1657.

Al-Suwaidi, M., Ward, B.C., Wilson, M.C., Hebda, R.J., Nagorsen, D.W., Marshall, D., et al., 2006. Late Wisconsinan Port Eliza cave sediments and biotic remains and their implications for human coastal migration, Vancouver Island, Canada. Geoarchaeology 21, 307–332.

Armstrong, J.E., 1975. Quaternary geology, stratigraphic studies and revaluation of terrain inventory maps, Fraser Lowland, British Columbia. Geological Survey of Canada, Ottawa, ON Paper 75-1A: pp. 377–380.

Armstrong, J.E., 1981. Post-Vashon Wisconsin Glaciation, Fraser Lowland, British Columbia. Geological Survey of Canada, Ottawa, ON, 34 pp.

Armstrong, J.E., Clague, J.J., 1977. Two major Wisconsin lithostratigraphic units in southwest British Columbia. Can. J. Earth Sci. 14, 1471–1480.

Armstrong, J.E., Crandell, D.R., Easterbrook, D.J., Noble, J.B., 1965. Late Pleistocene stratigraphy and chronology in southwestern British Columbia and northwestern Washington. Geol. Soc. Am. Bull. 76, 321–330.

Barrie, J.V., Conway, K.W., 2002. Rapid sea-level change and coastal evolution on the Pacific margin of Canada. Sed. Geol. 150, 171–183.

Blaise, B., Clague, J.J., Mathewes, R.W., 1990. Time of maximum late Wisconsinan Glaciation, west coast of Canada. Quatern. Res. 47, 140–146.

Bobrowsky, P.T., Rutter, N.W., 1992. The Quaternary geologic history of the Canadian Rocky Mountains. Géogr. phys. Quatern. 46, 5–50.

Clague, J.J., 1976. Quadra Sand and its relation to the late Wisconsin glaciation of southwest British Columbia. Can. J. Earth Sci. 13, 803–815.

Clague, J.J., 1980. Late Quaternary Geology and Geochronology of British Columbia. Part 1: Radiocarbon dates Geological Survey of Canada, Ottawa, ON. Paper 80–13, 28 pp.

Clague, J.J., 1981. Late Quaternary Geology and Geochronology of British Columbia. Part 2: Summary and Discussion of Radiocarbon-Dated Quaternary History. Geological Survey of Canada, Ottawa, ON. Paper 80–35, 41 pp.

Clague, J.J., 1983. Part 1: Radiocarbon dates Glacio-isostatic effects of the Cordilleran Ice Sheet, British Columbia, Canada. In: Smith, D.E., Dawson, A.G. (Eds.), Shorelines and Isostasy. Academic Press, London, pp. 321–343.

Clague, J.J., 1984. Quaternary Geology and Geomorphology, Smithers-Terrace-Prince Rupert Area, British Columbia. Geological Survey of Canada, Ottawa, ON, Memoir 413, 71 pp.

Clague, J.J., 1986. The Quaternary stratigraphic record of British Columbia—evidence for episodic sedimentation and erosion controlled by glaciation. Can. J. Earth Sci. 23, 885–894.

Clague, J.J., 1987. Quaternary stratigraphy and history, Williams Lake, British Columbia. Can. J. Earth Sci. 24, 147–158.

Clague, J.J., 1989. Cordilleran ice sheet. In: Fulton, R.J. (Ed.), Chapter 1 of Quaternary Geology of Canada and Greenland. Geological Survey of Canada, Ottawa, ON, pp. 40–43.

Clague, J.J., James, T.S., 2002. History and isostatic effects of the last ice sheet in southern British Columbia. Quatern. Sci. Rev. 21, 71–87.

Clague, J.J., Armstrong, J.E., Mathews, W.H., 1980. Advance of the late Wisconsinan Cordilleran Ice Sheet in southern British Columbia since 22,000 yr BP. Quatern. Res. 13, 322–326.

Clague, J., Harper, J.R., Hebda, R.J., Howes, D.E., 1982a. Late Quaternary sea levels and crustal movements, coastal British Columbia. Can. J. Earth Sci. 19, 597–618.

Clague, J.J., Mathewes, R.W., Warner, B.G., 1982b. Late Quaternary geology of eastern Graham Island, Queen Charlotte Islands, British Columbia. Can. J. Earth Sci. 19, 1786–1795.

Clague, J.J., Saunders, I.R., Roberts, M.C., 1988. Ice free conditions in southwestern British Columbia at 16 000 years BP. Can. J. Earth Sci. 25, 938–941.

Clague, J.J., Mathewes, R.W., Guilbault, J.P., Hutchinson, I., Ricketts, B.D., 1997. Pre-Younger Dryas resurgence of the southwestern margin of the Cordilleran ice sheet, British Columbia, Canada. Boreas 26, 261–278.

Clague, J.J., Mathewes, R.W., Ager, T.A., 2004. Environments of northwestern North America before the last glacial maximum. In: Madsen, D. (Ed.), Entering America: Northeast Asia and Beringia Before the Last Glacial Maximum. University of Utah Press, Salt Lake City, UT, 63–94.

Clague, J.J., Froese, D., Hutchinson, I., James, T.S., Simon, K.M., 2005. Early growth of the last Cordilleran ice sheet deduced from glacio-isostatic depression in southwest British Columbia, Canada. Quatern. Res. 63, 53–59.

Cosma, T.N., Hendy, I.L., Chang, A.S., 2008. Chronological constraints on Cordilleran Ice Sheet glaciomarine sedimentation from core MD02-2496 off Vancouver Island (Western Canada). Quatern. Sci. Rev. 27, 941–955.

Davis, N.F.G., Mathews, W.H., 1944. Four phases of glaciation with illustrations from southwestern British Columbia. J. Geol. 52, 403–413.

Edwards, B.R., Russell, J.K., Simpson, K., 2011. Volcanology and petrology of Mathews Tuya, northern British Columbia, Canada: constraints on, interpretations of the 730 ka Cordilleran paleoclimate. Bull. Volcanol, doi 10.1007/s00445-010-0418-z.

Edwards, B.R., Russell, J.K., Anderson, R.G., 2002. Subglacial phonolitic volcanism at Hoodoo Mountain volcano, northern Canadian Cordillera. Bull. Volcanol. 64, 254–272.

Flint, R.F., 1971. Glacial and Quaternary Geology. John Wiley & Sons, New York, 892 pp.

Friele, P.A., Clague, J.J., 2002. Readvance of glaciers in the British Columbia Coast Mountains at the end of the last glaciation. Quatern. Int. 87, 45–58.

Fulton, R.J., 1967. Deglaciation Studies in Kamloops Region, an Area of Moderate Relief, British Columbia. Geological Survey of Canada, Ottawa, ON, 36 pp.

Fulton, R.J., 1971. Radiocarbon Geochronology of Southern British Columbia. Geological Survey of Canada, Ottawa, ON, 28 pp.

Fulton, R.J., 1991. A conceptual model for growth and decay of the Cordilleran Ice Sheet. Géogr. phys. Quatern. 45, 281–286.

Fulton, R.J., Smith, G.W., 1978. Late Pleistocene stratigraphy of South-central British Columbia. Can. J. Earth Sci. 15, 971–980.

Fulton, R.J., Irving, E., Wheadon, P.M., 1992. Stratigraphy and paleomagnetism of Brunhes and Matuyama (>790 ka) Quaternary deposits at Merritt, British Columbia. Can. J. Earth Sci. 29, 76–92.

Fyles, J.G., 1963. Surficial Geology of Horne Lake and Parksville Map Areas, Vancouver Island, British Columbia. Geological Survey of Canada, Ottawa, ON, 142 pp.

Harder, M., Russell, J.K., 2007. Basanite glaciovolcanism at Langorse Mountain, northern British Columbia, Canada. Bull. Volcanol. 69, 329–340.

Hetherington, R., Barrie, J.V., Reid, R.G.B., MacLeod, R., Smith, D.J., 2004. Paleogeography, glacially induced crustal displacement, and late Quaternary coastlines on the continental shelf of British Columbia, Canada. Quatern. Sci. Rev. 23, 295–318.

Hickson, C.J., Moore, J.G., Calk, L., Metcalfe, P., 1995. Intraglacial volcanism in the Wells Gray—Clearwater volcanic field. Can. J. Earth Sci. 32, 838–851.

Hicock, S.R., 1990. Last interglacial Muir Point Formation, Vancouver Island, British Columbia. Géogr. phys. Quatern. 44, 337–340.

Hicock, S.R., Armstrong, J.E., 1981. Coquitlam Drift: a pre-Vashon Fraser glacial formation in the Fraser Lowland, British Columbia. Can. J. Earth Sci. 18, 1443–1451.

Hicock, S.R., Armstrong, J.E., 1983. Four Pleistocene formations in southwest British Columbia: their implications for patterns of sedimentation of possible Sangamonian to early Wisconsinan age. Can. J. Earth Sci. 20, 1232–1247.

Hicock, S.R., Armstrong, J.E., 1985. Vashon Drift: definition of the formation in the Georgia Depression, southwest British Columbia. Can. J. Earth Sci. 22, 748–757.

Hicock, S.R., Lian, O.B., 1995. The Sisters Creek Formation: Pleistocene sediments representing a nonglacial interval in southwestern British Columbia at about 18 ka. Can. J. Earth Sci. 32, 758–767.

Hicock, S.R., Lian, O.B., 1999. Cordilleran Ice Sheet lobal interactions and glaciotectonic superposition through stadial maxima along a mountain front in southwestern British Columbia. Boreas 28, 531–542.

Hicock, S.R., Hebda, R.J., Armstrong, J.E., 1982. Lag of the Fraser glacial maximum in the Pacific Northwest: pollen and macrofossil evidence from western Fraser Lowland, British Columbia. Can. J. Earth Sci. 19, 2288–2296.

Hicock, S.R., Lian, O.B., Mathewes, R.W., 1999. 'Bond cycles' recorded in terrestrial Pleistocene sediments of southwestern British Columbia, Canada. J. Quatern. Sci. 14, 443–449.

Howes, D.E., 1981. Late Quaternary sediments and geomorphic history of north-central Vancouver Island. Can. J. Earth Sci. 18, 1–12.

Howes, D.E., 1983. Late Quaternary sediments and geomorphic history of northern Vancouver Island, British Columbia. Can. J. Earth Sci. 20, 57–65.

Howes, D.E., 1997. Quaternary geology of Brooks Peninsula, 1997. In: Hebda, R.J., Haggarty, J.C. (Eds.), Brooks Peninsula; an Ice Age Refugium on Vancouver Island. B.C. Ministry of Environment, Lands and Parks, Victoria, British Columbia, pp. 3.1–3.19.

Huntley, D.H., Bobrowsky, P.T., Clague, J.J., 2001. Ocean drilling program leg 169S: surficial geology, stratigraphy and geomorphology of the Saanich Inlet area, southeastern Vancouver Island, British Columbia. Mar. Geol. 174, 27–41.

Jackson Jr., L.E., Clague, J.J. (Eds.), 1991. The Cordilleran Ice Sheet, In: Géogr. Phys. Quatern. 45, 261–377.

Jackson Jr., L.E., Phillips, F.M., Little, E.C., 1999. Cosmogenic ^{36}Cl dating of the maximum limit of the Laurentide ice sheet in southwestern Alberta. Can. J. Earth Sci. 36, 1347–1356.

James, T.S., Clague, J.J., Wang, K., Hutchinson, I., 2000. Postglacial rebound at the northern Cascadia subduction zone. Quatern. Sci. Rev. 19, 1527–1541.

Josenhans, H., Fedje, D., Pienitz, R., Southon, J., 1997. Early humans and rapidly changing Holocene sea levels in the Queen Charlotte Islands-Hecate Strait, British Columbia, Canada. Science 277, 71–74.

Kelman, M.C., Russell, J.K., Hickson, C.J., 2002. Effusive intermediate glaciovolcanism in the Garibaldi Volcanic Belt, southwestern British Columbia, Canada. In: Smellis, J.L., Chapman, M.G. (Eds.), Volcano Ice Interactions on Earth and Mars. Geological Society, London, pp. 195–211.

Kerr, F.A., 1934. Glaciation in northern British Columbia. R. Soc. Can. Trans. 28 (4), ser. 3, 17–31.

Kovanen, D.J., 2002. Morphologic and stratigraphic evidence for Allerod and Younger Dyras age glacier fluctuations of the Cordilleran Ice Sheet, British Columbia, Canada, and northwest Washington, USA. Boreas 31, 163–184.

Lacourse, T., Mathewes, R., Fedje, D.W., 2005. Late-glacial vegetation dynamics of the Queen Charlotte Islands and adjacent continental shelf, British Columbia, Canada. Palaeogeogr. Palaeoclimatol. Palaeoecol. 226, 36–57.

Lakeman, T.R., Clague, J.J., Menounos, B., 2008. Advance of alpine glaciers during final retreat of the Cordilleran ice sheet in the Finlay River area, northern British Columbia, Canada. Quatern. Res. 69, 188–200.

Lian, O.B., Hu, J., Huntley, D.J., Hicock, S.R., 1995. Optical dating studies of Quaternary organic-rich sediments from southwestern British Columbia and northwestern Washington State. Can. J. Earth Sci. 32, 1194–1207.

Lian, O.B., Barendregt, R.W., Enkin, R.J., 1999. Lithostratigraphy and paleomagnetism of pre-Fraser glacial deposits in south central British Columbia. Can. J. Earth Sci. 36, 1357–1370.

Lian, O.B., Mathewes, R.W., Hicock, S.R., 2001. Paleoenvironmental reconstruction of the Port Moody Interstade, a nonglacial interval in southwestern British Columbia at about 18 000 ^{14}C years BP. Can. J. Earth Sci. 38, 943–952.

Manley, W.F., Kaufman, D.S., 2002. Alaska PaleoGlacier Atlas. University of Colorado, Institute of Arctic and Alpine Research, Boulder, CO. http://instaar.colorado.edu/QGISL/akpaleoglacieratlas, v. 1.

Mathews, W.H., 1989. Development of Cordilleran landscapes during the Quaternary. In: Fulton, R.J. (Ed.), Chapter 1 of Quaternary Geology of Canada and Greenland. Geological Survey of Canada, Ottawa, ON, pp. 32–34.

Mathews, W.H., Rouse, G.E., 1986. An Early Pleistocene proglacial succession in south-central British Columbia. Can. J. Earth Sci. 23, 1796–1803.

Miller, R.F., Morgan, A.V., Hicock, S.R., 1985. Pre-Vashon fossil Coleoptera of Fraser age from the Fraser Lowland, British Columbia. Can. J. Earth Sci. 22, 498–505.

Muhs, D.R., Thorson, R.M., Clague, J.J., Mathews, W.H., McDowell, P.F., Kelsey, H.M., 1987. Pacific Coast and Mountain System. In: Graf, W.L. (Ed.), Geomorphic Systems of North America. Geological Society of America, Boulder, CO, pp. 517–581.

Nagorsen, D.W., Keddie, G., 2000. Late Pleistocene mountain goat (*Oreamnos america*nus) from Vancouver Island; biogeographic implications. J. Mammal. 81, 666–675.

Porter, S.C., Swanson, T.W., 1998. Radiocarbon age constraints on rates of advance and retreat of the Puget Lobe of the Cordilleran ice sheet during the last glaciation. Quatern. Res. 50, 205–213.

Ramsey, C.L., Griffiths, P.A., Fedje, D.W., Wigen, R.J., Mackie, Q., 2004. Preliminary investigations of a late Wisconsinan fauna from K1 cave, Queen Charlotte Islands (Haida Gwaii), Canada. Quatern. Res. 62, 105–109.

Riedel, J.L., Clague, J.J., Ward, B.C., 2010. Timing and extent of early marine isotope stage 2 alpine glaciation in Skagit Valley, Washington. Quatern. Res. 73, 313–373.

Ryder, J.M., 1976. Terrain Inventory and Quaternary Geology, Ashcroft, British Columbia. Geological Survey of Canada, Ottawa, ON. Paper 74–49, 17 pp.

Ryder, J.M., 1981. Terrain Inventory and Quaternary Geology, Lytton, British Columbia. Geological Survey of Canada, Ottawa, ON. Paper 79-25, 20 pp.

Saunders, I.R., Clague, J.J., Roberts, M.C., 1987. Deglaciation of Chilliwack River valley, British Columbia. Can. J. Earth Sci. 24, 915–923.

Spooner, I.S., Osborn, G.D., Barendregt, R.W., Irvine, E., 1995. A record of Early Pleistocene glaciation on the Mt. Edziza Plateau, northwestern British Columbia. Can. J. Earth Sci. 32, 2046–2056.

Spooner, I.S., Osborn, G.D., Barendregt, R.W., Irvine, E., 1996. A Middle Pleistocene (isotope stage 10) glacial sequence in the Stikine River valley, British Columbia. Can. J. Earth Sci. 33, 1428–1438.

Stumpf, A.J., Broster, B.E., Levson, V.M., 2000. Multiphase flow of the late Wisconsinan Cordilleran ice sheet in western Canada. Geol. Soc. Am. Bull. 112, 1850–1863.

Thorson, R.M., 1989. Glacio-isostatic response of the Puget Sound area, Washington. Geol. Soc. Am. Bull. 101, 1163–1174.

Waitt Jr., R.B., Thorson, R.M., 1983. The Cordilleran ice sheet in Washington, Idaho, and Montana. In: Porter, S.C. (Ed.), Late-Quaternary Environments of the United States, vol. 1, The Late Pleistocene. University of Minnesota Press, Minneapolis, MN, pp. 53–70.

Ward, B.C., Thomson, B., 2004. Late Pleistocene stratigraphy and chronology of lower Chehalis River Valley, southwestern British Columbia: evidence for a restricted Coquitlam Stade. Can. J. Earth Sci. 41, 881–895.

Ward, B.C., Wilson, M.C., Nagorsen, D.W., Nelson, D.E., Driver, J.C., Wigen, B., 2003. Port Eliza cave: North American west coast interstadial environment and implications for human migrations. Quatern. Sci. Rev. 22, 1383–1388.

Wilson, J.T., Falconer, G., Mathews, W.H., Prest, V.K., (Compilers), 1958. Glacial Map of Canada. Geological Association of Canada, Toronto, ON.

Chapter 45

Limits of Successive Middle and Late Pleistocene Continental Ice Sheets, Interior Plains of Southern and Central Alberta and Adjacent Areas

Lionel E. Jackson[1,*], Laurence D. Andriashek[2] and Fred M. Phillips[3]

[1] Geological Survey of Canada, Pacific Division, 625 Robson Street, Vancouver, British Columbia, Canada V6B 5J3
[2] E.R.C.B., Alberta Geological Survey, 4th Floor, 4999-98th Avenue, Edmonton, Alberta, Canada T6B 2X3
[3] Department of Earth and Environmental Science, New Mexico Tech, Sorcorro, New Mexico 87801, USA
*Correspondence and requests for materials should be addressed to Lionel E. Jackson. E-mail: lijackso@nrcan.gc.ca

45.1. INTRODUCTION

Jackson and Little (2004) concluded that the last glacial maximum (LGM) was the only time that the Rocky Mountain Foothills of Alberta, Canada were glaciated by a *continental* ice sheet (Fig. 45.1). They showed that stacked till sequences in this region can be related to belts of moraine and glacial limits dated to the last (Late Wisconsinan) glaciation (marine isotope stage (MIS) 2) by cosmogenic ^{36}Cl dating of erratics situated on these moraines. This conclusion was not novel: it was previously articulated a half century earlier by Horberg (1952, 1954) and further corroborated by Wagner (1966). It has also been documented in northern, central and south-central Alberta by radiocarbon dating of organic material in preglacial gravels underlying a single continental till in those areas (Liverman et al., 1989; Young et al., 1994, 1999).

These conclusions contrasted strongly with previous workers in Alberta (e.g. Stalker and Harrison, 1977) and in the western plains of the State of Montana, USA, where extensive areas of drift have been referred to the Illinoian Stage (last magnetic reversal to MIS 6; Fullerton et al., 2004). The Illinoian age was assigned by these workers-based relative dating techniques: no absolute dating had documented the Illinoian age designation for those areas.

Problems with the resolution of ages of drift sheets in this region are rooted in its geography and climate: in addition to being an international boundary, the 49th parallel roughly parallels the continental divide between drainage into Hudson Bay on its north side and the Gulf of Mexico on its south side (Fig. 45.1). Construction and correlation of glacial stratigraphy commonly employ examination and tracing of stratigraphy along river valleys. The Gulf of Mexico–Hudson Bay drainage divide causes glacial stratigraphical solitudes north and south of the 49th parallel. Further, the arid climate of the region during and between glaciations and abundant detrital coal and mineral carbonate content in glacial sediments have frustrated the use of radiocarbon dating of glacial maxima and interglacial or interstadial sediments (MacDonald et al., 1987). With the exception of palaeomagnetism (Barendregt, 1995; Verosub, 2000), which can assign glacial sediments to greater or lesser age than the last (Matuyama/Bruhnes) magnetic reversal (780 ka: Cande and Kent, 1992) or fortuitous preservation of tephras (Barendregt et al., 1991), confident absolute dating of glacial sediments dating from the Middle and Late Pleistocene (magnetically normal Brunhes Chron) has been problematic.

The first part of this chapter discusses new developments in cosmogenic dating of glacial limits on both sides of the International Boundary that further demonstrate a single glaciation of the Interior Plains southwest Alberta

---

This chapter is dedicated to the memory of Dr. David N. Proudfoot who died before his dissertation on the Pleistocene stratigraphy of the Medicine Hat area could be published in the wider geological literature. He is fondly remembered as a true friend of the Pleistocene.

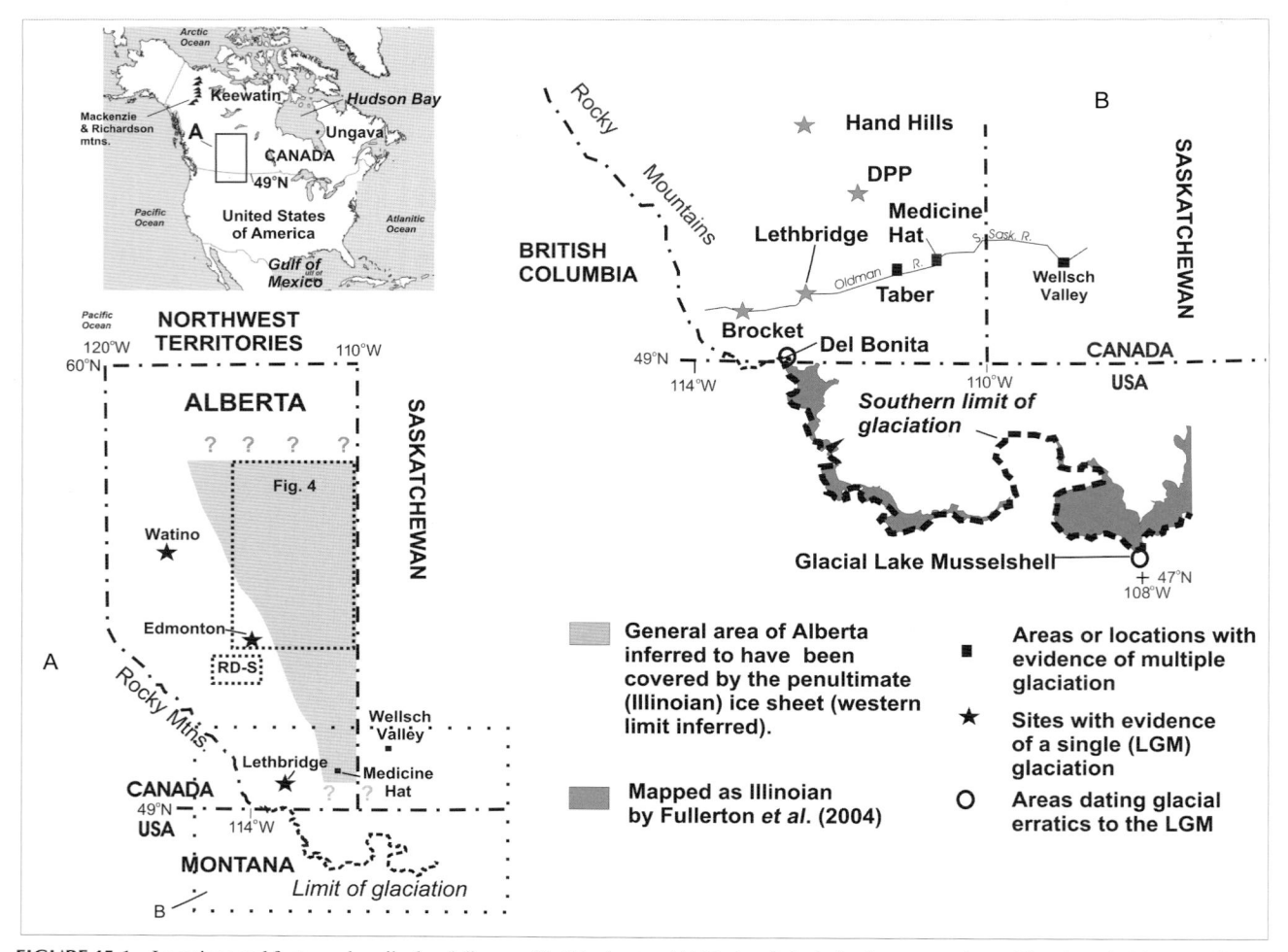

FIGURE 45.1 Locations and features described and discussed in this chapter. (A) Limits of glaciation by past continental ice sheets in Montana. Shaded area approximates the area of Alberta covered by the penultimate ice sheet to affect the region. Southern and northern limits lack data as indicated by question marks. Locations: DPP, Dinosaur Provincial Park; RD–S, Red Deer–Stettler area (discussed in Appendix C). (B) Detail of southern Alberta showing line of section in Fig. 45.3.

and adjacent Montana during the Late Wisconsinan Substage. The second part of this chapter reviews stratigraphy of glacial and non-glacial sediments within the buried Pleistocene valley systems of southern, central and northern Alberta. We show that this stratigraphy supports a single glaciation of much of Alberta. We also constrain the limits earlier glaciations of this region.

45.2. COSMOGENIC EXPOSURE DATING OF GLACIAL ERRATICS IN SOUTHERN ALBERTA AND NORTHERN MONTANA

Cosmogenic exposure dating of erratic boulders has been among the leading innovations in absolute dating of glacial limits and other glacial features (Zreda and Phillips, 2000). Although the absolute accuracy of cosmogenic ages can be debated, the great difference between the ages of the glaciations (about 100 ka) makes cosmogenic dating an appropriate and robust tool to discriminate between deposits of different glaciations. For example, cosmogenic ^{36}Cl dating of the Foothills erratics train has demonstrated that the Laurentide Ice Sheet coalesced with piedmont glaciers from Rocky Mountain ice caps during the Late Wisconsinan (Jackson et al., 1997). Crystalline erratics from the Canadian Shield (CS) that mark glacial limits or rest in end moraine systems along the Rocky Mountain front in the extreme southwest of Alberta also have yielded ^{36}Cl ages that date to the Late Wisconsinan (Jackson et al., 1999a, b; Jackson and Little, 2004). This glacial limit (C1 of Jackson and Little, 2004 and Jackson et al., 2008) marks the most extensive limit of a continental ice sheet in this region. Subsequent work in adjacent areas of Alberta and Montana (detailed below) supports this finding.

45.2.1. Cosmogenic Dating of the All-Time Glacial Limit in the Area of Del Bonita Upland

Del Bonita upland, which straddles the International Boundary, has an elevation of 1320 m above sea level (a.s.l.). It lies 50 km to the east of dated glacial limits in southwestern Alberta. It is underlain by Neogene non-glacial (Flaxville) gravel that predated glaciation. This gravel displays fine examples of periglacial modification including ice-wedge casts (Brierley, 1988). Its distinctive rose and olive clasts were derived from the Proterozoic Belt/Purcell Supergroup in the Rocky Mountains to the southwest. It contrasts with nearby glacial sediments derived from the dark argillaceous rock of the Interior Plains and crystalline rock of the CS. These factors make the glacial limits around Del Bonita upland the clearest to be found anywhere: hummocky terminal moraine surrounds the unglaciated upland to the east, north and west. The hummocky moraine is directly traceable to the C1 glacial limit of Jackson et al. (2008), Jackson and Little (2004), Little et al. (2001). Erratics resting in the terminal moraine that forms this all-time glacial limit on the northern margin of the Del Bonita upland were dated using cosmogenic ^{36}Cl exposure dating. Preliminary ages for these were published in Jackson and Phillips (2003). Descriptions and final ages for the erratics are presented in Tables 45.1 and 45.2 and Fig. 45.2. Methodology is described in Appendix A.

The nearest source for the dolostone erratics is in the headwaters of the Oldman River basin 200 km to the northwest in the Rocky Mountains. However, the most extensive montane piedmont lobe originating from the Oldman basin (M1 of Jackson and Little, 2004 and Jackson et al., 2008) terminated about 10 km west of Lethbridge which is 75 km northwest of Del Bonita (Fig. 45.1; Jackson et al. (1997, 2008). Deposits from the M1 advance clearly underlie those of the first advance of a continental ice sheet into the region (Jackson et al., 2008). The angular shape of these bedded dolostone blocks is in accord with supraglacial transport similar to the quartzite blocks of the Foothills erratics train (Jackson et al., 1997): had these blocks been deposited by a montane piedmont glacier then overrun by an ice sheet, they most likely would have been broken and smoothed by englacial or subglacial transport. Reconstruction of regional ice-flow for southern and central Alberta at the climax of the LGM is shown in Fig. 45.2 (after Shaw et al., 2010). The Bow River valley in the area

TABLE 45.1 Location and Description of Erratics Dated by the Cosmogenic ^{36}Cl Method in the Del Bonita area

Field Sample	Latitude N, Longitude W (degrees-minutes)	Elevation a.s.l. (m)	Lithology	Dimensions (m)/Shape	Sampling Method	Comments on Weathering
JJO01419	49-06.918, 112-51.313	1273	Gneiss	2 × 3 × 1.5/ ovoid	Drill and chisel	Sample taken from horizontal 100 cm² area along peak of erratic; surface smooth; no grus or spalling
JJO01420	49-03.252, 112-20.605	1197	Dolostone	4 × 4 × 3/ pyramidal	Hammer and chisel	Surface has 'alligator skin' surface; microrelief 0.5–1.0 cm
JJO01421	49-03.456, 112-20.188	1204	Bedded dolostone	2.5 × 3 × 1.5; tabular	Hammer and chisel	As above
JJO01423	49-03.587, 112-19.805	1208	Bedded dolostone	1.0 × 1.5 × 0.7/ tabular	As above	As above
JJO01425	49-06.640, 112-29.393	1196	Dolostone with solution cavities	2 × 2 × 1.5/ pyramidal	As above	Solution cavities common on sides of erratic
JJO01426	49-06.918, 112-51.313	1273	Granitoid	2 × 2 × 0.8/ cubic	Drill and chisel	Smooth, no grus or spalling
JJO01429	49-05.464, 112-45.275	1250	Granitoid	2 × 2 × 0.5/ tabular	Drill and chisel	Surface smooth; no grus or spalling
JJO01433	49-05.462, 112-44.764	1273	Pegmatitic granitoid	1 × 1 × 0.8	Drill and chisel	Surface rough due to large feldspars
JJO01435	49-0.871, 112-02.827	1059	Pebbly quartzite	3 × 1 × 1/ whaleback	Drill and chisel	Surface smooth (member of Foothills erratics train)

TABLE 45.2 Cosmogenic ^{36}Cl Ages

Field Sample	Zero Erosion Assumption		1 mm/ka Erosion Assumption		3 mm/ka Assumption	
	Age (years)	±	Age (years)	±	Age (years)	±
JJO01419	13,450	860	13,190	730	12,780	700
JJO01420	11,470	960	11,240	960	10,810	700
JJO01421	18,830	1710	17,710	1470	16,200	1180
JJO01423	10,900	660	10,490	1020	9840	840
JJO01425	25,680	2190	23,550	1730	21,080	1450
JJO01429	18,170	1030	17,560	890	16,680	890
JJO01433	19,260	1160	19,320	1140	19,460	1150
JJO01435	28,340	1720	27,250	1670	25,850	1440

of Calgary is the closest possible zone of convergence of a former Rocky Mountain trunk glacier with the Laurentide Ice Sheet. There, the Bow Valley glacier joined the Laurentide Ice Sheet (Jackson, 1980) in ice-flow that could have transported supraglacial dolostone blocks to their present location. Sources farther north in the North Saskatchewan River basin are also a possibility: trunk glaciers from these basins coalesced with the Laurentide Ice Sheet in a similar manner. Angular quartzite block 1435 is one of the Foothills erratics and lies east of the dolostone erratics. Its position is in accord with a Bow River (or North Saskatchewan) glacier joining southeastern coalescent flow between the Laurentide Ice Sheet and an Athabasca valley glacier along the Rocky Mountain Foothills (Fig. 45.2).

Glacial drift directly south and southwest of Del Bonita upland in the Milk River valley has been mapped as Illinoian (Karlstrom, 1999; Fullerton et al., 2004). The floor of the Milk River valley in that area is about 100 m lower than along the north side of Del Bonita upland. Jackson and Phillips (2003) argued that the most extensive glacial advance on the higher (northern) margin of Del Bonita upland would have been equally extensive on the lower southern side as well: the ground moraine south of Del Bonita upland was thus inferred to be Late Wisconsinan in age and not Illinoian. Subsequent work by Davis et al. (2006) (see below) is in accord with this assertion.

45.2.2. Cosmogenic Dating in the Lake Musselshell, Montana Area

Davis et al. (2006) determined cosmogenic ^{36}Cl surface exposure ages of 27 glacial erratic boulders deposited in ice-marginal Lake Musselshell of west-central Montana (Fig. 45.1). All yielded geomagnetically corrected ages within MIS 1 and 2 with 18 falling within the range of 20–10 ka. They concluded that the lake was dammed by the Laurentide Ice Sheet between 20 and 11.5 ka during the Late Wisconsinan Stage (MIS 2), rather than during the Illinoian Stage (pre-MIS 5). This ice-marginal lake was dammed by the margin of the last continental ice advance to reach this area. The lake area was mapped as Illinoian by Fullerton et al. (2004). It is approximately 25 km south of their Late Wisconsinan limit. The findings of Davis et al. further discount any glacial advances more extensive than the maximum Late Wisconsinan advance into southwestern Alberta and adjacent western Montana: other ice-marginal lakes west of Glacial Lake Musselshell such as glacial lakes Cut Bank and Great Falls occupy similar topographic environments. Although they were not dated by Davis et al., they are also likely the same age as Glacial Lake Musselshell based upon regional topographic considerations.

45.3. TESTING THE BARENDREGT–IRVING HYPOTHESIS: STRATIGRAPHY IN ALBERTA'S BURIED VALLEYS

The Late Wisconsinan age for the limit of glaciation in southwestern Alberta and western Montana is consistent with the model postulated by Barendregt and Irving (1998): deposits of successive North American continental ice sheets show a general pattern reflecting progressive westward expansion of the western margins of the continental ice sheets during the Pleistocene. This culminated in the development of an extensive ice-sheet-accumulation centre in the District of Keewatin during the Late Wisconsinan Stage (Barendregt and Duk-Rodkin, 2004). The balance of this chapter further tests this hypothesis by examining the subsurface stratigraphy of Alberta in order to

1. determine whether it corroborates the cosmogenic dating evidence for a single glaciation of much of Alberta;

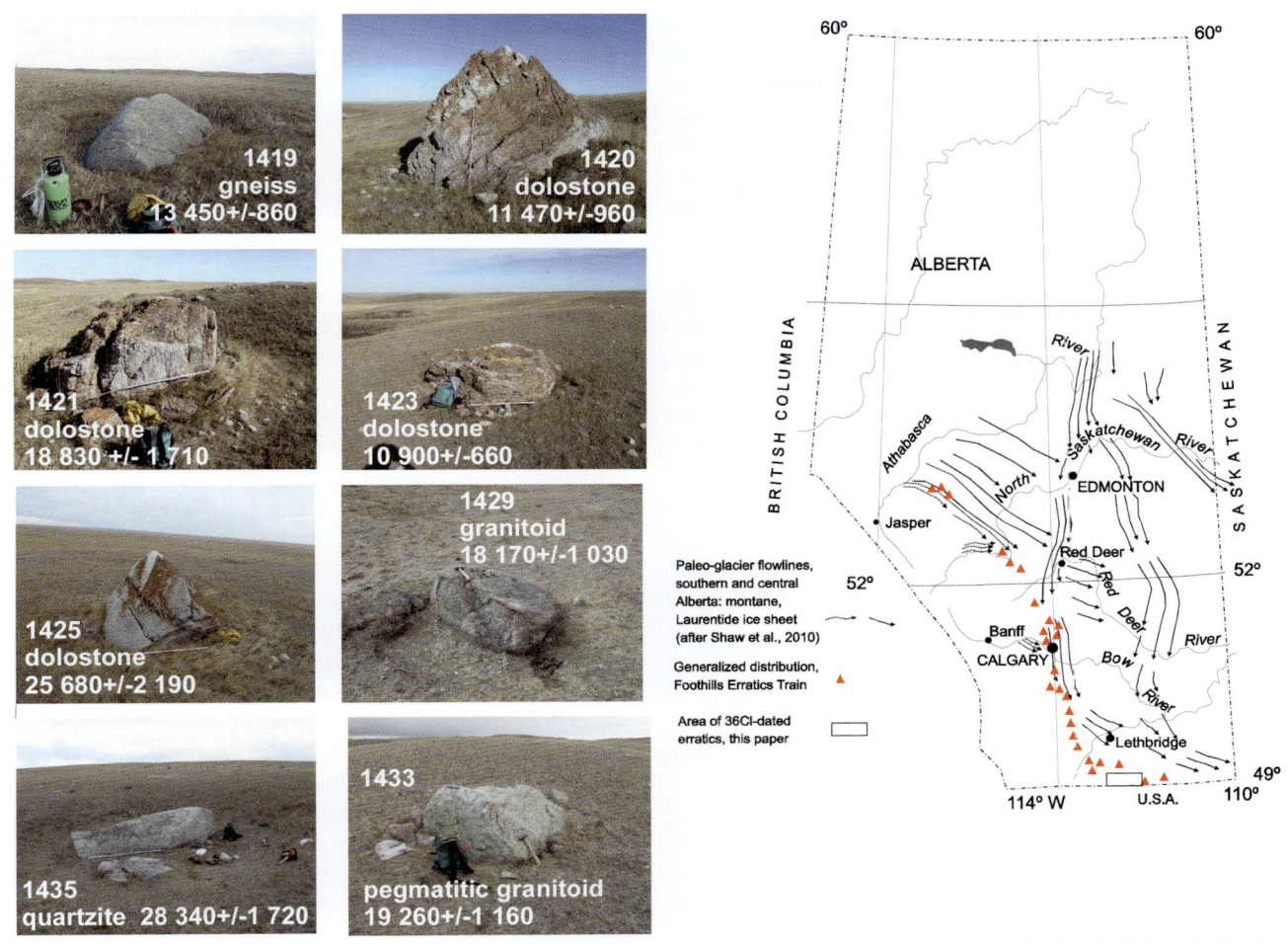

FIGURE 45.2 Erratics dated by the cosmogenic ^{36}Cl method (Tables 45.1 and 45.2). The ages of their deposition are clearly coinciding with the Late Wisconsinan Stage. The four dolostone erratics were derived from Devonian or Mississippian carbonate formations in the Rocky Mountains at least as far north as the Bow River Basin (measuring tape divisions are 10 cm). The quartzite erratic is a member of the Foothills erratics train and has its source in the headwaters of the Athabasca River Basin (Jackson et al., 1997). The angular shapes of these sedimentary blocks are in accord with supraglacial transport: they were transported along the zone of confluence between the Laurentide Ice Sheet and valley glaciers from the glacier complex that covered the Rocky Mountains in a similar way to the Foothills Erratics Train (Jackson et al., 1997; Young et al., 1999; Shaw et al., 2010). Ages shown assume a zero rate of erosion since deposition. Granitoid blocks originate in the Canadian Shield. Their smooth shapes and abraded surfaces are in accord with basal glacial transport following plucking (Benn and Evans, 1998, pp. 182–192).

2. determine the extent of limits of the penultimate or older ice sheets where evidence of multiple glaciation is present.

45.3.1. Stratigraphical Units in the Buried Valley Systems of Alberta

Buried valley systems cross Alberta from west to east (Stalker, 1961; Fravolden, 1963; Andriashek and Fenton, 1989). These carried drainage from the Rocky Mountains of southern and central Alberta and adjacent northern Montana down the regional slope towards Hudson Bay prior to glaciation by continental (CS-centred) ice sheets during the Pleistocene. They contain extensive fills of predominantly glacial diamicton, glacio-lacustrine and glacio-fluvial sediments laid down by incursion of one (or more) continental ice sheets. The fills of these valleys have been exposed in section as cliff banks by the cross-cutting post-glacial river courses (Stalker, 1963; Evans and Campbell, 1995) and have been exposed in the walls of gravel excavations. They have also been intersected by stratigraphical test wells related to ground water exploration and oil-sands development (Andriashek and Fenton, 1989). Investigation and interpretation of the sedimentary record contained in these buried valleys span almost 130 years since initial investigations by the Geological Survey of Canada began in the latter half of the nineteenth century. Understanding of this sedimentary record has evolved with progress in glacial sedimentology and absolute dating techniques. For example, Stalker (1963, 1969) was the first to systematically describe the best exposed Pleistocene sections in the Oldman–South Saskatchewan River Basin and correlate glacial sediments between sections in a systematic way. His work largely predated the revolution of insight into contemporary glacial sedimentary processes, particularly in the origin of till (*sensu lato*) that accelerated in the 1970s and continues to expand (e.g. Krüger, 1979, 1984; Lawson, 1981; Halderson and Shaw, 1982; Shaw, 1982; Dreimanis, 1988; Hicock et al., 1996; Benn and Evans, 1998; van der Meer et al., 2003; Meriano and Eyles, 2009). In this regard, Stalker commonly failed to clearly separate observation and interpretation. He routinely described glaciogenic diamictons as till with each serving as evidence for a separate glaciation. Naming and correlation of tills (alloformations) were done largely on the basis of appearance alone from type localities in the Red Deer–Stettler area (Stalker, 1960) to the Oldman River–South Saskatchewan River Basin hundreds of kilometres south (Fig. 45.1). This led to his conclusion that there is widespread evidence of multiple glaciations throughout southern Alberta (rather than only in the eastern part of that region as we argue below).

Another problem in interpreting the stratigraphy of buried valleys concerns stratigraphical nomenclature for gravelly and sandy sediments that overlie bedrock and underlie successions of glacial sediments in buried valley systems. These sediments are key to dating the first glacial incursions into the Interior Planes of Alberta because they lack any igneous or metamorphic rocks from the CS. There are two allostratigraphical names used for these sediments *Saskatchewan Gravel and Sand* (SGS) and *Empress Group (or Formation)* (EG).

In this chapter, we have discarded the term *Labuma Till* for the oldest (stratigraphically lowest) till containing clasts from the CS outside of the Red Deer–Stettler area. We also use *Empress Group* (after Christiansen, 1992; Evans and Campbell, 1995) for the time transgressive preglacial alloformation that underlies the succession of glacial and interglacial fill in buried valley systems. A discussion of the rationale for this revision of stratigraphy is presented in Appendices B and C.

45.3.2. Glacial Advance, Retreat and Re-Advance Sequence Stratigraphy in Alberta

As noted above, deposits interpreted as till in Alberta's buried valleys were regarded by some workers as marking a glacial event: evidence for up to four glaciations were purported to be recorded by some sections such as the Brocket Section near Pincher Creek Alberta (Stalker, 1963; Prest, 1970, p. 692). Since ca. 1980, analysis of the succession of lithofacies in vertical sequences of the Pleistocene sediments has dramatically revised interpretation of glacial events. The term 'till' has come to be used with considerable care and with genetic modifiers. Classification schemes have been devised for glaciogenic deposits based upon the precise knowledge of their origin (see Dreimanis, 1988, for a history of classification schemes to that time). In practice, such schemes have proven to have limited application for regional mapping and correlation of glacial geology and stratigraphy (e.g. Lundqvist, 1988; Jackson et al., 2008): different glacial environments can produce diamictic sediments with many overlapping properties that make them largely indistinguishable in the field (Dreimanis, 1988, his Appendix D; Levson and Rutter, 1988). Alberta investigators have used the term 'till' *sensu lato* after Dreimanis (1982): till is a sediment that has been transported and is subsequently deposited by or from glacier ice with little or no sorting by water. With regards to interpretation of sequences of glacial sediments in Alberta, a twofold classification scheme has generally been used: till is recognized to have originated from processes near or at the base of a glacier (shearing and comminution of underlying soft rock and sediment, lodgement and meltout of debris-laden basal ice) and processes associated with upper ice surfaces such as the sediment gravity flow onto land or into water and deposition of superglacial debris through melting of underlying ice.

As discussed above, the physiography of the Interior Plains of Alberta causes westward expanding ice sheets centred in the Ungava–Hudson Bay region and the District of Keewatin (Fig. 45.1) to initially advance against, or up the regional slope in opposition to trunk stream systems. Consequently, advancing ice-sheet margins create ice-dammed lakes as they advance and similarly pond flights of ice dammed lakes in their wake during retreat. This has led to a recognition of glacial advance and retreat cycles in sedimentary successions of the stratigraphical record (Alley and Harris, 1974; Proudfoot, 1985; Andriashek and Fenton, 1989; Jackson and Little, 2004). A similar repeating pattern is seen in the glacial stratigraphy of Saskatchewan (Christiansen, 1992). The nature of contacts between units is crucial in recognizing these sequences: sheared contacts are indicative of glacial overriding, contacts within sequences are conformable, and sequences usually terminate with a conformable contact grading upward into ice-proximal lake facies or are truncated by an erosional unconformity. This succession is seen in sections throughout the buried valley systems of Alberta (Figs. 45.3 and 45.4).

45.3.3. Discrimination of Advance and Retreat Sequences from Separate Glaciations

The advance and retreat sequences seen in the vertical succession of sediments in themselves cannot discriminate

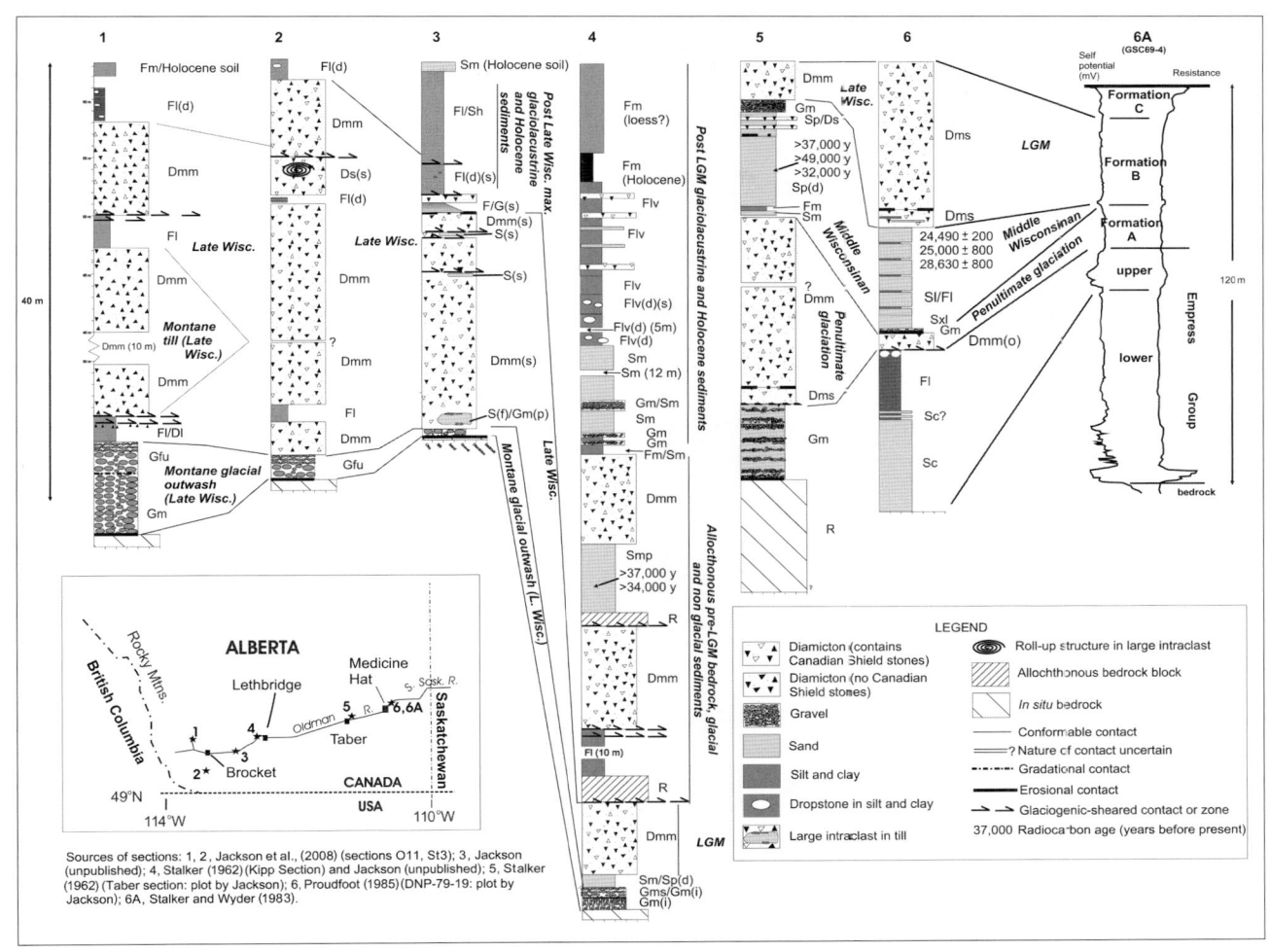

FIGURE 45.3 Succession of sediments in buried valleys within the ancestral Oldman–South Saskatchewan river systems west to east across southern Alberta. Sections west of Lethbridge (1–3) are composed of sediments dating entirely from the last glaciation. At Kipp (4) near Lethbridge and along the margin of the Lethbridge moraine (Eyles et al., 1999), allochthonous bedrock blocks and interglacial sediments are stacked upon the first (LGM) continental till to be deposited in the area. This stacking of allochthons may be related to the Lethbridge moraine which marks a stillstand or local re-advance of the Laurentide Ice sheet. Deposits of the last two glaciations and interglacial sediments occur in Taber and Medicine Hat areas (5, 6). Section 6A is an electric log of one of the most complete sections encountered in stratigraphical test drilling in the area (scale change). Facies codes modified from Benn and Evans (1998, p. 383): D, diamicton; G, gravel; S, sand; F, silt and clay; R, bedrock (sandstone and mudstone). Modifiers: mm, matrix supported and massive; ms, matrix supported & stratified; fu, upward fining; lv, fine laminations with rhythmites or varves; p, medium to very coarse and planar cross-bedded; l, fine laminations with minor sand; c, steeply dipping planar cross-bedding; xl, cross-laminated; (s), sheared; (f), fissile; (d), contains dropstones; (i), involuted (modified by past permafrost action); (o), oxidized.

between advances and re-advances within a glaciation (such as the Late Wisconsinan Substage (MIS 2)) and those separated by an extended non-glacial interval such the ca. 100 ka between MIS 2 and 6 (Lisiecki and Raymo, 2005). Three indicators of extended exposure between glaciations have been identified in this region. These may be present separately or co-occur.

45.3.3.1. Laterally Extensive Lag Gravel

The present landscape in the Interior Plains of Alberta is dominated by erosion. An interval of fluvial gravel or sand nested between glacial deposits in a single natural exposure or test well is not indicative an interglacial hiatus unless it contains plant or animal remains that are indicative of interglacial conditions (Table 45.3). However, if such a horizon is traceable over several kilometres along exposed sections and or between stratigraphical test wells, it defines an eroded land surface similar to the contemporary land surface that developed during the Holocene: drift deposited during the last glaciation has been incised by headward growth of small tributaries during this period. These tributaries have deposited gravelly lags in adjacent valley bottoms and on pediments along valley margins. Proudfoot (1985) interpreted the lag deposit that occurs between

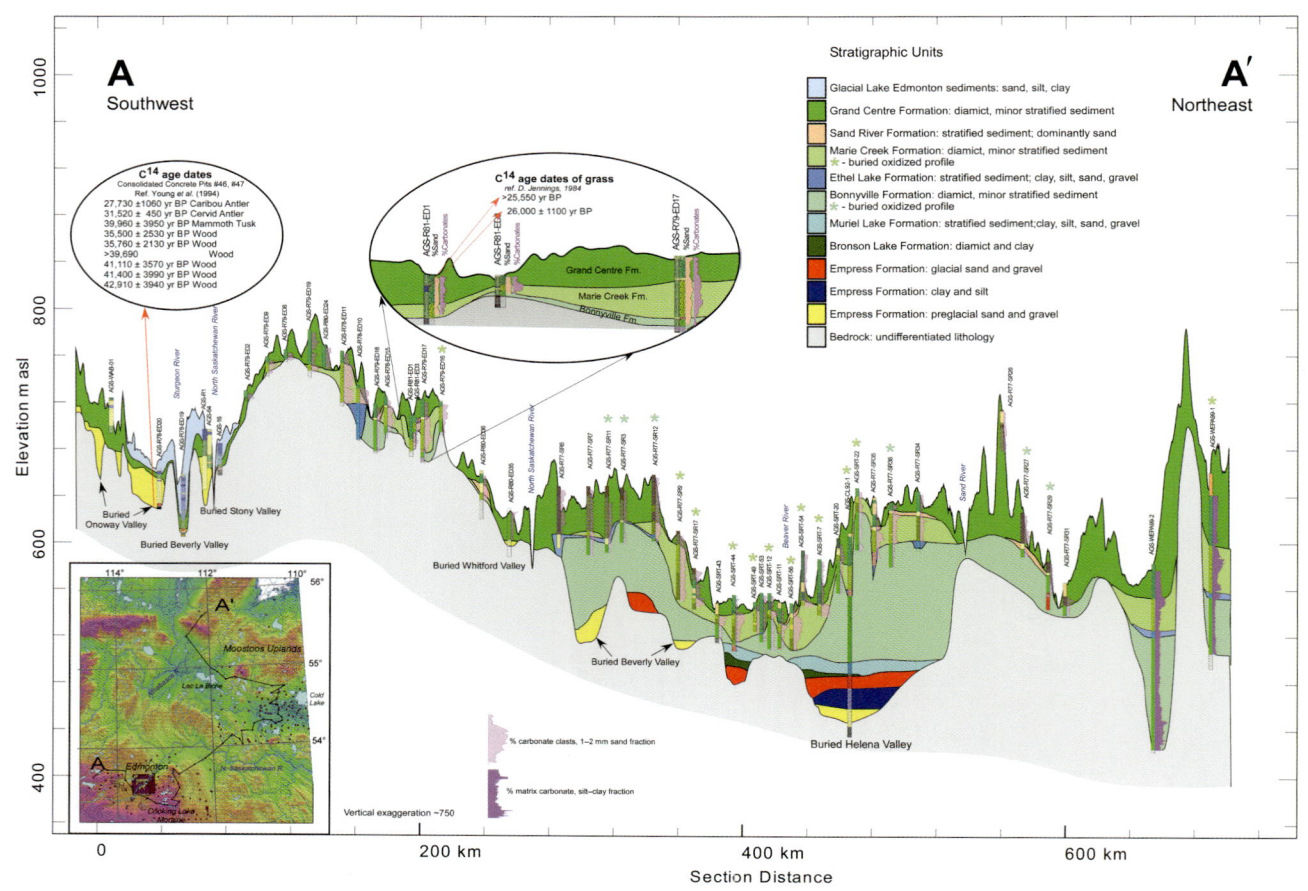

FIGURE 45.4 Architecture of the Quaternary lithostratigraphy in central and northeast Alberta highlighting formations associated with multiple glaciations and illustrating the distribution of time-transgressive preglacial deposits of the Empress Formation in buried valleys. Diamictons of major glacigenic units are differentiated and correlated primarily by differences in grain size and carbonate content in both the coarse-sand, and silt–clay fractions. Palaeo-weathered horizons on surfaces of buried diamicts indicate substantial periods of subaerial exposure prior to burial by deposits of subsequent glaciations.

glacial units A (below) and B (above) as having this origin (Fig. 45.3, sections 6 and 6A. He recognized this unit and the erosional unconformity that it overlies in natural exposures and bore-holes and traced its extent across the Medicine Hat area. Further, this gravel lag unit is overlain by the slack water deposits interpreted to mark damming of regional drainage caused by the advance of the last (Laurentide) ice sheet into the area.

45.3.3.2. Fossil-Bearing Sediments Dated to Known Non-Glacial Time Periods Between Glacial Advance Sequences

The slackwater sediments described above contain plant detritus and wood fragments radiocarbon dated between 24 and 39 ^{14}C ka BP (Table 45.3). These ages clearly fall within the non-glacial Middle Wisconsinan Substage which further corroborates an interglacial origin for the underlying topographically extensive lag gravel. Interglacial sediments equivalent to those in the Medicine Hat area cannot be traced farther west than the Taber area (49.9°N, 112.2°W; elevation ca. 750 m a.s.l.). There, sandy sediments containing wood that yielded non-finite radiocarbon ages underlie advance–retreat lithofacies sequence related to the Late Wisconsinan glaciation and overlie a series of older diamictons from a preceding advance–retreat sequence or sequences (Stalker, 1963). A similar horizon between advance sequences was intersected in a stratigraphical test well in the Elk Island area, east of Edmonton in central Alberta. Intersected grass and wood in lacustrine sediments were radiocarbon dated at 26 ± 1.1 and > 25.55 ^{14}C ka BP (Fig. 45.4; Jenner, 1984).

45.3.3.3. Oxidized Till

The interglacial climate of central and southern Alberta is semi-arid and vegetation is dominated by sparse short grasses or parkland. Soils developed under these conditions are typically shallow chernozemic or solenetzic soils in the south or grey wooded soils farther north. However,

TABLE 45.3 Summary of Sub-Till or Inter-Till Radiocarbon Ages with Significance for Demonstrating a Single Glaciation or Multiple Glaciation

Age (Radiocarbon Years BP)	Lab No.	Location	Material	Significance (S, Ages Support One Glaciation; M, Ages Support Two Glaciations)
Medicine Hat area				
24,490 ± 200	GSC 205	50°6.0′N, 110°39.7′W; field name 'Evil Smelling Bluff'	Plant fragments	M; dates advance sequence below LGM till; overlies unconformity above pre-LGM till
25,000 ± 800	GSC 1370	As above	Wood fragments	M; as above
28,630 ± 800	GSC 578	As above	As above	M; as above
37,900 ± 1100	GSC 144-2	50°8.9′N, 110°39.1′W; field name 'Galt Island Bluff'	Wood	M; as above; the second of two radiocarbon ages determined on the same sample. GSC 144-1 dated at 37,700 ± 1100
Taber area				
35,980 ± 1060	GSC 728	49°54.0′N, 112°12′W	Wood	M; dates fine sediments that underlie a succession of tills; overlies several older tills
Lethbridge area (Kipp)				
>30,000 >34,000 >25,000 >37,000	L433A L433B L455A L455B	49°43.7′N, 112°58.1′W	Wood humic Portion Wood humic portion	S; situated between tills but section contains bedrock blocks indicating that it has been thickened by glacial tectonism: host sediments are allochthonous
Dinosaur Provincial Park				
52,450 ± 1910 27,420 ± 240 >42,460	TO-2693, TO-1464 AEVC-1436C	ca. 50°46′N, 111°29′W	Pine fragments Gastropod shell Mammuthus tusk	S; from three separate localities; samples situated in gravel and sand lacking lithologies from the Canadian Shield; overlain by till containing clasts from the Canadian Shield.
Hand Hills and Wintering Hills				
Eight radiocarbon ages between 17,000 and 34,000 years (Young et al., 1999)		51°30′N–51°35′ 112°00′–112°30′W	Bone collagen	S; dated samples from a fossil prairie dog colony (Burns and McGillivary, 1989) underlying a single till with clasts from Canadian Shield; sediments hosting prairie dog colony lacks Canadian Shield lithologies
Edmonton				
29 finite ages 21,300–42,910; 4 non-finite ages (>39,690 to >41,220, Young et al., 1999)		Seven sites within area of 53°30′–54°00′N, 113°00′–114°00′W	Bone, tusk, wood	S; dated samples taken from gravel and sand lacking Canadian Shield lithologies or carbonate clasts (Empress Group) that underlie a till containing clasts from Canadian Shield. Dated bone and tusk material from horse, bison, mammoth, caribou, ground sloth, deer
Elk Island National Park area				
>25,550	S2159	53°39.9′N, 112°52.0′W	Wood fragments, grass	M; from sediments intersected at 11.5 m in a 35 m stratigraphical test well. Collected using a solid-stem auger Enclosing sediments lacustrine silt and clay lying between tills (Jenner, 1984)

Continued

TABLE 45.3 Summary of Sub-Till or Inter-Till Radiocarbon Ages with Significance for Demonstrating a Single Glaciation or Multiple Glaciation—cont'd

Age (Radiocarbon Years BP)	Lab No.	Location	Material	Significance (S, Ages Support One Glaciation; M, Ages Support Two Glaciations)
26,000 ± 1100	S2160	As above	As above	M; as above: depth 14.5 m
Watino area				
8 finite radiocarbon ages between 27,000 and 37,000 years; 2 non-finite ages (Liverman et al., 1989)		Watino section 55°42.8′N, 117°44.7′W Simonette section 56°08′N, 118°10.9′W	Wood (7 samples); organic detritus (3) sample	S; samples taken from gravels that lack any lithologies from the Canadian Shield. Radiocarbon ages are a compilation of those reported by Liverman et al. (1989), Westgate et al. (1971, 1972) and Reimchen (1968)

extended contact with the atmosphere during periods of interglacial duration results in oxidation of iron and manganese and translocation of gypsum and iron and manganese oxides in underlying till (Christiansen, 1992). In the subsurface of Saskatchewan, oxidized zones within till range in thickness from several metres to several tens of metres. The upper contacts of oxidized zones with the overlying unoxidized deposits are sharp and distinct, whereas the lower contacts with the underlying unoxidized deposits are gradational. The deepest oxidation occurs along joints that extend downward into unoxidized till. There, oxidation has progressed laterally from the joint surfaces. Christiansen (1992) has found these oxidized horizons to be laterally persistent in the natural exposures and exploratory wells in Saskatchewan. The lower till in the Medicine Hat area (Fig. 45.3, section 6 (below the lag gravel)) contains such oxidation features, whereas the Late Wisconsinan Till overlying it does not (Proudfoot, 1985). Two such oxidized zones in successive tills are correlated across east-central and north-eastern Alberta (see below).

45.3.4. Summary of Stratigraphical Evidence of Two or More Glaciations in Eastern Alberta

Successions of advance–retreat lithofacies sequences with evidence of intervening non-glacial periods described above are confined to the eastern part of Alberta. Figures 45.3 and 45.4 correlate glacial sediments in buried valleys across southern, central and northern Alberta. In southern Alberta, there is no bona fide evidence of multiple glaciation west of Taber (Fig. 45.3, section 5).

In east-central and north-east Alberta (Fig. 45.4), Andriashek and Fenton (1989) and Andriashek (2003) recognized two extended non-glacial intervals based on till oxidation that predate Late Wisconsinan glacial sediments. The stratigraphy of these deposits is correlated with the well-established multiple glaciation stratigraphy of Christiansen (1992) in adjacent Saskatchewan. Deposits of the stratigraphically highest oxidized till (Marie Creek Formation) are encountered up to an elevation of about 750 m a.s.l. in test wells.

45.3.5. Summary of Sites Spatially Constraining Limits of Pre-Late Wisconsinan Ice Sheets

Stratigraphical sequences that definitively indicate a single Late Wisconsinan glaciation of Alberta's Interior Plains have glacial sediments containing clasts from the CS which overlie sediments radiocarbon-dated to the Middle Wisconsin Stage that lack CS lithologies. Where direct radiocarbon dating of preglacial gravel is lacking, indirect assignment of overlying till(s) to the Late Wisconsinan Stage has employed tracing of till units to end moraines that have been dated using cosmogenic exposure dating.

45.3.6. Oldman River Basin

In the Rocky Mountain Foothills (western part of the basin ca. 49.3°N, 113.8°W; elevation ca. 1050 m a.s.l.), successions of diamictons near Brocket (Fig. 45.1) have been attributed to three or more glaciations (Stalker, 1963; Alley and Harris, 1974; Stalker and Harrison, 1977). These lack any oxidation zones, evidence of extensive erosion between glacial advances, or organic material that can be radiocarbon dated. They have been shown to be clearly related to belts of moraine that have been dated to the Late Wisconsinan by cosmogenic ^{36}Cl exposure dating of erratics (Jackson et al., 1999a,b, 2008; Jackson and Little, 2004). They were deposited by successive advances and re-advances of the Laurentide Ice Sheet and montane glaciers during the Late Wisconsinan. Moving eastward, a single thick till with multiple shear zones in (Fig. 45.3, section 3), correlates to the stack of Laurentide Tills in sections 1 and 2.

Farther to the east in the area of Lethbridge, Stalker (1963) reported finite Middle Wisconsinan radiocarbon ages between tills in the exposure at Kipp (Fig. 45.3, section 4; 49.7°N, 112.9°W; elevation ca. 900 m a.s.l.). However, reinvestigation of this exposure by one of the authors (L. E. J.) found that this radiocarbon-dated sequence is contained within stacked glaciotectonic allochthons that overlie the basal glacial-advance lithofacies. Such stacks of ice-thrusted sheets of bedrock and sediment are common in the Interior Plains of Alberta and Saskatchewan (Christiansen and Whitaker, 1976; Moran et al., 1980). The basal till in the Kipp section overlies outwash gravel from a montane glacial advance. Jackson et al. (2008) traced this till to the all-time limit of continental glaciation (their limit C1) which was dated as Late Wisconsinan by cosmogenic ^{36}Cl dating.

45.3.7. Dinosaur Provincial Park

This extensive badland area is located along the Red Deer River about 100 km northwest of Medicine Hat. The prairie surface is about 710 m a.s.l. (area of 50.8°N, 111.5°W). Buried valleys are extensively exposed in section in this area. Evans and Campbell (1995) described 29 Quaternary in there. Sections exposed Empress Group sediments overlain by a single till or tills lacking any indication of intervening breaks in deposition of interglacial duration. In two of the sections, the Empress Group yielded organic material dated to the Middle Wisconsinan and third produced a non-finite radiocarbon age (Table 45.3).

45.3.8. Hand Hills

Hand Hills are located 100 km northwest of Dinosaur Provincial Park (area of 51.6°N, 112.3°W). The summit of this upland is about 1000 m a.s.l. Like the Del Bonita upland, the Hand Hills upland is a relict of the surface of the Interior Plains at an earlier level of incision. The bedrock surface is overlain by Neogene fluvial sediments of Rocky Mountain provenance. These in turn are overlain by a single till. Seven Middle Wisconsinan radiocarbon ages (Table 45.3) have been determined on a fossil prairie dog colony that was established in the preglacial fluvial sediments prior to deposition of a single till (Burns and McGillivary, 1989; Young et al., 1994).

45.3.9. Edmonton

The prairie surface in the Edmonton metropolitan area ranges between about 650 m and 725 m a.s.l. A single advance lithofacies succession is present across the Edmonton area (Figs. 45.1 and 45.4; Shaw, 1982). It overlies gravel and sand of the Empress Group equivalent Empress Formation (EF). EF in the Edmonton area contains faunal material that has yielded 29 finite radiocarbon ages that fall within the non-glacial Middle Wisconsinan Substage (Young et al., 1994). Gravel and sand in EF are derived entirely from the sedimentary rocks of the Interior Plains and Rocky Mountains: EF deposition clearly predates incursion of a continental ice sheet into the Edmonton area.

45.3.10. Watino

A similar stratigraphy has been documented in cliff-bank exposures in the Watino area 360 km northwest of Edmonton (Fig. 45.1; Table 45.3). The sections are successions of gravel, sand, silt and clayey silt. The section lacks any CS clasts. Till and other glacial sediments have been eroded from the area. The elevation of the surrounding prairie surface, which is underlain by a continental till in this region, is no higher than ca. 560 m a.s.l. Eight finite Middle Wisconsinan radiocarbon ages have been determined on wood from fluvial sediments compositionally equivalent to EG (Westgate et al., 1971, 1972; Liverman et al., 1989). The sections clearly document the absence of continental glaciation of this region prior to the Late Wisconsinan.

45.4. DISCUSSION AND CONCLUSIONS

The age and distribution of glacial sediments in the buried valley systems of Alberta are in accord with a Late Wisconsinan age for the all-time most extensive continental glaciation in the Interior Plains of southern Alberta and adjacent Montana. Bona fide glacial sediments predating the Late Wisconsinan sediments are restricted to the eastern one-third to one-half of the province. Their maximum elevation is ca. 750 m a.s.l. (Fig. 45.1). The older glaciations responsible for these sediments occurred during the Illinoian Stage: palaeomagnetic polarity measurements in the Medicine Hat area and in Saskatchewan have proven to be magnetically normal referring them to the last 0.78 Ma of geologic time (Barendregt and Irving, 1998). The limited western extent of pre-Late Wisconsinan ice sheets is in accord with the Barendregt–Irving hypothesis.

A few details concerning flow patterns in the penultimate ice sheet can be inferred from existing petrologic and fabric data. Till pebble fabrics in the Marie Creek Till, the youngest of pre-Late Wisconsinan Tills in east-central Alberta, indicate a strong north–south orientation, implying flow from the north. On this basis, carbonates in the Marie Creek Till are inferred to source from the Devonian outcrops that fringe the Shield in northeast Alberta. No flow indicators have been measured for the pre-Late Wisconsinan Till at Medicine Hat. The carbonate and crystalline pebbles in this till could have come from the north or the northeast. However, the northwest-to-southeast flow of glacial ice into Alberta from Keewatin region that paralleled the Rocky Mountain Foothills during the climax of the last glaciation (Dyke and Prest, 1987; Shaw et al., 2010) was unique to the Late Wisconsinan Laurentide Ice Sheet as was coalescence of montane and continental ice (Jackson et al., 1997). Illinoian Keewatin ice centres were apparently of minor extent compared to that of the Late Wisconsinan: ice from this source never reached as far south and west as Watino until the Late Wisconsinan.

It follows that broad, ice-free corridors existed in Alberta during all glaciations except the Late Wisconsinan. These ice-free corridors extended north to the Arctic Ocean: the Late Wisconsinan was the only time that a continental ice sheet advanced into the Mackenzie and Richardson Mountains (Zazula et al., 2004; Fig. 45.1). It follows that plants and animals were free to move north and south through these ice-free corridors at the climaxes of all glaciations prior to the Late Wisconsinan. The coalescence of montane and continental ice at the climax of the Late Wisconsinan was a unique event.

It may be postulated that this unique event was a significant factor in the dramatic biogeographical changes at the end of the Pleistocene including extinctions. These contrast with the less dramatic changes associated with previous glacial terminations.

ACKNOWLEDGEMENT

Natural Resources Canada, Earth Sciences Sector contribution 20100063.

APPENDIX A. METHODS IN THE COSMOGENIC ^{36}CL DATING OF GLACIAL ERRATICS AROUND DEL BONITA UPLAND

Erratics were sampled along a traverse across end and ground moraine from east and north of the Del Bonita upland (Tables 45.1 and 45.2, Fig. 45.2). Latitude and longitude of sample erratics were determined by a global positioning system. Elevations were determined from 1:50,000 topographic maps for the area. Angles $\geq 5°$ from sampling sites to the summits of the local topographic were recorded to determine the degree of topographic blocking of the cosmic ray flux. The upper 2 cm of the erratics were sampled by hammer and chisel in the case of dolostone erratics and diamond drilling a matrix of cores and chiselling intervening rock between core holes. At least 500 g of rock was obtained from each erratic. Samples were obtained from flat surfaces and nearly level surfaces with the exception of pyramidal erratics where samples were taken from the upper 2 cm of apices.

In the laboratory, carbonate samples were dissolved in nitric acid and silicate samples in a mixture of nitric and hydrofluoric acids, ^{35}Cl-enriched carrier was added, chlorine was precipitated by the addition of AgNO$_3$, and the resulting AgCl purified of sulphur using standard techniques (Marrero, 2009). The ^{36}Cl/^{35}Cl ratio was measured at Lawrence Livermore National Laboratory by accelerator mass spectrometry and the Cl concentration simultaneously determined by isotope dilution mass spectrometry (Desilets et al., 2006). Surface exposure ages were calculated using a prototype of the CRONUS-Earth (Cosmic-Ray Produced Nuclide Systematics on Earth) on-line calculator. This calculator is based on the CHLOE (CHLOrine-36 Exposure) age calculator (Phillips and Plummer, 1996), but incorporates time-varying nuclide production rates and full uncertainty propagation. The geomagnetic scaling method of Lifton et al. (2008) was employed in the calculations. The elemental ^{36}Cl production rates were based on the calibration data set of Phillips et al. (2001) rescaled using the Lifton et al. (2008) method. The analyses from Meteor Crater were deleted from the calibration data set due to poor fit with the rest of the data. This rescaling yielded the following sea-level high-altitude production rates: 142 atoms ^{36}Cl (g K)$^{-1}$ year^{-1}, 69.0 atoms ^{36}Cl (g Ca)$^{-1}$ year^{-1} and 645 neutrons (g air)$^{-1}$ year^{-1}. Muon production was calculated according to Stone et al. (1998) and Evans (2001). Exposure ages were calculated for assumed rock–surface erosion rates ranging from 0 to 5 mm/ka.

APPENDIX B. BASAL GRAVEL PREDATING CONTINENTAL GLACIATION AND ITS TIME-TRANSGRESSIVE NATURE

The lowest deposits in buried preglacial valley systems are gravel and sand derived from Mesozoic sandstone units that underlie the Interior Plains and chemically and physically resistant quartz, quartzite and chert clasts from Palaeozoic or Precambrian sedimentary rocks of the Rocky Mountains. Although the gravel-size clasts in these deposits were ultimately derived from a terrain rich in carbonate rock in the Rocky Mountains, all carbonate clasts were destroyed by dissolution and abrasion before deposition in preglacial valleys of central and southeastern Alberta. Further, these gravels and sands lack any clasts derived from plutonic and high grade metamorphic rocks from the CS (clasts in this chapter refer to sand and larger rock and mineral grains except where otherwise stated). Two alloformational names have been commonly used for these sediments: 'SGS' (Stalker, 1968) and 'Empress Group' (Christiansen, 1992). The two have been widely used and correlated but Empress Group has proven to be the most robustly and precisely defined and widely correlated in southeastern and central Alberta (Proudfoot, 1985; Andriashek, 1988; Andriashek and Fenton, 1989). In this chapter, SGS is not used in favour of the Empress Group (or Formation: it has formational status in central Alberta; the two are used synonymously for the purposes of this chapter; Figs. 45.3 and 45.4). Our reasoning is that SGS, as defined by Stalker (1969), includes all gravel and sand units lacking clast lithologies derived from the CS and underlie the oldest till bearing clasts from the CS. By his definition, this unit includes gravel outwash from montane glaciers in the area of the Rocky Mountains: these gravels contain clasts of limestone and dolostone. Such gravel units can be found as far east as the greater Lethbridge area (Fig. 45.3, section 4). In contrast, Empress Group sediments lack carbonate clasts. They contrast with overlying glacial sediments that contain abundant carbonate clasts as well as silt and clay size particles in their matrices. This carbonate content is derived from the passage of glacial ice over the belt of Palaeozoic carbonate rock that outcrops along the CS from Manitoba to northern Alberta (Shetsen, 1984; Thorleifson and Garrett, 2000). Empress Group by definition lacks limestone and dolostone clasts and is clearly non-glacial except where it is clearly transitional into overlying glacial sediments. Jackson and Little (2004) and Jackson et al. (2008, p. 35) recognized this problem and abandoned SGS term. They instead assigned an arbitrary unit (1) to outwash gravel in the Rocky Mountain Foothills and specified that its deposition was coincident with an onset with the last cycle of montane glaciation. We reserve the term 'Empress Group' for siliceous gravel and sand units that underlie the first glacial sediments bearing clasts from the CS and carbonate clasts from the CS margin.

A notable aspect of the Empress Group is an upper member which fines upward into fine-grained, sediments. These have been interpreted to be slack water deposits resulting from damming of stream systems to the east by the margin of an advancing ice sheet (Proudfoot, 1985; Andriashek and Fenton, 1989; Evans and Campbell, 1995): all ice sheets pressing into central and southern Alberta from the east and north must advance up the regional gradient of the Canadian Interior Plains which progressively rise from the east and northeast to west and southwest. Magneto-stratigraphical investigations, radiocarbon dating and relative dating based upon vertebrate fossils have shown this member and the Empress Group as a whole to be time-transgressive: it decreases in age with increasing westward longitude: in Saskatchewan (Wellsch Valley: 50.5°N 107.9°W) at approximately 640 m, it is magnetically reversed (>780 ka) and contains Irvingtonian or Blancan fauna (Barendregt et al., 1991; Churcher, 2010, personal communication). By contrast, it contains fauna radiocarbon-dated to the Middle Wisconsinan in the Edmonton and Dinosaur Provincial Park areas.

The time-transgressive aspect of the upper member of the Empress Group is significant in that it suggests that progressively younger and progressively more extensive ice sheets affected the region with time in accord with the Barendregt–Irving hypothesis.

APPENDIX C. ELIMINATION OF 'LABUMA TILL' AND RELATED ALLOSTRATIGRAPHIC UNITS OUTSIDE OF THE RED DEER–STETTLER AREA

The Labuma Till alloformation has been widely correlated across central and southern Alberta (Alley and Harris, 1974; Stalker and Harrison, 1977). It was originally defined in the Red Deer–Stettler area of central Alberta (Stalker, 1960; Fig. 45.1) based upon its position directly above preglacial gravel and by its physical properties: dark grey colour, matrix rich in silt and clay and content of clasts (about one-third) from the CS. Its high content of expansive clay produces a characteristic prismatic peds on its surface in natural exposures. It was regarded as dating from the Early Pleistocene by Stalker. He correctly recognized that these properties were the result of continental ice travelling from the CS over previously unglaciated dark Cretaceous shales of the Interior Plains of Canada for the first time (Stalker, 1969, p. 9). However, if successive ice sheets were increasingly larger during the late Cenozoic (as concluded by Barendregt and Irving, 1998), each advance would create till with the characteristics of Labuma Till but each would date from different glaciations.

For this reason, we do not use 'Labuma Till' for regional correlation. In fact, it has not been used for regional correlation over the past 25 years (Proudfoot, 1985; Andriashek, 1988; Andriashek and Fenton, 1989; Jackson et al., 2008) as have Maunsel and Buffalo Lake tills also defined in the Red Deer–Stettler area by Stalker (1960).

REFERENCES

Alley, N.F., Harris, S.A., 1974. Pleistocene glacial lake sequences in the Foothills, Southwestern Alberta, Canada. Can. J. Earth Sci. 1, 1220–1235.

Andriashek, L.D., 1988. Quaternary Stratigraphy of the Edmonton Map Area, NTS 83H. EUB/AGS Earth Sciences Report 198804.

Andriashek, L.D., 2003. Quaternary geological setting of the Athabasca Oil Sands (in situ) Area, Northeast Alberta. EUB/AGS Earth Sciences Report 2002-03, 286pp.

Andriashek, L.D., Fenton, M.M., 1989. Quaternary stratigraphy and surficial geology of the Sand River area, 73L. Alberta Research Council, Alberta Geological Survey Bulletin 57, 154pp.

Barendregt, R.W., 1995. Paleomagnetic dating methods. In: Rutter, N.W., Catto, N.R. (Eds.), Dating methods for Quaternary Deposits, Geological Association of Canada GEOTEXT, 2, pp. 29–50.

Barendregt, R.W., Duk-Rodkin, A., 2004. Chronology and extent of Late Cenozoic ice sheets in North America: a magnetostratigraphic assessment. In: Ehlers, J., Gibbard, P.L. (Eds.), Quaternary Glaciations—Extent and Chronology. Part II. Developments in Quaternary Science 2b: North America. Elsevier, Amsterdam, pp. 1–7.

Barendregt, R.W., Irving, E., 1998. Changes in the extent of North American ice sheets during the late Cenozoic. Can. J. Earth Sci. 35, 504–509.

Barendregt, R.W., Thomas, F.F., Irving, E., Baker, J., Stalker, A. MacS, Churcher, C.S., 1991. Stratigraphy and paleomagnetism of the Jaw Face section, Wellsch Valley site, Saskatchewan. Can. J. Earth Sci. 28, 1353–1364.

Benn, D.I., Evans, D.J.A., 1998. Glaciers and Glaciation. Arnold, London, 734pp.

Brierley, J.A., 1988. A comparison of two soils from the Milk River Ridge, Southwest Alberta. Unpublished M.Sc. thesis, Department of Soil Science, University of Alberta, Edmonton, Canada.

Burns, J.A., McGillivary, W.B., 1989. A new prairie dog, *Cynomys churcherii*, from the late Pleistocene of southern Alberta. Can. J. Zool. 67, 2633–2639.

Cande, S.C., Kent, D.V., 1992. A new geomagnetic polarity timescale for the Late Cretaceous and Cenozoic. J. Geophys. Res. 97, 13917–13951.

Christiansen, E.A., 1992. Pleistocene stratigraphy of the Saskatoon Area, Saskatchewan, Canada: an update. Can. J. Earth Sci. 29, 1767–1778.

Christiansen, E.A., Whitaker, S.H., 1976. Glacial thrusting of drift and bedrock. In: Leggett, R.F. (Ed.), Glacial Till, Royal Society of Canada, Special Publication 12, pp. 121–130.

Davis, N.K., Locke, W.W. III, Pierce, K.L., Finkel, R.C., 2006. Glacial Lake Musselshell: Late Wisconsin slackwater on the Laurentide ice margin in central Montana, USA. Geomorphology 75, 330–345.

Desilets, D., Zreda, M., Almasi, P.F., Elmore, D., 2006. Determination of cosmogenic Cl-36 in rocks by isotope dilution: innovations, validation and error propagation. Chem. Geol. 233 (3–4), 185–195.

Dreimanis, A. (1982). INQUA-Commission on genesis and lithhology of Quaternary ceposits. Work group (1) - genetic classification of tills and criteria for their differentiation:progress report on activities 1977-1982, and definitions of glacigenic terms. In: Schlüchter, C. (Ed.), INQUA Commission on genesis and lithology of Quaternary deposits. Report on activities 1977-1982. Zurich, p. 12–31.

Dreimanis, A., 1988. Tills: their genetic terminology and classification. In: Goldthwait, R.P., Matsch, C.L. (Eds.), Genetic Classification of Glacigenic Deposits. A.A. Balkema, Rotterdam, pp. 17–83.

Dyke, A.S., Prest, V.K., 1987. Paleogeography of Northern North America 18,000–5000 Years Ago. Geological Survey of Canada Map, 1703A scale 1:12,500,000.

Evans, J.M., 2001. Calibration of the production rates of cosmogenic 36Cl from potassium. Unpublished, Ph.D. dissertation, Australia National University, Canberra, 142pp.

Evans, D.J.A., Campbell, I.A., 1995. Quaternary stratigraphy of the buried valleys of the lower Red Deer River, Alberta. J. Quatern. Sci. 10, 123–148.

Eyles, N., Boyce, J.I., Barendregt, R.W., 1999. Hummocky moraine; sedimentary record of stagnant Laurentide ice sheet lobes resting on soft beds; reply. Sediment. Geol. 129 (1–2), 169–171.

Fravolden, R., 1963. Bedrock channels of southern Alberta. Alberta Res. Council Bull. 12, 63–75.

Fullerton, D.S., Colton, R.B., Bush, C.A., 2004. Limits of mountain and continental glaciations east of the Continental Divide in northern Montana and northwestern North Dakota, U.S.A. In: Ehlers, J., Gibbard, P. L. (Eds.), Quaternary Glaciations—Extent and Chronology. Part II. Developments in Quaternary Science 2b: North America. Elsevier, Amsterdam, pp. 131–150.

Halderson, A., Shaw, J., 1982. The problem of recognizing melt-out till. Boreas 11, 261–277.

Hicock, S.R., Goff, J.R., Lian, O.B., Little, E.C., 1996. On the interpretation of subglacial till fabric. J. Sediment. Petrol. 66, 928–934.

Horberg, L., 1952. Pleistocene drift sheets in the Lethbridge region, Alberta, Canada. J. Geol. 60, 303–330.

Horberg, L., 1954. Rocky Mountain and continental Pleistocene drift sheets in the Waterton region, Alberta, Canada. Geol. Soc. Am. Bull. 65, 1093–1150.

Jackson, L.E. Jr., 1980. Quaternary history and stratigraphy of the Alberta portion of the Kananaskis Lakes map area. Can. J. Earth Sci. 17, 459–477.

Jackson, L.E. Jr., Little, E.C., 2004. A single continental glaciation of Rocky Mountain Foothills, south-western Alberta, Canada. In: Ehlers, J., Gibbard, P.L. (Eds.), Quaternary Glaciations—Extent and Chronology. Part II. Developments in Quaternary Science 2b: North America. Amsterdam, Elsevier, pp. 29–38.

Jackson, L.E., Phillips, F.M., 2003. Cosmogenic ^{36}Cl dating of the all time limit of glaciation, Del Bonita upland Alberta/Montana border and insights into changing extents and ice-flow patterns in successive continental ice sheets. In: Abstracts, 2003 Fall Meeting, American Geophysical Union H42C-1086.

Jackson, L.E. Jr., Phillips, F.M., Shimamura, K., Little, E.C., 1997. Cosmogenic 36Cl dating of the Foothills erratics train, Alberta, Canada. Geology 25, 195–198.

Jackson, L.E. Jr., Leboe, E.R., Little, E.C., Holme, P.J., Hicock, S.R., Shimamura, K., 1999a. Late Quaternary geology of the foothills from Calgary to the Alberta-Montana border. In: Field trip guidebook, 1999 Biennial Meeting of the Canadian Quaternary Association (CANQUA), p. 36, http://www3.telus.net/lejgeology/etrain/htmls/Resources.htm.

Jackson, L.E. Jr., Phillips, F.M., Little, E.C., 1999b. Cosmogenic ^{36}Cl dating of the maximum limit of the Laurentide Ice Sheet in southwestern Alberta. Can. J. Earth Sci. 36, 1347–1356.

Jackson, L.E. Jr., Leboe, E.R., Little, E.C., Holme, P.J., Hicock, S.R., Shimamura, K., et al., 2008. Quaternary stratigraphy and geology of the Foothills, southwestern Alberta. Geol. Surv. Can. Bull. 583, CD-ROM.

Jenner, D.B., 1984. The Late Quaternary geomorphology of Elk Island National Park, Central Alberta. Unpublished, M.Sc. thesis, Department of Geology, University of Alberta.

Karlstrom, E.T., 1999. Evidence for coalescence of Laurentide and Rocky Mountain glaciers east of Glacier National Park, Montana. Phys. Geogr. 20, 225–239.

Krüger, J., 1979. Structures and textures in till indicating subglacial deposition. Boreas 8, 323–340.

Krüger, J., 1984. Clasts with stoss-lee form in lodgement till: a discussion. J. Glaciol. 30, 241–243.

Lawson, D.E., 1981. Sedimentological characteristics and classification of depositional processes and deposits in the glacial environment. U.S. Army Cold Regions Research and Engineering Laboratory, Report 81-27.

Levson, V.M., Rutter, N.W., 1988. A lithofacies analysis and interpretation of depositional environments of montane glacial diamictons, Jasper Alberta, Canada. In: Goldthwait, R.P., Matsch, C.L. (Eds.), Genetic Classification of Glacigenic Deposits. A.A. Balkema, Rotterdam, pp. 117–140.

Lifton, N., Smart, D.F., Shea, M.A., 2008. Scaling time-integrated in situ cosmogenic nuclide production rates using a continuous geomagnetic model. Earth Planet. Sci. Lett. 268, 190–201.

Lisiecki, L.E., Raymo, M.E., 2005. A Pliocene–Pleistocene stack of 57 globally distributed benthic 18O records. Paleoceanography doi:10.1029/2004PA001071 PA1003.

Little, E.C., Jackson, L.E. Jr., James, T.S., Hicock, S.R., Leboe, E.R., 2001. Continental/Cordilleran ice interactions: a dominant cause of westward super-elevation of the last glacial maximum continental ice limit in southwestern Alberta, Canada. Boreas 30, 43–52.

Liverman, D.G.E., Catto, N.R., Rutter, N.W., 1989. Laurentide glaciation in west central Alberta: a single (Late Wisconsinan) event. Can. J. Earth Sci. 26, 266–274.

Lundqvist, J., 1988. Glacigenic processes, deposits and landforms. In: Goldthwait, R.P., Matsch, C.L. (Eds.), Genetic Classification of Glacigenic Deposits. A.A. Balkema, Rotterdam, pp. 3–16.

MacDonald, G.M., Beukens, R.P., Kieser, W.E., Vitt, D.H., 1987. Comparative radiocarbon dating of terrestrial plant macrofossils and aquatic moss from the 'ice-free corridor' of western Canada. Geology 15, 837–840.

Marrero, S.M., 2009. Chlorine-36 production rate calibration using shorelines from Pleistocene Lake Bonneville, Utah. Unpublished, M.Sc. thesis, New Mexico Institute of Mining & Technology, Socorro, 201pp.

Meriano, M., Eyles, N., 2009. Quantitative assessment of the hydraulic role of subglaciofluvial interbeds in promoting deposition of deformation till. Quatern. Sci. Rev. 28, 608–620.

Moran, S.R., Clayton, L., Hooke, R. LeB, Fenton, M.M., Andriashek, L.D., 1980. Glacier-bed landforms of the prairie region of North America. J. Glaciol. 25, 457–476.

Phillips, F.M., Plummer, M.A., 1996. CHLOE: a program for interpreting in-situ cosmogenic nuclide data for surface exposure dating and erosion studies. In: Radiocarbon (Abstracts, 7th International Conference on Accelerator. Mass Spectrometry) 38, pp. 98–99.

Phillips, F.M., Stone, W.D., Fabryka-Martin, J.T., 2001. An improved approach to calculating low-energy cosmic-ray neutron fluxes near the land/atmosphere interface. Chem. Geol. 175, 689–701.

Prest, V.K., 1970. Quaternary geology of Canada. In: Douglas, R.J.W. (Ed.), Geology and Economic Minerals of Canada, 838.

Proudfoot, D.N., 1985. A lithostratigraphic and genetic study of Quaternary sediments in the vicinity of Medicine Hat, Alberta. Unpublished Ph.D. dissertation, Department of Geology, University of Alberta, 248pp.

Reimchen, T.H.F. (1968). Pleistocene mammals from the Saskatchewan Gravels in Alberta. Unpublished M.Sc. thesis, Department of Geology, University of Alberta, 155 p.

Shaw, J., 1982. Melt-out till in the Edmonton area. Can. J. Earth Sci. 19, 1548–1569.

Shaw, J., Sharp, D., Harris, J., 2010. A flowline map of glaciated Canada based on remote sensing data. Can. J. Earth Sci. 47, 89–101.

Shetsen, I., 1984. Application of till pebble lithology to the differentiation of glacial lobes in southern Alberta. Can. J. Earth Sci. 21, 920–933.

Stalker, A.MacS., 1960. Surficial Geology of the Red Deer–Stettler map area. Alberta Geological Survey of Canada Memoir 306, 140pp.

Stalker, A.MacS., 1961. Buried valleys in central and southern Alberta. Geological Survey of Canada Paper 60–32, 13pp.

Stalker, A. MacS., 1963. Quaternary stratigraphy in southern Alberta. Geological Survey of Canada Paper 62–34, 52pp.

Stalker, A. MacS., 1968. Identification of Saskatchewan gravels and sands. Can. J. Earth Sci. 5, 155–163.

Stalker, A. MacS., 1969. Quaternary stratigraphy in southern Alberta, report II: sections near Medicine Hat. Geological Survey of Canada Paper 69–26, 28pp.

Stalker, A. MacS., Harrison, J.E., 1977. Quaternary glaciation of the Waterton-Castle River region of Alberta. Bull. Can. Petroleum Geol. 25, 882–906.

Stalker, A. MacS., Wyder, J.E., 1983. Borehole and outcrop stratigraphy compared with illustrations from the Medicine Hat area of Alberta. Geol. Surv. Can. Bull. 296, 28.

Stone, J.O.H., Evans, J.M., Fifield, L.K., Allan, G.L., Cresswell, R.G., 1998. Cosmogenic chlorine-36 production in calcite by muons. Geochim. Cosmochim. Acta 62, 433–454.

Thorleifson, L.H., Garrett, R.G., 2000. Lithology, mineralogy and geochemistry of glacial sediments overlying kimberlite at Smeaton, Saskatchewan. Geological Survey of Canada Bulletin 551, 40 p.

van der Meer, J.J.M., Menzies, J., Rose, J., 2003. Subglacial till: the deforming glacier bed. Quatern. Sci. Rev. 22, 1659–1685.

Verosub, K.L., 2000. Paleomagnetic dating. In: Noller, J.S., Sowers, J.M., Lettis. W.R. (Eds.), Quaternary Geochronology, Methods and Applications. AGU Reference Shelf 4. American Geophysical Union, Washington, DC, pp. 339–356.

Wagner, W.P., 1966. Correlation of Rocky Mountain and Laurentide glacial chronologies in southwestern Alberta, Canada. Unpublished Ph.D. thesis, University of Michigan, Ann Arbor, Michigan, 141 p.

Westgate, J.A., Fritz, P., Matthews, J.V. Jr., Kalas, L., Green, R., 1971. Sediments of mid-Wisconsin age in west-central Alberta; geochronology, insects, ostracodes, mollusca, and the oxygen isotope composition of molluscan shells. In: Geological Society of America, Abstracts with Programs, Rocky Mountain Section Annual Meeting, 3419.

Westgate, J.A., Fritz, P., Matthews, J.V. Jr., Kalas, L., Delorme, L.D., Green, R., et al., 1972. Geochoronology, and palaeoecology of mid-Wisconsin sediments in west-central Alberta. In: International Geological Congress, 24th Session, Abstracts. 380.

Young, R.R., Burns, J.A., Smith, D.G., Arnold, L.D., Rains, R.B., 1994. A single, late Wisconsin, Laurentide glaciation, Edmonton area and southwestern Alberta. Geology 22, 683–686.

Young, R.R., Burns, J.A., Rains, R.B., Schowalter, D.B., 1999. Late Pleistocene glacial geomorphology and environment of the Hand Hills region and southern Alberta, related to Middle Wisconsin fossil prairie dog sites. Can. J. Earth Sci. 36, 1567–1581.

Zazula, G.D., Duk-Rodkin, A., Schweger, C.E., Morlan, R.E., 2004. Late Pleistocene chronology of glacial Lake Old Crow and the north-west margin of the Laurentide Ice Sheet. In: Ehlers, J., Gibbard, P.L. (Eds.), Quaternary Glaciations—Extent and Chronology. Part II. Developments in Quaternary Science 2b: North America, 347–362.

Zreda, M.G., Phillips, F.M., 2000. Cosmogenic nuclide buildup in surficial materials. In: Noller, S., Sowers, J.M., Lettis, W.R. (Eds.), Quaternary Geochronology, Methods and Applications. AGU Reference Shelf 4. American Geophysical Union, Washington, DC, pp. 61–76.

Chapter 46

Magnetostratigraphy of Quaternary Sections in Eastern Alberta, Saskatchewan and Manitoba

René W. Barendregt

Department of Geography, The University of Lethbridge, 4401 University Drive, Lethbridge, Alberta, Canada T1K 3M4

46.1. INTRODUCTION

A record of continental glaciations has been developed from extensive exposures of Quaternary sediments which outcrop along major river valleys in western Canada. Although ages of Late Pleistocene glacial continental events in the study area are relatively well constrained, the age and extent of earlier glaciations are less well defined. Palaeomagnetism, tephras, palaeosols and fossils constrain the timing of some of these glaciations.

While at least 14 glaciations are recognised in the NW Cordillera of Canada, only about half that number can be documented in the Interior Plains of Canada (Barendregt and Duk-Rodkin, 2011). The magnetostratigraphical record suggests that cold conditions had a very different impact in western Canada than in northwest Canada (Barendregt and Irving, 1998; Barendregt and Duk-Rodkin, 2011; Duk-Rodkin and Barendregt, 2011).

46.2. STUDY SITES

The earliest evidence of Laurentide glaciation on the Canadian Prairies occurs in the Wellsch Valley, Regina and Saskatoon areas of south central Saskatchewan, and in the Medicine Hat area of southern Alberta (Fig. 46.1). Of the five sites discussed here, four are composites of multiple sections (Medicine Hat, Alberta, Wellsch Valley and Saskatoon, Saskatchewan and Gillam, Manitoba), and one is derived from a single outcrop (Wascana Creek, near Regina, Saskatchewan). For a number of sites in Saskatchewan where stratigraphic exposures reveal the upper units only, borehole data are used to describe the lower units. The composite sections reported here have been developed from recent work carried out in Saskatchewan, as well as from previously published and unpublished data. Section descriptions are based primarily on magnetostratigraphy, tephrochronology and fossil data and do not provide detailed facies descriptions (these are published in Christiansen, 1968, 1992; Stalker, 1969, 1976, 1982; Churcher and Stalker, 1988). Data presented here highlight extensive pre-Illinoian (pre-Saalian) deposits which are assigned to the (late) Matuyama Reversed Chron and the Brunhes Normal Chron (Fig. 46.2).

46.2.1. Medicine Hat, Alberta

At least three major till sheets, referred to as the Labuma, Maunsell and Buffalo Lake tills in the early literature, are present in eastern Alberta, and at the Medicine Hat sites. These tills have been correlated on the basis of lithostratigraphical characteristics, to the Saskatoon Group in Saskatchewan (Lower Floral Formation (Fm), Upper Floral Fm, and Battleford Fm, respectively). The Labuma till exhibits a distinctive dark colour, derived from Late Cretaceous shales which were incorporated into the glacial deposits laid down by the first continental glacier to reach the area. It forms the lowermost till at most sections throughout eastern Alberta, and near the outskirts of Medicine Hat, at the Galt Island section (Figs. 46.3 and 46.4) this till is underlain by normally magnetised preglacial sands, silts and clays which contain the Galt Island Tephra (0.43 ± 0.07 Ma; Westgate et al., 1978) as well as late Irvingtonian fossils. (This age of the Galt Island Tephra is considered a minimum age. The tephra has a mineralogical assemblage similar to that of the Wellsch Valley Tephra (0.78 ± 0.04 Ma) and both have a Cascadian source.) The Labuma till is overlain by interglacial sediments, which are Sangamon in age, based on dates obtained from organics and fossil bones and teeth (Stalker, 1976). These sediments are assumed to be the equivalent of the

FIGURE 46.1 Location map of Alberta, Saskatchewan and Manitoba study sites.

interglacial Riddell Member of the Floral Fm, in Saskatchewan. The interglacial deposits are overlain by three tills, separated in places by intertill beds, some of which are considered to mark interglacial or interstadial conditions. The tills are clearly Wisconsian in age, and the uppermost till can be traced westward across southern Alberta to sites where ^{36}Cl dates on erratics (Jackson et al., 1999) and radiocarbon dates on organics and bone have confirmed a Late Wisconsinan age (Stalker, 1969, 1976, 1982; Stalker and Wyder, 1983; Stalker and Vincent, 1993). Further refinement of the post-Sangamon deposits at Medicine Hat into early, middle and late Wisconsin ages has been suggested by previous workers (Stalker, 1976). At Manyberries, Alberta (60 km south of Medicine Hat), glacio-lacustrine sediments deposited by receding ice of the last glaciation contain the 12,000 ^{14}C ka BP Glacier Peak Tephra (Westgate, 1968), further supporting a Late Wisconsinan age for the surface till in southern Alberta.

All sediments at the Medicine Hat sections (including Galt Island) are normally magnetised (Barendregt, 1976; Barendregt and Stalker, 1978; Barendregt et al., 1977, 1988) and are assigned to the Brunhes Normal Chron. This assignment is supported by tephra and fossil data, as well as stratigraphical correlations with nearby sites at Wellsch Valley, Saskatchewan. The preglacial sediments in the Medicine Hat area are also normally magnetised, and based on the Galt Island Tephra and late Irvingtonian fossils found there (Fig. 3), fall within the early to mid Brunhes Chron (0.78–0.30 Ma; Marine Isotope Stages (MIS) 9–19).

46.2.2. Wellsch Valley Sections, near Swift Current, Saskatchewan

Preglacial, glacial and non-glacial deposits of Late Pliocene to Late Pleistocene age near Wellsch Valley, Saskatchewan (Figs. 46.1 and 46.5) have been described from surface exposures and borecores. Preglacial sediments at Wellsch Valley are reversely magnetised (Barendregt et al., 1991, 1998a,b) and contain the Wellsch Valley Tephra (0.78 ± 0.04 Ma) as well as early Irvingtonian fossils (Stalker, 1969, 1976, 1982; Szabo et al., 1973; Zymela et al., 1988; Stalker and Vincent, 1993; Stewart and Seymour, 1996), and therefore the sediments fall within the late Matuyama (1.0–0.78 Ma). These and underlying normally magnetised preglacial sediments are assigned to the Empress Group, which ranges from the late Gauss Normal Chron to the late Matuyama Reversed Chron (Figs. 46.2 and 46.5). The overlying glacial tills (five in total) are all normally magnetised and assigned to the Brunhes Normal Chron. These tills can be correlated to outcrop and borehole records near Regina and Saskatoon based on lithostratigraphical characteristics described in Christiansen (1968, 1992). Overlying the preglacial sediments is the upper till of the Dundurn Formation. This till is recognised through large parts of Saskatchewan. At the Wascana Creek site near Regina (Figs. 46.1 and 46.5), the Dundurn is overlain by glacio-lacustrine sediments containing the Wascana Creek Tephra, which is equivalent to the Lava Creek B Tephra in the American Midwest

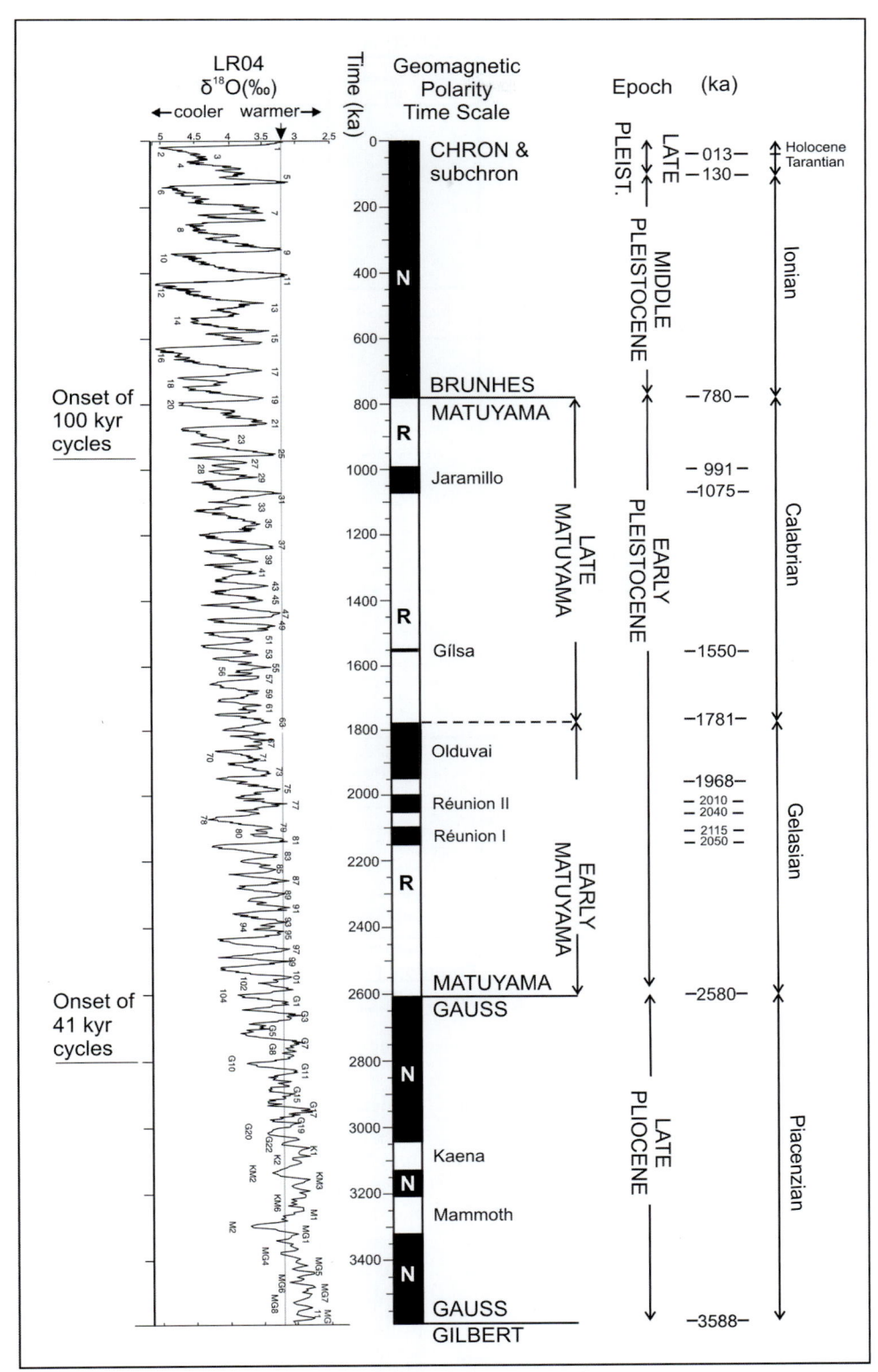

FIGURE 46.2 Geomagnetic polarity time scale (Cande and Kent, 1995) for LR04 benthic $\delta^{18}O$ palaeotemperature profile (Lisiecki and Raymo, 2005). Black/white intervals represent normal/reversed polarity. Marine isotope stages (MIS) are labelled on LR04 (even numbers represent glacials; odd numbers interglacials). MIS numbering scheme follows Ruddiman et al. (1986, 1989) and Raymo et al. (1989, 1992) from present to MIS 104, and Shackleton et al. (1995) in the Gauss Chron. Arrow marks Holocene mean $\delta^{18}O$ (Raymo, 1992).

FIGURE 46.3

SEDIMENTS	AGE	ISOTOPE STAGE	DATING TOOL	MAGNETIC POLARITY
post-glacial loess lacustrine silts & clays	<14,000	1	C^{14}	
Buffalo Lake (2 tills)	14,000 – 23,000	2		N
sands & gravel	?	?	–	?
Illinoian Till (Labuma)		6 8 (?)	pmag	N
Preglacial -sands, silts, and clays -Mitcheel Bluff equivalent		11 (?)	pmag	N
Wood-bearing sands		11 (?)	-C^{14} >39 Ka -pmag	N
Fluvial sands and silts -bone-bearing-		11 (?)	-ESR 0.23–0.41 Ma -F.T -pmag	N
Preglacial Gravels	?	?	?	?
Upper Cretaceous Foremost Fm.				

★★★ Galt Island Tephra ★★★ 0.43 ± 0.07 Ma.*

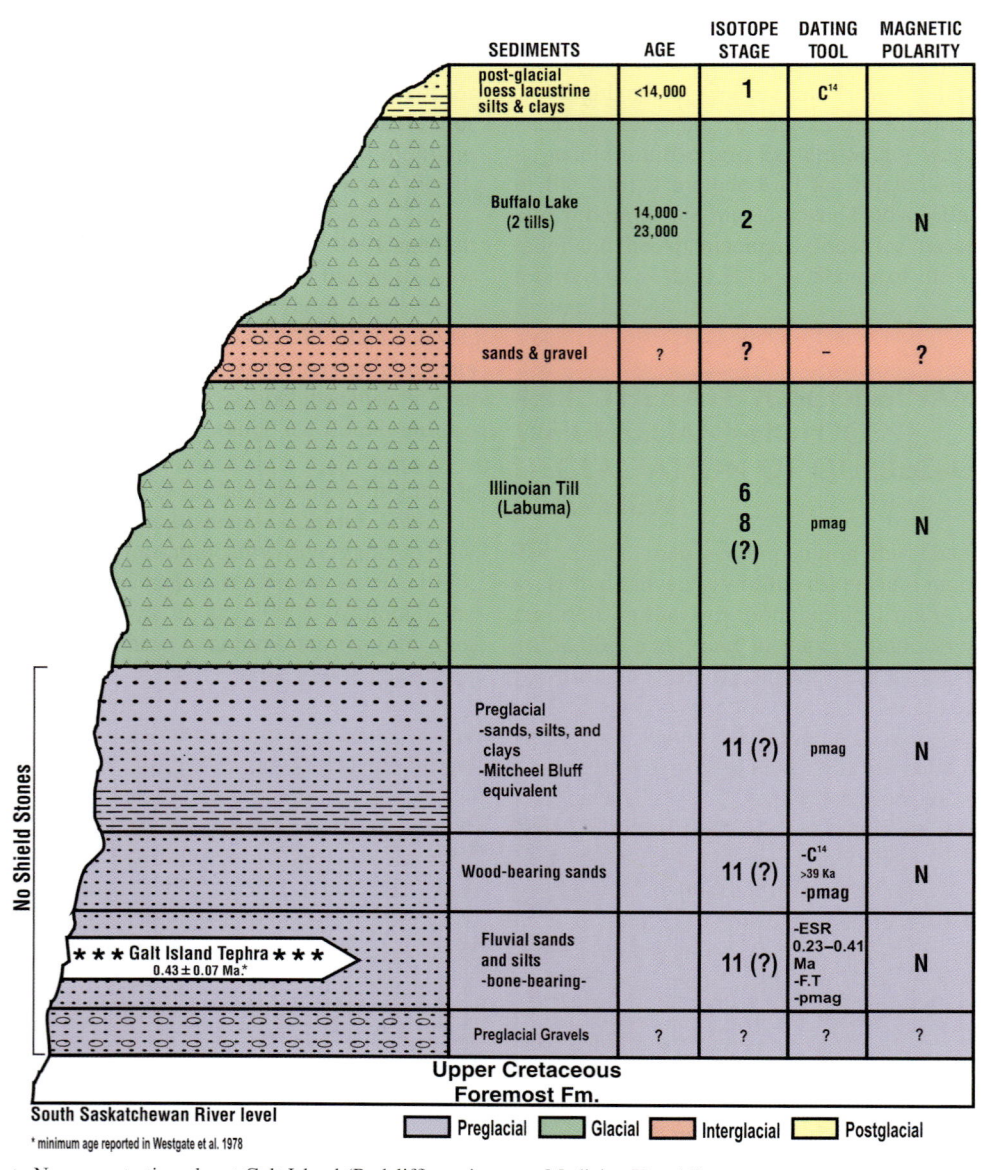

No Shield Stones

South Saskatchewan River level
* minimum age reported in Westgate et al. 1978

Legend: Preglacial | Glacial | Interglacial | Postglacial

FIGURE 46.3 Late Neogene stratigraphy at Galt Island (Redcliff) section near Medicine Hat, Alberta.

(0.639 ± 0.002 Ma; Lanphere et al., 2002). The association of the ash with sediments of the Dundurn deglaciation implies that this till was deposited during an MIS 16 glaciation. The Dundurn is overlain by the normally magnetised Warman Fm (MIS 12 or 14?) (Fullerton et al., 2004) which in turn is overlain by lower and upper tills of the Floral Fm (MIS 6 and 4), and by the Battleford Fm (MIS 2) at the surface.

46.2.3. Wascana Creek Section and Borecore near Regina, Saskatchewan

The Wascana Creek Tephra site is located 19 km northwest of Regina, Saskatchewan. Only the upper two-thirds of the Quaternary sediments outcrop along the west bank of the Wascana Creek valley. The lower third has been studied from borecores (Fig. 46.6). Outcrop and borecore records reveal four tills (Mennon, upper till of the Dundurn, Warman and Battleford formations) whose characteristics can be correlated with type sections to the north, near Saskatoon, Saskatchewan. Of note at this site is the presence of the Wascana Creek Tephra (0.639 ± 0.002 Ma; Lanphere et al., 2002) within glacio-lacustrine sediments associated with the Dundurn Fm and overlain by the Warman till. This tephra is equivalent to the Lava Creek B Tephra, a Pearlette ash from the Yellowstone area, and is found in a number of localities in the American Midwest (Westgate et al., 1977). It clearly dates the Upper till of the Dundurn Fm to MIS 16

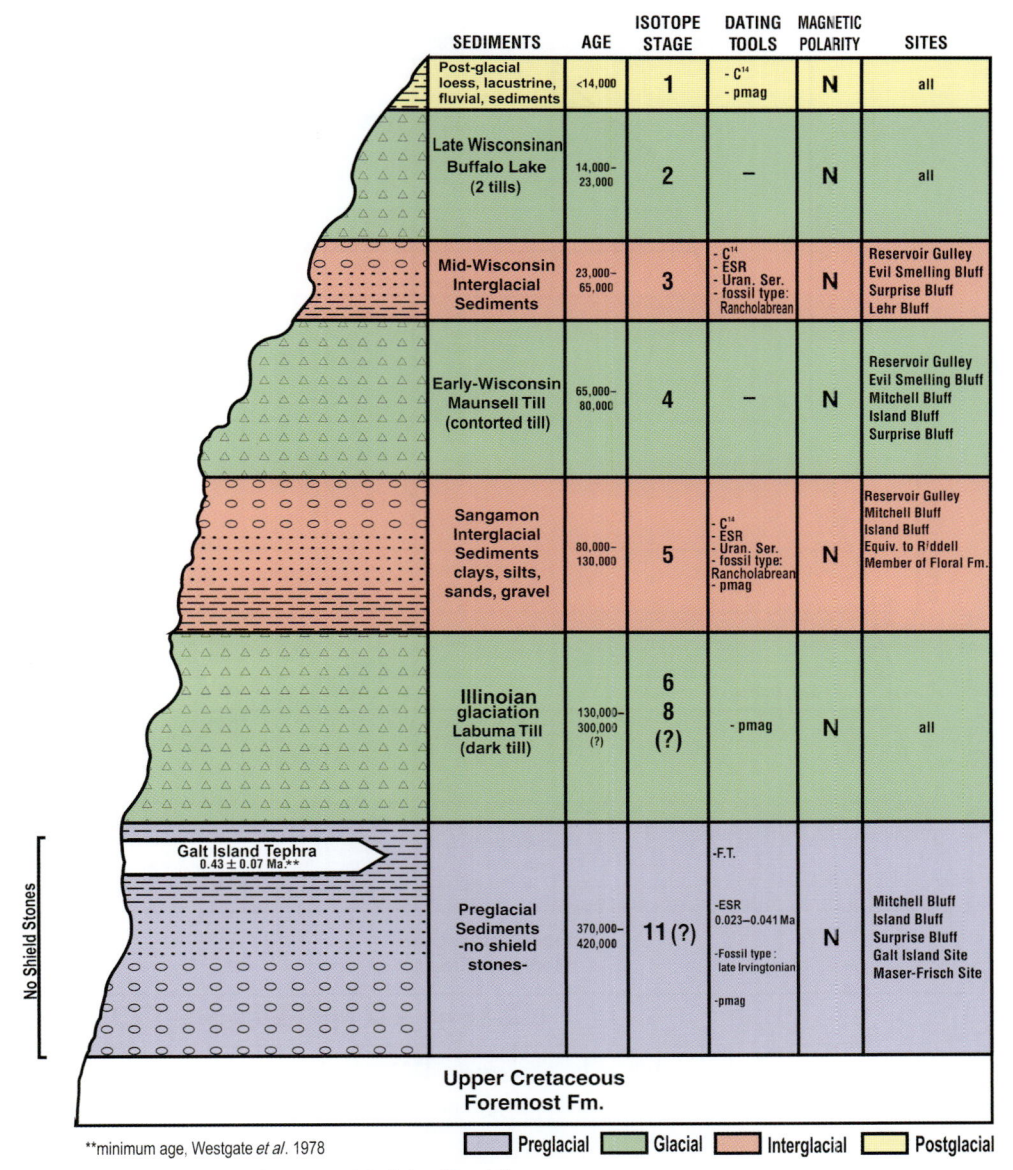

SEDIMENTS	AGE	ISOTOPE STAGE	DATING TOOLS	MAGNETIC POLARITY	SITES
Post-glacial loess, lacustrine, fluvial, sediments	<14,000	1	- C¹⁴ - pmag	N	all
Late Wisconsinan Buffalo Lake (2 tills)	14,000–23,000	2	–	N	all
Mid-Wisconsin Interglacial Sediments	23,000–65,000	3	- C¹⁴ - ESR - Uran. Ser. - fossil type: Rancholabrean	N	Reservoir Gulley Evil Smelling Bluff Surprise Bluff Lehr Bluff
Early-Wisconsin Maunsell Till (contorted till)	65,000–80,000	4	–	N	Reservoir Gulley Evil Smelling Bluff Mitchell Bluff Island Bluff Surprise Bluff
Sangamon Interglacial Sediments clays, silts, sands, gravel	80,000–130,000	5	- C¹⁴ - ESR - Uran. Ser. - fossil type: Rancholabrean - pmag	N	Reservoir Gulley Mitchell Bluff Island Bluff Equiv. to Riddell Member of Floral Fm.
Illinoian glaciation Labuma Till (dark till)	130,000–300,000	6 8 (?)	- pmag	N	all
Galt Island Tephra 0.43 ± 0.07 Ma.** Preglacial Sediments -no shield stones-	370,000–420,000	11 (?)	-F.T. -ESR 0.023–0.041 Ma -Fossil type: late Irvingtonian -pmag	N	Mitchell Bluff Island Bluff Surprise Bluff Galt Island Site Maser-Frisch Site
Upper Cretaceous Foremost Fm.					

**minimum age, Westgate et al. 1978   Preglacial Glacial Interglacial Postglacial

FIGURE 46.4 Late Neogene composite stratigraphy at Medicine Hat, Alberta.

and points to the antiquity of the underlying tills (lower till of the Dundurn Fm and till of the Mennon Fm). If the latter two tills represent separate glaciations, then the Mennon Fm would be expected to fall in MIS 20 or older and should be reversely magnetised. In fact, recent measurements (Barendregt et al., 2007, 2011) made from borecore sediments at Wascana Creek and Sutherland (see below) indicate that the Mennon indeed falls within the Matuyama Reversed Chron. The Mennon is underlain by reversely magnetised sediments (gravels, clays and sands) of the Empress Formation, and these sediments are probably of the same age as the preglacial sediments at Wellsch Valley (latest Matuyama). Beneath the reversely magnetised sediments of the Empress Fm, at the base of both the Wascana Creek and Sutherland borecores (described below), is an interval of normally magnetised sediments, also assigned to the Empress Fm (Fig. 46.7). This normal interval may be one of the Early Pleistocene normal subchrons within the Matuyama (i.e. Jaramillo or Olduvai), or may be Gauss age (>2.58 Ma).

46.2.4. Sutherland Borecore, Within City Limits of Saskatoon, Saskatchewan

Borecores drilled in the Sutherland district of the City of Saskatoon were collected from the reference sites for the Sutherland Group (Christiansen, 1968). The Sutherland Group (Fig. 46.7) includes all sediments between the

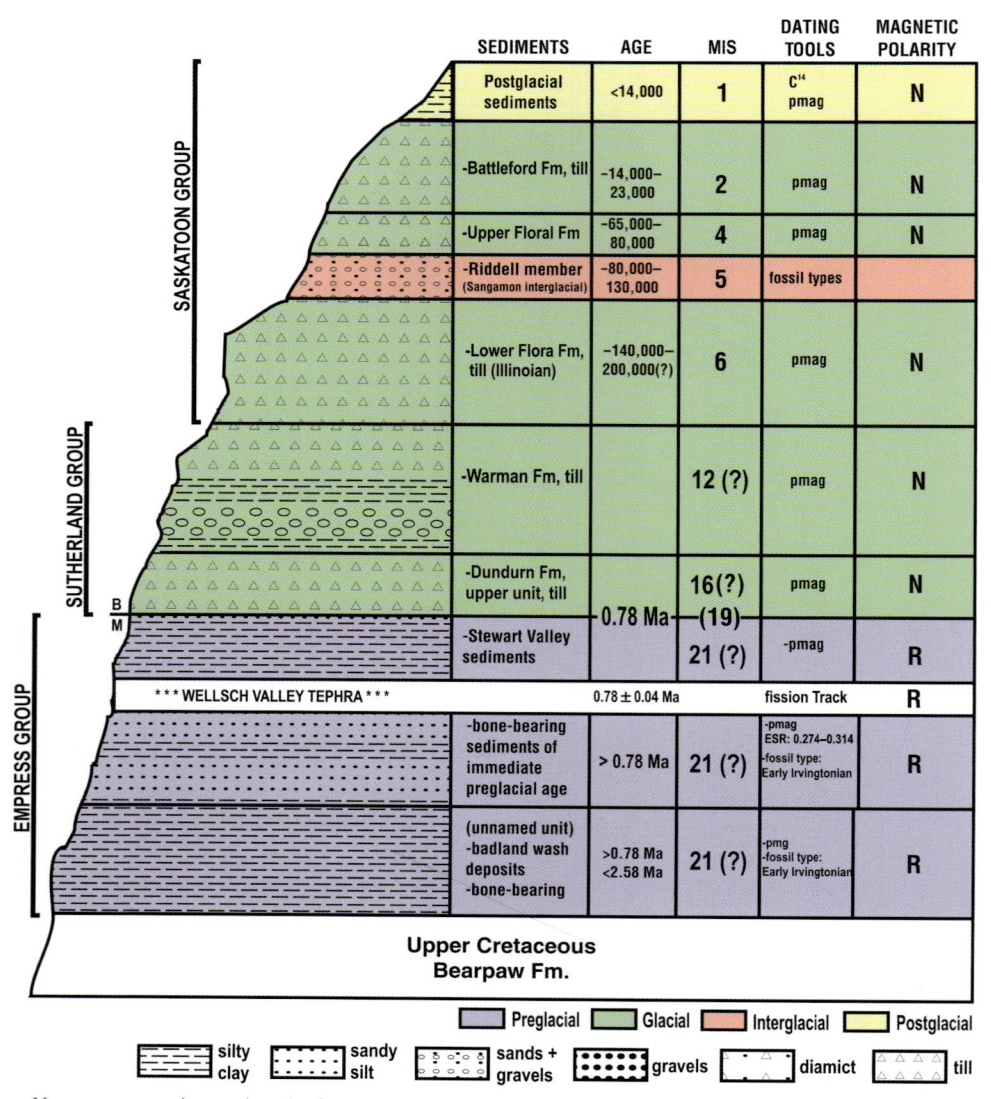

SEDIMENTS	AGE	MIS	DATING TOOLS	MAGNETIC POLARITY
Postglacial sediments	<14,000	1	C¹⁴, pmag	N
-Battleford Fm, till	~14,000–23,000	2	pmag	N
-Upper Floral Fm	~65,000–80,000	4	pmag	N
-Riddell member (Sangamon interglacial)	~80,000–130,000	5	fossil types	
-Lower Flora Fm, till (Illinoian)	~140,000–200,000(?)	6	pmag	N
-Warman Fm, till		12 (?)	pmag	N
-Dundurn Fm, upper unit, till	0.78 Ma	16(?)–(19)	pmag	N
-Stewart Valley sediments		21 (?)	-pmag	R
WELLSCH VALLEY TEPHRA	0.78 ± 0.04 Ma		fission Track	R
-bone-bearing sediments of immediate preglacial age	> 0.78 Ma	21 (?)	-pmag, ESR: 0.274–0.314, -fossil type: Early Irvingtonian	R
(unnamed unit) -badland wash deposits -bone-bearing	>0.78 Ma, <2.58 Ma	21 (?)	-pmg, -fossil type: Early Irvingtonian	R
Upper Cretaceous Bearpaw Fm.				

Groups (left axis): SASKATOON GROUP, SUTHERLAND GROUP, EMPRESS GROUP. B/M boundary marked at 0.78 Ma.

Legend: Preglacial | Glacial | Interglacial | Postglacial; silty clay, sandy silt, sands + gravels, gravels, diamict, till.

FIGURE 46.5 Late Neogene composite stratigraphy from borecores and outcrops at the Wellsch Valley sites, near Swift Current, Saskatchewan.

preglacial Empress Group and the Saskatoon Group (Illinoian and younger deposits). Results (Barendregt et al., 2007) from the first borecore were mixed, with a number of tills yielding incoherent magnetisations and therefore a second core was collected nearby. (incoherent magnetisation may result from insufficient water content of till slurries at time of deposition, preventing silt and clay-sized ferromagnetic minerals from becoming oriented in the earth's field.) While some of the units produced incoherent results in both cores, the Mennon, lower and upper units of the Dundurn and upper Floral Fms gave good results (Barendregt et al., 2007, 2011). The Mennon Fm (lowermost till in Saskatchewan) is reversely magnetised, while the Dundurn Fm (lower and upper unit tills) and the Floral Fm (upper till) are normally magnetized (Fig. 46.7). Results from the Warman and Floral Fm (lower till) were incoherent. Only the lower part of the Empress Fm was sampled, and it is normal.

Taken together, the records from Wellsch Valley and Wascana Creek, and Sutherland confirm a reversely magnetised till (late Matuyama) at the base of the Quaternary sequence, underlain by preglacial sediments which are likewise reversed, and based on the age and position of the Wellsch Valley Tephra, were probably deposited right up until the first continental (Laurentide) ice sheet arrived in this region of Saskatchewan.

46.2.5. Lower Nelson/Hayes River Area near Gillam, Manitoba

Thick sequences of both glacial and non-glacial sediments (Fig. 46.8) are exposed where major rivers in the Hudson

Chapter | 46 Magnetostratigraphy of Quaternary Sections in Eastern Alberta

Sediments	Age	MIS	Dating Tools	Magnetic Polarity
Post-glacial sediments -Regina clays	<14,000	1	C^{14} pmag	
Battleford Fm, till	14,000–23,000	2	pmag	N
Warman Fm, till	?	12 (?)	pmag	N
glacial lacustrine	0.639 Ma	15 / 16?	pmag / fission track	N(tephra) / N
till	?	16?	pmag	N
diamict	?	16?	pmag	N
till	?	16? (19)	pmag	N
Mennon Fm, till	?	20(?)	pmag	R
Empress Group	>0.78 Ma	21(?)	pmag	R / N
Upper Cretaceous Bearpaw Fm.				

Wascana Creek Tephra (0.639 ± 0.002 Ma) (=Lava Creek B Tephra in USA)

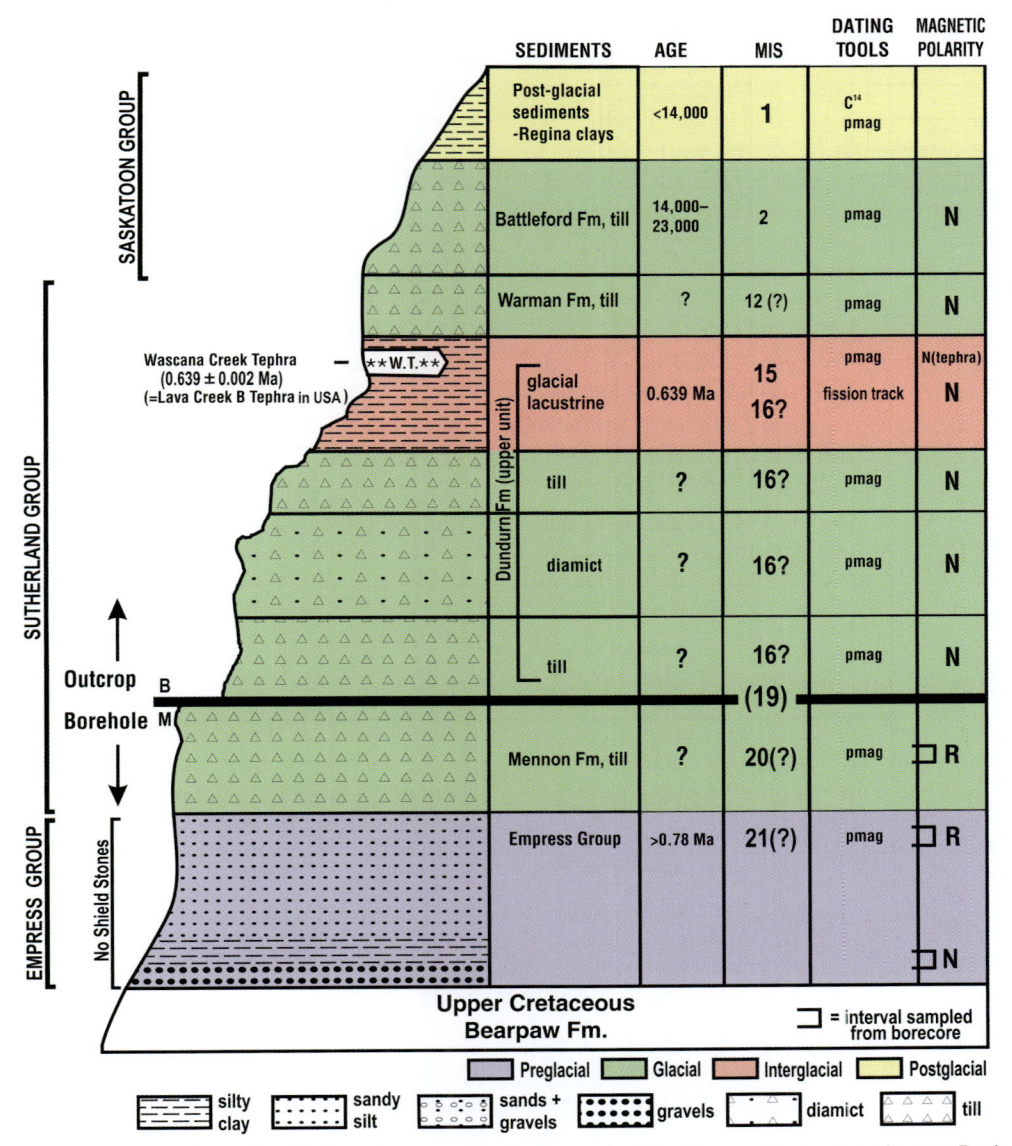

Groups: SASKATOON GROUP, SUTHERLAND GROUP, EMPRESS GROUP
Dundurn Fm (upper unit)
No Shield Stones

Legend: Preglacial | Glacial | Interglacial | Postglacial
silty clay | sandy silt | sands + gravels | gravels | diamict | till
⬜ = interval sampled from borecore

FIGURE 46.6 Late Neogene composite stratigraphy from borecore and outcrop records at the Wascana Creek valley site, near Regina, Saskatchewan.

Bay Lowlands are deeply incised through Quaternary deposits. Four tills have been identified on the basis of provenance, lithological composition, texture and colour. They are, from oldest to youngest: Sundance till, Amery till, Long Spruce till and Sky Pilot till. Sundance till is a sandy granitic till of northwestern (Keewatin) provenance, while the others are calcareous and of eastern (Labrador) provenance (Dredge and Nielsen, 1987; Roy, 1998).

Non-glacial beds have been identified between tills at a number of sites and have been correlated on the basis of their stratigraphic position relative to the tills and on palaeoecological criteria. They yield infinite radiocarbon dates, and on the basis of palaeogeographic and palaeoecological constraints, indicate an ice-free Hudson Bay. The top of the oldest till has a 2-m-thick leached zone containing a subarctic pollen assemblage, which is therefore interpreted as a truncated palaeosol. Amino acid racemization studies of shells contained in non-glacial sediments and in the Amery till suggest several marine inundations during the Middle to Late Pleistocene.

Six sections in the lower Nelson/Hayes River area were selected for palaeomagnetic study. These are Limestone, Henday, Sundance, Port Nelson, Stupart Creek and Echoing Creek. They were chosen because each contains an extensive glacial and non-glacial stratigraphy, which are well preserved in a major preglacial valley system (of Tertiary age). Collectively, the deposits span a considerable amount of time, based on the presence of multiple tills sheets (of

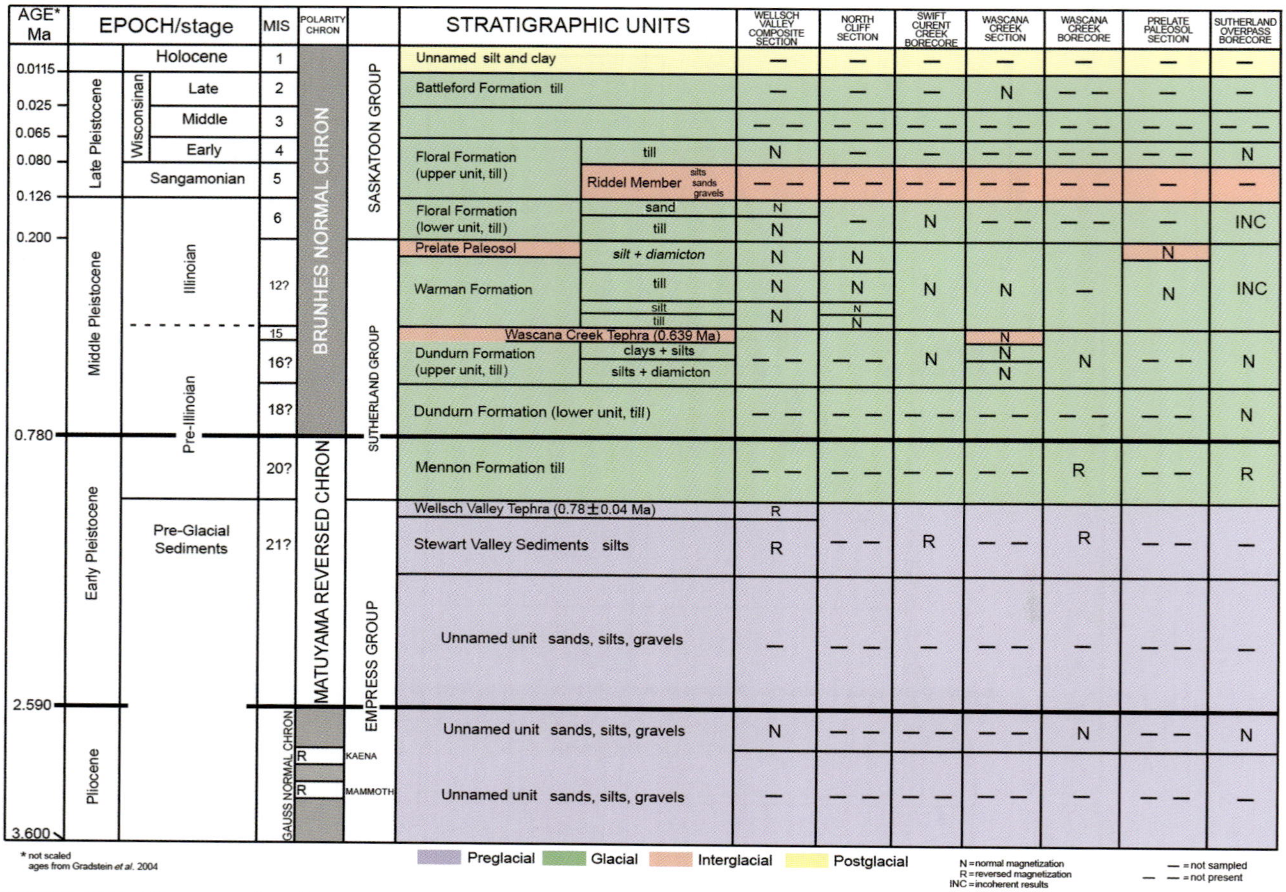

AGE* Ma	EPOCH/stage		MIS	POLARITY CHRON		STRATIGRAPHIC UNITS		WELLSCH VALLEY COMPOSITE SECTION	NORTH CLIFF SECTION	SWIFT CURRENT CREEK BORECORE	WASCANA CREEK SECTION	WASCANA CREEK BORECORE	PRELATE PALEOSOL SECTION	SUTHERLAND OVERPASS BORECORE
0.0115		Holocene	1	BRUNHES NORMAL CHRON	SASKATOON GROUP	Unnamed silt and clay		—	—	—	—	—	—	—
0.025	Late Pleistocene / Wisconsinan	Late	2			Battleford Formation till		—	—	—	N	— —	—	—
0.065		Middle	3					— —	—	—	—	— —	—	—
0.080		Early	4			Floral Formation (upper unit, till)	till	N	—	—	—	—	—	N
0.126		Sangamonian	5				Riddel Member silts sands gravels	— —	—	—	—	— —	—	—
0.200	Middle Pleistocene / Illinoian		6			Floral Formation (lower unit, till)	sand	N	—	N	—	— —	—	INC
							till	N	—					
			12?		SUTHERLAND GROUP	Prelate Paleosol	silt + diamicton	N	N	—	—	—	N	—
						Warman Formation	till	N	N	N	N	—	N	INC
			15				silt	N	N					
			16?			Wascana Creek Tephra (0.639 Ma)						N		
						Dundurn Formation (upper unit, till)	clays + silts	— —	—	N	N	N	N	N
			18?				silts + diamicton				N			
0.780		Pre-Illinoian				Dundurn Formation (lower unit, till)		— —	—	—	—	—	— —	N
	Early Pleistocene / Pre-Glacial Sediments		20?	MATUYAMA REVERSED CHRON		Mennon Formation till		— —	—	—	—	—	R	R
			21?			Wellsch Valley Tephra (0.78±0.04 Ma)		R						
						Stewart Valley Sediments silts		R	—	R	—	R	— —	—
					EMPRESS GROUP	Unnamed unit sands, silts, gravels		—	— —	—	—	— —	—	—
2.590	Pliocene			GAUSS NORMAL CHRON (KAENA, MAMMOTH)		Unnamed unit sands, silts, gravels		N	— —	—	—	N	— —	N
3.600						Unnamed unit sands, silts, gravels		—	— —	—	—	— —	—	—

* not scaled ages from Gradstein et al. 2004

Legend: Preglacial | Glacial | Interglacial | Postglacial
N = normal magnetization; R = reversed magnetization; INC = incoherent results; — = not sampled; — — = not present

FIGURE 46.7 Magnetostratigraphic correlation of Saskatchewan outcrop and borehole data from the Wellsch Valley, Wascana Creek and Sutherland sites.

differing provenance), the amount of time required to raise and lower sea level by substantial amounts, the considerable thicknesses of interglacial sediments, as well as extended periods of soil development. As no absolute dating techniques could be applied to the older sediments, palaeomagnetic measurements were made to determine whether reversely magnetised (Matuyama Chron; 2.6–0.78 Ma) sediments are present in the Hudson Bay Lowlands.

All units exhibit normal magnetisations (Barendregt et al., 2007, 2011), indicating that the Quaternary record in the Hudson Bay Lowlands can be assigned to the Brunhes Chron (<0.78 Ma) and falls within the Middle to Late Pleistocene age (Fig. 46.2). The magnetostratigraphy of the Hudson Bay Lowlands indicates that Brunhes Chron glaciations occurred repeatedly in this region.

46.3. DISCUSSION

The age of North American Interior Plains glaciations is less well constrained than Cordilleran glaciations, both in terms of age and extent (Barendregt and Duk-Rodkin, 2011; Duk-Rodkin and Barendregt, 2011). Only in the American Midwest (Minnesota, Nebraska, Iowa, Kansas and Missouri) have reversely magnetised continental tills been previously reported (Roy et al., 2004; Balco et al., 2005). Based on tephra dates, these reversely magnetised glacial sediments fall within the early as well as late Matuyama Reversed Chron (2.60–0.78 Ma). Barendregt et al. (2011) report the results of palaeomagnetic measurements made on borecore samples collected from south-central Saskatchewan. There a single reversely magnetised till (Mennon Fm) occurs at the base of the glacial sequence, and can be confidently assigned to the latest Matuyama (MIS 20), based on underlying and overlying tephras. At least 4 Middle Pleistocene (MIS 18?, 16?, 12?, 6?) and 2 Late Pleistocene (MIS 4 and 2) continental tills overlie the Mennon Fm, and all are normally magnetized. To the west, in Alberta, only a single (Late Wisconsinan) glaciation (MIS 2) covered most of the province, while areas to the east of the ~700 m contour line saw two or more glaciations. In south-central Saskatchewan, evidence for seven glaciations is reported. The earliest continental

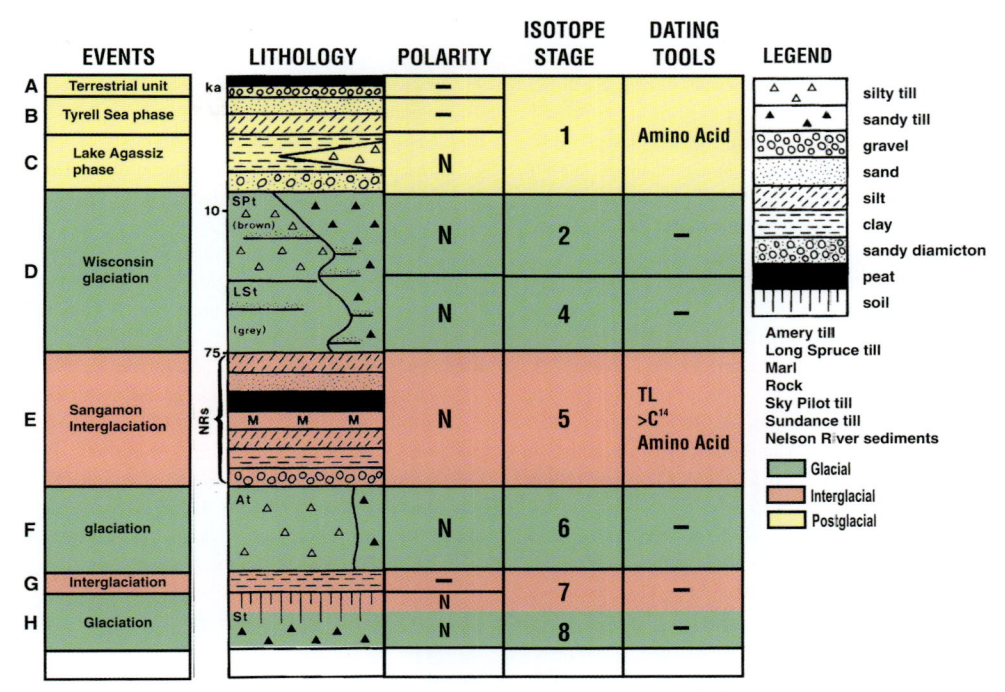

FIGURE 46.8 Late Neogene composite stratigraphy of Hudson Bay Lowland near Gillam, Manitoba.

glaciation in the southern Canadian Prairies occurred during the latest Matuyama Reversed Chron (> 0.78 Ma; MIS 20), based on palaeomagnetism, tephras and fossil data.

46.4. CONCLUSIONS

In 1998, Barendregt and Irving provided a summary of the magnetostratigraphical data for western Canada and the northwestern USA. In these summaries, distribution and extent of ice sheets in western North America are reconstructed, based on some 70 magnetostratigraphic records, and show marked differences between the Matuyama and Brunhes Chrons (see Fig. 46.3 in Barendregt and Duk-Rodkin, 2011, Chapter 32). During the Matuyama Chron, ice appears to have been largely absent from large areas of the southern prairie provinces in Canada and the adjacent states of Montana and North Dakota, as well as from much of the Arctic Islands. In the Late Matuyama a modest Keewatin ice centre formed, delivering ice as far distant as Banks Island, (Barendregt and Vincent, 1990; Barendregt et al., 1998b), North West Territories (NWT) and south-central Saskatchewan. In contrast, the Labrador/Hudson Bay ice centre delivered ice as far south as Kansas during both the Early and Late Matuyama (Roy et al., 2004; Balco et al., 2005). During the Brunhes Chron, ice caps appear for the most part to have been far more extensive than in the Matuyama, and only in the southern Midwestern states did Brunhes-age ice not reach previous limits. During the last major glaciation (Late Wisconsinan), ice cover was continuous from Atlantic to Pacific, with Cordilleran and Keewatin ice sheets in contact in western Alberta, and along the eastern margin of the Mackenzie Mountains in the NWT.

REFERENCES

Balco, G., Stone, J.O.H., Mason, J., 2005. The first glacial maximum in North America. Science 307, 222.

Barendregt, R.W., 1976. A detailed geomorphological survey of the Pakowki-Pinhorn Area, Southeastern Alberta. Ph.D. Dissertation, Department of Geography, Queen's University, 275 pp.

Barendregt, R.W., Duk-Rodkin, A., 2011. Chronology and extent of late Cenozoic ice sheets in North America: a magnetostratigraphic assessment. In: Ehlers, J., Gibbard, P.L., Hughes, P.D. (Eds.), Quaternary Glaciations—Extent and Chronology. Part IV: A Closer Look. Elsevier, Amsterdam.

Barendregt, R.W., Irving, E., 1998. Changes in the extent of North American ice sheets during the late Cenozoic. Can. J. Earth Sci. 35, 504–509.

Barendregt, R.W., Stalker, A.MacS., 1978. Characteristic magnetization of some Middle Pleistocene sediments from the Medicine Hat Area of Southern Alberta. Geological Survey of Canada Paper, 78-1A, pp. 487–488.

Barendregt, R.W., Vincent, J.-S., 1990. Late Cenozoic paleomagnetic record of Duck Hawk Bluffs, Banks Island, Canadian Arctic Archipelago. Can. J. Earth Sci. 27, 124–130.

Barendregt, R.W., Foster, J.H., Stalker, A. MacS, 1977. Paleomagnetic remanence characteristics of surface tills found in the Pakowki-Pinhorn area of southern Alberta. Geological Survey of Canada Paper, 77-1B.

Barendregt, R.W., Churcher, C.S., Stalker, A. MacS, 1988. Stratigraphy, paleomagnetism and vertebrate paleontology of Quaternary preglacial sediments at the Maser-Frisch Site, southeastern Alberta. Geol. Soc. Am. Bull. 100, 1824–1832.

Barendregt, R.W., Thomas, F.F., Irving, E., Baker, J., Stalker, A. MacS, Churcher, C.S., 1991. Stratigraphy and paleomagnetism of the Jaw Face section, Wellsch Valley site, Saskatchewan. Can. J. Earth Sci. 28, 1353–1364.

Barendregt, R.W., Irving, E., Christiansen, E.A., Sauer, E.K., Schreiner, B.T., 1998a. Stratigraphy and Paleomagnetism of Late Pliocene and Pleistocene sediments from Wellsch Valley and Swift Current Creek areas, southwestern Saskatchewan, Canada. Can. J. Earth Sci. 35, 1347–1361.

Barendregt, R.W., Vincent, J.-S., Irving, E., Baker, J., 1998b. Magnetostratigraphy of Quaternary and Late Tertiary sediments on Banks Island, Canadian Arctic Archipelago. Can. J. Earth Sci. 35, 147–161.

Barendregt, R.W., Duk-Rodkin, A., Enkin, R.E., Baker, J., Christiansen, E., Naeser, N.D., et al., 2007. New magnetostratigraphic data from central Saskatchewan, and from the Horton Plateau, Mackenzie District, NWT: evidence for Matuyama Chron glaciations in the Interior Plains of Canada. In: XVII I NQUA Congress, Cairns, Australia, 28 July–3 August, 2007.

Barendregt, R.W., Enkin, R.J., Christiansen, E.A., Tessler, D.L., 2011. Magnetostratigraphy of Late Neogene glacial, interglacial, and preglacial sediments in the Saskatoon and Regina areas, Saskatchewan. Studia Geophysica and Geodaetica 55 (in press).

Cande, S.C., Kent, D.V., 1995. Revised calibration of the geomagnetic timescale for the Late Cretaceous and Cenozoic. J. Geophys. Res. 100 (B4), 6093–6095.

Christiansen, E.A., 1968. Pleistocene stratigraphy of the Saskatoon area, Saskatachewan, Canada. Can. J. Earth Sci. 5, 1167–1173.

Christiansen, E.A., 1992. Pleistocene stratigraphy of the Saskatoon area, Saskatchewan, Canada: an update. Can. J. Earth Sci. 29, 1767–1778.

Churcher, C.S., Stalker, A. MacS, 1988. Geology and vertebrate paleontology of the Wellsch Valley site, Saskatchewan. unpublished manuscript.

Dredge, L.A., Nielsen, E., 1987. Glacial and interglacial stratigraphy, Hudson Bay Lowlands, Manitoba. In: Biggs, D.L. (Ed.), Geological Society of America Centennial Field Guide. North Central Section 3, 43–46.

Duk-Rodkin, A., Barendregt, R.W., 2011. The glacial history of Northwestern Canada. In: Ehlers, J., Gibbard, P.L., Hughes, P.D. (Eds.), Quaternary Glaciations—Extent and Chronology. Part IV: A Closer Look. Elsevier, Amsterdam.

Fullerton, D.S., Colton, R.B., Bush, C.A., 2004. Limits of mountain and continental glaciations east of the continental divide in northern Montana and northeastern North Dakota, USA. In: Ehlers, J., Gibbard, P.L. (Eds.), Quaternary Glaciations—Extent and Chronology. Part II: North America. Elsevier, Amsterdam.

Gradstein, F.M., Ogg, J.G., Smith, A.G., Agterberg, F.P., Bleeker, W., Cooper, R.A., 2004. A Geologic Timescale. Geological Survey of Canada, Miscellaneous Report 86, and 1 map.

Jackson, L.E., Jr., Phillips, F.M., Little, E.C., 1999. Cosmogenic ^{36}Cl dating of the maximum limit of the Laurentide Ice Sheet in southwestern Alberta. Can. J. Earth Sci. 36, 1347–1356.

Lanphere, M.A., Champion, D.E., Christiansen, R.L., Izett, G.A., Obradovich, J.D., 2002. Revised age for tuffs of the Yellowstone Plateau volcanic field—assignment of the Huckleberry Ridge Tuff to a new geomagnetic polarity event. Geol. Soc. Am. Bull. 114, 559–568.

Lisiecki, L.E., Raymo, M., 2005. A Pliocene–Pleistocene stack of 57 globally distributed benthic $\delta^{18}O$ records. Paleoceanography 20, PA1003. doi:10.1029/2004PA001071.

Raymo, M.E., 1992. Global climate change: a three million year perspective. In: Kukla, G.K., Went, E. (Eds.), Start of a Glacial, Nato ASI Series, Series I, Global Environmental Change 3, 207–223.

Raymo, M.E., Ruddiman, W.F., Backman, J., Clement, B.M., Martinson, D.G., 1989. Late Pliocene variation in Northern Hemisphere ice sheets and North Atlantic Deep Water circulation. Paleoceanography 4, 413–446.

Roy, Martin, 1998. Pleistocene stratigraphy of the Lower Nelson River Area: implications for the evolution of the Hudson Bay Lowland of Manitoba, Canada. Unpublished M.Sc. Thesis, Université de Quebec a Montreal, 220 pp.

Roy, M., Clark, P.U., Barendregt, R.W., Glasmann, J.R., Enkin, R.J., Baker, J., 2004. Glacial Stratigraphy and Paleomagnetism of late Cenozoic deposits of the north-central United States. Geol. Assoc. Am. Bull. 116, 30–41.

Ruddiman, W.F., Raymo, M.E., McIntyre, A., 1986. Matuyama 41,000-year cycles; North Atlantic Ocean and northern hemisphere ice sheets. Earth Planet. Sci. Lett. 80, 117–129.

Ruddiman, W.F., Raymo, M.E., Martinson, D.G., Clement, B.M., Backman, J., 1989. Pleistocene evolution of Northern Hemisphere climate. Paleoceanography 4, 353–412.

Shackleton, N.J., Hall, M.A., Pate, D., 1995. Pliocene stable isotope stratigraphy of site 846. Proc. Ocean Drill. Prog. Sci. Results 138, 337–353.

Stalker, A. MacS, 1969. Alberta report II: sections near Medicine Hat. Geological Survey of Canada, Paper, 69-26, 28 pp.

Stalker, A. MacS, 1976. Quaternary stratigraphy of the Southwestern Prairies. In: Mahaney, W.C. (Ed.), Quaternary Stratigraphy of North America. Hutchinson and Ross Inc., Stroudsberg, Pennsylvania, Dowden, pp. 381–407.

Stalker, A., 1982. Ice age deposits and animals from the southwestern part of Great Plains of Canada. Geological Survey of Canada, Miscellaneous Report, 31, (wall chart).

Stalker, A. MacS, Vincent, J.-S., 1993. Quaternary; subchapter 4K in Sedimentary Cover of the Craton in Canada. In: Stott, D.F., Aitken, J.D. (Eds.), Geological Survey of Canada, Geology of Canada, No. 5. pp. 466–482 (also Geological Society of America, The Geology of North America, v. D-1).

Stalker, A. MacS, Wyder, J.E., 1983. Borehole and outcrop stratigraphy compared with illustrations from the Medicine Hat area of Alberta. Geol. Surv. Can. Bull. 296, 28 pp. and maps.

Stewart, K.M., Seymour, K.L. (Eds.), 1996. Palaeoecology and Palaeoenvironments of Late Cenozoic Mammals: Tributes to the Career of C.S. (Rufus) Churcher. University of Toronto Press, Toronto, Canada 521 pp.

Szabo, B.J., Stalker, A. MacS, Churcher, C.S., 1973. Uranium-series ages of some Quaternary deposits near Medicine Hat, Alberta, Canada. Can. J. Earth Sci. 10, 1464–1469.

Westgate, J.A., 1968. Surficial geology: foremost-Cypress Hills area, Alberta. Alberta Research Councill Bulletin 22, 120.

Westgate, J.A., Christensen, E.A., Boellstorff, J.D., 1977. Wascana Creek Ash (Middle Pleistocene) in southern Saskatchewan: characterization, source, fission track age, paleomagnetism and stratigraphic significance. Can. J. Earth Sci. 14, 357–374.

Westgate, J.A., Briggs, N.D., Stalker, A. MacS, Churcher, C.S., 1978. Fission-track age of glass from tephra beds associated with Quaternary vertebrate assemblages in the southern Canadian Plains. Geol. Soc. Am. Abstr. 10, 514–515.

Zymela, S., Schwarcz, H.P., Grun, R., Stalker, A. MacS, Churcher, C.S., 1988. ESR dating of Pleistocene fossil teeth from Alberta and Saskatchewan. Can. J. Earth Sci. 25, 235–245.

Chapter 47

Late Pleistocene–Early Holocene Decay of the Laurentide Ice Sheet in Québec–Labrador

Serge Occhietti[1,2,*], M. Parent[3], P. Lajeunesse[4], F. Robert[5] and É. Govare[6]

[1] Département de géographie and Geotop, UQAM CP 8888 Centre-ville, Montréal, QC, Canada H3C 3P8
[2] CERPA, Université de Lorraine, Nancy, France
[3] Geological Survey of Canada, 490 rue de la Couronne, Québec, QC, Canada G1K 9A9
[4] Centre d'études nordiques & Département de géographie, Université Laval, Québec, QC, Canada G1V 0A6
[5] Groupe Omégalpha, Crabtree, Québec, QC, Canada J0K 1B0
[6] Hydro-Québec Équipement, Expertise Géomatique, 855 rue Sainte-Catherine Est, Montréal, QC, Canada H2L 4P5
*Correspondence and requests for materials should be addressed to Serge Occhietti. E-mail: serge.occhietti@gmail.com

47.1. INTRODUCTION

During the Wisconsinan Stage, the area of Québec–Labrador was completely covered by the Laurentide Ice Sheet (LIS), except for nunataks in the Torngat Mountains. For this reason, pre-Late Pleistocene events are only documented in scattered stratigraphical sections, and by erosional glacial marks. After the Last Glacial Maximum (LGM), the Eastern Sector of the LIS (ice from the Eastern Canadian Shield and the Appalachians) evolved from a single, predominant dispersal centre with subsidiary ice divides, into peripheral ice domes and masses which remained connected or became detached from the central dome. The latter was not a simple and stable dome-shaped ice mass, but an evolving ice mass. Thinning of the LIS through ablation, and mechanical drawdown along its margins as the result of diachronic ice streams in the St. Lawrence Corridor and Hudson Strait are the main features of Late-glacial ice-flow dynamics. In the areas south of the St. Lawrence Corridor, ice masses over the Appalachian uplands evolved from a glacier complex confluent with the LIS into separate local ice caps.

During part of the warm Bølling–Allerød phase, between 13.5 and 13.1 cal. ka BP, a series of ice-front positions marked the rapid retreat of the ice in the Appalachians of southern Québec. Between about 13 and 6 cal. ka BP, the ice mass over the Canadian Shield, north of the St. Lawrence Corridor, dissipated slowly. The deglaciation pattern includes the differentiation of an ice mass over the Hudson Bay, early deglaciation of the Labrador Highlands, a major ice flow in the Ungava Bay, and a roughly concentric ice retreat pattern in the southwest, south and southeast margins of the remnant main ice mass. By ca. 7.5 cal. ka BP, most of the remaining ice mass (about 0.55×10^6 km^2) had become stagnant. It fragmented to residual ice masses, located in the Labrador Trough and Nunavik, which finally disappeared ca. 6.0 cal. ka BP. In the northern Ungava and Labrador peninsulas, major glacial lakes in low-lying areas were dammed between ice and the tilted deglaciated land. Lowlands depressed by glacioisostatic loading were momentarily invaded by marine waters, mostly between 13 and 7 cal. ka BP.

47.2. OVERVIEW OF THE GLACIAL HISTORY OF QUÉBEC–LABRADOR

Observational evidence suggests a very erosional Illinoian Stage (s.l.) glaciation (Saalian Stage s.l.) over all the territory of Québec–Labrador. The evidence for pre-Wisconsinan events is limited, including rare exposures in the St. Lawrence Valley, tills and stratified deposits encountered in boreholes in the St. Lawrence Estuary (Occhietti et al., 1995) and the Appalachians of southern Québec (Shilts and Smith, 1986), and inter-till stratified units at one site in the Labrador Trough (Klassen et al., 1988). In the middle St. Lawrence Estuary, unpublished seismostratigraphical data provide evidence for tunnel valleys incised into pre-Illinoian deposits and bedrock. During the last glaciation (the Weichselian Stage which is the equivalent to the cold and cool phases of the Sangamonian and the Wisconsinan Stages in Canadian terminology, Fulton, 1984), all the territory was glaciated during the LGM or earlier, except for small areas in the Torngat Mountains of northern Labrador and Québec. For Wisconsinan glacial history, three main areas can be distinguished: the Appalachian area, the St. Lawrence Corridor (including the St. Lawrence Valley, Estuary and Gulf), and the larger area over the Canadian Shield (Laurentians to the south, Abitibi to the west, New Québec to the centre and north and Labrador to the east). Although each of the areas has a specific glacial style, the stratigraphical record of stadial and interstadial events in the St. Lawrence Corridor is critical because it was affected by both the Québec–Labrador ice dome and the Appalachian glacier complex.

Early ice sheet inception over the Laurentian highlands (including the Manicouagan Plateau, Occhietti, 1982) is evidenced by glacial striations indicating divergent ice flow to the northwest and the southeast (Parent et al., 1995; Veillette et al., 1999; Clark et al., 2000). Although the age of this early glacial phase is not known, we infer that ice began accumulating as early as Marine Isotope Substage 5d (Occhietti et al., 1996).

There is strong evidence for extensive ice cover on the eastern Canadian Shield during Marine Isotope Substage 5b. In the St. Lawrence Valley, early glacioisostatic subsidence followed by glacial lake inundation and glaciation is indicated by raised fluvial (Lotbiniere Sand), varved lake sediments (Deschaillons Varves), till (Levrard Till) and marine clay (La Pérade Clay) (Lamothe, 1989; Ferland and Occhietti, 1990; Occhietti et al., 1996). A coeval (?) Appalachian ice cap over southern Québec deposited a local till (Chaudière Till, McDonald and Shilts, 1971; Lamothe et al., 1992). Glacial cirques in highlands north of the middle St. Lawrence Estuary (Charlevoix area; J. Rondot, unpublished data; Govare, 1995) formed during repeated inception phases during the Pleistocene.

Stratigraphical data in the St. Lawrence Corridor and in the James Bay lowlands (Andrews et al., 1983) indicate that ice masses persisted in the Laurentians and New Québec during Marine Isotope Substage 5a.

From Marine Isotope Stage (MIS) 4 to the end of MIS 2, most of Québec and Labrador remained glaciated, with limited deglaciation episodes in the Appalachians (Lake Gayhurst episode), marine episode in the estuary and gulf of St. Lawrence ca. 35 ka (Gratton et al., 1984; Dionne and Occhietti, 1996), glacial lake in the upper St. Lawrence Valley (cf. Occhietti, 1989) and marine episodes in the James Bay lowlands (Andrews et al., 1983). The New Québec–Labrador ice mass formed an irregular dome whose limits changed through time. During glacial maxima, the dome extended over the Laurentians and the Labrador Trough, was connected to the Labrador highland glaciers and extended into Hudson Bay, the St. Lawrence Valley, Estuary and Gulf, and the Great Lakes basins and lowlands. Laurentide ice also flowed across the Appalachians uplands and highlands of Québec and New England, and was connected to regional ice caps in the Atlantic Provinces. This large and multi-origin Eastern Sector was the major dome of the LIS; it included all the related autochthonous and allochthonous ice masses over Hudson Bay–Québec–Labrador and the Appalachians (these ice masses were formerly referred to as the Labrador Sector of the LIS, Prest, 1984; Fulton, 1989).

47.3. THE LATE WISCONSINAN (LATE WEICHSELIAN)–HOLOCENE DEGLACIATION OF QUÉBEC–LABRADOR: OVERVIEW

The digital map accompanying this chapter comprises a systematic inventory of the eskers and moraines (Fig. 47.1) reported on the 143 topographical maps of Québec at the 1:250,000 scale, and in papers, public reports, published papers and special volumes. It includes glacial features of Labrador, compiled by R. Klassen and supplied by A. Moore (Geological Survey of Canada, Ottawa). The digital outlines of post-glacial glacioisostatic marine invasions and lakes (Fig. 47.2), extracted from the general maps of Dyke and Prest (1987), were kindly supplied by A. Moore.

Most of these data were published in a previous version of this chapter (Occhietti et al., 2004). The terrestrial features remain valid, for example, moraines, eskers, glacial lakes and post-glacial seas, but since the writing of the previous chapter, many studies have brought new regional field evidence (e.g. Lake Ontario Ice Stream (LOIS), Saint-Narcisse Moraine extension, deglaciation pattern along the coast of Hudson Bay), new approaches and syntheses (Clark et al., 2000; Dyke et al., 2003) and addressed the chronological problem of the timing of deglaciation in southern Québec (Richard and Occhietti, 2005). Nevertheless, the timing of the LIS decay during Holocene remains limited by the lack of reliable calendar ages. When significant, ^{14}C ages have been recalculated in calendar 1000 years (cal. ka BP) using the CALIB 6.0.1 programme (Reimer et al., 2009). However, in many cases, the 1σ error on ^{14}C ages is greater than 100 years and the 2σ range of the calculated calendar ages overlaps the time-interval of the mapped deglaciation isochrones (0.5–0.7 ka). For this reason, ages will be indicated either in cal. ka BP or, by default, in ^{14}C years (^{14}C ka BP). ^{14}C dates from marine shells are provided to give approximate ages; they are reservoir-corrected by a 0.4 ka mean value (for $\delta^{13}C = 0‰$); but the hard water and local reservoir effects are not corrected (cf. Occhietti and Richard, 2003).

The deglaciation of Québec can be subdivided into three major phases: (1) the St. Lawrence Ice Stream phase and its

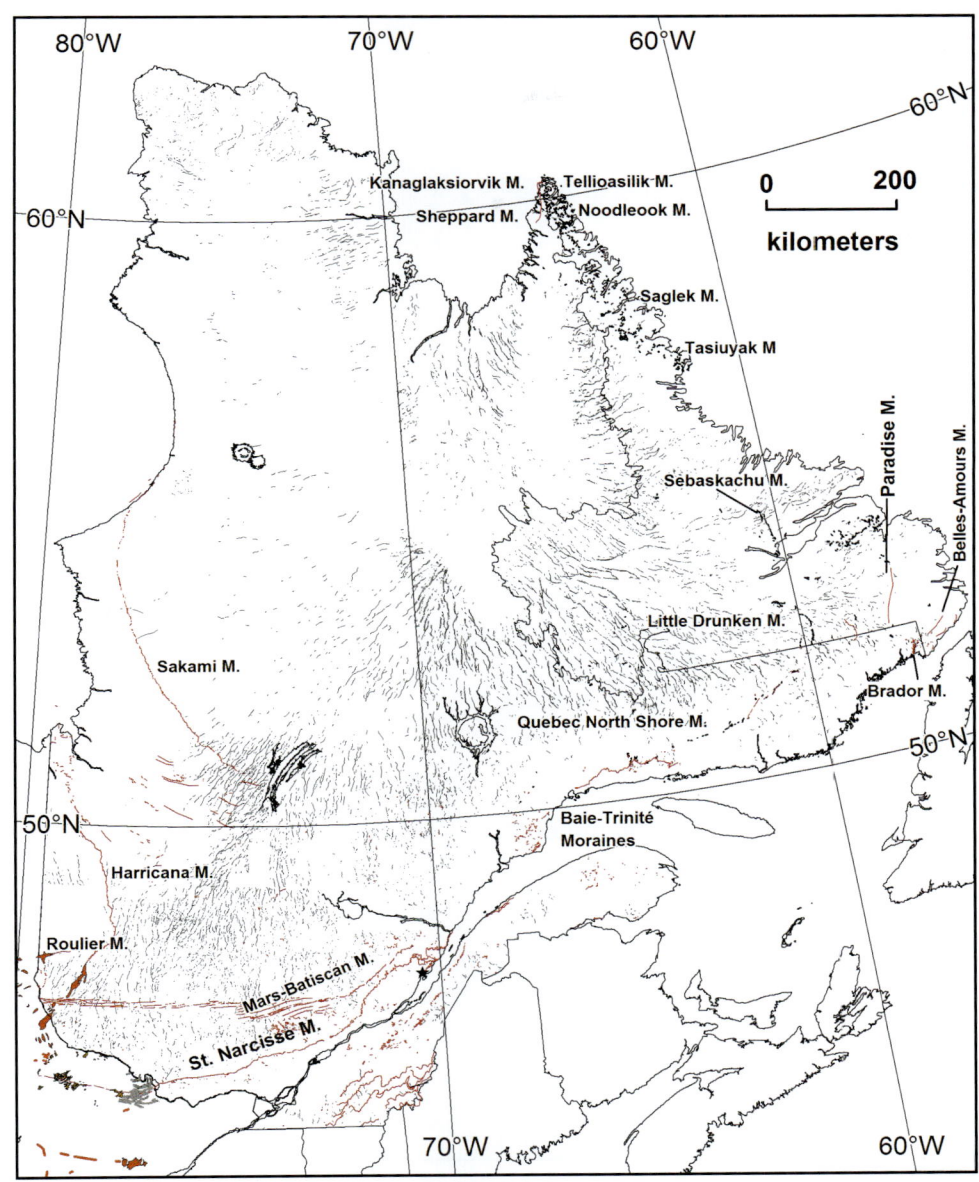

Chapter | 47 Decay of the Laurentide Ice Sheet in Québec

FIGURE 47.1 Compilation of Late Wisconsinan–Early Holocene moraines and eskers in Québec–Labrador, including a compilation by R.A. Klassen and A. Moore (Labrador area).

consequences (from about 16.0 to 13.0 cal. ka BP), (2) the Younger Dryas (YD) Chronozone (12.9–11.7 cal. ka BP; ca. 11–10 ^{14}C ka BP) and (3) the early Holocene deglaciation. In southeastern Québec, the margin of the LIS covered a physiographically diverse region: the Appalachian Uplands in southern Québec and Gaspé Peninsula, the St. Lawrence Corridor, and the deeply indented southern margin of the Laurentian Highlands. This landscape promoted deglacial styles that differed substantially from region to region. In central and northern Québec and in Labrador, a concentric thermo-latitudinal retreat pattern (retreat rate depending on the distance from the centres of ice mass or from the ice divides over New Québec–Labrador, and on the insolation) predominated, nevertheless with shifting centres of ice mass (Occhietti, 1982, 1983; Boulton and Clark, 1990; Parent et al., 1995, 1998; Clark et al., 2000). Glacial landform assemblages have been interpreted as broadly concentric zones comprising Zone 1, the outermost-characterised by extensive end moraines, hummocky moraine, ice thrust masses, and several generations of ice-flow lineaments; Zone 2, characterised by long eskers and contemporaneous ice-flow lineaments; and Zone 3, the

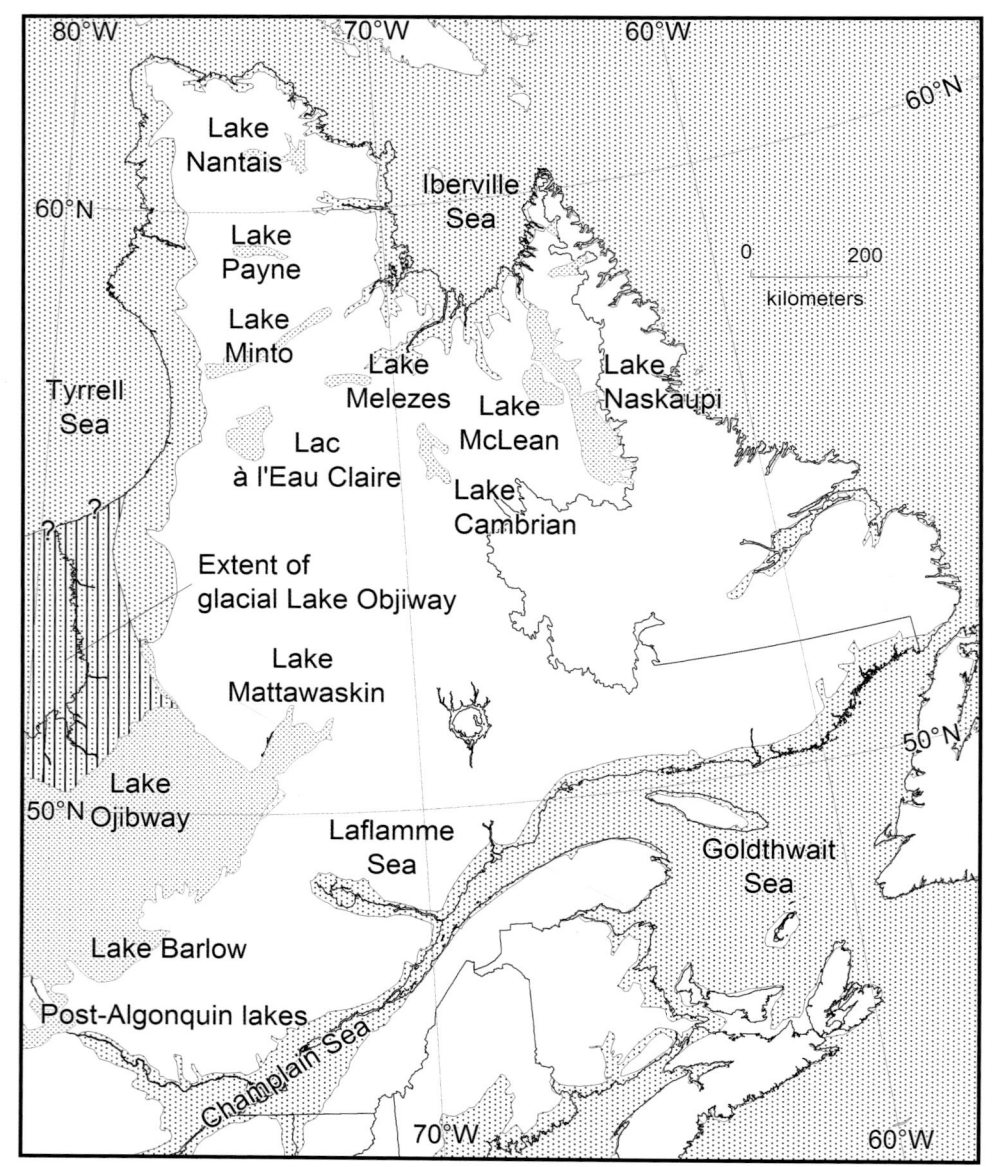

FIGURE 47.2 Diachronic outlines of the post-glacial marine invasions and lacustrine inundations in Québec–Labrador (mainly from a compilation by Dyke and Prest, 1987).

innermost, characterised by extensive ribbed moraine associated with drumlins and flutings (Prest et al., 1968). These assemblages have been interpreted as recording deposition in the marginal zones of a retreating sheet, and to become younger towards the geographical centre of glaciation. From the striation evidence, and new mapping of the landform record, which includes evidence of glacial overprinting (e.g. Kleman et al., 1995; Veillette et al., 1999; Clark et al., 2000), the record requires re-interpretation in terms of its relative age, context in the ice sheet, and ice-flow dynamics. Further, they provide a new basis for ice sheet modelling and palaeoclimatic reconstructions.

47.4. THE DEGLACIATION IN SOUTHERN QUÉBEC DURING LATE WISCONSINAN: FROM THE ST. LAWRENCE ICE STREAM TO THE EARLY PHASE OF YOUNGER DRYAS

The mode of deglaciation in the southern Québec part of the LIS was controlled by a series of semi-independent climatic and non-climatic factors (Occhietti et al., 2001). The global warming after the LGM continued with some minor climatic cooling from 18 cal. ka BP to the end of the Allerød ca. 12.9 cal. ka BP (see GRIP curve, Johnsen et al., 1992; Dansgaard et al., 1993; Grootes et al., 1993). The generally negative mass balance conditions for the LIS resulted in the gradual transition from a multidome-shaped ice sheet such as the Antarctic ice sheet (Lliboutry, 1965), to a much flatter plateau-like ice sheet, mostly in the marginal areas, much like the modern-day Greenland ice sheet (Lliboutry, 1965). This thinning of the ice sheet had several consequences in its marginal areas including: (1) increasing topographical control on ice-flow patterns, (2) development of ice streams, as in the Greenland and Antarctic ice sheets, (3) migration of ice divides and ice mass centres and (4) several secondary glacio-dynamic readjustments.

47.4.1. The St. Lawrence Ice Stream

Below a certain ice thickness and in conjunction with accelerated calving in the Gulf of St. Lawrence, a major northeast-trending ice stream formed within the ice sheet (Parent and Occhietti, 1999). Diachronously, between about 18 and 13 cal. ka BP, the head of the St. Lawrence Ice Stream migrated 1000 km along the axis of the St. Lawrence Corridor deep into the LIS. This major feature of the LIS was characterised by flow rates that were at least one order of magnitude higher than in adjacent ice masses, similar to present flow rates in Greenland (Lliboutry, 1965). This accelerated ablation regime (ice stream and iceberg calving) caused substantial thinning within the catchment area of the ice stream, particularly on the northwest flank of the Appalachian Uplands (Genes et al., 1981; Shilts, 1981; Lowell, 1985) and along the southern margin of the Laurentian Highlands (Fournier, 1998). Instead of the expected concentric marginal ablation pattern in the southeastern margin of the LIS (concentric to the New Québec Dome), the St. Lawrence Ice Stream favoured the progressive isolation of Appalachian ice masses in New Brunswick, Nova Scotia and northern Maine, in the Gaspé Peninsula and in the Notre Dame Mountains of southern Québec. Instability in the ice masses generated by the St. Lawrence Ice Stream strengthened the role of regional topographical control on glacial dynamics and increased the sensitivity of ice margins to climatic fluctuations.

47.4.2. The Lake Ontario Ice Stream

While the SLIS was funnelling ice flow towards the NE through the lower St. Lawrence Estuary and discharging large volumes of ice into the Gulf of St. Lawrence (Parent and Occhietti, 1999), another ice stream had formed almost coevally in the upper St. Lawrence Valley and was discharging ice in the Lake Ontario basin (Ross et al., 2006; Ross and Parent, 2007). This ice stream (Fig. 47.3), which may be informally called the LOIS, caused major flow line reorientations and glacial thinning in the St. Lawrence Valley upstream of Montreal and thus further contributed to dynamically disconnect the Appalachian ice masses from the main ice sheet. The southwest-flowing LOIS deposited the distinctive Oka Till, which lies above a till sheet (Argenteuil Till) which had been deposited by ice flowing towards the SE during the LGM or earlier (Ross et al., 2006). In this new model, flow line readjustments that followed deposition of the Oka Till led to deposition of the Fort-Covington Till in the upper St. Lawrence Valley, thus resolving a longstanding controversy on the timing and significance of regional deglacial events (McClintock and Stewart, 1965; Prest, 1977; Prest and Hode-Keyser, 1977; Dreimanis, 1977, 1985; Carl, 1978; Clark and Karrow, 1983). The Oka Till and Fort-Covington Till succession clearly postdates the LGM; the Fort-Covington Till, which was deposited during the Lake Candona episode (Ross et al., 2006), is therefore a truly deglacial unit.

In summary, two fast-flowing ice streams which largely controlled deglacial dynamics in southern Québec discharged large ice volumes into two distinct water bodies, the Goldthwait Sea in the lower St. Lawrence Valley and Glacial Lake Iroquois in the Lake Ontario basin.

47.4.3. The Lake Candona Episode

Ice retreat following the LOIS episode allowed Lake Iroquois to expand towards the NE and to become confluent with Glacial Lake Vermont which had also expanded north of the Lake Champlain Valley. This coalescent water body (Fig. 47.4) has been named Lake Candona on the basis of a characteristic freshwater ostracod, *Candona subtriangulata*, present in generally low abundance in its sediments (Parent and Occhietti, 1988, 1999; de Vernal et al., 1989; Rodrigues, 1992). In addition to ice-contact and ice-proximal glaciolacustrine features formed at the margin of the Appalachian uplands (Parent and Occhietti, 1988), the occurrence of *Candona*-bearing glaciolacustrine fine-grained sediments below Champlain Sea sediments constitutes the key method of delineating the position of the ice margin in the St. Lawrence and Ottawa valleys at the time of the Champlain Sea incursion.

The aerial extent of Glacial Lake Candona at the time of marine incursion, as shown in Fig. 47.4, can be expected to increase somewhat as new boreholes in the St. Lawrence and Ottawa valleys provide new micro-faunal records of the glaciolacustrine–glaciomarine transition. Our current reconstruction, based on evidence presented in Parent and Occhietti (1988, 1999), Rodrigues (1992), Naldrett (1988), Ross et al. (2006), indicates that the ice margin had retreated northward at least to the latitude of Montreal in the St. Lawrence Valley. While the ice-proximal, northeastern shorelines of Lake Candona now lie at elevations reaching 230 m a.s.l., its southwestern shores now lie well below the level of modern Lake Ontario. Current reconstructions along the Appalachian and Adirondack piedmonts indicate that water levels fell by 50–60 m at the

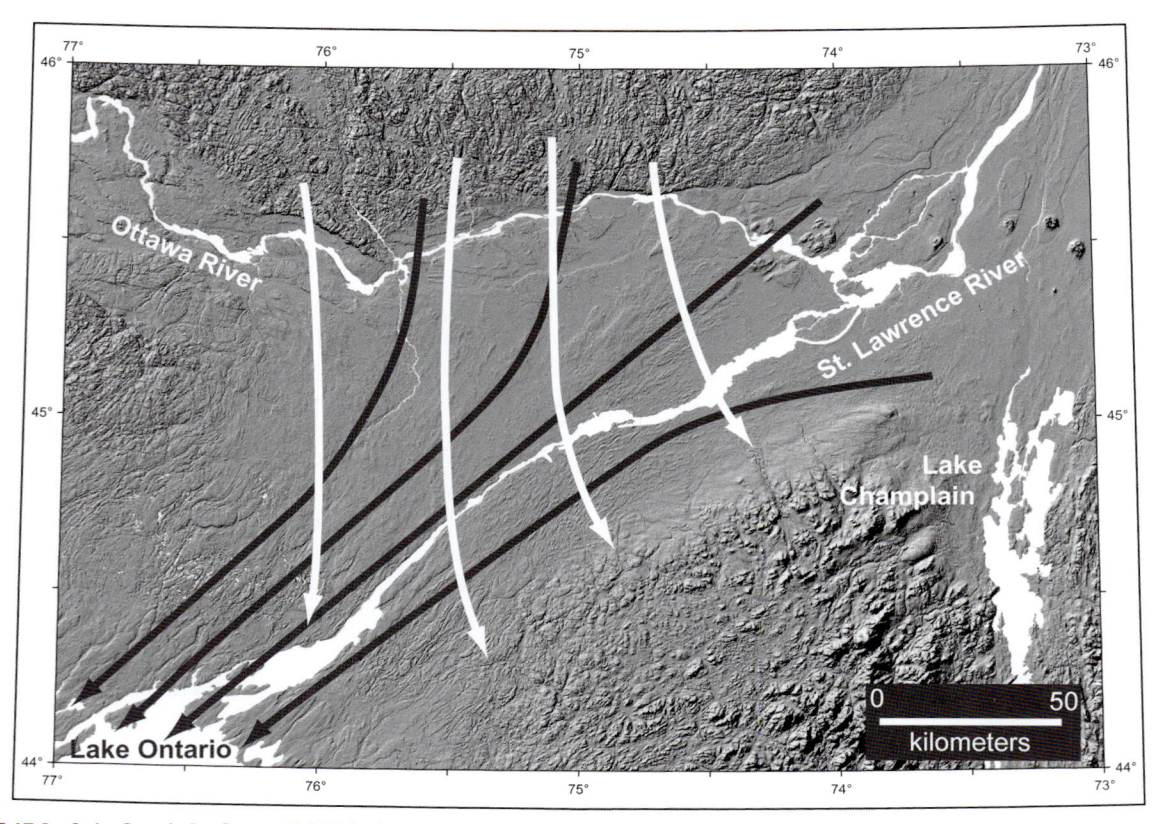

FIGURE 47.3 Lake Ontario Ice Stream (LOIS) in the upper St. Lawrence Valley (black arrows) and ultimate ice-flow reorientation (white arrows) prior to deglaciation and glacial Lake Candona (from Ross and Parent, 2007).

time that Lake Candona stopped spilling over towards the Hudson Valley and drained suddenly towards the Goldthwait Sea (Parent and Occhietti, 1988), an arm of the Atlantic Ocean which had already invaded the St. Lawrence Estuary. While the drainage route of Lake Candona towards the Goldthwait Sea is likely concealed below marine sediments and remains to be elucidated, the drainage event is thought to be coeval with the Champlain Sea incursion (Parent and Occhietti, 1988, 1999) ca. 13 cal. ka BP (Richard and Occhietti, 2005).

47.4.4. Deglaciation in the Appalachian Sector: Overview

Mostly allochthonous ice masses over the Appalachians were progressively isolated from the New Québec–Labrador ice mass, at first as a result of an ice-flow reversal and then by a marine embayment which cut them off from the main ice sheet (Fig. 47.5). This twofold sequence of events occurred in rapid succession between about 16 and 13.7 cal. ka BP from the Gaspé Peninsula to the Bois Francs region (Dionne, 1977; Lebuis and David, 1977; Locat, 1977; Allard and Tremblay, 1981; Genes et al., 1981; Shilts, 1981; Martineau and Corbeil, 1983; Chauvin et al., 1985; David and Lebuis, 1985; Lowell, 1985; Newman et al., 1985; Lowell et al., 1986, 1990; Rappol, 1993). This separation, initially caused by the dynamics of the St. Lawrence Ice Stream, led to the formation of independent ice domes in the Gaspé Peninsula, in New Brunswick, in northern New England and adjacent Québec (Lowell et al., 1986; Stea et al., 1998). The latter dome became increasingly autonomous as a result of recession through coastal Maine and ice flow towards the St. Lawrence at its northern margin. The eastern part of this dome remained connected with the New Brunswick one (Rappol, 1989) while its western part was characterised by rapid ablation and by flow convergence towards the Hudson–Champlain Valley (Connally and Sirkin, 1973; Connally, 1982; Hughes et al., 1985) and the upper St. Lawrence Valley. This late differentiation of the ice mass over the Appalachians was the result of topographical, glacio-dynamic and climatic factors.

At the height of the Bølling warm interval (ca. 14.65–14 cal. ka BP), the overall budget must have been strongly negative. The proposed mode of deglaciation in the Appalachians is that of a low mountain ice complex. The high ridges of the White Mountains and of the Notre Dame Mountains progressively separated ice masses on their northwestern flank in Québec from those on their southeastern flank in Maine (Genes et al., 1981; Shilts, 1981; Lowell, 1985). Four sectors can be recognised on the northern flank

Chapter | 47 Decay of the Laurentide Ice Sheet in Québec

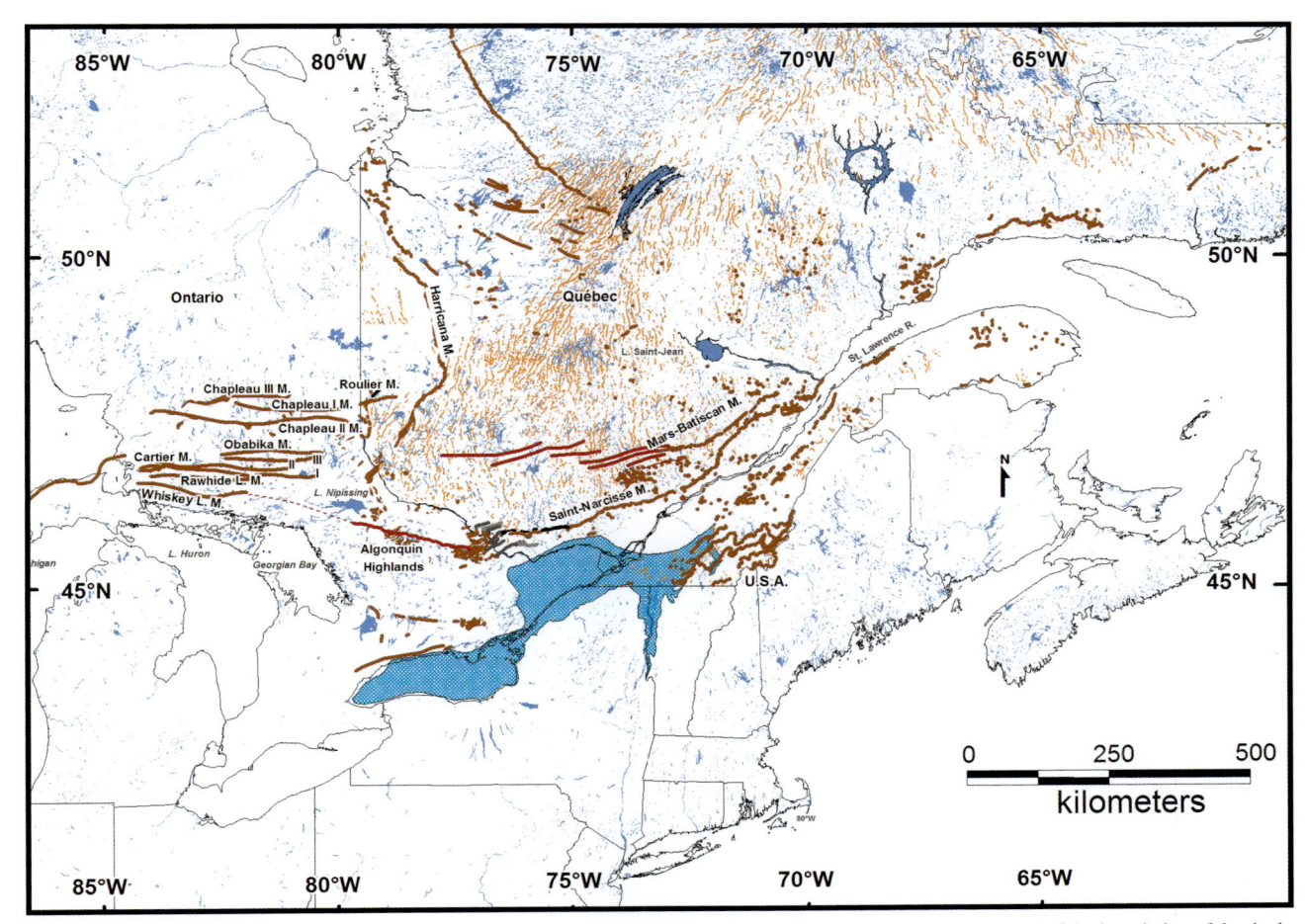

FIGURE 47.4 Glacial Lake Candona, as a result of the coalescence of lakes Iroquois, Memphremagog and Vermont and of the inundation of the deglaciated St. Lawrence–Ottawa River lowlands.

of the mountains: (1) southwestern ice masses which were still connected with the main LIS (see Thompson et al., 1999 for deglaciation in New Hampshire, south of the Québec border); (2) the Bois Francs residual ice cap, which was characterised by widespread downwasting and stagnation (Parent and Occhietti, 1988; LaSalle and Chapdelaine, 1990); (3) the Lower St. Lawrence River area which was affected by Late-glacial flow towards the St. Lawrence and then by downwasting; and (4) the Gaspé Peninsula (Gaspésie) which acted like an isolated ice cap and downwasted lately (Richard et al., 1997).

47.4.5. The Western Arm of Goldthwait Sea: Early Deglaciation Along the Southern Coast of the Lower Estuary of St. Lawrence

During deglaciation, thinning of the ice along the Appalachian piedmont favoured early deglaciation and incursion of the western arm of the Goldthwait Sea between the main ice sheet and the piedmont. This mode of glacial retreat does not require the development of a calving bay across the axis of the corridor; on the contrary, it implies calving parallel to the axis and to the shore. The upstream extent from the Gaspé Peninsula to the opposite side of the Saguenay mouth occurred by 12.0 ^{14}C ka BP (from shells, Dionne and Occhietti, 1996). The major water gap of the Chaudière Valley adds further local complications in the overall deglaciation pattern. At that time, the other areas of Québec were still glaciated (Fig. 47.5).

47.4.6. Dynamics of the Appalachian Ice Masses During the Chignecto Ice-Flow Phase Between About 15.6 and 14 cal. ka BP: Deglaciation of the Baie des Chaleurs Area

In the Atlantic Provinces, the Ice-Flow Phase 4 or Chignecto Phase records a glacio-dynamic event which occurred between about 15.6 and 14 cal. ka BP and is correlated to the Port Huron Event in the Great Lakes area (Stea et al., 1998). At that stage, the dome over the

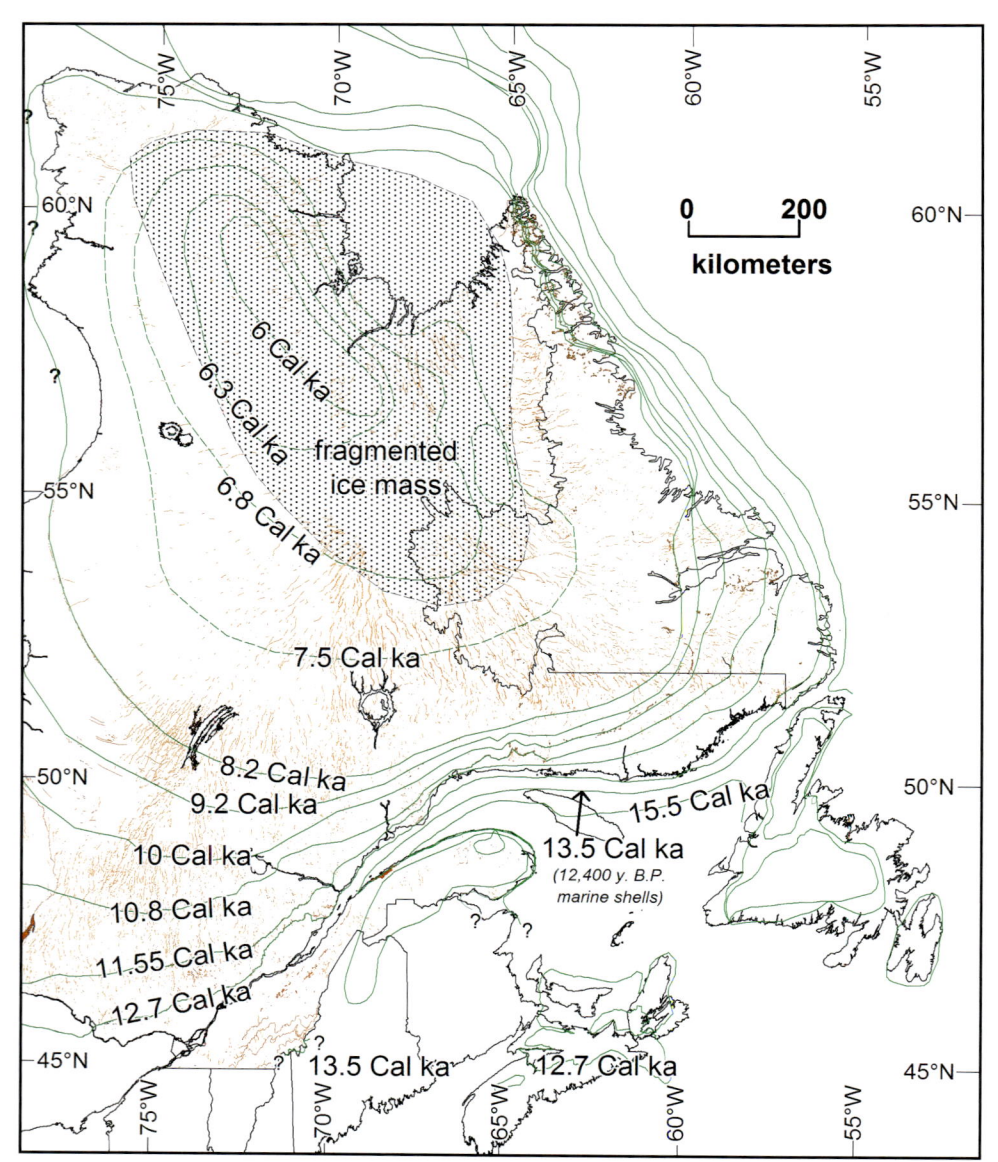

FIGURE 47.5 Tentative model of the ice-front retreat of the Laurentide Ice Sheet, in Québec–Labrador, between 13.5 and 6 cal. ka BP. Zone in grey: area from Clark et al. (2000) where pre-late ice retreat forms seem to be preserved, that is, extension of residual cold-based ice or stagnant ice. After the fragmentation of this ice mass, ice disappeared ca. 6 cal. ka BP (see Dyke et al., 2003, for current model).

southeastern and central part of the Gulf of St. Lawrence separated into several interconnected, radially flowing ice masses. At the end of this phase, the Baie des Chaleurs was deglaciated, although ice remained over New Brunswick and the Gaspé Peninsula.

47.4.7. The Major Northward Appalachian Ice-Flow Reversal

A Late-glacial reversal in ice flow is well documented by striations and rat-tails in the eastern part of the southern Appalachians of Québec (Lamarche, 1971, 1974; Lortie and Martineau, 1987; Rappol, 1993). This reversal was a powerful event as it caused a northward ice flow beyond the St. Lawrence River (Lanoie, 1995; Fournier, 1998; Paradis and Bolduc, 1999), and related glacial striations are observed almost as far south as the Québec/Maine boundary (Shilts, 1981). Shilts (1981) suggested an ice divide, the Québec Ice Divide, which is the equivalent to the late phase of a moving North Maine Ice Divide defined later by Lowell and Kite (1986) and Lowell et al. (1986). The Appalachian reversal is considered as a major event and seems to have been a re-equilibration event of the Appalachian ice masses. The ice over southern Québec and northern Maine flowed towards thinner ice in the middle and upper estuary. This event could be pre-Bølling in age. A tentative 16.5–15 cal. ka BP age bracket is proposed and the event could be more or less coeval of the Anticosti Island Phase or the Chignecto Phase. Heavy snow falls on the Maine–New Hampshire–southern Québec ice mass may have amplified the strength of the reversal.

47.4.8. Re-equilibration of the Maine–New Hampshire and Southern Québec Ice Mass and Strong Ice Stream in the Middle Estuary

From the striation record, Blais (1989) and Shilts (1997) identified northeastward ice flow in the upper middle Chaudière Valley subsequent to northward ice flow. The change in ice-flow direction is related to drawdown towards the St. Lawrence middle estuary documented by north-northeast to northeast striations in the Bois Francs, the Appalachian piedmont, an area north of the St. Lawrence River, and on islands and the southern coast of the middle estuary (Dionne, 1972). Inland of western and central Charlevoix, glacial striae record eastward ice flow convergent towards the estuary (Lanoie, 1995; Fournier, 1998). The ice-flow patterns indicate unstable ice and rapid thinning of Appalachian ice over southern Québec. The following deglaciation events were influenced by ice thinning, emergence of nunataks in the Appalachians and differential ice-surface lowering, as described by Syverson (1995) in historic deglaciation of Burroughs Glacier in Alaska.

47.4.9. Rapid Disintegration of the Southern Québec Appalachian Ice Masses During a Part of the Bølling–Allerød Warm Episode

At the start of the final deglaciation, Appalachian ice in southern Québec occupied three settings. In the west, it was connected to the Hudson–Lake Champlain lobe of the LIS. In the central part, it terminated at the Saint-Maurice lobe and was fed by the LIS. The eastern part, from Bois Francs to Bas du Fleuve, was initially connected to the LIS but thinned rapidly during a part of the Bølling–Allerød warm episode (Parent and Occhietti, 1999). Recessional moraines comprising the Frontier, Dixville–Ditchfield, Sutton-Cherry River-East-Angus-Megantic, Mont Ham and Saint-Ludger, and Ulverton–Tingwick morainic belts in the southern Appalachians of Québec (Fig. 47.6) were described by McDonald (1966, 1967, 1968, 1969), Shilts (1970, 1981), Prichonnet (1984) and Parent (1987) (see Parent and Occhietti, 1999). These ice-frontal features are transverse to the structural trend of the Appalachians. The ice margin outlined by the moraines is lobate and topographically influenced, with lobes in valleys and re-entrants on adjacent uplands and ice surface slopes gentle. Such morainic belts do not occur in northwestern Maine, where ice was probably stagnant.

Local ice readvances are identified on the Appalachian in the Bas du Fleuve (Rappol, 1993), in the Rimouski area (Rappol, 1993; Hétu, 1998) where the Neigette Readvance (Fig. 47.6) is dated at 12.4 ^{14}C ka BP (marine shell age). Later, a local readvance or reactivation of the Laurentide ice occurred in the Chaudière Valley, named the Beauce Event (Occhietti et al., 2001) and formerly related to the Highland Front Moraine System (Gadd et al., 1972a,b). The local recessional Saint-Sylvestre Moraine constructed later is tentatively related to the Ulverton–Tingwick morainic belt.

The age of the last morainic belt on the southwestern Appalachian piedmont, the Ulverton–Tingwick Moraine, is established from the number of glacial varves at the Rivière Landry section (Parent, 1987; Parent and Occhietti, 1999). The ice front receded from the moraine ca. 11.25–11.15 ^{14}C ka BP (13.2 cal. ka BP), more than 120 years prior to the opening of the St. Lawrence Valley to marine waters. Glacial Lake Candona (Parent and Occhietti, 1988; or Lake St. Lawrence of Rodrigues, 1992) inundated valleys in the Appalachians (Parent and Occhietti, 1988) until about 13 cal. ka BP.

47.4.10. The Ultimate Deglaciation Along the Appalachian Piedmont and Champlain Sea Transgression

Along the northeastern Appalachian piedmont, the ultimate episode of Laurentide ice retreat is the Saint-Raphael Moraine (Fig. 47.6). After the retreat from this position, the western arm of Goldthwait Sea extended rapidly between the Appalachian piedmont and the retreating LIS margin upstream up to the Bois Francs piedmont where the Saint-Maurice lobe front still abutted against the piedmont. Residual ice persisted on the Bois Francs plateau and in the northern side of the middle estuary. Then, the opening of the St. Lawrence Valley to the marine waters of Champlain Sea (Fig. 47.7) occurred by thinning of the LIS ice front along the piedmont and by overflow of the lacustrine waters of Lake Candona through a spillway incised into the ice and parallel to the piedmont and along the area of Lake Chaudière–Etchemin (Occhietti et al., 2001). As recorded in the Rivière Landry section (Parent, 1987) and in the western basin of Champlain Sea (Rodrigues, 1992), an early phase of mixing fresh and marine waters is recorded at the base of the Champlain Sea sediments. The age of the marine incursion is close to 11.1 ± 0.1 ^{14}C ka BP (13 cal. ka BP) (Occhietti and Richard, 2003).

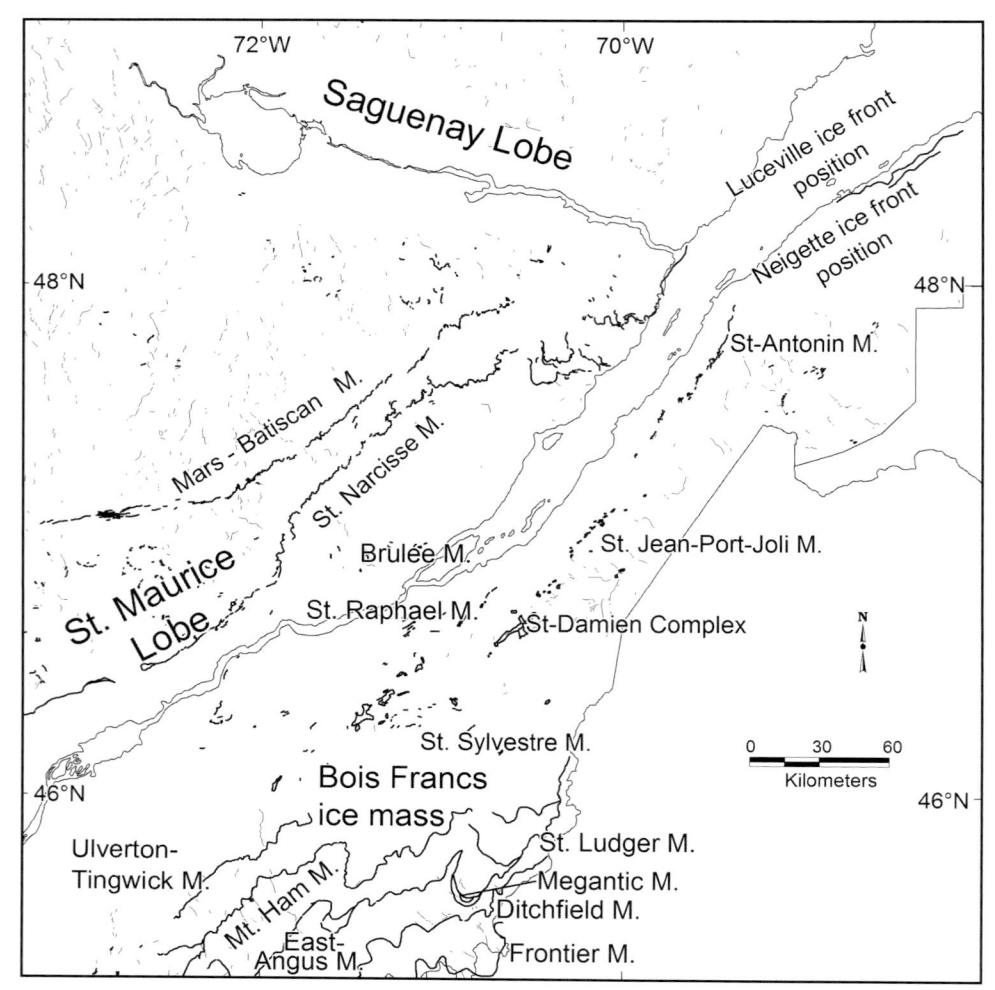

FIGURE 47.6 Moraines and eskers in southern Québec.

47.4.11. Deglaciation in the St. Lawrence Valley and Middle Estuary

The style of deglaciation in the St. Lawrence Valley is poorly known because many deglacial features were either buried or reworked as the result of marine inundation. The northward pattern of glacial retreat implies that glacial isostatic rebound began earlier along the Appalachian piedmont than along the southern edge of the Laurentians. Glacial thinning generated by the St. Lawrence Ice Stream also caused early crustal unloading along the axis of the St. Lawrence Corridor and a delayed response from east to west (Dionne, 1977, 1988; Lebuis and David, 1977; Locat, 1977; Parent, 1987; Dionne and Coll, 1995).

47.4.12. Deglaciation Along the Southern Margin of the Laurentian Highlands

The southern margin of the Laurentian Highlands includes the Mont Tremblant Highlands, the Saint-Maurice Valley, the Parc des Laurentides Highlands and the Saguenay fjord. This physiographic context favoured faster ice-marginal retreat in areas down-glacier (south) of the highlands while arcuate lobes were maintained in the two main valleys (Occhietti, 1980; Govare, 1995). The St. Lawrence Ice Stream interfered with this topographical context because flow convergence towards the estuary from the eastern flank of the Parc des Laurentides Highlands (Charlevoix) resulted in accelerated thinning. On the North Shore of the middle and upper estuary, ice lobes persisted in major valleys and readvanced later (Occhietti, 1980; Govare, 1995; Dionne and Occhietti, 1996), modifying significantly the mode of deglaciation of the St. Lawrence Valley.

47.5. THE YOUNGER DRYAS

The YD is a cold phase which lasted in northern hemisphere from about 12.9 to 11.7 cal. ka BP (ca. 11.0–10.0 ^{14}C ka BP). In Québec, the Saint-Narcisse Moraine (Figs. 47.1, 47.4, 47.6 and 47.7) was traditionally correlated with this

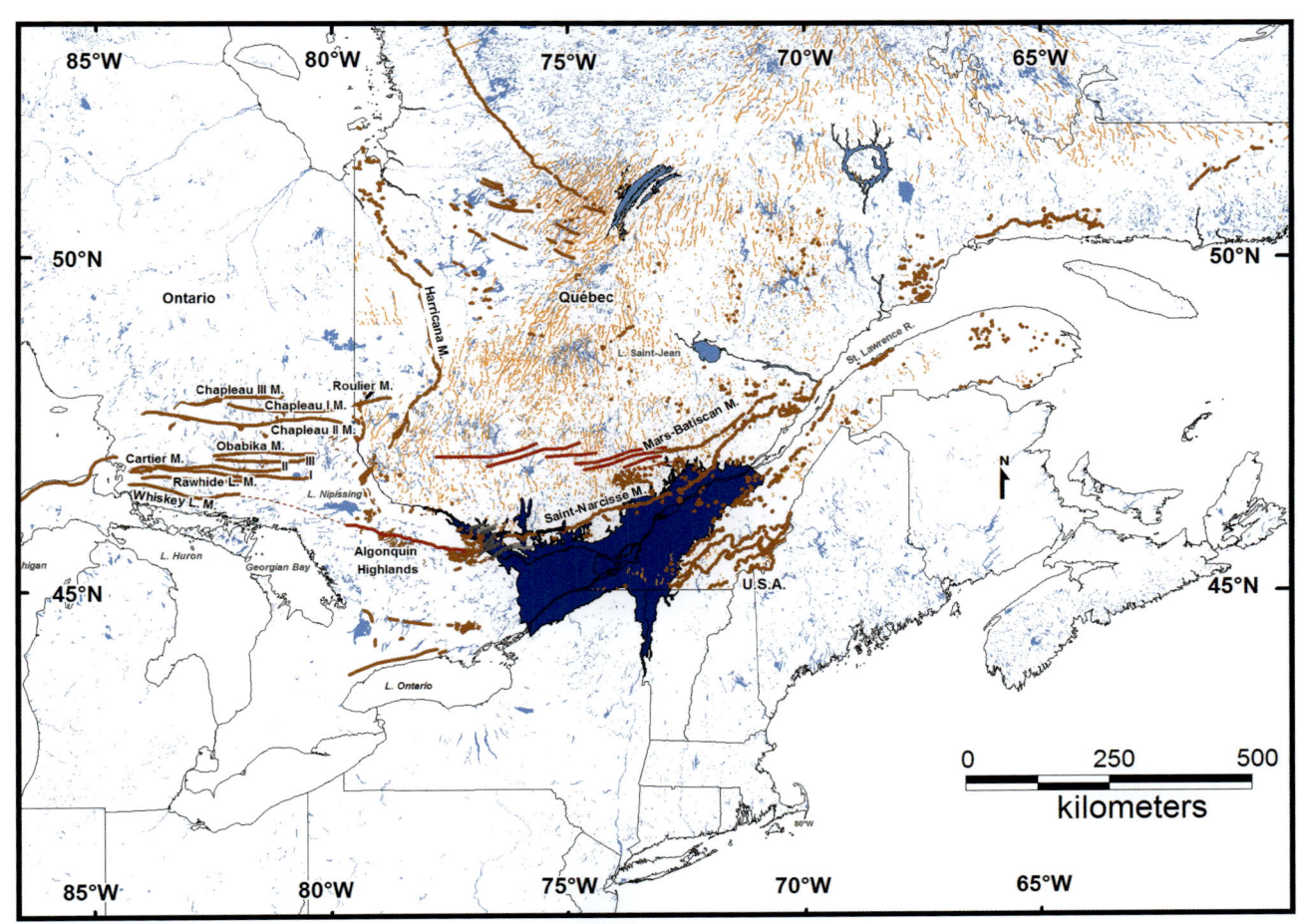

FIGURE 47.7 Diachronic maximum extent of Champlain Sea, between 13.1 and 10.8 cal. ka BP.

episode (LaSalle and Elson, 1975; Occhietti, 1980), but north of the Saint-Narcisse Moraine, a new outline of ice-front features (Mars-Batiscan Moraine, Figs. 47.1, 47.4, 47.6 and 47.7), initially named the Batiscan moraine (Bolduc, 1995), was compared by this author to one of the Salpausselkä moraines of Finland (Sauramo, 1929; Saarnisto and Saarinen, 2001). This ice-front feature was also previously noted in the Charlevoix area, close to the Mars River (Govare, 1995). Robert (2001) has shown that several outlines of glacial features, younger than the Saint-Narcisse Moraine, can be followed as far 100 km west of the Batiscan moraine, at least until the Saint-Maurice River.

The Saint-Narcisse Moraine episode, between about 12.8 and 12.2 cal. ka BP (10.8 and 10.5 ^{14}C ka BP), is related to the first major cold episode of the YD (Occhietti, 2007). The younger morainic systems, such as the Mars-Batiscan and unnamed intermediate moraines, result probably from the late cold episodes of YD, between 12.2 and 11.5 cal. ka BP.

The Saint-Narcisse Moraine is an almost continuous feature, extending more than 750 km from the Ottawa River area (Occhietti, 2007) to the Saguenay fjord (Dionne and Occhietti, 1996) (Figs. 47.1, 47.4, 47.6 and 47.7). This length is equivalent to the distance between Bergen, on the North Sea coast of Norway, and Stockholm, on the Baltic coast of Sweden.

It continues to the west in Ontario (Daigneault and Occhietti, 2006). East of the Saguenay, for 50 km along the cliffs of the upper North Shore, glacial deposits intercalated in marine sediments are related to the same event (Dionne and Bernatchez, 2000).

Some undated moraines are correlated with the Saint-Narcisse Moraine: the outer Baie-Trinité Moraine on the North Shore of the St. Lawrence Estuary (Vincent, 1989), the Bradore Moraine close to the Québec–Labrador limit and the Belles Amours Moraine in Labrador. The last three moraines are not a part of the following discussion. The outline of the Saint-Narcisse Moraine can be subdivided in several lobes (Gatineau, Saint-Maurice, central Charlevoix, Saguenay) and re-entrants (Mont Tremblant, Parc des Laurentides). A topographical influence is implied from this general setting: the gentle slope of the frontal zone of the

LIS favoured the ice convergence towards the relative depressions of the Laurentian Highlands and locally to the post-glacial sea. This general outline is related both to the climatic cause and to the topographical context. The western limit, at lower latitude than the eastern limit (45°45′N as compared to 48°10′N) and at a greater distance from the New Québec dome, is characterised by discontinuous features with dominant meltwater deposits. In the Saint-Maurice River area, the moraine consists of large bodies of glacial, glaciomarine and proglacial deposits. In the eastern area, in Charlevoix, the moraine is in fact a series of 30 small concentric frontal ridges (Rondot, 1974) which indicate a more active ice than in the western part. These local ice dynamics are partly related to the proximity of the ice centre and mainly to a steeper slope of the LIS margin which is the consequence of the fast ablation generated earlier by the St. Lawrence Ice Stream.

The Saint-Narcisse Moraine indicates a global readvance or at least a stabilisation of the LIS and that the flow lines of the LIS margin were maintained for several centuries (Figs. 47.5 and 47.8).

The late YD episode is indicated by the Mars-Batiscan moraine and other parallel ice-front features. These features emphasise the different styles of ice retreat, with dominant meltwater features (eskers, ice margin trains) in the western part and ice-front rims in the eastern part.

In the Québec City area, fossiliferous diamictons, with *Balanus hameri* fragments, have also been correlated to a YD glacial readvance (the Saint-Nicolas readvance, LaSalle and Shilts, 1993). The first author thinks that the Saint-Nicolas readvance represents an ice-flow event in Champlain Sea. The climatic significance of this local episode is not demonstrated. It could record an early minor cold pulse of YD age, ca. 11.0 ^{14}C ka BP (age from shells) or a local re-equilibration of the glacier.

47.6. DEGLACIATION OF THE CANADIAN SHIELD AREA (QUÉBEC AND LABRADOR) DURING THE EARLY HOLOCENE

The deglaciation of the Canadian Shield in Québec and Labrador is described by Vincent (1989) and interpreted by Clark et al. (2000) and Dyke et al. (2003). Previous studies at the scale of the LIS gave also a good overview of the ice retreat in the area (Prest et al., 1968; Bryson et al., 1969; Prest, 1969; Dyke and Prest, 1987, 1989). The general deglaciation style reflects thermo-latitudinal retreat with topographical control, and shifting of ice centres and divides (Figs. 47.5, 47.8 and 47.9). Glacial flow line reconstructions indicate large, complex and evolving ice masses of the Eastern Sector located over Hudson Bay–Québec–Labrador. As the Hudson Bay ice became dynamically independent, a dome over Québec–Labrador (New Québec or New Québec–Labrador Dome) dissipated slowly. In addition to ablation by surface melting, ice loss to the Eastern Sector was the result of mechanical, physical processes along its margins, including

(1) an ice stream in the Hudson Strait which induced convergence of ice flow towards Ungava Bay and Hudson Strait and

(2) accelerated ice flow in the south, towards the Great Lakes, and to the west, towards an extended Lake Agassiz.

On the eastern border, LGM outflow was restricted to the main valleys of the northern Labrador Highlands (Torngat Mountains) which were progressively deglaciated during late Wisconsinan and Early Holocene. In southern Labrador, a concentric retreat pattern predominated, due to the termination of the ice sheet margin on land, unaffected by ice streaming and calving. All these factors, including a negative budget of climatic origin, re-equilibration of the ice masses (rapid ice flow and thinning in response to the physical loss of ice at the margins), accelerated deglaciation of the Labrador Highlands and the development of ice streams, explain the shape and dynamics of the Eastern Sector at the eve of the Holocene and the subsequent mode of ice sheet decay.

47.7. ICE RETREAT IN THE SOUTHERN MARGIN OF THE NEW QUÉBEC–LABRADOR DOME: THE LAURENTIANS

North of the YD moraines (Saint-Narcisse and the Mars-Batiscan-Saint-Maurice morainic systems), the marginal retreat history of the ice sheet is well known in the southwestern part (the Lake Temiscaming–Ottawa River area), and poorly known in its central part, between the Ottawa and Manicouagan Rivers. Along the northern shore of the St. Lawrence lower Estuary and Gulf (the middle and lower Côte Nord) and in southern Labrador, controversy remains on the age of regionally extensive morainic systems (North Shore, Bradore, Belles Amours, Paradise and Little Drunken moraines). Although they outline marginal positions of New Québec–Labrador ice, their age and correlation are unknown.

47.7.1. Deglaciation in the Western Part of the Laurentians and Lake Temiscaming Area

West of Ottawa River, the southern margin of the Hudson Bay ice mass over the James Bay lowlands and northern Ontario and Manitoba, was dissipating rapidly, with successive readvances or surges in Lake Superior (Marquette readvance, ca. 11.6 cal. ka BP (10.05 ^{14}C ka BP; Lowell

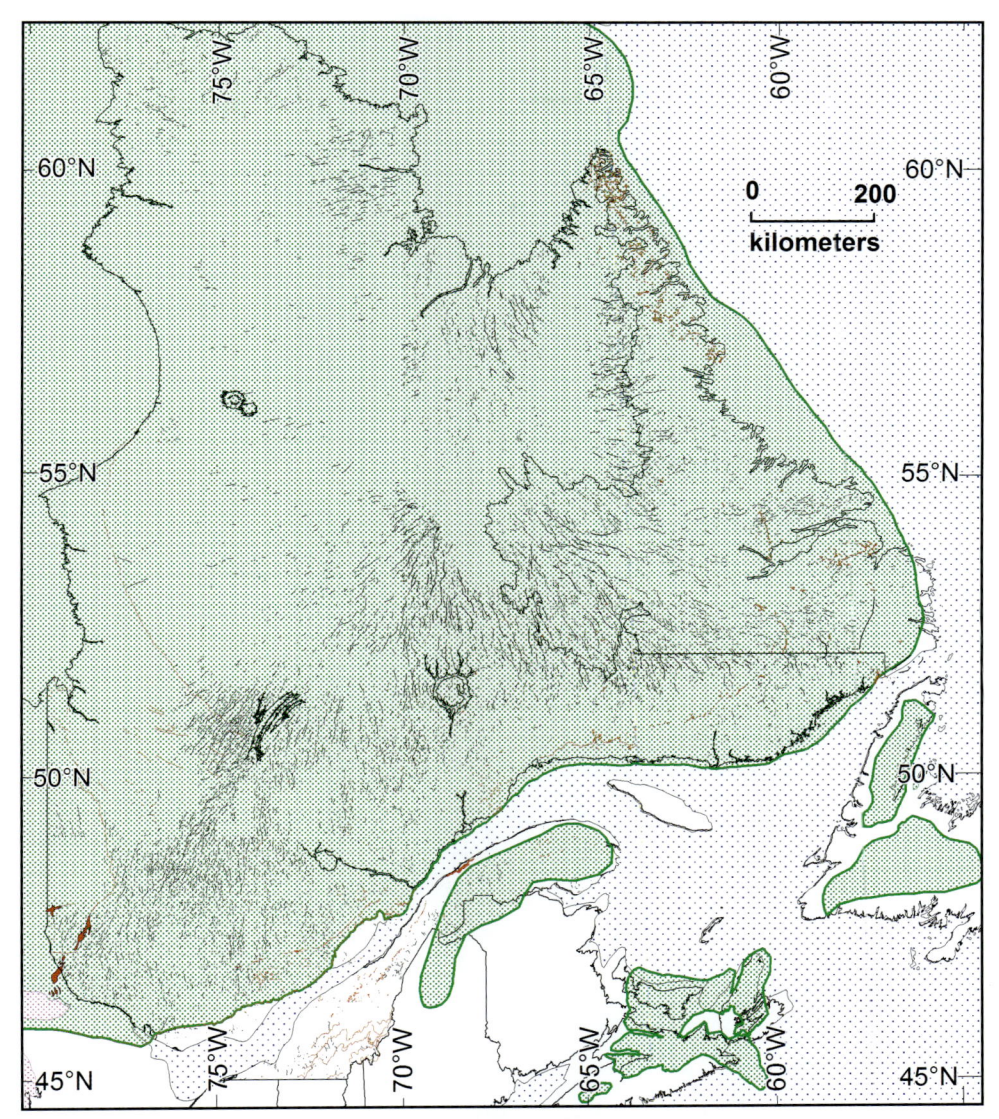

Chapter | 47 Decay of the Laurentide Ice Sheet in Québec

FIGURE 47.8 Estimated limits of the glaciers during the Saint-Narcisse Moraine episode, ca. 12.7–12.4 cal. ka BP, during the main early phase of the Younger Dryas.

et al., 1999), in Lake Agassiz (several secondary lobes and the Marquette readvance, Dredge and Cowan, 1989), and by the Cochrane non-climatic ice surges in Lake Ojibway, ca. 9.4 cal. ka BP (8.4 ^{14}C ka BP). In the upper Ottawa River and Lake Temiscaming area, uncoupling along the suture between the Hudson Bay ice and the New Québec ice is evidenced by converging ice-flow indicators (Veillette, 1986), by the diachronous Harricana Interlobate Moraine (Vincent and Hardy, 1977, 1979), and by a continuous glacial lake inundation which followed the ice-front retreat and partly resulted from the glacioisostatic lowering at the front of the ice sheet.

During deglaciation, southwest-flowing New Québec ice and southeast-flowing Hudson ice converged towards the Harricana Interlobate Moraine (Hardy, 1976), which was originally described by Wilson (1938) as 'a moraine between two ice-sheets'. This interpretation of the moraine is disputed (Brennan et al., 1996; Brennan and Shaw, 1996). According to these authors, the Harricana features are a mega-esker, deposited prior to the deglaciation and during a short episode. From the west southwest–east northeast orientation of ice front during the deglaciation on the Laurentian Plateau in Québec (Simard, 2003; Simard et al., 2003), on the eastern side of the Ottawa River, and the lack of visible interlobate features between the Lake McConnell Moraine and the southern limit of the Harricana Moraine, it seems that the two pre-existing models do not apply in the southern Temiscaming area. The Harricana Moraine

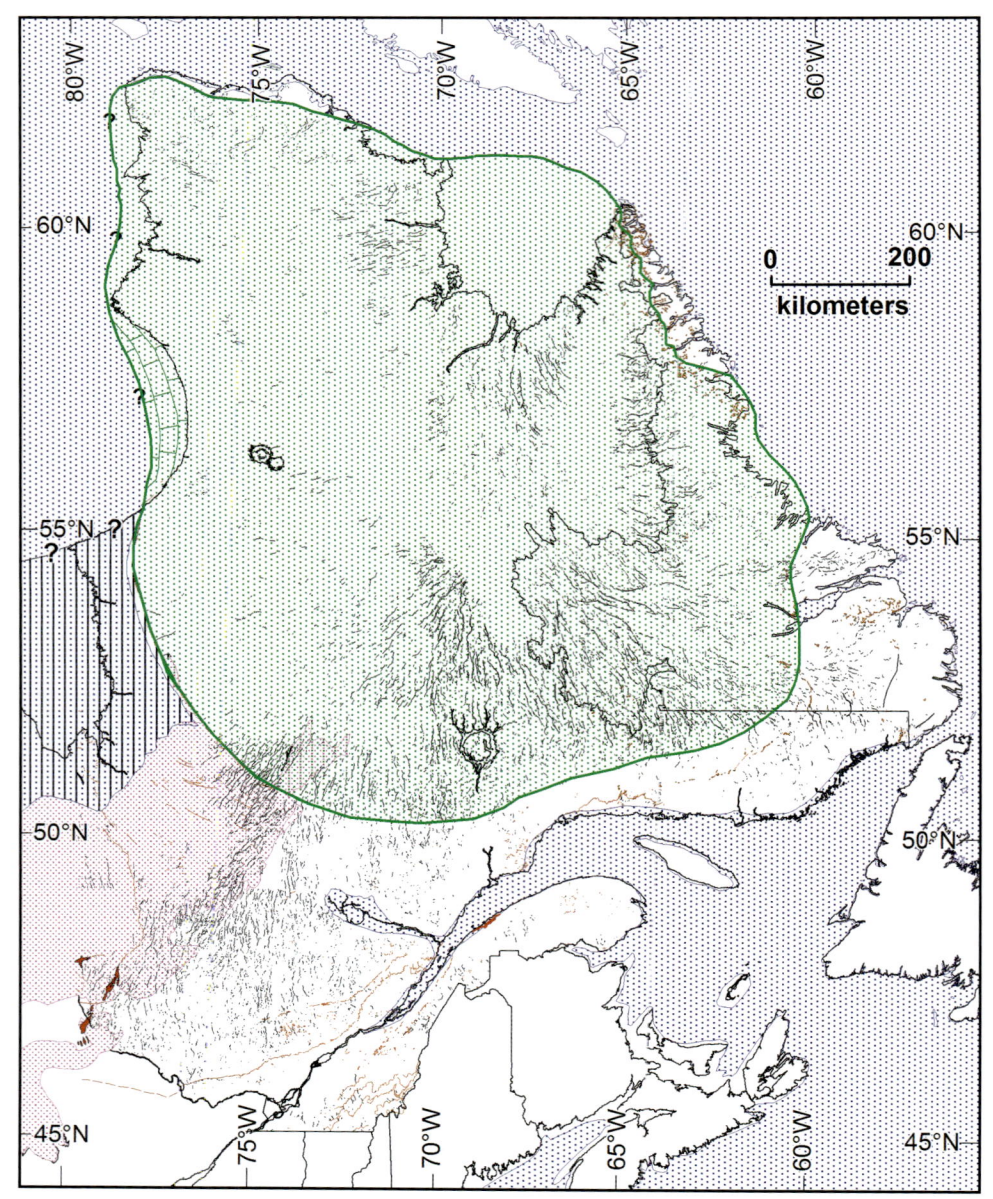

FIGURE 47.9 Estimated limits of the glacier in New Québec–Labrador during the Sakami Moraine episode, ca. 8.2 cal. ka BP, after the opening of the Hudson Strait and outflow of glacial lakes Agassiz and Ojibway. The area of this dome is of the order of 1.2×10^6 km².

is probably a composite feature, interlobate converging deposits and tunnelled-meltwater deposits.

The southwestern margin of the New Québec Dome retreated initially in a north-northeasterly direction (Veillette, 1997a; Simard et al., 2003). The southern region on the west side of the Harricana Moraine became ice free first (Veillette, 1983, 1986, 1988). The Laverlochère Moraine (Veillette, 1983, 1986, 1988) was built on the borders of an ice lobe in the northern Lac Témiscamingue trough. West of the Harricana Interlobate Moraine and north of Lake Temiscaming, the Roulier Moraine (Vincent and Hardy, 1977, 1979) was formed as a result of a halt or as a re-equilibration moraine.

Glacial lakes abutted the ice margin in low-lying western and northern areas (Vincent and Hardy, 1977, 1979; Veillette, 1983, 1988). The earliest glacial lake phase is related to the northeastern extension of the Post-Algonquin glacial lake from the Great Lakes basin (Harrison, 1972) (Fig. 47.2). Possibly during the Sheguiandah and certainly during the Korah Phase, the lake was still dammed by ice blocking drainage down Ottawa River in the Mattawa area, and extended later northeast of Temiscaming along a re-entrant in the ice front along the Harricana Interlobate Moraine (Vincent and Hardy, 1977, 1979; Veillette, 1988; Figs. 47.1 and 47.4). When ice withdrew from the Ottawa River Valley, water levels dropped, and glacial Lake Barlow (Wilson, 1918) occupied the Lake Temiscaming basin. Maximum glacial Lake Barlow levels in the area east of Lac Témiscamingue were at about 300 m and rise to the northeast to about 380 m in the vicinity of the present Hudson Bay watershed where the Harricana Moraine crosses it (Veillette, 1988). The northeastern tilt of glacial Lake Barlow water planes confirms that ice in New Québec was thicker than in Hudson Bay (Hillaire-Marcel et al., 1980).

The chronology of deglaciation in the Mattawa and Temiskaming areas has been reassessed. Anderson et al. (2001) re-sampled basal lake deposits. Terrestrial plant debris yielded an AMS age of 9.45 ± 0.05 ^{14}C ka BP (CAMS-46195) which shows that original dates from total organic matter sampled at the same depth are too old by 2000 years (hardwater effect). This result (Lewis et al., 2005, 2006) is confirmed by the reassessment of the position of the western extent of the Saint-Narcisse Moraine in a more southerly location (Robert, 2001; Occhietti, 2001; in Bhiry et al., 2001; Daigneault and Occhietti, 2006) than previously interpreted. The nearly west–east ice-front outline of the LIS on the Laurentian Plateau between about 11.55 and 11.15 cal. ka BP (10.0–9.7 ^{14}C ka BP) is another field evidence of a late deglaciation in the area.

47.7.2. Laurentian Highlands Between Lake Temiscaming and the Manicouagan River

The characteristic deglacial deposits of this understudied area are short segments of end moraines, isolated ice-contact deposits, eskers and outwash trains (Parry, 1963; Hardy, 1970; Denis, 1974; Lamothe, 1977; Pagé, 1977; Tremblay, 1977; Occhietti, 1980). In this moderate relief area higher summits first became ice free. The pattern of deglaciation is mostly the result of thinning of a wide marginal zone of the ice sheet. In the western area, the ice-front outline changes from a nearly west–east outline observed by Simard et al. (2003) to a curved outline related to the southwest limit of the New Québec Dome. From west to east, the rate of ice retreat decreases, the more distant ice margin from the ice centre is thinner and melts faster. In the Lac Saint-Jean area, Tremblay (1971), Dionne (1973) and LaSalle and Tremblay (1978) have shown that the ice moved in a southerly direction except along the Saguenay fjord, where ice flow was generally southeasterly along the axis of this depression (Dionne and Occhietti, 1996).

As the ice retreated from the uplands south of Lake Saint-Jean, small glacial lakes were dammed and De Geer moraines were built. A late ice lobe occupied the Lake Saint-Jean depression and ice-contact materials were deposited at its receding margin, among them is the Metabetchouane Moraine (LaSalle and Tremblay, 1978).

As the ice receded northwesterly up the Saguenay Valley and into the Lake Saint-Jean basin, marine waters from the Gulf of St. Lawrence extended over lower lying deglaciated land. This arm of the Goldthwait Sea was referred to as the Laflamme Sea by Laverdière and Mailloux (1956). Marine limit lies at 167 m at the mouth of Rivière Saguenay and at about 167 and 198 m south and north of Lake Saint-Jean, respectively (LaSalle and Tremblay, 1978). The oldest age (marine shells) for marine invasion and deglaciation is 10.25 ± 0.35 ^{14}C ka BP (Gif-424) south of Lake Saint-Jean. It seems too old by several centuries. Following deglaciation of the Lake Saint-Jean basin, the ice retreated northward leaving behind numerous eskers and fluted landforms.

47.7.3. Québec North Shore

The Quaternary history of the coastal fringe north of the Gulf of St. Lawrence between the mouth of Saguenay Fjord and the Québec/Labrador border is relatively well known, but few data are available for farther inland. During the LGM, south-flowing ice of the New Québec–Labrador Dome covered the Québec North Shore and extended in the Gulf of St. Lawrence (cf. Grant, 1989), without apparently overriding the eastern tip of Anticosti Island (Gratton et al., 1984).

Ice retreated generally northwestward during deglaciation. The eastern extremity of the North Shore and the headland in the Baie-Trinité region were probably the first areas to become ice free. Moraines were built in the Baie-Trinité area between 13.5 and 9.0 ^{14}C ka BP (Dredge, 1976, 1983). The outer Baie-Trinité Moraine is correlated with the Saint-Narcisse Moraine.

The marine submergence (Fig. 47.2) in the Estuary and Gulf of St. Lawrence, east of Québec City, was named Goldthwait Sea by Elson (1969) and studied by Dionne (1977), Hillaire-Marcel (1979) and Dubois (1980). Marine limit varies from 150 m in the Québec/Labrador border area (de Boutray and Hillaire-Marcel, 1977) to 130–145 m along the North Shore (Dubois et al., 1984) and near the mouth of the Saguenay Fjord (Dionne and Occhietti, 1996). The age (from marine shells) for deglaciation of the Québec North Shore comprises between 10.9 ± 0.14 ^{14}C ka BP (GSC-2825) (site in Labrador near the Québec border) and 10.23 ± 0.18 ^{14}C ka BP (Gif-3770) at Rivière Romaine, 9.14 ± 0.2 ^{14}C ka BP (GSC-1337) at Rivière Moisie and 9.97 ± 0.13 ^{14}C ka BP (QU-574) at Rivière Manicouagan. The upstream part of the North Shore was deglaciated prior to 11.1 ^{14}C ka BP (from marine shells, Dionne and Bernatchez, 2000) and reglaciated during the YD.

Moraines and ice-contact fans observed offshore Sept-Îles and Mingan about 20 km south of the North Shore Moraine were deposited at ~11.0–10.8 ^{14}C ka BP (from shells, Lajeunesse et al., 2008). These landforms were constructed during the YD cold period and could correspond to an eastward submarine extent of the early YD Saint-Narcisse Morainic System.

As ice retreated farther north onto the Canadian Shield, the Québec North Shore Moraine (Dubois and Dionne, 1985), more than 800 km long, was deposited between Rivière Manicouagan and south of Lake Melville in Labrador (Dubois, 1979, 1980; Fulton and Hodgson, 1979). The moraine includes segments in the Rivière Manicouagan area (Sauvé and LaSalle, 1968), the Lac Daigle Moraine of Dredge (1976, 1983), the Manitou-Matamec Moraine of Dubois (1976, 1977, 1979, 1980), the Little Drunken Moraine of Fulton and Hodgson (1979) and the Aguanus-Kenamiou Moraine of Dionne and Dubois (1980). The moraine is considered by Dubois and Dionne (1985) to represent a halt of the ice sheet during a cooler climatic phase. The age of the moraine, still uncertain, is apparently of 9.7–9.5 ^{14}C ka BP (Dubois and Dionne, 1985), and a correlation with the Mars-Batiscan Moraine of late YD is still very tentative. The Goldthwait Sea was in contact with the moraine locally in the Rivière Moisie and Rivière Manicouagan areas.

The ice margin retreated northward and northwestward from the Québec North Shore Moraine towards central New Québec leaving eskers and fluted landforms. According to Dubois (1980), the Goldthwait Sea in the middle North Shore area attained its inland limit (128–131 m) at about 9.5 ^{14}C ka BP (marine shells).

47.8. DEGLACIATION OF CENTRAL AND SOUTHERN LABRADOR

During deglaciation, ice flow was generally northeastward in central Labrador, eastward in the Lake Melville area and southeastward in southeastern Labrador. Flow was topographically controlled by the Mealy Mountains, around which flow trends part, and by Lake Melville, where a calving bay may have existed (Fulton and Hodgson, 1979). Major end moraines were built during the retreat phase. If Laurentide Ice covered all of southeastern Labrador and joined with the Newfoundland Ice Cap, then the Bradore, Belles Amours and Paradise moraines (Fig. 47.1; Grant, in Vincent, 1989; King, 1985) are probably Late Wisconsinan retreat features. Many other moraines were built during Holocene. The longest of these is the Sebaskachu–Little Drunken Moraine System (Blake, 1956; Fulton and Hodgson, 1979), which could be the extension of the Québec-North Shore System of Dubois and Dionne (1985).

Later retreat towards central New Québec was marked largely by construction of numerous eskers and fluted landforms. Glaciofluvial landforms generally trend in the last principal directions of ice flow defined by glacially streamlined landforms. There are, however, notable exceptions that occur principally in central and western Labrador. In the central Smallwood Reservoir, a large esker system with prominent tributaries entering from the northwest crosscuts ice-flow trends, suggesting major change in subglacial hydrology during the late deglaciation. The change may have been in response to evolution in the glacial lakes of northern Labrador or to subglacial discharge towards Lake Melville. Although free drainage to the Atlantic prevented major glacial lake development in western Labrador, glacial lakes of uncertain extent were dammed against the watershed divide; their full extent and evolution are not well known.

Atlantic Ocean waters submerged glacioisostatically depressed coastal areas of central and southern Labrador during deglaciation. Marine limit has been traced by Fulton (1986a,b) at the Geological Survey of Canada (Fig. 47.2).

Along the coast, north of the Strait of Belle Isle, it may have reached 150 m, between Sandwich Bay and Lake Melville 113 m (Rogerson, 1977) and northeast of Lake Melville 85 m (Hodgson and Fulton, 1972). In the Lake Melville area, marine limit increases in elevation inland from about 75 m on the outer coast to 150 m west of the lake (Fitzhugh, 1973). The maximum age and elevation of marine inundation decreases northwards along the Labrador Coast, and the 0 m hinge line is located at the northernmost tip of the Labrador Peninsula (Løken, 1962a,b).

On the basis of radiocarbon ages of shells, the area north of the Straits of Belle Isle was ice free by at least 10.9 ± 0.14 ^{14}C ka BP (GSC-2825), the coastal area east of Kanairiktok River by 10.275 ± 0.225 ^{14}C ka BP (GX-6345), central Lake Melville by 7.97 ± 0.09 ^{14}C ka BP (TO-200), and the west end of Lake Melville by 7.6 ± 0.1 ^{14}C ka BP (GSC-2970). The age determinations of 9.64 ± 0.17 ^{14}C ka BP (GSC-3067), 10.55 ± 0.29 ^{14}C ka BP (S1-3139) and 10.24 ± 1.24 ^{14}C ka BP (S1-1737) on lake sediments provide minimum ages (in the case of no hard water effect) for the deglaciation of the upper St. Paul River area, the southern Mealy Mountains and the eastern end of Lake Melville, respectively. They also provide a minimum age for the construction of the Paradise Moraine. Farther inland, a date of 6.46 ± 0.2 ^{14}C ka BP (GSC-1592) on peat provides a minimum age for the deglaciation of upper Churchill River area and a 6.5 ± 0.1 ^{14}C ka BP (GSC-3241) date for lake sediments, a minimum age for the deglaciation of central Labrador.

At Hopedale and Nain, on the central Labrador coast, glacier ice remained until ca. 7.6 ± 0.2 and 8.5 ± 0.2 ^{14}C ka BP, respectively (Awadallah and Batterson, 1990; Clark and Fitzhugh, 1990).

47.9. DEGLACIATION ON THE WESTERN SIDE OF THE NEW QUÉBEC DOME

47.9.1. Québec Clay Belt and James Bay Area

East of the Harricana Interlobate Moraine ice flow was first towards the south–southwest and swung progressively to the west (Hardy, 1976). In the La Grande Rivière area, superposed drumlins and striae (Lee et al., 1960) show that New Québec Ice was free to move in a more westerly direction following separation of the ice masses (Vincent, 1977). West of the Harricana Interlobate Moraine, Veillette (1986) has shown Late-glacial southeasterly flow.

As New Québec and Hudson ices retreated, glacial Lake Ojibway (Fig. 47.2) was dammed between the ice front and the drainage divide to the south (Coleman, 1909; Vincent and Hardy, 1977, 1979). The maximum lake limit rises northward from about 380 m to more than 450 m. The maximum depth of Lake Ojibway was more than 500 m on the east coast of James Bay. Lake waters flooded the lowlands east of James Bay as far as the Sakami Moraine (Figs. 47.1 and 47.2) (Hardy, 1976) and as far north as Kuujjuarapik (Hillaire-Marcel, 1976). Glacial Lake Obijway became separated from glacial Lake Barlow with the emergence of a sill (Vincent and Hardy, 1977, 1979). This sill was then the lowest point on the drainage divide, which was displaced south by isostatic tilting of the crust. Large channels on the divide and wide and deeply incised Rivière Kinojévis Valley were cut by overflowing glacial lake waters.

Extensive fields of De Geer moraines were built in glacial Lake Ojibway. The varve chronology established for lakes Barlow and Ojibway (Antevs, 1925; Hughes, 1965; Hardy, 1976) indicates 2110 years of glacial lacustrine sedimentation prior to the drainage of Lake Ojibway into Tyrrell Sea. Using this, Hardy (1976) calculated the rate of ice retreat which increased from about 320 m/year southeast of James Bay to 900 m/year in the La Grande Rivière area.

Late-glacial ice surges occurred into glacial Lake Ojibway. Hardy (1976) has indicated three Cochrane surges in the lowlands southeast of James Bay which came from Hudson ice. The maximum Cochrane I and II surges occurred 300 and 75 years before glacial Lake Ojibway drained. These surges presumably formed at the southern terminus of a short-lived ice stream that formed in the central James Bay region (Parent et al., 1995; Veillette, 1997a,b).

When New Québec Ice had retreated to the approximate position of the Sakami Moraine, marine waters from Hudson Strait penetrated Hudson and James Bays and flooded the isostatically depressed lowlands. Opening to the sea led to sudden drainage of glacial lakes Ojibway and Agassiz and to formation of the Sakami Moraine (Hardy, 1976), a major feature extending inland over a distance of 630 km from the coast of Hudson Bay to southern Lac Mistassini (Figs. 47.1, 47.4, and 47.9). Hillaire-Marcel et al. (1981) considered the Sakami Moraine as a re-equilibration moraine which results from the stabilisation of the ice front when the glacier grounded after drainage of Lake Ojibway. This drainage led also to a slight readvance in the Lac Mistassini area (Bouchard, 1980), recognised as the Waconichi ice advance by DiLabio (1981). The sudden drainage of glacial Lake Ojibway and the submergence by the Tyrrell Sea are recorded on hills in the Lake Ojibway basin by upper and lower wave washing limits (Norman, 1939; Hardy, 1976). The massive freshwater outflow from lakes Agassiz and Ojibway, through the Hudson Straight to the Labrador Sea and eastern North Atlantic Ocean, forced the 8.2 cal. ka BP cooling (Barber et al., 1999). From mean ^{14}C ages, and using sub-regional estimated deltaRs, the calendar ages of this event and the Sakami Moraine are close respectively to 8.4 and 8.2 cal. ka BP. In order to avoid any arbitrary interpretation, the 8.2 cal. ka BP isochrone (Figs. 47.5 and 47.9) will correspond to the Sakami Moraine, but original ^{14}C ages from shells will be indicated in the text.

The Tyrrell Sea followed the retreating ice front after construction of the Sakami Moraine. Marine limit is at about 198 m in the southern part of its basin (Hardy, 1976) and rises northwards up to 315 m in the Kuujjuarapik area (Hillaire-Marcel, 1976). In La Grande Rivière area, marine limit decreases eastwards (inland) from 270 m in the area of Sakami Moraine to 246 m farther up river (Vincent, 1977). Extensive swarms of De Geer moraines, many of which overlie drumlins and eskers, were built east of the Sakami Moraine. Rates of ice retreat averaged 217 m/y (Vincent, 1977). In the Lac Mistassini area, the relatively shallow glacial Lake Mattawaskin (Bouchard, 1986) followed the retreating ice front (Fig. 47.2). Rates of ice recession in this lake basin were estimated at 220–260 m/year (Bouchard, 1980).

An approximate age of ca. 8.0 ^{14}C ka BP from marine shells (Hardy, 1976) and from concretions (Hillaire-Marcel, 1976) was assigned to the Sakami Moraine and the Tyrrell Sea invasion. This age corresponds closely with the age proposed for the drainage of glacial lakes Agassiz and Ojibway and for the incursion of Tyrrell Sea on the west side of Hudson Bay (Dredge and Cowan, 1989; Dyke and Dredge, 1989). Using this age and the varve chronology, the area west of Lac Témiscamingue was deglaciated 10.0 ^{14}C ka BP; the height of land, at the Québec–Ontario border, was deglaciated by 9.2 ^{14}C ka BP; while the Cochrane I reached its maximum at 8.3 ^{14}C ka BP and Cochrane II 8.0 ^{14}C ka BP. East of the Sakami Moraine, based on the De Geer moraine chronology (Vincent, 1977), the Tyrrell Sea reached its eastern limit in the La Grande Rivière area about 7.5 ^{14}C ka BP. Farther inland, the oldest radiocarbon age of 6.6 ± 0.1 ^{14}C ka BP (B-9516) was obtained on basal gyttja (Richard, 1995).

47.9.2. Area East of Hudson Bay

Ice from central New Québec and central Ungava Peninsula (Nunavik) flowed generally westwards and southwestwards into Hudson Bay (Hillaire-Marcel, 1979; Gray and Lauriol, 1985; Bouchard and Marcotte, 1986; Veillette and Roy, 1995; Parent and Paradis, 1997; Veillette, 1997a,b) as far as the Belcher Islands (Jackson, 1960). When Hudson ice finally separated from New Québec ice in Hudson Bay, only the land area west of the Sakami Moraine in the Kuujjuarapik area, and the extreme northwestern part of Ungava Peninsula, was ice free (Figs. 47.5 and 47.9).

The location of ice-stillstand features in the Nastapoka Hills, a hill range that follows the arc-shaped coastline of eastern Hudson Bay, at ages (from shells) ranging between 7.85 and 7.61 ^{14}C ka BP (Lajeunesse and Allard, 2002, 2003a) suggests that the position of the New Québec ice margin was much further to the east from what was reported in previous models. This position occurred at the same time or shortly after (within the age of the standard error of a radiocarbon date, i.e. first century) the stillstand that deposited the Sakami Moraine (Lajeunesse, 2008). By 7.7 ^{14}C ka BP, during the Nastapoka Hills stillstand, the ice margin extended in an arc-like shape to Manitounouk Strait and Petite-Rivière-de-la-Baleine, where contemporaneous submarine ice-contact sediments and landforms are identified (Zevenhuizen, 1996). From there, the ice-terminus slightly curved northwards to the Nastapoka Hills near the Nastapoka River mouth, where moraines and emerged ice-contact submarine fans occur (Lajeunesse and Allard, 2002). Other raised ice-contact submarine fans located along the Nastapoka Hills north of Nastapoka River indicate that the ice margin extended up to Inukjuak (Lajeunesse, 2008). From Inukjuak, the ice front continued in the shallow water zone (<80 m) that extends from Inukjuak to the Ottawa Islands. The 7.43 ± 0.18 ^{14}C ka BP age from Ottawa Islands (GSC-706; Andrews and Falconer, 1969) provided from shells found on the surface of deltaic deposits, 17 m below marine limit, indicate a rather late deglaciation of western Ungava. North of Ottawa Islands, in the northern section of eastern Hudson Bay, the ice front curved northeastwards to Cape Smith where oldest age of 8.04 ± 0.11 ^{14}C ka BP; (UQ-761; Gray and Lauriol, 1985), redated at ca. 6.8 ^{14}C ka BP (Lauriol and Gray, 1987) was obtained. From the Cape Smith area, the ice front continued northwestwards to Mansell Island and then curved to the Ivujik area. The occurrence of ice-stillstand deposits and landforms in eastern Hudson Bay that show similar ages with the Sakami Moraine deposits indicate that the eastern part of Hudson Bay was deglaciated very rapidly following the drainage of Lake Agassiz–Ojibway. In fact, no geological evidence for a stillstand of the ice margin offshore in the eastern sector of the bay associated with the 8.0 ^{14}C ka BP position reported by Dyke et al. (2003) has been observed by geophysical investigations on the seafloor. The rapid ice retreat in this sector could be attributed to the thinning of the LIS over Hudson Bay and the development of a dense network of crevasses on its surface after the formation of the James Bay Ice Stream and before the lake drainage (Lajeunesse and St-Onge, 2008). The densely crevassed glacial ice, which might have formed an ice shelf at that time (Lajeunesse and St-Onge, 2008), probably broke up very rapidly after the lake drainage, allowing the ice margin to retreat to the eastern coast of Hudson Bay within a few decades or a century and without depositing any major ice-marginal landforms.

As the ice margin retreated to the east, the Tyrrell Sea covered the newly deglaciated areas and swarms of De Geer moraines were built at the ice front. The marine limit decreases northwards from about 315 m to possibly as low as 105 m east of Povungnituk. From there, it rises northerly to about 170 m near Hudson Strait (Gray and Lauriol, 1985). Marine limit also declines inland from 248 to 196 m along Nastapoka River (Allard and Seguin, 1985), and from 158 m on the Ottawa Islands (Andrews and Falconer, 1969) to 105 m east of Povungnituk (Gray and Lauriol, 1985). Rates of uplift, as measured in the Lac Guillaume-Delisle (Richmond Gulf) area, were between 7 and 10 m/century at the time of deglaciation (Hillaire-Marcel, 1976; Allard and Seguin, 1985; Lajeunesse and Allard, 2003b). According to Hillaire-Marcel (1976), by 7.0 ^{14}C ka BP the rate of uplift was 6.5 m/century and it decreased linearly to the present rate of 1.1 m/century. However, a relative sea level curve from the Nastapoka River area shows an inflection between 6.0 and 5.0 ^{14}C ka BP where rates decreased to 4 m/century and then to 1.6 m/century in the past 1000 years (Lajeunesse and Allard, 2003b).

East of marine limit, ice continued its retreat towards central Ungava Peninsula in the north or towards central New Québec farther south. Beyond the Tyrrell Sea limit, a shallow glacial lake was formed in the Lac à l'Eau Claire area (Fig. 47.2) between an uplifted sill and the ice front (Allard and Seguin, 1985). At the head of Rivière aux Mélèzes, glacial Lake Minto formed between the Hudson Bay–Ungava Bay drainage divide and the receding ice front (Lauriol, 1982; Lauriol and Gray, 1983).

47.10. DEGLACIATION IN THE NORTHERN AREAS OF QUÉBEC–LABRADOR

47.10.1. Ungava Peninsula (Nunavik)

Both northerly flow into southern Ungava Bay, from central New Québec, and flow towards Hudson Bay, western Hudson Strait, and western Ungava Bay, from a central north–south ice divide on Ungava Peninsula are recorded (Gray and Lauriol, 1985; Bouchard and Marcotte, 1986).

Ice flowed westwards and northwards from this ice-flow centre, called the Payne centre by Bouchard and Marcotte (1986), apparently coalesced with ice moving northeastwards in Hudson Bay and eastward in Hudson Strait (flow in offshore area from Andrews and Falconer, 1969; Shilts, 1980; Laymon, 1984; Gray and Lauriol, 1985). According to Clark et al. (2000), some of these ice-flow features correspond to an earlier phase of ice dynamics, and the pattern of the final deglaciation is characterised by a large mass of cold-based ice which was progressively fragmented to residual local ice masses, before the complete deglaciation. Instead of cold-based ice, we would infer passive stagnant ice such as the Bois Francs ice in the Appalachians of southern Québec.

In Hudson Strait, major submarine moraines lie subparallel to the northeastern part of the Ungava Coast (MacLean et al., 1992). Although their age is not known, they appear to define the northern limit of Late-glacial ice on the Wakeham Bay-Baie Héricart region ca. 8.5 ^{14}C ka BP, or a little earlier, stratigraphically overlying and deforming stratified glaciomarine sequences that floor the central part of the Strait (Maclean et al., 1992).

The northwestern and northern extremities of the Ungava Peninsula were deglaciated first (Fig. 47.9). Marine waters submerged the Ungava coast along Hudson Strait where marine submergence generally decreases from west to east (167 m at Cape Wolstenholme, Matthews, 1967; 138 m near Diana Bay, Gray et al., 1980), and from north to south (170 m on Charles Island and 120 m at the head of Deception Bay, Gray and Lauriol, 1985). Rates of uplift at about 8.0 ^{14}C ka BP were estimated at 7.9 m/century (Matthews, 1967). The timing of deglaciation of the southern Hudson Strait shore is subject of controversy. Most researchers agree that some areas were ice free at least 7.97 ± 0.25 ^{14}C ka BP (GSC-672, from shells), but some, on the basis of three 'older' radiocarbon age determinations, postulate earlier deglaciation. A 10.45 ± 0.25 ^{14}C ka BP (1-488; Matthews, 1966, 1967) age may in fact be erroneous since shells collected by Lauriol from the immediate vicinity of the Deception Bay site originally sampled by Matthews, were dated at 7.13 ± 0.1 ^{14}C ka BP (GSC-3947). Notwithstanding this, two other age determinations, the oldest of which is 9.8 ± 0.22 ^{14}C ka BP (Beta-11121), have been obtained from the Deception Bay area by dating *in situ Portlandia arctica* and *Nuculana minuta* shells collected in glacial marine sediments overlying till (Gray and Lauriol, 1985; Lauriol and Gray, 1987). This conflicts with evidence on Meta Incognita Peninsula of Baffin Island which requires that ice extended to the mouth of Hudson Strait until 8.6 ^{14}C ka BP or later (Andrews, 1989). Miller et al. (1988) have suggested that a late readvance of ice from New Québec Labrador across Hudson Strait, between 9.0 and 8.2 ^{14}C ka BP could account for the late presence of ice on southern Baffin Island. This corresponds to the Gold Cove readvance onto Baffin Island observed by Kaufman et al. (1993) and dated to 9.9–9.6 ^{14}C ka BP (see Fig. 11 in Clark et al., 2000). Upon retreat of the ice front towards the interior of Ungava Peninsula, glacial lakes were dammed between the ice front and higher ground on the Hudson Bay–Hudson Strait and Hudson Bay–Ungava Bay drainage divides (Prest et al., 1968; Prest, 1970). Of these lakes, the best documented is glacial Lake Nantais (Lauriol and Gray, 1987).

Standing water bodies extending far inland in the lower parts of the valleys of Rivière aux Mélèzes (Gray and Lauriol, 1985), Rivière Caniapiscau (Drummond, 1965) and of the George, Shepherd and Koroc Rivers require significant Late-glacial ice masses remaining in Ungava Bay when interior of Québec and Labrador were largely ice free. Eastern Hudson Strait was deglaciated by 9.1 ^{14}C ka BP on the basis of a radiocarbon age on shells collected from a sea-bed core (9.12 ± 0.48 ^{14}C ka BP, GSC-2946). The Iberville Sea (Laverdière and Bernard, 1969) followed the retreating ice front. On the west coast of Ungava Bay, marine limit rises from 138 m near Diana Bay to 195 m in upper Rivière aux Mélèzes drainage basin (Gray and Lauriol, 1985). On Akpatok Island, 75 km offshore, marine limit is much lower (58–74 m; Løken, 1978). According to radiocarbon ages by Allard et al. (1989) on the southeastern coast of the Ungava Bay and other ages on the western coast, all Ungava Bay coast became ice free more or less simultaneously, by about 7.0 ^{14}C ka BP (7.9 cal. ka BP).

47.10.2. Northeastern Québec and Northern Labrador

During the LGM, the New Québec–Labrador ice flow was east-northeasterly from central New Québec and easterly from Ungava Bay (Ives, 1957, 1958; Løken, 1962a) towards the Labrador Shelf over northeastern Québec and northern Labrador (Saglek Glaciation in the Torngat Mountains; Andrews, 1963). North of Fraser River, coastal summits and much of the northern Torngat Mountains and adjacent coastal forelands were not overtopped by ice (Ives, 1978). Clark (1984) illustrated that ice from west of the Labrador watershed crossed the Torngat Mountains as outlet glaciers extending to fiord mouths, leaving large nunataks. In part, the evidence is based on differential weathering zones delimited by lateral moraines (the Koroksoak and higher weathering zones). The probable upper limit of Late Wisconsinan ice is recorded in several local studies, and the regional limits of ice cover are not well constrained (Andrews, 1963; Clark, 1984; Evans, 1984; Evans and Rogerson, 1986). Generally, the upper glacial trimline of Saglek Glaciation declines eastward from the watershed divide to the sea and decreases northwards in overall elevation. Between Fraser River and Okak Bay, ice reached maximum elevations of about 700 m a.s.l. (Andrews, 1963). Mount Thoresby and Man O'War Peak are the most southerly nunataks (Andrews, 1963; Johnson, 1969). The glacial limit in the Saglek Fiord area is 615 m (Smith, 1969). In the Ryans Bay region, Clark (1984) stated that the ice passing through the Torngat Mountains did not reach elevations of more than 800 m on the drainage divide. In summit areas lying above the Saglek Glaciation level, cirque glaciers or small local ice caps existed independent of the LIS (Ives, 1960a; Løken, 1962a; Clark, 1984; Evans, 1984; Evans and Rogerson, 1986; Bell et al., 1987). The extent of grounded glacier ice on the Labrador Sea continental shelf is the subject of controversy. Initial studies by Clark (1984) and the studies of Evans (1984), Evans and Rogerson (1986) and Rogerson and Bell (1986) indicated that the Late Wisconsinan ice only extended to fjord mouths. In opposition, Josenhans et al. (1986) and Clark and Josenhans (1986) stated that grounded glacier ice extended well offshore to near the shelf edge. The interpretative differences may be reconciled by low shear strength of subglacial deposits on the shelf.

The deglacial history of the rugged coastal areas of Labrador is complex, is characterised by local readvances, and has been discussed for various locations by Ives (1958, 1960a), Løken (1962b, 1964), Tomlinson (1963), Andrews (1963), Johnson (1969), Clark (1984), Evans (1984) and Evans and Rogerson (1986). At the limit reached by the Late Wisconsinan ice, extensive systems of lateral moraines and kame complexes were built (Saglek Moraines of Ives, 1976). Several end and lateral moraines mark positions of halts or local readvances of the ice front during retreat from the Saglek Moraines towards an ice mass located west of the Labrador Sea/Ungava Bay watershed. Notable examples are the Tasiuyak Moraines in the Fraser River–Okak Bay area (Andrews, 1963) and the well-correlated Noodleook, Two Loon and Kangalaksiorvik (=Sheppard) moraines of Løken (1962b, 1964) on the Torngat Peninsula. Andrews (1977) suggested that the Kangalaksiorvik Moraines may be equivalent to the moraines of Cockburn age elsewhere in Arctic Canada (Andrews and Ives, 1978) which are dated about 8.0 ^{14}C ka BP. In their northernmost 100 km, the Kangalaksiorvik (Sheppard) moraines contain abundant indicator erratics derived from the Labrador Trough. They demonstrate net eastwards to northeastwards ice flow over 100s of kilometres across Ungava Bay and onto the northernmost Labrador Peninsula. They are interpreted to reflect a major, Late-glacial readvance after ca. 9.0 ^{14}C ka BP (Løken, 1964) that extended to ca. 280 m a.s.l. on the northern Peninsula. In contrast, there is no evidence for onshore ice flow in southeastern Ungava Bay (Allard et al., 1989).

Ice-flow indicators between the southwestern Torngat Mountains and lower Rivière George (Matthew, 1961), and along the east coast of Ungava Bay clearly indicate a Late-glacial reversal of ice flow into Ungava Bay which is interpreted as the result of drawdown towards a calving bay. The area affected by drawdown extended rapidly southwards to the Schefferville area of central Labrador–Québec (Ives, in Vincent, 1989; Parent and Paradis, 1994, 1997; Parent et al., 1995, 1996; Veillette et al., 1999).

As a result of glacial isostatic depression, both ice-free and newly deglaciated, low-lying coastal areas were submerged. Generally, marine limit progressively decreases in elevation northwards, from 93 m south of Okak Bay (Andrews, 1963) to 16 m on Killinek Island (Løken, 1964). The oldest Holocene age so far obtained on the north coast is 9.82 ± 0.07 ^{14}C ka BP (TO-305, from shells). The presence of tilted shorelines in northernmost Labrador truncated by a 15 m high horizontal shoreline is considered by Løken (1962b) as recording an early Holocene transgression and also suggests that thick continental ice did not overlie the northern tip of Labrador.

During deglaciation of the Torngat Mountains, east of the drainage divide, numerous small ephemeral glacial lakes were ponded in tributary valleys, blocked by ice tongues. As the ice margin receded westward from the divide, larger and longer-lived glacial lakes were created. Lakes in tributary valleys of Rivière Alluviaq drained into a fjord south of Iron Strand (Ives, 1957), whereas other lakes, in the Rivière Koroc basin, drained towards Saglek Bay (Ives, 1958), and later into Nachvak Fiord via Palmer River with continued deglaciation and westward marginal retreat. The largest lakes were glacial lakes Naskaupi and McLean (Ives, 1960a,b; Matthew, 1961; Barnett and Peterson, 1964; Barnett, 1964, 1967; Peterson, 1965; Fig. 47.2), which extended over large areas in the upper Rivière George and Rivière à la Baleine drainage basins. These lakes were dammed between the watershed divide on the east, the southwestwards retreating main body of New Québec ice on the west and Late-glacial ice in Ungava Bay (Prest, 1970, 1984). Glacial Lake Naskaupi cut a series of well-defined strandlines, some of which are incised into bedrock. Fine-grained glacial lake deposits, however, are virtually non-existent.

Glacial Lake McLean, in upper Rivière à la Baleine basin, was separate from Lake Naskaupi but drained into it by a channel west of Lac de la Hutte Sauvage. Both lakes finally drained into the Iberville Sea when ice in Ungava Bay had retreated sufficiently to allow free northwards drainage.

When the ice receded from the Saglek Moraines is unknown. From marine shell dates, the minimum age for deglaciation of the coastal areas is of about 9 ^{14}C ka BP. The deglacial history of the central Labrador coast has been used to infer that the glacial lakes were fully established only after 7.6 ± 0.2 ^{14}C ka BP, and that Late-glacial ice in Ungava Bay collapsed about 7.0 ^{14}C ka BP (Clark and Fitzhugh, 1990). Abandoned moraines and lichen-kill areas adjacent to present cirque glaciers and glacier-free cirques provide a record of Neoglacial expansion of glaciers in the Torngat Mountains (McCoy, 1983; Clark, 1984; Evans, 1984; Evans and Rogerson, 1986).

47.11. THE LATEST DEGLACIATED AREAS IN NUNAVIK, CENTRAL NEW QUÉBEC AND WESTERN LABRADOR

Since Low (1896) first recognised central Labrador–Ungava Peninsula as one of the final centres of ice disintegration, controversy has surrounded the situation of residual ice centres of the Eastern Sector. As portrayed in Wilson et al. (1958), Ives (1960a), Prest et al. (1968) and Prest (1969), net patterns of glacially streamlined landforms and eskers indicate that New Québec Ice flowed radially from a horseshoe-shaped area extending from northwest of Lake Delorme in the west to northern Smallwood Reservoir in the east. In the area enclosed by the horseshoe, ice flow was convergent northwards towards Ungava Bay, whereas outside ice flow was broadly radial outwards. Whether the ice divide, determined from the landforms, was a stable feature of Late Wisconsinan ice or whether its location fluctuated considerably or whether it existed at all (e.g. Kirby, 1961, 1961) has only recently been addressed (Clark et al., 2000). Indicator erratics derived from bedrock of the Labrador Trough have been used to constrain net glacier flowpaths and key aspects of glacial history. For example, Hughes (1964) and Richard et al. (1982) suggest that the ice divide must at one time have been situated well to the northeast of its final position because indicator erratics were transported from areas east of the assumed ice divide. Ives (1960c) identified evidence for transport from areas west of it.

In common with Shield terranes elsewhere in Canada (e.g. Klassen and Thompson, 1993; Parent et al., 1996; Veillette et al., 1999), the erosional record of striations has been proved to provide a comprehensive basis for reconstructing much of Wisconsinan glacial history, not only the Late glacial. Along with the evidence for net distances and directions of ice flow provided by indicator erratics, striations show a complex glacial history for Labrador–Québec (Henderson, 1959; Ives, 1960a,b,c; Kirby, 1961a,b; Hughes, 1964; Bouchard and Martineau, 1985; Bouchard and Marcotte, 1986; Veillette, 1986; Klassen and Thompson, 1993; Veillette and Roy, 1995; Parent et al., 1996; Veillette et al., 1999) and indicate that the depositional record of landforms does not define a simple integrated record of ice flow in the marginal areas of a decaying ice sheet (e.g. Dyke and Prest, 1987; Kleman et al., 1995; Clark et al., 2000). From the relative ages shown by striations, the ice divide determined from the glacial landform record is the product of multiple, distinct ice-flow events. It does not reflect a single glacial configuration of the Eastern Sector but is the integrated product of multiple glacial events (Klassen and Thompson, 1993; Veillette et al., 1999; Clark et al., 2000).

Field investigations in the Schefferville area of the Labrador Trough by Perrault (1955), Grayson (1956), Henderson (1959), Ives (1959, 1960a,b,c, 1968, 1979), Kirby (1961a,b) and Derbyshire (1962) brought conclusive evidence for the presence of small ice remnants in the low-lying basins of Howells River (Kivivic ice divide; just west of Schefferville) and Swampy Bay River valleys (north–northwest of Schefferville). Other authors, basing their arguments on glacial ice-flow indicators and landforms (Low, 1896; Hughes, 1964; Laverdière, 1967; Richard et al., 1982) and on the intersection of projected strandline tilt directions of glacial lakes as an indicator of the position of the maximum ice thickness (Ives, 1960b; Harrison, 1963; Barnett, 1964; Barnett and Peterson, 1964), proposed that the final ice masses disintegrated in the Schefferville area, near the southern base of the 'horseshoe-shaped' ice divide. Much controversy has ensued between the different authors (Ives, 1968; Bryson et al., 1969; Laverdière, 1969a,b; Laverdière and Guimont, 1982), but it is likely that there were numerous retreat centres in low-lying basins where discrete ice masses finally melted (Clark et al., 2000; Dyke et al., 2003).

Ages from lake sediment cores have been used to date the final disappearance of ice. In the Lac Delorme area, in the northern part of the Labrador Trough, the ice stagnated locally after 6.3 ^{14}C ka BP (6.32 ± 0.18 ^{14}C ka BP, GSC-3094) and disappeared in lower parts ca. 5.6 ^{14}C ka BP (Richard et al., 1982). These ages are compatible with 6.2 ± 0.1 ^{14}C ka BP (GSC-3644; King, 1985) in the Lac Stakel area. In Nunavik, a basal date of 5.23 ± 0.185 ^{14}C ka BP (Richard, 1981) gives the minimum age of local deglaciation (see also Gagnon and Payette, 1981). Apart perhaps from small remnant ice masses in depressions, it is probably safe to assume that glacier ice in the Québec–Labrador interior had completely melted by 6.0 cal. ka BP.

47.12. PENDING QUESTIONS ON THE DEGLACIATION OF QUÉBEC–LABRADOR

There are two types of problems which limit the complete model of deglaciation in Québec–Labrador, problems related to chronology and the need for geological field evidence.

The first chronological problem is related to ages measured on marine shells, in the St. Lawrence Valley, Gulf and Estuary (Champlain and Goldthwait seas), Lake Saint-Jean lowland (Laflamme Gulf) and the other post-glacial areas inundated by post-glacial marine waters (Tyrrell and Iberville seas, Labrador coast). Mean deltaR values (local marine reservoir value -400 y of mean oceanic reservoir) in the different marine and coastal areas of Canada and Greenland are measured by McNeely et al. (2006) on living shells collected prior to the nuclear bomb testing. On the whole, these McNeely et al. (2006) deltaR values can be applied as the deltaR values of the post-glacial basins to calculate calendar ages of the late Pleistocene–Holocene shells collected in marine deposits. They are not reliable for closed basins (i.e. Champlain Sea, Laflamme Gulf and every former bay of the inundated areas) as cross-dating of shells-vegetal debris from Champlain Sea revealed highly variable deltaR values, from 1500 to 350 years (Occhietti and Richard, 2003; Richard and Occhietti, 2005). Finally, as already stated, the 1 sigma error of most of the pre-AMS datations on marine shells is over 100 y, this means that a systematic programme of redating marine selected shells is necessary to get more reliable calendar ages and improve the deglaciation timing of Québec–Labrador.

The second chronological problem is the question of the age of the moraines on the North Shore of the St. Lawrence lower Estuary and Gulf and in southern–southeastern Labrador. Radiocarbon evidence for the deglacial history of southeastern Labrador is summarised and discussed by King (1985). Evidence for tundra conditions in southeastern Labrador during the glacial maximum is represented by analyses of fine-grained organic carbon in marine cores (Vilks and Mudie, 1978). The potential for contamination by old carbon makes the radiocarbon dates suspect. From dated lake sediment dates, King (1985) places the Late Wisconsinan ice margin east of the Paradise Moraine, with the maximum limits possibly at the Bradore Moraine and Belles Amours Moraines. The interpretation of extensive ice cover in southeastern Labrador is supported by Grant (1992) who places the Late Wisconsinan ice limit offshore, in the Strait of Belle Isle, consistent with the reconstruction of Dyke and Prest (1987). According to Grant (1992), the Bradore and Belles Amours moraines formed about 12.6 ka through regional readvance of the ice margin. The Paradise Moraine is thought to have formed ca. 10 ka, possibly as the result of regional climatic cooling during glacial retreat. The Little Drunken Moraine and the Kenamu Moraine are represented as segments of the Québec North Shore Moraine, and they are estimated to have formed about 9.0 ^{14}C ka BP; the Sebaskachu Moraine to the north of Lake Melville is inferred to be younger, forming about 8.0 ^{14}C ka BP. The Little Drunken Moraine is associated with glacial drawdown into Lake Melville; no major terminal moraines occur west of it, in the Labrador interior.

The third chronological problem is the age of the final disappearance of the LIS in northern Québec, in the Ungava Peninsula. ^{10}Be cosmogenic exposure ages on boulders of 6.8 ± 0.2 ka (Carlson et al., 2007) are older than the 6 cal. ka BP (from a ^{14}C age) related to the end of the ultimate ice masses. From the ^{10}Be age, Carlson et al. (2007) put forward a very rapid deglaciation of the New Québec–Labrador ice mass after the 8.2 ka cold event. As for any dating method, this new model has to be confirmed by field evidence.

There are large areas of Québec–Labrador that have never been mapped or simply explored, mostly because of the lack of mineral interest and the inaccessibility (cf. the general map of Quaternary deposits of Canada, Fulton, 1995). This is the case of most of the Laurentians between the Saint-Narcisse Moraine and the Sakami Moraine. Current work in the middle reaches of the Saint-Maurice River shows that there is a high potential for the discovery of new significant evidence about the LIS front retreat. The two interglacial deposits at Wabush (Klassen et al., 1988) are a prime example.

There is a problem that is not only unresolved but scarcely looked at that concerns the age, origin and timing of the glacial lakes in the interior, notably Naskaupi, McLean and Koroc, but also the existence of others in the Caniapiscau River basin (glacial Lake Minto, etc.). The problem is compounded by the INSTAAR models on the southern Baffin/western Hudson Strait that focus on the geological record of southeastern Baffin Island and are difficult to reconcile with the Québec–Labrador record. The problem is also linked to the glacial model applied to the Labrador highlands, especially to the Torngat mountains. The nunatak theory favours the absence of ice over the mountain tops. Gangloff (1983) shows that tors and felsenmeers are not necessarily indicative of unglaciated areas.

Another problem, currently in the process of resolution, concerns the acquisition of ice-flow data, and the construction of a comprehensive ice-flow record for much of the last glaciation. The existence of an early centre for ice sheet accumulation and dispersal on the Manicouagan uplands, north of the St. Lawrence Estuary and Gulf (considered by Occhietti, 1982 and others), has never been adequately accounted for through modelling.

47.13. CONCLUSIONS

During the Wisconsinan, the area of Québec–Labrador was completely covered by the LIS, except maybe for nunataks in the Torngat Mountains of northeastern Québec–northern Labrador. For this reason, pre-Upper Pleistocene events are only represented in scattered stratigraphical sections, mostly in southern Québec, and by erosional glacial marks. Early Wisconsinan ice-flow patterns indicate the LIS could have been initiated in highland regions north of the St. Lawrence Estuary and Gulf. After the LGM, the Eastern Sector of the LIS evolved from a single, predominant dispersal centre with subsidiary ice divides, into peripheral ice domes and masses which remained connected or not to the central dome. The central dome was not a simple and stable dome-shaped ice mass, but an evolving ice mass. The general model of deglaciation is still evolving as new field evidence becomes available.

Thinning of the LIS through ablation and through mechanical drawdown along its margins as the result of diachronic ice streams in the St. Lawrence Corridor and Hudson Strait are the main features of Late-glacial ice-flow dynamics. In the areas south of the St. Lawrence Corridor, ice masses over the Appalachian uplands evolved from a glacier complex confluent with the LIS into separate local ice caps. During a part of the warm Bølling–Allerød phase, a series of ice-front positions mark a fast retreat of the ice front in the Appalachians of southern Québec, between 13.5 and 13.1 cal. ka BP. The ice mass over the Canadian Shield, north of the St. Lawrence Corridor, dissipated slowly, between about 13 and 6 cal. ka BP. The deglaciation pattern includes the differentiation of an ice mass over the Hudson Bay, early deglaciation of the Labrador Highlands, a major change of ice flow from the southern part of the Ungava Bay towards the Hudson Strait and a very roughly concentric ice retreat pattern in the southwest, south and southeast margins of the remnant main ice mass. A series of concentric moraines (Saint-Narcisse, Mars-Batiscan and intermediate moraines), built between about 12.7 and 11.5 cal. ka BP, are related to the YD. During the early Holocene, the Harricana Interlobate Moraine is thought to record the separation of Hudson Bay ice from the ice mass over Québec and western Labrador. The Sakami Moraine, built ca. 8.2 cal. ka BP, is related to the drainage of Lake Ojibway and Lake Agassiz towards the Hudson Strait and grounding of the ice front. In the northern Ungava and Labrador peninsulas, major glacial lakes in low-lying areas were dammed between the ice masses and the tilted deglaciated land. Lowlands depressed by glacioisostasy were momentarily invaded by marine waters. Fossils from these post-glacial seas (Goldthwait, Champlain, Laflamme, Tyrrell and Iberville seas) give approximate ages of the deglaciation episodes, between 13 and 7 ka. The last glacial ice masses were situated in the Labrador Trough and Nunavik and finally disappeared ca. 6.8–6.0 cal. ka BP.

By 11.55 cal. ka BP (10 ^{14}C ka BP), the LIS was still a major ice mass of more than 5×10^6 km^2; it took about 5500 calendar years to dissipate completely. Taking into account the latent heat of ice (80 cal./g to transform ice at 0 °C to water), the LIS decreased the rate of global climatic warming during Early Holocene. The loss of heat energy due to phase change was reinforced by the albedo. The LIS was in disequilibrium with the interglacial insolation conditions and generated a major effect on the world's climatic system.

ACKNOWLEDGEMENTS

The compilation of maps and documents and the preparation of the digital maps of glacial features of Québec were technically and financially supported by the Département de Géographie, Université du Québec à Montréal (UQAM). The Glacial Landforms Map of Labrador and parts of adjacent Québec (compilation of R. A. Klassen) were supplied in digital format by A. Moore, Geological Survey of Canada (Ottawa). Research on the field and by air photography was supported by the Natural Sciences and Engineering Council of Canada (S. Occhietti). A special thanks is addressed to Hélène Boisvert and Claire Gagnon who supported firmly the two first authors.

REFERENCES

Allard, M., Seguin, M.K., 1985. La déglaciation d'une partie du versant hudsonien québécois; bassin des rivières Nastapoca, Sheldrake et à l'Eau Claire. Géogr. Phys. Quatern. 39, 13–24.

Allard, M., Tremblay, G., 1981. Observations sur le Quaternaire de l'extrémité orientale de la péninsule de Gaspé, Québec. Géogr. Phys. Quatern. 35, 105–125.

Allard, M., Fournier, A., Gahé, É., Seguin, M.K., 1989. Le Quaternaire de la côte sud-est de la baie d'Ungava. Québec nordique. Géogr. Phys. Quatern. 43 (3), 325–336.

Anderson, W., Lewis, M., Mott, R., 2001. AMS-revised radiocarbon ages at Turtle Lake, North Bay-Mattawa area, Ontario: implications for the deglacial history of the Great Lakes Region. In: 27e rencontre scientifique annuelle de l'Union géophysique canadienne, con-jointement avec la 58e conférence de neige de l'Est. Université d'Ottawa, Canada.

Andrews, J.T., 1963. End Moraines and Late-Glacial Chronology in the Northern Nain Okak Section of the Labrador Coast. Geogr. Ann. 45A, 158–171.

Andrews, J.T., 1977. Status of Late Quaternary correlation <125,000 BP along the eastern Canada seaboard-latitude 45°N to 82°N. IGCP Project 73/1/24. Quaternary Glaciations in the Northern Hemisphere, vol. 4. 180–195.

Andrews, J.T., 1989. Quaternary geology of the northeastern Canadian Shield. In: Fulton, R.T. (Ed.), Quaternary Geology of Canada and Greenland, Geological Survey of Canada, Geology of Canada, 1, pp. 276–302 (also: Geological Society of America, The Geology of North America, K-1).

Andrews, J.T., Falconer, G., 1969. Late glacial and post-glacial history and emergence of the Ottawa Islands, Hudson Bay, N.W.T.: evidence on the deglaciation of Hudson Bay. Can. J. Earth Sci. 6, 1263–1276.

Andrews, J.T., Ives, J.D., 1978. "Cockburn" nomenclature and the Late Quaternary history of the Eastern Canadian Arctic. Arct. Alpine Res. 10, 617–633.

Andrews, J.T., Shilts, W.W., Miller, G.H., 1983. Multiple deglaciation of the Hudson Bay Lowlands, Canada, since deposition if the Missinaibi (last inter-glacial?) Formation. Quatern. Res. 6, 167–183.

Antevs, E., 1925. Retreat of the Last Ice Sheet in Eastern Canada. Geological Survey of Canada, Memoir 146, 142 pp.

Awadallah, S.A., Batterson, M.J., 1990. Comment on "Late deglaciation of the central Labrador coast and its implications for the age of glacial lakes Naskaupi and McLean and for prehistory", by P.U. Clark and W.W. Fitzhugh. Quatern. Res. 34, 372–373.

Barber, D.C., Dyke, A.S., Hillaire-Marcel, C., Jennings, A.E., Andrews, J.T., Kerwin, M.W., et al., 1999. Forcing of the cold event of 8,200 years ago by catastrophic drainage of Laurentian lakes. Nature 400, 344–348.

Barnett, D.M., 1964. Some aspects of the deglaciation of the Indian House Lake Area, with particular reference to the former proglacial lakes. Unpublished MSc thesis, McGill University, Montréal, 175 pp.

Barnett, D.M., 1967. Glacial Lake McClean and its relationships with Glacial Lake Naskaupi. Geogr. Bull. 9, 96–101.

Barnett, D.M., Peterson, J.A., 1964. The significance of glacial Lake Naskaupi 2 in the deglaciation of Labrador–Ungava. Can. Geogr. 8, 173–181.

Bell, T., Rogerson, R.J., Klassen, R.A., Dyer, A., 1987. Acoustic survey and glacial history of Adam Lake, outer Nachvak Fiord, northern Labrador. In: Current Research, Part A, Geological Survey of Canada, Paper, 87-IA 101–110.

Bhiry, N., Dionne, J.-C., Clet, M., Occhietti, S., Rondot, J. (Eds.), 2001. Stratigraphy of the Pleistocene Units on Land and Below the St. Lawrence Estuary, and Deglaciation Pattern in Charlevoix, 64th Annual Reunion of the North East Friends of the Pleistocene. Québec City, QC, Canada, Field Guide, 124 pp.

Blais, A., 1989. Lennoxville glaciation of the middle Chaudiere and Etchemin valleys, Beauce region, Quebec. [MSc Memoir]: Carleton University, Ontario, Canada, 124 pp.

Blake, W., Jr., 1956. Landforms and topography of the Lake Melville area, Labrador, Newfoundland. Geographical Bulletin 9, 75–100.

Bolduc, A.M., 1995. Landforms in the Laurentians of southern Quebec: implications for the deglaciation history of the Laurentide Ice Sheet. In: CANQUA-CGRG Joint Meeting, St. John's, Newfoundland, Program, Abstracts and Fieldguides, CA5.

Bouchard, M.A., 1980. Late Quaternary geology of the Témiscamie Area, Central Québec, Canada. Unpublished PhD thesis, McGill University, Montreal, 284 pp.

Bouchard, M.A., 1986. Géologie des dépôts meubles de la région de Témiscamie (territoire du Nouveau-Québec). Ministère de l'Énergie et des Ressources, Rapport, MM-83-03, 88 pp.

Bouchard, M.A., Marcotte, C., 1986. Regional glacial dispersal patterns in Ungava, Nouveau-Québec. In: Current Research, Part B, Geological Survey of Canada, Paper 86-1B, 295–304.

Bouchard, M.A., Martineau, G., 1985. Southeastward ice flow in central Quebec and its paleogeographic significance. Can. J. Earth Sci. 22, 1536–1541.

Boulton, G.S., Clark, C.D., 1990. A highly mobile Laurentide ice sheet revealed by satellite images of glacial lineations. Nature 346, 813–817.

Brennan, T.A., Shaw, J., 1996. The Harricana glaciofluvial complex, Abitibi region, Quebec: its genesis and implications for meltwater regime and ice-sheet dynamics. Sed. Geol. 102, 221–262.

Brennan, T.A., Shaw, J., Sharpe, D.R., 1996. Regional-scale meltwater erosion and deposition patterns, northern Quebec, Canada. Ann. Glaciol. 22, 85–92.

Bryson, R.A., Wendland, W.M., Ives, J.D., Andrews, J.T., 1969. Radiocarbon isochrones and the disintegration of the Laurentide Ice Sheet. Arct. Alpine Res. 1, 1–14.

Carl, J.D., 1978. Ribbed moraine–drumlin transition belt, St. Lawrence Valley, New York. Geology 6, 562–566.

Carlson, A.E., Clark, P.U., Raisbeck, G.M., Brook, E.J., 2007. Rapid Holocene deglaciation of the Labrador sector of the Laurentide Ice Sheet. J. Climate 20, 5126–5133.

Chauvin, L., Martineau, G., LaSalle, P., 1985. Deglaciation of the Lower St. Lawrence region, Québec. In: Borns Jr., H.W., Lasalle, P., Thompson, W.B. (Eds.), Late Pleistocene History of Northeastern New England and Adjacent Quebec: Geological Society of America, Special Paper 197, pp. 111–123.

Clark, P.U., 1984. Glacial geology of the Kangalaksiorvik-Abloviak region, northern Labrador, Canada. Unpublished PhD thesis, University of Colorado, Boulder, 240 pp.

Clark, P.U., Fitzhugh, W.W., 1990. Late deglaciation of the central Labrador coast and its implications for the age of glacial lakes Naskaupi and McLean and for prehistory. Quatern. Res. 34, 296–305.

Clark, P.U., Josenhans, H., 1986. Late Quaternary land-sea correlations, northern Labrador and Labrador Shelf. In: Current Research, Part B, Geological Survey of Canada, Paper 86-1B, pp. 171–178.

Clark, P.U., Karrow, P.F., 1983. Till stratigraphy in the St. Lawrence Valley near Malone, New York: revised glacial history and stratigraphic nomenclature. Geol. Soc. Am. Bull. 94, 1308–1318.

Clark, C.D., Knight, J.K., Gray, J.T., 2000. Geomorphological reconstruction of the Labrador Sector of the Laurentide Ice Sheet. Quatern. Sci. Rev. 19, 1343–1366.

Coleman, A.P., 1909. Lake Ojibway; last of the great glacial lakes. Ontario Bureau of Mines, 18th Annual Report, 18 (1), pp. 284–293.

Connally, G.C., 1982. Deglacial history of western Vermont. In: Larson, J.G., Stone, B.D. (Eds.), Late Wisconsinan Glaciation of New England, A Proceeding Volume of the Symposium: Late Wisconsinan Glaciation of New England, Held at Philadelphia, Pennsylvania, March 13, 1980. Kendall/Hunt Publishing Company, Dubuque, pp. 183–193.

Connally, G., Sirkin, L.A., 1973. The Wisconsinian history of the Hudson–Champlain lobe. The Wisconsinian stage. Geological Society of America, Memoir 136, pp. 47–69.

Daigneault, R.A., Occhietti, S., 2006. Les moraines du massif Algonquin, Ontario, au début du Dryas récent, et corrélation avec la Moraine de Saint-Narcisse. Géogr. Phys. Quatern. 60, 103–118.

Dansgaard, W., Johnsen, S., Clausen, H., Dahl-Jensen, D., Gundestrup, N., Hammer, C., et al., 1993. Evidence for general instability of past climate from a 250-kyr ice core record. Nature 364, 218–220.

David, P.P., Lebuis, J., 1985. Glacial Maximum and Deglaciation of Western Gaspé, Québec, Canada. Geological Society of America, Special Paper, 197, pp. 85–109.

de Boutray, B., Hillaire-Marcel, C., 1977. Aperçu géologique du substratum et des dépôts quaternaires dans la région de Blanc-Sablon, Québec. Géogr. Phys. Quatern. 32, 207–215.

De Vernal, A., Goyette, C., Rodrigues, C.G., 1989. Contribution palynostratigraphique (dynokystes, pollens et spores) à la connaissance de la Mer de Champlain:coupe de St-Césaire, Québec. Can. J. Earth Sci. 26, 2450–2464.

Denis, R., 1974. Late Quaternary Geology and Geomorphology in the Lake Maskinongé Area, Quebec. Uppsala Universitet Naturgeografiska Institutionen, Report 28, 125 pp.

Derbyshire, E., 1962. The deglaciation of the Howells River valley and the adjacent parts of the watershed region, Central Labrador–Ungava. University of McGill, McGill Sub-Arctic Research Papers 14, 23 pp.

DiLabio, R.N.W., 1981. Glacial dispersal of rocks and minerals at the south end of Lac Mistassini, Quebec, with special reference to the Icon dispersal train. Geological Survey of Canada Bulletin 323, 46 pp.

Dionne, J.-C., 1972. Le Quaternaire de la région de Rivière-du-Loup/Trois-Pistoles, côte sud du Saint-Laurent. Environnement Canada, Centre de recherches forestières des Laurentides, Québec, Rapport d'information, Q-F-X-27, 95 pp.

Dionne, J.-C., 1973. La dispersion des cailloux ordoviciens dans les formations quaternaires, au Saguenay/Lac Saint-Jean, Québec. Rev. Géogr. Montréal 27, 339–364.

Dionne, J.-C., 1977. La mer de Goldthwait au Québec. Géogr. Phys. Quatern. 31, 61–80.

Dionne, J.-C., 1988. Holocene relative sea-level fluctuations in the St. Lawrence estuary, Québec, Canada. Quatern. Res. 29, 233–244.

Dionne, J.-C., Bernatchez, P., 2000. Les erratiques de dolomie sur le rivage des Escoumins, Côte Nord de l'estuaire maritime du Saint-Laurent, Québec. Atlantic Geol. 36, 117–129.

Dionne, J.-C., Coll, D., 1995. Le niveau marin relatif dans la région de Matane (Québec), de la déglaciation à nos jours. Géogr. Phys. Quatern. 49, 363–380.

Dionne, J.-C., Dubois, J.-M., 1980. Le complexe morai-nique frontal d'Aquanus-Kénamiou, Basse Côte Nord du Saint-Laurent (résumé). In: Résumés et programmes, 4e Colloque sur le Quaternaire du Québec, Québec, 13.

Dionne, J.-C., Occhietti, S., 1996. Aperçu du Quater-naire à l'embouchure du Saguenay, Québec. Géogr. Phys. Quatern. 50, 5–34.

Dredge, L.A., 1976. Moraines in the Godbout-Sept-Iles area, Quebec North Shore. Report of Activities, Part C, Geological Survey of Canada, Paper 76-IC, pp. 183–184.

Dredge, L.A., 1983. Surficial geology of the Sept-Iles area, Quebec North Shore. Geological Survey of Canada, Memoir 408, 40 pp.

Dredge, L.A., Cowan, W.R., 1989. Quaternary geology of the southwestern Canadian Shield. In: Fulton, R.T. (Ed.), Quaternary Geology of Canada and Greenland, Geological Survey of Canada, Geology of Canada, 1, pp. 214–235 (also: Geological Society of America, The Geology of North America, K-1).

Dreimanis, A., 1977. Correlation of Wisconsin glacial events between the eastern Great Lakes and the St. Lawrence Lowlands. Géogr. Phys. Quatern. 31, 37–51.

Dreimanis, A., 1985. Till stratigraphy in the St. Lawrence Valley near Malone, New York: revised glacial history and stratigraphic nomenclature: discussion. Geol. Soc. Am. Bull. 96, 155–156.

Drummond, R.N., 1965. Glacial geomorphology of the Cambrian Lake Area, Labrador–Ungava. Unpublished PhD thesis, McGill University, Montréal, 222 pp.

Dubois, J.-M., 1976. Levé préliminaire du complexe morainique du Manitou-Matamek sur la Côte Nord de l'estuaire maritime du Saint-Laurent; dans Report of Activities, Partie B. Commission géologique du Canada, Étude, 76-1B, 89–93.

Dubois, J.-M., 1977. La déglaciation de la Côte-Nord du Saint-Laurent: analyse. Géogr. Phys. Quatern. 31, 229–246.

Dubois, J.-M., 1979. Télédétection, cartographie et interprétation des fronts glaciaires sur la Côte Nord du Saint-Laurent entre le lac Saint-Jean et le Labrador. Département de géographie, Université de Sherbrooke, Bulletin de recherche, 42, 33 pp.

Dubois, J.-M., 1980. Environnements quaternaires et évolution littorale d'une zone côtière en émersion en bordure du bouclier canadien: la moyenne Côte-Nord du Saint-Laurent, Québec. PhD, Université d'Ottawa, 754 pp.

Dubois, J.-M., Dionne, J.-C., 1985. The Québec North Shore Moraine System: a major feature of Late Wisconsin deglaciation. In: Borns Jr., H.W., Lasalle, P., Thompson, W.B. (Eds.), Late Pleistocene History of Northeastern New England and Adjacent Quebec, Geological Society of America, Special Paper 197, pp. 125–133.

Dubois, J.-M., Desmarais, G., Brouillette, D., Perras, S., 1984. Géologie des formations en surface de la Mer de Goldthwait sur la Côte-Nord du Saint-Laurent. Commission Géologique du Canada, DP-1045.

Dyke, A.S., Dredge, L.A., 1989. Quaternary geology of the northwestern Canadian Shield. In: Fulton, R.T. (Ed.), Quaternary Geology of Canada and Greenland, Geological Survey of Canada, Geology of Canada, 1, pp. 189–214 (also: Geological Society of America, The Geology of North America, K-1).

Dyke, A.S., Prest, V.K., 1987. Late Wisconsinan and Holocene history of the Laurentide Ice Sheet. Géogr. Phys. Quatern. XLI (2), 237–263.

Dyke, A.S., Prest, V.K., 1989. Late Wisconsinan and Holocene retreat of the Laurentide Ice Sheet. Geological Survey of Canada, Map 1702A.

Dyke, A.S., Moore, A., Robertson, L., 2003. Deglaciation of Canada. Geological Survey of Canada Open File 1574. Thirty-two digital maps at 1:7,000,000 scale with accompanying digital chronological database and one poster (two sheets) with full map series. http://ess.nrcan.gc.ca/ercc-rrcc/proj4/theme1/act1_f.php.

Elson, J.A., 1969. Late quaternary marine submergence of Quebec. Rev. Géogr. Montréal XXIII, 247–258.

Evans, D.J.A., 1984. Glacial geomorphology and chronology in the Selamiut Range/Nachvak Fiord area, Torngat Mountains, Labrador. Unpublished MSc thesis, Memorial University of Newfoundland, St. John's, 138 pp.

Evans, D.J.A., Rogerson, R.J., 1986. Glacial geomorphology and chronology in the Selamiut Range/Nachvak Fiord area, Torngat Mountains, Labrador. Can. J. Earth Sci. 23, 66–76.

Ferland, P., Occhietti, S., 1990. L'Argile de la Pérade: nouvelle unité marine antérieure au Wisconsinien supérieur, vallée du Saint-Laurent, Québec. Géogr. Phys. Quatern. 44, 159–172.

Fitzhugh, W., 1973. Environmental approaches to the prehistory of the north. J. Wash. Acad. Sci. 63, 39–53.

Fournier, M., 1998. Stratigraphie des dépôts quaternaires et modalités de déglaciation au Wisconsinien supérieur dans le Charlevoix occidental, Québec. [MSc memoir], Université du Québec à Montréal, 147 pp.

Fulton, R.J., 1984. Summary: Quaternary stratigraphy of Canada/Sommaire: stratigraphie Quaternaire au Canada. In: Fulton, R.J. (Ed.), Quaternary Stratigraphy of Canada, 1–5 contribution to IGCP project 24, Geological Survey of Canada, Paper, 84–10.

Fulton, R.J., 1986a. Surficial geology, Red Wine River, Labrador, Newfoundland. Geological Survey of Canada, Map, 1621A, scale 1:500,000.

Fulton, R.J., 1986b. Surficial Geology, Cartwright, Labrador, Newfoundland. Geological Survey of Canada, Map, 1620A, scale 1:500,000.

Fulton, R.J. (Ed.), 1989. Quaternary geology of Canada and Greenland, Geological Survey of Canada (Ottawa). 839 p., 5 maps: Late Wisconsinan and Holocene retreat of the Laurentide Ice Sheet, by A.S. Dyke and V.K. Prest, GSC map 1702A:- Paleogeography of northern North America 18,000–12,000 years ago, GSC map 1703A (3 sheets):- Status of Quaternary geology mapping in Canada with bibliography, GSC map 1704A.

Fulton, R.J., 1995. Matériaux superficiels du Canada. Geological Survey of Canada, Map, 1880A.

Fulton, R.J., Hodgson, D.A., 1979. Wisconsin glacial retreat, Southern Labrador. Current Research, Part C, Geological Survey of Canada, Paper 79-1C, 17–21.

Gadd, N.-R., Lasalle, P., MacDonald, B.C., Shilts, W.W., Dionne, J.-C., 1972a. Géologie et géomorphologie du Quaternaire dans le sud du Québec. In: 24ème congrès international de géologie, Montréal, livret-guide d'excursion C-44, 74 pp.

Gadd, N.R., Lasalle, P., MacDonald, B.C., Shilts, W.W., 1972b. Deglaciation of Southern Quebec. Commission Géologique du Canada, paper, 71–47, 19 pp., carte, 10–1971.

Gagnon, R., Payette, S., 1981. Fluctuations holocènes de la limite des forêts de mélèze, rivière aux Feuilles, Québec nordique: une analyse macrofossile en milieu tourbeux. Géogr. Phys. Quatern. 35, 57–72.

Gangloff, P., 1983. Les Fondements géomorphologiques de la Théorie des Paléonunataks: Le Cas des Monts Torngats. Z. Geomorphol. 47, 109–136.

Genes, A.N., Newman, W.A., Brewer, T.B., 1981. Late Wisconsinan glaciation models of northern Maine and adjacent Canada. Quatern. Res. 16, 48–65.

Govare, É., 1995. Paléoenvironnements de la région de Charlevoix, Québec. PhD de l'Université de Montréal, 429 pp.

Grant, D.R., 1989. Quaternary geology of the Atlantic Appalachian region of Canada. In: Fulton, R.J. (Ed.), Quaternary Geology of Canada and Adjacent Greenland: Geological Survey of Canada, 1, 393–440, (Geological Society of America, Geology of North America, K-1).

Grant, D.R., 1992. Quaternary geology of St. Anthony–Blanc Sablon area, Newfoundland and Quebec. Geological Survey of Canada, Memoir, 427, 60 pp.

Gratton, D., Gwyn, Q.H.J., Dubois, J.M., 1984. Les paléoenvironnements sédimentaires au Wisconsinien moyen et supérieur, île d'Anticosti, Golfe du Saint-Laurent, Québec. Géogr. Phys. Quatern. 38, 229–242.

Gray, J.T., Lauriol, B., 1985. Dynamics of the Late Wisconsin Ice Sheet in the Ungava Peninsula interpreted from geomorphological evidence. Arct. Alpine Res. 17, 289–310.

Gray, J., de Boutray, B., Hillaire-Marcel, C., Lauriol, B., 1980. Postglacial emergence of the west coast of Ungava Bay, Quebec. Arct. Alpine Res. 12, 19–30.

Grayson, J.T., 1956. The post-glacial history of vegetation and climate in the Labrador–Quebec Region as determined by Palynology. Unpublished PhD thesis, University of Michigan, 252 pp.

Grootes, M., Stuiver, M., White, J.W.C., Johnsen, S., Jouzel, J., 1993. Comparison of oxygen isotope records from the GISP2 and GRIP Greenland ice cores. Nature 366, 552–554.

Hardy, L., 1970. Géomorphologie glaciaire et post-glaciaire de Saint-Siméon à Saint-François d'Assises. Maîtrise de l'Université Laval, 112 pp.

Hardy, L., 1976. Contribution à l'étude géomorphologique de la portion québécoise des basses terres de la Baie James. PhD de l'Université McGill, 264 pp.

Hardy, L., 1977. La déglaciation et les épisodes lacustres et marins sur le versant québécois des basses terres de la Baie James. Géogr. Phys. Quatern. 31, 261–273.

Harrison, D.A., 1963. The tilt of the abandoned lake shorelines in the Wabush-Shabogamo Lake area, Labrador. Geographical Studies in Labrador, Annual Report 1961–1962; McGill Sub-Arctic Research Paper 15, pp. 14–22.

Harrison, J.E., 1972. Quaternary Geology of the North Bay-Mattawa Region. Geological Survey of Canada, Paper 71–26, 37 pp.

Henderson, E.P., 1959. A glacial study of central Quebec–Labrador. Geological Survey of Canada Bulletin 50, 94 pp.

Hétu, B., 1998. La déglaciation de la région de Rimouski, Bas-Saint-Laurent (Québec): Indices d'une récurrence glaciaire dans la Mer de Goldthwait entre 12 400 et 12 000 BP. Géogr. Phys. Quatern. 52, 325–347.

Hillaire-Marcel, C., 1976. La déglaciation et le relèvement isostatique sur la côte est de la baie d'Hudson. Cahiers de géographie du Québec 20, 185–220.

Hillaire-Marcel, C., 1979. Les mers post-glaciaires du Québec: quelques aspects. thèse de doctorat d'état non publiée, Université Pierre et Marie Curie, Paris VI, 1, 293 pp., 2, 249 pp.

Hillaire-Marcel, C., Grant, D.R., Vincent, J.-S., 1980. Comment and reply on "Keewatin Ice Sheet-re-evaluation of the traditional concept of the Laurentide Ice Sheet" and "Glacial erosion and ice sheet divides, northeastern Laurentide Ice Sheet, on the basis of the distribution of limestone erratics". Geology 8, 466–468.

Hillaire-Marcel, C., Occhietti, S., Vincent, J.-S., 1981. Sakami moraine, Quebec: a 500-km-long-moraine without climatic control. Geology 9, 210–214.

Hodgson, D.A., Fulton, R.J., 1972. Site description, age and significance of a shell sample from the mouth of the Michael River, 30 km south of Cape Harrison, Labrador. Report of Activities, Part B, Geological Survey of Canada, Paper 72-1B, pp. 102–105.

Hughes, O.L., 1964. Surficial geology, Nichicun-Kaniapiskau map-area, Quebec. Geol. Surv. Can. Bull. 106, 20 pp.

Hughes, O.L., 1965. Surficial geology of part of the Cochrane District, Ontario, Canada. In: Wright, H.E., Frey, D.G. (Eds.), International Studies on the Quaternary, Geological Society of America, Special Paper, 84, pp. 535–565.

Hughes, T., Borns, H.W., Fastook, J.L., Hyland, J.S., Kite, J.S., Lowell, T.V., 1985. Models of glacial reconstruction and deglaciation applied to Maritime Canada and New England. In: Borns Jr., H.W., Lasalle, P., Thompson, W.B. (Eds.), Late Pleistocene History of Northeastern New England and Adjacent Quebec, Geological Society of America, Special Paper, 197 139–150.

Ives, J.D., 1957. Glaciation of the Torngat Mountains, Northern Labrador. Arctic 10, 67–87.

Ives, J.D., 1958. Glacial geomorphology of the Torngat Mountains, Northern Labrador. Geogr. Bull. 12, 47–75.

Ives, J.D., 1959. Glacial drainage channels as indicators of late-glacial conditions in Labrador–Ungava: a discussion. Cahiers de géographie du Québec 3, 57–72.

Ives, J.D., 1960a. The deglaciation of Labrador–Ungava, an outline. Cahiers de géographie du Québec 3, 323–343.

Ives, J.D., 1960b. Former ice-dammed lakes and the deglaciation of the middle reaches of the George River, Labrador–Ungava. Geographical Bulletin 14, 44–70.

Ives, J.D., 1960c. Glaciation and deglaciation of the Helluva Lake area, central Labrador Ungava. Geographical Bulletin 15, 46–64.

Ives, J.D., 1968. Late-Wisconsin events in Labrador–Ungava: an interim commentary. Canadian geographer 12, 192–203.

Ives, J.D., 1976. The Sagiek moraines of northern Labrador: a commentary. Arctic and Alpine Research 8, 403–408.

Ives, J.D., 1978. The maximum extent of the Laurentide Ice Sheet along the east coast of North America during the last glaciation. Arctic 31 (1), 24–53.

Ives, J.D., 1979. A proposed history of permafrost development in Labrador–Ungava. Géographie physique et Quaternaire 33, 233–244.

Jackson, G.D., 1960. Belcher Islands, Northwest Territories. Geological Survey of Canada, Paper 60–20, 13 pp.

Johnsen, S.J., Clausen, H.B., Dansgaard, W., Fuhrer, K., Gundestrup, N., Hammer, C.U., et al., 1992. Irregular glacial interstadials recorded in a new Greenland ice core. Nature 359, 312–313.

Johnson Jr., J.P., 1969. Deglaciation of the central Nain-Okak Bay section of Labrador. Arctic 22, 373–394.

Josenhans, H.W., Zevenhuizen, J., Klassen, R.A., 1986. The Quaternary geology of the Labrador Shelf. Can. J. Earth Sci. 23, 1190–1213.

Kaufman, D.S., Miller, G.H., Stravers, J.A., Andrews, J.T., 1993. Abrupt early Holocene (9.9–9.6 ka) ice stream advance at the mouth of Hudson Strait, Arctic Canada. Geology 21, 1063–1066.

King, G.A., 1985. A standard method for evaluating radiocarbon dates of local deglaciation: application to the deglaciation history of southern Labrador and adjacent Quebec. Géogr. Phys. Quatern. 39, 163–182.

Kirby, R.P., 1961a. Deglaciation in central Labrador–Ungava as interpreted from glacial deposits. Geographical bulletin 16, 4–23.

Kirby, R.P., 1961b. Movements of ice in Central Labrador–Ungava. Cahiers de géographie du Québec 5 (10), 205–218.

Klassen, R.A., Thompson, F.J., 1993. Glacial history, drift composition, and mineral exploration, central Labrador. Geol. Surv. Can. Bull. 435, 76.

Klassen, R.A., Matthews, J.V.J., Mott, R.J., Thompson, F.J., 1988. The stratigraphic and paleobotanical record of Interglaciation in the Wabush region of western Labrador (abstract). In: Climatic Fluctuations and Man 3, Annual Meeting of the Canadian Committee on Climatic Fluctuations. January 28–29, Ottawa, pp. 24–26.

Kleman, J., Borgström, I., Hättestrand, C., 1995. Evidence for a relict glacial landscape in Quebec–Labrador. Palaeogeogr. Palaeoclimatol. Palaeoecol. 111, 217–228.

Lajeunesse, P., 2008. Early Holocene deglaciation of the eastern coast of Hudson Bay. Geomorphology 99, 341–352.

Lajeunesse, P., Allard, M., 2002. Sedimentology of an ice-contact glaciomarine fan complex, Nastapoka Hills, eastern Hudson Bay, northern Québec. Sed. Geol. 152, 201–220.

Lajeunesse, P., Allard, M., 2003a. The Nastapoka drift belt, eastern Hudson Bay: implications of a stillstand of the Quebec–Labrador ice margin in the Tyrrell Sea at 8 ka BP. Can. J. Earth Sci. 40, 65–76.

Lajeunesse, P., Allard, M., 2003b. Late Quaternary deglaciation, glaciomarine sedimentation and glacioisostatic recovery in the Rivière Nastapoka area, eastern Hudson Bay, northern Québec. Géogr. Phys. Quatern. 57, 65–83.

Lajeunesse, P., St-Onge, G., 2008. The subglacial origin of Lake Agassiz–Ojibway final outburst flood. Nat. Geosci. 1, 184–188.

Lajeunesse, P., St-Onge, G., Duchesne, M., Occhietti, S., 2008. Late-Wisconsinan grounding lines of the Laurentide Ice Sheet margin in northwestern Gulf of St. Lawrence. In: Quebec 2008: 400 Years of Discoveries. Joint Meeting of the Geological Association of Canada, Mineralogical Association of Canada, Society of Economic Geologists and the Society for Geology Applied to Mineral Deposits May 26–28, 2008. Québec City, QC.

Lamarche, R.Y., 1971. Northward moving ice in the Thetford Mines area of southerne Quebec. Am. J. Sci. 271, 383–388.

Lamarche, R.Y., 1974. Southeastward, northward and westward ice movement in the Asbestos area of southern Quebec. Geol. Soc. Am. Bull. 85, 465–470.

Lamothe, M., 1977. Les dépôts meubles de la région de Saint-Faustin-Saint-Jovite, Québec. Cartographie, sédimentologie et stratigraphie. Maîtrise de l'Université du Québec à Montréal, 118 pp.

Lamothe, M., 1989. A new framework for the Pleistocene stratigraphy of the central St. Lawrence lowland, southern Québec. Géogr. Phys. Quatern. 43, 119–129.

Lamothe, M., Parent, M., Shilts, W.W., 1992. Sangamonian and Early Wisconsinan events in the St. Lawrence Lowland and Appalachians of southern Québec, Canada. In: Clark, P.U., Lea, D.D. (Eds.), The Last Inter-Glacial Transition in North America, Geological Society of America, Special Paper, 270, pp. 171–184.

Lanoie, J., 1995. Les écoulements glaciaires du Wisconsinien supérieur en Charlevoix occidental. [M.Sc. mémoire]: Université du Québec à Montréal, 83 pp.

LaSalle, P., Chapdelaine, C., 1990. Review of late-glacial and Holocene events in the Champlain and Goldthwait Seas areas and arrival of man in eastern Canada. In: Lasca, N.P., Donahue, J. (Eds.), Archeological Geology of North America, Geological Society of America, Centennial Special volume, 4, 1–19.

LaSalle, P., Elson, J.A., 1975. Emplacement of the Saint-Narcisse Moraine as a climatic event in eastern Canada. Quatern. Res. 5, 621–625.

LaSalle, P., Shilts, W.W., 1993. Younger Dryas-age readvance of Laurentide ice into the Champlain Sea. Boreas 22, 25–37.

LaSalle, P., Tremblay, G., 1978. Dépôts meubles Saguenay–Lac Saint-Jean. Ministère des Ressources naturelles, Rapport géologique, 191, 61 pp.

Lauriol, B., 1982. Géomorphologie quaternaire du sud de l'Ungava. Paléo-Québec 15, 174 pp.

Lauriol, B., Gray, J.T., 1983. Un lac glaciaire dans la région du lac Minto-Nouveau Québec. Journal Canadien des Sciences de la Terre 10, 1488–1492.

Lauriol, B., Gray, J.T., 1987. The decay and disappearance of the Late Wisconsinan Ice Sheet in the Ungava Peninsula, northern Quebec, Canada. Arct. Alpine Res. 19, 109–126.

Laverdière, C., 1967. Sur le lieu de fonte sur place de la calotte glaciaire de Scheffer. Géogr. Can. 11, 87–95.

Laverdière, C., 1969a. The Scheffer Ice-Sheet: a reply to Ives' comments. Can. Geogr. 13 (3), 269–283.

Laverdière, C., 1969b. Le retrait de la calotte glaciaire de Scheffer: du Témiscamingue au Nouveau-Québec. Rev. Géogr. Montréal 23, 233–246.

Laverdière, C., Bernard, C., 1969. Sur quelques néochronymes (Mer d'Iberville); dans Le vocabulaire de la géomorphologie glaciaire (Ve article). Rev. Géogr. Montréal 23, 355–358.

Laverdière, C., Guimont, P., 1982. Le réservoir du Caniapiscau, étude du milieu physique. Rapport interne, Société de développement de la baie James, 125 pp.

Laverdière, C., Mailloux, A., 1956. État de nos connaissances d'une transgression marine post-glaciaire dans les régions du haut Saguenay et du lac Saint-Jean. Rev. Can. Géogr. 10, 201–220.

Laymon, C., 1984. Glacial geology of western Hudson Strait with reference to Laurentide Ice Sheet dynamics (abstract). In: Abstracts with Program, Geological Society of America, 16, Annual Meeting, Reno, Nevada. 571.

Lebuis, J., David, P.P., 1977. La stratigraphie et les événements du quaternaire de la partie occidentale de la Gaspésie. Géogr. Phys. Quatern. 31, 275–296.

Lee, H.A., Eade, K.E., Heywood, W.W., 1960. Surficial geology, Sakami Lake (Fort George-Great Whale Area, New Quebec). Geological Survey of Canada, Map, 52–1959, scale 1:506,880.

Lewis, C.F.M., Blasco, S.M., Gareau, L., 2005. Glacial isostatic adjustment of the Laurentian Great Lakes Basin: using the empirical record of strandline deformation for reconstruction of Early Holocene paleolakes and discovery of a hydrologically closed phase. Géogr. Phys. Quatern. 59, 187–210.

Lewis, C.F.M., Anderson, T.W., Gareau, P.L., Karrow, P.F., Mott, R.J., Rodrigues, C.G., 2006. Outburst floods to Champlain Sea from glacial Lake Algonquin during the Younger Dryas cold event. In: Geological Association of Canada, Montréal Annual Meeting, Abstract, 3188.

Llibourty, L., 1965. Traité de glaciologie, Tome II, Glaciers Variations du climat Sols gelés. Masson, Paris, pp. 429–1040.

Locat, J., 1977. L'émersion des terres dans la région de Baie-des-Sables/Trois-Pistoles, Québec. Géogr. Phys. Quatern. 31, 297–306.

Løken, O.H., 1962a. On the vertical extent of glaciation in northeastern Labrador–Ungava. Can. Geogr. 6, 106–119.

Løken, O.H., 1962b. The late-glacial and postglacial emergence and the deglaciation of northernmost Labrador. Geogr. Bull. 17, 23–56.

Løken, O.H., 1964. A study of the late and postglacial changes of sea level in northernmost Labrador. Report to Arctic Institute of North America, Montréal, 74 pp.

Løken, O.H., 1978. Postglacial tilting of Akpatok Island, Northwest Territories. Can. J. Earth Sci. 15, 1547–1553.

Lortie, G., Martineau, G., 1987. Les systèmes de stries glaciaires dans les Appalaches du Québec. Québec, ministère de l'Énergie et des Ressources, DV 85–10, 45 pp.

Low, A.P., 1896. Report on exploration in the Labrador Peninsula along the East Main, Koksoak, Hamilton, Manicouagan and portions of other rivers in 1892-93-94-95. Geological Survey of Canada, Annual Report, 1895, 8, Part L, 387 pp.

Lowell, T.V., 1985. Late Wisconsin ice-flow reversal and deglaciation, northwestern Maine. In: Borns Jr., H.W., Lasalle, P., Thompson, W.B. (Eds.), Late Pleistocene History of Northeastern New England and Adjacent Quebec, Geological Society of America, Special Paper, 197 71–83.

Lowell, T.V., Kite, J.S., 1986. Glaciation style of northwestern Maine. In: Kite, J.S., Lowell, T.V., Thompson, W.B. (Eds.), Contributions to the Quaternary of Northern Maine and Adjacent Canada, Maine Geological Survey Bulletin, 37 53–68.

Lowell, T.V., Becker, D.A., Calkin, E., 1986. Quaternary stratigraphy in northwestern Maine: a progress report. Géogr. Phys. Quatern. 40, 71–84.

Lowell, T., Kite, J.S., Calkin, E., Halter, E.F., 1990. Analysis of small-scale erosional data and a sequence of late Pleistocene flow reversal, northern New England. Geol. Soc. Am. Bull. 102, 74–85.

Lowell, T.V., Larson, G.J., Hughes, J.D., Denton, G.H., 1999. Age verification of the lake Gribben forest bed and the Younger Dryas advance of the Laurentide Ice Sheet. Can. J. Earth Sci. 36, 383–393.

Maclean, B., Vilks, G., Deonarine, B., 1992. Depositional environments and history of Late Quaternary sediments in Hudson Strait and Ungava Bay: further evidence from seismic and biostratigraphic data. Géogr. Phys. Quatern. 46 (3), 311–329.

Martineau, G., Corbeil, P., 1983. Réinterprétation d'un segment de la moraine de Saint-Antonin, Québec. Géogr. Phys. Quatern. 37, 217–221.

Matthew, E.M., 1961. Deglaciation of the George River basin Labrador–Ungava. In: Andrews, J.T., Matthew, E.M. (Eds.), Field Research in Labrador–Ungava, Annual Report 1959–1960, McGill Sub-Arctic Research Paper, 11 29–45 (Geographical Branch, Department of Mines and Technical Surveys, Ottawa, Geographical Paper 29, pp. 17–29.).

Matthews, B., 1966. Radiocarbon dated postglacial land uplift in Northern Ungava, Canada. Nature 211 (5054), 1164–1166.

Matthews, B., 1967. Late Quaternary land emergence in northern Ungava, Quebec. Arctic 20, 176–202.

McClintock, P.P., Stewart, D.P., 1965. Pleistocene geology of the St-Lawrence lowland. In: New York State Museum and Science Service, Bulletin, 394 152.

McCoy, W.D., 1983. Holocene glacier fluctuations in the Torngat Mountains, Northern Labrador. Géogr. Phys. Quatern. 37, 211–216.

McDonald, B.C., 1966. Surficial geology, Richmond-Dudswell, Quebec. Geological Survey of Canada, carte, 4–1966.

McDonald, B.C., 1967. Surficial geology. Sherbrooke-Orford-Memphremagog, Quebec. Geological Survey of Canada, carte, 5–1966.

McDonald, B.C., 1968. Deglaciation and differential postglacial rebound in the Appalachian region of southeastern Quebec. J. Geol. 76, 664–677.

McDonald, B.C., 1969. Surficial Geology of La Patrie-Sherbrooke area, Quebec, including Eaton River watershed. Geological Survey of Canada, Paper, 67–52, 21 pp.

McDonald, B.C., Shilts, W.W., 1971. Quaternary stratigraphy and events in southeastern Quebec. Geol. Soc. Am. Bull. 82, 682–698.

McNeely, R., Dyke, A.S., Southon, J.R., 2006. Canadian Marine Reservoir Ages, Preliminary Data Assessment. Geological Survey of Canada, Open File 5049, pp. 3, 1 CD-ROM.

Miller, G.H., Andrews, J.T., Stravers, J.A., Laymon, C.A., 1988. The Cockburn Readvance in northeastern Canada: a Younger Dryas style regional climatic oscillation during the last deglaciation (abstract). In: Program and Abstracts, American Quaternary Association, 10th Biennial Meeting, Amherst, 139.

Naldrett, D.L., 1988. The late glacial-early glaciomarine transition in the Ottawa Valley: evidence for a glacial lake? Géogr. Phys. Quatern. 42, 171–179.

Newman, W.A., Genes, A.N., Brewer, T., 1985. Pleistocene geology of northeastern Maine. In: Borns Jr., H.W., Lasalle, P., Thompson, W.B. (Eds.), Late Pleistocene History of Northeastern New England and Adjacent Quebec, Geological Society of America, Special Paper, 197, 59–71.

Norman, G.W.H., 1939. The south-eastern limit of Glacial Lake Barlow-Ojibway in the Mistassini Lake region, Quebec. Transactions of the Royal Society of Canada Section, IV, 59–65.

Occhietti, S., 1980. Le Quaternaire de la région de Trois-Rivières-Shawinigan, Québec. Contribution à la paléo-géographie de la vallée moyenne du Saint-Laurent et corrélations stratigraphiques: Université du Québec, Paléo-Québec, 10, 227 pp.

Occhietti, S., 1982. Synthèse lithostratigraphique et paléo-environnements quaternaires au Québec méridional. Hypothèse d'un centre d'englacement wisconsinien au Nouveau-Québec. Géogr. Phys. Quatern. 36 (1–2), 15–49.

Occhietti, S., 1983. Laurentide ice sheet: oceanic and climatic implications. Palaeogeogr. Palaeoclimatol. Palaeoecol. 44, 1–22.

Occhietti, S., 1989. Quaternary geology of St. Lawrence Valley and adjacent Appalachian subregion. In: Fulton, R.T. (Ed.), Quaternary Geology of Canada and Greenland, Geological Survey of Canada, Geology of Canada, 1, pp. 350–379, (also: Geological Society of America, The Geology of North America, K-1).

Occhietti, S., 2001. Deglaciation of the middle Estuary and Charlevoix: an overview. In: Bhiry, N., Dionne, J.-C., Clet, M., Occhietti, S., Rondot, J. (Eds.), Stratigraphy of the Pleistocene units on land and below the St. Lawrence Estuary, and deglaciation pattern in Charlevoix, 64th annual Reunion of the North East Friends of the Pleistocene, Québec City, QC Canada, Field Guide, pp.1–20, Chapter 1.

Occhietti, S., 2007. The Saint-Narcisse morainic complex and early Younger Dryas events on the southern margin of the Laurentide Ice Sheet. Géogr. Phys. Quatern. 61, 5–34.

Occhietti, S., Richard, P.J.H., 2003. Effet réservoir sur les âges ^{14}C de la Mer de Champlain à la transition Pléistocène–Holocène: révision de la chronclogie de la déglaciation au Québec méridional. Géogr. Phys. Quatern. 57 (2–3), 115–138.

Occhietti, S., Long, B., Boespflug, X., Sabeur, N., 1995. Séquence de la transition Illinoien–Sangamonien: forage IAC-91 de l'île aux Coudres, estuaire moyen du Saint-Laurent, Québec. Can. J. Earth Sci. 32, 1954–1967.

Occhietti, S., Balescu, S., Lamothe, M., Clet, M., Cronin, T., Ferland, P., et al., 1996. Late Stage 5 glacio-isostatic sea in the St. Lawrence Valley, Canada and United States. Quatern. Res. 45, 128–137.

Occhietti, S., Chartier, M., Hillaire-Marcel, C., Cournoyer, M., Cumbaa, S.L., Harington, C.R., 2001. Paléoenvironnements de la Mer de Champlain dans la région de Québec, entre 11 300 et 9750 BP: le site de Saint-Nicolas. Géogr. Phys. Quatern. 55, 23–46.

Occhietti, S., Govare, É., Klassen, R., Parent, M., Vincent, J.-S., 2004. Late Wisconsinan–Early Holocene deglaciation of Québec–Labrador. In: Ehlers, J., Gibbard, P.L. (Eds.), Quaternary Glaciations—Extent and Chronology, Part II. Elsevier Publisher, North America, pp. 243–273.

Pagé, P., 1977. Les dépôts meubles de la région de Saint-Jean-de-Matha—Sainte-Émilie-de-l'Énergie, Québec, cartographie, sédimentologie et stratigraphie. Maîtrise de l'Université du Québec à Montréal, 118 pp.

Paradis, S.J., Bolduc, A.M., 1999. Mouvement glaciaire vers le nord sur le piémont laurentien dans la région de Québec. Geological Survey of Canada, Current Research, 1999-D, 1–7.

Parent, M., 1978. Géologie du quaternaire de la région de Stoke-Watopek, Québec. Thèse de maîtrise, Université de Sherbrooke, département de géographie, 206 pp.

Parent, M., 1987. Late Pleistocene stratigraphy and events in the Asbestos-Valcourt region, southern Quebec. PhD thesis, Western Ontario University, London, 320 pp.

Parent, M., Occhietti, S., 1988. Late Wisconsinan deglaciation and Champlain Sea invasion in the St. Lawrence valley, Québec. Géogr. Phys. Quatern. 42, 215–246.

Parent, M., Occhietti, S., 1999. Late Wisconsinian deglaciation and glacial lake development in the Appalachians of southeastern Québec. Géogr. Phys. Quatern. 53, 117–135.

Parent, M., Paradis, S.J., 1994. Géologie des formations superficielles Région de la Petite rivière de la Baleine, Québec nordique. Geological Survey of Canada, Open File, 2643, 48 pp., 2 maps.

Parent, M., Paradis, S.J., 1997. Mouvements glaciaires polyphasés dans la région d'Ashuanipi (23 F, 23 C ET 23 G/W), Moyen-Nord québécois. Ministère des Ressources naturelles du Québec, Séminaire d'information sur la recherche géologique, Programme et résumés, DV 97–03, 34.

Parent, M., Paradis, S.J., Boisvert, É., 1995. Ice flow patterns and glacial transport in the eastern Hudson Bay region: implications for the late Quaternary dynamics of the Laurentide Ice Sheet. Can. J. Earth Sci. 32 (12), 2057–2070.

Parent, M., Paradis, J.S., Doiron, A., 1996. Palimpsest glacial dispersal trains and their significance for drift prospecting. J. Geochem. Explor. 56, 123–140.

Parent, M., Occhietti, S., Lamothe, M., 1998. The Late Pleistocene record of eastern Canada—a history of shifting glacial outflow centers, surging ice streams and extensive marine incursions. In: Geological Association of Canada/Mineralogical Association of Canada/Association professionelle de géologues et géophysiciens du Québec, Joint Meeting, May 18–19–20, Québec 1998. Abstract volume, 141.

Parry, J.T., 1963. The Laurentians: a study in geomorphological development. Unpublished PhD thesis, McGill University, Montreal, 222 pp.

Perrault, G., 1955. Geology of the western margin of the Labrador Trough, Part 1: general geology of the western margin of the Labrador Trough, Part II: the Sokoman Iron Formation, Part III: some data on iron silicate minerals occurring in iron rich sedimentary rocks. Unpublished PhD thesis, University of Toronto, Toronto, 300 pp.

Peterson, J.A., 1965. Deglaciation of the White Gull Lake area, Labrador–Ungava. Cahiers de géographie du Québec 9, 183–196.

Prest, V.K., 1969. Retreat of Wisconsin and recent ice in North America. Geological Survey of Canada, Map, 1257A, scale 1:5,000,000.

Prest, V.K., 1970. Quaternary geology of Canada. In: Douglas, R.J.W. (Ed.), fifth ed. Geology and Economic Minerals of Canada, Geological Survey of Canada, Economic Geology Report, vol. 1. 676–764.

Prest, V.K., 1977. General stratigraphic framework of the Quaternary in Eastern Canada. Géogr. Phys. Quatern. 31, 7–14.

Prest, V.K., 1984. The Late Wisconsinan Glacier Complex. In: Fulton, R.J. (Ed.), Quaternary Stratigraphy of Canada—A Canadian Contribution to IGCP Project 24, Geological Survey of Canada, Paper, 84–10 21–36.

Prest, V.K., Hode-Keyser, J., 1977. Geology and Engineering Characteristics of Surficial Deposits, Montreal Island and Vicinity, Quebec. Geological Survey of Canada, Paper 75–27.

Prest, V.K., Grant, D.R., Rampton, V.N., 1968. Glacial map of Canada. Geological Survey of Canada, Map, 1253A, scale 1:5,000,000.

Prichonnet, G., 1984. Réévaluation des systèmes morainiques du sud du Québec. Commission géologique du Canada. Commission géologique du Canada, étude 83–29, 20 pp.

Rappol, M., 1989. Glacial history and stratigraphy of the northwestern New Brunswick. Géogr. Phys. Quatern. 43, 191–206.

Rappol, M., 1993. Ice flow and glacial transport in Lower St. Lawrence, Québec. Geological Survey of Canada, Paper 90–19, 28 pp.

Reimer, P.J., Baillie, M.G.L., Bard, E., Bayliss, A., Beck, J.W., Blackwell, P.G., et al., 2009. CALIB 6.1. Radiocarbon 51, 1111–1150.

Richard, P.J.H., 1981. Paléophytogéographie postglaciaire en Ungava, par l'analyse pollinique. Collection PALÉO-QUÉBEC 13, p. 153.

Richard, P.J.H., 1995. Le couvert végétal du Québec–Labrador il y a 6000 ans BP: essai. Géogr. Phys. Quatern. 49 (1), 117–140.

Richard, P.J.H., Occhietti, S., 2005. ^{14}C chronology for ice retreat and inception of Champlain Sea in the St. Lawrence Lowlands, Canada. Quatern. Res. 63, 353–358.

Richard, P.J.H., Larouche, A., Bouchard, M.A., 1982. Âge de la déglaciation finale et histoire postglaciaire de la végétation dans la partie centrale du Nouveau-Québec. Géogr. Phys. Quatern. 36, 63–90 R.-A..

Richard, P.J.H., Veillette, J.-J., Larouche, A.C., Hétu, B., Gray, J.T., Gangloff, P., 1997. Chronologie de la déglaciation en Gaspésie: nouvelles données et implications. Géogr. Phys. Quatern. 51 (2), 163–184.

Robert, F., 2001. Photo-interprétation à grande échelle et système d'information géographique: outils de reconstitution du retrait glaciaire dans les Laurentides Application aux régions de La Tuque et de la Gatineau. MSc thesis, Université du Québec à Montréal, 129 pp.

Rodrigues, C.G., 1992. Successions of invertebrate microfossils and the late Quaternary deglaciation of the Central St Lawrence Lowland, Canada and United States. Quatern. Sci. Rev. 11, 503–534.

Rogerson, R.J., 1977. Glacial geomorphology and sediments of the Porcupine Strands, Labrador, Canada. Unpublished PhD thesis, Macquarie University, 276 pp.

Rogerson, R.J., Bell, T., 1986. The Late-Wisconsin maximum in the Nachavak Fiord area of northern Labrador (abstract). In: Abstracts, 15th Arctic Workshop, Boulder, pp. 57–60.

Rondot, J., 1974. L'épisode glaciaire de Saint-Narcisse dans Charlevoix, Québec. Rev. Géogr. Montréal 28, 375–388.

Ross, M., Parent, M., 2007. Quaternary stratigraphy and late glacial dynamics in the Lower Laurentians and adjacent lowlands. In: CANQUA Biennial Meeting, Ottawa, June 2007, Fieldtrip Guidebook. 39 pp.

Ross, M., Parent, M., Benjumea, B., Hunter, J., 2006. The Late Quaternary stratigraphic record northwest of Montreal: Regional ice sheet dynamics, ice stream activity and early deglacial events. Can. J. Earth Sci. 43, 461–485.

Saarnisto, M., Saarinen, T., 2001. Deglaciation chronology of the Scandinavian ice sheet from the Lake Onega Basin to the Salpausselkä End Moraines. Glob. Planet. Change 31, 387–405.

Sauramo, M., 1929. The Quaternary geology of Finland. Bulletin de la Commission Géologique de Finlande 86, 110.

Sauvé & LaSalle, 1968. Notes sur la géologie glaciaire de la région de Manic 2. Naturaliste Can. 95, 1293–1300.

Shilts, W.W., 1970. Pleistocene Geology of the LacMe-Gantic region, Southeastern Quebec, Canada. PhD Université de Syracuse, 154 pp.

Shilts, W.W., 1980. Flow patterns in the central North American ice sheet. Nature 286, 213–218.

Shilts, W.W., 1981. Surficial geology of the lac Megantic area, Quebec. Commission géologique du Canada, mémoire, 397, 102 pp., carte 1497A.

Shilts, W.W., 1997. Erosional and depositional stratigraphy of the Appalachians of southeastern Quebec. In: 8th Biennial Congress Canadian Quaternary Association (CANQUA), Montréal, Québec, Canada, Programme—Abstracts, 72.

Shilts, W.W., Smith, S.L., 1986. Stratigraphic setting of buried gold-bearing sediments, Beauceville area, Quebec. Geological Survey of Canada, Current Research, Part B, 86-1A, 271–278.

Simard, J., 2003. Mode de retrait glaciaire dans les Laurentides au début de l'Holocène: transect de 500 km entre Manouane et le Témiscamingue. MSc thesis, Université du Québec à Montréal, 157 pp.

Simard, J., Occhietti, S., Robert, F., 2003. Retrait de l'inlandsis sur les Laurentides au début de l'Holocène : transect de 600 km entre le Saint-Maurice et le Témiscamingue, Québec. Géographie physique et Quaternaire 57 (2–3), 189–204.

Smith, P.A.W., 1969. Glacial geomorphology of the Saglek Fjord area of northeast Labrador; in Field Research in Labrador–Ungava. McGill Sub-Arctic Research, Paper 24, pp. 115–123.

Stea, R.R., Piper, D.J.W., Fader, G.B.J., Boyd, R., 1998. Wisconsinan glacial and sea-level history of Maritime Canada and the adjacent continental shelf: a correlation of land and sea events. Geol. Soc. Am. Bull. 110, 821–845.

Syverson, K.M., 1995. The ability of ice-flow indicators to record complex, historic deglaciation events, Burroughs Glacier, Alaska. Boreas 24, 232–244.

Thompson, W.B., Fowler, B.K., Dorion, C.C., 1999. Deglaciation of the northwestern White Mountains, New Hampshire. Géogr. Phys. Quatern. 53 (1), 59–77.

Tomlinson, R.F., 1963. Pleistocene evidence related to glacial theory in northeastern Labrador. Can. Geogr. 7, 83–90.

Tremblay, G., 1971. Glaciation et déglaciation dans la région Saguenay–Lac Saint-Jean, Québec, Canada. Cahiers de géographie du Québec 15, 467–494.

Tremblay, G., 1977. Géologie du quaternaire de la région de Rawdon—Laurentides—Shawbridge—Sainte-Agathe. Ministère des Richesses naturelles, DP-551, 1 map.

Veillette, J.J., 1983. Déglaciation de la vaIIée supérieure de l'Outaouais, le lac Barlow et le sud du lac Ojibway, Québec. Géographie physique et Quaternaire 37, 67–84.

Veillette, J.J., 1986. Former southwesterly ice flows in Abitibi-Timiskaming region: implications for the configuration of the Late Wisconsinan ice sheet. Can. J. Earth Sci. 23, 1724–1741.

Veillette, J.J., 1988. Déglaciation et évolution des lacs proglaciaires Post-Algonquin et Barlow au Témiscamingue, Québec et Ontario. Géogr. Phys. Quatern. 42, 7–31.

Veillette, J.J., 1997a. Surficial geology, Angliers, Ville-Marie & Lac Simard, Quebec. Commission géologique du Canada. Open File, 871, 3 maps.

Veillette, J.J., 1997b. Le Rôle d'un courant de glace tardif dans la déglaciation de la Baie James. Géogr. Phys. Quatern. 51 (2), 141–161.

Veillette, J.J., Roy, M., 1995. The spectacular cross-striated outcrops of James Bay, Quebec. Current Research, Geological Survey of Canada, 1995-C, pp. 243–248.

Veillette, J.J., Dyke, A.S., Roy, M., 1999. Ice-flow evolution of the Labrador Sector of the Laurentide Ice Sheet: a review, with new evidence from northern Quebec. Quatern. Sci. Rev. 18, 993–1019.

Vilks, G., Mudie, J., 1978. Early deglaciation of the Labrador Shelf. Science 202, 1181–1183.

Vincent, J.-S., 1977. Le Quaternaire récent de la région du cours inférieur de La Grande Rivière, Québec. Commission géologique du Canada, Étude, 76–19, 20 pp.

Vincent, J.S., 1989. Quaternary geology of the southeastern Canadian Shield. In: Fulton, R.T. (Ed.), Quaternary Geology of Canada and Greenland, Geological Survey of Canada, Geology of Canada, 1, pp. 249–275 (also: Geological Society of America, The Geology of North America, K-1).

Vincent, J.-S., Hardy, L., 1977. L'évolution et l'extension des lacs glaciaires Barlow et Ojibway en territoire québécois. Géogr. Phys. Quatern. 31, 57–372.

Vincent, J.-S., Hardy, L., 1979. The evolution of glacial lakes Barlow and Ojibway, Quebec and Ontario. Geol. Surv. Can. Bull. 316, 18 pp.

Wilson, M.E., 1918. Timiskaming County, Quebec. Geological Survey of Canada, Memoir 103, 197 pp.

Wilson, J.T., 1938. Glacial geology of part of northwestern Quebec. R. Soc. Can. Trans. 4 (32), 49–59.

Wilson, J.T., Falconer, G., Mathews, W.H., Prest, V.K., 1958. Glacial Map of Canada. Geological Association of Canada, scale 1:3 801 600.

Zevenhuizen, J., 1996. Late Quaternary and surficial geology of southeaster Hudson Bay. MSc thesis, Dalhousie University, 215 pp.

Chapter 48

The Appalachian Glacier Complex in Maritime Canada

Rudolph R. Stea[1,*], Allen A. Seaman[2], Toon Pronk[2], Michael A. Parkhill[3], Serge Allard[2] and Daniel Utting[4]

[1] Stea Surficial Geology Services, 851 Herring Cove Road, Halifax, Nova Scotia, Canada B3R 1Z1
[2] New Brunswick Department of Natural Resources, Fredericton, New Brunswick, Canada E3B 5H1
[3] New Brunswick Department of Natural Resources, Bathurst, New Brunswick, Canada E2A 3Z1
[4] Nova Scotia Department of Natural Resources, 1701 Hollis Street, Halifax, Nova Scotia, Canada B3J 2T9
*Correspondence and requests for materials should be addressed to Rudolph (Ralph) Stea. E-mail: ralphstea@eastlink.ca

48.1. INTRODUCTION

The Canadian Maritime Provinces include New Brunswick (NB), Nova Scotia (NS) and Prince Edward Island (PEI) and lie east of the U.S. State of Maine, and south of the Province of Québec (Fig. 48.1). The main topographic features in the region are the Laurentian Channel, a prominent 500-m deep U-shaped submarine valley within the Gulf of St. Lawrence, and highland promontories, including the Chic-Choc Mountains (>1000 m a.s.l.) of Québec, the Miramichi Highlands (>500 m a.s.l.) of NB and the Cape Breton Highlands (>500 m a.s.l.) of NS. The purpose of this chapter is to present a comprehensive review and update of the Quaternary history of the Maritime Provinces from the pre-Wisconsinan through to the last glacial pulse during the Younger Dryas. The data for this chapter have been compiled from field data collected in the region over the past two centuries and synthesised into provincial-scale glacial landform maps (Prest, 1973; Rampton et al., 1984; Stea et al., 1992a). Regional summary papers by Rampton et al. (1984), Grant and King (1984), Grant (1989) and King (1996) are considered benchmark works upon which this work is based.

At the turn of the past century, when the first geological maps were being completed in the Maritimes, a controversy emerged about the nature of the last major glaciation in the area that still resonates today. Was the ice local, originating in Appalachian upland areas or was it part of the Laurentide Ice Sheet (LIS) with its centre in Québec (Chalmers, 1895; Prest, 1896; Bailey, 1898; Goldthwait, 1924). These two opposing models of glaciation for the Maritimes have come to be known as the 'minimum' and 'maximum' models. The maximum model invokes the LIS covering the entire region during the last glacial maximum (LGM) (~20 ka) (Flint, 1971; Mayewski et al., 1981; Dyke et al., 2002). The minimum model evolved from the recognition of autonomous local glaciers in Maritime Canada (Chalmers, 1895; Flint, 1951; McNeill and Purdy, 1951), into a concept of limited glaciation with highland nunataks and glaciers terminating just offshore of the land areas during the LGM (Grant, 1977; Dyke and Prest, 1987; Grant, 1989, 1994). The minimum model was found to be invalid when the inner shelf and terrestrial ice margins defined in that model formerly thought to be ~20–21 ka were radiocarbon dated in the Bay of Fundy at ~13 ^{14}C ka BP (Stea and Wightman, 1987) and on the inner Scotian shelf at ~15 ^{14}C ka BP (King, 1996). The maximum extent of LGM glaciers was defined at the edge of the continental shelf (Mosher et al., 1989). The model that has emerged to counter both extreme models is one of local ice divides situated over Maritime Canada collectively termed the Appalachian Glacier Complex (AGC), large enough to cross local highlands and effectively prevent ingress of the LIS over most of the region (Prest and Grant, 1969; Grant, 1989; Stea et al., 1989, 1998; Stea, 2004; Shaw et al., 2006). Unfortunately, the minimum model is still being used for calibration of continent-wide numerical models of the last glaciation (e.g. Marshall et al., 2002; Tarasov and Peltier, 2004).

Past studies have emphasised the role of ice streams in the Bay of Fundy and Laurentian Channel in the ablation of the LIS and in the formation of local ice divides (e.g. Mayewski et al., 1981; Belknap et al., 1989; Grant, 1989). An important regional synthesis of Atlantic Canada has been recently completed by Shaw et al. (2006) in which they further develop the ice stream model of the development and evolution of Appalachian divides during the LGM.

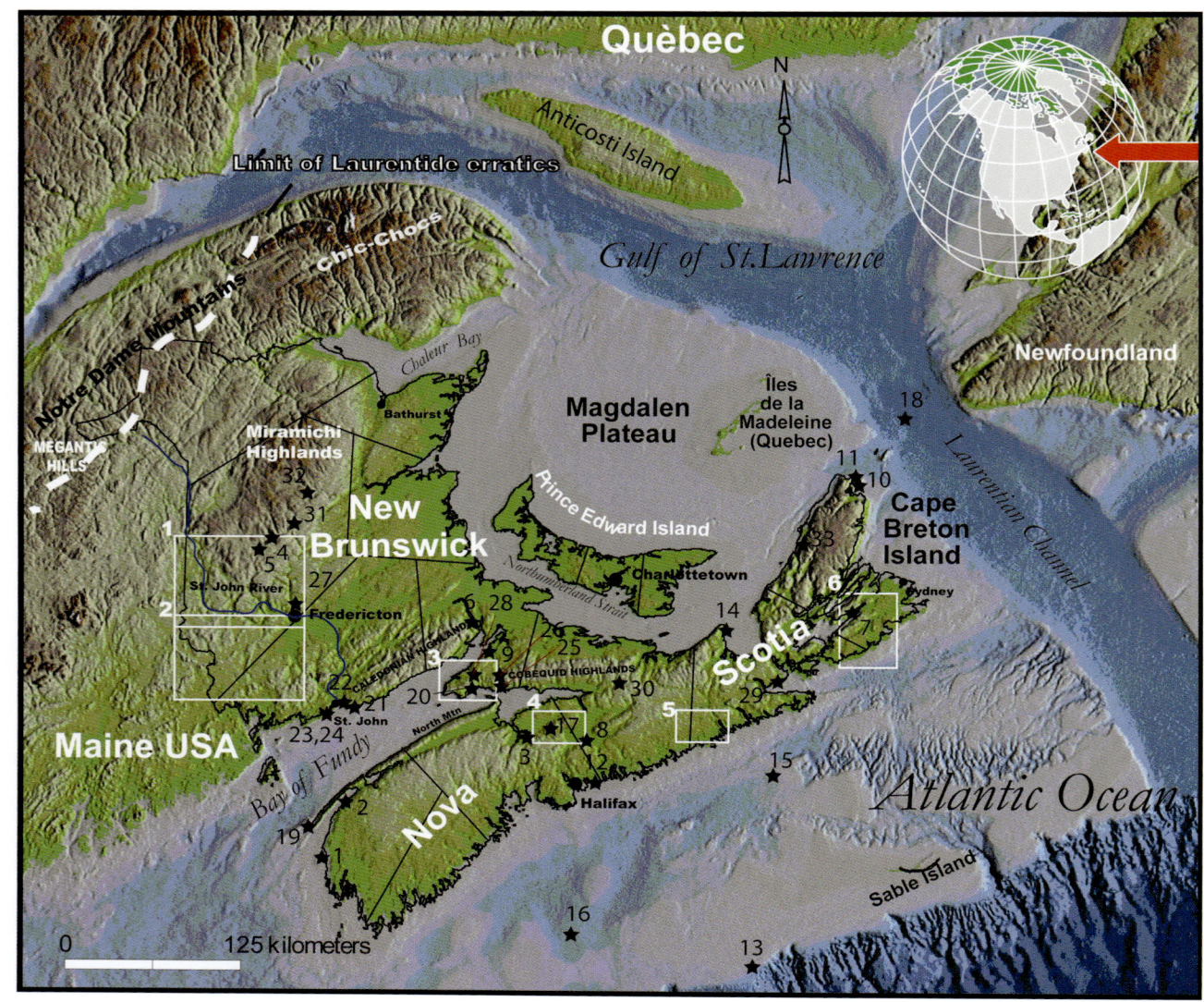

FIGURE 48.1 Location of the Maritime Provinces of Canada, main physiographic features, type areas and type sections. Boxes-type areas. 1, Northumberland Phase; 2, Caledonia Phase; 3, Shulie Lake Phase; 4, Chignecto Phase; 5, Escuminac Phase; 6, Scotian Phase. Stars-type sections. 1, Salmon River; 2, Gilbert Cove; 3, Miller Creek Quarry; 4, Half Moon section; 5, Sisson Brook; 6, Hillsborough; 7, East Bay; 8, East Milford; 9, Joggins; 10, Bay St. Lawrence; 11, Dingwall; 12, West Lawrencetown; 13, Core 90-015-1; 14, Ballantynes Cove; 15, Core 79-011-C; 16, Core 88-010-007; 17, Rawdon; 18, Core HU90-028-10; 19, Brier Island Bog; 20, Spencers Island; 21, Mispec Bay; 22, St. John; 23, Sheldon Point; 24, Taylors Island; 25, Leak lake; 26, Gilbert Lake; 27, Killarney lake; 28, Shulie Lake; 29, Collins Pond; 30, Millbrook; 31, Todd Mountain; 32, Joe Lake; 33, Cormier Lake.

The glacial history of the Maritimes is complex. Evidence for pre-Late Wisconsinan events is fragmentary, and the glacial striation record attributed to the Late Wisconsinan, extremely varied. Bedrock surfaces in the Maritimes reveal palimpsest erosional forms including cross-striated outcrops (cf. Veillette et al., 1999), 'bevelled' facets (Prest, 1983; Grant, 1989) and multi-stossed outcrops (Stea and Finck, 1984). The relationships between older, weathered, striations on lee-side surfaces and freshly striated surfaces have been termed 'erosional stratigraphy' (term suggested by W. W. Shilts, personal communication, 1986; Stea et al., 1992b, 1998). Flow events defined by discrete, regionally mappable trends of striations and large-scale streamlined glacial landforms of the same relative age are termed ice-flow phases (cf. Rampton et al., 1984). Glacial flowlines based on field data are utilised in the reconstruction of former ice sheets. As part of this compilation, type areas and type stratigraphical sections for each glacial phase and interglacial period are defined. Type area maps show the distribution and cross-cutting relationships of streamlined glacial landforms (drumlins/fluting) and small-scale glacier erosional features (striae/roches moutonnées) and are based on new digital elevation model imagery, created from topographic data (e.g. Fisher et al., 2006) and elevation data obtained from NASA's Shuttle Radar Topography Mission (SRTM http://dds.cr.usgs.gov/srtm/version2_1/SRTM3/North_America/). Type sections are defined and briefly described for till sheets formed during each flow phase and interglacial period.

48.2. PRE-ILLINOIAN RECORD

Occurrences of intensely weathered bedrock (saprolites) are common in the Maritimes (e.g. Gauthier, 1980; McKeague et al., 1983; Rutherford and Thacker, 1988; Broster and Seaman, 1991; Lamothe, 1992; Parkhill, 1994; O'Beirne-Ryan and Zentilli, 2006). Weathering of granite has been described as definitely pre-Wisconsinan and probably of pre-Pleistocene age (Wang et al., 1986). Symons et al. (1996) have shown that weathered gossan in northern NB began forming no earlier than the beginning of the Pleistocene, and no later than 1 Ma.

O'Beirne-Ryan and Zentilli (2006) describe three distinct weathering profiles in granites of the South Mountain in NS, a relithified clay-rich saprolite beneath a Carboniferous sedimentary sequence; a pre-Triassic profile; and a pre-Pleistocene arenaceous saprolite beneath Pleistocene till. The widespread preservation of preglacially weathered bedrock in the Maritimes indicates that previous glaciations in this area were not highly erosional.

The oldest Quaternary deposits in Maritime Canada are thought to be the iron-cemented Bridgewater and Mabou Conglomerates, assigned by various authors to the Tertiary to early Pleistocene (Prest et al., 1972; Grant, 1989). These are tillites which rest directly on striated bedrock and are overlain by Late Quaternary tills.

A pre-Illinoian interglacial interval (Yarmouthian) (250–300 ka) is indicated by amino acid racemisation dates from a shell-bearing diamicton in south-western NS (Wehmiller et al., 1988). The diamicton hosting the shells (Little Brook Till) is thought to be Illinoian because a correlative unit is found beneath Sangamonian marine beds at the reference section of Salmon River (Figs. 48.1 and 48.2; Grant, 1980a; Stea et al., 1992b; Stea, 2004).

Lack of evidence for pre-Illinoian glaciations and interglacial periods in the Maritime Provinces is puzzling, considering the preservation of unconsolidated Early Cretaceous outliers in lowland fault basins beneath a Wisconsinan sediment sequence (e.g. Stea and Pullan, 2001) and the offshore record of glaciations spanning 1 Ma (Piper et al., 1994). Karst topography in many lowland areas provides ample opportunity for the preservation of older glacial and non-glacial deposits, and these have been made accessible by gypsum mining and shoreline erosion. Nonetheless, the Quaternary record appears to begin at the penultimate glaciation. A possible explanation for the lack of deposits may be the development and persistence of local, cold-based ice caps in the early Pleistocene.

48.3. NORTHUMBERLAND PHASE (MIS 6; ~190–130 KA)

Stratigraphical evidence for an Illinoian glaciation (MIS 6) is found in gypsum quarries in central NS where an indurated till (Miller Creek Till; Williams et al., 1982) with upper soil development is found beneath wood and peat beds of the last interglacial (MIS 5; Stea, 2004). Cobequid Highland erratics within the Miller Creek Till suggest emplacement by an ice sheet from the west to north.

In the previous stratigraphical model for NB (Seaman, 2004), all known glacial phenomena were assigned to the Wisconsinan (MIS 4–2). This was based on the interpretation of Lamothe (1992) that a non-finite radiocarbon date (>36 ka; GSC-4199) for charcoal from the Half Moon sediments indicated that they were of Mid-Wisconsinan age (Figs. 48.1 and 48.2A; Table 48.1). To confirm this interpretation, a new charcoal sample was analysed using AMS technology. The new date (>51.7 ka; ULA-1337) indicates that these fluvial sediments are significantly older, probably of Sangamonian Interglacial age (Table 48.1; Seaman, 2009). The reinterpretation of the Half Moon Sediments as Sangamonian in age bumps the underlying 'lower till' of Lamothe (1992) back to the penultimate glaciation, presumed to be the Illinoian (Seaman, 2009).

Seaman (2004) correlated the 'lower' till with indicators of an early eastward flow identified by Rampton et al. (1984) across NB. A similar early eastward ice-flow event was defined in NS (Chalmers, 1885, 1895; Grant, 1994) and assigned to the early part of the Caledonia Phase, inferred to be of Early Wisconsinan age (Stea et al., 1998; Stea, 2004). In light of this new data, we reinterpret this probably Illinoian ice-flow phase as a separate flow event 'Northumberland Phase' (Seaman, 2009). Evidence for the east to east–southeast flow of the Northumberland Phase is found across the Maritime Provinces.

Rampton et al. (1984) inferred a source area in the Megantic Hills of Quebec and the White Mountains of New England for the early eastward flow now assigned to the Northumberland Phase. Subsequently, based on work in southern Quebec, Elson (1987) presented evidence for an axis of accumulation in the northern Appalachian Mountains, and inferred the eastward flow to be of mid-Wisconsinan age. Rappol (1988, 1989) suggested that this axis may have been the source area for the Northumberland flow pattern. Although evidence for eastward outflow from an ice divide in New England is strong, the location of this early divide remains to be confirmed by workers in that area (e.g. Kite et al., 1982).

48.3.1. Type Area (West-Central NB)

The type area for the Northumberland Phase is west-central NB (Fig. 48.3). This area exhibits large-scale east–

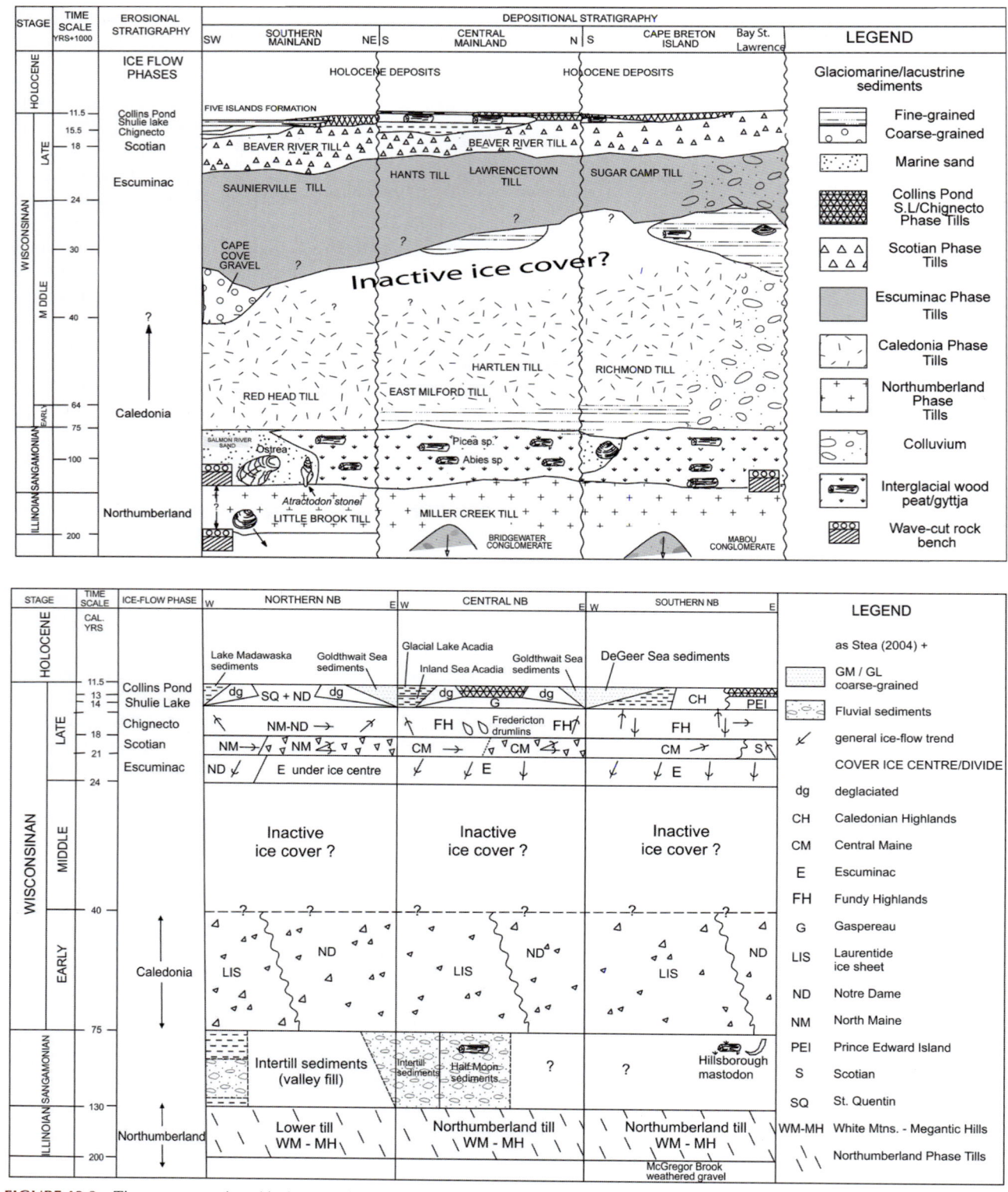

FIGURE 48.2 Time–space stratigraphical summaries of New Brunswick (A) and Nova Scotia (B) including ice-flow phases, Pleistocene stratigraphical units and age designations.

Chapter | 48 The Appalachian Glacier Complex in Maritime Canada

TABLE 48.1 Compilation of Radiocarbon, Dosimetric and Uranium Series Ages for Sediments in NS and NB

Section	Date	Type	CAL	Material/deposit	LAbN	Source
Pre-Sangamonian (MIS 7) pre-Northumberland Phase						
(2) Gilbert Cove	>250,000	AAR		shell in till	–	Wehmiller et al. (1988)
Sangamonian Interglaciation (MIS 5) pre-Caledonia Phase						
(3) Miller Creek Quarry	>52,000	^{14}C		Wood under till	GSC-2694	Stea et al. (1992a)
(4) Half Moon Pit	>51,700	^{14}C		Charcoal in sand under till	ULA-1337	Seaman (in press)
(6) Hillsborough	51,500±1270	^{14}C		Wood from coprolite	GSC-2467	Harington et al. (1993)
(6) Hillsborough	>43,000	^{14}C		Peat	GSC-1680	Harington et al. (1993)
(7) East Bay	86.9±12 ka	U/Th		Wood/silty clay (Unit 2)	UQT-109	de Vernal et al. (1986)
(7) East Bay	62.5±5	U/Th		Wood in sand (Unit 3)	UQT-177	de Vernal et al. (1986)
(7) East Bay	126.400±15 ka	U/Th		Wood in peat (Unit 1)	UQT-175	de Vernal et al. (1986)
(8) East Milford Quarry	>50,000	^{14}C		Wood under till	GSC-1642	Mott and Grant (1985)
(8) East Milford Quarry	84.0±6.5 ka	U/Th		Wood under till	UQT-185	de Vernal et al. (1986)
(8) East Milford Quarry	74.9±6.5 ka	ESR		Dental enamel	–	Godfrey-Smith et al. (2003)
Mid-Wisconsinan Recession (MIS 3)						
(10) Bay St. Lawrence	21,920±150	^{14}C		Shell in glacio-marine	TO-246	Grant (1994)
(10) Bay St. Lawrence	36,730±285	^{14}C		Shell in glacio-marine	LYON-2125	S. Occhietti personal communication (2005)
(10) Bay St. Lawrence	41,725±415	^{14}C		Shell in glacio-marine	LYON-2124	S. Occhietti personal communication (2005)
(11) Dingwall	32,700±560	^{14}C		Wood in till	GSC-3381-2	Grant (1994)
Late Wisconsinan Maximum (MIS 2) Escuminac Phase						
(13) 90-015-1	20,780±170	^{14}C	24.1–24.6	Shell in debris flow	TO-2077	Baltzer et al. (1994)
Late Wisconsinan Readvance (MIS 2) Scotian Phase						
(15) 79-011-C	14,850±170	^{14}C	17.9–18.5	Shell in Emerald Silt A	–	King (1996)
(16) 88-010-007	17,450±155	^{14}C	20–20.3	Shell In Emerald Silt	Beta-27229	Piper and Fehr, 1991
Late Wisconsinan Readvance Chignecto Phase						
(18) HU90-028-10	13,650±80	^{14}C	15.4–15.8	Shell-till tongue margin	OS-4865	Josenhans and Lehman (1999)
(19) Brier Island Bog Lake	13,200±130	^{14}C	14.9–15.3	Shell in sand	GSC-4431	Stea and Mott (1998)
(20) Spencers Island	13,400±300	^{14}C	14.9–15.8	Shell in outwash delta	BETA13728	Stea and Wightman (1987)

Continued

TABLE 48.1 Compilation of Radiocarbon, Dosimetric and Uranium Series Ages for Sediments in NS and NB—cont'd

Section	Date	Type	CAL	Material/deposit	LAbN	Source
(21) Mispec Bay	14,400±530	^{14}C	15.9–17.6	Shell in ice-contact delta	GSC-2573	Rampton et al. (1984)
(22) St. John	13,325±500	^{14}C	14.4–15.9	Shell in delta	GSC-7	Gadd (1973)
(23) Sheldon Point	13,200±200	^{14}C	14.7–15.4	Shell in moraine	GSC-965	Gadd (1973)
(23) Sheldon Point	12,650±175	^{14}C	13.9–14.5	Shell in moraine	GX-12971	Nicks (1988)
(24) Taylors Island,	12,040±150	^{14}C	13.3–13.6	Barnacle in raised beach	BGS 1010	Seaman et al. (1993)
Late Wisconsinan Readvance (Older Dryas?) Shulie Lake Phase						
(25) Leak Lake	12,260±160	^{14}C	13.9–14.4	Wood basal lake core	TO-3971	Stea and Mott (1998)
(26) Gilbert Lake	11,360±112	^{14}C	13–13.4	Wood basal lake core	TO-807	Stea and Mott (1998)
(23) Sheldon Point	11,620±250	^{14}C	12.9–13.3	Shell in moraine	BGS 1223	Nicks (1988)
(27) Killarney Lake	11,180±120	^{14}C	12.9–13.1	Wood in basal lake core	TO-3001	Levesque et al. (1993)
Late Wisconsinan Readvance (Younger Dryas) Collins Pond Phase						
(29) Collins Pond (top)	10,900±100	^{14}C	12.8–12.9	Peat under till	GSC-4475	Mott and Stea (1993)
(30) Millbrook Pipeline	10,900±110	^{14}C	12.8–13.0	Wood under till	GSC-6435	Stea and Mott (2005)
(31) Todd Mountain	11,500±150	^{14}C	13.2–13.5	Organic silt under till	GSC-4277	Lamothe (1992)
(5) Sisson Brook	11,030±60	^{14}C	12.9–13.0	Twigs in till	TO-13589	Seaman and McCoy (2008)
(5) Sisson Brook	10,990±60	^{14}C	12.9–13.0	Twigs in till I	TO-13588	Seaman and McCoy (2008)
(32) Joe Lake	11,180±90	^{14}C	12.9–13.1	Wood in gyttja	TO-2303	Mayle et al. (1993)
(33) Cormier Lake	9970±80	^{14}C	11.4–11.4	Wood in gyttja	BETA-61401	Stea and Mott (1998)

Locations in Fig. 48.1. Compiled and condensed from much larger databases in Rampton et al. (1984), Grant (1989), King (1996), Stea and Mott (1998) and Seaman (in press).

southeast trending bedrock fluting that was not readily apparent prior to the availability of the high-resolution, high-altitude images. Remnants of an old east–southeast trending dispersal train from the Carboniferous Carlisle Formation red beds have also been noted in this area. The relative antiquity of this dispersal event is indicated by the observation of pebbles of the reddish till reworked into the overlying regional yellowish-brown till (Seaman, 2000).

48.3.2. Type Section (Sisson Brook, NB)

Lamothe (1992) correlated the 'lower till' of the Half Moon section with a similar till underlying fluvial deposits in the Saint John River valley area in northwestern NB (Fig. 48.2A). This 'lower till' has an east–west fabric (Rampton et al., 1984; Rappol, 1989; Lamothe, 1992) and contains clasts of Deboullie Syenite from a source area

Chapter | 48   The Appalachian Glacier Complex in Maritime Canada

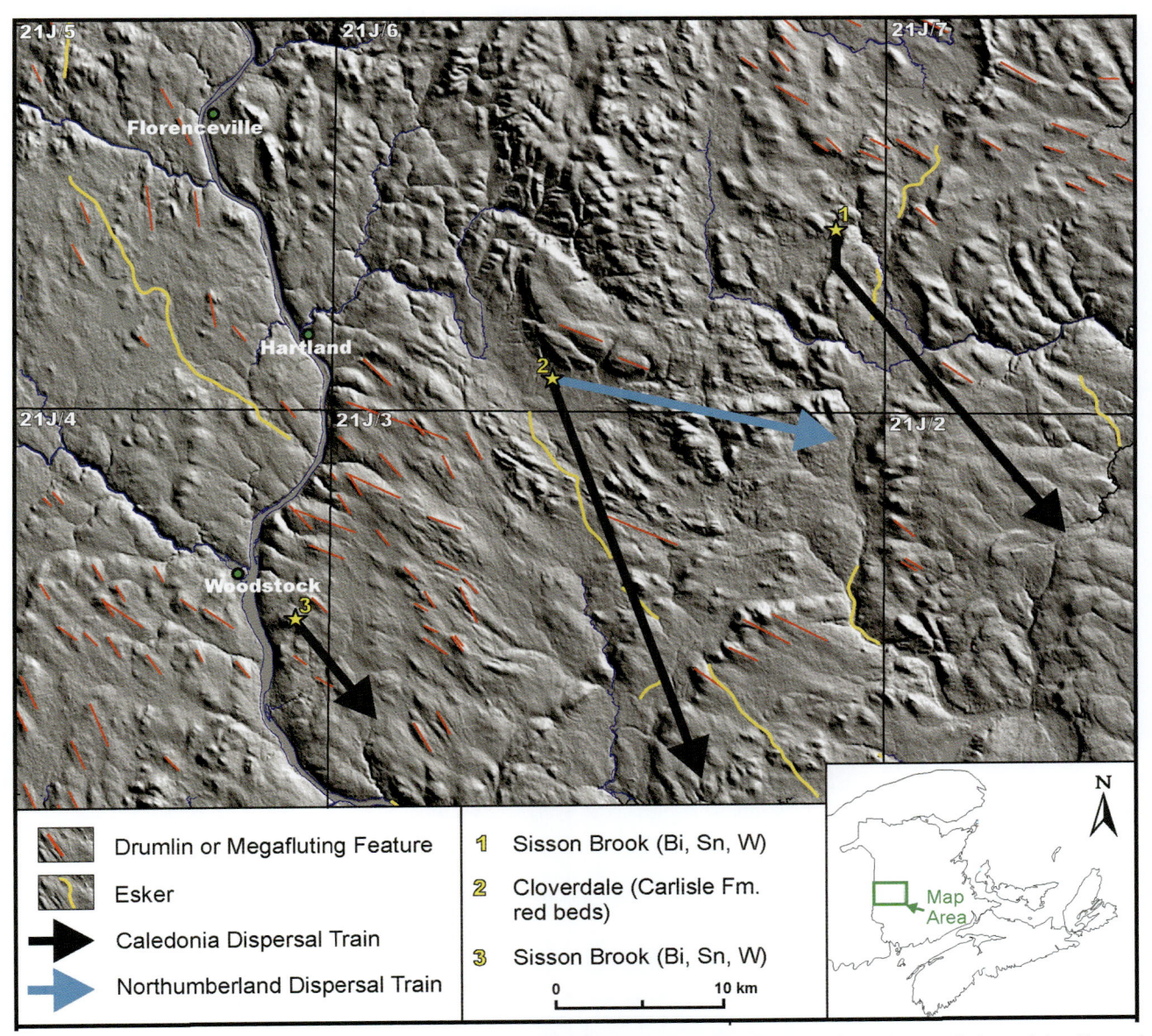

FIGURE 48.3 Type area for the Northumberland Flow Phase as demonstrated by east–southeastward-trending palimpsest fluting and drumlins, and dispersal trains from mineral deposits at Sisson Brook. Caledonia Phase flow pattern is also imprinted on the landscape.

in northern Maine that confirms eastward dispersal (Rappol, 1988, 1989). The 'lower till' is now referred to informally as the 'Northumberland Till' in NB. A reference section for the Northumberland Till is in a mineral exploration trench at Sisson Brook (Figs. 48.2A and 48.12), previously described by Seaman and McCoy (2008). This site lies approximately 15 km to the north of the eastern end of the above noted dispersal train. The Northumberland Till was exposed at the western end of the trench, where it appeared to occur as a remnant wedge in a narrow northeast–southwest trending valley between the relatively flat surface of the mineralised zone to the east and a small hill to the west. At the eastern end of the wedge, the Northumberland Till lies directly upon the mineralised bedrock surface. It is overlain by what is now interpreted to be a Caledonia Phase till, of Early Wisconsinan age.

48.3.3. Margin of the Northumberland Phase Glacier

The outer margins of the Northumberland glacier lay in the offshore. Its northern part probably fed into an ice stream in the Laurentian Channel in the Gulf of St. Lawrence (Fig. 48.4).

48.4. SANGAMONIAN INTERGLACIAL (MIS 5; 128–75 KA)

Organic deposits (peat and wood) that underlie Wisconsinan glacial deposits have been assigned to the Sangamonian Interglacial (MIS 5; e.g. Fulton, 1984; Mott and Grant, 1985; de Vernal et al., 1986; Stea et al., 1992b). More than 30 of these pre-Wisconsinan sites are found throughout NS and four sites are located in NB (Grant, 1989; Figs. 48.1 and 48.2A, B). During this interval, the climate fluctuated

Chapter | 48   The Appalachian Glacier Complex in Maritime Canada

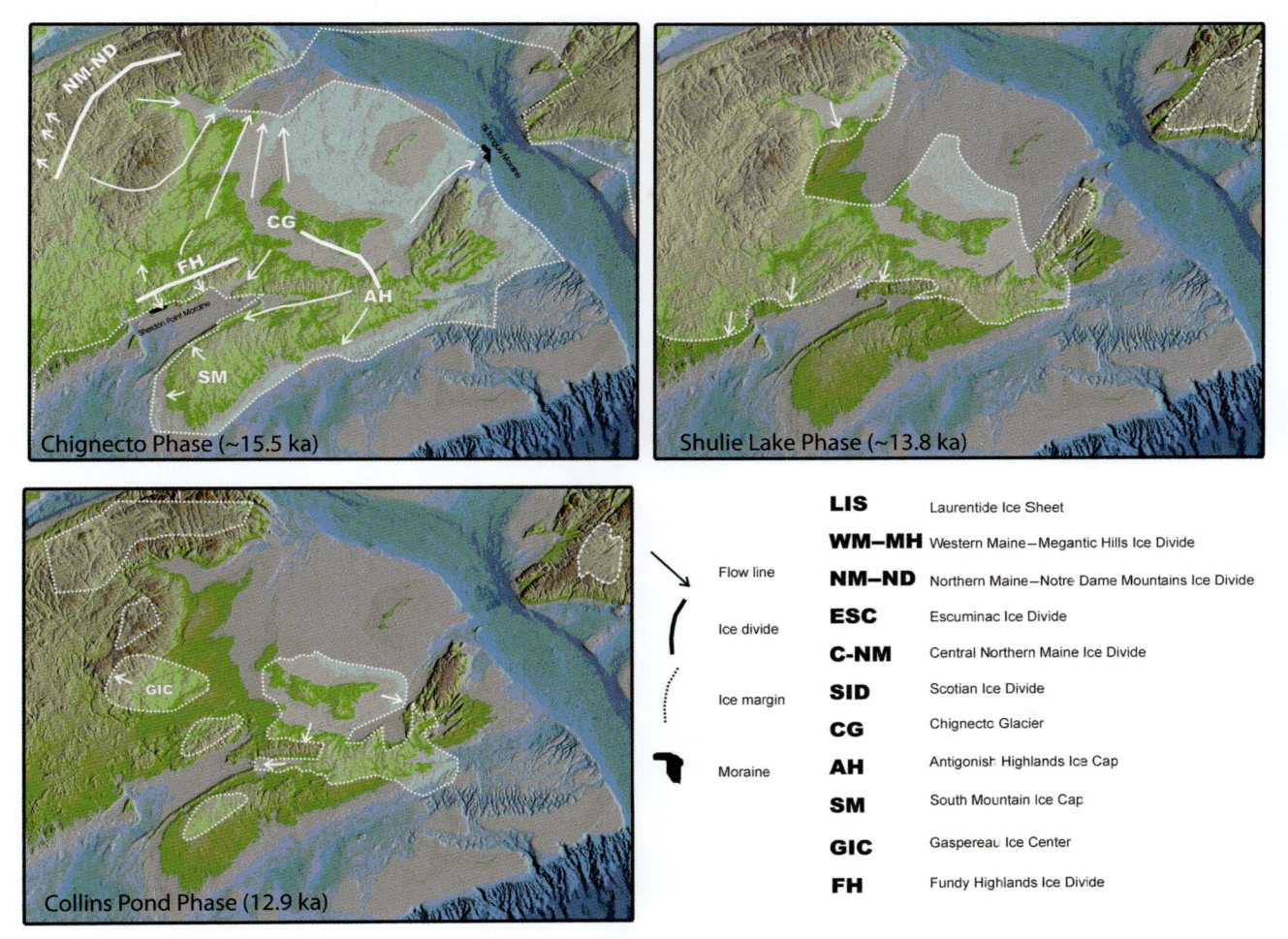

FIGURE 48.4 Evolution of the Appalachian Glacier Complex from the Illinoian (MIS 6) to the Late Wisconsinan in (MIS 2) in Maritime Canada. The location of hypothetical ice centres and divides, and flow lines are shown. The extent of the glacio-isostatic Goldthwait and DeGeer Seas are not defined. Flow data is largely derived from striae and landform compilations by Prest (1973), Rampton et al. (1984), Seaman (1989), Stea et al. (1992a) and Grant (1994).

considerably with an early climatic optimum (MIS 5e), followed by less temperate cycles (MIS 5d–5a) that culminated in regional glaciation (MIS 4–2). The early climatic optimum of the Sangamonian was characterised by sea level 4–6 m higher than at present (Grant, 1980a). This higher sea level cut a shoreline whose remnants are flat, wave-cut rock benches upon which non-glacial and glacial sediments have been deposited (Fig. 48.2A, B).

The Half Moon sediments (Fig. 48.2A), originally interpreted as Middle Wisconsinan age, are now assigned to the Sangamonian stage based on a new date of >51.7 ^{14}C ka BP (ULA-1337—Table 48.1). The Hillsborough mastodon find in NB and its associated sediments were assigned to the latter part of the Sangamonian (MIS 5a) by Harington et al. (1993), after originally being interpreted as a Late Wisconsinan animal that died in a mid-Wisconsinan mire (Schroeder and Arseneault, 1978) or Early to Middle Wisconsinan age (Foisy and Prichonnet, 1991).

48.4.1. Type Section (East Bay, NS)

The type section for organic beds spanning the last interglacial in Maritime Canada is considered to be East Bay, NS (Fig. 48.1). At the site, a woody peat layer is overlain by organic silty clay and then a gravelly sand with peat inclusions (Mott and Grant, 1985; de Vernal et al., 1986). This organic sequence is overlain by glaciolacustrine clay and then till. In the organic sequence, three palynostratigraphical units (Units 1–3) were differentiated. The woody peat is characterised by taxa indicative of climatic conditions warmer than present and forests containing abundant white pine and other thermophilous hardwood taxa (Unit 1). The organic silty clay spectra of Unit 2 suggest climate similar to the present and the peat inclusions in gravel record spectra typical of boreal coniferous to woodland and tundra during a stratigraphically younger interval.

Uranium series age dates confirm a Sangamonian age for the East Bay beds (Causse and Hillaire-Marcel, 1986), but the youngest unit (Unit 3) also produced some Middle Wisconsinan ages (Table 48.1). de Vernal et al. (1986) proposed an extended interglacial interval with fluctuating climate and no glacial interruptions until the Middle Wisconsinan. The Middle Wisconsinan U/Th dates, however, are considered minimum ages because of the strong possibility of post-depositional U migration in the sites with the youngest dates (Stea et al., 1992b). Godfrey-Smith et al. (2003) have presented optical luminescence and electron spin resonance dating of sediments and mastodon bone and dental enamel at the reference section in East Milford, which have confirmed a Sangamonian age for sub-till organic beds (Table 48.1; Fig. 48.2A). They show that a mastodon was living in a boreal/tundra forest transition environment at the end of the last interglacial (MIS 5a; ~75 ka; Table 48.1).

48.5. EARLY–MIDDLE WISCONSINAN CALEDONIA PHASE (75–50?KA)

The oldest Wisconsinan ice-flow patterns on land in Maritime Canada are southeastward (Caledonia Phase, Fig. 48.4). The Caledonia Phase was initially defined by Rampton and Paradis (1981a,b,c) to identify early eastward to southeastward ice movements in south-eastern NB. However, as noted above, at least part of the early eastward flow is now attributed to the Northumberland Phase of the Illinoian Glaciation (MIS 6). Rampton et al. (1984) inferred that the southeastward flow of the Caledonia Phase was of Early to Middle Wisconsinan age, since striae with this trend were observed on the Sangamonian wave-cut bench in the upper Bay of Fundy area.

The Caledonia Phase glacier crossed highlands of 300 m a.s.l. in southern NB and northern NS and deposited overconsolidated, matrix-rich tills found in stratigraphical sections throughout NS (Fig. 48.2A) below as many as three Late Wisconsinan till sheets (Grant and King, 1984; Stea, 2004). These 'mature' basal tills have been linked to the regional Caledonia southeastward flow pattern through till fabric and provenance studies (Nielsen, 1976; Alcock, 1984; Stea and Finck, 1984; Stea et al., 1986, 1992b; Graves and Finck, 1988; Grant, 1994; McClenaghan and DiLabio, 1995, 1996; Stea and Pe-Piper, 1999).

The timing of onset of the last glacial cycle in eastern Canada is uncertain (e.g. Clark et al., 1993). The age of the lowermost of the Wisconsinan tills was assigned to the Early–Middle Wisconsinan (MIS 4–3) because:

1. The first Wisconsinan till sheet lies directly on the interglacial (MIS 5) Sangamonian organic beds and the interglacial rock platform (Grant, 1980b).

2. Varved glaciolacustrine sediments conformably overlie interglacial organic lacustrine sediments in lowland basins of NS and are overlain conformably by till, suggesting a continuum of deposition from the interglacial (MIS 5) to the glacial period (MIS 4) (Stea et al., 2003; Stea and Mott, 2005).

3. The Bay St. Lawrence section records the retreat phase of a major glaciation during the Middle Wisconsinan (50–35 ka; Stea et al., 1992b; Grant, 1994).

Foisy and Prichonnet (1991) and Grant (1994) presented evidence that glaciers developed in some highland areas such as the Caledonian Highlands of NB and the Cape Breton Highlands before coalescing with regional ice sheets in the Early Wisconsinan.

Rampton et al. (1984), Stea et al. (1998) and Stea (2004) attributed the Caledonia Phase in south-eastern NB and NS to flow from the Gaspereau Ice Centre in central NB based on a lack of striae or streamlined landforms indicating SE ice flow in east-central NB (Fig. 48.4). Seaman (2006), however, has since shown that the Gaspereau Ice Centre was active only during the Younger Dryas Stadial at the end of the Late Wisconsinan. Also, the Gaspereau Ice Centre in central NB cannot be the source for early southeastward ice flow noted in western and northern NB that is now correlated with the Caledonia Phase (Seaman, 2009). The source area therefore lies to the northwest, in either the LIS or the North Maine–Notre Dame Ice Divide of Rampton et al. (1984) which straddled the Chic-Choc Mountains of Gaspé.

Is the Caledonia Phase of Laurentide origin? The following points support arguments against a Laurentide centre for most of this regional southeastward ice flow. These arguments apply to not only the Early Wisconsinan but also the configuration during the LGM as they are based on erratic dispersal on highland areas and the assumption that similar flow paths were used.

1. A lack of striae or dispersal trains indicating the passage of a regional, erosive, southeastward-flowing LIS from Québec across northeastern NB (Rampton et al., 1984; Pronk et al., 1989; Lamothe, 1992; Parkhill and Doiron, 2003).

2. Laurentide erratics from the Precambrian rocks of the Laurentian Highlands of Quebec, to the north of the St. Lawrence River, are known only from the area of the Saint John River valley in the northwestern part of the province (Fig. 48.1; Rampton et al., 1984; Pronk et al., 1989; Rappol, 1989; Lamothe, 1992).

Charbonneau and David (1993) concluded that the LIS crossed the western part of the Gaspé Peninsula of Quebec, to the north, during the Early Wisconsinan. In contrast, Olejczyk and Gray (2007) provided detailed petrological data for erratics on the eastern Gaspé mountains, concluding that the LIS never crossed northern Gaspé during the Wisconsinan. It is therefore suggested that the Caledonia

Phase involved the combined efforts of the LIS, to the west, and the Notre Dame Ice Divide over Gaspé, to the east.

The coasts of eastern NB, northern NS and Cape Breton have been extensively surveyed by both bedrock and Quaternary geologists, and Canadian Shield erratics have not been reported. If Laurentide ice centred in Quebec had crossed the region during the Wisconsinan, abundant erratics would be expected as they are found in many areas of central Canada 300–500 km south of the shield margin (Prest et al., 2001). A petrologic and geochemical study of erratics from a type section with a Caledonia Phase Till sheet along the Atlantic coast near Halifax, for example, confirmed an Appalachian origin of all crystalline erratics found in tills there (Stea and Pe-Piper, 1999). Prest and Nielsen (1987), however, not only reported an anorthosite boulder in western PEI and suggested an Archean Canadian Shield source but also noted that the anorthosite was found with Devonian-age Appalachian granite erratics. Recent geological mapping has documented occurrences of anorthosite in the highlands of southern NB (Barr et al., 2001), west of PEI, and along the flowlines of the Northumberland Phase glacier.

Many Laurentide erratics are reported from northwestern NB and these probably entered via LIS flow down transverse valleys through the Notre Dame Mountains (Parkhill, 2005; Wilson et al., 2005). Much of the Laurentide basal debris transported by the LIS was probably diverted eastward down the Laurentian Channel to the north (e.g. Pronk et al., 1989).

48.5.1. Type Area (South-Western NB)

The type area for the Caledonia Phase is in southwestern NB (Fig. 48.5). This area is characterised by southeastward trending streamlined bedrock hills and drumlins and also by a number of southeastward to south-southeastward trending glacial dispersal trains with a magnitude in the order of 20–30 km (Fig. 48.5; Seaman, 2000, in press). This flow trend was attributed to the St. Stephen, Tay and Woodstock flow patterns by Rampton et al. (1984), which they assigned to the Late Wisconsinan. However, striae mapping in this area has shown that this was an early ice-flow event (Seaman, 1991; Seaman et al., 1993), comparable to the Caledonia flow pattern of Rampton et al. (1984) to the east. Subsequently, Allard (2003) and Allard and Pronk (2003) have interpreted the till in southwestern NB as the Caledonia Till. A local characteristic of this till is 'hybridisation' by reworking by later ice flows, particularly that towards the south–southwest, without significant displacement (Allard, 2003; Allard and Pronk, 2003).

48.5.2. Type Section (Joggins, NS)

The type section for the Caledonia Phase ice flow is at Joggins, NS, located at the border of NS and NB, and features three tills, from bottom to top, coloured red, grey and yellow (Wickenden, 1941; Prest et al., 1972). Stea et al. (1986) completed the most detailed study of the section to date and found that the lower red till (McCarron Brook Till) contains pebble lithologies (igneous, volcanic erratics and purple argillite) derived from the Caledonian Highlands to the west and northwest, deposited during the Caledonia Phase flow. The middle grey till (Joggins Till) contains limestone and coal clasts and was deposited by (Escuminac Phase) south/southwestward ice flow. The upper yellowish till (Shulie Lake Till) with sideritic sandstone clasts was originally thought to be formed by southwestward flow, but recent bedrock mapping (Ryan et al., 1990) suggests that the source of the sandstone clasts is to the south indicating northward flow (Scotian Phase). Fabric studies of the Shulie Lake Till suggest that it was later reworked by SW ice flows (Chignecto/Shulie Lake Phases).

48.5.3. Margin of the Caledonia Phase Glacier

Caledonia Phase tills in NS are generally thicker and denser than Late Wisconsinan tills and have a substantial far-travelled erratic component, suggesting to some that the Early Wisconsinan glaciation was the most extensive in eastern Canada (Grant, 1977; Grant and King, 1984). North Mountain (Fig. 48.1) basalt erratics found at Sable Island Bank (King, 1970) indicate that a southeastward flow may have indeed extended out at least to the outer banks and probably farther (Fig. 48.1). The limit of relict iceberg furrowing on the outer banks occurs in water depths of 300–600 m and indicates that ice extended out to the shelf/slope topographic break (Dodds and Fader, 1986). Debris-flow deposits, interfingering with ice proximal glacio-marine deposits related to a series of tidewater ice margins, have been found on the Scotian Slope and the lowest or earliest feature has been interpreted as Early Wisconsinan (Mosher et al., 1989; Table 48.1) based on a sedimentation rate extrapolation. Scott et al. (1989) interpreted an MIS 4 age for subglacial channel fills mapped on Sable Island. Huppertz and Piper (2009) present evidence for a glacier on the Grand Banks south of Newfoundland during MIS 4. Ice rafting records in the North Atlantic also indicate that glaciers existed in the Laurentian Channel during the Early–Middle Wisconsinan (Bond and Lotti, 1995).

48.6. MID-WISCONSINAN RETREAT PHASE (MIS 3; ~50–25 KA)

Dredge and Thorleifson (1987) concluded from evidence available at that time that the southern margin of the Middle Wisconsinan LIS lay along Anticosti Island; the eastern margin lay along the coast of Labrador, Newfoundland;

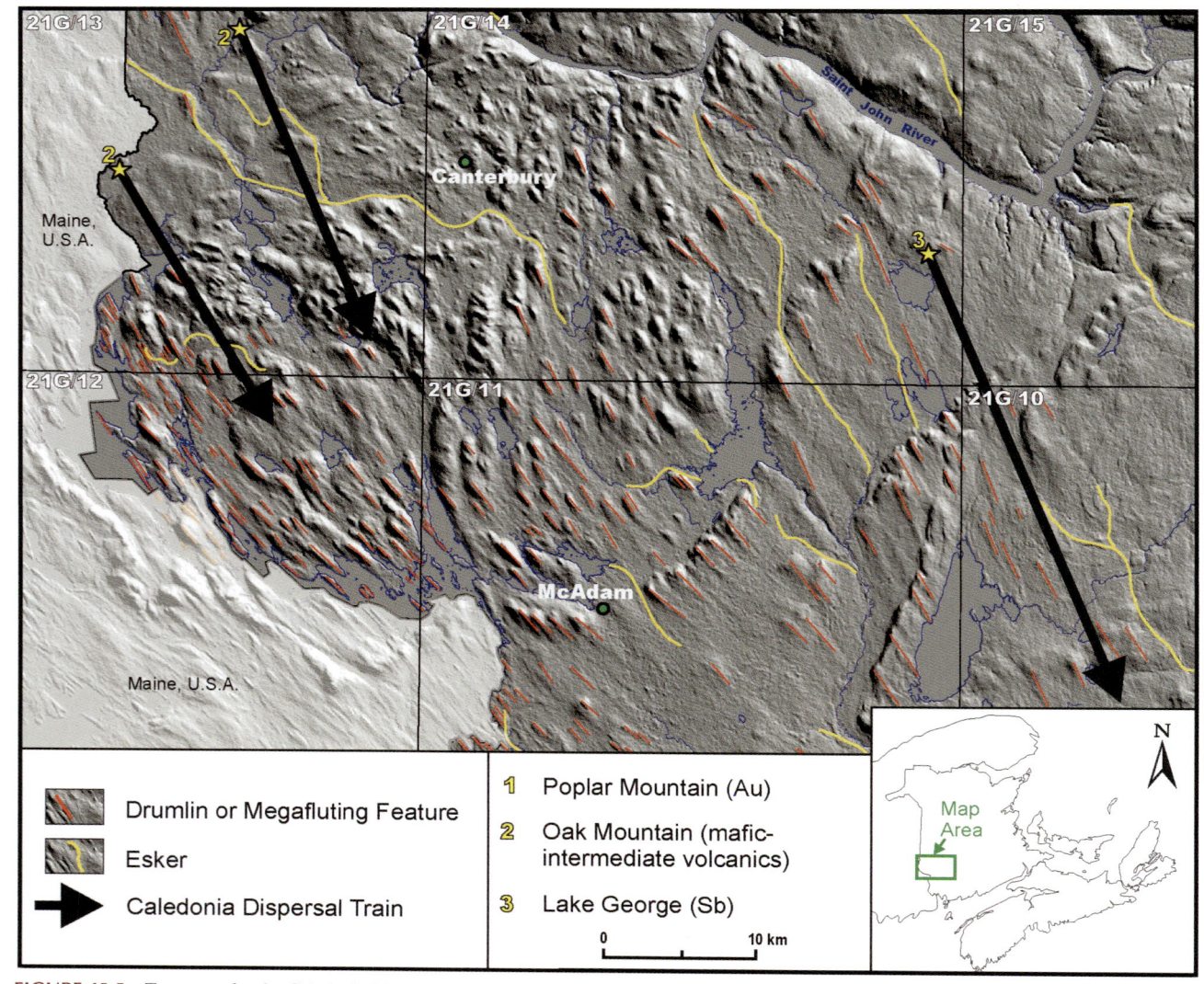

FIGURE 48.5 Type area for the Caledonia Phase (west-central NB).

and grounded ice extended only as far as the Atlantic coast. Apart from controversial evidence of sub-till oxidation zones, the terrestrial record for NS and NB (Fig. 48.2) indicates that much of the land mass was covered by ice from local (Appalachian) ice caps, although some coastal areas may not have been glaciated (Stea, 2004). As presented earlier, Mid-Wisconsinan radiocarbon dates on sub-till organic beds are considered spurious and the beds are now assigned to the last interglacial. The one exception is the Bay St. Lawrence section in northern Cape Breton which is now considered the type section for the Mid-Wisconsinan in Atlantic Canada.

48.6.1. Type Section (Bay St. Lawrence, NS)

This important Quaternary section is nearly 50 m in elevation at its highest point and features a Sangamonian to Late Wisconsinan stratigraphical sequence (Fig. 48.2B; Newman, 1971; de Vernal et al., 1983; Grant, 1994; Stea et al., 1992b, 1998). A wave-cut bedrock platform 5 m above MSL is at the base of the section. This platform is found throughout the Maritime Provinces and was interpreted as a sea-level marker during the warmest substage of the last interglacial (MIS 5e; Grant, 1980a). Directly above the bedrock platform is a gravel-bed overlain by a continuous peat layer. Above the peat lies a stony diamicton either colluvial or till in origin (Grant, 1994). Stratigraphically above the lower diamicton at an elevation of 20–30 m above MSL is a glacio-marine sand unit with shells consisting of alternating, coarse gravelly sand and fine sand units. Shells from this sand unit were radiocarbon dated at 36 and 41 ^{14}C ka BP (S. Occhietti, personal communication, 2005) and 21 ^{14}C ka BP (Grant, 1994) and also produced ages of

~40–60 ka using amino acid racemisation ratios (Table 48.1; Stea et al., 1992b). Above this is another colluvial deposit capped with a till deposit.

Based on the radiocarbon dates, sedimentology and elevation above sea level, the marine sand deposit is interpreted as an ice proximal glacio-marine deposit formed on an isostatically depressed landscape following a major glaciation (Caledonia Phase; MIS 4). There are no raised marine intertill deposits like it anywhere in Atlantic Canada and its formation is likely due to the proximity of the section to the Laurentian Channel calving bay which during MIS 3 had migrated southward into the Cape Breton Channel (Fig. 48.4).

48.7. LATE WISCONSINAN ESCUMINAC PHASE (MIS 2; 25–20 KA)

Striated outcrops in NB, NS and PEI record a shift in flow from southeastward during the Caledonia Phase to southward and southwestward during the Escuminac Phase (Rampton et al., 1984; Stea, 2004; Fig. 48.4). This flow is assigned a Late Wisconsinan age because it is the first regional ice flow postdating the Caledonia Phase and can be traced out to the edge of the continental shelf where the margin was dated at ~21 ka (Mosher et al., 1989; Stea et al., 1998).

During the Escuminac Phase, ice flowed southward across Chignecto Bay and the Bay of Fundy (Rampton et al., 1984; Stea, 2004; Shaw et al., 2006) and radiated north skirting the Chaleur Bay trough and over the Magdalen Islands, perhaps merging with the Laurentian Channel ice stream (see also Pronk et al., 1989; Occhietti, 1989). Rampton et al. (1984) originally assigned south–southwestward ice-flow patterns in southern NB to the Chignecto Phase with a restricted (minimum model) ice margin on the Chignecto Peninsula, but Stea and Finck (1984) and Stea et al. (1989) recognised this flow pattern on highlands of northern NS and as far south as the Atlantic coast, prompting Stea et al. (1998) to name this flow pattern the Escuminac Phase. The Chignecto Phase was then reassigned to a more restricted flow pattern during deglaciation.

Goldthwait (1924) attributed the major southward flow over mainland NS to a lobe of Laurentide ice from eastern Québec crossing the Laurentian Channel called the Acadian Bay Lobe. As discussed earlier, the absence or lack of reports of Canadian Shield erratics over wide areas of northern NS and Cape Breton imply a Maritime ice divide rather than Laurentide ice. This idea is reinforced by the radiating ice-flow trends clearly derived from an ice divide on the Magdalen Shelf (Fig. 48.4). However, it appears that the AGC was confluent with an ice stream emanating from the LIS in the Laurentian Channel (Fig. 48.4) based on abundant anorthosite and granulite erratics in the mid-Gulf region (Loring and Nota, 1973; Stea, 2001).

Parent and Dubois (1990) and Josenhans and Lehman (1999) describe tills formed by ice stemming from Escuminac Divide, on the Magdalen Islands and in the Laurentian Channel. In much of NB, the Escuminac Phase is interpreted to be represented by southward to southwestward (and possibly westward) ice flow from the Escuminac Ice Centre (Broster et al., 1997; Seaman, 2009). However, northwestern NB may have been affected by parallel flow from the Notre Dame Ice Divide. The original concept of the Escuminac ice centre was radiating flow from a centre on the Magdalen Plateau. This incorporated NW-trending striae (Tracadie flow pattern; Wilson et al., 2005) found in eastern NB, south of Chaleur Bay (Rampton et al., 1984; Stea et al., 1998). This single dome concept is now considered glaciologically inconsistent, and more likely, the two areas were linked by a divide responsible for regional southward flow across PEI, NS and NB. The dividing line between NB and PEI derived flows in NS is found in the Bridgewater area where red mud making up the Lawrencetown Till drumlins with Carboniferous sources to the northeast, changes into grey till drumlins with north-northwest Paleozoic grey-bed sources (Grant, 1963). The LGM reconstructions of Stea (2004) and Shaw et al. (2006) and this study are very similar, but our recent reconstruction has the Escuminac ice divide further north to satisfy evidence of southward movement on the north coast of PEI (Prest, 1973; Catto, 1998). The Tracadie flow pattern, mentioned earlier, is isolated and may have formed during an initial growth phase in the Mid-Wisconsinan (Fig. 48.4).

Ice outflow from the Escuminac Ice Divide during the Escuminac Phase was likely in the form of rapidly flowing ice streams or ice 'currents', as first envisioned by Grant (1976). Rapid ice stream flow may be inferred from the properties of the drumlin Lawrencetown Till which has a consistently high erratic content, as if dilution with local debris along its flow path was suppressed (Grant, 1976; Finck and Stea, 1995). Clark (1987), for example, suggested that rapid ice sheet flow inhibits basal mixing, and produces long dispersal trains. The Lawrencetown Till is associated with relatively narrow drumlin fields and streamlined drift in low-lying areas of NS, consistent with ice stream-generated landforms inferred elsewhere (Patterson, 1998). Drumlin orientations and Escuminac Phase striae suggest that the ice streams converged into inter-bank channels on the outer shelf, drawn by calving of the ice margin in deeper water (Piper, 1991; Shaw et al., 2006)

No till has been associated to date with the Escuminac Phase in NB, but Escuminac Phase flow is represented by widespread south–southwest trending glacial striae, which cross-cut Caledonia flow indicators. Locally, the Escuminac flow has reworked the older Caledonia till, generating a hybrid till with Caledonia geochemical and physical characteristics but an Escuminac fabric (e.g. Allard, 2003; Allard and Pronk, 2003).

48.7.1. Type Area (Eastern Shore NS)

A type area for the Escuminac Phase is the Liscomb/Indian River drumlin field (Grant, 1963; Stea and Brown, 1989) along the Eastern Shore of NS (Figs. 48.1 and 48.6). Palimpsest drumlin forms in this area show a shift in flow from southeastward during the Caledonia Phase to south-southeastward during the Escuminac Phase. This flow shift is also marked by crossing glacial striae on local outcrops.

48.7.2. Type Section (West Lawrencetown, NS)

The type section for the Escuminac Phase is at West Lawrencetown, NS, east of Halifax (Figs. 48.1 and 48.7). It is a section through a SE-trending drumlin found along the coast revealing three compositionally distinct till units (Grant, 1963; Nielsen, 1976). Stea and Pe-Piper (1999) conducted a detailed study of boulder whole rock geochemistry and till fabrics and defined the source and genesis of these till units. The lowermost Hartlen Till with a strong SE fabric and distinctive North Mountain basalt erratics was formed by Caledonia Phase flow. The overlying red Lawrencetown Till contains a suite of granitoid erratics derived from a source in the Cobequid Highlands due north of the section and was formed by Escuminac Phase ice flow. A thin till unit at the top of the section is a hybrid till formed by reworking of the Lawrencetown Till by the penultimate Scotian Phase ice flow (Graves and Finck, 1988). A LIDAR (Light Detection and Ranging) survey of the type section region reveals subtle features of the landscape that were previously obscured by vegetation (Fig. 48.7; Utting, 2009). Based on the LIDAR, the surface morphology of the drumlin is oriented towards the southeast (155°) and not the south, as previously thought based on topographic maps and aerial photographs (Stea and Brown, 1989). The surface form of drumlin matches fabric in upper Beaver River–Lawrencetown hybrid till, which is significant because previously this till was interpreted as an ice-divide till, which only draped the surface of the drumlins (Stea and Pe-Piper, 1999). The Scotian Phase glacier remoulded the surface of the drumlin and may have formed the drumlin itself, suggesting on the Atlantic coast the Scotian Ice Divide featured areas of fast ice flow or ice streams.

48.7.3. Margins of the Escuminac Phase Glacier

Escuminac Phase ice streams likely crossed the inner Scotian Shelf to the outer banks, as they appeared to have crossed the 300-m high Cobequid and Caledonian Highlands (Stea et al., 1986; Broster et al., 1997) and there are no major topographic barriers to prevent their passage. NS-derived metamorphic erratics have been noted in glacigenic deposits sampled at the shelf/slope margin (Mosher et al., 1989). The distinctive Carboniferous red mud that makes up the Lawrencetown Till along the Atlantic Coast can be traced in cores across the Scotian Shelf to the shelf edge (Cok, 1970; Stanley et al., 1972; Hill, 1981; Amos and Miller, 1990). Based on this correlation, a Late Wisconsinan age is assumed for the Escuminac Phase because it represents the last major shelf-crossing glacial event (Fig. 48.4). The age of the Late Wisconsinan calving ice margin at the shelf edge is between 18 and 21 ^{14}C ka BP, as determined by shell dates from piston cores in marginal glacio-marine deposits, interfingering with debris-flow deposits relating to these ice margins (Mosher et al., 1989; Piper, 1991; Baltzer et al., 1994; Table 48.1). Throughout most of the Escuminac Phase and later flow phases, the Magdalen Plateau was an ice divide or rise bypassed by ice streams in the Laurentian Channel, Cape Breton and Chaleur Bay troughs (Fig. 48.4). Cold-based ice located over this region may explain thin till cover over PEI (Prest, 1973) and the style of glaciotectonic deformation on the Magdalen Islands (Dredge et al., 1992). Josenhans and Lehman (1999) demonstrated that the Laurentian Channel was filled with ice prior to ~ 15 ^{14}C ka BP, thus defining the maximum extent of Late Wisconsinan ice streams within the Laurentian Channel.

48.8. LATE WISCONSINAN SCOTIAN PHASE (20–17 KA)

During deglaciation, the Escuminac Ice Divide was gutted by drawdown and calving bay migration into the Laurentian Channel. Local trough-controlled ice streams were 'switched on'. In NS, this drawdown resulted in the formation of the Scotian Ice Divide and ice flow essentially reversed from south to northwards. In NB, ice sheet deflation resulted in a province-wide eastward to northeastward ice flow with a source in the North Maine Ice Divide of Rampton et al. (1984) and a hypothetical Central Maine Ice Centre (Seaman, 2009; Fig. 48.4). Earlier models of the Scotian Ice Divide (e.g. Stea et al., 1989) show a relatively simple arcuate divide over NS, but a more recent conceptual model by Shaw et al. (2006) shows a much more complex triple-junction configuration over northern NS. This model (we believe) is too complex and does not fit the field data, which includes a strong north–northwestward flow through the Northumberland Strait merging with streams out of Chaleur Bay (Fig. 48.4). These models can be reconciled if we consider a sequential development of the Scotian Ice Divide resulting from the westward migration of a calving bay in the Laurentian Channel whereby the first phase of flow is northeastward and then switching to northwestward as thinning of Escuminac ice progressed, and the Chaleur Bay ice stream became dominant

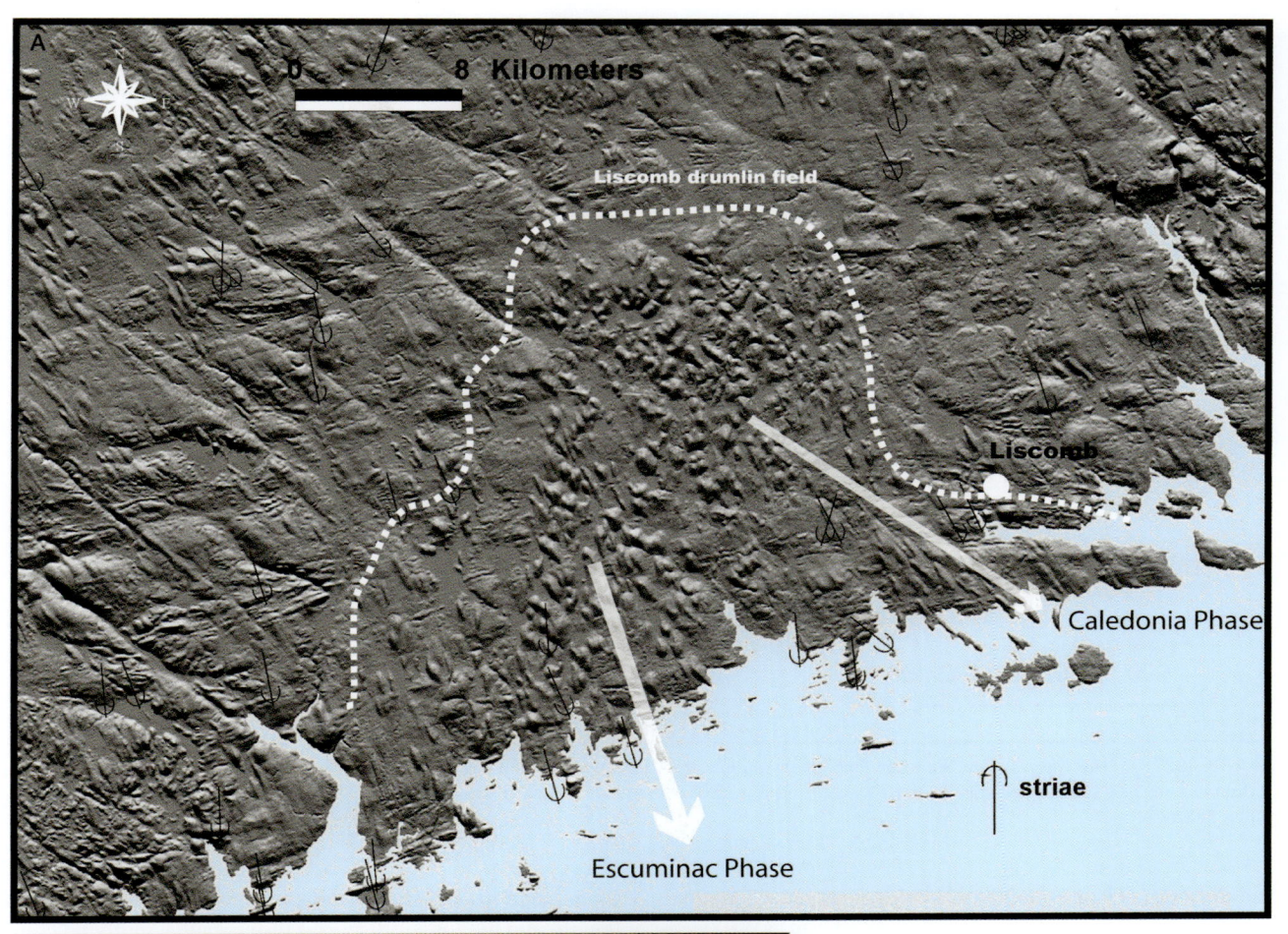

Chapter | 48 The Appalachian Glacier Complex in Maritime Canada

FIGURE 48.6 (A) Digital terrain model of the type area for the Escuminac Phase (location—Fig. 48.1; central mainland NS) showing drumlins and striae (plotted). White arrows show main drumlin/fluting trends. Striae from Stea et al. (1992b). (B) Reference striae site in NB showing crossing Escuminac/Caledonia Phase striae.

Site AS-FV-03, west of Fredericton in west-central NB: Large SSE trending rat-tails (pen-151°-Caledonia Phase) cross-cut by SSW trending striae (compass-191°- Escuminac Phase).

(Fig. 48.4). Our sequential model explains crossing northeast and northwest landforms and striae in northern NS (Fig. 48.7; Stea and Finck, 1984; Grant, 1994; McClenaghan and DiLabio, 1995). It is important to note that northward flow from the Scotian Ice Divide in both configurations crossed highlands in southern Cape Breton

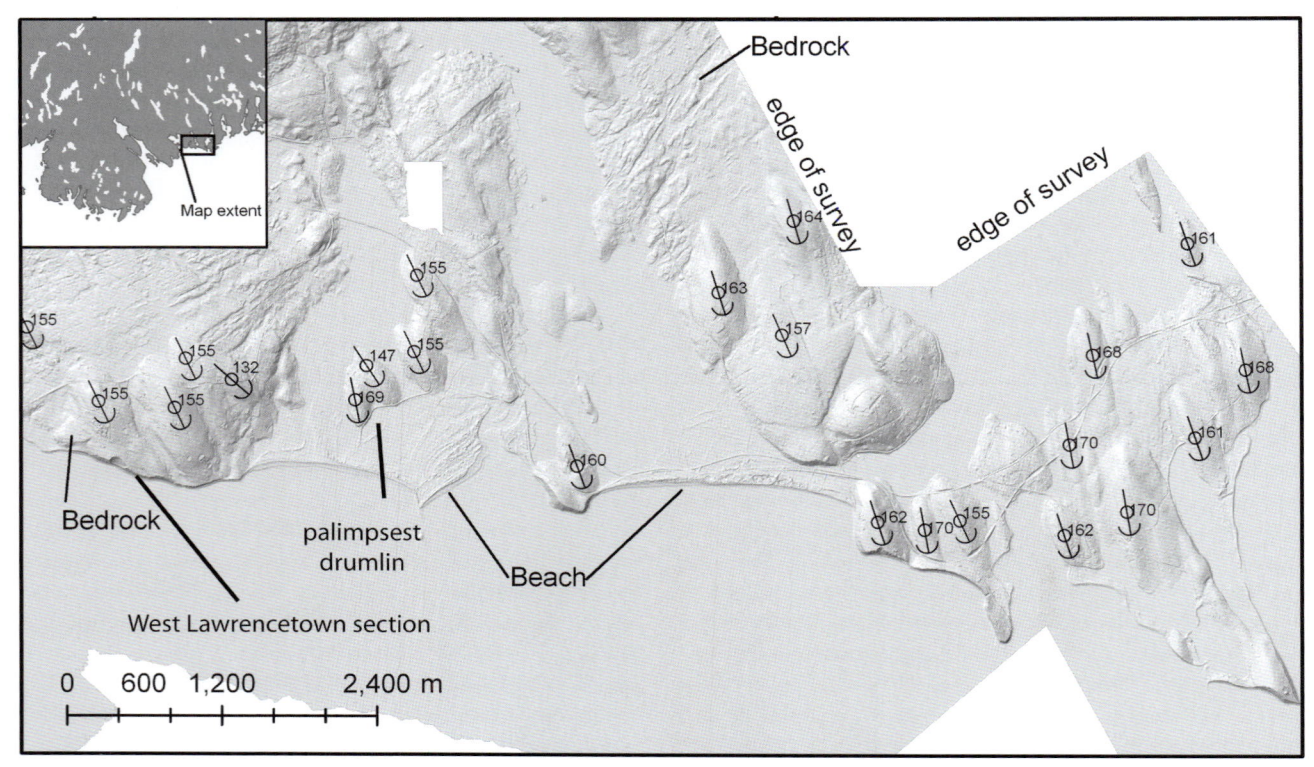

FIGURE 48.7 LIDAR images of the West Lawrencetown section in the Chezzetcook drumlin field just west of Halifax (Fig. 48.1). Two dominant trends of drumlins are 147–155° (Caledonia–Scotian Phases?) and 170–180° (Escuminac Phase). Reoriented (palimpsest) drumlins are found in the field (after Utting, 2009).

and northern NS of 200–300 m in elevation (McClenaghan and DiLabio, 1995, 1996; Stea et al., 2006). Stea (1995) calculated an ice thickness of ~1 km for the Scotian Ice Divide using a modified Nye equation based on the minimum thickness of ice required to cross these highlands. The Mid-Wisconsinan ice dome off Cape Breton Island, proposed by Grant (1977), was probably part of the Late Wisconsinan Scotian Ice Divide whose northwestward flow pattern can be traced across Chedabucto Bay to mainland NS (Fig. 48.4).

The provenance of surface tills in sections along the Georges Bay coast of northern NS and southern Cape Breton indicates a northeastward Scotian Phase flow (Stea et al., 1989; McClenaghan and DiLabio, 1995, 1996). The distinctive, clast-dominated Beaver River Till which overlies both Escuminac and Caledonia Phase tills along the Atlantic coast of NS was deposited by southward and southeastward ice flow from the Scotian Ice Divide (Grant, 1976; Grant and King, 1984; Graves and Finck, 1988).

In west-central, southwestern and south-eastern NB, the Scotian flow pattern is represented by striae and local hybrid tills. However, in central, eastern and northern NB, a number of dispersal trains have been identified, suggesting that a Scotian till was deposited. While dispersal of individual boulders of up to 30 km has been noted, dispersal trains tend to be short, from commonly <0.5–4 km in length, and the longer dispersal trains may actually be compound, reflecting multiple bedrock sources (Lamothe, 1992; Parkhill and Doiron, 2003). Dispersal trains are an order of magnitude shorter than those associated with the Caledonia Phase. This could reflect the slow ice velocities in proximity to the divides, significantly shorter time interval associated with the Scotian Phase, limited addition of newly eroded material to reworked Caledonia till, or a combination of these three factors. Also, the relatively long distance of dispersal of individual clasts may reflect early eastward transport during the Northumberland Phase.

48.8.1. Type Area (South-Eastern Cape Breton Island)

The type area for the Scotian Phase is in south-eastern Cape Breton (Fig. 48.8). Here, ice streams emanating from an early Scotian Phase Divide configuration produced pronounced northeastward-trending elongate, streamlined landforms (drumlins and fluting). This ice stream corridor shows topographically channelled northeastward flow and converging flowlines indicating an area of accelerating flow (Stokes and Clark, 2003).

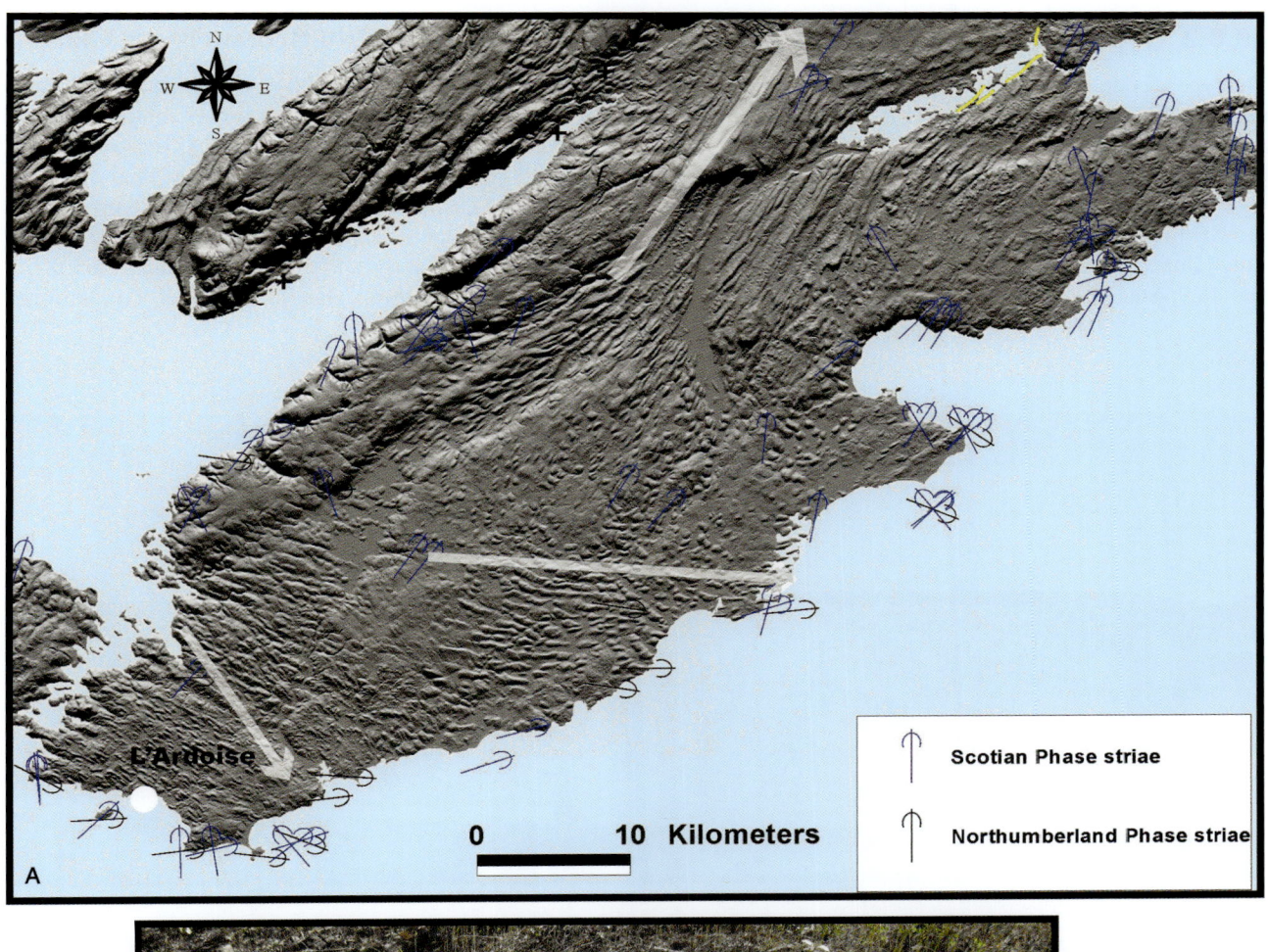

Chapter | 48 The Appalachian Glacier Complex in Maritime Canada

FIGURE 48.8 (A) Digital terrain model of the type area for the Scotian Phase (Fig. 48.1; Cape Breton Island, NS) showing drumlins and striae (plotted). White arrows show main drumlin/fluting trends. Striae from Grant (1994), and McClenaghan and DiLabio (1996). Late SE-trending striae-tentatively assigned to Chignecto Phase omitted from diagram. (B) Crossing striation sets at Belledune NW of Bathurst (Fig. 48.1) 1–096° (Scotian Phase-Baie des Chaleurs flow pattern) 2–056° (Chignecto Phase-Belledune flow pattern) 3–177°(Shulie Lake Phase?) after Wilson et al. (2005).

48.8.2. Type Section (Ballantynes Cove NS)

The type section for Scotian Phase flow is located near Ballantynes Cove near Antigonish, NS. The pebble lithology and provenance of three tills in a 30-m section along the coast of Georges Bay record shifting ice centres beginning with SE flow (Lower Till, Unit 1—Caledonia Phase) to NW Flow (Middle Till, Unit 2—Scotian Phase 1) followed by E flow (Upper Till, Unit 3—Collins Pond Phase?; Figs. 48.1 and 48.9; Stea et al., 1989).

48.8.3. Margins of the Scotian Phase Glaciers

Stea et al. (1998) and Stea (2004) placed the Scotian Phase glacier margin at the series of submerged moraines termed the Scotian Shelf End Moraine Complex (King, 1969, 1996; King and Fader, 1986; Fig. 48.4). Scotian Phase ice-flow patterns, and the Scotian Shelf End Moraine Complex, were linked by moraine orientations and lithological studies of morainal deposits at sea and tills on land (Stea, 1995; King, 1996; Stea et al., 1998). The Beaver River Till deposited

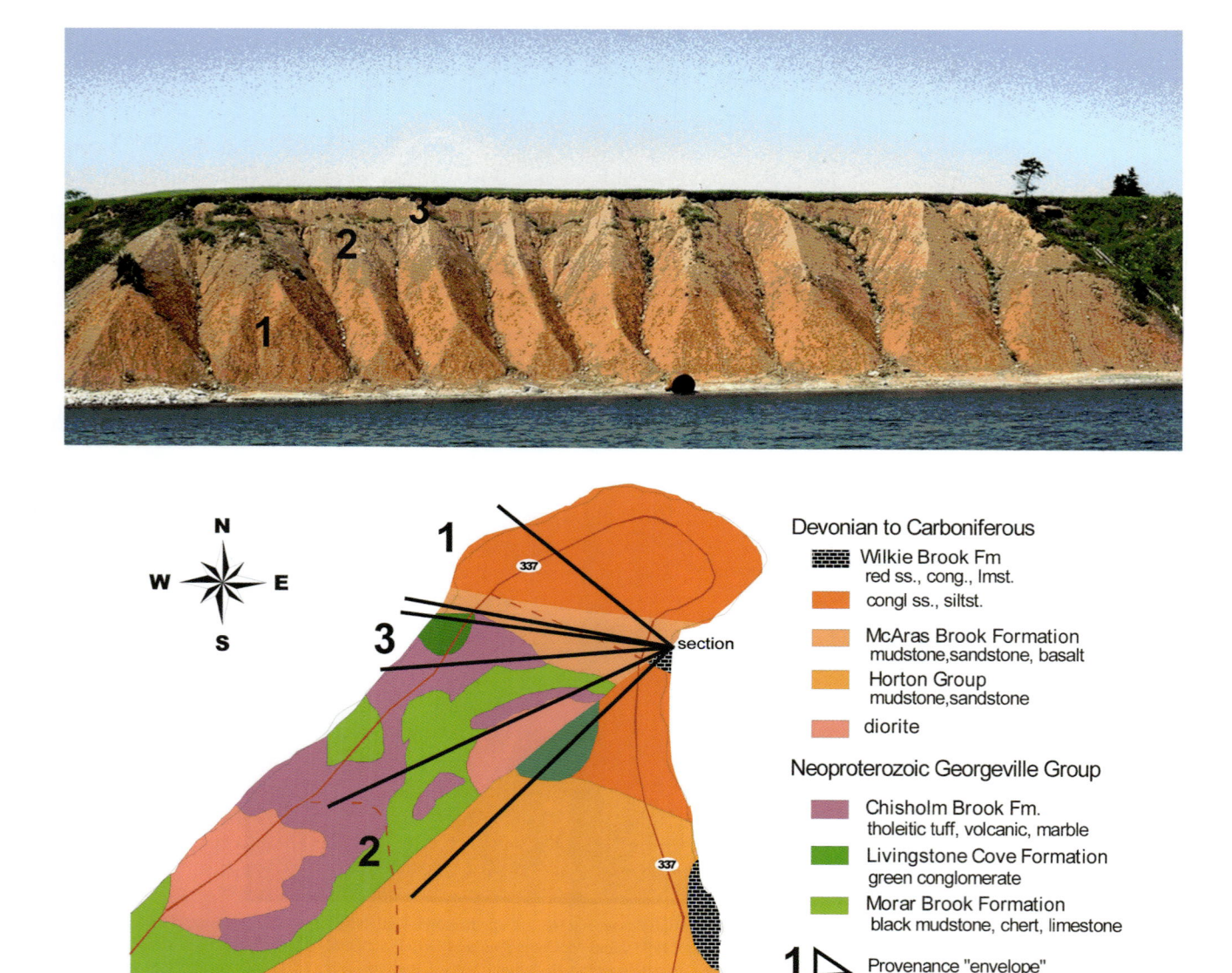

FIGURE 48.9 Type section of Scotian Phase ice flow in northern mainland Nova Scotia (Fig. 48.1—14, Ballantynes Cove). Till Unit 1—Caledonia Phase source area west/northwest. Till Unit 2—Scotian Phase source area southeast. Till Unit 3—post-Scotian Phase source area west (Stea et al., 1989). A ships boiler in the foreground is 4 m high for scale.

under the Scotian Ice Divide consists almost entirely of local metamorphic rocks, as do the off-shore moraine diamictons which can be differentiated lithologically from erratic-rich Caledonia and Escuminac Phase tills (Stea, 1995; Stea et al., 1998).

In order to satisfy symmetry requirements and to explain long sediment records in the outer shelf basins and evidence of local ice rises acting as centres of outflow (Gipp and Piper, 1989; Piper and Fehr, 1991), Stea et al. (1998) suggested that this southern margin of the Scotian Ice Divide on the inner Scotian Shelf may have been buttressed by a short-lived ice shelf, extending out to the shelf edge. An alternative possibility is that grounded ice of the Scotian Phase extended out to the edge of the shelf (Shaw et al., 2006). The location of the divide is likely over NS itself and not offshore as depicted in Shaw et al. (2006) model as the Scotian Phase Beaver River Till shows southward dispersal and northward shelf-based ice-flow indicators have not been documented along the Atlantic coast. The Scotian Shelf End Moraine Complex formed between 17 ka in the northeast and 20 ka in the south-west (King and Fader, 1988; Gipp and Piper, 1989; Piper and Fehr, 1991; King, 1996). Keigwin and Jones (1995) note a peak in ice-rafted debris production in Scotian Slope cores at around 18 ka, without a concomitant decrease in ^{18}O values, suggesting a cooling-related ice advance that may relate to the Scotian Phase.

48.9. CHIGNECTO PHASE (15.9–14.7 KA)

Striae on bedrock exposures along the lowlands that border the NS coast of the Bay of Fundy record an intermittent shifting of ice flow from northeastward to northwestward during the Scotian Phase and finally west- and southwestward during the Chignecto Phase (MacNeill, in Prest et al., 1972; Stea and Finck, 1984; Fig. 48.4). As discussed earlier, the term 'Chignecto Phase' was originally assigned by Rampton et al. (1984) for a restricted SW ice flow at LGM, which is no longer tenable, so rather than abandon the term Stea et al. (1998) decided to use it for a deglacial SW phase of ice flow with a margin similar to that presented in Rampton et al. (1984), equivalent to their 'Bantalor' Phase and 'phase 3' of Foisy and Prichonnet (1991). During this time interval, the North Maine–Notre Dame Ice Divide traversed the northwestern part of NB, generating the northwestward ice-flow trend observed in the Edmundston area (e.g. Rappol, 1989; Parkhill, 2005; Wilson et al., 2005). In southern NB, the centre of outflow migrated southeastward from the Central Maine Ice Centre to the Fundy Highlands Ice Divide, overlying the Caledonian Highlands to the east and the southern part of the St. Croix Highlands to the west (Fig. 48.4). In central NB, northward flow from this ice divide generated scattered northwest to north–northeast trending striae, and probably reoriented the Fredericton drumlins (Seaman, 2009). Further north, interaction between northward flowing ice from the Fundy Ice Divide and southeast flowing ice from the North Maine–Notre Dame Ice Divide may have maintained an ice stream in the Curventon–Bathurst valley (Fig. 48.4).

A suite of streamlined glacial landforms and eskers also define that late westward flow pattern in the lowlands adjacent to the Minas Basin and the Annapolis Valley in NS (Fig. 48.4; Stea et al., 1992a; Rivard et al., 2007). During the Chignecto Phase, small ice caps formed over southern NS (South Mountain Ice Cap—MacNeill and Purdy, 1951), the Northumberland Strait area (Chignecto Glacier—Chalmers, 1895; Northumberland Strait Ice Centre—Rampton and Paradis, 1981a), the Antigonish Highlands (Myers and Stea, 1986) and Cape Breton Highlands (Grant, 1994). These were a result of calving bay ingress into the Bay of Fundy and Cape Breton trough and drawdown or deflation of the Scotian Ice Divide (Fig. 48.4).

48.9.1. Type Area (Central-Mainland, NS)

The type area for the Chignecto Phase is considered to be central mainland NS in the vicinity of the Rawdon Hills (Figs. 48.1 and 48.10). On this horst block highland plateau, the last set of glacial striae inscribed on slate bedrock surfaces trend westward (Fig. 48.10). In the lowlands north of the Rawdon Hills drumlins, fluting and eskers, parallel the latest westward flow trend (Stea and Finck, 1984).

48.9.2. Type Section (Rawdon, NS)

The type section for the Chignecto Flow Phase is located at Rawdon, NS (Figs. 48.1 and 48.10), where a road cut revealed two tills, a lower red-brown till with Cobequid erratics (Hants Till) overlain by a grey silty till with local slate clasts and a strong westward till fabric parallel to local west-trending striae (Stea and Finck, 1984). This upper till was termed the Rawdon till after the type locality (Williams et al., 1985).

48.9.3. Margins of the Chignecto Phase Glaciers

On the Atlantic coast of NS, ice flow during the Chignecto Phase was southwestward from a centre in the Antigonish Highlands (Myers and Stea, 1986; Stea et al., 1992a; Fig. 48.4). The SW margin of the Antigonish Highlands glacier was probably on the inner Scotian Shelf, represented by NW–SE oriented, submarine moraines, perpendicular to the Chignecto Phase flow pattern (Stea et al., 1996, 1998; Fig. 48.4). This offshore margin is estimated to be around 15.1 ka, based on age dates of the offshore lowstand shoreline truncating the moraines and glacio-marine sediment

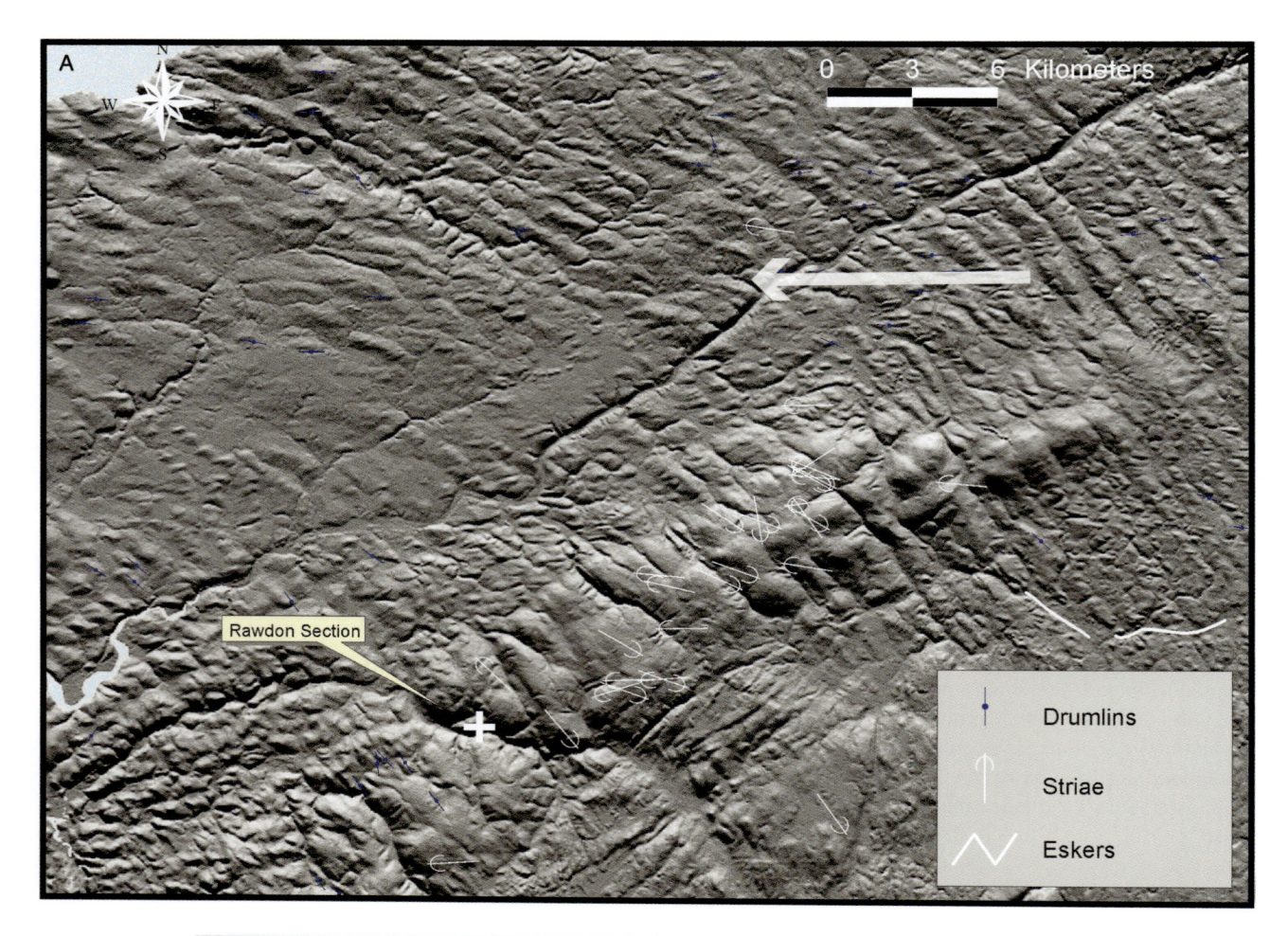

FIGURE 48.10 (A) Type area of the Chignecto Phase (Fig. 48.1; central mainland NS). Note strong SE terrain fabric northeast of the Rawdon Hills, probably a structural feature, as ridges are cored by bedrock. (B) Type section—Rawdon—a road cut where Rawdon Till formed by westward ice flow overlies the Hants Till.

with the moraines (Stea et al., 1996). A glacier lobe or ice stream emanating from both the Antigonish Highlands and highland ice caps in Cape Breton flowed northwards and terminated in 200 m water depth at the mouth of the Cape Breton trough in the Laurentian Channel leaving a distinctive 'till tongue' marking its readvance (Fig. 48.4; Josenhans and

Lehman, 1999) dated at around 13.2 ^{14}C ka BP (Table 48.1). The Antigonish Highland Glacier and possibly a cap on the Cobequid Highlands fed ice into the southern lowlands of the Minas Basin–Annapolis Valley terminating most likely at Margaretsville, in the middle of the Annapolis Valley where ice-contact deposits related to a westward flowing lobe of ice are in contact with outwash and an outwash delta (Hickox, 1962; Stea and Mott, 1998; Stea, 2004; Rivard et al., 2007). While the age of this delta is not directly known because it is lacking in fossils, a raised marine deposit at Brier Island Bog, a few kilometres south of the delta, produced shell ages of 13.2 ^{14}C ka BP (Stea and Mott, 1998).

The Chignecto Glacier, another one of the residual ice centres from the Scotian Ice Divide, was located somewhere in the PEI, Northumberland Strait area, and flowed southwestward into Chignecto Bay (Fig. 48.4). Ice margins for the Chignecto Glacier were most likely along the north shore of the Minas Basin and are represented by outwash delta deposits termed the Five Islands Formation and Minas Terrace (Swift and Borns, 1967; Wightman, 1980). These deposits have been dated around 13 ^{14}C ka BP using *in situ* *Portlandia arctica* shells in the deltaic sediments (Stea and Wightman, 1987) and AMS (wood) basal sediment dates from kettle lakes at the margin (Stea and Mott, 1998). The outwash fans were previously thought to relate to the Late Wisconsinan glacier maximum as in the 'minimum model' (Grant, 1977).

In southern NB, the margin of the Chignecto Phase/Fundy Ice Divide was essentially defined by moraines and deltas along the northern shore of the Bay of Fundy. With calving into the rising glacio-isostatic De Geer Sea (Grant, 1989), the Bay of Fundy was largely ice free by the middle of the Chignecto Phase. The Chignecto ice margin existed as a tidewater margin resulting in the formation of a system of impressive moraines and ice-contact deltas extending approximately parallel to the shoreline from the eastern side of Saint John Harbour to Campobello Island in the west (Figs. 48.1 and 48.4; Gadd, 1973; Rampton et al., 1984; Seaman et al., 1993; Seaman, 2004). This system continues westward into Maine as the Pond Ridge moraine (e.g. Borns et al., 2004). Radiocarbon dates on marine shells from deltas and moraines range from 14.4 to 12.6 ^{14}C ka BP (Table 48.1). The magnitude of the Pennfield-Utopia delta complex and a wide range of radiocarbon dates from the Sheldon Point Moraine indicate that the ice margin remained close to this location for most of the Chignecto Phase (Nicks, 1988; Seaman et al., 1993) while the remainder of NB remained under glacial cover. Discounting a couple of radiocarbon dates that Lamothe (1992) considered to be questionable, the available data suggest that the waters of the Goldthwait Sea in the Gulf of St. Lawrence did not encroach into the current onshore area of northeastern or eastern NB until after 13 ^{14}C ka BP (14.7 cal. ka BP), near the termination of the Chignecto Phase.

48.10. SHULIE LAKE PHASE (~13.8 KA)

A short-lived readvance of the Chignecto Glacier called the Shulie Lake Phase was proposed by Stea et al. (1986). They defined an ice margin on the Chignecto Peninsula north of the Minas Basin marked by the pinch-out of a till sheet (Shulie Lake Till) and the formation of a cross-valley moraine (Gilbert Lake Moraine) in the Parrsboro Gap (Fig. 48.11). The moraine was dated between 13 and 11.6 ^{14}C ka BP (Fig. 48.11; Stea and Mott, 1998; Table 48.1) based on lake core bottom, AMS radiocarbon dates from Leak Lake south of the moraine and from Gilbert Lake which is dammed just behind the Gilbert Lake Moraine. It is likely that this readvance correlates with a minor readvance in NB at the end of the Chignecto Phase marked by glacial over-riding of the Sheldon Point and St. George moraines (Nicks, 1988; Seaman, 2009) and the formation of the Pineo Ridge moraine of Eastern Maine. (Thompson et al., 2007) equivalent to the Older Dryas Chronozone of Europe (12–11.8 ^{14}C ka BP; ~14 cal. ka BP; Lowe and Walker, 1998). A late southward flow in NE NB mapped by Parkhill (1994) is tentatively placed in this phase. The flow crosses Chaleur Bay likely terminating in a terrestrial margin south of Bathurst, implying lowering of the glacio-isostatic Goldthwait Sea (Grant, 1989). Indeed, shallow raised marine features mapped in western PEI (Prest, 1973) were dated to 12.4 ^{14}C ka BP (GSC-101), suggesting that isostatic rebound was well underway just preceding the Shulie Lake Phase.

The Shulie Lake Phase preceded a short-lived climatic cooling known as the Killarney Oscillation (11.2–10.9 ^{4}C ka BP—Levesque et al., 1993).

48.10.1. Type Area and Section (western NS)

The type area and type sections for the Shulie Lake Phase are located on the Chignecto Peninsula in NS between Chignecto Bay and the Minas Basin. Surficial mapping in this region revealed a distinct terrain boundary defined by the pinch-out of the Shulie Lake Till with strongly rolling and incised terrain south of the boundary and flatter terrain with ribbed moraine north of the boundary (Fig. 48.11; Stea et al., 1986). The Shulie Lake Till, a sandy stony till facies, outcrops north of the boundary while south of the boundary, a reddish-brown silty till (Eatonville Till) is found. Soil development is significantly greater on the Eatonville Till south of the margin (Wang et al., 1986). The Shulie Lake terrain boundary is considered an 'indistinct' glacier margin without development of an end moraine, analogous to the Olean Drift Boundary in Pennsylvania (Crowl and Sevon, 1980).

At the Shulie Lake Phase limit, a backhoe excavation revealed the Shulie Lake Till overlying the Eatonville Till and this is considered the type section for this advance (Fig. 48.1; Stea et al., 1986).

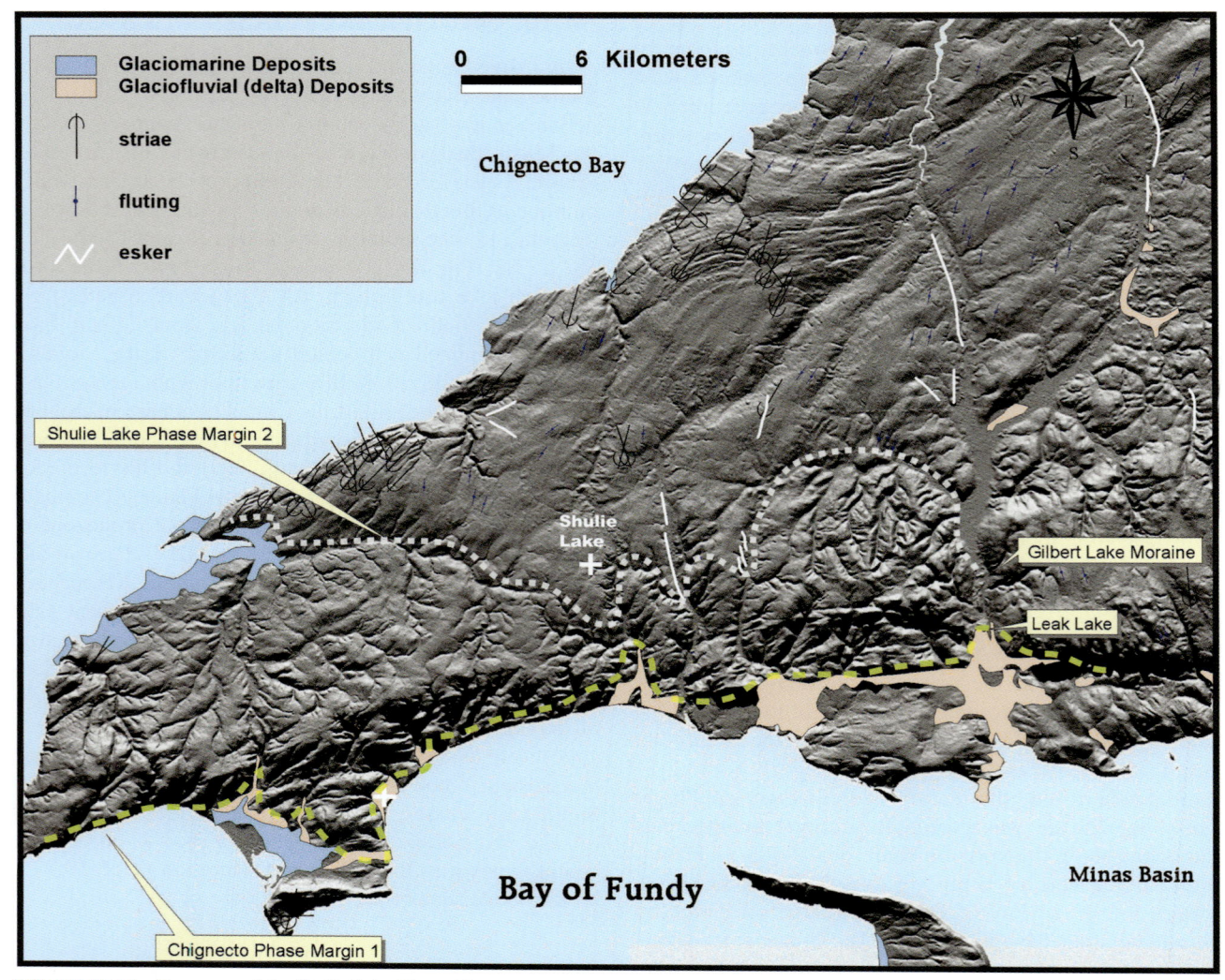

FIGURE 48.11 Type area of the Shulie Lake Phase (Fig. 48.1; west-mainland NS).

48.11. COLLINS POND PHASE (YOUNGER DRYAS CHRONOZONE—12.9–11.7 KA)

A prolonged period of climatic warming and ice retreat post-dating the Shulie Lake Phase resulted in the dissipation of most glacier ice in Maritime Canada and a rapid fall in sea level (Mott, 1994; Stea et al., 1998). This was interrupted by an abrupt and pronounced phase of climatic cooling, dated just before 11.0 ^{14}C ka BP that strongly affected the terrestrial landscape, its vegetation cover and the nature of sedimentation and fauna of adjacent oceanic basins (Mott et al., 1986). Various authors had proposed that glaciers formed and were reactivated during this time (Borns, 1966; Grant, 1989; Stea and Mott, 1989, 1998, 2005; Lamothe, 1992; Grant, 1994; King, 1994; Mott and Stea, 1993; Seaman, 2006), based on ~30 sites with glacigenic deposits overlying organic beds. Direct evidence for an ice readvance during the Younger Dryas occurs at Collins Pond, NS (Stea and Mott, 1989, 2005; Mott and Stea, 1993) and in central NB (Lamothe, 1992; Seaman, 2006; Seaman and McCoy, 2008) where glacial till overlies deformed peat beds. Stea and Mott (1998) also describe glacial lake sediments in south-west Cape Breton that overlie an 11 ka peat bed. Offshore, Piper and Fehr (1991), King (1994) and Stea et al. (1996) described a distinctive seismic marker horizon in glacio-marine sediments formed during the Younger Dryas Chronozone.

In central NB, some of the morainal deposits originally attributed to their Millville/Dungarvon phase, ca. 12.7 ka, by Rampton et al. (1984), are now known to relate to the western margins of the Gaspereau Ice Centre during the Younger Dryas, ca. 10.7 ka (Seaman, 2009). In southwestern NB, the morainal systems that marked the Millville/Dungarvon phase boundary have been identified as disintegration moraine (Seaman et al., 1993), formed by downwasting following the Shulie Lake Phase. Therefore, the name 'Millville/Dungarvon phase' is retired. In addition, the late and contemporaneous Plaster Rock and Chaleur Phases of Rampton et al. (1984), inferred to relate to events ca. 12.2 ka, are now interpreted to more likely represent Younger Dryas events. However, this interpretation remains to be confirmed.

48.11.1. Type Section (Collins Pond, NS)

The Collins Pond site is considered the type section of glacigenic deposits of the Younger Dryas Chronozone in eastern Canada (Stea and Mott, 1989, 2005). At the site, sections through two drumlins revealed peat and wood under a compacted till. The peat layer was dated from 12.1 to 10.9 ^{14}C ka BP (~14–12.9 cal. ka BP). A reference section in NB for the Collins Pond Phase is in the same mineral exploration trench at Sisson Brook used as a type section for the Northumberland Phase (Figs. 48.1 and 48.2A and 48.12). At this site, the Collins Pond till overlies and deforms an Allerød organic silty clay horizon and a lower Caledonia till. Till fabric trends and deformation structures of the Collins Pond phase till suggest deposition by a glacier flowing towards the west to northwest (Fig. 48.12; Seaman and McCoy, 2008) attributed to the Gaspereau Ice Centre.

48.11.2. Margins of the Collins Pond Phase Glaciers

The margins of the Younger Dryas glacier complex were not previously known, largely because of the lack of till-buried organic sites, that would allow for correlation of a till sheet with ice-marginal deposits (Stea and Mott, 1998). This situation changed, however, in the summer of 1999, when the Maritimes and Northeast Pipeline Company excavated a continuous, 3-m trench across northern mainland NS, to host the Sable Island gas pipeline (Stea and Mott, 2005). A paleosol, with preserved soil A-horizon (peat and wood), was found buried under 2–10 m of surface till at six sites over a wide area of the pipeline route. Till fabric analyses in the surface till sheet from three sites indicate a strong fabric parallel to regional late south- and southwestward glacial lineations. Radiocarbon dates on wood from the buried paleosol cluster around 10.9 ka (Stea and Mott, 2005; Table 48.1). The regional till sheet overlying the soil can be traced to ice-marginal deposits near the Cobequid Highlands to the south, including ice-dammed glaciolacustrine sediments overlying peat found along the coasts of northern NS and south-west Cape Breton Island (Fig. 48.4). The source of this southward flowing ice must be near PEI, the site of other Late Wisconsinan ice centres of the AGC.

FIGURE 48.12 Reference section of Younger Dryas till at Sisson Brook, NB (Fig. 48.1).

The margins of the Collins Pond Phase glaciation in NB are poorly constrained. Ice margins illustrated in Fig. 48.4 are hypothetical, based on the following limits:

1. Glacial Lake Acadia in the central Saint John River valley and Lake Madawaska in the upper Saint John River valley remained open during the Younger Dryas (Rampton et al., 1984);
2. Goldthwait Sea marine submergence features in eastern NB were not over-ridden by Younger Dryas glaciers (Rampton et al., 1984; Seaman, 2009);
3. Joe Lake in central NB (Mayle et al., 1993) was not glaciated during the Younger Dryas; and
4. Southwestern NB was deglaciated prior to the Younger Dryas and remained deglaciated (Seaman, 2006; Dickinson, 2008).

48.12. DISCUSSION: PROBLEMS OF THE QUATERNARY IN MARITIME CANADA

The formation and reactivation of both highland and lowland ice caps during the Younger Dryas in NS is a key to the understanding of the initiation of the AGC during the Wisconsinan. During the YD, positive mass balance had to be maintained by high snowfall and cooler temperatures because these small glaciers developed near sea level. Albedo-feedback cooling and formation of blocking high pressure cells over the LIS in the high latitude regions forced the jet stream and storm tracks southward over the Maritime-Appalachian region (e.g. Isarin and Renssen, 1999). Moisture-laden air masses from the Gulf of Mexico followed the northeast jet stream path and produced heavy snowfall accumulations, nourishing the incipient glaciers. This process likely occurred in the initiation of glaciation during MIS 4 and the reorganisation of glaciers during MIS 3.

The concepts of LIS incursion in Maritime Canada, prevalent in the past, were based partly on an assumption that the region was too far south for the development of ice sheets. Stratigraphical sections in Maritime Canada spanning the MIS 5/4 transition and the Allerød/Younger Dryas are similar, showing the development of ice advance glacial lakes before till cover. Glaciers that dammed these lowland basins had to be formed in local highlands, because of the position of outlets relative to regional ice flow (Stea and Mott, 2005). Initial outflow of Appalachian highland glaciers during MIS 4 was defined by striae and local tills (Grant, 1994).

After the Caledonia Phase, the formation of calving bays in the Laurentian Channel and Bay of Fundy (Fig. 48.4) left an ice mass stranded over the Magdalen Plateau during MIS 3. With renewed cooling during MIS 2 and by the climatic processes described above, this remnant ice mass was reactivated into the Escuminac Ice Divide. Indeed, the presence of the plateau can be used as an argument against Laurentide incursion during the LGM and in the many glaciations before, as one might expect ice streams to have eroded swaths into the soft red beds that make up the geology of that prominent physiographic region. The development of the Escuminac centre may have been hastened by the freezing of glacial lakes dammed against surrounding uplands, in the mode described by Denton and Hughes (1981).

Ice streams in the lowland and marine channels bordering the Maritime Provinces and sea-level rise controlled ice sheet configurations during the LGM and during regional deglaciation (Mayewski et al., 1981; Grant, 1989; Stea et al., 1998; Shaw et al., 2006). The major flow shifts during the Scotian Phase were in all likelihood initiated by sea-level rise and enhanced ice streaming. However, it is interesting to note that computer modelling of the LGM and deglaciation, using the glacial/dynamic ice sheet model at the University of Maine, was unable to reproduce the Scotian Phase configuration no matter what factors were used for ice streaming (J. Fastook, personal communication, 2005). Climatic inputs for this model were rather rudimentary, so if local precipitation increases were factored in a different result may accrue.

48.13. CONCLUSIONS

In conclusion, we will list some of the major problems addressed in this chapter and some that need further study.

1. Record of glaciations in eastern Canada. A stratigraphical record of glaciations and interglacial periods prior to MIS 6 in Maritime Canada appears to be absent.
2. Extent and duration of LIS during the Wisconsinan. The nature of the interaction between the LIS and the Notre Dame Ice Divide during the Caledonia Phase. Provenance of western PEI and northwest NB gneissic, metamorphic and anorthosite boulders-Appalachian or Laurentide?
3. The nature and action, if any, of the Mid-Wisconsinan glacial cover.
4. Age and correlation of flow patterns. In many areas, flow directions during several phases were similar. An example is SE flow along the Atlantic shore (Fig. 48.7) where Caledonia, Escuminac and Scotian flows in many areas have a similar trajectory. Till sections and stratigraphy are an invaluable tool in sorting out these events, but problems still remain. For example, were SE-trending drumlins on eastern shore of NS formed during the deglacial Scotian Phase, rather than Caledonia Phase?
5. Inception, development of glaciers in Maritime Canada. The extent of highland glaciation in advance of the main southeast flow of the Caledonia Phase. The climatic and dynamic factors that produced local ice divides over Maritime Canada. Were ice streams the controlling

factor or did climatic factors also play a role? Our empirical model for the Scotian Phase glacier features a divide extending on to the continental shelf and with ice thickness of 1 km or more. Is this physically plausible?

REFERENCES

Alcock, W., 1984. A sedimentologic and stratigraphic study of Wisconsin tills in an area of the Cobequid Highlands, north-western Nova Scotia. M.Sc. thesis, University of Toronto, 255 pp.

Allard, S., 2003. Geochemistry, lithology, and depositional history of the Rollingdam (NTS 21 G/06) map area till, southwestern NB. New Brunswick Department of Natural Resources and Energy; Minerals, Policy and Planning Division, Open File, 2003-2, 218 pp.

Allard, S., Pronk, T., 2003. Glacial history of southwestern New Brunswick: challenges to drift prospecting. In: Carroll, B.M.W. (Ed.), Current Research 2002, New Brunswick Department of Natural Resources; Minerals, Policy and Planning Division, Mineral Resource Report 2003-4, pp. 1-11.

Amos, C.L., Miller, A.A.L., 1990. The Quaternary stratigraphy of southwest Sable Island Bank, eastern Canada. Geol. Soc. Am. Bull. 102, 915–934.

Bailey, L.W., 1898. Report on the geology of south-western Nova Scotia. Geological Survey of Canada, Annual Report, 9, pt. M, pp. 30–60.

Baltzer, A., Cochonat, P., Piper, D.J.W., 1994. In situ geotechnical characterization of sediments on the Nova Scotian slope, eastern Canadian Continental margin. Mar. Geol. 120, 291–308.

Barr, S.M., White, C.E., Venugopal, D.V., Hamilton, M.A., Stirling, J.A.R., 2001. Petrology and age of the lower Coverdale high-Ti, P, and V gabbro-anorthosite complex and associated granite, Moncton area. Abstracts Volume; Annual Review of Activities, New Brunswick, Minerals and Energy Division, Report. 2001-1, pp. 3–6.

Belknap, D.F., Shipp, R.C., Kelley, J.T., Schnitker, D., 1989. Depositional sequence modelling of late Quaternary geologic history, west central Maine. In: Tucker, R.D., Marvinney, R.G. (Eds.), Studies in Maine Geology, vol. 5, Quaternary Geology. Maine Geological Survey, Augusta, pp. 29–46.

Bond, G.C., Lotti, R., 1995. Iceberg discharges into the North Atlantic on millennial time scales during the last glaciation. Science 267, 1005–1010.

Borns, H.W., 1966. The geography of paleo-Indian occupation in Nova Scotia. Quaternaria 8, 49–57.

Borns Jr., H.W., Doner, L.A., Dorion, C.C., Jacobsen Jr., G.L., Kaplan, M.R., Kreutz, K.J., et al., 2004. The deglaciation of Maine, U.S.A. In: Ehlers, J., Gibbard, P.L. (Eds.), Quaternary Glaciations—Extent and Chronology, Part II: North America. Developments in Quaternary Science vol. 2b. Elsevier, Amsterdam, pp. 89–109.

Broster, B.E., Seaman, A.A., 1991. Glacigenic rafting of weathered granite: Charlie Lake, New Brunswick. Can. J. Earth Sci. 28, 649–654.

Broster, B.E., Munn, M.D., Pronk, A.G., 1997. Inferences of glacial flow from till clast dispersal, Waterford area, New Brunswick. Géogr. Phys. Quatern. 51, 29–39.

Catto, N.R., 1998. Comparative study of striations and basal till clast fabrics, Malpeque-Bedeque region, Prince Edward Island, Canada. Boreas 27 (4), 251–274.

Causse, C., Hillaire-Marcel, C., 1986. Géochemie des familles U et Th dans la matière organique fossile des dépôts interglaciares et interstadiares de l'est et du nord du Canada; potentiel radiochronologique; dans recherches en cours. Partie B; Commission geologique du Canada, Étude 86-lB, pp. 11–18.

Chalmers, R., 1885. Preliminary report on the surface geology of New Brunswick. Geological and Natural History Survey of Canada, Annual Report, I, Part GG, 58 pp.

Chalmers, R., 1895. Report on the surface geology of eastern New Brunswick, north-western Nova Scotia and a Portion of Prince Edward Island. Geological Survey of Canada, Annual Report, 1894, 1 (7), pt. m, 144 pp.

Charbonneau, R., David, P., 1993. Glacial dispersal of rock debris in central Gaspesie, Québec, Canada. Can. J. Earth Sci. 30, 1697–1707.

Clark, U., 1987. Subglacial sediment dispersal and till composition. J. Geol. 95, 527–541.

Clark, U., Clague, J.J., Curry, B.B., Dreimanis, A., Hicock, S., Miller, G.H., et al., 1993. Initiation and development of the Laurentide and Cordilleran ice sheets following the last glaciation. Quatern. Sci. Rev. 12, 79–114.

Cok, A.E., 1970. Morphology and surficial sediments of the eastern half of the Nova Scotia shelf. Unpublished Ph.D. dissertation, Dalhousie University, 261 pp.

Crowl, G.H., Sevon, W.D., 1980. Glacial border deposits of Late Wisconsinan age in northeastern Pennsylvania. Pennsylvania Department of Environmental Resources General Geology Report 71, 23 pp.

de Vernal, A., Richard, P., Occhietti, S., 1983. Palynologie et paléoenvironnements du Wisconsinien de la région de la Baie Saint-Laurent. Ile du Cap Breton, Geogr Phys Quatern; Proc NS inst.sc; T Roy Soc Can 37, 307–322.

de Vernal, A., Causse, C., Hillaire-Marcel, C., Mott, R.J., Occhietti, S., 1986. Palynostratigraphy and Th/U ages of Upper Pleistocene interglacial and interstadial deposits on Cape Breton Island, Eastern Canada. Geology 14, 554–557.

Denton, G.H., Hughes, T.J., 1981. The Last Great Ice Sheets. John Wiley and Sons, Inc., Toronto, Ontario, 484 pp.

Dickinson, P.J., 2008. Geomorphological processes and the development of the lower Saint John River human landscape. Unpublished Ph.D. thesis, University of New Brunswick, 203 pp.

Dodds, D.J., Fader, G.B.J., 1986. A combined seismic reflection profiler and sidescan Sonar System for deep ocean geological surveys. In: Merklinger, H. (Ed.), Proceedings of the 12th International Congress on Acoustics. Plenum Press, New York, pp. 169–179.

Dredge, L.A., Thorleifson, L.H., 1987. The middle Wisconsinan history of the Laurentide ice sheet. Géogr. Phys. Quatern. 41 (2), 215–236.

Dredge, L., Mott, R.J., Grant, D.R., 1992. Quaternary stratigraphy, paleoecology & glacial geology, Iles de la Madelaine, Québec. Can. J. Earth Sci. 29, 1981–1996.

Dyke, A.S., Prest, V.K., 1987. Late Wisconsinan and Holocene history of the Laurentide ice sheet. In: Fulton, R.J., Andrews, J.T. (Eds.), The Laurentide Ice Sheet, Géographie physique et Quaternaire 41 237–264.

Dyke, A.S., Andrews, J.T., Clark, P.U., England, J.H., Miller, G.H., Shaw, J., et al., 2002. The Laurentide and Innuitian Ice Sheets during the Last Glacial Maximum. Quatern. Sci. Rev. 21 (1–3), 9–31.

Elson, J.A., 1987. West-southwest dispersal of pillow-lava boulders, Philipsburg—Sutton region, Eastern Townships, Quebec. Can. J. Earth Sci. 24, 985–991.

Finck, W., Stea, R.R., 1995. The compositional development of tills overlying the South Mountain Batholith. Nova Scotia Department of Natural Resources; Mines and Minerals Branch, Paper 95–1, 51 pp.

Fisher, B.E., Poole, J.C., McKinnon, J.S., 2006. Shaded relief images derived from a 25 metre digital elevation model of the Province of Nova Scotia. Nova Scotia Department of Natural Resources, DP ME 56.

Flint, R.F., 1951. Highland centres of former glacial outflow in northeastern North America. Geol. Soc. Am. Bull. 62, 21–38.

Flint, R.F., 1971. Glacial and Quaternary Geology. John Wiley and Sons, London, 892 pp.

Foisy, M., Prichonnet, G., 1991. Reconstruction of glacial events in southeastern New Brunswick. Can. J. Earth Sci. 28, 1594–1612.

Fulton, R.J., 1984. Summary: Quaternary stratigraphy of Canada. In: Fulton, R.J. (Ed.), Quaternary Stratigraphy of Canada—A Canadian Contribution to IGCP Project 24, Geological Survey of Canada Paper 84-10, pp. 1–5.

Gadd, N.R., 1973. Quaternary geology of southwest New Brunswick with particular reference to Fredericton area. Geological Survey of Canada Paper 71–34, 31 pp.

Gauthier, C. 1980. Decomposed granite, Big Bald Mountain area, New Brunswick. In Current Research, Part B. Geological Survey of Canada, Paper 80-1B, pp. 277–282.

Gipp, M.R., 1994. Late Wisconsinan deglaciation of Emerald Basin, Scotian Shelf. Can. J. Earth Sci. 31, 554–566.

Gipp, M.R., Piper, D.J.W., 1989. Chronology of Late Wisconsinan glaciation, Emerald Basin, Scotian Shelf. Can. J. Earth Sci. 26, 333–335.

Godfrey-Smith, D.I., Grist, S.W., Stea, R.R., 2003. Dosimetric and radiocarbon chronology of a pre-Wisconsinan mastodon fossil locality at East Milford, Nova Scotia, Canada. Quatern. Sci. Rev. 22, 1353–1360.

Goldthwait, J.W., 1924. Physiography of Nova Scotia. Geological Survey of Canada, Memoir 140, 179 pp.

Grant, D.R., 1963. Pebble lithology of the tills of south-east Nova Scotia. M.Sc. thesis, Dalhousie University, Halifax, Nova Scotia, 235 pp.

Grant, D.R., 1976. Tills of the Appalachian Province. In: Leggett, R.F. (Ed.), Glacial Till, an Inter-Disciplinary Study, The Royal Society of Canada, Special Publication, 12 52–57.

Grant, D.R., 1977. Glacial style and ice limits, the Quaternary stratigraphic record & changes of land and ocean level in the Atlantic Provinces, Canada. Géogr. Phys. Quatern. 31 (3–4), 247–260.

Grant, D.R., 1980a. Quaternary sea-level change in Atlantic Canada as an indication of crustal delevelling. In: Mörner, N.-A. (Ed.), Earth Rheology, Isostasy and Eustasy. John Wiley and Sons, London, pp. 201–214.

Grant, D.R., 1980b. Quaternary stratigraphy of southwestern Nova Scotia: glacial events and sea level changes. Geological Association of Canada, Annual Meeting (Halifax), Field Trip Guidebook. 63 pp.

Grant, D.R., 1989. Quaternary geology of the Atlantic Appalachian region of Canada. In: Fulton, R.J. (Ed.), Chapter 5 in Quaternary Geology of Canada and Adjacent Greenland, 393–440 Geological Survey of Canada, Geology of Canada no. 1 (also Geological Society of America, The Geology of North America, K-l).

Grant, D.R., 1994. Quaternary Geology, Cape Breton Island. Geol. Surv. Can. Bull. 482, 159 pp.

Grant, D.R., King, L.H., 1984. A stratigraphic framework for the Quaternary history of the Atlantic Provinces. In: Fulton, R.J. (Ed.), Quaternary Stratigraphy of Canada—A Canadian Contribution to IGCP Project 24, Geological Survey of Canada Paper 84–10, pp. 173–191.

Graves, R.M., Finck, W., 1988. The provenance of tills overlying the eastern part of the South Mountain Batholith, Nova Scotia. Marit. Sed. Atlantic Geol. 24, 61–70.

Harington, C.R., Grant, D.R., and Mott, R.J. 1993. The Hillsborough, New Brunswick, mastodon and comments on other Pleistocene mastodon fossils from Nova Scotia. Canadian Journal of Earth Sciences, 30, pp. 1242–1253.

Hickox, Jr., C.F., 1962. Pleistocene geology of the central Annapolis Valley, Nova Scotia. Nova Scotia Department of Mines Memoir, 5, 36 pp.

Hill, R., 1981. Detailed morphology and Late Quaternary sedimentation on the Nova Scotian Slope south of Halifax. Unpublished Ph.D. thesis, Dalhousie University, Halifax, 331 pp.

Huppertz, T.J., Piper, D.J.W., 2009. The influence of shelf-crossing glaciation on continental slope sedimentation, Flemish Pass, eastern Canadian continental margin. Mar. Geol. 265, 67–85.

Isarin, R.F.B., Renssen, H., 1999. Reconstructing and modelling late Weichselian climates; the Younger Dryas in Europe as a case study. Earth Sci. Rev. 48 (1–2), 1–38.

Josenhans, H., Lehman, S., 1999. Late glacial stratigraphy and history of the Gulf of St. Lawrence, Canada. Can. J. Earth Sci. 36, 1327–1345.

Keigwin, L.D., Jones, G.A., 1995. The marine record of deglaciation from the Continental margin off Nova Scotia. Paleoceanography 10, 973–985.

King, L.H., 1969. Submarine end moraines and associated deposits on the Scotian Shelf. Geol. Soc. Am. Bull. 80 (1), 83–96.

King, L.H., 1994. Proposed Younger Dryas glaciation of the eastern Scotian Shelf. Can. J. Earth Sci. 31, 401–417.

King, L.H., 1996. Late Wisconsinan ice retreat from the Scotian Shelf. Geol. Soc. Am. Bull. 108, 1056–1067.

King, L.H., Fader, G.B.J., 1986. Wisconsinan glaciation of the continental shelf of south-east Atlantic Canada. Geol. Surv. Can. Bull. 363, 72 pp.

King, L.H., Fader, G.B.J., 1988. A comparison between the Late Wisconsinan history of south-west and north-east Emerald Basins. Geological Survey of Canada Open File Report 2060, 12 pp.

King, L.H. (1970). Surficial geology of the Halifax-Sable Island map area. Marine Science Branch, Geological Survey of Canada Paper 1, 16 pp.

Kite, J.S., Lowell, T.V., Nicholas, G.P., 1982. Quaternary studies in the upper St. John River basin, Maine and New Brunswick. Guidebook for the 1982 NBQUA field trip 1982, 54 pp.

Lamothe, M., 1992. Pleistocene stratigraphy and till geochemistry of the Miramichi Zone, New Brunswick. Geol. Surv. Can. Bull. 433, 58.

Levesque, A.J., Mayle, F.E., Walker, I.R., and Cwynar, L.C. 1993. A previously unrecognized late-glacial cold event in eastern North America. Nature, 361, 623–626.

Loring, D.H., Nota, D.J.G., 1973. Morphology and sediments of the Gulf of St. Lawrence. Fish. Res. Board Can. Bull. 182, 147.

Lowe, J.J., Walker, M.J.C., 1998. Reconstructing Quaternary environments. Longman, Harlow, UK 446 pp.

Marshall, S.J., James, T.S., Clarke, G.K.C., 2002. North American Ice Sheet reconstructions at the Last Glacial Maximum. Quatern. Sci. Rev. 21 (1–3), 175–192.

Mayewski, A., Denton, G.H., Hughes, T.J., 1981. Late Wisconsin Ice Sheets in North America. In: Denton, G.H., Hughes, T.J. (Eds.), The Last Great Ice Sheets. John Wiley & Sons, New York, 484 pp.

Mayle, F.E., Levesque, A.J., Cwynar, L.C., 1993. Accelerator-mass-spectrometer ages for the Younger Dryas event in Atlantic Canada. Quatern. Res. 39, 355–360.

McClenaghan, M.B., Dilabio, R.N.W., 1995. Till geochemistry and its implications for mineral exploration: south-eastern Cape Breton Island, Nova Scotia, Canada. Quatern. Int. 20, 107–122.

McClenaghan, M.B., DiLabio, R.N.W., 1996. Ice flow history and glacial dispersal patterns, south-eastern Cape Breton Island, Nova Scotia: implications for mineral exploration. Can. J. Earth Sci. 33, 351–362.

McKeague, J.A., Grant, D.R., Kodama, H., Beke, G.J. and Wang, C. 1983. Properties and genesis of soil and the underlying gibbsite bearing saprolite, Cape Breton Island, Canada; Canadian Journal of Earth Sciences, v. 20, no. 1, p. 37–38.

McNeill, R.H. & Purdy, C.A. (1951). A local glacier in the Annapolis - Cornwallis Valley (abstract). Proceedings and transactions of the Nova Scotian Institute of Science, 23 (1), 111.

Mosher, D.C., Piper, D.J.W., Vilks, G.V., Aksu, A.E., Fader, G.B.J., 1989. Evidence for Wisconsinan glaciations in the Verrill Canyon area, Scotian Slope. Quatern. Res. 31, 27–40.

Mott, R.J., 1994. Wisconsinan late-Glacial environmental change in Nova Scotia: a regional synthesis. J. Quatern. Sci. 9, 155–160.

Mott, R.J., Grant, D.R.G., 1985. Pre-Late Wisconsinan paleoenvironments in Atlantic Canada. Géogr. Phys. Quatern. 39, 239–254.

Mott, R.J., Stea, R.R., 1993. Late-Glacial (Allerød/Younger Dryas) buried organic deposits, Nova Scotia, Canada. Quatern. Sci. Rev. 12, 645–657.

Mott, R.J., Grant, D.R.G., Stea, R.R., Occhietti, S., 1986. Late-glacial climatic oscillation in Atlantic Canada equivalent to the Allerød–Younger Dryas event. Nature 323 (6085), 247–250.

Myers, R.A., Stea, R.R., 1986. Surficial mapping results: Pictou, Guysborough and Antigonish Counties, Nova Scotia. Mines and Minerals Branch, Nova Scotia Department of Mines and Energy Report, 86–1, pp. 189–194.

Newman, W.A., 1971. Wisconsin glaciation of northern Cape Breton Island, Nova Scotia. Unpublished Ph.D. thesis, Syracuse University, Syracuse, New York, 117 pp.

Nicks, L., 1988. The study of the glacial stratigraphy and sedimentation of the Sheldon Point Moraine, St John, New Brunswick. M.Sc. thesis, Dalhousie University Halifax, Nova Scotia, 171 pp.

Nielsen, E., 1976. The composition and origin of Wisconsinan tills in mainland Nova Scotia. Unpublished Ph.D. thesis, Dalhousie University, Halifax, Nova Scotia, 256 pp.

O'Beirne-Ryan, A.M., Zentilli, M., 2006. Weathering of Devonian monzogranites as recorded in the geochemistry of saprolites from the South Mountain Batholith, Nova Scotia, Canada. Atlantic Geol. 153, 153–159.

Occhietti, S., 1989. Quaternary geology of St. Lawrence Valley and adjacent Appalachian subregion. In: Fulton, R.J. (Ed.), Quaternary Geology of Canada and Greenland, 350–389 Geological Survey of Canada, Geology of Canada, 1.

Olejczyk, P., Gray, J.T., 2007. The relative influence of Laurentide and local ice sheets during the last glacial maximum in the eastern Chic-Chocs Range, northern Gaspé Peninsula, Quebec. Can. J. Earth Sci. 44 (11), 1603–1625.

Parent, M., Dubois, J.M.M., 1990. Les dépôts glaciares et marins des Îls de la Madelaine (Québec), temoins d'une englaciation rapide dans les appalaches septentrionales au Pléistocène superieur. Canadian Quaternary Association-American Quaternary Association, First Joint Meeting; Programme and Abstracts 28. P. 123.

Parkhill, M.A., 1994. Surficial geology and till geochemistry of the Nepisiguit Lakes (NTS 21 O/7) and California Lake (NTS 21 O/8) map areas; Gloucester, Northumberland, Restigouche, and Victoria counties, New Brunswick. New Brunswick Department of Natural Resources and Energy, Minerals and Energy Division Geoscience Report, 94–3, 275 pp.

Parkhill, M.A., 2005. Till geochemistry of the Kedgwick, Gounamitz River, States brook, and Menneval map areas (NTS 21 O/11, 12, and 13). New Brunswick Department of Natural Resources and Energy, Minerals and Energy Division Open File, 2005–4.

Parkhill, M.A., Doiron, A., 2003. Quaternary geology of the Bathurst Mining Camp and implications for base metal exploration using drift prospecting. In: Goodfellow, W.D., McCutcheon, S.R., Peter, J.M. (Eds.), Massive Sulphide Deposits of the Bathurst Mining Camp, New Brunswick and Northern Maine, Economic Geology Monograph 11 631–660.

Patterson, C.J., 1998. Laurentide glacial landscapes; the role of ice streams. Geology 26, 643–646.

Piper, D.J.W., 1991. Surficial geology and physical properties 7: paleooceanography and paleo-glaciology. Geological Survey of Canada East Coast Atlas: Scotian Shelf. Atlantic Geoscience Center, p. 123.

Piper, D.J.W., Fehr, S.D., 1991. Radiocarbon chronology of Late Quaternary sections on the inner and middle Scotian Shelf. Current Research, Part E, Geological Survey of Canada Paper 91-1E, pp. 321–325.

Piper, D.J.W., Mudie, J., Aksu, A.E., Skene, K.I., 1994. A 1 Ma record of sediment flux south of the Grand Banks used to infer the development of glaciation in south-eastern Canada. Quatern. Sci. Rev. 13, 23–27.

Prest, V.K., 1973. Surficial deposits of Prince Edward Island. Geological Survey of Canada Map, 1366A.

Prest, W. H. 1896. Glacial succession in central Lunenburg, Nova Scotia; Proceedings and Transactions of the Nova Scotia Institute of Science, Halifax, 9, Session 1895-96, 158–170.

Prest, V.K., 1983. Canada's heritage of glacial features. Geological Survey of Canada, Miscellaneous Report 28, 89 pp.

Prest, V.K., Grant, D.R., 1969. Retreat of the last ice sheet from the Maritime Provinces—Gulf of St. Lawrence region. Geological Survey of Canada Paper 69–33, 15 pp.

Prest, V.K., Nielsen, E., 1987. The Laurentide ice sheet and long distance transport. In: Kujansuu, R., Saarnisto, M. (Eds.), INQUA Till Symposium, Finland 1985 Geological Survey of Finland Special Paper, 3, pp. 91–102.

Prest, V.K., Grant, D.R., MacNeill, R.H., Brooks, I.A., Borns, H.W., Ogden III, J.G., et al., 1972. Quaternary geology, geomorphology and hydrogeology of the Atlantic Provinces. In: 24th International Geological Congress, Excursion Guidebook A61-C61, 79 pp.

Prest, V.K., Donaldson, J.A., Mooers, H.D., 2001. The omar story: The role of omars in assessing glacial history of west-central North America. Géogr. Phys. Quatern. 54, 257–270.

Pronk, A.G., Bobrowsky, T., Parkhill, M.A., 1989. An interpretation of the late Quaternary glacial flow indicators in the Baie des Chaleurs region, northern New Brunswick. Géogr. Phys. Quatern. 43 (2), 79–190.

Rampton, V.N., Paradis, S., 1981a. Quaternary geology of Amherst map area (21 H), New Brunswick. New Brunswick Department of Natural Resources, Mineral Development Branch, Map Report 81-3, 36 pp.

Rampton, V.N., Paradis, S., 1981b. Quaternary geology of Moncton map-area (21I). New Brunswick Department of Natural Resources Map Report 81–2, 31 pp.

Rampton, V.N., Paradis, S., 1981c. Quaternary geology of Woodstock map area (21 J), New Brunswick. New Brunswick Department of Natural Resources, Mineral Development Branch, Map Report 81–1, 37 pp.

Rampton, V.N., Gauthier, R.C., Thibault, J., Seaman, A.A., 1984. Quaternary geology of New Brunswick. Geological Survey of Canada Memoir 416, 77 pp.

Rappol, M., 1988. Glacial dispersal of Deboullie syenite, northern Maine, into western New Brunswick. Current Research, Part B Geological Survey of Canada Paper 88–1B, 49-53.

Rappol, M., 1989. Glacial history and stratigraphy of northwestern New Brunswick. Géogr. Phys. Quatern. 43, 191–206.

Rivard, C., Deblonde, C., Boivin, R., Bolduc, A., Paradis, S.J., Paradis, D., et al., 2007. Canadian groundwater inventory: hydrogeological atlas of the Annapolis Valley. Geological Survey of Canada Open File, 5541, 130 pp.

Rutherford, G.K., Thacker, D.J., 1988. Characteristics of two mafic saprolites and their associated soil profiles in Canada. Can. J. Soil Sci. 68, 223–231.

Ryan, R.J., Boehner, R.C., Deal, A.J., Calder, J.H., 1990. Cumberland Basin Geology Map, Amherst, Springhill and Parrsboro, Cumberland County, Nova Scotia [21H/08, 21H/09 and 21H/16][1:50,000]. Nova Scotia Department of Natural Resources Map, ME 1990-012.

Schroeder, J., Arseneault, S., 1978. Discussion of a karst in the Hillsborough gypsum, New Brunswick. Géogr. Phys. Quatern. 32 (3), 249–261.

Scott, D.B., Boyd, R., Douma, M., Medioli, F.S., Yuill, Y., Leavitt, E., et al., 1989. Holocene relative sea-level changes and Quaternary glacial events on a Continental shelf edge: Sable Island Bank. In: Scott, D.B., Pirazzoli, A., Honig, C.A. (Eds.), Late Quaternary Sea-Level Correlations and Applications. Kluwer Academic Publishers, Dordrecht, pp. 105–119.

Seaman, A.A., 1989. Glacial striae trends in New Brunswick: a compilation. New Brunswick Department of Natural Resources and Energy, Minerals and Energy Division Open File Report 89-34, 136 pp.

Seaman, A.A., 1991. Complex glacial ice-flow events in New Brunswick with particular reference to the Oromocto Lake area, and implications for drift prospecting. New Brunswick Department of Natural Resources, Mineral Resources Branch, Geoscience Report 89–2, 48 pp.

Seaman, A.A., 2000. Glacial dispersal in west-central New Brunswick. Atlantic Geol. 36, 71–72.

Seaman, A.A., 2004. Late Pleistocene history of New Brunswick, Canada. In: Ehlers, J., Gibbard, P.L. (Eds.), Quaternary Glaciations—Extent and Chronology, Part II: North America. Developments in Quaternary Science vol. 2b. Elsevier, Amsterdam, pp. 151–167.

Seaman, A.A., 2006. A new interpretation of Late Glacial history of central New Brunswick: The Gaspereau Ice Center as a Younger Dryas ice cap. In: Martin, G.L. (Ed.), Geological Investigations in New Brunswick for 2005, New Brunswick Department of Natural Resources, Mineral Resources Report 2006–3, pp. 1–36.

Seaman, A.A. 2009. The Appalachian Glacier complex, and the Middle to Late Pleistocene history of west-central New Brunswick, Canada. In: Martin, G.L. (Ed.), Geological Investigations in New Brunswick for 2008, New Brunswick Department of Natural Resources; Minerals, Policy and Planning Division, Mineral Resource Report 2009-2, pp. 66–140.

Seaman, A.A., McCoy, S.M., 2008. Multiple Wisconsinan till in the Sisson Brook exploration trench of Geodex Minerals Ltd., York County, west-central New Brunswick. In: Martin, G.L. (Ed.), Geological Investigations in New Brunswick for 2005, New Brunswick Department of Natural Resources, Mineral Resources Report 2008–1, pp. 1–34.

Seaman, A.A., Broster, B.E., Cwynar, L.C., Lamothe, M., Miller, R.F., Thibault, J.J., 1993. Field guide to the Quaternary geology of southwestern New Brunswick. New Brunswick Department of Natural Resources and Energy, Mineral Resources, Open File Report 93-1, 102 pp.

Shaw, J., Piper, D.J.W., Fader, G.B.J., King, E.L., Todd, B.J., Bell, T., et al., 2006. A conceptual model of the deglaciation of Atlantic Canada. Quatern. Sci. Rev. 25, 2059–2208.

Stanley, D.J., Swift, D.J.P., Silverberg, N., James, N.P., Sutton, R.G., 1972. Late Quaternary progradation and sand spillover on the outer continental margin off Nova Scotia, south-east Canada. Smithson. Contrib. Earth Sci. 8, 88 pp.

Stea, R.R., Feetham, M., Pullan, S.E., Ostrom W., Baechler, L and. Ryan, R. J., 2006. Geology and economic potential of the glacial clay and sand deposits of Inverness County, Southwest Cape Breton Island, Nova Scotia, Department of Natural Resources, Economic Geology Series 2006, 1, 110p.

Stea, R. R., and Brown, Y. 1989. Variation in drumlin orientation, form and stratigraphy relating to successive ice flows in southern and central Nova Scotia; Sedimentary Geology, v. 62, p. 223–240.

Stea, R. R., 2001. Late-glacial stratigraphy and history of the Gulf of St. Lawrence: Discussion, Canadian Journal of Earth Sciences. 38, p. 479-482.

Stea, R.R., 1995. Late Quaternary glaciations and sea-level change along the Atlantic coast of Nova Scotia. Unpublished Ph.D. dissertation; Dalhousie University, Halifax, 407 pp.

Stea, R.R., 2004. The Appalachian glacier complex in Maritime Canada. In: Ehlers, J., Gibbard, P.L. (Eds.), Quaternary Glaciations—Extent and Chronology, Part II North America. Elsevier, Amsterdam, pp. 213–232.

Stea, R.R., Finck, P.W., 1984. Patterns of glacier movement in Cumberland, Colchester, Hants and Pictou Counties, northern Nova Scotia. Current Research, Part A; Geological Survey of Canada Paper 84-lA, pp. 477–484.

Stea, R.R., Mott, R.J., 1989. Deglaciation environments and evidence for glaciers of Younger Dryas age in Nova Scotia, Canada. Boreas 18, 169–187.

Stea, R.R., Mott, R.J., 1998. Deglaciation of Nova Scotia; Stratigraphy and chronology of lake sediment cores and buried organic sections. Géogr. Phys. Quatern. 41, 279–290.

Stea, R.R., Mott, R.J., 2005. Younger Dryas glacial advance in the southern Gulf of St. Lawrence, Canada: analogue for ice inception? Boreas 34, 345–362.

Stea, R.R., Pe-Piper, G., 1999. Using whole-rock geochemistry to locate the source of igneous erratics from drumlins on the Atlantic coast of Nova Scotia. Boreas 28, 308–325.

Stea, R.R., Pullan, S.E., 2001. Hidden Cretaceous basins in Nova Scotia. Can. J. Earth Sci. 38, 1335–1354.

Stea, R.R., Wightman, D.M., 1987. Age of the Five Islands Formation, Nova Scotia and the deglaciation of the Bay of Fundy. Quatern. Res. 27, 211–219.

Stea, R.R., Finck, P.W., Wightman, D.M., 1986. Quaternary geology and till geochemistry of the Western part of Cumberland County, Nova Scotia (sheet 9). Geological Survey of Canada, Paper 85–17, 58 pp.

Stea, R.R., Turner, R.G., Finck, P.W., Graves, R.M., 1989. Glacial dispersal in Nova Scotia: a zonal concept. In: DiLabio, R.N.W., Coker, W.B. (Eds.), Drift Prospecting, Geological Survey of Canada Paper 89–20, pp. 155–169.

Stea, R.R., Conley, H., Brown, Y., (Compilers), 1992a. Surficial geology of the Province of Nova Scotia. Nova Scotia Department of Natural Resources Map 92–1, Scale 1:500,000.

Stea, R.R., Mott, R.J., Belknap, D.F., Radtke, U., 1992b. The Pre-Late Wisconsinan Chronology of Nova Scotia, Canada. In: Clark, P.U., Lea, P.D. (Eds.), The Last Interglaciation/Glaciation Transition in North America, Geological Society of America, Special Paper 270, pp. 185–206.

Stea, R.R., Boyd, R., Costello, O., Fader, G.B.J., Scott, D.B., 1996. Deglaciation of the inner Scotian Shelf, Nova Scotia: correlation of terrestrial and marine glacial events. In: Andrews, J.T., Bergsten, H.H., Jennings, A.E. (Eds.), Late Quaternary Palaeoceanography of the North Atlantic Margins, Geological Society Special Publication 111, pp. 77–101.

Stea, R.R., Piper, D.J.W., Fader, G.B.J., Boyd, R., 1998. Wisconsinan glacial and sea-level history of Maritime Canada, a correlation of land and sea events. Geol. Soc. Am. Bull. 110 (7), 821–845.

Stea, R.R., Pullan, S.E., Feetham, M., 2003. Mesozoic and Cenozoic stratigraphy of the lowlands of southwestern Cape Breton Island, (NTS 11 F 11, 11 F 14). Nova Scotia Department of Natural Resources Report, 2003–1, pp. 103–126.

Stokes, C.R., Clark, C.D., 2003. The Dubawnt Lake palaeo-ice stream: evidence for dynamic ice sheet behaviour on the Canadian Shield and insights regarding the controls on ice-stream location and vigour. Boreas 32, 263–279.

Swift, D.J.P., Borns Jr., H.W., 1967. A raised fluviomarine outwash terrace, north shore of the Minas Basin, Nova Scotia. J. Geol. 75 (6), 693–710.

Symons, D.T.A., Lewchuk, M.T., Boyle, D.R., 1996. Pliocene–Pleistocene genesis for the Murray Brook and Heath Steele Au-Ag gossan ore deposits, New Brunswick, from paleomagnetism. Can. J. Earth Sci. 33, 1–11.

Tarasov, L., Peltier, W.R., 2004. A geophysically constrained large ensemble analysis of the deglacial history of the North American ice-sheet complex. Quatern. Sci. Rev. 23 (3–4), 359–388.

Thompson, W.B., Borns Jr., H.W., Hall, B., 2007. Extrapolation of the Littleton-Bethlehem (Older Dryas) and Pineo Ridge Moraine systems across New Hampshire and Maine. In: Geological Society of America Northeastern Section—42nd Annual Meeting March 2007 Abstracts with Programs, 39–1, p. 55.

Utting, D.J., 2009. LIDAR-based glacial geology of the Halifax Metropolitan Area. Mineral Resources Branch, Report of Activities 2008, Nova Scotia Department of Natural Resources, Report ME 2009–1, pp. 129–137.

Veillette, J.J., Dyke, A.S., Roy, M., 1999. Ice flow evolution of the Labrador sector of the Laurentide ice sheet: a review, with new evidence from northern Quebec. Quatern. Sci. Rev. 18, 993–1019.

Wang, C., Stea, R.R., Ross, G.J., Holmstrum. D., 1986. Age estimation of the Shulie Lake and Eatonville tills in Nova Scotia by pedogenic development. Can. J. Earth Sci. 23, 115–119.

Wang, C., Ross, G.J., and Rees, H.W. 1981. Characteristics of residual and colluvial soils developed on granite and of the associated pre-Wisconsinan landforms in north-central New Brunswick. Canadian Journal of Earth Sciences, 18, 487–494.

Wehmiller, J.F., Belknap, D.F., Boutin, B.S., Mirecki, J.E., Rahaim, S.D., York, L.L., 1988. A review of the aminostratigraphy of Quaternary molluscs from United States Atlantic Coastal Plain sites. In: Easterbrook, D.J. (Ed.), Dating Quaternary Sediments, Geological Society of America Special Paper 227 69–110.

Wickenden, R.T.D., 1941. Glacial deposits of part of northern Nova Scotia. Transactions of the Royal Society of Canada 143–149, Section IV.

Wightman, D.M., 1980. Late Pleistocene glaciofluvial and glaciomarine sediments on the north side of the Minas Basin, Nova Scotia. Unpublished Ph.D. thesis, Dalhousie University, Halifax, Nova Scotia, 426 pp.

Wilson, R.A., Parkhill, M.A., Carroll, J.I., 2005. New Brunswick Appalachian transect: bedrock and Quaternary geology of the Mount Carleton—Restigouche River area. Field Trip B8. In: Geological Association of Canada Mineralogical Association of Canada—Canadian Society of Petroleum Geologists—Canadian Society of Soil Sciences, Joint Meeting Halifax, May 2005. Atlantic Geoscience Society Special Publication 34, 94 pp.

Williams, G.L., Fyffe, L.R., Wardle, R.J., Colman-Sadd, S.P., Boehner, R. C. & Watt, J.A. (1985). Lexicon of Canadian Stratigraphy, Volume VI, Atlantic Region, Canadian Society of Petroleum Geologists, Yarmouth NS. Can, 572 pp.

Chapter 49

Stratigraphical Record of Glacials/Interglacials in Northwest Canada

Alejandra Duk-Rodkin[1],* and René W. Barendregt[2]

[1]*Geological Survey of Canada, 3303-33rd St. NW, Calgary, Alberta, Canada T2L 2A7*
[2]*Department of Geography, The University of Lethbridge, 4401 University Drive, Lethbridge, Alberta, Canada T1K 3M4*
*Correspondence and requests for materials should be addressed to Alejandra Duk-Rodkin E-mail: Alejandra.Duk-Rodkin@NRCan-RNCan.gc.ca

49.1. INTRODUCTION

The glacial record of north-western (NW) North America is one of the most extensive of any preserved worldwide. This record was left by Cordilleran, montane, continental and plateau glaciers. The oldest glacial sediments deposited by Cordilleran glaciations are found near the Yukon/Alaska border (Canada/USA). Regional scale glaciations (Cordilleran and continental) commenced in NW Canada and east central Alaska between 2.9 and 2.6 million years ago (Ma). Multiple glaciations have been recorded in the mountains (Cordilleran and montane ice) and in the interior plains (ice caps and continental ice sheets) of northern Canada. Cordilleran and plateau ice cap glaciations occurred repeatedly throughout the Late Pliocene to Late Pleistocene. Local montane glaciers were widespread throughout glacial periods; however, they were more susceptible to local changes of climate and moisture supply. They often extended only relatively short distances from their source, leaving only loess deposits beyond their terminus. Only during the Late Pleistocene is a continental (Laurentide) ice sheet record documented in NW Canada.

New data are presented here to complement the record of glaciations (Duk-Rodkin et al., 2004). The summary of northwestern Canadian glacial history is presented here with more detailed stratigraphical descriptions and some additional data from the Fort Selkirk (Yukon), Smoking Hills (northern Interior Plains), Mount Edziza (northern British Columbia) and Merritt (southern British Columbia) areas.

The wide variety of surficial deposits exposed across the northern Canadian mainland provides a record of glaciations which span the Late Pliocene to Late Pleistocene. Glacial tills were deposited across the northern Canadian Shield and northern Interior Plains during a succession of plateau glaciations and a single continental glaciation (Figs. 49.1 and 49.2). Retreats of these ice sheets were accompanied by the development of large glacial lakes and deposition of glacio-lacustrine silts and clays across large areas within the northern Interior Plains.

Post-glacial erosion and downcutting of streams, in part related to post-glacial isostatic rebound, removed surficial deposits from large areas of the Canadian Shield in Nunavut and the Northwest Territories. Glacial deposits are absent across most of northern and west-central Yukon Territory because this area remained unglaciated (Fig. 49.2). Old Crow Basin occurs within this unglaciated region and was the depositional centre for extensive glacio-lacustrine sediments resulting from damming of rivers to the east by the Late Pleistocene continental (Laurentide) glaciation.

The record of Cordilleran glaciations is mostly found in the Yukon Territory, while continental and plateau glacial records are found in the Northwest Territories (NWT) and Banks Island (Fig. 49.1). Before the development of the first Cordilleran glaciation in NW Canada, tidewater glaciers are thought to have formed as early as the Late Miocene in southeastern Alaska (Denton and Armstrong, 1969; Lagoe et al., 1993). Stratigraphical sequences described for Cordilleran and/or montane glaciations in the central Yukon indicate that glaciation commenced there only in the Late Pliocene. The record of the first Cordilleran glaciation in the Latest Pliocene (late Gauss) between 2.7 and 2.8 Ma (Marine Isotope Stage (MIS) G10–G6) in the Yukon is reported in Duk-Rodkin and Barendregt (1997), Froese et al. (2000) and Duk-Rodkin et al. (2001, 2004); and is summarised in the stratigraphical correlations of Fig. 49.3. Multiple glaciations spanning the Matuyama Reversed Chron (2.58–078 Ma) and Brunhes Normal Chron (0.78 Ma–present) were deposited after this first glaciation and are documented in the magnetostratigraphy of tills, outwash and loess deposits. The most complete record is found in the Tintina Trench of central Yukon. However, other sites such as those in the Klondike Plateau, Fort Selkirk and Mackenzie Mountains (NWT) provide records which are complementary to those of the Tintina Trench (Figs. 49.1 and 49.3).

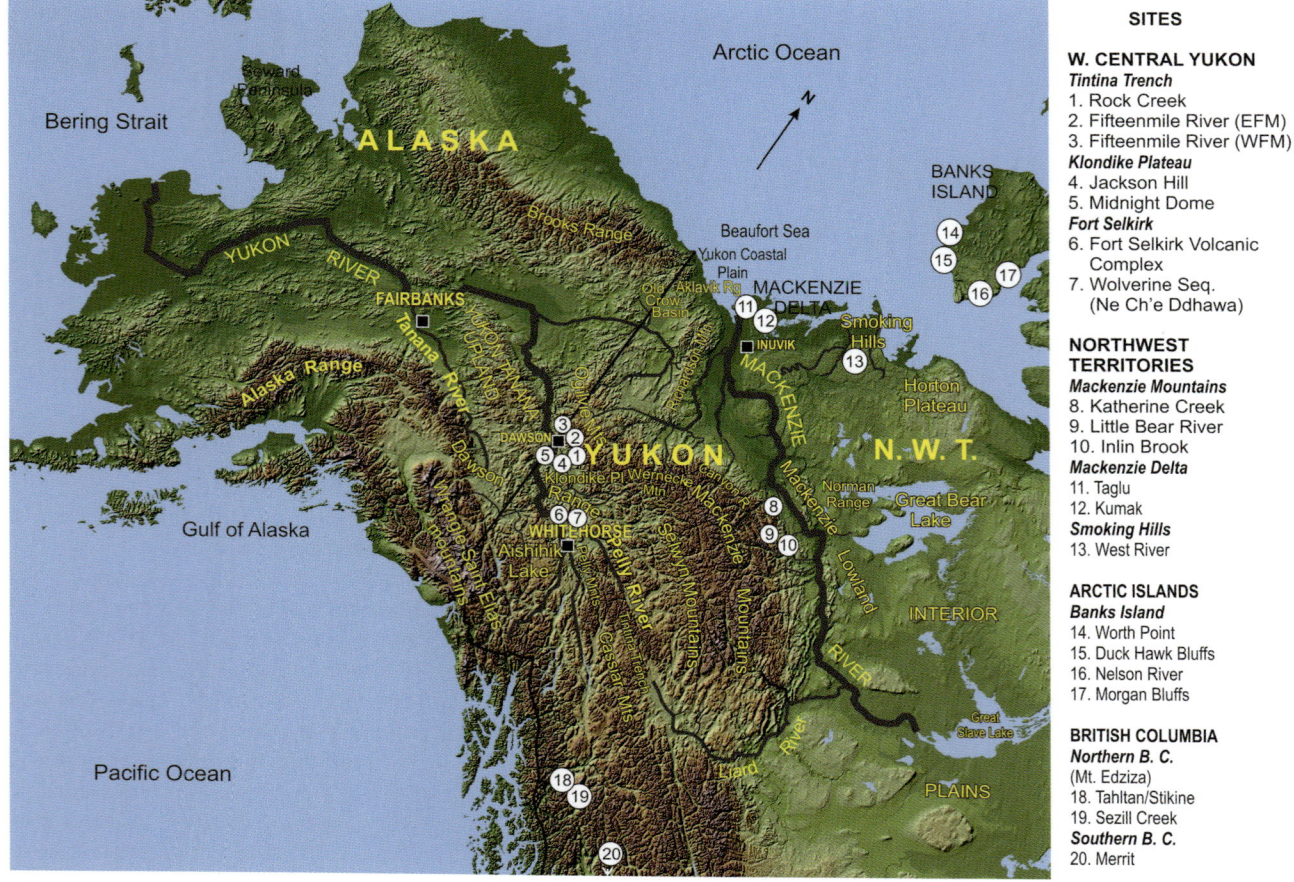

FIGURE 49.1 Relief map of NW North America and locations of sites described in text.

SITES

W. CENTRAL YUKON
Tintina Trench
1. Rock Creek
2. Fifteenmile River (EFM)
3. Fifteenmile River (WFM)
Klondike Plateau
4. Jackson Hill
5. Midnight Dome
Fort Selkirk
6. Fort Selkirk Volcanic Complex
7. Wolverine Seq. (Ne Ch'e Ddhawa)

NORTHWEST TERRITORIES
Mackenzie Mountains
8. Katherine Creek
9. Little Bear River
10. Inlin Brook
Mackenzie Delta
11. Taglu
12. Kumak
Smoking Hills
13. West River

ARCTIC ISLANDS
Banks Island
14. Worth Point
15. Duck Hawk Bluffs
16. Nelson River
17. Morgan Bluffs

BRITISH COLUMBIA
Northern B. C.
(Mt. Edziza)
18. Tahltan/Stikine
19. Sezill Creek
Southern B. C.
20. Merrit

Northern Interior Plains ice was present in northwest Canada, but the available record is neither as complete nor as well defined as that of the Cordilleran. In the continental Arctic, older diamicts have been reported from the Taglu and Kumak cores in the Mackenzie Delta (Fig. 49.3; Dallimore and Matthews, 1997). In the Smoking Hills area, there are as many as four tills of local plateau (Horton Plateau) origin separated by up to four silty clay beds (Fig. 49.3; Duk-Rodkin and Barendregt, unpublished field notes, 2004). The palaeomagnetic measurements of these tills and intertill beds indicate a sequence of glacial events within the Matuyama Chron (Early Pleistocene). The lowest till overlies reversely magnetised preglacial sands and gravels (presumably the late Beaufort Formation) containing large ice-wedge casts. Evidence of continental glaciation extending back to the Early Late Pleistocene is found on Banks Island at 1.6 Ma (Vincent et al., 1984; Barendregt and Vincent, 1990; Vincent, 1990; Fyles et al., 1994; Barendregt et al., 1998). Based on the Banks Island data, the first Keewatin-centred glaciation postdates Cordilleran glaciation in the Yukon by about 1 Ma.

49.2. WEST-CENTRAL YUKON (TINTINA TRENCH, KLONDIKE PLATEAU, FORT SELKIRK)

49.2.1. Tintina Trench

A long stratigraphical record extending from Late Cretaceous to Middle Pleistocene has been preserved along the walls of the Tintina Trench where it is exposed in modern landslide scars or along slopes (Duk-Rodkin et al., 2001, 2004). Stratigraphical sites are located on both sides of the trench along a distance of approximately 100 km near the Yukon/Alaska border (Figs. 49.1 and 49.2). At the Trench sites (sites 1–3; Figs. 49.1–49.6), tills and/or outwash gravels conformably overlie Pliocene preglacial gravels, which are unconformably underlain by faulted alluvial fan deposits of Miocene age (Duk-Rodkin et al., 2001).

Sections in the Trench show well-exposed beds below and above the Pliocene/Miocene unconformity at all three sites. Strata beneath the unconformity are composed of faulted alluvial fan deposits and have been grouped into

Chapter | 49 Stratigraphical Record of Glacials/Interglacials in Northwest Canada

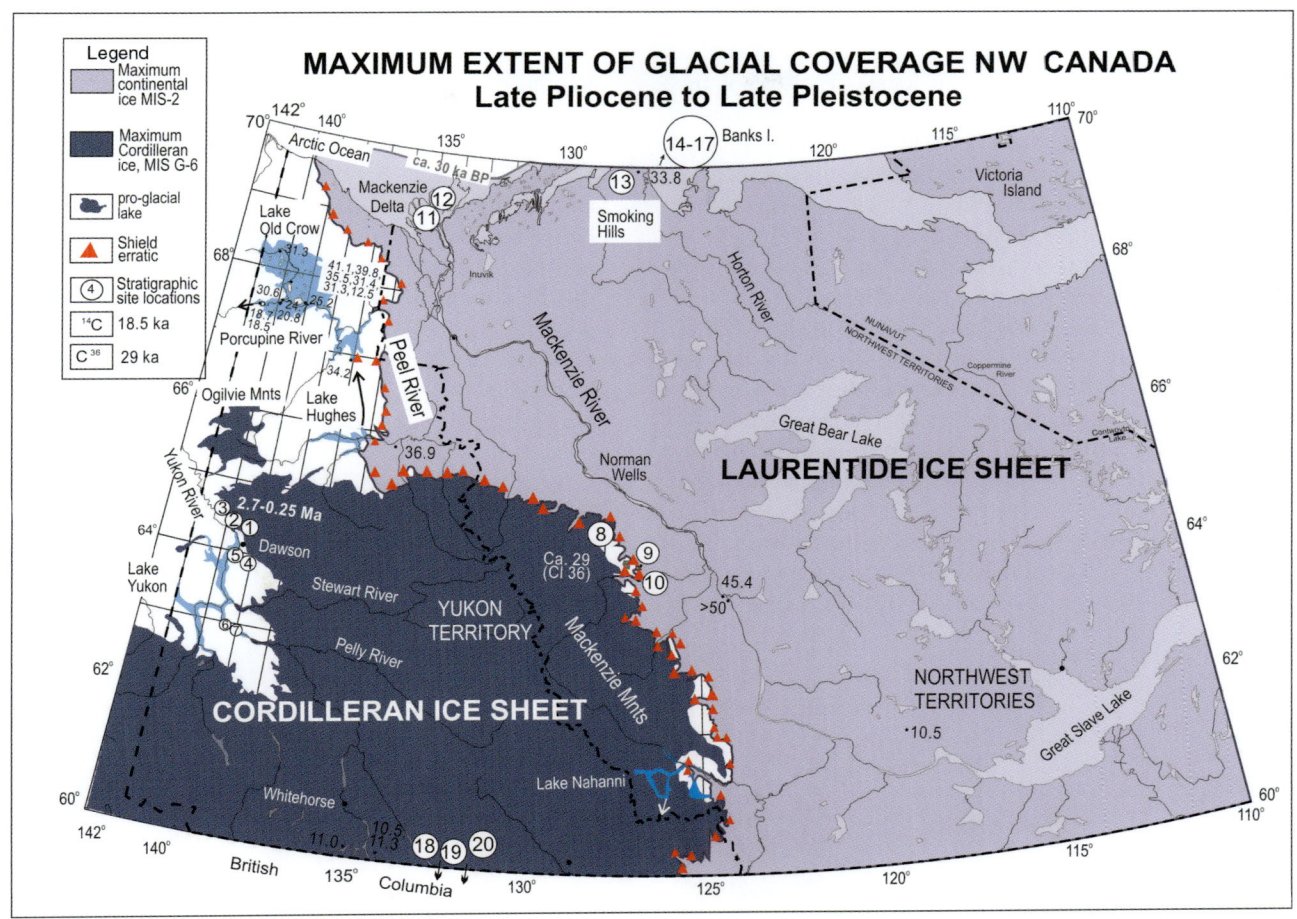

FIGURE 49.2 Maximum extent of glaciations in Northwest Canada. The maximum extent of Cordilleran ice occurred during the Late Pliocene, MIS G6, while the maximum ice extent of continental ice occurred during the Late Pleistocene (MIS 2).

two parts: a lower recessive succession of claystone and very coarse grained, pebbly sandstone with coal beds (Eocene–Cretaceous age) and an upper succession of interbedded conglomerate, sandstone and organics (peat) with in most places indurated silt beds at the top of the strata (Miocene age; Duk-Rodkin et al., 2001).

During the glaciations recorded in the trench, Cordilleran ice advanced along the trench to the west, merging with local glaciers in the Southern Ogilvie Mountains and transporting erratic pebbles from both Cordilleran (continental divide: Selwyn/Wernecke mountains) and Ogilvie Mountain (local) sources. Depending on the locality within the Trench, glacial clast lithologies may be either entirely local in origin or of both local and mixed provenance, but in all cases, sites reveal similar stratigraphical sequences of tills with palaeosols, outwash and loess. Collectively they record seven pre-Middle Pleistocene glacial events (Fig. 49.3). These sections have a normal–reverse–normal–reverse–normal–reverse–normal palaeomagnetic sequence extending from late Gauss to late Middle Pleistocene (Barendregt et al., 2010; Duk-Rodkin et al., 2010).

1. *Rock Creek site* (RC site 1, Figs. 49.1–49.4) in the Tintina Trench reveals beds above the unconformity which contain records of both Cordilleran and local glaciations (Duk-Rodkin et al., 2010). This site exposes preglacial gravels consisting of a debris flow deposit (unit 1a) overlain by a thin lacustrine bed (unit 1b). The shallow lacustrine deposit contains *Polemonium* sp. pollen and is normally magnetised. *Polemonium* sp. first appears in the upper Miocene (Muller, 1981) and is present in this region today. It is present in the Lost Chicken beds (ca. 2.9 Ma) of the Yukon and Alaska (White et al., 1999) which are thought to be correlative with this unit. Preglacial sediments are capped by outwash gravels (unit 2) and sand, deposited by glaciers associated with the first regional glaciation (Figs. 49.2–49.4). These deposits and the underlying unit 1 are normally magnetised. Outwash deposits only document this first glaciation at this site in the Tintina Trench and in the Klondike Plateau to the south. The outwash is overlain by a reversely magnetised till (unit 3) with a luvisolic palaeosol developed at its upper contact. This palaeosol is similar to soils

associated with the southern limit of boreal forest vegetation today. The pebble lithologies of the till are of mixed origin (Cordilleran and local). The age of this deposit has been bracketed between 1.97 and 2.6 Ma (Early Pleistocene, Matuyama Reversed Chron). Unit 4 is a very coarse deposit enriched in clay content as a result of a prolonged period of moderate weathering and translocation of surface clays. It is a thick glacial mudflow deposit for which no polarity was determined; however, it is overlain by a normally magnetised outwash so must fall within the early Matuyama Chron (Figs. 49.3 and 49.4). Unit 5 consists of an outwash and a till. The till has a weathering horizon developed at its surface suggesting that it may represent the lowermost part of a truncated luvisol. There is no weathering horizon between the outwash and the till and therefore both are considered to be part of the same glacial event. A normal polarity was obtained from the outwash, indicating an Olduvai age based on its stratigraphical position (Figs. 49.3 and 49.4). Unit 6 is reversely magnetised outwash gravel assigned to the middle Matuyama, based on its stratigraphical position between the Olduvai and Jaramillo subchrons. Unit 7 comprises crude to well-stratified outwash gravels. Silty sand beds have normal polarity and are assigned to the Jaramillo normal subchron. They occur between reversely magnetised units above and below, assigned to the late Matuyama Chron. Unit 8 sediments are outwash sands and silts containing pockets of deformed material. Silts have reversed polarity and are underlain by normally magnetised outwash deposits (Jaramillo) and overlain by normal sediments (Brunhes Chron), and can therefore be assigned to the latest Matuyama Chron. Unit 9 is the uppermost glacial outwash at RC site and has a luvisolic palaeosol developed at its surface (comparable to a Wounded Moose palaeosol; Smith et al., 1986). Unit 9 is overlain by loess (unit 11, described below) and underlain by outwash (unit 8). Clasts within the soil exhibit clay skin coatings. The normal polarity recorded in fine-grained sand lenses is assigned to the Brunhes Normal Chron. Unit

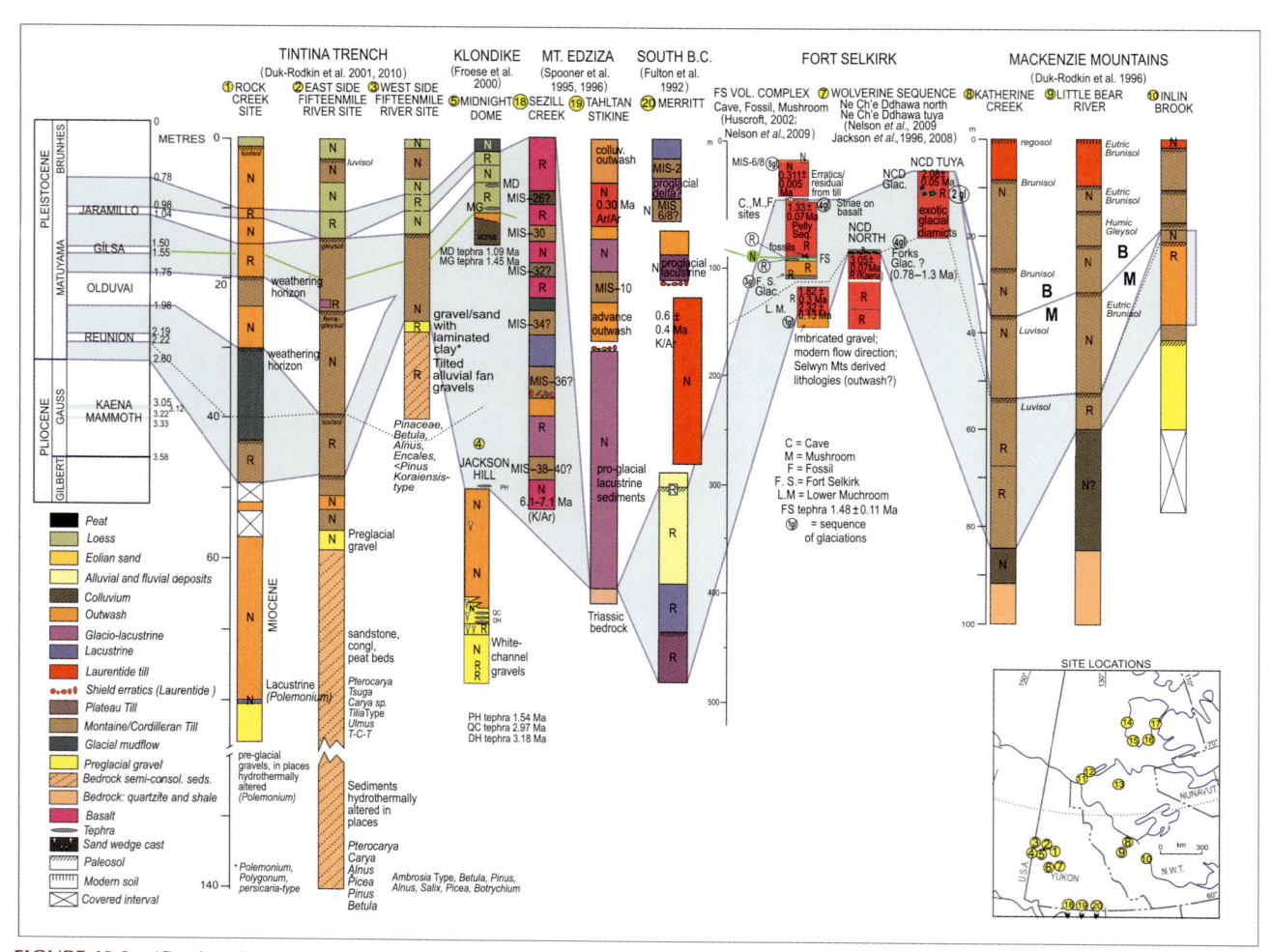

FIGURE 49.3 (Continued)

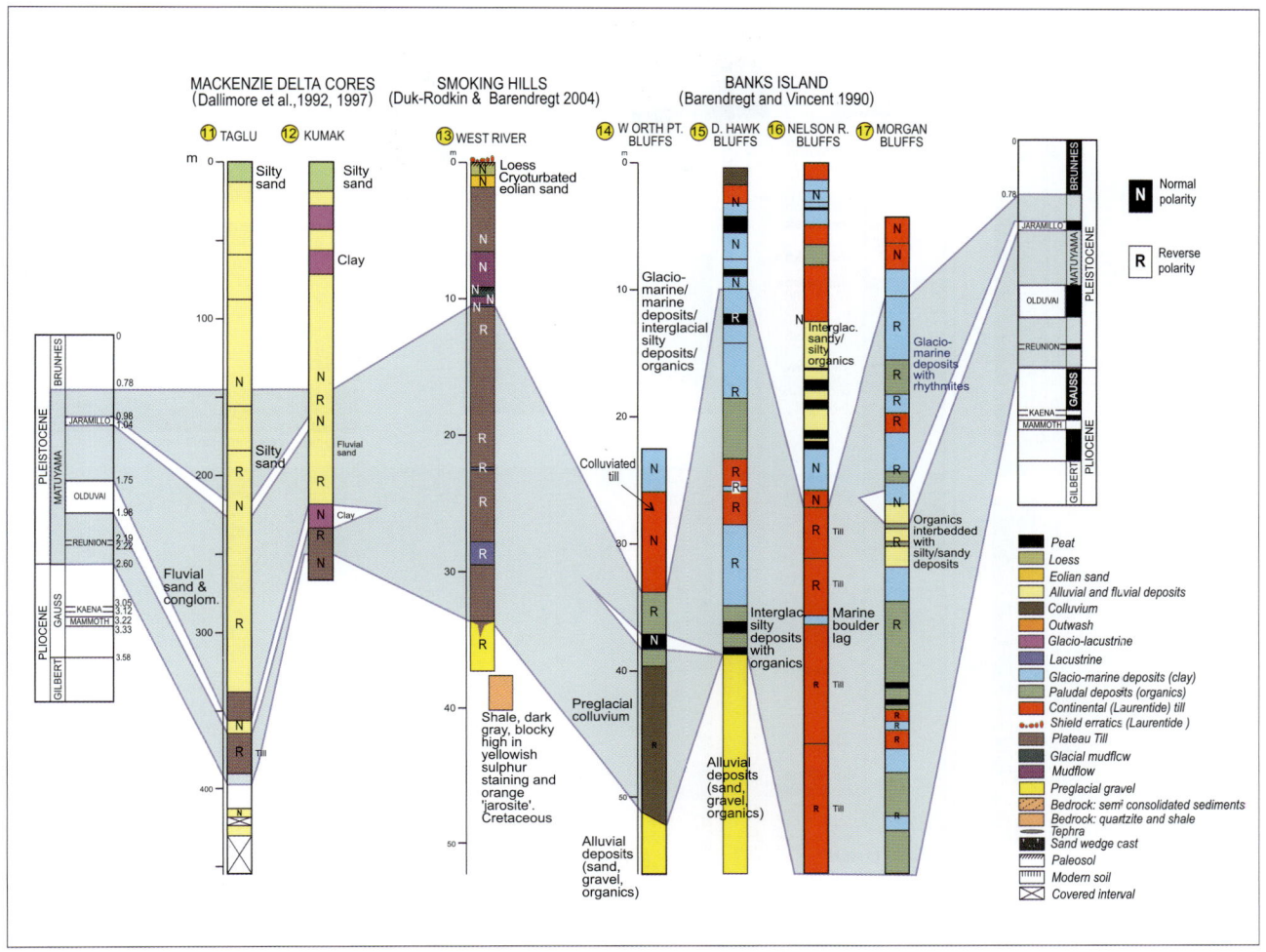

Chapter | 49 Stratigraphical Record of Glacials/Interglacials in Northwest Canada

FIGURE 49.3 Late Pliocene to Late Pleistocene stratigraphical correlations of Cordilleran and Northern Interior Plains sites in Northwest Canada. Light blue colour band identifies Reversed Matuyama Chron stratigraphy, while Normal Gauss and Normal Brunhes Chron strata lie below and above, respectively. (For interpretation of the references to colour in this figure legend, the reader is referred to the Web version of this chapter.)

11 forms the uppermost loess unit (<20 cm) at RC. This regionally extensive unoxidised surface loess lacks structure, suggesting little or no reworking. It is normally magnetised and overlies Reid till (MIS 6/8; Westgate et al., 2008), and is therefore younger than 126 ka, probably of McConnell age (<23 ka, MIS 2).

2. *Fifteenmile River* (EFR site 2, Figs. 49.1–49.3 and 49.5) stratigraphy outcrops in a Holocene landslide scar on the north side of the Tintina Trench, 4.0 km east of Fifteenmile River (Figs. 49.3 and 49.5). The EFR outcrops expose a package of Plio–Pleistocene preglacial and glacial sediments separated unconformably from the underlying tilted Miocene beds. The latter are predominantly alluvial deposits laid down following extensional faulting in the trench. Glaciers from the continental divide area did not reach this portion of the trench, and only deposits of local provenance (Ogilvie Mountains) are present. Unit 1 exposes preglacial alluvial sand and gravel deposits that overlie tilted alluvial Miocene deposits. These preglacial sediments contain a palaeosol (feragleysol) at its upper contact. The palaeosol is partially truncated and deformed (Fig. 49.5, unit 1b) and in places incorporated into the overlying till, forming a wavy upper contact. Fine-grained beds are normally magnetised and correlated to the Gauss Normal Chron. Unit 1b overlies the Miocene/Pliocene unconformity and reveals the same polarity (and palaeomagnetic directions) as the overlying (unit 2) glacial deposits. This relationship is also seen at RC site. Unit 2 is a till with an interbed of silty fine sand, which is normally magnetised and is correlated to the latest Gauss

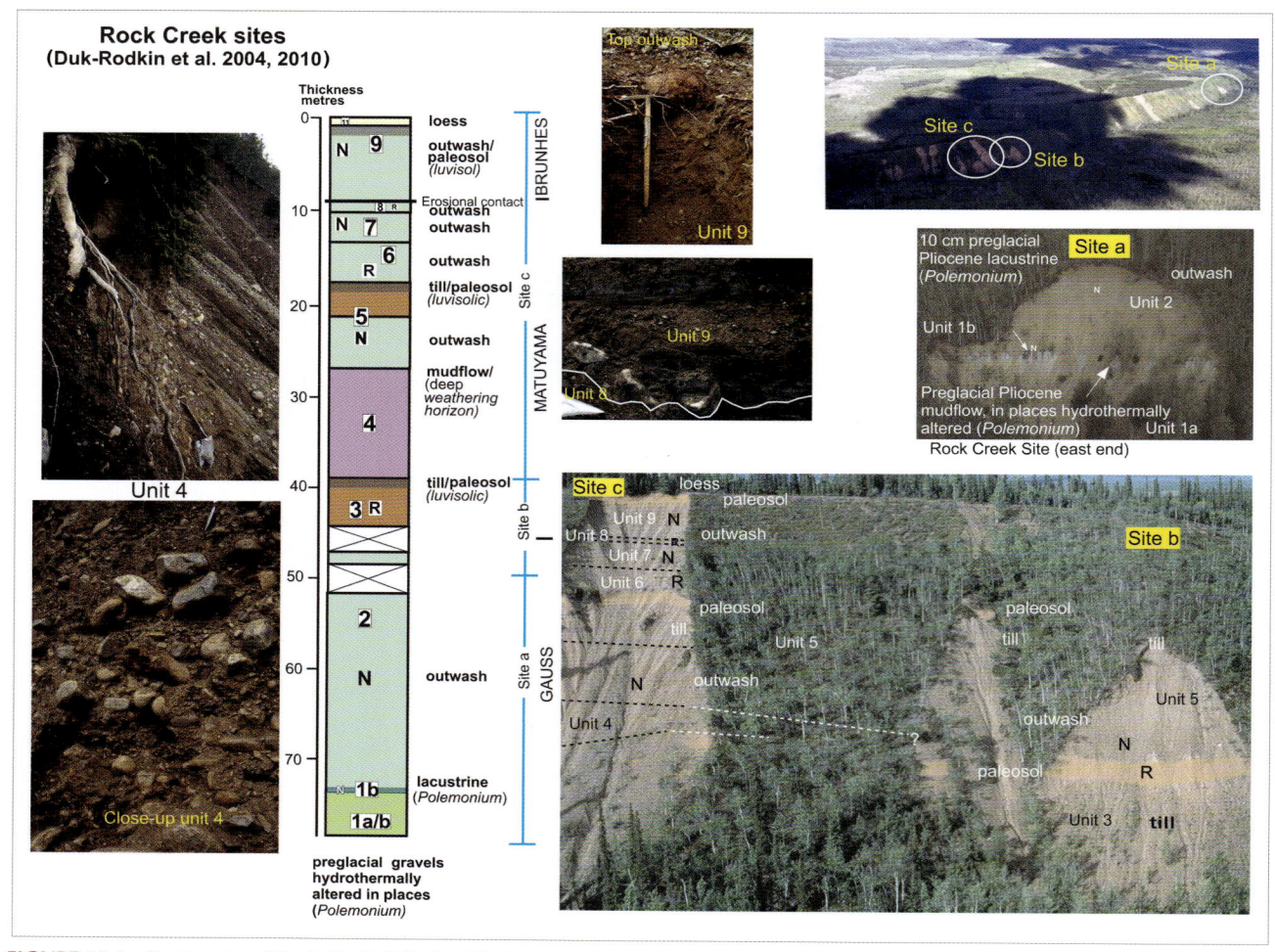

FIGURE 49.4 Stratigraphy of Rock Creek (RC) site, Tintina Trench, Yukon. Stratigraphy extends from Late Pliocene (Gauss Normal Polarity Chron) preglacial and glacial deposits to Late Pleistocene loess cover at top of sections. The RC stratigraphy is a composite of three sites which lie in close proximity to each other. Unit numbers in the text correspond to numbers on left of stratigraphical column. Note hydrothermally altered preglacial sediments at base of strata.

Normal Chron (2.7–2.58 Ma) based on its stratigraphical location above alluvial deposits of unit 1 and its comparable polarity and palaeomagnetic directions (Barendregt et al., 2010). The weathering horizon at the top of the unit has characteristics of a truncated luvisol. Unit 3 is interpreted as a till with a well-developed luvisol at its surface, and has a reversed polarity assigned to the early Matuyama Chron (2.58–1.97 Ma). It occurs between the Gauss Normal Chron (unit 2) and the Olduvai subchron (unit 5) and is stratigraphically similar to unit 3 at RC site. Unit 5 is a till, with a gleysolic palaeosol developed in the top of the till. It forms a sharp contact with lacustrine unit 6 above. The base and upper parts of the unit 5 till have normal polarity, and because it occurs between reversely magnetised sediments of the early Matuyama (unit 3) and a pre-Jaramillo reversed till (unit 6), is assigned to the Olduvai subchron (1.97–1.78 Ma). Unit 6 comprises a basal lacustrine clay subunit, overlain by till with melt-out characteristics. The top of the unit exhibits a gleysolic weathering horizon that formed under poorly drained conditions and is highly disturbed, probably due to cryoturbation. The lacustrine subunit and till are reversely magnetised, and unit 6 is assigned to the early late Matuyama Chron (1.78–1.05 Ma; pre-Jaramillo) because it lies stratigraphically between sediments assigned to the Olduvai and Jaramillo subchrons (Figs. 49.3 and 49.5). Unit 8 is a silt and fine sand deposit with mottled appearance. The lower contact is sharp. Samples collected for palaeomagnetic measurements were also used for pollen identification. Only one sample yielded some identifiable pollen grains at

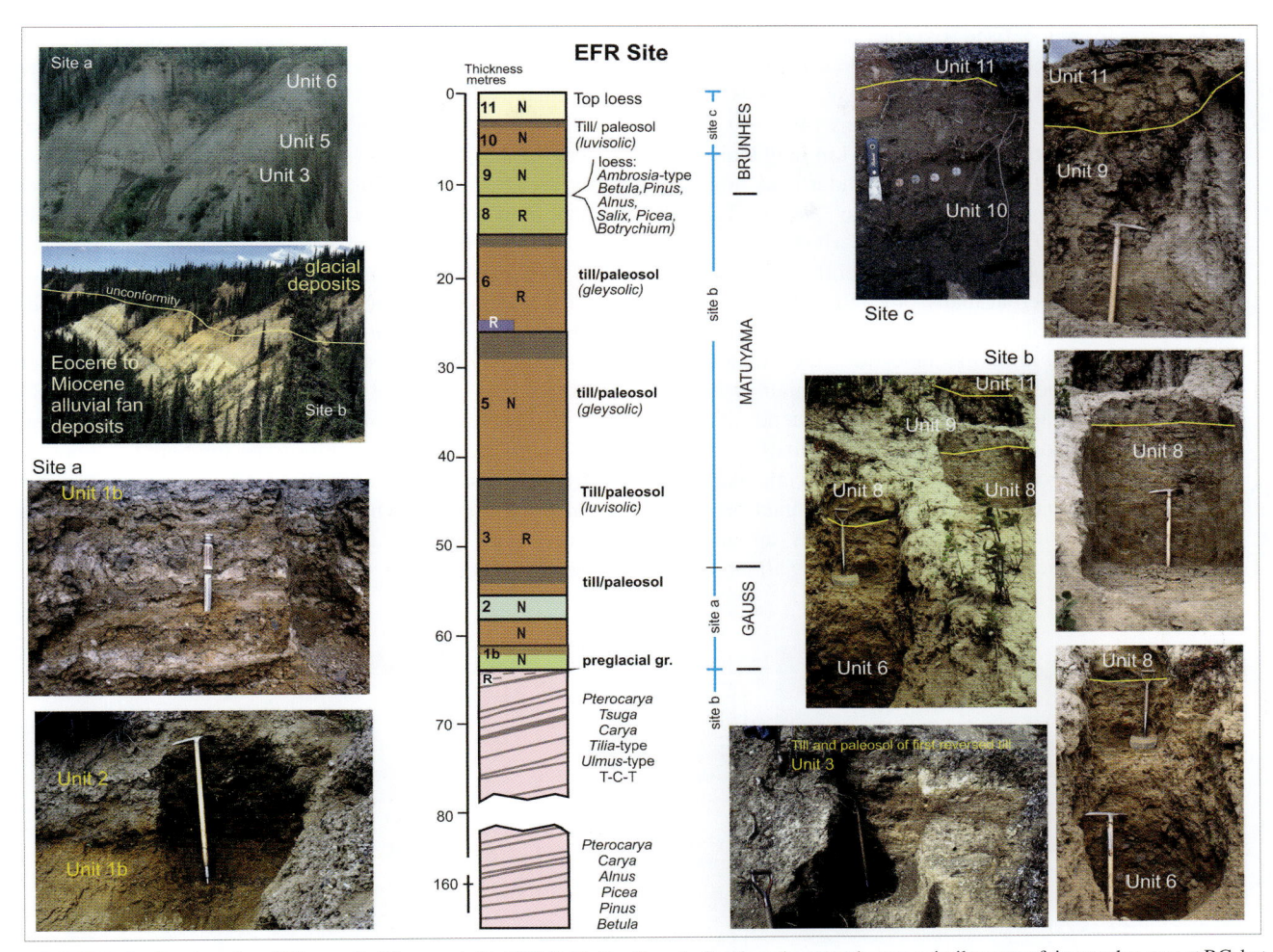

FIGURE 49.5 Stratigraphy of Fifteenmile River, east side (EFM), Tintina Trench. Stratigraphy extends over a similar span of time as that seen at RC, but includes an MIS 8/6 till near the top. It is a composite of two sites located about 100 m apart. Note deformation and incorporation of unit 1b into the overlying till.

the transition to unit 9. There were 51 grains identified revealing the following: *Ambrosia*-type 33%, *Betula* 29%, *Pinus* 20%, *Alnus* 6%, *Salix* 6%, *Picea* 4% and *Botrychium* 2%. Unit 8 is reversely magnetised loess and is assigned to the latest Matuyama. This assignment is based on its stratigraphical position between Jarmillo and Brunhes age deposits, and comparable palaeomagnetic directions for unit 8 at RC and WFM sites (Barendregt et al., 2010). Unit 9 is a silt and fine sand deposit with mottled appearance and minor stratification marked by fine clay beds. Unit 9 is normally magnetised loess, assigned to the early Brunhes based on stratigraphical position (Fig. 49.5) and similarities to the record found at WFR (Barendregt et al., 2010). It is reasonable to suppose that the boundary between units 8 and 9 at the EFR and WFR sites is correlative with the boundary between units 1 and 2 at Midnight Dome

(MD) in the Klondike area (Froese et al., 2000) and that this stratigraphical contact marks the Brunhes–Matuyama (B/M) boundary. The Brunhes/Matuyama boundary (0.78 Ma) falls within marine oxygen isotope stage (MIS) 19, an interglacial period (Fig. 49.5). Pollen obtained from the base of unit 9 reveals *Pinus*, *Ambrosia*-type, *Betula*, *Alnus*, *Salix*, and *Picea* and suggests deposition during an interglacial. Unit 10 is a diamicton (till) with a silty-clay matrix and 20% clasts up to 15 cm in size. It forms a sharp contact with underlying unit 9. Unit 10 has a normal polarity, is assumed to have been deposited by the Reid glaciation and is correlated with unit 10 at WFR (Fig. 49.5). Unit 11 is a 0.2-m-thick, massive silt and fine sand bed loess which corresponds to the regionally extensive unoxidised surface loess that discontinuously covers much of the western Yukon. Its massive characteristic suggests little or no reworking.

It is normally magnetised, overlies unit 10 at EFR and WFR sites, and is therefore considered to be the Late Pleistocene (<23 ka) McConnell loess.

3. *Fifteenmile River* (WFR site 3, Figs. 49.1–49.3 and 49.6) stratigraphy outcrops along a 1-km long Holocene landslide scar reactivated within an older landslide, and is located on the north side of the Tintina Trench, approximately 1.0 km west of the river (Figs. 49.2 and 49.6). The exposure reveals several older landsliding events, also seen at many other localities, and may relate to extensional faulting along the trench in the Late Miocene. Beds within the landslide have undergone minor rotation and lowering but have remained intact. The WFR stratigraphy exposes a 14 m sequence of tilted Miocene alluvial deposits, overlain by 26 m of conformable and horizontally stratified glacial deposits described at two sites (Fig. 49.6). The contact between these two depositional sequences forms an angular unconformity which is clearly visible along all landslide scars in the Trench. Approximately 200 m of partially exposed and poorly preserved Tertiary (and older) strata extend to creek level. The stratigraphy above the unconformity at WFR site consists of Pliocene preglacial gravel and laminated sand, overlain by till and loess units. This stratigraphical sequence is less complete than that seen at EFR site but is equivalent to the lower units at RC and EFR. Unit 1 has preglacial gravel and sand with minor laminated clay beds which overlie tilted alluvial Miocene deposits and are overlain by the first till (Fig. 49.6). A silty fine sand bed near the unconformity is reversely magnetised and probably belongs to either the Kaena (3.12–3.05 Ma) or the Mammoth (3.33–3.22 Ma) reversed subchrons within the Gauss Normal Chron (Late Pliocene). This age assignment is supported by the presence of *Polemonium*, *Polygonum* and *persicaria*-type pollen which is generally thought to have a maximum age of Late Miocene, and by recently dated sediments in the Klondike Plateau which are

FIGURE 49.6 Stratigraphy of Fifteenmile River, west side (WFM), Tintina Trench. It is exposed in a major landslide scar and extends over a similar time span as that of EFR. The contacts between units 3 and 6 were not studied due to the steep terrain and were observed only through field glasses. WFM is a composite of two transects located about 70 m apart. Note blocks of preglacial sediments incorporated in the lowermost till.

thought to be equivalent, and contain the Quartz Creek tephra (3.00 ± 0.33 Ma; Sandhu et al., 2000). The pollen indicates a cool/cold alpine climate. Unit 2 is exposed in a 16-m vertical outcrop where study was limited to the basal portion. The exposure may contain sediments from other glaciations, and additional polarity histories may also be present. Unit 2 is a till, which in places has incorporated blocks of underlying preglacial sediments. The fine sand and silty-clay beds at the base may be outwash deposits. The till appears to be capped by a palaeosol which is discontinuously preserved. The normal polarity is assigned to the upper Normal Gauss Chron based on stratigraphical position and polarities recorded in units above and below (Figs. 49.3 and 49.6). Units 7–9 are loess deposits recording three polarities (N–R–N), suggesting a considerable time span. They are assigned (from bottom to top), to the Jaramillo subchron (N), latest Matuyama Chron (R) and early Brunhes Chron (N) and are correlated with units 8 and 9 at EFM and at RC. At EFR and WFR, these units are overlain by the middle Brunhes Reid till (MIS 8 or 6). The extensive loess sequence at WFR indicates that glaciers from the Ogilvie Mountains most likely did not reach this part of the Tintina Trench. The upper part of unit 9 exhibits cryoturbation indicating periglacial activity during this time. Unit 10 is a till with a basal glacio-lacustrine component. Both till and lacustrine sediments are normally magnetised and are assigned to the Brunhes Normal Chron. This till is most probably the late Middle Pleistocene (Reid age) till (MIS 6 or 8) and is correlated with the uppermost till at EFM. Unit 11 is a 40-cm-thick unoxidised regional loess cover, which lacks structure, suggesting little or no reworking. It is normally magnetised, overlies the assumed Reid age till, and is thought to be Late Pleistocene (<23 ka, (MIS) 2) McConnell loess.

49.2.2. Klondike Gold Fields

The Lower Klondike River Valley has yielded an extensive late Neogene stratigraphy and chronology. Two sites (MD and Jackson Hill (JH), sites 4 and 5; Figs. 49.1–49.3, 49.7 and 49.8) have preserved a stratigraphical record that complements most of the late Neogene geologic record in the Tintina Trench (Duk-Rodkin et al., 2010).

1. *JH* (site 4, Figs. 49.3 and 49.7) records upper preglacial gravels (White Channel) and outwash gravels related to the first glaciation in west-central Yukon which occurred in the Late Pliocene (ca. 2.65 Ma; Froese et al., 2000). The White Channel Gravels are of local provenance (Klondike Plateau gold fields; Fig. 49.1), while the glacial outwash gravels are of Cordilleran origin. The lower part of the White Channel Gravels reveals a reversed/normal/reversed/normal magnetic polarity sequence assigned to the Gilbert/Gauss Epochs (4.18–3.05 Ma; Figs. 49.3 and 49.7). The upper reversed/normal sequence within the White Channel Gravels contains ice-wedge casts which provide clear evidence for the onset of a cold climate (Fig. 49.7). The upper White Channel Gravels interfinger with Cordilleran outwash, and sediments are normally magnetised throughout this zone of mixed sedimentation (Fig. 49.7). Elsewhere in the gold fields, two tephras were obtained from this strata (Dago Hill and Quartz Creek) which have fission-track ages of 3.18 and 2.97 Ma, respectively (Sandhu et al., 2000; Westgate et al., 2002).

2. *MD* (site 5; Figs. 49.3 and 49.8) preserves Middle–Early Pleistocene basal fluvial gravels, outwash gravels, and loess deposits along the north side of the lower Klondike River Valley. The loess sequence has a reversed/normal/reversed/normal magnetisation which intermittently spans the late Matuyama and Brunhes Chrons (1.45 Ma–present) and includes the Jaramillo normal subchron (0.99–1.05). This sequence contains two tephras, one in the lower reversed loess which was deposited during a warm interval (1.4 Ma Mosquito Gulch tephra) and the other in the upper part of the same reversed unit deposited during a cold event (1.09 Ma, MD tephra; Froese et al., 2000; Froese and Westgate, 2001). This loess sequence reveals a short interval at its base containing pine pollen, followed by cold and very cold (glacial) conditions (colluviated loess, or yedoma).

49.2.3. Fort Selkirk

Fort Selkirk sites in the west-central Yukon preserve an extensive late Neogene record of volcanic, glacial and interglacial events (Figs. 49.1–49.3). The stratigraphy outcrops along the Yukon River (Fig. 49.1) and is depicted in four composite sites (Figs. 49.9–49.12) labelled (from older to younger): (1) Wolverine sequence (Ne Ch'e Ddhawa north and Ne Ch'e Ddhawa tuya, Fig. 49.9); (2) Fort Selkirk Volcanic Complex: Lower Mushroom site (Fig. 49.10); (3) Fort Selkirk Volcanic Complex: composite of Mushroom, Cave and Fossil sites (Fig. 49.11) and (4) Fort Selkirk Volcanic Complex: Black Creek flows (Tip, Angel and Pillow Point sites, Fig. 49.12). In total, five glacial events are recorded from sediments which are only locally preserved beneath, within and above lava flows and hyaloclastite complexes. Four of these glaciations fall within the Matuyama Reversed Chron and one within the Brunhes Normal Chron. The ages of these events span from 1.83 to 0.30 Ma. The deposits have been dated by K–Ar, Ar–Ar, fission-track, palaeomagnetism

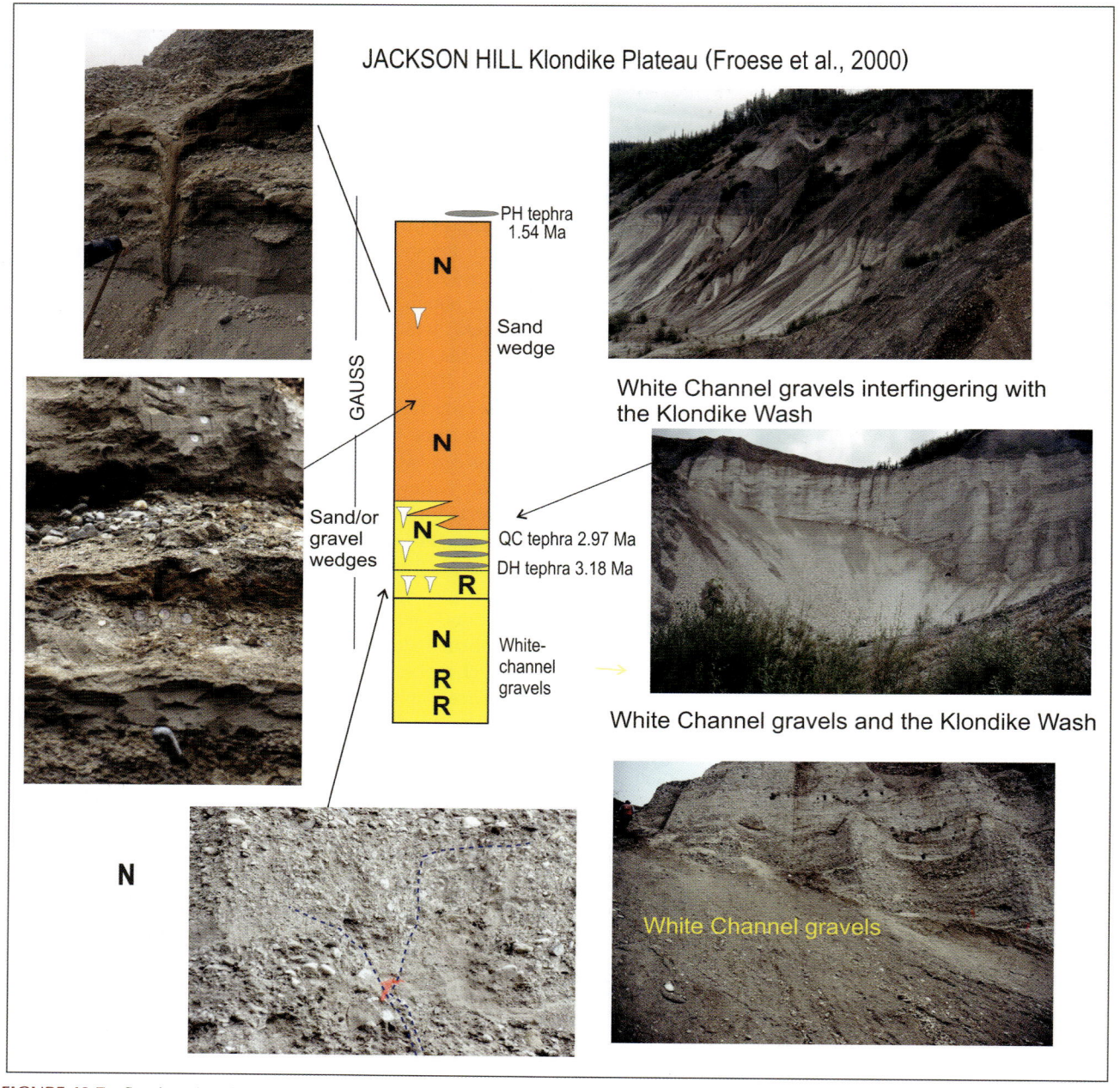

FIGURE 49.7 Stratigraphy of Jackson Hill is exposed in a large placer mine outcrop near Dawson townsite in the Klondike Valley. The stratigraphy reveals preglacial White Channel gravels which interfinger with the Klondike outwash near the upper contact (MIS G10). Both preglacial gravels and outwash have sand and gravel ice-wedge casts developed at various palaeo-surfaces.

and fossil evidence. In addition, geomorphological evidence of glaciations, such as striae on lava beds, exotic glacial pebbles within lavas, and the presence of pillow lavas as indicators of eruption through ice, have also been used (Jackson et al., 1996; Huscroft et al., 2004; Nelson et al., 2009).

Wolverine sequence (Ne Ch'e Ddhawa north and Ne Ch'e Ddhawa tuya, Figs. 49.3 and 49.9) is composed of three basalts flow (total thickness ~ 60 m). The lower flow (4.34 ± 0.06 Ma) is reversely magnetised and falls within the Upper Gilbert, the middle flow (3.21 ± 0.07 Ma) is also reversed and falls within the Mammoth subchron of the early Gauss, while the third flow (3.05 ± 0.07 Ma) is reversed and falls within the Kaena subchron of the early Gauss. The upper flow is overlain by a reversely magnetised till (Forks Glaciation, Jackson et al., 2010) considered to be of late

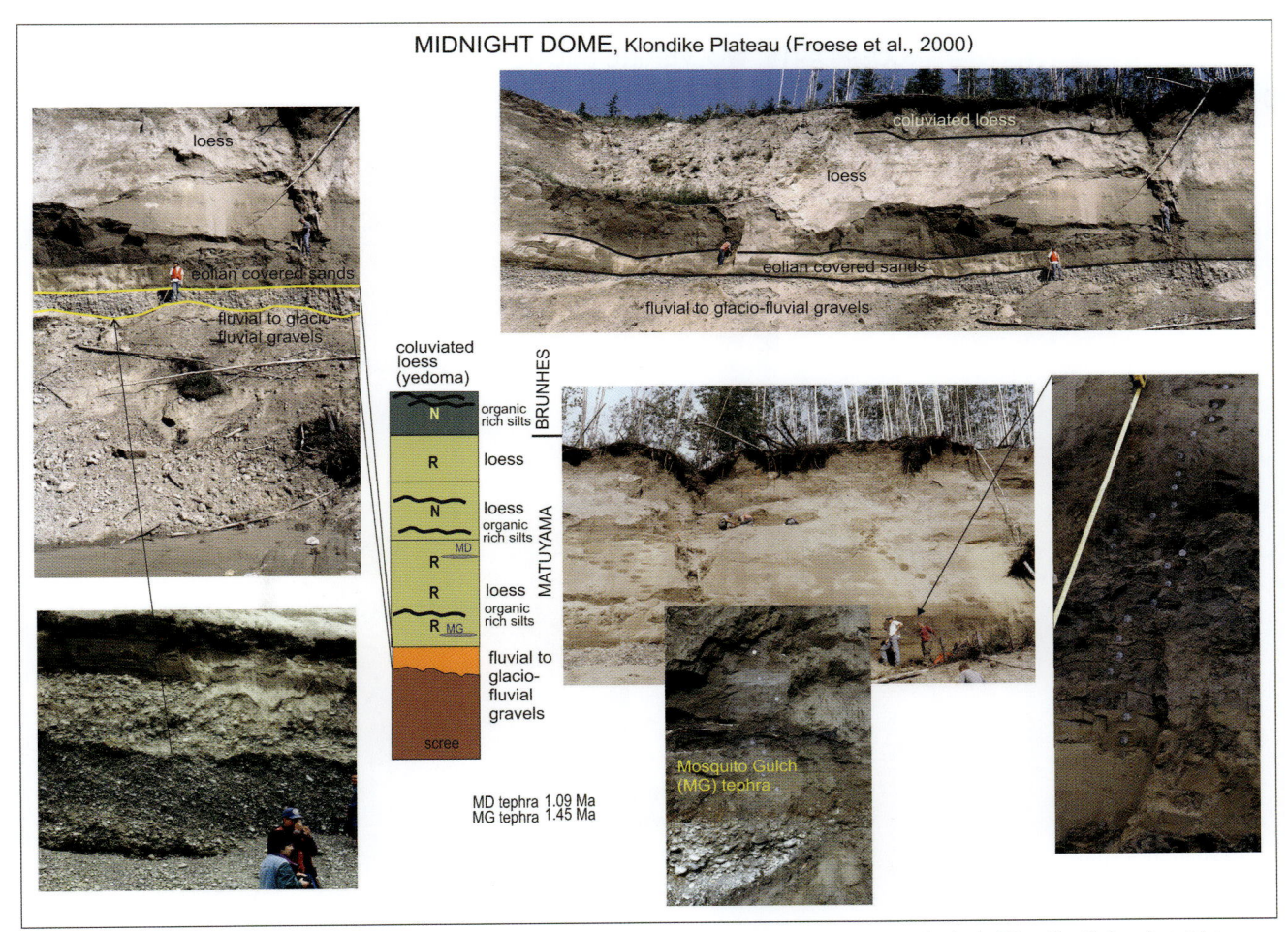

FIGURE 49.8 Stratigraphy of Midnight Dome (MD) is exposed in a placer mine outcrop near Dawson townsite in the Klondike Valley. Late Matuyama to early Brunhes age deposits are exposed at MD. The stratigraphy reveals loesses developed under both glacial and interglacial conditions, and contains organic rich silt horizons (palaeosols) and ice-wedge casts.

Matuyama age (Fig. 49.3). Stratigraphically above, and some 300 m to the south of Ne Ch'e Ddhawa north is the *Ne Ch'e Ddhawa tuya site* (Figs. 49.3 and 49.9). The tuya is the product of subglacial eruption and contains blocks of exotic glacial diamicts (Fig. 49.9). The upper part of the Tuya has an age of 2.08 ± 0.05 Ma and is reversely magnetised, and therefore this glaciation falls within the early Matuyama (Jackson et al., 2008; Nelson et al., 2009).

Fort Selkirk Volcanic Complex: Lower Mushroom site (Figs. 49.3 and 49.10) consists of a basalt flow underlain by outwash gravel. Two Ar/Ar ages for the reversely magnetised basalt (2.32 ± 0.13 and 1.83 ± 0.3 Ma) provide a minimum age for the underlying outwash. This glaciation may correspond to the one which produced the NCD Tuya, or may be an earlier glaciation.

Fort Selkirk Volcanic Complex: Composite of Mushroom, Cave and Fossil sites (Figs. 49.3 and 49.11) is a sedimentary and volcanic sequence occurring between two glacial events (Fort Selkirk Glaciation and an unnamed glaciation). The Fort Selkirk Glaciation consists of till overlain by outwash, both reversely magnetised, and both considered to belong to the same glaciation, based on their nearly identical palaeomagnetic directions. Reversely magnetised interglacial deposits overlie the outwash and contain a tephra (1.48 ± 0.11 Ma, also reversed), and a short interval of normally magnetised silt (Gilsa subchron, Froese and Westgate, 2001). A 14-m-thick basalt (1.33 ± 0.07 Ma) overlies the interglacial deposits, reveals striations at its upper surface, and is overlain by discontinuous drift of a younger glaciation (<1.33 Ma), perhaps equivalent to the Forks Glaciation at NCD North.

Fort Selkirk Volcanic Complex: Black Creek flows (Tip, Angel and Pillow Point sites; Figs. 49.3 and 49.12) consist

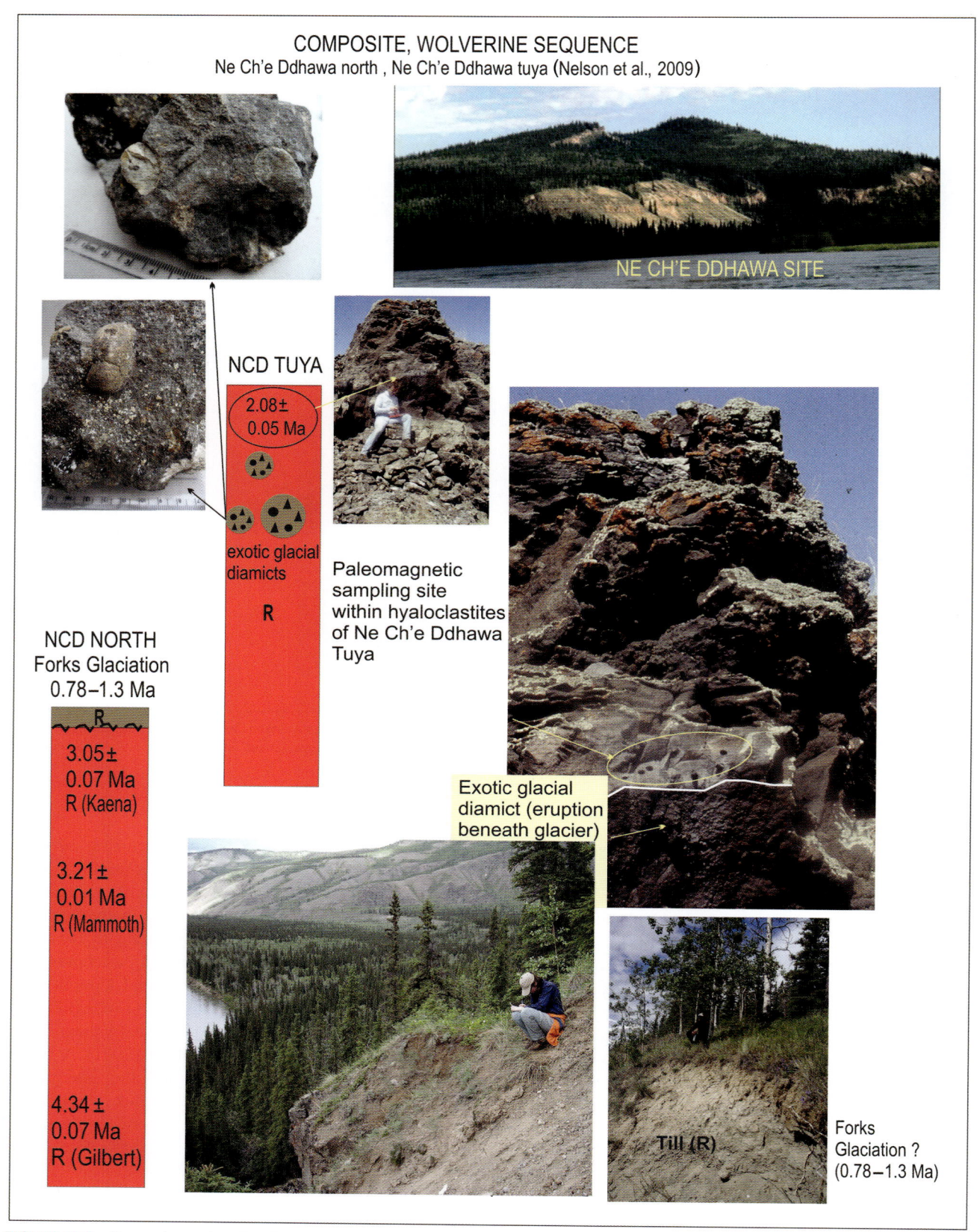

FIGURE 49.9 Composite stratigraphy of the Wolverine sequence is described at two sites: NCD Tuya and NCD North. NCD Tuya exposes diamictites (till) incorporated into lava during a subglacial eruption and provides an accurate age for this glaciation (2.14 ± 0.14 Ma). NCD North exposes a till (latest Matuyama) overlying a 3.05 ± 0.07 Ma basalt.

FIGURE 49.10 Lower Mushroom site (Fort Selkirk volcanic complex) exposes a basalt (1.83 ± 0.3 and 2.32 ± 0.13 Ma) overlying outwash (?) gravels which were deposited during the Gauss (G10) or earliest Matuyama (MIS 100) Chrons.

of a series of five lava flows exposed along the Yukon River downstream from Fort Selkirk and are discontinuously overlain by outwash and loess. The uppermost basalt has an Ar/Ar age of 0.311 ± 0.005 Ma, and all flows are normally magnetised. The outwash deposit is assumed to be the late Middle Pleistocene Reid glaciation (MIS 8/6) and is overlain by loess of the latest Pleistocene (McConnell Glaciation, MIS 2; Huscroft, 2002; Huscroft et al., 2004; Nelson et al., 2009).

49.3. NORTHWEST TERRITORIES

49.3.1. Mackenzie Mountains

Katherine Creek (KC) section (Fig. 49.13) consists of five montane tills, capped by one Laurentide till. These tills overlie 8 m of colluvium developed above a bedrock pediment. It is difficult to differentiate between the colluvium and the lowest montane till because the till clasts are subangular and glaciers travelled only short distances from

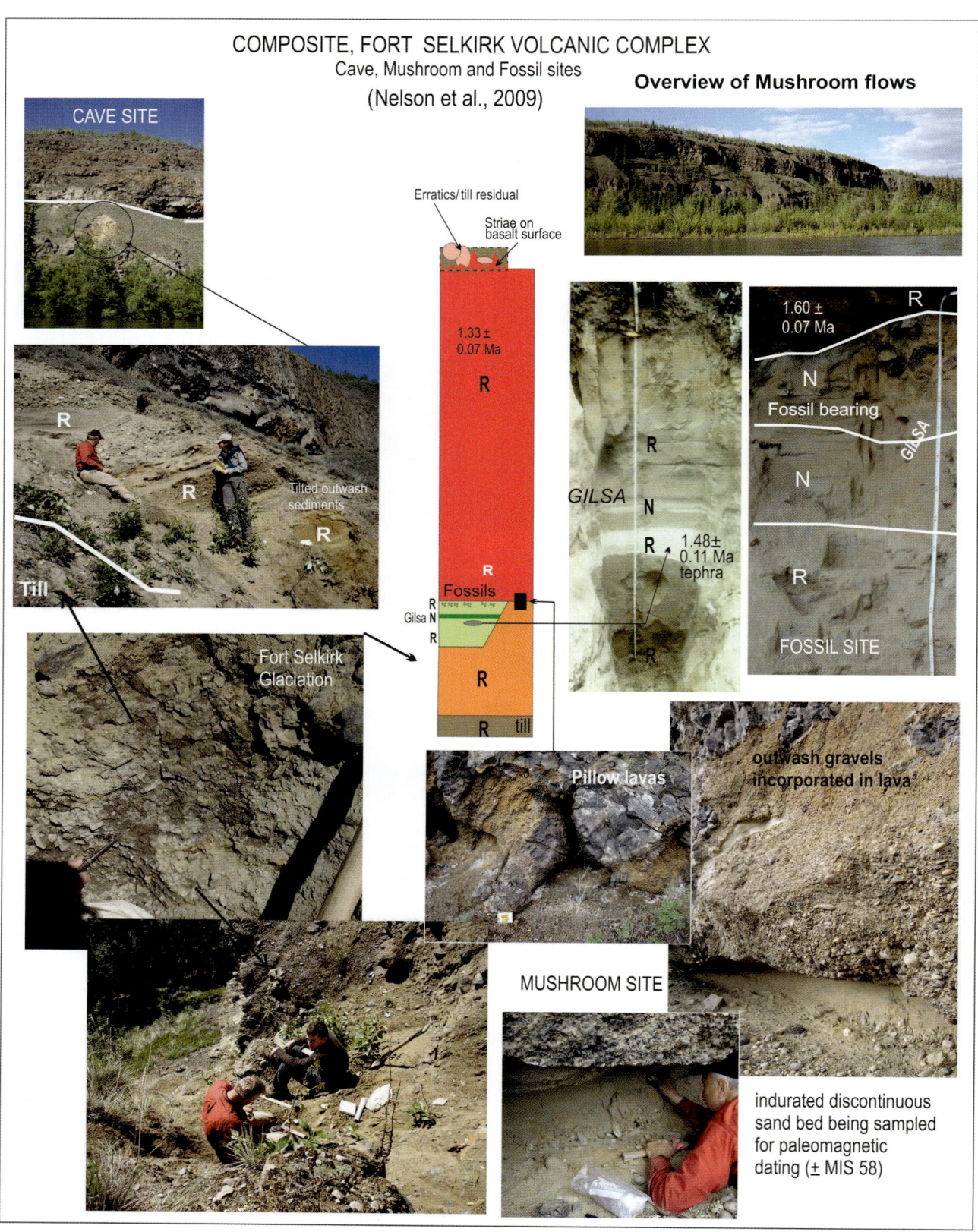

FIGURE 49.11 Cave, upper Mushroom and Fossil sites of the Fort Selkirk volcanic complex expose glacial and interglacial deposits underlying a reversed basalt (1.33 ± 0.07 Ma). The basalt has been overridden by ice, as evidenced by glacial striae and residual erratics. Reversed till and outwash deposits at the base of the exposure are post-Olduvai (< 1.78 Ma) but younger that the Gilsa subchron (1.6–1.55 Ma). Interglacial deposits above the till and outwash contain a reversed–normal–reversed polarity sequence and a tephra (1.48 ± 0.11 Ma). The normal sediments are assigned to the Gilsa subchron and are overlain by reversely magnetised sediments containing Blancan fossils.

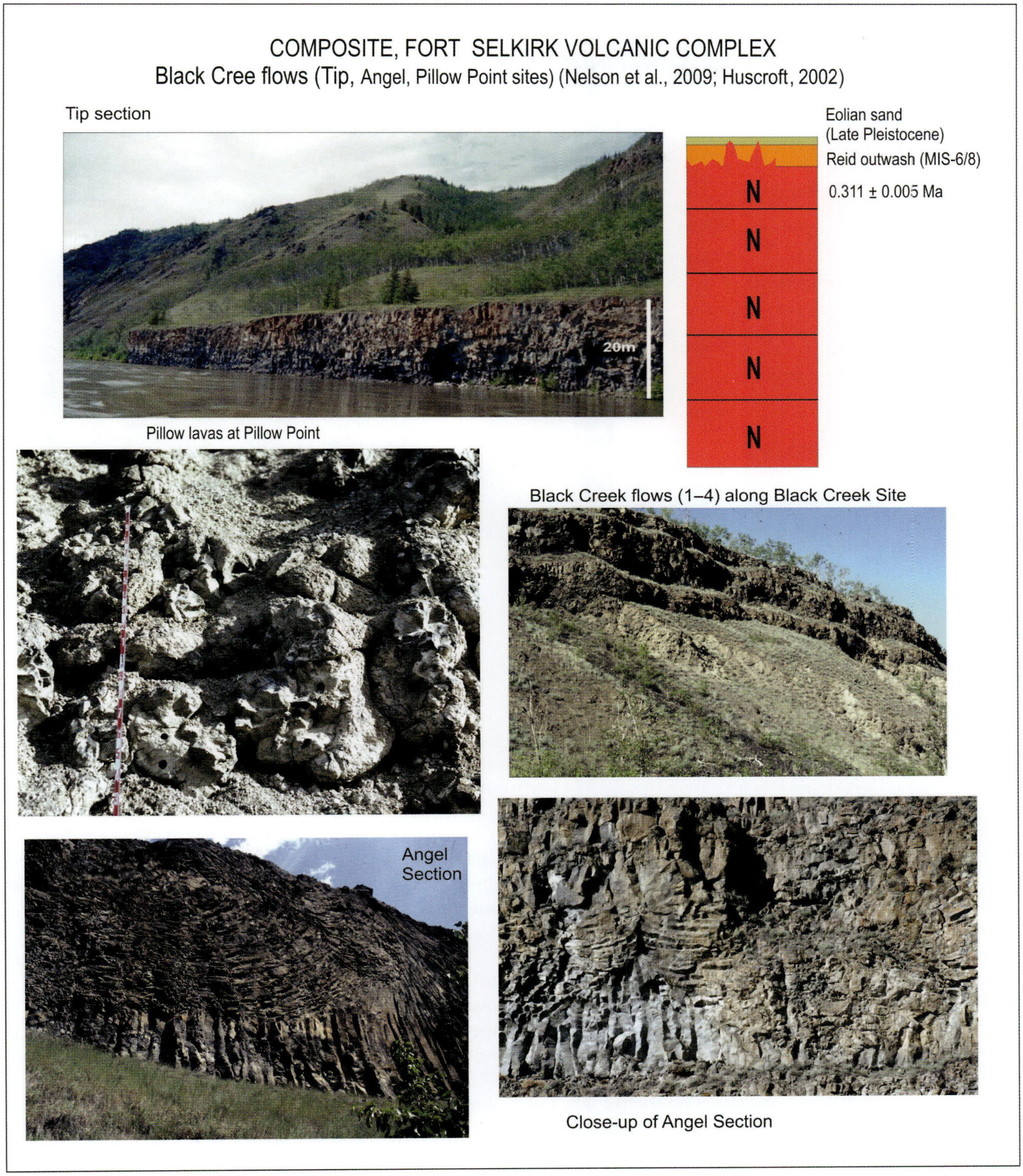

COMPOSITE, FORT SELKIRK VOLCANIC COMPLEX
Black Cree flows (Tip, Angel, Pillow Point sites) (Nelson et al., 2009; Huscroft, 2002)

FIGURE 49.12 Black Creek outcrop (Fort Selkirk volcanic complex) exposes five basalt flows, and the upper flow (0.311 ± 0.005 Ma) is overlain by outwash of the Reid glaciation (MIS 6/8).

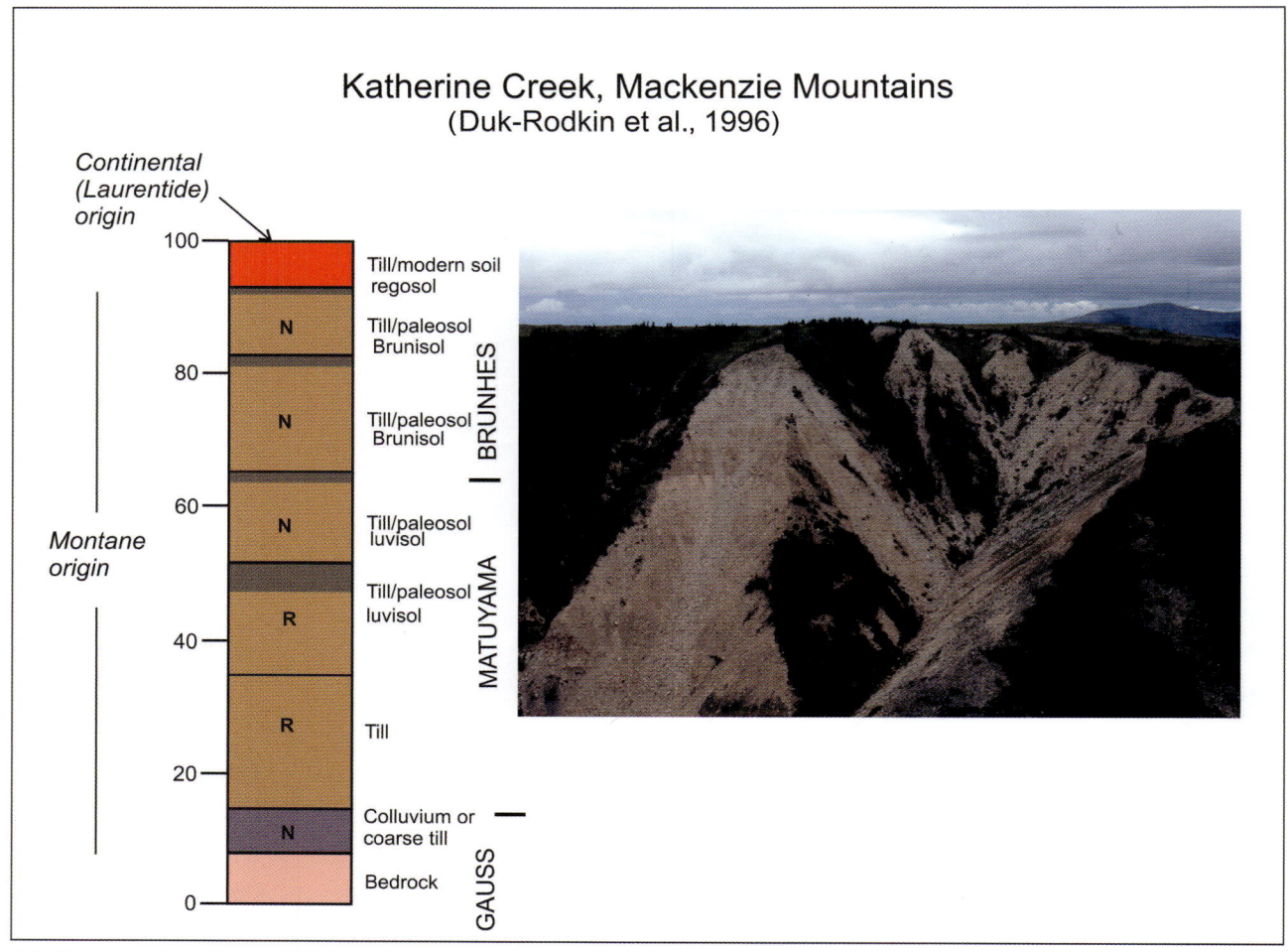

FIGURE 49.13 Katherine Creek (KC) site exposes a series of montane (local) tills of both Matuyama and Brunhes age. These tills are underlain by normally magnetised colluvium over bedrock and are overlain by a Late Wisconsin (Laurentide) glacial till at the surface.

their mountain sources. This colluvial unit is also present at the Little Bear (LB) site where it reaches thicknesses of over 20 m. In Duk-Rodkin et al. (1996), this first deposit overlying the bedrock pediment was considered to be a colluvium. At both the KC and LB sections, this lowest unit is normally magnetised, and if this unit is in fact a till it may correlate with the earliest glacial deposits in the Yukon, which are also normally magnetised, and assigned to the latest Gauss Normal Chron. Above the normally magnetised colluvial unit are two reversed tills (pre-Olduvai), thought to be laid down by a single glaciation, as there is no sharp contact or palaeosol between the two. A luvisol is developed on the upper reversed till. The overlying till is normally magnetised and assigned to the Olduvai (1.95–175 Ma). It contains ice cast pseudomorphs developed at its surface, as well as a luvisolic B horizon. Two normally magnetised tills were deposited above the Olduvai till, and these are assigned to the Brunhes Normal Chron (<0.78 Ma; MIS 16?, 12? or 8?). The surface till is continental (Laurentide) and considered to be Late Pleistocene (MIS 2). Chlorine 36 ages of 28–26 ka obtained from boulders near LB site indicate that the glacier had receded from its maximum position by that time (Duk-Rodkin et al., 1996).

LB River section (Fig. 49.14) comprises five montane tills overlying a colluvial deposit. The entire sequence is overlain by a till of Laurentide origin (Fig. 49.14). As in the KC section, three tills (normal/reversed/normal polarity) in the lower half of the outcrop span the late Gauss (2.70–2.58 Ma) to early Matuyama (2.58–1.98 Ma). The three overlying montane tills are coarse-textured, and only the last till provided a (normal) polarity. The tills are assigned to the Brunhes Normal Chron, and the uppermost Montane till is assumed to have been deposited by a Reid equivalent MIS 8/6 glaciation. The palaeosols at LB indicate a progressive deterioration of climate. The palaeosol developed on the lowermost colluvium exhibits a thick weathering horizon, while at the top of the section, a thin Eutric Brunisol is identified (Fig. 49.14).

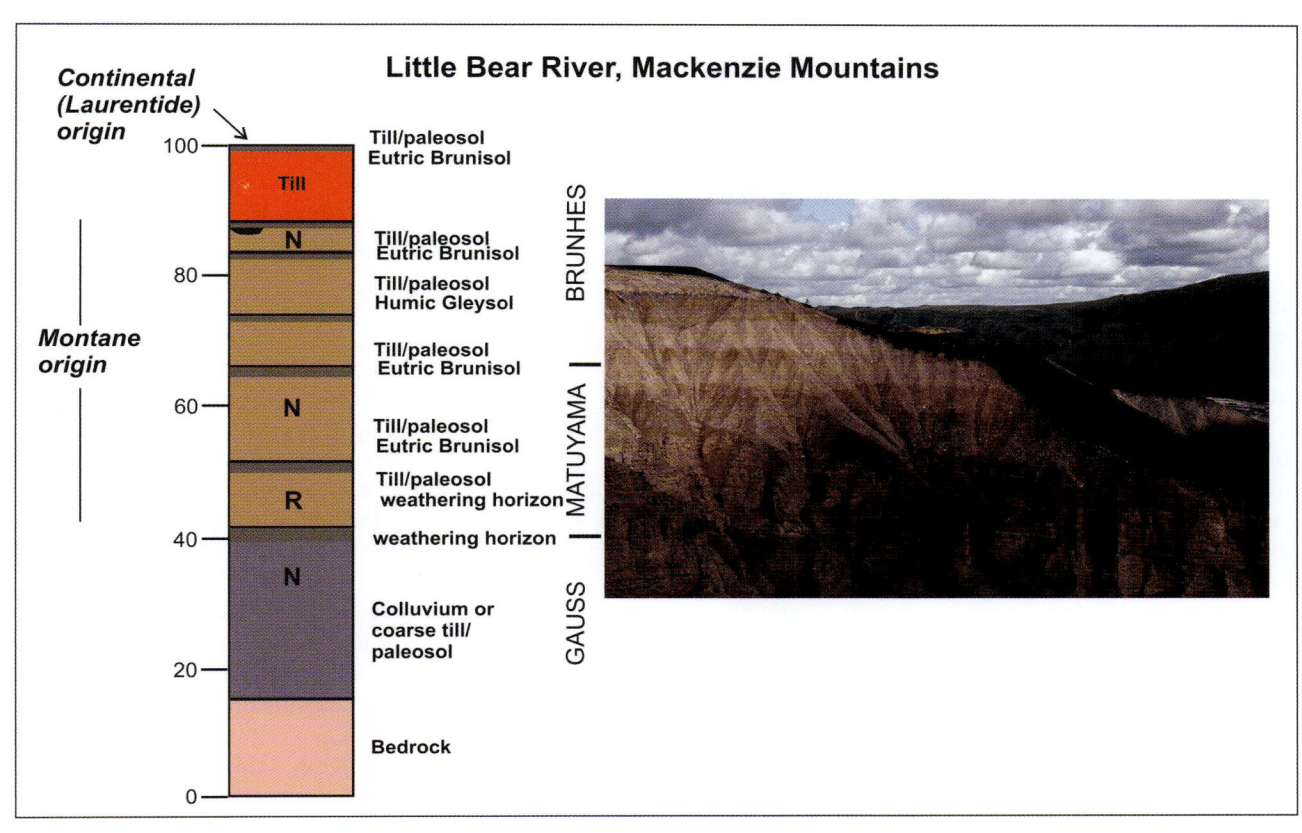

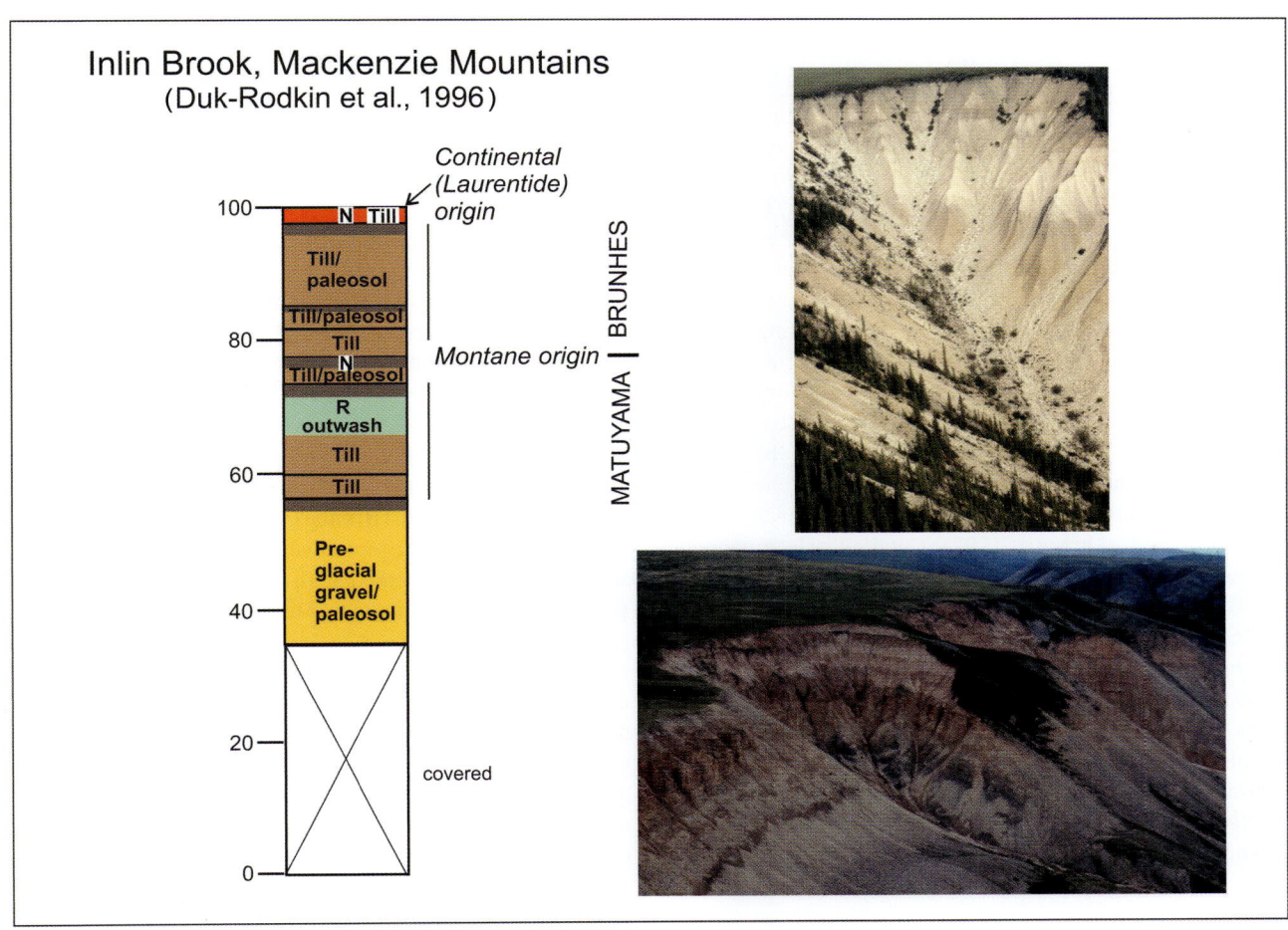

Chapter | 49   Stratigraphical Record of Glacials/Interglacials in Northwest Canada

FIGURE 49.14 Little Bear (LB) and Inlin Brook (IB) are two exposures near the KC site and reveal a similar stratigraphy to that of KC.

Inlin Brook (IB) section (Fig. 49.14) has a less complete stratigraphy than the KC and LB sites and is considered as a complimentary section. It consists of a preglacial gravel deposit of Tertiary age at the base, overlain by tills and outwash deposits capped by a single continental (Laurentide) till. There are five distinctive glacial units (Fig. 49.14), four of which have a palaeosol developed at their upper contact (detailed study of the palaeosols has not been carried out at this locality). Two tills and outwash at the base of the section are reversely magnetised and are assigned to the early Matuyama (2.58–1.98 Ma). These are overlain by a till and palaeosol of normal polarity (Olduvai 1.95–177 Ma), which in turn are overlain by two tills and palaeosols for which no polarity was obtained. However, they are assumed to be correlative with the Brunhes age tills at the KC and LB sites. The three sections were correlated on the basis of lithology, palaeomagnetism and palaeosol properties. For the younger strata, chlorine dates obtained from surface boulder erratics were used in the correlation. A formal stratigraphical nomenclature has been developed for the deposits of this region (Duk-Rodkin et al., 1996). The sequence of glacial tills and palaeosols marks a long record of glacial–interglacial cycles, while soil properties for the two oldest palaeosols point to climatic conditions much warmer and wetter than today.

49.3.2. Mackenzie Delta

Glacial and non-glacial deposits have been recorded in the Mackenzie Delta area from the Taglu, Kumak and Unipkat boreholes, which extend to 450, 250 and 95 m depth, respectively (Figs. 49.3 and 49.15). Only the Taglu and Kumak boreholes contain diamictons thought to be of glacial origin and fall within the early Matuyama Reversed Chron (Dallimore and Matthews, 1997). The cores represent mostly fluvial, glaciofluvial and marine deposits consisting of intercalated sands/silts/clays and lesser amounts of gravels. Cross-beds and foresets are common, indicating deltaic deposition. The diamictons are located at the base of the cores and are non-glacial in origin below the Gauss/Matuyama boundary (2.58 Ma), and Pleistocene in age above the boundary, based on the presence of *Artemisia* (Dallimore and Matthews, 1997). Detailed palaeomagnetic measurements of all three cores were carried out by Wang and Evans (1997). Pollen analysis of the Kumak borecore was carried out by Jette, and macrofossils were studied by Matthews and are reported in Dallimore and Matthews (1997).

The Taglu borecore (Fig. 49.15) reveals a sequence of 35 m (450–415 m) of interbedded clay and silt with occasional gravel zones interbedded with silty sand beds. It is overlain by a diamicton of about 25 m thickness, with some core gaps above and below the diamicton. The diamicton extends from 415 to 390 m depth and is normally magnetised (Gauss) (Dallimore, 1992). The samples are rich in reworked pollen, indicating floral characteristics of a climate warmer than today. However, one sample contains pollen which suggests a climate as cold as today. The diamicton also contains reworked pollen which was derived from Eocene to Jurassic bedrock. This diamicton is clearly associated with non-glacial conditions. Above this diamicton is a 15 m (390–375 m) thick silty/clay bed containing recycled pollen from Eocene to Jurassic rocks. It is reversely magnetised and assigned to the Kaena or Mammoth subchron. This silty/clay bed is overlain by a 35-m-thick diamicton and sand unit with normal polarity (Gauss) at its base, and reversed polarity (earliest Matuyama) at the top of the unit. This diamicton extends from 375 to 340 m with the reversely magnetised sediments extending from 351 to 360.5 m. These sediments contain no plant macrofossils but do contain shells. This second diamicton, which spans the Gauss–Matuyama boundary, contains a great admixture of pollen types revealing herbaceous-shrub tundra vegetation. This mix suggests cold (glacial?) conditions for the second diamicton (Dallimore and Matthews, 1997). Both diamicton units are predominantly clay-textured, with varying amounts of granules and pebbles. No Shield erratics were found in the diamicton. This second diamicton (reversed) is underlain and overlain by normally magnetised sediments which are assigned to the Gauss Normal Chron and the Olduvai normal subchron, respectively (Fig. 49.15). Overlying this second diamicton are 80 m (340–260 m) of conglomerate with silty sand beds and sandy gravel beds containing organic debris, wood and occasional shells (to the top). Several oxidised horizons in the upper segment of this core interval may represent palaeowater table levels. Polarity for this interval indicates a reversed/normal/reversed/normal/reversed sequence, where the normal intervals are assigned to the Reunion and Olduvai, respectively (Fig. 49.15). This interval is overlain by about 115 m (260–145) of sandy/silt to silty/sand beds some with organic detritus, wood and occasional shells. This core interval records, from bottom to top, an R–N–R polarity sequence which is assigned to the late Matuyama (post-Olduvai), Jaramillo and latest Matuyama, respectively. The overlying sediments (145–0 m) are all normally magnetised (Brunhes Normal Chron) and from bottom to top consist of massive sand with organic detritus, silty beds grading to sand with minor laminations, clayey silt and ice-rich organic silt of terrestrial and marine origin to the top of the core.

The Kumak borecore, at its base, reveals a normally magnetised diamicton (262–225 m) consisting of massive pebbly/silty/clay (Fig. 49.15). This lowest segment of the core (Late Pliocene) contains abundant pine, spruce, fir and hemlock pollen (Dallimore and Matthews, 1997) which is typical of Pliocene deposits in this region and points to a climate warmer than present. A sharp contact separates this

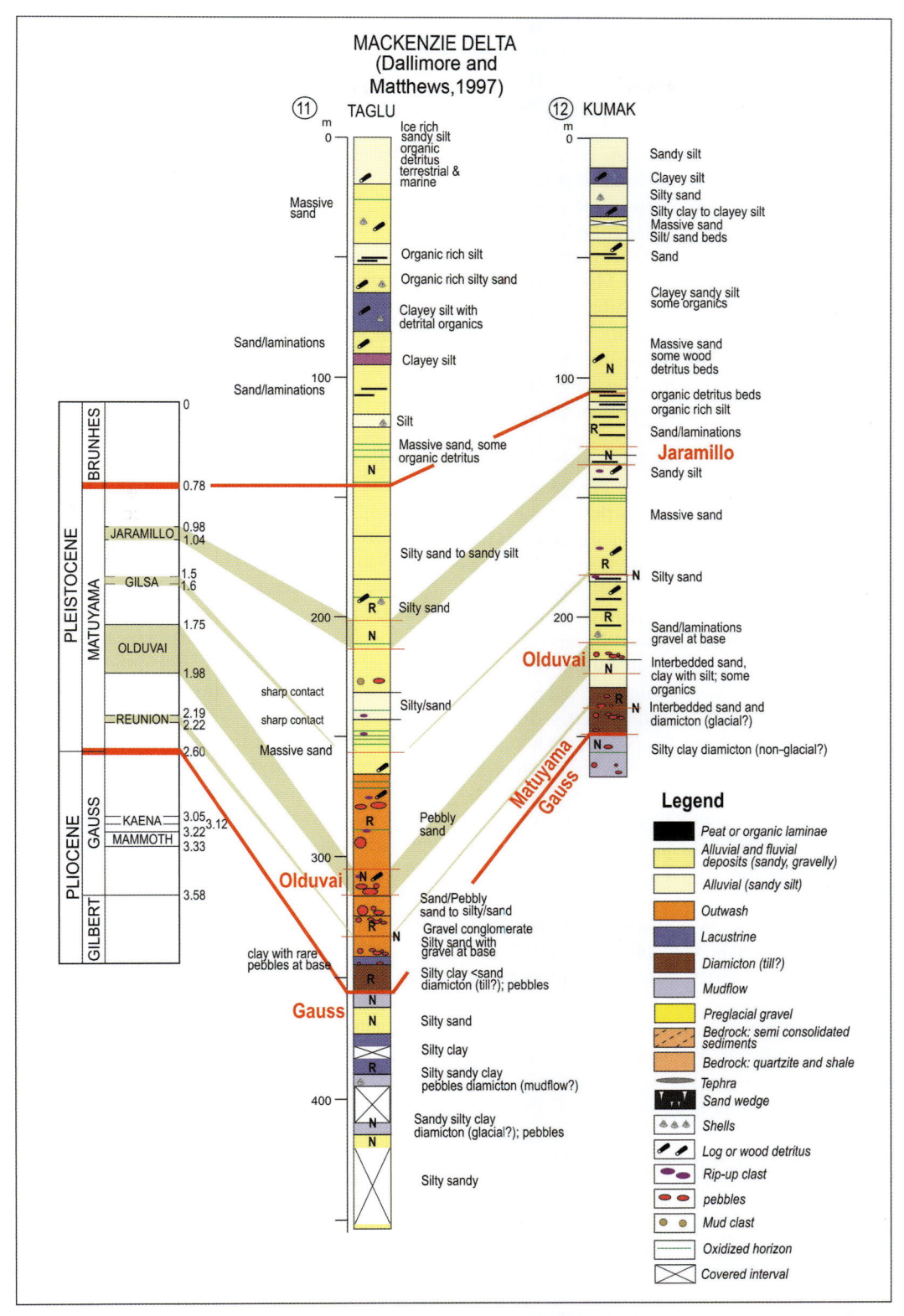

Chapter | 49 Stratigraphical Record of Glacials/Interglacials in Northwest Canada 679

FIGURE 49.15 Taglu (450 m) and Kumak (260 m) cores collected from the Mackenzie Delta expose a continuous sequence of sediments which span the late Gauss, Matuyama and Brunhes Chrons and include most, if not all, of the subchrons. While the sedimentology of the cores has not been fully studied, a rich pollen assemblage has been obtained, which suggests that the Mackenzie Delta may have been impacted by early glaciations only (earliest Matuyama to Gilsa; 2.58–1.6 Ma). Sandy pebbly diamicts and reworked pollen near the base of the cores are thought to mark glacial conditions.

lower normally magnetised (Gauss age) diamict from the pebbly reversely magnetised (earliest Matuyama) upper diamict. The pebbly diamicton changes to sand interbedded with clayey silt beds containing little or no organic detritus (250–225 m) and reworked pollen. This pollen mix derived from the upper diamicton contains a high percentage of spruce and Polypodiaceae, a good indicator of poorly preserved, rebedded pollen. The low quantity of pollen and organics detritus in addition to the sedimentological characteristics of the diamicton suggests glacial conditions or the initial stages of a cooling episode (earliest Matuyama, MIS 100?) (Dallimore and Matthews, 1997). Pebble lithologies were not available, precluding an estimate of the provenance of the diamicts. The next core segment (225–215 m) reveals silt interbedded with sand and clay, and has no record of organic detritus. The pollen record near the base of this core segment shows a decrease of spruce and alder with fluctuations of Cupressaceae (juniper, *Thuya*, etc.), *Abies* (fir), and pine, a low percentage of herbs, and an increase of grasses and sedges. Further up the core segment, pollen indicates a change to colder climate, with increasing spruce, grasses, sedges and birch, and a decrease of pine pollen. The lower part of the core (262–215 m) records an N–R–N polarity sequence assigned to the late Gauss, early Matuyama and Olduvai (Late Pliocene to Early Pleistocene), respectively. The core segment between 215 and 174 m is predominantly reversely magnetised but contains a short interval of normally magnetised sediments assigned to the Gilsa subchron (1.5–1.55 Ma). This core segment consists of sand with laminations and gravel at the base. At a depth of 183 m, these sediments are high in *Picea* (spruce) pollen but contain low amounts of pollen from shrubs and herbs, indicating generally cool conditions, with temperatures similar to today. The next core segment (175–149 m) consists of reversely magnetised massive sands, which have a generally low pollen count, and contain three oxidised horizons which may indicate palaeowater table conditions. Above these horizons, a sample rich in pollen (*Picea*, *Abies*, *Sambucus*) indicates conditions warmer that today. The core segment extending from 149 to 99 m consists of sandy/silt deposits with laminations, rip-up clasts and wood detritus at the base, organic silt beds increasing to the top containing reworked pollen. The sediments reveal a reversed/normal (Jaramillo)/reversed/normal polarity sequence (assigned to the late Matuyama, Jaramillo, latest Matuyama and early Brunhes). Between 99 and 74 m, the core sediments are composed of massive sand with wood detritus and an oxidised horizon changing to clayey sandy/silt with organics. They are normally magnetised (Brunhes). Pollen here consists of a high percentage of *Picea*, little *Pinus*, and some reworked Cupressaceae. From 75 to 49 m, sediments are clayey sandy silt, and contain organics. They are normally magnetised (Brunhes) and contain mostly spruce pollen, very little pine, and about 10% *Betula*, and *Alnus*. The uppermost part of the Kumak core (49–0 m) consists of silty clay to clayey silt with some massive sandy beds containing wood detritus and shells, and abundant spruce pollen.

49.3.3. Smoking Hills

An exposure along the West River, a tributary to the Horton River near the coast of the northern mainland (Figs. 49.1–49.3, 49.16), reveals a sequence of multiple tills separated by silty clay deposits, some of which contain organic material. The surface till was deposited by continental (Laurentide) ice, and a lag of boulders and cobbles of Canadian Shield (Keewatin) provenance litter the landscape. The Laurentide till is covered discontinuously by massive loess (Duk-Rodkin and Barendregt, field notes 2004; Fig. 49.16). The glacial sequence overlies preglacial sands and gravels probably correlative with the Late Tertiary (Pliocene) Beaufort Formation or the Olduvai age Worth Point Formation (Vincent, 1990). The preglacial deposits are underlain by Upper Cretaceous shales. The glacial sequence consists of three reversely magnetised tills separated by two silty clay deposits (also reversed). The tills contain predominantly chert and quartzite pebble lithologies and minor conglomerate (<1%). The silty/clay deposits may be remnants of interglacial deposits. Pollen analysis is currently being carried out for these units. The upper reversed till is truncated by two silty/clay to clayey/silt and minor sand deposits and is overlain by a thin lacustrine unit containing minor mudflow lenses and minor organics. The lacustrine unit has a palaeosol developed at its surface which is overlain by overbank deposits containing organic and wood detritus. The units above the upper reversed till are all normally magnetised. A 13-cm diameter log (*Larix* sp.) was found within the upper unit, suggesting that climatic conditions at the time of deposition of this unit were similar to those presently seen some 500 km to the south. Overlying the organic rich interglacial deposits is a till which is normally magnetised and contains pebble lithologies of a much greater variety than lower units. These include dolomite, siltstone, sandstone, chert, quartzite and gneiss. The till is overlain by cryoturbated aeolian sand and loess with a well-developed soil and scattered Laurentide glacial erratics. Pebble lithologies of all the subsurface tills are of local origin only, supporting the hypothesis that the Horton ice cap (Duk-Rodkin et al., 2004; Barendregt and Duk-Rodkin, 2011) was restricted to the local uplands only. The normally magnetised till exhibits a larger variety of lithologies which may indicate ice provenance from a greater region, including perhaps Banks Island. It is important to note that no shield lithologies were found in any of the subsurface tills.

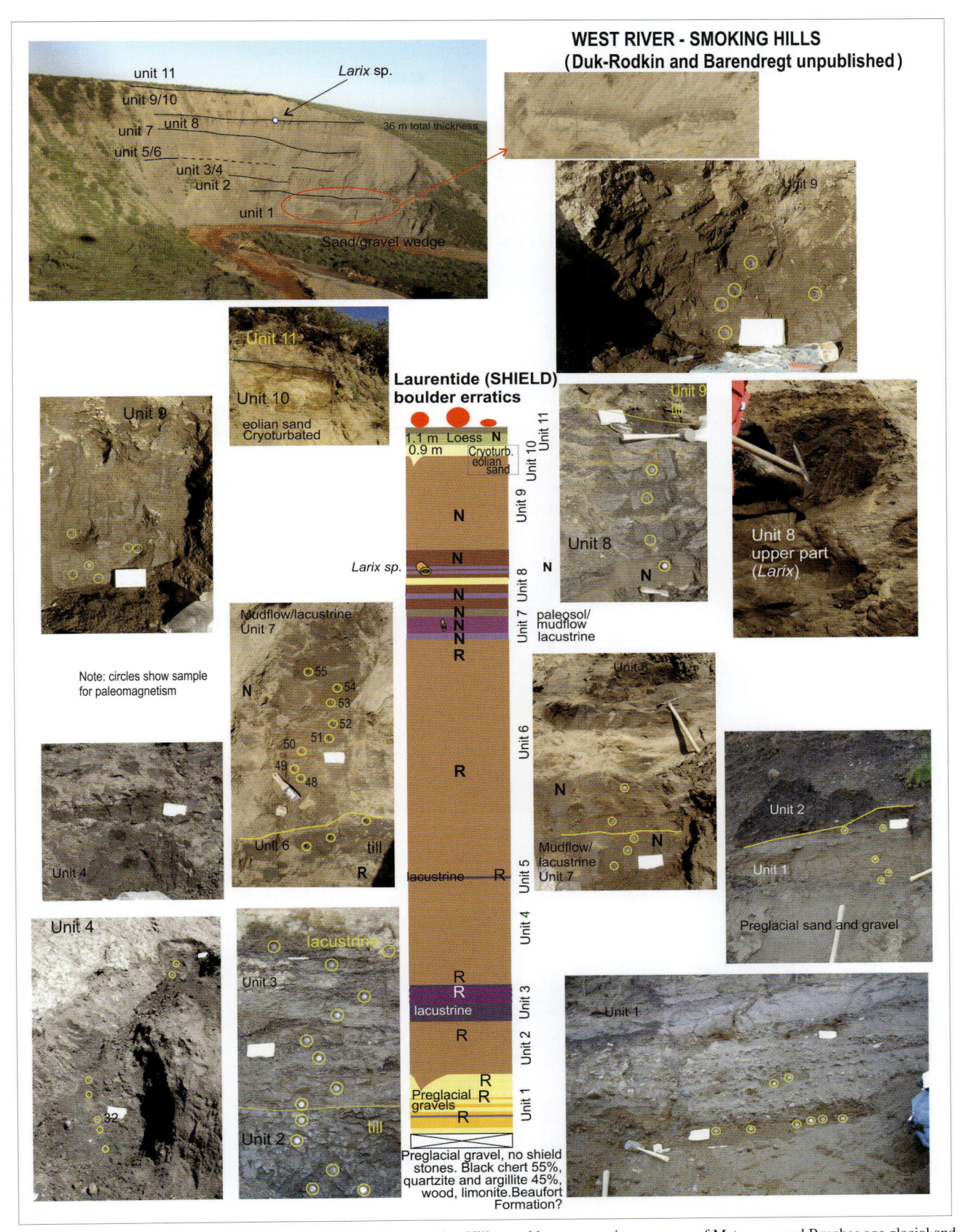

Chapter | 49 Stratigraphical Record of Glacials/Interglacials in Northwest Canada

FIGURE 49.16 West River (tributary of Horton River) in the Smoking Hills provides an extensive exposure of Matuyama and Brunhes age glacial and interglacial deposits, overlying reversely magnetised preglacial deposits which may be late Beaufort Formation or younger (possibly Worth Point equivalent?). The preglacial sediments contain ice-wedge casts suggesting cold conditions, and these were preserved by a glacial till which was deposited shortly after their formation. Only the surface till (Late Wisconsin) contains Shield erratics. Pollen analysis of interglacial deposits is currently in progress.

49.3.4. Banks Island

Evidence of older glacial deposits is recorded in sections along the southern, western and eastern coasts of Banks Island (Figs. 49.3 and 49.17–49.20). These deposits extend back to the Early Pleistocene (Vincent et al., 1984; Barendregt and Vincent, 1990; Vincent, 1990; Barendregt and Irving, 1998). Preglacial (Worth Point Formation) sediments are found above the Beaufort Formation which occurs at the base of two sections (Figs. 49.17 and 49.18) and are assigned to the Olduvai Subchron (1.95–1.77 Ma). These fluvial, aeolian and colluvial deposits contain buried peat and a palynological record indicating open larch-dominated forest-tundra vegetation (Matthews et al., 1986; Vincent, 1990). Overlying the preglacial Worth Point Formation is a series of tills, collectively referred to the Banks Glaciation(s), which may represent as many as four separate glaciations during the interval 1.77–1.05 Ma. They are overlain by interglacial deposits, including peat and soil, which contain the Jaramillo Subchron (1.05–0.99 Ma at the Morgan Bluffs Site; Fig. 49.19). These deposits in turn are overlain by a reversely magnetised till of late Matuyama age (unnamed glaciation), and interglacial silt deposits and soil developed during the Brunhes/Matuyama transition. The Early Pleistocene glacial limit is marked by deposits of the Banks Glaciation(s), which covered most of the terrain except for the NW part of the island (Vincent, 1983). During this time, the Bernard and Plateau Tills of northeastern Banks Island were also deposited. During the Matuyama Reversed Chron, sediments on Banks Island provide a record of at least two and probably as many as five glaciations and two interglaciations.

Evidence for two middle Pleistocene glaciations is found on Banks Island, an unnamed glaciation, and the Thomsen Glaciation, separated by an extensive suite of interglacial sediments (unnamed interglaciation). These two glaciations occurred between the Brunhes/Matuyama boundary (0.78 Ma) and 0.13 Ma (Vincent, 1983; Barendregt et al., 1998). They are underlain by deposits of an unnamed interglacial which spans the Brunhes/Matuyama boundary (Fig. 49.3), and is overlain by marine transgressive sediments and organic beds of the interglacial Cape Collinson Formation (Vincent, 1992). These glacio-marine sediments contain *in situ* shells which yielded dates of >37 ka (GSC-3698) and a U/Th age estimate of 85.9–130.0 ± 11.9 ka (UQT-143) (Vincent, 1992). These deposits are therefore

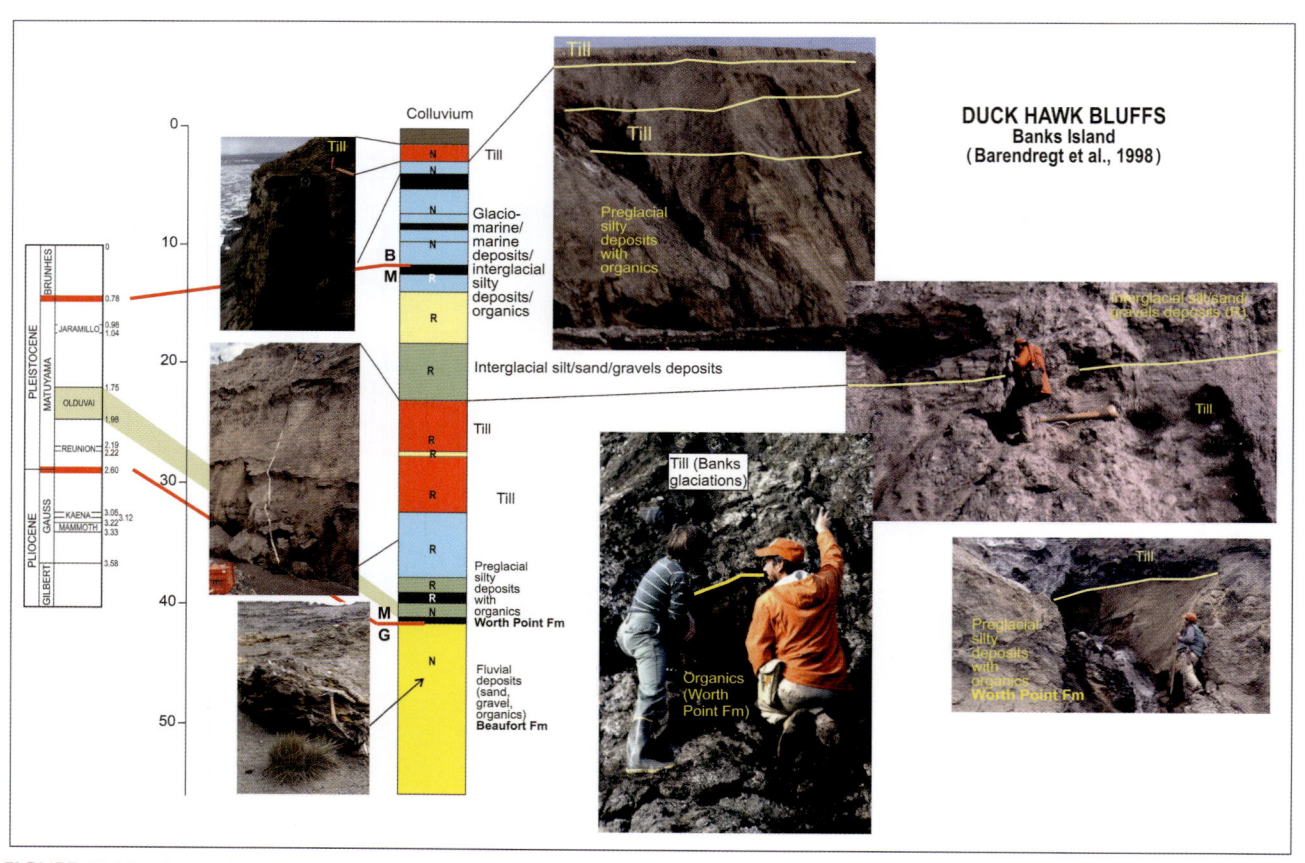

FIGURE 49.17 Composite stratigraphy at Duck Hawk Bluffs sections indicates preglacial (Olduvai) sediments overlain by a sequence of two reversed tills and one normal till, as well as glacio-marine and interglacial deposits.

Chapter | 49 Stratigraphical Record of Glacials/Interglacials in Northwest Canada

WORTH POINT BLUFFS Banks Island (Barendregt et al., 1998)

Pliocene gravels and Pleistocene till/paludal/marine sediments

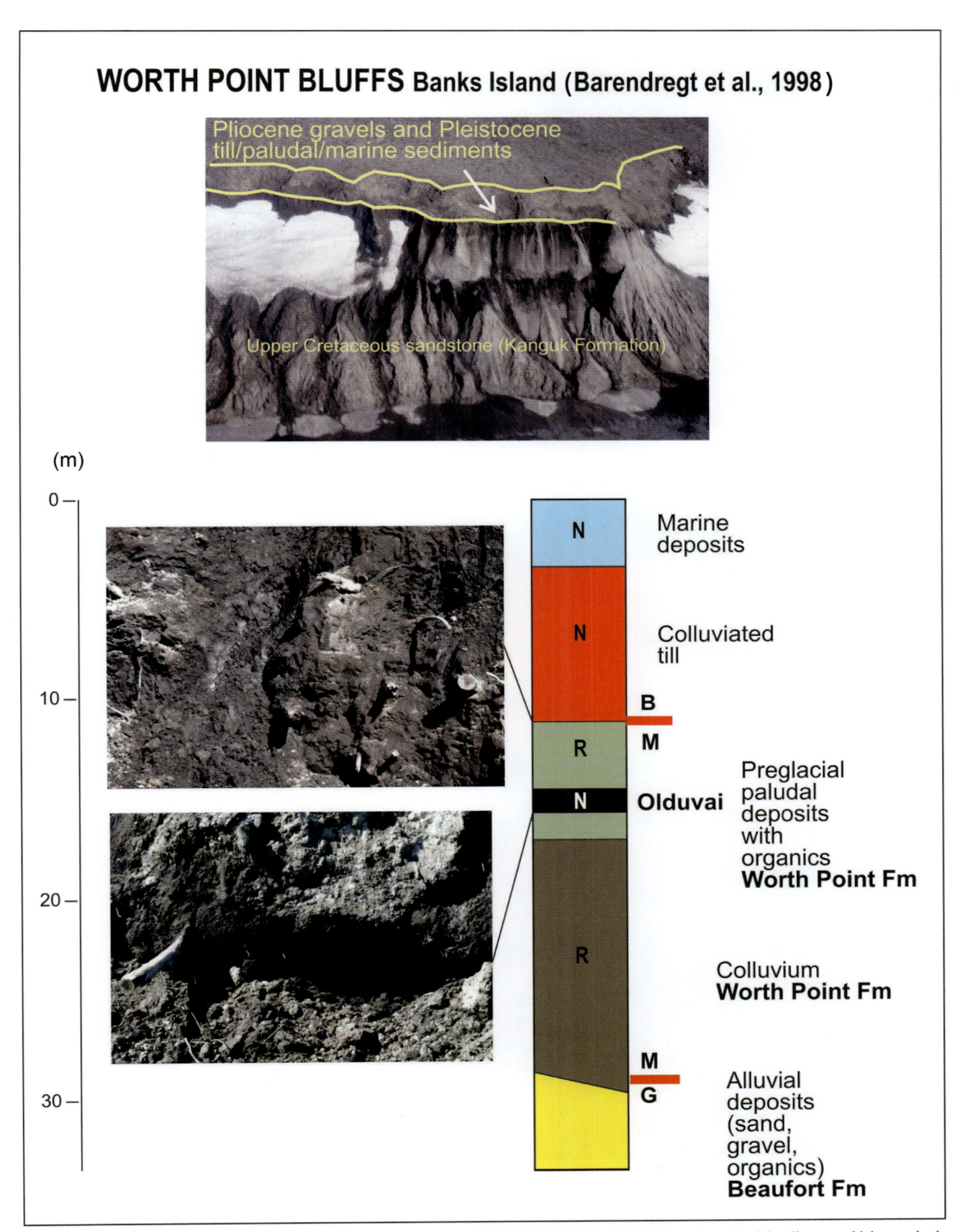

Upper Cretaceous sandstone (Kanguk Formation)

N	Marine deposits
N	Colluviated till
	B
R	M
N	Olduvai — Preglacial paludal deposits with organics **Worth Point Fm**
R	Colluvium **Worth Point Fm**
	M
G	Alluvial deposits (sand, gravel, organics) **Beaufort Fm**

FIGURE 49.18 Worth Point stratigraphy exposes (at its base) an extensive sequence of reversely magnetised preglacial sediments which contain the Olduvai subchron (1.97–1.78 Ma) which is overlain by a Brunhes age till and marine deposits.

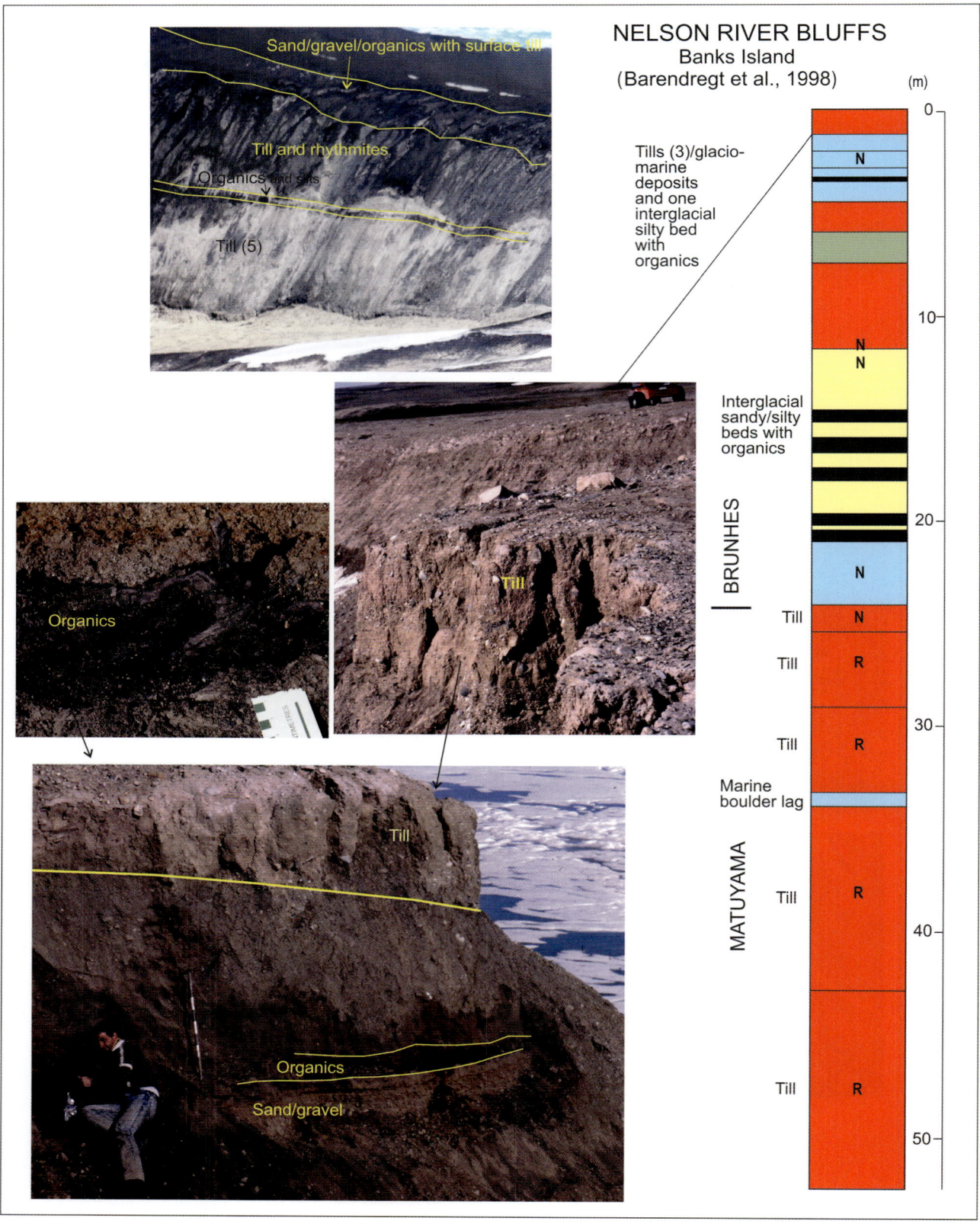

FIGURE 49.19 Nelson River Bluffs expose an extensive sequence of reversely magnetised tills overlain by normally magnetised interglacial deposits and three tills.

FIGURE 49.20 The composite stratigraphy at Morgan Bluffs exposes a long sequence of marine and paludal sediments and three tills (all reversed). A short normal interval occurs near the middle of the section and is assigned to the Jaramillo. The two upper tills are normally magnetised and assigned to the Brunhes.

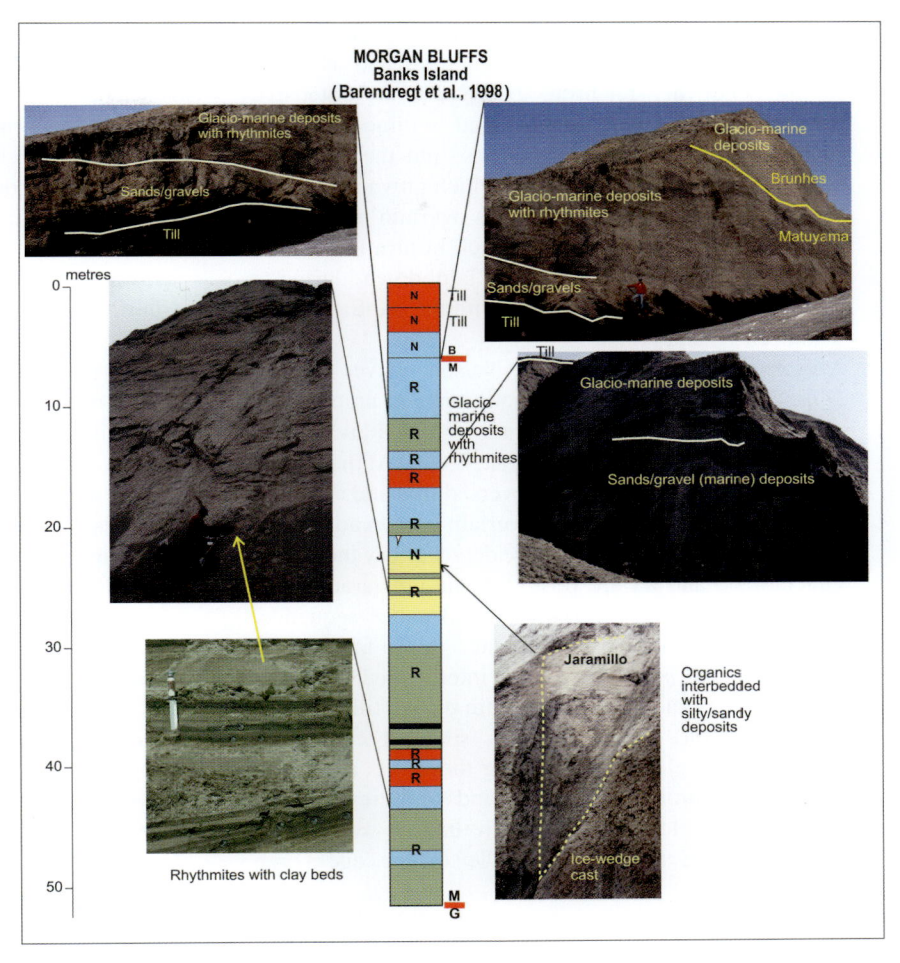

considered to be of Sangamonian age (80–130 ka) and the underlying till of the Thomsen Glaciation is pre-Sangamonian, and most likely of Middle Pleistocene age (250 ka (MIS) 8?). Apart from the stratigraphy, there is no geomorphological evidence for these two Middle Pleistocene glaciations. However, it is thought that the Thomsen Glaciation marks the northern extent of Middle Pleistocene-ice in NW Canada. It reached the south-central part of the island and must have travelled along the eastern coast, cutting across the north-east of the island, and bordered along the Plateau Hills to the west, south and east. Thomsen Glaciation deposits were over-ridden along their middle-eastern extent by Late Pleistocene ice (Early Wisconsinan Amundsen Glaciation; Vincent, 1983). This limit was considered to be of Late Wisconsinan age by various authors, and this age has recently been verified by England et al., 2009.

1. *Duck Hawk Bluffs.* These sites contain 11 lithostratigraphical units (Figs. 49.3 and 49.17; Matthews et al., 1986; Vincent, 1990). Reversely magnetised aeolian sediments and associated pond deposits assigned to the Worth Point Formation overlie fluvial facies of the Pliocene (Gauss Normal Chron) Beaufort Formation. These are overlain by marine sediments, and two tills (pre-Jaramillo age Banks Glaciations), all of which are reversed. Above these deposits are reversely magnetised interglacial colluvial, fluvial and aeolian deposits (Late Matuyama) (Vincent, 1992). These are overlain by marine and woody peat deposits which record periglacial and interglacial conditions, and at their base record the Brunhes/Matuyama boundary, and at their upper contact are overlain by Late Pleistocene till (Amundson Glaciation) and Holocene deposits.

2. *Worth Point Bluffs.* At Worth Point (Figs. 49.3 and 49.18), three units occur above Pliocene fluvial coastal plain sediments of the Beaufort Formation. The lowest unit is composed of preglacial colluvium and paludal deposits which contain a rich palaeoflora and fauna (Matthews et al., 1986; Vincent, 1990), indicating that climatic conditions were sufficiently warm to allow conifer growth. The unit is reversely magnetised, except

for a 2-m-thick subunit comprising paludal (pond) deposits, which is normally magnetised and assigned to the Olduvai subchron, based on an extensive sequence of reversely magnetised sediments which overlie this unit at other localities, and the generally warmer than modern conditions which prevailed during the Early Pleistocene. This unit is overlain by periglacially reworked till (normal), which in turn is overlain by sands and gravels (normal) laid down during a marine transgression associated with the pre-Sangamon Thomsen Glaciation.

3. *Nelson River Bluffs.* The Nelson River Bluffs (Fig. 49.19) expose 13 lithostratigraphical units which are assigned to the Matuyama (post-Olduvai) and Brunhes Chrons (Barendregt et al., 1998). At their base, the sequence consists of two reversed tills and a marine boulder lag, which in turn is overlain by two additional reversed tills. The four tills were deposited by the Banks Glaciations, and all are post-Olduvai–pre-Jaramillo in age. These tills are overlain by a Brunhes sequence, comprising a till, marine sands, thick interglacial sediments, and *in situ* peat units interbedded in a peri-marine delta sequence. It is overlain by two tills (Thomsen Glaciation) separated by a silt bed containing organics (interglacial). Overlying these tills are a few metres of rythmites and deltaic sands and gravels associated with deglaciation, which in turn are overlain by the Amundsen till of Latest Pleistocene age (Late Wisconsinan).

4. *Morgan Bluffs Site (composite).* At these bluffs, 13 lithostratigraphic units have been recognised from outcrops at many sections (Fig. 49.20). The first five units record the first glaciation(s) (Banks Glaciations including the Bernard Till) and associated high and low sea-stand deposits. They are reversely magnetised and pre-date the Jaramillo (1.05–0.99 Ma). These units are overlain by interglacial deposits, which are also reversely magnetised, but contain a normally magnetised large ice-wedge cast near its upper contact, which is assigned to the Jaramillo (Barendregt et al., 1998). These units are in turn overlain by reversely magnetised till and associated high and low sea-stand deposits (post-Jaramillo, Latest Matuyama). The three upper units are normally magnetised tills associated with the Thomsen and Amundsen Glaciations.

49.4. NORTHERN BRITISH COLUMBIA

1. *Sezill Creek.* This site is located in a Plio–Pleistocene volcanic complex (Fig. 49.21). A basalt unit forming the base of this section is dated between 6.1 and 7.1 Ma. It is overlain by a diamicton (till?) and sandy sediments which have a reversed polarity (Gilbert–Gauss? early Matuyama?). These are overlain by a boulder lag, followed by a diamict and overlying rythmites. Up section is a coarse to medium diamict (MIS 34?) containing striated clasts (intrusive and metamorphic megaclasts), and isolated pockets of openwork gravels (till). The glacial event that deposited this till was regional in scope, based on clast lithologies. Continuing up section, two tills are preserved between three basalt flows bracketed between a 1.1 Ma (Pyramid Formation) and a 1.0 Ma (Ice Peak Formation(s); Souther, 1992; Spooner et al., 1995). The lower till contains exotic megaclasts, rare pyroclastic debris, and several erosional contacts which may represent several glacial events. The upper till (MIS 30) interfingers with lacustrine tuff at a section approximately 5 km to the east of Sezill Creek section, indicating that the Pyramid Formation basalt is coeval with the glacial event that deposited this lower till. The top of the section is covered by a thick colluvial unit which is overlain by a reversed (post-Jaramillo) basalt.

2. *Tahltan/Stikine River.* The sediments recorded at the Tahltan Canyon and Stikine River sites are glacial and non-glacial in origin. They overlie bedrock of Triassic age. A composite stratigraphy of the two sites is presented in Fig. 49.22. The deposits have been preserved beneath Pleistocene basalt flows from Mount Edziza and have yielded normal polarity (Brunhes Chron) (Spooner et al., 1996). These sediments record a regional glaciation from the Coast Mountains of British Columbia. The initial ice blocked the Stikine River, forming a glacial lake and glaciofluvial delta. Basalts were deposited on fluvial as well as glaciofluvial sediments. The basalt flow (K–Ar age of 0.30 ± 0.03 Ma) at the top of the section contains hyaloclastic/palagonite minerals indicating that it was deposited in water, or in association with ice. The deposits beneath the basalt consist of horizontally bedded, cross-bedded, massive and laminated sands interbedded with thin diamictons.

49.5. SOUTHERN BRITISH COLUMBIA
49.5.1. Merritt Sections (Composite)

Deposits exposed along Lily Lake Road, and the Coldwater River in the vicinity of Merritt, British Columbia (Fulton et al., 1992), contains evidence of five glaciations and six interglaciations (Fig. 49.23). At the base of the sequence, proglacial lacustrine sediments and overlying glacio-lacustrine sediments are separated by a Chernozemic palaeosol and are reversely magnetised. The intervening soil indicates that these lacustrine sediments represent two separate glaciations. These units are

FIGURE 49.21 Sezill Creek section near Mount Edziza, northern B.C. exposes a sequence of basalts and interbedded glacial sediments. The base of the section exposes a normally magnetised basalt (6.1–7.1 Ma) overlain by three tills separated by glaciofluvial and glacio-lacustrine deposits. These sequences are overlain by four basalts units and interbedded tills spanning the Jaramillo and latest Matuyama.

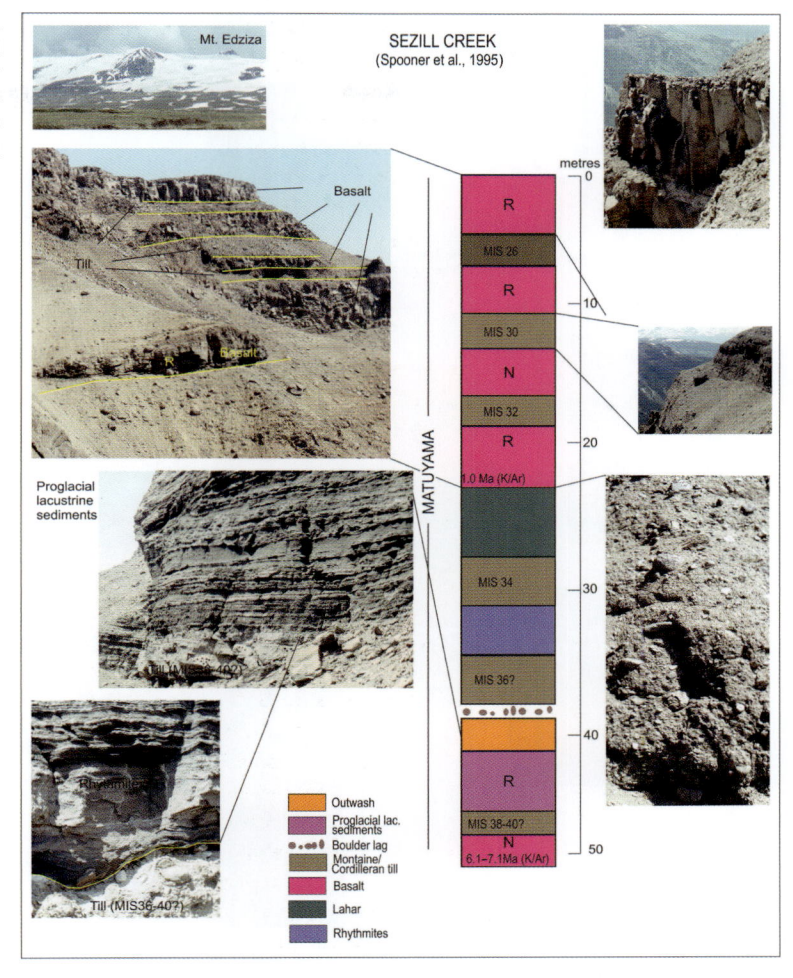

unconformably overlain by interglacial sediments containing a tephra (>670 ka) at its base and a palaeosol near the top. The Brunhes/Matuyama boundary occurs directly above the palaeosol. The palaeosol and underlying sediments are reversely magnetised, while the uppermost interglacial sediments are normally magnetised. The latter are overlain by a suite of glacial deposits and are normally magnetised. The age of these glacial deposits cannot be further constrained at this time, but it is possible that this glaciation occurred relatively soon after the eruption of a 0.6 ± 0.4 Ma normally magnetised valley basalt in the Merritt basin because the boulder lag (Fig. 49.23) is made up of material derived from this basalt, and the basalt is striated. The basalt was deposited during an interglacial period. The suite of glacial sediments containing the basalt boulder lag was succeeded by at least two additional glaciations, a penultimate glaciation, and the Late Pleistocene (Wisconsin) Fraser Glaciation.

49.6. REGIONAL CORRELATIONS

The NW Canadian glacial/interglacial record of the Late Cenozoic is preserved in a large number of sites which have been studied by various authors. The most complete stratigraphical record has been obtained from sites in the Yukon, northern Interior Plains and British Columbia. These sites (Fig. 49.1, sites 1–5 and 8–10) record multiple Cordilleran and/or montane glaciations, as well as periods of soil formation and loess deposition. The glacial/interglacial deposits overlie a preglacial stratigraphy, which in most cases is marked by a regional unconformity. Other sites record multiple plateau and/or continental glacial deposits overlying preglacial fluvial stratigraphy, and these records are less extensive than that seen in the cordillera (Fig. 49.1, sites 11–17). A third group of sites (Fig. 49.1, sites 6, 7, 18–20) provides a record of glaciations (tills) preserved between lava flows.

FIGURE 49.22 The composite of the Tahltan and Stikine sections reveals a sequence of Brunhes age glacial sediments: glacio-lacustrine and glaciofluvial beds and a till. In addition, one basalt (0.30 Ma) bed occurs between outwash units near the surface of the outcrop.

Sites within the Tintina Trench (Figs. 49.3–49.6) contain the most complete Late Cenozoic stratigraphy, and other study sites are compared to this group. The Tintina Trench sites provide a robust magnetostratigraphy with clear polarity chrons and subchrons which can be reliably correlated with the global Geomagnetic Polarity Timescale (Fig. 49.24). Other sites provide radiometric ages (tephra, basalt, organics) and exposure dates (^{36}Cl) that have further refined the ages of these deposits. When this magnetostratigraphy is combined with sites throughout NW Canada, an extensive lithostratigraphy is developed which can be correlated with the marine isotope record and provides mappable evidence of the extent and timing of glaciations.

In the Tintina Trench, 10 glacial units and seven non-glacial units span a substantial portion of the Quaternary record, and this record is evidently more extensive than those found at other sites in NW Canada. Data from the Mackenzie Mountains and Banks Island in the NWT reveals two additional glaciations within the Brunhes Chron which are not found elsewhere (Fig. 49.3; Duk-Rodkin and Barendregt, 2004).

49.7. LATE PLIOCENE PREGLACIAL SEDIMENTS AND THE FIRST GLACIATION (MIS G6)

49.7.1. The Cordillera

Pliocene and older Tertiary unconsolidated preglacial sediments are found in many sections in the Northern Cordillera, but few of these sediments have been dated. Along the Tintina Trench and on the northern slopes of the Klondike Plateau (upper White Channel Gravels), excellent exposures of preglacial gravels occur. They are largely composed of sands and gravels deposited as alluvial fans, and locally occur as mudflow deposits and

Chapter | 49 Stratigraphical Record of Glacials/Interglacials in Northwest Canada

MERRITT B.C.
(composite) Fulton et al., 1992

- Glacio-lacustrine
- Till (MIS 2)
- Proglacial lacustrine
- Till (Brown drift; MIS 6/8?)

- proglacial delta?
- proglacial lacustrine
- Boulder lag

Basalt 0.6 ± 0.4 Ma

N / R B/M

Outwash

R Pelecypod fossils (freshwater shells)

Coutlee Interglacial sediments

Tephra > 670 ka

Sub-Coutlee silts
R Glacio-lacustrine

Paleosol A (Chernozem?)

Coldwater silts
R Proglacial-lacustrine

(Early B.C. glaciation/Matuyama age)

Outwash — lacustrine — Boulder lag — Uppermost Coutlee Interglacial sediments

Brunhes / Matuyama
Paleosol B (Brunisol)

interglacial sediments

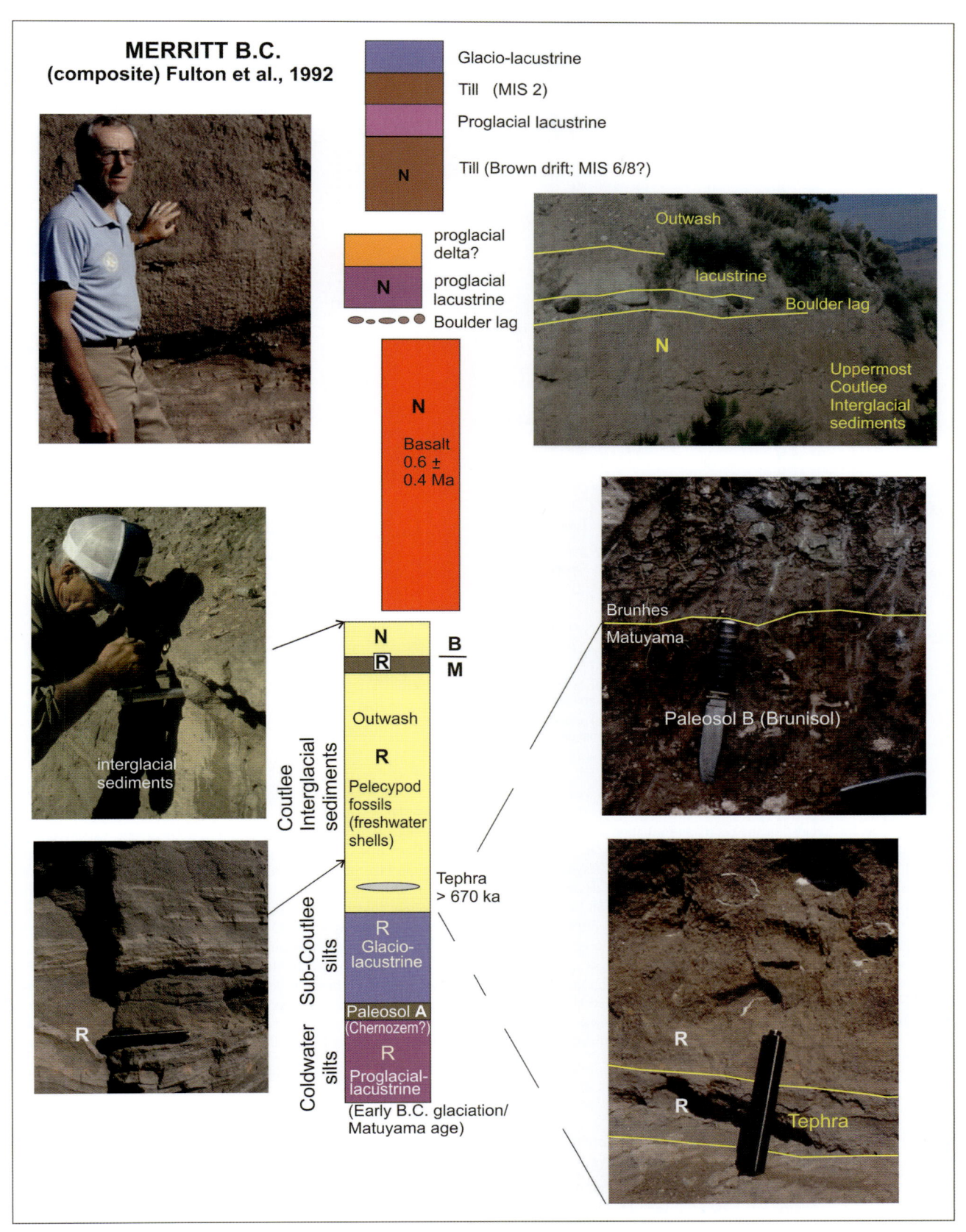

R

R Tephra

FIGURE 49.23 The composite of the sections near Merritt, in southern B.C. expose from bottom to top, a sequence of reversely magnetised glacio-lacustrine and interglacial sediments containing a tephra (>0.670 Ma) and palaeosol. Above these deposits is a normal basalt (0.60±0.4 Ma) which is overlain by outwash deposits and two tills (all normal) representing up to three glaciations.

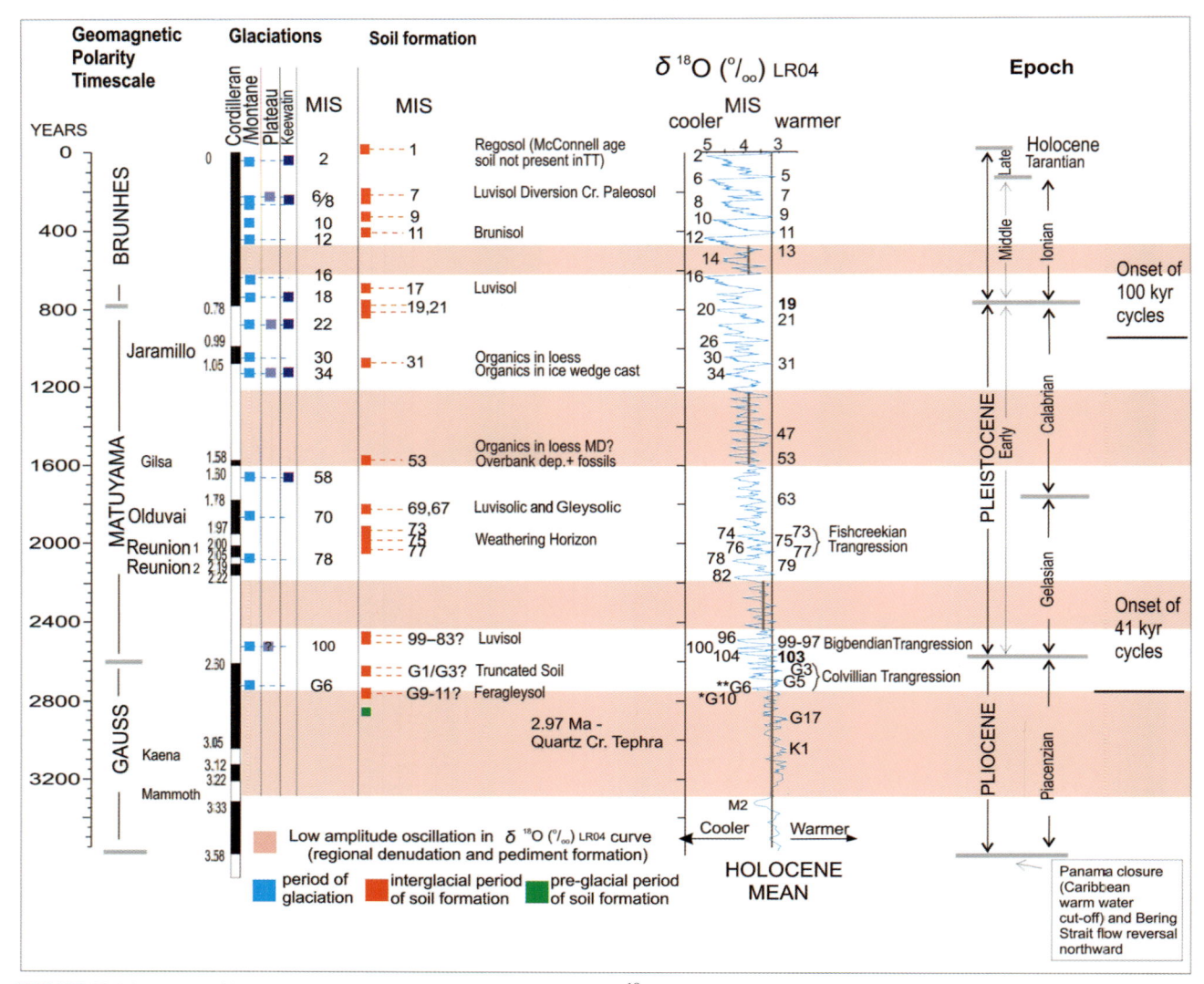

FIGURE 49.24 Zones of low amplitude oscillation in the LR04 composite $\delta^{18}O$ record during the Late Pliocene and Pleistocene are considered to be periods of regional denudation and pediment formation, and are compared to the glacial/interglacial record of NW Canada. Geomagnetic polarity timescale is based on (Cande and Kent, 1995; Gradstein et al., 2004) and composite $\delta^{18}O$ LR04 marine isotopic record (relative palaeotemperature) is obtained from multiple deep ocean cores (Lisiecki and Raymo, 2005). The base of the Pleistocene (2.58 Ma) and new subdivisions of the Pliocene and Pleistocene follow the recently ratified convention described in Gibbard et al. (2010). Black and white areas are normal and reversed polarity, respectively. Marine Isotope Stages (MISs) are labelled on LR04 (even numbers represent colder (glacial) and odd numbers warmer (interglacial) conditions). MIS numbering scheme follows Ruddiman et al. (1986, 1989), Raymo et al. (1989) and Raymo (1992) from present to MIS 104, and Shackleton et al. (1995) in the Gauss Chron. Vertical line marks Holocene mean $\delta^{18}O$ (Raymo 1992). Suggested correlation of glacial deposits to cold stages in the marine isotopic record (blue squares) is shown to the right of Geomagnetic Polarity Timescale. There are three types of glacial regimes shown (Cordilleran/montane, Plateau and Keewatin glaciations). Suggested correlation of soil forming periods to warm stages in the marine isotopic record is shown with red squares. G18 (2.97 Ma)—First Cordilleran glaciation in Yukon; *G10—Worldwide cooling commences; **G6—First global glacial event coincides with final closure of Panama Isthmus, start of Chinese loess/palaeosol sequences, and appearance of genus Homo; MIS 25–21—Marks onset of pronounced glacial/interglacial cyclicity. (For interpretation of the references to colour in this figure legend, the reader is referred to the Web version of this chapter.)

ponded silts, clays and minor sands. In the Tintina Trench, they occur above a major Miocene–Pliocene unconformity which separates tilted Miocene sands and conglomerates from horizontal Pliocene preglacial fluvial deposits. The tilted beds contain *Pterocarya* and *Tsuga* pollen grains which argues for a Miocene and older age for the sediments and indicates warm climatic conditions (Duk-Rodkin et al., 2001) while the preglacial conformable beds above have an assemblage with *Polemonium* pollen grains indicative of cooler conditions. The preglacial deposits in the Tintina Trench have been assigned to MIS G19 because the *Polemonium* assemblage argues for a cool climate, and after stage G19 the isotopic record reveals a strong cooling trend leading to the first glaciation (G18) in the area (Fig. 49.3). These deposits are correlated with the nearby Klondike Wash stratigraphy (22 km to the south) reported by Froese et al. (2000) who provided evidence for the interfingering of White Channel gravels and the first glacial outwash in the region (Klondike Wash) during the late Gauss. These beds are considered to be equivalent to the preglacial White Channel gravels based on the presence there of the 3.0 ± 0.33 Ma Quartz Creek tephra (Sandhu et al., 2000), pollen (late Miocene or younger) and normal polarity sediments (of probable Gauss age, Froese et al., 2000; Duk-Rodkin et al., 2001).

In the absence of tephra dates in the Tintina Trench, the reversed polarity of unit 1a at WFR cannot be assigned with confidence to Kaena, Mammoth or Gilbert, but may be equivalent to the Kaena age upper White Channel gravels reported in Froese et al. (2000). The sediments are older than 3.18 Ma (base of upper Gauss normal) and older than the preglacially ponded deposits at RC which contain *Polemonium* pollen and a normal Gauss polarity and therefore are broadly correlative to the upper White Channel gravels.

At the Fort Selkirk volcanic complex, the Lower Mushroom site contains gravels which may be associated with the first glaciation and is overlain by basalt dated at 2.32 ± 0.13 Ma (Westgate et al., 2001). This gravel deposit is thought to be glacial in origin, based on clast lithologies, although this cannot be said with certainty. If it is glacial in origin, then this first glaciation in the area likely occurred around or before 2.3 Ma (Fig. 49.10).

In the Mackenzie Mountains, preglacial deposits are found at the KC and LB sections (Figs. 49.3, 49.13 and 49.14). At these sites, the preglacial deposits are mostly coarse subangular gravel with minor fine sand and silt developed on bedrock pediments (Duk-Rodkin et al., 1996). These deposits (colluvium/diamicton/till?) have a weathering horizon at its upper contact and are normally magnetised (Gauss Normal Chron) and may be equivalent to the preglacial sediments at the base of RC (unit 1b). Alternatively, it may be correlated to the till at RC (RC unit 2). At the IB section, the preglacial gravels are thick alluvial fan deposits (Palaeocene, Summit Creek Formation) which overlie bedrock.

49.7.2. Northern Interior Plains

In the northern Interior Plains of Canada, three separate regions, each with multiple sites, record glacial deposits laid down by local ice caps. The glacial history of the Mackenzie Delta region is characterised on the basis of two long cores (Taglu and Kumak). In the Taglu core, silty sand sediments (normal) at the base of the core are considered to be preglacial fluvial deposits. They are followed by a sequence of mudflow, lacustrine, fluvial sand and diamicton deposits. This diamict is non-glacial in origin and is normally magnetised (Gauss Chron). At the base of the Kumak core, normally magnetised silty clay diamicton (mudflow) contains pollen of warmer/preglacial conditions, and is therefore assigned to the late Gauss Chron. This non-glacial diamicton grades to a glacial diamicton at the Gauss/Matuyama boundary (Fig. 49.15).

In sharp contrast to the Late Pliocene preglacial sediments described for the Mackenzie Delta cores and the Smoking Hills, the Banks Island record reveals an Early Pleistocene age for the preglacial sediments underlying the first till (Worth Point, Fig. 49.18), based on a suite of normally magnetised paludal deposits contained within a much more extensive silt, sand and aeolian sequence, which is reversed. The first glaciation on Banks Island (Barendregt et al., 1998) is post-Olduvai (late Matuyama, MIS 58?).

49.8. EARLIEST MATUYAMA GLACIATIONS (2.2–2.6 MA, MIS 100 OR 98 OR 96)

49.8.1. The Cordillera

At the Tintina Trench RC and EFR sites, a reversely magnetised till (unit 3) represents an early Matuyama glaciation (the second glaciation to affect this region). Morphologic evidence of this early Matuyama glaciation is recorded on two terrace outwash remnants in the lower Klondike Valley which are loess covered, and these loesses are reversely magnetised (Froese et al., 2000) (see Dawson map, Duk-Rodkin, 1996). The till of this second glaciation is overlain by thin beds of fine sands and silts and a well-developed luvisol from which palaeomagnetic samples were collected. Like the till parent material below, the soil yielded a reversed polarity. The soil forming period has been tentatively correlated with the Bigbendian marine transgression (Fig. 49.24) in Alaska, and with isotope stage 97. At the WFR section, unit 3 may or may not be present. The extensive vertical outcrop precluded detailed assessment.

This early Matuyama glaciation may or may not be present in the Mackenzie Mountains as there is only one pre-Olduvai reversed till recorded there (Figs. 49.3, 49.13 and 49.14). Deposits of this glaciation may also be present at the base of Sezill Creek section in northern British Columbia, but this cannot be stated with certainty (Figs. 49.3 and 49.21).

49.8.2. Northern Interior Plains

In the northern Interior Plains (Fig. 49.3), evidence for an early Matuyama glaciation is found in the Mackenzie Delta (reversely magnetised diamicts in the lower part of the Taglu and Kumak cores, below the Reunion normal subchron, Fig. 49.15). In the Mackenzie Delta cores, the diamicton marking the first glaciation in this region extends across the Gauss/Matuyama boundary, changing its characteristics from non-glacial to glacial, and from normal to reversed polarity. In the Smoking Hills area, along the West River, one of the three exposures described in Duk-Rodkin et al. (2004) has yielded much additional information. The West River exposure reveals preglacial gravels with wood detritus and other organic debris at the base. These gravels may be the Late Pliocene Beaufort, or may be considerably younger (Worth Point equivalent). The gravels yield a reversed polarity, most likely the Mammoth or Kaena subchron within the Gauss Chron. This preglacial deposit contains sand-wedges developed at its upper contact indicating cold (periglacial/cryogenic) conditions following deposition of the host sediments. The same ice-wedges are, in places, filled with diamicton (till) (Fig. 49.16) which is reversely magnetised. The preservation of the underlying ice-wedge cast argues for till deposition fairly soon after the formation of the ice wedge, which otherwise would probably have been destroyed by subaerial processes. This event likely records the first glaciation on the northwest continental Arctic coast. This event may be correlative with the first glacial event in the Mackenzie Delta cores, and with sediments of the second glaciation in the Cordilleran stratigraphy (MIS 100?, Fig. 49.3). It is unlikely that ice from the Horton Plateau Icecap (Smoking Hills; Duk-Rodkin et al., 2004) reached the Mackenzie Delta because of the distance between the two sites (>400 km) and the very low ice gradient which would have existed. Further, the presence of the Horton Plateau Ice cap would preclude continental (Laurentide) ice from reaching the Mackenzie Delta region, and therefore the glacial diamicts at the bottom of Taglu and Kumak cores are almost certainly of local origin. The only possible local source would be the Aklavik Range, where geomorphologic evidence suggests the presence of pre-Late Pleistocene piedmont glaciers (Duk-Rodkin and Hughes, 1992).

49.9. EARLY MATUYAMA GLACIATION (1.98–2.15 MA, MIS 78)

49.9.1. The Cordillera

At the Tintina Trench RC site, an extensive mudflow deposit of glacial origin (unit 4; Fig. 49.4) overlies deposits of two older glaciations. The mudflow has a silty clay matrix and was further enriched in clays as a result of a prolonged period of moderate weathering and translocation of surface clays from a >2-m-thick regosolic soil developed at its surface. No polarity was obtained from these sediments, but it is overlain by a normally magnetised outwash (Olduvai subchron, unit 5) and underlain by a reversely magnetised (early Matuyama) till/luvisolic palaeosol (unit 3, MIS 100 and Bigbendian marine transgression). The mudflow (unit 4) is tentatively correlated with the Fishcreekian marine transgression in Alaska (MIS 78). This unit may be equivalent to the pre-Olduvai reversely magnetised till present at any of the Mackenzie Mountains sites: KC, LB, IB.

49.9.2. Northern Interior Plains

The early Matuyama glaciation is represented in the Taglu core by reversely magnetised sediments composed predominantly of coarse gravel conglomerate (outwash?) which are underlain by normally magnetised sediments assigned to the Reunion normal subchron. In the Kumak core, this unit is a diamicton occurring between normally magnetised sediments of the Reunion and Olduvai. There were no macrofossils recovered from this core interval, and it was not possible to determine whether these sediments are glacial or non-glacial in nature.

At the Smoking Hills section, this unit may be equivalent to one of the upper two reversely magnetised tills (Fig. 49.16).

49.10. OLDUVAI GLACIATION (1.75–1.98 MA, MIS 70?)

49.10.1. The Cordillera

At the Tintina Trench RC and EFR sites, a suite of normally magnetised glacial sediments and overlying palaeosol (unit 5) occur between the underlying glacial mudflow (unit 4) and an overlying outwash (unit 6) which is reversely magnetised and thus limits the age of unit 5 to the Olduvai subchron. Unit 5 consists of an outwash overlain by a till on which a weathering horizon was developed. The sharp unconformity between the weathering horizon and the overlying outwash may suggest truncation of what was at one time a fully developed palaeosol. Evidence also exists in the Mackenzie Mountains for a probable Olduvai age glaciation (Duk-Rodkin et al., 1996) on which a well-developed luvisol is found; both till and soil are normally magnetised.

49.10.2. Northern Interior Plains

In the Mackenzie Delta cores, deposits of Olduvai age include fluvial sediments (Taglu core) and marine clays (Kumak core) indicating interglacial conditions. In the Smoking Hills, normally magnetised interglacial lacustrine deposits with a palaeosol overlain reversed tills may be of Olduvai age. Likewise, paludal (preglacial Worth Point) deposits on Banks Island (Fig. 49.18) are normally magnetised and assigned to the Olduvai subchron (Barendregt et al., 1998). Macrofossils and pollen of these deposits indicate non-glacial conditions. The first glaciation to affect Banks Island developed shortly after the end of the Olduvai normal subchron.

49.11. LATE MATUYAMA GLACIATIONS (1.06–1.78 MA, MIS 58, 34)

49.11.1. The Cordillera

Reversely magnetised outwash and diamicton (till) deposits occurring between the Olduvai and Jaramillo subchrons are found at the RC and EFR sites, respectively (unit 6; Figs. 49.3–49.6). The palaeosol developed on the till at EFR is a gleysol and is also reversely magnetised. Both units are post-Olduvai in age. The age between 1.78 and 1.06 Ma represents one of the MISs in the range of 60–58 or 38–34. The most likely candidate is MIS 58 because the till in all likelihood correlates with a reversely magnetised till and outwash at Fort Selkirk occurring directly below the Gilsa subchron (an interglacial, falling somewhere between MIS 55 and 51; Fig. 49.11).

At the MD site on the Klondike Plateau, the upper Matuyama is represented by an extensive loess stratigraphy deposited under both glacial and interglacial conditions (at least three separate interglaciations; Fig. 49.8). The lower part of the exposure contains reversely magnetised loess with interglacial pollen, the Mosquito Gulch tephra (1.45 Ma), loesses deposited under glacial conditions (MIS 34?), another tephra (MD, 1.09 Ma), and ice-wedge casts in which a palaeosol (cryosol) is developed. In the upper half of the MD section is a unit of normally magnetised loess (Jaramillo, MIS 29?) in which a luvisol is developed, and is overlain by reversed loess, normal loess (B/M boundary marked by colluvial organic silt or yedoma), and finally by colluviated loess (yedoma) at the top (normal, Brunhes age) (Froese et al., 2000).

At Fort Selkirk Cave, Mushroom and Fossil sites reversely magnetised outwash and till (Fig. 49.11) were deposited during MIS 58, and overlying reversely magnetised interglacial deposits (MIS 57) contain the Fort Selkirk tephra (1.48 ± 0.11 Ma) which in turn are overlain by normally magnetised silts (Gilsa subchron, 1.58–1.60 Ma), in turn overlain by reversely magnetised post-Gilsa interglacial paludal deposits containing fossils (MIS 55?) (Jackson et al., 1996; Nelson et al., 2009).

At the Sezill Creek site (Mount Edziza, B.C., Fig. 49.20), reversely magnetised glacial deposits at the base of the exposure are underlain by a diamicton (till?), which in turn is underlain by 6.1–7.1 Ma basalt. The glacial deposits and diamict are overlain by a series of undated deposits (glacio-lacustrine, outwash gravels, diamicton (till?), glacio-lacustrine and two further diamicts (tills?)). This sequence is capped by a reversely magnetised basalt (1.0 Ma), and three interbedded tills and basalts for which only polarity data is available. Each basalt and till pair is assumed to be of similar age, and the pairs are assigned to MIS 32, 30 and 26, respectively. At the Sezill Creek site, there are two additional glacial packages beneath the reversely magnetised basalt (1.0 Ma) and above the lowermost till. There are no records of more than two glaciations between the Olduvai and Jaramillo subchrons in the stratigraphy of northwest Canada, possibly due to the low amplitude and uniform fluctuation of climate during the time interval between MIS 56–37 (Fig. 49.24), and therefore these two glaciations probably occurred after MIS 37 (1.22 Ma) and before 1.0 Ma and are most likely MIS 36 and 34 cold events. The lower most diamicton (till?) at the Sezill Creek section may be correlative to MIS 58 or may correspond to a much older (pre-Olduvai) glacial event.

Finally, at the Merritt section in British Columbia (Fig. 49.23), there are two sets of glacio-lacustrine deposits separated by a palaeosol (all reversed) which may be Late Matuyama, based on a minimum age of 0.670 Ma on an overlying tephra, and may therefore have been deposited by glaciations which occurred between the Olduvai and Jaramillo (MIS 78?, 70?, 58?, 34?).

49.11.2. Interior Plains

Post-Olduvai–pre-Jaramillo age sediments are present in the cores of the Mackenzie Delta (Fig. 49.15). However, it is not possible to determine with certainty whether these sediments are glacial or non-glacial in nature. The bulk of the evidence would suggest that the sediments are non-glacial. In the Smoking Hills, two reversed upper tills underlain by reversed glacio-lacustrine (?) sediments and overlain by a normal polarity sequence may correspond to MIS 58 and/or MIS 34 (Fig. 49.15).

On Banks Island, reversely magnetised preglacial paludal deposits containing the Olduvai subchron (Worth Point Formation at the Worth Point site, Fig. 49.18) are assigned to the late Matuyama. One of the two reversed tills at Duck Hawk Bluffs, one of the four reversed tills at Nelson River and one of the two reversed tills at Morgan Bluffs may correspond to MIS 58 and/or 34 (Fig. 49.3).

49.12. JARAMILLO SUBCHRON GLACIATION (0.99–1.05 MA, MIS 30)

49.12.1. The Cordillera

At the Tintina Trench RC site, a normally magnetised outwash (unit 7; Fig. 49.4) is underlain by reversely magnetised sediments. The upper contact of the outwash forms an unconformity with blocks of sediment that are reversed (unit 8). At WFR, this period of time is represented by loess (unit 7) which is not found at EFR.

Normally magnetised loess deposits of Jaramillo age occur in the Klondike Plateau at the MD site (Fig. 49.8). The loess deposits contain pollen and a palaeosol (luvisol) in the middle part of the loess unit. They are overlain by a diamict containing reworked pockets of the palaeosol and are also considered to be of Jaramillo age. Both loess and diamict are thought to have been deposited during interglacial conditions (MIS 31?). Above the diamict is a massive loess, which is reversely magnetised (Froese et al., 2000).

Within the Fort Selkirk Volcanic complex, Cave, Mushroom and Fossil sites contain glacial erratics and striae over a basalt surface (1.33 ± 0.07 Ma). It is assumed that the glacial event that left the erratics and the striae on this surface basalt corresponds to one of MIS 36, 34, 30?

At the Mount Edziza Sezill Creek site (Fig. 49.20), Jaramillo age deposits are thought to be represented by a diamicton (till?) (MIS 30?) overlain by a normal basalt (undated) occurring within a sequence of reversed basalts and interbedded diamictons (tills?).

49.12.2. Interior Plains

The Jaramillo subchron occurs in interglacial massive sand and silt with reworked organics, wood detritus and shells, in the Taglu and Kumak cores in the Mackenzie Delta.

The West River site in the Smoking Hills has several metres of interglacial sediments separated by a palaeosol, all of which are normally magnetised. It is not certain whether this normal interval is of Jaramillo age or is entirely within the Brunhes Normal Chron. Pollen analysis of this unit is still ongoing.

On Banks Island, Jaramillo age deposits occur only at the Morgan Bluffs Sites (Barendregt et al., 1998) where normal sediments occur within ice-wedge casts. The cryogenic polygons formed in reversely magnetised interglacial sediments deposited immediately before the Jaramillo (MIS 32) and the infilling occurred during the Jaramillo (MIS 31).

49.13. LATEST MATUYAMA GLACIATION (0.99–0.78 MA, MIS 20 OR 22)

49.13.1. The Cordillera

At the Tintina Trench RC site, post-Jaramillo outwash gravels (unit 8) and loess at both EFR and WFR (unit 8) are reversely magnetised and are overlain by normally magnetised (Brunhes) loess (unit 9) containing pollen. At RC, unit 8 is a remnant of glacial deposits (pockets of outwash sediment) preserved intermittently along an erosional and discontinuous contact. On the Klondike Plateau at the MD site, reversely magnetised sediments overlie Jaramillo age interglacial deposits (Froese et al., 2000). These deposits are discontinuous organic silts and diamicts with interbedded organics. They are interglacial and were probably deposited during (MIS 21?).

At the Sezill Creek site near Mount Edziza, post-Jaramillo reversed basalt and interbedded diamicton (till?) extend to the top of the section.

At the Merritt sites in southern British Columbia, there are reversely magnetised proglacial lacustrine and glacio-lacustrine deposits separated by a palaeosol (Chernozem). They are stratigraphically below reversely magnetised interglacial sediments containing a tephra dated at >0.67 Ma. These interglacial deposits have a palaeosol at their upper contact which is also reversed, and are overlain by normally magnetised interglacial deposits marking the Brunhes/Matuyama boundary. The reversely magnetised interglacial sediments are thought to be latest Matuyama, while the age of the underlying reversely magnetised glacio-lacustrine deposits are less well constrained.

49.13.2. Interior Plains

In the Mackenzie Delta cores, the latest Matuyama is represented by silty sand in the Taglu core and by organic rich silt interbedded with clay laminations in the Kumak core (Fig. 49.15).

The West River site in the Smoking Hills contains three reversed tills which could be assigned to any of the glacial intervals within the Matuyama Chron (Fig. 49.16), and it may be that the top reversed till is late Matuyama in age.

On Banks Island, the Morgan Bluffs Sites are the only sites with Jaramillo age deposits and also reveal post-Jaramillo (reversed) glacial sediments which can confidently be assigned to the latest Matuyama. This latest Matuyama sequence contains a till which is underlain and overlain by glacio-marine sediments (all reversed). The till was most probably deposited during MIS 22 or 20?

49.14. EARLY BRUNHES GLACIATIONS (0.78–0.40 MA, MIS 18, 16 AND 12?)

49.14.1. The Cordillera

The Brunhes–Matuyama boundary is present in most stratigraphical records in NW Canada. In the Tintina Trench at the RC site, the top normally magnetised outwash (unit 9, Fig. 49.4) unconformably overlies reversely magnetised outwash (latest Matuyama). Loess of equivalent age occurs at EFR and WFR sections. Significant cold periods are recorded worldwide for MIS 16 and 12, and it is likely that the outwash and loess was deposited at this time. The loess deposits of unit 9 do not reveal any clear sedimentary break with unit 8. Polarity changes which are not accompanied by stratigraphical boundaries are commonly reported for loess.

Stratigraphically, these early Brunhes deposits (unit 9) are bracketed between two interglacials, a lower one which is represented by the pollen assemblage found at EFR site and an upper one which is represented by a palaeosol ("Wounded Moose" type luvisol). The lower contact occurs at MIS 19 (the Brunhes/Matuyama boundary) and the upper contact falls somewhere in the range of MIS 17–11 (strong warm spikes in the oceanic δO^{18} record). The 1.2-m-thick luvisol which was developed on the outwash at RC is a typical Wounded Moose type palaeosol described by Tarnocai and Schweger (1991). The pollen record in unit 9 reflects interglacial conditions (MIS 19; Fig. 49.5). The assemblage includes *Ambrosia*-type, *Betula*, *Pinus*, *Alnus*, *Salix*, *Picea* and *Botrychium*.

The Brunhes/Matuyama boundary (MIS 19) thus occurs within a thick loess sequence at both the EFR and WFR sites, while at RC site, the boundary is marked by an unconformity which presumably indicates that interglacial conditions prevailed here, as elsewhere. The Brunhes/Matuyama boundary occurs worldwide at MIS 19, an interglacial period.

At the MD site in the Klondike Plateau, the Brunhes/Matuyama boundary occurs within colluviated organic silt (Froese et al., 2000).

In the Mackenzie Mountains, several normally magnetised tills may represent glaciations during MIS 18 or 16.

At the Merritt site in British Columbia, normally magnetised proglacial lacustrine and outwash sediments, and a boulder lag composed of the same lithology as the underlying flow (0.6 ± 0.4 Ma) suggest that at least one early Brunhes glaciation may be present.

49.14.2. Interior Plains

In the Mackenzie Delta cores, the early Brunhes is marked by fluvial deposits containing much organic detritus (Fig. 49.15).

At West River site in the Smoking Hills, the normally magnetised upper suite of sediments awaits further pollen analysis. On Banks Island, the Brunhes/Matuyama boundary occurs in interglacial deposits, while at the Nelson River site, this boundary is marked by an unconformity between two till units.

49.15. LATE BRUNHES GLACIATIONS (0.40–0.015 MA, MIS 10–2)

The normally magnetised surface till at the EFR and WFR sites (unit 10) forms a generally thin veneer and is of local provenance. During the Reid Glaciation (MIS 6 or 8) and subsequent glaciations, neither local nor Cordilleran ice reached the RC area. The normally magnetised palaeosol (brunisol) developed on this till is thinner than that seen at the Reid type locality, and also thinner than the underlying palaeosols at these sections. The Reid glaciation was as extensive as the first local glaciation in the area, and in both cases these glaciers extended from the Fifteenmile Valley and reached the Yukon River, blocking it and forming a glacial lake (Duk-Rodkin, 1996; Fig. 49.4). The Late Pleistocene glaciations (MIS 2) in this area were not as extensive as earlier glaciations, and did not reach the Tintina Trench.

Recent exposure dating in the Aishek map area (Ward et al., 2008) has yielded ca. 50 ka ages for glacial erratics that were previously estimated to be "pre-Reid" (pre-Middle Pleistocene). This opens the possibility that a "Reid" designation may include more than a single glaciation. These young ages suggest a later advance of local ice. Therefore, while the Tintina Trench unit 10 (Reid) glaciation is assigned to MIS 8/6, some of the so-called Reid deposits may be as young as MIS 4 (Ward et al., 2008).

At the Fort Selkirk Volcanic complex, the Black Creek flows comprise a series of five basalt flows for which the upper flow has a date of 0.311 ± 0.005 Ma. They are overlain by Reid Glaciation outwash (MIS 6 or 8) and Latest Pleistocene loess.

Equivalents to the MIS 6 or 8 glaciations are almost certainly present in the Mackenzie Mountains, where up to 3 normal tills occur beneath the Late Pleistocene Laurentide glacial deposits at the surface.

At the Mount Edziza, Tahltan/Stikine sites, a sequence of normally magnetised glacial sediments are overlain by basalt dated by Ar–Ar at 0.30 Ma. The till has been assigned to MIS 10 (Spooner et al., 1996).

At the Merritt site, a normally magnetised till and proglacial lacustrine sediment sequence has tentatively been interpreted as being MIS 6 or 8 (Fulton et al., 1992). These deposits are overlain by glacial deposits of the Late Pleistocene (Wisconsin) Fraser glaciation.

49.15.1. Interior Plains

In the West River site, in the Smoking Hills, there is one normally magnetised till beneath the Late Pleistocene (Wisconsin) Laurentide till (Fig. 49.15). At the Banks Island sites up to two tills occur beneath the Late Pleistocene deposits (Fig. 49.3).

49.16. LATEST BRUNHES GLACIATIONS (32–12 KA, MIS 2)

49.16.1. The Cordillera

Late Pleistocene loess deposits are found at the Tintina Trench sites, Klondike Plateau sites and Fort Selkirk sites (Fig. 49.3). The Mackenzie Mountain sites are all capped by Late Wisconsin continental (Laurentide) tills containing glacial erratics transported from the Shield terrain of Canada.

49.16.2. Interior Plains

The Smoking Hills are covered by a discontinuous lag of pebbles, cobbles and boulders of Shield origin, as well as by loess (Late Wisconsin). All glacial sediments beneath the late Wisconsin loess in the Smoking Hills are of local origin, indicating that only the Latest Pleistocene Laurentide Ice Sheet reached the northern Interior Plains. The Late Pleistocene (Wisconsin) Laurentide Ice Sheet had a very profound effect upon the Canadian landscape, including the reversal of flow of major rivers and/or the rerouting of many of the major rivers and their tributaries during retreat of the ice sheet (Duk-Rodkin and Hughes, 1992, 1994; Duk-Rodkin et al., 2004).

49.17. CONCLUSIONS

The Tintina Trench stratigraphy (RC, EFR and WFR) contains one of the most extensive records of preglacial, glacial and interglacial sediments in northwest Canada. Perhaps its most notable attribute is the large number of glaciations which occurred within the Latest Pliocene and Early Pleistocene. Seven of the eight glaciations documented here fall within the Matuyama Reversed Chron and provide a snapshot of deteriorating Late Pliocene climate conditions and the series of regional glaciations which followed. In addition to the nine glaciations and six interglaciations defined for the Tintina Trench sites, the trench contains an overlying till and loess sequence which are assigned to the late Brunhes. Other Cordilleran sites (sites 4–10, 18–20, Fig. 49.1) exposing a less complete Plio–Pleistocene stratigraphical record have been correlated to the sites described in the Tintina Trench. The large moisture supply from the north Pacific and Arctic Ocean and the onset of global cooling around 3.0 Ma (Figs. 49.1, 49.2 and 49.24) were key elements in the formation of the largest Cordilleran ice sheets in early Late Pliocene, in the Tintina Trench and Klondike areas, as well as in parts of Alaska. During cold conditions (glacial periods), the region experienced the build-up of both local and regional ice sheets and saw the deposition of extensive loess sheets beyond glacier margins. The extent of regional and local ice varied temporally and spatially. During warmer conditions (interglacial periods) fluvial, alluvial and colluvial processes predominated, and weathering of landscape surfaces produced a variety of soils which today are preserved as palaeosols (G10, Fig. 49.24).

In the topographically lower regions of NW Canada, the first glaciations appear to have occurred somewhat later. Rising sea level led to the opening of the Bering Strait around about 3.0–3.5 Ma (Dowsett et al., 1994; Robinson, 2009), and heat transfer across the Bering Strait and into the Arctic Ocean may have warmed and moderated the climate along the continental Arctic and southern Arctic Islands. The build-up of ice on the ocean surface was probably quite restricted and seasonal (Dowsett et al., 1994). Recent work (Spielhagen et al., 1997; Marincovich, 2000; Gusev et al., 2009) has suggested a relatively warm and largely ice-free Arctic in the earliest Pleistocene and this probably accounts for the somewhat later development of ice in the Mackenzie Delta and Mackenzie Mountains (<2.6 Ma) as well as in the Smoking Hills (<2.6–1.6? Ma). Initiation of glaciation appears to have been considerably later (<1.8 Ma, post-Olduvai) on Banks Island.

Magnetostratigraphy has been used extensively to establish a correlation both within and between study sites reported here. Chronological data obtained from tephrochronology, Ar/Ar, K/Ar, ^{14}C and ^{36}Cl, as well as fossils, pollen and palaeosols, have been used where available. This has greatly facilitated the assignment of glacial and interglacial sedimentary units found at the 20 major study sites (Fig. 49.1) to the Geomagnetic Polarity Timescale and, where possible, to the δ^{18}O marine isotopic record. In this manner, a systematic comparison was made with records from the Cordilleran (Klondike Plateau, Fort Selkirk, Mackenzie Mountains, British Columbia) and Northern Interior Plains sites (Mackenzie Delta, Smoking Hills and Banks Island).

ACKNOWLEDGEMENTS

The authors are thankful for the thorough review provided by James N. White and the Elsevier series editors. Useful comments and photographs were supplied by authors of previous work and have been incorporated in this stratigraphic correlation project (R. Fulton, L.E. Jackson, Jr. D.G. Froese, I. Spooner and C. Huscroft).

REFERENCES

Barendregt, R.W., Duk-Rodkin, A., 2011. Chronology and extent of late Cenozoic ice sheets in North America: a magnetostratigraphic assessment. In: Ehlers, J., Gibbard, P.L. (Eds.), Quaternary Glaciations Extent and Chronology, Part II. Elsevier, Amsterdam, this volume.

Barendregt, R.W., Irving, E., 1998. Changes in the extent of North American ice sheets during the late Cenozoic. Can. J. Earth Sci. 35, 504–509.

Barendregt, R.W., Vincent, J.S., 1990. Late Cenozoic paleomagnetic record of Duck Hawk Bluffs, Banks Island, Canadian Arctic Archipelago. Can. J. Earth Sci. 27, 124–130.

Barendregt, R.W., Vincent, J.-S., Irving, E., Baker, J., 1998. Magnetostratigraphy of Quaternary and Late Tertiary sediments on Banks Island, Canadian Arctic Archipelago. Can. J. Earth Sci. 35, 47–161.

Barendregt, R.W., Enkin, R.J., Duk-Rodkin, A., Baker, J., 2010. Paleomagnetic evidence for multiple Late Cenozoic glaciations in the Tintina Trench of west central Yukon, Canada. Can. J. Earth Sci. 47, 987–1002.

Cande, S.C., Kent, D.V., 1995. Revised calibration of the geomagnetic polarity timescale for the Late Cretaceous and Cenozoic. J. Geophys. Res. B Solid Earth Planets 100, 6093–6095.

Dallimore, S.R., 1992. Taglu and Kumak borehole data. Geological Survey of Canada Open file 2561.

Dallimore, S.R., Matthews, J.V., 1997. The Mackenzie Delta Borehole Project. Environmental studies Research Funds Report, No. 135. Calgary, 1 CD-ROM.

Denton, G.H., Armstrong, R.L., 1969. Miocene–Pliocene glaciations in southern Alaska. Am. J. Sci. 267, 1121–1142.

Dowsett, H., Thompson, J., Barron, R., Cronin, T., Fleming, F., Ishman, R., et al., 1994. Joint investigations of the Middle Pliocene climate. Global Planet. Change 9, 169–195.

Duk-Rodkin, A., 1996. Surficial geology, Dawson, Yukon Territory. Geological Survey of Canada, Open File 3288, 1 map, Scale 1:250 000.

Duk-Rodkin, A., Barendregt, R.W., 1997. Gauss and Matuyama glaciations in the Tintina Trench. Canadian Quaternary Association, Abstracts, Montreal, 22.

Duk-Rodkin, A., Hughes, O.L., 1992. Pleistocene montane glaciations in the Mackenzie Mountains, Northwest Territories. Géogr. Phys. Quatern. 46, 69–83.

Duk-Rodkin, A., Hughes, O.L., 1994. Tertiary–Quaternary drainage of the pre-glacial Mackenzie Basin. Quaternary International: Tertiary Quaternary Boundaries, 22/23, 221–241.

Duk-Rodkin, A., Barendregt, R.W., Tornacai, C., Philips, F.M., 1996. Late Tertiary to late Quaternary record in the Mackenzie Mountains, Northwest Territories, Canada: stratigraphy, paleosols, paleomagnetism, and chlorine-36. Can. J. Earth Sci. 33, 875–895.

Duk-Rodkin, A., Barendregt, R.W., White, J.M., Singhroy, V.H., 2001. Geologic evolution of the Yukon River: implications for placer gold. In: Duk-Rodkin, A., Patyk-Kara, N.G. (Eds.), Quaternary International: Quaternary Placer Deposits, 82,, 5–31

Duk-Rodkin, A., Barendregt, R.W., Froese, G.D., Weber, F., Enkin, R., Smith, R., et al., 2004. Timing and extent of Plio-Pleistocene glaciations in north-western Canada and east-central Alaska. In: Ehlers, J., Gibbard, P.L. (Eds.), Quaternary Glaciations—Extent and Chronology, Part II, Elsevier North America, Amsterdam, 313–345.

Duk-Rodkin, A., Barendregt, R.W., White, J.M., 2010. An extensive late Cenozoic terrestrial record of multiple glaciations preserved in the Tintina Trench of west central Yukon: stratigraphy, paleomagnetism, paleosols, and pollen. Can. J. Earth Sci. 47, 1003–1028.

England, J.H., Furze, M.F.A., Doupe, J.P., 2009. Revision of the NW Laurentide Ice Sheet: implications for paleoclimate, the northeast extremity of Beringia, and Arctic Ocean sedimentation. Quatern. Sci. Rev. 28, 1573–1596.

Froese, D.G., Westgate, J.A., 2001. Klondike Goldfields stratigraphy. In: Froese, D.G., Duk-Rodkin, A., Bond, J.D. (Eds.), Heritage Branch, Government of the Yukon, Canada, Occasional Papers in Earth Sciences 2, 44–47.

Froese, D.G., Barendregt, R.W., Enkin, R.J., Baker, J., 2000. Paleomagnetic evidence for multiple late Pliocene-early Pleistocene glaciations in the Klondike area, Yukon Territory. Can. J. Earth Sci. 37, 1–15.

Fulton, R.J., Irving, E., Wheadon, P.M., 1992. Stratigraphy and paleomagnetism of Brunhes and Matuyama (>790 ka) Quaternary deposits at Merritt, British Columbia. Can. J. Earth Sci. 29, 76–92.

Fyles, J.G., Hills, L.V., Mathews Jr., J.V., Barendregt, R.W., Baker, J., Irving, E., et al., 1994. Ballast Brook and Beaufort Formations (Late Tertiary) on northern Banks Island, 1227 Arctic Canada. Quatern. Int. 22-23, 141–171.

Gibbard, P.L., Head, M.J., Walker, M.J.C., subcommission on Quaternary Stratigraphy, 2010. Formal ratification of the Quaternary system/period and the Pleistocene Series/Epoch with a base at 2.58 Ma. J. Quatern. Sci. 25, 96–102.

Gradstein, F.M., Ogg, J.G., Smith, A.G., Agterberg, F.P., Bleeker, W., Cooper, R.A., 2004. A geologic timescale (2004). Geological Survey of Canada, Miscellaneous Report 86, and 1 map.

Gusev, E.A., Andreeva, I.A., Anikina, N.Y., Bondarenko, S.A., Derevyanko, L.G., Iosifidi, A.G., et al., 2009. Global Planet. Change 68, 115–131.

Huscroft, C.A., 2002. Late Cenozoic history of the Yukon River Valley, Fort Selkirk to the Stevenson Ridge map area (115 I/13, 14 & J/13, 14, 15). M.Sc. thesis, Simon Fraser University, Burnaby, BC.

Huscroft, C.A., Ward, B.C., Barendregt, R.W., Jackson Jr., L.E., 2004. Pleistocene volcanic damming of Yukon River, and age of the Reid glaciation, west central Yukon. Can. J. Earth Sci. 41, 151–164.

Jackson Jr., L.E., Barendregt, R.W., Baker, J., Irving, E., 1996. Early Pleistocene volcanism and glaciation in central Yukon: a new chronology from field studies and paleomagnetism. Can. J. Earth Sci. 33, 904–916.

Jackson, L.E., Huscroft, C.A., Ward, B.C., Villeneuve, M., 2008. Age of volcanism of the Wolverine Volcanic Center, West-Central Yukon Territory, Canada and its implications for the history of Yukon River. EOS Trans. AGU 89 (53) Fall Meeting Supplement, abstract V41D-2110.

Lagoe, M.B., Eyles, C.H.N., Hale, C., 1993. Timing of the Late Cenozoic tidewater glaciation in the far North Pacific. Geol. Soc. Am. Bull. 105, 1542–1560.

Lisiecki, L.E., Raymo, M., 2005. A Pliocene–Pleistocene stack of 57 globally distributed benthic $\delta^{18}O$ records. Paleoceanography 20, PA1003.

Marincovich Jr., L., 2000. Central American paleogeography controlled Pliocene Arctic Ocean molluscan migrations. Geology 28, 551–554.

Matthews Jr., J.V., Mott, R.J., Vincent, J.-S., 1986. Preglacial and interglacial environments of Banks Island: pollen and macrofossils from Duck Hawk Bluffs and related sites. Géogr. Phys. Quatern. 40, 279–298.

Muller, J., 1981. Fossil pollen records of extant angiosperms. Bot. Rev. 47 (1), 1–142. doi:10.1007/BF02860537.

Nelson, F.E., Barendregt, R.W., Villeneuve, Mike, 2009. Stratigraphy of the Fort Selkirk Volcanogenic complex in Central Yukon, and its paleoclimatic significance: Ar/Ar and Paleomagnetic Data. Can. J. Earth Sci. 46, 381–401.

Raymo, M.E., 1992. Global climate change: a three million year perspective. In: Kukla, G.J., Went, E. (Eds.), Start of a Glacial, NATO ASI Series I, Global Environmental Change, vol. 3. 207–223.

Raymo, M.E., Ruddiman, W.F., Backman, J., Clement, B.M., Martinson, D.G., 1989. Late Pliocene variation in Northern Hemisphere ice sheets and North Atlantic Deep Water Circulation. Paleoceanography 4, 413–446.

Robinson, M.M., 2009. New quantitative evidence of extreme warmth in the Pliocene Arctic. Stratigraphy 6, 265–275.

Ruddiman, W.F., Raymo, M.E., McIntyre, A., 1986. Matuyama 41,000-year cycles: North Atlantic Ocean and northern hemisphere ice sheets. Earth Planet. Sci. Lett. 80, 117–129.

Ruddiman, W.F., Raymo, M.E., Martinson, D.G., Clement, B.M., Backman, J., 1989. Pleistocene evolution of Northern Hemisphere climate. Paleoceanography 4, 353–412.

Sandhu, A.S., Westgate, J.A., Preece, S.J., Froese, D.G., 2000. Glass fission-track ages of late Cenozoic distal tephra beds in the Klondike district, Yukon Territory. In: Emond, D.S., Weston, L.H. (Eds.), Yukon Exploration and Geology 2000. Exploration and Geological Services Division, Yukon, Indian and Northern Affairs, Canada, pp. 247–256.

Shackleton, N.J., Hall, M.A., Pate, D., 1995. Pliocene stable isotope stratigraphy of ODP site 846. Proc. Ocean Drilling Program Sci. Results 138, 337–356.

Smith, C.A.S., Tarnocai, C., Hughes, O.L., 1986. Pedological investigations of Pleistocene drift surfaces in the central Yukon. Géogr. Phys. Quatern. 40, 69–77.

Souther, J.G., 1992. The late Cenozoic Mount Edziza Volcanic Complex, British Columbia. Geological Survey of Canada Memoir, 420.

Spielhagen, R.F., Bonani, G., Eisenhauer, A., Frank, M., Frederichs, T., Kassens, H., et al., 1997. Arctic Ocean evidence for late Quaternary initiation of northern Eurasian ice sheets. Geology 25, 783–786.

Spooner, I.S., Osborn, D.G., Barendregt, R.W., Irving, E., 1995. A record of Early Pleistocene glaciation on the Mount Edziza Plateau, Northwestern British Columbia. Can. J. Earth Sci. 32, 2046–2056.

Spooner, I.S., Osborn, D.G., Barendregt, R.W., Irving, E., 1996. A Middle Pleistocene (Isotope stage 10) glacial sequence in the Stikine River valley, British Columbia. Can. J. Earth Sci. 33, 1428–1438.

Tarnocai, C., Schweger, C.E., 1991. Late Tertiary and Early Pleistocene paleosols in northwestern Canada. Arctic 44, 1–11.

Vincent, J.-S., 1983. La geologie du Quaternaire et la geomorphologie de L'ile Banks, Arctique Canadien. Geological Survey of Canada, Mémoire 405, 118pp.

Vincent, J.-S., 1990. Late Tertiary and Early Pleistocene deposits and history of Banks Island, southern Canadian Archipelago. Arctic 43, 339–363.

Vincent, J.-S., 1992. The Sangamonian and Early Wisconsin glacial in the western Canadian Arctic. In: Clark, P.U., Lea, P.D. (Eds.), The Last Interglacial Transition in North America, 233–252, Geological Society of America, Special Paper 270.

Vincent, J.-S., Morris, W.A., Occhietti, S., 1984. Glacial and non glacial sediments of Matuyama paleomagnetic age on Banks Island, Canadian Arctic Archipelago. Geology 12, 139–142.

Wang, Y., Evans, M.E., 1997. Paleomagnetism of Canadian Arctic permafrost; Quaternary magnetostratigraphy of the Mackenzie Delta. Can. J. Earth Sci. 34, 135–139.

Ward, B.C., Bond, J.D., Froese, D., Jensen, B., 2008. Old Crow tephra (140 ± 10 ka) constrains penultimate Reid glaciation in central Yukon Territory. Quatern. Sci. Rev. 27, 1909–1915.

Westgate, J.A., Preece, S.J., Froese, D.G., Walter, R.C., Sandhu, A.S., Schweger, C.E., 2001. Dating Early and Middle (Reid) Pleistocene Glaciations in Central Yukon by tephrochronology. Quatern. Res. 56, 335–348.

Westgate, J.A., Sandhu, A.S., Preece, S.J., 2002. Age of the gold-bearing White Channel Gravel, Klondike district, Yukon. Yukon Exploration and Geology, Geological Fieldwork, pp. 241–250.

Westgate, J.A., Preece, S.J., Froese, D.G., Pearce, N.J.G., Roberts, D., Demurod, J.A., et al., 2008. Changing ideas on the identity and stratigraphic significance of the Sheep Creek tephra beds in Alaska and the Yukon Territory, northwestern North America. Quatern. Int. 178, 183–209.

White, J.M., Ager, T.A., Adam, D.P., Leopold, E.B., Liu, G., Jette, H., et al., 1999. Neogene and Quaternary quantitative palynostratigraphy and paleoclimatology from sections in Yukon and adjacent NWT and Alaska. Geol. Surv. Can. Bull. 543, 30pp.

Chapter 50

The Greenland Ice Sheet During the Past 300,000 Years: A Review

Svend Funder[1,*], Kristian Kjellerup Kjeldsen[1], Kurt Henrik Kjær[1] and Colm Ó Cofaigh[2]

[1] Centre for GeoGenetics, Natural History Museum, University of Copenhagen, Øster Voldgade 5-7, DK-1350 Copenhagen K, Denmark
[2] Department of Geography, Durham University, South Road, Durham DH1 3LE, United Kingdom
*Correspondence and requests for materials should be addressed to Svend Funder. E-mail: svf@snm.ku.dk

50.1. INTRODUCTION

The Greenland ice sheet's response to temperature change is a major issue in the debate surrounding future climate warming (Lenton et al., 2008). This chapter provides a brief review of the response of the ice-sheet margin to climate change over the past 300 ka (1000 years)—the period for which adequate evidence exists. An emphasis in this chapter is on field observations from the continental shelf surrounding Greenland and from the ice-free land, with the aim being to increase knowledge and understanding of the ice sheet's history in order to inform predictions of its likely future sensitivity to environmental change (cf. Alley et al., 2010). The reconstructions presented here are an updated version of previous reviews (Funder, 1989; Funder and Hansen, 1996; Funder et al., 2004), which contain references to older work. Among the new advances, the increasing application of cosmogenic isotope surface exposure dating of bedrock surfaces and glacial erratics has been a key development which has added significantly to our knowledge of past ice-margin behaviour (Zreda et al., 1999; Håkansson et al., 2007a,b, 2009; Roberts et al., 2008, 2009; Rinterknecht et al., 2008; Möller et al., 2010).

The Greenland ice sheet is composed of two separate domes (Fig. 50.1). The southern dome with ca. 15% of the total ice volume is a climatically sensitive, highland ice cap, resting on bedrock ca. 500 m a.s.l. Much of the northern dome is grounded below present sea-level (Bamber et al., 2001). As indicated below, this sector of the ice sheet has been relatively stable during the climate changes of the Quaternary. A detailed summary of the ice sheet's physical characteristics has been given in Dahl-Jensen et al. (2009). The expansion and shrinking of the ice sheet has been controlled not only by climate, but also by such physiographic factors such as the dimensions of fjords and cross-shelf troughs which act as the main drainage conduits for the ice sheet during periods of ice sheet advance. These range from the Melville Bugt area of northwest Greenland where deep and wide troughs extend inshore across the shelf to the present ice margin, to the wide, shallow and physiographically complex shelf of northeast Greenland (Fig. 50.1). The importance of these troughs and the fast-flowing ice streams that have shaped them has recently been emphasised by several studies (Evans et al., 2009; Roberts et al., 2009; Dowdeswell et al., 2010; Winkelmann et al., 2010; Larsen et al., 2010).

In the text below, all ages are expressed as thousand calendar years before present (ka BP). Ages, which appear as C14 years in the original papers, have been calibrated using the mean of the extremes in the 2 sigma range in the CALIB version 6.0 conversion programme (Reimer et al., 2009). If not already reservoir corrected, C14 ages on marine organisms have been corrected with -550 years ($\Delta R = 150$) for north Greenland and east Greenland north of Denmark Strait (areas dominated by polar water from the Arctic Ocean) and -400 years ($\Delta R = 0$) for the rest of Greenland (dominance of Atlantic water) (Mörner and Funder, 1990).

50.2. SAALIAN STAGE (FIG. 50.2, MIS ≥ 6, CA. 300–130 KA BP)

In the Scoresby Sund area, east Greenland, maximum ice cover was achieved during the Scoresby Sund glaciation in Marine Isotope Stage (MIS) 6, and this was the last time that the ice sheet overrode the Jameson Land peninsula and coastal mountains. This interpretation was based on optically stimulated luminescence (OSL)-dating of aolian and glaciofluvial sediments overlying till on high mountain plateaux, supported by ice-rafted detritus (IRD) evidence in cores from the outer shelf, as well as recent exposure dating (Funder et al., 1998; Nam and Stein, 1999; Adrielsson and Alexanderson, 2005; Håkansson et al., 2009).

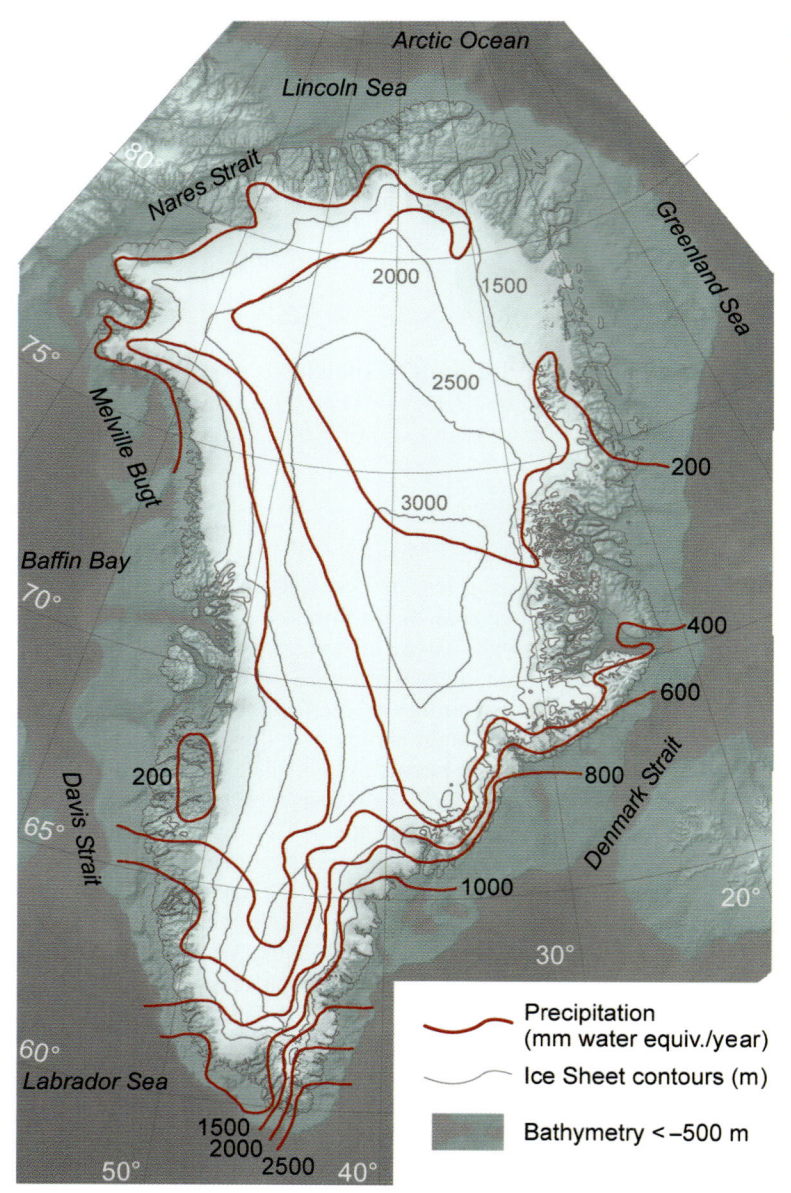

FIGURE 50.1 Greenland with ice sheet topography, shelf bathymetry (from IBCAO, 2008), annual precipitation over the ice sheet (from Ohmura and Reeh, 1991) and oceanic place names mentioned in the text.

In northwest Greenland till from two glaciations, the Agpat and Narsaarsuk glaciations, pre-dates the Eemian Stage interglacial (~MIS 5e), as shown by OSL dating and biostratigraphy of raised marine sediments between till-beds in the Wolstenholme Fjord area and Smith Sund. Both were more extensive than the Weichselian, and were referred to as Saalian Stage in age by Kelly et al. (1999).

In the Sisimiut area of West Greenland, exposure dating of bedrock and glacial erratics has recently indicated that the coastal mountain tops were ice free during the Weichselian and overridden by ice for the last time during the Saalian (Rinterknecht et al., 2008; Roberts et al., 2009). Along this coast, from 69°30′ to 61°N, high resolution seismic studies in the 1970s showed the presence of several sets of submarine moraine systems on the shelf (Zarudski, 1980). From correlation with weathering boundaries on coastal mountains, Kelly (1985) provisionally proposed that the outermost system, the Hellefisk moraines, located at the shelf edge and on the outer shelf, were Saalian Stage in age (Fig. 50.2). By counting down, the younger Fiskebanke moraines on the inner shelf and along cross-shelf troughs were then assigned to the Last Glacial Maximum (LGM; Sisimiut glaciation). Subsequently, some authors have suggested that the Hellefisk moraines may actually date to the LGM, and the Fiskebanke moraines to the Late Weichselian Substage, the Younger Dryas Stadial (van

FIGURE 50.2 Saalian proposed ice margins and ice flow, modelled Eemian minimum ice sheet (Letréguilly et al., 1991), location of problematic ice cores used in the reconstruction, and place names referring to Saalian and Eemian key localities. *Note*: The Eemian ice sheet reconstruction is based to some extent on erroneous assumptions of presence/absence of Eemian ice in ice cores. This has led to larger than present ice distribution in some areas. The Renland ice cap was formed in the late Eemian as a local ice cap (Johnsen et al., 1992), and the Dye 3 area was ice covered during the Eemian (see text).

Tatenhove et al., 1996; Weidick et al., 2004; Roberts et al., 2009). However, the shelf moraines themselves have not been dated directly, and it is not clear as to whether they mark several distinct advance–retreat events or one advance followed by recessional stages. Modelling of the extent and thickness of partly floating and partly grounded ice on the continental shelf cannot be done with a precision high enough to correlate individual moraines with ice thickness on land (cf. van Tatenhove et al., 1996). Important to this debate are recent preliminary results from marine studies in southwest Greenland which suggest that here the ice sheet reached the shelf edge, or even beyond, during the Early Weichselian Substage, thus indicating an age or a maximum age for the Hellefisk moraines (Seidenkrantz et al., 2010). We therefore regard the precise age of the Hellefiske moraines as an open question, but pending direct dating control from the shelf we have retained the original age-model.

50.3. THE EEMIAN STAGE (FIG. 50.2, MIS 5E, 135–115 KA BP)

The sedimentology of the Eemian Stage (Langelandselv interglacial) marine sediment in Scoresby Sund suggests that, in spite of the thick Saalian ice sheet, the relative sea-level history was characterised by dominance of eustasy over isostasy in the early Eemian (Funder et al., 1998). This is contrary to the Holocene and must be explained by the rapidity of sea-level rise at the end of the Saalian and/or that the melting of the Greenland ice sheet was leading the global ice melting.

The ice sheet's response to the Eemian warmth has been modelled from climate data from ice cores and climate models (Cuffey and Marshall, 2000; Huybrechts, 2002; Tarasov and Peltier, 2003; Greve, 2005; Lhomme et al., 2005; Otto-Bliesner et al., 2006). All the models agree that the southern ice dome was the most sensitive, and either disappeared entirely or had shrunk to a local ice cap over the mountains at the southeast coast (Fig. 50.2). The models show an ice sheet that was reduced to between one-fourth and two-thirds of its present size (Huybrechts, 2002; Otto-Bliesner et al., 2006). However, dating of the basal ice in the Dye 3 ice core (Fig. 50.2) has shown that ice has been continuously present over this locality for at least ca. 450 ka BP (Willerslev et al., 2007). Further, this core site is located in the saddle between the southern and northern domes. Collectively, this implies that the models generally overestimate the magnitude of Eemian ice sheet melting. Figure 50.2 shows the model of Letréguilly et al. (1991), the only model which has ice over Dye 3.

50.4. THE WEICHSELIAN STAGE (MIS 5D-1, CA. 115–11.7 KA BP)

50.4.1. Early Weichselian Substage (MIS 5d-a, 115–75 ka BP)

Only the Scoresby Sund area has a detailed record of glaciation dating to the Early Weichselian Substage. In this region, terrestrial investigations have shown that warm-based glaciers advanced for several hundred kilometres through the fjords and reached the inner shelf. After this glacier retreat was followed by a marine transgression, and subsequent readvance and retreat (the Aucellaelv and Jyllandselv stades, referred to MIS 5d and 5c, 115–105 and 93–85 ka BP, Funder et al., 1998).

In northwest Greenland, a post-Eemian and pre-LGM till bed is referred to the Kap Abernathy stade, which may date from Early–Mid Weichselian times and show a mid-fjord advance of outlet glaciers (Kelly et al., 1999). As noted above, preliminary results from offshore southwest Greenland have indicated maximum glaciations in this area in MIS 4 (Seidenkrantz et al., 2010).

50.4.2. Last Glacial Maximum, Ca. 24–16 ka BP, Fig. 50.3

Figure 50.3 shows our reconstruction of the LGM ice-sheet margin around Greenland. The line is drawn on the basis of observations, often undated, of tangible evidence such as moraines and fresh glacially sculptured landforms from the shelf and coastal mountains, and interpolated between these control areas. The absence of data from many parts of the shelf leaves evidence from land as the only clue to ice distribution on the shelf. This opens the possibility that the ice sheet was actually more extensive during the LGM. Thus, the LGM margin in Fig. 50.3 should be regarded as a 'minimum' in some areas (see below). The reconstruction implies spatial variations in the response of the ice sheet to forcing factors during the LGM. Below, we briefly describe the different regimes clockwise from north to south.

50.4.3. North Greenland

Recent onshore field work has indicated that a partly grounded ice shelf existed in the Lincoln Sea between Canada and Greenland, nourished from both Greenland and Canada and buttressed against stationary multiyear sea ice in the Arctic Ocean (Möller et al., 2010; Larsen et al., in press). An IRD record from the nearby Arctic Ocean indicates that the build-up of ice on North Greenland began at ca. 30 ka BP, and OSL and exposure dates show that disintegration began at 16 ka BP (Larsen et al., in press). At the mouth of Independence Fjord, marine cores and shallow seismics show erosional troughs and covered by Holocene glacimarine sediments, and indicate that ice debouching from the large Independence Fjord system extended onto the mid-outer shelf. Evidence from a sediment core from the adjacent shelf slope indicates that the ice-sheet advance in this area culminated at ca. 25–20 ka BP (Nørgaard-Pedersen et al., 2009).

50.4.4. North and Central East Greenland

The continental shelf is at its widest and shallowest offshore of northeast Greenland (Fig. 50.1). High-resolution multibeam swath bathymetry and shallow seismic data from both troughs and shallow intervening banks in this region show a number of glacial landforms, both subglacial and ice-marginal, including mega-scale glacial lineations and a belt of recessional moraines (Evans et al., 2009; Winkelmann et al., 2010). The streamlined subglacial bedforms indicate that fast-flowing, warm-based ice streams occupied the cross-shelf troughs in this region, whilst moraines on adjacent banks provide evidence for the former presence of grounded, but non-streaming, ice. Glacimarine debris flows on the continental slope are suggested by Evans et al. (2009) to be compatible with an interpretation that the ice sheet extended beyond these features to the shelf edge where the debris flows were sourced by sediment dumped from the ice margin, although Wilken and Mienert (2006) argued that the debris flows could have been supplied from other

Chapter | 50 The Greenland Ice Sheet During the Past 300,000 Years: A Review

FIGURE 50.3 Reconstructed LGM ice sheet margin, ice flow, ice-sheet free areas, dates for break-up of shelf-bound ice and names of key localities.

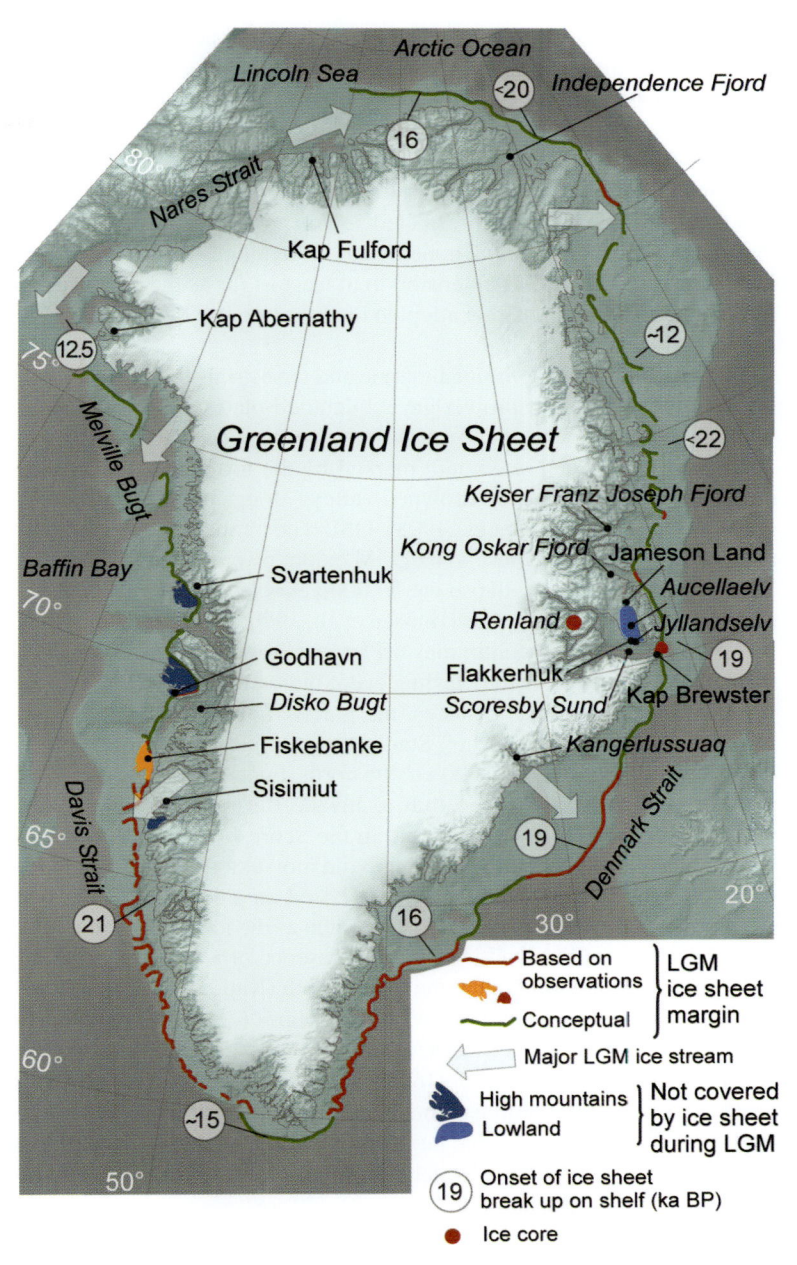

sources. From their freshness, the submarine features are dated to the LGM.

In our reconstruction (Fig. 50.3), the LGM ice sheet limit is placed on the mid-shelf at the belt of recessional moraines, although we note that this is a minimum position, given that the outer edge of the subglacial lineations was not mapped by Evans et al. (2009) or Winkelmann et al. (2010). Interestingly, low weathering limits on adjacent coastal mountains indicate that the ice sheet here was thin (Landvik, 1994; Hjort, 1997). This could be consistent either with a mid-shelf position or alternatively with low-gradient ice streams extending further across the shelf.

Five hundred kilometres to the south, a large mid-shelf moraine in a bathymetric trough extending across the shelf from Kejser Franz Joseph Fjord was interpreted to represent either the maximum LGM ice-marginal position or a recessional position during deglaciation (Evans et al., 2002). On the basis of dated sediment core records from the submarine channels which showed the channels were last active during regional deglaciation at 17–18 ka BP, Ó Cofaigh et al.

(2004) argued for extensive glaciation of the shelf in this region during the LGM, with the ice sheet extending to the outermost shelf and possibly the shelf edge. Till units overlain by Holocene deglacial sediments extend from the mid-shelf moraine of Evans et al. (2002) across the outer shelf and provide indirect support for this interpretation. Resolution of this issue requires dated sediment cores from the outer continental shelf but nonetheless the available data indicate a *minimum* LGM extension to at least as far as the mid-shelf. This is the position adopted in our reconstruction shown in Fig 50.3.

In Scoresby Sund, combined marine and terrestrial studies during the PONAM project gave a detailed picture not only of the attainment of the LGM but also provided the duration of the glacial maximum marked by the formation of glacier-dammed lakes in drowned valleys alongside the glacier margin (Elverhøj et al., 1998). OSL dates imply that the glacier was already close to its LGM extent by ca. 60 ka BP and, until its final collapse ca. 50,000 years later, only oscillated over short distances (Hansen et al., 1999; Hansen, 2001). The thin and discontinuous till and the low-gradient surface slope left by this long lasting event indicates that, in contrast to the previous advances, the glacier at this time was cold based (the Flakkerhuk Stadial), similar to present low-gradient glacier outlets 600 km the north (Funder et al., 1998). The ice sheet was too thin to invade the Renland mountain plateau in the interior, and in the Scoresby Sund basin a thin low-gradient ice stream did not extend far beyond the fjord mouth. This view was based on a combination of evidence from different fields, such as the ice core record in the local Renland ice cap, the presence of a large submarine terminal moraine at the fjord mouth (Kap Brewster Moraine), a distinct and low weathering limit (<100 m a.s.l.) on land at the fjord mouth, and low marine limits at the coast and in the outer fjord (Johnsen et al., 1992; Dowdeswell et al., 1994; Funder and Hansen, 1996; Funder et al., 1998). This would leave large areas of upland Jameson Land ice free during the Weichselian to serve as an ice-age refugium for plants (Funder, 1979). More recently, however, this view of a restricted LGM ice sheet in Scoresby Sund has been challenged by Håkansson et al. (2007b, 2009), who from a set of ca. 50 Be10 and Al36 dates, suggested that the uplands of Jameson Land and the Kap Brewster plateau were not ice free, but rather covered by a cold-based, local ice cap, and the Flakkerhuk stade ice stream actually reached the outer shelf. The exposure ages are rather evenly distributed over a range from ca. 200 to ca. 11 ka BP with discordant Be and Al ages. All boulders on Jameson Land were thought to have been brought into the area during the Saalian, and the spread of ages reflected stepwise transport from the interior fjords and repeated shielding/exhumation by the waxing and waning of a local polar ice cap over Jameson Land. The so-called driftless area (Möller et al., 1994), where these samples were taken from, consists of Mesozoic sandstone and shale, which has been deeply weathered by fluvial erosion and subaerial denudation. Isolated tors, 3–4 m high, give a measure of the lowering of the terrain surface since the Saalian, and the spread of ages may therefore also reflect a long exhumation history, rather than the presence of polar ice, which would have preserved the Saalian terrain. On Kap Brewster, the three dated boulders could have been shielded from exposure by cold-based ice. Indeed, Mangerud and Funder (1994) presented arguments to show that a local ice cap covered this area possibly until Late Weichselian times. In Fig. 50.3, we have therefore retained the restricted LGM ice-sheet margin.

To the north of this, at Kong Oscar Fjord, an undated submarine fjord-mouth moraine occupies a position similar to that of the Kap Brewster Moraine and has been interpreted by Hubberten et al. (1995) to mark the LGM position in this fjord system.

The break-up of the ice sheet along east Greenland's northern and central coast is dated by the timing of maximum IRD flux and terrigenous material to the adjacent ocean, the change to coarser grain sizes and the isotopic evidence for meltwater events in marine sediment cores (Stein et al., 1996; Evans et al., 2002). These, in combination, indicate that break-up began at ca. 19 ka BP. In the north, the break-up may have been delayed, and did not begin until ca. 11.5 ka BP (Landvik, 1994).

This evidence from the coast of north and central east Greenland indicates a starvation-regime where outlet glaciers advanced through the fjords to the inner or mid-shelf while the passive ice-sheet margin in the south was more restricted, increasing its coverage on the shelf towards the north. This situation mirrors that of the present where the Scoresby Sund area is the part of Greenland where the ice-sheet margin is farthest away from the coast, ca. 300 km. Northwards from here, the width of the ice-free land gradually decreases (Fig. 50.1). Today, south of Scoresby Sund, there is a dramatic south–north decrease in precipitation over the ice sheet (Fig. 50.1), and this would have been enhanced during the LGM, when the passage of cyclones through Denmark Strait was obstructed by both the Greenland and Iceland ice sheets (Funder and Hansen, 1996; Andrews, 2008), causing precipitation starvation along the northeast Greenland coast, decreasing from south to north.

50.4.5. Southern East Greenland

In contrast, to the south of Scoresby Sund, the ice-sheet margin extended to the shelf edge, up to 200 km from the coast. In the cross-shelf trough at Kangerlussuaq, basal till and streamlined subglacial bedforms in the trough and turbidite sedimentation at the adjacent shelf break show that a fast-flowing ice stream moved through the trough and was bordered by more passive ice on the shelf (Jennings et al., 2002, 2006; Andrews, 2008; Dowdeswell et al., 2010). Two hundred and fifty kilometres further to the south, terminal moraines on the shelf break were bathymetrically mapped by Sommerhoff (1981), showing that the ice-sheet margin reached the shelf break over this long distance. At Kangerlussuaq, the ice margin had reached the shelf edge before 31 ka BP and began to retreat at 17 ka BP (Andrews et al., 1997, 1998; Jennings et al., 2006). A change of sedimentation rates in a deep sea core 600 km from the coast to the south show initial deglaciation beginning at ca. 16 ka BP (Kuijpers et al., 2003).

The glacial regime along this coast was therefore very different from that north of Scoresby Sund with the build-up of ca. 200,000 km^2 of shelf-bound ice. Presently, the margin of the ice sheet and its satellite local ice caps reach close to the coast along this ca. 1500 km coastline, which has Greenland's highest precipitation (Fig. 50.1). The record suggests that the change from LGM to present took place mainly on the shelf where the ice build-up became an integral part of the ice sheet's accumulation area. This is supported by the generally low isostatic uplift of the ice-free land (Funder and Hansen, 1996), and is also suggested by modelling from isostatic uplift (Long et al., 2008a).

50.4.6. Southern and Central West Greenland

Detailed studies of the glacial history of the west Greenland shelf are still wanting, but ongoing studies may soon provide better-dated records from this area (Ó Cofaigh et al., 2010; Seidenkrantz et al., 2010). As mentioned above, submarine moraine-belts have been identified intermittently from ca. 69°30′N to ca. 61°N. Kelly (1985) provisionally correlated the inner system, the Fiskebanke moraines, with weathering limits at ca. 800 m a.s.l. on the adjacent coast and interpreted the moraines as LGM in age (Sisimiut Stadial). Later studies have tended to downgrade the significance of weathering limits and shift the LGM margin to the shelf break (van Tatenhove et al., 1996; Weidick et al., 2004; Roberts et al., 2009). However, recent exposure dating on the coastal mountains in the Sisimiut area has indicated that there the weathering limits do indeed reflect the Weichselian Stage maximum ice thickness (Rinterknecht et al., 2008; Roberts et al., 2009). We therefore retain Kelly's (1985) LGM ice limit at the Fiskebanke moraines on the inner shelf (pending a precise dating of the moraines), while acknowledging that it may be a minimum as it does not preclude thin ice extending to the shelf break (cf. Roberts et al., 2009). The LGM line in Fig. 50.3 is based on Roksandic's (1979) division between an older extensive glaciations-phase and a younger one with glacial deposition restricted mainly to the cross-shelf troughs. The younger phase includes areas on the inner shelf where glacial erosion has exposed bedrock. To the north and south, we have matched this to correspond with Holtedahl's (1970) outer limit for the erosional marginal channel that runs along the coast from ca. 60° to ca. 70°N. An implication of this interpretation is that this part of the LGM ice margin was characterised by ice streams that effectively drained the ice sheet, but that some inter-trough areas may have been at least partially ice free.

Today, Disko Bugt drains some of the largest outlets from the Greenland Ice Sheet. Moraines on Disko Island show that an ice stream occupied the wide embayment, but current terrestrial evidence suggests that the ice stream only reached the inner shelf and left high coastal mountains ice free as seen from the weathering boundary outside the moraines (Godhavn Stadial; Ingólfsson et al., 1990; Weidick and Bennike, 2007). If these moraines mark the LGM limit, the ice stream must have persisted well into the Holocene since its youngest moraines at the mouth of the bay were not abandoned until ca. 10 ka BP, but the main part of the embayment was ice free at 10.2 ka BP (Ingólfsson et al., 1990; Lloyd et al., 2005).

Only at Greenland's southern tip where the shelf is narrow did thick ice build-up over the shelf and coastal area, as seen from modelling of isostatic uplift (Bennike et al., 2002). This is similar to conditions on the neighbouring southeast coast.

To date, there is only limited chronological control on the LGM ice sheet in West Greenland. Recent exposure dates from the Sisimiut area have suggested that the thinning began at 21 ka BP, but some ice streams may have traversed the shelf up until ca. 10 ka BP (Roberts et al., 2009). In contrast, at the southernmost tip of Greenland initial deglaciation took place at ca. 15 ka BP, and proceeded rapidly to reach the present ice-margin location at ca. 10 ka BP (Bennike et al., 2002; Sparrenbom et al., 2006a,b).

50.4.7. Northern West Greenland

From the low weathering limits and the occurrence of undisturbed pre-Weichselian marine sediment, Kelly (1985) suggested that glaciation was restricted, and fjord glaciers on the Svartenhuk Peninsula did not reach the shelf, thereby leaving adjacent mountains free of ice-sheet ice. To the north, recent results from ice cores indicate that the Weichselian ice sheet surface was 600 m higher than today at Camp Century (Vinther et al., 2009). To accommodate this, we have drawn our ice margin ca. 200 km further out in Baffin Bay. The shelf in Melville Bugt is traversed by Greenland's widest and deepest cross-shelf trough (Fig. 50.1), and this was possibly a drainage channel for a major ice stream during the LGM. To the north of this, at the mouth of Smith Sund, Blake's (1999) ice-thickness calculation and the substantial amount of isostatic uplift (Funder and Hansen, 1996; Blake, 1999; Bennike, 2002), together indicate that a thick ice stream in Smith Sund possibly reached the 600 m depth contour in northern Baffin Bay. This ice stream was coalescent with ice from Ellesmere Island (Blake, 1999; England et al., 1999). By 12.5 ka BP, the ice stream had retreated from both ends of Nares Strait (Fig. 50.4). At the northern end, moraines indicate that glacier fronts had withdrawn to fjord mouths at ca. 11.2 ka BP (Fig. 50.4, the Kap Fulford stade, Kelly and Bennike, 1992), and in the south, southern Smith Sund had been transgressed by the sea at 12.5 ka BP (Knudsen et al., 2008). An implication of the ice-stream-dominated regime suggested here for west and northwest Greenland is that the sea ice along the coast of Davis Strait and Baffin Bay was in sufficient motion to allow the escape of calving ice from the ice streams.

50.4.8. LGM Glacial Regimes

In our reconstruction, the LGM ice sheet occupied an area of ca. 2.7 Ma km^2, which is ca. 65% larger than at present. Approximately, two-thirds of this ice sheet was on the shelf and one-third over the currently ice-free land. The areal

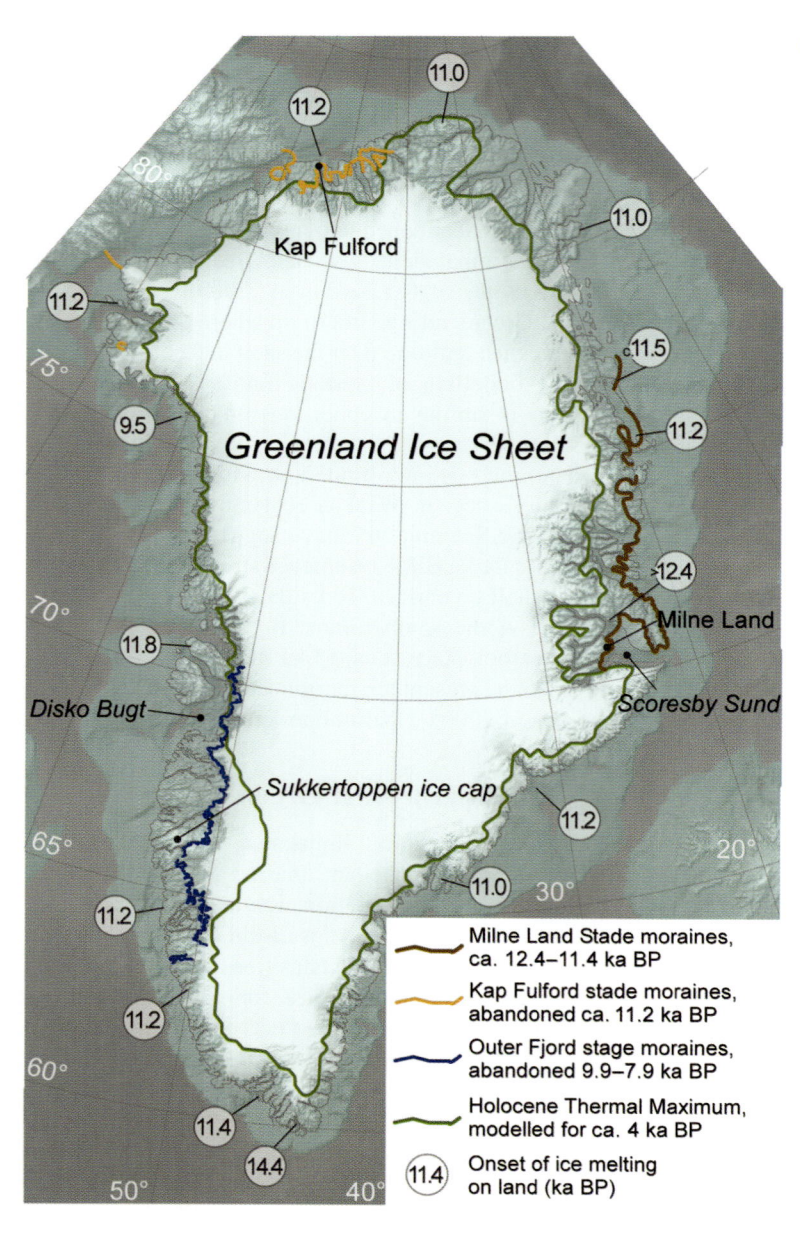

FIGURE 50.4 Deglaciation. Date for onset of melting of land-based ice, prominent moraine stages and modelled minimum extent of ice sheet (from Simpson et al., 2009).

coverage in the reconstruction is slightly smaller than Simpson et al.'s (2009) preferred model, which with an estimated area of 2.9 Ma km^2 is also based to large extent on field observations (the main difference is our smaller shelf-bound ice in northeast Greenland). As described above, the reconstructed ice margin suggests that the response of the ice margin to general climate change in different areas was modified by regional differences in local climate and physiography. In the north, on the Arctic Ocean coast, accumulation and ablation were probably low, but permanent stationary sea ice allowed the outlet glaciers to build a thin ice shelf along the coast. Along the northeastern coast (ca. 81–71°N), ice streams from north to south apparently became less vigorous as a consequence of decreased cyclone passages and precipitation, caused by the restriction of the Denmark Strait. Further south (ca. 68–51°N), the LGM was characterised by major ice build-up on the shelf, where the growing ice sheet may have attracted the precipitation from cyclones that now hit the coastal ice-sheet margin. In southern and central West Greenland (ca. 51–68°N), we interpret the lobate moraines in the troughs and lateral moraines on the banks to indicate a regime that was dominated by fast-flowing ice streams that calved into the Davis Strait and the Labrador Sea. In the northwest, some of Greenland's largest ice streams may have occupied the Nares Strait and Melville Bugt. An implication of the ice-stream-dominated regime is that the sea ice in Baffin Bay and Davis Strait, contrary to the Arctic Ocean, was in sufficient motion to transport away calved icebergs.

50.5. DEGLACIATION

50.5.1. Ice Sheet Retreat from the Continental Shelf (ca. 17–11.7 ka BP)

The timing of initial retreat of the ice sheet from its LGM position on the continental shelf varies from ca. 17 to ca. 11.5 ka BP. To some extent, this variation is undoubtedly an artefact arising from the use of different proxies with different sensitivity, but it does indicate that the break-up of the ice sheet on the continental shelf was driven not only by temperature, but also by other factors such as sea-level change and ocean warming, shelf bathymetry and drainage and sea ice conditions. Moreover, the initial loss was achieved largely by calving rather than melting. Evidence for the latter are the giant iceberg plough marks, which can be seen on the sea floor down to depths greater than 1000 m depth (Syvitski et al., 2001; Kuijpers et al., 2007). These combined factors may also account for the apparent lack of an ice-marginal response to the dramatic climate changes occurring in the atmosphere above Greenland during the Late Weichselian (Rasmussen et al., 2007; Kobashi et al., 2008; Steffensen et al., 2008; Vinther et al., 2009). First to become ice free were the Scoresby Sund area and Greenland's southern tip. The outer Scoresby Sund fjord basin was ice free before 12.4 ka BP, when fjord glaciers again advanced to the mid-fjord Milne Land moraines (Hall et al., 2008, 2010; Kelly et al., 2008). In southernmost Greenland, lake coring and modelling from isostasy indicates that the main phase of melting was between 14 and 12 ka BP and the area was deglaciated by the end of the Weichselian (Bennike et al., 2002).

50.5.2. The Younger Dryas Stadial (12.8–11.7 ka BP), and the Preboreal Oscillation (PBO 11.5–11.4 ka BP)

Although the Younger Dryas is recorded by a prominent signal in the Greenland ice core record of lowered temperatures ending with abrupt warming (Steffensen et al., 2008), interestingly, there is no uniform evidence for an associated ice-marginal response. In Scoresby Sund, fjord and valley glaciers advanced during the Milne Land Stage, but the advance was apparently already underway in Allerød times before the onset of Younger Dryas Stadial cooling, and the subsequent retreat started before the end of the Younger Dryas (Hall et al., 2008, 2010; Kelly et al., 2008). On the Kangerlussuaq shelf and in southernmost Greenland, the ice-sheet margin was apparently in retreat (Jennings et al., 2006; Sparrenbom et al., 2006b), and on the nearshore shelf of southeast Greenland, the Younger Dryas came and went without leaving a signal of change (Kuijpers et al., 2003).

During the abrupt warming at the end of the Younger Dryas period at 11.7 ka BP, mean annual air temperatures over the ice sheet rose ca. 10 °C in ca. 60 years (Steffensen et al., 2008; Walker et al., 2009). However, none of the few available records from the land and the nearshore shelf mentioned has a signal of these abrupt changes, neither the initial cooling, nor the abrupt warming at the termination. One explanation of this may be that in the high Arctic even the dramatic temperature rise did not cause temperatures to rise above the melting point because the warming was mainly in the winter temperatures (Denton et al., 2005; Hall et al., 2010).

In most areas, the ice margin lingered on close to the present coastline until the end of the PBO at 11.4 ka BP when a subsequent phase of abrupt warming occurred (Björck et al., 1997; Kobashi et al., 2008; Rasmussen et al., 2008). Figure 50.4 shows a selection of dates for the regional onset of ice-margin retreat from the coastal areas based on C14 dates from marine molluscs and lake sediments from coastal lakes (from compilations by Funder and Hansen, 1996; and Bennike and Björck, 2002, supplemented with newer results from Jennings et al., 2006; Sparrenbom et al., 2006b; Hall et al., 2008; Long et al., 2008a; Wagner et al., 2010; Larsen et al., in press).

Radiocarbon dating of this time interval is precarious because of C14 plateaux and, for marine organisms, because of the varying reservoir effects as a consequence of changing ocean circulation (Björck, 2007; Cao et al., 2007). Nonetheless, the lack of a uniform ice-marginal response and the consistently younger ages for warming in lakes and marine environments indicate that the Younger Dryas event and the abrupt warming at its end did not have a profound influence on the ice-sheet margin as a whole, as otherwise expected and modelled (Weidick et al., 2004; Roberts et al., 2009; Simpson et al., 2009). This shows that the response of the ice margin to temperature change is not just a direct function of temperatures, and it underlines the necessity of including field observations in modelling. Also the neighbouring ice sheet on Svalbard lacks distinct traces of Younger Dryas cooling (Mangerud and Landvik, 2007).

50.5.3. Disappearance of Ice on Land (ca. 11–7 ka BP)

From Scoresby Sund in the south equivalents of the Milne Land moraines have been traced ca. 800 km to the north, and the youngest moraines in this moraine belt, probably dating to PBO, apparently mark the last ice-age 'flicker' along the ice-sheet's eastern margin (Funder et al., 2004) (Fig. 50.4). After this, the ice margin everywhere was in retreat. This retreat was driven by surface melting, and calving from fjord glaciers (Funder and Hansen, 1996), and the speed and amplitude the retreat varied greatly from area to area.

In the south of Greenland, and in areas with dense spacing of fjord basins, such as the Nuuk Fjord complex, interior fjords close to the present ice-sheet margin were already ice free by 9.5–10 ka BP. This has been shown by dating the onset of organic sedimentation in lakes adjacent to the fjords (Fredskild, 1983; Weidick et al., 2004; Bennike and Sparrenbom, 2007). At the same time, some ice streams persisted on the inner shelf until ca. 10 ka BP (Weidick, 1975; Roberts et al., 2009). In areas with scant fjord-drainage such as the Kangerluussuaq area in West Greenland, the present ice-margin position may not have been attained until ca. 6 ka BP (Van Tatenhove et al., 1996). In the fjords of east and north Greenland, the outlet glaciers retreated through the fjords, while local ice caps in some areas persisted on adjacent plateaux, and the present day positions were attained at ca. 8 ka BP (Bennike, 2008). The Nares Strait, today one of the major connections between the Arctic Ocean and the North Atlantic, remained blocked by the coalescent Greenland and Innuitian ice sheets until ca. 10 ka BP (Zreda et al., 2010).

The melting was most marked in West Greenland where the largest land-areas were uncovered. During the three millennia from ca. 11 to ca. 8 ka BP, numerous lateral and terminal moraines were formed along the fjords and valleys, since the ages of the moraines are scattered evenly over this time interval, they are generally interpreted as marking topographically conditioned still-stands that lack climatic significance (Kelly, 1985; Weidick and Bennike, 2007). Exceptions to this are the Fjord Stage moraines (Weidick, 1968; Weidick and Bennike, 2007). This moraine belt is composed of individual fjord and valley moraines correlated from ca. 70°30′ to ca. 63°30′N by their relation to raised marine sediments (Fig. 50.4), and dated to the period 9.9–7.9 ka BP (Weidick and Bennike, 2007). Based on evidence in Disko Bugt, Long et al. (2006) suggested that these moraines did not record a uniform ice-sheet marginal response but were a function of the interplay between topography and ice-stream dynamics. However, Weidick and Bennike (2007) have argued that since the moraine belt at Jakobshavn Isfjord had taken 2000 years to form it should mark a climatically conditioned slowing down or cessation of the deglaciation process, ending with the 8.2 ka BP cold event. On Disko Island, outlet glaciers from the local ice cap advanced to sea-level at ca. 10 ka BP, almost immediately after this large embayment became ice free (Disko stade, Ingólfsson et al., 1990). Recently, an early Holocene readvance, within the interval 9.6–6.3 ka BP, has been recorded for a major valley glacier in North Greenland (Möller et al., 2010).

50.5.4. Holocene Thermal Maximum (Ca. 8–5 ka BP)

Ice-marginal recession continued behind the present ice margin for some millennia during the Holocene Thermal Maximum (HTM; Fig. 50.4). This is seen from C14-dates on organic material reworked into historical moraines or brought to the ice surface by movement along shear planes (Bennike and Weidick, 2001; Weidick et al., 2004; Weidick and Bennike, 2007). In Disko Bugt and in southern Greenland, this type of evidence indicates that the ice-sheet margin retreated at least 15–20 km behind the present margin (Weidick et al., 2004; Weidick and Bennike, 2007).

Somewhat larger estimates on the extent of HTM ice melting have been derived from the relative sea-level curves from Disko Bugt, Sisimiut and South Greenland (Long et al., 2006, 2008b; Sparrenbom et al., 2006a,b). These show a transgression, which, at least to some extent, was caused by subsidence created by readvance of the ice-sheet margin. The amount of loading indicates that the ice margin may have been as much as 40–80 km behind its present position in West Greenland (Fig. 50.4) (Fleming and Lambeck, 2004; Simpson et al., 2009). GPS and tide gauge evidence support this conclusion showing that while most of Greenland is still being uplifted, West Greenland is currently subsiding, indicating that this area reacted most vigorously to the HTM warming and neoglacial cooling (Tarasov and Pelletier, 2002; Khan et al., 2008).

50.5.5. Neoglacial and Later (≤5 ka BP)

In general terms, after the HTM the ice sheet began to grow again until it reached a maximum during the Little Ice Age in the late 1800s or early 1900s. However, the precise timing of the start of this regrowth is unknown and different estimates vary by several millennia. For West Greenland, this was discussed by Long et al. (2008b), who noted that the late Holocene history of relative sea-level is determined not only by changes in ice load but also by factors unrelated to the Greenland ice sheet. From different parts of the country, different proxies with different sensitivities record the climate cooling. In the north lakes register cooling from ca. 5 ka (Schmidt et al., 2010), they became permanently ice covered, and polar desert expanded, but summer temperatures were still higher than the present until ca. 4 ka BP (Blake et al., 1982; Funder and Abrahamsen, 1988), and from ca. 3.5 ka local ice caps were formed on bare land (Madsen & Thorsteinson, 2001). In northeast and east Greenland, summer temperatures began to fall at ca. 5 ka BP, glaciers were advancing at 4.5 ka BP and sea ice increased at 4.7 ka BP (Wagner et al., 2000; Bennike and Weidick, 2001; Klug et al., 2009). In southeast Greenland, vegetation declined from ca. 6 ka BP, and temperatures dropped significantly at ca. 4 ka BP (Jakobsen et al., 2008); in west Greenland, it appears to have been later. In South Greenland, glacier readvance began at 4.8 ka BP (Nørgaard-Pedersen and Mikkelsen, 2009), but the local ice cap, Sukkertoppen ice cap, had already advanced at ca. 7 ka BP (compilation by Kelly and Lowell, 2009). To the north, on land and in the fjords in central West Greenland, cooling did not become apparent until ca. 3.2 ka BP, as summarised by Ren et al. (2009) and Seidenkrantz et al. (2008), while some lakes record initial cooling already at 6.5 ka BP (Bennike et al., 2010). In conclusion, the different proxies suggest that ice-margin retreat during HTM changed to slow advance at ca. 6 ka BP, picking up speed at ca. 4 ka BP.

In most areas, the Little Ice Age moraines from the 1800s or early 1900s mark the maximum of late Holocene glaciation. In a few areas, moraines testify to an earlier more advanced position of the ice margin (Bennike and Sparrenbom, 2007; Forman et al., 2007; Weidick and Bennike, 2007).

50.6. SUMMARY

The maximum growth of the Greenland ice sheet during the mid- and late Quaternary probably occurred during the Saalian Stage (MIS 6, ca. 188–135 ka BP) when the ice sheet overrode coastal mountains and likely reached the shelf break in all areas. In the warm climate of the Eemian Stage (MIS 5e, ca. 135–115 ka BP), the ice sheet shrank to ca. 70% of its present size, and especially the area of the southern dome was reduced, but by the end of the Eemian local ice caps began to form on some high mountain plateaux.

The Early Weichselian is only recorded in east and northwest Greenland. In east Greenland, it began with the large-scale advance of warm-based ice streams in the fjords (MIS 5d and B, 115–105 and 93–85 ka BP). Already by 60 ka BP, the low-gradient polar glaciers had reached close to their LGM distribution in east Greenland, which currently the best evidence suggests was restricted to low-gradient ice streams that only reached the inner or mid-shelf.

In southeast and southwest Greenland, however, thick ice built up on the shelf to form its own accumulation area. The contrasting behaviour of the ice margin in the south and northeast is also seen today and is a function of the dramatic decline in precipitation at the northern end of the Denmark Strait. For the West Greenland coast, we propose an LGM regime dominated by fast-flowing ice streams in cross-shelf troughs, with the passive ice margin closer to the coast—although the evidence for this is sparse and leaves room for other interpretations as well. In northern West Greenland recent evidence suggest that, during the LGM, large ice streams, possibly the largest in Greenland, transported ice into Baffin Bay where sea ice was sufficiently mobile to allow ice bergs to be transported away. In contrast, on the north Greenland coast new evidence suggests that permanent stationary sea ice allowed an ice shelf to develop between Canada and Greenland.

The LGM ice sheet covered an area of ca. 2.7 km^2, that is, 65% larger than at present, and its break-up began between ca. 20 and ca. 16 ka BP. By ca. 11.5 ka BP, the shelf as well as areas in the south and on the east coast had been cleared of ice, an area of ca. 0.8 million km^2. There is no clear evidence that the dramatic climate changes, which are so clearly recorded in ice cores, affected the ice margin, possibly because factors other than temperature played a larger role in the break-up of the ice sheet on the continental shelf.

Disappearance of the ice on land apparently only gathered speed after the PBO (ca. 11.4 ka BP). Deglaciation was a slow process that proceeded at different rates in different areas. By ca. 8 ka BP, the present distribution of ice had been achieved in most areas, and by ca. 6 ka BP, the ice may have reached its Holocene minimum. According Simpson et al.'s (2009) model, the ice sheet at this time was reduced by ca. 60,000 km^2 or ca. 4% compared to the present, mainly in West Greenland and in the saddle between the northern and southern domes. By 6 ka BP, the first signs of cooling are seen, and the ice sheet began slowly growing. The Neoglacial ice sheet reached its maximum during the Little Ice Age in the late 1880s or early 1900s, although some areas may have seen a maximum earlier.

REFERENCES

Adrielsson, L., Alexanderson, H., 2005. Interactions between the Greenland Ice Sheet and the Liverpool Land coastal ice cap during the last two glaciation cycles. J. Quatern. Sci. 20, 269–283.

Alley, R.B., Andrews, J.T., Brigham-Grette, J., Clarke, G.K.C., Cuffey, K. M., Fitzpatrick, J.J., et al., 2010. History of the Greenland Ice Sheet: paleoclimatic Insights. Quatern. Sci. Rev. 29, 1728–1756.

Andrews, J.T., 2008. The role of the Iceland Ice Sheet in the North Atlantic during the late Quaternary: a review and evidence from Denmark Strait. J. Quatern. Sci. 23, 3–20.

Andrews, J., Thang, P., Smith, I.M., Preston, R., Cooper, T., Jennings, A.E., 1997. Spatial and temporal patterns of iceberg rafting (IRD) along the East Greenland margin, ca. 68°N, over the last 14 cal. ka. J. Quatern. Sci. 12, 1–13.

Andrews, J.T., Cooper, T.A., Jennings, A.E., Stein, A.B., Erlenkeuser, H., 1998. Late Quaternary iceberg-rafted detritus events on the Denmark Strait-Southeast Greenland continental slope (c. 65°N) related to North Atlantic Heinrich events? Mar. Geol. 149, 211–228.

Bamber, J.L., Ekholm, S., Krabill, W.B., 2001. A new, high-resolution digital elevation model of Greenland fully validated with airborne laser altimeter data. J. Geophys. Res. B 106, 6733–6745.

Bennike, O., 2002. Late Quaternary history of Washington Land, North Greenland. Boreas 31, 260–272.

Bennike, O., 2008. An early Holocene Greenland whale from Melville Bugt, Greenland. Quatern. Res. 69, 72–76.

Bennike, O., Björck, S., 2002. Chronology of the last deglaciation of Greenland. J. Quatern. Sci. 17, 211–219.

Bennike, O., Sparrenbom, C.J., 2007. Dating of the Narssarssuaq stade in southern Greenland. Holocene 17, 279–282.

Bennike, O., Weidick, A., 2001. Late Quaternary history around Nioghalvfjerdsfjorden and Jøkelbugten, North-East Greenland. Boreas 30, 205–227.

Bennike, O., Björck, S., Lambeck, K., 2002. Estimates of South Greenland late-glacial ice limits from a new relative sea level curve. Earth Planet. Sci. Lett. 197, 171–186.

Bennike, O., Anderson, N.J., McGowan, S., 2010. Holocene palaeoecology of southwest Greenland inferred from macrofossils in sediments of an oligosaline lake. J. Paleolimnol. 43, 787–798.

Björck, S., 2007. Younger Dryas oscillation, global evidence. Encyclopedia of Quaternary Science Elsevier, Amsterdam, 1985–1993.

Björck, S., Rundgren, M., Ingólfsson, Ö., Funder, S., 1997. The Preboreal oscillation around the Nordic Seas: terrestrial and lacustrine responses. J. Quatern. Sci. 12, 445–465.

Blake, W., 1999. Glaciated landscapes along Smith Sund, Ellesmere Island, Canada and Greenland. Ann. Glaciol. 28, 40–46.

Blake, W., Boucherle, M.M., Fredskild, B., Janssens, J.A., Smol, J.P., 1982. The geomorphological setting, glacial history and Holocene development of Kap Inglefield Sø, Inglefield Land, North-West Greenland. Medd. Gronl. Geosci. 27, 42pp.

Cao, L., Fairbanks, R.G., Mortloc, R.A., Risk, M.J., 2007. Radiocarbon reservoir age of high latitude North Atlantic surface water during the last deglacial. Quatern. Sci. Rev. 26, 732–742.

Cuffey, K.M., Marshall, S.J., 2000. Substantial contribution to sealevel rise during the last interglacial from the Greenland ice sheet. Nature 404, 591–594.

Dahl-Jensen, D., Bamber, J., Bøggild, C.E., Buch, E., Christensen, J.H., Dethloff, K., et al., 2009. The Greenland Ice Sheet in a Changing Climate: Snow, Water, Ice and Permafrost in the Arctic (SWIPA). Arctic Monitoring and Assessment Programme (AMAP), Oslo, 115pp.

Denton, G.H., Alley, R.B., Comer, G.C., Broecker, W.S., 2005. The role of seasonality in abrupt climate change. Quatern. Sci. Rev. 24, 1159–1182.

Dowdeswell, J.A., Uenzelmann-Neben, G., Whittington, R.J., Marienfeld, P., 1994. The Late Quaternary sedimentary record in Scoresby Sund, East Greenland. Boreas 23, 294–310.

Dowdeswell, J.A., Evans, J., Cofaigh, C.Ó., 2010. Submarine landforms and shallow acoustic stratigraphy of a 400 km-long fjord-shelf-slope transect, Kangerlussuaq margin, East Greenland. Quatern. Sci. Rev. 29, 3359–3369.

Elverhøj, A., Dowdeswell, J.A., Funder, S., Mangerud, J., Stein, R., 1998. Glacial and oceanic history of the polar north Atlantic margins: an overview. Quatern. Sci. Rev. 17, 1–10.

Evans, J., Dowdeswell, J.A., Grobe, H., Niessen, F., Stein, R., Hubberten, H.W., et al., 2002. Late Quaternary sedimentation in Kejser Franz Joseph Fjord and the continental margin of East Greenland. In: Dowdeswell, J.A., O´Cofaigh, C. (Eds.), Glacier-Influenced Sedimentation on High-Latitude Continental Margins. Geological Society, London, pp. 149–179, Special Publication 203.

Evans, J., Ó.Cofaigh, C.A., Dowdeswell, J.A., Wadhams, P., 2009. Marine geophysical evidence for former expansion and flow of the Greenland Ice Sheet across the north-east Greenland continental shelf. J. Quatern. Sci. 24, 279–293.

Fleming, K., Lambeck, K., 2004. Constraints on the Greenland Ice Sheet since the Last Glacial Maximum from sea-level observations and glacial-rebound models. Quatern. Sci. Rev. 23, 1053–1077.

Forman, S.L., Marín, L., van der Veen, C., Tremper, C., Csatho, B., 2007. Little Ice Age and neoglacial landforms at the Inland Ice margin, Isunguata Sermia, Kangerlussuaq, west Greenland. Boreas 36, 341–351.

Funder, S., 1979. Ice-age plant refugia in East Greenland. Palaeogeogr. Palaeoclimatol. Palaeoecol. 28, 279–295.

Funder, S., 1989. Development of climate, glaciation, and oceanographis circulation (Greenland). Chapter 13 In: Fulton, R.J. (Ed.), Quaternary Geology of Canada and Greenland, 783–786, Geological Survey of Canada, Geology of Canada 1 (also Geological Society of America, The Geology of North America K-1).

Funder, S., Abrahamsen, N., 1988. Palynology in a Polar desert, eastern North Greenland. Boreas 17, 195–207.

Funder, S., Hansen, L., 1996. The Greenland ice sheet—a model for its culmination and decay during and after the last glacial maximum. Bull. Geol. Soc. Denmark 42, 137–152.

Funder, S., Hjort, C., Landvik, J.Y., Nam, S.I., Reeh, N., Stein, R., 1998. History of a stable ice margin—east Greenland during the Middle and Upper Pleistocene. Quatern. Sci. Rev. 17, 77–125.

Funder, S., Jennings, A., Kelly, M., 2004. Middle and Late Quaternary glacial limits in Greenland. In: Ehlers, J., Gibbard, P. (Eds.), Quaternary Glaciations. Extent and Chronology II. Elsevier, Amsterdam, pp. 425–430.

Greve, R., 2005. Relation of measured basal temperatures and the spatial distribution of the geothermal heat flux for the Greenland ice sheet. Ann. Glaciol. 42, 424–432.

Håkansson, L., Graf, A., Strasky, S., Ivy-Ochs, S., Kubik, P.W., Hjort, C., et al., 2007a. Cosmogenic Be-10-ages from the Store Koldewey island, NE Greenland. Geogr. Ann. 89A, 195–202.

Håkansson, L., Briner, J.P., Alexanderson, H., Aldahan, A., Possnert, G., 2007b. 10Be ages from coastal northeast Greenland constrain the extent of the Greenland Ice Sheet during the Last Glacial Maximum. Quatern. Sci. Rev. 26, 2316–2321.

Håkansson, L., Alexanderson, H., Hjort, C., Möller, P., Briner, J.P., Aldahan, A., et al., 2009. Late Pleistocene glacial history of Jameson Land, central East Greenland, derived from cosmogenic ^{10}Be and ^{26}Al exposure dating. Boreas 38, 244–260.

Hall, B., Baroni, C., Denton, G., Kelly, M.A., Lowell, T., 2008. Relative sea-level change, Kjove Land, Scoresby Sund, East Greenland: implications for seasonality in Younger Dryas time. Quatern. Sci. Rev. 27, 2283–2291.

Hall, B., Baroni, C., Denton, G., 2010. Relative sea-level changes, Schuchert Dal, East Greenland, with implications for ice extent in late-glacial and Holocene times. Quatern. Sci. Rev. 29, 3370–3378.

Hansen, L., 2001. Landscape and coast development of a lowland fjord margin following deglaciation, East Greenland. Geogr. Ann. 83A, 131–144.

Hansen, L., Funder, S., Murray, A.S., Mejdahl, V., 1999. Luminescence dating of the last Weichselian glacier advance in East Greenland. Quatern. Geochronol. 18, 179–190.

Hjort, C., 1997. Glaciation, climate history, changing marine levels and the evolution of the northeast water polynya. J. Mar. Syst. 10, 23–33.

Holtedahl, O., 1970. On the morphology of the west greenland shelf with general remarks on the "marginal channel" problem. Mar. Geol. 8, 155–172.

Hubberten, H.W., Grobe, H., Jokat, W., Melles, M., Niessen, F., Stein, R., 1995. Glacial history of east Greenland explored. EOS Trans. Am. Geophys. Union 76, 353, 356.

Huybrechts, P., 2002. Sea-level changes at the LGM from icedynamic reconstructions of the Greenland and Antarctic ice sheets during the glacial cycles. Quatern. Sci. Rev. 21, 203–231.

IBCAO, 2010. International bathymetric chart of the Arctic Ocean, Version 2.23 http://www.ngdc.noaa.gov/mgg/bathymetry/arctic/.

Ingólfsson, O., Frich, P., Funder, S., Humlum, O., 1990. Paleoclimatic implications of an early Holocene glacier advance on Disko Island, West Greenland. Boreas 19, 297–311.

Jakobsen, B.H., Fredskild, B., Pedersen, J.B.T., 2008. Holocene changes in climate and vegetation in the Ammassalik area, East Greenland, recorded in lake sediments and soil profiles. Danish J. Geogr. 108, 21–50.

Jennings, A.E., Grönvold, K., Hilberman, R., Smith, M., Hald, M., 2002. High-resolution study of Icelandic tephras in the Kangerlussuaq Trough, southeast Greenland, during the deglaciation. J. Quatern. Sci. 17, 747–757.

Jennings, A.E., Hald, M., Smith, L.M., Andrews, J.T., 2006. Freshwater forcing from 722 the Greenland Ice Sheet during the Younger Dryas: evidence from Southeastern 723 Greenland shelf cores. Quatern. Sci. Rev. 25, 282–298.

Johnsen, S.J., Clausen, H.B., Dansgaard, W., Gundestrup, N.S., Hansson, M., Jonsson, P., et al., 1992. A 'deep' ice core from East Greenland. Medd. Gronl. Geosci. 29, 22 pp.

Kelly, M., 1985. A review of the Quaternary geology of western Greenland. In: Andrews, J.T. (Ed.), Quaternary Environments Eastern Canadian Arctic, Baffin Bay and Western Greenland. Allen and Unwin, Boston, pp. 461–501.

Kelly, M., Bennike, O., 1992. Quaternary geology of western and central North Greenland. Rapport Grønlands geoogiske Undersøgelser 153, 34pp.

Kelly, M.A., Lowell, T.V., 2009. Fluctuations of local glaciers in Greenland during latest Pleistocene and Holocene time. Quatern. Sci. Rev. 28, 2088–2106.

Kelly, M., Funder, S., Houmark-Nielsen, M., Knudsen, K.L., Kronborg, C., Landvik, J.Y., et al., 1999. Quarternary glacial and marine environmental history of north-west Greenland: a review and reappraisal. Quatern. Sci. Rev. 18, 373–392.

Kelly, M.A., Lowell, T.V., Hall, B.L., Schaefer, J.M., Finkel, R.C., Goehring, B.M., et al., 2008. A Be-10 chronology of lateglacial and Holocene mountain glaciation in the Scoresby Sund region, east Greenland: implications for seasonality during lateglacial time. Quatern. Sci. Rev. 27, 2273–2282.

Khan, S.A., Wahr, J., Leuliette, E., van Dam, T., Larson, K.M., Franci, O., 2008. Geodetic measurements of postglacial adjustments in Greenland. J. Geophys. Res. 113, 16. doi:10.1029/2007JB004956, B02402.

Klug, M., Bennike, O., Wagner, B., 2009. Repeated short-term bioproductivity changes in a coastal lake on Store Koldewey, northeast Greenland: an indicator of varying sea-ice coverage? Holocene 19, 653–663.

Knudsen, K.L., Stabell, B., Seidenkrantz, M.-S., Eiríksson, J., Blake, W., 2008. Deglacial and Holocene conditions in northernmost Baffin Bay: sediments, foraminifera, diatoms and stable isotopes. Boreas 37, 346–376.

Kobashi, T., Severinghaus, J.P., Barnola, J.M., 2008. $4\pm1.5\,°C$ abrupt warming 11,270 yr ago identified from trapped air in Greenland ice. Earth Planet. Sci. Lett. 268, 397–407.

Kuijpers, A., Troelstra, S.R., Prins, M.A., Linthout, K., Akhmetzhanov, A., Bouryak, S., et al., 2003. Late Quaternary sedimentary processes and ocean circulation changes at the Southeast Greenland margin. Mar. Geol. 195, 109–129.

Kuijpers, A., Dalhoff, F., Brandt, M.P., Hümbs, P., Schot, T., Zotova, A., 2007. Giant iceberg plow marks at more than 1 km water depth offshore West Greenland. Mar. Geol. 246, 60–64.

Landvik, J.Y., 1994. The last glaciation of Germania Land and adjacent areas, northeast Greenland. J. Quatern. Sci. 9, 81–92.

Larsen, N.K., Kjær, K.H., Funder, S., Möller, P., van der Meer, J., Schomacker, A., Linge, H., Darby, D. (2010). Extensive ice shelf glaciation in northern Greenland during the Last Glacial Maximum. Quatern. Sci. Rev. 29, 3399–3414.

Lenton, T.M., Held, H., Kriegler, E., Hall, J.W., Lucht, W., Rahmstorf, S., et al., 2008. Tipping elements in the Earth's climate system. Proc. Natl. Acad. Sci. USA 105, 1786–1793.

Letréguilly, A., Reeh, N., Huybrechts, P., 1991. The Greenland ice sheet through the last glacial-interglacial cycle. Palaeogeogr. Palaeoclimatol. Palaeoecol. Global Planet. Change Sect. 90, 385–394.

Lhomme, N., Clarke, G.K.C., Marshall, S.J., 2005. Tracer transport in the Greenland Ice Sheet: constraints on ice cores and glacial history. Quatern. Sci. Rev. 24, 173–194.

Lloyd, J.M., Park, L.A., Kuijpers, A., Moros, M., 2005. Early Holocene palaeoceanography andd eglacial chronology of Disko Bugt, West Greenland. Quatern. Sci. Rev. 24, 1741–1755.

Long, A.J., Roberts, D.H., Dawson, S., 2006. Early Holocene history of the west Greenland Ice Sheet and the GH-8.2 event. Quatern. Sci. Rev. 25, 904–922.

Long, A.L., Roberts, D.H., Simpson, M.J.R., Dawson, S., Milne, G.A., Huybrechts, P., 2008a. Late Weichselian relative sea-level changes and ice sheet history in southeast Greenland. Earth Planet. Sci. Lett. 272, 8–18.

Long, A.J., Woodroffe, S.A., Dawson, S., Roberts, D.H., Bryant, C.L., 2008b. Late Holocene relative sea level rise and the Neoglacial history of the Greenland ice sheet. J. Quatern. Sci. 24, 345–359.

Madsen, K.N., Thorsteinsson, T., 2001. Textures, fabrics and meltlayer stratigraphy in the Hans Tausen ice core, North Greenland—indications of late Holocene ice cap generation? Copenhagen, Danish Polar Center. Medd. Gronl. Geosci. 39, 97–114.

Mangerud, J., Funder, S., 1994. The interglacial–glacial record at the mouth of Scoresby Sund, East Greenland. Boreas 23, 349–358.

Mangerud, J., Landvik, J.Y., 2007. Younger Dryas cirque glaciers in western Spitsbergen: smaller than during the Little Ice Age. Boreas 36, 278–285.

Möller, P., Hjort, C., Adrielsson, L., Salvigsen, O., 1994. Glacial history of interior Jameson Land, East Greenland. Boreas 23, 320–348.

Möller, P., Larsen, N.K., Kjær, K., Funder, S., Schomacker, A., Linge, H., 2010. Early to middle Holocene valley glaciations on northernmost Greenland. Quatern. Sci. Rev. 29, 3379–3398.

Mörner, N.-A., Funder, S., 1990. C-14 dating of samples collected during the NORQUA 86 expedition, and notes on the marine reservoir effect. In: Funder, S. (Ed.), Late Quaternary Stratigraphy and Glaciology in the Thule Area. Northwest Greenland, 57–59. Meddelelser om Grønland, Geoscience.

Nam, S.-I., Stein, R., 1999. Late Quaternary variations in sediment accumulation rates and their paleoenvironmental implications: a case study from the East Greenland Continental Margin. GeoRes. Forum 5, 223–240.

Nørgaard-Pedersen, N., Mikkelsen, N., 2009. 8000 year marine record of climate variability and fjord dynamics from Southern Greenland. Mar. Geol. 264, 177–189.

Nørgaard-Pedersen, N., Mikkelsen, N., Kristoffersen, Y., 2009. Late glacial and Holocene marine records from the Independence Fjord and Wandel Sea regions, North Greenland. Polar Res. 27, 209–221.

Ó Cofaigh, C., Dowdeswell, J.A., Evans, J., Kenyon, N.H., Taylor, J., Mienert, J., et al., 2004. Timing and significance of glacially-influenced mass-flow activity in the submarine channels of the Greenland Basin. Mar. Geol. 207, 39–54.

Ó Cofaigh, C., Dowdeswell, J.A., Kilfeather, A.A., Jennings, A.E., Evans, J., Noormets, R., et al., 2010. Palaeo-ice streams on the west Greenland continental margin during the last glacial cycle. Geophys. Res. Abstr. 12, EGU2010-4959.

Ohmura, A., Reeh, N., 1991. New precipitation and accumulation maps for Greenland. J. Glaciol. 37, 140–148.

Otto-Bliesner, B.L., Marshall, S.J., Overpeck, J.T., Miller, G.H., Hu, A., CAPE Last Interglacial Project Members, 2006. Simulating warm-Arctic climate and ice sheet sensitivity for the last interglaciation. Science 311, 1751–1753.

Rasmussen, S.O., Vinther, B.M., Clausen, H.B., Andersen, K.K., 2007. Early Holocene climate oscillations recorded in three Greenland ice cores. Quatern. Sci. Rev. 26, 1907–1914.

Reimer, P.J., Baillie, M.G.L., Bard, E., Bayliss, A., Beck, J.W., Blackwell, P.G., et al., 2009. Intcal09 and marine09 radiocarbon age calibration curves, 0–50,000 years cal BP. Radiocarbon 51, 1111–1150.

Ren, J., Jiang, H., Seidenkrantz, M.S., Kuijpers, A., 2009. A diatom-based reconstruction of Early Holocene hydrographic and climatic change in a southwest Greenland fjord. Mar. Micropaleontol. 70, 166–176.

Rinterknecht, V., Gorokhovich, Y., Schaefer, J., Caffee, M., 2008. Preliminary ^{10}Be chronology for the last deglaciation of the western margin of the Greenland Ice Sheet. J. Quatern. Sci. 24, 270–278.

Roberts, D.H., Long, A.J., Schnabel, C., Simpson, M., Freeman, S., 2008. The deglacial history of the southeast sector of the Greenland ice sheet during the Last Glacial maximum. Quatern. Sci. Rev. 27, 1505–1516B.

Roberts, D.H., Long, A.J., Schnabel, C., Davies, B.J., Xu, S., Simpson, M.J.R., et al., 2009. Ice sheet extent and early deglacial history of the southwestern sector of the Greenland Ice Sheet. Quatern. Sci. Rev. 25–26, 2760–2773.

Roksandic, M.M., 1979. Geology of the continental shelf off West Greenland between 61°15′N and 64°00′N: an interpretation of sparker seismic and echo sounder data. Grønlands Geol. Unders. 92, 15pp.

Schmidt, S., Wagner, B., Heiri, O., Klug, M., Bennike, O., Melles, M., 2010. Chironomids as indicators of the Holocene climatic and environmental history of two lakes in Northeast Greenland. Boreas 10.1111/j.1502-3885.2010.00173.x. ISSN 0300-9483.

Seidenkrantz, M.-S., Roncaglia, L., Fischel, A., Heilmann-Clausen, C., Kuijpers, A., Moros, M., 2008. Variable North Atlantic climate seesaw patterns documented by a late Holocene marine record from Disko Bugt, West Greenland. Mar. Micropaleontol. 68, 66–83.

Seidenkrantz, M.-S., Kuijpers, A., Aagaard-Sørensen, S., Andersson, S., Lindgren, H., Ploug, J., et al., 2010. Glacial ocean circulation and shelf edge glaciation offshore SW Greenland during the past 75,000 years. Geophys. Res. Abstr. 12, EGU2010-4721.

Simpson, M.J.R., Milne, G.A., Huybrechts, P., Long, A.J., 2009. Calibrating a glaciological model of the Greenland ice sheet from the Last Glacial Maximum to present-day using field observations of relative sea level and ice extent. Quatern. Sci. Rev. 28, 1631–1657.

Sommerhoff, G., 1981. Geomorphologische Prozesse in der Labrador - und Imingersee. Ein Beitrag zur submarinen Geomorphologie einer subpolaren Meeresregion.. Polarforschung 51, 175–191.

Sparrenbom, C.J., Bennike, O., Björck, S., Lambeck, K., 2006a. Holocene relative sea-level changes in the Qaqortoq area, southern Greenland. Boreas 35, 171–187.

Sparrenbom, C.J., Bennike, O., Björck, S., Lambeck, K., 2006b. Relative sea-level changes since 15 000 cal. yr BP in the Nanortalik area, southern Greenland. J. Quatern. Sci. 21, 29–48.

Steffensen, J.P., Andersen, K.K., Bigler, M., Clausen, H.B., Dahl-Jensen, D., Fischer, H., et al., 2008. High-resolution Greenland ice core data show abrupt climate change happens in few years. Science 321, 680–684.

Stein, R., Nam, S., Grobe, H., Hubberten, H., 1996. Late Quaternary glacial history and short-term ice-rafted debris fluctuations along the east Greenland Continental Margin. In: Andrews, et al., (Ed.), Late Quaternary Paleoceanography of the North Atlantic Margins, Geological Society Special Publication, 111, pp. 135–151.

Syvitski, J.P.M., Stein, A.B., Andrews, J.T., 2001. Icebergs and the Sea Floor of the East Greenland (Kangerlussuaq) continental Margin. Arct. Antarct. Alpine Res. 33, 52–61.

Tarasov, L., Peltier, W.R., 2003. Greenland glacial history, borehole constraints, and Eemian extent. J. Geophys. Res. 108 (B3), 2143.

van Tatenhove, F.G.M., van der Meer, J.J.M., Koster, E.A., 1996. Implications for deglaciation chronology from new AMS age determinations in central west Greenland. Quatern. Res. 45, 245–253.

Vinther, B.M., Buchardt, S.L., Clausen, H.B., Dahl-Jensen, D., Johnsen, S.J., Fisher, D.A., et al., 2009. Holocene thinning of the Greenland ice sheet. Nature 461, 385–388.

Wagner, B., Melles, M., Hahne, J., Niessen, F., Hubberten, H.-W., 2000. Holocene climate history of Geographical Society Ø, East Greenland. Evidence from lake sediments. Palaeogeogr. Palaeoclimatol. Palaeoecol. 160, 45–68.

Wagner, B., Bennike, O., Cremer, H., Klug, M., 2010. Late Quaternary history of the Kap Mackenzie area, northeast Greenland. Boreas 39, 492–504.

Walker, M., Johnsen, S., Rasmussen, S.O., Popp, T., Steffensen, J.-P., Gibbard, P., et al., 2009. Formal definition and dating of the GSSP (Global Stratotype Section and Point) for the base of the Holocene using the Greenland NGRIP ice core, and selected auxiliary records. J. Quatern. Sci. 24, 3–17.

Weidick, A., 1968. Observations on some Holocene glacier fluctuations in West Greenland. Medd. Grønland 165 (6), 202pp.

Weidick, A., 1975. Quaternary geology of the area between Frederikshabs Isblink and Ameralik. Geological Survey of Greenland Report 70, 22pp.

Weidick, A., Bennike, O., 2007. Quaternary glaciation history and glaciology of Jakobshavn Isbræ and the Disko Bugt region. West Greenland: a review Geological Survey of Denmark and Greenland Bulletin 14, 77pp.

Weidick, A., Kelly, M., Bennike, O., 2004. Late Quaternary development of the southern sector of the Greenland ice sheet, with particular reference to the Qassimiut lobe. Boreas 33, 284–299.

Wilken, M., Mienert, J., 2006. Submarine glacigenic debris flows, deep-sea channels and past ice-stream behaviour of the East Greenland continental margin. Quatern. Sci. Rev. 25, 784–810.

Willerslev, E., Cappellini, E., Boomsma, W., Nielsen, R., Hebsgaard, M.B., Brand, T.B., et al., 2007. Ancient biomolecules from deep ice cores reveal a forested southern Greenland. Science 317, 111–114.

Winkelmann, D., Jokat, W., Jensen, L., Schenke, H.-W., 2010. Submarine end moraines on the continental shelf off NE Greenland—implications for Lateglacial dynamics. Quatern. Sci. Rev. 29, 1069–1077.

Zarudzki, E.F.K., 1980. Interpretation of shallow seismic profiles over the continental shelf in West Greenland between latitudes 64 and 69 30'N. Grønlands Geol. Unders. 100, 58–61.

Zreda, M., England, J., Phillips, F., Elmore, D., Sharma, P., 2009. Unblocking of the Nares Strait by Greenland and Ellesmere ice-sheet retreat 10,000 years ago. Nature 398, 139–142.

Chapter 51

Pleistocene Glaciations in Southern Patagonia and Tierra del Fuego

Andrea Coronato[1,2,*] and Jorge Rabassa[1,2]

[1] CONICET-CADIC. B. Houssay 200, 9410 Ushuaia, Tierra del Fuego, Argentina
[2] Universidad Nacional de la Patagonia San Juan Bosco, Sede Ushuaia, Darwin y Canga, 9410 Ushuaia, Tierra del Fuego, Argentina
*Correspondence and requests for materials should be addressed to A. Coronato. E-mail: acoronato@cadic-conicet.gob.ar

51.1. INTRODUCTION

Continental and alpine-type glaciations of different ages and extent are both very well represented in the landscapes of southern Patagonia and Tierra del Fuego. Morphological evidence of ancient glaciations is widespread in the mountains and also in the lowlands at the present Atlantic coast close to the Magellan Straits and Bahía Inútil (Fig. 51.1). Till deposits interbedded with basalt flows indicate the occurrence of glaciation even before the Pliocene–Pleistocene boundary (ca. 2.6 Ma).

One of the unsolved questions regarding Quaternary climatic changes concerns when the cooling periods started, how extended they were, and how much time they took. There are several main reasons why frequent glaciations have occurred in southern Patagonia since the latest Miocene. They include factors such as latitudinal position, the existence of the permanent South Pacific Anticyclone, the westerlies belt, the Andean chains as a rather low mountain barrier, volcanic activity, tectonic uplift, the presence of a large number of tectonically developed, longitudinal and transverse valleys, the steep precipitation gradient towards the eastern slope and the persistent arid conditions in the Patagonian tablelands. Palaeoclimatic studies should be focused to explain why even the older glaciations, such as the Greatest Patagonian Glaciation (GPG, ca. 1 Ma), are so well preserved, their landform morphology almost untouched by erosion. A detailed overview of the physical geography of Patagonia is given by Coronato et al. (2008a,b,c).

Some of the first observations of glacial boulders found scattered over the landscape were made 200 years ago by Darwin (1842) on his trip along the Beagle Channel and the Santa Cruz River (Fig. 51.1). Many other scientists identified glacial deposits and landforms in Patagonia, but the first serious and extensive mapping was done by Caldenius (1932), who mapped in detail four glacial boundaries east of the Patagonian Andes, along 410,000 km^2, from latitude 41°S to Cape Horn and surrounding islands, at the southern tip of South America (56°S). Several authors have been interested in the distribution of glaciation in Patagonia, including Auer (1956), Mercer (1976 and many later works), Heusser (1989), Rabassa and Clapperton (1990), Clapperton (1993), Lowell et al. (1995); recently Coronato and Rabassa (2007a,b), Rabassa (2008), Glasser et al. (2008) and Glasser and Ghiglione (2009). These authors provided regional investigations of glaciations all over Patagonia. For a complete modern review of Late Cenozoic Patagonian glaciations, see Rabassa (2008). Reconstruction and modelling of the Patagonian Ice Sheet during the Last Glaciation have been presented by Hulton et al. (2002), Hubbart et al. (2005) and Kilian et al. (2007), among others.

Pliocene and pre-Pliocene glacial deposits have been investigated by Rabassa and Coronato (2009), Wenzens (1999, 2000, 2006) and Lagabrielle et al. (2010), whereas the relationship of the Pleistocene glaciations with the Pampean biostratigraphical stages was presented by Rabassa et al. (2005) and, more recently, by Rabassa and Coronato (2009). Holocene glaciations, including the Little Ice Age, have been recently studied by Glasser et al. (2004), Wenzens (2005) and Kilian et al. (2007), among others.

This chapter is an updated review based on its first version published by Coronato et al. (2004a,b). However, the study area comprises between 46°S (Lago Buenos Aires/Carreras) and the Beagle Channel (54°55′S), whereas the northern Patagonian region is presented by Martínez et al. (2011). The Patagonian glacial limits are depicted following Caldenius (1932) and later authors but they were redrawn using Digital Terrain Models (90 m; freely available on http://srtm/csi.cgiar.org) and based on new field information. The application of dating methods has improved the glacial geology knowledge during the past decades,

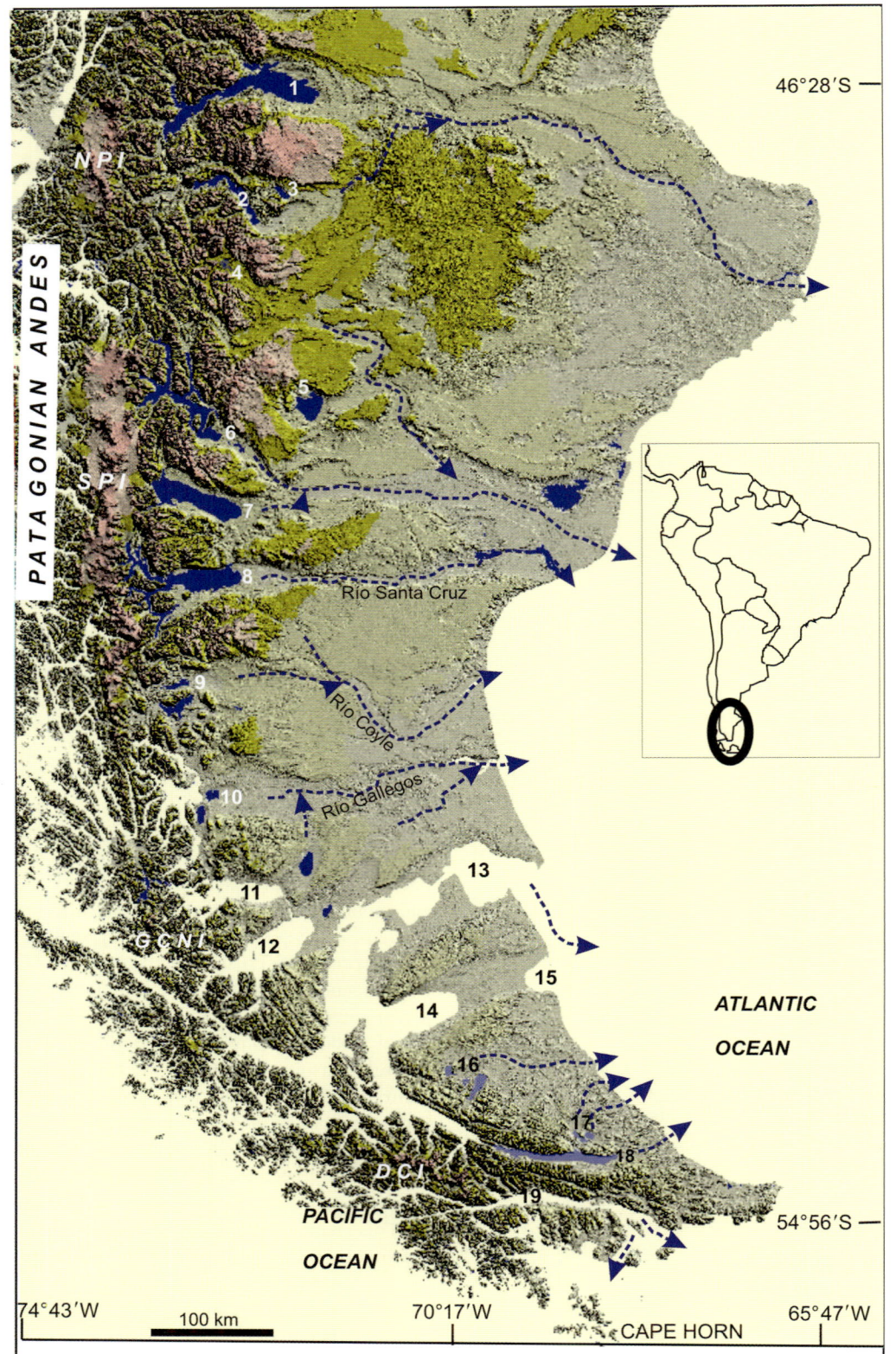

FIGURE 51.1 *Location map of the southern Patagonia region.* The main ancient eastward-flowing glacier lobes are today occupied by major lakes or river valleys. The dashed lines show the position of ancient outwash lines, most of them totally occupied by modern rivers. Considering the position of the sea coast at ca. −120 m a.s.l. during the Last Glacial Maximum, the mouth of these ancient rivers is still unknown. They must be located on the present Atlantic continental shelf. The map is based on Shuttle Radar Terrain Model (90 m resolution) 22-22, 22-23, 23-22, 23-23 and 23-24, freely available on http://srtm/csi.cgiar.org.

most of them based on $^{40}Ar-^{39}Ar$, K–Ar, cosmogenic isotopes, mainly ^{10}Be, ^{26}Al, ^{3}He and ^{36}Cl, radiocarbon and tephrochronology techniques.

This chapter contributes to the understanding of the palaeo-Patagonian ice fields as a unique source of ice from which outlet glaciers flowed to the east and west, although each of them behaved slightly differently depending on the local conditions, a factor that gave rise to differing capabilities of landscape sculpturing.

Following Coronato et al. (2004a,b), the GPG, firstly identified by Mercer (1976) in the Viedma and Argentino lakes (Fig. 51.1) and later recognised and dated in the Lago Buenos Aires (Singer et al., 2004a,b and other papers by this group) and the Río Gallegos valley (Meglioli, 1992; Ton-That et al., 1999), is largely considered the key to which the glaciations that followed are referred. This means that both the pre-GPG and post-GPG boundaries are given in the digital maps. The Last Glacial Maximum (LGM) and its re-advance phases during overall glacial retreat (LGM-RP1 to RP3) are also shown. The comparison to local nomenclature as well as the available chronological data is given in several tables.

Papers published after Coronato et al. (2004a,b) are particularly considered in the present work. Although an extensive reference list is provided, a more complete bibliography list is given by Rabassa (2008, and other papers cited).

during Marine Isotope Stage (MIS) 6 and within the period between MIS 18 and 8–10 (Table 51.1). The present chronostratigraphy of glacial deposits in the area includes the Telken moraine VII (1016–1168 ka), the Telken moraines I–VI (760–1016 ka), the Deseado moraines I–III (?–760 ka), the Moreno moraines I–III (109–?), the Fenix moraines I–IV (16–25 ka) and the Menucos moraine (<15.8 ka). Turner et al. (2005) demonstrated a rapid glacier retreat at 15–16 ka (calendar ages), followed by stability and a complete deglaciation phase at ca. 12.8 cal. ka based on stratigraphy and radiocarbon ages on exposures located in the present Chilean sector. Later, Douglass et al. (2006) refined the LGM and Late-glacial chronology by calculating ^{10}Be and ^{26}Al surface exposure age on morainic boulders. They found six glacial advances between 22.7 and 14.4 ka (Table 51.2 for detailed chronology) in phase with the Antarctic Cold Reversal signal.

Although none of these authors found ages older than 1 Ma (older than the GPG) in this area, the presence of frontal moraines older than that age has been proposed by Malagnino (1995) from the north-eastern corner of the Argentine sector of the lake. This is presented here as a pre-GPG limit. Malagnino (1995) defined the so-called Chipanque moraines based on geomorphology and stratigraphy, but he was unable to determine their absolute age. However, the stratigraphical position allowed him to postulate an age range of between 7 and 4.6 Ma or 3.5 and 2.3 Ma.

51.2. THE SOUTHERN PATAGONIAN PALAEO-ICE-LOBES

51.2.1. The Lago Buenos Aires–Carreras (Argentina–Chile) Lobe (46°16′–46°56′S; 71°09′–72°50′W)

Since the work of Ton-That et al. (1999), in which a maximum age for the GPG was indicated at 1.15 Ma, further research has been undertaken in the area. It has mainly been devoted to establishing the chronology of the glacial advances previously mapped by Caldenius (1932), Malagnino (1995) and others. Singer et al. (2004a) and Kaplan et al. (2004) describe 19 morainic suites that relate to the frontal position of a piedmont lobe flowing from the palaeo-Northern Patagonian Icefield (NPI) between 1.2 Ma and 16 ka BP. The occurrence of lava flows interbedded with till or stratigraphically related to moraines has provided absolute numerical ages for the glacial limits to be determined. Radiocarbon dates from glaciolacustrine deposits, and $^{40}Ar/^{39}Ar$ dates on basalts, have constrained the detailed timing of Pleistocene glaciations in the Lago Buenos Aires area (Tables 51.1 and 51.2). Later, Kaplan et al. (2005) determined the age of two glacial advances using cosmogenic isotopes measurements. These advances occurred

51.2.2. The Lagos Pueyrredón-Posadas Lobe (47°08′–47°30′S; 71°47′–72°33′W)

A recent paper by Hein et al. (2009) reports the detailed mapping of four glacial boundaries, almost coincident with those of Caldenius (1932), named as Gorra de Poivre (Initioglacial), Cañadón Caracoles (Daniglacial), Hatcher (Gotiglacial) and Río Blanco (Finiglacial). The glacial chronology is mainly referred to the glacial model of Lago Buenos Aires proposed by Singer et al. (2004a,b) and Kaplan et al. (2004, 2005). Refined evidence is presented by Hein et al. (2009) for the Middle Pleistocene glaciations, based on ^{10}Be concentrations from outwash cobbles deposited on very extensive outwash terraces. Thus, the major glacial advance, which deposited the Hatcher moraines, has been referred to ca. 260 ka (i.e. MIS 8). These data help to constrain the wide range of Moreno III moraines of the Lago Buenos Aires (Table 51.1), which have been dated as older than 190 ka (Singer et al., 2004a,b). The age of the Río Blanco glaciation was determined at ca. 27–25 ka, which is similar to that obtained from the Lago Buenos Aires Fénix moraines (Singer et al., 2004a,b). Glaciation during MIS 4 is represented in neither of the palaeo-ice lobes at the surface. Similarly, evidence of moraines dating to MIS 6 has not been found yet in this lobe.

TABLE 51.1 Quaternary Chronostratigraphy of the Main East-Flowing Palaeoglacier Lobes from 46°S to 54°S

Time (ka)	Geomagnetic Polarity Time Scale / Chron and subchron	Lake Buenos Aires–Carreras (after 2,3)	Lakes Pueyrredón–Posadas (after 4)	Lake Viedma (after 5,6)	Lake Argentino (after 7,8,9)	Río Gallegos valley (after 10, 2)	Magellan Straits (after 10)	Inútil–San Sebastián depression (after 10,11)	Caldenius glacial limits (1932)
0	Last Interglacial / BRUNHES				See table 2				Finiglacial
200		Moreno I and II moraines			Glaciation 2	Río Turbio Drift	? Primera Angostura and Punta Delgada Drift ?	? Lagunas Secas and San Sebastián Drift ?	Gotiglacial
400		Moreno III moraine	Hatcher m.	Glaciations 4 to 2	Glaciation 3				?
600			Caracoles moraines	?	Glaciation 4				Daniglacial
800	BRUNHES / MATUYAMA	Deseado moraines		?		Glen Cross Drift	Cabo Virgenes Drift	Río Cullen Drift	
1000	R / MATUYAMA	Telken I–VI moraines	Gorra de Poivre moraines	Glaciation 5	Till 1	Bella Vista Drift	Sierra de los Frailes Drift	Pampa de Beta Drift	Initioglacial
1200	J	Telken VII till → Pre-GPG Chipanque moraines (5)		→ Pre-GPG Glaciations 8 to 6	→ Pre-GPG Cerro del Fraile Tills 6 to 1			→ Pre-GPG Río Grande Drift	

δ¹⁸O(‰) (1) — ← cooler / warmer → — MIS stages labeled 1–35

(1) Lisiecki and Reimo (2005); (2) Singer et al. (2004a); (3) Kaplan et al. (2005); (4) Hein et al., (2009); (5) Malagnino (1995); (6) Wenzens (2006 and previous papers); (7) Schellman (1988); (8) Singer et al. (2004b); (9) Strelin (1995); (10) Meglioli (1992); (11) Walther et al. (2007).

The ages are based on uncalibrated radiocarbon dates and on cosmogenic isotope measurements providing minimum exposure ages on boulders. The dashed lines and question marks indicate chronological uncertainties.

TABLE 51.2 Comparison of the Chronostratigraphy of Southern Patagonian Glaciation During MIS 2 (Late Wisconsinan, Late Weichselian) of the Major Palaeoglacier Lobes from 46°S to 54°S

ka BP	Lago Buenos Aires–Carreras (after 1)	Lagos Pueyrredón–Posadas (after 2)	Lago Viedma (after 3)	Lago Argentino region (after 4)	Magellan Straits Inútil–San Sebastián depression (after 5)	Lago Fagnano lobes (after 6)	Caldenius (1932)	
10	Fachinal moraines		Río Condor m. / A° Guanaco m.	Herminita m. / P. Bandera m.			Post Finiglacial	NEOGLACIAL
12			Río Guanaco moraines		Stage E	C submerged m. ?		YD
14	Menucos Moraine	Columna m.					Finiglacial	ACR
16	Fenix I Moraine							LATE GLACIAL
18	Fenix II Moraines	Río Blanco moraines			Stage D ?	B. submerged m. ?		
20	Fenix III–V moraines		Lake Viedma moraines	Moraine 1a and 1b	Stage C	Tolhuin–Yehuin m. ? / Río Fuego m. ?	?	LAST GLACIAL MAXIMUM
25			?	?				
30			?					

(1) Douglass et al. (2005, 2006); (2) Hein et al. (2009, 2010); (3) Wenzens (1999, 2003); (4) Schellman et al. (1988), Strelin et al. (2008); Ackert et al. (2005); (5) McCulloch et al. (2005a); (6) Coronato et al. (2006, 2008a), Waldmann et al. (2010). A° = creek; m=moraines.

The ages are based on uncalibrated radiocarbon dates and on cosmogenic isotope measurements providing minimum exposure ages on boulders. The dashed lines and question marks indicate chronological uncertainties.

51.2.3. The Belgrano and Burmeister Lake Lobes (47°47′–47°57′S; 72°04′–72°18′W)

Wenzens (2003) proposed that these lakes formed part of the same glacial outlet which flowed south-eastwards from Mount San Lorenzo (3700 m a.s.l.) and joined the palaeo-ice lobe of the Río Chico which flowed north-eastwards from the Sangra Mountains. The LGM moraines have been identified several kilometres east of the present heads of the lakes (Wenzens, 2005). The ice lobes flowed between the Belgrano and Strobel tablelands; the present Río Belgrano valley should had been the spillway around the time of the LGM. Moraines closing the lakes were referred to Late-glacial re-advances.

51.2.4. The Lago San Martín–Lago Tar Lobe (48°55′–49°15′S; 71°54′–72°31′W)

Four glacial limits have previously been recognised in this area (Coronato et al., 2004a,b, for a reference list). Later, LGM moraines were defined surrounding Lago Tar and the southern coast of Lago San Martín (Wenzens, 2003, 2005). Radiocarbon dating on basal organic matter in a drainage channel between two morainic arcs indicates that the oldest Late-glacial advance occurred before 13.1 ka BP and the youngest Late-glacial advance occurred prior to 9.3 ka BP (Wenzens, 1999). Late-glacial moraines have been mapped enclosing the two branches of Lago San Martín and enclosing both shores of the central lake branch. They were interpreted as representing a third Late-glacial re-advance at ca. 10.5–9.5 ^{14}C ka BP (Wenzens, 2003).

Further east, in the extra-Andean tableland region, several morainic arc and spillway evidence have been mapped. They are placed on the tablelands, in higher positions than those of the valley floor. They were interpreted as indicating glaciations of Miocene age (Wenzens, 2006), herein mapped as pre-GPG events and undetermined age moraines for those of middle topographic position. Moreover, northern Lago Cardiel ancient glacial limits and spillways were recognised by Wenzens (2006) as evidence of the eastern frontal position of very ancient piedmont glaciers.

51.2.5. The Lago Viedma Lobe Area (49°24′–49°54′S; 70°46′–72°50′W)

A set of morainic arcs surrounding the head of the lake and further south, in the Río Guanaco valley were recognised by Wenzens (1999) as marking the position of ice during the LGM limit. An extensive drumlin and megaflute field occurs in the valley bottom between the LGM moraines and the lake heads. Late-glacial ice limits have been identified along the north and south coast of Lago Viedma, the Cóndor and the Guanaco valleys. Based on radiocarbon dating, Wenzens (1999, 2005) proposed that these glaciers advanced three times between 14 and 10 ka BP (Table 51.2), the younger probably in partially equivalent to the Younger Dryas Stadial (12.9–11.7 ka BP).

Pre-GPG glacial evidence, stratigraphically related to lava flows, has been pointed out in detail in Coronato et al. (2004a,b). Around 76 km east of the LGM limit, pre-GPG relict moraines have been recognised by Wenzens (2000) (Table 51.1). Moraines of undetermined ages are also herein mapped based on Wenzens (1999).

51.2.6. The Lago Argentino Lobe (50°09′–50°28′S; 70°53′–72°08′W)

Several different terminal moraines have been mapped by different authors along a 130 km-long transect, indicating the occurrence of at least nine separate glacial advances since the Late Pliocene to the Weichselian Late-glacial (Coronato et al., 2004a, and Rabassa, 2008, for a complete reference list). Since then, new published work has focused on the pre-GPG glaciations and the Late-glacial advances. Singer et al. (2004b) performed palaeomagnetic analyses and obtained ^{40}Ar/^{39}Ar and unspiked K–Ar dates from the Cerro del Fraile sequence, where eight till units interbedded with lava flows are exposed. Piedmont glaciations of the tableland region occurred in the period 2.1–1.0 Ma, herein considered as pre-GPG events. These glaciations could not be mapped as morainic arcs in the present map, but a violet dot indicates its location, south of Lago Argentino.

In relation to the Late-glacial advances, Strelin and Malagnino (2000) proposed that three events, termed as Punta Bandera I (the oldest) to III (the youngest), occurred on the northern and southern coast of Lago Argentino. Later, Strelin and Denton (2005) offered refined radiocarbon dating of this sequence and proposed that the three glacial advances occurred between 11.7 and 10.3 ka BP and were interpreted as representing responses to the southern hemisphere Late-glacial cooling recorded in Antarctica. Becker et al. (2005) obtained cosmogenic isotope ages suggesting that the moraines were deposited in several phases between 15.4 and 11.9 ka. By contrast, cosmogenic isotope studies on the most prominent moraine boulders indicate younger ages. Based on ^{10}Be and ^{36}Cl measurements of 16 boulders, Becker et al. (2005) obtained a weighted mean age of 11.1 ka for the maximum Late-glacial advance to the Puerto Banderas moraines, thus implying that the moraines are younger than the Antarctic Cold Reversal.

In the Cachorro glacial valley, a southern ancient ice lobe which joined the Lago Argentino palaeoglacier, a sequence of four frontal morainic suites has been mapped by Lovecchio et al. (2008). Although no dating was available, they were assigned to the Punta Banderas sequence based on geomorphological characteristics. These moraines were tentatively shown on the accompanying maps because of problems of poor recognition in the Digital Terrain Models used for mapping.

51.2.7. The Río Coyle Valley Lobe (50°49′–51°16′S; 70°56′–72°36′W)

Five glacial limits were recognised based on geomorphology along the upstream valley and the heads of the Sarmiento-del Toro lakes (Coronato et al., 2004a, for a complete list of previous authors). Pre-GPG moraines were assumed to be equivalent to those at Lago Argentino (Strelin et al., 1999), and both ancient morainic systems are located on the tableland region, east of the mountains and at ca. 400 m a.s.l. No further work has been undertaken on the Quaternary glacial processes in this area.

51.2.8. The Río Gallegos Valley Lobe (51°27′–52°03′S; 70°37′–72°26′W)

The upper stream of the Río Gallegos valley flowed along the Late and Middle Pleistocene moraines, but the younger morainic arcs are located in the Chilean lakes Balmaceda and Pinto, close to the Última Esperanza sound in the Pacific Ocean. The older glacial landscape has been mapped here by several authors (Coronato et al., 2004a, for complete reference list), but Meglioli (1992), Ton-That et al. (1999) and Singer et al. (2004a) constrained the age of the GPG event. Dates obtained on lava flows by ^{39}Ar–^{40}Ar incremental heating have yielded an age of 1.16 Ma (Singer et al., 2004a) for these features. The GPG has been named here the Estancia Bella Vista Glaciation. Moraines and huge erratic boulders occur on top of Miocene sedimentary rocks in the northern side of the valleys at 290–300 m a.s.l. Morainic arcs developed to the north-east, although they are assumed to be of 'Initioglacial' age (Caldenius, 1932). Their absolute age is still unknown; they could belong to the GPG, as well as to a pre-GGP glaciation, because the latter occurs in the Lago Argentino and Río Coyle valleys to the north. Here, they are mapped as moraines of undetermined age.

Early Pleistocene drumlins, megaflutes and drumlinoid forms have been described by Ercolano et al. (2004) in the ancient Río Gallegos valley resulting from subglacial deposition during the GPG advance (1.1–1.05 Ma). The dry climate of the region since the Early Pleistocene and the fact that younger glaciations never overrode the GPG have contributed to the excellent preservation of these glacial landforms.

^{10}Be measurements on boulders placed on top of the GPG moraines were performed by Kaplan et al. (2007). They found discrepancies between the cosmogenic ages of 106–124 ka (without erosion and with erosion correction, respectively) and the well-dated tills interbedded with lava flows. As it also occurred further south, in northern Tierra del Fuego moraines, these authors postulated that cosmogenic nuclide dating cannot be used reliably as a chronometer of glacial deposits older than the LGM in these areas, as is the case in the Lago Buenos Aires area (Kaplan et al., 2005; Douglass et al., 2006). They hypothesised that Quaternary geomorphological processes operated differently towards the south, mainly due to wetter climate and sea proximity during interglacial times (Kaplan et al., 2007).

Two post-GPG glacial limits have been mapped by Caldenius (1932) and Meglioli (1992) in the region, although no detailed chronological constraint is yet available for these deposits. Their assumed age is presented in Table 51.1.

According to Sagredo et al. (personal communication), a piedmont ice lobe, formed by the coalescence of outlet glaciers flowing from the South Patagonian Icefield (SPI), covered the head of this valley during the LGM, after which the Balmaceda and Pinto lakes were formed. Based on geomorphology, stratigraphy and radiocarbon dating, these authors proposed that the LGM termination began shortly before 17.5 cal. ka BP, in phase with the Lago Buenos Aires sequence (Kaplan et al., 2004) and the Magellan Straits area (McCulloch et al., 2005a,b). Deglaciation was interrupted by a stabilisation phase which ended at 16.3 cal. ka BP, similar to in the Lago Buenos Aires area (Douglass et al., 2006), and the intermediate Lago Pinto complex was formed. A final advance was radiocarbon dated to between 14.6 and 12.8 cal. ka BP in the Última Esperanza sound area. This late cool episode was attributed to the Antarctic Cold Reversal in phase with the sequence at Buenos Aires and Argentino lakes and the Magellan Straits glacial advances.

51.2.9. The Seno Skyring Lobe (52°08′S–71°07′W; 52°48′S–72°40′W)

Four glacial advances have been recognised along this palaeo-ice lobe (Caldenius, 1932; Meglioli, 1992, among others), although their proposed chronology was not coincident. The GPG was not recognised as morainic systems, but till remnants were observed in the higher hills. The accompanying map shows the limits of the three Middle Pleistocene glaciations recognised in detail by Meglioli (1992). Its chronostratigraphical relationships are shown in Table 51.1. No further information has been presented from this area since that work. By contrast, Kilian et al. (2007) proposed a Late-glacial/Holocene ice retreat based on geomorphology, sedimentology, seismic, pollen studies, tephrochronology and radiocarbon dating. A non-dated LGM limit (Moraine A) is coincident with the morainic arcs mapped as post-GPG II and III by Meglioli (1992), but the morainic limit closing the present sound (Moraines B and C) is coincident with Meglioli's LGM position. From this limit, the beginning of the ice retreat was identified between 18.3 and 17.5 cal. ka BP. Then, a more rapid glacier retreat occurred until around 15–14 cal. ka BP. Much slower retreat then took place from 14 to 11 cal. ka BP during both the Antarctic Cold Reversal and the Younger Dryas stadials. Moraines D and E occur over 20 km to the west. Late Holocene glacial advances have also been recognised between AD 1220 and 1910 and mapped as F Moraines. These authors postulate a distinct reaction, the Andean palaeo-mountain ice sheet (the Gran Campo Nevado icefield) being distinctly more climate sensitive here of than the Southern Patagonian Icefield.

51.2.10. The Seno Otway Lobe (52°03′S–70°11′W; 73°12′S–71°54′W)

The glacial limits of this palaeo-ice lobe presented here follow those of Caldenius (1932) and Meglioli (1992) for the Middle Pleistocene Glaciations. They are termed as post-GPG I, II and III by Coronato et al. (2004a) (Table 51.1). New information was provided by Kilian et al. (2007) who recognised the same morainic limits presented in northern Skyring sound. They have defined the B and C Moraine suites as the limit of the present sound head. Because no submerged moraines were found by using echo-sounding profiles between the north-eastern shoreline and the small islands in the south-western section, it is assumed that the D Moraines are located very close to the Andes and are probably submerged. The glaciological behaviour of the palaeoglacier that occupied this sound was interpreted in a similar manner to those of the Skyring and Magellan palaeoglaciers.

51.2.11. The Magellan Straits Lobe (51°46′S–69°12′W; 53°34′S–70°32′W)

The glacial history of the largest palaeoglacier in southern South America has been described and mapped beginning with the GPG (1.1–1.05 Ma), followed by three Middle Pleistocene glaciations (post-GPG I to III, after Coronato et al., 2004a), the LGM and three Late-glacial advances (Clapperton et al., 1995; McCulloch et al., 2005a,b). Details of pre-LGM glacial advances have been fully described in Coronato et al. (2004a,b).

Recently, it has been proposed that the GPG was not a single glacial advance (Bockheim et al., 2009). Sand-wedge casts developed in two till units in the Tres de Enero site (51°49′S, 69°24′W), south-west of Río Gallegos, previously recognised as part of the de los Frailes drift (Meglioli, 1992) or 'Initioglacial' (Caldenius, 1932), reveal that this extensive glacial advance would have occurred during different stadials or even in two different glacial events (Table 51.1).

The tundra environment, with strong aeolian cold conditions, allowed the development of sand wedges in a lower basal till unit; after erosional processes cut their upper portions, a new glacial advance deposited the upper basal till unit in which a new set of sand wedges was also formed. This indicates that not one but at least four strong cold climatic periods have occurred close to the present Atlantic Ocean coast 1 million years ago or earlier, two of them in the form of true glacial advances and two as the development of tundra environments.

Till, erratic boulders and glaciofluvial deposits have been recognised 20 km offshore on the present Atlantic Ocean continental shelf and the present easternmost outlet of the Magellan Straits, by oceanographic and seismic studies performed by Mouzo (2005a,b). Pleistocene till and outwash deposits forming submerged tablelands gently sloping to the east were recognised 20 km offshore on Tertiary rocks. Also, a south-eastwards-trending submerged glacial valley, 125 km long, was identified as the ancient outwash channel (see dashed line in the Magellan Straits east mouth in Fig. 51.1). Although chronological control is still not available for these submerged glacial deposits, it could be possible that they belong to different glaciations. Nevertheless, based on this evidence, it can be postulated that sometime in the Pleistocene, the Magellan palaeoglacier was part of an extensive piedmont glacier that covered southern Patagonia and northern Tierra del Fuego. During recessional phases, the main outwash channel terminated close to the southern Bahía San Sebastián palaeoglacier.

On the western side of the Magellan Straits (Chile), new detailed mapping focused on the LGM and Late-glacial landforms and palaeoglacier limits was published by Bentley et al. (2005) and McCulloch et al. (2005a). Based on geomorphology, radiocarbon dating and ^{10}Be concentration measurements on glacial boulders, four glacial limits described the LGM and its recession, and one limit was referred to the last, Middle Pleistocene Glaciation or post-GPG III (following Coronato et al., 2004a) (Table 51.2).

The LGM limits are coincident with those mapped by previous authors, but the chronology has been constrained. McCulloch et al. (2005a) proposed that the LGM limit (stage B) occurred after 31.2 cal. ka BP and culminated at 25.2–23.1 cal. ka BP but was followed by a less extensive advance (stage C) before 22.4–20.3 cal. ka BP. A considerably less extensive Stage D culminated before 17.7–17.6 cal. ka BP. Subsequently a rapid and widespread glacial retreat occurred. The last advance (E) dammed glacial lakes and spanned 15.5–11.7 cal. ka BP coinciding with the Antarctic Cold Reversal. At the end of this phase, the lake drained catastrophically to the Pacific Ocean but between the LGM limit and the previous one (the post-GPG III limit) a terrestrial valley existed. The present marine environment was fully installed during the Early Holocene times, ca. 8.2–7.4 ^{14}C ka BP (McCulloch et al., 2005b).

51.3. THE FUEGIAN PALAEO-ICE-LOBES

The Isla Grande de Tierra del Fuego is the main island of the Fuegian Archipelago. Its separation from continental South America is the consequence of glacial overdeeping of major Andean faults during the Pleistocene glaciations and the later marine flooding of glacial troughs during the Holocene. Quaternary glaciations affected this island along the Magellan Straits, the Bahía Inútil–Bahía San Sebastián depression, the Lago Fagnano, the Carbajal valley, the Beagle Channel and small tributary or hanging valleys.

The map accompanying this chapter clearly shows that five main glaciations have been mapped in the northern area but only the last two were recognised in the mountainous southern area. Since the overview by Coronato et al. (2004b), some new work has been done by several authors, but gaps already exist north and south of the Río Grande region (the 'Driftless area', *sensu* Meglioli, 1992) and in the eastern Atlantic coast.

51.3.1. The Inútil–San Sebastián Bays Lobe

The classic glacial model, proposed by Caldenius (1932) and later refined by Meglioli (1992), is presented on the accompanying map. Apart from this, several new contributions have appeared. Submerged till, glaciofluvial deposits and erratic boulders have been recognised by seismic investigations and coring in front of the Bahía San Sebastián coast (Mouzo, 2005a,b). He interpreted this glacial evidence as the result of a huge piedmont glacier which covered the area during the Early Pleistocene, following Caldenius' work. The easternmost till position was found 30 km offshore, whereas boulders occurred at 90 km. He denies the existence of the two frontal morainic sets identified by Isla and Schnack (1995), arguing that they reflect both structural control and megadunes development. Exploratory studies of palaeomagnetism on well-exposed basal till along Punta Sinaí cliffs were performed by Walther et al. (2007). They proposed a normal, Brunhes-age polarity for this till, thus providing the first chronological data for the post-GPG I moraines (*sensu* Coronato et al., 2004b) in Tierra del Fuego (Table 51.1). Based on the stratigraphical position of paraglacial fans and raised marine beaches, Bujalesky et al. (2001) postulated that the Sierras de San Sebastián and Lagunas Secas glacial limits should correspond to MIS 10 and 6, respectively.

An extensive erratic boulder field overlying this moraine and along the cliff shore was interpreted as rock avalanche supraglacial debris produced sometime along the southern margin of the palaeoglacier when it flowed from the Cordillera Darwin, where the allochthonous boulder lithology occurs (Coronato et al., 1999).

Boulders belonging to post-GPG I and II moraines were sampled for cosmogenic isotopes measurements in order to obtain a chronostratigraphical constraint on the glaciations on the southern margin of Bahía San Sebastián. The data obtained indicate that glacial erratics have a much shorter exposure period than the assumed age of the landforms in which they are found (Kaplan et al., 2007). The ages obtained relate the clasts to the LGM, which reached its frontal position 70 km to the west. Intensive weathering and complex geomorphological processes throughout the Pleistocene have been proposed to explain the discrepancy between assumed and obtained ages (Kaplan et al., 2007). A recent paper by Evenson et al. (2009) also presented extremely young cosmogenic ages for these boulders confirming the discrepancy between boulder exposure and landform age. According with Meglioli (1992) and Coronato et al. (1999), the ages explain the concentration of boulders as a supraglacial rock avalanche in the glacier accumulation area.

McCulloch et al. (2005a,b) studied the western side of this depression, the Bahía Inútil region. Here they proposed the same glacial model developed for the Magellan Straits during the LGM and Late-glacial times (Table 51.2). As the palaeoglaciers which covered these areas belong to the same outlet glacier, the glaciological behaviour was understood under the same framework conditions. The D-phase moraines are marked by stillstand episodes which built up three distinct limits defined by moraines and meltwater channels (Bentley et al., 2005). The glacial lobe here had retreated into the Whiteside Channel sometime before ca. 17.5–16.6 cal. ka BP (McCulloch et al., 2005a). A minimum age for the proglacial lake Stage E provided by a basal peat bog radiocarbon date in Isla Dawson refers to 10.3–10 ^{14}C ka BP, but the lake phase is thought to have persisted until 7.2 ^{14}C ka BP. Glacial stages D and E demonstrate that the glacial behaviour here was again controlled by the Antarctic Cold Reversal before the complete disappearance of the ice.

51.3.2. The Lago Lynch and Lago Blanco Lobes (53°42′–54°09′S; 68°27′–69°41′W)

These are small lakes located in the Andean footslope of the Chilean region of Tierra del Fuego, north of Lake Fagnano. The glacial morphology presented here was firstly proposed by Meglioli (1992) and included in the map by Coronato et al. (2004b). Because of the lakes and terminal moraine emplacement direction, it was assumed that the area was glaciated by different glaciers flowing from the northern slope of the Fuegian Andes range. Terminal morainic arcs close the head of the lakes but also ancient terminal topography is placed northwards close to the foot of isolated tablelands and hills. These frontal moraines were interpreted as belonging to the GPG and post-GPG I glaciations (*sensu* Coronato et al., 2004b; according to Meglioli, 1992).

This was based on comparison with the glacial morphology and sedimentary characteristics of till along the Bahía Inútil–Bahía San Sebastián lobe. Neither updated detailed surveys nor absolute numerical ages are available for this region.

51.3.3. The Lago Fagnano Lobe (54°09′–54°39′S; 66°49′–68°43′W)

The Lago Fagnano, placed along a transform plate boundary in central Tierra del Fuego, was one of the main glacial axes along which ice flowed eastwards from the Darwin Cordillera mountain ice sheet. The LGM and the pre-LGM limits have been recognised in this region by previous authors (full reference list in Coronato et al., 2004b). The accompanying map shows the LGM Late-glacial frontal positions along the Lago Fagnano and San Pablo, Fuego and Ewan valleys, based on Coronato et al. (2002, 2008b,c, 2009) and Waldmann et al., 2010. The post-GPG III limit is only presented in the Río Fuego valley because it is the clearest evidence in the region.

Recent surveys have demonstrated that the Fagnano glacial outlet was fed by almost 50 alpine-type glaciers flowing from the northern and southern mountain ranges. Close to the ablation zone, it was joined laterally by alpine glaciers flowing from the southern mountains (Coronato et al., 2009). The most conspicuous moraines are developed at the head of the lakes and were defined as the Tolhuin moraines. These features did not develop at the maximum extent of the LGM, but during the first re-advance during an equivalent of Stage C in the Bahía Inútil lobe (after McCulloch et al., 2005a). An ancient proglacial lake was also reported in this area by Coronato et al. (2002). Beyond this proglacial environment, an outwash plain with kettles and lateral moraines enclose the valley. Frontal moraines are not well developed, perhaps due to a very low gradient of the ice at its terminus (Coronato et al., 2009). The Lateglacial advances have been found in Lago Fagnano by seismic and coring activities. Waldmann et al. (2010) have recognised a series of five elongated crested topographic highs by seismic stratigraphy and interpreted them as either a sequence of lateral moraines or as medial moraines forming by the junction of the Fagnano palaeoglacier flowing from the west with other glaciers advancing from the southern mountains. A standstill position, forming a proglacial lake with calving during the recession to the west, was interpreted here by Waldmann et al. (2008) from seismic stratigraphy. Towards the west, the terminal moraines and semicircular crested structures have been recognised by seismic studies and interpreted as the last glacial re-advances or standstills before complete ice disintegration. Although the chronological evidence is sparse, the advancing phases of the Lago Fagnano palaeoglacier were compared to Stages D and E in the Bahía Inútil lobe and are again attributed to the influence of the Antarctic Cold Reversal (Waldmann et al., 2010). Table 51.2 shows a preliminary chronostratigraphical model based on the available data.

A glacial lobe flowed from the Fagnano palaeoglacier towards the north and north-east, over the Fuego and Ewan valleys. Along the Río Fuego valley, four morainic arcs have been distinguished. A thermoluminescence age of 25.7 ka was obtained in the latero-frontal LGM moraine in the Río Fuego valley (Coronato et al., 2008b, Table 51.2). The moraine was correlated with Stage B in the Bahía Inútil lobe (following McCulloch et al., 2005a). The uppermost morainic arc, which surrounds Lago Yehuin, was interpreted as a Stage C moraine (Coronato et al., 2008b).

Three morainic arcs have been recognised in the Río Ewan valley, the two outermost were interpreted as belonging to the LGM and the inner, that surrounds Lago Chepelmut, as the Stage C moraine (Coronato et al., 2008c), based on topographical, morphological and sedimentological characteristics. A basal radiocarbon date of 9.2 ka BP, obtained for a peat bog developed into the second morainic arc (Coronato et al., 2008c), reveals that ice was far from the area during the early Holocene. No other datable evidence has been found to constrain the chronology of the glacial advances in the area.

51.3.4. The Beagle Channel Lobe (54°50′S; 66°29′–68°33′W)

The LGM and pre-LGM boundaries (here termed the post-GPG III, after Coronato et al., 2004b) have been indicated on the accompanying map. No further papers have been published on the area to date. However, Antonione (2006, unpublished) surveyed the Carbajal-Harberton drumlin field in the Estancia Harberton area. He distinguished drumlin types based on their sedimentary composition as rock drumlins, basal till-cored drumlins and glaciofluvial and glaciolacustrine sedimentary-core drumlins and postulated that the drumlin-field emplacement was conditioned by the joining of the palaeo-Beagle glacier with the tributary Carbajal glacier in the widest section of the trough with horst and graben tectonic behaviour.

Recently, Rabassa et al. (2008) have identified the location of marine deposits on the southern coast of the Beagle Channel (Chile). Tree trunks and marine shells from here have been radiocarbon dated to 41.7 ^{14}C ka BP and also older than 46.1 ^{14}C ka BP. Although the dates are close to the confidence limit of the radiocarbon method, they clearly indicate that the marine environment in which they were deposited did not belong to the Holocene Marine Transgression, which is well known in the area (Bujalesky et al., 2004). A probable Sangamonian age (last interglacial) was suggested for these deposits but how they were preserved from the subsequent LGM advance is not yet certain.

51.4. FINAL COMMENTS

During recent years, glacial geology and palaeoclimatic changes have been intensively researched in the eastern flank of the Andes and in Tierra del Fuego. Much attention has been focused in the Early Pleistocene glaciations. However, the timing of glacier advances around the time of the LGM and re-advance phases during the Late-glacial has been improved due to the application of combination of methods such as radiocarbon dating, cosmogenic isotope analyses, $^{40}Ar/^{39}Ar$ dating of lava flows, sedimentology, tephrochronology and palynology. In addition to improved geochronology, the geomorphological knowledge has been expanded, mainly by the application of Digital Terrain Models as mapping tools. The latter offer much improved precision for regional mapping. Seismic, coring and stratigraphical analyses in sediments performed in glacial lakes and fjords are also contributing to the understanding of the Late-glacial climate and glacier re-advances of the main palaeoglacier lobes.

Different glaciological behaviour in the Patagonian Pleistocene ice sheets has been recognised through palaeo-ice-modelling, and the Northern, Southern, Gran Campo Nevado and Cordillera Darwin icefields have been described as the main sources of palaeoglaciers. Moreover, the timing and how and where they separated from each other are still under research.

The cold stages, MIS 32–2 are almost all represented by glacial morphology or palaeoenvironmental records in the region; only MIS 4 is less well understood or represented. No full evidence of this cold period has been found so far in most of the Patagonian glacial valleys. This has become an important paramount research challenge.

The concept that the natural system was not forced by the same mechanisms all around Patagonia is being increasingly accepted by the scientific community. In this way, some efforts have been achieved towards understanding the effects and interplay of both the weathering/geomorphological processes and the atmospheric dynamics.

The scientific question mostly actively debated today is whether the Antarctic Cold Reversal forced the main Late-glacial cooling in Southern Patagonia, while the Younger Dryas Stadial event affected the northern Patagonian Andes instead. In addition, the shifting of the westerlies during the Pleistocene, as the key process for humidity-snow delivery and the increase of glacial accumulation areas, is one of the targets of the present palaeoenvironmental studies being developed in the region.

Patagonia could be considered one of the best regions in the Southern Hemisphere where Quaternary palaeoclimates can be studied and interpreted, but much further multidisciplinary and intra-regional research is still required to achieve a full understanding of the problem.

ACKNOWLEDGEMENTS

This chapter is humbly devoted to the bi-centenary of Charles Darwin's birthday, who made the first geological observations in Patagonia. We also devote this chapter to Carl Caldenius, the Swedish geologist who was the first to map Patagonian glaciations in extraordinary detail 75 years ago and tried to correlate terminal moraines and ancient lakes with the past global climatic changes. We are deeply grateful to the two reviewers' suggestions which largely improved a preliminary version of this chapter.

REFERENCES

Antonione, G., 2006. La genesi dei drumlin Tardo Pleistocenici nell´ analisi geomorfologica di Estancia Harberton (Terra del Fuoco, Argentina). Degree Thesis, Facoltá di Scienze Matematiche, Fisiche e Naturali, Universitá degli Studio di Trieste, Italy. Unpublished.

Auer, V., 1956. The Pleistocene of Fuego-Patagonia. Part I: the Ice and Interglacial Ages. Ann. Acad. Sci. Fennicae Ser. A III Geol. Geogr. 45, 1–226.

Becker, R., Ackert, R., Singer, B., Douglass, D., Caffe, M., Kurz, M., Mickelson, D., Rabassa, J., 2005. ^{10}Be and ^{36}Cl Surface Exposure Age of Puerto Banderas Moraine, Lago Argentino, Argentina, 50°S. Geological Society of America, Abstracts with Programs, Washington, DC.

Bentley, M., Sugden, D., Hulton, N., McCulloch, R., 2005. The landforms and pattern of deglaciation in the Strait of Magellan and Bahía Inútil, southernmost South America. Geogr. Ann. 87A (2), 313–334. Oxford, Blackwell Publishing Ltd.

Bockheim, J., Coronato, A., Ponce, F., Ercolano, B., Rabassa, J., 2009. Relict sand wedges in southern Patagonia and their stratigraphic and paleoenvironmental significance. Quatern. Sci. Rev. 28 (13–14), 1188–1199.

Bujalesky, G., Coronato, A., Isla, F., 2001. Ambientes glacifluviales y litorales Cuaternarios de la región del Río Chico, Tierra del Fuego, Argentina. Rev. Asoc. Geol. Argentina 56 (1), 73–90.

Bujalesky, G., Coronato, A., Roig, C., Rabassa, J., 2004. Holocene differential tectonic movements along the Argentine sector of the Beagle Channel (Tierra del Fuego) inferred from marine palaeoenvironments. Geosur 2004 International Symposium. Boll. Geof. T.y. App. 45 (2), 235–238.

Caldenius, C., 1932. Las Glaciaciones Cuaternarias en Patagonia y Tierra del Fuego. Min. Agric. y Gan. de la Nac. Dir. Gen. Minas y Geol. 1–148, Buenos Aires.

Clapperton, C., 1993. Nature and environmental changes in South America at the Late Glacial Maximum. Palaeogeogr. Palaeoclimatol. Palaeoecol. 101, 189–208.

Clapperton, C., Sudgen, D., Kaufman, D., McCulloch, R., 1995. The last glaciation in Central Magellan Strait, Southernmost Chile. Quatern. Res. 44, 133–148.

Coronato, A., Rabassa, J., 2007a. Mid-Quaternary Glaciations in the Southern Hemisphere. In: Elias, S. (Ed.), Encyclopedia of Quaternary Science, vol. 2. Elsevier, Amsterdam, pp. 1051–1056.

Coronato, A., Rabassa, J., 2007b. Late Quaternary glaciations in South America. In: Elias, Scott (Ed.), Encyclopedia of Quaternary Science vol. 2. Elsevier, Amsterdam, pp. 1101–1108.

Coronato, A., Roig, C., Rabassa, J., Meglioli, A., 1999. Erratic boulder field of Pre-Illinoian age at Punta Sinaí, Tierra del Fuego, Southernmost South America. In: XV INQUA Congress, pp. 47–48. Durban, South Africa, Abstract.

Coronato, A., Roig, C., Mir, X., 2002. Geoformas glaciarias de la región oriental del Lago Fagnano, Tierra del Fuego, Argentina. In: Cabaleri, N., Cingolani, C., Linares, E., López de Luchi, M., Ostera, H., Panarello, H. (Eds.), XV Congreso Geológico Argentino CD-Rom, Article No. 24, 6pp. (El Calafate).

Coronato, A., Martínez, O., Rabassa, J., 2004a. Pleistocene glaciations in Argentine Patagonia, South America. In: Ehlers, J., Gibbard, P.L. (Eds.), Quaternary Glaciations—Extent and Chronology, Part III. Elsevier, Amsterdam, pp. 49–67.

Coronato, A., Meglioli, A., Rabassa, J., 2004b. Glaciations in the Magellan Straits and Tierra del Fuego, Southernmost South America. In: Ehlers, J., Gibbard, P.L. (Eds.), Quaternary Glaciations—Extent and Chronology, Part III. Elsevier, Amsterdam, pp. 45–48.

Coronato, A., Coronato, F., Mazzoni, E., Vázquez, M., 2008a. Physical geography of Patagonia and Tierra del Fuego. In: Rabassa, J. (Ed.), Late Cenozoic of Patagonia and Tierra del Fuego. Development in Quaternary Sciences, vols. 11, 3. Elsevier, Amsterdam, pp. 13–56.

Coronato, A., Ponce, F., Seppälä, M., Rabassa, J., 2008b. Englazamiento del valle del río Fuego durante el Pleistoceno tardío, Tierra del Fuego, Argentina. In: XVII Congreso Geológico Argentino, Actas, pp. 1194–1195, (San Salvador de Jujuy).

Coronato, A., Ponce, F., Rabassa, J., Seppälä, M., 2008c. Evidencias morfológicas del englazamiento del valle del río Ewan, Tierra del Fuego, Argentina. In: XVII Congreso Geológico Argentino, Actas, pp. 1196–1197, (San Salvador de Jujuy).

Coronato, A., Seppälä, M., Ponce, F., Rabassa, J., 2009. Glacial geomorphology of the Pleistocene Lake Fagnano ice lobe, Tierra del Fuego, southern South America. Geomorphology 112, 67–81.

Darwin, C., 1842. On the distribution of the erratic boulders and on the contemporaneous unstratified deposits of South America. Trans. Geol. Soc. Lond. 6, 415–431.

Douglass, D., Singer, B., Kaplan, M., Michelson, D., Caffee, M., 2006. Cosmogenic nuclide surface exposure dating on boulders on last-glacial and late glacial moraines and paleoclimatic implications. Quatern. Geochronol. 1, 43–58.

Ercolano, B., Mazzoni, E., Vázquez, M., Rabassa, J., 2004. Drumlins y formas drumlinoides del Pleistoceno Inferior en Patagonia Austral, Provincia de Santa Cruz. Rev. Asoc. Geol. Argentina 59 (84), 771–777, (Buenos Aires).

Evenson, E., Burkhart, P., Goose, J., Baker, G., Jacofsky, G., Meglioli, A., et al., 2009. Enigmatic boulder trains, rock avalanches and the origin of "Darwin Boulders", Tierra del Fuego. GSA Today 19 (12), 4–10.

Glasser, N., Ghiglione, M., 2009. Structural, tectonic and glaciological controls on the evolution of fjord landscapes. Geomorphology 105, 291–302.

Glasser, N., Harrison, S., Winchester, V., Aniya, M., 2004. Late Pleistocene and Holocene paleoclimate and glacier fluctuations in Patagonia. Global Planet. Change 43, 79–101.

Glasser, N., Jansson, C., Harrrinson, S., Kleman, J., 2008. The glacial geomorphology and Pleistocene history of South America between 38°S and 56°S. Quatern. Sci. Rev. 27, 365–390.

Hein, A., Hulton, N., Dunai, T., Scnabel, Ch., Kaplan, M., Naylor, M., et al., 2009. Middle Pleistocene glaciation in Patagonia dated by cosmogenic-nuclide measurements on outwash gravels. Earth Planet. Sci. Lett. 286, 184–197.

Hein, A., Hulton, N., Dunai, T., Sudgen, D., Kaplan, M., Xu, Sh., 2010. The chronology of the last Glacial Maximum and deglacial events in central Argentine Patagonia. Quatern. Sci. Rev. 29, 1212–1227.

Heusser, C., 1989. Climate and chronology of Antarctica and adjacent South America over the past 30,000 yr. Palaeogeogr. Palaeoclimatol. Palaeoecol. 76, 31–37.

Hubbart, A., Hein, A., Kaplan, M., Hulton, N., Glasser, N., 2005. A modelling reconstruction of the Last Glacial Maximum ice sheet and its deglaciation in the vecinity of the Northern Patagonian Icefield, South America. Geogr. Ann. 87A (2), 375–392.

Hulton, N., Purves, R., McCulloch, R., Sudgen, D., Bentley, M., 2002. The Last Glacial Maximum and deglaciation in southern South America. Quatern. Sci. Rev. 21, 233–241.

Isla, F., Schnack, E., 1995. Submerged moraines offshore northern Tierra del Fuego, Argentina. In: Quaternary of South America & Antarctic Peninsula, vol. 9. A.A. Balkema, Rotterdam, pp. 205–222.

Kaplan, M., Ackert, R., Singer, B., Douglass, D., Kurz, M., 2004. Cosmogenic nuclide chronology of millennial-scale glacial advances during the O-isotope stage 2 in Patagonia. Geol. Soc. Am. Bull. 116 (3), 308–321.

Kaplan, M., Douglass, D., Singer, B., Ackert, R., Mc Caffee, M., 2005. Cosmogenic nuclide chronology of pre-last glacial maximum moraines at Lago Buenos Aires, 46°S, Argentina. Quatern. Res. 63, 301–315.

Kaplan, M., Coronato, A., Hulton, N., Rabassa, J., Kubik, P., Freeman, S., 2007. Cosmogenic nuclide measurements in southernmost South America and implications for landscape change. Geomorphology 87, 284–301.

Kilian, R., Schneider, C., Koch, J., Fesq-martin, M., Biester, D., Casassa, G., et al., 2007. Paleoecological constraints on Late Glacial and Holocene ice retreat in the Southern Andes (53°S). Global Planet. Change 59, 49–66.

Lagabrielle, Y., Scalabrino, B., Suarez, M., Ritz, J., 2010. Mio-Pliocene glaciations of Central Patagonia: new evidence and tectonic implications. Andean Geol. 37 (2), 276–299.

Lisiecki, L.E., Raymo, M.E., 2005. A Pliocene- Pleistocene stack of 57 globally distributed benthic delta 18O records. Paleoceanography 20. PA 1003, 1–17.

Lovecchio, J., Strelin, J., Astini, A., 2008. Cronología de las morenas del Cachorro, Lago Argentino, provincia de Santa Cruz. In: Actas del XVII Congreso Geológico Argentino III, pp. 713–714, (San Salvador de Jujuy).

Lowell, T., Heusser, C., Andersen, B., Moreno, P., Hauser, A., Heusser, L., et al., 1995. Interhemispheric correlation of Late Pleistocene Glacial Events. Science 269, 1541–1549.

Malagnino, E., 1995. The discovery of the oldest extra-Andean glaciation in the Lago Buenos Aires basin, Argentina. Quaternary of South America & Antarctic Peninsula 9, A.A. Balkema Publishers, Rotterdam, 69–83.

Martínez, O., Coronato, A., Rabassa, J. 2011. Pleistocene Glaciations in Northern Patagonia, Argentina: An updated Review. Developments in Quaternary Science. 15, 729–734. Elsevier, Amsterdam.

McCulloch, R., Fogwill, C., Sudgen, M., Bentley, M., Kubik, P., 2005a. Chronology of the last glaciation in Central Strait of Magellan and Bahía Inútil. Geogr. Ann. 87A, 289–312.

McCulloch, R., Bentley, M., Tipping, M., Clapperton, C., 2005b. Evidences for late glacial dammed-lakes in the Central Straits of Magellan, southern Chile. Geogr. Ann. 87A, 335–362.

Meglioli, A., 1992. Glacial geology of Southernmost Patagonia, the Strait of Magellan and Northern Tierra del Fuego. Ph.D. Dissertation, Lehigh University, Bethlehem, USA, Unpublished.

Mercer, J., 1976. Glacial history of Southernmost South America. Quatern. Res. 6, 125–166.

Mouzo, F., 2005a. Estructura somera y cubierta sedimentaria Plio-Pleistocena en la Plataforma Continental al noreste de Tierra del Fuego. In: XVI Congreso Geológico Argentino, CD-Rom, Full Article.

Mouzo, F., 2005b. Límite de las glaciaciones Plio-Pleistocenas en la Plataforma Continental al noreste de Tierra del Fuego. In: XVI Congreso Geológico Argentino, CD-Rom, Full Article.

Rabassa, J., 2008. Late Cenozoic glaciations in Patagonia and Tierra del Fuego. In: Rabassa, J. (Ed.), Late Cenozoic of Patagonia and Tierra del Fuego. Developments in Quaternary Sciences, vol. 11. Elsevier, Amsterdam, pp. 151–204.

Rabassa, J., Clapperton, C., 1990. Quaternary glaciations of the Southern Andes. Quatern. Sci. Rev. 9, 153–174.

Rabassa, J., Coronato, A., 2009. Glaciations in Patagonia and Tierra del Fuego during the Ensenadan stage/age (Early Pleistocene–earliest Middle Pleistocene). Quatern. Int. 210, 18–36.

Rabassa, J., Coronato, A., Salemme, M., 2005. Chronology of the Late Cenozoic Patagonian Glaciations and their correlation with biostratigraphic units of the Pampean Region (Argentina). J. South Am. Earth Sci. 20 (1–2), 81–103.

Rabassa, J., Gordillo, S., Ocampo, C., Rivas Hurtado, P., 2008. The southernmost evidence for an interglacial transgression (Sangamon?) in South America. First record of upraised Pleistocene marine deposits in Isla Navarino (Beagle Channel, Southern Chile). Geol. Acta 6 (2), 251–258.

Sagredo, E., Moreno, P., Villa-Martínez, R., & Kaplan, M. (personal communication). Glacial fluctuations of the Seno Última Esperanza piedmont lobe during the last glacial–interglacial Transition, SW Patagonia, Chile (manuscript).

Schellmann, G., 1988. Jungkänozoische Landschaftsgeschichte Patagoniens (Argentinien). Andine Vorlandvergletscherungen, Talentwicklung und marine Terrasen. Ess. Geog. Arb. 29, 1–218.

Singer, B., Ackert, R., Guillou, H., 2004a. ^{40}Ar/^{39}Ar and K–Ar chronology of Pleistocene glaciations in Patagonia. Geol. Soc. Am. Bull. 116 (2), 434–450.

Singer, B., Brown, L., Rabassa, J., Guillou, H., 2004b. ^{40}Ar/^{39}Ar ages of Late Pliocene and Early Pleistocene geomagnetic and glacial events in Southern Argentina. AGU Geophysical Monograph "Timescales of the Internal Geomagnetic Field". pp. 175–190.

Strelin, J., Denton, G., 2005. Las morenas de Puerto Bandera, Lago Argentino. In: Actas del XVI Congreso Geológico Argentino, III, 129–134.

Strelin, J., Malagnino, E., 2000. Late glacial history of Lago Argentino, Argentina and age of the Puerto Banderas moraines. Quatern. Res. 54, 339–347.

Strelin, J., Re, G., Keller, R., Malagnino, E., 1999. New evidence concerning the Plio-Pleistocene landscape evolution of southern Santa Cruz region. J. South Am. Earth Sci. 12, 333–342.

Ton-That, T., Singer, B., Mörner, N., Rabassa, J., 1999. Datación de lavas basálticas por ^{40}Ar/^{39}Ar y geología glacial de la región del Lago Buenos Aires. Rev. Asoc. Geol. Argentina 54 (4), 333–352.

Turner, K., Fogwill, R., McCulloch, R., Sudgen, D., 2005. Deglaciation of the Eastern flank of the North Patagonian Icefield and associated continental scale lake diversions. Geogr. Ann. 87 (A), 363–374.

Waldmann, N., Ariztegui, D., Anselmetti, F., Austin, J., Dunbar, R., Moy, C., et al., 2008. Seismic stratigraphy of Lago Fagnano sediments (Tierra del Fuego, Argentina)—a potencial archive of paleoclimatic change and tectonic activity since the late glacial. Geol. Acta 6 (1), 101–110.

Waldmann, N., Ariztegui, D., Anselmetti, F., Coronato, A., Austin, J., 2010. Geophysical evidence of multiple glacier advances in Lago Fagnano (54°S, southernmost Patagonia) since the LGM. Quatern. Sci. Rev. 29 (9–10), 1188–1200.

Walther, A., Rabassa, J., Coronato, A., Tassone, A., Vilas, J.F., 2007. Paleomagnetic study of glacigenic sediments from Tierra del Fuego. In: Geosur 2007, Abstract volume, p. 174.

Wenzens, G., 1999. Fluctuations of outlet and valley glaciers in the Southern Andes (Argentina) during the past 13,000 years. Quatern. Res. 51, 238–247.

Wenzens, G., 2000. Pliocene piedmont glaciation in the Río Shehuen valley, Southwest Patagonia, Argentina. Arctic Antarct. Alpine Res. 32 (1), 46–54.

Wenzens, G., 2003. Comment on: "The Last Glacial Maximum and deglaciation in southern South America" Quatern. Sci. Rev. 22, 751–754.

Wenzens, G., 2005. Glacier advances east of the Southern Andes between the Last Glacial Maximum and 5000 BP compared with lake terraces of the endorrheic Lago Cardiel (49°S, Patagonia, Argentina). Z. Geomorphol. 49, 433–454.

Wenzens, G., 2006. Terminal moraines, outwash plains, and lake-terraces in the vicinity of Lago Cardiel (49°S, Patagonia, Argentina)—evidence of Miocene Andean Foreland Glaciations. Arctic Antarct. Alpine Res. 38 (2), 276–291.

Chapter 52

Pleistocene Glaciations in Northern Patagonia, Argentina: An Updated Review

Oscar Martínez[1,*], Andrea Coronato[2,3] and Jorge Rabassa[2,3]

[1] Universidad Nacional de la Patagonia San Juan Bosco, Sede Esquel, Ruta 259, km 16,5, (9200) Esquel, Chubut, Argentina
[2] CONICET-CADIC, B. Houssay 200, 9410 Ushuaia, Tierra del Fuego, Argentina
[3] Universidad Nacional de la Patagonia San Juan Bosco, Sede Ushuaia, Darwin y Canga, 9410 Ushuaia, Tierra del Fuego, Argentina
*Correspondence and requests for materials should be addressed to Oscar Martínez. E-mail: oscarm@unpata.edu.ar

52.1. INTRODUCTION

The area of northern Patagonia extends between latitude 39°30′S (central Neuquén Province) and latitude 46°S (the boundary between the Chubut and Santa Cruz provinces). The eastern limits of the mountain ice sheet, which formed here during each of the Patagonian Pleistocene glaciations, have been drawn according to the proposed positions by other authors who worked in the area. This chapter is also an update of Coronato et al. (2004). These contributions were adapted, when it was considered necessary, to new evidence from the analysis of terrain digital models (SRTM 90 m) and higher definition satellite imagery which are available nowadays. For the spatial and chronological ordering of the glacial limits, the boundaries of the Great Patagonian Glaciation (GPG; Mercer, 1976) have been used as the main references. The GPG occurred at around 1 Ma, in the Early Pleistocene (Ton-That et al., 1999; cf. Rabassa, 2008) in the glacial sequence of Lago Buenos Aires (latitude 46°30′S) and in the Río Gallegos Valley (52°S; Meglioli, 1992; Singer et al., 2004). Following Coronato et al. (2004), ancient glaciations were named as Pre-GPG, while the Middle Pleistocene Glaciations have been termed Post-GPG 1, 2 and 3 and, finally, the Last Glacial Maximum (LGM). Likewise, the glaciostratigraphical schemes, defined long ago for the Lago Nahuel Huapi area (latitude 41°S) and for the Esquel Valley (latitude 43°S) (Flint and Fidalgo, 1964, 1969), have been very important, since they provide a highly reliable mapping of the GPG moraines and/or glacial drifts. This is, thanks to the occurrence in the region of a regional drainage deepening event of tectonic nature, close to the Early Pleistocene/Middle Pleistocene boundary. This uplift event allowed the preservation of sediments and landforms of the older glaciations over the divides.

A maximum number of five major glacial events have been identified, some of which would be including more than one glacial advance. Although the occurrence of glaciogenic units older than the GPG cannot be ruled out (especially among the sequences found in the southern portion of the area), this group of moraines is that which is usually located in the easternmost positions, that is, at the greatest distance from the Andean mountain front. Not all these five glacial events are clearly represented in each of the glacial valley sequences described below, and in some cases, as a consequence of their younger age and better preservation, the Last Glaciation moraines are those that provide greater certainties concerning the local glaciostratigraphical scheme and the subsequent regional correlation.

52.2. THE RÍO MALLEO VALLEY (LATITUDE 39°37′30″S; LONGITUDE 71°17′W)

This sequence has been modified from Turner (1973), Rabassa et al. (1987, 1990) and Rabassa (1993, 2008).

The Río Malleo Valley begins in Lago Tromen and extends for over 40 km south-eastwards. Different drift units are irregularly distributed along the valley and have been ordered chronologically (Rabassa et al., 1987, 1990) by applying geomorphological, palynological and palaeomagnetic criteria. In particular, the chronology relies on the numerical dating of basalts that can be spatially related to the glacigenic deposits. Thus, it has been possible to correlate these glaciogenic units with different and successive positions of glacial still stand phases, as follows: Paso Tromen Drift/Puesto de Paja Drift, with the GPG (Early Pleistocene), San Huberto Drift with a pre-Penultimate Glaciation (ca. late Early or Middle Pleistocene), Criadero de Zorros Drift with the Penultimate Glaciation (late

Middle Pleistocene, ca. 0.2 Ma) and the Mamuil Malal Drift with the Last Glaciation (Late Pleistocene, 0.1–0.01 Ma).

52.3. HEADS OF THE RÍO LIMAY AND LAGO NAHUEL HUAPI (LATITUDE 41°8′S; LONGITUDE 71°8′W)

This sequence has been modified from Flint and Fidalgo (1964), Kodama et al. (1985, 1986), Rabassa and Evenson (1989, 1996) and Rabassa and Clapperton (1990). A general overview is given in Rabassa (2008).

The glacial stratigraphy defined for this region corresponds more closely to the glacial sequence established by Porter (1981) for the Chilean slope of the Andes at this latitude and to the global model of the marine isotope stratigraphy. Rabassa and Evenson (1996) suggested that the Pichileufú Drift (i.e. the GPG Drift), the oldest unit, includes the products of at least three different glacial advances which could correspond to several glaciations, all of which preceded a fluvial canyon cutting event during the Early Pleistocene (Rabassa and Clapperton, 1990). They proposed that the El Cóndor Drift could be subdivided into two glaciogenic units that would correspond to the Post-GPG 1 and Post-GPG 3 glaciations (both of pre-Late Pleistocene age). These glaciations also occur at different elevations, in relation to the 'Last Glaciation'. This fact results from tectonic uplift and/or fluvial incision intermediate between each glacial event, which demonstrates their relatively older ages. The Last Glaciation is represented by the Nahuel Huapi Drift, with its Nahuel Huapi I and II Stadials (Marine Isotope Stages (MIS) 2 and 4). No evidence of Late-glacial readvances has been confirmed in this region, based on absolute chronologies.

52.4. THE PRE-ANDEAN VALLEY OF EL MAITÉN (LATITUDE 42°10′S; –LONGITUDE 71°10′S)

This sequence has been modified from Miró (1967), Flint and Fidalgo (1969), González Díaz and Andrada de Palomera (1995, 1996) and González Díaz (1993a,b).

Several problems have been encountered in the establishment of a stratigraphical correlation between the glaciogenic deposits of this area because no absolute dates are available and because of the differing opinions of the authors who have worked in the region. These authors have neither followed a single stratigraphical scheme nor identified the same number of marginal moraines. The pre-Andean longitudinal valley of El Maitén has been successively occupied by Pleistocene glaciations. This important ice body was fed by two ice lobes, the Epuyén Valley Lobe (Fig. 52.1) and the Cholila Valley Lobe, which flowed from the east and south-east, respectively. However, it is possible that during the oldest glaciations, ice may have flowed from the north-western side. Smaller transverse valleys that cross the Sierra Oriental de El Maitén (Fig. 52.1) allowed the oldest ice sheet, corresponding to the GPG, to reach its maximum extent at around latitude 70°40′W. Immediately towards the west, the Post-GPG 1 deposits are found occurring as entrenched sedimentary bodies at lower topographical levels. These younger hills today act as water divides, separating the basins that drain towards the Pacific Ocean from those that run towards the Atlantic Ocean. At the latitude of Lago Epuyén, between this and the Río Chubut, the best-preserved moraines of the area are found (Fig. 52.1). They have been studied in detail by Miró (1967). They would correspond with the Post-GPG 2, Post-GPG 3 and the Last Glaciation advances.

52.5. VALLEY OF ESQUEL AND RÍO CORINTOS (LATITUDE 43°S; LONGITUDE 71°W)

These sequences have been modified from Andrada de Palomera and González Díaz (1996), Andrada de Palomera (2002), Martínez (2002) and Martínez (2005a,b).

The present 16 de Octubre Valley was the pathway regularly followed by successive ice advances from the southwest which occupied the mountain passes of Esquel, in the north, and Súnica, some kilometres towards the south-east, in the Río Corintos Valley. The easternmost moraines, which today occur as highly degraded landforms, reached their maximum extent at longitude 70° 49′W in the pre-Andean Valley of the Río Tecka. These deposits do not correspond with the oldest and largest mountain ice sheet of the GPG. Instead, the former ice sheet of that age is represented by moraines which occur further west, above 1000 m a.s.l. These moraines have been identified east of the Esquel Airport and along the mountain sides east and north of Laguna Súnica. Thus, the innermost and topographically highest morainic systems should be correlated with the GPG and the easternmost moraines, deeply entrenched in the valleys, would correspond to the Post-GPG 1 advance. The Post-GPG 2 moraine is very well preserved closing the northernmost extreme Esquel Valley. The Post-GPG 3 moraines (Penultimate Glaciation) are represented by three belts, over 5 km in length, that enclose Laguna Súnica, and by the moraines located west of the Esquel Airport. South-west of Laguna Súnica, a moraine 2 km long and the remnants of a second, more distal one that corresponds to the Last Glaciation event are found. The Esquel Valley is blocked at its eastern portion by an elongated morainic arc of the same age, although today it appears highly degraded by

FIGURE 52.1 Satellite image of the Sierra Oriental de El Maitén showing moraines and outwash plains (Martínez and Coronato, 2008).

glaciofluvial and glaciolacustrine erosion. Between these Last Glaciation moraines and those beyond, important terraced glaciofluvial deposits occur which reach the Río Tecka Valley. This stream and, further north, the Río Chubut acted as glaciofluvial drainage spillways towards the Atlantic Ocean, during at least the last three major glaciations. The moraines of the Penultimate and the Last Glaciation dammed the melt waters during ice recession. These lakes were interconnected and occupied a large part of the Cordilleran environment until they ultimately drained towards the Pacific Ocean when the water level overtopped the surface of the ice bodies located towards the west.

52.6. THE RÍO HUEMUL VALLEY (LATITUDE 43°30′S; LONGITUDE 71°10′W)

This sequence has been modified from Lapido et al. (1989), Haller et al. (2003), Haller (1979) and Martínez (2002, 2005a,b).

Lapido et al. (1990) studied the stratigraphy of the glaciogenic deposits that occur in the transverse valley east of the town of Corcovado and west of the Río Tecka Valley. They defined three glacial events for this area which consist of (i) isolated deposits that they named the 'Caquel Drift', that occurs along the mountain slopes above 1000 m a.s.l.;

(ii) the 'Tecka Drift', represented by two or more terminal moraines, located in the eastern margin of the Río Tecka; and (iii) the 'Mallín Grande Drift', equivalent to the Last Glaciation event. The latter is represented by two, well-preserved, morainic arcs with their corresponding glaciofluvial plains that extend eastwards and adjacent glaciolacustrine deposits towards the west. Recent studies (Martínez, 2002, 2005a,b) have suggested a probably similar age for the 'Caquel Drift' and the regional GPG event. The 'Tecka moraines' could be equivalent to the Post-GPG 1 advance and the 'Mallín Grande Drift' included the Post-GPG 3 and Last Glaciation deposits.

52.7. THE RÍO CORCOVADO VALLEY (LATITUDE 43°45′S; LONGITUDE 71°20′W) AND LAGO GENERAL VINTTER (LATITUDE 43°55′S; LONGITUDE 71°25′W)

These sequences have been modified from Haller (1979), Haller et al. (2003), Martínez (2002, 2005a,b) and Martínez et al. (2009).

The Lago General Vintter, which straddles the Argentina–Chile border, drains towards the Pacific Ocean through the Río Corcovado. In Argentinian territory, this stream wanders aimlessly between morainic arcs, before turning west, at latitude 43°30′S. A frontal moraine dams the lake (Fig. 52.2), and together with a second that occurs immediately to the east, it represents the Last Glaciation deposits (Martínez et al., 2009). Two very well-preserved morainic arcs, equivalent to these units and deposited by a glacier advancing from the north along the Río Corcovado Valley, are developed. Eastwards, highly eroded glacial deposits, corresponding to the three older systems, are found. The two westernmost ones occur between the Río Corcovado and the Río Putrachoique Valleys. They are considered to represent the Post-GPG 3, the Post-GPG 2 and Post-GPG 1 events, respectively.

52.8. THE RÍO PICO VALLEY (LATITUDE 44°10′S; LONGITUDE 71°20′W)

This local sequence has been modified from Beraza and Vilas (1989) and Lapido (2000).

Four end moraine groups were distinguished at the Río Pico Valley by Beraza and Vilas (1989). They determined that all these units showed normal polarity magnetisation and correspond to the Brunhes Chron (younger than 0.78 Ma). Thus, the different drifts termed as I (outer), II, III and IV (inner) are all younger than 0.78 Ma and Middle or Late Pleistocene in age. In a more extensive work, Lapido (2000) identified five drift units, two undifferentiated tills and terraced glaciofluvial deposits. An inner unit mapped by this author is the 'Las Mulas Drift' which is represented by dissected marginal and frontal moraines and glaciofluvial deposits that coincide with the 'Drift IV' and 'Drift III' of Beraza and Vilas (1989), equivalent to the LGM in this contribution. The 'Tremenhau Drift' is a more external morainic arc, and it coincides with 'Drift III' of Beraza and Vilas (1989) and is tentatively equivalent to the Penultimate Glaciation or Post-GPG 3 event. The 'Cherque Drift' occurs in an external position and comprises a complex and voluminous glaciogenic unit that probably includes deposits of several glaciations. There is good spatial superposition between these moraines and the 'Drift I' unit mapped by Beraza and Vilas (1989). The authors believe that it is appropriate to correlate this unit with the Antepenultimate Glaciation or the Post-GPG 2 event. Actually, the 'Drift I' includes another, even more external group of moraines, in a higher position, which are tentatively correlated to the Post-GPG 1 event. Finally, the 'Baguales Drift', of Pliocene–Pleistocene age (Lapido, 2000), which was not considered by Beraza and Vilas (1989), is represented by both morainic and glaciofluvial deposits that are found closing the valley at higher elevations (1300–1000 m a.s.l.). This drift, which usually appears as being strongly eroded, is tentatively correlated with the GPG but it probably includes older deposits.

52.9. THE RÍO APELEG VALLEY (LATITUDE 44°30′S; LONGITUDE 71°20′W)

This sequence has been modified from Ramos (1981), Ploszkiewicz (1987) and Lapido (2000).

The moraines that cross this valley were deposited by a glacial lobe derived from the Río Frías Valley (west to the present Argentina–Chile international border), when the ice divide was located further west. Most of the moraines corresponding to the Pleistocene glacial fronts are today found on Chilean territory. Although they may be clearly identified and precisely mapped in the images, it is not possible to define a glaciostratigraphical scheme because of the lack of specific studies in this area. Therefore, the glacial boundaries presented in this chapter are proposed based upon the spatial correlation with those sequences located immediately north and south, and they should be considered as tentative.

52.10. THE SEQUENCE OF THE LA PLATA AND FONTANA LAKES (LATITUDE 45°S; LONGITUDE 71°10′W)

This local sequence has been modified from Ramos (1981), Ploszkiewicz (1987) and Lapido (2000).

Initially, two glacial advances were described by Ramos (1981). The 'Río Moro Till' is represented by basal moraines that occur to the north and south of the lakes, at over 1000 m a.s.l. These moraines were deposited by a

FIGURE 52.2 Frontal moraines damming the lake of Lago General Vintter, which straddles the Argentina–Chile border.

mountain ice sheet the frontal position of which was further east, in the Post-GPG 3 or older advance limits. The last-glacial advance confined the lakes between morainic deposits. Ramos (1981) showed that the 'Last Glaciation' would be represented by the 'Fontana Till', deposited around the head of Lago Fontana and on its southern margins. The absence of specific studies in the area does not allow a more complete glacial map to be drawn.

52.11. THE EL COYTE (LATITUDE 45°15′S; LONGITUDE 71°15′W), HEADS OF THE RÍO MAYO (LATITUDE 45°30′S; LONGITUDE 71°15′W) AND LAGO BLANCO (LATITUDE 45°55′S; LONGITUDE 71°15′W) REGIONS

These sequences have been modified from Beltramone (1991) and Dal Molin and González Díaz (2002).

Three drift units, that indicate three glacial still stand positions, occur in the westernmost and more depressed portions of these valleys. Based mainly in what was proposed by Beltramone (1991), these units may be correlated with the Last Glaciation (La Elvira Drift), with the Penultimate Glaciation or Post-GPG 3 event (Río Mayo Drift) and with the pre-Penultimate Glaciation or Post-GPG 2 episode (Ricardo Rojas Drift). In addition, two outermost, more eroded and less continuous glacial fronts have been identified, which usually emerge above the glaciofluvial deposits of the glaciations already mentioned, which would correspond to the Post-GPG 1 and the GPG events, respectively. It should not be excluded that some of the drift deposits, included in this section or adjacent to these last units, would instead represent pre-GPG glaciations.

REFERENCES

Andrada de Palomera, R.P., 2002. Geomorfología del valle de Esquel y alrededores de las lagunas Willimanco, Zeta y Carao, noroeste del Chubut. In: XV Congreso Geológico Argentino El Calafate. CD-ROM. Artículo No. 052, 6pp.

Andrada de Palomera, R.P., González Díaz, E.F., 1996. Geomorfología de la zona comprendida entre las localidades de Leleque y Esquel, Noroeste del Chubut, Argentina. In: XIII Congreso Geológico Argentino and III Congreso de Exploración de Hidrocarburos, pp. 333–343. Actas IV, Buenos Aires.

Beltramone, C., 1991. Estratigrafía glacial del valle de Río Mayo, Provincia de Chubut, Argentina. In: Congreso Geológico Chileno, vol.1, pp. 58–61, Santiago, Chile. Resúmenes Expandidos.

Beraza, L., Vilas, J., 1989. Paleomagnetism and relative age from Pleistocenic end moraines at Río Pico valley, Patagonia, Argentina. In: International Geological Congress, vol. 28(1), pp. 128–129, Washington, DC.

Coronato, A., Martínez, O., Rabassa, J., 2004. Pleistocene glaciations in Argentine Patagonia, South America. In: Ehlers, J., Gibbard, P.L. (Eds.), Quaternary Glaciations—Extent and Chronology. Part III. Elsevier, Amsterdam, pp. 49–67.

Dal Molin, C.N., González Díaz, E.F., 2002. Geomorfología del área comprendida entre el Río Senguerr y el Lago Blanco, sudoeste de la Provincia del Chubut. In: Cabaleri, N., Cingolani, C.A., Linares, E., López de Luchi, M.G., Ostera, H.A., Panarello, H.O. (Eds.), Actas del XV Congreso Geológico Argentino CD-ROM, Artículo 284, El Calafate, 6pp.

Flint, R., Fidalgo, F., 1964. Glacial Geology of the East Flank of the Argentine Andes between Latitude 39°10′S and Latitude 41°20′S. Geol. Soc. Am. Bull. 75, 335–352.

Flint, R., Fidalgo, F., 1969. Glacial drift in the Eastern Argentine Andes between latitude 41°10′S and latitude 43°10′S. Geol. Soc. Am. Bull. 80, 1043–1082.

González Díaz, E.F., 1993a. Nuevas determinaciones y mayores precisiones en las localizaciones de los términos glaciarios del "Inicio" y "Daniglacial" en el sector de Cushamen (Chubut), Noroeste del Chubut. In: XII Congreso Geológico Argentino and II Congreso de Exploración de Hidrocarburos, pp. 48–55. Actas VI, Mendoza.

González Díaz, E.F., 1993b. Mapa Geomorfológico del Sector de Cushamen (NO de Chubut): reinterpretación genética y secuencial de sus principales geoformas. In: XII Congreso Geológico Argentino, pp. 56–65. Actas VI, Mendoza.

González Díaz, E.F., Andrada de Palomera, R.P., 1995. Los "sistemas de morenas terminales" de Caldenius al sur de la localidad de Ñorquinco, sudoeste de la Provincia de Río Negro. Rev. Asoc. Geol. Arg 50 (1–4), 212–218.

González Díaz, E.F., Andrada de Palomera, R.P., 1996. Nueva propuesta genética y evolutiva geomórfica de la "pampa" de Gualjaina, NO del Chubut extrandino. In: XIII Congreso Geológico Argentino and III Congreso de Exploración de Hidrocarburos, pp. 221–230. Actas IV, Buenos Aires.

Haller, M., 1979. Descripción Geológica de la Hoja 44 a-b, Trevelin, Provincia de Chubut. Servicio Geológico Minero, Buenos Aires unpublished report.

Haller, M., Lech, R.R., Martínez, O.A., Meister, C.M., Page, S.M., 2003. Hoja Geológica 4373IV/III, Trevelin, Provincia del Chubut. Programa Nacional de Cartas Geológicas de la República Argentina, 1:250.000. Servicio Geológico Nacional, Buenos Aires.

Kodama, K.P., Evenson, E.B., Clinch, J.M., Rabassa, J., 1985. Anomalous geomagnetic field behaviour recorded by glacial sediments from Northwestern Patagonia, Argentina. J. Geomagn. Geoelectr. 37, 1035–1050.

Kodama, K.P., Rabassa, J., Evenson, E.B., Clinch, J.M., 1986. Paleomagnetismo y edad relativa del Drift Pichileufú en su área tipo, San Carlos de Bariloche, Río Negro. Rev. Asoc. Geol. Arg. 41 (1–2), 165–178.

Lapido, O., 2000. Carta Geológica de la República Argentina, Escala 1:250.000. Gobernador Costa 4572-II/I. Edición Cartográfica PreliminarServicio Geológico Nacional, Buenos Aires.

Lapido, O., Beltramone, C., Haller, M., 1989. Glacial deposits on the Patagonian Cordillera at latitude 43°30′South. In: Rabassa, J. (Ed.), Quaternary of South America and Antarctic Peninsula vol. 6. A.A. Balkema, Rotterdam, pp. 257–266.

Lapido, O., Beltramone, C., Haller, M., 1990. Glacial deposits on the Patagonian Cordillera at latitude 43°30′ South. Quaternary of South America and Antartic Península. 6, 257–266.

Martínez, O., 2002. Geomorfología y geología de los depósitos glaciarios y periglaciarios de la región comprendida entre los 43° y 44° lat. Sur y 70°30′ y 72° long. Oeste, Chubut, República Argentina. Universidad Nacional de la Patagonia-San Juan Bosco, unpublished doctoral thesis, Comodoro Rivadavia and Esquel.

Martínez, O., 2005a. Incisión fluvial y glaciaciones durante el Pleistoceno a los 43°lat. S., noroeste de la Provincia de Chubut. In: XVI Congreso Geológico Argentino La Plata, abstracts.

Martínez, O., 2005b. Geomorfología y geología de los depósitos glaciarios y periglaciarios de la región comprendida entre los 43° y 44° lat. Sur y 70°30′ y 72° long. Oeste, Chubut, República Argentina. Naturalia Patagónica, extended abstract, Universidad Nacional de la Patagonia-San Juan Bosco, 2, 1, 108–112. Comodoro Rivadavia.

Martínez, O., Coronato, A., 2008. The Late Cenozoic fluvial deposits of Argentine Patagonia. In: Rabassa, J. (Ed.), The Late Cenozoic of Patagonia and Tierra del Fuego. Developments in Quaternary Sciences vol. 11. Elsevier, Amsterdam, pp. 205–226.

Martínez, O., Gosse, J., Yang, G., 2009. Las Morenas Frontales de la Ultima Glaciación en el Lago General Vintter (Provincia de Chubut): Edades Absolutas y su Correlación con las Secuencias Glacigénicas de la Región. Resumen. In: IV Congreso Argentino de Cuaternario y Geomorfología La Plata, Abstracts, 375pp.

Meglioli, A., 1992. Glacial geology of Southernmost Patagonia, the Strait of Magellan and Northern Tierra del Fuego. Ph.D. Dissertation, Lehigh University, Bethlehem, U.S.A. Unpublished.

Mercer, H., 1976. Glacial History of southernmost South America. Quatern. Res. 6, 125–166.

Miró, R., 1967. Geología glaciaria y preglaciaria del Valle de Epuyén. Rev. Asoc. Geol. Arg. 22 (3), 177–202.

Ploszkiewicz, J.V., 1987. Descripción Geológica de la Hoja 47c, Apeleg, Provincia de Chubut. Dirección Nacional de Minería y Geología, Boletín No. 204. Buenos Aires.

Porter, S.C., 1981. Pleistocene Glaciation in the southern Lake District of Chile. Quatern. Res. 16, 263–292.

Rabassa, J., 1993. Geología del Cuaternario del valle del río Malleo. Provincia de Neuquén. Réplica. Rev. Asoc. Geol. Arg. 47 (3), 344–346.

Rabassa, J., 2008. Late Cenozoic glaciations in Patagonia and Tierra del Fuego. In: Rabassa, J. (Ed.), Late Cenozoic of Patagonia and Tierra del Fuego. Developments in Quaternary Science, vol. 11. Elsevier, Amsterdam.

Rabassa, J., Clapperton, C.M., 1990. Quaternary glaciations of the Southern Andes. Quatern. Sci. Rev. 9, 153–174.

Rabassa, J., Evenson, E.B., 1989. Revisión de la estratigrafía glacial del área del Lago Nahuel Huapi. In: XI Congreso Geológico Argentino Resúmenes. San Juan.

Rabassa, J., Evenson, E.B., 1996. Reinterpretación de la estratigrafía glaciaria de la región de San Carlos de Bariloche (Prov. de Río Negro, Argentina). In: XIII Congreso Geológico Argentino y III Congreso de Exploración de Hidrocarburos, p. 237. Actas IV.

Rabassa, J., Evenson, E.B., Schlieder, G., Clinch, J.M., Stephens, G., Zeitler, P., 1987. Edad Pre-Pleistoceno superior de la Glaciación El Cóndor, valle del río Malleo, Neuquén, Argentina. In: X Congreso Geológico Argentino, San Miguel de Tucumán, 1987, pp. 261–263. Actas III.

Rabassa, J., Evenson, E.B., Clinch, J.M., Schlieder, G., Zeitler, P., Stephens, G., 1990. Geología del Cuaternario del valle del río Malleo, Provincia del Neuquén. Rev. Asoc. Geol. Arg. 45 (1–2), 55–68. Buenos Aires.

Ramos, V.A., 1981. Descripción geológica de la Hoja 47ab Lago Fontana. Serv. Geol. Nac., Bol. Boletín 183, Buenos Aires.

Singer, B., Ackert, R., Guillou, H., 2004. $^{40}Ar/^{39}Ar$ and K-Ar chronology of Pleistocene glaciations in Patagonia. Geol. Soc. Am. Bull. 116, 434–450.

Ton-That, T., Singer, B., Mörner, N., Rabassa, J., 1999. Datación de lavas basálticas por $^{40}Ar/^{39}Ar$ y geología glacial de la región del Lago Buenos Aires, provincia de Santa Cruz, Argentina. Rev. Asoc. Geol. Arg. 54 (4), 333–352.

Turner, J.C., 1973. Descripción Geológica de la Hoja 37a, b, Junín de los Andes. Provincia de Neuquén. Servicio Nacional Minero Geológico Boletín, No. 138. Buenos Aires.

Chapter 53

The High-Glacial (Last Glacial Maximum) Glacier Cover of the Aconcagua Group and Adjacent Massifs in the Mendoza Andes (South America) with a Closer Look at Further Empirical Evidence

Matthias Kuhle

Department of Geography and High Mountain Geomorphology, Geographisches Institut, University of Göttingen, Goldschmidtstr. 5, D-37077 Göttingen, Germany

53.1. INTRODUCTION

This volume offers the author an opportunity to present the empirical basis of his reconstruction of glacier areas in the southern Andes (Kuhle, 2004) in a more comprehensible way. This is provided by means of 77 photographs taken in the field which can be download from the website http://booksite.elsevier.com/9780444534477/. Thus the results of the glacial limits, which had already been summarised in 2004, are intended to be substantiated empirically so that further examples and key sites are more easily understood.

Fig. 53.1A (Kuhle, 2004, Fig. 1) is included again here. It shows the unchanged result of further detailed investigations presented here, that is, of the global Last Glacial Maximum (LGM in the terms of Ehlers and Gibbard, 2003)-glacier cover of the Aconcagua group and adjacent massifs in the Mendoza Andes (South America). The digital maps of the LGM glaciation included on the enclosed DVD naturally also remain unchanged.

The glaciation map in Fig. 53.1A is overlain on the same satellite image which indicates the localities and directions of the photographs in Fig. 53.1B. Accordingly, the empirical observations in the field, marked and explained in the photographs, are unambiguous with regard to their positions in reference to the reconstruction of the glacier areas. This record of the local observations and their interpretation allows cross-checking of the evidence and approval or rejection of the interpretations.

In any case, the observations made possible by the photographs not only allow 'a closer look' but also contribute to transparency as a precondition for objections or confirmations.

Despite the fact that a subtropical arid mountain area is concerned here, the climate of which is extremely hostile with regard to glaciation, a valley glaciation of the 'ice stream network' type has been reconstructed. The author thus asks the reader to exercise some understanding of the numerous panoramas.

53.2. RESULTS

High-Glacial (LGM) ice termini have been recognised up to a level below 1300 m since 1929 on the western slope of the Aconcagua massif (Brüggen, 1929). Now the author presents corresponding findings from the eastern slope. In the lower parts of the Rio de las Cuevas and Rio de las Vacas, a minimum thickness of the valley glaciers of 1020 m has been reconstructed by means of morainic material with erratics at points where, until 1984, former researchers had postulated the ends of the LGM glacier tongues (Fig. 53.1A, Profile I). The author has related these ice termini (Stages 1–5) to the Late-Glacial levels of the Last Glacial stage (i.e. Wisconsinan, Weichselian, etc., cf. Kuhle, 2004, Figs. 3 and 6) (cf. Fig. 53.1B). The largest thickness of pre-modern valley glaciers was reached in those regions where the glacier surface was on the level of the snow line (3380–3435 m a.s.l.). In the Aqua Relincho valley below the eastern flank of the Cerro Aconcagua, in the feeding area of the glacier system, the minimum thickness reached was 834 m (Fig. 53.1A, Profile II, Nos. 3, 2 and 1) (cf. Kuhle, 2004, Fig. 10, Nos. 3, 2 and 1; Figs. 6–9). Five Late-Glacial glacier levels have also been found on the western slope (Kuhle, 2004, Fig. 11). In the Valle del (Rio) Aconcagua (Fig. 53.1A), an ice thickness between 850 and 500 m has been reconstructed by polish lines and ground-moraines of the lowest distinct trough profile, 3.5–4 km upvalley from Guardia Vieja. This corresponds to the moraine 18.75 km downvalley as the matching High-Glacial deposit at 1220 m a.s.l. The lowest glacier margin lay at ~1220 m a.s.l. (on the western slope) in the Valle del (Rio) Aconcagua (Fig. 53.1A, cf. Kuhle, 2004, Fig. 1).

In the Rio Mendoza valley (E-slope), a glacier terminus has been reconstructed based on moraines with striated boulders at 1870 m a.s.l. (cf. Figs. 53.1A and 53.1B; Kuhle, 2004, Figs. 1, 2 and 12). In the eastern flank of the Cerros del Chacay, a glacier reached as far down as 2060 m a.s.l. (cf. Figs. 53.1A and 53.1B; Kuhle, 2004, Fig. 13). Thus, in the Cerro Aconcagua massif and the adjacent massifs, that is, the Cerro Tupungato and Cerro Juncal massif, an ice stream network (Fig. 53.1A) existed with partly up to 112.5-km-long ice streams and a vertical range of up to 5150 m. The climatic snow line (ELA) ran around 3200 m which amounts to a depression of 1400 m and a drop in annual temperature of ca. 8.4 °C, especially during the summer months.

ACKNOWLEDGEMENTS

The author is very grateful to the editors Jürgen Ehlers and Philip Gibbard for their reworking and correction of the text.

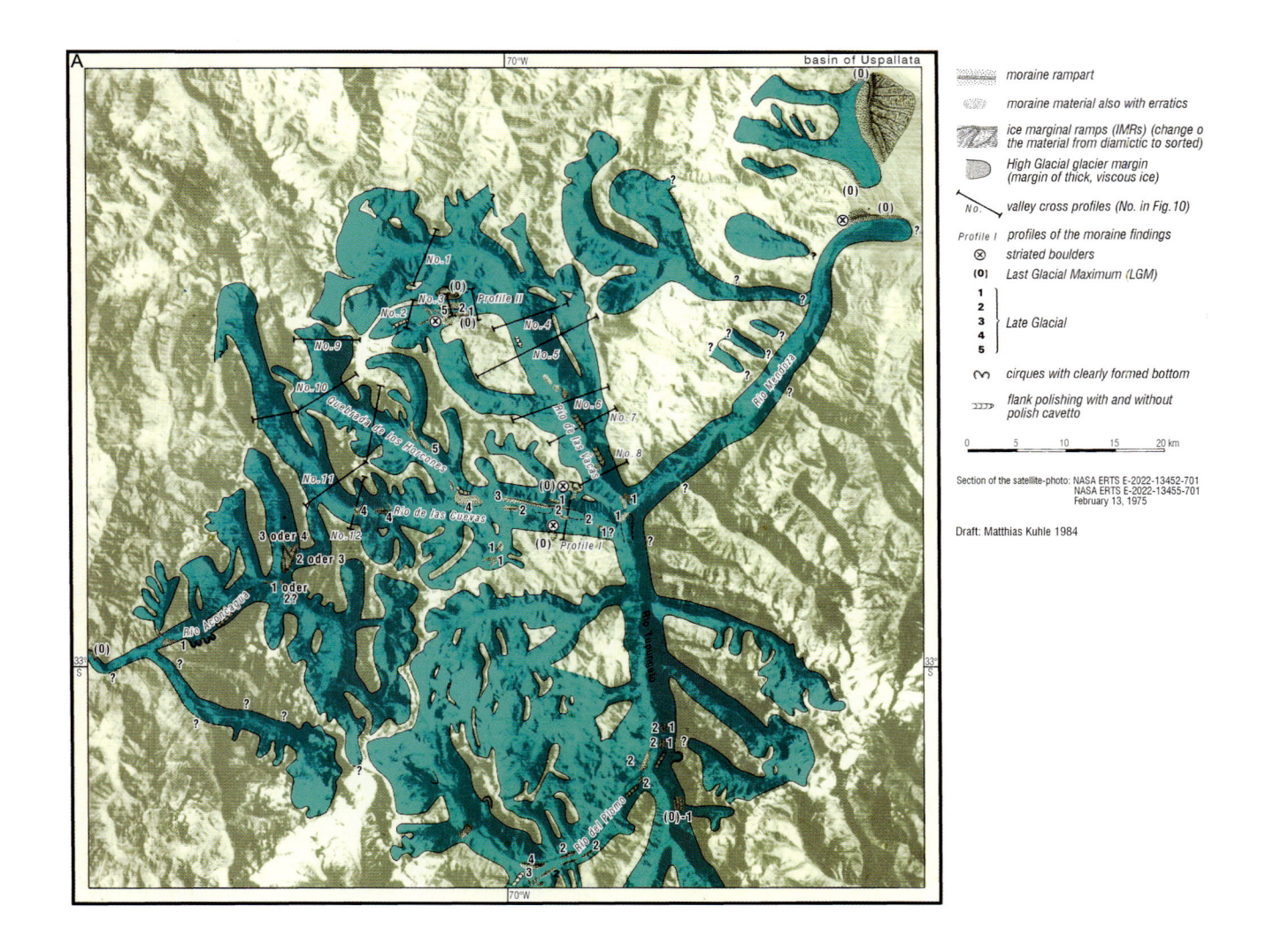

Chapter | 53   The High-Glacial (Last Glacial Maximum) Glacier Cover of the Aconcagua Group

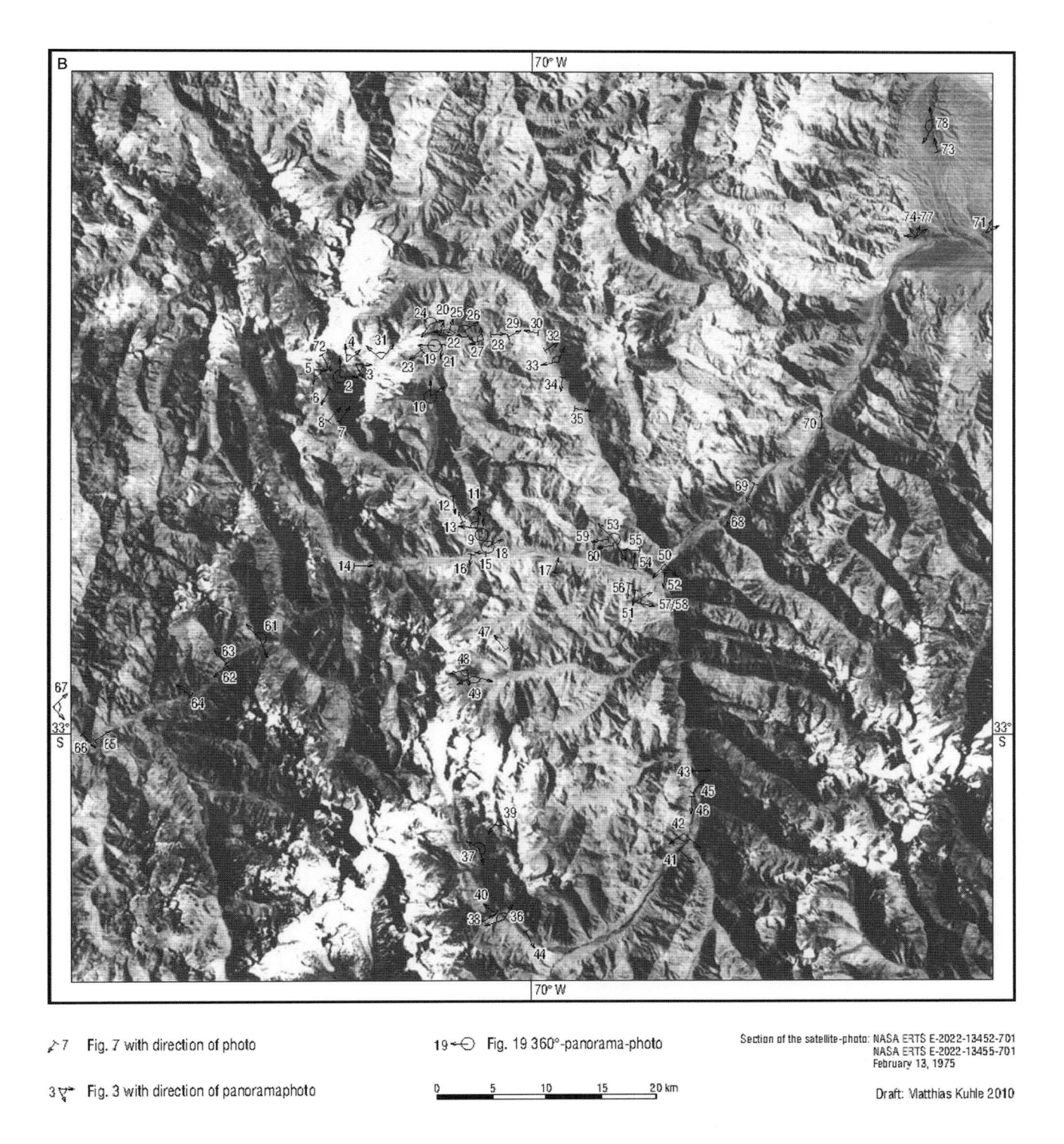

↗7 Fig. 7 with direction of photo

3▽ Fig. 3 with direction of panoramaphoto

19←◯ Fig. 19 360°-panorama-photo

Section of the satellite-photo: NASA ERTS E-2022-13452-701
NASA ERTS E-2022-13455-701
February 13, 1975

0   5   10   15   20 km

Draft: Matthias Kuhle 2010

FIGURE 53.1   (A) Satellite image map after Kuhle (2004), showing the global Last Glacial Maximum (LGM) ice cover of the Aconcagua group, Argentina. For valley cross profiles, see Kuhle (2004; Fig. 10). (B) Aconcagua group and adjacent massifs, Argentina: satellite image showing the locations of photographs 2–78 (download files from the website http://booksite.elsevier.com/9780444534477/).

REFERENCES

Brüggen, J., 1929. Zur Glazialgeologie der chilenischen Anden. Geol. Rundsch. 20, 1–35, Berlin.

Ehlers, J., Gibbard, P.L., 2003. Extent and chronology of glaciations. Quatern. Sci. Rev. 22, 1561–1568.

Kuhle, M., 2004. The Last Glacial Maximum (LGM) glacier cover of the Aconcagua group and adjacent massifs in the Mendoza Andes (South America). In: Ehlers, J., Gibbard, P.L. (Eds.), Quaternary Glaciation—Extent and Chronology. Part III: South America, Asia, Africa, Australia, Antarctica. Development in Quaternary Science, vol. 2c. Elsevier B.V., Amsterdam, pp. 75–81.

Chapter 54

The Pleistocene Glaciations of Chile

Stephan Harrison[1,]* and Neil F. Glasser[2]

[1]College of Life and Environmental Sciences, University of Exeter, Penryn, Cornwall, United Kingdom
[2]Institute of Geography and Earth Sciences, Aberystwyth University, United Kingdom
*Correspondence and requests for materials should be addressed to Stephan Harrison. E-mail: Stephan.harrison@exeter.ac.uk

54.1. INTRODUCTION

Reconstructing the timing and extent of Quaternary glaciations in Chile (and South America more generally) has the potential to shed light on the nature of inter-hemispheric climate changes during this period and the forcing factors which may have been in operation. Of special interest to palaeoclimatologists is assessments of the timing of local glacial maxima (local LGMs) and the status, extent and timing of rapid climate change events in the Late-Glacial such as the Younger Dryas Chronozone (YDC) and Antarctic Cold Reversal (ACR).

This chapter provides an overview of the Quaternary glaciations of Chile. An earlier review (Harrison, 2004) discussed the scientific research on Chilean glaciations up to that point; the present review will not revisit much of this earlier work but concentrate on the large amount of recent research that has been published, especially in the arid Andes of northern and central Chile and in southern Chile. The chapter builds on a number of important recent reviews of the Late Pleistocene and Holocene glacial history of South America, especially those provided by Rodbell et al. (2009) and Zech et al. (2009).

Over the past few decades, there has been an enormous expansion in the amount of research on the Quaternary history of Chile. Covering 38° of latitude and interacting with important climatic systems such as the southern westerlies and South American Summer Monsoon, Chile has assumed considerable importance as a location where the patterns and timing of Southern Hemisphere climatic events can be reconstructed and where models of global climate change can be tested.

The Andes form one of the world's great mountain ranges and the extreme altitudinal, latitudinal and climatic variability of the Chilean Andes provides the setting for a remarkable range of glacial environments. These include glaciers with low accumulation and discharge rates in high arid mountains glaciers to lacustrine and tidewater calving glaciers which are among the most dynamic on earth.

54.2. THE ANDES

In northern Chile (between 18°S and 30°S), precipitation is low (commonly, 200–300 mm a^{-1}), the contemporary equilibrium line altitude (ELA) is at high altitudes and glaciers and small ice caps exist only on volcanic summits lying above 6000 m a.s.l. (Ammann et al., 2001). In central Chile (defined as lying between 30°S and 40°S), the ELA falls below 4000 m associated with increased precipitation and higher seasonality, and glaciers are common on the higher peaks. South of 37°S, the average altitude of the Andes decreases sharply to ca. 2000 m a.s.l. and glaciers are only found on peaks rising above 4000 m. In southern Chile (between 40°S and 55°S), the average altitude of the Andes lies between 1500 and 3000 m. South to 46°S, the ELA falls to 1800 m a.s.l., reflecting still higher precipitation. Records of atmospheric circulation across southern South America show strong inter-annual, inter-decadal and inter-centennial variability (Glasser et al., 2004). Climatic changes in South America are associated directly or indirectly through long-term (Ma) mountain-range uplift (Hartley, 2003); atmospheric teleconnections, with large-scale atmospheric/oceanic forcing such as the El Niño-Southern Oscillation (ENSO) (Aceituno, 1988; Allan et al., 1995; Diaz and Markgraf, 2000); the temperature gradient between tropical and extra-tropical region; the sea surface temperatures of the South Atlantic and South Pacific Oceans, and the circum-Antarctic ocean circulation (Villalba et al., 1997; Lamy et al., 2004).

Below 46°S, the climate changes to cool maritime and here are located the two last major stores of temperate ice on earth. The North Patagonian Ice cap (NPI) lies between 46°30′S and 47°30′S and covers an area of 3900 km². Its outlet glaciers are among the most active in the world and include the lowest latitude tidewater calving glacier, Glaciar San Rafael. The larger South Patagonian Ice cap covers an area of some 14,000 km² between 48°30′S and 50°30′S. Both ice masses owe their form and extent to the operation of the southern westerlies, interacting with the Humboldt Current, whose vigorous circulation and high precipitation accounts for the high ablation and accumulation rates of the glaciers, their mass balance gradients and outlet velocities. Precipitation ranges from ca. 3000 mm per year on the western seaboard to 10,000 mm per year at 700 m a.s.l. on the ice caps (Aniya and Enomoto, 1986). A marked climatic gradient exists over the icefields, with precipitation totals decreasing markedly to the east of the topographic divide (Harrison, 2010). The location and dynamics of the southern westerlies are important components of the Southern Hemisphere climate, and changes in this system may provide information on the nature and timing of inter-hemispheric climate variations and linkages. The southern westerlies vary in response to changes in the pressure gradients between the subtropical high and subpolar systems. During Late Pleistocene glaciations, it is hypothesised that the position of the westerlies moved north to between 45°S and 50°S, creating the wet/cool conditions that allow glaciation of the Chilean Lake District at 41°S (Hubbard, 1997; Denton et al., 1999a,b; McCulloch et al., 2000). This movement of the pressure systems northwards reduced precipitation between 50°S and 55°S thus inhibiting glacier expansion during glaciations in those latitudes (Hulton et al., 1994). It is assumed that the westerlies returned to their present position during the Late-Glacial, and the reconstruction of their variations is possible through reconstructions of glacier fluctuations and palaeoecological records. However, Zech et al. (2008) follow (Markgraf et al., 1992) by suggesting that this view has been disputed, and these issues are discussed later.

54.3. METHODS

Geomorphological mapping has been widely used to establish former ice limits (e.g. Caldenius, 1932, recently updated and augmented by Glasser et al., 2008). A number of dating techniques have previously been applied to moraine systems in Chile and their associated glacigenic deposits to help reconstruct the Quaternary history of the country. These techniques include radiocarbon (^{14}C) dating of moraines, peat bogs and lacustrine deposits in Patagonia (e.g. Mercer, 1965, 1968, 1976; Aniya, 1995; Denton et al., 1999b; Hajdas et al., 2003; McCulloch et al., 2005a,b). However, in the dry Andes with few opportunities for radiocarbon dating of organic material, accurate dating of glacial deposits has only recently been achieved with the introduction of Cosmogenic Radionuclide Dating (CRN) to these regions (e.g. Zech et al., 2006, 2007, 2009). CRN dating has been used in Patagonia (e.g. Kaplan et al., 2004, 2005, 2007; Fogwill and Kubik, 2005; McCulloch et al., 2005a,b; Douglass et al., 2006). Recently, and for the first time in South America, OSL dating has been used to date glacier fluctuations and in combination with CRN to support glacier chronologies (Winchester et al., 2005; Glasser et al., 2006; Harrison et al., 2008). ^{40}Ar/^{39}Ar and K–Ar dating of lava flows interbedded with glacigenic and glaciofluvial deposits has been employed by several workers (Wenzens, 2000, 2006b; Singer et al., 2004). Numerical ice-sheet modelling experiments have also significantly increased our understanding of the extent and dimensions of the former ice sheets and their interaction with former climates (e.g. Hulton et al., 1994, 2002; Sugden et al., 2002; Hubbard et al., 2005).

54.4. THE MAXIMUM GLACIATIONS

While there has been much new research on the glacial history of Chile since 2004, most of it has concentrated (as before) on the Southern Andes, south of 35°. The sensitivity of the southern Andes to climate changes has meant that the most complete evidence for the Pleistocene maximum glaciation (occurring in the middle Pleistocene) has been found here. Much of this evidence comes from glacial sediments deposited to the east of the Andes. In southern Chile, the terminal positions of piedmont lobes from glaciers flowing to the west from the Andes at this time are located, in the main, below sea level.

The conditions suitable for the development of ice caps in southern Chile have probably existed for at least 14 Ma. Mercer (1983; cited in Clapperton, 1993) believed that ice sheets had existed in west Antarctica since the Late Miocene. These ice masses and their ice shelves produce cold, saline bottom water which surfaces as the cold Humboldt Current. This runs northwards up the Pacific coast of South America and cools the moisture-laden westerlies, creating the snow accumulations that nourish the Patagonian ice caps (Clapperton, 1993).

54.4.1. The Chilean Lake District

Early research in the Chilean Lake District between the Andes and the Pacific Ocean at 41–43°S identified four drift sheets on the basis of weathering characteristics of tills, clasts and moraine morphology (Porter, 1981). These drift sheets were taken to represent four glaciations and were assigned local names; the oldest to youngest being the Caracol, Rio Llico, Santa Maria and Llanquihue drifts (see Fig. 54.1). The oldest drift (Caracol) records a less extensive glaciation than does the Rio Llico and demonstrates that Andean glaciers in this sector flowed westwards as a continuous ice front. All the drifts pre-date the Last Glacial Maximum (LGM), although age data from them is patchy. Wood from the Santa Maria Till has given an age of $57.8 \pm 2.3/3.2$ ^{14}C ka BP, but Clapperton (1993) suggests that, in view of the possibility of contamination, it may be much older.

FIGURE 54.1 Data on Quaternary glacial extent in the western Chilean Lake District. The background is a pan-sharpened Landsat-7 image draped over a hillshade of the SRTM 90 m DEM. (A) Glacial geomorphology adapted from the mapping of Glasser et al. (2008). (B) Information on glacial extent from published dates, divided into those that provide maximum ages for glacier advance (crosses) and those that provide minimum ages for glacier recession (stars). The ages and sources of information are provided on the labels, which also indicate whether these represent a single date or the mean age of more than one date. This figure was prepared by Ingo Wolff.

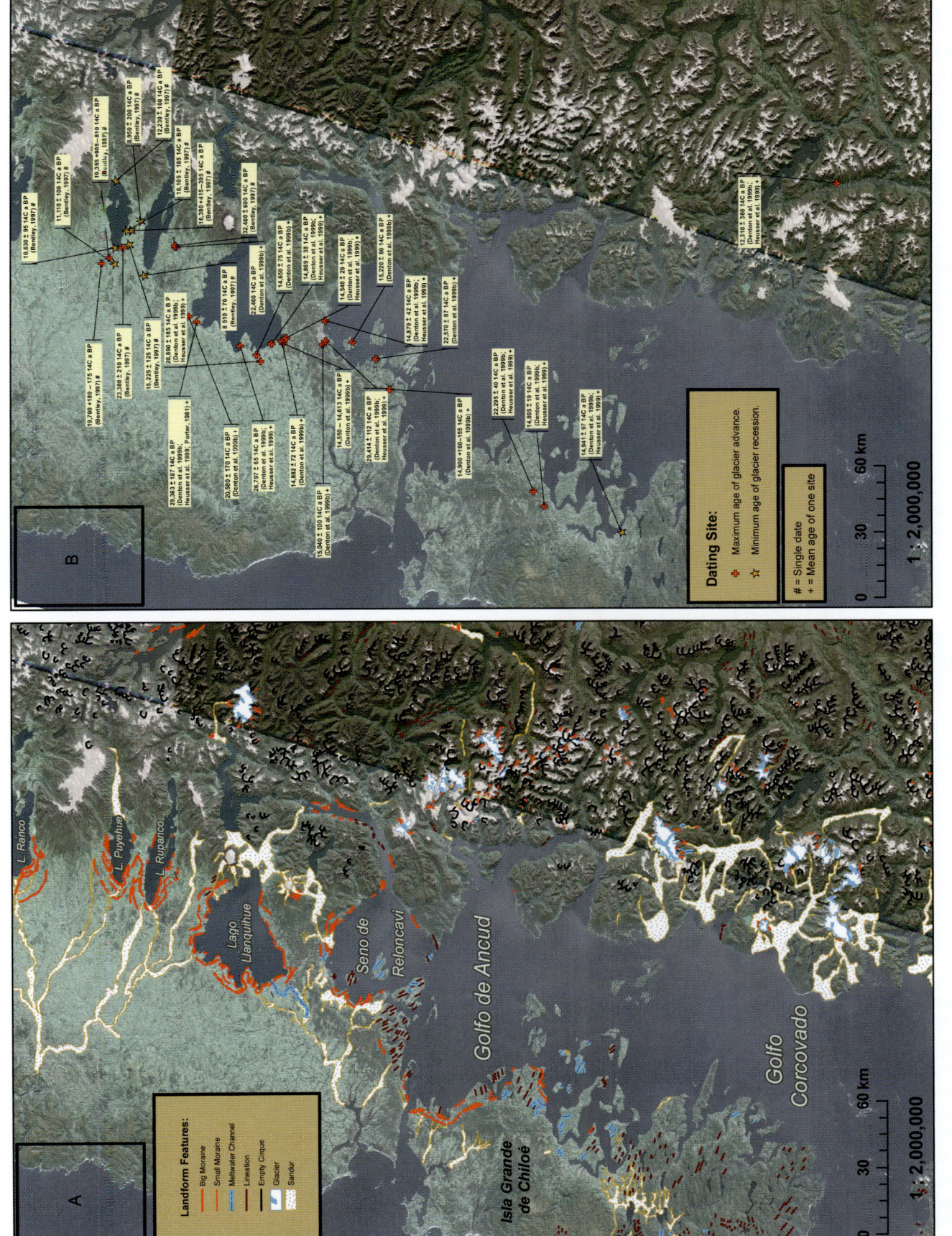

54.4.2. The Patagonian Icefields

Current understanding of the terrestrial extent of the former Patagonian Ice Sheets owes much to the pioneering study of Caldenius (1932) who first mapped moraine systems to the east of the contemporary icefields. Caldenius (1932) distinguished four separate moraine belts, and he concluded from their state of preservation that the three inner moraine systems were relatively young. In accordance with the stages of the last Weichselian Glaciation in northern Europe, Caldenius named the three moraine limits (from inner to outer) the 'Finiglacial', the 'Gotiglacial' and the 'Daniglacial'.

The Finiglacial moraines were correlated to the LGM and the Daniglacial and the Gotiglacial moraine systems to the middle Pleistocene (Mörner and Sylwan, 1989). The fourth (outermost) moraine system, termed the 'Initioglacial' by Caldenius (1932), is still poorly constrained in age, but is thought to be between 1.1 and 2.3 Ma in age (Mörner and Sylwan, 1989; Singer et al., 2004). Rabassa et al. (2005) have argued from a synthesis of records that the oldest known Patagonian glaciations took place between approximately 7 and 5 Ma but moraine systems from these glaciations are not preserved and their occurrence is inferred mainly on stratigraphic and sedimentological grounds. Rabassa et al. (2005) suggest that a minimum of eight glaciations occurred in the Middle–Late Pliocene (Marine Isotopic Stages 54–82). The 'Great Patagonian Glaciations' developed between 1.168 and 1.016 Ma (MIS 30–34; Early Pleistocene).

Outside the LGM Fenix moraines at Lago Buenos Aires (Fig. 54.2), the Moreno moraines have been dated to 140–150 ka (Kaplan et al., 2004). To the east and outside these are another set of moraines dated to < 1016 ka but > 760 ka (Singer et al., 2004). The maximum extent of glaciation is dated to ca. 1100 ka (Singer et al., 2004). These ages are contested, however; Wenzens (2006a) has argued on geomorphological grounds (mainly the relationship between the moraines and dated outwash terraces) that both the Fenix and Moreno moraines are in fact LGM in age.

54.4.3. The Magellan Straits

There is evidence for pre-LGM glacial advances in the Magellan Straits area and Bahía Inútil at Punta Dungeness, Punta Angostura and Segunda Angostura (see Fig. 54.3). The moraine belts are some 50 km apart and 10 km wide. Recent work has suggested that the Segunda Angostura moraine is a composite feature, possibly formed by three different glacier advances. Using relative age determinations based upon weathering of moraines and surface clasts, Porter (1989) correlated three sets of moraines on both sides of the Magellan Strait. Shells incorporated into the Primera Angostura moraine yielded infinite radiocarbon ages of greater than 47.0 ^{14}C ka BP (Clapperton, 1993). The advanced weathering of these moraines compared to the younger (probably Last Glaciation age) Segunda Angostura moraines suggests that they may be of pre-Last Glaciation age and perhaps are as old as 140 ka. On the basis of this estimate, Porter (1989) argued for an age of between 0.74 and 1.2 Ma for the moraines at Punta Dungeness. Since there is evidence of more than one glaciation at Segunda Angostura, Clapperton (1993, p. 357) believes 'that the drifts of possibly all major Quaternary glaciations are present along Magellan Strait'. Later work (Clapperton et al., 1995) argues that the maximum extent of glacial ice occurred during the early part of the last glacial cycle, possibly before 30,000 years ago.

54.5. THE EARLY/MIDDLE WEICHSELIAN/ WISCONSINAN GLACIATION

54.5.1. Dry Andes

In the dry Andes of northern and central Chile (30°S), Zech et al. (2006, 2007) have used CRN to date extensive glacial advances in the Cordon de Dona Rosa to 42/39 ka BP. The central Andes are an important region for Quaternary research, as they are situated in the transition zone between the tropical and mid-latitude atmospheric systems (Zech et al., 2006). However, until recently, the absence of organic material for dating has restricted the opportunities for glacier reconstruction (Ammann et al., 2001). Zech et al. (2006) used CRN on moraines dated to around 32 ka and speculate that these pre-LGM glacier advances were driven by increased precipitation in association with a northward movement of the southern westerlies in response to increased Antarctic sea ice extent (Mosola and Anderson, 2006).

54.5.2. The Chilean Lake District

In this region, large moraine complexes are common on the western sides of the major lakes (Fig. 54.1). Commonly three ridges and associated outwash deposits are found and have been regarded as Last Glaciation in age (Porter, 1981; Mercer, 1983); the largest and most extensive of these moraines are the outermost in the sequence. Glacial deposits and landforms assigned to Marine Isotope Stage (MIS) 4 and 3 have been assigned the chronostratigraphical term 'Llanquihue' by Clapperton (1993) since the moraine sequence in the vicinity of Lago Llanquihue in the Chilean Lakes District of southern Chile is the most closely dated sequence from the country. The stage Llanquihue I records glacier advances from 70 to 65 ka; Llanquihue II covers the period from 28 to 18 ka and Llanquihue III from 15 to 14 ka. The latter two advances are described in the next section. Radiocarbon dating of a log from peat deposits overlain by tills in the outermost moraine complex of

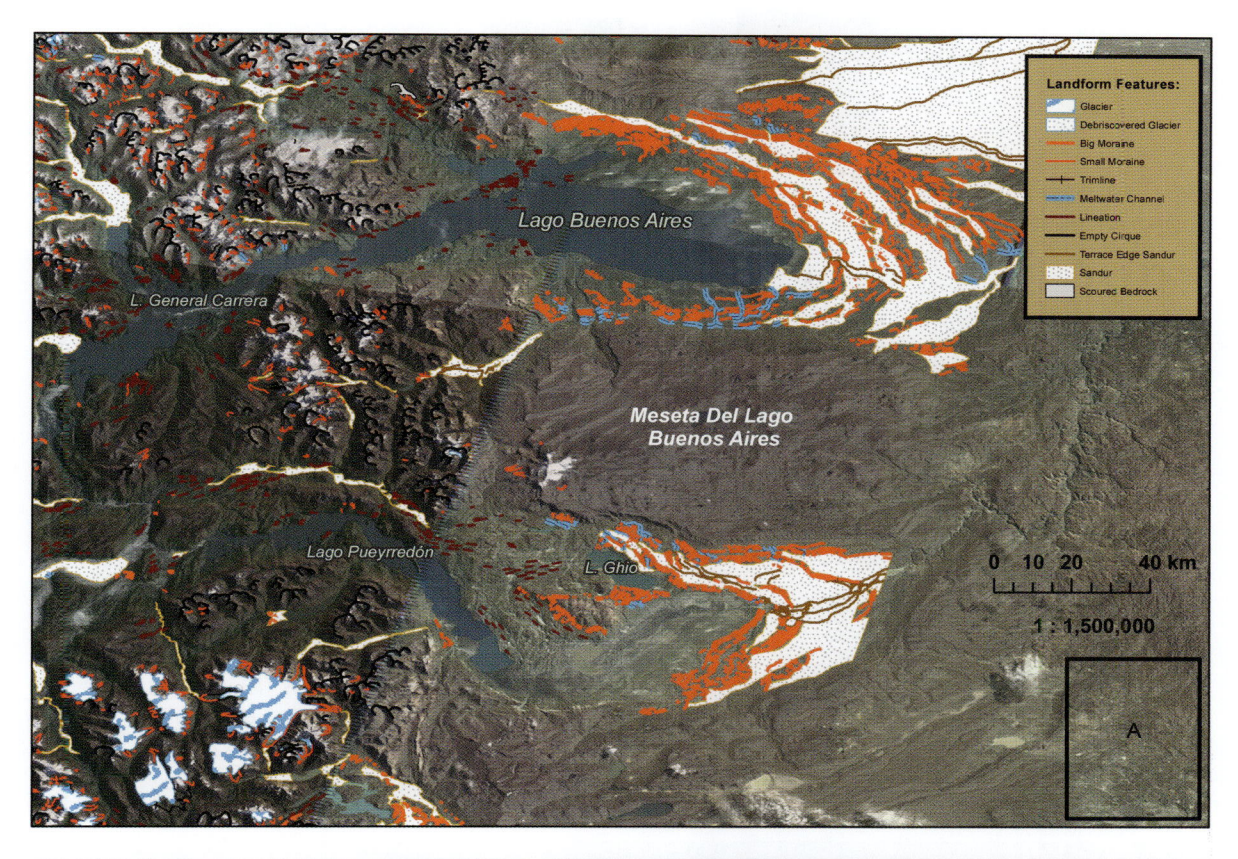

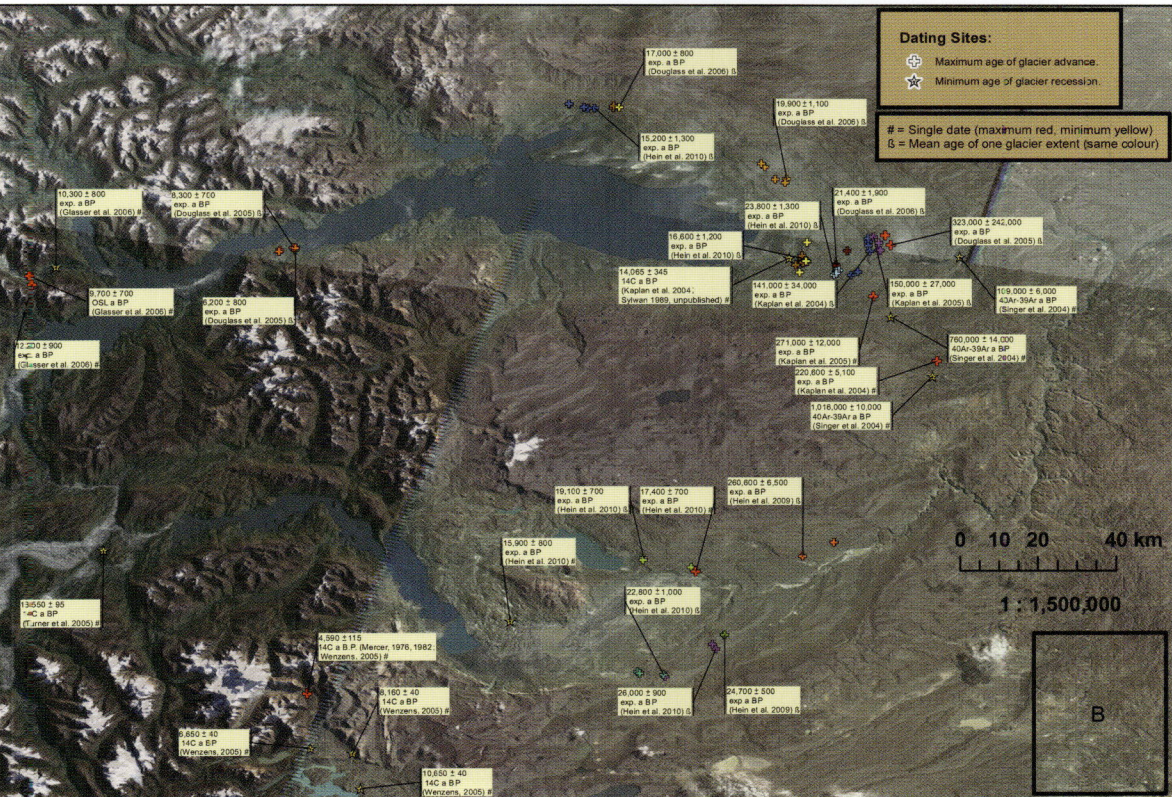

FIGURE 54.2 Data on Quaternary glacial extent in the Lago Buenos Aires–Lago Pueyrredon area. The background is a pan-sharpened Landsat-7 image draped over a hillshade of the SRTM 90 m DEM. (A) Glacial geomorphology adapted from the mapping of Glasser et al. (2008). (B) Information on glacial extent from published dates, divided into those that provide maximum ages for glacier advance (crosses) and those that provide minimum ages for glacier recession (stars). The ages and sources of information are provided on the labels, which also indicate whether these represent a single date or the mean age of more than one date. This figure was prepared by Ingo Wolff.

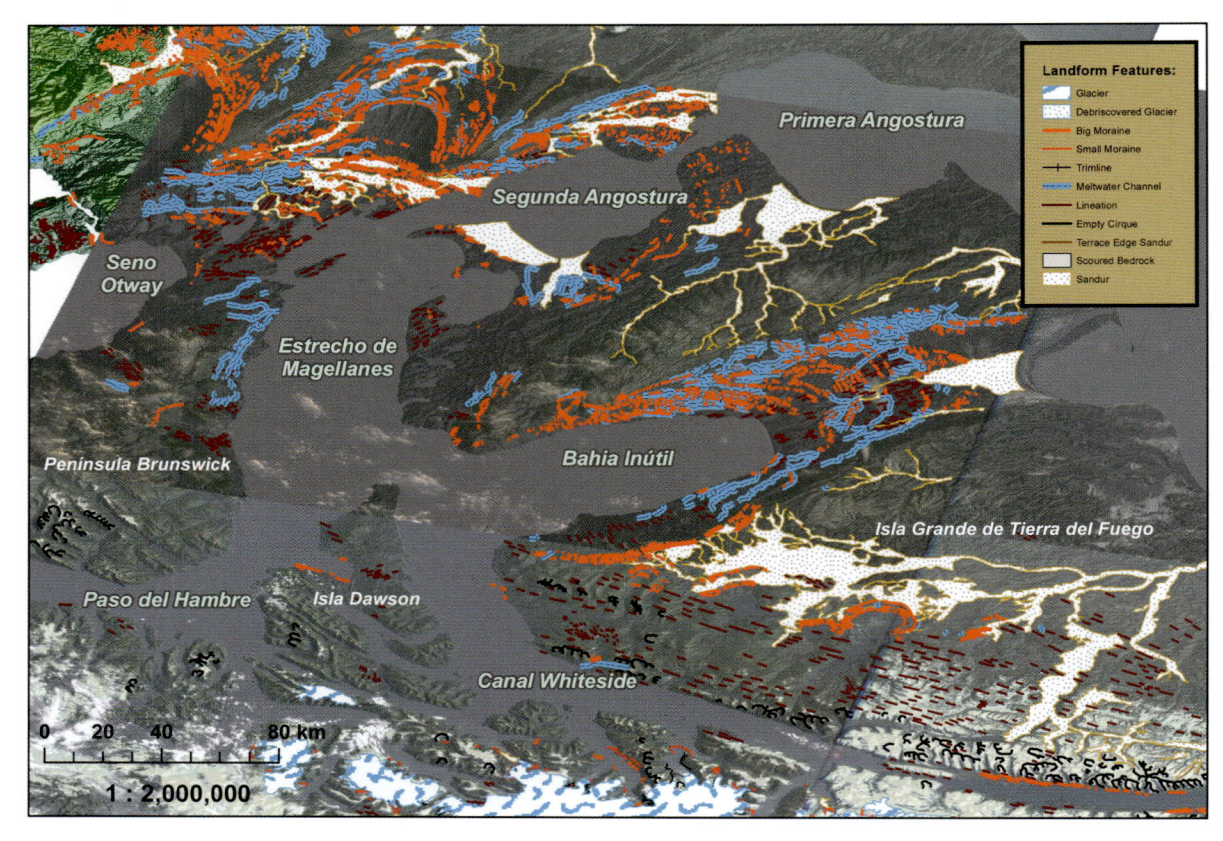

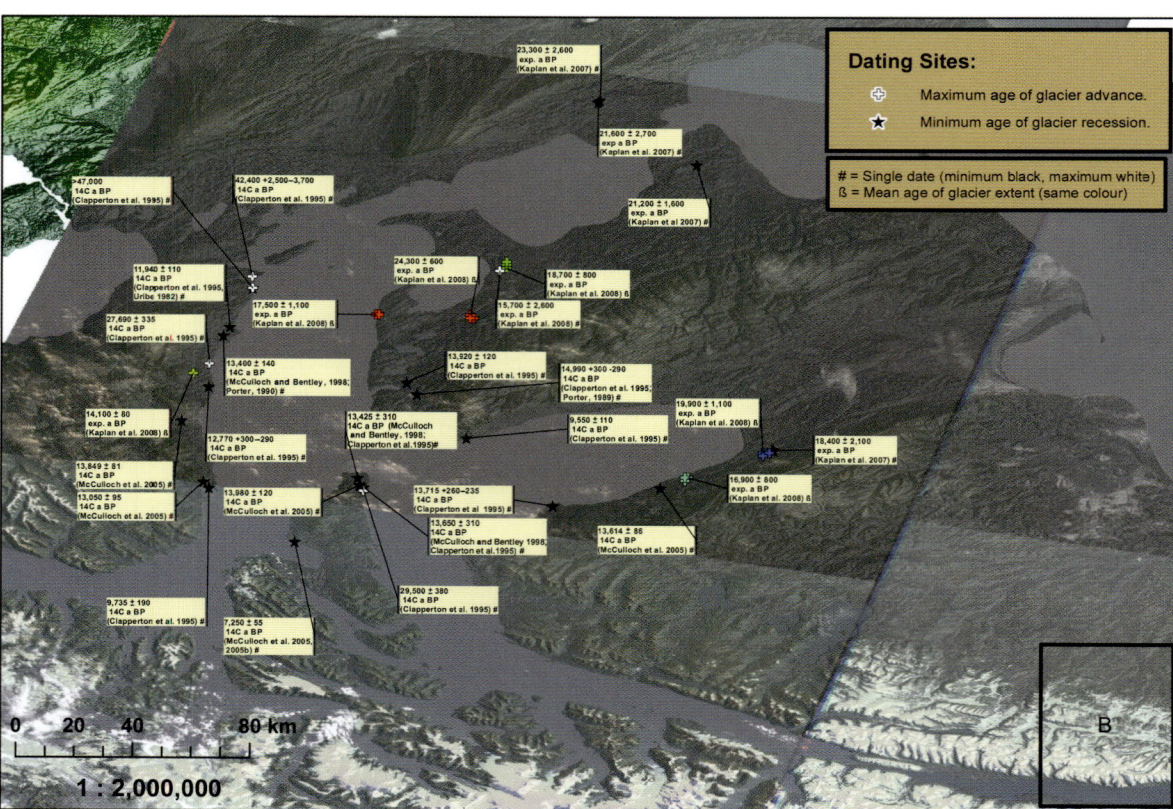

FIGURE 54.3 Data on Quaternary glacial extent in the Magellan Strait region. The background is a pan-sharpened Landsat-7 image draped over a hill-shade of the SRTM 90 m DEM. (A) Glacial geomorphology adapted from the mapping of Glasser et al. (2008). (B) Information on glacial extent from published dates, divided into those that provide maximum ages for glacier advance (crosses) and those that provide minimum ages for glacier recession (stars). The ages and sources of information are provided on the labels, which also indicate whether these represent a single date or the mean age of more than one date. This figure was prepared by Ingo Wolff.

Llanquihue I near Puerto Varas has shown them to be older than 39 ^{14}C ka BP. Other evidence from the northern part of the region suggests that the glaciers during the Last Glaciation reached their maximum positions before about 40 ka BP. Zech et al. (2008) argue that this indicates drier conditions during the course of the last glaciation from 40 to 18 ka and a southwards movement of the westerlies over this time period.

On the basis of the oxygen isotope record from deep sea cores, Mercer (1983) concluded that Llanquihue I moraines probably represented a glacier advance at about 73 ka and were therefore of MIS 4 age. Further evidence that the Llanquihue I moraines pre-date the LGM was presented by Heusser (1981) from radiocarbon dating of peat from kettle holes inside the outermost Llanquihue I moraine on Isla Chiloé. The peat deposits (at Taiquemo, 8 km west of Quemchi) were dated to 42.7 ± 1.2 ka, and this date supports the pattern of glaciation and moraine development suggested by Mercer (1983) and Laugenie (1984) from the Lago Llanquihue area, although this date is probably a minimum age. Evidence for interstadial conditions occurring between cold phases of the Last Glaciation is also available from the Chilean Lake District. Pollen analysis of cores from peat basins to the west of the glaciation limit shows that climatic conditions varied considerably from >43 to 28 ka Heusser (1981, 1983). Between 45 or so and 28 ka, conditions in the region were warm and dry. Following 28 ka, pollen taxa show the onset of very cold conditions reflecting the climatic deterioration of the LGM.

54.5.3. The Patagonian Icefields

Marking the eastern end of Lago Buenos Aires, moraines stratigraphically within the limits of the 'Greatest Patagonian Glaciation' of 1.1 Ma are a series of pre-LGM moraines. These represent glacial advances dated to 140–150 ka (MIS 6) (Kaplan et al., 2005). Further south in the Lago Pueyrredón valley, Hein et al. (2009) dated moraines from the 'Hatcher Glaciation' to 260 ka (MIS 8).

54.6. THE LAST GLACIAL MAXIMUM AND LATE-GLACIAL

With the introduction of radiometric dating techniques, many of the moraines of Chile which were assumed to have been constructed during the LGM (MIS 2) have been shown to span a much longer period of the Pleistocene (Clapperton, 1993). However, many moraine complexes (especially those from drier northern Chile) are undated. Nonetheless, moraines of probable LGM age are found throughout the country and, especially in the Chilean Lake District and in Patagonia, are chronologically reasonably well constrained.

54.6.1. Northern and Central Chile

High snowlines (above 5000 m a.s.l.) and aridity throughout much of the Pleistocene have meant that the high mountains and volcanoes of the northern Andes display relatively little evidence for glacier advance during the LGM. Earlier work (e.g. Clapperton, 1993) reported that some summits above 5000 m a.s.l. in northern Chile were probably unglaciated throughout the Pleistocene and only the highest summits (above 6000 m) may have LGM-age moraines. More recent studies have demonstrated considerable climatic shifts in northern Chile at this time.

In the Cordon de Dona Rosa, glacier advances are dated to the Late-Glacial (ranging from 19 to 15 ka BP) and to 18–13 ka BP in the Encierro valley (Zech et al., 2006; Fig. 54.4). Climate reconstructions from northern Chile suggest that increased winter precipitation may have triggered the early LGM glacial advances; with increasing aridity following the LGM glaciers became more restricted (Stuut and Lamy, 2004; Maldonado et al., 2005).

54.6.2. Southern Chile

After the Great Patagonian Glaciations, 14–16 cold (glacial/stadial) events alternated with corresponding warm (interglacial/interstadial) equivalents (Rabassa et al., 2005). They argue that the LGM occurred between 25 and 16 ka (MIS 2; Late Pleistocene) and that two readvances (or still stands) took place during the Late-Glacial (15–10 ^{14}C ka BP). During the Quaternary, the Patagonian icefields expanded and contracted in response to this climatic forcing a number of times (Heusser, 2003; Harrison, 2004; Sugden et al., 2005). At times they coalesced to form the much larger Patagonian Ice Sheet. Modelling studies (Hulton et al., 1994, 2002) and evidence from marine sediment cores (Lamy et al., 2004; Kaiser et al., 2007) are in close agreement that the regional ice maximum coincided with an ~6 °C sea surface temperature lowering in the southeast Pacific off southern Chile.

54.6.3. The Chilean Lake District

Mercer (1976) provided the first radiocarbon dates for the terminal moraines around the largest lake basin, at Lago Llanquihue, demonstrating that the innermost moraines belonged to the last glaciation and that there had been at least three, and probably four, prior advances. The glacial maximum occurred at ca. 19.45 ka, and a final advance reached the western shore of Lago Llanquihue at 13 ka. Porter (1981) confirmed and refined the chronology with many more radiocarbon dates and demonstrated that the advances occurred before 30 ka, between 20 and 19 ka, and shortly after 13 ka. More recent work on the timing of Llanquihue glacial advances has yielded a more comprehensive picture of Lake District glaciation (Bentley, 1997; Denton et al.,

FIGURE 54.4 Early LGM lateral moraines from the Encierro valley in the arid Andes of Chile. Picture by Zech.

1999b; Anderson et al., 1999). Denton et al. (1999b) dated LGM positions to the interval 29.4 to 14.55 ^{14}C ka BP with a maximum extent at ca. 21 ^{14}C ka BP. The high-resolution radiocarbon chronology of Lowell et al. (1995) identified at least six glacier advances during the later part of the last glaciation. Radiocarbon dates on three glacier lobes (Llanquihue, Seno Reloncavi, Castro) have established that advances occurred at least once before 35, at 29.2, 26.9, 23.1 and 21 ka, and between 14.5 and 14.7 ka (Fig. 54.1). Unfortunately, no data exist for the ages of the glacier advances east of the Andes in this sector.

Work in the region by Lowell et al. (1995) and Denton et al. (1999b) supported by 450 radiocarbon dates has refined the chronostratigraphy of MIS 2 in this region (see Fig. 54.1). The moraine chronology shows that full glacial conditions existed almost without interruption throughout the period from 29.4 to 14.5 ^{14}C ka BP, although pollen evidence from Isla Grande de Chiloe suggests that full glaciation did not commence there before about 26 ^{14}C ka BP. During the advances of the LGM, this work suggests that the snowline was depressed by about 1000 m compared to today.

The ^{14}C dates in the northern Chilean Lake District indicate that the piedmont lobes reached their maximum extent during the LGM sometime between 29.4 and 14.5 ^{14}C ka BP (Bentley, 1997; Denton et al., 1999b). The southern lobes also advanced to maximum positions some time before 49.9 ^{14}C ka BP (Denton et al., 1999b). There is therefore reasonable certainty that the large arcuate moraines formed in front of glaciers 1–13 represent a number of separate advances, most recently during the LGM. The expansion of glaciers into the Chilean lake basins occurred during an interval of known global cooling, but it has also been suggested that the advance was driven largely by a northwards migration of the polar front, which resulted in a substantial increase in precipitation in the Lake District (Heusser, 1989, 1990; Hulton et al., 1994; Lamy et al., 2004).

From a synthesis of key proxy records, McCulloch et al. (2000) concluded that there was a sudden rise in temperature that initiated deglaciation of the Patagonian Ice Sheet synchronously over 16° of latitude at 14.6–14.3 ^{14}C ka BP (17.5–17.15 ka). There was a second step of warming in the Chilean Lake District at 13–12.7 ^{14}C ka BP (15.65–15.35 ka), which saw temperatures rise to close to modern values. A third warming step, particularly clear in southern Chile, occurred ca. 10 ^{14}C ka BP (11.4 ka). Following the initial warming, there was a lagged response in precipitation as the westerlies, after a delay of ca. 1.6 ka, migrated from their northern glacial location to their present latitude, which was attained by 12.3 ^{14}C ka BP (14.3 ka) (McCulloch et al., 2000). Recent work in the Chilean Lake District (Massaferro et al., 2010) used pollen and chironomid records to argue that the latter part of the LGM (at 20–17.6 ka) was characterised by extreme cold and wet conditions, followed by gradual warming until

16.8 ka. Noticeable cooling started after 14 ka, correlating with the ACR and this intensified from 13.5 to 11.5 ka, coincident with the end of the ACR and YDC.

54.6.3.1. North Patagonian Icefield Region

East of the North Patagonian Icefield, glacial lineations indicate that ice discharge was concentrated into large, fast-flowing, topographically determined outlet glaciers (Glasser et al., 2005, 2008; Glasser and Jansson, 2005). Dates for advances of these eastern outlet glaciers have been obtained from CRN dating (Kaplan et al., 2004, 2005, 2006; Turner et al., 2005; Douglass et al., 2006), using $^{40}Ar/^{39}Ar$ and K–Ar dating (Singer et al., 2004), ^{14}C dating (Wenzens, 2006b) and using a combination of CRN and OSL dating (Glasser et al., 2006). The area at the eastern end of Lago General Carrera/Lago Buenos Aires in Argentina is one of the most comprehensively dated in this area of South America. Working eastward from the lake shore, the youngest moraines here (the Menucos moraines) have been dated to 14.4 ka (Douglass et al., 2006). The next moraine system to the east, the Fenix moraines, has been dated to the LGM at between 15 and 23 ka (Kaplan et al., 2004). To the south of here recent work by Hein et al. (2010) in the Lago Pueyrredòn Valley used CRN to date boulders on Rio Blanco moraines to 27–25 ka.

54.6.3.2. South Patagonian Icefield Region

The terminal moraines at the heads of Lago Viedma and Lago Argentino have been dated to the LGM by Wenzens (1999, 2005). Further west, at Puerto Bandera, the prominent moraines that protrude into Lago Argentino have been dated to between 10.39 and 13.0 ^{14}C ka BP (Strelin and Malagnino, 2000). Recession from the Puerto Bandera moraines is indicated by dates of 6.74 ± 0.13 and 10.0 ± 0.14 ^{14}C ka BP from within this limit (Mercer, 1968). To the south, around the margins of the contemporary South Patagonian Icefield (SPI) in the Torres del Paine area, Marden and Clapperton (1995) defined the LGM extent at a point only ca. 70 km from the contemporary icefield. Younger moraines, located closer to the contemporary SPI, are dated to 13.2 ± 0.8 ka (Fogwill and Kubik, 2005) and between 9.1 and 12.7 ^{14}C ka BP (Marden and Clapperton, 1995). Other moraines in the region, although stratigraphically similar to LGM moraines, remain undated (Fig. 54.5).

54.6.3.3. The Magellan Straits and Bahía Inútil (53–55°S)

Much of the research in this region by researchers such as Coronato et al. (1989), Coronato (1995), Clapperton (1993), Clapperton et al. (1995), Benn and Clapperton (2000) and

FIGURE 54.5 Undated moraines to the east of the SPI. Photograph by Glasser.

McCulloch and Davies (2001) in southernmost Patagonia was reviewed by Harrison (2004). Large moraine complexes, consisting of multiple moraine sets and associated glacial lineations, are developed in the areas occupied by former outlet glacier in Seno Skyring, Seno Otway, Estrecho Magallenes and Bahía Inútil. The LGM in this area is reasonably well established at 23–25 ka for a terminal moraine in the Magellan Strait (McCulloch et al., 2005a, b) and 18–20 ka in Bahía Inútil (McCulloch et al., 2005a, b; Kaplan et al., 2007). A deglacial recessional stage of the LGM (between 10.315 and 12.7 ^{14}C ka BP) on Isla Dawson has also been reported by Rabassa et al. (1986) (Fig. 54.3).

Recently, Kaplan et al. (2008) used CRN, amino acid dating and reviewed earlier ^{14}C dating from this region of southernmost Patagonia. CRN dates on moraines mapped by Bentley et al. (2005) show glacial advances at 24.6 ± 0.9 ka, 18.5 ± 1.8, and 17.6 ± 0.2 in the Magellan Straits and 20.4 ± 1.2 and 17.3 ± 0.8 ka in Bahía Inútil (Kaplan et al., 2008). Late-Glacial moraines dated to 15.5 and 11.7 ka by radiocarbon dating are located at the head of the Straits of Magellan and Bahía Inútil (McCulloch et al., 2005a,b). Kaplan et al. (2008) argue that the last glaciation in southernmost Patagonia began around 31 ka, with the maximum ice extent at 25 ka and with major ice recession occurring after 19 ka. Following the LGM, ice recession was very rapid in response to warming. They compare LGM and Late-Glacial events from the Magellan Straits area with reconstructed glaciations from Lago Buenos Aires and show that in both regions 4–5 moraine belts were constructed between 25 and 16 ka. The timing of deglaciation differs, however; the Magellan Straits were largely ice-free by 17 ka, while to the north at Lago Buenos Aires, ice was still extensive at 15–14 ka (Fig. 54.3).

54.7. GLACIATION DURING THE YOUNGER DRYAS CHRONOZONE AND ANTARCTIC COLD REVERSAL

Assessing the global significance of the Younger Dryas cold event is of considerable interest to palaeoclimatologists wishing to test hypotheses on the timings and forcings of global climate change. There are several views on the nature of climate change in southern South America during the last Glacial–Interglacial transition, although the final pattern is likely to be more complex than any of them. The first recognises the dominance of Antarctic climate signals in the palaeoclimate record and thus the existence of an antiphase relationship between Northern and Southern Hemisphere climates at this time (Blunier et al., 1998). Similar views support this bipolar see-saw relationship driven by changes in deep-water formation at high latitudes (Broecker, 1998). The second main view argues that there is an in-phase synchronous relationship between the hemispheres (e.g. Denton et al., 1999a,b) suggesting that the trigger for climate change at this time was atmospheric rather than oceanic in origin.

It is clear that resolution of these issues is crucial to our understanding of inter-hemispheric climatic teleconnections during the Pleistocene. The Younger Dryas in the Northern Hemisphere was associated with a strong April–July insolation maximum, a period of Milankovitch cyclicity unfavourable for ice-sheet growth (e.g. Berger, 1990). In the Southern Hemisphere, the winters at this time would have been relatively shorter than now, and therefore, the accumulation phases of glacier mass balance might well have been less effective. This could have tended to reduce the impact of Younger Dryas forcing on ice cap and glacier growth. The role of Milankovitch cyclicity should therefore be borne in mind when discussing the likelihood of Younger Dryas glacier expansion in Chile.

In southern South America, however, identifying climate changes during the YDC is hampered by the presence of other cooling events at around this time, notably the ACR, a period of lower temperatures recorded in Antarctic ice cores, at 14.8–12.7 ka BP (Blunier et al., 1997, 1998; Raynaud et al., 2000; Blunier and Brook, 2001; Ahn and Brook, 2008).

Recent studies in the Torres del Paine region of Patagonia (Moreno et al., 2009) have clarified the relationship between Late-Glacial cold events in this region. Using CRN and ^{14}C dating, they show that the Late-Glacial maximum readvance in southwest Patagonia occurred at 14.8–12.6 ka at the time of the ACR EPICA Dome C record (Stenni et al., 2003). Like other researchers in the region, they found no evidence for an YDC glacial event and suggest that the pattern and timing of Late-Glacial climate change in southern South America is latitudinally dependent (cf. Sugden et al., 2005) and argue that the position of the Antarctic Polar Front determines the strength of Late-Glacial cooling signals.

The geomorphological and palaeoecological evidence for a general Younger Dryas glacier advance in Chile is therefore equivocal (Markgraf, 1993). Numerous, undated moraines exist within the LGM limit and outside Neoglacial limits, and this has tempted some workers to suggest that they represent glacial advances coeval with the Younger Dryas (e.g. Marden and Clapperton, 1995). Similarly, there is some pollen core evidence of climate changes at this time, although other studies find no evidence for these changes. The evidence discussed here comes from the Chilean Lake District, various locations around the North and SPIs and from Tierra del Fuego.

The evidence to show cooling from palaeoecological studies in the vicinity of the Chilean Lake District and the North Patagonian Icefield is certainly ambiguous. Some workers show a climate shift during the Pleistocene–

Holocene transition (e.g. Heusser, 1993), although other workers do not (Ashworth et al., 1991; Markgraf, 1991, 1993; Lumley and Switsur, 1993; Bennett et al., 2000).

54.7.1. The Chilean Lake District

Little geomorphological evidence has yet been reported from the Chilean Lake District which can be attributed to Younger Dryas glacier advances. However, at around 41°S, a glacial advance at Lago Mascadi has been dated using radiocarbon dating to between 11.4 and 10.15 ± 0.09 ^{14}C ka BP (Hajdas et al., 2003), preceding the Younger Dryas by around 550 years. Generally, though, the glacier/climate record in this region is (as elsewhere) partly obscured by the presence of deep lakes into which the glaciers must have advanced. These created the conditions for glacier calving and this partly decoupled the glacier fluctuations from climate change. As a result, glacier expansion during short-lived episodes of climate deterioration may have been attenuated by an increase in calving fluxes.

The main evidence for a Younger Dryas climatic reversal in this region comes from palaeoecological reconstructions by Heusser and Streeter (1980), Heusser (1984) and Heusser et al. (1999). The latter work demonstrates that from 12.2 to 10.0 ^{14}C ka BP, there was an expansion of cold-tolerant species and an opening of the forest canopy and suggests that this reflected a temperature decrease in the order of 2–3 °C. This argument is countered by work which suggests that changes in some of the indicator species used by Heusser et al. (1999) may be related more to changes in soil and groundwater levels than to changes in climate (Markgraf, 1989). In addition, the fossil beetle records do not support the evidence for a cooling in this period (Hoganson and Ashworth, 1992; Ashworth and Hoganson, 1993). Near Alerce in the Chilean Lake District at 41°S, Heusser and Streeter (1980) employed palynology to show reduced temperature and increased precipitation at around 10.4 ^{14}C ka BP.

54.7.2. The Patagonian Icefields

Although large moraine systems lying outside the Holocene glacier limits exist around the margins of the NPI and SPI, very few of these have been dated. The only information about the likely climatic conditions that prevailed during the Younger Dryas period around the NPI comes from pollen analysis and radiocarbon-dated organic sequences in small lakes on the Taitao Peninsula, to the west of the icefield. These results (Lumley and Switsur, 1993; Bennett et al., 2000) show that there is no evidence for a climatic deterioration in the region during the YDC. Indeed, Bennett et al. (2000, p. 326) argue that 'the YDC was in fact a period of stable, or possibly slightly increasing, temperatures'. This supports earlier results by Hoganson and Ashworth (1992) who used beetle remains to suggest that no cooling occurred during this time in the region. In addition, in southern Patagonia, White et al. (1994) presented evidence for considerable warming at 10.0 and 12.8 ^{14}C ka BP. Further south in the Chilean Channels, Ashworth et al. (1991) report similar findings. However, Massaferro and Brooks (2002) used chironomid data from the Taitao Peninsula to show cooling during the YDC.

The geomorphological evidence for glacier expansion during the YDC is similarly equivocal and hampered by the absence of well-dated moraine sequences in the outlet valleys. The only moraines dated from this time are from the Exploradores valley of the NPI where Glasser et al. (2005) used CRN dating and OSL methods to date a significant glacial advance between 12.5 and 9.6 ka.

South of the NPI, there is conflicting evidence for the nature of climate change during Younger Dryas times. Recent retreat of Glaciar Tempano (48°45'S) exposed organic deposits dated to 11.07 ± 0.16 and 11.1 ± 0.17 ^{14}C ka BP (Mercer, 1976). This suggests that there was no Younger Dryas advance of the glacier beyond its twentieth century limit. However, a number of workers have identified large moraines lying within the limits of the LGM and around the margins of the SPI (e.g. Caldenius, 1932; Mercer, 1976; Clapperton, 1983), and it may be that these represent a glacier advance during the Younger Dryas. At the present, few of these moraines have been closely dated. Clapperton (2000) hypothesises that they may correlate with moraine stages in the Chilean Lakes region (although no Younger Dryas moraines have been found here) and the Magellan Straits. Marden (1997) dates end moraines in the region to between 11.88 and 9.18 ^{14}C ka BP. At Glaciar Grey on the southern side of the SPI, a glacial advance dated by tephrochronology occurred between 12.01 and 9.18 ± 0.12 ^{14}C years BP (McCulloch et al., 2000). Ackert et al. (2008) use CRN to date the Puerto Bandero moraine on the eastern flank of the SPI to 10.8 ± 0.5 ka, near the end of the YDC, contemporaneous with the highest shoreline of Lago Cardiel. They suggest that these events show that increased precipitation during YDC times rather than cooling drove glacier behaviour.

54.7.3. The Magellan Straits

Ice had retreated into several of the upland valleys of the Cordillera Darwin by 13.3 ^{14}C ka BP (Clapperton et al., 1995; McCulloch and Bentley, 1998). Between 12,700 and 10.3 ^{14}C ka BP, the ice readvanced to the north up the Straits of Magellan, and the culmination of this is constrained by tephrochronology to 12.01 and 10.3 ^{14}C ka BP. This period spans the ACR (McCulloch et al., 2000). Palaeoenvironmental information covering this time period has been gathered by Accelerator Mass Spectrometry

(AMS) dated pollen and diatom cores from the Straits of Magellan. The evidence to show significant climate change at the time of the ACR or YDC is ambiguous; between 12.3 and 10.3 ^{14}C ka BP, there were numerous and short-lived changes between heathland and grassland but these variations in vegetation do not necessarily represent temperature changes. Slightly wetter conditions are indicated by the transitions to heathlands and these increases in precipitation might have been sufficient to trigger glacier advances. At 10.3 ^{14}C ka BP, the arrival of *Nothofagus* signals a rise in temperature; after this, a period of aridity occurred until ca. 8.5 ^{14}C ka BP.

54.8. DISCUSSION

While a number of patterns may be discerned in the timing and extent of glaciations in Chile. The most complete evidence of Pleistocene glacier fluctuations and associated palaeoenvironmental reconstructions comes from the well-dated moraine sequences and palaeoecological information covering the LGM.

In the arid mountains of northern and central Chile, the limited available evidence suggests an 'early' LGM at 42–39 ka (Zech et al., 2009; see Fig. 54.4). In southern Chile, there appears to be a regionally synchronous LGM at around 27–25 ka which is also generally in-phase with the global LGM (Schaefer et al., 2006), followed by a readvance in several locations which is probably related to the ACR at 14.8–12.7 ka.

After this, Chile warmed synchronously and deglaciation followed. This pattern is nearly global in extent and Denton (2000) argues that this may reflect changes in oceanic deep-water production. However, as McCulloch et al. (2000) note, the evidence shows that there was no time lag between deglaciation in the Northern Hemisphere and southern South America. Glaciation in the Chilean Lake District is probably associated with, and partly driven by, northward movement of the southern westerlies belt (e.g. Moreno, 1997; Denton et al., 1999b) concomitant with northward expansion of Antarctic sea ice (e.g. Heusser, 1989). In Magellan Straits area, Benn and Clapperton (2000) argue on the basis of geomorphological evidence that permafrost was present at sea level during the LGM. The capacity of such cold air masses to hold moisture would be relatively reduced, and as a result, glacier expansion may have been limited by accumulation. McCulloch et al. (2000) suggest that after deglaciation, it took some 2500 years for the southern westerlies to move south to their present position in the vicinity of 50°S. They further suggest that this southerly movement of the westerlies occurred following reorganisation of the oceans after the last termination. This took place after a delay of 1600 years, took about 2500 years to complete and the delay coincides with the Heinrich 1 iceberg discharge event in the North Atlantic (Denton, 2000). They hypothesise that this event suppressed thermohaline circulation and reorganisation of the oceans occurred only after this was restarted, allowing the southerly movement of the westerlies from the latitudes of the Chilean Lake District. McCulloch et al. (2000, p. 415) conclude that 'This scenario explains the ca. 1.6 ka wet transitional period following the initial warming in the Chilean Lake District and the precipitation peaks in the Taitao and Magellan areas some 2.5 ka after the initial warming'. However, it should be noted that there is debate over the position of the southern westerlies at this time. Markgraf (1989) and Markgraf et al. (1992) argue that the southern westerlies were positioned further south during the LGM. Markgraf and Kenny (1997) further suggest that the southern westerlies did not vary their latitudinal position seasonally as much as at present and were located around 43–45°S. In addition, glacier–climate modelling studies from northern Chile by Kull and Grosjean (2000) suggest that an equatorward displacement of the westerlies cannot account for the Late Pleistocene glaciation in the region.

54.8.1. Some Problems

Harrison (2004) identified a number of unresolved problems which had to be addressed before a more complete understanding of Pleistocene glaciations in Chile could be achieved. Six years later, at least five pressing issues can be identified; several of them (the status of Younger Dryas-age and Neoglacial glacier advances) have direct relevance to understanding the nature of climate linkages between the hemispheres during cold climate episodes.

54.8.1.1. Glaciation of Western Patagonia

With the exception of three mid-Holocene radiocarbon dates obtained from two western outlet glaciers of the SPI (Mercer, 1970) and research on the moraines of the San Rafael Glacier on the NPI, very few moraines older than late Holocene have been dated on the western flanks of the Patagonian icefields. In addition, and with few exceptions (Glasser et al., 2006; Harrison et al., 2008), the bulk of the research on moraines on the eastern side of the icefields has employed radiocarbon dating on organic material within moraine sequences or lake deposits (Wenzens, 1999), or CRN dating on boulders deposited on moraine surfaces (Ackert et al., 2008). It is known that both of these techniques have limitations in the Patagonian context; radiocarbon dating only usually produces minimum ages, while use of CRN is restricted to the drier eastern flanks of the icefields with limited vegetation cover and soil development to interfere with the isotopic signal. Hence, no CRN dates have been successfully obtained from the wetter and heavily vegetated western side of the Patagonian Icefields.

This, combined with the more restricted occurrence of moraines along the western margins of the icefields (most western outlet glaciers terminate as calving glaciers and evidence is therefore offshore in fjords), has meant that very few attempts have been made to date the Late Quaternary expansion of the icefields in the west, and this must be seen as a significant omission in our attempts to reconstruct their behaviour and the trends of climate change in southern South America.

West of the Andes and in the vicinity of the NPI, there is geomorphological evidence in the form of terminal moraines along the eastern and southern arms of Lago Presidente Rios for an independent ice mass on the Taitao Peninsula. Heusser (2002, 2003) also mapped these moraines and suggested that they formed sometime after 14.355 ^{14}C ka BP (based on a minimum date from nearby Laguna Stibnite provided by Lumley and Switsur, 1993). The arrangement of the moraines indicates that a locally nourished ice cap developed on the peninsula, which was entirely independent of the nearby NPI. The existence and preservation of these moraines at this latitude and this close to sea level is difficult to explain for two reasons:

(a) If the Taitao ice cap is LGM or pre-LGM in age, then this implies severely restricted expansion of glaciers (<10 km of expansion) from the NPI at the LGM. This is difficult to reconcile with dated geomorphological evidence elsewhere (e.g. the Chilean Lake District immediately to the north) where large west-flowing outlet glaciers developed at the LGM. It is also difficult to imagine a climatological setting where temperature and precipitation allowed the growth of a substantial ice cap on the low-lying Taitao Peninsula but little or no expansion of existing glaciers in the high-accumulation area of the Andes currently occupied by the NPI. If the Taitao ice cap is LGM in age, one possible explanation is that outlet glaciers from the LGM Patagonian Ice Sheet were prevented from over-running the Taitao Peninsula because of vigorous ice flow and associated iceberg calving in the deep NE–SW Elefantes Channel, which separates the peninsula from the mainland. This would allow a separate LGM ice mass to develop on the peninsula.

(b) If the Taitao ice cap is not LGM or pre-LGM in age, then it must have formed some time after the LGM. In this case, it most likely dates from the ACR or from glacier expansion during the YDC (13.3–12.0 cal ka BP; 11.4–10.2 ^{14}C ka BP; Hajdas et al., 2003). Glacier expansion in this area during the YDC is, however, incompatible with palaeoecological evidence derived from lake cores on the peninsula that suggests no significant cooling occurred in the YDC (Bennett et al., 2000), although YDC cooling has been reported from palaeoecological records further south on Tierra del Fuego (Heusser and Rabassa, 1987). Indeed, a radiocarbon-dated pollen sequence from nearby Laguna Stibnite on the Taitao Peninsula provides evidence for an early deglaciation (before 14.0 ka BP) and no evidence for a YDC climatic reversal in this region of Chile (Lumley and Switsur, 1993). Evidence from the chironomid (midge) assemblage record is equivocal, with changes in assemblages during the Late-Glacial and Holocene indicating that the climate may have become cooler and drier during the YDC at nearby Laguna Stibnite (Massaferro and Brooks, 2002), but not ~300 km to the north at Laguna Facil (Massaferro et al., 2005).

Further dates are clearly required from the Lago Presidente Rios moraines in order to determine their age and palaeoclimatological significance. Based on the geomorphological record, it seems most likely that the LGM Patagonian Ice Sheet was prevented from over-running the Taitao Peninsula because of the deep NE–SW Elefantes Channel. This would allow a separate LGM ice mass to develop on the peninsula at this time without the need to invoke glacier advances during the YDC.

54.8.1.2. Undated Moraine Systems

As has been seen, most of the available information on the glaciations of Chile comes from the Chilean Lake District, the Patagonian Icefields and the Magellan Straits region. Elsewhere, for instance, in the arid north of the country, very little is known of the glacial history despite the pioneering work of Zech and co-workers. Consequently, large gaps exist in the mapped extents of the ice limits. However, large moraine complexes exist in most of the mountainous regions of Chile and in many piedmont locations; most are undated.

54.8.1.3. Ice Sheet Trimlines

Glacial trimlines marking the upper limits of the PIS are evident in most mountain regions of Patagonia. Despite this, almost no information is available on their age and only in three cases have they been used to assess the thickness of former ice masses. Fieldwork by SH in the Nef valley in 1998 and in the Leones valley in 2000 identified lateral moraines some 700 m upslope of the late Holocene glacier limits and terminal moraines 10 km or so downvalley of these. The age and climatic significance of these glacial systems are not known at present, but it is clear that they reflect a considerable reduction in ELA below present values. Recent work in the Rio Chacabuco Valley has identified and mapped glacial trimlines marking the upper limits of the former PIS and used CRN to date these limits (Boex et al., 2010). When complete, this will be the first detailed assessment of the thickness of the PIS at times in the past and will add considerably to our understanding and modelling of the dynamics of this ice mass.

54.8.1.4. What Is the Record of Glacier Fluctuations During Neoglaciation?

Glasser et al. (2004) reviewed evidence for Holocene glacier advances in Patagonia. From this, they concluded that during the early Holocene (10.0–5.0 ^{14}C ka BP), atmospheric temperatures east of the Andes were about 2 °C above modern values in the period 8.5–6.5 ^{14}C ka BP. The period between 6.0 and 3.6 ^{14}C ka BP appears to have been colder and wetter than present, followed by an arid phase from 3.6 to 3.0 ^{14}C ka BP. From 3.0 ^{14}C ka BP to the present, there is evidence of a cold phase, with relatively high precipitation. West of the Andes, the available evidence points to periods of drier than present conditions between 9.4–6.3 ^{14}C ka BP and 2.4–1.6 ^{14}C ka BP. Holocene glacier advances in Patagonia began around 5.0 ^{14}C ka BP, coincident with a strong climatic cooling around this time (the Neoglacial interval). Glacier advances can be assigned to one of three time periods following a 'Mercer-type' chronology, or one of four time periods following an 'Aniya-type' chronology. The 'Mercer-type' chronology has glacier advances 4.7–4.2 ^{14}C ka BP; 2.7–2.0 ^{14}C ka BP and during the Little Ice Age (LIA). The 'Aniya-type' chronology has glacier advances at 3.6, 2.3, 1.6–1.4 ^{14}C ka BP and during the LIA (also see Porter, 2000; Harrison et al., 2007). They concluded that these chronologies should be regarded as broad regional trends, as there are also dated examples of glacier advances outside these time periods.

54.8.1.5. The Non-Linear Response of Glaciers to Climate Forcing

Hein et al. (2010) suggested that the small difference in timing of the LGM ice extent between the area to the east of the NPI and the Chilean Lake District may reflect the partial decoupling of fast-flowing outlet glaciers from the climate record. This non-linear response of Patagonian glaciers to climate forcing was identified by Harrison (2004) as a potentially important issue for climate reconstructions from glacial extents, especially in Chile with a wide range of glacial environments driven by a range of dynamic variables (e.g. arid high mountain glaciers with high sublimation rates and low precipitation; lacustrine and tidewater calving glaciers on icefields with high mass fluxes; high-velocity palaeoglaciers flowing into regions of high aridity during Pleistocene glaciations).

The assumption that glacier behaviour reflects a linear response of the glaciers to climate forcing is questioned by the behaviour of certain glaciers whose characteristics mean that their response to first-order climate forcing is partly obscured by second-order controls. An interesting example of this is the modelling study of Kaplan et al. (2009), who demonstrated that the drainage morphology of the southern Andes changed radically from a non-glaciated to a glaciated landscape through the Quaternary. Glacial modification of the mountains caused changes in topography that meant that successive glaciations decreased in extent through time. Thus, the extent of Quaternary glaciations is not necessarily purely climatically driven; landscape modification by glacial erosion also plays a role.

In addition, there are difficulties in identifying the glacier/climate signal from calving glaciers. The proximity of the Pacific Ocean to the Andes mountains in much of southern Chile means that during periods of glacier expansion, many of the glacier termini ended in the sea and therefore experienced tidewater calving. Freshwater calving also affected many of the Pleistocene glaciers in the Chilean Lake District and on the eastern side of the icefields which terminated in deep freshwater lakes (e.g. Warren and Aniya, 1999). It is well known that calving processes established second-order controls on glacier behaviour and therefore partly decouple and obscure the glacier/climate signal (e.g. Van der Veen, 2002). Since much of our geomorphological evidence for climate change in Chile is derived from the position and ages of moraines associated with glaciers which are calving or have calved in the past, the problem of assigning climatic inference to oscillations of calving fronts is a serious one. The picture is further complicated by the different calving responses in freshwater and tidewater. Further, many of the climatic reconstructions from the Chilean Lake District and the Magellan Straits area are partly based upon dates of the fluctuations of piedmont lobes with low ice gradients. It has been suggested that the termini of such ice masses may oscillate in response to variations in bed conditions rather than climatic inputs (Glasser and Jansson, 2005). These factors mean that considerable caution should therefore be exercised when reconstructing palaeoclimates from the positions of moraines.

54.9. CONCLUSIONS

Chile's position in the Southern Hemisphere straddling 38° of latitude and its possession of one of the longest and most complete terrestrial records of glaciation in the world means that the country has become a critical global location for testing models of climate change. While considerable gaps remain in our knowledge of the timing and extent of glaciation in Chile and of the drivers behind them, it is clear that the behaviour of the southern westerlies and the Humboldt current are crucial variables. There does not appear to be a Younger Dryas signal in the southern Chile palaeoecological record, yet there are glacier advances in this region which span both Younger Dryas and ACR times. This may represent a return to southerly latitudes of the southern westerlies. Finally, it is clear that there are a large number of problems in the Pleistocene glacial history of Chile and these will only be resolved by a combination of accurate dating and multi-proxy geomorphological, climate modelling and palaeoecological approaches.

ACKNOWLEDGEMENTS

We thank NERC, Leverhulme Trust, the Royal Geographical Society and Raleigh International for supporting our research in Patagonia over the past 20 years. Ingo Wolff kindly gave us permission to use his database of published ice limits in Patagonia.

REFERENCES

Aceituno, P., 1988. On the functioning of the Southern oscillation in the South American sector, Part I: surface climate. Mon. Weather Rev. 116, 505–524.

Ackert Jr., R.P., Becker, R.A., Singer, B.S., Kurz, M.D., Caffee, M.W., Mickelson, D.M., 2008. Patagonian glacier response during the Late Glacial–Holocene transition. Science 321, 392–395.

Ahn, J., Brook, E.J., 2008. Atmospheric CO_2 and climate on millennial time scales during the last glacial period. Science 322, 83–85.

Allan, R., Lindesay, J., Parker, D., 1995. El Nino Southern Oscillation and Climatic Variability. CSIRO, Collingwood, Australia.

Ammann, C., Jenny, B., Kammer, K., Messerli, B., 2001. Late Quaternary glacier response to humidity changes in the arid Andes of Chile. Palaeogeogr. Palaeoclimatol. Palaeoecol. 172, 313–326.

Andersen, B., Denton, G.H., Lowell, T.V., 1999. Glacial Geomorphologic Maps of Llanquihue Drift in the Area of the Southern Lake District, Chile. Geogr. Ann. 81A, 155–166.

Aniya, M., 1995. Holocene glacial chronology in Patagonia: Tyndall and Upsala Glaciers. Arct. Alpine Res. 27, 311–322.

Aniya, M., Enomoto, H., 1986. Glacier variations and their causes in the Northern Patagonia Icefield, Chile, since 1944. Arct. Alpine Res. 18, 307–316.

Ashworth, A.C., Hoganson, J.W., 1993. The magnitude and rapidity of the climate change marking the end of the Pleistocene in the mid-latitudes of South America. Palaeogeogr. Palaeoclimatol. Palaeoecol. 101, 263–270.

Ashworth, A.C., Markgraf, V., Villagran, C., 1991. Late Quaternary climatic history of the Chilean Channels based on fossil pollen and beetle analyses, with an analysis of the modern vegetation and pollen rain. J. Quatern. Sci. 6, 279–291.

Benn, D.I., Clapperton, C.M., 2000. Glacial sediment landform associations and paleoclimate during the Last Glaciation, Strait of Magellan, Chile. Quatern. Res. 54, 13–23.

Bennett, K.D., Haberle, S.G., Lumley, S.H., 2000. The last Glacial–Holocene transition in Southern Chile. Science 290, 325–328.

Bentley, M.J., 1997. Relative and radiocarbon chronology of two former glaciers in the Chilean Lake District. J. Quatern. Sci. 12, 25–33.

Bentley, M., Sugden, D., Hulton, N., McCulloch, R., 2005. The landforms and pattern of deglaciation in the Strait of Magellan and Bahía Inutil, southernmost South America. Geogr. Ann. 87A, 313–334.

Berger, W.H., 1990. The Younger Dryas cold spell—a quest for causes. Palaeogeogr. Palaeoclimatol. Palaeoecol. 89, 219–237.

Blunier, T. Brook, E.J., 2001. Timing of millennial-scale climate change in Antarctica and Greenland during the last glacial period. Science 291, 109–112.

Blunier, T., Schwander, J., Stauffer, B., Stocker, T.F., Dällenbach, A., Indermühle, A., et al., 1997. Timing of the Antarctic Cold Reversal and the atmospheric CO_2 increase with respect to Younger Dryas event. Geophys. Res. Lett. 24, 2683–2686.

Blunier, T., Chappellaz, J., Schwander, J., Dallenbach, A., Stauffer, B., Stocker, T.F., et al., 1998. Asynchrony of Antarctic and Greenland climate during the last glacial period. Nature 394, 739–743.

Boex, J., Harrison, S., Fogwill, C.J., Schnabel, C., 2010. The 3-D geometry of the North Patagonian icefield from glacial to interglacial conditions. Geophys. Res. Abstr. 12, EGU2010-11816.

Broecker, W.S., 1998. Paleocean circulation during the last deglaciation: a bipolar seesaw? Paleoceanography 13, 119–121.

Calcenius, C.C., 1932. Las glaciaciones cuaternarios en la Patagonia y Tierra del Fuego. Geogr. Ann. 14, 1–64.

Clapperton, C.M., 1983. The glaciation of the Andes. Quatern. Sci. Rev. 2 (3), 83–155.

Clapperton, C.M., 1993. Quaternary Geology and Geomorphology of South America. Elsevier, Amsterdam.

Clapperton, C.M., 2000. Interhemispheric synchroneity of Marine Oxygen Isotope Stage 2 glacier fluctuations along the American cordilleras transect. J. Quatern. Sci. 15, 435–468.

Clapperton, C.M., Sugden, D.E., Kauffman, D., McCulloch, R.D., 1995. The last glaciation in central Magellan Strait, southernmost Chile. Quatern. Res. 44, 133–148.

Coronato, A.M.J., 1995. The last Pleistocene glaciation in tributary valleys of the Beagle Channel, Fuegan Andes, South America. In: Quaternary of South America and the Antarctic Peninsula. Balkema, Amsterdam, pp. 153–171.

Coronato, A., Rabassa, J., Serrat, D., Marti, C., 1989. Drumlins en el Canal Beagle; fotointerpretacion y parametros morfometricos. In: Symposium Argentine Teledeteccion, Resumenes. Universidad Nacional del Sur, Bahia Blanca, Argentina.

Denton, G.H., 2000. Does an asymmetric thermohaline oscillator drive 100,000-yr glacial cycles? J. Quatern. Sci. 15, 301–318.

Denton, G.H., Heusser, C.J., Lowell, T.V., Moreno, P.I., Anderson, B.G., Heusser, L.E., et al., 1999a. Interhemispheric linkages of palaeoclimate during the last glaciation. Geogr. Ann. 81A, 107–153.

Denton, G.H., Lowell, T.V., Heusser, C.J., Schlüchter, C., Andersen, B.G., Heusser, L.E., et al., 1999b. Geomorphology, stratigraphy & radiocarbon chronology of Llanquihue drift in the area of the southern Lake district, Seno Reloncavi & Isla Grande de Chiloe, Chile. Geogr. Ann. 81A, 155–166.

Diaz, H.F., Markgraf, V., 2000. El Niño and the Southern Oscillation. Cambridge University Press, Cambridge, MA.

Douglass, D.C., Singer, B.S., Kaplan, M.R., Ackert, R.P., Mickelson, D. M., Caffee, M.W., 2005. Evidence of early Holocene glacial advances in southern South America from cosmogenic surface-exposure dating. Geology 33 (3), 237–240.

Douglass, D.C., Singer, B.S., Kaplan, M.R., Mickleson, D.M., Caffee, M.W., 2006. Cosmogenic nuclide surface exposure dating of boulders on last-glacial and late-glacial moraines, Lago Buenos Aires, Argentina: interpretative strategies and paleoclimate implications. Quat. Geochronol. 1, 43–58.

Fogwill, C.J., Kubik, P.W., 2005. A glacial stage spanning the Antarctic cold reversal in Torres del Paine (51°S), Chile, based on preliminary cosmogenic exposure ages. Geogr. Ann. 87A, 403–408.

Glasser, N.F., Jansson, K.N., 2005. Fast-flowing outlet glaciers of the Last Glacial maximum Patagonian Icefield. Quatern. Res. 63, 206–211.

Glasser, N.F., Harrison, S., Winchester, V., Aniya, M., 2004. Late Pleistocene and Holocene palaeoclimate and glacier fluctuations in Patagonia. Glob. Planet. Change 43, 79–101.

Glasser, N.F., Jansson, K.N., Harrison, S., Rivera, A., 2005. Geomorphological evidence for variations of the North Patagonian Icefield during the Holocene. Geomorphology 71, 263–277.

Glasser, N.F., Harrison, S., Ivy-Ochs, S., Duller, G.A.T., Kubik, P., 2006. Evidence from the Rio Bayo valley on the extent of the North Patagonian icefield during the Late Pleistocene–Holocene transition. Quatern. Res. 65, 70–77.

Glasser, N.F., Jansson, K., Harrison, S., Kleman, J., 2008. The glacial geomorphology and Pleistocene history of southern South America between 38°S and 56°S. Quatern. Sci. Rev. 27, 365–390.

Hajdas, I., Bonani, G., Moreno, P.I., Ariztegui, D., 2003. Precise radiocarbon dating of Late-Glacial cooling in mid-latitude South America. Quatern. Res. 59, 70–78.

Harrison, S., 2004. The Pleistocene glaciations of Chile. In: Ehlers, J., Gibbard, P.L. (Eds.), Pleistocene Glaciations: Extent and Chronology, Part III: South America, Asia, Africa, Australasia, Antarctica. Elsevier, Amsterdam, pp. 89–103.

Harrison, S., 2010. Patagonia. In: Singh, V.P., Singh, P., Haritashya, U.K. (Eds.), Encyclopaedia of Snow, Ice and Glaciers. Springer, London.

Harrison, S., Winchester, V., Glasser, N.F., 2007. The timing and nature of recession of outlet glaciers of Hielo Patagónico Norte, Chile, from their Neoglacial IV (Little Ice Age) maximum positions. Glob. Planet. Change 59, 67–78.

Harrison, S., Glasser, N.F., Haresign, E., Winchester, V., Warren, C.R., Duller, G.A.T., et al., 2008. Glaciar Leon, Chilean Patagonia: Late Holocene chronology and geomorphology. Holocene 18, 643–652.

Hartley, A.J., 2003. Andean uplift and climate change. J. Geol. Soc. Lond. 160, 7–10.

Hein, A.S., Hulton, N.R.J., Dunai, T.J., Schnabel, C., Kaplan, M.R., Naylor, M., et al., 2009. Middle Pleistocene glaciation in Patagonia dated by cosmogenic-nuclide measurements on outwash gravels. Earth Planet. Sci. Lett. 286, 184–197.

Hein, A.S., Hulton, N.R.J., Dunai, T.J., Sugden, D.E., Kaplan, M.R., Xu, S., 2010. The chronology of the Last Glacial Maximum and deglacial events in central Argentinian Patagonia. Quatern. Sci. Rev. doi:10.1016/jquascirev.2010.01.020.

Heusser, C.J., 1981. Palynology of the last interglacial glacial cycle in mid-latitudes of southern Chile. Quatern. Res. 16, 293–321.

Heusser, C.J., 1983. Quaternary pollen record from Laguna de Tagua, Chile. Science 219, 1429–1432.

Heusser, C.J., 1984. Late Glacial–Holocene Climate of the lake District of Chile. Quatern. Res. 22, 77–90.

Heusser, C.J., 1989. Southern westerlies during the last glacial maximum. Quatern. Res. 31, 423–425.

Heusser, C.J., 1990. Chilotan piedmont glacier in the Southern Andes during the last glacial maximum. Rev. Geol. Chile 17, 3–18.

Heusser, C.J., 1993. Late-glacial of southern South America. Quat. Sci. Rev. 12, 345–350.

Heusser, C.J., 1999. 14 C age of glaciation in Estrecho de Magallanes—Bahía Inútil, Chile. Radiocarbon 41 (3), 287–293.

Heusser, C.J., 2002. On glaciation of the southern Andes with special reference to the Peninsula de Taitao and adjacent Andean Cordillera. J. S. Am. Earth Sci. 15, 577–589.

Heusser, C., 2003. Ice Age Southern Andes: A Chronicle of Paleoecological Events. Elsevier, Amsterdam, 240 pp.

Heusser, C.J., Rabassa, J., 1987. Cold climatic episode of Younger Dryas age in Tierra del Fuego. Nature 328, 609–611.

Heusser, C.J., Streeter, S., 1980. A temperature and precipitation record of the past 16,000 years in Southern Chile. Science 210, 1345–1347.

Heusser, C.J., Heusser, L.E., Lowell, T.V., 1999. Palaeoecology of the southern Chilean Lake District-Isla Grande de Chiloe during middle-late Llanquihue glaciation and deglaciation. Geogr. Ann. 81A, 231–284.

Hoganson, J.W., Ashworth, A.C., 1992. Fossil beetle evidence for climatic change 18,000–10,000 years B.P. in south-central Chile. Quatern. Res. 37, 101–116.

Hubbard, A., 1997. Modelling climate, topography and palaeoglacier fluctuations in the Chilean Andes. Earth Surf. Proc. Landf. 22, 79–92.

Hubbard, A., Hein, A.S., Kaplan, M.R., Hulton, N.R.J., Glasser, N.F., 2005. A modelling reconstruction of the late glacial maximum ice sheet and its deglaciation in the vicinity of the Northern Patagonian Icefield, South America. Geogr. Ann. 87A, 375–391.

Hulton, N.R.J., Sugden, D.E., Payne, A., Clapperton, C.M., 1994. Glacier modelling and the climate of Patagonia during the Last Glacial Maximum. Quatern. Res. 42, 1–19.

Hulton, N.R.J., Purves, R.S., McCulloch, R.D., Sugden, D.E., Bentley, M.J., 2002. The Last Glacial maximum and deglaciation in Southern South America. Quatern. Sci. Rev. 21, 233–241.

Kaiser, J., Lamy, F., Arz, H.W., Hebbeln, D., 2007. Dynamics of the millennial-scale sea surface temperature and Patagonian Ice Sheet fluctuations in southern Chile during the last 70 kyr (ODP Site 1233). Quatern. Int. 161, 77–89.

Kaplan, M.R., Ackert, R.P., Singer, B.S., Douglass, D.C., Kurz, M.D., 2004. Cosmogenic nuclide chronology of millenial-scale glacial advances during O-isotope Stage 2 in Patagonia. Bull. Geol. Soc. Am. 116, 308–321.

Kaplan, M.R., Douglass, D.C., Singer, B.S., Ackert, R.P., Caffee, M.W., 2005. Cosmogenic nuclide chronology of pre-last glacial maximum moraines at Lago Buenos Aire, 461 S, Argentina. Quatern. Res. 63, 301–315.

Kaplan, M.R., Singer, B.S., Douglass, D.C., Ackert, R.P., Caffee, M.W., 2006. Comment on: cosmogenic nuclide chronology of pre-last glacial maximum moraines at Lago Buenos Aires, 461 S, Argentina. [Letter to the Editor]Quatern. Res. 66, 367–369.

Kaplan, M.R., Coronato, A., Hulton, N.R.J., Rabassa, J.O., Kubik, P.W., Freeman, S.P.H.T., 2007. Cosmogenic nuclide measurements in southernmost South America and implications for landscape change. Geomorphology 87, 284–301.

Kaplan, M.R., Fogwill, C.J., Sugden, D.E., Hulton, N., Kubik, P.W., Freeman, S., 2008. Southern Patagonian glacial chronology for the Last Glacial period and implications for Southern Ocean climate. Quatern. Sci. Rev. 27, 284–294.

Kaplan, M.R., Hein, A.S., Hubbard, A., Lax, S.M., 2009. Can glacial erosion limit the extent of glaciation? Geomorph. 103 (2), 172–179.

Kull, C., Grosjean, M., 2000. Late Pleistocene climate conditions in the north Chilean Andes drawn from a climate-glacier model. J. Glaciol. 46, 622–632.

Lamy, F., Kaiser, J., Ninnemann, U., Hebbeln, D., Arz, H.W., Stoner, J., 2004. Antarctic timing of surface water changes off Chile and Patagonian ice sheet response. Science 304, 1959–1962.

Laugenie, J.C., 1984. Le dernier cycle glaciaire Quaternaire et la construction des nappes fluviatiles d'avant pays dans les Andes Chiliennes. Bull. Assoc. Fr. Etude Quatern. 1–3, 139–145.

Lowell, T.V., Heusser, C.J., Anderson, B.G., Moreno, P.I., Hauser, A., Heusser, L.E., et al., 1995. Interhemispheric correlation of Late Pleistocene glacial events. Science 269, 1541–1549.

Lumley, S.H., Switsur, R., 1993. Late Quaternary chronology of the Taitao Peninsula, southern Chile. J. Quatern. Sci. 8, 161–165.

Maldonado, A., Betancourt, J.L., Latorre, C., 2005. Pollen analyses from a 50,000-yr rodent midden series in the southern Atacama Desert (25°30′S). J. Quatern. Sci. 20, 493–507.

Marden, C.J., 1997. Late-glacial fluctuations of south Patagonian Icefield, Torres del Paine National park, southern Chile. Quatern. Int. 38 (39), 61–68.

Marden, C.J., Clapperton, C.M., 1995. Fluctuations of the South Patagonian Icefield during the last glaciation and the Holocene. J. Quatern. Sci. 10, 197–210.

Markgraf, V., 1989. Reply to C.J. Heusser's 'Southern Westerlies during the Last Glacial Maximum'. Quatern. Res. 31, 426–432.

Markgraf, V., 1991. Late Pleistocene environmental and climatic evolution in southern South America. Bamb. Geograph. Schriften 11, 271–281.

Markgraf, V., 1993. Younger Dryas in southernmost South America—an update. Quatern. Sci. Rev. 12, 351–355.

Markgraf, V., Kenny, R., 1997. Character of rapid vegetation and climate change during the late-glacial in southernmost South America. In: Huntley, B. et al., (Ed.), Past and Future Rapid Environmental Changes: The Spatial and Evolutionary Responses of Terrestrial Biota. Springer-Verlag, Berlin, pp. 81–90.

Markgraf, V., Dodson, J.R., Kershaw, A.R., McGlore, M.S., Nicholls, N., 1992. Evolution of late Pleistocene and Holocene climates in the circum-South Pacific land areas. Climate Dyn. 6, 193–211.

Massaferro, J., Brooks, S.J., 2002. Response of chironomids to Late Quaternary environmental change in the Taitao Peninsula, southern Chile. J. Quatern. Sci. 17, 101–111.

Massaferro, J., Brooks, S.J., Haberle, S.G., 2005. The dynamics of chironomid assemblages and vegetation during the Late Quaternary at Laguna Facil, Chonos Archipelago, Southern Chile. Quatern. Sci. Rev. 24, 2510–2522.

Massaferro, J.I., Moreno, P.I., Denton, G.H., Vandergoes, M., Dieffenbacher-Krall, A., 2010. Chironomid and pollen evidence for climate fluctuations during the Last Glacial Termination in NW Patagonia. Quatern. Sci. Rev. 28, 517–525.

McCulloch, R.D., Bentley, M.J., 1998. Late Glacial ice advances in the Strait of Magellan, Southern Chile. Quatern. Sci. Rev. 17, 775–787.

McCulloch, R.D., Davies, S.J., 2001. Late-glacial and Holocene palaeoenvironmental change in the central Strait of Magellan, southern Patagonia. Palaeogeogr. Palaeoclimatol. Palaeoecol. 173, 143–173.

McCulloch, R.D., Bentley, M.J., Purves, R.S., Hulton, N.R., Sugden, D.E., Clapperton, C.M., 2000. Climatic inferences from glacial and palaeoecological evidence at the last glacial termination, southern South America. J. Quatern. Sci. 15, 409–417.

McCulloch, R.D., Bentley, M.J., Tipping, R.M., Clapperton, C.M., 2005a. Evidence for Late-Glacial ice dammed lakes in the Central Strait of Magellan and Bahía Inútil, Southernmost South America. Geogr. Ann. 87A, 335–362.

McCulloch, R., Fogwill, C., Sugden, D., Bentley, M., Kubik, P., 2005b. Chronology of the Last Glaciation in Central Strait of Magellan and Bahía Inutil, Southernmost South America. Geogr. Ann. 87A, 289–312.

Mercer, J.H., 1965. Glacier variations in Southern Patagonia. Geogr. Rev. 55, 390–413.

Mercer, J.H., 1968. Variations of some Patagonian glaciers since the Late-Glacial. Am. J. Sci. 266, 91–109.

Mercer, J., 1970. Variations of some Patagonian glaciers since the late glacial II. Am. J. Sci. 269 (1), 1–25.

Mercer, J.H., 1976. Glacial history of southernmost South America. Quatern. Res. 6, 125–166.

Mercer, J.H., 1982. Holocene glacier variations in southern South America. Striae 18, 35–40.

Mercer, J.H., 1983. Cenozoic glaciation in the Southern Hemisphere. Annu. Rev. Earth Planet. Sci. 11, 99–132.

Moreno, P.I., 1997. Vegetation and climate near Lago Llanquihue in the Chilean Lake District between 20200 and 9500 14 C yr BP. J. Quatern. Sci. 12, 485–500.

Moreno, P.I., Kaplan, M.R., Francois, J.P., Villa-Martinez, R., Moy, C.M., Stern, C.R., et al., 2009. Renewed glacial activity during the Antarctic cold reversal and persistence of cold conditions until 11.5 ka in southwestern Patagonia. Geology 37, 375–378.

Mörner, N.-A., Sylwan, C., 1989. Magnetostratigraphy of the Patagonian moraine sequence at Lago Buenos Aires. J. S. Am. Earth Sci. 2, 385–390.

Mosola, A.B., Anderson, J.B., 2006. Expansion and rapid retreat of the West Antarctic Ice Sheet in eastern Ross Sea: possible consequences of over-extended ice streams? Quatern. Sci. Rev. 25, 2177–2196.

Porter, S.C., 1981. Pleistocene glaciations in the southern Lake District of Chile. Quatern. Res. 16, 263–292.

Porter, S.C., 1989. Character and ages of Pleistocene drifts in a transect across the Strait of Magellan. Quatern. S. Am. Antarct. Peninsula 7, 35–50.

Porter, S.C., 2000. Onset of neoglaciation in the southern hemisphere. J. Quatern. Sci. 15, 395–408.

Rabassa, J., Heusser, C., Stuckenrath, R., 1986. New data on Holocene Sea transgression in the Beagle Channel: Tierra del Fuego, Argentina. Quatern. S. Am. Antarct. Peninsula 4, 291–309.

Rabassa, J., Coronato, A.M., Salemme, M., 2005. Chronology of the Late Cenozoic Patagonian glaciations and their correlation with biostratigraphic units of the Pampean region (Argentina). J. S. Am. Earth Sci. 20, 81–103.

Raynaud, D., Barnola, J.-M., Chappellaz, J., Blunier, T., Indermuhle, A., Stauffer, B., 2000. The ice record of greenhouse gases: a view in the context of future changes. Quatern. Sci. Rev. 19, 9–17.

Rodbell, D.T., Smith, J.A., Mark, B.G., 2009. Glaciation in the Andes during the Late-glacial and Holocene. Quatern. Sci. Rev. 28, 2122–2165.

Schaefer, J.M., Denton, G.H., Barrell, D.J.A., Ivy-Ochs, S., Kubik, P.W., Andersen, B.G., et al., 2006. Near-synchronous interhemispheric termination of the Lastglacial maximum in mid-latitudes. Science 312, 1510–1513.

Singer, B.S., Ackert, R.P., Guillou, H., 2004. 40Ar/39Ar and K–Ar chronology of Pleistocene glaciations in Patagonia. Geol. Soc. Am. Bull. 116, 434–450.

Stenni, B., et al., 2003. A late-glacial high-resolution site and source temperature record derived from the EPICA Dome C isotope records (East Antarctica). Earth Planet. Sci. Lett. 217, 183–195.

Strelin, J.A., Malagnino, E.C., 2000. Late-glacial history of Lago Argentino, Argentina, and age of the Puerto Bandera Moraines. Quat. Res. 54, 339–347.

Stuut, J.-B.W., Lamy, F., 2004. Climate variability at the southern boundaries of the Namib (south-western Africa) and Atacama (northern Chile) coastal deserts during the last 120,000 yr. Quatern. Res. 62, 301–309.

Sugden, D.E., Hulton, N.R.J., Purves, R.S., 2002. Modelling the inception of the Patagonian ice sheet. Quatern. Int. 95–96, 55–64.

Sugden, D.E., Bentley, M.J., Fogwill, C.J., Hulton, N.R.J., McCulloch, R.D., Purves, R.S., 2005. Late-glacial glacier events in southernmost

South America: a blend of 'Northern' and 'Southern' hemispheric climate signals? Geogr. Ann. 87A, 273–288.

Turner, K.J., Fogwill, C.J., Mcculloch, R.D., Sugden, D.E., 2005. Deglaciation of the eastern flank of the North Patagonian Icefield and associated continental-scale lake diversions. Geogr. Ann. 87A, 363–374.

Van der Veen, C.J., 2002. Calving glaciers. Prog. Phys. Geogr. 26, 96–122.

Villalba, R., Cook, E.R., D'Arrigo, R.D., Jacoby, G.C., Jones, P.D., Salinger, M.J., et al., 1997. Sea-level pressure variability around Antarctica since AD 1750 inferred from subantarctic tree-ring records. Climate Dyn. 13, 375–390.

Warren, C.R., Aniya, M., 1999. The calving glaciers of southern South America. Glob. Planet. Change 22, 59–77.

Wenzens, G., 1999. Fluctuations of outlet and valley glaciers in the Southern Andes (Argentina) during the past 13,000 years. Quatern. Res. 51, 238–247.

Wenzens, G., 2000. Pliocene piedmont glaciation in the Rio Shehuen Valley, Southeast Patagonia, Argentina. Arct. Antarct. Alpine Res. 32, 46–54.

Wenzens, G., 2005. Glacier advances east of the Southern Andes between the Last Glacial Maximum and 5000 BP compared with lake terraces of the endorrheic Lago Cariel (49¹S, Patagonia, Argentina). Z. Geomorphol. 49, 433–454.

Wenzens, G., 2006a. Comment on: cosmogenic nuclide chronology of pre-last glacial maximum moraines at Lago Buenos Aires, 46¹ S, Argentina. [Letter to the Editor]. Quatern. Res. 66, 364–366.

Wenzens, G., 2006b. Terminal moraines, outwash plains, and lake terraces in the vicinity of Lago Cardiel (49¹ S; Patagonia, Argentina)—evidence for Miocene Andean foreland glaciations. Arct. Antarct. Alpine Res. 38, 276–291.

White, J.W.C., Cias, P., Figge, I.A., Kenny, R., Markgraf, V., 1994. A high-resolution record of atmospheric CO_2 content from carbon isotopes in peat. Nature 367, 53–156.

Winchester, V., Harrison, S., Bailey, R., 2005. A 2.5 kyr luminescence date for a terminal moraine in the Leones valley, Southern Chile. J. Glaciol. 51, 186–188.

Zech, R., Kull, C., Veit, H., 2006. Late Quaternary history in the Encierro Valley, northern Chile (29°S), deduced from ^{10}Be surface exposure dating. Palaeogeogr. Palaeoclimatol. Palaeoecol. 234, 277–286.

Zech, R., Kull, C., Veit, H., 2007. Exposure dating of Late Glacial and pre-LGM moraines in the Cordillera Dona Rosa, Northern Chile (31°S). Clim. Past 3, 1–14.

Zech, R., May, J.-H., Kull, C., Ilgner, J., Kubik, P., Veit, H., 2008. Timing of the Late Quaternary Glaciation in the Andes from ~15 to 40°S. J. Quat. Sci. 23, 635–647.

Zech, R., Smith, J., Kaplan, M.R., 2009. Chronologies of the Last Glacial Maximum and its termination in the Andes (10–55°S) based on surface exposure dating. In: Vimeux, F. et al., (Ed.), Past Climate Variability in South America and Surrounding Regions. Developments in Palaeoenvironmental Research, 14 Springer, New York, pp. 61–87.

Chapter 55

Late Quaternary Glaciations in Bolivia: Comments on Some New Approaches to Dating Morainic Sequences

Klaus Heine

Geographisches Institut der Universität Regensburg, Universitätsstr. 31, D-93040 Regensburg, Germany

55.1. INTRODUCTION

The Bolivian part of the Andes extends between 14°S and 23°S (Fig. 55.1). This area belongs to the 'arid diagonal', reaching from the Atacama desert in Peru to the dry regions of Argentina in the south. The higher mountains of the Bolivian Andes were glaciated in the Pleistocene. Age determination of the Pleistocene glaciations of the Bolivian Andes is dependent on records from sites in the eastern Cordillera and of isolated volcanoes (Sajama). The first notes on Pleistocene glaciations in Bolivia were published by Hettner (1889), Hoek and Steinmann (1905) and Hauthal (1906, 1911). Based on geomorphological and geological observations, the authors described (i) at least two glaciations, similar to the European Würm(ian) and Riss(ian) glaciations; (ii) a massive glacier retreat during recent times documented by recessional moraines; and (iii) an asymmetry of the modern and Pleistocene snow-line (equilibrium line altitude: ELA). Troll (cf. 1927, 1929, 1968) and Troll and Finsterwalder (1935) presented evidence for three Pleistocene glacier advances and asserted that an older glaciation (similar to the European Günz/Mindel(ian) glaciations) could clearly be distinguished from two younger glaciations (compared to the European Rissian and Würmian events) by geomorphological evidence. Further, they identified three groups of recessional moraines. The moraine groups of all glaciations, as well as those related to the late glacial glacier retreat, were observed in all valleys of the eastern Cordillera Real between the modern glaciers and the Altiplano. Troll and Finsterwalder (1935) had already stressed the importance of detailed morphostratigraphical fieldwork. Only many decades later, problems of the Quaternary glacial chronostratigraphy were pursued by Argollo (1980), Clapperton (1981), Müller (1985), Graf (1975, 1979, 1981), Gouze et al. (1986) and Lauer and Rafiqpoor (1986) and, more recently, by Seltzer (1990, 1992), Heine (1996), Clapperton et al. (1997), Klein et al. (1999) and Argollo and Mourguiart (2000), among others. At most sites, the ages of the moraines were determined by ^{14}C dating of organic sediments behind moraine walls in the valleys (cf. Seltzer, 1991, 1992). It remained unclear whether the local last glacial maximum (LGM) occurred synchronously with, prior to or following the global LGM (24–18 ka) (Zech et al., 2008). For more details, see Heine (1996), Mark et al. (2004), Rodbell et al. (2009) and La Frenierre et al. (2011).

In recent years, surface exposure dating (SED) using cosmogenic nuclides was used as a valuable method for absolute dating of moraines, as well as for constraining the ages of glacier retreat. Smith et al. (2008), Rodbell et al. (2009) and La Frenierre et al. (2011) have summarised these efforts for Bolivia, Peru and Ecuador. The results were used to reconstruct local LGM ELAs and to determine the average lowering of the ELA from LGM to modern to ice-age temperature and precipitation changes (cf. Ramage et al., 2005). However, the interpretation of an SED dataset from single moraines is not always straightforward. Reuther et al. (2006) provided a discussion of factors that result in 'too old' and 'too young' ages from morainic boulders. At nearly all their sites in Bolivia, Smith et al. (2005a,b,c, 2008, 2009) implicitly equated the exposure ages of boulders on moraine crests with the age of the moraine. For example, they described Late-glacial moraines in the Zongo Valley at ~3400 m a.s.l. only based on SED results. Yet, the relation between the apparent exposure age and the age of the moraine deposition or even the time of the glacial advance referred to can be complex (Reuther et al., 2006). While the normal distribution of scatter of SED datasets from a single moraine with a favourable lithology can be explained by analytical uncertainties alone, biased or polymodal

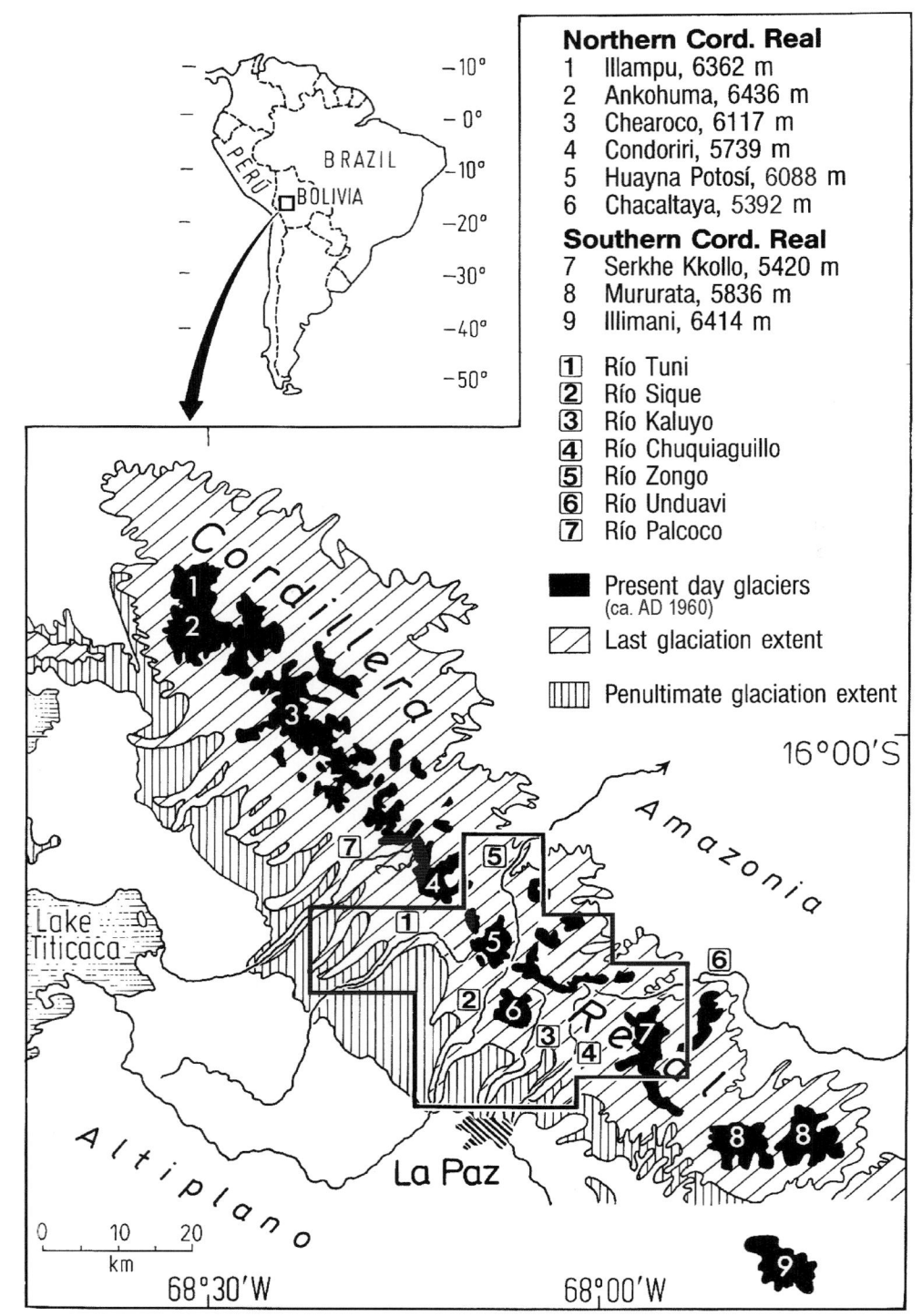

FIGURE 55.1 Map showing the area of investigation. The extent of the glaciation in the Bolivian Andes (shown here as presented by Lauer and Rafiqpoor, 1986) must be revised. The penultimate glaciation extent could only partly be identified in the field; the last glaciation extent mainly belongs to the pre-LGM glaciations. The area of Fig. 55.2 is indicated.

distributions can indicate the influence of geomorphological processes. This may arise from multiple glacial advances, complex moraine stratigraphy, post-depositional overprinting of the moraines, spalling of rock surface or inheritance. As Reuther et al. (2006) demonstrate, any of these processes is potentially important for the interpretation of the boulder age but they are difficult to quantify. With respect to the Bolivian glacial chronostratigraphy, SED datasets have frequently been discussed by applying different age models to the exposure ages obtained with different calculations for the scaling system (cf. Smith et al., 2008; Zech et al., 2008). More recently, developed scaling systems yield much younger ages compared to calculations used by Smith et al. (2005a,b,c), but ultimately, local calibration studies will be necessary to validate the exposure-age calculations (Zech et al., 2008).

Here, the author shows that the interpretation of age distributions should be reached against the background, not only of systematic methodological uncertainties but also of detailed fieldwork on a case-to-case basis. A consistent glacial chronology is provided by combining geomorphological evidence with all available SED and ^{14}C dating (cf. Smith et al., 2008; Rodbell et al., 2009). At least glacier advances of three glaciations are represented and show that the SED ages published by many authors must be interpreted in the light of the glacial, periglacial and palaeoclimatic setting, together with the time they have been exposed to moraine degradation. Detailed geomorphological mapping of the landforms and an understanding of the depositional processes are required to interpret an age distribution for a single moraine and to choose the right approach for determining the moraine age (cf. Reuther et al., 2006).

55.2. GLACIAL GEOMORPHOLOGICAL SEQUENCES OF THE CORDILLERA REAL

Apart from comprehensive mapping of glacial landforms and sediments of late Quaternary age and the palaeopedological observations in the field and the laboratory, ^{14}C and SED ages complement the studies. Although the development of SED during recent years and their first applications in the central Andes reveal a huge potential for establishing a sound glacial chronostratigraphy (cf. Zech et al., 2008, 2009b), in the following special attention is paid to the importance of the glacial setting and the climatic boundary conditions for the interpretation of moraine ages and their palaeoclimatic implications. The processes that affect the moraine sequences, groups and numbers of moraine ridges and moraine morphology are mainly determined by the climatic conditions that persisted during the glacial phase. LGM-age moraines in the arid climate of the Bolivian Altiplano and the adjacent slopes of the Cordilleras have undergone very little degradation of the morainic surface since deposition, whereas in the more humid regions of the eastern Andean mountain ranges, moraines have been remarkably degraded by denudation processes (cf. Reuther et al., 2006). Therefore, it is supposed that *moraine successions* of different valleys resemble each other in their main features, no matter whether they face the Altiplano or they are orientated to the Oriente. Further, periglacial phenomena, such as bedded slope deposits and rock glaciers, provide a better understanding of the glacial chronology (for further details, see Heine, 1995, 2011). Bedded slope deposits (stratified screes, Francou, 1990) result from permafrost. The lowermost occurrence of permafrost is bound to ~ 0 °C mean annual temperature. By combining LGM-age permafrost evidence with LGM-age moraines, not only temperature and ELA depressions can be estimated for the late Quaternary but also a distinction can be made between pre-LGM and LGM/post-LGM-age moraines.

For the first time, a detailed map of the glacial history from the tropical Andes is presented here (Fig. 55.2). This map shows the extent and age of the glaciation in a selected area of the Cordillera Real of Bolivia, based on geomorphological mapping in the field, relevant ^{14}C ages and SED, weathering characteristics and periglacial features (bedded slope deposits). The hierarchical order of our discussion is as follows: (i) the relative chronostratigraphy of the geomorphological forms of the glaciated valleys (the glaciated valleys evolved through a sequence of 'stages' characterised by certain processes and forms, cf. Troll and Finsterwalder, 1935), (ii) periglacial slope deposits on moraines and hillslopes and (iii) relevant ^{14}C and SED ages.

By using this approach, the author does not follow Rodbell et al. (2009) and Smith et al. (2008) who included a great number of ages from the literature and disregarded only ages that had been questioned regarding accuracy by the authors themselves, to develop LGM, Late-glacial and Holocene chronologies. Rodbell et al. (2009) and Smith et al. (2008) failed to discuss the accuracy of the ages they used with respect to the overall geomorphological setting.

55.3. PRE-WEICHSELIAN GLACIATION

Glacial limits in the region were mapped in the Cordillera Real and in some additional valleys in the Cordillera Quimsa Cruz, together with the occurrence of both active (above ca. 4600 m a.s.l.) and inactive bedded slope deposits (between ca. 4600 and 3950 m a.s.l.) on moraine ridges (Heine, 1995, 1996, 2000). The inactive bedded slope deposits developed during the LGM and during earlier cold periods. The occurrence of inactive bedded slope deposits on moraine slopes shows that the morainic ridges formed before the LGM and are older than ~ 21 ka. By using the occurrence of inactive bedded slope deposits on moraine ridges, a distinction can easily be made between (i) moraines older than the LGM and (ii) moraines of LGM and younger age. The results suggest that the pre-Würmian (pre-Wisconsinan) glacial limits of Lauer and Rafiqpoor (1986) require revision (Fig. 55.1). At least three pre-~ 21 ka glacier advances are represented by moraine ridges

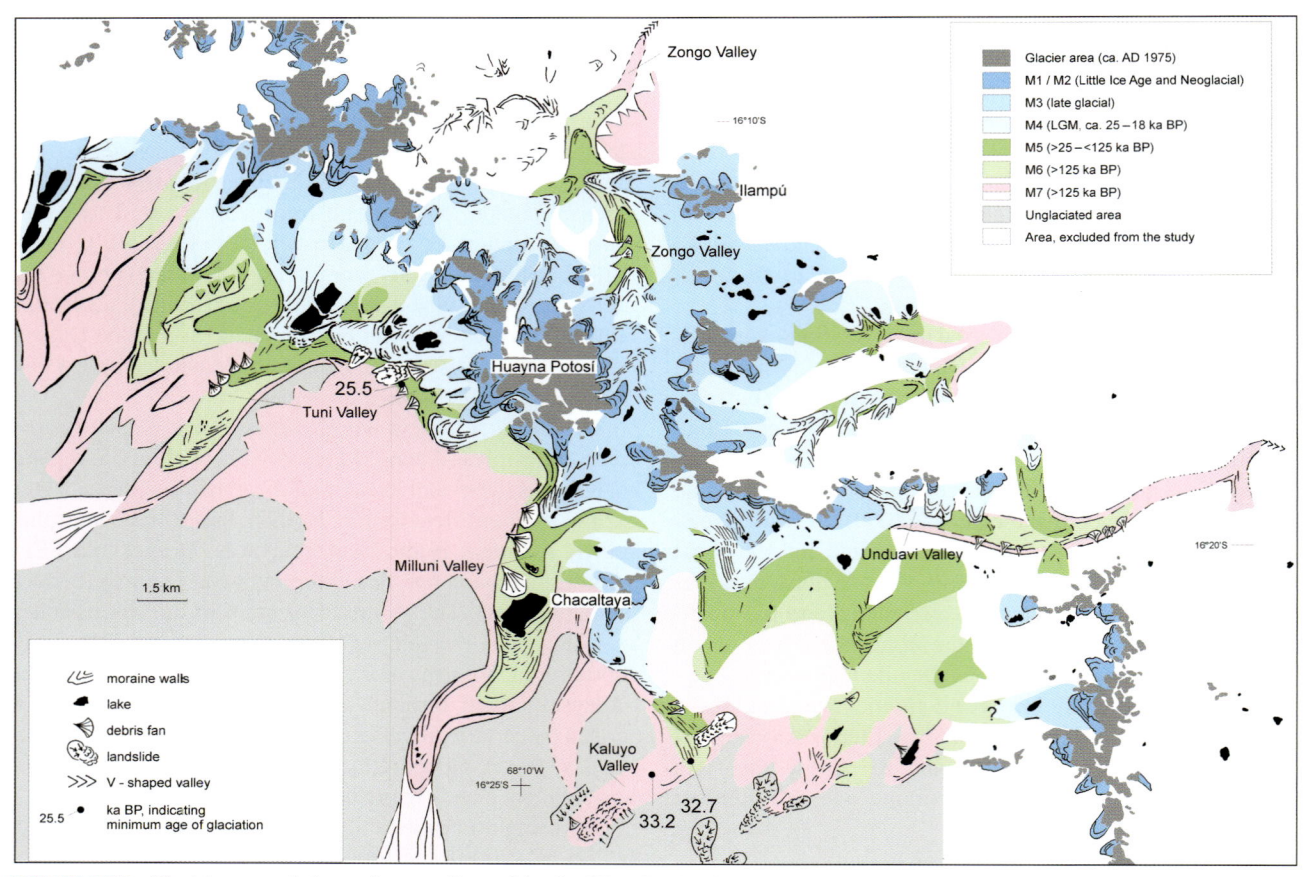

FIGURE 55.2 Glacial geomorphology of some valleys of the Cordillera Real (Milluni, Zongo, Kaluyo, Tuni, Unduavi). Neoglacial and Little Ice Age moraines, after Jordan (1991), supplemented. Additional observations of the Río Zongo Valley by Müller (1985). The moraines of groups M5–M7 were previously thought to be of LGM age. The revised glacial chronology shows that the moraines of group M4 are actually of LGM age and the moraines of group M5 are older than 25 ka.

in different valleys of the Andes (Fig. 55.2). In the Zongo Valley, the oldest lateral and end moraines (group M7) are found at 2800–2850 m a.s.l. On the basis of the weathering of the till, the morphology of the moraine ridges, the mean lower limits of the terminal moraines and the occurrence of relict stratified slope deposits, the M7 and M6 moraines could be as old as (or older than) Marine Isotope Stage (MIS) 6 and 8, respectively (Heine, 2004).

The pre-LGM age of M5 tills and moraine ridges is corroborated by the glacial sequence found in the Río Zongo, the Río Kaluyo, the Río Tuni and the Río Unduavi Valleys. Stratified slope deposits occur on the M5 moraine ridges down to 3950 m a.s.l. in the Zongo Valley. In contrast, the M4 moraine ridges that terminate at about 4120 m a.s.l. have not been influenced by periglacial processes and do not show bedded slope deposits. In the Kaluyo Valley, sharp-crested moraine ridges of the M4 group without stratified slope deposits reach down to 4380 m a.s.l., and in the Tuni Valley, moraines of group M4 down to around 4400 m a.s.l. While the moraines of groups M3 and M4 lack bedded slope deposits below 4300 m a.s.l., the M5, M6 and M7 moraines are covered in all valleys by stratified slope deposits even in lower altitudes and thus record an age of $>\sim 21$ ka (Heine, 1995, 2000, 2004). Further, lacustrine sediments, displaced by landslides, intercalate with glacial deposits in the Tuni Valley and have been dated to ~ 25.5 ka (Heine, 1996, 2004). In many places, alluvial fans cross the moraine ridges of groups M5–M7. Organic material from these fans in the Kaluyo Valley has yielded radiocarbon ages of roughly 28 and 27 ^{14}C ka BP (~ 33.2 and ~ 32.7 ka). It should be noted that landslides have redeposited at least part of the supposedly LGM till deposits from the valley flanks to the valley floor. Observations by the author have shown that the till in these landslides is of pre-Weichselian age and is intermixed with younger organic material from the valley floor. In many cases, the radiocarbon ages of these organic deposits have been used to date the tills. This explains why a number of young radiocarbon dates of till in combination with peat have been obtained from low positions in the valleys (cf. Graf, 1979; Villarroel and Graf, 1979). Consequently, they present no evidence for large-scale glacier advances during the LGM (Heine, 1995).

Although direct age control concerning the pre-LGM glacial history on the Bolivian Altiplano is extremely poor (Heine, 2004; Mark et al., 2004; La Frenierre et al., 2011), the first ^{10}Be SED (Smith et al., 2005a) provided additional evidence for pre-LGM glacial advances. Smith et al. (2005a) presented ages from boulders of two M7 moraine crests of the Milluni Valley (ca. 16°24′S, 68°11.5′W; ~4600 m a.s.l.). The ^{10}Be exposure ages fall within the last glacial cycle (34–23 ka). Reasoning that post-depositional erosion of the moraine matrix and of the boulder surface is the controlling process of the age distribution leads to the 'oldest-age' method (Reuther et al., 2006). This means that the oldest exposure age can be considered to be a good estimate for the minimum deposition age of the moraine, unless it is a clear outlier or there are stratigraphical inconsistencies (Zech et al., 2008). If large, tall boulders are sampled, the scatter in the dataset can be minimised since such boulders on the moraine crest are less likely to have been exhumed from till sediment following deglaciation or toppled under gravitational forces (periglacial phenomena documented by bedded slope deposits). The size of the boulders dated by Smith et al. (2005a: Group C) on the pre-LGM moraine crests of the Milluni Valley was relatively small (height between 0.4 and 1.2 m) so that sampling of small boulders to increase the dataset will not improve its quality (Reuther et al., 2006). It is suggested that the oldest ^{10}Be exposure age (~34 ka) represents the minimum age of the moraine. The ^{10}Be exposure ages corroborate the pre-LGM age of the M7 moraines by Heine (1996, 2004). Besides the SED of the M7 moraines, Smith et al. (2005a) dated boulders from the M3 moraine crests about 8 km up-valley in the Milluni Valley, which gave Würmian (Wisconsinan) Late-glacial ages between 17.14 and 8.8 ka (Group B of Smith et al., 2005a). The sequence of the moraine groups in the Milluni Valley (Fig. 55.2) shows at least three more moraine groups between the terminal moraines M3 and M7. The distance between the SED-dated M3 and M7 moraine groups is about 8 km, and the moraine groups M4–M6 show pronounced moraine walls and a great number of recessional moraines. It seems unlikely that the M4–M6 moraines were laid down by glacier fluctuations within only a few thousand years after the LGM, during the Late-glacial phase. Based on the moraine sequences (M1–M7) of the different valleys, the strong degree of weathering (cf. Leinweber et al., 1996) of the M7 morainic material, the periglacial bedded slope deposits covering the M7 till, and the alluvial fans that pre-date the local LGM glacier advance and post-date the M7 moraines, it is suggested that the local LGM glacier advance did not reach as far down-valley as was postulated by Smith et al. (2005a). Moreover, based on detailed stratigraphies of glacial and fluvial sequences, Troll (1929) and Troll and Finsterwalder (1935) could already show that the tall lateral moraines (M7) did not represent the last glacial cycle (Würmian) but, at least, the penultimate glacial cycle (Rissian). Contrary to these and other earlier discussions, Smith et al. (2005a) only rely upon their ^{10}Be exposure ages.

Since similar lateral moraines in the Peruvian Junin region, dated by Smith et al. (2005a,c: Group D), show populations of exposure ages that span tens of thousands to hundreds of thousands of years, the oldest Bolivian M7 moraines are thought to be of pre-penultimate glaciation age. These moraines are equated with the Group D moraines in neighbouring Peru. In the latter, Smith et al. (2005c) distinguished two major age clusters from 10 dated moraines, with ranges of 440–325 and 280–50 ka or 530–380 and 315–51 ka, respectively. According to different estimates of boulder erosion and surface uplift, they concluded that low long-term rates of geomorphological processes would help explain the extraordinary preservation of these older moraines. In the Peruvian Cordillera Blanca, Farber et al. (2005) ^{10}Be-dated the oldest moraines at the lowest elevations to greater than 400 ka, thereby corroborating the Bolivian observations. Recently, Smith and Rodbell (2010) presented a detailed glacial chronology from the southern Peruvian Cordillera Blanca also based on ^{10}Be-dated boulders; their dating indicated that glaciations culminated at ca. 65, 32 and 18–15 ka and that a larger advance occurred prior to 65 ka.

Müller (1985) and Troll and Finsterwalder (1935) described 'Hochflächenmoränen' of an old glaciation (moraines of high plains (Altiplano moraines)). Valleys with flat bottoms were incised into these high-plain moraines and were later occupied by glaciers of younger glaciations. Figure 55.3 shows an ancient morainic relief north of the Lake Titicaca. The age of these moraines is unknown. Since these glacial forms and sediments show that during glaciation the upper Sorata Valley was not incised that deep and/or the upper Sorata Valley was filled with glacier ice, the Altiplano moraines document glacial conditions that were different from those during the younger Pleistocene. From Peru, Smith et al. (2005c) and Farber et al. (2005) have also reported very old (greater than 800 ka) glacial moraines shaped like those in Bolivia.

55.4. THE LAST GLACIAL MAXIMUM

Until recently, some authors have assumed that glacier advances occurred at the end of MIS 3 around 30 ka ^{14}C BP (~34.7 ka) in the Bolivian Andes (e.g. Seltzer, 1992; Argollo and Mourguiart, 1995; Seltzer et al., 2002). However, the evidence is ambiguous. It appears that the Bolivian glaciers advanced and reached their LGM extent between 25 and 15 ka ^{14}C BP (~32 and ~18 ka) (Jordan et al., 1993; Heine, 1996; Fig. 55.4). Only a few bracketing radiocarbon dates exist. Based exclusively on radiocarbon dates, it is not yet clear whether the LGM glaciation culminated

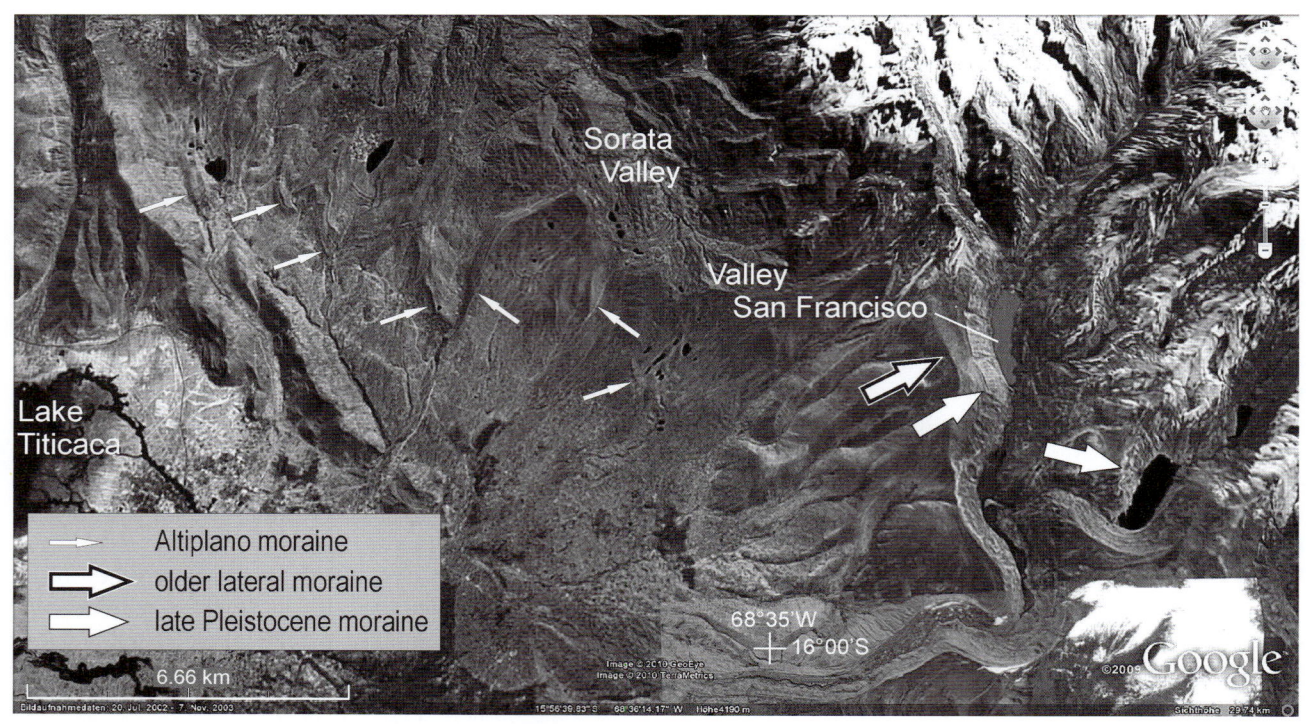

FIGURE 55.3 Satellite image of the Altiplano east of Lake Titicaca, showing the 'Altiplano moraines' and Pleistocene moraines of San Francisco Valley. The Altiplano moraines were thought to be of pre-penultimate glaciation by Troll (1929) and Troll and Finsterwalder (1935) and of penultimate glaciation by Lauer and Rafiqpoor (1986). They are definitely older according to geomorphological evidence and SED of similar moraines from Peru (see text). Note that, in the San Francisco Valley, the Late Pleistocene moraines, dated by SED (Zech et al., 2007), overlie the older, lateral, tall moraines.

around 20 ka ^{14}C BP (~24 ka) (Jordan et al., 1993; Heine, 1995, 1996), between 16 and 12 ka ^{14}C BP (~19 and ~14 ka) (Argollo, 1980; Seltzer, 1992; see critical discussion in Heine, 1996) or between 14 and 13 ka ^{14}C BP (~17 and ~15.7 ka) (Clapperton et al., 1997; Clapperton and Seltzer, 2001; Mark et al., 2004). Timing of the local LGM with SED is still also ambiguous, since Smith et al. (2005a,b,c, 2008, 2009) and Zech et al. (2007, 2008) published their data (cf. La Frenierre et al., 2011). Smith et al. (2005a) dated the local LGM moraines to 34–15 ka. Zech et al. (2007, 2008) recalculated the extensive datasets of Smith et al. (2005a,b,c) and suggested that the local LGM in the Bolivian tropical Andes occurred between 25 and 20 ka, that is, roughly synchronous with the global temperature minimum. However, the recalculated ages of Smith et al. (2005a) refer to boulders on the M7 moraines from the Milluni Valley (Figs. 55.2 and 55.5). The SED of M7 moraines by Smith et al. (2005a, 2008) in the Milluni Valley (Fig. 55.5) yielded ages between 34 and 15 ka (Smith et al., 2005a) and recalculated between 32 and 15 ka (Smith et al., 2008), respectively. As shown above, these surface exposure ages do not record the ages of the moraine deposition. The SED ages are younger than the deposition of the M7 moraines because of degrading of the high-moraine ridges since they were abandoned by the downwasting glacier. Moraine degradation was active for tens of thousands of years, when severe periglacial climatic conditions with intense gelifluction processes (cf. Troll and Finsterwalder, 1935) existed during the cold stages (e.g. MIS 4 and 2). During these stages, the lower limit of permafrost and of processes causing erosion and active bedded slope deposits was depressed by about 800 m (Heine, 1995). Apparently, boulders on the M7 moraine ridges emerged after the sediments were eroded. Besides boulder exhumation, secondary rotation and erosion of the boulder surface might also have occurred, together with upheaval processes that could have brought boulders to the surface (Reuther et al., 2006). The SED age distribution from the M7 moraine boulders scatters very much, thereby supporting the hypothesis that the ages are too young.

Zech et al. (2007) ^{10}Be-dated moraines in the Valle San Francisco in the Cordillera Real using the Lifton et al. (2005) calculation scheme. They yielded exposure ages of boulders from an outer moraine at 4670 m altitude between 24.1 and 15.3 ka and from the dominant inner lateral moraine between 19.3 and 17.2 ka. Hypothetically, the outer moraine can be correlated with the M5 group and the inner moraine with the M4 group.

According to the author's investigations, the LGM climate was characterised by cooler (more than 4.8 °C temperature depression in ~4000 m altitude: Heine, 1995) and

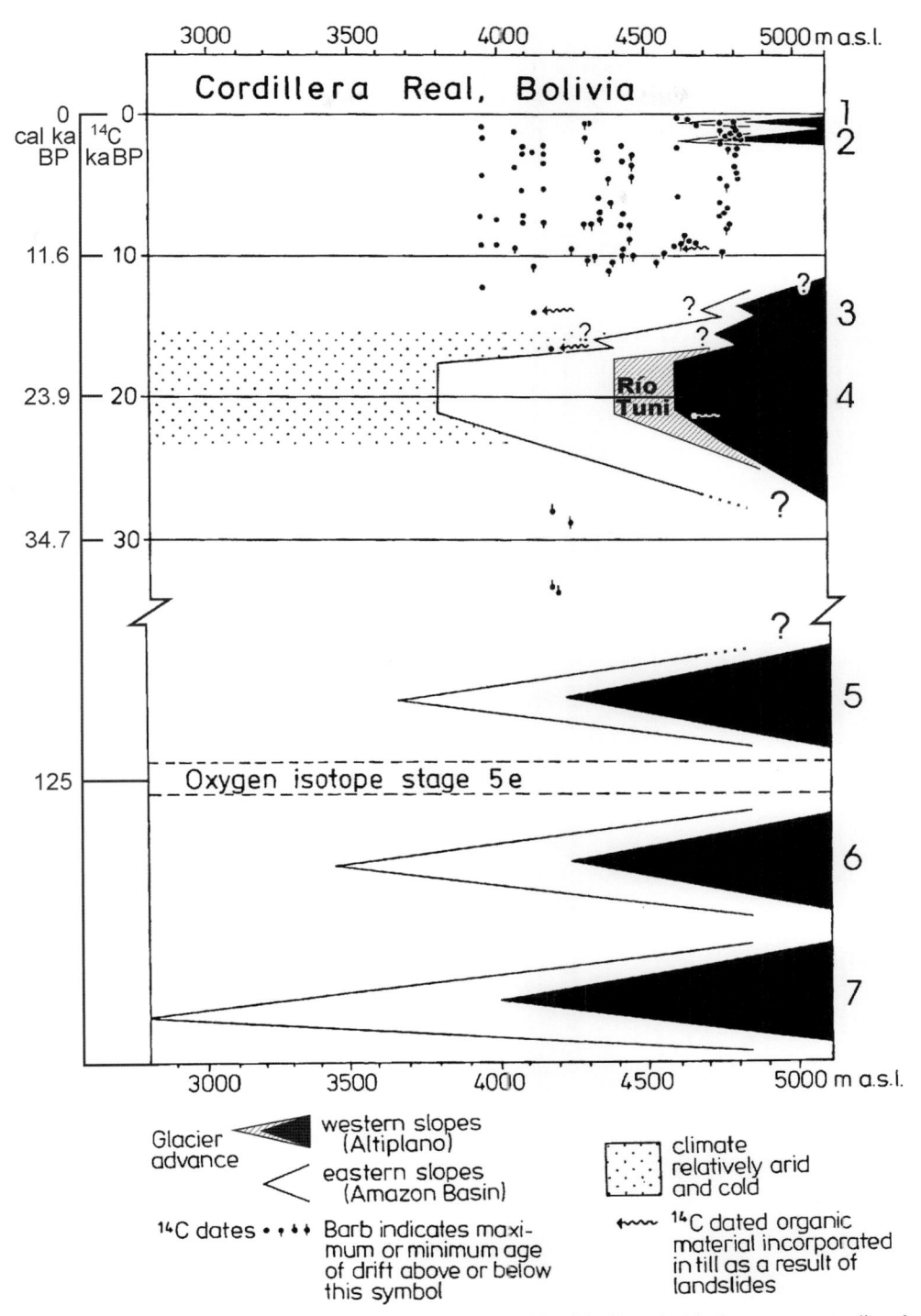

FIGURE 55.4 Glacial stratigraphy of the Cordillera Real, Bolivia. Note that the LGM and the Late-glacial advances are not radiocarbon dated. Relevant radiocarbon dates have been compiled from various sources (cf. Heine, 1995, 2004). Surface exposure dating of moraines will improve the chronostratigraphy if methodological uncertainties are carefully evaluated (see text).

drier conditions compared to that of today (Heine, 2000). The local LGM ELA was over 4750/4800 m a.s.l. in the western part of the Cordillera Real. Bedded slope deposits of pre-Holocene age occur below 4800 m a.s.l. Further, the steep climatic gradients (precipitation, humidity) from the Amazon Basin in the east to the Atacama Desert in the west, and from the lowlands of Amazonia to the Cordilleran peaks, exceeding 6000 m elevation, have resulted in

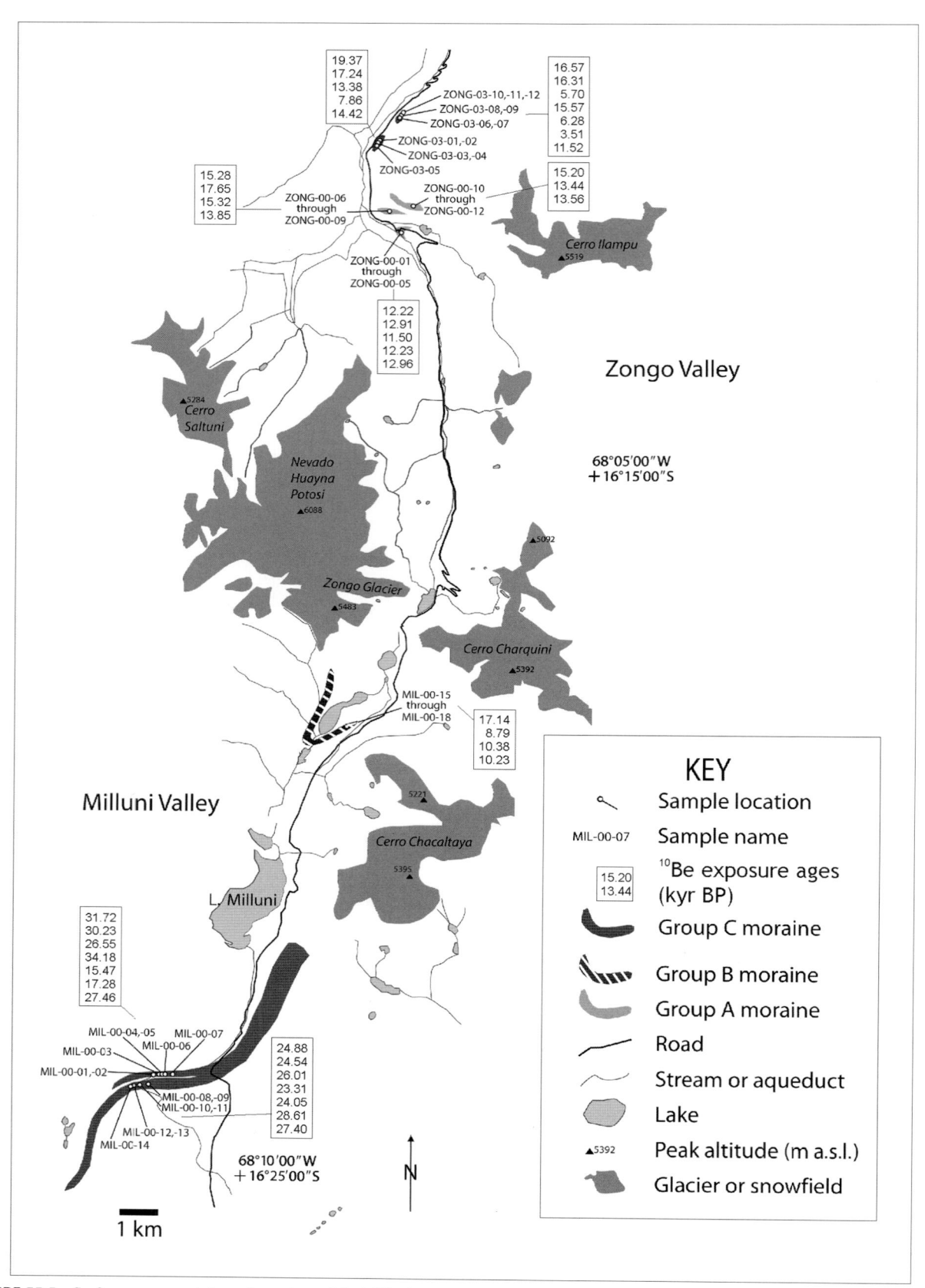

FIGURE 55.5 Surface exposure dating of moraines of the Milluni and Zongo Valleys (after Smith et al., 2005a).

pronounced differences in cloudiness, precipitation, radiation and temperature not only today (Jordan, 1991) but also during the Late Pleistocene. Although no precise dates can be given for the LGM-glacial advances, which might have occurred around 25–20 ka, it can be shown that the extent of the LGM glaciers was very small because of the dry climate (cf. Thompson et al., 1998). Many moraine ridges, formerly associated with the LGM glacier advances, are definitely older. Hence, LGM snow-line reconstructions (e.g. Klein et al., 1999; Porter, 2001; Seltzer, 2001; Smith et al., 2005b) are of limited value for the Bolivian Andes (Heine, 1995, 1996, 2000). Even cosmogenic dating of erratics from moraines provides limited absolute dating with sufficient accuracy to identify deposits of the local LGM. Moreover, palaeoclimatic implications derived from a reconstructed shift of the ELA suffer from the fact that ELA changes are dependent on temperature changes as well as of other climatic parameters (Müller, 1985; Kaser, 1995; Kaser et al., 1996; Kull et al., 2008).

Figure 55.2 gives examples for the extent of the LGM glaciers based on geomorphological mapping, palaeosols, radiocarbon dates, the distribution of active and inactive bedded slope deposits, as well as an evaluation of published radiocarbon and SED dates from the Bolivian Andes. Figure 55.6 illustrates an example for the reconstruction of an LGM glacier in the Bolivian Andes in the Zongo Valley. The local LGM ELA was at ca. 4670 m altitude (Heine, 2004; Fig. 55.6). Based only on SED, Smith et al. (2005a) suggested that LGM glaciers extended to ~4300 m in the Milluni Valley and to ~3000 m in the Zongo Valley, that

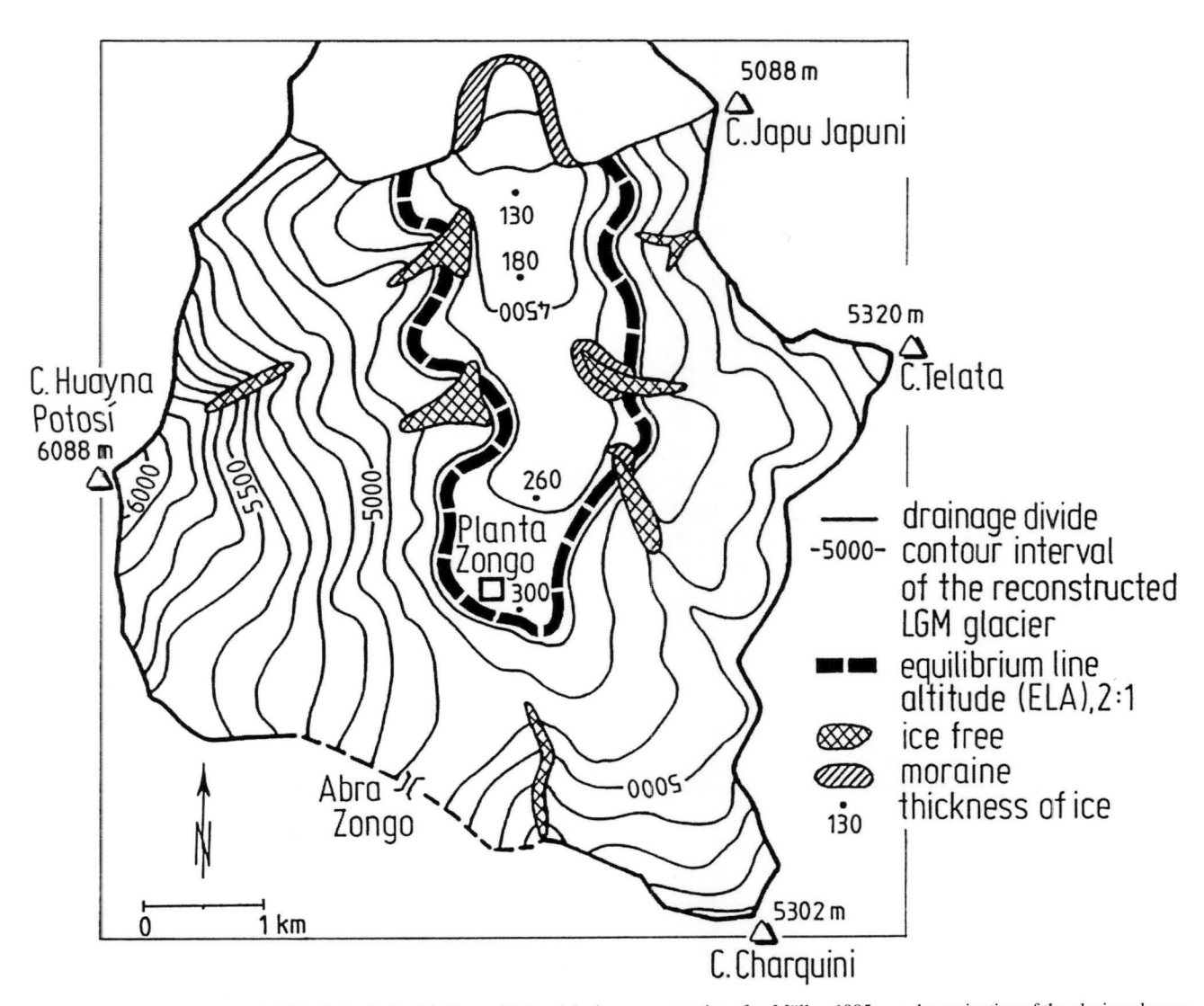

FIGURE 55.6 Reconstructed LGM glacier in the Río Zongo Valley (glacier reconstruction after Müller, 1985; age determination of the glacier advance after Heine, 1995). See also Fig. 55.2. The method used for this reconstruction is discussed by Müller (1985, pp. 84–104) and Gross et al. (1978).

is, the LGM glacier terminus extended further down-valley than Müller (1985) and Heine (1995, 1996) had suggested. Smith et al. (2005a) estimated that ELA depression of the local LGM was 300–600 m in the Milluni Valley and 800–1000 m in Zongo Valley. In the Zongo area, the modern ELA is not at ca. 5150 m altitude as Smith et al. (2005a) conclude but rises from ~4650 m on south-facing slopes in the eastern ranges (Ilampú, 16°11.5′S, 68°06.3′W) to ~5230 m altitude on north-east-facing slopes at the Huayna Potosí in the western Zongo catchment (16°15′S, 68°08.6′W) (Jordan, 1991). The local LGM ELA was not depressed to ~4200 m in the Zongo Valley (Smith et al., 2005a), but only to ~4650–4700 m altitude in the upper Milluni Valley and to ~4300 m in Ilampú tributary valley (cf. Fig. 55.2: M4 moraines in the Zongo Valley originated on two different mountains: Huayna Potosí and Ilampú). Based on the detailed modern ELA calculations by Jordan (1991), an LGM ELA depression for the Zongo catchment can be estimated at ~500–400 m in the upper valley; only 9 km to the north-east, the local LGM ELA depression is ~350–400 m, that is, only half of the value recorded by Smith et al. (2005a). The same is true for the Milluni Valley. Similar to the modern ELA (Jordan, 1991), the LGM ELA experienced great differences in altitude over small horizontal distances between the eastern mountain ranges and the central mountains of the Cordillera Real.

Research on tropical glacial fluctuations by Kaser (1995, 1999), Kaser et al. (1996) and Wagnon et al. (1999) has shown that, in the tropics, the ELA reacts generally less sensitive in comparison to that on mid-latitude glaciers, yet, more strongly, if forced by a change in temperature; major glacier tongues as well as small glaciers react sensitively on increasing ablation, whereas the reaction to long-term dynamic forcing is weak (Kaser, 1995). Therefore, in a place like Bolivia, LGM palaeotemperature estimates should not be based on LGM glacier reconstructions, as was done by Smith et al. (2005a). The distribution of inactive bedded slope deposits, which can be dated by their occurrence on moraine ridges and inactive rock glaciers, is more appropriate for an assessment of palaeotemperatures. Heine (1995) calculated, based on the shift of the lower limit of bedded slope deposits, an LGM temperature depression of more than 4.8 °C. This interpretation is corroborated by glacier–climate modelling: temperature depression during full-glacial times was more likely to exceed 5 °C (Zech et al., 2008; Kull et al., 2008).

The Late Pleistocene lake-level fluctuations of Lake Titicaca and other palaeolakes on the Bolivian Altiplano are recorded in their lacustrine sediments and landforms. Although it is very difficult to determine the precise age of the sediments and landforms (for problems related to radiocarbon dating, cf. Geyh et al., 1999; Argollo and Mourguiart, 2000), two Late Pleistocene/early Holocene highstands in lake levels have been recognised: the Minchín phase prior to 73–30 ka (Fornari et al., 2001) and the Tauca phase. The latter was dated by Argollo and Mourguiart (2000) to 14.0–11.5 ^{14}C ka BP (~17 and ~13.5 ka), by Fornari et al. (2001) to 16–12 ka and by Geyh et al. (1999) to 10.8–9.2 ^{14}C ka BP (~12.7 and ~8.5 ka). A recent study applying ^{14}C and U/Th dating techniques could not corroborate the existence of the Minchin wet period (Placzek et al., 2006). Shorelines from deep palaeolakes on the Altiplano were dated to 120–98 ka (called Ouki), 18–14 ka (Tauca) and 13–11 ka (Coipasa). Only shallow lakes existed between 98 and 18 ka (Zech et al., 2007). Both high lake-level stands have been correlated with undated glacier advances by many authors (e.g. Clapperton et al., 1997; Baker et al., 2001). Because of correlation uncertainties and dating problems of the lake-level highstands, neither the derived ages of the LGM glacier advances nor the extent of the LGM glaciers, as shown by many authors, can as yet be confirmed for the Bolivian Andes.

55.5. DEGLACIATION

There has also been much debate concerning the deglaciation from the LGM positions in Bolivia (Figs. 55.2 and 55.4). The numerous moraine ridges suggest that a number of glacier oscillations occurred during the deglaciation period (Jordan et al., 1993; Heine, 1995, 1996, 2000, 2004). However, the precise age of the different recessional stages is not known. This was also ascertained in the review of Rodbell et al. (2009) that highlighted the need for improved age control. In the past, some moraine groups have been correlated with the Tauca high lake-level stand of Lake Titicaca and of the Uyuni Salar area. However, there is no evidence for a synchronous occurrence of high lake levels and terminal moraines. Until now, because of the lack of age determinations, neither the Antarctic Cold Reversal (ACR) nor the Younger Dryas (YD) Chronozone can be correlated with any glacier advances or retreats (Heine, 1996; Heine and Geyh, 2002). The conspicuous M3 moraines are of Late-glacial age (>11.0 ^{14}C ka BP, ~>12.9 ka). The Late-glacial is defined here as the interval between ~18 and 11.6 ka.

Late-glacial ages for the M3 moraines are corroborated by SED ages from the Milluni Valley published by Smith et al. (2005a). The authors dated small boulders (0.3–0.4 m high) from a hummocky end moraine complex (Fig. 55.2: M3) and yielded ^{10}Be ages of 17.14, 10.38, 10.23 and 8.79 ka (Fig. 55.5). Smith et al. (2005a) used these ages for their glacial chronology and suggested that the construction of this group of recessional moraines occurred by 18 ka. The distribution of boulder ages and the morphology of the sampling sites, showing that the shape of the moraines were unstable because of dead-ice melting, indicate that moraine stabilisation continued over several thousand years during the Late-glacial period. The SED of M3 moraine group to ~18 ka is documented by only one ^{10}Be age from the Milluni Valley. A true age of the M3 moraines cannot be given. More recently, Smith et al. (2008) presented an age for these recessional moraines (Group B) between 20 and 15 ka. A correlation of the M3 moraine with recessional moraines of the San Francisco Valley ca. 60 km to the north-west and moraines of the Cochabamba area (Zech et al., 2007) is not possible. Zech et al. (2007) ^{10}Be-dated significant readvances during the Late-glacial to around 16 ka and between 13 and 11 ka and discussed the previously published SED Late-glacial chronologies from the Bolivian Cordillera Real.

In the Zongo Valley, Smith et al. (2005a) also ^{10}Be-dated boulders on four moraines. The ages were recalculated (Smith et al., 2008; Rodbell et al., 2009), and all were found to be younger than 18 ka, even at altitudes of ~3400 m a.s.l. (Smith et al., 2008). Smith et al. (2005a, 2008) correlated the upper moraines (M3, M4 and M5? in Fig. 55.2) with their Group A (14–11 ka) and the lower two (M5? and M6 in Fig. 55.2) with Group B (20–15 ka) from the Milluni Valley and from the Peruvian Junin region. Unfortunately, Smith et al. (2005a,b,c) presented data by latitude and longitude for the Zongo Valley sample sites that did not match the moraines shown in Fig. S1 of their paper. Therefore, the dated moraines of Smith et al. (2005a) can only be hypothetically correlated with the chronostratigraphy presented herein (cf. Fig. 55.5): The ^{10}Be ages of ZONG-00-01 through ZONG-00-05 represent moraine group M3 of a glacier that originated on the Ilampú and did not represent the Zongo Valley glacier; the ages of ZONG-00-06 through ZONG-00-09 represent the M4 moraine group of the Ilampú glacier; the ages of ZONG-00-10 through ZONG-00-12 represent moraine group M4 (or M5?) of the Ilampú glacier; the ages of ZONG-03-01 through ZONG-03-05 represent the M5 moraine group of the Ilampú glacier; the ages of ZONG-03-06 through ZONG-03-12 represent moraine group M6 of the combined Zongo Valley/Ilampú glacier. The ^{10}Be ages of the M3 and M4 moraines show a normal distribution of scatter and date these moraine groups to the Late-glacial to about 18 ka for M4 and 13 ka for M3.

The ^{10}Be ages of the lower moraines are widely scattered (Fig. 55.5). The interpretation of the SED dataset by Smith et al. (2005a) is ambiguous and inconsistent with the observations of Müller (1985) and Heine (1995, 2004). According to the moraine sequence of the Zongo Valley (Figs. 55.2 and 55.5) and the relative chronology of the moraines, slope deposits, alluvial fans and weathering properties of the tills, the ^{10}Be ages refer to moraine group M6 and pre-date the last glacial cycle. This is also corroborated by the low altitude of ~3500–3400 m a.s.l. It is thought that many processes, such as degradation of the moraine ridges, secondary rotation of the boulders, boulder-surface erosion, erosion of the fine moraine matrix and shielding of the boulder surface, influenced the ages and that the rate of the different processes was not uniform, even on the same moraine ridge. The shape of the M6 moraine ridges records a pre-LGM age. It is definitely impossible that these moraines represent post-LGM glacier movements in the Zongo Valley, as Smith et al. (2005a, 2008) deduced.

Many authors (e.g. Clapperton et al., 1997; Zech et al., 2007) discussed the correlation of the Late-glacial glacier advances/retreats with lake-level fluctuations of the Altiplano lakes (Lake Titicaca, Lake Poopo and Salar de Uyuni). Zech et al. (2007) observed evidence for Late-glacial glacier advances synchronous with the lake transgression phases (ca. 13.3 and 12–10 ^{14}C ka BP, i.e. ca. 16 and 14–11.6 ka) and therefore referred that glaciers are/were mainly more moisture- than temperature-sensitive and that precipitation became the limiting factor for glacial advances. Based on the SED of Bolivian moraines (Zech et al., 2007, 2009a), Kull et al. (2008) concluded that the precipitation sensitivity of the glaciers was corroborated by glacier–climate modelling and explained the observed synchrony of the glacial advances on the Bolivian Altiplano. Although the synchrony between glacier movements and lake-level changes can be postulated, the accuracy of the glacial history derived from SED ages is still equivocal. As shown above, contrary to the ^{10}Be-dated moraines and the ELA reconstructions based thereon (cf. Smith et al., 2005a,b,c), LGM glacier extent of the tropical Andes appear to be smaller than previously thought (Heine, 2011), but with large intra-regional variance (cf. Mark, 2008).

55.6. THE HOLOCENE

For the Holocene, Neoglacial and Little Ice Age (LIA) glacier advances have been recorded (Seltzer, 1992; Heine, 1995, 1996, 2004). The M2 moraines were dated to a Neoglacial age (ca. 3.5–1.5 ^{14}C ka BP); the M1 moraine sequences are of LIA age (ca. AD 1350–1850). Heine's (1996) oldest minimum-limiting radiocarbon age of the M2 moraines is 1565 ± 85 ^{14}C a BP (~260–650 AD), and according to the results of Müller (1985), the Neoglacial M2 moraines are younger than ~3.5 ^{14}C ka BP (~3.8 ka).

Recently, Rodbell et al. (2009) summarised the Würmian Late-glacial and Holocene glaciations of the tropical Andes and concluded that the spatial–temporal pattern of Holocene glaciation exhibits tantalising but incomplete evidence for an early to mid-Holocene ice advance(s) in many regions, but not in the arid subtropical Bolivian Andes. This is because here the moraines deposited during the past millennium record the most extensive advance of the Holocene. Rodbell et al. (2009) did not refer to Müller's (1985) or Heine's (1996) observations and concluded that there is no evidence for expanded ice cover that exceeded the maximum ice extent of the past millennium. Müller (1985) mapped moraines in detail, used lichenometry to date 'sub-recent' moraine groups and concluded that the Neoglacial glaciers advanced only, if at all, a few hundred metres further than the LIA glaciers. She remarked that many glaciers might have overrun their Neoglacial (M2) moraines during the LIA and referred to Röthlisberger (1986) who presented several sites from the Peruvian Cordillera Blanca where stacked Neoglacial moraines were dated by 15 ^{14}C ages from wood and soils to the Neoglacial since <3.5 ^{14}C ka BP. Often the maximum Neoglacial/LIA glacial extent was reached at first during the LIA. Moraines that correlate with the LIA of the Northern Hemisphere are seen in all presently glacierised mountain ranges (Jordan, 1991); most of these date to within the past 450 years (Rodbell et al., 2009).

On the southern side of Huayna Potosí, a lateral moraine dammed a small stream. Behind the moraine wall (16°17′51″S; 68°09′17″W; ~4775 m a.s.l.), lacustrine silty clays (0.2 m) were deposited starting at about 1565 ± 85 ^{14}C a BP (~260–650 AD), whereas rapid peat growth (1 m) began no earlier than 360 ± 55 ^{14}C a BP (~1446–1642 AD). These observations show that basal peat ages are of little value for yielding true moraine ages. Nevertheless, they corroborate the results of Seltzer (1992) that show at least two phases of Neoglacial advances of near equal extent. In their synthesis of Neoglacial glaciations in the tropical Andes, Rodbell et al. (2009) concluded that the most extensive Holocene glaciation in the Cordillera Real occurred just prior to the LIA. They also show that there was an apparent lack of moraines that pre-date the past millennium in the Cordillera Real and further suggested that perhaps aridity precluded glaciers in these regions from advancing earlier. Since the LIA glacial advances were in many places as extensive as or even more extensive than earlier Neoglacial advances in Bolivia, they had not been identified with coring and dating of sediments from inside the morainic arcs and by using SED and lichenometric dating. To resolve these specific chronological problems, only the stratigraphical methods used by Röthlisberger (1986) will provide bracketing ages for moraines and will improve upon the present state of the moraine chronology. It is worthwhile noting that radiocarbon ages of many slope deposits from many glaciated valleys record solid evidence for slope processes that represent the environmental conditions during the Holocene since about ~8.0 ^{14}C ka BP (~9 ka). Periglacial slope deposits were active in an altitude of 4300 m a.s.l. only during the LIA. At least three periglacial debris layers could be identified, being younger than 645 ± 60 ^{14}C a BP (~1269–1413 AD) and 510 ± 60 ^{14}C a BP (1296–1485 AD) and recording lower temperatures during the past ~700 years. During the Neoglacial, not only (periglacial) slope and alluvial fan deposits moved at higher rates but peat also developed more rapidly since ~3.5 ka (relevant maximum-limited ^{14}C ages from slope deposits: 3170 ± 70 (1614–1271 BC), 2705 ± 70 (1023–771 BC), 2320 ± 80 (752–194 BC), 2315 ± 70 (748–195 BC); from peat: 4445 ± 85 (3353–2915 BC) and 2290 ± 100 (753–98 BC)). The data from slope deposits and peat bogs can help in the unravelling of the Holocene glacial chronology. Similar to the observation that the moraine sequences do not show moraine groups of Holocene age predating the Neoglacial (cf. Müller, 1985; Jordan, 1991; Seltzer, 1992), there is no evidence of an intensification of geomorphological processes between ~8.0 and ~3.5 ka. Palaeohydrological evidence has illustrated a consistent overall pattern of aridity from the Late Pleistocene through the middle Holocene with wetter condition starting around 3.4 ka, while the wettest period began 2.3 ka (Abbott et al., 2003; Rowe and Dunbar, 2004). Only during the interval between roughly 3.5 ka and the LIA is there evidence for an increase in precipitation, followed by a reduction of the LIA temperatures. This is recorded by the depression of the lower limit of periglacial processes and dead ice in low areas (cf. Hurlbert and Chang, 1984). Thus, the first phase of the Neoglacial glacier advances was caused mainly by higher precipitation, whilst during the second phase, the LIA, it was caused by lower temperatures and higher precipitation compared to modern times. This is confirmed by the shape of the LIA glacier tongues that can be reconstructed from the lateral moraines. The moraine roots from which the lateral walls begin are the same as those of the oldest LIA moraines (and the buried M2 moraines?) as well as for the youngest LIA moraines. Decreasing precipitation during the LIA resulted in shorter glacier tongues without an increase of the ELA. Only the post-LIA moraines represent an increase of the ELA (cf. Jordan, 1991). These climatic conditions, a humid and cold first part of the LIA and a drier and cold second part after ca. AD 1700, are confirmed by Thompson et al. (1986) from the Quelccaya ice core and by Liu et al. (2005) from the Sajama ice core.

Jordan (1991) mapped all modern glaciers with their 'sub-recent' moraines in the Bolivian Andes and established a characteristic sequence of moraine groups for the LIA, disregarding the many annual moraine walls of the glacier foreland. On the youngest moraines (I), lichens are absent (apart from microscopically detected lichens, Müller, 1985); ~10 flowering plants grow on the moraines which formed in AD 1922–1927 (Müller, 1985). Moraine group II, III and IV shows regular lichen growth and ~60 flowering plants and was dated to the mid-nineteenth century to the early LIA, perhaps around AD 1700 (personal communication, Jordan, 1991). The material of the oldest moraine group is dark because of the cover with the lichens *Aspicilia* sp. and *Sporastatia* sp. Weak soil development was observed and the deposition was thought to be early LIA (Müller, 1985) or older. These moraines are situated immediately beside the group IV moraines or they were covered by group IV moraines.

Rabatel et al. (2008) dated moraines of 15 glaciers in the Bolivian Eastern Cordillera by lichenometry and presented a detailed chronology of glacier advances and retreats during the LIA, ignoring the comprehensive work of Müller (1985) and Jordan (1991). Using the shape of moraines (size, height, slope of the outer and inner side), the continuity of ridges on the proglacial margin, any evidence that moraines have (or have not), removed previous deposits, and the position of moraines along the glacier foreland, Rabatel et al. (2008) established a sequence of 10 moraines. The lichen measurements reveal that Bolivian glaciers reached their maximal extent during the second half of the seventeenth century (Rabatel et al., 2008) or rather AD 1630–1680 (Jomelli et al., 2009). This glacier maximum coincides with the Maunder minimum of solar irradiance and with observations by the author of rock glaciers that developed inside the Neoglacial moraines (M2). Glaciers began retreating at varying rates in the Bolivian Andes after AD 1740. The glacial retreat was moderate but continuous until about AD 1870 (Rabatel et al., 2008).

A correlation of the moraine sequence of Rabatel et al. (2008) with the sequence presented by Jordan (1991) is difficult. In many glacier forelands (e.g. Yaypuri/Jankho Loma glaciers, Southern Charquini glacier), Jordan's (1991) moraines I, II, III and IV equate to the moraines M9, M6, M3 and M1, respectively, of Rabatel et al. (2008). Yet, the moraines M6–M10 of the Western Huayna Potosí glacier, as shown in the figure of Rabatel et al. (2008), are definitely younger than indicated by the authors themselves. They were deposited after December 1907 as is clearly documented by a photo (Jordan, 1991). Moraines M9 and M10, dated to ca. AD 1860/1870 and ca. AD 1910 by Rabatel et al. (2008), must have developed during the 1920s according to the interpretation of many photographs of the first half of the past century (Müller, 1985). Therefore, the interpretations from the moraine ages, presented by Rabatel et al. (2008) regarding the velocity of glacier recession, glacier volume, ELA changes, etc. should be revised.

Notwithstanding these critical remarks, a synthesis of the research on LIA environmental conditions and glacier fluctuations (e.g. Müller, 1985; Jordan, 1991; Heine, 1995, 1996; Rabatel et al., 2008) confirms the local LIA chronology and proves that glacier response in Bolivia during the LIA was mainly driven by a combination of cooler temperatures (caused by solar energy modulation) and regional changes in precipitation (Rabatel et al., 2008).

55.7. MAPPING GLACIAL LIMITS

The occurrence of inactive bedded slope deposits, overlying the pre-LGM moraines (groups M5–M7), as well as the relative dating of these inactive slope deposits by moraine sequences, indicates that the altitudinal belt with active bedded slope deposits was depressed by about 800 m during the LGM (Heine, 1995). Bedded slope deposits are characteristic phenomena of the periglacial altitudinal belt and are active in Bolivia today above ca. 4800 m a.s.l. Since the lower limit of the active stratified slope deposits depends on the thermal regime (mean annual temperature), variations of the lower limit of inactive bedded slope deposits reflect changes in mean annual temperature. The last period that saw extensive periglacial processes and the formation of bedded slope deposits down to 4000 m a.s.l. occurred during the formation of the M4 moraine group. On the basis of stratigraphical correlations, it must be attributed at least to the LGM. Pre-LGM moraines have been identified by mapping the occurrence of inactive bedded slope deposits on different tills and moraine ridges (Heine, 1995, 2000). Many moraine groups, hitherto thought to be of LGM age, are covered by stratified slope deposits down to elevations of about 4000 m a.s.l. These 'buried' moraines are definitely older than the LGM.

A major problem arises from the frequent occurrence of landslides in the Bolivian Andes. Till of pre-LGM age can be redeposited by landslides onto younger peat on the valley floor. The radiocarbon dates from peat overridden by landslide material (redeposited till) cannot give correct ages for the glacier advances. This must be taken into account for future work on the glacial chronostratigraphy.

55.8. REMAINING QUESTIONS

The Bolivian glacial chronology suffers from a lack of absolute dating, although much progress has been achieved since 2004 (Ehlers and Gibbard, 2004). The precise ages of the LGM maximal glacier extent and of the Late-glacial advances are not yet known despite many efforts in dating with new methods (SED, cf. Smith et al., 2005a,b,c, 2008) (Fig. 55.4). Using many of the published ^{14}C and SED ages (cf. SED ages of boulders from the M7 moraines of the Milluni Valley and from the M3–M6 moraines of the Zongo Valley) to establish a glacial chronology, but without evaluating the sound and consistent quality of these data, ambiguous results are achieved (Rodbell et al., 2008, 2009; Smith et al., 2008, 2009). Only multiproxy studies, combined geomorphological and geochronological evidence, can solve the chronological problems. In future, more reliable age determinations of moraines must be correlated to robust well-calibrated palaeoclimate evidence and the results of sound geomorphological mapping of moraine sequences, shape of moraines, internal structure, weathering properties, (relict) talus, active/inactive slope deposits (solifluction-mantled slopes), intercalation of forms and sediments (glacial–fluvial–colluvial), cycles of glacial and paraglacial sedimentation (cf. Ballantyne,

2002) and the like. Because of the steep climatic gradient from the Amazon Basin to the western Altiplano and the deeply incised valleys heading to the Amazon Basin, the terminal moraines of the different glacier advances are located up to 800 m (e.g. M4, during the LGM) lower in the east than in the west of the Cordillera Real (Figs. 55.2 and 55.4). The M7 terminal moraines (pre-penultimate glaciation and/or older) are accordingly found up to 1200 m lower in the east. Apart from the dating problems, there are still many palaeoclimatic problems to be solved. When and why did the LGM climate become more arid in comparison to that of today? Is there any evidence for an ACR- and/or YD-age glacier oscillation in Bolivia? Did the Neoglacial and LIA glaciers advance and retreat synchronously in Bolivia, Peru, Ecuador and Colombia on the one hand and in Chile and Argentina on the other, or were there differences north and south of the Equator (cf. Zech et al., 2008)?

Further, it remains unclear why the oldest glacial advances (M7 and M6 moraine groups) extended so far down-valley in the eastern cordilleras from Bolivia to Ecuador (Heine, 2011; Fig. 55.2). The age of these ice advances is still very uncertain.

ACKNOWLEDGEMENTS

I would like to thank the *Deutsche Forschungsgemeinschaft* (DFG) and the *Volkswagen-Stiftung* for financial support of my research in South America, and M.A. Geyh for numerous absolute age determinations and discussions. I am indebted to many persons for assistance in the field and the laboratory. I would like to thank P. Gibbard and P. Hughes for critically reviewing the chapter.

REFERENCES

Abbott, M.B., Wolfe, B.B., Wolfe, A.P., Seltzer, G.O., Aravena, R., Mark, B.G., et al., 2003. Holocene paleohydrology and glacial history of the central Andes using multiproxy lake sediment studies. Palaeogeogr. Palaeoclimatol. Palaeoecol. 194, 123–138.

Argolo, B.J., 1980. Los Pie de Montes de la Cordillera Real entre los Valles de la Paz y de Tuni, Estudio Geológico, Evolución Plio-Cuaternaria. Ph. D. thesis, Universidad Mayor de San Andrews, La Paz. 86 pp.

Argolo, B.J., Mourguiart, P., 1995. Palaeohidrología de los últimos 25,000 años en los Andes bolivianos. Bulletin de l'Institut francais d'Études Andines 24, 551–562.

Argolo, B.J., Mourguiart, P., 2000. Late Quaternary climate history of the Bolivian Altiplano. Quatern. Int. 72, 37–51.

Baker, P.A., Seltzer, G.O., Fritz, S.C., Dunbar, R.B., Grove, M.J., Tapia, P.M., Cross, S.L., Rowe, H.D., Broda, J.P., 2001. The history of South American tropical precipitation for the past 25,000 years. Science 291, 640–643.

Ballantyne, C.K., 2002. Paraglacial geomorphology. Quatern. Sci. Rev. 21, 1935–2017.

Clapperton, C.M., 1981. Quaternary glaciations in the Cordillera Blanca, Perú, and the Cordillera Real, Bolivia. Revista Centro Interamericano de Fotointerpretación C.I.A.F., Bogotá, 6, 93–111.

Clapperton, C.M., Seltzer, G., 2001. Glaciation during marine isotope stage 2 in the American Cordillera. In: Markgraf, V. (Ed.), Interhemispheric Climate Linkages. Academic Press, San Diego, pp. 173–181.

Clapperton, C.M., Clayton, J.D., Benn, D.J., Marden, C.J., Argollo, J., 1997. Late Quaternary glacier advances and palaeolake highstands in the Bolivian Altiplano. Quatern. Int. 38 (39), 49–59.

Ehlers, J., Gibbard, P.L. (Eds.), 2004. Quaternary Glaciations—Extent and Chronology, Part III. In: Developments in Quaternary Science, 2 Elsevier, Amsterdam, p. 380.

Farber, D.L., Hancock, G.S., Finkel, R.C., Rodbell, D.T., 2005. The age and extent of tropical alpine glaciation in the Cordillera Blanca, Peru. J. Quatern. Sci. 20 (7–8), 759–776.

Fornari, M., Risacher, F., Féraud, G., 2001. Dating of paleolakes in the central Altiplano of Bolivia. Palaeogeogr. Palaeoclimatol. Palaeoecol. 172, 269–282.

Francou, B., 1990. Stratification mechanisms in slope deposits in high subequatorial mountains. Permafrost Periglac. Process. 1, 249–263.

Geyh, M.A., Grosjean, M., Núñez, L., Schotterer, U., 1999. Radiocarbon reservoir effect and the timing of the Late-Glacial/Early Holocene humid phase in the Atacama Desert (Northern Chile). Quatern. Res. 52, 143–153.

Gouze, P., Argollo, J., Saliège, J.-F., Servant, M., 1986. Interprétation paléoclimatique des oscillations des glaciers au cours des 20 derniers millénaires dans les régions tropicales: exemple des Andes boliviennes. CR 'Académie Sci. Paris II 303, 219–223.

Graf, K., 1975. Estudios Geomórficos en los Andes y el Altiplano Bolivianos. Soc. Geol. Boliviano 21, 3–23.

Graf, K., 1979. Untersuchungen zur rezenten Pollen- und Sporenflora in der nördlichen Zentralkordillere Boliviens und der Versuch einer Auswertung von Pollen aus postglazialen Torfmooren. Habilitationsschrift Phil. Fak. II, Universität Zürich, Juris, Zürich, 104 pp.

Graf, K., 1981. Palynological investigations of two post-glacial peat bogs near the boundary of Bolivia and Peru. J. Biogeogr. 8, 353–368.

Gross, G., Kerschner, H., Patzelt, G., 1978. Methodische Untersuchungen über die Schneegrenze in alpinen Gletschergebieten. Z. Gletscherk. Glazialgeol. 12, 223–251.

Hauthal, R., 1906. Quartäre Vergletscherung der Anden in Bolivien und Peru. Z. Gletscherk. 1906, 230.

Hauthal, R., 1911. Reisen in Bolivien und Peru, ausgeführt 1908. Wiss. Veröff. Ges. Erdkunde Leipzig 7, 1–247.

Heine, K., 1995. Bedded slope deposits with respect to the Late Quaternary Glacial sequence in the high Andes of Ecuador and Bolivia. In: Slaymaker, O. (Ed.), Steepland Geomorphology. Wiley & Sons, Chichester, pp. 257–278.

Heine, K., 1996. The extent of the last glaciation in the Bolivian Andes (Cordillera Real) and palaeoclimatic implications. Z. Geomorphol. 104, 187–202, Suppl.-Band.

Heine, K., 2000. Tropical South America during the Last Glacial Maximum: evidence from glacial, periglacial and fluvial records. Quatern. Int. 72, 7–21.

Heine, K., 2004. Late Quaternary glaciations of Ecuador. In: Ehlers, J., Gibbard, P.L. (Eds.), Quaternary Glaciations—Extent and Chronology, Part III, (Developments in Quaternary Science 2). Elsevier, Amsterdam, pp. 165–169.

Heine, K., 2011. Late Quaternary glaciations of Ecuador. In: Ehlers, J., Gibbard, P.L., Hughes, P.D. (Eds.), Quaternary Glaciations—Extent and Chronology, Part IV—A Closer Look. Elsevier, Amsterdam.

Heine, K., Geyh, M.A., 2002. Neue ^{14}C-Daten zur Jüngeren Dryaszeit in den ecuadorianischen Anden. Eiszeit. Ggw 51, 33–50.

Hettner, A., 1889. Dritter Bericht von Herrn Alfred Hettner über seine Reisen in Peru und Bolivien. Verh. Ges. Erdkunde Berlin 16, 269–276.

Hoek, H., Steinmann, G., 1905. Erläuterungen zur Routenkarte der Expedition Steinmann, Hoek, v. Bistram in den Anden von Bolivien 1903–04. Petermanns Geogr. Mitt. 52, 1–13 25–31.

Hurlbert, S.H., Chang, C.C.Y., 1984. Ancient Ice Islands in Salt Lakes of the Central Andes. Science 224, 299–302.

Jomelli, V., Favier, V., Rabatel, A., Brunstein, D., Hoffmann, G., Francou, B., 2009. Fluctuation of glaciers in the tropical Andes over the last millennium and palaeoclimatic implications: a review. Palaeogeogr. Palaeoclimatol. Palaeoecol. 281, 269–282.

Jordan, E., 1991. Die Gletscher der bolivianischen Anden. Erdwissenschaftliche Forschung XXIII Steiner, Stuttgart, 365 pp.

Jordan, E., Reuter, G., Leinweber, P., Alfaro, H., Condo, A., Geyh, M.A., 1993. Pleistocene moraine sequences in different areas of glaciation in the Bolivian Andes. Zentralbl. Geol. Paläontol. Teil I 1993 (1/2), 455–470.

Kaser, G., 1995. Some notes on the behaviour of tropical glaciers. Bulletin de l'Institut francais d'Études Andines 24, 671–681.

Kaser, G., 1999. A review of the modern fluctuations of tropical glaciers. Glob. Planet. Change 22, 93–103.

Kaser, G., Georges, Ch., Ames, A., 1996. Modern glacier fluctuations in the Huascarán-Chopicalqui massif of the Cordillera Blanca, Perú. Z. Gletscherk. Glazialgeol. 32, 91–99.

Klein, A.G., Seltzer, G.O., Isacks, B.L., 1999. Modern and last local glacial maximum snowlines in the Central Andes of Peru, Bolivia, and Northern Chile. Quatern. Sci. Rev. 18, 63–84.

Kull, C., Imhof, S., Grosjean, M., Zech, R., Veit, H., 2008. Late Pleistocene glaciation in the Central Andes: temperature versus humidity control—a case study from the eastern Bolivian Andes (17°S) and regional synthesis. Glob. Planet. Change 60, 148–164.

La Frenierre, J., In Huh, K., Mark, B.G., 2011. Quaternary glaciations—extent and chronology: Ecuador, Peru, Bolivia. In: Ehlers, J., Gibbard, P.L., Hughes, P.D. (Eds.), Quaternary Glaciations—Extent and Chronology, Part IV—A Closer Look. Elsevier, Amsterdam.

Lauer, W., Rafiqpoor, M.D., 1986. Die jungpleistozäne Vergletscherung im Vorland der Apolobamba-Kordillere (Bolivien). Erdkunde 40, 125–145.

Leinweber, P., Jordan, E., Schulten, H.-R., 1996. Molecular characterization of soil organic matter in Pleistocene moraines from the Bolivian Andes. Geoderma 72, 133–148.

Lifton, N.A., Bieber, J.W., Clem, J.M., Duldig, M.L., Evenson, P., Humble, J.E., et al., 2005. Addressing solar modulation and long-term uncertainties in scaling secondary cosmic rays for in situ cosmogenic nuclide applications. Earth Planet. Sci. Lett. 239 (1–2), 140–161.

Liu, K.B., Reese, C.A., Thompson, L.G., 2005. Ice-core pollen record of climatic changes in the central Andes during the last 400 years. Quatern. Res. 64 (2), 272–278.

Mark, B.G., 2008. Tracing tropical Andean glaciers over space and time: some lessons and transdisciplinary implications. Glob. Planet. Change 60, 101–114.

Mark, B.G., Seltzer, G.O., Rodbell, D.T., 2004. Late Quaternary glaciations of Ecuador, Peru and Bolivia. In: Ehlers, J., Gibbard, P.L. (Eds.), Quaternary Glaciations Extent and Chronology, Part III: South America, Asia, Africa, Australia, Antarctica. Elsevier, Amsterdam, pp. 151–163.

Müller, C., 1985. Zur Gletschergeschichte in der Cordillera Quimsa Cruz (Depto. La Paz, Bolivien). Ph.D. thesis, University of Zürich, Zürich, 188pp.

Placzek, C., Quade, J., Patchett, P.J., 2006. Geochronology and stratigraphy of late Pleistocene lake cycles on the southern Bolivian Altiplano: implications for causes of tropical climate change. Geol. Soc. Am. Bull. 118 (5), 515–532.

Porter, S.C., 2001. Snowline depression in the tropics during the Last Glaciation. Quatern. Sci. Rev. 20, 1067–1091.

Rabatel, A., Francou, B., Jomelli, V., Naveau, P., Grancher, D., 2008. A chronology of the Little Ice Age in the tropical Andes of Bolivia (16°S) and its implications for climate reconstruction. Quatern. Res. 70 (2), 198–212.

Ramage, J.M., Smith, J.A., Rodbell, D.T., Seltzer, G.O., 2005. Comparing reconstructed Pleistocene equilibrium-line altitudes in the tropical Andes of central Peru. J. Quatern. Sci. 20, 777–788.

Reuther, A.U., Ivy-Ochs, S., Heine, K., 2006. Application of surface exposure dating in glacial geomorphology and the interpretation of moraine ages. Z. Geomorphol. Suppl.-vol. 142, 335–359.

Rodbell, D.T., Smith, J.A., Mark, B.G., 2009. Glaciation in the Andes during the Lateglacial and Holocene. Quatern. Sci. Rev. 28, 2165–2212.

Röthlisberger, F., 1986. 10 000 Jahre Gletschergeschichte der Erde. Sauerländer, Aarau, Frankfurt/Main, Salzburg, 416 pp.

Rowe, H.D., Dunbar, R.B., 2004. Hydrologic-energy balance constraints on the Holocene lake-level history of lake Titicaca, South America. Climate Dyn. 23, 439–454.

Seltzer, G.O., 1990. Recent glacial history and paleoclimate of the Peruvian–Bolivian Andes. Quatern. Sci. Rev. 9, 137–152.

Seltzer, G.O., 1991. La regresión del hielo en Peru y en Bolivia desde el Pleistoceno tardió. Bol. Soc. Geológica Boliviana 26, 13–32.

Seltzer, G.O., 1992. Late Quaternary glaciation of the Cordillera Real, Bolivia. J. Quatern. Sci. 7, 87–98.

Seltzer, G.O., 2001. Late Quaternary glaciation in the tropics: future research directions. Quatern. Sci. Rev. 20, 1063–1066.

Seltzer, G.O., Rodbell, D.T., Baker, P.A., Fritz, S.C., Tapia, P.M., Rowe, H.D., et al., 2002. Early warming of tropical South America at the last glacial–interglacial transition. Science 296, 1685–1686.

Smith, J.A., Rodbell, D.T., 2010. Cross-cutting moraines reveal evidence for North Atlantic influence on glaciers in the tropical Andes. J. Quatern. Sci. 25, 243–248.

Smith, J.A., Seltzer, G.O., Farber, D.L., Rodbell, D.T., Finkel, R.C., 2005a. Early local last glacial maximum in the Tropical Andes. Science 308, 678–681.

Smith, J.A., Seltzer, G.O., Rodbell, D.T., Klein, A.G., 2005b. Regional synthesis of the last glacial maximum snowlines in the tropical Andes, South America. Quatern. Int. 138–139, 145–167.

Smith, J.A., Finkel, R.C., Farber, D.L., Rodbell, D.T., Seltzer, G.O., 2005c. Moraine preservation and boulder erosion in the tropical Andes: interpreting old surface exposure ages in glaciated valleys. J. Quatern. Sci. 20 (7–8), 735–758.

Smith, J.A., Mark, B.C., Rodbell, D.T., 2008. The timing and magnitude of mountain glaciation in the tropical Andes. J. Quatern. Sci. 23 (6–7), 609–634.

Smith, J.A., Lowell, T.V., Caffee, M.W., 2009. Lateglacial and Holocene cosmogenic surface exposure age glacial chronology and geomorphological evidence for the presence of cold-based glaciers at Nevado Sajama, Bolivia. J. Quatern. Sci. 24, 360–372.

Thompson, L.G., Mosley-Thompson, E., Dansgaard, W., Grootes, P.M., 1986. The little ice age as recorded in the stratigraphy of the tropical Quelccaya ice cap. Science 234, 361–364.

Thompson, L.G., Davis, M.E., Mosley-Thompson, E., Sowers, T., Henderson, K.A., Zagorodnov, V.S., et al., 1998. A 25,000-year tropical climate history from Bolivian ice cores. Science 282, 1858–1864.

Troll, C., 1927. Forschungen in den zentralen Anden von Bolivien und Peru. Petermanns Geogr. Mitt. 73, 41–43.

Troll, C., 1929. Die Cordillera Real. Z. Ges. Erdkunde Berlin 1929 (7/8), 279–312.

Troll, C., 1968. The Cordilleras of the Tropical Americas. Aspects of Climatic, Phytogeographical and Agrarian Ecology. Colloquium Geographicum 9, 15–56.

Troll, C., Finsterwalder, R., 1935. Die Karten der Cordillera Real und des Talkessels von La Paz (Bolivien) und die Diluvialgeschichte der zentralen Anden. Petermanns Geogr. Mitt. 81, 393–399 445–455.

Villarroel, C., Graf, K., 1979. Zur Entstehung des Talkessels von La Paz/Bolivien und Umgebung. Geogr. Helv. 1979, 43–48.

Wagnon, P., Ribstein, P., Kaser, G., Berton, P., 1999. Energy balance and runoff seasonality of a Bolivian glacier. Glob. Planet. Change 22, 49–58.

Zech, R., Kull, C., Kubik, P.W., Veit, H., 2007. LGM and late glacial glacier advances in the Cordillera Real and Cochabamba (Bolivia) deduced from 10Be surface exposure dating. Clim. Past 3, 623–635.

Zech, R., May, J.-H., Kull, C., Ilgner, J., Kubik, P.W., Veit, H., 2008. Timing of the late Quaternary glaciation in the Andes from 15 to 40°S. J. Quatern. Sci. 23, 635–647.

Zech, J., Zech, R., May, J., Kubik, P.W., Veit, H., 2009a. Heinrich I and Younger Dryas glaciation in the Central Andes. In: AGU Fall Meeting 2009. Abstracts.

Zech, J., Zech, R., Kubik, P.W., Veit, H., 2009b. Glacier and climate reconstruction at Tres Lagunas, NW Argentina, based on 10Be surface exposure dating and lake sediment analyses. Palaeogeogr. Palaeoclimatol. Palaeoecol. 284, 180–190.

Chapter 56

Ecuador, Peru and Bolivia

Jeff La Frenierre*, Kyung In Huh and Bryan G. Mark

Department of Geography, The Ohio State University, Derby Hall, North Oval Mall, Columbus, Ohio 43210, USA

*Correspondence and requests for materials should be addressed to Jeff La Frenierre. E-mail: la-frenierre.1@osu.edu

56.1. INTRODUCTION

The development of an accurate glacial chronology for sites in Ecuador, Peru and Bolivia is important for understanding factors that force glacial change in the Andes as well as the linkages between tropical and global climate systems (Klein et al., 1999; Hastenrath, 2009). Because tropical mountains tend to experience high precipitation but minimal thermal seasonality, tropical glaciers exhibit different mass-balance behaviour than glaciers at higher latitudes. Their distinctive ablation zone environments make tropical glacier termini highly responsive to climatic variation (Kaser and Osmaston, 2002; Smith et al., 2008). As such, the glacial geomorphology of tropical mountains can serve as a strong indicator of past climatic fluctuation (Seltzer, 1990; Heine, 2000; Licciardi et al., 2009).

Mark et al. (2004) reviewed the state of palaeoglacial knowledge for Ecuador, Peru and Bolivia in the first edition of this volume. This chapter synthesises information from that review with more recent research and provides a more comprehensive cartographic overview of mapped moraines in this portion of the Andes. The past 5 years have been rich in both new moraine dating studies and studies that revisit previously published age estimates in key localities. Further, two comprehensive overview papers have been published in the past 2 years: Smith et al. (2008) examined palaeoglaciation in Ecuador, Peru and Bolivia from Marine Isotope Stage (MIS) 5e (132–117 ka) through the Younger Dryas chron (12.9–11.7 ka), while Rodbell et al. (2009) focused on Andean glacial chronologies referencing the Late-glacial and Holocene (16.7 ka–present). Readers are referred to these two papers for a more detailed discussion than that which is presented here.

All relevant features of palaeoglacial chronologies included on maps within the literature described in this chapter have been digitised and incorporated into an ArcGIS geodatabase. This database includes 1550 individual mapped moraines and 350 sampling locations, as well as the present glacial extent (generally as represented on the original maps) and the location of lakes utilised for sediment coring. Maps produced from this geodatabase accompany this chapter. The original maps were georeferenced using the coordinate grid provided on the map and/or by matching features on Google Earth imagery. Individual features were then hand-digitised and attributed based on information both on the maps and in the text. The shaded relief background utilised on the maps created for this chapter is derived from 30-m resolution digital elevation models (DEMs) produced by the Advanced Spaceborne Thermal Emission and Reflection radiometer (ASTER) optical sensor (ASTER Global DEM is a product of the Japanese Ministry of Economy, Trade and Industry (METI) and NASA). Some ASTER DEMs are prone to 'pitting'—artificial sinks in the elevation model. Where pitting is pervasive, 90-m resolution DEMs produced by the NASA Shuttle Radar Topography Mission (SRTM) have been used instead. In some instances, moraines digitised from the original maps have been shifted slightly to better match features identifiable on Google Earth imagery or the topographic surfaces generated from the DEMs. While every effort has been made to represent reality as closely as possible, the complexities of georeferencing and the uncertain ground accuracy of the original maps do not allow use of this geodatabase at small mapping scales.

The literature reviewed in this chapter primarily utilises radiocarbon or surface exposure dating (SED) methods (Fig. 56.1). In both cases, adjustment of calculated values is necessary in order to estimate the true calendar age of the sample. Because the production rate of ^{14}C has varied through time, radiocarbon dates must be calibrated using dendrochronologically derived curves while scaling factors must be applied to surface exposure dates calculated from cosmogenic radio-nuclides (CRN) such as ^{10}Be, ^{36}Cl and ^{3}He when samples are obtained at low latitudes and/or high elevations. Because there are multiple scaling methods in use, a range of age interpretations can exist for single feature. At this time, no single universally accepted scaling method has been identified, though the CRONUS-Earth

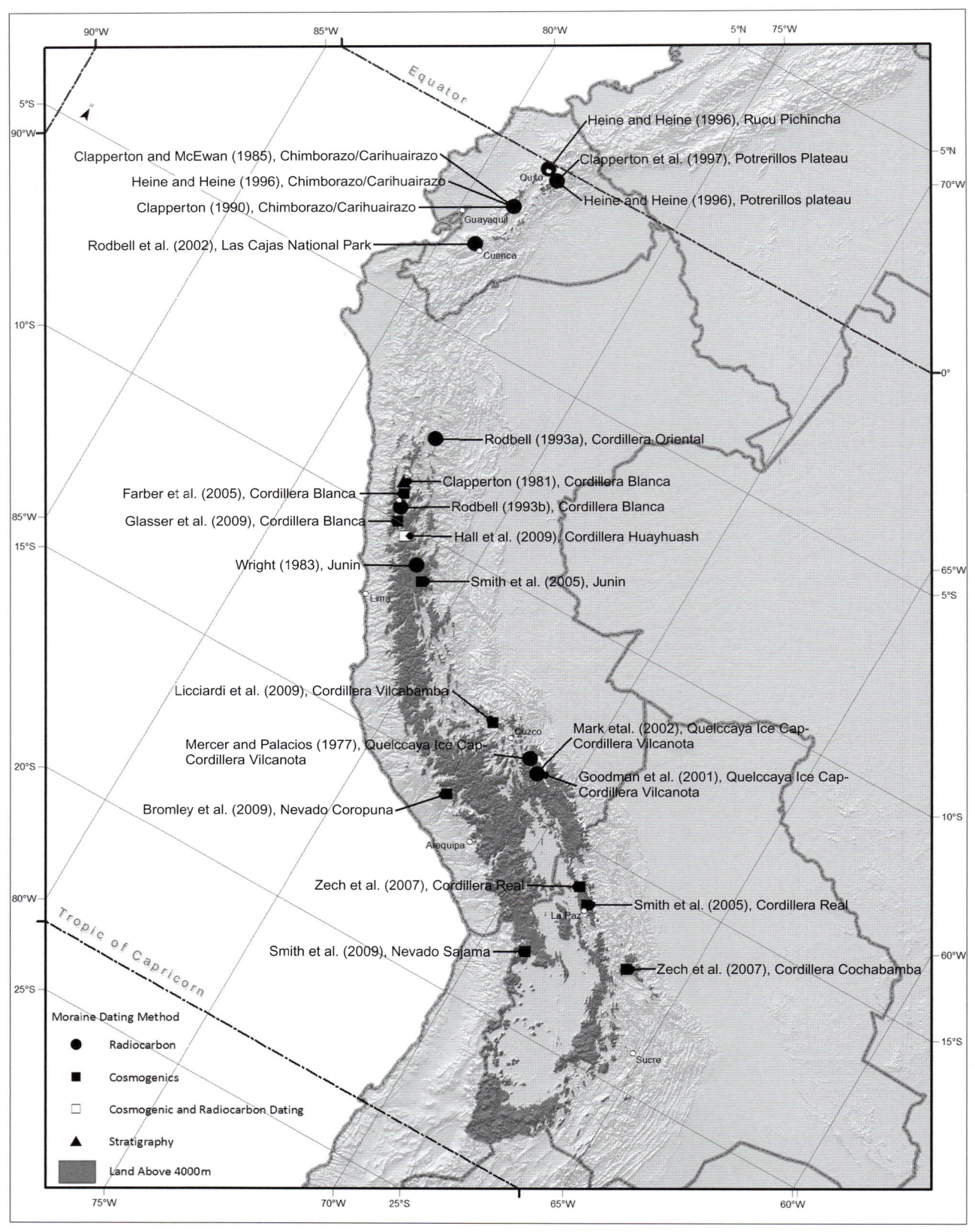

FIGURE 56.1 Geographical locations and moraine dating methods of the literature reviewed in this chapter.

Project initiative is aimed at synthesising existing scaling methods and producing consistent calibration dates (Balco et al., 2008).

All radiocarbon dates reported in this chapter are calibrated years before present (cal. ka BP, with BP as AD 1950) unless otherwise indicated. Rodbell et al. (2009) provided calibrated radiocarbon dates for numerous Andean sites where these dates were not initially reported, and these are used here. In other cases, calibration calculations are made using Calib version 5.0 (Stuiver and Reimer, 1993). Readers are referred to the original literature for the original ^{14}C dates. CRN dates reported in this chapter are those provided in the original literature, though instances where revised calculations have been made by other authors are noted. Rodbell et al. (2009) have calibrated the results of several tropical Andean SED studies using the CRONUS calculator and include a useful table of correlated values for several of the papers reviewed here.

In both this chapter and the accompanying ArcGIS geodatabase, glacial chronologies are defined by their correlative MISs (Fig. 56.2). Based on sea-level lowstands, the global Last Glacial Maximum (LGM) is defined herein as 22–19 ka (Yokoyama et al., 2000); in the tropical Andes, the LGM has been situated at 21 ± 2 ka (Mark et al., 2005; Smith et al., 2008). The transition between MIS 2 and MIS 1 occurs between 11.4 and 12.1 ka (Martinson et al., 1987), and the base of the Holocene has recently been placed at 11.7 ka using the Greenland NGRIP ice core (Walker et al., 2009). Following the convention of Rodbell et al., 2009, the Late-glacial is defined in this chapter as 16.7–11.7 ka.

56.2. GEOGRAPHICAL SETTING

The Andes span the entire western flank of the South American continent. Within Ecuador, Peru and Bolivia, the Andes stretch for over 3000 km and contain an estimated 2650 km^2 of glacial ice, $\sim$95% of the glaciers presently existent within the tropical latitudes (Kaser and Osmaston, 2002). This portion of the Andes comprises a collection of individual ranges that are generally grouped into two parallel cordilleras—the Cordillera Oriental and the Cordillera Occidental—with a high plateau between the two. In southern Peru and Bolivia, this plateau, known as the Altiplano, is more than 200 km wide. In Ecuador, the two cordilleras generally reach elevations between 4000 and 4400 m but are punctuated by seven stratovolcanoes in excess of 5000 m and an eighth (Volcán Chimborazo) that reaches 6000 m (Smith et al., 2008). In Peru and Bolivia, extensive sections of each cordillera exceed 5000 m, with numerous individual peaks topping 6000 m, particularly in the ranges of the Cordillera Occidental, while the Altiplano averages 3500–4000 m (Smith et al., 2008). At present, glacial limits in the tropical Andes are $\sim$5000 m, though terminus elevations are somewhat lower in the Cordillera Blanca (4700 m; Mark, 2008), Cordillera Huayhuash (4800 m; Hall et al., 2009) and Cordillera Vilcanota (4600 m; Mercer and Palacios, 1977) of Peru. The Quelccaya Ice Cap in southern Peru is the only remaining such glacial feature in the Earth's tropical latitudes.

56.3. TERTIARY AND EARLY PLEISTOCENE GLACIATIONS (>MIS 19; >790 KA)

There is scant documented evidence of tropical Andean glaciers from this Early in geologic history. It is estimated that the Bolivian Andes were no more than 50% of their present altitude 10 Ma and that the Peruvian Andes have attained most of their present elevation in only the past 5–6 Ma (Smith et al., 2008), suggesting that the range may have been of insufficient altitude to support glaciation at this latitude prior to these dates.

Clapperton (1983) described the 1962 research of Earnest Dobrovolny, who identified a pair of glacial tills near La Paz, Bolivia that attest to very old palaeoglaciations. One, a 7-m thick till, is overlain by an ignimbrite formation dated to $\sim$3.25 Ma. The second till is highly oxidised, yet buried well beneath two other tills associated with the most recent two glacial cycles. The depth of the oxidation suggests a considerable lag between the deposition of this till and those above it. No specific glacial stadial is hypothesised for this material.

56.4. MIDDLE PLEISTOCENE GLACIATIONS (MIS 19–6; 790–132 KA)

Heine (1995) and Heine and Heine (1996) describe the sequence of moraines on Rucu Pichincha, Ecuador (0.2°S, 78.5°W; Fig. 56.3), and classify these into seven distinct groups, M1–M7, sequentially numbered from oldest to youngest. Groups M1–M3 pre-date the last glacial cycle (MIS 4–2), though precise dating control is lacking. M1, the oldest, is underlain by a lava formation dated to greater than 900 ka. On Rucu Pichincha, no limiting age for these moraines has been determined, however correlative sequences have been identified on the Potrerillos Plateau (0.3°S, 78.2°W; Fig. 56.4A and B), where a lava flow inside the valley carved by the M1 glaciation is dated to >150–180 ka provides a minimum-limiting age for the M1 and M2 sequences (Heine, 1995; Smith et al., 2008). On Rucu Pichincha, M1 moraines occur at 3550–3600 m, with M2 moraines located in close proximity. M3 moraines are only identified in two locations, at elevations as low as 3700 m.

On the Chimborazo–Carihuairazo massif (1.4°S, 78.8°W; Fig. 56.5A and B), there are no moraines interpreted as dating to prior to MIS 2; however, there is other geomorphic evidence for earlier and potentially more

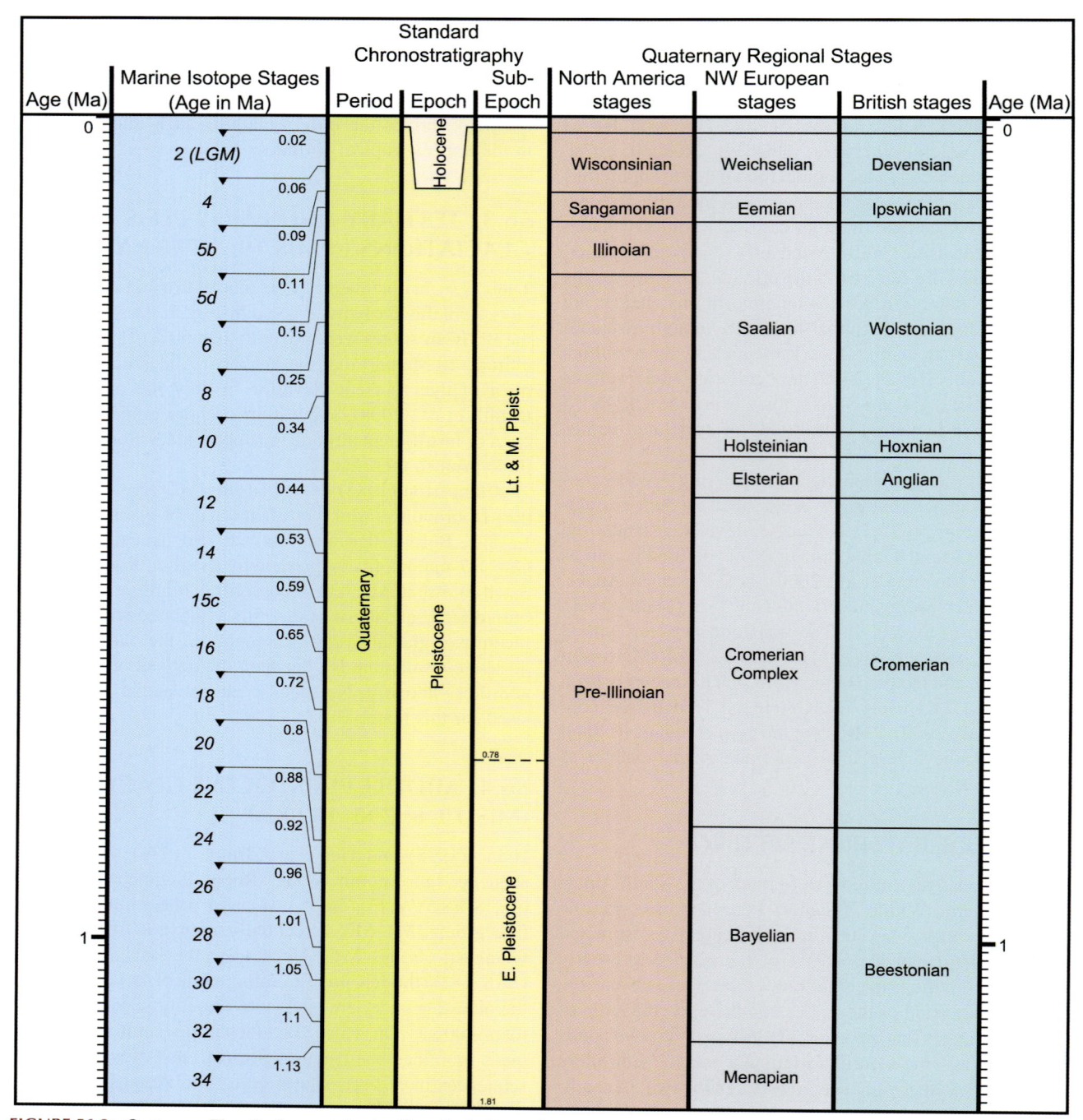

FIGURE 56.2 Quaternary Time Scale, created using Time Scale Creator© A. Lugowski and J. Ogg.

extensive glacial events. Clapperton and McEwan (1985) described lateral moraines at 3600 m in the Rio Mocha Valley between the two volcanoes that represent the local MIS 2 maximum extent of glaciation, but note that these moraines appear too small for the deeply eroded glacial valley in which they have been deposited, suggesting an earlier glaciation. The authors also described the presence of oxidised tills outside the limits of MIS 2 moraines. Clapperton (1986) (reported in Smith et al., 2008) placed these tills as low as 2750 m and noted that this is the lowest altitude at which glacial deposits have been identified in Ecuador. Clapperton (1990) described the presence of highly weathered glacial deposits inter-bedded with volcanic material on Carihuairazo and suggested that these may date from as early as MIS 8 based on rates of weathering and oxidation relative to a separate till constrained by a minimum-limiting

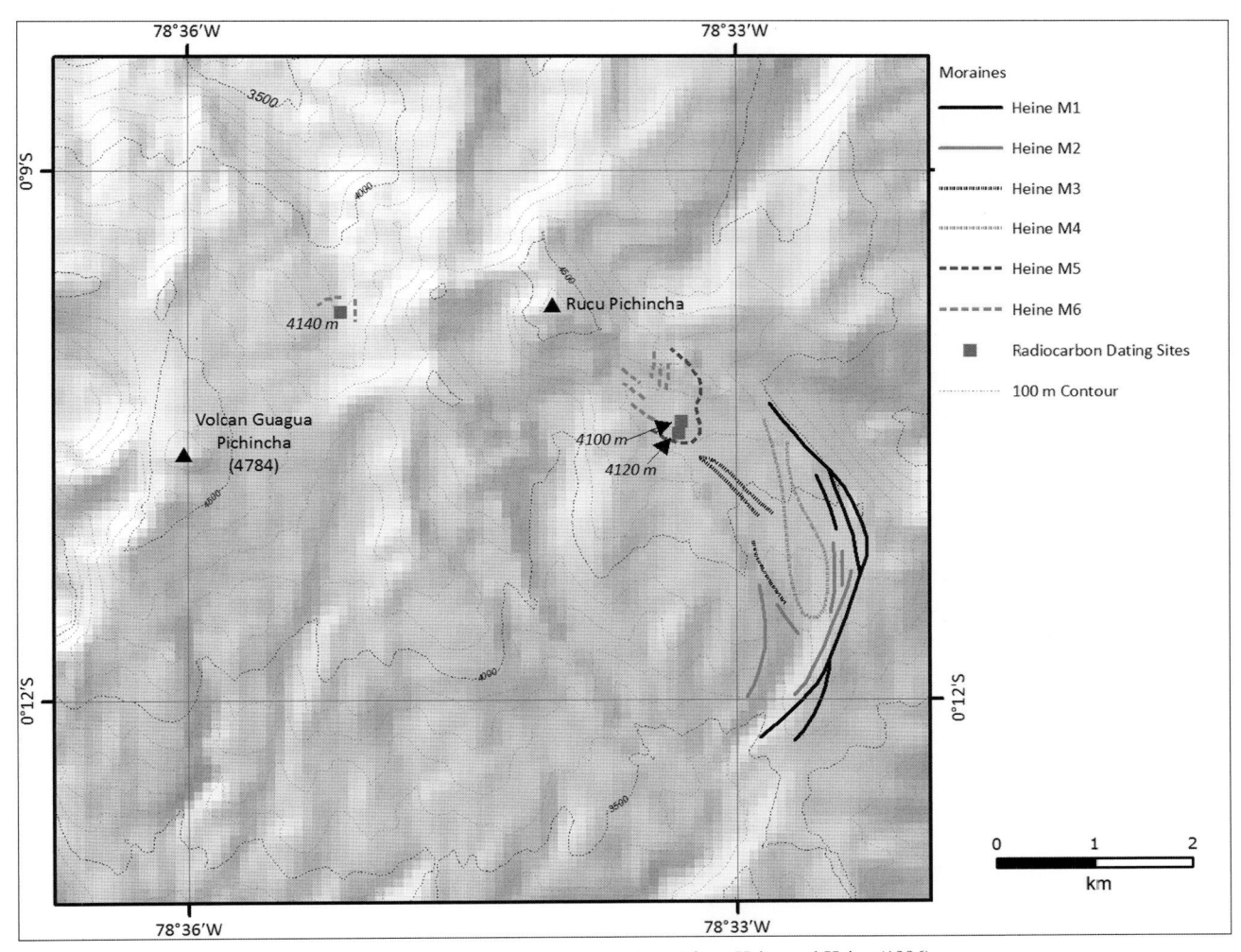

FIGURE 56.3 Quaternary glacial features on Rucu Pichincha, Ecuador. Adapted from Heine and Heine (1996).

^{14}C date of >43 ka. Lack of dating control makes the age of this till uncertain.

Advancing the work of Clapperton (1981), Rodbell (1993a) measured rock weathering and soil development, and used radiocarbon dating from lake/bog sediments to define four pre-Holocene moraine sequences in the Cordillera Blanca (9.3°S, 77.5°W; Fig. 56.6A–F). The oldest of these four sequences, the Cojup, was assigned one of three potential date ranges based on either extrapolation of rock weathering or Profile Development Index (PDI) of moraine soils, the youngest being 46–29 ka, while the oldest was >4.3 Ma. Farber et al. (2005) revisited the age classifications of Rodbell, using ^{10}Be SED to improve the age precision of these features. Based on this work, the authors suggested a date range of 440–120 ka and speculated that the wide range of dates associated with Cojup boulders in their study may actually be indicative of two separate glacial advances. Only small isolated remnants of Cojup moraines are found in the Cordillera Blanca, at elevations as low as 3650 m (Rodbell, 1993a).

Further evidence of Middle Pleistocene glaciation in the tropical Andes is found in the Lima Conglomerate, a unit deposited in sediment fans along Peru's Pacific Coast littoral. Le Roux et al. (2000) identified distinct cycles of coarse and fine material within the conglomerate and suggested that they related to variations in river discharge resulting from the release of meltwater and glacial debris during low sea-level stands at the beginning of interglacial stades. Specific cycles within the unit were tentatively identified with lowstands during MIS 16, 12 and 6.

56.5. LATE PLEISTOCENE GLACIATIONS (MIS 5–2; 132–11.7 KA)

56.5.1. MIS 5e–3 (132–35 ka)

There is little documented evidence in Ecuador for glacial events during this portion of the Pleistocene. Clapperton et al. (1997) briefly mentioned the existence of moraines at 3000 m on the west flank of the Potrerillos Plateau and

2700 m on the east flank, which he suggested pre-dates the global LGM. These were not linked to any of the M1–M6 moraine sequences that Heine and Heine (1996) describe in this region. On the north flank of Carihuairazo, Clapperton (1990) noted the presence of a glacial till constrained by a minimum-limiting ^{14}C date of >43 ka. Clapperton suggested that this till dates either from a glacial advance at MIS 4 (~70 ka) or perhaps even MIS 6 (before 132 ka).

Among his four pre-Holocene moraine classifications for the Cordillera Blanca, Peru (Fig. 56.6A–F), Rodbell (1993a) designated Rurec moraines as pre-global LGM; however, he was uncertain about their estimated age. Based on the method of data analysis (extrapolation of weathering rates, linear or logarithmic extrapolation of morainic soil PDI values) and the assumed age of seemingly related moraine groups elsewhere in Peru, four possible age ranges were suggested: (1) 32–20.5 ka, (2) 46.5–27 ka, (3) 132–75 ka or (4) 591–228 ka. Later SED analysis of the Rurec moraines (Farber et al., 2005) confirmed that they were distinct from and older than the Laguna Baja moraine sequence that Rodbell (1993a) believed represented the global LGM throughout the Cordillera Oriental in northern Peru (including the Cordillera Blanca), but placed their age at 29–20.5 ka, and thus also within MIS 2. Recent recalculation of the CRN data from Farber et al. (2005) by Bromley et al. (2009) further revise the age of the Rurec moraines to ~20.5 ka.

Extensive evidence of pre-MIS 2 glaciations exists in the mountains surrounding the Junin Plain in central Peru (11.0°S, 76.2°W; Fig. 56.7A and B). Wright (1983) classified the older of two distinct glacial sequences as the Rio Blanco phase and determined a minimum-limiting age of ~42 ka based on a radiocarbon date obtained from Laguna Coochachuyco. Wright further noted that the topographic contrast between the Rio Blanco phase and the younger Punrun phase was very similar to differences between the Bull Lake and Pinedale glaciations in the Rocky Mountains. More recently, Smith et al. (2005) used SED to date the oldest apparent moraines on the east slope of the Junin Plain, slightly south of the area mapped by Wright. This study

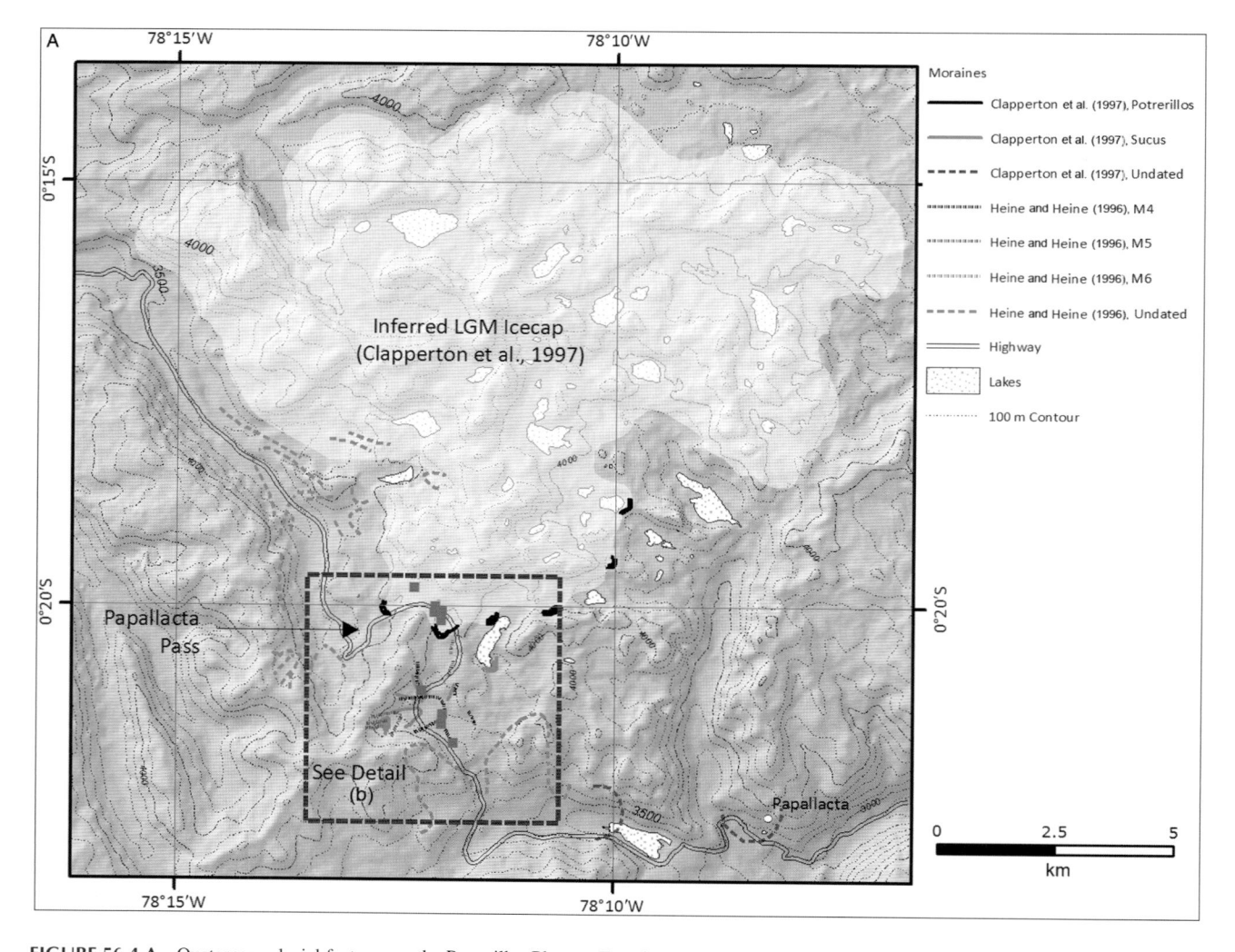

FIGURE 56.4 A Quaternary glacial features on the Potrerillos Plateau, Ecuador. Adapted from Heine and Heine (1996) and Clapperton et al. (1997).

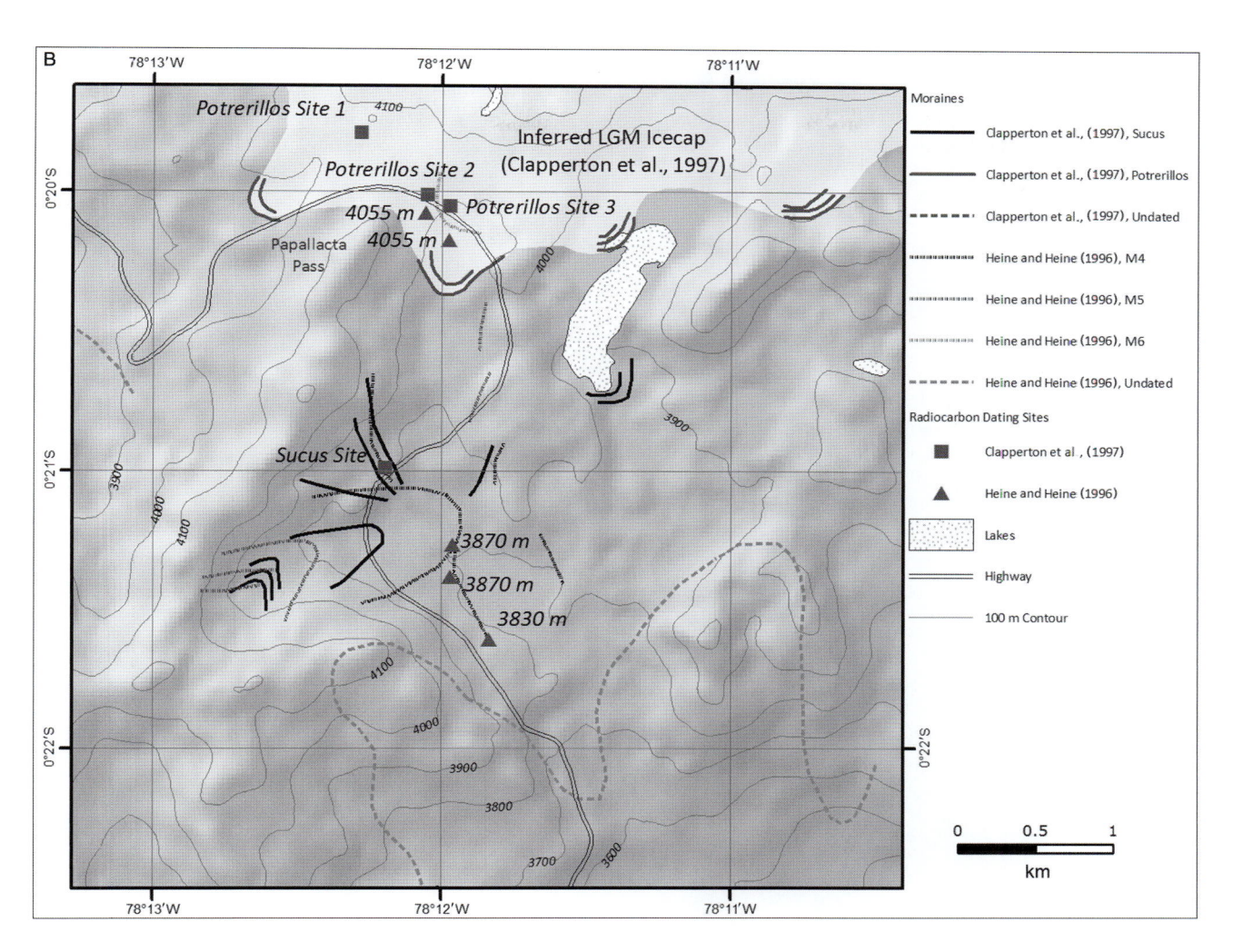

FIGURE 56.4 B (Continued).

indicated that these oldest moraines were at least 67 ka, and possibly much older.

The pre-MIS 2 moraines of the Junin Plain are consistent with the findings of Mercer and Palacios (1977), who mapped moraines in the Cordillera Vilcanota (14.0°S, 71.0°W; Fig. 56.8), farther south in Peru's Cordillera Oriental. Here, moraines slightly up-valley from the most distal moraines were constrained by a minimum-limiting radiocarbon date of ~41.5 ka, while soil development profiling suggested that these moraines could be 115–70 ka, or even older (Goodman et al., 2001). The most distal moraines, located at 3600 m and more than 16 km from present-day glacier termini, were undated. At the same time, glaciers on the south side of the Cordillera Vilcanota merged with glaciers emanating from the Quelccaya Ice Cap, covering an extensive area above 4500 m.

Pre-MIS 2 moraines are also identified on Nevado Coropuna (15.5°S, 72.6°W; Fig. 56.9) by Bromley et al. (2009). Here, adjacent to an MIS 2 moraine sequence designated C-I, the authors noted but could not date a series of moraines that are markedly more weathered than those of the C-I group. In Quebrada Huayllaura, these pre-C-I moraines are located 3 km farther down-valley (at 4150 m) than the C-I group, indicating that this pre-LGM glaciation was more extensive than anything that occurred subsequently.

56.5.2. MIS 3–2 (35–16.7 ka)

The identity and location of LGM moraines in Ecuador has been an issue of contention. Heine (1995) and Heine and Heine (1996) placed the M4 moraine sequence identified on both Rucu Pichincha (Fig. 56.3) and the Potrerillos Plateau (Fig. 56.4A and B) at the global LGM. The M4 moraines, which occur at 3800–3900 m, are so dated based on their relative position, weathering and oxidation

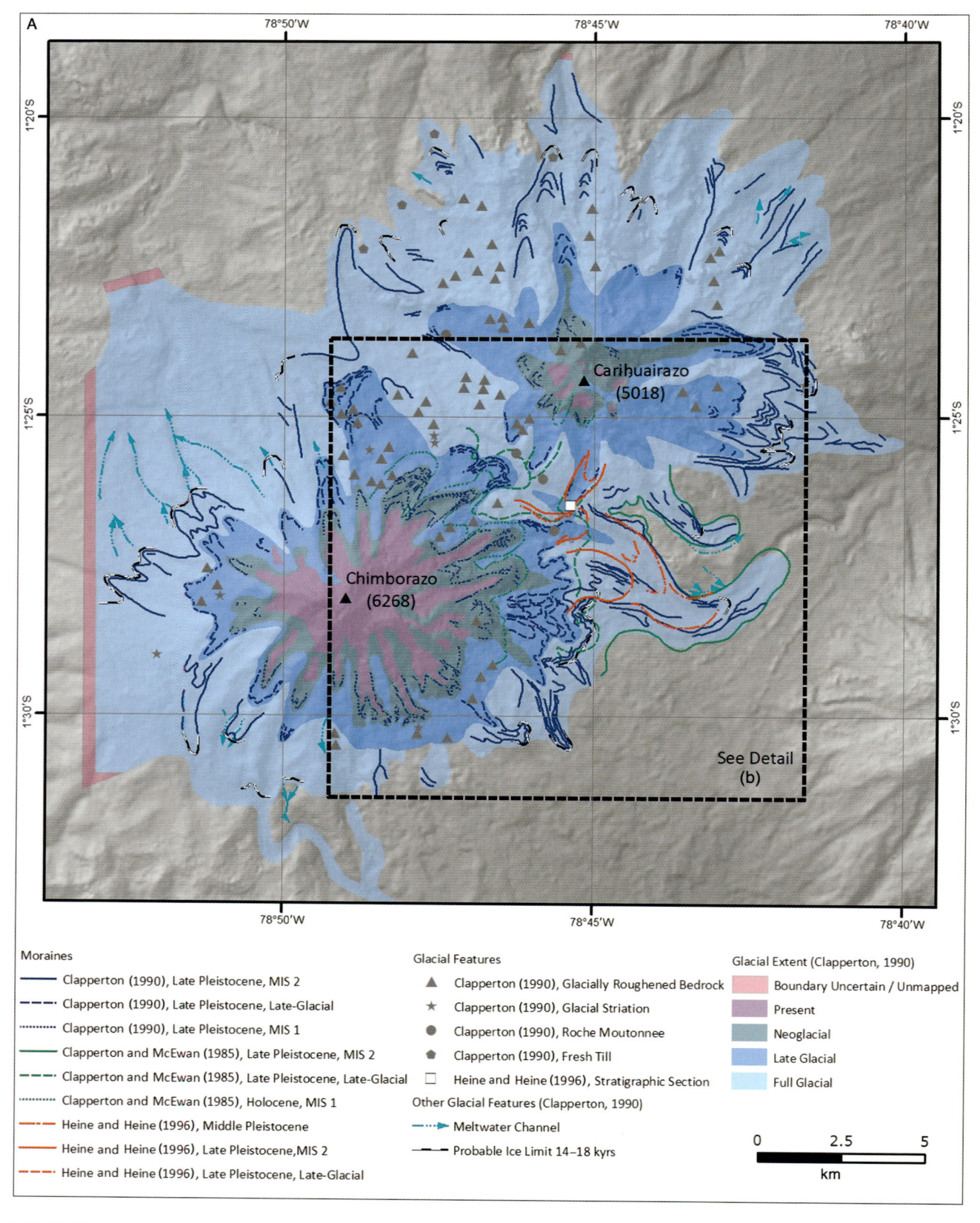

FIGURE 56.5 A Quaternary glacial features on Volcán Chimborazo and Carihuairazo, Ecuador. Adapted from Clapperton and McEwan (1985), Clapperton (1990) and Heine and Heine (1996).

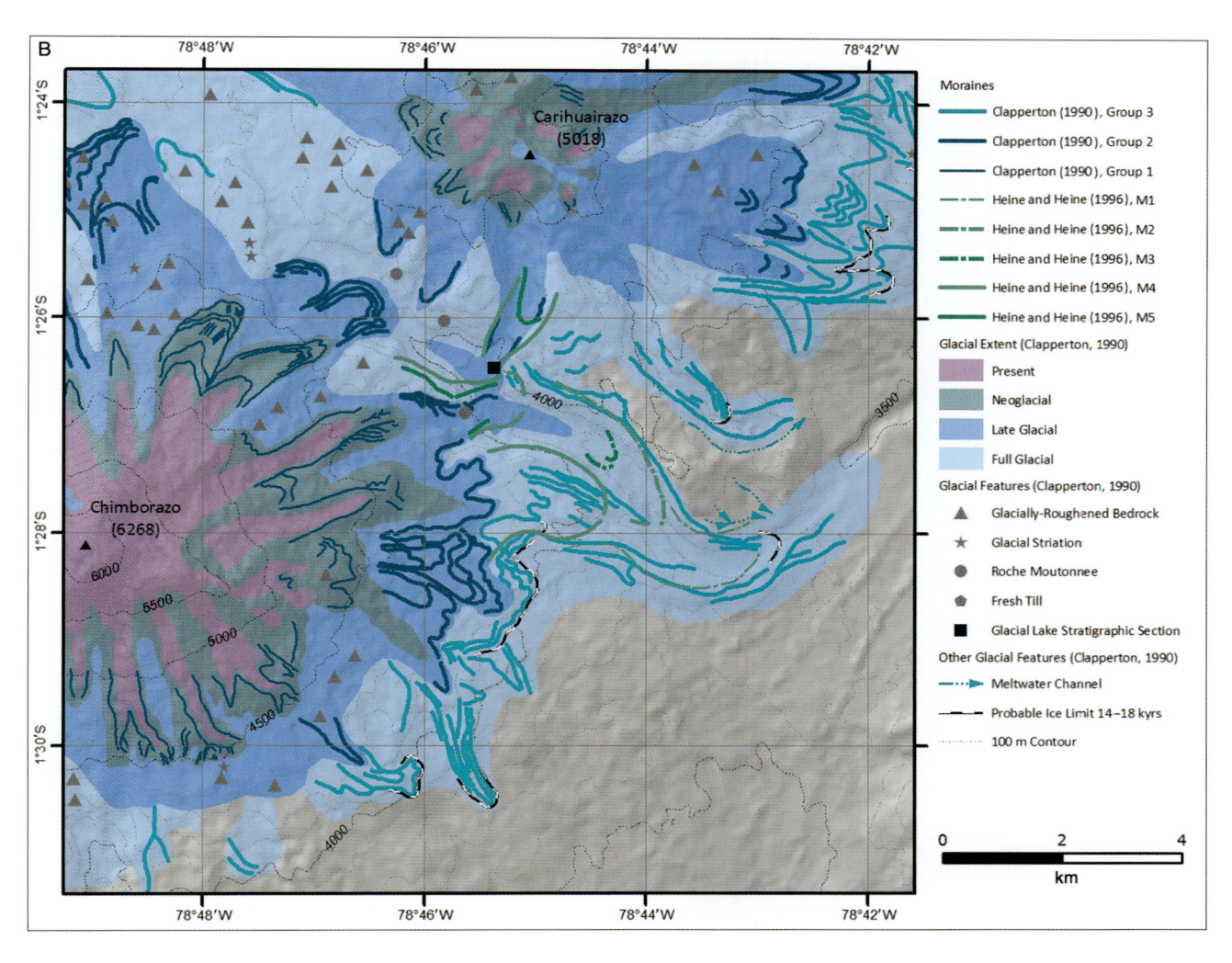

FIGURE 56.5 B (Continued).

compared to M5 moraines, which are constrained on Rucu Pichincha by a minimum-limiting age of ~15.4 ka. Clapperton et al. (1997), however, identified analogous moraines at 3850 m on the Potrerillos Plateau as belonging to a Late-glacial event he termed the Sucus Glacier Advance and proposed as being associated with an 800 km² ice cap atop the plateau. Clapperton et al. (1997) obtained a minimum-limiting radiocarbon date of ~15.5 ka for the Sucus Advance. On the Potrerillos Plateau, M4 moraines could only be constrained by a minimum-limiting radiocarbon age of ~14.2 ka (Heine and Heine, 1996).

It is difficult to reconcile the results obtained by both research teams. The relatively young minimum-limiting age obtained for M4 moraines does not necessarily preclude them from being associated with the 22–19 ka LGM; however, ~15.5 ka minimum-limiting age for the Sucus Advance suggests the possibility that the associated moraines may be older and not of Late-glacial origin. Further, Rodbell et al. (2009) noted the potential for mistaken interpretation in Clapperton et al. (1997) based on the possibility that what they interpret as primary glacial tills are in fact solifluyed tills originating from upslope moraines. Rodbell et al. (2009) also noted that the presence of a Late-glacial ice cap at 4300 m at this latitude is inconsistent with evidence from Las Cajas National Park, located at a similar altitude but 2° further south, showing it to have been ice free by 15.6 ka at the latest.

Clapperton and McEwan (1985) provided a relative sequence of Quaternary moraines in the Rio Mocha watershed, located between the cones of the Chimborazo–Carihuairazo massif (Fig. 56.5A and B), designating groups 1–3. Group 3 moraines are the oldest and most extensive on the massif and occur as low as 3600 m in the Rio Mocha watershed. Though no locally collected datable material

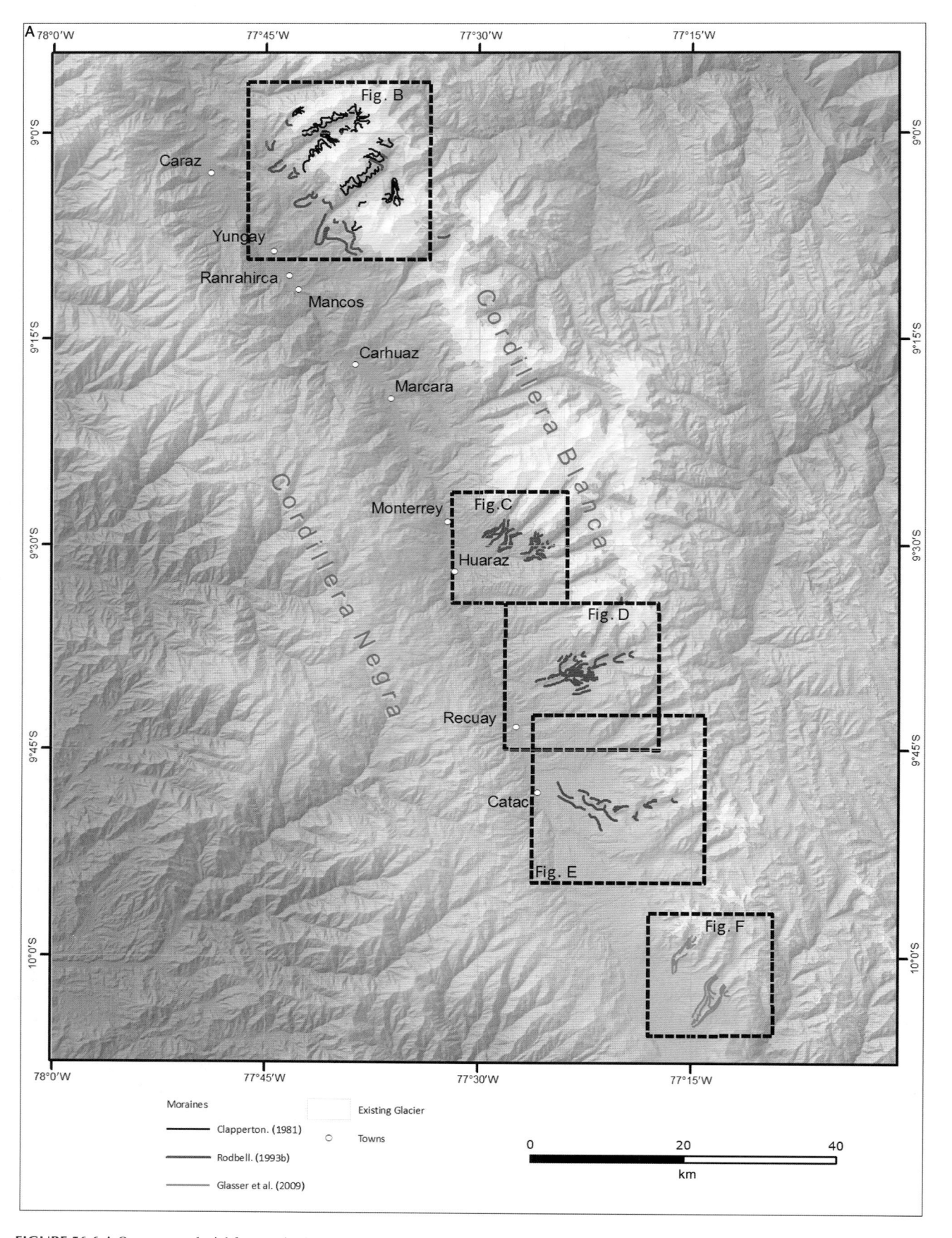

FIGURE 56.6 A Quaternary glacial features in the Cordillera Blanca, Peru. Adapted from Rodbell (1993a), Mark et al. (2004) and Glasser et al. (2009).

constrains the ages of Group 3 moraines, the authors relate these moraines to a till on the northern flanks of Carihuairazo constrained by a maximum-limiting ^{14}C age of ~35,500. Clapperton (1990) extended the mapping of the three moraine groups to the entire Chimborazo–Carihuairazo massif, estimated the spatial extent of the glaciations associated with each moraine group and differentiated two subgroups within Group 3: an inner set of low moraines (5–10 m high) and outer set of high moraines (100–200 m high). Outer moraines descend to 3400 m on the eastern flanks of Carihuairazo and 3800 m on the western flanks of both cones. Inner moraines generally terminate 200–300 m higher in altitude. The structure of Group 3 moraines between Carihuairazo and Chimborazo suggested to Clapperton (1990) that a continuous icefield existed between the two mountains, fed both by glaciers flowing off the two volcanoes as well as its own accumulation zone if/when the local snowline was depressed below 4400 m.

During MIS 2, a 400 km^2 ice cap appears to have existed along the crest of Southern Ecuador's Cordillera Occidental in what is now Las Cajas National Park (Rodbell et al., 2002) (2.8°S, 79.2°W; Fig. 56.10). The authors suggested that this MIS 2 ice cap covered ground above 2800 m; however, Clapperton (1986) limited MIS 2 glaciation in the region to 3200 and 3800 m on the range's southeast and northwest flanks, respectively. Clapperton (1986) (reported in Smith et al., 2008) noted that glaciation appears to have occurred as low as 2800 m in the past but links the tills that provide evidence of this to the same weathered tills that exist beyond the limit of Group 3 moraines on the Chimborazo–Carihuairazo massif, and that have been suggested as originating during either MIS 4 or 6 glaciations. Moraine dates are not tightly constrained in the Las Cajas area; however, radiocarbon data obtained from the Chorreras Valley by Rodbell et al. (2002) indicate that the area was ice free by ~14 ka.

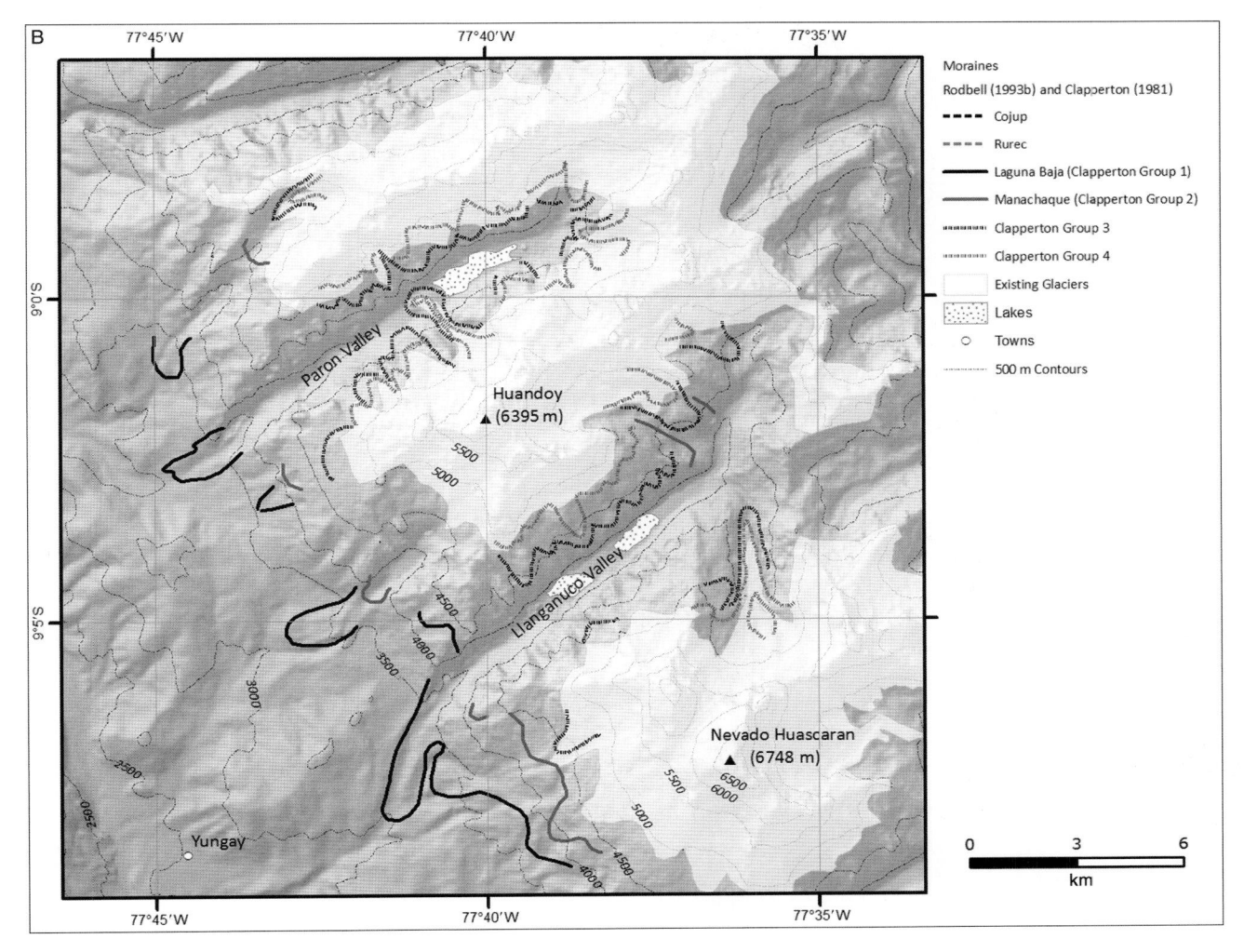

FIGURE 56.6 B

Rodbell (1993b) provided the northernmost palaeoglacial documentation in Peru's segment of the Cordillera Oriental (7.7°S, 77.5°W; Fig. 56.11) and dated all existent moraines in the area to the global LGM, suggesting that the MIS 2 glaciation was at least as extensive as earlier glaciations. These were classified as Laguna Baja moraines based on their location relative to the eponymous lake, and this has become the type-name for moraines their age both here and in the Cordillera Blanca. Because recent research indicates that, in the Cordillera Blanca, the Laguna Baja sequence is in fact not representative of the greatest MIS 2 glacial expansion (Farber et al., 2005; Bromley et al., 2009) (discussed below), there is now uncertainty about the age classification (though not the relative relationship) of the moraines here.

In the Cordillera Blanca (Fig. 56.6A–F, the Rurec moraines that Rodbell (1993a) initially believed to pre-date MIS 2 have been determined to be representative of the global LGM (Farber et al., 2005) and are dated ~20.5 ka (Bromley et al., 2009). Terminal moraines of the Rurec group are located between 3400 and 3800 m, ~5 km down-valley from the oldest Laguna Baja moraines that Rodbell (1993a) believed represented the global LGM. Thus Rurec moraines appear to mark the most extensive MIS 2 glacial advance in the range.

Laguna Baja moraines are located between 3550 and 4250 m in the Cordillera Blanca. Rodbell (1993a) considered this sequence to be equivalent to the Group 1 (MIS 2) moraines identified by Clapperton (1981) and tentatively dated them to either 30–25 or 14.8–14 ka based on their likely correlation with MIS 2 moraines of uncertain date in the Cordillera Vilcanota reported by Mercer and Palacios (1977). As with the Rurec moraines, Farber et al. (2005) reported a revised date of 16.5 ka for the Laguna Baja sequence based on SED investigations. Though Bromley et al. (2009) did not specifically address the age of the

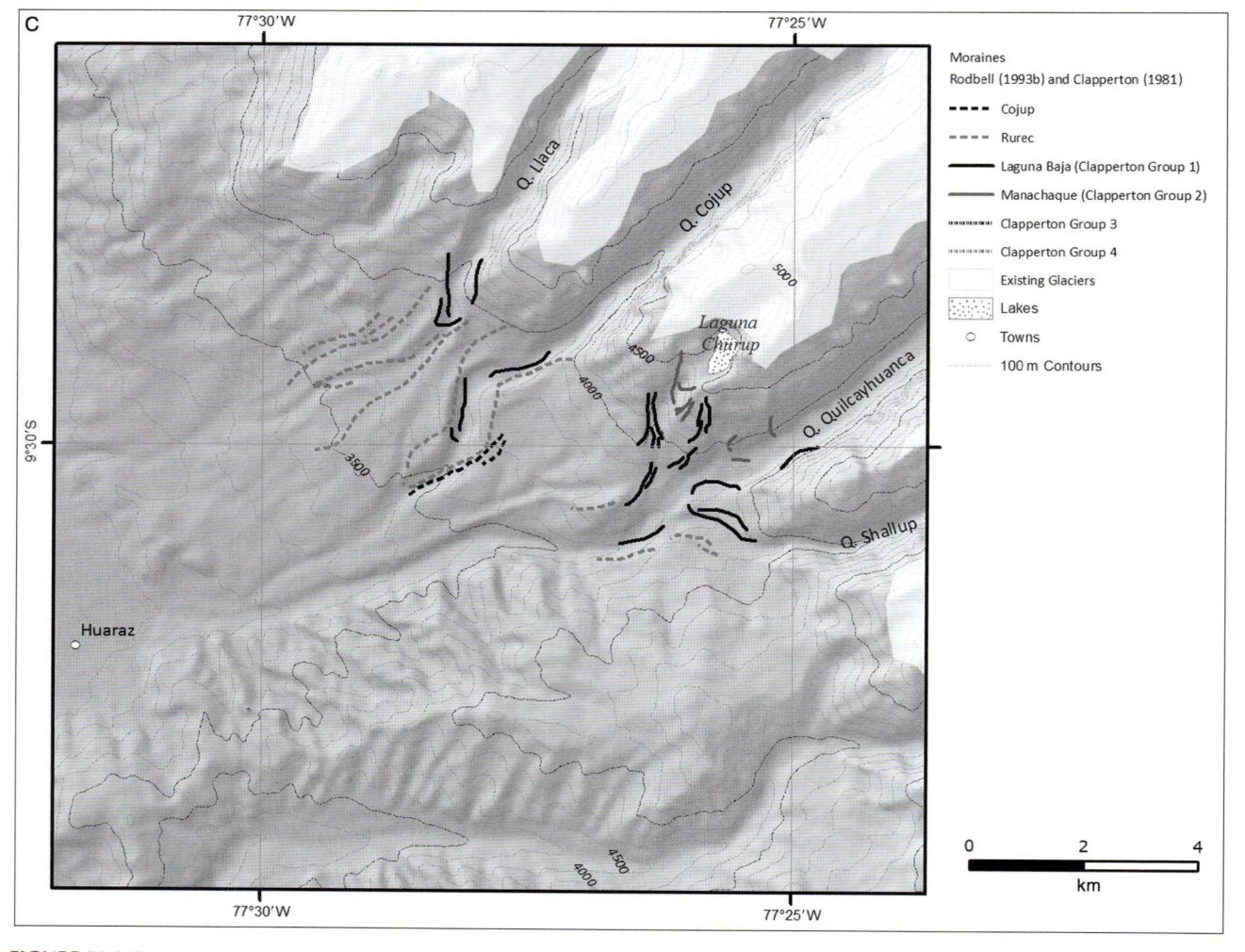

FIGURE 56.6 C

Laguna Baja sequence, it should be noted that their additional downward revision of the Rurec dates based upon CRONUS calculator recalibration of the Farber et al. (2005) CRN data suggests the possibility that the Laguna Baja sequence may also be younger, and may be of Lateglacial origin.

A recent SED study from the Juellesh and Tuco Valleys at the southern end of the Cordillera Blanca (Fig. 56.6F) found no evidence of global LGM moraines (Glasser et al., 2009). Because the oldest moraines identified in the study were approximately concurrent with the Younger Dryas chron, the authors posited that the local climate was too arid between 25 and 18 ka to support extensive glacial advancement.

In the Cordillera Huayhuash (10.3°S, 76.8°W; Fig. 56.12), Hall et al. (2009) found very little evidence of early MIS 2 glaciations and ascribe this to the morphology of the range, where steep valley slopes inhibit preservation of moraines from one glacial period to the next. Nonetheless, SED of exposed bedrock at 3800 m, the lowest elevation at which glacial features are found in the Jahuacocha Valley, provides a minimum-limiting age of 26 ka, suggesting that deglaciation from MIS 2 maxima may have occurred no later than this. A second set of MIS 2 moraines, which the authors classify as Stage IV, is dated at ~22–20 ka, coeval with the global LGM. These are inferred to represent a local ice advance just prior to the onset of an extended period of retreat.

The cordilleras on either side of the Junin Plain (Fig. 56.7A and B) exhibit ample evidence of MIS 2 glaciation. Wright (1983) reported that most areas above 4500 m in the western Junin Plain and surrounding cordillera were ice-scoured, while glacier termini on the eastern portion of the plain occurred down to 4200 m. Moraines from MIS 2 were classified as Punrun phase moraines, yet limiting age dates for the Punrun were confounded by somewhat

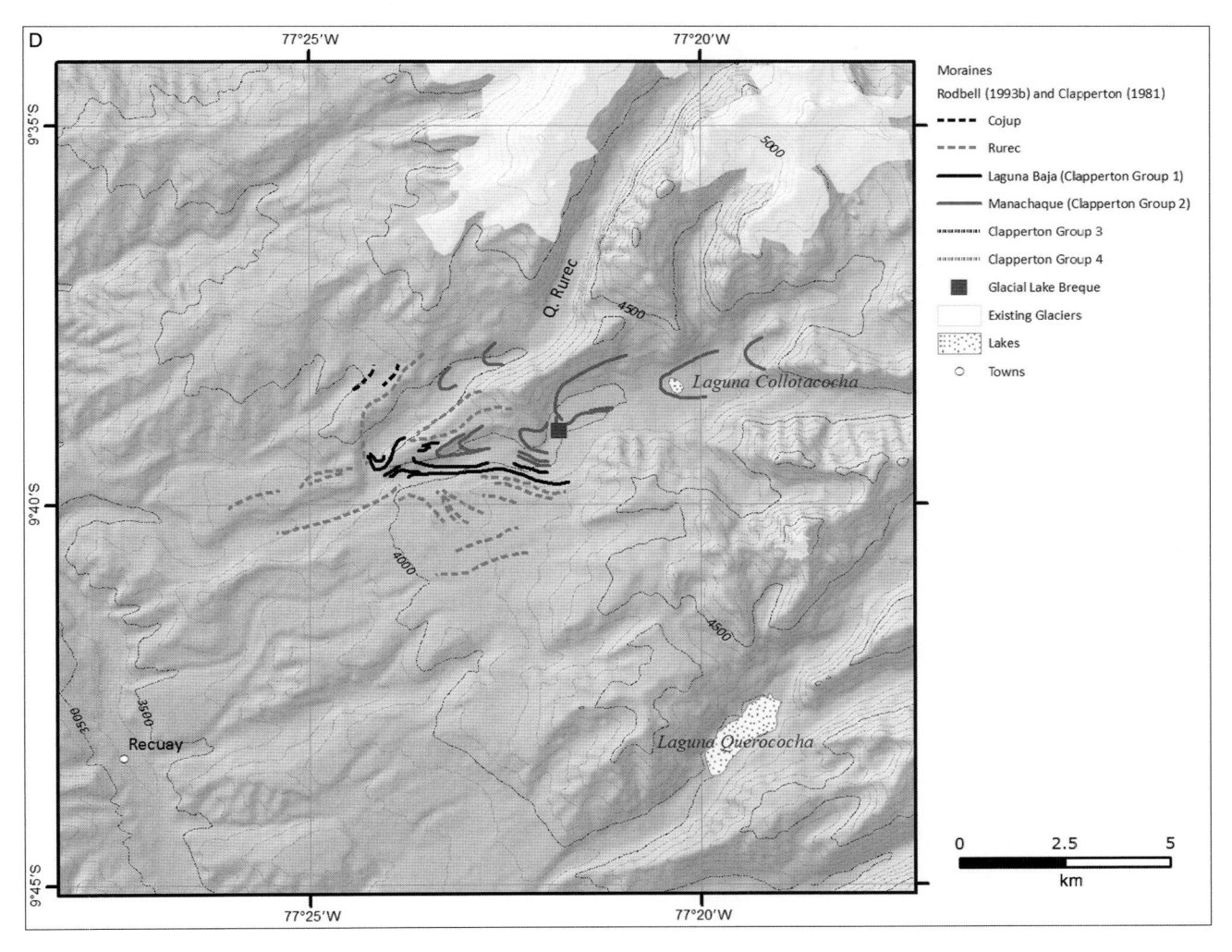

FIGURE 56.6 D

conflicting radiocarbon evidence. A sediment core from Lago Junin, on the floor of the plain, revealed a glacial till constrained by maximum–minimum dates of 24.0 and 12.0 ka, respectively. However, a radiocarbon date obtained from a sediment core at the south end of Laguna Huatacocha, ~40 km closer to the glacier's source, indicated that ice had retreated past this point by 13.5 ka.

Smith et al. (2005) examined moraines on the east side of the Junin Plain, overlapping the earlier mapping work of Wright (1983) in the Calcalcocha Valley. The authors classified four groups of moraines (A–D), with Group C being representative of the local LGM and thus equivalent to the Punrun sequence. Their SED results indicate that the local glacial maxima occurred as early as 34 ka, with recession commencing no later than 21 ka. Though the 34 ka date was later revised downward to 31 ka (Smith et al., 2008), the key conclusion remained that the local LGM at the Junin Plain significantly pre-dated the global LGM at 22–19 ka.

Zech et al. (2007) recalculated the CRN data of Smith et al. (2005) using a different calibration method, and determined that the local LGM was 25–22 ka, in accord with work from the same study in Bolivia's Cordillera Real. Bromley et al. (2009) also recalculated the Smith et al. (2005) dataset using the CRONUS calculator and agreed that the local LGM dates had been overestimated. Smith et al. (2005) identified another MIS 2 morainic sequence, classified as Group B and dated to 20–16 ka. Bromley et al. (2009) recalibrated these dates using the CRONUS calculator and estimated a mean age of 16.1 ± 1.7 ka. On the map accompanying this chapter (Fig. 56.7B), the Group D moraines of Smith et al. (2005), dated at >67 ka, appear to correlate with the MIS 2 Punrun phase of Wright (1983). This is not the interpretation of the authors, however; instead this apparent spatial relationship is due only to the different scales and possible slight spatial inaccuracies of the original maps.

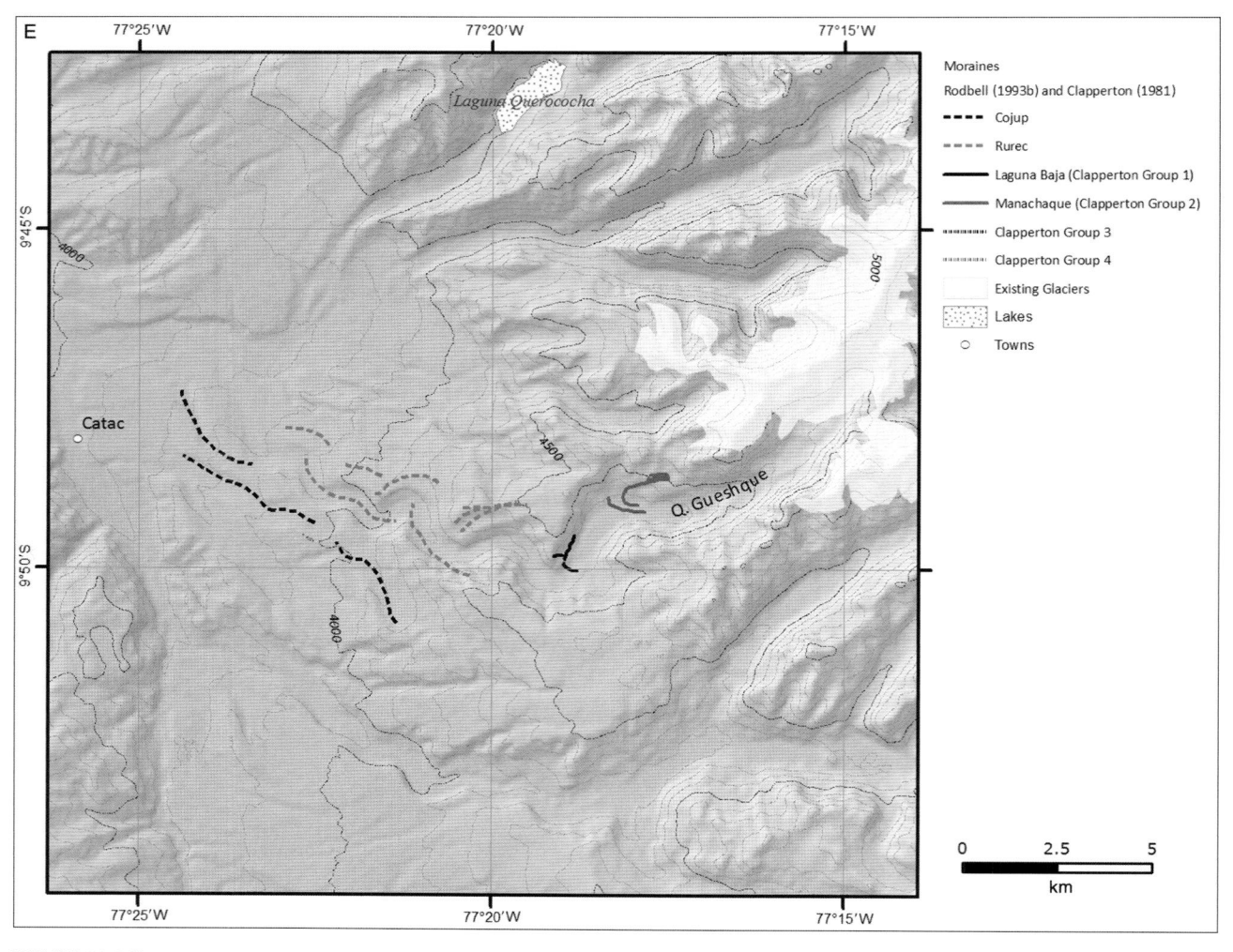

FIGURE 56.6 E

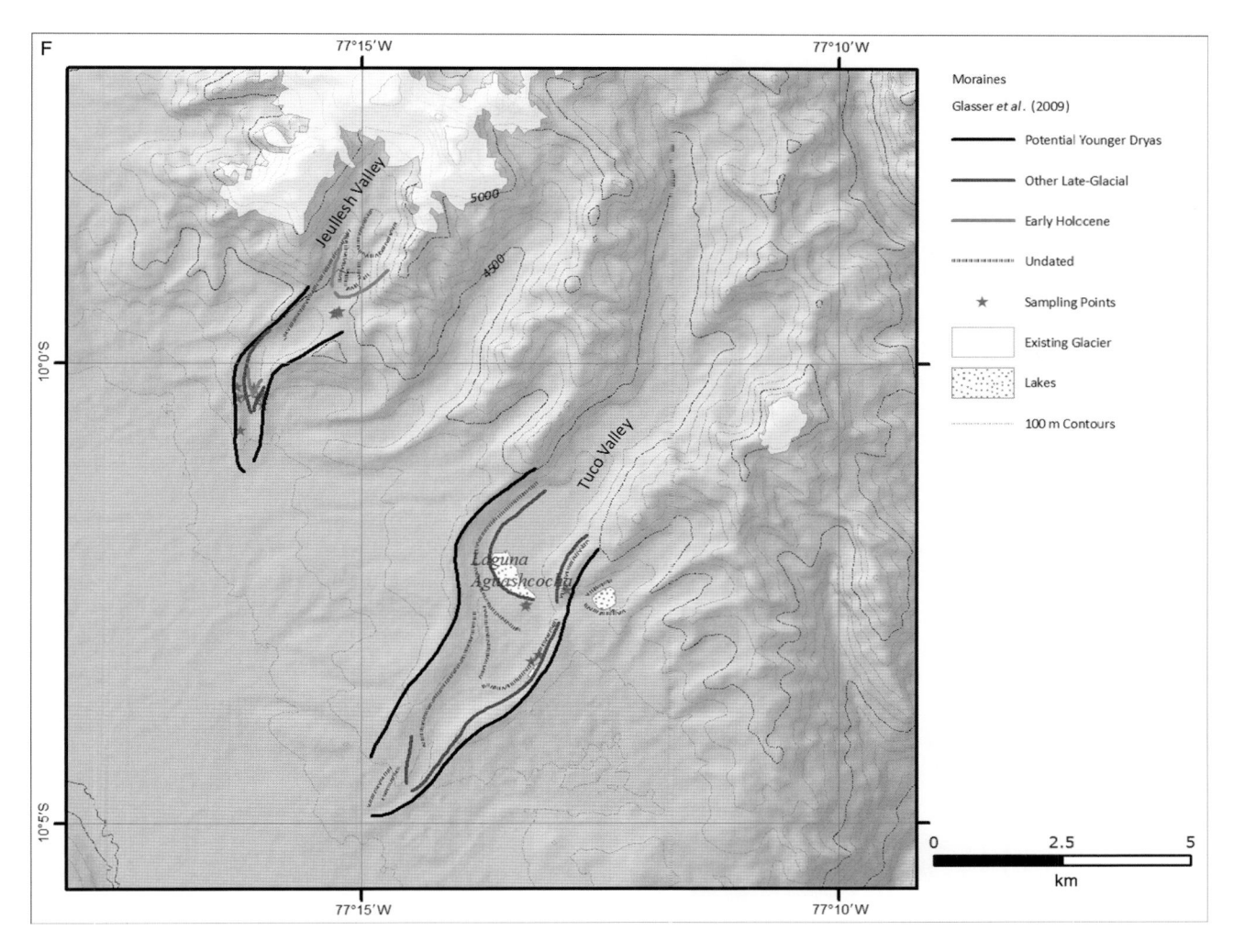

FIGURE 56.6 F

On the northwest flank of the Cordillera Vilcanota (14.0°S, 71.0°W; Fig. 56.8), a cluster of well-nested moraines between 4000 and 4350 m in the Upismayo and Jalacocha Valleys was determined to be of MIS 2 origin based on maximum-limiting radiocarbon date of ~34.4–31.7 ka and a minimum-limiting radiocarbon date of 16.8 ka (Mercer and Palacios, 1977). These moraines have subsequently been classified as the U3 sequence (Mark et al., 2004). Goodman et al. (2001) combining soil development profiling and radiocarbon dating of lake/bog sediments, calculated a 16.65 ka minimum-limiting age for the U3 sequence and argued that these moraines were the result of a distinct glacial advance that culminated at this time, noting that such a chronology was contemporaneous with other sites in the tropical Andes as well as the Heinrich Event One evidenced by ice-rafted debris in the North Atlantic. Mark et al. (2002) examined the U3 moraines and noted that, while well-nested, the moraines were also cross-cutting and thus suggestive of multiple glacial advances that happened to terminate at the same location. Further, the authors obtained a radiocarbon minimum-limiting date from nearby Laguna Casercocha in a sediment core of ~20 ka, which they argued bolstered their hypothesis that the U3 sequence represents two different MIS 2 advances.

Mercer and Palacios (1977) mapped moraines in the vicinity of the Quelccaya Ice Cap (14.0°S, 70.9°W; Fig. 56.13) and defined three glacial phases: Huancané I, Huancané II and Huancané III. Huancané III moraines are the oldest of these but are considered to be of Late-glacial origin based on a minimum-limiting age of ~14 ka. Goodman et al. (2001) identified a series of moraines ~14 km southwest of the present ice face and obtained a minimum-limiting radiocarbon date of 25.5 ka. Their age and distance from the ice face indicate that these moraines originated from a different advance than those of the Huancané III group.

A distinct group of moraines on Nevado Coropuna, a glacier-capped stratovolcano at the southern end of Peru's Cordillera Occidental (Fig. 56.9) were estimated at 21.4–20.7 ka based on SED by Bromley et al. (2009), using the CRONUS calculator scaling method. These moraines, classified C-I, occur at 4600 m and 4350 m on the west and north sides of the mountain, respectively, and are interpreted as representative of an LGM advance on the mountain. Based on recalibration of surface exposure data from Smith et al. (2005) and Farber et al. (2005), the authors argued that the C-I sequence is concurrent with the Group C (Junin Plain) and Rurec (Cordillera Blanca) moraine sequences.

Smith et al. (2005) found synchrony between the glacial phases of in Bolivia's Cordillera Real (16.3°S, 68.2°W; Fig. 56.14A and B) and those of the Junin Plain region of central Peru, and use the same moraine classification (Group C for the MIS 2 local LGM, Group B for Late-glacial). As with the Junin Plain moraines, Zech et al. (2007) recalculated the Group C dates of Smith et al. (2005) and estimated an age range of ~25–22 ka here as well. In the Cordillera Real, Smith et al. (2005) studied a pair of valleys on either side of Nevado Huayna Potosi (Fig. 56.14B) and found distinctly different glacial behaviour during each sequence. In the west-facing Milluni Valley, the lowest Group B moraines are located about 8–10 km up-valley from the local LGM Group C moraines, while in the steep, narrower, east-facing Zongo Valley, no Group C moraines are identified while Group B moraines appear to have progressed at least 1000 m lower in altitude and are the oldest in the valley. The authors attributed the difference in Group B extent to increased snowfall, debris-cover on the glacier surface, and topographic shading in the Zongo Valley having allowed the maintenance of ice at lower elevations.

Approximately 60 km northwest of the Milluni Valley, Zech et al. (2007) used SED to estimate the age of moraines at the mouth of the Valle San Francisco (Fig. 56.14A) as

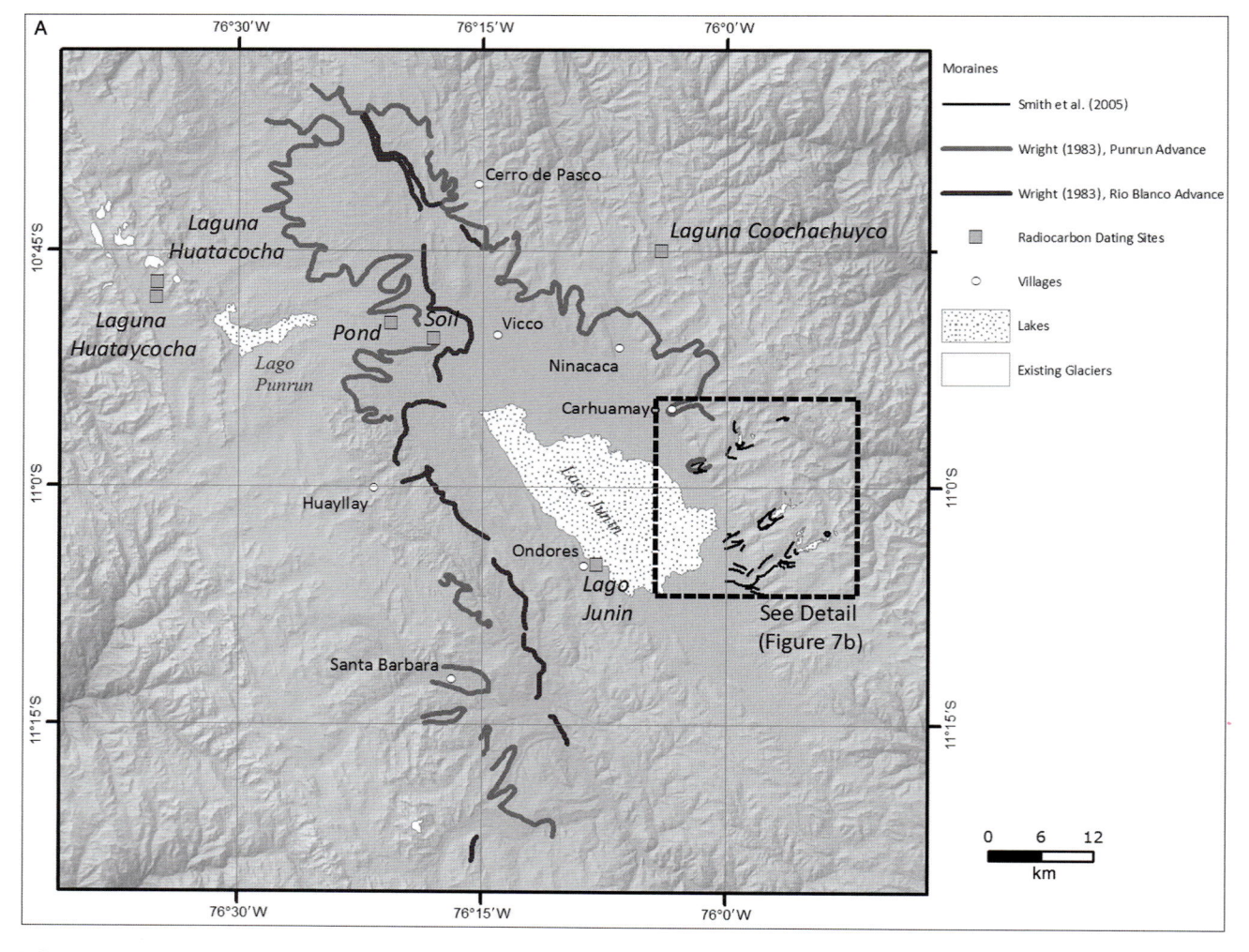

FIGURE 56.7 A Quaternary glacial features adjacent to the Junin Plain, Peru. Adapted from Wright (1983) and Smith et al. (2005).

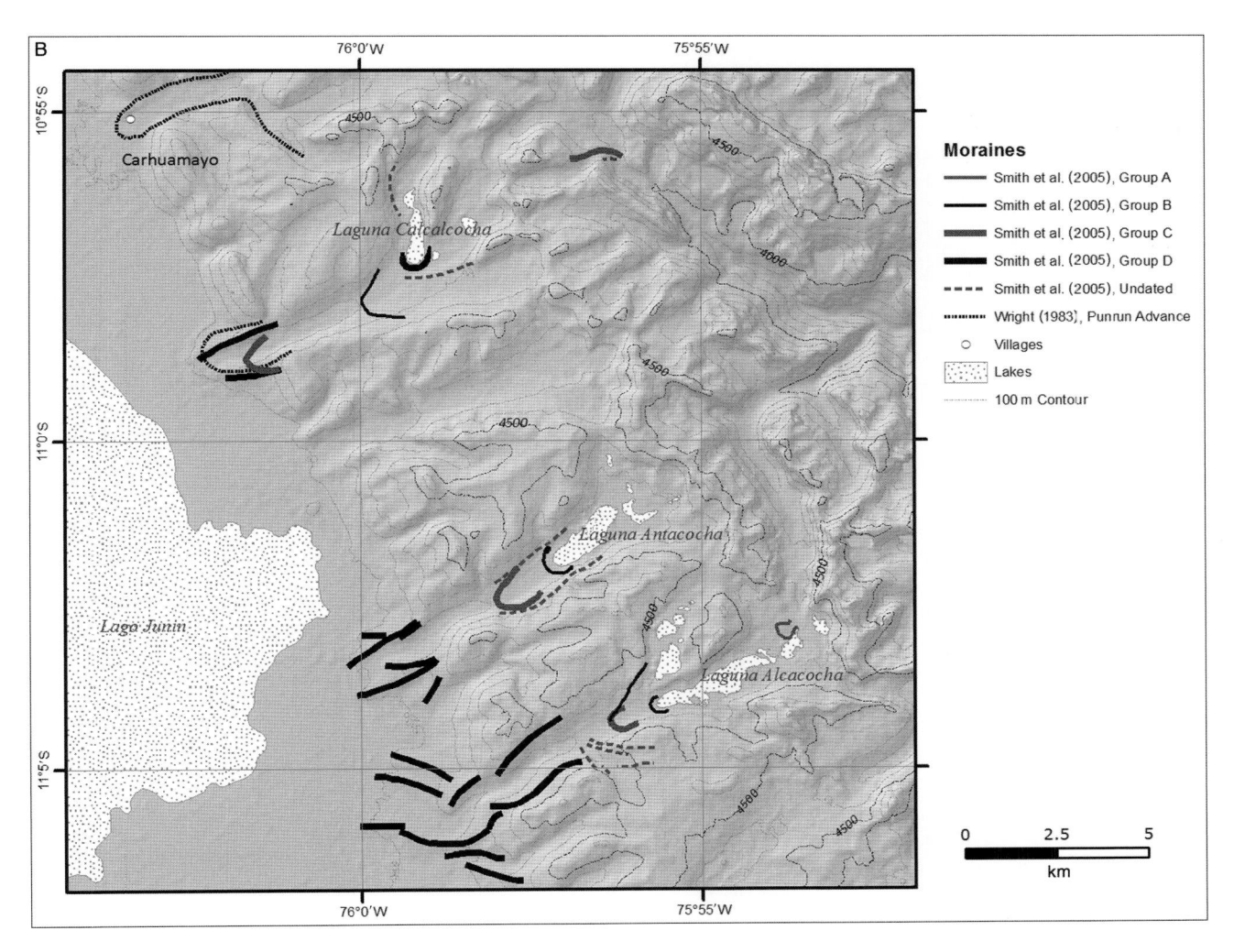

FIGURE 56.7 B (Continued).

~25–22 ka, or coeval with the Group C moraines of Smith et al. (2005). Zech et al. (2007) also identified a local glacial advance at ~15 ka, which could be equivalent to the Group B sequence. Farther south in the Cordillera Cochabamba (17.2°S, 66.5°W; Fig. 56.15), Zech et al. (2007) estimated an identical ~25–22 ka range for the local LGM.

56.5.3. Late-glacial (16.7–11.7 ka)

In addition to their disagreement over the global LGM or Late-glacial provenance of Potrerillos Plateau moraines located at 3800–3900 m, Heine and Heine (1996) and Clapperton et al. (1997) provided differing opinions as to Late-glacial conditions, particularly whether or not there is evidence of cooling that correlates with the Younger Dryas chron in the Ecuadorian glacial record. Heine and Heine described M5 and M6 moraines on both Rucu Pichincha (Fig. 56.3) and the Potrerillos Plateau (Fig. 56.4A and B). M5 moraines are constrained by a minimum-limiting age of ~15.4 ka (Rucu Pichincha) and ~14.8 ka (Potrerillos Plateau), and in both locales occur between 4100 and 4200 m. M6 moraines occur inside M5 moraines at roughly the same elevation in both locales and are age-bracketed by dates of ~13 ka (obtained from an underlying peat) and 8.2 ka (obtained from an overlying tephra).

Based on the above chronology, Heine and Heine (1996) suggested the following Late-glacial scenario for Rucu Pichincha and the Potrerillos Plateau: after the local LGM, glaciers re-advance by 15.4 ka (Rucu Pichincha) or 14.8 ka (Potrerillos Plateau) to deposit M5 moraines. As M5 glaciers recede, lakes (Rucu Pichincha) and peat bogs (Potrerillos Plateau) form behind M5 moraines, but these desiccate as the climate dries at ~13 ka (Rucu Pichincha)

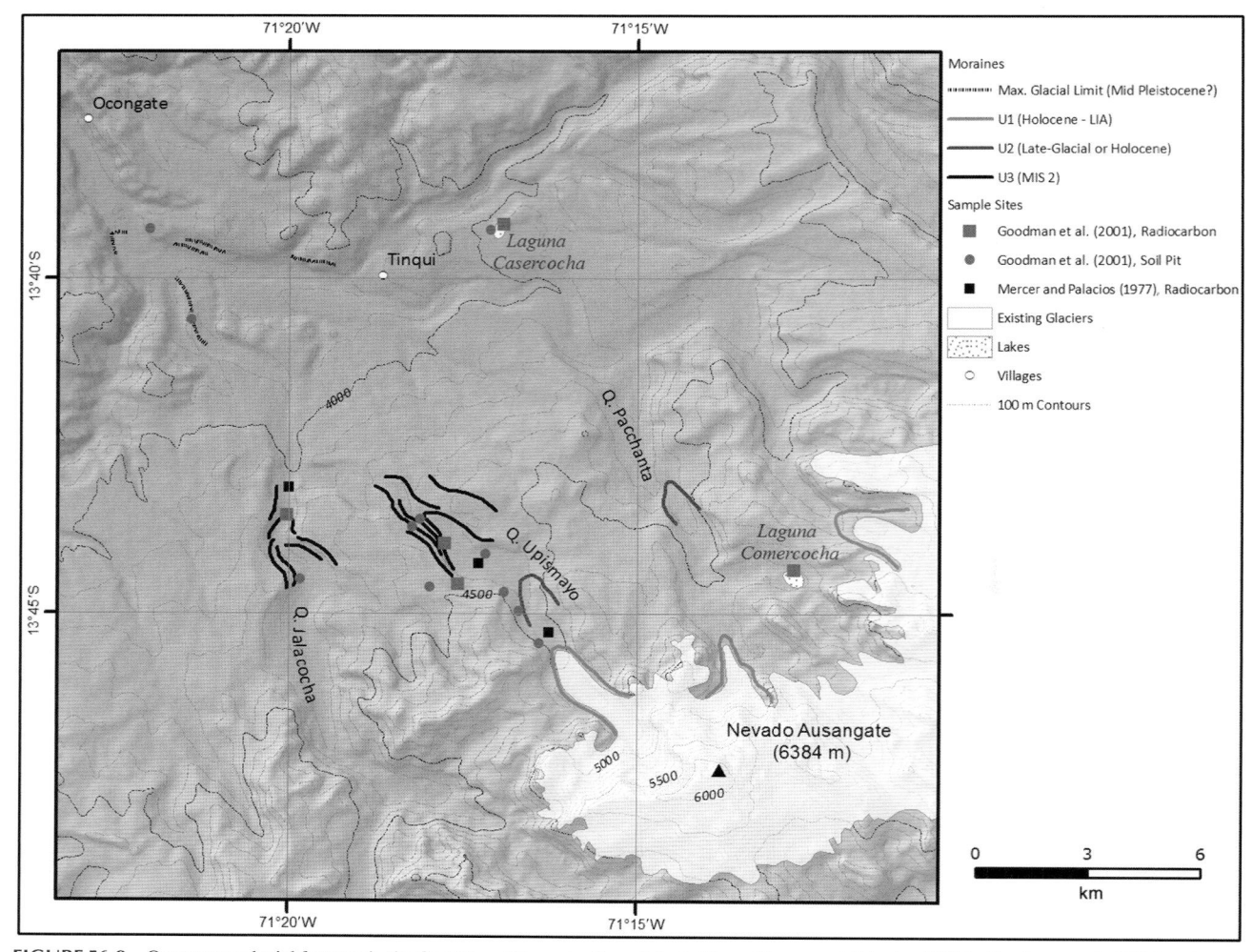

FIGURE 56.8 Quaternary glacial features in the Cordillera Vilcanota, Peru. Adapted from Mercer and Palacios (1977), Goodman et al. (2001) and Mark et al. (2002).

and ~13.3 (Potrerillos Plateau). Glaciers advance once more to deposit M6 moraines sometime between 12.5 and 9 ka.

Clapperton et al. (1997) identified moraines correlating to the M6 sequence as evidence of the 'Potrerillos Glacier Advance', which they associate with the redevelopment of a smaller (140 km^2) ice cap atop the Potrerillos Plateau. Because this advance is constrained by a maximum-limiting age of 12.9 ka and is evidenced by the burial of paramó vegetation between 12.71 and 12.26 ka BP in one sediment core, the authors argued it was coeval with the onset of the Younger Dryas chron in the Northern Hemisphere. Heine and Heine (1996) argued that the onset of M6 glaciation occurred 400–500 years after the beginning of Younger Dryas cooling in Europe and that, without explanation of this lag, a correlation could not be made. However, the calibrated radiocarbon dates derived from the original Heine and Heine (1996) ^{14}C dates (~12.85 ka, per Rodbell et al., 2009) are in very close proximity to the 12.9 ka date generally accepted as the onset of the Younger Dryas chron (Rasmussen et al., 2006). As noted earlier, however, the presence of any ice cap at the altitude/latitude combination present at the Potrerillos Plateau is inconsistent with observed glacial behaviour elsewhere in the Ecuadorian Andes (Rodbell et al., 2009). Both research groups agree that both Rucu Pichincha and the Potrerillos Plateau were deglaciated no later than the Early Holocene (~11 ka per Clapperton et al., 1997 and no later than 8.2 ka per Heine and Heine, 1996), and have remained ice free ever since.

Late-glacial moraines on the Chimborazo–Carihuairazo massif were classified by Clapperton and McEwan (1985) and Clapperton (1990) as belonging to Group 2 (Fig. 56.5A and B). In the Rio Mocha drainage, these moraines occur as low as 3900 m and indicate that glaciers from both Chimborazo and Carihuairazo were confluent during this period as well. Clapperton and McEwan (1985) identified the site of a former glacial lake created by the damming

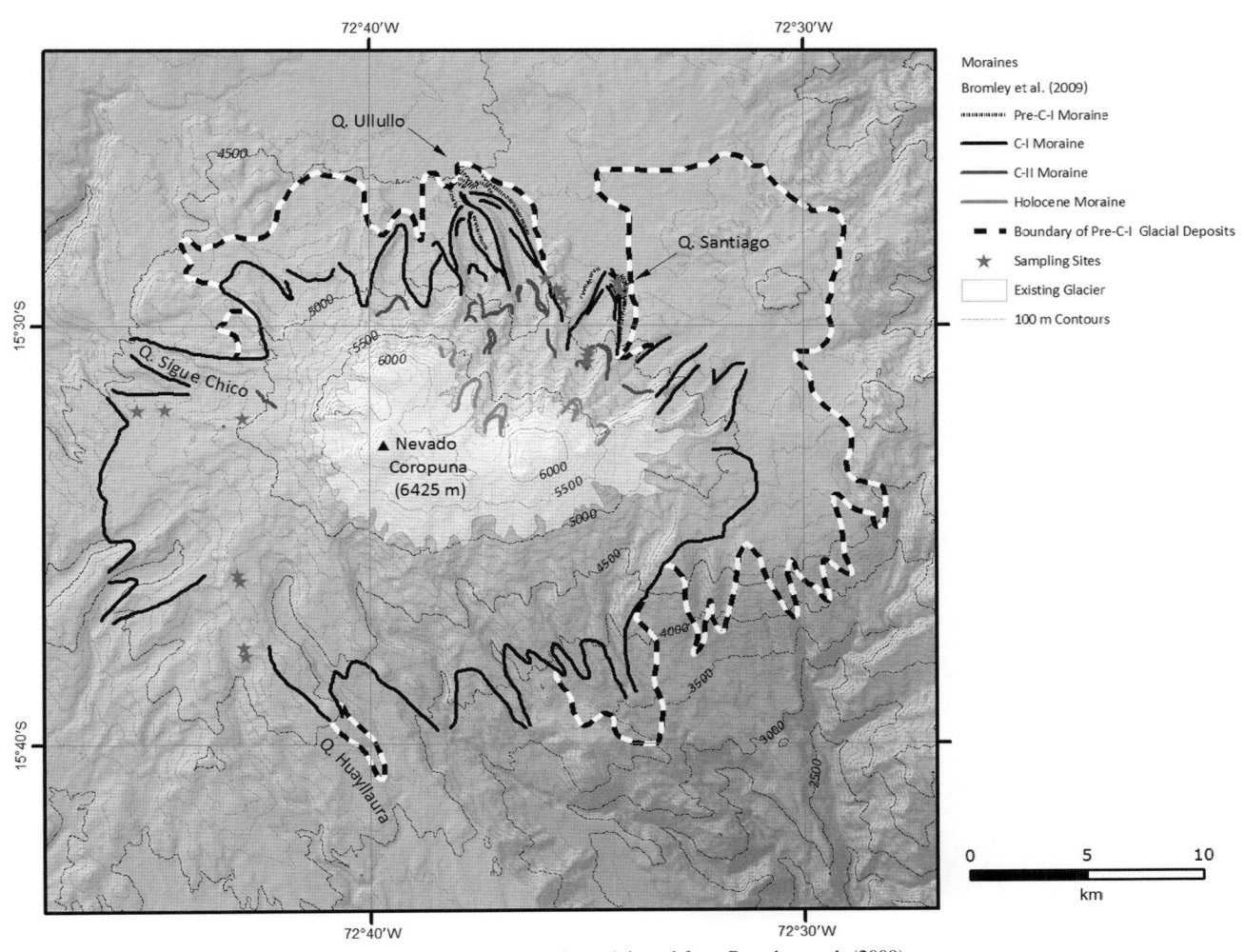

FIGURE 56.9 Quaternary glacial features on Nevado Coropuna, Peru. Adapted from Bromley et al. (2009).

of the Rio Mocha by a Group 2-era lobe of the modern-day Reschreiter Glacier on Chimborazo. Radiocarbon dates obtained from a pair of peat layers exposed at this site were calculated as ~13.2 and ~12.7 ka, respectively. The authors interpreted this as being the result of multiple lacustrine-peat bog transitions associated with glacial terminus variations and correlate the advance responsible for the lake with a contemporaneous glacial stage identified by Mercer and Palacios (1977) at the Quelccaya Ice Cap—interpreted as coincident with the Younger Dryas chron.

Heine and Heine (1996) countered the assumption that the dates obtained from the two peat layers were indicative of the dates of the glacial advance, noting that more than 8 m of lacustrine sediments underlie the two peat layers and that the lake must have been present, based on local sedimentation rates, at least 2.5 ka years prior to the dates calculated by Clapperton and McEwan (1985). Further, Heine and Heine (1996) posited that a moraine was the lake-forming feature rather than the glacier itself and that the lake would have been present even after the glacier had retreated. Instead, the authors hypothesised that the moraine responsible for the formation of this lake was associated with the M5 glacier advance that they identified at Rucu Pichincha and the Potrerillos Plateau approximately 14.8–15.6 ka. Rodbell and Seltzer (2000) reported visiting the site of this glacial lake in 1993 and stated that they agree with the interpretation of Heine and Heine (1996) that a direct relationship between lake fluctuations and variation in ice margins is questionable.

Radiocarbon evidence from the northern Cordillera Oriental in Peru (Fig. 56.11) indicates that glaciers had retreated past Laguna Baja (3600 m) by ~14 ka, representing a greater than 50% reduction in areal extent relative to their MIS 2 maxima (Rodbell, 1993b). There is evidence of glacial re-expansion above 3800 m between ~14.2 and 12.0 ka; however, final deglaciation was rapid, with the area being ice free between 12.0 and 11.1 ka.

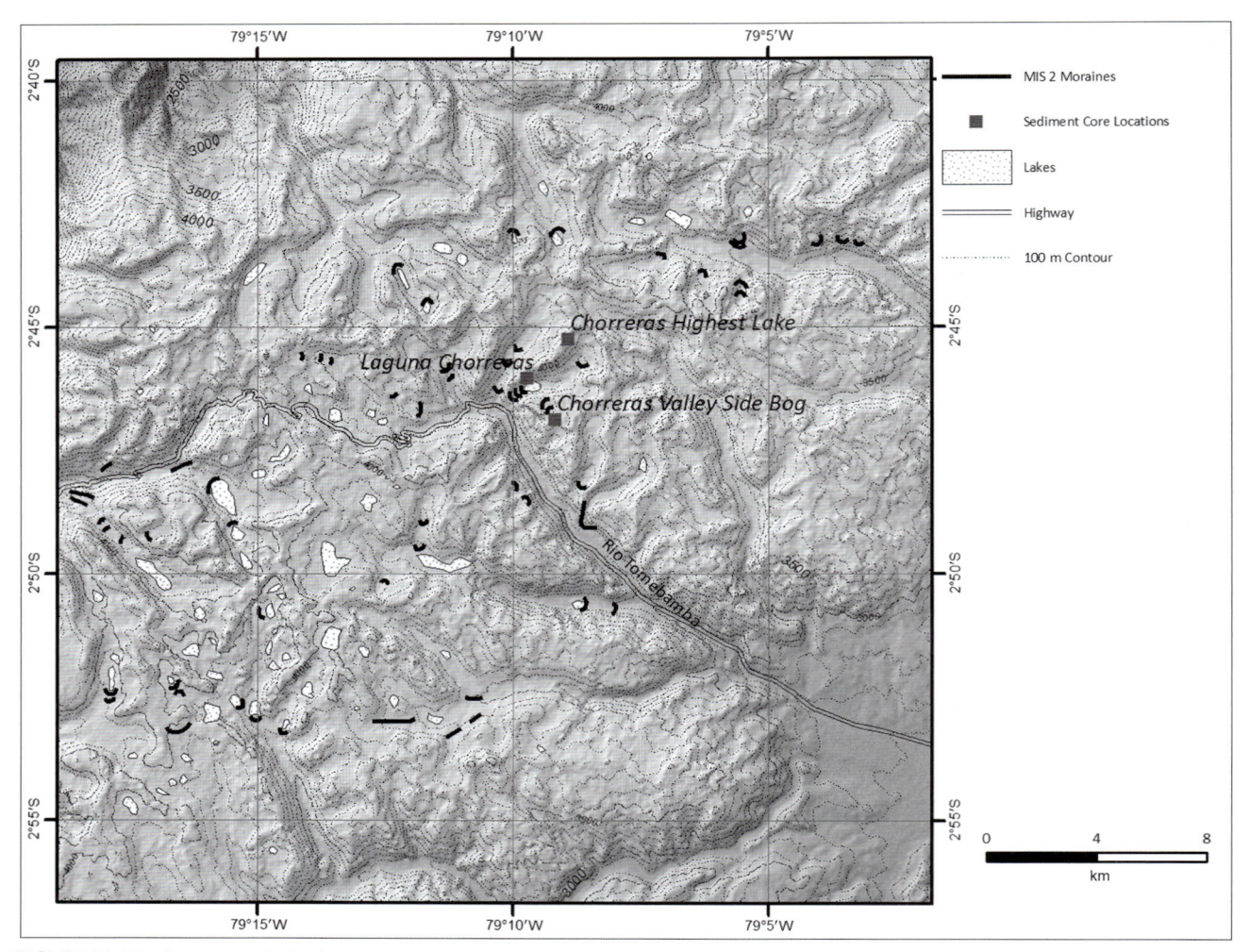

FIGURE 56.10 Quaternary glacial features in Las Cajas National Park, Ecuador. Adapted from Rodbell et al. (2002).

There is strong geomorphologic evidence for a Lateglacial ice advance in the Cordillera Blanca (Fig. 56.6A–F). Rodbell (1993a) defined the Manachaque moraine sequence (equivalent to the Group 2 moraines of Clapperton, 1981) and dated these features to ~13.5–10.3 ka. Manachaque moraines are located between 3800 and 4400 m and are distributed over a distance of 10 km down-valley from the oldest Holocene moraines.

Rodbell and Seltzer (2000) examined a lacustrine sediment sequence from a former glacial lake (Glacial Lake Breque, Fig. 56.6D) and were able to closely bracket a glacial advance that culminated at 12.9 ka and was immediately followed by rapid ice retreat. Because this retreat occurred despite strong evidence of regional cooling at this time, the authors inferred a sharp decrease in precipitation that starved glaciers of mass even though temperatures were reduced. This evidence of glacial retreat commencing at 12.9 ka, which the authors also found present at the Quelccaya Ice Cap, would appear to contradict other research (Clapperton and McEwan, 1985; Rodbell, 1993a; Clapperton et al., 1997) finding glacial advances concurrent with the Younger Dryas chron in the northern hemisphere. The authors, however, did not eliminate the possibility that the drying conditions were associated with a global scale event during the Younger Dryas chron. Instead, they noted that the cool and moist conditions that led to a glacial advance prior to 12.9 ka indicate markedly different conditions in the tropical Andes than those that define the proceeding Bølling-Allerød interstadial in the northern hemisphere.

Glasser et al. (2009) confirmed the presence of moraines dating from immediately prior to or early in the Younger Dryas chron in the southern Cordillera Blanca (Fig. 56.6F). In the Juellesh Valley, they dated a moraine at 12.4 ± 0.6 ka, while in the Tuco Valley, the oldest, most exterior lateral moraine returned a date of 12.5 ± 1.1 ka. The authors remarked upon the potential coincidence of these moraines with the Younger Dryas chron but did not

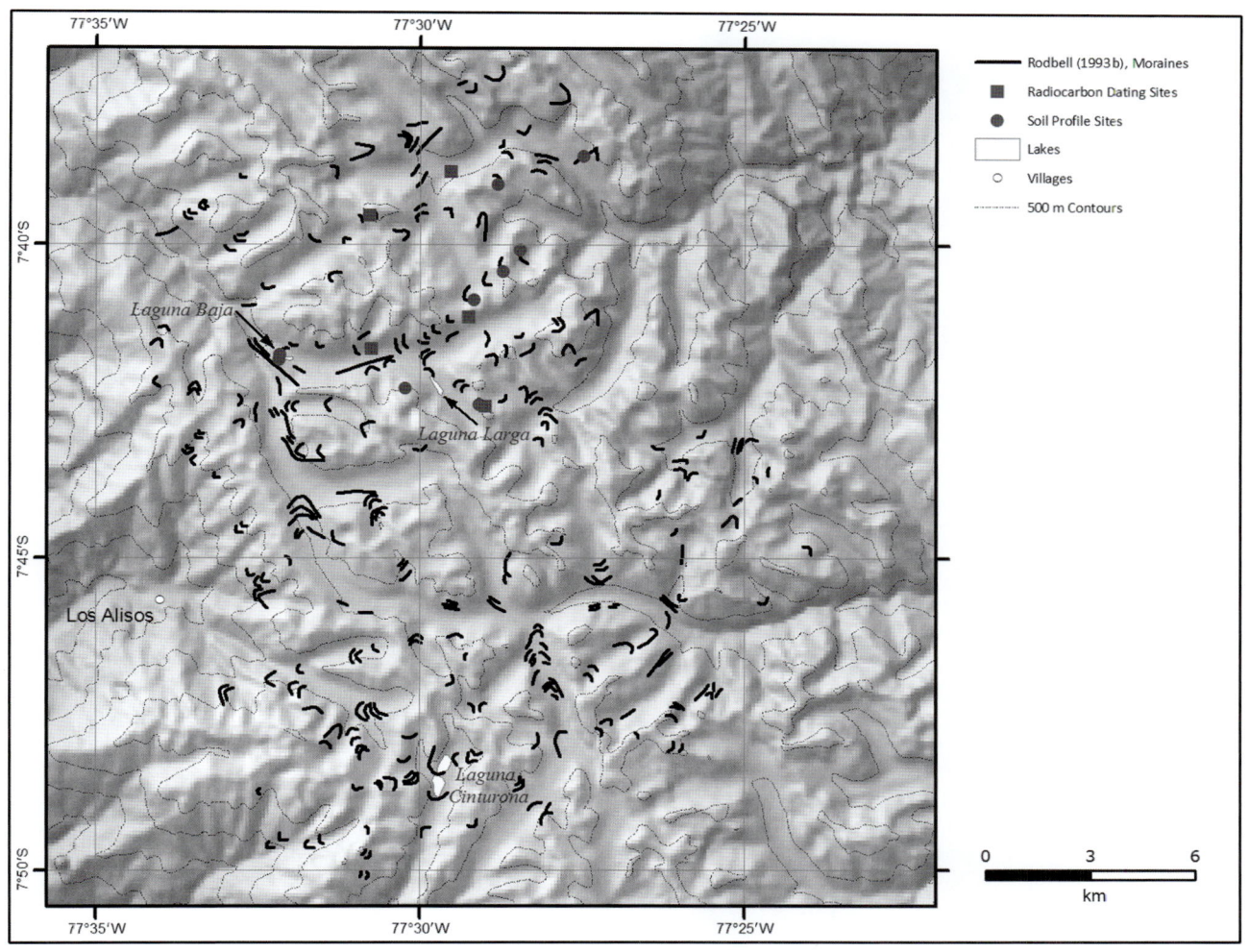

FIGURE 56.11 Quaternary glacial features in the Cordillera Oriental, northern Peru. Adapted from Rodbell (1993b).

explicitly link them, noting that the separation of temperature from precipitation as a forcing mechanism for tropical glacier change is difficult. A pair of nested moraines located ~1 km farther up the Juellesh Valley were dated at ~10.8 and ~9.7 ka, respectively, suggesting either a somewhat subdued rate of retreat from the ~12.4 ka position followed by an extended stillstand or possibly even a minor re-advance.

In the Cordillera Huayhuash (Fig. 56.12), at least two, and possibly three, sets of Late-glacial moraines have been identified at elevations as low as 4000 m (Hall et al., 2009). Using a combination of SED and radiocarbon dating, the authors placed Stage III moraines at 14–13 ka, and related them to Manachaque moraines in the Cordillera Blanca (Rodbell, 1993a) and Group A moraines of the Junin Plain region (Smith et al., 2005). While Stage II moraines, dated at 10–9 ka, were the next prominent set of moraines, the authors also identified a Stage IIa represented by smaller moraines immediately down-valley from the larger Stage II features. Stage IIa moraines were dated at ~12–11.5 ka, which the authors suggested potentially represent an advance synchronous with the Younger Dryas chron. They noted, however, that the ages of the more dominant Stage II and Stage III moraines are not synchronous with the Younger Dryas.

The existing literature from the Junin Plain (Fig. 56.7A and B) describes very little in the way of Late-glacial ice extent and behaviour. Smith et al. (2005) defined a Group A moraine classification and dated these moraines to ~15–12 ka; however, they mapped only one such moraine, in the Alcacocha Valley.

In the Upismayo Valley of the Cordillera Vilcanota (Fig. 56.8), Mark et al. (2002) described a group of moraines 4 km up-valley from the late-MIS 2 U3 group, which Mark et al. (2004) classified as the U2 sequence. Mark et al. (2002) were unable to define a precise date range for the U2 moraines, determining only maximum/minimum-limiting ages of 12.25 and 5.05 ka, respectively. The

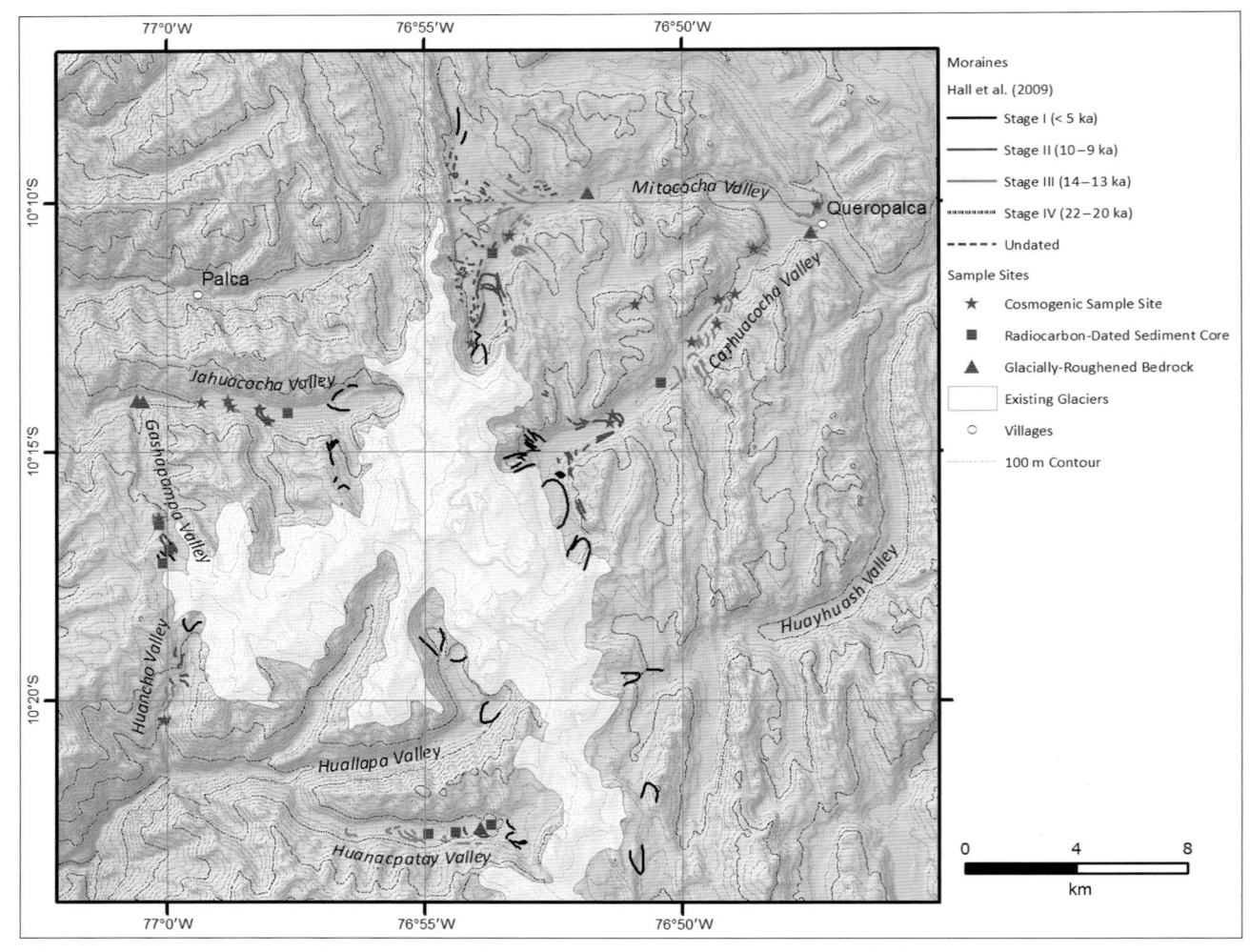

FIGURE 56.12 Quaternary glacial features in the Cordillera Huayhuash, Peru. Adapted from Hall et al. (2009).

authors placed greater weight on the lower end of this range and suggested that U2 moraines were most likely associated with a Middle Holocene advance.

Mercer and Palacios (1977) described three distinct terminal moraine sequences in the Rio Huancané Valley, west of the Quelccaya Ice Cap (Fig. 56.13), and placed two of these, Huancané III and Huancané II, within the Late-glacial period. Huancané III is the oldest and is constrained by a minimum-limiting radiocarbon date obtained from a peat sample of 14.2 ka. While this date is only a minimum limit, the authors believed that the ice cap retreated from the Huancané III moraines very near to these limiting dates. Based on additional radiocarbon dates, the authors argued that the Huancané II sequence resulted from a glacial advance that was in progress by 13.3 ka and over by ~12.9 ka. Rodbell and Seltzer (2000) agreed with this Quelccaya chronology, and their research in the Cordillera Blanca identified a glacial advance constrained by very similar dates, leading them to conclude that glaciers in both central and southern Peru varied in near 'lockstep' and that this was indicative of at least regional climatic forcing at that time.

Goodman et al. (2001) studied the Huancané II moraines and calculated minimum-limiting radiocarbon dates of 13.1 and 12.8 ka based on their own coring analysis. They also obtained a minimum-limiting date of 10.1 ka for moraines fronting Laguna Churuyo, ~8 km southwest of the present ice face. The location of these moraines, however, is suggestive of the Huancané III sequence. Mark et al. (2002) also examined the Huancané sequences, obtained from a peat sample a minimum-limiting radiocarbon date of 14.3 ka for the Huancané III sequence. The authors also mentioned the 12.8 ka radiocarbon date obtained from Laguna Paco Cocha by Rodbell and Seltzer (2000) and noted that this indicates that the Quelccaya Ice Cap has not been farther than 1 km from its current position anytime in the past 12.8 ka years.

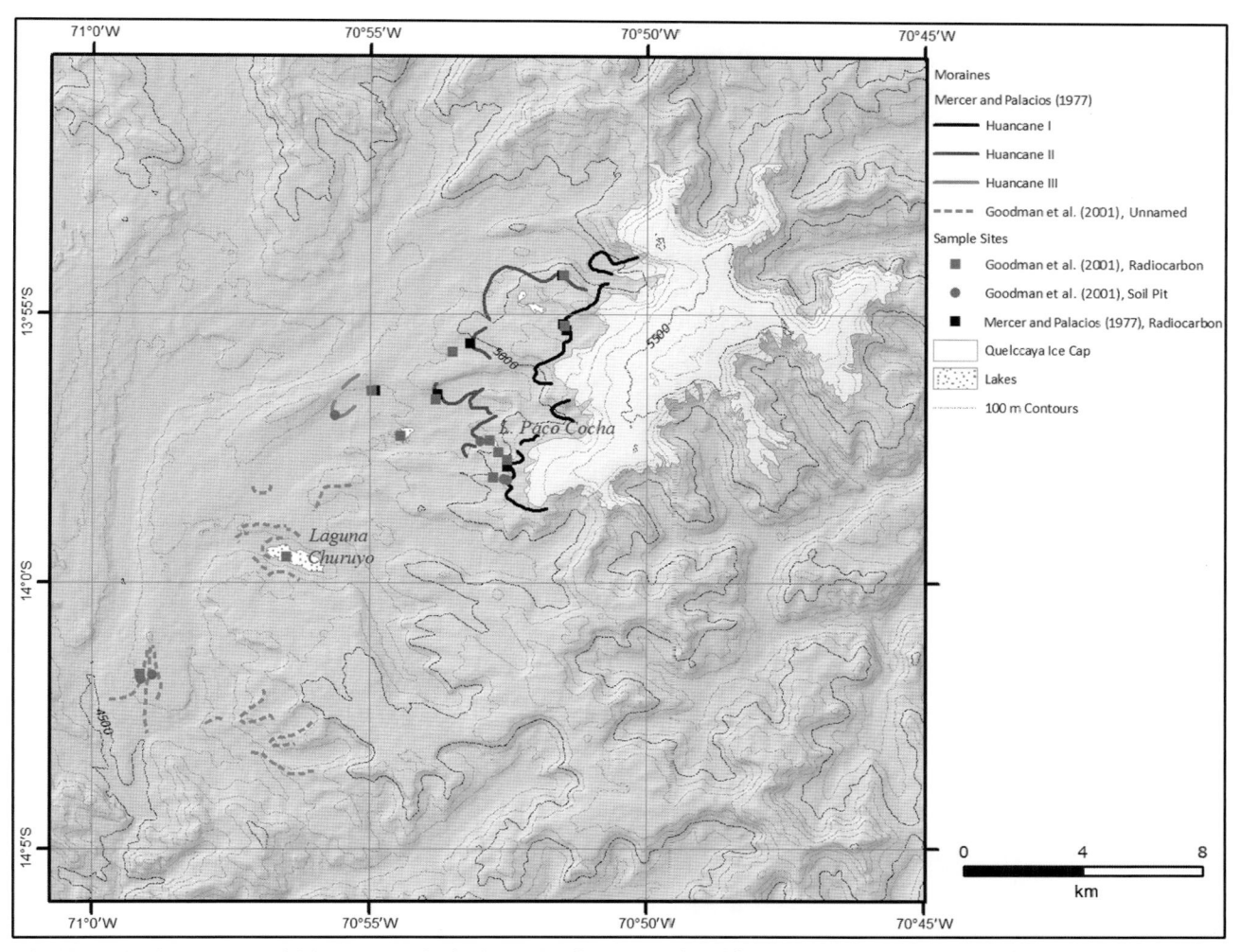

FIGURE 56.13 Quaternary glacial features near the Quelccaya Ice Cap, Peru. Adapted from Mercer and Palacios (1977), Goodman et al. (2001) and Mark et al. (2002).

On Nevado Coropuna (Fig. 56.9), a group of moraines classified as the C-II sequence are constrained by surface exposure dates of 14.2 and 10.6 ka, with a mean age of 13.0 ± 1.6 ka (Bromley et al., 2009). In Quebradas Ullullo and Santiago, C-II moraines terminate 3.5 km up-valley from the C-I sequence, and on the north side of the mountain, they terminate at 4700 m, 350 m higher than the C-I group. The authors tentatively suggested that the C-II group was equivalent to the Laguna Baja sequence in the Cordillera Blanca, based on the fact that the Laguna Baja age had been revised downward to 16.5 ka by Farber et al. (2005) and that the revised Rurec sequence dates of Farber et al. (2005) had themselves been overestimated due to the SED scaling method they had applied.

A key point made by Bromley et al. (2009) was that while the exact dates of the C-I and C-II sequences may be uncertain due to the variability associated with the use of competing calibration procedures, the 6- to 7-k year difference between the two phases is consistent with similar global LGM to Late-glacial recessional timelines identified at sites in New Zealand, North America and Europe. This, they argued, suggests that in this portion of the Andes, these glacial cycles were in synchrony with global events and that the variation in specific reported dates may be more a function of imprecise dating methods.

As with the Junin Plain, Smith et al. (2005) identified Group A moraines in the Zongo Valley of the Cordillera Real, Bolivia (Fig. 56.14B), and suggested a date range of ~ 15–12 ka based on SED.

In the Cordillera Cochabamba (Fig. 56.15), Zech et al. (2007) dated prominent lateral moraines in both the Valle Huara Loma and the Valle de Rio Suturi to 13.3–11.0 ka, again broadly consistent with advances identified elsewhere in the tropical Andes at roughly the same time as the

Younger Dryas chron. The authors also dated the most exterior, lower-elevation moraine in the Valle de Rio Suturi to 16.9 ka, though they admitted difficulty in reconciling this date with the ages of older moraines identified in the Valle Haura Loma and noted that the particular boulder from which the sample was obtained had clear signs of erosion, suggesting an example of cosmogenic underestimation.

Nevado Sajama (18.1°S, 68.9°W; Fig. 56.16), a 6500-m stratovolcano in Bolivia's Cordillera Occidental, is among the southernmost glacierised peaks in the tropical Andes. Smith et al. (2009) applied SED methods to classify four groups of moraines on the mountain, one of which dates from the Late-glacial. Ground Moraine 1, with minimum elevations between 4350 and 4635 m, is composed of previously englacial or supraglacial materials and is broadly dated to between 16.9 and 11.3 ka. However, lateral and terminal moraines associated with Ground Moraines 1 (and Holocene Ground Moraine 2) are composed of two different morainic structures, tall bounding moraines and short cross-cutting moraines, which confounds age estimation attempts. The tall moraines lack the boulders necessary for SED analysis; however, the authors hypothesised that these are relic landforms rather than the products of a Late-glacial or MIS 2 event while the short cross-cutting moraines were proposed to be the result of past cold-based ice, a glacier type not previously identified in the tropical Andes. Because of the intermixture of these moraine types and the lack of differentiation on the original Smith et al. (2009) map, they are not dated in Fig. 56.16. In any case, the authors argue that the most extensive local glacial advance appears to have occurred during the Late-glacial period rather than synchronous with the global LGM, since the outermost moraines appear no older than 16.9 ka, though they do not rule out the possibility of an earlier MIS 2 advance that has yet to be identified in Bolivia's Cordillera Occidental.

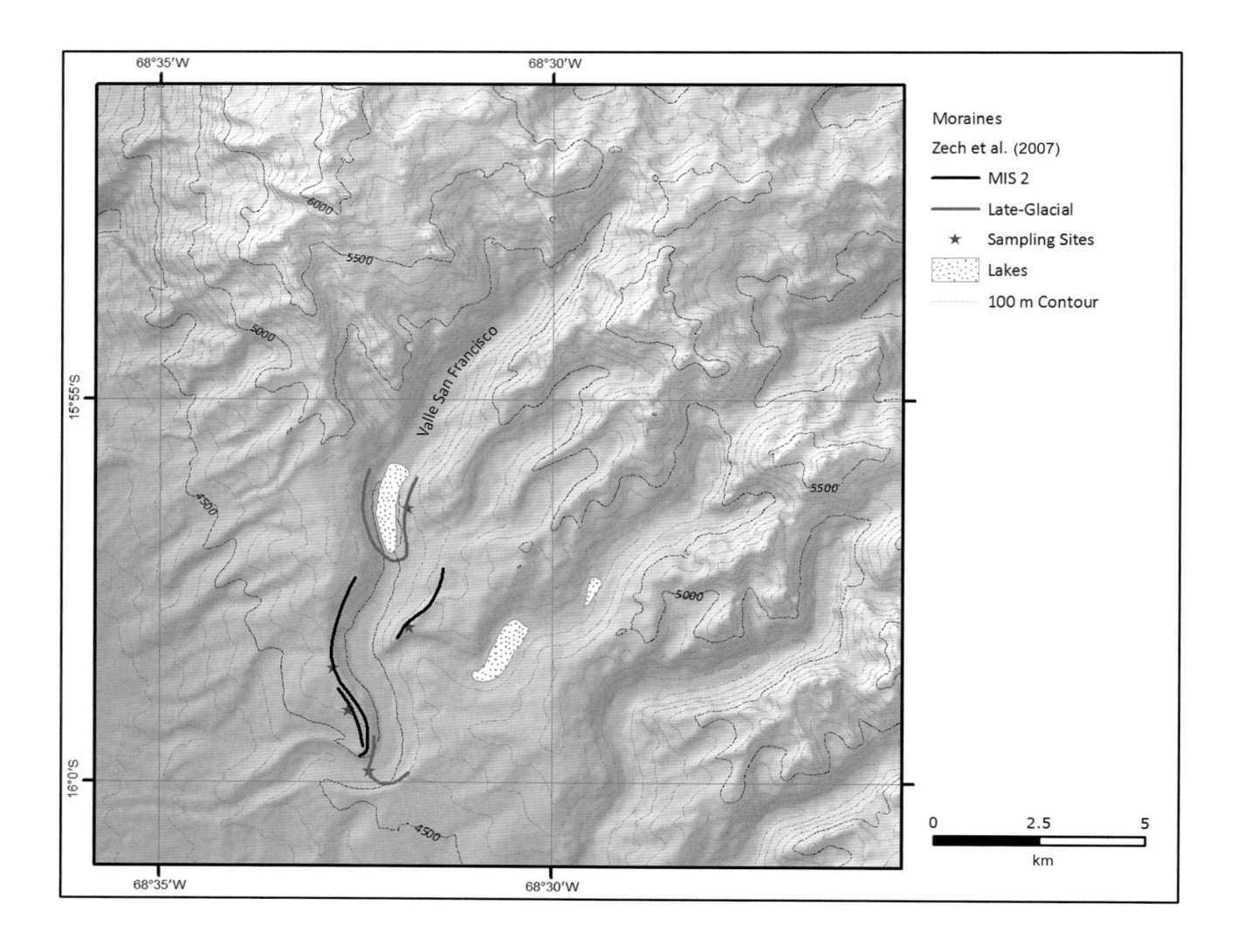

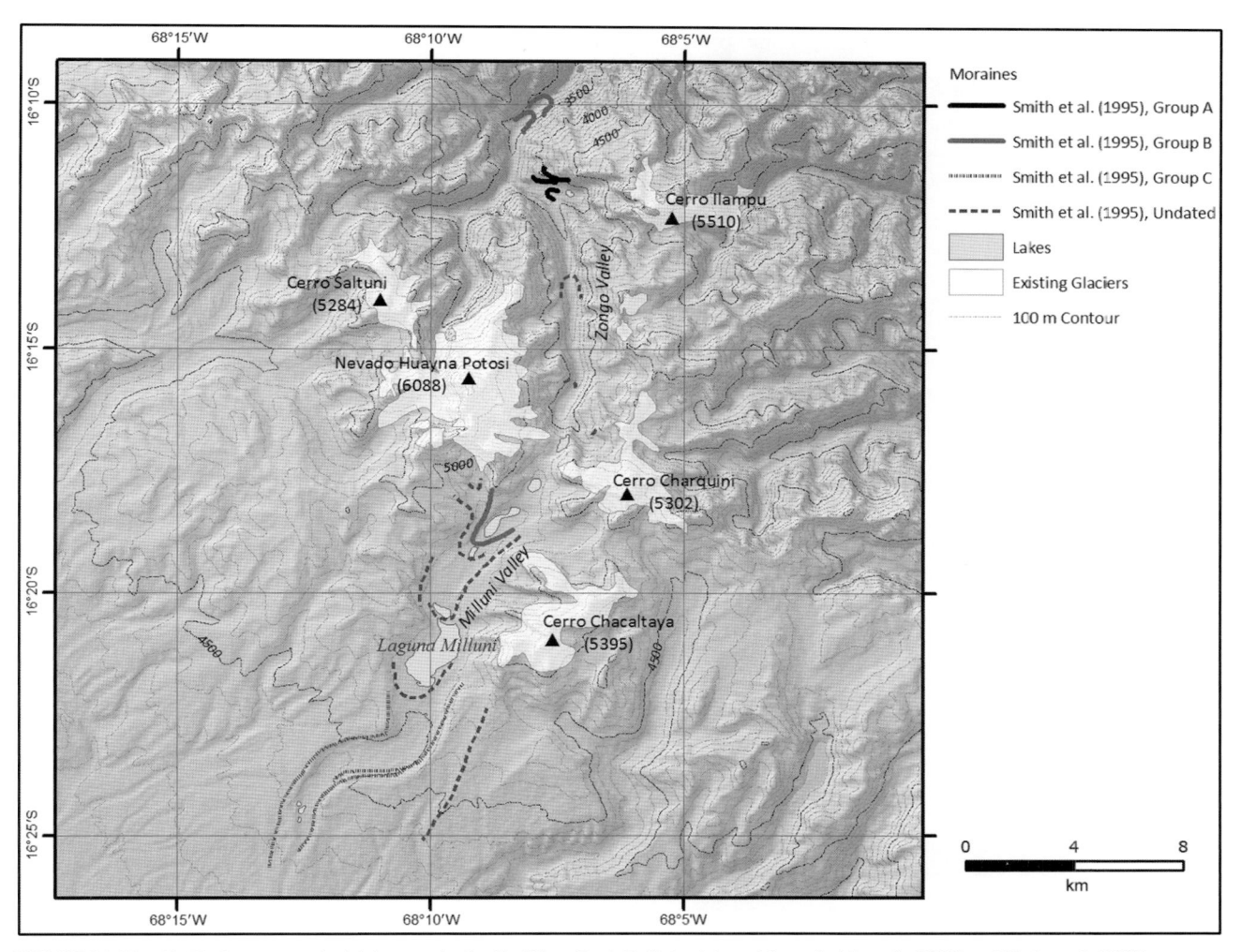

FIGURE 56.14 (A, B) Quaternary glacial features in the Cordillera Real, Bolivia. Adapted from Smith et al. (2005) and Zech et al. (2007).

56.6. MIS 1 (HOLOCENE) GLACIATIONS (11.7 KA–PRESENT)

Groups of superimposed moraines above 4300 m on the Chimborazo–Carihuairazo massif are classified as Holocene Group 1 moraines by Clapperton and McEwan (1985) and Clapperton (1990) (Fig. 56.5A and B). These are interpreted as the result of multiple, similarly sized Neoglacial (>6 ka) re-advances.

In the Cordillera Blanca (Fig. 56.6A–F), Clapperton (1981) classified Group 3 and Group 4 moraines as representative of Holocene ice-face fluctuations, while Rodbell (1993a) identified four separate Holocene moraine groups in the range. In the Juellesh Valley, Glasser et al. (2009) dated a largely unaltered moraine 1.5 km below the present-day glacier terminus at 7.6 ka and suggested that this date is broadly coincident with the 8200-year cooling event identified in Greenland and elsewhere. No such advance is identified in the Cordillera Huayhuash (Fig. 56.12); however, the Stage I moraines described by Hall et al. (2009) indicate a minor Middle Holocene advance between ~5.6 and 3.5 ka.

In the Cordillera Callejon, west of the Junin Plain (Fig. 56.7A), Wright (1983) noted that, while the current ice coverage is approximately 6 km^2, it appears to have been as much as 22 km^2, 'in the recent past'. No specific dates are provided, however.

In the Cordillera Vilcabamba (13.4°S, 72.6°W; Fig. 56.17), northwest of Cuzco, Peru, Licciardi et al. (2009) obtained surface exposure dates for two Holocene moraine sequences. The oldest and most extensive of the two sequences yielded an estimated date of 8.6 ± 0.3 ka which is similar in time frame to the advance described above in the Cordillera Blanca. The authors also obtained a date of 200 ± 20 years (i.e. CE 1810 ± 20) on an inner moraine in the Rio Blanco Valley. From a procedural

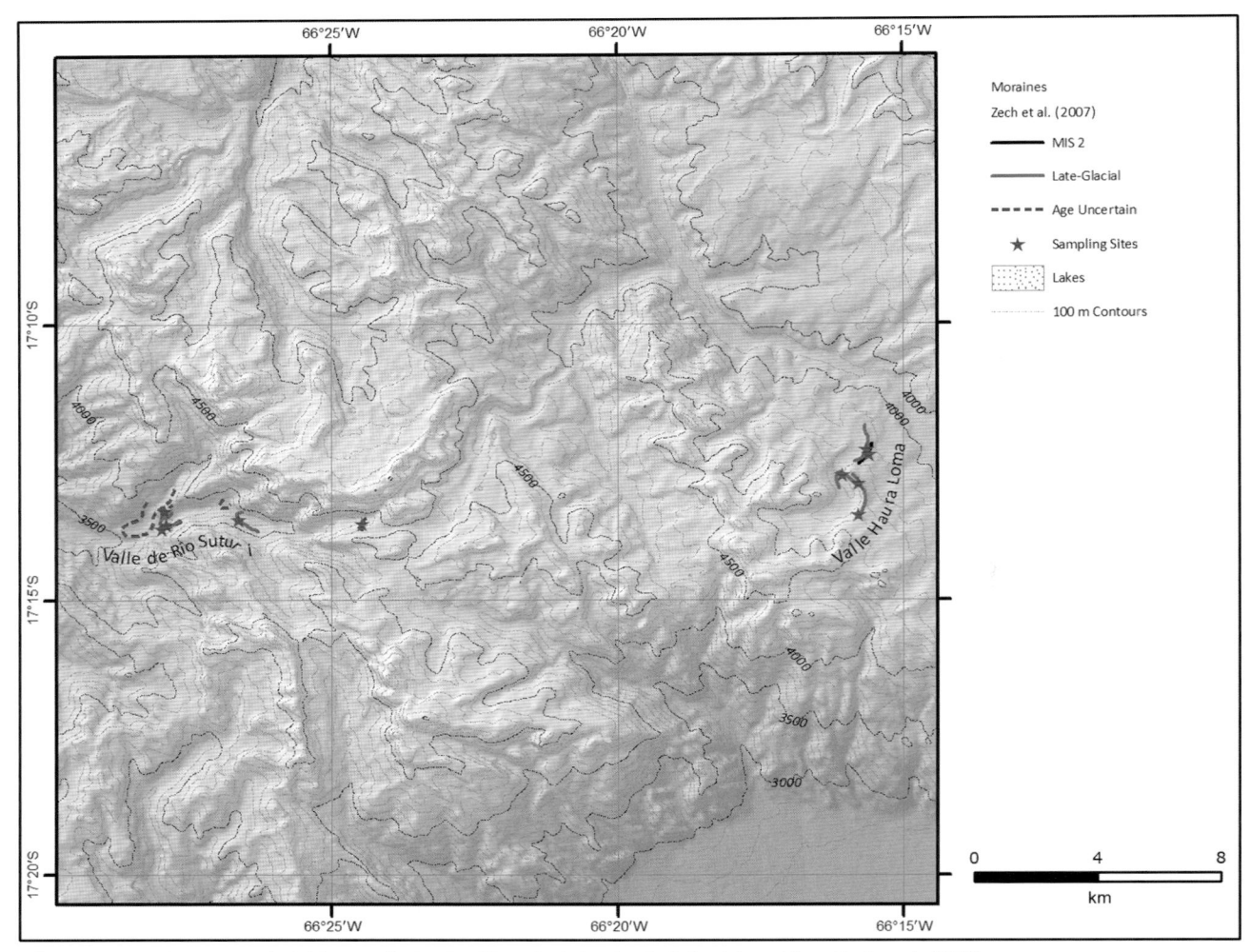

FIGURE 56.15 Quaternary glacial features in the Cordillera Cochabamba, Bolivia. Adapted from Zech et al. (2007).

standpoint, this is interesting because it shows that ^{10}Be SED can be used on morphological features of even historic age. From a climatological standpoint, these results indicate that a glacial advance occurred here that was broadly contemporaneous with the northern hemisphere Little Ice Age (LIA; CE 1300–1860) and that the present-day glacier has receded nearly 2 km since this time.

Compared to lichenometry studies of recent Holocene moraines elsewhere in the tropical Andes (e.g. Rodbell, 1992; Jomelli et al., 2009), which identify LIA maxima at CE 1630 ± 30 in the Cordillera Blanca and $\sim$CE 1669–1690 in the Cordillera Oriental of Bolivia, the LIA here appears to lag by over a century. While this may be due to imprecision associated with both lichenometry and SED, the authors believed that there were in fact differences in the timing of LIA glacial maxima within these regions. The authors also noted that the LIA maximum in the Cordillera Vilcabamba postdates a transition from wetter to drier conditions recorded in Quelccaya ice cores but is coincident with an abrupt shift towards nutrient-rich, oxygen-depleted ocean water recorded in marine sediment cores obtained at the same latitude off the Peruvian coast (Licciardi et al., 2009).

In the Cordillera Vilcanota (Fig. 56.8), Holocene moraines have been classified as the U1 sequence (Mark et al., 2004). Goodman et al. (2001) obtained a radiocarbon date of 390 years from the highest peat bog they cored in the Upismayo Valley, indicating that all moraines located up-valley from this point are LIA in origin. A minimum-limiting radiocarbon date of 2.9 ka was obtained for a moraine $\sim$2 km down-valley from the present ice face in the same study. Goodman et al. (2001) concluded that the LIA was the most extensive Late Holocene glacial advance both here and at the Quelccaya Ice Cap.

The Huancané I sequence described by Mercer and Palacios (1977) at Quelccaya (Fig. 56.13) is of Holocene provenance, and moraines of this sequence occur less than 1 km from the 1977 face of the ice cap. The authors noted

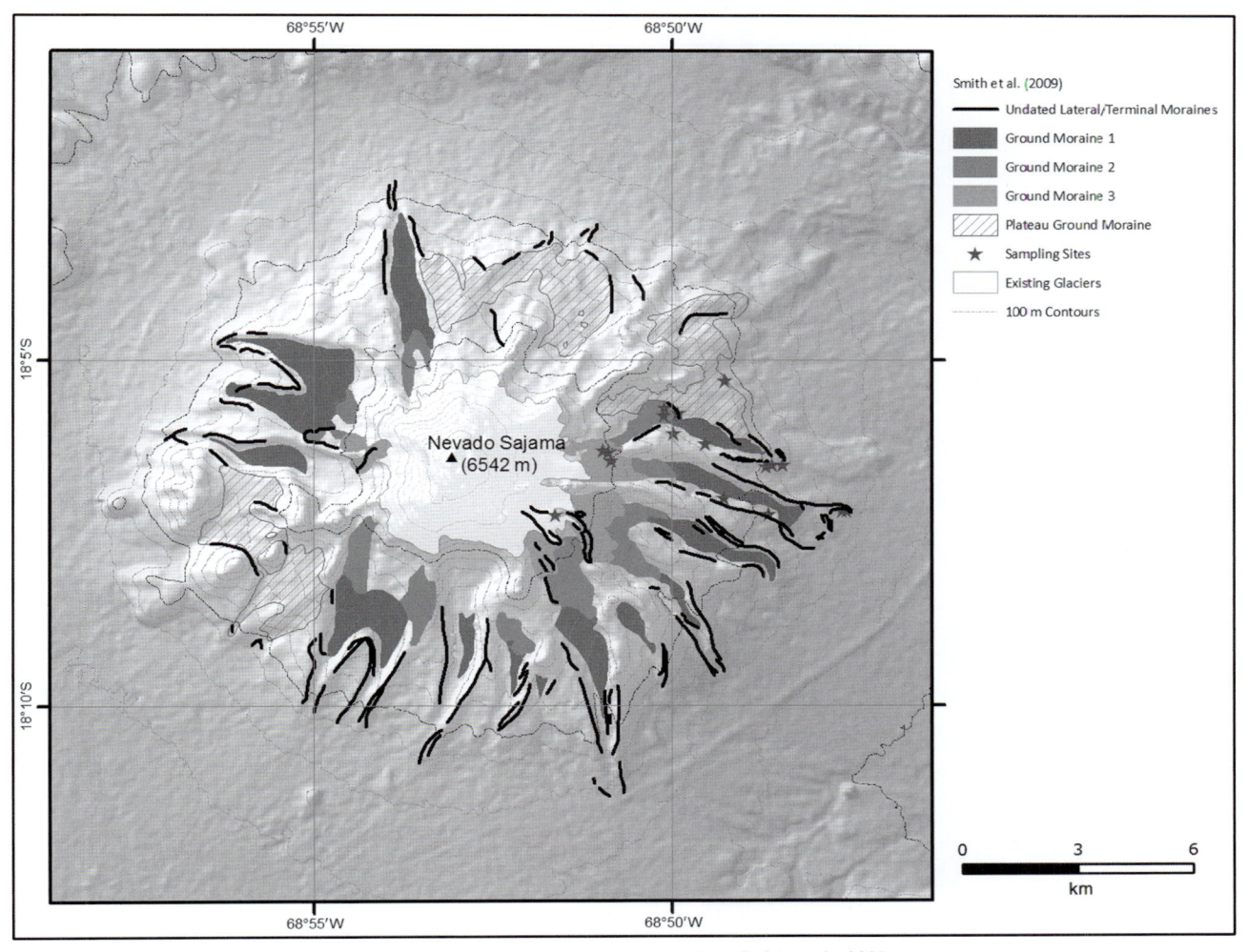

FIGURE 56.16 Quaternary glacial features on Nevado Sajama, Bolivia. Adapted from Smith et al. (2009).

that the Huancané I moraines contain a great deal of bulldozed peat, and one apparently stratigraphically intact peat deposit returned a radiocarbon date of 750–850 years. Mark et al. (2002) calculated a maximum-limiting age for the Huancané I moraines of 300 (+202/−300) years, indicating that this moraine sequence marks the LIA extent of the ice cap. The authors also noted that peat exposed at the present ice margin has been radiocarbon dated to ∼2.8 ka, implying that the ice cap has in fact been smaller during the Holocene than it is at present.

In the Cordillera Cochabamba (Fig. 56.15), Zech et al. (2007) dated one recessional moraine to 7.0 ka but noted that they have been unable to corroborate this date with other evidence.

Three of the four ground moraine classifications identified by Smith et al. (2009) on Nevado Sajama date from the Holocene (Fig. 56.16). Ground Moraine 2 occurs in 12 of the 15 valleys radiating from the mountain, with minimum elevations ranging from 4585 to 4925 m. Like the Lateglacial Ground Moraine 1, these are composed of previously englacial and/or supraglacial material. Using SED, the authors estimate an age of 7.1–4.0 ka for this sequence. Ground Moraine 3 has been identified in eight of Sajama's valleys, at elevations ranging from 4745 to 5000 m. The structure of this sequence appears similar to that of a rock glacier in that they lack lateral moraines; however, they are not related to rock glaciers currently present on the mountain. These are estimated to date from 4.7 to 3.3 ka. A final group, the Plateau Ground Moraines, are found only on high-elevation, low angle plateaus separated from the main massif by secondary peaks and extensive bedrock exposures. Because of this spatial characteristic and their lack of associate lateral and terminal moraines, these are interpreted as being the result of small, cold-based ice caps. Plateau moraines are estimated to date from 7.0 to 6.5 ka.

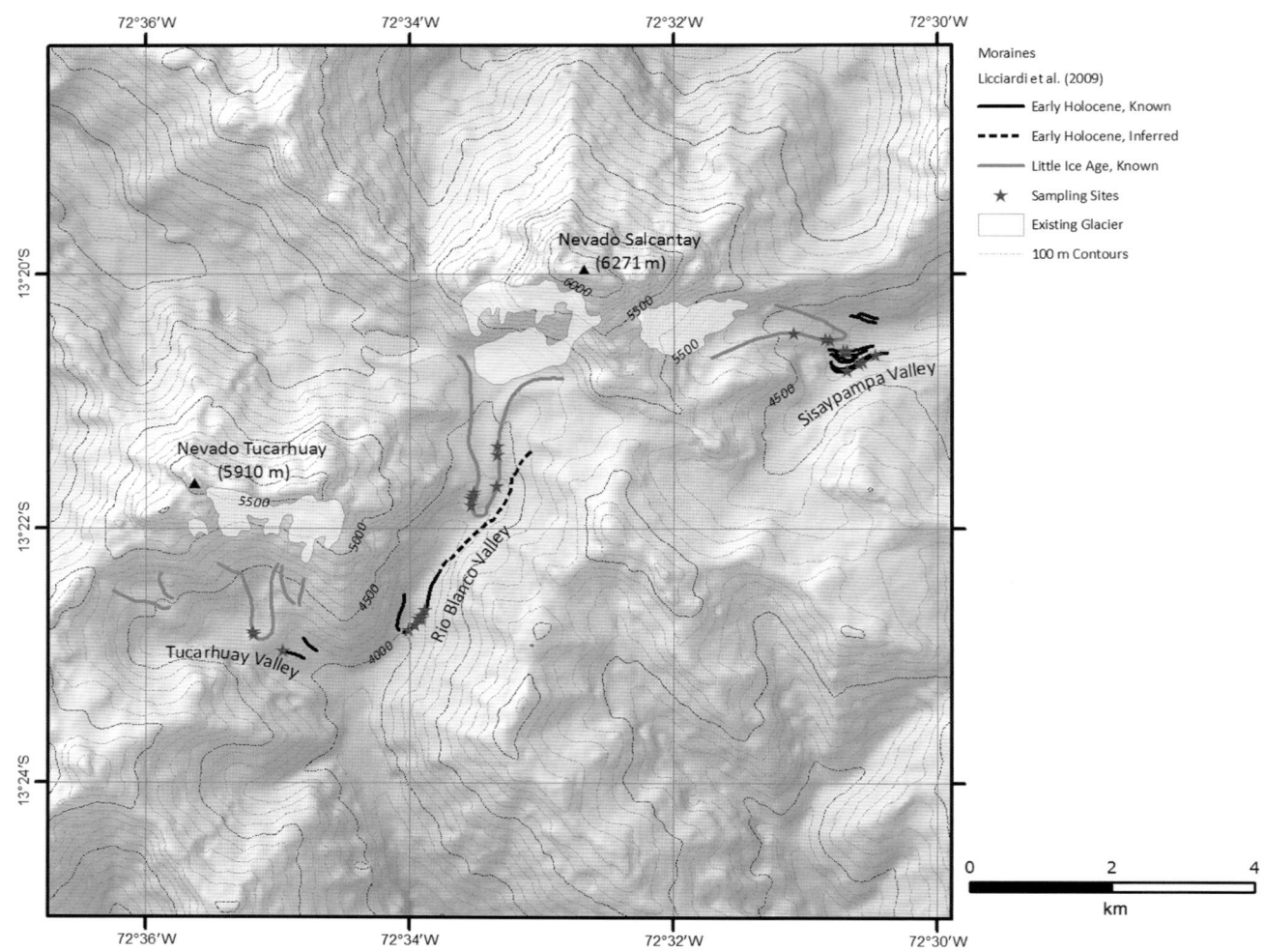

FIGURE 56.17 Quaternary glacial features in the Cordillera Vilcabamba, Peru. Adapted from Licciardi et al. (2009).

Smith et al. (2009) noted that the Holocene glacial sequences identified on Sajama are potentially synchronous with Holocene advances identified in the Cordillera Blanca by Rodbell (1992). Further, they suggested that the Holocene advances appear to coincide with lower lake levels at Lake Titicaca, on the Peruvian–Bolivian Altiplano. While such lowstands have previously been interpreted as representing a warm/dry climate period, the authors hypothesise that while it may have been drier at this time, perhaps it was not as warm, thus allowing glacial advance contemporaneous with reduced regional lake levels. These Holocene advances, however, are not coeval with any Holocene glacial advances reported from the Cordillera Vilcabamba (Licciardi et al., 2009). There, Licciardi et al. (2009) found that Early Holocene glacial behaviour did closely correlate with variation in Titicaca's lake levels, suggesting that both are forced by some combination of increased precipitation and decreased temperature.

56.7. CONCLUSIONS

The Quaternary glacial chronology of the tropical Andes has come into better focus over the past 5 years as numerous additional studies highlighting both new sites and utilising new dating techniques have been published. Examining the literature, it is now possible to identify broadly synchronous glacial advances within the region:

1. An MIS 2 advance that culminates between 25 and 20 ka (with emphasis on the lower end of this range, coeval with the global LGM) is marked by moraine sequences of the Rurec in the Cordillera Blanca (20.5 ka) (Rodbell, 1993a; Farber et al., 2005; Bromley et al., 2009); Stage IV in the Cordillera Huayhuash (22–20 ka) (Hall et al., 2009); Group C (and Punrun) in the Junin Plain and Cordillera Real (25–22 ka) (Wright, 1983; Smith et al., 2005; Bromley et al., 2009); and C-I on Nevado Coropuna (Bromley et al., 2009).

2. A second MIS 2 glacial advance that culminates ∼16 ka is evidenced by moraines of the M5 on Rucu Pichincha and the Potrerillos Plateau (∼15.4 ka) (Heine and Heine, 1996); Laguna Baja in the Cordillera Blanca (16.5 ka) (Rodbell, 1993a; Farber et al., 2005); Group B in the Junin Plain and Cordillera Real (16.1 ka) (Smith et al., 2005; Zech et al., 2007; Bromley et al., 2009); and probably some portion of the U3 moraines in the Cordillera Vilcanota (16.9 ka) (Mercer and Palacios, 1977; Goodman et al., 2001; Mark et al., 2002).
3. Finally, there is strong evidence for a glacial advance either just prior to or coeval with the Younger Dryas chron, including the Potrerillos Advance on the Potrerillos Plateau (12.9 ka) (Clapperton et al., 1997); the Breque Moraine in the central Cordillera Blanca (12.9 ka) (Rodbell and Seltzer, 2000); Juellesh and Tuco Valley moraines in the southern Cordillera Blanca (12.5 ka) (Glasser et al., 2009); Stage III in the Cordillera Huayhuash (14–13 ka) (Hall et al., 2009); Huancané II moraines below the Quelccaya Ice Cap (13.3–12.9 ka) (Mercer and Palacios, 1977; Rodbell and Seltzer, 2000; Goodman et al., 2001; Mark et al., 2002); and C-II moraines on Nevado Coropuna (13.0 ka) (Bromley et al., 2009).

Not every dated moraine sequence fits neatly within this timeline, however, and numerous problems must be resolved before any assumptions can be made about palaeoglaciation in the tropical Andes. First, identification and acceptance of a preferred SED scaling method that best accounts for the variations inherent to CRN decay at low latitudes and high altitudes is necessary for SED studies to be compared with one another and studies using other dating methodologies. Second, an improved understanding of the relationship between precipitation and temperature forcing mechanisms of glacial behaviour is needed, both between the inner tropical Andes (Ecuador) and the outer tropical Andes (Peru and Bolivia), as well as between the Cordilleras Occidental and Oriental in all three countries. Third, because of its potential importance in explaining climatic linkages between the northern hemisphere and the tropics, a better understanding of glacial behaviour during the Younger Dryas chron (12.9–11.7 ka) is necessary. The presence of evidence for glacial advance during the Younger Dryas chron in the tropical Andes has been an issue of particular contention, and while the interpretation of Rodbell and Seltzer (2000) is supported by evidence from several Peruvian sites, much uncertainty remains. Fourth, additional investigations of Holocene glacial behaviour in the region are needed, especially with regard to a potential signal from the northern hemisphere 8200-year event identified at select sites in Peru. Finally, despite the number of new studies, additional research is needed at sites where the palaeoglacial chronology remains unexamined or poorly understood, especially in the Ecuadorian Andes.

REFERENCES

Balco, G., Stone, J.O., Lifton, N.A., Dunai, T.J., 2008. A complete and easily accessible means of calculating surface exposure ages or erosion rates from Be-10 and Al-26 measurements. Quatern. Geochronol. 3, 174–195.

Bromley, G.R.M., Schaefer, J.M., Winckler, G., Hall, B.L., Todd, C.E., Rademaker, K.M., 2009. Relative timing of last glacial maximum and late-glacial events in the central tropical Andes. Quatern. Sci. Rev. 28, 2514–2526.

Clapperton, C.M., 1981. Quaternary glaciations in the Cordillera Blanca, Peru and the Cordillera Real, Bolivia. Revista Centro Interamericano de Fotointerpretacion C.I.A.F., Bogata, Columbia, pp. 93–111.

Clapperton, C.M., 1983. The glaciation of the Andes. Quatern. Sci. Rev. 2, 83–84, IN1–IN2, 85–155.

Clapperton, C.M., 1986. Glacial geomorphology, Quaternary glacial sequence and paleoclimatic inferences in the Ecuadorian Andes. In: Gardiner, V. (Ed.), International Geomorphology 1986, Part II. Wiley, London, pp. 843–870.

Clapperton, C.M., 1990. Glacial and Volcanic Geomorphology of the Chimborazo–Carihuairazo Massif, Ecuadorian Andes. Trans. R. Soc. Edinburgh Earth Sci. 81, 91–116.

Clapperton, C.M., McEwan, C., 1985. Late Quaternary moraines in the Chimborazo Area, Ecuador. Arctic Alpine Res. 17, 135–142.

Clapperton, C.M., Hall, M., Mothes, P., Hole, M.J., Still, J.W., Helmens, K.F., et al., 1997. A Younger Dryas icecap in the equatorial Andes. Quatern. Res. 47, 13–28.

Farber, D.L., Hancock, G.S., Finkel, R.C., Rodbell, D.T., 2005. The age and extent of tropical alpine glaciation in the Cordillera Blanca, Peru. J. Quatern. Sci. 20, 759–776.

Glasser, N.F., Clemmens, S., Schnabel, C., Fenton, C.R., McHargue, L., 2009. Tropical glacier fluctuations in the Cordillera Blanca, Peru between 12.5 and 7.6 ka from cosmogenic 10Be dating. Quatern. Sci. Rev. 28, 3448–3458.

Goodman, A.Y., Rodbell, D.T., Seltzer, G.O., Mark, B.G., 2001. Subdivision of glacial deposits in southeastern Peru based on pedogenic development and radiometric ages. Quatern. Res. 56, 31–50.

Hall, S.R., Farber, D.L., Ramage, J.M., Rodbell, D.T., Finkel, R.C., Smith, J.A., et al., 2009. Geochronology of Quaternary glaciations from the tropical Cordillera Huayhuash, Peru. Quatern. Sci. Rev. 28, 2991–3009.

Hastenrath, S., 2009. Past glaciation in the tropics. Quatern. Sci. Rev. 28, 790–798.

Heine, K., 1995. Late Quaternary glacier advances in the Ecuadorian Andes: a preliminary report. In: Rabassa, J., Salemme, M. (Eds.), Quaternary of South America and Antarctic Peninsula. A.A. Balkema, Rotterdam, pp. 1–22.

Heine, K., 2000. Tropical South America during the Last Glacial Maximum: evidence from glacial, periglacial and fluvial records. Quatern. Int. 72, 7–21.

Heine, K., Heine, J.T., 1996. Late glacial climatic fluctuations in Ecuador: glacier retreat during younger dryas time. Arctic Alpine Res. 28, 496–501.

Jomelli, V., Favier, V., Rabatel, A., Brunstein, D., Hoffmann, G., Francou, B., 2009. Fluctuations of glaciers in the tropical Andes over the last millennium and palaeoclimatic implications: a review. Palaeogeogr. Palaeoclimatol. Palaeoecol. 281, 269–282.

Kaser, G., Osmaston, H., 2002. Tropical Glaciers. Cambridge University Press, Cambridge, UK.

Klein, A.G., Seltzer, G.O., Isacks, B.L., 1999. Modern and last local glacial maximum snowlines in the Central Andes of Peru, Bolivia, and Northern Chile. Quatern. Sci. Rev. 18, 63–84.

le Roux, J.P., Tavares Correa, C., Alayza, F., 2000. Sedimentology of the Rímac-Chillón alluvial fan at Lima, Peru, as related to Plio-Pleistocene sea-level changes, glacial cycles and tectonics. J. South Am. Earth Sci. 13, 499–510.

Licciardi, J.M., Schaefer, J.M., Taggart, J.R., Lund, D.C., 2009. Holocene glacier fluctuations in the Peruvian Andes indicate northern climate linkages. Science 325, 1677–1679.

Mark, B.G., 2008. Tracing tropical Andean glaciers over space and time: some lessons and transdisciplinary implications. Global Planet. Change 60, 101–114.

Mark, B.G., Seltzer, G.O., Rodbell, D.T., Goodman, A.Y., 2002. Rates of deglaciation during the Last Glaciation and Holocene in the Cordillera Vilcanota-Quelccaya Ice Cap Region, Southeastern Perú. Quatern. Res. 57, 287–298.

Mark, B.G., Seltzer, G.O., Rodbell, D.T., 2004. Late Quaternary glaciations of Ecuador, Peru and Bolivia. In: Ehlers, J., Gibbard, P.L. (Eds.), Quaternary Glaciations—Extent and Chronology Part III. Elsevier, Amsterdam, pp. 151–163.

Mark, B.G., Harrison, S.P., Spessa, A., New, M., Evans, D.J.A., Helmens, K.F., 2005. Tropical snowline changes at the last glacial maximum: a global assessment. Quatern. Int. 138–139, 168–201.

Martinson, D.G., Pisias, N.G., Hays, J.D., Imbrie, J., Moore, T.C., Shackleton, N.J., 1987. Age dating and the orbital theory of the ice ages: development of a high-resolution 0 to 300,000-year chronostratigraphy. Quatern. Res. 27, 1–29.

Mercer, J.H., Palacios, M.O., 1977. Radiocarbon dating of the last glaciation in Peru. Geology 5, 600–604.

Rasmussen, S.O., Andersen, K.K., Svensson, A.M., Steffensen, J.P., Vinther, B.M., Clausen, H.B., et al., 2006. A new Greenland ice core chronology for the last glacial termination. J. Geophys. Res. A Space Phys. 111, D06102.

Rodbell, D.T., 1992. Lichenometric and radiocarbon dating of Holocene glaciation, Cordillera Blanca, Peru. Holocene 2, 19–29.

Rodbell, D.T., 1993a. Subdivision of Late Pleistocene moraines in the Cordillera Blanca, Peru, based on rock-weathering features, soils, and radiocarbon dates. Quatern. Res. 39, 133–143.

Rodbell, D.T., 1993b. The timing of the last deglaciation in Cordillera Oriental, Northern Peru, based on glacial geology and lake sedimentology. Geol. Soc. Am. Bull. 105, 923–934.

Rodbell, D.T., Seltzer, G.O., 2000. Rapid ice margin fluctuations during the Younger Dryas in the Tropical Andes. Quatern. Res. 54, 328–338.

Rodbell, D.T., Bagnato, S., Nebolini, J.C., Seltzer, G.O., Abbott, M.B., 2002. A Late Glacial-Holocene Tephrochronology for Glacial Lakes in Southern Ecuador. Quatern. Res. 57, 343–354.

Rodbell, D.T., Smith, J.A., Mark, B.G., 2009. Glaciation in the Andes during the Lateglacial and Holocene. Quatern. Sci. Rev. 28, 2165–2212.

Seltzer, G.O., 1990. Recent glacial history and paleoclimate of the Peruvian–Bolivian Andes. Quatern. Sci. Rev. 9, 137–152.

Smith, J.A., Seltzer, G.O., Farber, D.L., Rodbell, D.T., Finkel, R.C., 2005. Early local last glacial maximum in the tropical Andes. Science 308, 678–681.

Smith, J.A., Mark, B.G., Rodbell, D.T., 2008. The timing and magnitude of mountain glaciation in the tropical Andes. J. Quatern. Sci. 23, 609–634.

Smith, C.A., Lowell, T.V., Caffee, M.W., 2009. Lateglacial and Holocene cosmogenic surface exposure age glacial chronology and geomorphological evidence for the presence of cold-based glaciers at Nevado Sajama, Bolivia. J. Quatern. Sci. 24, 360–372.

Stuiver, M., Reimer, P.J., 1993. Extended 14C database and revised CALIB radiocarbon calibration program. Radiocarbon 35, 215–230.

Walker, M., Johnsen, S., Rasmussen, S.O., Popp, T., Steffensen, J.-P., Gibbard, P., et al., 2009. Formal definition and dating of the GSSP (Global Stratotype Section and Point) for the base of the Holocene using the Greenland NGRIP ice core, and selected auxiliary records. J. Quatern. Sci. 24, 3–17.

Wright, H.E. Jr., 1983. Late-Pleistocene glaciation and climate around the Junin Plain, Central Peruvian Highlands. Geogr. Ann. A Phys. Geogr. 65, 35–43.

Yokoyama, Y., Lambeck, K., De Deckker, P., Johnston, P., Fifield, L.K., 2000. Timing of the last glacial maximum from observed sea-level minima. Nature 406, 713–716.

Zech, R., Kull, C., Kubik, P.W., Veit, H., 2007. LGM and Late Glacial glacier advances in the Cordillera Real and Cochabamba (Bolivia) deduced from 10Be surface exposure dating. Climate Past 3, 623–635.

Late Quaternary Glaciations of Ecuador

Klaus Heine

Geographisches Institut der Universität Regensburg, Universitätsstr. 31, D-93040 Regensburg, Germany

57.1. INTRODUCTION

The glaciations of the Ecuadorian Andes (Fig. 57.1) have been widely discussed for over 100 years (Jordan et al., 2003; cf. Reiss and Stübel, 1886; Wolf, 1892; Meyer, 1904, 1907; Sauer, 1971; Hastenrath, 1981; Clapperton, 1985, 1987a,b, 1990; Clapperton and McEwan, 1985; Lauer and Rafiqpoor, 1986; Heine, 1993, 1995a,b, 2000; Mark et al., 2004; Smith et al., 2008). Clapperton (1990) developed a glacial chronology of Ecuador, which pointed out the sometimes fundamental errors of former studies and which provided a foundation for later research. Jordan and Hastenrath (1998) mentioned 13 mountains in the wetter Eastern Cordillera with glaciers and four glacier-capped mountains in the drier Western Cordillera (altogether up to 100 small glaciers with ~ 100 km^2 area). J.T. Heine (1993, 1995a) and K. Heine (1995a,b, 2004) studied the glacial history of the late Quaternary of the region. Their correlation of moraine suites between valleys and volcanoes is based on an assessment of the position (altitude) of the moraines and tills on eight volcanoes, the geomorphological character of the moraines, the degree of weathering (palaeosols) of the tills, the tephrostratigraphical position of the tills and the ^{14}C ages obtained from many sections. The authors have contributed new evidence to the discussion of the age of the local last glacial maximum (LLGM) glacier extent, the late-glacial advances and the glacial (marine isotope stage (MIS) 2) aridity in the tropical Andes (Heine, 2000, 2004). Glacial chronologies, based on surface exposure dating, had not yet been published for the Ecuadorian Andes (cf. also Smith et al., 2008; Rodbell et al., 2009). Little progress has been made since the publication of Heine (2004). The occurrence of inactive bedded slope deposits as evidence of former permafrost conditions has also been studied as a means of characterising the last glacial maximum (LGM)-temperature depression and to distinguish between pre- and post-LGM moraines in the field (Heine, 1995b).

57.2. BEDDED (STRATIFIED) SLOPE DEPOSITS ON MORAINES IN ECUADOR

Apart from comprehensive mapping of glacial landforms and sediments of late Quaternary age and the palaeopedological and tephrostratigraphical observations in the field and the laboratory, ^{14}C and some U/Th ages complement the studies. Special attention has been paid to periglacial phenomena, such as bedded slope deposits and rock glaciers, to provide a better understanding of the glacial chronology (Heine, 1995b). The occurrence of bedded slope deposits indicates that permafrost was active in the region. The lowermost occurrence of permafrost is bound to the ~ 0 °C mean annual temperature. By combining LGM-permafrost evidence with that of LGM moraines, temperature and equilibrium-line altitude (ELA) depressions can be estimated for the late Quaternary time.

Stratification mechanisms in slope deposits in high equatorial mountains are summarised by Francou (1990) and Lautridou and Francou (1992). The term *stratified slope deposit* refers to detrital materials which show an alternation of beds arranged by distinct grain-size sorting (Dewolf, 1988; Francou, 1990). Francou (1990) distinguishes between stratified screes and deposits of *grèzes-litées* type. Stratified screes show a slope angle up to 30°, a marked heterogeneity of material and the possible intervention of gravity-induced mechanisms as factors producing the bedding. Deposits of *grèzes-litées* type display slope angles between less than 28–30° and more than 5° and a greater homogeneity in the granule and sand fraction. Bedding is generated by solifluction sheets (<20 cm thick) which move at the speed of a few centimetres per year. Solifluction involves needle ice, frost creep and gelifluction. Sorting is caused by frost heaving. Coarse particles are concentrated at the front of the sheets, and then they are buried by the advancing fine layer. Coarse-grained sheets are active in areas with permafrost. In the central highland of

FIGURE 57.1 Location map. 1, Cayambe; 2, Rumiñahui; 3, Cotopaxi; 4, Illiniza.

Ecuador, permafrost occurs on volcanoes above 5200–5300 m a.s.l. (Fig. 57.2). In relict slope deposits at lower altitudes, sedimentary characteristics allow the recognition of features which indicate the origin of the bedding (Francou, 1990). Therefore, variations in the elevation of the lower limit of bedded slope deposits can be used as permafrost and palaeotemperature indicators. They allow the separation of temperature from precipitation as forcing mechanism for tropical glacier change (La Frenierre et al., 2011).

57.3. PRE-LGM GLACIATION (OLDER THAN 30 KA)

The Quaternary glacial sequence, mean lower limits of the terminal moraines and the ages of the glacial advances in the Ecuadorian Andes are shown in Table 57.1. The estimated mean altitudes of the lower limits of the moraine groups, based on dated moraine sequences in combination with inactive bedded slope deposits on moraines and their reconstructed lower limit, show that the postulated extent (e.g. Schubert and Clapperton, 1990) of the LLGM glacier advances is far too great. Heine (1995a,b) presents a well-documented account of seven glaciations varying in age from over 150–180 ka (determined by fission-track dating) to the Little Ice Age (Fig. 57.3).

The moraines of the M I and M II glaciations are easily distinguished on the Pichincha volcano by deformed varved clays and tephra layers. Although palaeosols do not allow the differentiation of the two moraine groups, the arrangement of the M I and M II lateral moraines in many places records two distinct ice advances. The M I and M II tills are highly weathered from the top to the base, whereas the M III till is less weathered, commonly only to 2–4 m depth. The M I and M II moraines might be as old as (or older than) MISs 6 and 8, respectively.

On the basis of re-examination of Clapperton's Ecuadorian sections on the northern flanks of the Chimborazo–Carihuarazo volcanoes (Heine 1995a,b, 2000), a tentative correlation is suggested of the author's stratigraphy with the till/moraine sequence of Clapperton (1990): M I is equivalent to GL 1 of Clapperton (1990), M II equates with GL 2 and M III equates with GL 3. If correct, then the GL 3 till is not of LGM age (<35 and >12 ^{14}C ka BP; ~40.1–13.9 cal.ka BP), as suggested by Clapperton (1990), but older (?MIS 4), and the GL 2 and GL 1 tills are both of pre-Wisconsinan age (cf. Heine, 1995a).

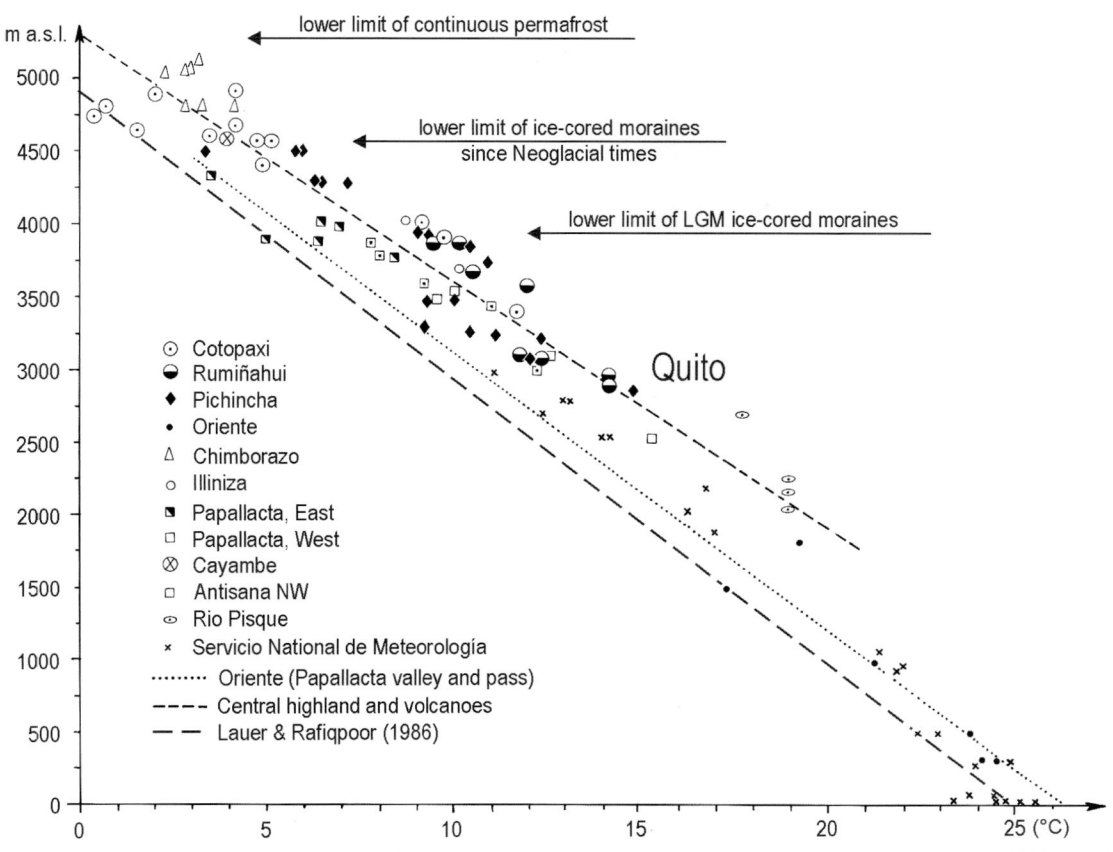

Chapter | 57 Late Quaternary Glaciations of Ecuador

FIGURE 57.2 Soil temperatures in Ecuador. According to the measurements, permafrost occurs in the eastern Andes above ~4800 m a.s.l. and in the central Andes above ~5200–5300 m a.s.l. The data by Lauer and Rafiqpoor (1986) refer to the humid eastern slopes of the Antisana volcano and show a lapse rate of 0.52 °C/100 m. The author's soil temperature data which were obtained in AD 1990/1991 show an increase of about 0.5 °C in heights of 4000–4500 m a.s.l. compared to the data of Lauer and Rafiqpoor (1986).

The pre-LGM age of the M III, M II and M I tills and/or moraines is corroborated by the glacial sequences from the Rucu Pichincha and Guagua Pichincha (Fig. 57.4). The M III moraines had already been formed when the Guagua Pichincha volcano was partly destroyed by explosive activity about 30 ka ago. At the Rucu Pichincha volcano, the deposits of the glaciations M I–M VI are separated by tephra layers, varves, periglacial sediments and palaeosols. The age of these deposits is constrained not only by geomorphological, sedimentological and pedological evidence but also by some radiometrically dated strata (Heine, 1995a).

Glacial sequences from the Cayambe, Illiniza volcanoes and from the Antisana/Papallacta valley also include moraines of seven glaciations, thus presenting further evidence for the proposed glacial stratigraphy.

57.4. THE LAST GLACIAL MAXIMUM

The glacial-chronological research by Heine and Heine (1996) and Heine (1995a,b, 2000) suggested that the maximum glacial extent in Ecuador was not in-phase with classic mid-latitude records and the record of global ice volume in the marine isotope record (where the LGM occurs at ca. 21 ka). At the end of MIS 3, the Ecuadorian glaciers advanced and reached their maximum extent between 30.0 and 25.0 ^{14}C ka BP (~34.75–27.9 cal. ka BP; Fig. 57.3). Between about 25.0 and 16.0 ^{14}C ka BP (~27.9–19.2 cal. ka BP), the volume of the glaciers was reduced and ice-cored moraines replaced the M IV glacier tongues in many valleys. The occurrence of ice-cored moraines—and in some places rock glaciers—record a shift towards greater aridity after ca. 25.0 ^{14}C ka BP (~27.9 cal. ka BP). The occurrence of inactive bedded slope deposits overlying the pre-LGM moraines (groups I–III) and on the early-LGM moraine IV indicates that the altitudinal belt with active bedded slope deposits was depressed by about 700–800 m during the LGM. This depression corresponds to a temperature decline of ~4.2–4.8 °C (taking a lapse rate of 0.6 °C/100 m; Fig. 57.2). The combined geomorphological and geochronological evidence clearly indicates that the LGM-temperature depression reached 4–5 °C in the tropics. Moreover, the position of the permafrost evidence (bedded slope deposits, rock glaciers) constrains not only

TABLE 57.1 The Quaternary Glacial Sequence in the Ecuadorian Andes (after Heine, 2000)

		age and duration	lower limit (mean value) in m		Schubert & Clapperton (1990)	
		^{14}C kyr ka BP BP	western and central Cordillera	eastern Cordilleran slopes	lower limit West East	age
M VII	Little Ice Age Neoglacial	> 0,01 < 3,5	4800-4500	4400-3900	4800 - ~4100 4600 - 3900	Neoglacial interval
M VI	Early Holocene and/or Late Pleistocene	> 8,2 < 10,6	4500-4200	4050	4400 - 3800	Late-Glacial stade
M V	Late Pleistocene	> 13 < LGM	4100-4000	3800-3700	3700 - 3600	Last Gaciation (late)
M IV	Late Glacial Maximum (LGM)	> 20 < 30	4000-3800	3750-3500	3900 - 3000	Last Glaciation (main)
M III	Wisconsin (stage 4?)	> 49,5	3900-3800	?		
M II	Pre Wisconsin (stage 6?)	> 125	3700	3200-3000	3500 - 2900	Last Glaciation (early)
M I	Pre Wisconsin (stage 6?, stage 8?)	> 150 –180	3500	2800	3500 - 2750	Penultimate Glaciation

MIS, marine isotope stage.

a temperature depression of about 4.5 °C during the LLGM but also a palaeoclimatic change towards greater aridity. In the Ecuadorian Andes, the shifting balance between temperature and precipitation changes supported glacier development and expansion early in MIS 3/MIS 2 transition (cf. Smith et al., 2005a). These conclusions are corroborated by Smith et al. (2005a) from Peru and Bolivia (cf. Heine, 2011). These observations tie in with the message that the LGM sea-surface temperatures in the tropics were indeed 5 °C lower than today (Anderson and Webb, 1994) and that the Amazon Basin (source of the moisture) was more arid (Heine, 2000). A minor readvance occurred before 13.0 ^{14}C ka BP (~15.5 cal. ka BP). The melting of the last glacial stage glaciers was completed before 13.0 ^{14}C ka BP (Heine, 1995a,b, 2000).

57.5. DEGLACIATION

There has been much debate about the deglaciation from the LGM positions in Ecuador (cf. J.T. Heine, 1993, 1995). Apart from the data presented by Heine and Heine (1996) for a glacier advance at the end of the Younger Dryas (YD) period between 10.5 and 9.0 ^{14}C ka BP (~12.56–10.25 cal. ka BP), Heine and Geyh (2002) presented an evaluation and extensive discussion of the late-glacial/early Holocene radiocarbon ages and conclude that an advance, represented by tills and moraines in the Papallacta Pass area, occurred during the latter half of the YD Chronozone and the early Holocene (Fig. 57.5). The glacial deposits of the Papallacta Pass site are both underlain and overlain by tephras, peat and other deposits rich in organic matter. These deposits were first invoked by Heine (1995a) and Heine and Heine (1996) to establish a glacial stratigraphy for the YD/early Holocene transition period. The 29 radiocarbon dates published later by Clapperton et al. (1997), and the 29 radiocarbon dates presented by Heine (1995a) and Heine and Geyh (2002), have been used to date a drier period of glacier retreat during the first half of the YD and a period of more humid conditions and glacier advance from the second half of the YD to the early Holocene. Thus the glacial history during the late-glacial/Holocene transition (Termination 1 of the North Atlantic region) is fairly well established in Ecuador (Fig. 57.6).

57.6. THE HOLOCENE

After the late YD/early Holocene ice advance (moraine group M VI), Andean climate seems to have become warmer than present day between 8.9 and 3.3 cal. ka BP (Niemann and Behling, 2008); the glaciers advanced only during the Neoglacial/Little Ice Age. A short summary of the discussion is presented by Heine (1995a). Until today, no further detailed research has been undertaken with regard to the Neoglacial and Little Ice Age glaciations of the Ecuadorian Andes. According to historical sources

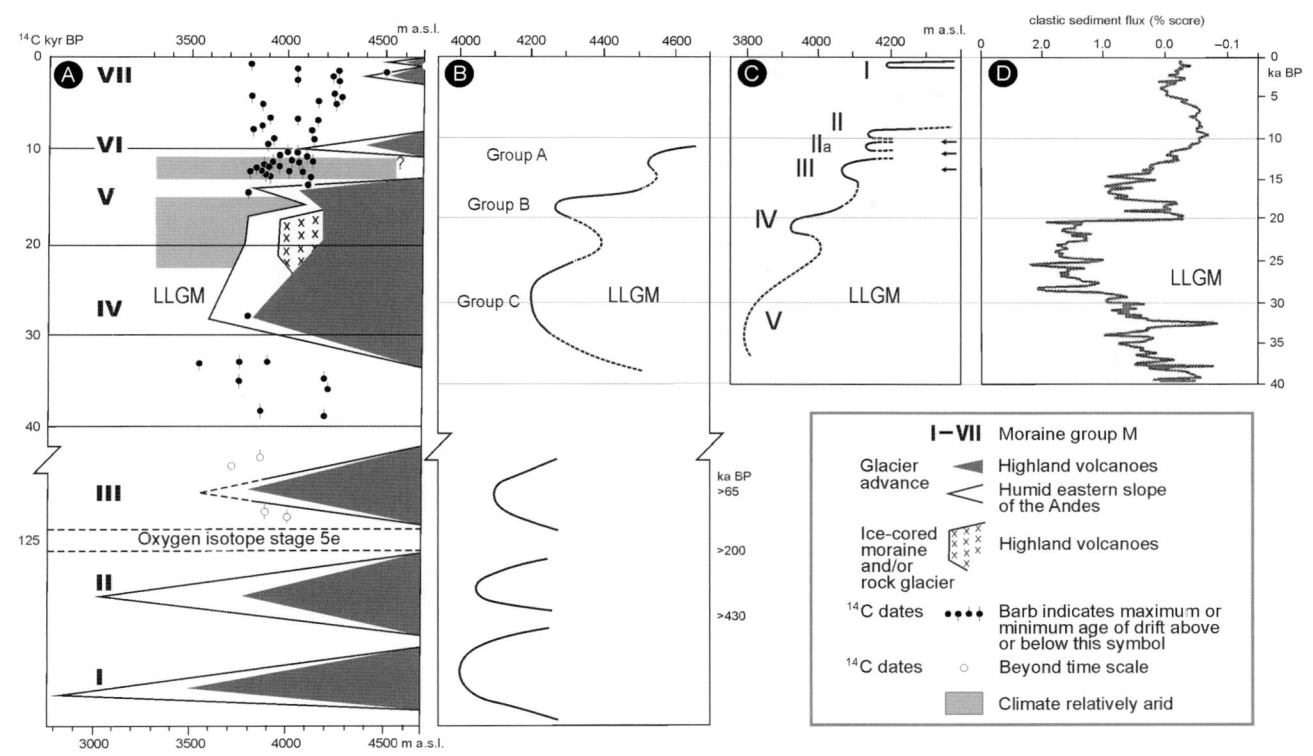

FIGURE 57.3 Glacial stratigraphy in the Ecuadorian Andes (A) compared to proxy records from Peru. (B) Quaternary glacial stages in the central Peruvian Andes (after Smith et al., 2005a,c), moraine group C demarcates the ice limit of the LLGM in the region. (C) Late Quaternary glacial stages (I–V) of the Cordillera Huayhuash, Peru, according to Hall et al. (2009), cf. also Licciardi et al. (2009). Late glacial/early Holocene glacier advances in Bolivia (arrows) after Zech et al. (2009). (D) Stacked record of sediment flux as proxy for the extent of ice cover in the region from southern Ecuador to the Cordillera Real, Bolivia (after Rodbell et al., 2008).

discussed and interpreted in detail by Hastenrath (1981), snowline depression was greater between AD 1740 and 1800 than since AD 1800, indicating that the maximum Little Ice Age glacier extension occurred before AD 1800 and was broadly correlative with glacial records in Colombia (Bartels, 1970; Schubert, 1989; Chapter 58) and Venezuela (Rull and Schubert, 1989; Schubert and Vivas, 1993; Chapter 59). Jomelli et al. (2009) used lichenometry to date Ecuadorian terminal moraines: the Little Ice Age moraines of volcanoes with summits above 5700 m had their maximum extent AD 1720 ± 16, whereas volcanoes with heights below 5400 m show maximum glacial advances around AD 1830 ± 14. On the Quelccaya ice cap in Peru, snow accumulation was greatest between ca. AD 1500 and 1700 (Thompson et al., 1986), indicating also a period favourable for glacier advances during the first half of the Little Ice Age.

57.7. MAPPING GLACIAL LIMITS

Currently inactive bedded slope deposits overlie the pre-LGM moraines (group M I–M III) and the early-LGM moraine M IV, as noted above. Dating of these inactive slope deposits, using tephrostratigraphically correlated sequences, has shown that the altitudinal belt with active bedded slope deposits was depressed by about 700–800 m during the LGM (Heine, 1995b). Bedded slope deposits are characteristic of the periglacial altitudinal zone and are only active in Ecuador above ca. 4800 m a.s.l. today. Since the lower limit of the bedded slope deposits forming processes depends on the thermal regime (mean annual temperature), changes in the altitude of the lower limit of relict-bedded slope deposits represent changes in mean annual temperature. The last period that underwent extensive periglacial processes and the formation of bedded slope deposits down to 4000 m a.s.l. occurred during the formation of moraine group M IV and, based on stratigraphical correlations, must be attributed to the LGM, at least.

By mapping the occurrence of inactive bedded slope deposits on different tills and moraine slopes, pre-LGM moraines have been identified (Heine, 1995a,b, 2000). Many moraine suites, hitherto assumed to be of LGM age, are covered by stratified slope deposits down to elevations of about 4000 m a.s.l. These moraines, overlain by bedded slope deposits, definitely pre-date the global LGM.

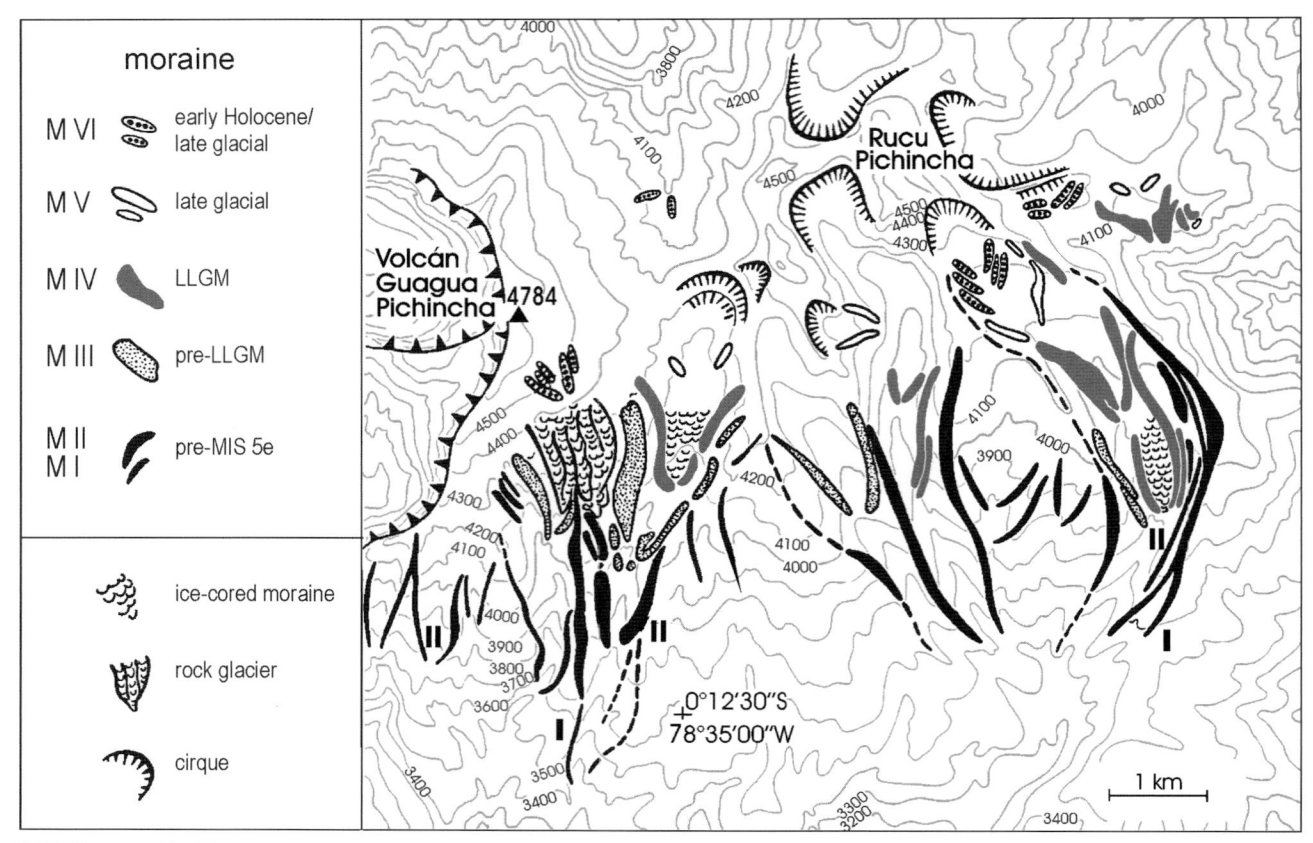

FIGURE 57.4 Glacial geomorphology of the Pichincha volcanoes, Ecuador. The moraines of groups M I–M III were previously thought to be of LGM age. Chronostratigraphical investigations and bedded slope deposits clearly show that the moraines of group M IV are actually of LGM age (after Heine, 1995b).

57.8. OPEN QUESTIONS

Recently, most authors who evaluate and comment the glacial chronologies from the tropical Andes exclusively adhere to physical dating methods, such as radiocarbon dating (^{14}C), accelerator mass spectrometry radiocarbon dating (AMS ^{14}C), thermoluminescence (TL) dating, optically stimulated luminescence (OSL) dating and surface exposure dating techniques (cf. Smith et al., 2005b, 2008; La Frenierre et al., 2011). Nevertheless, by combining geomorphological, (palaeo)pedological, tephrostratigraphical and limnological chronologies with other proxy data (palynology, archaeology) and by evaluating all available absolute-age determinations, reliable glacial chronostratigraphies can be critically assessed. A combination of different palaeoenvironmental proxies can assist in overcoming the problem of establishing an Ecuadorian glacial sequence. This is because the Ecuadorian glacial chronology suffers from a lack of numerical dating using different dating methods (^{14}C, OSL, surface exposure dating), the only well-dated (by ^{14}C) glacier advance being the late YD/early Holocene advance (Heine and Heine, 1996; Heine and Geyh, 2002). However, much work is still required to elucidate the precise duration and magnitude of the climatic reversal during the YD Chronozone (cf. van 't Veer et al., 2000; Seltzer, 2001). The precise ages of the LLGM maximal glacier extent and of the late-glacial ice advances are not yet known (Fig. 57.3; cf. Smith et al., 2005b). Smith et al. (2005b) stated that no LGM moraines have been definitely identified and dated in the Ecuadorian Andes. This comment does not consider all the observations and conclusions based on an assessment of the position (altitude) of the moraines and tills on more than eight volcanoes, the geomorphological character of the moraines, the degree of weathering (palaeosols) of the tills, the tephrostratigraphical position of the tills and the ^{14}C ages obtained from many sections and correlated tephrostratigraphically and geomorphologically. From these proxies, J.T. Heine (1993,1995), Heine (1995a,b), Heine and Heine (1996) and Heine and Geyh (2002) have been able to establish a robust late Quaternary glacial chronology.

The present synthesis of the research in Ecuador yields the following outline of the late Quaternary glacial history: The first phase of the LLGM seems to have been more humid than that during the late LLGM. This is indicated by the maximum glacier extension (M IV) of the LLGM.

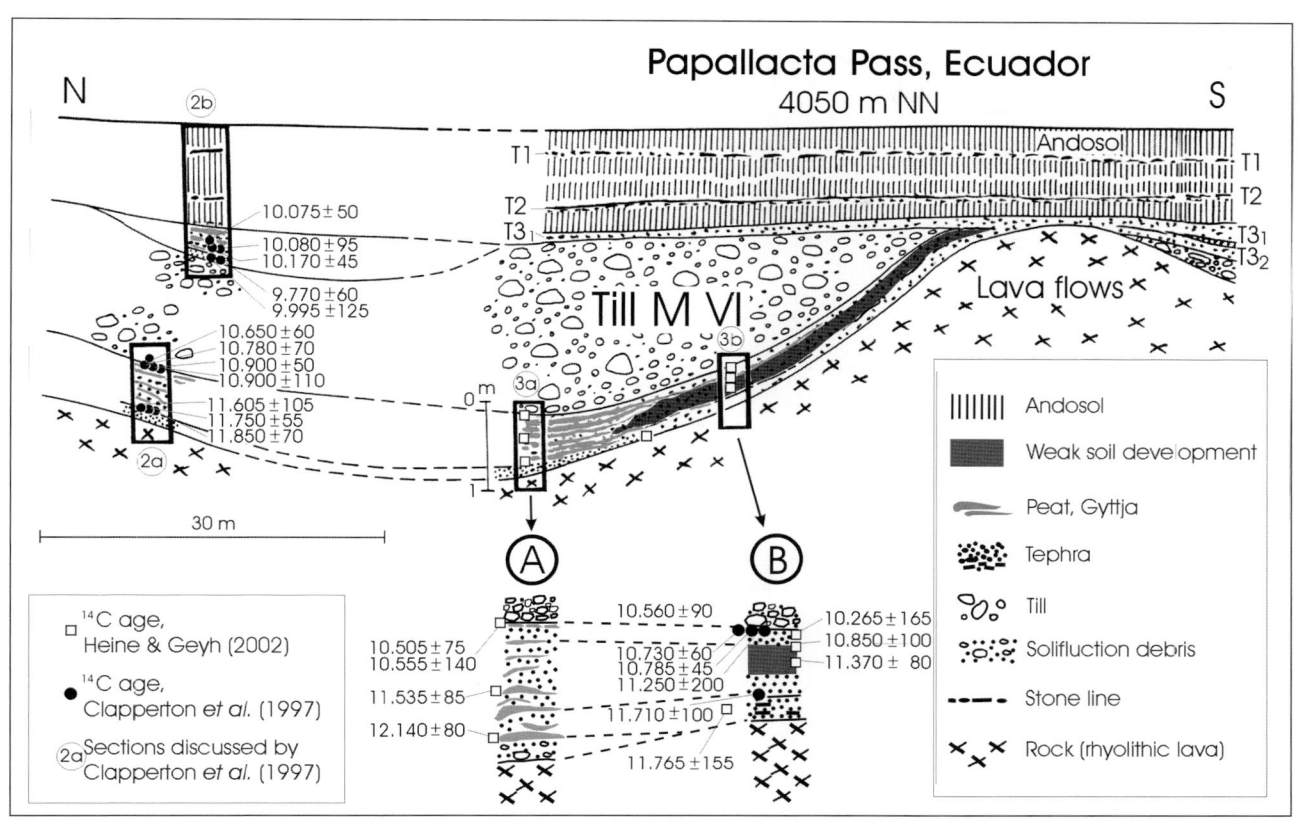

FIGURE 57.5 Combined sections of the Papallacta Pass area. The glacial Younger Dryas/early Holocene stratigraphy is discussed in detail by Heine and Geyh (2002).

The LLGM ice-cored moraines and rock glaciers developed when the M IV glaciers had receded in response to the contemporaneous greater aridity. This is corroborated by findings from Peru (Smith et al., 2008) suggesting that the LLGM was closer to ca. 30 ka than 21 ka (i.e. before the global LGM as inferred from the marine isotope record). It also shows that glaciers reached their maxima as early as 35–29 ka and were receding during the global LGM (~25–17 ka). Rodbell et al. (2008) report clastic sediment fluxes into tropical (Ecuadorian) lakes that record the extent of ice cover in the region (cf. Fig. 57.3); cosmogenic radionuclide exposure ages from moraines in Peru and Bolivia (cf. Smith et al., 2005a) and stacked composite lake-core records indicate that the expansion of ice cover appeared at least by 40 ka, and the LLGM culminated between 30 and 20 ka (and corresponds with glacier advance M IV in Ecuador). It also demonstrates that the interval between 20 and 18 ka appears to have been marked by much reduced ice extent. This was followed after 18 ka by initiation of an interval of expanded ice cover that lasted until ~14 ka (corresponding to glacier advance MV). This glacial chronology is inconsistent with the exposure ages reported by Bromley et al. (2009), who remark that their ages, constraining the LGM maximum and the end of the glacier extent, range between 25.3 and 24.5 ka and between 21.1 and 16.7 ka, respectively (depending on the cosmogenic production rate scaling model applied).

The Ecuadorian glacial sequence shows that the sensitivity of the mass balance of the late Quaternary Ecuadorian (and Peruvian/Bolivian) glaciers to changes in precipitation is not restricted to semi-arid regions but can also be traced in the humid tropical Andes. The writer agrees with Smith et al. (2005a) that increased precipitation and persistent cool temperatures favoured glacier expansion during the MIS 3/MIS 2 transition and caused glacial culmination early in MIS 2. Yet, contrary to Smith et al. (2005a), it is suggested that not warming but moisture deficiency initiated deglaciation ~22–21 ka. The depression of the lower permafrost limit by 700–800 m during the global LGM, indicated by rock-glacier formation and bedded slope deposits, is clear evidence that the Ecuadorian glacier retreat was not a response to warming during the global LGM, but was caused by severe aridity (Fig. 57.7). Likewise, according to marine sediment-based reconstructions, more aridity was associated with glacial conditions along the coast of Ecuador, caused by more La Niña-like conditions, coupled with a northward shift of the intertropical zone during glacial periods (Rincón-Martínez et al., 2010).

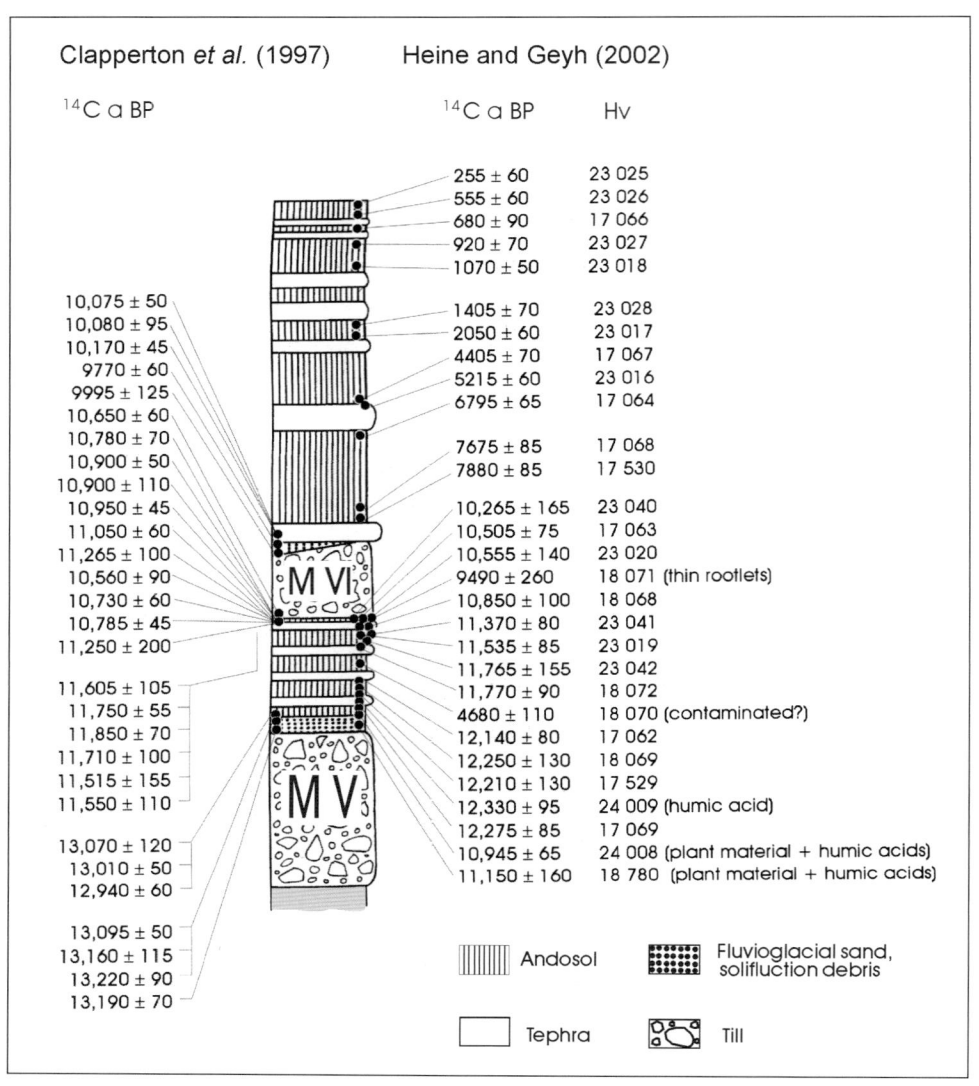

FIGURE 57.6 Schematic section of the Papallacta Pass glacial sequence with radiocarbon dates (after Heine and Geyh, 2002). The radiocarbon dates on the left side after Clapperton et al. (1997).

It is instructive to compare the Ecuadorian late-glacial/early Holocene glacier record to regional late-glacial/early Holocene glacier fluctuations in Peru (Glasser et al., 2009; Hall et al., 2009; Licciardi et al., 2009; Fig. 57.3). Evidently, the robust YD/early Holocene chronology from Ecuador is confirmed by Rodbell and Seltzer (2000). These authors reported retreating ice fronts during the YD in Peru, whilst Licciardi et al. (2009) reported glacier advances during the early Holocene between ∼9.9 and ∼8.0 ka and the Little Ice Age from Peru (cf. Hall et al., 2009). The late YD/early Holocene short but pronounced glacier advance (M VI) provides insight into the regional palaeoclimatic trends and forcing processes, suggesting climate linkages between the tropics and the North Atlantic region (Licciardi et al., 2009), as well as the southern hemisphere. The Antarctic Cold Reversal (ACR) and the YD Stadial lasted from ∼14.3 to 13.5 ka and 12.6 to 11.5 ka, respectively. Smaller-scale climatic oscillations may correlate with the ACR and the YD and with the multiple moraine sets dated between ∼14 and ∼8 ka in Peru (Hall et al., 2009; Licciardi et al., 2009). During the late-glacial and early Holocene, the alternation of advancing and retreating glaciers can be attributed to influences of both, the ACR and the YD events. The glacier movements agree closely with the timing of the ACR and the YD and, thus, may be attributed to past fluctuations of (i) sea-surface temperatures and the intertropical convergence zone migration in the eastern Pacific (cf. Lea et al., 2000) and (ii) the tropical Atlantic (cf. Licciardi et al., 2009) sea-surface temperatures. In the Ecuadorian Andes, it is thought there was a greater influence of the YD fluctuation, while in the Andes of Bolivia (and Peru) a greater influence of the ACR caused the glacier movements. This

FIGURE 57.7 Reconstruction of temperature and precipitation trends in the Ecuadorian Andes (3000–4000 m a.s.l.) based on glacial and periglacial evidence since 25 ka BP. (A) Climate phases as reconstructed by glacial and periglacial proxies. (B) Phases with glacier advances (moraines) and ice-cord moraines/rock glaciers. (C) and (D) Sajama ice core, Bolivia: accumulation record of ice (the data presented are deviations from the mean of two combined records) and oxygen isotopes ($\delta^{18}O$ [‰]) after Thompson et al. (1998). (e) Vostok ice core, Antarctica: δD (‰) after Jouzel et al. (1987, 2007).

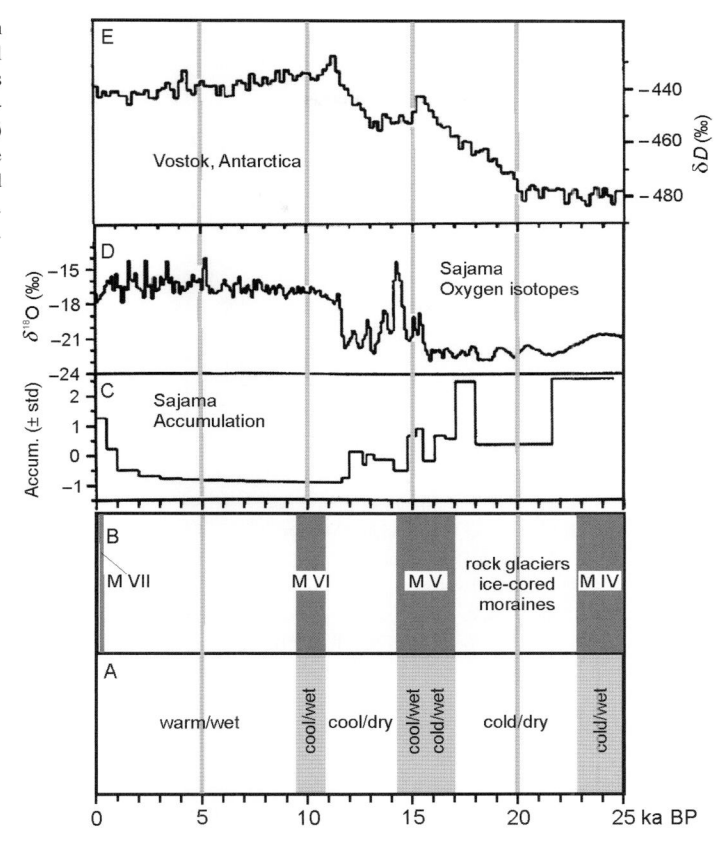

hypothesis is consistent with the moraine records of the tropical Andes, which suggest that the influence from Antarctica is more stringent in the south (Bolivia, Chile) than in the north (Ecuador, Colombia) and that the influence from the Arctic is more obvious in the north than in the southern areas.

The Neoglacial moraines record glacier fluctuations that are not dated with high precision. Although the glacial Little Ice Age chronology from Ecuador corresponds to the records from Peru, presented by Licciardi et al. (2009), who obtained exposure ages tightly clustered around a mean of AD 1810 ± 20, the data do not exclude the possibility of asynchronously advancing glaciers in Ecuador, Peru and Colombia during the Little Ice Age (AD 1300–1860). As in the Ecuadorian Andes, Neoglacial moraines that were deposited between the early Holocene and the Little Ice Age moraine groups were not observed and dated in the tropical Andes of Peru. However, by comparing multiple proxy datasets from regional lake and ice cores, Hall et al. (2009) proposed a glacial readvance at ~5 ka and additional minor readvances in the latest Holocene.

Further, it remains unclear why the oldest glacial advance periods, during which considerably longer palaeoglaciers (group M I and M II) extended far down-valleys in the eastern cordilleras (Table 57.1). Dating of these advances by palaeopedological and geomorphological methods has yielded ages that pre-date the penultimate glacial stage (i.e. Illinoian, Saalian, Rissian Stages). This dating also requires more research.

ACKNOWLEDGEMENTS

I thank the *Deutsche Forschungsgemeinschaft* (DFG) and the *Volkswagen-Stiftung* for financial support of my research in South America and M. A. Geyh for numerous absolute-age determinations and discussions. I am indebted to many persons for assistance in the field and the laboratory. I would like to thank Philip Gibbard and Philip Hughes for critically reviewing the chapter.

REFERENCES

Anderson, D.M., Webb, R.S., 1994. Ice-age tropics revisited. Nature 367, 23–24.

Bartels, G., 1970. Geomorphologische Höhenstufen in der Sierra Nevada de Santa Marta (Kolumbien). Gießener Geogr. Schriften 21, 1–56.

Bromley, G.R.M., Schaefer, J.M., Winckler, G., Hall, B.L., Todd, C.E., Rademaker, K.M., 2009. Relative timing of last glacial maximum and late-glacial events in the central tropical Andes. Quatern. Sci. Rev. 28, 2514–2526.

Clapperton, C.M., 1985. Significance of a late-glacial readvance in the Ecuadorian Andes. Quatern. S. Am. Antarct. Peninsula 3, 149–158.

Clapperton, C.M., 1987a. Glacial geomorphology, Quaternary glacial sequence and palaeoclimatic inferences in the Ecuadorian Andes. In: Gardiner, V. (Ed.), International Geomorphology 1986 Part II. J. Wiley & Sons, Chichester, pp. 843–870.

Clapperton, C.M., 1987b. Maximal extent of late Wisconsin glaciation in the Ecuadorian Andes. Quatern. S. Am. Antarct. Peninsula 5, 165–179.

Clapperton, C.M., 1990. Glacial and volcanic geomorphology of the Chimborazo-Carihuairazo Massif, Ecuadorian Andes. Trans. R. Soc. Edinburgh Earth Sci. 81, 91–116.

Clapperton, C.M., McEwan, C., 1985. Late Quaternary moraines in the Chimborazo area, Ecuador. Arct. Alpine Res. 17, 135–142.

Clapperton, C.M., Hall, M., Mothes, P., Hole, M.J., Still, J.W., Helmens, K.F., et al., 1997. A Younger Dryas icecap in the Equatorial Andes. Quatern. Res. 47, 13–28.

Dewolf, Y., 1988. Stratified slope deposits. In: Clark, M.J. (Ed.), Advances in Periglacial Geomorphology. Wiley, New York, pp. 113–149.

Francou, B., 1990. Stratification mechanisms in slope deposits in high subequatorial mountains. Permafrost Periglac. Process. 1, 249–263.

Glasser, N.F., Clemmens, S., Schnabel, C., Fenton, C.R., McHargue, L., 2009. Tropical glacier fluctuations in the Cordillera Blanca, Peru between 12.5 and 7.6 ka from cosmogenic ^{10}Be dating. Quatern. Sci. Rev. 28, 3448–3458.

Hall, S.R., Farber, D.L., Ramage, J.M., Rodbell, D.T., Finkel, R.C., Smith, J.A., et al., 2009. Geochronology of Quaternary glaciations from the tropical Cordillera Huayhuash, Peru. Quatern. Sci. Rev. 28, 2991–3009.

Hastenrath, S., 1981. The glaciation of the Ecuadorian Andes. Balkema, Rotterdam.

Heine, J.T., 1993. A reevaluation of the evidence for a Younger Dryas climatic reversal in the tropical Andes. Quatern. Sci. Rev. 12, 769–779.

Heine, J.T., 1995a. Comments on C.M. Clapperton's "Glacier readvances in the Andes at 12,500–10,000 yr BP: implications for mechanism of late-glacial climatic change (JQS 8, 197–215)" J. Quatern. Sci. 10, 87–88.

Heine, K., 1995. Late Quaternary glacier advances in the Ecuadorian Andes: a preliminary report. Quatern. S. Am. Antarct. Peninsula 9, 1–22.

Heine, K., 1995b. Bedded slope deposits with respect to the late Quaternary glacial sequence in the high Andes of Ecuador and Bolivia. In: Slaymaker, O. (Ed.), Steepland Geomorphology. J. Wiley & Sons, Chichester, pp. 257–278.

Heine, K., 2000. Tropical South America during the last glacial maximum: evidence from glacial, periglacial and fluvial records. Quatern. Int. 72, 7–21.

Heine, K., 2004. Late Quaternary glaciations of Ecuador. In: Ehlers, J., Gibbard, P.L. (Eds.), Quaternary Glaciations—Extent and Chronology Part III. Developments in Quaternary science, 2 Amsterdam, Elsevier, pp. 165–169.

Heine, K., 2011. Late Quaternary glaciations in Bolivia: comments on some new approaches to dating morainic sequences. In: Ehlers, J., Gibbard, P.L., Hughes, P.D. (Eds.), Quaternary Glaciations—Extent and Chronology, Part IV. Elsevier, Amsterdam.

Heine, K., Geyh, M.A., 2002. Neue ^{14}C-Daten zur Jüngeren Dryaszeit in den ecuadorianischen Anden. Eiszeit. Gegenwart 51, 33–50.

Heine, K., Heine, J.T., 1996. Late glacial climatic fluctuations in Ecuador: glacier retreat during the Younger Dryas time. Arct. Alpine Res. 28, 496–501.

Jomelli, V., Favier, V., Rabatel, A., Brunstein, D., Hoffmann, G., Francou, B., 2009. Fluctuation of glaciers in the tropical Andes over the last millennium and palaeoclimatic implications: a review. Palaeogeogr. Palaeoclimatol. Palaeoecol. 281, 269–282.

Jordan, E., Hastenrath, S., 1998. Glaciers of Ecuador. United States Geological Survey, Washington, Professional Paper, 1386-I, pp. 131–150.

Jordan, E., Cáceres, B., Francou, B., Ungerechts, L., 2003. Die Glazialforschungen Hans Meyers aus heutiger Sicht. Wertung der wissenschaftlichen Leistungen Meyers in den Hochanden von Ekuador aus aktueller Sicht und Ausblick auf die geographischen Forschungsergebnisse der vergangenen 100 Jahre. Wiss. Alpenvereinshefte 37, 159–193.

Jouzel, J., Lorius, C., Petit, J.R., Genthon, C., Barkov, N.I., Kotlyakov, V.M., et al., 1987. Vostok ice core: a continuous isotope temperature record over the last climatic cycle (160,000 years). Nature 329, 403–408.

Jouzel, J., Masson-Delmotte, V., Cattani, O., Dreyfus, G., Falourd, S., Hoffmann, G., et al., 2007. Orbital and millennial Antarctic climate variability over the past 800,000 years. Science 317, 793–797.

La Frenierre, J., In Huh, K., Mark, B.G., 2011. Quaternary glaciations—extent and chronology: Ecuador, Peru, Bolivia. In: Ehlers, J., Gibbard, P.L., Hughes, P.D. (Eds.), Quaternary Glaciations—Extent and Chronology, Part IV. Elsevier, Amsterdam.

Lauer, W., Rafiqpoor, M.D., 1986. Geoökologische Studien in Ecuador. Erdkunde 40, 68–72.

Lautridou, J.P., Francou, B., 1992. Present-day periglacial processes and landforms in mountain areas. Permafrost Periglac. Process. 3, 93–101.

Lea, D.W., Pak, D.K., Spero, H.J., 2000. Climatic impact of late Quaternary equatorial pacific sea surface temperature variations. Science 289, 1719–1724.

Licciardi, J.M., Schaefer, J.M., Taggart, J.R., Lund, D.C., 2009. Holocene glacier fluctuations in the Peruvian Andes indicate northern climate linkages. Science 325, 1677–1679.

Mark, B.G., Seltzer, G.O., Rodbell, D.T., 2004. Late Quaternary glaciations of Ecuador, Peru and Bolivia. In: Ehlers, J., Gibbard, P.L. (Eds.), Quaternary Glaciations—Extent and Chronology Part III: South America, Asia, Africa, Australia, Antarctica. Elsevier, Amsterdam, pp. 151–163.

Meyer, H., 1904. Die Eiszeit in den Tropen. Geogr. Ztschr. 10, 593–600.

Meyer, H., 1907. In den Hoch-Anden von Ecuador. Reisen und Studien, Reimer, Berlin, 551 pp.

Niemann, H., Behling, H., 2008. Late Quaternary vegetation, climate and fire dynamics inferred from the El Tiro record in the southeastern Ecuadorian Andes. J. Quatern. Sci. 23, 203–212.

Reiss, W., Stübel, A., 1886. Skizzen aus Ecuador. Asher & Co., Berlin.

Rincón-Martínez, D., Lamy, F., Contreras, S., Leduc, G., Bard, E., Saukel, C., et al., 2010. More humid interglacials in Ecuador during the past 500 kyr linked to latitudinal shifts of the equatorial front and the intertropical convergence zone in the eastern tropical Pacific. Paleoceanography 25, PA2210. doi:10.1029/2009PA001868, 2010.

Rodbell, D.T., Seltzer, G.O., 2000. Rapid ice margin fluctuations during the Younger Dryas in the tropical Andes. Quatern. Res. 54, 328–338.

Rodbell, D.T., Seltzer, G.O., Mark, B.G., Smith, J.A., Abbott, M.B., 2008. Clastic sediment flux to tropical Andean lakes: records of glaciation and soil erosion. Quatern. Sci. Rev. 27, 1612–1626.

Rodbell, D.T., Smith, J.A., Mark, B.G., 2009. Glaciation in the Andes during the late glacial and Holocene. Quatern. Sci. Rev. 28, 2165–2212.

Rull, V., Schubert, C., 1989. The Little Ice Age in the tropical Venezuelan Andes. Acta Cient. Venez. 40, 71–73.

Sauer, W., 1971. Geologie von Ecuador. Borntraeger, Berlin, Stuttgart.

Schubert, C., 1989. Glaciaciones cuaternarias en el norte de América del Sur. Memorias VII Congreso Geológico Venezolano 3, pp. 1304–1317.

Schubert, C., Clapperton, C.M., 1990. Quaternary glaciations in the northern Andes (Venezuela, Colombia and Ecuador). Quatern. Sci. Rev. 9, 123–135.

Schubert, C., Vivas, L., 1993. El Cuaternario de la Cordillera de Mérida, Andes Venezolanos. Universidad de los Andes/Fundación POLAR, Mérida, Venezuela.

Seltzer, G., 2001. Paleoclimates of the Central Andes. Workshop report. Tucson, Arizona, 11–16 January 2001. PAGES News 9 (1), 16.

Smith, J.A., Seltzer, G.O., Farber, D.L., Rodbell, D.T., Finkel, R.C., 2005a. Early local last glacial maximum in the tropical Andes. Science 308, 678–681.

Smith, J.A., Seltzer, G.O., Rodbell, D.T., Klein, A.G., 2005b. Regional synthesis of the last glacial maximum snowlines in the tropical Andes, South America. Quatern. Int. 138–139, 145–167.

Smith, J.A., Finkel, R.C., Farber, D.L., Rodbell, D.T., Seltzer, G.O., 2005c. Moraine preservation and boulder erosion in the tropical Andes: interpreting old surface exposure ages in glaciated valleys. J. Quatern. Sci. 20 (7–8), 735–758.

Smith, J.A., Mark, B.C., Rodbell, D.T., 2008. The timing and magnitude of mountain glaciation in the tropical Andes. J. Quatern. Sci. 23 (6–7), 609–634.

Thompson, L.G., Mosley-Thompson, E., Dansgaard, W., Grootes, P.M., 1986. The Little Ice Age as recorded in the stratigraphy of the tropical Quelccaya ice cap. Science 234, 361–364.

Thompson, L.G., Davis, M.E., Mosley-Thompson, E., Sowers, T.A., Henderson, K.A., Zagorodnov, V.S., et al., 1998. A 25,000-year tropical climate history from Bolivian ice cores. Science 282, 1858–1864.

van 't Veer, R., Islebe, G.A., Hooghiemstra, H., 2000. Climatic change during the Younger Dryas chron in northern South America: a test of the evidence. Quatern. Sci. Rev. 19, 1821–1835.

Wolf, T., 1892. Geografía y geología del Ecuador. Brockhaus, Leipzig.

Zech, J., Zech, R., May, J., Kubik, P.W., Veit, H., 2009. Heinrich I and Younger Dryas glaciation in the Central Andes. AGU Fall Meeting 2009. Abstracts.

Chapter 58

Quaternary Glaciations of Colombia

Karin F. Helmens

Department of Physical Geography and Quaternary Geology, Stockholm University, 106 91 Stockholm, Sweden

58.1. INTRODUCTION

The Andes in Colombia are situated in northernmost South America between latitudes ca. 1°N and 11°N. The Colombian Andes consist of three north–south to north-east–south-west orientated mountain chains named the Cordilleras Occidental, Central and Oriental or the Western, Central and Eastern Cordilleras, respectively (Fig. 58.1). The Eastern Cordillera rises sufficiently high to support glaciers on its northern portion, where the Sierra Nevada del Cocuy (Fig. 58.1) and adjacent ranges are ice covered over a total length of ca. 35 km. The Central Cordillera encompasses a series of glacier-capped stratovolcanoes, including the Nevado del Huila and the Ruíz–Tolima volcanic massif. Glaciers are also found on the highest peaks and ridges of the Sierra Nevada de Santa Marta in northernmost Colombia. The total area of glacier ice in Colombia measured from Landsat images from the early 1970s has been determined at 104 km^2 (Hoyos-Patiño, 1998). A significant decline in total area of the tropical glaciers in the Sierras Nevadas del Cocuy and Santa Marta and the Ruíz–Tolima massif of no less than 50% has been recorded from the 1950s to 2003, based on historical information and satellite (ASTER and Landsat) data (Morris et al., 2006).

The total area of glaciated terrain was substantially enlarged during the Pleistocene. Estimates by Thouret et al. (1996) indicate that ca. 26,000 km^2 of the Colombian Andes would bear evidence of Pleistocene glaciation, amounting to ca. 7.5% of its total surface area. Despite the fact that only a minor portion of the formerly glaciated mountains has been studied, the investigations of the Eastern and Central Cordilleran morainic sequences (van der Hammen et al., 1980/1981; Helmens, 1988; Thouret et al., 1996; Helmens et al., 1997b) are characterised by their great detail, and provided one of the best-dated Late Pleistocene glacial records for the Andes (Clapperton, 2000). Chronologies are based on the radiocarbon dating of organic-rich sediments and palaeosols, the accumulation of organic matter in depressions and the soil cover being favoured by the cool and humid climate of the northern high Andes. In addition, the study of glaciofluvial sediment in the inter-montane basin of Bogotá (Eastern Cordillera; Fig. 58.1) has provided evidence on the timing and intensity of older glaciations in the region (Helmens et al., 1997a). Palynological studies in the Colombian Andes have been numerous and have provided an important tool for comparison with the glacial records (e.g. van der Hammen et al., 1980/1981; Helmens and van der Hammen, 1994; Helmens and Kuhry, 1995; Thouret et al., 1996).

This chapter focuses on the glacial record of the Eastern Cordillera, in particular, the high plain of Bogotá region. Detailed maps are presented for the Bogotá mountains and the Sierra Nevada del Cocuy that show morainic sequences and interpreted glacial limits in type areas, and type localities for radiocarbon dating. The areas mapped outside the type areas are small and for their glacial morphology and former glacial limits, the reader is referred to the original literature. A comparison is made with the Late Pleistocene glacial record of the Ruíz–Tolima volcanic massif in the Central Cordillera. For studies on the morainic sequence of the Sierra Nevada de Santa Marta, Raasveldt (1957) should be consulted. The present chapter is modified after Helmens (2004). It includes photographs of geological evidence and references to more recent studies and discusses in more detail problems/limitations and suggestions for further study.

58.2. EARLY AND MIDDLE PLEISTOCENE GLACIATIONS RECORDED BY GLACIOFLUVIAL SEDIMENT IN THE BOGOTÁ BASIN

The tectonic basin of Bogotá and direct surroundings (Eastern Cordillera; Figs. 58.1 and 58.2) holds a long sedimentary record that extends from the Holocene into the Pliocene. The sediments have been divided into 16 lithostratigraphical units and mapped over a total area of nearly 2000 km^2 (e.g. van der Hammen et al., 1973; Helmens, 1990; Helmens and van der Hammen, 1994). The record

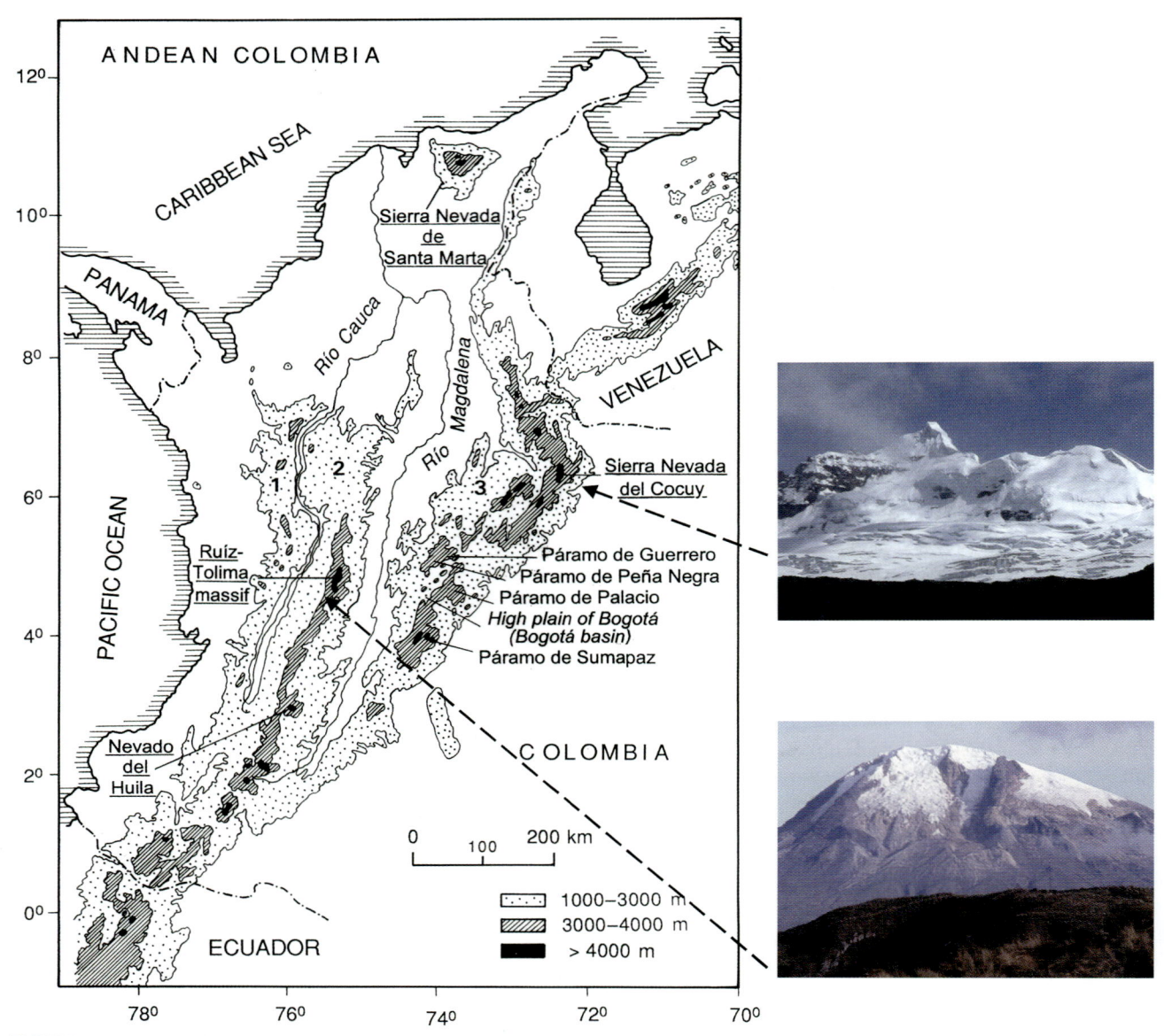

FIGURE 58.1 Andean Colombia showing the Western (1), Central (2) and Eastern Cordilleras (3). The sites mentioned in the text are indicated; those underlined are currently glaciated (see photographs).

includes seven biozones based on fission-track dated palynological evidence (e.g. van der Hammen et al., 1973; Kuhry and Helmens, 1990; Hooghiemstra and Ran, 1994; Wijninga, 1996; van der Hammen and Hooghiemstra, 1996). Glacial deposits are found in the higher mountains (see below). Glaciers did not descend into the Bogotá basin, but thick sequences of glaciofluvial sediments, derived from these ice tongues, have accumulated in its outer valleys.

Where the Bogotá basin is bordered by high mountains, such as along its north-eastern margin below the Páramo de Palacio, a distinct lithological transition is observed from morainic deposits at high elevations towards the glaciofluvial sediment in the basin (Figs. 58.2 and 58.3). The coarse and angular boulders of which the moraines are composed (Río Chisacá Formation) pass into rounded boulders and gravel in a series of large coalescing outwash fans directly at the foot of the formerly glaciated mountain slopes (Río Siecha Formation). The latter deposits grade, within the Bogotá basin, into thick sequences of well-rounded gravel (Río Tunjuelito Formation), which in their turn grade into a series of sand and gravel units away from the main rivers that enter the basin. The sand and gravel units alternate with, and in places truncate, more fine-grained sediment beds of a mostly lacustrine origin (Subachoque Formation). Palynological evidence from the fine-grained lacustrine intercalations in the Subachoque Formation, from thin organic beds in the Río Tunjuelito

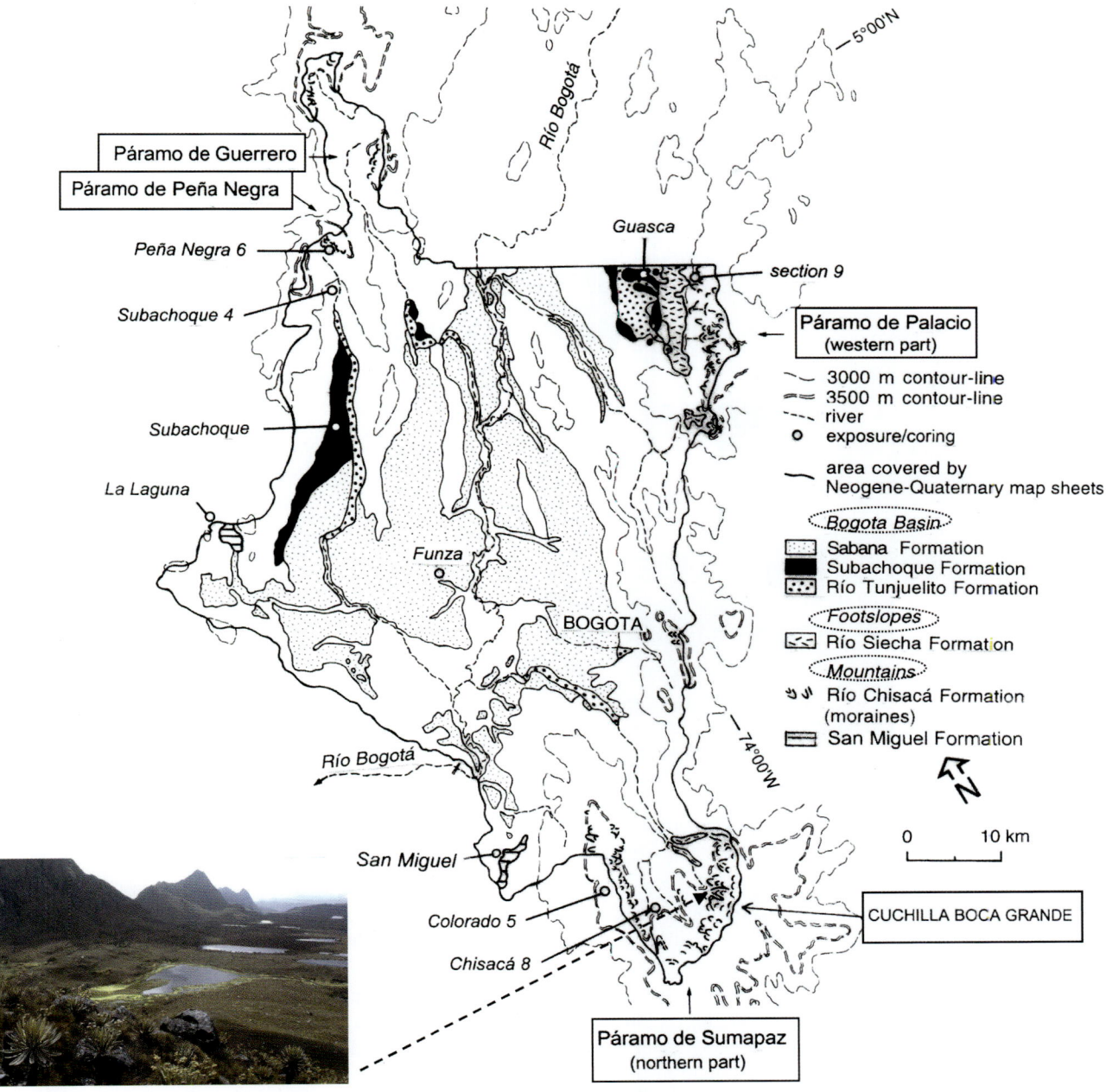

Chapter | 58 Quaternary Glaciations of Colombia

FIGURE 58.2 The tectonic basin of Bogotá and surrounding mountains (Eastern Cordillera in Fig. 58.1), showing the surface distribution of lithostratigraphical units (including moraines) and the location of exposures/boreholes, mentioned in the text (based on Helmens, 1990). The photograph shows the glacial morphology in the Cuchilla Boca Grande area (Fig. 58.5).

Formation and from sequences of up to 150 m thick of clays interbedded with peaty and sandy sediment in the deeper parts of the central basin, that have been correlated with the type Subachoque, reflect the changing climatic conditions of the Pleistocene. The pollen assemblages from these sediments indicate alternately Andean forest vegetation representing interglacial (or interstadial) conditions, and treeless, tropic-alpine vegetation, called *paramo* in the northern Andes, that indicate glacial (stadial) conditions (van der Hammen et al., 1973; Kuhry and Helmens, 1990; Hooghiemstra and Ran, 1994). The sand and gravel interbeds in the Subachoque Formation, and the gravels of the Tunjuelito Formation, are interpreted to represent the coldest intervals of the Pleistocene when the surrounding mountains were glaciated. The glaciers caused the outer valleys of the Bogotá basin to be infilled by glaciofluvial sediment, restricting the Bogotá Lake to the central part of the basin (van der Hammen et al., 1973; Helmens, 1990). Radiocarbon dates from the Tunjuelito Formation and

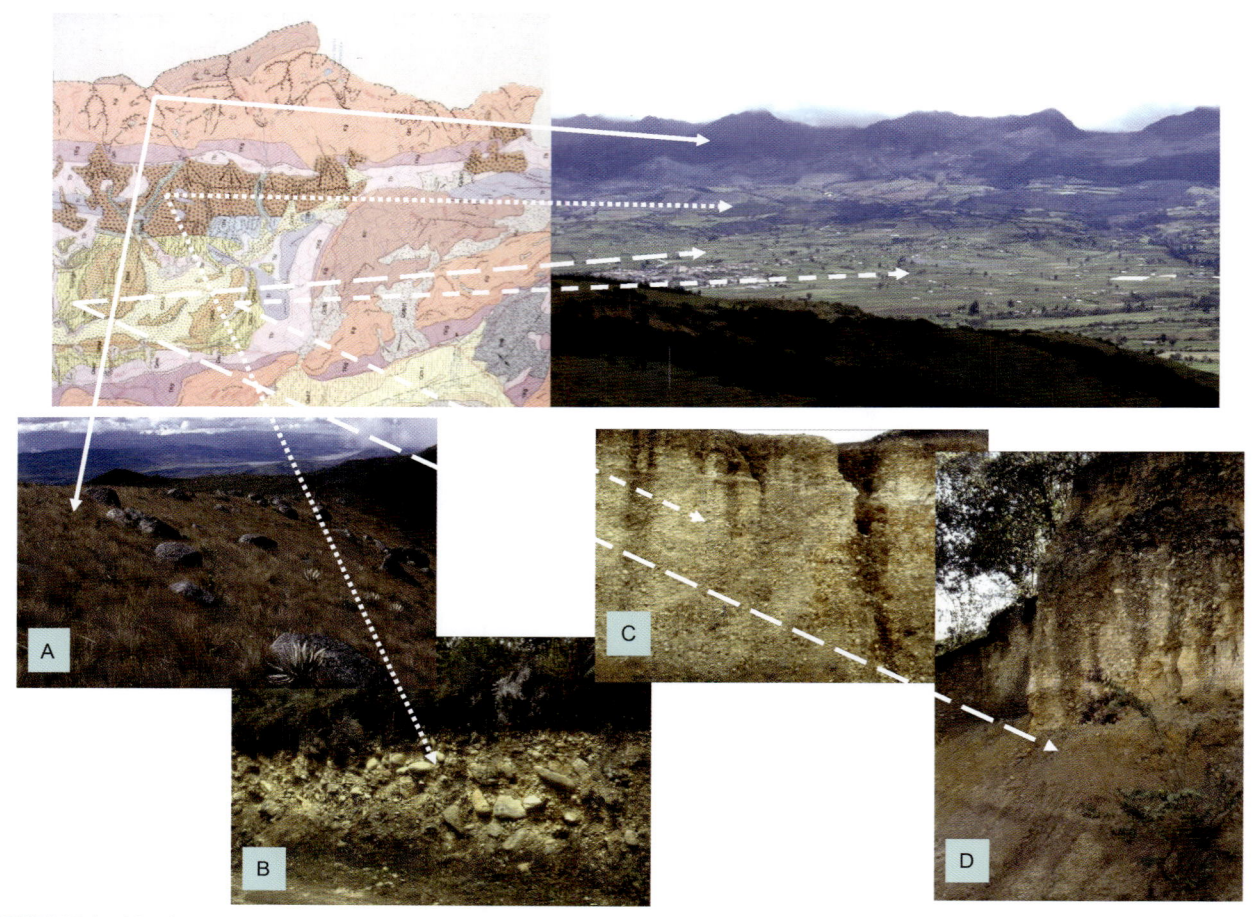

FIGURE 58.3 Map fragment (Helmens, 1990) and photographs illustrating the Río Chisacá (A: moraines), Río Siecha (B: outwash), Río Tunjuelito (C: glaciofluvial) and Subachoque Formations (D: glaciofluvial sediments alternating with lake sediments) along the western slopes of the Páramo de Palacio and in the adjacent Bogotá basin (Fig. 58.2).

from organic-rich sediments and palaeosols found associated with the moraines of the Río Chisacá Formation suggest synchrony between the deposition of gravels in the Bogotá basin and glacial events in the mountains through the Late Pleistocene (van der Hammen et al., 1980/1981; Van der Hammen, 1986).

A detailed geochronology for the sediment infill of the outer parts of the Bogotá basin is presented by Helmens et al. (1997a). The study focussed on two major outcrops in the Subachoque Formation and the sediments of the underlying lacustrine Guasca Member of the Upper Tilatá Formation (Guasca and Subachoque sections in Fig. 58.2). The two outcrops that expose a total thickness of ca. 75 m were sampled at ca. 10 cm intervals for palaeomagnetic polarity studies. Interbedded tephras, derived from the Ruíz–Tolima volcanic massif in the Central Cordillera (Fig. 58.1), were sampled for fission-track dating. The correlation with the geomagnetic polarity reference timescale (Cande and Kent, 1995) presented in Helmens et al. (1997a; Fig. 58.4) takes into account discontinuities in sedimentation as well as the error limits of the fission-track dates. The sediments of the Guasca Member yielded fission-track dates of ca. 3 and 2.9 Ma, and their polarities are assigned to the upper part of the Gauss Chron. The Subachoque Formation yielded fission-track dates of 2.5, 1.7 and 1 Ma, and its magnetisation indicates that it ranges from the Matuyama to the Brunhes Chron.

58.2.1. Onset of Glaciations near the Pliocene–Pleistocene Boundary (2.6 Ma)

Magnetic polarity and fission-track dating places the first glaciation in the Bogotá mountains, as recorded at the base of the Subachoque Formation by a sudden influx of glaciofluvial sand and gravel into the outer valleys of the Bogotá basin, near the Gauss/Matuyama magnetic reversal at 2.6 Ma (Helmens et al., 1997a; Fig. 58.4). This dating coincides closely with the base of the Quaternary as recently formalised by the International Union of Geological Sciences (IUGS; Gibbard et al., 2010). At the lower boundary of the

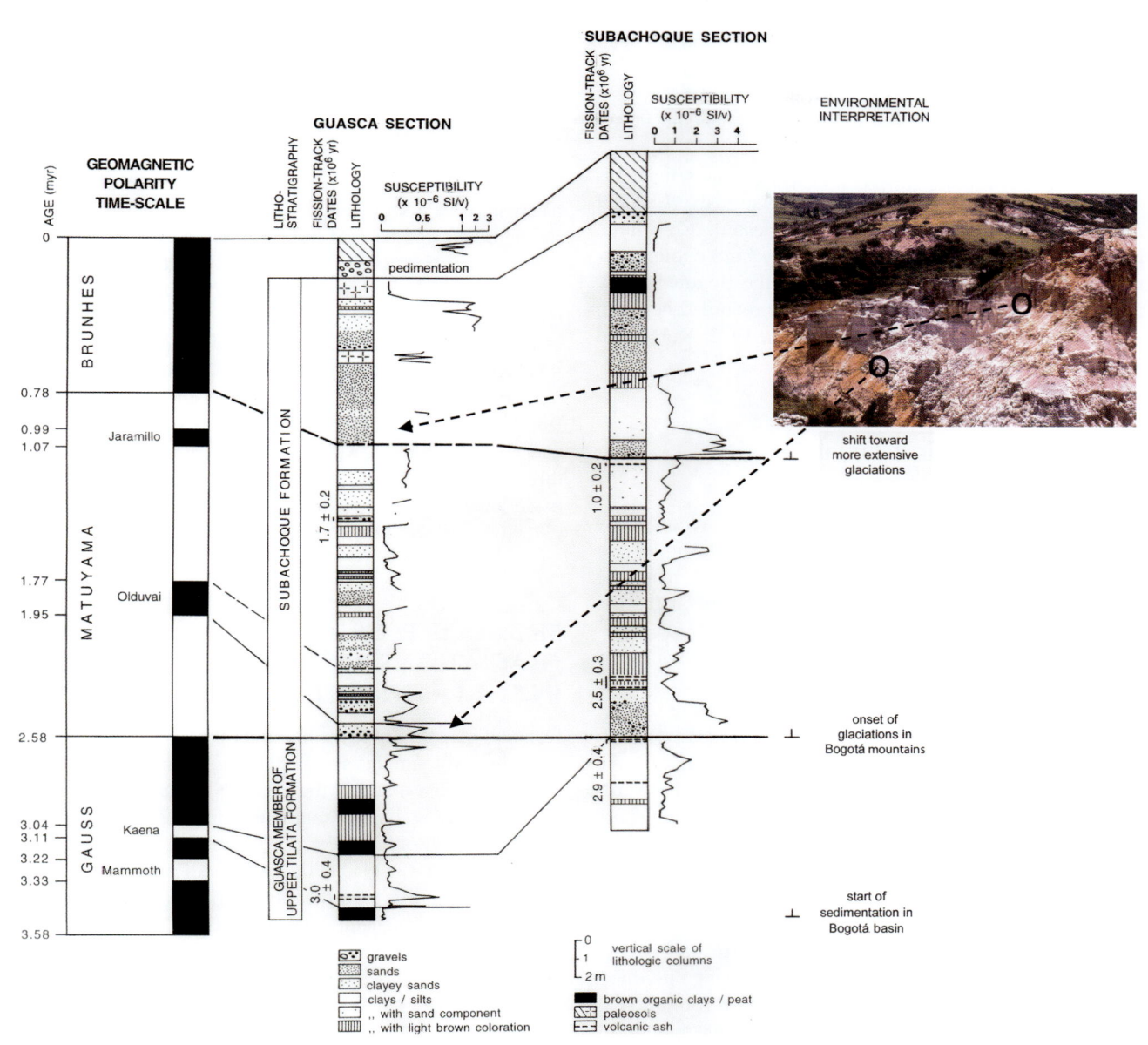

FIGURE 58.4 The Late Pliocene–Quaternary environmental record of the Guasca Member of the Upper Tilatá Formation and the Subachoque Formation in the marginal valleys of the Bogotá basin, based on magnetostratigraphy, fission-track chronology, lithology and magnetic susceptibility of the Guasca and Subachoque sections (Fig. 58.2). The geomagnetic reversal chronology is based on Cande and Kent (1995). The vertical scale of lithological columns is linear. The sand and gravel interbeds in the Subachoque Formation represent glaciofluvial sediment derived from glaciers in the higher mountain ranges surrounding the Bogotá basin (Helmens et al., 1997a).

Quaternary at 2.6 Ma, a global-scale cooling is recorded that led to a major expansion of Northern Hemisphere ice sheets and the initiation of a pattern of glacial–interglacial cycles that has dominated the world's climate to the present day (Pillans and Naish, 2004; Walker and Lowe, 2007).

The long Funza pollen record from the central Bogotá basin (Fig. 58.2) records a considerable lowering of the regional forest limit at the base of the Subachoque Formation, which also indicates sudden cooling (Hooghiemstra and Ran, 1994). The base of the Subachoque sediments in the Funza II borehole has been fission-track dated as slightly younger than 2.7 ± 0.6 Ma (Andriessen et al., 1993). In addition, widespread downslope movement of old tropical weathering product is recorded at the base of the Quaternary (van der Hammen et al., 1973; Helmens, 1990). These slope deposits are characterised by weathered clasts in a strongly reddish/orange coloured clayey matrix (San Miguel Formation in Fig. 58.2; Helmens, 1990). They are found on hill and mountain slopes in the Bogotá area, downslope from deeply weathered Late Cretaceous sandstones. The base of the San Miguel Formation has been tightly bracketed by fission-track dates on tephra between 2.8 ± 0.2 and 2.5 ± 0.3 Ma (San Miguel section in Fig. 58.2; P. A. M. Andriessen, K. F. Helmens and R. W. Barendregt, unpublished data).

The marine oxygen isotope record indicates that episodes of global cooling preceded the Quaternary (Lisiecki and Raymo, 2005). The Colombian Eastern Cordillera, however, might have lacked sufficiently high enough terrain to support glaciers at that time. In the Bogotá region, major tectonic uplift by some 2000 m is recorded between ca. 6 and 3 Ma (Van der Hammen et al., 1973; Kuhry and Helmens, 1990; Wijninga, 1996).

58.2.2. Increase in Glacial Activity since ca. 0.8 Ma

A shift towards more extensive glaciations is recorded by lithology and magnetic susceptibility (MS) in the upper part of the Subachoque Formation and is dated near the Matuyama/Brunhes magnetic Chron boundary at 0.8 Ma (Helmens et al., 1997a).

The detailed lithological description of the Subachoque Formation by Helmens et al. (1997a) has revealed that the sand and gravel interbeds become distinctly more coarse-grained in the upper part of the formation (Fig. 58.4). Helmens et al (1997a) also made MS measurements, despite the fact that their field sampling was initially directed towards studies of palaeomagnetic remanence, concentrating only on fine-grained lithologies. As a result there are gaps in the MS record corresponding to the coarsest units (Fig. 58.4). Nevertheless, the data obtained seems to indicate a further increase in MS amplitude in the upper part of the Subachoque Formation, compared to the increase recorded at the base of the formation. The change in lithology and MS signal in the upper part of the formation is dated at ca. 0.8 Ma. Apparently, periods with conditions colder than those prevailing during the Matuyama Chron resulted in more extensive glaciations (and a larger influx of magnetite-rich sediment into the Bogotá basin), with rapidly aggrading floodplains leaving a distinct series of coarse-grained sand and gravel beds. Additionally, episodes with conditions warmer than during the Matuyama Chron, and higher evaporation and evapotranspiration rates (Hooghiemstra, 1984; Kuhry, 1991), resulted in lower lake levels and, in the orographically dry Guasca valley, in periods of soil formation (Fig. 58.4).

A shift towards higher magnitude climate oscillations is also recorded in the upper part of the Funza pollen sequence and has been placed at ca. 0.8 Ma based on fission-track dating and land–sea correlation (Andriessen et al., 1993; Hooghiemstra et al., 1993; Hooghiemstra and Ran, 1994). This change has been found associated with a distinct change in the frequency of oscillations from 41 to 100 ka climate cycles (Hooghiemstra et al., 1993).

58.2.3. Conclusions and Limitations

The geological and palaeoecological data from the Bogotá basin shows clear evidence of a major change in the environment near the base of the Quaternary at 2.6 Ma. Major cooling occurred and a long period of alternately glacial and interglacial conditions succeeded (e.g. van der Hammen et al., 1973; Hooghiemstra and Ran, 1994; Helmens et al., 1997a). Climate changes and glaciations were primarily driven by astronomical influences, as suggested by comparison with the deep-sea oxygen isotope record (e.g. Ruddiman et al., 1986; Lisiecki and Raymo, 2005; Walker and Lowe, 2007). Although glaciations are recorded throughout the Quaternary, the glacial record of the Bogotá basin lacks the detail to allow for a detailed correlation of glacial events with the marine isotope stratigraphy.

58.3. LATE PLEISTOCENE AND HOLOCENE GLACIATIONS RECORDED BY MORAINES IN THE EASTERN CORDILLERA

Detailed mapping of moraines and till beds has been undertaken in the mountains surrounding the high plain of Bogotá (ca. latitude 5°N) and in the Sierra Nevada del Cocuy (ca. lat. 6°N; Fig. 58.1). The mapping, together with the radiocarbon dating of organic-rich sediments and palaeosols associated with the glacial deposits/landforms, has resulted in a high-resolution record of glacier fluctuations for the last ca. 45 ^{14}C ka BP. The highest peak in the Cocuy mountains is the Ritacuba Blanca at 5490 m and the modern snowline is above 4700 m a.s.l. (Helmens et al., 1997b). The mountains near Bogotá reach to almost 4000 m altitude. Although currently ice free, a distinct glacial landscape of cirques, U-shaped valleys and glacially scoured, relatively flat areas dotted with glacial lakes, combined with numerous morainic ridges, characterise mountain ranges, the higher parts of which exceed 3600 m a.s.l. (Helmens, 1988).

Maximum glaciation in the Colombian Eastern Cordillera pre-dated the Late Weichselian (Late Wisconsinan) glacial maximum. One (possibly two) major glacier advance(s) of Middle Weichselian age: a twofold glaciation maximum for the Late Weichselian (referred to as an early and late stadial of the global last glacial maximum: LGM) and a minor advance of early Late-Glacial age are presented, based on the Bogotá record. Evidence from the Sierra Nevada del Cocuy is discussed which additionally indicates a glacial event of most probable Younger Dryas (YD) Chron age and a series of glacier limits for the

Holocene. The small scale of the digital maps inhibits the reproduction of the morainic ridges. However, several maps are included in this chapter, which illustrate in detail the morainic sequence in both the Bogotá and Cocuy mountains together with inferred glacial limits, as well as the location of sites that have provided radiocarbon-dating control for the different glacial events recognised.

The glacial sequence discussed below is compared with the palynological record from the Eastern Cordillera. Numerous pollen records have been obtained from sediment and peat accumulations in deep tectonic basins, weathering depressions in sandstone and, in the high altitude areas, in glacial basins. Syntheses of palynological evidence for the radiocarbon-dated interval 25 ^{14}C ka BP to the present day have been published by van der Hammen et al. (1980), Kuhry (1988) and Helmens and Kuhry (1995).

Finally, the late Quaternary glacial record of the Eastern Cordillera is compared with results obtained from a detailed study of moraines on the Ruíz–Tolima volcanic massif in the Central Cordillera (Thouret et al., 1996).

58.4. MOUNTAIN RANGES NEAR BOGOTÁ

Glacial landforms in the high mountain ranges of the Páramo de Guerrero, the Páramo de Peña Negra, the westernmost part of the Páramo de Palacio and the northernmost part of the Páramo de Sumapaz (Figs. 58.1 and 58.2) have been mapped in detail by Helmens (1988). The mapping especially focussed on the mountain slopes that drain towards the high plain of Bogotá (Helmens, 1990). The main morainic lobes are shown in Fig. 58.2.

Based on general morphology and degree of modification by erosion and denudation, the maps by Helmens (1988) differentiate four morainic complexes, defined from oldest to youngest as morainic complexes 1–4. All four morainic complexes are well-represented on the slopes below the Cuchilla de Boca Grande in the Páramo de Sumapaz (Figs. 58.2 and 58.5). In addition, the latter area includes a site with 'older glacial deposits'. These deposits, which no longer show a morainic topography, have been encountered beyond the sequence of moraines (van der Hammen et al., 1980; Helmens, 1990).

Absolute chronological control for the glacial sequence in the Bogotá mountains is provided by a series of radiocarbon dates (uncalibrated) from peat, lake sediments and palaeosols from borehole cores and exposures in the formerly glaciated areas (van der Hammen et al., 1980; Helmens, 1988, 1990; Helmens and Kuhry, 1995; Helmens et al., 1997b). The palaeosols are formed in volcanic ash or ash-dominated slope deposits. Aluminium–humus complexes have protected organic matter against microbial degradation, and the organic material accumulated in considerable quantities in the soil profile. This stable humus has been successfully used to radiocarbon date palaeosol sequences in the area (e.g. Fölster and Hetsch, 1978; Guillet et al., 1988). Combined dating results on palaeosols and peaty or lacustrine sediments further illustrate the reliability of the palaeosol dates (Helmens and Kuhry, 1995). A histogram of dating results on soil humus from the volcanic ash–palaeosols in the Bogotá area (Fig. 58.6) indicates that intervals of soil formation occurred throughout the region over the last 50 ^{14}C ka BP (van der Hammen, 1981; Helmens and Kuhry, 1995). Because of their widespread occurrence, palaeosols have been tentatively used as stratigraphical markers for the glacial sequence (Helmens and Kuhry, 1995).

58.4.1. Middle Weichselian Glacial Advance(s)

The glacial record of the Bogotá area provides evidence for major glaciation(s) during the Middle Weichselian Substage. A major glacier advance occurred here before 31 ^{14}C ka BP, and an even more extensive advance took place before 38 ^{14}C ka BP. The glacial events are represented by morainic complex 1 and the older glacial deposits, respectively.

Morainic complex 1 consists of the oldest morainic landforms identified. These moraines are strongly subdued and only preserved locally. They occur at elevations some 200 m below the moraines of complex 2 which represent an early stadial of the LGM. A palaeosol directly overlying morainic complex 1 yielded a radiocarbon date of 30.93 ± 0.42 ^{14}C ka BP (Chisacá 8 in Fig. 58.2; Helmens, 1990). A similar date of 31.35 ± 0.5 ^{14}C ka BP has been obtained from a palaeosol overlying sediments developed within sediments interpreted as reworked sediment of morainic complex 1. The latter sediments compose the upper part of a terrace sequence found exposed at the foot of the Cuchilla Boca Grande (section 190 in Fig. 58.5; Helmens, 1990; Helmens and Kuhry, 1995). Remnants of moraines of complex 1 were encountered in the apical zone of a similar terrace situated some distance to the north. The reworked sediment of morainic complex 1 in section 190 is separated from an underlying series of older glacial deposits by a volcanic ash bed overlain by organic-rich clays that have yielded a radiocarbon date of $38.1 + 2.5/-1.9$ ^{14}C ka BP. The older glacial deposits in the lower part of the terrace sequence include a thick layer of basal till. These rather compact clays scattered with rock fragments are underlain by rounded gravels and boulders most probably of glaciofluvial origin and overlain by solifluction deposits that probably represent reworked supraglacial till. Slightly reworked supraglacial till has also been found in another valley several kilometres downvalley from morainic

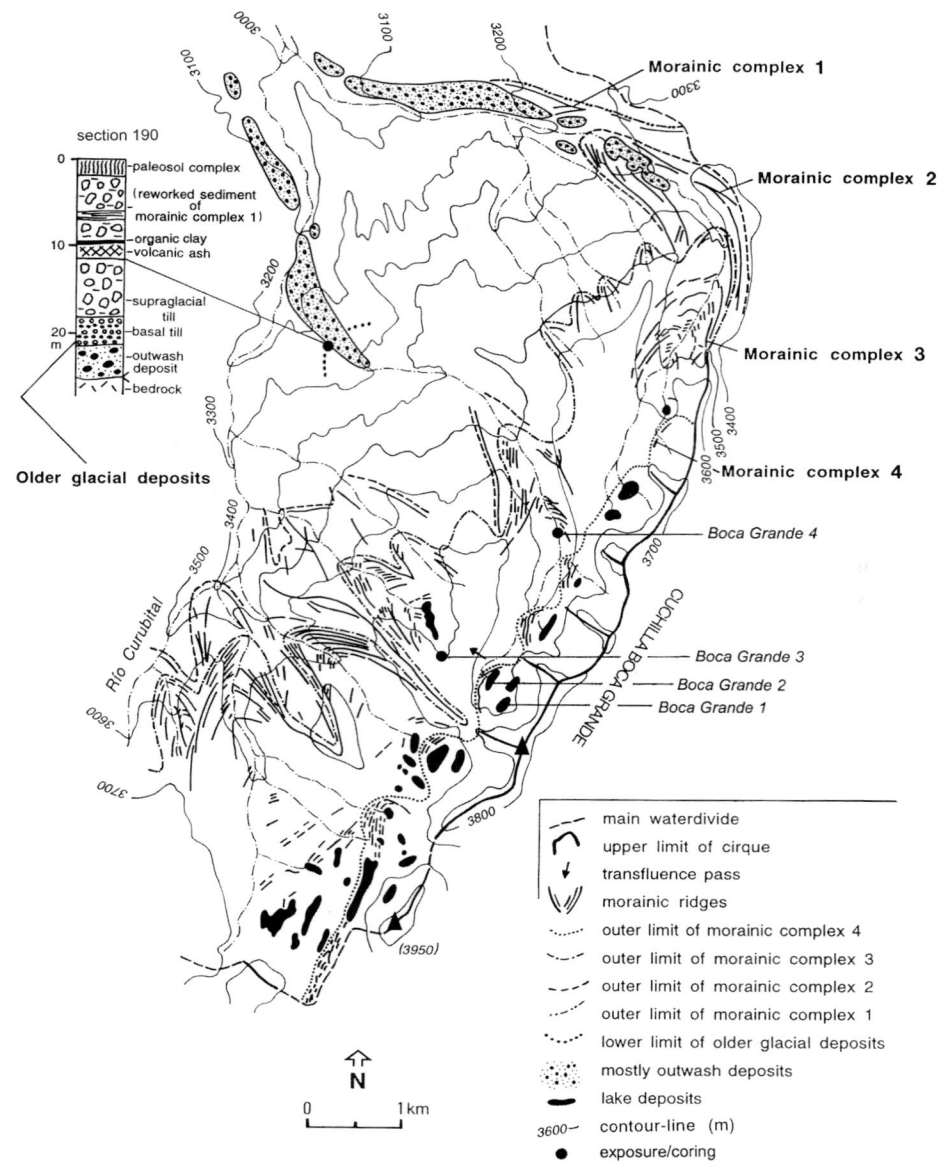

FIGURE 58.5 The late Quaternary glacial sequence in the Bogotá mountains (Cuchilla Boca Grande, Páramo de Sumapaz; Fig. 58.2). The older glacial deposits pre-date 38 ka and possibly post-data 43 ^{14}C ka BP. Morainic complexes 1–4 have been dated to the time interval ca. 36–12.5 ^{14}C ka BP (Helmens et al., 1997b). The sites of exposures/boreholes mentioned in the text are indicated.

complex 1 (Subachoque 4 in Fig. 58.2; van der Hammen et al., 1980). The surface here, at an elevation of 2850 m a.s.l., is characterised by concentrations of large erratic boulders. A palaeosol developed on the till has been dated to $35.8+1.1/-0.9$ ^{14}C ka BP. Moreover, a possible maximum age of 43 ^{14}C ka BP for the older glacial deposits is indicated by a palaeosol developed in the Bogotá area during this time interval (Fig. 58.6) that is represented in neither of the above-mentioned sequences of the older glacial deposits (Helmens and Kuhry, 1995).

An early timing for maximum glaciation during the Last Glacial stage in the Colombian Andes was first indicated by van der Hammen et al. (1980/1981) and van der Hammen (1986). Their interpretation was based on limited absolute dating control for the morainic sequence of the Sierra Nevada del Cocuy, combined with evidence on glacial deposits, glaciofluvial gravels, palaeosols and pollen from the Bogotá area. Van der Hammen et al. (1980/1981) correlated the older glacial events with stadial periods of Middle Weichselian age, characterised in the palynological and lake-level records from the area by cold and distinctly humid conditions. More limited ice advances during the even colder Late Weichselian were ascribed to prevailing drier climatic conditions in the Colombian Andes.

FIGURE 58.6 Histogram of radiocarbon dates of soil humus from volcanic ash–palaeosols in the Bogotá area (Helmens and Kuhry, 1995).

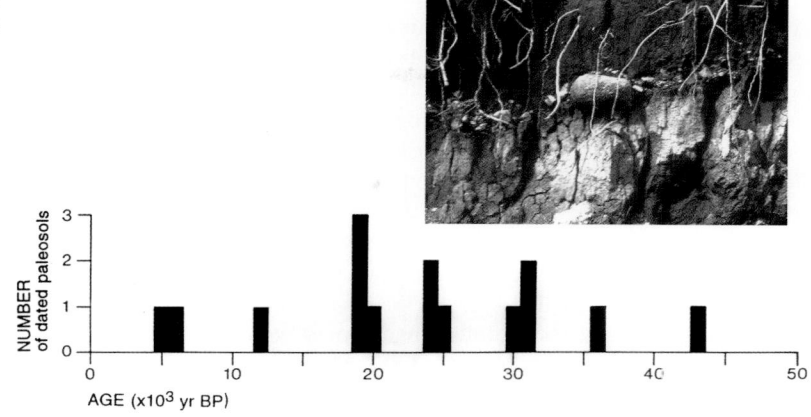

Additional glacial geomorphological/geological, palaeoecological and absolute chronological data later available for the Bogotá area (Helmens and Kuhry, 1986; Helmens, 1988, 1990) supported this interpretation and placed the most extensive glaciations recognised before 38 ^{14}C ka BP and probably after 43 ^{14}C ka BP, and between 36 and 31 ^{14}C ka BP (Helmens and Kuhry, 1995).

An early timing for the last maximum glacier advance has also been reported from the Ecuadorian Andes where radiocarbon dates from peat interbedded between tills yield ages in the range of 33 to >43 ^{14}C ka BP (Clapperton, 1987). Evidence for major glacial events pre-dating the last global ice volume maximum at ca. 18 ^{14}C ka BP in mountain ranges around the world is summarised and discussed in Gillespie and Molnar (1995).

58.4.2. Late Weichselian Glaciation Maxima

In the Bogotá mountains, the Late Weichselian global glacial maximum (LGM) consisted of two glacial maxima separated by an interval of glacier retreat. The glacial advances are marked by morainic complexes 2 and 3. The glaciers had retreated from complex 2 just before 19.5 ^{14}C ka BP, but re-advanced shortly afterwards close to this limit and deposited morainic complex 3, from where they subsequently receded just before 15.5 ^{14}C ka BP.

The youngest of the two LGM morainic complexes, complex 3, shows the most impressive morainic morphology of the different morainic complexes recognised in the Bogotá mountains. The arcuate, multiple ridge system rises tens of metres above the valley floors, and the related maximum ice extent can be continuously traced throughout the mountain ranges studied. Its distribution shows that limits of former glacier systems were influenced by local climatological conditions (Helmens, 1988). It appears that glaciers reached further downvalley in areas with high orographic precipitation, most significantly on the wet, eastern slopes of the Cordillera, where morainic complex 3 is found at the low elevation of 3100 m a.s.l. In the rain shadow created by large topographic barriers, complex 3 occurs at elevations as high as 3750 m a.s.l. A rain-shadow effect is to some extent demonstrated in the area of the Cuchilla Boca Grande (Fig. 58.5). Northwards, a decrease in elevation in combination with a narrowing of the water divide gives rise to an increasing influence of moisture-bearing winds from the eastern slopes of the Cordillera. Moraines of complex 3 reach down to elevations of ca. 3400 m a.s.l. in the northern part of the Cuchilla Boca Grande area, but only to about 3600 m a.s.l. in the more southerly parts. The moraines of complex 2 partially enclose the moraines of complex 3 but reach ca. 100–150 m farther downvalley. Morainic complex 2, however, has been distinctly more affected by erosional and denudational processes, is generally incomplete, and displays more subdued ridges than the sharp crests of morainic complex 3.

Two radiocarbon dates are available for the retreat of glaciers from morainic complex 3, that is, 15.51±0.19 ^{14}C ka BP (Colorado 5 in Fig. 58.2; Helmens et al., 1997b) and 14.66±0.28 ^{14}C ka BP (Boca Grande 3 in Fig. 58.5; Helmens, 1988) obtained from basal lake sediments enclosed by the moraines. Minimum ages for complex 2 of 19.19±0.12 and 18.13±0.17 ^{14}C ka BP, obtained from organic-rich sediments found overlying glaciofluvial gravel directly behind the moraines of complex 2 (Peña Negra 6 in Fig. 58.2; Helmens, 1988), are in accordance with a radiocarbon date of 19.37±0.23 ^{14}C ka BP for a palaeosol found on top of the complex (section 9 in Fig. 58.2; Khobzi, oral communication, in Helmens, 1988). The maximum age for morainic complex 2 most probably is not older than 23.5 ^{14}C ka BP (Helmens and Kuhry, 1995). This age corresponds to a period of widespread formation of organic-rich soils in the Bogotá area

(Fig. 58.6); the period is not represented in the dated palaeosol sequence developed on the morainic complex, nor is it expressed in the sediment sequence behind the moraines.

The time interval between 21 and 14 ^{14}C ka BP was originally defined in the palynological record as the Fúquene Stadial (van Geel and van der Hammen, 1973). Pollen evidence from the inter-montane Fúquene Lake, located at ca. 100 km NE of Bogotá, reflects a lowering in temperature in the order of 8 °C compared to present conditions for this interval. The Fúquene Stadial was followed by the Susacá Interstadial (14–13 ^{14}C ka BP) when temperatures rose to values ca. 4 °C lower than at present (van der Hammen and Vogel, 1966; Kuhry, 1988). More recently, a pollen record obtained from a series of lake sediments, collected from an ecologically highly sensitive situation on the western slopes of the Eastern Cordillera (La Laguna in Fig. 58.2), has revealed that cold, glacial conditions did not persist throughout the Fúquene Stadial, but instead were interrupted by a distinct interval of climate warming around ca. 18 ^{14}C ka BP (Helmens et al., 1996). High pollen frequencies during the Holocene for Andean forest elements reflect the present location of the La Laguna site within the Andean forest belt. During the Late Weichselian, the pollen record shows that the Andean forest was replaced by open, treeless, paramo vegetation, indicating that under the influence of considerable climate cooling, the forest limit had descended to well below the elevation of the La Laguna site. A more detailed inspection of the Late Weichselian sequence, however, reveals several intervals, including around 18 ^{14}C ka BP, with a clear increase in tree pollen. The application of well-studied quantitative relationships of modern vegetation cover and pollen rain (Grabandt, 1980, 1985) translates into an important upward shift in the upper Andean forest limit for the 18 ^{14}C ka BP interval.

Figure 58.7 shows the climate record obtained from the La Laguna site over the past 30 ka. The upper Andean forest limit throughout the Eastern Cordillera corresponds to the 9 °C annual isotherm, both in wet and relatively dry areas (Kuhry et al., 1993). Altitudinal shifts in the forest limit are translated into temperature values using thermal lapse rates with elevation, which at present are consistently in the order of 0.66 °C/100 m (van der Hammen and Gonzalez, 1963; Kuhry, 1988) and which were similar throughout the past 25 ka (Bakker, 1990; Kuhry et al., 1993). The chronology of the La Laguna core is based on 10 radiocarbon dates from macrofossils and bulk peaty samples. The La Laguna pollen record indicates a lowering of the forest

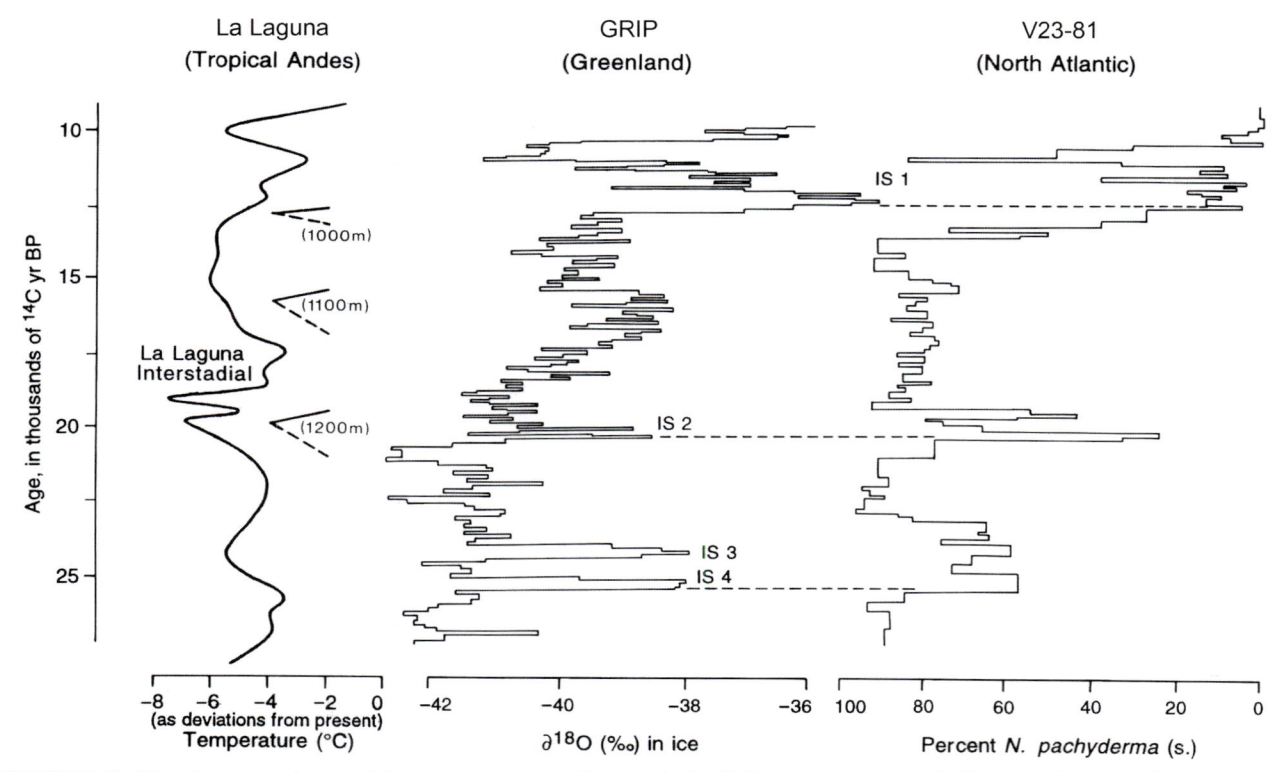

FIGURE 58.7 The climate record obtained from the La Laguna pollen record (Fig. 58.2) compared with the GRIP ice core from Greenland (Dansgaard et al., 1993) and the North Atlantic Ocean V23-81 core (Bond et al., 1993). The correlation between the Greenland and North Atlantic cores is according to Bond et al. (1993); the timescale follows the radiocarbon chronology of the marine record (from Helmens et al., 1996). Intervals of glacier expansion recorded in the Bogotá mountains (the average lowering in glacier-front positions given in brackets) have been added directly to the right of the La Laguna temperature curve (Helmens et al., 1997b).

limit by 1100–900 m beneath its present elevation at ca. 3300 m a.s.l. for the stadial intervals directly preceding and following the 18 ka interval, implying a drop in mean annual temperatures of 8–6 °C. Glaciers advanced downvalley during the stadial periods to elevations of ca. 3350 m a.s.l. (morainic complex 2) and 3500 m a.s.l. (complex 3), reflecting a lowering of the ice front by ca. 1200–1100 m compared than present. Between 19.5 and 17 ^{14}C ka BP, the ice front retreated, extensive soil formation took place (Fig. 58.6) and the upper Andean forest limit shifted to elevations similar to that of the La Laguna site (2900 m a.s.l.). At the same time, temperatures rose considerably to values only 3–4 °C lower than those in the present interglacial period. The interval between 19.5 and 17 ka has been defined as the La Laguna Interstadial (Helmens et al., 1996). Following Kuhry (1988), the preceding stadial period (21–19.5 ^{14}C ka BP) and the following stadial period (17–14 ^{14}C ka BP) are termed the Early Fúquene Stadial and the Late Fúquene Stadial, respectively.

Recently, Mark and Helmens (2005) have reconstructed 23 palaeo-glacier surfaces on the basis of the Early Fúquence (early LGM) morainic complex 2 using digital elevation data in a geographical system (ArcGIS^{TM}). Equilibrium-line altitudes (ELAs) have been reconstructed using the area–altitude balance ration (AABR) method. By incorporating the full-glacier hypsometry, the AABR method accounts for the area distribution of glacier mass relative to the ELA. The method allows non-standard glacier shapes to be accommodated (Furbish and Andrews, 1984), as well as non-linear vertical mass-balance gradients. Tropical glaciers show a stronger vertical mass-balance gradient than those in mid-high latitudes (Kaser, 2001), and both the topography and geomorphological evidence suggest that glaciers in the Bogotá mountains took on non-standard forms (i.e. not a standard alpine glacier form). The glaciers are best described as poorly developed valley glaciers or ice caps, rather than distinct U-shaped valley forms (Helmens, 1988). The glaciers were flat, and in several places, glaciers overtopped adjacent valley walls and spread out laterally across surrounding slopes. The average ELA for all palaeo-glaciers was reconstructed at 3488 m a.s.l. with a standard deviation of 182 m. The overall lowering in ELA from early LGM to modern of ca. 1300 m is of similar magnitude to the pollen-based reconstructed lowering in forest limit. This provides multi-proxy evidence for significant climate cooling during the LGM in the northern tropical Andes. Notwithstanding the fact that the overall average lowering in ELA at ca. 1300 m a.s.l. can only be explained by a considerable lowering in temperature, the data by Mark and Helmens (2005) do show a considerable amount of intra-regional variance in LGM ELA. This has been ascribed to topography and its indirect effect on precipitation, cloudiness and/or glacier form, with lower headwall elevations being correlated to larger accumulation areas and lower ELAs (cf. also Mark et al., 2005).

In Fig. 58.7, the temperature record of the La Laguna core and the glacial record are compared with high-resolution climate records obtained from the Greenland ice cores (Dansgaard et al., 1993) and North Atlantic Ocean sediments (Bond et al., 1993). The ice-core and marine records show several abrupt temperature shifts to markedly warm interstadial periods for the Last Glacial period, including during the LGM around 20 ka (Greenland Interstadial 2). The response of vegetation in the circum-Atlantic tropics to the large, relatively short-lived (centuries to millennia) climate shifts recorded in the North Atlantic region is further explored in Hessler et al. (2010) by comparing pollen records from both marine and terrestrial sediment cores including the La Laguna sequence near Bogotá.

58.4.3. Glacial Advance of Weichselian early Late-Glacial Age

The Bogotá mountains were deglaciated at ca. 12.5 ^{14}C ka BP. It is at this time that glaciers had retreated from the youngest morainic complex 4. Complex 4 includes a distinct system of winding morainic ridges that stand a few metres high. The moraines are only found on high mountain ridges with high orographic precipitation, where at elevations of ca. (3300) 3500–3700 m a.s.l., they generally enclose well-defined cirques.

Basal lake sediments in a cirque basin, inside morainic complex 4, have yielded a radiocarbon date of 12.76 ± 0.16 ^{14}C ka BP (Boca Grande 1 in Fig. 58.5; Helmens, 1988). A date of 14.46 ± 0.17 ^{14}C ka BP, obtained from lake sediments below till, directly behind the moraines of complex 4 (Boca Grande 2 in Fig. 58.5; Helmens, 1988), appears to correspond to the time when ice associated with morainic complex 3 was ablating at this site, having begun to disappear from a nearby site by 14.66 ± 0.28 ^{14}C ka BP (Boca Grande 3 in Fig. 58.5; Helmens, 1988). The pre-existing lacustrine sediments were apparently not eroded during the subsequent re-advance, when glacier ice just overtopped the rock threshold of the adjacent cirque, depositing morainic complex 4 close to the riegel's base. A rough maximum age for complex 4 is provided by a radiocarbon date of 13.71 ± 0.08 ^{14}C ka BP obtained from a peat lamina in lake sediments that were found underlying glaciofluvial sands and gravels in front of the complex (Boca Grande 4 in Fig. 58.5; Helmens, 1988).

The time interval during which morainic complex 4 was most probably formed correlates with the La Ciega Stadial defined in the palynological record. This stadial represents a short but intense cold spell, between 13 and 12.5 ^{14}C ka BP, with temperature depletion estimated at about 6 °C compared to present conditions. This stadial was followed by the relatively warm Guantiva Interstadial (12.5–11 ^{14}C ka BP) when temperatures rose considerably to values only 1–2 °C lower than present (van der Hammen and Gonzalez, 1965; van Geel and van der Hammen, 1973; Kuhry, 1988).

Following the widespread deglaciation that began at ca. 14 ^{14}C ka BP, mountain glaciers on different continents, as well as segments of the large continental ice sheets, are recorded to have re-advanced during the early part of the Late-Glacial period. While moraines of this episode are seldom closely dated, they seem to have formed during the time interval ca. 13–12 ^{14}C ka BP (Clapperton, 1995, 2000).

58.4.4. Limitations and Suggestions for Further Study

The radiocarbon-dated glacial and pollen records from the Bogotá region provide a coherent picture of glacier fluctuations and vegetation changes during the Late Pleistocene. The glacial record is relatively well-dated when compared to many other glacial records in South America (Clapperton, 2000). Nevertheless, more absolute dates are needed, especially when attempting a correlation with the millennium-scale Heinrich events or Dangaard/Oeschger (D/O) climate cycles of the North Atlantic region. Datable organic matter is abundant in the area, both in sediments and the tephra-rich soil cover. A key area for further study might be the central and southern parts of the Páramo de Sumapaz (Fig. 58.2), where maps by Brunnschweiler (1981), as well as 'Google Earth' images, show many well-preserved moraines. The area is, however, remote, and in addition, the special political status of the Sumapaz region makes it difficult to access.

58.5. SIERRA NEVADA DEL COCUY

The currently glaciated mountain range of the Sierra Nevada del Cocuy is ca. 300 km north-east of Bogotá (Fig. 58.1). A map of the region, covering some 1250 km^2 of glaciated and formerly glaciated terrain, which shows the ice cover in the late 1950s and a sequence of six drift bodies with accompanying moraines, was constructed by van der Hammen et al. (1980/1981). This map was based on earlier reconnaissance by Gonzalez et al. (1966). Figure 58.8 gives a simplified version of part of the map by van der Hammen et al. (1980/1981), which covers the extensively glaciated western slopes of the Cocuy range and which includes the type areas for the drifts that they numbered 3–6 (old to young, respectively). The type area for the still older drift 2 is situated along the steep and sparsely glaciated eastern slopes of the Sierra Nevada del Cocuy mountains (Fig. 58.9). The glacial origin of landforms and deposits of drift 1 is uncertain. Recently, Helmens et al. (1997b) presented a detailed map of the morainic sequence along the Ríos San Pablín and Cóncavo (Figs. 58.8 and 58.10) and have defined the still unnamed moraines in the upper parts of drifts 3 and 5. Here, new detailed maps are presented of the morainic sequence of the youngest drifts 5 and 6, in their type areas along the Río Bocatoma and in the vicinity of the Laguna de la Sierra (Figs. 58.8, 58.11 and 58.12). They allow a further subdivision of the moraines in the lower part of drift 5. The maps in Figs. 58.10–58.12, which are based on aerial photographs taken in the early 1980s, additionally distinguish an extensive zone of glacially scoured bedrock adjacent to the retreating ice front.

Only limited dating control is available for the moraines of the Sierra Nevada del Cocuy. Extrapolation of a radiocarbon date from sediments cored in Laguna Ciega (Laguna Ciega III in Fig. 58.8; van der Hammen et al., 1980/1981) yields a possible age of 24.5–27.0 ^{14}C ka BP for the beginning of lake sedimentation on the lower part of drift 3 following glacier retreat from the Cóncavo moraines, renamed the Lower Cóncavo moraines by Helmens et al. (1997b). No dates are available for the still older Río Negro moraines of drift 2, which like the Lower Cóncavo moraines, are only found very locally. Although van der Hammen et al. (1980/1981) use these data combined with those from the Bogotá area, including palynological evidence, to suggest extensive glaciation during the Middle Weichselian, additional dating control is required to correlate the older moraines in the Cocuy mountains with the Middle Weichselian moraines and older glacial deposits in the mountains near Bogotá.

Glacier retreat from the Upper Lagunillas moraines in the upper part of drift 4 took place shortly before ca. 12.5 ka. The radiocarbon date of 12.32 ± 0.1 ^{14}C ka BP was obtained from a thin peat horizon overlying basal, laminated minerogenic lake sediments at a site directly behind the moraines (Lagunillas V in Figs. 58.8 and 58.11; Gonzalez et al., 1966). The dates of 24.5–27.0 and 12.5 ^{14}C ka BP obtained at the Laguna Ciega III and Lagunillas V sites seem to place both the Lower Lagunillas moraines in the lower part of drift 4 (van der Hammen et al., 1980/1981) and the Upper Cóncavo moraines in the upper part of drift 3 (Helmens et al., 1997b) at the time of the LGM. Helmens et al. (1997a) have noted the striking similarity in morphology displayed by the moraines in the Cocuy and Bogotá mountains. The Lower Lagunillas moraines, which rise to 100 m above the valley floors (Fig. 58.10), and the late-LGM morainic complex 3 of the Bogotá mountains show a most impressive morainic morphology. The arcuate, multiple ridge systems are enclosed by the Upper Cóncavo moraines and the early-LGM morainic complex 2, which are distinctly more subdued and are incomplete. Like the Late-Glacial morainic complex 4, the Upper Lagunillas moraines include winding ridges of much lesser height. Along the Río San Pablín, on the western slopes of the Sierra Nevada del Cocuy, the Upper and Lower Lagunillas moraines reach downvalley to elevations of ca. 3600 and ca.

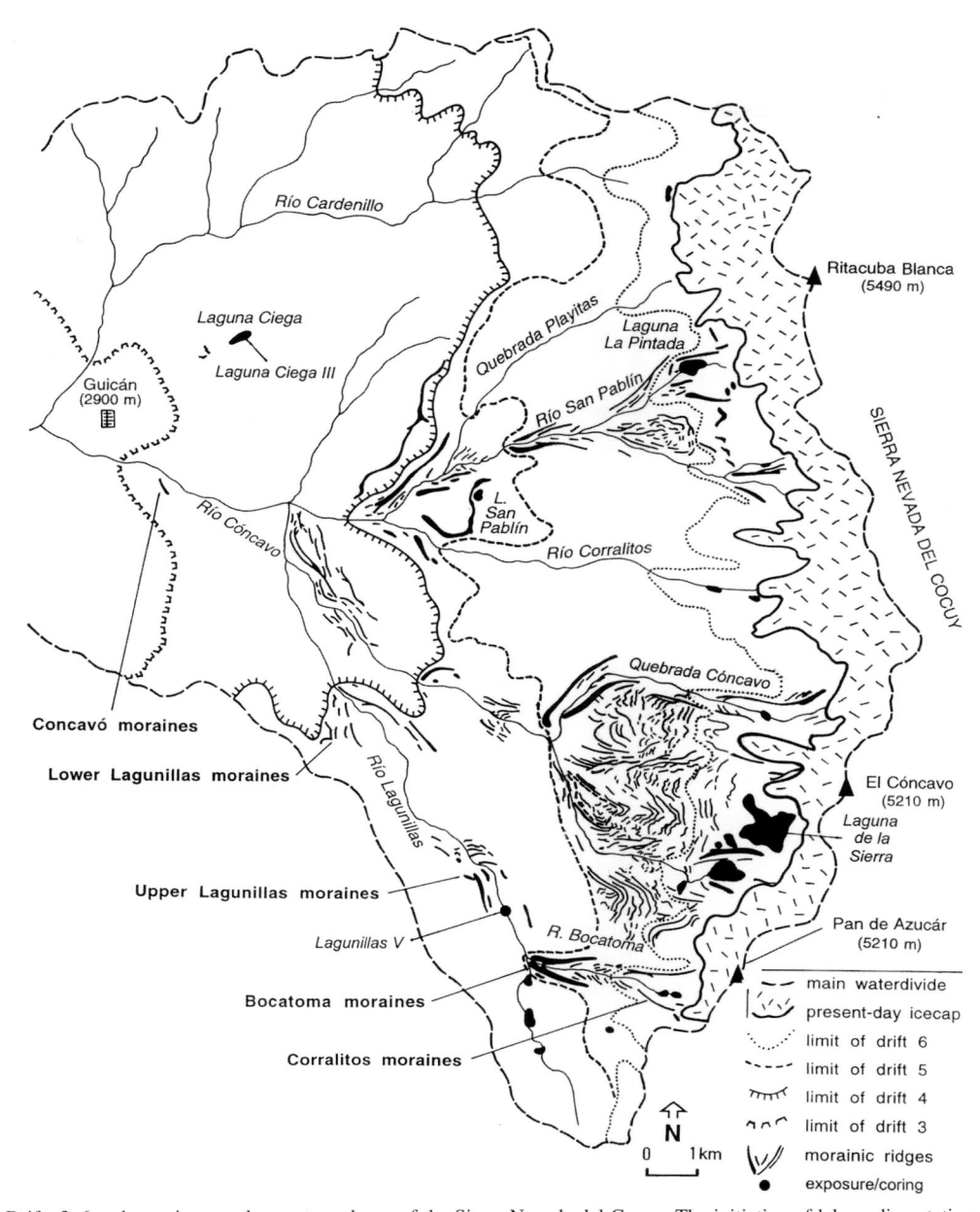

FIGURE 58.8 Drifts 3–6 and moraines on the western slopes of the Sierra Nevada del Cocuy. The initiation of lake sedimentation on the lower part of drift 3 and on the upper part of drift 4 has been dated at ca. 24.5–27 and 12.5 ^{14}C ka BP, respectively (based on van der Hammen et al., 1980/1981). The positions of exposures/boreholes mentioned in the text are indicated.

3400 m a.s.l. and the Upper and Lower Cóncavo moraines to ca. 3300 and ca. 2900 m a.s.l., respectively. The Río Negro moraines on the eastern slopes of the Cocuy range occur at elevations of ca. (2200) 2600–2800 m a.s.l.

58.5.1. Younger Dryas Glacial Advance

A glacial event corresponding to the YD Stadial climatic oscillation of the northern North Atlantic region, between ca. 11 and 10 ka, has been proposed by Gonzalez et al. (1966) and van der Hammen et al. (1980/1981) based on the Cocuy record. The event is marked by the Bocatoma moraines in the lower part of drift 5, renamed the Lower Bocatoma moraines by Helmens et al. (1997b). The massive, lobate morainic complex (photograph in Fig. 58.10) displays ridges which are distinctly sharper and fresher looking than the morainic crests of drift 4. The moraines reach down on the western slopes of the Cocuy range to elevations of ca. 3800–4000 m a.s.l. They indicate a maximum lowering of the glacier fronts by some 700 m compared to present conditions.

Detailed mapping of the Lower Bocatoma moraines suggests that they represent two glacial advances, in which

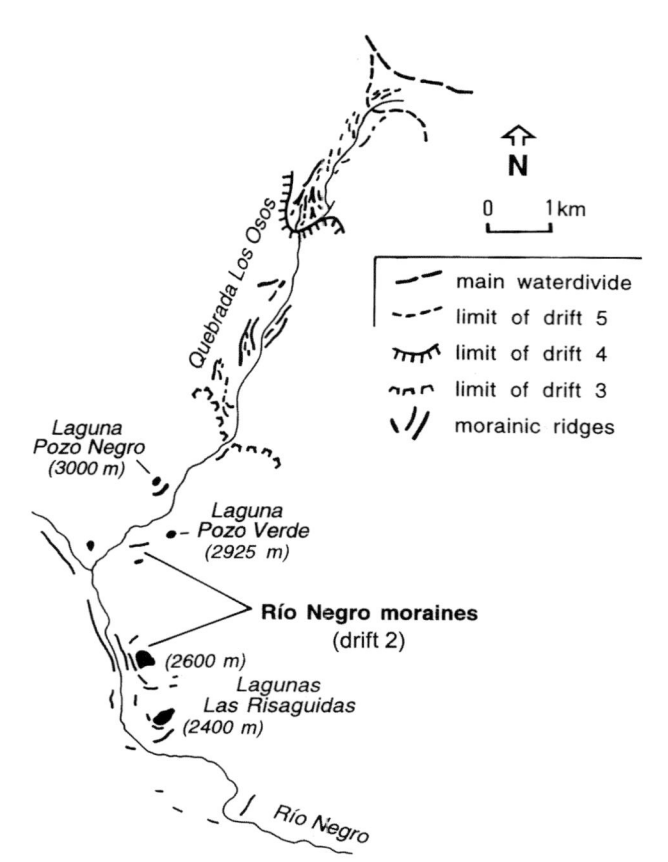

FIGURE 58.9 Drifts 2–4 and moraines on the eastern slopes of the Sierra Nevada del Cocuy (van der Hammen et al., 1980/1981).

the moraines of the younger advance in part have buried the moraines formed by the older advance (indicated as, respectively, outer to outermost moraines in Figs. 58.10–58.12) and a minor re-advance or distinct still-stand that left an inner moraine or lobe of moraines. As Gonzalez et al. (1966) mentioned, the individual morainic ridges that make up the Lower Bocatoma moraines do not show a noticeable difference in intensity of weathering, suggesting only a slight age difference between their periods of formation.

A YD age for the Lower Bocatoma moraines is suggested by a radiocarbon-dated sediment sequence deposited in front of the moraines and a pollen record obtained from basal lake sediments accumulated directly behind the (outer) moraines (Lagunillas V and XI, respectively, in Fig. 58.11; Gonzalez et al., 1966). The Lagunillas V section is exposed in a major body of valley-floor sediments enclosed by the Upper Lagunillas moraines in the upper part of drift 4. Ice should have retreated from the Upper Lagunillas moraines and sandy, laminated lake sedimentation had begun at the section-site shortly before 12.32 ± 0.1 ^{14}C ka BP. Around the level of that date, the sediments in the section become more clayey with intercalations of peat. After ca. 11.9 ^{14}C ka BP, peat was deposited. Sandy intercalations in this peat layer have been dated between 11.35 ± 0.14 and 10.03 ± 0.09 ^{14}C ka BP and might indicate that a glacier occurred nearby, behind the Lower Bocatoma moraines. Palynological evidence from the Lagunillas XI section suggests glacial retreat from the outer morainic ridges of the Lower Bocatoma moraines and the start here of sandy/clayey, laminated lake sedimentation near the transition from the Late-Glacial to the Holocene.

An attempt to provide the Lower Bocatoma moraines with additional chronological control has resulted in a rough minimum age for the corresponding glacial event. The base of a peat deposit found overlying sand directly behind the inner ridges of the Lower Bocatoma moraines on the slopes below the Laguna de la Sierra yielded a radiocarbon date of 8.54 ± 0.26 ^{14}C ka BP (Cóncavo II in Fig. 58.12; Helmens, 2004). The sample dates the start of formation of the slope bog at the site (currently dominated by Espeletia–*Carex*–*Sphagnum* vegetation). Since there is a time lag before a slope bog can develop on deglaciated terrain, the date of ca. 8.5 ^{14}C ka BP does not necessarily provide a precise timing for glacier retreat from the moraines. The date, however, suggests that ice retreated from the Lower Bocatoma moraines shortly after 10 ^{14}C ka BP, because during the time interval ca. 10–8.5 ^{14}C ka BP, mostly peat accumulated at the Lagunillas V site.

A cooling event in the YD Chronozone, the El Abra Stadial (van Geel and Van der Hammen, 1973), is distinguished in the palynological record. The El Abra Stadial has been recognised in many pollen records of the Eastern Cordillera and has been dated to the time interval ca. 11 to 10–9.5 ^{14}C ka BP (Kuhry et al., 1993). New evidence obtained by van't Veer et al. (2000) also indicates the lowest temperatures between ca. 11 and 10.5 ^{14}C ka BP; cool and dry conditions are interpreted after ca. 10.5 ^{14}C ka BP, extending into the earliest Holocene. The maximum temperature decline during the El Abra Stadial, compared to late Holocene conditions, is estimated in the order of 3–4 °C (Kuhry et al., 1993; van't Veer et al., 2000). Although the YD-age assignment for the Lower Bocatoma moraines requires supporting radiocarbon dates, the available evidence on the glacial sequence of the Cocuy mountains, in combination with that provided by the palynological record of the Eastern Cordillera, strongly suggests that the moraines were formed at the time of the YD cooling event.

Strong evidence for glacier expansion in the tropical Andes during the YD Stadial was presented by Clapperton et al. (1997). At the Papallacta Pass in the Eastern Cordillera of Ecuador, multiple radiocarbon dates from macrofossils, peat and gyttja above and below till indicate that a ca. 140 km^2 ice cap developed between ca. 11 and 10 ^{14}C ka BP on ground with a mean elevation of 4200 m where none exists now. Palynological evidence from the organic beds indicates cold and humid conditions just prior to this so-called Potrerillos glacial advance, and a gradual warming to temperatures slightly higher than modern values following the retreat. Radiocarbon dating evidence from the same area

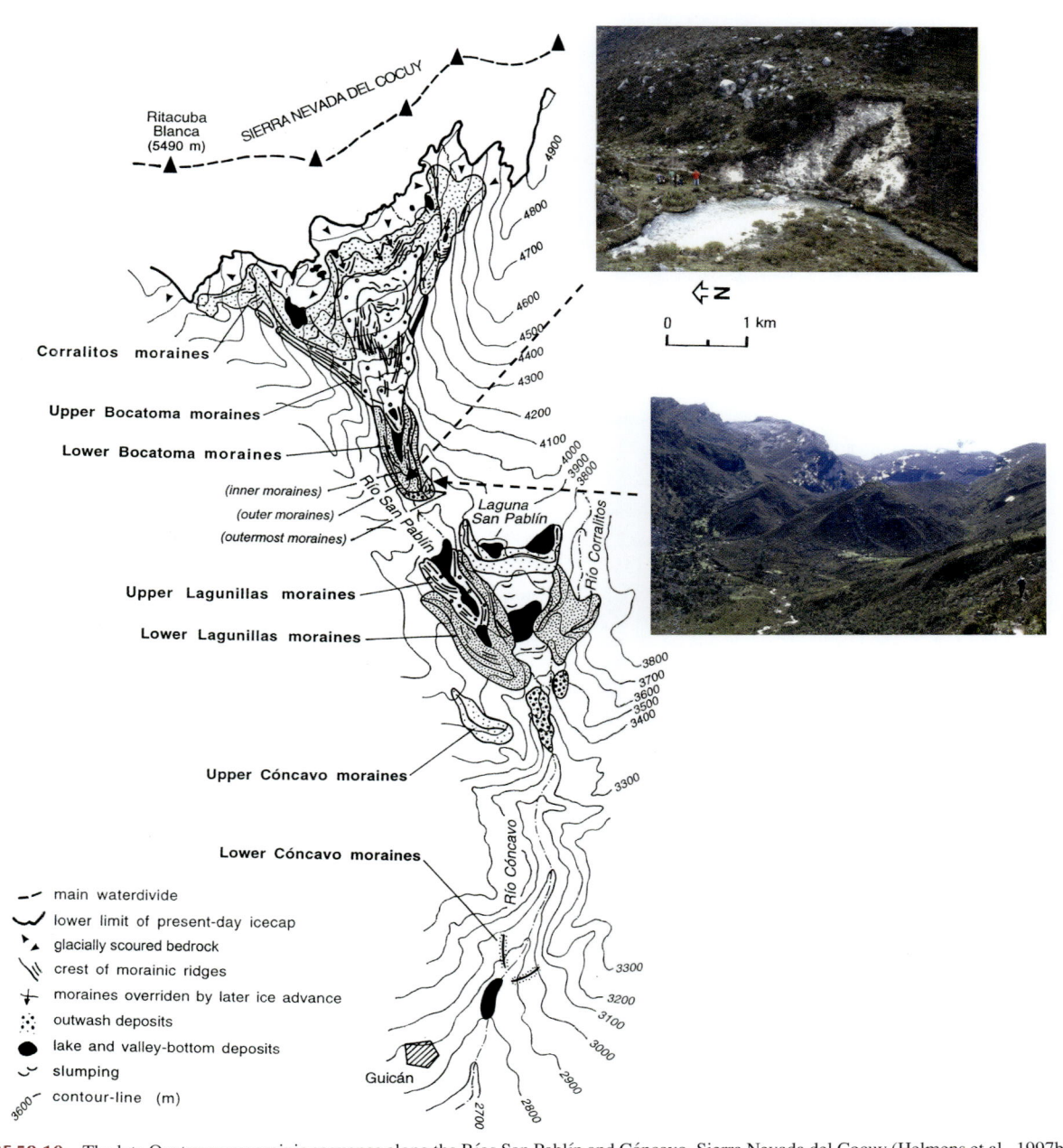

FIGURE 58.10 The late Quaternary morainic sequence along the Ríos San Pablín and Cóncavo, Sierra Nevada del Cocuy (Helmens et al., 1997b, modified from van der Hammen et al., 1980/1981; Fig. 58.8). The subdivision of the Lower Bocatoma moraines (into outermost, outer and inner moraines) is based on the present chapter. The photographs show the Lower Bocatoma moraines and minerogenic lake sediments enclosed by the moraines.

has been used by Heine and Heine (1996), Heine and Geyh (2002) and Heine (2004) to place the glacier expansion during the late YD/early Holocene. The composite section shown in Heine (2004) of the Papallacta Pass glacial sequence, however, shows that the radiocarbon dates at ca. 8 ^{14}C ka BP, that were used to propose an early Holocene age, are from a stratigraphical unit higher in the sequence than the peat bed dated by Clapperton et al. (1997) at ca. 10 ^{14}C ka BP, that directly overlies the Potrerillos Till.

It seems therefore that the numerical ages obtained by Clapperton et al. (1997) more tightly bracken the glacier advance, placing it during the YD Chronozone.

58.5.2. Holocene Glacier Fluctuations

Numerous morainic ridges in the Cocuy mountains indicate that glaciers were more extensive during earlier parts of the Holocene than at present. The Corralitos moraines, which

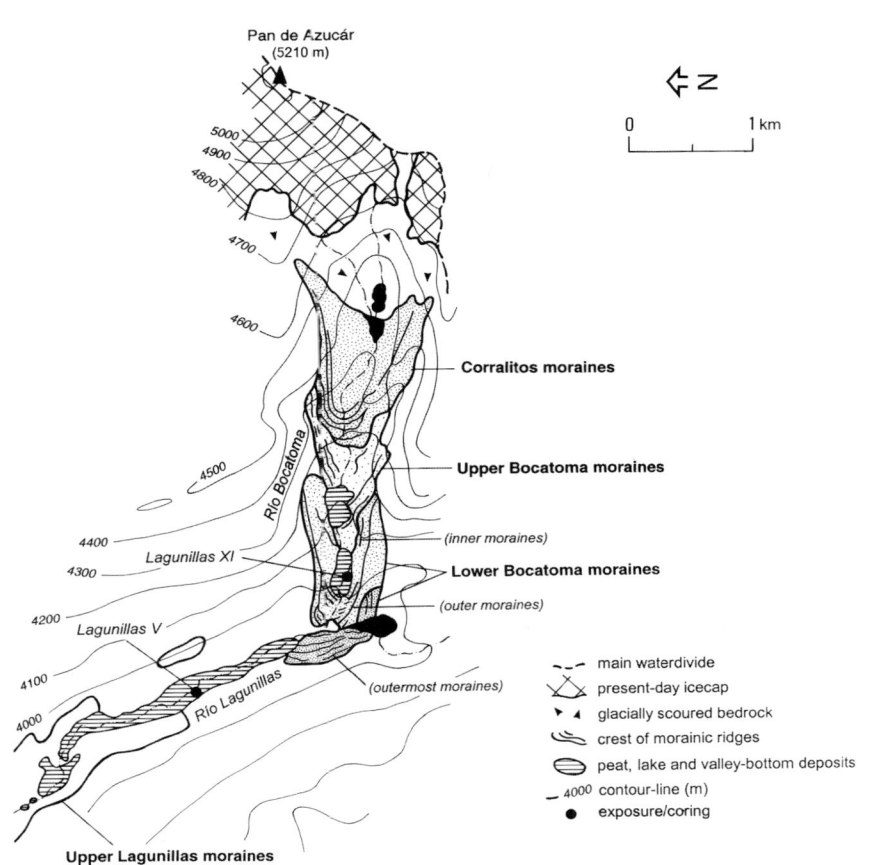

FIGURE 58.11 The Late-Glacial–Holocene morainic sequence along the Río Bocatoma, Sierra Nevada del Cocuy (modified from van der Hammen et al., 1980/1981; Fig. 58.8). The Lower Bocatoma moraines are probably of Younger Dryas Chronozone age (11–10 ^{14}C ka BP), whereas the Corralitos moraines were formed during the Little Ice Age. The sites of exposures/boreholes mentioned in the text are indicated.

comprise the youngest drift 6, most probably formed during the later part of the past millennium. The lower limit of drift 6 is found on the western, ice-covered slopes of the Cocuy range between ca. 4200 and 4500 m altitude. During maximum advance, glaciers extended some 200–400 m lower than at present.

The Corralitos moraines are very conspicuous and complete, and are vegetation free. Radiocarbon dates obtained for similar moraines in other parts of the Colombian Andes (Herd, 1982; van der Hammen, 1984) place their time of formation within the past ca. 500 years during the Little Ice Age (van der Hammen et al., 1980/1981). Glaciers may still have been near the outer ridges of the Corralitos moraines as late as ca. AD 1850. At this time, Ancízar on a pilgrimage through the region reported 4150 m as the lowest elevation for a glacier terminus in the Cocuy mountains (cited in Hoyos-Patiño, 1998). Glacial retreat during the latter part of the past century has been significant. For example, a photograph taken by E. Kraus in 1938 shows glacier ice in contact with Laguna de la Sierra in the uppermost part of drift 6 (van der Hammen et al., 1980/1981).

Aerial photographs taken in 1981 (Fig. 58.12) indicate that the ice had disappeared from the lake and that glacier margins have retreated considerably to elevations of ca. 100–150 m above the lake. During this recent retreat, glaciers in the Cocuy range have exposed an extensive zone of glacially scoured bedrock which is covered by only very little till.

On the upper part of drift 5, aerial photographs very clearly depict still another series of moraines, which have been defined the Upper Bocatoma moraines by Helmens et al. (1997a). The moraines are similar in appearance, but much smaller in size than those of the Late-Glacial Lower Bocatoma moraines in the lower part of drift 5. No radiocarbon dates are available to constrain the timing of formation of the Upper Bocatoma moraines. They might have formed in the early Holocene from short still-stands during glacier retreat following the YD cooling event.

The palynological record of the Eastern Cordillera shows slightly elevated temperatures between ca. 7 and 3 ^{14}C ka BP (Kuhry, 1988). No distinct short episodes of cooling are represented.

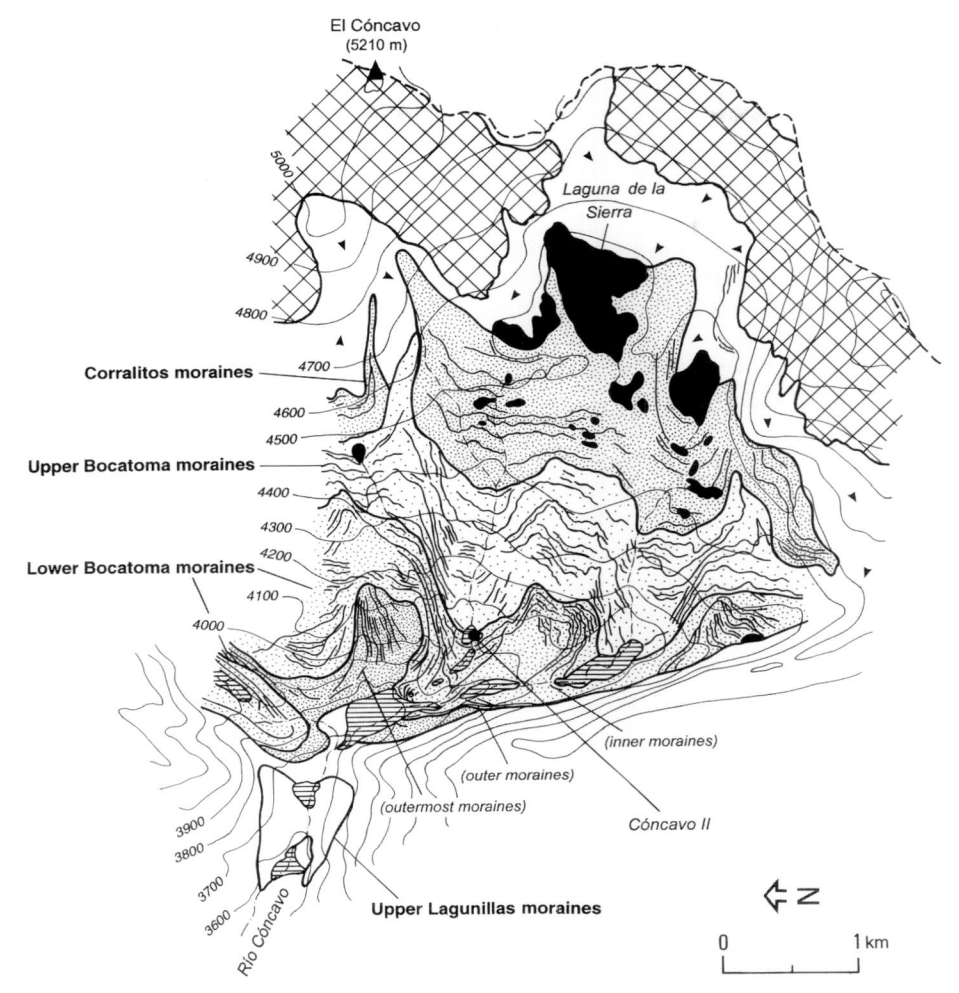

FIGURE 58.12 The Late-Glacial–Holocene morainic sequence in the vicinity of the Laguna de la Sierra, Sierra Nevada del Cocuy (modified from van der Hammen et al., 1980/1981; Fig. 58.8). The Lower Bocatoma moraines are probably of Younger Dryas Chronozone age (11–10 ^{14}C ka BP), whereas the Corralitos moraines were formed during the Little Ice Age. The sites of exposures/boreholes mentioned in the text are indicated. For legend, see Fig. 58.11.

58.5.3. Limitations and Suggestions for Further Study

The well-preserved and abundant moraines of the Sierra Nevada del Cocuy have been mapped in great detail. However, absolute dates are scarce and the morainic sequence is in need of detailed chronology. In 1994, the author visited the area, together with R.W. Rutter and P. Kuhry, in order to collect organic materials to date the Late-Glacial moraines. Access to field sites, however, was restricted to the main valleys of the Ríos Lagunillas, Cóncavo and San Pablín (Fig. 58.8), because of local warfare. Thick sequences of laminated lake sediments were found exposed at several localities directly behind moraines (photograph in Fig. 58.10), but the organic content of the sediments was too low to provide reliable radiocarbon dates. In addition, palaeosol sequences here proved to be thin and fragmented on the steep slopes of the impressive morainic ridges (Fig. 58.10). The author's original target had been minor tributary valleys, in size more or less comparable as in the Bogotá mountains, where, on smaller moraines, soils could be expected to be thicker and more differentiated and organic matter might have started to accumulate behind moraines before lakes were drained. The Cocuy range also holds an impressive Holocene moraine sequence in need of absolute dating and is additionally of great interest to study present glacier-front retreat, since over 50% of total glacier area has been lost during the past 50 years (Morris et al., 2006).

58.6. COMPARISON WITH THE GLACIAL RECORD OF THE CENTRAL CORDILLERA

A detailed study of the morainic sequence on the Ruíz–Tolima volcanic complex in the Central Cordillera (Fig. 58.1; ca. latitude 5°N; highest peak, Nevado del Ruíz at 5400 m a.s.l.) has been made by Thouret et al. (1996). An absolute chronology for a series of morainic belts is provided by radiocarbon dating and inter-site correlation using tephra/palaeosol/peat stratigraphies.

Although the late Río Recío moraines have not been adequately dated, stratigraphical evidence appears to indicate that the moraines here were formed before 28 and after 42 ^{14}C ka BP. This information suggests that glaciers were more extensive during the Middle, rather than during the Late Weichselian, as in the Eastern Cordillera. Peat directly overlying the still older early Río Recío moraines has yielded radiocarbon dates of ca. 49 to up to 53 ^{14}C ka BP. Since no evidence of an interglacial soil has been discovered in the locally more than 10-m-thick sediment-soil sequence on the early Río Recío moraines, it is assumed that the relatively fresh moraines were formed during the early part of the Wisconsinan period. The Río Recío moraines occur at elevations of ca. (2900) 3200–3300 m a.s.l.

The Murillo moraines, which represent two phases of glacier expansion down to ca. 3300–3400 and to ca. 3400–3600 m a.s.l., have been dated as older than 16 ^{14}C ka BP and probably younger than 28 ^{14}C ka BP. Using palynological evidence, the glacier advances are correlated with stadial intervals between ca. 28 and 21 ^{14}C ka BP (early Murillo moraines) and ca. 21 and 14 ^{14}C ka BP (late Murillo moraines). An independent radiocarbon chronology is required for the early Murillo moraines in order to support the correlation proposed by Thouret et al. (1996) or, as an alternative, to propose a correlation of the early and late Murillo moraines with the Early (21–19.5 ^{14}C ka BP) and Late (17–14 ^{14}C ka BP) Fúquene Stadials, respectively, of the Eastern Cordillera (Kuhry, 1988; Helmens et al., 1996). The early Murillo moraines form the most voluminous glacial deposits in the Ruíz–Tolima massif.

Peat near the base of the up to 4-m-thick sediment-soil sequence overlying the late Otún moraines (at ca. 3800–4000 m altitude) has yielded a radiocarbon date of ca. 8.8 ^{14}C ka BP. The overlying stratigraphical sequence lacks a widespread tephra, dated between ca. 11.5 and 10.8 ^{14}C ka BP which occurs in the sequence on the early Otún moraines (3600–3800 m) of early Late-Glacial age. The late Otún moraines have been correlated by Thouret et al. (1996) with the YD Stadial of the northern North Atlantic region.

An early Holocene age for the Santa Isabel moraines (4150–4300 m a.s.l.) has been interpreted on the basis of a radiocarbon date of ca. 7.5 ka obtained from a palaeosol which forms the oldest stratigraphical unit overlying the moraines. The youngest moraines on the Ruíz–Tolima volcanic complex occur at ca. 4300–4600 m altitude; they only bear a few centimetres thick entisol (inner Ruíz moraines), with a thin tephra bed at the base (outer Ruíz moraines). Peat found at the base of the outermost ridge of these unweathered and vegetation-free moraines has yielded a date of ca. 500 ^{14}C ka BP (Herd, 1982).

58.7. SUMMARY

The onset of glaciations in the Colombian Andes, as recorded by the start of glaciofluvial sedimentation in the inter-montane Bogotá basin, is dated near the Gauss/Matuyama polarity reversal at 2.6 Ma, with episodes of increased glacial activity occurred since ca. 0.8 Ma. Moraines and till beds in the higher parts of the Andes record a series of glacier fluctuations of late Quaternary age. Radiocarbon dates of organic-rich sediments and palaeosols found associated with glacial landforms/deposits in mountain ranges exceeding 3600 m altitude in the Eastern Cordillera, in combination with evidence provided by the radiocarbon-dated palaeosol sequence in the region, place glacial events between probably 43 and 38, 36–31, 23.5–19.5, 18.0–15.5, 13.5–12.5 ka, and most probably 11–10 ka (all ages are ^{14}C ka BP). Independent chronologies for the glacial and palynological records of the Eastern Cordillera suggest a close match between the stadials characterised by low upper Andean forest limits and glacier advances in the surrounding high-mountain ranges. Major glacier advances during the Middle Weichselian (middle Wisconsinan) seem to have responded to cool and humid climatic conditions. The Late Weichselian global LGM is recorded as a twofold glaciation maximum just before 19.5 and 15.5 ^{14}C ka BP, with glaciers advancing some 1200–1100 m below their present limits; during the cooling events, the forest limit was depressed by 1100–900 m, implying a decline in mean annual temperature of ca. 8–6 °C, respectively. Interstadial conditions prevailed around 18 ka, when temperatures rose considerably to values up to 4 °C higher than during the preceding and following stadial periods. Mountain ranges below 4000 m altitude were deglaciated at ca. 12.5 ka following a Late-Glacial advance of cirque glaciers. YD-age cooling is well registered in the palynological record; glaciers in the highest parts of the Andes seem to have responded to the cooling by extending to elevations some 700 m lower than at present. The glacial record registers a high climatic variability in the northern Tropical Andes during the late Quaternary period.

ACKNOWLEDGMENTS

Research by the author on the glaciations of the Eastern Cordillera of Colombia has been supported by grants of the Dutch Foundation for the Advancement of Tropical Research (WOTRO; doss. nr. W77-103), the Natural Sciences and Engineering Research Council of Canada ('Canada International Fellowship') and the National Geographic Society, USA. (grant 4871-92). Thanks are due to the following persons (placed in alphabetical order) who have contributed in an important way to the research over the years: Dr. P. A. M. Andriessen (Free University, Amsterdam, The Netherlands), Dr. R. W. Barendregt

(University of Lethbridge, Canada), Dr. R. J. Enkin (Geological Survey of Canada, Pacific Geoscience Centre Subdivision), Dr. P. Kuhry (Stockholm University, Sweden), Dr. W. G. Mook (University of Groningen, The Netherlands), Dr. N. W. Rutter (University of Alberta, Edmonton, Canada), Dr. T. van der Hammen has died and Dr. K. van der Borg (University of Utrecht, The Netherlands).

REFERENCES

Andriessen, P.A.M., Helmens, K.F., Hooghiemstra, H., Riezebos, P.A., Van der Hammen, T., 1993. Absolute chronology of the Pliocene-Quaternary sediment sequence of the Bogotá area, Colombia. Quatern. Sci. Rev. 12, 483–501.

Bakker, J.G.M., 1990. Tectonic and climatic controls on Late Quaternary sedimentary processes in a neotectonic intramontane basin (The Pitalito basin, South America). Unpublished Ph.D dissertation, Agricultural University of Wageningen.

Bond, G., Broecker, W.S., Johnsen, S.J., McManus, J., Labeyrie, L., Jouzel, J., et al., 1993. Correlations between climate records from North Atlantic sediments and Greenland ice. Nature 365, 143–147.

Brunnschweiler, D., 1981. Glacial and periglacial form systems of the Colombian Quaternary. Rev. CIAF 6 (1–3), 53–56.

Cande, S.C., Kent, D.V., 1995. Revised calibration of the geomagnetic polarity timescale for the Late Cretaceous and Cenozoic. J. Geophys. Res. 100, 6093–6095.

Clapperton, C.M., 1987. Maximal extent of Late Wisconsin glaciation in the Ecuadorian Andes. Quatern. S. Am. Antarct. Peninsula 5, 165–179.

Clapperton, C.M., 1995. Fluctuations of local glaciers at the termination of the Pleistocene: 18–8 ka BP. Quatern. Int. 28, 41–50.

Clapperton, C.M., 2000. Interhemispheric synchroneity of Marine Oxygen Isotope Stage 2 glacier fluctuations along the American cordilleras transect. J. Quatern. Sci. 15, 435–468.

Clapperton, C.M., Hall, M., Mothes, P., Hole, M.J., Helmens, K.F., Kuhry, P., et al., 1997. A Younger Dryas Icecap in the Equatorial Andes. Quatern. Res. 47, 13–28.

Dansgaard, W., Johnsen, S.J., Clausen, H.B., Dahl-Jensen, D., Gundestrup, N.S., Hammer, C.U., et al., 1993. Evidence for general instability of past climate from a 250-kyr ice-core record. Nature 364, 218–220.

Fölster, H., Hetsch, W., 1978. Paleosol sequences in the Eastern Cordillera of Colombia. Quatern. Res. 9, 238–248.

Furbish, D.J., Andrews, J.T., 1984. The use of hypsometry to indicate long-term stability and response of valley glaciers to changes in mass transfer. J. Glaciol. 30, 199–211.

Gibbard, P., Head, M.J., Walker, M.J.C., The Subcommission on Quaternary Stratigraphy, 2010. Formal ratification of the Quaternary System/Period and the Pleistocene Series/Epoch with a base at 2.58 Ma. J. Quatern. Sci. 25, 96–102.

Gillespie, A., Molnar, P., 1995. Asynchronuous maximum advances of mountain and continental glaciers. Rev. Geophys. 33 (3), 311–364.

Gonzalez, E., van der Hammen, T., Flint, R.F., 1966. Late Quaternary glacial and vegetational sequence in Valle de Lagunillas, Sierra Nevada del Cocuy, Colombia. Leidse Geol. Meded. 32, 157–182.

Grabandt, R.A.J., 1980. Pollen rain in relation to arboreal vegetation in the Colombian Cordillera Oriental. Rev. Palaeobot. Palynol. 29, 65–147.

Grabandt, R.A.J., 1985. Pollen rain in relation to vegetation in the Colombian Cordillera Oriental. Unpublished Ph.D dissertation, University of Amsterdam.

Guillet, B., Faivre, P., Mariotti, A., Khobzi, J., 1988. The ^{14}C dates and $^{13}C/^{12}C$ ratios of soil organic matter as a means of studying the past vegetation in intertropical regions: examples from Colombia (South America). Palaeogeogr. Palaeoclimatol. Palaeoecol. 65, 51–58.

Heine, K., 2004. Late Quaternary glaciations of Ecuador. In: Ehlers, J., Gibbard, P.L. (Eds.), Quaternary Glaciations—Extent and Chronology, Part III: South America, Asia, Africa, Australasia, Antarctica. Elsevier, Amsterdam, pp. 165–169.

Heine, K., Geyh, M.A., 2002. Neue 14C-Daten zur Jüngeren Dryaszeit in den ecuadorianischen Anden. Eiszeit. Ggw 51, 33–50.

Heine, K., Heine, J.T., 1996. Late Glacial climatic fluctuations in Ecuador: glacier retreat during Younger Dryas time. Arct. Alpine Res. 28, 496–501.

Helmens, K.F., 1988. Late Pleistocene glacial sequence in the area of the high plain of Bogotá (Eastern Cordillera, Colombia). Palaeogeogr. Palaeoclimatol. Palaeoecol. 67, 263–283.

Helmens, K.F., 1990. Neogene-Quaternary Geology of the High Plain of Bogotá, Eastern Cordillera, Colombia (Stratigraphy, Paleoenvironments and Landscape Evolution). Dissertationes Botanicae, vol. 163. J. Cramer, Berlin-Stuttgart.

Helmens, K.F., 2004. The Quaternary glacial record of the Colombian Andes. In: Ehlers, J., Gibbard, P.L. (Eds.), Quaternary Glaciations–Extent and Chronology, Part III: South America, Asia, Africa, Australasia, Antarctica. Elsevier, Amsterdam, pp. 115–134.

Helmens, K.F., Kuhry, P., 1986. Middle and Late Quaternary vegetational and climatic history of the Páramo de Agua Blanca (Eastern Cordillera, Colombia). Palaeogeogr. Palaeoclimatol. Palaeoecol. 56, 291–335.

Helmens, K.F., Kuhry, P., 1995. Glacier fluctuations and vegetation change associated with Late Quaternary climatic oscillations in the Andes near Bogotá, Colombia. Quatern. S. Am. Antarct. Peninsula 9, 117–140.

Helmens, K.F., van der Hammen, T., 1994. The Pliocene and Quaternary of the high plain of Bogotá (Colombia): a history of tectonic uplift, basin development and climatic change. Quatern. Int. 21, 41–61.

Helmens, K.F., Kuhry, P., Rutter, N.W., Van der Borg, K., De Jong, A.F.M., 1996. Warming at 18,000 yr B.P. in the Tropical Andes. Quatern. Res. 45, 289–299.

Helmens, K.F., Barendregt, R.W., Enkin, R.J., Bakker, J., Andriessen, P.A.M., 1997a. Magnetic polarity and fission-track chronology of a Late Pliocene-Pleistocene paleoclimatic proxy record in the Tropical Andes. Quatern. Res. 48, 15–28.

Helmens, K.F., Rutter, N.W., Kuhry, P., 1997b. Glacier fluctuations in the Eastern Andes of Colombia (South-America) during the last 45,000 radiocarbon years. Quatern. Int. 38 (39), 39–48.

Herd, D.G., 1982. Glacial and volcanic geology of the Ruiz-Tolima volcanic complex Cordillera Central, Colombia. Publicaciones geológicas especiales del INGEOMINAS, 8 (Bogotá, Colombia), pp. 1–48.

Hessler, I., Dupont, L., Bonnefille, R., Behling, H., González, C., Helmens, K.F., et al., 2010. Millennial-scale changes in vegetation records from tropical Africa and South America during the last glacial. Quatern. Sci. Rev. 29, 2882–2899. doi:10.1016/j.quascirev.2009.11.029.

Hooghiemstra, H., 1984. Vegetational and climatic history of the high plain of Bogotá, Colombia: a continuous record of the last 3.5 million years. Dissertationes Botanicae, vol. 79. J. Cramer, Vaduz. 368pp.

Hooghiemstra, H., Ran, E.T.H., 1994. Late Pliocene-Pleistocene high resolution pollen sequence of Colombia: an overview of climatic change. Quatern. Int. 21, 63–80.

Hooghiemstra, H., Melice, J.L., Berger, A., Shackleton, N.J., 1993. Frequency spectra and paleoclimatic variability of the high-resplution 30–1450 ka Funza I pollen record (Eastern Cordillera, Colombia). Quatern. Sci. Rev. 12, 141–156.

Hoyos-Patiño, F., 1998. Glaciers of South-America—Glaciers of Colombia. In: Williams Jr., R.S., Ferrigno, J.G. (Eds.), Satellite Image Atlas of Glaciers of the World, U.S. Geological Survey Professional Paper 1386-I, pp. I11–I30.

Kaser, G., 2001. Glacier-climate interaction at low latitudes. J. Glaciol. 47, 195–204.

Kuhry, P., 1988. Palaeobotanical-palaeoecological studies of tropical high Andean peatbog sections (Cordillera Oriental, Colombia). Dissertationes Botanicae, vol. 116. J. Cramer, Berlin-Stuttgart.

Kuhry, P., 1991. Comparative paleohydrology in the Andes of Colombia. In: Abstracts of the XIII INQUA Congress, Beijing, China, p. 179.

Kuhry, P., Helmens, K.F., 1990. Neogene-Quaternary biostratigraphy and paleoenvironments. In: Helmens, K.F. (Ed.), Neogene-Quaternary Geology of the High Plain of Bogotá, Eastern Cordillera, Colombia (Stratigraphy, Paleoenvironments and Landscape Evolution). Dissertationes Botanicae, vol. 163 J. Cramer, Berlin-Stuttgart, pp. 89–132.

Kuhry, P., Hooghiemstra, H., Van Geel, B., Van der Hammen, T., 1993. The El Abra stadial in the Eastern Cordillera of Colombia (South America). Quatern. Sci. Rev. 12, 333–343.

Lisiecki, L.E., Raymo, M.E., 2005. A Pliocene-Pleistocene stack of 57 globally distributed benthic $\delta^{18}O$ records. Paleoceanography 20, PA1003.

Mark, B.G., Helmens, K.F., 2005. Reconstruction of glacier equilibrium-line altitudes for the Last Glacial Maximum on the High Plain of Bogotá, Eastern Cordillera, Colombia: climatic and topographic implications. J. Quatern. Sci. 20 (7–8), 789–800.

Mark, B.G., Harrison, S.P., Spessa, A., New, M., Evans, D.J.A., Helmens, K.F., 2005. Tropical snowline changes at the last glacial maximum: a global assessment. Quatern. Int. 138–139, 168–201.

Morris, J.N., Poole, A.J., Klein, A.G., 2006. Retreat of tropical glaciers in Colombia and Venezuela from 1984 to 2004 as measured from ASTER and Landsat images. In: 63rd Eastern Snow Conference. Newark, Delaware U.S.A. 2006, pp. 181–191.

Pillans, B., Naish, T., 2004. Defining the Quaternary. Quatern. Sci. Rev. 23, 2271–2282.

Raasveldt, H.C., 1957. Las glaciaciones de la Sierra Nevada de Santa Marta. Revista de la Academia Colombiana de Ciencias Exactas, Físicas y Naturales 9 (38), 469–482.

Ruddiman, W.F., Raymo, M.E., McIntyre, A., 1986. Matuyama 41,000-year cycles: North Atlantic Ocean and Northern Hemisphere ice sheets. Earth Planet. Sci. Lett. 80, 117–129.

Thouret, J.-C., van der Hammen, T., Salomons, B., 1996. Paleoenvironmental changes and stades of the last 50,000 years in the Cordillera Central, Colombia. Quatern. Res. 46, 1–18.

van der Hammen, T., 1981. Glaciales and glaciaciones en el Cuaternario de Colombia: paleoecología y estratigrafía. Rev. CIAF 6, 635–638.

van der Hammen, T., 1984. Datos sobre la historia de clima, vegetación y glaciación de la Sierra Nevada de Santa Marta. In: van der Hammen, T., Ruiz, P.M. (Eds.), Studies on Tropical Andean Ecosystems, vol. 2. J. Cramer, Vaduz, pp. 561–580.

van der Hammen, T., 1986. La Sabana de Bogotá y su lago en el PLeniglacial Medio. Caldasia 15, 249–262.

van der Hammen, T., Gonzalez, E., 1963. Historia de clima y vegetación del Pleistoceno Superior y del Holoceno de la Sabana de Bogotá. Boletín Geológico (Ingeominas, Bogotá) 11 (1–3), 189–266.

van der Hammen, T., Gonzalez, E., 1965. A Late Glacial and Holocene pollen diagram from Ciénaga del Visitador (Dept. Boyacá, Colombia). Leidse Geol. Meded. 32, 193–201.

van der Hammen, T., Hooghiemstra, H., 1996. Chronostratigraphy and correlation of the Pliocene and Quaternary of Colombia Quatern. Int. 40, 81–91.

van der Hammen, T., Vogel, J.C., 1966. The Susacá Interstadial and the subdivision of the Late Glacial. Geol. Mijnbouw 45 (2), 33–35.

van der Hammen, T., Werner, J.H., Van Dommelen, H., 1973. Palynological record of the upheaval of the Northern Andes: a study of the Pliocene and Lower Quaternary of the Colombian Eastern Cordillera and the early evolution of its High-Andean biota. Rev. Palaeobot. Palynol. 16, 1–122.

van der Hammen, T., Dueñas, H., Thouret, J.C., 1980a. Guía de excursión—Sabana de Bogotá. In: Primer seminario sobre el Cuaternario de Colombia, Bogotá-Colombia 25 al 29 de Agosto, 1980, Bogotá, 8 pp.

van der Hammen, T., Barelds, J., De Jong, H., De Veer, A.A., 1980b. Glacial sequence and environmental history in the Sierra Nevada del Cocuy (Colombia). Palaeogeogr. Palaeoclimatol. Palaeoecol. 32, 247–340.

van Geel, B., van der Hammen, T., 1973. Upper Quaternary vegetational and climatic sequence of the Fúquene area (Eastern Cordillera, Colombia). Palaeogeogr. Palaeoclimatol. Palaeoecol. 14, 9–92.

van't Veer, R., Islebe, G.A., Hooghiemstra, H., 2000. Climatic change during the Younger Dryas chron in northern South America: a test of the evidence. Quatern. Sci. Rev. 19, 1821–1835.

Walker, M., Lowe, J., 2007. Quaternary science 2007: a 50 year retrospective. J. Geol. Soc. Lond. 164, 1073–1092.

Wijninga, V.M., 1996. Paleobotany and palynology of Neogene sediments from the high plain of Bogotá (Colombia)—evolution of the Andean flora from a paleoecological perspective. PhD thesis, University of Amsterdam, 370pp.

Chapter 59

Late Quaternary Glaciations in the Venezuelan (Mérida) Andes

Volli Kalm[1,*] and William C. Mahaney[2]

[1]*Department of Geology, Institute of Ecology and Earth Sciences, University of Tartu, Ravila 14a, Tartu 50411, Estonia*
[2]*Quaternary Surveys, 26 Thornhill Avenue, Thornhill, Ontario, Canada L4J 1J4*
*Correspondence and requests for materials should be addressed to Volli Kalm. E-mail: volli.kalm@ut.ee

59.1. INTRODUCTION

The Venezuelan (Mérida) Andes are elongated mountain ranges in the north-western part of the country, consisting of two north-east to south-west trending cordilleras—the Mérida and Sierra de Perija—between 7°30′N and 10°10′N and 69°20′W and 72°30′W (Schubert, 1979; Schubert and Vivas, 1993). The glacial geology of the Venezuelan Andes has been studied by numerous researchers for over a century, starting with Sievers (1886) and Jahn (1925, 1931), continuing with Royo y Gómez (1956, 1960), Schubert (1970, 1972, 1974a,b, 1979, 1984, 1998) and Mahaney et al. (2000a, 2001, 2007a, 2008, 2010a,b), among others. Reviews of the history of glacial geology and Pleistocene stratigraphy in the Venezuelan Andes are available in a number of publications (Schubert, 1979, 1998; Bradley et al., 1985; Schubert and Vivas, 1993; Mahaney and Kalm, 1996; Mahaney et al., 2000a,b, 2010a,b; Rull, 2005). All early workers described glacial features down to near 3000 m a.s.l. elevation and postulated two stages of the last (Mérida, Wisconsinan) glaciation, an older stage reaching to < 3000 m and a younger one reaching 3500–3600 m a.s.l. (Table 59.1). The earliest published map of glaciers in the Sierra Nevada de Mérida was based on field work by Jahn in 1910 (Jahn, 1925), showing the distribution of glaciers in the Sierra Nevada de Mérida until maps were published by Schubert (1972, 1980, 1984). The total glacier-covered area in the Mérida Andes was approximately 600 km^2, while at the later stage of the Mérida Glaciation, the mountain chain had an approximate glacier area of 200 km^2 (Schubert, 1998).

59.2. EARLY AND MIDDLE PLEISTOCENE

Schubert and Vivas (1993) concluded that the Mérida Andes reached elevations similar to those today at the end of the Tertiary and therefore must have been affected by glaciations characteristic of the Quaternary Period. However, with the exception of sites on Mesa del Caballo, there is very little evidence of glacial deposition prior to that of the Mérida (Wisconsinan, Weichselian) Glaciation in the Mérida Andes. This has been explained by supposing that the Andes reached a sufficient elevation to generate glaciers very late (Vivas, 1974; Giegengack and Grauch, 1975), or that the rate of uplift was sufficiently high so that each glaciation could eliminate the evidence of the former (Schubert, 1979; Schubert and Vivas, 1993). The latter interpretation is supported by the fact that basement uplift accelerated at the end of the Late Oligocene, and the most intensive uplift and erosion of the Mérida Andes was from Late Miocene (ca. 10 Ma) to recent (Kohn et al., 1984; Pindell et al., 1998).

There are few reports available suggesting the presence of old, possibly pre-Mérida age glacial diamictons in the Mérida Andes. Schubert (1979) reported an old, highly weathered Middle Mucujún Moraine in the lower part of Páramo de La Culata at 2600 m a.s.l. Mahaney and Kalm (1996) described two different types of deeply weathered till layers in the ridge-like landform of Mesa del Caballo near El Pedregal (sections LAG1–3) and presumed their pre-Mérida age. The Mesa del Caballo ridge comprises superficial tills believed to represent, at depth, Middle to Late Pleistocene sediments displaced from their original position in front of the Mucuchache Valley by movement of the Boconó fault (Audemard et al., 1999, 2005). Mahaney et al. (2000b) report a contorted unit of weathered and oxidised gravel and till under the 81-ka (an optically stimulated luminescence (OSL) date) old sandy interlayer in the RF3 section at Pueblo Llano in the central Mérida Andes. Although Bezada et al. (1995) refer to an occasional 300 ka thermoluminescence (TL) date from pre-global Last Glacial Maximum (LGM) sediments, the possible pre-Mérida age of mentioned tills and respective glaciation(s) is not

TABLE 59.1 Glacial Stratigraphy of the Mérida Andes Compiled from Different Sources (Indicated in the Table with Numbers in Square Brackets): [1], Mahaney et al. (2000a); [2], Salgado-Labouriau et al. (1988); [3], Schubert and Vivas (1993); [4], Salgado-Labouriau and Schubert (1976); [5], Mahaney et al. (2008); [6], Rinaldi (1996); [7], Rull et al. (2005); [8], Salgado-Labouriau et al. (1977); [9], Mahaney and Kalm (1996); [10], Mahaney et al. (2004); [11], Schubert and Rinaldi (1987); [12], Mahaney et al. (2009); [13], Mahaney et al. (2000b); [14], Schubert and Valastro (1980); [15], Mahaney et al. (2001); [16], Dirszowsky et al. (2005); [17], Rull (2005); [18], Mahaney et al. (2010a,b); [19], Schubert and Vaz (1987); [20], Bezada et al. (1995)

	Glaciations and cold phases / Warm periods	Ice margin, m a.s.l. (5000, 4000, 3000)	Age, key sites and references [n] - reference no in Table caption	MIS	Age, ka
Holocene	Little Ice Age (LIA)	4900 ● ● 4200	Humboldt Glacier (1); LIA≈<500 yr BP, Lake El Suero (2), Humboldt Massif, [1]	1	0
	Early Neoglacial	● 4150			
	Miranda Warm Phase		3.6–2.5 ka, glaciofluvial terrace (3), Páramo de Miranda [2,3]		5
	La Culata Cold/Dry Phase		6.0–5.3 ka, all. terr. at La Culata (4), Rio Mucujun valley [3,4]		
	Miranda Warm Phase		9.4–6.3 ka, glaciofluvial terrace (3), Páramo de Miranda [2,3]		10
Late Wisconsin	Younger Dryas / Mucubají Warm Ph.	3800 / 4000	12.4 ka, Lake El Suero (2), Lake Mucubají catchm. (5), [5,6,7]; 14.1–13.8 cal. ka, Lake Mucubají (6), [8]; 15.7–14.1 cal. ka, peat layers, El Pedregal fan complex (7), [8,9,10]	14.7	15
	Late Mérida Stade	2900 / 3200	LGM at ca 21.5 cal. ka, El Pedregal fan complex (7), Lake Mucubají (6) and Lake Coromoto (8), [9, 10, 11, 12]; 27.4–22.2 cal. ka, peat layers, El Pedregal fan complex (7), [9]	2	20 / 25
Middle Wisconsin	El Pedregal Interstade		31–26 ka OSL, section RF3 in La Canoa v. at Pueblo Llano (9), [13]; 50.6–33.7 ka ¹⁴C, 2nd alluvial terrace at Tuñame (10), Rio Motatán [3, 14]; 47.5–45.9 ka TL, 2nd all. terr. at Timotes (11) and Mesa Grane (12), Rio Motatán [3, 14]; >61–48 ka ¹⁴C, section PED5 (7 peat layers) in El Pedregal fan complex (7), [15, 16, 17]	27.7 / 3	30 / 40 / 50 / 60
	Early Mérida Stade	2600 / 2800	ca 90–60 ka OSL, Lower basal till bed in section PED5 (7) in El Pedregal fan complex [9, 15, 16, 18]; End moraine in Los Zerpas (13) valley at Los Frailes [9, 18]; Contorted glaciotectonized glaciofluvial gravel and till in section RF3 (9) in La Canoa valley at Pueblo Llano [9, 13].	4 / 74	70 / 80
Early Wisconsin				5a-d	90 / 100
Eem (Sang.)				5e	115 / 120 / 130
Illinoian	Pre-Mérida glaciation	3500	169.7–147.8 ka TL, 4th alluvial terrace at Timotes (11), Rio Motatán [3,19]; ? ka, 'Younger' pre-Mérida weathered till, sections LAG2 and LAG3 (14) at Mesa del Caballo [9]; ? ka, 'Older' pre-Mérida weathered till, section LAG1 (14) at Mesa del Caballo [9]; ? ↓ 300 ka TL, pre-LGM sediments [20]	6	140 / 160 / 180 / 186

The numbers in round brackets indicate location of sites/sections in Fig. 59.1. Note the changes in the time scale factor.

directly determined. The only dated evidence of earlier glaciations is the 4th alluvial terrace in Rio Motatán valley near Timotes, interpreted as a valley fill that occurred during glacial advance. TL dating of alluvial sand yielded an age of 147.8 and 169.7 ka (Schubert and Vaz, 1987), which potentially can be related to pre-Eemian Middle Pleistocene glaciation.

59.3. LATE PLEISTOCENE

Although Van der Hammen and Hooghiemstra (2003) have identified interglacial (Palo Blanco or Eemian interglacial) lacustrine sediments recovered from Lake Fuquene in the Eastern Cordillera of Colombia, and while these sediments are inferred to correspond to marine isotope stage (MIS) 5, there is no evidence of interglacial sediments recorded in the Mérida Andes, some 450 km north-east of Fuquene.

59.3.1. Mérida (Wisconsinan, Weichselian) Glaciation

The Late Pleistocene Mérida Glaciation was first defined by Schubert (1974a,b). Traditionally, moraine evidence has been used to differentiate an early stage, with ice reaching about 2600–2800 m a.s.l., from a later stage with ice reaching 2900–3500 m a.s.l. (Schubert, 1979; Bezada, 1990; Schubert and Vivas, 1993; Bezada et al., 1995; Mahaney and Kalm, 1996; Mahaney et al., 2000a). Recently, chronological data became available (Mahaney et al., 2001; Dirszowsky et al., 2005), that set an upper limiting age (>60 ka) for the Early Mérida Stage, indicating considerable duration of the interstadial (from ~60 to 27–26 ka) between the two stages of the Mérida Glaciation.

59.3.1.1. Early Mérida Stadial

The Mérida Glaciation began with a poorly constrained early stage that left either till buried by the LGM deposits, dispersed patches of glacial diamicton eroded by runoff and/or intact end and lateral moraines such as exist in the Santo Domingo Valley. According to Schubert (1974b), evidence of the Early Mérida Stage is located in the Sierra de Santo Domingo and Páramo de La Culata, although later investigations (Bezada, 1990; Mahaney and Kalm, 1996; Mahaney et al., 2000b) have not revealed Early Mérida Till in the Páramo de La Culata. Field mapping and stratigraphical analysis has shown that there are only a few drainage basins where the early stage of Mérida Glaciation can be studied, either buried under Late Mérida Till and associated deposits or on the surface. One site is at Los Zerpa (for site location, see Fig. 59.1) where ice from Victoria, Aguila and Los Zerpa drainage basins coalesced and flowed north-east along the Santo Domingo River. The second locality is at La Canoa in the western Mérida Andes, where the type section of the Mérida Glaciation was first documented by Mahaney et al. (2000b). Recently, Mahaney et al. (2010a), based on the data from PED5 section near El Pedregal, at 3600 m a.s.l., provide evidence for the growth and demise of an ice cap over the northern flank of the eastern Mérida Andes during the Early Mérida Stage of the last glaciation. The question of age controls for the growth of the Early Mérida Stage ice is impossible to ascertain but is assumed to be younger than MIS 5 or approximately <90 ka BP. In the case of the PED5 section, the minimum age of Early Mérida till is known to be ~60 ka BP by ^{14}C AMS dating (Mahaney et al., 2001; Dirszowsky et al., 2005).

59.3.1.2. El Pedregal Interstadial

The PED5 section near El Pedregal (Fig. 59.1) clearly represents materials of both Early and Late Mérida origin as well as sediments spanning the full interstadial period (Dirszowsky et al., 2005). On the basis of the stratigraphical context, palaeodepositional environment and a series of ^{14}C dates spanning most of the 40 m sequence and ranging from approximately 12 to >60 ka, the whole Mérida Glaciation has been defined at the PED5 Section and adjacent sites (Dirszowsky et al., 2005, 2006). In the PED5 Section, interstadial conditions are represented by a minimum of 8 m of predominantly lacustrine material with seven buried peat layers (Peat I–VII, ^{14}C AMS ages between >61 and 48 ^{14}C ka BP) and related interstadial palaeosols (Mahaney et al., 2001; Dirszowsky et al., 2005, 2006). Although exact correlations are considered tentative, there is some indication that the lowermost Peat VII (^{14}C AMS ages 56.9, 58.4, >59.8 and >60.7 ka ^{14}C ka BP) corresponds to a particularly pronounced warm event (IS number 19, in Dansgaard et al., 1993) that occurred near the end of the Early Wisconsinan/Mérida Stadial between 66.5 and 70 ka BP (Dirszowsky et al., 2005). If that holds true, the El Pedregal Interstadial occurred at approximately the same time as the Colombian Bachué Interstadial, which began in the Early/Middle Fuquenian transition at approximately 60 ka (Van der Hammen, 1995; Rull, 2005). Rull (2005) recorded typical interstadial pollen in the Peat VII and considered this particular layer equivalent to the whole Pedregal Interstadial. Later, Rull (2006) reinterpreted the Peat VII 'event' to be the Pedregal Interstadial 1, and the entire assemblage of PED5 peat layers to be the El Pedregal Interstadial Complex. During the time of accumulation of Peat VII with maximum tree pollen, the PED5 site (3600 m a.s.l.) was occupied by ecosystems (intermediate Páramo/Superpáramo communities) which today grow from 4000 m a.s.l. upwards (Rull, 2005). The mid-Wisconsinan palaeosols (Spodosols) in the Mérida Andes are sandy pedons showing strong podsolic soil morphologies in different degrees of evolutionary development that do not have equivalent

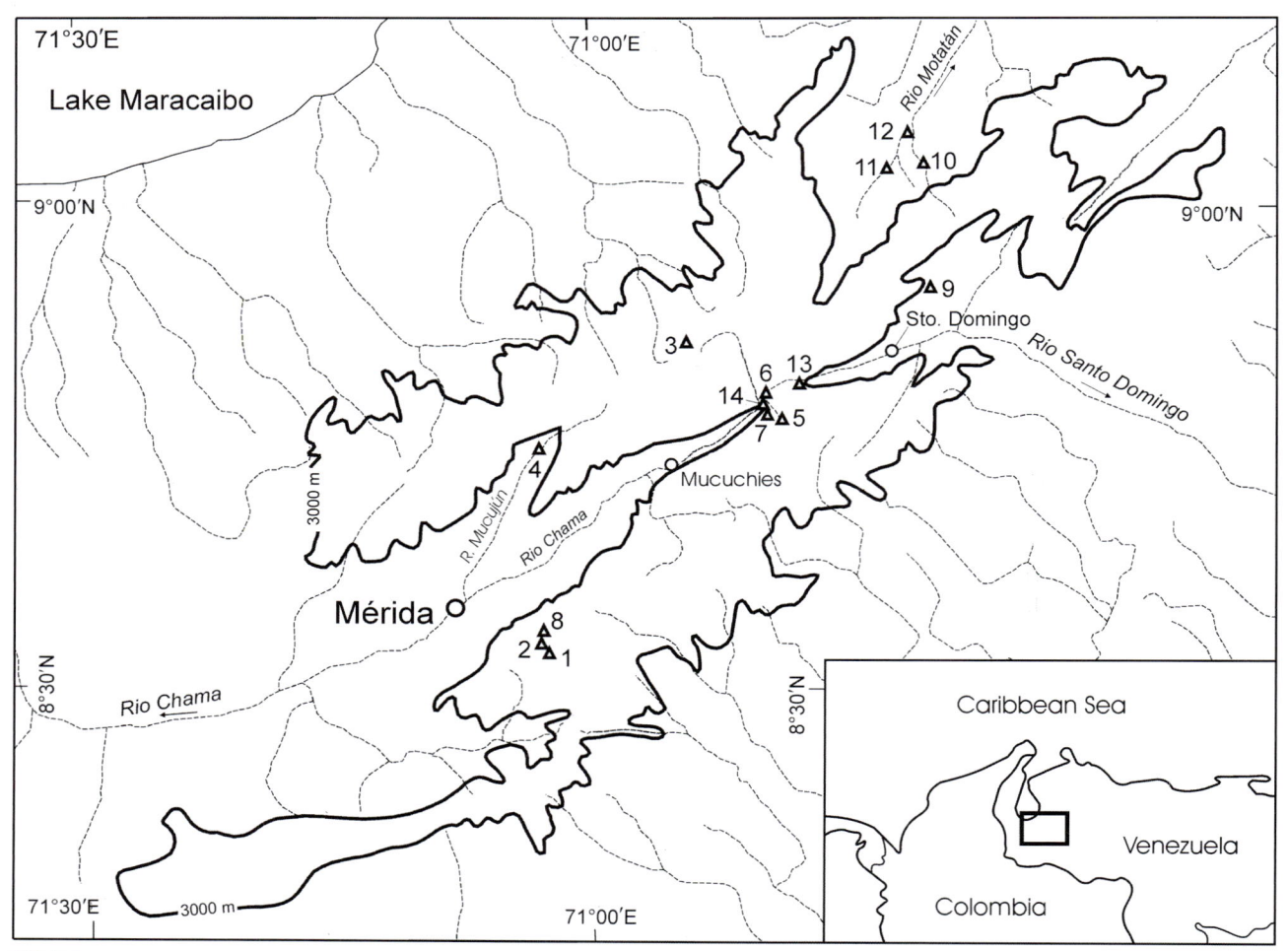

FIGURE 59.1 Location map showing central Mérida Andes. The black solid line indicates areas above 3000 m a.s.l.; dashed lines show river network. Specific sites/sections mentioned in the text and Table 59.1 are indicated with numbers: (1), Pico Humboldt and Humboldt Glacier; (2), Lake El Suero in Humboldt Massif; (3), glaciofluvial terrace, Páramo Miranda; (4), alluvial terrace at La Culata, Rio Mucujún; (5), Lake Mucubají catchment; (6), Lake Mucubají; (7), El Pedregal fan complex (sections PED1-5); (8), Lake Coromoto; (9), end moraine in La Canoa valley at Pueblo Llano (section RF3); (10), 2nd alluvial terrace at Tuñame, Rio Motatan; (11), 2nd and 4th alluvial terraces at Timotes, Rio Motatan; (12), 2nd alluvial terrace at Mesa Grande, Rio Motatan; (13), end moraine at Los Zerpas; (14), sections LAG1, LAG2 and LAG3 at Mesa del Caballo.

surface pedons in the Holocene environment of the El Pedregal area (Mahaney et al., 2001).

Peaty layers with ^{14}C dated wood (ages between 50.6 and 33.7 ^{14}C ka BP) in the 2nd alluvial terrace in the valley of Rio Motatán at Tuñame (Schubert and Valastro, 1980; Schubert and Vivas, 1993) also indicate a relatively warmer and wetter climate. Together with the PED5 Section, the series of peaty layers indicates an interstadial environment, although with fluctuating climatic conditions from >61 to 33.7 ^{14}C ka BP. Mahaney et al. (2000b) reported OSL ages of 31–26 ka from sandy silt overlain by glacially deformed sand in the RF3 section (2800 m a.s.l.) in La Canoa valley at Pueblo Llano and calibrated ^{14}C ages of 27.4 ka BP age of a peat in the PED4 section at El Pedregal (Mahaney and Kalm, 1996), thus extending the record of interstadial conditions to the immediate onset of the Late Mérida Stadial at ca. 25 ka (Schubert and Clapperton, 1990).

59.3.1.3. Late Mérida Stadial

Traditionally, the latter stadial of the Mérida Glaciation is estimated to have occurred between 25 and 13 ka ^{14}C BP culminating at around 18 ka ^{14}C ka BP (uncalibrated ages), as indicated by radiocarbon ages and correlation with the Cordillera Oriental of Colombia (Schubert, 1974b; Schubert and Valastro, 1974; Schubert and Clapperton, 1990; Mahaney and Kalm, 1996; Rull, 2005). If calibrated, the culmination of the Late Mérida Stage is placed at ca. 21.5 ka BP (henceforth in the text, the ^{14}C ages are given in calibrated; OxCal v4.1.5, Bronk Ramsey, 2009 format). Earlier and recent chronological data from El Pedregal allow refining the onset and culmination of the last glacial advance in central Mérida Andes at an elevation of ca. 3600 m a.s.l. Although the precise position of dated samples is not known, Schubert and Rinaldi (1987) report a

22.9 cal. ka BP age (calibrated from 19.08 ± 0.82 ^{14}C ka BP) buried organic sediments 2 m above the base and 19.7 cal. ka BP age (calibrated from 16.5 ± 0.29 ^{14}C ka BP) 3 m below the top of the 40-m thick El Pedregal fan complex. Later, Mahaney and Kalm (1996) dated several clay-rich peat layers from below and above the rhythmically bedded glaciolacustrine silt and clay representing the ice advance during the LGM. They obtained ^{14}C ages of 22.9 and 22.2 cal. ka BP (calibrated from 19.2 ± 0.35 and 18.6 ± 0.3 ^{14}C ka BP, section PED1) from the pre-LGM peat layers, and these ages nicely conform with the data of Schubert and Rinaldi (1987) and set the maximum limiting age for the LGM at ca. 22.9 cal. ka BP. The reconstructed glacier equilibrium-line altitude (ELA) depression for the LGM in the Mérida Andes was $\sim$1100 m (Lachniet and Vazquez-Selem, 2005), with the ELA depression ranging between 850 and 1420 m in different areas (Stansell et al., 2007).

Upstream of Lake Mucubají (3600 m a.s.l.) and Lake Coromoto (3000 m a.s.l.), a distinct series of recessional moraines located between 3200 and 3800 m a.s.l. (Schubert, 1972, 1974a; Salgado-Labouriau et al., 1977), record several stillstands that occurred following the LGM. Several post-LGM peat strata in sections PED1 and PED1A at El Pedregal fan complex yielded ^{14}C ages of 17.8, 15.6 and 14.1 cal. ka BP (calibrated from 14.57 ± 0.45, 12.95 ± 0.45 and 12.17 ± 0.2 ^{14}C ka BP, Mahaney and Kalm, 1996; Mahaney et al., 2004). This together with the earlier data of Giegengack and Grauch (1976) and Salgado-Labouriau et al. (1977) provides evidence for climate fluctuations with periodic interruptions of glaciolacustrine sedimentation at ca. 17.8, 15.6, 15.0 and 14.1 cal. ka BP. Stansell et al. (2005) concluded that on the wetter south-eastern side of the Mérida Andes, glaciers had significantly retreated by 15.7 cal. ka BP, followed by several minor glacial advances and retreats between 14.9 and 13.8 cal. ka BP with at least one major glacial readvance between 13.8 and 10.0 cal. ka BP.

59.3.1.4. Younger Dryas Stadial

Deposits of push moraine, outwash and glaciolacustrine sediments, recovered from the upstream area of Lake Mucubají ($\sim$3800 m a.s.l.) and at Lake El Suero ($\sim$4000 m a.s.l.) in the Humboldt Massif, document the latest Pleistocene advance of the ice in the Mérida Andes. Underlying peats provide maximum limiting ages (13.66, 13.64, 13.29 and 12.40 cal. ka BP; Mahaney et al., 2008) on till and outwash. In upper Mucuñque Valley at 3800 m, a succession of glaciolacustrine and glaciofluvial sediment yielded a 'burnt layer' (Mahaney et al., 2008), ^{14}C dated to $\sim$12.9 cal. ka BP, tentatively correlated with the 'black mat', a carbonaceous coating of ejecta considered to have originated from the break-up of a comet over the Laurentide Ice Sheet. Fallout from this airburst/impact may have initiated the onset of the Younger Dryas (YD) Stadial (Mahaney et al., 2010b). In both the Mucuñque and Humboldt Massifs, the YD glacial advance nearly reached earlier Late-glacial ice positions (Mahaney et al., 2008). However, although Stansell et al. (2005) argued for an ice advance into the lower Mucuñque Valley, the magnitude of the advance was contested by Mahaney et al. (2007b). This glacial advance is approximately synchronous with the YD cold event of Europe and the North Atlantic Region (ca. 12.9–11.7 cal. ka BP). Rull (1999) and Rull et al. (2005) correlate the YD readvance with the cold phases of the Mucubají (III) and the Páramo Miranda of Rinaldi (1996).

59.4. HOLOCENE GLACIER FLUCTUATIONS

According to Schubert's (1998) estimation, at the end of the last glaciation, the Mérida Andes had an approximate glacier area of 200 km^2 (Schubert, 1998). By 1952, the total glacier area in the mountain chain was 2.9 km^2, by 1991, ca. 2.0 km^2 (Schubert, 1998) and by 2003, the glacier area had decreased to 0.29 km^2 (Morris et al., 2006).

Pollen records have shown that vegetation and climate have remained similar to today through most of the Holocene (Salgado-Labouriau et al., 1988, 1992; Rull, 1999), but some minor cold events occurred at ca. 6.0–5.3 ^{14}C ka BP (La Culata cold/dry Phase; Salgado-Labouriau and Schubert, 1976) and in the eleventh to fourteenth centuries (Piedras Blancas Cold Phase; Salgado-Labouriau, 1989) and relatively warm conditions occurring ca. 9.4–6.3 and 3.6–2.5 ^{14}C ka BP (Miranda Warm Phases; Salgado-Labouriau et al., 1988; Schubert and Vivas, 1993). The Piedras Blancas Cold Phase has been correlated with the Little Ice Age (LIA) (Rull et al., 1987; Salgado-Labouriau, 1989) and with moraines present at upper elevations of the Humboldt massif (Schubert, 1972, 1998). Mahaney et al. (2000a, 2008, 2009) dated the moraine/outwash complexes and related soils of Late-glacial/YD to LIA from the Humboldt Massif at >4000 m a.s.l. Dated buried peats (0.97 ± 0.26 and 1.58 ± 0.09 ^{14}C ka BP) at Lake El Suero (ca. 4200 m a.s.l.) indicate less erosion and possibly warmer and wetter periods of short duration at two different times in the later Holocene around 1.48 and 0.92 cal. ka BP (Mahaney et al., 2000a). Assertions by Stansell et al. (2005) that LIA moraines exist in the upper Mucuñque Valley has been contested by Mahaney et al. (2007b) who showed that the upper valley is dominated principally by talus and rockfall and is lacking in any LIA moraines. The only LIA moraines occur on the Humboldt Massif and below Pico Bolivar.

REFERENCES

Audemard, F., Pantosti, D., Machette, M., Costa, C., Okumura, K., Cowan, H., et al., 1999. French investigation along the Mérida section of the Boconó fault (central Venezuelan Andes), Venezuela. Tectonophysics 308, 1–21.

Audemard, F., Romero, G., Rendon, H., Cano, V., 2005. Quaternary fault kinematics and stress tensors along the southern Caribbean from fault-slip data and focal mechanism solutions. Earth Sci. Rev. 69, 181–233.

Bezada, M., 1990. Geologia glacial de Cuaternario de la región de Santo Domingo—Pueblo Llano—Las Mesitas (Estado Mérida y Trujillo). PhD thesis, Instituto Venezolano de Investigaciones Cientificas, 245 pp.

Bezada, M., Kalm, V., Mahaney, W.C., 1995. Pleistocene glaciation in the Venezuelan Andes. In: Abstracts, XIV INQUA International Congress, August 3–10, 1995. Freie Universität, Berlin, p. 28.

Bradley, R.S., Yuretich, R.F., Salgado-Laboriau, M., Weingarten, B., 1985. Late Quaternary paleoenvironmental reconstruction using lake sediments from the Venezuelan Andes: preliminary results. Z. Gletscherk. Glazialgeol. 21, 97–106.

Bronk Ramsey, C., 2009. Bayesian analysis of radiocarbon dates. Radiocarbon 51, 337–360.

Dansgaard, W., Johnsen, S.J., Clausen, H.B., Dahl-Jensen, D., Gundestrup, N.S., Hammer, C.U., et al., 1993. Evidence for general instability of past climate from a 250-kyr ice-core record. Nature 364, 218–220.

Dirszowsky, R.W., Mahaney, W.C., Hodder, K.R., Milner, M.W., Kalm, V., Bezada, M., et al., 2005. Lithostratigraphy of the Mérida (Wisconsinan) glaciation and Pedregal interstade, Mérida Andes, northwestern Venezuela. J. S. Am. Earth Sci. 19, 525–536.

Dirszowsky, R.W., Mahaney, W.C., Kalm, V., Beukens, R., 2006. Comment on "A Middle Wisconsin interstadial in the northern Andes" by Valentí Rull. J. S. Am. Earth Sci. 21, 310–314.

Giegengack, R., Grauch, R.I., 1975. Quaternary geology of the central Andes, Venezuela: a preliminary assessment. Boletín de Geología, Publ. Especial Venez. 7 (1), 241–283.

Giegengack, R., Grauch, R.I., 1976. Late Cenozoic climatic stratigraphy of the Venezuelan Andes. Boletín de Geología, Publ. Especial 7 (2), 1187–1200.

Jahn, A., 1925. Observaciones glaciológicas en los Andes Venezolanos. Cultura Venezolana, Caracas 64, 265–280.

Jahn, A., 1931. El deshielo de la Sierra Nevada de Mérida y sus causas. Cultura Venezolana, Caracas 110, 5–15.

Kohn, B., Shagam, R., Banks, P., Burkley, L., 1984. Mesozoic–Pleistocene fission track ages on rocks of the Venezuelan Andes and their tectonic implications. Geol. Soc. America Memoir. 162, 365–384.

Lachniet, M.S., Vazquez-Selem, L., 2005. Last Glacial Maximum equilibrium line altitudes in the circum-Caribbean (Mexico, Guatemala, Costa Rica, Colombia, and Venezuela). Quatern. Int. 138 (139), 129–144.

Mahaney, W.C., Kalm, V., 1996. Field Guide for the International Conference on Quaternary Glaciation and Paleoclimate in the Andes Mountains. Quaternary Surveys, Toronto, 79 pp.

Mahaney, W.C., Milner, M.W., Sanmugadas, K., Kalm, V., Bezada, M., Hancock, R.G.V., 2000a. Late Quaternary deglaciation and Neoglaciation of the Humboldt Massif, northern Venezuela. Z. Geomorphol. Suppl. 122, 209–226.

Mahaney, W.C., Kalm, V., Bezada, M., Hütt, G., Milner, M.W., 2000b. Stratotype for the Mérida Glaciation at Pueblo Llano in the northern Venezuelan Andes. J. S. Am. Earth Sci. 13, 761–774.

Mahaney, W.C., Milner, M.W., Russell, S., Kalm, V., Bezada, M., Hancock, R.G.V., et al., 2001. Paleopedology of Middle Wisconsinan/Weichselian paleosols in the Mérida Andes, Venezuela. Geoderma 104, 215–237.

Mahaney, W.C., Dirszowsky, R.W., Milner, M.W., Menzies, J., Stewart, A., Kalm, V., et al., 2004. Quartz microtextures and microstructures owing to deformation of glaciolacustrine sediments in the Northern Venezuelan Andes. J. Quatern. Sci. 19, 23–33.

Mahaney, W.C., Dirszowsky, R., Milner, M.W., Harmsen, R., Finkelstein, S., Kalm, V., et al., 2007a. Soil stratigraphy and plant–soil interactions on a Late Glacial–Holocene terrace sequence, Sierra Nevada National Park, northern Venezuelan Andes. J. S. Am. Earth Sci. 23, 46–60.

Mahaney, W.C., Dirszowsky, R.W., Kalm, V., 2007b. Comment on "Late Quaternary deglacial history of the Mérida Andes, Venezuela" by N.D. Stansell, et al.. J. Quatern. Sci. 22, 1–5.

Mahaney, W.C., Milner, M.W., Kalm, V., Dirszowsky, R.W., Hancock, R.G.V., Beukens, R.P., 2008. Evidence for a Younger Dryas glacial advance in the Andes of northwestern Venezuela. Geomorphology 96, 199–211.

Mahaney, W.C., Kalm, V., Kapran, B., Milner, M.W., Hancock, R.G.V., 2009. A soil chronosequence in Late Glacial and Neoglacial moraines, Humboldt Glacier, northwestern Venezuelan Andes. Geomorphology 109, 236–245.

Mahaney, W.C., Kalm, V., Menzies, J., Milner, M.W., 2010a. Reconstruction of the Early Mérida, pre-LGM Glaciation with comparison to Late Glacial Maximum till, northwestern Venezuelan Andes. Sed. Geol. 226, 29–41.

Mahaney, W.C., Kalm, V., Krinsley, D.H., Tricart, P., Schwartz, S., Dohm, J., et al., 2010b. Evidence from the northwestern Venezuelan Andes for extraterrestrial impact: the black mat enigma. Geomorphology 116, 48–57.

Morris, J.N., Poole, A.J., Klein, A.G., 2006. Retreat of tropical glaciers in Colombia and Venezuela from 1984 to 2004 as measured from ASTER and Landsat images. In: Proceedings of 63rd Eastern Snow Conference, pp. 181–191, Newark, Delaware, USA.

Pindell, J.L., Higgs, R., Dewey, J.F., 1998. Cenozoic palinspastic reconstruction, paleogeographic evolution and hydrocarbon setting of the northern margin of South America. Soc. Sed. Geol., Special Publ. 58, 45–86.

Rinaldi, M., 1996. Evidencias del evento frío Younger Dryas en los Andes Venezolanos. In: Abstracts. International conference on Quaternary Glaciation and Palaeoclimate in the Andes mountains. Universidad Pedagógica Experimental Libertador, Mérida, Venezuela, p. 14.

Royo y Gómez, J., 1956. Quaternary in Venezuela. In: Stratigraphical Lexicon of Venezuela (English Edition), Dirección de Geología Venezolana, Boletín de Geología, Publicacíon Especial 1, pp. 468–478.

Royo y Gómez, J., 1960. Glaciarismo pleistoceno en Venezuela. Asoc. Venez. de Geología, Mineria y Petró., Boletín Informativo 2 (11), 333–357.

Rull, V., 1999. Palaeoclimatology and sea-level history in Venezuela. New data, land-sea correlations, and proposals for future studies in the framework of the IGBP-PAGES project. Interciencia 24 (2), 92–101.

Rull, V., 2005. A Middle Wisconsin interstadial in the Northern Andes. J. S. Am. Earth Sci. 19, 173–179.

Rull, V., 2006. Reply to a comment by Dirszowsky, R.W., Mahaney, W.C., Kalm, V., and Beukens, R. on 'A Middle Wisconsin interstadial in the northern Andes' (Rull, V., 2005. JSAES 19, 173-179). J. S. Am. Earth Sci. 21, 315–317.

Rull, V., Salgado-Labouriau, M.L., Schubert, C., Valastro, S., 1987. Late Holocene temperature depression in the Venezuelan Andes. Palynological evidence. Palaeogeogr. Palaeoclimatol. Palaeoecol. 60, 109–121.

Rull, V., Abbott, M.B., Polissar, P.J., Wolfe, A.P., Bezada, M., Bradley, R.S., 2005. 15,000-yr pollen record of vegetation change in the high altitude tropical Andes at Laguna Verde Alta, Venezuela. Quatern. Res. 64, 308–317.

Salgado-Labouriau, M.L., 1989. Late Quaternary climatic oscillations in the Venezuelan Andes. Biol. Int. 18, 12–14.

Salgado-Labouriau, M.L., Schubert, C., 1976. Palynology of Holocene peat bogs from Central Venezuelan Andes. Palaeogeogr. Palaeoclimatol. Palaeoecol. 19, 147–156.

Salgado-Labouriau, M.L., Schubert, C., Valastro, S., 1977. Paleoecologic analysis of a Late Quaternary terrace from Mucubají, Venezuelan Andes. J. Biogeogr. 4, 313–325.

Salgado-Labouriau, M.L., Rull, V., Schubert, C., Valastro, S., 1988. The establishment of vegetation after Late Pleistocene deglaciation in the Páramo de Miranda, Venezuelan Andes. Rev. Palaeobot. Palynol. 55, 5–17.

Salgado-Labouriau, M.L., Bradley, R.S., Yuretich, R., Weingarten, B., 1992. Paleoecological analysis of the sediments of lake Mucubají, Venezuelan Andes. J. Biogeogr. 19, 317–327.

Schubert, C., 1970. Glaciation of the Sierra de Santo Domingo, Venezuelan Andes. Quaternaria 13, 225–246.

Schubert, C., 1972. Geomorphology and glacier retreat in the Pico Bolivar area, Sierra Nevada de Mérida, Venezuela. Z. Gletscherk. Glazialgeol. 8, 189–202.

Schubert, C., 1974a. Late Pleistocene glaciation of Páramo de la Culata, north-central Venezuelan Andes. Geol. Rundsch. 63, 516–538.

Schubert, C., 1974b. Late Pleistocene Mérida Glaciation, Venezuelan Andes. Boreas 3, 147–152.

Schubert, C., 1979. Glacial sediments in the Venezuelan Andes. In: Schlüchter, C. (Ed.), Moraines and Varves. Origin, Genesis, Classification, Proceedings of an INQUA Symposium on Genesis and Lithology of Quaternary Deposits, Zurich, 10–20 September, 1978, pp. 43–49.

Schubert, C., 1980. Contribucíon de Venezuela al inventario mundial de glaciares. Boletín de la Soc. Venez. de Ciencias Naturales. 34, 267–279.

Schubert, C., 1984. The Pleistocene and recent extent of the glaciers of the Sierra Nevada de Mérida, Venezuela. Erdwissensch. Forsch. 18, 269–278.

Schubert, C., 1998. Glaciers of Venezuela. In: Williams Jr., R.S., Ferrigno, J.G. (Eds.), Satellite Image Atlas of Glaciers of the World—South America. United States Government Printing Office, Washington, DC, pp. 1–10. USGS Professional Paper 1386-I.

Schubert, C., Clapperton, C.M., 1990. Quaternary glaciations in the northern Andes (Venezuela, Colombia & Ecuador). Quatern. Sci. Rev. 9, 123–135.

Schubert, C., Rinaldi, M., 1987. Nuevos datos sobre la cronología del estadio tardío de la Glaciación Mérida, Andes Venezolanos. Acta Cient. Venez. 38, 135–136.

Schubert, C., Valastro, S., 1974. Late Pleistocene glaciation of páramo de La Culata, north-central Venezuelan Andes. Geol. Rundsch. 63, 516–538.

Schubert, C., Valastro, S., 1980. Quaternary Esnujaque Formation, Venezuelan Andes: preliminary alluvial chronology in a tropical mountain range. Z. Dtsch. Geol. Ges. 131, 927–947.

Schubert, C., Vaz, J.E., 1987. Edad termoluminiscente del complejo aluvial cuaternario de Timotes, Andes venezolanos. Acta Cient. Venez. 38, 285–286.

Schubert, C., Vivas, L., 1993. El Cuaternario del la Cordillera de Mérida, Andes Venezolanos. Universidad de Los Andes / Fundacion POLAR, Mérida, Venezuela, 345 pp.

Sievers, W., 1886. Über Schneeverhältnisse in der Cordillere Venezuelas. Mitteilungen der Geogr. Geselsch. München 1885, 54–57.

Stansell, N.D., Abbott, M.B., Polissar, J., Wolfe, A.P., Bezada, M., Rull, V., 2005. Late Quaternary deglacial history of the Mérida Andes, Venezuela. J. Quatern. Sci. 20, 801–812.

Stansell, N.D., Polissar, P.J., Abbott, M.B., 2007. Last glacial maximum equilibrium-line altitude and paleo-temperature reconstructions for the Cordillera de Mérida, Venezuelan Andes. Quatern. Res. 67, 115–127.

Van der Hammen, T., 1995. La última glaciación en Colombia (Glaciación Cocuy, Fuquense). Anál. Geogr. 24, 69–88.

Van der Hammen, T., Hooghiemstra, H., 2003. Interglacial–glacial Fuquene-3 pollen record from Colombia: an Eemian to Holocene climate record. Glob. Planet. Change 36, 181–199.

Vivas, L., 1974. Estudio geomorfológico de la cuenca superior de la quebrada Tuñame. Revista Geogr. (Universidad de los Andes) 11, 69–112.

Chapter 60

Costa Rica and Guatemala

Matthew S. Lachniet* and Alex J. Roy[†]

Department of Geoscience, University of Nevada, Las Vegas, 4505 Maryland Parkway, Las Vegas, Nevada 89154, USA
*Correspondence and requests for materials should be addressed to Matthew S. Lachniet E-mail: matthew.lachniet@unlv.edu

60.1. INTRODUCTION

The highest peaks in Guatemala and Costa Rica were glaciated to altitudes as low as ca. 3100 m a.s.l. during the late Quaternary (Orvis and Horn, 2000; Lachniet and Seltzer, 2002b). The presence of glacial landforms in Costa Rica has long been known, based on the pioneering studies of Weyl (1956, 1962), Hastenrath (1973) and others (Bergoeing, 1978; Barquero and Ellenberg, 1982/1983, 1986; Shimizu, 1992). In Costa Rica, several valley glaciers existed around Cerro Chirripó (3819 m a.s.l.) and other high peaks in the Cordillera de Talamanca (9°30′ N, 83°30′ W). More recently, ice limits for glaciations in Valle Morrenas were mapped and described (Horn, 1990; Orvis and Horn, 2000, 2005), to complement a suite of radiocarbon ages from basal sediments of tarns (Horn et al., 2005) within the upper Morrenas Valley. Other efforts described the geomorphology (van Uffelen, 1991; Kappelle and van Uffelen, 2005) and ice limits in the Cordillera de Talamanca in general (Lachniet et al., 2005; Lachniet, 2007). Vestiges of possible glaciation are also present on Cerro de la Muerte (3475 m a.s.l.; Lachniet and Seltzer, 2002b; Lachniet et al., 2005; Lachniet, 2007).

Research on the glaciation in Guatemala has been more sparse due largely to a contracted civil war in that country. The glacial geological observations of Anderson (Anderson, 1969a,b; Anderson et al., 1973) and Hastenrath (1973, 1974) provided the first detailed look at glaciation in the Sierra Cuchumatanes, and ice limits were established by the orientations of moraines on the high limestone plateau. After Hastenrath's pioneering work, little research was completed in the Cuchumatanes until recently (Lachniet and Vazquéz-Selem, 2005; Lachniet, 2007; Roy and Lachniet, 2010), when moraine limits and glacial geology were mapped in greater detail, based on the availability of aerial stereophotographs, topographic maps, and field access after the ending of the civil war in 1996.

The present contribution will focus primarily on the glaciation and ice limits of Guatemala, as previous publications have discussed the glaciation in Costa Rica more thoroughly (Horn, 1990; Orvis and Horn, 2000, 2005; Lachniet and Seltzer, 2002a,b; Lachniet, 2004, 2007; Horn et al., 2005; Lachniet and Vazquéz-Selem, 2005; Lachniet et al., 2005). Ice limits for Costa Rica were published in GIS format with the previous edition of this work (Lachniet, 2004). For details on the glacial geology of Costa Rica, the reader is referred to these previous publications.

60.2. GLACIAL GEOLOGY

The Sierra los Cuchumatanes (15.5°N, 91.5°W) is a high-altitude, low-relief limestone plateau (Blake, 1934; Enjalbert, 1967) which contains the highest non-volcanic point in Central America at 3837 m a.s.l. (Bundschuh et al., 2007). The Sierra los Cuchumatanes is located on the southern margin of North America, just north of the Motagua fault zone, making this location the most southerly glaciated site in North America (despite suggestion to the contrary; Owen et al., 2003). Relief is typically less than 400 m, reaching a height of 3837 m a.s.l. at El Torre peak, and with glaciated valley floors reaching below 3400 m a.s.l. (Fig. 60.1). Clear cockpit-karst features, as well as dolines and other solutional landforms, are also apparent near the main glaciated plateau, which is somewhat of a rarity at this high-altitude and cool tropical environment. Evidence of former glaciation in the Sierra was noted during reconnaissance visits and geologic mapping (Enjalbert, 1967; Anderson, 1969a,b; Anderson et al., 1973; Hastenrath, 1974) and demonstrated the former presence of an ice cap on the main karstic limestone plateau. These studies showed the presence of lateral, terminal and recessional moraines, as well as glacially scoured bedrock, broad U-shaped valleys in a low-relief environment, and an outwash plain in the Llanos de San Miguel Valley. Although

---

[†] Present address: Department of the Environment, Wetlands and Waterways Program, 1800 Washington Boulevard, Baltimore, Maryland 21230-4170, USA.

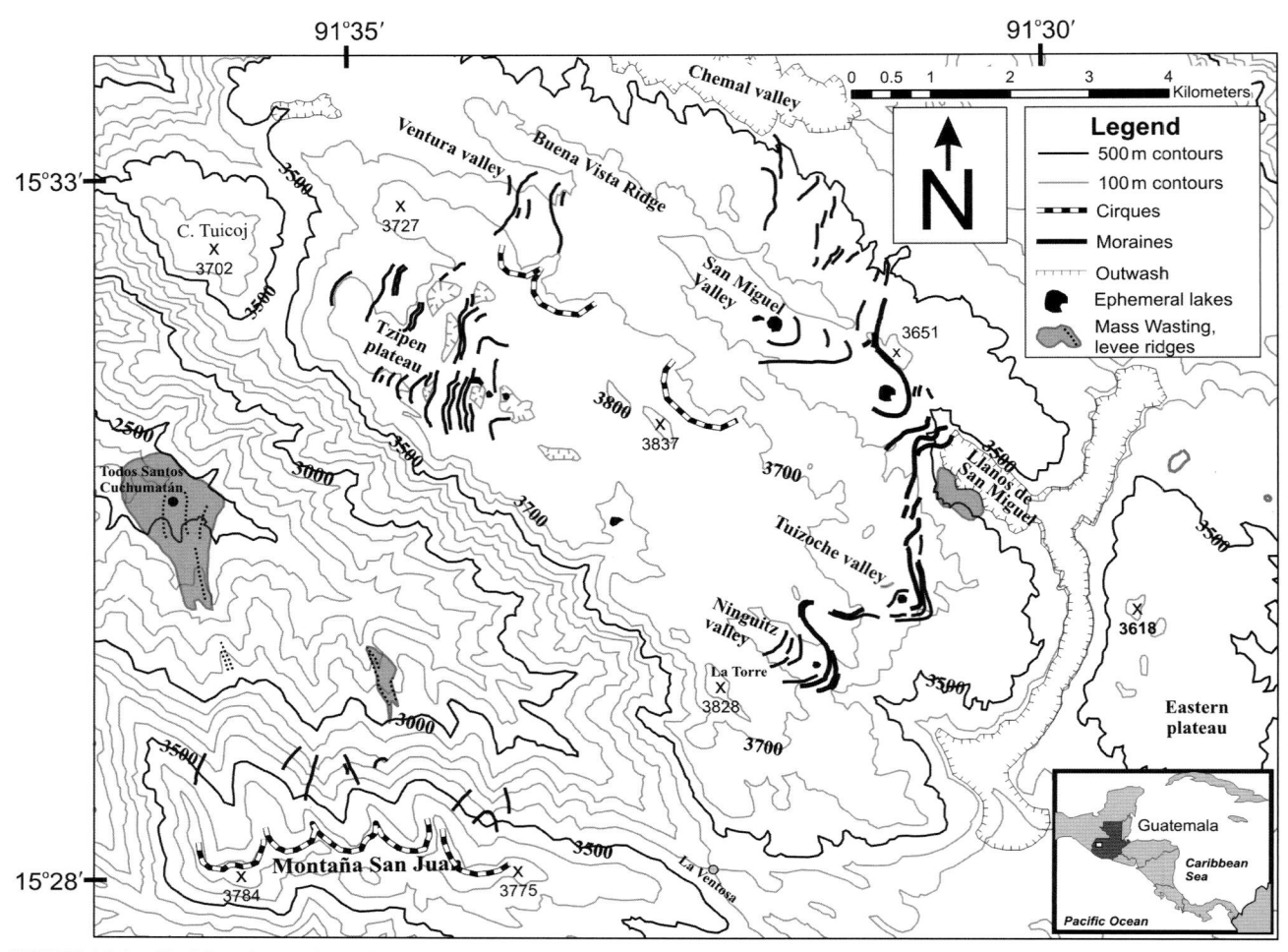

FIGURE 60.1 Glacial geology and moraine limits of the Sierra los Cuchumatanes, including Montaña San Juan. Inset is location of Guatemala.

no radiometric ages for the glaciation exist, the slightly weathered till indicated a late Wisconsin age (Anderson, 1969b), and the moraine ages may be correlative to the Mexican glacial sequence at Iztaccihuátl, located 850 km northwest of the Cuchumatanes.

Our recent field work and stereophotographic analysis allowed for the more detailed and precise mapping of lateral, terminal and recessional moraines; outwash plains; ephemeral moraine-dammed lakes; and other glacial landforms to construct a map of the greatest extent of ice during the last local glacial maximum (LLGM). Moraines marking sub-maximal ice limits are also present in many valleys (Fig. 60.1). Because an absolute chronology of glaciation has not yet been developed, the timing of the LLGM and subsequent smaller ice phases remains unknown.

Our mapping was concentrated on the eastern plateau, including the San Miguel, Tuizoche and Ninguitz Valleys (Fig. 60.1) along with reconnaissance field exploration of the Ventura Valley, the Buena Vista Ridge and the Tzipen plateau. Our mapping delineated moraine sequences and glacial landforms, with prominent nearly continuous ice-cap marginal moraines located along the eastern edge of the plateau. The largest moraine segments on the eastern plateau have peaked crests while some with rounded crests—interpreted as a possibly older sequence—are located just down-valley but in contact with the sharper-crested moraines. Some moraine segments contain multiple crests, suggesting that ice advanced to nearly the same terminal position on multiple occasions. In the San Miguel Valley, lateral/terminal moraines are well developed and impound ephemeral lake basins (Fig. 60.2). On the Tzipen plateau, a sequence of smaller-relief (<15 m) rounded and nested moraines are found up-valley of the maximal ice limits. These moraines were interpreted as recessional moraines. Separated from the main plateau by a fault zone, the nearby Montaña San Juan (3784 m a.s.l.) comprises an east–west trending ridge flanked by a group of north-facing cirques separated by arêtes. The cirque headwalls at ~3650 m a.s.l. lead down-valley to small lateral and end moraine segments.

FIGURE 60.2 Two views of the lateral/end moraine sequence and moraine-dammed lake in San Miguel Valley. (A) Up-valley; (B) down-valley. Stars indicate location of viewpoints for the respective photographs.

60.3. ICE CAP AND VALLEY GLACIER RECONSTRUCTION

The LLGM ice extent (Fig. 60.3) was reconstructed using moraine limits along the eastern and northern parts of the plateau, and by the transition from U- to V-shaped valleys along the western plateau. The southern ice margin is defined by the distinct fault-controlled escarpment, and ice limits on Montaña San Juan were defined by extrapolation of lateral moraines. The ice thickness was determined by matching to the glacial geology and by estimation of realistic basal shear stresses (Roy and Lachniet, 2010). Area analysis shows that the plateau region of the Sierra de los Cuchumatanes supported an ice cap ~40 km^2, with additional ice cover over the Montaña San Juan. Possible nivation basins and other areas of perennial snow cover during the late Quaternary may have been present on other areas above ca. 3500 m a.s.l. that do not show evidence of valley glaciation.

60.4. EQUILIBRIUM LINE ALTITUDES

The late Quaternary equilibrium line altitudes in the Guatemalan highlands were estimated separately for (1) the main ice cap and (2) the valley glaciers on Montaña San Juan, using the area–altitude balance ratio (AABR) method (Osmaston, 2005). Using a typical 'tropical' balance ratio of 5.0, the ELAs were 3670 m a.s.l. for the ice cap and 3470 m a.s.l. for the valley glaciers. The ~200 m lower altitude of the equilibrium line of the valley glaciers on Montaña San Juan may be due to their northerly aspect in valleys that are deeper and shadier relative to the broad plateau. As the altitude of the modern 0 °C isotherm, taken as approximately equivalent to the modern equilibrium line, is 4840 ± 230 m, the ELA depression associated with the LLGM ice limits in Guatemala was 1170 m for the ice cap and 1370 m for the valley glaciers. If interpreted in terms of temperature along using the modern environmental lapse rate of -5.3 °C km^{-1}, LLGM climate was ~6.2–7.3 °C cooler than modern at ca. 3600 m altitude.

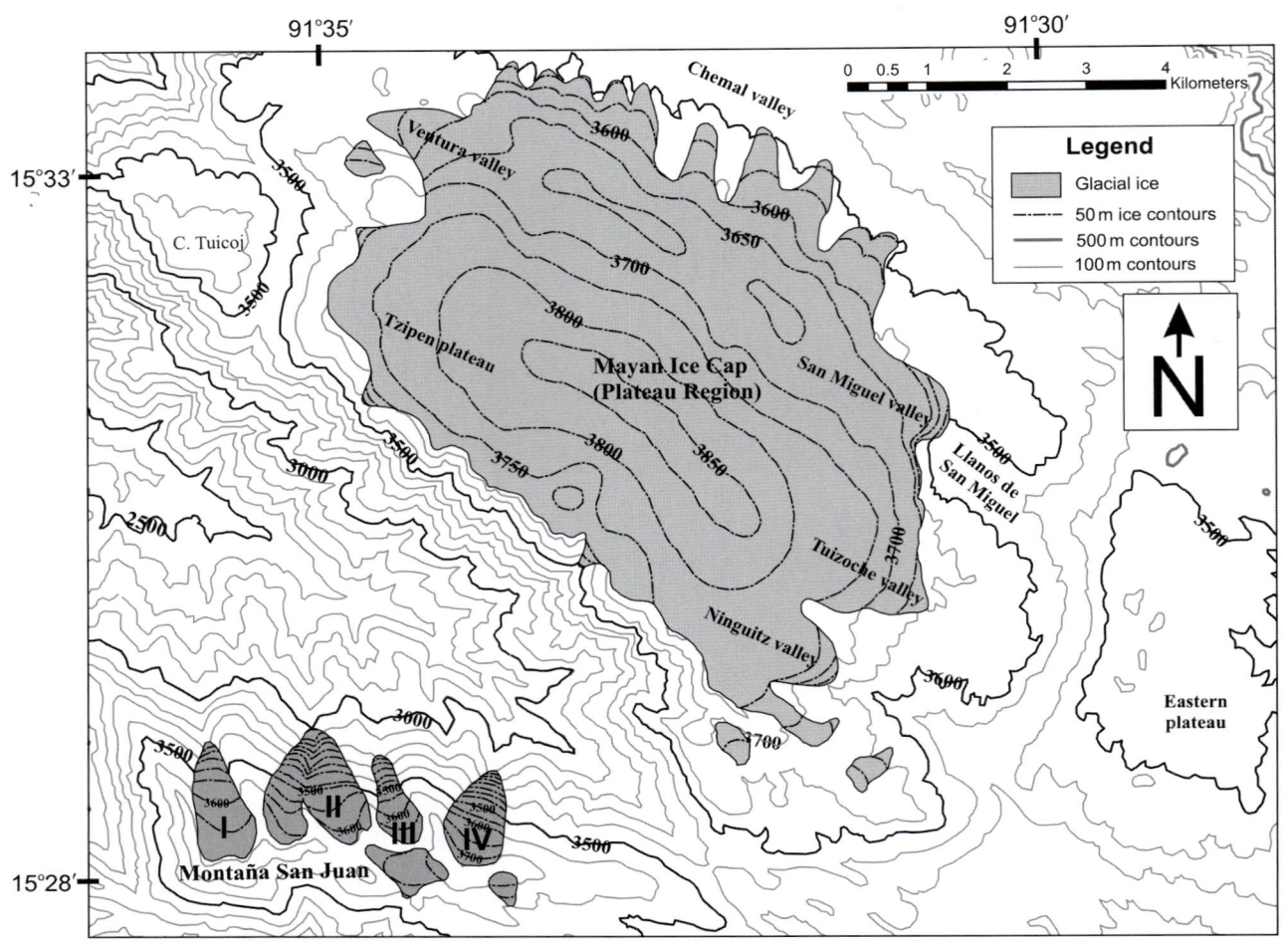

FIGURE 60.3 Last local glacial maximum ice limits of the Sierra los Cuchumatanes, including Montaña San Juan. The ice cap was associated with an equilibrium line altitude of 3670 and the valley glaciers of ~3470 m a.s.l.

60.5. DISCUSSION

The estimated Guatemalan ELAs of 3500–3600 m a.s.l. are similar to those from Costa Rica (Orvis and Horn, 2000; Lachniet and Seltzer, 2002a,b). Thus, the LLGM ELA appears to have been spatially consistent over Central America and should be considered a calibration target for general circulation models. A key focus of future work should be to generate absolute chronologies of glaciation in Costa Rica and Guatemala. Currently, only minimum limiting ages are available for Costa Rica glaciation, and no ages for Guatemala, despite extensive searches for organic material associated with moraine emplacement in both locations by multiple research groups. The absence of organic material associated with moraines and other glacial features is suggestive of a paucity of vegetation during late Quaternary glaciation.

Despite being located near active volcanic centres of Guatemala, soils forming on glacial diamicts comprising the LLGM moraines lack visible ash layers, and glacial sediments exposed in quarries and shallow wells within the glaciated area similarly lacked ash layers. The silicic Los Chocoyos Ash is a widespread ash layer in Guatemala that has been identified in terrestrial and marine stratigraphic successions and dated to ~84 ka by correlation to marine isotope stratigraphy (Drexler et al., 1980; Ledbetter, 1985; Rose et al., 1999). The origin of the Los Chocoyos Ash is at Lago Atitlán, ~95 km southeast of our study area (Koch and McLean, 1975), and it is found throughout the Guatemalan highlands (Rose et al., 1987). The Los Chocoyos Ash reached latitudes of $>25°$N in the Gulf of Mexico, and the Sierra Cuchumatanes lies within the area covered by >10 cm of ash (Drexler et al., 1980). An age of >40 ^{14}C age ka BP, based on ^{14}C dating, for the I2–I5 ashes (Rose et al., 1999) overlying the Los Chocoyos Ash in the Atitlán area may represent an infinite ^{14}C age but is likely younger than 75 ka based on the lack of fission tracks on the I3 ash layer (Rose et al., 1987). The apparent absence of ash layers associated with the LLGM deposits in our study area suggests that glaciation is younger than 84 ka and possibly younger than ca. 40 ^{14}C ka BP. This inference is consistent with the apparently fresh appearance of the LLGM moraines.

Application of cosmogenic nuclide dating of moraines is hampered in Costa Rica by observations of surface weathering and spallation of morainal boulders (Orvis and Horn, 2000) in the wet and humid environment and by chemical dissolution of the limestone moraine boulders and bedrock features in Guatemala (Roy and Lachniet, 2010). Application of luminescence dating of buried sediments may be applicable in both locations, and radiocarbon dating of basal lake sediments in moraine- and bedrock-dammed ephemeral lake basins in Guatemala should be pursued. A tentative correlation to ^{36}Cl-dated moraines on Iztaccíhuatl, central Mexico, suggests that the LLGM moraines in Guatemala may date to between 20.0 and 17.5 ka (Hueyatlaco-1 glaciation) or 17.0–14.0 ka (Hueyatlaco-2; Vazquez-Selem and Heine, 2004).

ACKNOWLEDGEMENTS

The authors would like to thank the collaboration and field work participation of Sr. Gerónimo Pablo Ramírez, of La Ventosa, for his enthusiastic support and curiosity in the Sierra los Cuchumatanes, without which our field work would not have been possible. We also thank the support of the Alcaldia Municipal of Todos Santos Cuchumatán for permission to complete the field work in this area. This research was partially supported by the University of Nevada, Las Vegas. Much of the research on Guatemala was part of the M.S. thesis of Alex Roy at UNLV.

REFERENCES

Anderson, T.H., 1969a. First evidence for glaciation in Sierra Los Cuchumatanes Range, northwestern Guatemala. Geol. Soc. Am. Spec. Pap. 121, 387.

Anderson, T.H., 1969b. Geology of the San Sebastian Huehuetenango Quadrangle, Guatemala. University of Texas, Austin.

Anderson, T.H., Burkart, B., Clemons, R.E., Bohnenberger, O.H., Blount, D.N., 1973. Geology of the Western Altos Cuchumatanes, Northwestern Guatemala. Geological Society of America Bulletin. 84, 805–826.

Barquero, J., Ellenberg, L., 1982. Geomorfologfa del piso alpino del Chirripo en la Cordillera de Talamanca, Costa Rica. Revista Geográfica de América Central. 17 (18), 293–299.

Barquero, J., Ellenberg, L., 1986. Geomorphologie der alpinen Stufe des Chirripo in Costa Rica. Eiszeit. Ggw 36, 1–9.

Bergoeing, J.P., 1978. Modelado Glaciar en la Cordillera de Talamanca. Costa Rica. Informe Semestral del Instituto Geografico Nacional, Ministerio de Obras Publicas y Transportes Julio/Diciembre, pp. 33–44.

Blake, S.F., 1934. New Asteraceae from Guatemala and Costa Rica collected by A.F. Skutch. Journal of the Washington Academy of Sciences. 24, 432–443.

Bundschuh, J., Birkle, P., Finch, R.C., Day, M., Romero, J., Paniagua, S., et al., 2007. Geology-related tourism for sustainable development. In: Bundschuh, J., Alvarado, G.E. (Eds.), Central America: Geology, Resources and Hazards. Taylor & Francis, London, pp. 1015–1098.

Drexler, J.W., Rose, W.I., Sparks, R.S.J., Ledbetter, M.T., 1980. The Los Chocoyos Ash, Guatemala: a major stratigraphic marker in Middle America and in three ocean basin. Quatern. Res. 13, 327–345.

Enjalbert, H., 1967. Les montagnes calcaires du Mexique et du Guatemala. Annales de Geographie 76, 25–59.

Hastenrath, S., 1973. On the Pleistocene glaciation of the Cordillera de Talamanca, Costa Rica. Zeitschrift für Gletscherkunde und Glazialgeologie. 9, 105–121.

Hastenrath, S., 1974. Spuren pleistozaener Vereisung in den Altos de Cuchumatanes, Guatemala. Traces of Pleistocene glaciation in the Altos de Cuchumatanes. Eiszeitalter und Gegenwart 25, 25–34.

Horn, S.P., 1990. Timing of deglaciation in the Cordillera de Talamanca, Costa Rica. Climate Res. 1, 81–83.

Horn, S.P., Orvis, K.H., Haberyan, K.A., 2005. Limnología de las lagunas glaciales en el páramo del Chirripó, Costa Rica. In: Kappelle, M.S.P.H. (Ed.), Páramos de Costa Rica. Editorial INBio, San Jose, CR., Santo Domingo de Heredia, pp. 161–181.

Kappelle, M., van Uffelen, J.G., 2005. Los suelos de los páramos de Costa Rica. In: Kappelle, M., Horn, S.P. (Eds.), Páramos de Costa Rica. Editorial INBio, Santo Domingo de Heredia, pp. 147–159.

Koch, A.J., McLean, H., 1975. Pleistocene tephra and ash-flow deposits in the volcanic highlands of Guatemala. Geol. Soc. Am. Bull. 86, 529–541.

Lachniet, M.S., 2004. Late Quaternary glaciation of Costa Rica and Guatemala, Central America. In: Ehlers, J., Gibbard, P.L. (Eds.), Quaternary Glaciations—Extent and Chronology. Part III: South America, Asia, Africa, Australasia, Antarctica. Elsevier, Amsterdam, pp. 135–138.

Lachniet, M.S., 2007. Glacial geology and geomorphology. In: Bundschuh, J., Alvarado, G. (Eds.), Central America: Geology, Resources, and Hazards. Taylor & Francis, London, pp. 171–182.

Lachniet, M.S., Seltzer, G.O., 2002a. Erratum: late Quaternary glaciation of Costa Rica [modified]. Geol. Soc. Am. Bull. 114, 921–922.

Lachniet, M.S., Seltzer, G.O., 2002b. Late Quaternary glaciation of Costa Rica. Geol. Soc. Am. Bull. 114, 547–558.

Lachniet, M.S., Vazquéz-Selem, L., 2005. Last glacial maximum equilibrium line altitudes in the circum-Caribbean (Mexico, Guatemala, Costa Rica, Colombia, and Venezuela). Quatern. Int. 138–139C, 129–146.

Lachniet, M.S., Seltzer, G.O., Solís, S.L., 2005. Geología, geomorfología y depósitos glaciares en los páramos de Costa Rica. In: Kappelle, M., Horn, S.P. (Eds.), Páramos de Costa Rica. Editorial INBio, Santo Domingo de Heredia, p. 1293.

Ledbetter, M.T., 1985. Tephrochronology of marine tephra adjacent to Central America. Geol. Soc. Am. Bull. 96, 77–82.

Orvis, K., Horn, S., 2000. Quaternary glaciers and climate on Cerro Chirripo, Costa Rica. Quatern. Res. 54, 24–37.

Orvis, K.H., Horn, S.P., 2005. Glaciares cuaternarios y clima del Cerro Chirripó, Costa Rica. In: Kappelle, M., Horn, S.P. (Eds.), Páramos de Costa Rica. Editorial INBio, Santo Domingo de Heredia, pp. 185–213.

Osmaston, H., 2005. Estimates of glacier equilibrium line altitudes by the area × altitude, area × altitude balance ratio, and the area × altitude balance index methods and their validation. Quatern. Int. 238–239, 22–32.

Owen, L.A., Finkel, R.C., Minnich, R., Perez, A., 2003. Extreme southern margin of late Quaternary glaciation in North America: timing and controls. Geology 31, 729–732.

Rose, W.I., Newhall, C.G., Bornhorst, T.J., Self, S., 1987. Quaternary silicic pyroclastic deposits of Atitlán Caldera, Guatemala. J. Volcanol. Geoth. Res. 33, 57–80.

Rose, W.I., Conway, F.M., Pullinger, C.R., Deino, A., McIntosh, W.C., 1999. An improved age framework for late Quaternary silicic eruptions in northern Central America. Bull. Volcanol. 61, 106–120.

Roy, A.J., Lachniet, M.S., 2010. Late Quaternary glaciation and equilibrium line altitudes of the Mayan Ice Cap, Guatemala, Central America. Quatern. Res. 74, 1–7.

Shimizu, C., 1992. Glacial landforms around Cerro Chirripo in Cordillera de Talamanca, Costa Rica. J. Geogr. Jpn 101, 615–621 (in Japanese).

van Uffelen, J.G., 1991. A geological/geomorphological and soil transect study of the Chirripo Massif and adjacent areas, Cordillera de Talamanca, Costa Rica. Unpublished report to Centro Agronomico Tropical de Investigacion y Ensefianza Agricultural, University Wageningen, Ministerio de Agricultura y Ganaderia de Costa Rica, Universidad Nacional, Heredia.

Vazquez-Selem, L., Heine, K., 2004. Late Quaternary glaciation of México. In: Ehlers, J., Gibbard, P.L. (Eds.), Quaternary Glaciations— Extent and Chronology. Part III. Elsevier, Amsterdam, pp. 233–242.

Weyl, R., 1956. Eiszeitliche Gletscherspuren in Costa Rica (Mittelamerika). Z. Gletscherk. Glazialgeol. 3, 317–325.

Weyl, R., 1962. Glaciares Pleistocenos en la zona tropical de Centroamerica. Sonderdruck aus Alemania—La Revista de la Republica Federaf' 2. Jahrgang—Heft no. 4, unknown.

Chapter 61

Late Quaternary Glaciation in Mexico

Lorenzo Vázquez-Selem[1],* and Klaus Heine[2]

[1]Instituto de Geografía, Universidad Nacional Autónoma de México, Ciudad Universitaria, 04510 México, D.F., Mexico
[2]Geographisches Institut, Universität Regensburg, Universitätsstr.31, D-93040 Regensburg, Germany
*Correspondence and requests for materials should be addressed to Lorenzo Vázquez-Selem. E-mail: lselem@igg.unam.mx

61.1. INTRODUCTION

The glacial history of the central Mexican volcanoes has been researched extensively, but both the limits and ages of the Wisconsinan late-glacial and early Holocene glaciations remain poorly understood in many places. There is evidence of glaciation on 13 volcanoes in central Mexico (Fig. 61.1). Deposits of five late-glacial and Holocene glaciations have been recognised (White, 1962, 1987; Heine, 1975, 1988, 1994; White et al., 1990; Vázquez-Selem, 1991, 1997, 2000). Age assignments for these deposits have been qualitative and based on stratigraphical position of the moraines and tills, and their relation to periglacial deposits, palaeosols, debris flows, fluvial gravels, fluvial sands, loess-like so-called 'toba' sediments, lava flows, ignimbrite deposits and tephra. In addition, radiocarbon dating of tephra layers, palaeosols, peats and gravels has contributed to the late Quaternary glacial chronostratigraphy, as well as cosmogenic ^{36}Cl exposure ages. Yet, the dated sections containing glacial deposits from different volcanoes of central Mexico do not include evidence for late Quaternary glacier advances that occurred between 25 and 18 ^{14}C ka BP.

Here, the authors report on stratigraphical sections relevant to the timing of the late Quaternary glacial sequences from the Iztaccíhuatl massif, the Nevado de Toluca volcano and the La Malinche, Cofre de Perote and Tancítaro volcanoes (Fig. 61.1), and revise earlier proposed stratigraphies (White and Valastro, 1984; Heine, 1994). From field observations, tephra studies, laboratory analyses, ^{14}C dates, ^{36}Cl exposure ages and literature estimates, the writers reconstruct glacier advances that occurred during late Quaternary time.

61.2. IZTACCÍHUATL

The Iztaccíhuatl volcanic massif rises south-east of Mexico City to heights of 5286 m a.s.l. (Fig. 61.1). It occurs in the central part of the Sierra Nevada, a north–south orientated mountain range that forms the south-eastern margin of the Valley of Mexico. The glaciated summit of Iztaccíhuatl ranks it third highest of the Mexican volcanoes after Popocatépetl, situated at the southern end of the Sierra Nevada and the Pico de Orizaba to the east. The Iztaccíhuatl shows a complex history of cone construction comprising a number of coalescing and superimposed central volcanoes. The eruptive products are calc-alkaline andesites and dacites, comprising viscous flows and flow breccias with minor intercalated pyroclastic material. The cone construction began prior to 0.9 Ma and continued until the late Pleistocene (Nixon, 1989). The evolution of Iztaccíhuatl is divided into two main phases represented by rocks of the Older Volcanic Series (>0.6 Ma) and Younger Volcanic Series (<0.6 Ma).

The chronology of late Quaternary glacial moraines and tills identified by White (1962, 1987) on the western flanks of Iztaccíhuatl has been extended to glacial deposits mapped on the eastern side of the volcano by Heine (1975) and Nixon (1989), and on the northern flanks (Téyotl peak) by Vázquez-Selem (1991, 1997). More recently, Vázquez-Selem (2000) elaborated a new late Quaternary glacial chronology for Iztaccíhuatl (Fig. 61.2). Here, the authors refer to this chronology (Vázquez-Selem and Phillips, 1998; Vázquez-Selem, 2000; Table 61.1).

Morphostratigraphy, tephrochronology and 94 cosmogenic ^{36}Cl exposure ages provide the basis for this chronology of the Iztaccíhuatl volcano. Age estimates of the different glacier advances are based upon three or more ^{36}Cl exposure ages, in some cases supplemented by radiocarbon ages associated to tephras. The total uncertainty of ^{36}Cl age estimates should be in the range of 5–10% (Zreda and Phillips, 1994; Zreda et al., 1994). Exposure ages are stated in ^{36}Cl thousand of years before present (hereafter ^{36}Cl ka) which are considered equivalent to calibrated (cal) ka BP.

The chronology, together with a study of equilibrium line altitudes (ELAs) of past glaciers and their palaeoclimatic implications, reveals evidence for marked climatic change in the Mexican tropics during the late Quaternary.

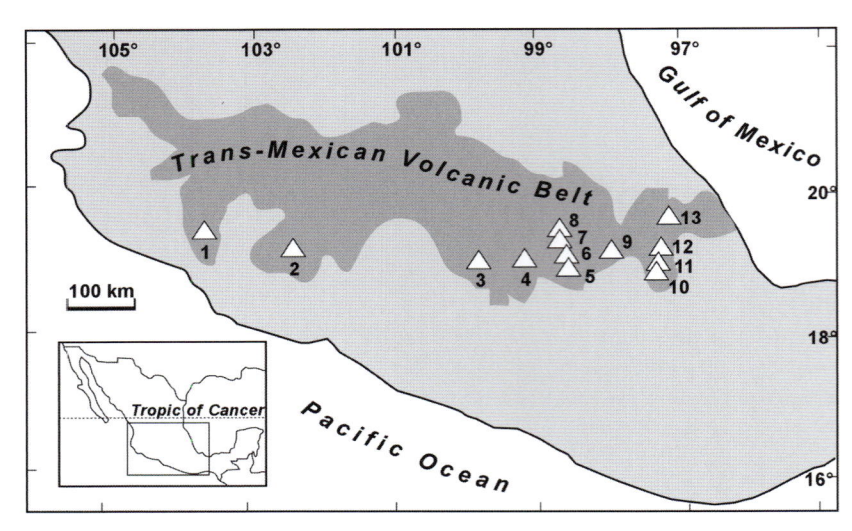

FIGURE 61.1 Location of volcanoes in central Mexico with evidence of glaciation and sites mentioned in the text. The three highest volcanoes currently support small glaciers. 1. Nevado de Colima (4180 m a.s.l.). 2. Tancítaro (3842 m a.s.l.). 3. Nevado de Toluca (4690 m a.s.l.). 4. Ajusco (3952 m a.s.l.). 5. Popocatépetl (5452 m a.s.l.). 6. Iztaccíhuatl (5286 m a.s.l.). 7. Telapón (4090 m a.s.l.). 8. Tláloc (4120 m a.s.l.). 9. La Malinche (4461 m a.s.l.). 10. Sierra Negra (4600 m a.s.l.). 11. Pico de Orizaba (5675 m a.s.l.). 12. Las Cumbres (3950 m a.s.l.). 13. Cofre de Perote (4282 m a.s.l.).

The oldest glacial deposits unequivocally identified on Iztaccíhuatl (Nexcoalango, as in White, 1962) have been dated at 205–175 ^{36}Cl ka, that is, they probably correspond to an advance during early Marine Isotope Stage 6 (MIS 6, Illinoian). Two lava flows from northern Iztaccíhuatl dated in ~120 and ~140 ^{36}Cl ka partially cover Nexcoalango glaciated areas. The next advance recorded (Hueyatlaco-1) peaked at 21–20 ^{36}Cl ka. The lack of evidence for glacial advances between ~175 and 21 ^{36}Cl ka suggests that Hueyatlaco-1 advance was more extensive than any other post-Nexcoalango event. It is also possible that post-Nexcoalango volcanic activity of Iztaccíhuatl obliterated part of the glacial record. However, the possibility of glacial advances prior to Hueyatlaco-1 is suggested by deposits older than the local last glacial maximum (LLGM) on other volcanoes in central Mexico.

The Hueyatlaco-1 (~21–17.5 ^{36}Cl ka) is the most extensive glacial advance of the late Pleistocene and represents the LLGM. Glaciers had a mean terminus elevation of 3390 m and mean ELA of 3940 ± 130 m (THAR = 0.4). Recession was apparently in progress by 17.5 ^{36}Cl ka.

The Hueyatlaco-2 (~17–14.5 ^{36}Cl ka) advance was underway when pumice from Popocatépetl volcano fell around 17.1 cal. ka BP. Massive lateral moraines actively developed shortly before or during the period 15–14 ^{36}Cl ka, when glaciers reached a mean terminus elevation of 3500 m and a mean ELA of 4040 ± 130 m (THAR = 0.4). Short before 14 ^{36}Cl ka glaciers started receding from their maximum positions. Up to three Hueyatlaco-2 recessional moraines formed at 3400–3800 m a.s.l. (end positions) between 14.5 and 13.5 ka, probably indicating a period of marked climatic instability.

Available data suggest that the first half of the Younger Dryas Chronozone (12.9–11.7 ka) is coeval with the final recession of Hueyatlaco-2 glaciers, while the second half may coincide with a small re-advance (Milpulco-1). Therefore, it appears that both recession and advance of glaciers of Iztaccíhuatl were time transgressive with respect to the Younger Dryas Chronozone.

Milpulco-1 advance peaked around 12 ^{36}Cl ka, with glacier termini at a mean elevation of 3810 m and the mean ELA at 4240 ± 60 m (THAR = 0.4). Recessional moraines developed between ~11 and ~10 ^{36}Cl ka. By 10 ^{36}Cl ka, a major recession was apparently in progress and continued at least until ~9.5 ^{36}Cl ka.

The maximum expansion of the Milpulco-2 glaciers occurred shortly before 8 ^{36}Cl ka, with mean glacier termini at 4050 m a.s.l. and mean ELA at 4420 ± 120 m (THAR = 0.4). Up to five recessional moraines developed between ~8 and ~7 ^{36}Cl ka. Talus rock glaciers formed on shaded slopes probably as low as 3900 m a.s.l. By 7 ^{36}Cl ka, all areas below 4300–4400 m a.s.l. were ice free. All deglaciated areas and moraines were mantled by a pumice deposit from Popocatépetl at 5.7 cal. ka BP.

There is no evidence of glacier expansion between the early-to-mid Holocene advance (Milpulco-2) and the late Holocene advance (Ayoloco). Moraines of the latter post-date the deposition of a pumice from the Popocatépetl volcano dated to ca. 1 ka BP (Siebe et al., 1996). At the maximum of Ayoloco, represented by conspicuous massive moraines, the mean elevation of glacier termini was 4510 m a.s.l. and the mean ELA 4715 ± 75 m (THAR = 0.4). Up to four Ayoloco recessional moraines developed, the youngest of which was formed during the first half of the twentieth century. The distribution of talus rock glaciers and protalus ramparts indicates that periglacial conditions with abundant snowfall prevailed elevations as low as 4260 m a.s.l. at least during the peak of the advance.

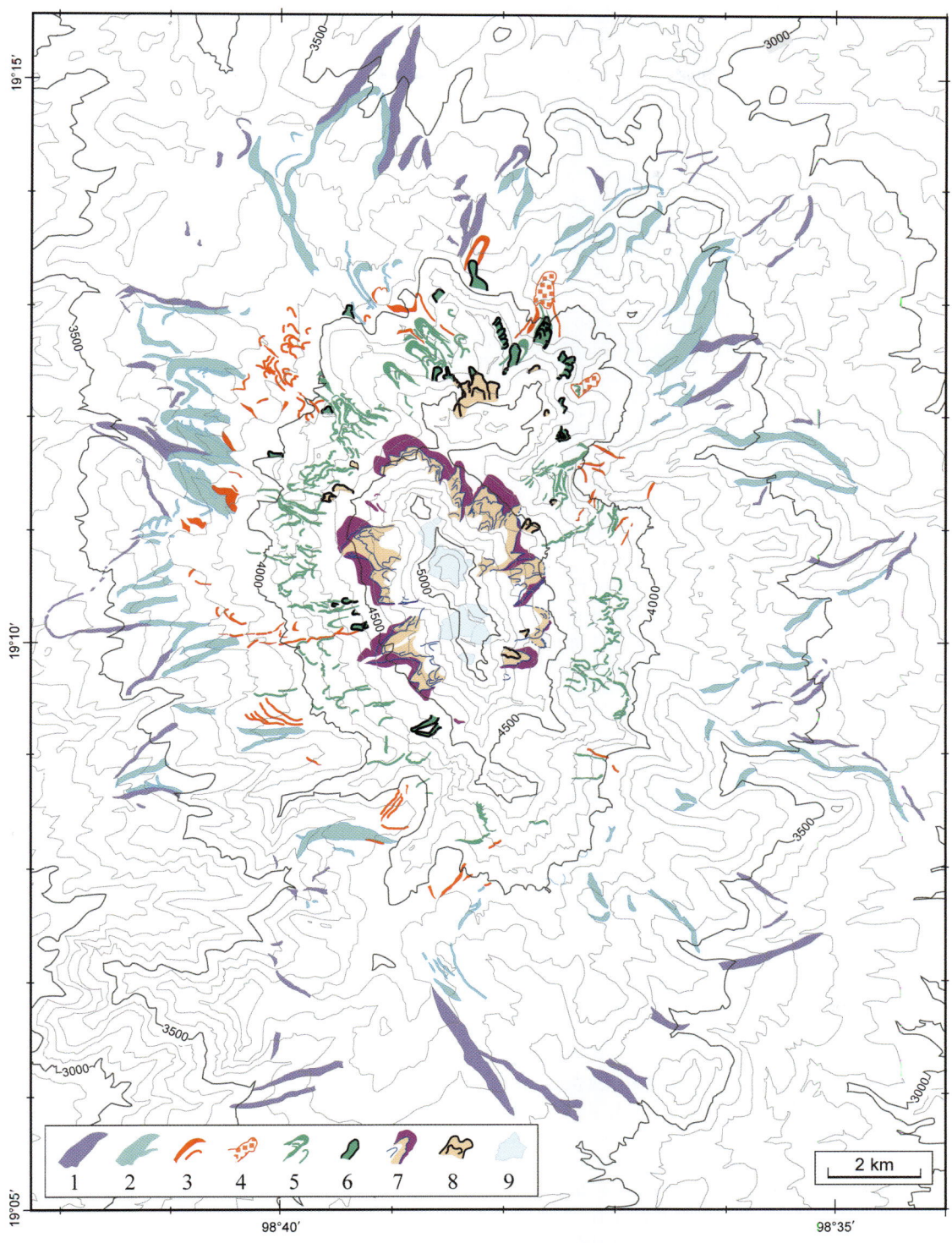

Chapter | 61 Late Quaternary Glaciation in Mexico

FIGURE 61.2 The late Pleistocene and Holocene glacial features of Iztaccíhuatl massif. 1. Hueyatlaco-1 moraine. 2. Hueyatlaco-2 moraine. 3. Milpulco-1 moraine. 4. Milpulco-1 rock glacier. 5. Milpulco-2 moraine. 6. Milpulco-2 rock glacier. 7. Ayoloco moraine. 8. Ayoloco rock glacier. 9. Glacier in AD 1983 (modified after Vázquez-Selem, 2000).

TABLE 61.1 Glacial Chronology of the Iztaccíhuatl Volcano (Modified from Vázquez-Selem, 2000)

Glacial advance	Age (^{36}Cl ka)	Glaciated Area (km²)	Mean Altitude of Glacier Terminus (m a.s.l.)	Mean ELA THAR = 0.4 (m a.s.l.)
Modern (AD 1960)	0.04	1.3	4860 ± 130	4970 ± 90
Ayoloco	<1	7.6	4510 ± 110	4720 ± 70
Milpulco-2 (recessional)	7.8–7.3	–	4180 ± 80	4530 ± 100
Milpulco-2 (main advance)	~8.3	29	4050 ± 120	4420 ± 120
Milpulco-1	12–10	46	3810 ± 80	4240 ± 60
Hueyatlaco-2 recessional	14.5–13.5	–	–	–
Hueyatlaco-2	17–14.5	101	3500 ± 190	4040 ± 130
Hueyatlaco-1	21–17.5	131	3390 ± 160	3940 ± 130
Nexcoalango	~205–175	>200	~3000	?

61.3. NEVADO DE TOLUCA

The Nevado de Toluca volcano rises to 4690 m a.s.l. and is situated about 80 km west of Mexico City (Fig. 61.1). It is a strongly eroded polygenetic central stratovolcano with moderate slopes and a summit crater 0.5 × 1.5 km in size. The volcano largely comprises dacitic lava flows but its lower slopes are mantled by a thick spread of volcanoclastic deposits, including pyroclastic flow deposits, lahars and pumice beds (Bloomfield and Valastro, 1977; Cantagrel et al., 1981). Despite several highly explosive eruptions during the late Pleistocene (Macías, 2005), it shows clear evidence of past glaciation and periglacial activity, both before and after the major eruptions (Heine, 1976a,b, 1994).

The summit area of Nevado de Toluca volcano includes several moraine systems (Fig. 61.3). Heine (1976a,b, 1988, 1994) presented detailed descriptions of the moraines and rock glaciers of the north-western, northern and south-eastern slopes. Additional field observations and ^{14}C and ^{36}Cl dating by both authors have led to a slightly revised glacial chronology (Table 61.2).

Tephrostratigraphic investigations document that the moraines of the different groups can be attributed to the following late Quaternary intervals: M I, older than 32 ^{14}C ka BP; M II, older than 11.6 ^{14}C ka BP; M III, younger than 11.6 ^{14}C ka BP. Several pyroclastic deposits are useful for stratigraphical correlations. The Lower Toluca Pumice (LTP), which is only found on the north-eastern slopes of the volcano, has an age of about 24–25 ^{14}C ka BP (Bloomfield and Valastro, 1977; Newton and Metcalfe, 1999; Caballero et al., 2001) or ca. 21.7 ^{14}C ka BP (Capra et al., 2006) and does not mantle the M II moraines. A much thicker deposit, the Upper Toluca Pumice (UTP) beds, mantled most of the mountain at 11.6 ^{14}C ka BP (Bloomfield and Valastro, 1977; 10.5 ^{14}C ka BP according to Arce et al., 2003), including M II moraines and rock glaciers. In many places, the UTP beds rest on a strong unconformity that is developed on palaeosols. Two organic-rich horizons that underlie the UTP were dated, one at the Nevado de Toluca (99°48.5′W, 19°08′N) giving an age of 11.37 ± 0.175 ^{14}C ka BP (Hv 20242, 13.404–13.086 cal. ka BP, 1 sigma), and one at Ajusco volcano (99°16′W, 19°13′N), yielding an age of 11.730 ± 0.110 ^{14}C ka BP (Hv 24529, 13.701–13.455 cal. ka BP, 1 sigma). These dates are in good agreement with other dates (Heine, 1994; Caballero et al., 2001) but are older than that suggested by Arce et al. (2003) of 10.445 ± 0.095 ^{14}C ka B.P (12.608–12.161 cal. ka BP). Cosmogenic ^{36}Cl ages of 7.4 ± 0.3 and 7.0 ± 0.3 ^{36}Cl ka reported by Arce et al. (2003) correspond to moraines deposited after the UTP eruption (M III in Fig. 61.3), thus supporting their formation during the early-to-mid Holocene.

The rock glaciers can be grouped into four systems: the oldest groups correspond to moraine groups M II and M III; the M IV group developed during the early to mid Holocene (based on an unpublished ^{36}Cl age of 7.5 ^{36}Cl ka); and the M V group during Little Ice Age times.

Using all the available data, the glacial chronostratigraphy for Nevado de Toluca volcano is proposed in Table 61.2.

61.4. LA MALINCHE VOLCANO

La Malinche volcano is a Tertiary–Quaternary andesitic stratovolcano and rises to 4461 m a.s.l. It is an isolated mountain located 60 km east of Iztaccíhuatl, between the basins of Puebla/Tlaxcala and Oriental (Fig. 61.1). Unlike Iztaccíhuatl, it has been active during the late Pleistocene and Holocene (Heine, 1975; Castro-Govea and Siebe,

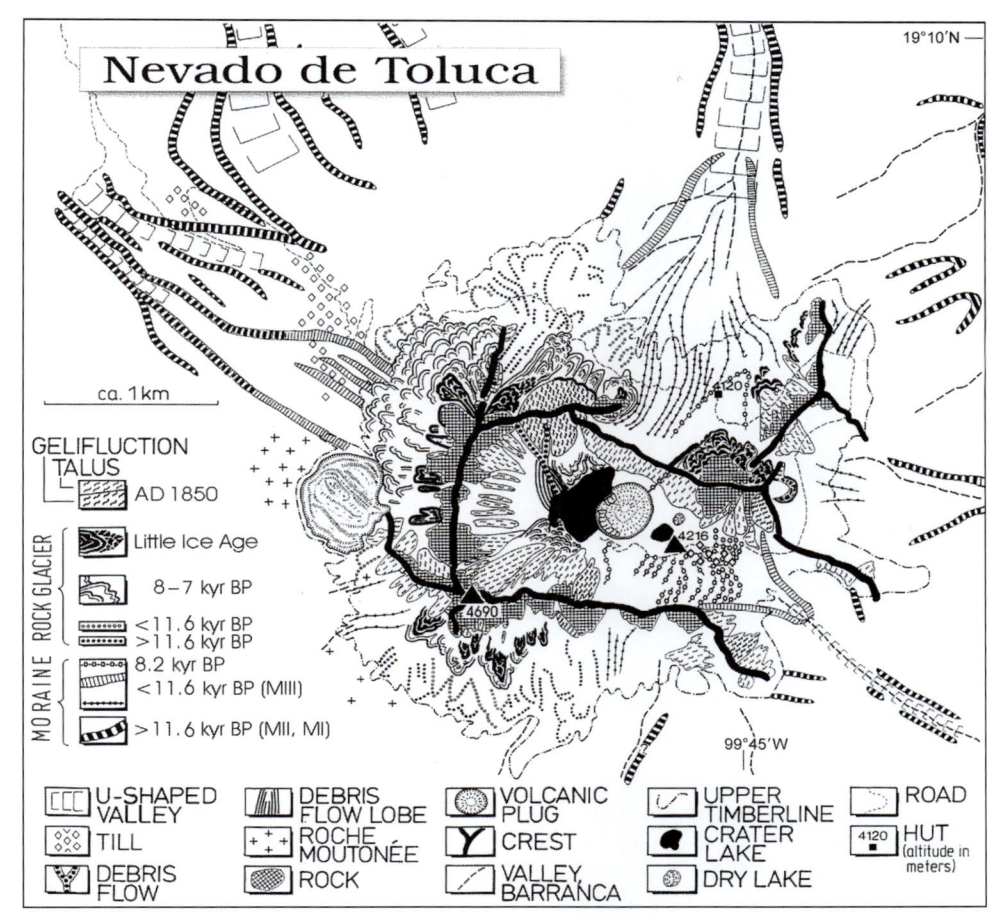

Chapter | 61 Late Quaternary Glaciation in Mexico

FIGURE 61.3 Glacial morphology of Nevado de Toluca volcano (modified after Heine, 1994).

TABLE 61.2 Glacial Stratigraphy of the Nevado de Toluca Volcano

M V rock glacier	Little Ice Age
M IV moraine/till/rock glacier	Younger than 11.6 ^{14}C ka BP; ^{36}Cl dating suggests 8–7 ka
M III moraine/till/rock glacier	Younger than 11.6 ^{14}C ka BP, most probably late Younger Dryas/early Holocene
Upper Toluca Pumice	ca. 11.6 ^{14}C ka BP
Palaeosol (fBo2)	ca. 11–12 ^{14}C ka BP
M II 2 moraine/till	Older than 11.6 ^{14}C ka BP, younger than ca. 17 ^{14}C ka BP
Gelifluction layer	Older than 17 ^{14}C ka BP, younger than 21.7 ^{14}C ka BP
M II 1 moraine/till	Older than ca. 17 ^{14}C ka BP, younger than ca. 21.7 ^{14}C ka BP
Lower Toluca Pumice	ca. 21.7 ^{14}C ka BP
Palaeosol (fBo1)	ca. 28–21.7 ^{14}C ka BP
Varved clays, reworked till, gelifluction deposits	Older than 25 ^{14}C ka BP
M I till	Older than ca. 40 ^{14}C ka BP, probably MIS 4 and/or older

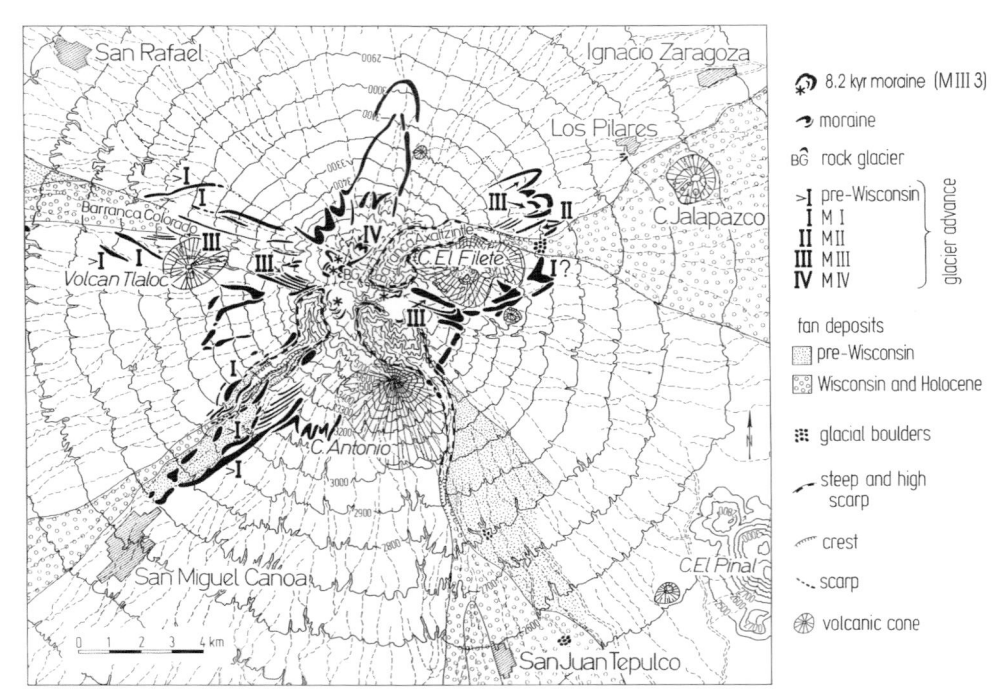

FIGURE 61.4 Glacial morphology of La Malinche volcano (modified after Heine, 1994).

2007). Radial stream drainage has produced many shallow narrow ravines leading down from the summit area. Several glaciations produced moraine ridges around the volcano. Heine (1975, 1994) developed the record of late Pleistocene and Holocene glaciations on Malinche as his standard (Fig. 61.4). The oldest glaciation, M I, is dated to 36–32 ^{14}C ka BP. Moraines of the next glaciation, M II, were limited in earlier publications (e.g. Heine, 1975) to a very short period around 12 ^{14}C ka BP, but were subsequently dated at ~16–12 ^{14}C ka BP (~19– 14 cal. ka BP; Heine, 1994). The moraines of group M III are dated between 12 and 8.5 ^{14}C ka BP by means of tephra correlation. Based on extensive field observations and sedimentation rate calculations of the loess-like toba deposits, as well as on an estimation of the time needed for soil development, the glacial advance of the M III group moraines was estimated to have occurred after ~10 ^{14}C ka BP (~11.5 cal. ka BP). The moraines of the M III group show three individual advances: M III 1, M III 2 and M III 3. During the Neoglacial between 3 and 2 ^{14}C ka BP and during the Little Ice Age, periglacial deposits (rock glaciers, gelifluction layers) developed.

61.5. AJUSCO VOLCANO

Ajusco is a group of eroded volcanic peaks rising to 3937 m a.s.l. along the south-western divide of Cuenca de México, ca. 60 km west of Iztaccíhuatl. It is an aggregate of late Miocene to late Pliocene andesitic dacite lava flows, tuffs, breccias and ignimbrites (White, 1987; White et al., 1990). The glacial chronology of Ajusco is based on a few radiocarbon ages that admittedly represent only minimum ages (White and Valastro, 1984). Radiocarbon dating provides minimum ages of about 27 ^{14}C ka BP for the Marqués glaciation and about 25 ^{14}C ka BP for the Santo Tomás glaciation (White, 1987). However, based on soil development, White et al. (1990) admitted the possibility that the moraines of these two advances could be much older, even pre-Wisconsinan in age. The Albergue glaciation is dated by White and Valastro (1984) to the period 15–8 ^{14}C ka BP. White (1987) generally placed it between 16 and 10 ^{14}C ka BP. White and Valastro (1984) assigned an age of <2 ^{14}C ka BP to moraines of Neoglaciation I and II, despite the fact that these moraines lie ca. 1000 m lower than comparable moraines of the Iztaccíhuatl massif.

61.6. CITLALTÉPETL

The Citlaltépetl or Pico de Orizaba is the highest mountain in Mexico and lies 140 km due east of Iztaccíhuatl (Fig. 61.1). Its altitude has never been accurately determined and references range from 5675 to 5700 m a.s.l. Pico de Orizaba is a stratovolcano which developed in three stages, and has had recent volcanic activity. Its lithology is dominated by andesites and dacites (Hoskuldsson and Robin, 1993; Palacios and Vázquez-Selem, 1996). According to Heine (1975, 1988), several late Pleistocene and early Holocene glacier advances are comparable to the late

Quaternary stratigraphical succession of the La Malinche volcano. Only relative age-dating methods have been applied to determine age differences in the moraine sequence.

61.7. COFRE DE PEROTE

Cofre de Perote (4282 m) is a Quaternary compound volcano located 50 km to the north of Citlaltépetl. Here the youngest lavas were emplaced from 250–200 ka and the edifice has remained inactive ever since, except for two collapses, ca. 40 and 13–11 ^{14}C ka (Carrasco-Núñez et al., 2010). Evidence of glaciation was previously reported by Lorenzo (1964) and Heine (1975). Recent mapping (Fig. 61.5) and ^{36}Cl dating have revealed extensive valley glaciers between ∼20 and ∼14 ^{36}Cl ka, with mean termini at 3390 ± 70 m a.s.l. and mean ELA at 3650 ± 40 m (THAR = 0.4) and 3620 ± 60 m (the maximum altitude of lateral moraines; Lachniet and Vázquez-Selem, 2005). Deglaciation here took place from 14 to 11 ^{36}Cl ka, probably contributing to the youngest collapse event of the edifice (Carrasco-Núñez et al., in press). Small undated moraines located at the foot of two cirques at 4000–3850 m a.s.l., probably formed during the terminal Pleistocene.

Cerro Las Cumbres (3950 m), located between Cofre de Perote and Citlaltépetl, has clear glacial morphology. Assuming an ELA of ∼3650 m a.s.l. as in Cofre de Perote, small glaciers must have existed on this mountain during the LLGM (∼20–14 ka).

61.8. TANCÍTARO

Tancítaro peak (3842 m) is an extinct andesitic stratovolcano located in west-central Mexico. It was built between ∼800 and ∼240 ka (Ownby et al., 2007). The upper parts of the mountain

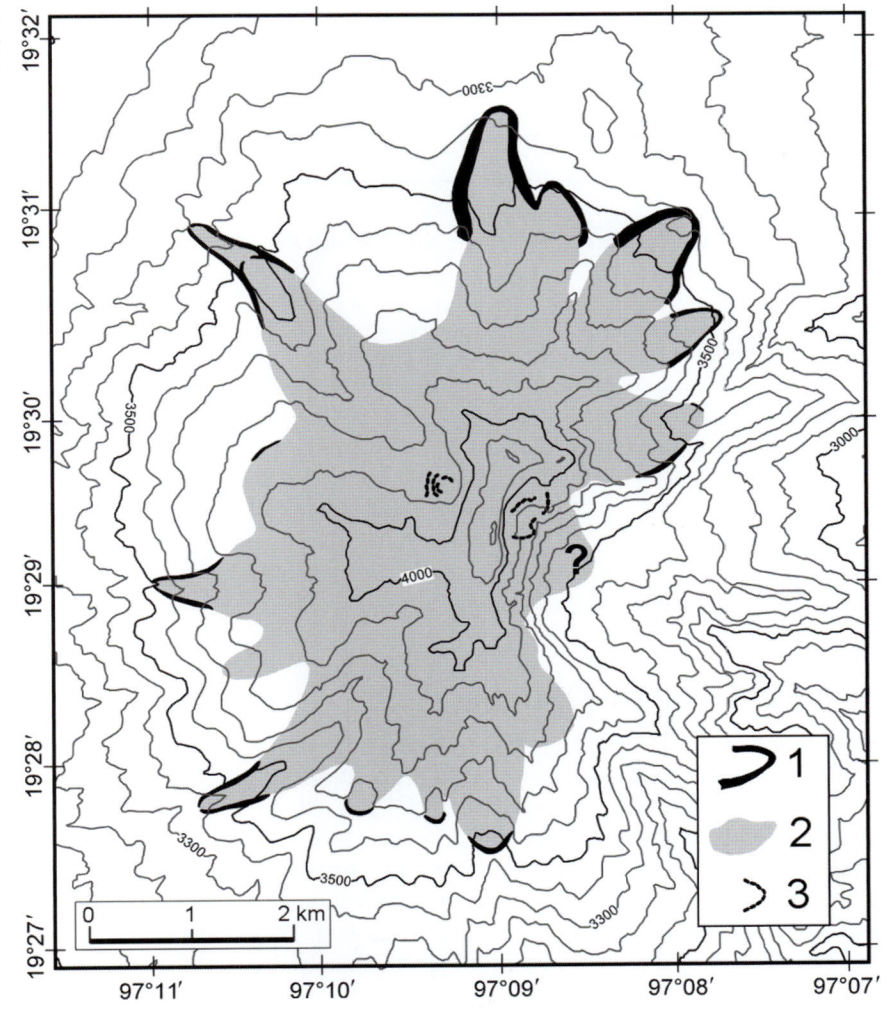

FIGURE 61.5 Glacial features of Cofre de Perote volcano. 1. Moraines 20–14 ^{36}Cl ka. 2. Glacier area 20–14 ^{36}Cl ka. 3. Moraines terminal Pleistocene (?).

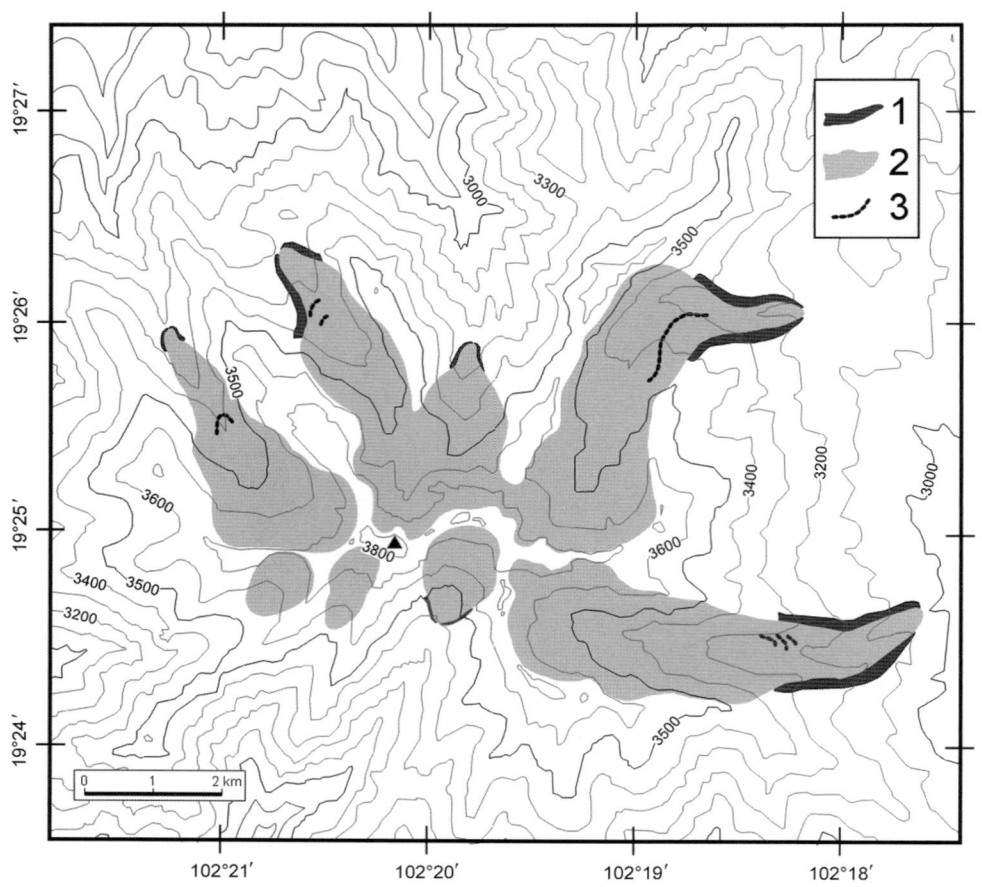

FIGURE 61.6 Glacial features of Tancítaro volcano. 1. Moraines 19–14 ^{36}Cl ka. 2. Glacier area 19–14 ^{36}Cl ka. 3. Recessional moraines ~13.5 ^{36}Cl ka.

show clear evidence of glaciation (Fig. 61.6). During the maximum advance, glaciers on the three main valleys ended at 3200–3100 m with an ELA of 3390 ± 50 m (THAR = 0.4) and 3380 ± 40 m (maximum altitude of lateral moraines; Lachniet and Vázquez-Selem, 2005). Unpublished ^{36}Cl dates indicate an age of ~19–14 ^{36}Cl ka for this phase, formation of small recessional moraines at ~13.5 ^{36}Cl ka and deglaciation between 13.5 ka and 11.6 ^{36}Cl ka. The mountain apparently lacks evidence of a younger glacial advance.

61.9. NEVADO DE COLIMA

Nevado de Colima (4180 m) is an extinct late Pleistocene stratovolcano located in the western part of the Trans-Mexican Volcanic Belt, only 85 km from the Pacific Ocean. Glacial landforms were first described there by Lorenzo (1961), but no further work has been conducted in the area. According to Lorenzo moraines of two different advances on the northern side suggest a glacier reaching ~3500, then ~3650 m. They are probably related to the phase of maximum advance recorded on Tancítaro, located 135 km to the west.

61.10. CERRO POTOSÍ

Pleistocene glaciation on the mountains of northern Mexico has long been considered a possibility, in particular on the highest peaks of Sierra Madre Oriental to the south of Monterrey, where several mountain ranges reach 3700 m (Lorenzo, 1964; Heine, 1975). Recent work has confirmed that Cerro Potosí (24°53′N, 100°14′W), a limestone mountain peaking at 3715 m, supported at least one glacier on its northern side (and probably a small ice cap on the summit area; Fig. 61.7). The main moraines indicate a glacier reaching 3200 m, with a length of ~1.4 km, an area of ~0.4 km^2, and an ELA around 3550 m. Four small inner (<3 m high) moraines are present near the cirque headwall between 3560 and 3600 m.

Although dating is pending, it is worth noticing that the ELA value of ~3550 m at Cerro Potosí is compatible with the values of 3400–3950 m mentioned above for central Mexico at 21–14 ka and with the presence of glaciers between 3850 and 3100 m in northern New Mexico (35°N) just before ~14 ka (Armour et al., 2002).

FIGURE 61.7 Glacial features of Cerro Potosí, north-eastern Mexico. 1. Moraines, maximum advance (LGM?). 2. Glacier area (LGM?). 3. Inner moraines (terminal Pleistocene?).

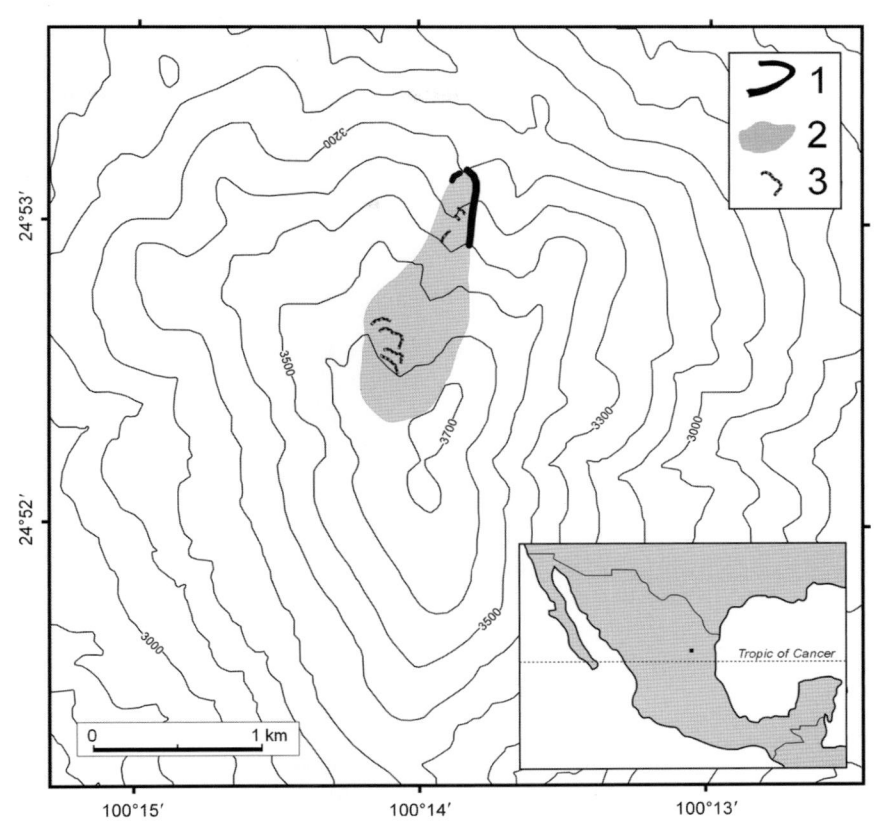

61.11. PRE-WEICHSELIAN (PRE-WISCONSINAN)

The oldest glacial deposits are dated to $\sim 205–175$ ^{36}Cl ka (Vázquez-Selem, 2000) on Iztaccíhuatl, that is, during the early MIS 6 (Illinoian Stage). They indicate the most extensive glaciation reached ~ 3000 m. We assume that some of the oldest tills mentioned by White et al. (1990; Nexcoalango advance on Iztaccíhuatl; till-like deposits on Ajusco) and by Heine (1975, 1988; M I advance) may correspond to this advance.

The possibility of late Pleistocene glacial advances prior to the LLGM during MIS 4 and/or MIS 3 is suggested by the chronologies of La Malinche, Ajusco and Nevado de Toluca. However, these advances lack precise dating.

61.12. WEICHSELIAN/WISCONSINAN GLACIAL MAXIMUM (LLGM) AND HOLOCENE

Table 61.3 presents the suggested correlations between the glacial chronology of Iztaccíhuatl (Vázquez-Selem, 2000) and those from Ajusco (White and Valastro, 1984), La Malinche and Nevado de Toluca (Heine, 1988, 1994). The authors offer no correlations with records for Popocatépetl and Pico de Orizaba (Heine, 1988) because of the poor chronometric controls. Correlations are based in principle on numerical ages as presented by the respective authors for each mountain. However, the writers comment on some of these ages and suggest alternative correlations.

The 'standard Mexican glacial sequence', elaborated by White (1987) and White and Valastro (1984), requires revision. It is based only on seven radiocarbon dates, some of which are minimum ages. The authors interpret some of the field evidence of White and Valastro (1984) on Ajusco differently: the Marques till is ignimbrite and as such is of volcanoclastic rather than glacial origin. Some moraines mentioned and mapped by White and Valastro (1984) are walls and ridges consisting of debris layers and caused by erosion. Therefore, correlations between the sequences of Ajusco and the well-dated Iztaccíhuatl glacier advances (Vázquez-Selem, 2000) seem problematic.

The La Malinche glacial sequence shows probable equivalences with the Iztaccíhuatl. Although Heine (1994) does not report a numerical age for the M I moraines, they could be equivalent to the Nexcoalango deposits of Iztaccíhuatl. The M II deposits correspond well with the Hueyatlaco moraines. The Hueyatlaco-1 advance is dated to 20–17.5 ^{36}Cl ka, when surface warming started in Antarctica and a rise in atmospheric CO_2 and in global sea level occurred (Rahmstorf, 2002). The M III 1 and M III 2 moraines of Malinche match the Milpulco-1 moraines of

TABLE 61.3 Correlation of Glacial Chronologies of Central Mexican volcanoes

Iztaccíhuatl[a] (^{36}Cl a)	La Malinche[b] (^{14}C a BP)	Nevado de Toluca[c] (^{14}C a BP)	Ajusco[d] (^{14}C a BP)
Ayoloco (Little Ice Age, <1000)	M V rock glacier (Little Ice Age)	M V rock glacier (Little Ice Age)	
	M IV rock glaciers (ca. 3500–2000)		Neoglaciation I and II (<2000)
Milpulco-2 (8300–7300)	M III 3 (>5750)	M IV rock glacier (8000–7000 ^{36}Cl a)	
Milpulco-1 (12,000–10,000)	M III 1 and M III 2 (10,000–9000)	M III moraines and rock glaciers (<11,500)	
Hueyatlaco-2 (17,000–14,500) Hueyatlaco-1 (21,000–17,500)	M II (>15,000–12,000)	M II (>11,500)	Albergue (<16,000–>10,000)
	M I (36,000–32,000)		Santo Tomás (>25,080) Marqués (>27,190)
Nexcoalango (~190,000 ^{36}Cl a)	Pre-Wisconsinan	M I (>40,000)	Till-like deposits

The chronology of Iztaccíhuatl is based on ^{36}Cl dating; the other chronologies are based on ^{14}C, except where indicated. Correlations assume ^{36}Cl ages are equivalent to calibrated ^{14}C ages (a = annum).
[a] Modified from Vázquez-Selem (2000).
[b] Modified from Heine (1988, 1994).
[c] Modified from Heine (1994).
[d] Modified from White and Valastro (1984) and White (1987).

Iztaccíhuatl. They proceeded approximately synchronously with the abrupt warming (11.5 cal. ka BP) of the North Atlantic at the end of the YD. M III 3 moraines (ca. 9–8.5 ^{14}C ka BP) from La Malinche are dated ca. 1 ka older than Milpulco-2 moraines. However, a careful examination of chronometric evidence from La Malinche shows (Vázquez-Selem, 2000) that the M III 3 moraines of La Malinche volcano are probably coeval with Milpulco-2 moraines of Iztaccíhuatl and could represent the 8.2 cal. ka event (Heine and Vázquez-Selem, 2002). This event has been linked to a meltwater-induced weakening of the thermohaline circulation in the North Atlantic with strong influence in the tropics (Alley et al., 1997; Cheng et al., 2009). The M IV glaciation on La Malinche is represented by rock glaciers, yet a revision of the stated timing might be necessary. As on the Nevado de Toluca, the M IV rock glaciers could represent a late phase of the M III 3 deposits (8–7 ka). The M V glaciation is probably represented on La Malinche by a talus rock glacier. Although its age is uncertain, it can be ascribed to the Ayoloco advance of Iztaccíhuatl (Vázquez-Selem, 2000). There is a clear correlation between the chronologies of Iztaccíhuatl (Vázquez-Selem, 2000) and La Malinche (Heine, 1988, 1994) for the Hueyatlaco-1, Hueyatlaco-2, Milpulco-1, Milpulco-2 and Ayoloco glaciations.

Moraines mapped by Heine (1994) on the Nevado de Toluca, such as the M II, probably correspond to the Hueyatlaco-1 and Hueyatlaco-2 moraines of Iztaccíhuatl. The M III moraines are probably equivalent to those of Milpulco-1. The M III late advance moraines of Heine (1994; M III 3 in Heine, 1988) can be correlated with the Milpulco-2 moraines of Iztaccíhuatl. This is based on their similar morphology and elevation and on published (Arce et al., 2003) and unpublished ^{36}Cl exposure ages from Nevado de Toluca. The same comments made for rock glaciers M IV and M V of Malinche can be applied to the deposits on Nevado de Toluca.

61.13. OPEN QUESTIONS

The chronology of Pleistocene glaciations on the high mountains of Mexico suffers from a lack of absolute dating from several volcanoes. Nevertheless, the glacial sequence presented by Vázquez-Selem (2000) for Iztaccíhuatl can be used as the standard glacial sequence of Mexico today for MIS 2 (late Wisconsinan) and MIS 1 (Holocene).

Little is known about the glacier advances prior to MIS 2. Pre-Wisconsinan (MIS 6) glacial deposits have been identified and dated on Iztaccíhuatl but their extent and timing on other mountains remains unclear.

The chronology of Iztaccíhuatl proposed by Vázquez-Selem (2000) and supported by records from other mountains suggests a maximum MIS 2 extent for glaciers around 21–20 ^{36}Cl ka (Hueyatlaco-1), thus near the end of the global last glacial maximum (LGM: 26.5 to 20–19 ka), as defined by Clark et al. (2009). In common with several mountain areas of the western United States (Licciardi et al., 2004), glaciers on Iztaccíhuatl re-advanced after the LLGM, forming the Hueyatlaco-2 moraines. However, widespread deglaciation at ~17 ka, recorded in the western United States (Licciardi et al., 2004), did not occur in central Mexico. Instead here glaciers remained close to their maximum positions until ~15 ka and marked deglaciation occurred only after 14 ka. On mountains near the Pacific Ocean (Tancítaro) and the Gulf of Mexico (Cofre de Perote) there is no clear evidence of a double glacial advance, but rather of a maximum stable position between 20 and 14 ^{36}Cl ka. This and other particularities of the late Pleistocene climate in central Mexico are probably related to the variable influence of moisture and temperature, including changes in the sources of moisture and the effects of meltwater input into the Gulf of Mexico.

ELA differences along a W–E transect through central Mexico at 21–14 ka probably reflect moisture gradients. The mean ELA was at ~3400 m a.s.l. on a low mountain near the Pacific (Tancítaro); at 3950–4050 m a.s.l. on a high mountain in the interior (Iztaccíhuatl); and ~3650 m a.s.l. near the Gulf of Mexico (Cofre de Perote). This pattern suggests increased moisture supply from the Pacific, thus supporting the hypothesis of stronger influence of the westerlies during the LGM and Wisconsinan late glacial in the region (Bradbury, 1997). This is consistent with the idea of a relatively dry climate in the central highlands (Metcalfe et al., 2000; Lozano-García et al., 2005).

It remains uncertain why the much colder Younger Dryas Chronozone is less clearly associated to a glacial advance in central Mexico than the 8.2 cal. ka event. If both, the Younger Dryas and the 8.2 cal. ka event, have similar pattern and forcing mechanisms, as suggested by Barber et al. (1999), a tentative explanation is that during the first half of the Younger Dryas the climate in the region was cold and too dry, thus keeping the ELAs at higher elevation than during the second half, coeval with the Milpulco-1 advance. The Milpulco-2 glaciation could have resulted not only from a temperature decrease (as the precedent advances), but also from higher precipitation than today, as suggested by palaeolimnological evidence from central Mexico (Caballero et al., 2002). On the other hand, the Younger Dryas could have been relatively warm in the region due to the diversion of glacial meltwater away from the Mississippi River and the Gulf of Mexico, as suggested by Nordt et al. (2002) for south-central Texas. More research effort is necessary to solve the problems that arise from the late-glacial–early Holocene climatic history. The clue to understanding the mechanisms of climate change in Mexico during the Pleistocene/Holocene transition lies in the North Atlantic–North America–Gulf Stream–Gulf of Mexico region and the palaeo-moisture conditions in Mesoamerica (Bradbury, 1997; Bradbury et al., 2001).

Finally, the identification of glacial landforms of probable late Pleistocene age in north-eastern Mexico (Cerro Potosí) opens new avenues for research in that region, potentially filling the geographical gap in the records of glaciation between the south-western United States and central Mexico.

61.14. SUMMARY

The high stratovolcanoes of central Mexico (~19°N) have a record of late Quaternary glaciation relevant for the northern tropics. The glacial chronology is based mainly on glacial landforms of Iztaccíhuatl (5282 m, 19°10′N, 98°40′W), Nevado de Toluca (4558 m, 19°08′N, 99°45′W), and La Malinche (4461 m, 19°14′N, 98°00′W), supplemented by the recently studied records of Tancítaro (3842 m, 19°25′N, 102°19′W) and Cofre de Perote (4282 m, 19°29′N, 97°09′W). Morphostratigraphy, tephrochronology and cosmogenic ^{36}Cl exposure ages provide the chronological support. On Iztaccíhuatl, the most extensive recorded advance (Nexcoalango) occurred during MIS 6 at ~190 ka and reached ~3000 m a.s.l. Iztaccíhuatl lacks clear evidence of Wisconsinan glaciation prior to 21 ka BP. The local late Pleistocene glacial maximum occurred around the global LGM. A first, pulse (Hueyatlaco-1 advance) peaked at 21–17.5 ka and a second one (Hueyatlaco-2) at ~17–14.5 ka. Valley glaciers reached ca. 3400–3500 m. Recessional moraines developed from 14.5 to 13.5 ka, followed by a rapid glacier retreat at 13–12 ka. Glaciers peaked again at ~12 ka reaching ca. 3800 m (Milpulco-1), and built recessional moraines until ~10 ka. Between ~8.5 and 7.5 ka glaciers formed small but distinctive moraines above 4000 m (Milpulco-2). No evidence of glacier expansion was found between the Milpulco-2 deposits and the massive moraines of <1 ka (Ayoloco) located at 4300–4700 m. The ELAs of glaciers for the five Wisconsinan advances on Iztaccíhuatl were located at 1030, 930, 730, 550 and 250 m, respectively, below the modern (AD 1960) ELA. Correlations between the glacial record of Iztaccíhuatl and other glacial sequences are presented. Evidence of Pleistocene glaciation recently found on Cerro Potosí (3715 m, 24°53′N, 100°14′W) in north-eastern Mexico is compatible with ELAs in central Mexico at 21–14 ka. The overall pattern of glaciation in the region is similar to those of mid-latitude North America and tropical South America, thus supporting the general synchroneity of major climatic events. A temperature decrease of 5–9 °C estimated from Hueyatlaco-1 ELA supports marked cooling over tropical land and oceans during the LLGM. Lower ELAs near the Pacific Ocean support the hypothesis of increased moisture supply from the westerlies during the LLGM. Available evidence indicates glacier expansion during the second half of the Younger Dryas Chronozone. An early-to-mid Holocene cold event (8.2 cal ka event) in the North Atlantic is coeval with Milpulco-2 advance, and the Ayoloco advance is contemporaneous with the Little Ice Age.

ACKNOWLEDGEMENTS

Research by L. V. S. was funded by NSF Grant 9422220, NASA-ESSFP Grant 1995-GlobalCh00006, CONACYT Grants G28528T and 50780-F and UNAM-PAPIIT Grants INI01199 and IN111206.

Work from K. H. was funded by the *Deutsche Forschungsgemeinschaft* (DFG), grant numbers He 722/1–9, 16, 17 and 31. We thank M.A. Geyh (Hannover) for numerous ^{14}C age determinations and discussions and F.M. Phillips (New Mexico Tech) for his advice on ^{36}Cl dating. We are indebted to many persons for assistance in the field and in the laboratory.

REFERENCES

Alley, R.B., Mayewski, P.A., Sowers, T., Stuiver, M., Taylor, K.C., Clark, P.U., 1997. Holocene climatic instability: a prominent, widespread event 8200 yr ago. Geology 25, 483–486.

Arce, J.L., Macías, J.L., Vázquez-Selem, L., 2003. The 10.5 ka Plinian eruption of Nevado de Toluca volcano, Mexico: stratigraphy and hazard implications. Geol. Soc. Am. Bull. 115, 230–248.

Armour, J., Fawcett, P.J., Geissman, J.W., 2002. 15 k.y. paleoclimatic and glacial record from northern New Mexico. Geology 30, 723–726.

Barber, D.C., Dyke, A., Hillaire-Marcel, C., Jennings, A.E., Andrews, J.T., Kervin, M.W., et al., 1999. Forcing the cold event of 8,200 years ago by catastrophic drainage of Laurentide lakes. Nature 400, 344–348.

Bloomfield, K., Valastro, S., 1977. Late Quaternary tephrochronology of Nevado de Toluca volcano, central Mexico. Overseas Geol. Miner. Resour. 46, 1–15.

Bradbury, J.P., 1997. Sources of glacial moisture in Mesoamerica. Quatern. Int. 43 (44), 97–110.

Bradbury, J.P., Grosjean, M., Stine, S., Sylvestre, F., 2001. Full and late glacial lake records along the PEP 1 transect: their role in developing interhemispheric paleoclimate interactions. In: Markgraf, V. (Ed.), Interhemispheric Climate Linkages. Academic Press, San Diego, pp. 265–291.

Caballero, M., Macías, J.L., Lozano-García, S., Urrutia-Fucugauchi, J., Castañeda-Berrnal, R., 2001. Late Pleistocene-Holocene volcanic stratigraphy and palaeoenvironment of the upper Lerma basin, Mexico. Int. Assoc. Sedimentol. Spec. Publ. 30, 247–261.

Caballero, M., Ortega, B., Valadez, F., Metcalfe, S., Macias, J.L., Sugiura, Y., 2002. Sta. Cruz Atizapan: a 22-ka lake level record and climatic implications for the late Holocene human occupation in the upper Lerma Basin, Central Mexico. Palaeogeogr. Palaeoclimatol. Palaeoecol. 186, 217–235.

Cantagrel, J.M., Robin, C., Vincent, P., 1981. Les grandes étapes d'évolution d'un volcan andesitique composite: exemple du Nevado de Toluca. Bull. Vulcanol. 44, 177–188.

Capra, L., Carreras, L.M., Arce, J.L., Macías, J.L., 2006. The lower Toluca Pumice: a ca. 21,700 yr B.P. Plinian eruption of Nevado de Toluca volcano, México. In: Siebe, C., Macías, J.L., Aguirre-Díaz, G.J. (Eds.), Neogene-Quaternary Continental Margin Volcanism: A Perspective from Mexico, 155–173. Geological Society of America Special Paper 402, Penrose Conference Series.

Carrasco-Núñez, G., Siebert, L., Díaz-Castellón, R., Vázquez-Selem, L., Capra, L., 2010. Evolution and hazards of a long-quiescent compound shield-like volcano: Cofre de Perote, Eastern Trans-Mexican Volcanic Belt. J. Volcanol. Geotherm. Res. 197, 209–224.

Castro-Govea, R., Siebe, C., 2007. Late Pleistocene-Holocene stratigraphy and radiocarbon dating of La Malinche volcano, Central Mexico. J. Volcanol. Geotherm. Res. 162, 20–42.

Cheng, H., Fleitmann, D., Edwards, R.L., Wang, X., Cruz, F.W., Auler, A. S., et al., 2009. Timing and structure of the 8.2 kyr B.P. event inferred from δ^{18}O records of stalagmites from China, Oman, and Brazil. Geology 37, 1007–1010.

Clark, P.U., Dyke, A.S., Shakun, J.D., Carlson, A.E., Clark, J., Wohlfarth, B., et al., 2009. The last glacial maximum. Science 325, 710–714.

Heine, K., 1975. Studien zur jungquartären Glazialmorphologie mexikanischer Vulkane— mit einem Ausblick auf die Klimageschichte. Das Mexiko-Projekt der Deutchen Forschungsgemeinschaft VII. Steiner, Wiesbaden, 178pp.

Heine, K., 1976a. Auf den Spuren der Eiszeit in Mexiko. Nat. Museum 106, 289–298.

Heine, K., 1976b. Blockgletscher- und Blockzungen-Generationen am Nevado de Toluca,, Mexiko. Die Erde 107, 330–352.

Heine, K., 1988. Late Quaternary glacial chronology of the Mexican volcanoes. Die Geowissenschaften 6, 197–205.

Heine, K., 1994. The late-glacial moraine sequences in Mexico: is there evidence for the Younger Dryas event? Palaeogeogr. Palaeoclimatol. Palaeoecol. 112, 113–123.

Heine, K., Vázquez-Selem, L., 2002. Das 8,2 ka-Ereignis in Mexiko: Gletscherverhalten und klimatische Folgerungen. Schriftenreihe Dt. Geol. Ges. 21, 154.

Hoskuldsson, A., Robin, C., 1993. Late Pleistocene to Holocene eruptive activity of Pico de Orizaba, eastern Mexico. Bull. Volcanol. 55, 571–587.

Lachniet, M.S., Vázquez-Selem, L., 2005. Last glacial maximum equilibrium line altitudes in the circum-Caribbean (Mexico, Guatemala, Costa Rica, Colombia, and Venezuela). Quatern. Int. 138–139, 129–144.

Licciardi, J.M., Clark, P.U., Brook, E.J., Elmore, D., Sharma, P., 2004. Variable responses of western U.S. glaciers during the last deglaciation. Geology 32, 81–84.

Lorenzo, J.L., 1961. Notas sobre la geología glacial del Nevado de Colima. Boletín del Instituto de Geología. UNAM. Mexico 61, 77–92.

Lorenzo, J.L., 1964. Los Glaciares de México. Instituto de Geofísica, U.N. A.M., México, 124pp.

Lozano-García, S., Sosa-Najera, S., Sugiura, Y., Caballero, M., 2005. 23,000 yr of vegetation history of the upper Lerma, a tropical high-altitude basin in Central Mexico. Quatern. Res. 64, 70–82.

Macías, J.L., 2005. Geología e historia eruptiva de algunos de los grandes volcanes activos de México. Bol. Soc. Geol. Mex. LXVII, 379–424.

Metcalfe, S.E., O'Hara, S.L., Caballero, M., Davies, S.J., 2000. Records of late Pleistocene—Holocene climatic change in Mexico—a review. Quatern. Sci. Rev. 19, 699–721.

Newton, A.J., Metcalfe, S.E., 1999. Tephrochronology of the Toluca Basin, central Mexico. Quatern. Sci. Rev. 18, 1039–1059.

Nixon, G.T., 1989. The geology of Iztaccíhuatl volcano and adjacent areas of the Sierra Nevada and Valley of Mexico. The Geological Society of America, Special Paper 219, 58pp.

Nordt, L.C., Boutton, T.W., Jacob, J.S., Mandel, R.D., 2002. C_4 plant productivity and climate-CO_2 variations in south-central Texas during the late Quaternary. Quatern. Res. 58, 182–188.

Ownby, S., Delgado Granados, H., Lange, R.A., Hall, C.M., 2007. Volcan Tancítaro, Michoacán, Mexico, ^{40}Ar/^{39}Ar constraints on its history of sector collapse. J. Volcanol. Geotherm. Res. 161, 1–14.

Palacios, D., Vázquez-Selem, L., 1996. Geomorphic effects of the retreat of Jamapa glacier, Pico de Orizaba volcano (Mexico). Geogr. Ann. 78A, 19–34.

Rahmstorf, S., 2002. Ocean circulation and climate during the past 120,000 years. Nature 419, 207–214.

Siebe, C., Abrams, M., Macías, J.L., Obenholzner, J., 1996. Repeated volcanic disasters in prehispanic time at Popocatépetl, central Mexico: past key to the future? Geology 24, 399–402.

Vázquez-Selem, L., 1991. Glaciaciones del Cuaternario Tardío en el Volcán Téyotl, Sierra Nevada. Investigaciones Geográficas. Boletin del Instituto de Geografía, UNAM, Mexico, Vol. 22, 22–45.

Vázquez-Selem, L., 1997. Late Quaternary glaciations of Téyotl volcano, central Mexico. Quatern. Int. 43 (44), 67–73.

Vázquez-Selem, L., 2000. Glacial Chronology of Iztaccíhuatl Volcano, central México. A Record of Environmental Change on the Border of the Tropics. Unpublished Ph.D. Thesis, Arizona State University, 257pp.

Vázquez-Selem, L., Phillips, F.M., 1998. Glacial chronology of Iztaccíhuatl volcano, central Mexico, based on cosmogenic ^{36}Cl exposure ages and tephrochronology. American Quaternary Association. Program and Abstracts of the 15th Biennial Meeting, Puerto Vallarta, Mexico, 5–7 September 1998 174.

White, S.E., 1962. Late Pleistocene glacial sequence for the west side of Iztaccíhuatl, Mexico. Geol. Soc. Am. Bull. 73, 935–958.

White, S.E., 1987. Quaternary glacial stratigraphy and chronology of Mexico. Quatern. Sci. Rev. 5 (1–4), 201–206.

White, S.E., Valastro Jr., S., 1984. Pleistocene glaciation of volcano Ajusco, central Mexico, and comparison with standard Mexican glacial sequence. Quatern. Res. 21, 21–35.

White, S.E., Reyes-Cortes, M., Ortega Ramírez, J., Valastro, S., 1990. El Ajusco: Geomorfología volcánica y acontecimientos glaciales durante el Pleistoceno superior y comparación con las series glaciales mexicanas y las de las Montañas Rocallosas. Colección Científica, Serie Arqueología. INAH, México, D.F. 77pp.

Zreda, M.G., Phillips, F.M., 1994. Surface exposure dating by cosmogenic chlorine-36 accumulation. In: Beck, C. (Ed.), Dating in Exposed and Surface Contexts. University of New Mexico Press, Albuquerque, pp. 161–183.

Zreda, M.G., Phillips, F.M., Elmore, D., 1994. Cosmogenic ^{36}Cl accumulation in unstable landforms. 2. Simulations and measurements on eroding moraines. Water Resour. Res. 30, 3127–3136.

Late Pleistocene Glaciation of the Hindu Kush, Afghanistan

Stephen C. Porter

Department of Earth and Space Sciences, University of Washington, Seattle, Washington 98195-1360, USA

Study of Pleistocene glaciation in the Hindu Kush has been hampered by remote, difficult terrain and political instability that has limited or prevented access to most of the key glaciated areas of the range. Published reports are few, and maps are small in scale and based on limited field data. Proposed chronologies are based on inferred correlations with other regions rather than radiometric ages. As a consequence, the most reliable estimates of the extent and character of Pleistocene glaciers now available are based on remote-sensing data, supplemented by a few published reports from ground-based surveys.

Porter (1985) used eighteen 1:500,000-scale LANDSAT images to map the inferred extent of Late Pleistocene glaciers in Afghanistan (Fig. 62.1). The interpretation was based primarily on glacial-erosional landforms, especially valley morphology, and secondarily on moraines. The downvalley limit of a former valley glacier could generally be inferred by noting the change from typical characteristics of a glaciated valley (parabolic cross profile, truncated interfluve spurs, cirque(s) at valley head) to those of a non glaciated valley (steep nonparabolic cross profiles, interlocking interfluve spurs). This change is often abrupt and can be mapped with an accuracy of 1–2 km. Uplands indented by cirques are distinct from those lying below the glacial-age snowline, which lack cirques. In some major valleys, moraines and (or) valley trains and outwash heads are resolvable on the images and were used to map former ice limits.

In the absence of dating control, and based on only very limited field reports, it was assumed that downvalley limits of mapped glaciers were those of the last glaciation (i.e. marine isotope stage 2). Although older, more-extensive ice limits likely exist, as noted in adjacent mountain belts (e.g. Porter, 1970; Owen et al., 1997), because of the steep terrain in all but the largest valleys, such glacial limits probably lie relatively close to those of the last glaciation.

Late Pleistocene glaciers in the Hindu Kush headed in peaks that range in altitude from ca. 4650 to 6400 m. The longest descended as low as 2600 m in the Panjshir Valley. The smallest mapped glaciers were < 1 km long, whereas the largest were 70 km long; 14 glaciers reached or exceeded lengths of 50 km. The longest glaciers (50–70 km) flowed down north-trending valleys. The largest in south-trending valleys reached a length of 50 km, but most were < 25 km long.

Based on the mapped extent of glaciers and their altitude distributions, the full-glacial snowline, or equilibrium line altitude (ELA) in the Hindu Kush lay between ca. 4000 and 4600 m, some 1000 ± 100 m below the estimated altitude of the modern snowline (von Wissmann, 1959; Grötzbach and Rathjens, 1969; Desio, 1975).

REFERENCES

Desio, A., 1975. Notes on the Pleistocene of central Badakhshan (north-east Afghanistan). In: Desio, A. (Ed.), Scientific Results of the Italian Expedition to the Karakoram (K2) and Hindu Kush, vol. 3, 339–422.

Grötzbach, E., Rathjens, C., 1969. Die heutige und die jungpleistozäne Vergletscherung des Afghanischen Hindu-Kusch. Z. Geomorph. Suppl. 8, 58–75.

Owen, L.A., Bailey, R.M., Rhodes, E.J., Mitchell, W.A., Coxon, P., 1997. Style and timing of glaciation in the Lahul Himalaya, northern India: a framework for reconstructing late Quaternary palaeoclimatic change in the western Himalayas. J. Quatern. Sci. 12, 83–109.

Porter, S.C., 1970. Quaternary glacial record in Swat Kohistan, West Pakistan. Geol. Soc. Am. Bull. 81, 1421–1446.

Porter, S.C., 1985. Extent of late Pleistocene glaciers in Afghanistan based on interpretation of LANDSAT imagery. In: Agrawal, D.P., Kusumgar, S., Krishnamurthy, R.V. (Eds.), Climate and Geology of Kashmir and Central Asia: The Last Four Million Years. Current Trends in Geology. Today and Tomorrow's Printers and Publishers, New Delhi, pp. 191–195.

von Wissmann, H., 1959. Die heutige Vergletscherung und Schneegrenze in Hochasien. Abhandlungen der Akademie der Wissenschaften und der Literatur (Mainz), Mathematich-Naturwissenschaftliche Klasse 14, 5–307.

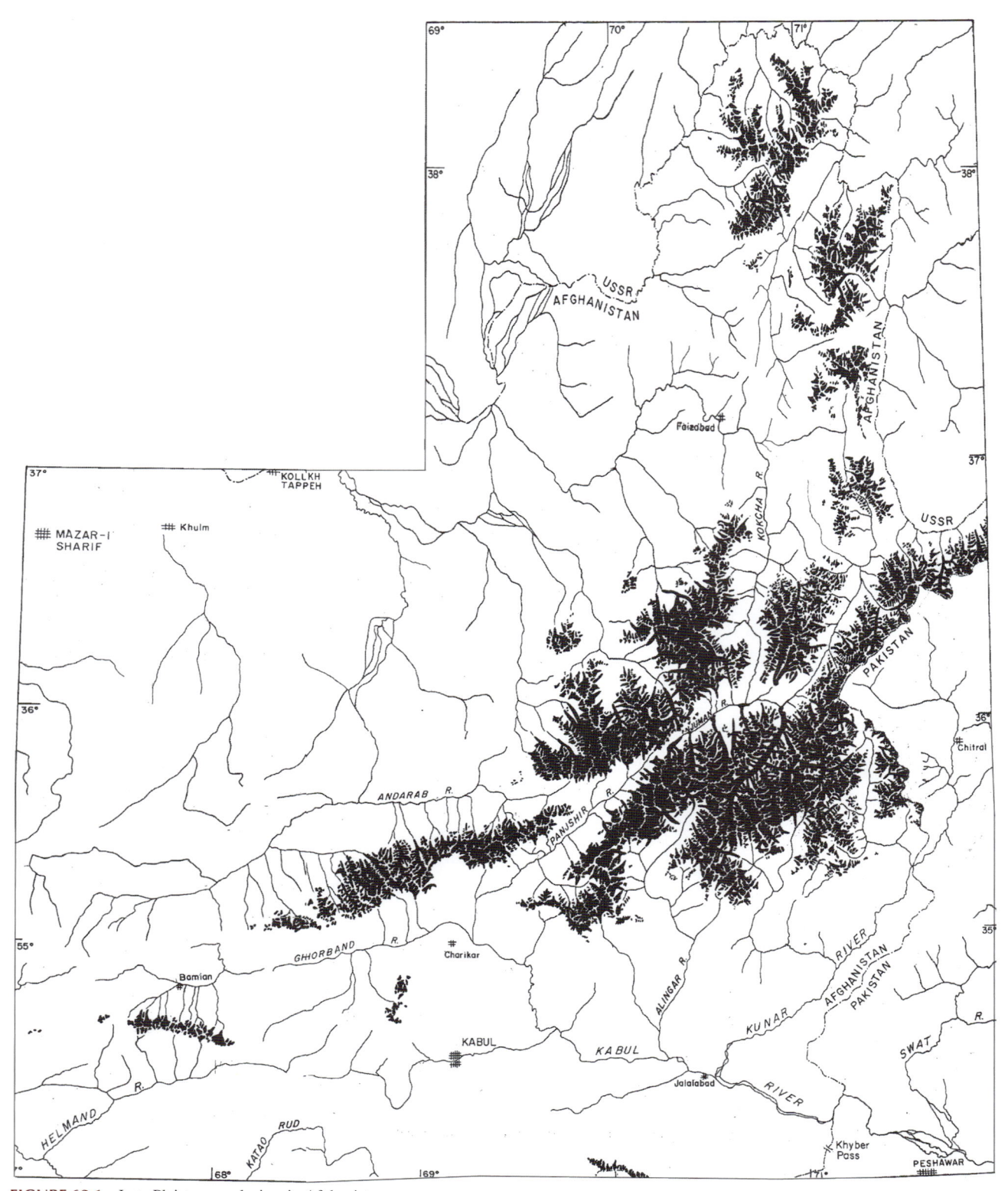

FIGURE 62.1 Late Pleistocene glaciers in Afghanistan.

Chapter 63

Late Pleistocene Glaciations in North-East Asia

O.Yu. Glushkova

North-eastern Interdisciplinary Research Institute FEB RAS, 16 Portovaya St., 685000 Magadan, Russia, E-mail: glushkova@neisri.ru

63.1. INTRODUCTION

The vast territory of north-east Asia includes several climatic zones and is orographically diverse. Accordingly, it offers great opportunities in studying Late Pleistocene glaciations in various orographic and climatic contexts. However, most of this territory remains poorly investigated. Therefore, the comparative geomorphological analysis of glacial features should be the basic method for assimilating existing data on the mode, aerial extent and stages of Late Pleistocene glaciation in diverse parts of north-east Asia.

The reconstruction of palaeoclimates is based on the results of palynological and carpological studies of correlative deposits. Published results of ^{14}C, cosmogenic exposure ^{36}Cl, optically stimulated luminescence [OSL] and infrared stimulated luminescence [IRSL] dating have been employed for chronological subdivision and correlation of the glacial events.

63.2. RELIEF AND CLIMATE

The major orographic features of the region are the Verkhoyansky Range, the Chersky Mountain System, the Koryak Upland, ridges and plateaus of the Okhotsk–Chukotka Mountain System, forming the Pacific–Arctic Watershed. The relief foundation is composed by low mountains ca. 800–1500 m high, above which rise mountain massifs 2000–3000 m high. The plain relief is spread in coastal areas (Yano-Indigirka, Kolyma, Lower Anadyr and others) as well as in large intermontane depressions.

The Verkhoyansky, Chersky, Suntar-Khayata ranges and the Koryak Upland contain modern glaciers. They are mostly small cirques, rarely cirque-valley and valley glaciers.

The study area is situated in several climatic zones: arctic desert, arctic tundra, tundra and forest-tundra. The tundra and forest-tundra zones are humid, with cold summers, snowy winters and extremely severe weather. Two basic climate types are distinguished by winter temperatures: the extreme continental climate, with very frosty winters with the mean January temperature of $-32°$ C, and the moderate continental climate, with mean January temperature higher than $-32°$ C (the Okhotsk and Bering seas coasts).

The climate of ice ages in north-east Asia is known from a series of key sections. The integrated study of bottom sediments of Lake Elgygytgyn, Chukotka, (cores PG-1351 and LZ-1024) permitted to trace changes in vegetation and climate within the past 350,000 years (Matrosova, 2009). Nine stages of regional restructuring of the vegetation cover have been distinguished (Fig. 63.1). Five warm and four cold periods have been inferred. Dating of lacustrine sediments (Forman et al., 2007; Juschus et al., 2007) allowed correlation of the pollen zones with marine isotope stages (MIS) 1–10. The two coolings identified in the lower part of the sections are correlated with the Samarovo (MIS 8) and Taz (MIS 6) glacial stages of West Siberia (Stratigraphy of USSR, 1982). Several pollen zones relate to two major coolings of the Late Pleistocene. Climate reconstructions based on pollen assemblages suggest a similarity between glacial climates of Middle and Late Pleistocene.

The palynological record of Lake Elikchan-4 close to the Okhotsk-Kolyma Watershed, reflects vegetation changes within the past 70 ka. Two herbaceous pollen zones reflecting the deep cooling of the climate are identified by pollen spectra. They correlate with glacial events of the Late Pleistocene (MIS 4 and MIS 2; Lozhkin et al., 2008).

63.3. LATE PLEISTOCENE GLACIATIONS

Two Late Pleistocene glacial complexes are reliably identified in north-east Asia. They are clearly different in their geomorphological positions and correlate with the Zyrianka and Sartan glaciations of Siberia. Mostly mountainous areas were glaciated in the Late Pleistocene. Only on the Chukchi

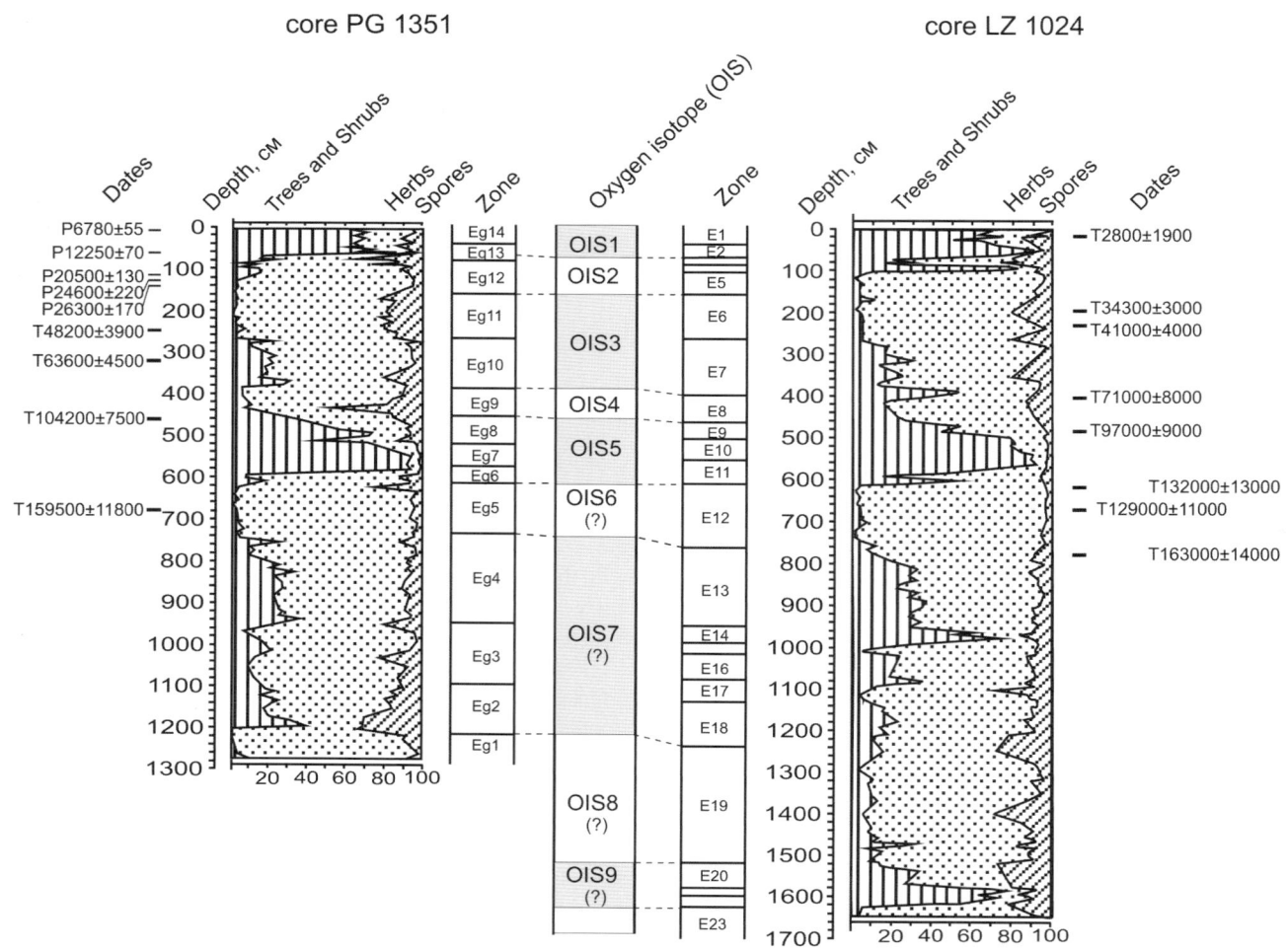

FIGURE 63.1 Correlation of pollen zones (borehole cores PG-1351 and LZ-1024) of Elgygytgyn Lake with marine isotope stages: P—^{14}C ages (Forman et al., 2007), T—infrared stimulated luminescence ages (Juschus et al., 2007).

Peninsula, glaciers advanced onto the coast and emerged shelf of the Bering Sea.

Early and Middle Pleistocene glaciations left only sparse remnants of morainic formations making reconstructions of glacial events of that time a hard task. However, in the recent time Middle Pleistocene glacial complexes have been found, preserved on the western slope of the Verchoyansky Range, which will be discussed below.

Traces of the last two glaciations are found in widespread morainic and glacio-aqueous formations. In many areas, large valley glaciers, over 100 km long, are reconstructed for MIS 4 time when the Zyrianka glaciers expanded onto intramontane depressions and submontane plains (Fig. 63.2). In some areas, reticulate networks or ice caps developed. Voluminous piedmont glaciers of Malaspina type are reconstructed for areas adjacent to the coasts of the Bering Sea (Lower Anadyr Lowland) and the Sea of Okhotsk (the Yama Lowland, Gizhiga Plain).

The Sartan glaciation (MIS 2) was localised in a few isolated areas (Fig. 63.3) in the Chersky Mountain System, in the axial part of the Verkhoyansky Range, in mountain systems of the Pacific–Arctic Watershed, on the Koryak Upland. Valley glaciers 10–20 km long and numerous cirque glaciers prevailed.

One of the important peculiarities of Late Pleistocene ice ages in north-east Asia is the fact that the Sartan glaciation was two to three times less extensive than the Zyrianka glaciation. However, in western territories (Verkhoyansky, Chersky and Suntar-Khayata ranges), the Sartan glaciation was not much smaller, while in the east, the Zyrianka glaciation area exceeded that of the Sartan manifold. The observed disproportion is explained by the fact that, during the Sartan Ice Age, the rise of the snow line left low mountains in the eastern part of the region below the snow accumulation area. In the western alpine mountains, the Sartan rise of the snow line reduced accumulation area insignificantly.

In coastal lowlands beyond mountain systems, no traces of Late Pleistocene glaciers have been found. Continuous accumulation of periglacial sediments has been described

Chapter | 63   Late Pleistocene Glaciations in North-East Asia

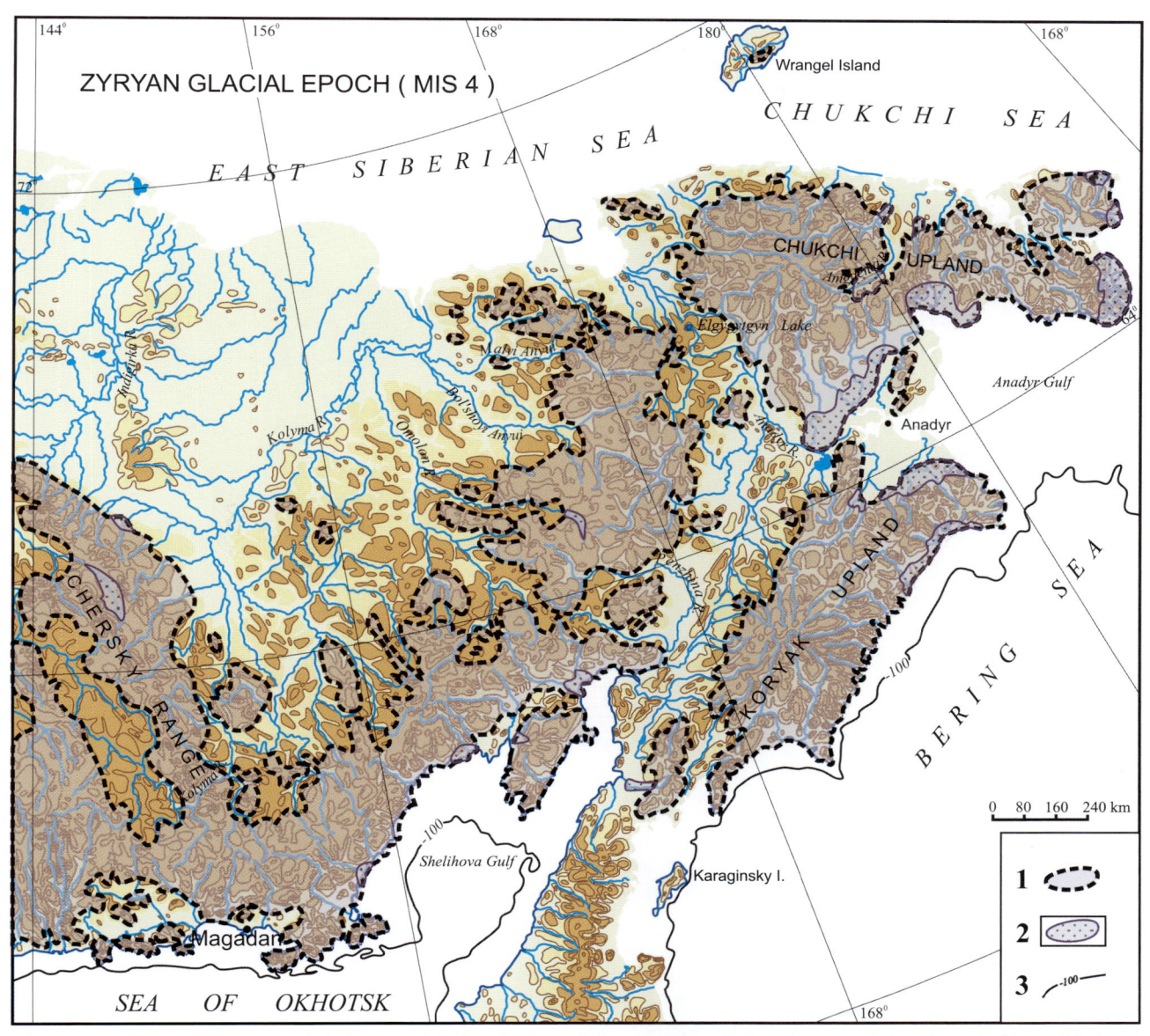

FIGURE 63.2 Zyryan Glaciation (MIS-4). 1—area of glaciers; 2—large piedmont and foot glaciers; 3—assumed shoreline at the Zyryan Maximum.

for the Karginsky interstadial and the Sartan Ice Age (MIS 3 and 2). The sedimentological studies showed that in the interval between 60 and 12 ka, the environment here had not radically changed. The vegetation cover was mosaic arctic tundra with tundra-steppe communities dominating. The climate was continental, with the summer no colder than it is today, but with colder winters. Very fast environmental changes occurred during Pleistocene–Holocene transition (Sher et al., 2005).

The visions of ages and types of Late Pleistocene glaciations based on comprehensive studies of glacier complexes in some basic districts of north-east Asia: Verkhoyansky, Chersky, West Chukotka and Lower Anadyr (Glushkova, 1994).

63.3.1. Verkhoyansky Range

Many researchers believe that the largest in the Verkhoyansky Range was the Zyrianka glaciation. This view is based on the fact that the basal moraine of this age is widely spread in the Verkhoyansky Range foothills and on the right banks of the Lena and the Aldan Rivers. Traces of older glaciations are believed to have been destroyed by the Zyrianka glaciation (Kind, 1975; Velichko, 1993).

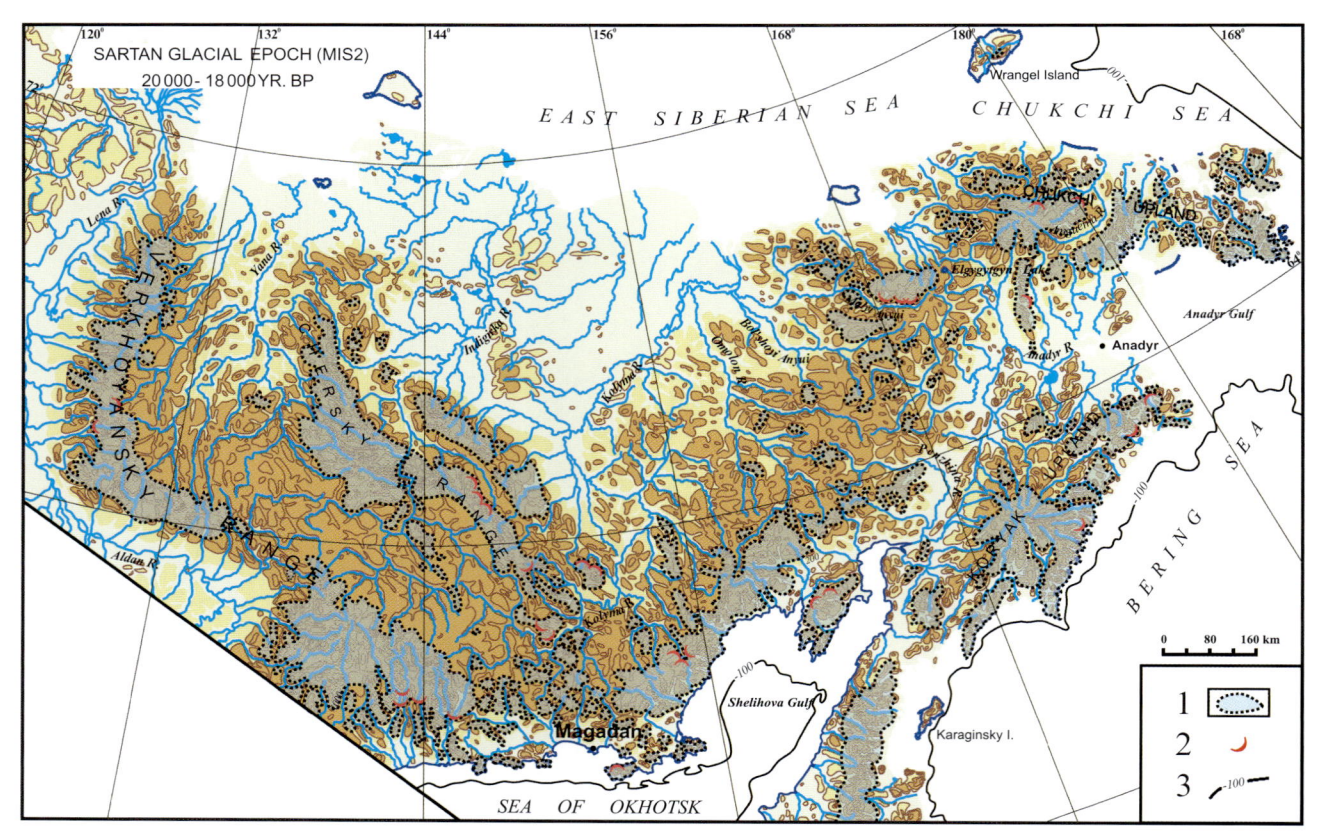

FIGURE 63.3 Sartan Glaciation (MIS-2). 1—area of glaciers; 2—large ridges of terminal moraine; 3—assumed shoreline at the Sartan Maximum.

According to the published data, the Verkhoyansky Range area was also subject to glaciation in the Sartan time, and its scale was but a little smaller than that of the Zyrianka. The palaeogeographic reconstruction for the time of the last glacial maximum, based on the analysis of the relic glacial morphosculpture and its relations with the dated correlative sediments (Kolpakov and Belova 1980), imply that the Verkhoyansky Range was heavily glaciated. Ice limits were traced by marginal landforms, readily identifiable on submontane plains of the south-western Verkhoyansky Range. It is assumed that the central, highest part of the range hosted an ice cap, with outlet glaciers crossing the lower mountains.

Publications on this problem have been analysed by Zamoruyev, who came to the conclusion that all observed glacial features belong to the last glacial maximum (Zamoruyev, 2004).

However, recently published data contradict the above conclusions. Stauch et al. (2007) and Siegert et al. (2007) report the IRSL ages of cover sediments on moraines within the submontane valleys of Tumara and Dyanyshka rivers 107 ± 10–123 ± 9.5 ka and 135 ± 9–157 ± 11 ka, respectively. Based on this, the authors argue that the largest glaciation in the Verkhoyansky Range is older than it has been believed before and that the results obtained 'do not permit to reconstruct any significant glaciation for Verkhoyanye in the last glacial maximum' (Siegert et al., 2007, p. 5). Thus, there are two antipodal views of the character of the Late Pleistocene glaciation in the Verkhoyansky Range.

63.3.2. The Chersky Range System

In the Chersky mountain chain, the greatest Late Pleistocene glaciers of north-east Asia developed. Here the largest present-day glacier system is also found. We consider the mode and extent of the glaciations using the two exemplary areas the Ulakhan-Chistay and the Bol'shoy Annachag Ranges Ridges.

63.3.2.1. The Ulakhan-Chistay Range

One hundred and nine modern glaciers of total area of 85.0 km^2 exist in this range 2000–3000 m high (Catalogue of Glaciers of the USSR, 1981). Glaciers are of cirque-hanging, cirque-valley and valley types. Most of them are found in the Buordakh Massif; the largest glaciers up to 8.9 km long are concentrated around Mt. Pobeda (3147 m a.s.l.).

Mostly valley glaciers, reticulate networks or ice caps are reconstructed. The largest valley glaciers (Buordakh, Tirekhtyakh and others) existed in the highest (2500–3000 m a.s.l.) central part of the Ulakhan-Chistay Ridge. During the glaciation maximum, they reached 60–100 km in length (Nekrasov and Sheinkman, 1974). Their maximum advance is indicated by large end moraines within the Momskaya Depression at 700–800 m a.s.l., that is, by 800–1000 m below the snouts of modern glaciers. In the south-eastern part of the Momskaya Depression, spacious amphitheatres are observed, containing serried of contiguous terminal moraine. The frontal part of the outer amphitheatre with the arc perimeter about 30 km is 22 km from the foot of the Ulakhan-Chistay Ridge (Glushkova, 2001).

Traces of the Zyrian glaciation have been found only in soft sediment sections in the Bugchan Depression. The moraine of the two age generations overlaps the alluvium thickness to 200 m thick. In the natural exposure in the mouth of the Bugchan Creek, at the depth of 30–33 m, in the moraine thickness, the ^{14}C age of 40.2 ± 2.7 and 40.8 ± 2.5 ^{14}C ka BP has been fixed as well as two dates: >42 and >48 ^{14}C ka BP. The data from the spore-and-pollen analysis show a certain similarity between the palynological complex of the lower moraine and the complex of Upper Quarternary (Zyrian) deposits on the Indigirka River (Anderson and Lozhkin, 2002).

63.3.2.2. Bol'shoi Annachag Range

The Bol'shoy Annachag Ridge is located 370 km south-east from the Ulakhan-Chistay Ridge. It is averages ca. 2000–2200 m a.s.l. (Fig. 63.4). The geomorphological analysis allowed the reconstruction of the Sartan glaciation boundaries as well as the basic parameters, movement routes, and the dynamic of development for the largest glaciers of the district, Vos'miozyorny and Jack London (Lozhkin et al., 1995; Glushkova 2001). At the glacial Lake Jack London, Lake El'gennya, Lake Sosednee, morphology of form glacial complexes has been studied comprehensively and bottom sediments have been drilled, yielding a series of radiocarbon dates between 22.0 and 1.8 ^{14}C ka BP. The Vos'miozyorny Glacier, descending from the western slope of the Bol'choy Annachag Ridge, was 21 km long in its maximum phase. Its movement has been fixed by a spacious terminal-moraine range, blocking the El'gennya River valley with the formation of the lake, named similarly and 4.3 km long. The range goes 13 km from the north to the south and almost all over has a narrow crest, 10–20 m wide. The ridge height increases from 60–70 m in the south to 200 m in the north. The Jack London Glacier descended from the eastern slope of the ridge along the Purga Creek valley and filled the intermontane trough.

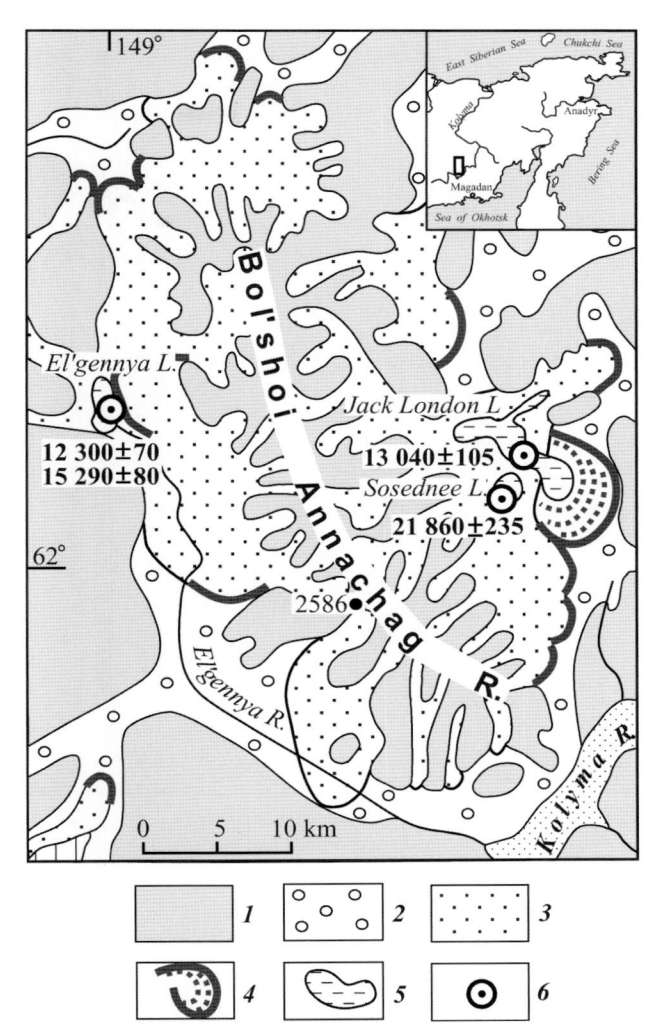

FIGURE 63.4 Extent of Sartan and Zyryanka moraines in Bolshoy Annachag range. 1—unglaciated terrains; 2—Zyryanka moraines; 3—Sartan moraines; 4—Sartan terminal moraines: main and stadial (dotted line); 5—glacial lakes; 6—studied section.

63.3.3. North-Western Chukotka

Chronology of Western Chukotka's glacial complexes still remains disputable. There are hypotheses of two Middle Pleistocene and two Late Pleistocene glaciations (Verkhovskaya, 1986; Ivanov, 1986). Terminal moraines are most informative for reconstructing glacial events. However, terminal moraine dates have still remains single. Morphologically, two generations of Late Pleistocene glacial features are distinguished (Fig. 63.5). There are moraines preserved in intermontane basins, on submontane plains, and along the rim of the Chaun Lowland. Glaciers that descended from the southern slopes of the Ilirney

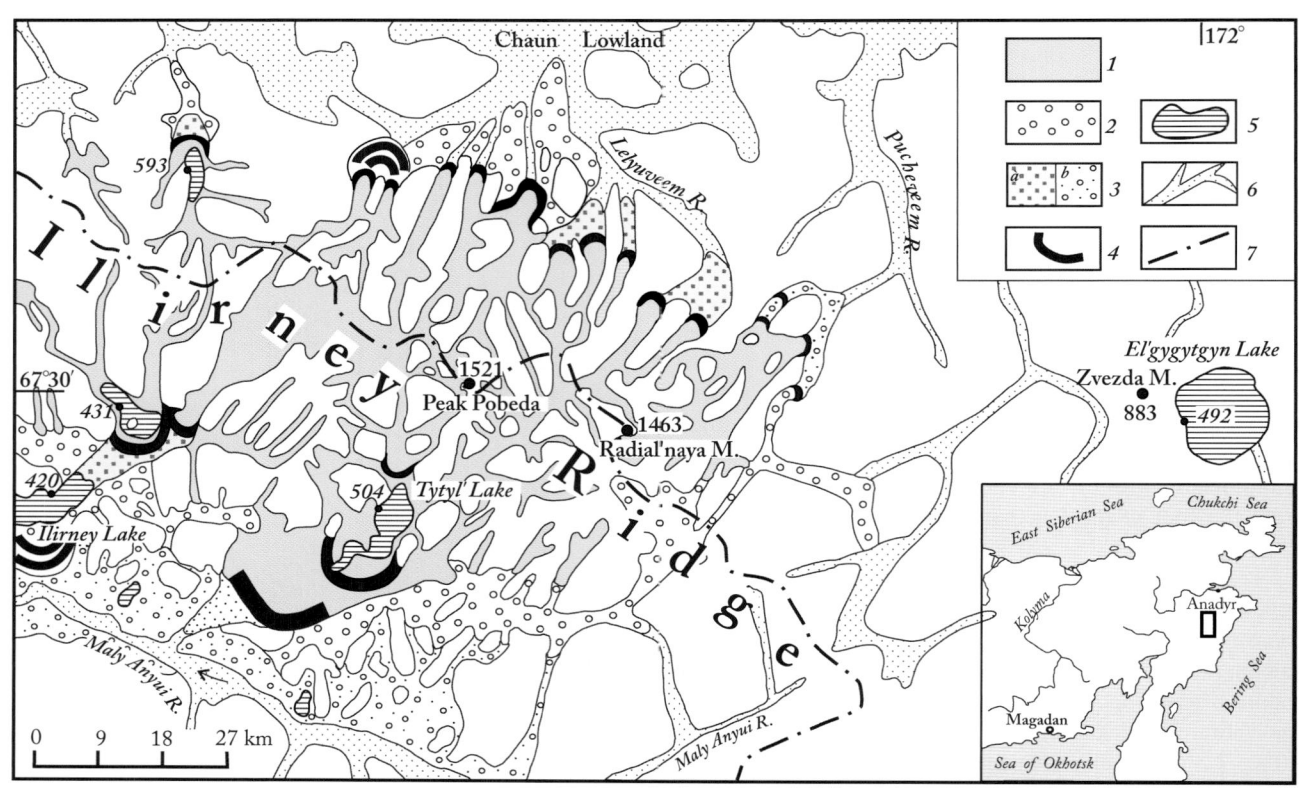

FIGURE 63.5 Late Pleistocene glacial complexes in Ilirney range. Area of glaciers: 1—Sartan; 2—Zyryanka; 3—fluvioglacial deposit: a—Sartan, b—Zyryanka; 4—end moraines; 5—glacial lakes; 6—river valleys; 7—watershed line.

Range advanced into the broad valley of the Maly Anyui River. Erosional glacial features are represented by cirques and troughs (Glushkova, 2001).

In the Zyrianka time, two glaciated regions and a few isolated glacial centres appeared in north-western Chukotka. The northern Region includes the Pyrkanay Mountains and the Rauchuan Range. The largest southern region includes the Chuvan Mountains, the Ilirney Range and mountain massifs along upper Bol'shoy Anyui and Maly Anyui rivers. Glacial formations are represented by poorly preserved terminal moraines, the hill-range relief of the basic moraine in valley bottoms. Zyrianka glaciers not only filled most valleys but also flowed over low watershed saddles from one basin into another. The largest and longest glaciers of produced a dendritic pattern on the southern slope of the Ilirney Range: the Tytyl'vaam Glacier was 60 km long; the Ilirney, 65 km long. The north slope glaciers were 25–40 km long.

The Sartan glaciations were two to three times smaller than the Zyrianka one. It is dispersed in 10 isolated mountain centres, attached to the central, most elevated parts of the Rauchuan, Anyui and Ilirney ridges. In mountain massifs of the northern part of the region, only short and passive valley and cirque glaciers emerged. In the Rauchua River valley, the glacier was 25 km long. The glacier discharged into a small intermontane basin. The arcs of four contiguous terminal moraines are well topographically expressed; they have a length of 3–6 km, width of 0.3–1.7 km and height of 40–80 km.

63.3.4. The Lower Anadyr Regions

63.3.4.1. The Northern Slope of the Koryak Range

Steep (20–24°) morainic hummocks and ridges built of coarse rubbly materials are observable in valleys of alpine-type mountains on the northern slope (Fig. 63.6). The ridges, 0.8–1.5 km long and 20–40 m high, have sharp crests. The largest terminal-moraine complex is located at the southern margin of the Lower Anadyr Lowland, 60 km from the source of the Nygchekveem River (Glushkova and Gualtieri, 1998). Its perimeter is about 11 km; its width varies from 4–6 km; the height of individual hills is 120–140 m. Deep troughs and kettles are frequent; lakes, of various shapes are, numerous. At different sites of the moraine, a series of samples was taken for the ^{36}Cl cosmogenic analysis. Average age values vary from 15 to 16 ka. In the Gytgykai River valley, fragments of six terminal-moraine ridges are preserved. Four of them block the large flowing Lake Mainits,

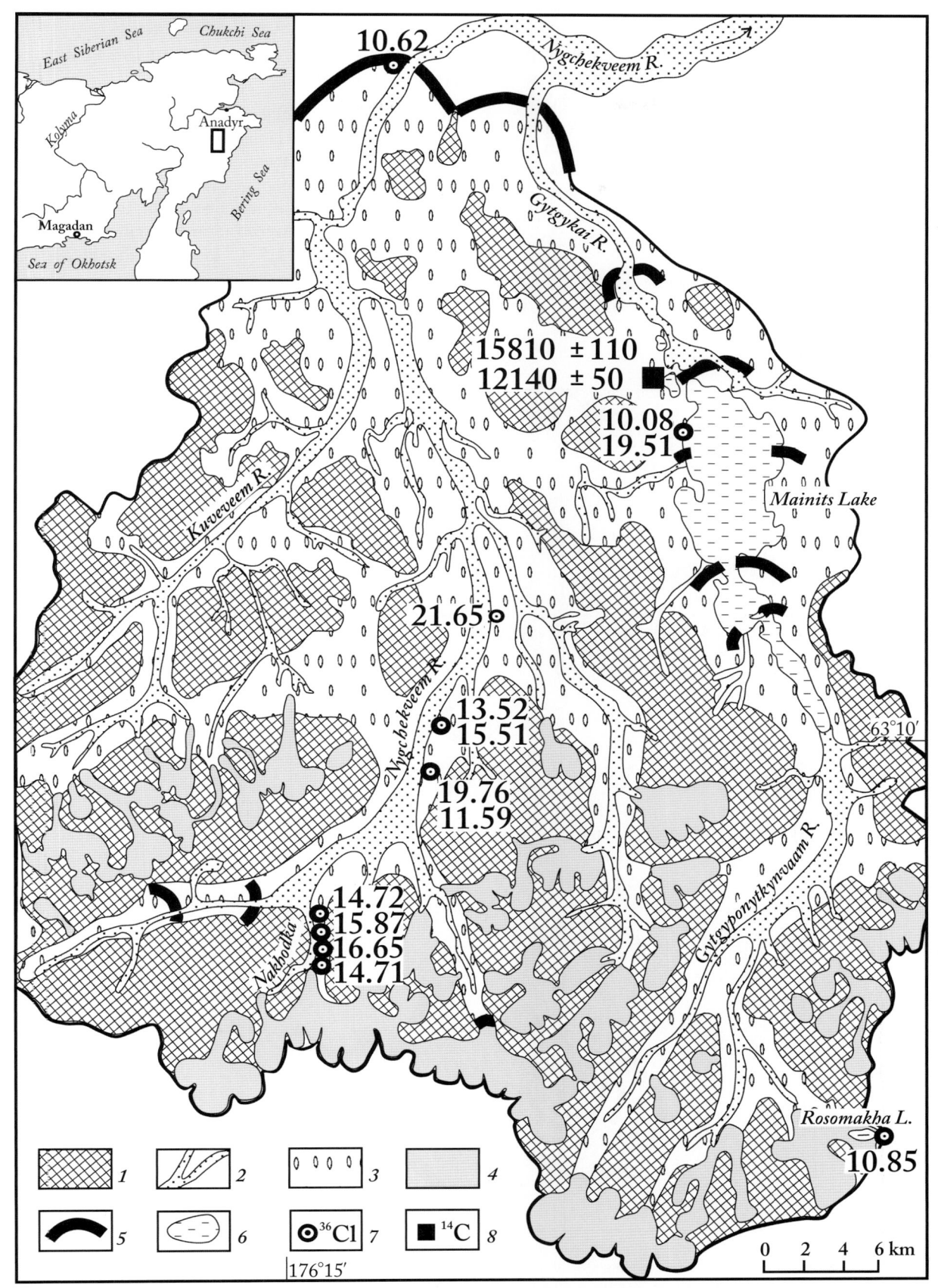

FIGURE 63.6 Area of LGM glacial complexes on northern slope of Koryak Upland. 1—unglaciated terrain; 2—river valleys; 3—Sartan moraines; 4—Holocene moraines; 5—end morainic ridges; 6—glacial lakes; 7—sample locations ^{36}Cl; 8—sample locations ^{14}C.

about 20 km long, up to 5 km wide and up to 130 m deep. The morainic ridge, damming Lake Mainits from the north is 1.5–2.0 km wide and over 3 km long. From boulders exposed on one of the hillcrests, two ^{36}Cl dates have been obtained, 10.08 ± 0.85 and 19.51 ± 2.27 ka. Besides, 29 dates within the range of 1.38–15.81 ^{14}C ka BP have been obtained from bottom sediments of Lake Patricia, located between the terminal-moraine ridges, damming Lake Mainits (Lozhkin et al., 2008). They indicate that the lake appeared during the final phase of the last ice age.

63.3.4.2. North-Western Part of the Lower Anadyr Lowland

In this region, large glacial complexes with numerous and various Zyrianka glacial landforms and correlative deposits have preserved. Characteristic for this area is the wide spread of ridge and hill-sink landscapes. Very large terminal-moraine ridges, basal moraine relief and glacial lakes prevail (Fig. 63.7). Arched terminal-moraine ridges reach 20–30 km at length, 1.5–4.5 km at width and 20–60 m at height.

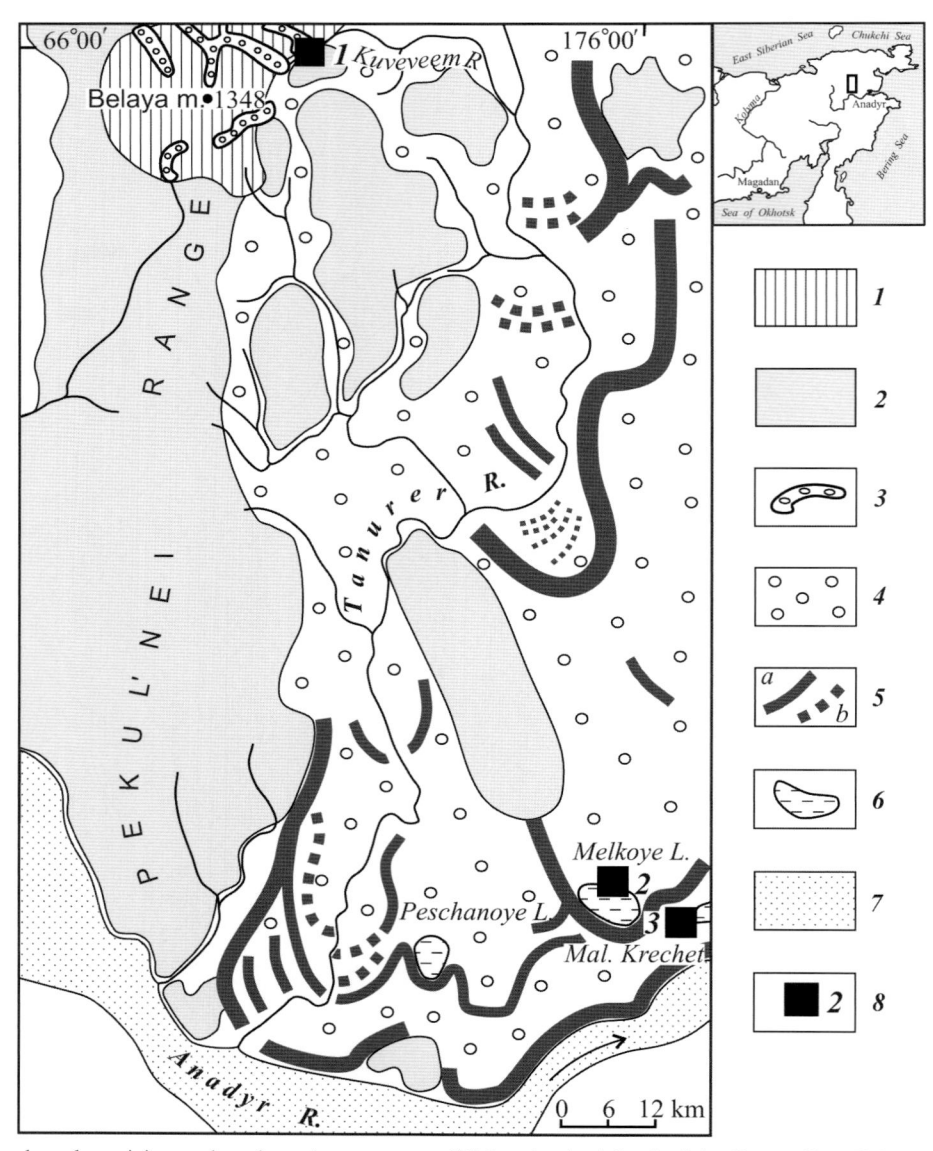

FIGURE 63.7 Zyryanka end morainic complexes in north-western part of Nizhne-Anadyrsk Lowland. 1—Center of Late Pleistocene glaciation in Pekul'ney Range; 2—unglaciated terrains; 3—Sartan moraines; 4—Zyryanka moraines; 5—end morainic ridges: a—main, b—stadial; 6—glacial lakes; 7—valley of Anadyr River; 8—studied sections: a—Kuveveem River, b—Melkoe Lake, c—Maly Krechet Lake.

Chapter | 63 Late Pleistocene Glaciations in North-East Asia

The analysis of glacial formations, in the basin of the middle and lower Tanurer and Kanchalan rivers, showed that the northern part of the Lower Anadyr Lowland was inundated by glaciers of the first Late Pleistocene glaciation (MIS 4). Palaeogeographic reconstructions showed that the Tanurer Glacier had been one of the longest in north-east Asia, about 30 km long. It had been formed from mountain–valley glaciers descending from the Pekul'nei Ridge and the Chukotka Upland. In the lowland, it transformed into an enormous Malaspina-type foot glacier. Observations in the area of small elevations on the lowland surface showed that, in the middle part of the glacier, ice had been

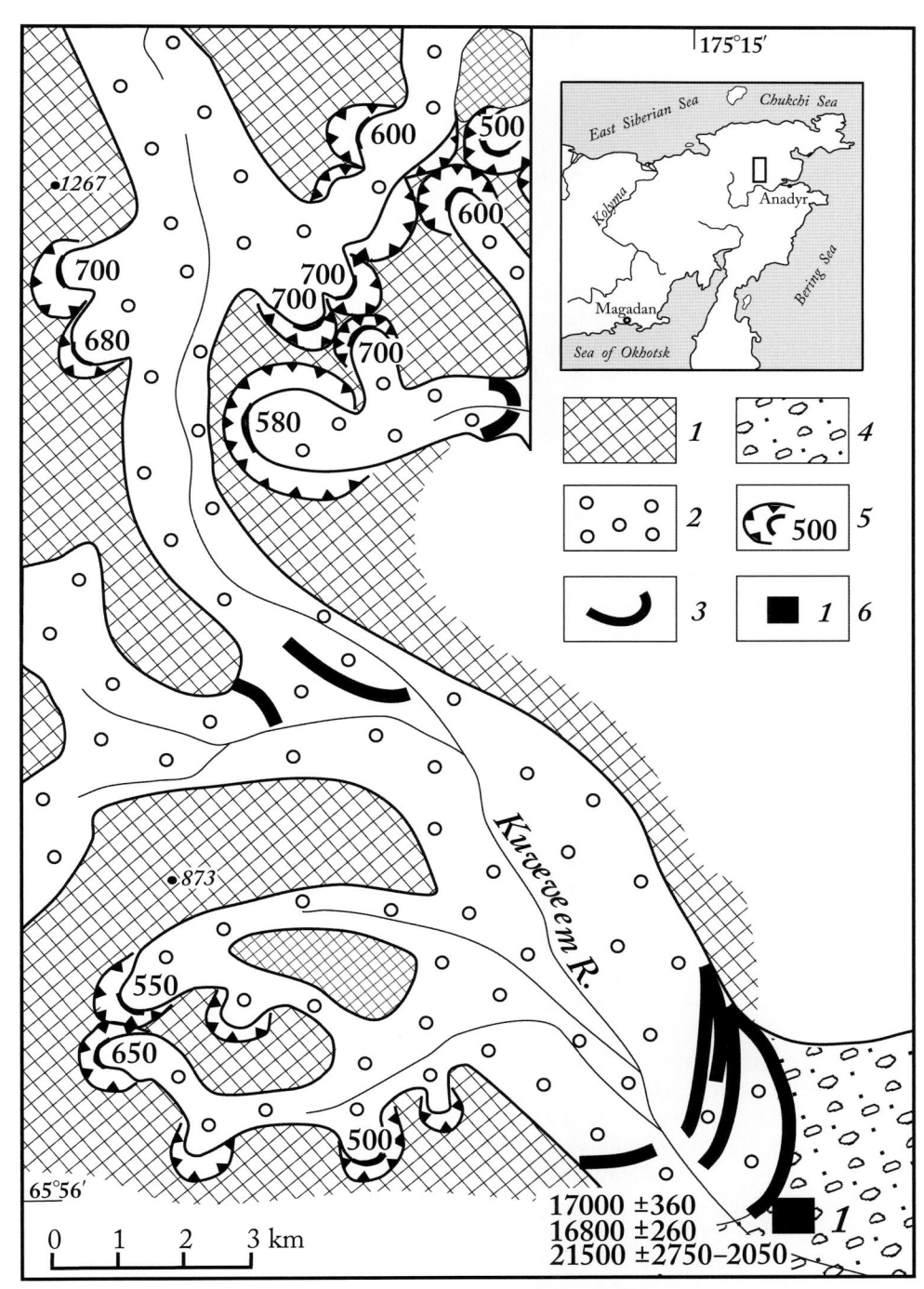

FIGURE 63.8 Sartan glacial complex in the Pekylney Range. 1—mountain slopes; 2—moraines; 3—end morainic ridges; 4—outwash; 5—glacial cirque; 6—studied sections.

no less than 150–200 m thick. In the basin of the middle and lower Tanurer River, seven large marginal terminal-moraine belts and seven low banks and ranges often drawn together in between, which testifies to the long and dynamic glacier maximum.

The time of glaciation development is witnessed by materials from drilling bottom sediments of the glacial Maly Krechet Lake, included in the outer glacier complex, approaching the Anadyr River. Palynological analysis and radiocarbon dates (44–46 ^{14}C ka BP) indicate the emergence of Maly Krechet Lake at the beginning of the Late Pleistocene interglacial, or on the border of MIS 4 and 3 (Lozhkin et al., 2005).

In the upper Tanurer River basin, on the eastern slope of the Pekul'nei Ridge, a glacial complex formed during the Sartan glaciations has been formed. It has a limited spread in the upper parts of valleys. Its marginal moraines seldom protrude to foothills. Glacial formations of the complex are most impressive in the basin of the Kuviveem River, the right tributary to Tanurer River (Fig. 63.8). Altitudes of the bottoms of 28 glacier cirques are 600–700 m a.s.l. Three contiguous steep morainic arcs 15–140 m high block the river valley some 20 km from the source, indicating the maximum ice advance. The sandur fringing the outer moraine has yielded two radiocarbon dates, 16.86 ± 0.26 ^{14}C ka BP and 21.5 ± 2.75 ^{14}C ka BP (Brigham-Grette et al., 2003).

The example of the Kuviveem Glacier testifies to a limited extent of the last glaciations in Chukotka. Glaciers were concentrated in the mountains, never exceeded 10–20 km in length and rarely advanced into the foothills.

63.4. CONCLUSION

The comparative analysis of Late Pleistocene glacial events in north-east Asia's districts, varying in height and distance from the seashore, shows the following. The first Late Pleistocene glaciations (the Zyrianka) occupied vast territories in all mountain systems of north-east Asia. In many areas, valley glaciers over 100 km long advanced onto piedmont plains and formed large piedmont glaciers.

The Sartan glaciation was by two to three times less extensive than the previous glaciation. It localised mostly in a few regions of the highest mountains. Valley glaciers, 10–20 km long, and numerous cirque glaciers prevailed. The difference in glaciations extent can be explained by evolution of temperature regime and humidity through the Late Pleistocene. However, another significant cause of this phenomenon was apparently the orographic factor on uplands with maximum heights under 1500 m a.s.l. As a result of the snow line rise, vast areas occurred in the zone of negative mass balance. The available data also indicate that in north-east Asia there are no traces of the Late Pleistocene ice sheets and domes suggested by some researchers (Grosswald and Hughes, 1995).

The important task of further studying of numerous and various exposures of glacial activity is to determine the age, geomorphological position and depositing conditions of glacial forms. The data on the Late Pleistocene glacial formations in north-east Asia may be of great significance for palaeoclimatic and palaeoglaciological reconstructions as well as for correlating glaciations of north-eastern Siberia and Alaska.

ACKNOWLEDGEMENT

I thank Prof. V. Astachov for editing the English version of this chapter.

REFERENCES

Anderson, P.M., Lozhkin, A.V., 2002. Late Quaternary Vegetation and Climate of Siberia and the Russian far East (Palynological and Radiocarbon Database). NESC FEB RAS, Magadan, 369pp.

Brigham-Grette, J., Gualtierie, L.V., Glushkova, O.Y., Hamilton, T.D., Mostoller, D., Kotov, A., 2003. Chlorine-36 and ^{14}C chronology support a limited last glacial maximum across central. Chukotka, north-eastern Siberia, and no Beringian ice sheet. Quatern. Res. 59, 386–398.

Catalogue of Glaciers of the USSR (1981). vol. 17, 19. Leningrad, Gidrometeoizdat, 88pp. (in Russian).

Forman, S., Pierson, J., Gomez, J., Brigham-Grette, J., Nowaczyk, N.R., Melles, M., et al., 2007. Luminescence geochronology for sediment from Lake El'gygytgyn, northeast Siberia, Russia: constraining the timing of paleoenvironmental events for the past 200 ka. J. Paleolimnol. 37, 77–88.

Glushkova, O.Yu., 1994. Paleogeography of Late Pleistocene glaciation of the North-Eastern Asia. In: Proceedings International Conference on Arctic Margins, Anchorage, Alaska, August 1992, pp. 339–344.

Glushkova, O.Yu., 2001. Geomorphological correlation of late Pleistocene glacial complexes of Western and Eastern Beringia. Quatern. Sci. Rev. 20, 405–417.

Glushkova, O.Yu., Gualtieri, L., 1998. Features of late Pleistocene glaciation of the northern Koryak uplands. In: Simakov, K.V. (Ed.), Environmental Changes in Beringia During the Quaternary. NEISRI FEB RAS, Magadan, pp. 112–132, (in Russian).

Grosswald, M.G., Hughes, T.J., 1995. Paleoglaciology's grand unsolved problem. J. Glaciol. 41, 313–332.

Ivanov, V.F., 1986. Quaternary Deposits of Coastal Eastern Chukotka. Far Eastern Science Center, Academy of Sciences of the USSR, Vladivostok, 138pp. (in Russian).

Juschus, O., Preusser, F., Melles, M., Radtke, U., 2007. Appling SAR-IRSL methodology for dating fine-grained sediments from Lake El'gygytgyn, north-eastern Siberia. Quatern. Geochronol. 2, 187–194.

Kind, N.V., 1975. Glaciations in the Verkhojansk Mountains and their place in the radiocarbon chronology of the Siberian late Antropogene. Biul. Peryglac. 2, 41–54, (in Russian).

Kolpakov, V.V., Belova, A.P., 1980. Radiokarbon dating in the glacial region of Verkhoyan'ye and its framing. In: Kind, N.V. (Ed.),

Geochronology of the Quaternary Period. Nauka Press, Moskau, pp. 230–235.

Lozhkin, A.V., Anderson, P.M., Eisner, W.R., Rovako, L.G., Hopkins, D.M., Brubaker, L.B., et al., 1995. New palinological and radiocarbon data on vegetation evolution in western Beringia in late Pleistocene and Holocene. In: Bychkov, Yu.M., Lozhkin, A.V. (Eds.), Evolution of Climate and Vegetation in Beringia in Late Cenozoic. NEISRI FEB RAS, Magadan, pp. 5–24.

Lozhkin, A.V., Anderson, P.M., Brown, T.A., Vazhenina, L.N., Glushkova, O.Yu., Kotov, A.N., et al., 2005. Glaciation of the Anadyr lowland (as inferred from lake sediments). In: Pages from Quaternary History of Northeast Asia. NEISRI FEB RAS, Magadan, pp. 4–22, (in Russian).

Lozhkin, A.V., Anderson, P.M., Matrosova, T.V., Solomatkina, T.B., 2008. Experience of study of lacustrine pollen records for reconstruction of natural environment in Quaternary. Vestnik NESC FEB RAS 1, 24–32, (in Russian).

Matrosova, T.V., 2009. Reconstruction of vegetation and climates during the last 350 thousand years in northern Chukotka (According to palynologic evidences of El gygytgyn lake). Vestnik NESC FEB RAS 2, 23–30, (in Russian).

Nekrasov, I.A., Sheinkman, V.S., 1974. Stage of late Pleistocene glaciers reduction in the Cherski mountain system. DokladyAS USSR, 219, N 2, pp. 421–424, (in Russian).

Sher, A.V., Kuzmina, S.A., Kuznetsova, T.V., Sulerzhitsky, L.D., 2005. New insights into Weichselian environment and climate of the East Siberian Arctic, derived from fossil insects, plants, and mammals. Quatern. Sci. Rev. 24, 533–569.

Siegert, K., Stauch, G., Lehmkuhl, F., Sergeenko, A.I., Diekmann, B., Popp, S., et al., 2007. Development of glaciation in the Verkhoyansk range and his piedmont in the Pleistocene: results of new researches. Regional Geol. Metallogeny 30–31, 222–228 (in Russian).

Stauch, G., Lehmkuhl, F., Frechen, M., 2007. Luminescence chronology from the Verkhoyansk Mountains (North- Eastern Siberia). Quatern. Geochronol. 2, 255–259.

Stratigraphy of USSR, 1982. Quaternary, vol. 1. Moscow, Nedra, 443pp. (in Russian).

Velichko, A.A., 1993. Evoultion of landscapes and climates of the Northern Eurasia. Late Pleistocene—Holocene. Elements of Prognosis. I. Regional Paleogeography. Nauka, Moscow, 102pp. (in Russian).

Verkhovskaya, N.B., 1986. Pleistocene in Chukotka, palinostratigraphy and the major paleogeographic events. Russian Academy of Sciences, Vladivostok, 112pp. (in Russian).

Zamoruyev, V., 2004. Quaternary glaciation of North-East Asia. In: Ehlers, J., Gibbard, P.L. (Eds.), Quaternary Glaciations-Extent and Chronology Part III. Elsevier, Amsterdam, pp. 321–323.

Chapter 64

Extent and Timing of Quaternary Glaciations in the Verkhoyansk Mountains

Georg Stauch* and Frank Lehmkuhl

Department of Geography, RWTH Aachen University, Templergraben 55, 52056 Aachen, Germany
*Correspondence and requests for materials should be addressed to Georg Stauch. E-mail: gstauch@geo.rwth-aachen.de

64.1. INTRODUCTION

The extent and timing of Quaternary glaciations in the Verkhoyansk Mountains have been debated since the beginning of the twentieth century. The remoteness of these vast mountain areas and harsh climate conditions has constrained Quaternary research for many centuries. The location of the mountain system at the eastern edge of the Atlantic-influenced Eurasian continent is a unique situation in which to study not only regional climate effects but also the influence of Quaternary ice sheets in the north of the continent. The Verkhoyansk Mountains stretch for about 1200 km from the Laptev Sea coast to central Yakutia in an s-shaped manner (Fig. 64.1). Maximum elevations of 2959 m a.s.l. are reached in the south-eastern part of the mountains, in an area known as the Suntar Chajata. However, in the northern parts, most summits are around 1400 and 2000–2200 m a.s.l. in the central part. Harsh climate conditions are characteristic of the area and the so-called 'cold pole' of the Earth is situated at the village of Oimjakon, which is just 100 km away from the eastern branch of the mountain system. The mean annual temperature is at most stations below −10 °C with mean temperatures in January of −40 and up to 20 °C in July. Present-day precipitation varies considerably in the area as a consequence of orographic effects. In the valley of the Lena River in the west of the region, climate stations record values of around 200 mm a year, while on the western side of the mountains up to 700 mm are recorded. On the eastern side of the Verkhoyansk Mountains, values drop again to 130 mm (Lydolph, 1977; Murzin, 2003). At present, only two parts of the Verkhoyansk Mountains contain glaciers. In the northern branch, some isolated glaciers with a maximum length of 3.5 km exist (Stauch, 2006). In the high parts of the Suntar Chajata, an area of more than 200 km is currently glaciated (Koreisha, 1991; Ananicheva and Krenke, 2005).

64.2. VERKHOYANSK MOUNTAINS

Early reports of past extensive glaciations have been made by several Russian researchers such as V. Obruchev and V.N. Saks in the first half of the twentieth century. However, detailed research on the extent and timing of the Quaternary glaciations began in the 1960s with the work of V.N. Kind, V.V. Kolpakov and A.P. Belova and several other scientists. This work resulted in different models of Quaternary glaciations in the Verkhoyansk Mountains. Most research has been undertaken in the central Verkhoyansk Mountains, especially in the valley of the Tumara River (Fig. 64.1). Small terminal moraines in the main valleys of the central Verkhoyansk Mountains and in the western foreland have been dated by Kind (1975) on the basis of their relative position as remnants of retreating gLGM (global Last Glacial Maximum; ca. 21 cal.ka BP, 18 ^{14}C ka BP) glaciers and Early Holocene advances (Fig. 64.2). Three large terminal moraines in the western foreland have been dated by radiocarbon to 29–15 ^{14}C ka BP (Kind, 1975; Kolpakov, 1979; Kolpakov and Belova, 1980). According to Kind (1975), a further terminal moraine, marking an even larger glacial extent, dates to 33–30 ^{14}C ka BP. This moraine had been named Karginsk after the Karginsk Interstadial of the Siberian stratigraphy and had not been identified in other area of north-eastern Russia. Two more extensive moraines are only preserved as ground moraine with no terminal ridges at the surface. They have been attributed to the early part of the last glacial cycle and the previous glacial cycle (Kind et al., 1971; Kind, 1975).

Several maps have been published in the past few decades regarding the extent of glaciations mainly during the gLGM. According to Grosswald and Hughes (2002), north-eastern Russia was completely covered by a large ice sheet with maximum ice thickness of more than 2000 m east of the Verkhoyansk Mountains. This ice sheet

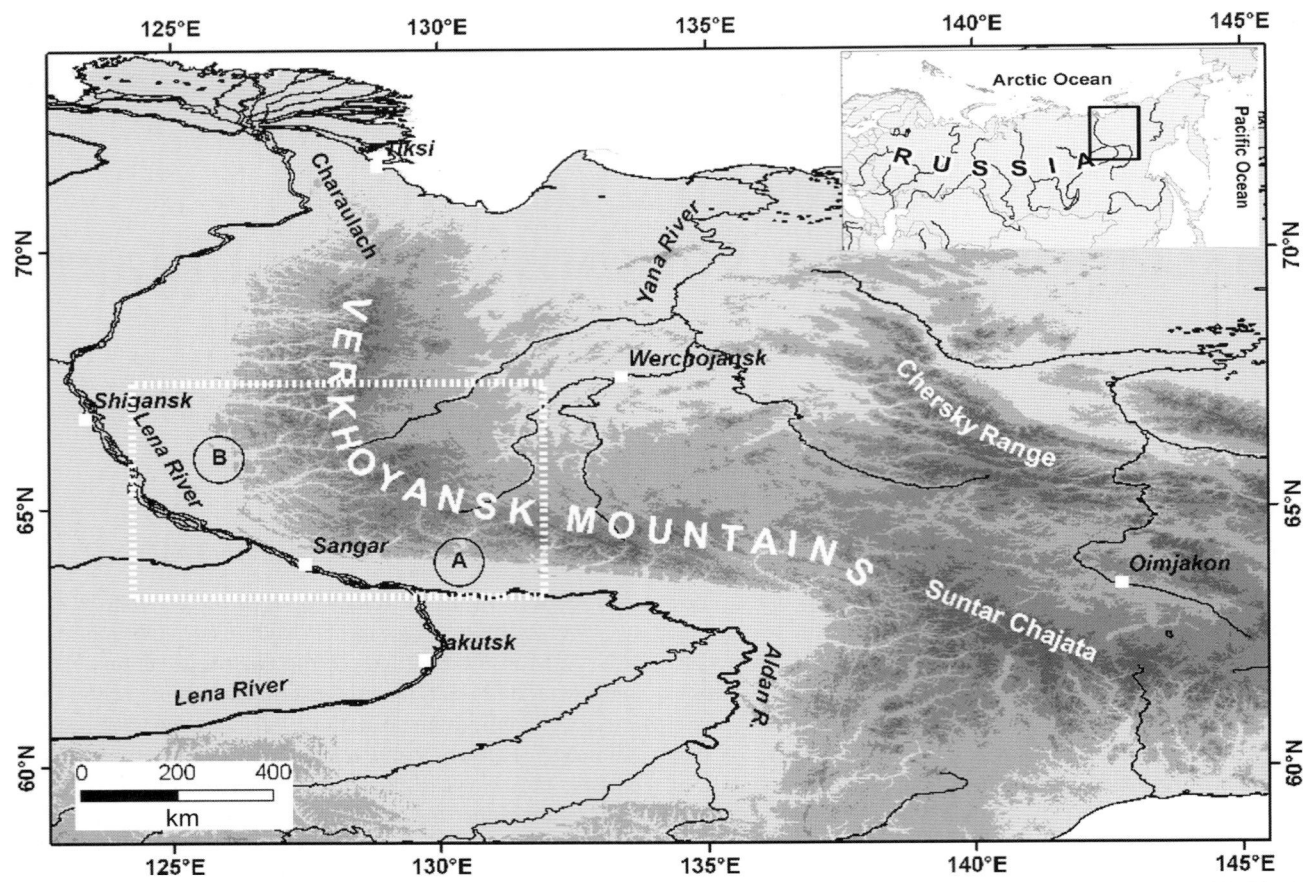

FIGURE 64.1 The Verkhoyansk Mountains (the hatched box indicates the central Verkhoyansk Mountains: (A) Tumara River; (B) Djanushka River).

was part of a large pan-Arctic Ice Sheet covering most of the Arctic regions. This hypothesis has been rejected by several studies in recent years (e.g. Velichko and Spasskaya, 2002; Brigham-Grette et al., 2003; Svendsen et al., 2004; Spielhagen et al., 2005). Other reconstructions show mountain glaciations of much smaller extent (e.g. Arkhipov et al., 1986; Zamoruyev, 2004). However, all maps indicated glaciers with a length of more than 100 km on the western side of the Verkhoyansk Mountains stretching from the high mountain areas down to the foothills and out in the forelands.

Research from the start of the twenty-first century, based on a re-evaluation of the morphological landforms in combination with IRSL (infrared-stimulated luminescence) dating of overlying aeolian sediments as well as glacial sediments, resulted in a different chronology of glaciations in the Verkhoyansk Mountains. According to these new results, glaciations in the central Verkhoyansk Mountains were either very limited or non-existent during the past 50 ka (Fig. 64.2). The last glaciation which left traceable morphological landforms occurred before 50 ka (Stauch et al., 2007). Terminal moraines (termed I; Fig. 64.3) related to this glaciation have been deposited inside of the mountain system and are relatively small compared to older glacial landforms. These results correspond with studies from the Charaulach Mountains in the northernmost part of the Verkhoyansk Mountains close to the delta of the Lena River. This mountain chain has not been glaciated at least during the past 60 ka (Schirrmeister et al., 2002; Hubberten et al., 2004). Glacial deposits at the mountain front of the Verkhoyansk Mountains are generally much larger in size. In the Tumara River valley, terminal moraines (II) form impressive arcs with diameters of 20 km. Aeolian sediments covering these terminal moraines have been dated to 19.2 and 46.8 ka (Stauch et al., 2007) but on the basis of their relative position, the underlying glacial sediments were deposited prior to 50 ka. A third set of terminal moraines further downstream of the rivers is even larger. Similar glacial landforms can be observed in many river valleys in the western foreland of the Verkhoyansk Mountains. IRSL ages indicate formation around 80–90 ka ago. At the Tumara River, one further terminal moraine is preserved (IV), while two more former glacier terminus positions can be reconstructed in the Djanushka River valley (IV and V). Preservation of both moraines is

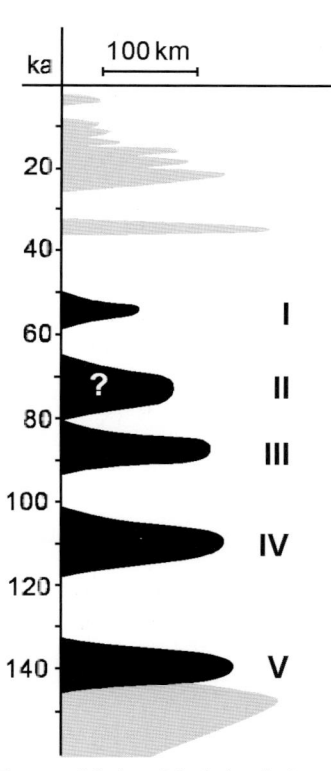

FIGURE 64.2 Extent and timing of glaciations in the central Verkhoyansk Mountains (grey: according to Kind et al., 1971; black: Stauch et al., 2007).

only sketchy as a result of fluvial erosion caused by the younger glacial advances and in the case of the outermost moraine V, the Lena River. According to the IRSL ages obtained from the Tumara River, moraine IV gives an age of 100–120 ka, while the sediments of the outermost moraine at the Djanushka River (V) have been deposited around 135 ka. However, only two IRSL ages are at present available from this moraine V (Stauch et al., 2007). Similar successions of terminal moraines have been mapped in several neighbouring valleys on the western side of the Verkhoyansk Mountains (Stauch, 2006; Stauch and Lehmkuhl, 2010). In contrast, terminal moraines on the eastern side of the mountains are smaller and closer to the higher part of the Verkhoyansk Mountains. Late Quaternary glaciers must have been much smaller on this side and all terminal moraines have been deposited inside the valleys of the mountains and not on the piedmont.

These results indicate a primary moisture source from western direction during times of glaciations and, therefore, a similar situation as today. This is also confirmed by the orientation of cirques and cirque floor elevations. Nearly 500 cirques have been mapped in the central Verkhoyansk Mountains on the basis of satellite images. Cirques located at lower elevations are generally found on the western side, while cirques on the eastern side are ~500 m higher. The average cirque floor elevation is around 1400 m a.s.l. (Stauch, 2006).

64.3. REGIONAL COMPARISON

Comparing these results with recent studies from areas west and east of the Verkhoyansk Mountains shows some similar trends in the timing of glaciations but also some striking differences. Moisture-bearing winds from the west have been strongly influenced by development of the large ice sheets in the northern part of the Eurasian continent. Svendsen et al. (2004) reconstructed the extent of the Eurasian Ice Sheet for four time periods in the Late Pleistocene. The two segments of the ice sheet, the Scandinavian Ice Sheet and the Barents–Kara Ice Sheet, developed opposite trends (Svendsen et al., 2004; Chapter 28). While the western sector became progressively larger during this period, the eastern sector became smaller in size. The trend of the Barents–Kara Ice Sheet is therefore comparable to the development of mountain glaciations in the Verkhoyansk Mountains. This is further confirmed by the development of mountain glaciations in the Polar Urals (Fig. 64.4). An age of around 20 ka has been obtained for a terminal moraine only 1 km from the present ice margin, while moraines further away from the mountains have an age of 50–60 ka (Mangerud et al., 2008). A similar age has been given by Astakhov and Mangerud (2007) for river terraces related to glaciers from the Putorana Plateau.

East of the Verkhoyansk Mountains, only few results concerning the extent and timing of Quaternary glaciations in the Russian Far East are available (Chapter 63). A recent summary highlighted two areas of more detailed studies, the Pekulney Mountains in the Anadyr Uplands and the Koryak Mountains (Fig. 64.4) on the eastern side of the Eurasian continent (Stauch and Gualtieri, 2008). Three suites of terminal moraines have been mapped and dated in the Pekulney Mountains (Brigham-Grette et al., 2003). The youngest of these sets has been termed the Kuveveem Moraines. Cosmogenic isotope dating revealed a formation during marine isotope stage (MIS) 2. Remnants of two older glaciations have been identified at the Tanyurer River. The upper moraine has been termed the Chumyveem moraine. Cosmogenic isotope dating of erratic boulders has yielded ages between 41 and nearly 70 ka. The second moraine (Anadyr River moraine) has been dated on the basis of morphological criteria. Brigham-Grette et al. (2003) assumed Middle to Early Pleistocene age for this moraine. In the Koryak Mountains, a study by Gualtieri et al. (2000) presented evidence for at least two Pleistocene glaciations. A younger glaciation occurred again during MIS 2, while cosmogenic isotope data from the lower Anadyr depression, north of the mountains, indicate a glaciation before 50 ka. No terrestrial evidence has yet been found relating to the early part of the last glacial cycle (Weichselian Stage). However, Brigham-Grette (2001) and Brigham-Grette et al. (2001) suggested, on the basis of glaciomarine sediments, a rapid glaciation of Chukotka Peninsula directly after the last interglacial.

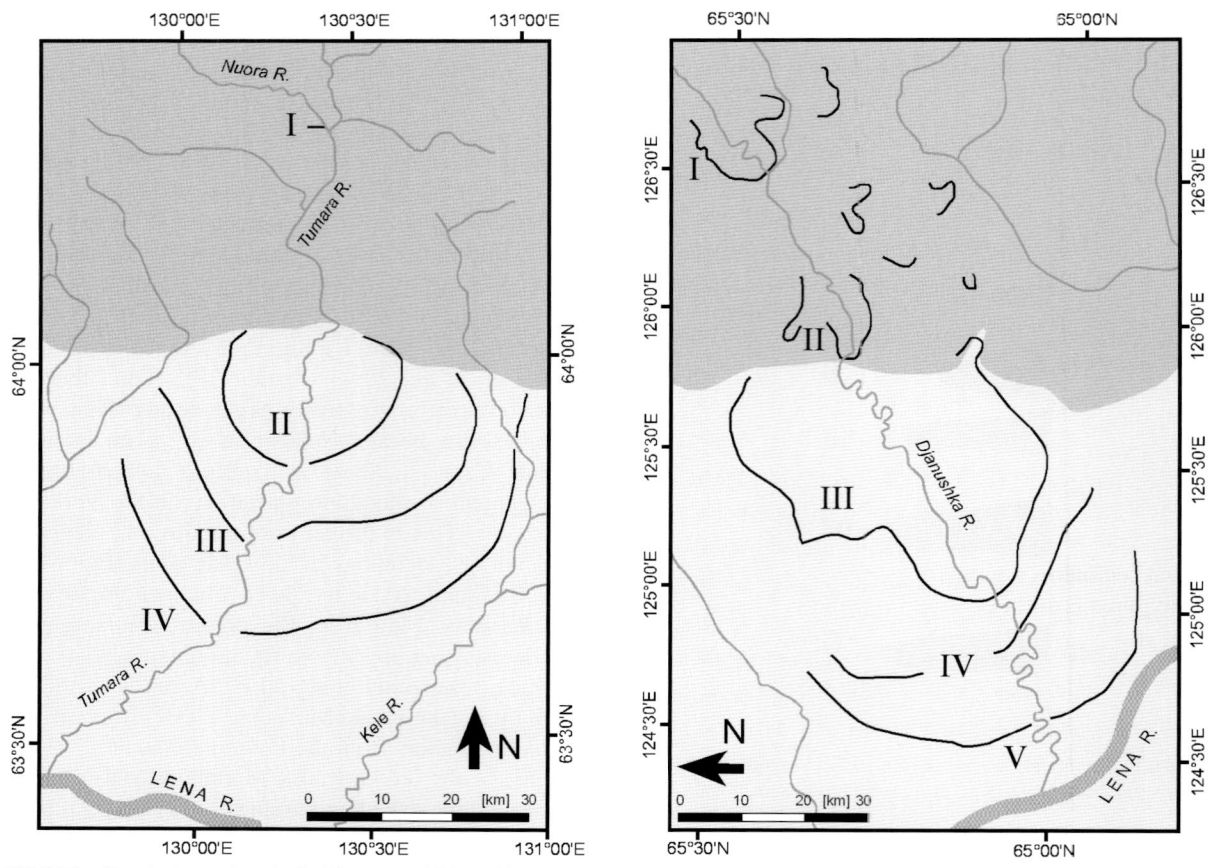

FIGURE 64.3 Terminal moraines in the Tumara and Djanushka valley (modified from Stauch et al., 2007).

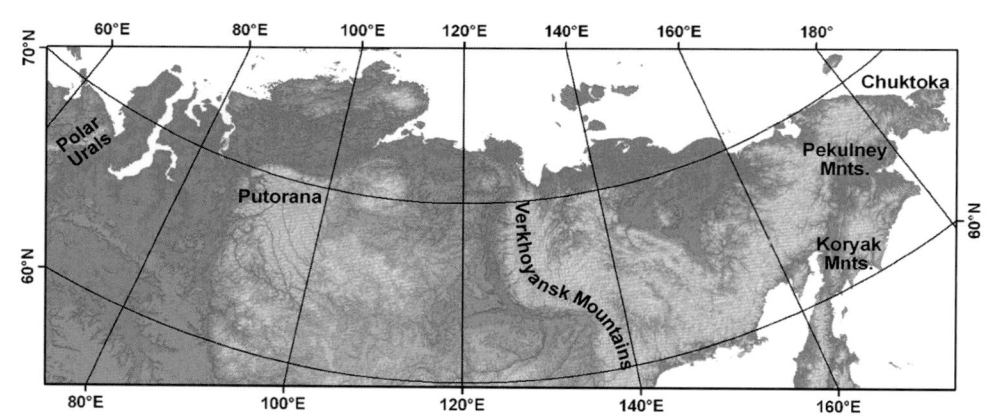

FIGURE 64.4 Northern Asia and selected mountain areas.

64.4. CONCLUSION

Quaternary glaciation in the central Verkhoyansk Mountains became more limited in extent throughout the Late Quaternary. The largest extent of ice has so far been attributed to the penultimate glacial cycle (Saalian Stage) prior to the Eemian interglacial. Further glaciations occurred at around 100–120, 80–90 and before 50 ka. It seems that there was no glaciation in the area during the gLGM (MIS 2). However, whether the dated terminal moraines during the early part of the glacial cycle represent individual glaciations with a considerable ice retreat between, or whether they are remnants of glacier oscillations, remains one of the unanswered questions. The extent and timing of glaciations in the Verkhoyansk Mountains shows similar trends to mountain glaciations further west and the Barents–Kara Ice Sheet (Chapters 27 and 28). However, there have been more glaciations identified in the Verkhoyansk area during the early part of the last glacial cycle. East of the Verkhoyansk Mountains, a different pattern is recognisable. In these areas, glaciation also decreased in size progressively through the last glacial cycle, but prominent terminal moraines were developed during MIS 2. The asynchronous formation of glaciers on the western and eastern slopes of the Verkhoyansk Mountains is the result of different moisture sources. In the west of the Eurasian continent, the growth of the Scandinavian Ice Sheet blocked the moisture-bearing air masses from entering the far east of Siberia and prevented the growth of glaciers in the Verkhoyansk Mountains during the MIS 2, despite very low temperatures. In contrast, on the easternmost side of the continent, moisture from the Pacific supported the development of glaciers in the Pekulney and Anadyr Mountains during MIS 2 (Stauch and Gualtieri, 2008; Stauch and Lehmkuhl, 2008). However, there are still large areas in north-eastern Russia with large mountain systems where no dating results for glacial sediments are available. Further research in this interesting area is still needed.

REFERENCES

Ananicheva, M.D., Krenke, A.N., 2005. Evolution of climatic snow line and equilibrium line altitudes in the North-Eastern Siberian Mountains (20th Century). Ice Climate News 6, 3–6.

Arkhipov, S.A., Isayeva, L.L., Bespaly, V.G., Glushkova, O., 1986. Glaciations of Siberia and North-East USSR. Quatern. Sci. Rev. 5, 463–474.

Astakhov, V., Mangerud, J., 2007. The Geochronometric age of Late Pleistocene terraces on the Lower Yenisei. Doklady Earth Sci. 416, 1022–1026.

Brigham-Grette, J., 2001. New perspectives on Beringian Quaternary paleogeography, stratigraphy, and glacial history. Quatern. Sci. Rev. 20, 15–24.

Brigham-Grette, J., Hopkins, D.M., Ivanov, V.F., Basilyan, A., Benson, S.L., Heiser, P.A., et al., 2001. Last interglacial (isotope stage 5) glacial and sea level history of coastel Chukotka Peninsula and St. Lawrence Island, western Beringia. Quatern. Sci. Rev. 20, 419–436.

Brigham-Grette, J., Gualtieri, L.M., Glushkova, O.Yu., Hamilton, T.D., Mostoller, D., Kotov, A., 2003. Chlorine-36 and 14C chronology support a limited last glacial maximum across central Chukotka, north-eastern Siberia, and no Beringian ice sheet. Quatern. Res. 59, 386–398.

Grosswald, M.G., Hughes, T.J., 2002. The Russian component of an Arctic ice sheet during the Last Glacial Maximum. Quatern. Sci. Rev. 21, 121–126.

Gualtieri, L., Glushkova, O., Brigham-Grette, J., 2000. Evidence for restricted ice extent during the last glacial maximum in the Koryak Moutains of Chukotka, far eastern Russia. Geol. Soc. Am. Bull. 112, 1106–1118.

Hubberten, H.-W., Andreev, A., Astakhov, V.I., Demidov, I., Dowdeswell, J.A., Henriksen, M., et al., 2004. The periglacial climate and environment in northern Eurasia during the last glaciation. Quatern. Sci. Rev. 23, 1333–1357.

Kind, N.V., 1975. Glaciations in the Verkhojansk Mountains and their place in the radiocarbon chronology of the Late Pleistocene Arthropogene. Biul. Perygl. 24, 41–54.

Kind, N.V., Kolpakov, V., Suleržicku, L.D., 1971. To the problem of the age of glaciations in the Verkhoyansk area. Proc. USSR Acad. Sci. Geol. Ser. 10, 135–144. (in Russian).

Kolpakov, V.V., 1979. Glacial and periglacial relief of the Verkhojansk glacial region and new radiocarbon datings. In: Muzins, A.I. (Ed.), Regional Geomorphology of Newly Developed areas. Geographical Society of the USSR, Moskau, pp. 83–97.

Kolpakov, V.V., Belova, A.P., 1980. Radiocarbon dating in the glacial region of Verkhoyan'ye and its framing. In: Ivanova, I.K., Kind, N.V. (Eds.), Geochronology of the Quaternary period. Nauka Press, Moskau, pp. 230–235.

Koreisha, M.M., 1991. Glaciation of the Verkhoyansk-Kolyma Region. Academy of Sciences of the USSR, Moskau Soviet Geophysical Committee (in Russian).

Lydolph, P.E., 1977. Climates of the Soviet Union. World Survey of Climatology 7. Elsevier, Amsterdam.

Mangerud, J., Gosse, J., Matiouchkov, A., Dolvik, T., 2008. Glaciers in the Polar Urals, Russia, were not much larger during the Last Global Glacial Maximum than today. Quatern. Sci. Rev. 27, 1047–1057.

Murzin, A.M., 2003. Glaciers of Yakutia. Nauka i technika w Jakutii (Science and techniques in Yakutia) 2 (5), 102–107. (in Russian).

Schirrmeister, L., Siegert, C., Kuznetsova, T., Kuzmina, S., Andreev, A., Kienast, F., et al., 2002. Paleoenvironments and paleoclimatic records from permafrost deposits in the Arctic region of Northern Siberia. Quatern. Int. 89, 97–118.

Spielhagen, R.F., Erlenkeuser, H., Siegert, C., 2005. History of freshwater runoff across the Laptev Sea (Arctic) during the last deglaciation. Global Planet. Change 48, 187–207.

Stauch, G., 2006. Jungquartäre Landschaftsentwicklung im Werchojansker Gebirge, Nordost Sibirien (Late Quaternary landscape development in the Verkhoyansk Mountains, North-Eastern Siberia). Aachener Geographische Arbeiten 45, Aachen (in German, English Summary), 1–179.

Stauch, G., Gualtieri, L., 2008. Late Quaternary glaciations in North-Eastern Russia. J. Quatern. Sci. 23, 545–558.

Stauch, G., Lehmkuhl, F., 2008. Dry climate conditions in Northeast Siberia during the MIS2. In: Kane, D.L., Hinkel, K.M. (Eds.), Proceedings of the Ninth International Conference on Permafrost, University of Alaska Fairbanks June 29–July 3, 2008. University of Fairbanks, Fairbanks,1701–1704.

Stauch, G., Lehmkuhl, F., 2010. Quaternary glaciations in the Verkhoyansk Mountains, northeast Siberia. Quatern. Res. 74, 145–155.

Stauch, G., Lehmkuhl, F., Frechen, M., 2007. Luminescence chronology from the Verkhoyansk Mountains (North-Eastern Siberia). Quatern. Geochronol. 2, 255–259.

Svendsen, J.I., Alexanderson, H., Astakhov, V.I., Demidov, I., Dowdeswell, J.A., Funder, S., et al., 2004. Late Quaternary ice sheet history of northern Eurasia. Quatern. Sci. Rev. 23, 1229–1271.

Velichko, A.A., Spasskaya, I., 2002. Climatic change and the development of landscapes. In: Shahgedanova, M. (Ed.), The Physical Geography of Northern Eurasia. Oxford University Press, Oxford, pp. 36–69.

Zamoruyev, V., 2004. Quaternary glaciation of north-eastern Asia. In: Ehlers, J., Gibbard, P.L. (Eds.), Quaternary Glaciations—Extent and Chronology, Part III: South America, Asia, Africa, Australasia, Antarctica. Elsevier, Amsterdam, pp. 321–323.

Chapter 65

Glaciation in the High Mountains of Siberia

Vladimir S. Sheinkman

Institute of the Earth Cryosphere, Malygina Street, 86, Russian Academy of Sciences, Siberian Branch, Tyumen 62500, Russia
V.B. Sochava Institute of Geography, Ulan-Batorskaya Street, 1, Russian Academy of Sciences, Siberian Branch, Irkutsk 664033, Russia

65.1. INTRODUCTION

The high mountain belt surrounding Siberia on the south and east (here referred to as the 'Siberian Mountains') provides important insights into the nature of glaciation in High Central and north-eastern Asia. They represent a wide range of environments, where numerous modern glaciers, as well as ancient glacial relics, occur together with well-expressed permafrost. Nevertheless, for a long time, glaciological and palaeoglaciological phenomena from many mountain regions of the belt had not been studied in detail. However, as early as end of the nineteenth century, Kropotkin (1876), a famous Russian researcher, first collected material to test his ideas on the glacial theory in Siberia.

Reliable reconstructions of the Pleistocene glaciation were only based on a superficial geological survey, which, towards the end of the 1960s, had been completed and eliminated the last blank areas from the geological maps. However, lack of knowledge inhibited a clear understanding of ancient glaciations in this major continental climatic regime, which occurred in the heart of the Eurasian continent. This was also hampered by the fact that most of the modern glaciers in the Siberian Mountains were only studied in any detail in the latter half of the twentieth century. Until today, the discussion revolved around the long-standing debate between adherents and opponents of Voeikov (1881), a famous nineteenth century Russian climatologist, who claimed that it was impossible for great glaciers to form in the heart of Siberia because of its dry climate. From the very beginning of the debate, many famous scientists took sides, with such celebrities as Cherskii (1882) and Berg (1938) supporting Voeikov's point of view, whereas Obruchev (1951), a renowned Siberian researcher, did not agree. The debate continues, and the discussion is as topical today, as ever.

The first palaeoglaciological reconstructions were generally influenced by the classical European model and did not always take into account the specific environment of the Siberian Mountains. This was because the study of Quaternary phenomena was largely initially focused on North-western Siberia (Sax, 1953; Arkhipov, 1989), where the environment resembles that of North Europe. In addition, study of modern glaciers in Siberia began in the more accessible and most humid West Altai with its relatively mild climate. It is often assumed that all glaciers, if they are of the same morphological type, as well as characteristic landforms yielded by their activity, are similar to one another in their appearance. Consequently, Alpine glaciological and palaeoglaciological concepts derived from Europe were widely transferred to the mountainous regions of Siberia without consideration of the thermodynamics of the glaciers in this region. For this reason, the similarity of the European and Siberian glaciation schemes were thought, for a long time, to form a firm basis for subsequent investigations. Even after discovery of large modern glaciers under the extra-continental climate conditions in north-eastern Siberia, the first discoverers of these phenomena still accepted Voeikov's concept on the basis of the modern climatic situation. However, they assumed that humid environments must have existed during the Pleistocene glacial epochs to explain the significant advances of ancient glaciers (Vas'kovskii, 1963; Grave et al., 1964). As has been demonstrated in a recent overview (Astakhov, 2006), this concept is still perpetuated by some modern authors.

All in all, the problem is objective: its key component is inseparably connected with the history of study of the Siberian Mountains. The first summary investigation of Siberian glaciers based on a regular field study was published in the mid-twentieth century by Tronov (1949), a famous Russian glaciologist. He initially focused on the Altai; the similar work on Eastern and north-eastern Siberia did not appear until the mid-1960s (Preobrazhenskii, 1960; Koreisha, 1963). This was simply because north-eastern Siberian is a very remote place.

Moreover, in spite of the fact that permafrost phenomena in Siberia have been noticed for a long time, only in the late 1920s were they catalogued and described by Sumgin

(1927), the founder of geocryology in Russia. This also hampered the determination of the specific distribution of Siberian glaciers.

The main evidence relating to the glaciation in Siberia was only presented in the 1970–1980s. Then, together with detailed evaluation of geological material, numerous data were collected during geocryological mapping and a general survey of both glaciers and icings (naleds) of the former USSR (Nekrasov, 1976; Sheinkman, 1987; Koreisha, 1991; Kotlyakov, 1997).

All these data demonstrated that the old palaeoglaciological schemes required correction, since the Quaternary events in the Siberian Mountains did not coincide with their European equivalents. Further, some of the landforms and sediments previously considered to be of glacial origin were found to have been formed by non-glacial processes. Overall, glaciations in western Eurasia (the Alpine scheme) were found to be fundamentally different to those of the mountainous regions of Central and north-eastern Asia (Sheinkman, 1992, 1995, 2002a, 2002c, 2008; Zamoruev, 1995; Astakhov, 2006). An important point is that glaciological and geocryological investigations throughout Siberia (Vinogradov, 1966–1981; Yershov, 1988–1989) demonstrated that high moisture availability and, in turn, high snowfall are *not* favourable conditions to the formation of cryogenic ice. On the contrary, a continental climate inhibits the development of glaciers. In addition, the atmospheric circulation throughout the Quaternary was characterised by strongly continental conditions over Siberia (Sheinkman and Barashkova, 1991; Borzenkova, 1992; Gavrilova, 1998; Sheinkman and Antipov, 2007).

Both glacial and permafrost processes have occurred in the Siberian Mountains under this continental climatic regime. However, unfortunately, the original lack of exchange between glacier and permafrost researchers resulted in some major interpretational disagreements. Some investigators, who promoted the view that permafrost environments prevailed in Siberia during the Quaternary, underestimated the role of glaciers (e.g. Danilov, 1978; Tomirdiaro, 1980). Whereas others (e.g. Grosswald and Hughes, 2002) ignored the permafrost phenomena and postulated giant ice sheets covering most of Siberia. After having studied both modern and ancient glaciation, as well as permafrost, along the entire mountain belt surrounding Siberia, for many years, the present author does not agree with either of these extreme points of view. Today, it is obvious that glaciers in the Siberian Mountains *can* develop under conditions of low snow accumulation. The reason is simple: because temperatures are so cold over the glacier surfaces. Consequently, despite significant summer melting, a sufficient part of melt water is repeatedly frozen on glacier surfaces all over the ablation season. This forms superimposed ice, both at the beginning of ablation period and throughout that period; this process noticeably supplements the mass budget of glaciers.

Glaciation, as an integrated phenomenon, must be then seen in Siberia as a development of different glacial and cryogenic ice agents, which can shape the valley morphology either individually or in combination (Sheinkman, 1993, 1995, 2007). In order to reconstruct and assess the events across the region during the Quaternary, the study of former glaciation must include the interaction of both geocryological and glacial phenomena, together with their evolution through time as an integrated process. Such systems were earlier termed (Sheinkman, 2004, 2007) 'cryoglacial systems' (CGS).

A further important problem involves understanding the timing of the Pleistocene glacial events. Quaternary deposits in mountain valleys are often unsuitable for traditional radiometric dating (^{14}C, U-series, etc.), and the application of dosimetric dating methods (thermoluminescence (TL), optically stimulated luminescence (OSL), etc.) has been so far been limited in mountain regions. There are also significant problems in interpreting those dates that do exist for the deposits of this region.

Modelling experiments have been carried out during the past decades (Sheinkman, 1995, 2002a,b,c; Shlukov and Sheinkman, 2007; Sheinkman et al., 2009). They were able to demonstrate that (a) the contradictions are solvable, (b) the estimated ages for Quaternary deposits in mountain regions in cases when the traditional techniques are inapplicable and (c) randomness in dating by processing larger numbers of measurements instead of isolated cases of age determinations can be reduced. To avoid errors, careful geological interpretation of data has been combined with rigorous insights into the dating procedures. They resulted in numerous dates obtained from the representative sections and made it possible to show the main glacial events in the Siberian Mountains and to correlate them with chronostratigraphical schemes.

65.2. METHODS

Standard methods used to reconstruct former glaciation from landforms and deposits are not always applicable in the Siberian Mountain areas. This is because of the interdependency of glacial and permafrost processes that interacted intimately in the region throughout the Quaternary. The system geographical approach is based on Sochava's concept (1974, 1978) and has been applied by the present author to avoid such a confrontation. It also helps to deal with the development of ice phenomena as a single entity in every form of their interaction. This enables the identification of the dominant processes in such a system and also helps to define the different types of the cryospheric system for each stage of the glacial–interglacial process during the Pleistocene.

The onset of a cold, dry continental climate in the heart of Siberia began during the late Pliocene, promoting the development of permafrost under arid conditions (cryo-aridisation). The 'cold component' of glaciation appears then as the parameter connecting all ice bodies and phenomena. Some of them were completely controlled by this phenomenon, whilst others were partially regulated at least, by it. As a result, the ice bodies and phenomena are closely related elements of a single whole, designated in our work as the CGS.

Glaciers in the Siberian Mountains are important, but they are not the only members of the CGS which itself includes different types of ice forms. Some non-glacier agents of the permafrost regime are also very active. Icings (naleds), for example, are not infrequently comparable with glaciers both in volume of ice and in the volume of geological work they can achieve, although the processes involved are markedly different (Alekseev, 1987; Sheinkman, 1987, 1993, 1995, 2007; Romanovskiy, 1993). Under these conditions, the original glacial landforms and deposits generally retain their main features but they can be significantly modified in detail because the geological work of the CGS has been realised under the control of glaciers, icings and ground ice in combination.

In order to apply the system geographical approach efficiently, the main units of the systems and their modes of geological work differentiated have first to be defined (Sochava, 1978; Christopherson, 1998). Environmental diversity in the Siberian Mountains ranges from relatively humid to extremely continental, and from relatively warm to extremely cold conditions (the coldest place in the Northern Hemisphere occurs in north-eastern Siberia). Therefore, diverse typical modes of the CGS occur in the mountains. A few main types of the CGS have been distinguished (see below) so the differences in their temperature regime are used as a basis on which to typify the phenomena. This is because the principal characteristic of CGS is their cold storage capacity. The main indicators of glacial environments used herein are as follows: (a) temperature regime of glaciers and surrounding rocks, (b) appearance or disappearance of the certain types of icings and ground ice in the proglacial and periglacial areas and (c) permafrost characteristics, as a background parameter, of the non-glaciated zone.

Establishing the timing of the glacial events was achieved using a special TL method of regional saturated standards and developed in the 1970s on the base of the research at the Russian Plain and in the 1980s and later in Siberia (Matrosov et al., 1979; Sheinkman, 1995, 2002a,b,c; Shlukov et al., 2001; Shlukov and Sheinkman, 2007; Perevalov et al., 2009). In other techniques, the saturated samples are prepared by laboratory irradiation in a radioactive field much stronger than the *in situ* field, which are not equivalent (Matrosov et al., 1979; Shlukov et al., 1993). Fast laboratory saturation of minerals for hours or days is by no means equivalent to natural saturation that takes thousands of years. The fast saturation produces an additional induced signal that causes significant distortion of ages and yields most of the debates (Shlukov et al., 1993; Sheinkman, 2002a,b). Generally, TL dating is a good tool but the results of the technique associated with induced TL signals should be taken with due caution.

The testing of the ages obtained in our study has been carried out in the different sites reliably dated earlier by ^{14}C and U/Th methods in Russia and elsewhere. In general, the ages obtained were practically identical to those obtained by radiometric methods despite some scatter, which is common to any age method (Sheinkman, 2002a, b,c; Sheinkman et al., 2009).

Remains dated by radiometric methods are rare in the Siberia Mountains. The results obtained here have been compared to the ages obtained by the so-called 'varve-cyclite' (VC) method indicated initially as the 'nano-cyclite' technique (Afanas'ev, 1991). This is a modification of the varve counting analysis that relies on the identification of subtle changes in the case of deposition of the silt and clay fraction in water basins stable over the studied time span. Here, deposition is sensitive to gravity perturbations, and thickness variations of unit sediment layers reflect interaction of planets in the solar system, which is the main forcing mechanism. In this process, deposition of coarser particles in periods of a stronger gravity field determines the unit layer thickness, whereas thinner layers of finer particles are deposited in a lower gravity field. This rhythmic pattern reflects variations of the gravity field, and therefore, its chronology can be calculated by using astronomical methods.

65.3. CHARACTERISTICS CONTROLLING THE DEVELOPMENT OF GLACIATION

The Siberian Mountains consist of numerous ranges which were characterised by many similar features. They reach from the Altai through the Trans-Baikal Mountains to the Chukchi Upland (Fig. 65.1). The highest peaks are found in the Altai, many of which exceed 3000 m above sea level (a.s.l.) and some peaks even above 4000 m a.s.l. (the highest is Mount Belukha, 4506 m a.s.l.). In the Sayan Ranges, many mountains reach 3000 m a.s.l., and some of them rise to almost 3500 m a.s.l. The highest peaks of the Trans-Baikal Mountains are close to 3000 m a.s.l. (the highest is Peak BAM, 3072 m a.s.l.); those in north-east Siberia reach a similar altitude, some of them 3000 m a.s.l. However, the Chukchi Upland is lower, close to the 2000 m a.s.l.

Overall, the Siberian Mountains show a certain morphological uniformity, and, being in a relatively homogeneous environment, they have been subject to a similar erosional

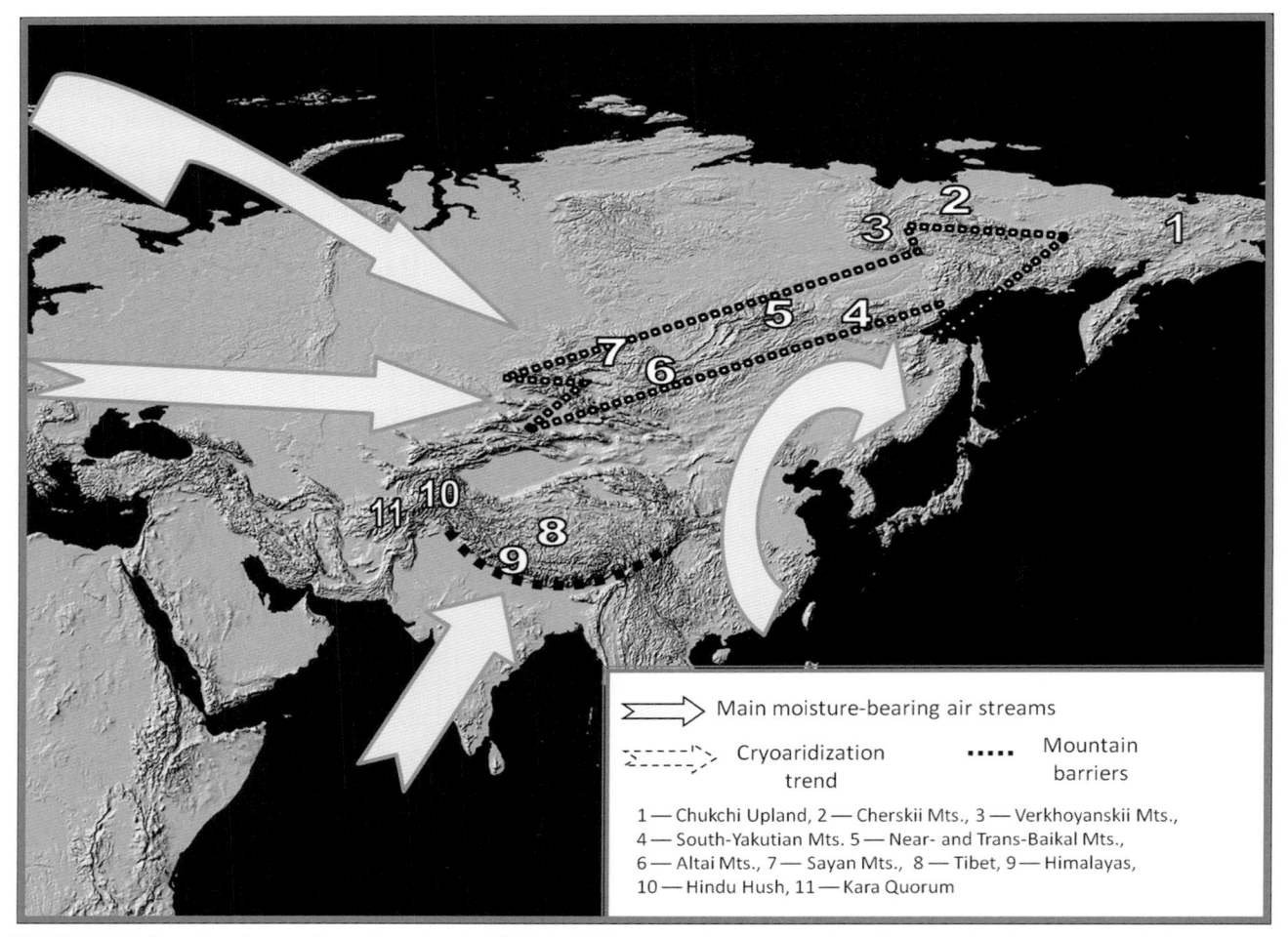

FIGURE 65.1 Location of mountainous formations in Siberia and influence on these on the main moisture-bearing air streams.

history. At present, most of the chain is influenced by of the Siberian Anticyclone, the main feature of which is a continental environment with low winter and high summer air temperatures. As a consequence of the dominance of westerly winds, precipitation reaches inland only with barometric depressions from the Atlantic and North-west Arctic, warmed by the Gulf Stream. Monsoons from the Indian Ocean are blocked by the ranges of Hindu-Hush, Kara-Quorum and Himalaya, whereas moisture from the Pacific either is blocked by the Gobi Desert or meets opposed westerly air currents and turns to the Pacific. The cryoaridisation trend reflects this increase in climatic continentality and enhanced cooling of the Siberian Mountain terrain (Fig. 65.1).

The south-western part of the Siberian Mountains is still reached by the air masses providing significant precipitation. This is because the warm Gulf Stream of the North Atlantic reaches through the Arctic Ocean to the coast of North-western Siberia, and air mass from there can cross the West-Siberian Plain and moisten the south-western branches of the Siberian Mountains. As a result, the High West Altai receives about 2000 mm annually (mm/a) of atmospheric precipitation that is very much in respect to Siberia. However, far inland of Siberia, the precipitation decreases sharply. About 1000 mm/a is typical for the high north-east Altai, West Sayan and the north-western part of the East Sayan (Revyakin, 1981), although further to the East, as well as to the inland part of the Altai-Sayan terrain, the precipitation decreases to only a few hundreds of mm/a. In those parts of the mountain belt, within an area of many thousands of square kilometres, the precipitation is of 250–400 mm/a in the foothills, increasing to some 500–700 mm/a in the high mountainous zone. In the intermountain depressions, precipitation can decrease to 100–200 mm/a.

The mean annual air temperatures along the Siberian Mountain chain are everywhere below zero; falling from -3 to $-6\,°C$ in the south-west to -15 to $-17\,°C$ in the north-east. As a consequence, permafrost occurs everywhere. It appears as of high-temperature permafrost in the south-west of the Siberian Mountains within valley bottoms at middle and low altitudes. At altitudes over 2000 m a.s.l. in the Altai and further towards the north-east, the permafrost becomes more strongly developed, turning into low-temperature mode so that, only in the Altai, the piedmont terrain is characterised by discontinuous permafrost where it grades into continuous modes at the same height towards the Near-Baikal area and upwards in the Altai Ranges (Shats, 1978; Yershov, 2002; Gorbunov and Severskii, 2007). Thus, cryogenic ice ranges from seasonal forms in the low Altai area to substantial long-lived perennial bodies within the mountainous terrain of north-eastern Siberia.

The occurrence of polygonal ice wedges (PIW) clearly indicates the severity of that low-temperature permafrost as these wedges require for their origin a mean annual temperature of some $-3\,°C$ in rocks, at the depth of null fluctuations, and twice as low as those in air. Figure 65.2 shows that most of the studied surface area includes polygons with ice wedges. They have been mapped from the north in the Angara Valley, within which only ancient PIW pseudomorphs occur (e.g. Slagoda and Medvedev, 2004) to Baikal Lake and the Amur River. Following the finds of PIW in the foothills of the Sayan Ranges (Osadchii, 1982; Yamskikh and Yamskikh, 1999) and in north-eastern China (Tong, 1993), the ice-wedge polygon limit has been drawn (Vasil'chuk, 2004) skirting the Angara River, Lake Baikal and the Amur River from the south. The most southern PIW occur in North Mongolia at latitude $50°30'N$ (Sheinkman and Krivonogov, 2005; Sheinkman, 2008). In the author's opinion, all these findings show separate areas with PIW, whereas their main area is situated north of the Angara–Baikal–Amur line, mapped by Vtyurin and Koreisha (Vtyurin, 1975; Kotlyakov, 1997).

Overall, it is obvious that even under the present-day interglacial conditions, the low-temperature permafrost with PIW covers a vast area in Eurasia from the East Arctic in the north to the Central Mongolian steppe in the south, and to the Altai Mountains in the south-west. Correspondingly, the Siberian Mountains must have been characterised by deeply frozen permafrost during the Pleistocene cold stages when the freezing increased significantly. These areas would have been much drier than today too, because precipitation would have been intercepted by ice sheets in north-west Eurasia.

Ablation only occurs today on the glaciers of the Siberian Mountains during short frost-free seasons but is characterised by intensive melting, because of the warm summers associated with the continental climate. The ablation period ranges from 75 to 120 days on the Altai glaciers, to

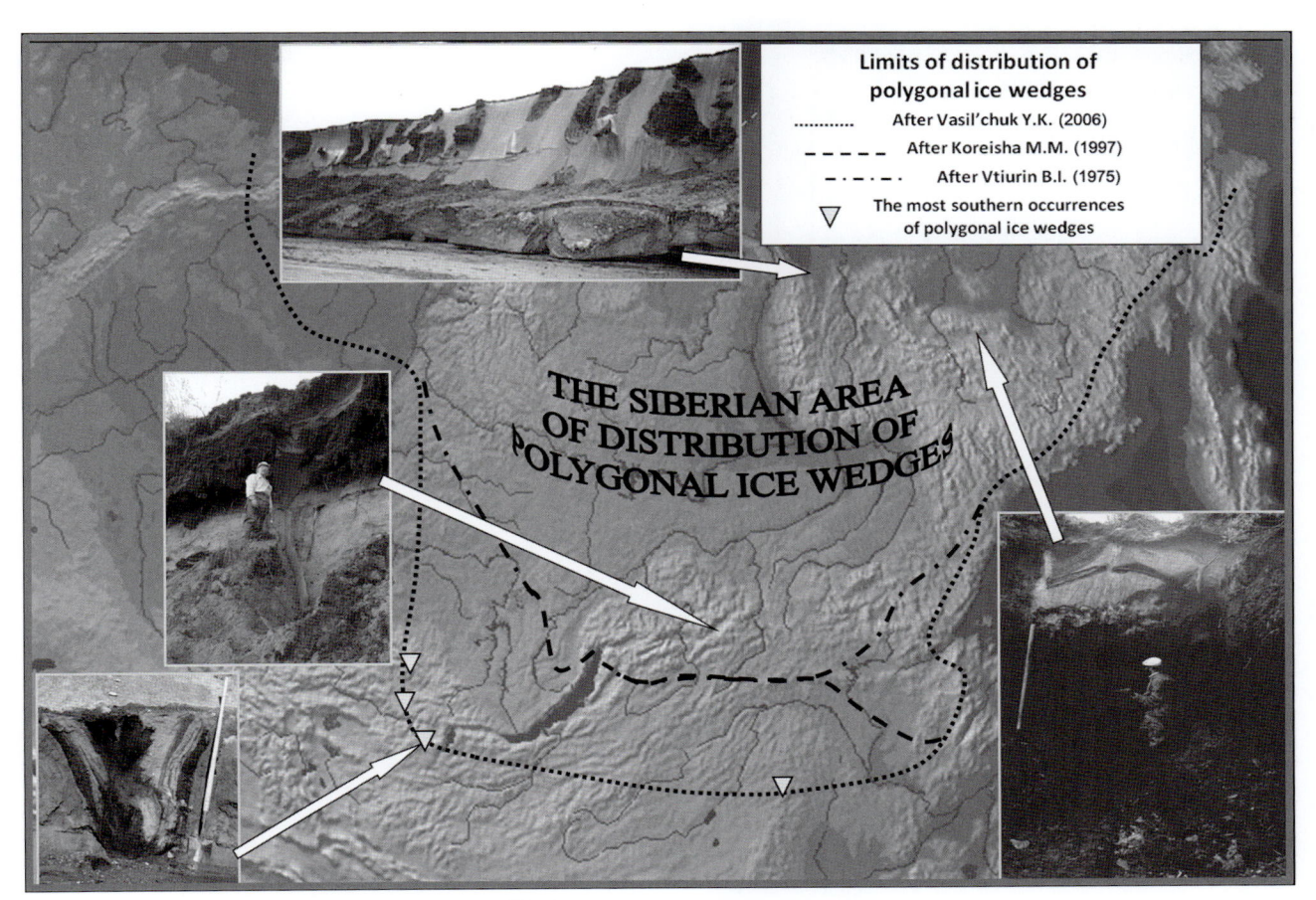

FIGURE 65.2 Distribution of polygonal ice wedges in Siberia.

50–60 days in north-eastern Siberia. Water from melted snow and ice not only flows away from the glaciers, taking heat away, but also repeatedly freezes on their cold surface as superimposed ice. This explains the frequent snow loss over the glacier surface yet positive mass balance (Fig. 65.3). Hence, empirical formulae required to calculate ablation/accumulation at the glaciers through snow thawing at the equilibrium line are not acceptable in Siberia. These formulae (e.g. the Krenke–Khodakov's formula: Krenke, 1982) do not take into account the repeated freezing of ice onto the cold surface of glaciers and must be corrected on the basis of mass-balance research (Koreisha, 1991).

Significant modern glaciation occurs in two-opposing area of the Siberian Mountains, supported either by very low annual air temperatures (the north-east) or by relatively abundant precipitation (the south-west). In the coldest area of north-eastern Siberia, the longest glaciers flow down from the mountains at about 3000 m a.s.l. These glaciers reach 10 km in length, whereas the total glaciation here occupies more than 350 km^2 (Sheinkman, 1987, 2007; Koreisha, 1991). Under a similar climate in the Chukchi Upland, but with mountains of lesser height, there are only a few small glaciers. In the Trans-Baikal Mountains, where many peaks reach close to 3000 m a.s.l., and in the Sayan Ranges, where some reach to almost 3500 m a.s.l., the adverse environment also allows the development of a many small glaciers. The latter occupy an overall area of about 20 and 30 km^2, respectively (Vinogradov, 1966–1981). The most extensive modern glaciation is restricted to the highest (up to 4506 m a.s.l.) snowbound Altai where the glaciers at times exceed 10 km in length and cover, altogether an area of about 800 km^2 (Narozhnii and Nikitin, 2003).

65.4. FUNCTION AND STRUCTURE OF THE CGS

It should be noted that any CGS is spatially bounded by an area where interaction of glacial and permafrost processes appear according to the range of the system—local,

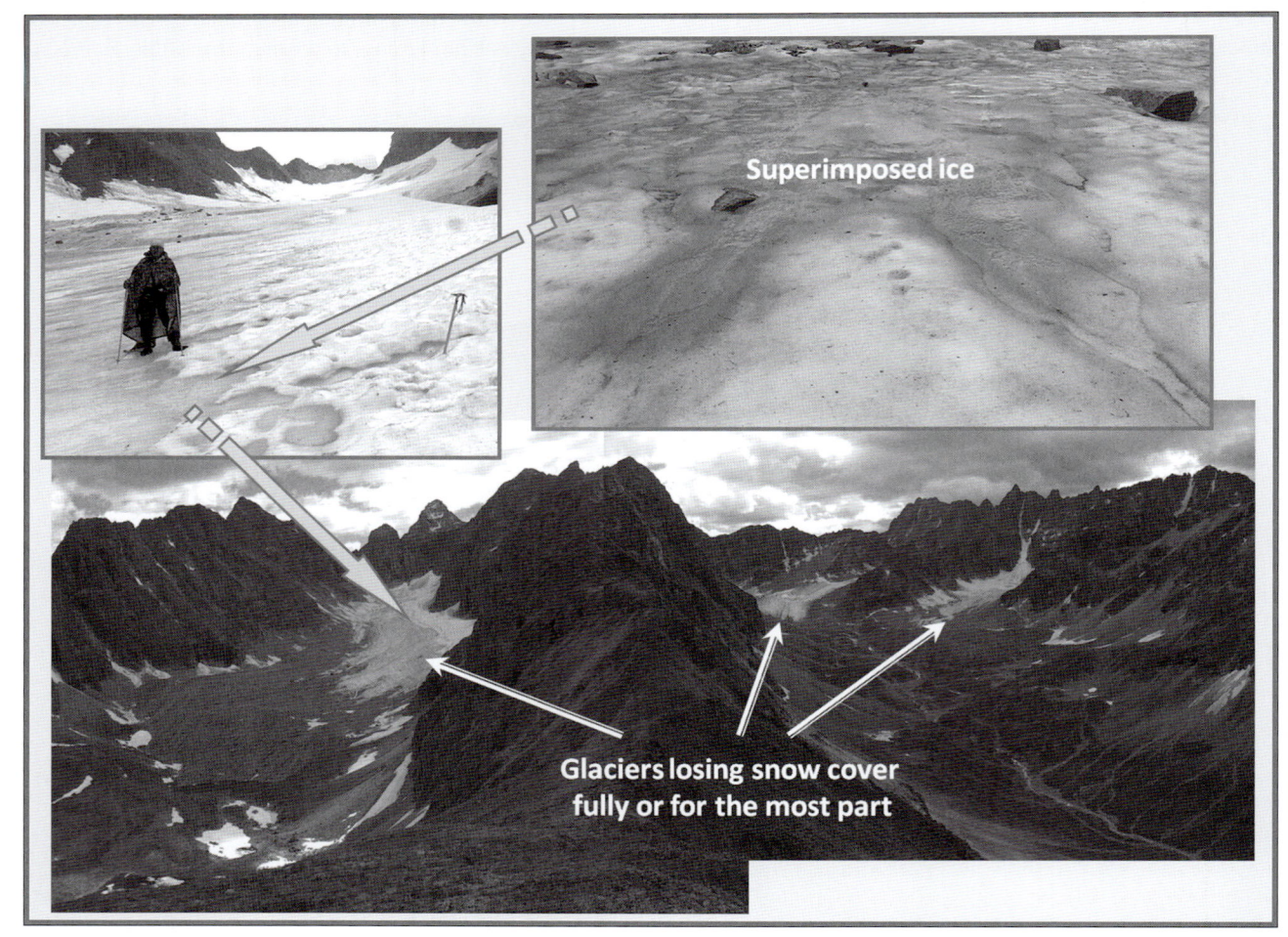

FIGURE 65.3 Trans-Baikal region: superimposed ice development on the Kodar Glaciers showing the loss of snow cover over the whole glacier, or over most parts.

regional and so on. With regard to the temporal boundaries of CGS, this is defined by the interaction of glacial and permafrost processes through time, and includes successions of geological events. This is well expressed for many cryogenic elements of the CGS by the appearance of environments that former glaciers produce. To formulate the definition 'CGS' is possible, thus, as follows. A CGS includes all ice bodies, as well as sediments connecting them, which together form a complete whole within the topography and interact through mass and energy exchange.

Development of any Siberian CGS follows certain common patterns. To form the CGS, moisture must first be moved to the Siberian inland via atmospheric circulation. In order to reach the Siberian interior, moisture must be derived only either from the Atlantic, or from the Northwestern Arctic, warmed by the Gulf Stream. As stated above (Fig. 65.1), only a little moisture reaches most areas in the Siberian Mountains, and this moisture supply must have decreased during the Pleistocene cold stages, when the Gulf Stream did not operate and when, in north-western Eurasia, ice sheets intercepted significant part of the remaining moisture. This mechanism caused a continental climatic environment to develop in the Siberian Mountains throughout the Quaternary (Sheinkman and Barashkova, 1991; Sheinkman, 2007). Therefore, the limited volume of moisture that reached the interior Siberian regions may have changed little through the Quaternary apart from minor variations. Another parameter that has remained almost constant among the CGS' characteristics is the altitude of the mountains, which rose gradually from the Early to Late Pleistocene but have not significantly altered during the glaciation stages, when the main glacial advances occurred (Sheinkman, 2002a,b,c, 2008). This means that the state of the Siberian CGS is mainly determined by cold storage acquired first from the atmosphere and, second, by the additional quantity of cold storage resulting from freezing of the substrate, as the permafrost process embraces all components of the CGS.

Usually, it is assumed that the greater the cold storage in any geosystem, the lesser is the energy and its effect of erosion/accumulation processes. Under the limited precipitation regime over most of the Siberian Mountains, annual run-off volume per unit area is low. However, by contrast, significant cold storage results in active erosion/accumulation processes driven by the different ice agents. Even in relation to glaciers, investigations of Serebryannii et al. (1989) have demonstrated that greater geological work is carried out by frozen cold-base glaciers that are slowly moving, than by warm glaciers that move more rapidly.

At times, glaciers dominated the structure of the CGS and filled all the valleys, and other CGS' elements played then a subordinate role. At other times, the glaciers retreated and the non-glacial elements took over, actively reworking landforms and rocks left by the glaciers. The most dominant role of glaciers in the Siberian Mountains occurred during the Pleistocene cold stages when they increased significantly in size. They were mainly valley glaciers, and only became reticular forms during their maximal advance in the Late Pleistocene when they reached inter- and sub-mountain depressions and lowlands, and formed large piedmont ice fields. Even then, the glaciers were still connected with valleys and acted as members of the mountain-valley CGS. This is clearly shown by appearance or absence of different moraines outlining the former glaciers (Fig. 65.4). There have been, however, attempts to model giant ice sheets in the Siberian Mountains (e.g. in Butvilovskii, 1993; Grosswald and Hughes, 2002), based on Alpine schemes and calculations. However, the author considers that these models are flawed because they ignore the arguments provided above. It should be underlined that any CGS in Siberia represents a combined whole, which reflects only a certain type of glaciation that must be recognised. There are different dominants in structure of each types of CGS, and they do not remain the same relics. This is because not the same energy components participate in such a process, and there is difference even in the case of glaciers. Their effect is usually estimated through kinetic energy of moving ice. However, the Siberian glaciers move slowly and are frozen to their bed (cold-based). Slow movement of the glaciers at the frozen glacier–bed interface causes the removal of masses of bedrock, and, as mentioned above, glaciers erode the valley more intensively than high-speed moving glaciers on the thawed ground base (Serebryannii et al., 1989).

The Siberian CGS also utilise actively thermodynamic energy components, by dint of transitions of the CGS' substance from solid rock (ice) into melt (water) and back again. Icings are an obvious case. They are a product of migration of subsurface water streams (also the supplied by run-off of glaciers) from the zone of positive temperatures in thawed rocks into the zone of negative temperatures at the day surface. Icings are related to repeated congelation ice (in terms of Shumskii, 1955) and can be located directly on glaciers, abutted against glaciers or located at some distance from glaciers. No motion of these ice agents is observed, although they can achieve intensive geological work. Weathering is sharply increased within 'icing glades' (widened parts of valleys where vegetation clearings form along icings), and streams are deflected by icings. The result is a specific effect that causes widening of the icing glades and filling of the valleys with particular deposits that is designated as 'icing alluvium'. In the course of time, wide and levelled icing glades develop in the valleys downstream of glaciers, and the former trough morphology consequently changes (Fig. 65.5) in many details (Alekseev, 1987; Sheinkman, 1987, 1993; Romanovskiy, 1993). Only when taking into account the different modes of energy, using by the CGS elements, can we develop adequate understanding of the specificity of glaciation in the Siberian Mountains.

FIGURE 65.4 West Sayan Range, the left-hand upper reach of the Yenisei River. A typical Pleistocene moraine complex is present in the bottom of the valley, whilst no moraines are found at the top of the slopes and close to the watershed.

Some note should also be made regarding perennial snow forms, which sometimes are to be included in structure of CGS. These CGS elements can yield pseudo-glacier and pseudo-icing landforms that must be taken into consideration, especially in regions that were not subjected to any Pleistocene glaciation. The reason is that in spite of low snow accumulation in most area of the Siberian Mountains, sometimes wind redistribution of snow causes formation of large snow-banks (Fig. 65.6). For example, precipitation in the Chuckchi Upland is of only about 300 mm/a, but large wind-generated snow-banks occur at altitudes much lower than the moraines of the maximal glaciers and yield pseudo-glacial relief that sometimes confuses researchers.

65.5. TYPES AND OCCURRENCE OF THE CGS

In the Siberian Mountains, a few types of CGS may be distinguished based on their temperature regime (Sheinkman, 2004). This all-purpose approach, forming the basis of many classifications of glaciers and permafrost (Kotlyakov, 1984; Yershov, 2002), is representative (Sheinkman, 2007). In comparison with the previous (Sheinkman, 2004), the renewed classification of the CGS, based on the data obtained in the past years, has been used in the present chapter. Let us consider the main units of this classification to apprehend principal distinctive features of Siberian mode of glaciation because confusions occur so far through ignoring such a difference.

First of all, it should be underlined that the 'warm CGS' and 'moderately warm CGS' (Fig. 65.7), which are similar to those considered by traditional Alpine models (Sheinkman, 2004), have never occurred in Siberia. Simple ways to use background resources by CGS have been presented in this mode of glaciation: moisture, evaporated from the ocean and transferred by air stream to the continent, passes a relatively short distance through the land terrain and falls on a *predominantly unfrozen bed*. In the case of the warm CGS, there is no permafrost under the glaciers, and in the

FIGURE 65.5 North-eastern Siberia: different perennial icings in the bottom of the Pleistocene troughs.

case of moderately warm CGS, permafrost occurs only in the high glacial accumulation zone, but the tongues of the glaciers are permanently in isometric state, with 0 °C temperature. Glaciation develops then only as classical totality of nival-glacial objects, which are a result of high snow accumulation compensating active ablation (Fig. 65.7).

As opposed to this, forming the Siberian glaciers *is always imposed on development of permafrost* that is generated earlier as a result of cryoaridisation. Principally, other types of CGS designated as the 'moderately cold CGS' and 'cold CGS' (Figs. 65.8 and 65.9) can be distinguished, and much more complicated ways to exploit the background resources have been used by the CGS in this case. In such a situation, moisture evaporated from the ocean has to overcome a great distance and loses a significant part of its volume along the way through the continent. Consequently, in the Siberian Mountains, amounts of atmospheric precipitation are very small. However, precipitation totals are sufficient to form glaciers (cold and dry continental glaciers) because, already in the West Altai, ice bodies accumulate a big volume of cold storage *in situ*, at the earth surface, and realise it to supply glaciers by superimposed ice. The difference between two cold types of CGS is that, in the case of cold CGS, continuous permafrost spreads for a great distance away from glaciers, whereas in the case of moderately cold CGS, only discontinuous permafrost occurs downstream of glaciers. Correspondingly, different types of icings and ground ice interact with glaciers and are included into CGS' composition (Figs. 65.8 and 65.9).

The moderately cold CGS are currently widespread in the Altai and extend to the right bank of the Upper Yenisei where indicators of the stronger freezing appear (PIW, etc.). Although the Altai, with significant evidence of glaciation, was previously considered as the more humid and warmer area (Sheinkman, 2004), recently obtained data confirm that all glaciers here occur above the boundary of continuous permafrost areas. This does not contradict with Voeikov's concept in its climatic essence, as Voeikov did not take in account feeding the glaciers by superimposed ice, when solar energy has been used in thawing ice that existed at the beginning of the ablation period and then for thawing newly formed portions of superimposed ice.

FIGURE 65.6 Different perennial snow-banks: A—as a snow belt at the top of the slope, and B—as a separate ice body in the former glacial cirque in the West Sayan Range; C—as a separate ice body at the top of a slope in the Kuznetskii Alatau; D—as a large surface cover body in the bottom of a valley in the Chukchi Upland.

The low and middle mountains in the Altai are a typical area with the discontinuous permafrost (Shats, 1978; Sheinkman, 1993, 2007). The Altai glaciers, therefore, were previously related to the warm firn zone (Mikhalenko, 2007) and designated as cold only within upper layers by Nikitin et al. (1986). Later it became clear that these observations were incorrect and new evidence (Aizen et al., 2006; Gorbunov and Severskii, 2007) has shown that continuous permafrost replaces island and sporadic permafrost everywhere in the High Altai and its glaciers are deeply frozen.

In the accumulation zone of a glacier at the slopes of Mount Belukha, in the West Altai, in the layer of null fluctuations, a temperature of −16 °C has been recorded, and even at the bed of the glacier, at the depth of 170 m, the temperature was −14 °C (Aizen et al., 2006). Taking into account some 200 m thickness of the glaciers and even conceding their advance to the altitudes close to some 2000 m, that is, the lowest limit of the continuous permafrost, the glaciers must be entirely frozen. This is made even more likely given that the thickness of permafrost at altitudes above 2000 m is of 300–400 m (Shats, 1978), which obviously exceeds thickness of the Altai glaciers. It is confirmed also by the fact that thawed run-off from the glaciers has been stopped in late autumn and caught to a considerable extent by near-glacier icings. In the Central Altai, there are representative temperature records on the glaciers in the Aktru Valley. Temperatures of ca. −4 °C have been recorded in the lower tongue part of the 100 m thick Small Aktru Glacier and of −8 to −9 °C at the output from its accumulation zone. At the neighbouring 60-m thick Vodopadnii Glacier, temperature as low as −15 °C has been recorded (Narozhnii, 1993). Taking into account the fact that the glaciers are frozen more weakly than the surrounding rocks (Nekrasov, 1976) and that the temperatures are low enough, they indicate that glaciers in the Central Altai occur over continuous permafrost. Large icings intercept a significant part of the glacial run-off that has been stopped in late autumn.

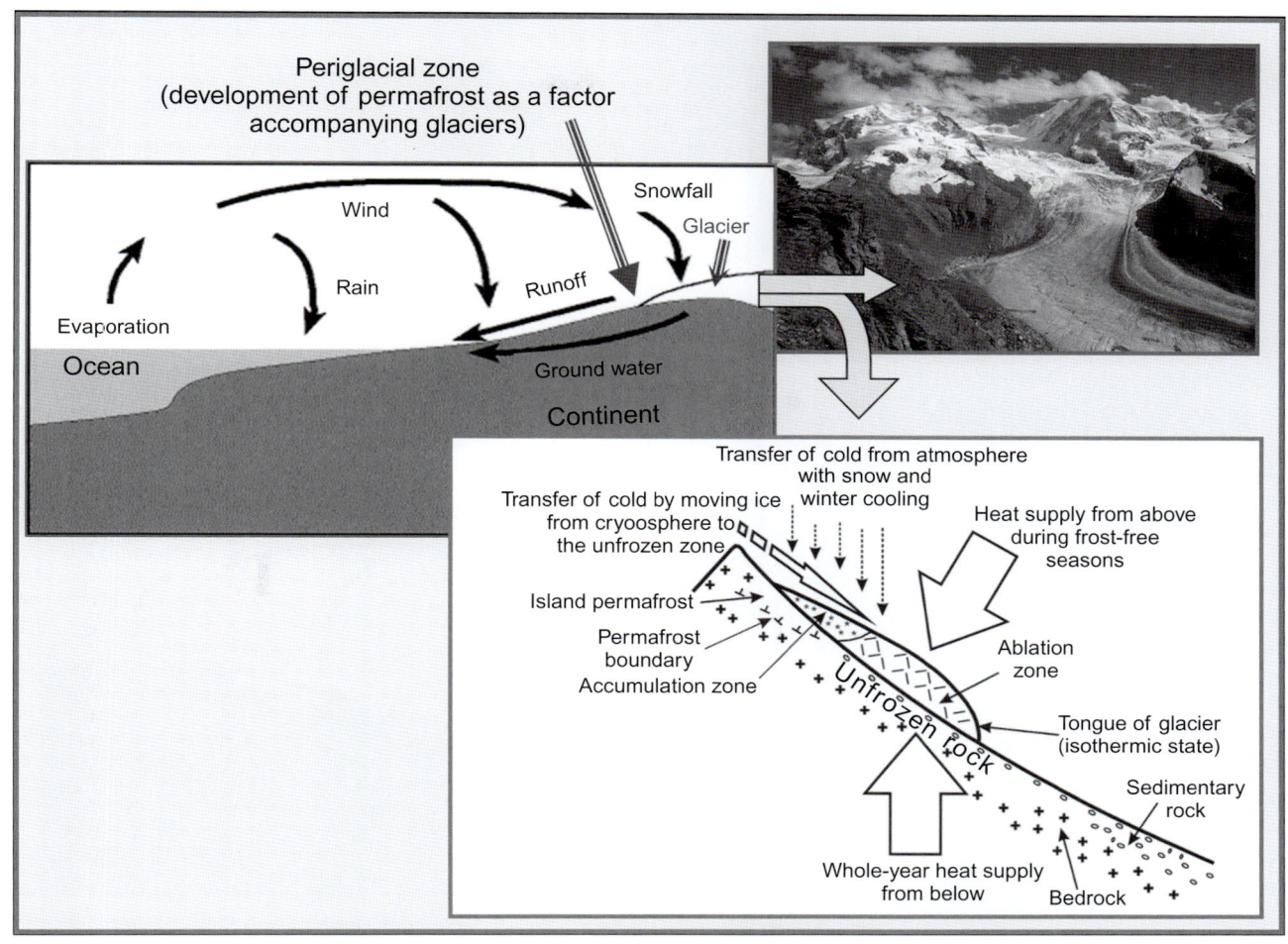

FIGURE 65.7 Structure and development of the warm CGS (cryo-glacial systems).

In the south-east Altai, freezing increases but the features of CGS remains yet as the moderately cold type; indicators of the more severe environments do not appear here. However, it has been noted (Fukui et al., 2007) that in this region, in front of the Sofiiskii Glacier, polygonal structures with ice wedges are able to exist, though these are an indicator of a stronger freezing. However, in the past, the same authors (Mikhailov et al., 2006) had demonstrated temperature curves through those rough debris rocks and shown that the seasonal thawing layer is about 3 m thick and the rock temperatures are of -3 to $-4\,°C$, and no concrete PIW are found. Indeed, these temperatures are low, but they can provide a PIW only in finely dispersed sediments, whereas in rough debris, it requires temperatures of -5 to $-7\,°C$ or even lower (Romanovskii, 1977). The polygons noted above are usually a result of a weaker freezing, more so since neither PIW nor their relics have been found throughout the Altai region.

The next unit is the 'cold CGS' (Fig. 65.9) that is widespread today from the right-bank upper reaches of the Yenissei River to the Chukchi Upland. This CGS varies in cold storage, rock thawing and ablation intensity. Therefore, southern and northern subtypes can be distinguished. As a result of relatively high ablation, in the Near-Baikal region at latitudes of 51–57°N, there are only a few tens of small cirque glaciers and a few of valley glaciers up to 2 km in length, restricted to the peaks of about 3000 m a.s.l. (Fig. 65.3). In contrast, on the slopes of comparable mountains in north-eastern Siberia, at latitudes of 63–65°N, glaciers have developed large enough valley and even dendritic forms. Nevertheless, well-expressed PIW have accompanied the cold CGS everywhere and, since they are present in the foothills in the East Sayan, become wide-presented in the Trans-Baikal region (Figs. 65.2 and 65.10). Radiocarbon dates from sections with these PIW have established that the PIW originated at about 20 ^{14}C ka, that is, during MIS 2, and gradually developed until present through the Holocene optimum (Sheinkman, 2007, 2008). All of this underlines the constancy of the environmental conditions in the region under very cold and dry conditions. Temperature evidence also confirms this situation. The Trans-Baikal glaciers

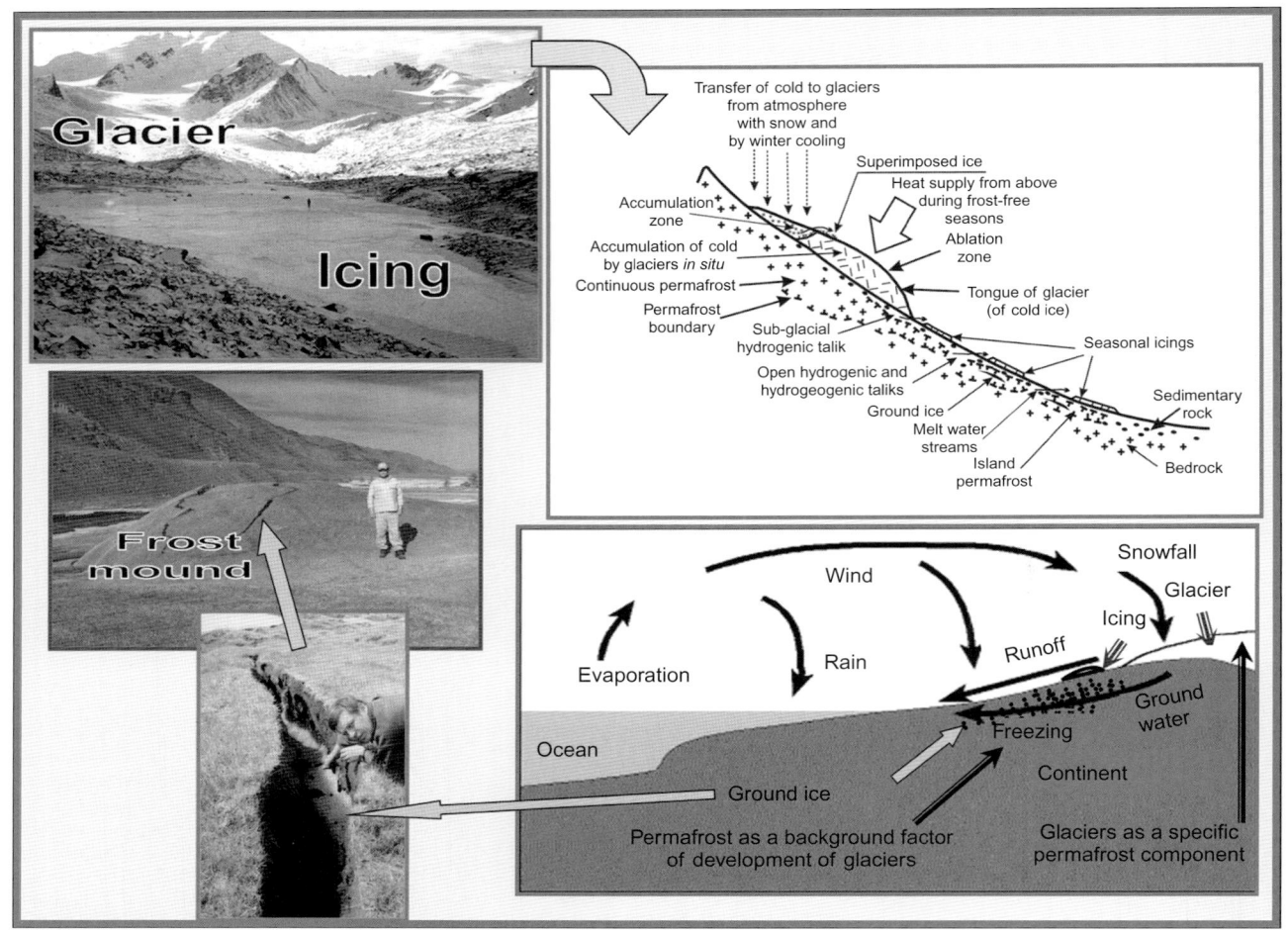

FIGURE 65.8 Structure and development of the moderately cold CGS (cryo-glacial systems).

temperatures are of ca. −7 °C, and of −10 to −12 °C in the surrounding rocks (Shesternev and Sheinkman, 2008), whereas in the similar environments in north-eastern Siberia, temperatures of −7 to −10 °C are recorded in the glaciers and −14 °C in the surrounding rocks (Grave et al., 1964; Sheinkman, 1987). However, because of the shorter 1.5–2 month ablation period (because of the proximity to the Arctic Circle), the glaciers in north-eastern Siberia, advancing from the mountains at a height of some 3000 m, reach up to 10 km in length and cover a significantly greater area than in the Trans-Baikal region. Large perennial icings and frequent PIW in front of the modern glaciers, as well as along the glacial troughs supplement the picture in north-eastern Siberia (Fig. 65.10).

Overall, in the cold type of CGS, the ground is deeply frozen and many ice agents are caused by continuous low-temperature permafrost. Run-off from glaciers is restricted to the warm season, and a significant part of this water is intercepted by near-glacier icings. Reworking by icings in the glacial valleys is very intense. Major glacial and fluvioglacial landforms (including lateral and terminal moraines, high terraces, etc.) become, as a result, clearly expressed since most of the small-scale landforms have been removed by icing erosion (Fig. 65.5). At present, the first ice agents reflecting very low snow accumulation and deep freezing of rocks (frequent PIW, perennial icings, frost mounds, etc.) occur in the Siberian Mountains in the East Sayan Ranges. In contrast, relict Late Pleistocene landforms of the same type (former icings, ice-wedge pseudomorphs, etc.) occur in the periglacial zone on the left bank of the upper reaches of the Yenissei River, in the inner parts of the West Sayan (Fig. 65.11). However, they are absent from the Altai (Sheinkman, 1993, 1995, 2007).

The 'super-cold' CGS may be also designated. These CGS do not occur today, but their relics (dense ice-wedge network, traces of very strong freezing, etc.) are expressed in the Pleistocene troughs in north-eastern Siberia. Development of icings in such a situation was limited. This is because of the great cold storage of the CGS: the groundwater supply routes for the icings were largely frozen; thus, large icings could form mainly in front of the glaciers, where water was supplied by the ice melting.

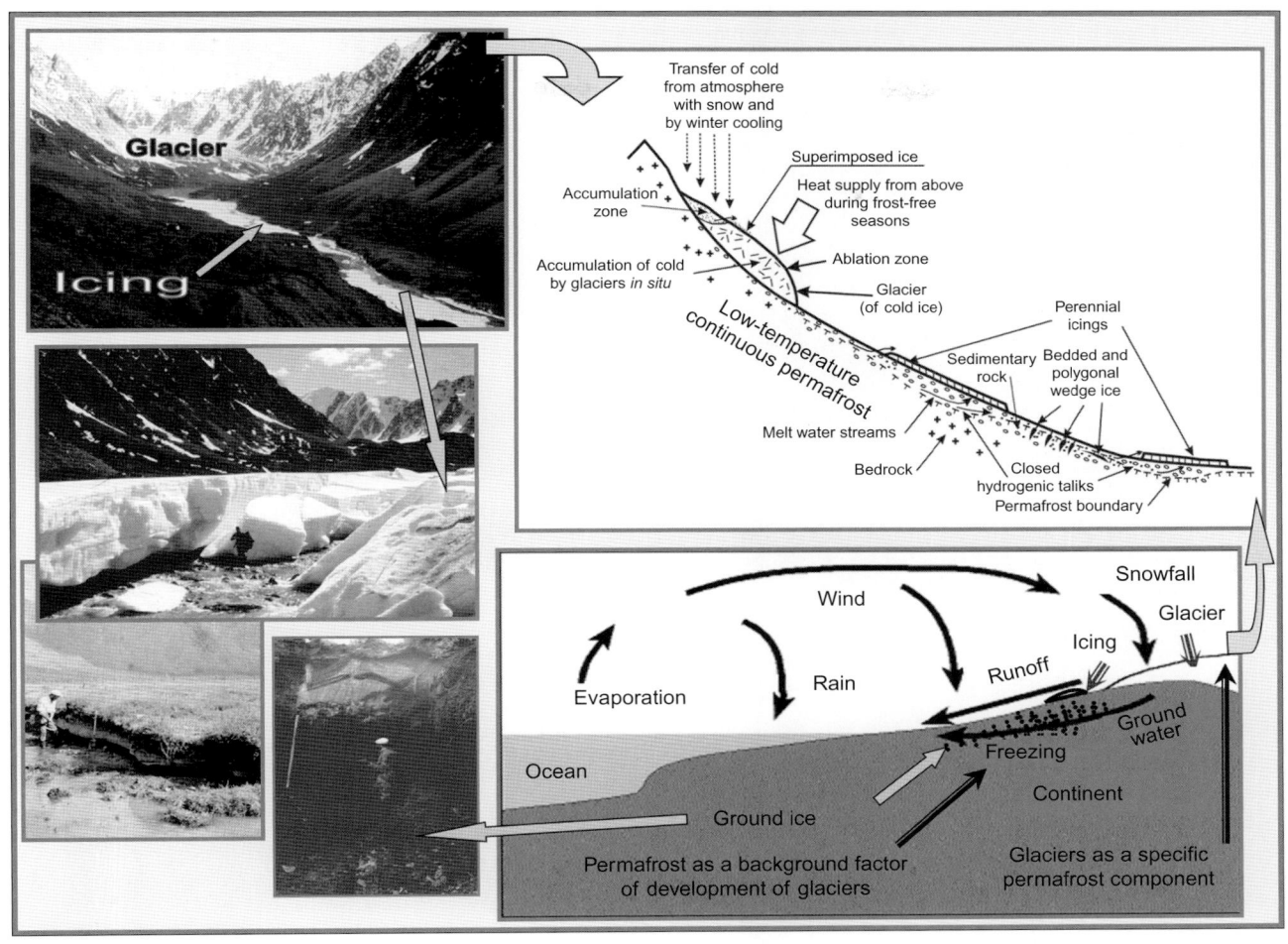

FIGURE 65.9 Structure and development of the cold CGS (cryo-glacial systems).

The different types of the CGS during the Pleistocene changed the distribution of ice volume in different parts of the Siberian Mountains. Today, the Altai Ranges support the largest glaciers, surrounded by relatively small seasonal icings. However, during the Late Pleistocene, these glaciers advanced and reached 70 km in length. In north-eastern Siberia, the present-day glaciers are less extensive than in the Altai, but they are surrounded by giant icings, the largest of which occupy areas up to 100 km^2. During the Late Pleistocene, the north-eastern Siberian glaciers were twice as long as those in the Altai, although they were initiated from mountains more than 1000 m lower than the Altai ranges, because the greater cold storage of the Quaternary CGS in the Siberian Mountains provided larger glaciation.

Some notes should be added concerning the glaciers that fall under the Pacific influence—in Kamchatka and Koryak Mountains where the CGS resemble those in the West Altai and include significant glaciation. Being a cold mountainous region where annual air temperatures are below zero and the frost-free season is less than 100 days, the Kamchatka Peninsula, together with the Koryak area (with the highest peaks of 4688 and 2453 m a.s.l., respectively), receives relatively abundant precipitation—ca. 1000 and 700 mm/a, respectively. This provides significant glaciation: about 400 glaciers cover more than 800 km^2 in the Kamchatka and more than 1000 glaciers cover over 200 km^2 in the Koryak region, that is, this is comparable to the Altai. However, during the Pleistocene cold stages, the shelf sea around the peninsula and the continent was frozen so that all this area became a part of the giant frozen continent. Therefore, Pleistocene glaciers in the region reached only a few tens of kilometres in length, again resembling the situation in the Altai Ranges.

65.6. TIMING OF GLACIATION AND RESULTS OF THE DATING

The Alpine glacial model was usually used to form the basis for understanding the glacial history of Siberia. However, deep freezing of rocks was not originally considered. Today, the most representative sections reflecting glacial

FIGURE 65.10 Trans-Baikal region, foothills of the Kodar Range. Development of polygonal ice wedges on the periphery of the maximal Pleistocene glacier advance.

events in the Siberian Mountains have to be interpreted in order to understand how and when glaciation developed in relation to the cold CGS, that is, on the basis of independently determined local sequences.

Sections in the valley of the Chagan-Uzun River, the main tributary of the Chuia River, in turn, an upper reach of the Ob' River, were taken as references for modern study (Fig. 65.12). Large modern glaciers occur in the upper reaches of the Chagan-Uzun River, as well as perfectly preserved relics of older ice streams retracing the glacial history along the valley. In front of the glaciers and along glacial troughs, permafrost phenomena clearly prove that the CGS are inherented from the cold conditions that existed during both in the Quaternary and today. Although this type of CGS is moderately cold, it is sufficiently representative. Outcrops in this valley preserve an uninterrupted Quaternary glacial–interglacial stratigraphical sequence. No other such complete sections with glacial deposits in the Siberian Mountains are known, and this uniqueness has given rise to intense discussions since the 1960s (e.g. reviews in Sheinkman, 2002a,b,c, 2008). The Pleistocene ice streams emanating from high in the mountains terminated in the Chuiskii and Kuraiskii intermountain depressions and formed foot (piedmont) glaciers around basins than fill them. The form of these glaciers is now marked by terminal frontal moraines that block troughs that begin high in the surrounding mountains. The Chagan-Uzun valley in its lower reach is also blocked by clear frontal moraines left during the maximum glacier advance (Fig. 65.12).

There are two very representative outcrops that extend along the valley for a few kilometres. Altogether they comprise a common sedimentary sequence in which all members are closely related. The first outcrop is in the middle reach of the Chagan-Uzun River, along its right-hand tributary—the Chagan River. Here, 150-m sections of grey glacial–interglacial facies overlying red-coloured preglacial deposits are exposed. A second outcrop with a similar thickness of glacial deposits occurs in the lower reach of the Chagan-Uzun valley and reveals where glacial

FIGURE 65.11 West Sayan Range, the left-hand upper rich of the Yenisei River. Pseudomorphs formed by polygonal ice wedges in the body of the Pleistocene basal moraines.

sediments and other associated formations indicate the maximal advance of the Chagan-Uzun glacier (Figs. 65.12–65.14).

The end moraines in the lower reaches of the Chagan-Uzun River were often assigned a Late Pleistocene age based on geological evidence (e.g. Popov, 1972). However, other authors (e.g. Deviatkin, 1965) argued that the moraines should be assigned a Middle Pleistocene age since that was universally accepted in West-Siberian schemes, which inherit the classical Alpine models. The grey sections exposed in cliffs along the middle course of the Chagan-Uzun River were also either attributed to the Late Pleistocene or divided into a Late Pleistocene unit and older tills in which case the latter were considered to be of Middle or even Early Pleistocene age (e.g. Markov, 1978).

The previous TL dating in the Chagan-Uzun Valley was mostly applied to glacial deposits (Il'ichev et al., 1973). However, there have been some problems with this approach. In addition to age distortion arising from laboratory radioactive saturation of the samples (see above), the pre-genetic memory of minerals in the sediments was interpreted from the measured residual TL signal in moraines of present-day glaciers. It is incompatible with the interpretations of the glacial geology. Ancient and modern glacial sediments are buried in quite different conditions and by different mechanisms. Moreover, it is impossible to estimate the pre-genetic memory of buried minerals in most moraines since they are not exposed to the sunlight, whereas the contents of radiogenic elements responsible for the background radiation and its effect on the TL signal are distributed unevenly across the valley (Sheinkman, 1995; Sheinkman, 2002a,b,c; Sheinkman et al., 2009). Therefore, dating determinations obtained by this method are inevitably biased and uncorrectable. Besides, they were not tested by comparison with dates obtained by more reliable methods.

In our case, sampling has been carried out very thoroughly, based on the latest developments and extensive experience of in dating in Siberia (Sheinkman et al., 2009). Also, the methods applied allowed distortions to be avoided, as noted above, and testing using a range of

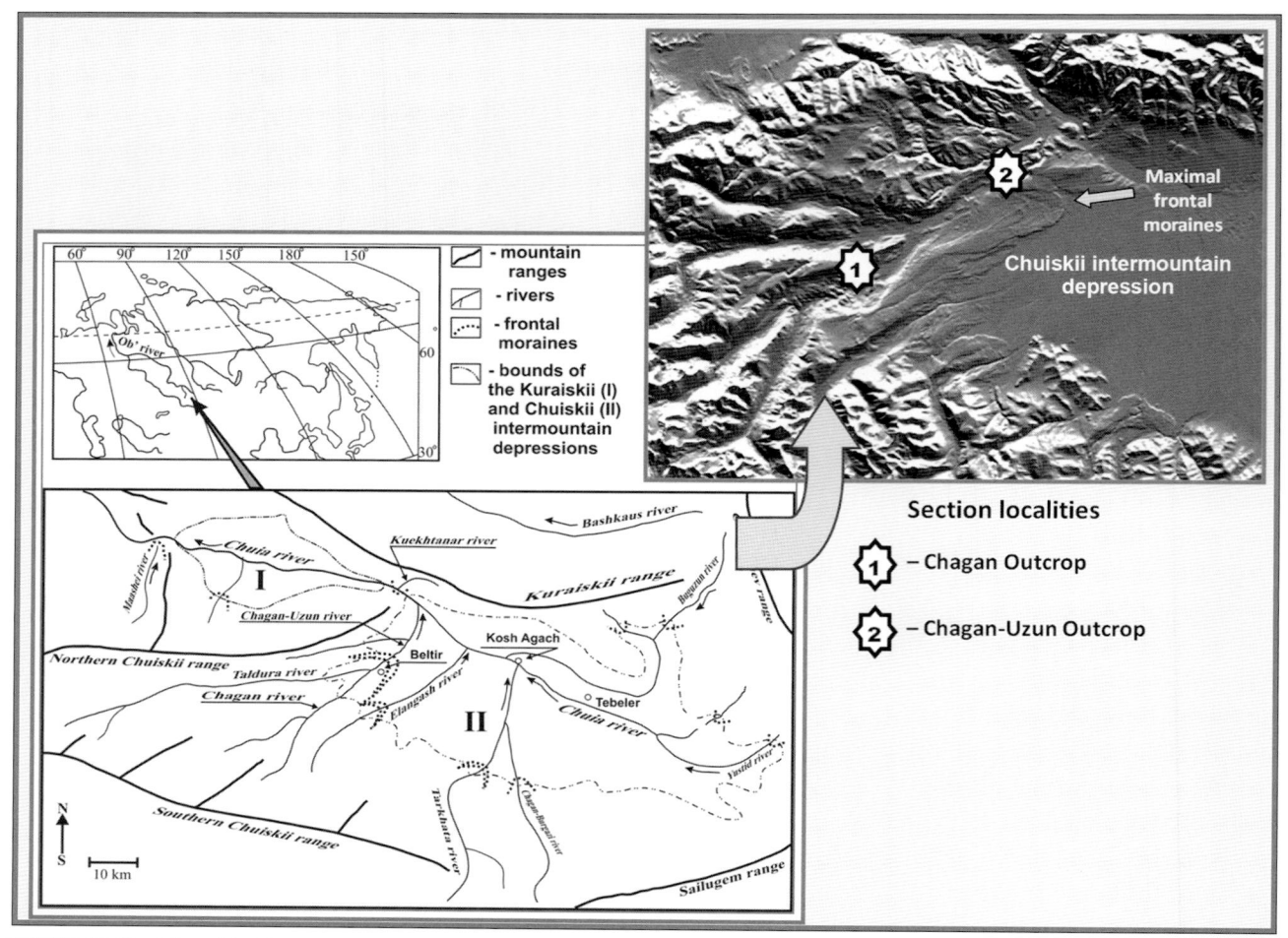

FIGURE 65.12 Map of the Upper Chuya basin and position of the dated outcrops.

reliable dating techniques has been applied. Consequently, the ages obtained by Il'ichev et al. (1973), which are much older than those found in our research, are to be considered to be incorrect.

In general, glacial and glaciofluvial deposits suitable for TL analysis are extremely rare in the region being restricted to fine-grained ablation sand buried upon glacier surfaces, as sand near large boulders washed by melt water, or bottom sediments of small shallow lakes within glaciers. Therefore, in this study, the present author preferred to determine most ages of glacial sediments by correlating deposits rather than by directly dating the moraines. This was also because all the samples could not be processed because of methodological limitations.

The VC measurements were also carried out in lacustrine-glacial pelitic silt. Occasionally, VC measurements revealed seasonal and annual rhythms in glaciolacustrine deposits. These patterns indicate the duration of seasonal freezing in lakes, and shed light on some related problems. The VC data also showed a particular depositional mechanism in lacustrine sections in the Chagan-Uzun Valley for which estimated sedimentation rates varied from a few millimetres to tens of centimetres per year. They showed deposition of pelitic silt alternating with fine sand interbeds at 1–1.5 cm per year, as typically found in many overflowing mountain lakes (Sheinkman, 1997).

In relation to dating, in general, VC analysis requires continuous sections: the greater the number of continuous layers analysed, the greater the accuracy of age estimates, while breaks of a few hundreds of years may cause considerable uncertainty. Depending on this, the VC age of the deposits has been determined either to thousands or to hundreds of years or even to years and seasons.

The grey-colour deposits in the Chagan-Uzun Valley originated in different ways and record advance and retreat cycles of Pleistocene glaciers. An exposure of Quaternary deposits over 10 km long on the left side of the Chagan Valley is especially spectacular. Several blocks in the longest Kyzyl-Jar section and in the Tedash section upstream have been dated (Figs. 65.12 and 65.13).

The uppermost strata in the Chagan exposure involve alternation of the basal till and glaciofluvial gravel (8–12 m

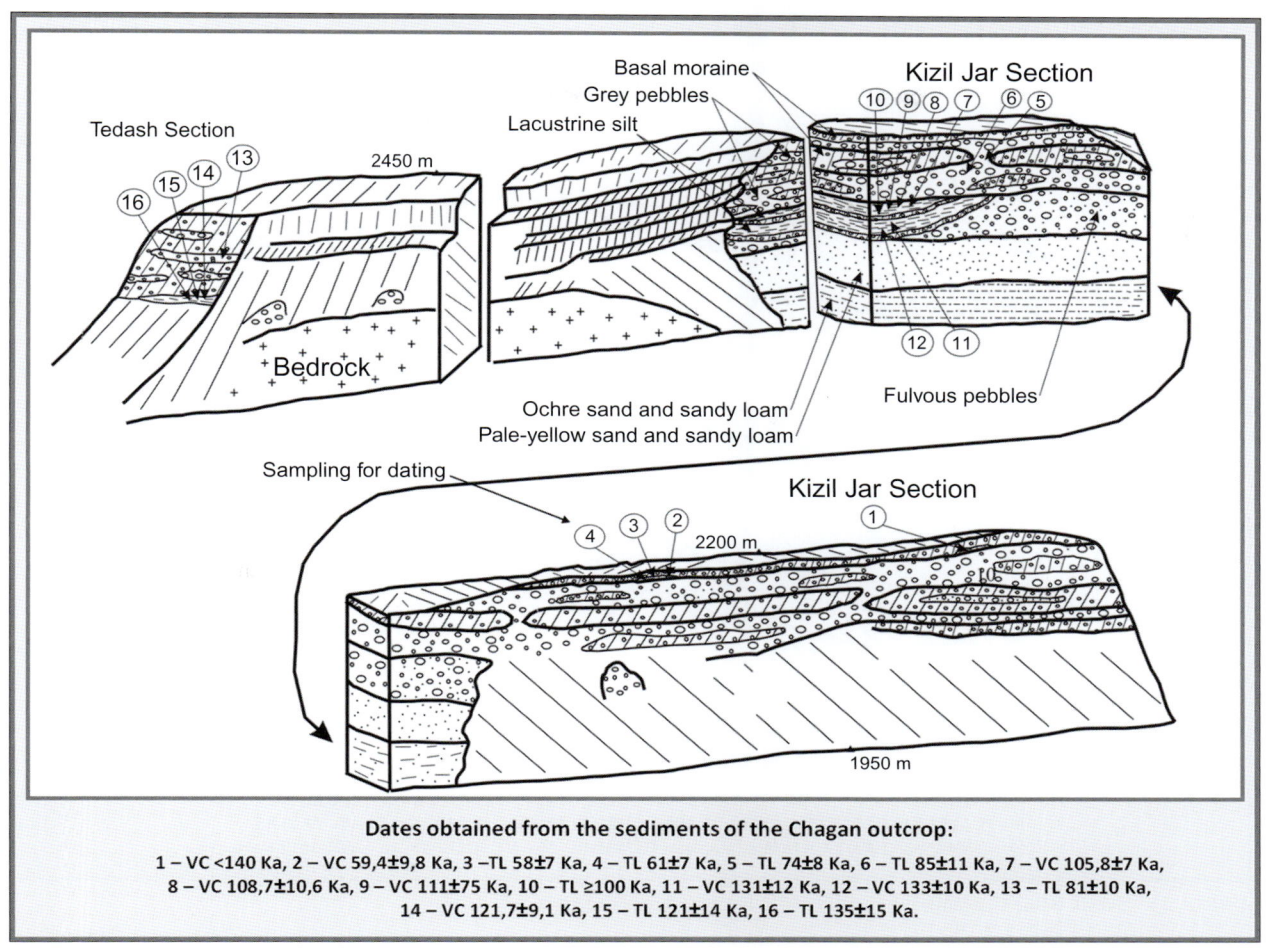

FIGURE 65.13 Sketch of the Chagan outcrop: dated layers and obtained ages.

Dates obtained from the sediments of the Chagan outcrop:
1 – VC <140 Ka, 2 – VC 59,4±9,8 Ka, 3 – TL 58±7 Ka, 4 – TL 61±7 Ka, 5 – TL 74±8 Ka, 6 – TL 85±11 Ka, 7 – VC 105,8±7 Ka, 8 – VC 108,7±10,6 Ka, 9 – VC 111±75 Ka, 10 – TL ≥100 Ka, 11 – VC 131±12 Ka, 12 – VC 133±10 Ka, 13 – TL 81±10 Ka, 14 – VC 121,7±9,1 Ka, 15 – TL 121±14 Ka, 16 – TL 135±15 Ka.

thick, occasionally up to 20 m) that periodically grade into syngenetic glaciolacustrine and fluvio-lacustrine deposits. Glacial erosion and deposition processes sometimes cause sequence discontinuities. Massive blocks of overlapping basal till and interglacial deposits may contain interbedded material such as boulders, pebbles, sand and silt. Nevertheless, together the sections record the glacial–interglacial cycles from three to six units in different parts of the Chagan exposure.

The uppermost till forms a 5–8- to 10–12-m thick continuous uniform mantle upon a peneplain surface. Syngenetic deposits periodically overlie or are included into this basal till. Those that were deposited in water environments and are suitable for the dating have been dated several times. The first VC date of <140 ka was obtained for a small lens of glaciolacustrine silt in the section at Kyzyl-Jar-1 block (point 1; Fig. 65.13). Here a VC age of 59.4±9.8 ka was obtained for equivalent sediments in the Kyzyl-Jar-2 block and two TL ages (58±7 and 61±7 ka) for syngenetic sands (points 2–4; Fig. 65.13).

The following cycle in the Chagan exposure is difficult to distinguish. Often massive blocks of similar overlapping basal tills cut by glaciofluvial pebble lenses occur to depths 70–80 m below the surface, for example, in the Tedash section, or gravels occur as massive blocks cut with till lenses. Such blocks occur in the Kyzyl-Jar section (Fig. 65.13). TL ages of 79±8 and 85±11 ka were obtained from glaciofluvial pebbles in the Kyzyl-Jar-3 block (points 5 and 6; Fig. 65.13) and similar gravels in the Tedash section (point 13; Fig. 65.13) showed about the same age: 81±10 ka. A few values of the same order were obtained from equivalent sequences in the neighbouring catchments of the Taldura Valley (Fig. 65.12) incised into the same peneplain.

Two lenses of glaciolacustrine pelitic silts (10–12 and ~20 m thick) separated by 10–12 m of basal till occur. They underlie an 80 m thick sequence of alternating tills and gravels exposed in the middle of the Chagan exposure. The lower lens is underlain by another till unit of broadly similar thickness (Fig. 65.13).

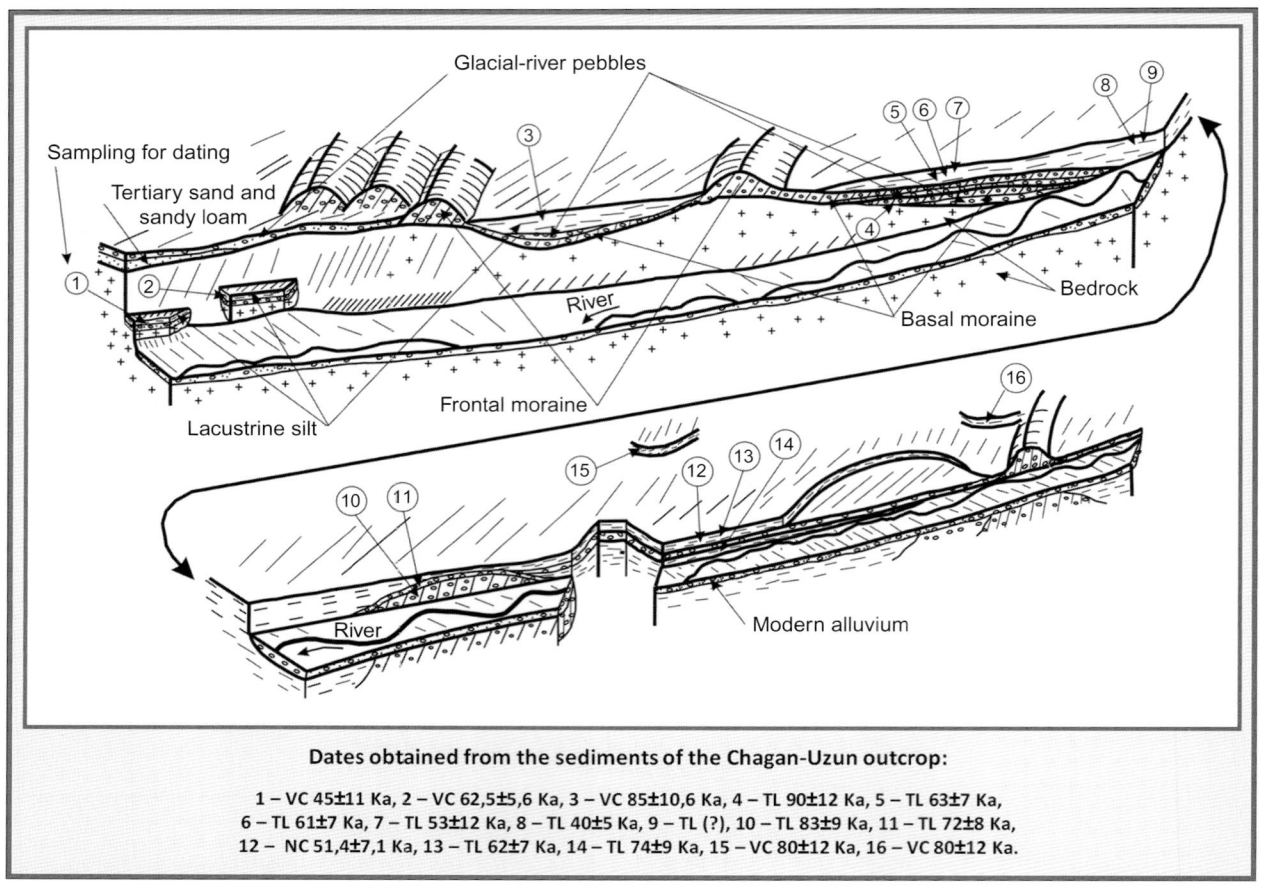

FIGURE 65.14 Sketch of the Chagan-Uzun outcrop: dated layers and obtained ages.

Dates obtained from the sediments of the Chagan-Uzun outcrop:

1 – VC 45±11 Ka, 2 – VC 62,5±5,6 Ka, 3 – VC 85±10,6 Ka, 4 – TL 90±12 Ka, 5 – TL 63±7 Ka, 6 – TL 61±7 Ka, 7 – TL 53±12 Ka, 8 – TL 40±5 Ka, 9 – TL (?), 10 – TL 83±9 Ka, 11 – TL 72±8 Ka, 12 – NC 51,4±7,1 Ka, 13 – TL 62±7 Ka, 14 – TL 74±9 Ka, 15 – VC 80±12 Ka, 16 – VC 80±12 Ka.

Three VC dates of 105.8 ± 7, 108.7 ± 10.6 and 111 ± 75 ka (the poor accuracy of the latter is caused by discontinuous bedding) and a TL age of ≥ 100 ka were obtained from the upper lens of lacustrine silts (Kyzyl-Jar-4 block, points 7–10; Fig. 65.13), and VC ages of 131.7 ± 12 and 133 ± 10 ka from the lower lens (Kyzyl-Jar-5 block, points 11 and 12; Fig. 65.13). Equivalent glaciolacustrine silts beneath a till block yielded a VC age of 121.7 ± 9.1 ka and TL ages of 121 ± 14 and 135 ± 15 ka (cited down section, points 14–16 in Fig. 65.13), and a TL age of 81 ± 10 ka was obtained for the gravel that cuts the overlying till (point 13 in Fig. 65.13). Samples from the red-coloured section at the base of the Chagan-Uzun exposure gave the saturated TL signal. Unfortunately, there were no sediments suitable for VC dating, and therefore, the exact age of the red-colour formation could not be estimated.

The section in the lower reaches of the Chagan-Uzun River is exposed in a deep canyon (Figs. 65.12 and 65.14). The age of the distal moraine marking the maximum glacier extent here could not be determined to a better accuracy than just Early Zyryanian Stage (i.e. Early Weichselian, Early Wisconsinan) because it was impossible to separate the chronometer minerals from impurities. However, for the bottom sediments of a lake that emerged before the proximal terminal moraine (point 3; Fig. 65.14), a VC age of 85 ± 10.6 ka was obtained. These sediments are preserved upstream either buried or within isolated fragments of high terraces along the valley sides (up to 60 m and higher above the present river surface). Two further VC ages (both 80 ± 12 ka, points 15 and 16; Fig. 65.14) were determined for the upper strata in these terraces, and a TL age of 74 ± 9 ka (point 13; Fig. 65.14) shows the presence of buried deposits of the same generation.

Another moraine complex, overlying a big riegel (rock step), occurs about 1 km upstream from the terminal moraine and is associated with another larger and a more intricately structured glacier tongue basin. The riegel is overlain by basal till interbedded with pebble beds that resemble the equivalent strata in the Chagan exposure. Upstream, the moraine complex is buried by lacustrine sands and pelitic silts, often cut with pebble beds, which belong to a terrace a few tens of metres above the present river level. A TL date of 90 ± 12 ka was obtained for a sand lens here at base of the basal till (point 4; Fig. 65.14). However, rocks within the glacier tongue basin were found to be much younger (points 5–8, respectively; Fig. 65.14) since they gave TL ages of 63 ± 7 and 61 ± 7 ka for the lower, 53 ± 6 ka for the central and 40 ± 5 ka for the upper parts of the lacustrine unit. A TL age of 83 ± 9 ka was obtained for a sand lens in a fragment of a till from a site in the centre of the glacier tongue basin beneath glaciolacustrine and glaciofluvial deposits, while a pebble bed overlying the till gave an age of 72 ± 8 ka (Fig. 65.14; points 10 and 11). Samples from the middle of a 20-m-thick section of a lake terrace upstream, above the pebble bed, gave a TL age of 62 ± 7 ka and a VC age of 51.4 ± 7.1 ka, and a sample from below the pebbles gave a TL age of 74 ± 9 ka (points 12–14; Fig. 65.14).

The upstream part of the Chagan-Uzun glacier tongue basin is dammed by a moraine overlying lacustrine deposits ploughed by a glacier up the valley from the trough floor. Here, a ^{14}C age of 25.3 ± 0.6 ka was obtained from these deposits (Markov, 1978). Unfortunately, it is impossible to judge its reliability or context, since there is no description of the sampling. The sample was most likely from younger superimposed deposits often encountered in glacier tongue basins. Therefore, this ^{14}C age can only be used as an indication that the lake deposits are older, a conclusion that is supported by the current author's evidence.

Several TL, ^{14}C and VC dates from units in the Chagan Valley, beyond the glacier tongue basin, have revealed a Sartan Stage (Late Weichselian, Late Wisconsinan) section produced by glaciers that never reached the peneplain (Sheinkman, 1992, 1995, 2002a). Note that a thorough study of the glacial setting showed a non-glacial origin of the sediments from the Chagan-Uzun Valley below the terminal moraine, where some authors expected to find traces of older glaciers (Deviatkin, 1965).

Glacial deposits occasionally encountered in the Chuya Valley are related to Chagan-Uzun syngenetic glaciers that descended along other tributaries of the river. These deposits are found, for instance, at the end of the Kuekhtanar Valley (Fig. 65.12) which was occupied by a short but voluminous glacier that, unlike that in the Chagan-Uzun Valley, reached and barred the Chuya Valley which itself was unglaciated at that time. The motion of glaciers (the Kuekhtanar glacier, together with the small neighbours and a glacier that moved in the opposite direction from the North Chuya Ridge) that barred the Chuya Valley gave rise to a number of ephemeral lakes (Sheinkman, 1997). They produced wave-cut cavities in the outer cirque slope of the terminal moraine that blocked the Chagan-Uzun Valley and left a series of lake terraces. VC ages from silt sediments of the terraces preserved in fragments in the water gap (points 1 and 2; Fig. 65.14) are of 62.2 ± 5.6 ka (from a higher terrace) and 45 ± 11 ka (from a lower terrace).

TL ages of Sartan (Late Weichselian) age, for example, 14.5 ± 1.5 and 13 ± 1.5 ka, were obtained from the uppermost section of glacial/glaciofluvial/glaciolacustrine deposits exposed in the mouth of the Kuekhtanar River along the Chuya Valley. They comprise the bottom sands of a proglacial lake formed before the youngest moraine. The average TL signal in samples from the underlying section indicates a Zyryanian (Early and Middle Weichselian) age of the older moraines. However, according to Butvilovskii (1993), the ^{14}C ages from the mouth of the Kuekhtanar River, the entire glacial complex is of Sartan age. Though, this timing may be biased by an underestimation of ^{14}C ages obtained near the temporal limit for this dating method. Dating of subaerial periglacial deposits that are ubiquitous in the Chuya basin (cryogenic-aeolian cover loam and fan deposits) places them within the Sartan Glacial event, the coldest period of the Pleistocene. This is confirmed by TL determinations.

65.7. PALAEOGEOGRAPHICAL INTERPRETATION

It should be remembered that the interpretation of Quaternary events in Siberia has been long viewed from the perspective of events reconstructed in Western Siberia that were related to those described from Western Europe. This model was revised, updated, generally confirmed by evidence from the north-west of the region, but contradicted by observations from inland Siberia. Such a dissimilarity has been explained by different combination of allochthonous and autochthonous glacial processes which act in each case in different ways.

Following on from the author's own analysis (Sheinkman, 2008), the overall progressive climate cooling through the Quaternary resulted in maximum glaciation in north-western Siberia much earlier than in north-eastern Siberia, because this reflects respective reaction of different CGS to the cooling.

In north-western Siberia, glaciation developed in the form of a great ice sheet (Svendsen et al., 2004), which was initiated as a warm CGS existing under the influence of the Atlantic Ocean. When that CGS went through a climatic cooling, the firn line, under conditions with primarily abundant snowfalls, descended rapidly and enabled glaciers to form. At first, these glaciers prevented freezing of the ice bed and frozen rocks were located in the periglacial zone only. Such a situation occurred until the cryo-hygrotic (cold and humid) phase of glaciation had been replaced by the cryo-xerotic (cold and dry) phase (Grichuk and Grichuk, 1960; Velichko, 1981). Once the cryo-xerotic phase became well established, the glaciers became deeply frozen, and permafrost formed also under the glaciers. The strongest cryo-arid conditions occurred during the Late

Pleistocene, when glaciation in north-west Siberia was lesser in extent than in previous cold periods of the Pleistocene.

In inner Asia, continental climatic conditions were stable throughout the Quaternary, and the cooling was associated with a colder, dryer climate. Cryo-arid environments have prevailed in the Siberian Mountains throughout the Quaternary. The largest phase of glaciation occurred in this region during the early Late Pleistocene and was represented only by the cold CGS both initially and subsequently. However, even during the most severe cooling during the latest period of the Pleistocene, glaciers reduced in size, even in north-eastern Siberia. Nevertheless, severely dry and cold conditions were accompanied by the most extensive distribution of permafrost of the Late Pleistocene.

Recently, significant glacier advance in the early Late Pleistocene has been demonstrated by independent evidence from boreholes in the sedimentary fill of Lake Baikal (Karabanov et al., 1998). Repeated dating of the representative sediments in the foothills of the Verkhoyanskii Mountains (Kind, 1974; Stauch et al., 2007) has also supported this chronology. The SPECMAP marine isotope record (Imbrie et al., 1984) has been increasingly used as a temporal reference for this sequence.

It could be argued that the Siberian Mountains are highly sensitive to environmental changes, and the Milankovitch rhythms (Milankovitch, 1969, translated from Milankovitch, 1941) must be well pronounced there. A good basis for correlating the glaciation events with these rhythms in the light of the marine isotope record of global ice volume has been provided by the evidence from the Chuya basin, since it is a typical continental mountain region of Siberia, and has the most robust glacial geochronology of the region. It must be noted that, unlike the continuous records from ice cores, lake-bottom sediments, loess sequences, etc., the record in the glacial deposits is frequently incomplete because of repeated depositional breaks. Moreover, although there is no alternative to TL method in mountain environments, it can provide an accuracy of no better than $\pm 10\%$, which gives an uncertainty of ± 10–15 ka for the early Late Pleistocene. Therefore, the tens of age determinations available may even be insufficient for detailed correlations. Nevertheless, despite some inevitable scatter of ages, a quite well-pronounced pattern is observed in the Chuya basin. The coincidence is not fortuitous because inland continental regions are the most sensitive to changes in insolation. Thus, it is no surprise that when these results are compared with the SPECMAP global ice volume record, as they reflect the main orbitally driven climate cycles.

The ages of the end moraines in the lower reaches of the Chagan-Uzun River, where the maximum advance of the 70 km long glacier is marked by a succession of moraine ridges, have been interpreted by correlation with ages of 80 ± 12 to 85 ± 11 ka obtained from the lake-bottom sediments in the related moraine-dammed basin. The same buried sediments have given an age of 74 ± 9 ka. The river on the valley floor is locally eroding till deposits correlated to 83 ± 9 to 90 ± 12 ka sediments overlain by pebble beds dated to 72 ± 8 ka. A number of dates from 40 ± 5 to 62 ± 7 ka were obtained for lacustrine deposits of the second glacier tongue basin located upstream which owes its existence to another moraine complex. (Note that VC analysis of both dated lake sequences shows freezing seasons typical of interglacial time.) Thus, the proximal moraine of the first complex must have been formed at about 85–90 ka ago and the other at about 60–65 ka ago. These results suggest that the events occurred during the Early Weichselian Substage or MIS 5b (85–95 ka ago) and MIS 4 (60–70 ka ago), whereas the moraine-dammed lakes that pre-dates these moraines formed in the time corresponding to the Early Weichselian interstadial (MIS 5a; 70–85 ka ago) and the Middle Weichselian (MIS 3; 25–60 ka ago). Hence the distal maximal moraine may have formed during MIS 5d (105–115 ka ago).

More agreement is found in dating the Chagan exposure. The oldest till underlies 135 ± 15 ka lacustrine deposits and is therefore older. Hence, the lacustrine deposits overlying the till may be correlated to the last interglacial (Mikulino, Eemian Stage; MIS 5e) and the till itself must represent the preceding late Saalian cold Stage (i.e. MIS 6). The age of the younger till interbedded between two lacustrine units of 100–110 and 130–135 ka lie within the interval 110–120 ka in the Kyzil-Jar section and <120 ka in the Tedash section where the underlying sediments were dated to 121–135 ka. Thus, the till can be equated with the Early Weichselian MIS 5d and the overlying deposits with MIS 5c. These lenses result from deposition in lakes between moraines, and, on the basis of the VC evidence, reflect a freezing regime similar to that in the existing proglacial lakes. The reverse magnetisation in these deposits (Markov, 1978) therefore presumably corresponds to the Blake excursion, the parameters of which agree at least with the ages of the lower lacustrine unit (Harland et al., 1990). According to these data, the MIS 5d-age till at Chagan can be correlated with the distal moraine in the terminal moraine complex in the Lower Chagan-Uzun Valley.

Short et al. (1991) showed that the amplitudes of the SPECMAP cycles (Imbrie et al., 1984) in Central Eurasia do not fit the mean summer air temperatures because of specific filtering of the Milankovitch cycles by local conditions. At the same time, glaciation in Central Siberia directly depends on summer temperatures (Koreisha, 1991; Sheinkman, 2007, 2008), whilst winter-temperature variations are less important for the cold glaciers of those regions (these temperatures are very low). However, even minor variations in summer temperatures change ablation

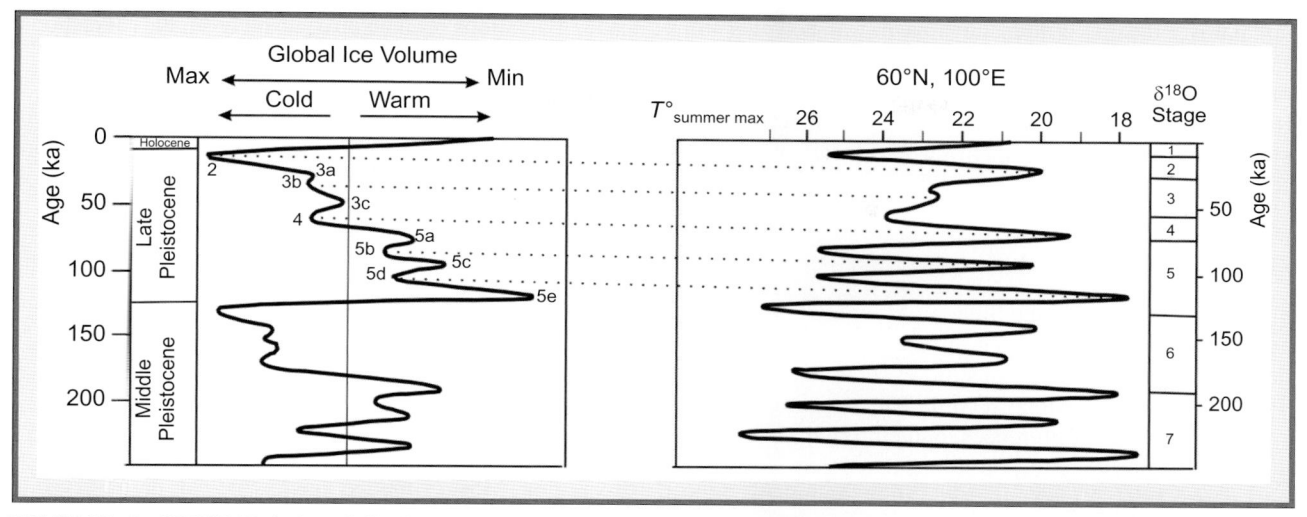

FIGURE 65.15 SPECMAP rhythms (left) after Imbrie et al. (1984) and continental-filtered Milankovitch cycles in Siberia (right) after Short et al. (1991).

and ice volume. In the region of about 60°N, 100°E (close to the study area), the strongest last interglacial warming (115–130 ka ago) followed a peak of highest summer temperatures. The ensuing earliest Weichselian cooling of MIS 5d (105–115 ka ago) was associated with a relatively small increase in global ice volume. However, in inland Siberia, this stadial occurred after a strong summer temperature fall, coinciding with the transition to the cooling from the strongest last interglacial warming (Fig. 65.15). This led to an unusual climatic combination of relatively high humidity from the eastward propagation of the Gulf Stream and the onset of cooling with lowest summer temperatures. Such a situation triggered glaciation, which was accompanied by progressive cooling and drying of climate and mountain building, and a specific combination of autochthonous and allochthonous glacial processes. It culminated in the earliest Late Pleistocene. Similar climatic combinations occurred earlier, but then the Siberian Mountains were insufficiently high to intrude above the snowline. As opposed to the West-Siberian environments, it did not produce glacier advances like those in the Late Pleistocene. Therefore, the till correlated with the late Saalian Substage (MIS 6) was deposited at the end of the previous 100-ka cycle, in its driest time, by a glacier that cannot have reached maximum advance and whose traces were buried under the deposits of larger Late Pleistocene glaciers.

The climatic warming during MIS 5c was moderate. It was followed by the long cold stadial of MIS 5b (85–95 ka ago). Although the summer temperature depression during MIS 5b was much less than during the MIS 5d, the glaciers in this stage overran successive moraines of the terminal complex that barred the lower Chagan-Uzun Valley. Nevertheless, a considerable increase in summer temperatures during the MIS 5b caused the glaciers to retreat to about one-third of their length during MIS 5c and produced glacial lakes with sediment records 10–12 m thick (e.g. at Chagan). The glaciers retreated again during the relatively long warm MIS 5a interstadial (70–85 ka ago; Fig. 65.15), and left the first glacier tongue basin in the lower Chagan-Uzun. In the middle part of the valley, however, oscillations of glaciers complicated the formation of the basal moraine, the age of which can be inferred from the cutting pebbles dated from 74 ± 8 to 85 ± 11 ka. MIS 4 (60–70 ka ago) is marked by the uppermost till at Chagan which is equated to sediments dated to 60 ± 7 ka. This event was associated with considerable cooling and followed a fall in summer temperatures almost to the Late Pleistocene minimum. Therefore, the Chagan-Uzun glacier in the middle part of the valley was still able to reach the watershed peneplain and approached the maximum-advance moraines in depressions. The glaciers that barred the Chuya Valley were also substantial. The dammed Chuya produced a lake that abraded the outer slope of the distal moraine in the lower course of the Chagan-Uzun, and provided lacustrine deposition dated to 62.2 ± 5.6 ka.

During the Middle Weichselian (MIS 3; 25–60 ka ago), the Chagan-Uzun glacier retreated to the upper part of the valley leaving a large glacier tongue basin. Lags in this retreat were accompanied by deposition of glaciofluvial gravels and lacustrine silts, and the short cold excursion of the MIS 3b (40–45 ka ago) may have allowed glacial readvance. At least, such an advance is indicated by 45 ± 11 ka lacustrine deposits produced by repeated damming of the lower Chuya Valley by glaciers, and by a terminal moraine overlying the tongue basin deposits.

The Sartan Stage (Late Weichselian) complex of deposits corresponds to MIS 2 during which the Gulf Stream had already been blocked. Despite of the strong cooling, low humidity forced the glaciers up-valley and widespread permafrost developed in the Siberian

Mountains. Freezing in the Chuya basin was however limited: deposition of subaerial loam and permafrost phenomena occurred under a strongly cold and arid climate. However, there are no traces of PIW, which were found as the pseudomorphs after them far to the east of Altai, in the left-bank upper reaches of the Yenicei River.

65.8. CONCLUSIONS

In general, glaciation in the Siberian Mountains was highly sensitive to climate change but different regions apparently showed clear responses to specific events recorded in the SPECMAP marine isotope record. The magnitude of the present-day climatic differences found through the Siberian Mountains is similar to the variations that occurred (Berger and Loutre, 1991; Kutzbach et al., 1998; Velichko, 2009) during the Pleistocene cold stages. This is because the general circulation pattern of the atmosphere did not radically change during the Quaternary, in this part of the world, and this allows comparison to respective environments.

Despite some uncertainty, the obtained results reveal the main features of the Late Pleistocene history of continental mountainous regions in Siberia and their response to climate change. It has also showed a real possibility for the determination of the ages of deposits, at least for the Late Pleistocene sediments. Of course, at the current stage of knowledge, the timing of glaciations in mountain valleys can only outline a fragmentary sequence of past events. It requires large- or at least medium-scale collections of dated sediment samples to provide a statistically representative filtered data set. This requirement can be met only by the application and improvement of new dating techniques.

Overall, the glaciations in the Siberian Mountains were of mountain-valley type and associated with advance and retreat of glaciers apparently corresponding to the main Milankovitch cycles. The lowering of temperatures, which mainly controlled the development of glaciation under a cold and dry climate, provided close interaction of glaciers, icings and ground ice, the totality of which clearly displayed the environmental changes. This is because the relationship between the different ice agents is a reliable indicator of environmental changes.

Glaciers in the Siberian Mountains did not reach the final form of their development, the ice sheet. Instead, they were confined to their troughs, and only during the maximal advance did the ice streams reach the piedmont areas and form fan-shaped piedmont glaciers at the foot of the mountains. Mostly, the glaciers were a few tens of kilometres in length, although the largest of them in north-eastern Siberia stretched over 150 km in length.

The conditions for the glacier advance were the most favourable in the earliest Late Pleistocene rather than in the Early or Middle Pleistocene, as in western Siberia.

Later, glaciations became progressively smaller in response to strong cooling and drying in accordance with the main events of the marine oxygen isotope record. As a result, the Early Zyryanian (Early Weichselian) glaciation in the Siberian Mountains produced twice as much ice volume as the Sartan (Late Weichselian) Glaciation, which was restricted to the high mountains.

Overall, throughout the Pleistocene, glaciers developed on to permafrost, since the latter is a background environmental factor—appearing in the region much earlier than the first glaciers. Cold and dry conditions first promoted permafrost development, and only when the mountains were uplifted into ionosphere could the glaciers be initiated. Freezing involves any rock, as well as glaciers themselves, and all of this imparts specific features to the glaciation in this region. Most of the ice agents that can be observed in the Siberian Mountains were much more frozen during the Pleistocene cold stages. As a result, glacial sediments and landforms underwent significant reworking by other nonglacial processes of the CGS, among which icings were the most active.

All in all, comparison with European (Alpine) glacial chronology requires special adaptations for the Siberian Mountains, because of the very different environments. Moreover, it should be remembered that the interpretation of events in Siberia has long been viewed in the light of the West-Siberian model that stems from the reconstructions viewed from a European perspective. Although this model has recently been confirmed by new information from West Siberia, studies of glacial environments in the Siberian Mountains show that special adaptions must be applied to inner Siberia, since its continental conditions were quite different from those in West Siberia.

REFERENCES

Afanas'ev, S.L., 1991. Nano-Cyclic Method to Determine Geological Ages by Micro-Layers, Varves and Salt Inter-Layers. Rosvuznauka of Quaternary deposits, Moscow 218pp (in Russian).

Aizen, V.B., Aizen, E.M., Joswiak, D.R., Fujita, K., Takeuchi, N., Nikitin, S.A., 2006. Climatic and atmospheric circulation pattern variability from ice-core isotope/geochemistry records (Altai, Tien Shan and Tibet). Annals of Glaciology 43, 49–60.

Alekseev, V.R., 1987. Icings. Nauka, Novosibirsk (in Russian, English summary).

Arkhipov, S.A., 1989. Pleistocene chronostratigraphy of Northern Siberia. Geol. Geophys. 7, 13–22.

Astakhov, V.I., 2006. About chronostratigraphical units of the Upper Pleistocene. Geol. Geophys. 47 (11), 1207–1220.

Berg, L.S., 1938. Foundations of climatology. Utchpedgiz, Leningrad 456pp (in Russian).

Berger, A., Loutre, V.F., 1991. nsolation values for the climate of the last 10 million years. Quatern. Sci. Rev. 10, 297–317.

Borzenkova, I.I., 1992. Climate Change in the Cainozoic. Gidrometeoizdat, St-Petersburg, 248p (in Russian, English summary).

Butvilovskii, V.V., 1993. Palaeogegraphy of the Last Glaciation and Holocene of Altai: Event-Catastrophic Model. Tomsk State University, Tomsk, 252p (in Russian).

Cherskii, I.D., 1882. To the question of about ancient glaciers in East Siberia. Proc. East-Siberian Div. Imperial Russ. Geogr. Soc. 12 (4–5), 28–62 (in Russian).

Christopherson, R.W., 1998. Elemental Geosystems. Prentice Hall, Inc, Gustav Fischer, Stuttgart, 534pp.

Danilov, I.D., 1978. Polar Lithogenesis. Nedra, Moscow, 238pp (in Russian, English summary).

Deviatkin, E.V., 1965. Cainozoic Deposits and Neo-Tectonics of the South-eastern Altai. Nauka, Moscow, 244pp (in Russian, English summary).

Fukui, K., Fujii, Y., Mikhailov, N., Ostanin, O., Iwahana, G., 2007. The lower limit of mountain permafrost in the Russian Altai Mountains. Permafrost Periglacial Process. 18 (2), 129–136.

Gavrilova, M.K., 1998. Climate of Cold Regions in the World. Russian Academy of sciences, Siberian Branch, Yakutsk, 206pp (in Russian).

Gorbunov, A.P., Severskii, E.V., 2007. East Altai: pecularities of high geocryological zonality. Cryosphere World XI (4), 15–19 (in Russian, English summary).

Grave, N.A., Gavrilova, M.K., Gravis, G.F., Katasonov, E.M., Klyukin, N.K., Koreisha, M.M., et al., 1964. Freezing of the Earth Surface and Glaciation of the Suntar-Khaiata Range. Nauka, Moscow, 144pp (in Russian, English summary).

Grichuk, M.P., Grichuk, V.P., 1960. About Periglacial Vegetation at the USSR Area: Periglacial Phenomena at the USSR Area. Moscow State University, Moscow, pp. 66–100 (in Russian).

Grosswald, M.G., Hughes, T.J., 2002. The Russian component of an Arctic Ice Sheet during the Last Glacial Maximum. Quatern. Sci. Rev. 21, 121–146.

Harland, W.B., Armstrong, R.L., Cox, A.V., Craig, L.E., Smith, A.G., Smith, D.G., 1990. A Geologic Time Scale, 1989. Cambridge University Press, New York, 262 pp.

Il'ichev, V.A., Kulikov, O.A., Faustov, S.S., 1973. New Data of Palaeomagnetic and Thermoluminescence Researches of the Chagan Section (Mountain Altai): Chronology of the Pleistocene and Climatic Stratigraphy. Russian Geographical Society, Leningrad, pp. 252–257 (in Russian, English summary).

Imbrie, J., Hays, J.D., Martinson, D.G., McIntyre, A., Mix, A.C., Morley, J.J., et al., 1984. The orbital theory of Pleistocene climate: support from a revised chronology of the marine $\delta^{18}O$ record. In: Berger, A., Imbrie, J., Hays, J., Kukla, G., Saltsman, B. (Eds.), Milankovich and Climate, Part I. Reidel, Boston, pp. 269–305.

Karabanov, E.B., Prokopenko, A.A., Williams, D.F., Colman, S.M., 1998. Evidence from Lake Baikal for Siberian glaciation during oxygen-isotope substage 5d. Quatern. Res. 50, 44–55.

Kind, N.V., 1974. Late Quaternary Geochronology According to Isotopes. Nauka, Moscow, 256pp.

Koreisha, M.M., 1963. Modern Glaciation of the Suntar-Khaiata Range. Glaciology 11, 105pp (in Russian, English summary).

Koreisha, M.M., 1991. Glaciation of Verkhojan-Kolimski Region. Russian Academy of Sciences, Moscow, 144pp (in Russian, English summary).

Kotlyakov, V.M. (Ed.), 1984. Glaciological Glossary. Gidrometeoizdat, Leningrad, 528pp (in Russian, English summary).

Kotlyakov, V.M. (Ed.), 1997. World Atlas of Snow and Ice Resources. Russian Academy of Sciences, Moscow, vol. I, 392pp; vol. II, Book 1, 264pp, Book 2, 270pp (in Russian, English summary).

Krenke, A.N., 1982. Mass-Exchange in Glacier Systems in the USSR Area. Gidrometeoizdat, Leningrad, 288pp (in Russian English summary).

Kropotkin, A.P., 1876. Study of Glacial Period. Proceedings of Imperial Russian Geographical Society in General Geography Vol. 7. St-Petersburg, 770pp (in Russian).

Kutzbach, J., Gallimore, R., Harrison, S., Behling, P., Selin, R., Laarif, F., 1998. Climate and biome simulations for the past 21,000 years. Quatern. Sci. Rev. 17, 473–506.

Markov, K.K. (Ed.), 1978. The Section of the Latest Deposits of the Altai. Moscow State University, Moscow, 208pp (in Russian).

Matrosov, I.I., Chistiakov, V.K., Pogorelov, Y.L., 1979. Research of Thermoluminescence of Geologic Materials. Tomsk State University, Tomsk, 114pp (in Russian).

Mikhailov, N.N., Ostatnin, O.V., Fukui, K., Fujii, E., 2006. Experience in use of automatic temperature loggers in the High Altai Mountains. Geography and Nature Management in Siberia, No. 8. Altai University, Barnaul, pp. 134–146 (in Russian).

Mikhalenko, V.N., 2007. Inner Structure of Glaciers in Non-Polar Regions. URSS, Moscow, 315pp (in Russian English summary).

Milankovitch, M., 1969. Canon of Insolation and the Ice-Age Problem (Kanon der Erdbestrahlungen und seine Anwendung auf das Eiszeitenproblem. Belgrade 1941). Israel Program for Scientific Translations, Jerusalem, 633pp.

Narozhnii, Y.K., 1993. Temperature regime in the active layer of the Aktru Glaciers. Glaciol. Siberia 4 (19), 140–150 Tomsk, Tomsk State University (in Russian).

Narozhnii, Y.K., Nikitin, S.A., 2003. Modern Glaciation of Altai at the Turn of the XIX Century. Data of Glaciological Studies, No. 95. Russian Academy of Sciences, Moscow, pp. 93–101 (in Russian, English summary).

Nekrasov, I.A., 1976. Cryolithozone of North-eastern and Southern Siberia and regularities of its development. Yakutsk, 246pp (in Russian, English summary).

Nikitin, S.A., Vesnin, A.V., Menshikov, V.A., Negoduiko, A.G., Smutkin, A.G., 1986. Temperature regime of the Small Aktru Glacier in the ablation period. Glaciol. Siberia 3 (18), 81–84, Tomsk, Tomsk State University (in Russian).

Obruchev, V.A., 1951. Events of the Glacial Period in Northern and Central Asia: Selected Works on Geography of Asia, vol. 3. Geografgiz, Moscow, pp. 49–128 (in Russian).

Osadchii, S.S., 1982. Repeated Ice Wedges in the East Sayan and Their Palaeogeographic Significance: Late Pleistocene and Holocene of the Southern East Siberia. Nauka, Novosibirsk, pp. 146–155 (in Russian).

Perevalov, A.V., Sheinkman, V.S., Lutoev, V.P., Tsidenov, A.B., 2009. Dosimetric properties of natural quartz and possibility of its use to determine age of Quaternary deposits: radioactivity and radioactive elements in the environment of the Man Habitation. In: Materials of the III International Conference, Tomsk, pp. 440–443.

Popov, V.E., 1972. About possibility to apply geomorphologic criterion to determine age of the Quaternary deposits in the Chagn-Uzun basic section in the Altai. Glaciology of the Altai, No 7. Tomsk State University, Tomsk, pp. 104–114 (in Russian).

Preobrazhenskii, V.S., 1960. Kodar glacier region (Trans-Baikalia). Glaciology, No 4. Russian Academy of Sciences, Moscow, 74pp (in Russian, English summary).

Revyakin, V.S., 1981. Natural Ice of the Altai-Sayan Mountain Country. Gidrometeoizdat, Leningrad, 288pp (in Russian, English summary).

Romanovskii, N., 1977. Formation of Repeated Ice Wedges. Nauka, Novosibirsk, 215pp (in Russian, English summary).

Romanovskiy, N.N., 1993. Foundations of Cryogenesis in Lithosphere. Moscow University, Moscow, 336pp (in Russian, English summary).

Sax, V.N., 1953. The Quaternary in the Soviet Arctic. Gidrometeoizdat, Leningrad, 627pp (in Russian).

Serebryannii, L.R., Orlov, V.A., Solomina, O.N., 1989. Moraines as a Source of Glaciological Information. Results of Research on the International Geophysical Projects. Russian Academy of Sciences, Moscow (in Russian, English summary).

Shats, M.M., 1978. Geocryological Conditions of the Altai-Sayan Mountain Country. Nauka, Novosibirsk, 104pp (in Russian, English summary).

Sheinkman, V.S., 1987. Glaciology and Palaeoglaciology of the Chersky Mountain System and Adjacent Area of the Northeast of USSR. Results of Research on the International Geophysical Projects. Russian Academy of Sciences, Moscow, 154pp (in Russian, English summary).

Sheinkman, V.S., 1992. About the character of the ancient glaciation in the Siberian Mountains. Proc. Russ. Geogr. Soc. 124 (2), 158–164 (in Russian, English summary).

Sheinkman, V.S., 1993. About Interpretation of the Evidence of the Ancient Glaciation in the Siberian Mountains. Data of Glaciological Studies, No. 77. Russian Academy of Sciences, Moscow, pp. 114–120 (in Russian, English summary).

Sheinkman, V.S., 1995. Experience of Palaeoglaciological Research in the Siberian Mountains. Data of Glaciological Studies, No. 79. Russian Academy of Sciences, Moscow, pp. 111–118 (in Russian, English summary).

Sheinkman, V.S., 1997. Glaciogenic Dammed Lakes in the Siberian Mountains: Causes and Factors of Origin and Development. Data of Glaciological Studies, No. 82. Russian Academy of Sciences, Moscow, pp. 43–50 (in Russian, English summary).

Sheinkman, V.S., 2002a. Age diagnostics of glacial deposits of the Mountain Altai, testing the dating results at the Dead Sea sections and palaeoglaciological data interpretation. Data of Glaciological Studies, No. 93. Russian Academy of Sciences, pp. 17–25 (in Russian, English summary).

Sheinkman, V.S., 2002b. Late Pleistocene invasion of Palaeo-Dead Sea into the Lower Zin valley, the Negev Highlands, Israel. EGS Stephan Mueller Special Publication Series 2, 113–122.

Sheinkman, V.S., 2002c. Testing the S-S technique of TL dating on the Dead Sea sections, its use in the Altai Mountains and palaeogeographic interpretation of results. Archeol. Ethnogr. 2 (10), 22–37.

Sheinkman, V.S., 2004. Quaternary Glaciation in the High Mountains of Central and North-east Asia. In: Ehlers, J., Gibbard, P.L. (Eds.), Quaternary Glaciations—Extent and Chronology, Part III: South America, Asia, Africa, Australia, Antarctica. Elsevier, Amsterdam, pp. 325–335.

Sheinkman, V.S., 2007. Peculiarity of Glaciation in the High Mountains of Siberia. Data of Glaciological Studies, No. 102. Russian Academy of sciences, Moscow, pp. 54–64.

Sheinkman, V.S., 2008. Quaternary Glaciation in the Siberian Mountains: General Regularities, Data Analysis: Data of Glaciological Studies, No. 105. Russian Academy of Sciences, Moscow, pp. 51–72 (in Russian, English summary).

Sheinkman, V.S., Antipov, A.N., 2007. Baikal palaeoclimatic record: discussion questions of possible correlation with ancient glaciations of the Siberian Mountains. Geogr. Nat. Res. 1, 5–13 (in Russian, English summary).

Sheinkman, V.S., Barashkova, N.K., 1991. The Pleistocene Glaciation of the Siberian Mountains and atmospheric circulation. Glaciers-Ocean-Atmosphere interaction. IAHS Publication, 208, pp. 415–423.

Sheinkman, V.S., Krivonogov, S.K., 2005. Peculiarities of the Pleistocene Cryogenesis in Mongolia and Altai. Materials of the Third Conference of Cryologists of Russia, vol. 1. Moscow State University, Moscow, pp. 206–213 (in Russian, English summary).

Sheinkman, V.S., Antipov, A.N., Shlukov, A.I., 2009. Absolute dating of the Quaternary complexes: problems and possible solutions. Fundamental Problems of the Quaternary. Russian Academy of Sciences, Siberian Branch, Novosibirsk, pp. 625–628 (in Russian, English summary).

Shesternev, D.M., Sheinkman, V.S., 2008. Glaciers and Cryogenc Ice in the High Mountains of the Kodar Range (Trans-Baikalia) and Modern Climate Change. Data of Glaciological Studies, vol. 105. Russian Academy of Sciences, Moscow, pp. 178–182 (in Russian, English summary).

Shlukov, A.S., Sheinkman, V.S., 2007. Dating the highest Sediments of the Dead Sea Late Pleistocene Precursor by new T-T technique. Quatern. Int. 382, 167–168 XVII INQUA Congress 2007, Cairns, Australia.

Shlukov, A.I., Shakhovets, S.A., Voskovskaya, L.T., Lyashenko, M.G., 1993. A criticism of standard TL dating technology. Nucl. Instrum. Methods Phys. Res. 73, 373–381.

Shlukov, A.I., Usova, M.G., Voskovskaya, L.T., Shakhovets, S.A., 2001. New dating techniques for Quaternary sediments and their application on the Russian Plain. Quatern. Sci. Rev. 20, 875–878.

Short, D.A., Mengel, J.G., Crowlev, T.J., Hyde, W.T., North, G.R., 1991. Filtering of Milankovich Cycles by Earth's geography. Quatern. Res. 35, 157–173.

Shumskii, P.A., 1955. Foundations of Structural Ice Science. Russian Academy of Sciences, Moscow, 492pp (in Russian, English summary).

Slagoda, E.A., Medvedev, G.I., 2004. Palaeocryogenic formations, stratigraphy and geoarcheology of Quaternary deposits of Near-Baikal Siberia. Cryosphere World VIII (1), 18–28 (in Russian, English summary).

Sochava, V.B., 1974. Geotopology as a Part of the Geosystem Teaching: Topologic Aspects of the Geosystem Teaching. Nauka, Novosibirsk, pp. 3–86 (in Russian, English summary).

Sochava, V.B., 1978. Introduction into the Geosystem Teaching. Nauka, Novosibirsk, 319pp (in Russian, English summary).

Stauch, G., Lemkuhle, F., Frechen, M., 2007. Luminescence chronology from Verkhoyansk Mountains (North-Eastern Siberia). Quatern. Geochronol. 2, 255–259.

Sumgin, M.I., 1927. Soil Permafrost in the USSR. Far East Observatory. Vladivostok, 372pp (in Russian).

Svendsen, J.I., Alexanderson, H., Astakhov, V.I., Demidov, I., Dowdeswelle, J.A., Funder, S., et al., 2004. Late Quaternary ice sheet history of northern Eurasia. Quatern. Sci. Rev. 23, 1229–1271.

Tomirdiaro, S.V., 1980. Loess-Ice Formation of East Siberia in the Late Pleistocene and Holocene. Nauka, Moscow, 184pp (in Russian, English summary).

Tong, B., 1993. Ice wedges in northeastern China. In: Proceedings of the 6th International Conference on Permafrost, vol. 1, Guangzhou, Southern China University of Technology Press, 617–621.

Tronov, M.V., 1949. Features of Glaciation of the Altai. Geografgiz, Moscow, 376pp (in Russian).

Vas'kovskii, A.P., 1963. Features of stratigraphy of anthropogenic deposits of the Utmost North-eastern Asia. Materials on Geology and Minerals of the North-Eastern USSR, Vol. 16, Magadan Publishing Company, Magadan, pp. 24–53 (in Russian).

Vasilchuk, Y.K., 2004. Southern limit of distribution of repeated ice wedges in Eurasia. Kryosphere World VIII (3), 34–51 (in Russian, English summary).

Velichko, A.A., 1981. To the question of sequence and principal structure of climatic rhythms of the Pleistocene. In: Velichko, A.A., Grichuk, V.P. (Eds.), Questions of Palaeogeography of the Pleistocene Glacial and Periglacial Areas. Nauka, Moscow, pp. 220–246 (in Russian, English summary).

Velichko, A.A. (Ed.), 2009. Palaeoclimates and palaeoenvironments of extra-tropical regions of Northern Hemisphere. In: Late Pleistocene—Holocene. Atlas-Monograph GEOPS, Moscow, 120pp (in Russian, English summary).

Vinogradov, O.N. (Ed.), 1966–1981. Catalogue of Glaciers of the USSR. Gidrometeoizdat, Leningrad (in Russian, English summary).

Voeikov, A.I., 1881. Climatic conditions of the present and past glacial phenomena. Proc. Mineral. Soc. Series 2 (Pt. 16), 21–90 St. Petersburg (in Russian).

Vtyurin, B.S., 1975. Underground Ice of the USSR. Nauka, Moscow, 214pp (in Russian, English summary).

Yamskikh, A.F., Yamskikh, A.A., 1999. Dynamics of the Todza ice-dammed lake (Upper Yenisei River, Southern Siberia) during the Late Pleistocene and Holocene. Science Reports of Tohoku University, 7th Series (Geography) 49 (2), 143–259. Sendai, Japan.

Yershov, E. (Ed.), 1988–1989. Geocryology of the USSR, vols. 1–5. Nedra, Moscow (in Russian, English summary).

Yershov, E.D., 2002. General Geocryology. Moscow State University, Moscow, 684pp (in Russian, English summary).

Zamoruev, V.V., 1995. About the main questions of Quaternary glaciation in mountainous regions. Geography and Natural Resources, No. 4. Russian Academy of Sciences, pp. 142–148 (in Russian, English summary).

Late Quaternary Glaciation of Northern Pakistan

Ulrich Kamp[1,*] and Lewis A. Owen[2]

[1]Department of Geography, The University of Montana, Missoula, Montana 59812-5040, USA
[2]Department of Geology, University of Cincinnati, Cincinnati, Ohio 45221-0013, USA
*Correspondence and requests for materials should be addressed to Ulrich Kamp. E-mail: ulrich.kamp@umontana.edu

66.1. INTRODUCTION

The mountains of northern Pakistan constitute one of the most glaciated regions outside of the Polar realms (Fig. 66.1). Yet our knowledge of the timing and extent of glaciation is still in its infancy. This is due largely to logistical and political problems, and in recent years, it has become increasingly more difficult to work in the region because of insurgencies within Pakistan and the adjacent countries. Since the publication of Ehlers and Gibbard (2004) that included the first version of this review (Kamp and Haserodt, 2004), several new studies have been undertaken on the Quaternary glaciation of this region (e.g. Spencer and Owen, 2004; Meiners, 2005; Seong et al., 2007). We review these studies in this contribution, building on the earlier article of Kamp and Haserodt (2004), and those of Derbyshire and Owen (1997) and Kamp and Haserodt (2002). We focus on the Late Quaternary glacial history of the Hindu Kush, Karakoram and Himalaya in northern Pakistan to highlight the new insights into our understanding of nature and dynamics of glaciation in this remote and inaccessible part of the world.

In their review, Kamp and Haserodt (2004) described the Quaternary glaciation systematically by region. We retain this structure to provide a comprehensive overview of the region. In particular, Kamp and Haserodt (2004) highlighted how many studies misused climatostratigraphical terms such as Last Glacial Maximum (LGM) when assigning ages to glacial successions and how many of these studies failed to provide adequate documentation of key data essential for assessing and recalculating indices such as equilibrium line altitudes (ELAs). Other reviews on Quaternary glaciation that include northern Pakistan and the adjacent ranges in the Himalayan–Tibetan orogen also touched on these issues including the reviews of Derbyshire and Owen (1997), Owen et al. (1998, 2002a, 2008) and Lehmkuhl and Owen (2005). Other significant reviews on the glacial geology of the Himalaya, which are particularly relevant to the understanding of landscape evolution and palaeoenvironmental change in northern Pakistan, include those of Benn and Owen (1998) on the role of the south Asian summer monsoon and the mid-latitude westerlies in Himalayan glaciation, Benn and Owen (2002) on Himalayan glacial sedimentary systems, Benn and Lehmkuhl (2000) and Owen and Benn (2005) on mass-balance and ELAs for Himalayan glaciers.

Study of the Quaternary glaciation in Pakistan has a long history. Of particular note is the early work of Drew (1875) in the Indus Valley and Dainelli (1922, 1934, 1935) who attempted to correlate the glaciation in the Karakoram with Penck and Bruckner's (1909) fourfold glaciation scheme for the European Alps. Cotter (1929), Visser (1928), Trinkler (1932), Misch (1935) and Troll (1938) extended these studies into the Indus Valley region. The first modern studies were undertaken by Schneider (1959, 1969) in the Karakoram and by Porter (1970) in Swat. Since the 1980s, there has been a plethora of studies, firstly with the work by the International Karakoram Project in Hunza (Miller, 1984) and then with increased interest in the Nanga Parbat syntaxis and the possible role of glaciation and fluvial erosion in driving and/or controlling topography (Brozovic et al., 1997; Zeitler et al., 2001).

Over the past decade, there has been much focus on applying newly developing dating techniques such as optically stimulated luminescence (OSL) and terrestrial cosmogenic nuclide (TCN) surface exposure methods to define the timing of glaciation throughout the region (Phillips et al., 2000; Richards et al., 2000; Owen et al., 2002b,c; Spencer and Owen, 2004; Seong et al., 2007).

These techniques have greatly improved the glacial chronologies for the region, which were sparse because of the lack of radiocarbon ages due to the paucity of organic matter in glacial and associated sediment in high mountain environments. One of the key aims of these recent studies has been to test the hypothesis of Benn and Owen (1998), which argues that glaciation throughout the Himalaya and Tibet may be asynchronous because of the differing spatial and temporal variations of the two major climatic systems, the mid-latitude westerlies and the south Asian monsoon, that drive glaciation in the region. Moreover, once the timing and extent of glaciation can be defined, over time periods that span and exceed those covered by radiocarbon dating, the newly developing geomorphic paradigms that suggest glaciers influence uplift essentially by denudational unloading and can control topography can now begin to be tested.

The glaciated mountains of northern Pakistan include the Hindu Kush, Karakoram and Himalaya. These mountains constitute the greatest topography on our planet, with the Central Karakoram, focused around K2—the world's second highest mountain, containing the greatest concentration of 7000 and 8000 m peaks on our planet. The mid-latitude westerlies and the south Asian monsoon that dominate the climate vary on Milankovitch timescales (thousands of years) linked to changes in Northern Hemisphere insolation and shorter-term variability (years to decades) attributed to changes within the climate system such as variation in Eurasian snow cover, the El Nino–Southern Oscillation and tropical sea surface temperatures (Benn and Owen, 1998). Most of the region experiences a pronounced summer precipitation maximum, reflecting moisture advected northwards from the Indian Ocean by the south Asian summer monsoon. Summer precipitation declines sharply from south to north across the Greater Himalaya into the Karakoram and Hindu Kush. There is, however, a winter precipitation maximum in the extreme west, in the Hindu Kush, influenced by winter westerly winds that bring moisture from the Mediterranean, Black and Caspian seas. The seasonal distribution of precipitation creates mostly glaciers of the summer accumulation type, with maxima in accumulation and ablation occurring more or less simultaneously during the summer (Benn and

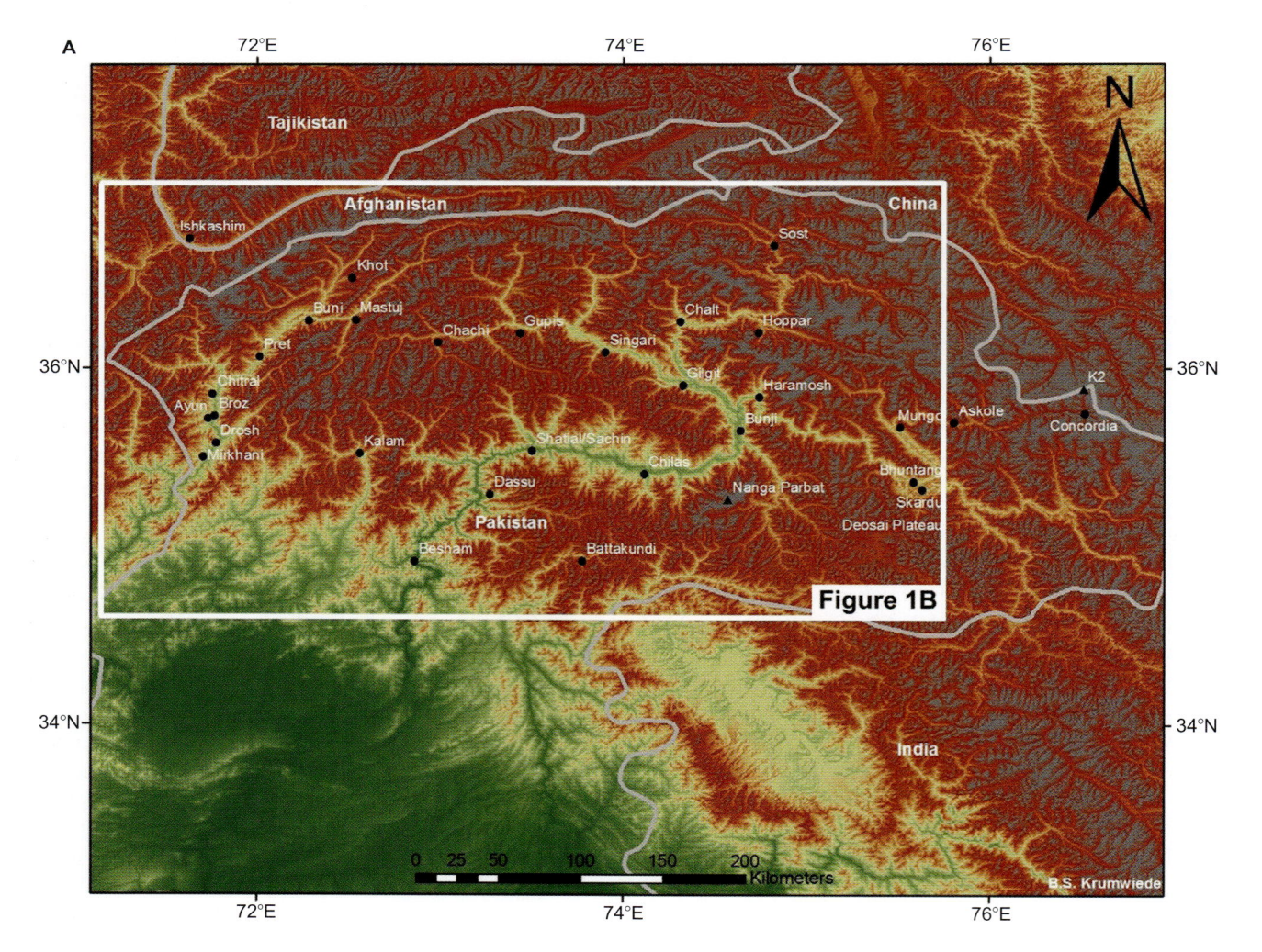

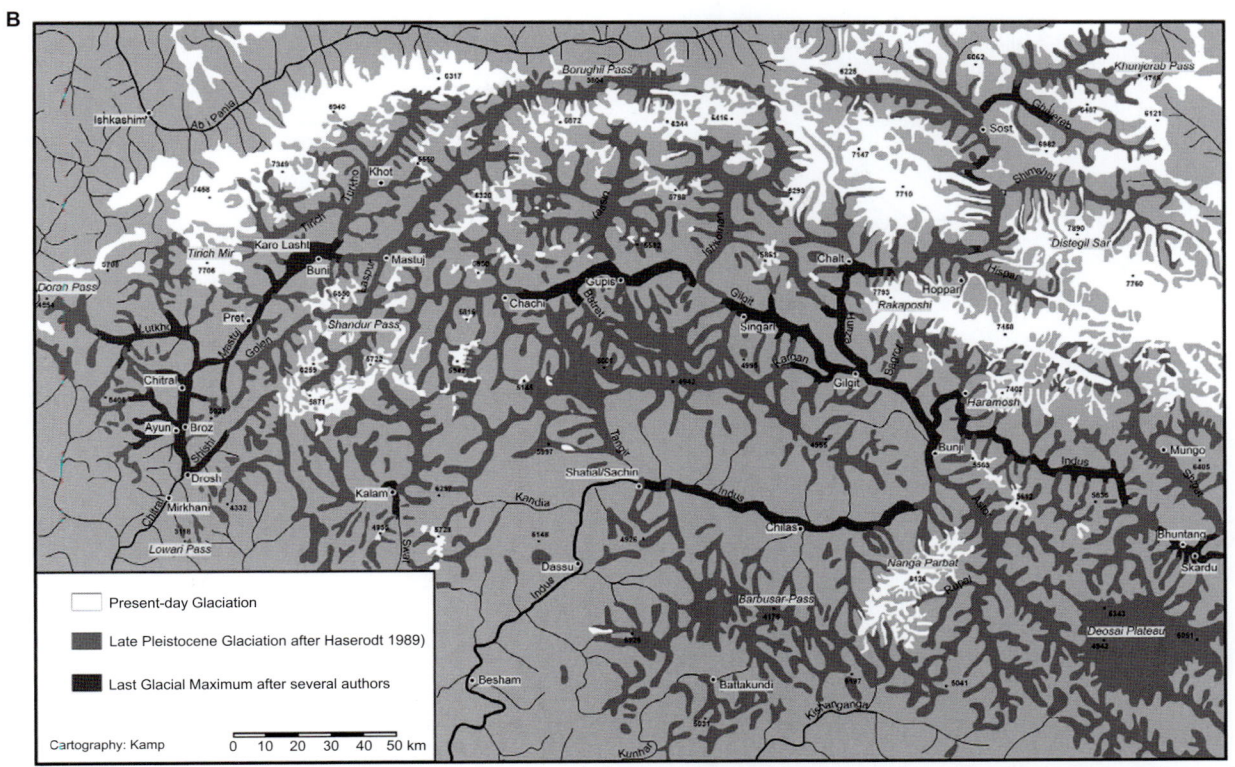

FIGURE 66.1 (A) Locations of main study areas in northern Pakistan. (B) Locations of main study areas in the high mountains of northern Pakistan and the extent of the present and former glaciation in the northern Pakistan mountains (except Central Karakoram). Adapted from Haserodt (1989) and Kamp (1999).

Owen, 1998). Understanding the variations in the two major climatic systems and appreciating the importance of the monsoon seasonality in controlling glaciation are essential for understanding and modelling the nature of Quaternary glaciation in the region (Rupper et al., 2009).

66.2. HINDU KUSH

66.2.1. Chitral

The Quaternary glaciation of Chitral was first described by Haserodt (1989; Fig. 66.2). Using a combination of remote sensing, field mapping and OSL dating, Kamp (1999, 2001a,b) built on the work of Haserodt (1989) to develop the first comprehensive glacial history for Chitral. The OSL ages presented in Kamp (1999, 2001a,b) were later refined to produce a revised chronology for the region (Owen et al., 2002b; Kamp et al., 2004). These studies presented evidence for multiple valley glaciation in the region, which strongly controlled the landscape development, specifically producing thick valley fills and extensive alluvial fans during times of deglaciation.

Evidence for three major glaciations is present along the Chitral Valley. The oldest, the penultimate glacial stage is preserved only at a few locations at very high altitudes. In contrast, there is abundant evidence for the Drosh glacial stage throughout the region, and this represents the local LGM, which occurred during marine isotope stage (MIS) 3. During the next glaciation, the Drosh glacial stage, Chitral was covered by a wide ice-stream network including a trunk glacier that was $\gg 500$ m thick and >270 km long. This glacier advanced to the village of Drosh at 1300 m a.s.l. During deglaciation, extensive alluvial/debris flow fans formed by paraglacial processes, constituting the Broz Formation (upper terrace). Incision and deposition of mainly fluvial sediments led to the development of the Ayun Formation (middle terrace). The next glaciation, the Pret glacial stage (occurring during MIS 2/Early Holocene), was characterized by a trunk glacier that was only ~ 500 m thick and ~ 200 km long. This terminated at 1670 m a.s.l. near the village of Pret. Upon deglaciation, extensive alluvial fans formed by paraglacial action, constituting the Urghuch Formation (lower terrace). During the early Holocene (Sonoghar glacial stage), tributary valley glaciers advanced into the Chitral Valley. In the middle to late Holocene, tributary glaciers advanced four times, during the Shandur I and II and the Barum I and II glacial stages.

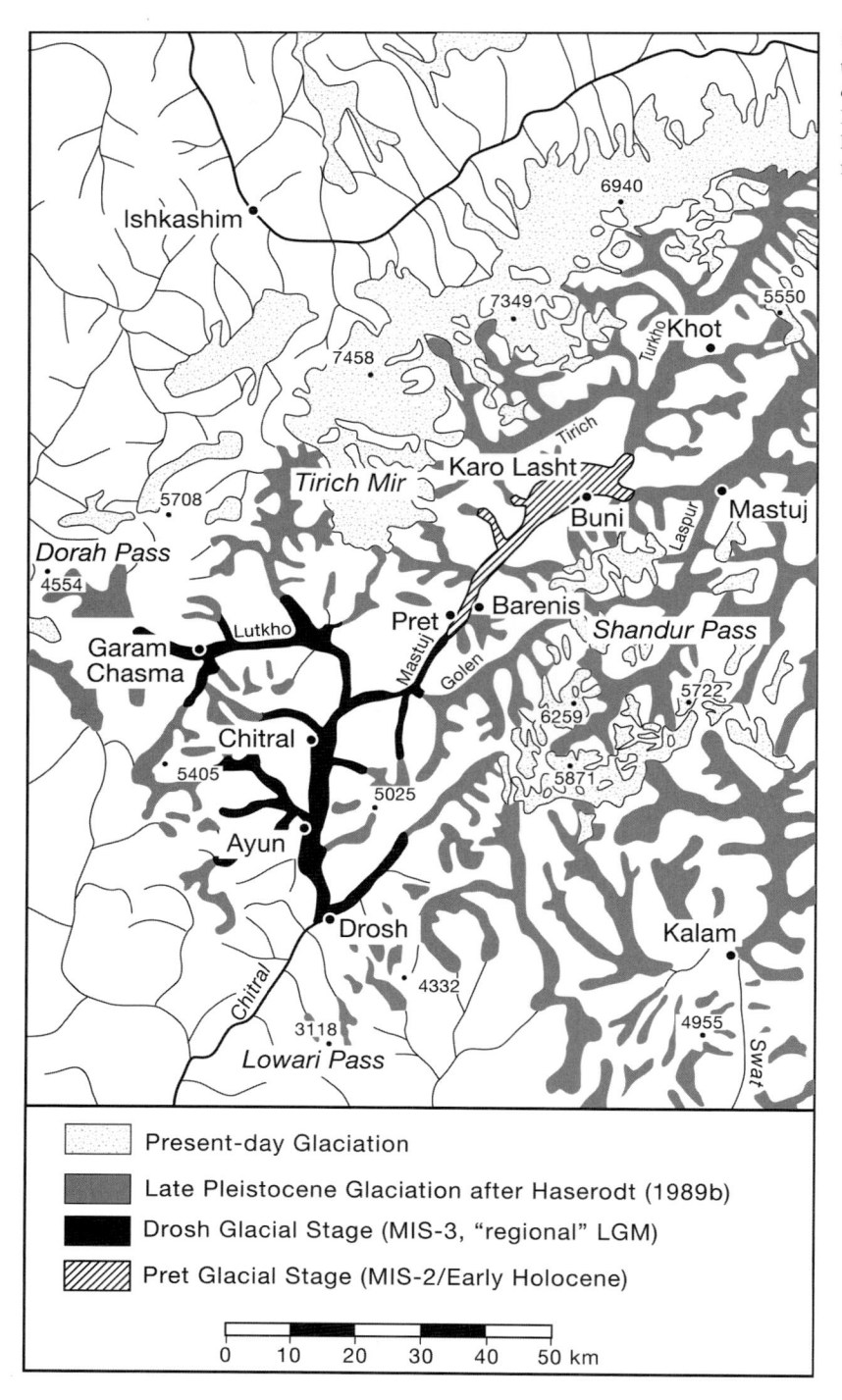

FIGURE 66.2 Reconstruction of the extent of glaciation in Chitral during the Drosh and Pret glacial stages of Kamp (1999, 2001a,b), Owen et al. (2002b) and Kamp et al. (2004a,b) compared to the extent of Late Pleistocene Glaciation of Haserodt (1989). Adapted from Owen et al. (2002b) and Kamp et al. (2004a,b).

Kuhle (2001) argued that there is evidence for a much longer and much thicker (~1500 m) valley glacier during the local LGM that terminated at the village of Mirkhani at 1050 m a.s.l. On the basis of this, Kuhle (2001) reconstructed an extensive ice-stream network but did not have any numerical dating to support a last glacial age for his ice-stream network.

66.2.2. Ghizar

The Ghizar region is located east of Chitral and stretches from Gilgit to the Shandur Pass (3700 m a.s.l.), which forms the drainage divide between the Chitral and Gilgit rivers. Haserodt (1989) argued that a relatively restricted glacier network existed in this region during the last glacial and postulated that the Ghizar glacier system advanced from the Shandur Pass to below the village of Chachi at 2800 m a.s.l. If correct, then numerous tributary glaciers would have extended up to 100 km in length and would have advanced into the main Ghizar–Gilgit Valley. In contrast, according to Owen (1989) and Derbyshire and Owen (1990), a trunk glacier extended down the Gilgit Valley to its confluence with the Hunza Valley during the Yunz and Borit Jheel glacial stages (see below for description of these stages). Kuhle (2001) argued for a very extensive ice-stream network in Chitral that spilled over the Shandur Pass into the Ghizar region, which eventually advanced into the Indus Valley to join with glaciers from Nanga Parbat. However, Kuhle (2001) provided no chronological control to define the age of this glaciation and his field evidence is somewhat ambiguous.

66.2.3. Swat-Kohistan

Evidence for three major glaciations is present in the Swat Valley of Kohistan (Porter, 1970; Fig. 66.3). The evidence for the earliest glaciation, the Laikot glacial stage, is only preserved at a few locations, at high altitudes. The penultimate glaciation, Gabral (I–II) glacial stage, is represented by well-developed moraines throughout the Swat Valley. Abundant evidence for the youngest glaciation, the Kalam (I–III) glacial stage, is present throughout the valleys down to 2200 m a.s.l. near the town of Kalam. A number of limited local glacier advances occurred in the Holocene, which are represented by well-formed moraines near the heads of the tributary valleys.

Many of the moraines and their associated terraces are capped with loess, which Porter (1970) showed increased in thickness with the age of the moraine or terrace that they capped. Using thermoluminescence (TL) methods, Owen et al. (1992) dated some of the loess deposits and showed that the age of the Laikot glacial stage was probably early last glacial (MIS 4) and that the Gabral glacial stage likely pre-dated 20 ka and possibly was coincident with the global LGM (Fig. 66.3). In addition, Owen et al. (1992) showed that the Kalam glacial stage was younger than 7 ka and was possibly Neoglacial. However, Richards et al. (2000), using improved OSL methods, re-dated the successions and showed that the glacial stages were older than proposed by Owen et al. (1992), with an age of 77 ka for the Gabral II glacial stage (MIS 4) and 38 ka for the Kalam I glacial stage (MIS 3).

66.3. KARAKORAM
66.3.1. Hunza Valley

The Hunza Valley has a long history of study. This began in the late 1970s as a consequence of the construction of the Karakoram Highway, which links northern Pakistan with China across the Himalaya and Karakoram. Numerous Chinese publications were produced on the glacial geomorphology and history, which culminated in the International Karakoram Project in 1980 involving Pakistani, Chinese and British scientists (Batura Glacier Investigation Group, 1976, 1979; Zhang and Shi, 1980; Miller, 1984).

Derbyshire et al. (1984) presented the first comprehensive study of glacial chronology of the upper Hunza Valley, which was supported by TL and radiocarbon ages (Figs. 66.4 and 66.5). They showed that scattered erratics on high benches represented an extensive glaciation, the Shanoz glacial stage. This glaciation was followed by an extensive valley glaciation, the Yunz glacial stage (>139 ka, MIS 6). Several glacier advances occurred during the last glacial. This included the Borit Jheel glacial stage (65 ka, MIS 4) when a large valley glacier extended down the Hunza Valley into the Gilgit and Indus valleys to at least the mouth of the Astor Valley at 1300 m a.s.l. (Fig. 66.5). This was followed by a more restricted valley glaciation during the Ghulkin I glacial stage (<47 ka, MIS 3). Subsequently, during the Ghulkin II glacial stage (probably MIS 2), tributary valley glaciers advanced into the Hunza Valley. Glaciers advanced several times during the Holocene including the Batura glacial stage (Neoglacial), the Pasu I glacial stage (Little Ice Age, LIA) and the Pasu II glacial stage (historical).

Haserodt (1984) provided documentation of glacial landforms in the Shimshal Valley to the east of the Hunza Valley. Notably, he described historical (<1 ka BP) glacier advances that led to the formation of what he called a typical great lateral moraine near Mulungutti Glacier. Several studies (e.g. Kick, 1985) showed a general glacier retreat since ~ 1910–1920, for example, at Biafo, Chogo-Lungma, Hispar, Minapin and Hoppar glaciers.

Haserodt (1989) also recognized three major glaciations in the Hunza Valley but rejected the view that a valley glacier extended all the way down the Hunza Valley to the town of Gilgit. Rather, Haserodt (1989) argued that moraines near the towns of Chalt, Nager and Khaibar (~ 2700 m a.s.l.) were produced by tributary valley glaciers. In contrast, Kuhle (2005) reconstructed an extensive >1750-m-thick Hunza ice-stream network for the local LGM, when Hunza Glacier joined Gilgit Glacier and contributed to the supply of Indus Glacier. Meiners (2005) supported this view and showed Bar Glacier on the south side of the Greater Karakoram Range was connected to Hunza Glacier.

Owen (1989), Shroder (1984) and Shroder et al. (1989) refined the chronologies and glacier reconstructions of Derbyshire et al. (1984). They argued that Batura Glacier during the Shanoz glacial stage advanced ~ 250 km to Chilas in the Indus Valley and that during the Yunz glacial stage, the glacier extended farther down to Shatial at ~ 850 m a.s.l.

Richards et al. (2000) provided the OSL ages for moraines produced by Indus Glacier at Shatial formed

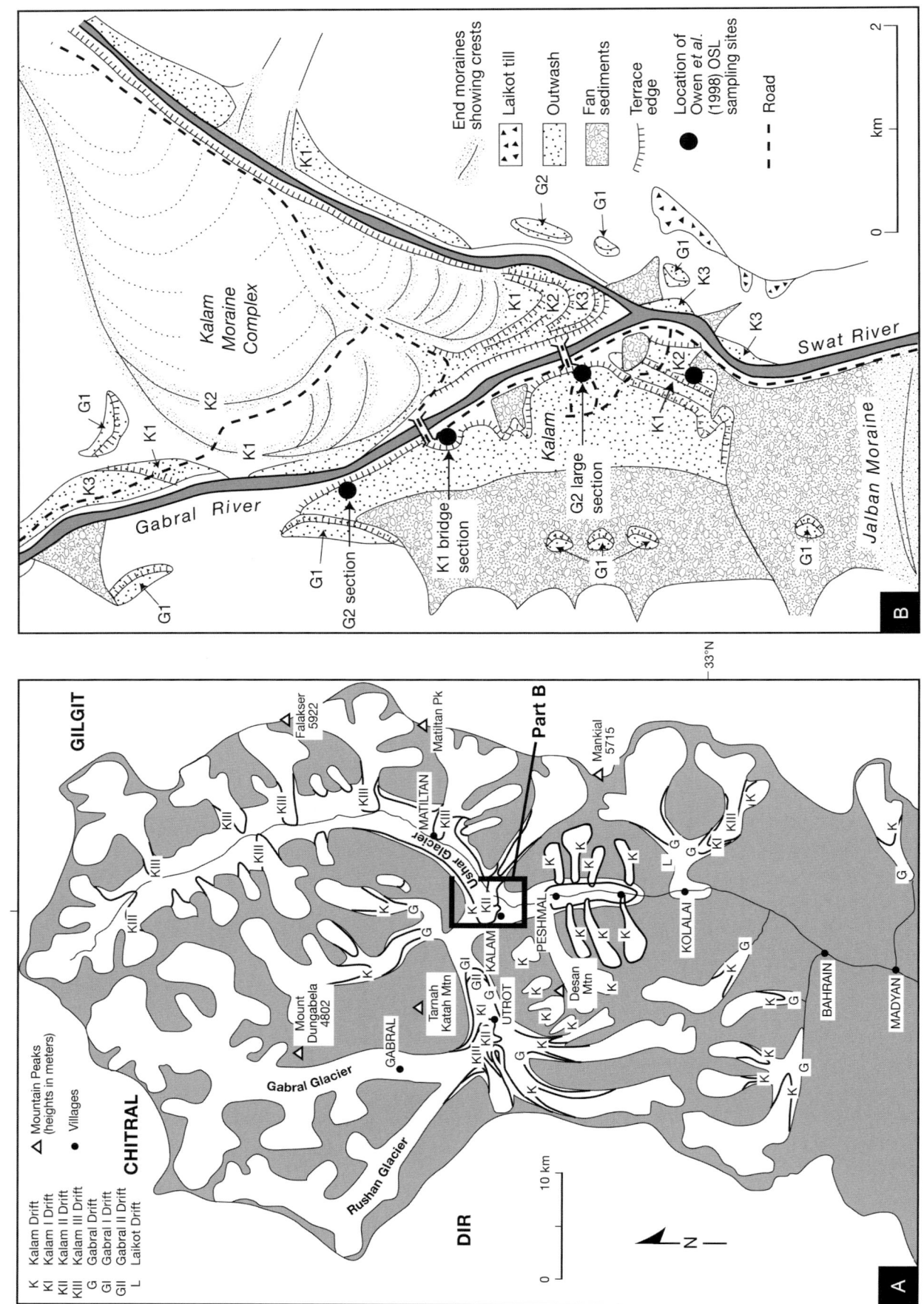

FIGURE 66.3 (A) Reconstruction of former glacier extent in the upper Swat Valley after Porter (1970) and (B) the geomorphology of region around Kalam village showing the locations of sampling sites for the OSL dating of Owen et al. (1998).

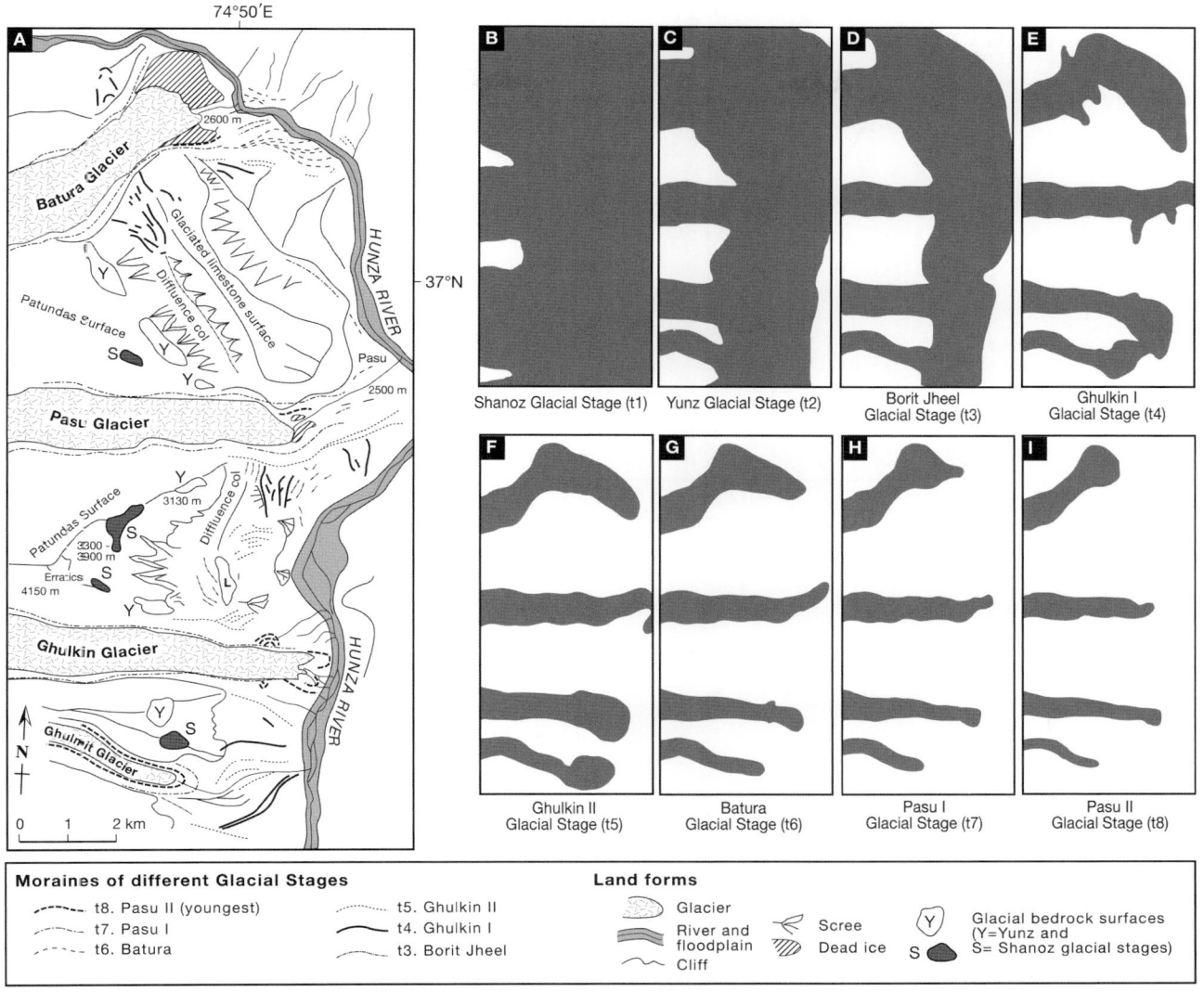

FIGURE 66.4 (A) Distribution of glaciers and moraines in the upper Hunza Valley and (B)–(I) reconstruction of glacier extent for area covered in (A) for different glacial stages. Adapted from Derbyshire et al. (1984) and Owen et al. (2002c).

during the Yunz glacial stage at ~60 ka (MIS 3) and showed that the Borit Jheel glacial stage occurred around 27 ka (MIS 2) with an Indus Glacier that ended at Chilas. Owen (1988a, 1996) examined the landforms and valley fills around the confluence of the Hunza and Gilgit rivers and argued that the evidence showed that glaciers advanced from the Hunza and Gilgit valleys to the confluence at Gilgit during the Yunz glacial stage. However, during the Borit Jheel glacial stage, only Hunza Glacier advanced far enough to reach the confluence (Figs. 66.6 and 66.7). This glacier blocked the Gilgit Valley to form a large glacial lake, which Owen (1988a, 1996) called Glacial Lake Gilgit (Fig. 66.7).

Owen et al. (2002c) presented ages on moraines in the upper Hunza Valley based on TCN dating: ~50 ka (MIS 3) for the Borit Jheel glacial stage, ~23 ka for the Ghulkin I glacial stage (MIS 2), ~16.5 ka for the Ghulkin II glacial stage and ~10 ka for the Batura glacial stage. Spencer and Owen (2004) compared the TCN ages, from boulders on moraine ridges produced by Owen et al. (2002c), with OSL dates from glaciogenic sediments from associated moraines and concluded that both dating techniques gave concordant results.

The understanding and knowledge of the physics of TCN dating have been advancing rapidly over the past few years. Researchers are beginning to appreciate the uncertainty in scaling production rates for TCN, particularly for high-altitude and high-latitude regions such as the Himalaya. Owen et al. (2008), therefore, reassessed the dating of glacial successions throughout the Himalaya

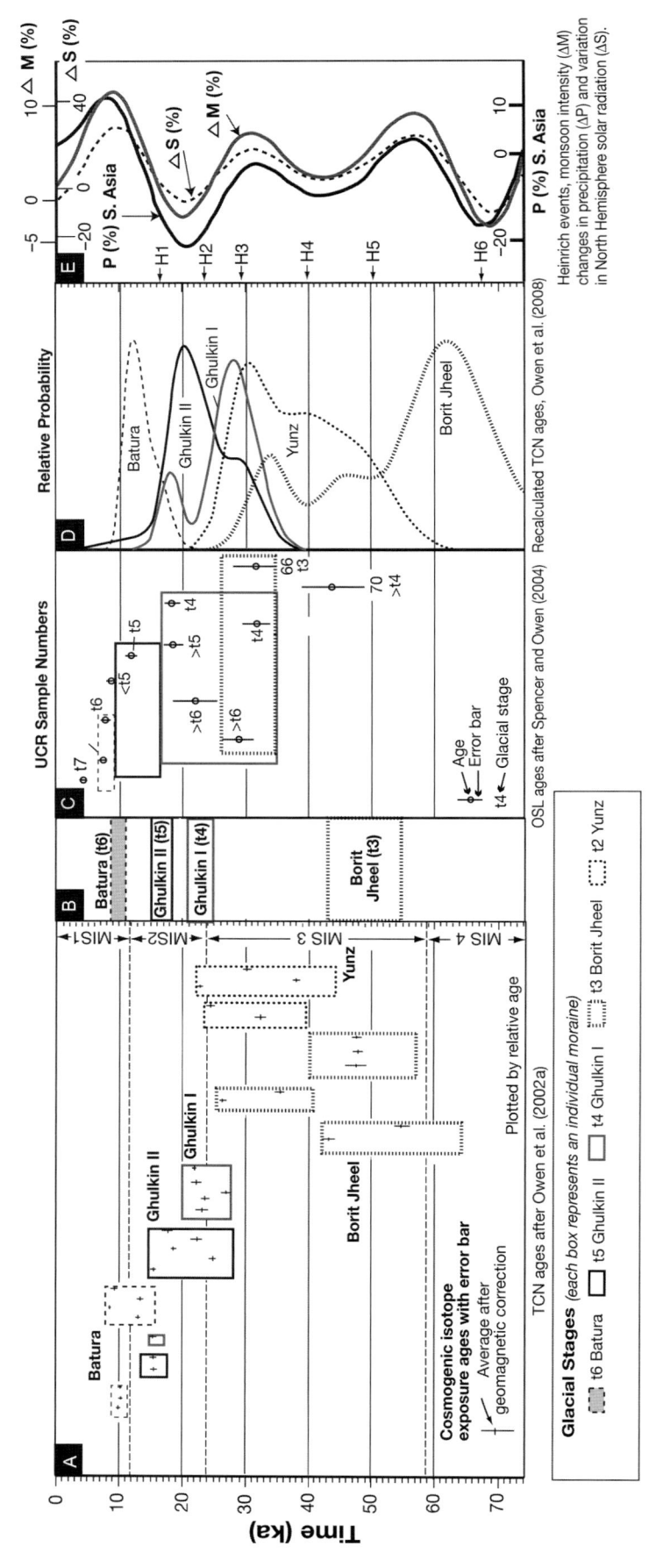

FIGURE 66.5 Glacial chronology for the upper Hunza Valley summarized by Owen et al. (2008), in which they recalculated the ^{10}Be ages from the original publication using new production rates and scaling models. (A) TCN ages for boulders on moraines. The TCN ages are plotted along the x-axis according to their geographical locations and their relative ages. Each box encloses ages for samples from the same moraine. The dashed horizontal lines represent the boundaries of the marine isotope stages (MISs). t_2–t_7 refer to the glacial stages. (B) Timing of glacial stages. (C) OSL ages for glaciogenic sediments associated with moraines (original data from Spencer and Owen, 2004). The relative age is highlighted by '*t*', such that >*t* is for sediment below a moraine, <*t* is for sediment above a moraine and *t* is for sediment within the moraines. (D) Probability density functions of TCN ^{10}Be ages for each glacial stage recalculated by Owen et al. (2008). (E) Simulated monsoon pressure index (ΔM percentage, grey line) for the Indian Ocean and simulated changes in precipitation (ΔP percentage, black line) and variations in Northern Hemisphere solar radiation (ΔS dashed line). H1–H6 show the timing of Heinrich events during the past 70 ka.

FIGURE 66.6 Reconstruction of the extent of glaciation during the Borit Jheel glacial stage according to Derbyshire et al. (1984) and the current extent of glaciation.

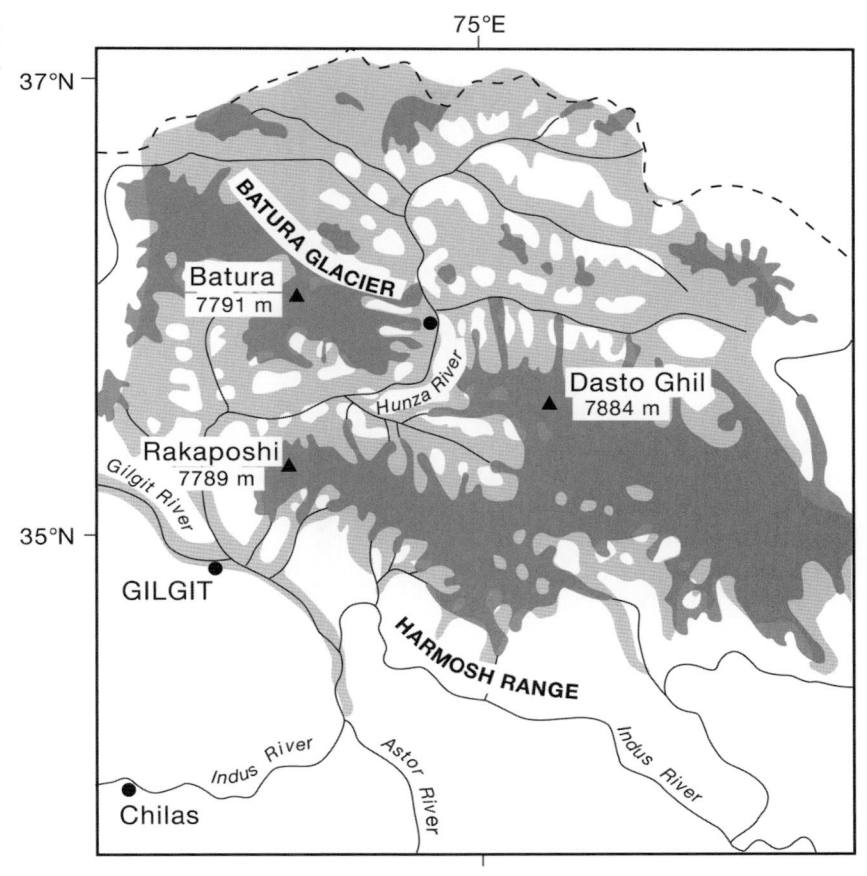

and Tibet. They highlighted the importance of the chronology in the Hunza Valley, showing that it is one of the few areas where both TCN and OSL ages have been obtained on the same glacial systems. In a reassessment of the Hunza data, Owen et al. (2008) showed that recalculating the previously published TCN ages of Owen et al. (2002c) makes the glacial stages a little older, by a few thousand years, as compared to the original data. However, Owen et al. (2008) showed that in combination with the OSL ages presented by Spencer and Owen (2004), the timing of glaciation is similar to that suggested in Owen et al. (2002a) (Fig. 66.5). Owen et al. (2002a) argued that glaciers in Hunza were responding both to changes in insolation-controlled monsoon precipitation that allowed glaciers to advance during the early Holocene and MIS 3 and Northern Hemisphere cooling that allowed glaciers to advance during the global LGM.

66.3.2. Rakaposhi and Haramosh

Rakaposhi and the Haramosh Massif represent a significant mountain mass sandwiched between the Greater Himalaya and Nanga Parbat, and the Karakoram. Evidence for Quaternary glaciation in the Haramosh Massif was first presented by Wiche (1958), who argued that Mani and Baskei glaciers during the last glaciation only advanced by ~3–4 km from their present-day positions. Wiche (1958) believed the large latero-frontal moraines within the valleys of Haramosh were formed during the Late Glacial. In contrast, Haserodt (1989) reconstructed a large glacier that advanced down the Mani–Baskei–Khaltoro valleys and terminated near Sassi at 1500 m a.s.l. In addition, Mani Glacier still reached down to 2100 m a.s.l. during the Late Glacial. Haserodt (1989) regarded the large latero-frontal moraines as Holocene landforms. Owen (1988a) and Meiners (2001) described a >1000-m-thick Late Glacial glacier network south of the Haramosh Massif, which supplied an existing Indus Glacier (Fig. 66.8).

Derbyshire et al. (1984) briefly described some of the Holocene moraines on the northern slopes of Rakaposhi and sited radiocarbon ages for them. Haserodt (1989) described numerous large moraines throughout Rakaposhi, which he attributed to tributary glaciers that advanced during the local LGM. These included moraines produced by Jaglot Glacier that advanced and crossed the Hunza Valley to 1750 m a.s.l. and those formed by Bagrot Glacier that

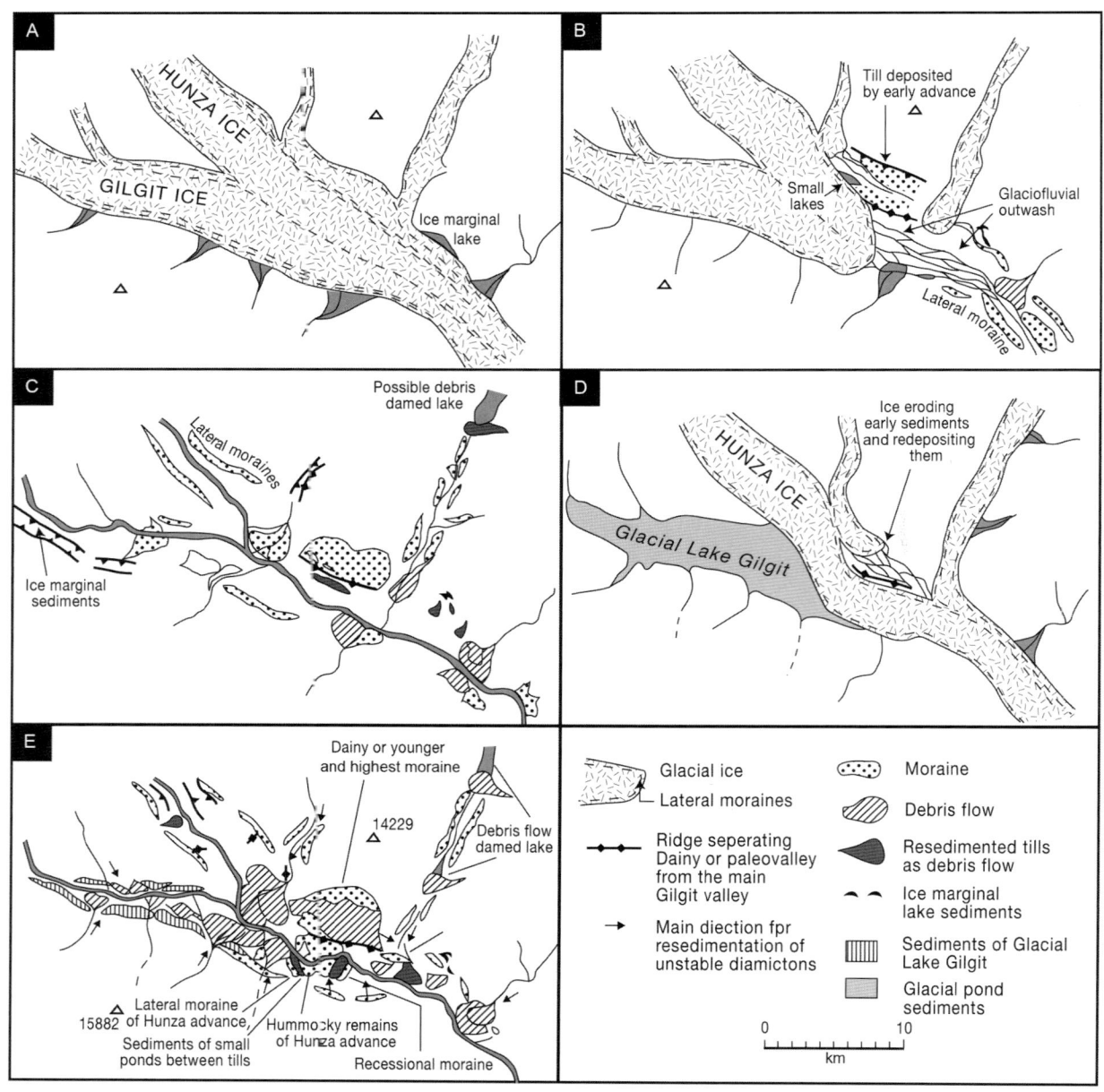

FIGURE 66.7 Sequence of events at the confluence of the Hunza and Gilgit rivers. Adapted from Owen (1988a, 1989).

extended ~25 km to Oshikandaz at 1600 m a.s.l. in the Gilgit Valley. Röthlisberger and Geyh (1985) argued that Burchee Glacier in upper Bagrot Valley advanced between 3.6 and 3.2 ^{14}C ka BP to form distinct moraines, which their lower parts dated to 0.585 and 0.190 ^{14}C ka BP. Haserodt (1989) also presented tree-ring dates at the same moraines that dated to 0.260–0.290 ka, with similar ages on the large moraines associated with Barpu, Jaglot and Hoppar glaciers. Further, Haserodt (1989) argued that smaller moraines at the inner flank of the large latero-frontal moraines likely represented an advance that occurred between 1880 and 1920.

66.3.3. Central Karakoram

The Central Karakoram includes the high mountains focused around K2 (8611 m a.s.l.), Gasherbrum I (8068 m a.s.l.), Broad Peak (8047 m a.s.l.), Gasherbrum II (8035 m a.s.l.) and the main glaciated valleys that drain into the Indus, including the Braldu, Shigar and Hushe. An impressive intermontane basin, the Skardu Basin, is present at the confluence of the Shigar and Indus rivers. This region has had a long history of study, from the early work of Drew (1875) and Dainelli (1922) to Owen (1988b, 1989), Hewitt (1961, 1989, 1998, 1999, 2005), Cronin (1989), Cronin

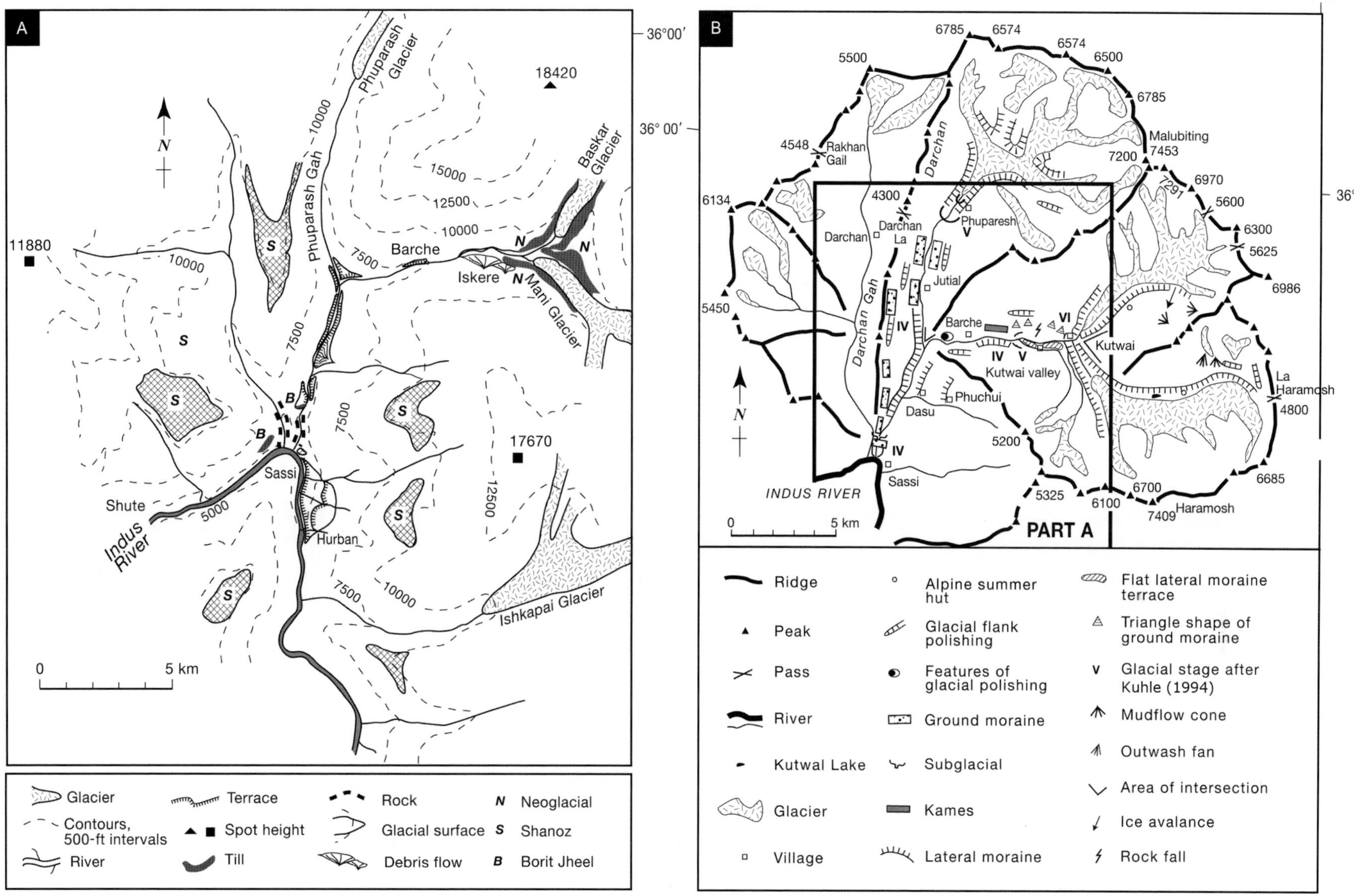

FIGURE 66.8 Geomorphic maps for the Haramosh Valley after (A) Owen (1988a) and (B) Meiners (2001).

et al. (1989), Haserodt (1989), Kuhle (1998, 2004, 2005) and Seong et al. (2007, 2008a,b, 2009a,b).

Drew (1875) and Dainelli (1922) highlighted the presence of glacial landforms, alluvial fans and lake beds along the Indus and its tributaries. The first detailed mapping was undertaken by Cronin (1989) and Owen (1988b, 1989) who focused their studies on the Skardu Basin and the lower reaches of the Shigar Valley (Fig. 66.9). Cronin (1989), Owen (1988b, 1989) and Haserodt (1989) argued that the Indus Valley between the western end of the Skardu Basin and the confluence with the Gilgit Valley was virtually ice free during the last glacial with only tributary glaciers reaching down to the Indus Valley. In contrast, Kuhle (1998, 2004, 2005) reconstructed a 125,000-km^2-large Indus-Karakoram ice-stream network for the last glaciation ($\sim$60–18 ka) and calculated ice thicknesses of 2400 m in the Shigar Valley and $\sim$1500 m in the Skardu Basin. Kuhle (1998, 2004, 2005) argued that the ice-stream network merged into an Indus Glacier that reached down to Sachin at $\sim$800 m a.s.l.

Based on TCN dating of boulders on moraines and glacially eroded surfaces, Seong et al. (2007) developed the first chronostratigraphy based on numerical dating for Quaternary glaciations in the Central Karakoram (Fig. 66.10). Seong et al.'s (2007) glacial chronology included four glacial stages: the Bunthang (>0.7 Ma), the Skardu (MIS 6 or older), the Mungo (MIS 2) and the Askole (Holocene) glacial stages. During each glacial stage, the glaciers advanced several times. These advances are not well defined for the oldest glacial stages, but during the Mungo and Askole glacial stages, glacier advances likely occurred at $\sim$16, $\sim$11–13, $\sim$5 and $\sim$0.8 ka. The extent of glaciation in this region became increasingly more restricted over time. In the Braldu and Shigar valleys, glaciers advanced >150 km during the Bunthang and Skardu glacial stages, whilst glaciers advanced >80 km beyond their present positions during the Mungo glacial stage. In contrast, glaciers advanced a few kilometres from present ice margins during the Askole glacial stage.

Kuhle (2008) argued that the work of Seong et al. (2007) was not valid because the TCN dating method cannot be applied to glacial deposits in high altitudes. However, as Seong et al. (2009a) pointed out, TCN can be applied at high altitudes, and although there is some uncertainty associated with the techniques, it is not great enough to change the interpretations of the ages they presented in their paper. Further, Seong et al. (2009a) expanded on the problems associated with using former ELAs that have been reconstructed from glacial geologic evidence to correlate glacial successions within and across mountain belts. Seong et al. (2009a) argued that this and the misinterpretation of glacial and non-glacial landforms and sediments have lead to the erroneous reconstructions of ice sheets across Tibet, as proposed by Kuhle (1988b, 1995, 2001, 2004, 2006).

The problems associated with the misidentification of glacial and non-glacial landforms and methods to reconcile some of these issues have been discussed by Owen and Derbyshire (1988), Owen (1991) and Hewitt (1998, 1999) specifically for the Karakoram and by Benn and Owen (2002) for the larger Himalayan region. Figure 66.9 illustrates this by highlighting some of the landforms in the Skardu Basin, which have been assigned different origins in different studies.

In the past three decades, there has been much interest in the nature of recent and historical glacier fluctuations (Mayewski and Jeschke, 1979; Mayewski et al., 1980; Goudie et al., 1984; Diolaiuti et al., 2003; Hewitt, 2005; Meiners, 2005; Copland et al., 2009). Hewitt (2005) showed that many glaciers in the highest parts of the Central Karakoram began expanding and thickening in the late 1990s and called this the 'Karakoram Anomaly'. He also found exceptional numbers of glacier surges. Hewitt (2005) explained this anomaly with the 'elevation effect', that an increase in temperatures and precipitation at higher elevations is a result of global climate change. Meiners (2005) described the retreat of Kukuar and Baltar glaciers since 1915 by 8 km. She also described older moraines, which she assigned to the Neoglacial and the last glacial, but does not have any numerical ages to confirm the timing of glaciation. Copland et al. (2009) delivered the first synthesis of glacier velocities across the Central Karakoram based on feature-tracking calculations using field measurements and satellite imagery from 2005 to 2007. Their results showed that all of the glaciers were active, even when debris of often >1 m covered the terminus. Basal sliding was identified as the dominant motion mechanism, and velocities appeared to be influenced by local conditions. Compared with results from Hewitt et al. (1989), Copland et al. (2009) showed that velocities of $\sim$100–150 ma^{-1} at Biafo Glacier had changed little over the past $\sim$20 years. Diolaiuti et al. (2003) described Liligo Glacier, a tributary of Baltoro Glacier, as of surge type and calculated an advance of at least 1450 m between 1986 and 1997. Copland et al. (2009) identified several surge-type glaciers and described terminus advances of several hundred metres per year and the displacement of trunk glaciers as surge-type glaciers pushed into them.

66.4. WESTERN HIMALAYA

66.4.1. Nanga Parbat

Nanga Parbat is one of the most rapidly rising massifs in the Himalayan–Tibetan orogen and as such has attracted considerable attention from the geological community (Zeitler et al., 2001). The earliest studies of the glacial geology of Nanga Parbat were undertaken by Wadia (1932/1933) who examined the valleys from Gilgit to Nanga

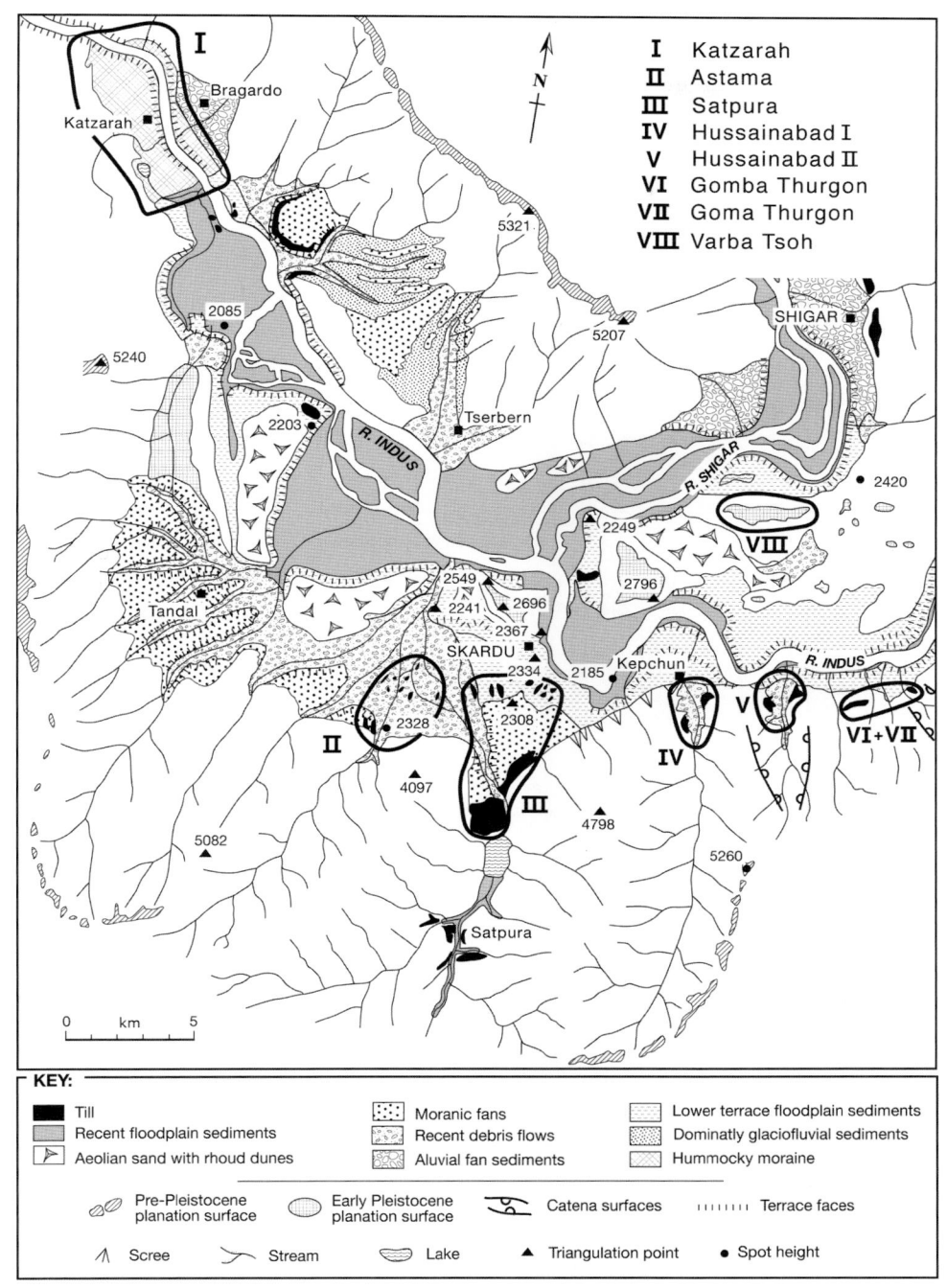

FIGURE 66.9 Geomorphology of the Skardu Basin (after Owen et al., 1989). Hewitt (1998, 1999) argues that some of the mapped moraines are long run-out landslides; these are highlighted by the bold circles.

Parbat and Finsterwalder (1936, 1937) on the massif itself. Modern studies were undertaken by Scott (1992) and Owen et al. (2000) who identified two major glaciations at Nanga Parbat, which they correlated with the Yunz and the Borit Jheel glacial stages in the Hunza Valley, and two advances in larger tributaries such as the Rupal,

Rakhiot and Rama valleys that were correlated to the Ghulkin I and II glacial stages (Fig. 66.11). Scott (1992) and Owen et al. (2000) also described evidence for an early Holocene, Batura glacial stage and two historical glacial stages, Pasu I and II. Scott (1992) and Owen et al. (2000) argued that Astor Glacier did not advance into the Indus

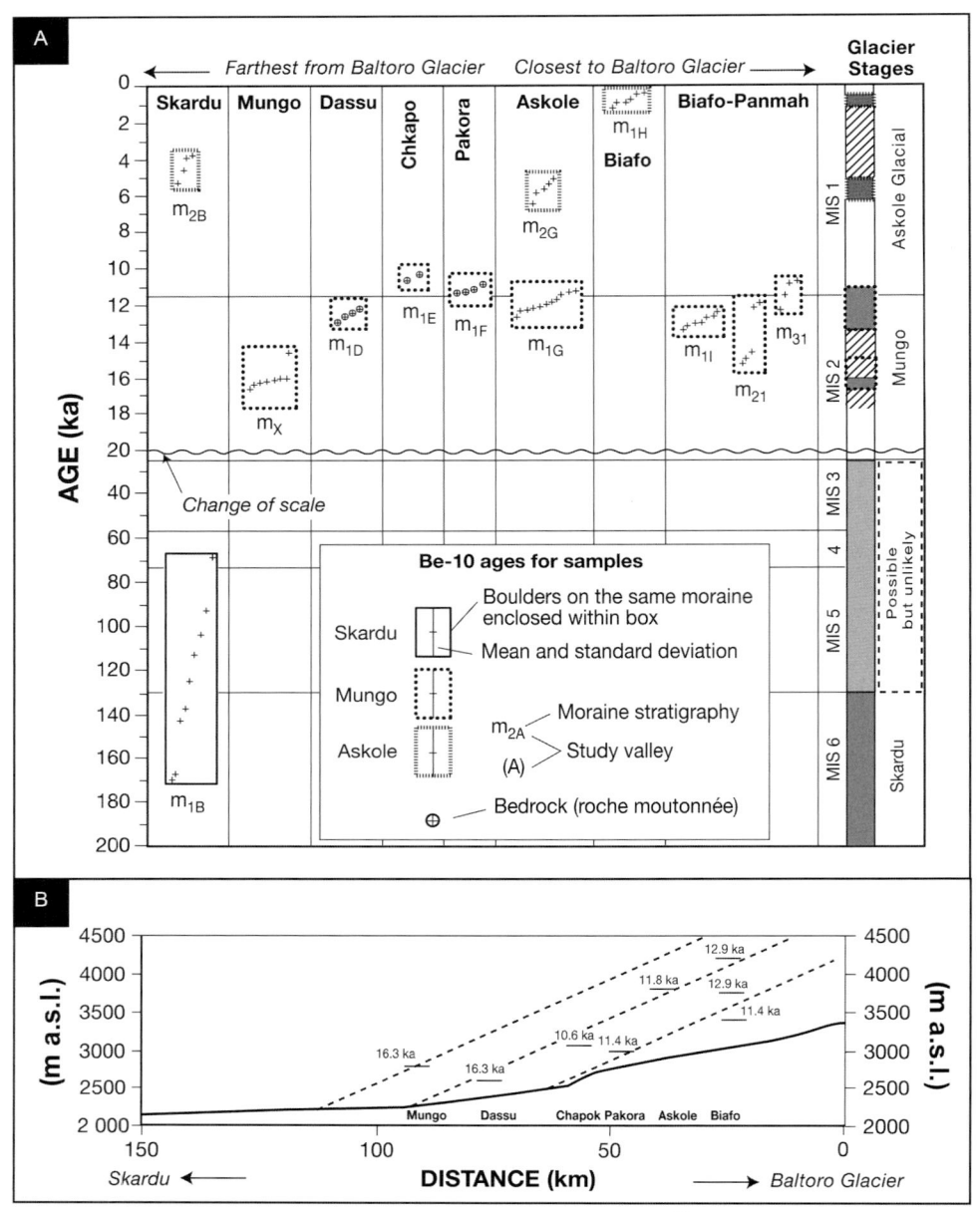

FIGURE 66.10 Chronostratigraphy of the Central Karakoram (after Seong et al., 2007). (A) ^{10}Be TCN surface exposure dates for boulders on the moraines dated from study areas in the Central Karakoram (after Seong et al., 2007). Each box represents one moraine with the data points representing individual ^{10}Be ages. (B) Distribution of moraines formed during Mungo glacial stage in the Central Karakoram (after Seong et al., 2007). The dashed lines are schematic profiles of the main trunk glacier during the Mungo glacial stage. The profiles are based on the present gradient of the Baltoro Glacier and three altitudinally distinct former ice limits are apparent.

Valley, but the Indus Valley was glaciated during the local LGM.

Phillips et al. (2000) used cosmogenic ^{3}He dating and defined the timing of the two last glacial advances to 56 and 35 ka (during MIS 3). Phillips et al. (2000) rejected the idea of a Late Glacial (MIS 2) advance because of regional aridity but identified several Holocene glacier advances. In contrast, Richards et al. (2000) argued on the basis of OSL dating that a glacier existed at the confluence of the Indus and Astor valleys at ~27 ka (MIS 2). Moreover, Owen et al. (2000) showed that after the local LGM, at least two phases of glacial advance or stagnation occurred, and that glaciers advanced at least three times in the Holocene. In contrast, Kuhle (2005) reconstructed an extensive 800–1400-m-thick Late Glacial ice-stream network around Nanga Parbat, but his evidence is ambiguous.

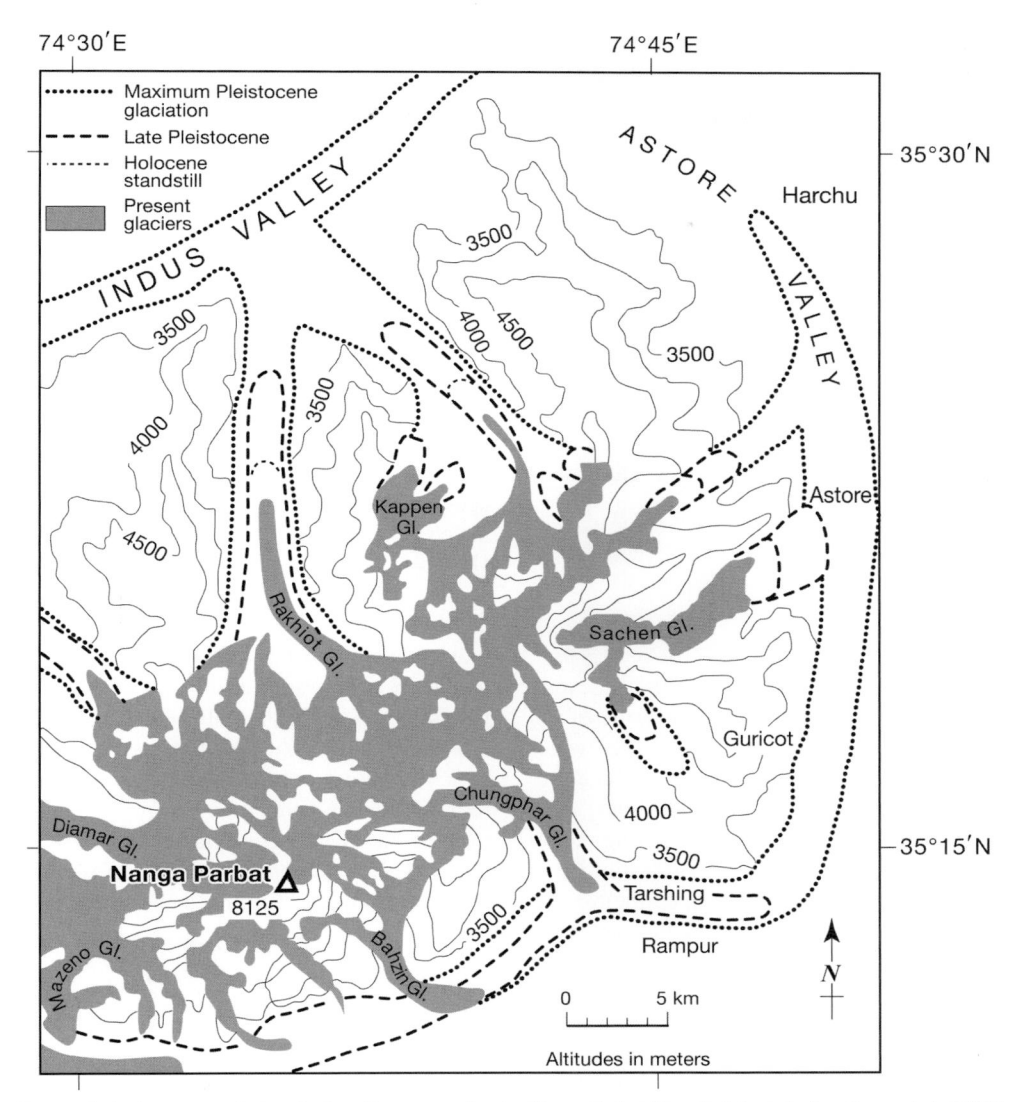

FIGURE 66.11 Proposed glacier extents during the late Quaternary for the Nanga Parbat Massif. Adapted after Owen et al. (2000).

66.4.2. Kaghan

The only glacial geologic work that has been undertaken in Kaghan was by Haserodt (1989). Haserodt (1989) identified evidence for two glacial advances. The older glaciation produced glaciers that advanced to at least Battakundi at 2620 m a.s.l., which Haserodt (1989) correlated with the Wurmian glaciation in Europe. During the younger glaciation, glaciers advanced down the Kaghan Valley to below Battakundi, and tributary glaciers advanced to the town of Kaghan at 2022 m a.s.l.

66.4.3. Deosai Plateau

The Deosai Plateau has not been studied in any great details. However, Haserodt (1989) argued that the Deosai Plateau and its surrounding mountains were extensively glaciated during the last glacial, when 25-km-long Tukshan Glacier advanced into the ice-free Indus Valley. Later, Kuhle (2004) reconstructed an ~24-km-wide ice cap on the Deosai Plateau with outlet glaciers reaching down into the surrounding valleys and connecting the ice cap with the Indus-Karakoram ice-stream network.

66.5. INDUS VALLEY

There is much contention over whether a trunk valley glacier existed in the Indus Valley and, if one did, the timing of its advance down the Indus Valley. Loewe (1924) argued that the Indus Valley was ice free during the last glacial. In contrast, Trinkler (1932) argued that a glacier filled the Indus Valley reaching the foreland for the Himalaya. Later, Desio and Orombelli (1971) proposed that a large glacier existed in the Indus Valley, and Owen (1988a) argued that the Indus Valley had been filled to Shatial at 800–850 m a.s.l. where an impressive latero-frontal moraine complex exists. In contrast, Haserodt (1989) believed that the lower Gilgit and the Indus valleys were ice free and even areas of the lower Hunza Valley had never been glaciated. Haserodt (1989) believed that glacial deposits in the Indus Valley were simply formed by tributary glaciers that advanced from Nanga Parbat. However, the existence of an Indus Glacier is now generally accepted (e.g. Derbyshire et al., 1984; Kuhle, 1988a; Owen, 1989; Shroder et al., 1993; Richards et al., 2000). Richards et al. (2000) provided OSL ages that suggested the Indus Valley was glaciated during MIS 2.

66.6. CONCLUSIONS

The study of the Quaternary glacial geology of the high mountains of northern Pakistan is still in its infancy despite this region being one of the most glaciated outside of the polar realms and the importance of glaciation as the water source for the vast population that lives in and adjacent to the region. Studies in the past few years, however, have highlighted the potential to apply newly developing dating techniques such as TCN and OSL to define the timing of glaciation, and remote sensing to help map and monitor glacier changes. However, these new techniques are time intensive and very costly. Complete study and coverage of the entire region is therefore unlikely to be undertaken in the coming years. Moreover, ground-truth and sample collection is difficult at present, greatly hindered by the dangerous political environment. Remote sensing, however, has the potential to be used to examine very recent glacier fluctuations, which is particularly important for predicting the likely future changes that the region will experience as the world's climate continues to warm due to human activity. In addition, digital elevation modelling, computer modelling and simulations are helping to analyse mountain topography and its impact on glaciation, and vice versa (e.g. Bishop et al., 2001, 2002, 2003; Bolch et al. 2005; Bolch and Kamp, 2006). This will lead on to improved use of remote sensing methods to define the nature, style and timing of glaciation in this inaccessible and remote region.

The latest glacial geologic studies confirm the summaries in earlier reviews, including those of Derbyshire and Owen (1997), Kamp and Haserodt (2002, 2004), Owen et al. (2002a, 2008), Lehmkuhl and Owen (2005), and Owen (2009). The new studies during the past few years have shown that glaciation in this region was more extensive prior to the last glacial cycle and that the local LGM in the region occurred early in the last glacial cycle and was not coincident with the global LGM. The timing of the local LGM is not well defined, but it is likely this occurred in most regions during MIS 3, a time of higher low latitude insolation and increased monsoon intensity. The increased precipitation falling as snow at high altitudes and increased cloudiness result in positive glacier mass balances (Owen et al., 2002c, 2005; Rupper et al., 2009). The new studies are showing that significant early Holocene glacier advances, during any time of increased insolation and monsoon enhancement, have occurred in most areas.

The glacial geologic record also shows, although not well defined, multiple glacial advances during the latter part of the Holocene and the Neoglacial. Glaciers likely advanced during the LIA, but few studies have numerical ages on moraines attributed to the LIA. In general, glaciers have retreated since the beginning of the twentieth century. However, in some regions, recent studies are beginning to show that certain glaciers have begun to stabilize and/or advance during the past few years (Bagla, 2009). This may be analogous to the growth of glaciers under enhanced monsoon condition during time of increased low latitude insolation. The past century has experienced increased monsoon precipitation in Pakistan (Treydte et al., 2006), which would likely enhance glacier growth. Improved remote sensing techniques promise to provide new insights into the nature of recent glacier fluctuations, which in turn should help in the understanding of the importance of the different climatic forcing factors for glaciation in this region.

ACKNOWLEDGEMENTS

The authors thank Brandon Krumwiede for comments on this chapter and drafting Fig. 66.1A, and Tim Phillips for drafting Figs. 66.2–66.11.

REFERENCES

Bagla, P., 2009. No sign yet of Himalayan meltdown, Indian report finds. Science 326, 924–925.

Batura Glacier Investigation Group, 1979. The Batura Glacier in the Karakoram Mountains and its variations. Sci. Sinica 22, 959–974.

Batura Glacier Investigation Group, 1976. Investigation Report on the Batura Glacier in the Karakoram Mountains, the Islamic Republic of Pakistan (1974–1975). Batura Investigation Group, Engineering Headquarters, Peking, 123 pp.

Benn, D., Lehmkuhl, F., 2000. Mass balance and equilibrium-line altitudes of glaciers in high-mountain environments. Quatern. Int. 65 (66), 15–29.

Benn, D., Owen, L.A., 1998. The role of the Indian summer monsoon and the mid-latitude westerlies in Himalayan glaciation: review and speculative discussion. J. Geol. Soc. 155, 353–363.

Benn, D., Owen, L.A., 2002. Himalayan glacial sedimentary environments: a framework for reconstructing and dating the former extent of glaciers in high mountains. Quatern. Int. 97 (98), 3–25.

Bishop, M.P., Bonk, R., Kamp, U., Shroder, J.F., 2001. Terrain analysis and data modeling for alpine glacier mapping. Polar Geogr. 25, 182–201.

Bishop, M.P., Shroder, J.F., Bonk, R., Olsenholler, J., 2002. Geomorphic change in high mountains: a western Himalayan perspective. Glob. Planet. Change 32, 311–329.

Bishop. M.P., Shroder, J.F., Colby, J.D., 2003. Remote sensing and geomorphometry for studying relief production in high mountains. Geomorphology 55, 345–361.

Bolch, T., Kamp, U., 2006. Glacier mapping in high mountains using DEMs, Landsat and ASTER data. In: Proceedings of the eighth International Symposium on High Mountain Remote Sensing Cartography (HMRSC) March 20–27, 2005, La Paz, Bolivia, Grazer Schriften der Geographie und Raumforschung, 41, Graz, 13–24.

Bolch, T., Kamp, U., Olsenholler, J., 2005. Using ASTER and SRTM DEMs for studying geomorphology and glaciers in high mountain areas. In: Oluic, M. (Ed.), New Strategies for European Remote Sensing, Proceedings of the 24th Annual Symposium European Association of Remote Sensing Laboratories (EARSeL), May 25–27, 2004, Dubrovnik, Croatia. Millpress, Rotterdam, pp. 119–127.

Brozovic, N., Burbank, D.W., Meigs, A.J., 1997. Climatic limits on landscape development in the northwestern Himalaya. Science 276, 571–574.

Copland, L., Pope, S., Bishop, M.P., Shroder, J.F., Clendon, P., Bush, A., et al., 2009. Glacier velocities across the central Karakoram. Ann. Glaciol. 50, 41–49.

Cronin, V.S., 1989. Structural setting of the Skardu intermontane basin, Karakoram Himalaya, Pakistan. Geol. Soc. Am. Spec. Pap. 232, 183–201.

Cronin, V.S., Johnson, W.P., Johnson, N.M., Johnson, G.D., 1989. Chronostratigraphy of the upper Cenozoic Bunthang sequence and possible mechanisms controlling base level in Skardu intermontane basin, Karakoram Himalaya, Pakistan. Geol. Soc. Am. Spec. Pap. 232, 295–309.

Dainelli, G., 1922. Studi sul glaciale. Spedizione Italiane de Filippi nell' Himalaia, Caracorum e Turchestan Cinese (1913–1914). Seriae II 5.3.

Dainelli, G., 1934. Studi sul Glaciale: Spedizone Italian de Filippi nel'Himalaia, Caracorum e Turchestan Cinese (1913–1914). Ser. II., 3: Zanichelli, Bologna, 658 pp.

Dainelli, G., 1935. La Series dei Terreni: Spedizione Ilaliana De Fillippi nell'Himalaia, Caracorum e Turchestan Cinese (1913–1914). Ser. II, 2: Zanichelli, Bologna, 230 pp.

de Cotter, G., 1929. The erratics of the Punjab. Rec. Geol. Sur. India 61, 327–336.

Derbyshire, E., Owen, L.A., 1990. Quaternary alluvial fans in the Karakoram Mountains. In: Rachocki, A.H., Church, M. (Eds.), Alluvial Fans: A Field Approach. John Wiley and Sons, Chichester, pp. 27–53.

Derbyshire, E., Owen, L.A., 1997. Quaternary glacial history of the Karakoram mountains and northwest Himalayas: a review. Quatern. Int. 38 (39), 85–102.

Derbyshire, E., Li, J., Perrot, F.A., Shuying, Xu., Waters, R.S., 1984. Quaternary glacial history of the Hunza Valley, Karakoram Mountains, Pakistan. In: Miller, K.J. (Ed.), The International Karakoram Project 2. Cambridge University Press, Cambridge, pp. 456–495.

Desio, A., Orombelli, G., 1971. Preliminary note on the presence of a large valley glacier in the middle Indus Valley (Pakistan) during the Pleistocene. Atti della Accademia Nazionale dei Lincei 51, 387–392.

Diolaiuti, G., Pecci, M., Smiraglia, C., 2003. Liligo Glacier, Karakoram, Pakistan: a reconstruction of the recent history of a surge-type glacier. Ann. Glaciol. 36, 168–172.

Drew, F., 1875. The Jummoo and Kashmir Territories: A Geographical Account. Indus Publications, Karachi, Pakistan.

Ehlers, J., Gibbard, P. (Eds.), 2004. Quaternary Glaciations, Extent and Chronology, Part III: South America, Asia, Africa, Australasia, Antarctica. In: Development in Quaternary Science, 2c Elsevier, Amsterdam, 380 pp.

Finsterwalder, R., 1936. Die Formen der Nanga-Parbat-Gruppe. Zeit. Gesell. Erdkunde Berlin 71, 321–341.

Finsterwalder, R., 1937. Die Gletscher des Nanga Parbat. Glaziologische Arbeiten der Deutschen Himalayas-Expedition 1934 und ihre Ergebnisse. Zeit. Glet. 25, 57–108.

Goudie, A.S., Jones, D.K.C., Brunsden, D., 1984. Recent fluctuations in some glaciers of the western Karakoram mountains, Pakistan. In: Miller, K.J. (Ed.), The International Karakoram Project, 2. Cambridge University Press, Cambridge, pp. 411–455.

Haserodt, K., 1984. Aspects of the actual climatic conditions and historic fluctuations of glaciers in Western Karakoram. J. Cent. Asia 7, 77–94.

Haserodt, K., 1989. Zur pleistozänen und postglazialen Vergletscherung zwischen Hindukusch, Karakorum und Westhimalaya. In: Haserodt, K. (Ed.), Hochgebirgsräume Nordpakistans im Hindukusch Karakorum und Westhimalaya. Beiträge und Materialien zur Regionalen Geographie, 2 Technical University Berlin, Berlin, pp. 181–233.

Hewitt, K., 1961. Karakoram glaciers and the Indus. Indus 2, 4–14.

Hewitt, K., Wake, C.P., Young, G.J., David, C., 1989. Hydrological investigations at Biafo Glacier, Karakoram Himalaya, Pakistan: an important source of water for the Indus River. Ann. Glaciol. 13, 103–108.

Hewitt, K., 1989. The altitudinal organisation of Karakoram geomorphic processes and depositional environments. Z. Geomorphol. N. F. Suppl.-Bd 76, 9–32.

Hewitt, K., 1998. Catastrophic landslides and their effects on the Upper Indus streams, Karakoram Himalaya, northern Pakistan. Geomorphology 26, 47–80.

Hewitt, K., 1999. Quaternary moraines vs. catastrophic avalanches in the Karakoram Himalaya, northern Pakistan. Quatern. Res. 51, 220–237.

Hewitt, K., 2005. The Karakoram anomaly? Glacier expansion and the 'elevation effect', Karakoram Himalaya. Mt. Res. Dev. 25, 332–340.

Kamp, U., 1999. Jungquartäre Geomorphologie und Vergletscherung im östlichen Hindukusch, Chitral, Nordpakistan. Berliner Geographische Studien, 50 Technical University Berlin, Berlin, 254 pp.

Kamp, U., 2001a. Die jungquartäre Vergletscherung Chitrals im östlichen Hindukusch, Pakistan. Z. Gletscherk. Glazialgeol. 36, 81–106.

Kamp, U., 2001b. Jungquartäre Terrassen und Talentwicklung in Chitral, östlicher Hindukusch. Z. Geomorphol. 45, 453–475.

Kamp, U., Haserodt, K., 2002. Quartäre Vergletscherung im Hindukusch, Karakorum und West-Himalaya, Pakistan: ein Überblick. E&G Quatern. Sci. J. 51, 93–113.

Kamp, U., Haserodt, K., 2004. Quaternary glaciations in the high mountains of northern Pakistan. In: Ehlers, J.J., Gibbard, P.L. (Eds.), Quaternary Glaciations Extent and Chronology, Part III: South America, Asia, Africa, Australasia, Antarctica. Development in Quaternary Science, 2c Elsevier, Amsterdam, pp. 293–311.

Kamp, U., Haserodt, K., Shroder, J.F., 2004. Quaternary landscape evolution in the eastern Hindu Kush, Pakistan. Geomorphology 57, 1–27.

Kick, W., 1985. Geomorphologie und rezente Gletscheränderungen in Hochasien. In: Hartl, W., Engelschalk, W. (Eds.), Geographie,

Naturwissenschaft und Geisteswissenschaft, Regensburger Geographische Schriften, 19/20, pp. 53–77, Institut für Geographie, Regensburg.

Kuhle, M., 1988a. The Pleistocene glaciation of Tibet and the onset of ice ages: an autocycle hypothesis. GeoJournal 17, 581–596.

Kuhle, M., 1988b. Geomorphological findings on the built-up of Pleistocene glaciation in southern Tibet and on the problem of inland ice. GeoJournal 17, 457–512.

Kuhle, M., 1995. Glacial isostatic uplift of Tibet as a consequence of a former ice sheet. GeoJournal 37, 431–449.

Kuhle, M., 1998. Reconstruction of the 2.4 million km^2 Late Pleistocene ice sheet on the Tibetean plateau and its impact on the global climate. Quatern. Int. 45 (46), 71–108.

Kuhle, M., 2001. The maximum ice age (LGM) glaciation of the Central- and South Karakorum: an investigation of the heights of its glacier levels and ice thicknesses as well as lowest prehistoric ice margin positions in the Hindukush, Himalaya and in East-Tibet on the Minya Konka-massif. GeoJournal 54, 109–396.

Kuhle, M., 2004. The Pleistocene glaciation in the Karakoram-mountains: reconstruction of past glacier extensions and ice thicknesses. J. Mt. Sci. 1, 3–17.

Kuhle, M., 2005. Glacial geomorphology and ice ages in Tibet and the surrounding mountains. Island Arc 14, 346–367.

Kuhle, M., 2006. The past Hunza glacier in connection with a Pleistocene Karakorum ice stream network during the Last Ice Age (Würm). In: Kreutzmann, H. (Ed.), Karakoramin Transition—Culture Development and Ecology in the Hunza Valley. Oxford University Press, Karachi, Pakistan, pp. 24–48.

Kuhle, M., 2008. Correspondence on Quaternary glacier history of the central Karakarum by Yeong Bae Seong, Lewis A. Owen, Michael P. Bishop, Andrew Bush, Penny Clendon, Luke Copland, Robert Finkel, Ulrich Kamp and John F. Shroder Jr.. Quatern. Sci. Rev. 27, 1655–1656.

Lehmkuhl, F., Owen, L.A., 2005. Late Quaternary glaciation of Tibet and the bordering mountains: a review. Boreas 34, 87–100.

Loewe, F., 1924. Die Eiszeit in Kaschmir, Baltistan und Ladakh. Zeit. Gesell. Erdkunde Berlin 42–53.

Mayewski, P.A., Jeschke, P.A., 1979. Himalayan and Trans-Himalayan glacier fluctuations since AD 1812. Arct. Alpine Res. 11, 267–287.

Mayewski, P.A., Pregent, G.P., Jeschke, P.A., Ahmad, N., 1980. Himalayan and Trans Himalayan glacier fluctuations and the South Asian Monsoon record. Arct. Alpine Res. 12, 171–182.

Meiners, S., 2001. The post to late glacial valley reconstruction on the Haramosh north side (Mani, Baska and Phuparash valleys). GeoJournal 54, 429–450.

Meiners, S., 2005. The glacial history of landscape in the Batura Muztagh, NW Karakoram. GeoJournal 63, 49–90.

Miller, K.J. (Ed.), 1984. The International Karakoram Project (1+2). Cambridge University Press, Cambridge 412 pp. +635 pp.

Misch, P., 1935. Ein gefalteter junger Sandstein im Nordwest-Himalaya und sein Gefuge. Festschrift zum 60 geburstag von Hans Stille. Verlag, Stuttgart.

Owen, L.A., 1988a. Wet-sediment deformation of Quaternary and recent sediments in the Skardu Basin, Karakoram Mountains, Pakistan. In: Croot, D. (Ed.), Glaciotectonics: Forms and Processes. Balkema, Rotterdam, pp. 123–148.

Owen, L.A., 1988b. Terraces, uplift and climate, Karakoram Mountains, Northern Pakistan. Unpublished PhD thesis, University of Leicester, 399 pp.

Owen, L.A., 1989. Terraces, uplift and climate in the Karakoram Mountains, northern Pakistan: Karakoram intermontane basin evolution. Z. Geomorphol. 76, 117–146.

Owen, L.A., 1991. Mass movement deposits in the Karakoram Mountains: their sedimentary characteristics, recognition and role in Karakoram landform evolution. Z. Geomorphol. 35, 401–424.

Owen, L.A., 1996. Quaternary lacustrine deposits in a high-energy semi-arid mountain environment, Karakoram Mountains, northern Pakistan. J. Quatern. Sci. 11, 461–483.

Owen, L.A., 2009. Latest Pleistocene and Holocene glacier fluctuations in the Himalaya and Tibet. Quatern. Sci. Rev. 28, 2150–2164.

Owen, L.A., Benn, D.I., 2005. Equilibrium-line altitudes of the Last Glacial Maximum for the Himalaya and Tibet: an assessment and evaluation of results. Quatern. Int. 138 (139), 55–78.

Owen, L.A., Derbyshire, E., 1988. Glacially deformed diamictons in the Karakoram Mountains, northern Pakistan. In: Croot, D. (Ed.), Glaciotectonics: Forms and Processes. Balkema, Rotterdam, pp. 149–176.

Owen, L.A., White, B.J., Rendell, H., Derbyshire, E., 1992. Loess silt deposits in the western Himalayas: their sedimentology, genesis and age. Catena 19, 493–509.

Owen, L.A., Derbyshire, E., Fort, M., 1998. The Quaternary glacial history of the Himalaya. Quatern. Proc. 6, 91–120.

Owen, L.A., Scott, C.H., Derbyshire, E., 2000. The Quaternary glacial history of Nanga Parbat. Quatern. Int. 65 (66), 63–79.

Owen, L.A., Finkel, R.C., Caffee, M.W., 2002a. A note on the extent of glaciation throughout the Himalaya during the global Last Glacial Maximum. Quatern. Sci. Rev. 21, 147–157.

Owen, L., Kamp, U., Spencer, J., Haserodt, K., 2002b. Timing and Style of Late Quaternary Glaciations in the Eastern Hindu Kush, Chitral, Northern Pakistan: a review and revision of the glacial chronology based on new optically stimulated luminescence dating. J. Quatern. Sci. 97 (98), 41–55.

Owen, L.A., Finkel, R.C., Caffee, M.W., Gualtieri, L., 2002c. Timing of multiple late Quaternary glaciations in the Hunza valley, Karakoram Mountains, Northern Pakistan: defined by cosmogenic radionuclide dating of moraines. Geol. Soc. Am. Bull. 114, 593–604.

Owen, L.A., Finkel, R.C., Barnard, P.L., Ma, H., Asahi, K., Caffee, M.W. et al., 2005. Climatic and topographic controls on the style and timing of Late Quaternary glaciation throughout Tibet and the Himalaya defined by $_{10}Be$ cosmogenic radionuclide surface exposure dating. Quaternary Sci. Rev.24, 1391–1411.

Owen, L.A., Caffee, M.W., Finkel, R.C., Seong, B.Y., 2008. Quaternary glaciations of the Himalayan-Tibetan orogen. J. Quaternary Sci. 23, 513–532.

Penck, A., Bruckner, E., 1909. Die Alpen im Eiszeitalter. Tauchnitz, Liepzig.

Phillips, W.M., Sloan, V.F., Shroder, J.F., Sharma, P., Clarke, M.L., Rendell, H.M., 2000. Asynchronous glaciation at Nanga Parbat, northwestern Himalayas Mountains, Pakistan. Geology 28, 431–434.

Porter, S.C., 1970. Quaternary glacial record in Swat Kohistan, West Pakistan. Bull. Geol. Soc. Am. 81, 1421–1446.

Richards, B.W., Owen, L.A., Rhodes, E.J., 2000. Timing of Late Quaternary glaciations in the Himalayas of northern Pakistan. J. Quatern. Sci. 15, 283–297.

Röthlisberger, F., Geyh, M.A., 1985. Glacier variations in Himalayas and Karakorum. Z. Gletscherk. Glazialgeol. 21, 237–249.

Rupper, R., Roe, G., Gillespie, A., 2009. Spatial patterns of Holocene glacier advance and retreat in Central Asia. Quatern. Res. 72, 337–346.

Schneider, H.J., 1959. Zur diluvialen Geschichte des NW-Karakorum. Mitt. Geogr. Ges. München 44, 201–216.

Schneider, H.J., 1969. Minapin—Gletscher und Menschen im NW-Karakorum (Erläuterungen zur Expeditionskarte 1: 50.000). Erde 100, 266–286.

Scott, C.H., 1992. Contemporary sediment transfer in the Himalayan glacial system: implications for the interpretation of the Quaternary record. Unpublished PhD thesis, University of Leicester, 352 pp.

Seong, Y.B., Owen, L.A., Bishop, M.P., Bush, A., Clendon, P., Copland, L., et al., 2007. Quaternary glacial history of the Central Karakoram. Quatern. Sci. Rev. 26, 3384–3405.

Seong, Y.B., Owen, L.A., Bishop, M.P., Bush, A., Clendon, P., Copland, L., et al., 2008a. Rates of fluvial bedrock incision within an actively uplifting orogen: Central Karakoram Mountains, northern Pakistan. Geomorphology 97, 274–286.

Seong, Y.B., Owen, L.A., Bishop, M.P., Bush, A., Clendon, P., Copland, P., et al., 2008b. Reply to comments by Matthias Kuhle on "Quaternary glacial history of the central Karakoram" Quatern. Sci. Rev. 27, 1656–1658.

Seong, Y.B., Bishop, M.P., Bush, A., Clendon, P., Copland, L., Finkel, R., et al., 2009a. Landforms and landscape evolution in the Skardu, Shigar and Braldu Valleys, Central Karakoram. Geomorphology 103, 251–267.

Seong, Y.B., Owen, L.A., Caffee, M.W., Kamp, U., Bishop, M.P., Bush, A., et al., 2009b. Rates of basin-wide rockwall retreat in the K2 region of the Central Karakoram defined by ^{10}Be terrestrial cosmogenic nuclides. Geomorphology 107, 254–262.

Shroder, J.F., 1984. Batura glacier terminus, 1984, Karakorum Himalayas. Geol. Bull. Univ. Peshawar 17, 119–126.

Shroder, J.F., Khan, M.S., Lawrence, R.D., Madin, I.P., Higgins, S.M., 1989. Quaternary glacial chronology and neotectonics in the Himalayas of northern Pakistan. In: Malinconico, C. (Ed.), Geology of the Western Himalayas, Geological Society of America, Special Paper 232 pp. 275–294 Geological Society of America.

Shroder, J.F., Owen, L.A., Derbyshire, E., 1993. Quaternary glaciation of the Karakoram and Nanga Parbat Himalayas. In: Shroder, J.F. (Ed.), Himalayas to the Sea: Geology Geomorphology and the Quaternary. Routledge, London, pp. 132–158.

Spencer, J.Q., Owen, L.A., 2004. Optically stimulated luminescence dating of Late Quaternary glaciogenic sediments in the upper Hunza valley: validating the timing of glaciation and assessing dating methods. Quatern. Sci. Rev. 23, 175–191.

Treydte, K.S., Schleser, G.H., Helle, G., Frank, D.C., Winiger, M.W., Haug, G.H., et al., 2006. The Twentieth Century was the wettest period in northern Pakistan over the past millennium. Nature 440, 1179–1182.

Trinkler, E., 1932. Geographische Forschungen im westlichen Zentralasien und Karakorum-Himalaya. Wissenschaftliche Ergebnisse der Dr. Trinklerschen Zentralasien-Expedition (1). Reimer/Vohsen, Berlin, 133 pp.

Troll, C., 1938. Reliefenergie und Vergletscherung in der Nanga Parbat-Gruppe. Zeit. Glet. 26, 303–307.

Visser, P.C., 1928. Von den Gletschern am obersten Indus. Zeit. Glet. 16, 169–229.

Wadia, D.N., 1932/33. Note on the geology of Nanga Parbat (Mt. Diamir), and adjoining portions of Chilas, Gligit District, Kashmir. Rec. Geol. Surv. India 66, 212–234.

Wiche, K., 1958. Die österreichische Karakorum-Expedition 1958. Mitt. Geogr. Ges. Wien 10, 280–294.

Zeitler, P.K., Meltzer, A.S., Koons, P.O., Craw, D., Hallet, B., Chamberlain, C.P., et al., 2001. Erosion, Himalayan geodynamics and the geomorphology of metamorphism. GSA Today 11, 4–8.

Zhang, X., Shi, Y., 1980. Changes of the Batura Glacier in the Quaternary and recent times. Professional Papers on the Batura Glacier, Karakoram Mountains, Beijing, 173–190.

Chapter 67

Quaternary Glaciation of Northern India

Lewis A. Owen

Department of Geology, University of Cincinnati, Cincinnati, Ohio 45221, USA, E-mail: lewis.owen@uc.edu

67.1. INTRODUCTION

The mountains of Northern India contain impressive glacial landforms and sediments that provide abundant evidence for Late Quaternary glaciation. The main mountain ranges comprise some of the greatest on our planet, including the Greater Himalaya, Pir Panjal, Zanskar Range, Ladakh Range and Karakoram (Fig. 67.1). Undertaking research in these mountains is logistically and politically difficult. This has hindered the systematic reconstruction of the extent and timing of glaciation in this region. However, new studies have begun to yield important insights into the complex nature of glaciation in this region during the past few years. Slowly, regional correlations and reconstructions of glaciation are beginning to be developed (Owen et al., 2006; Dortch et al., 2010; Hedrick et al., 2010). This has been aided by the application of newly developing numerical dating methods such as optically stimulated luminescence (OSL) and terrestrial cosmogenic nuclide (TCN) surface exposure dating (e.g. Owen et al., 2006; Dortch et al., 2010; Hedrick et al., 2010). These new methods have been particularly useful because in most of the study areas, the lack of organic matter within sediments in the high-mountain environments inhibits the use of radiocarbon dating. Further, many of the glacial successions within the region are far older than the age range to which radiocarbon dating can be applied.

Despite the recent work, it is still not possible at this time to produce a comprehensive and systematic reconstruction of the extent of Late Quaternary glaciation throughout all of Northern India. Moreover, constructing a simple map of the glaciation of Northern India for any particular time in the Late Quaternary is still not possible. Further, quantifying the degree of glaciation, for example, by using equilibrium-line altitudes (ELAs), is difficult because of the problems associated with calculating present and former ELAs for mountain glaciers. Some of these problems are discussed in more detail, with specific reference to the Himalaya, in Benn and Lehmkuhl (2000), Owen and Benn (2005) and Benn et al. (2005).

This contribution builds on the review of Owen (2004) that was published in the first edition of Ehlers and Gibbard (2004). As in Owen (2004), individual mountain ranges and valley systems are reviewed in detail with the aim of providing a framework for future regional syntheses. This contribution also builds on the reviews of Derbyshire and Owen (1997), Owen et al. (1998, 2002a, 2008), Owen (2009) and Lehmkuhl and Owen (2005).

67.2. REGIONAL SETTING

The mountains of Northern India are tectonically active and are the consequence of the continued northward collision of the Indian and Asian continental lithospheric plates, and intense fluvial and glacial erosion and weathering. The mountains trend NW and progressively rise in elevation from south to north to the highest peaks above 8000 m a.s.l. in the Karakoram (Fig. 67.1). The climate is dominated by the Southeast Asian summer monsoon and the mid-latitude westerlies. Benn and Owen (1998) provide a summary of the climatic setting and the importance of these climatic systems in driving glaciation in this region. A strong climatic gradient exists across Northern India, with the region becoming more arid towards the north, and striking microclimatic differences occur within individual valley systems. The southern mountain ranges—the Pir Panjal, Greater Himalaya and southernmost Zanskar Range—receive most of their precipitation from the SE Asian summer monsoon during the summer, while the northern ranges of Zanskar and the Ladakh Range and the Karakoram receive somewhat limited precipitation during the winter, supplied by the mid-latitude westerlies (Fig. 67.2). The strong regional climatic gradient and the large range of altitudes control the distribution of vegetation, which from south to north and the lowest to highest elevations vary from subtropical forest, coniferous forest, semi-arid steppe, alpine meadows to barren snowfields (Schweinfurth, 1968; Singh and Singh, 1987). Moreover, the strong climatic gradients, regionally and altitudinally, produce markedly different forms and rates of weathering on landforms of the same

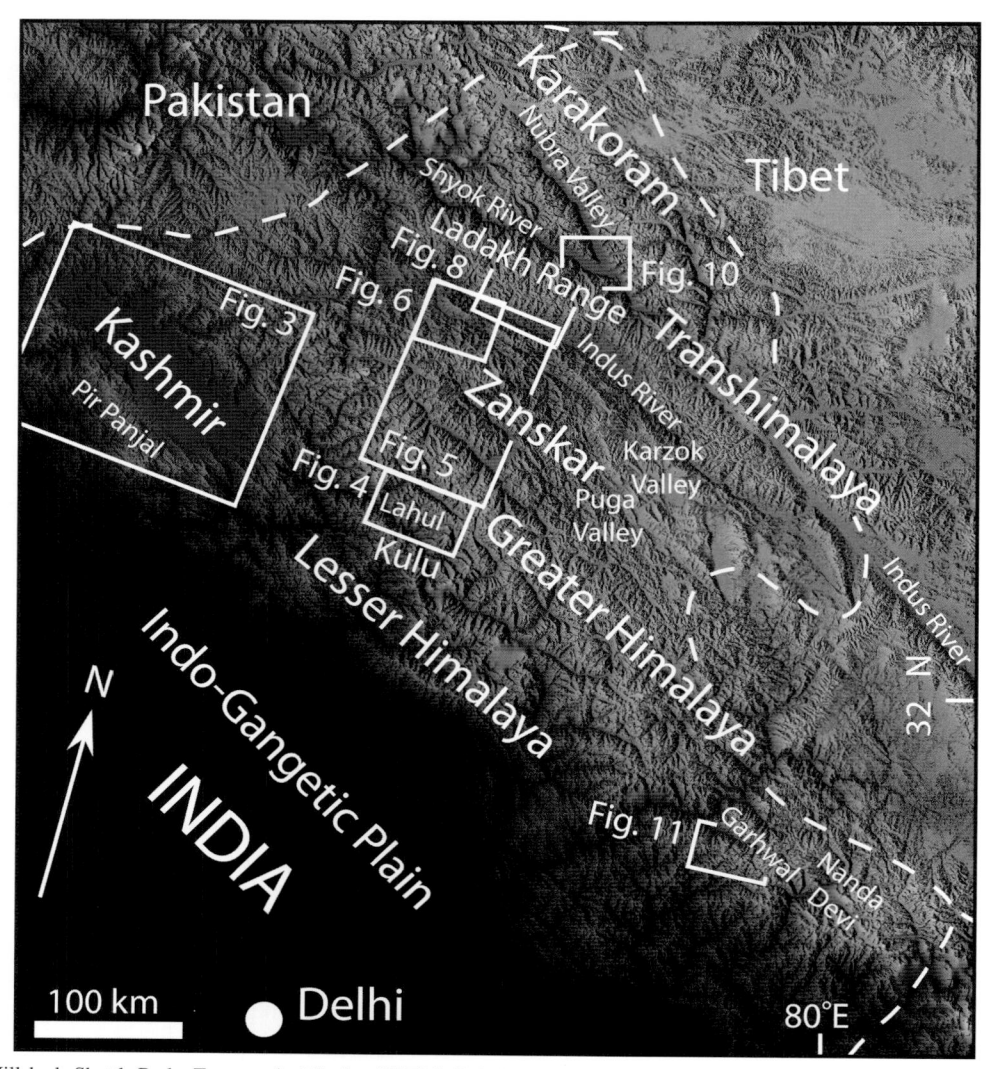

FIGURE 67.1 Hillshade Shuttle Radar Topography Mission (SRTM) digital elevation model (DEM) of northern India showing the locations of the main study areas discussed within the text.

age, both across the region and within individual valley systems (Burbank and Fort, 1985; Owen et al., 1997).

The glaciers in Northern India are complex, varying from sub-polar types at very high altitudes in the Karakoram to maritime types in the Greater Himalaya (Benn and Owen, 2002). Most glaciers have very high-altitude source areas (>5500 m a.s.l.) and may extend to altitudes as low as 1500 m a.s.l. Most glaciers are of summer accumulation type, are steep and of the high-activity type (Benn and Owen, 2002). Many glaciers are debris mantled as a consequence of the abundant supply of rock from the adjacent steep slopes by rock fall and avalanche processes (Owen et al., 1995; Benn and Owen, 2002). The dominance of supraglacial debris results in glacial sediments that can easily be misidentified as mass movement deposits and similarly mass movement deposits look like glacial sediments. The difficulty in distinguishing glacial from non-glacial landforms often within landforms that have undergone intense mass wasting, as well as fluvial and glacial erosion, results in them losing their diagnostic morphologies, which in turn has led to erroneous reconstructions of glacial extents (Derbyshire, 1983; Fort and Derbyshire, 1988; Owen, 1988, 1991, 1993; Hewitt, 1999; Benn and Owen, 2002).

67.3. KASHMIR

No new research has been undertaken on the Quaternary glaciation of Kashmir since the review of Owen (2004). This is because of continued civil unrest in the region. Nevertheless, the early research undertaken on Quaternary geology of the Kashmir Basin and adjacent Greater Himalaya and Pir Panjal Mountains were landmarks in the study of

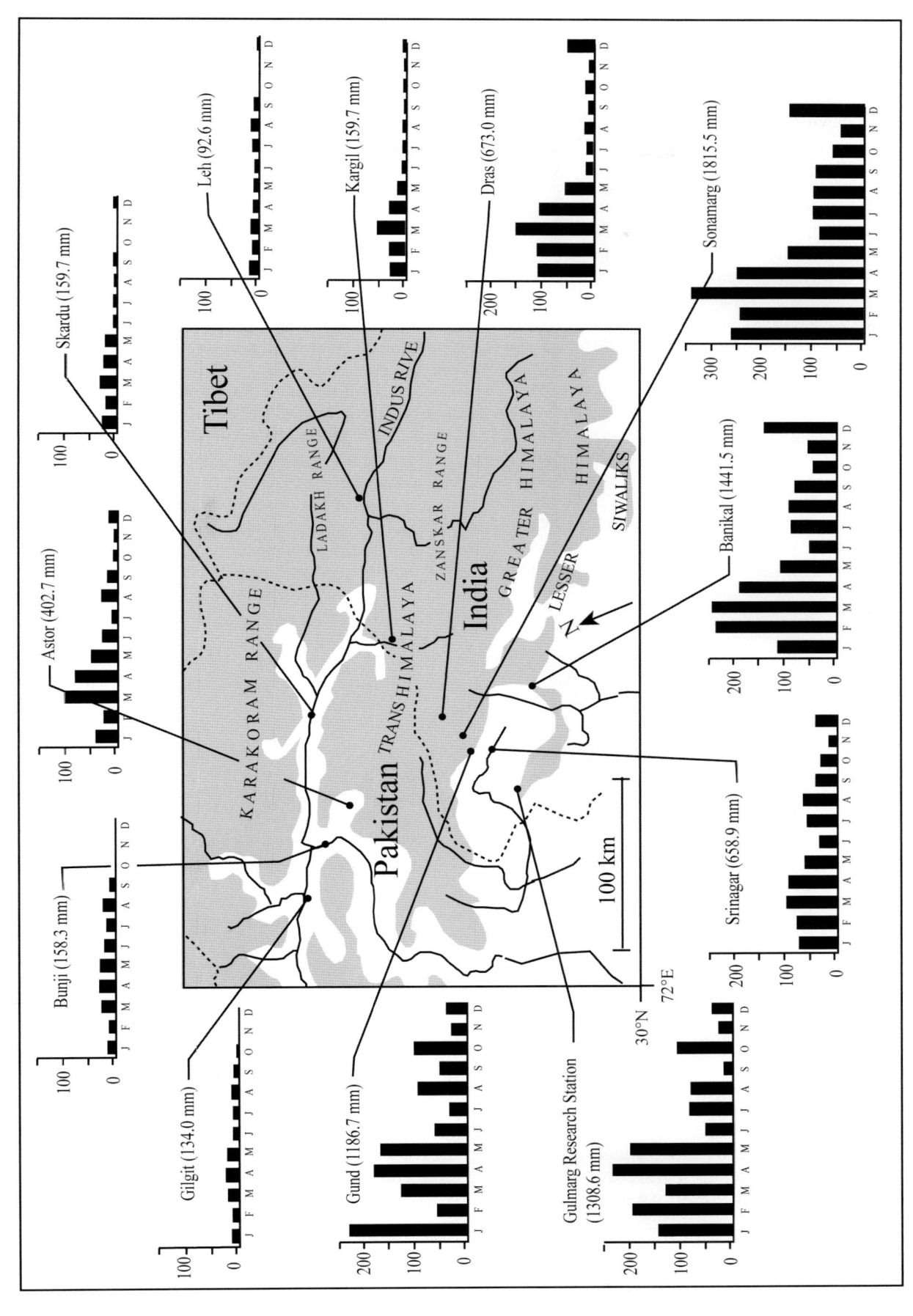

FIGURE 67.2 Regional variation in precipitation (in mm) across the Himalaya in NW India and Pakistan (adapted from Owen and England, 1998). The total annual precipitation is shown in parentheses for each location.

the Quaternary geology of the Himalaya. This includes the work of Grinlinton (1928) and De Terra and Paterson (1939) on lacustrine and glacial sediments, the more recent interdisciplinary study of the Quaternary history by Agrawal (1992), and the glacial geological studies of Holmes (1988, 1993) and Holmes and Street-Perrott (1989).

Holmes (1988, 1993) and Holmes and Street-Perrott (1989) reinterpreted much of De Terra and Paterson (1939) work, particularly moraines as mass movement deposits, showing that the lower altitudinal limit of glaciation was much higher (above 2200 m a.s.l.) than previously believed (Fig. 67.3). Holmes (1988, 1993) and Holmes and Street-Perrott (1989) calculated the maximum ELA depression to be ~800 m and highlighted that there is a strong modern ELA gradient across the region rising northeastwards from ~3900 to 4700 m a.s.l. In contrast, Holmes (1988, 1993) and Holmes and Street-Perrott (1989) showed that the former ELA gradient was in the reverse direction. This apparent reversal of ELA gradient was attributed to rapid differential uplift of the Pir Panjal along the southern edge of the Kashmir basin, which they argued tectonically uplifted the glacial landforms to give the apparent impression of a former reversed gradient. Holmes (1988, 1993) and Holmes and Street-Perrott (1989) highlighted that there is evidence for two major phases of ice advance in Kashmir, but they did not describe these in any detail or provide any numerical ages on the landforms.

67.4. LAHUL HIMALAYA

The Lahul Himalaya is located in the Greater Himalaya immediately north of the Pir Panjal in the Kulu area of Himachal Pradesh. The geomorphology and glacial geology of the region have been examined and discussed by Owen et al. (1995, 1996, 1997, 2001), Coxon et al. (1996) and Adams et al. (2009). On the basis of morphostratigraphy, Owen et al. (1995, 1996, 1997) recognized five glacial stages: Chandra (oldest), Batal, Kulti and Sonapani I and II (youngest). Owen et al. (1997) dated deltaic sediments using OSL methods to determine that they were formed during the Batal glacial stage at ~40 ka. Reinterpretation of the morphostratigraphic relationships of these deltaic deposits and a comprehensive programme of TCN dating showed the Batal glacial stage to be significantly

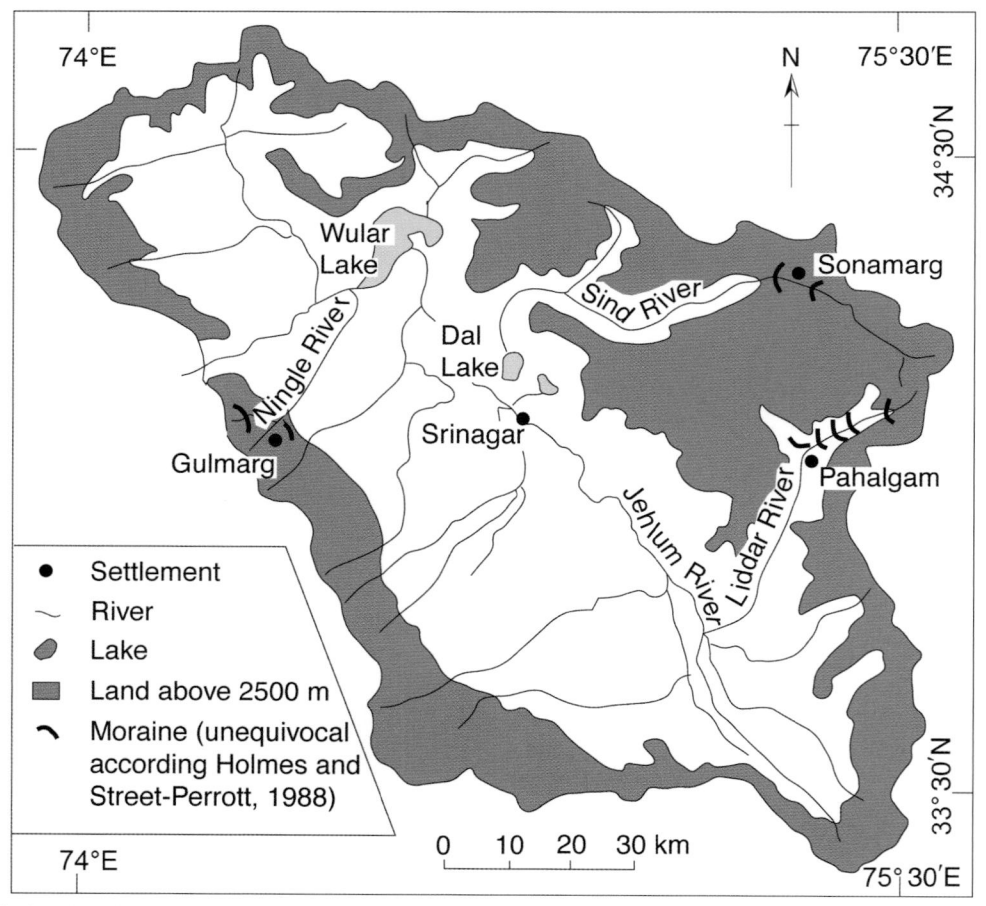

FIGURE 67.3 The location of the main end moraines in Kashmir (adapted from Holmes and Street-Perrott, 1989).

younger (~12–15.5 ka: Late Glacial Interstadial; Owen et al., 2001; Fig. 67.4). Owen et al. (2001) were also able to show that the Kulti glacial stage occurred in the early Holocene (~10–11.4 ka). Further, Owen et al. (1996) showed by comparing photographs of glaciers taken by Walker and Pascoe (1907) in 1905 that the Sonapani II glacial stage occurred during the latter part of the nineteenth century. The Chandra and Sonapani I glacial stages have still to be dated.

67.5. ZANSKAR

Zanskar lies between the Greater Himalaya and the Indus valley in Northern India. The first glacial geological study on the region was by Dainelli (1922), and Osmaston (1994) undertook the first modern study. Osmaston (1994) recognized evidence for four glaciations, the Kilima (oldest), sTongde, Tepuk and Drung-drung (Fig. 67.5). Damm (1997, 2006), Taylor (1999), Mitchell et al. (1999) and Taylor and Mitchell (2000) extended this earlier work into adjacent valleys.

Taylor and Mitchell (2000) mapped glacial successions in Zanskar and correlated the glacial landforms with the glacial successions in the Lahul Himalaya, replacing the Osmaston's glacial stage names with those of Owen et al. (1995, 1996, 1997) for the Lahul Himalaya (Fig. 67.6). Thus Taylor and Mitchell (2000) replaced the Kilima (oldest), sTongde, Tepuk and Drung-drung stages with the Chandra, Batal, Kulti and Sonapani glacial stages, respectively. Using OSL methods, Taylor and Mitchell (2000) showed that glaciers during their Batal glacial stage reached its maximum at 78.0 ± 12.3 ka and existed until at least 40 ka. The Kulti glacial stage landforms of Taylor and Mitchell (2000) is represented by distinct sets of well-preserved moraine ridges within 10–20 km of the present ice margins (Mitchell et al., 1999; Taylor and Mitchell, 2000). Taylor and Mitchell (2000) presented infrared stimulated luminescence (IRSL) ages of between 10 and 16 ka on glaciofluvial sediments associated with moraines of their Kulti glacial stage, suggesting that this glaciation may be coincident with the global Last Glacial Maximum (LGM) or might have occurred slightly later in the Late Glacial. Mitchell et al. (1999) and Taylor and Mitchell (2000) attributed sharp-crested moraines within 2 km of the present glaciers to the Sonapani glacial stage, but no numerical ages were provided for the moraines. However, Taylor and Mitchell (2000) suggested that the youngest sets of moraine ridges are coincident with the Little Ice Age (LIA). Figure 67.6 shows the extent of glaciation and the location and results of the luminescence dating. The TCN dating undertaken by Owen et al. (2001), however, showed that the Kulti and Batal Stages in the Lahul Himalaya are significantly younger than the corresponding glacial stages named by Taylor and Mitchell (2000) for Zanskar. This highlights the difficulty in correlating between mountain ranges in the Himalaya based on morphostratigraphy. Damm (2006) mapped the Markha valley and the northern Nimaling Mountains (Northern Zanskar Range) and produced a chronology that included eight glacial stages. Damm (2006) argued that the two oldest glaciations probably occurred prior to the last glacial cycle, with the following three glacial stages occurring during the global LGM, the Late Glacial and the early Holocene. Large moraine complexes in the vicinity of the present glaciers were assigned to the Neoglacial and historical time.

To examine the transition in pattern and timing of glaciation between Lahul and Ladakh, Hedrick et al. (2011) mapped moraines and dated them using ^{10}Be methods in the Puga and Karzok valleys of Zanskar. Hedrick et al. (2011) showed that in the Puga valley, glaciers advanced >10 km at ~115 ka and <10 km at ~40 ka, ~3.3 ka and ~0.5 ka. Whereas in the Karzok valley, glaciers advanced >10 km during marine isotope stage (MIS) 9 or older, <5 km at ~80 ka and <2 km at ~3.6 ka. Hedrick et al. (2011) argued that moraines in the Puga and Karzok valleys broadly correlate with previous studies in the Zanskar Range, but the paucity of data for many of the glacial stages across the Zanskar region makes the correlations tentative (Fig. 67.7). There was no evidence for an early Holocene glaciation in the Puga and Karzok valleys, which is in stark contrast to many regions of the Himalaya. Moreover, Hedrick et al. (2011) highlighted that the similarity between the glacial records in the Puga and Karzok study areas suggests that the transition to Lahul style glaciation is to the south of the Karzok valley and suggest that this geographical transition is abrupt.

67.6. LADAKH RANGE

The Ladakh Range is located in the Transhimalaya of Northern India and its valleys drain southward into the Indus and northward into the Shyok. Drew (1875), Dainelli (1922), Fort (1978, 1981, 1983), Burbank and Fort (1985), Osmaston (1994), Brown et al. (2002) and Owen et al. (2006) provide basic descriptions of the glacial geomorphology in Ladakh. Fort (1978, 1981, 1983) and Burbank and Fort (1985) presented the first modern framework for the Quaternary glacial history of Ladakh Range. These studies described the evidence for former glaciers filling tributary valleys down to the Indus valley along the Ladakh Range and partially filling tributary valleys along the northern slopes of the Zanskar Range along the Indus. Fort (1978) calls this glaciation the "Leh Stage". Burbank and Fort (1985) suggested that these moraines at the mouths of tributary valleys represent the local LGM and are probably coincident with the global LGM and had a maximum ELA depression of between 900 and 1000 m. Burbank and Fort (1985) also describe sets of moraines at higher altitudes

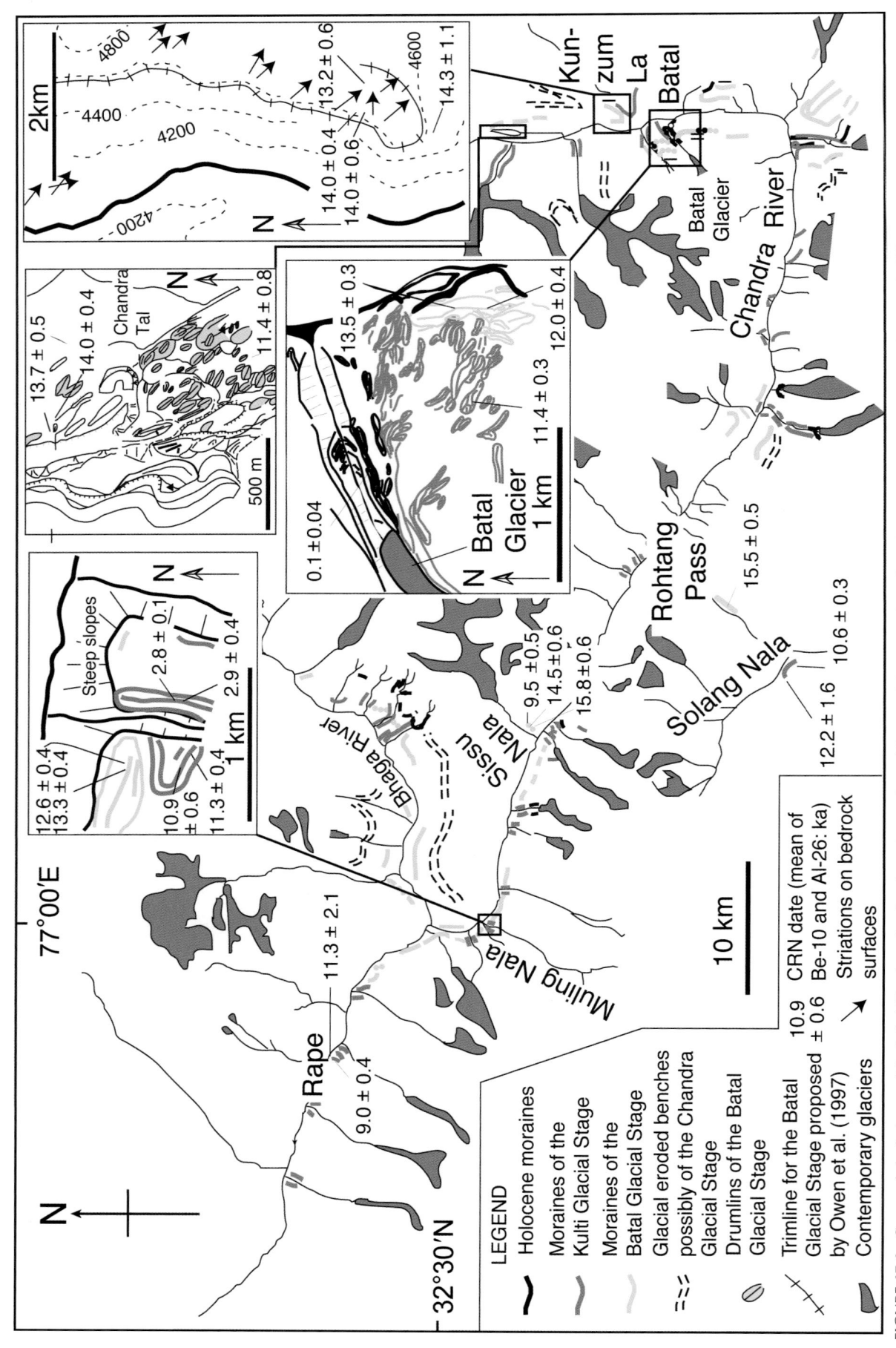

FIGURE 67.4 Geomorphological map of the Lahul Himalaya showing the main moraines with terrestrial cosmogenic radionuclide (CRN) ages (adapted from Owen et al., 2001).

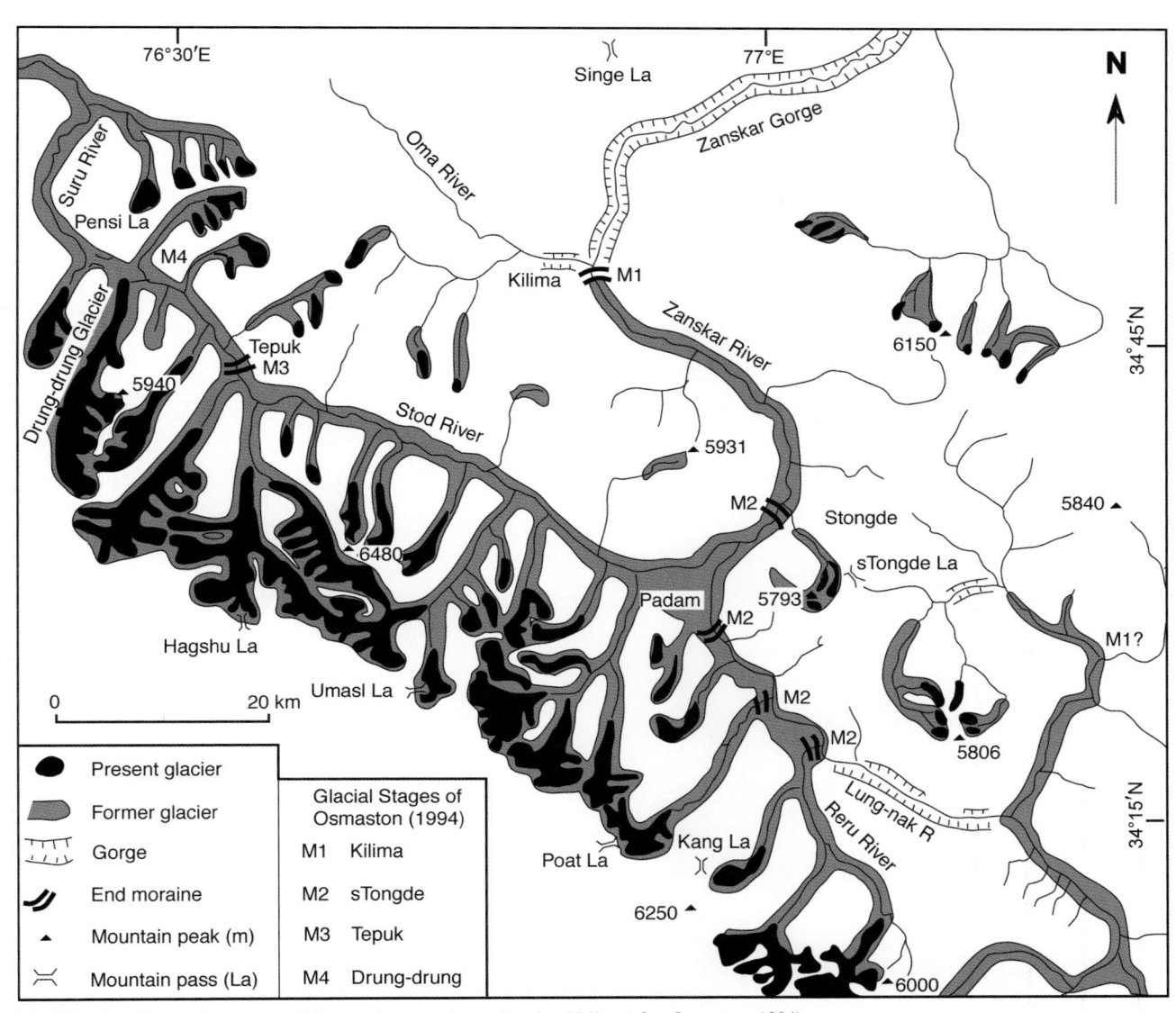

FIGURE 67.5 Extent of present and former glaciation in the Zanskar Valley (after Osmaston, 1994).

within these valleys, which they called "recessional moraines", and they also recognized moraines within 1 km of the present glaciers that they assigned to the Neoglacial.

Burbank and Fort (1985), however, had no numerical ages on these moraines to confirm their assigned ages. The first numerical ages on these moraines were provided by Brown et al. (2002) using ^{10}Be TCN dating on four glacial boulders on a Leh glacial stage moraine north of Leh, providing an age of 90 ± 15 ka. Owen et al. (2006) recalculated these ages of Brown et al. (2002) using new production rates and improved scaling models and showed the range from between 79 and 225 ka.

Owen et al. (2006), however, showed, using ^{10}Be TCN dating of moraine boulders and alluvial fan sediments, that the timing of glaciation in the Ladakh Range was much older than proposed by any of the previous studies. Moreover, Owen et al. (2006) recognized two additional glacial stages, the Bazgo and Khalling glacial stages (Fig. 67.8). The timing of five glacial stages was defined to older than 430 ka (Indus Valley glacial stage), the penultimate glacial cycle or older (Leh glacial stage), early part of the last glacial cycle (Kar glacial stage), middle of the last glacial cycle (Bazgo glacial stage) and early Holocene (Khalling glacial stage) (Fig. 67.9).

Owen et al. (2006) highlighted that like other areas in the interior of the Himalayan–Tibetan orogen, a pattern of progressively more restricted glaciation over each progressive glacial cycle was evident in the Ladakh Range. Owen et al. (2006) argued that this likely reflected indicating a progressive reduction in the moisture supply necessary to sustain glaciation. One possible explanation for

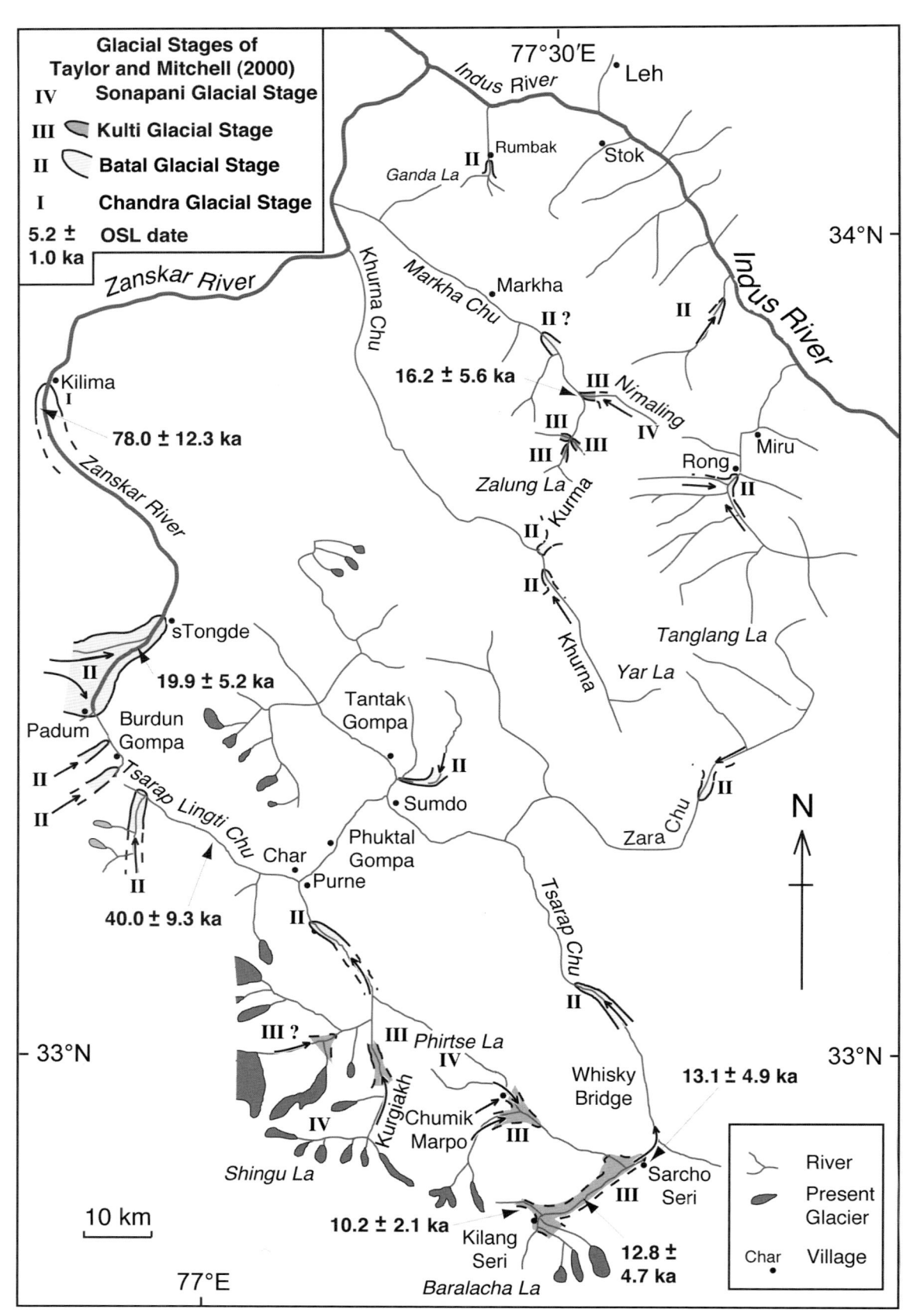

FIGURE 67.6 Extent of glaciation in Zanskar showing the sampling locations and results of OSL dating (adapted from Taylor, 1999; Mitchell et al., 1999; Taylor and Mitchell, 2000).

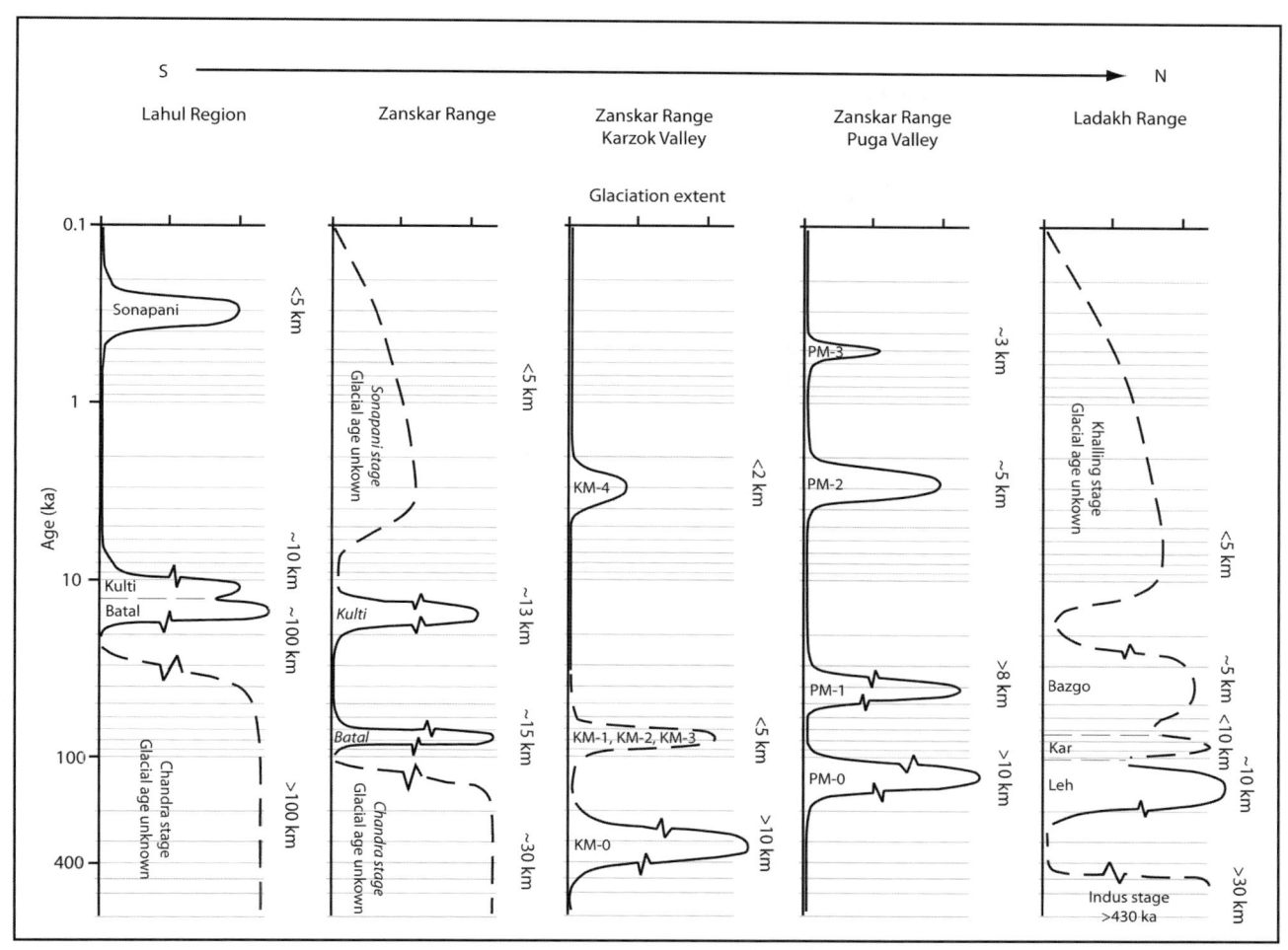

FIGURE 67.7 Comparison of timing and extent of glaciation in Lahul, Zanskar and Ladakh (after Hedrick et al., 2011). Lahul data are from Owen et al. (1997, 2001), Zanskar data from Taylor and Mitchell (2000) and Hedrick et al. (2010), and Ladakh data from Brown et al. (2002) and Owen et al. (2006). The extent of glaciation is relative with the approximate extent beyond the present ice margin listed to the right of the advance curve. Dashed lines indicate a glacial advance of undated age. Glacial stages shown in italics were assigned by Taylor and Mitchell (2000) by relative dating to Owen et al.'s (1997, 2001) chronology for Lahul. These stage names must be revised in light of the discussion presented in Owen et al. (2002b).

this is that uplift of Himalayan ranges to the south and/or of the Karakoram Mountains to the west of the region may have effectively blocked moisture supply by the south Asian summer monsoon and mid-latitude westerlies, respectively. Owen et al. (2006) also suggested that this pattern of glaciation might reflect a trend of progressively less extensive glaciation in mountain regions that has been observed globally throughout the Pleistocene.

The Quaternary sediments on the northern side of the Ladakh Range were first described by Phartiyal et al. (2005). Phartiyal et al. (2005) examined a diamict deposit, which they interpreted as a landslide mass, and a thick ($\sim$20 m) deposit of lacustrine sediments at the Nubra–Shyok confluence that they dated to $24,970 \pm 550$ radiocarbon years BP. Deformation structures within the sediments were attributed to tectonic events. Pant et al. (2005) and Dortch et al. (2010) later reinterpreted the diamict to be a moraine. Dortch et al. (2010) extended the glacial geological study of Owen et al. (2006) to the northern side of the Ladakh Range, where they recognized three glacial stages (Deshkit 1, Deshkit 2 and Dishkit 3 glacial stages) in the Nubra and Shyok valleys (Fig. 67.10). Using ^{10}Be TCN, Dortch et al. (2010) dated these to $\sim$45 ka (Deshkit 1 glacial stage), $\sim$81 ka (Deshkit 2 glacial stage) and $\sim$144 ka (Deshkit 3 glacial stage). Dortch et al. (2010) calculated a mean ELA depression of $\sim$290 m for the Deshkit 1 glacial stage but did not determine ELA depressions for other glacial stages because it was not possible to make accurate reconstructions of glaciers belong to those stages. Dortch et al. (2010) suggested that comparison of glaciation in the Nubra and Shyok valleys with glaciations in the adjacent Central Karakoram of northern Pakistan and northern side of the Ladakh Range of Northern India showed that glaciation was broadly synchronous across the region

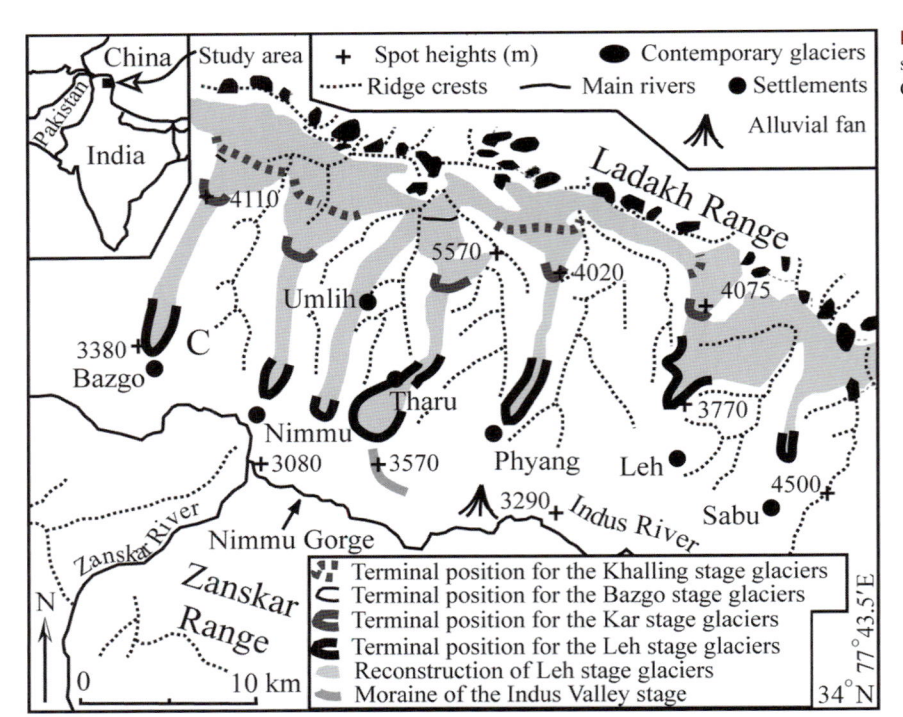

FIGURE 67.8 Glacial landforms along the southern side of the Ladakh Range (adapted from Owen et al., 2006).

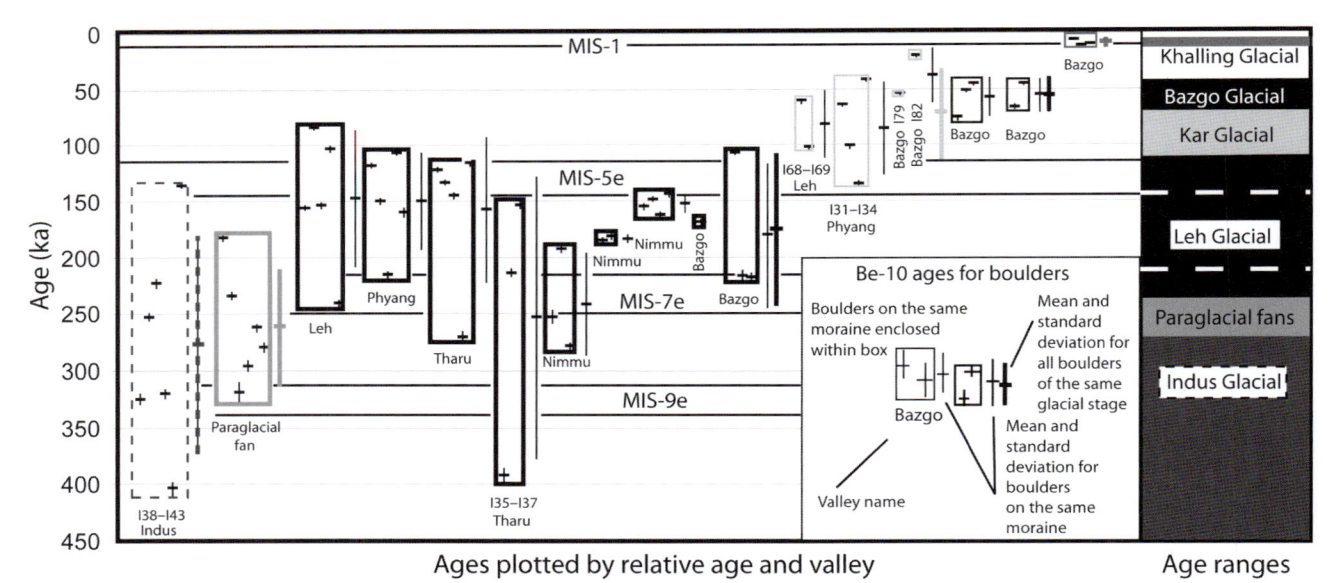

FIGURE 67.9 Plot of ^{10}Be ages for boulders on moraines for the Ladakh Range (after Owen et al., 2006) illustrating the type of geochronologic data that are beginning to be developed for selected regions of northern India. The coloured boxes surround data obtained from moraines of a given stage in the same valley. The minimum age of the alluvial fan that was dating using a TCN depth profile is also plotted. The durations of interglacials are bracketed by grey lines. The column on the right illustrates our best estimates for ages for each glacial stage. The uncertainty in assigning an age to the Leh glacial stage is highlighted by the stripes, and the dashed black lines show our best estimate for the age of this glacial stage. (For interpretation of the references to colour in this figure legend, the reader is referred to the Web version of this chapter.)

during MIS 6. However, the extent of glaciation differed greatly, with more extensive glaciation in the Karakoram than the morphostratigraphically equivalent glaciation on the northern slopes of the Ladakh Range. Dortch et al. (2010) argued that this highlights the strong contrast in the extent of glaciation across ranges in the Himalayan–

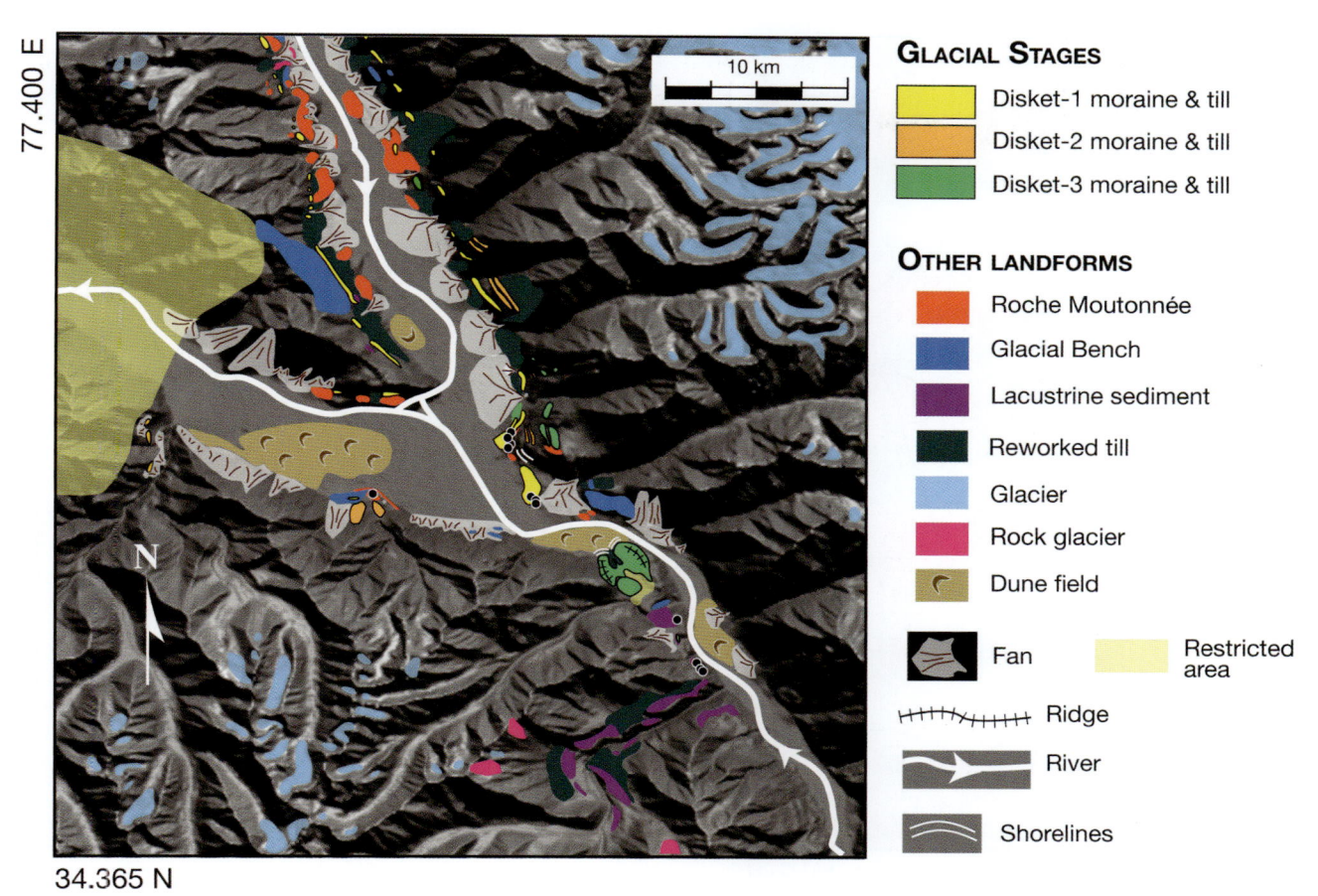

FIGURE 67.10 Hillshade SRTM DEM showing glacial and non-glacial landforms and contemporary glaciers of the Nubra–Shyok confluence (adapted from Dortch et al., 2010).

Tibetan orogen and that it necessitates caution when correlating glacial successions within and between mountain ranges.

67.7. GARHWAL

Garhwal is part of the Greater Himalaya and stretches from the eastern border of Nepal across Nanda Devi, Kedarnath, Yamnotri to Gangotri. Sharma and Owen (1996) provided the first study of the glacial successions in Garhwal. Focusing on the valleys between the Gangotri glaciers and along the upper Bhagirathi, Sharma and Owen (1996) recognized a major valley glaciation, the Bhagirathi glacial stage, and two minor advances, the Shivling and Bhujbas glacial advances (Fig. 67.11). Using OSL methods, Sharma and Owen (1996) dated the maximum extent of glaciation during the Bhagirathi glacial stage to ~63 ka and argued that this glaciation persisted into the early Holocene. They reconstructed a major trunk valley glacier that extended ~40 km from the contemporary Gangotri glacier to Jhala at an altitude of ~2300 m a.s.l. This study showed that there was a regional ELA depression of ~640 m during the Bhagirathi glacial stage. Sharma and Owen (1996) showed that Shivling glacial advance occurred during the middle Holocene and had an ELA depression of between 40 and 100 m. Sharma and Owen (1996) used a combination of historical documentation and simple dendrochronology to show that the Bhujbas glacial advance occurred ~200–300 years BP and was equivalent to the LIA, with an ELA depression of between 20 and 60 m. IPCC (2007) highlights Gangotri Glacier as a rapidly retreating glacier in the Himalaya and suggests that by the year 2035 all glaciers within the Himalaya will have melted away. Unfortunately, the original sources to these data were not provided to assess this statement, and the work of Sharma and Owen (1996) and recent observations (Bagla, 2009) suggest that glaciers are not retreating as quickly are suggested by the IPCC (2007) report. Barnard et al. (2004a) reassessed the work of Sharma and Owen (1996) and dated glacial and paraglacial landforms (fans, terraces and associated moraines) using ^{10}Be TCN in the Bhagirathi Valley. The new study and dating showed that fan and terrace formation is intimately

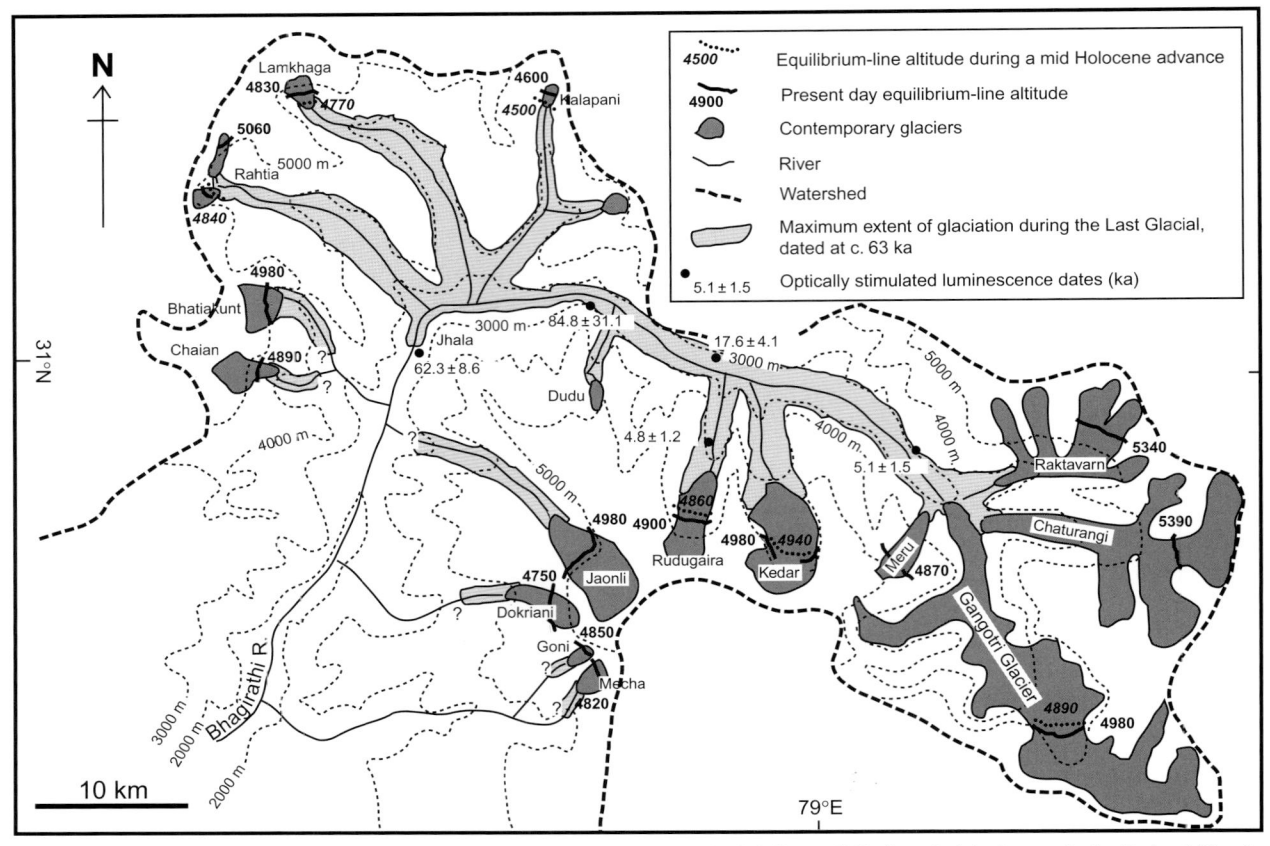

FIGURE 67.11 Reconstruction of ice extents for the Bhagirathi glacial stage, and the Shivling and Bhujbas glacial advances in the Garhwal Himalaya. The location and results of OSL dating are also shown (adapted from Sharma and Owen, 1996).

related to glaciation and that fluctuations in glacial and associated environments during times of climatic instability cause rapid sediment transfer and re-sedimentation of glacial landforms. The TCN ages confirmed that the Bhagirathi glacial stage of Sharma and Owen (1996) dates to ~63–11 ka, the Shivling glacial stage dates to ~5 ka and the Bhujbas glacial stage dates to ~200–300 years BP. In addition, Barnard et al. (2004a) identified two new glacial stages, the Kedar stage dated to ~7 ka and the Gangotri glacial stage dated to ~1 ka.

The only study undertaken in NE Garhwal was by Barnard et al. (2004b) who used ^{10}Be TCN to date moraines, fans and river and strath terraces in the Gori Ganga Valley of Nanda Devi. Barnard et al. (2004b) showed that moraines date to the Late Glacial Interstadial, early Holocene, late Holocene and LIA. The study also showed that fans and river terraces within the Gori Ganga Valley developed rapidly by debris flow and flood processes during periods of deglaciation. This highlighted the importance of climatic controls on landscape evolution and suggests a strong monsoonal control on the dynamics of earth surface processes in this region.

67.8. CONCLUSIONS

Knowledge of the Quaternary glacial history of Northern India is still in its infancy, but new studies are beginning to provide data on regional variability. This is showing marked contrasts in the extent and timing of glaciation between mountain ranges in Northern India. In the Lahul Himalaya, for example, glaciation was very extensive during the Late Glacial with an extensive valley glacier system filling the main trunk valleys for >100 km (Owen et al., 1997, 2001). In contrast, in Zanskar and Ladakh, glaciers only advanced a few kilometres from their present positions during the Late Glacial (Owen et al., 2006; Hedrick et al., 2010). As Owen et al. (2005) suggested, this likely reflects temporal and spatial variability in the south Asian monsoon and mid-latitude westerlies, and regional precipitation gradients. Owen et al. (2005) also pointed out that in zones of greater aridity, the extent of glaciation has become increasingly restricted throughout the Late Quaternary leading to the preservation of old glacial landforms, as is the case in the Ladakh Range. In contrast, in regions that are very strongly influenced by the monsoon, the preservation potential of pre-Late Glacial moraine successions is generally poor, as is the case in NE Garhwal. This is possibly because Late Glacial and Holocene glacial advances may have been more extensive than early glaciations and hence may have destroyed evidence of earlier glaciations. Further, Owen et al. (2005) argue that the intense denudation, mainly by fluvial and mass movement processes, which characterize these wetter environments, results in rapid erosion and re-sedimentation of glacial and associated landforms, which also contributes to their poor preservation potential.

The glacial geological research projects that have been undertaken generally have concentrated on relatively small and accessible areas of the Himalaya, and few studies have defined glacial chronologies using numerical dating. The new studies, however, allow us to begin to test regional patterns of glaciation. Most seem to support the view of Benn and Owen (1998) that on millennial time scales, glacier oscillations in the Himalaya reflect periods of positive mass-balance coincident with times of increased insolation and intensified monsoon activity in the Himalaya. In particular, glaciation throughout the Himalaya was probably most extensive during the early part of the last glacial cycle and likely during MIS 3, a time of high insolation. Further, early Holocene glacial advances in many ranges in the Himalaya suggest that the insolation maximum and increased precipitation helped drive glaciation at this time. It is quite noteworthy that glaciation was limited to advances within ~10 km of the present ice fronts during the global LGM in many regions and that there is no unequivocal evidence for a glacial advance during the Younger Dryas Stade. The rapid and possibly differential tectonic uplift of some mountain ranges such as the Pir Panjal and/or the Greater Himalaya may have significantly influenced the style of glaciation throughout the region such as in Kashmir and the Ladakh Range. However, quantifying rates of surface uplift and relating this to glaciation has not been adequately achieved and such links will need to be explored in future studies.

ACKNOWLEDGEMENTS

The author thanks Tim Phillips for drafting many of the figures and Jason Dortch and Kate Hedrick for their discussions on the glacial geology of Northern India and suggestions on this contribution.

REFERENCES

Adams, B., Dietsch, C., Owen, L.A., Caffee, M., Spotila, J., Haneberg, B., 2009. Exhumation and incision history of the Lahul Himalaya, northern India, based on (U-Th)/He thermochronometry and terrestrial cosmogenic nuclide dating techniques. Geomorphology 107, 285–299.

Agrawal, D.P., 1992. Man and Environment in India Through Ages. Book and Books, New Delhi.

Bagla, P., 2009. No sign yet of Himalayan meltdown, Indian report finds. Science 326, 924–925.

Barnard, P.L., Owen, L.A., Finkel, R.C., 2004a. Style and timing of glacial and paraglacial sedimentation in a monsoonal influenced high Himalayan environment, the upper Bhagirathi Valley, Garhwal Himalaya. Sed. Geol. 165, 199–221.

Barnard, P.L., Owen, L.A., Sharma, M.C., Finkel, R.C., 2004b. Late Quaternary (Holocene) landscape evolution of a monsoon-influenced high Himalayan valley, Gori Ganga, Nanda Devi, NE Garhwal. Geomorphology 61, 91–110.

Benn, D.I., Lehmkuhl, F., 2000. Mass balance and equilibrium-line altitudes of glaciers in high-mountain environments. Quatern. Int. 65 (66), 15–30.

Benn, D.I., Owen, L.A., 1998. The role of the Indian summer monsoon and the midlatitude westerlies in Himalayan glaciation; review and speculative discussion. J. Geol. Soc. Lond. 155, 353–364.

Benn, D., Owen, L.A., 2002. Himalayan glacial sedimentary environments: a framework for reconstructing and dating the former extent of glaciers in high mountains. Quatern. Int. 97 (98), 3–25.

Benn, D.I., Owen, L.A., Osmaston, H.A., Seltzer, G.O., Porter, S.C., Mark, B., 2005. Reconstruction of equilibrium-line altitudes for tropical and sub-tropical glaciers. Quatern. Int. 138 (139), 8–21.

Brown, E.T., Bendick, R., Bourles, D.L., Gaur, V., Molnar, P., Raisbeck, G.M., et al., 2002. Slip rates of the Karakoram fault, Ladakh, India, determined using cosmic ray exposure dating of debris flows and moraines. J. Geophys. Res. 107, B9, 2192, 7-1–7-8.

Burbank, D.W., Fort, M.B., 1985. Bedrock control on glacial limits; examples from the Ladakh and Lanshan Ranges, north-western Himalaya. Indian J. Glaciol. 31, 143–149.

Coxon, P., Owen, L.A., Mitchell, W.A., 1996. A late Quaternary catastrophic flood in the Lahul Himalayas. J. Quatern. Sci. 11 (6), 495–510.

Dainelli, G., 1922. Studi sul glaciale. Spedizione Italiane de Filippi nell' Himalaia, Caracorum e Turchestan Cinese (1913–1914). Seriae II 5.3.

Damm, B., 1997. Vorzeitliche und aktuelle Vergletscherung des Markhatales und der nördlichen Nimaling-berge Ladakh (Nordindien). Z. Gletscherk. Glazialgeol. 33, 133–148.

Damm, B., 2006. Late Quaternary glacier advances in the upper catchment area of the Indus River (Ladakh and Western Tibet). Quatern. Int. 145–155, 87–99.

De Terra, H., Paterson, T.T., 1939. Studies on the Ice Age in India and Associated Human Cultures, 493. Carnegie Institute of Washington Publications, Washington, 354 pp.

Derbyshire, E., 1983. The Lushan dilemma: Pleistocene glaciation south of the ChangJiang (Yangste River). Z. Geomorphol. 27, 445–471.

Derbyshire, E., Owen, L.A., 1997. Quaternary glacial history of the Karakoram mountains and northwest Himalayas: a review. Quatern. Int. 38 (39), 85–102.

Dortch, J.M., Owen, L.A., Caffee, M.W., 2010. Quaternary glaciation in the Nubra and Shyok confluence, northernmost Ladakh, India. Quatern. Res. 74, 132–144.

Drew, F., 1875. The Jummoo and Kashmir Territories: A Geographical Account. Indus Publications, Karachi, Pakistan.

Ehlers, J., Gibbard, P.L. (Eds.), 2004. Quaternary Glaciations, Extent and Chronology, Part III: South America, Asia, Africa, Australasia, Antarctica. In: Development in Quaternary Science, 2c Elsevier, Amsterdam, 380 pp.

Fort, M., 1978. Observations sur la géomorphologie du Ladakh. Bull. 'Assoc. Géographes Fr. 452, 159–175.

Fort, M., 1981. Un exemple de milieu périglaciaire sec d'altitude: le versant Tibetain de la chaîne Himalayenne. Recherches Géographiques à Strabourg 16–17, 169–178.

Fort, M., 1983. Geomorphological observations in the Ladakh area (Himalaya). Quaternary evolution and present dynamics. In: Gupta, V.J. (Ed.), Contribution to Himalayan Geology, 2. Inhustan Publishing Corporation, India, pp. 39–58.

Fort, M., Derbyshire, E., 1988. Some characteristics of tills in the Annapurna Range, Nepal. In: Chen, E. (Ed.), Proceedings of the Second Conference on the Palaeoenvironment of East Asia from the Mid-Tertiary. Geology, Sea Level Changes, Palaeoclimatology and Palaeobotany vol. 1. University of Hong Kong, Hong Kong, pp. 195–214.

Grinlinton, A., 1928. The former glaciation of the East Liddar Valley, Kashmir. Mem. Geol. Soc. India 49, 100 pp.

Hedrick, K.A., Seong, Y.B., Owen, L.A., Caffee, M.C., Dietsch, C., 2011. Towards defining the transition in style and timing of Quaternary glaciation between the monsoon-influenced Greater Himalaya and the semi-arid Transhimalaya of Northern India. Quatern. Int. 236, 21–33.

Hewitt, K., 1999. Quaternary moraines vs catastrophic avalanches in the Karakoram Himalaya, northern Pakistan. Quatern. Res. 51, 220–237.

Holmes, J.A., 1988. Pliocene and Quaternary environmental change in Kashmir, northwest Himalayas. Unpublished Ph.D. thesis, University of Oxford, 552 pp.

Holmes, J.A., 1993. Present and past patterns of glaciation in the northwest Himalaya: climatic, tectonic and topographic controls. In: Shroder, J.F. (Ed.), Himalaya to the Sea: Geology, Geomorphology and the Quaternary. Routledge, London, pp. 72–90.

Holmes, J.A., Street-Perrott, F.A., 1989. The Quaternary glacial history of Kashmir, North-West Himalaya: a revision of De Terra and Paterson's sequence. Z. Geomorphol. 76, 195–212.

IPCC (Intergovernmental Panel on Climate Change), 2007. Working Group II: in climate change impacts, adaptation and vulnerability. http://www.ipcc-wg2.org/.

Lehmkuhl, F., Owen, L.A., 2005. Late Quaternary glaciation of Tibet and the bordering mountains: a review. Boreas 34, 87–100.

Mitchell, W.A., Taylor, P.J., Osmaston, H., 1999. Quaternary geology in Zanskar, NW Indian Himalaya: evidence for restricted glaciation and preglacial topography. J. Asian Earth Sci. 17, 307–318.

Osmaston, H., 1994. The geology, geomorphology and Quaternary history of Zangskar. In: Osmaston, H. (Ed.), Buddhist Himalayan Villages in Ladakh. University of Bristol Press, Bristol, pp. 1–35.

Owen, L.A., 1988. Terraces, Uplift and Climate, Karakoram Mountains, Northern Pakistan. Unpublished Ph.D. thesis, University of Leicester, 399 pp.

Owen, L.A., 1991. Mass movement deposits in the Karakoram Mountains: their sedimentary characteristics, recognition and role in Karakoram landform evolution. Z. Geomorphol. 35, 401–424.

Owen, L.A., 1993. Glacial and non-glacial diamictons in the Karakoram Mountains. In: Croots, D., Warren, W. (Eds.), The Formation and Deformation of Glacial Deposits. Balkema, Rotterdam, pp. 9–29.

Owen, L.A., 2004. The Late Quaternary glaciation of Northern India. In: Elhers, J., Gibbard, P.L. (Eds.), Extent and Chronology of Glaciations, volume III: South America, Asia, Africa, Australasia, Antarctica. Development in Quaternary Science, 2c Elsevier, Amsterdam, pp. 201–210.

Owen, L.A., 2009. Latest Pleistocene and Holocene glacier fluctuations in the Himalaya and Tibet. Quatern. Sci. Rev. 28, 2150–2164.

Owen, L.A., Benn, D.I., 2005. Equilibrium-line altitudes of the Last Glacial Maximum for the Himalaya and Tibet: an assessment and evaluation of results. Quatern. Int. 138 (139), 55–78.

Owen, L.A., England, J., 1998. Observations on rock glaciers in the Himalayas and Karakoram Mountains of northern Pakistan and India. Geomorphology 26, 199–213.

Owen, L.A., Benn, D.I., Derbyshire, E., Evans, D.J.A., Mitchell, W., Sharma, M., et al., 1995. The geomorphology and landscape evolution of the Lahul Himalaya, Northern India. Z. Geomorphol. 39, 145–174.

Owen, L.A., Benn, D.I., Derbyshire, E., Evans, D.J.A., Mitchell, W.A., Richardson, S., 1996. The Quaternary glacial history of the Lahul Himalaya, Northern India. J. Quatern. Sci. 11, 25–42.

Owen, L.A., Mitchell, W., Lehmkuhl, F., Bailey, R.M., Coxon, P., Rhodes, E., 1997. Style and timing of Glaciation in the Lahul Himalaya, northern India. J. Quatern. Res. 12, 83–110.

Owen, L.A., Derbyshire, E., Fort, M., 1998. The Quaternary glacial history of the Himalaya. Quatern. Proc. 6, 91–120.

Owen, L.A., Gualtieri, L., Finkel, R.C., Caffee, M.W., Benn, D.I., Sharma, M.C., 2001. Cosmogenic radionuclide dating of glacial landforms in the Lahul Himalaya, Northern India: defining the timing of Late Quaternary glaciation. J. Quatern. Sci. 16, 555–563.

Owen, L.A., Finkel, R.C., Caffee, M.W., 2002a. A note on the extent of glaciation throughout the Himalaya during the global Last Glacial Maximum. Quatern. Sci. Rev. 21, 147–157.

Owen, L.A., Gualtieri, L., Finkel, R.C., Caffee, M.W., Benn, D.I., Sharma, M.C., 2002b. Reply—cosmogenic radionuclide dating of glacial landforms in the Lahul Himalaya, Northern India: defining the timing of Late Quaternary glaciation. J. Quatern. Sci. 17, 279–281.

Owen, L.A., Finkel, R.C., Barnard, P.L., Ma, H., Asahi, K., Caffee, M.W., et al., 2005. Climatic and topographic controls on the style and timing of Late Quaternary glaciation throughout Tibet and the Himalaya defined by ^{10}Be cosmogenic radionuclide surface exposure dating. Quatern. Sci. Rev. 24, 1391–1411.

Owen, L.A., Caffee, M., Bovard, K., Finkel, R.C., Sharma, M., 2006. Terrestrial cosmogenic surface exposure dating of the oldest glacial successions in the Himalayan orogen. Geol. Soc. Am. Bull. 118, 383–392.

Owen, L.A., Caffee, M.W., Finkel, R.C., Seong, B.Y., 2008. Quaternary glaciations of the Himalayan–Tibetan orogen. J. Quatern. Sci. 23, 513–532.

Pant, R.K., Phadtare, N.R., Chamyal, L.S., Juyal, N., 2005. Quaternary deposits in Ladakh and Karakoram Himalaya: a treasure trove of the paleoclimate records. Curr. Sci. 88, 1789–1798.

Phartiyal, B., Sharma, A., Upadhyay, R., Ram-Awatar, Sinha, A.K., 2005. Quaternary geology, tectonics and distribution of palaeo- and present flivio/glacio lacustrine deposits in Ladakh, NW Indian Himalaya-a study based on field observations. Geomorphology 65, 241–256.

Schweinfurth, U., 1968. Vegetation of the Himalaya. In: Law, B.C. (Ed.), Mountains and Rivers of India, 21st International Geographical Congress, Aligarh Muslim University, India 110–136.

Sharma, M.C., Owen, L.A., 1996. Quaternary glacial history of NW Garhwal Himalayas. Quatern. Sci. Rev. 15, 335–365.

Singh, J.S., Singh, S.P., 1987. Forest vegetation of the Himalaya. Bot. Rev. 53, 80–192.

Taylor, P.J., 1999. The Quaternary glacial history of the Zanskar Range, north-western Indian Himalaya. Unpublished Ph.D. thesis, University of Luton.

Taylor, P.J., Mitchell, W.A., 2000. The Quaternary glacial history of the Zanskar Range, north-west Indian Himalaya. Quatern. Int. 65 (66), 81–99.

Walker, H., Pascoe, E.N., 1907. Notes on certain glaciers in Lahaul. Rec. Geol. Surv. India 35, 139–147.

Chapter 68

The High Glacial (Last Ice Age and Last Glacial Maximum) Ice Cover of High and Central Asia, with a Critical Review of Some Recent OSL and TCN Dates

Matthias Kuhle

Department of Geography and High Mountain Geomorphology, Geographisches Institut, University of Göttingen, Goldschmidtstr. 5, D-37077 Göttingen, Germany

68.1. PREFACE

Since 2004 the state of research has not changed very much; so only a brief overview is given in this chapter. There are, however, dates in the newest literature that conflict with the reconstruction of the glacial limits, the chronological classification of the glacial features and even the existence of a Pleistocene ice sheet. A closer look on these dates and their accompanying arguments is given in this chapter (Section 68.4).

The empirical database for the glacier map collected from 1973 to 2002 was already presented in the first publication (Kuhle, 2004a). This also applies to the literature that was cited in the same article, which was the basis of the ca. 40 maps on a scale of 1:1 million. In Section 68.3 of this chapter, references are given to the empirical database of the glacier map which has been added in the meantime, that is, since 2002 (Fig. 68.1). Accordingly, only those papers which include previously unpublished empirical field and laboratory data are cited here.

Since a few map sheets were incorrectly reproduced in Ehlers and Gibbard (2004a), the digital maps in the present volume have been corrected. Due to an error, they were based on a provisional map at a scale of 1:3,000,000 from the middle of the 1990s instead of the final version on a scale of 1:1,000,000 from the year 2002.

68.2. INTRODUCTION: THE STATE OF RESEARCH TO 1973 IN RELATION TO THE AUTHOR'S OBSERVATIONS

A synopsis of earlier results and views on the Pleistocene glaciation of High and Central Asia has been compiled by von Wissmann's (von Wissmann, 1959; for Iran, see also Bobek, 1937). More recently, the former ice cover of Tibet was discussed by Shi and Wang (1979). Their view is reflected in the CLIMAP map (Cline, 1981). According to those authors, glaciation was limited to a 10–20% ice cover of the mountain areas and plateaux of Tibet. In the "Quaternary Glacial Distribution Map of Qinghai-Xizang (Tibet) Plateau" by Shi Yafeng, Li Binyuan, Li Jijun, Cui Zhijiu, Zheng Benxing, Zhang Quingsong, Wang Fubao, Zhou Shangzhe, Shi Zuhui, Jiao Keqin and Kang Jiancheng (cf. Shi et al., 1992), this view is largely repeated. However, there is one notable alteration: Zhou Shangzhe conceded that a Pleistocene ice sheet had existed in the Animachin region. Consequently, the map in Shi et al. (1992) includes a plateau glacier of 400 × 300 km.

Time and again, from since the turn of the twentieth century, individual researchers, such as von Loczy (1893), Oestreich (1906), Handel-Mazzetti (1927), Dainelli and Martinelli (1928), Norin (1932), De Terra (1934) and others, have described former ice-marginal positions scattered throughout the high regions of Asia. According to the author's calculations, these works reflect an equilibrium line altitude (ELA) depression of over 1000 m and suggest, locally at least, a significantly more extensive glacier cover than envisaged by von Wissmann. Other early researchers, such as Huntington (1906), Tafel (1914), Odell (1925), Prinz (1927), Trinkler (1932) and Zabirov (1955), found evidence for formerly only slightly larger glaciers which reflected only a few hundred metres of ELA depression.

The author has carried out about 40 expeditions and research visits to High Asia since 1973 (Fig. 68.1),

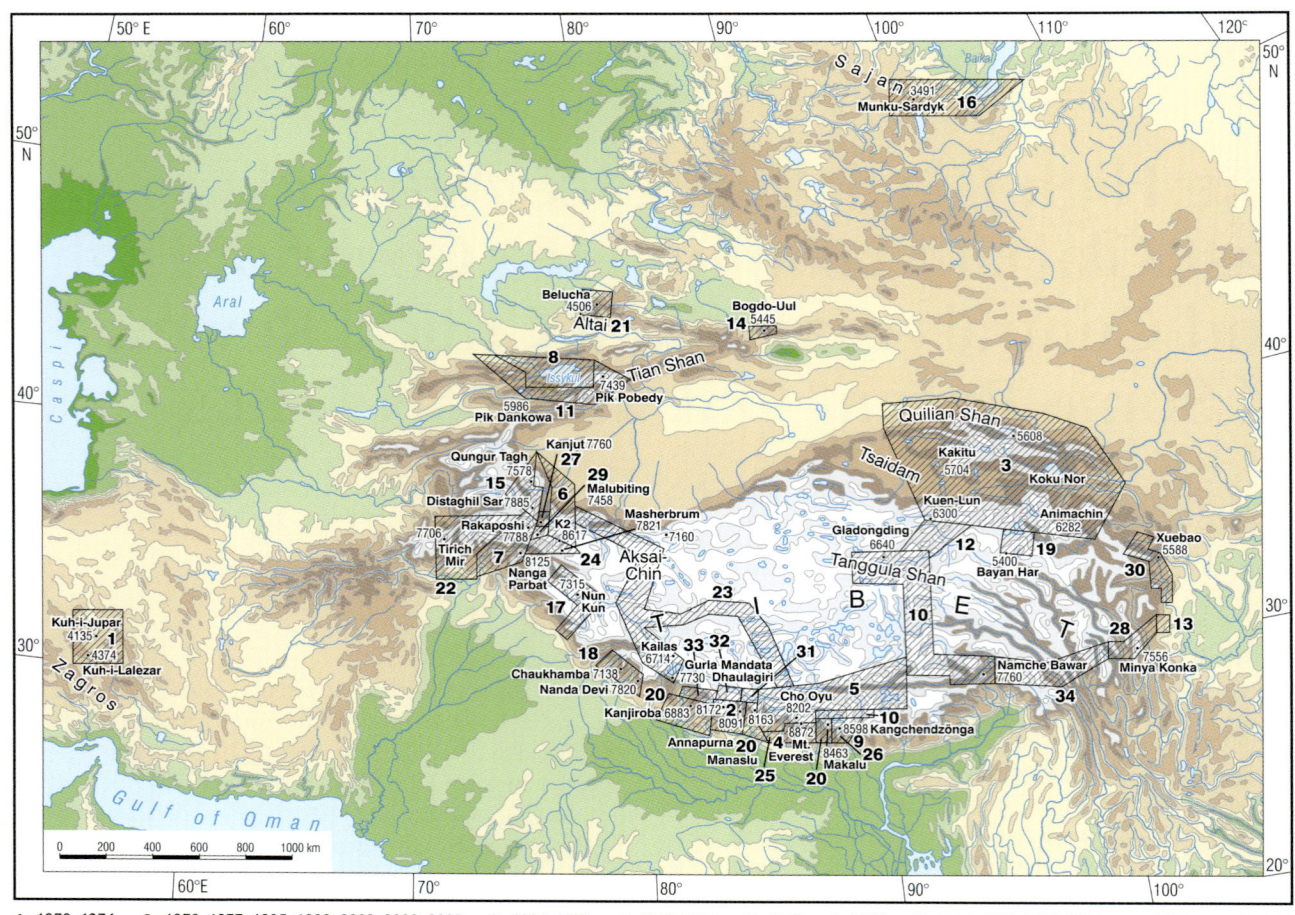

FIGURE 68.1 Areas investigated by the author in High and Central Asia 1973–2009 (cf. Fig. 68.9).

1: 1973, 1974   2: 1976, 1977, 1995, 1998, 2000, 2002, 2005   3: 1981, 1998   4: 1982, 2003   5: 1984   6: 1986   7: 1987, 1992, 1995, 2000   8: 1988   9: 1988/89
10: 1989   11: 1991   12: 1991   13: 1984, 1991   14: 1986, 1992   15: 1992, 2006   16: 1993   17: 1993, 1996   18: 1993, 2004   19: 1994   20: 1994/95, 2000, 2004, 2007, 2008
21: 1995   22: 1995   23: 1996   24: 1997, 2006   25: 1998, 2008   26: 1999   27: 1999, 2000   28: 2000   29: 2000   30: 2002   31: 2004, 2007   32: 2005
33: 2009   34: 2009

Draft: M. Kuhle (2009)

some of which extended to several months, with the purpose of reconstructing the extent of Pleistocene glaciation. The results of those expeditions have allowed the reconstruction of glacial areas for all of Tibet and parts of Central Asia. Reconstructions are supported by data from some earlier authors (see above) but are in glaring contrast to the negligible ice cover suggested by the CLIMAP team (Cline, 1981) and by the "Quaternary Glacial Distribution Map of Qinghai-Xizang (Tibet) Plateau" (Shi et al., 1992). According to the present author's reconstruction, during the Last Ice Age, the Tibetan ice sheet covered approximately 2.4 million km^2 and is estimated to have had a central thickness of about 2 km (Fig. 68.9). At the plateau margins, ice discharged through the surrounding mountains as steep outlet glaciers (e.g. Kuhle, 1987a, 1988a, 1989, 1998, 2005b, 2005c, 2009, 2010).

While the author's research results are based on classical glacial-geological field investigations (Kuhle, 1974–2010; see also part of the references in Kuhle, 2004a), several authors including Colgan et al. (2006), Seong et al. (2008a) and Owen et al. (2008) have recently chosen a different approach. Their investigations are largely based on optically stimulated luminescence (OSL) and terrestrial cosmogenic nuclide (TCN) dates, which are discussed below.

Since the first publication of the results of the INQUA Extent and Chronology of Quaternary Glaciations project in 2004, the author has carried out further research expeditions to High and Central Asia (Fig. 68.1) so that the regional data and information have increased (Kuhle, 2003, 2004b, 2005a,b,c, 2006a,b, 2007a,b, 2008, 2009, 2010). These data, which are not yet fully published, confirm and corroborate the overall picture of the High- and Central-Asian glacier extent (Section 68.3).

68.3. GLACIAL LIMITS: REFERENCES TO THE NEW EMPIRICAL GLACIAL MAP DATABASE SINCE 2002

New regional results, based on expeditions between 2002 and 2009 and the corresponding field and laboratory data on which the glacier map and the glacial chronology are based, have been published in detail elsewhere (see below). The new publications are listed here together with the corresponding investigation areas (cf. Fig. 68.1).

No. 1: Zagros: Kuh-i-Jupar, Kuh-i-Lalezar and Kuh-i-Hezar Massifs (Kuhle, 2007b)

No. 2: Western Annapurna Himal S-slope; lower and middle Thak Khola (Kuhle, 2005c, 2007a)

No. 4: Khumbu and Khumbakarna Himalaya (Kuhle, 2005b,c. 2007a)

No. 6: Karakoram N-slope (Kuhle, 2005a,c)

No. 7: W-Karakoram: Hispar Valley, Destighil Sar-Massif, Spantik-Malubiting Massif W-side; Husainabad Valley, Bar Valley, Batura-Massif, Rakaposhi-Massif, Hunza Valley, Gilgit Valley, Indus Valley with side-valleys (Kuhle, 2006b, 2007a, 2008, 2009)

No. 15: NW-Karakoram: Chapursan Valley, Shimshal Valley (Kuhle, 2006b, 2007a, 2008, 2009)

No. 18: Garwhal Himalaya: Gangotri Valley, Bargarathi Valley, Nandakini Valley, Mandakini Valley, Ala-knanda Valley, Nanda Devi-Massif, Kamet Massif (Kuhle, 2006a)

No. 29: Muztagh Karakoram (Kuhle, 2004b, 2006b, 2008, 2010)

No. 31: Damodar Himal (Kuhle, 2010)

No. 32: S-Tibet, Mustang, uppermost Thak Khola (Kuhle, 2005c, 2006a)

No. 33: Tibetan Himalaya: Dolpo, Kanjiroba Himal and W-Dhaulagiri-N-face, upper Barbung Khola (Kuhle, 2010)

No. 34: SE-Tibet with the glacial gorges of Saluen (or Nu Jiang), Mekong (or Lanoang Jiang), Jangtsekian (or Jinsha Jiang) and Jalunkiang which are eroded backward into the plateau (Kuhle, 2010).

68.4. DATING METHODS AND EMPIRICAL EVIDENCE FOR THE GENERAL OVERESTIMATION BY NUMERICAL DATING IN HIGH ASIA

The Quaternary geological method and the morphological features applied by the author are explained in detail by means of photographs in the volume *Quaternary Glaciations—Extent and Chronology, Part III: South America, Asia, Africa, Australia, Antarctica* (Kuhle, 2004a).

Here, the chronological classification based on the author's field and laboratory investigations supplemented by ^{14}C-dates, as well as the OSL and TCN determinations from different authors, is discussed against the glacial-geological context of the maximum past glaciation of High and Central Asia (i.e. of Tibet, Himalaya, Karakoram, Tian Shan, etc.).

68.4.1. Examples and Key Sites from the Himalaya and South Tibet

Until recently, age control of the glacial features mapped in the field has been largely based on radiocarbon dating. ^{14}C dates do exist, for example, from the Khumbu Himalaya, that is, Mount Everest and Cho Oyu-S-slopes (Figs. 68.2, 68.3, 68.8; Kuhle, 1986, 1987a,b, 2005b). They can be considered as being representative for the monsoon-facing Himalaya slope (Table 68.1). The data were obtained in connection with a large-scale and detailed glaciological analysis of the area (Kuhle, 1988a,b, 1990b, 1991, 1998, 1999, 2001a,b, 2002a). In the meantime, diverse publications that have included OSL and TCN dates have become available from this area (e.g. Finkel et al., 2003; Owen et al., 2008) and are summarised and compared. A disadvantage of these overviews is that they did not carry out a detailed glacial geomorphological analysis of the area. Obviously, these authors' work has been focused on the vicinity of the sampled sites. Consequently, the broader geomorphological setting has been somewhat neglected.

For instance, in the tributary valleys of the Ngozumpa Drangka, a western parallel valley of the Khumbu and Tshola Drangka, only ca. 10 km from the sampling sites, referred to by Finkel et al. (2003) and Owen et al. (2008) (cf. Fig. 68.2, 68.3, 68.8; see below), the author has obtained six ^{14}C dates of between 2.1 and 4.2 ka for the Neoglacial Stages V (Nauri Stage), VI (Older Dhaulagiri Stage) and VII (Middle Dhaulagiri Stage) (cf. Table 68.1) (Kuhle, 1986, 1987a: 408; Kuhle, 1987b, 2005b). The evidence is glaciologically consistent with sequences of end moraines and corresponds to a rising ELA from old to young (cf. Fig. 68.2). According to the relative chronology, these three Neoglacial stages correspond to an ELA depression of between 560 and 280 m (Kuhle, 2005b). Based on its chronology and its state of preservation, the moraine in the main valley near the Periche settlement (Fig. 68.3: V between Profile 12 and Profile 16) can be classified as belonging to Stage V (Table 68.3; Kuhle, 2005b). However, for this Periche Stage, with an ELA depression of no more than 300–350 m, Finkel et al. (2003) have provided TCN dates that put it into the Last High Glacial (23 ± 3 ka) and a slightly weaker glacier re-advance at about 16 ± 2 ka. Thus, compared with the ^{14}C data, the TCN evidence gives higher ages by a factor of 6.5 on average.

Finkel et al. (2003) assign the allegedly oldest and most extensive former glaciation of the Khumbu area at

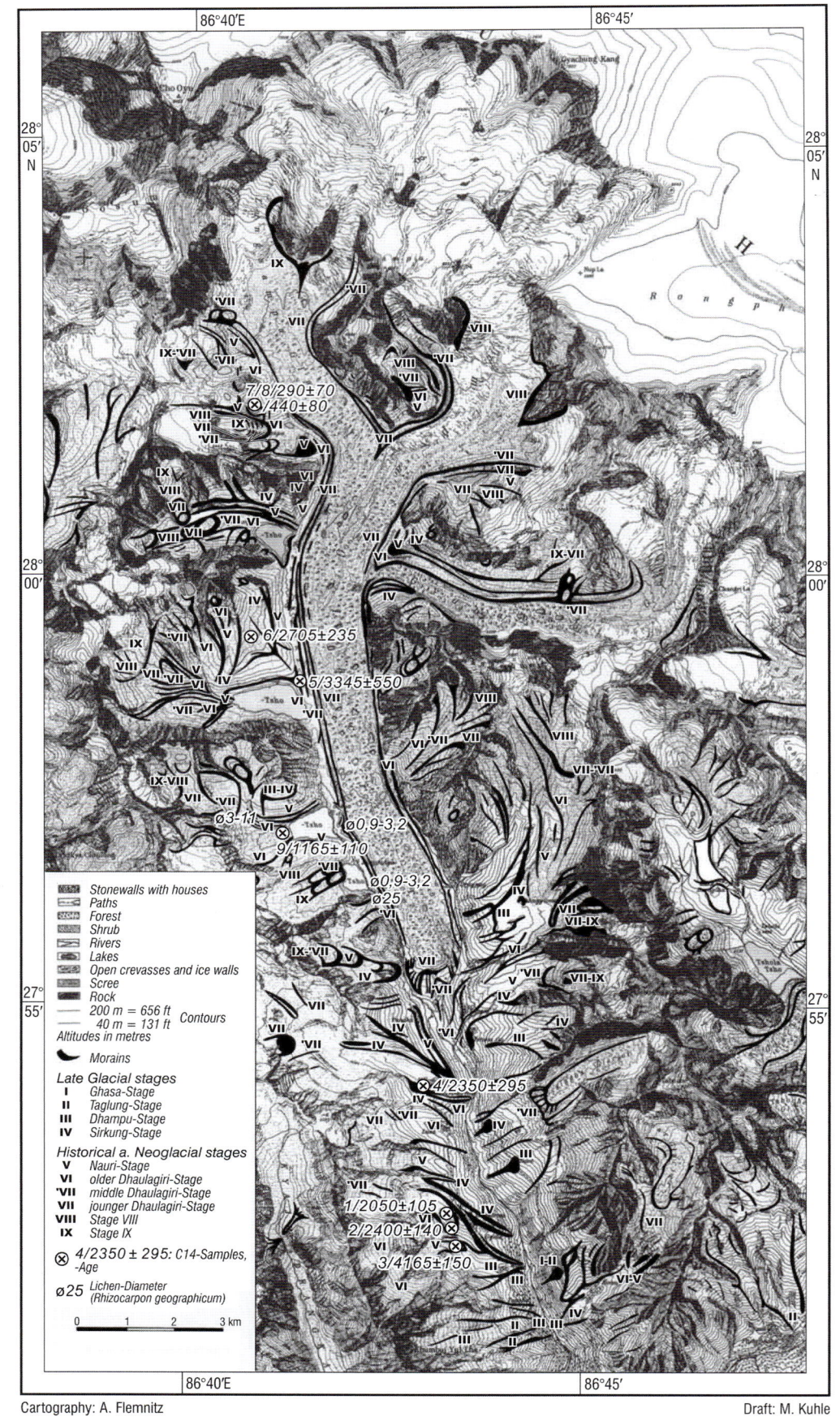

FIGURE 68.2 Late Glacial, Neoglacial and historical glacier stages on the south slope of Cho Oyu with the Ngozumpa glacier system and the Ngozumpa Drangka. Reconstruction of the glacier stages, after Kuhle, 1986, 1987b (cf. Table 68.1; Figs. 68.3 and 68.8). Basic topographic map: Khumbu Himal 1:50,000, Schneider (1978).

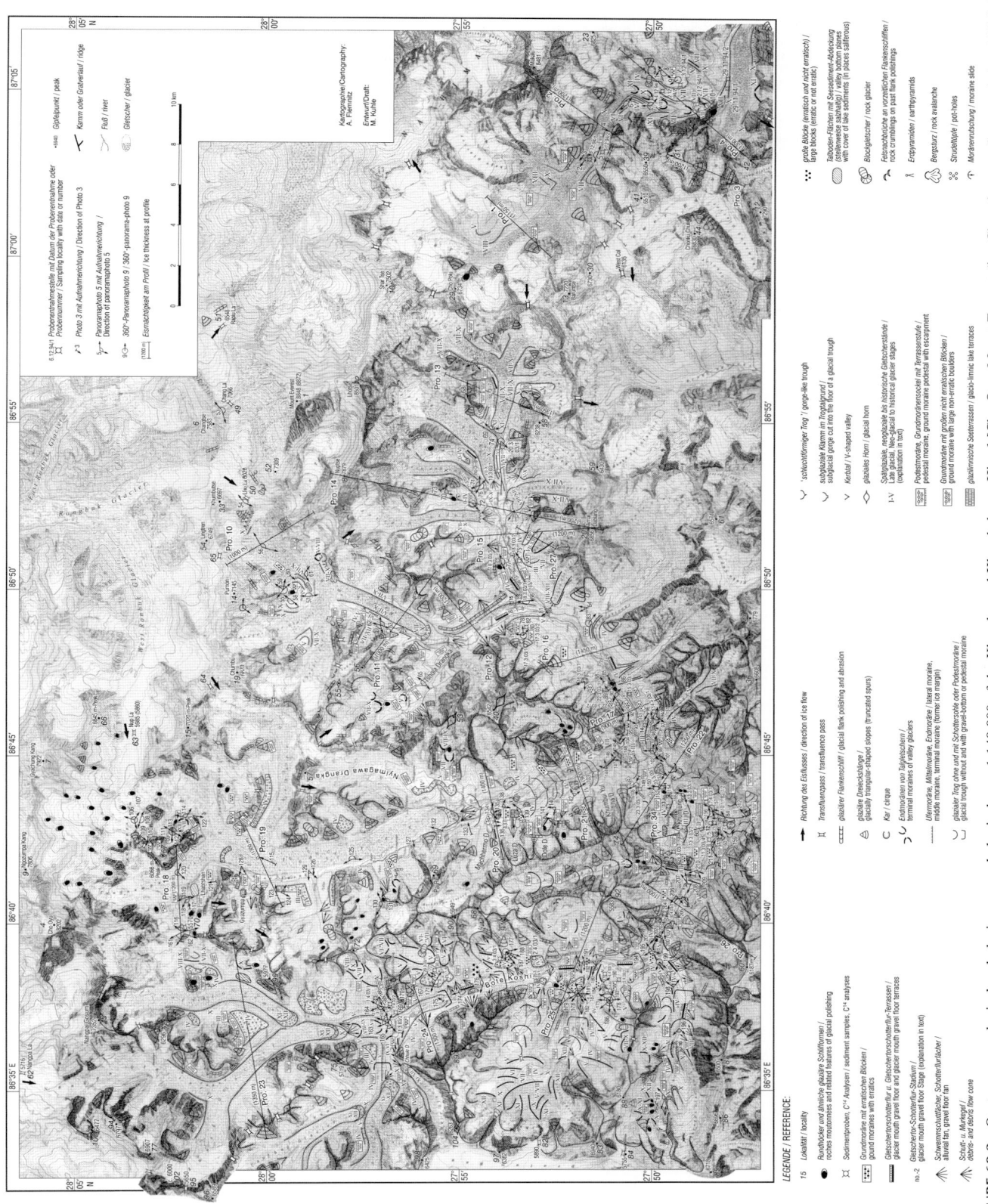

FIGURE 68.3 Quaternary geological and glaciogeomorphological map 1:140,000 of the Khumbu and Khumbakarna Himal [Cho Oyu, Mount Everest (i.e. Chogolungma, Sagarmatha) and Makalu massifs] in the Central Himalaya with sampling, photograph and profile localities, after Kuhle (2005b) (cf. Table 68.1, Nos. 10–15; Fig. 68.1, No. 4; Figs. 68.2 and 68.8). Basic topographic map: Khumbu Himal 1:50,000, Schneider (1978).

TABLE 68.1 ^{14}C-Datings from the Khumbu Himalaya After Kuhle (1986, 1987b) (Fig. 68.1: No. 4)

No.	Material	Altitude (m a.s.l.)	Locality	Depth of Sample (m)	Underlying Substratum	Recent Vegetation Vover	δ^{13}C (‰)	^{14}C Age (Years Before 1950)	Comments
1	Soil	4410	Adjacent valley of Dole (Dole Drangka), end moraine; 27°52′32″N/86°43′21″E	0.3	Moraine, unconsolidated rock—metamorphosed greywacke	Cyperacae turf	−26.0	2050 ± 105	See Fig. 68.2, No. 1
2	Peat	4400	Adjacent valley of Dole, tongue basin; 27°52′21″N/86°43′24″E	0.25	Gneiss gravel	Alpine turf and moist alpine scrub	−24.8	2400 ± 140	See Fig. 68.2, No. 2
3	Muds of acid alpine moor soil	4230	Adjacent valley of Dole, end moraine; 27°52′10″N/86°43′30″E	0.5	Glacial till—gneiss and metamorphosed greywacke	Moist alpine scrub	−24.8	4165 ± 150	See Fig. 68.2, No. 3
4	Humus soil and peat	4440	Machhermo Khola (Drangka), lateral to end moraine; 27°54′06″N/86°43′03″E	0.2	Moraine material, sand with blocks of gneiss	See No.2	−23.8	2350 ± 295	See Fig. 68.2, No. 4
5	Soil	4910	Lateral depression right of the Ngozumpa glacier; 27°58′45″N/86°41′30″E	0.12	Glaciofluvial sand, sand bar material	See No. 1	−25.1	3345 ± 550	See Fig. 68.2, No. 5
6	Peat of hummocks	5230	340-m-high pedestal ground moraine terrace right of the Ngozumpa glacier; 27°59′16″N/86°41′01″E	0.6	Coarse moraine blocks of greywacke and gneiss	Alpine turf	−24.5	2705 ± 235	See Fig. 68.2, No. 6

Sampling localities: see Figs. 68.2 and 68.3.

up-valley morainic boulders at the exit of the Tsola Khola (Tshola Drangka). These boulders are thought to lie at an altitude above 4500 m a.s.l. Here, TCN determinations give ages between 86 and 33 ka. However, the recent tongue of the Tshola glacier is situated at 4500 m a.s.l. (cf. Fig. 68.3: on the left above Profile 12) (Kuhle, 2005b), so that—as shown by the map of glacier reconstruction in the Khumbu area by Owen et al. (2008)—this locality was overridden by ice up to at least 4800 m a.s.l. during the Periche Stage (cf. Kuhle, 2005b).

The poor description of the localities, topography and geomorphological connections in Finkel et al. (2003) allows two different interpretations. Either the samples were really taken at an altitude of over 4500 m a.s.l. Then there is no other possibility than that they are simultaneous with the Periche Stage. If this is correct, then there would already be a variation of the TCN data by the factor ~4 within this stage (cf. Fig. 68.3; Kuhle, 2005b). Or, if the dated samples were taken at least 300 m higher, that is, above 4800 m a.s.l., they might represent an older event. Assuming the TCN age of 86 ka given by Finkel et al. (2003) represents an overestimation by a factor of 6.5, the age of these moraines is reduced to a Late Glacial age, that is, about ~13 ka. This would therefore correspond without difficulty to Kuhle's Late Glacial glacier stage IV (Sirkung Stage), established glaciosedimentologically as well as geomorphologically for this locality (Kuhle, 1987a, 2005b). Consequently, the Last Ice Age (Last Glacial Maximum, LGM) glacier extent, suggested by Finkel et al. (2003) for the Khumbu area, does not match their age determinations. The equivalent glacier terminus would have reached further down-valley than that of their LGM Periche Stage (see above), and even that would be the *youngest* Late Glacial stage. The LGM advance must have reached a maximum position much further down the valley, in agreement with the present author's opinion (Kuhle, 1987a, 1998, 2001b, 2005b).

However, instead of discussing any geomorphological alternative for the formation of the landforms and sediments down-valley, Finkel et al. (2003) and Owen et al. (2008) categorically declare that evidence of a larger glaciation does not exist in the Khumbu area "because intense erosion and slope instability have destroyed much of the glacial evidence" (Finkel et al. 2003: 562). As Kuhle (1987a,b, 1998, 2001b, 2005b) has demonstrated, their interpretation is contradicted by the empirical evidence (cf. Figs. 68.3 and 68.8). At 4850 m a.s.l., that is, ~750 m up-slope in the same valley cross-profile of the alleged LGM glacial terminus at Periche (Finkel et al., 2003; Fig. 1), large-scale till cover with erratic boulders (Fig. 68.3: between Profile 12 and Profile 16: Nos. 78, 80; Fig. 68.5: Nos. 18, 19) (cf. Kuhle, 2005b: Photo 78–80) is found on the orographic right-hand valley slope. These tills extend 30 km down-valley on both sides of the thalweg (Figs. 68.3–68.8; cf. Kuhle, 1987a, 2005b). In addition, glacially polished landforms (Fig. 68.3 and 68.8) (cf. Kuhle, 2001b, 2004a, 2005b) and potholes, situated in the valley flanks high above the thalweg (Figs. 68.3 and 68.8) (Kuhle, 2004a), provide evidence in support of this Pleistocene valley glaciation reaching ca. 3000 m further down-valley. These features were formed by outlet glaciers from southern Tibet (Fig. 68.8 and 68.9 on the right-hand side of Mount Everest). The lowered ELA is also reflected by corresponding cirques (Kuhle, 1987a). The accompanying lowest ice-marginal position of the glacier tongue of the Imja–Dudh-Koshi parent glacier has been reconstructed in the confluence area of the Inkhu Khola at 27°28′30″N/86°43′20″E at ca. 900 m a.s.l. (Fig. 68.8; Kuhle, 2005b; cf. the findings from the E-adjacent Kangchendzönga massif in Kuhle, 1990b). No dates are available, but due to the state of preservation and extent of these morainic deposits and polished landforms—in the most erosion-intensive mountains on earth—it is unlikely that this ice margin could be older than the Last Glacial period.

Finkel et al. (2003) and Owen et al. (2008), however, ignore the evidence for this much more extended LGM

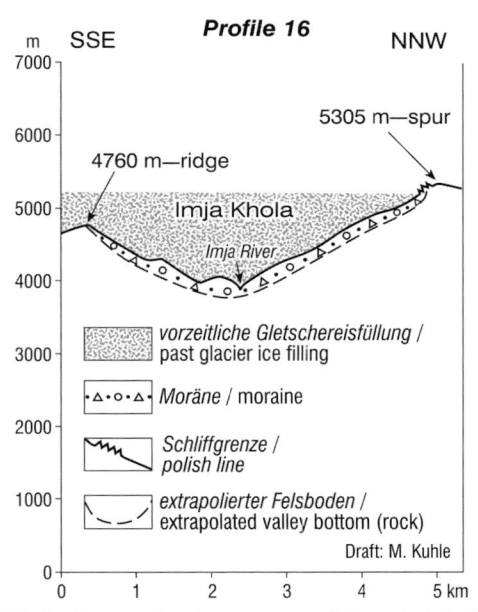

FIGURE 68.4 Cross section (not exaggerated) from the left side down-valley across the lower Imja Khola (Drangka) with the 4760-m ridge to the 5305-m spur with its glacier ice filling reconstructed for the Last Glacial period (ca. 60–18 ka = Stage 0; Table 68.3) (Kuhle, 2005b: Photograph 83 from behind ▲ below No. 36 up to the right margin). On the orographic right side is an upper abrasion limit at 5250 m; on the orographic left, the rock ridges have been glacigenetically rounded and abraded, that is, they have been overflowed. The entire valley cross-profile is buried by ground moraine. According to the abrasion limit, a maximum past ice thickness of ca. 1450 m can be inferred up to the rock ground of the valley bottom. The glacier surface lay ca. 1500 m above the simultaneous high-glacial ELA. The present-day Imja river is cut into the ground moraine (locality: Fig. 3, Profile 16) (after Kuhle, 2005b).

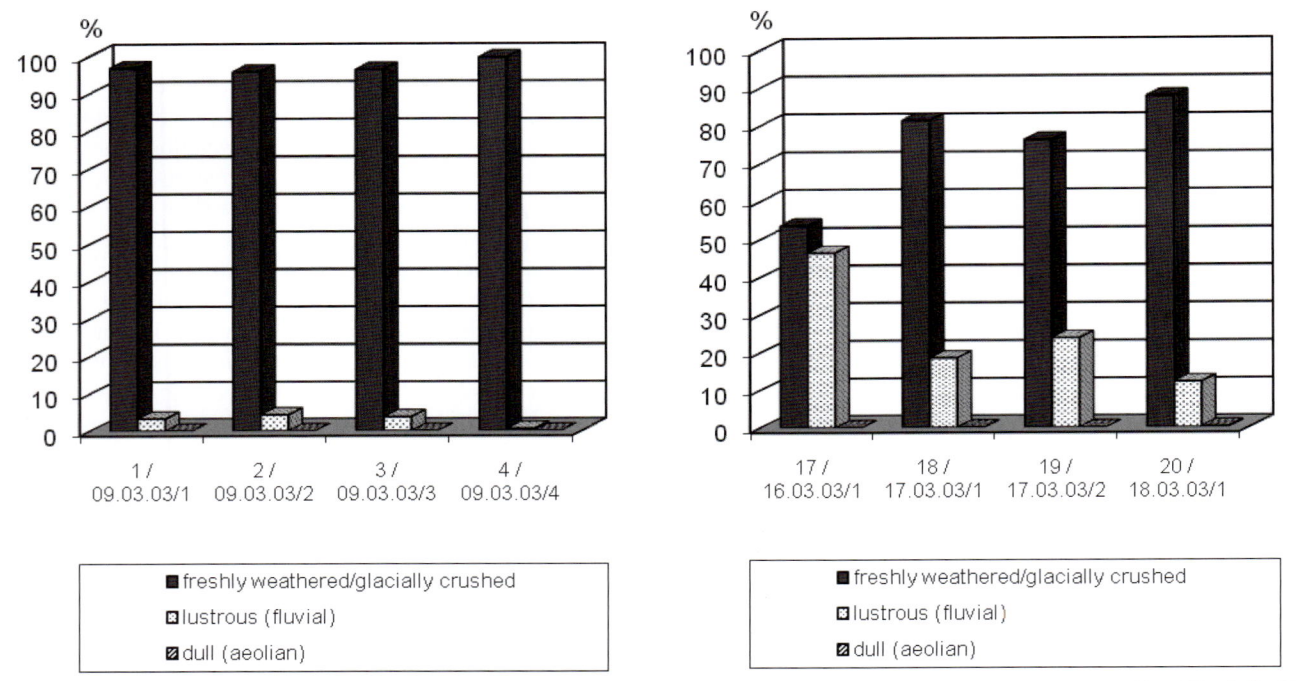

FIGURE 68.5 Morphometric quartz-grain analysis of eight representative samples from the Khumbu Himalaya with the Bote Koshi and Dudh Koshi Nadi (sampling locality Fig. 68.3) (after Kuhle, 2005b). Sampling: M. Kuhle.

glaciation and the corresponding ^{14}C dates and observations published since 1986 (among others Kuhle, 1986, 1987a,b, 1988a,b, 1998, 1999, 2001a,b, 2002a, 2004a, 2005b). Further, Owen et al. (2008) in their summary paper are of the opinion that N of the main ridge of the Himalaya, an ELA depression of merely 300 m resulted in even more limited glaciation during the LGM than they had reconstructed for the south slope (ibid., Figs. 13A and 4D). However, the Periche Stage on the south slope, which they classify as belonging to the LGM (see above), also shows an ELA depression of 300–350 m.

But also here, N of the Himalaya main ridge, in addition to many glacial-geological findings (see below), the available ^{14}C dates suggest a different interpretation (Fig. 68.1, Nos. 2, 5, 10, 23, 31–33; Kuhle, 1988a,b, 1991, 1998, 2001a, 2002a). However, this is also completely ignored by the authors cited above: in the area of the Tsangpo bend in Southern Tibet, 1 km down-valley from the Ganga Bridge (29°18′N/94°21′E, at 3090 m a.s.l.), a last strong glacier advance was reconstructed for about and after 9.82 ± 0.35 ^{14}C ka BP (Table 68.2). Moreover, there are dates from wood from an 80-m-high exposure (conifer trunks which were taken from 2 m above the base) which were analysed in 1989 (Kuhle, 1991). The eight ^{14}C samples were taken from the lower part of the exposure (32 m high) and show ages of up to 48.58 ^{14}C ka BP. The sequence of data in Table 68.2 is listed from top to bottom of the exposure. Varved clays 8 m thick (Kuhle, 1991) overlie the dated basal lacustrine sands. They provide evidence of an ice-dammed lake in this lower section of the Tsangpo valley at only ~3000 m a.s.l. Accordingly, this ice-dammed lake formed either at or after 48.58 ^{14}C ka BP and disappeared before 9.82 ka ^{14}C BP. Therefore, it may belong to the LGP or LGM (Table 68.3, glacier stage 0) up to the Late Glacial (Table 68.3, glacier stages I–IV). The varve counts showed the lake to have existed for approximately 1000 years. It was situated (Fig. 68.1, No. 10 left-hand side of Namche Bawar) between the ice complexes I2 and I3 (Kuhle, 1998, 2004a) and was dammed up by the Nyang Qu glacier (Linchi Chu glacier). The latter, which was an outlet glacier of the ice-sheet complex I2, flowed down the Nyang Qu (Linchi Chu) valley and eventually reached the Tsangpo valley (Kuhle, 1998), 17.5 km down-valley from the exposure (see above). Till and lateral moraines (Kuhle, 1991) confirm that the glacier turned into the Tsangpo valley. Thus the ice-dammed lake came into being. On the one hand, these ^{14}C dates indicate the timing of the ice sheet in Central Tibet during the LGP or LGM. At the same time, these findings demonstrate that this region that is entirely glacier-free today, north of the Himalaya, once must have been covered by an immense ice sheet, extending even to the lowest parts of the Tsangpo valley at about 3000 m a.s.l., for which an ELA depression of at least 700–800 m was necessary.

Owen et al. (2008) envisage an ELA depression of only 100–150 m for the Mount Everest N-slope, while Burbank and Kang (1991) assume ca. 400 m. These limit is impossible, because the ice margin Burbank and Kang (1991) refer

FIGURE 68.6 Photograph taken from the orographic left flank of the Dudh Koshi Nadi, on the right side of the Handi Khola-exit in the Lukla settlement at ca. 2800 m a.s.l., approximately 800 m above the thalweg of the Dudh Koshi Nadi (27°41′02″N/86°44′45″E; Fig. 8 left above (north) of Handi, K.), facing E, showing a 2-m-deep excavation exposure of ground moraine in the Würmian (Weichselian) LGP ground moraine pedestal. The exposed moraine material (■) consists of more or less round-edged and facetted augen-gneiss boulders up to extensions of 1.5 m (○), embedded into a strongly condensed matrix, which is rich in clay and associated with edged, less metamorphic components of schist. Analogue photograph M. Kuhle, 9 March 2003.

to lies only 200 m lower than the current glacier terminus (Kuhle, 2007a). In their investigation area of only 8 km in length, that with regard to their method is much too small for the statements they make, the authors classify the lowest former ice-marginal position they have found as belonging to marine isotope stage (MIS) 6 (Middle Pleistocene). Their interpretation is based on relative degrees of weathering. However, the allegedly high degree of weathering of the tourmaline granite boulders on the moraine probably results from hydrothermal processes that had already taken place in the solid bedrock (Heydemann and Kuhle, 1988). Further, Burbank and Kang do not consider the five to seven lower end moraines and ice-marginal positions situated far beyond and down-valley from their small study area, which have been mapped in the neighbouring areas (cf. Fig. 68.8; Kuhle, 1987a, 1988a,b, 1989, 1990b). In addition, the authors fail to consider the erratic boulders high above the thalweg, and the thick ground moraine covers present tens of kilometres down the Rongbuk valley, for example, at 4350 m a.s.l. (Kuhle, 1988b) and 3950 m a.s.l. in the

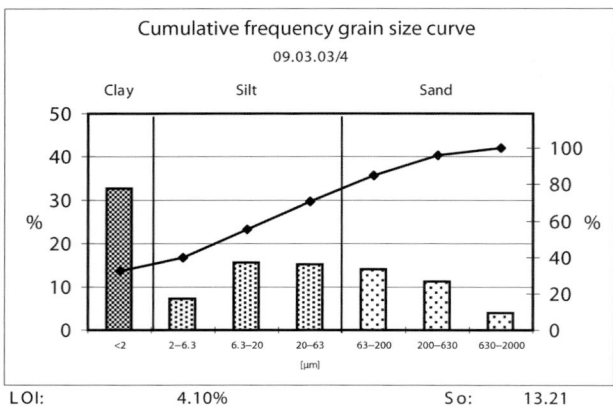

FIGURE 68.7 Fine material matrix of an orographic left ground moraine pedestal at 2480 m a.s.l. ca. 780 m above the Dudh Koshi river, taken from an exposure 3 m below the surface. The primary maximum with 33% is in the clay, and the secondary peak with 16% is in medium silt. The bimodal course of the cumulative curve of the fine-material matrix (i.e. inclusive of the large components as the boulders and the skeleton portion, trimodal course of the curve of the entire material) is typical of till (ground moraine) matrix. Here, the very high clay portion is typical of the moraine, especially in the high mountain relief. It is a decametre-thick accumulation of loose rock with numerous embedded, edged, round-edged and facetted boulders up to metres in size. The sorting coefficient calculated according to Engelhardt (1973, p. 133) amounts to So = 13.21 (So = $\sqrt{Q_3/Q_1}$). This is an extremely high value, which also speaks in favour of moraine; the loss of ignition (LOI) amounts to 4.1%. Locality: 27°40′20″N/86°43′20″E, N above the Surke settlement (Surket); Fig. 8: left above (north) of Khari Khola (Fig. 5: No. 4) (after Kuhle, 2005b).

lower Rongbuk valley and its continuation, the Dzakar Chu. Similar spreads also occur in the valley chamber of Kadar in the Pum Qu, the upper Arun valley, a further continuation of Rongbuk and Dzakar Chu, very high above the thalweg, which lies at only just 3700 m a.s.l. (Kuhle, 1991).

On the basis of his ice-marginal position chronology, the author has classified the ice-marginal position in the upper Rongbuk valley as belonging to the oldest Neoglacial stage (Nauri Stage V, ca. 4.0–4.5, i.e. 5.5 ^{14}C ka BP, middle Holocene; cf. Table 68.3) (Kuhle, 1988b). This is in strong contrast to Burbank and Kang (1991) who place the limit in the Middle Pleistocene. A snowline depression of only 100–150 m for and since the glacial maximum of MIS 6 (Middle Pleistocene) compared to the present snowline—as postulated by the authors—is not possible.

68.4.2. Example and Key Sites from Central Tibet

The publication of Colgan et al. (2006) concerning TCN dates from the Tanggula Shan in Central Tibet (Fig. 68.1, No. 10) is a similarly prominent example. The authors give a TCN age of ~32 ka for a moraine only 3 km from the modern glacier front. If one applies the above extracted factor of overestimation between 4 and 6.5, an age of ~5–8 ka is the result. That would be a more realistic age and well in accordance with the Quaternary geological findings of Kuhle (1991).

In the highest mountain group of the Tanggula Shan, that is, the Geladaindong massif, situated 50 km to the NW (ibid.: Fig. 43, No. 1), remnants of dead ice are found at some distance (up to ca. 100 m) from the present ice margin of the Geladaindong (Gladongding) east glacier (ibid.: Photo 1 and 5), indicating the rapid current back-melting of the glacier tongue. Up to ca. 1–1.5 km from this glacier tongue, no alpine meadow vegetation is encountered, although it is common in the further surroundings. The same applies to the forefields of neighbouring glaciers. This demonstrates the intensive glacier retreat of the Tanggula Shan glaciers—not only today but also during historical times. These obviously historical moraines lie only a few kilometres apart, and their altitudes differ by only some 100–200 m from the Holocene Neoglacial maximum stage glacier margins (Nauri Stage V; cf. Table 68.3) formed ca. 4.0–5.5 ^{14}C ka BP (Kuhle, 1991). The area close to the N-Tanggula Shan pass, where Colgan et al. (2006) collected their TCN samples, shows the same extremely small vertical distances of current, historical to Neoglacial glacier stages and also the same Quaternary geological characteristics (Kuhle, 1991).

Apart from this glaciological and vegetational evidence of the historical to Neoglacial age of the features, the TCN dates imply an ELA depression of about 100 m for the LGM. This would be a global singularity that would have to be explained. N and S of the Tanggula pass till sheets and glacial landforms are found over a distance of over 100 km that prove a widespread former glaciaton (ice sheet) of the Tibetan plateau in this area (Kuhle, 1991, 1995). Unfortunately, these findings were not discussed by Colgan et al. (2006).

68.4.3. Example and Key Sites of the Semi-Arid Western Margin of Tibet

For the dry Karakoram extreme age, overestimations by OSL and TCN dates have been reported, that are in disagreement with the Quaternary geological evidence.

Spencer and Owen (2004) have compared OSL and TCN dates from the Hunza valley (NW Karakoram). From here, there were two previous ^{14}C-determinations which give an indication for the calibration. They come from an end moraine of the Minapin glacier and provide two ^{14}C ages of 325 ± 60 and 830 ± 80 ^{14}C a BP (Derbyshire et al., 1984; cited after Spencer and Owen, 2004). These ages are classified as belonging to a glacial stage called Pasu I (t7). From the same stage (t7), Spencer and Owen (2004) have now obtained a much older OSL date of 8.4 ± 0.9 and 4.3 ± 0.4 ka on the Batura glacier situated

60 km away. Because the moraines of the Pasu I Stage occur in the close vicinity of the recent glacier tongues, Spencer and Owen consider the OSL ages as too old and the ^{14}C data as correct. The present author agrees with them. Consequences should be drawn with regard to other OSL data which are not—as in this case—checked by ^{14}C-dates. The rejected OSL dates are apparently consistent with the relative chronology of the other OSL determinations given by Spencer and Owen (2004). They also match with Kuhle's dates if the results are divided by 10, a calibration suggested by the comparison of the youngest dates with the radiocarbon dates. The TCN dates are more coherent than those of OSL. The next older stage (t6) of the chronology, according to the TCN data, should be 9.0–10.8 ka old.

Stage (t4), only some kilometres from the recent glacier tongue, is interpreted by Spencer and Owen (2004) as marking the LGM limit. Its OSL age is 31.5 ± 5 ka and the TCN age 21.8–25.7 ka. Considering the younger OSL and TCN dates checked by the radiocarbon dates, the present author assumes that the ages are highly overestimated.

Thus, the TCN dating of Seong et al. (2008a) from the E-adjacent K2-Baltoro area in the Central Karakoram requires correction. Firstly, it should be pointed out that their TCN dates are inconsistent. According to these dates in the Hunza valley (Spencer and Owen, 2004), the glacier advance allegedly amounted to only a few kilometres during the LGM, which is dated here to $\sim$24 ka TCN. In the Baltoro area, situated only 30–70 km to the SE, the LGM glaciers should have advanced over 100 km. This advance gives a TCN age of only 16 ka. If correct, this date should be referred to as being from the middle *Late Glacial* (Heinrich Event 1).

Unfortunately, many dates have been carried out without consideration of the local, as well as regional Quaternary geological context. Taylor and Mitchell (2002) write with regard to a paper on the Lahul Himalaya: "In

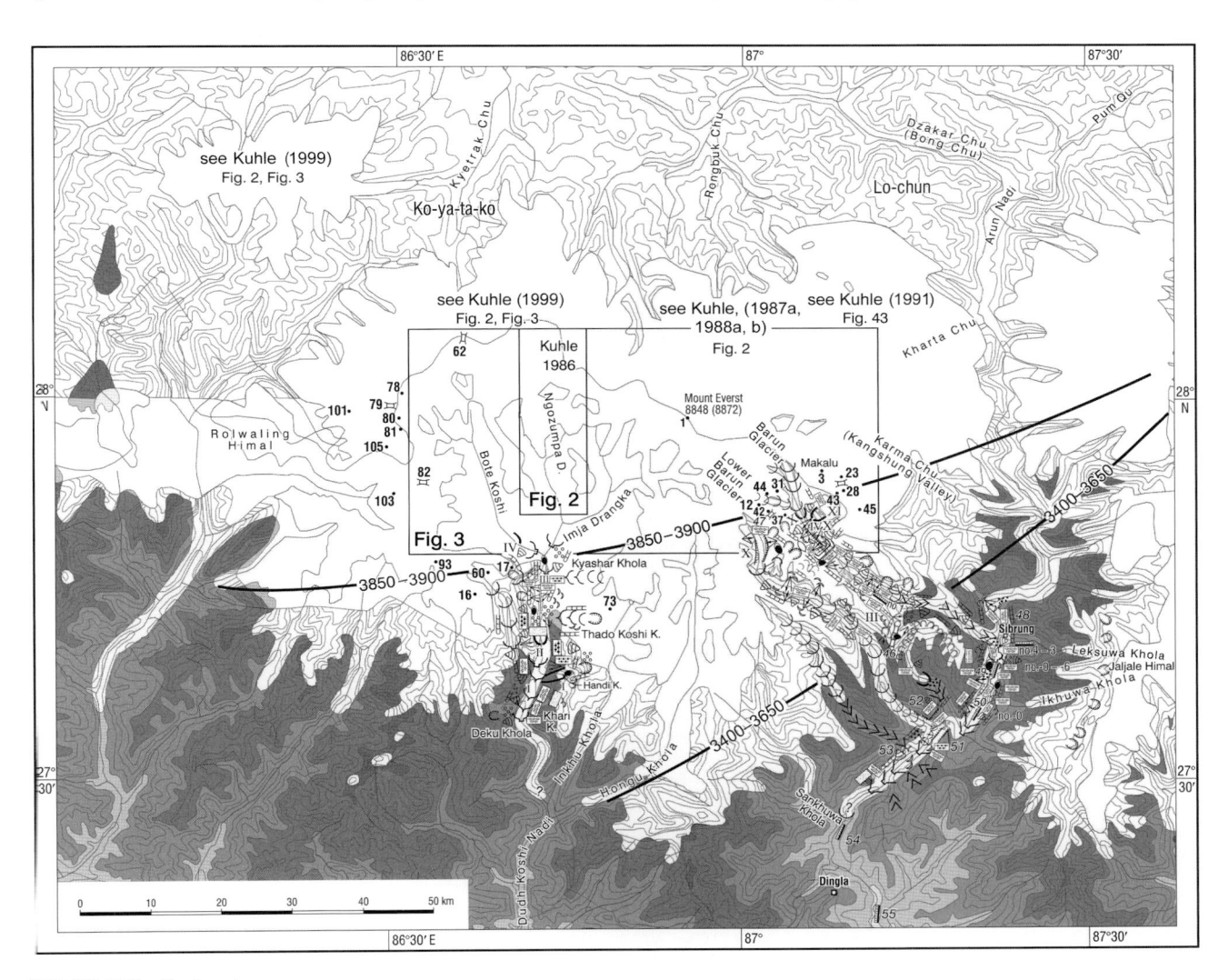

FIGURE 68.8 Continued.

LEGENDE/REFERENCE:

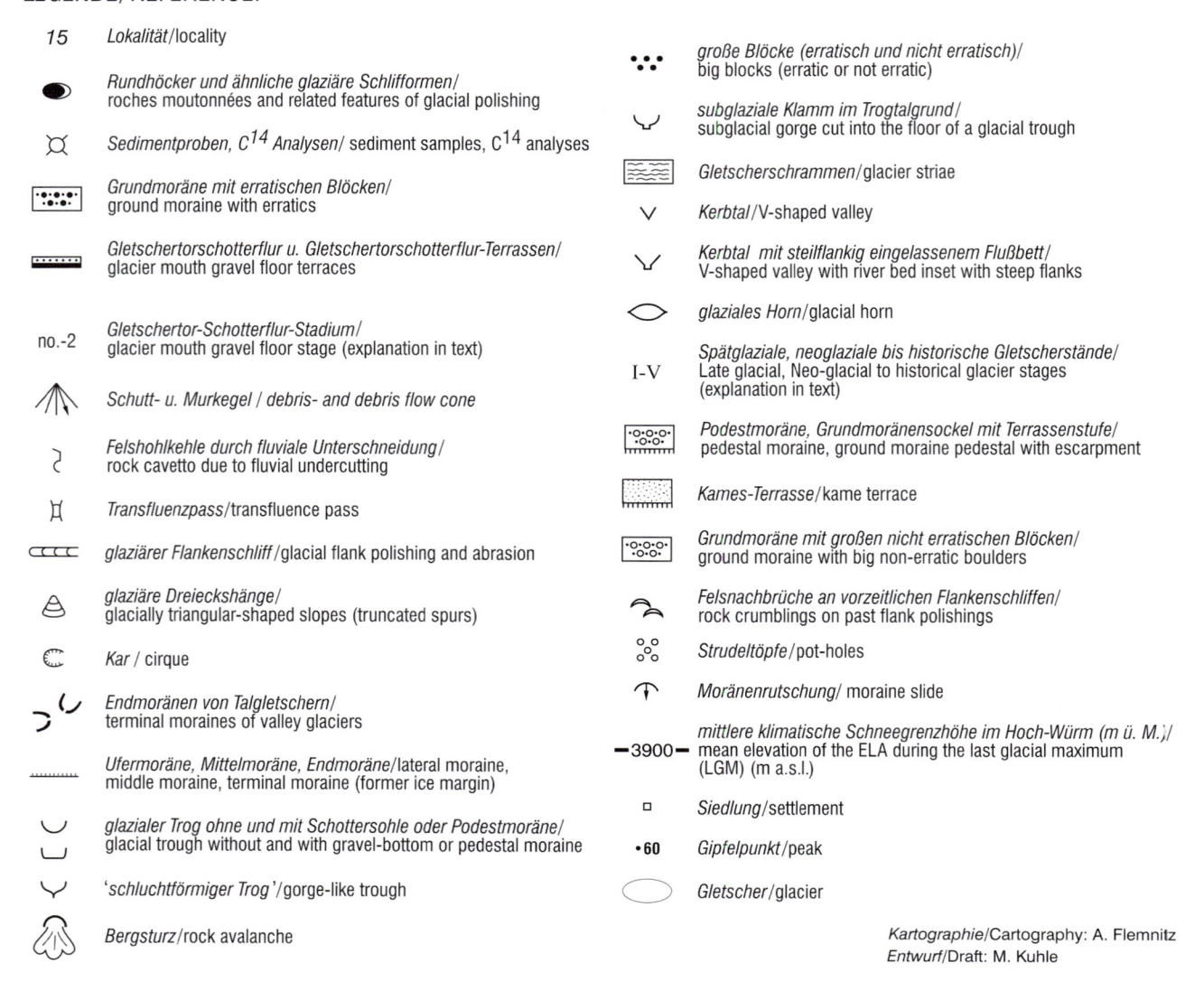

- 15   Lokalität/locality
- Rundhöcker und ähnliche glaziäre Schlifformen/ roches moutonnées and related features of glacial polishing
- Sedimentproben, C^{14} Analysen/ sediment samples, C^{14} analyses
- Grundmoräne mit erratischen Blöcken/ ground moraine with erratics
- Gletschertorschotterflur u. Gletschertorschotterflur-Terrassen/ glacier mouth gravel floor terraces
- no.-2   Gletschertor-Schotterflur-Stadium/ glacier mouth gravel floor stage (explanation in text)
- Schutt- u. Murkegel / debris- and debris flow cone
- Felshohlkehle durch fluviale Unterschneidung/ rock cavetto due to fluvial undercutting
- Transfluenzpass/transfluence pass
- glaziärer Flankenschliff/glacial flank polishing and abrasion
- glaziäre Dreieckshänge/ glacially triangular-shaped slopes (truncated spurs)
- Kar / cirque
- Endmoränen von Talgletschern/ terminal moraines of valley glaciers
- Ufermoräne, Mittelmoräne, Endmoräne/lateral moraine, middle moraine, terminal moraine (former ice margin)
- glazialer Trog ohne und mit Schottersohle oder Podestmoräne/ glacial trough without and with gravel-bottom or pedestal moraine
- 'schluchtförmiger Trog'/gorge-like trough
- Bergsturz/rock avalanche
- große Blöcke (erratisch und nicht erratisch)/ big blocks (erratic or not erratic)
- subglaziale Klamm im Trogtalgrund/ subglacial gorge cut into the floor of a glacial trough
- Gletscherschrammen/glacier striae
- Kerbtal/V-shaped valley
- Kerbtal mit steilflankig eingelassenem Flußbett/ V-shaped valley with river bed inset with steep flanks
- glaziales Horn/glacial horn
- I-V   Spätglaziale, neoglaziale bis historische Gletscherstände/ Late glacial, Neo-glacial to historical glacier stages (explanation in text)
- Podestmoräne, Grundmoränensockel mit Terrassenstufe/ pedestal moraine, ground moraine pedestal with escarpment
- Kames-Terrasse/kame terrace
- Grundmoräne mit großen nicht erratischen Blöcken/ ground moraine with big non-erratic boulders
- Felsnachbrüche an vorzeitlichen Flankenschliffen/ rock crumblings on past flank polishings
- Strudeltöpfe/pot-holes
- Moränenrutschung/ moraine slide
- −3900−   mittlere klimatische Schneegrenzhöhe im Hoch-Würm (m ü. M.)/ mean elevation of the ELA during the last glacial maximum (LGM) (m a.s.l.)
- Siedlung/settlement
- •60   Gipfelpunkt/peak
- Gletscher/glacier

Kartographie/Cartography: A. Flemnitz
Entwurf/Draft: M. Kuhle

FIGURE 68.8 Quaternary geological and glaciogeomorphological map 1:700,000 of the Khumbu and Khumbakarna Himal [Cho Oyu, Mt. Everest (i.e. Chogolungma, Sagarmatha) and Makalu massifs] in the Central Himalaya. See Fig. 68.1: No. 4; Figs. 68.2, 68.3 and **68.9** (after Kuhle, 2004a, 2005b).

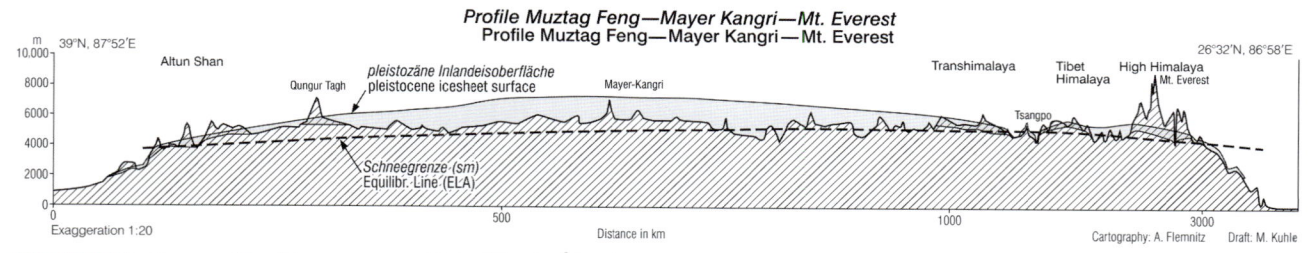

FIGURE 68.9 Cross-profile of the reconstructed 2.4 million km² ice sheet, and marginal ice stream networks, covering the Tibetan plateau, from Qungur Tagh north-slope (East Pamir plateau) in the north-west to Mt. Everest (Chogolungma) south slope in the south-east. A synthesis of the author's empirical data in addition to the mapwork 1:1,000,000 of High Asia (Fig. 68.1: Nos. 15, 6, 23, 5, 4).

TABLE 68.2 ^{14}C-Dating of the Tibetan Inland Ice in Its Southern Central Part at the Locality of Fig. 68.1: No. 10 Below; 29°18′N/94°21′E (Kuhle, 1991, Fig. 43: No. 38; Kuhle, 2004a, Fig. 37) Lower Tsangpo Valley, Up-Valley of the Nyang Qu (Linchi Chu) Confluence, at a Distance of Ca. 1 km from the Ganga Bridge on the Orographic Right-Hand Bank of the Tsangpo River

Sample no.	Sample Material	Taking of the Sample	Sample Depth	Sample Location	Sample Environment	δ^{13}C (‰)	Conv. ^{14}C-Age (Before Stadium (Stage) 1950)	Stadium (Stage)
26.9.89/13 L-Hv 17657	Tree trunk	Exposure in limnic sands	48 m below surface	29°18′N/94°21′E lower Tsangpo Valley; 3090 m a.s.l.	In ice-dammed lake sediments 42 m below root horizon, highest level with wood	−25.6	24,040 ± 450	Last High Glacial Maximum, LGM; Würm: 0
26.9.89/14 L-Hv 17659	Tree trunk	Exposure in limnic sands	50 m below surface	29°18′N/94°21′E lower Tsangpo Valley; 3088 m a.s.l.	In ice-dammed lake sediments 44 m below root horizon, second highest level with wood	−24.3	495 ± 85	?
26.9.89/9 L-Hv 17658	Tree trunk	Exposure in limnic sands	56 m below surface	29°18′N/94°21′E lower Tsangpo Valley; 3082 m a.s.l.	In ice-dammed lake sediments 50 m below root horizon, third highest level with wood	−23.8	1575 ± 75	?
26.9.89/8 L-Hv 17656	Tree trunk	Exposure in limnic silt	63 m below surface	29°18′N/94°21′E lower Tsangpo Valley; 3075 m a.s.l.	In ice-dammed lake sediments 57 m below root horizon, fourth highest wood level	−25.9	16,350 ± 580	Late Glacial Ghasa-stage: I
26.9.89/7 L-Hv 17655 cf./6/4	Tree trunk	Exposure in limnic sands	76 m below surface	29°18′N/94°21′E lower Tsangpo Valley; 3062 m a.s.l.	In ice-dammed lake sediments 70 m below root horizon, fifth highest wood level	−26.1	48,580 ± 4660−2930	Last High Glacial Maximum; Würm: 0
26.9.89/6 L-Hv 17652 cf./7/4	Tree trunk	Exposure in limnic sands	76 m below surface	29°18′N/94°21′E lower Tsangpo Valley; 3062 m a.s.l.	In ice-dammed lake sediments 70 m below root horizon, fifth highest wood level	−26.2	45,700 ± 3050−2200	Last High Glacial; Würm: 0
26.9.89/4 L-Hv 17653 cf./7/6	Tree trunk	Exposure in limnic sands	76 m below surface	29°18′N/94°21′E lower Tsangpo Valley; 3062 m a.s.l.	In ice-dammed lake sediments 70 m below root horizon, fifth highest wood level	−27.2	43,130 ± 2500−1750	Last High Glacial Maximum; Würm: 0
26.9.89/1 L-Hv 17654	Tree trunk	Exposure in limnic sands	78 m below surface	29°18′N/94°21′E lower Tsangpo Valley; 3060 m a.s.l.	In ice-dammed lake sediments 72 m below root horizon, sixth highest wood level	−24.3	9820 ± 350	Late Glacial Sirkung-stage: IV = older than 9820

Sample type: tree trunk. ^{14}C-analyses: M. A. Geyh, Lower Saxony State Soil Research Office, Hanover, Germany; sampled by M. Kuhle (after Kuhle, 1997, 1998).

TABLE 68.3 Glacier Stages of the Mountains in High Asia, That Is, in and Surrounding Tibet (Himalaya, Karakoram, E-Zagros and Hindukush, Sajan Mountains, E-Pamir, Tien Shan with Kirgisen Shan and Bogdo Uul, Quilian Shan, Kuenlun with Animachin, Nganclong Kangri, Tanggula Shan, Bayan Har, Gangdise Shan, Nyainquentanglha, Namche Bawar, Minya Gonka) from the Pre-Last High Glacial (Pre-LGM) to the Present-Day Glacier Margins and Corresponding Sanders (Glaciofluvial Gravel Fields and Gravel Field Terraces) with Their Approximate Age (After Kuhle, 1974–2009; Fig. 68.1)

Glacier Stage	Gravel Field (Sander)	Approximated Age (YBP)	ELA Depression (m)
−I = Riß (pre-Last High Glacial Maximum)	No. 6	150,000–120,000	ca. 1400
0 = Würm (last High Glacial maximum)	No. 5	60,000–18,000	ca. 1300
I–IV = Late Glacial	No. 4–No. 1	17,000–13,000 or 10,000	ca. 1100–700
I = Ghasa Stage	No. 4	17,000–15,000 (older than 12,870)	ca. 1100–900
II = Taglung Stage	No. 3	15,000–14,250	ca. 1000
III = Dhampu Stage	No. 2	14,250–13,500	ca. 800
IV = Sirkung Stage	No. 1	13,500–13,000	ca. 700
V–VII = Neo-Glacial	No. −0 to No. −2	5500–1700 (older than 1610)	ca. 300–80
V = Nauri Stage	No. −0	5500–4000 (4165)	ca. 150–300
VI = older Dhaulagiri Stage	No. −1	4000–2000 (2050)	ca. 100–200
VII = middle Dhaulagiri Stage	No. −2	2000–1700 (older than 1610)	ca. 80–150
VII–XI = historical glacier stages	No. −3 to No. −6	1700–0 (=1950)	ca. 80–20
VII = younger Dhaulagiri Stage	No. −3	1700–400 (440 resp. older than 355)	ca. 60–80
VIII = Stage VIII	No. −4	400–300 (320)	ca. 50
IX = Stage IX	No. −5	300–180 (older than 155)	ca. 40
X = Stage X	No. −6	180–30 (before 1950)	ca. 30–40
XI = Stage XI	No. −7	30–0 (=1950)	ca. 20
XII = Stage XII = recent resp. present glacier stages	No. −8	+0 to +30 (1950–1980)	ca. 10–20

dismissing previous OSL dates with little explanation, the work of Owen et al. (2001) raises serious questions as to the accuracy of existing chronologies based on OSL dating and whether comparison between these and the new chronology can be made."

The explanation for the fact that Seong et al. (2008a) are forced to associate the LGM with much too young dates is that they did not find the lower moraines. A further reason is that the only glacial features (in this case tills) they dated on the Karpochi Rock further down-valley (Seong et al., 2008a) were of an average age of 125 ka and, therefore, were classified as belonging to MIS 6. The apparent absence of real LGM moraines was explained by Seong et al. (2008a): "The lack of evidence for a glacial advance during the early part of the last glacial cycle is likely due to the poor preservation potential in this high energy environment as paraglacial and post-glacial processes easily rework and destroy glacial landforms. (. . .) it is likely that glacial, paraglacial and post-glacial processes destroyed the evidence of former glacial deposits".

Without exception, all authors who have worked in the Karakoram and Himalaya are of the opinion that this is the area with possibly the most extreme erosion rates (cf. Owen et al., 2008; Seong et al., 2008a; Hewitt, 2009). Thus, it is improbable that moraines *older* than those dating from the LGM are preserved on the extremely steep and erosion-intensive valley flanks. The author considers this opinion to be correct, without reservation. Even so, LGM moraines do exist. By detailed mapping of very high-lying moraines, glacier polish and erratic boulders, Kuhle (1988a,d, 1989, 1997, 2001b, 2006b) has been able to reconstruct the extensive LGM ice-stream network in the Baltoro and Hunza region (Fig. 68.10). Its longest parent glacier, the Indus main glacier, flowed down to below 870 m a.s.l. The accompanying indicators, with their glacial–physical interpretations, that do not show any alternative and the glacial-

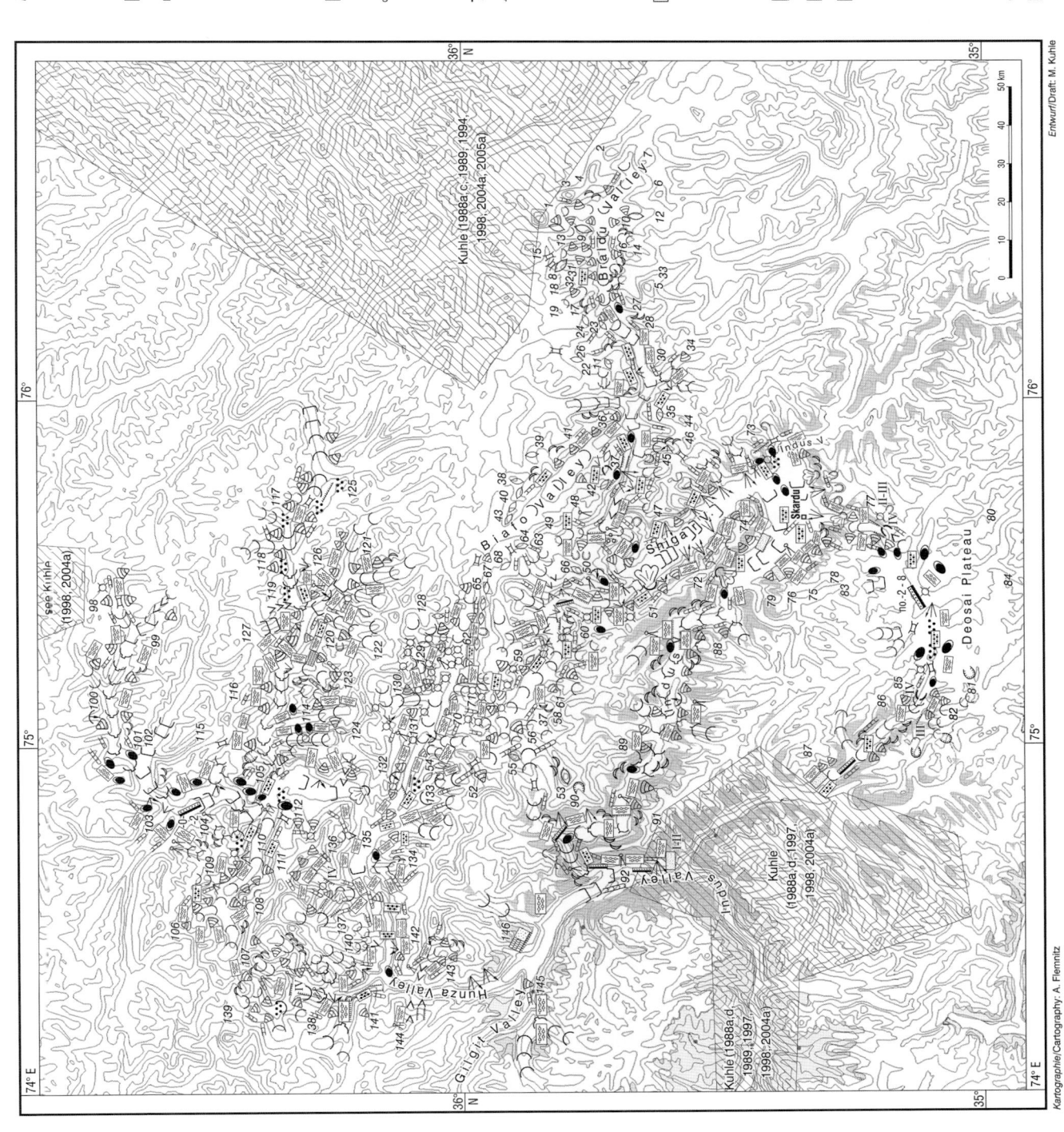

FIGURE 68.10 Quaternary geological and glaciogeomorphological map of the Karakoram and Deosai plateau between the upper Hunza Valley in the N, the Central Karakoram with the K2 and Baltoro Glacier in the E, the Gilgit- and lower Hunza Valley in the NW and the lower Indus Valley with the Nanga Parbat massif in the SW; with sampling, photograph and profile localities after Kuhle (2001b, 2004b, 2006b) (Fig. 68.1: Nos. 6, 7, 15, 22, 24, 27, 29).

geologically established classification into the LGM, have been confirmed by Hewitt (2009). Spencer and Owen (2004) and Seong et al. (2008a), however, disregard these findings. At the same time, they do not offer an alternative explanation for the reconstructed glacial genesis (Kuhle, 2008). Their interpretation is entirely based on the allegedly absolute, but geomorphologically and palaeoclimatically unlikely results of their OSL and TCN dates.

In this connection, the palaeoclimatologically dubious nature of these alternative conclusions must also be mentioned: according to Seong et al. (2008a), the extent of the Baltoro glacier (Braldu valley) during the LGM would have required an ELA depression of ∼700 m. The Hunza area, however (cf. in Fig. 68.10), situated only ∼50 km to the W, was supposed to have had a maximum High Glacial ELA depression of only 150 m. This is especially out of place, because the Hunza valley glaciers were nourished by the precipitation from the west that means by winter precipitation. The nourishing areas of the Baltoro glacier (Braldu valley area) situated to the E, profit from the current summer monsoon precipitation. During the glacial periods, however, the monsoon system had collapsed. Accordingly, the Baltoro area (Braldu valley area) had a *still smaller* ELA depression during the Ice Age than the Hunza (valley) area. Evidence for the contrary (Seong et al., 2008a) are solely the TCN dates that are neither consistent with the geological evidence nor in agreement with the ^{14}C dates for High Asia (see above).

Already the wide range of TCN ages of between 170 and 70 ka that are from boulders on the Karpochi Rock seems to be surprising. The Karpochi Rock is an isolated hill, present in the middle of the 10 km wide Skardu basin of the middle Indus valley, so that neither debris flows, landslides nor rock falls could go down to it. Considering the average age of the boulders of 125 ka, and reducing this number by a factor of 4 (see above), an age of ∼31 ka results, that is, a reduction from MIS 6 to MIS 2. If one takes the youngest age of 70 ka, (divided by 4) the resulting date of ∼18 ka is in good agreement with the Late Glacial age of ca. 15.0 ka BP to which the author has classified the moraines of the Karpochi Rock (Kuhle, 2001b).

Unfortunately, a major part of the discussion on the former glaciation of High Asia is based on verbal accusations. Seong et al. (2008b) are mistaken, when they write: "Kuhle (2008) highlights, for example, that he had (...) recognized the till on Karpochi rock (Kuhle, 2001b). He failed to note, however, that earlier authors had also mapped these landforms, including Drew (1873), Cronin (1982) and Owen (1988), and others". Kuhle (2008) *exclusively refers to the temporal classification* of the till; it is not suggested in any way that he had "recognised" it. In Kuhle (2001b), Owen is cited with two articles in this context (Owen 1988a,b) and discussed in three different passages (Kuhle 2001b). The article by Cronin (1982, unpublished) is not cited by Seong et al. (2008a) instead, they cite Cronin (1989) and this same article is also cited by Kuhle (2001b). Drew's paper (1873) is of minor significance, mainly because it was preceded by the fundamental work of Godwin-Austen (1864), which is cited by Kuhle but not by Seong et al. (2008a). Further, these authors neglect to include the equally relevant works of Lydekker (1881, 1883), Oestreich (1906) and Norin (1925) in their reference list. Therefore, the accusation of insufficient, neglected citation is unfounded and appears somewhat strange considering the fact that Seong et al. (2008a) totally ignore Kuhle (2001b) and other relevant works of Kuhle (1994; 2006b).

The above overall picture also demonstrates that in the Karakoram the OSL and TCN age determinations show considerable inconsistencies. Apparently, underestimation of ages does not occur, while the factor of overestimation of the real age may lie between 4 and 10. There are more examples of this kind the discussion of which would go beyond the scope of this chapter. However, the examples cited should be sufficient to demonstrate that a glacial chronology cannot be based entirely on numerical dates alone. It has to be based on the field evidence.

Therefore, in this chapter, the glacier stage chronology has been and will continue to be referred to the 12–14 steps (Table 68.3) between the Last Glacial Maximum of MIS 3–2 and the recent glacier margins between 1950 and 1980 (Kuhle, 1998). They have been placed not only into ^{14}C-dates in Southern Tibet and the Himalaya (see above) but also in different areas of High Asia as in the Karakoram, Kuenlun, Animachin, Nanshan, Kakitu, Tienshan, etc. (cf. Kuhle, 2004a). This provides a consistent overall picture, additionally ensured by ELA reconstructions, for the LGM glaciation, Late Glacial events, Neoglacial advances and historical glacier stillstands.

68.5. THE METHODOLOGICAL CONSEQUENCES OF THE COMPARISONS

Methodology is the basis of science but is seldom discussed, as if something that is continuously applied must automatically be correct. Obviously, OSL and TCN dating do not originate in Quaternary geology but derive from other disciplines such as nuclear physics, astrophysics and chemistry. Their application is caused by the desire to establish a timescale for the Quaternary stratigraphical subdivision. They do not replace Quaternary geological analysis but must be applied according to its conditions to be useful, that is, they must be methodologically acceptable. This demands an appropriate relationship between Quaternary geological and geomorphological evidence and the chemico-physical measuring methods. As demonstrated by the examples above, in High Asia, this connection is still missing up to now. Not only for this reason is it astonishing that the two age determination techniques, OSL and TCN dates, have gained such great acceptance during the past 10 years. The second major problem is that these methods do not have a physically well-established calibration, a problem that because of the high altitude in High Asia is an extreme source of error.

This methodological mistake is so fatal because not only is the dating technique mistaken for the autochthonous method of the Quaternary geology itself, but also—as exemplified above—it eventually begins to replace field work. Geological and geomorphological evidence tends to be neglected where it is not in agreement with the dates. In more recent papers on the glacial history of High Asia, there is a tendency towards substituting Quaternary geology, large-scale geomorphologically detailed analysis in the field, the reconstruction of glacial limits, the extent of the recent glaciation, the reconstruction of the past ELA, the establishment of the recent ELA and of the ELA depression as control for the small- and large-scale comparison, the state of preservation of glacial indicators, sedimentological analyses, etc.—sometimes even the information is missing whether the dated sample comes from a moraine, landslide, rock fall, debris flow, etc. It is just referred to as "diamicton". The information often is restricted to the locality, the height (frequently imprecise), the process of dating and the measured age of a sample (with TCN often not even the boulder sizes and rock types are given). The glaciological discussion is restricted to giving the locality of a glacier, for example, the "Tanggula Glacier" in Central Tibet (with no mention of the catchment area, terminus and glacier type). Further, a TCN date: "(the) age of 31 900 ± 3400 ^{10}Be years comes from a boulder on an end moraine located about 3 km from the terminus of Tanggula Glacier" (Colgan et al., 2006: 337) and a conclusion is stated: "the limited extent of the Tanggula Glacier suggests that it is unlikely that the Tanggula Shan were extensively glaciated at the LGM as suggested by Kuhle (1998)" (ibid., 338). With this style of information, quickly summarised from numerous similar publications, Owen et al. (2008) come to the conclusion, "Now it is generally accepted that a large ice sheet did not cover the Tibetan Plateau, at least not during the past few glacial cycles".

Those researches in this manner have lost sight of the classic glacial-geological method as centre of his special technique and defer to the unrelated authority of numerical dating technique even in Quaternary geological journals. However, for High Asia, this dating technique is wrongly calibrated, whereas in astrophysical journals, these conclusions with regard to ages would not be accepted. Until now, in the papers cited by Owen et al. (2008), the authors have never managed or only tried to disprove the evidently glaciological indicators in favour of a large-scale LGM glaciation of High Asia, that is, to classify them as being of a different origin (cf. Kuhle, 1974–2010 and references in Kuhle, 2004a). The authors fully content themselves with dates for their sample material by which they can place it into the High Glacial or an even older cycle—when, according to my chronology, it must be attributed to the Post- and Late Glacial period. All other indicators not based on uncalibrated datings and speaking in favour of a considerably more extensive glaciation are simply being disqualified and dismissed without any argumentation.

The authors concede that the features described by Kuhle and others might be relics of a much older glaciation. Strangely enough, it has been agreed that the High Asian areas (Karakoram, Himalaya) belong to the most extreme areas of removal and, accordingly, the large-scale preservation of morainic deposits, traces of polishing, subglacial potholes in high positions, etc. from glacial cycles older than LGM and LGP, is not very likely (cf. Spencer and Owen, 2004; Owen et al., 2008; Seong et al., 2008a). Consequently, an explanation is required, despite the high erosion rates, to account for why these indicators were preserved in large numbers. If the phenomena presented by Kuhle are thought to be non-glaciogenic in origin, that should also be specified. Thus far, this has not been done.

If the presented glacial morphological indicators of a High Asian glaciation can be refuted, accordingly, this cannot be carried out cursorily but only by detailed field investigations. Numerical dating methods alone are insufficient.

68.6. PROBLEMS: MAGNETIC FIELD EXCURSIONS—ON THE ASTROPHYSICAL SOURCES OF ERROR OF THE TCN TECHNIQUE

Radiocarbon dating is a well-calibrated procedure. However, its limited chronological range (maximum ca. 60 ka BP), and especially the rareness of organic material available for dating in the morainic deposits of High Asia, restricts its applicability. Consequently, different dating methods are highly desirable. But the available methods have their own problems.

The main problem with OSL dating relies on the incomplete bleaching of the material before the deposition, which is a necessary condition. The glacial dynamics in the form of a sub-, intra-, para-, that is, latero- and supraglacial transport of sediments, inevitably results in a mixture of bleached, incompletely bleached and unbleached material. Strong redeposition in a mountainous landscape adds to the problem. The mixing ratio remains unknown. In addition, the intensity of bleaching per unit of time varies because of the differing roughness of the grain surfaces. Because the measuring technique is only calibrated for the case of complete bleaching, and the samples contain unknown proportions of unbleached material, the age determination by OSL leads to an unknown overestimation of the true age (cf. Owen et al., 2008). A further factor of OSL calibration is the portion of cosmic rays, to which the already deposited material has been exposed. This portion, however, can only be estimated and is—as will be illustrated below in the context of the TCN technique—underestimated. This must lead to an additional overestimation of the age of samples. In a summary article, Owen et al. (2008) discuss the problems connected with the OSL method. As they put it: it "(...) is extremely difficult to assess the validity of many published OSL ages" (ibid., 518). Consequently, their article is focussed on TCN dates instead, setting aside the OSL ages altogether.

The TCN technique, based, for instance, on Beryllium-10, at first sight seems to be suitable for the dating of moraines. The material is infinitely available and uncertainties resulting from factors such as weathering, toppling, inheritance and shielding possibly may be taken in account or kept within reasonable limits by taking a great number of samples. Currently, the question for suitable scaling factors which especially in low and high latitudes—and thus in the High Asian area—are reliable is being heavily discussed. These factors are strained with great uncertainties of up to 30% differences among the present scaling models (cf. Owen et al., 2008). The production of TCN in the rocks depends on the total amount of cosmic rays which hit the surface. Unfortunately, the intensity of cosmic radiation is not homogeneous but subject to temporal fluctuations in dependence on variations of the geomagnetic field and the solar magnetic activity. These fluctuations of intensity have different causes:

(a) turbulence in the heliosphere, created by the following three factors: (1) by the solar wind varying in the solar activity cycle, (2) by coronal mass ejections (CMEs) and (3) by the structure of the interplanetary magnetic field;
(b) the position of the sun/heliosphere in the interstellar medium;
(c) the solar activity over larger periods of time;
(d) the intensity of the terrestrial magnetic field and its topology;
(e) the constant/variation of the interstellar sources creating cosmic radiation, for example, super novae, Wolf-Rayet stars.

These past variations of cosmic-ray flux occurring over a quite long time period can approximately be reconstructed by comparing variations of ^{14}C and ^{10}Be cosmogenic isotopes in the earth's atmosphere. Their chronology has been stored in fossil wood, ice cores and marine sediments. The minima of magnetic activities (i.e. periods of a weak geomagnetic field as well as minima of solar activity) generally coincide with maxima of cosmic radiation, as shown by correlation analyses. The latter coincide with the terrestrial cold phases (Dergachev et al., 2007; Usoskin and Kovaltsov, 2008). Unfortunately, the reconstruction of cosmic-ray flux via cosmogenic isotopes can only be regarded as an average value that does not correspond with the amount of *local* TCN productivity. The values of the local TCN productivity are subject to strong variations. They can only be measured by ground-based neutron monitors (NMs). Without such data being available for High Asia, TCN dates are prone to errors of unknown size.

Without going too far into detail, the following conclusions can be drawn:

1. Both OSL and TCN techniques are dependent on the amount of cosmic radiation. Up to now, the temporal and spatial fluctuations of the intensity of this radiation cannot be reliably calculated. Both techniques tend to an overestimation of the ages. They cannot be used for a mutual calibration because they depend on the same factor. Only the ^{14}C-dating can provide an independent age control for both techniques. This alone is not dependent on the direct, that is, local cosmic-ray intensity, but only indirectly on the production of ^{14}C isotopes in the atmosphere, for which a comparatively reliable chronology is available.
2. Even if OSL and TCN techniques *generally* provide data that are too old, they could possibly provide a *relative* chronology of glacial stages. Therefore, this relative chronology does not permit the correctness of absolute ages. Chronology and absolute age do not establish a connection. In addition, because the OSL technique is based on the dominant influence and factor of uncertainty of incomplete bleaching, that causes higher ages, and this influence has nothing to do with the chronology, it has to be expected that the OSL data may fluctuate more widely within the relative chronology than the TCN data.

68.7. THE QUESTION OF ARIDITY

When Owen et al. (2008) claim "Now it is generally accepted that a large ice sheet did not cover the Tibetan Plateau, at least not during the past few glacial cycles", they also repeat the old legend of "too great dryness" in High Asia that is invoked to make an extended glaciation impossible. As summarised by Owen et al. (2008), this interpretation comes from Klute (1930) and von von Wissmann (1959), both of whom never worked in High Asia. Their general interpretations are considered by Owen et al. (2008) as "detailed work". At the same time, they cite the reconstruction of a Tibetan ice sheet by Kuhle as "based on field observations and extrapolation of large ELA depressions (>1000 m) from the margins of Tibet into the interior regions". Precisely the same interpretation, that is, that the glaciation of the Tibetan plateau was only deduced from a mistaken extrapolation of the ΔELA is also repeated by Seong et al. (2008b).

Seong et al. (2008b) claim that, in Kuhle's reconstruction of a Tibetan ice sheet, "(...) the ΔELA was (...) erroneously extrapolated across the Tibetan plateau to argue for an ice sheet at the LGM". Moreover, Seong et al. (2008b) refer to the results of Kuhle's sedimentological and glacial morphological investigations cursorily as "(...) rather equivocal field evidence, including exotic boulders and eroded landforms (...)", which Kuhle apparently used "(...) to hypothesise that an ice sheet existed over Tibet during the last glacial". By contrast, Seong et al. refer to the works of those other authors whose views reinforce their own opinions as "extensive studies (...) who present glacial geologic evidence that shows that an ice sheet could not have existed on the Tibetan plateau during at least the last two glacial cycles". The fact is that, up till now, neither Seong et al. nor any of the authors cited by them, have even done so much as *suggest* an alternative explanation for any of the glacial indicators referred to by Kuhle. Therefore, the labelling of Kuhle's findings (which are documented in ~1700 printed photographs and photographic panoramas) as "equivocal field evidence", on the basis of which Kuhle supposedly "extrapolated" and "hypothesised", is not acceptable. The *only* argument brought forward by Seong et al. and authors cited by them who oppose the existence of a Tibetan ice sheet during the LGM is their OSL and TCN dates, which lack reliable calibration.

It is also unclear why Owen et al. (2008) and Seong et al. (2008b) object to the "extrapolation". Naturally, when the Tibetan plateau was covered by an ice sheet, the position of the ELA can *only* be constructed from the marginal areas with the lowest ice-marginal positions and not in the centre, where there are no ice margins during a total glaciation. Therefore, the ELA *can only be extrapolated* to the central plateau. The existence of a former ice sheet on the Tibetan plateau during the LGM has been demonstrated in detail using precisely the same evidence that is valid for the reconstruction of Pleistocene ice sheets in North America and Europe, that is, due to the widespread occurrence of a till cover with glacigenic, partly striated erratics and glacially polished landforms, etc. (Kuhle, 2004a).

In this context, it is symptomatic that Owen et al. (2008) ignore Kuhle's field evidence, based on numerous sediment analyses and detailed mapping, instead they recommend "remote sensing" as the adequate technique for the reconstruction of the Ice Age ice cover in High Asia (Owen et al., 2008).

The argument offered against a glaciation of Tibet, that there was a lack of precipitation during the glacial periods, so that the region was too arid to build up an ice sheet like this, is based—as can be shown easily—on a consequence caused by the Tibetan ice. Thus, the climatological necessity of the Tibetan ice even is confirmed.

As Owen et al. (2008) rightly point out the strongly reduced precipitation during the LGM was caused by the breakdown of the summer monsoon. The collapse of the monsoon circulation is well documented (cf. summary description in Kuhle, 2002b). There is no doubt that the existence of the summer monsoon requires an ice-free Tibetan plateau. Accordingly, Tibet is the high-lying subtropical heating face causing the thermodynamics which produce the monsoon. Thus, the only acceptable explanation as to the breakdown of the monsoon during the High Glacial is that this heating face did not exist at that time. This is best explained by a perennial snow cover (cf. discussion in Kuhle, 2002b). In the course of decades and centuries, a perennial snow cover would develop into a glaciation. This is similar to the way lowland ice sheets formed elsewhere. Due to local glacier advances down to the plateau, the Tibetan mountains would have also contributed to the glaciation, similar to that in the mountains of Scandinavia.

For the build-up of a glaciation in this manner, only a small amount of precipitation is necessary, combined with annual mean temperatures below -6 to $-8\,°C$. Even today, large parts of the plateau surface are situated above the permafrost limit (cf. Fig. 68.9). Accordingly, they show an annual mean temperature below -4 to $-8\,°C$ (Kuhle, 1985, 1990a, 1997). With a continental LGM cooling of the northern hemisphere by only ~8 to $-12\,°C$ up to $-16\,°C$, the entire plateau surface would lie above the snowline. This is also valid under cold arid conditions. Model calculations (Kuhle et al., 1989; Kuhle, 1997) have shown that, in combination with low temperatures, precipitation of only 100 mm/a is sufficient to develop within 10,000 years an ice sheet ~1000 m thick in Tibet. The amount of precipitation necessary to initiate a glaciation like this clearly remains under the present amount of precipitation (cf. Owen et al., 2008). In addition, the development of large glacial period lakes in the Qaidam Basin, Tarim Basin and in the Gobi (Tengger) immediately to the north of the Tibetan plateau reflects an increasing humidity towards the N (Chen and Bowler, 1986; Pachur and Wünnemann, 1995; Rhodes et al., 1996; Wünnemann et al., 1998). Climate model calculations show that under the conditions of glacial-period temperatures in Tibet, a permanent ice sheet must have formed (cf. discussion in Kuhle, 2002b).

Finally, it should be kept in mind that according to ^{14}C dates by Kashiwaya et al. (1991), Van Campo and Gasse (1993), Gasse et al. (1996) and Avouac et al. (1996), the lakes on the plateau are *all younger than the LGM*. The only reasonable explanation of this is that they formed after the decay of an extensive ice sheet.

68.8. CONCLUSIONS

For several years, the OSL and TCN dating of morainic material has come into fashion in High Asian investigations. These techniques, however, lack reliable calibration methods. The intensity of the cosmic-ray flux is being modulated by the terrestrial and solar magnetic fields. In the past, these magnetic fields have shown considerable secular fluctuations. Especially for High Asia, these variables still cannot be converted into corresponding local TCN-production rates. There are reasons that point to the fact that the ages determined up till now are in general clearly overestimated. This is confirmed by (1) conventional radiocarbon dates, and (2) it is especially stable because of the independently acquired glacial chronology. This has been recorded on the basis of the classical Quaternary geological (glacial geomorphological) methods on a small and large scale. Evidence given for a much more extended LGM glaciation of High Asia that resulted from these investigations, however, is rejected by some authors based on poorly calibrated numerical dates. The OSL and TCN dates in High Asia provide an inconsistent and inhomogeneous picture and, accordingly, are implausible. Moreover, the climatological and glacial-geological consequences of an ice-free Tibetan plateau during the LGM are difficult to reconcile with the field evidence. It is much more difficult to explain that a Tibetan ice sheet did *not* exist during the LGM, than the opposite way around. If the glaciation of Tibet is to be refuted, this should not be done on the basis of OSL and TCN dates but by field investigations.

ACKNOWLEDGEMENTS

I thank Volker Bothmer (Astrophysical Institute, University of Göttingen) for important discussions and references to the literature, and I like to express my thanks to Jürgen Ehlers and Philip Gibbard for their careful reworking of this chapter.

REFERENCES

Avouac, J.-Ph., Dobremez, J.-F., Bourjot, L., 1996. Palaeoclimatic interpretation of a topographic profile across middle Holocene regressive shorelines of Longmu Co (Western Tibet). Palaeogeogr. Palaeoclimatol. Palaeoecol. 120, 93–104.

Beer, J., 2000. Neutron monitor records in broader historical context. Space Sci. Rev. 93, 107–119.

Benestad, R.E., 2006. Solar Activity and Earth's Climate. Springer, Berlin/Heidelberg/New York.

Bobek, H., 1937. Die Rolle der Eiszeit in Nordwestiran. Z. Gletscherk. 25, 130–183.

Bothmer, V., Zhukov, A., 2007. The Sun as the prime source of space weather. In: Bothmer, V., Daglis, I.A. (Eds.), Space Weather—Physics and Effects. Springer, Berlin/Heidelberg/New York, pp. 31–102.

Burbank, D.W., Kang, J.C., 1991. Relative dating of Quaternary Moraines, Rongbuk Valley, Mount Everest, Tibet: implications for an ice sheet on the Tibetan Plateau. Quatern. Res. 36, 1–18.

Chen, K., Bowler, J.M., 1986. Late Pleistocene evolution of salt lakes in the Qaidam basin, Qinghai province, China. Palaeogeogr. Palaeoclimatol. Palaeoecol. 54, 87–104.

Cline, R. (Ed.), 1981. Climap Project: Seasonal Reconstructions of the Earth's Surface at the Last Glacial Maximum. Geological Society of America, New York, pp. 1–18.

Colgan, P.M., Munroe, J.S., Shangzhe, Zhou, 2006. Cosmogenic radionuclide evidence for the limited extent of last glacial maximum glaciers in the Tanggula Shan of the central Tibetan Plateau. Quatern. Res. 65, 336–339.

Cronin, V.S., 1982. Physical and magnetic polarity stratigraphy of the Skardu basin, Baltistan, northern Pakistan. Unpublished MA Thesis, Dartmouth College, New Hampshire, pp. 1–226 (unpublished).

Cronin, V.S., 1989. Structural setting of the Skardu intermontane basin, Karakoram Himalaya, Pakistan. In: Manlinconico, L.L., Lillie, R.J. (Eds.), Tectonics of the Western Himalayas. Geological Society of America Special Paper 232 eine Ortsangabe, konnte ich nicht finden, pp. 183–201.

Dainelli, G., Martinelli, O., 1928. Le condizione fisiche attuali. Risultati Geologici e Geografici, Relazione scientifiche della Speditizione Italiana De Filippi nell'Himalaya, Caracorum e Turchestan Cinese (1913–1914), Ser.II, Bd.IV, Bologna.

Derbyshire, E., Li, Jijun, Perrott, F.A., Shuying, Xu., Waters, R.S., 1984. Quaternary glacial history of the Hunza Valley, Karakoram Mountains, Pakistan. In: Miller, K. (Ed.), International Karakoram Project, pp. 456–495, Cambridge University Press, Cambridge.

Dergachev, V.A., Dmitriev, P.B., Raspopov, O.M., Jungner, H., 2007. Cosmic ray flux variations, modulated by the solar and terrestrial magnetic fields, and climate changes. Part 2: the time interval from 10,000 to 100,000 years ago. Geomagn. Aeron. 47 (1), 109–117.

de Terra, H., 1934. Physiographic results of a recent survey in Little Tibet. Geogr. Rev. 24, 12–41 1934, Karte 1:506 880 (Rupschu-Panggong-Tschangtschenmo) von A. Gul Khan.

Drew, F., 1873. Alluvial and lacustrine deposits and glacial records of the upper Indus basin; Part 1 Alluvial deposits. Geol. Soc. Lond. Q. J. 29, 449–471.

Dunai, T.J., 2000. Scaling factors for production rates of in situ produced cosmogenic nuclides: a critical reevaluation. Earth Planet. Sci. Lett. 176, 157–169.

Dunai, T.J., 2001. Influence of secular variation of the geomagnetic field on production rates of in situ produced cosmogenic nuclides. Earth Planet. Sci. Lett. 193, 197–212.

Ehlers, J., Gibbard, P., 2004. Quaternary Glaciations - Extent and Chronology. Part III: South America, Asia, Africa, Australasia, Antarctica. Development in Quaternary Science 2c Elsevier B.V., Amsterdam, pp. 1–380.

Engelhardt, W.v., 1973. Die Bildung von Sedimenten und Sedimentgesteinen. In: Engelhardt, W.v., Füchtbauer, H., Müller, G. (Eds.), Sediment-Petrologie, vol. III, 1–378.

Finkel, R.C., Owen, L.A., Barnard, P.L., Caffee, M.W., 2003. Beryllium-10 dating of Mount Everest moraines indicates a strong monsoon influence and glacial synchroneity throughout the Himalaya. Geology 31, 561–564.

Gasse, F., Fontes, J.Ch., Van Campo, E., Wei, K., 1996. Holocene environmental changes in Bangong Co basin (Western Tibet). Part 4: discussion and conclusions. Palaeogeogr. Palaeoclimatol. Palaeoecol. 120, 79–92.

Godwin-Austen, H.H., 1864. On the glaciers of the Muztagh Range. Proc. R. Geogr. Soc. 34, 19–56.

Handel-Mazzetti, F.H., 1927. Das nordostbirmanische-westyünnanische Hochgebirgsgebiet. Karsten-Schenck, Vegetationsbilder 17, 7–8.

Hewitt, K., 2009. Catastrophic rock slope failures and late Quaternary developments in the Nanga Parbat—Haramosh Massif, Uper Indus basin, northern Pakistan. Quatern. Sci. Rev. 28 (11–12), 1055–1069.

Heydemann, A., Kuhle, M., 1988. The petrography of Southern Tibet—results of microskopic and X-ray analyses of rock samples from the 1984 expedition area (Transhimalaya to Mt. Everest N Slope). GeoJournal 17 (4), 615–625. Tibet and High-Asia, Results of the Sino-German Joint Expeditions (I), Eds: Kuhle, M.; Wang Wenjing).

Huntington, A., 1906. Pangong, a glacial lake in the Tibetan Plateau. J. Geol. 14, 599–617.

Kashiwaya, K., Yaskawa, K., Yuan, B., Liu, J., Gu, Z., Cong, S., et al., 1991. Paleohydrological processes in Siling-Co (Lake) in the Qing-Zang (Tibetan) Plateau based on the physical properties of its bottom sediments. Geophys. Res. Lett. 18 (9), 1779–1781.

Klute, F., 1930. Verschiebung der Klimagebiete der letzten Eiszeit. Peterm. Mittl. Ergänzungsh 209, 166–182.

Kovaltsov, G.A., Usoskin, I.G., 2007. Regional cosmic ray induced ionization and geomagnetic field changes. Adv. Geosci. 13, 31–35.

Kuhle, M., 1982a. Was spricht für eine pleistozäne Inlandvereisung Hochtibets? In: Sonderband: Die Chinesisch/Deutsche Tibet-Expedition 1981, Braunschweig-Symposium vom 14.-16.04.1982. Sitzungsberichte und Mitteilungen der Braunschweigischen Wissenschaftlichen Gesellschaft 6, pp. 68–77.

Kuhle, M., 1982b. Erste Chinesisch-Deutsche Gemeinschaftsexpedition nach NE-Tibet und in die Massive des Kuen-Lun-Gebirges 1981. Ein Expeditions- und vorläufiger Forschungsbericht. In: Hagedorn, H., Giessner, K. (Eds.), Tagungsber. und wiss. Abhandlungen des 43. Deutschen Geographentages Mannheim 1981. Steiner, Wiesbaden, pp. 63–82.

Kuhle, M., 1983. A new expedition in Tibet—a contribution of climatology and high mountain research. In: Universitas (A Survey of Current Research), vol. 25(10), pp. 59–63 Wissenschaftliche Verlagsgesellschaft M.B.H., Stuttgart.

Kuhle, M., 1985. Permafrost and periglacial indicators on the Tibetan Plateau from the Himalaya Mountains in the south to the Quilian Shan in the north (28-40°N). Z. Geomorphol. N.F. 29 (Bd.2), 183–192.

Kuhle, M., 1986. Former glacial stades in the mountain areas surrounding Tibet—in the Himalayas (27-29°N: Dhaulagiri-, Annapurna-, Cho Qyu-, Gyachung Kang-areas) in the south and in the Kuen Lun and Quilian Shan (34-38°N: Animachin, Kakitu) in the north. In: Joshi, S.C., Haigh, M.J., Pangtey, Y.P.S., Joshi, D.R., Dani, D.D. (Eds.), Nepal-Himalaya—Geo-Ecological perspectives 437–473 H. R. Publishers, Delhi.

Kuhle, M., 1987a. Subtropical mountain- and highland-glaciation as Ice Age triggers and the waning of the glacial periods in the Pleistocene. GeoJournal 14 (4), 393–421.

Kuhle, M., 1987b. Absolute Datierungen zur jüngeren Gletschergeschichte im Mt. Everest-Gebiet und die mathematische Korrektur von Schneegrenzberechnungen. In: Hütteroth, W.-D. (Ed.), Tagungsbericht und wissenschaftliche Abhandlungen des 45. Deutschen Geographentages Berlin 1985. Steiner, Stuttgart, pp. 200–208.

Kuhle, M., 1987c. The problem of a Pleistocene inland glaciation of the northeastern Qinghai-Xizang-Plateau. In: Hövermann, J., Wenjing, W. (Eds.), Reports on the Northeastern Part of Quinghai-Xizang (Tibet)-Plateau by the Sino-German Scientific Expedition 1981. Science Press, Beijing, China, pp. 250–315.

Kuhle, M., Jacobsen, J.P., 1988. On the Geoecology of Southern Tibet - Measurements of Climate Parameters including Surface- and Soil-Temperatures in Debris, Rock, Snow, Firn and Ice during the South Tibet- and Mt. Everest Expedition in 1984. GeoJournal 17 (4, Tibet and High-Asia, Results of the Sino-German Joint Expeditions (I), Eds: Kuhle, M.; Wang Wenjing), 597–615.

Kuhle, M., 1988a. The Pleistocene glaciation of Tibet and the onset of Ice Ages—an autocycle hypothesis. GeoJournal 17 (4), 581–596. (Tibet and High-Asia. Results of the Sino-German Joint Expeditions (I), Eds: Kuhle, M.; Wang Wenjing).

Kuhle, M., 1988b. Geomorphological findings on the build-up of Pleistocene glaciation in southern Tibet, and on the problem of inland ice. Results of the Shisha Pangma and Mt. Everest Expedition 1984. GeoJournal 17 (4), 457–513 (Tibet and High-Asia, Results of the Sino-German Joint Expeditions (I), Eds: Kuhle, M.; Wang Wenjing).

Kuhle, M., 1988c. Zur Geomorphologie der nivalen und subnivalen Höhenstufe in der Karakorum-N-Abdachung zwischen Shaksgam-Tal und K2 Nordsporn: Die quartäre Vergletscherung und ihre geoökologische Konsequenz. In: Becker, H. (Ed.), Tagungsbericht und wissenschaftliche Abhandlung des 46. Deutschen Geographentag 1987 München. Steiner, Stuttgart, pp. 413–419.

Kuhle, M., 1988d. Letzteiszeitliche Gletscherausdehnung vom NW-Karakorum bis zum Nanga Parbat (Hunza-Gilgit- und Indusgletschersystem). In: Becker, H. (Ed.), Tagungsbericht und wissenschaftliche Abhandlungen des 46. Deutschen Geographentages München 1987. AK-Sitzung "Neueste physisch-geographische Forschungsergebnisse aus Hochtibet und angrenzenden Gebieten" Steiner, Stuttgart, pp. 606–607.

Kuhle, M., 1989. Die Inlandvereisung Tibets als Basis einer in der Globalstrahlungsgeometrie fußenden, reliefspezifischen Eiszeittheorie. Peterm. Geogr. Mittl. 133 (4), 265–285.

Kuhle, M., 1990a. The cold deserts of high Asia (Tibet and contiguous mountains). GeoJournal 20 (3), 319–323.

Kuhle, M., 1990b. New data on the Pleistocene glacial cover of the southern border of Tibet: the glaciation of the Kangchendzönga Massif (8585 m, E-Himalaya). GeoJournal 20, 415–421.

Kuhle, M., 1991. Observations supporting the Pleistocene inland glaciation of high Asia. GeoJournal 25 (2/3), 133–233 (Tibet and High Asia, Results of the Sino-German Joint Expeditons (II), Eds: Kuhle, M.; Xu Daoming).

Kuhle, M., 1994. Present and Pleistocene glaciation on the north-western margin of Tibet between the Karakorum Main Ridge and the Tarim Basin supporting the evidence of a Pleistocene inland glaciation in Tibet. GeoJournal 33 (2/3), 133–272 (Tibet and High Asia, Results of the Sino-German and Russian-German Joint Expeditions (III), Ed: Kuhle, M.).

Kuhle, M., 1995. Glacial isostatic uplift of Tibet as a consequence of a former ice sheet. GeoJournal 37 (4), 431–449.

Kuhle, M., 1997. New findings concerning the Ice Age (Last Glacial Maximum) glacier cover of the East-Pamir, of the Nanga Parbat up to the Central Himalaya and of Tibet, as well as the age of the Tibetan Inland Ice. GeoJournal 42 (2–3), 87–257 (Tibet and High Asia. Results of Investigations into High Mountain Geomorphology, Paleo- Glaciology and Climatology of the Pleistocene (Ice Age Research) IV, Ed: Kuhle, M.).

Kuhle, M., 1998. Reconstruction of the 2.4 million qkm Late Pleistocene Ice Sheet on the Tibetan Plateau and its impact on the global climate. Quatern. Int. 45/46, 71–108 (Erratum: 47/48, 173–182 (1998) included).

Kuhle, M., 1999. Reconstruction of an approximately complete Quaternary Tibetan Inland Glaciation between the Mt. Everest- and Cho Oyu Massifs and the Aksai Chin. - A new glaciogeomorphological southeast-northwest diagonal profile through Tibet and its consequences for the glacial isostasy and Ice Age cycle. GeoJournal 47 (1–2), 3–276 (Kuhle M. (ed.), Tibet and High Asia (V), Results of Investigations into High Mountain Geomorphology, Paleo-Glaciology and Climatology of the Pleistocene).

Kuhle, M., 2001a. Reconstruction of outlet glacier tongues of the Ice age South-Tibetan ice cover between Cho Oyu and Shisha Pangma as a further proof of the Tibetan Inland Ice Sheet. Polarforschung 71 (3), 79–95.

Kuhle, M., 2001b. The maximum ice age (LGM) glaciation of the Central- and South Karakorum: an investigation of the heights of its glacier levels and ice thicknesses as well as lowest prehistoric ice margin positions in the Hindukush, Himalaya and in East-Tibet on the Minya Konka-massif. GeoJournal 54(2–4), 55(1), 109–396 (Tibet and High Asia (VI): Glaciogeomorphology and Prehistoric Glaciation in the Karakorum and Himalaya. Kluwer Academic Publishers, Dordrecht/Boston/London).

Kuhle, M., 2002a. Outlet glaciers of the Pleistocene (LGM) south Tibetan ice sheet between Cho Oyu and Shisha Pangma as potenial sources of former mega-floods. In: Martini, P., Baker, V.R., Garzón, G. (Eds.), Flood and Megaflood Processes and Deposits: Recent and Ancient Examples, Special Publication of the International Association of Sedimentologists (IAS) vol. 32, pp. 291–302. Blackwell Science, Oxford.

Kuhle, M., 2002b. A relief-specific model of the ice age on the basis of uplift-controlled glacier areas in Tibet and the corresponding albedo increase as well as their positiv climatological feedback by means of the global radiation geometry. Climate Res. 20, 1–7.

Kuhle, M., 2003. The former Glaciation of High- and Central Asia and its climatic Impact. Comments on the INQUA-COG-GLACIATION MAP 1:1 Mio. XVI INQUA Congress in Reno Nevada USA, July 23–30, 2003, Abstract Volume, 70.

Kuhle, M., 2004a. The High Glacial (Last Ice Age and LGM) ice cover in High and Central Asia. In: Ehlers, J., Gibbard, P.L. (Eds.), Quaternary Glaciations—Extent and Chronology, Part III: South America, Asia, Africa, Australia, Antarctica. Development in Quaternary Science 2c Elsevier B.V., Amsterdam, pp. 175–199.

Kuhle, M., 2004b. Past glacier (Würmian) ice thickness in the Karakoram and on the Deosai Plateau in the catchment area of the Indus river. E&G Quatern. Sci. J. 54, 95–123.

Kuhle, M., 2005a. The maximum Ice Age glaciation between Karakoram Main Ridge (K2) and the Tarim Basin and its influence on the global energy balance. J. Mountain Sci. 2 (1), 5–22 (Institute of Mountain Hazards and Environment, Chinese Academy of Science, Science Press, Beijing).

Kuhle, M., 2005b. The maximum Ice Age (Würmian, Last Ice Age, LGM) glaciation of the Himalaya—a glaciogeomorphological investigation of glacier trim-lines, ice thicknesses and lowest former ice margin positions in the Mt. Everest-Makalu-Cho Oyu massifs (Khumbu and Khumbakarna Himal) including informations on late-glacial, neoglacial and historical glacier stages, their snow-line depressions and ages. GeoJournal 62 (3–4), 191–650 (Tibet and High Asia(VII): Glaciogeomorphology and Former Glaciation in the Himalaya and Karakorum, Ed: Kuhle, M.).

Kuhle, M., 2005c. Glacial geomorphology and ice ages in Tibet and surrounding mountains. The Island Arc 14(4), 346–367 (Blackwell Publishing Asia Pty Ltd).

Kuhle, M., 2006a. The reconstruction of the Ice Age glaciation of the Himalaya and high Asia by Quaternary geological and glaciogeomorphological methods. In: Saklani, P.S. (Ed.), Himalaya (Geological Aspects), vol. 4. Satish Serial Publishing House; SSPH, New Delhi, pp. 181–214, ISBN 81-89304-14-3.

Kuhle, M. (2006b). The Past Hunza Glacier in Connection with a Pleistocene Karakorum Ice Stream Network during the Last Ice Age (Würm). In: *Karakoram in Transition*. (Ed: Kreutzmann, H.) Saijid, A., Oxford University Press, Karachi, Pakistan, ISBN-13: 978-0-19-547210-3, 24–48.

Kuhle, M., 2007a. Critical approach to the methods of glacier reconstruction in High Asia (Qinghai-Xizang (Tibet) Plateau, West Sichuan Plateau, Himalaya, Karakorum, Pamir, Kuenlun, Tienshan) and discussion of the probability of a Qinghai-Xizang (Tibetan) inland ice. J. Mountain Sci. 4 (2), 91–123 (June 2007) (Science Press, Beijing, Springer, Germany, ISSN 1672–6316, Ed: Yu Dafu, C.).

Kuhle, M., 2007b. The Pleistocene Glaciation (LGP and pre-LGP, pre-LGM) of SE-Iranian Mountains exemplified by the Kuh-i-Jupar, Kuh-i-Lalezar and Kuh-i-Hezar Massifs in the Zagros. Polarforschung 77 (2–3), 71–88 (Erratum/Clarification concerning Figure 15: 78(1–2), 83, 2008).

Kuhle, M., 2008. Correspondence to on-line-edition (doi.101016/jj.quascirev.2007.09.015 Elsevier) of Quaternary Science Reviews (QSR) article. In: Seong, Yeong Bae, Owen, Lewis A., Bishop, Michael P., Andrew, Bush, Penny, Cendon, Luke, Copland, Robert, Finkel, Ulrich, Kamp, Shroder Jr., John F. (Eds.), Quaternary Glacier History of the Central Karakorum. Quaternary Science Reviews vol. 27. Elsevier, Amsterdam, pp. 1655–1656.

Kuhle, M., 2009. The former glaciation of High-(Tibet) and Central Asia and it's global climatic impact—an Ice Age theory with a remark on potential warmer climatic cycles in the future. In: Krugger, M.I., Stern, H.P. (Eds.), New Permafrost and Glacier Research. Nova Science Publisher, Inc., New York, pp. 163–235.

Kuhle, M., 2010. New indicators of a former Tibetan ice sheet and an ice stream network in the surrounding mountain systems: new observations and dating on the SE-, S- and W-margin of Tibet from expeditions in 2004–2009. In: Program and Extended Abstracts of the 25th Himalaya-Karakoram-Tibet Workshop in San Francisco 2010, pp. A104–A105.

Kuhle, M., Herterich, K., Calov, R., 1989. On the Ice Age glaciation of the Tibetan highlands and its transformation into a 3-D model. GeoJournal 19 (2), 201–206.

Lal, D., 1991. Cosmic ray labeling of erosion surfaces; in situ nuclide production rates and erosion models. Earth Planet. Sci. Lett. 104, 429–439.

Lifton, N.A., Bieber, J.W., Clem, J.M., Duldig, M.L., Evenson, P., Humble, J.E., et al., 2005. Addressing solar modulation and long-term uncertainties in scaling secondary cosmic rays for in situ cosmogenic nuclide applications. Earth Planet. Sci. Lett. 239, 140–161.

Lydekker, R., 1881. Geology of part of Dardistan, Baltistan and neighbouring districts. Rec. GsuI. 14, 1–56.

Lydekker, R., 1883. The geology of Kashmir and Chamba Territories. Mem. Gsul. 22, 3838–3841.

Mursula, K., Usoskin, I.G., Kovaltsov, G.A., 2001. Long-term cosmic ray intensity vs. Solar proxies: a simple linear relation does not work. In: Proceedings of ICRC 2001, 3838, Copernicus Gesellschaft.

Norin, E., 1925. Preliminary notes on the late Quaternary glaciation of the Northwest Himalaya. Geogr. Ann. 7, 166–194.

Norin, E., 1932. Quaternary climatic changes within the Tarim Basin. Geogr. Rev. 22, 591–598 (New York).

Odell, N.E., 1925. Oberservation on the rocks and glaciers of Mount Everest. Geogr. J. 66, 289–315.

Oestreich, K., 1906. Die Täler des nordwestlichen Himalaya. Petermanns Geographische Mitteilungen, Ergänzungsband 1–106.

Owen, L.A., 1988a. Terraces, Uplift and Climate, in the Karakoram Mountains, northern Pakistan. Unpublished Ph.D. thesis, Department of Geography, University of Leicester, UK (unpublished).

Owen, L.A., 1988b. Wet-sediment deformation of Quaternary and recent sediments in the Skardu Basin, Karakoram mountains, Pakistan. In: Grott, D.G. (Ed.), Glaciotectonics. Forms and Processes. Balkema, Rotterdam.

Owen, L.A., Gualtieri, L., Finkel, R.C., Caffee, M.W., Benn, D.I., Sharma, M.C., 2001. Cosmogenic radionuclide dating of glacial landforms in the Lahul Himalaya, northern India: defining the timing of Late Quaternary glaciation. J. Quatern. Sci. 16, 555–563.

Owen, L.A., Caffee, M.W., Finkel, R.C., Seong, Y.B., 2008. Quaternary glaciation of the Himalayan-Tibetan orogen. J. Quatern. Sci. 23, 513–531.

Owen, L.A., Robinson, R., Benn, D.I., Finkel, R.C., Davis, N.K., Chaolu Y., Putkonen, J., Dewen L., Murray, A.S., 2009. Quaternary glaciation of Mount Everest. Quatern. Sci. Rev. 28, 1412–1433.

Pachur, H.J., Wünnemann, B., 1995. Lake evolution in the Tengger Desert, northwestern China, during the last 40,000 years. Quatern. Res. 44, 171–180.

Pigati, J.S., Lifton, N.A., 2004. Geomagnetic effects on time-integrated cosmogenic nuclide production with emphasis on in situ ^{14}C and ^{10}Be. Earth Planet. Sci. Lett. 226, 193–205.

Prinz, G., 1927. Beiträge zur Glazilogie Zentralasiens. Mitt. Jahrb. Kgl. Ungar. Geolog. Anst., 25, Budapest, Berlin, pp. 1–133.

Rhodes, T.E., Gasse, F., Ruifen, Lin, Fontes, J.Ch., et al., 1996. A Late Pleistocene—Holocene lacustrine record from Lake Manas, Zunggar (Northern Xinjiang, Western China). Palaeogeogr. Palaeoclimatol. Palaeoecol. 120, 105–118.

Russell, C.T., 2007. The coupling of the solar wind to the Earth's magnetosphere. In: Bothmer, V., Daglis, I.A. (Eds.), Space Weather—Physics and Effects. Springer, Berlin/Heidelberg/New York, pp. 103–130.

Seong, Y.B., Owen, L.A., Bishop, M.P., Bush, A., Clendon, P., Copland, L., et al., 2008a. Quaternary glacial history of the Central Karakoram. Quatern. Sci. Rev. 26 (25–28), 3384–3405.

Seong, Y.B., Owen, L.A., Bishop, M.P., Bush, A., Clendon, P., Copland, L., et al., 2008b. Reply to comments by Matthias Kuhle (2008) on "Quaternary glacier history of the Central Karakorum" by Yeong Bae Seong, Lewis A. Owen, Michael P. Bishop, Andrew Bush, Penny Clendon, Luke Copland, Robert Finkel, Ulrich Kamp, John F. Shroder Jr.. Quatern. Sci. Rev. 27, 1655–1656.

Shi, Y., Wang, J.-T., 1979. The fluctuations of climate, glaciers and sea level since Late Pleistocene in China: Sea level, ice and climatic change. In: Proceedings Canberra Symposium.

Shi, Y., Zheng, B., Li, S., 1992. Last glaciation and maximum glaciation in the Qinghai-Xizang (Tibet) Plateau: a controversy to M. Kuhle's ice sheet hypothesis. Z. Geomorphol. N.F. 84, 19–35 (Berlin, Stuttgart).

Shi Yafeng (Scientific advisor), Li Binyuan, Li Jijun (Chief editors), Cui Zhijiu, Zheng Benxing, Zhang Qingsong, Wang Fubao, Zhou Shangzhe, Shi Zuhui, Jiao Keqin & Kang Jiancheng (Editors)., 1991. Quaternary Glacial Distribution Map of Qinghai-Xizang (Tibet) Plateau, Scale 1:3,000,000, Science Press, Beijing.

Spencer, J.Q., Owen, L.A., 2004. Optically stimulated luminescence dating of Late Quaternary glaciogenic sediments in the upper Hunza valley: validating the timing of glaciation and assessing dating methods. Quatern. Sci. Rev. 23, 175–191.

Stone, J.O., 2000. Air pressure and cosmogenic isotope production. J. Geophys. Res. 105, 23753–23759.

Stozhkov, Yu.I., 2007. What can be extracted from data on the concentrations of Be-10 and C-14 natural radionuclides? Bull. Lebedev Phys. Inst. 34 (5), 135–141.

Tafel, A., 1914. Meine Tibetreise. Eine Studienreise durch das nordwestliche China und durch die innere Mongolei in das östliche Tibet. 2 Bde. Karte, Stuttgart.

Taylor, P.J., Mitchell, W.A., 2002. Comment: cosmogenic radionuclide dating of glacial landforms in the Lahul Himalaya, northern India: defining the timing of late Quaternary glaciation. J. Quatern. Sci. 17 (3), 277–281.

Trinkler, E., 1932. Geographische Forschungen in westlichen Zentralasien und Karakorum-Himalaya. Wiss. Ergeb. Dr. Trinklerschen Zentralasien-Expedition.

Usoskin, I.G., Kovaltsov, G.A., 2008. Cosmic rays and climate of the Earth: possible connection. C. R. Geosci. 340, 441–450.

Van Campo, E., Gasse, F., 1993. Pollen- and diatom-inferred climatic and hydrological changes in Sumix Co Basin (Western Tibet) since 13,000 yr B.P. Quatern. Res. 39, 300–313.

von Loczy, L., 1893. Die wissenschaftlichen Ergebnisse der Reise des Grafen Bla Széchéniy in Ostasien 1877–1880. 3 Abschnitt. Geologie 1, 307–836.

von Wissmann, H., 1959. Die heutige Vergletscherung und Schneegrenze in Hochasien mit Hinweisen auf die Vergletscherung der letzten Eiszeit. Akademie der Wissenschaften und der Literatur in Mainz, Abhandlungen der Mathematisch-Naturwissenschaftlichen Klasse 14, 1103–1407.

Wünnemann, B., Pachur, H.J., Li, J., Zhang, H., 1998. Chronologie der pleistozänen und holozänen Seespiegelschwankungen des Gaxun Nur/Sogo Nur und Baijian Hu, Innere Mongolei, Nordwestchina. Peterm. Geogr. Mittl. 142 (3+4), 191–206.

Wang, H., 2001. Effects of glacial isostatic adjustment since the late Pleistocene on the uplift of the Tibetan Plateau. Geophys. J. Int. 144, 448–458.

Zabirov, R.D., 1955. Oledenenie Pamira. GeografGIZ, Moskau.

Map

Khumbu Himal 1:50 000, 1978: Nepal-Kartenwerk der Arbeitsgemeinschaft f. vergleichende Hochgebirgsforschung Nr.2; Schneider, E.; Second Edition 1978; Freytag-Berndt u. Artaria, Vienna.

Chapter 69

The Extent and Timing of Late Pleistocene Glaciations in the Altai and Neighbouring Mountain Systems

Frank Lehmkuhl[1,*], Michael Klinge[2] and Georg Stauch[1]

[1]Department of Geography, RWTH Aachen University, Templergraben 55, 52056 Aachen, Germany
[2]Kischenangerweg 10, 37139 Adelebsen / Lödingsen
*Correspondence and requests for materials should be addressed to Frank Lehmkuhl. E-mail: flehmkuhl@geo.rwth-aachen.de

69.1. INTRODUCTION

We present first the extent of Late Pleistocene ice in the Altai (the Russian, Mongolian and Chinese parts) and add results concerning the Pleistocene glaciations in Western Mongolia, such as the Khangai Mountains and the mountain ranges surrounding Lake Hovsgol (Hövsgöl, Khovgul Nuur) in northern Mongolia (Fig. 69.1). Secondly, we present the current state of research concerning the timing of glacier advances in this region. The Altai Mountain system and the Khangai Mountains are situated at the northern border of Central Asia between the Siberian taiga in the north and the steppe to desert steppe regions in the south. Regarding their different geomorphological settings and climate conditions, the Altai Mountains (82–95°E; 46–52°N) can be divided into the western Russian Altai, the eastern Mongolian Altai and the southern Chinese Altai. In the north-east, the Altai Mountains are bounded by the Sayan Mountains. The south-eastern Altai extends into the Gobi Altai. East of the Altai, the so-called Valley of Great Lakes leads over to the Khangai Mountains. The Khangai Mountains trend from north-west to south-east and are located in central Mongolia (96–103°E; 46–50°N, Fig. 69.1), approximately 100 km east of the Altai Mountains. Lake Hovsgol occurs in this region, approximately 300 km north of the Khangai Mountains (Fig. 69.1). From here, Prokopenko et al. (2009) published new results from a long sediment core containing palaeoclimate records of ca. 1 Ma from a drilling in the lake-bottom sediments.

The extent of the Last Glacial Maximum (LGM) in the eastern Altai and the Khangai is often clearly determinable by the surface morphology of glacial sediments using air photographs, satellite images and topographic maps. However, intensive fieldwork is required to examine glacial relics and sediments, especially in the steep and narrow valleys of the western Altai, and in few regions of the Mongolian Altai and Khangai, respectively. The Altai Mountains are divided into the territories of four different countries: Russia, Mongolia, China and Kazakhstan. Therefore, research work in the Altai is always limited to a part of these mountains. The different regions are discussed separately in the literature, and up till today, there is no comprehensive of the entire Altai Mountains massif, apart from the authors' previous work (Lehmkuhl et al., 2004).

In several areas of this region, the extent and especially the timing of glaciers during the LGM are still a matter of debate. Some relative chronologies exist for mountain glacier fluctuations, but the timing is poorly understood because of the lack of absolute ages for moraines. However, field evidence in Mongolia and Russia show that there is a considerable difference between the Late Pleistocene moraines (of the last glacial cycle = local Last Glacial Maximum, lLGM) and their connecting outwash plains, and those of the penultimate glaciation or Late Glacial to Holocene moraines.

The underlying topography and the climatic conditions control the modern and Pleistocene glaciations. In the western and southern Altai, the alpine relief is dominated by steep V-shaped valleys. At the margins of the Russian Altai and in Mongolia, the highest parts of the mountains comprise flat mountain tops, which are shaped by a major planation surface rising from 2000 m a.s.l. in the north-western Altai to between 3000 and 3500 m a.s.l. on the eastern fringe of the Altai. The summits, which stretch above the main mountain surface, reach elevations of over 4000 m a.s.l., including mountains up to 4506 m a.s.l., including

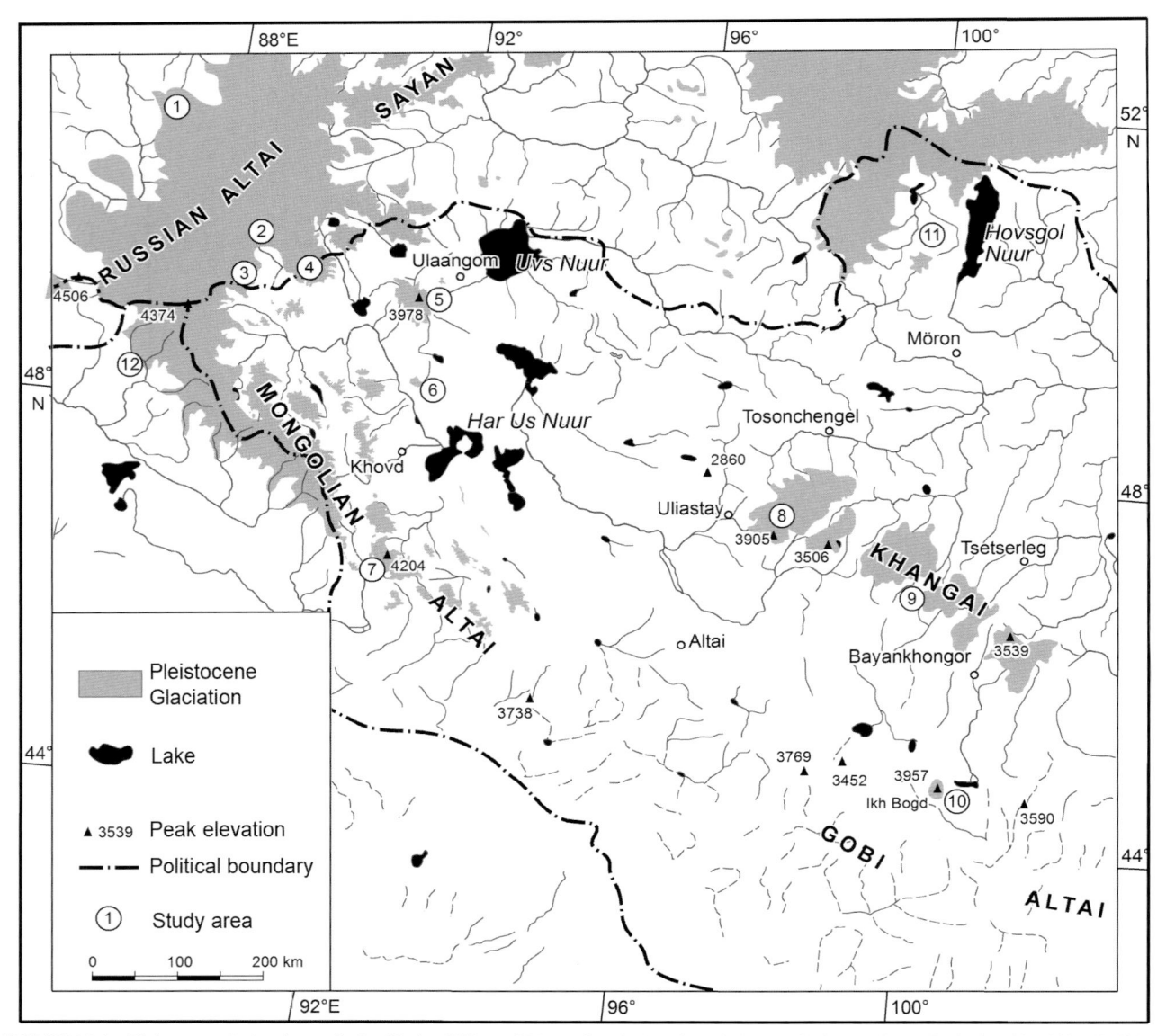

FIGURE 69.1 Pleistocene glaciated areas in the Altai and Khangai Mountains, the Russian Altai: compiled according to Budvylovski (1993), Kotljakov et al. (1997) and the authors' observations. For the numbers of the study areas, see Table 69.1.

the Belucha and the Tavan Bogd (4374 m a.s.l.). In contrast, intramontane basins with widespread alluvial fans, fanglomerats and lakes are dividing the various isolated mountain systems of the eastern Altai. The mountain systems of this region, especially in Mongolia, comprise wide and extensive flat surfaces in their central parts. Here, they have been dissected by deep glacial cirques and U-shaped glacial troughs in the higher parts of the mountains. At the outer margins of the mountains, these flat surfaces give way to broad flat ridges. In contrast, in the central part of the Khangai, three main planation surfaces (peneplains) can be recognised (Klimek, 1980); in the northern part of the Altai, two to three main planation surfaces also occur. As a result of the more active fault scarps and tectonics, they are divided into smaller patches.

The present-day continental climatic conditions can be shown by large temperature variations with winter temperatures below $-20\ °C$ and summer temperatures up to more than $20\ °C$. The highest amount of annual rainfall occurs in the western Altai, which is influenced by the westerlies (Lydolph, 1977). The eastern part of the Altai is situated in a rain shadow and is consequently more arid. In the north-western Russian Altai, the estimated amount of precipitation exceeds 1500 mm/a in the mountains but is reduced to less than 300 mm/a in the south-eastern region. (Komitet Geodesii I Kartografii CCCP, 1991). In western

Mongolia, the precipitation is about 200 mm at 2000 m a.s.l. but is estimated to be greater than 300 mm in the higher mountain areas, whilst precipitation is less than 50 mm/a in the Valley of Great Lakes. Further to the east, in the Khangai Mountains, the annual precipitation rises to over 400 mm/a, whereas in the basins, it is below 100 mm/a (Barthel, 1983; Academy of Sciences of Mongolia and Academy of Sciences of USSR, 1990). The distribution of glaciers and other altitudinal belts, such as periglacial environments or the distribution of forest, is controlled by these general climatic conditions (Lehmkuhl, 2008). Böhner and Lehmkuhl (2005) and Klinge et al. (2003) show results of climate modelling, and the distribution of the snowline and timberline for Central and High Asia or the Altai Mountains, respectively.

69.2. EXTENT OF PRESENT AND LATE PLEISTOCENE GLACIATIONS

Even today, many recent glaciers exist in the Altai. Information about the modern glaciations in the Altai and Khangai Mountains are given by the Academy of Sciences of Mongolia and Academy of Sciences of USSR (1990), Klinge (2001), Klinge et al. (2003), Kotljakov et al. (1997), Lehmkuhl (1999) and Lehmkuhl et al. (2004). The present glaciation of the eastern Altai is restricted to the highest peaks in the central parts of the various mountain systems. Plateau glaciers, cirque glaciers, many isolated ice patches and several smaller valley glaciers occur. The size of the currently glaciated area is greater than 900 km^2 in the Russian Altai (Bussemer, 2000), about 200 km^2 in the Chinese Altai (Shi, 1992) and about 850 km^2 in the Mongolian Altai (Klinge, 2001). In the Khangai Mountains, exclusively, only one small glacier occurs in the westernmost part at the Otgon Tenger Uul (Fig. 69.1 and Table 69.1: No. 8; Fig. 69.5: No. 34).

69.2.1. Russian Altai

The extent of Pleistocene glaciations in the north-western Russian Altai is still debated. According to Rudoi (2002), for example, the Pleistocene ice reached the foothill of the Altai, close to Gorno Altaisk (about 300 m a.s.l.) in the Katun valley, whereas Budvylovski (1993) identified the terminal moraines about 100 km further upstream. Bussemer (2000) compiled the various descriptions concerning the extent of the last maximum glaciation in the north-western Russian Altai in the Russian literature. Figure 69.2 shows the extent of Pleistocene ice according to Budvylovski (1993) and Rudoi (2002). The figure includes the occurrence of ice-dammed lakes in the Altai Mountains and remnants of catastrophic floods in the mountain foreland (Baker et al., 1993; Carling et al., 2002, 2009; Rudoi, 2002). The authors' observations indicate that there is no evidence for these ice-dammed lakes in Western Mongolia. However, the information concerning the Pleistocene glaciated area in the Russian Altai varies between 32,000 and 35,000 km^2 (Bussemer, 2000). On the basis of the maps compiled and the author's own observations, it is estimated that the extent of Pleistocene glacial ice in the Russian Altai was at least 80,000 km^2 (Lehmkuhl et al., 2004).

Most researchers argue that valley glaciers of a limited ice cap reached at least Lake Teleski at 430 m a.s.l. in the northernmost parts of the Russian Altai (Fig. 69.1: No. 1; e.g. Baryshnikov, 1992; Budvylovski, 1993; Bussemer, 2000). This indicates a Pleistocene ELA of lower than 2000 m a.s.l. (Table 69.1). The ice extent was much greater in the western parts of the Altai and more restricted in the eastern regions. The Pleistocene glacial landforms here include mainly cirques, U-shaped valleys, hanging glaciers and small ice caps. In the ablation areas, different types of moraines are present. The southern limit of this small ice sheet can be shown north of the Chuja Basin (Fig. 69.1 and Table 69.1: Nos. 2, 3 and Fig. 69.3). Several moraines here reach the mountain foothills surrounding this tectonic basin. Here, different morainic stages can be mapped as a consequence of the geomorphology, the weathering and the overlying strata, especially in the south-western part close to the settlement Beltir. This is one of the areas where valley glaciers extended down to the basin and dammed the main rivers.

69.2.2. Mongolian Altai

Four areas can be distinguished in which there is geological evidence for multiple Pleistocene glaciations in the Mongolian Mountains: the Khentey, the Khangai and the Mongolian Altai. In addition, there are traces of Pleistocene glaciations in the mountains regions surrounding the Lake Khovgul (Hovsgol Nuur) and the Darhad Basin in Northern Mongolia (Fig. 69.1: No. 11; Devjatkin, 1981; Kotljakov et al., 1997; Gillespie et al., 2008). The Pleistocene glaciated area in the Mongolian Altai (western Mongolia) is calculated to be about 28,750 (Devjatkin, 1981) and 20,700 km^2 (Klinge, 2001), respectively. Florensov and Korzhnev (1982) published a map showing the extent of a substantial Pleistocene ice sheet in the Mongolian Altai. This could not be confirmed by other authors (Klinge, 2001; Lehmkuhl et al., 2004).

On the basis of geomorphological criteria, the degree of erosion and the preservation of evidence, Devjatkin (1981) has differentiated three different Pleistocene glacial advances in Mongolia. The most extensive glaciation occurred during the Middle Pleistocene, whilst both the younger moraines belong to the Late Pleistocene. This chronological classification was confirmed by the correlation of fluvial and alluvial sediments with the different terminal

TABLE 69.1 Estimation of Modern and Pleistocene Snowlines (ELA/GEI) and Snowline Depressions for the Russian Altai (1–3), Mongolian Altai (4–6), Khangai Mountains (8–9), Gobi Altai (10), Northern Mongolia (Mountains Surrounding Darhad Basin = 11) and Kanas Lake in the Chinese Altai (12)

Pleistocene Ice Margin	Elevation (m a.s.l.)	Catchment area (Distance to Pleistocene Ice Margin)	Peak Elevation (m a.s.l.)	Modern Ice Margin (m a.s.l.)	Modern ELA (m a.s.l.)	Pleistocene ELA (Depression)
1. Lake Telisker (Bija River)	424	Chulyshman Plateau (ca. 100 km, no modern glaciers)	3110	–	>3100	1800 (ca. 1300)
2. Chuja Basin (N)	ca. 1750	North-Chuja (60 km)	3936	2350–2700	3150–3300	2800 (−350 to 500)
3. Chuja Basin (S)	2100–2200	South-Chuja (ca. 10 km)	3936	2800–2900	3300–3350	3000–3050 (−300)
4. Ikh Türgen (N-Seite)	1900	Ikh Türgen (ca. 30 km)	4029	3100	3500	3000 (−500)
5. Turgen–Kharkhiraa	1900–2200	Turgen/Kharkhiraa (different aspects, ca. 40 km)	3966	2900–3200	3500	2900 (−600)
6. Tsambagarav Uul	2200–2800	Mongolian Altai (central part) (max. 80 km)	4208	3300–3600	3700	3200 (−500)
7. Munk Khairkhan	2300–2900	Mongolian Altai (southern part) (25 km)	4204	3300–3500	>3700	3200 (−500)
8. Otgon Tenger (Khangay)	2100–2500	Western Khangai (45 km)	3905	3400–3600	3750	3000 (−750)
9. N and S slope of Khangay	2100–2300	Central Khangai (44 km)	3514	–	>3900	2800 (>1000)
10. Ikh Bogd	3900	Gobi Altai (northern aspect)	3980	–	>4000	3800–3900 (?)
11. Darhad Basin (Gillespie et al., 2008)	ca. 1600	Darhad Basin (W of Lake Hovsgol)	2900–3200	–	>3200	2100–2400 (900–1200)
12. Kanas Lake (Xu et al., 2009)	1340	Chinese Altai	4374	2416	3300[a]	2800–2900 (>500?)[a]

For the locations, see Fig. 69.1.
[a] The authors' calculations.

moraine stages and the radiometric dating of overlying peat, fossil soils and aeolian sediments (Lehmkuhl, 1999; Grunert et al., 2000; Klinge, 2001; Lehmkuhl and Lang, 2001). The correlation between the different Late Quaternary glacial advances in the Altai and Khangai Mountains and the glaciations in Northern Asia (Arkhipov et al., 1986) and Europe is shown in Table 69.2. During several expeditions to western Mongolia between 1996 and 1999, detailed glacio-geomorphological fieldwork was undertaken by the authors at sites in the Khangai and Mongolian Altai. In western Mongolia, several terminal-moraine systems were separated on the basis of geomorphological criteria. These include surface morphology, weathering and different types of soil development. Two types of carbonate crusts, which develop under arid climate conditions on the underside of stones, can be differentiated: the younger one of Holocene origin has a bright white colour and mostly lies in its initial position; an older one is light yellow in colour and has been weathered and turned upside down by fluvial or periglacial processes. Old carbonate crusts are often distributed within an older till (Table 69.2).

Although at least a few Pleistocene glaciers reach the intramontane basins in the northernmost part of the Altai, they are restricted to the central part of the mountains in the Mongolian Altai. For example, the detailed glacial geomorphology of the Turgen–Kharkhiraa Mountains (Fig. 69.1 and Table 69.1: No. 5) is presented by Lehmkuhl (1998). According to Grunert et al. (2000), the widespread aeolian mantles in the Turgen–Kharkhiraa Mountains were accumulated mainly during the Weichselian Late Glacial or

Chapter | 69 The Extent and Timing of Late Pleistocene Glaciations in the Altai

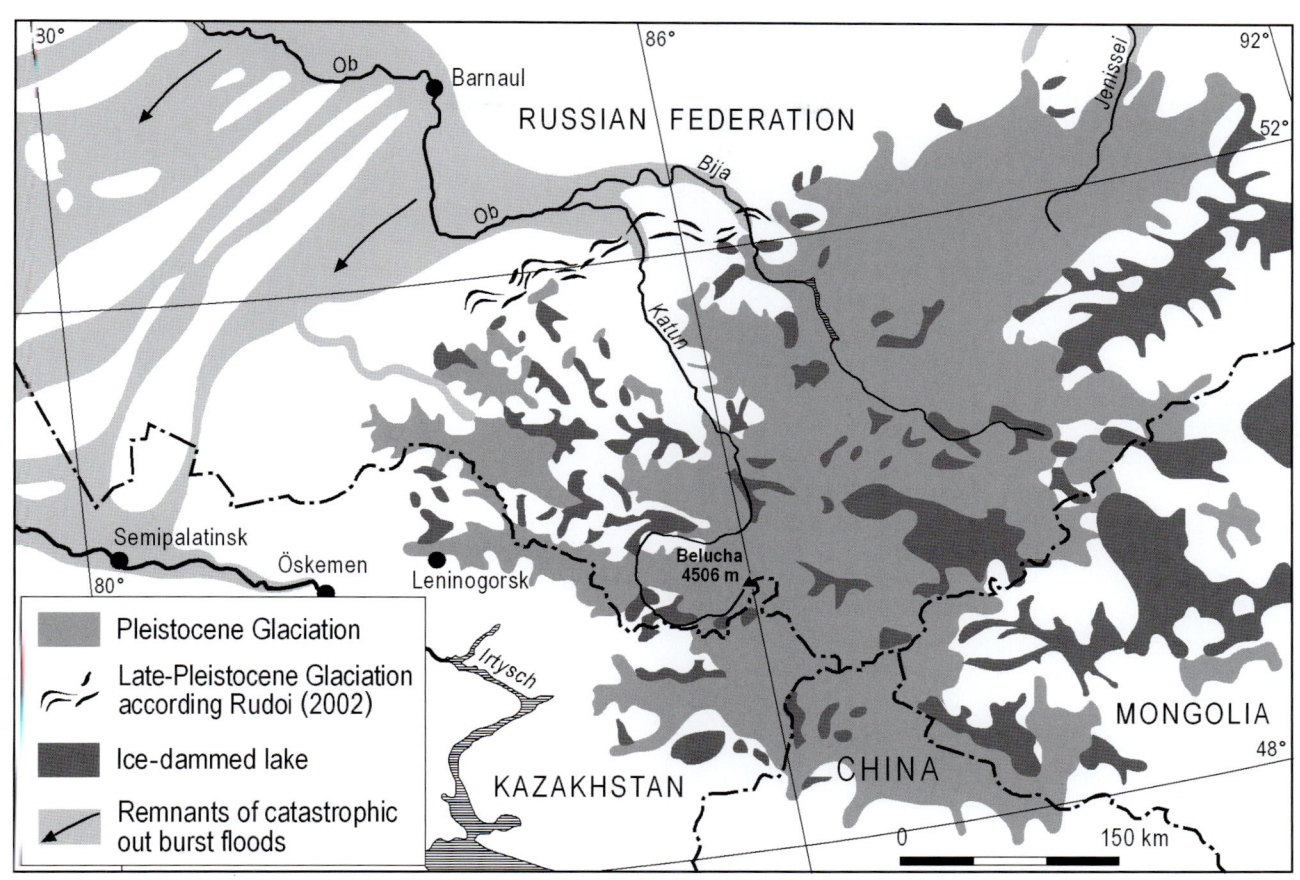

FIGURE 69.2 Extent of Pleistocene glaciations in the Russian Altai, according to Budvylovski (1993) and Rudoi (2002).

Early Holocene. However, in some regions, interstadial loess-like silt overlies different older terrace sequences and the latter ones can be correlated to ice margins. Results from luminescence dating confirm the morphostratigraphy and suggest maximum ice advances during the two maximum cooling periods of the last glacial cycle equivalent to the marine isotope stages (MIS) 4 and 2.

In the southern Mongolian Altai, local Pleistocene mountain glaciations in the Munkh Khairkhan Mountains (Fig. 69.1: No. 7) yield moraine ages of Middle- to Late Pleistocene age. Late Glacial moraines are rare and indistinct in the Mongolian Altai (Klinge, 2001), whilst there are more of them in the Khangai (Richter, 1961; Devjatkin, 1981; Florensov and Korzhnev, 1982; Lehmkuhl, 1998). This morphostratigraphical system (Table 69.2) is confirmed by luminescence dating of overlying aeolian sediments that give ages of 57 and 17 ka BP (Lehmkuhl and Lang, 2000; Klinge, 2001). The clearly identifiable Late Pleistocene terminal moraines M_1 derived from large valley glaciers. Most of them are marked by two moraine ridges, which indicate two main glacial periods during the Late Pleistocene (M_{1a}: 14/15–32 ka BP, M_{1b}: 50–70 ka BP). The moraine stages can be combined with glacial–fluvial outwash plains and terraces, which are connected with the pediments in the basins. In some areas throughout the whole region, relics of older, deeply weathered moraines (M_2) can be found not far from the last glacial ice margins. They probably belong to a slightly more extensive Middle Pleistocene glaciation.

69.2.3. Khangai

Six centres of Pleistocene glaciations in the Khangai Mountains covering an area of approximately 12,900 km^2 were shown by Florensov and Korzhnev (1982) and Lehmkuhl (1998). Richter (1961) and Klinge (2001) described Pleistocene glaciers and terminal moraines from the western Khangai, near Otgon Tenger Uul (Fig. 69.1: No. 8). Klimek (1980) reported two glacial events from the Last Glaciation of the southern side of the Khangai in the Tsagan-Turutuin-Gol drainage basin. Evidence concerning the Pleistocene glaciations in the upper catchment area of Baydragiyn Gol, in the central part of the Khangai, is given by Lehmkuhl (1998), Lehmkuhl and Lang (2000) and Walther (1998). Regarding the timing of the glaciation during the last glacial cycle, up till now only Lehmkuhl and

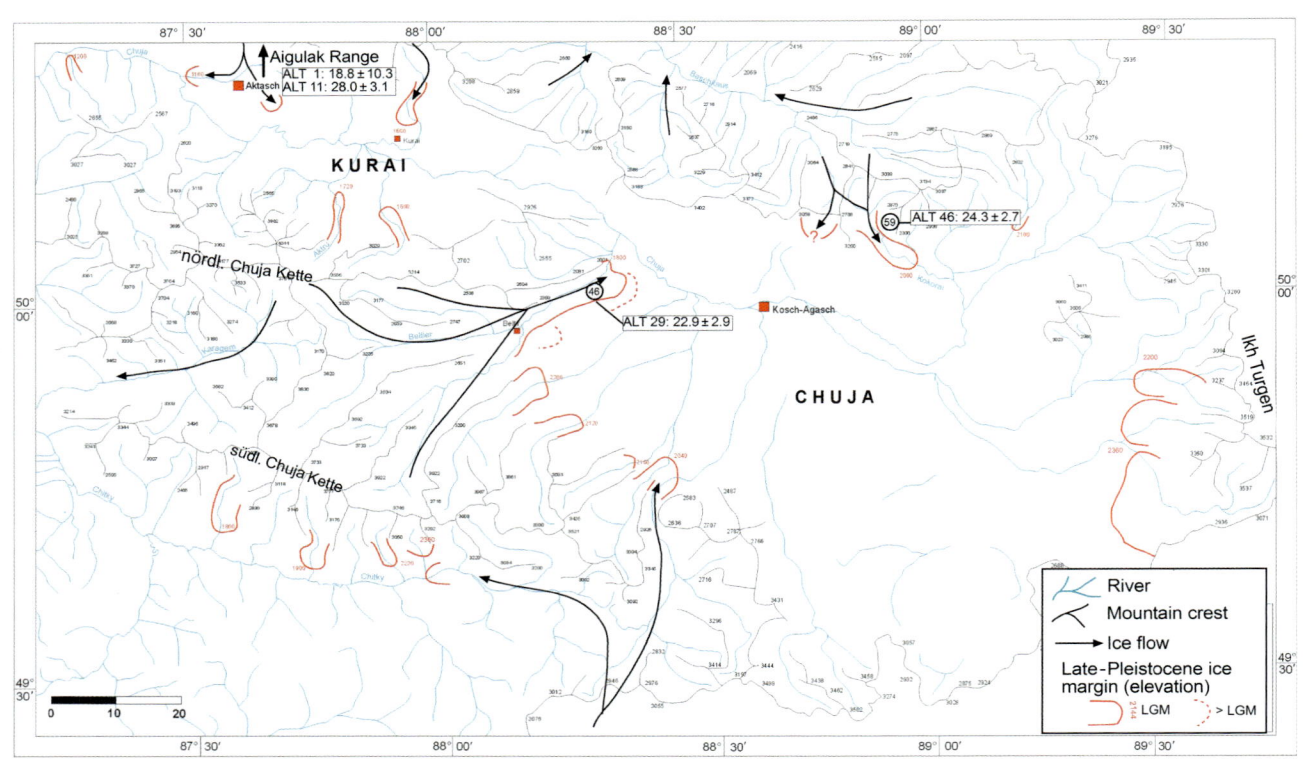

FIGURE 69.3 Extent of selected ice margins surrounding the Chuja Basin (compiled according to Budvylovski, 1993 and the authors' observations, see Fig. 69.1, No. 2/3). Modified according to Lehmkuhl et al. (2004).

Lang (2001) have published one luminescence age of about 22 ka for a sand deposit overlying the terrace which is connected to the Last Glacial ice margin.

69.3. PRESENT AND LATE PLEISTOCENE ELA RECONSTRUCTIONS

Palaeoclimatic reconstructions based on the limits of former glaciers often make use of estimates of the associated equilibrium-line altitudes (ELAs). The equilibrium line marks the position where, over a period of 1 year, accumulation of snow and ablation is exactly balanced. There is a very close connection between the ELA and local climate, particularly solid precipitation and air temperatures. The ELA is sensitive to perturbations in either of these two variables and rises in response to decreasing snowfall and/or increasing frequency of positive air temperatures, and vice versa. Fluctuations in the ELA, therefore, provide an important indicator of glacier response to climate change and allow reconstructions of former climates and the prediction of future glacier behaviour (Benn and Lehmkuhl, 2000).

Frenzel (1959) reconstructed the Pleistocene snowline in Eurasia. He includes the Russian references and shows a steep west–east increase of about 1500 m from about 1250 to 2750 m a.s.l. for the Pleistocene snowlines in the Russian Altai. A map of the Pleistocene glaciations in southern Siberia, presented by Kotljakov et al. (1997), based on information of Grosswald (1980), shows an increase in altitude of the Pleistocene snowline from 1800 m in the north-western Russian Altai to 2600 m a.s.l. in the central Mongolian Altai. It then decreases again to 2600 m a.s.l. in the south-eastern Mongolian Altai. In contrast, Klinge (2001) calculates a steady altitudinal increase of the Pleistocene snowline from 2900 m a.s.l. in the centre to more than 3100 m a.s.l. in the south-eastern Mongolian Altai. For the Turgen–Kharkhiraa Mountains (91.5°E, 49.6°N), which are situated in the northern Mongolian Altai, Kotljakov et al. (1997) propose a Pleistocene snowline of about 2700 m a.s.l. Differing from this, Lehmkuhl (1998, 1999) and Grunert et al. (2000) report a Pleistocene snowline of 2900 to 3000 m a.s.l. for the Turgen–Kharkhiraa. These differences between the snowline altitudes of 200 and 300 m can be arise from unequal snowline calculation methods (Benn and Lehmkuhl, 2000).

For the Pleistocene glaciation in the southern Altai region, which is part of Kazakhstan and China, Kotljakov et al. (1997) and Shi (1992) reported glacial snowline altitudes between 2350 and 2500 m a.s.l. In this study, the authors use the detailed description of the Pleistocene glaciated area by Kotljakov et al. (1997). For the Chinese

TABLE 69.2 Stratigraphy of Late and Middle Quaternary Glacier Advances in the Altai and Khangai Mountains (Lehmkuhl et al., 2004)

Terminal Moraine Stage	Glacier Advances	Oxygen Isotope Stage	Siberian Classification	Alpine Classification	Pedological Environment of the Specific Till
M_{LIA}	Sixteenth to middle of nineteenth centuries AC	1		Little Ice Age	Fresh till without any sediment and soil cover
M_0	10–15 ka BP	1		Late Glacial	Aeolian cover bed and Kastanozem soil with carbonate layer
M_{1a}	15/20–32 ka BP	2	Sartan	Late Würm	Aeolian cover bed and castanoseme soil with carbonate layer
		3	Karginsky Interstadial		Higher lake levels
M_{1b}	50–70 ka BP	4	Early Zyrianka	Early Würmian	Aeolian cover bed and castanoseme soil with carbonate layer; erosion, older carbonate crusts are common
		5	Kazantsevo Interglacial	Eemian	
M_2	>132 ka BP	>5	Taz, Somarova	Rissian	Intensively eroded; relics
		>9/16	Shaitan	Middle- to Early Pleistocene	No glacial deposits

Altai, the extent of Pleistocene glacial ice is estimated to be about 23,000 km².

To show the variations in the ice extent, the modern and Late Pleistocene ELAs for this area were presented in two cross sections by Lehmkuhl et al. (2004). They mapped the modern and Late Pleistocene ice extent and named the latter event, the lLGM. This was done to contrast it with the global Last Glacial Maximum (gLGM) which occurred during MIS 2 because there are only a few numerical dates that can be related to the timing of the terminal moraines (Lehmkuhl and Owen, 2005). The ELAs occur at a relatively low altitude in the more humid outermost ranges of the Russian Altai but rise towards the central part of Mongolia (Fig. 69.1 and Table 69.1). The limited extent of present and Pleistocene glaciers in the eastern part of the Russian Altai and the Mongolian Altai is the result of reduced precipitation from west to east. This results in a rise of present and Pleistocene ELAs towards the east. However, this general increase of the ELAs towards the more arid regions was more evident during glacial times (Table 69.1). The present ELA rises about >1000 m from the north-western part of the Russian Altai (<2600–3100 m a.s.l.; Table 69.1: No. 2) towards the south-east (southern part of the Mongolian Altai, about 3700 m a.s.l.; Table 69.1: No. 7). The Late Quaternary ELA is calculated to be below 2000 m a.s.l. in the north-western mountains ranges of the Russian Altai (Table 69.1: No. 1) and >3100 m a.s.l. in the southern parts of the Mongolian Altai (Table 69.1: No. 7). In the Khangai Mountains, the present ELA occurs at about 3800–3900 m a.s.l. (Table 69.1: Nos. 8 and 9; Lehmkuhl, 1998; Klinge, 2001), whereas in the Chinese Altai, the Late Pleistocene ELA can be estimated at elevations between 2800 and 3000 m. The Pleistocene ELA is between 2700 and 2800 m a.s.l. in the Khangai Mountains.

Klinge (2001) calculated the modern ELAs and the distribution of ice on 8 different slope aspects for 29 mountain massifs in western Mongolia (Fig. 69.4 and Table 69.3). The map in Fig. 69.4 shows the influence of humidity on the present ELAs in the different mountain massifs and the overall ELA of the Mongolian Altai. A general increase from low to high latitude can be seen resulting from the lowering of temperature especially in the eastern part of western Mongolia. Above the western and central Altai, the snowline changes its latitudinal (W–E) direction into a longitudinal direction, which is caused by the maximum rainfall in the western mountain ranges.

The resulting ELA depressions are higher than 1000 m in the wettest parts of the western Russian Altai and between 800 and 400 m in the eastern part of the Russian Altai and in the Mongolian Altai, respectively (Fig. 69.5

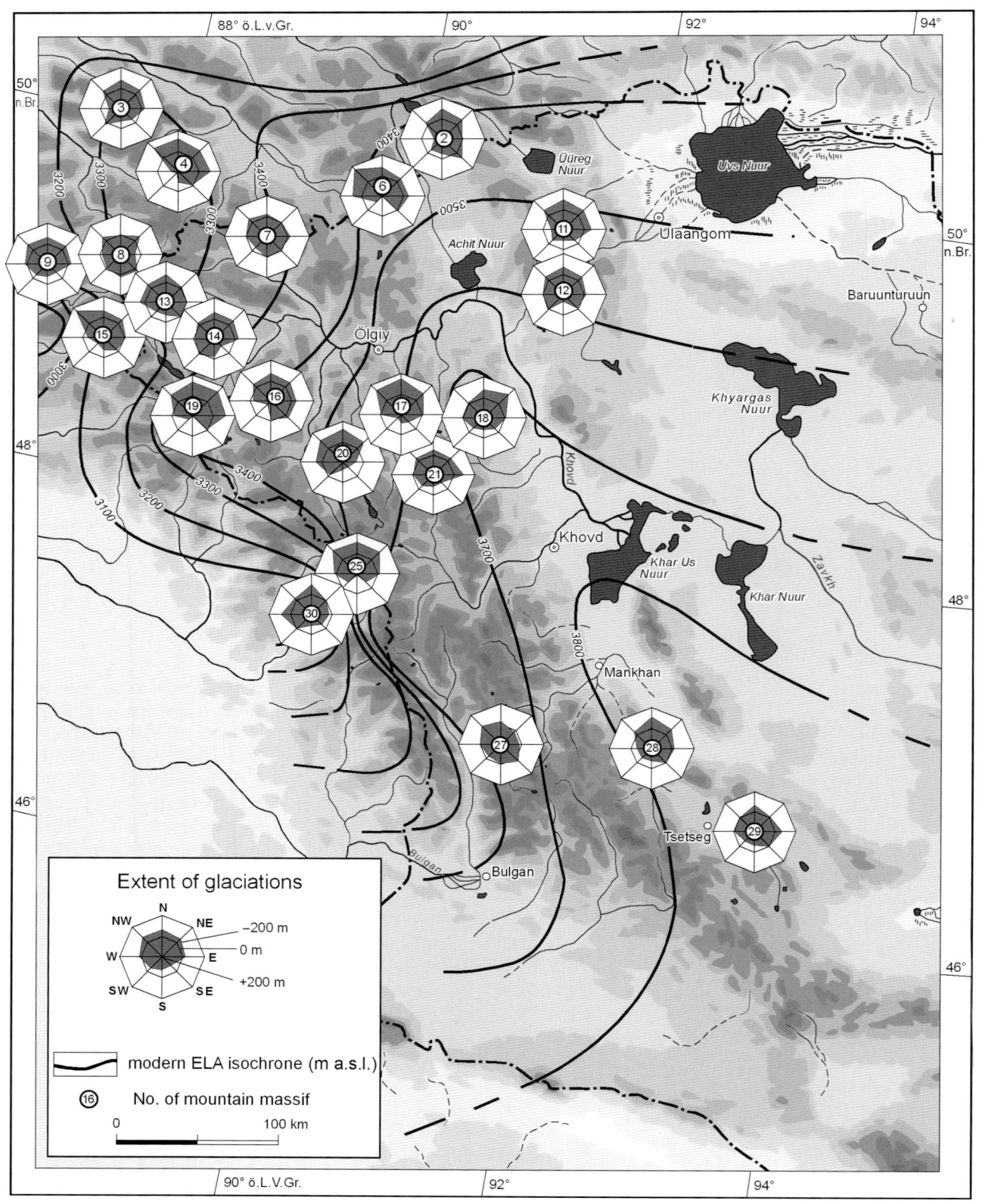

FIGURE 69.4 Map of modern ELAs and ice extent of 29 mountain regions in the Mongolian Altai and adjacent areas (modified according to Klinge, 2001). For the numbers of the mountain massifs, refer to Table 69.3.

Chapter | 69 The Extent and Timing of Late Pleistocene Glaciations in the Altai

TABLE 69.3 Estimates of Modern and Pleistocene Snowlines (ELA/GEI) and Snowline Depressions for the Mongolian Altai and Adjacent Areas (for the Areas, See Figs. 69.4 and 69.5)

No	Area	Longitude E	Latitude N	Modern Snowline (m a.s.l.)	Snowline LGM (m)	Snowline Depression (m)
1	Russian Altai, N	90°00′	50°30′	3195	2807	389
2	Munkhu Khairkhan Uul	90°09′	50°16′	3461	2954	507
3	Russian Altai, NW	87°38′	49°49′	3298	2894	405
4	Russian Altai, main ridge	88°05′	49°45′	3322	2894	428
5	Talbu Aur	89°19′	49°58.5′	3314		
6	Ikh Turgen Uul	89°45′	49°48′	3450	2971	480
7	Sarschetami	88°45′	49°28′	3231	2913	318
8	Tavan Bogd	87°50′	49°09′	3400	3019	381
9	Main ridge, west of Tavan Bogd	87°22′	49°5.5′	3061		
10	Nagon Nurankhel	89°27′	49°21′	3328		
11	Turgen	91°20′	49°41′	3587	3006	581
12	Kharkhiraa	91°23′	49°34′	3537	3006	531
13	Arschanu Ikhe Ula	88°14′	48°58.7′	3358	3053	305
14	Khoumunu Njuruk Ula	88°27′	48°40′	3220	2986	234
15	Tavan Bogd, S	87°51′	48°57′	3153	2918	235
16	Tsengel Khairkhan	89°05′	48°38′	3418	3115	303
17	Khrumuni Khairkhan Uul/Tolbo	90°10′	48°38.5′	3538	3030	508
18	Tsambagarav Uul	90°50′	48°40′	3737	3163	574
19	Main ridge, south-west of Tsengel Khairkhan Uul	80°36′	48°21′	3506	2969	537
20	Khrebet Gurban Duschiin Nuru	89°50′	48°20′	3511	3047	465
21	Sair, south of Tsambagarav Uul	90°33′	48°02′	3625	3184	441
22	Khokh Serkhiyn Nuruu, N	90°57′	47°57.5′	3673	3171	502
23	Chinese Altai, N	89°30′	47°51′	3291		
24	Khokh Serkhiyn Nuruu, S	90°48′	48°10′	3689	3171	518
25	Main ridge, west of Khovd, N	87°22′	49°05.5′	3514	3041	473
26	Main ridge, west of Khovd, S	90°45′	47°15′	3625	3041	584
27	Munkh Khairkhan	91°40′	47°05′	3680	3183	497
28	Munkh Tsast/Baltarin Nuruu/Ekhen Burkhad Uul	92°44′	46°58′	3835	3174	661
29	Sutai/Tsasamu Bogda Ula	93°35′	46°37′	3838	3162	675
30	Chinese Altai, S	89°55′	47°30′	3057		
31	Munkh Khairkhan, N	91°18′	47°20′		3124	
32	Main ridge, south of Tsengel Khairkhan	89°11′	48°14′		2906	
33	Changai, Otgon Tenger	97°33′	47°34.5′	3703	3007	696
34	Changai, north of Otgon Tenger	98°00′	48°00′		2738	

Continued

TABLE 69.3 Estimates of Modern and Pleistocene Snowlines (ELA/GEI) and Snowline Depressions for the Mongolian Altai and Adjacent Areas (for the Areas, See Figs. 69.4 and 69.5)—cont'd

No	Area	Longitude E	Latitude N	Modern Snowline (m a.s.l.)	Snowline LGM (m)	Snowline Depression (m)
35	Mid-Changai, W	99°54′	47°38′		2822	
36	Mid-Changai, E	100°05′	47°09′		2826	

Data adapted from Klinge (2001, p. 83).

and Tables 69.1 and 69.3). This is a result of the greater aridity in the Mongolian Altai (Lehmkuhl et al., 2004). In the Khangai, further to the east, the ELA depression is again over 1000 m (Table 69.1: No. 9). This may be the consequence of a strong monsoonal influence, which is also evident further east in the mountains of Northern China, such as the Qinling Shan or Wutai Shan (Lehmkuhl and Rost, 1993).

The extent of present and Late Quaternary glaciations in the Mongolian mountains (Khangai and Gobi Altai) towards the southern part of the Tibetan Plateau was published by Lehmkuhl (1998). Even though there are no modern glaciers in the central part of the Khangai and Gobi Altai, Lehmkuhl and Lang (2001) calculated that the Pleistocene ELA was between 2700 and 2800 m a.s.l. in the Khangai. Lehmkuhl (1998) suggested that limited Quaternary glaciers were also present in the Gobi Altai in the vicinity of the Ikh Bogd (3957 m a.s.l.; Fig. 69.1: No. 10) during glacial times. This implies that the ELA depression in the Khangai and Gobi Altai was approximately 1000 m, that is, similar to that on the northern slopes of the Qilian Shan in north-eastern Tibet (Lehmkuhl and Rost, 1993).

69.4. TIMING OF PLEISTOCENE GLACIATIONS IN THE RUSSIAN ALTAI AND WESTERN MONGOLIA

Only a few published results constrain the dating of Late Pleistocene glaciations in this vast region. In the eastern part of the Russian and Mongolian Altai, there is evidence of two to three major Pleistocene glaciations (Devjatkin, 1981; Florensov and Korzhnev, 1982). However, the extent of two of the main stages is similar because the terminal moraines are close together (Lehmkuhl, 1998; Klinge, 2001). Here, Russian scientists established a Pleistocene stratigraphy for this region and used the local names from the Siberian stratigraphy (cf. Arkhipov et al., 1986; Lehmkuhl, 1998; Table 69.2). According to this stratigraphy, the Last Glaciation is divided into two glacial periods, the Sartan Glaciation (corresponding to the MIS 2) and the Early Zyrianka Glaciation (MIS 4), interrupted by the Karginsky Interstadial (MIS 3). Remnants of these events can be separated on the basis of their differing degrees of weathering of sediments and the connection with different gravel spreads and terraces (Lehmkuhl, 1998). In contrast, Lehmkuhl (1998), Klinge (2001) and Lehmkuhl et al. (2004) differentiated two major Pleistocene glaciations in the Altai Mountains on the same bases (Table 69.2).

Four luminescence ages determined from sand and silt layers within till sediments in the Russian Altai have been published Lehmkuhl et al. (2007). These dates range from 28 to 19 ka and are related to the Last Glacial ice margins (Fig. 69.3: ALT11, 28.0 ± 3.1 ka; ALT29, 22.9 ± 2.9 ka; ALT46, 24.3 ± 2.7 ka; ALT1, 18.8 ± 10.3 ka). In addition, Reuther et al. (2006) showed that glacier-dammed floods down the Ob River date to 15.8 ± 1.8 ka (^{10}Be), suggesting that Late Weichselian (MIS 2) glaciers were close to their maximum extents at this time.

Based on luminescence dating of overlying aeolian cover beds, huge alluvial fans and lake-level variations, Grunert et al. (2000) support the view that there are two major ice advances equivalent to MIS 2 and 4 in the Turgen–Kharkhiraa Mountains (Fig. 69.1: No. 5). For the Munkh Khairkhan Mountains (Fig. 69.1: No. 7), Klinge (2001) published luminescence dating of overlying aeolian sediments (17 and 57 ka BP) post-dates the glaciations. They confirm the local morphostratigraphy and also suggest maximum ice advances in the MIS 2 and 4. In addition, Lehmkuhl and Lang (2001) published a luminescence age of 21.7 ± 2.2 ka for a sand deposit overlying the terrace which is related to the Last Glacial ice margin several hundred metres for the Khangai Mountains (Fig. 69.1: No. 9) nearby. Because this aeolian or fluvial deposit is deposited immediately on top of the glaciofluvial gravel, this can be taken as minimum age for this ice advance in the southern Khangai (Fig. 69.1: No. 9), suggesting an equivalent to MIS 2.

Gillespie et al. (2008) presented the first results concerning the timing and extent of Late Pleistocene glaciers around the Darhad Basin (Fig. 69.1: No. 11). Here, the glaciers advanced to near their maximum positions at least

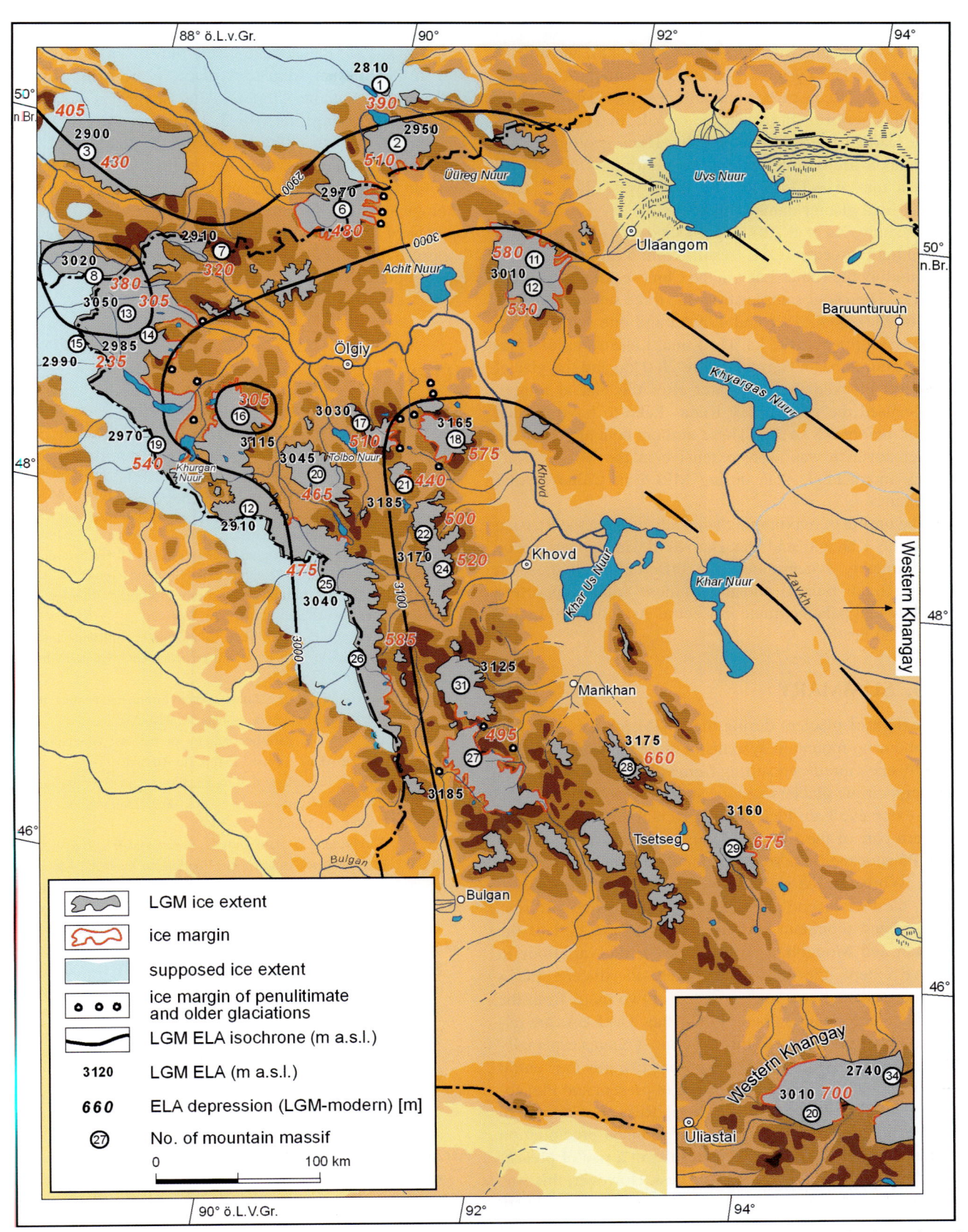

FIGURE 69.5 Map of Pleistocene ice extent and LGM ELAs in the Mongolian Altai and adjacent areas (modified after Klinge, 2001). For the numbers of the mountain massif, refer to Table 69.3.

three times, twice during the Zyrianka glaciation (MIS 2, at about 17–19 ka and MIS 4, at about 35–53 ka), and at least once earlier. The Zyrianka glaciers were smaller than their predecessors. Radiocarbon and luminescence dating of lake sediments confirms the existence of palaeolake highstands in the Darhad Basin before ~35 ka. Here, geological evidence and ^{10}Be cosmic-ray exposure dating of the drift till suggest that at ~17–19 ka, the basin was filled at least briefly by a glacier-dammed lake.

Recently, Xu et al. (2009) published first optically stimulated luminescence (OSL) results from terminal moraines surrounding the Kanas Lake in the Chinese Altai (Fig. 69.1: No. 12). Their four samples from the three sets of moraines date to 28.0, 34.4, 38.1 and 49.9 ka and one sample from outwash sediment to 6.8 ka. The Kanas Lake occupies a glacially scoured basin formed before 28.0 ka, and the glacial moraines blocked the meltwater after the glacial retreat. This caused the present-day Kanas Lake to have formed at least before 6.8 ka BP. These authors conclude that the glacier advance was more extensive in the MIS 3 (Karginsky Interstadial) than during MIS 2 in Chinese Altai Mountains.

The timing of maximum glacial advances in the Russian Altai, the Khangai and the Darhad Basin appears to be approximately synchronous, but different from those in Siberia (Lake Baikal) and the Chinese Altai, supporting the view that palaeoclimate in Central Asia differed in different regions.

69.5. SUMMARY

The extent of glaciers during the lLGM for the Altai and neighbouring mountain systems is more or less identified. Pleistocene moraines resulting from valley glaciers, plateau glaciers, cirque glaciers and ice streams can be found. The most extensive Late Pleistocene glaciation occurred in the western part of the Russian Altai. However, the extent of Late Pleistocene ice is still a matter of debate in the Russian Altai. Whilst an ice sheet formed in the western Russian Altai, the Pleistocene glaciers have been restricted to several isolated mountain systems in the eastern part of the Russian and Mongolian Altai. To show the variations in the ice extent, the modern and Late Pleistocene ELAs and the glaciated area are presented in two maps from the Russian and Mongolian part of the Altai. The ELAs are relatively low in the more humid outermost ranges especially in the Russian part of the Altai and rise towards the central part of Mongolia. The limited extent of present and Pleistocene glaciers in the eastern part of the Russian Altai and in the Mongolian Altai is the result of reduced precipitation gradient from west to east. This results in a rise of present and Pleistocene ELAs towards the east. However, this was more pronounced during the Pleistocene than today. There is an essential lack of absolute dating of glacial sediments in this region. Nevertheless, on the basis of the present knowledge, most Late Pleistocene glacier advances in Mongolia and in the Russian Altai took place in the MIS 2 and 4. OSL dates from the Chinese Altai suggest a maximum ice advance during the MIS 3. However, the knowledge concerning the exact timing and the fluctuations of Late Quaternary glaciation within the glacial cycles is still very small in this vast area. The application of modern geomorphological and sedimentological techniques, and the development of new dating techniques, especially CRN surface exposure dating, has opened up the possibility of regional and temporal correlations across this huge area.

REFERENCES

Academy of Sciences of Mongolia and Academy of Sciences of USSR, 1990. National Atlas of the Peoples Republic of Mongolia. In: Ulaan Bataar, Moscow, 144pp.

Arkhipov, S.A., Isayeva, L.L., Bespaly, V.G., Glushkova, O., 1986. Glaciation of Siberia and North-East USSR. Quatern. Sci. Rev. 5, 463–474.

Baker, V.R., Benito, G., Rudoi, A.N., 1993. Palaeohydology of late Pleistocene superflooding, Altai Mountains, Siberia. Science 259, 348–350.

Barthel, H., 1983. Die regionale und jahreszeitliche Differenzierung des Klimas in der Mongolischen Volksrepublik. Stud. Geogr. 34, 3–91.

Baryshnikov, G.J., 1992. Cenozoic development of relief of mountainous regions and adjacent areas. 181pp (in Russian).

Benn, D.I., Lehmkuhl, F., 2000. Mass balance and equilibrium-line altitudes of glaciers in high mountain environments. Quatern. Int. 65 (66), 15–29.

Böhner, J., Lehmkuhl, F., 2005. Climate and environmental change modelling in Central and High Asia. Boreas 34, 220–231.

Budvylovski, V.V., 1993. Palaeogeography of the last Glaciation and Holocene in the Altai. Tomsk, 251pp. (in Russian).

Bussemer, S., 2000. Jungquartäre Vergletscherung im Bergaltai und in angrenzenden Gebirgen—analyse des Forschungsstandes. Mitt. Geogr. G. München Band 85, 45–64.

Carling, P.A., Kirkbride, A.D., Parnachov, S., Borodavko, P.S., Berger, G.W., 2002. Late Quaternary catastrophic flooding in the Altai Mountains of south-central Siberia: a synoptic overview and an introduction to flood deposit sedimentology. Spec. Publ. Int. Assoc. Sedimentol. 32, 17–35.

Carling, P.A., Martini, P., Herget, J., Parnachov, S., Borodavko, P., Roberts, B., 2009. Megaflood sedimentary valley fill—Altai Mountains, Siberia. In: Burr, D., Carling, P., Baker, V. (Eds.), Megaflooding on Earth and Mars. Cambridge University Press, Cambridge, pp. 243–264.

Devjatkin, E.V., 1981. Cenozoic of Inner Asia. Nauka, Moscow, 196pp. (in Russian).

Florensov, N.A., Korzhnev, S.S., 1982. Geomorphology of Mongolian People Republic. Joined Soviet-Mongolian scientific research geological expeditions. Transactions, 28. Moscow (in Russian).

Frenzel, B., 1959. Die Vegetations- und Landschaftszonen Nord-Eurasiens während der letzten Eiszeit und während der postglazialen Wärmezeit. I. Teil: Allgemeine Grundlagen. Abh. math.-nat. Kl. d. Akad. Wiss. u. Lit. 13, 937–1099.

Gillespie, A.R., Burke, R.M., Komatsu, G., Bayasgalan, A., 2008. Late Pleistocene glaciers in Darhad Basin, northern Mongolia. Quatern. Res. 69, 169–187.

Grosswald, M.G., 1980. Late Weichselian Ice Sheet of Northern Eurasia. Quatern. Res. 13, 1–32.

Grunert, J., Lehmkuhl, F., Walther, M., 2000. Palaeoclimatic evolution of the Uvs Nuur Basin and adjacent areas (Western Mongolia). Quatern. Int. 65 (66), 171–192.

Klimek, K., 1980. Major physico-geographical features of the southern slope of the Khangai Mountains. Geogr. Stud. 136, 9–13. (Polish Academy of Sciences, Institute of Geography and Spatial Organisation).

Klinge, M., 2001. Glazialgeomorphologische Untersuchungen im Mongolischen Altai als Beitrag zur jungquartären Landschafts- und Klimageschichte der Westmongolei. Aachener Geogr. Arb. 35, Aachen, 125pp.

Klinge, M., Böhner, J., Lehmkuhl, F., 2003. Climate patterns, snow- and timberline in the Altai Mountains, Central Asia. Erdkunde 57, 296–308.

Komitet Geodesii I Kartografii CCCP, 1991. Atlas Altaiskowo kraja. Moskwa, 36pp. (in Russian).

Kotljakov, V.M., Kravzova, V.I., Dreyer, N.N., 1997. World Atlas of snow and ice resources. Moscow, 392pp.

Lehmkuhl, F., 1998. Quaternary Glaciations in Central and Western Mongolia. In: Owen, L.A. (Ed.), Mountain glaciations, Quaternary Proceedings, Wiley, Chichester, vol. 6, pp. 153–167.

Lehmkuhl, F., Lang, A., 2001. Geomorphological investigations and luminescence dating in the southern part of the Khangay and the Valley of the Gobi Lakes (Mongolia). Journal of Quaternary Science (JQS?) 16, 69–87.

Lehmkuhl, F., 1999. Rezente und jungpleistozäne Formungs- und Prozeßregionen im Turgen-Charchiraa, Mongolischer Altai. Die Erde 130, 151–172.

Lehmkuhl, F., 2008. The kind and distribution of mid-latitude periglacial features and alpine permafrost in Eurasia. In: Kane, D.L., Hinkel, K.M. (Eds.), Proceedings of the Ninth International Conference on Permafrost, University of Alaska Fairbanks, June 29–July 3, 1031–1036.

Lehmkuhl, F., Lang, A., 2001. Geomorphological investigations and luminescence dating in the southern part of the Khangay and the Valley of the Gobi Lakes (Mongolia). J. Quatern. Sci. 16, 69–87.

Lehmkuhl, F., Owen, L.A., 2005. Late Quaternary glaciation of Tibet and the bordering mountains: synthesis and new research. Boreas 34, 87–100.

Lehmkuhl, F., Rost, K.T., 1993. Zur pleistozänen Vergletscherung Ostchinas und Nordosttibets. Petermanns Geogr. Mitt. 137, 67–78.

Lehmkuhl, F., Klinge, M., Stauch, G., 2004. The extent of Late Pleistocene Glaciations in the Altai and Khangai Mountains. In: Ehlers, J., Gibbard, P.L. (Eds.), Quaternary Glaciations—Extent and Chronology, Part III: South America, Asia, Africa, Australia, Antarctica. Elsevier, Amsterdam, pp. 243–254.

Lehmkuhl, F., Frechen, M., Zander, A., 2007. Luminescence chronology of fluvial and aeolian deposits in the Russian Altai (Southern Siberia). Quatern. Geochronol. 2, 195–201.

Lydolph, P.E., 1977. Climates of the Soviet Union. World Survey of Climatology, vol. 7, 443pp.

Prokopenko, A.A., Kuzmin, M.I., Li, H.C., Woo, K.-S., Catto, N.R., 2009. Lake Hovsgol basin as a new study site for long continental paleoclimate records in continental interior Asia: general context and current status. Quatern. Int. 205, 1–11.

Reuther, A.U., Herget, J., Ivy-Ochs, S., Borodavko, P., Kubik, P.W., Heine, K., 2006. Constraining the timing of the most recent cataclysmic flood event from ice-dammed lakes in the Russian Altai Mountains, Siberia, using cosmogenic in situ ^{10}Be. Geology 34 (11), 913–916.

Richter, H., 1961. Probleme der Eiszeitlichen Vergletscherung des Changai. Verhandl. d. deutsch. Geogr. Köln 33, 400–407.

Rudoi, A.N., 2002. Glacier-dammed lakes and geological work of glacial superfloods in the Late Pleistocene, Southern Siberia, Altai Mountains. Quatern. Int. 87, 119–140.

Shi, Y., 1992. Glaciers and glacial geomorphology in China. Zeitsch. Geomorphol. N.F. Suppl. Bd. 86, 51–63.

Walther, M., 1998. Paläoklimatische Untersuchungen zur jungpleistozänen Landschaftsentwicklung im Changai-Bergland und in der nördlichen Gobi (Mongolei). Petermanns Geogr. Mitt. 142, 205–215.

Xu, X., Yang, J., Dong, G., Wang, L., Miller, L., 2009. OSL dating of glacier extent during the Last Glacial and the Kanas Lake basin formation in Kanas River valley, Altai Mountains, China. Geomorphology 112, 306–317.

Chapter 70

Quaternary Glaciations: Extent and Chronology in China

Zhou Shangzhe[1,*], Li Jijun[2], Zhao Jingdong[3], Wang Jie[2] and Zheng Jingxiong[1]

[1] *School of Geographical Science, South China Normal University, Guangzhou, China*
[2] *Department of Geography, Lanzhou University, Lanzhou, China*
[3] *State Key Laboratory of Cryospheric Sciences, Cold and Arid Regions Environmental and Engineering Research Institute, Chinese Academy of Sciences, Lanzhou, China*
*Correspondence and requests for materials should be addressed to Zhou Shangzhe. E-mail: zhsz@lzu.edu.cn

70.1. DISTRIBUTION OF QUATERNARY GLACIAL REMNANTS ON THE QINGHAI-TIBET PLATEAU

Quaternary glaciations of China have been investigated since the 1950s. Glacial erosional and depositional features are found widespread on the Qinghai-Xizang (Tibet) Plateau, in the Tianshan Mountains and the Altai Mountains in western China. They are also found in the Taibai Mountain, Changbai Mountain and on the island of Taiwan in eastern China.

The Qinghai-Tibetan Plateau (Fig. 70.1) covers an area of 2,500,000 km^2 (25–40°N, 73–103°E) and lies at an average altitude of 4500 m a.s.l. It consists of planation surfaces, basins and mountain ranges trending generally from west to east. Most notable among the mountain ranges are the Karakoram Mountains, the Kunlun Mountains, the Qilian Mountains, the Tanggula Mountains, the Nyainqentanglha Mountains, the Gangdise Mountains, the Hengduan Mountains and the Himalayas, which constitute the principal parts of highland Asia. Of the 96 mountain peaks higher than 7000 m in the world, 95 are gathered in and around the Qinghai-Tibetan Plateau. The Qinghai-Tibetan Plateau hence became known as the 'roof of the world', and it is the source of many of the greatest rivers of Asia, including the Yangtze, Huang He, Brahmaputra, Mekong, Salween, Indus and the Ganges. Consequently, the plateau is also referred to as the Asian water tower. As a result of the influence of the south Asian monsoon, the south-eastern part of the Qinghai-Tibetan Plateau has much more precipitation than the north-western part. The catalogue of glaciers in China comprises 46,377 mountain glaciers, with a total area of 59,425.18 km^2 in western China (Shi Yafeng, 2008), of which 36,790 with 49,873.44 km^2 are distributed on Qinghai-Tibetan Plateau. Others are found in the Tianshan Mountains and Altai Mountains. The Qinghai-Tibetan Plateau and the bordering mountains contain most of the mountain glaciers in the middle and lower latitudes of the world.

In the early twentieth century, the present and Quaternary glaciers in this region were studied by western scientists (Hedin, 1909, 1922; Huntington, 1906; De Fillipi, 1915; Younghuband, 1926; Trinkler, 1930; Word, 1934; Heim, 1936; Wissmann, 1937). Later, Quaternary glaciations in the Tianshan Mountains and Qilian Mountains were investigated by some Chinese scientists (Huang, 1944; Huang and Cheng, 1945; Liu Zengqian, 1946; Weng and Lee, 1946). A systematic and continuous study on these issues was carried out in conjunction with investigations on existing glaciers since 1958 by Chinese geographers (Xia Kairu, 1960; Liu Zechun et al., 1962; Li Jijun, 1963; Zheng Benxing and Shi Yafeng, 1976, Shi Yafeng et al., 1982; Derbyshire et al., 1984). The first monograph on recent and Quaternary glaciations in Tibet was published in 1986 (Li Jijun et al., 1986), and the first Quaternary glacial distribution map of the Qinghai-Tibetan Plateau (Fig. 70.2) was published in 1991 (Li Bingyuan et al., 1991). In the past decade, a significant achievement has been made with the application of various chronological techniques to date the glacial sediments (Zhou Shangzhe et al., 2001, 2002a,b, 2007; Owen et al., 2002, 2005, 2006, 2009; Lehmkuhl and Owen, 2005; Seong et al., 2007, 2009). In 2006, a monograph on glaciations and environmental variations in China was published (Shi Yafeng et al., 1982).

The extent and chronology of Quaternary glaciations has been a central theme of glacial geological investigation for the past 50 years. Although the Qinghai-Tibetan Plateau is the highest in the world, Quaternary glaciations were largely restricted to the higher mountain ranges rising above

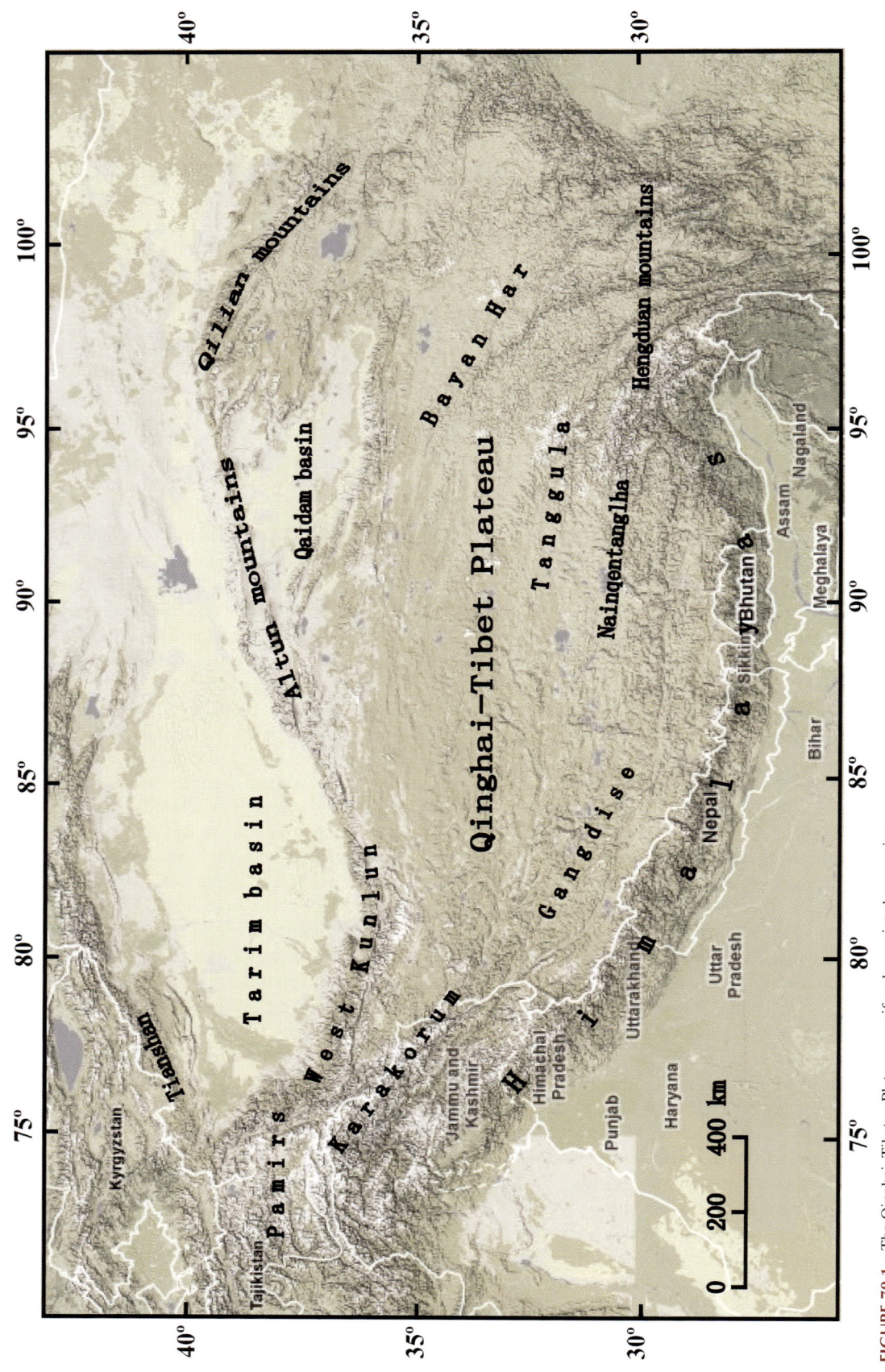

FIGURE 70.1 The Qinghai–Tibetan Plateau massif and associated mountains.

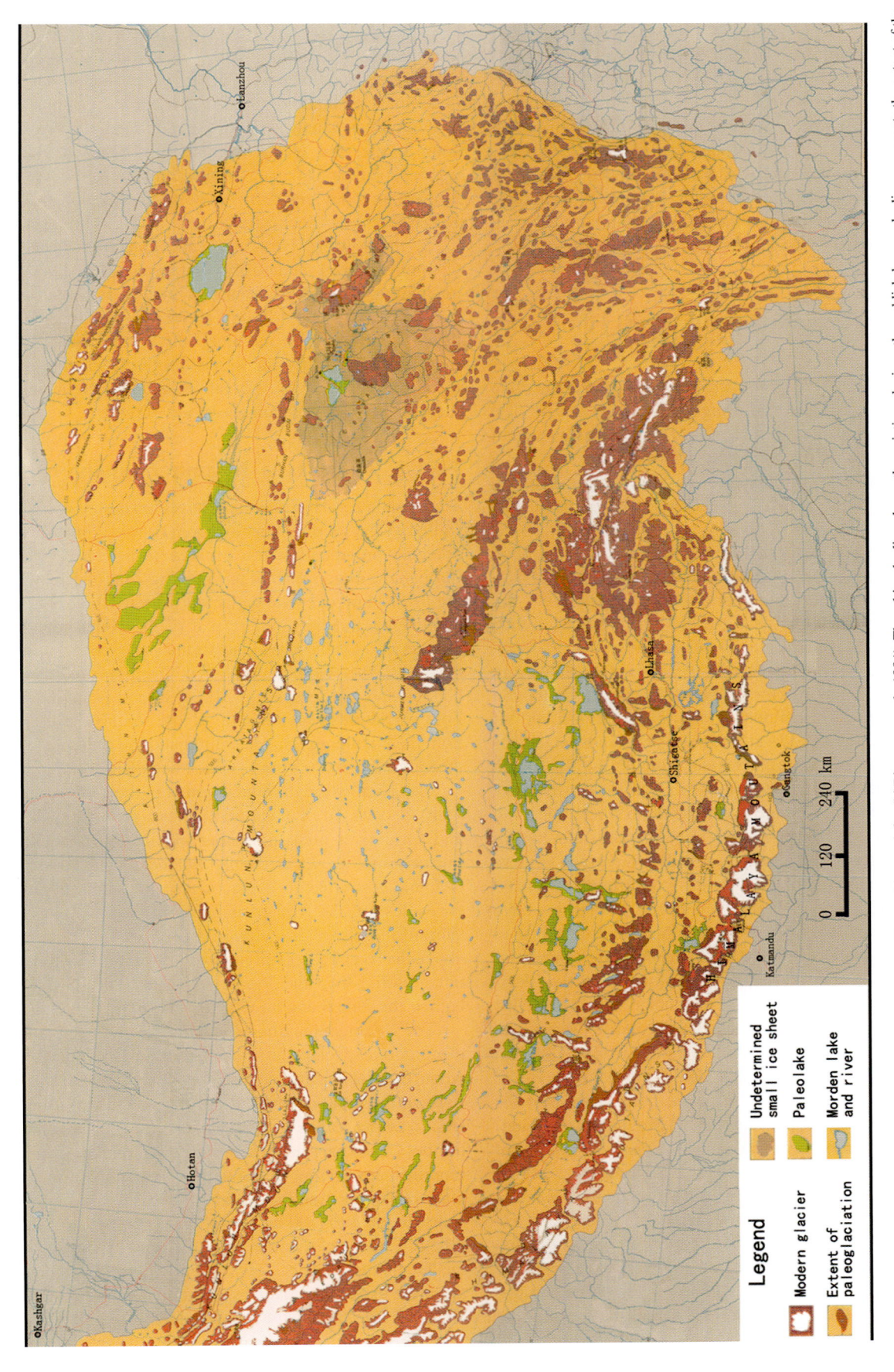

FIGURE 70.2 Quaternary glacial distribution map of Qinghai–Tibetan Plateau (after Li Bingyuan et al., 1991). The white shading shows the existing glaciers, the reddish-brown shading represents the extent of the most extensive Quaternary glaciations and the brown-shaded area shows the location of a small former ice sheet (the dimensions of which are as yet undetermined). The blue and green are lakes and grasslands, respectively. (For interpretation of the references to colour in this figure legend, the reader is referred to the Web version of this chapter.)

the plateau. The limiting factor preventing the development of a coalescing ice sheet covering the entire plateau was probably precipitation. On the basis of the geographical position and climatic conditions, five major areas of Quaternary glaciation can be distinguished on the plateau: the Karakoram, Kunlun and western Himalayas region; the Himalayas and the Nyainqentanglha region; the Hengduan and the Bayan Har region; the Qilian Mountains Region; and the Qiangtang Plateau region.

70.1.1. Karakoram–Kunlun–West Himalayas Region

The Karakoram, Kunlun and western Himalayas constitute the western margin of the Qinghai-Tibetan Plateau. The climate of these mountain ranges is dominated by winter precipitation delivered by westerly air currents.

The Karakoram stretches NW–SE 500 km between China and Kashmir. As the highest mountains in the plateau with a mean elevation of 6500 m a.s.l., the Karakoram has the most extensive modern glaciers, of which there are 3563 with a total area of 6262 km^2 distributed in the Chinese part. This is despite the climate here being comparatively dry (600 mm precipitations near the equilibrium-line altitude (ELA)). The Yinsugaiti glacier (42 km long, 379.97 km^2), located north-west of the K2 peak in the Shaksgam River Basin, is one of the biggest glaciers in China (Su Zhen, 2006). Its terminus is found at an altitude of 4000 m a.s.l., and its ELA at 5420 m a.s.l. Chinese scientists have investigated Quaternary glaciations in the Shaksgam Valley and the Chang Chenmo River Valley, the tributary of the Shayok River, on the north-western slope of the main range, and in the Hunza River Valley on the south-eastern slope of the Karakoram main range. The moraines in these valleys indicate that the glaciers have repeatedly advanced during the Quaternary. Four Pleistocene glaciations have been recognised in this region, which include the Shanoz Glaciation, an intermediate Glaciation, the penultimate glaciation [marine isotope stage (MIS) 6] and the Last Glaciation (Weichselian), as well as Neoglaciation and Little Ice Age glaciation during the Holocene. Of these, the penultimate glaciation was the most extensive (Su Zhen, 2006).

The Kunlun Mountains stretches from 75°E to 105°E along 33–37°N. It partially forms the north margin of the Qinghai–Tibetan Plateau. The Kunlun Mountains is the longest mountain range in central Asia. It consists of a series of mountain peaks with 7924 modern glaciers covering an area of 12,538.63 km^2. They are mainly distributed in its western section where most of the peaks reach to above 6500 m a.s.l., although the climate there is very dry. Based on the distribution of glacial sediments, three Quaternary glaciations have been distinguished in the west Kunlun Mountains, namely the Kunlun Glaciation, the penultimate glaciation and the Last Glaciation. The west Kunlun slopes, with a steep gradient northwards to the Tarim Basin. Thus, the Quaternary glacial sediments on the northern slope are present mainly in tributary valleys of Yurungkax River. On the south slope, the glacial sediments cover the piedmont region, and the most extensive Kunlun Glaciation was much larger than today, reaching the northern shore of the Gozha Lake (Fig. 70.3; Zheng Benxing et al., 1991).

In the western Himalayas, the glaciation near Polan County has been studied in great detail. Four Quaternary glaciations have been identified in this area. The oldest moraine occurs about 200 m above the river bed. Further upstream from this moraine, three additional sets of moraines are readily identified, representing the penultimate glaciation, Last Glaciation and the Neoglaciation, respectively (Zhou Shangzhe et al., 2004).

70.1.2. Himalayas–Nyainqentanglha region

The Himalayas–Nyainqentanglha region is climatically controlled by the south Asian monsoon. Humid air masses bring abundant precipitation into this area from the Bay of Bengal through the large valley of the Yarlung Zangbo River. There are 19,118 km^2 of existing glaciers in total in both the Nyainqentanglha and the Chinese part of the Himalayas. In the Himalayas, Quaternary glaciations have been studied in detail in the Qomolangma–Xixiabangma regions (Shi Yafeng and Liu Tungsheng, 1964; Zheng Benxing, 1988). Four glaciations were recognised: the Xixiabangma Glaciation, the Nieniexiongla Glaciation and the two substages of the Qomolangma (Everest) Glaciations (Fig. 70.4).

The Nieniexiongla Glaciation was the most extensive, with glaciers from tributary valleys joining each other and forming a great piedmont glacier during that time. Further, another set of deeply weathered glacial sediment is found restricted to some ridges and summits between the valleys of the Xixiabangma north slope. These 'ridge-summit moraines' are 6200 m a.s.l. high and some 600 m above the river beds, implying that this moraine was formed much earlier than the Nieniexiongla moraine. It has been named the Xixiabangma Glaciation.

The Nyainqentanglha Mountain was the local centre of glaciation, especially in its east section (south-east Tibet) where the glaciers are fed by extensive moisture transported through the great channel of the Brahmaputra Valley. The Boduizangbo Valley is a crucial area for the understanding of the Quaternary glaciations of this region. Numerous moraines are present in the valley. The most prominent is a set of lateral moraine which extends discontinuously at a height of 800–500 m above the river bed to Guxiang village (Fig. 70.5), some 100 km from the headwater of the Guanxing Glacier. This lateral moraine implies that the most extensive advance of the glaciers covered a distance of 100 km. It is also the oldest moraine preserved

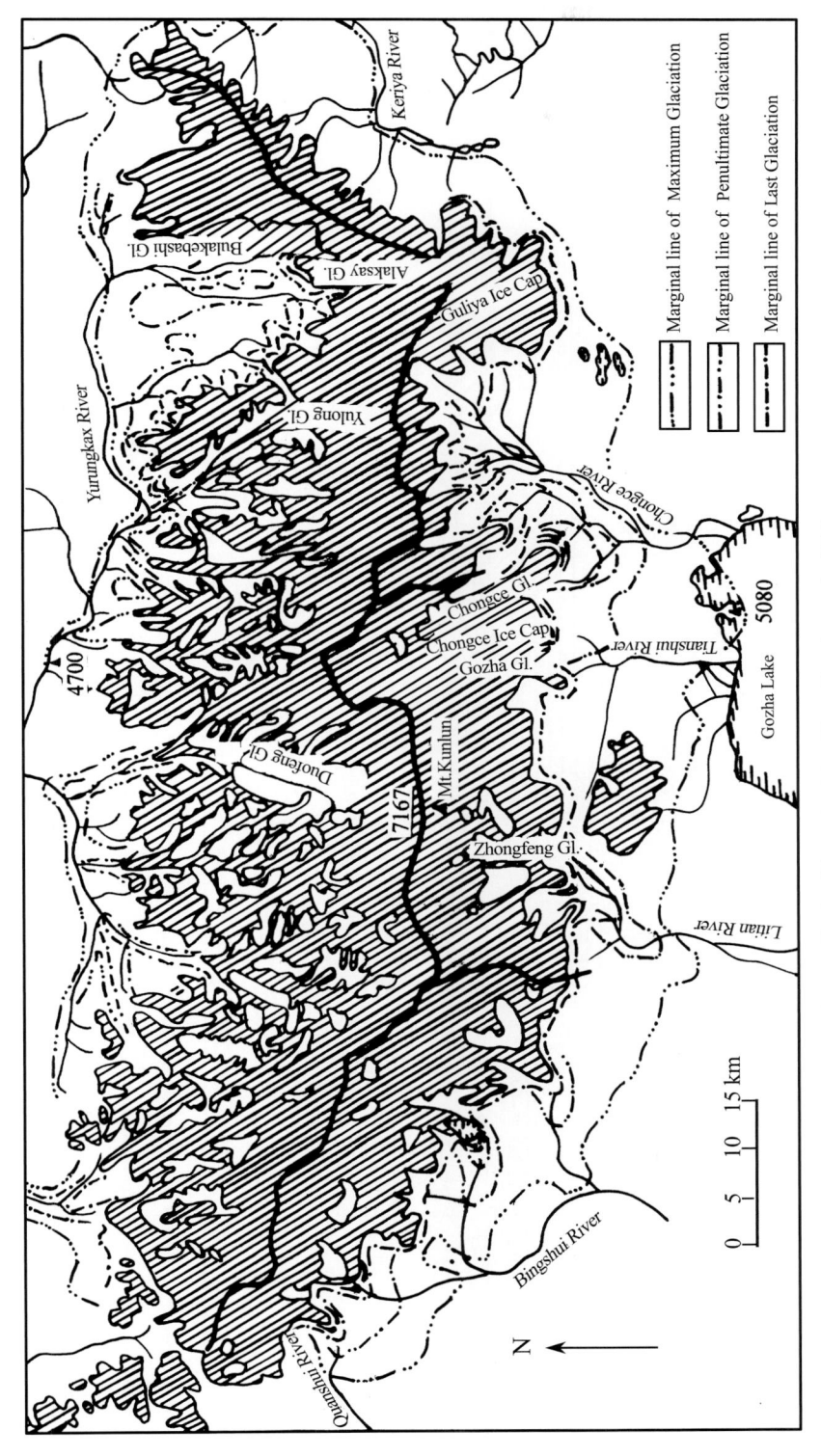

FIGURE 70.3 Distribution of glacial deposits and landforms in west Kunlun Mountains (after Zheng Benxing et al., 1991).

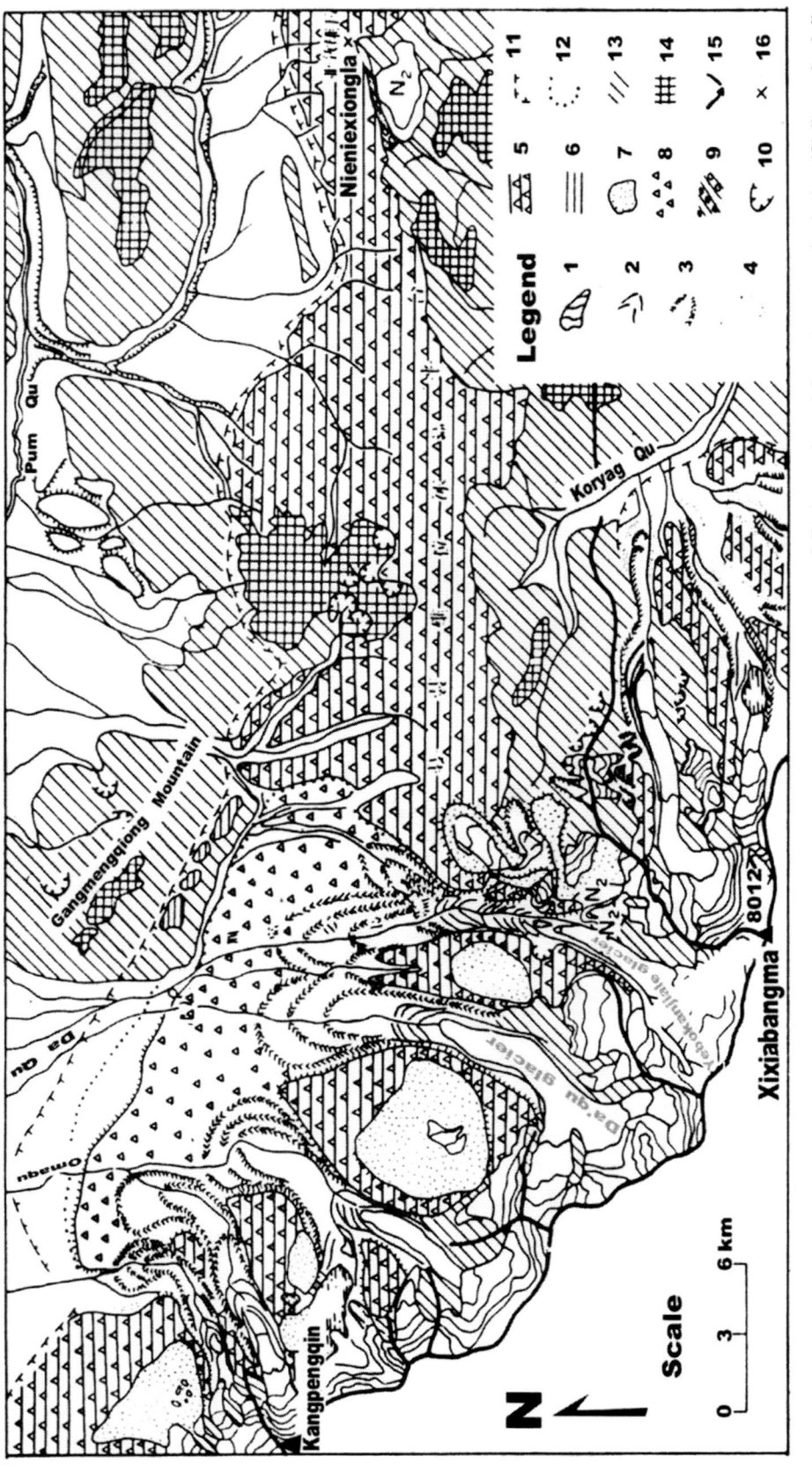

FIGURE 70.4 Geomorphological map of the north slope of Mount Xixiabangma (after Li Jijun et al., 1986).

1. Modern glacier
2. End moraine of Neoglaciation
3. End moraine of Qomolongma glaciation (2) (last glacial)
4. End and lateral moraine of Qomolangma glaciation (1) (Penultimate)
5. Moraine platform of Nieniexiongla glaciation (last but two)
6. Glacial erosion platform of Nieniexiongla glaciation
7. Highest moraine platform of Xixabangma glaciation (last but third)
8. Hummocky moraine plain of Qomolangma glaciation (1)
9. Gongba conglomerate of Lower Pleistocene
10. Cirque
11. Lower limit of glacier of Nieniexiongla glaciation
12. Lower limit of glacier of Qomolangma glaciation
13. Mountains
14. Planation surface on the mountain top
15. Ridge and peak
16. Pass

FIGURE 70.5 Lateral moraines of the Guxiang and Baiyu Glaciations (after Zhou Shangzhe et al., 2010).

in the basin. During that time, the glaciers joined each other in the main valley to form a network of valley glaciers (Fig. 70.6). This glacial period was termed the Guxiang Glaciation by Li Jijun et al. (1986).

Another smaller lateral moraine extends at a height of 300–100 m along the right side of the Boduizangbo River. It ends at the Baiyu Gou outlet, just touching the large end moraine from the Baiyu Gou. The latter is 260–300 m high and a completely preserved end moraine. Not far from this site, another large end moraine arc can be found at the mouth of the Zhuxi Gou, which resembles that at the mouth of the Baiyu Gou in scale (Fig. 70.7). Both moraines, and the smaller lateral moraine from the main valley, undoubtedly represent another glacial advance. This glacial advance was termed the 'Baiyu Glaciation' by Li Jijun et al. (1986). During this glaciation, the large branch glaciers also joined in the main valley to form a composite glacier, 80 km in length reaching from the Guanxing Glacier to the Baiyu Gou (Figs. 70.5 and 70.8). As well as these moraines from the last cold stage, Holocene Neoglacial and Little Ice Age moraines are also present in front of the modern glaciers.

The N–S-orientated Hengduan Mountains and the NW–SE-orientated Bayan Har Mountains comprise the eastern part of the Qinghai-Tibetan Plateau. Both the south-west and the south-east monsoons influence the climate of this area. Despite the scarcity of modern glaciers (1600 km^2 in total) in this region, Quaternary glaciation was rather extensive. Evidence indicates that ice caps and even small ice sheets developed. For example, between the Jinsha River (upper reaches of the Yangtze River) and Yalongjiang Rivers, a relatively large and well-preserved palaeoplanation surface with a mean height of 4700 m a.s.l. is present on which glacial sediments and glacial landforms (e.g. *roches moutonnées* and rock drumlins) are ubiquitous. Near Daocheng County (Sichuan Province), the ancient Daocheng ice cap covered an area of over 3000 km^2. A contemporaneous Xinlong ice cap, covering an area of about 2000 km^2, formed nearby. The Queer Mountain, through which the Sichuan–Tibetan highway passes, is a region of typical Alpine-type glaciations. This area is characterised by all kinds of glacial geomorphological features, such as troughs, cirques, glacial lakes, hanging valleys, end moraines, lateral moraines and glaciofluvial terraces. Field investigations also demonstrated the existence of a small Quaternary ice sheet (covering an area of ca. 50,000 km^2) in the source area or the Yellow River, extending down from the Bayan Har Mountains (5268 m a.s.l.) to the piedmont region of these mountain ranges (Zhou Shangzhe and Li Jijun, 1998). Streamlined landforms such as *roches moutonnées* and whale-back drumlins are quite typically present. Moreover, till deposits and erratics (including striated rocks) are found to be distributed on both flanks of the mountain range, up to 60 km away from their source area at the Bayan Har summit. The till deposits and granite erratics are also found near the town of Huashixia, some 150 km north of the Bayan Har summit, as well as to the west of Gyarling Lake. This ice sheet formerly covered the source area of the Yellow River. Based on the geological evidence, four glacial stages can be identified in this area: the Huanghe Glaciation (MIS 12), the Yematan Glaciation (MIS 6), the Galala Glaciation (MIS 4) and the Bayan Har Glaciation (MIS 2).

The former two were the most extensive that have been identified so far on the Qinghai-Tibetan Plateau. Although there are different ideas about its limits and scale (Zhou Shangzhe and Li Jijun, 1998; Stroeven et al., 2009), all researchers agree that an extensive glaciation had occurred in this region.

70.1.3. Qilian Mountain Region

The Qilian Mountains stretch about 800 km in a north-west–south-eastwards direction along the north-eastern margin of the Qinghai-Tibetan Plateau. In winter, the climate of this region is controlled by the Mongolian High Pressure cell. Present-day precipitation is 700–800 mm in the east and 300 mm in the west near the ELA (4500 m a.s.l.). It is highly seasonal and mainly falls in the summer, caused by invasions of the south-westerly and south-easterly monsoon. Modern glaciers cover a total area of 1972 km^2. Three Pleistocene glaciations are distinguished in the Qilian Mountains. Glacial geomorphological features, such as trough valleys and cirques, are widely distributed, and moraines, till hills and outwash terraces can be found in many valleys. End moraines and lateral moraines, representing different glacial episodes, are found at varying altitudes. A complete sequence of Quaternary glaciations has been established based on the evidence found in the Bailang River Valley (38°52′–39°10′N, 99°15′–99°28′E) on the northern slope of the Qilian Mountains. Four sets of Pleistocene moraines have been recognised in this valley (Zhou Shangzhe et al., 2002a). The oldest moraine is located at the top of a piedmont hill, Zhonglianggan, which is 2996 m a.s.l. and 500 m above the Bailang River level. The morainic sediment was derived from glacier Nos. 14 and 16 of Zoulangnanshan Mountains (the peak altitude here is 5121 m a.s.l., and the modern ELA is 4500 m a.s.l.). The till of the moraine was dated to 463 ka (MIS 12) by electron spin resonance (ESR). Another large moraine, the Changgousi moraine, and its contemporary outwash terrace (130 m above the river level) were dated by ESR to 135 and 130 ka, respectively. The base of the loess, which lies 18 m thick on the outwash terrace, was dated using thermoluminescence (TL) to 142 ka. The TL date of the loess should not be older than the ESR dates of the underlying outwash and moraine, but there is potential error of the measurement (about 10%). Those dates imply that glaciation occurred during both MIS 12 and 6

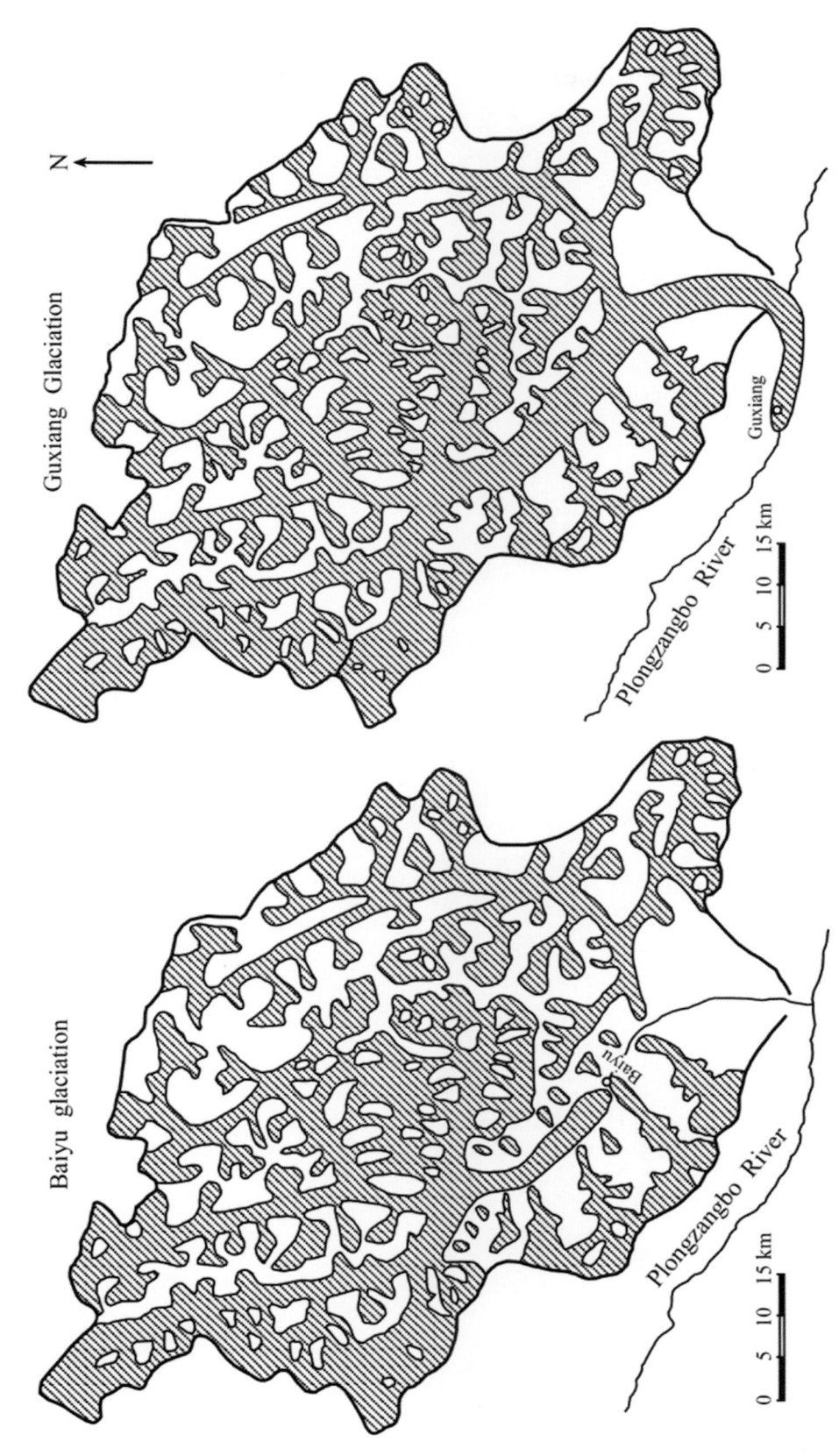

FIGURE 70.6 The dissected glaciers of the Guxiang and Baiyu Glaciations (after Li Jijun et al., 1986).

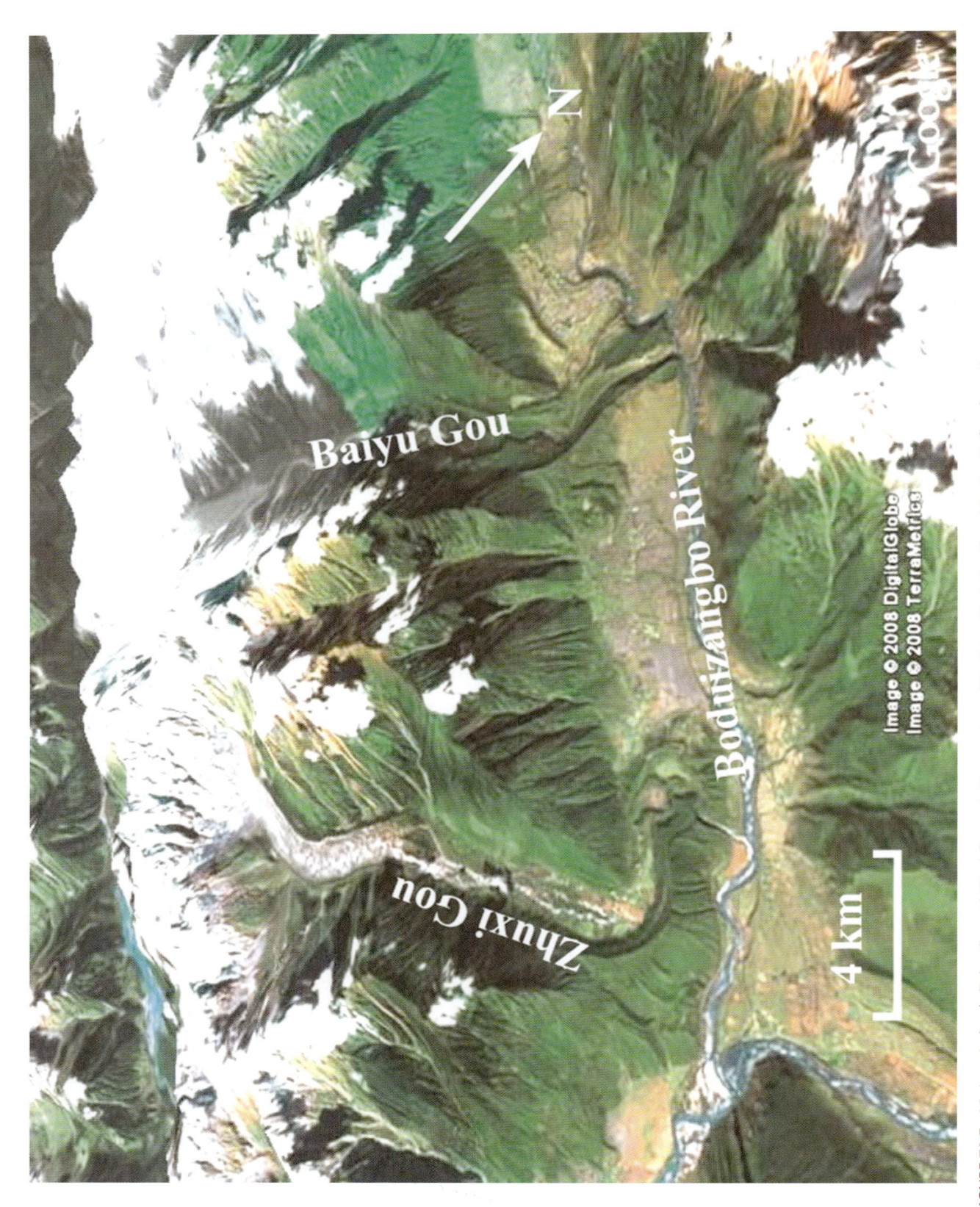

FIGURE 70.7 Modern glaciers in Baiyu Gou and Zhuxi Gou and older moraines of the Baiyu Glaciation (from http://earth.Google.com).

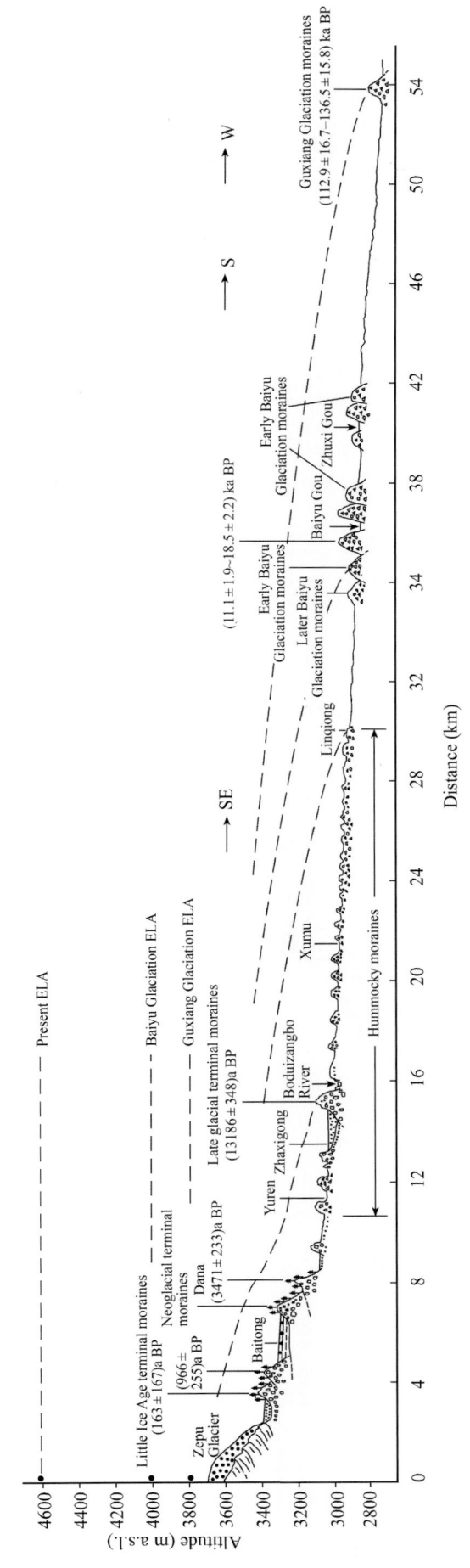

FIGURE 70.8 Sketch map showing the advances of the glaciers in Boduizangbo Valley in the late Quaternary (after Zhou Shangzhe et al., 2010).

in the Qilian Mountains (Zhou Shangzhe et al., 2002a). The other two sets of moraines are distributed in a fan-like form on the steep slope beyond the valley. They are thought to represent the glaciations during MIS 4 and 2. The moraines of the Holocene Neoglaciation and the Little Ice Age are also present in the valley (Fig. 70.9).

70.1.4. Hinterland of the Plateau

The hinterland of the Qinghai–Tibetan Plateau is termed the Qiangtang Plateau (30–37°N, 80–91°E and 5000 m in altitude). Here, the modern annual precipitation is only 100–150 mm, and this region is also the coldest part of the Qinghai-Tibetan Plateau. In this region, some relatively gentle mountains rise to 6000–6200 m a.s.l. above the high plateau. They are covered by modern ice caps. Quaternary glaciations once extended beyond the mountains and formed piedmont glaciers. However, the extent of glaciation was quite limited due to the aridity of the region. It seems that the moisture supply from the south-westerly and south-easterly monsoon was equally insufficient then, as it is today. Overall, it appears that the eastern part of the Qinghai-Tibetan Plateau was more favourable for the development of Pleistocene glaciers than the western part. If the modern glacial area is compared with that glaciated during the Pleistocene, the ratio in the eastern part of the plateau is greater than in the western part, indicating that the eastern part has undergone more pronounced environmental changes during the Quaternary than the western part of the Qinghai-Tibetan Plateau.

70.2. OTHER MOUNTAINS

70.2.1. Tianshan Mountains

The Tianshan Mountains between 40°16′–44°35′N and 69°00′–95°00′E is situated in dry central Asia. A stretch of 1500 km of this mountain range lies within China. The highest peak, the Tomur, reaches 7435 m a.s.l., and there are more than 40 peaks above 6000 m a.s.l. Climatically, the Tianshan Mountains is influenced by the prevailing westerly winds. Modern glaciers cover an area of 9225 km^2 of these mountains. The modern ELAs vary between 3500 and 4500 m a.s.l. from west to east as a consequence of the eastwards decrease in precipitation. Evidence for Quaternary glaciations has been studied at the southern slope of the Tomur, at the source of the Urumqi River and on the Bogeda Mountain (Huang, 1944; Cui

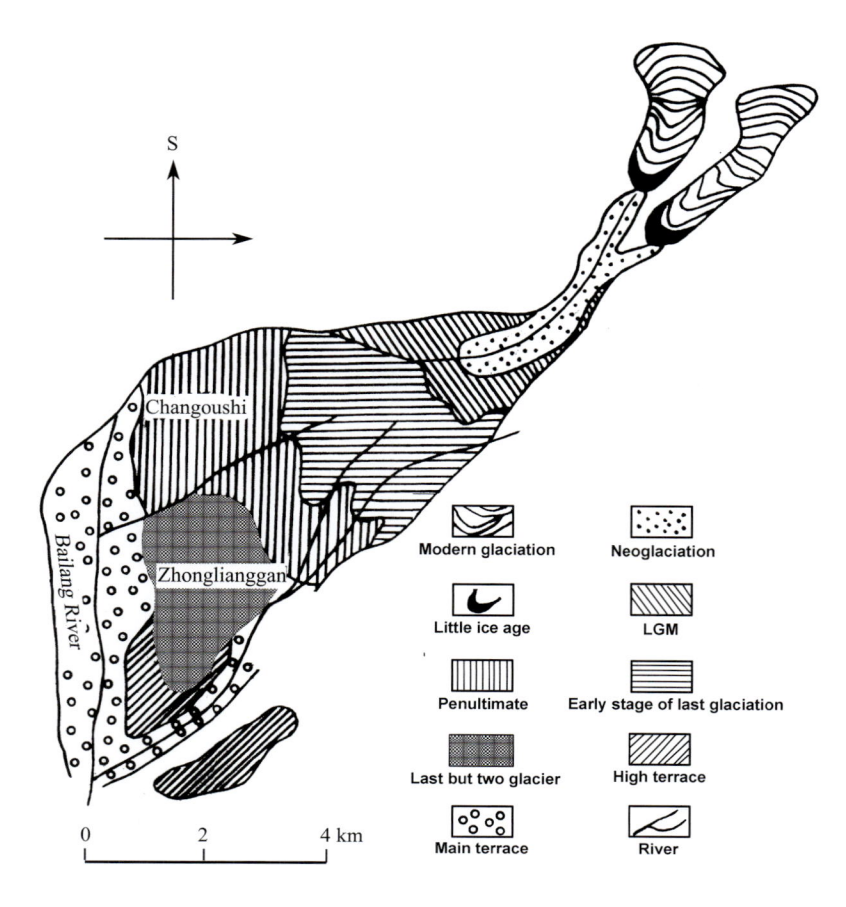

FIGURE 70.9 Glacial geomorphological map of the Bailang River area (Qilian Mountains) (after Zhou Shangzhe et al., 2002a).

Zhijiu, 1981; Wang jingtai, 1981; Zhou Shangzhe et al., 2002b; Zhao Jingdong et al., 2006, 2009, 2010a,b). The source area of the Urumqi River is just where the Tianshan Mountains Glaciology Station of the Chinese Academy of Sciences is located, so the problem of Quaternary glaciation has been studied here in most detail. Near Wangfeng Daoban, 12 km from the Glacier No. 1 terminus, there are two moraines that were probably formed during the Last Glacial period: the upper and lower Wangfeng moraines. Additionally, a remnant of a U-shaped valley at an altitude of 3300–3400 m a.s.l. is present, some 200–300 m above the river level. Glacial deposits are still preserved in this high valley and are attributed to the High Wangfeng Glaciation. These are thought to be the oldest glacial deposits in the Urumqi River Valley area. This High upper Wangfeng moraine was dated to 471 and 460 ka by Yi Chaolu and Zhou Shangzhe, respectively, by ESR (Zhou Shangzhe et al., 2001). These dates are very close to that of the Zhonglianggan Glaciation in the Qilian Mountains, implying the earliest glaciation in the middle section of Tianshan Mountains probably occurred during MIS 12. Moraines from Last Glacial have been dated using AMS ^{14}C and ESR to MIS 2 and 4 by Yi Chaolu et al. (1998, 2002). ESR dating of the outwash terraces in this valley shows a glacial advance occurred during MIS 6 (Zhou Shangzhe et al., 2002b, 2004). In front of the current glacier No. 1, Little Ice Age and Neoglacial moraines are preserved. They have been dated by lichenometry (Chen Jiyang, 1989).

70.2.2. Altai Mountains

The Altai Mountains is another large mountain range in central Asia, lying in the boundary area of China, Mongolia, Russia and Kazakhstan. It stretches about 1650 km with a NW–SE direction, and about 500 km is within China, and consists of a series of peaks above 4000 m a.s.l. and the highest mountain reaches 4374 m a.s.l. The Altai Mountains is also climatically influenced by westerly winds. It is the only mountain range in China that drains into the Arctic Ocean. There are 280 km^2 of contemporary glaciers in the Chinese part of the Altai Mountains. The equilibrium line of the glaciers is between 2950 and 3000 m. The largest glacier is the Kanas glacier which covers an area of 30.13 km^2 and a length of 10.8 km. Quaternary glacial deposits are well preserved in the Kanas Valley, where three Pleistocene and two Holocene glaciations have been recognised. The three Pleistocene glaciations were termed the Burqin Glaciation, the Kanas Glaciation and the Last Glaciation.

The Last Glaciation probably includes advances during both MIS 4 and 2, and perhaps even MIS 3 (Cui Zhijiu et al., 1992; Xu Xiangke et al., 2009). The earliest, Burqin Glaciation almost covered the mountains and reached down to 1000 m a.s.l. on the Burqin plain outside the mountains. The Penultimate Glaciation in this region was characterised by dendritic valley glaciers, and the contemporaneous Kanas glacier was 114 km long (Fig. 70.10).

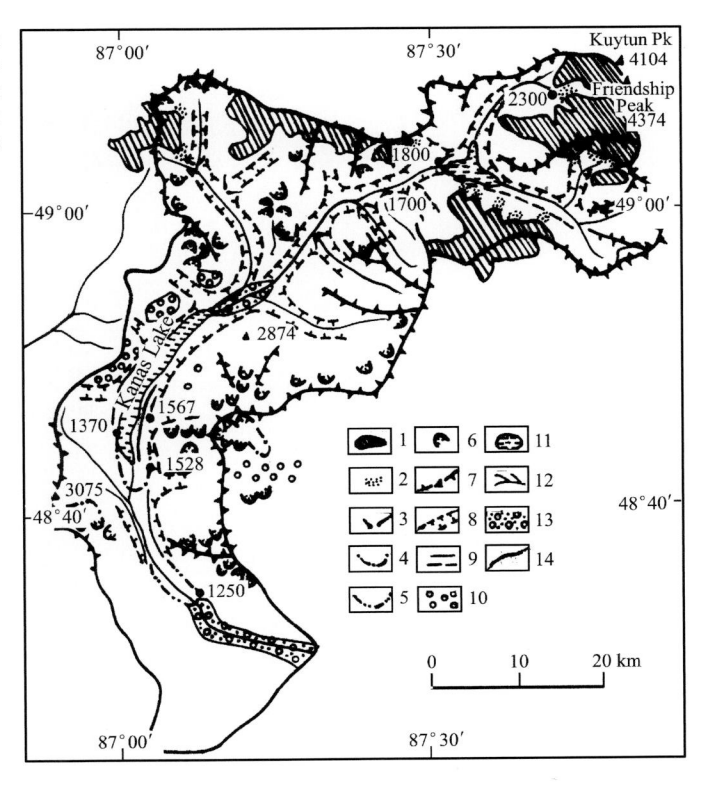

FIGURE 70.10 Glacial geomorphological map of the Kanas River in the Altai area. (Cui Zhiju et al., 1992). 1, modern glacier; 2, Little Ice Age moraine ridge; 3, Neoglaciation moraine; 4, Last Glaciation moraine; 5, penultimate glaciation moraine; 6, cirque; 7, arête and horn; 8, palaeo-trough; 9, lateral moraine; 10, glacial boulders of the Buerjin Glaciation; 11, lake; 12, river; 13, lake sediment; and 14, watershed.

70.2.3. Central Mountains of Taiwan

The Central Mountains of Taiwan at 23–24°N, 121°E, with a maximum altitude of 3970 m a.s.l., are the highest in eastern China. Research work by Japanese scientists in the early twentieth century and, more recently, by Chinese colleagues suggests that two discrete glaciations occurred on these mountains during the Late Pleistocene (Cui Zhijiu et al., 2002). These glaciations can be correlated with the Late and the Early Weichselian (i.e. MIS 2 and 4), respectively.

70.2.4. Changbai Mountain

The Changbai Mountain is the highest mountain (2570 m a.s.l.) in north-eastern China (42°N, 128°°E), on the border between China and Korea. Because of its higher latitude and abundant precipitation, glaciers developed around the volcanic massif during the Last Glacial Maximum (LGM; 20.0 ± 2.1 ka) and during the Late Glacial (11.3 ± 1.2 ka), dated by optically stimulated luminescence (OSL) (Zhang Wei et al., 2008). However, its relatively low elevation has been a limiting factor for glacial development.

70.2.5. Taibai Mountain

The Taibai is the highest mountain (3767 m a.s.l.) in Shanxi province, north-western China (33°7′N, 107°45′E). Glaciers developed here during the LGM because of its high elevation. Although precipitation was limited, the ELA during the LGM is about 3500 m a.s.l.

70.3. THE SEQUENCE OF PLEISTOCENE GLACIATIONS IN CHINA

The marginal mountains of the Qinghai-Tibetan Plateau exhibit well-preserved Quaternary glacial sequences. The oldest glacial sediments are preserved here on high terraces or platforms, indicating a complex glacial geomorphological and sedimentological system. Detailed analysis of this system will help to elucidate both the processes of formation of these glacial sequences and their chronology. To date, three to four Pleistocene glaciations have been recognised (Table 70.1). In recent years, great strides have been made in establishing the chronology of the glacial sequence using radiometric dating techniques.

70.3.1. The Earliest Glaciations in Some Mountain Ranges

Great attention has being paid to the earliest glacial sediments in the Qinghai-Tibetan Plateau because this issue is associated with the tectonic uplift of the plateau. So far, some elementary conclusions can be drawn that the earliest glaciations probably occurred during MIS 18–16 in Kunlun Mountains, Himalayas, Nyainqentanglha Mountains and Hengduan Mountains. In Qilian Mountains and Tianshan Mountains, it probably occurred during MIS 12. This assessment is based on the following evidence (Zhou Shangzhe et al., 2006).

The Kunlun Glaciation was suggested based on the moraines at the Kunlun Pass region (34°30′N, 94°00′E) of the eastern Kunlun Mountains. Here, the moraines are

TABLE 70.1 Local Pleistocene Glacial Sequences in China

Region	Last Glaciation (MIS 4–2)	Penultimate Glaciation (MIS 6)	≥Antepenultimate Glaciation (≥MIS 12)
Everest	Rongbu Temple	Jilong Temple	Nieniexongla
Gangdisi	Gangrenboqi	Selong	Gangdisi
Nianqingtanggula	Hailong	Yebaguo	Liuhuangshan
Karakoram	Qingchengmo 2	Qingchengmo 1	Lanake
Tanggula	Basicuo	Zhajazangbu	Tanggula
West Kunlun	Bingshuigou	Quanshuigou	Kunlun
Bayan Har	Galala	Yematan	Huanghe
Southeast Tibet	Baiyu	Guxiamng	
Hengduan Mountains	Zhuqing	Rongbacha	Daocheng
Qilian Mountain	LG	Changgousi	Zhonglianggan
Tianshan Mountains	Uper Wangfeng	Lower Wangfeng	High Wangfeng
Taiwan	LG		
Changbai Mountain	LG		

situated on the ridges (5000–5100 m a.s.l.) of two sides of the Kunlun Pass (4700 m a.s.l.), below which is lacustrine sediments dating from the Brunhes/Matayama (B/M) magnetic boundary based on palaeomagnetic data. The overlying moraine was dated at 710 ka using ESR (Cui Zhijiu et al., 1998b). Additionally, the basal ice of the Guliya ice core of 309.8 m deep in west Kunlun Mountains was dated to 760 ka (Thompson et al., 1997), suggesting that glaciers began to occur in Kunlun Mountains after B/M boundary and have never since disappeared from the western Kunlun. In the Himalayas, the earliest glaciation was thought to be the Xixiabangma Glaciation. However, the subsequent Nieniexiongla Glaciation was more extensive and seems comparable with the Kunlun Glaciation. Unfortunately, both glaciations are still undated. The Gongba conglomerate, a glaciofluvial sediment thought to be synchronous with the Xixiabangma moraine, has been palaeomagnetically dated to 2.3–1.9 Ma (Qian Fang, 1991), but this is disputable. Recently, the earlier Tingri moraine in Rongbuk Valley on the northern slope of Mount Everest was dated to 330 ka BP (Owen et al., 2009). However, the relationship between the Tingri moraine and the Xixiabangma/Nieniexiongla moraines is uncertain because they are in different areas and are separated by a long distance. The oldest moraines (deeply weathered) in western Nyainqentanglha are found in high positions in the Cangxung-Yangpachen Valley. This glaciation was named "Ningzhong Glaciation". These moraine and associated glaciofluvial sediments were dated by ESR to 678–593 ka (Zhao Xitao et al., 2002; Wu Zhonghai et al., 2003).

The earliest glaciation in the Hengduan Mountains, the Daocheng Glaciation, is characterised by a set of deeply weathered moraines. These moraines are located at an altitude of 3800 m a.s.l. Thick red clays are present, with a SiO_2/Al_2O_3 ratio of 2.42 and a SiO_2/R_2O_3 ($Al_2O_3 + Fe_2O_3$) ratio of 2.35, implying that they were deposited sufficiently early to experience deep weathering under moist, warm interglacial conditions at a much lower altitude than today. The moraine sediments are derived from the high planation surface (4600–4800 m a.s.l.) to the north where there are no current glaciers. This moraine was dated at 560 ka by ESR. Further, the oldest moraine in Yulong Mountains (5596 m a.s.l.), lying in the south Hengduan Mountains in Yunnan Province, was ESR dated to 593 ka (Zheng Benxing, 2000) and 500–600 ka (Guo Zhengtang et al., 2001). The Yulong Mountains is the southernmost mountain where modern maritime glaciers occur in China (Table 70.2).

The traces of the earliest glaciation in the Qilian Mountains, and in the middle section of the Tian Shan, are found in the Zhonglianggan, the High Wangfeng (described above) and the Qingshantou moraines. Closely matching ESR dates have been obtained from those moraines in different areas and glaciofluvial deposits in high landscape positions, implying the existence of MIS 12 glaciation in western China. This is supported by other evidence from the Indus Valley (Owen et al., 2006).

Although the earliest glaciations have been reconstructed in some places, it is still difficult to map its extent for the whole Qinghai-Tibetan Plateau based on those segments of glacial sediments.

TABLE 70.2 The Published Data for Earliest Glaciations in Qinghai-Tibetan Plateau and Its Adjacent Mountains

Mountain	Glaciation Name	Latitude (N)	Longitude (E)	Altitude (m)	Dates (ka)	References
Himalayas	Xixiabangma	27°00′	87°00′	6200	?	Zheng et al. (1976)
Kunlun Mountains	Kunlun				ESR 710, 760	Cui et al. (1998); Thompson et al. (1997)
Karakoram	Bunthang	36°10′	75°00′	4700	MAG>700	Seong et al. (2007)
Nyainqentanglha	Ningzhong	30°20′	90°04′	5000	ESR 678, 593	Zhao et al. (2002); Wu et al. (2003)
Yulong Mountains	Yunshanping	27°30′	101°15′	3240	ESR 500–600	Zheng (2000); Guo et al. (2001)
Hengduan Mountains	Daocheng	29°05′	101°15′	3800	ESR 571	Zhou et al. (2005)
Qilian Mountains	Zhonglianggan	39°00′	99°20′	2996	ESR 463	Zhou et al. (2002a)
Tianshan Mountains	Upper Wangfeng	43°05′	86°50′	3450	ESR 471 and 460	Zhou et al. (2001)
Tianshan Mountains	Qingshantou	41°41′	80°06′	3327	ESR 440	Zhao et al. (2009)
Altai Mountains	Burqin	48°40′	87°10′	1000	TL 476	Cui et al. (2006)

70.3.2. MIS 6 Glaciation

Few glacial advances during MIS 10 and 8 have been confirmed by physical dating in the Qinghai-Tibetan Plateau and its adjacent mountains. However, MIS 6 glaciation has been identified in many mountain ranges using various techniques. These data are listed in Table 70.3. The MIS 6 glaciation was more extensive than that during the LGM.

70.3.3. Last Glaciation

The moraines and erratics of the Last Glaciation are widely distributed on the Qinghai-Tibetan Plateau and the bordering mountain ranges. These moraines are well preserved and have only weakly weathered surfaces. In recent years, a lot of chronometric data have been obtained by various methods including cosmogenic radionuclide (CRN), OSL, ESR and ^{14}C. It is not possible to list all the dates here, but some of the important results are briefly summarised. The data indicate that the local glacial maximum on the Qinghai-Tibetan Plateau and the bordering mountains occurred during the early part of the Last Glacial cycle, in most areas during mid-MIS 3. However, around the time of the global LGM (18–24 ka), Himalayan Glaciation was very restricted in extent, generally extending < 10 km from contemporary ice margins. During MIS 3, increased insolation strengthened the South Asian summer monsoon so that the glaciers received much more snowfall and advanced, while during MIS 2 lower insolation produced a weaker summer monsoon and lower snowfall, restricting the glacial advance (Owen et al., 2002). This interpretation seems to suggest that in the Himalayas, the global warm phases (interglacials) are more advantageous for glacier development than global glacials. However, recent data also show glacial advances in Hengduan Shan during Heinrich Event 1 and the Younger Dryas Stadial, both towards the end of the last cold stage (Strasky et al., 2009).

70.3.4. The Holocene Glaciations

The Holocene glaciations are marked by several sets of very complete terminal moraines in front of contemporary glacier tongues. The moraines represent Neoglacial and Little Ice Age glacial advances. There are still great differences in vegetation development between both sets of moraines, and the vegetation on Neoglacial moraines is much better developed than on Little Ice Age moraines. A lot of these moraines have been dated using ^{14}C and other methods. The Neoglacial moraines were dated to 4–3 ka BP in the western and eastern Kunlun, Tanggula, Nyainqentanglha and eastern Tianshan Mountains. In the source of Urumqi River, eastern Tianshan Mountains, the Little Ice Age moraines in the source of Urumqi River were dated using lichenometry to 1871–1538 AD (Chen Jiyang 1989). In many areas, the Little Ice Age moraines are quite fresh and are interpreted as having formed just hundreds of years ago. New data show that glacial advances also occurred in Early–Middle Holocene in the Gongga Mountain and Karola Pass regions (Owen et al., 2005).

70.4. ELAs DISTRIBUTIONS ON THE PLATEAU AND THE BORDERING MOUNTAINS

While it is difficult to reconstruct the entire extent of Early–Middle Pleistocene glaciations for the whole Qinghai-Tibetan Plateau because of the scarcity of surviving glacial landforms and deposits, this is possible for the Last Glaciation. In 1991, a first 'Quaternary glacial distribution map of Qinghai-Tibetan Plateau' was published .This was followed in 2006 by the 'Quaternary glacial distribution map of China', published together with the monograph 'The Quaternary glaciations and environmental variations in China' (Shi Yafeng et al., 2006). In fact, these maps only show the extent of the Last Glaciation, while local

TABLE 70.3 The Chronological Data for MIS 6 Glaciations

Mountain	Glaciation Name	Latitude (N)	Longitude (E)	Altitude (m)	Dates (ka)	References
Karakorom	Yunz	36°00′	74°20′	3000 TL	139	Derbyshire et al. (1984)
Karakoram	Skardu	36°10′	75°00′	4500	CRN 70–170	Seong (2007)
Kunlun Mountains	Quanshuigou	35°30′	81°30′	5300	TL 206	Zheng Benxing et al., 1991
SE Tibet	Guxiang	30°05′	95°35′	2700	CRN 112–137	Zhou et al. (2007)
Qilian Mountains	Changgousi	39°00′	99°15′	2600	ESR 130–142	Zhou et al. (2002a)
Tanggula	Zhajiazangbo	32°50′	92°00′	5100	CRN 107–181	Owen et al. (2005)
Tianshan Mountains	Low Wangfeng	43°14′	87°00′	3400	ESR 170–185	Zhao et al. (2006)
Tianshan Mountains	The fifth set till	41°41′	80°63′	2850	ESR 134.4–219.7	Zhao et al. (2009)

Chapter | 70 Quaternary Glaciations: Extent and Chronology in China

illustrations for sequence of glaciations in some places were appended on the maps. The ELA distribution pattern for Last Glacial was reconstructed based on well-preserved cirques. The ELA in the Last Glaciation followed a similar spatial pattern to the present in that it rose from the margin towards the inner plateau and from east to west, displaying an asymmetrical concentric distribution with the Qiangtang Plateau as the centre. However, when compared with today, the palaeo-ELA in the Last Glaciation was much lower in the marginal areas than in the centre. The difference reaches 1000 m in the east margin, while in the centre it is less than 300 m. This implies that the difference in extent of glaciations between the Last Glaciation and the present-day in the marginal areas is much larger than in the centre (Fig. 70.11). The ratio between the glaciated areas of the Last Glaciation and the present glaciated area reaches 20 or more in the eastern part of the plateau. In Bayan Har Mountains, Hengduan Mountains and the eastern Nyainqentanglha Mountains, ice caps of some thousands of square kilometres were present during the Last Glaciation, but today no

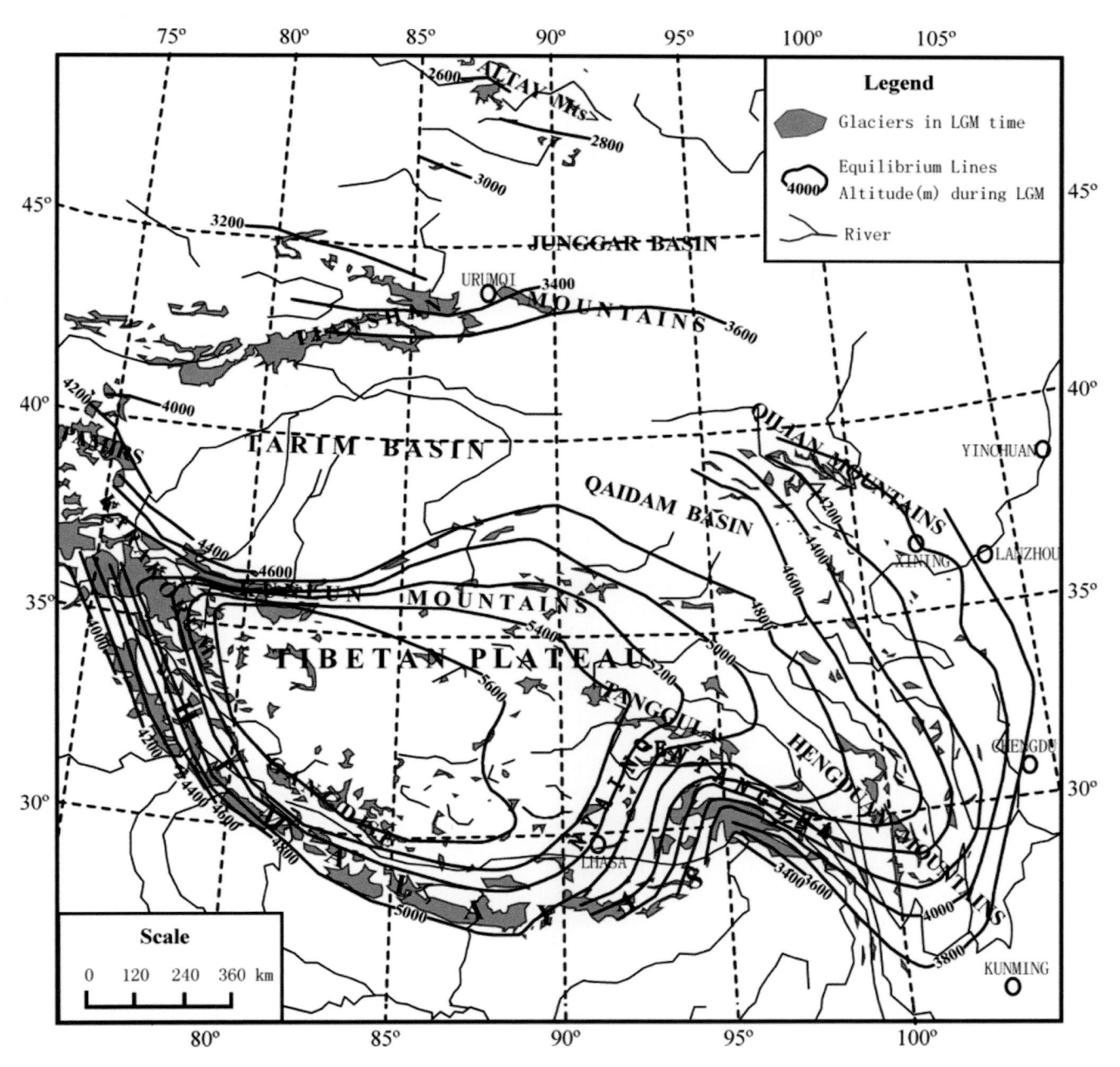

FIGURE 70.11 Distribution of glaciers and equilibrium line altitudes (ELAs) in western China during the Last Glacial Maximum (after Shi Yafeng et al., 2006).

contemporary glaciers exist within these areas. However, in the Karakoram Mountains, Kunlun Mountains, western Himalaya and Qiangtang Plateau, the ratios of Last Glaciation glaciated areas to modern glaciers are much smaller. This situation reflects the Asian monsoonal climate, which in combination with the plateau topography determines glacier development.

70.5. THE GLACIATIONS AND THE UPLIFT OF THE QINGHAI-TIBETAN PLATEAU

The Quaternary glaciations of the Qinghai-Tibetan Plateau are considered to be the combined result of global climatic deterioration and local tectonic uplift. The available evidence suggests that the plateau began to be uplifted intermittently from an extensive planation surface in the late Pliocene (Li Jijun et al., 1979). There are two tectonic uplift events which are very pivotal to glacier formation. They are called the Kunlun-Huanghe and Gonghe movements. The Kunlun-Huanghe movement was suggested based on a sequence of lacustrine sediment at the Kunlun Pass. The lacustrine Qiangtang Formation, thought to be associated with the lake sediment in the Qaidam Basin, palaeomagnetically dated to 2.5 Ma at the base and 1.1 Ma at the top, has been uplifted to 4700 m a.s.l. A fossiliferous freshwater deposit (including *Nelumbo nucifera* Gaertn, *Typhalatifolia*, *Rhizoma Scirpi yagarae*, *Potamogeton*) has been found both in the Qiangtang Formation and in a core drilling at Dabsan Lake at the centre of the Qaidam Basin. The sediment in the Dabsan core is 800 m deep and has an altitude of 2000 m a.s.l. This shows that the Kunlun Pass has been uplifted about 2700 m since 1.1 Ma. This uplift continued until 0.6 Ma (Cui Zhijiu et al., 1998a,b; Li Jijun and Fang Xiaomin, 1999). Further, the rapid tectonic uplift since 1.1 Ma has displaced the till deposits of the Kunlun Glaciation. Based on its distribution at different altitudes today, the Kunlun Pass must have been uplifted by about 1700 m since the Kunlun Glaciation. More geological evidence has been found to be related to the Kunlun-Huanghe movement (Cui Zhijiu et al., 1998b). Hence, this movement has been confirmed as the tectonic uplift of the Qinghai-Tibetan Plateau vital for triggering glacier formation. As mentioned above, the earliest glaciation in the plateau occurred during MIS 18–16, which corresponds with the late Kunlun-Huanghe movement.

The Gonghe movement is considered to responsible for the tectonic uplift of the plateau in the Late Pleistocene. It has been suggested that it caused intense incision of the Huanghe River since the Last Interglacial, which cut down into the Gonghe Formation (a set of fluvial–lacustrine sediments of Late Pleistocene age) and the underlying bedrock, resulting in the spectacular 500–600 m deep Longyang Gorge (Li Jijun et al., 1996). The significance of the Gonghe movement for the glacial development seems obvious. For example, uplifted mountains (4000–4600 m), east of the plateau, passed beyond the ELA and resulted in glaciations during the Last Glacial. No earlier glaciations were found in those mountains. So it is reasonable to think that the glaciations in Qinghai-Tibetan Plateau and the bordering mountains are the combined results of global glacial–interglacial cycles and local tectonic uplift (Fig. 70.12).

70.6. THE PROBLEM OF ICE COVER

The hypothesis that a continuous ice sheet covered the whole of Qinghai-Tibetan Plateau (2.5×10^6 km^2) during the Last Glaciation, advocated by Kuhle (2004), can be disputed based on following evidence.

1. Glacial deposits of three to four Pleistocene glaciations have been found in many places. The older moraines were usually retained on piedmonts or higher abandoned valleys. If they had been covered by an ice sheet during the Last Glaciation, they would have been destroyed. In particular, the U-shaped valleys, the cirques and the moraines of the Last Glaciation all of which indicate that glaciation was of limited extent.
2. Drillings in contemporary lakes on the Qinghai-Tibetan Plateau have revealed that some lacustrine deposits comprise a successive sedimentary history representing a major part of the Pleistocene, implying that those lakes were not covered and destroyed by an ice sheet.
3. Many periglacial remnants formed during the Last Glaciation and penultimate glaciation, including ice-wedge casts and involutions have been found in many places, showing that these places were subaerially exposed under a periglacial environment and not covered by an ice sheet.

70.7. CONCLUSIONS

1. In the Quaternary cold stages, glaciers expanded in the higher mountains of the Qinghai-Tibetan Plateau. However, glaciations were largely restricted to mountain glaciers, with limited ice caps. No continuous ice sheet was formed. In eastern China, just a few of the highest peaks were found to have developed glaciers in the Last Glaciation.
2. Four discrete Pleistocene glaciations have been identified on the Qinghai-Tibetan Plateau and the bordering mountains. These four main Pleistocene glaciations are correlated with MIS 18–16, 12, 6 and 4–2. Glaciations in mountains that are lower than 4500 m a.s.l. in the eastern Qinghai-Tibetan Plateau only occurred during the Last Glaciation.

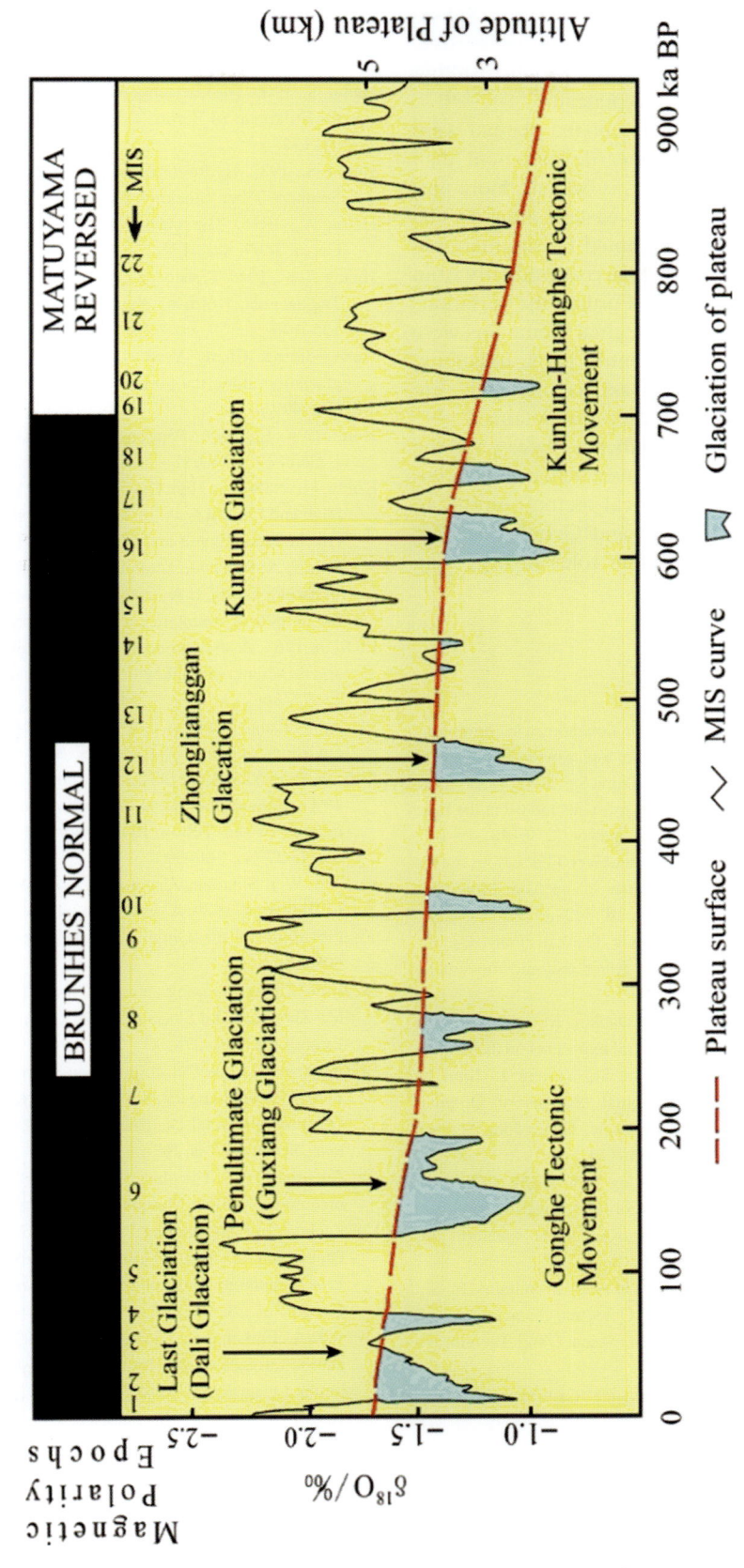

FIGURE 70.12 Coupling between the uplift of the Qinghai-Xizang Plateau and former glaciations (after Zhou Shangzhe et al., 2006).

3. The Kunlun Glaciation was the most extensive. Subsequent glaciations have been successively less extensive, probably caused by increasingly arid climates resulting from the Quaternary uplift of the plateau.
4. The eastern mountainous margins of the plateau were much more favourable for glacier development than the central and western parts. This is because more precipitation was brought into these eastern areas by the monsoons and effectively intercepted by the marginal mountains. The ELAs on the plateau rise gradually from the marginal to the central areas. Further, the variations in glacial extent in the eastern marginal mountains were larger than in the western and central parts, indicating that the glaciers in the western and central areas are more stable than those in the eastern mountains.
5. The glaciations in the Qinghai-Tibetan Plateau and the bordering mountains were the consequences of a combination between climate and local tectonic uplift. In particular, the Kunlun-Huanghe and Gonghe tectonic uplifts have played very important roles in triggering glaciations in high Asia.

REFERENCES

Chen, Jiyang, 1989. A preliminary researches on lichenometric chronology of Holocene glacial fluctuations and on other topics in the headwater of Urumqi River, TianShan Mountains. Sci. China Ser. B 32, 1487–1500.

Cui Zhijiu, 1981. Kinds and features of glacial moraine and till at the head of Urumqi River Tian Shan. J. Glaciol. Geocryol. 3, 36–48.

Cui Zhijiu, Yi Chaolu, Yan Jinfu, 1992. Quaternary Glaciations in the Halasi River catchment and its surroundings in the Altai Mountains in Xingjiang. China. J. Glaciol. Geocryol. 14, 342–351.

Cui Zhijiu, Wu Yongqiu, Liu Gengnian, Ge Daokai, Pang Qiqing, Xu Qinghai, 1998a. On Kunlun-Yellow River tectonic movement. Sci. China Ser. D 41, 592–600.

Cui Zhijiu, Wu Yongqiu, Liu Gengnian, Ge Daokai, Xu Qinghai, Pang Qiping, Yin Jiarun, 1998b. Records of natural exposures on the Kunlun Shan Pass of Qinghai-Xizang highroad. In: Shi Yafeng, Li Jijun, Li Bingyuan (Eds.), Uplift and environmental changes of Qinghai-Xizang (Tibetan) Plateau in the late Cenozoic. Guangdong Science and Technology Press, Guangzhou, pp. 81–114.

Cui Zhijiu, Yang Chienfu, Liu Gengnian, Zhang Wei, Wang Shin, Sung Quocheng, 2002. The Quaternary glaciation of Shesan Mountain in Taiwan and glacial classification in monsoon areas. Quatern. Int. 97–98, 147–153.

Cui Zhijiu, Su Zhen, 2006. Quaternary glaciers in Altay Mountains. In: Shi Yafeng, Cui Zhijiu, Su Zhen, (Eds.), The Quaternary Glaciations and Environmental Variations in China. Hebei Science and Technology Press, Shijiazhuang, pp. 503–519.

De Fillipi, 1915. Expedition to the Karakoram and Central Asia, 1913–1914. Geogr. J. 46, 85–105.

Derbyshire, E., Li Jijun, Perrott, F.A., Xu Shuying, Waters, R.S., 1984. Quaternary glacial history of the Hunza Valley, Karakoram Mountains, Pakistan. In: Miller, K. (Ed.), International Karakoram Project. Cambridge University Press, Cambridge, pp. 456–495.

Guo Zhengtang, Yao Xiaofeng, Zhao Xitao, Wei Lanying, 2001. A tropical paleosol at high elevation in the Yulong Mountains and its implication on the uplift of the Tibetan Plateau. Chin. Sci. Bull. 46, 69–72.

Hedin, S., 1909. Die Gletscher des Mustaghta. Zeitschrift der Gesellschaft für Erdkunde zu Berlin. Trans-Himalaya. Macmillan, New York, 30, 94–135.

Hedin, S., 1922. Hydrography, Geography and Geomorphology of Tibet, Southern Tibet. Stockholm, 5, 493–600.

Heim, A., 1936. The glaciation and solifluction of Minya Gongkar. Geogr. J. 87, 444–454.

Huang, T.K., 1944. Pleistocene morainic and non-morainic deposits in the Taglag area, North of Agsu, Sinkiang. Bull. Geol. Soc. China 24, 125–146.

Huang Jiqing, Cheng Yuqi, 1945. On the occurrence of Pleistocene glaciation along the northern margin of the Tarim basin. Sci. Rec 1, 492–499.

Huntington, E., 1906. Pangong: a glacial lake in the Tibetan Plateau. J. Geol. 14, 599–617.

Kuhle, M., 2004. The High Glacial (Last Ice Age and LGM) ice cover in the High and Central Asia. In: Ehlers, J., Gibbard, P.L. (Eds.), Quaternary Glaciations—Extent and Chronology. Part III: South America, Asia, Africa, Australasia, Antarctica. Elsevier, pp. 175–199.

Lehmkuhl, F., Owen, L.A., 2005. Late Quaternary glaciation of Tibet and the bordering mountains: a review. Boreas 34, 87–100.

Li Jijun, 1963. A study concerning the recent mountain geomorphic ages and the Quaternary glaciations of Nanshan. J. Lanzhou Univ. Nat. Sci. 13, 77–86.

Li Jijun, Fang Xiaomin, 1999. Uplift of the Tibetan Plateau and environmental change. Chin. Sci. Bull. 44, 2117–2124.

Li Jijun, Fang Xiaomin, Ma Haizou, Zhu Junjie, Pan Baotian, Chen Huailu, 1996. Geomorphological and environmental evolution in the upper reaches of the Yellow River during the late Cenozoic. Sci. China Ser. D 39 (4), 380–390.

Li Jijun, Wen Shixuan, Zhang Qingsong, Wang Fubao, Zheng Benxing, Li Bingyuan, 1979. A discussion on the period, amplitude and type of the uplift of the Qinghai-Xizang Plateau. Sci. China Ser. A 22, 608–616.

Li Jijun, Zheng Benxing, Yang Xijin, Xie Yingqin, Zhang Linyuan, Ma Zhenghai, Xu Shuying, 1986. Glaciers of Xizang. Science Press, Beijing, pp. 194–276.

Li Bingyuan, Li Jijun, Cui Zhijiu, Zheng Benxing, Zhang Qingsong, Wang Fubao, Zhou Shangzhe, Shi Zuhui, Jiao Keqin, Kang Jiancheng, 1991. Quaternary Glacial Distribution Map of Qinghai-Xizang (Tibet) Plateau. Science Press, Beijing, China.

Liu Zengqian, 1946. The glacial landforms of Datong River Basin in south piedmont of Qilian Mountains. Geol. Rev. 11, 247–252.

Liu Zechun, Liu Zhengzhong, Wang Fubao, 1962. A comparison of the Quaternary glaciers development patterns nearby the Qomolangma Peak, Hantengri Peak and the Tuanjie Peak of Qilian Mountains. Acta Geogr. Sin. 28, 19–33.

Owen, L.A., Finkel, R.C., Caffee, M.W., 2002. A note on the extent of glaciation throughout the Himalaya during the global Last Glacial Maximum. Quatern. Sci. Rev. 21, 147–157.

Owen, L.A., Finkel, R.C., Barnard, P.L., Ma, H., Asahi, K., Caffee, M.W., Derbyshire, E., 2005. Climatic and topographic controls on the style and timing of Late Quaternary glaciation throughout Tibet and the Himalaya defined by ^{10}Be cosmogenic radionuclide surface exposure dating. Quatern. Sci. Rev. 24, 1391–1411.

Owen, L.A., Caffee, M.W., Bovard, K.R., Finkel, R.C., Sharma, M.C., 2006. Terrestrial cosmogenic nuclide surface exposure dating of the oldest glacial successions in the Himalayan orogen: Ladakh range, northern India. Geol. Soc. Am. Bull. 118, 383–392.

Owen, L.A., Robinson, R., Benn, D.I., Finkel, R.C., Davis, N.K., Yi Chaolu, Putkonen, J., Li, Dewen, Murray, A.S., 2009. Quaternary glaciation of Mount Everest. Quatern. Sci. Rev. 28, 1412–1433.

Qian Fang, 1991. The age and formation environment of Gongba conglomerate at Tingri County, Tibet. In: Research Center of the Quaternary Glaciation and Environment & Committee of Quaternary Research in China (Eds.). Quaternary Glacier and Environment in West China. Science Press, Beijing, pp. 285–291.

Seong, Y.B., Owen, L.A., Bishop, M.P., Bush, A., Clendon, P., Copland, L., Finkel, R., Kamp, U., Shroder Jr., J.F., 2007. Quaternary glacial history of the central Karakoram. Quatern. Sci. Rev. 26, 3384–3405.

Seong, Y.B., Owen, L.A., Yi Chaolu, Finkel, R.C., Schoenbohm, L., 2009. Geomorphology of anomalously high glaciated mountains at the northwestern end of Tibet: MuztagAta and Kongur Shan. Geomorphology 103, 227–250.

Shi Yafeng, Cui Zhijiu, Su Zhen, 2006. The Quaternary Glaciations and Environmental Variations in China. Hebei Science and Technology Press, Shijiazhuang.

Shi Yafeng, Liu Tungsheng, 1964. Preliminary report on the Mount Shisha Pangma scientific expedition. Chin. Sci. Bull. 10, 928–938.

Shi Yafeng, Cui Zhijiu, Zheng Benxing, 1982. Inquiry on ice ages in Mount Xixiabangma Region. In: Shi Yafeng, Liu Dongsheng (Eds.), Reports of Scientific Expedition in Mount Xixiabangma Region Science Press, Beijing, pp. 155–176.

Shi Yafeng, 2008. Concise glacier inventory of China. Shanghai Popular Science Press, Shanghai.

Strasky, S., Graf, A.A., Zhao Zhizhong, Kubik, P.W., Baur, H., Schlüchter, C., Wieler, R., 2009. Late Glacial ice advances in southeast Tibet. J. Asian Earth Sci. 34, 458–465.

Stroeven, A.P., Hättestrand, C., Heyman, J., Harbor, J., Li, Y.K., Zhou, L.P., et al., 2009. Landscape analysis of the Huang He headwaters, NE Tibetan Plateau— Patterns of glacial and fluvial erosion. Geomorphology 103, 212–226.

Su Zhen, 2006. Quaternary glaciers in the Karakorum Mountains. In: Shi Yafeng, Cui Zhijiu, Su Zhen (Eds.), The Quaternary Glaciations and Environmental Variations in China. Hebei Science and Technology Press, Shijiazhuang, pp. 241–264.

Thompson, L.G., Yao, T., Davis, M.E., Henderson, K.A., Mosley-Thompson, E., Lin, P.N., et al., 1997. Tropical climate instability: the last glacial cycle from a Qinghai-Tibetan ice core. Science 276, 1821–1825.

Trinkler, E., 1930. The Ice Age on the Tibetan Plateau and in the adjacent regions. Geogr. J. 75, 225–232.

Wang jingtai, 1981. Ancient glaciers at the head of Urumqi River, Tianshan. J. Glaciol. Cryopedol. 3 (Special Issue), 57–63.

Weng Wenbo, Lee, T.S., 1946. A preliminary study on the Quaternary glaciation of Nanshan. Bull. Geol. Soc. China 26, 163–171.

Wissmann, H.V., 1937. The Pleistocene glaciation in China. Bull. Geol. Soc. China 17, 145–168.

Word, F.K., 1934. The Himalaya east of the Tsangpo. Geogr. J. 84, 369–397.

Wu Zhonghai, Zhao Xitao, Jiang Wang, Wu Zhenhan, Zhu Dagang, 2003. Dating of Pleistocene glacial deposits in the southeast piedimont of Nianqingtanggula Shan. Glaciol. Geocryol. 25 (3), 272–274.

Xia Kairu, 1960. Preliminary observation on the ancient glaciations and the characteristics of the periglacial geomorphology on the north section of the Qilian Mountains. Acta Geogr. Sin. 26, 165–180.

Xu Xiangke, Yang Jianqiang, Dong Guocheng, Wang Liqiang, Miller, L., 2009. OSL dating of glacier extent during the Last Glacial and the Kanas Lake basin formation in Kanas River valley, Altai Mountains, China. Geomorphology 112, 306–317.

Yi Chaolu, Liu Kexin, Cui Zhijiu, 1998. AMS dating on glacial tills at the source area of the Urumqi River in the Tianshan mountains and its implications. Chin. Sci. Bull. 43, 1749–1751.

Yi Chaolu, Jiao Keqing, Liu Kexin, He Yuanqing, Ye Yuguang, 2002. ESR dating of the sediments of the Last Glaciation at the source area of the Urumqi River, Tian Shan Mountains, China. Quatern. Int. 97-98, 141–146.

Younghuband, F.E., 1926. The problem of the shakasgam valley. Geogr. J. 68, 225–230.

Zhang Wei, Niu Yunbo, Yan Ling, Cui Zhijiu, Li Chuanchuan, Mu Kehua, 2008. Late Pleistocene glaciation of the Changbai Mountains in northeastern China. Chin. Sci. Bull. 53, 2672–2684.

Zhao Xitao, Wu Zhonghai, Zhu Dagang, Hu Daogong, 2002. Quaternary glaciations in the west Nyaiqentanglha Mountains. Quatern. Sci. 22, 424–433.

Zhao Jingdong, Zhou Shangzhe, He Yuanqing, Ye Yuguang, Liu Shiyin, 2006. ESR dating of glacial tills and glaciations in the Urumqi River headwaters, Tianshan Mountains, China. Quatern. Int. 144, 61–67.

Zhao Jingdong, Liu Shiyin, He Yuanqing, Song Yougui, 2009. Quaternary glacial chronology of the Ateaoyinake River Valley, Tianshan Mountains, China. Geomorphology 103, 276–284.

Zhao Jingdong, Liu Shiyin, Wang Jie, Song Yougui, Du Jiankuo, 2010a. Glacial advances and ESR chronology of the Pochengzi Glaciation, Tianshan Mountains. China. Sci. China Earth Sci. 53, 403–410.

Zhao Jingdong, Song Yougui, King, J.W., Liu Shiyin, Wang Jie, Wu Min, 2010b. Glacial geomorphology and glacial history of the Muzart River valley, Tianshan range. China. Quatern. Sci. Rev. 29, 1453–1463.

Zheng Benxing, 1988. Quaternary glaciation of Mount Qomolangma-Xixiabangma Region. Geojournal 17, 525–543.

Zheng Benxing, Jiao Keqin, Li Shijie, Ma Qiuhua, 1991. The evolution of Quaternary glaciers and environmental change in the west Kunlun Mountains, West China. In: Research Center of the Quaternary Glaciation and Environment & Committee of Quaternary Research in China (Eds.), Quaternary Glacier and Environment in West China. Science Press, Beijing, pp. 15–23.

Zheng Benxing, 2000. Quaternary glaciations and alacier evolution in the Yulong Monut, Yunan. J. Glaciol. Geocryol. 22, 53–61.

Zheng Benxing, Shi Yafeng, 1976. On Quaternary glaciation of Mount Qomolangma area. In: Comprehensive Scientific Expedition to the Qinghai-Xizang Plateau, Chinese Academy of Sciences (Ed.), Reports of Qomolangma area Scientific Expedition (1966–1968), Quaternary Geology. Science Press, Beijing, pp. 29–62.

Zhou Shangzhe, Li Jijun, 1998. Quaternary glaciation sequence in Bayan Har Mountains. Quatern. Int. 45 (46), 133–142.

Zhou Shangzhe, Yi Chaolu, Shi Yafeng, Ye Yuguang, 2001. Study on the glacial advance of MIS-12 in western China. J. Geomech. 7, 321–327.

Zhou Shangzhe, Wang Jie, Xu Liubing, Wang Xiaoli, Colgan, P.M., Mickelson, D.M., 2010. Glacial advances in southeastern Tibet during late Quaternary and their implications for climatic changes. Quatern. Int. 218, 58–66.

Zhou Shangzhe, Li Jijun, Zhang Shiqiang, 2002a. Quaternary glaciation of the Bailang River Valley, Qilian Shan. Quatern. Int. 97 (98), 103–110.

Zhou Shangzhe, Jiao Keqin, Zhao Jingdong, Zhang Shiqiang, Cui Jianxin, Xu Liubing, 2002b. Geomorphology of the Urumqi River Valley and the uplift of the Tianshan Mountains in Quaternary. Quatern. Sci. China Ser. D 45 (11), 961–968.

Zhou Shangzhe, Li Jijun, Zhang Shiqiang, Zhao Jingdong, Cui Jianxin, 2004. Quaternary Glaciations in China. In: Ehlers, J., Gibbard, P.L. (Eds.), Quaternary Glaciations—Extent and Chronology. Part III. Elsevier, Amsterdam, pp. 105–113.

Zhou Shangzhe, Wang Xiaoli, Wang Jie, Xu Liubing, 2006. A preliminary study on timing of the oldest Pleistocene glaciation in Qinghai-Tibetan Plateau. Quatern. Int. 154-155, 44–51.

Zhou Shangzhe, Xu Liubing, Colgan, P.M., Mickelson, D.M., Wang Xiaoli, Wang Jie, et al., 2007. Cosmogenic ^{10}Be dating of Guxiang and Baiyu glaciations. Chin. Sci. Bull. 52, 1387–1393.

Zhou Shangzhe, Xu Liubing, Cui Jianxin, Zhang Xiaowei, Zhao Jingdong, 2005. Geomorphologic evolution and environmental changes in the Shaluli Mountain region during the Quaternary. Chin. Sci. Bull. 50, 52–57.

Late Pleistocene and Early Holocene Glaciations in the Taiwanese High Mountain Ranges

Margot Böse* and Robert Hebenstreit

Freie Universität Berlin, Institute of Geographical Sciences, Malteserstraße 74-100, 12249 Berlin, Germany
*Correspondence and requests for materials should be addressed to Margot Böse. E-mail: m.boese@fu-berlin.de

71.1. INTRODUCTION

The study of the former glacier extent in the high mountain range of Taiwan, with its unique location as an isolated high-altitude area in monsoonal Asia at the junction of the Eurasian continent and the western Pacific Ocean, provides useful high-altitude terrestrial proxy data which can mediate between the climate archives in marine sediments (Thunell and Miao, 1996; Kiefer and Kienast, 2005) and continental deposition (An, 2000; Porter, 2001a; Zhou et al., 2001). Knowledge of former glaciations in Taiwan also provides a meridional link between the Pleistocene glaciation areas of the Japanese Islands (Ono, 1991; Kerschner and Heuberger, 1997; Ono et al., 2003, 2005), Borneo (Koopmans and Stauffer, 1967; Hope, 2004) and New Guinea (Löffler, 1972, 1980; Brown, 1990; Prentice et al., 2005).

The general reconstruction of the Pleistocene and Holocene palaeoclimate of Taiwan is mainly based on palynological evidence from low-altitude terrestrial sediments (Liew et al., 2006a,b) and on marine sediment proxy data from the surrounding seas (Huang et al., 1997; Wei, 2002; Xiang et al., 2007).

The Taiwanese high mountain range is not glaciated today. The present (theoretical) equilibrium line altitude (ELA) is calculated at 3950 ± 100 m a.s.l. (Hebenstreit, 2006), and thus it lies around the altitude of the highest peaks of the island (*Yushan* 3952, *Hsueh Shan* 3884 m a.s.l., Fig. 71.1). The presence of 62 summits (Porter, 2001b) and an area of about 450 km² (calculation bases on SRTM-3, Farr et al., 2007) over 3000 m a.s.l. suggest that the conditions may have been favourable for the development of true glaciation in many of the Taiwanese mountain ranges during the last glacial cycle.

The first reports on glacial landforms and sediments in the Taiwanese high mountain range were published in the 1930s (Kano, 1932; Tanaka and Kano, 1934; Panzer, 1935). The former ELA calculations were based on these observations or on extrapolations from Japan and China (Shi, 1992; Ono and Naruse, 1997; Porter, 2001b; Ono et al., 2003). Recent field studies in *Yushan* (3952 m a.s.l.), *Hsueh Shan* (3884 m a.s.l.) and *Nanhuta Shan* (3742 m a.s.l.) have provided a more detailed view on these phenomena (Chu et al., 2000; Yang, 2000; Cui et al., 2002; Hebenstreit and Böse, 2003; Hebenstreit et al., 2006; Siame et al., 2007; Hebenstreit et al., 2011). The *Nanhuta Shan* is the best studied area to date (Fig. 71.1). Here, glacial landforms and deposits have been described and mapped in detail. Sediment analysis undertaken from key localities includes grain-size distribution, lithology, clast morphology and sediment fabric. Glacial and glaciofluvial sediments, as well as glacial boulders, have been dated using optically stimulated luminescence (OSL) and surface exposure dating (SED) on *in situ* produced cosmogenic ^{10}Be, respectively.

Potential glacial phenomena have also been reported from *Hohuan Shan* (3416 m a.s.l.) and the *Shangyang Shan* (3602 m a.s.l.)–*Sancha Shan* (3496 m a.s.l.) in north-central and southern Taiwan, respectively (Chu, 2009).

The present morphodynamic activity in Taiwan is dominated by extraordinarily effective erosional processes. In particular, mass movements and fluvial transport generate one of the highest erosion rates in mountain ranges worldwide (Dadson et al., 2003). These processes are induced by fast terrain uplift of averaged 5 mm/yr (Lundberg and Dorsey, 1990), in combination with effective rock destruction by earthquakes and the availability of high precipitation rates, especially during large-scale precipitation events like typhoons (Hartshorn et al., 2002; Chen, 2006). The consequence is that glacial deposits in low valley positions are sparsely preserved.

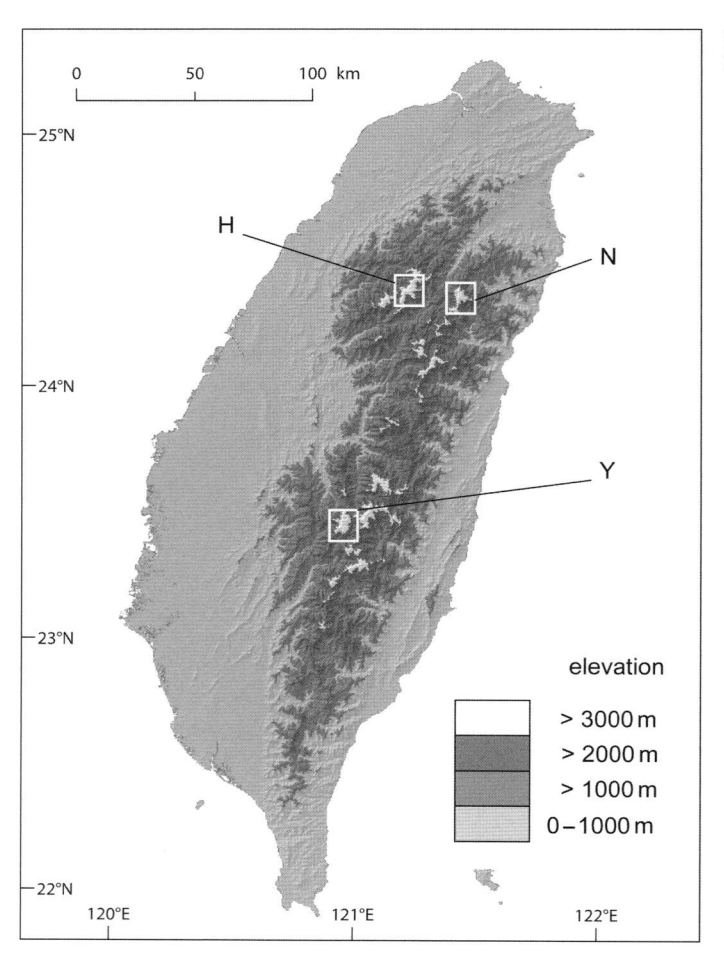

FIGURE 71.1 Location of study areas in Taiwan: N, Nanhuta Shan; H, Hsueh Shan; Y, Yushan. Data source is SRTM-3 (Farr et al., 2007).

In strong contrast to these processes, the present and recent geomorphological activity in the central (uppermost) parts of the studied mountain massifs is minor and limited to little frost weathering at the highest peaks and episodic fluvial transport with a tendency to accumulation (Böse, 2006; Klose, 2006). This allows the preservation of glacial landforms and deposits. Nevertheless, backward erosion has reached the upper parts of several valleys already, causing an ongoing destruction of the fossil glacial relief.

71.2. LATE PLEISTOCENE GLACIATIONS

The central parts of many mountain ranges above 3000 m a.s.l. are dominated by large erosional glacial landforms, indicating a high elevation of the former glacier surface in the accumulation area. This implies a wide extent of mountain glaciation during the last glacial cycle. Unfortunately, corresponding glacial sediments of the maximum stage are usually badly preserved owing to the strong post-glacial and modern-day erosion. Remnants are hard to find or inaccessible in the dense vegetation of the steeply sloped lower valley sections. Therefore, a detailed reconstruction of the glacier configuration during the maximum glacier extent in Taiwan has not so far been possible, and age control of this stage is still very limited.

71.2.1. Nanhuta Shan

Wide U-shaped valleys, with over-deepened floors, alpine ridges (*arêtes*), rock bars, glacial knobs (*roches moutonnées*) and subglacial gorges are common above 3000 m a.s.l. in this mountain range in north-east Taiwan. Glacial trimlines indicate the former presence of a small plateau glacier, which almost completely covered the central *Nanhuta Shan* massif (3742 m a.s.l.) during the maximum-stage glaciation (Fig. 71.2). Only the highest summits and parts of the mountain ridges rose above the ice surface. Huge valley glaciers radiated down from the plateau into the surrounding valleys. The ice thickness reached at least 150 m in the upper sections (Hebenstreit and Böse, 2003).

A diamicton, which is preserved further in the West in the *Nanhu Xi* (Nanhu Valley) at 2250 m a.s.l., is the only known potential deposit of this stage in *Nanhuta Shan*. The sediment characteristics, such as sorting, fabric and

FIGURE 71.2 Overview on Lower Valley (centre) and Upper Valley (left), central Nanhuta Shan. Seen from the cirque below North Peak (cf. Fig. 71.3), 24.374°N, 121.442°E, 3592 m a.s.l., facing S, 22 Mar 2000. The flanks of this U-shaped valley have been abraded nearly up to the Main Peak (3742 m a.s.l.). The trimline at about 3500–3600 m a.s.l. (– – –) is marked by a distinct difference in the rock surface structure. This indicates an almost complete ice infill in the valley. The glacier originated on a small plateau called the saddle and the ice outflow was over a rock bar (rb) to the W into the Nanhu Xi (→) during a pre-Holocene stage. Well-preserved end moraine ridges, which cross the valley bottom at 3380 m a.s.l. (bbm1, 2, 3) and erratic boulders deposited on the rock ridge between the Upper and the Lower Valley (e) at 3500 m a.s.l., belong to an Early Holocene (ca. 10 ka) stage, when the glacier did not longer reach as far as the rock bar (adapted from Hebenstreit and Böse, 2003; Hebenstreit et al., 2011).

clast shape, indicate a glacial origin for this material (Hebenstreit et al., 2006). The sediment is undated so far, but it is morphostratigraphically older than the Late glacial and Holocene sediments dated in the central part of the massif above 3000 m a.s.l.

71.2.2. Hsueh Shan

Various glacial landforms have been reported from *Hsueh Shan* (3884 m a.s.l.) in Central North Taiwan. They include cirques, alpine ridges, glacial rock thresholds, polished and scratched rock surfaces, and moraine ridges (Yang, 2000; Cui et al., 2002; Hebenstreit and Böse, 2003).

A sediment ridge was deposited at 3050 m a.s.l. on the steep right-hand slope of the *Qi Jia Wan Xi* (NE-Valley) about 500 m above the present valley bottom. The ridge is about 10 m high and parallel to the valley. It has a convex shape towards the valley with a depression on its upslope side. It is composed of a deeply brownish weathered diamicton with rounded or sub-rounded, often faceted or polished schist and quartzite boulders with diameters of at least 40 cm embedded in a matrix with a high content of silt. Most of the slate clasts have been broken by frost weathering.

Two samples from the matrix material on the slopeward side of the ridge, at about 5 m below the crest, gave OSL ages of 51 ± 10 and 56 ± 4 ka (Hebenstreit and Böse, 2003). An additional thermoluminescence (TL) age of 44.25 ± 3.27 ka, termed the *Shanzhuang* stage (Cui et al., 2002), is reported from the depression on the upslope side of the moraine. The sediment in the depression may include some slope debris, which would explain the younger age. Although the origin of this deposit may be more complex and is not yet clear in detail, Hebenstreit and Böse (2003) classified it as a dislocated lateral moraine, because its position on the steep slope and its shape exclude an origin as a pure landslide deposit or an end moraine, as suggested by Cui et al. (2002). However, it marks the minimum extent of the glacier down to 3050 m a.s.l.

71.2.3. Yushan

Glacial evidence from *Yushan* (3952 m a.s.l.), the highest summit in Taiwan, was reported early in the past century (Panzer, 1935). U-shaped and hanging valleys, alpine ridges, cliffs and glacial horns are clear legacies of a former glaciation in this area. Here, a diamicton deposit, described by Panzer in the *Lao Nong Xi* (NE-Valley) at 3140 m a.s.l., has been partly destroyed by a recent landslide. Hebenstreit et al. (2006) agree with the classification of this massive diamicton as a glacial till. Although the deposit is still undated, in the light of dating results from *Nanhuta Shan* in a similar altitude (Section 71.3.1), a Late Pleistocene or even Late glacial age is more likely than a Middle Pleistocene age (~Rissian), which was suggested by Panzer (1935).

Further downvalley, a several hundred square metre plateau, called *Ba Tong Guan* (2850 m a.s.l.) with smooth hills contrasts with the rough surface of the surrounding area which was created by fluvial erosion and landslide activity. The present destruction of the plateau from both the north and the south shows that this landform is ancient and inactive. Erratic boulders are spread on the plateau surface and support the idea of its glacial origin. Judging from the dimensions and position of the plateau, and the U-shaped cross-section of the valley here, the plateau does not represent the maximum extent of a former glacier. However, the lowest position and age of the glacier terminus are still unknown (Hebenstreit et al., 2006).

71.3. LATE GLACIAL AND EARLY HOLOCENE GLACIATIONS

71.3.1. Nanhuta Shan

The central *Nanhuta Shan* massif provides good conditions for sediment preservation, because of its unique topography. Two parallel, north–south striking glacially eroded and over-deepened basins, called the Upper Valley and the Lower Valley, represent a sediment trap and the source of the *Nanhu Xi* (Nanhu Valley), whose drainage is orthogonal to the west (Figs. 71.2, 71.3 and 71.4). However, the sediments deposited in these basins correspond neither in dimension nor in position to the large valley forms in which they are imbedded. They must therefore belong to a younger phase of activity in terms of their morphostratigraphical setting. Besides modern accumulations (Klose, 2006), several fossil deposits, including prominent diamictons, are preserved. Three types of glacial deposits have been described and dated in the *Nanhuta Shan* area:

1. *Schist moraines* are heterogeneous sediment complexes with rapid horizontal and vertical changing lithofacies, which accumulated in the northern parts of both basins. Their structure varies from massive and matrix-supported diamictons to zones which contain mainly small schist plates, usually not larger than 0.5 m, or finer units with gravel concentrations. Metre-size boulders are concentrated in the massive units and dispersed on their surfaces or in the upper sediment layers. The whole complex appears fresh and unweathered. The assemblage of sediment parameters, including striated boulders, clast shape, grain size, sorting and fabric, indicates a glacial origin for these deposits (Hebenstreit et al., 2006). The morphology of the sediment complex is amorphous as is typical for latero-frontal ice-marginal positions (Benn and Owen, 2002). Moreover, post-sedimentary degradation (periglacial, nival, surface outwash, fluvial erosion and re-sedimentation) has modified the morphology and the structure of the upper sediment layers. However, a lateral moraine which corresponds to a small cirque is clearly recognisable at least in the Lower Valley (Fig. 71.3). Four samples from the finer units and the matrix of the schist moraines have yielded OSL ages of between 7.3 ± 0.9 and 9.7 ± 0.7 ka (Hebenstreit et al., 2006).

2. *Big-Boulder Moraines* are characterised by their specific composition (Hebenstreit et al., 2006). This type of glacial deposit consists of metre-sized rounded quartzite and slate boulders with very little sediment matrix. It forms well-defined and stable ridges in the Lower Valley and the SE-Valley (*He Ping Nan Xi*). Because of their mixed lithology, the boulder shape and the shape of the landform itself, these deposits are clearly identifiable as moraines (Hebenstreit and Böse, 2003).

In the Lower Valley at 3380 m a.s.l., three staggered big-boulder moraine ridges, each about 10 m high, indicate a repeated oscillation of the glacier, which had originated on a small plateau-like plain (*the saddle*) south-east of the Main Peak (3742 m a.s.l.) where the present elevation is 3550 m a.s.l. (Fig. 71.2). Two boulders from the middle (second oldest) ridge were sampled for ^{10}Be exposure dating (Hebenstreit et al., 2011), because the first (oldest) one did not have any suitable quartzite boulders. They gave ages of 9.22 ± 1.13 and 10.06 ± 1.22 ka.

Another big-boulder moraine crosses the valley bottom at the end of a flat valley section at 3180 m a.s.l. in the SE-Valley (*He Ping Nan Xi*, Fig. 71.5). Here, there is a ridge about 12 m high consisting of metre-sized slate and meta-sandstone (quartzite) boulders. A sample from a quartzite boulder on the crest of the ridge gave an exposure

FIGURE 71.3 Glacial evidences in the Lower and Upper Valley, Nanhuta Shan, 24.370°N, 121.443°E, 3375 m a.s.l. (at location Nan-VII), facing NE. 12 Mar 2002. A lateral (schist) moraine was accumulated by the small cirque glacier (C) running down southward from the North Peak (1). The big-boulder moraines (bbm1 and bbm2) were accumulated by a northward outlet tongue of the plateau glacier (coming from the right in the picture). The schist moraines in the Upper Valley (Nan-II and Nan-III) belong to the Early Holocene glacier advance, too. During a pre-Holocene stage of glaciation, the ice outflow resulted into the Nanhu Xi (3) crossing a rock bar. An accumulation of metre-size erratic boulders on a rock shoulder (×) marks the minimum level of the ice surface during that stage (adapted from Hebenstreit et al., 2006).

age of 13.61 ± 1.45 ka (Hebenstreit et al., 2011). This age corresponds to an OSL age determination from glaciofluvial sediments in the former glacier basin of 11.1 ± 2.5 ka (Hebenstreit et al., 2006).

3. *Erratic boulders* are widespread and frequent in the study area (Hebenstreit and Böse, 2003; Siame et al., 2007). They mainly comprise abundant quartzite boulders, but also include some slate boulders, that have preserved striations on their surfaces and offer clear proof of their glacial transport (Fig. 71.6). However, there are only a few that are suitable for exposure dating, most being too small, potentially eroded, previously buried or not in a suitable position. Two carefully selected quartzite boulders on the middle ridge, between the Upper and the Lower Valley, were dated. This ridge represents a lateral position of the former outlet glacier into the Upper Valley. The samples from here yielded ages of 8.80 ± 1.12 and 9.77 ± 1.09 ka, which is in good agreement with those from the big-boulder moraines in the Lower Valley (Hebenstreit et al., 2011).

71.3.2. Hsueh Shan

Cui et al. (2002) reported three TL ages, which were derived from a sediment section 1500 m from the main peak of *Hsueh Shan* at about 3300 m a.s.l. Here, the sediment is associated with moraine ridges and yielded ages in three depths of 18.26 ± 1.25, 14.28 ± 1.13 and 10.68 ± 0.84 ka, respectively.

71.4. EQUILIBRIUM LINE ALTITUDES AND PALAEOCLIMATIC CORRELATIONS

Since no evidence for a Little Ice Age (LIA) glacier advance has yet been reported from the Taiwanese

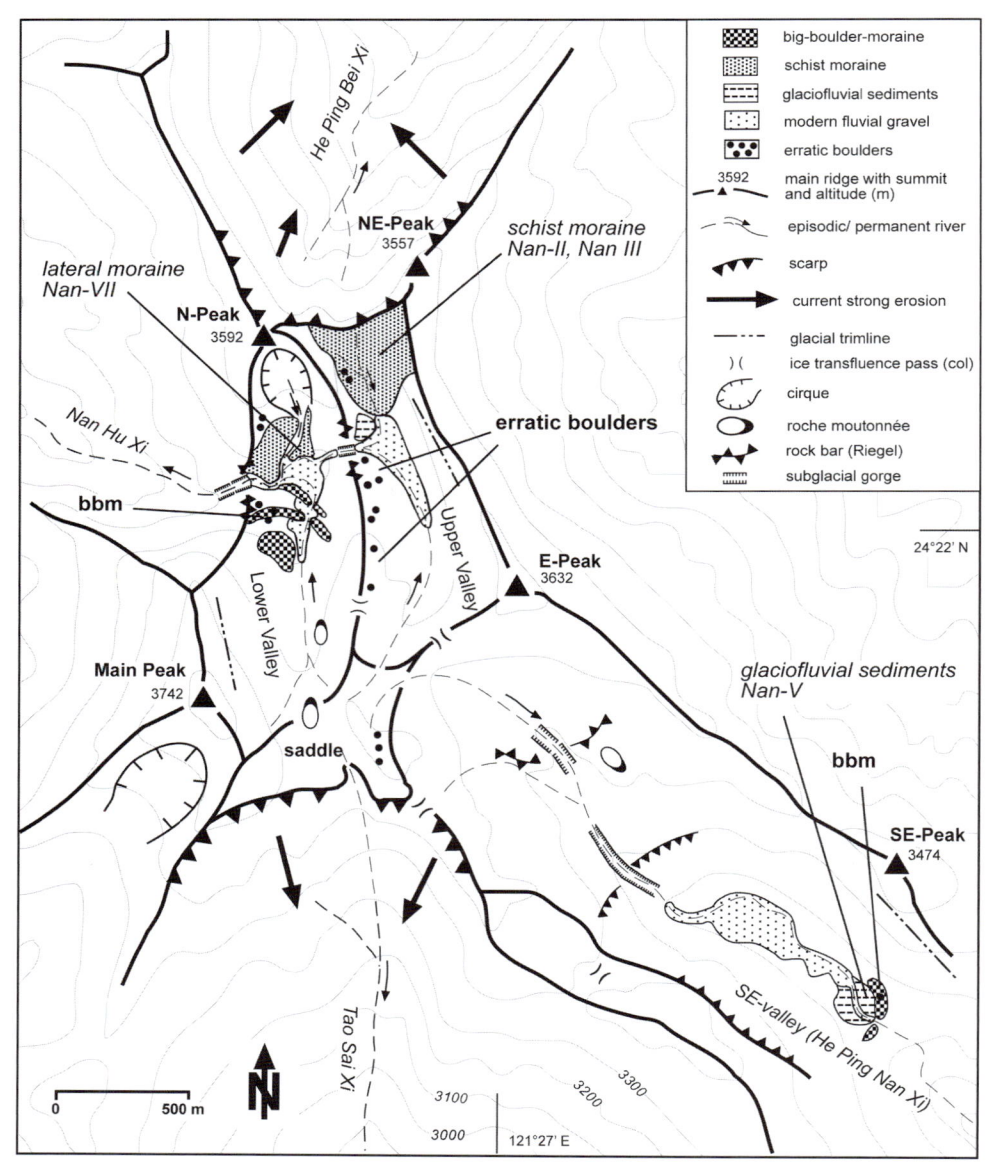

FIGURE 71.4 Geomorphological sketch map of central Nanhuta Shan with sampling sites of Late glacial/Early Holocene deposits for surface exposure dating (bold, Siame et al., 2007; Hebenstreit et al., 2011) and OSL (italic, Hebenstreit et al., 2006). bbm, big-boulder moraine.

mountains, all ELA depressions (ΔELA) are given with respect to the present (theoretical) ELA. This is calculated at 3950 ± 100 m a.s.l. (Hebenstreit, 2006) using the empirical equation for the relationship of summer temperatures and annual total precipitation from Ohmura et al. (1992) and a 30-year (1971–2000) database (Klose, 2006) from *Pei Shan* (3845 m, north peak of *Yushan*). Other authors calculate the present ELA to 4354 m a.s.l. (Yang, 2000) or even 4600 m a.s.l. (Ono et al., 2005). The latter altitude is derived using only winter precipitation.

Late Pleistocene and Holocene ELAs have been reconstructed for Taiwan (Hebenstreit, 2006) in most cases using the terminal to summit altitudinal method (TSAM), established by Louis (1955). Since this method gives values that are slightly too high, so that the resulting ΔELAs can be taken as minimum values.

Datable sediments and features which determine the glacial-ice maximum are still very sparsely known for the maximum stage of Taiwanese glaciation. Consequently, the reconstruction of the ELA is difficult. Data from *Hsueh Shan* (Cui et al., 2002; Hebenstreit and Böse, 2003) indicate a marine isotope stage (MIS) 3 or 4 age, which corresponds with the idea that mountain glaciations in East Asia were stronger in the early last glacial than during the global Last Glacial Maximum (LGM) (Ono and Naruse, 1997). Cui et al. (2002) suggest an ELA at 3400 m a.s.l. for that Early

FIGURE 71.5 SE-Valley (He Ping Nan Xi), Nanhuta Shan, about 24.357°N, 121.455°E, 3300 m a.s.l., facing SE, 21.03.2000. The U-shaped valley was almost completely ice filled during the maximum stage of the glaciation, which is indicated by scoured flanks and rock cliffs (– – –). The ice thickness reached 150 m. The valley bottom is very flat here and probably over-deepened by glacial erosion. A younger (Late glacial) stage is documented by an end moraine ridge (big-boulder moraine), which crosses the valley at about 3180 m a.s.l. (bbm). Meltwater-related sediments behind the moraine (Nan-V) represent the subsequent glacier retreat. The asymmetry of the valley cross-section is due to the bedrock structure (adapted from Hebenstreit and Böse, 2003; Hebenstreit et al., 2006).

last glaciation, not including mountain uplift. Assuming an average uplift rate of 5 mm/yr (Lundberg and Dorsey, 1990), an ELA at 3200 m a.s.l. would result. Depending on the interpretation of the moraine origin, Hebenstreit (2006) calculates the ELA in *Hsueh Shan* to 2775–3195 m a.s.l. for MIS 3–4 using the maximum elevation of lateral moraines for the lower value and including 5 mm/yr uplift. This results in an ΔELA of 700–1125 m. Due to the lack of data from *Nanhuta Shan* and *Yushan*, the ΔELA for a pre-Late glacial stage can be only given as >900 and >600 m, respectively.

These values correspond widely with ΔELA values of >1000 m for the LGM in the Japanese mountains (Ono et al., 2005) and also with the mean depression of 900 ± 135 m for the last glaciation in tropical and subtropical areas (Porter, 2001b). These values especially correspond with those from Mount Kinabulu/Borneo (905 ± 150 m) and Papua New Guinea (900–1000 m), even though recent calculations for West Papua have yielded an LGM ELA depression of only 600 m (Prentice et al., 2005).

A more detailed view on the Late glacial and Early Holocene glaciation is possible. Both OSL and exposure dating result from *Nanhuta Shan* point independently to a glacial advance in the Taiwanese high mountain area around 10 ka, called the Nanhuta glacier advance (Hebenstreit et al., 2006; Siame et al., 2007; Hebenstreit et al., 2011). The three staggered moraines in the Lower Valley show an oscillation of the glacier terminus during that time, whereas it is still unknown whether they mark backwasting positions or repeated advances. Nevertheless, they do indicate unstable climatic conditions during the earliest Holocene in the Taiwanese mountains. The ELA at the time was 3440 m a.s.l., allowing for 50 m uplift, and the corresponding ELA depression was about 510 m (Hebenstreit, 2006). This is in contrast to previously reported

FIGURE 71.6 Erratic boulder. Nanhuta Shan, valley divide between Upper and Lower Valley, 24.368°N, 121.445°E, about 3500 m a.s.l. The well-rounded boulder shows typical glacial striations and a polished surface on its underside. It consists of hard quartzite, which occurs as bedrock some hundred metres from this location, while the bedrock here is slate. Similar erratics are found at several locations in this area. They reach several metres in size. (The boulder was turned over for the photo; adapted from Hebenstreit and Böse, 2003.)

assumptions, which related the glacial sediments in the central *Nanhuta Shan* to the LGM with a derived ELA at 3500–3600 m a.s.l. (Tanaka and Kano, 1934; Shi, 2002; Ono et al., 2005).

The moraine in the SE-Valley probably marks an earlier glaciation stage at 13.6 ± 1.5 ka, because it is located 200 m lower than the moraines in the Lower Valley. An OSL age of 11.1 ± 2.5 ka from glaciofluvial sediments in the former glacier basin represents the subsequent glacier retreat from this position, in this scenario. The ELA during that time was about 3340 m a.s.l. and the corresponding ELA depression about 610 m (Hebenstreit et al., 2011).

A precise correlation of the Nanhuta glacial advance with the millennial timescale of the Early Holocene climate events, obtained from ice cores and regional terrestrial or marine datasets, is difficult. Since erratic boulders and boulders on moraine crests represent the onset of the glacier retreat and thus the end of a cold period (Gosse, 2005), a correlation with the Younger Dryas (YD) Stadial event (Alley et al., 1993; Rasmussen et al., 2006; Vinther et al., 2006) seems possible. This is especially likely in the light of the dating uncertainties of the determined exposure ages and the relatively deep ELA depressions.

Pollen assemblages from the Toushe basin (in Central Taiwan, 650 m) show a significant cold period between 13 and 11.6 ka that is indicated by sharp increase of *Salix* and *Gramineae* pollen. This could well represent the YD event (Liew et al., 2006a). Evidence from the neighbouring marine environment suggests that the cold period (YD) ended here at 11.7 ka (Xiang et al., 2007).

Other possible correlations with short timescale Holocene anomalies obtained from ice-core records are the 9.95, the 9.3 and the 8.2 ka events (Rasmussen et al., 2007). A corresponding cold episode in Taiwan between 9.6 and 9.4 ka is marked by a rise of *Salix*, *Ilex* and *Symplocos* pollen (Liew et al., 2006a). High-resolution records of planktonic foraminifera and oxygen isotopes from the middle Okinawa trough exhibit millennial-scale oscillations during the Early Holocene, with cold events at about 10.6, 9.6 and 8.2 ka BP caused by variations in the influence of the Kuroshio current (Jian et al., 2000; Xiang et al., 2007). Additional rapid sea-level rise (meltwater pulse 1C) at 9.5–9.2 ka in the East China Sea, Yellow Sea and South China Sea (Liu et al., 2004) may have supported glacier growth in the Taiwanese mountains by providing more atmospheric moisture in the area.

71.5. CONCLUSIONS

The geomorphological evidence of glacial landforms and sediments reported by various authors from the high mountain range of Taiwan at altitudes above 3000 m a.s.l. corresponds widely with each other. It confirms the concept of repeated, multi-stage former glacier advances in the Taiwanese high mountain range (Tanaka and Kano, 1934; Panzer, 1935; Yang, 2000; Cui et al., 2002; Hebenstreit and Böse, 2003; Hebenstreit et al., 2006; Siame et al., 2007; Hebenstreit et al., 2011).

Nevertheless, glacial sediments in the central *Nanhuta Shan* do not represent the maximum of the last glacial cycle. These sediments have been dated using two independent methods (OSL and ^{10}Be exposure dating), yielding Late glacial (latest Pleistocene) or Earliest Holocene ages, around 10 ka, called the Nanhuta glacier advance. This advance had an ELA at 3440 m a.s.l. and an ELA depression of 510 m. In the SE-Valley, there is evidence of a possible Late glacial glacier advance between 15 and 12 ka. This advance had an ELA ca. 3340 m a.s.l. and an ELA depression of 610 m. Consequently, the glacier extent must have been much wider during the maximum (pre-Late glacial) stage of the last glacial cycle, as indicated by large-scale erosional landforms in the central parts of *Nanhuta Shan*, *Hsueh Shan* and *Yushan* above 3000 m a.s.l. Although reliable terminal deposits of this stage are unknown, luminescence dating from *Hsueh Shan* suggests an age of MIS 3 or 4, where the ELA was likely lower than 3000 m a.s.l. and the ELA depression >1000 m.

REFERENCES

Alley, R.B., Meese, D.A., Shuman, C.A., Gow, A.J., Taylor, K.C., Grootes, P.M., et al., 1993. Abrupt increase in Greenland snow accumulation at the end of the Younger Dryas event. Nature 362, 527–529.

An, Z., 2000. The history and variability of the East Asian paleomonsoon climate. Quatern. Sci. Rev. 19 (1–5), 171–187.

Benn, D.I., Owen, L.A., 2002. Himalayan glacial sedimentary environments: a framework for reconstructing and dating the former extent of glaciers in high mountains. Quatern. Int. 97–98, 3–25.

Böse, M., 2006. Geomorphic altitudinal zonation of the high mountains of Taiwan. Quatern. Int. 147 (1), 55–61.

Brown, I.M., 1990. Quaternary glaciations of New Guinea. Quatern. Sci. Rev. 9 (2), 273–280.

Chen, H., 2006. Controlling factors of hazardous debris flow in Taiwan. Quatern. Int. 147 (1), 3–15.

Chu, H.T., 2009. Glacial landforms and relicts in the high mountains of Taiwan, European Geosciences Union General Assembly 2009. Geophys. Res. Abstr. Vienna, p. EGU2009-7083.

Chu, H.T., Sonic, W.L., Kuo, Y.C., 2000. New evidences of late Quaternary glacial relicts in the Nanhutashan area, Taiwan. In: The Symposium on Taiwan Quaternary & Workshop of the Asia Paleoenvironmental Change Project, Keelung, p. 18.

Cui, Z., Yang, C., Liu, G., Zhang, W., Wang, S., Sung, Q., 2002. The Quaternary glaciation of Shesan Mountain in Taiwan and glacial classification in monsoon areas. Quatern. Int. 97–98, 147–153.

Dadson, S.J., Hovius, N., Chen, H., Dade, W.B., Hsieh, M.-L., Willett, S.D., et al., 2003. Links between erosion, runoff variability and seismicity in the Taiwan orogen. Nature 426 (6967), 648–651.

Farr, T.G., Rosen, P.A., Caro, E., Crippen, R., Duren, R., Hensley, S., et al., 2007. The Shuttle Radar Topography Mission. Rev. Geophys. 45, 1–33.

Gosse, J.C., 2005. The contribution of cosmogenic nuclides to unraveling alpine paleoclimate histories. In: Huber, U.M., Bugmann, H.K.M., Reasoner, M.A. (Eds.), Global Change and Mountain Regions—An Overview of Current Knowledge. Advances in Global Change Research, 23. Springer, Dordrecht, pp. 39–49.

Hartshorn, K., Hovius, N., Dade, B.W., Slingerland, R.L., 2002. Climate-driven bedrock incision in an active mountain belt. Science 297, 2036–2038.

Hebenstreit, R., 2006. Present and former equilibrium line altitudes in the Taiwanese high mountain range. Quatern. Int. 147, 70–75.

Hebenstreit, R., Böse, M., 2003. Geomorphological evidence for a Late Pleistocene glaciation in the high mountains of Taiwan dated with age estimates by optically stimulated luminescence (OSL). Z. Geomorphol. N. F. Suppl. Bd. 130, 31–49.

Hebenstreit, R., Böse, M., Murray, A., 2006. Late Pleistocene and early Holocene glaciations in Taiwanese mountains. Quatern. Int. 147 (1), 76–88.

Hebenstreit, R., Ivy-Ochs, S., Kubik, P.W., Schlüchter, C., Böse, M., 2011. Lateglacial and early Holocene surface exposure ages of glacial boulders in the Taiwanese high mountain range. Quatern. Sci. Rev. 30 (3–4), 298–311.

Hope, G.S., 2004. Glaciation of Malaysia and Indonesia, excluding New Guinea. In: Ehlers, J., Gibbard, P.L. (Eds.), Quaternary Glaciations—Extent and Chronology, Part III: South America, Asia, Africa, Australasia. Developments in Quaternary Science. Elsevier, Amsterdam, pp. 211–214.

Huang, C.-Y., Liew, P.-M., Meixun, Z., Chang, T.-C., Kuo, C.-M., Chen, M.-T., et al., 1997. Deep sea and lake records of the Southeast Asian paleomonsoons for the last 25 thousand years. Earth Planet. Sci. Lett. 146 (1–2), 59–72.

Jian, Z., Wang, P., Saito, Y., Wang, J., Pflaumann, U., Oba, T., et al., 2000. Holocene variability of the Kuroshio Current in the Okinawa Trough, northwestern Pacific Ocean. Earth Planet. Sci. Lett. 184 (1), 305–319.

Kano, T., 1932. Preliminary notes on the morphology of the high mountain lands of Formosa. Geogr. Rev. Jpn 8 (3 & 6), 196–202, pp. 505–520.

Kerschner, H., Heuberger, H., 1997. New results on the Pleistocene Glaciation of the Japanese alps (Honshu) and the "Hettner Stein" problem—a preliminary report. Z. Gletscherk. Glazialgeol. 33 (1), 1–14.

Kiefer, T., Kienast, M., 2005. Patterns of deglacial warming in the Pacific Ocean: a review with emphasis on the time interval of Heinrich event 1. Quatern. Sci. Rev. 24 (7–9), 1063–1081.

Klose, C., 2006. Climate and geomorphology in the uppermost geomorphic belts of the Central Mountain Range, Taiwan. Quatern. Int. 147 (1), 89–102.

Koopmans, B.N., Stauffer, P.H., 1967. Glacial phenomena on Mount Kinabulu, Sabah. Malays. Geol. Surv. (Borneo Region) Bull. 8, 25–35.

Liew, P.-M., Huang, S.-Y., Kuo, C.-M., 2006a. Pollen stratigraphy, vegetation and environment of the last glacial and Holocene—a record from Toushe Basin, central Taiwan. Quatern. Int. 147 (1), 16–33.

Liew, P.M., Lee, C.Y., Kuo, C.M., 2006b. Holocene thermal optimal and climate variability of East Asian monsoon inferred from forest reconstruction of a subalpine pollen sequence, Taiwan. Earth Planet. Sci. Lett. 250 (3–4), 596–605.

Liu, J.P., Milliman, J.D., Gao, S., Cheng, P., 2004. Holocene development of the Yellow River's subaqueous delta, North Yellow Sea. Mar. Geol. 209 (1–4), 45–67.

Löffler, E., 1972. Pleistocene Glaciation in Papua and New Guinea. Z. Geomorphol. N. F. Suppl. Bd. 13, 32–58.

Löffler, E., 1980. Neuester Stand der Quartärforschung in Neuguinea. Eiszeit. Ggw 30, 109–123.

Louis, H., 1955. Schneegrenze und Schneegrenzbestimmung. Geogr. Taschenb. 1954/55, 414–418.

Lundberg, N., Dorsey, R.J., 1990. Rapid Quaternary emergence, uplift, and denudation of the Coastal Range, eastern Taiwan. Geology 18 (7), 638–641.

Ohmura, A., Kasser, P., Funk, M., 1992. Climate at the equilibrium line of glaciers. J. Glaciol. 38 (130), 397–411.

Ono, Y., 1991. Glacial and periglacial paleoenvironments in the Japanese Islands. Quatern. Res. (Daiyonki Kenkyu) 30 (2), 203–211.

Ono, Y., Naruse, T., 1997. Snowline elevation and eolian dust flux in the Japanese Islands during Isotope Stages 2 and 4. Quatern. Int. 37, 45–54.

Ono, Y., Shiraiwa, T., Dali, L., 2003. Present and last-glacial Equilibrium Line Altitudes (ELAs) in the Japanese high mountains. Z. Geomorphol. N. F. Suppl. Bd. 130, 217–236.

Ono, Y., Aoki, T., Hasegawa, H., Dali, L., 2005. Mountain glaciation in Japan and Taiwan at the global Last Glacial Maximum. Quatern. Int. 138–139, 79–92.

Panzer, W., 1935. Eiszeitspuren auf Formosa. Z. Gletscherk. 23 (1–3), 81–91.

Porter, S.C., 2001a. Chinese loess record of monsoon climate during the last glacial-interglacial cycle. Earth Sci. Rev. 54 (1–3), 115–128.

Porter, S.C., 2001b. Snowline depression in the tropics during the Last Glaciation. Quatern. Sci. Rev. 20 (10), 1067–1091.

Prentice, M.L., Hope, G.S., Maryunani, K., Peterson, J.A., 2005. An evaluation of snowline data across New Guinea during the last major glaciation, and area-based glacier snowlines in the Mt. Jaya region of Papua, Indonesia, during the Last Glacial Maximum. Quatern. Int. 138–139, 93–117.

Rasmussen, S.O., Andersen, K.K., Svensson, A.M., Steffensen, J.P., Vinther, B.M., Clausen, H.B., et al., 2006. A new Greenland ice core chronology for the last glacial termination. J. Geophys. Res. 111 (D06102), 1–16.

Rasmussen, S.O., Vinther, B.M., Clausen, H.B., Andersen, K.K., 2007. Early Holocene climate oscillations recorded in three Greenland ice cores. Quatern. Sci. Rev. 26 (15–16), 1907–1914.

Shi, Y., 1992. Glaciers and glacial geomorphology in China. Z. Geomorphol. N. F. Suppl. Bd. 86, 51–63.

Shi, Y., 2002. Characteristics of late Quaternary monsoonal glaciation on the Tibetan Plateau and in East Asia. Quatern. Int. 97–98, 79–91.

Siame, L., Chu, H.T., Carcaillet, J., Bourles, D., Braucher, R., Lu, W.C., et al., 2007. Glacial retreat history of Nanhuta Shan (north-east Taiwan) from preserved glacial features: the cosmic ray exposure perspective. Quatern. Sci. Rev. 26 (17–18), 2185–2200.

Tanaka, K., Kano, T., 1934. Glacial topographies in the Nankotaizan Mountain group in Taiwan. Geogr. Rev. Jpn 10 (3), 169–190.

Thunell, R.C., Miao, Q., 1996. Sea surface temperature of the Western Equatorial Pacific Ocean during the Younger Dryas. Quatern. Res. 46 (1), 72–77.

Vinther, B.M., Clausen, H.B., Johnsen, S.J., Rasmussen, S.O., Andersen, K.K., Buchardt, S.L., et al., 2006. A synchronized dating of three Greenland ice cores throughout the Holocene. J. Geophys. Res. 111 (D13102), 1–11.

Wei, K.-Y., 2002. Environmental changes during the late Quaternary in Taiwan and adjacent seas: an overview of recent results of the past decade (1990–2000). West. Pac. Earth Sci. 2 (2), 149–160.

Xiang, R., Sun, Y., Li, T., Oppo, D.W., Chen, M., Zheng, F., 2007. Paleoenvironmental change in the middle Okinawa Trough since the last deglaciation: evidence from the sedimentation rate and planktonic foraminiferal record. Palaeogeogr. Palaeoclimatol. Palaeoecol. 243 (3–4), 378–393.

Yang, C.F., 2000. The study of glacial relicts in Shesan Peak during the Last Glaciation. PhD thesis, National Taiwan University, Taipei.

Zhou, W., Head, M.J., Deng, L., 2001. Climate changes in northern China since the late Pleistocene and its response to global change. Quatern. Int. 83–85, 285–292.

Late Quaternary Glaciations in Japan

Takanobu Sawagaki[1,*] and Tatsuto Aoki[2]
[1] Faculty of Environmental Earth Science, Hokkaido University, Sapporo, Japan
[2] School of Regional Development Studies, Kanazawa University, Ishikawa, Japan
*Correspondence and requests for materials should be addressed to Takanobu Sawagaki. E-mail: sawagaki@ees.hokudai.ac.jp

72.1. INTRODUCTION

Today no glaciers are found in Japan. Despite this, late Quaternary glacial landforms and deposits are preserved in the high mountain ranges in the central to northern part of Japan. Although the total distribution of glaciated mountains in Japan is still debated, the spatial extent of the glaciated areas is estimated at nearly 800 km^2 (Iozawa, 1966, 1979; Yonekura et al., 2001). The latitudinal distribution of these glaciated areas ranges from 35°N to 45°N, at altitudes decreasing from 2500 m a.s.l. in the south to 1000 m a.s.l. in the north.

A general summary of Quaternary glaciations in the Japanese high mountains and results of the studies up to 1980 were given by Ono (1980). Since then, additional geomorphological evidence has been obtained from the Japanese Alps on Honshu Island and the Hidaka Mountains on Hokkaido Island. These studies have been summarised by Aoki (2002), Aoki and Hasegawa (2003) and Sawagaki et al. (2003). Thus, this chapter will review the recent knowledge of the Japanese Alps and the Hidaka Mountains for the Quaternary glaciations in Japan.

The Japanese Alps include high mountain ranges on the main island of Japan (Honshu), composed of three north–south-trending ranges, the summits of which range between 2900 and 3200 m a.s.l. These three ranges are officially called the Hida, Kiso and Akaishi Mountains and are commonly referred to as the Northern, Central and Southern Japanese Alps, respectively (Fig. 72.1). Quaternary glaciation was first recognised in Japan in the Northern Japanese Alps in the 1900s (Yamazaki, 1902). Since then, a number of glacial geomorphological studies have been conducted in this area.

The Hidaka Mountain Range is situated in the northernmost island of Japan, extending in an NNE–SSE direction from almost 42°N to 43°N at heights of some 1500–2000 m a.s.l. This is a key area for the Quaternary glaciation in Japan where the glaciated features were first noted in this area from the 1950s onwards (Iozawa, 1966).

In the 1960s, Iozawa conducted a systematic interpretation of aerial photographs and field research from which he presented several maps showing the distribution of glacial landforms in the country (Iozawa, 1966). Since then, these maps have been used as valuable references for glaciation studies and have been repeatedly modified by a number of new observations and interpretations from subsequent investigations. Consequently, the authors have attempted to incorporate the most recent results into Iozawa's maps in the digital maps in this volume.

Geomorphological evidence indicates that small mountain glaciers formed glacial troughs and cirques with moraines and till. These features are mainly concentrated on north-east- to east-facing slopes because of drifting snow supplied by westerly winds. Ono (1980, 1981, 1982, 1984, 1988) pointed out that the orientation of former glaciers was controlled by two decisive factors. The first factor was the morphological alignment of the high mountains, reflecting the intensive tectonic activity of the eastern margin of the Pacific Plate island arc. The second factor is the strong winter monsoon which supplied snow to the mountains from the west. Low insolation on shaded north- and east-facing slopes is also likely to have been an important factor.

Glacial landforms in the two mountain ranges are generally classified into two groups: those of younger forms and those of older forms. The younger forms show well-preserved fresh features and are always confined to the areas most favourable for snow accumulation. The older landforms have been dissected and extend into lower altitudes than the younger ones. Early studies in the 1950s and 1960s considered generally that the Quaternary glaciation in Japan could be subdivided into two stages, correlated with the Rissian and Würmian cold stages in European Alps. However, by the 1970s, these two events had been reinterpreted as being more probably correlated with the early and late stadials of the last cold stage (Weichselian, Wisconsinan, Valdaian), respectively.

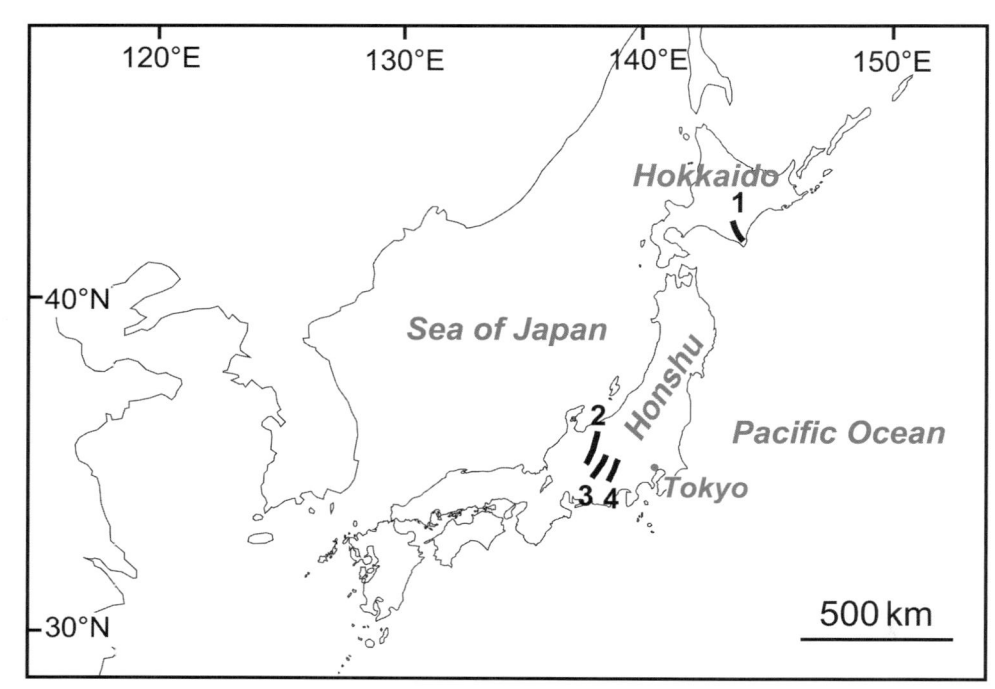

FIGURE 72.1 Japanese high mountain ranges. 1, Hidaka Mountain Range; 2, Hida Mountain Range (Northern Japanese Alps); 3, Kiso Mountain Range (Central Japanese Alps); and 4, Akaishi Mountain Range (Southern Japanese Alps).

Despite early attempts to correlate the Japanese glacial record with glacial chronostratigraphical frameworks elsewhere in the world, moraines and glacial deposits in Japan have been rarely studied. This is because of their limited distribution, poor preservation and dense vegetation cover. Even in these well-studied areas, it is still difficult to determine the maximum glacial limits owing to a lack of distinctive glacial terminal landforms. In many cases, these landforms are located on steep valley-side slopes, several tens to hundred metres above the present riverbeds, and where steep gorges disrupt the continuity of river terraces in their lower reaches. Further, some uncertainties still remain in distinguishing the true glacial landforms and sediments from other similar features, such as landslide or volcanic landforms commonly found in Japanese mountain areas. This latter issue remains the most controversial problem for the glacial studies in Japan.

While few direct datings of glacial sediments using the ^{14}C method have been undertaken, many marker tephras are available for chronological studies, owing to the active volcanic eruptions during the Quaternary. Recently, new numerical dating techniques, such as cosmogenic radionuclide and optically stimulated luminescence, can be applied to date glacial deposits and eroded rock surfaces. These studies are revealing that the younger event corresponds to several glacial advances that occurred during marine isotope stage (MIS 2) and latter half of MIS 3. The older event is correlated with glacial advances during the period between the early MIS 3 and MIS 5a. It is notable that the glacial expansion in the early part of the last cold stage was more extensive than during the latter event.

Some sediments from the older glacial event are buried by tephras that date from 90–120 to 95–130 ka, possibly indicating that they may represent the penultimate glaciation.

The recent glacial chronological results are summarised in Fig. 72.2. As indicated in this figure, the maximum glacial extent during the last cold stage in the Japanese Alps and the Hidaka Mountains is dated to MIS 4 and the middle of MIS 3, respectively. These periods were warmer than the coldest interval (MIS 2). It is thus notable that the glaciers in the Japanese high mountains were less extensive during the Last Glacial Maximum (LGM, 24–18 ka) and became smaller than those in the warmer period. This is clearly out-of-phase with the main continental ice sheets in the northern hemisphere. Consequently, the digital map does not indicate the glaciated areas at the LGM but the maximum glacial extent during an early stadial in the last cold stage.

72.2. MIDDLE PLEISTOCENE GLACIATIONS

72.2.1. Glacial Limits

The glacial advances older than the last cold stage are recognised in the four valleys in the Northern Japanese Alps. In the Kitamatairi valley, Mount Shirouma, Koaze et al. (1974) identified the lowest moraine at 1275 m a.s.l. Similarly, at the bottom of the Ohtsumetazawa valley, east of Mount Kashima-Yarigatake, about 13 km south from Mount

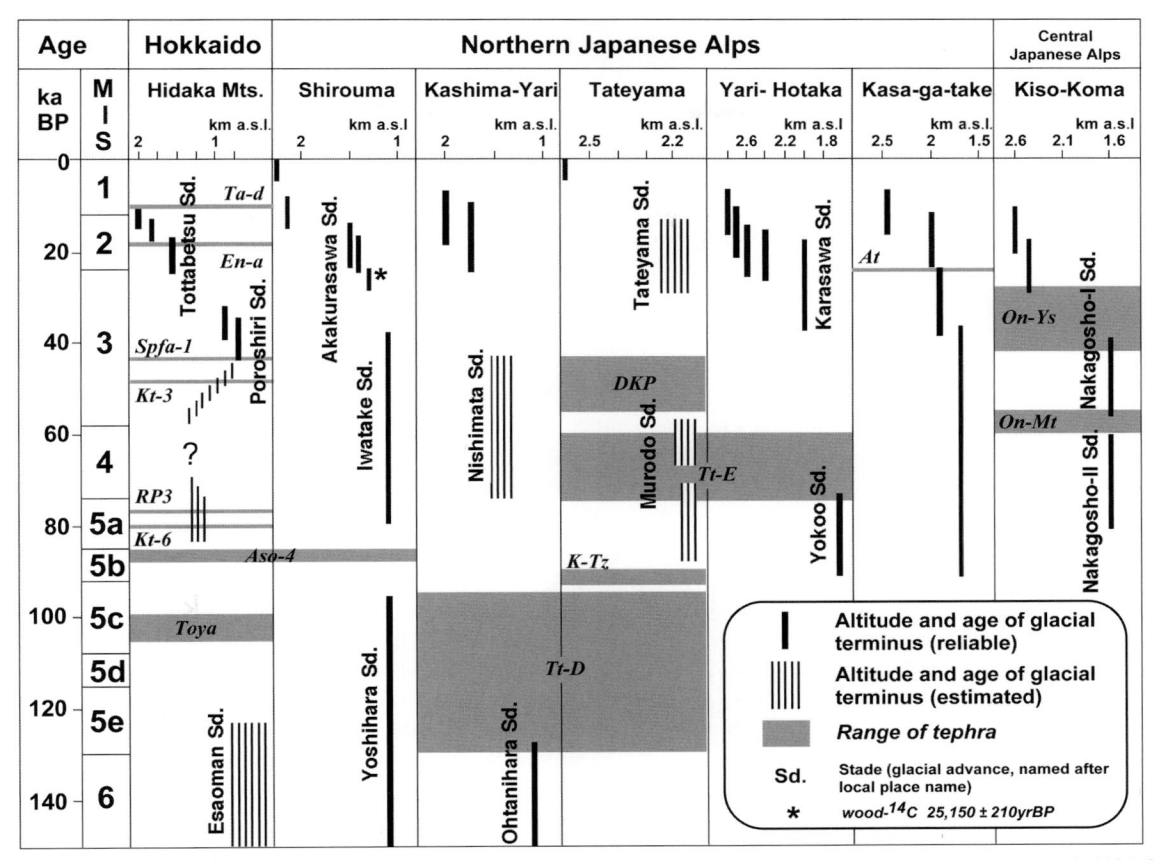

FIGURE 72.2 Spacing and timing of glacial advances in Japanese high mountain ranges. (modified from Iwata, 2001; Yonekura et al., 2001; Sawagaki et al., 2003).

Shirouma, the lowest terminal moraine occurs at 1150–1250 m a.s.l. (Ito and Shimizu, 1987). In addition, Hasegawa (1996a) suggested that the penultimate glaciation in the Hidarimata Valley of the Gamata river, east of Mount Kasagatake, reached to around 1500 m a.s.l.

The penultimate cold stage (MIS 6) in the northern Hidaka Mountains is identified by Iwasaki et al. (2000a,b) and is named the Esaoman Glaciaiton. The glacier advance during the Esaoman Glaciation possibly reached down to 750 m a.s.l., some 100 m lower than during the last cold stage.

72.2.2. Morphological Features

Glacial landforms during penultimate cold stage (Esaoman Glaciations; MIS 6) in the Northern Japanese Alps are represented by glaciated valleys and glacigenic and fluvioglacial deposits filling these valleys. The lowest terminal moraine in the Kitamatairi valley (Koaze et al., 1974; Ito and Shimizu, 1987) is covered by the Aso-4 volcanic ash (85–90 ka: Machida and Arai, 1992). In addition, the Tateyama-D pumice (hereafter, Tt-D, 100–130 ka: Machida and Arai, 1992) was discovered in the lowest part of the tephra layer that directly overlies the till of the lowest terminal moraine in the Ohtsumetazawa valley (Ito and Masaki, 1987).

In the Mount Chogatake region, Ito (1983) identified cirques, glaciated valleys and moraines which he also correlated with MIS 6. The lowest moraine occurs at around 1550 m a.s.l. in the Honzawa Valley, a branch of the Karasugawa River, east of the Mount Chogatake. However, some problems associated with this identification are discussed later in Section 72.2.4.

The Esaoman Glaciations are represented by buried glacial troughs and glacigenic sediments filling them (Iwasaki et al., 2000a). Three marker tephras, the Toya tephra (100–107 ka), the Kt-6 tephra (80 ka) and the RP-3 tephra (77 ka), are interstratified within the non-glacial debris flow sediments within local tributary valleys. These local sediments are overlain by deformation till and underlain by fluvioglacial sediments (or possibly meltout till) accompanying glaciotectonically deformed structures.

72.2.3. Key Sites

The key site for the penultimete glaciation in the Japanese Alps is eastern flank of the Mount Shurouma, where has the well-developed glacial landforms which dated by the marker tephra and carbon dating (Koaze et al., 1974; Ito and Shimizu, 1987; Kariya, 2000).

The critical sites for the late Quaternary glaciations in the Hidaka Mountains are the places where glacial landforms are well correlated with the key marker tephras, because tephras are particularly significant for the correlation and chronology in this area. The tephras Kt-6, a marker for MIS 5a, and the Toya, a marker for MIS 5e–d, are found in the Tottabetsu valley (Iwasaki et al., 2000a).

72.2.4. Problems

Within the drainage area of the Karasugawa River, located on the eastern side of Mount Chogatake, Ito (1983) identified the glacial landforms correlated to MIS 6, based on the stratigraphical succession of river terraces, identified as outwash deposits, and the Tt-D pumice overlying these deposits. However, Oguchi et al. (2003) suggested that Ito's correlation of these terraces was wrong because the sequence is confused by displacement due to tectonic activity. Moreover, Tomita et al. (2010) discussed the cirque-like hollow and moraine-like mounds identified by Ito (1983) as the product of large-scale slope failure rather than glaciers.

72.3. LATE PLEISTOCENE GLACIATIONS

72.3.1. Glacial Limits

The glacial landforms formed during the last cold stage in the Japanese Alps are distributed from the northern Japanese Alps to the southern Japanese Alps. As indicated in Iozawa (1979) and the digital map in this volume, small cirque floors occur at altitudes between 2400 and 2700 m a.s.l. and short U-shaped valleys extend to below 1000 m a.s.l. The glacial landforms in the Northern Japanese Alps are extensive, while the glaciated areas in the Central and Southern Japanese Alps are very restricted.

Early studies in the 1950s and 1960s pointed out that the formation of glacial landforms and glacial sediments in the Hidaka Mountains can be divided into two periods (Minato and Hashimoto, 1954a,b; Hashimoto and Kumano, 1955; Hashimoto and Minato, 1955; Minato and Ijiri, 1966). These periods were termed the Poroshiri Glacial Stage and the Tottabetsu Glacial Stage, thought to have corresponded to the Rissian and the Würmian cold stages in the European Alps, respectively. Later, Hashimoto et al. (1972) re-examined the glacial landforms and glacial sediments in the Hidaka Mountains and suggested that these were formed by several glaciations during the last cold stage. On the basis of this interpretation, they redefined the two glacial periods as the Poroshiri Stadial and the Tottabetsu Stadial, respectively. Since then, these stadials have been correlated with the Older and Younger Stadials of the Japanese Alps. Consequently, the Poroshiri Stadial is attributed to the largest glaciation during the last cold stage in this area, and the Tottabetsu Stadial to a smaller glacial readvance (Fig. 72.2).

Until the 1970s, it was believed that the growth of glaciers in the Hidaka Mountains was restricted to the vicinity of cirques. However, in the 1980s, Yanagida et al. (1982), Schlüchter et al. (1985a,b) and Shibano et al. (1988) found moraine ridges on valley floors, and these authors presumed that glaciers formed these moraines during the Poroshiri Stadial. They demonstrated that the glaciers in this area during the last cold stage were larger than was thought prior to 1970.

Schlüchter et al. (1985b) concluded that the maximum glacial expansion during the last cold stage in the Hidaka Mountains (the Poroshiri stadial) reached down to an altitude of 750 m a.s.l. However, since no description was given in their paper, their argument in support of this conclusion cannot confirmed. Instead, Iwasaki et al. (2000a) and Nakamura et al. (2000) dated the Older Stadial by finding the Spfa-1 tephra in a typical terminal moraine at 850 m a.s.l. on the valley floor in the northern Hidaka Mountains. This tephra is conformably overlain by its supraglacial till and indicates that the glacier reached its maximum just after the fall of Spfa-1 tephra, namely at 40–42 ka (MIS 3). Thus, Iwasaki et al. (2000a) contested the interpretation of Schlüchter et al. (1985b). Iwasaki et al. (2000a) also argued that this advance might belong to the preceding cold period of MIS 4, and the glaciation would be established around 85 ka, as described below.

The Kt-6 tephra (80 ka) is the marker for the glacier advance during the early part of the last cold stage. However, very little is known about the glacial fluctuations during this period. Kt-6 pumice grains, in the proglacial fluvioglacial sediments at just beyond the cirque floor, indicate that a glacier had already fluctuated in the vicinity of the cirque floor even during the relatively warm period of MIS 5a. This glaciation had possibly begun during the MIS 5b substage or earlier.

According to the palynological record from Hokkaido Island (Ooi et al., 1997), both marine isotope substages 5b and 5d were extremely cold and dry in the region, while the MIS 5c was warm, and saw the spread of lowland forest. Therefore, the glacial advance of MIS 5 probably occurred during marine isotope substage 5b. There is no evidence regarding whether the glacier had existed during the cold MIS 5d. Moreover, a glacier fluctuation towards the maximum expansion of MIS 3 (described above) is not at all evident. Consequently, as in the Japanese Alps, the maximum glacial extension in the Hidaka Mountains occurred before the LGM and probably during MIS 4 and these glaciations possibly persisted between MIS 5b and MIS 3.

The Younger Stadial in the Hidaka Mountains (the Tottabetsu Stadial) has been dated by the En-a tephra, which is found almost throughout fluvioglacial sediments at some localities close to cirque floors in the northern part of the mountains (Ono and Hirakawa, 1975b; Iwasaki et al., 2000a,b; Nakamura et al., 2000). This glacier advance probably lasted through the LGM and continued into the Late glacial. The glacier terminus lay at 1250 m a.s.l. during the most extensive stage, and the glacier advance was restricted to the cirque floor and its immediate surroundings (Iwasaki et al., 2000a,b). After the maximum glacial advance during the Tottabetsu Stadial, there were at least four short ice-equilibrium stillstands or glacier readvances represented by the series of terminal moraines on the cirque floors (Iwasaki et al., 2000b). These moraines can be correlated to the substadials of the Younger Stadial in the Japanese Alps.

The several lines of evidence presented above suggest that the glaciers in the Hidaka Mountains around the time of the LGM had a terminus somewhere on the glacial accumulation terrace surface and cirque floor and formed a fluvioglacial terrace in the valley. The glacier had probably retreated to the cirque floor from its maximum expansion around 40–42 ka (MIS 3). There is no direct geomorphological evidence concerning the extent or fluctuation of the maximum advancing (MIS 3) to LGM (MIS 2). Slope deposits, in which the En-a tephra separates two sheets of fluvioglacial sediments, probably indicate that the glacier had already retreated long before the deposition of the En-a tephra.

72.3.2. Morphological Features

Based on morphostratigraphical analysis and radiocarbon dating, four glacial advances during the last cold stage have been identified in the Northern Japanese Alps; Substadial 1 occurred around MIS-4, and three glacial advances (Substadials 2–4, from older to younger) occurred around MIS-2 (Koaze et al., 1974; Ito, 1982; Ito and Vorndran, 1983; Ito and Masaki; 1987; Hasegawa, 1992, 1999). In the Central and Southern Japanese Alps, however, only three glacial advances occurred during the last cold stage: Substadial 1, Substadials 2–3 (not clearly subdivided, Aoki, 2000a) and Substadial 4 (e.g. Kobayashi, 1966; Yanagimachi, 1983; Aoki, 1994, 2000a; Kanzawa and Hirakawa, 2000).

Based on the distribution of the glacial landforms, the glaciers in the Japanese Alps were most extensive during Substadial 1. The size of these glacial landforms decreases from Substadial 1 to Substadial 4. The distirbution and the size of the glacial landforms of the Substadial 4 are strictly controlled by the local-scale landforms (such as altitudinal distance from the ridge, slope aspect) and microclimate (such as drifting snow, radiation). Consequently, the equilibrium altitude of these small glaciers did not correspond to the regional climatic snowline altitude (Aoki, 2002; Aoki and Hasegawa, 2003).

The summit area of the Hidaka Mountains is characterised by a number of deglaciated cirques. Although these features suggest that glaciers had been distributed through the mountains, glacial landforms become insignificant towards the southern part of the mountains, because their altitude decreases towards the south. The cirques are distributed through the mountains mainly on the east side of the mountain ridge (Ono and Hirakawa, 1975a), and U-shaped valleys, indicating past glacier expansions, are seen in some eastwards-directed tributaries. Glacial and fluvioglacial sediments are also preserved on the valley bottom, especially in the northern part of the mountains (e.g. Shibano et al., 1988). In contrast, glacial and fluvioglacial sediments in the central and southern part of the Hidaka Mountains have not been studied in detail and their chronology has not yet been established.

72.3.3. Key Sites

72.3.3.1. Mount Shirouma, Northern Japanese Alps

Around the Mount Shirouma, located in the northernmost of the Northern Japanese Alps, the glacial landforms are the most expanded area in the Japanese Alps (Iozawa, 1966, 1974, 1979). Koaze et al. (1974) found the evidence for the Substadial 1 advance on Mount Shirouma. They showed that the U-shaped valleys and the lowest terminal moraines provide evidence for the Older Stadial, which is the period of maximum glacial expansion in the Japanese Alps during the last cold stage. The glacial landforms of this stadial are more weathered than those of the Younger Stadial (Substadials 2–4). Koaze et al. (1974) also obtained a radiocarbon date of 25 ^{14}C ka BP for the Substadial 2 terminal moraine on the eastern slope of Mount Shirouma.

72.3.3.2. Range Between Mount Yari and Mount Hotaka, Northern Japanese Alps

In the Yari-Hotaka Range of the southern part of the Northern Japanese Alps, Ito and Masaki (1989) argued that the maximum advance of the Substadial 1 glaciation occurred during MIS 4. Their argument was based on their discovery of the Tateyama-E pumice (hereafter, Tt-E, 60–75 ka: Machida and Arai, 1992), which fell during MIS 4 on the lowest terminal moraine of the Substadial 1 advance. Further evidence for a Substadial 1 glaciation was recognised on the Tateyama volcano of the Northern Japanese Alps based on till stratigraphy, lava flows and other volcanic deposits (Fukai, 1975). Substadial 1 glaciation occurred between the eruption that caused the Tt-E and Tamadono lava flow, which is stratigraphically underlain by the Daisen-Kurayoshi pumice (DKP, 45–55 ka: Machida and Arai, 1972). Recently, Kawasumi (2000) argued that the

Substadial 1 glaciation continued from MIS 5b into MIS 3, based on the succession of tills, Kikai-Tozurahara tephra (K-Tz, 90–95 ka: Machida and Arai, 1992), Tt-E pumice and Tamadono lava. However, a terminal moraine, produced during the Substadial 1 glaciation, has yet to be discovered and thus the maximum age of the glaciation has not been resolved.

In addition, Hasegawa (1992) found the Aira-Tanzawa pumice (AT, 24–25 ka: Machida and Arai, 1992) on Substadial 2 ground moraine on the north-western slope of Mount Kasagatake, near the Yari-Hotaka range. Moreover, Aoki (2003) used cosmogenic ^{10}Be dating to show that the Substadial 4 terminal moraines, existing north of Mount Kasagatake, date from 10 to 11 ka. This age corresponds to the Early Holocene and suggests that the glacial advance occurred well after the LGM in the Japanese Alps.

72.3.3.3. Mount Kiso-Komagatake and Mount Kumazawa, Central Japanese Alps

In the Central Japanese Alps, the marker tephra called the Ontake-Mitake scoria (On-Mt, 43–55 ka: Machida and Arai, 1992) was found on the Substadial 1 terminal moraine on the eastern slope of Mount Kiso-Komagatake and Mount Kumazawa (Kobayashi, 1966; Ono and Shimizu, 1982; Yanagimachi, 1983). Consequently, the maximum glacial extension in the Japanese Alps occurred before the LGM and could be correlated to MIS 4 (prior to 55–43 ka), and these glaciations possibly continued from MIS 5b to MIS 3.

The glacial landforms of the Younger Substadial (Substadials 2–4) are relatively well developed in the Northern Japanese Alps. The subdivision of the Younger Stadial in the Central Japanese Alps is also not clear in comparison to those in the Northern Japanese Alps. For example, Aoki (1994, 2000a) reconstructed the glacial fluctuation in the Central Japanese Alps using the relative dating method of measuring weathering-rind thicknesses (WRT) of the moraine clasts. He found that the time interval between Substadials 2 and 3 advances was too short to be detected with WRT methods. However, Substadials 3 and 4 can be clearly differentiated. Moreover, Aoki (1994, 2000a) suggested that Substadial 4 should be correlated with the last cold stage based on statistical analysis of the growth rate of the weathering rind. The WRT method shows that Substadial 3 terminal moraines occur extensively in the Central Japanese Alps (Aoki, 2000a). A part of these moraines has been dated to 16–18 ka by the cosmogenic ^{10}Be method (Aoki, 2000b). This shows that the Substadial 3 advance should represent the latter part of the LGM event.

72.3.3.4. Mount Senjo, Southern Japanese Alps

In the Southern Japanese Alps, Substadial 1 and two younger substadials were recognised in the Mount Senjo, the northern part of the Southern Japanese Alps (Kanzawa and Hirakawa, 2000), but their ages are unknown because of a lack of volcanic ashes and organic material for radiocarbon dating from the sediments. However, Kanzawa and Hirakawa (2000) suggest that the two younger stadials should be assigned to the LGM and the Late glacial, based on relative weathering criteria. From these results, it is likely that glacial advances were synchronous throughout the Japanese Alps. Substadial 1 occurred between MIS 5b and MIS 3 (probably during MIS 4). During the Younger Stadial, Substadial 2 occurred before 24 ka, Substadial 3 occurred between 20 and 18 ka and Substadial 4 between 11 and 10 ka.

72.3.3.5. Hidaka Mountains

The key sites for the late Quaternary glaciations in the Hidaka Mountains are the places where glacial landforms can be easily correlated with the key marker thephras, because tephras are particularly significant for the correlation and chronology in this area. In fact, seven marker tephras are found within the glacial- and non-glacial sediments in the northern part of the Hidaka Mountains (Kasugai et al., 1968; Ono and Hirakawa, 1974; Nakamura et al., 2000). Since the 1980s, these tephras have been well dated and assigned to MISs. Of these marker tephras, the Ta-d, En-a, Spfa-1, Kt-3, RP3, Kt-6 and Toya are dated to 9, 17–18, 40–42, 50, 77, 80 and 100–107 ka, respectively. This is achieved by ^{14}C dating, their relationship to marine terrace sequences and the ocean-core stratigraphy of the Japan Sea (e.g. Umezu, 1986; Katsui et al., 1988; Katoh, 1994; Yanagida, 1994; Katoh et al., 1995; Machida, 1995; Nakajima et al., 1996; Shirai et al., 1997).

For example, the En-a tephra (17–18 ka) allows the glacier extent and fluctuations around the so-called LGM (MIS 2) to be examined very precisely. Ono and Hirakawa (1975a,b) confirmed the interpretation of Hashimoto et al. (1972) by finding En-a tephra from outwash sediments at the bottom of the Nanatsu-numa cirque. The other tephras are also useful for determining the same issues. Especially, Spfa-1 tephra (40–42 ka) in a terminal moraine at 850 m a.s.l. on the valley floor of the Esaoman-Tottabetsu river indicates the maxmum glacier extent of the Older Stadial in this area (Iwasaki et al., 2000a; Nakamura et al., 2000).

72.3.4. Problems

Numerous workers have discussed the glaciation of the Hidaka Mountains, and some important conclusions have been described on the glacier fluctuations and their glaciological characters during the last cold stage. However, no firm history of glacial fluctuations has yet been provided. For example, Schlüchter et al. (1985b) and Shibano et al. (1988) presented different interpretations of the maximum glacial expansion during the Older Stadial. The main reason for their disagreement is the scarcity of field evidence of glacial sediments and dated materials. Recently, Liu and Ono (1997) and Liu et al. (1998a,b) made numerical reconstructions of flow conditions, equilibrium line altitudes (ELA) and the characteristics of mass balance of the glaciers during the last cold stage. Although they used a precise digital elevation model, their reconstruction of the maximum glacial expansion during the last cold stage is doubtful. It is unclear, whether landforms have been correctly interpreted because their study lacked proper field control. Thus, verification from field evidence is required to substantiate their results.

To overcome this problem, a series of chronological studies have been conducted during the past two decades, on the basis of intensive field survey on glacial landforms, sediments and some marker tephras (Iwasaki et al., 2000a,b; Nakamura et al., 2000; Sawagaki et al., 2003). These studies were concentrated on the two valleys in the northern part of the Hidaka Mountains. The two previously known stadials of the last cold stage, the Poroshiri and Tottabetsu, were reassessed, and three stages of glacier advances were reconstructed for the last cold stage (MIS 5a, MIS 3 and MIS 2). In addition, one older glaciation probably during the penultimate cold stage (MIS 6) was also identified.

It remains uncertain whether the chronological results, presented above, are applicable to the glaciated landforms in the central and southern part of the Hidaka Mountains. However, it has been confirmed that some cirques in the southern part of the mountains commonly have two significant step-like features at their bottom. This means that the younger and the older stadials during the last cold stage were possibly also recognisable in this region. Consequently, further investigation in the central and southern part of the mountains is expected to reveal the glacial history for the whole of the Hidaka Mountains.

72.4. HOLOCENE GLACIATION

There is also some evidence that suggests that glaciation occurred in parts of the Northern Japanese Alps during the Holocene, but overall, it is not clear whether significant glaciation took place during this period. For example, Koaze and Okazawa (1983) obtained a radiocarbon date of 7.8 ^{14}C ka BP from a deformed peat layer within a push moraine at a site near Mount Shirouma. In the Kuranosuke cirque on Mount Tateyama, there is a perennial ice mass called the Kuranosuke Snow Patch. Glaciological analysis of this ice mass suggests that it has an internal structure similar to that of a cirque glacier (Yamamoto and Yoshida, 1987). Radiocarbon dating of plant debris from a dirt layer within the ice mass indicates that the ice was formed at least 1000 years ago (Yoshida et al., 1990). These results show that the Kuranosuke Snow Patch is a fossil glacial ice mass.

Further, Hasegawa (1996b) found distinct cirque-like hollows inside the Substadial 4 glacial landforms around Mount Kasagatake and suggested that they had been formed by Holocene glaciation. If stronger evidence for such a glacial advance was found, this advance would be referred to Substadial 5.

REFERENCES

Aoki, T., 1994. Chronological study of glacial advances based on the Weathering-Rind thickness of moranic gravels in the northern part of the Central Japan Alps. Geogr. Rev. Jpn 64A, 601–618, (in Japanese with English abstract).

Aoki, T., 2000a. Late Quaternary equilibrium altitude in the Kiso Mountain Range, Central Japan. Geogr. Rev. Jpn 73B, 105–118.

Aoki, T., 2000b. Chronometry of the glacial deposits based on the ^{10}Be exposure age method—a case study in Senjojiki cirque and Nogaoki cirque, northern part of the Kiso mountain range. Quatern. Res. (Daiyonki-Kenkyu) 39, 189–198, (in Japanese with English abstract).

Aoki, T., 2002. Classification and characteristics of the distribution of glacial landforms in the Japanese Alps. J. Geogr. (Chigaku Zassi) 111, 498–508, (in Japanese with English abstract).

Aoki, T., 2003. Younger Drias glacial advances in Japan datad with in situ produced cosmogenic radionuclides. Trans. Jpn. Geomorph. Union 24, 27–39.

Aoki, T., Hasegawa, H., 2003. Late Quaternary glaciations in the Japanese Alps: controlled by sea level changes, monsoon oscillations and topography. Zeit. für Geomorphologie, Suppl. Bd. 130, 195–215.

Fukai, S., 1975. Formation and age of glacial landforms in the northern Japanese Alps. In: Shiki, M. (Ed.), Problems on the Glacial Period in Japan. Kokon Shoin, Tokyo, pp. 1–14.

Hasegawa, H., 1992. Glacial geomorphology and late Pleistocene glacial fluctuations in the Uchikomi valley, Northern Japanese Alps. Geogr. Rev. Jpn 65A, 320–338, (in Japanese with English abstract).

Hasegawa, H., 1996a. Glacial and periglacial landforms around Mount Kasaga-take, Northern Japanese Alps. Geogr. Rev. Jpn 69A, 75–101, (in Japanese with English abstract).

Hasegawa, H., 1996b. Late Quaternary glaciation and periglaciation in the southern part of the northern Japanese Alps. Unpublishued Ph.D. Dissertation, Meiji University, Tokyo (in Japanese).

Hasegawa, H., 1999. Glacial geomorphology in the Ichinomata-dani Valley, Northern Japanese Alps. Q. J. Geogr. 51, 114–124, (in Japanese with English abstract).

Hashimoto, S., Kumano, S., 1955. Zur Gletschertopo-graphie im Hidaka-Gebirge, Hokkaido, Japan. J. Geol. (Chishitsugaku Zassi) 61, 208–217, (in Japanese with German abstract).

Hashimoto, S., Minato, M., 1955. Quaternary geology of Hokkaido, 1st Report: on the ice-ages and post glacial age of the Hidaka mountain-range. J. Fac. Sci. Hokkaido Univ. IV (9), 7–20.

Hashimoto, W., Ono, Y., Taira, K., Makino, Y., Masuda, H., 1972. New knowledge of the geologic history of the Quaternary deposits in the Hidaka Range and the western part of the Tokachi Plain. Prof. Junichi Iwai Memorial Volume, Tohoku University, Sendai, Japan, 259–275.

Iozawa, T., 1966. Glacial landforms in Japan. Geography (Chiri) 11 (3), 24–30, (in Japanese).

Iozawa, T., 1974. Aero-observation on glacial landforms. Geography (Chiri) 19 (2), 38–50, (in Japanese).

Iozawa, T., 1979. Bird's Eye View of the Japanese Alps. Kodansha, Tokyo, 190 pp. (in Japanese).

Ito, M., 1982. Glacial geomorphology around the Migimata, Gamata Valley, Northern Japanese Alps. J. Geogr. (Chigaku Zassi) 91, 20–103, (in Japanese with English abstract).

Ito, M., 1983. Glacial landforms and accumulation terraces around Mount Chogatake, the Northern Japanese Alps, Central Japan. Geogr. Rev. Jpn 56, 35–49, (in Japanese with English abstract).

Ito, M., Masaki, T., 1987. Glacial landforms and the height of Pleistocene orographic snowline around Mount Harinoki and Mount Renge, Northern Japanese Alps. Ann. Tohoku Geogr. Assoc. 39, 247–267 (in Japanese with English).

Ito, M., Masaki, T., 1989. Chronological significance of the late Pleistocene maker tephra on the lowest terminal moraine in the Yari-Hotaka Range, Northern Japanese Alps. Geogr. Rev. Jpn 62A, 438–447, (in Japanese with English abstract).

Ito, M., Shimizu, F., 1987. Chronological significance of the discovery of the Late Pleistocene marker tephra on the terminal moraine in the Matsukawa Kitamata Valley, the eastern part of Mount Shirouma, northern Japanese Alps. J. Geogr. (Chigaku Zassi) 96, 11.

Ito, M., Vorndran, G., 1983. Glacial geomorphology and snowlines of younger Quaternary around the Yari-Hotaka mountain range, Northern Alps, central Japan. Polarforschung 53, 75–89.

Iwasaki, S., Hirakawa, K., Sawagaki, T., 2000a. Late Quaternary glaciation in the Esaoman-Tottabetsu Valley, Hidaka Range, Hokkaido, Japan. J. Geogr. (Chigaku Zassi) 109, 37–55, (in Japanese with English abstract).

Iwasaki, S., Hirakawa, K., Sawagaki, T., 2000b. Late Quaternary glaciation and its chronology in the Totta-betsu Valley, Hidaka Range. Geogr. Rev. Jpn 73A, 498–522, (in Japanese with English abstract).

Iwata, S., 2001. Introduction to Japanese Alps. Alpine geomorphology in central Japan: glaciation and peri- glaciation of the Japanese Alps and Mount Fuji. In: Matsuoka, N. Ono, Y. (Eds.), A guidebook of post-conference field excursion B7/B8, Fifth International Conference on Geomorphology, Tokyo, Japan, pp. 6–14.

Kanzawa, K., Hirakawa, K., 2000. Glaciation and accumulation terracing in the Yabusawa Valley, Mount Senjogatake in the Southern Japanese Alps, since the Last Glacial Substade. Geogr. Rev. Jpn 73A, 124–136, (in Japanese with English abstract).

Kariya, Y., 2000. New radiocarbon ages from the Pleistocene –Holocane deposits in the Shirouma-dake district, central Honshu, Japan. In: Proceedings of the generan meeting of the Association of Japanese Geographers, 58 68–69, (in Japanese).

Kasugai, A., Kimura, H., Kosaka, T., Matsuzawa, I., Nokawa, K., 1968. On so-called "Obihiro volcanic sand" distributed in Tokachi Plane. Earth Sci. (Chikyu Kagaku) 22, 137–146 (in Japanese with English abstract).

Katoh, S., 1994. The age of the En-a pumice-fall deposit and the paleoclimate around its age. Geogr. Rev. Jpn 67A, 45–54, (in Japanese with English abstract).

Katoh, S., Yamagata, K., Okumura, K., 1995. AMS-14C dates of Late Quaternary tephra layers erupted from the Shikotsu and Kuttara Volcanoes. Quatern. Res. (Daiyonki-Kenkyu) 34, 309–313, (In Japanese with English abstract).

Katsui, Y., Yokoyama, I., Okada, H., Abiko, T., Mutoh, H., 1988. Kuttara: Hiyori-yama, Volcanic geology, active history, activity and disaster prevention measures of the Kuttara: Mount Hiyoriyama. Report of Study on Volcanos in Hokkaido, Comitee for Prevention of Disasters of Hokkaido, Hokkaido, 99 pp. (in Japanese).

Kawasumi, T., 2000. Ages of glaciations in the early substadial of the Late Pleistocene, based on glacial and volcanic deposits in Tateyama, Hida Mountains, Central Japan. Geogr. Rev. Jpn 73A, 26–43, (in Japanese with English abstract).

Koaze, T., Sugihara, S., Shimizu, F., Utsunomiya, Y., Iwata, S., Okazawa, S., 1974. Geomorphological studies of Mount Shirouma and its surroundings, Central Japan. Sundai Hist. Rev. 35, 1–186, (in Japanese with English abstract).

Koaze, T., Okazawa, S., 1983. Holocene push moraine in the Mt. Asahi, northern Ushiro-Tateyama Mountains. Program and abstracts of the annual meeting of Japanese Association for Quaternary Research, 13, 110–111.

Kobayashi, K., 1966. Significance of the Ikenotaira Interstadial indicated by moraines on the Mount Kumazawa of the Kiso Mountain Range, Central Japan. J. Fac. Liberal Arts Sci. Shinshu Univ. 1, 97–113.

Liu, D., Ono, Y., 1997. New method for reconstruction of glacial landforms and ELA (Equilibrium Line Altitudes) by using photogrammetric workstation: example of the Hidaka Range, Hokkaido. Trans. Jpn. Geomorph. Union 18, 365–387, (in Japanese with English abstract).

Liu, D., Ono, Y., Naruse, R., 1998a. A reconstruction of alpine glaciers during the Last Glacial Maximum in the Hidaka Range, central Hokkaido. J. Jpn Soc. Snow Ice (Seppyo) 60, 47–54, (in Japanese with English abstract).

Liu, D., Ono, Y., Naruse, R., 1998b. Characteristics of mass balance of reconstructed glaciers during the Last Glaciation in the Hidaka Range, Hokkaido. Trans. Jpn. Geomorph. Union 19, 91–106, (in Japanese with English abstract).

Machida, H., 1995. Frequency and periodicity of explosive volcanic activity, and climatic changes. In: Koizumi, I., Yasuda, Y. (Eds.), Civilization and Environment—The Cycle of the Earth and Civilization-, 1. Asakura Shoten, Tokyo, pp. 116–127.

Machida, H., Arai, F., 1972. Daisen_kurayoshi Pumice: stratigraphy, chronology, distribution and implication to Late Pleistocene events in Central Japan. J. Geogr. (Chigaku-Zassi) 88, 313–330, (in Japanese).

Machida, H., Arai, F., 1992. Atlas of Tephra in and Around Japan. University of Tokyo Press, Tokyo, 276 pp. (in Japanese).

Minato, M., Hashimoto, S., 1954a. Zur Karbildung im Hidaka-Gebirge, Hokkaido, Japan. Proc. Imperial Acad. 30, 106–108, (in Japanese).

Minato, M., Hashimoto, S., 1954b. Poroshiri glaciation, Tottabetsu glaciation and Poroshiri-Tottabetsu inter-glacial stage. J. Geogr. (Chigaku-zashi) 60, 460.

Minato, M., Ijiri, S., 1966. Japanese Archipelago, second ed. Iwanami-Shinsho, Tokyo, 221 pp. (in Japanese).

Nakajima, T., Kikkawa, K., Ikehara, K., Katayama, H., Kikawa, E., Joshima, M., et al., 1996. Marine sediments and late Quaternary stratigraphy in the southeastern part of the Japan Sea-concerning the timing of dark layer deposition. J. Geol. Soc. Jpn 102, 125–138, (in Japanese with English abstract).

Nakamura, Y., Hirakawa, K., Iwasaki, S., Sawagaki, T., 2000. Late Quaternary tephras in the Tokachi Plain and the Hidaka Range, with special reference to glacial sediments. Quatern. Res. (Daiyonki-Kenkyu) 39, 33–44, (in Japanese with English abstract).

Oguchi, T., Aoki, T., Matsuta, N., 2003. Identification of an active fault in the Japanese Alps from DEM-based hill shading. Comput. Geosci. 29, 885–891.

Ono, Y., 1980. Glacial and periglacial geomorphology in Japan. Prog. Phys. Geogr. 4, 149–160.

Ono, Y., 1981. Mass balance of the Pleistocene alpine glacier in the Japanese Alps. Annual Report of Institute of Geosciences, University of Tsukuba 7, pp. 35–38.

Ono, Y., 1982. Reconstruction of the amount of snowfall in the Last Glacial Age by the glacial landforms and the sea level fluctuations. Quatern. Res. (Daiyonki-Kenkyu) 21, 229–243, (in Japanese with English abstract).

Ono, Y., 1984. Last Glacial paleoclimate reconstruction from glacial and periglacial landforms in Japan. Geogr. Rev. Jpn 57B, 87–100.

Ono, Y., 1988. Last Glacial snowline altitude and paleoclimate of the Eastern Asia. Quatern. Res. (Daiyonki-Kenkyu) 26, 271–280, (in Japanese with English abstract).

Ono, Y., Hirakawa, K., 1974. The Pleistocene tephra in the western and southern part of Tokachi Plain. Quatern. Res. (Daiyonki-Kenkyu) 13, 35–47, (in Japanese with English abstract).

Ono, Y., Hirakawa, K., 1975a. Glacial and periglacial morphogenetic environments around the Hidaka Range in the Würm Glacial Age. Geogr. Rev. Jpn 48, 1–26, (in Japanese with English abstract).

Ono, Y., Hirakawa, K., 1975b. Geological significance of the discovery of the Eniwa-a Pumice-fall deposit in the Hidaka Range. J. Geol. Jpn 81, 333–334, (in Japanese).

Ono, Y., Shimizu, C., 1982. Glacial landforms around Mt. Kisokomagatake. Abstr. Assoc. Jpn. Geogr. (Chiri-Gakkai-Yokoushu) 21, 60–61, (in Japanese).

Ooi, N., Tsuji, S., Danhara, T., Noshiro, S., Ueda, Y., Minaki, M., 1997. Vegetation change during the early last Glacial in Haboro and Tomamae, northwestern Hokkaido, Japan. Rev. Palaeobot. Palynol. 97, 79–95.

Sawagaki, T., Iwasaki, S., Nakamura, Y., Hirakawa, K., 2003. Late Quaternary glaciation in the Hidaka Mountain range, Hokkaido, northernmost Japan: its chronology and deformation till. Zeit. für Geomorphologie Suppl. Bd. 130, 237–262.

Schlüchter, C., Heuberger, H., Horie, S., 1985a. Evidence for valley glaciation in Hidaka Mountains, Hokkaido, I. upper Satsunai Valley. Proc. Jpn Acad. 61B, 433–436.

Schlüchter, C., Kerschner, H., Horie, S., 1985b. Evidence for valley glaciation in Hidaka Mountains, Hokkaido, II. Poroshiri, Esaoman-Tottabetsu Area. Proc. Jpn Acad. 61B, 437–440.

Shibano, A., Sawaguchi, S., Koaze, T., 1988. Glacial landforms in the Hachinosawa river, Satsunai river source, Hidaka Range. Abstr. Assoc. Jpn. Geogr. 33, 104–105, (in Japanese).

Shirai, M., Tada, T., Fujioka, K., 1997. Identification and chronostratigraphy of middle to upper Quaternary marker tephras occurring in the Anden Coast based on comparison with ODP cores in the Sea of Japan. Quatern. Res. (Daiyonki-Kenkyu) 36, 183–196, (in Japanese with English abstract).

Tomita, K., Kariya, Y., Sato, G., 2010. Cirque-like and terminal moraine-like featuers caused by large-scale slope failure on the eastern side of Mount Chogatake, ther southern Hida Mountains, central Japan. Quatern. Res. (Daiyonki-Kenkyu) 49, 11–22.

Umezu, Y., 1986. 14C ages of the Eniwa-a pumice-fall deposit (En-a) and the Tarumae-d pumice fall deposit (Ta-d). Ann. Tohoku Geogr. Assoc. 39, 141–143, (in Japanese).

Yamamoto, K., Yoshida, M., 1987. Impulse radar sounding of fossil ice within the Kuranosuke perennial snow patch, central Japan. Ann. Glaciol. 9, 218–220.

Yamazaki, N., 1902. Glaciers, didn't they exist in Japan? J. Geol. Soc. Jpn 9, 361–369, 390–398, (in Japanese).

Yanagida, M., 1994. Age of the Shikotsu pumice fall-1 deposit. Quatern. Res. (Daiyonki-Kenkyu) 33, 205–207, (in Japanese).

Yanagida, M., Shimizu, C., Nakano, M., 1982. Glacial landforms of downstream side at the Mt. Poroshiri North cirque in the Hidaka Range. Graduate School of Geography Research, Komazawa University, Tokyo, Japan, 12, 15–25, (in Japanese).

Yanagimachi, O., 1983. Glacial fluctuations and chronology during the Last Glacial Age in the northern part of the Kiso Mountain Range, Central Japan. J. Geogr. (Chigaku Zassi) 92, 12–172, (in Japanese with English abstract).

Yonekura, N., Kaizuka, S., Nogami, M., Chinzei, K., 2001. Regional Geomorphology of the Japanese Islands vol. 1, Introduction to Japanese Geomorphology. University of Tokyo Press, Tokyo 349 pp. (in Japanese).

Yoshida, M., Yamamoto, K., Higuchi, K., Iida, H., Ohata, T., Nakamura, T., 1990. First discovery of fossil ice of 1000 1700 years BP in Japan. In: Higuchi, K. (Ed.), Study on Structures and Formations of the Oldest Fossil Ice Mass in Japan (Kuranosuke-sawa, Northern Japanese Alps), 9–14. Report of the Grants-in-Aid for Scientific Research from the Ministry of Education, Culture, Sports, Science and Technology, 63302016.

Chapter 73

The Glaciation of the South-East Asian Equatorial Region

M.L. Prentice[1], G.S. Hope[2,*], J.A. Peterson[3] and Timothy T. Barrows[4]

[1] Indiana Geological Survey, Indiana University, 611 North Walnut Grove, Bloomington, Indiana 47405-2208, USA
[2] Department of Archaeology and Natural History, CAP, Australian National University, Canberra 0200, Australia
[3] Centre for GIS School of Geography and Environmental Science, Monash University, Clayton 3800, Australia
[4] School of Geography, The University of Exeter, Exeter, Devon EX4 4RJ, United Kingdom
*Correspondence and requests for materials should be addressed to G.S. Hope. E-mail: geoffrey.hope@anu.edu.au

73.1. INTRODUCTION

This chapter discusses mountain glaciation in equatorial south-east Asia, an area of mostly continental islands within 10° of the equator. The mountainous islands extend from the west in Sumatra at 3°N and Borneo (Kalimantan) at 6°N for 5900 km to the east to New Guinea at 9°S and span three countries, Malaysia (Sabah), Indonesia and Papua New Guinea (PNG) (Fig. 73.1). In general terms, they occupy both sides of the major collision zone between the Asian and Indo-Australian plates, the latter moving northwards since the Late Mesozoic (Hall, 2001). The mountains thus formed are young and, in most cases, still actively rising, with widespread volcanism and igneous emplacement as well as uplifted Tertiary and Mesozoic marine sediments, ultramafics and metamorphics. The climates are wet and relatively aseasonal, as the zone largely lies within the Asian Intertropical Convergence Zone (ITCZ) which is very broad

FIGURE 73.1 Location map for western mountain areas Sumatra to Sulawesi.

in the western Pacific. Glaciers are only found in the western side of New Guinea in Papua province of Indonesia. There, the glaciers have retreated strongly since the early 1970s as the snowline has risen to about 4900 m a.s.l. (Peterson and Peterson, 1994; Quarles van Ufford and Sedgwick, 1998; Prentice and Hope, 2007; Prentice and Glidden, 2010).

The legacy of Quaternary glaciation in the region has been documented in several places, including earlier editions of Ehlers and Gibbard (2004) in which (Hope, 2004) reviewed the evidence for Indonesia and Malaysia outside New Guinea and Peterson et al. dealt with PNG. Papua Province (western New Guinea) was omitted from those reviews, but the Pleistocene snowlines for the whole of New Guinea were discussed by Prentice et al. (2005) and recent changes in ice extent by Hastenrath (2009), Porter (2001), Mark et al. (2005) and Prentice and Hope (2007). Previous workers have interpreted and mapped glacial landforms and past ice extent with the aid of air photos, but so far, a comprehensive overview based on orthorectified (i.e. scale-consistent) satellite images has been lacking. Here, we present the results of simple spatial queries to estimate likely palaeo-ice extents. Detailed contour maps exist for PNG but do not exist (or are unavailable) for the high peaks of Malaysia and Indonesia. The Shuttle Radar Topography Mission (SRTM) digital elevation model has provided the first reliable height data and has been used in concert with Google Earth. However, it should be noted that there seems to be a significant offset between SRTM altitudes and elevations determined by survey around Mount Jaya.

73.1.1. Mount Kinabalu and Western Indonesian Mountains

Table 73.1 lists mountains in Borneo, Sumatra, Java, Lombok and Sulawesi which are high enough to have experienced nival conditions during the Pleistocene. Of these eight mountains, only Mount Kinabalu in Borneo was definitely glaciated.

Mount Kinabalu, at 4101 m a.s.l., is the highest mountain between the subtropical southern Himalaya in northern Yunnan and the equatorial peaks of New Guinea. No other mountains in Malaysia approach its height. It lies near the northern tip of Borneo (Sabah), rising steeply from the coastal plain. It is an isolated hornblende adamellite and adamellite porphyry dome, thrust above the Miocene sandstones and shale of the Crocker Range, with associated ultramafic serpentinites to the south up to 3100 m (Lee and Choi, 1996). Several dates indicate that the granites cooled at 7 Ma and post-dated the collision of the South China Sea plate, the Sulu Plate and Eurasian plate which also created north-west Borneo (Cottam et al., 2010). Jacobson (1986) quotes a long-term tectonic uplift rate of 0.5 m/1000 years.

The granite area currently exposed is about 150 km^2 in extent with an area of 19.3 km^2 above 3000 m a.s.l. The summit crest consists of an eastern and western plateau at 3900 m with several isolated peaks rising 50–100 m above it. The plateaux are almost separated by Lows Gully, a very deep fault-controlled valley of the northwards-flowing Sungai Panataran. Very steep slopes or cliffs border the summit plateaux, limiting ice accumulation below 3800 m. The plateaux and peaks are remarkable in being mostly bare rock, having only minimal vegetation and soil (MacKinnon et al., 1996). Koopmans and Stauffer (1968) proposed that, at the Last Local Glacial Maximum (LLGM), ice covered 5.5 km^2 as a thin ice cap on the eastern (and highest) plateau and a thicker cover on the western plateau. Small valley glaciers developed and fed ice into Lows Gully, giving its eastern arm a 1000-m-deep U-shaped valley with a pile of rock debris, interpreted as a moraine, in the main valley at about 2900 m (Fig. 73.2).

TABLE 73.1 Mountains Exceeding 3400 m a.s.l. in Western Indonesia and Malaysia

Mountain	Summit Height (m)		Location	Latitude, Longitude	Area (ha) Above 3450 m	Area (ha) Above 2900 m
Kemiri–Bandahara-Leseur	3407–3381	U	Aceh, North Sumatra	3°45′N, 97°11′E 3°33′N, 97°28′E	0	6173
Kinabalu	4101	I	Sabah, Malaysia	6°05′N, 116°33′E	550	1930
Kerinci	3805	V	West Sumatra	1°42′S, 101°16′E	185	1335
Slamet	3428	V	Central Java	7°14′S, 109°14′E	0	220
Semuru	3676	V	East Java	8°06′S, 112°55′E	10	120
Rinjani	3726	V	Lombok Is	8°14′S, 116°28′E	125	1050
Rantemario	3440	U	South Sulawesi	3°23′S, 120°02′E	0	1500

Origin: U, ultramafic; I, intrusive; V, volcanic.

FIGURE 73.2 Glacial limits inferred by Koopmanns, re-drawn using a high-resolution DEM from Mike Cottam.

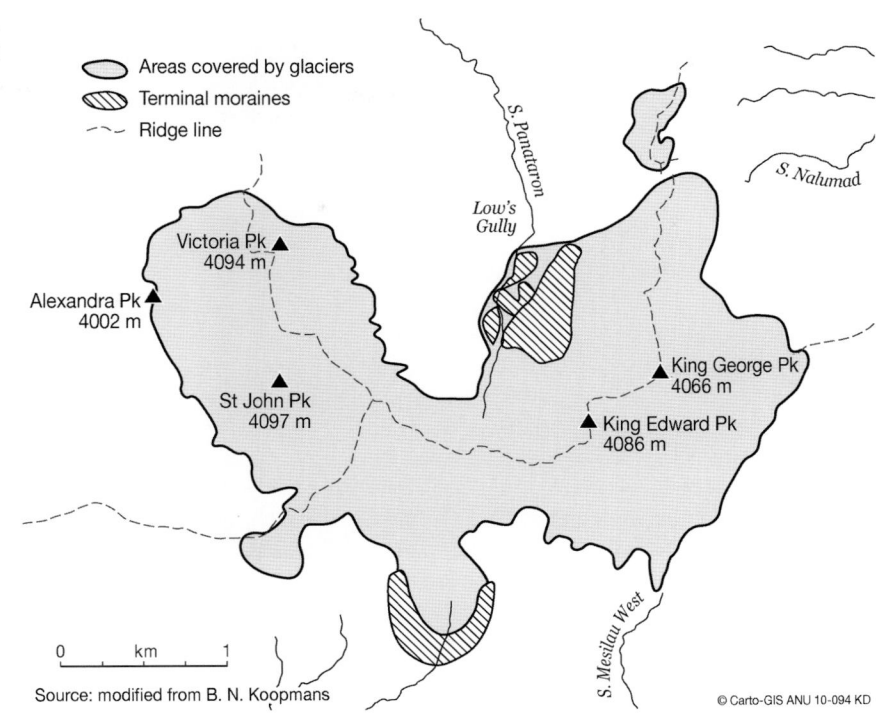

A possible arcuate terminal moraine of poorly sorted blocks occurs on the northern slope near Paka Cave at 3100 m a.s.l. The most prominent cirque has been identified on the eastern plateau between King George and North Peak and on the western faces of King Edward Peak and King George Peak (Smith and Lowry, 1968). Two steep-sided recesses at the base of Lows Peak and the lip of Lows Gully near Ugly Sister peak may be small cirques.

The very steep slopes below 3600 m prevent a geomorphological assessment of the position of the LLGM snowline or equilibrium-line altitude (ELA), but Koopmans and Stauffer (1968) estimated an ELA of 3660 m a.s.l. The stripping of the summit plateau suggests that the snowline lay below the plateau altitude. Smoothing and plucking of the summit peaks indicate that plateau ice was up to 80 m thick. Koopmans and Stauffer (1968) suggested that the northern and western slopes accumulated thicker ice, due to shading and daily cloud build-up. Although Lows Gully has been proposed as having been filled by over 1000 m of ice (Lee and Choi, 1996), it has not yet been shown that it carried a substantial ice stream during the LLGM (Fig. 73.2).

Several mountains above 3000 m a.s.l. occur elsewhere in western Indonesia (Table 73.1). North Sumatra has substantial areas of ultramafic and volcanic peaks and plateaux above 3000 m a.s.l., including Mount Leseur, Mount Bandahara and Mount Kemiri which are all 3400–3440 m a.s.l. in height. van Beek (1982) argues that an area of about 100 km^2 was glaciated in the Mount Leseur massif in north Sumatra (3°45′N) with a Pleistocene snowline at 3100 m. This is currently regarded as unlikely in that the regional snowline is unlikely to have been below 3450 m and was possibly as high as 3650 m a.s.l.

The other high mountains of Sumatra, Java and Lombok are active stratovolcanoes with 15 peaks above 3050 m a.s.l. The tallest volcano is Mount Kerinci at 3805 m a.s.l., in central Sumatra. Based on the SRTM, a total area of 2.4 km^2 lay above the possible Pleistocene snowline at 3450 m a.s.l. on Mounts Kerinci, Semuru and Rinjani. Although the volcanic mountains may have exceeded the Pleistocene snowline, their activity and the cinder cones from which they are made wiped out most traces of glaciation. They, and perhaps some slightly lower volcanoes, may also have gained or lost height during the Holocene. For example, the 2755 m a.s.l. Mount Tamboran, on Sumbawa Island, east of Lombok, experienced a huge eruption in April 1815, with an estimated eruptive loss of 150–180 km^3 (Zen and Ganie, 1992). This eruption removed the summit that is reported as having been around 4000 m in height, leaving a caldera and large crater lake. Mount Rantemario, 3440 m a.s.l., is the highest mountain of Sulawesi (Kilmaskossu and Hope, 1985), formed from ultramafics and sediments, and it lay below the maximum ice extent. A total of about 73.5 km^2 above 2900 m a.s.l. reflects the probable extent of alpine environments and former regular but ephemeral snow falls in western Indonesia.

FIGURE 73.3 Summit plateau with St. John Peak (4097 m). Photo SE Asia Research Group, Royal Holloway.

73.2. WESTERN NEW GUINEA (PAPUA, INDONESIA)

In contrast to the equivocal evidence from the west, clear glacial landforms are abundant across the main island of New Guinea. In Papua (also known as Irian Jaya), formerly glaciated summits are confined to the central ranges which extend from the Weyland Mountains in the west to the Star Mountains in the east (Fig. 73.4). An almost unbroken line of crests above 3500 m (the Snow Mountains or Maoke Range) extends from Mount Idenberg to Mount Mandala, and there are more isolated mountains on the northern flank of the range. A structure of Tertiary limestones folded over Mesozoic clastic sediments and tuffs occurs throughout and extends into PNG. Intrusions of granidiorite are present, for example, Grasberg 4200 m a.s.l. (now removed by mining) and Mount Antares 4170 m a.s.l. In Table 73.2, we break the Maoke Range into western, central and eastern blocks at lower points in the range.

Only limited surveys of glacial geomorphology have been undertaken in Papua, principally around Mount Jaya (Prentice et al., 2005) and two other high peaks, Mount Trikora (M. L. Prentice, unpublished) and Mount Mandala (Verstappen, 1952, 1964; Visser and Hermes, 1962). An estimate of 1400 km^2 for the total glaciated area given by Hope and Peterson (1975) was based on planimetry of the 3500 m contour of the 1:1,000,000 ONC chart on the assumption that the LLGM snowline in Papua matched the snowline in PNG. Reconstruction and analysis of palaeo-glacier morphology on Mount Jaya suggest a much higher ELA of ca. 3700–4050 m a.s.l. during the LLGM for this area and a probable lower ELA to the east of 3400–3600 m a.s.l. (Mark et al., 2005).

Figure 73.4 shows the 90 m SRTM for the area. Analysis of this DEM and new, higher, ELA values determined by Prentice et al. (2005) around Mount Jaya suggests that the glacierised area of Papua lies in the range of 2850–3350 km^2. This is more than twice the previous estimate. Substantial ice fields formed along the range crests with relatively steep, short glaciers up to 250 m thick emerging from ice domes (Prentice et al., 2005). On the Maoke Range (Snow Mountains), the general distribution of glaciers displays a north–south asymmetry due to the precipitous southern fall wall of the range over which ice-falls fed narrow valley glaciers in steep-sided valleys (e.g. see Charlesworth, 1957, p. 91) that may have reached uncharacteristically low altitudes (Hope and Peterson, 1975). To the north, glaciers extended out 4–6 km on to high plateaux such as the Kemabu and East Baliem. Ice streams follow fault-controlled valleys for up to 20 km to the west (Prentice, unpublished). Substantial areas above 2900 m altitude extend beyond the ice limits along the range. These would have been an alpine area of occasional snow falls at maximum temperature lowering.

Figure 73.5A is a sketch map of an area north of the highest peaks on Mount Jaya (Mount Carstensz). Figure 73.5B shows some of this ice in 1942 and moraines to the north.

The modern ice fields above 4650 m on Mount Jaya have rapidly declined from about 11 km^2 in 1942 to 2.4 km^2 by 2000, representing about an 80% decrease in ice area (Prentice and Hope, 2007). Meanwhile, other ice fields in Papua have disappeared, reflecting mainly a long-term warming trend (Hope et al., 1973; Klein and Kincaid, 2008).

The glaciation of the more isolated mountain areas mentioned in Table 73.2 is unmapped, as high-resolution images are not currently available. They will be important in indicating the limits of glaciation, but ELA may differ

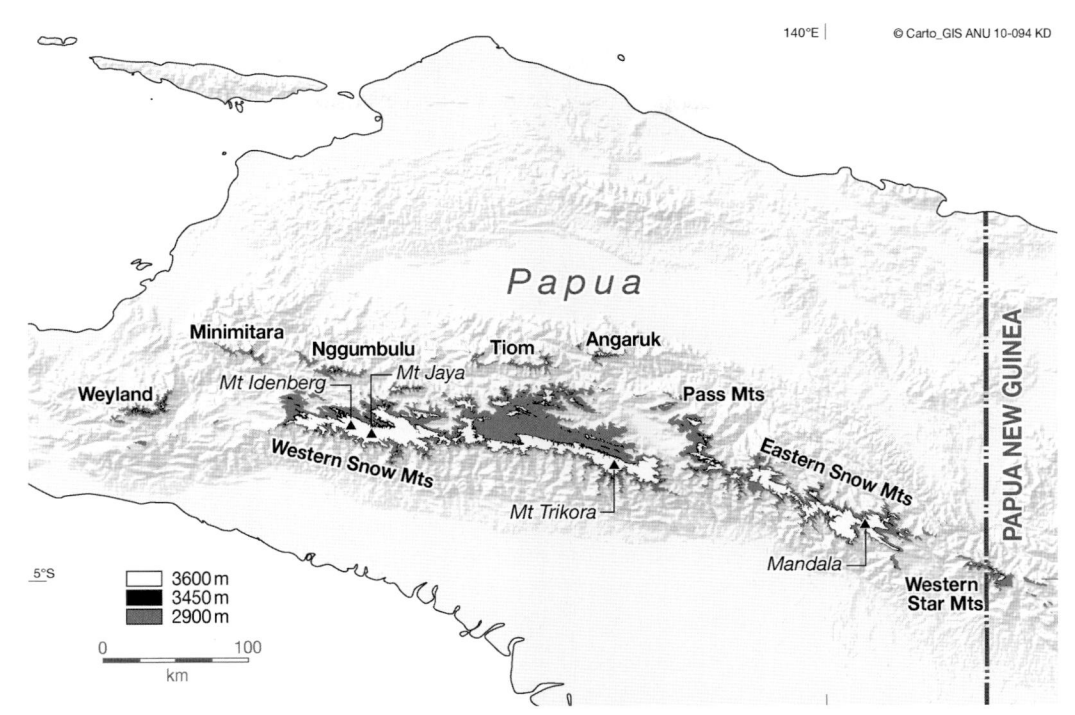

FIGURE 73.4 Maximum extent of nival and glacial environments in Papua. Digital terrain model of Snow Mountains and outliers above 2900, 3450 and 3600 m a.s.l.

from the main range due to greater reliance on precipitation from the north, rather than the Arafura Sea to the south.

73.3. EASTERN NEW GUINEA (PAPUA NEW GUINEA)

The central ranges continue into PNG but become less continuous and broader, with discrete high areas of former glaciation. Other isolated ranges in the north (Bismarck Ranges) and north-east (Finnesterre–Saruwaged) and south-east (Owen Stanley Ranges) are structurally separate (Löffler, 1982). The Owen Stanley Ranges extend south-east to Mount Suckling (3676 m a.s.l.) and experience more seasonal rainfall than the Central Range. Former glaciation was recognised early in PNG (e.g. Detzner, 1921; Reiner, 1960; Shepherd, 1966) and has been documented in several reviews (e.g. Bik, 1972; Löffler, 1972; Hope and Peterson, 1975; Brown, 1990; Peterson et al., 2001, 2004; Prentice et al., 2005). It has become clear that at least 20 mountains that were not too steep to retain snow cover, and were over 3500 m a.s.l. or so, carry evidence of former glaciation which has been assumed to reflect the LGM. The ice cap on Mount Giluwe was the most extensive area of glaciation (170–188 km^2) followed by Mount Wilhelm (~107 km^2; Löffler, 1972, 1982; Peterson et al., 2004). Löffler (1972) placed the LLGM ELA (snowline) at 3500–3550 m a.s.l. in central PNG, rising to between 3600 and 3700 m a.s.l. at higher latitude in the Saruwaged and Owen Stanley Ranges. The locations of the main glaciated areas are shown in Fig. 73.6. Other isolated high mountains are listed by Löffler (1972), who provides sketch maps of glacial features based on field and air photograph studies (Table 73.3).

Where broad accumulation areas existed, ice caps formed as on Mount Giluwe and Mount Albert Edward. These ice caps left mammilated surfaces dotted with tarns and moraines marking scattered tongues of ice extending out from the ice cap. Peterson et al. (2004) review the extent of the main ice areas using satellite images and estimate former ice areas that are slightly less than Löffler's (1972) estimates. Their mapping differs slightly from that of Löffler (1972) (Fig. 73.7).

Small mountain glaciers are typical of the steep peaks such as Mount Wilhelm. Valley glaciers extended from deep, completely glacierised valleys to form complex lateral and end moraines at 3100–3400 m a.s.l. (Fig. 73.8). These moraines commonly exhibit two to three individual crests. Retreat moraines within the valleys by contrast are minor features and support a general picture of steady retreat with few re-advances. As Prentice et al. (2005) point out, the existing chronology is based on dates from organics which provide only minimum-limiting ages for ice advances. Hence, the assumption that the largest moraines date to the LGM remains untested except for some preliminary cosmogenic ages noted below.

TABLE 73.2 Mountain Areas in Western New Guinea (Papua, Indonesia)

Mountain	Summit Height (m)		Location	Latitude, Longitude	Area (ha) Above 3600 m	Area (ha) Above 3450 m	Area (ha) Above 2900 m
Weyland	3860	S	South-east of Nabire	3°53.5′S, 135°44.86′E	246	3413	27,131
Minimitara	3750	S	North-east of Nabire	3°31.1′S, 136°24.5′E	4	181	10,024
Western Snow Mountains—Mount Jaya	4967	S, I	Paniai to Jila Pass	4°4.75′S, 137°9.7′E	108,247	151,073	292,580
Nggumbulu	3885	? U	North and north-west of Beoga	3°37.9′S, 136°55.3′E	1211	3308	29,444
Central Snow Mountains—Trikora	4760	S	Jila Pass to Baliem Gorge	4°15.8′S, 138°40.6′E	99,094	142,374	335,605
Tiom Mountains	3915	S	North of Tiom	3°46.4′S, 138°17.3′E	9056	25,143	141,589
Pass Mountains—Angaruk	3962	S	North-east of Wamena	3°58.6′S, 139°11.0′E	54	1483	19,595
Eastern Snow Mandala	4628	S	Baliem Gorge to Ok Sibil	4°42.5′S, 140°17.3′E	89,581	132,898	313,799
Western Star Mountains	3910	I	Ok Sibil to PNG border	4°53.7′S, 140°54.2′E	290	746	7308
Papua Total km²					3077.8	4606.2	11,770.8

U, ultramafic; I, Intrusive; S, sediments?, not known.

73.4. MIDDLE PLEISTOCENE GLACIATIONS

The only evidence for Mid-Pleistocene glaciation in New Guinea comes from Mount Giluwe, which was volcanically active over the past 1 million years or so until about 200 ka. The presence of palagonite on the mountain is taken to indicate that eruptions took place under ice (Blake and Löffler, 1971; Löffler et al., 1980). Palagonite near the east peak dates back to more than 650 ka and may signify the first permanent snow or ice on the mountain. More conclusive evidence of glaciation is represented by an inferred till that is interbedded between two lava flows dated at 290–320 ka (Löffler et al., 1980). Extensive breccia containing palagonite occurs on the north side of the mountain and is believed to represent an eruption under a thick cover of ice (Blake and Löffler, 1971). The breccia is dated at 285–300 ka, approximately the same time as the lava flows (Löffler et al., 1980). No moraines or other glacial landforms can be attributed to these early glaciations, not least because the mountain continued to be volcanically active. The oldest recognisable moraines occur on the south-west side of the mountain outside the limits of the last glaciation and are termed the Mengane glaciation (Barrows et al., in press). Some of the boulders on these moraines are highly weathered, and the enclosing valleys are incised. Exposure dating using ^{36}Cl reveals that these moraines are 130–160 ka in age.

73.4.1. Key Sites

Mount Giluwe is unusual in having been an active volcano through the Mid-Pleistocene. The dome-like structure has been very sensitive to retreat and an excellent site for ice accumulation. Volcanics at high altitudes have also been located on Mount Bangeta but have not been examined for palagonite. In Papua, the extensive high areas beyond Late Pleistocene glacial limits on the Kemabu Plateau and north of Mount Trikora also deserve more attention to assess the possibility of earlier glaciation. Hope et al. (1993) noted sequences of up to five outwash terraces along the east Baliem River north of Mount Trikora. The youngest terrace is less than 34 ka in age, so the older ones may extend back to Mid-Pleistocene times.

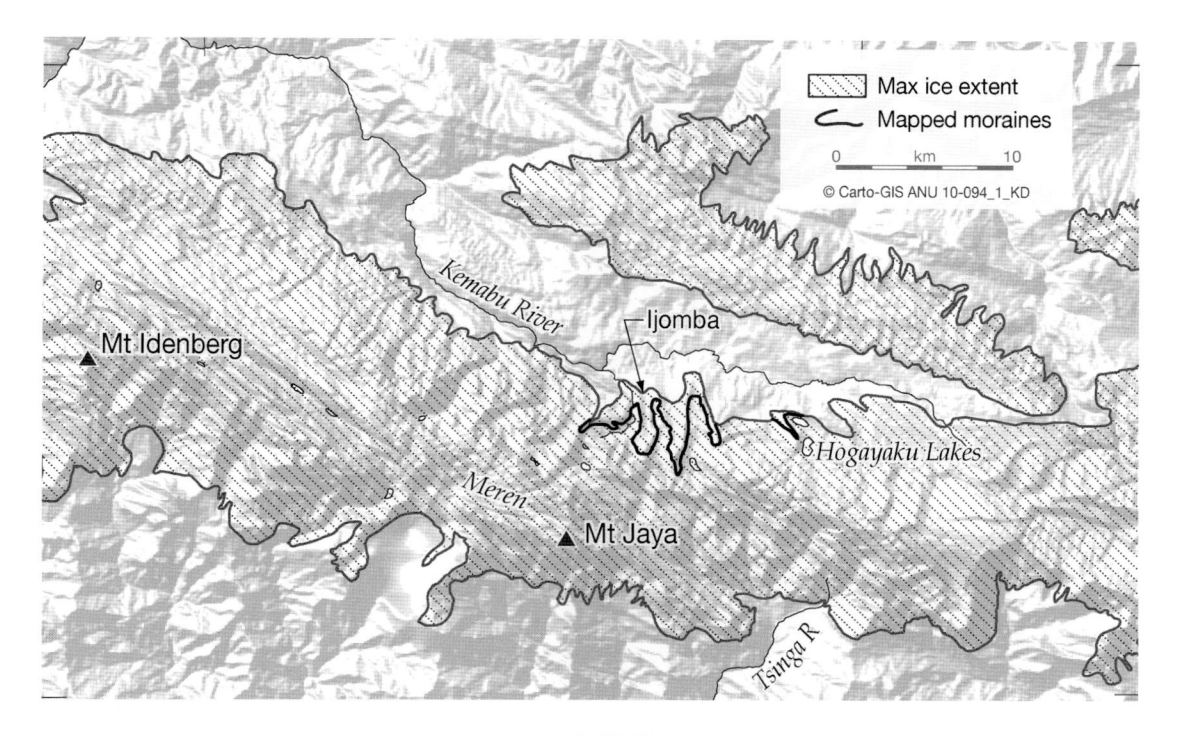

Chapter | 73 The Glaciation of the South-East Asian Equatorial Region

FIGURE 73.5 (A) Map of ice limits on Mount Jaya showing the limits of glaciers which developed down to terminii at 3650 m. (B) USAF 1942 trimetrogon air photo from Ngga Pulu ice cap to the Kemabu Plateau (the centre of (A)).

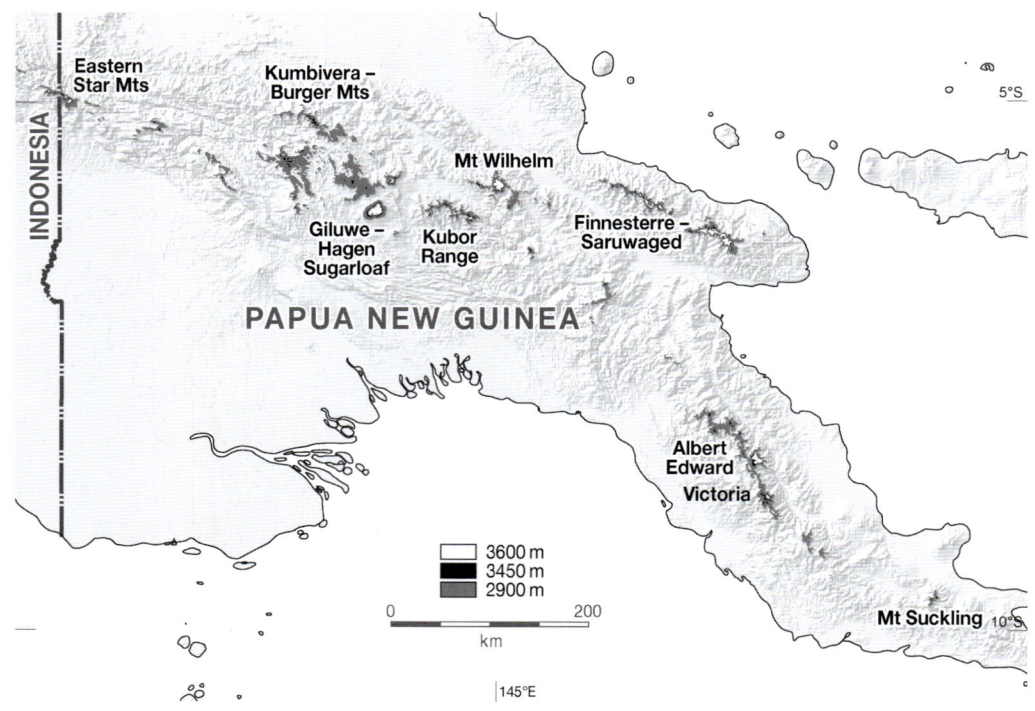

FIGURE 73.6 DEM of maximum nival and glacial areas in Papua New Guinea above 2900, 3450 and 3600 m a.s.l.

TABLE 73.3 Mountain Areas in Papua New Guinea

Mountain	Summit Height (m)		Location	Latitude, Longitude	Area (ha) Above 3600 m	Area (ha) Above 3450 m	Area (ha) Above 2900 m
Eastern Star Mountains	4210	S, I	Telefomin to PNG border	4°59.7'S, 141°5.0'E	1015	2953	21,307
Kumbivera	3750	S	South-east of Telefomin	5°11.5'S, 141°52.7'E	785	5581	179,435
Giluwe–Hagen Sugarloaf	4368	V	Wabag-Mount Hagen	6°2.6'S, 143°53.2'E	9594	18,091	132,003
Kubor Ra	4060	S	South of Wahgi Valley	6°3.6'S, 144°36.9'E	2019	6128	55,246
Wilhelm	4510	I	North of Kundiawa	5°46.8'S, 145°1.8'E	6479	9745	46,371
Finnesterre–Saruwaged	4040	S	47 km north of Lae	6°18.7'S, 147°5.4'E	12,856	24,249	94,475
Victoria–Albert Edward	4072	M, S	75 km north of Port Moresby	8°53.6'S, 147°32.0'E	6553	15,719	135,595
Mount Suckling	3676	U	East of Port Moresby	9°40.0'S 149°0.5'E	7	312	6099
*PNG Total km²					393.1	827.8	6705.3

U, ultramafic; I, intrusive; V, volcanic; S, sediments; M, metamorphic. Additional PNG Mountains mentioned by Löffler (1972): Burgers Mountains, Wamtakin, Sugarloaf, Udon.
*Includes minor areas.

FIGURE 73.7 North eastern plateau of Mount Giluwe. Large moraines, retreat termini and small parallel cross-valley moraines marking ice cap edges are visible.

FIGURE 73.8 Pindaunde Valley of Mount Wilhelm with deep cirque lakes and a prominent lateral moraine that has multiple crests near the terminus.

73.4.2. Problems

Evidence for Mid-Pleistocene glaciation has probably been overridden or removed by erosion on most mountains. Highly weathered possible tills and outwash sheets have been noted outside the existing limits of glaciation on the Kemabu Plateau and the East Baliem valleys in Papua (Hope and Peterson, 1976; Hope et al., 1993) but not in PNG or on Mount Kinabalu. Much more field work needs to be carried out. It has been suggested (Löffler, 1972) that some rapidly rising areas such as the Saruwaged Range may not have a long history of glaciation.

73.5. LATE PLEISTOCENE GLACIATIONS

For Mount Kinabalu, only one study has been carried out that provides a minimum age for deglaciation of the summit area. Flenley and Morley (1978) cored the Sacrificial Pool on the summit plateau and found 72 cm of organic-rich sandy clay. A near basal date of $10,393 \pm 130$ cal. yr BP (SRR-234) was obtained from 70 to 72 cm. The possible moraines mapped by Koopmans and Stauffer (1968) have not been dated. The bulk of mapping and dating has been focussed on large, well-preserved moraines and retreat basins at a range of altitudes across New Guinea. Prentice et al. (2005) list the available dates from the base of organic infills of basins within glaciated areas from across New Guinea. These lie in the range of 14,000–16,000 cal. yr BP.

From an earlier version of these data, Hope and Peterson (1975) inferred a progressive retreat from around 16,000 cal. yr BP yr and the disappearance of ice below 4500 m a.s.l. by ca. 9500 cal. yr BP. But Prentice et al. (2005) pointed out that the large, sharp-edged lateral and terminal moraines are often composite with three to five ridges involved, and the possibility of multiple ages needs to be directly determined. T.T. Barrows (unpublished data) has found more extensive weathering on boulders on the outer ridges than on the terminal moraines in the Pindaunde valley on Mount Wilhelm and is currently obtaining exposure ages for these.

73.5.1. Glacial Limits—Early Würmian and LGM

For most of the New Guinea mountains, only one phase of glaciation has been detected. ELA estimates are 4050 m a.s.l. for Mount Jaya and 3450–3600 m a.s.l. for the other mountains, possibly slightly higher in the Saruwaged Mountains. Terminal moraines typically occur at 3100–3500 m a.s.l. but a few extend to 2900 m a.s.l., reflecting large ice accumulation areas. There is now evidence from at least two widely separated areas, Mount Giluwe and Mount Trikora, that a more complex Late Pleistocene glacial history occurred.

On Mount Giluwe, small frontal moraines overlie large down-valley laterals indicating a later advance of relatively thick ice over an earlier glacial (Fig. 73.9). Through relative stratigraphy and exposure dating, Barrows et al. (in press) have subdivided the youngest moraines inside the Mengane glaciation limits into the Komia glaciation and the Tongo glaciation. The Komia glaciation has very limited extent and has mostly been overridden by the later Tongo glaciers. Exposure ages put this advance between 60 and 65 ka (Barrows et al., in press). The ensuing Tongo glaciation is far more extensive and well preserved. The maximum extent of ice is exposure dated at ~ 19 ka, and all ice was gone from the mountain by 11.5 ka (Barrows et al., in press). The exposure ages are corroborated by radiocarbon ages from the base of tarns and mires at several locations. Ages for retreat are 15,500–17,000 cal. yr BP in the main valleys at 3550 m a.s.l., while the highest site at 4170 m a.s.l. on the main summit is deglaciated by 11,250–11,900 cal. yr BP.

North of Mount Trikora, Papua, cross-cutting of older moraines clearly occurred, suggesting at least two Würmian Stage advances of similar extent (Fig. 73.11). Reconnaissance mapping in the Lake Purmoree Valley (Fig. 73.10) which channelled ice northwards from a crest at 4300 m a.s.l. indicates that there are several significant moraine systems. A major moraine system defines the valley downvalley from the lake with relief of > 100 m. We refer to this system as the Pur 14 sequence. Moraine sequence Pur 12 constitutes separate, significantly different up-valley moraines. The Pur 14 moraine sequence features large right and left lateral moraine walls with a nearly constant surface elevation of 3400 m a.s.l. The moraine walls stand 150 m in relief, have steep proximal slopes and are single crested. This moraine system is well preserved. Moraine sequence Pur 13 includes retreat moraines between Pur 14 and Pur 12 sequences.

In situ cosmogenic ^{10}Be and ^{26}Al exposure ages are available for 13 different boulders distributed on these moraines (Fink et al., 2003). Five boulders spread across the Pur 12 moraine sequence have post-LGM exposure ages. Boulders on moraine sequences Pur 14 and 13 exhibit a range of exposure ages that, on average, are older than those on Pur 12. Four exposure ages date to the LGM.

Based on this, we speculate that maximum ice extent, represented by the Pur 14 sequence, was achieved during the LGM. At this time, the ice front reached as low as 3100 m a.s.l. In valleys immediately to the east of Purmoree Valley, the moraine sequences are probably correlative to Pur 14 cross-cut older, large moraine systems. This suggests that, prior to the LGM, ice extent was as large or perhaps larger than during the LGM.

Sediment sequences from Mount Jaya (Prentice et al., 2005), Tari (Haberle, 1998), Sirunki (Walker and Flenley, 1979), Mount Wilhelm (Hope, 1976) and Mount Albert Edward (Hope, 2009) support the evidence for a

FIGURE 73.9 Glaciated area on Mount Giluwe, PNG.

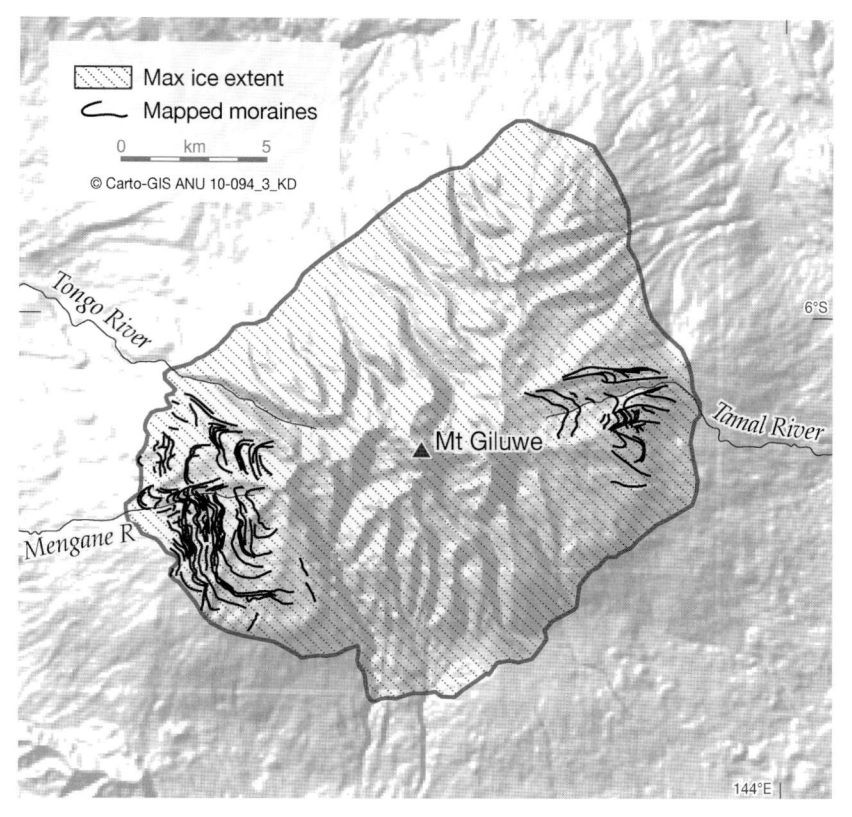

FIGURE 73.10 Ice extent near Mount Trikora showing mapping in the Purmoree Valley, also the site of surface exposure dating.

glacial cooling about 65 ka and two cool periods, one from ca. 30–25 ka and the other from 22–18 ka. A notably low treeline of ca. 2200 m marked the LGM and may reflect low CO_2 concentrations rather than low temperatures.

73.5.2. Key Sites

The areas north of Mountains Jaya and Trikora, where ice debouched onto high plateaux, provide very good opportunities for improving the detail of ice movement and

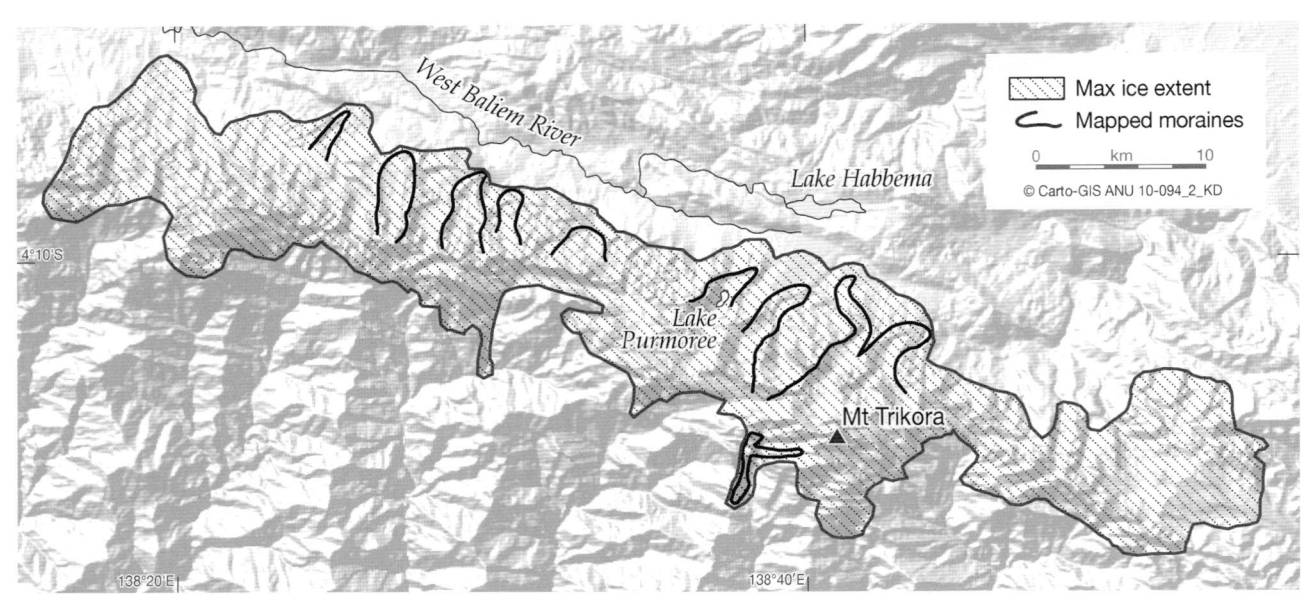

FIGURE 73.11 North of Mount Trikora, Papua. The older glacial valley flowing to the right has been cut off by a more recent ice flow to the left.

constructing a detailed chronology. Mount Giluwe is also a key site because of the very good preservation of moraines and a complex evolution in ice mass balances. It is likely that Mount Albert Edward will also prove to have a great deal of detail for comparison with the more equatorial mountains. Mount Kinabalu will repay detailed mapping and cosmogenic dating due to its very isolated location.

73.5.3. Problems

Considerable further work is required with strong chronological control to determine the pattern of glacial advance and retreat across the region. Glacier models that provide ELAs tied by time and location may eventually define the Pleistocene glacial climates and allow testing of causal hypotheses. For example, at present, the regions with relatively high LGM ELA, Mount Jaya, the Saruwaged Range and Mount Albert Edward and possibly Mount Kinabalu, are explained by quite different processes. Mount Jaya may have lain in a rainshadow caused by the drying of the shallow Arafura shelf. The Saruwaged Range and Mount Kinabalu are subject to rapid tectonic uplift, so their glacial limits may have been lower. Mount Albert Edward, at 8°S, may reflect a latitudinal upward trend in the regional snowlines, or the effects of lower precipitation.

73.6. CONCLUSIONS

The Asian equatorial high mountains are amongst the wettest on the planet and have probably been this way through the Quaternary. Some of the New Guinea glaciers were nourished by snow-bearing winds from across the West Pacific Warm Pool and others from snow-bearing winds from across the Coral Sea. The Arafura Sea was greatly affected by changing sea level. Similarly, the mountains of Sumatra and Borneo must have been affected by the exposure of the Sunda shelf but also remained relatively moist. The modelled ELAs for the LLGM are 3450–3900 m a.s.l., and the differences probably reflect differential moisture gradients. They may also lie behind the glacial advances that preliminary evidence points to as having occurred during marine isotope stages (MIS) 3–4 that seem to have had different mass balance relationships than those around 19 ka. The youngest exposure ages from Mount Trikora are consistent with the Younger Dryas cold period that is so prominent in the GISP 2 record. A small advance on the southern side of Mount Jaya has also been dated from around this time ($13,740 \pm 210$ cal. yr BP; Hope and Peterson, 1975). These sites would possibly have experienced increased precipitation at this time as the Arafura shelf flooded but indicate that temperatures remained cold.

Neoglacial advances have only been documented on Mount Jaya with four tills dating from >3000 cal. yr BP to present, the youngest probably reflecting the little ice age (Hope and Peterson, 1976). However, Late Holocene interbedded peats and sands on the summit area of Mount Wilhelm (Hope, 1976) suggest that permanent snow and ice probably affected all areas above 4400 m a.s.l.

The reviews of tropical glaciation such as those of Mark et al. (2005) and Hastenrath (2009) emphasise the variability of palaeo-ELA estimates within single regions. Their figures show that the Asian tropical glaciers are relatively wet and equable, leading to generally low ELA values, but with a surprising variability. Clearly, an intensive effort to improve mapping and dating will be necessary to relate the Asian equatorial glaciation to that of other areas and resolve causal mechanisms. Such an effort is justified by the fact that ice changes are clearly climate related and hence provide an important control for other palaeoenvironmental records, such as pollen, speleothems and diatoms, from lower altitudes.

ACKNOWLEDGEMENTS

The authors owe a debt to the first major researcher of New Guinea ice history, Ernst Löffler and thank him for his assistance over many years. Permission to work in Papua was provided by LIPI and KLH and sponsored by PT Freeport and Universitas Cenderawasih. Accelerator mass spectrometry has been carried out by David Fink, ANSTO and Keith Fifield, ANU. Mike Cottam, SE Asia Research Group, Royal Holloway, kindly provided a DEM and information about Mount Kinabalu. Kay Dancey and Jenny Sheehan undertook the DEM analysis and drew the figures. We thank our respective institutions and the Australian Research Council for providing major funding over the past 40 years. Prentice's contribution is based upon work supported in part by the National Science Foundation, Paleoclimate Program and PT Freeport Indonesia.

REFERENCES

Barrows, T.T., Prentice, M.L., Hope, G.S., Fifield, L.K., Tims, S, in press. Late Pleistocene glaciation of Mt Giluwe, Papua New Guinea. Quatern. Sci. Rev.

Bik, M.J.J., 1972. Pleistocene glacial and periglacial landforms on Mt. Giluwe and Mt. Hagen, western and southern highlands districts, Territory of Papua and New Guinea. Z. Geomorphol. 16, 1–15.

Blake, D.H., Löffler, E., 1971. Volcanic and glacial landforms on Mount Giluwe, Territory of Papua and New Guinea. Geol. Soc. Am. Bull. 82, 1605–1614.

Brown, I.M., 1990. Quaternary glaciations of New Guinea. Quatern. Sci. Rev. 9, 273–280.

Charlesworth, J.K., 1957. The Quaternary Era. Arnold, London.

Cottam, M., Hall, R., Sperber, C., Armstrong, R., 2010. Pulsed emplacement of the Mount Kinabalu granite, northern Borneo. J. Geol. Soc. Lond. 167, 49–60.

Detzner, H., 1921. Vier Jahre Under Kannibalen. Scherl, Berlin.

Ehlers, J., Gibbard, P.L., 2004. Quaternary glaciations—extent and chronology, part III: South America, Asia, Africa, Australasia, Antarctica. Developments in Quaternary Science vol. 2c. Elsevier, Amsterdam.

Fink, D., Peterson, J., Prentice, M.L., Hope, G.S., 2003. The Last Glacial Maximum and deglaciation events based on 10-Be and 26-Al exposure ages from the Mt. Trikora Region, Irian Jaya, Indonesia. In: Abstract, INQUA 2003 Conference Proceedings, Reno.

Flenley, J.R., Morley, R.J., 1978. A minimum age for the deglaciation of Mount Kinabalu, East Malaysia. Mod. Quatern. Res. S. east Asia 4, 57–61.

Haberle, S.G., 1998. Late Quaternary vegetation change in the Tari Basin, Papua New Guinea. Palaeogeogr. Palaeoclimatol. Palaeoecol. 137, 1–24.

Hall, R., 2001. Cenozoic reconstructions of SE Asia and the SW Pacific: Changing patterns of land and sea In Metcalfe, I. Smith, J.M.B. Morwood, M. and Davidson, I. (Eds.), Faunal and Floral Migrations and Evolution in SE Asia- Australasia. Balkema, Lisse, pp 35–56.

Hastenrath, S., 2009. Past glaciation in the tropics. Quatern. Sci. Rev. 28, 790–798.

Hope, G.S., 1976. The vegetational history of Mt. Wilhelm, Papua New Guinea. J. Ecol. 64, 627–664.

Hope, G.S., 2004. Glaciation of Malaysia and Indonesia, excluding New Guinea. In: Ehlers, J., Gibbard, P.L. (Eds.), Quaternary Glaciations—Extent and Chronology, Part III: South America, Asia, Africa, Australasia, Antarctica. Developments in Quaternary Science vol. 2c. Elsevier, Amsterdam, pp. 211–214.

Hope, G.S., 2009. Environmental change and fire in the Owen Stanley Ranges, Papua New Guinea. Quatern. Sci. Rev. 28, 2261–2276.

Hope, G.S., Peterson, J.A., 1975. Glaciation and vegetation in the high New Guinea mountains. Bull. R. Soc. NZ 13, 153–162.

Hope, G.S., Peterson, J.A., 1976. Palaeoenvironments. In: Hope, G.S., Peterson, J.A., Radok, U., Allison, I. (Eds.), The Equatorial Glaciers of New Guinea. A.A. Balkema, Rotterdam, pp. 173–205.

Hope, G.S., Peterson, J.A., Mitton, R., 1973. Recession of the minor ice fields of Irian Jaya. Z. Glestcherkunde Glazialgeol. IX, 73–87.

Hope, G.S., Flannery, T.F., Boeardi, N., 1993. A preliminary report of changing Quaternary mammal faunas in subalpine New Guinea. Quatern. Res. 40, 117–126.

Jacobson, S., 1986. Kinabalu Park. Sabah Parks Publication 7. Kota kinabalu.

Kilmaskossu, M.St.E., Hope, G.S., 1985. A mountain research program for Indonesia. Mt. Res. Dev. 5, 339–348.

Klein, A.G., Kincaid, J.L., 2008. On the disappearance of the Puncak Mandala ice cap, Papua. J. Glaciol. 54, 195–198.

Koopmans, B.N., Stauffer, P.H., 1968. Glacial phenomena on Mount Kinabalu, Sabah. Borneo Region Malays. Geol. Surv. Bull. 8, 25–35.

Lee, D., Choi, T., 1996. Geology of Kinabalu. In: Wong, K.M., Phillipps, A. (Eds.), Kinabalu, Summit of Borneo. second ed. Revised Sabah Society and Sabah Parks, Kota Kinabalu, 544 pp.

Löffler, E., 1972. Pleistocene glaciation in Papua New Guinea. Z. Geomorphol. N. F. Suppl 13, 32–58.

Löffler, E., 1982. Pleistocene and present-day glaciations. In: Gressitt, J.L. (Ed.), Biogeography and Ecology of New Guinea. Junk, The Hague, pp. 39–56.

Löffler, E., Mackenzie, D.E., Webb, A.W., 1980. Potassium-argon ages from some of the Papua New Guinea highlands volcanoes, and their relevance to Pleistocene geomorphic history. J. Geol. Soc. Aust. 26, 387–397.

MacKinnon, K., Hatta, G., Halim, H., Mangalik, A., 1996. The Ecology of Kalimantan. Periplus Edition (HK) Ltd., Singapore, 872 pp.

Mark, B.G., Harrison, S.P., Spessa, A., New, M., Evans, D.J.A., Helmens, K.F., 2005. Tropical snowline changes at the last glacial maximum: a global assessment. Quatern. Int. 138–139, 168–201.

Peterson, J.A., Peterson, L.F., 1994. Ice retreat from the neoglacial maxima in the Puncak Jayakesuma, Republic of Indonesia. Z. Gletscherk. Glazialgeol. 30, 1–9.

Peterson, J.A., Hope, G.S., Prentice, M., Hantoro, W., 2001. Mountain environments in New Guinea and the late Glacial Maximum "warm seas/cold mountains" enigma in the West Pacific Warm Pool region. In: Kershaw, P., David, B., Tapper, N., Penny, D., Brown, J. (Eds.), Bridging Wallace's Line. Advances in GeoEcology, 34 Catena Verlag, Reiskirchen, pp. 173–187.

Peterson, J.A., Chandra, S., Lundberg, C., 2004. Landforms from the Quaternary glaciation of Papua New Guinea: an overview of ice extent during the LGM. In: Ehlers, J., Gibbard, P.L. (Eds.), Quaternary Glaciations—Extent and Chronology, Part III: South America, Asia, Africa, Australasia, Antarctica. Developments in Quaternary Science vol. 2c. Elsevier, Amsterdam, pp. 313–320.

Porter, S.C., 2001. Snowline depression in the tropics during the Last Glaciation. Quatern. Sci. Rev. 20, 1067–1091.

Prentice, M.L., Glidden, S., 2010. Glacier crippling and the rise of the snowline in western New Guinea (Papua Province, Indonesia) from 1972 to 2000. Terra Australis 32, 257–271.

Prentice, M.L., Hope, G.S., 2007. Climate of Papua. In: Marshall, A.J., Beehler, B.M. (Eds.), The Ecology of Papua. Periplus Editions, Singapore, pp. 177–195.

Prentice, M.L., Hope, G.S., Maryunani, K., Peterson, J.A., 2005. An evaluation of snowline data across New Guinea during the last major glaciation and area-based glacier snowlines in the Mt. Jaya region of Papua, Indonesia, during the LGM. Quatern. Int. 138–139, 93–117.

Quarles van Ufford, A., Sedgwick, P., 1998. Recession of the equatorial Puncak Jaya glaciers (1825 to 1995), Irian Jaya (Western New Guinea), Indonesia. Z. Gletscherk. Glazialgeol. 34 (2), 131–140.

Reiner, E., 1960. The glaciation of Mount Wilhelm, Australian New Guinea. Geogr. Rev. 50, 491–503.

Shepherd, M., 1966. Glacial geomorphology of the eastern Star Mountains. (Unpublished Honours thesis)University of Sydney, Sydney.

Smith, J.M.B., Lowry, J.B., 1968. Further exploration and observations on Mount Kinabalu East. Malay Nat. J. 22, 29–40.

van Beek, C.G.G., 1982. A geomorphological and pedological study of the Gunung Leseur National Park, North Sumatra, Indonesia. Utrechtse Geografische Studies, 26 Wageningen Agricultural University, Wageningen, 187 pp.

Verstappen, H.Th., 1952. Luchtfotostudies over het Centrale Bergland van Nederlands Nieuw-Guinea. Tijdschrift Van Het Koninklijk Nederlandsch Aardrijkskundig Genootschap Series 2 69, pp. 336–363, 425–431.

Verstappen, H.Th., 1964. The geomorphology of the Star Mountains. Nova Guinea NS Geol. 5, 101–155.

Visser, W.A., Hermes, J.J., 1962. Geological results of the exploration for oil in Netherlands New Guinea. Koninklijk Nederlands Geologisch Mijnbouwkundig Genootschap Verhandelingen, Geologische Serie 20, pp. 1–256.

Walker, D., Flenley, J., 1979. Late Quaternary vegetational history of the Enga Province of upland Papua New Guinea. Philos. Trans R. Soc. Lond. B 286, 265–344.

Zen, M.T., Ganie, B.M., 1992. Tambora 1815 eruption. In: Degens, E.T., Wong, H.K., Zen, M.T. (Eds.), The sea off Tambora, Mitteilungen aus dem Geologisch-Paläontologischen Institut der Universität Hamburg, 70, 173–185.

Chapter 74

The Glaciation of Australia

Eric A. Colhoun[1] and Timothy T. Barrows[2,*]

[1] Earth Sciences, The University of Newcastle, Callaghan, NSW 2308, Australia
[2] School of Geography, University of Exeter, Exeter EX4 4RJ, United Kingdom
*Correspondence and requests for materials should be addressed to Timothy T. Barrows. E-mail: t.barrows@exeter.ac.uk

74.1. INTRODUCTION

Despite low latitude and altitude, Australia was glaciated numerous times during the Pleistocene. Most mountains are located along the east of the continent with the highest peaks in the Snowy Mountains of the south-east. The mountains of Tasmania are lower in altitude but the vast majority of glaciation was concentrated there because of its higher latitude. Tasmania has been partially glaciated many times during the Pleistocene, the earliest occurring around 1 million years ago and the youngest during marine isotope stage (MIS) 2. The earliest glaciations were the most extensive, and later ice advances were successively smaller. Least is known about the age of the earliest glaciations and associated climates. The application of exposure-age dating is revealing unprecedented detail about the timing and extent of glaciation during the past few hundred thousand years. This chapter presents an overview of the Pleistocene glaciations of Australia.

74.2. SNOWY MOUNTAINS

The Snowy Mountains experience a temperate–subtropical climate with precipitation mostly in spring and winter which arrives from westerly low-pressure systems. Mt Kosciuszko is the tallest mountain reaching 2228 m a.s.l. as part of a larger massif generally over 1800 m a.s.l. The highest mountains are the coldest part of the continent indicated by a mean annual temperature of 4.5 °C at Thredbo Crackenback station (1957 m a.s.l.). Precipitation is also high on the Kosciuszko massif with 2056 mm/year at Charlotte Pass (1755 m a.s.l.; Bureau of Meteorology, 2010).

The extent of glaciation over mainland Australia has been a highly controversial topic since the action of glaciers was first recognised in the Snowy Mountains during the nineteenth century (David et al., 1901). Early models envisaged three stages of glaciation (e.g. Browne and Vallance, 1957). The first phase was an ice cap with ice area of more than 400 km². The second phase was dominated by valley glaciation, with a glacier extending down the Snowy River valley for 22 km. The final phase was an episode of cirque glaciation. However, later research showed that only the highest peaks were glaciated by a cirque phase of glaciation (Galloway, 1963).

Unequivocal evidence for glaciation in the Snowy Mountains is restricted to about 12–15 km² (Barrows et al., 2001). The style of glaciation was of the form of niche glaciers and short valley glaciers. The largest glacier was at least 1.7 km in length and formed on the flanks of Mt Twynam (2196 m a.s.l.) extending down the Snowy River valley from the cirque above Blue Lake to the moraines below Headley Tarn. Within the Blue Lake valley, there are four distinct terminal moraines. Valley heads nearby were only occupied by short valley glaciers or niche glaciers. The most important occurred on the lee side of the Main Range in the Pounds Creek, Club Lake, Lake Albina and leeward of Mt Kosciuszko at Lake Cootapatamba.

The terminal moraines below Blue Lake can be separated on the basis of exposure dating using the cosmogenic nuclide ^{10}Be (Fig. 74.1; Barrows et al., 2001). The outermost moraine has a mean age of 59.3 ± 5.4 ka and represents an ice advance (the Snowy River advance) during MIS 4. Galloway (1963) was prescient in estimating that the maximum glacial extent was about middle Würm in age. The next oldest advance occurs late in MIS 3 at 32.0 ± 2.5 ka (the Hedley Tarn advance). This is the oldest evidence for cold conditions leading into the last glacial maximum (LGM), and corresponds to high volumes of river runoff on the Riverine Plain (Page et al., 1991). Ice advanced during the LGM at 19.1 ± 1.6 ka (the Blue Lake advance), and after a brief still stand at 16.8 ± 1.4 ka (the Twynam advance), the Snowy Mountains were completely deglaciated before commencement of the Holocene. Barrows et al. (2001) divided the first advance into the Early Kosciuszko glaciation and grouped the last three advances into the Late Kosciuszko glaciation (Fig 74.1).

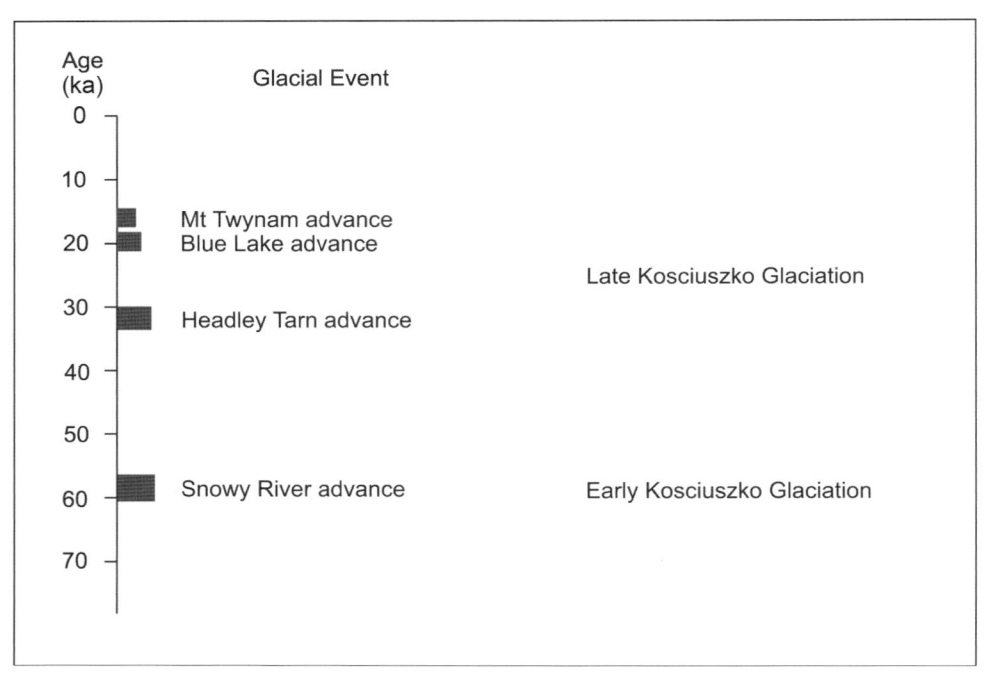

FIGURE 74.1 Timeline showing the age of ice advances in the Snowy Mountains against time (Barrows et al., 2002).

Evidence for glaciation older than the Late Pleistocene in south-eastern mainland Australia is ambiguous. Early workers assumed that rounded valley heads and valley floor debris represented widespread glaciation throughout the Snowy Mountains and into Victoria (e.g. Browne and Vallance, 1957). However, most of this evidence can be satisfactorily explained in terms of natural weathering of the landscape particularly of granites, and by the operation of periglacial processes that formed extensive coarse clastic debris on the higher mountain slopes and summit areas (Galloway, 1963).

Cirque floors on the Kosciuszko massif lie between 1875 and 2040 m a.s.l., with an average height of 1970 m a.s.l. (Galloway, 1963, 1965), approximately 700 m higher than cirques in northern Tasmania. Galloway (1965) inferred that the orographic snowline was approximated by the cirque floor level and was 600–700 m lower than at present. Without a change in precipitation, this would correspond to a summer mean temperature ~6 °C colder than present. However, there is abundant evidence that mainland Australia was drier during the LGM, and that summer temperatures were as much as 9 °C colder (Galloway, 1965). Consequently, precipitation could have been half present values at the LGM equilibrium line altitude (ELA). It is likely that conditions were wetter, and not as cold, during the earlier ice advances, but temperature and precipitation values were not as high as modern values until after final deglaciation.

74.3. TASMANIA

Tasmania lies within the westerly wind belt which brings abundant precipitation to the generally north–south oriented mountain ridges of western Tasmania, the Central Highlands and high Central Plateau. These mountains and plateaux extend from 600 to 1500 m in altitude and hosted cirque, valley and ice cap glaciation numerous times during the Pleistocene. Mapped landforms and till sheets have been subdivided on the basis of clast weathering, radiocarbon dating, palaeomagnetism, uranium-series dating and, most recently, exposure dating using cosmogenic nuclides (Fig. 74.2). For convenience, we discuss glaciation in three sections, the Early, Middle and Late Pleistocene. The Brunhes–Matuyama boundary is used to separate the Early from the Middle Pleistocene, and Termination II (MIS 6/MIS 5 boundary) separates the Middle from the Late Pleistocene.

74.3.1. Early Pleistocene

The most extensive ice cover of Tasmania occurred during the Early Pleistocene, when an ice sheet formed on the Central Plateau and Central Highlands extended westwards to the West Coast Range (Fig. 74.3). In the upper Pieman Valley, it extended to Success Creek at 170–200 m a.s.l. and 60 km upstream (33 km inland) from the coast. Large outlet glaciers extended northwards from the ice sheet along the Forth and Mersey valleys, and to the south and west along

Chapter | 74   The Glaciation of Australia

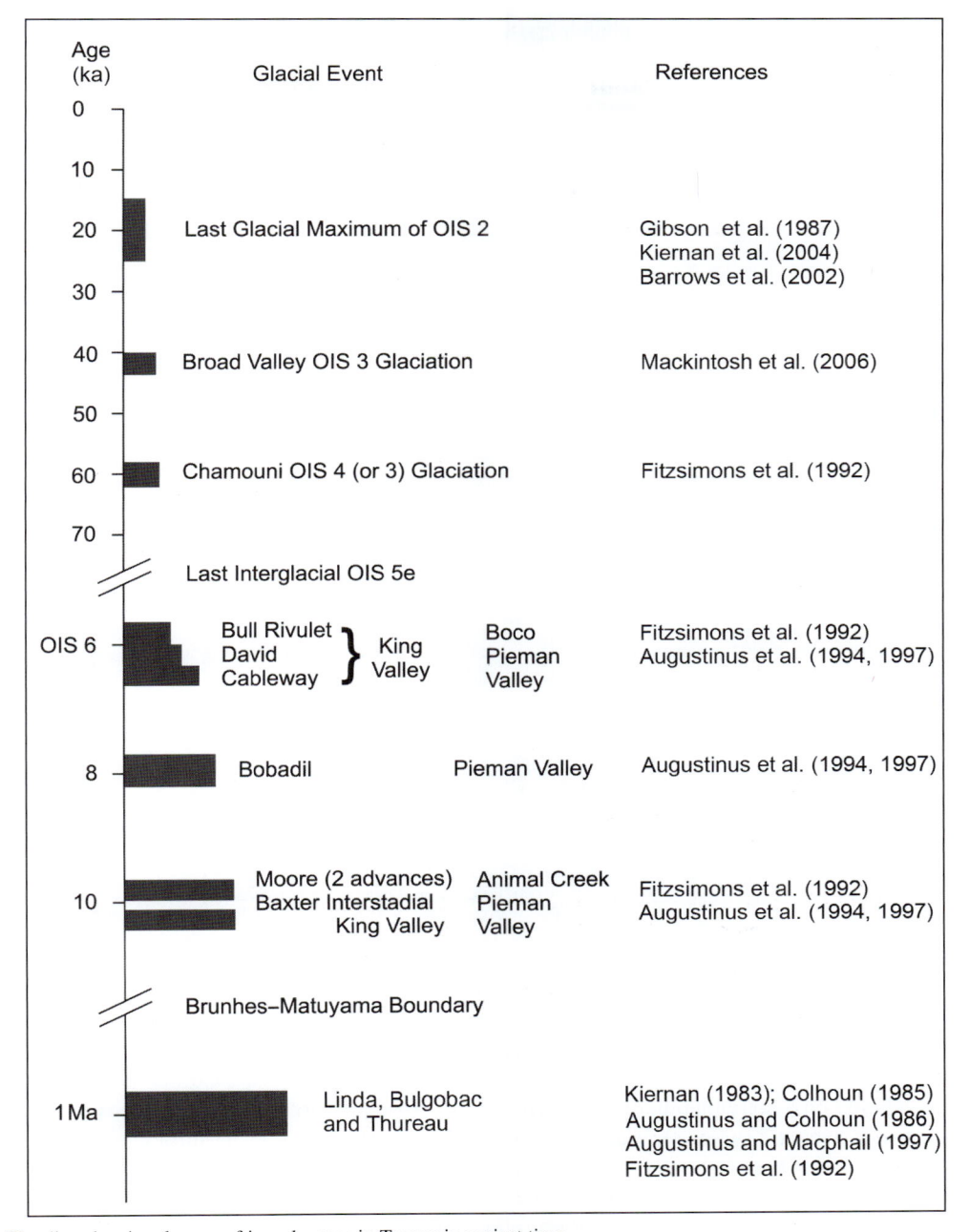

FIGURE 74.2 Timeline showing the age of ice advances in Tasmania against time.

the upper Derwent, Franklin and King valleys. This ice sheet glacial system was first recognised by Arndell Lewis as a period of ancient glaciation, but its antiquity was dismissed by others who thought it was of Würmian/Wisconsinan age (Lewis, 1945; Jennings and Banks, 1958).

Examination of landforms and deposits in the Linda Valley in the West Coast Range during the late 1970s and early 1980s confirmed that this glaciation did not occur during the last cold stage (Kiernan, 1983; Colhoun, 1985). The ice sheet covered approximately 7000 km^2 and was named the Linda Glaciation. Mapping in the Pieman Valley of part of this system, there referred to as the Bulgobac Glaciation, demonstrated there was also an older ice advance at Que River (Fig. 74.3; Augustinus and Colhoun, 1986). Although the extent of Early Pleistocene ice limits is broadly known in Tasmania, the number of ice advances and stages of glaciation have not yet been fully differentiated. At present, it appears certain that multiple Early Pleistocene and perhaps even Late Pliocene glaciations occurred, but refinement of the number, extent and ages of the phases of glaciation awaits further research.

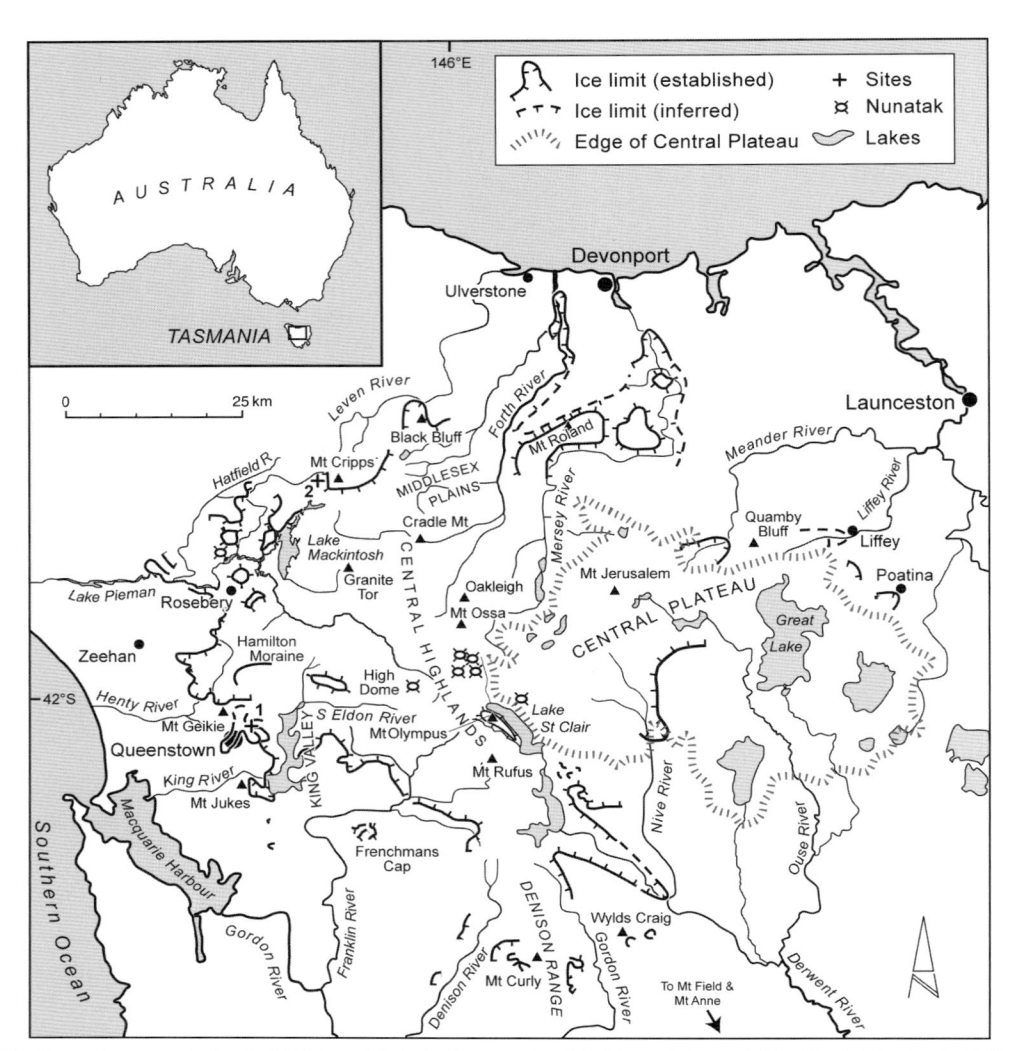

FIGURE 74.3 Maximum ice extent during the Middle and Early Pleistocene (modified from Colhoun, 2004). Site 1 is the Lynda Valley and Site 2 is the upper Pieman Valley.

Currently, assignment of the Linda Glaciation to the Early Pleistocene is based on a combination of weathering criteria, palaeomagnetic data and pollen stratigraphy. The glacial till, ice-proximal sands and gravels and outwash deposits associated with the Linda Glaciation are strongly chemically weathered and contain erratics of a variety of igneous rocks that are highly decomposed. The most widely distributed erratic is dolerite of Jurassic age. Study of the mean thicknesses of weathering rinds and estimation of the time elapsed since deposition led Kiernan to believe the Linda age glacial deposits were over 600,000 years old (Kiernan, 1983).

The end moraine marking the limit of the Linda Glaciation near Gormanston was deposited on its ice-distal side into a glacial lake in the upper part of the valley. In the lake, a sequence of varved clays were interstratified with flow-till units. The magnetic polarity of sediment samples from the varved clays was determined and found to be reversed (Fig. 74.3; Barbetti and Colhoun, 1988). Palaeomagnetism was extended to varved lake deposits associated with till and ice-contact deposits throughout the area of the West Coast Range, Pieman, Forth and Franklin valleys. It was found that the strongly chemically weathered glacial deposits containing dolerite clasts with mean weathering rind thicknesses of 57–90 mm and sometimes as much as 300 mm always had reversed polarity. On this basis, the Linda Glaciation and Bulgobac correlate were assigned a pre Brunhes–Matuyama age of >783,000 years and placed in the Early Pleistocene (Spell and McDougall, 1992; Pollington et al., 1993; Augustinus et al., 1995; Colhoun et al., 2010).

The exact ages of the Early Pleistocene glaciations are not known, but a section in a tributary of the Upper Linda Valley shows a soil horizon within alluvial fan deposits that contains pollen of Late Pliocene age. The alluvial deposits marginally abut and are stratigraphically older than the adjacent Linda age gravels (Macphail et al., 1995). In addition, a section in the Pieman Valley on the divide between Marionoak and Huskisson rivers showed glacial silts belonging to the Bulgobac Glaciation overlying organic-rich silts with normal magnetization. The organic silts contain pollen of numerous extinct taxa known from the Late Pliocene. The normal magnetization suggests that the silts may represent the Jaramillo event, the first phase of normal magnetization before the Brunhes/Matuyama reversal. The Jaramillo has an age of 0.99–1.07 Ma. If this interpretation is correct, it suggests the Linda Glaciation occurred around 1 Ma ago, but it could be older (Augustinus and Macphail, 1997).

74.3.2. Middle Pleistocene

The terminal locations of the large Middle Pleistocene outlet glaciers occur mainly within the major valley systems of the Pieman and King rivers in the west, the Forth and Mersey north of the Central plateau and Derwent, Gordon and Franklin south of it (Fig. 74.3). In addition, a piedmont expansion of ice from Cradle Mountain reached Middlesex Plains, and large cirque and short valley glaciers occurred in some southern mountain ranges. However, although multiple advances of outlet glaciers have been recognised in many valleys and areas of piedmont ice on adjacent plains, the extent of deposits has not been sufficiently well differentiated to define the areas covered during each advance of ice. A rough estimate of around 4000–5000 km^2 indicates up to five times more Middle Pleistocene than Late Pleistocene ice.

Because of difficulty imposed by the heavily forested terrain and access to remote mountains, the pattern of multiple glaciations and their dating has been determined in only a few areas. The palaeomagnetic direction in all Middle Pleistocene glacial lake sediments is normal, but more refined differentiation of stages is based on mapped spatial patterns, stratigraphy, uranium-series dating of associated deposits and exposure dating of glacial boulders on moraine ridges.

Most of the Middle Pleistocene glacier advances were initially identified on the basis of mean thicknesses of weathering rinds developed on erratic dolerite clasts in the glacial deposits (Kiernan, 1983, 1989, 1990, 1991, 1992, 1995; Kiernan and Hannan, 1991). The average weathering rind thicknesses of 5–15 mm for Middle Pleistocene deposits is much less than in Early Pleistocene deposits. This suggests a long time elapsed between the Early and the Middle Pleistocene glaciations which occurred much closer in time to the Late Pleistocene ice advances.

The clearest dated Middle Pleistocene sequence occurs in the upper part of the Pieman Valley where three ice advances younger than the Bulgobac Glaciation and older than the Last Glaciation have been identified as the Animal Creek, Bobadil and Boco Glaciations (Augustinus et al., 1994, 1995, 1997; Augustinus, 1999a,b). Weathering rind thicknesses on dolerite erratics in Boco deposits (7.1 mm rinds) are half the relative age of Bobadil deposits (13.1 mm) which are at least an order of magnitude younger than Bulgobac deposits (130–240 mm). Also, post-depositional chemical weathering of andesite erratics shows a decrease in density with increase in age (Boco, 2.46–2.6 g/cm^3; Bobadil, 2.22–2.24 g/cm^3; Bulgobac, 2.05–2.25 g/cm^3).

In the middle King Valley, a complex sequence of Middle Pleistocene deposits has been determined by morphostratigraphical mapping (Fitzsimons and Colhoun, 1991; Fitzsimons et al., 1992). The Middle Pleistocene glacial deposits are attributed to two ice advances within one glaciation, the Moore Glaciation (Fitzsimons et al., 1990). The advances are separated by cold climate organic-rich interstadial deposits at Baxter Rivulet, for which an amino-acid racemisation date suggests a minimum age of MIS 10 (B. J. Pillans, personal communication, 1987 in Fitzsimons et al., 1992). These deposits are underlain by a till comparable in age to the Linda (and Bulgobac) glacial deposits, which is locally referred to as the Thureau Glaciation. This till is highly weathered, and associated glacial lake deposits show reversed magnetic polarity and are considered to be Early Pleistocene in age.

Further north in the King Valley, three ice advances occur: the Cableway, David and Bull Rivulet. The oldest Cableway deposits have mean dolerite weathering rind thicknesses of 7.1 mm which is identical with the Boco deposits in the Pieman Valley. Deposits of the David ice advance comprise till, gravel and plastically deformed silts and clays up to 80 m thick. It is the most extensive Middle Pleistocene deposit in the King Valley. No weathering rind data are available, but the geographic and stratigraphic dispositions of the sequence make it younger than the Cableway. There is no exposure in the Bull Rivulet moraine, but it is part of an ice limit that occurs at the mouth of the Linda Valley, which contains dolerite erratics with mean weathering rind thicknesses of 5–10 mm (Kiernan, 1983). Kiernan correctly identified this moraine as of pre-Last Interglacial age and referred it to a Comstock Glaciation. This identification is confirmed by all three ice advances in the Middle King Valley being older than the organic-rich Smelter Creek pollen site which has a minimum amino-acid racemisation age of MIS 5 (B. J. Pillans, personal communication, 1988 in Fitzsimons et al., 1992; Colhoun et al., 1992). Thus, the threefold sequence of Middle Pleistocene ice advances in the middle King Valley occurred during MIS 6.

Initial semi-quantitative ages for the Pieman Valley Middle Pleistocene ice advances were based on uranium/thorium measurements on iron pan and ferricrete horizons formed within a till, sand, silt and gravel sediment sequence in Core AK 1 in the Bulgobac Valley. The ferricrete horizons were formed by chemical weathering and deposition of iron during the weathering intervals succeeding periods of deposition of till and associated sand and gravel deposits. The age measurements are therefore minimal for the preceding glacial deposits and inferred glaciations. The results show the Boco Glaciation is older than $78+23/-20$ ka, the Bobadil older than $178+20/-18$ ka and the Animal Creek more than $240+45/-35$ ka old (Augustinus and Colhoun, 1986; Augustinus et al., 1994; Augustinus, 1999a,b). Because the ferricretes were probably formed under interglacial conditions, the most likely ages of the glaciations are MIS 6, MIS 8 and MIS 10 (Fig. 74.2).

Until recently, it has been thought that the large moraines occurring at the exits of short valleys and cirque basins in western Tasmania marked MIS 2 glaciation limits, the prime example being the Hamilton Moraine in the West Coast Range (Lewis, 1945; Colhoun, 1985). Exposure dating indicates that this is not so and that at least some large moraines were formed during more than one glaciation preceding the Last Glaciation (Barrows et al., 2002). It is likely that the Hamilton Moraine west of Lake Margaret formed between 190 and 220 ka (Barrows et al., 2002; Fink, personal communication, 2009 in Colhoun et al., 2010). Four ages from the moraine fall within MIS 7 and one within MIS 8. Since they are likely to be minimum values, the dates suggest the ice last advanced and constructed most of the moraine during MIS 8.

74.3.3. Late Pleistocene

The best preserved and most well-known record of glaciation is attributed to the Late Pleistocene. During the last glaciation, most of the ice was confined to an ice cap that covered much of the western Central Plateau and the Central Highland ridge with short outlet glaciers extending into the Forth and Mersey valleys in the north and the upper Derwent in the south (Fig. 74.4). A small ice cap was formed on the crest of the Tyndall Range in the West Coast Range and outlet glaciers extended to west, north and south. Elsewhere, cirque and short valley glaciers formed in individual mountain ranges. The area covered by ice is estimated at 1085 km^2 (Colhoun et al., 1996).

All the deposits are weakly chemically weathered with mean rind thicknesses on dolerite erratics of approximately 1 mm and rarely exceeding 2–3 mm (Kiernan, 1983), clearly differentiating them from those of the Middle Pleistocene advances but not allowing any differentiation within the Late Pleistocene. Very few sites have suitable exposure for radiocarbon dating. The best known is the site at Dante Rivulet which has been radiocarbon dated to 18.8 ± 0.5 and 19.1 ± 0.17 ^{14}C ka BP ($\sim$22.3–22.5 cal. ka BP; Gibson et al., 1987; Colhoun and Fitzsimons, 1996).

The application of exposure dating using cosmogenic nuclides has led to a revolution in the knowledge of the timing of glaciation in the Late Pleistocene. Barrows et al. (2002) sampled 18 moraines in eight areas to better define the timing of ice advance during the LGM. They found that the ice attained its maximum extent between 22 and 17 ka, with a mean of around 19–20 ka. Ice retreated rapidly after 18 ka and there is little evidence for extensive ice after 16 ka though there is evidence for survival of a small glacier in the Rhona Valley of the Denison Range until 14.2 ka (McMinn et al., 2008). There is no evidence for significant ice advance in Tasmania during the Younger Dryas Chronozone. It is unlikely that there was any ice in Tasmania after around 14–13 ^{14}C ka BP, because pollen diagrams from many sites throughout western and central Tasmania indicate climate had warmed and wet forest vegetation was expanding extensively.

Exposure dating has also revealed the presence of significant ice advance during MIS 3 (Barrows et al., 2002; Mackintosh et al., 2006). The best evidence for MIS 3 glaciation occurs at Mt Field where the Broad Valley glacier extended 4 km down valley beyond the MIS 2 ice limits. Two large dolerite boulders known as the Griffith and Taylor boulders occur on the small outwash plain distal to the fourth moraine beyond the MIS 2 ice limit. They have been dated using ^{36}Cl to 44.1 ± 2.2 and 41 ± 2 ka BP (Mackintosh et al., 2006). Boulders from other moraine ridges and glaciated surfaces have also been dated from other sites outside MIS 2 ice limits to between 24 and 53 ka, suggesting cirque and valley glacier formation during MIS 3 (Barrows et al., 2002).

Unlike on the mainland, there is no directly dated evidence for glaciation during MIS 4 in Tasmania. Until recently, it was thought that there were two advances of ice within the last glaciation, one that occurred during the LGM and an earlier advance in the King Valley which extended to Chamouni where it was radiocarbon dated to >48.7 ^{14}C ka BP (Fitzsimons et al., 1992). Current evidence suggests glaciations during MIS 2, MIS 3 and possibly MIS 4 at Chamouni.

74.4. CLIMATE

Climate was the principle control on the distribution of the glacial systems. Today, Tasmanian weather is dominated by easterly moving low-pressure systems. The precipitation gradient from the West Coast Range (3600–2000 mm) across the Central Highlands (2500–2000 mm) and Central Plateau (1500–1000 mm) to the Midlands (800–600 mm) is very steep. Eastern Tasmania lying in the rain shadow of the mountains is relatively dry and had no glaciers except on the Ben Lomond Plateau (1200–1575 m a.s.l.) in the north-east. There is also a strong contrast between the mean lowland temperatures of 11–12 °C and the mountain and plateaux

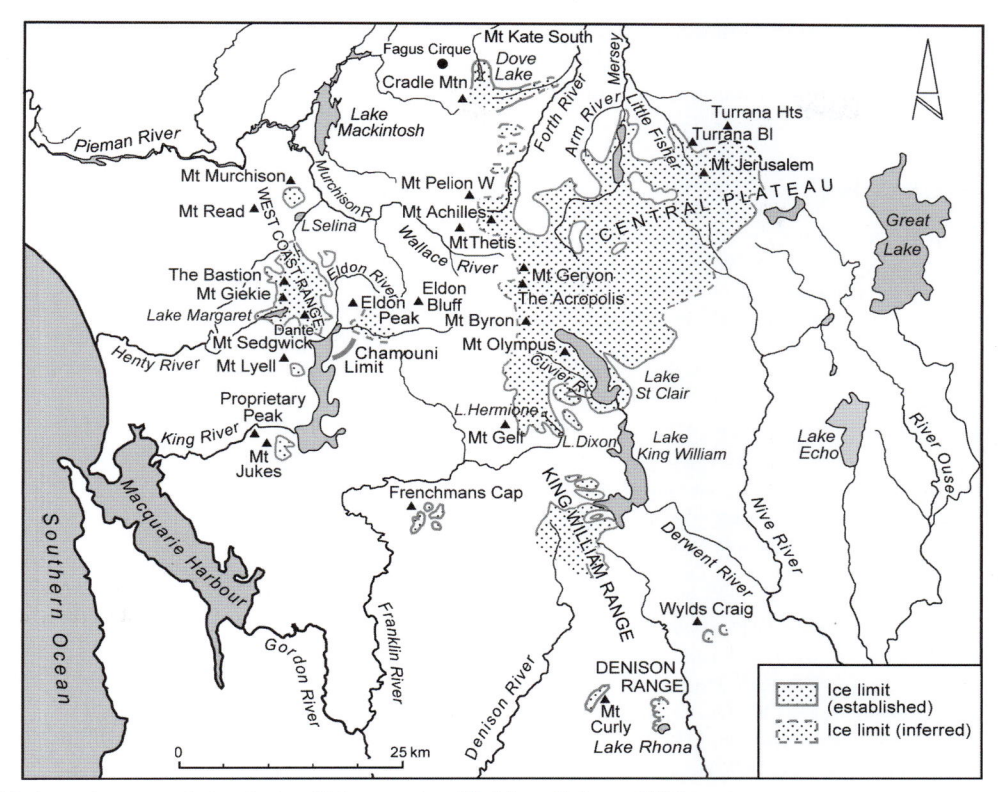

FIGURE 74.4 Maximum ice extent during the late Pleistocene (modified from Colhoun, 2004).

areas where, at tree line (~1000 m), mean temperatures are 4–6 °C (Bureau of Meteorology, 2010).

During the Pleistocene, the regional snowline rose from 610 m a.s.l. in south-west Tasmania to 1220 m a.s.l. over the Central Plateau and at least 1350–1400 m a.s.l. on Ben Lomond in the north-east, mirroring modern precipitation patterns. The snowline gradient is consistent with the occurrence of glaciers at low altitude in the south-west mountains and a thinning ice cap across the northern and eastern Central Plateau (Davies 1967; Peterson, 1968; Peterson and Robinson, 1969). The above snowline estimates refer mostly to the best preserved glacial landforms of the Late Pleistocene but historically pre-date differentiation into Early, Middle and Late Pleistocene landforms. Estimates made for the most recent and limited glaciation of MIS 2 comparing glacier ELAs from the West Coast Range with the mean level of atmospheric freezing using a lapse rate of 0.65 °C/100 m (Colhoun 1985; Nunez and Colhoun, 1986) give an average temperature depression of 6.5 °C and a snowline depression of around 1000 m. However, higher values up to 10 °C colder for 1500 m snowline depression have been suggested for the Cradle Mountain glacial system using mean summer temperatures (Thrush, 2008). The proximity of suggested MIS 3 from MIS 4 ice limits to the MIS 2 limits indicates that temperature and snowline depression during earlier stages of the Last Glaciation have been only slightly greater. Also, since a small decrease in mean temperature can cause considerable ice expansion, temperature depression during the Middle and Early Pleistocene ice advances was probably not more than 1–2 °C colder than during MIS 2 (Colhoun et al., 2010).

74.5. CONCLUSIONS

During the past 35 years, advances in glacial studies in south-eastern Australia have contributed to partially resolving the complexity of glacial events that affected the region during the Pleistocene and have begun to provide a robust chronology of the events. Currently, it has been ascertained that all known evidence for Pleistocene glaciation in the Snowy Mountains is of last glaciation age but there were at least two distinct periods of glaciation during MIS 2 and MIS 4 termed the Early and Late Kosciusko glaciations. In Tasmania, the Pleistocene glaciation was more complex with multiple ice advances recorded from the Early, Middle and Late Pleistocene. The earliest Pleistocene advances are thought to have been about 1 million years old but may be older. Middle Pleistocene ice advances occurred during MIS 10, MIS 8 and MIS 6. During the Late Pleistocene, ice was probably present during MIS 4, MIS 3 and MIS 2. Deglaciation from the LGM was complete by around

14 ka, and there is no evidence of the Antarctic Cold Reversal or Younger Dryas events causing renewed ice advances either in Tasmania or in the Snowy Mountains region.

ACKNOWLEDGEMENTS

We thank Sue Rouillard of The University of Exeter and Olivier Rey-Lescure of The University of Newcastle, NSW, Australia for invaluable assistance in drafting the maps and diagrams. Figures 74.3 and 74.4 have been updated from figures published in Colhoun (2004) by Elsevier Press, Amsterdam.

REFERENCES

Augustinus, P.C., 1999a. Reconstruction of the Bulgobac Glacial System, Pieman river basin, Western Tasmania. Aust. Geogr. Stud. 37 (1), 24–36.

Augustinus, P.C., 1999b. Dating the Late Cenozoic glacial sequence, Pieman River basin, western Tasmania, Australia. Quatern. Sci. Rev. 18, 1335–1350.

Augustinus, P.C., Colhoun, E.A., 1986. Glacial history of the upper Pieman and Boco valleys, western Tasmania. Aust. J. Earth Sci. 33, 181–191.

Augustinus, P.C., Macphail, M.K., 1997. Early Pleistocene stratigraphy and timing of the Bulgobac Glaciation, Western Tasmania, Australia. Palaeogeogr. Palaeoclimatol. Palaeoecol. 128, 253–267.

Augustinus, P.C., Short, S.A., Colhoun, E.A., 1994. Pleistocene stratigraphy of the Boco Plain, western Tasmania. Aust. J. Earth Sci. 41, 581–591.

Augustinus, P.C., Pollington, M.J., Colhoun, E.A., 1995. Magnetostratigraphy of the Late Cenozoic glacial sequence, Pieman River basin, western Tasmania. Aust. J. Earth Sci. 42, 509–518.

Augustinus, P.C., Short, S.A., Heijnis, H., 1997. Uranium/thorium dating of ferricretes from mid-to late Pleistocene glacial sediments, western Tasmania, Australia. J. Quatern. Sci. 12 (4), 295–308.

Barbetti, M., Colhoun, E.A., 1988. Reversed magnetisation of glaciolacustrine sediments from Western Tasmania. Search 19 (3), 151–153.

Barrows, T.T., Stone, J.O., Fifield, L.K., Cresswell, R.G., 2001. Late Pleistocene Glaciation of the Kosciuszko Massif, Snowy Mountains, Australia. Quatern. Res. 55, 179–189.

Barrows, T.T., Stone, J.O., Fifield, L.K., Cresswell, R.G., 2002. The timing of the last glacial maximum in Australia. Quatern. Sci. Rev. 21, 159–173.

Browne, W.R., Vallance, T.G., 1957. Notes on some evidences of glaciation in the Kosciusko region. Proc. Linn. Soc. NSW 82, 125–144.

Bureau of Meteorology, 2010. Climate data online. http://www.bom.gov.au/climate/data/index.shtml, accessed 06.10.

Colhoun, E.A., 1985. The Glaciations of the West Coast Range, Tasmania. Quatern. Res. 24, 39–59.

Colhoun, E.A., 2004. Quaternary glaciations of Tasmania and their ages. In: Ehlers, J., Gibbard, P.L. (Eds.), Quaternary Glaciations—Extent and Chronology, Part III: South America, Asia, Africa, Australasia, Antarctica. Developments in Quaternary Science, 2 Elsevier, Amsterdam, pp. 353–360.

Colhoun, E.A., Fitzsimons, S.J., 1996. Additional radiocarbon date from Dante Outwash Fan, King Valley, and dating of The Late Wisconsin glacial maximum in Western Tasmania. Pap. Proc. R. Soc. Tas. 130 (1), 81–84.

Colhoun, E.A., van de Geer, G., Fitzsimons, S.J., 1992. Late Quaternary organic deposits at Smelter Creek and vegetation history of the Middle King Valley, western Tasmania. J. Biogeogr. 19, 217–227.

Colhoun, E.A., Hannan, D., Kiernan, K., 1996. Late Wisconsin glaciation of Tasmania. Pap. Proc. R. Soc. Tas. 130 (2), 33–45.

Colhoun, E.A., Kiernan, K., Barrows, T.T., Goede, A., 2010. Advances in Quaternary Studies in Tasmania. In: Pillans, B., Bishop, P. (Eds.), Australian landscapes. Geological Society, London, Special Publication 346, Geological Society of London, London, pp. 165–183.

David, T.W.E., Helms, R., Pittman, E.F., 1901. Geological notes on Kosciusko, with special reference to evidences of glacial action. Proc. Linn. Soc. NSW 26, 26–74.

Davies, J.L., 1967. Tasmanian landforms and Quaternary climates. In: Jennings, J.N., Mabbutt, J.A. (Eds.), Landform Studies from Australia and New Guinea. ANU Press, Canberra, pp. 1–25.

Fitzsimons, S.J., Colhoun, E.A., 1991. Pleistocene glaciations of the King Valley, western Tasmania, Australia. Quatern. Res. 36, 135–156.

Fitzsimons, S.J., Colhoun, E.A., van de Geer, G., 1990. Middle Pleistocene glacial stratigraphy at Baxter Rivulet, western Tasmania, Australia. J. Quatern. Sci. 5, 17–27.

Fitzsimons, S.J., Colhoun, E.A., van de Geer, G., Pollington, M., 1992. The Quaternary geology and glaciation of the King Valley. Geol. Surv. Bull. 68, 1–57 Tasmanian Department of Mines, Hobart.

Galloway, R.W., 1963. Glaciation in the Snowy Mountains: a re-appraisal. Proc. Linn. Soc. NSW 88, 180–198.

Galloway, R.W., 1965. Late Quaternary climates in Australia. J. Geol. 73, 603–618.

Gibson, N., Kiernan, K.W., Macphail, M.K., 1987. A fossil bolster plant from the King River, Tasmania. Pap. Proc. R. Soc. Tas. 121, 35–42.

Jennings, J.N., Banks, M.R., 1958. The Pleistocene glacial history of Tasmania. J. Glaciol. 3 (24), 298–303.

Kiernan, K., 1983. Weathering evidence for an additional glacial stage in Tasmania. Aust. Geogr. Stud. 21, 197–220.

Kiernan, K., 1989. Multiple glaciation of the Upper Franklin Valley, Western Tasmania Wilderness World Heritage Area. Aust. Geogr. Stud. 27, 208–233.

Kiernan, K., 1990. The extent of late Cainozoic glaciation in the Central Highlands of Tasmania. Arct. Alpine Res. 22, 341–354.

Kiernan, K., 1991. Glacial history of the upper Derwent Valley, Tasmania. NZ J. Geol. Geophys. 34, 157–166.

Kiernan, K., 1992. Glacial geomorphology and the Last Glaciation at Lake St Clair. Pap. Proc. R. Soc. Tas. 126, 47–57.

Kiernan, K., 1995. A reconnaissance of the geomorphology and glacial history of the upper Gordon River Valley, Tasmania. Tasforests 7, 51–76.

Kiernan, K., Hannan, D., 1991. Glaciation of the upper forth river catchment, Tasmania. Aust. Geogr. Stud. 29 (1), 155–173.

Kiernan, K., Fifield, L.K., Chappell, J., 2004. Cosmogenic nuclide ages for Last Glacial Maximum moraine at Schnells Ridge, Southwest Tasmania. Quatern. Res. 61, 335–338.

Lewis, A.N., 1945. Pleistocene glaciation in Tasmania. Pap. Proc. R. Soc. Tas. 1944, 41–56.

Mackintosh, A.N., Barrows, T.T., Colhoun, E.A., Fifield, L.K., 2006. Exposure dating and glacial reconstruction at Mt. Field, Tasmania, Australia, identifies MIS 3 and MIS 2 glacial advances and climatic variability. J. Quatern. Sci. 21 (4), 363–376.

Macphail, M.K., Colhoun, E.A., Fitzsimons, S.J., 1995. Key periods in the evolution of the Cenozoic vegetation and flora in Western Tasmania: the Late Pliocene. Aust. J. Bot. 43, 505–526.

McMinn, M.S., Kiernan, K., Fink, D., 2008. Cosmogenic nuclide dating in the Denison Range, Southwest Tasmania. In: Cohen, T., Household, I.

(Eds.), Program and Abstracts. Australian and New Zealand Geomorphology Group, 13th Conference Queenstown Tasmania, 63.

Nunez, M., Colhoun, E.A., 1986. A note on air temperature lapse rates on Mt. Wellington, Tasmania. Pap. Proc. R. Soc. Tas. 120, 11–15.

Page, K.J., Nanson, G.C., Price, D.M., 1991. Thermoluminescence chronology of late Quaternary deposits on the Riverine Plain of south-eastern Australia. Aust. Geogr. 22, 14–23.

Peterson, J.A., 1968. Cirque morphology and Pleistocene ice formation conditions in southeastern Australia. Aust. Geogr. Stud. 6, 67–83.

Peterson, J.A., Robinson, G., 1969. Trend surface mapping of cirque floor levels. Nature 222, 75–76.

Pollington, M.J., Colhoun, E.A., Barton, C.E., 1993. Palaeomagnetic constraints on the ages of Tasmanian glaciations. Explor. Geophys. 24, 305–310.

Spell, T.L., McDougall, I., 1992. Revisions to the age of the Brunhes-Matuyama boundary and the Pleistocene geomagnetic polarity timescale. Geophys. Res. Lett. 19, 1181–1184.

Thrush, M.N., 2008. The Pleistocene glaciations of the Cradle Mountain Region, Tasmania. Unpublished PhD thesis, University of Newcastle, New South Wales, Australia.

Chapter 75

Quaternary Glaciers of New Zealand

D.J.A. Barrell
GNS Science, Private Bag 1930, Dunedin, New Zealand, E-mail: d.barrell@gns.cri.nz

75.1. INTRODUCTION

A small mountainous landmass in an oceanic setting amidst the southern mid-latitude westerlies, New Zealand has a remarkable footprint of Quaternary glaciations preserved in landforms and near-surface deposits. The moraine record begins at the margins of modern glaciers and extends outwards geographically and backwards in time to the last glaciation and beyond. The record becomes increasingly fragmentary into the Middle and Early Pleistocene. The merging of outwash plains, or the relative geomorphological positions and morphologies of moraines in specific catchments, provides a basis for regional correlation of glacial sequences. Many localities show a progressive decrease in the extents of ice from Middle Pleistocene to Late Pleistocene glaciations. Excellent examples of late-glacial to Holocene moraines are preserved in some valleys near the highest parts of the Southern Alps. Improved dating technologies, especially surface exposure methods, have opened new avenues for quantifying the history of Quaternary glaciers in New Zealand.

Astride an active plate boundary in the south-western Pacific Ocean, the islands of New Zealand lie in the mid-latitude westerly wind belt. The axial ranges of New Zealand, best expressed in the Southern Alps of the South Island, reflect ongoing uplift in proximity to the plate boundary. This tectonic regime began in the mid-Cenozoic, but became more convergent in the latest Miocene, initiating mountain-building and leading eventually to land elevations that intersected glacial-phase snowlines. Today, higher parts of the Southern Alps also intersect interglacial snowlines, producing modern valley glaciers and ice fields.

The history of scientific discovery of New Zealand glaciations is outlined by Gage (1985) and Suggate (1990, 2004). Former ice extents have been identified largely from the morphological record of glaciogenic landforms. Relative chronologies have emerged from the stratigraphy of glaciogenic deposits or morpho-stratigraphy of landforms, aided by the degree of weathering and stratigraphy of younger cover-beds such as loess. The marine oxygen isotope timescale (MIS) provides numerical limits on past interglaciation and glaciation events, to which the relative chronologies may be correlated, commonly by 'counting-back' methods. Direct numerical chronologies have relied largely on radiocarbon dating (^{14}C), supplemented recently by luminescence dating techniques and increasingly by surface exposure dating (SED) targeting *in situ* produced cosmogenic isotopes such as ^{10}Be.

Suggate (1990) provides a more detailed introduction to New Zealand glaciations. This chapter presents an update of the state of knowledge regarding Quaternary glacier fluctuations in New Zealand.

The glacial sequence is subdivided into Early Pleistocene versus Middle to Late Pleistocene and Holocene events, in accord with the recent redefinition of the base of the Pleistocene at ~2.6 million years ago (Ma). The Early Pleistocene ended ~780 thousand years ago (ka) and was succeeded by the Middle Pleistocene spanning ~780 to ~125 ka, including MIS stages 19–6. The Late Quaternary extends from ~125 ka to the present day and comprises the Late Pleistocene and Holocene (MIS stages 5–1). This chapter utilises calendrical timescales (ka) rather than uncorrected isotopic timescales, such as radiocarbon (e.g. ^{14}C kyr). In this chapter, the term till is used in a general sense for all materials deposited in direct association with glacier ice.

75.2. EARLY PLEISTOCENE GLACIERS

Ross Glaciation—New Zealand's earliest known Late Cenozoic ice advance—is recognised from till and glacial lake beds interbedded with fluviatile gravel in the north Westland area of the South Island (Gage, 1945, 1961; Bowen, 1967). These sediments have undergone considerable tectonic tilting and folding. The type locality is at 170°49′E, 42°54′S (Fig. 75.1), but ice limits are not known and the extent of the Ross Glaciation cannot be delineated. Suggate (1990) inferred an age of 2.4–2.5 Ma from biostratigraphical evidence and by correlation to the Wanganui Basin sedimentary sequence (see below). Subsequently,

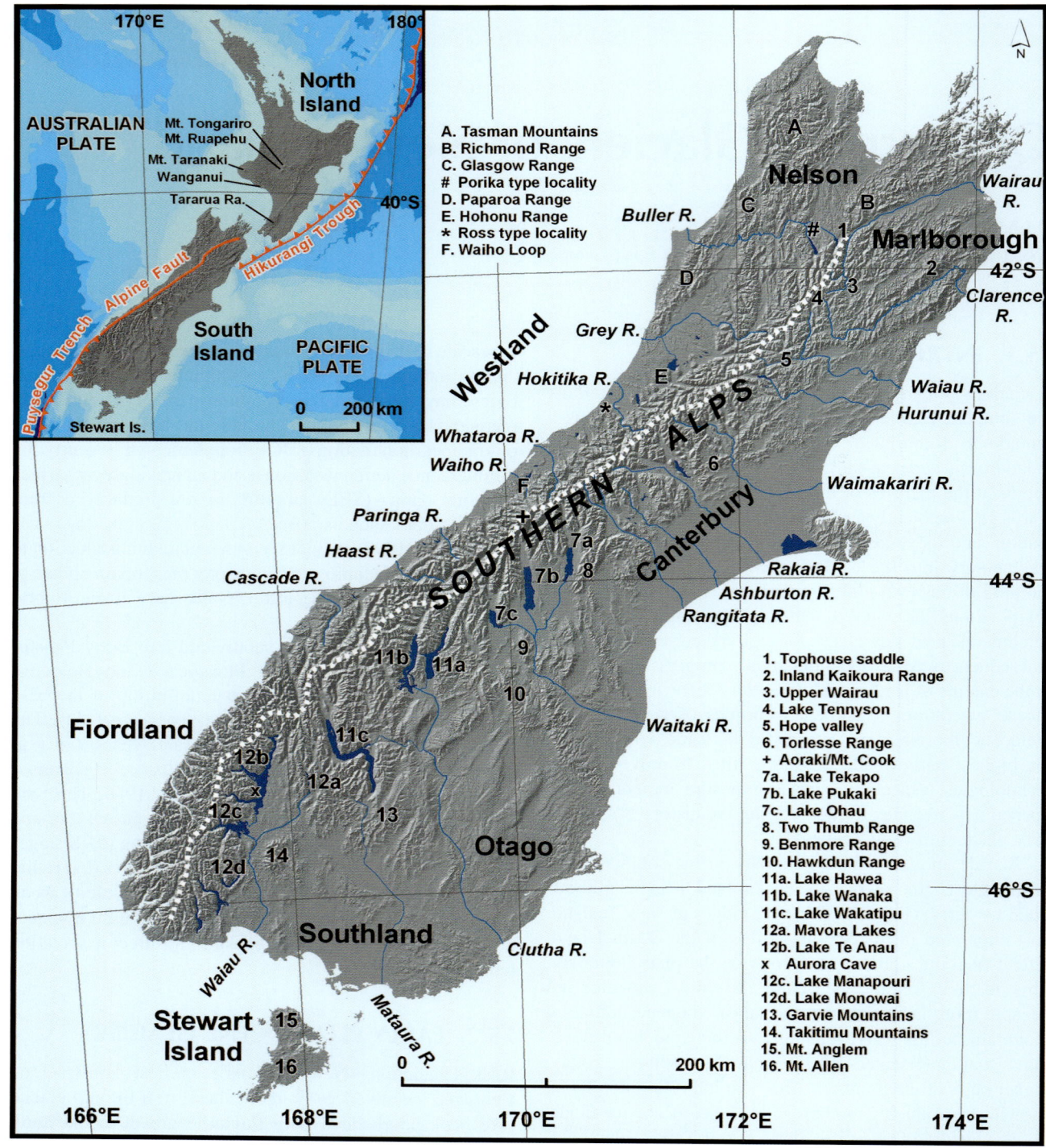

FIGURE 75.1 Plate tectonic setting of New Zealand and locations of features. The white dotted line denotes the Main Divide.

the age ranges of diagnostic fossils reported by Gage (1945) and Couper and McQueen (1954), from deposits stratigraphically beneath the Ross glacial sequence, have progressively been refined. Currently accepted age ranges (Beu et al., 2004) constrain the Ross glacial deposits to no better than Late Pliocene or younger. An Early Pleistocene age for the Ross Glaciation deposits is likely but more specific estimates are tentative. At the type area, Fitzsimons et al. (1996) reported a reverse magnetic polarity, consistent with an Early Pleistocene age during the Matuyama Chron.

Porika Glaciation (type locality 172°36′E, 41°47′S; Fig. 75.1) is known from glacial deposits resting on basement rocks or Pliocene gravel in south-east Nelson (Mildenhall and Suggate, 1981). Pollen of extinct taxa within glacially deformed lake silts indicates an age no younger than the Gelasian Stage (i.e. >1.8 Ma).

Porika glacial deposits are generally considered to be younger than Ross glacial deposits, based on biostratigraphical considerations. An age of 2.2–2.1 Ma assigned to the Porika Glaciation is based on correlation to the marine oxygen isotope record and is compatible with biostratigraphical indicators (Suggate, 1990). As with the Ross Glaciation, the specific age estimate is tentative. At the type area, Fitzsimons et al. (1996) reported a reverse magnetic polarity, consistent with an Early Pleistocene age.

Several other stratigraphical exposures of glaciogenic deposits north-east of the Ross type area were correlated to the Ross Glaciation by Bowen (1967). However, Fitzsimons et al. (1996) reported that these exposed deposits all have normal magnetic polarities, except for one with transitional polarity. These deposits probably represent glacial events that differ in age from those of the Ross type area. Whether they were formed during the Late Pliocene normal Gauss Chron, during normal subchrons in the Early Pleistocene, or subsequently, is unknown. Further research is desirable.

75.3. GAPS IN THE EARLY PLEISTOCENE GLACIAL RECORD

There is no clear stratigraphical or geomorphological record of glaciations in New Zealand during the latter part of the Early Pleistocene. This represents a period of at least 1 million years following the Porika Glaciation, assuming that its assigned age (above) is correct. The apparent gap is due, most likely, to uplift and erosion, as the configuration of ranges, basins and drainage systems evolved to their present forms. Indeed, if not for serendipitous stratigraphical exposures, we would not know of the Ross Glaciation. Porika deposits presumably have survived in a localised area due to unique combinations of favourable tectonic and geomorphological circumstances.

A spectacular record of Late Cenozoic sea-level fluctuation is represented in stratigraphical exposures of coastal Wanganui (Wanganui Basin) in the western North Island (e.g. Saul et al., 1999). In this sequence of cyclic sedimentary packages, each package is interpreted as a lowstand–highstand succession representing a glacial–interglacial glacio-eustatic fluctuation. These deposits span a period of few records of Early Pleistocene onland glaciation in New Zealand but do not shed direct light on the presence or absence of glaciers in New Zealand during the Early Pleistocene.

75.4. MIDDLE AND LATE PLEISTOCENE (AND HOLOCENE) GLACIERS

Onland records of glaciations during the past three-quarters of a million years differ from place to place in length and completeness. This may be attributable to local tectonic settings and rates of geomorphological activity.

Local names for New Zealand glaciations and interglaciations are correlated to MIS stages (Table 75.1). In addition, local Quaternary stratigraphical names have been erected in different regions or catchments, as a way of overcoming long-standing difficulties in quantifying ages for non-marine Quaternary deposits. This approach enabled a succession of relative ages to be established in areas where there is clear superposition of deposits or cross-cutting relationships between landforms (Suggate, 1965). These local stratigraphies are correlated between different areas, with varying degrees of confidence.

^{14}C dating has been the chronometrical 'mainstay' for Late Pleistocene deposits, despite limitations that include its short effective time window, the possibility of organic materials having been reworked from older deposits and, particularly in high rainfall regions, contamination by younger carbon introduced from plant roots or organic leachates. Methods such as luminescence and SED are being used increasingly for age control. Palynology has been useful within some regions, but regional climatic gradients and their effects on vegetation types limit their interregional value as a correlation tool.

From 1993 to 2010, a nationwide update of 1:250,000 scale geological maps was undertaken by GNS Science, a research institute owned by the New Zealand Government. This Quarter-million scale mapping ('QMAP') project took the bold step of mapping the ages of Quaternary deposits in terms of MIS stages, with stage numbers prefixed by 'Q'.

These ages would be little more than guesswork were it not for an important suite of morphological and stratigraphical relationships between the Grey and Hokitika rivers (Fig. 75.1). There, slow uplift has preserved a morpho-stratigraphical flight of proximal glacial outwash aggradation terraces, separated by interglacial raised shorelines and beach deposits (Suggate, 1965, 1992; Suggate and Waight, 1999). Assuming an approximately steady uplift rate, these sequences provide a good foundation for correlation to MIS stages. An important consideration is that the outwash terraces can be traced to nearby moraine sequences. Poorer preservation of landforms with increasing elevation means that confident correlations cannot be extended back beyond MIS 8 (Suggate, 1990, 2004). The Greymouth–Hokitika area is a key reference locality for correlation of Middle to Late Pleistocene glacial sequences in New Zealand.

Expanding upon earlier work (Suggate, 1965), R.P. Suggate progressively compiled maps of the limits of ice advances in New Zealand, at a scale of 1:250,000.

TABLE 75.1 New Zealand Names for Glaciations and Interglaciations (Suggate, 1990; Suggate and Waight, 1999)

MIS Stage	Glaciation	Interglaciation	Approximate Age (cal. ka)
1		Aranui	0–11.5 (by definition, Aranui includes LGIT, and Spans 0–18)
2	'Last Glacial/Interglacial transition—LGIT' (including 'Late-glacial' events – e.g. Antarctic Cold Reversal (ACR)) Late Otira		11.5–18 18–30
3	Mid-Otira		30–50
4	Early Otira		~65
5		Kaihinu	Little direct age control in NZ—refer to published MIS stage boundaries
6	Waimea		
7		Karoro	
8	Waimaunga		
9		(Not named)	
10	Nemona		
11		(Not named)	
12	Kawhaka		
(Preceding glaciations and interglaciations not assigned formal names)			
	Porika		>1.8 Ma, ?<2.6 Ma
	Ross		?<2.6 Ma

That compilation, completed in 2000, drew upon publications, interpretation of large-scale topographic maps and consultation with colleagues, but ultimately relied considerably upon personal judgement (Suggate, 2004).

Building upon the work of Suggate, this update comprises revisions and additions arising from the QMAP project, other published research and additional compilation and interpretation by the present author. The compilation includes ice limits in the North Island, which were not part of the Suggate (2004) paper. Also featured is a map of the extent of ice during the last glacial maximum (LGM), at ~20 ka (Fig. 75.2). Although keyed in to ice limits defined by well-known moraines, moraines in general are few and far between, and for the most part, the ice extent map reflects my interpretation of general geomorphology as well as consideration of the altitudes and slope aspects at which LGM ice was likely to have formed. This map represents inferred extents of LGM ice that I hope will motivate further investigation and in turn be tested and refined by future research and dating.

In the ensuing subsections, the Middle to Late Pleistocene, and Holocene, extents of glaciers in New Zealand are summarised on a region by region basis.

75.4.1. Physiographic and Climatic Setting

It is thought that the New Zealand landscape was largely in its present configuration by the Middle Pleistocene. In the central North Island, glaciation affected the highest volcanoes as well as the highest parts of the Tararua Range in the south. In the South Island, ice was extensive on the Southern Alps during glaciations and extended onto the margins of adjoining forelands. Major glaciers flowed either west or east from the Main Divide of the Southern Alps. The higher ranges west and east of the Southern Alps also carried ice during glaciations, as did the highest peaks of Stewart Island, the southernmost of New Zealand's main islands.

The prevailing westerly circulation interacts with the axial ranges, producing an orographic precipitation pattern. Voluminous precipitation west of the Main Divide contrasts with much drier conditions to the east (Henderson and Thompson, 1999). Modern glaciers are fed largely by snow brought by prevailing westerly winds. In the South Island, snowfall diminishes rapidly east of the divide, and snowlines rise eastwards. Some snow is also brought by southerly winds.

The extent to which the present-day climate regime prevailed during glacial cycles is an area of active research

Chapter | 75 Quaternary Glaciers of New Zealand

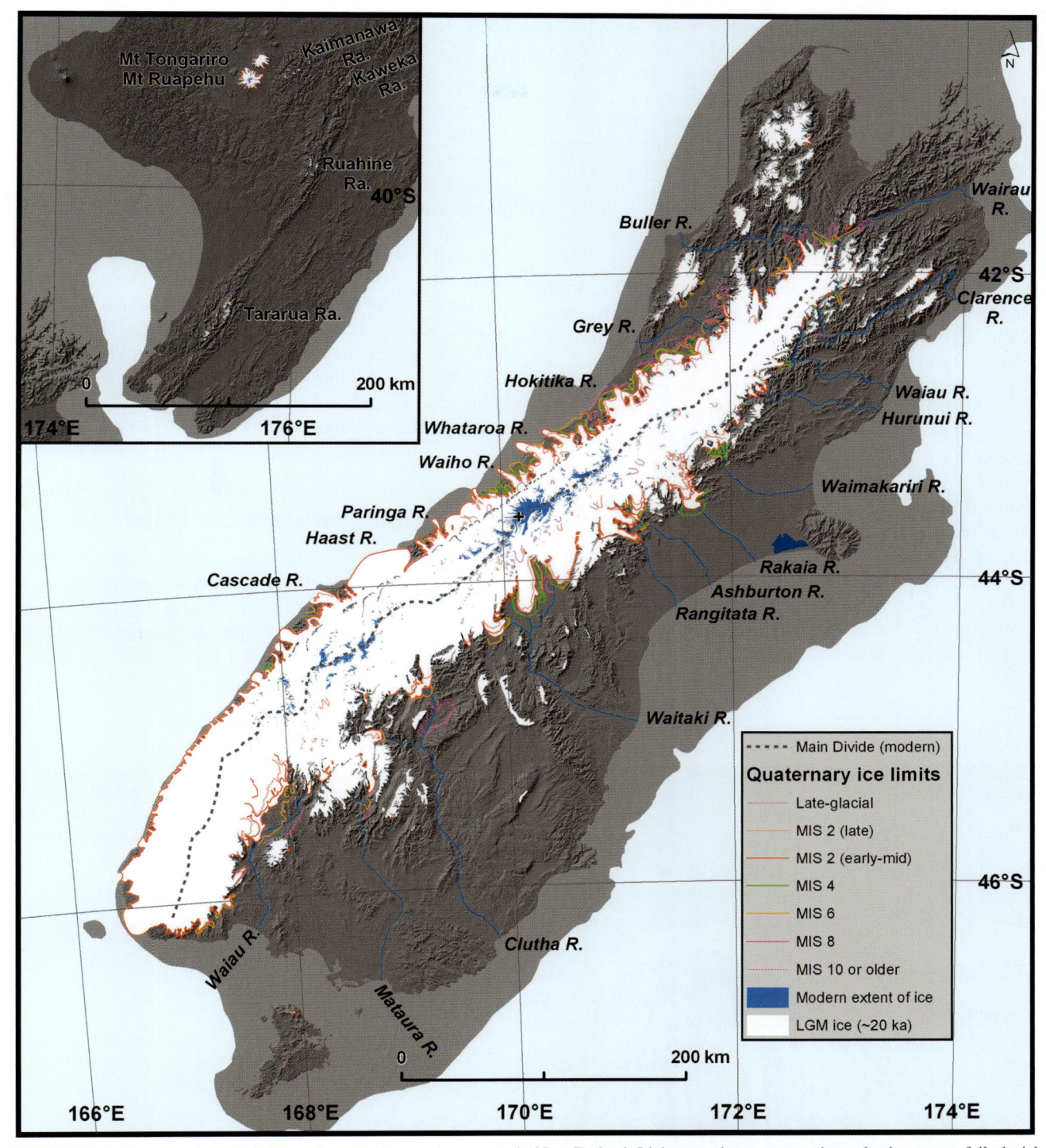

FIGURE 75.2 Interpretation of Middle to Late Pleistocene ice extents in New Zealand. Light grey denotes approximate land extent at full-glacial (−125 m) sea level.

(e.g. Alloway et al., 2007). For the past extents of glaciers, it is commonly assumed that advances either side of the divide were broadly synchronous on decadal to centennial timescales.

75.4.2. A Note on the Interglacial Record

In addition to the Greymouth–Hokitika sequence, several other South Island localities display morphological and/or stratigraphical relationships between glacial and interglacial sequences and may be fruitful targets for future research aimed at refining New Zealand glacial chronology. Localities include the uplifted terraces of the Fiordland coast (Turnbull et al., 2007, 2010 and references therein) and intercalated glacial–interglacial deposits beneath the subsiding eastern margin of the Canterbury Plains (Suggate, 1958; Brown and Wilson, 1988; Browne and Naish, 2003).

75.4.3. North Island

Only Mt Ruapehu, a large active volcano and the North Island's highest mountain, carries glaciers today (Krenek, 1959). The central North Island volcanoes Ruapehu and Tongariro had been built up sufficiently high by the Late Pleistocene to have developed valley glaciers (Mathews, 1967; McArthur and Shepherd, 1990; Townsend et al., 2008; Lee et al., 2010). Most are thought to have formed during MIS 2 and perhaps MIS 4 (Fig. 75.3).

Based on elevation and aspect, localised LGM glaciers may possibly have formed on parts of the North Island axial ranges (Fig. 75.2). Beyond doubt is the presence of Late Pleistocene ice-age glaciers on the Tararua Range (Brook, 2009a and references therein).

To the west, the Mt Taranaki volcano most likely attained its present altitude in the Holocene. No evidence for past glaciers has been found (Townsend et al., 2008).

75.4.4. North-Western South Island

Although ice free today, during glaciations, the Tasman Mountains (Fig. 75.1) carried minor ice caps from which glaciers extended down many mountain valleys (Fig. 75.2). Terminal moraines are few, only sparse remnants of aggradation terraces survive and inferred ice limits are tentative (Thackray et al., 2009). Recent studies have placed some constraints on the extents of last glaciation ice, both MIS 2 and MIS 4 (McCarthy et al., 2008) and on the late MIS 2 ice retreat (Shulmeister et al., 2001, 2005).

Highest parts of the Richmond Range have U-shaped valleys and till remnants of uncertain age (Rattenbury et al., 2006). Near the west coast, higher ranges have glacial topography in the form of south-east-facing cirques and localised U-shaped valleys, such as in the Glasgow Range and the Paparoa Range (Figs. 75.1 and 75.4).

75.4.5. North-Eastern South Island

In the ranges of central to eastern Marlborough, scattered cirque-like features near the crests of the highest peaks suggest local ice cover during glaciations. Cirque-like and moraine-like features are best developed in the Inland Kaikoura Range (Bacon et al., 2001; Fig. 75.1).

FIGURE 75.3 Paired lateral moraines (?MIS 2) of the Waihohonu valley descend away from the Tongariro summit plateau. The post-glacial cone of Ngauruhoe is in the foreground (D.L. Homer, GNS Science CN3011/4).

FIGURE 75.4 Glacier-scoured bedrock topography in the Paparoa Range (D. L. Homer, GNS Science CN35866/9).

75.4.6. Southern Alps—The North-Eastern Valleys

Ice occupied the headwaters of the east-draining Wairau, Clarence, Waiau and Hurunui rivers during glaciations (Figs. 75.1 and 75.2). Limits of successive ice advances are much farther apart in the Wairau valley than in the west-draining Buller valley. This may reflect the numerous high-altitude catchments draining to the lower reaches of the former Wairau Glacier. Confluence, or not, of tributary ice may have controlled glacier length. The most recent coalescence of ice across the Tophouse saddle between the Buller and Wairau glacier systems was during the Middle Pleistocene, probably MIS 6. Two sets of older moraine and outwash gravels were mapped by Johnston (1990). Severe weathering of the oldest is the basis for its assignment to MIS 10.

The maximum MIS 2 moraine is identified by its position at the head of the main low-level aggradation surface of the Wairau valley. This moraine was sourced from the Rainbow River tributary. The uppermost reaches of the Wairau valley lie farther upstream in a separate glacial basin. There, older subdued moraines lie 10 km or more downvalley of the MIS 2 limits, implying much larger glaciers during previous glaciations (McCalpin, 1992b). These moraines are assigned to MIS 6. Moraines attributed to MIS 2 have well-defined morphologies. Just inside the inner margin of the main belt of these moraines, McCalpin (1992b) presented radiocarbon dates bracketing a deposit interpreted as till, indicating ice advance sometime between ∼14.7 and ∼10.8 ka. The site is much farther downvalley than would be expected for late-glacial ice. If correct, it implies a need for substantial revision of the glacial chronology around the northern end of the Main Divide. Future research should investigate whether this deposit is landslide debris rather than till, noting that a large landslide lies immediately upstream, and a strand of the active Awatere Fault passes through the site.

In the upper Clarence valley, a set of moraines impounding Lake Tennyson was judged by McCalpin (1992a) to be MIS 2 to early MIS 1, based on weathering-rind dating and indirect ^{14}C evidence (Fig. 75.5). On geomorphological grounds, I follow Suggate (1965, 2004) in assigning them all to MIS 2. Limits of older glacial advances in the upper Clarence valley are poorly defined.

In the Waiau glacier system, terminal moraines attributed to MIS 2 are well preserved in the main northern valley. Farther downstream, features such as broad valley floors and deltaic deposits indicate that ice was more extensive during older glaciations (Clayton, 1968). Weathered till, assessed as MIS 8 (Rattenbury et al., 2006), near Edwards Pass between the Waiau and Clarence valleys, lies about 20 km downvalley of the MIS 2 termini. In the western part of the catchment, glaciers from several tributaries coalesced in the Hope valley. Deposits of fluvial gravel, till and lake silt are shown by luminescence dating to extend back to at least MIS 6, but moraines of only MIS 2 age are preserved (Cowan, 1990; Rother et al., 2010).

To the south, the Hurunui glacier system has well-defined MIS 2 moraines and prominent outwash aggradation terraces (Powers, 1962; Suggate, 1965). Valley form implies that ice was more extensive during older advances,

FIGURE 75.5 Lake Tennyson is impounded by a large moraine complex (left) of probable MIS 2 age (D. L. Homer, GNS Science CN22138/10H).

which presumably were responsible for high river terraces farther down the Hurunui valley, but ice limits are ill defined. Attempts to relate these terraces to interglacial marine shorelines at the coast are not convincing owing to tectonic warping and uplift in the middle to lower reaches of the valley (Suggate, 2004).

75.4.7. Southern Alps West of Main Divide—North of 43°30′S

North Westland is regarded as the type area for Middle to Late Pleistocene glaciations in New Zealand on account of detailed studies by Suggate (1965) and Suggate and Waight (1999). Nearly all the New Zealand glaciations and interglaciations (Table 75.1) are named after localities in this region. The Greymouth–Hokitika glacial terrace sequence has been linked via outwash terrace profiles north-east through to the Nelson Lakes area (Suggate, 1965; Johnston, 1990; Challis et al., 1994; Nathan et al., 2002). Correlations south-west to sequences in southern Westland are not so clear, but improvements in the confidence of correlations have come about through loess and soil stratigraphical studies, and through numerical dating (Lowell et al., 1995; Denton et al., 1999; Almond et al., 2001; Preusser et al., 2005; Suggate and Almond, 2005).

Throughout most of this region, as many as five major Middle to Late Pleistocene glacial advances are widely recognised, with their ages distinguished by relative positions, landform preservation or degrees of weathering. In some areas, ice from adjoining valley systems coalesced across low divides; in others, outwash aggradation terraces merged. In both circumstances, correlations may be extended to the wider region. Valleys that are isolated by gorges lacking terraces or by truncation at the coast contain glacial sequences whose relative ages stand alone, even though wider correlations are uncertain.

The Alpine Fault, marking the local boundary between the Australian and Pacific plates, runs along the western foot of the Southern Alps (Fig. 75.1). Right-lateral rates of strike-slip displacement are up to 26 mm/year. On timescales of the order of 10^5 years, this implies several-kilometre lateral dislocations between moraines on the western foreland and their formative glacier catchments in the Southern Alps.

Middle Pleistocene ice advances during MIS 10, 8 and 6 are recorded by patchy remnants of moraine and outwash terraces. They are best preserved in the north or north-east sectors of the main valleys, reflecting progressive north-east migration of the lowland along the Alpine Fault. These advances appear to have been more extensive than those of the Late Pleistocene.

There is a better preserved landform record of Late Pleistocene ice advances. Moraines formed during the LGM (~MIS 2) are easily defined on morphological grounds, being bordered on their inner margins by prominent moraine walls and abandoned glacial troughs occupied by post-glacial lakes or broad river floodplains. Recent work has shown the LGM record to be more complicated than originally thought, with at least three major episodes of ice advance, to full-glacial extent, between ~29 and 19 ka, thus spanning late MIS 3 and MIS 2 (Suggate and Almond, 2005). Organic sediments interbedded within till or outwash deposits constrain the timings of ice advances and concomitant outwash aggradation (Suggate, 1965; Moar, 1980; Challis et al., 1994; Lowell et al., 1995; Denton et al., 1999; Hormes et al., 2003, 2004; Suggate and Moar, 2004; Suggate and Almond, 2005).

More dissected or topographically subdued remnants of moraine or outwash outboard of the LGM moraines are assigned to MIS 4. MIS 4 glacial deposits closely follow a cooling that marked the end of MIS 5a. Although the climatic history of MIS 3 is not well established (Moar and Suggate, 1996), no evidence of early- or mid-MIS 3 ice advance has been found preserved in north Westland. In south Westland, loess stratigraphy led Almond et al. (2001) to postulate a mid-MIS 3 advance to near the full-glacial extent. However, pending further research, these postulated mid-MIS 3 moraines and outwash terraces were provisionally included within MIS 4 by Cox and Barrell (2007).

Cirques and U-shaped valleys abound above 900 m a.s.l. in the Hohonu Range, a granitic massif rising from the western foreland (Fig. 75.1). Cirques also lie on the south-eastern sides of other nearby isolated hills at similar elevations. Rarely are moraines found in association with these features, which presumably were last occupied by ice during MIS 2.

The prominent arcuate moraine known as the Waiho Loop has attracted much scientific attention (Fig. 75.1). About 1.5 km upstream of the loop, a veneer of wood-bearing till is plastered on granite bedrock at Canavans Knob (Mercer, 1988; Denton and Hendy, 1994). The current best age estimate for emplacement of the Canavans Knob till, attesting to advance of Franz Josef Glacier over this site, is ~13.1 ka (Turney et al., 2007). However, it remains unclear whether this till was emplaced by the same glacier event that formed the Waiho Loop. Ages of boulders on the Waiho Loop determined by SED methods indicate deposition at ~10.5 ka according to Barrows et al. (2007), but this interpretation has been debated (Applegate et al., 2008; Barrows et al., 2008; Tovar et al., 2008). Reconnaissance investigation of detrital clast lithologies led Tovar et al. (2008) to hypothesise a non-climatic landslide event that caused ice to advance and form the Waiho Loop, a concept explored further by Shulmeister et al. (2009). For now, the Waiho Loop, hitherto regarded as a classic expression of late-glacial climatic reversal in New Zealand, a notion supported by paleotemperature estimates derived from coupled mass balance-glacier flow models (Anderson and Mackintosh, 2006), is of uncertain status. Nonetheless, other remnants of moraine upstream of the Southern Alps range front (Fig. 75.2), for example, in the Whataroa River catchment, lie in positions commensurate with a late-glacial age (Cox and Barrell, 2007). The landform records of late-glacial climatic events in New Zealand, and associated ice limits, will benefit from further research.

Moraine sequences close to many of the modern glaciers represent mid to late Holocene ice limits. Ages have been estimated by a variety of methods including ^{14}C dating, lichenometry and dendrochronology (Gellatly et al., 1988; Suggate, 1990; McKinzey et al., 2004).

75.4.8. Southern Alps West of Main Divide—43°30 to 44°30′S

Farther north, the continental shelf is broad and all the LGM glaciers probably terminated onland during glacio-eustatic sea-level lowstands. South of about 43°30′S, the continental shelf is narrower and, particularly south of 44°S, some LGM glaciers may have had tidewater-calving termini. The most direct indication of marine influence is at Paringa River (Fig. 75.1), where marine shells were deposited in a fiord-like setting at ~16 ka (Suggate, 1968; Simpson et al., 1994). Mostly, moraines are poorly preserved, and only MIS 2 limits can be estimated with any confidence (Rattenbury et al., 2010).

A spectacular sequence of lateral moraines lies on the north-east side of the Cascade valley (Fig. 75.1). Sutherland et al. (2007) report SED ages indicating major moraine-forming events at ~80, ~60 ka, possibly ~30 and at ~20 ka. South of 44°30′S is the Fiordland region (see later Section 75.4.10).

75.4.9. Southern Alps—East of Main Divide—42°50′S to 45°S

This region comprises the eastern side of the central part of the Southern Alps, including their highest part in the Aoraki/Mt Cook area (Figs. 75.1 and 75.2), and extends south-west to the Fiordland region. The central Southern Alps contain many present-day glaciers, unlike the north-eastern sector (see above) which, for the most part, appears to have lacked Holocene glaciers.

75.4.9.1. The Waimakariri–Rakaia–Ashburton–Rangitata Glacier Systems

These four major glacier systems are linked through the coalescing of outwash fans that form the Canterbury Plains (Suggate, 1963). Within the mountains, ice margins joined locally, adding support to correlations indicated by the outwash surfaces.

Previous studies of the Waimakariri system (Gage, 1958), the Rakaia system (Soons, 1963; Soons and Gullentops, 1973) and upper Ashburton and Rangitata systems (Mabin, 1984, 1987; Oliver and Keene, 1989, 1990) provide a morphological framework that remains largely accepted

FIGURE 75.6 This view upstream along the left-lateral moraines of the Rangitata valley shows the contrast between younger, sharper moraines attributed to MIS 2 (left) and subdued older moraines (right) assigned to MIS 4 (D. L. Homer, GNS Science CN35958/20).

today (Fig. 75.6), apart from adjustments to the ages assigned to parts of the moraine and outwash sequences (Cox and Barrell, 2007; Forsyth et al., 2008). The largest departure is the marked reduction in the inferred downstream extent of older ice advances from the Waimakariri catchment. This chapter follows Forsyth et al. (2008) in only showing ice limits that are constrained unambiguously by deposits or morphology (Fig. 75.2). For example, I consider that interpreting poorly sorted coarse bouldery deposits as till, as did Gage (1958) at Otarama, is problematic in the narrow steep gorge of a large river, where sheer rock bluffs or steep tributary streams provide abundant potential sources of rock slide or debris flow material. The contrasting interpretations shown here in Fig. 75.2 versus Fig. 6 of Suggate (1990) represent alternative hypotheses that beg further research.

A number of ^{14}C ages constrain, in particular, the retreat of glaciers from MIS 2 limits (McGlone, 1995; Fitzsimons, 1996; McGlone et al., 2004; Burrows, 2005 and references therein). Reconnaissance dating in progress using luminescence and SED is expected to refine the glacial histories of these catchments (J. Shulmeister, personal communication, 2009). There is little direct age control on moraines and associated ice limits judged to be MIS 4 or older, and age estimates are based on relative positions and degrees of weathering (Fig. 75.6).

Moraines of late-glacial to Holocene age are preserved in the heads of many valleys in these catchments. At a few localities, ages have been estimated using methods such as lichenometry and weathering rinds, or more directly by ^{14}C (Birkeland, 1982; Gellatly et al., 1988; McGlone, 1995; Burrows, 2005) and ^{10}Be (Ivy-Ochs et al., 1999). Elsewhere, ages are based on geomorphological considerations (Cox and Barrell, 2007).

75.4.9.2. The Upper Waitaki Glacier Systems

During glaciations, three major valley glaciers drained from snowfields in the highest part of the Southern Alps into the Mackenzie Basin, a large intermontane depression. LGM troughs of these glaciers are occupied by lakes Tekapo, Pukaki and Ohau. The dry rain-shadow climate of the Mackenzie Basin has aided the preservation of spectacular moraine and outwash sequences. The excellent preservation is reflected by a general congruence of interpretation of the moraine sequence and ice limits by previous workers (e.g. Speight, 1963; NZ Geological Survey, 1973; Oborn, 1978; Maizels, 1989; Suggate, 1990; Cox and Barrell, 2007 and references therein). Moraines attributable to the LGM ($\sim$MIS 2) are easily recognised by their morphologies. There is little direct control on ages of belts of older moraine farther out, and age estimates for these are tentative. Generally speaking, ice extents have been progressively smaller during later glaciations (Fig. 75.2).

Late-glacial to Holocene moraines are common in the headwaters of these catchments and in adjacent ranges. The best known moraine sequences lie in the Pukaki catchment (e.g. Porter, 1975; Birkeland, 1982; Gellatly et al., 1988; Suggate, 1990; McGlone, 1995; Schaefer et al., 2009 and references therein).

Extensive dating of upper Waitaki moraines using SED methods is in progress, building on preliminary work by Schaefer et al. (2006). Increasing precision in the SED method has provided a detailed chronological picture of the mid- to late-Holocene moraines surrounding the modern glaciers near Aoraki/Mt Cook (Schaefer et al., 2009). A further improvement has been the quantification of a local production rate for *in situ* produced Beryllium-10 (Putnam et al., 2010). This production rate implies that many previously calculated ^{10}Be ages are between 12% and 14% too young. Much improved confidence in the age estimations of the moraine sequences in the upper Waitaki catchment can be expected to emerge in publications in the near future.

75.4.9.3. The Upper Clutha Glacier System

The upper Clutha glacial sequence (McKellar, 1960) may be one of the most comprehensive in New Zealand (McSaveney et al., 1992; Turnbull, 2000). In most of the eastern Southern Alps glacier systems, each advance extended a shorter distance downvalley than the previous one, but nowhere more so than the upper Clutha, where Middle Pleistocene glaciers were as much as twice as long as MIS 2 glaciers.

Uranium–thorium (U/Th) dating of Middle Pleistocene components of the glacial sequence has yielded an age framework for regional correlation of moraine/outwash sequences (Turnbull, 2000). However, this framework is not entirely satisfactory because uncertainties of the age measurements are large, and median, or in some cases, older-bound, ages are used as a basis for correlation to MIS stages. For example, the age of the glaciogenic Lindis Formation is constrained by a U/Th age of 413 ± 85 ka for a carbonate concretion in lake silt just inside the Lindis moraine limits. Ideally, one would take the younger bound (328 ka; early MIS 9) as a minimum and adopt MIS 10 as a minimum age for Lindis Formation glacial deposits. In contrast, the median age (early MIS 11) was taken as a minimum, and Lindis Formation was thus assigned to the preceding MIS 12 glacial stage (Turnbull, 2000).

Distal outwash gravel remnants, lacking preserved aggradation surfaces, are at three locations cemented by coarsely crystalline travertine that returned U/Th ages of 430 ± 108, 455 ± 115 ka and 420 ± 10 ka. These ages are minima for the host outwash gravel. Correlation of these gravel remnants to Lowburn Formation glacial deposits, at least 15 km farther upstream, is based on topographic profiling. The older bound of the oldest finite age (i.e. 570 ka; late MIS 15) appears to have been the basis for Turnbull (2000) assigning Lowburn Formation to the preceding glacial stage, MIS 16. While the assigned age is not necessarily wrong, the younger-bound values of the same ages (between 310 and 340 ka—MIS 9) could equally well be used to assign Lowburn Formation to MIS 10 or older. As Lindis Formation could justifiably be assigned to the Nemona Glaciation (MIS 10), then assignment of older Lowburn Formation to the Kawhaka Glaciation (MIS 12) is justifiable, compatible with the U/TH ages and more in keeping with the approach used in the Greymouth–Hokitika area.

Within the Late Pleistocene sequence, till regarded as having been deposited during the Albert Town advance overlies an older outwash gravel that is cemented by travertine with a U/Th age of 98 ± 35 ka. This being a maximum age for the Albert Town till, the till is thus constrained to be younger than the older bound of the travertine age (i.e. 133 ka—late MIS 6). An age of MIS 4 was adopted by Turnbull (2000) although younger (e.g. early MIS 2—favoured here based on surface form and soil development) or older (e.g. MIS 5b, 5d) ages should not be ruled out. Farther inboard is a moraine belt, which at Lake Wanaka comprises at least two ice limits. An MIS 2 age for this moraine belt is compatible with glacial lakes Wanaka and Hawea impounded at its inner margin and is consistent with a minimum age of ~ 18.3 ka for the onset of ice retreat provided by ^{14}C dating of peat on a degradational terrace beside the Lake Hawea outlet channel (McKellar, 1960).

The western part of the Clutha catchment, drained by the Kawarau River and including the large LGM glacier trough occupied by Lake Wakatipu, has a sequence of moraines and outwash terraces that has been correlated with the upper Clutha glacial sequence by valley and terrace profiling (Turnbull, 2000). Outwash from the Wakatipu basin also extended southwards into the Mataura River catchment.

Heights of moraine remnants mapped around Lake Wakatipu (Turnbull, 2000 and references therein) imply glacier surface gradients in the main Wakatipu valley ranging from 0.003 (MIS 2) to 0.01 (MIS 4–10). However, this valley is dominated by ice-smoothed bedrock, with moraines preserved only as scattered discontinuous remnants. Better preserved are trim-line features, marking the upper limit of more freshly sculpted bedrock. In the main valley, this transition climbs upstream at about 100 m/4 km (gradient of 0.02) and I suggest that it demarcates the approximate level of MIS 2 ice. Evidence for ice gradients of this order comes from glacier systems where lateral moraines are preserved continuously. Examples include the expanded Franz Josef Glacier, Westland, during MIS 2 (gradient 0.03); Tasman Glacier, Canterbury, in the late Holocene (gradient 0.03); and Pukaki Glacier, Canterbury, during MIS 2 (gradient 0.015). For this chapter, I reconstructed the MIS 2 Wakatipu glacier starting at the Kingston terminus, and progressing upstream guided by trim-line heights (Fig. 75.7). This reconstruction, which is just one interpretation, implies higher ice levels than hitherto proposed, and relatively steep ice gradients (as much as 0.05) on distributary ice lobes such as Arrowtown and South Von (Fig. 75.7).

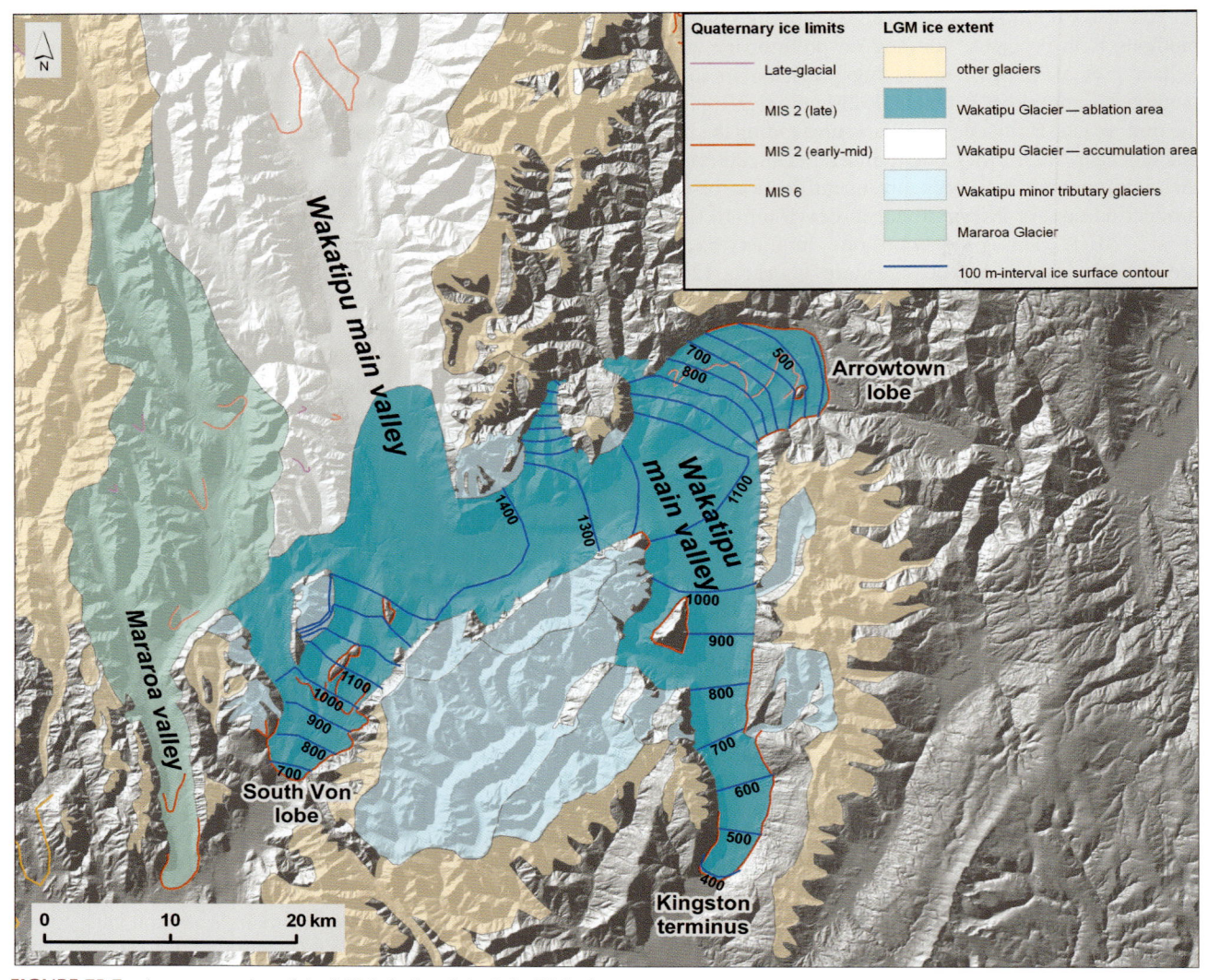

FIGURE 75.7 A reconstruction of the LGM glacier of the Lake Wakatipu area.

This interpretative scenario makes it possible for an MIS 2 age to be assigned to the South Von moraines of the true right of the middle reaches of the Wakatipu valley system. In turn, this allows more satisfactory connections to the moraines and outwash of the Mararoa Glacier (see below). Although the mapping of tills and the attendant controls on ice limits by Turnbull (2000) are robust, the proposed upvalley correlations are questionable. My reconstruction represents one alternative way of correlating moraines around this topographically complex valley system. In Fig. 75.7, I have inferred a boundary between the accumulation and ablation areas, assuming an equilibrium line altitude (ELA) of about 1500–1600 m on the main valley glacier. As depicted, the accumulation area ratio (accumulation area vs. ablation area) is about 0.65, providing some corroboration of the worth of this scenario.

In summary, I favour a glacier flow model approach to ice reconstruction and also a more sequential 'counting-back' approach to age assignment, via a more circumspect interpretation of direct ages such as by U/Th derived by valley profiling from the upper Clutha area (see above; Fig. 75.2). Overall for the Clutha catchment, this chapter offers alternative hypotheses to those summarised by Turnbull (2000). It is my hope that these alternatives will highlight uncertainties and encourage further research on the Clutha glacial sequences.

Modern glaciers in the Clutha catchment are small, and little work has been done on their post-glacial moraine sequences.

75.4.9.4. The Mararoa Glacier

The mountains between lakes Wakatipu and Te Anau–Manapouri are drained by the Mararoa River, in the middle reaches of which lie the South and North Mavora lakes, impounded by moraine (Fig. 75.8). These moraines were assigned to MIS 4 by Turnbull (1980, 2000), while MIS 2 moraines were mapped farther upvalley. It is unlikely that these narrow lakes in this large catchment would have survived subsequent infilling by sediment if they had formed as long ago as MIS 4. Following Suggate (2004), the Mavora moraines and associated outwash deposits are assigned to MIS 2 (Fig. 75.2).

75.4.10. Fiordland

The Fiordland massif forms the south-western corner of the South Island. It is for the most part formed in hard crystalline plutonic and gneissic rocks, and is deeply dissected by steep-sided U-shaped valleys headed by innumerable cirques, with an array of lakes and fiords around the margins of the massif. The massif is highest at its northern end and there carries numerous present-day cirque glaciers (Augustinus, 1992).

On its western (seaward) side, the massif has no coastal lowland and the seafloor descends steeply down the continental slope. Many glaciers would have reached the sea at the times of glacial maxima/sea-level minima, with glacier trough morphologies and submarine outwash fan features evident in fiord and continental slope bathymetries (e.g. Barnes, 2009). MIS 2 ice limits are depicted schematically (Fig. 75.2), and no earlier limits are easily recognisable. Little is known of late-glacial or Holocene ice limits because relatively few moraines are preserved in the mid to upper parts of the catchments.

To the south, a broad continental shelf flanks Fiordland and relief is gentler. Several large glacial valleys, their troughs now occupied by lakes, are bordered by moraine and outwash sequences which locally are intersected by uplifted coastal terraces (Turnbull and Uruski, 1995; Turnbull et al., 2010 and references therein).

To the east, Fiordland glaciers drained onto a valley and basin foreland. The most extensive and best preserved moraine and outwash sequences are adjacent to the two northernmost lakes, Te Anau and Manapouri (Fig. 75.9). From there, correlations are attained farther south via outwash surfaces in the Waiau valley. The Te Anau glacial deposits (McKellar, 1973; Turnbull, 1985, 1986) are subdivided into four main groups, comprising oldest to youngest, the Moat Creek Advance (Early Pleistocene?), the Whitestone Advance (MIS 6?), Ramparts 1 and 2 Advances (MIS 4?) and Marakura 1 and 2 Advances (MIS 2). No direct age control has been reported. Having reviewed the mapping on morphological grounds, I suspect that the Ramparts Advance is more likely to be early MIS 2 or perhaps late MIS 3, and that MIS 4 is either missing from the terminal moraine sequence or forms an inner part of the Whitestone moraines. This interpretation is reflected in Fig. 75.2 and leads to a satisfactory accord between the Ramparts (MIS 2) moraine/outwash of Te Anau and the Mavora Lakes moraine and outwash in the Mararoa valley (see above).

Fluvial deposits and speleothems in Aurora Cave on the western side of Lake Te Anau (Williams, 1996; Fig. 75.1)

FIGURE 75.8 A view north-west up the Mararoa valley, occupied by the Mavora Lakes. The moraine impounding the downstream lake (bottom left) is considered more likely to be MIS 2 rather than MIS 4 (D. L. Homer, GNS Science CN6490/24H).

FIGURE 75.9 A view from the rugged mountains of Fiordland south-east across the Manapouri glacial trough. The lake is studded with islands of ice-shorn bedrock and enclosed by moraines and outwash middle left to right. The Takitimu Mountains lie to the distant right (D. L. Homer, GNS Science CN10405/5).

attest to episodes of sediment aggradation (stadials) interspersed with periods of calcite deposition (interstadials). These features were interpreted to reflect glacier expansion/reduction and correlated to a set of benches on the valley side above the cave, interpreted as kame terraces marking glacier margin positions.

Williams (1996) suggested correlations between these benches and terminal moraines farther downvalley, based on similarities of heights above sea level. However, moraine morphological features east of Lake Te Anau, such as elongated ice-scoured bedrock highs, along with a rise in height of the Marakura-1 moraine belt from its terminus around to its left-lateral margin, indicate that the Te Anau glacier flowed to the south-east, with an ice surface gradient of about 100 m/4.5 km (grade of 0.02). Topographic map contours imply similar gradients on the higher, most continuous, of the benches (kame terraces or lateral moraines) above Aurora Cave. On this basis, the 560–570 m a.s.l. bench at Aurora Cave may be extrapolated at 0.02 grade to meet the Marakura-1 terminal moraine (~280 m a.s.l.) near Te Anau town, 14 km south-east of the cave, while the 660–700 m a.s.l. bench may be extrapolated 17 km south-east to the Ramparts terminal moraines (~300 m a.s.l.).

In contrast, Williams suggested correlations of these two benches to the Moat Creek and Whitestone moraines, an interpretation that is difficult to reconcile with glacier gradients and the contrasting morphologies of the well-preserved benches versus these eroded, dissected and subdued moraines. My preferred estimates of terminal moraine ages imply an MIS 2 age for the 560–570 m bench, and a likely age of MIS 2 to early MIS 3 for the 660–700 m bench. The close association inferred by Williams (1996) between cave deposits and the lower-level benches (<560 m a.s.l.) requires reconsideration. SED of the benches is in progress and may clarify this issue (P. Williams, personal communication, 2009).

Although not persuaded that the cave records offer direct connections to ice surface height or correlation to terminal moraines, I concur that the speleothem record reflects phreatic water levels and associated subaqueous and subglacial sedimentation. The periods of interstadial calcite deposition recorded and dated to high precision in Aurora Cave may well represent a unique record of times of ice minima, a parameter captured nowhere else in the glacial sequences of New Zealand.

75.4.11. South Island—Central and Eastern Ranges

Several of the highest mountain ranges to the east of the Southern Alps axial range carry geomorphological indicators of former ice cover. These include south-east-facing cirques and cirque-like basins, and over-widened, in some cases U-shaped, upper reaches of main valleys in the ranges. In most cases, it is likely that any such ice-related features were last occupied during MIS 2, though undoubtedly most of these features were initiated during earlier glaciations (e.g. Brook et al., 2008). Notable examples include the Torlesse Range (43°15′S), southern Two Thumb Range

(43°50′S), Benmore Range (44°25′S), Hawkdun Range (44°45′S), Garvie Mountains (45°30′S) and Takitimu Mountains (43°15′S).

75.4.12. Stewart Island

Stewart Island rises from a shallow area of the continental shelf and would have comprised hills surrounded by a coastal plain at times of lower sea levels. No glaciers exist today, but small cirques lie on the south-east sides of two of the highest peaks. Mt Anglem (46°45′S; ~980 m a.s.l.) has two cirques, each containing a small pond impounded by moraine (Allibone and Wilson, 1997 and references therein). Assigned a Holocene age by Turnbull and Allibone (2003), an MIS 2 age seems more likely, judging by a lack of well-preserved moraines farther downslope. The cirque moraines imply an ELA of about 750 m for the former glaciers. At Mt Allen (47°05′S; 750 m a.s.l.), a well-defined moraine ridge, probably also MIS 2, represents an ELA of ~600 m (Allibone and Wilson, 1997; Brook, 2009b).

75.5. CONCLUSIONS

New Zealand has a remarkable footprint of Quaternary glaciations preserved in landforms and near-surface deposits. Tectonic deformation, uplift and erosion have resulted in, at best, a fragmentary survival of older parts of the record, but the Late Quaternary record, right through to the present day around modern glaciers, is excellent. The advance of knowledge had, for some time, been slowed by a paucity of quantitative age control. Technologies such as SED have re-energised research, and ever-improving chronologies are enhancing the relevance and usefulness of glacier moraines as climatic indicators. There remains immense potential for further refinement and quantification of the Quaternary glacial history of New Zealand.

ACKNOWLEDGEMENTS

It has been a privilege to build upon the work of Pat Suggate. I thank Pat for the advice and insights he has provided over recent years. Many other colleagues have contributed to my understanding of glaciers, in particular, Trevor Chinn and George Denton. I am grateful for the assistance of Delia Strong with GIS data manipulation. The chapter benefited greatly from review comments by George Denton and Andrew Mackintosh.

REFERENCES

Allibone, A.H., Wilson, S., 1997. Evidence of glacial activity at Mt Allen, southern Stewart Island, New Zealand. NZ J. Geol. Geophys. 40, 151–155.

Alloway, B.V., Lowe, D.J., Barrell, D.J.A., Newnham, R.M., Almond, P.C., Augustinus, P.C., et al., 2007. Towards a climate event stratigraphy for New Zealand over the past 30,000 years (NZ-INTIMATE Project). J. Quatern. Sci. 22, 9–35.

Almond, P.C., Moar, N.T., Lian, O.B., 2001. Reinterpretation of the glacial chronology of south Westland. NZ J. Geol. Geophys. 44, 1–15.

Anderson, B., Mackintosh, A., 2006. Temperature change is the major driver of late-glacial and Holocene glacier fluctuations in New Zealand. Geology 34, 121–124.

Applegate, P.J., Lowell, T.V., Alley, R.B., 2008. Comment on "Absence of cooling in New Zealand and the adjacent ocean during the Younger Dryas Chronozone" Science 320, 746d.

Augustinus, P.C., 1992. Outlet glacier trough size-drainage area relationships, Fiordland, New Zealand. Geomorphology 4, 347–361.

Bacon, S.N., Chinn, T.J., Van Dissen, R.J., Tillinghast, S.F., Goldstein, H.L., Burke, R.M., 2001. Paleo-equilibrium line altitude estimates from late Quaternary glacial features in the Inland Kaikoura Range, South Island, New Zealand. NZ J. Geol. Geophys. 44, 55–67.

Barnes, P.M., 2009. Postglacial (after 20 ka) dextral slip rate of the offshore Alpine fault, New Zealand. Geology 37, 3–6.

Barrows, T.T., Lehman, S.J., Fifield, L.K., De Deckker, P., 2007. Absence of cooling in New Zealand and the adjacent ocean during the Younger Dryas Chronozone. Science 318, 86–89.

Barrows, T.T., Lehman, S.J., Fifield, L.K., De Deckker, P., 2008. Response to Comment on "Absence of cooling in New Zealand and the adjacent ocean during the Younger Dryas Chronozone" Science 320, 746e.

Beu, A.G., Alloway, B.V., Cooper, R.A., Crundwell, M.P., Kamp, P.J.J., Mildenhall, D.C., et al., 2004. Pliocene, Pleistocene, Holocene (Wanganui Series). Chapter 13. In: Cooper, R.A. (Ed.), The New Zealand Geological Timescale, Institute of Geological & Nuclear Sciences Monograph 22, 197–228.

Birkeland, P.W., 1982. Subdivision of Holocene glacial deposits, Ben Ohau Range, New Zealand, using relative-dating methods. Geol. Soc. Am. Bull. 93, 433–449.

Bowen, F.E., 1967. Early Pleistocene glacial and associated deposits of the West Coast of the South Island, New Zealand. NZ J. Geol. Geophys. 10, 164–181.

Brook, M.S., 2009a. Lateral moraine age in Park Valley, Tararua Range, New Zealand. J. R. Soc. NZ 39, 63–69.

Brook, M.S., 2009b. Glaciation of Mt Allen, Stewart Island (Rakiura): the southern margin of LGM glaciation in New Zealand. Geogr. Ann. 91A, 71–81.

Brook, M.S., Kirkbride, M.P., Brock, B.W., 2008. Temporal constraints on glacial valley cross-profile evolution: Two Thumb Range, central Southern Alps, New Zealand. Geomorphology 97, 24–34.

Brown, L.J., Wilson, D.D., 1988. Stratigraphy of late Quaternary deposits of the northern Canterbury Plains, New Zealand. NZ J. Geol. Geophys. 31, 305–335.

Browne, G.H., Naish, T.R., 2003. Facies development and sequence architecture of a Late Quaternary fluvial-marine transition, Canterbury Plains and shelf, New Zealand: implications for forced regressive deposits. Sed. Geol. 158, 57–86.

Burrows, C.J., 2005. Julius Haast in the Southern Alps. Canterbury University Press, Christchurch, New Zealand, 215 pp.

Challis, G.A., Johnston, M.R., Lauder, W.R., Suggate, R.P., 1994. Geology of the Lake Rotoroa area, Nelson. Scale 1:50,000. Institute of Geological & Nuclear Sciences Geological Map 8. GNS Science, Lower Hutt, New Zealand.

Clayton, L., 1968. Late Pleistocene glaciations of the Waiau valleys, north Canterbury. NZ J. Geol. Geophys. 11, 753–767.

Couper, R.A., McQueen, D.R., 1954. Pliocene and Pleistocene plant fossils of New Zealand and their climatic interpretation. NZ J. Sci. Technol. 35B, 398–420.

Cowan, H.A., 1990. Late Quaternary displacements on the Hope Fault at Glynn Wye, North Canterbury. NZ J. Geol. Geophys. 33, 285–293.

Cox, S.C., Barrell, D.J.A., (compilers), 2007. Geology of the Aoraki area. Institute of Geological & Nuclear Sciences 1:250,000 Geological Map 15. GNS Science, Lower Hutt, New Zealand.

Denton, G.H., Hendy, C.H., 1994. Younger Dryas age advance of Franz Josef glacier in the Southern Alps of New Zealand. Science 264, 1434–1437.

Denton, G.H., Heusser, C.J., Lowell, T.V., Moreno, P.I., Andersen, B.G., Heusser, L.E., et al., 1999. Interhemispheric linkage of paleoclimate during the last glaciation. Geogr. Ann. 81A, 107–153.

Fitzsimons, S.J., 1996. Late-glacial and early Holocene glacier activity in the Southern Alps, New Zealand. Quatern. Int. 38/39, 69–76.

Fitzsimons, S.J., Pollington, M., Colhoun, E., 1996. Palaeomagnetic constraints on the ages of glacial deposits in north-western South Island, New Zealand. Z. Geomorphol. 105, 7–20.

Forsyth, P.J., Barrell, D.J.A., Jongens, R., (compilers), 2008. Geology of the Christchurch area. Institute of Geological & Nuclear Sciences 1:250,000 Geological Map 16. GNS Science, Lower Hutt, New Zealand.

Gage, M., 1945. The Tertiary and Quaternary geology of Ross, Westland. Trans. R. Soc. NZ 75, 138–159.

Gage, M., 1958. Late Pleistocene glaciation of the Waimakariri valley, Canterbury New Zealand. NZ J. Geol. Geophys. 1, 123–155.

Gage, M., 1961. On the definition, date, and character of the Ross Glaciation, Early Pleistocene, New Zealand. Trans. R. Soc. NZ 88, 631–637.

Gage, M., 1985. Glaciation in New Zealand—the first century of research. Quatern. Sci. Rev. 4, 189–214.

Gellatly, A.F., Chinn, T.J.H., Roethlisberger, F., 1988. Holocene glacier variations in New Zealand: a review. Quatern. Sci. Rev. 7, 227–242.

Henderson, R.D., Thompson, S.M., 1999. Extreme rainfalls in the Southern Alps of New Zealand. J. Hydrol. NZ 38, 309–330.

Hormes, A., Preusser, F., Denton, G., Hajdas, I., Weiss, D., Stocker, T.F., et al., 2003. Radiocarbon and luminescence dating of overbank deposits in outwash sediments of the Last Glacial Maximum in North Westland, New Zealand. NZ J. Geol. Geophys. 46, 95–106.

Hormes, A., Preusser, F., Schlüchter, C., 2004. Dating the main north Westland glacial advance of the late Otira Glaciation: the Raupo section. Reply to Suggate and Moar (2004). NZ J. Geol. Geophys. 47, 156.

Ivy-Ochs, S., Schlüchter, C., Kubik, P.W., Denton, G.H., 1999. Moraine exposure dates imply synchronous Younger Dryas glacier advances in the European Alps and in the Southern Alps of New Zealand. Geogr. Ann. 81A, 313–323.

Johnston, M.R., 1990. Tophouse, Sheet N29AC. Geological Map of New Zealand 1:50, 000. Department of Scientific & Industrial Research, Wellington, New Zealand.

Krenek, L.O., 1959. Changes in the glaciers of Mount Ruapehu in 1955. NZ J. Geol. Geophys. 2, 643–653.

Lee, J.M., Bland, K.J., Townsend, D.B., Kamp, P.J.J., (compilers), 2010. Geology of the Hawke's Bay area. Institute of Geological & Nuclear Sciences 1:250,000 Geological Map 8. GNS Science, Lower Hutt, New Zealand.

Lowell, T.V., Heusser, C.J., Anderson, B.G., Moreno, P.I., Hauser, A., Huesser, L.E., et al., 1995. Interhemispheric correlation of late Pleistocene glacial events. Science 269, 1541–1549.

Mabin, M.C.G., 1984. Late Pleistocene glacial sequence in the Lake Heron basin, mid-Canterbury. NZ J. Geol. Geophys. 27, 191–202.

Mabin, M.G.C., 1987. Early Aranuian sedimentation in the Rangitata Valley, mid Canterbury. NZ J. Geol. Geophys. 30, 87–90.

Maizels, J.K., 1989. Differentiation of late Pleistocene terrace outwash deposits using geomorphic criteria: Tekapo valley, South Island, New Zealand. NZ J. Geol. Geophys. 32, 225–242.

Mathews, W.H., 1967. A contribution to the geology of the Mount Tongariro massif, North Island, New Zealand. NZ J. Geol. Geophys. 10, 1027–1039.

McArthur, J.L., Shepherd, M.J., 1990. Late Quaternary glaciation of Mt Ruapehu, North Island, New Zealand. J. R. Soc. NZ 20, 287–296.

McCalpin, J.P., 1992a. Glacial and postglacial geology near Lake Tennyson, Clarence River, New Zealand. NZ J. Geol. Geophys. 35, 201–210.

McCalpin, J.P., 1992b. Glacial geology of the upper Wairau valley, Marlborough, New Zealand. NZ J. Geol. Geophys. 35, 211–222.

McCarthy, A., Mackintosh, A., Rieser, U., Fink, D., 2008. Mountain glacier chronology from Boulder Lake, New Zealand, indicates MIS 4 and MIS 2 ice advances of similar extent. Arct. Antarct. Alp. Res. 40, 695–708.

McGlone, M.S., 1995. Late glacial landscape and vegetation change and the Younger Dryas climatic oscillation in New Zealand. Quatern. Sci. Rev. 14, 867–881.

McGlone, M.S., Turney, C.S.M., Wilmshurst, J.M., 2004. Late-glacial and Holocene vegetation and climatic history of the Cass Basin, central South Island, New Zealand. Quatern. Res. 62, 267–279.

McKellar, I.C., 1960. Pleistocene deposits of the upper Clutha valley, Otago, New Zealand. NZ J. Geol. Geophys. 3, 432–460.

McKellar, I.C., 1973. Te Anau-Manapouri District, 1:50, 000. New Zealand Geological Survey Miscellaneous Series Map 4. Department of Scientific & Industrial Research, Wellington, New Zealand.

McKinzey, K.M., Lawson, W., Kelly, D., Hubbard, A., 2004. A revised Little Ice Age chronology of the Franz Josef Glacier, Westland, New Zealand. J. R. Soc. NZ 34, 381–394.

McSaveney, M.J., Thomson, R., Turnbull, I.M., 1992. Timing of relief and landslides in Central Otago, New Zealand. In: Bell, D.H. (Ed.), Landslides/Glissements de Terrain: Proceedings of the Sixth International Symposium, February 1992, Christchurch, New Zealand, vol. 2. Balkema, Rotterdam, pp. 1451–1456, Christchurch, New Zealand.

Mercer, J.H., 1988. The age of the Waiho Loop terminal moraine, Franz Josef Glacier, Westland. NZ J. Geol. Geophys. 31, 95–100.

Mildenhall, D.C., Suggate, R.P., 1981. Palynology and age of the Tadmor Group (late Miocene-Pliocene) and Porika Formation (early Pleistocene), South Island, New Zealand. NZ J. Geol. Geophys. 24, 515–528.

Moar, N.T., 1980. Late Otiran and early Aranuian grassland in central South Island, New Zealand. NZ J. Ecol. 3, 4–12.

Moar, N.T., Suggate, R.P., 1996. Vegetation history from the Kaihinu (Last) Interglacial to the present, West Coast, South Island, New Zealand. Quatern. Sci. Rev. 15, 521–547.

Nathan, S., Rattenbury, M.S., Suggate, R.P., (compilers), 2002. Geology of the Greymouth area. Institute of Geological & Nuclear Sciences 1:250,000 Geological Map 12. GNS Science, Lower Hutt, New Zealand.

New Zealand Geological Survey, 1973. Quaternary Geology—South Island, 1:1, 000, 000. New Zealand Geological Survey Miscellaneous Series Map 6. Department of Scientific and Industrial Research, Wellington, New Zealand.

Oborn, L.E., 1978. Waitaki catchment. In: Suggate, R.P., Stevens, G.R., Te Punga, M.T. (Eds.), The Geology of New Zealand. Government Printer, Wellington, New Zealand, pp. 608–611.

Oliver, P.J., Keene, H.W., 1989. Mount Somers—Sheet K36 AC & Part Sheet K35. Geological Map of New Zealand 1:50,000. Department of Scientific & Industrial Research, Wellington, New Zealand.

Oliver, P.J., Keene, H.W., 1990. Clearwater—Sheet J36 BD and Part Sheet J35. Geological Map of New Zealand 1:50,000. Department of Scientific & Industrial Research, Wellington, New Zealand.

Porter, S.C., 1975. Equilibrium-line altitudes of late Quaternary glaciers in the Southern Alps, New Zealand. Quatern. Res. 5, 27–47.

Powers, W.E., 1962. Terraces of the Hurunui River. NZ J. Geol. Geophys. 5, 114–129.

Preusser, F., Andersen, B.G., Denton, G.H., Schlüchter, C., 2005. Luminescence chronology of late Pleistocene glacial deposits in North Westland, New Zealand. Quatern. Sci. Rev. 24, 2207–2227.

Putnam, A.E., Schaefer, J.M., Barrell, D.J.A., Vandergoes, M., Denton, G.H., Kaplan, M.R., et al., 2010. In situ cosmogenic ^{10}Be production-rate calibration from the Southern Alps, New Zealand. Quatern. Geochronol. 5, 392–409.

Rattenbury, M.S., Townsend, D.B., Johnston, M.R., (compilers), 2006. Geology of the Kaikoura area. Institute of Geological & Nuclear Sciences 1:250,000 Geological Map 13. GNS Science, Lower Hutt, New Zealand.

Rattenbury, M.S., Jongens, R., Cox, S.C., (compilers), 2010. Geology of the Haast area. Institute of Geological & Nuclear Sciences 1:250,000 Geological Map 14. GNS Science, Lower Hutt, New Zealand.

Rother, H., Shulmeister, J., Rieser, U., 2010. Stratigraphy, optical dating chronology (IRSL) and depositional model of pre-LGM glacial deposits in the Hope Valley, New Zealand. Quatern. Sci. Rev. 29, 576–592.

Saul, G., Naish, T.R., Abbott, S.T., Carter, R.M., 1999. Sedimentary cyclicity in the marine Pliocene-Pleistocene of the Wanganui basin (New Zealand): Sequence stratigraphic motifs characteristic of the past 2.5 m.y. Geol. Soc. Am. Bull. 111, 524–537.

Schaefer, J.M., Denton, G.H., Barrell, D.J.A., Ivy-Ochs, S., Kubik, P.W., Andersen, B.G., et al., 2006. Near-synchronous interhemispheric termination of the Last Glacial Maximum in mid-latitudes. Science 312, 1510–1513.

Schaefer, J.M., Denton, G.H., Kaplan, M., Putnam, A., Finkel, R.C., Barrell, D.J.A., et al., 2009. High-frequency Holocene glacier fluctuations in New Zealand differ from the northern signature. Science 324, 622–625.

Shulmeister, J., McKay, R., Singer, C., McLea, W., 2001. Glacial geology of the Cobb Valley, northwest Nelson. NZ J. Geol. Geophys. 44, 55–62.

Shulmeister, J., Fink, D., Augustinus, P.C., 2005. A cosmogenic nuclide chronology of the last glacial transition in North-West Nelson, New Zealand—new insights in Southern Hemisphere climate forcing during the last deglaciation. Earth Planet. Sci. Lett. 233, 455–466.

Shulmeister, J., Davies, T.R., Evans, D.J.A., Hyatt, O.M., Tovar, D.S., 2009. Catastrophic landslides, glacier behaviour and moraine formation—a view from an active plate margin. Quatern. Sci. Rev. 28, 1085–1096.

Simpson, G.D.H., Cooper, A.F., Norris, R.J., 1994. Late Quaternary evolution of the Alpine Fault Zone at Paringa, South Westland, New Zealand. NZ J. Geol. Geophys. 37, 49–58.

Soons, J.M., 1963. The glacial sequence in part of the Rakaia valley, Canterbury, New Zealand. NZ J. Geol. Geophys. 6, 735–756.

Soons, J.M., Gullentops, F.W., 1973. Glacial advances in the Rakaia valley, New Zealand. NZ J. Geol. Geophys. 16, 425–438.

Speight, J.G., 1963. Late Pleistocene historical geomorphology of the Lake Pukaki area, New Zealand. NZ J. Geol. Geophys. 6, 160–188.

Suggate, R.P., 1958. Late Quaternary deposits of the Christchurch metropolitan area. NZ J. Geol. Geophys. 1, 103–122.

Suggate, R.P., 1963. The fan surfaces of the central Canterbury Plain. NZ J. Geol. Geophys. 6, 281–287.

Suggate, R.P., 1965. Late Pleistocene geology of the northern part of the South Island. New Zealand. New Zealand Geological Survey Bulletin 77.

Suggate, R.P., 1968. The Paringa Formation, Westland, New Zealand. NZ J. Geol. Geophys. 11, 345–355.

Suggate, R.P., 1990. Late Pliocene and Quaternary glaciations of New Zealand. Quatern. Sci. Rev. 9, 175–197.

Suggate, R.P., 1992. Differential uplift of middle and late Quaternary shorelines, northwest South Island, New Zealand. Quatern. Int. 15/16, 47–59.

Suggate, R.P., 2004. South Island, New Zealand: ice advances and marine shorelines. In: Ehlers, J., Gibbard, P.L. (Eds.), Quaternary Glaciations—Extent and Chronology, Part III. Elsevier, Amsterdam, pp. 285–291.

Suggate, R.P., Almond, P.C., 2005. The Last Glacial Maximum (LGM) in western South Island, New Zealand: implications for the global LGM and MIS 2. Quatern. Sci. Rev. 24, 1923–1940.

Suggate, R.P., Moar, N.T., 2004. Dating the main north Westland glacial advance of the late Otira Glaciation: the Raupo section. Comment on Hormes et al., 2003. NZ J. Geol. Geophys. 47, 155.

Suggate, R.P., Waight, T.E., 1999. Geology of the Kumara-Moana area. Scale 1:50,000. Institute of Geological & Nuclear Sciences Geological Map 24. GNS Science, Lower Hutt, New Zealand.

Sutherland, R., Kim, K., Zondervan, A., McSaveney, M., 2007. Orbital forcing of mid-latitude Southern Hemisphere glaciation since 100 ka inferred from cosmogenic nuclide ages of moraine boulders from the Cascade Plateau, southwest New Zealand. Geol. Soc. Am. Bull. 119, 443–451.

Thackray, G.D., Shulmeister, J., Fink, D., 2009. Evidence for expanded middle and late Pleistocene glacier extent in northwest Nelson, New Zealand. Geogr. Ann. 91A, 291–311.

Tovar, D.S., Shulmeister, J., Davies, T.R., 2008. Evidence for a landslide origin of New Zealand's Waiho Loop moraine. Nat. Geosci. 1, 524–526.

Townsend, D., Vonk, A., Kamp, P.J.J., (compilers), 2008. Geology of the Taranaki area. Institute of Geological & Nuclear Sciences 1:250,000 Geological Map 7. GNS Science, Lower Hutt, New Zealand.

Turnbull, I.M., 1980. Walter Peak (West), Sheet E42 AC. Geological Map of New Zealand 1:50,000. Department of Scientific & Industrial Research. Wellington, New Zealand.

Turnbull, I.M., 1985. Te Anau Downs, Sheet D42AC & Part Sheet D43. Geological Map of New Zealand 1:50,000. Department of Scientific & Industrial Research, Wellington, New Zealand.

Turnbull, I.M., 1986. Snowdon, Sheet D42BD & Part Sheet D43. Geological Map of New Zealand 1:50,000. Department of Scientific & Industrial Research, Wellington, New Zealand.

Turnbull, I.M., (compiler), 2000. Geology of the Wakatipu area. Institute of Geological & Nuclear Sciences 1:250,000 Geological Map 18. GNS Science, Lower Hutt, New Zealand.

Turnbull, I.M., Allibone, A.H., (compilers), 2003. Geology of the Murihiku area. Institute of Geological & Nuclear Sciences 1:250,000 Geological Map 20. GNS Science, Lower Hutt, New Zealand.

Turnbull, I.M., Uruski, C.I., 1995. Geology of the Monowai-Waitutu Area. Scale 1:50,000. Institute of Geological & Nuclear Sciences Geological Map 19. GNS Science, Lower Hutt, New Zealand.

Turnbull, I.M., Sutherland, R., Beu, A., Edwards, A.R., 2007. Pleistocene glaciomarine sediments of the Kisbee Formation, Wilson River, southwest Fiordland, and some tectonic and paleoclimatic implications. NZ J. Geol. Geophys. 50, 193–204.

Turnbull, I.M., Allibone, A.H., Jongens, R., (compilers), 2010. Geology of the Fiordland area. Institute of Geological & Nuclear Sciences 1:250,000 Geological Map 17. GNS Science, Lower Hutt, New Zealand.

Turney, C.S.M., Roberts, R.G., de Jonge, N., Prior, C., Wilmshurst, J.M., McGlone, M.S., et al., 2007. Redating the advance of the New Zealand Franz Josef Glacier during the Last Termination: evidence for asynchronous climate change. Quatern. Sci. Rev. 26, 3037–3042.

Williams, P.W., 1996. A 230 ka record of glacial and interglacial events from Aurora Cave, Fiordland, New Zealand. NZ J. Geol. Geophys. 39, 225–241.

Chapter 76

Quaternary Glaciations of the Atlas Mountains, North Africa

Philip D. Hughes[1,*], C.R. Fenton[2] and Philip L. Gibbard[3]

[1] *Quaternary and Geoarchaeology Research Group, Geography, School of Environment and Development, The University of Manchester, Manchester M13 9PL, United Kingdom*
[2] *NERC Cosmogenic Isotope Analysis Facility, Scottish Universities Environmental Research Centre, Scottish Enterprise Technology Park, Rankine Avenue, East Kilbride, Glasgow G75 OQF, United Kingdom*
[3] *Cambridge Quaternary, Department of Geography, Downing Place, University of Cambridge, Cambridge CB2 3EN, United Kingdom*
*Correspondence and requests for materials should be addressed to Philip D. Hughes. E-mail: philip.hughes@manchester.ac.uk

76.1. INTRODUCTION

The Atlas Mountains contain the highest mountains of North Africa and some of the highest mountains that surround the Mediterranean basin, several of which exceed 4000 m a.s.l. The mountains stretch over 1500 km from central Morocco to northern Tunisia and are commonly subdivided into the High, Middle and Saharan Atlas (Fig. 76.1). The mountain chain is situated at the Africa–Eurasia plate boundary and has been uplifted and deformed together with its Palaeozoic basement, during the Mesozoic and Cenozoic. The highest peaks are formed in Palaeozoic basement granites and extrusive lavas such as basalt, andesite and rhyolite, whilst elsewhere the mountains are formed in uplifted and deformed Mesozoic carbonate rocks (Dresch, 1941; Pique, 2001; Pouclet et al., 2007). The highest peaks are situated at the south-western end of the range in the High Atlas and culminate in the Jbel Toubkal massif (4167 m a.s.l.). North of the Atlas proper are the Rif and Tell mountains which border the Mediterranean Sea and extend from northern Morocco to Tunisia. They are a folded range consisting of carbonates, flysch and crystalline basement rocks (Pique, 2001).

Glacial and periglacial activity has occurred in North and West Africa on numerous occasions. The oldest recorded is that of the Ordovician ice sheet, which existed over what was then western Gondwanaland (Ghienne, 2003). More recent, Quaternary glacial and periglacial features are in evidence throughout the Atlas Mountains (Hughes et al., 2004) like in many other mountains surrounding the Mediterranean basin (Hughes et al., 2006) and also other high mountains across Africa (Osmaston and Harrison, 2005; Mark and Osmaston, 2008). However, no modern glaciers exist although periglacial activity is thought to be still active above 2000 m (Couvreur, 1966; Robinson and Williams, 1992).

Climatically, the Atlas Mountains experience humid conditions in the northern section bordering the Mediterranean Sea and progressively drier conditions towards the southwest. The climate is strongly seasonal with dry summers and precipitation in winter (Delannoy, 1971). In the north, in regions bordering the Mediterranean, such as the Djurdjura massif in the Greater Kabylie area of the Algerian Tell, precipitation exceeds 2000 mm (Vita-Finzi, 1969, p. 54). At Ifrane, in the Middle Atlas, Morocco, mean annual precipitation (1961–1990) is also relatively high at 1118 mm (World Meteorological Organisation, 1998). In contrast, at Midelt, in the lee of the Middle Atlas at the foot of the northern-eastern High Atlas, the 1957–1989 annual precipitation average is just 208 mm (World Meteorological Organisation, 1998), although totals vary considerably year on year from < 100 to nearly 500 mm (Rhanem, 2009). Precipitation values from the highest parts of the High Atlas are rare, although Messerli (1967) noted mean annual precipitation totals of ca. 800 mm on the north-western flanks of the Toubkal massif. The eastern flanks of the Atlas range, draining into the Sahara region, are particularly arid with annual precipitation as low as 200 mm in the south-east foothills of Jbel Toubkal in the High Atlas (Messerli, 1967; Rhanem, 2009).

76.2. THE HIGH ATLAS

The highest mountains in North Africa exist in this region, including the Jbel Toubkal (4165 m a.s.l.), Irhil M'Goun (4071 m a.s.l.) and Jbel Ayachi (3751 m a.s.l.). Glacial

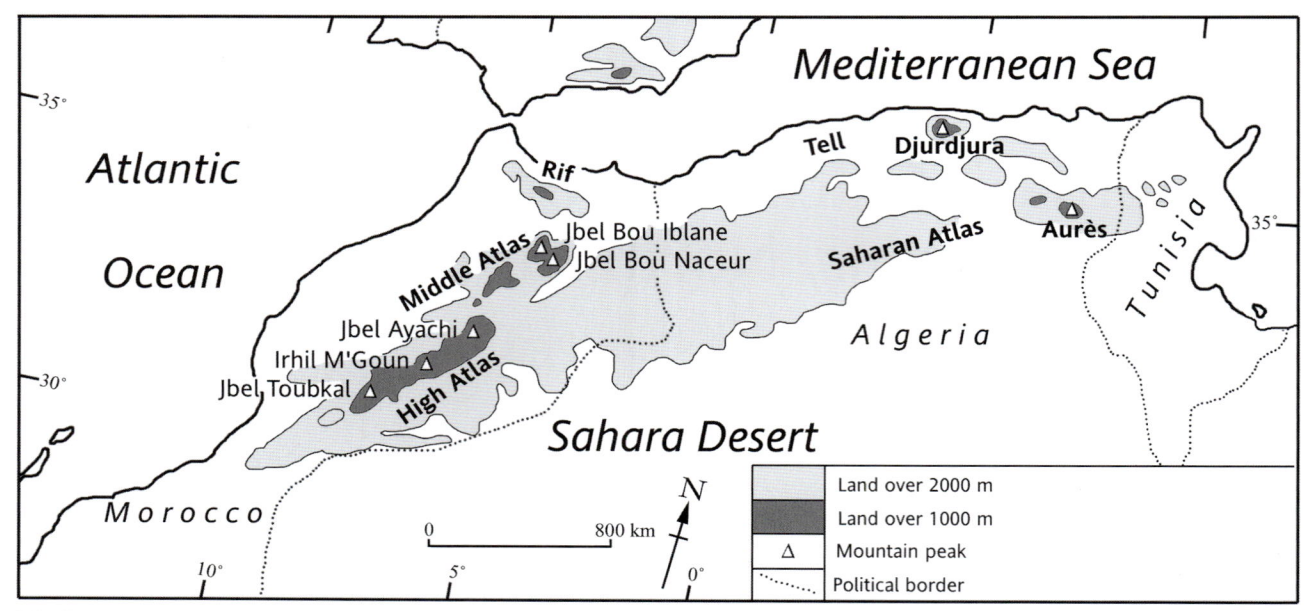

FIGURE 76.1 Location map of the Atlas Mountains indicating mountain areas mentioned in the text.

features such as cirques, troughs, roche moutonée, riegels and moraines have been noted in all these massifs (Dresch, 1941, 1949; Heybrock, 1953; Mensching, 1953; Wiche, 1953; Awad, 1963; Beaudet, 1971).

The largest glaciers formed in the Toubkal massif, where valley glaciers emanated from a central ice field which formed between the two highest summits, Toubkal (4167 m a.s.l.) and Ouanuokrim (4067 m a.s.l.). The northern outlet glacier draining this ice field extended nearly 10 km to an altitude of just 2000 m a.s.l., ca. 1 km downvalley of the shrine of Sidi Chamarouch. Valley glaciers also formed on the northern slopes of Aksoual and Bou Iguenouane. Well-defined cirques are present in the upper catchments around Toubkal and on the northern slopes of Aksoual and Bou Iguenouane and these cirques contain moraines (Figs. 76.2 and 76.3). In fact, at least three separate moraine units are present in these mountains, widely separated in altitude with some moraines present at elevations as low as 2000 m a.s.l. (Fig. 76.4).

Dresch (1941) estimated that the Pleistocene snowline was situated at 3600–3800 m a.s.l., whilst later workers have derived slightly lower estimates at 3400–3500 m a.s.l. (Mensching, 1953) and at 3300–3400 m a.s.l. (Awad, 1963). However, most researchers have underestimated the extent of glaciation in the Toubkal area and overestimated the altitude of Pleistocene snowlines. Dresch (1941) and Mensching (1953) both recognised clear moraines in the higher valleys but underestimated the extent of glaciation in some valleys where snowlines are as low as 3000 m a.s.l. and sometimes lower. In addition, some features have been misidentified. For example, Dresch (1941) interpreted large accumulations of boulder debris in some glaciated valleys as rock glaciers, whereas these features better resemble landslides as a result of rock-slope failure. Nevertheless, these coarse boulder deposits, like at Arroumd (also spelt Aremd or Aremdt), are closely associated with glaciation since rock-slope failures and resultant landslides are common in the Mediterranean mountains and frequently occur in overdeepened glacial valleys (Figs. 76.5 and 76.6). Furthermore, it is possible that rock-slope failures produced a catastrophic landslide onto the surface of a retreating glacier in this valley. This is supported by the presence of lateral moraines bounding rock-slope-failure debris on both sides of the valley (Fig. 76.6). Cosmogenic nuclide exposure ages from boulders on both the surfaces of the landslide debris and the lateral moraines will help test this scenario further. A similar mechanism of paraglacial glacier–landslide interaction has been suggested for the formation of the barrage at Lac D'Ifni on the southern slopes of Toubkal (Celerier and Charton, 1923).

Until recently, the age of the High Atlas moraines was a matter of speculation. However, a programme of cosmogenic isotope analyses is now underway at the UK Natural Environment Research Council Cosmogenic Isotope Analysis Facility, at East Kilbride in Scotland. This dating programme is focusing on dating glacial landforms in the Toubkal massif and is applying ^{10}Be analyses to quartz veins and ^{36}Cl analyses to andesite boulders to provide estimates of exposure history for the moraines in this area. Preliminary ages are calculated using the Dunai (2001) scaling scheme in the CRONUS-Earth online calculator (Balco et al., 2008; wrapper script version 2.2, main calculator 2.1, constants 2.2.1 and muons 1.1). Three ^{10}Be ages from boulders on the highest and stratigraphically youngest

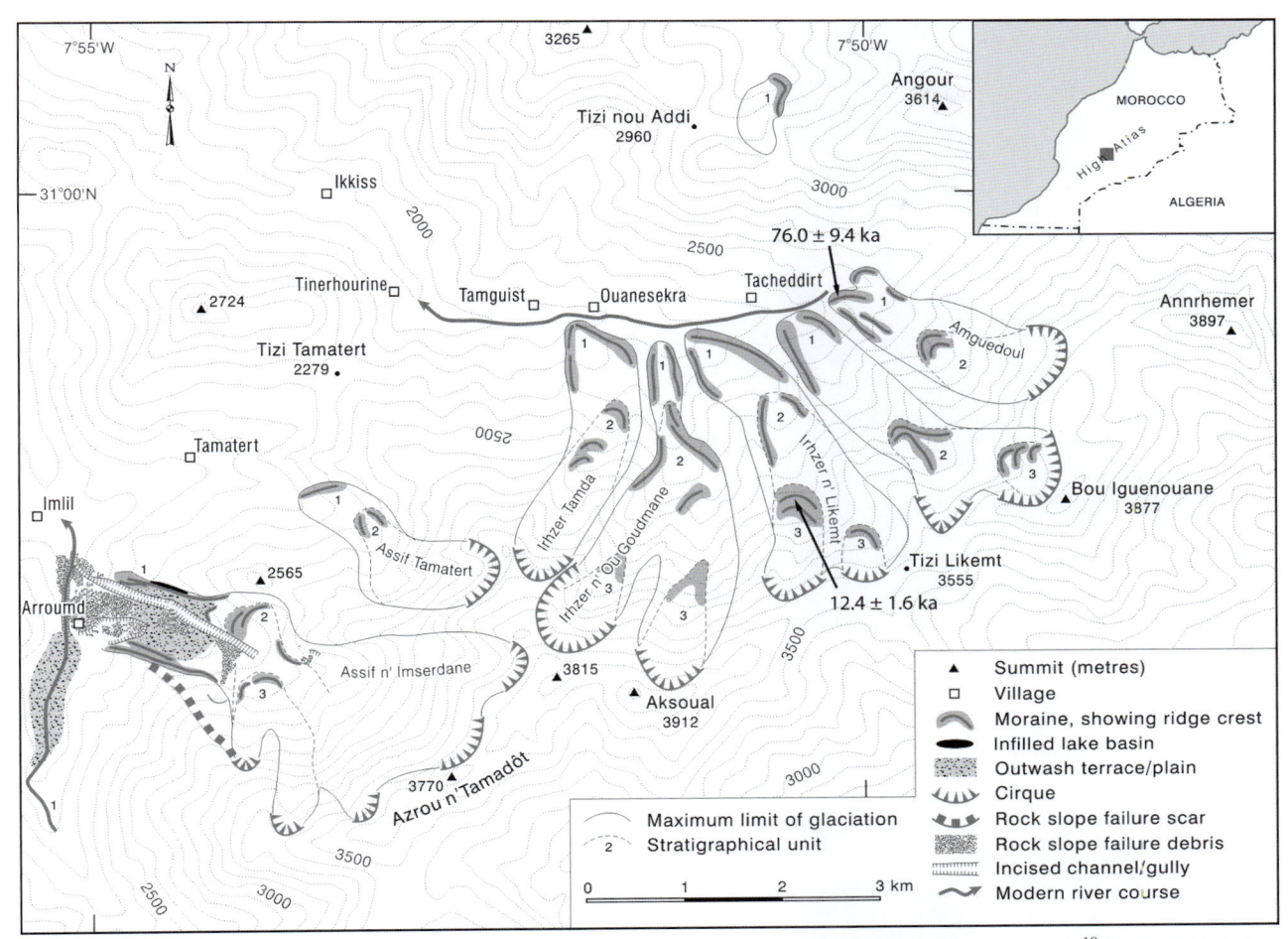

FIGURE 76.2 Glacial geomorphological map of the northern slopes of Aksoual (3912 m a.s.l.), High Atlas, Morocco, with ^{10}Be ages indicated.

FIGURE 76.3 Cirque moraines (Units 2 and 3) on the northern slopes of Bou Iguenouane (3877 m a.s.l.) in the High Atlas, Morocco (see Fig. 76.2 for location).

FIGURE 76.4 Lateral moraines at ca. 2000–2200 m a.s.l. on the northern slopes of Aksoual. The flat light-coloured area to the right of the moraine crest is an infilled basin containing several metres accumulation of lake sediments.

FIGURE 76.5 Landslide deposits resulting from a rock-slope failure in a glaciated valley near Arroumd in the High Atlas (see Fig. 76.2 and 76.6 for location).

moraines in the Azib Mzik and Irhzer Likemt valleys (Figs. 76.2 and 76.7; glacial unit 3; 12.4 ± 1.6, 12.2 ± 1.5 and 11.1 ± 1.4 ka) suggest that glaciers formed moraines in the High Atlas during the African Humid Period (14.8-5.5 ka). Interestingly, these preliminary ages fall within, or overlap within error, the Younger Dryas (12.9–11.7 ka)—a well-known cold reversal in the North Atlantic. It must be noted that despite being the highest moraines in the Azib Mzik valley, the moraines are very low (ca. 2000–2100 m a.s.l.) compared with the peak altitudes in this area, and lower than the equivalent moraines in the Irhzer Likemt (2900–3000 m a.s.l.). This may be because of strong local topoclimatic controls on glacier development at this particular site. Nevertheless, it does raise the question: 'how large were equivalent glaciers elsewhere in the nearby mountains and did Holocene glaciers survive in the highest cirques?' Another ^{10}Be age, this time from the lowest and stratigraphically oldest moraines (Fig. 76.2; glacial unit 1), yielded an age of 76.0 ± 9.4 ka (uncorrected for erosion). Finally, a ^{10}Be age from an intermediate moraine ridge, between the

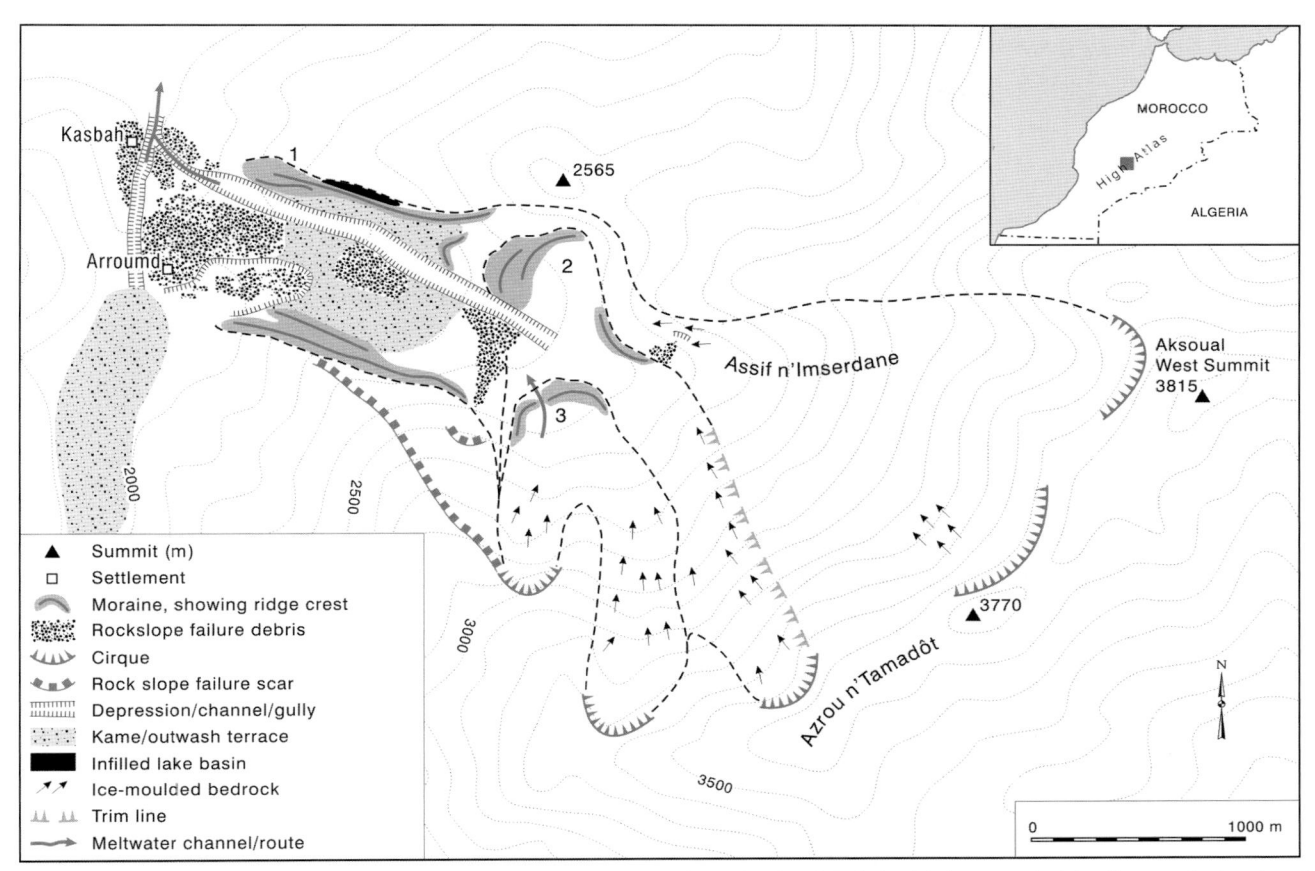

FIGURE 76.6 Geomorphological map of the Arroumd valley (Assif n'Imserdane) showing the relationship between glacial sediment units (numbered 1–3) and rock-slope-failure deposits.

lowest and the highest moraines in the Azib Mzik valley, yielded an exposure age of 24.4 ± 3.0 ka (Fig. 76.7). These preliminary results demonstrate that there is a large difference in exposure history between the oldest and youngest moraines in the High Atlas. The preliminary evidence of separate stratigraphical units yielding exposure ages of ca. 76, 24 and 12 ka suggest evidence for at least three glacial events widely separated in time. However, additional dating is needed to be confident of the timings of glaciations, and eventually, this will provide a platform from which to better understand the relationship between glacier expansions and the climatic state of the Sahara.

Several hundred kilometres north-east of the Toubkal massif, the glacial features of the Irhil M'Goun and Jbel Ayachi area were recorded by Wiche (1953) and Awad (1963). Wiche (1953) mapped the cirques and associated moraines on the northern slopes of both the highest peak of Irhil M'Goun (4070 m a.s.l.) and on nearby Jbel Ouaougoulzat (3770 m a.s.l.). On Jbel Ayachi too, cirques are well developed (Awad, 1963), although no detailed glacial geomorphological mapping has been done in this area.

Periglacial features are widespread in the High Atlas, and features such as solifluction lobes, rasentrappen, thufurs, polygons, stone stripes and felsenmeer are active today above ca. 2000 m a.s.l. (Couvreur, 1966). Frosts are frequent, even in modern times, where in winter, minimum temperatures at 2000 m a.s.l. are often in the range 0 to $-10\ °C$ and can fall to $-20\ °C$ (Robinson and Williams, 1992). Blockfields are widespread on Jbel Toubkal, and frost-shattered bedrock forms extensive talus slopes which supply the lower valleys with huge amounts of debris. For example, the valley containing Lac D'Infi (2312 m a.s.l.), to the east of Toubkal, is choked with thick accumulations of debris. On Toubkal, Chardon and Riser (1981) reported an active rock glacier in the western cirque above the Toubkal (formerly Neltner) refuge. However, the feature could equally be interpreted as a rock-slope-failure deposit, and similar features occur at a very wide range of altitudes on Toubkal. Relict rock glaciers have been reported elsewhere in the High Atlas, such as on Irhil M'Goun and Jbel Ouaougoulzat (Wiche, 1953) and at Arroumd and several other sites near Toubkal (Dresch, 1941). However, in the Toubkal area at least, many of these 'rock glaciers' may in fact be better interpreted as landslides resulting from rock-slope failures—possibly in association with glacier retreat (Celerier and Charton, 1923).

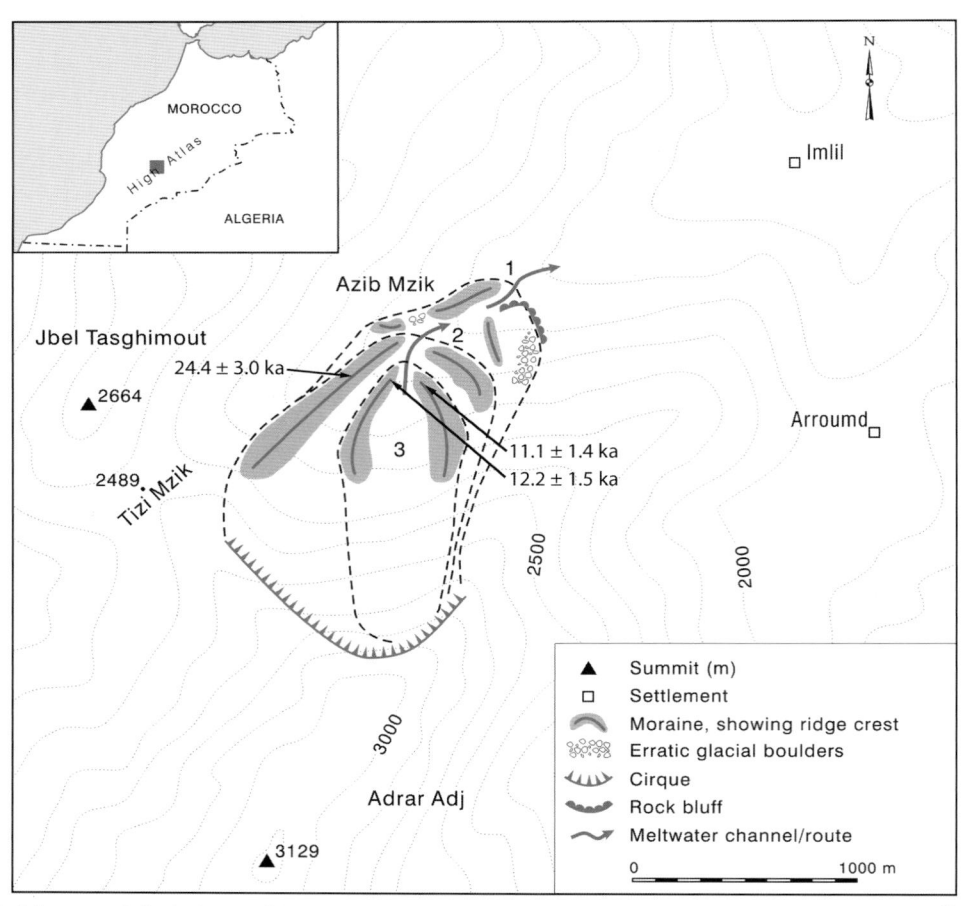

FIGURE 76.7 Glacial geomorphological map of the Azib Mzik area on the northern slopes of Adrar Adj (3129 m a.s.l.) with ^{10}Be ages indicated.

Even today, in some sheltered cirques on Toubkal, snow fields are perennial, although the true snowline lies slightly above the highest peaks, probably at ca. 4200 m a.s.l. (Messerli, 1967). A notable snowfield exists below the northern cliffs of the Tazaghart plateau. This feature was mapped as 'névé permanent' on French maps dating from the 1940s and a large snowfield was observed in July 2008, although whether it is still truly perennial is unclear. The size and nature of this feature in recent centuries is also unclear and it would not be implausible to hypothesise the presence of small niche glaciers at this site and others around the High Atlas during the 'Little Ice Age', as has been found in other mountains around the Mediterranean basin (e.g. González Trueba et al., 2008; Hughes, 2010).

76.3. THE MIDDLE ATLAS AND THE RIF

The Middle Atlas lies to the north-east of the High Atlas in central Morocco. Here, the highest peaks include Jbel Bou Iblane (3340 m a.s.l.) and Jbel Bou Naceur (3310 m a.s.l.), both of which show evidence of former glaciation (Dresch and Raynal, 1953; Raynal et al., 1956; Awad, 1963; Beaudet, 1971). Periglacial features are also widely observed in the Middle Atlas as in the High Atlas further south. Stone polygons, solifluction features and rock glaciers are described on Bou Iblane and Jbel Bou Naceur by numerous workers (Raynal, 1952; Dresch and Raynal, 1953; Awad, 1963). Further details on the extensive work of René Raynal in this region can be found in Joly (2002).

The former regional snowline in the Middle Atlas is estimated to have been at ca. 2800 m a.s.l. during the most extensive glacial phase (Awad, 1963). According to Awad (1963), the most remarkable collection of glacial troughs in the Atlas Mountains is to be found on the southern and eastern slopes of Jbel Bou Iblane. Terminal and lateral moraines are widely preserved and extend down to ca. 2400–2500 m a.s.l. (Raynal et al., 1956). Awad (1963) also stated that rock glaciers are especially abundant between 2100 and 2500 m in the eastern valleys of the Bou Naceur massif, although, given the doubtful interpretations of rock glaciers made in earlier literature by other workers in the High Atlas, caution must be given to these claims until further research has been undertaken. Further north, the Rif mountains are thought to have lain beneath the regional snowline, although, according to Mensching (1960), there is evidence for former perennial snow patches and, again, rock glaciers on the highest mountain Tidirhin-Kette (2456 m a.s.l.).

76.4. ALGERIAN ATLAS

Glacial and periglacial features have been noted in the Djurdjura massif of the Algerian Tell (Barbier and Cailleux, 1950; Büdel. 1952; Tihay, 1972; Tihay, 1973) and in the Aurès massif of the Saharan Atlas (Ballais, 1983). In the Djurdjura massif (2308 m), to the north-west, cirques, U-shaped valleys and terminal moraines are all in evidence, according to Barbier and Cailleux (1950). Glacial deposits extend to exceptionally low altitudes for this latitude, reaching as low as 750 m a.s.l. on the northern slopes and 1270 m a.s.l. on the west. The formation of glaciers was probably aided by snow accumulation in dolines—a topographical situation also identified as important for former glaciation elsewhere in the Mediterranean (Hughes et al., 2007), and even today, snow lies in such dolines throughout the summer above 2000 m a.s.l. This probably arises from high precipitation of this area, which exceeds 1500 mm, of which most falls during the winter months. The Pleistocene snowline here, during the most extensive glacial phase, was as low as 1900 m a.s.l., although Büdel (1952) puts the snowline slightly higher at 2100 m a.s.l. As is the case today, this probably results from the influence of maritime air masses from the nearby western Mediterranean.

South-east of the Djurdjura massif, in the Aurès massif of the Saharan Atlas, Ballais (1983) noted the presence of moraines on Jbel Ahmar Khaddou (2017 m a.s.l.) and Jbel Mahmel (2321 m) above 1600 m a.s.l. However, none were noted on the highest peak, Jbel Chélia (2326 m a.s.l.). On the Jbel Mahmel, two phases of glaciation are evident. It would appear that conditions in this region were marginal to glaciation and glaciers only formed in exceptional topographic localities. This is no doubt due to the proximity of the arid Sahara region and low precipitation levels compared with the Djurdjura massif.

There has been very little glacial research in Algeria since the early 1980s. The extent and characteristics of the former glaciers remain very unclear. In fact, the same is true for the Middle Atlas and the Rif as well as large areas of the High Atlas, with the exception of the Toubkal massif—although even here there are still a lot of unanswered questions as to the extent and timing of past glaciations.

76.5. PALAEOGLACIERS IN NORTH AFRICA: THEIR IMPORTANCE FOR UNDERSTANDING PALAEOCLIMATES

The former glaciers of the Atlas Mountains in North Africa are strategically positioned for understanding Pleistocene cold-stage atmospheric circulation between the North Atlantic and the Mediterranean Sea. However, former glaciers in the Atlas Mountains have the added advantage of being in close proximity to the largest desert on earth—the Sahara. The nature of the Sahara Desert in the past, especially during Pleistocene cold stages, has been the subject of debate. Traditionally, cold stages were considered to have been drier than interglacials as a result of a cooler global atmosphere with lower moisture capacity and a large amount of the Earth's hydrological budget being locked up in large ice sheets (Flint, 1971). Thus, deserts such as the Sahara were considered to have expanded during cold stages and retracted during interglacials. However, 'to equate glaciations with aridity and interglacials with an increase in desert rainfall is to oversimplify' (Williams et al., 1998). Indeed, there is widespread evidence that the end of the last cold stage over the Sahara region was characterised by a much wetter climate than present, during a time known as the African Humid Period (14.8–5.5 ka). For much of this interval, the Sahara was almost entirely covered by vegetation, whilst the mid and late Holocene saw major desert expansion and associated dust flux over the Saharan region. In fact, moisture supply to the Sahara region has fluctuated throughout the Pleistocene, and DeMenocal (2008) indentified several 'wet' phases during the last glacial cycle.

Important palaeoclimatic questions remain unanswered regarding the sources of moisture supply during glaciations. For example, did Atlantic depressions deliver moisture equitably between latitudes 30°N and 45°N during cold-stage glaciations, or were there major latitudinal differences? The position of the Polar Front during cold stages at the latitude of southern Portugal (Ruddimann and McIntyre, 1981) may suggest that cyclogenesis occurred at this latitude. If so, then one might expect Atlantic depressions to track through the Straits of Gibraltar delivering moisture to both North Africa and Iberia (cf. Florineth and Schlüchter, 2000; Fig. 76.8). In the Alboran Sea, just east of the Straits of Gibraltar, Cacho et al. (1999) have shown that north-westerly winds were dominant and strengthened during stadials associated with Heinrich Events in the North Atlantic within the last cold stage and it is possible that this reflects a southerly shift in the track of depressions at these times. The extent to which Atlantic depressions penetrated North Africa may be reflected by a marked increase in glacier snowlines between the Middle and High Atlas, for example. Alternatively, there may be no differences in snowlines at particular times, and if so, this may suggest that spatial variations in glacier snowlines were controlled by precipitation rather than summer temperatures.

Another major question is 'what were the relative roles of Atlantic depressions and the West African Monsoon in delivering moisture to the Atlas Mountains (and perhaps southern Iberia)?' Various hypotheses have been put

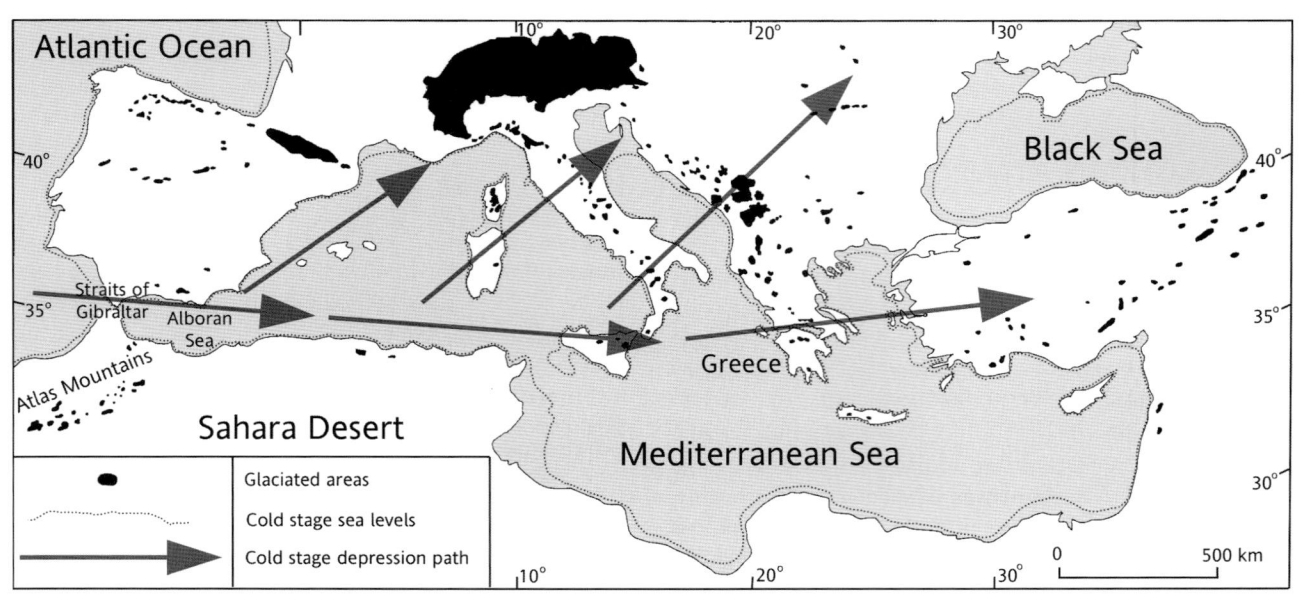

FIGURE 76.8 Distribution of Pleistocene glaciation around the Mediterranean basin and the postulated track of depressions during Pleistocene cold stages (Florineth and Schlüchter, 2000).

forward to explain the African humid periods. One hypothesis is that the West African Monsoon was strengthened when sea-surface temperatures in the North Atlantic were much cooler than present, such as during the Late Pleistocene/Early Holocene transition when the vast Laurentide ice sheet was decaying and releasing cold meltwater into the North Atlantic. This resulted in a steep thermal gradient between the North Atlantic Ocean and the Western Sahara, thereby enhancing the West African Monsoon (Renssen et al., 2003)—a situation promoted by cold upwelling in the Atlantic off the coast of North Africa at the end of cold phases such as the Younger Dryas (12.9–11.7 ka) and the 8.2 ka Event (Adkins et al., 2006). This produced a greater latitudinal temperature gradient between the tropics and the mid-latitudes over Africa, which effectively forced the West African Monsoon further north over the current Sahara region. Another condition thought to be favourable to this mechanism during the Late Pleistocene/early Holocene transition was the peak in solar radiation on the precessional orbital cycle at ca. 10.5 ka (DeMenocal et al., 2008). This effectively intensified tropical heating and thus further amplifying the temperature contrasts between the North Atlantic Ocean and the Tropics causing the West African Monsoon to penetrate far into North Africa, and possibly into southern Europe (DeMenocal, 2008, DeMenocal et al., 2008). As noted earlier, DeMenocal (2008) found evidence for this mechanism occurring several times during the last cold stage, and it is likely that this fluctuation in the West African Monsoon also occurred during other cold stages.

In addition to providing an environmental proxy adjacent to the largest desert region on Earth, the glacial record of the High Atlas is also crucial for understanding the capacity of these mountain areas to support refugial populations of plant and animal species during Pleistocene cold stages. The mountainous peninsulas of southern Europe, as well as the mountains of North Africa, are thought to have hosted biotic refugia during Pleistocene cold stages (Hewitt, 2000). It is from these refugial centres that plants and animals colonised Europe during subsequent interglacials. The presence of glaciers in these proposed refugial centres during Pleistocene cold stages has important bearing on moisture supply as well as temperature, and large glaciers in the uplands would have limited the availability of 'temperate' refugial sites in topographically sheltered sites such as valley floors. Moreover, the strategic position of the High Atlas, in close proximity to routes of Palaeolithic human migration from Africa to Europe (Gibert et al., 2003, Finlayson et al., 2006), highlights the importance of understanding environmental conditions these mountains during Pleistocene cold stages.

76.6. CONCLUSIONS

The glacial history of the Atlas Mountains is important for several reasons. First, glaciers have shaped the high mountain landscape we see today and also affected geomorphological processes long after glacier retreat, such as through valley oversteepening, which caused subsequent major landslides and alluvial fans that are still active today. Second, evidence of past glaciers also tell us a great deal about past climate in the Atlas Mountains, which is of major significance given the strategic position of these mountains for understanding palaeo-atmospheric circulation, bordering the North Atlantic Ocean and the Mediterranean Sea. Third, the Atlas Mountains also border the Sahara—the largest modern desert region on Earth. Palaeoglaciers effectively act as palaeotemperature and palaeoprecipitation gauges and documenting their presence, size and altitude at different times during the Quaternary is of crucial importance for understanding palaeoenvironmental conditions in the Sahara region—especially moisture supply. Finally, the palaeoclimate record available from evidence of past glaciers in these mountains is important in providing an understanding of the environmental conditions endured by plant and animal species, including early humans, during Pleistocene cold stages. Recent progress towards establishing detailed geochronologies for the different glaciations will help constrain the palaeoclimatic and wider palaeoenvironmental interpretations that can be drawn from this record (Hughes and Woodward, 2008). Eventually, this data can be combined with other glacial records around the Mediterranean basin and used to help in our understanding of the climatological, geomorphological and biological evolution of southern Europe and North Africa.

ACKNOWLEDGEMENTS

This research was funded by a Thesiger-Oman International Fellowship to Philip Hughes from the Royal Geographical Society (with IBG) and NERC awards for cosmogenic isotope support (NERC CIAF Allocation Nos. 9038.1007 and 9070.1009).

REFERENCES

Adkins, J., DeMenocal, P., Eshel, G., 2006. The 'African Humid Period' and the record of marine upwelling from excess 230^{Th} in Ocean Drilling Program Hole 658 C. Paleoceanography 21, PA4203. doi:10.1029/2005PA001200.

Awad, H., 1963. Some aspects of the geomorphology of Morocco related to the Quaternary climate. Geogr. J. 129, 129–139.

Balco, G., Stone, J.O., Lifton, N.A., Dunai, T.J., 2008. A complete and easily accessible means of calculating surface exposure ages or erosion rates from ^{10}Be and ^{26}Al measurements. Quat. Geochronol. 3, 174–195.

Ballais, J.-L., 1983. Moraines et glaciers quaternaires des Aurès (Algerie). In: 108ème Congrès national de Sociétés savantes, Grenoble, 1983, Géographie, 291–303.

Barbier, A., Cailleux, A., 1950. Glaciare et périglaciaire dans le Djurdjura occidental (Algérie). CR Séances 'Académie Sci. Paris 231, 365–366, Juillet-Décembre 1950.

Beaudet, G., 1971. Le Quaternaire Marocain: édat des études. Rev. Géogr. Maroc 20, 3–56.

Büdel, J., 1952. Bericht über klimamorphologische und Eiszeitforschungen in Niederafrika. Erdkunde 6, 104–132.

Cacho, I., Grimalt, J.O., Pelejero, C., Canals, M., Sierro, F.J., Flores, J.A., et al., 1999. Dansgaard-Oeschger and Heinrich event imprints in Alboran Sea palaeotemperatures. Palaeoceanography 14, 698–705.

Celerier, J., Charton, A., 1923. Un lac d'origine glaciarire dans le Haut Atlas (Le lac d'Ifni). Hespéris 3, 501–513.

Chardon, M., Riser, J., 1981. Formes et processus géomorphologiques dans le Haut-Atlas marocain. Rev. Géogr. Alpine 69, 561–582.

Couvreur, G., 1966. Les formations périglaciares du Haut Atlas central Marocaine. Rev. Géogr. Maroc 10, 47–50.

Delannoy, H., 1971. Aspects du climat de la région de Marrakesh. Rev. Géogr. Maroc 20, 69–106.

DeMenocal, P.B., 2008. Africa on the edge. Nat. Geosci. 1, 650–651.

DeMenocal, P.B., Ortiz, J., Guilderson, T., Adkins, J., Sarnthein, M., Baker, L., et al., 2008. Abrupt onset and termination of the African Humid Period: rapid climate responses to gradual insolation forcing. Quatern. Sci. Rev. 19, 347–361.

Dresch, J., 1941. Recherches sur l'évolution du relief dans le Massif Central du Grand Atlas le Haouz et le Sous. Arrault et Cie, Maitres Imprimeurs, Tours. 653 pp.

Dresch, J., 1949. Sur des formations de remblaiement continental et la présence de formes glaciaires dans le Haut Atlas calcaire. CR Sommaire Séances Soc. Géol. Fr. 9–10, 169–171.

Dresch, J., Raynal, R., 1953. Les formes glaciaires et périglaciaires dans le Moyen Atlas. CR Sommaire Séances Soc. Géol. Fr. 11/12, 195–197.

Dunai, T., 2001. Influence of secular variation of the magnetic field on production rates of in situ produced cosmogenic nuclides. Earth Planet. Sci. Lett. 193, 197–212.

Finlayson, C., et al., 2006. Late survival of Neanderthals at the southernmost extreme of Europe. Nature 443, 850–853.

Flint, R.F., 1971. Glacial and Quaternary Geology. Wiley, New York, 892 pp.

Florineth, D., Schlüchter, C., 2000. Alpine evidence for atmospheric circulation patterns in Europe during the last glacial maximum. Quatern. Res. 54, 295–308.

Ghienne, J.-F., 2003. Late Ordovician sedimentary environments, glacial cycles, and post-glacial transgression in the Taoudeni Basin, West Africa. Palaeogeogr. Paleoclimatol. Palaeoecol. 189, 117–145.

Gibert, J., Gibert, L., Iglesias, A., 2003. The Gibraltar Strait: a Pleistocene door of Europe? Hum. Evol. 18, 147–160.

González Trueba, J.J., Martín Moreno, R., Martínez de Pisón, E., Serrano, E., 2008. 'Little Ice Age' glaciation and current glaciers in the Iberian Peninsula. Holocene 18, 551–568.

Hewitt, G.M., 2000. The genetic legacy of the Quaternary ice ages. Nature 405, 907–913.

Heybrock, W., 1953. Eiszeitliche Gletscherspuren und heutige Schneeverhältnisse im Zentralgebiet des Hohen Atlas. Z. für Gletscherk. Glazialgeol. 2, 317–321.

Hughes, P.D., 2010. Little Ice Age glaciers in Balkans: low altitude glaciation enabled by cooler temperatures and local topoclimatic controls. Earth Surf. Process. Land. 35, 229–441.

Hughes, P.D., Woodward, J.C., 2008. Timing of glaciation in the Mediterranean mountains during the last cold stage. J. Quatern. Sci. 23, 575–588.

Hughes, P.D., Gibbard, P.L., Woodward, J.C., 2004. Quaternary glaciation in the Atlas Mountains, North Africa. In: Ehlers, J., Gibbard, P.L. (Eds.), Quaternary Glaciation—Extent and Chronology. Volume 3: Asia, Latin America, Africa, Australia, Antarctica. Elsevier, Amsterdam, pp. 255–260.

Hughes, P.D., Woodward, J.C., Gibbard, P.L., 2006. Glacial history of the Mediterranean mountains. Prog. Phys. Geogr. 30, 334–364.

Hughes, P.D., Gibbard, P.L., Woodward, J.C., 2007. Geological controls on Pleistocene glaciation and cirque form in Greece. Geomorphology 88, 242–253.

Joly, F., 2002. Hommage à René Raynal (1914–2002). Géomorphologie 8, 269–271.

Mark, B.G., Osmaston, H.A., 2008. Quaternary glaciation in Africa: key chronologies and climatic implications. J. Quatern. Sci. 23, 589–608.

Mensching, H., 1953. Morphologische Studien im Hohen Atlas von Marokko. Würzburger Geographische Arbeiten H.1, 104 pp.

Mensching, H., 1960. Bericht und Gedanken zur Tagung der Kommission für Periglazialforschung in der IGU in Marokko vom 19. bis 31. Oktober, 1959. Z. Geomorphol. N.F. 4, 159–170.

Messerli, B., 1967. Die eiszeitliche und die gegenwartige Vertgletscherung im Mittelemeerraum. Geogr. Helv. 22, 105–228.

Osmaston, H.A., Harrison, S.P., 2005. The late Quaternary glaciation of Africa: a regional synthesis. Quatern. Int. 138, 32–54.

Pique, A., 2001. Geology of Northwest Africa. Beiträge Zur Regionalen Geologie Der Erde, Band 29. Gebrüder Borntraeger.

Pouclet, A., Aarab, A., Fekkak, A., Benharraf, M., 2007. Geodynamic evolution of the northwestern Paleo-Gondwanan margin in the Morocco Atlas at the PreCambrian-Cambrian boundary. Geol. Soc. Am. Spec. Publ. 423, 27–60.

Raynal, R., 1952. Quelques examples de l'action du froid et le neige sur les formes du relief au Maroc. Notes Marocaines 2, 14–18.

Raynal, R., Dresch, J., Joly, F., 1956. Deux exemples régionaux de glaciation quaternaire au Maroc: Haut Atlas Oriental, Moyen Atlas Septentrional. In: IV Congrès INQUA, Rome-Pisa, pp. 108–117.

Renssen, H., Brovkin, V., Fichefet, T., Goosse, H., 2003. Holocene climate instability during the termination of the African Humid Period. Geophys. Res. Lett. 30 (4), 1184. doi:10.1029/2002GL016636, 2003.

Rhanem, M., 2009. L'Alfa (Stipa tenacissima L.) dans las plaine de Midelt (haut bassin versant de la Moulouya, Maroc)—elements de climatologie. Physio Géo Géogr. Phys. Environ. 3, 1–20.

Robinson, D.A., Williams, R.B.G., 1992. Sandstone weathering in the High Atlas, Morocco. Z. Geomorphol. 36, 413–429.

Ruddimann, W.F., McIntyre, A., 1981. The North Atlantic Ocean during the last deglaciation. Palaeogeogr. Palaeoclimatol. Palaeoecol. 35, 145–214.

Tihay, J.-P., 1972. Modelés cryonival et glaciaire dans la haute montagne Algérienne: l'exemple de la chaîne du Djurdura (Grande Kabylie). Rev. Géogr. Montréal 26, 447–463.

Tihay, J.-P., 1973. Note sur quelques paléoformes <periglaciaires> observées en Algérie orientale. Mediterranée 13 (2), 37–47.

Vita-Finzi, C., 1969. The Mediterranenan Valleys. Geological Changes in Historical Times. Cambridge University Press, London, 140 pp.

Wiche, K., 1953. Klimamorphologische und talgeschichtliche Studien im M'Goungebiet. Mitt. Der Geographischen Ges. Wien 95, 4–41.

Williams, M., Dunkerley, D., De Decker, P., Kershaw, P., Chappell, J., 1998. Quaternary Environments, second ed. Arnold, London, 329 pp.

World Meteorological Organisation, 1998. 1961–1990 Global Climate Normals. National Climatic Data Center, Asheville, NC, US Electronic resource. (CD-ROM).

Quaternary Glacial Chronology of Mount Kenya Massif

William C. Mahaney

Quaternary Surveys, 26 Thornhill Avenue, Thornhill, Ontario, Canada L4J 1J4

The 1:250,000 scale map showing moraine distributions on Mount Kenya is based primarily on Plates 1–5 in Mahaney (1990) which summarize the Mount Kenya Quaternary History Project. Mount Kenya, one of the principal mountains of East Africa, rises to 5199 m a.s.l., and contains a record of Quaternary events reaching as far back as the Plio-Pleistocene boundary. In this sense, Mount Kenya stands alone, not only as a major mountain, exceeded in altitude by Kilimanjaro, which is much younger, but also much higher than its nearest rivals, Elgon and the Aberdares, and closer in elevation to the Virungas and the Ruwenzori (Fig. 77.1). Mount Kenya has undergone many glaciations since the crystallization of the main volcanic plug 2.64 Ma (Everden and Curtis, 1965) revised to ~2.71 Ma (Veldkamp et al., 2007); as well, it contains a wealth of interglacial palaeosols that have provided much important information on palaeoenvironments between glaciations.

For just over a century, Mount Kenya has attracted the attention of glacial geologists and ecologists who have been keen to refine the chronology of episodic glaciation and climatic change (Gregory, 1894; Baker, 1967; Hastenrath, 1984; Mahaney, 1990), which are the underpinnings necessary to understand the dispersal of remnant flora and fauna (Coe, 1967; Harmsen, 1989) and genesis of soils and palaeosols. After Gregory, Mackinder (1900) climbed Batian (5199 m a.s.l.) and offered new insights into the biogeography of the mountain, but unfortunately, most of his specimens were lost on the way home. After several years of scientific inactivity, Eric Nilsson (1935) in 1927 and 1932 mapped moraines in Gorges Valley near the Nithi Waterfall (Liki moraines of Mahaney, 1990); his observations (Nilsson, 1931) of an 'early till', below the 3200-m contour in the Nithi Catchment on the eastern flank, were not followed up until Mahaney carried out detailed investigations starting in 1983.

Nilsson's observations are particularly important as he was the first to describe the moraine sequence around Harris Tarn, on the north-eastern flank of the mountain, into which the Kolbe Glacier descended in the first few decades of the twentieth century. The Kolbe Glacier has since melted away and will soon be followed by many of the other small glaciers on the mountain (Mahaney, 1990; Mizuno, 2005). Moreover, Nilsson was the first to surmise that since the moraine sequence on Mount Kenya correlated closely with Mount Kilimanjaro, Mount Elgon and the Ruwenzori Mountains, it was quite likely that all resulted from the same climatic change. Nilsson's maps of the eastern flank of the mountain are among the most accurate ever made on any of the East African mountains.

At about the same time (1934), Troll and Wien began glaciological investigations on Mount Kenya which resulted in a photogrammetric survey and flow measurements (Troll and Wien, 1949). Just after WWII, Spink (1945, 1949) began monitoring glacier recession, a series of observations that have continued on and off (Hastenrath, 1984; Mahaney, 1990) up to the present day. In 1947, Zeuner (1949) studied soil frost structures and loess on the mountain, focusing attention on needle ice development, a peculiar process involving diurnal freezing in the tropical mountains. Subsequent work by Mahaney (1987, 1990) showed, among other things, that loess exists in the surface of almost all soils on Mount Kenya and, in particular, that it is represented by quartz-rich horizons in the epipedons of interglacial palaeosols.

During the IGY, a number of reconnaissance studies in glaciology, glacial geology, ecology, meteorology and hydrology were undertaken in various catchments on the mountain and briefly documented by Baker (1967). Baker also mapped the bedrock and published the first detailed bedrock map of the mountain which proved of considerable use to the Mount Kenya Quaternary History Project initiated in 1976. Between 1981 and 1986, the U.K. Geological Survey began a 5-year mapping project, starting on the northern flank of the mountain including the northern

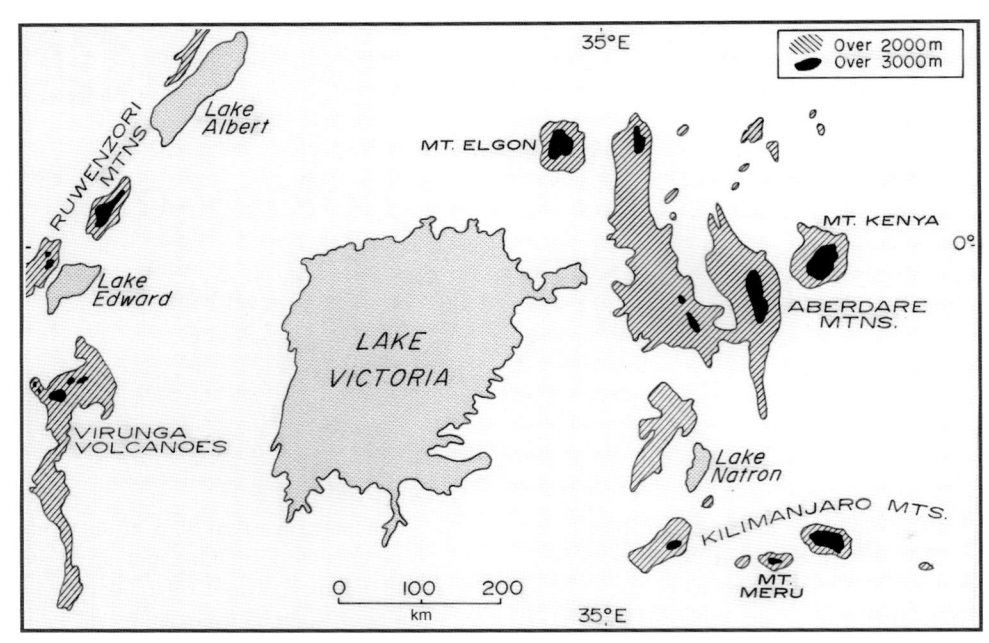

FIGURE 77.1 Location map showing major mountain areas in East Africa.

catchments. Surficial geological studies conducted under the aegis of this project (Charsley, 1989) provided new information on the age of outwash fans emplaced during times of expanded glacier activity on the mountain.

The Central Plateau of East Africa, of which Mount Kenya is a part, is an area of intense tectonic activity that has waxed and waned since the early Miocene (Mahaney, 1990). Following fractures as old as Cretaceous, a system of rift valleys formed in the Late Tertiary with enormous lakes; major faulting led to down dropping of several rift valleys, while others remained several hundred metres above sea level. Because the Rift Valleys formed from fractures in the crust, immense lava flows and volcanoes, such as Mount Kenya (5199 m a.s.l.), formed aligned with major fracture zones. Not all mountain massifs are volcanic as some formed as erosional remnants (Cherangani Hills north of Mount Kenya) and others from regional faulting of meta-sedimentary rocks (Mount Elgon on the Uganda–Kenya border).

These lava flows, overlain by loess, are principally trachyandesite. They form the base of the Mount Kenya Volcanic Series, which in the early literature is described as being of probable Miocene/Pliocene age. Mahaney et al. (2011) reported $^{39}Ar/^{40}Ar$ dates ($\sim$5.2 to $\sim$5.5 Ma) and reversed magnetizations which establish a latest Miocene to earliest Pliocene age for two flows on the west flank of the mountain near Mweiga (0°18′S; 37°48′E). Palaeosols interbedded with the lavas indicate generally dry climatic conditions during the Late Miocene, punctuated with humid events during the Pliocene and Quaternary. The succession depicts a relatively long and complex weathering history, initially of lava plus loess during a 300-k interval, followed by loess deposition during the Pliocene and Quaternary, The palaeosols were episodically deflated when surfaces were devoid of vegetation and wind systems intensified during periods of climatic deterioration. Such weathering histories within palaeosol profiles are also documented on nearby Mount Kenya, where well-weathered lower palaeosol horizons developed on Matuyama-age tills are overlain by much younger, less weathered horizons developed on Brunhes-age loess. The geochronology of Late Miocene lavas reported here provides maximum ages for weathering histories of palaeosols formed in the xeric tropical highland climate at $\sim$1800 m a.s.l.

The succession of volcanic rock which forms the Mount Kenya Massif amounts to a thickness of $\sim$6200 m overlying the African Precambrian basement series. Estimates of erosion of the summit vary from over 2000 (Gregory, 1921) to $<$1000 m (Mahaney, 1990). The mountain itself has been rather stable since the time of its formation in the late Pliocene with faulting restricted to minor displacements of only tens of metres.

A full discussion of the climate on Mount Kenya is given by Mahaney (1990). While not known with precision, the MAT at 3000 m a.s.l. is estimated at close to 10 °C; at 4000 m a.s.l. at +3 °C which conforms to a lapse rate of 6.5 °C/km (International Standard Atmosphere). Data on precipitation collected during the IGY show the wettest drainages to be on the west and south, fed by thunderstorms. The eastern flank of the mountain is dry above 3000 m altitude, sufficiently so to allow the ingress of adders thriving in the grasslands there.

Below the 3000-m contour, the climate becomes wet with precipitation well over 1 m/year and, during the monsoon (April–July), it is exceedingly humid. The northern slopes are often under anticyclonic activity and very dry. In effect, this climatic distribution led to the development of the largest modern glaciers on the western flank where they would be least starved of moisture, the smallest glaciers restricted to the north. The ancient Pleistocene climatic regime was quite different, with the largest glaciers on the eastern side of the mountain, in Gorges Valley. Presumably, during each glaciation, there may have been an upward shift in the monsoon, which fed moisture in much greater quantity onto the upper eastern slopes.

Along with Quaternary climatic changes, pollen evidence indicates that vegetation belts shifted by as much as 1500 m (Coetzee, 1967). Each descent of the timberline displaced the Afroalpine flora into a larger area, perhaps twice its present extent. Under a mean annual temperature 7 °C colder than today, the freezing limit came very close to the 3000-m elevation level. Palaeosol evidence for timberline shifts, as indicated by the lower altitude at which buried organic-rich Afroalpine epipedons are found, is 2600 m. With present timberline located at ~3200 m a.s.l., the floristic shift may have been considerably less.

Moraine positions, outlined on the Mount Kenya map and taken largely from Mahaney (1990, Plates 1–5), are based on geologic mapping and stratigraphic data compiled into the chronology shown in Fig. 77.2. However, the map shading approximates the terminal positions of the older glaciation (~2900–3000 m a.s.l.), and the younger Liki terminal limits at ~3200 m a.s.l. The area of the glaciers shown in summit area is from maps more than three decades old, and the present surface area is <1.0 km^{-2} or about half. The glacier to the right, the Lewis–Gregory, has been separated into two very small systems for the past 20 years.

The oldest glaciations, Gorges, Lake Ellis and Naro Moru, are known only from the eastern and south-eastern flanks of the mountain. Gorges Glaciation, which includes the Olduvai subchron, is older than 1.9 Ma with an indeterminate upper limit at <1.7 Ma. The younger Lake Ellis Glaciation is older than 0.78 Ma judging by its reversed remnant magnetism (Barendregt and Mahaney, 1988) and younger than the Olduvai subchron. With advanced weathering rinds on surface clasts, loss of steep slopes and sharp moraine crests and thick, well-weathered palaeosols, the Gorges and Lake Ellis palaeosols document the longest weathering histories in the tropical mountains (Mahaney, 1990). Both the Gorges and Lake Ellis moraines contain well-weathered tills at depth, with surface horizons of loess yielding normal remnant magnetism indicating a much younger Brunhes age. This suggests a period of loess stripping followed by renewed aeolian input during a subsequent glaciation (see Mahaney et al., 1997).

The interpretation of moraine ages in lower Gorges Valley and attempted revision of the chronology based on isotopic ages put forward by Shanahan and Zreda (2000) are clearly at odds with ages assigned by Mahaney (1990), the latter based on field geological relationships, weathering and palaeomagnetism. In summary, the Liki I moraines described by Mahaney (1990) could not possibly have 400 ka ages as proposed by Shanahan and Zreda (2000), the latter based on mean concentrations of ^{36}Cl from three boulders, hardly a reliable population of data. Gorges Valley has 300-m high cliffs that extend well above reconstructed glacier surfaces, and quite likely the isotopic age of the boulders results from exposure to the atmosphere over ~300–400 kyr, followed by mass wasting onto the ice during the Liki Glaciation (<90 ka). Clearly, attempts by Shanahan and Zreda (2000), Mark and Osmaston (2008) and Osmaston and Harrison (2005) to limit glaciation on Mount Kenya to 400 ka are based on flawed field examination of deposits and read too much into too few isotopic dates. Further, their interpretation flies in the face of weathering and palaeomagnetic data both of which indicate the oldest Gorges tills were emplaced during the Olduvai subchron, clearly a difference of 1.5 million years. Rearranging the chronology of the mountain, based on too few sporadic cosmogenic ^{36}Cl dates on boulders that likely began accumulating cosmic Cl long before a particular glaciation occurred, creates considerable confusion that hardly adds to the true time-stratigraphic framework of the mountain (Mahaney, 1990).

Younger tills of Naro Moru age are known from excavations in Gregory's left lateral moraine in Teleki Valley and from surface occurrences in Gorges Valley (see Mount Kenya Map; Fig. 77.1). As with older stratigraphic deposits, till of Naro Moru age is well weathered and contains palaeosols (formerly called pre-Teleki) that contain elevated levels of metahalloysite, taken to record well-developed leaching during time of formation and hence a strong and more humid climate. Still younger palaeosols, those associated with Teleki (cf. Illinoian in North America; Saale in Europe) tills and loess, are known from several drainages on the mountain: Teleki, Gorges and Hobley, also situated beyond the Liki moraines on the flanks of Ithanguni, a 3894 m a.s.l. parasitic cone 10 km to the northeast of Mount Kenya. As with the Teleki tills, most older drifts probably lie buried underneath younger sediments awaiting excavation by erosion at some future time.

The post-Teleki palaeosol at site TV23 (Mahaney, 1990) is also the type section for the Teleki till. When Gregory (1894) reported the pre-last glacial age of the lateral moraine in which TV23 is situated, he could not know the wealth of information that lay recorded in weathered deposits below the surface. The ~5-m deep weathered section at TV 23 is actually a pedostratigraphic complex of

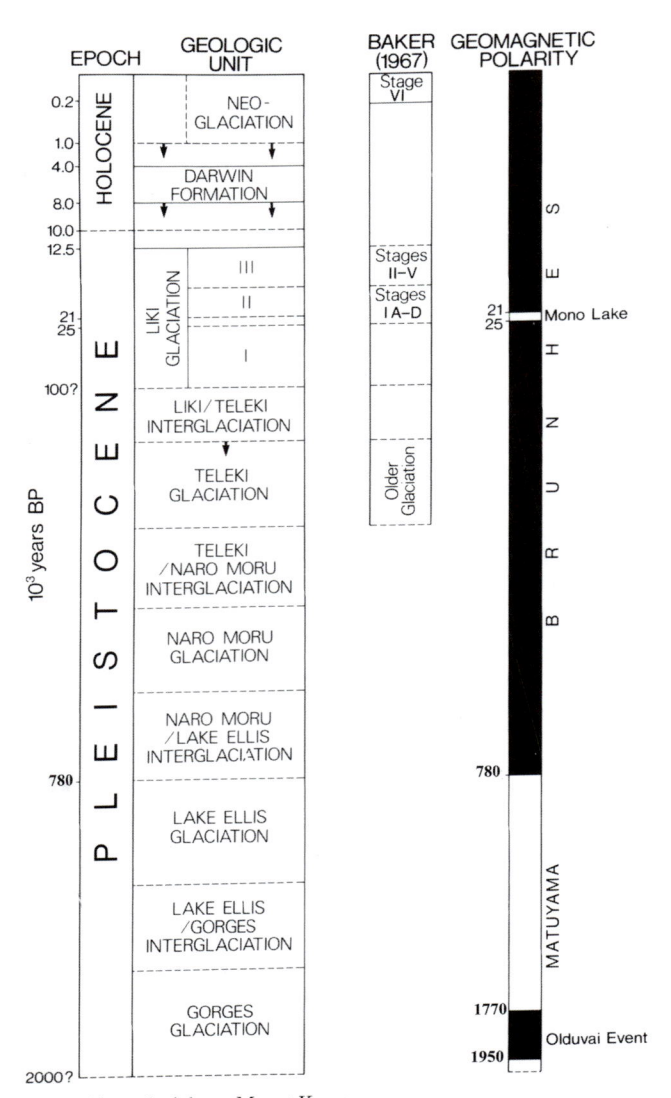

FIGURE 77.2 Chronology of glaciations and interglacials on Mount Kenya.

post-Naro Moru palaeosol underlying the post-Teleki palaeosol with the latter still exposed to the subaerial atmosphere. Thus, the surface palaeosol is relict; the buried system is fossil (stratigraphy with detailed mineralogy is on pp. 108–109; Mahaney, 1990). The percentage of altered volcanic glass, reduced amounts of metahalloysite and greatly increased concentrations of gibbsite argue for a strong Teleki–Liki Interglacial, with high soil water and aggressive leaching conditions. While greater clay is present in the post-Naro Moru palaeosol than in the post-Teleki palaeosol, it may be that aggressive leaching translocated some of it downward during the last interglacial.

The Liki Glaciation (=Wisconsinan, Weichselian) is well represented in most major catchments on the mountain as low as 3200 m a.s.l. Liki I, considered coeval with oxygen isotope stage 4 is present in only a few drainages—Hinde and Gorges valleys—although the age has been disputed by Shanahan and Zreda (2000) who argue for a middle Pleistocene range of 355–420 ka, an assessment that does not fit the landform morphology of the valley. As previously stated, such old ages could result from rocks exposed to cosmogenic isotopes for a long period of time finally toppling off high mountain ridges to ride 'piggy back' on the surface of the Liki ice. It is also possible, however, as Shanahan and Zreda report, that Liki I in Gorges Valley could be younger and belong to the Last Glacial Maximum which would correlate with the Liki II substage.

On the flanks of Gregory's older moraine in Teleki Valley (Gregory, 1894), a succession of Brunhes-age deposits at Site TV61 (Mahaney, 1990), of almost completely weathered sediments, was studied to reconstruct its magnetic history. In the upper horizons of the pedostratigraphic profile, a suite of undeformed colluvial and slope wash sediment showed an intermediate polarity indicating a short-lived excursion ^{14}C dated between 21,500 ± 430 and 24,520 ± 450 year BP (Mahaney, 1990). These sediments almost perfectly represent the Mono Lake excursion, first described at Summer Lake, Oregon by Verosub (1984).

At Two Tarn, near the headwaters of the Burguret Catchment, a ^{14}C date on organic gyttja of 33,840 ± 2810 (Gak-11270) from recovered sediment taken by piston core provides a minimum age on till which most certainly belongs to the Liki I stade. In still other drainages, such as Teleki and Hausberg, Liki moraines were washed away at the end of the last glaciation leaving only isolated moraine remnants. The few Liki II moraines in Teleki and Liki North valleys have yielded ^{14}C ages of between 14 and 15 ka, and cores on Liki II moraines in Teleki Valley have yielded ages of 10.3–2.2 ka, age extrapolations using rates of sedimentation below the dated beds to till yield approximate ages of 14.5–15.5 ka for the tills.

The glacial limits of Hastenrath (1984) shown on his Mount Kenya map are either possible Liki I moraines or bedrock outcrops. Recessional moraines mapped by Baker (1967) above the Liki II limits in most valleys are bedrock outcrops, there being almost no recessional moraines above 3200 m a.s.l.; at ~4000 m a.s.l., in most drainages, cross-valley moraines belong to the recessional stillstand of Liki ice that occurred during the Late Glacial. Attempts to test the possibility that these moraines were emplaced by ice advances either during the Younger Dryas, or in the early Holocene, resulted in a search for buried soils, bog sediment or weathered bedrock underneath the moraine. As described by Mahaney (1990), a search for buried evidence failed to support a re-advance hypothesis. Thus, these moraines are considered to represent stillstand at the end of the Liki Glaciation. To add to this, Shanahan and Zreda (2000) have used ^{36}Cl to date boulders on the Liki III moraines in upper Gorges and Teleki valleys with equivocal results: an average of 10.2 ka for Teleki moraines and 14.1 ka for Gorges moraines with rather wide standard deviations. Conclusive proof of a *Younger Dryas* origin for these moraines remains elusive despite assertions to the contrary by Shanahan and Zreda (2000). Their YD moraines could just as easily be the product of glacial stillstand during retreat phases of the Late Glacial.

Glacier positions during the early and middle Holocene are impossible to reconstruct with precision. Glaciers either retreated to their cirque positions or disappeared entirely from the mountain. Despite various hypotheses of ice advances (Perrott, 1982; Johansson and Holmgren, 1985; Mark and Osmaston, 2008) from the western cirques during the middle Holocene, no unequivocal evidence has been found to indicate expanded ice (Mahaney, 1987). All moraines above the rock glaciers in Teleki Valley (Mahaney, 1980), and below the Neoglacial deposits, carry similar rock weathering ratios and rock rinds and have in them palaeosols of a similar kind (Mahaney, 1990, Chapter 7), all correlated with the Liki III substage. If there were differences in age, as, for example, on moraines below the Neoglacial limit to Naro Moru Tarn, there would certainly be variations in rock weathering features and soil expression. No middle Holocene tills have been found (Mahaney, 1990) despite pollen evidence indicating colder conditions (Coetzee, 1967). The interpretation (Mark and Osmaston, 2008) of recovered ^{14}C dated lake core material as lying below glacial sediment is of dubious quality since there is no database indicating the 'glacial sediment' is indeed glacial. In past examples, sediment with glacially crushed quartz has been taken to represent glacial advances (Johansson and Holmgren, 1985) when in fact glacially crushed quartz can be found in Mount Kenyan loessic beds (Mahaney, 1987, 1990).

The onset of Neoglaciation is a tenuous boundary to establish in the sediment record. Evidence for a change in alluvial regime comes from the TV4 soil in Upper Teleki Valley directly in front of the massive valley side lobate-shaped rock glaciers of Liki III age. As shown by Mahaney (1990), fine textured silty sandy sediment older than 1940 ± 120 year BP (Gak 8273) was replaced with coarser pebbly sandy material that probably records renewed glacier activity and increased meltwater run-off from the high cirques.

Two advances of Neoglaciation are known on Mount Kenya: Tyndall with an indeterminate minimum age of ~1 ka and Lewis considered coeval with the Little Ice Age (500–100 years). Moraines and outwash of both advances are dated principally by relative criteria, including weathering rinds, presence/absence of different species of lichens, and by a lichen growth curve for *Rhizocarpon* section *Rhizocarpon* (Mahaney, 1990). The maximum age on the Tyndall moraines and outwash can be approximated from changes in stream regimen but is otherwise impossible to date with recoverable organic material for radiocarbon.

REFERENCES

Baker, B.H., 1967. Geology of the Mount Kenya Area. Geological Report no. 79. Kenya Geological Survey, Nairobi, 78 pp.

Barendregt, R.W., Mahaney, W.C., 1988. Paleomagnetism of selected Quaternary sediments on Mount Kenya, east Africa; a reconnaissance study. J. Afr. Earth Sci. 7, 219–225.

Charsley, T.J., 1989. Composition and age of older outwash deposits along the northwestern flank of Mount Kenya. In: Mahaney, W.C. (Ed.), Quaternary and Environmental Research on the East African Mountains. Balkema, Rotterdam, pp. 165–174.

Coe, M.J., 1967. The Ecology of the Alpine Zone of Mount Kenya. Junk, The Hague, 136 pp.

Coetzee, J.A., 1967. Pollen analytical studies in East and Southern Africa. Palaeoecol. Afr. 3, 1–146.

Everden, J.F., Curtis, G.H., 1965. The potassium-argon dating of late Cenozoic rocks in East Africa and Italy. Curr. Anthropol. 6, 343–385.

Gregory, J.W., 1894. The glacial geology of Mount Kenya. Q. J. Geol. Soc. 56, 205–222.

Gregory, J.W., 1921. The Rift Valleys and Geology of East Africa. Seeley, Service, London.

Harmsen, R., 1989. Recent evolution and dispersal of insects on the East African mountains. In: Mahaney, W.C. (Ed.), Quaternary and Environmental Research on East African Mountains. Balkema, Rotterdam, pp. 245–256.

Hastenrath, S., 1984. The Glaciers of Equatorial East Africa. Reidel, Dordrecht, 353 pp.

Johansson, L., Holmgren, K., 1985. Dating of a moraine on Mount Kenya. Geogr. Ann. 67A, 123–128.

Mackinder, H.J., 1900. A journey to the summit of Mount Kenya, British East Africa. Geogr. J. 15, 453–486.

Mahaney, W.C., 1980. Late Quaternary rock glaciers, Mount Kenya, East Africa. J. Glaciol. 25, 492–497.

Mahaney, W.C., 1987. Dating of a moraine on Mount Kenya: a discussion. Geogr. Ann. 69A, 359–363.

Mahaney, W.C., 1990. Ice on the Equator. Wm Caxton Press, Ellison Baym, Wisconsin, 386 pp.

Mahaney, W.C., Barendregt, R.W., Vortisch, W., 1997. Relative ages of loess and till in two Quaternary paleosols in Gorges Valley, Mount Kenya, East Africa. J. Quatern. Sci. 12, 61–72.

Mahaney, W.C., Barendregt, R.W., Villeneuve, M., Dostal, J., Hamilton, T., Milner, M.W., 2011. Upper Neogene volcanics and interbedded paleosols near Mount Kenya. In: Out of Africa—3.8 Ga of Earth History, Geological Society Special Volume London, in press.

Mark, B.G., Osmaston, H.A., 2008. Quaternary glaciation in Africa: key chronologies and climatic implications. J. Quatern. Sci. 23, 589–608.

Mizuno, K., 2005. Glacial fluctuation and vegetation succession on Tyndall Glacier, Mt. Kenya. Mt. Res. Dev. 25, 68–75.

Nilsson, E., 1931. Quaternary glaciation and pluvial lakes in British East Africa and Abyssinia. Geogr. Ann. H13, 241–248.

Nilsson, E., 1935. Traces of ancient changes of climate in East Africa. Geogr. Ann. 1–2, 1–21.

Osmaston, H.A., Harrison, S.P., 2005. The Late Quaternary glaciation of Africa: a regional synthesis. Quatern. Int. 138–139, 32–54.

Perrott, R.A., 1982. A high altitude pollen diagram from Mount Kenya, its implications of the history of glaciation. In: Coetzee, J.A., van Zinderen Bakker, E.M. (Eds.), Palaeoecology of Africa, 14. Balkema, Rotterdam, pp. 77–83.

Shanahan, T.M., Zreda, M., 2000. Chronology of Quaternary glaciations in East Africa. Earth Planet. Sci. Lett. 177, 23–42.

Spink, P.C., 1945. Further notes on the Kibo inner crater and glaciers of Kilimanjaro and Mount Kenya. Geogr. J. 106, 210–216.

Spink, P.C., 1949. The equatorial glaciers of East Africa. J. Glaciol. 1, 277–283.

Troll, C., Wien, K., 1949. Der Lewisgletcher am Mount Kenya. Geogr. Ann. 31, 257–274.

Veldkamp, A., Buis, E., Wijbrans, J.R., Olago, D.O., Boshoven, E.H., Marée, M., et al., 2007. Late Cenozoic fluvial dynamics of the River Tana, Kenya, an uplift dominated record. Quatern. Sci. Rev. 26, 2897–2912.

Verosub, K.L., 1984. Mono Lake geomagnetic excursion found at Summer Lake, Oregon. Geology 12, 643–646.

Zeuner, F.E., 1949. Frost soils on Mount Kenya and the relation of frost soils to aeolian deposits. J. Soil Sci. 1, 20–30.

Chapter 78

Glaciation in Southern Africa and in the Sub-Antarctic

Kevin Hall[1,2,*] and Ian Meiklejohn[3]

[1] Department of Geography, Geoinformatics and Meteorology, University of Pretoria, Pretoria 0002, South Africa
[2] Department of Geography, University of Northern British Columbia, 3333 University Way, Prince George, Canada BC V2N 4Z9
[3] Department of Geography, Rhodes University, Grahamstown 6140, South Africa
*Correspondence and requests for materials should be addressed to Kevin Hall. E-mail: hall@unbc.ca

78.1. INTRODUCTION

The two regions have been consolidated into one Chapter owing to, as will be noted in each section, the overall paucity of suitable new information since the last edition (Hall, 2004a,b).

78.2. GLACIATION IN SOUTHERN AFRICA

As presented previously (Hall, 2004a) and subsequently by Osmaston and Harrison (2005), the evidence in southern Africa of Quaternary glaciations has been both weak and controversial despite the topographical possibility of areas such as Lesotho, the Natal Drakensberg and parts of the Eastern Cape Mountains having experienced glaciation of some kind. Subsequent to that report, and as identified in Mark and Osmaston (2008), a more rigorous undertaking (Mills, 2006) has provided new information strongly indicative of moraines (Fig. 78.1) within several valleys of the Lesotho-South Africa border: in the Tsatsa-La-Mangaung, Sekhokong and Leqooa valleys (Fig. 78.2). Although individual attributes of this study (Mills, 2006), such as clast orientation (p. 313), clast shape (p. 317) or thin section analyses (p. 319) cannot provide information uniquely diagnostic of a glacial origin, the totality of the data coupled with morphology and location strongly suggests that the observed ridges are moraines. An implication of these being moraines would be that they have experienced little modification to form during the Holocene. Given a temperature depression of 6 °C (Mills, 2006, p. 396), then small niche/cirque glaciers could have existed in the eastern Drakensberg if there were adequate precipitation (see below cf. Nel and Sumner, 2008) and protection from solar radiation.

The evidence (Mills, 2006, p. 398) strongly suggests that "glaciers... were climatologically restricted to a few sites, where microtopographic (e.g. slope gradient and aspect) and particularly microclimatic factors (e.g. late-lying snow cover, snowblow and solar radiation) played a major role". In other words, the evidence does not suggest widespread glaciation (Mills, 2006) and hence is not in conflict with periglacial findings elsewhere within this general area (e.g. Boelhouwers et al., 2002; Boelhouwers and Meiklejohn, 2002; Sumner, 2004). Consequent upon the work of Mills (2006), other material (Mills and Grab, 2005; Carr et al., 2009; Mills et al., 2009) expands and provides further detail regarding possible palaeoclimatic reconstruction and glacial implications.

Concurrent with the above works directly associated with glaciation, other studies have provided ancillary, if sometimes conflicting, information. Nel and Sumner (2008) discuss rainfall and temperature attributes for the suggested glacial area and conclude that the precipitation extrapolations may be somewhat inflated and that this may have an impact on the proposed glaciation. Mills (2006, p. 375) accepts this in respect to widespread glaciation but shows that the small glaciers she identifies may have been sustained locally by the addition of snowblow and avalanching. From a different perspective, Boelhouwers et al. (2002) provide information regarding periglacial blockstreams in Lesotho which argue against widespread glaciation of this area, but their data would not be in significant conflict with more recent reconstructions of Mills (2006). Boelhouwers and Meiklejohn (2002) present much of the temperature, precipitation and periglacial evidence that appear to provide a (so far) convincing argument that widespread glaciation did not occur and that earlier

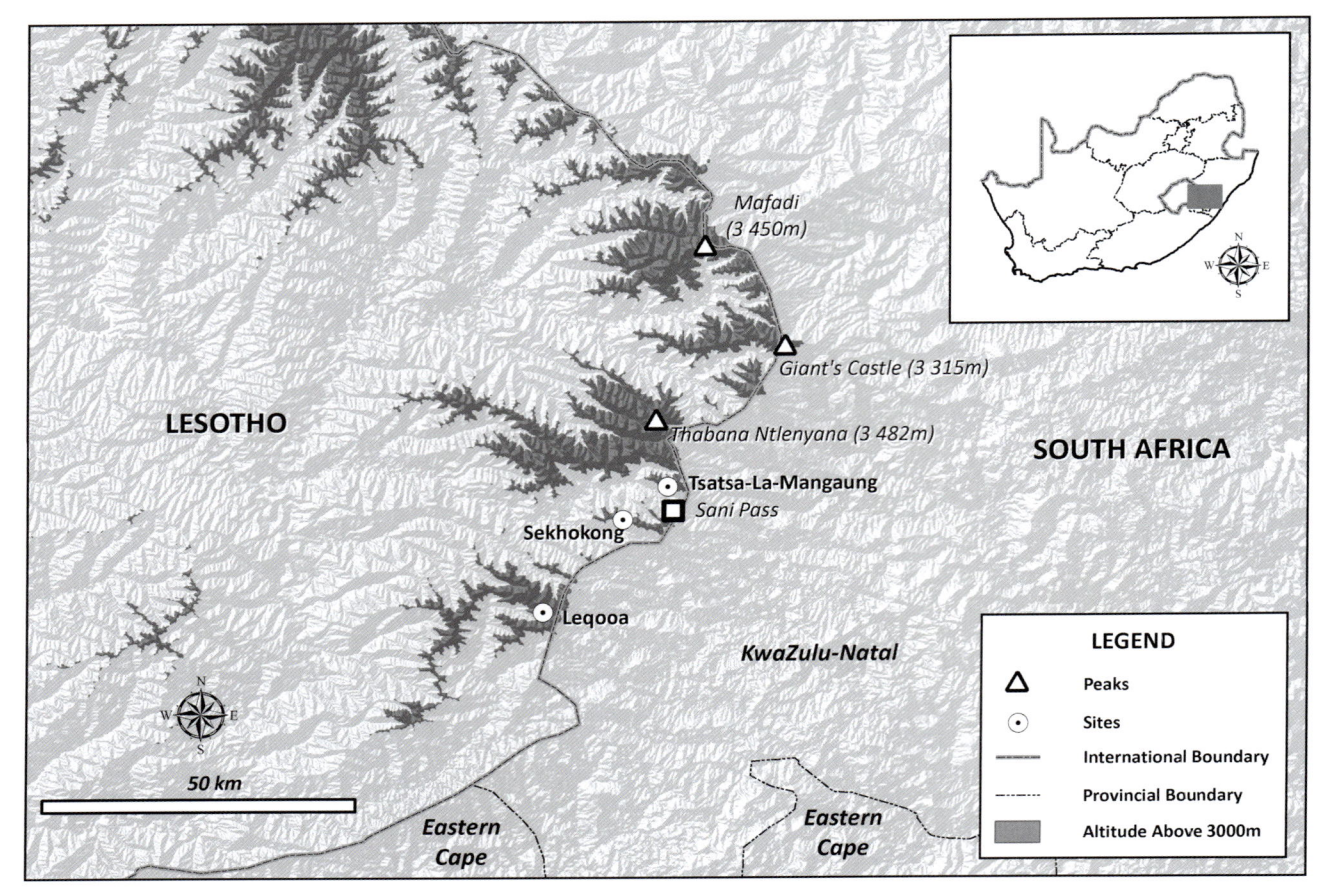

FIGURE 78.1 The location of the three proposed glaciers at the Lesotho–South Africa border.

supposed glacial evidence for the Eastern Cape Drakensberg (e.g. Lewis, 1994) is unsupported by either field observations or proxy data; that temperatures would have needed to drop by 19–24 °C to facilitate glaciation in this area appears unrealistic.

Thus, the recent advances (notably Mills, 2006) indicate that local, topographically specific, small-scale glaciers may well have existed at a number of locations. At the same time, periglacial evidence argues against any notions of a more widespread glacial event within the Lesotho–Drakensberg region. As limited as these advances may appear within the big scheme of things, they are, nevertheless, significant for this region. This is the more so as the now determined nature of the glaciation does not create the same degree of conflict (as was previously the case) with other climatic or periglacial information such that the two come closer together and, in so doing, provide a better picture of Quaternary conditions for this area. Hall (2010) recently presented an argument that the small glacial valleys, which maintained a V-shape, identified by Mills (2006), might be explained by the presence of very thin but cold ice.

78.3. THE SUB-ANTARCTIC

Sadly there appears to have been little in the way of new work in this interesting but complex region, despite it being an area critical for understanding of the Southern Ocean circulation and climate (Hall, 2009). Hall (2009) provides a synthesis of Holocene glacial history and bemoans, as noted here for the Quaternary, the (p. 2226) "lack of sufficient data, particularly chronological information, (which) makes it difficult to synthesize the Holocene history of sub-Antarctic glaciers". That said, a number of new advances have been made for Marion Island, Heard Island and South Georgia.

In respect of Marion Island, Boelhouwers et al. (2008) have synthesised the new information. Key here was the finding of new glacial evidence (e.g. Nel, 2001) coupled with the loss of the island ice cap since ca. 1982 (Sumner et al., 2004) which has opened up new areas for investigation. Concurrent with these findings has been the dating of more lava flows (McDougall et al., 2001) which shows that, contrary to earlier thoughts based upon lava dates available at that time (e.g. Hall, 1983), volcanism and lava outpourings were not limited to interglacial periods. Based on the

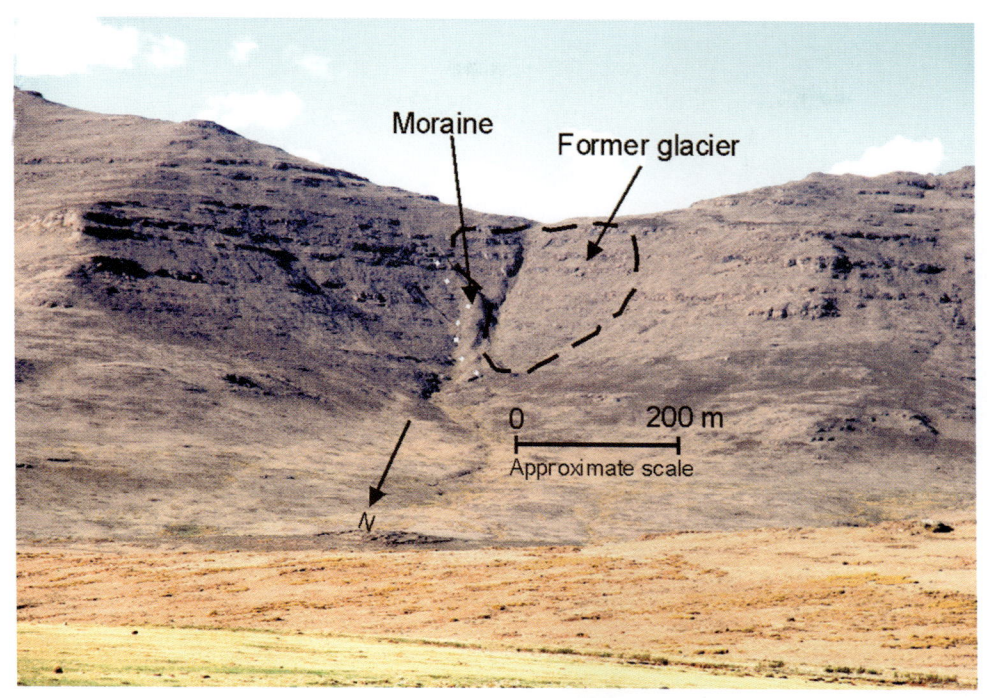

FIGURE 78.2 Photograph of the proposed glacier site at Tsatsa-La-Mangaung showing the moraine and the rough outline of the former glacier.

above, coupled with fieldwork in 2009, the new reconstruction of the Quaternary ice cover for Marion Island (Fig. 78.3) is presented in Hall et al. (2011). The new findings, including evidence for Holocene glaciation as well (Boelhouwers et al., 2008; Hall et al., 2011), indicate that some of the glaciers were much larger than was previously thought, that the hypothesis for faulting and volcanism due to tectonism resulting from rapid deglaciation (Hall, 1982) was in error, that there were nunataks providing refugia for various organisms, and that new information allowed the reconstruction of glaciers for the southern part of the island (Hall et al., 2011). Although, from a glacial perspective, the proposed changes (Hall et al., 2011) are not particularly significant, they do have ramifications for the geology, the topographic evolution of the island, and for explaining some of the biological findings that required nunatak refugia (e.g. Chown and Froneman 2008, p. 354).

Balco (2007) identifies that Heard Island may have experienced a substantial marine ice cap with ice as much as 50–80 km from the present shoreline. Based on off-shore observations of features resulting from possible glacial erosion (troughs) and deposition (moraines), this would extend the ice at least 50 km from the present coast and necessitate erosion to a depth of ca. 180 m below the glacial sea level, with the palaeo-grounding line at 120 m below the Last Glacial Maximum (LGM) sea level. Based on these observations, Balco (2007, p. 3) suggests the ice was ca. 135 m thick at its margin and, hence, several hundred metres thick at its centre. To achieve this, ice thickness would necessitate significantly increased precipitation and/or that ice loss by calving was less effective than at present (Balco, 2007)— the latter being favoured and facilitated by increased sea ice that would have decreased wave energy.

For South Georgia, as noted previously (Hall, 2004b; Fig. 5), Clapperton (1990) had suggested that the LGM ice extended out to the edge of the continental shelf but more recent work by Bentley et al. (2007) indicated rather that (p. 674) "glacier advance at the LGM was either: (i) restricted to the inner fjords; or (ii) extended to the outer fjords or shelf but without leaving an onshore record. This latter possibility seems unlikely, so the suggestion of restricted glaciation ... is supported by our mapping". The more restricted glaciation is explained by precipitation being limited due to the extensive sea ice (Bentley et al., 2007) and hence constraining glacier growth towards the edge of the continental shelf. Subsequently, Graham et al. (2008) utilized marine echo-sounding to map the geomorphology of the entire continental shelf around South Georgia at high resolution (5 m cell sizes with 1 m vertical resolution and a positional accuracy of 10 m) and identified what are thought to be large glacially eroded troughs that link to the present fjords and thus suggest ice cover did indeed extend much further off-shore than was thought by Bentley et al. (2007). Further, there are also (what are interpreted as) terminal, lateral and recessional moraines marking former positions of the ice, which extend out to the shelf edge north of South Georgia (Graham et al., 2008, p. 16). Interestingly, Graham et al. (2008) do not see a necessary conflict with the findings of

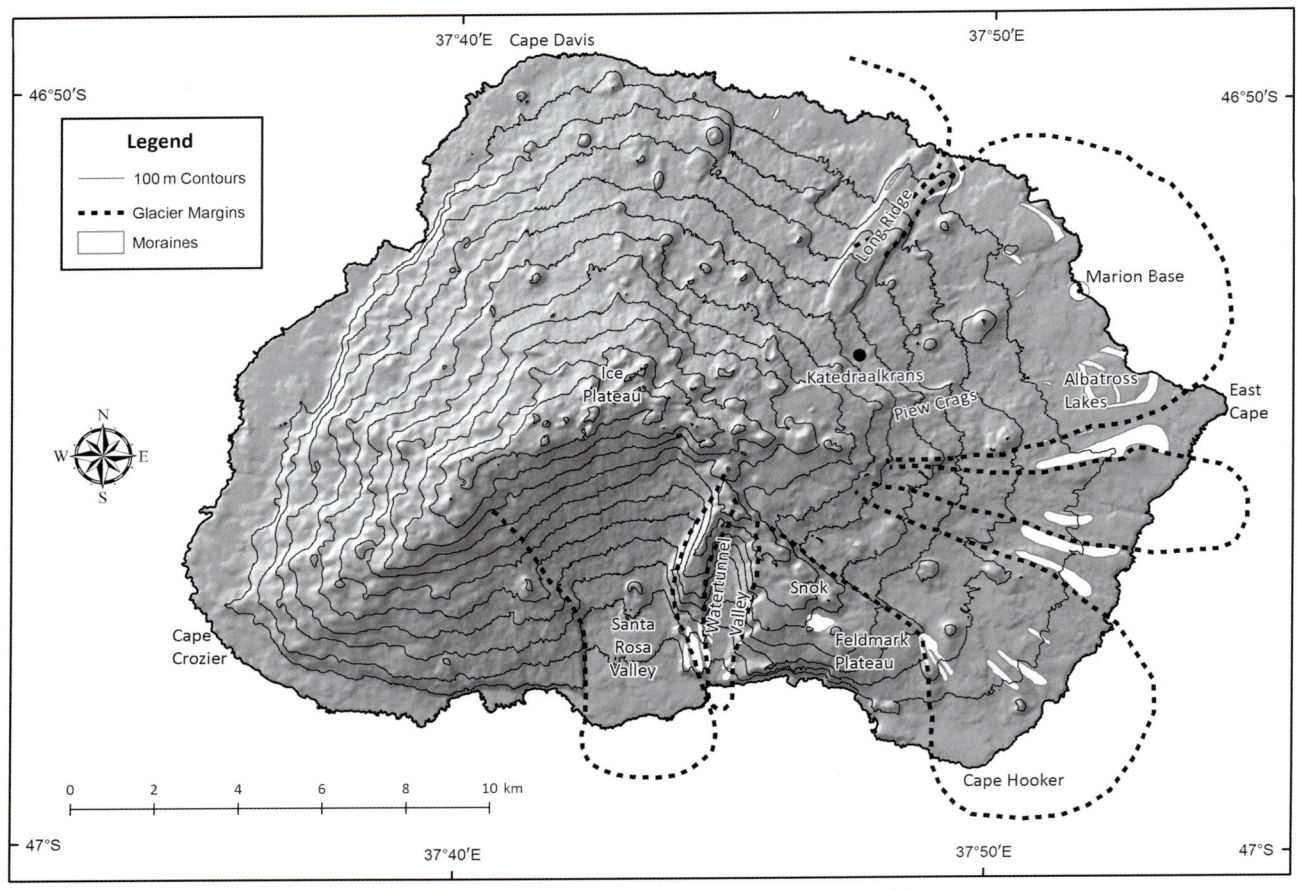

FIGURE 78.3 Reconstruction of glaciers on sub-Antarctic Marion island based on newly available evidence.

Bentley et al. (2007) as they argue (p. 17) that "it could be considered that terrestrially mapped moraines are in fact stillstand or readvance margins of a post-LGM ice sheet, formed subsequent to the maximum extension of the ice sheet onto the shelf". Thus, as with Heard Island and Marion Island (above), it appears that on-going research indicates a (perhaps) a wider, more extensive ice extent than had been previously thought.

With respect to the Îles Crozet, Camps et al. (2001) note a glacial episode that produced U-shaped valleys but offer no idea of the ice distribution. Giret et al. (2003) also refer to periods of glaciation on L'île de La Possession and L'île de l'Est of Îles Crozet but also offer no information regarding ice distribution. On Îles Kerguelen, there is a substantial amount of material regarding Holocene glacial fluctuations (e.g. Frenot et al., 1993; Frenot, 1996; Frenot et al., 1997; Berthier, 2008) but none (it appears) that directly relates, and advances knowledge of, Quaternary glaciation. For Macquarie Island, there appears to be no new information that indicates any form of glaciation, other than the few small, localised glaciers, referred to previously (Hall, 2004b, p. 341). Last, for the Falkland Islands, there seems to be no works that extend the glaciation beyond the cirques and associated upper reaches of their valleys (ca. 2.7 km long glaciers) as expounded by Clapperton and Sugden (1976). Wilson et al. (2008), in a discussion regarding cosmogenic isotope dating of stone runs, show that (p. 471) "sampled surfaces have not been covered for long periods of time by ice or sediment ... our data clearly demonstrate that there have been no glaciers in the Prince's Street valley in the last 700–800 kyr;" the arguments of Clapperton and Sugden (1976) being further validated. Other studies of environmental change (e.g. Clark et al., 1998; Wilson et al., 2002) also fail to indicate any evidence that would contradict these earlier findings.

REFERENCES

Balco, G., 2007. A surprisingly large marine ice cap at Heard Island during the Last Glacial Maximum? U.S. Geological Survey and National Academies; USGS OF-2007-1047, Extended Abstract 147, pp. 1–4.

Bentley, M.J., Evans, D.J.A., Fogwill, C.J., Hansom, J.D., Sugden, D.E., Kubik, P.W., 2007. Glacial geomorphology and chronology of deglaciation, South Georgia, sub-Antarctic. Quatern. Sci. Rev. 26, 644–677.

Berthier, É., 2008. Recul des glaciers de montagne: que nous apprennent les satellites? La Météo. 63, 32–39.

Boelhouwers, J.C., Meiklejohn, K.I., 2002. Quaternary periglacial and glacial geomorphology of southern Africa: review and synthesis. S. Afr. J. Sci. 98, 47–55.

Boelhouwers, J., Holness, S., Meiklejohn, I., Sumner, P., 2002. Observations on a blockstream in the vicinity of Sani Pass, Lesotho Highlands, southern Africa. Permafrost Periglac. Process. 13, 251–257.

Boelhouwers, J.C., Meiklejohn, K.I., Holness, S.D., Hedding, D.W., 2008. Geology, geomorphology and climate change. In: Chown, S.L., Froneman, P.W. (Eds.), The Prince Edward Islands: Land-Sea Interactions in a Changing Ecosystem. SUN Press, Stellenbosch, pp. 65–96.

Camps, P., Henry, B., Prévot, M., Faynot, L., 2001. Geomagnetic palaeo-secular variation recorded in Plio-Pleistocene volcanic rocks from Possession Island (Crozet Archipelago, southern Indian Ocean). J. Geophys. Res. 106 (B2), 1961–1971.

Carr, S.J., Lukas, S., Mills, S.C., 2009. Glacier reconstruction and mass-balance modeling as a geomorphic and palaeoclimatic tool. Geophys. Res. Abstr. 11, 1360–1362.

Chown, S.L., Froneman, P.W., 2008. The Prince Edward Islands: Land-Sea Interactions in a Changing Ecosystem. SUN Press, Stellenbosch 450 pp.

Clapperton, C.M., 1990. Quaternary glaciations in the Southern ocean and Antarctic Peninsula area. Quatern. Sci. Rev. 9, 229–252.

Clapperton, C.M., Sugden, D.E., 1976. The maximum extent of glaciers in part of West Falkland. J. Glaciol. 17, 73–77.

Clark, R., Huber, U.M., Wilson, P., 1998. Late Pleistocene and environmental change at Plaza Creek, Falkland Islands, South Atlantic. J. Quatern. Sci. 13, 95–105.

Frenot, Y., 1996. Conséquences des variations climatiques récentes sur l'évolution des marges glaciaires en milieu subantarctique. Comité écologie et gestation du patrimoine, Notes, ENV-SRAE-93083, 29 pp..

Frenot, Y., Gloaguen, J.-C., Picot, G., Bougère, J., Benjamin, D., 1993. *Azorella selago* Hook. Used to estimate glacier fluctuations and climatic history in the Kerguelen Islands over the last two centuries. Oecologia 95, 140–144.

Frenot, Y., Gloaguen, J.-C., Van de Vijver, B., Beyens, L., 1997. Datation de quelques sediments tourbeux holocènes et oscillations glaciaires aux Îles Kerguelen. Cr. Acad. Sci. III-vie 320, 567–573.

Giret, P.A., Weis, D., Zhou, X., Cottin, J.-Y., Tourpin, S., 2003. Les Îles Crozet. Géologues 137, 15–23.

Graham, A.C.C., Fretwell, P.T., Larter, R.D., Hodgson, D.A., Wilson, C.K., Tate, A.J., et al., 2008. A new bathymetric compilation highlighting extensive paleo-ice sheet drainage on the continental shelf, South Georgia, sub-Antarctica. Geochem. Geophys. Geosyst. 9 (7), 1–21.

Hall, B.L., 2009. Holocene glacial history of Antarctica and the sub-Antarctic islands. Quatern. Sci. Rev. 28, 2213–2230.

Hall, K., 1982. Rapid deglaciation as an initiator of volcanic activity: an hypothesis. Palaeogeogr. Palaeoclimatol. Palaeoecol. 29, 243–259.

Hall, K., 1983. A reconstruction of the Quaternary ice cover on Marion Island. In: Oliver, R.L., James, P.R., Jago, J.B. (Eds.), Antarctic Earth Science. Australian Academy of Science, Canberra, pp. 461–464.

Hall, K., 2004a. Glaciation in southern Africa. In: Ehlers, J., Gibbard, P.L. (Eds.), Quaternary Glaciations—Extent and Chronology. Part III: South America, Asia, Africa, Australasia, Antarctica. Elsevier, Amsterdam, pp. 337–338.

Hall, K., 2004b. Quaternary glaciation of the sub-Antarctic Islands. In: Ehlers, J., Gibbard, P.L. (Eds.), Quaternary Glaciations—Extent and Chronology. Part III: South America, Asia, Africa, Australasia, Antarctica. Elsevier, Amsterdam, pp. 339–345.

Hall, K., 2010. The shape of glacial valleys and implications for southern African glaciation. S. Afr. Geogr. J. 92, 35–44.

Hall, K., Meiklejohn, I., Bumby, A., 2011. Marion Island (sub-Antarctic) volcanism and glaciation: new findings and reconstructions. Antarct. Sci. 23, 155–163.

Lewis, C.A., 1994. Protalus ramparts and the altitude of the local equilibrium-line during the Last Glacial stage in Bokspruit, East Cape Drakensberg, South Africa. Geogr. Ann. A 76, 37–48.

Mark, B.G., Osmaston, H.A., 2008. Quaternary glaciation in Africa: key chronologies and climatic implications. J. Quatern. Sci. 23, 589–608.

McDougall, I., Verwoerd, W.V., Chevallier, L., 2001. K-Ar geochronology of Marion Island, Southern Ocean. Geol. Mag. 138, 1–17.

Mills, S.C., 2006. The origin of slope deposits in the southern Drakensberg, eastern Lesotho. Unpublished Ph.D., School of Geography, Archaeology and Environmental Studies, University of the Witwatersrand, 459pp.

Mills, S.C., Grab, S.W., 2005. Debris ridges along the southern Drakensberg escarpment as evidence for Quaternary glaciation in southern Africa. Quatern. Int. 129, 61–73.

Mills, S.C., Grab, S.W., Carr, S.J., 2009. Recognition and palaeoclimatic implications of late Quaternary niche glaciation in eastern Lesotho. J. Quatern. Sci. 24, 647–663.

Nel, W., 2001. A spatial inventory of glacial, periglacial and rapid mass movement forms on parts of Marion Island: implications for Quaternary environmental change. Unpublished M.Sc., Geography Department, University of Pretoria, South Africa.

Nel, W., Sumner, P., 2008. Rainfall and temperature attributes on the Lesotho-Drakensberg escarpment edge, southern Africa. Geogr. Ann. A 90, 97–108.

Osmaston, H.A., Harrison, S.P., 2005. The Late Quaternary glaciation of Africa: a regional synthesis. Quatern. Int. 139, 32–54.

Sumner, P.D., 2004. Geomorphic and climatic implications of relict open-work block accumulations near Thabana-Ntlenyana, Lesotho. Geogr. Ann. Ser. A 86, 289–302.

Sumner, P., Meiklejohn, I., Boelhouwers, J., Hedding, D., 2004. Climate change melts Marion's snow and ice. S. Afr. J. Sci. 100, 395–398.

Wilson, P., Clark, R., Birnie, J., Moore, D.M., 2002. Late Pleistocene and Holocene landscape evolution and environmental change in the Lake Sulivan area, Falkland Islands, South Atlantic. Quatern. Sci. Rev. 21, 1821–1840.

Wilson, P., Bentley, M.J., Schnabel, C., Clark, R., Xu, S., 2008. Stone run (block stream) formation in the Falkland Islands over several cold stages, deduced from cosmogenic isotope (^{10}Be and ^{26}Al) surface exposure dating. J. Quatern. Sci. 23, 461–473.

Index

Note: The letters 'f' and 't' following the locators refer to figures and tables respectively.

A

AAR. *See* Amino-acid racemisation
Aberdeen Ground Formation, 62
　Cromerian Complex and, 61
Accelerator mass spectrometry (AMS), 98–99, 378, 450
　in Ecuador, 808
　in Magellan Straits, 749–750
Aconcagua group, 735–738, 737f
ACR. *See* Antarctic Cold Reversal
Advanced Spaceborne Thermal Emission and Reflection Radiometer (ASTER), 3, 4f, 375, 773
Afghanistan, Hindu Kush Mountains of, in Late Pleistocene, 863–864, 864f
Africa. *See also* Kenya, Mount
　East, 1076f
　　Central Plateau of, 1076
　North
　　Atlas Mountains of, 1065–1074
　　paleoglaciers and palaeoclimates in, 1071
　in Pleistocene, 7
　southern, 1081–1086
African Humid Period, 1071
Agassiz, Lake, 501–504
AGC. *See* Appalachian Glacier Complex
Agnano-Monte Spina Tephra (AMST), 218
Agpat, 699–700
Agri, 400–401
Ahklun Mountains, 440
　in Early Wisconsinan, 441f
　in Late Wisconsinan, 441f
　in Pleistocene, 440, 441f
AHVC. *See* Alban Hills Volcanic Complex
Ajusco, 854
Akaishi Mountains, 1013, 1014f
Aksoual, 1066, 1067f, 1068f
Aktru Valley, 892
Akulovo, 337–339, 341f, 342f, 344f
Aladagllar, 396
Alaska, 11–12, 423–425
　cosmogenic exposure ages in, 430t
　in Early Weichselian, 430t
　in Early Wisconsinan, 429
　glacial limits in, 428–437
　in Late Wisconsinan, 429–437, 430t
　moraines in, 437
　in Neogene, 7–8
　penultimate glaciation in, 429
　in Pleistocene, 428–437
　Tertiary in, 429
Alaska Palaeo-Glacier Atlas (APG Atlas), 427–448
　Early Wisconsinan and, 428f
　GIS and, procedures, 427–428
　Late Wisconsinan and, 428f
　Pleistocene and, 428f
　regional updates for, 437–443
Alaska Range, 441–442
Alban Hills Volcanic Complex (AHVC), 213–214
Albert Edward, Mount, 1027
Albert Town, 1057
Alberta, 576f, 592f
　basal gravel predating in, 586–587
　buried valleys of, 578–585
　　stratigraphical units in, 579–580
　discrimination of advance and retreat sequences in, 580–584, 581f
　fossil-bearing sediments in, 582
　glacial advance, retreat and readvance in, 580
　in Late Pleistocene, 575–590
　in Late Wisconsinan, 578–579, 579f, 584
　laterally extensive lag gravel in, 581–582
　LGM of, 575, 581f
　LIS in, 576, 581f
　magnetostratigraphy for, 591–600
　in Middle Pleistocene, 575–590
　oxidized till in, 582–584
　radiocarbon ages in, 583t
　stratigraphical evidence in, 584
Albina, Lake, 1037
Aldan River, 867–868
Alesund/Tolsta, 272–273
Aleutian Range, 442
Algerian Atlas, 1071
Algerian Tell, 1065
Allentown, Pennsylvania, 524–527
Allerød, 137, 792
　in Québec-Labrador, 609
Alpes provençales, 117–118
Alpine Fault, 1054
Alpine glaciations
　in Austria, 15–18
　in Germany, 149
　in Siberia, 895–896
Alpine-Himalayan, 305–307
Alps
　in Early Pleistocene, 8
　East, 24f
　French, 117–126

　　EMC in, 119, 121f
　　end-moraines in, 119
　　key sites in, 119, 122
　　in Late Pleistocene, 119–124, 121f
　　in Middle Pleistocene, 117–119, 118f, 120f
　　morphological features of, 119, 122
　in Germany, 149
　Julian, 389
　Southern, 1053–1054
　　Main Divide and
　　　east of, 1058–1059
　　　north of, 1054–1055
　　　west of, 1055
　Transylvanian, 305
Altai Mountains, 887, 893, 993
　in China, 981
　ELA of, 972–976
　in Late Pleistocene, 967–980
　LGM of, 967
　of Mongolia, 969–971
　　ELA in, 974f, 975t
　　LGM of, 977f
　　in Pleistocene, 977f
　　timing of, 976–978
　in Pleistocene, 968f
　of Russia, 969, 971f
　　ELA in, 970t
　　in Pleistocene, timing of, 976–978
　stratigraphy of, 973t
Ältere Deckenschotter, 15, 18, 165, 167, 168
Amino-acid racemisation (AAR), 66–67
AMS. *See* Accelerator mass spectrometry
AMST. *See* Agnano-Monte Spina Tephra
Anadyr River, 874
Anadyr Uplands, 879
Anatolian Plateau, volcanoes of, 400–401
Ancylus Lake, 113
Andell Saddle, 38
Andes. *See also specific South American countries*
　of Bolivia, 757–772, 758f, 775
　of Chile
　　in Early Weichselian, 742–745
　　LGM of, 745–748
　　in Middle Weichselian, 742–745
　　in Wisconsinan, 742–745
　of Ecuador, 775
　in Middle Pleistocene, 9
　of Peru, 775

Andes. *See also specific South American countries* (cont.)
  of Pleistocene Chile, 739–740
Andøya-Andfjord, 291–294, 294f, 295f
Anglia, East, 63–64, 64f, 68, 79–81
  tills of, 66–67
  in Wolstonian, 83–84
Anglian, 63, 66–67, 66f
  Great Britain in, 78, 79–82
    offshore areas and, 82
  in Happisburgh Formation, 67
  in Midlands, 81
  in North Sea, 82
  traditional model of, 78–89
  Wales in, 81–82
Animal Creek, 1041
Antarctic Cold Reversal (ACR)
  in Bolivia, 766
  in Chile, 748–750
Antarctica
  in Cenozoic, 7
  in Late Pleistocene, 9
  North Sea and, 261–262
Antares, Mount, 1026
Anyticyclone, 885–886
Apennines. *See* Italian Apennines
APG Atlas. *See* Alaska Palaeo-Glacier Atlas
Appalachians
  in Chignecto Ice-Flow, 607–608
  piedmont, 609
  in Québec-Labrador, 606–607
    ice-flow reversal in, 608–609
Appalachian Glacier Complex (AGC)
  in Illinoian, 639f
  in Late Wisconsinan, 639f
  in Maritime Canada, 631–660
Apuseni Mountains, 305–307, 313–317, 316t
  ELA of, 320f
Aqua Relincho Valley, 735–736
Ararat, Mount, 400–401
Arc, 119
Arctic Ice Sheet, 877–878
Arctic Ocean, 702
  Siberia and, 886
ArcView, 3
Ardleigh Terrace, 62
Argentina. *See also* Patagonia
  in Plio-Pleistocene, 8
Ariège catchment, 127–128, 129–132, 135
  in Pleistocene, 131f
Armchair-like cirque, 416f
Arroumd Valley, 1069f
Ashburton, 1055–1056
Asia. *See also specific countries and regions*
  central, 944f, 955t
    glacial limits of, 945
    LGM of, 943–966
    methodological consequences and comparisons for, 958–959
  high, 944f
    dating methods for, 945–958
    glacial limits of, 945
    glacier stages in, 956t
    LGM of, 943–966
    methodological consequences and comparisons for, 958–959
  north-east
    in Early Pleistocene, 866
    in Late Pleistocene, 865–876
    in Middle Pleistocene, 866
    relief and climate of, 865
  south-east, 1023–1036
Asian Intertropical Convergence Zone (ITCZ), 1023–1024
ASTER. *See* Advanced Spaceborne Thermal Emission and Reflection Radiometer
Athens, 489–491
Atlanta Formation, 555, 558
Atlas Mountains, 1066f
  of Algeria, 1071
  high, 7, 1065–1070, 1067f, 1068f
  middle, 1070–1071
  of North Africa, 1065–1074
Augustowian, 300–301
Aurès, 1071
Aurora Cave, 1060
Australasia. *See also* New Zealand
  in Plio-Pleistocene, 8
Australia, 11–12, 1037–1046
  ELA of, 1038
  in Late Pleistocene, 1037–1038
  LGM of, 1038
Austria, 15–28, 16f
  alpine glaciations in, 15–18, 17f
  in Bühl, 24–25
  chronology in, 22–23, 23f, 25–26
  in Daun, 26
  drainage systems of, 17f
  in Egesen, 26
  glacial development in, 18–21, 19f
  in Gschnitz, 25
  ice decay in, 23–24
  last interglacial-glacial cycle in, 22
  overdeepened Valleys in, 21–22, 21f
  in Steinach, 25
  tectonic activity in, 18
Ayoloco, 850, 851f, 857–858
Azib Mzik Valley, 1066–1069, 1070f

B

Ba Tong Guan, 1006
Bacton Green, 63–64
Bad Ischl, 24–25
Baffin Island, 705–706
Baginton-Lillington Gravel, 82
Bahía Inútil, 747–748
Baie des Chaleurs, 607–608
Baikal, Lake, 902
Bailang River, 992f
Baiyu Gou, 987f, 988, 990f
Ballantynes Cove, 648
Baltic depression, 48
Baltic Elster, 50
Baltic River, 149
Baltic Sea, 96–97
Baltija, 224–226, 242
Baltoro, 958
Baltuja-Braslav, 34

Bandahara, Mount, 1025
Bangeta, Mount, 1028
Banks Island, 424f, 682–686
Barendregt-Irving hypothesis, 578–585
Barents Sea, 361–372, 363f
  deglaciation of, 368
  in Early Miocene, 7–8
  in Early Pleistocene, 364–365
    early part of, 364
    later part of, 364–365
    seismic stratigraphy and chronology of, 364
  in Early Weichselian, 366–367
  Kara Sea and, 366
  in Late Pleistocene, 9–10, 365
  in Late Pliocene, 363
  in Late Weichselian, 361f, 365, 367–370, 368f
  in Middle Pleistocene, 365
  in Middle Weichselian, 366–367
  in Neogene, 363–364
  north-west Russia and, 366
  in Palaeogene, 363–364
  physiography of, 362–363
  sediments in, 362–363
  in Svalbard, 368–370, 369f
  Weichselian in, 365
  Western, 366–367
Barents Sea Ice Sheet, 879
  in Late Pliocene, 70
  LGM of, 286f
Barnum Formation, 510
Barra-Donegal Fan, 60–61
Basement, 65, 84
Batura, 952–953
Baumkirchen, 22
Bavaria, 43
Bavarian Forest, 171
Bay of Biscay, 266
Bayan Har Mountains, 988
Bayerischer Wald, 171
Be, 227
Beaded Valleys, 526f, 527
Beagle Channel, 723, 724–725
Bear Island, 362f, 363
Belarus
  in Berezina, 30–31
  in Dnieper, 31
  glacial limits in, 32f
  in Narev, 30, 300–301
  in Oka, 405–408
  in Pleistocene, 29–36, 29t, 30f, 31f
  in Poozerian, 33–34
  in Pripyat', 31–33
  Sozh in, 32–33
  in Varyazh, 30
Belgrano, 720
Belomorsk ice flow, 356
Belovežje bed, 95
Belukha, Mount, 885, 892
Ben Lomond, 1043
Berezina, 30–31
Bergen-Hardanger, 293f
BGS. *See* British Geological Survey

Bhagirathi Valley, 939–940, 940f
Biba-Boko, 145
Biber, 163, 167
Big Fork Formation, 510
Big-boulder moraines, in Nanhuta Shan, 1006
BIS. *See* British Ice Sheet
Bismarck Range, 1027
Bistra Malrului, 308–309, 309f
Bistritlei Mountains, 311
Bjerka Moraine, 291–293
Bjørnøyrenna, 362, 367–368
Black Creek, 675, 675f
Black Forest, 170–171
Black Sea, 400
    Turkey and, 400
Black Sea Mountains, 400
Bled-Radovljica, 387–388, 388f
    in Late Pleistocene, 389f
Bléone Valley, 117–118
Bloody Canyon, 452, 454f
Blue Lake, 1037
Bobadil, 1041
Boco, 1041
Boduizangbo Valley, 984–988, 991f
Bogotá, 821f, 823f, 825–830, 826f
    in Early Pleistocene, 819–824
    in Middle Pleistocene, 819–824
Böhener Feld, 167
Bohinj, 386, 387–388, 390, 390f
Bolar Mountains, 397–398
Bolivia, 773–802, 774f, 797f, 798f, 799f
    ACR in, 766
    Andes of, 757–772, 758f, 775
    deglaciation of, 766–767
    glacial limit mapping in, 769
    in Holocene, 767–769, 797–800
    in Late Pleistocene, 777–796
    LGM of, 761–766
    in Middle Pleistocene, 775–777
    moraines in, 757–772
    in pre-Weichselian, 759–761
    in pre-Wisconsinan, 759–760
    in Tertiary, 775
    in Würmian, 767–768
Bølling, 291, 356–357
Bølling-Allerød, 792
    in Québec-Labrador, 609
Bol'shoi Annachag Range, 869
    Sartan in, 869f
    Zyrianka in, 869f
Bolshoi Shar, 331
Bœlthav, 56f
Borit Jheel, 917f
Bosies Bank, 274
Bote Koshi, 950f
Bou Iguenouane, 1066, 1067f
Bovbjerg, 54
Bovec basin, 387f, 391
Bozeat till, 66–67
Brahmaputra River, 981
Bresse River, 117–118
Brest chronostratigraphical unit, 30
Bridlington Member, 84
British Columbia, 566f, 569t
    Cordilleran Ice Sheet in, 563–564, 564f
    erosion and deposition in, 567
    growth and decay of, 564–567
    stratigraphical record and chronology of, 568–570
    crustal deformation in, 567
    deglaciation of, 565, 567f, 570
    in Early Pleistocene, 568
    LGM of, 570f
    in Middle Pleistocene, 568
    northern, 686
    in Pleistocene, 563–574
    in Sangamonian, 568
    sea-level in, 567f
    southern, 686–687
    volcanoes in, 568
British Geological Survey (BGS), 262
British Ice Sheet (BIS), 59–74, 265
    in Middle Pleistocene, 62, 65–66
    in North Sea, 62, 261–262
    Northern Hemisphere and, 70
    in Scotland, 60–61, 70
    shelf-edge expansion of, 68–69
    SIS and, 273–274
    in Wales, 62–63, 70
    Witch Ground and, 274
British Isles, 60f, 69f
    continental shelf of, 60–61
    Cromerian Complex in, 61
    in Early Pleistocene, 59–74
    glaciation in, 68–69
    in Late Saalian, 63
    long-term sedimentary archives in, 59
    in Middle Pleistocene, 59–74
    mountain glaciation in, 70
    North Sea and, 61–62
    Northern Hemisphere and, 70
    proto-Thames terrace sequences in, 62–63
    tills in, 65–66
Briton's Lane Formation, 63–64, 67, 68
Broad Valley, 1042
Brooks Range, 437–438
    in Early Wisconsinan, 437–438
    in Late Wisconsinan, 438
Brørup, 53, 108
    in Finland, 107–108
Browerville Formation, 509, 510
Broz Formation, 911
Brumunddal, 282
Brunhes/Matuyama palaeomagnetic reversal, 8
    Austria and, 22
    British Isles and, 61
    China and, 994–995
    Cordilleran Ice Sheet and
        early, 694–695
        latest, 695
    Illinois and, 473
    Klondike and, 669
    Linda and, 1040
    North America and, 419–421, 423f
    North Sea and, 264–265
    Northern Interior Plains and
        early, 695
        late, 693
        latest, 696
    northern Missouri and, 558
    Tasmania and, 1038, 1040–1041
    Wisconsin and, 539–541
Bucher Schneckenmergel, 167
Buena Vista Ridge, 844
Bugchan Depression, 869
Bühl, 24–25
Bulgobac, 1039, 1040–1041
Bull Rivulet, 1041
Burmiester Lake, 720
Buzul Mountains, 394–396
Byrranga Mountains, 378
Bytham River, 62, 67

C

Cabanes basin, 136
Cableway, 1041
Cachorro, 720–721
Calderone, 217
Caledonia, 640–641, 642f
    margins of, 641
Callimani Mountains, 311
Campo Felice, 211, 213f
Canada. *See also specific provinces*

    GeoBase and, 3
    LIS in, 591
    in Neogene, 7–8
    printed maps in, 3–4
Canadian Shield (CS), 576, 602
    deglaciation of, 612
    in northwest Canada, 661
Candona, Lake, 605–606
Canteen member, 469–470
Calpaltlânei Mountains, 313
Cape Breton Island, 646–647
Cape Mountains, 1081
Cape Sabler-type, 378
Capra-Buha Massif, 313
Capron Member, 541–542
Caquel Drift, 731–732
Caracol Drift, 740
Carihuairazo, 775–777, 781f
Carol, 128, 129f
    in Middle Pleistocene, 129–132
    moraines in, 134–135
Carpathian Mountains, 41–42
Carpathians
    of Romania, 305–322, 306f, 307t
        age assignment problems with, 317–318
        arguments with, 308–310
        Eastern, 310–313, 312t, 319f
        ELA in, reconstruction of, 318, 319f
        glacier orientations in, 317
        methodological aspects for, 307–308
        setting of, 305–307
        Southern, 313, 314t, 319f
    southern, 313, 314t, 319f
    of Ukraine, mountain glaciation in, 413–414, 415f, 416f
Cascade Range, 534–535, 563
    Sierra Nevada and, 458

Caucasus
  Central, 144–146
  Eastern, 146
  in Late Pleistocene, 141–148, 143t
  Main Range of, 144
  Minor, 146
  Western, 144
  in Würmian, 141–148, 142f, 143t
Cayambe, 804f
Ceahla1u Massif, 311
Celtic Sea, 88
Cenozoic. *See also* Late Cenozoic
  Antarctica in, 7
  Atlas Mountains in, 1065
  Carpathians in, 305–307
  climate change in, 7
  New Zealand in, 1047
  North America in, 419–426
  Southern Hemisphere in, 7
Central Asia, 944f, 955t
  glacial limits of, 945
  LGM of, 943–966
  methodological consequences and comparisons for, 958–959
Central Caucasus, 144–146
Central Finland end-moraine (CFEM), 112f
Central Karakoram, 918–920, 922f
Central Minnesota, 509–510
Central Mountains, of Taiwan, 994
Central Plateau, East Africa, 1076
Central Plateau, Tasmania, 1038, 1043
Central Russia, 328
Central Siberian uplands, 328–329
Central Taurus Mountains, 396–398
Central Tibet, 952
Cerro Las Cumbres, 855
Cerro Potosí, 856
  LGM of, 857f
Cerros del Chacay, 736
CFEM. *See* Central Finland end-moraine
CGS. *See* Cryo-glacial systems
Chagan, 898–899, 899f
  in Eemian, 902
Chagan-Uzun, 896–897, 900, 900f, 901, 902
  in Late Pleistocene, 897
  TL and, 897
Chambaran Plateau, 117
Champlain Sea, 609, 611f
Changbai Mountain, 994
Changgousi, 988–992
Charaulach Mountains, 878–879
Charnogora Ridge, 416f
Chaun Lowland, 869–870
Chelmos, Mount, 178f
Chersky Mountains, 865, 868–869
  in Sartan, 866
Cheshire Plain, 88
Chic-Choc Mountains, 631
Chignecto, 650f
  Appalachians in, 607–608
  margins of, 649–651
  Maritime Canada in, 649–651
Chile
  ACR in, 748–750

Andes of
  in Early Weichselian, 742–745
  LGM of, 745–748
  in Middle Weichselian, 742–745
  in Wisconsinan, 742–745
Lake District of, 740–741, 741f
  LGM of, 745–748
  in Wisconsinan, 742–745
  in YD, 749
moraines in, 751
neoglaciation in, 752
non-linear responses to climate forcing in, 752
northern and central, 745
Patagonian Ice Sheets in, 742
  in Wisconsinan, 745
in Pleistocene, 739–756
  Andes of, 739–740
  maximum glaciations in, 740–742
  methods for, 740
in Plio-Pleistocene, 8
southern, 745
trimlines in, 751–752
in YD, 748–750
Chimborazo-Carihuairazo, 790–791, 804
China, 981–1002
  Altai Mountains in, 981
  Brunhes/Matuyama palaeomagnetic reversal and, 994–995
  ELA of, 997f
  in Holocene, 996
  Karakoram Mountains in, 981, 984
  Kunlun in, 984
  LGM of, 996, 997f
  in Middle Pleistocene, 8–9
  in Pleistocene, 994–996, 994t
    earliest mountain glaciations in, 994–995
    last glaciations in, 996
  West Himalaya in, 984
Chitral, 911–912, 912f
Cho Oyu, 946f
Chogatake, Mount, 1015, 1016
Cholila Valley, 730
Chronozones, 12
Chucelná, 40–41
Chudskoe ice flow, 356
Chuia River, 896
Chuiskii, 896
Chukchi, 865–866, 888, 892f, 893–894
Chukotka, 869–870, 873–874, 879
Chusovskoye, Lake, 323–324
Chuvan Mountains, 870
Chuya Valley, 901
Cinca catchment, 134
Citlaltépetl, 854–855
Clarence River, 1053
CLIMAP, 11
Clio, Mount, 394–396
Club Lake, 1037
Clutha, 1057–1058
Coast Mountains, 563–564, 570
Cofre de Perote, 849, 855, 855f
Col du Pourtalet, 133–134
Colchester Formation, 62

Coldwater River, 686–687
Collins Pond, 652–654
Colombia, 819–838, 820f
  ELA of, 829
  in Holocene, 824–825
  in Late Pleistocene, 824–825
  in Late Weichselian, 827–829
  LGM of, 827
  in Middle Weichselian, 825–827
  in Plio-Pleistocene, 822–824
  in Weichselian, 829–830
  in YD, 831
Columbia member, 555
Combe d'Ain Lake, 122
Cook Inlet area, 442
Coon Creek, 509
Cootapatamba, Lake, 1037
Copper Falls Formation, 543, 544–545
Coquitlam, 570
Cordilleras, Colombia and, 819
Cordillera Blanca, 778, 784–785, 787f, 792, 797
  in YD, 792, 793
Cordillera Callejon, 797
Cordillera Cochabamba, 788–789, 799, 798f
  in YD, 795–796
Cordillera Huauyhuash, 785, 793, 794f
Cordillera Occidental, 788, 796
  Colombia and, 819
Cordillera Oriental, 791, 793f
  Colombia and, 819
Cordillera Real, 760f, 762, 763, 763f, 788, 795, 797f
Cordillera Vicabamba, 797–798, 800f
Cordillera Vilcanota, 798
Cordilleran Ice Sheet, 428–437, 565f
  Aleutian Range and, 442
  in British Columbia, 563–564, 564f
    erosion and deposition in, 567
    growth and decay of, 564–567
    stratigraphical record and chronology of, 568–570
  Brunhes/Matuyama palaeomagnetic reversal and
    early, 694–695
    latest, 695
  in Early Wisconsinan, 440f
  in Jaramillo, 694
  in Late Pleistocene, 665f
  in Late Pliocene, 665f
  in Late Wisconsinan, 440f
  in Matuyama
    earliest, 691
    early, 692
    late, 693
    latest, 694
  in northern Patagonia, 730–731
  in northwest Canada, 661, 663–669, 688–691
  in Olduvai, 692
  in Pleistocene, 440f
  southern margin of, 443
  in western Washington, 531–533
Cordon de Dona Rosa, 742, 745
Corralitos, 834

Index

Cosmogenic radionuclide (CRN), 128, 773–775
    in Ariège catchment, 129
    in Central Asia, 943–966
    in Chile, 740, 748
    in Del Bonita upland, 577–578, 577t, 586
    in high Asia, 943–966
    in Lower Anadyr Lowland, 870–872
    magnetic field excursions and, 959–960
    in Montana, 578
    in Musselshell, Lake, 578
    in north-east Asia, 865
    in northern India, 929
    in northern Missouri, 558–559
    in northern Montana, 576–578
    in northern Pakistan, 909
    in southern Alberta, 576–578
    in Tibet, 953
    in Wisconsin, 537
    in Zanskar, 933
Costa Rica, 843–848
    ELA of, 845
    geology of, 843–844
    ice cape and valley reconstruction of, 845
    in LLGM, 844, 845
Cotopaxi, 304f
Cotswold escarpment, 62
Courbiére, 136
Craig Basins, 61–62
Crest zone cirques, 137
CRN. See Cosmogenic radionuclide
Cromer Forest-bed Formation, 61, 67
Cromer Ridge, 63–64, 68
    deglaciation of, 81
Cromer Tills, 63
Cromerian
    in Latvia, 222–223
    Netherlands in, 248–249
    in North Sea, 265
Cromerian Complex, 49–50, 67, 68
    Aberdeen Ground Formation and, 61
    British Isles in, 61
    Don Glaciation and, 342–343
    Netherlands and, 248–249, 345
Cromwell Formation, 509, 510
Crown Point-Port Bruce, 491
Cryo-glacial systems (CGS), 884, 885, 893f, 894f, 895f
    in Siberia, 888–890
        glaciation timing and dating, 895–901
        palaeogeographical interpretation of, 901–904
        types and occurrences of, 890–895
CS. See Canadian Shield
Cuchilla Boca Grande, 825–826
Czech Silesia, 37–38
    in Elsterian, 40–41
    in Saalian, 41
Czechia
    continental glaciation in, 37–38
    in Elsterian, 38–41
    mountain glaciation in, 41–42
    in Pleistocene, 37–46
    in Saalian, 41

D

Dagda, 226
Dainava, 237–238
Dainelli, 909
Damp Century, 705–706
Daniglacial, 742
Dansgaard-Oeschger, 283
Dante Rivulet, 1042
Danube River, 37–38, 165
Daocheng, 995
Darhan Basin, 969
    in Late Pleistocene, 976–978
Dashwood Drift, 568
Daumantai Formation, 234–235, 236f
Daun, 26
David, 1041
DCW. See Digital Chart of the World
De Geer moraines, 617
Dead ice, 356–357
Deckenschotter, 164–165
    Ältere, 15, 18, 165, 167, 168
    Jüngere, 18, 165, 168–169
        Hochterrasse and, 169
    Mittlere, 168
Dedegöl Mountains, 398–400
Deglaciation
    of Baie des Chaleurs, 607–608
    of Barents Sea, 368
    of Bolivia, 766–767
    of British Columbia, 565
    of Cofre de Perote, 855
    of Cromer Ridge, 81
    of CS, 612
    of Early Weichselian, 378
    of East Anglia, 81
    of Ecuador, 806
    of Estonia, 99
    of Finland, 112f
    of Greenland Ice Sheet, 706f, 707–709
    of Labrador, 616
    of Laurentian Highlands, 610, 615
    of Laurentians, 612–616
    of Middle Weichselian, 110
    of New Québec Dome, 617–618
    of North Sea, 275
    of Norway, 289
    of Pyrenees, 136–137
    of Québec-Labrador, 604–610, 612
        latest areas of, 621
        northern areas of, 618–621
        pending questions on, 621–622
    of St. Lawrence, 610
    of Taymyr Peninsula, 378
Del Bonita upland, 577–578, 577t, 586
Delta River Valley, 441
DEM. See Digital elevation model
Denmark, 52f
    in Drenthe, 50
    in Early Pleistocene, 8, 49
    in Early Weichselian, 53
    in Eemian, 51
    in Elsterian, 50
    in Late Pleistocene, 48f, 49, 51–57
    in Late Saalian, 51
    in Late Weichselian, 47f, 54–56, 57
    in Middle Pleistocene, 48f, 49–50
    in Middle Weichselian, 47f, 53, 57
    in Neogene, 49
    North Sea and, 270–271
    Norway and, 283–284
    OSL in, 156
    in Pleistocene, 47–58
    in Saalian, 47f, 50–51
    SIS in, 53
    TL in, 156
    valleys in, 49
    in Warthe, 50–51, 51f
Deosai Plateau, 923
Derwent Valley, 1038–1039
Des Moines Lobe, 545
Devensian, 78, 86–89. See also Late Devensian
Devil's Hole, 61, 61f
Devils Postpile National Monument, 452
Diachronic classification system, 489–490
Digital Chart of the World (DCW), 1–2, 5f, 6f
Digital elevation model (DEM), 930f
    of PNG, 1030f
Digital maps, 1–6, 2f
Digital Terrain Elevation Data (DTED), 2
Dinosaur Provincial Park, 585
Disko Bugt, 705
Disna Lake, 33–34
Djanushka River, 878–879, 878f, 880f
Djurdjura, 1071
Dnieper, 346f, 351–352, 412f
    Belarus in, 31
    Estonia in, 97–98
    features of, 409
    glacial limits of, 409–412
    ice-marginal zone and, 412
    key sites in, 413
    Oka and, 406–408
    Pleistocene in, 413–414
    Plio-Pleistocene in, 411f
    Russia in, 408–413
    Ukraine in, 406–408, 407f, 408–413
Dogger Bank, 270–271, 272–273
Dömnitz, 153
Donau, 163, 167
Donian, 8, 341–343
Drenthe
    Denmark in, 50
    Estonia in, 97
    Great Britain in, 84–85
    Netherlands in, 85, 252
    north Germany in, 153
    Ukraine in, 413
Drenthe-Warthe, 97
Driftless Area, in Wisconsin, 538f, 541, 548
Drosh, 911, 912f
Dry Weather Bed, 507
DTED. See Digital Terrain Elevation Data
Duck Hawk Bluffs, 682f, 685
Dudh Kosh Nadi, 950f, 951f
Dzakar Chu, 950–952
Dzulkija, 237–238

E

Eagle Bend Formation, 510
Early Fúquence, 829
Early Holocene
    Hsueh Shan in, 1007
    Iceland in, 206
    Nanhuta Shan in, 1006–1007, 1007f
    Québec-Labrador in, 603f
    Taiwanese high mountain ranges in, 1003–1012
    Turgen-Kharkhiraa Mountains in, 970–971
Early Mérida, 837
Early Miocene
    Barents Sea in, 7–8
    Norway in, 7–8
Early Pleistocene, 7, 8–9
    Barents Sea in, 364–365
        Early part of, 364
        Later part of, 364–365
        seismic stratigraphy and chronology of, 364
    Bogotá basin in, 819–824
    British Columbia in, 568
    British Isles in, 59–74
    Denmark in, 49
    Devil's Hole in, 61f
    Estonia in, 95
    Fladen Ground in, 61f
    German Alpine Foreland in, 167
    Iceland in, 201–203
    Latvia in, 221
    Lithuania in, 234–237
    mountain glaciation in, 70
    Netherlands in, 248f
    New Zealand in, 1047–1049
        gaps in, 1049
    north Germany in, 150
    North Sea in, 264–265
    north-east Asia in, 866
    Northern Hemisphere in, 7
    northern Missouri in, 553–562, 558
    Ohio in, 513–514
    Pennsylvania in, 524–527
    Poland in, 300–301
    in Río Gallegos Valley, 721
    Ross in, 1047–1048
    Sierra Nevada in, 450–452
    Slovenia in, 385–388
        glacial limits in, 385–386
    Tasmania in, 1038–1041, 1040f
    Venezuela in, 835–837
    Verkhoyansk Mountains in, 879
    Wisconsin in, 537–541
        ice extent and chronology for, 539–541
Early Pleniglacial, 122–124, 240
Early Portage Mountain, 568
Early Ugandi, 97
Early Valdaian, 98
Early Weichselian
    Alaska in, 430t
    Barents Sea in, 366–367
    Chilean Andes in, 742–745
    deglaciation of, 378
    Denmark in, 53
    Estonia in, 98
    Finland in, 107–110
    Greenland Ice Sheet in, 702
    Kara Sea Ice Sheet in, 333f
    North Sea in, 269–270
    northern Russia in, 331–332
    Norway in, 282–287
    NTZ in, 380
    Siberia in, 902
    Svalbard in, 367
    Taymyr Peninsula in, 377–378
Early Wisconsinan
    Ahklun Mountains in, 441f
    Alaska in, 429
    Alaska Range in, 441–442
    APG Atlas and, 428f
    Brooks Range in, 437–438
    Cordilleran Ice Sheet in, 440f
    Maritime Canada in, 640–641
    Seward Peninsula in, 439f
    Wisconsin in, 543–544
East Africa, 1076f
    Central Plateau of, 1076
East Alps, 24f
East Anglia, 63–64, 64f, 68, 79–81
    tills of, 66–67
    in Wolstonian, 83–84
East Bay, 639–640
East European Plain, 337–362, 340f, 354f
    Akulovo and, 337–339, 341f, 342f, 344f
    in Dnieper, 351–352
    in Donian, 341–343
    in Ikorets, 345–346
    in Kamenka, 351
    in Krasikovo, 339–341, 343f
    in Likhvin, 349, 350f
    in Likovo, 337
    in Mikulino, 352, 353f
    in Moscow, 351–352
    in Muchkap, 344–345, 348f
    in Oka, 346–349
    in Okatovo, 341, 344f, 345f
    in Pechora, 349–351
    in Pleistocene, 338f
    in Roslavl', 347f
    in Setun', 341
    in Valdaian, 352–357
East Jylland, 55f
Eastern Black Sea Mountains, 400
Eastern Cape Mountains, 1081
Eastern Carpthathians, 310–313, 312t, 319f
Eastern Caucasus, 146
Eastern England, 63
Eburonian, 247
Ecuador, 773–802, 774f, 777f, 779f, 792f, 804f, 807f
    Andes of, 775
    deglaciation of, 806
    glacial limits mapping of, 807
    in Holocene, 797–800, 806–807
    late glaciations of, 803–814
    in Late Pleistocene, 777–796
    LGM of, 779–781, 805–806
    in Middle Pleistocene, 775–777
    pre-LGM glaciation in, 804–805
    in pre-Wisconsinan, 804
    soil temperatures in, 805f
    stratified slop deposits in, 803–804
    temperature/precipitation trends in, 811f
    in Tertiary, 775
    volcanoes in, 808f
    in YD, 789
Edmonton, 585
Edziza, Mount, 661, 687f
Eemian, 8, 10f
    Ariéga catchment in, 129–132
    Austria in, 22
    Chagan in, 902
    Denmark in, 51
    Estonia in, 95, 98
    Greenland Ice Sheet in, 701–702
    Holocene and, 155
    Latvia in, 224
    north Germany in, 155–156
    northern Russia in, 375
    Scandinavian Mountains in, 107
    Severnaya Zemlya islands in, 375
    Southern Germany in, 170
    Taymyr Peninsula in, 375
    Venezuela in, 837
Egesen, 26
žeimena, 237–239
El Abra, 832
El Cóndor Drift, 730
El Coyte, 733
El Maitén, 730, 731f
El Niño-Southern Oscillation (ENSO), 739
El Pedregal, 837–838
El Suero, Lake, 839
ELA. *See* Equilibrium-line altitude
Elbe Valley, 155
Electron spin resonance (ESR), 373–375
    in Qilian Mountains, 988–992
El'gennya, Lake, 869
Elgin, 489–491
Elgygytgyn, Lake, 865, 866f
Elikchan-4, Lake, 865
Ellis, Lake, 1077
Elsterian, 8. *See also* Pre-Elsterian
    Czech Silesia in, 40–41
    Czechia in, 38–41
    Denmark in, 50
    Estonia in, 95
    Finland in, 105
    Germany in, 149
    Latvia in, 223
    Netherlands in, 249–251, 250f
        drainage history of, 251, 252f
    north Germany in, 150–152
    North Sea in, 249–251, 266–267
        glacial limits in, 266
        key sites of, 267
        morphological features of, 266–267
    northern Bohemia in, 38–39
    northern Moravia in, 40–41
    Šluknov Hilly land in, 39f
žemaitija, 238–239
EMC. *See* External moraine complex

Empress Formation, 580, 582f, 587
En-a, 1016–1017
Encierro Valley, 746f
End-moraines, 156–157
    in Finland, 111
    in French Alps, 119
England, eastern, 65
English Midlands. See Midlands
Enguri river, 145
Enns Valley, 19–21, 20f
ENSO. See El Niño-Southern Oscillation
Equilibrium-line altitude (ELA), 9–10
    of Altai, 972–976
    of Andes, 739
    of Apuseni Mountaofs, 320f
    of Australia, 1038
    of Bolivian Andes, 757
    of Cerro Potosí, 856
    of Chofa, 997f
    of Colombia, 829
    of Costa Rica, 845
    of Ecuador, 303
    of Garhwal, 939–940
    of Greece, 195
    of Guatemala, 845
    of high Asia, 943
    of Hofdu Kush, 863
    of Italian Apennofes, 214, 216f, 217f
    of Iztaccíhuatl, 849–850
    of Kashmir, 932
    of Ladakh Range, 937–939
    of Makanaka, 464
    of Mexico, 859
    of Mongolian Altai, 974f, 975t
    of Mount Everest, 950–952
    of Mount Jaya, 1026
    of northern India, 929
    of northern Pakistan, 909
    at Recess Peak, 457
    of Romanian Carpathians, 307–308
        reconstruction of, 318, 319f
    of Russian Altai, 970t
    of Sierra Nevada, 452, 459–460, 459f
    of Taiwanese high mountain ranges, 1003, 1007–1010
Erciyes, Mount, 401
Erratic boulders, in Nanhuta Shan, 1007, 1010f
Esaoman 1015
Escuminac, 643–644, 645f
    margins of, 644
Esquel Valley, 729, 730–731
ESR. See Electron spin resonance
Estonia, 96f
    Baltic Sea and, 96–97
    deglaciation of, 99
    in Early Pleistocene, 95
    in Elsterian, 95
    Haanja ice-marginal zone in, 99–100
    in Holsteinian, 96–97
    in Late Pleistocene, 98–101
    in Late Weichselian, 100f
    LGM of, 99
    in Middle Pleistocene, 95–98
    Otepää ice-marginal zone in, 100–101
    Palivere ice-marginal zone in, 101
    Pandivere ice-marginal zone in, 101
    in Pleistocene, 95–104
    in Saalian, 97–98
    Sakala ice-marginal zone in, 100–101
    till sheets in, 95
    in Weichselian, 98–101
Etna, Mount, 214–216
Eurasion Ice Sheet, 879
Europe
    in Late Pleistocene, 9–10
    in Middle Pleistocene, 9
European Continental Margin, 60–61
Evans Creek, 534
Everest, Mount, 947f, 954f
    ELA of, 950–952
External moraine complex (EMC), 117
    in French Alps, 119, 121f
    in Jura Mountains, 121f

F

FAD. See First Appearance Datum
Falun member, 509
Farcalu-Mihailec Massif, 310–311
Fårop, 54
Fedje Till, 281
Felicianova, 224
Fenix, 742
Fennoscandian border zone, 49, 50
    Estonia and, 95
    Middle Weichselian and, 110
Fennoscandian Ice Sheet (FIS), 265
    in Germany, 149
    North Sea and, 261–262
Feursteinlinie, 37
Fifteenmile River, 665, 667f, 668, 668f
Finchley-Hendon area, 80
Finiglacial, 742
Finland, 98, 105–116
    deglaciation of, 112f
    in Early Weichselian, 107–110
    in Elsterian, 105
    end-moraines in, 111
    ice-lobes in, 112f
    ice-marginal zones in, 111–114
    Lapland of, 105, 107
    in Late Pleistocene, 107–114
    in Late Weichselian, 106f, 111
    in Middle Pleistocene, 105–107
    in Middle Weichselian, 110–111
    in pre-Holsteinian cold stages, 105
    in Saalian, 105–107
    SIS in, 105, 108f
    till sheets in, 110–111
    in Weichselian, 107–114
Fiordland, 1059–1060
First Appearance Datum (FAD), 62–63
First Cromer Till, 63–64
First Salpausselkä, 113
Firth Approach, 61
FIS. See Fennoscandian Ice Sheet
Fjøsangerian, 282
Fladen Ground
    in Early Pleistocene, 61f
    in Late Saalian, 62
Fleuve Manche, 85–86, 266
Flint line, 37
Fljótsdalur, 200, 203
Floral Formation, 591–592
Foix-Cadirac, 129–132
Fontana Lake, 732–733
Formations, 164. See also specific formations
Fort Selkirk, 669–673, 691
    Black Creek site at, 671–673, 675f
    Lower Mushroom site at, 671, 673f
    Mushroom and Fossil sites at, 674f, 693
Forth Valley, 1038–1039
Frankfurt Formation, 157–158
Franklin Valley, 1038–1039
Franz Josef, 1057
Fraser Ice Sheet, 531–532, 531f, 534, 568, 569
French Alps, 117–126
    EMC in, 119, 121f
    end-moraines in, 119
    key sites in, 119, 122
    in Late Pleistocene, 119–124, 121f
    in Middle Pleistocene, 117–119, 118f, 120f
    morphological features of, 119, 122
Frýdlant Hilly land, 38, 41
Fuegian Palaeo-ice-lobes, 723–725
Fulton member, 555
Funkley Formation, 510
Fúquence, 829

G

Gait Island, 592, 594f
Galala, 988
Gállego catchment, 128, 130f
    moraines in, 133–134
Ganga Bridge, 955t
Ganges River, 981
Gangotri, 939–940
Gârbova-Baiul Mountains, 311–313
Garfield Heights, 516, 516f
Garhwal, 939–940
Gauss Normal Chron, 673–675, 678
Gauss/Matuyama magnetic reversal, 8
    Austria and, 15
    British Isles and, 60–61
    Iceland and, 201–202
Gave de Pau catchment, 135
Gave d'Ossau catchment, 135
Geladaindong, 952
Geneva, Lake, 122
GeoBase, 3
Geographic Information System (GIS), 1, 427–428
Geologische Orgeln, 168–169
Geomagnetic polarity timescale, 420f, 593f
Germany
    Alpine Foreland of
        in Early Pleistocene, 167
        LGM of, 164f, 167f
        stratigraphical systems of, 166t
    northern
        in Dömnitz, 153
        in Early Pleistocene, 150
        in Eemian, 155–156

Germany (cont.)
    in Elsterian, 150–152
    in Holsteinian, 152
    in Late Weichselian, 156
    in Middle Saalian, 153
    in Older Saalian, 153–155
    in Pleistocene, 149–162
    in Saalian, 153
    in Saalian Complex, 152–153
    in Weichselian, 156–158
    in Younger Saalian, 155
  southern
    in Eemian, 170
    LGM of, 170
    low mountain range glaciation in, 170–171
    morainic amphitheatres in, 169
    in Pleistocene, 163–174
    regional descriptions of, 165–166
    stratigraphical results and concepts with, 163–165
    in Würmian, 170
Ghizar, 912–913
Ghulkin, 920–922
Giant Mountains, 171
Giants Ridge, 510
Gilead, Mount, 515
Gillam, Manitoba, 596–598
Giluwe, Mount, 1027, 1028, 1031f, 1032
Girrabet frontal system, 135
GIS. *See* Geographic Information System
Giumaalu Mountains, 311
Glaciaire de la Dombes, 117–118
Glasford Formation, 475
Glaven Valley, 63–64, 68
Glazov, 344
Goldthwait Sea, 607
Gondwanaland, 1065
Gonghe, 998
Gorges, 1077
Goring Gap, 67–68
Gotiglacial, 742
GPG. *See* Great Patagonian Glaciation
Gran Sasso Massif, 217
Grand Marias, 495
Great Britain
    in Anglian, offshore areas and, 82
    glacial evidence in, 75–77, 76f
    in Late Devensian, 86–89, 87f
    LGM of, 88
    maximum glacial limits in, 77f
    North Sea and, 77–78
    in Pleistocene, 75–94
    in Weichselian, 86–89
    in Wolstonian, 82–86
        miscellaneous glacial deposits in, 84–86
Great Interglacial, 169
Great Lakes. *See* Michigan
Great Patagonian Glaciation (GPG), 8, 729
Greater Himalaya, 929–930
Greater Kabylie, 1065
Greece, 175–198, 176f, 177t
    carbonate formation in, 183
    ELA of, 195

LGM of, 183–185
palaeoclimate and, 193
stratigraphical framework for, 186–192
uranium-series ages in, 183–185, 191f
Green Bay Lobe, 546–547, 546f
Green River Lowland, 484
Green stimulated luminescence (GSL), 373–375
Greenland, 8, 12
    disappearance of ice on land, 708
    north and central eastern, 702–704
    North Sea and, 261–262
    northern, 702
    northern west, 705–706
    southern and central west, 705
    southern east, 704–705
Greenland Ice Sheet, 59, 279, 399f, 700f
    deglaciation of, 706f, 707–709
    in Early Weichselian, 702
    in Eemian, 701–702
    HTM in, 708–709
    in Late Pliocene, 70
    LGM of, 702, 703f
        regimes for, 706–707
    Michigan and, 492f
    neoglacial and later, 709
    in past 300,000 years, 699–714
    in Preboreal, 707–708
    in Saalian, 699–701, 701f
    in Weichselian, 702–707
    in YD, 707–708
Greymouth-Hokitika, 1051–1052
Grèzes-litées, 803–804
Grodno highland, 31, 32
Grönenbacher Feld, 168–169
Großes Interglazial, 15–18
Group A moraines, 795
Group C moraines, 788
Group D moraines, 761
Gruda-Ozerskaya, 33
Gschnitz, 25
    Austria in, 25
GSL. *See* Green stimulated luminescence
GTOPO30 terrain model, 2
Guatemala, 843–848
    ELA of, 845
    geology of, 843–844
    ice cape and valley reconstruction of, 845
    in LLGM, 844
Gulbene, 227
Gulf of Finland, 98
Gulf of Riga, 98
Gulf Stream, in North Atlantic, 886
Günz, 18, 163, 167
Guxiang, 987f, 989f
Gyarling Lake, 988
Gytgykai River Valley, 870–872

H

Haanja ice-marginal zone, in Estonia, 99–100
Half Moon, 639
Hamilton Moraine, 1042
Hand Hills, 585

Happisburgh Formation, 63–64, 66f, 67, 68
    in Anglian, 67
    in Pleistocene, 78
Haramosh, 917–918, 919f
Harricana Interlobate Moraine, 613–614, 617
Haslach, 163, 168
Hawaii
    in Late Pleistocene, 464f
    LGM of, 463
    in Pleistocene, 463–466
    volcanoes in, 463f
Hawk Creek Formation, 507
Hayes River, 596–598
He Ping Nan Xi, 1006–1007, 1009f
Hebrides Shelf, 60–61
    dropstones in, 68
Heinrich Event, 787
Hellefisk, 700
Helsinki, 107
Henderson Formation, 507
Hengduan Mountains, 988, 995
Hercynian Mountains, 41–42
Hersey Member, 538, 539–541
Hida Mountains, 1013, 1014f
Hidaka Mountains, 1013, 1014f, 1016
    in Late Pleistocene, 1018
    LGM of, 1016, 1017
    in Tottabetsu, 1016–1017
High Asia, 944f
    dating methods for, 945–958
    glacial limits of, 945
    glacier stages in, 956t
    LGM of, 943–966
    methodological consequences and comparisons for, 958–959
High Atlas, 7, 1065–1070, 1067f, 1068f
Hilgard, 457
Himalaya Mountains, 909, 910–911
    Alpine, 305–307
    Greater, 929–930
    keys sites in, 945–952
    Khumbu, 945, 947f, 948t, 950f, 954f
    LGM of, 949
    Lahul, 932–933, 934f, 937f
    precipitation in, 931f
    Trans-Himalaya, 933–935
    West, in China, 984
    western, 920–923
Himalayas-Nyainqentanglha, 984–988
Hindu Kush Mountains
    of Afghanistan, in Late Pleistocene, 863–864, 864f
    of Pakistan, 910–913
Hochflächenmoräen, 761
Hochschwab, 20
Hochstand, 23
Hochterrasse, 18, 165
    Jüngere Deckenschotter and, 169
Hohonu Range, 1055
Hohuan Shan, 1003
Holocene, 7. *See also* Early Holocene
    Bolivia in, 767–769, 797–800
    China in, 996
    Colombia in, 824–825

Ecuador in, 797–800, 806–807
Eemian and, 155–156
Italian Apennines in, 211
　　neoglaciation in, 217–218
Iztaccíhuatl in, 850, 851f
Japan in, 1019
Kattegate Ice Streams and, 56
La Malinche in, 852–854
Laguna de la Sierra in, 835f
Mexico in, 857–858
Mount Olympus in, 179
New Zealand in, 1049–1061
northern Minnesota in, 510
Peru in, 797–800
Québec-Labrador in, 602–604
Sierra Nevada del Cocuy in, 833–834, 834f
southern Africa in, 1081
sub-Antarctic in, 1082
Taurus Mountains in, 396
Upper Waitaki in, 1056
Venezuela in, 839
Holocene Ground Moraine, 796
Holocene Thermal Maximum (HTM), 708–709
Holsteinian, 50, 51, 105
　Estonia in, 95, 96–97
　Latvia in, 223
　north Germany in, 152
Holy Hill Formation, 544–545
Hope Valley, 1053
Horn Mountains, 439–440
Horton PLateau, 423–425
Hotaka, Mount, 1017–1018
Hoxnian
　Great Britain in, 78
　Midlands in, 81
Hozon Valley, 122
Hsueh Shan, 1004f, 1005–1006
　in Early Holocene, 1007
HTM. *See* Holocene Thermal Maximum
Huancané, 794, 798–799
Huang He River, 981
Huanghe, 988
Huayna Potosí, 768
Hudson Bay, 618
Hudson Strait, 619
Hueyatlaco-1, 850, 851f, 857–858
Hueyatlaco-2, 850, 851f
Humboldt Massif, 839
Hunza Valley, 913–917, 915f, 916f, 952–953, 957f
Hurunui, 1053–1054
Hurunui River, 1053
Hyaloclastites, 201f

I

Iceland, 202f, 203f
　in Early Holocene, 206
　in Early Pleistocene, 201–203
　Gauss/Matuyama magnetic reversal and, 201–202
　glacial limits in, 206–207
　in Late Pleistocene, 203–204
　LGM of, 204–206, 205f
　in Middle Pleistocene, 201–204
　in Miocene, 200
　neoglaciation in, 206
　in Pleistocene, 199–210
　quality of data for, 206–207
　SIS and, 202–203
　in Tertiary, 200–201, 204f, 207
　volcanoes in, 201f
Iceland Ice Sheet, 201–203
Ice-marginal zones
　Dnieper and, 412
　in Finland, 111–114
　in Late Weichselian, 111–114
　in Latvia, 226–228
　of Mokoritto, 379f
　in Taymyr Peninsula, 379–380
Ice-rafted debris (IRD), 7, 279, 283, 287
　in British Isles, 60–61
　in Denmark, 53
　in Great Britain, 86–88
Icicle Creek, 534–535
Idenberg, Mount, 1026
židin1i, 222–223
IGME. *See* Institute for Geological and Mineral Exploration
I1htiyars1ahap, Mount, 396
I1kiyaka Mountains, 394–396
Ikorets, 345–346
Îles Crozet, 1084
Ilirney Range, 869–870
Illinoian, 8, 474–484. *See also* Late Illinoian; Pre-Illinois Episode
　AGC in, 639f
　age and paleoclimate of, 476
　fluvial, lacustrine and aeolian sediments of, 475–476
　glacial history and diamictons of, 474–475
　Ohio in, 515
　Québec-Labrador in, 601–602
　radiocarbon ages in, 479t
　Wisconsin in, 537–538, 541–542
　Wisconsinan in, 476–484, 482f
　　aeolian and lacustrine deposits of, 484
Illinois, 467–488, 468f, 469f
　bedrock of, 471f
　key sites in, 472f
Illniza, 804f
IMC. *See* Internal moraine complex
Imja Khola, 949f
Independence Formation, 510
Indus River Valley, 909, 913–915, 923–924, 981
Infrared stimulated luminescence (IRSL), 865
　in Verkhoyansk Mountains, 868, 878–879
Infrared-radiofluorescence (IR-RF), 152–153
Inland Kaikoura Range, 1052
Inlin Brook, 677f, 678
Inn Valley, 18–19, 19f
　LGM of, 165
　overdeepening in, 21–22
Innere Jungenmoräne, 170
Institute for Geological and Mineral Exploration (IGME), 175
Internal moraine complex (IMC)
　in French Alps, 119–124
　in Isère Valley, 122
　in Jura Mountains, 119–124
International Commission on Stratigraphy, 12
International Karakoram Project, 909
Inútil-San Sebastián Bays, 723
Ipswichian, 78
IRD. *See* Ice-rafted debris
Irhil M'Goun, 1065–1066, 1069
Irhzer Likemt Valley, 1066–1069
Irish Sea Glaciation, 81–82
IR-RF. *See* Infrared-radiofluorescence
IRSL. *See* Infrared stimulated luminescence
Isar Valley, 19f
Isère Valley, 122
Isla Grande de Tierra del Fuego, 723
Italian Apennines
　ELA of, 216f, 217f
　in Holocene, 211
　　neoglaciation in, 217–218
　in Late Pleistocene, 211, 212f, 214–217
　LGM of, 215f
　in Middle Pleistocene, 211–220, 212f
ITCZ. *See* Asian Intertropical Convergence Zone
Iztaccíhuatl, 849–851, 857, 858–859
　ELA of, 850
　glacial chronology of, 852t
　in Holocene, 850, 851f
　in Late Pleistocene, 851f
　in YD, 850

J

Jack London, Lake, 869
Jackson Hill, 669, 670f
Jahuacocha Valley, 785
Japan, 1013–1022
　in Holocene, 1019
　in Late Pleistocene, 10, 1016–1019
　　glacial limits in, 1016–1017
　　key sites in, 1017–1018
　　morphological features of, 1017
　in Middle Pleistocene, 1014–1016
　　glacial limits in, 1014–1015
　　key sites in, 1015–1016
　　morphological features of, 1015
Jaramillo, 61
　Cordilleran Ice Sheet in, 694–695
　Northern Interior Plains in, 694
　in Tasmania, 1040–1041
Jaya, Mount, 1026, 1029f, 1032–1034
Jbel Ayachi, 1065–1066, 1069
Jbel Bou Iblane, 1070
Jbel Bou Naceur, 1070
Jbel Toubkal, 1065–1066
Jítrava Saddle, 37–38
Jizerské hory mountain belt, 38, 39f, 41–42
Joggins, 641
Juellesh Valley, 785
Julian Alps, 389
Jüngere Deckenschotter, 18, 165, 168–169
　Hochterrasse and, 169
Junin Plain, 785–786
Jura Mountains, 117–126, 124f
　EMC in, 121f
　end-moraines in, 119

Jura Mountains (cont.)
    IMC in, 119–124
    in Late Pleistocene, 119–124, 121f, 123f
    in Middle Pleistocene, 117–119, 118f, 120f
    morphological features of, 119

K

Kaagvere, 99–100
Kaçkar, Mount, 400
Kaghan, 923
Kaldabrun1a, 226–227
Kalviai, 234–237, 236f, 238f
Kamchatka Mountains, 893–894
Kame terraces, 24–25
Kamenka, 351
Kanas Lake, 976–978
Kanas River, 993f
Kanchalan River, 873–874
Kangerlussuaq, 704–705
Kansan, 553–554
Kap Fulford, 705–706
Kara Sea, 380f
    Barents Sea and, 366
    Severnaya Zemlya islands and, 377
Kara Sea Ice Sheet, 879
    Barents Sea and, 363
    in Early Weichselian, 333f
    in northern Russia, 375–377
    in Severnaya Zemlya islands, 375–377
    in Taymyr Peninsula, 375–377
Karagöl, Mount, 400
Karakoram Mountains, 910–911, 913–920, 929–930, 957f
    Central, 918–920, 922f
    in China, 981, 984
Karasugawa River, 1016
Karavanke Mountains, 389
Karginsky, 330, 330f, 877
    in Late Pleistocene, 866–867
Karpochi Rock, 956, 958
Karuküla, 96–97
Kashima-Yarigatake, Mount, 1014–1015
Kashmir, 930–932, 932f
Kaskaskia River, 475
Katherine Creek, 673–676, 676f
Kattegate Ice Streams, 51–53, 54, 54f
    in Holocene, 56
Kauvonkangas, 110
Kavus1s1ahap Mountains, 396
Kazakhstan, 972–973
Kazantzevo, 375
Keewatin Ice Centre, 423–425, 471–473, 538, 539–541
    in British Columbia, 564
    in Wisconsin, 537
Kejser Franz Joseph Fjord, 703–704
Kelling outwash plain, 63–64
Kelnase, 98
Kemabu Plateau, 1029f
Kemiri, Mount, 1025
Kenya, Mount, 7, 1075–1080, 1078f
    volcanoes and, 1076
Kerinci, Mount, 1025

Kesgrave Sands and Gravels/Formation. See Proto-Thames terrace sequences
Kettle Moraine, 547–548
Kewaunee Formation, 546, 548
Khangai, 971–972
    LGM of, 967
    in Pleistocene, 968f
Khangai Mountains, 973t
Khumbakarma, 947f, 954f
Khumbu Himalaya, 945, 947f, 948t, 950f, 954f
    LGM of, 949
Kidderminster, 81
Kieler Formation, 541
Kihnu Island, 98
Kilimanjaro, Mount, 7, 1075
Kinabalu, Mount, 1024–1025, 1032
King Valley, 1038–1039, 1041
Kinnickinnic Member, 539–541
Kirtisho, 145
Kiso Mountains, 1013, 1014f
Kiso-Komagatake, Mount, 1018
Kitamatairi Valley, 1014–1015
Kleiner Arbersee, 42
Klintholm Ice Stream, 51–54
Klondike gold fields, 669
Klondike Plateau, 693
Kodar, 888f, 896f
Kodori river, 144
Konakhovka, 344
Kong Oscar Fjord, 704
Koruldashi, 145
Koryak Range, 893–894
    LGM of, 871f
    northern slope of, 870–872
Koryak Upland, 865
Kosciuszko, Mount, 1037
Kozi Grzbiet, 300–301
Králický Sne1žník, 41–42
Krasikovo, 339–341, 343f
Kravar1e, 40–41
Krkonoše Mountains, 41–42
    in Late Weichselian, 42, 43f
    in pre-Weichselian, 42
    in Weichselian, 42–43
Krušné hory mountain belt, 41–42
Krzna Stadial, 301
KSIS. See Kara Sea Ice Sheet
Kt-6, 1015, 1016
Kuekhtaran River, 901
Kumak, 682, 683–684
Kumazawa, Mount, 1018
Kunlun, 8–9, 984, 985f, 994–995
Kuraiskii, 896
Kurenurme, 99–100
Kurzeme, 223–224
Kuskokwim Mountains, 439–440
Kuveveem, 874, 879
Kvichak Bay, 442
Kyzyl-Jar, 898

L

La Canoa Valley, 838
La Culata, 839
La Laguna, 828–829, 828f

La Malinche, 849, 852–854, 857–858
    glacial morphology of, 854f
    in Holocene, 852–854
    in Late Pleistocene, 852–854
La Niña, 809
La Plata Lake, 732–733
Labe River, 38, 42
Labrador. See also Québec-Labrador
    deglaciation of, 616
Labrador Trough, 621
Labradorean Ice Centre, 471–473
    in Wisconsin, 537
Labuma Till, 587
LAD. See Last Appearance Datum
Ladakh Range, 929–930, 933–939, 937f, 938f
    ELA of, 937–939
Ladoga ice flow, 356
Lago Argentino, 720–721
Lago Blanco, 723–724
Lago Buenos Aires, 717, 742, 743f
Lago Fagnano, 724
Lago General Vintter, 732, 733f
Lago Lynch, 723–724
Lago Nahuel Huapi, 730
Lago Pueyrredon, 743f
Lago San Martín, 720
Lago Tar, 720
Lago Viedma, 720
Lagos Pueyrredón, 717–719
Laguna Baja, 784–785
Laguna Ciega, 830
Laguna de la Sierra, 835f
Lahul Himalaya, 932–933, 934f, 937f
Lakes. See also specific lakes
    in Belarus, 33–34
    in Czechia, 38
    in northern Moravia, 41
    SRTM and, 3
Lake Candona Episode, 605–606
Lake Michigan Lobe, 545f
Lake Ontario Ice Stream (LOIS), 605, 606f
LANDSAT, 375
    Hindu Kush and, 863
    in Romanian Carpathians, 307–308
Landshut-Neuötting Hoch, 165
Lapland, of Finland, 105, 107
Las Cajas National Park, 792f
Las Mulas Drift, 732
Last Appearance Datum (LAD), 62–63
Last Glacial Maximum (LGM), 10f, 11–12. See also Local last glacial maximum
    of Aconcagua group, 735–738, 737f
    of Alberta, 575, 581f
    of Altai, 967
    of Australia, 1038
    of Austria, 23
    of Barents Sea Ice Sheet, 286f
    of Beagle Channel, 724
    of Blue Lake, 1037
    of Bolivia, 761–766
        Andes of, 757
    of British Columbia, 570f
    of Central Asia, 943–966
    of Cerro Potosí, 857f

of Chile, 739
    Andes of, 745–748
    Lake District of, 740, 745–748
of China, 996, 997f
of Colombia, 827
of Czechia, 42
of Ecuador, 779–781, 805–806
of Encierro Valley, 746f
of Estonia, 99
of German Alpine Foreland, 164f, 167f
of Great Britain, 88
of Greece, 183–185
of Greenland Ice Sheet, 702, 703f
    regimes for, 706–707
of Hawaii, 463
of Hidaka Mountains, 1016, 1017
of high Asia, 943–966
of Iceland, 204–206, 205f
of Inn, 165
of Italian Apennines, 214, 215f
of Khangai, 967
of Khumbu Himalaya, 949
of Koryak Range, 871f
of Lake Waktipu, 1058f
of Latvia, 224–225
of Lithuania, 241–242
of Llanquihue Drift, 745
of Mendoza Andes, 735–738
of Mongolian Altai, 977f
of New Zealand, 1050
of North Island, 1052
of northern Pakistan, 909
of northern Patagonia, 729
of northern Russia, 331–332
of Norway, 281–282, 287–294, 288f
of Pyrenees, 132–133
of Québec-Labrador, 619–620
of Río Zongo Valley, 765f
of Scandinavia, 280f
of SIS, 236f
of South Georgia, 1083–1084
of Southern Alps, 1054–1055
of Southern Germany, 170
of southern Patagonia, 717
of sub-Antarctic, 1083
of Taiwanese high mountain ranges, 1008–1009
of Tibet, 953
of Turkey, 394–396, 396f
of Venezuela, 835–837, 838–839
of Verkhoyansk Mountains, 868, 877
of Zanskar, 933
Late Cenozoic
    Great Britain in, 77
    New Zealand in, 1049
    Tintina Trench in, 689
Late Devensian, 81
    Great Britain in, 86–89, 87f
Late Illinoian, Pennsylvania in, 523, 524
Late Mérida, 838–839
Late Nemunas, 234f, 235t
Late Neogene
    Gait Island in, 594f
    Medicine Hat, Alberta in, 595f

    Wascana Creek, Saskatchewan in, 597f
    Wellsch Valley, Saskatchewan in, 596f
Late Pleistocene, 9–10
    Alberta in, 575–590
    Altai Mountains in, 967–980
    Australia in, 1037–1038
    Barents Sea in, 365
    Bled-Radovljica in, 389f
    Bolivia in, 777–796
    Caucasus in, 141–148, 143t
    Chagan-Uzun River in, 897
    Chukotka in, 869–870
    Colombia in, 824–825
    Cordilleran Ice Sheet in, 665f
    Darhan Basin in, 976–978
    Denmark in, 48f, 49, 51–57
    Ecuador in, 777–796
    Estonia in, 98–101
    Finland in, 107–114
    French Alps in, 119–124, 121f
    Hawaii in, 464f
    Hidaka Mountains in, 1018
    Hindu Kush, Afghanistan in, 863–864, 864f
    Iceland in, 203–204
    Italian Apennines in, 211, 212f, 214–217
    Iztaccíhuatl in, 851f
    Japan in, 1016–1019
        glacial limits in, 1016–1017
        key sites in, 1017–1018
        morphological features of, 1017
    Jura Mountains in, 119–124, 121f, 123f
    Karginsky in, 866–867
    La Malinche in, 852–854
    Latvia in, 224–226
    Lithuania in, 239–244
    Mexico in, 859
    Mount Tymphi in, 190f
    Nevado de Colima in, 856
    New Zealand in, 1049–1061, 1051f
    north-east Asia in, 865–876
    Northern Interior Plains in, 665f
    northern Russia in, 330–332, 375
    northwest Canada in, 661, 690f
    Ohio in, 516–518
    Peru in, 777–796
    PNG in, 1032–1034
        glacial limits in, 1032–1033
        key sites in, 1033–1034
    Poland in, 302
    Québec-Labrador in, 601–630
    Ruíz-Tolima in, 819
    Severnaya Zemlya islands in, 375
    Siberia in, 889, 902–903
    Sierra Nevada in, 455–458
    SIS in, 108f
    Slovenia in, 388–391
        glacial limits in, 388–389
        key sites of, 390–391
    Southern Alps in, 1054–1055
    Subachoque Formation in, 823f
    Taiwanese high mountain ranges in, 1003–1012
    Tasmania in, 1042, 1043f

    Taymyr Peninsula in, 375
    Venezuela in, 837–839
    Verkhoyansk Mountains in, 879
    Wellsch Valley in, 592–594
    western Washington in, 534–535
    Wisconsin in, 542–548
    Yushan in, 1006
Late Pleniglacial, 122–124
Late Pliocene
    Barents Sea Ice Sheet in, 70
    Barents Sea in, 363
    Cordilleran Ice Sheet in, 665f
    Greenland Ice Sheet in, 70
    Northern Interior Plains in, 665f
    northwest Canada in, 661, 663f, 688–691, 690f
    Siberia in, 885
Late Saalian
    British Isles in, 63
    Denmark in, 51
    Fladen Ground in, 62
Late Tertiary, 149
Late Ugandi, 97
Late Valdaian, 99
Late Vistulian, 302
Late Weichselian
    Barents Sea in, 361f, 365, 367–370, 368f
    Bavaria in, 43
    Colombia in, 827–829
    Denmark in, 47f, 54–56, 57
    Estonia in, 99, 100f
    Finland in, 106f, 111
    ice-marginal zones in, 111–114
    Krkonoše Mountains in, 42, 43f
    Lithuania in, 240–241
    north Germany in, 156
    North Sea in, 271f, 272f
        key sites of, 272–273
        maximum glaciation limits of, 270–271
        morphological features of, 271–272
    Norway in, 281–282, 287–294
    NTZ in, 381
    Siberia in, 903–904
    southern Patagonia in, 719t
    Svabard in, 369f
    Turgen-Kharkhiraa Mountains in, 970–971
    Witch Ground in, 271–272
Late Wisconsinan, 419, 539f
    AGC in, 639f
    Ahklun Mountains in, 441f
    Alaska in, 429–437, 430t
    Alaska Range in, 442
    Alberta in, 578–579, 579f, 584
    APG Atlas and, 428f
    Brooks Range in, 438
    Cordilleran Ice Sheet in, 440f
    Lake Musselshell in, 578
    Maritime Canada in, 643–644
    Minnesota in, 499
    Montana in, 575–576, 578–579, 579f
    in northeastern Pennsylvania, 53
    Ohio in, 515–516, 517
    Pennsylvania in, 523–524
    Québec-Labrador in, 592f, 602–604

Late Wisconsinan (cont.)
   Scotian in, 644–649
   Seward Peninsula in, 439f
   southern Patagonia in, 719t
   Wisconsin in, 544–547
      landforms and palaeoglaciology for, 547–548
Latorit1ei Mountains, 313
Latvia, 225f
   in Early Pleistocene, 8, 221
   in Eemian, 224
   in Elsterian, 223
   glacial limits in, 226–228
   in Holsteinian, 223
   ice-marginal zones in, 226–228
   in Late Pleistocene, 224–226
   in Middle Pleistocene, 222–224
   in Middle Weichselian, 222–224
   in Pleistocene, 221–230, 222f
   in Saalian, 223–224
   in Weichselian, 224–226
Lauenburg Clay, 50
   in north Germany, 152
Laurentian Highlands, deglaciation of, 610, 615
Laurentians, deglaciation of, 612–616
Laurentide Ice Sheet (LIS), 59, 70, 475f
   in Alberta, 576, 581f
   in Canada, 591
   Illinoian and, 478–483
   Illinois and, 467
   Lake Musselshell and, 578
   in Maritime Canada, 631
   Minnesota and, 499
   in northern Missouri, 553
   in northwest Canada, 661
   in Ohio, 513f
   in Pennsylvania, 521–522
   in Québec-Labrador, 601–630, 608f
   Rocky Mountains and, 577–578
   Sierra Nevada and, 448–450
Laurentide Ice Sheets, 9, 11–12
Lauze, 136
Laya-Adzva moraines, 332–333
Leaota Mountains, 313
Lebed Formation, 323–324
Lech River, 165
Lech Valley, 19f
Leh, 933–935
Lena River, 867–868, 878–879
Leseur, Mount, 1025
Lesotho, 1081, 1082f
Le2tiža, 223
LGM. See Last Glacial Maximum
LIA. See Little Ice Age
Likhvin, 96–97, 345, 349, 350f
   Central Russia in, 328
Liki, 1078, 1079
Likovo, 337
Lilleboelt advance, 51
Lilly Lake Road, 686–687
Lincoln Sea, 702
Lind, Lake, 509
Linda, 1039, 1040

Brunhes/Matuyama palaeomagnetic reversal and, 1040
Lindis Formation, 1057
Linkuva, 227
LIS. See Laurentide Ice Sheet
Lithostratigraphy, 163–165
Lithuania, 232f
   Belarus and, 33
   Daumantai Formation and, 234–235
   in Early Pleistocene, 8, 234–237
   in Late Pleistocene, 239–244
   in Late Weichselian, 240–241
   LGM of, 241–242
   in Middle Pleistocene, 237–239
      early part of, 238
      late part of, 238–239
   in Middle Weichselian, 239–244
   in Pleistocene, 231–246
   statigraphical comparison in, 237
Little Bear, 676, 677f
Little Ice Age (LIA), 137
   in Bolivia, 767
   in China, 996
   Taiwanese high mountain ranges in, 1007–1008
   Venezuela in, 839
Liwiecian, 301
Llanos de San Miguel Valley, 843–844
Llanquihue, 740, 742–745
   LGM of, 745
LLGM. See Local last glacial maximum
Local last glacial maximum (LLGM), 803, 808–809
   Costa Rica in, 844, 845
   Guatemala in, 844
   Mexico in, 857–858
   in Mount Kinabalu, 1024
   in Sierra los Cuchumatanes, 846f
   Venezuela in, 845
Loess
   Austria and, 15
   in Ohio, 517
   in Ukraine, 413
   in Wisconsin, 547
   Wisconsinian and, 476–477
LOIS. See Lake Ontario Ice Stream
Loisach Valley, 19f
L1omnica Valley, 42
Los Chocoyos Ash, 846–847
Los Zerpa, 837
Louis, Mont, 134–135
Loveland Formation, 508–509
Lowburn Formation, 1057
Lower Anadyr regions, 870–874
   northwestern-part of, 872–874
Lower Klondike River, 699
Lower Mushroom site, at Fort Selkirk, 671, 673f
Lower Nelson River, 596–598
Lower Ob Valley, 329f
Lowestoft Formation, 63, 66–67, 80, 83
Lows Gully, 1025
L'viv lobe, of Oka, 406
Lyntupy highland, 33

M

Mackenzie Basin, 1056
Mackenzie Delta, 678–680, 677f, 679f, 691
Mackenzie Mountains, 423–425, 563, 673–678, 676f, 691
Macubají, Lake, 839
Magdeburg-Bernburg-Halle, 153
Magellan Straits, 722, 742, 744f, 747–748
   YD in, 749–750
Main Advance, 51–53, 54f
Main Divide, 1048f, 1050, 1053
   Southern Alps and
      east of, 1058–1059
      north of, 1054–1055
      west of, 1055
Main Range, of Caucasus, 144
Main Stationary Line (MSL), 54
Maine-New Hampshire, 609
Mainitis, Lake, 870–872
Makanaka, 463–464
   ELA of, 464
Malaspina, 866, 873–874
Malaya Kheta, 330f
Malaysia, 1023–1036, 1024t
Maly Anyui river, 869–870
Maly Krechet Lake, 874
Mandala, Mount, 1026
Manicouagan River, 615
Manitoba, 592f
   magnetostratigraphy for, 591–600
Mansi Till, 323–324
Maoke Range, 1026
Maramures1 Mountains, 310
Mararoa, 1058–1059, 1059f
Marathon Formation, 538, 539–541
Marengo Moraine, 478–483
Margaret, Lake, 1042
Marie Creek Formation, 584
Marion Island, 1082–1083, 1084f
Marionoak river, 1040–1041
Maritime Canada, 632f
   AGC in, 631–660
   in Chignecto, 649–651
   in Collins Pond, 652–654
      margins of, 653–654
   in Early Wisconsinan, 640–641
   in Late Wisconsinan, 643–644
   LIS in, 631
   in Middle Wisconsinan, 640, 641–643
   in Northumberland, 633–637
   in PIE, 633
   in pre-Wisconsinan, 631
   in Sangamonian, 638–640
   sediment ages in, 635t
   Shulie Lake in, 651, 652f
   time-space stratigraphical summaries for, 634f
   in YD, 631, 652–654
Markhida moraine, 331
Marly Drift, 63
   East Anglia and, 80
Marquette, 495

Matuyama. *See also* Brunhes/Matuyama palaeomagnetic reversal; Gauss/Matuyama magnetic reversal
   Cordilleran Ice Sheet in
      earliest, 691
      early, 692
      late, 693
      latest, 694
   New Zealand in, 1047–1048
   Northern Interior Plains in, 692
      early, 695
      late, 693
      latest, 696
Mauna Kea, 463–464
Mauna Loa, 463
Mavroneri catchment, 179–181
Maximalstand, 23
Maximum likelihood classification (MLC), 182–183
McCredie Formation, 555
McGee Mountain, 450–451
McLean, Lake, 620
Mecklenburg-Vorpommern, 156
Medford Member, 541
Medicine Hat, Alberta, 591–592
   in Late Neogene, 595*f*
Medininkai, 237*f*, 238–239
Mediterranean basin, in Pleistocene, 1071, 1072*f*
Mega-scale glacial lineations (MSGLs), 49, 268, 271–272
Mekong River, 981
Melville Bugt, 705–706
Menapian, 247–248
   North Sea in, 265
Mendoza Andes, LGM of, 735–738
Mercan Mountains, 401
Mérida, 837. *See also* Venezuela
Merkine3, 239–244
Merrill Member, 543
Merritt, British Columbia, 686–688, 689*f*
Mersey Valley, 1038–1039
Mesa del Caballo, 835
Mesozoic, 63
   Atlas Mountains in, 1065
   Carpathians in, 305–307
   in south-east Asia, 1023–1024
Metropolis Formation, 470
Mexico
   ELA of, 859
   glacial chronologies of, 858*t*
   in Holocene, 857–858
   in Late Pleistocene, 859
   in LLGM, 857–858
   in pre-Weichselian, 857
   in pre-Wisconsinan, 857
   volcanoes of, 849–862, 850*f*
   in Weichselian, 857–858
   in Wisconsinan, 857–858
   in YD, 850, 859
Meyer Lake Formation, 510
Michigan
   Greenland Ice Sheet and, 492*f*
   major glacial phases in, 490–495
   radiocarbon dates for, 493*t*
   Wisconsinan in, 489–498
Michigan Subepisode, 478*f*, 490–491
Middle Atlas, 1070–1071
Middle Esterian Advance, 50
   Estonia and, 95
Middle Lithuanian ice-marginal zone, 243
Middle Macujún Moraine, 835–837
Middle Pleistocene, 8–9
   Alberta in, 575–590
   Ariéga catchment in, 129–132
   Banks Island in, 682–686
   Barents Sea in, 365
   BIS in, 62
   Bogotá basin in, 819–824
   Bolivia in, 775–777
   British Columbia in, 568
   British Isles in, 59–74
   Carol Valley in, 129–132
   Chukotka in, 869–870
   Denmark in, 48*f*, 49–50
   Eastern England in, 63
   Ecuador in, 775–777
   Estonia in, 95–98
   Finland in, 105–107
   French Alps in, 117–119, 118*f*, 120*f*
   Great Britain in, 78
   Iceland in, 201–204
   Italian Apennines in, 211–220, 212*f*
   Japan in, 1014–1016
      glacial limits in, 1014–1015
      key sites in, 1015–1016
      morphological features of, 1015
   Jura Mountains in, 117–119, 118*f*, 120*f*
   Latvia in, 222–224
   Lithuania in, 237–239
      early part of, 238
      late part of, 238–239
   Midlands in, 63, 65*f*
   Mongolia in, 969–970
   Mount Tymphi in, 190*f*
   New Guinea in, 1028–1032
      key sites in, 1028–1031
   New Zealand in, 1049–1061, 1051*f*
      interglacial record in, 1051–1052
      physiographic and climate setting in, 1050–1051
   North Sea in, 265–266
   north-east Asia in, 866
   northern Missouri in, 553–562, 558
   northern Russia in, 323–329, 373–375
   Ohio in, 514–516
   Peru in, 775–777
   Pieman Valley in, 1041–1042
   Poland in, 301–302
   Pyrenees in, 128–132
   Severnaya Zemlya islands in, 373–375
   Siberia in, 328*f*
   SIS in, 65–66, 108*f*
   Slovenia in, 385–388
      glacial limits in, 385–386
   Southern Alps in, 1054–1055
   Tasmania in, 1040*f*, 1041–1042
   Taymyr Peninsula in, 373–375
   Venezuela in, 835–837
   Verkhoyansk Mountains in, 879
   Wisconsin in, 541–542
      ice extent and chronology for, 541–542
   Yushan in, 1006
Middle Pleistocene Transition (MPT), 59, 70
Middle Saalian, 155
Middle Ugandi, 97
Middle Valdaian, 98–99
Middle Weichselian
   Barents Sea in, 366–367
   Chilean Andes in, 742–745
   Colombia in, 825–827
   deglaciation of, 110
   Denmark in, 47*f*, 53, 57
   Estonia in, 98–99
   Fennoscandian border zone and, 110
   Finland in, 110–111
   Lithuania in, 239–244
   Norway in, 282–287
   NTZ in, 381
   Siberia in, 903
Middle Wisconsinan, Maritime Canada in, 640, 641–643
Midlands, 64*f*
   in Anglian, 81
   in Middle Pleistocene, 63, 65*f*
   in Wolstonian, 82–83
Midnight Dome, 669, 671*f*
Mija cirque, 320*f*
Mikulino, 352, 353*f*
Milankovitch variations, 7
   in Siberia, 902
Milk River, 578
Millbrook Till, 515
Milluni Valley, 761, 764*f*, 766–767
Milpulco-1, 850, 851*f*, 857–858
Milpulco-2, 850, 851*f*, 857–858
Minas Basin, 651
Mindel, 163
Mindelian, 15–18
Minford Silts, 513–514
Minnesota, 499–512
   central, 509–510
   geologic map of, 502*f*
   lithostratigraphy of, 499–510, 502*t*, 503*t*, 504*f*
   northeast, 508*f*, 510
   northern, 508*f*, 510
   paleogeography of, 501*f*
   in Pleistocene, 500*f*
   Red River Valley in, 501–506
   southeast, 508–509
   southwest, 507–508
   stratigraphical columns in, 505*f*, 506*f*, 507*f*
   Twin Cities, 509
Minnesota River Valley, 507
Minor Caucasus, 146
Minsk highland, 31, 32
Miocene, 7. *See also* Early Miocene
   Iceland in, 200
   New Zealand in, 1047
Missouri, 556*t*
   chronology of, 558–559
   in Early Pleistocene, 553–562

Missouri (cont.)
    early work in, 553–554
    glacial boundaries in, 559
    in Middle Pleistocene, 553–562
    stratigraphy and lithologies for, 554–557, 555*f*
Mittlere Deckenschotter, 168
MLC. *See* Maximum likelihood classification
Moat Creek, 1059, 1060
Moberly Formation, 555, 558
Mokoritto, ice-marginal zone of, 379*f*
mole runs, 3
Momskaya Depression, 869
Møn, Young Baltic Ice Streams in, 56*f*
Mondsee, 22
Mongolia, 887
    Altai of, 969–971
        ELA of, 974*f*, 975*t*
        LGM of, 977*f*
        in Pleistocene, 977*f*
            timing of, 976–978
    in Middle Pleistocene, 969–970
    north, 887
    Zyrianka in, 976
Mono Lake, 452, 456*f*
Montana, 576*f*
    CRN in, 578
    in Late Wisconsinan, 575–576, 578–579, 579*f*
Montaña San Juan, 844, 845
Moraines. *See also specific moraines or moraine types*
    in Alaska, 437
    in Bolivia, 757–772
    in Carol catchment, 134–135
    in Chile, 751
    in Gállego catchment, 133–134
    in Pyrenees, 137
    in Têt catchment, 134–135
    in Urals Valley, 334*f*
    in YD, 289
Moravian Gate, 41
Moray Firth Approach, 61
Morgan Bluffs, 685*f*, 686
Morocco, 1065–1074
Moscow, 351–352
    Estonia in, 97–98
Mounds Gravel, 470
Mount Kenya
    neoglaciation on, 1079
    in Weichselian, 1078
    in Wisconsinan, 1078
    in YD, 1079
Mount Olympus, in Pleistocene, 181*t*
Mount Tymphi
    in Late Pleistocene, 190*f*
    in Middle Pleistocene, 190*f*
Mountain glaciation
    in British Isles, 70
    in Czechia, 41–42
    in Early Pleistocene, 70
    in Southern Germany, 170–171
    in Turkey, 395*t*

    in Ukrainian Carpathians, 413–414, 415*f*, 416*f*
MPT. *See* Middle Pleistocene Transition
MSGLs. *See* Mega-scale glacial lineations
MSL. *See* Main Stationary Line
Muchalat River Drift, 568
Muchkap, 344–345, 348*f*
Mulligan Lake Formation, 510
Muntele Mare Mountains, 317
Murillo, 836
Murukta glaciation, 330–331
Mushroom and Fossil sites, at Fort Selkirk, 674*f*, 693
Musselshell, Lake, CRN in, 578
Mystery Cave, 509

N

Nanga Parbat, 920–922, 923*f*, 957*f*
Nanhu Xi, 1004–1005
Nanhuta Shan, 1003, 1004–1005, 1004*f*, 1005*f*, 1008*f*
    big-boulder moraines in, 1006
    in Early Holocene, 1006–1007, 1007*f*
    erratic boulders in, 1007, 1010*f*
    OSL in, 1009–1010
    schist moraines in, 1006
Nano-cyclite, 885
Narev, in Belarus, 30, 300–301
Naro Moru, 1077
Narsaarsuk, 699–700
Nashwauk Formation, 510
Näsijärvi-Jyväskylä, 111
Natal Drakensberg, 1081
NATMAP, 419
Nauri, 952
Naust Formation, 279–281, 281*f*
Nebraska, 507–508
Nebraskan, 553–554
Nelson River Bluffs, 684*f*, 686
Nemona, 1057
Nemunas, 239–244
Nenana River Valley, 441
Neogene, 7. *See also* Late Neogene
    Alaska and, 7–8
    Austria and, 15
    Barents Sea in, 363–364
    Canada and, 7–8
    Denmark in, 49
Netherlands
    in Cromerian, 248–249
    in Cromerian Complex, 248–249, 345
    in Drenthe, 85, 252
    in Early Pleistocene, 248*f*
    in Elsterian, 249–251, 250*f*
        drainage history of, 251, 252*f*
    in Pleistocene, 247–260
    in pre-Elsterian, 247–249
    in Saalian, 251–255, 254*f*
        drainage during, 255
Neutron monitors (NMs), 960
Neva, 356–357
Nevado Coropuna, 788, 795
Nevado de Colima, 856

Nevado de Toluca, 849, 852, 853*f*, 858
    glacial stratigraphy of, 853*t*
Nevado Sajama, 796, 799, 799*f*
New Brunswick. *See* Maritime Canada
New Guinea, 1023–1036
    in Middle Pleistocene, 1028–1032
    key sites in, 1028–1031
New Québec Dome, 614
    deglaciation of, 617–618
New Ulm Formation, 507–509, 509–510
New Zealand, 1047–1064, 1048*f*, 1050*t*
    in Cenozoic, 1047
    in Early Pleistocene, 1047–1049
        gaps in, 1049
    in Holocene, 1049–1061
    in Late Cenozoic, 1049
    in Late Pleistocene, 1049–1061, 1051*f*
    LGM of, 1050
    in Matuyama, 1047–1048
    in Middle Pleistocene, 1049–1061, 1051*f*
        interglacial record in, 1051–1052
        physiographic and climate setting in, 1050–1051
    in Miocene, 1047
    volcanoes in, 1052
Ngauruhoe, 1052*f*
Ngozumpa Drangka, 945, 946*f*
Nidanian, 301
Niedersachsen, 152, 154
Niederterasse, 165
    in Würmian, 170
Niederterrasse, 18, 22
Nieniexiongla, 984
Ninguitz Valley, 844
Nizhne-Anadrsk Lowland, in Zyrianka, 871*f*
NMs. *See* Neutron monitors
Noguera Ribagorçana catchment, 134, 137
    deglaciation of, 136
    in Würmian, 133*f*
North Africa
    Atlas Mountains of, 1065–1074
    paleoglaciers and palaeoclimates in, 1071
North America. *See also specific regions or countries*
    Brunhes/Matuyama palaeomagnetic reversal and, 419–421, 423*f*
    in Cenozoic, 419–426
    in Late Pleistocene, 9
    in Middle Pleistocene, 9
    paleomagnetic data sites in, 422*t*
    polarity chrons and, 421*f*
    tectonic activity in, 425
North Atlantic
    in Early Pleistocene, 8
    Gulf Stream in, 886
    ocean currents in, 200*f*
North Germany
    in Dömnitz, 153
    in Early Pleistocene, 150
    in Eemian, 156–156
    in Elsterian, 150–152
    in Holsteinian, 152
    in Late Weichselian, 156

in Middle Saalian, 155
in Older Saalian, 153–155
in Pleistocene, 149–162
in Saalian, 153
in Saalian Complex, 152–153
in Weichselian, 156–158
in Younger Saalian, 155
North Island, 1049, 1052
    LGM of, 1052
North Lithuanian ice-marginal zone, 243–244
North Mongolia, 887
North Sea, 82, 262f, 263f, 273f
    Anglian and, 82
    BIS in, 62
    British Isles and, 61–62
    deglaciation of, 275
    in Early Pleistocene, 264–265
    in Early Weichselian, 269–270
    in Elsterian, 249–251, 266–267
        glacial limits in, 266
        key sites of, 267
        morphological features of, 266–267
    Great Britain and, 77–78
    in Late Weichselian, 271f, 272f
        key sites of, 272–273
        maximum glaciation limits of, 270–271
        morphological features of, 271–272
    in Middle Pleistocene, 265–266
    in Pleistocene, 261–278
        evidence techniques and methods, 264
    in pre-Elsterian, 265–266
    in Saalian, 251–252, 253–255, 268
        glacial limits in, 268
        key sites of, 268
        morphological features of, 268
    SIS in, 62
    tunnel valleys in, 266, 267f
    in Weichselian, 255–256, 268–275
    in Wolstonian, 85f
North Sea Drift Formation, 63, 64f, 65
    Norway and, 79–80
North Sea Fan, 279–281
North Taymyr ice-marginal zone (NTZ), 379–380
    in Early Weichselian, 380
    in Late Weichselian, 381
    in Middle Weichselian, 381
North Westland, 1054
North-central Wisconsin, 543f
    in Early Pleistocene, 538
North-east Asia
    in Early Pleistocene, 866
    in Late Pleistocene, 865–876
    in Middle Pleistocene, 866
    relief and climate of, 865
Northeast Minnesota, 508f, 510
Northeastern Pennsylvania, 526f
    late Wisconsinan in, 527
North-eastern South Island, 1052
Northern Bohemia, 37–38
    in Elsterian, 38–39
    in Saalian, 41
Northern Cordilera, 424f
Northern Europe, till sheets in, 8

Northern Hemisphere. *See also specific regions*
    BIS and, 70
    British Isles and, 70
    in Early Pleistocene, 7
    in Saalian, 48
    in Weichselian, 48
    in YD, 748, 790
Northern India, 929–942
    precipitation in, 931f
    regional setting of, 929–930
    SRTM of, 930f
Northern Interior Plains, 424f, 661–662, 692
    Brunhes/Matuyama palaeomagnetic reversal and
        early, 694–695
        late, 693
        latest, 696
    in Jaramillo, 694
    in Late Pleistocene, 665f
    in Late Pliocene, 665f
    in Matuyama, 692
        early, 692
        late, 693
        latest, 694
    in Olduvai, 692
Northern Minnesota, 508f, 510
Northern Missouri, 556t
    chronology of, 558–559
    in Early Pleistocene, 553–562
    early work in, 553–554
    glacial boundaries in, 559
    in Middle Pleistocene, 553–562
    stratigraphy and lithologies for, 554–557, 555f
Northern Montana, CRN in, 576–578
Northern Moravia, 37–38
    in Elsterian, 40–41
    in Saalian, 40–41, 40f
Northern Pakistan, 909–928, 911f
    ELA of, 909
    LGM of, 909
Northern Patagonia
    LGM of, 729
    in Pleistocene, 729–734
Northern Patagonian Ice cap (NPI), 747, 751
Northern Rocky Mountain Trench, 568
Northern Russia, 323–336, 324f, 325t, 373–384
    in Eemian, 375
    Kara Sea Ice Sheet in, 375–377
    in Late Pleistocene, 330–332, 375
    LGM of, 331–332
    in Middle Pleistocene, 323–329, 373–375
    OSL in, 331, 331f
    in Saalian, 323–324, 375
Northumberland, Maritime Canada in, 633–637
Northwest Canada, 662f
    Cordilleran Ice Sheet in, 688–691
    in Late Pleistocene, 690f
    in Late Pliocene, 663f, 688–691, 690f
    in Olduvai, 692
    regional correlations in, 689
    stratigraphical record of, 661–698
North-West Russia, Barents Sea and, 366

Northwest Territories, 673–686
North-western South Island, 1052
Norway, 279–298
    deglaciation of, 289
    Denmark and, 283–284
    Early Miocene and, 7–8
    in Early Weichselian, 282–287
    in Late Weichselian, 281–282, 287–294
    LGM of, 281–282, 287–294, 288f
    in Middle Weichselian, 282–287
    North Sea Drift and, 79–80
    in pre-Weichselian, 279
    in Weichselian, 281–282
    Western, 291
    in YD, 292f
Norwegian Channel, 60–61, 264–265, 279–281, 289–291, 293f
    North Sea and, 270
Norwegian Elster Advance, 50
Norwegian gravel, 50
Norwegian Saale Advance, 50
Notec-Randow Urstrom, 158
Nova Scotia. *See* Maritime Canada
Novogrudok highland, 31, 31f, 32
Nozon Valley, 124f
NPI. *See* Northern Patagonian Ice cap
NTZ. *See* North Taymyr ice-marginal zone
Nunataks, 289
Nunavik, 621
Nyainqentanglha Mountain, 984–988
Nyang Qu, 950, 955t
Nygchekveem River, 870–872

O

Oadby Till, 63–64, 66f, 67, 82–83
Oak Formation, 470
Ob' River, 896
Obergünzburg, 169
Ocean Drilling Program (ODP), 363
October Revolution Island, 375–377, 376f
Odderade, 107–108
Oder Formation, 157–158
ODP. *See* Ocean Drilling Program
Odra, north Germany and, 152
Odra fluvial terrace, 41
Odranian, 301
Ohau Lake, 1056
Ohio
    in Early Pleistocene, 513–514
    glacial limits in, 514f
    in Illinoian, 515
    in Late Pleistocene, 516–518
    in Late Wisconsinan, 515–516, 517
    LIS in, 513f
    lithologic units in, 515t
    in Middle Pleistocene, 514–516
    in Pleistocene, 513–520
    in Tertiary, 513
    in Wisconsinan, 517
Ojibway, Lake, 617
Oka
    in Belarus, 405–408
    Dnieper and, 406–408
    L'viv lobe of, 406

Oka (cont.)
    in Ukraine, 405–408, 406f
Oka Glaciation, 346–349
Okanogan Centre Drift, 568
Okatovo, 341, 344f, 345f
Okhotsk-Chukotka Mountain System, 865
Okhotsk-Kolyma Watershed, 865
Older Saalian
    in Estonia, 97
    in north Germany, 153–155
Oldman River, 577–578, 581f, 584–585
Olduvai
    Cordilleran Ice Sheet in, 692
    Northern Interior Plains in, 692–693
    northwest Canada in, 693
Olduvai Subchron, 203
Oligocene, 7
Olympic Mountains, 531, 535, 563
Olympus, Mount, 177–181, 180f
    glacial sedimentary sequences on, 179–181
    records of, 195
Onega-karelia ice flow, 356
Ontario, 489–490
Ontario, Lake, 605
Open StreetMap (OSM), 3, 6f
Optically stimulated luminescence (OSL), 66–67, 98, 100–101
    in Barents Sea, 366–367
    in Central Asia, 943–966
    in Denmark, 156
    in Finland, 108–109
    in Garhwal, 939–940
    in Greenland Ice Sheet, 702
    in high Asia, 943–966
    for Illinoian, 476
    in Illinois, 467
    in Lithuania, 234
    in Nanhuta Shan, 1009–1010
    in north-east Asia, 865
    in northern India, 929
    in northern Pakistan, 909
    in northern Russia, 331, 331f
    in Pyrenees, 128
    in Siberia, 884
    in Tibet, 952–953
    in Venezuela, 835–837, 838
    in Wisconsin, 537
Optically stimulated luminescence/thermoluminescence (OSL/TL), 95
Øresund, 56
Orsha-Mogilov pLateau, 32
Osa-Eidfjord moraines, 292f
Oshmyany highland, 31, 32
OSL. See Optically stimulated luminescence
Oslofjord, 289–291, 290f
OSL/TL. See Optically stimuLated luminescence/thermoluminescence
OSM. See Open StreetMap
Ostrobothnia, Finland, 98
Otepää ice-marginal zone, in Estonia, 100–101
outwash cones, 33
Overdeepened Valleys, in Austria, 21–22, 21f
Owen Stanley Range, 1027

P

Pacific-Arctic Watershed, 865, 866
Pakistan. See also Northern Pakistan
    Hindu Kush Mountains in, 910–913
    precipitation in, 931f
Palaeoclimate
    Greece and, 193
    of PIE, 474
Palaeogene, Barents Sea in, 363–364
Palivere ice-marginal zone, in Estonia, 101
Palo Blanco, 837
Pandivere ice-marginal zone, in Estonia, 101
Pangaea, 49
Papallacta Pass, 810f
Paparoa Range, 1053f
Papua, Indonesia, 1026–1027
Papua New Guinea (PNG), 1023–1036, 1027f, 1028t, 1030t
    DEM of, 1030f
    in Late Pleistocene, 1032–1034
        glacial limits in, 1032–1033
        key sites in, 1033–1034
Paramo, 820–822
Páramo de Guerrero, 825
Páramo de Palacio, 820–822
Páramo de Peña Negra, 825
Páramo de Sumapaz, 825
Paso Tromen Drift, 729–730
Pasu I, 952–953
Patagonia, western, 750–751
Patagonian Ice Sheets
    in Chile, 742
    of Chile, in Wisconsinan, 745
    YD in, 749
Pavlof Bay, 442
Pearl Formation, 476
Pebble fraction, 34
Pech de Varilles, 129–132
Pechora, 349–351
Pekul'nei Ridge, 873–874
Pekulney Mountains, 879
Pekylney Mountains, in Sartan, 873f
Penck, Albrecht, 163
Peneios River, 179
Pennfield-Utopia delta complex, 651
Pennsylvania, 521–530, 521f
    in Early Pleistocene, 524–527
    glacial limits in, 523
        dating of, 523–524
    in Late Illinoian, 523, 524
    in Late Wisconsinan, 523–524
    northeastern, 526f, 527
Penultimate glaciation, 20
    in Alaska, 429
    in Altai Shan, 993
    in northern Patagonia, 729–730
Peoria Formation, 508–509, 547
Peoria Silt, 478, 484
Peräpohjola, 107–108
Permafrost, 890
permafrost, 883–884
Permian rifting, 49
Peru, 761, 773–802, 774f, 794f, 795f, 800f
    Andes of, 775

    in Holocene, 797–800
    in Late Pleistocene, 777–796
    in Middle Pleistocene, 775–777
    in Tertiary, 775
Piatra Mare Massif, 311
Pichincha volcanoes, 808f
Pico de Orizaba, 854–855, 857
PIE. See Pre-Illinois Episode
Piedras Blancas, 839
Pieman Valley, 1039, 1041
    in Middle Pleistocene, 1041–1042
Pierce Formation, 509, 538, 539–541
Pietroasa Valley, 309f
Pietu2 Lietuvos, 242–243
Pindaunde Valley, 1031f
pingos, ASTER, 3
Pir Panjal, 929–932
PIW. See Polygonal ice wedges
Plan de la Sagne, 122
Pleistocene, 7. See also Early Pleistocene; Late Pleistocene; Middle Pleistocene
    Africa in, 7
    Ahklun Mountains in, 440, 441f
    Alaska in, 428–437
    Altai in, 968f
    APG Atlas and, 428f
    Ariéga catchment in, 131f
    Belarus in, 29–36, 29f, 30f, 31f
    British Columbia in, 563–574
    Chile in, 739–756
        Andes of, 739–740
        maximum glaciations in, 740–742
        methods for, 740
    China in, 994–996, 994t
        earliest mountain glaciations in, 994–995
        last glaciations in, 996
    Cordilleran Ice Sheet in, 440f
    Czechia in, 37–46
    Denmark in, 47–58
    Dnieper in, 413–414
    East European Plain in, 338f
    Estonia in, 95–104
    Great Britain in, 75–94
    Happisburgh Formation in, 78
    Hawaii in, 463–466
    Iceland in, 199–210
    Khangai in, 968f
    Lapland in, 107
    Latvia in, 221–230, 222f
    Lithuania in, 231–246
    Mediterranean basin in, 1071, 1072f
    Minnesota in, 500f
    Mongolian Altai in, 977f
        timing of, 976–978
    Mount Olympus in, 179, 181t
    Mount Tymphi in, 183
    Netherlands in, 247–260
    north Germany in, 149–162
    North Sea in, 261–278
        evidence techniques and methods, 264
    northern Patagonia in, 729–734
    Ohio in, 513–520
    Poland in, 299–300, 300f
    Puget Lobe in, 533f

Pyrenees in, 127f
Russian Altai in, timing of, 976–978
Sahara Desert in, 1071
Seward Peninsula in, 439f
Siberia in, 889, 891f
southern Germany in, 163–174
southern Patagonia in, 715–728
Tierra del Fuego in, 715–728
Turkey in, 394–396
Ukraine in, 405–418, 407f
Wisconsin in, 537, 542f
Witch Ground in, 269f
Pleniglacial, 122–124, 133, 240
Pliocene. *See also* Late Pliocene
Sierra Nevada in, 450–452
Plio-Pleistocene, 7–8
Colombia in, 822–824
in Dnieper, 411f
PNG. *See* Papua New Guinea
Pogeda Mountain, 992–993
Pohjanmaa, 109–110
Poland, 299–304, 299f
in Early Pleistocene, 8, 300–301
in Late Pleistocene, 302
in Middle Pleistocene, 301–302
in Pleistocene, 299–300, 300f
Polar Front, 1071
Polar Urals, 333–334, 333f, 879
Polycyclic push moraines, 40–41
Polygonal ice wedges (PIW), 893
in Siberia, 887, 887f
Pomeranian, 356–357
Poozerian, Belarus in, 33–34
Popocatétl, 857
Porika, 1049
Poroshiri, 1016
Port Huron, 491–492
Poruba Gate, 41
Potrerillos Plateau, 779f, 789–796
Pounds Creek, 1037
Powers Bluff, 539–541
Pozhizhevs'ka, 414
Prangli Island, 98
Preboreal
Greenland Ice Sheet in, 707–708
Iceland in, 206
Norway in, 291
Pre-Cromerian Complex, 337
Pre-Elsterian
Netherlands in, 247–249
North Sea in, 265–266
Pre-Holsteinian, 105
Pre-Illinois Episode (PIE), 467–474, 553
age and paleoclimate of, 474
drainages of, 470f
fluvial, lacustrine and aeolian sediments of, 473–474
glacial history of, 471–473
Maritime Canada in, 633
pre-glacial alluvium in, 469–470
Pre-Weichselian
Bolivia in, 759–761
Krkonoše Mountains in, 42
Mexico in, 857

Norway in, 279
Šumava Mountains in, 42
Pre-Wisconsinan, 509
Bolivia in, 759–760
Ecuador in, 804
in Illinoian, 601–602
Maritime Canada in, 631
Mexico in, 857
Prince Edward Island. *See* Maritime Canada
Pripyat', 31–33
Proto-Thames terrace sequences, 68
in British Isles, 62–63
in East Anglia, 80
Pseudo-moraine, 524–527, 525f
Ps'ol River, 409, 412f
Pudasjärvi, 109–110
Pudasjärvi-Taivalk-Hossa, 111
Pueblo Llano, 838
Puesto de Paja Drift, 729–730
Puget Lobe, 532, 532f
in Pleistocene, 533f
Pukaki, 1056, 1057
Pulvernieki, 223
Punta Bandera I, 720
Putorana Plateau, 334, 366
Pyrenees, 127–140
crest zone cirques in, 137
deglaciation of, 136–137
LGM of, 132–133
in Middle Pleistocene, 128–132
north slope of, 129–132
deglaciation in, 136–137
in Pleistocene, 127f
southern slope of, 128–129
in Würmian, 132–137

Q

Qilian Shan, 988–992, 992f, 995
Qinghai-Tibetan Plateau, 8–9, 981–992, 982f, 983f
earliest glaciations in, 994
glaciations and uplift of, 998, 999f
hinterland of, 992
ice cover on, 998
snowline distributions on, 996–998
tectonic activity in, 998
Qingshantou, 995
QMAP. *See* Quarter-million scale mapping
Qomolangma-Xixabangma, 984
Quarter-million scale mapping (QMAP), 1049
Que River, 1039
Québec North Shore, 615–616
Québec-Labrador, 604f, 610f, 614f
Appalachians in, 606–607
ice-flow reversal in, 608–609
Bølling-Allerød in, 609
deglaciation of, 604–610, 612
latest areas of, 621
northern areas of, 618–621
pending questions on, 621–622
in Early Holocene, 603f
glacial history of, 601–602
in Holocene, 602–604
in Illinoian, 601–602

in Late Pleistocene, 601–630
in Late Wisconsinan, 602–604, 603f
LGM of, 619–620
LIS in, 601–630, 608f
in Sangamonian, 601–602
in Weichselian, 602–604
in Wisconsinan, 601–602
in YD, 610–612
Quebrada Huayllaura, 779
Queen Charlottte Range, 563, 567, 570
Queer Mountain, 988
Quelccaya Ice Cap, 790–791, 792, 795f, 806–807
Qungur, 954f

R

Ra moraines, 289–291
Rainier, Mount, 534
Rakaia, 1055–1056
Rakaposhi, 917–918
Rangitata, 1055–1056, 1056f
Rantemario, Mount, 1025
Rauchuan Range, 870
Ravni Laz, 388f
Rawdon, 649
Recess Peak, 457
Red Altmark Till, 155
Red Deer-Stettler area, 587
Red River Valley, 501–506
Regina, Saskatchewan, 594–595
Regional subsidence, 49
Reid, 695
Retezat Mountains, 318
Rhine River, 165, 165f, 169
North Sea and, 82
Rhône, 119
Ricardo Rojas Drift, 733
Richmond Range, 1052
Riesengebirge, 171
Rif Mountains, 1070–1071
Rift Valleys, 1076
Rinjani, Mount, 1025
Río Apeleg Valley, 732
Río Chisacá Formation, 820–822, 822f
Río Corcovado Valley, 732
Río Corintos Valley, 730–731
Río Coyle Valley, 721
Rio de las Cuevas, 735–736
Rio de las Vacas, 735–736
Río Frías Valley, 732
Río Gallegos Valley, 721
Rio Huancané Valley, 794
Río Huemul Valley, 731–732
Río Kaluyo Valley, 760
Río Limay, 730
Río Malleo Valley, 729–730
Río Mayo, 733
Rio Mendoz Valley, 736
Río Moro, 732–733
Rio Motatán, 838
Río Pico Valley, 732
Río Recío, 836
Río San Pablín, 833f
Río Siecha Formation, 820–822, 822f

Río Tecka Valley, 730–731
Río Tuni Valley, 760
Río Tunjuelito Formation, 820–822, 822f
Río Unduavi Valley, 760
Río Zongo Valley, 765f
Rioni river, 146
Risbury Formation, 81
Rissian, 22, 42, 165
  gravel fields in, 169
  Yushan in, 1006
Ristinge, 51–54, 156
River Falls Formation, 509, 541
Rivière aux Mélèzes, 619
Rize Mountains, 400
Robein Member, 478
Roches moutonnés, 988
Rock Creek, 663, 666f
Rocky Mountains, 564, 576
  LIS and, 577–578
Rödschitz, 26
Romanian Carpathians, 305–322, 306f, 307t
  age assignment problems with, 317–318
  arguments with, 308–310
  Eastern, 310–313, 312t, 319f
  ELA of, reconstruction of, 318, 319f
  glacier orientations in, 317
  methodological aspects for, 307–308
  setting of, 305–307
  Southern, 313, 314t, 319f
Rona Wedge, 60–61
Rongbuk Valley, 950–952
Rõngu, 98
Rosenthal end-moraine, 156–157
Roslavl', 345, 347f
Ross, 1047–1048, 1049
Roxana Formation, 476–477, 478, 508–509, 544
RP-3, 1015
Ruapehu, Mount, 1052
Rucu Pichinacha, 777f, 790
Ruíz-Tolima, 819, 836
Rumiñahui, 804f
Russia, 11–12
  Altai of, 969, 971f
    ELA of, 970t
    in Pleistocene, timing of, 976–978
  central, 328
  in Dnieper, 408–413
  northern, 323–336, 324f, 325t, 373–384
    in Eemian, 375
    Kara Sea Ice Sheet in, 375–377
    in Late Pleistocene, 330–332, 375
    LGM of, 331–332
    in Middle Pleistocene, 323–329, 373–375
    OSL in, 331, 331f
    in Saalian, 323–324, 375
  north-west, Barents Sea and, 366
Russian Plain, 97

S

Saale-Elbe, 150, 152, 153
Saalian, 8. See also Late Saalian; Older Saalian; Younger Saalian
  Czech Silesia in, 41
  Czechia in, 41

Denmark in, 47f, 50–51
Estonia in, 97–98
Finland in, 105–107
Germany in, 149
Greenland Ice Sheet in, 699–701, 701f
Latvia in, 223–224
Middle, 155
Netherlands in, 251–255, 254f
  drainage during, 255
north Germany in, 151, 152–154
North Sea in, 251–252, 253–255, 268
  glacial limits in, 268
  key sites of, 268
  morphological features of, 268
northern Bohemia in, 41
Northern Hemisphere in, 48
northern Moravia in, 40–41, 40f
northern Russia in, 323–324, 375
Severnaya Zemlya islands in, 375
Taymyr Peninsula in, 375
Sachsen, 152
Sacrificial Pool, 1032
Saglek Moraines, 620
Sahara Desert, 1071
Saint-Narcisse Moraine, 611, 613f
Sakala ice-marginal zone, 100–101
Sakami Moraine, 614f, 617
Sakeni River, 144
Salamanca, 521–522
Salpausselkäs, 111
Salthouse outwash plain, 63–64
Salween River, 981
Salza Valley, 20
Salzach, 19, 21
Samarovo, 328, 375
  Siberia in, 373–375
San Huberto Drift, 729–730
San Miguel Valley, 844, 845f
San Rafael, 739–740
Sancha Shan, 1003
Sandiras, Mount, 398, 399f
Sangamonian, 489–490, 516–517
  Beagle Channel in, 724–725
  British Columbia in, 568
  Maritime Canada in, 638–640
  Québec-Labrador in, 601–602
Sangaste Till, 95
Sanian 1, 301
Sanian 2, 301
Santa Isabel, 836
Santa Maria Drift, 740
Santo Domingo Valley, 837
Sarala, 110
Sartan, 865–866, 868f
  in Bol'shoi Annachag Range, 869f
  in Chersky Mountain System, 866
  Pekylney Range in, 873f
  Siberia in, 903–904
  Zyrianka and, 870
Saskatchewan, 592f
  magnetostratigraphy for, 591–600
Saskatchewan Gravel and Sand (SGS), 580
Saskatoon, Saskatchewan, 595–596

Sauk Centre Formation, 510
Saum Formation, 510
Savala, 98–99
Sawmill Canyon, 455
Sayan Ranges, 887, 888, 890f, 892f, 897f
Scandinavia. See also specific countries and regions
  LGM of, 280f
  mountains of, in Eemian, 107
Scandinavian Ice Sheet (SIS), 59, 279, 281, 282f, 285f, 356
  Barents Sea and, 363, 366
  BIS and, 273–274
  in Denmark, 53
  in Finland, 105, 108f
  Great Britain and, 80
  Iceland and, 202–203
  in Late Pleistocene, 108f
  LGM of, 286f
  in Middle Pleistocene, 65–66, 108f
  Netherlands and, 253
  in North Sea, 62
  tills from, 65
  Verkhoyansk Mountains and, 879
  in Weichselian, 283f
Schefferville, 621
Scheldt River, 82
Schist moraines, in Nanhuta Shan, 1006
Schleswig-Holstein, 50–51
Schwarzwald, 170–171
Scoresby Sund, 699, 704
Scotian, 647f, 648f
  in Late Wisconsinan, 644–649
  margins of, 648–649
Scotland, 68, 273f
  BIS in, 60–61, 70
Second Comer Till, 63–64
Second Salpausselkä, 113
SED. See Surface exposure dating
Sediment exposure dating. See Surface exposure dating
Seisdon Formation, 81
Seisdon-Stourbridge Channel, 81
Selinsgrove, Pennsylvania, 524–527
Semiahmoo Drift, 568
Semuru, Mount, 1025
Senjo, Mount, 1018
Seno Otway, 722
Seno Skyring, 721–722
Setun', 341
Severnaya Zemlya islands, 367, 373–384, 380f
  in Eemian, 375
  Kara Sea and, 375–377
  in Late Pleistocene, 375
  in Middle Pleistocene, 373–375
  in Saalian, 375
  in Weichselian, 381–382
Seward Peninsula, 438
  in Early Wisconsinan, 439f
  in Late Wisconsinan, 439f
  in Pleistocene, 439f
Sezill Creek, 686, 687f, 693
SGS. See Saskatchewan Gravel and Sand
Shanzhuang, 1005–1006

Sheringham Cliffs Formation, 63–64, 67, 68
Sherwin, 451, 453f
Shetland Island, 270
Shimshal Valley, 913
Shirouma, Mount, 1014–1015, 1017–1018
Shirta, 375
Shooks Formation, 510
Shulie Lake, 651, 652f
Shurouma, Mount, 1015–1016
Shuttle Radar Topography Mission (SRTM), 3, 5f, 6f, 773
    of northern India, 930f
Shvencionys highland, 33
Šiaures Lie3tuvos, 243–244
Siberia
    Alpine glaciations in, 895–896
    CGS in, 888–890
        glaciation timing and dating, 895–901
        palaeogeographical interpretation of, 901–904
        types and occurrences of, 890–895
    Early Weichselian in, 902
    glaciation development characteristics in, 885–888
    high mountains of, 883–908, 886f
    in Late Pleistocene, 889, 902–903
    in Late Pliocene, 885
    in Late Weichselian, 903–904
    methods in, 884–885
    in Middle Pleistocene, 328f
    in Middle Weichselian, 903
    Milankovitch variations in, 902
    PIW in, 887, 887f
    in Pleistocene, 889, 891f
    Samarovo in, 373–375
    in Sartan, 903–904
    TL in, 902
Sidestrand Unio Bed, 67
Sidi Chamarouch, 1066
Sierra de Perija, 835
Sierra los Cuchumatanes, 843–844, 844f
    LLGM of, 846f
Sierra Madre Oriental, 856
Sierra Nevada
    California, 447–462, 449f
        glacial advances in, 450–458
    cosmic-ray exposure date reliability for, 458
    in Early Pleistocene, 450–452
    ELA of, 459–460, 459f
    glacial advances in, 451t
    lake records and, 458
    in Late Pleistocene, 455–458
    paraglacial deposit dating in, 458–459
    in Pliocene, 450–452
    in YD, 458
Sierra Nevada de Mérida. *See* Venezuela
Sierra Nevada de Santa Marta, 819
Sierra Nevada del Cocuy, 819, 830–835, 831f, 832f, 833f
    in Holocene, 833–834, 834f
SIS. *See* Scandinavian Ice Sheet
Sisimiut, 700–701
Sisson Brook, 636–637
    in YD, 653f

16 de Octubre Valley, 730–731
Sjœlland, 53–54
Skagerrak, 270
Skagerrak-Kattegat Trough, 48–49, 53
Skalisti Range, 146
Skamnellian, 187–188, 190–191, 193
Skardu Basin, 921f
Skœrumhede Group, 53
Skovbjerg, 51
SLIS. *See* St. Lawrence Ice Stream
Slovenia, 385–392, 386f
    in Early Pleistocene, 385–388
        glacial limits in, 385–386
    in Late Pleistocene, 388–391
        glacial limits in, 388–389
        key sites of, 390–391
    in Middle Pleistocene, 385–388
        glacial limits in, 385–386
Šluknov Hilly land, Elsterian and, 39f
Smith, Geoffrey, 177–179
Smith Sund, 705–706
Smoking Hills, 680, 681f
Sne1žné jámy Cirques, 42
Snežnik Mountains, 389
Snow Mountains, 1027f
Snowy Mountains, 11–12, 1037–1038
Soc1a, 386–387
    in Late Pleistocene, 389
Sog1anli Mountains, 398
Sokli, Finland, 108
Sonora Junction, 452
Sorata Valley, 761
Sosednee, Lake, 869
South Dakota, 507–508
South Georgia, LGM of, 1083–1084
South Island, central and eastern, 1060–1061
South Lithuanian ice-marginal zone, 242–243
South Patagonian Ice cap (SPI), 739–740, 747, 747f
South Polish, 408
South Tibet, keys sites in, 945–952
South-east Asia, 1023–1036
South-east England, 79–81
Southeast Minnesota, 508–509
South-eastern Taurus Mountains, 394–396
Southern Africa, 1081–1086
Southern Alberta, CRN in, 576–578
Southern Alps, 1053–1054
    Main Divide and
        east of, 1058–1059
        north of, 1054–1055
        west of, 1055
Southern Carpathians, 313, 314t, 319f
Southern Germany
    in Eemian, 170
    LGM of, 170
    low mountain range glaciation in, 170–171
    morainic amphitheatres in, 169
    in Pleistocene, 163–174
    regional descriptions of, 165–166
    stratigraphical results and concepts with, 163–165
    in Würmian, 170
Southern Hemisphere. *See also specific regions*

    in Cenozoic, 7
    in Late Pleistocene, 9
    in Plio-Pleistocene, 8
    in YD, 748
Southern Lapland Marginal Formations, 112f
Southern Patagonia, 716f
    in Late Weichselian, 719t
    in Late Wisconsinan, 719t
    LGM of, 717
    palaeo-ice-lobes in, 717–722, 718t, 719t
    in Pleistocene, 715–728
Southern Wisconsin
    in Early Pleistocene, 538
    in Middle Pleistocene, 541
Southwest Minnesota, 507–508
Sozh, Belarus in, 32–33
Speleotherms, 192–193
Spfa-1, 1016
SPI. *See* South Patagonian Ice cap
SRTM. *See* Shuttle Radar Topography Mission
St. Elias Mountains, 425
St. Francis Formation, 510
St. George's Channel, 82
St. John Peak, 1026f
St. Lawrence, 622
    deglaciation of, 610
St. Lawrence Estuary, 607
    tunnel Valleys in, 601–602
St. Lawrence Ice Stream (SLIS), 605
Staudenplatte, 167
Staufenberg, 167
Steinach, 25
Stewart Island, 1061
Steyr-Krems drainage system, 20
Stikine River, 686, 688f
Stockport Formation, 81
Stratigraphical terminology, 11–12
Subachoque Formation, 820–822, 822f
    in Late Pleistocene, 823f
Sub-Antarctic, 1082–1084
    LGM of, 1083
Suckling, Mount, 1027
Sudbury Formation, 62, 68
Sudetes mountain belt, 37–38
Sula Sgeir Trough Mouth Fan, 60–61
Sulawesi, 1023f
Sumatra, 1023–1036, 1023f
Šumava Mountains, 41–42
    in pre-Weichselian, 42
    in Weichselian, 43
Sundsøre Advance, 51–53, 56–57, 284
Sunrise Member, 509
Sunshine Mountains, 439–440
Suntar Chajata, 877
Suntar-Khayata Range, 865
Superior Lobe, 545
Süphan, Mount, 401
Surface exposure dating (SED), 111, 757–759, 773–775
    in New Zealand, 1047
Sutherland Borecore, Saskatchewan, 595–596
Svalbard
    Barents Sea and, 366–367, 368–370, 369f
    in Late Weichselian, 369f

Svartenhuk Peninsula, 705–706
Swarte Bank, 62
Swat Valley, 914f
Swat-Kohistan, 913
Swift Current, Saskatchewan, 592–594
Swift River Valley, 442
Swine Creek, 515
Switzerland, 8

T
Taffoni, 129
Tagh, 954f
Taglu, 678, 679f
Tahltan Canyon, 686, 688f
Tahoe, 453
Taibai Mountain, 994
Taitao ice cap, 751
Taiwan
    Central Mountains of, 994
    high mountain ranges of, 1004f
        in Early Holocene, 1003–1012
        ELA of, 1007–1010
        in Late Pleistocene, 1003–1012
        LGM of, 1008–1009
        in LIA, 1007–1008
    in Late Pleistocene, 10
Tamboran, Mount, 1025
Tampen, 274
Tamula, Lake, 99–100
Tancítaro, 849, 855–856, 856f
Tanggula Shan, 952
Tannheim, 168
Tanurer River, 873–874
Tanyurer River, 879
Taranaki, Mount, 1052
Tasman Mountains, 1052
Tasmania, 1038–1042, 1039f
    Brunhes/Matuyama palaeomagnetic reversal and, 1040–1041
    climate of, 1042–1043
    in Early Pleistocene, 1038–1041, 1040f
    in Jaramillo, 1040–1041
    in Late Pleistocene, 1042, 1043f
    in Middle Pleistocene, 9, 1040f, 1041–1042
Tateyama volcano, 1017–1018
Tatra Mountains, 414
Taurus Mountains
    central, 396–398
    south-eastern, 394–396
    of Turkey, 393–400
    western, 398–400
Taymyr Peninsula, 373–384, 374f, 380f
    deglaciation of, 378
    in Early Weichselian, 377–378
    in Eemian, 375
    ice-marginal zones in, 379–380
    Kara Sea Ice Sheet in, 375–377
    in Late Pleistocene, 375
    in Middle Pleistocene, 373–375
    in Saalian, 375
Taz, 329, 375
Tecka Drift, 731–732
Tectonic activity
    in Austria, 18
    in North America, 425
    in Qinghai-Tibetan Plateau, 998
Tekapo Lake, 1056
Teleki Valley, 1078–1079
Temiscaming, Lake, 612–615
Tennyson, Lake, 1053, 1054f
Tepsankumpu, 107
Terek River, 146
Terminal Moraine, 523
Terrestrial cosmogenic nuclide. See Cosmogenic radionuclide
Tertiary, 7
    Alaska in, 429
    Bolivia in, 775
    Ecuador in, 775
    Iceland in, 200–201, 204f, 207
    Ohio in, 513
    Peru in, 775
    in south-east Asia, 1023–1024
    Venezuela in, 835
Têt catchment
    deglaciation of, 136
    moraines in, 134–135
    in Würmian, 134f
Thames, 79–81, 82. See also Proto-Thames terrace sequences
THAR. See Toe-to-headwall altitude ratio
Thermoluminescence (TL), 42
    in Barents Sea, 366–367
    Chagan-Uzun River and, 897
    in Denmark, 156
    in Ohio, 515
    in Qilian Mountains, 988–992
    in Siberia, 884, 885, 902
    for Tobol, 373–375
    in Ukraine, 408
    in Venezuela, 835–837
Thrussington Till, 66–67, 82–83
Thüringen, 152
Tianshan Mountains, 981, 992–993
Tianshu, 8–9
Tibet, 955t. See also Qinghai-Tibetan Plateau
    central, 952
    in Middle Pleistocene, 8–9
    south, 945–952
    too great aridity prejudice and, 960–961
    western, 952–958
Tibetan Plateau, 960–961
Tibles1 Mountains, 311
TIC. See Total inorganic carbon
Tierra del Fuego, 715–728
Tiglian C4c, 247
Tills. See also specific tills
    in British Isles, 65–66
    of East Anglia, 66–67
    in Estonia, 95
    in Finland, 110–111
    in Pleistocene, 8
    from SIS, 65
Tintina Trench, 662–669, 691
    in Late Cenozoic, 688
Tioga, 455, 456f
Titicaca, Lake, 762f, 766, 799
Tjøme-Hvaler moraine, 289–291
Tjörnes, 200, 203, 204
TL. See Thermoluminescence
Tobol, 373–375
Toe-to-headwall altitude ratio (THAR), 308, 318
Tomur, 992–993
Tongariro, 1052, 1052f
Torlesse Range, 1060
Torngat Mountains, 620
Torres del Plaine, 748
Torun-Ebersvalde, 158, 302
Total inorganic carbon (TIC), 456f
Tottabetsu, 1016–1017
Tottenhill, 66f, 83–84
Toubkal, 1066, 1069
Toya, 1015
Trade River Formation, 509
Trans-Baikal Mountains, 885, 888, 888f, 893–894, 896f
Trans-Himalaya, 933–935
Transylvanian Alps, 305
Treene, 154–155
Tremenhau Drift, 732
Trikora, Mount, 1028, 1032, 1033–1034, 1033f, 1034f
Tromsø-Lyngen, 291–293
Tsangpo Valley, 950, 955t
Tsatsa-La-Mangaung, 1083f
Tuco Valley, 785
Tuizoche Valley, 844
Tukshan, 923
Tumara River, 878–879, 878f, 880f
Tuñame, 838
Tunisia, 1065–1074
Tunnel Valleys, 61, 80–81, 82, 84–85
    in North Sea, 266, 267f
    in St. Lawrence Estuary, 601–602
    in Witch Ground, 266–267
Turgeliai, 237–238
Turgen-Kharkhiraa Mountains, 970–971, 976
Turkey, 393–404
    Black Sea and, 400
    LGM of, 396f
    mountain glaciation in, 395t
    Taurus Mountains of, 393–400
Twin Cities, 509
Two Creeks Forest Bed, 546
Two Rivers-Onaway, 492–495
Two Tarn, 1079
Two Thumb Range, 1060
Twynam, Mount, 1037
Tymphi, Mount, 181–183, 182f, 184f, 188f
    geomorphological map of, 189f
    records of, 195
    uranium-series ages for, 191f
Tymphian, 187–188, 192, 193
Tyrrel Sea, 617
Tzipen plateau, 844

U
Ugandi, 97–98
    Early, 97
    Late, 97
    Middle, 97

Index

Ukraine
    Carpathians of, mountain glaciation in, 413–414, 415f, 416f
    in Dnieper, 406–408, 407f, 408–413
    in Drenthe, 413
    loess in, 413
    in Oka, 405–408, 406f
    in Pleistocene, 405–418, 407f
Ulakhan-Chistay Range, 868–869
Uludag1, 401
Umlach River, 169
Ungava Peninsula, 619
United States Geological Survey (USGS), 2, 3–4
Úpa Valley, 42
Upper Chuya basin, 898f
Upper Clutha, 1057–1058
Upper Pleniglacial 122–124
Upper regional unconformity (URU), 362–363
Upper Rhine Graben (URG), 164
Upper Tilatá Formation, 823f
Upper Waitaki, 1056–1057
Upper Würmian, 25f
Urals Valley, moraines in, 334f
Uranium-series ages
    in Greece, 183–185, 191f
    for Mount Tymphis, 191f
URG. *See* Upper Rhine Graben
Urghuch Formation, 911
URU. *See* Upper regional unconformity
Urumqi River, 992–993
U-series, 8–9
USGS. *See* United States Geological Survey
Utena, 240f

V

Vadaian, 10f
Vâlcan Mountains, 313
Valdaian, 352–357
    Early, 98
    Estonia in, 98–99
    Japan in, 1013
    Late, 99
    Middle, 98–99
Valdema2rpils 227–228
Vale of York, 38
Valgjärve, in Estonia, 98
Valira catchment, 128–129
Valle San Francisco, 762
Valley of Mexico, 849
Vancouver Island Range, 563–564
Varve-cyclite (VC), 885, 898, 900
Varyazh, 30
Vashon, 570
Vatnajökull Ice Sheet, 199, 200
VC. *See* Varve-cyclite
Vector Map Level 1 (VMAP1), 2–3, 6f
Venezuela, 835–842, 838f
    in Early Pleistocene, 835–837
    in Eemian, 837
    glacial stratigraphy of, 836t
    in Holocene, 839
    in Late Pleistocene, 837–839
    LGM of 835–837, 838–839
    in LIA, 839
    in LLGM, 845
    in Middle Pleistocene, 835–837
    in Tertiary, 835
    in Weichselian, 837–839
    in Wisconsinan, 837–839
    in YD, 839
Ventura Valley, 844
Vepsovo, 356–357
Verdon Valley, 117–118
Verkhoyansk Mountains, 865, 867–868, 877–882, 878f, 879f
    in Early Pleistocene, 879
    in Late Pleistocene, 879
    LGM of, 868, 877
    in Middle Pleistocene, 879
    regional comparison with, 879
    SIS and, 879
    in Weichselian, 879
Vicdessos, 136
Vidurio Lietuvos, 243
Vienna Basin, 18
Vikos Canyon, 181
Vindžu1nai, 234–235, 236f, 238f
Vistulian, 302
Vla1deasa Mountains, 315–317
Vlasian, 187–188, 191–192, 193
VMAP1. *See* Vector Map Level 1
Voidomatis River, 181–182, 182f, 193–195, 194f
Volcán Chimborazo, 775–777, 781f
Volcanoes
    of Anatolian Plateau, 400–401
    in British Columbia, 568
    in Ecuador, 808f
    in Hawaii, 463f
    in Iceland, 201f
    in Japan, 1017–1018
    of Mexico, 849–862, 850f
    Mount Kenya and, 1076
    in New Zealand, 1052
    in western Indonesia, 1025
Volkovysk highland, 31f
Vorskla River, 409
Vorstoßschotter, 22
Võrtsjärv, Estonia in, 99
Vos'miozyorny, 869

W

Wacken, 51
Waiau River, 1053
Waiho Loop, 1055
Waihohonu Valley, 1052f
Waimakariri, 1055–1056
Wairau River, 1053
Wakatipu, Lake, 1057
    LGM of, 1058f
Walcott Till, 63–64
Wales
    in Anglian, 81–82
    BIS in, 62–63, 70
Walworth Formation, 538, 541–542
Wanganui Basin, 1047–1048, 1049
Warren House Till, 65, 68
Warsaw-Berlin, 302
Wartanian, 301–302
Warthe
    Denmark in, 50–51, 51f
    Middle Saalian and, 155
Wascana Creek, Saskatchewan, 594–595
    in Late Neogene, 597f
Washington, 531–536
    Cordilleran Ice Sheet in, 531–533
    in Late Pleistocene, 534–535
    in YD, 533
Watino, 585
Wausau Member, 538
Wee Bankie, 274
Weichselian, 8, 10f. *See also* Early Weichselian; Late Weichselian; Middle Weichselian; Pre-Weichselian
    Barents Sea in, 365
    Belarus in, 34
    Colombia in, 829–830
    Estonia in, 98–101
    Finland in, 107–114
    Germany in, 149
    Great Britain in, 86–89
    Greenland Ice Sheet in, 702–707
    Japan in, 1013
    Krkonoše Mountains in, 42–43
    Latvia in, 224–226
    Mexico in, 857–858
    Mount Kenya in, 1078
    north Germany in, 156–158
    North Sea in, 255–256, 268–275
    Northern Hemisphere in, 48
    Norway in, 281–282
    Patagonian Ice Sheets in, 742
    Québec-Labrador in, 602–604
    Severnaya Zemlya islands in, 381–382
    SIS in, 283f
    umava Mountains in Š, 43
    Venezuela in, 837–839
    Verkhoyansk Mountains in, 879
Wellsch Valley, Saskatchewan, 592–594
    in Late Neogene, 596f
    magnetostratigraphic correlation for, 598f
West African Monsoon, 1071–1072
West Coast Range, 1038–1039
West Himalaya, in China, 984
West Lawrencetown, 644
west of, 1055
West Polissian, 406–409
    glacial limits of, 409–412
    key sites in, 413
West River, 680, 681f
West Walker River, 452
Western Barents Sea, 366–367
Western Caucasus, 144
Western Indonesia, 1024–1025, 1024t
Western New Guinea, 1026–1027
Western Norway, 291
Western Taurus Mountains, 398–400
Western Tibet, key sites in, 952–958
Western Washington, 531–536
    Cordilleran Ice Sheet in, 531–533
    in Late Pleistocene, 534–535
    in YD, 533
Western Wisconsin, 543f

Western Wisconsin (cont.)
  in Early Pleistocene, 538
  in Middle Pleistocene, 541–542
Westlynn drift, 568
Weybourne Town Till, 63–64
Whidbey Formation, 533
White Sea, 366
Whitestone, 1060
Whitney, Josiah, 448–450
Wiconsinan, Wisconsin in, 538
Wilhelm, Mount, 1027, 1031*f*
Windischgarsten basin, 20–21
Wirt Lake Formation, 510
Wisconsin, 537–552, 540*f*, 544*f*
  Driftless Area in, 538*f*, 541, 548
  in Early Pleistocene, 537–541
    ice extent and chronology for, 539–541
  in Early Wisconsinan, 543–544
  in Illinoian, 537–538, 541–542
  in Late Pleistocene, 542–548
  in Late Wisconsinan, 544–547
    landforms and palaeoglaciology for, 547–548
  in Middle Pleistocene, 541–542
    ice extent and chronology for, 541–542
  north-central, 543*f*
    in Early Pleistocene, 538
  in Pleistocene, 537, 542*f*
  southern
    in Early Pleistocene, 538
    in Middle Pleistocene, 541
  western, 543*f*
    in Early Pleistocene, 538
    in Middle Pleistocene, 541–542
  in Wisconsinan, 538, 542–548
Wisconsinan, 8, 10*f*. *See also* Early Wisconsinan; Late Wisconsinan; Middle Wisconsinan; Pre-Wisconsinan
  Chilean Andes in, 742–745
  Chilean Patagonian Ice Sheets in, 745
  in Illinoian, 476–484, 482*f*
    aeolian and lacustrine deposits of, 484
  Japan in, 1013
  Mexico in, 857–858
  in Michigan, 489–498
  Mount Kenya in, 1078
  Ohio in, 517
  Québec-Labrador in, 601–602
  Venezuela in, 837–839
  Wisconsin in, 542–548

Witch Ground, 265–266
  BIS and, 274
  in Late Weichselian, 271–272
  in Pleistocene, 269*f*
  tunnel Valleys in, 266–267
Wolston Formation, 63–64, 66–67, 82–83
Wolstonian
  East Anglia in, 83–84
  Great Britain in, 78, 82–86
    miscellaneous glacial deposits in, 84–86
  Midlands in, 82–83
  North Sea in, 85*f*
Wolverine, 670–671, 672*f*
Woodville Member, 538
Worth Point Bluffs, 683*f*, 685–686
Wroxham Crag, 61, 68
Würmian, 19–20, 22, 42, 163
  Ariège catchment in, 129
  Bolivia in, 767–768
  Caucasus in, 141–148, 142*f*, 143*t*
  in Niederterasse, 170
  Noguera Ribagorçana catchment in, 133*f*
  PNG in, 1032–1033
  Pyrenees in, 128, 132–137
  Southern Germany in, 170
  Šumava Mountains in, 43
  in Têt catchment in, 134*f*
  Upper, 25*f*

X
Xixiabangma, 984, 986*f*

Y
Yangtze River, 981
Yaran-Musyur, 332–333
Yari, Mount, 1017–1018
YD. *See* Younger Dryas
Yematan, 988
Yenissei River, 332*f*, 890*f*, 893–894, 897*f*
Yermak PLateau, 363
Yosemite Valley, 448–450
Young Baltic Ice Streams, 51–53
  in East Jylland, 55*f*
  in Bœlthav, 56*f*
  in Møn, 56*f*
Younger Dryas (YD), 26
  Barents Sea in, 369–370
  Chile in, 748–750
  Chilean Lake District in, 749
  Colombia in, 824–825, 831
  Cordillera Blanca in, 792–793
  Cordillera Cochabamba in, 795–796
  Ecuador in, 789
  Finland in, 113–114
  Greenland Ice Sheet in, 707–708
  Iztaccíhuatl in, 850
  in Magellan Straits, 749–750
  Maritime Canada in, 631, 652–654
  Mexico in, 850, 859
  Mount Kenya in, 1079
  Northern Hemisphere in, 748, 790
  Norway in, 289, 291, 292*f*
  Patagonian Ice Sheets in, 749
  Pyrenees in, 137
  Québec-Labrador in, 610–612
  Sierra Nevada in, 458
  Sisson Brook in, 653*f*
  Southern Hemisphere in, 748
  Venezuela in, 839
  western Washington in, 533
Younger Saalian
  Estonia in, 97–98
  north Germany in, 155
Yukon, 423–425, 563
  west-central, 662–673
Yukon-Koyukuk region, 438–439
Yushan, 1004*f*, 1006

Z
Zanskar, 929–930, 933, 935*f*, 936*f*, 937*f*
Zaogros Mountains, 394
Zeitz, 153–154
Zenda Formation, 541–542
Zhonglianggan, 988–992, 995
Zhuxi Gou, 988, 990*f*
Zittau Basin, 38
Zongo Valley, 757–760, 764*f*, 767, 795
Zorros Drift, 729–730
Zusamplatte, 167
  stratigraphical implications of, 168
Zyrianka, 865–866, 867*f*, 869
  in Bol'shoi Annachag Range, 869*f*
  in Mongolia, 976
  Nizhne-Anadrsk Lowland in, 871*f*
  Sartan and, 870